Methodicum Chimicum

A Critical Survey of Proven Methods
and Their Application in Chemistry,
Natural Science, and Medicine

Editor-in-Chief

Friedhelm Korte

Volume Editors

H. Aebi
H. Batzer
E. Baumgartner
K.-H. Büchel
G. Büyük
J. Falbe
R. Gompper
M. Goto
U. Hasserodt

F. Korte
E. Kutter
H. Machleidt
K. Niedenzu
G. Ohloff
T. Peterzylka
H. Zimmer
F. Zymalkowski

Academic Press New York · San Francisco · London   1975

Georg Thieme Publishers Stuttgart

Volume 6

C-N Compounds

Edited by

F. Zymalkowski

Contributions from

G. Baumeyer, Tübingen
E. V. Brown, Lexington,
  Kentucky
L. Caglioti, Rom
G. F. Cainelli, Bologna
B. Camerino, Milano
O. Červinka, Prag
W. Ehrenstein, Leverkusen
G. Entenmann, Stuttgart
M. Ferles, Prag
J. Goerdeler, Bonn
W. Hänsel, Freiburg
R. Haller, Freiburg

J. Hoyer, Freiburg
J. Keck, Biberach
J. K. Korobizina, Leningrad
U. Kraatz, Freising-
  Weihenstephan
G. Krüger, Biberach
K. Lehmann, Freiburg
S. Linke, Wuppertal-
  Elberfeld
H. Meier, Tübingen
G. Paolucci, Bologna
M. Regitz, Kaiserslautern
L. L. Rodina, Leningrad
G. Rosini, Bologna
H. J. Roth, Bonn

K. Schank, Saarbrücken
H. Scherer, Kaiserslautern
W. Schneider, Freiburg
H. Schönenberger,
  München
G. Simchen, Stuttgart
H. Stamm, Karlsruhe
W. Sucrow, Berlin
B. Unterhalt, Marburg
W. Walter, Hamburg
E. Wehinger, Wuppertal-
  Elberfeld
H. Wollweber, Wuppertal-
  Elberfeld
A. Zobáčová, Prag

104 tables

Academic Press New York · San Francisco · London   1975

Georg Thieme Publishers Stuttgart

Editors

In this handbook are listed many registered trade marks, trade names, etc. (although specific reference to this fact is not always made in the text), BIOS and FIAT reports, patents, and methods of manufacture and application. The editors and publishers draw attention to the fact that the legal situation must be carefully checked before commercial use is made of these. Only a selection of industrially produced apparatus and appliances is mentioned. This in no way implies a reflection on the products not included in this volume.

Journal titles are abbreviated chiefly in accordance with the system of Chemical Abstracts.

© 1975 Academic Press Inc., New York · San Francisco · London. – Typesetting: Filmsatz Stauffer + Cie., Basel. – Text Processing and Computer-controlled Photo-Typesetting of Index: Siemens TEAM Programme System. Register Typesetting: CRT Photo-Typesetting Equipment of Satz AG, Zürich. – Printed in Germany by Grammlich, Pliezhausen.

Library of Congress Catalog Card Number: 74-21580
ISBN 0-12-460706-3 (Academic Press)
ISBN 3-13-505301-6 (Thieme)

Preface of the Series

The METHODICUM CHIMICUM is a short critical description of chemical methods applied in scientific research and practice. It is particularly aimed at chemists as well as scientists working in associated areas including medicine who make use of chemical methods to solve their 'interrelated' problems.

Considering the present development of science and the necessity for concise and unambiguous information, the series provides a guide to rapid and reliable detection of the method suitable for the solution of the problem concerned. Thus, particular emphasis is placed on the description of proved procedures whereby a complete and exhaustive compilation of all reported methods and also a detailed description of experimental techniques have been deliberately omitted. Newer methods as well as those which have not yet been reported in review articles are treated more extensively, whereas conventional methods are dealt with cincisely. Biological procedures which, in specific cases, are more useful for characterizing substances than chemical or physical methods, will be discussed in the analytical volume. The interrelated methods and concepts which are constantly gaining importance will be fully discussed in the third 'Specific Part'.

The METHODICUM CHIMICUM is comprised of three parts. The first, the 'General Part' consists of Volumes 1, 2 and 3. Volume 1 (Analytical Methods) is concerned with chemical, physical, and biological analytical methods including those necessary for the elucidation of structures of compounds.

Volume 2 (Planning of Syntheses) contains a review on fundamentals, principles, and models with particular respect to the concepts and applications of theoretical chemistry essential to the practically working scientist.

Volume 3 (Types of Reactions) is designed to illustrate the scope and utility of proved working techniques and syntheses.

The second part (Vols. 4–8), which is particularly devoted to 'Systematic Syntheses', deals with proved methods for syntheses of specific compounds. These procedures are classified according to functional groups linked together in the last step of reaction.

Volume 4 (Syntheses of Skeletons) describes the construction of hydrocarbons and heterocyclic compounds.

Volume 5 the formation of C–O-bonds, Volume 6 the formation of C–N-bonds, Volume 7 the syntheses of compounds containing main group elements, and Volume 8 compounds containing transition metal elements.

The third 'Special Part' (Volume 9–11) is concerned with the chemical aspects connected with the formulation of a question or problem.

Volume 9 deals with nonmetallic synthetic fibers and synthetic materials as well as their additives, Volume 10 with synthetic compounds and Volume 11 with natural products and natural occuring compounds.

All Volumes should not contain more than 900 printed pages. They are intended to give the chemist and any person working in fields related of chemistry a sufficient answer to his problem. Selected review articles or important original works are cited for the sake of detailed information.

We wish to thank the Georg Thieme Verlag, Stuttgart, for making possible the realization of the basic concept of METHODICUM CHIMICUM and for the excellent presentation of the work.

Bonn, September 1974   Friedhelm Korte

Preface of Volume 6

Volume 6 of Methodicum Chimicum concerns itself with methods for the formation of carbon-nitrogen bonds. The material is arranged according to the resultant substance class; the suitable methods are described and their efficiency and limitations are compared.

The many new developments in the field of organic nitrogen chemistry have necessitated a selection which, in spite of all attempts at objectivity, may be influenced by the personal experiences and view-points of the individual authors. Heteroaromatic nitrogen compounds are described in Volume 4 of Methodicum Chimicum. In volume 6 only examples for the syntheses of heteroaromatic nitrogen compounds which make use of generally applicable methods for the formation of nitrogen-carbon bonds are included. It was not possible to avoid mentioning some methods for the preparation of organic nitrogen compounds which are based on the formation of nitrogen-nitrogen bonds or transformation of other organo-nitrogen compounds.

Possibilities for the further reactions as well as preparations have only been included when either the reaction products belong to a class of compounds which is not described elsewhere in M.C. or the products are so labile (even **explosive!**) that they are not isolated during organic syntheses but are always reacted further *in situ*.

A reference work is only of use when the reader can find what he seeks rapidly. Thus, much effort has been made in the clear structure of the individual chapters, in the extensive lists of contents, in the many formula schemes, and also in the tabular surveys to ensure that the search is as easy as possible. Although well-proven methods are described more briefly than new developments, references to the original literature are always given. The literature has been covered up to 1971.

I would like to thank all contributors to this volume for their willing cooperation and the punctual submission of their manuscripts in spite of the short dead-lines and other duties. Dr. Simchen, Stuttgart, helped me especially by editing the extensive and relatively heterogeneous chapter on *N*-functional carboxylic acid derivatives. The editor-in-chief, Prof. Korte, Munich, arranged many important contacts and gave many valuable hints. The publishers are to be thank for the technical preparation of the volume and the understanding cooperation during its preparation.

July 1975   *F. Zymalkowski*

Table of Contents

Table of Contents

1 Organic Nitro Compounds

Contributed by

G. Baumeyer
Tübingen

1.1 Preparation of aromatic *C*-nitro compounds by substitution

A variety of factors influence the substitution of aromatic compounds by the nitro group. In addition to nitrating agent and temperature the reactivity of the compound concerned plays the decisive role.

As nitration is initiated by the nitronium cation acting as the attacking group, aromatic compounds rich in π-electrons (aromatic compounds with electron-donating substituents or five-membered heteroaromatic compounds) may be described as readily nitratable. Aromatic compounds with few π-electrons (those with electron-withdrawing substituents or six-membered heteroaromatic compounds) are difficult to nitrate. (For details concerning the reactivity of aromatic and heteroaromatic compounds, see refs. [1, 168]).

The directing effect of already existing substituents on the aromatic compound on the direction of the substitution of an aromatic H-atom by the nitro group follow from the familiar empirically found rules for the electrophilic (secondary) substitution of aromatic systems [1, 168]. Substituents of *first order* direct the nitro group predominantly into the *o*- and *p*-positions. These include those substituents which due to their (predominant) +M and/or +I effect increase the basicity (*e.g.*, OH, NH$_2$, NHR, NR$_2$, O$^\ominus$, Alkyl) and the halogens. Substituents of 'second order' direct the entering substituent mainly into the *m*-position. These are substituents which reduce the reactivity of the aromatic compound (*e.g.*, CO—R, COOH, COOR, CN, No$_2$, $\overset{\oplus}{N}$R$_4$), *i.e.*, groups with a predominant —M and/or —I effect, except the halogens (cf. p. 4).

Promotion of nitration by the solvent diminishes in the order (*Houben-Weyl*, Vol. X/1, p. 472):

sulfuric acid > nitric acid > nitromethane > glacial acetic acid > 1,4-dioxane > water

The active agent during the nitration is the nitronium ion, NO$_2^\oplus$, which is formed from nitrating agents of the general formula X—NO$_2$.

The readiness with which X$^\ominus$ is split off leads to the following series having a progressively diminishing nitration action [1]:

Nitronium ion	$\overset{\oplus}{N}O_2$
Nitracidium ion	H$_2$O$^\oplus$—NO$_2$
Nitronium chloride	Cl—NO$_2$
Dinitrogen pentoxide	O$_3$N—NO$_2$
Acetyl nitrate	H$_3$C—CO—O—NO$_2$
Nitric acid	HO—NO$_2$
Ethyl nitrate	H$_5$C$_2$—O—NO$_2$

1.1.1 Replacement of hydrogen

1.1.1.1 Readily nitrated aromatic compounds

Electrophilic aromatic substitutions, including replacement of a hydrogen atom by the nitro group, are facilitated by excess π-electrons. This simple empirical rule explains the behavior of many aromatic compounds during nitration that is described in what follows.

In biphenyl the participation of the second phenyl ring in the mesomerism of the first leads to enhanced electrophilic reactivity (relative to benzene). However, this applies only if there are no substituents in the *o,o'*-position which bring about a stronger rotation of the ring planes [1].

When nitrating acid or nitric acid (admixed with acetic anhydride or glacial acetic acid) act on naphthalene, predominantly *1-nitronaphthalene* (95% at 40–45°) together with 4–5% of the 2-isomer are formed [2, 3]. The introduced nitro group deactivates (see p. 8) the aromatic compound, but this deactivation does not extend or extends only little to a fused ring, as is demonstrated by the relatively ready nitration of naphthalene to form *1,5*- and *1,8-dinitronaphthalene* [4].

Azulene can likewise be nitrated under mild conditions (nitric acid-acetic anhydride mixtures). Here the influence of temperature is very marked [5, 6]:

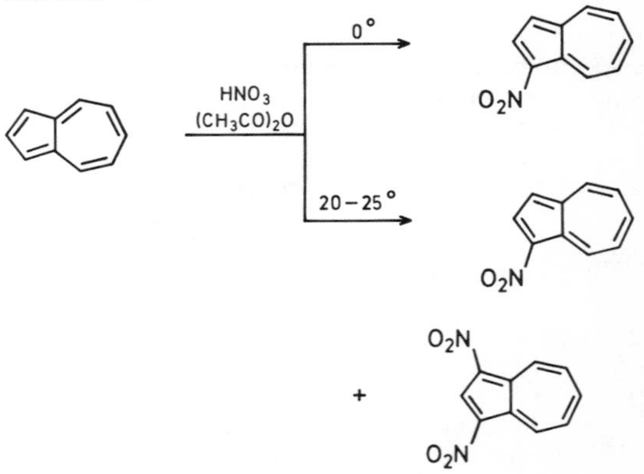

[1] *W. Seidenfaden, D. Pawellek* in *Houben-Weyl*, Methoden der organischen Chemie, Bd. X/1, p. 477, 488, Georg Thieme Verlag, Stuttgart 1971.

[2] *H.E.Fierz-David, R. Sponagel*, Helv. Chim. Acta **26**, 98 (1943).

[3] *M.J.S.Dewar, T.Mole*, J. Chem. Soc. (London) 1441 (1956); *A.A. Spryskow, I.K. Barvinskaja*, Zh. Org. Khim. **4**, 191 (1968); C.A. **68**, 104298 (1968).

[4] *H.H. Hodgson, J. Walker*, J. Chem. Soc. (London) 1346 (1933).

Monoalkylbenzenes, too, are much more reactive than benzene[7] and can be nitrated under mild conditions. Table 1 shows the dependence on nitrating agent and temperature of the isomer ratio obtained in the case of toluene.

In principle the same mild conditions are used for nitrating higher alkylaromatic compounds, though the size of the alkyl group plays a part[8]. Thus, nitration of, for example, 1-phenyl- or 1-benzyl-adamantane[9], cumene (>30° with acetyl nitrate[10]), and paracyclophanes[11, 12] yields practically only ring-nitrated derivatives.

Table 1. Nitration toluene with various nitrating agents (cf. *Houben-Weyl*, Vol. X/1, p. 516)

Nitrating agent	Temperature [°C]	o-Nitrotoluene %	m-Nitrotoluene %	p-Nitrotoluene %
85% Nitric acid	−30	55.6	2.7	41.7
	60	57.5	4.0	38.5
23.8% Nitric acid, 58.7% sulfuric acid, 17.5% water	50	62.2	4.3	33.5
5.4% Nitric acid, 71.4% sulfuric acid, 21.5% water	60	58.2	5.5	36.2
90% Nitric acid, Amberlite IR-120	50	41	—	45
90% Nitric acid, aromatic sulfonic acid	50	51.0	3.6	45.4

kylated benzenes are likewise readily nitrated. Their reactivity is dependent on the number, nature, and position of the alkyl substituents, and these parameters also influence the possibility of side-reactions occurring. Nitration of 1,3-dialkyl-substituted benzene proceeds relatively readily and uniformly, and generally goes to the trinitro stage. Thus *1,3-dimethyl-2,4,6-trinitrobenzene* is obtained from 1,3-xylene, while *1,3,5-trimethyl-2,4-dinitrobenzene* is obtained in high yield during nitration of mesitylene in liquid hydrogen fluoride with sodium nitrate at 5° (see *Houben-Weyl*, Vol. X/1, pp. 528, 533, 825).

If the *alkylbenzenes* are *monohalogenated* on the nucleus then nitration can ensue under relatively mild conditions just as in the case of the unsubstituted alkylbenzenes. Polyhaloalkylbenzenes require more energetic conditions (see *Houben-Weyl*, Vol. X/1, p. 552).

The nature and position of a halogen atom on the ring can reduce the *m*-directing effect of a trifluoromethyl group strongly. A fluorine or chlorine atom in the *para* or *meta* position with respect to the trihalomethyl group on the ring will determine where the nitro group enters.

3-Chlorotoluene is more readily and uniformly dinitrated than 2-chlorotoluene (see *Houben-Weyl*, Vol. X/1, p. 557).

A functional group in an alkyl substituent can influence the reaction in dependence on its nature and distance from the aromatic nucleus. As the

Mononitration of *dialkylated benzenes* with a mixed acid containing 10–20% water using approximately theoretical stoichiometric proportions also succeeds without difficulty. The entering nitro group seeks the proximity of the smaller alkyl group; the displacement of the alkyl group by the nitro group which occurs in certain polyalkylbenzenes can be suppressed by adopting suitable measures. Unless they are sterically hindered polyal-

[8] *C.K. Ingold*, Structure and Mechanism in Organic Chemistry, p. 258, Cornell University Press, Ithaka 1953;
J.R. Knowles, R.O.C. Norman, G.K. Radda, J. Chem. Soc. (London) 4885 (1960);
H.C. Brown, W.H. Bonner, J. Amer. Chem. Soc. 76, 605 (1954);
K. Le Roi Nelson, H.C. Brown, J. Amer. Chem. Soc. 73, 5605, 5607 (1951).

[9] *H. Stetter, J. Weber, C. Wulff*, Chem. Ber. 97, 3488 (1964);
F.N. Stepanov, E.I. Dikolenko, G.I. Danilenko, Zh. Org. Khim. 2, 640 (1966); Engl.: 640; C.A. 65, 8782h (1966).

[10] *G. Vavon, A. Callier*, Bull. Soc. Chim. France [4] 41, 357 (1927);
L. Bert, P.C. Dorier, Bull. Soc. Chim. France [4] 41, 1170 (1927).

[11] *D.J. Cram, R.A. Reeves*, J. Amer. Chem. Soc. 80, 3094 (1958).

[12] *D.J. Cram, R.H. Bauer, N.L. Allinger, R.A. Reeves, W.J. Wechter, E. Heilbronner*, J. Amer. Chem. Soc. 81, 5977 (1959).

[5] *Houben-Weyl*, Methoden der organischen Chemie, Bd. V/2, Herstellung carbocyclischer konjugierter π-Elektronensysteme, Georg Thieme Verlag, Stuttgart (in preparation).

[6] *A.G. Anderson jr., R. Scotoni jr., E.J. Cowles*, J. Org. Chem. 22, 1193 (1957).

[7] *C.K. Ingold, F.R. Shaw*, J. Chem. Soc. (London) 2918 (1927);
W.W. Jones, M. Russell, J. Chem. Soc. (London) 921 (1947);

distance grows the effect diminishes and the resulting isomer ratio approaches that in the case of an alkylaromatic compound.

Table 2. Nitration of substituted alkyl aromatic compounds

Side-chain substituent	Nitrating agent	see *Houben-Weyl*, Vol. X/1, p. 516
Nitro group	conc. HNO$_3$, nitrating acid	536
Carbonyl group	HNO$_3$	544
Carboxy group		545
Halogens	HNO$_3$	537
Hydroxy and alkoxy groups	HNO$_3$	542
Amino group	HNO$_3$ or mixtures with H$_2$SO$_4$, CH$_3$COOH	549

Nitration of *halogenated aromatic compounds* proceeds diversely. Substituent halogens hinder nitration because their −I effect reduces the overall basicity (−I effect > −M effect in the ground state). In the positively charged δ-complex the +M effect predominates, and so halogens are *o-p* directing[168] (for the *o-p* isomer distribution see *Houben-Weyl*, Vol. X/1, p. 500).

Dihalobenzenes can be nitrated approximately quantitatively (using calculated or slightly more mixed acid containing 5−10% water and slightly elevated temperature).

m-Dihalo derivatives are nitrated the most readily, *o*-compounds less so, *p*-compounds least readily. Nitration of 2,4-difluoro-1,3,5-trimethylbenzene with sodium nitrate in anhydrous hydrogen fluoride proceeds very smoothly[171].

The +M effect of the *hydroxy group* promotes aromatic nitration and directs the entering nitro group into the *o* and *p* positions. Nitration proceeds even in strongly diluted aqueous nitric acid if it contains nitrous acid; if not, an induction period is observed (formation of minor amounts of nitrous acid)[13].

Nitration of phenols may ensue spontaneously and **violently.** It is therefore appropriate to stir well and to check the temperature constantly during addition of the nitrating agent. Nitrate ions cata-

lyze the further nitration beyond 2-nitrophenol[14] to *2,4-* and *2,6-dinitrophenol*[15]. 80−90% nitric acid/glacial acetic acid or mixed acid with cooling is employed for nitrating nitrophenols themselves further. 2-Nitrophenol is nitrated to a mixture of *2,4-* and *2,6-dinitrophenol*, while with 3-nitrophenol the hydroxy group makes the nitro group enter the *o* or *p* position[16]. Nitrating phenol with a 50% excess of nitric acid invariably leads to some *2,4-dinitrophenol*[17].

Phenol may be nitrated also with tetranitromethane[18] to give equal parts 2- and 4-nitro derivative; a similar result is obtained with dinitrogen tetraoxide[19]. On passing nitrous gases into a solution of pyrogallol in ether one nitro group is introduced into the aromatic nucleus[172]. Nitration of phenol esters and ethers, too, proceeds smoothly (*Houben-Weyl*, Vol. X/1, p. 565).

Pyrocatechol and resorcinol (*o*- and *m*-dihydroxybenzenes) can be nitrated very readily. Depending on the reaction conditions used, nitric acid of varying concentration is employed[20].

[13] *A. Klemenc, E. Ekl*, Monatsh. Chem. *39*, 641 (1918);
S. Veibel, Ber. *63*, 1577 (1930);
S. Veibel, Z. Physik. Chem. (Leipzig) B *10*, 22 (1930).

[14] *F.W. Henle*, Anleitung für das organisch-chemische Praktikum, p. 39, Akademische Verlagsgesellschaft, Leipzig 1927;
L. Gattermann, Die Praxis des organischen Chemikers, 41. Aufl., p. 214, Walter de Gruyter Verlag, Berlin 1962;
H.E. Fierz-David, L. Blangey, Grundlegende Operationen der Farbenchemie, 8. Aufl., p. 142, Springer Verlag, Wien 1952;
F. Arnall, J. Chem. Soc. (London) *125*, 811 (1924);
Houben-Weyl, Methoden der organischen Chemie, Bd. VI/1c, Kap. Herstellung von Phenolen (in preparation);
K. Beaucourt, E. Hämmerle, J. Prakt. Chem. [2] *120*, 185 (1928).

[15] *J.J. Rinkes*, Rec. Trav. Chim. Pays-Bas *45*, 846 (1926);
D.H. Olson, P.W. Storms, J. Org. Chem. *31*, 1469 (1966).

[16] *A.F. Holleman, G. Wilhelmy*, Rec. Trav. Chim. Pays-Bas *21*, 436 (1902).

[17] *R.A. Slavinskaya*, Zh. Obshch. Khim. *27*, 1160 (1957); Engl.: 1243; C.A. *52*, 2793 (1958).

[18] *S. Skraup, W. Beifuss*, Ber. *60*, 1074 (1927);
K. Auwers, Ber. *35*, 455 (1902).

[19] *P. Podkowka, A. Tarnawski*, Monatsh. Chem. *90*, 179 (1959);
P.F. Frankland, R.C. Farmer, J. Chem. Soc. (London) *79*, 1356 (1901);
K. Auwers, Ber. *35*, 456 (1902);
H. Wieland, Ber. *54*, 1776 (1921);
L.B. Haines, H. Adkins, J. Amer. Chem. Soc. *47*, 1419 (1925);
L. Monti, Gazz. Chim. Ital. *67*, 628 (1937); Chem. Zentr. I, 3034 (1938);
P.P. Schorygin, A.V. Topchiev, Zh. Obshch. Khim. *8*, 986 (1938); Chem. Zentr. I, 2174 (1939).

As it is readily oxidized to *p*-benzoquinone, hydroquinone can be nitrated only in the form of its *O*-derivatives. Under overenergetic conditions oxidation nonetheless occurs due to hydrolysis. In tri- and polyphenols nitration may be made difficult in dependence on the position of the hydroxy groups; this effect is alleviated by partial *O*-acylation. During dinitration of phloroglucinol partial oxidation occurs under the drastic conditions required (*Houben-Weyl*, Vol. X/1, p. 597 ff.).

Aromatic compounds containing three and more hydroxy groups are nitrated with dilute nitric acid at room temperature only where quinone formation is excluded (*e.g.*, phloroglucinol = 1, 3, 5-trihydroxybenzene). In other cases two hydroxy groups need to be protected.

3-Alkoxyphenols are readily nitrated under mild conditions, *viz.* glacial acetic acid as diluent and solvent, efficient cooling[21]. The nitro group enters mainly in the *o*-position to the free hydroxy group. Alkyl-*m*-methoxyphenols react very **violently** with nitric acid.

Benzoyl nitrate in carbon tetrachloride or acetonitrile likewise lends itself for nitrating phenols and their derivatives; solutions of metal nitrates[22] in sulfuric acid, too, can be used for carrying out nitrations but offer no advantages.

During nitration of *halogenated phenols*[23], which ensues as readily as that of the halogenfree parent compounds, the directing effect of the halogen does not manifest itself. However, replacement of the *o*- or *p*-halogen by the nitro group can occur (more readily of Br than of Cl).

Alkylphenols are nitrated more readily than phenol, especially 3-alkylphenols. On the other hand, dinitration and formation of decomposition products are also favored, and occur even at −15° and using 65% nitric acid admixed with glacial acetic acid. Di- and polynitration of monoalkylphenols, above all of methylphenols, often proceed **violently**, so that it is appropriate to work in a solvent (*e.g.*, glacial acetic acid, carbon tetrachloride,

chloroform) and to use moderately diluted nitric acid in order to restrain the reaction[24]. 4-Cyclopentyl-, cyclohexyl-, and cyclooctylphenols[25], too, are readily dinitrated.

Dialkyl- and *polyalkylphenols* very readily react; not only nitration but side-reactions such as dinitration, resinification, *etc.*, are facilitated as a result[26]. These side-reactions are reduced by keeping the concentration of the nitric acid low and using moderately elevated temperature. Where di- or polynitration of dialkyl- or polyalkylphenols is desired it is possible to perform it under mild conditions[27].

On polynitrating polyalkylphenols with substituents in the *o*- and *p*-position a predominant simultaneous displacement of an alkyl substituent by the nitro group takes place[27].

In phenol ethers the substitution site is influenced as a function of the ether group. The ready nitration of this substance class often makes it difficult to arrest the reaction at the mononitro product stage[28].

Ring-halogenated alkylphenols also are readily nitrated[29]. A free hydroxy group determines the site of entry of the nitro group. Nitration of aromatic *aldehydes* is promoted by the presence of a hydroxy group including in the form of its derivatives. The same is true also if the *hydroxy aldehyde* contains an additional chlorine atom on the ring (*Houben-Weyl*, Vol. X/1, p. 600).

[20] *W. Seidenfaden, D. Pawellek* in *Houben-Weyl*, Methoden der organischen Chemie, Bd. X/1, p. 588, Georg Thieme Verlag, Stuttgart 1971.

[21] *H.H. Hodgson, H. Clay*, J. Chem. Soc. (London) 2775 (1929);
W. Seidenfaden, D. Pawellek in *Houben-Weyl*, Methoden der organischen Chemie, Bd. X/1, p. 592, Georg Thieme Verlag, Stuttgart 1971.

[22] *L. Spiegel, H. Haymann*, Ber. 59, 202 (1926).

[23] *W. Seidenfaden, D. Pawellek* in *Houben-Weyl*, Methoden der organischen Chemie, Bd. X/1, p. 569, Georg Thieme Verlag, Stuttgart 1971.

[24] DDRP. 12398 (1956), Erf.: *G. Fricke, R. Kaltofen*; Chem. Zentr. 11434 (1957).

[25] *A.R. Abdurasuleva, F.K. Shadieva*, Zh. Obshch. Khim. 29, 4021 (1959); Engl.: 3980; C.A. 54, 20940e (1960);
A.R. Abdurasuleva, N.G. Koralnik, Zh. Obshch. Khim. 30, 1635 (1960); Engl.: 1633; C.A. 55, 1512i (1961);
Brit. P. 861792 (1958), British Oxygen Co., Ltd., Inv.: *W.O. Jones*; C.A. 55, 18666c (1961).

[26] *M.S. Carpenter, W.M. Easter, T.F. Wood*, J. Org. Chem. 16, 586 (1951);
K. Ley, Eu. Müller, Chem. Ber. 89, 1402 (1956).

[27] *W. Seidenfaden, D. Pawellek* in *Houben-Weyl*, Methoden der organischen Chemie, Bd. X/1, p. 583, Georg Thieme Verlag, Stuttgart 1971.

[28] *W. Seidenfaden, D. Pawellek* in *Houben-Weyl*, Methoden der organischen Chemie, Bd. X/1, p. 584, 825, Georg Thieme Verlag, Stuttgart 1971.

[29] *P.W. Robertson*, J. Chem. Soc. (London) 93 T, 788 (1908);
G.M. Robinson, J. Chem. Soc. (London) 109, 1078 (1912);
R. v. Walther, W. Zipper, J. Prakt. Chem. [2] 91, 364 (1915);
R. v. Walther, K. Demmelmeyer, J. Prakt. Chem. [2] 92, 107 (1915).

Molten 1- and 2-naphthols react with liquid dinitrogen tetraoxide with **explosive** violence. Solutions in ether afford *2,4-dinitro-1-naphthol* (50%) and *1,6-dinitro-2-naphthol* (81%) respectively [30]. By contrast, the corresponding naphthol ethers furnish, for example, *1-methoxy-4-nitronaphthalene* and *1-ethoxy-4-nitronaphthalene* with liquid dinitrogen tetraoxide at −10 to −20°.

Acyl groups make nitration difficult; the *meta : ortho* ratio is strongly dependent on temperature [31]. *Ring-alkylated* aromatic ketones are more readily nitrated than the unsubstituted parent compounds [32], *p*-alkylated derivatives afford uniform products. Thus, nitration of 4-methylacetophenone gives *4-methyl-3-nitroacetophenone*. *C*-acylphenols are readily nitrated, severe reaction conditions bring about the entry of several nitro groups or a replacement of the acyl group by the nitro group [33], *e.g.*:

4'-Methoxy-3'-nitropropiophenone; 90%

2,4-Dinitroanisole; 64%

Anthraquinone is difficult to nitrate (see p. 9) [34], ring-alkylated anthraquinones [35] and hydroxyanthraquinones [36] are nitrated much more readily.

Because of the −I/−M effect of the carboxy group nitration of aromatic carboxylic acids is made difficult. By contrast, alkylated aromatic carboxylic acids are dinitrated with excess nitric acid. Alkali metal nitrates produce substantially only mononitration [37].

Trinitromethane (25°/pyridine) furnishes a good yield of *1-nitro-3-azulenecarboxylic acid* from 3-azulenecarboxylic acid [164].

During nitration of *hydroxybenzoic acids* the inhibiting effect of the carboxy group is entirely eliminated. The resinification that arises on nitrating phenols occurs to only a small extent or not at all [38].

Introducing several nitro groups directly into hydroxybenzoic acids can be performed without difficulty [38].

Table 3 shows that the nitration of *arylamines* is dependent on many different factors.

Nitration of *aminophenols* is readily accomplished if the amino or amino and hydroxy groups are acylated, or the hydroxy is alkylated [39]. Dinitration of *p*-aminophenols is achieved by using excess nitric acid. Both dinitration and polynitration are facilitated particularly where the two functional groups are in the *m*-position to each other. Nitration of carbonyl group containing aromatic amines takes place even under mild conditions [38].

Aromatic N-acyl and *N-alkyl amino acids* are readily nitrated [40]; *phenolic amino acids* particularly so [41].

Aromatic *sulfonic acids,* too, are easily nitrated if electron-donating substituents are present (*Houben-Weyl*, Vol. X/1, p. 635 ff.).

Azo compounds can be readily nitrated, and with the usual agents, to give preferentially *p*-substitution.

Aromatic compounds with substituents containing boron, silicon, or other heteroatoms can be fairly readily nitrated. Admittedly, a lesser or greater portion of the heteroatom may be split off (see p. 12) [42].

Furan, as a π-electron-rich heterocycle, decomposes **violently** on nitration in bulk [43]. Alkylated

[30] *J. Schmidt*, Ber. *33*, 3244 (1900).
[31] *G.T. Morgan, J.E. Moss*, J. Soc. Chem. Ind. *42*, T 461 (1923); Chem. Zentr. I, 1028 (1924);
G.T. Morgan, L.P. Walls, J. Soc. Chem. Ind. *49*, T 15 (1930); Chem. Zentr. I, 1132 (1930);
R.L. Jenkins, R.McCullough, C.F. Booth, Ind. Eng. Chem. *22*, 31 (1930).
[32] *W. Seidenfaden, D. Pawellek* in *Houben-Weyl*, Methoden der organischen Chemie, Bd. X/1, p. 610, Georg Thieme Verlag, Stuttgart 1971.
[33] *L. Szekeres, G. Fodor*, Acta Chem. Acad. Sci. Hung. *1*, 391 (1951); Chem. Zentr. 4674 (1953).
[34] *R.J. Gillespie, D.J. Millen*, Quart. Rev. Chem. Soc. *2*, 277 (1948).
[35] *V.Y. Fain, V.L. Plakidin*, Zh. Obshch. Khim. *31*, 1588 (1961); Engl.: 1476; C.A. *55*, 24699a (1961).
[36] DRP. 163 042, 170 728 (1904), Farb. Bayer; Fortschritte der Teerfabrikation und verwandter Industriezweige *8*, 268, 250 (1905); Chem. Zentr. II, 1062 (1905).

[37] *T.R. Govindachari, S. Rajappa, V. Sudarsanam*, Indian J. Chem. *1*, 247 (1963);
D. Peltier, Bull. Soc. Sci. Bretagne *31*, 7 (1956); C.A. *52*, 9016h (1958);
A. Giacolone, Gazz. Chim. Ital. *65*, 840 (1935); Chem. Zentr. I, 3137 (1936);
L. Vanscherpenzeel, Rec. Trav. Chim. Pays-Bas *20*, 149 (1901);
W. Findeklee, Ber. *38*, 3553 (1905);
M. Kerfanto, D. Raphalen, Compt. Rend. *258*, 6441 (1964).
[38] *W. Seidenfaden, D. Pawellek* in *Houben-Weyl*, Methoden der organischen Chemie, Bd. X/1, p. 629, 633, 679, Georg Thieme Verlag, Stuttgart 1971.
[39] *W. Weigel*, J. Prakt. Chem. [4] *4*, 79 (1956).

Table 3. Nitration of arylamines *(Houben-Weyl,* Vol. X/1, p. 642 ff.)

H_5C_6—R R =	Experimental conditions	Orientation of the nitro group relative to R
—NH$_2$ —NH—Alkyl —N (Alkyl)$_2$	Dissolved in at least fivefold sulfuric acid, preferably below 0°, using nitric acid or nitrating acid	Predominantly in *m*-position
—NH—Alkyl	In acid solution with sodium nitrite or dinitrogen trioxide	*p*-Orientation; *o*-position if *p*-position is occupied
—N (Alkyl)$_2$	Nitrous acid or dinitrogen trioxide	*o*- and *p*-orientation
—NH—CO—Alkyl —NH—CO—Aryl	With nitric acid With nitric acid-glacial acetic acid With nitric acid-sulfuric acid With nitric acid-acetic anhydride	*o*- and *p*-orientation *o*-Position strongly preferred
—N⟨ Acyl Alkyl	With nitric acid With nitric acid-glacial acetic acid With nitric acid-sulfuric acid	*o*- and *p*-positions preferred
—NH—SO$_2$—Aryl	Preferably in aqueous medium or in organic solvents with dilute nitric acid	*p*-Position is unambiguously preferred; if it is occupied *o*-substitution
—N$^{\oplus}$(Alkyl)$_3$	With usual nitrating agents	Predominantly *m*-position

furans are generally nitrated in acetic anhydride with concentrated or fuming nitric acid[43, 44].

Unlike xanthone, 1-hydroxyxanthone can be reacted to form *1-hydroxy-2-nitroxanthone*[45] under mild conditions.

With its aromatic character *thiophene* stands between benzene and furan and is therefore nitrated readily; thus benzo[b]thiophene gives *3-nitrobenzo[b]thiophene*[62] (50%). Selenophene behaves similarly[46].

Pyridine contains few π-electrons and hence is difficult to nitrate (see p. 9). By contrast, nitration in position 2 or 4 succeeds readily in *pyridine N-oxide.* This same technique is recommended for substituted pyridines; a strongly activating substituent makes the N-oxide detour unnecessary. Six-membered heteroaromatic compounds with two nitrogen atoms in the ring behave similarly[46].

The behavior of thiazole illustrates that an abundance of π-electrons is no unambiguous criterion of ready nitration (see p. 10).

Copper nitrate/acetic anhydride is a conservative nitrating agent used for substances which suffer substantial cleavage with conventional agents *(Houben-Weyl,* Vol. X/1, p. 768 ff.).

1.1.1.2 Reluctantly nitrated aromatic compounds

The following section discusses compounds with few π-electrons (aromatic compounds with electron-withdrawing substituents, six-membered heteroaromatic compounds) which make the attack by the nitronium cation difficult as a result of the slight basicity of the nucleus.

Nitro groups already present on the ring make an electrophilic secondary substitution difficult, including nitration. Thus, dinitration of benzene on reaction with nitrating acid takes place to

[40] *F.G. Mann, J.H. Turnbull,* J. Chem. Soc. (London) 754 (1951);
H. Rupe, Ber. *30,* 1097 (1897);
F. Reverdin, Ber. *40,* 2444 (1907);
F. Reverdin, E. Deletra, Ber. *39,* 972 (1906);
A. Einhorn, E. Uhlfelder, Justus Liebigs Ann. Chem. *371,* 176 (1909);
USP. 3 177 247 (1962), DuPont, Inv.: *J.M. Ross;* C.A. *63,* 545b (1965);
W. Theilacker, R. Hopp, Justus Liebigs Ann. Chem. *669,* 85 (1963).

[41] *J.L. Simonsen, M.G. Rau,* J. Chem. Soc. (London) *111,* 220 (1917);
C.S. Gibson, J.L. Simonsen, M.G. Rau, J. Chem. Soc. (London) *111,* 69 (1917).
[42] *R.T. Hawkins, W.J. Lennarz, H.R. Snyder,* J. Amer. Chem. Soc. *82,* 3053 (1960);
L.I. Zakharkin, V.I. Stanko, A.I. Klimova, Zh. Obshch. Khim. *35,* 394 (1965); Engl.: 393; C.A. *62,* 13165d (1965);
V.I. Stanko, A.V. Bobrov, Zh. Obshch. Khim. *35,* 2003 (1965); Engl.: 1994; C.A. *64,* 6674h (1966);
J.L. Speier, J. Amer. Chem. Soc. *75,* 2930 (1953);
F.S. Kipping, N.W. Cusa, J. Chem. Soc. (London) 1088 (1935);
F.S. Kipping, J.C. Blackburn, J. Chem. Soc. (London) 2200 (1932);
E.A. Chernyshev, M.E. Dolgaya, A.D. Petrov, Izv. Akad. Nauk SSSR 1425 (1960); C.A. *55,* 1424h (1961);
Fr. P. 1 205 994 (1958); US. P. 3 020 302 (1959), Union Carbide Corp., Erf.: *D.L. Bailey, R.M. Pike;* Chem. Zentr. 6586 (1962); 21350 (1963);
Fr. P. 1 303 018 (1961), Union Carbide Corp; C.A. *58,* 9137h (1963);
T. Hashimoto, J. Pharm. Soc. Jap. *80,* 730 (1960); C.A. *54,* 24480i (1960).

only a subordinate extent. By choosing more energetic nitration conditions, for instance, doubling of the sulfuric acid and a reaction temperature of just above 100°, dinitration of the benzene can be obtained in one step[47]. Further nitration of *m*-dinitrobenzene to *1,3,5-trinitrobenzene* (71%) succeeds by prolonged heating to 110° with a mixture of 60% oleum and anhydrous nitric acid[48].

The in part marked electronegativeness of the halogens makes the nitration of *halogenated aromatic compounds* difficult by comparison to that of the unsubstituted parent compound (in the ground state −I effect > +M effect). Practically no difference is observed during nitration with acetyl nitrate.

However, as in the positively charged δ-complex the +M effect of the halogens exceeds their −I effect they direct the entering nitro group into the *o*- and *p*-positions.

Dinitration of halobenzenes can be achieved by more energetic conditions and the use of appropriate amounts of nitrating agent. Predominantly the *2,4-dinitro-1-halo compound* and a little *2,6-dinitro-1-halo compound* are obtained (*Houben-Weyl*, Vol. X/1, p. 504).

During nitration of halobenzenes with dinitrogen tetroxide the yield increases as the halogen concerned is found higher up in the periodic table[49].

The more drastic conditions required for mononitrating tri- and tetrahalobenzenes consist in using either a large excess of concentrated nitric acid (up to 100%) or close to anhydrous nitrating acid with almost molar amounts of nitric acid at slightly elevated temperature. Dinitration of trifluorinated benzene also requires fuming nitric acid and sulfuric acid. A mixture of boron trifluoride and nitric acid is recommended for nitrating aromatic polyfluoro compounds[162].

While monoalkyl aromatic compounds are readily nitrated the entering nitro group inhibits the further reaction. 2,4,5- and 2,3,4-trinitrotoluene do not afford the tetranitro compound even with oleum-containing nitric acid at 140°[50].

Free positions in alkylbenzenes with multiple halogen atoms on the ring are deactivated in dependence on the number and type of the halogen; nitration requires more energetic conditions[51]. Halogen atoms in the side-chain as well as on the ring are an additional hindrance.

5-Bromo-1,3-di-tert-butylbenzene is relatively resistant to attempted nitration; it requires 100% nitric acid in acetic acid/sulfuric acid and elevated temperature to give *6-bromo-2,4-di-tert-butyl-1-nitrobenzene*. Twice as much acid and more drastic conditions afford *2-bromo-4,6-di-tert-butyl-1,3-dinitrobenzene*[52].

As a second order substituent the carbonyl function makes nitration more difficult because of its −M effect and the nitro group is directed into the *m*-position. Although this general property presupposes drastic nitration conditions, aldehydes of the benzene series are nitrated with pure nitric acid or anhydrous nitrating acid at below 10° in order to avoid oxidation to carboxylic acid[53]. Ring-halogenated aromatic aldehydes require at least the same strong nitrating agents (*Houben-Weyl*, Vol. X/1, p. 601).

[43] *R. Marquis*, Ann. Chim. (Paris) [8] 4, 196 (1905); Bull. Soc. Chim. France [3] 29, 276 (1903);
J.J. Rinkes, Rec. Trav. Chim. Pays-Bas 49, 1118, 1169 (1930); 50, 590 (1931);
B.T. Freure, J.R. Johnson, J. Amer. Chem. Soc. 53, 1142 (1931).
[44] *H.B. Hill, G.R. White*, Amer. Chem. J. 27, 193 (1902);
K. Kawabe, T. Suzui, M. Iguchi, J. Pharm. Soc. Jap. 80, 62 (1960); C.A. 54, 12091f (1960);
Fr. P. 1330332 (1962), Italfarmaco S.p.A.; C.A. 60, 2894e (1964); 59, 3895f (1963);
Jap. P. 62: 15635, 10688, 10689 (1959), Inv.: *R. Ueno*; C.A. 59, 9986d (1963);
USSR. P. 130045 (1959), Inv.: *S. Hillers, K. Venters, N. Saldábols, R. Yu. Kalnberg*; C.A. 55, 5530g (1961);
K. Venters, S. Hillers, Dokl. Akad. Nauk SSSR 137, 83 (1961); C.A. 55, 19907b (1961);
Y. Arata, T. Obashi, K. Aoki, M. Koseki, K. Sakai, J. Pharm. Soc. Jap. 76, 211 (1956); C.A. 59, 3895 (1963);
K. Venters, S.P. Korshunov, L.I. Vereshchagin, R.L. Bol'shedvorskaya, D.O. Lolya, S. Hillers, Khim. Geterotsikl. Soedin. 616 (1965); C.A. 64, 5028d (1966).
[45] *H. Haase*, J. Prakt. Chem. [4] 20, 161 (1963).

[46] *W. Seidenfaden, D. Pawellek* in *Houben-Weyl*, Methoden der organischen Chemie, Bd. X/1, p. 704, 706, 732, Georg Thieme Verlag, Stuttgart 1971.
[47] *H. Goldschmidt*, Chemiker-Ztg. 37, 642 (1913).
[48] *L. Desvergnes*, Chim. Ind. (Paris) 25, No. 3, 291 (1931);
L.G. Radcliffe, A.A. Pollitt, J. Soc. Chem. Ind. 40, 45 T (1921).
[49] *A. Schaarschmidt, H. Balzerkiewicz, J. Gante*, Ber. 58, 499 (1925).
[50] *P. Debeule*, Bull. Soc. Chim. Belg. 42, 27 (1933); Chem. Zentr. I, 2675 (1933).
[51] *W. Seidenfaden, D. Pawellek* in *Houben-Weyl*, Methoden der organischen Chemie, Bd. X/1, p. 552, Georg Thieme Verlag, Stuttgart 1971.
[52] *W. Rundel*, Chem. Ber. 96, 636 (1963).
[53] *O. Bayer* in *Houben-Weyl*, Methoden der organischen Chemie, Bd. VII/1, p. 406, Georg Thieme Verlag, Stuttgart 1954.

Anthraquinone is even more difficult to nitrate than benzoic acid or nitrobenzene[54] (it substitutes preferentially in the 1-position). Phenanthraquinone reacts more easily (*Houben-Weyl,* Vol. X/1, p. 614).

Although the nitration of aromatic carboxylic acids requires more drastic conditions, the use of excess nitrating agent at 185° leads to dinitration, *e.g.,* formation of *3,5-dinitrobenzoic acid*[55] from benzoic acid. Aromatic dicarboxylic acids require even more drastic conditions. Nitration of ring-halogenated benzoic acids ensues under approximately the same conditions as that of the non-halogenated parent compound. Di- and polynitration requires very energetic conditions which are avoided by using alkali metal nitrates (*Houben-Weyl,* Vol. X/1, p. 624).

Phenolic monocarboxylic acids are readily nitrated; using energetic conditions leads to replacement of the carboxy group by nitro. The nitro group enters particularly easily into alkyl- or halogen-substituted hydroxybenzoic acids (*Houben-Weyl,* Vol. X/1, p. 629ff.).

Of *O*-heteroaromatic compounds xanthone is difficult to nitrate because of the deactivating effect of the carbonyl function.

Six-membered heteroaromatic compounds containing nitrogen as the heteroatom contain few π-electrons. The consequent relatively low electron density of the C-2, C-4, and C-6 atoms makes electrophilic substitution difficult[56]. Pyridine is comparable with dinitrobenzene in its behavior toward nitrating agents. Drastic methods such as the use of anhydrous mixed acid and temperatures as high as 300° are required to bring about reaction to *3-nitropyridine* (5%)[57]. A preliminary oxidation to *pyridine N-oxide* facilitates the nitration (see p. 8) and affords *2-nitropyridine N-oxide* (8%) and *4-nitropyridine N-oxide* (72%)[60]. 2,2'-Bipyridyl is another compound that can be converted into *4,4'-dinitro-2,2'-bipyridyl via* the *N*-oxide[58].

Quinoline is nitrated only on the nitrogenfree ring with nitric acid[59], the nitro group enters both rings

as a function of the temperature when quinoline *N*-oxide is used[60]. At 65–70° 67% *4-nitroquinoline 1-oxide* and at 0–10° 5- and *8-nitroquinoline 1-oxides* are obtained.

5-Nitroquinoline; 35%

+

8-Nitroquinoline; 43%

With benzoyl nitrate partial oxidation occurs[61]

78%

+

7%

As has been shown, the substituent does not enter the heteroaromatic ring but the benzene ring during electrophilic substitution of condensed six-membered *N*-heteroaromatic compounds.

A lower π-electron density than benzene also characterizes heteroaromatic compounds with two nitrogen atoms. Attempts to nitrate pyrimidine, pyridazine, and pyrazine were therefore fruitless[63] and it required the presence of electron-donating substituents or the use of the *N*-oxides to bring about a satisfactory reaction (*Houben-Weyl,* Vol. X/1, p. 732ff.).

Oxadiazoles are among the heteroaromatic compounds with few π-electrons. Nonetheless, nitration of, *e.g.,* 1,2,4-oxadiazoles, even when the 3- and 5-positions are free, does not take place. The same is true of the sydnones. As a result phenyl-

[54] *R.J. Gillespie, D.J. Millen,* Quart. Rev. Chem. Soc. *2,* 277 (1948).
[55] DDRP. 34329 (1961), Erf.: *H. Cassebaum;* C.A. *63,* 13161g (1965).
[56] *A. Albert,* Chemie der Heterocyclen, p. 42, Verlag Chemie, Weinheim/Bergstr. 1962.
[57] *F. Friedl,* Ber. *45,* 428 (1912); *E. Ochiai, K. Arima, M. Ishikawa,* J. Pharm. Soc. Jap. *63,* 79 (1943); C.A. *45,* 5151 (1951).
[58] *F.H. Case,* J. Org. Chem. *27,* 640 (1962); *I. Murase,* J. Pharm. Soc. Jap. *77,* 682 (1956); C.A. *52,* 9100 (1958).
[59] *W. Seidenfaden, D. Pawellek* in *Houben-Weyl,* Methoden der organischen Chemie, Bd. X/1, p. 720, Georg Thieme Verlag, Stuttgart 1971; Re the preparation of *4-nitroisoquinoline* (13%) from isoquinoline/nitric acid/acetic anhydride (100°/1 hr.) s. *J.W. Bunting, W.G. Meathrel,* Org. Prep. Proc. Int. *4,* 9 (1972).

1,2,4-oxadiazole is nitrated on the phenyl ring[64, 65]:

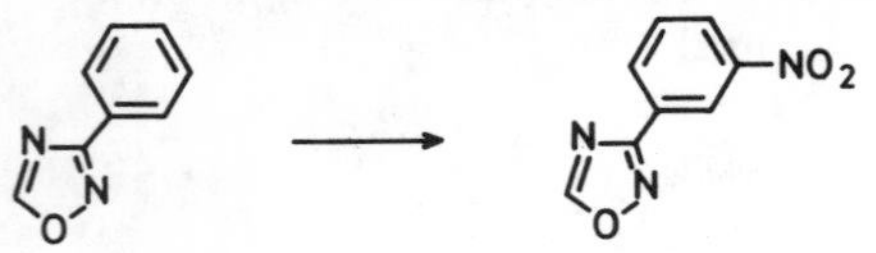

1-Phenylpyrazole behaves analogously. While here nitrating acid produces substitution in the aromatic ring[65, 66], acetyl nitrate at −5° affords *4-nitro-1-phenylpyrazole:*

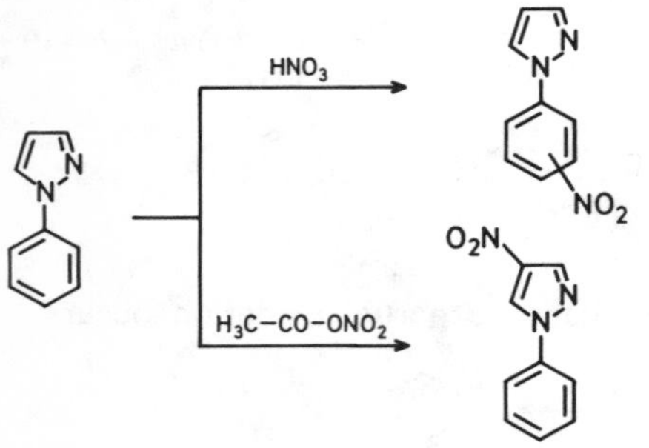

1,3-Thiazoles, too, although rich in π-electrons, are difficult to nitrate. Thus, 2-alkyl-1,3-thiazoles are hardly attacked, while 4-alkyl-1,3-thiazoles require elevated temperatures for success. 2-Aryl-1,3-thiazoles are nitrated on the aryl ring[67]. Thiadiazoles display an analogous reluctance to undergo nitration and only benzo-1,2,3-thiadiazoles are stable enough to produce a successful reaction (*Houben-Weyl*, Vol. X/1, p. 752 ff.).

1.1.1.3 Special nitrating agents

Nitrating benzene with nitric acid or nitrating acid affords mainly *nitrobenzene*[68]; dinitration is observed only when drastic conditions are employed. Iron(III) nitrate in acetic anhydride at 80° and sodium nitrate in hydrogen fluoride at 0°, too, convert benzene into nitrobenzene[67]. Calcium, strontium, and barium nitrate produce dinitration[67]. With nitronium tetrasulfate ($4SO_3$; N_2O_5, H_2O) benzene reacts vigorously to form nitrobenzene plus benzenesulfonic acid. Nitryl fluoride does not produce dinitration of benzene.

Perdeuterated benzene can be nitrated with dinitrogen pentoxide[69] in 56% yield or with nitrating acid (72%).

1.1.1.4 Side-reactions during nitration

The conditions under which nitration ensues and the properties of the substances being nitrated sometimes allow a number of side-reactions to occur.

Oxidation reactions are the principal reactions involved. For example, naphthacene is oxidized to *naphthacenequinone* by the action of nitric acid[70]. Phenols should be nitrated in the form of their esters or ethers because of possible oxidations (cf. p. 4 ff.). Aromatic aldehydes are usually not oxidized if working is performed at below 10° in anhydrous medium[71]. Using concentrated nitric acid is advisable also during nitration of aromatic alde-

[60] *E. Ochiai*, J. Org. Chem. *18*, 536, 549 (1953).

[61] *E. Ochiai, C. Kaneko*, Pharm. Bull. (Tokyo) *5*, 56 (1957); *7*, 191, 267 (1959).
W. Seidenfaden, D. Pawellek in *Houben-Weyl*, Methoden der organischen Chemie, Bd. X/1, p. 762, Georg Thieme Verlag, Stuttgart 1971.

[62] *K. Fries, E. Hemmecke*, Justus Liebigs Ann. Chem. *470*, 1 (1929);
K. Fries, H. Heering, E. Hemmecke, G. Siebert, Justus Liebigs Ann. Chem. *527*, 83 (1937);
D.A. Shirley, M.J. Danzig, F.C. Canter, J. Amer. Chem. Soc. *75*, 3278 (1953);
D.E. Boswell, J.A. Brennan, P.S. Landis, P.G. Rodewald, J. Heterocycl. Chem. *5*, 69 (1968).

[63] *S. Dixon, L.F. Wiggins*, J. Chem. Soc. (London) 3236 (1950);
B. Lythgoe, L.S. Rayner, J. Chem. Soc. (London) 2323 (1951).

[64] *G. Palazo, G. Corsi*, Gazz. Chim. Ital. *93*, 1196 (1963).

[65] *C. Moussebois, F. Eloy*, Helv. Chim. Acta *47*, 838 (1964).

[66] *I.L. Finar, R.J. Hurlock*, J. Chem. Soc. (London) 3024 (1957).

[67] *W. Seidenfaden, D. Pawellek* in *Houben-Weyl*, Methoden der organischen Chemie, Bd. X/1, p. 744, 748, 768, Georg Thieme Verlag, Stuttgart 1971.

[68] *H.E. Fierz-David, L. Blangey*, Grundlegende Operationen der Farbenchemie, 8. Aufl., p. 67, Springer Verlag, Wien 1952;
G. Schultz, J. Flachsländer, J. Prakt. Chem. [2] *66*, 156 (1902);
V. Migrdichian, Organic Synthesis, Vol. II, Chap. Aromatic Nitro and Nitroso Compounds, p. 1597, Reinhold Publishing Co., New York 1957;
K.J. Kobe, J.J. Mills, Ind. Eng. Chem. *45*, 287 (1953);
BIOS Final Rep. *1144*, 2; 3;
F.H. Cohen, J.P. Wibaut, Rec. Trav. Chim. Pays-Bas *54*, 409 (1935);
J.H. Simons, H.J. Passino, S. Archer, J. Amer. Chem. Soc. *63*, 608 (1941);
A.A. Spryskov, I.K. Barvinskaya, Zh. Obshch. Khim. *33*, 1885 (1963); Engl.: 1835; C.A. *59*, 11221b (1963).

[69] *A. Rieker, P. Ziemek*, Z. Naturforsch. *20b*, 640 (1965);
C. Bernasconi, W. Koch, H. Zollinger, Helv. Chim. Acta *46*, 1184 (1963).

[70] *S. Gabriel, E. Leupold*, Ber. *31*, 1277 (1898).

[71] *W. Seidenfaden, D. Pawellek* in *Houben-Weyl*, Methoden der organischen Chemie, Bd. VII/1, p. 406, Georg Thieme Verlag, Stuttgart 1954.

hydes containing two hydroxy groups. When nitrating acid is allowed to act on monohydroxy-benzaldehydes the formyl group may be exchanged against the nitro group. 3,4-Dihydroxy-benzaldehyde is completely oxidized by nitric acid (*Houben-Weyl*, Vol. X/1, p. 829/602).

Aromatic diamines and triamines, as free bases, are sensitive to oxidation. Free *o*- and *p*-aminophenols are oxidized to *quinones;* this is not possible with *m*-aminophenols (*Houben-Weyl*, Vol. X/1, pp. 643, 670).

In heterocycles containing a sulfur atom sulfoxide formation can arise in addition to nitration, *e.g.*[72]:

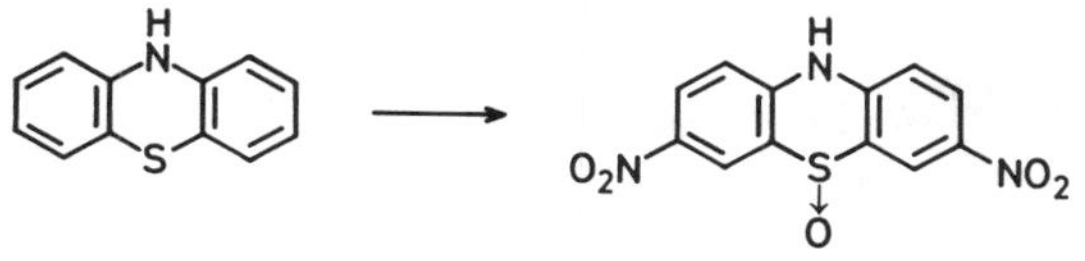

Sulfur atoms outside the ring may be split off during the oxidation[73], *e.g.:*

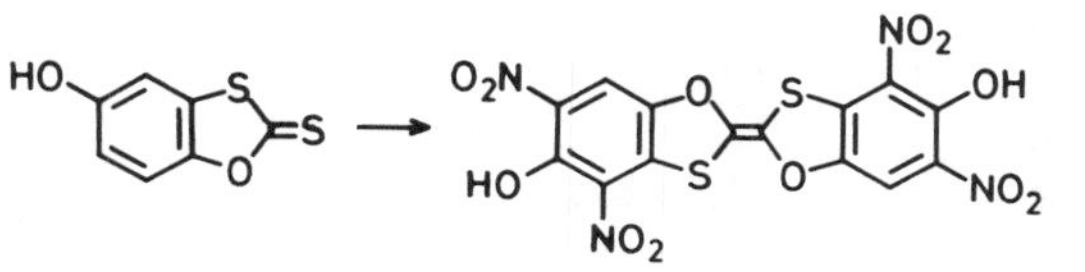

In the following pyrimidine derivatives oxidative cleavage of the heteroaromatic compound is observed[74]:

In addition to nitration an atom or a part of the molecule may be displaced. For a detailed description of the *interchange reactions,* above all in respect of their preparative utilization, see pp. 10, 11. *Decarboxylation* is another feature that is encountered in individual cases (see below).

Certain starting compounds are *sensitive to acid* (*e.g.*, indole, pyrrole) or polymerize in acid medium. Under these circumstances (*e.g.*, with pyrrole) the reaction is performed in acetic anhydride as solvent or, if the benzene ring is to be nitrated, the hetero ring is previously reduced (*e.g.*, indole)[75]. In neutral or alkaline medium alkyl nitrates are required to bring about a reaction. The acid-sensitive cyclopenta[b]thiapyran is nitrated with tetranitromethane[76].

[72] *J.S. Driscoll, R.H. Nealey*, J. Heterocycl. Chem. *2*, 272 (1965).
[73] *H. Fiedler*, Chem. Ber. *95*, 1771 (1962).
[74] *S. Gronowitz, A. Hallberg*, Acta Chem. Scand. *21*, 2296 (1967).

Other side-reactions may occur in addition to oxidation and polymerization, so that the desired product can be obtained only indirectly. For example, nitration of aminophenol in acetic anhydride proceeds anomalously because the solvent acetylates the amino group and the directing effect of the acylamino group is different from that of a free amino group (*Houben-Weyl*, Vol. X/1, p. 643). Another example of a 'detour reaction' is the nitration of furfural, whose aldehyde group needs to be converted into the acetal before carrying out a nitration in acid medium (*Houben-Weyl*, Vol. X/1, p. 691).

1.1.2 Replacement of functional groups

Under certain conditions it is possible to substitute nitro for halogen, alkyl, alkoxy, formyl, acyl or carboxy on aromatic compounds including during nitration of aromatic compounds with sulfo and amino groups. Often such exchange reactions represent undesired features of the reaction.

In aromatic halogen compounds the firmness[77] of the halogen atom bond is in the order Cl ≫ Br > I. A requisite for the replacement of halogen by NO_2 is the presence of further first order substituents. Displacement of the halogen may be accompanied by a simultaneous nitration of a different site of the molecule, so that dinitro compounds are formed, *e.g.*[78]:

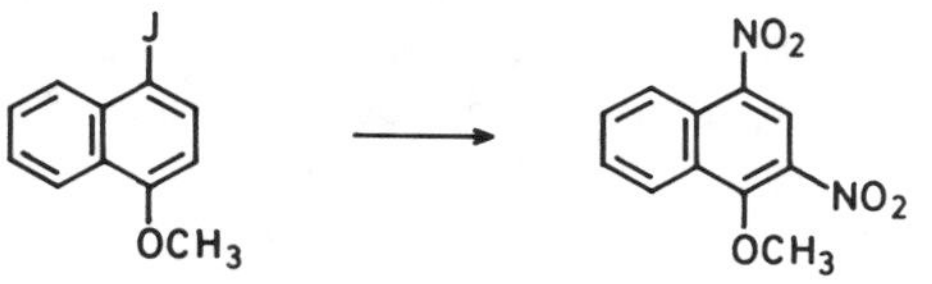

During replacement of alkyl by nitro the following decreasing tendency for the exchange to occur is observed: *tert* > *sec* > *prim*. A hydrogen atom may be replaced in addition to the alkyl group[79], *e.g.:*

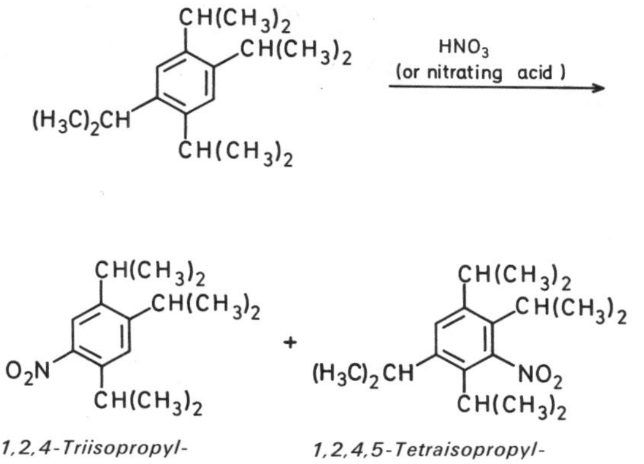

1,2,4-Triisopropyl-
5-nitrobenzene; 83%

1,2,4,5-Tetraisopropyl-
3-nitrobenzene; 15%

In dependence on the reaction conditions a hydrogen atom may be replaced at the same time as the formyl group in analogy to the behavior of alkylbenzenes. Some benzoic acids display the same behavior. Displacement of acyl groups is very much dependent on the reaction conditions (*Houben-Weyl*, Vol. X/1, p. 830f.).

Benzenesulfonic acids undergo replacement only if the sulfo group is activated by first order substituents; here again additional replacement of a hydrogen atom can take place, e.g.[80]:

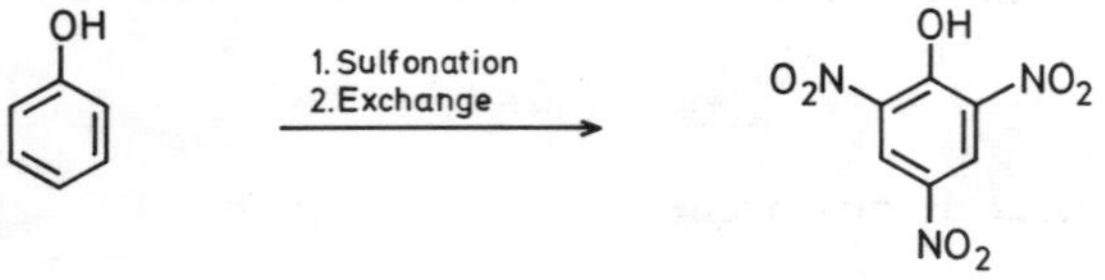

2,4,6-Trinitrophenol;
77% (Picric acid)

Exchanging amino against nitro *via* diazo is known as the Nitro-Sandmeyer reaction[81]:

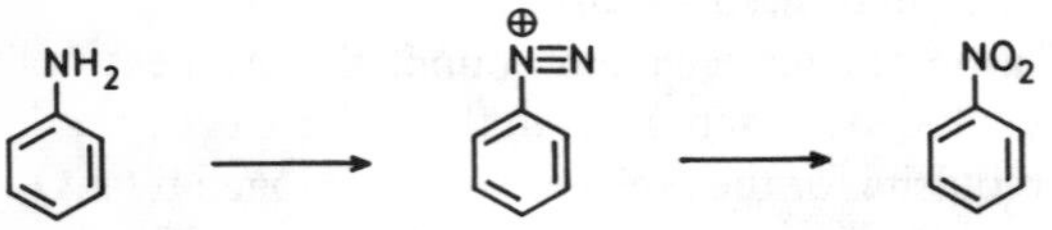

The replacement is performed in bulk (**Caution**) or in solution using alkali nitrites and, optionally, copper catalysts[81].

Direct and indirect conversion of an aromatic amine into the corresponding aromatic nitro compound is always advisable for positions which are difficult or impossible to fill by direct nitration of the aromatic compound. Substituents containing silicon and mercury, too, can be replaced[82].

In an interesting reaction *3-nitroisoxazoles* (30–63%)[183] are obtained from 1-bromo-2-alkynes and sodium nitrite in dimethylformamide by substitution, addition, and cyclization.

1.2 By oxidation
(cf. *Houben-Weyl*, Vol. X/1, pp. 843–854)

Oxidation is used for preparing aromatic nitro compounds mainly where usual nitration techniques do not lead to the desired aim.

In addition to *aromatic amines*, *N-arylhydroxylamines*, *quinone oximes*, and *aromatic nitroso* compounds are submitted to oxidation.

Aromatic amines are oxidized to the corresponding *nitro compounds* by peroxy acids:

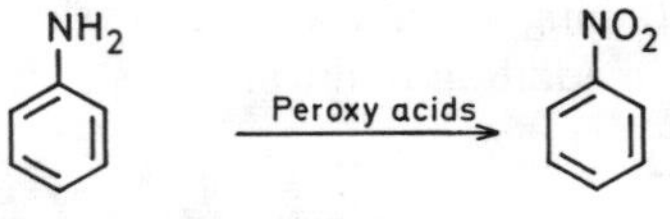

Thus, high yields of *dinitrobenzenes* are obtained on oxidizing the corresponding nitroanilines with ammonium peroxydisulfate in sulfuric acid in the presence of silver salts[83]. Caro's acid is less suitable as oxidizing agent and its possible use is restricted to the pyridine series.

Peroxyacetic acid, too, can be used for oxidizing amines to nitro compounds at the boiling point. Among other factors the position of the amino group in the molecule determines how easily it is oxidized to the nitro group.

Peroxytrifluoroacetic acid and peroxymaleic acid are other suitable oxidizing agents.

Exhaustive oxidation of *N*-arylhydroxylamines also leads to *nitro* compounds[84], e.g.:

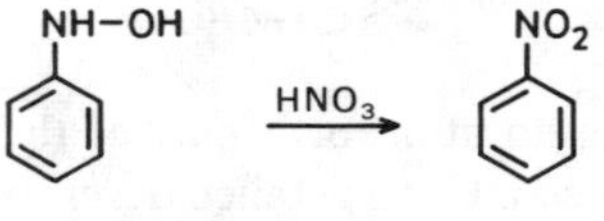

The reaction is of subordinate preparative importance because, in general, *N*-arylhydroxylamines are prepared by partial reduction of nitro compounds in the reverse reaction. Nitric acid, excess permanganate, atmospheric oxygen, or hydrogen peroxide are the agents used in the exceptional cases where this technique can be recommended.

[75] *W. Seidenfaden, D. Pawellek* in *Houben-Weyl*, Methoden der organischen Chemie, Bd. X/1, p. 706, 789, Georg Thieme Verlag, Stuttgart 1971.
[76] *R. Mayer, J. Franke*, J. Prakt. Chem. [4] *30*, 262 (1965).
[77] *W. Seidenfaden, D. Pawellek* in *Houben-Weyl*, Methoden der organischen Chemie, Bd. X/1, p. 821, Georg Thieme Verlag, Stuttgart 1971.
[78] *N. Campbell, N.M. Hasan*, Chem. & Ind. (London) 2056 (1966).
[79] *A. Newton*, J. Amer. Chem. Soc. *65*, 2434 (1943).
[80] DRP. 298 021 (1915);
H. Kast, Spreng- und Zündstoffe, p. 237, Friedr. Vieweg u. Sohn, Braunschweig 1921;
R. King, J. Chem. Soc. (London) *119*, 2105 (1921);
M. Marqueyrol, P. Carre, P. Loriette, Bull. Soc. Chim. France [4] *27*, 140 (1920).

[81] *W. Seidenfaden, D. Pawellek* in *Houben-Weyl*, Methoden der organischen Chemie, Bd. X/1, p. 836, Georg Thieme Verlag, Stuttgart 1971;
K. Holzach, Die aromatischen Diazoverbindungen, p. 214, Ferdinand Enke Verlag, Stuttgart 1947;
K.H. Saunders, The Aromatic Diazocompounds, 2. Ed., p. 293, E. Arnold & Co., London 1949;
E. Profft, Chemiker-Ztg. *74*, 455 (1950).
[82] *F.B. Deans, C. Eaborn*, J. Chem. Soc. (London) 498 (1957);
F.S. Kipping, J.C. Blackburn, J. Chem. Soc. (London) 2205 (1932);
Y. Ogata, M. Tscuchida, J. Org. Chem. *21*, 1065 (1956).
[83] *O.N. Witt, E. Kopetschni*, Ber. *45*, 1134 (1912).

Aromatic nitroso compound can be oxidized to *nitro* compounds analogously to the *N*-arylhydroxylamines[84]:

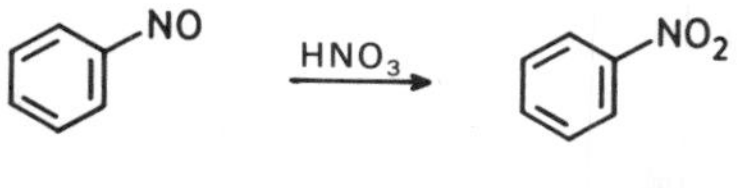

Alkaline hydrogen peroxide and peroxytrifluoroacetic acid can serve as the oxidizing agent in addition to nitric acid[84].

1.3 *N*-Nitration of aromatic compounds

During nitration of ring-alkylated aromatic monoamino compounds the amino group may be attacked in addition to the ring if the nitric acid used contains nitrogen oxides[86].

On reacting a solution of 2, 3, 5, 6-tetrachloraniline in glacial acetic acid with a twentyfold excess of nitric acid a 93% yield of *2, 3, 5, 6-tetrachloro-N, 4-dinitroaniline* is obtained[85]:

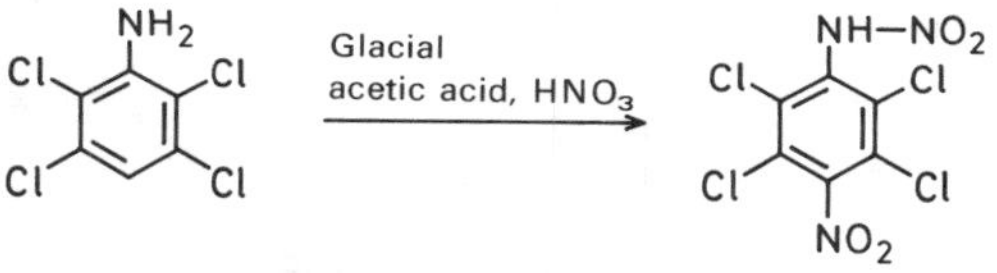

When nitrous gases or nitrous acid act on free aromatic amines arylnitramines are formed as well as other products[86]. Nitrating *N, N*-dimethylaniline in concentrated nitric acid causes replacement of a methyl group in addition to ring nitration with formation of *N-methyl-N, 4-dinitroaniline*[87]. Reacting pyridine bases or 5-nitroquinoline with nitronium tetrafluoroborate in acetonitrile affords *N-nitropyridinium* and *N-nitroquinolinium salts*[88]. *N*-Nitroarylamines themselves generally rearrange to *o-nitroarylamines* under acid catalysis (*e.g., N*-nitroaniline to *2-nitroaniline;* 93% of theory)[89]. This rearrangement is important only among heteroaromatic compounds; thus, *4-amino-5-bromo-3-nitropyridine* (96%) results from 5-bromo-4-(nitroamino)pyridine[90].

1-Nitropyrazole, prepared from pyrazole/nitric acid/acetic anhydride (to 84%), rearranges to *5-nitropyrazole* (200°)[166].

1.4 Transformation of aromatic nitro compounds

1.4.1 Reduction of aromatic nitro compounds

The importance of aromatic nitro compounds for organic synthesis is demonstrated by the possible reduction to a large number of aromatic nitrogen compounds (cf. the following scheme).

[84] *W. Seidenfaden, D. Pawellek* in *Houben-Weyl*, Methoden der organischen Chemie, Bd. X/1, p. 849, Georg Thieme Verlag, Stuttgart 1971.

[85] *A.T. Peters, F.M. Rowe, D.M. Stead*, J. Chem. Soc. (London) 372 (1943).

[86] *E. Bamberger*, Ber. *27*, 584 (1894); *E. Bamberger, E. Hoff*, Justus Liebigs Ann. Chem. *311*, 91 (1900); *H. Glaser* in *Houben-Weyl*, Methoden der organischen Chemie, Bd. XI/1, p. 117, Georg Thieme Verlag, Stuttgart 1957.

[87] *J. Glazer, E.D. Hughes, C.K. Ingold, A.T. James, G.T. Jones, E. Roberts*, J. Chem. Soc. (London) 2657 (1950).

[88] *C.A. Cupas, R.L. Pearson*, J. Amer. Chem. Soc. *90*, 4742 (1968).

[89] *W.N. White, J.T. Jolden*, Chem. & Ind. 138 (1962); *H.J. Shine*, Aromatic Rearrangements, p. 235, Elsevier Publ., Co., Amsterdam, London, New York 1967.

[90] *J.S. Wieczorek, T. Talik*, Roczniki Chem. *36*, 967 (1962); see a. *K. Lewicka, E. Plazek*, Roczniki Chem. *39*, 643 (1965).

[91] *W. Seidenfaden, D. Pawellek* in *Houben-Weyl*, Methoden der organischen Chemie, Bd. X/1, p. 875, Georg Thieme Verlag, Stuttgart 1971; *J. Becher*, Acta Chem. Scand. *26*, 1659 (1972); *A.F.M. Iqbal*, Helv. Chim. Acta. *55*, 798 (1972); *K. Srinivasan, K.G. Srinivasan, K.K. Balasubramanian, S. Swaminathan*, Synthesis 313 (1973); *T. Nishiwaki, T. Takahashi*, Synthesis 363 (1973).

[92] *W.E. Noland, J.H. Sellstedt*, J. Org. Chem. *31*, 345 (1966).

[93] *M.S. Newman, J. Blum*, J. Amer. Chem. Soc. *86*, 5600 (1964); *M.J.S. Dewar, A.R. Lepley*, J. Amer. Chem. Soc. *83*, 4560 (1961); *G. Briegleb*, Elektronen-Donator-Acceptor-Komplexe, Springer Verlag, Berlin, Göttingen, Heidelberg 1961; *G. Briegleb, J. Czekalla*, Angew. Chem. *72*, 401 (1960); *S.P. McGlynn*, Chem. Rev. *58*, 1113 (1958); *L.J. Andrews*, Chem. Rev. *54*, 713 (1954); *T.N. Hall, C.F. Poranski*, 'Polynitroaromatic Additions Compounds' in *H. Feuer*, The Chemistry of the Nitro and Nitroso Groups, Part 2, p. 329, Interscience Publishers, New York 1970; USP. 2 941 017 (1955), Standard Oil Co., Inv.: *F. Veatch, R.W. Foreman, J.A. Gecsy;* Chem. Zentr. 9898 (1962); USP. 2 941 019 (1956), Standard Oil Co., Inv.: *R.W. Foreman, A.F. Miller;* Chem. Zentr. 16846 (1962); *W. Seidenfaden, D. Pawellek* in *Houben-Weyl*, Methoden der organischen Chemie, Bd. X/1, p. 432, 440, Georg Thieme Verlag, Stuttgart 1958; *M.R. Crampton, V. Gold*, Chem. Commun. 256 (1965).

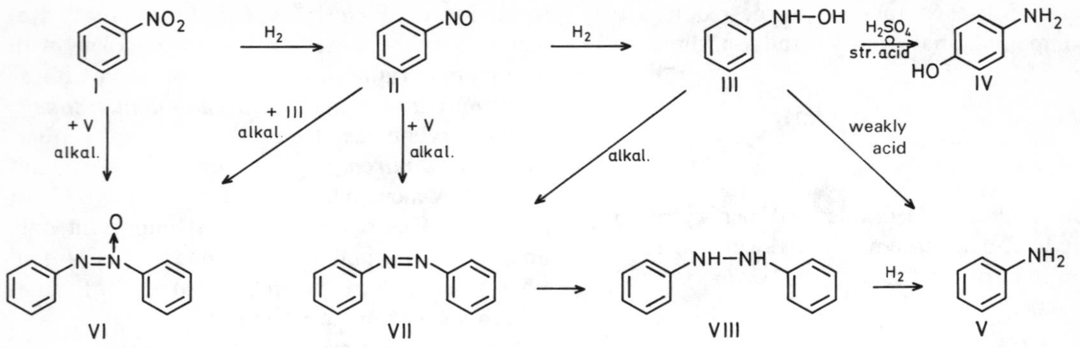

The pathway taken by the reduction is dependent on the reducing agent and the conditions (cf. the corresponding chapters in this volume).

1.4.1.1 Reduction without ring closure

The open-chain compounds formed on reducing nitro compounds are discussed in detail in *Houben-Weyl*, Methoden der organischen Chemie.

Table 4. Transformation of the nitro group by reduction

Reduction product	Houben-Weyl	
	Volume	Page
Arylhydroxylamines	X/1	1138
Arylamines	XI/1	360
Arylazoxy compounds	X/3	745
Diarylazo compounds	X/3	213
Hydrazo compounds	X/2	697
Aromatic nitroso compounds	X/1	1063

Nitrobenzene is used as an oxidizing agent in organic syntheses, *e.g.,* for the Skraup quinoline synthesis.

1.4.1.2 Reduction accompanied by ring closure

Where suitable substituents are present both intra- and intermolecular heterocyclic systems are formed during reduction of aromatic nitro compounds[91], *e.g.*[92]:

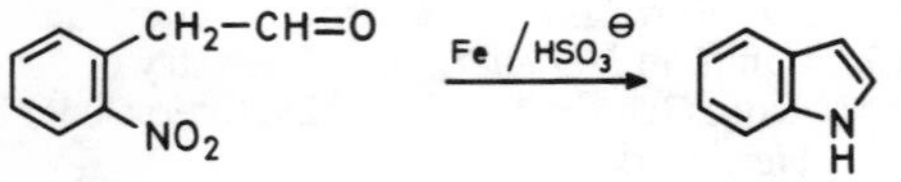

The reduction of suitable aromatic nitro compounds with trivalent organophosphorus compounds, too, offers a convenient and preparatively fertile access to many heterocycles[169].

1.4.2 Replacement of the nitro group

Exchange reactions of aromatic nitro groups can serve for preparing a variety of substance classes.

Table 5. Displacement of the nitro group by other substituents

Displacement of the nitro group by	Houben-Weyl	
	Volume	Page
Fluorine	V/3	166
Chlorine	V/3	750, 922
Bromine	V/4	296, 436
Compounds formed by interchange		
Alkyl ethers	VI/3	76
Diaryl disulfides	IX	65
Diaryl sulfides	IX	103
Aryl sulfides	IX	113
Arylsulfonic acids	IX	524
Arylamino compounds	XI/1	236

1.4.3 Charge transfer complexes

During reaction of certain aromatic nitro compounds with aromatic hydrocarbons electron acceptor-donor complexes (charge transfer complexes) may arise which can be employed for, *e.g.,* separating and identifiying the hydrocarbons. For details see ref.[93].

1.4.4 Photoreactions

Aromatic nitro compounds undergo photoreactions with light of various wavelengths.

In some cases an *aci*-nitro compound is postulated to be formed intermediately; the same compound is held to be responsible for the photochromism of certain aromatic nitro compounds[94].

1.5 Preparation of aliphatic *C*-nitro compounds

The entry of a nitro group into an aliphatic substituent group entails a far-reaching effect. Firstly, the

[94] *J. Weinstein, A.L. Bluhm, J.A. Sousa,* J. Org. Chem. *31,* 1983 (1966);
A.L. Bluhm, J. Weinstein, J.A. Sousa, J. Org. Chem. *28,* 1989 (1963);
J.A. Sousa, J. Weinstein, J. Org. Chem. *27,* 3155 (1962).

—I effect causes aliphatic nitro compounds to display a sometimes appreciably strong *CH* acidity (see p. 20); secondly, primary and secondary nitro compounds display nitro-*aci*-nitro tautomerism:

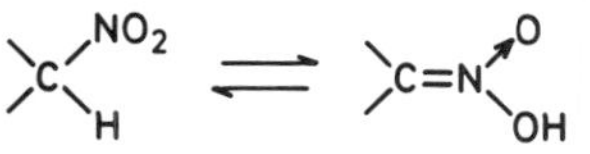

Normally the equilibrium is practically entirely on the side of the nitro compound, but stabilization of mesomerism of the *aci*-nitro form can favor the latter to make it present in the tautomerism equilibrium mixture in a measurable concentration.
Aliphatic nitro compounds are isomeric with the corresponding esters of nitrous acid, so that both substance classes can form side by side in certain S_N reactions of aliphatic halides in dependence on the nature of the reaction components (cf. p. 17).
On reacting dinitrogen tetroxide with olefins vicinal *nitronitrosooxyalkanes (nitronitrites)* are obtained (*Houben-Weyl*, Vol. X/1, p. 76)

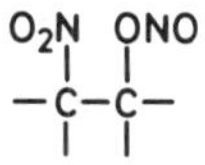

which can be isolated as nitro alcohols (following hydrolysis). Chloroalkyl nitrites are formed as side-products during addition of nitryl chloride to olefins[162] (for the rearrangement of nitrites to nitro compounds, see ref. [180]). During exchange reactions involving halogen atoms nitrites are formed along with nitro compounds (see p. 17).
Various different methods are available for preparing aliphatic nitro compounds. The ease with which the hydrogen is replaced is governed by its activation by other functional groups.

1.5.1 By substitution

1.5.1.1 Replacement of hydrogen

Replacement of a hydrogen atom in the alkane by a nitro group with pure nitric acid probably proceeds by a free-radical mechanism[95] (the paramagnetic monomeric NO_2 radical is considered to be the nitrating agent):

$$R-H + \circ NO_2 \longrightarrow R\circ + HNO_2$$
$$R\circ + \circ NO_2 \longrightarrow R-NO_2$$

The influence of the carbon skeleton on the compound being nitrated is expressed in the varying reactivity of primary, secondary, and tertiary *CH* bonds (primary : secondary : tertiary $\sim$ 1 : 3.46 : 4.43)[173].

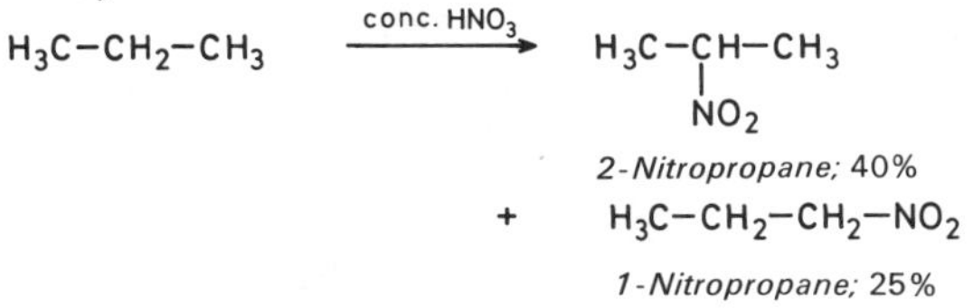

The difference between a secondary and a tertiary *CH* bond is illustrated by the nitration of decahydronaphthalene (*Houben-Weyl*, Vol. X/1, p. 17):

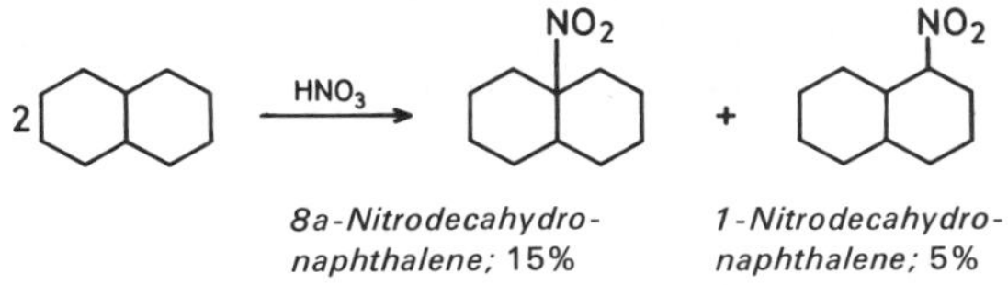

Thus nitration of alkanes or substituted alkanes almost invariably yields mixtures. Polynitrations can also ensue.

Nitration in the gas phase is primarily of technical interest and will therefore not be discussed in greater detail here[96].

For the liquid-phase nitration performed in the laboratory nitric acid, its derivatives and mixtures, and also nitrogen oxides may be employed (*Houben-Weyl*, Vol. X/1, pp. 13–44).
Nitration of compounds displaying sufficient *CH*-acidity with nitric acid in the cold is of only subordinate preparative interest. Malonic acid derivatives can be nitrated in good yield in this way[174]. Boiling *CH*-acid hydrocarbons with nitric acid under reflux to bring about nitration[97] leads to *C–C* cleavage in some cases, *e.g.*[98]:

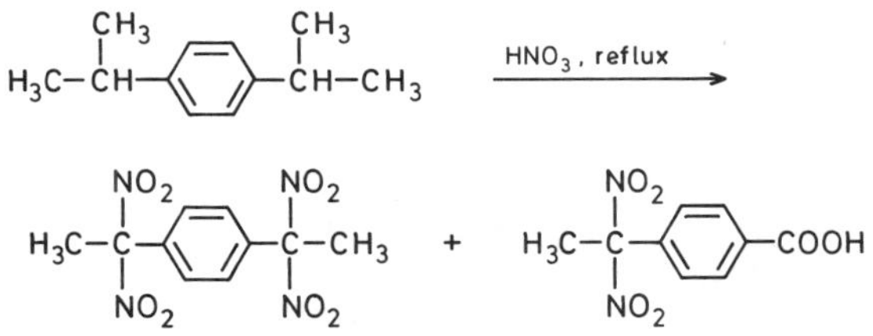

[95] *A.I. Titov*, Zh. Obshch. Khim. *19*, 1464 (1949); C.A. *44*, 1010 (1950).

[96] *H.G. Padeken, O. von Schickh, A. Segnitz* in *Houben-Weyl*, Methoden der organischen Chemie, Bd. X/1, p. 44, Georg Thieme Verlag, Stuttgart 1971; *Ullmann*, 3. Aufl. Bd. XIII, p. 13 (1961); Bd. XVI, p. 70 (1965).
[97] *R.A. Worstall*, J. Amer. Chem. Soc. *20*, 202 (1898); *21*, 210, 218 (1899); Amer. Chem. J. *22*, 164 (1899); Chem. Zentr. I, 926 (1898); I, 966 (1899).
[98] *J.E. McIntyre*, J. Chem. Soc. (London) 3540 (1964).

The yield of nitrations using hot nitric acid is often unsatisfactory.

Nitrating under pressure increases the yield and the speed of nitration[99]. With arylalkanes the aliphatic system may be nitrated side by side with the aromatic portion (see p. 3), *e.g.*[100]:

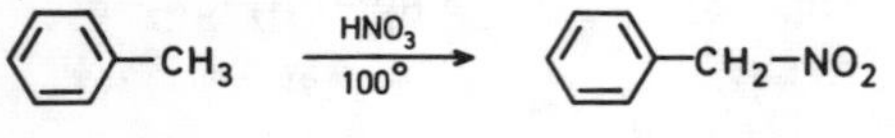

Side-chain nitration occurs only if the nitric acid contains oxides of nitrogen or by reacting, *e.g.*, toluene with nitric acid admixed with glacial acetic acid at elevated temperature but normal pressure[101] (radical mechanism→substitution in the side-chain; ionically→substitution in the aromatic nucleus).

Where the aromatic molecule contains several methyl groups it is possible to nitrate only one of them, *e.g.*[102]:

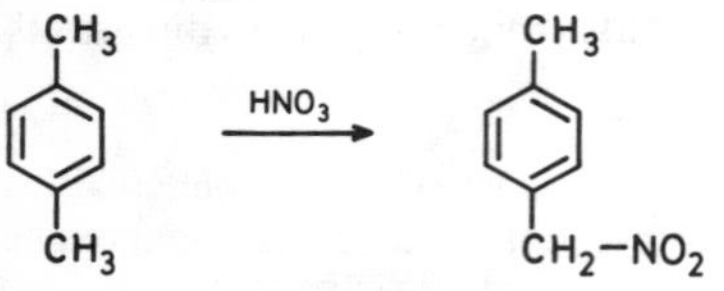

Functionally substituted aromatic alkyl compounds, too, can be nitrated on the alkyl group (*Houben-Weyl*, Vol. X/1, pp. 17, 19, 25, 30, 32), *e.g.*:

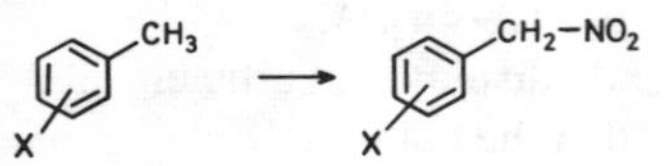

Using nitric acid mixtures during nitration of alkanes offers no or few advantages, but carboxylic acids are more readily nitrated with mixtures of nitric and sulfuric acids[175]. Nitric acid in glacial acetic acid and acetic anhydride admixed with nitric acid produce more ready nitration in carbonyl compounds than in pure alkanes[175]. Acetyl nitrate also can be used quite generally as a nitrating agent.

Like alkane dialkylamides, keto steroids, sulfones, and *N*-lithium compounds of primary amines, alkanes, cycloalkanes, and arylalkanes can be nitrated with esters of nitric and nitrous acid[103]. In strongly alkaline solutions (*e.g.*, potassium amide/liquid ammonia) alkyl nitrates nitrate ketones in the α-position[163].

Metal nitrates, too, represent potential nitrating agents[104]. According to the observed reactivity scale, no nitration takes place with alkali and ammonium nitrates, while of silver, mercury, aluminum, and bismuth nitrates the last-named exhibits the best results[105].

On reacting with dinitrogen tetraoxide or pentoxide a statistical attack[106] ensues on the alkane molecule.

Few examples of nitration with nitrogen monoxide (generally under pressure at 100–300°) have been described[107]. During working with dinitrogen tetraoxide the strong **explosiveness** displayed by its solutions in hydrocarbons needs to be guarded against. Here working reacting without pressure is limited by the boiling point of the compound being nitrated.

Nitration under pressure is of greater scope. Practically all hydrocarbons from pentane onwards can be nitrated in this way. Reacting toluene with dinitrogen tetraoxide leads to nitrotoluenes that are nitrated both in the side-chain and in the ring (*Houben-Weyl*, Vol. X/1, p. 44).

[99] *M. Konowalow*, Compt. Rend. *113*, 1052 (1891); *114*, 26 (1892); Ber. *28*, 1852 (1895).

[100] *M. Konowalow*, Ber. *28*, 1861 (1895); J. Russ. Phys.-Chem. Soc. *31*, 254 (1898); Chem. Zentr. I, 1237 (1899);
M.A. van Raalte, Rec. Trav. Chim. Pays-Bas *18*, 378 (1899);
DRP. 239 953, 246 381 (1910);
Monnet & Cartier, Paris, Société Chimique des Usines du Rhône, Erf.: *P. Gillard*, Chem. Zentr. II, 1564 (1911); I, 1677 (1912);
Fortschritte der Teerfabrikation u. verwandter Industriezweige *10*, 158, 159 (1913).

[101] *P.P. Schorygin, A.M. Sokolova*, J. Russ. Phys.-Chem. Soc. *62*, 673 (1930); Chem. Zentr. II, 2637 (1930);
R. Biedermann, A. Ledoux, Ber. *8*, 57 (1875);
R. Anschütz, E. Romig, Justus Liebigs Ann. Chem. *233*, 327 (1886).

[102] *M. Konowalow*, J. Russ. Phys.-Chem. Soc. *31*, 254 (1898), *37*, 530 (1905); Chem. Zentr. I, 1237 (1899); II, 817 (1905).

[103] USP. 2 883 433 (1957), DuPont, Inv.: *C.P. Spaeth*; C.A. *53*, 16025 (1959).

[104] *M. Konowalow*, J. Russ. Phys.-Chem. Soc. *33*, 393 (1901); Chem. Zentr. II, 580 (1901).

[105] *H.G. Padeken, O. von Schickh, A. Segnitz* in *Houben-Weyl*, Methoden der organischen Chemie, Bd. X/1, p. 37, Georg Thieme Verlag, Stuttgart 1971.

[106] *F. Asinger, G. Geiseler, W.D. Wirth*, Chem. Ber. *90*, 1987 (1957);
F. Asinger, K. Halcour, Chem. Ber. *94*, 83 (1961);
G.B. Bachman, H.B. Hass, J.V. Hewett, J. Org. Chem. *17*, 928 (1952);
G.B. Bachman, M.T. Atwood, M. Pollack, J. Org. Chem. *19*, 312 (1954);
A.I. Titov, N.V. Shchitov, Dokl. Akad. Nauk SSSR *81*, 1085 (1951); C.A. *46*, 7992 (1952);
A.I. Titov, Zh. Obshch. Khim. *19*, 258 (1949); C.A. *43*, 6584 (1949).

[107] USP. 2 883 432 (1957) ≡ DBP. 1 069 133, DuPont, Inv.: *C.P. Spaeth*; C.A. *53*, 16025 (1959).

Reactions with dinitrogen pentoxide also proceed *via* free radicals; dinitrogen tetraoxide is formed intermediately and this agent thus brings no preparative advantages (*Houben-Weyl*, Vol. X/1, p. 42).

1.5.1.2 Replacement of functional groups
(cf. *Houben-Weyl*, Vol. X/1, pp. 46–61)

Under certain conditions it may be advantageous to exchange a functional group against the nitro function. This technique offers the benefit relative to replacing hydrogen of introducing a nitro group into a definite position.

Replacement of bromine and iodine are most important here.

$$R-X + MNO_2 \longrightarrow R-NO_2 \text{ (or } R-O-NO) + MX$$

X = Br, J
M = Ag, Na, K

Esters of nitrous acid are formed as side-products in amounts prim < sec < tert. The yield of nitro compound increases inversely in the order tert < sec < prim haloalkane. Olefins are one product when secondary haloalkanes are reacted with silver nitrite and can react further to other products[108]. Tertiary haloalkanes give practically only nitrites with alkali nitrites[109]. Halogens of high atomic weight are replaced most easily[110].

$$F-CH_2-CH_2-CH_2-Br$$
$$F-CH_2-CH_2-CH_2-J \Bigg\} \longrightarrow F-CH_2-CH_2-CH_2-NO_2$$

Exchange with silver nitrate is possible also where the haloalkane concerned contains further functional groups. Dimethylformamide is a particularly suitable solvent for the reaction[109, 111]. The solu-

bility of sodium nitrite in this compound is increased by the addition of urea.

Re the exchange of other functional groups, see ref.[112].

1.5.2 Addition to *C–C* multiple bonds

Addition of pure nitric acid to olefins is of no significance in terms of the preparation of the lower-membered olefins[113] and proceeds *via* an addition-elimination mechanism[114]. On reacting 1, 1-diphenylethylene with nitric acid some of the initially formed addition product is isolated[170]:

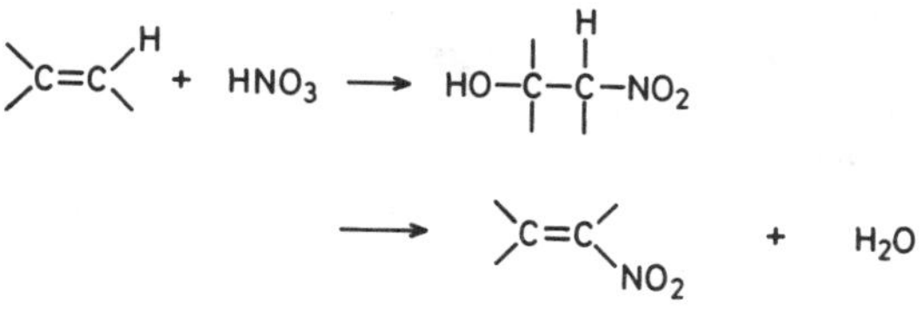

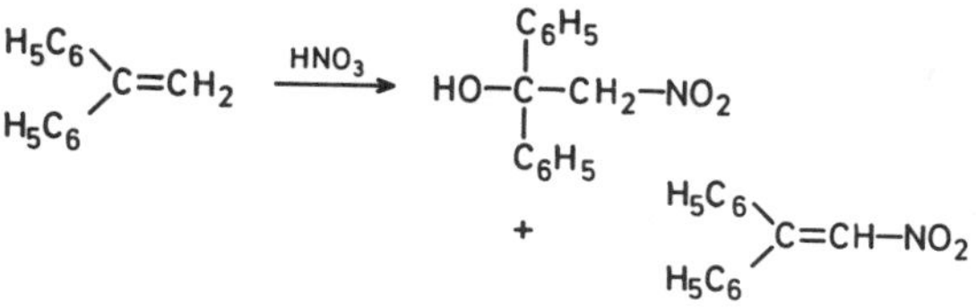

Reacting cinnamic acids with pure or fuming nitric acid affords either ring nitration or decarboxylation in dependence on the reaction conditions. As with dinitrogen trioxide α-nitration takes place accompanied by decarboxylation (*Houben-Weyl*, Vol. X/1, p. 64).

When olefins react with nitric acid/acetic anhydride mixtures acetyl nitrate is the active agent (Caution: acetyl nitrate is **explosive**)[115]:

$$(CH_3-CO)_2O + HNO_3 \rightarrow CH_3-CO-O-NO_2 + CH_3COOH$$

The addition is generally accompanied by other reactions. Thus, 2-alkenes give a mixture of *3-nitro-1-alkenes* and *3-nitro-2-alkyl acetates* as well as small amounts of *3-nitro-2-alkyl nitrates*[176]. If the reaction is conducted in the presence of tertiary amines practically no acetate is formed. On treating vinyl halides with nitrating acid oxidation often ensues in addition to nitration, *e.g.*[116]:

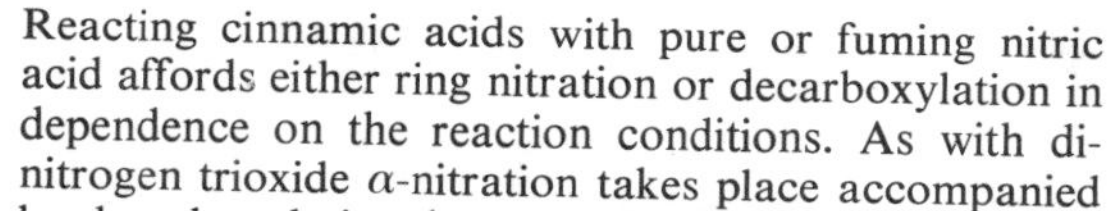

Chloronitroacetyl chloride

[108] *N. Kornblum, R.A. Smiley, H.E. Ungnade, A.M. White, B. Taub, S.A. Herbert jr.*, J. Amer. Chem. Soc. 77, 5528 (1955).
[109] *N. Kornblum, H.O. Larson, R.K. Blackwood, D.D. Mooberry, E.P. Oliveto, G.E. Graham*, J. Amer. Chem. Soc. 78, 1497 (1956).
[110] *E. Brackebusch*, Ber. 6, 1290 (1873);
F.L.M. Pattison, W.J. Cott, W.C. Howell, R.W. White, J. Amer. Chem. Soc. 78, 3484 (1956).
[111] *N. Kornblum, R.K. Blackwood*, J. Amer. Chem. Soc. 78, 4037 (1956).
[112] *H.G. Padeken, O. von Schickh, A. Segnitz* in *Houben-Weyl*, Methoden der organischen Chemie, Bd. X/1, p. 57, Georg Thieme Verlag, Stuttgart 1971.
[113] *L. Haitinger*, Justus Liebigs Ann. Chem. 193, 366 (1878); Monatsh. Chem. 2, 286 (1881);
H. Wieland, F. Rahn, Ber. 54, 1775 (1921);
A. Kekule, Ber. 2, 329 (1869);
T. Akestorides, J. Prakt. Chem. [2] 15, 62 (1877).
[114] *A. Michael, C.H. Carlson*, J. Amer. Chem. Soc. 57, 1268 (1935).

[115] *H.G. Padeken, O. von Schickh, A. Segnitz* in *Houben-Weyl*, Methoden der organischen Chemie, Bd. X/1, p. 33, Georg Thieme Verlag, Stuttgart 1971.
[116] *I.V. Martynov, Y.L. Kruglyak, S.P. Makarov*, Zh. Obshch. Khim. 33, 3382 (1963); Engl.: 3308.
[117] *J. Schmidt*, Ber. 35, 2323 (1902).

Hexafluoropropene reacts with 100% nitric acid/anhydrous hydrogen fluoride (nitryl fluoride) to a mixture of products[161]:

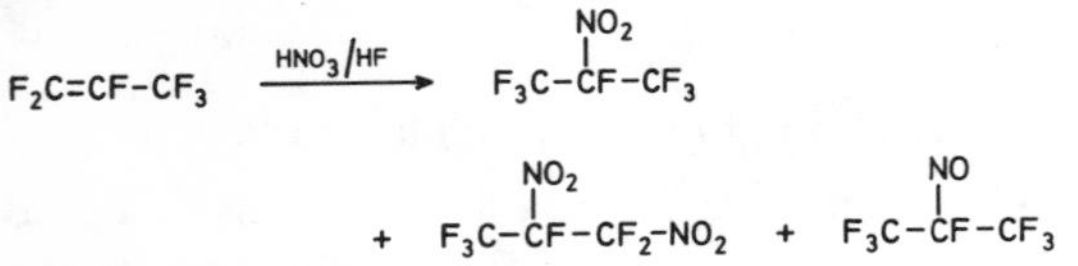

$$F_2C{=}CF{-}CF_3 \xrightarrow{\text{HNO}_3/\text{HF}} F_3C{-}\underset{\underset{\text{NO}_2}{|}}{C}{-}CF_3$$

$$+\; F_3C{-}\underset{\underset{\text{NO}_2}{|}}{C}{-}CF_2{-}NO_2 \;+\; F_3C{-}\underset{\underset{\text{NO}}{|}}{C}{-}CF_3$$

Nitrogen aoxides, too, can add onto *C–C* multiple bonds. Addition of dinitrogen trioxide leads to nitroso-nitro compounds ('pseudonitrosites')[117], which may dimerize[118] or can be transformed into α-nitro oximes. To carry out the reaction, either dinitrogen trioxide is introduced into a solution of the olefin, or a sodium nitrite solution layer is inserted under the olefin solution, and the dinitrogen trioxide is generated by acidifying whilst cooling[119].

Often the pseudonitrosites are not isolated, but the crude product is converted into the corresponding nitroolefin (*Houben-Weyl*, Vol. X/1, p. 333).

Dinitrogen tetraoxide, too, adds on to olefins to lead to a variety of products in dependence on the mechanism involved and the reaction conditions (*Houben-Weyl*, Vol. X/1, pp. 76–86):

Dinitrogen tetraoxide undergoes $1,\omega$-addition with conjugated dienes and trienes, *e.g.*[120]:

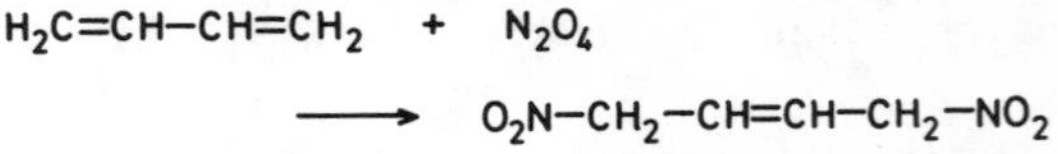

$$H_2C{=}CH{-}CH{=}CH_2 \;+\; N_2O_4$$
$$\longrightarrow O_2N{-}CH_2{-}CH{=}CH{-}CH_2{-}NO_2$$

1,4-Dinitro-2-butene; 36%

Dinitrogen pentoxide reacts with olefins in solution at lower temperatures. As the compound dissociates according to

$$N_2O_5 \; \rightleftharpoons \; NO_3^{\ominus} \;+\; NO_2^{\oplus}$$

esters of nitric acid are formed in addition to nitro compounds. It is advisable to work with rigorous exclusion of water[121].

In general, $1M$ solution of dinitrogen pentoxide in, for instance, dichloromethane, chloroform, or nitromethane at 0 to $-10°$ is added to the solution of the olefin cooled to -10 to $-30°$[122]. Tetraethylammonium nitrate is added to suppress undesired side-reactions.

During addition of nitrosyl chloride, which decomposes according to

$$NOCl \longrightarrow NO^{\oplus} \;+\; Cl^{\ominus}$$

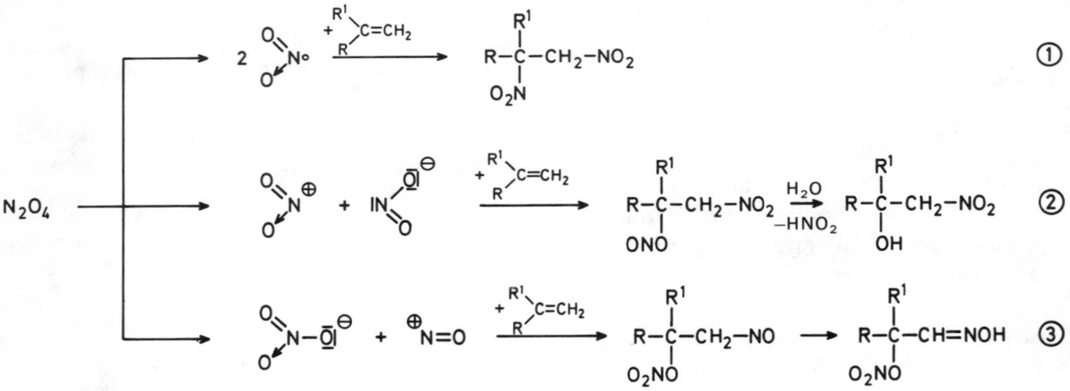

Oxidation of the vicinal nitronitrosooxyalkanes formed by pathway **2** leads to formation of some α-nitroalkanol nitrates during working up, while hydrolysis gives rise to nitroalcohols:

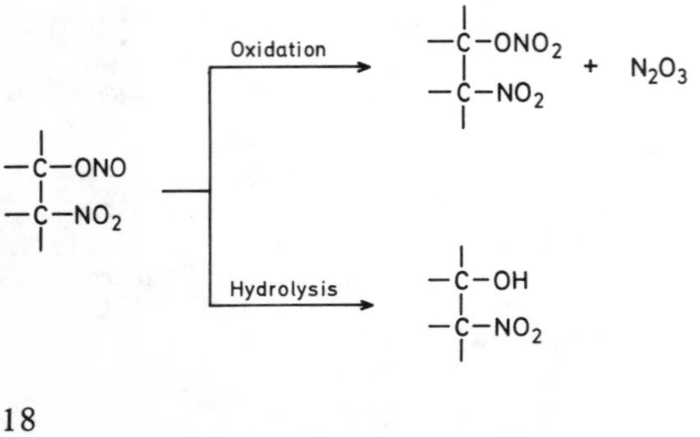

[118] *H. Wieland, E. Blümich*, Justus Liebigs Ann. Chem. *424*, 75 (1921).
[119] *A. Michael, G.H. Carlson*, J. Org. Chem. *4*, 169 (1939).
[120] *V.V. Perekalin, O.M. Lerner*, Dokl. Akad. Nauk SSSR *129*, 1303 (1959); C.A. *54*, 9729 (1960); USP. 2 314 615 (1943), I.C.I., Inv.: *R.G. Franklin, F.J. Wilkins;* C.A. *37*, 5188 (1943).
[121] *L.B. Haines, H. Adkins*, J. Amer. Chem. Soc. *47*, 1419 (1925).
[122] *T.E. Stevens, W.D. Emmons*, J. Amer. Chem. Soc. *79*, 6008 (1957).
[123] *H. Shechter, F. Conrad, A.L. Daulton, R.B. Kaplan*, J. Amer. Chem. Soc. *74*, 3052 (1952); *H. Petri*, Z. Anorg. Allgem. Chem. *257*, 180 (1948); *C.C. Price, C.A. Sears*, J. Amer. Chem. Soc. *75*, 3275 (1953).

oxidation of the primarily formed *α-chloro-1-nitrosoalkane* leads to the *nitro compound* (the primary addition of nitrosyl chloride follows the Markovnikov rule)[177]:

$$R-CH=CH_2 \xrightarrow{NOCl} \underset{\underset{Cl}{|}\ \underset{NO}{|}}{R-CH-CH_2}$$

$$\xrightarrow{NOCl} \underset{\underset{Cl}{|}\ \underset{NO_2}{|}}{R-CH-CH_2} + \underset{\underset{Cl}{|}\ \underset{Cl}{|}}{R-CH-CH_2}$$

Preparatively satisfactory yields are obtained with haloalkenes, while the reaction with olefins, and with nitro-, carboxy-, and alkoxyolefins is only sometimes successful. Working in the presence of oxygen leads to *chloronitro compounds* directly[178], e.g.:

$$H_3C-CH_2-CH=CH_2 + NOCl + O_2$$

$$\xrightarrow{N_2O_4} \underset{\underset{Cl}{|}\ \underset{NO_2}{|}}{H_3C-CH_2-CH-CH_2}$$

2-Chloro-1-nitrobutane

Addition of nitryl chloride to olefins affords chloroalkyl nitrites as well as chloronitro and dihalo compounds[123, 161]:

$$H_2C=CH_2 \xrightarrow{NO_2Cl} \underset{\underset{Cl}{|}}{H_2C-CH_2-NO_2}$$

$$+ \underset{\underset{Cl}{|}}{H_2C-CH_2-ONO} + \underset{\underset{Cl}{|}\ \underset{Cl}{|}}{H_2C-CH_2} + NO_2$$

In special cases tetranitromethane is reacted with olefins to add on a nitro group and form trinitromethane[124]. Arylpropenes are nitrated to *2-nitro-1-arylpropene* derivatives, and in some cases total addition of tetranitromethane to the olefin succeeds[125].

Tetranitromethane is the product obtained on reacting acetylene with nitrating acid, while *trinitromethane* is obtained when nitric acid is used alone[126]. It can be nitrated to tetranitromethane in a separate reaction[127].

Little is known about the addition of nitrogen oxides to acetylenes and almost nothing about that of dinitrogen trioxide[128]. Dinitrogen tetraoxide furnishes *dinitroalkenes* in moderate yield[129]:

$$-C\equiv C- + N_2O_4 \longrightarrow \underset{\underset{O_2N}{|}\ \underset{NO_2}{|}}{-C=C-}$$

Addition of nitrosyl chloride and nitryl chloride to acetylenes leading to *1-chloro-2-nitroalkenes* has likewise received little attention[130].

1.5.3 By oxidation of other nitrogen functions

While tertiary amines are oxidized to *nitroalkanes* with permanganate in relatively good yields (70–80%)[131], primary and secondary amines require the use or organic peroxy acids to afford substantial amounts of nitro compounds.

In general, chloronitrosoalkanes, nitrosonitroalkanes, and α-nitrosocarboxylic acids can be converted readily into the corresponding nitro compounds by oxidation. In addition to nitric acid a large number of alternative oxidizing agents may be employed: manganese(VII) oxide, chromic acid, lead(IV) oxide, oxygen, hydrogen peroxide, sodium hypochlorite, ozone, and trifluoroperoxyacetic acid (*Houben-Weyl*, Vol. X/1, p. 106 ff.).

Oximes are either formed *in situ* (e.g., by nitrating *CH*-acid compounds) or are used as such for the oxidation. Dinitrogen tetraoxide in ether is more suitable than nitric acid for oxidizing the *in situ* formed oximes. Oxidation of the oximes themselves with dinitrogen tetraoxide is known as the Ponzio reaction[132]:

$$4\ \underset{\underset{\overset{\|}{C}}{}}{H_3C-\overset{NOH}{C}-CH_3} \xrightarrow{3\ N_2O_4} 4\ \underset{\underset{NO}{|}}{H_3C-\overset{NO_2}{\underset{|}{C}}-CH_3}$$

$$+ 2\ H_2O + 2\ NO$$

$$\underset{}{H_5C_6-\overset{NOH}{\overset{\|}{C}}-C_6H_5} \xrightarrow{N_2O_4} \underset{\underset{NO_2}{|}}{H_5C_6-\overset{NO_2}{\underset{|}{C}}-C_6H_5}$$

[124] *A.T.Shulgin*, J.Med.Chem. *9*, 445 (1966); Experienta *20*, 366 (1964).

[125] *K.V. Altuchov, E.V. Racino, V.V. Perekalin*, Zh. Org. Khim. *5*, 2246 (1969).

[126] *K.J.P. Orton, P.V. McKie*, J. Chem. Soc. (London) *117*, 283 (1920);
K.F. Hager, Ind. Eng. Chem. *41*, 2168 (1949);
A. Wetterholm, Tetrahedron, Suppl. 155 (1963).

[127] *L. Schischkoff*, Justus Liebigs Ann. Chem. *119*, 247 (1861).

[128] *H. Wieland, E. Blümich*, Justus Liebigs Ann. Chem. *424*, 100 (1921).

[129] *H.H. Schlubach, W. Rott*, Justus Liebigs Ann. Chem. *594*, 59 (1955).

[130] *H.H. Schlubach, A. Braun*, Justus Liebigs Ann. Chem. *627*, 28 (1959).

[131] *N. Kornblum, R.J. Clutter*, J. Amer. Chem. Soc. *76*, 4494 (1954);
N. Kornblum, R.J. Clutter, W.J. Jones, J. Amer. Chem. Soc. *78*, 4003 (1956).

[132] *G. Born*, Ber. *29*, 90 (1896);
G. Ponzio, Gazz. Chim. Ital. *27*, I, 271 (1897); *28*, II, 269 (1898); J. Prakt. Chem. [2] *59*, 493 (1899).

While a number of oxidizing agents (Caro's acid[133], nitric acid[134], potassium hexacyanoferrate(III)[135], and potassium permanganate[136] oxidize oximes to nitro compounds in poor to moderate yield only, manganese(IV) oxide[137], ozone[138], and trifluoroperoxyacetic acid[179] furnish better results. The reaction is carried out in acetonitrile, which accelerates the rearrangement of the primarily formed *aci*-nitro compound to the more stable nitro compound and thus protects it from further oxidation.

1.5.4 Special procedures

Rearrangement of esters of nitrous and nitric acid leading to *nitro compounds*[139] is achieved by warming the esters in the presence of sulfuric acid or trifluoroacetic acid. Success is obtained also with *in situ* formed esters[140]:

$$H_9C_4-OH \xrightarrow[\text{2. } NaNO_2]{\text{1. } N_2O_4} H_9C_4-NO_2$$

Aromatic nitro compounds can be reduced to nitrocyclohexanes with sodium tetrahydroborate[141], *e.g.*, in methanol/water:

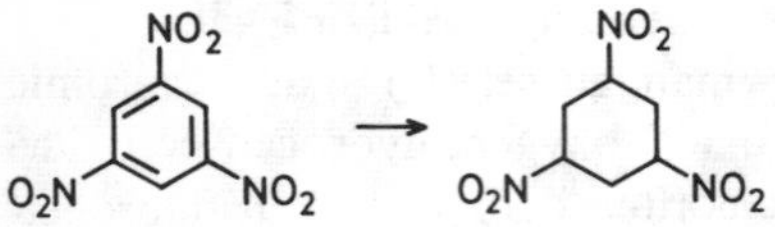

Sometimes, depending on the structure, only the cyclohexene stage is reached[142]. The requisite for a quantitative hydrogenation is the substitution of the benzene ring by electron-attracting groups, for instance, three nitro functions.

Good yields of defined 1-nitroalkanes are obtainable by eliminating carbon dioxide from carboxylic acid-nitric acid anhydrides[167].

Re the electrolytic dimerization of nitroalkanes to dinitro derivatives, see ref.[165].

1.5.5 *CH*-acidity of nitro compounds

The *CH*-acidity of a molecule is increased by the presence of a nitro group[143], *i.e.*, the formation of carbanions is facilitated. Accordingly, nitroalkanes may be regarded as strong *CH* acids which

can be ionized even with relatively weak bases. This follows, firstly, from the electron attraction by the nitro group (*i.e.* its −I effect) and, secondly, from the ability of the ion formed to delocalize the negative charge:

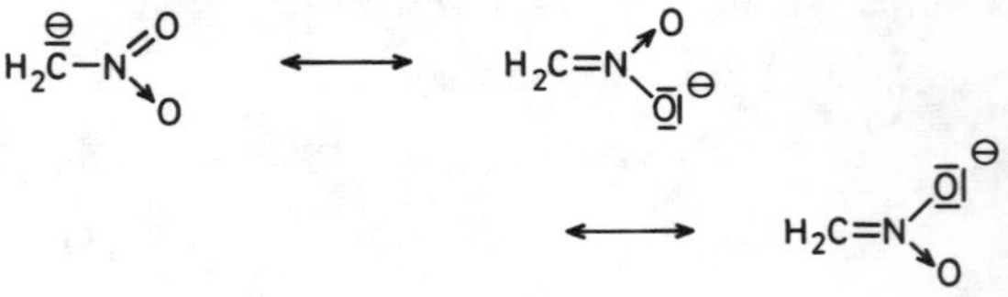

pK_a values form a quantitative measure of the acidity (K_a = dissociation constant of the acid in water[143, 144]). It follows that the acidity increases in the order

	H_3C-NO_2	$H_3C-CH_2-NO_2$	$\begin{array}{c}CH_3\\ \mid\\ CH-NO_2\\ \mid\\ CH_3\end{array}$
pK_a	10,2	8,3	7,7

and also increases as the number of nitro groups increases[144]:

	H_3C-NO_2	$H_2C(NO_2)_2$	$HC(NO_2)_3$
pK_a	10,2	3,6	0,17

Thus, all reactions which presuppose a certain *CH*-acidity in a molecule are promoted by the presence of a primary and secondary nitro group on the carbanionic carbon atom (cf. *Houben-Weyl*, Vol. X/1, pp. 130–166; 182–307).

1.5.6 *N*-Nitration of aliphatic compounds

Carboxamides are nitrated on the nitrogen with nitric acid in acetic anhydride at 0°, *e.g.*[147]:

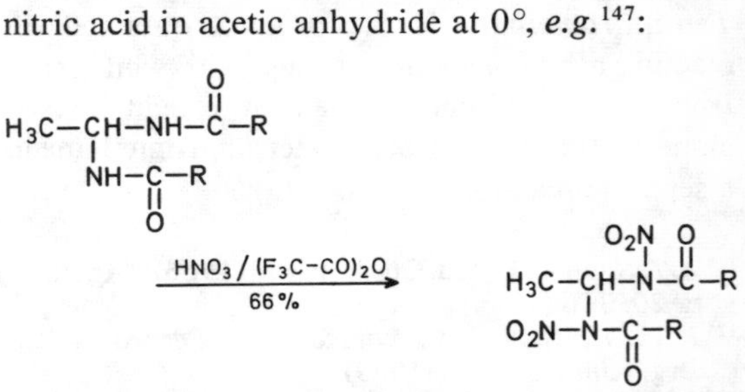

[133] *E. Bamberger, R. Seligman*, Ber. *35*, 3884 (1902); *E. Bamberger*, Ber. *33*, 1781 (1900).

[134] *G. Ponzio*, Gazz. Chim. Ital. *32*, II, 34 (1902).

[135] *M.O. Forster*, J. Chem. Soc. (London) *83*, 514 (1903).

[136] *T.M. Lowry, V. Steele*, J. Chem. Soc. (London) *107*, 1038 (1915).

[137] *L. Canonica*, Gazz. Chim. Ital. *77*, 92 (1947); C.A. *42*, 1885 (1948).

[138] *M. Freri*, Gazz. Chim. Ital. *66*, 23 (1936); *A.H. Riebel, R.E. Erickson, C.J. Abshire, P.S. Bailey*, J. Amer. Chem. Soc. *82*, 1801 (1960).

[139] *H.G. Padeken, O. von Schickh, A. Segnitz* in *Houben-Weyl*, Methoden der organischen Chemie, Bd. X/1, p. 127, Georg Thieme Verlag, Stuttgart 1971.

[140] *G.B. Bachman, N.W. Connon*, J. Org. Chem. *34*, 4121 (1969).

In the same way secondary amines can be nitrated on the nitrogen atom with nitric acid/acetic acid or nitric acid/acetic anhydride. Thus *N,N'-diethyl-N,N'-dinitroethylenediamine* is obtained from *N,N'*-diethylethylenediamine[148]:

$$H_5C_2-NH-CH_2-CH_2-NH-C_2H_5$$

$$\xrightarrow{HNO_3 \,/\, CH_3COOH} \quad H_5C_2-\overset{\overset{NO_2}{|}}{N}-CH_2-CH_2-\overset{\overset{NO_2}{|}}{N}-C_2H_5$$

Primary amines can be nitrated on the nitrogen by reacting their *N*-lithium compound derivative with alkyl nitrates[149]. 2-Cyano-2-propyl nitrate[150] is a particularly efficient nitrating agent for primary and secondary amines:

$$H_3C-\overset{\overset{ONO_2}{|}}{\underset{\underset{CH_3}{|}}{C}}-CN$$

Dinitrogen pentoxide successfully produces *N*-nitration of succinimide (62%)[151].

Hydrazines are not nitrated on the nitrogen[145,146]; for example, *N,N'*-dibenzoylhydrazines gives α,α-*dinitrobenzyl benzoate* with nitric acid/acetic anhydride[145].

1.6 Transformation reactions of aliphatic nitro compounds

Transformation reactions of aliphatic nitro compounds may be divided into two main classes:

1 Reactions with retention of the *C–N* bond
2 Reactions involving rupture of the *C–N* bond

A large numer of reducing agents have been used for reducing the nitro group. Amines[152,153,181] (see p. 494), hydroxylamines[154] (see p. 65), and oximes[155] (see p. 417) are among the products obtained. Under certain conditions the reduction of the nitro group is accompanied by ring closure[156,157,169,182].

The nitro group in nitroalkanes can be substituted. For example, the action of mineral acids on nitroalkanes leads to the formation of carboxylic acids[158]:

$$R-CH_2-NO \;+\; H_2O \;\xrightarrow{H^{\oplus}}\; \text{Carboxylic acids} \;+$$

$$NH_2OH$$

Preparation of aldehydes by the Nef reaction also starts from nitroalkanes[159], whose nitro group is eliminated as dinitrogen monoxide.

Condensation reaction of nitro compounds under the action of alkalis occasionally lead to heterocyclic systems with loss of nitrite[160].

[147] *R.G. Gafurov, A.G. Korepin, L.T. Eremenko,* Izv. Akad. Nauk SSSR 442 (1970);
J.A. Krimmel, J.J. Schmidt-Collerus, J.A. Young, G.E. Bohner, D.N. Gray, J. Org. Chem. *36*, 350 (1971);
H.F. Kauffmann, A. Burger, J. Org. Chem. *19*, 1662 (1954).

[148] Brit. P. 1 003 021 (1963), I.C.I., Inv.: *W.S. Lindsay, G.C. Mees;* C.A. *63*, 13077 (1965).

[149] *L.J. Winters, D.B. Learn, S.C. Desai,* J. Org. Chem. *30*, 2471 (1965).

[150] *W.D. Emmons, J.P. Freeman,* J. Amer. Chem. Soc. *77*, 4387 (1955);
J.P. Freeman, I.G. Shepard, Org. Synth. *43*, 83 (1963).

[151] *J. Runge, W. Treibs,* J. pr. [4] *15*, 223 (1965).

[152] *J. Malek, M. Cerny,* Synthesis 232 (1972);
H.B. Hass, E.F. Riley, Chem. Rev. *32*, 373 (1943).

[153] *R. Schröter* in *Houben-Weyl,* Methoden der organischen Chemie, Bd. XI/1, p. 360, Georg Thieme Verlag, Stuttgart 1957.

[154] *B. Zeeh, H. Metzger* in *Houben-Weyl,* Methoden der organischen Chemie, Bd. X/1, p. 1138, 1150, Georg Thieme Verlag, Stuttgart 1971.

[155] *H. Metzger* in *Houben-Weyl,* Methoden der organischen Chemie, Bd. X/4, p. 132, Georg Thieme Verlag, Stuttgart 1971;
H.G. Padeken, O. von Schick, A. Segnitz in *Houben-Weyl,* Methoden der organischen Chemie, Bd. X/4, p. 132, Georg Thieme Verlag, Stuttgart 1971.

[156] *H. Schnell, J. Nentwig* in *Houben-Weyl,* Methoden der organischen Chemie, Bd. XI/2, p. 542, 554, Georg Thieme Verlag, Stuttgart 1958.

[157] *G. Apel, H. Schwarz* in *Houben-Weyl,* Methoden der organischen Chemie, Bd. X/1, p. 439, Georg Thieme Verlag, Stuttgart 1971.

[158] *G. Apel, H. Schwarz* in *Houben-Weyl,* Methoden der organischen Chemie, Bd. X/1, p. 456, Georg Thieme Verlag, Stuttgart 1971.

[159] *O. Bayer* in *Houben-Weyl,* Methoden der organischen Chemie, Bd. VII/1, p. 272, Georg Thieme Verlag, Stuttgart 1954;
W.E. Noland, Chem. Rev. *55*, 137 (1955).

[160] *D.W.S. Latham, O. Meth-Cohn, H. Suchitzki,* Chem. Commun. 1040 (1972);
B.S. Thyagarajan, P.V. Gopalakrishnan, Tetrahedron *21*, 3305 (1965);

[141] *T. Severin, R. Schmitz,* Chem. Ber. *95*, 1417 (1962).

[142] *T. Severin, M. Adam,* Chem. Ber. *96*, 448 (1963).

[143] *F. Ebel,* Die Acidität der CH-Säuren, Georg Thieme Verlag, Stuttgart 1969.

[144] *E.S. Gould,* Mechanismus und Struktur in der organischen Chemie, p. 436, Verlag Chemie, Weinheim/Bergstr. 1962.

[145] *D. Harper, A.H. Lamberton,* J. Chem. Soc. [C], 1784 (1970).

[146] *I.S. Korsakova, I.S. Simonenko, S.S. Novikov,* Izv. Akad. Nauk SSSR 1202 (1970).

A.S. Sopova, V.V. Perekalin, V.M. Lebednova, Zh. Obshch. Khim, *34*, 2638 (1964); Engl.: 2659;

J. Skramstad, Acta Chem. Scand. *25*, 1287 (1971).

[161] *C.C. Price, C.A. Sears*, J. Amer. Chem. Soc. *75*, 3275 (1953).

[162] *P.L. Coe, A.E. Juges, J.C. Tatlov*, J. Chem. Soc. [C] 2323 (1966);

P.L. Coe, A.E. Jukes, J.C. Tatlov, Tetrahedron *24*, 5913 (1968).

[163] *H. Feuer, A.M. Hall, S. Golden, R.L. Reitz*, J. Org. Chem. *33*, 3622 (1968).

[164] *P.H. Donkas, T.J. Speaker*, J. Pharm. Sci. *60*, 184 (1971).

[165] *R. Pearson, W.V. Evans*, Trans. Electrochem. Soc. *84*, 173 (1943);

USP. 2 485 803 (1949), Inv.: *C.T. Balmer;* C.A. *44*, 2876 (1950);

C.F. Balmer, Ing. Eng. Chem. *44*, 317 (1952);

[166] *K.J. Klebe, C.L. Habraken*, Synthesis 294 (1973).

[167] *G.B. Bachman, T.F. Biermann*, J. Org. Chem. *35*, 4229 (1970).

[168] *H. Beyer*, Lehrbuch der organischen Chemie, 17. Aufl., p. 412, S. Hirzel Verlag, Stuttgart 1973; *Organicum*, 12. Aufl., p. 328, VEB Deutscher Verlag der Wissenschaften, Berlin 1973.

[169] *J.I.G. Cadogan*, Quart. Reviews *22*, 222 (1968); Synthesis 11 (1969).

[170] *H. Wieland, F. Rahn*, Ber. *54*, 1770 (1921).

[171] *G.C. Finger*, J. Amer. Chem. Soc. *73*, 149 (1951).

[172] *L. Barth*, Monatsh. Chem. *1*, 882 (1880).

[173] *F. Asinger, H.H. Vogel*, in *Houben-Weyl*, Methoden der organischen Chemie, Bd. V/1a, p. 626, Georg Thieme Verlag, Stuttgart 1970.

[174] *Houben-Weyl*, Bd. X/1, p. 15, 27, 32.

[175] *Houben-Weyl*, Bd. X/1, p. 13–44.

[176] *F.G. Bordwell, E.W. Garbisch*, J. Amer. Chem. Soc. *82*, 3588 (1960).

[177] The invariably co-formed 1,2-dichloroalkane is formed by addition of the chlorine liberated during the oxidation of 2-chloro-1-nitrosoalkane by nitrosyl chloride to unchanged olefin.

[178] *Houben-Weyl*, Bd. X/1, p. 89.

[179] *Houben-Weyl*, Bd. X/1, p. 119.

[180] *Houben-Weyl*, Methoden der organischen Chemie, Bd. X/1, p. 127, Georg Thieme Verlag, Stuttgart 1971.

1.7 Bibliography

G.A. Olah, S.J. Kuhn in *G.A. Olah*, Friedel-Crafts and Related Reactions, Vol. III/2, Chap. Nitration, p. 1393, Interscience Publishers, New York 1964.

T. Urbanski, Nitro-Compounds, Pergamon Press, Oxford 1964.

R.O.C. Norman, R. Taylor, Electrophilic Substitution in Benzenoid Compounds, p. 61, Elsevier Publishing Co., Amsterdam, London, New York 1965.

N. Kornblum, The Synthesis of Aliphatic and Alicyclic Nitro Compounds, Org. Reactions *12*, 101 (1962).

F.W. Stacey, J.F. Harries, Formation of Carbon-Hetero-Atom-Bonds by Free Radical Chain Additions to Carbon-Carbon Multiple Bonds, Org. Reactions *13*, 150, 224 (1963).

P.A.S. Smith, The Chemistry of Open-Chain Organic Nitrogen Compounds, Vol. II, p. 361, W.A. Benjamin, New York, Amsterdam 1966.

N.V. Sidgwick in *I.T. Millar, H.D. Springwall*, The Organic Chemistry of Nitrogen, 3. Ed., p. 361, Clarendon Press, Oxford 1966.

H. Feuer, The Chemistry of the Nitro- and Nitroso-Groups, Part 1 (1966), Part 2 (1970), Interscience Publishers, New York.

Ullman, Enzyklopädie der technischen Chemie, 3. Aufl., Band XIII, p. 13, 1961; Band XVI, p. 70, 1965, Urban & Schwarzenberg, München, Berlin.

B.T.F. Bedero, Encyclopedia of Explosives and Related Items, Picating Arsenal, USA 1961.

M.J. Astle, Industrial Organic Nitrogen Compounds, Reinhold Publishing Co., New York 1961.

W.W. Perekalin, Ungesättigte Nitro-Verbindungen, Gosschimisdat, Leningrad 1961.

---

[181] *T.L. Ho, C.M. Wong*, Synthesis 45 (1974).

[182] *D.W.S. Latham, O. Meth-Cohn, H. Suchitzki*, Chem. Commun. 41 (1973); J.C.S. Perkin Trans I 696 (1973);

D.W. Livingstone, G. Tennant, Chem. Commun. 96 (1973);

K. Wagner, H. Heitzer, L. Öhlmann, Chem. Ber. *106*, 640 (1973).

[183] *S. Rossi, E. Duranti*, Tetrahedron Lett. 485 (1973).

2 Nitroso Compounds

Contributed by

H. Meier
Chemisches Institut der Universität Tübingen,
Tübingen

2.1 Aromatic nitroso compounds

Aromatic nitroso compounds contain a nitroso group linked directly to a ring carbon atom of an aromatic or heteroaromatic system. In solution, melt, or in the gaseous state mainly green to blue *monomers* are present; by contrast, in the solid state the generally *colorless* dimers prevail. These aromatic bis-nitroso compounds, which thus represent N,N'-azo dioxides, normally have the *trans*-configuration*, but may be rearranged into the *cis*-form [1]:

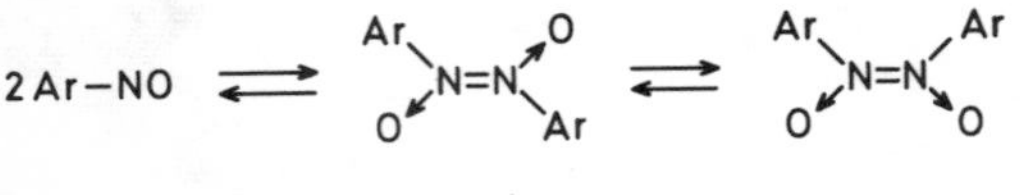

A structural speciality among monomeric *o*- and *p*-nitrosophenols and *o*- and *p*-nitrosoanilines is the tautomerism with the corresponding *quinone monooximes* and *quinonimine oximes* [2]. Aromatic 1,2-dinitroso compounds display no free nitroso groups but exist as *benzofurazan N-oxides* [3]:

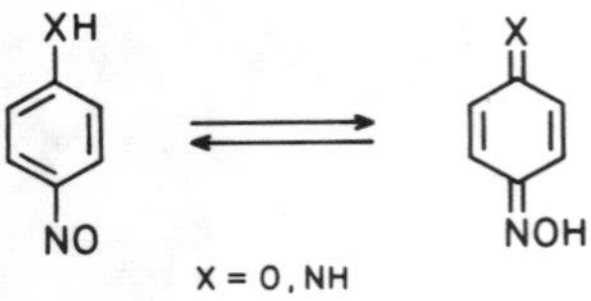

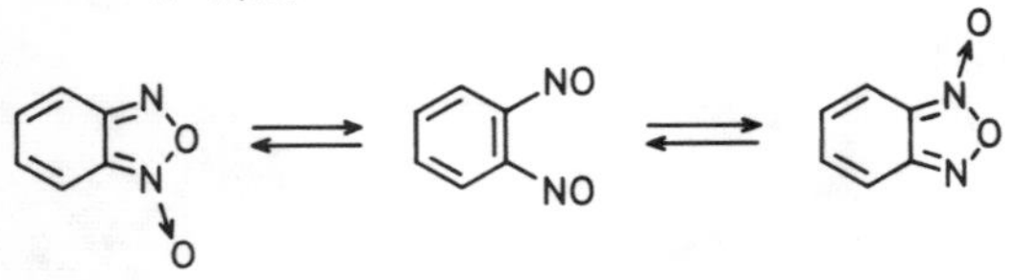

2.1.1 Preparation

2.1.1.1 Direct introduction of the nitroso group

Direct ring nitrosation, as an electrophilic substitution reaction, is generally applicable only to activated (electron-rich) aromatic compounds such as phenols and tertiary amines. Aromatic *hydroxy* compounds can be nitrosated with nitrous acid without difficulty; the electrophilic attack ensues preferentially in the *p*-position. *o*-Nitrosations are occasionally observed if the *p*-position is occupied.

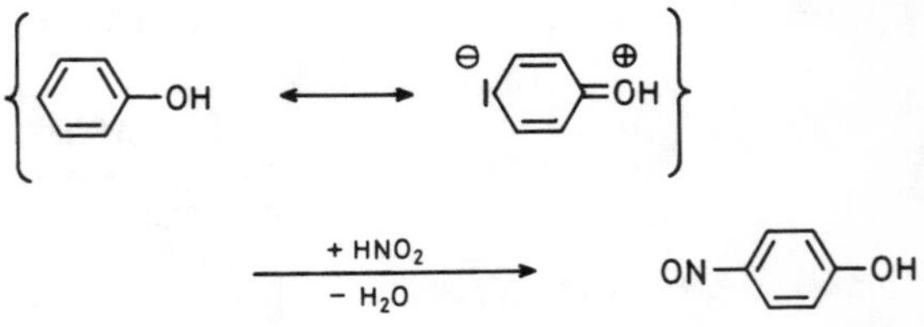

The following *techniques* have given good results:

1 Add hydrochloric, sulfuric, or acetic acid to the solution of the hydroxy compound in the calculated quantity of caustic soda or potash plus a small excess of alkali nitrite cooled to 0–10°.
2 Slowly add an aqueous or aqueous alkaline solution of phenol and sodium or potassium nitrite to the acid whilst cooling.
3 Dissolve or suspend the aromatic hydroxy compound in dilute acid and add concentrated nitrite solution drop by drop (or solid nitrite) whilst cooling.
4 Add the phenol dissolved in pyridine to nitrous acid prepared from sodium nitrite and concentrated sulfuric acid drop by drop whilst cooling.
5 Nitrosate in aqueous solution using a hydrolyzing metal nitrite such as aluminum or iron(III) nitrite (Particularly mild conditions).
6 Treat the phenol dissolved in alkali metal alcoholate or alkali hydroxide with ethyl, butyl, or pentyl nitrite.

---

*cis-Bis(nitroso)benzene and cis-bis(1-nitrosonaphthalene) form exceptions.

[1] *C.D. Darwin, D.C. Hodkin*, Nature *166*, 827 (1950);
B.G. Gowenlock, J. Trotman, J. Chem. Soc. (London) 4190 (1955);
W. Lüttke, Angew. Chem. *68*, 417 (1956);
W. Lüttke, Z. Elektrochemie *61*, 302, 976 (1957);
B.G. Gowenlock, W. Lüttke, Quart. Rev. Chem. Soc. *12*, 321 (1958).
[2] *H.H. Hudgson*, J. Chem. Soc. (London) *520* (1937);
P. Ramart-Lucas, M.M. Martynoff, Bull. Soc. Chim. France *15*, 571 (1948);
W.R. Vaughan, G.K. Finch, J. Org. Chem. *21*, 1201 (1956);
R.K. Norris, S. Sternhell, Australian J. Chem. *19*, 841 (1966);
H.H. Hudgson, J. Chem. Soc. (London) 520 (1937);
C.H. Sluiter, Rec. Trav. Chim. Pays-Bas *25*, 10 (1906);
A. Schors, A. Kraaijveld, E. Havinga, Rec. Trav. Chim. Pays-Bas *74*, 1243 (1955).

[3] *A.G. Green, F.M. Rowe*, J. Chem. Soc. (London) *110*, 2452 (1912);
M.O. Forster, M.F. Barker, J. Chem. Soc. (London) *113*, 1918 (1913);
D.L. Hammick, W.A.M. Edwardes, E.R. Steiner, J. Chem. Soc. (London) 3308 (1931);
A.R. Katritzky, S. Oksne, R.K. Harris, Chem. & Ind. (London) 990 (1961);
P. Diehl, H.A. Christ, F.B. Mallory, Helv. Chim. Acta *45*, 504 (1962).
[4] *H. Ficheroulle, R. Soule*, Mém. Poudres *37*, 339 (1955); C.A. *50*, 16 695 (1956).
[5] *H.H. Hodgson, E.A.C. Crouch*, J. Chem. Soc. (London) 221 (1943);
K.H. Meyer, S. Lenhardt, Justus Liebigs Ann. Chem. *398*, 79 (1913);
H. Weidel, J. Pollak, Monatsh. Chem. *21*, 29 (1900).

Table 1. Nitroso compounds by nitrosation of aromatic hydroxy compounds

Hydroxy compounds	Method	Reaction product	Yield [% of theory]	Ref.
Phenol	1	*4-Nitrosophenol*	72	12
	2		90	16
5-Bromo-2-cresol	1	*5-Bromo-4-nitroso-2-cresol*	~90	13
2-Naphthol	1	*1-Nitroso-2-naphthol*	—	14, 18
			80	
2-Phenanthrol	1	*1-Nitroso-2-phenanthrol*	—	15
2,4-Xylenol	3	*4-Nitroso-2,6-xylenol*	67	17
2,6-Di-*tert*-butylphenol	3	*2,6-Di-tert-butyl-4-nitrosophenol*	75	17
3-Fluorophenol	4	*3-Fluoro-4-nitrosophenol*	—	19
3-Chlorophenol	4	*3-Chloro-4-nitrosophenol*	80	20
2-Cresol	6	*4-Nitroso-2-cresol*	90	21
Hexylresorcinol	6	*4-Hexyl-6-nitrosoresorcinol*	—	22

Using these techniques, of which the first is the most frequently applied, allows one to nitrosate phenols, naphthols, anthranols, etc. Several nitroso groups can be introduced into *dihydroxy* and *trihydroxy compounds*. From resorcinol a 93% yield of *2,4-dinitrosoresorcinol* is obtained[4]. Certain *phenol ethers*, too, can be nitrosated using these procedures[5], ether cleavage is a frequent side-reaction[6]:

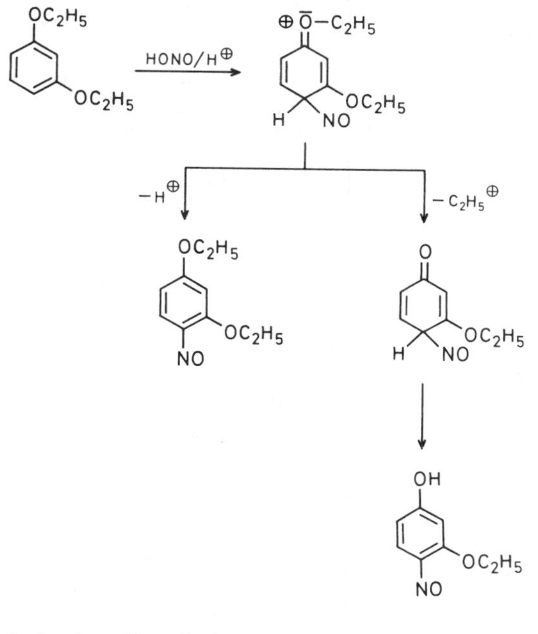

A further 'interfering reaction' is formation of nitro compounds due to the oxidizing effect of nitrous acid. Thus, *o*- and *m*-cresol can be smoothly nitrosated in the *p*-position to *4-nitrosoresorcinol* and *4-nitrosopyrocatechol*[7], but with *p*-cresol[7], as with 3,5-dichlorophenol[8], nitration to 3-nitro-*p*-cresol and 3,5-dichloro-4-nitrophenol respectively takes place.

Table 1 surveys nitrosations carried out using techniques 1–6.

With *p*-substituted phenols it is possible to obtain *o*-nitrosation in some cases by performing the reaction in the presence of heavy metal salts forming inner metal complexes with *o-nitrosophenols*[9]:

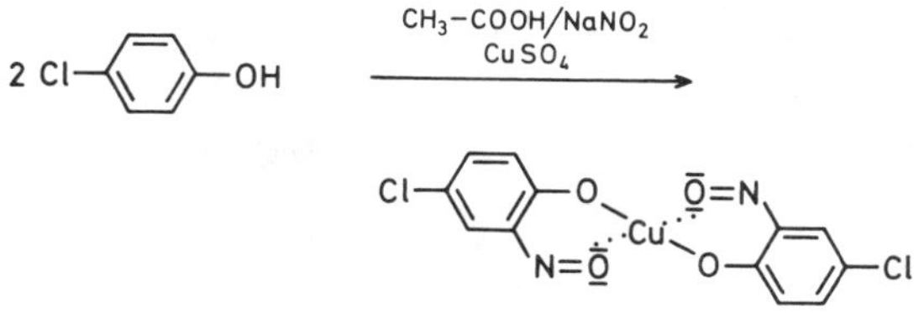

An efficient reagent for *o*-nitrosation — even if the *p*-position is free — is a mixture of hydroxylamine and hydrogen peroxide. This so-called *Baudisch reaction*[10] is performed in water, ethanol, or in the two-phase system water-petroleum ether in the presence of copper(II) salts or sodium penta-cyanoamineferrate(II). Phenols, alkylphenols, halophenols, and phenol carboxylic acids can be nitrosated in good yield in this way[11].

Starting from the aromatic hydrocarbons instead of from phenols in the Baudisch reaction leads to *o-nitrosophenols*[23] by oxidative nitrosation:

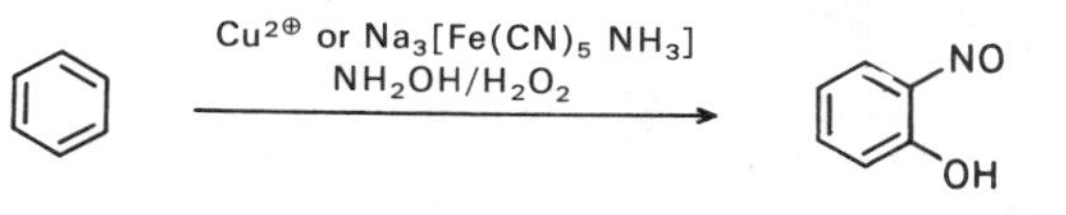

[6] *H.H. Hodgson, H. Clay*, J. Chem. Soc. (London) 1872 (1930);
A. Kraus, Monatsh. Chem. *12*, 368 (1891).
[7] *H.H. Hodgson, E.A.C. Crouch*, J. Chem. Soc. (London) 221 (1943);
H. van Erp, Rec. Trav. Chim. Pays-Bas *30*, 276 (1910);
M.J. Astle, W.P. Cropper, J. Amer. Chem. Soc. *65*, 2398 (1941);
H.A.J. Schoutissen, Rec. Trav. Chim. Pays-Bas *40*, 753 (1921).
[8] *H.H. Hodgson, J.S. Wignall*, J. Chem. Soc. (London) 2216 (1927);
R. Willstätter, G. Schudel, Ber. *51*, 786 (1918).
[9] *G. Cronheim*, J. Org. Chem. *12*, 7, 17 (1947).

Occasionally a carboxy group on the aromatic nucleus can be displaced by the nitroso group in addition to hydrogen[13]. Thus, when sodium nitrite is allowed to act on an aqueous alcoholic solution of 3, 5-dibromo-4-hydroxybenzoic acid a quantitative yield of *2,6-dibromo-4-nitrosophenol* is obtained[24]. While in the naphthalene series 3-hydroxy-2-naphthoic acid is nitrosated smoothly to *3-hydroxy-4-nitroso-2-naphthoic acid*[25], *nitrosating decarboxylation* occurs in 1-hydroxy-2-naphthoic acid and in 2-hydroxy-1-naphthoic acid with formation of *2-nitroso-1-naphthol*[26] and *1-nitroso-2-naphthol*[27] respectively.

The action of nitrous acid on primary aromatic amines generally leads to diazonium salts. Various *m* or *o* and *m* alkyl-substituted anilines are anomalies, and the following products are obtained, for example:

2,5-Xylidine ⟶ *4-Nitroso-2,5-xylidine*[28];

3-Anisidine ⟶ *4-Nitroso-3-anisidine*[29];

2,3-Dimethylaniline ⟶ *2,3-Dimethyl-4-nitrosoaniline*[28];

1,2,3,4-Tetrahydro-5-naphthylamine ⟶
        1,2,3,4-Tetrahydro-8-nitroso-5-naphthylamine[28];

Analogously, 4-amino-2-pyridinol is nitrated to *4-amino-3-nitroso-2-pyridinol*[30]. With nitrosylsulfuric acid in concentrated sulfuric acid 1-naphthylamine can be converted into *4-nitroso-1-naphthylamine* without diazotization[31].

Secondary aromatic amines form *N*-nitrosoamines with nitrous acid, and in this way ring nitrosation can be achieved *via* the detour of the Fischer-Hepp rearrangement (cf. p. 30). Independently of this method secondary aromatic amines can be frequently ring-nitrosated directly in fuming sulfuric acid, or in alcohol or glacial acetic acid saturated with hydrogen chloride, with solid sodium nitrite or nitrosylsulfuric acid[32]. The action of phosgene at 0–5° on a mixture of amine, nitrite, and methanol has given particularly good results in ring-nitrosating diarylamines directly[33].

Tertiary aromatic amines can be ring-nitrosated by the following procedures:

1 Add an equimolar amount or a small excess of an aqueous solution of alkali nitrite to the amine dissolved in aqueous or alcoholic hydrogen chloride at 5–10°.
2 Use an easily hydrolyzed ester of nitrous acid, *e.g.*, pentyl nitrite, in place of alkali nitrite when working in alcoholic hydrogen chloride.
3 Use tetranitromethane as nitrosating agent in aqueous alcoholic hydrochloric acid.
4 Introduce the nitroso group with $N_2O_3 \cdot BF_3$ in nitroethane as solvent at 0°.
5 Use nitrosylsulfuric acid in sulfuric acid as nitrosating agent.

[10] *O. Baudisch, H.S. Smith*, Naturwissenschaften *27*, 769 (1939).
[11] *K. Maruyama, I. Tanimoto, R. Goto*, J. Org. Chem. *32*, 2516 (1967);
G. Cronheim, J. Org. Chem. *12*, 7 (1947);
O. Baudisch, H.S. Smith, Naturwissenschaften *27*, 769 (1939).
[12] *N.J. Leonard, J.W. Curry*, J. Org. Chem. *17*, 1074 (1952).
[13] *H.H. Hudgson, F.H. Moore*, J. Chem. Soc. (London) 2036 (1926).
[14] *K. Lagozinski, D. Hardine*, Ber. *27*, 3075 (1894).
[15] *S. Marcinkiewicz, J. Green*, J. Chromatogr. *10*, 366 (1963).
[16] *A. Krassnowski*, Promysl. Org. Chim. *1*, 604 (1936);
Chem. Zentr. II, 3845 (1936).
[17] *V.V. Ershov, G.A. Zlobina, G.A. Nikiforov*, Izv. Akad. Nauk SSSR 1877 (1963); C.A. 60, 2809 (1964).
[18] USP. 3 051 750 (1960), Du Pont, Inv.: *W. Dettwyler*; C.A. 58, 7888 (1963);
Chem. Zentr. 43 2353 (1964).
[19] *H.H. Hudgson, D.E. Nicholson*, J. Chem. Soc. (London) 1268 (1940).
[20] *A. Kraaijeveld, E. Havinga*, Rec. Trav. Chim. Pays-Bas. *73*, 537, 546 (1954).
[21] *W. Borsche, A.D. Berkhout*, Justus Liebigs Ann. Chem. *330*, 95 (1904).
[22] Jap. P. 64/13 918 (1961), Inv.: *T. Teshigawara*; C.A. *62*, 7688 (1965).
[23] *O. Baudisch*, Naturwissenschaften *27*, 768 (1939);
O. Baudisch, Science *92*, 336 (1940);
O. Baudisch, J. Amer. Chem. Soc. *63*, 622 (1941);
G. Cronheim, J. Org. Chem. *12*, 1, 7 (1947);

O. Konecny, J. Amer. Chem. Soc. *77*, 5748 (1955);
K. Maruyama, I. Tanimoto, R. Goto, Tetrahedron Lett. 5889 (1966);
K. Maruyama, I. Tanimoto, R. Goto, J. Org. Chem. *32*, 2516 (1967).
G. Cronheim, J. Org. Chem. *12*, 7 (1947).
[24] *R.A. Henry*, J. Org. Chem. *23*, 648 (1958).
[25] *S.v. Kostanecki*, Ber. *26*, 2898 (1883);
F. Ullmann, R. Heisler, Ber. *42*, 4266 (1909);
R. Lantz, G. Mingasson, Bull. Soc. Chim. France [4] *49*, 1172 (1931);
W.G. Gulinow, Ukr. Khim. Zh. 7, 197 (1932); Chem. Zentr. II, 1434 (1933).
[26] *F. Reverdin, C. de la Harpe*, Ber. *26*, 1280 (1893);
R. Nietzki, A.L. Guitermann, Ber. *20*, 1274 (1887).
[27] *G. Heller*, Ber. *45*, 676 (1912).
[28] DRP. 519 729 (1929), I.G. Farben, Erf.: *K. Schirmacher, H. Schlichenmaier*; Fortschritte der Teerfabrikation u. verwandter Industriezweige *17*, 471 (1929).
[29] DRP. 561 424 (1931), I.G. Farben, Erf.: *L. Blangey*; Fortschritte der Teerfabrikation u. verwandter Industriezweige *19*, 695 (1931).
[30] *T. Talik, Z. Talik*, Roczniki Chem. *37*, 75 (1963).
[31] *L. Blangey*, Helv. Chim. Acta *21*, 1579 (1938).

Table 2. Aromatic nitroso compounds by nitrosation of tertiary aromatic amines

Tertiary aromatic amine	Method	Reaction product	Yield [% of theory]	Ref.
N,N-Dimethylaniline	1	*N,N-Dimethyl-p-nitrosoaniline*	90	[41]
	4		~100	[46]
	5		71	[47]
3-Bromo-*N,N*-dimethylaniline	1	*3-Bromo-4-nitroso-N,N-dimethylaniline*	—	[42]
N,N-Dimethyl-3-biphenylamine	1	*N,N-Dimethyl-6-nitroso-3-biphenylamine*	75	[43]
Triphenylamine	2	*4-Nitrosotriphenylamine*	79	[44]
N,N-Dimethyl-*m*-toluidine	3	*N,N-Dimethyl-4-nitroso-m-toluidine*	54	[45]

The first technique is by far the most commonly used. Variant **2** gives especially good results with weakly basic tertiary aromatic amines. Procedures **3–5** are generally limited to special cases.

Ring nitrosation generally takes place in the *p*-position with respect to the *tertiary* amino group, but there are exceptions, as is shown in the following example[34]:

5-[3-(Dimethylamino)-propyl]-10,11-dihydro-4-nitrosodibenz[b,f]azepine; 18%

2,5-Dimethyl-3-nitrosopyrrole[40]

1-Methyl-3-nitroso-2-phenylindole[48]; 85%

With *N,N*-dimethyl-α-phenyl-4-toluidine *o*-nitrosation also is observed[35]. Bulky substituents on the nitrogen of the amino group or in the *o*-position can prevent nitrosation[36]. In *tertiary* amines with two or three aryl substituents only *one* aromatic nucleus is nitrosated[37].

Table 2 surveys the various methods for preparing nitroso compounds of tertiary aromatic amines.

Some *heteroaromatic ring systems* such as pyrrole, pyrazole, imidazole, and indole can be ring-nitrosated directly. In others, such as pyridine and pyrimidine, this succeeds only if an amino or hydroxy group is present as a primary substituent[38]. The following are examples of direct ring nitrosation of heterocycles:

3-Chloro-1,5-dimethyl-4-nitrosopyrazole[31]; 64%

2,4-Dimethyl-4-nitroso-1-phenyl-2-pyrazolin-5-one[49]

5-Nitroso-2,4-diphenylpyrrole[39]; ~100%

7-Amino-2,6-dimethyl-3-nitroso-pyrazolo[1,5-a]pyrimidine[50]

2.1.1.2 Oxidation of aromatic nitrogen compounds

Primary aromatic amines can be oxidized to nitroso compounds with peracids in weakly acid solution. Nitro and azoxy compounds are often obtained as side-products. Table 3 surveys this preparative method (in relation to the oxidizing agent used).

Secondary aromatic amines, too, can be occasionally converted into nitroso compounds with peroxymonosulfuric acid in degradation reactions. From *N*-benzylaniline, *nitrosobenzene* is obtained side by side with *nitrobenzene, azoxybenzene,* and *benzoic acid*[51]. Cold, dilute permanganate solution degrades 3-Hydroxy-2-phenyl-3-indols to *2-nitrosobenzoic* acid[52].

Peracids oxidize *tertiary* aromatic amines such as *N*-pentachlorophenylpiperidine or 4-piperidinotetrachloropyridine not to amine oxides but to *pentachloronitrosobenzene*[60] and *4-tetrachloronitrosopyridine*[61] respectively.

Aromatic *hydroxylamines,* which are normally obtained by reduction from the corresponding nitro compounds, can be easily converted into nitroso compounds. Table 4 surveys this type of reaction using various different oxidizing media.

Table 3. Aromatic nitroso compounds by oxidation of primary aromatic amines

Amine	Oxidizing agent	Nitroso compound	Yield [% of theory]	Ref.
2-Nitroaniline	Peroxymonosulfuric acid	*1-Nitro-2-nitrosobenzene*	70	[53]
		1-Nitro-3-nitrosobenzene		
3-Nitroaniline	Peroxymonosulfuric acid	*3-Nitronitrosobenzene*	40	[54]
	Peracetic acid		54	[55]
4-Nitroaniline	Peroxymonosulfuric acid	*1-Nitro-4-nitrosobenzene*	35	[54]
	Peracetic acid		72	[55]
Pentafluoroaniline	Performic acid	*Pentafluoronitrosobenzene*	48	[56]
2,6-Difluoroaniline	Perbenzoic acid	*2,6-Difluoro-1-nitrosobenzene*	85	[57]
Aniline	Potassium permanganate/ formaldehyde	*Nitrosobenzene*	—	[58]
3'-Amino-5'-chloroacetophenone	Peroxydisulfuric acid	*5'-Chloro-3-nitrosoacetophenone*	—	[59]

Table 4. Aromatic nitroso compounds by oxidation of N-arylhydroxylamines

Hydroxylamine	Oxidizing agent	Nitroso compound	Yield [% of theory]	Ref.
N-Phenylhydroxylamine	$Na_2Cr_2O_7/H_2SO_4$	*Nitrosobenzene*	67	[66]
	Diethyl azodicarboxylate		89	[67]
N,N'-Dihydroxy-*m*-phenylenediamine	$FeCl_3$	*1,3-Dinitrosobenzene*	—	[68]
4-Hydroxylaminopyridine-1-oxide	$KMnO_4/H_2SO_4$	*4-Nitrosopyridine-1-oxide*	50	[69]
6-Hydroxylaminopurine	MnO_2	*6-Nitrosopurine*	21	[70]
1-Naphthylhydroxylamine	Ag_2O	*1-Nitrosonaphthalene*	67	[71]
1-Hydroxylaminoanthra quinone	$O_2(air)/NaOH$	*1-Nitrosoanthraquinone*	50	[72]
N,N'-Dihydroxy-4,6-dinitro-*m*-phenylenediamine	CrO_3	*1,3-Dinitro-4,6-dinitrosobenzene*	—	[73]

[32] *J. Houben,* Ber. *46,* 3984 (1913);
J.J. D'Amico, C.C. Tung, L.A. Walker, J. Amer. Chem. Soc. *81,* 5957 (1959);
USP. 3 504 034 (1967), Monsanto Chem. Co., Inv.: *R.L. Wright;* C.A. *72,* 112533 (1970);
DRP. 608 669 (1933), I.G. Farben, Erf.: *A. Zitscher, W. Seidenfaden;*
Chem. Zentr. I, 3046 (1935);
USP. 3 036 051 (1958), Monsanto Chemical Co., Erf.: *J.J. D'Amico;* C.A. *58,* 1413 (1963);
Fr. P. 783 088 (1933), I.C.I.;
Chem. Zentr. I, 1121 (1936);
J. Houben, W. Brassert, L. Ettinger, Ber. *42,* 2745, 2750 (1909).

[33] USP. 2 495 774 (1946), General Aniline a. Film Corp., Inv.: *S.M. Roberts;*
Chem. Zentr. I, 256 (1951).

[34] *B.A. Porai-Koshits, E.N. Sof'ina, I.Ya. Kvitko,* Zh. Obshch. Khim. *34,* 2094 (1964); C.A. *61,* 6990 (1964).

[35] *J. Limpricht,* Justus Liebigs Ann. Chem. *307,* 305 (1899).

[36] *T.C. van Hoek, P.E. Verkade, B.M. Webster,* Rec. Trav. Chim. Pays-Bas *77,* 559 (1958);
L.A. Walker, J.J. D'Amico, D.D. Mullins, J. Org. Chem. *27,* 2767 (1962);
P. Friedländer, Monatsh. Chem. *19,* 627 (1898);
W.J. Hickinbottom, J. Chem. Soc. (London) 946 (1933);
R.C. Fuson, Reactions of Organic Compounds, p. 106, John Wiley & Sons, New York 1962.

Appropriate oxidation of *quinone dioximes* can serve for preparing *aromatic dinitroso compounds, e.g.,* 1,4-dinitrosobenzene

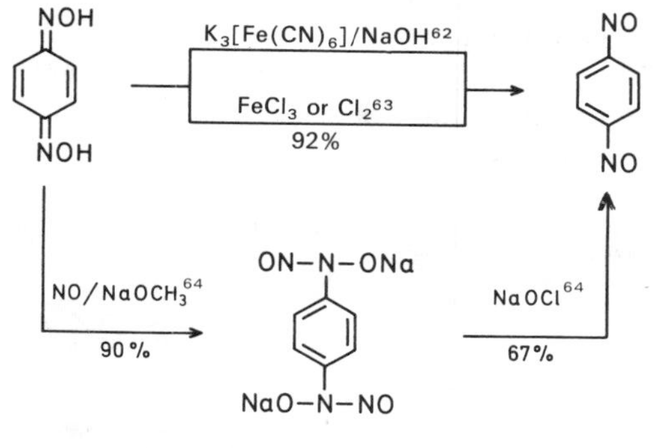

Finally, azomethines and their *N*-oxides, the *nitrones,* too, can be utilized for formation of aromatic *nitroso compounds* by oxidation[65], *e.g.:*

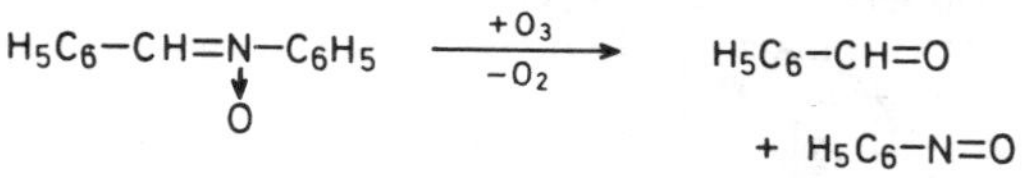

Nitrosobenzene; 30%

2.1.1.3 Reduction of aromatic nitro compounds

During reduction of aromatic nitro compounds success is seldom achieved in trapping the nitroso compound stage. In acid medium the reaction proceeds rapidly to *hydroxylamines* and *amines;* under basic conditions the aromatic nitroso compounds undergo condensation reactions. The best chance of obtaining the nitroso stage is over the neutral region at small redox potentials. Most reduction techniques described in the literature afford poor yields[74]. Reduction of 4-nitro-3,5-diphenylpyrazole[75] and of 2,4,6-trinitroazobenzene[76] are two examples:

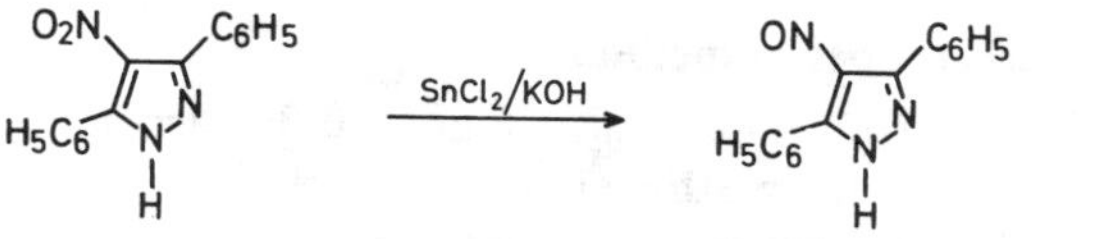

4-Nitroso-3,5-diphenyl-pyrazole; 84%

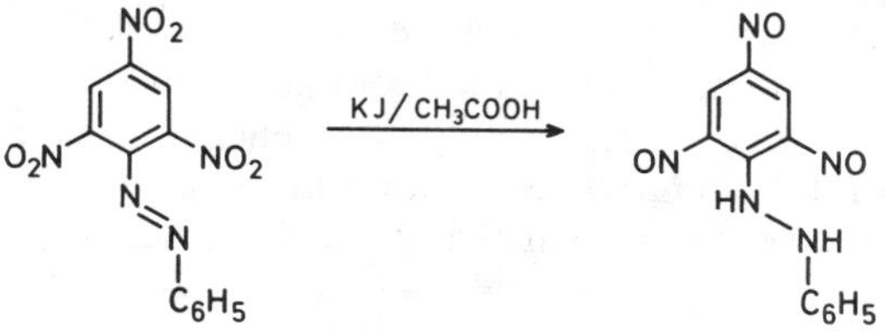

2,4,6-Trinitrosohydrazo-benzene

A number of aromatic nitro compounds display *intramolecular redox reactions* leading to nitroso compounds *via* displacement of oxygen, *e.g.:*

[37] *C. Cloëz,* Compt. Rend. *124,* 898 (1897).
[38] *A.I. Titov,* Zh. Obshch. Khim. *8,* 1483 (1938) Engl.: C.A. *33,* 4248 (1939);
  L. Gatterman, A. Skita, Ber. *49,* 499 (1916);
  A. Baeyer, Justus Liebigs Ann. Chem. *130,* 140 (1864);
  N. Chatak, S. Dutt, J. Indian Chem. Soc. *5,* 665 (1928);
  W. Traube, Justus Liebigs Ann. Chem. *331,* 73 (1904);
  F.E. King, T.J. King, J. Chem. Soc. (London) 943 (1947).
[39] *M.A.T. Rogers,* J. Chem. Soc. (London) 590 (1943);
  F. Angelico, E. Calvello, Gazz. Chim. Ital. I, *34,* 38 (1904).
[40] *J.M. Tedder, B. Webster,* J. Chem. Soc. (London) 3270 (1960).
[41] *L. Gattermann, T. Wieland,* Die Praxis der organischen Chemie, 41. Aufl., p. 271, Walter de Gruyter & Co., Berlin 1962.
[42] *D. Vorländer, E. Siebert,* Ber. *52,* 283, 288 (1919).
[43] *P. Groenewoud, R. Robinson,* J. Chem. Soc. (London) 1696 (1934).
[44] *J. Piccard, M. Kharasch,* J. Amer. Chem. Soc. *40,* 1074 (1918);
  J. Piccard, Helv. Chim. Acta *1,* 134 (1918); *J. Piccard,* Ber. *59,* 1653 (1926).
[45] *E. Schmidt,* Ber. *52,* 410 (1919).
[46] *G.B. Bachman, T. Hokama,* J. Amer. Chem. Soc. *79,* 4370, 4373 (1957).
[47] *J. Biehringer, W. Borsum,* Ber. *49,* 1405 (1916);
  G.A. Elliott et. al., J. Chem. Soc. (London) 1219, 1230 (1926).
[48] *N. Campbell, R.C. Cooper,* J. Chem. Soc. 1208 (1935);
  P.E. Verkade, J. Lieste, E.G.G. Werner, Rec. Trav. Chim. Pays-Bas *64,* 289 (1945).

[49] *L. Knorr,* Justus Liebigs Ann. Chem. *238,* 212 (1887).
[50] Jap. P. 65/24377 (1963), Shionogi a. Co., Inv.: *A. Takamizawa, S. Hayashi;* C.A. 64, 5110g (1966).
[51] *R. Hubner,* Ber. *35,* 731 (1902).
[52] *E. Fischer,* Ber. *29,* 2062 (1896).
[53] *W.J. Mijs, S.E. Hoekstra, R.M. Ulmann, E. Havinga,* Rec. Trav. Chim. Pays-Bas *77,* 746 (1958).
[54] *E. Bamberger, R. Hübner,* Ber. *36,* 3803 (1903).
[55] *J. D'Ans, A. Kneip,* Ber. *48,* 1136 (1915).
[56] *G.M. Brooke, J. Burdon, J.C. Tatlow,* Chem. & Ind. (London) 832 (1961).
[57] *L. Di Nunno, S. Florio, P.E. Todesco,* J. Chem. Soc. [C] 1433 (1970).
[58] *E. Bamberger, F. Tschirner,* Ber. *32,* 342 (1899).
[59] *J.H. Boyer,* Formation of the Nitroso Group and Its Reactions in *H. Feuer:* The Chemistry of the Nitro and Nitroso Groups I, p. 242, Interscience Publishers, John Wiley & Sons, New York 1969.
[60] *D.J. Berry, I. Collins, S.M. Roberts, H. Suschitzky, B.J. Wakefield,* J. Chem. Soc. [C] 1285 (1969).

4-Nitroaniline $\xrightarrow{H_2SO_4/SO_3}$ *2-Amino-5-nitrosophenol*
   + 3-Amino-6-nitrosophenol[77]

2-Nitrostyrene $\xrightarrow{H_2SO_4}$ *2'-Nitrosoacetophenone*[78]

1,8-Dinitronaphthalene $\xrightarrow{H_2SO_4/SO_3}$ *5-Nitroso-1-naphthol*[79]

In addition to being brought about by concentrated or fuming sulfuric acid, these processes can proceed photochemically:

2-Nitrobenzyl alcohol $\xrightarrow{h\nu}$ *2-Nitrosobenzaldehyde*[80]

2-Nitrobenzaldehyde $\xrightarrow{h\nu}$ *2-Nitrosobenzoic acid*[81]

4-Nitrobenzaldehyde $\xrightarrow{h\nu}$ *4-Nitrosobenzoic acid*[82]

2-Nitrotriphenylmethane $\xrightarrow{h\nu}$ *2-(Nitrosophenyl)benzhydrol*[83]

2.1.1.4 Special methods

Apart from direct nitrosation and nitrosating decarboxylation, there is a third method for introducing the nitroso group into aromatic ring systems by the agency of a substitution reaction. It involves treating the corresponding *organomercury*[84], *organotin*[85], or *organomagnesium*[86] compounds with nitrosyl chloride[84, 85], nitrosyl bromide[84], dinitrogen trioxide[54], or dinitrogen tetroxide[87]. The reactions of polymethylphenylmercury chloride[84] and arylmethyltin[85] with nitrosyl chloride have given particularly good results, *e.g.*:

Nitrosomesitylene

4-Nitrosoanisole

A further method of ring nitrosation is the Fischer-Hepp rearrangement[88] of the *N*-nitroso derivatives of secondary *N*-alkyl- and *N*-arylanilines. Nitrosation of secondary aromatic amines in general ensues as a reversible process on the nitrogen. In alcoholic hydrogen chloride or hydrogen bromide intermolecular cationotropic rearrangement to the 4-nitrosoanilines occurs[89]:

[66] *L. Gattermann, H. Wieland*, Die Praxis des organischen Chemikers, p. 158, Walter de Gruyter & Co., Berlin 1962.

[67] *E.C. Taylor, F. Yoneda*, Chem. Commun. 199 (1967).

[68] *F.J. Alway, R.A. Gortner*, Ber. *38*, 1899 (1905).

[69] *F. Parisi, P. Bovina, A. Quilico*, Gass. Chim. Ital. *90*, 903, 914 (1960);
E. Ochiai, H. Mitarashi, Itsuu Kenkyushu Nempo *13*, 19 (1963); C.A. *60*, 5336h (1964).

[70] *A. Giner-Sorolla*, J. Heterocycl. Chem. *7*, 75 (1970).

[71] *R. Willstätter, H. Kubli*, Ber. *41*, 1936 (1908).`

[72] *W.H. Beisler, L.W. Jones*, J. Amer. Chem. Soc. *44*, 2296, 2305 (1922).

[73] *J.H. Boyer* in *H. Feuer*, Formation of the Nitroso Groups and Its Reactions: The Chemistry of the Nitro and Nitroso Groups I, John Wiley & Sons, New York 1969.

[74] Ger. Pat. 192519; Erf.: *O. Dieffenbach*; C.A. *2*, 1765 (1908);
T. Zerevitinov, I. Ostromisslenski, Ber. *44*, 2402 (1911);
J. Meisenheimer, Ber. *36*, 4174 (1903);
Ger. Pat. 194811, Kalle and Co., C.A. *2*, 2304 (1908);
W.M. Lauer, M.M. Sprung, C.M. Langkamerer, J. Amer. Chem. Soc. *58*, 225 (1936);
M.F. Abidova, V.K. Pitsaris, A.S. Sultanov, L.Kh. Freidlin, Uzb. Khim. Zh. *7*, 60 (1963); C.A. *59*, 5049 (1963);
L.A. Errede, H.R. Davis, J. Org. Chem. *28*, 1430 (1963);
J. Meisenheimer, E. Hesse, Ber. *52*, 1173 (1919);
J. Meisenheimer, E. Patzig, Ber. *39*, 2526 (1906);
J. Meisenheimer, Justus Liebigs Ann. Chem. *323*, 205 (1902);
A. Wohl, W. Aue, Ber. *34*, 2442 (1901);
A. Wohl, Ber. *36*, 4135 (1903);
I.J. Rinks, Chem. Weekblad, II, 1961 (1914);
Jap. P. 4329 (1953), Inv.: *S. Kobayashi, Y. Aoyama*, C.A. *49*, 4712 (1955);
W.P. Norris, J. Amer. Chem. Soc. *81*, 4239 (1959);
E. Koerner, V. Gustorf, M.J. Jun, Z. Naturforsch. *20b*, 521 (1965);
E. Bamberger, R. Hubner, Ber. *36*, 3822 (1903).

[61] *S.M. Roberts, H. Suschitzky*, J. Chem. Soc. [C] 1537 (1968).

[62] *R. Nietzki, F. Kehrmann*, Ber. *20*, 613 (1887).

[63] *Y.S. Khishchenko, M.A. Makarov, G.A. Gareev, N.A. Cherkashina, A.S. Koptina*, Zh. Prikl. Khim. *42*, 2384 (1969); C.A. *72*, 31348 (1970).

[64] *R. Nietzki, F. Kehrmann*, Ber. *20*, 613 (1887);
P. Mehne, Ber. *21*, 729 (1888).

[65] *A.H. Riebel, R.E. Erickson, C.J. Abshire, P.S. Bailey*, J. Amer. Chem. Soc. *82*, 1801 (1960).

The rearrangement is suppressed if the *p*-position is occupied[89] or if the group R on the nitrogen is very large[90]. Secondary 1-nitrosonaphthylamines isomerize to the corresponding *1-nitroso-4-naph-thylamines*[89]. A particularly interesting example is the *Fischer-Hepp rearrangement* of *N,N'*-dimethyl-*N,N'*-dinitroso-*m*-phenylenediamine to *N,N'*-dimethyl-4,6-dinitroso-m-phenylenediamine[91].

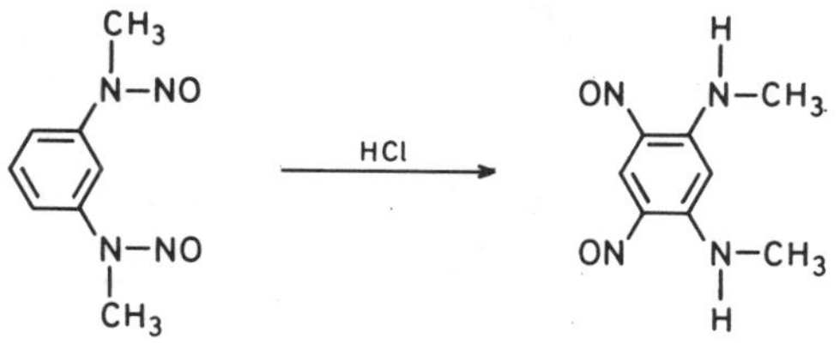

The *quinone monooximes*[92] obtained from quinones with hydroxylamine in weakly acid solution are in tautomeric equilibrium with the *nitrosophenols:*

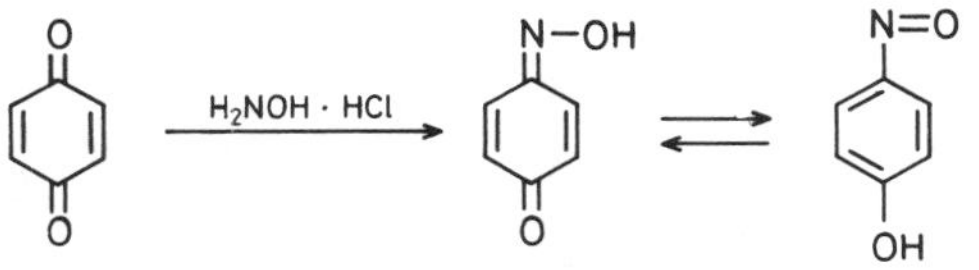

Finally, attention is drawn to the preparation of some heteroaromatic *C*-nitroso compounds by *ring closure reactions. 4-Nitrosopyrazoles* are formed from 1,2,3-trione 2-oximes and hydrazines[93]:

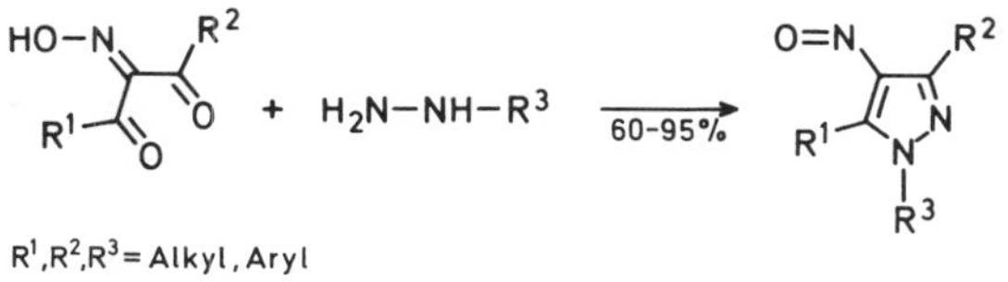

R¹,R²,R³ = Alkyl, Aryl

Heating amidinium salts of mesoxalonitrile oxime in high-boiling solvents such as 5-ethyl-2-methyl-pyridine leads to formation of *4,6-diamino-5-nitrosopyrimidines*[94]:

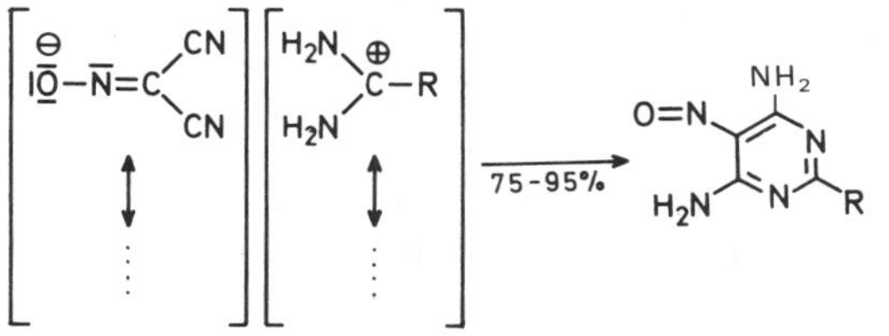

R = Alkyl, Aryl(−NH₂,−S−Alkyl)

Guanidine or 5-methylisothiourea may be used in place of the amidine[94, 95].

2.1.2 Transformation reactions and analytical hints

Ring substitutions of aromatic nitroso compounds lead to new nitroso compounds unless the nitroso group is attacked at the same time. During electrophilic substitution the entering substituent is directed to the *m*-position in monomeric nitroso compounds, but generally into the *p*-position in dinitroso compounds. *4-Nitrosophenols*[96] and *4-nitrosoanilines*[97] are often interconvertible:

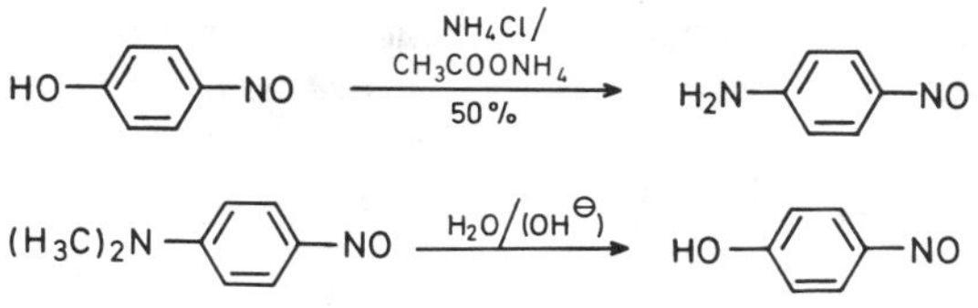

[75] *R. Hüttel, F. Büchele, P. Jochum,* Ber. 88, 1577 (1955).
[76] *C. Willgerodt,* Ber. 24, 592 (1891).
[77] *J. Horyna,* Collect. Czech. Chem. Commun. 24, 3579 (1959).
[78] *Y.S. Shabarov, S.S. Mochalov, I.P. Stepanova,* Dokl. Akad. Nauk SSSR 189, 1028 (1969); C.A. 72, 66523 (1970).
[79] DRP. 90 414 (1895), BASF; *Beilstein 7,* 732
[80] *E. Bamberger,* Ber. 51, 606 (1918).
[81] *G. Ciamician, P. Silber,* Ber. 34, 2040 (1901).
[82] *G.W. Wubbels, R.R. Havtala, R. Letsinger,* Tetrahedron Lett. 1689 (1970).
[83] *I. Tanasescu,* Bull. Soc. Chim. France 39, 1443 (1926).
[84] *A. Baeyer,* Ber. 7, 1638 (1874);
L.I. Smith, F.L. Taylor, J. Amer. Chem. Soc. 57, 2460 (1935).
[85] *E.H. Bartlett, C. Eaborn, D.R.M. Walton,* J. Chem. Soc. [C] 1717 (1970).

[86] *B. Oddo,* Gazz. Chim. Ital. 39, I, 659 (1909).
[87] *E. Bamberger,* Ber. 30, 506 (1897).
[88] *O. Fischer, E. Hepp,* Ber. 19, 2991 (1886);
O. Fischer, E. Hepp, Ber. 20, 1247, 247 (1887).
[89] *P.W. Neber, H. Rauscher,* Justus Liebigs Ann. Chem. 550, 182 (1942);
J. Glazer, E.D. Hughes, C.K. Ingold, A.T. James, E. Roberts, J. Chem. Soc. (London) 2657 (1950).
[90] *J. Willenz,* J. Chem. Soc. (London) 1677 (1955).
[91] *O. Fischer, E. Diepolder,* Justus Liebigs Ann. Chem. 286, 145 (1895).
[92] *H. Goldschmidt,* Ber. 17, 213 (1884).
[93] *S.F. Torf et al.,* Zh. Obshch. Khim. 32, 1740 (1962);
R. Hüttel, F. Büchele, P. Jochum, Ber. 88, 1577 (1955);
E.N. Padeiskaya, Vestn. Mosk. Univ., Ser. II Chim. 18, 69 (1963);
Chem. Zentr. 13 (1965);
USP. 2 510 724 (1950), United States Rubber Co., Inv.: *N.K. Sundholm;* C.A. 44, 8375i (1950);
G.L. McNew, N.K. Sundholm, Phytopathology 39, 721 (1949); C.A. 44, 781h, 4183g (1950);

Should resinification occur during the alkaline cleavage of *N,N-dialkyl-4-nitrosoaniline,* then it is advantageous to work with excess sodium hydrogen sulfite solution[98].

2.1.2.1 Oxidation

Aromatic nitroso compounds can be oxidized to the corresponding *nitro* compounds with nitric acid[99], with hydrogen peroxide in alkaline medium[100], with peroxytrifluoroacetic acid[101], and with other oxidizing agents such as Caro's acid, permanganate, hexacyanoferrate(III) (see p. 12).

2.1.2.2 Reduction

Aromatic nitroso compounds are reduced to *amines* by catalytic hydrogenation or by suitable reducing agents such as Raney nickel, sodium dithionite, tin-hydrochloric acid, or tin(II) chloride-hydrochloric acid (see p. 494).

Hydrogen iodide is a suitable selective reducing agent for reducing nitroso groups in the presence of nitro groups[102].

With metallic sodium, hydrogen sulfide, or hydrazobenzene, reduction of aromatic nitroso compounds stops at the *N-arylhydroxylamine* stage (see p. 65).

2.1.2.3 Condensation

A series of condensation reactions by nitroso compounds is reminiscent of aldehydes. Thus, water is split off with primary aromatic amines and (unsymmetrical) *azo compounds* are formed (see p. 133).

From 4-nitrosophenols or *N,N*-dialkyl-4-nitrosoanilines this technique leads to indophenol, indamine, phenazine, oxazine, and azapolymethine dyes.

While nitrosobenzene can be converted into a *benzene diazonium salt*[103] with hydroxylamine (see p. 175) and into *phenyl azide*[104] with hydroxylamine-*O*-sulfonic acid, azoxy compounds are formed with *N*-arylhydroxylamines (see p. 138).

In addition to primary amines, hydrazines, and hydroxylamines, *CH*-acid compounds such as nitroalkanes[105], arylacetonitriles[106], 2,4-diphenylpyrrole[107], flavanone[108], 2,5-piperazinedione[109], 3-indolol[110], and 3-thionaphthol[111] above all can be condensed with nitroso compounds. Condensation is performed in the presence of alkaline condensing agents such as caustic soda, sodium carbonate, sodium ethoxide, or sodium cyanide.

[97] *A. Bayer, M. Caro,* Ber. 7, 963 (1874);
J. Meisenheimer, Justus Liebigs Ann. Chem. *428,* 256 (1922);
K. Zeile, H. Meier, Hoppe-Seyler's Z. Physiol. Chem. *256,* 137 (1938).

[98] *J. v. Braun, K. Heider, E. Müller,* Ber. *51,* 737 (1918);
R. Munch, G.T. Thannhauser, D.L. Lottle, J. Amer. Chem. Soc. *68,* 1297 (1946).

[99] *C.K. Ingold,* J. Chem. Soc. *127,* 515, 517 (1925);
J. Meisenheimer, E. Hesse, Ber. *52,* 1173, 1175 (1918);
R. Kuhn, Ber. *70,* 1296 (1937);
R. Kuhn, Ber. *71,* 779 (1938).

[100] *E. Bamberger,* Ber. *33,* 120 (1900);
H.H. Hodgson, E. Kilner, J. Chem. Soc. *125,* 809 (1924);
H.H. Hodgson, H. Clay, J. Chem. Soc. 2777 (1929);
F.L. Gilbert, J. Chem. Soc. 2301 (1927).

[101] *J.H. Boyer, S.E. Ellzey,* J. Org. Chem. *24,* 2038 (1959);
E.C. Taylor, A. McKillop, J. Org. Chem. *30,* 3153 (1965).

[102] *J.H. Boyer, W. Schoen,* J. Amer. Chem. Soc. *78,* 423 (1956).

[103] *E. Bamberger,* Ber. *28,* 1218 (1895).

[104] *J.P. Anselme, N. Koga,* Chem. Commun. 443 (1970).

[105] *F. Kröhnke, H.H. Steuernagel,* Chem. Ber. *96,* 494 (1963).

[106] *P. Ehrlich, F. Sachs,* Ber. *32,* 2341 (1899);
T.S. Osdene, G.M. Timmis, Chem. & Ind. (London) 404 (1954).

[107] Brit.P. 562 756 (1942), I.C.I., Inv.: *E.P. Goodings, M.A.T. Rogers;* C.A. *40,* 3617 (1946).

[108] *D. Evans, I.M. Lockart,* J. Chem. Soc. [C] 711 (1966).

[109] *M. Augustin,* J. Prakt. Chem. [4] *32,* 158 (1966).

[110] *R. Pummerer, M. Goettler,* Ber. *42,* 4269 (1909);
R. Pummerer, H. Goettler, Ber. *43,* 1376 (1910).

[111] *R. Pummerer,* Ber. *43,* 1370 (1910).
DRP. 214 781 (1907), BASF; Chem. Zentr. II, 1603 (1909). Fortschritte der Teerfabrikation u. verwandter Industriezweige 9, 568 (1907);
J. Harley-Mason, F.G. Mann, J. Chem. Soc. (London) 410 (1942);
G.E. Dalgliesh, F.G. Mann, J. Chem. Soc. (London) 901 (1945).

USP. 2 832 785 (1958), Lakeside Laboratories Inc., Inv.: *J.H. Biel;* C.A. *52,* 15595 (1958);
USP. 2 751 395 (1956), May a. Baker Ltd., Inv.: *W.A. Freeman, R. Slack;* C.A. *51,* 2054 (1957);
F. Sachs, P. Alsleben, Ber. *40,* 664, 672 (1907);
R. Fusco, S. Rossi, Tetrahedron 3, 220 (1958).

[94] *E.C. Taylor, O. Vogl, C.C. Cheng,* J. Amer. Chem. Soc. *81,* 2442 (1959);
USP. 2 963 478 (1960), Smith, Kline a. French Laboratories, Inv.: *J. Weinstock;* C.A. *55,* 10483 (1961).

[95] Fr. P. 1 364 734 (1963), Laboratoires Lumière, Inst. Merieux; C.A. *62,* 572 (1965).

[96] *O. Fischer, E. Hepp,* Ber. *20,* 2475 (1887);
O. Fischer, E. Hepp, Ber. *21,* 684 (1888);
E.W. Malmberg, C.J. Hamilton, J. Amer. Chem. Soc. *70,* 2415 (1948);
O. Fischer, C. Dietrich, F. Weiss, J. Prakt. Chem. [2] *100,* 167 (1920).

In addition to the anticipated *azomethines*, this reaction, which bears the names of Ehrlich and Sachs[106], often leads to formation of *nitrones* (by virtue of the dehydrogenating effect of the nitroso compounds) (see p. 336). As the *Kröhnke* synthesis, the nitrone variant has achieved preparative importance for preparing aldehydes from alkyl halides[112] (see Vol. V).

In addition, nitroso compounds display a marked tendency to react with aromatic hydroxy or amino compounds with formation of *quinonoid* systems[113], *e.g.*:

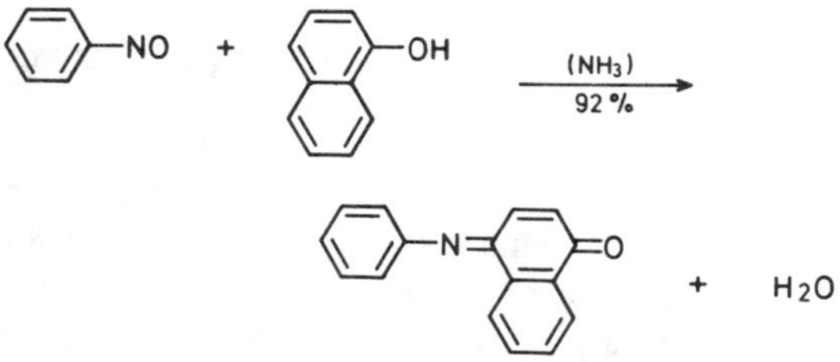

2.1.2.4 Addition

The $O{=}N$ double bond of the nitroso compounds is capable of undergoing a series of addition reactions with formation of *hydroxylamines* (see p. 55). Examples of reaction partners are olefins[114], dienes[119], acetylenes[115], allenes[116], ketenes[117], quinones[118], diazoalkanes[120], ylides[121], and organometallic compounds[122, 123, 119].

2.1.2.5 Photochemistry

Aromatic nitroso compounds absorb at ~ 7000 Å in the visible region ($n_N \rightarrow \pi^*$-transition). The *UV* bands of the nitroso group coincide with the absorption by aromatic compounds[124] (cf. Vol. 1, p. 258). While nitrosobenzene and other aromatic nitroso compounds are remarkably resistant to red light, they do display some characteristic photoreactions on *UV* irradiation, of which *nitroxide* formation[125]

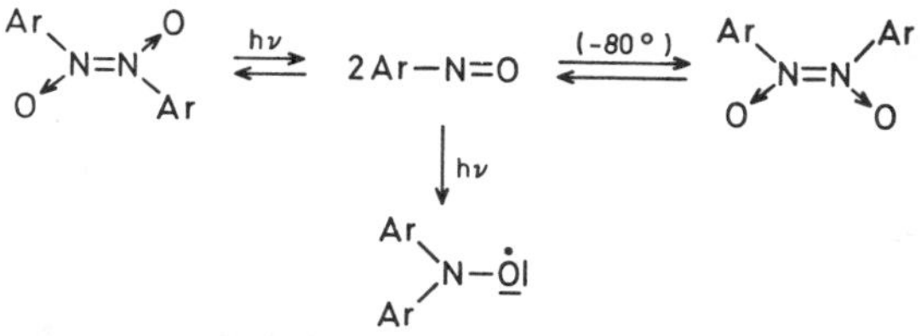

and photoreduction to *azoxy* compounds[126] may be mentioned here.

2.1.2.6 Analysis

The following spectroscopic data are useful for the rapid identification of aromatic nitroso compounds:

IR: (Vol. I, p. 268)	NO-valence vibrations of the monomers	$1538–1621$ cm^{-1}	Ref.: [127]
	trans-Dimers	$1253–1299$ cm^{-1}	[128]
	cis-Dimers	$1389–1397$ cm^{-1} ~ 1409 cm^{-1}	
UV: (Vol. I, p. 258)	$n_N \rightarrow \pi^*$-transition of the monomers	$680–760$ mμ (ε : $40–70$)	[129]
	Further transitions	~ 300 mμ (ε : ~ 5000) ~ 280 mμ (ε : $\sim 10,000$)	

[112] *F. Kröhnke*, Ber. *71*, 2584 (1938).
   F. Kröhnke, E. Börner, Ber. *69*, 2006 (1936).
[113] *L. Sander*, Ber. *58*, 824 (1925);
   H. Euler, Ber. *39*, 1035 (1906).
[114] *L. Alessandri*, Atti Accad. Naz. Lincei, Classe Sci. Fis. Mat. Natur. Rend. [5] *24*, 62 (1915);
   N.F. Hepfinger, C.E. Griffin, Tetrahedron Lett. 1361 (1963).
[115] *L. Alessandri*, Gazz. Chim. Ital. *52*, 193 (1922);
   L. Alessandri, Gazz. Chim. Ital. *54*, 426 (1924);
   L. Alessandri, Gazz. Chim. Ital. *55*, 729 (1925).
[116] *R.H. Howe*, J. Org. Chem. *33*, 2848 (1968).
[117] *H. Meier*, unpubl.
[118] *W. Gundel, R. Pummerer*, Justus Liebigs Ann. Chem. *529*, II (1937).
[119] *S.B. Needleman, M.C. Chang kno*, Chem. Rev. *62*, 405 (1962);
   G. Kresze, J. Firl, Fortschr. Chem. Forsch. II, 245 (1969).
[120] *H. v. Pechmann*, Ber. *28*, 855 (1895);
   H. v. Pechmann, Ber. *30*, 2461, 2871 (1897).
[121] *A.W. Johnson, R.B. La Count*, J. Amer. Chem. Soc. *83*, 417 (1961);
   DAS 1048568 (1957), BASF, Erf.: *G. Wittig, U. Schöllkopf, H. Pommer;*
   Chem. Zentr. 13313 (1959);
   A. Schönberg, K.H. Brosowski, Ber. *92*, 2602 (1959).
[122] *B. Buck, G. Köbrich*, Tetrahedron Lett. 1563 (1967);
   H. Wieland, Ber. *45*, 494 (1912);

   H. Wieland, Ber. *47*, 2113 (1914);
   H. Wieland, Ber. *48*, 1117 (1915);
   H. Wieland, Ber. *53*, 216 (1920);
   H. Wieland, Ber. *55*, 1802 (1922).
[123] *H. Wieland, A. Roseau*, Ber. *45*, 49 (1912);
   H. Wieland, K. Roth, Ber. *53*, 222 (1920).
[124] *K. Nakamoto, K. Suzuki*, J. Chem. Phys. *20*, 1971 (1952);
   K. Nakamoto, R. Rundle, J. Chem. Soc. *78*, 1116 (1956).
[125] *K. Maruyama, R. Tanikaga, R. Goto*, Bull. Chem. Soc. Japan *37*, 1893 (1964);
   A. Mackor, Th. Wajer, Th. de Boer, J. van Voorst, Tetrahedron Lett., 2115 (1966).
[126] *H. Mauser, H. Heitzer*, Z. Naturforsch. *20b*, 200 (1965).
[127] *W. Lüttke*, Z. Elektrochem. *61*, 302 (1957);
   W. Lüttke, Z. Elektrochem. *61*, 976 (1957);
   J. Mason, J. Dunderdale, J. Chem. Soc. (London) 754 (1956).
[128] *B.G. Gowenlock, W. Lüttke*, Quart. Rev. *12*, 321 1958).
[129] *K. Tabei, S. Nagakura*, Bull. Chem. Soc. Japan *38*, 965 (1965).

2.2 Aliphatic nitroso compounds

This section describes compounds which contain the nitroso group linked to an aliphatic carbon atom. Certain characteristic features distinguish these *C-nitroso compounds* from *N*-nitroso and *O*-nitroso compounds, especially their green to blue color in the monomeric state and their tendency to form colorless dimers:

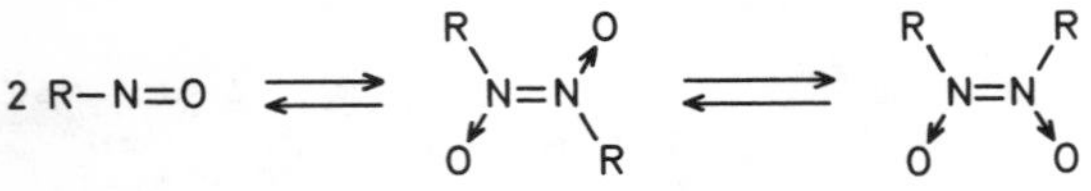

Primary and *secondary* dinitrosoalkanes form colorless solutions and some even form colorless melts. However, on heating the solutions or melts a blue color appears; rapid cooling regenerates the *dimeric* colorless *nitroso compounds* unchanged. *Tertiary* nitroso hydrocarbons pass over into the monomeric form much more readily and more completely in solution or in the molten state. *Geminal nitroso-nitro* compounds (pseudonitroles, **1**) are generally dimeric and colorless at room temperature.

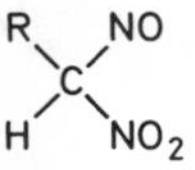

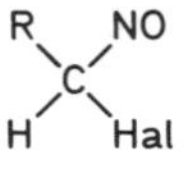

Geminal halonitroso compounds with the structure **2** are dimers at room temperature, while higher branched Type **3** compounds are monomeric even at low temperatures. Most of the dinitroso compounds prepared by conventional methods probably have the *trans configuration*. Intramolecular dinitroso compounds have an exclusively *cis* configuration **4**.
1, 2-Dinitrosoalkenes possess a furoxane structure **5**.

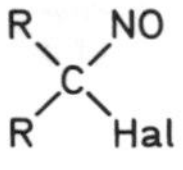

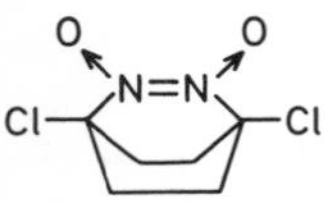

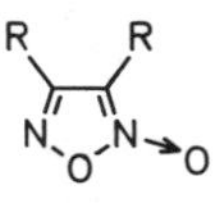

A characteristic property of the primary and secondary *C*-nitroso compounds is their ability of being readily transformed into the isomeric *oximes*:

2.2.1 Preparation

2.2.1.1 Replacement of hydrogen by the nitroso group

Direct nitrosation of hydrocarbons can be achieved as a free-radical gas-phase process[1] or by irradiation[2]:

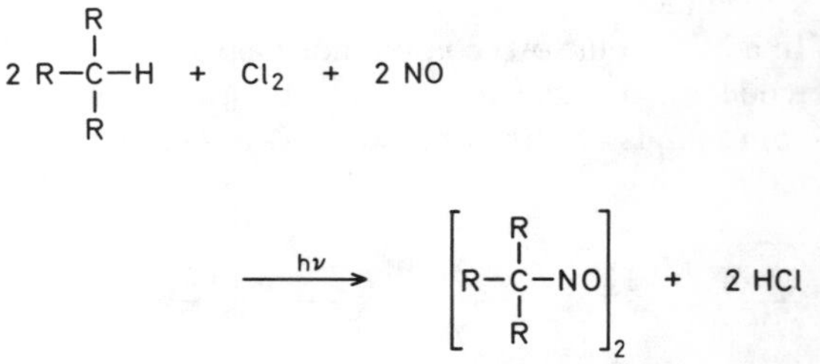

Depending on the hydrocarbon to be nitrosated, a gas mixture containing from 5 to 10 molecules nitrogen monoxide per molecule chlorine is used. The reaction is dependent on the wavelength of the irradiated light; the 380–420 mμ region is particularly effective[3]. When nitrosyl chloride is employed in place of the mixture of nitrogen monoxide and chlorine, then in general *oximes* and *geminal chloronitroso* compounds are formed. However, if the hydrogen chloride formed during the reaction is expelled with nitrogen[4], or if excess nitrogen monoxide is employed[5], *dinitroso* compounds are formed here as well.

Detailed investigations of the quantum yield[6] allow the postulate of a four-center mechanism for the photonitrosation with nitrosyl chloride.

Photonitrosation can be used for preparing both *primary* and *secondary nitroso* compounds or their dimers. In this way *bis(α-nitrosotoluene)* and *bis(nitrosocyclohexane)* are obtained in the pure form and in good yield from toluene and cyclohexane[7]. When heptane is made the starting material a mixture of position-isomeric *bis(nitrosoheptanes)* can be isolated[7]. During the photochemical reaction of nitrosyl chloride or of nitrogen monoxide/chlorine mixtures with *methyl* or *methylene* groups *nitroso* compounds are formed initially quite generally; a secondary dark reaction in the presence of formed or added hydrogen chloride causes them to undergo rearrangement into *oximes*. The behavior of the very reactive *methine* groups is still largely unsolved.
An alternative route for the photochemical nitrosation of alkanes, cycloalkanes, and alkylbenzenes containing an abstractable *primary* or *secondary* hydrogen starts from *alkyl nitrites,* for example, *tert*-butyl nitrite[8]. Irradiation at 0–5° with wavelengths of around 400 mμ leads to formation of

the corresponding *trans-dinitroso compounds* in yields of up to 80% side by side with a little oxime[9]. At even lower temperatures and with energy-richer radiation up to 30% of the *cis*-compound can be isolated[8, 10]:

$$(H_3C)_3C-O-NO \xrightarrow[-\overset{\cdot}{N}O]{h\nu} (H_3C)_3C-\overset{..}{\underset{..}{O}}\cdot$$

$$\xrightarrow[-R\cdot]{R-H} (H_3C)_3C-OH$$

$$2\ R\cdot\ +\ 2\ NO \longrightarrow$$

$$2\ R-NO \rightleftharpoons [R-NO]_2$$

Tertiary hydrogen atoms fundamentally react even more readily[11] in the manner described here. However, the tertiary nitroso compounds formed are largely present as monomers in solution and therefore are exposed to a number of secondary reactions such as attack by the *tert*-butoxy radicals.

Using a special irradiation technique by working in a soxhlet apparatus avoids this complication and *tertiary nitroso compounds,* too, can be synthesized in good yield by this technique[11].

Replacement of a hydrogen atom by the nitroso group in *activated C–H bonds* can be achieved also by a polar mechanism. Electron-attracting groups such as carbonyl, carboxy, ester, cyano, imino, nitro, and aryl functions are suitable activators.

Ketones may be nitrosated with nitrous acid, alkyl nitrites, dinitrogen trioxide, or nitrosyl chloride. Where a methine group that is in the α-position to the carbonyl group is attacked α-nitroso ketones are obtained. By contrast, if the reaction ensues on a methyl or methylene group, the initially formed nitroso compound generally quickly rearranges to the *oxime* and isolation of *α-nitroso ketones* is then possible under special conditions only[12].

The following Table 1 surveys the nitrosation of ketones in relationship to the nitrosation medium.

When dinitrogen trioxide acts on ethyl 2-methyl-3-oxo-3-phenylpropionate *ethyl 2-methyl-2-nitroso-3-oxopropionate* is formed quantitatively[18].

Table 1. Nitroso ketones by nitrosation of ketones

Ketone	Nitrosation method	Reaction product	M. p. [°C]	Yield [% of theory]	Ref.
3,3,7-Trimethyl-4-cyclohepten-1-one	Sodium nitrite/HCl	Bis-[3,3,7-trimethyl-7-nitroso-4-cyclohepten-1-one]	121–124	—	13
3,3,7-Trimethyl-4-cyclohepten-1-one	Pentyl nitrite/HCl	Bis-[3,3,7-trimethyl-7-nitroso-4-cyclohepten-1-one]	121–124	—	14
2,5-Dimethyl-3-hexanone	Ethyl nitrite/HCl	Bis-[2,5-dimethyl-2-nitroso-3-hexanone]	90	51	15
2,5-Dimethyl-3-hexanone	Ethyl nitrite/acetyl chloride	Bis-[2,5-dimethyl-2-nitroso-3-hexanone]	90	50	16
1,3-Diphenyl-1,3-propanedione	Dinitrogen trioxide/ether	Bis-[2-nitroso-1,3-diphenyl-1,3-propanedione]	125	50–60	17

[1] Belg. P. 649 637 (1963), Société Edison.
[2] *Eu. Müller, H. Metzger,* Chem. Ber. *87,* 1282 (1954); *88,* 165 (1966); DBP. 958 840 (1954); 966 055 (1955), Erf.: *Eu. Müller, H. Metzger, D. Fries;* C.A. *53,* 100 77 (1959).
[3] *H. Metzger, Eu. Müller,* Chem. Ber. *90,* 1179 (1957).
[4] *L.G. Donaruma,* J. Org. Chem. *23,* 1338 (1958).
[5] *Eu. Müller, G. Schmid,* Chem.Ber. *92,* 514 (1959).
[6] *Eu. Müller, G. Fiedler, J. Heiss,* Chem. Ber. *101,* 765 (1968).
[7] *Eu. Müller, H. Metzger,* Chem. Ber. *88,* 165 (1955).
[8] *A. Mackor, T.J. de Boer,* Rec. Trav. Chim. Pays-Bas *89,* 151 (1970).
[9] *A. Mackor, T.J. de Boer,* Rec. Trav. Chim. Pays-Bas *89,* 164 (1970).
[10] *A. Mackor, J.U. Veenland, T.J. de Boer,* Rec. Trav. Chim. Pays-Bas *88,* 1249 (1969).
[11] *A. Mackor, T.J. de Boer,* Rec. Trav. Chim. Pays-Bas *89,* 159 (1970).
[12] *O. Touster,* Org. Reactions *7,* 358 (1953).
[13] *A. v. Baeyer,* Ber. *27,* 1923 (1894).

While *acyclic* β-oxocarboxylic acid esters often fragment during nitrosation with loss of an acyl group[19], *cyclic* β-oxocarboxylic acid esters can be converted into *tertiary nitroso compounds* without a change in the carbon skeleton[20]. Diethyl-2,5-di-oxo-1,4-cyclohexanedicarboxylate with dinitrogen trioxide in ether affords *tetraethyl bis(1,4-dini-troso - 2, 5 - dioxo-1,4 - cyclohexanedicarboxylate* (25–51%)[21]. Substituted malonic acid diesters, like α-monosubstituted β-carbonyl compounds, tend to form oximes so readily that an ester is generally split off:

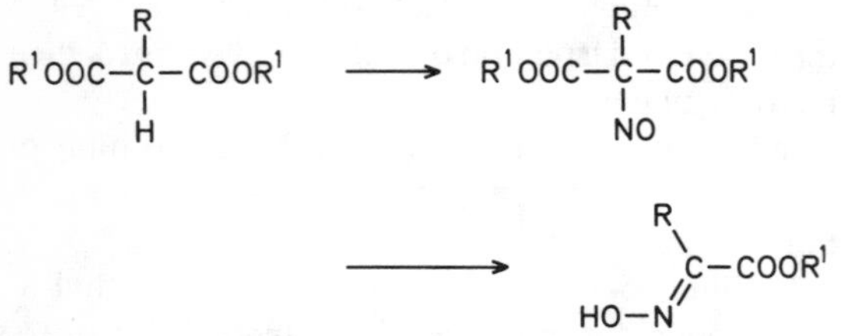

Ethyl tetrahydro-2-oxo-3-furoate forms an exception and with nitrous acid or dinitrogen trioxide can be nitrosated to *ethyl tetrahydro-3-nitroso-2-oxo-3-furoate*[22]:

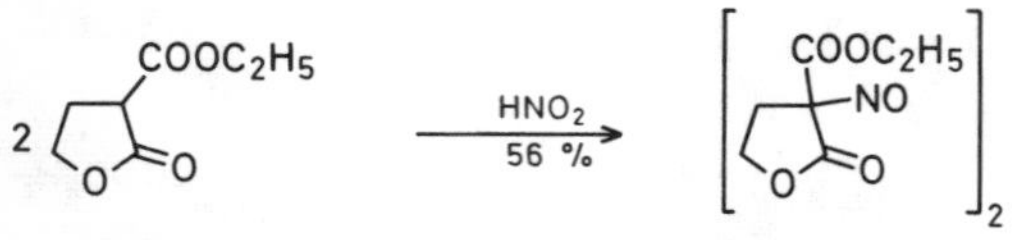

Primary and *secondary nitro compounds* likewise react with nitrous acid. Careful acidification of a solution of a primary nitroparaffin treated with al-kali nitrite furnishes a *nitrolic acid*[23], that of a secondary nitroparaffin a *pseudonitrole*[24]:

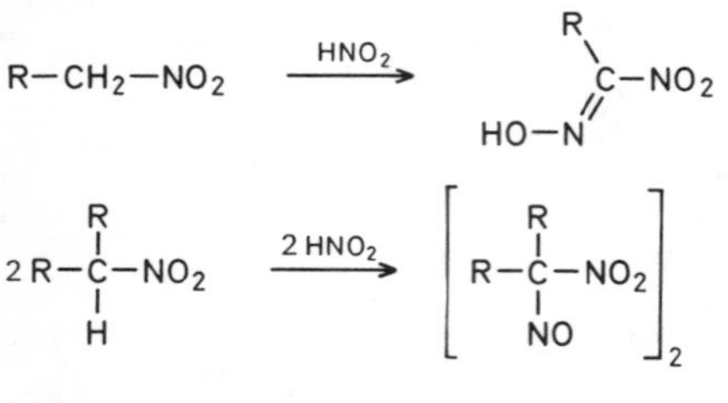

Tertiary nitro compounds do not react with nitrous acid.

Nitrosation of secondary nitro compounds to *pseudonitroles* is, as a rule, performed by allowing alkali nitrite to act on a solution of the nitro compound in caustic potash or soda followed by making this solution acid, *i.e.,* by allowing nitrous acid to act on the *aci*-nitro compound[25]:

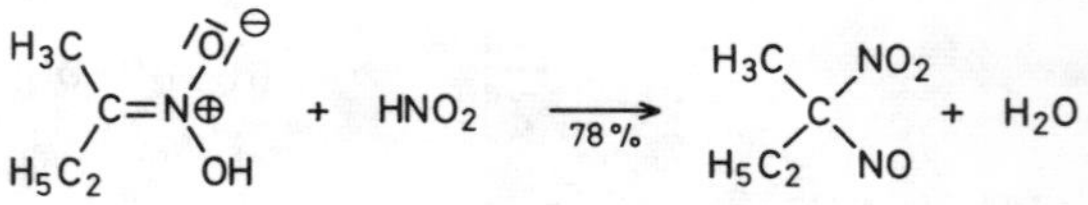

Pseudonitroles can be formed also when alkyl nitrites[26] or nitrosyl chloride[27] act on secondary nitro compounds (in suitable solvents).

2.2.1.2 Replacement of functional groups by the nitroso group

From polyfluoroiodoalkanes and polyfluorobro-moalkanes thermolysis or photolysis furnish *fluo-roalkyl radicals* and elementary iodine or bromine respectively. Working in the presence of nitrogen monoxide, which itself possesses a lone electron, results in the formation of *polyfluoronitrosoal-kanes*[28] in good yield (albeit at low conversions):

Such substitutions can be observed also on the *olefinic* carbon atom[29]:

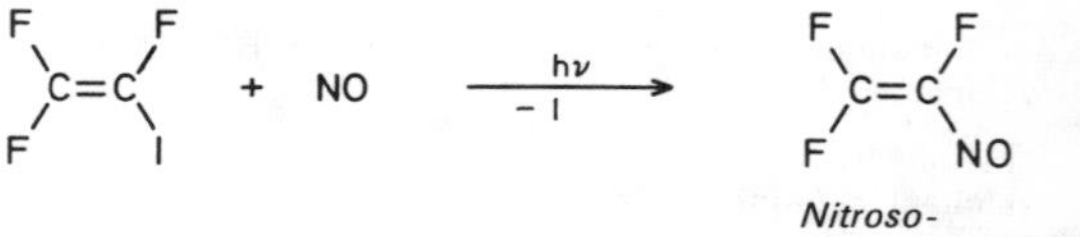

Nitroso-trifluoroethylene

[14] *A. v. Baeyer,* Ber. *28,* 646 (1895).
[15] *J.G. Aston, M.G. Mayberry,* J. Amer. Chem. Soc. *57,* 1888 (1935).
[16] *J.G. Aston, M.G. Mayberry,* J. Amer. Chem. Soc. *57,* 1888 (1935).
[17] *H. Wieland, S. Bloch,* Ber. *37,* 1524 (1904); *37,* 1530 (1904).
[18] *J. Schmidt, E. Aeckerle,* Justus Liebigs Ann. Chem. *398,* 255 (1913).
[19] *J. Schmidt, E. Aeckerle,* Justus Liebigs Ann. Chem. *398,* 251 (1913).

[20] *W. Dieckmann,* Ber. *33,* 581, 580, 594 (1900); *W. Dieckmann, A. Groeneveld,* Ber. *33,* 604 (1900).
[21] *H. Ebert,* Justus Liebigs Ann. Chem. *229,* 54 (1885).
[22] *W.W. Feofilaktow, A.S. Onitshenko,* Zh. Obshch. Khim. *9,* 304 (1939); C.A. *34,* 378 (1940); *H. Snyder, J.H. Andree, G.W. Cannon, C.F. Peters,* J. Amer. Chem. Soc. *64,* 2082 (1942); *W. Reppe,* Justus Liebigs Ann. Chem. *596,* 184 (1955).
[23] *V. Meyer,* Ber. *6,* 1492 (1873); *V. Meyer, J. Lochner,* Ber. 7, 670 (1874).
[24] *V. Meyer, J. Lochner,* Ber. 7, 670, 788, 1506 (1874).
[25] *V. Meyer, J. Lochner,* Ber. 7, 1508 (1874); *V. Meyer, J. Lochner,* Justus Liebigs Ann. Chem. *180,* 135 (1876).
[26] *N. Kornblum, R.K. Blackwood, D.D. Mooberry,* J. Amer. Chem. Soc. *78,* 1502 (1956).

Low haloalkanes, whose halogen atoms cannot be replaced directly by nitroso groups on irradiation, can be converted into *nitroso compounds via the organometallic compounds,* which in general can be prepared readily from them.

Dialkylmercury compounds are particularly suitable for this purpose. Nitrogen monoxide or nitrosyl chloride[30] serves as the nitrosating agent:

$$R-\underset{\underset{R}{|}}{CH}-Hg-\underset{\underset{R}{|}}{CH}-R \xrightarrow[-Hg]{\nabla \ or \ h\nu} 2 \ R-\underset{\bullet}{\underset{\underset{R}{|}}{CH}}$$

$$\xrightarrow{2 \bullet N=O} \left[R-\underset{\underset{R}{|}}{CH}-NO \right]_2$$

Nitrosoacetylenes[31] can be obtained from 1-dialkyn-1-ylmercury compounds at low temperatures with nitrosyl chloride; on warming these rearrange into *2-oxocarbonitriles*:

$$\underset{R-C\equiv C}{\overset{R-C\equiv C}{>}}Hg \xrightarrow[-HgCl_2]{2 \ NOCl} 2 \ R-C\equiv C-\bar{N}=\bar{O}$$

$$\xrightarrow{\nabla} 2 \ \underset{/O/}{\overset{R}{>}}C-C\equiv NI$$

In the aliphatic series an indirect route leads to *nitrosoalkanes* from *Grignard compounds.* *N-nitrosohydroxylamines* readily decomposed in the presence of nitrous acid arise as intermediates[32]:

N-Cyclohexyl-N-nitroso-hydroxylamine

$$\xrightarrow{-[NOH]} \tfrac{1}{2} \left[\text{(cyclohexyl)}-NO \right]_2$$

Bis-[nitrosocyclohexane]

When nitrosyl chloride acts on anhydrous *silver*[33], *mercury(II)*[34], or *lead(II)*[35] salts of *carboxylic acids, acyl nitrites* are formed which can be decarboxylated readily in the gas phase either thermally[33-35] or photolytically[36] to form *C-nitroso compounds* in good yield:

$$R-COOAg \xrightarrow[-AgCl]{NOCl} \underset{/O/}{\overset{R}{>}}C-\bar{O}-\bar{N}=\bar{O}$$

$$\xrightarrow[-CO_2]{\nabla \ or \ h\nu} R\bullet \ + \ I\dot{N}=\bar{O} \longrightarrow R-\bar{N}=O$$

R = Alkyl, Aryl

This nitrosating decarboxylation is closely related to the 'brominating decarboxylation' of silver salts of carboxylic acids commonly called the Hunsdiecker reaction[37]. While the Hunsdiecker degradation can be carried out in solution, disproportionation according to the equation

$$2 \ R-CO-O-NO \longrightarrow$$
$$R-CO-O-CO-R \ + \ N_2O_3$$

D.A. Barr, R.N. Haszeldine, J. Chem. Soc. (London) 1881 (1955); 3416 (1956);
Brit. P. 770619 (1953), Inv.: *R.N. Haszeldine;* C.A. *51,* 14790 (1957);
J. Banus, J. Chem. Soc. 3755 (1953);
J. Banus, Nature *171,* 173 (1953);
J. Mason, J. Dunderdale, J. Chem. Soc. (London) 754 (1956).
29 *C.E. Griffin, R.N. Haszeldine,* Proc. Chem. Soc. 369 (1959); 1398 (1960).
30 *P. Tarrant, D.E. O'Connor,* J. Org. Chem. *29,* 2012 (1964);
H.T.J. Chilton, B.G. Gowenlock, J. Chem. Soc. (London) 3232 (1953);
H.T.J. Chilton, B.G. Gowenlock, Nature *172,* 73 (1953);
H.T.J. Chilton, B.G. Gowenlock, J. Chem. Soc. (London) 3174 (1954);
W.A. Bryce, K.U. Ingold, J. Chem. Phys. *23,* 1968 (1955).
31 *E. Robson, J.M. Tedder,* Proc. Chem. Soc. 13 (1963).
32 *E. Müller, H. Metzger,* Chem. Ber. *89,* 396 (1956).
33 *L. Francesconi, Cialdea,* Gazz. Chim. Ital. *34 I,* 435 (1904);
J. Banus, J. Chem. Soc. (London) 3757, 3759 (1953);
R.N. Haszeldine, J. Jander, J. Chem. Soc. (London) 4172 (1953);
R.E. Banks, R.N. Haszeldine, M.K. McCreath, Proc. Chem. Soc. 64 (1961);
C.W. Taylor, T.J. Brice, R.L. Wear, J. Org. Chem. *27,* 1064 (1962);
W. Pritzkow, M. Nitzer, J. Prakt. Chem. [4] *25,* 69 (1964).

27 *N. Kornblum, R.K. Blackwood, D.D. Mooberry,* J. Amer. Chem. Soc. *78,* 1502 (1956).
28 *R.N. Haszeldine,* Nature *168,* 1029 (1951);
R.N. Haszeldine, J. Chem. Soc. (London) 275 (1953);
R.N. Haszeldine, Angew. Chem. *66,* 698 (1954);
J. Jander, R.N. Haszeldine, Naturwissenschaften *40,* 579 (1953);
J. Jander, R.N. Haszeldine, J. Chem. Soc. (London) 912 (1954);

predominates during thermal decomposition of the acyl nitrites in solution[38]. The acyl nitrite intermediate stage is reached also by allowing dinitrogen trioxide to act on *anhydrides* of *carboxylic acids.* Thus, from hexafluoroacetic anhydride *trifluoroacetyl nitrite* is formed whose pyrolysis affords *trifluoronitrosomethane* (85%)[39, 40]:

$$(F_3C-CO)_2O \xrightarrow{N_2O_3} F_3C-CO-O-NO$$

$$\xrightarrow{h\nu, \triangle} F_3C-NO + CO_2$$

In analogy to the nitrosating decarboxylation technique, *sulfinic acids* and their *salts,* too, can be nitrosated while splitting off sulfur dioxide. In this way *trichloronitrosomethane*[41] is formed from sodium trichloromethylsulfinate and nitrosyl chloride[41]:

$$Cl_3C-SO_2Na + NOCl$$

$$\longrightarrow Cl_3C-NO + SO_2 + NaCl$$

As has already been mentioned on p. 20, the acyl group in acyclic α-monosubstituted β-oxo-carboxylic acid esters can often be displaced by the nitroso group[42]:

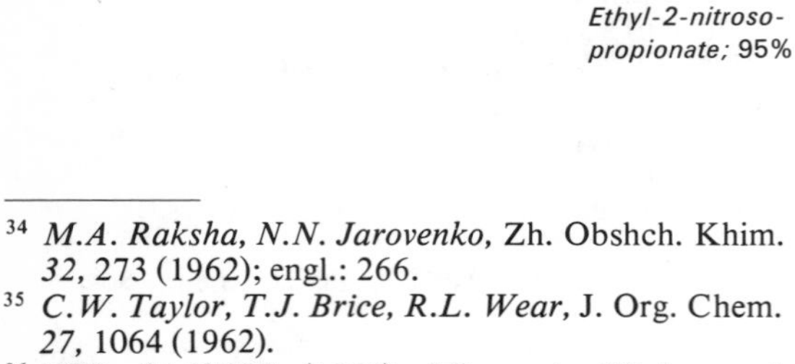

Ethyl-2-nitroso-propionate; 95%

The tendency to lose the acyl group increases in the sequence benzoyl, acetyl, formyl[43].

2.2.1.3 Addition reactions on double bonds

The usual nitrosating agents such as nitrosyl halides and nitrogen oxides add to $C{=}C$ and $C{=}N$ double bonds.

Addition of nitrosyl halides to *olefins* generally proceeds according to an electrophilic mechanism. The $NO^{\oplus}$ cation, as electrophilic agent, adds to the hydrogen-richer carbon atom and, accordingly, the halide anion to the carbon atom with fewer hydrogens (Markovnikov's rule), because in this way the *carbenium cation* with the *least energy* is formed intermediately:

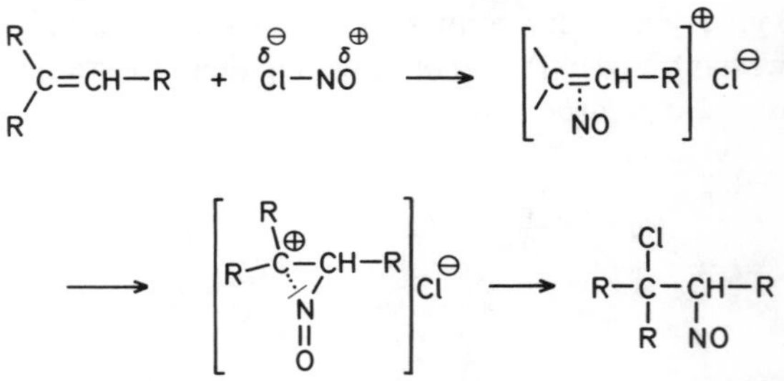

From 1, 2-disubstituted olefins with different substituents an *isomer mixture* is generally formed[44]. During addition of nitrosyl chloride to chloroethylene, 1, 1-dichloroethylene, or trichloroethylene with aluminum chloride catalysis the Markovnikov orientation is not followed[45]. The stereospecific *trans* addition corresponding to the A_E mechanism was confirmed with 1, 2, 3, 4, 5, 6, 7, 8-octahydronaphthalene but not with norbornene **7**[46]:

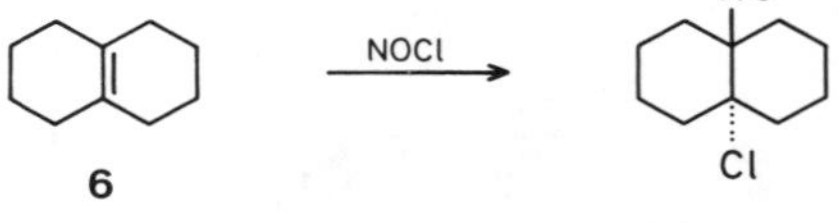

6

trans-4a-Chlorooctahydro-6a-nitrosonaphthalene

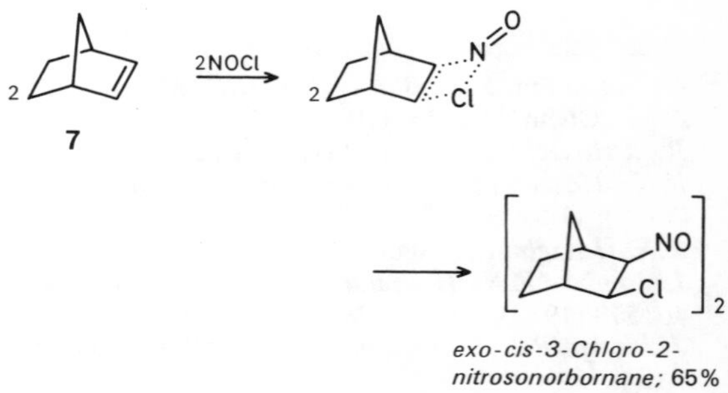

7

exo-cis-3-Chloro-2-nitrosonorbornane; 65%

[34] *M.A. Raksha, N.N. Jarovenko,* Zh. Obshch. Khim. *32,* 273 (1962); engl.: 266.

[35] *C.W. Taylor, T.J. Brice, R.L. Wear,* J. Org. Chem. *27,* 1064 (1962).

[36] USP. 3162590 (1964), Minnesota Mining and Manufg. Co., Inv.: *J.D. Park;* C.A. *62,* 9010 (1965).

[37] *H. Hunsdiecker, C. Hunsdiecker,* Ber. *75,* 291 (1942).

[38] *W. Pritzkow, M. Nitzer,* J. Prakt. Chem. [4] *25,* 69 (1964).

[39] *J.D. Park, R.W. Rosser, J.R. Lacher,* J. Org. Chem. *27,* 1462 (1962);
US Dept. Com., Office Tech. Serv., AD 418, 638 (1962), Inv.: *H.A. Brown, N. Knoll, D.E. Rice;* C.A. *60,* 14709 (1964);
H. Sutcliffe, J. Org. Chem. *30,* 3221 (1965).

Where the nitroso group adds to a tertiary carbon atom monomeric or dimeric *vicinal halonitroso compounds* are formed. By contrast, when the nitroso group becomes linked to a secondary or primary carbon atom either *α-halooximes* or *α-nitroso compounds* result in dependence on the reaction conditions; the latter, as a rule, only as dimers in the form of *bis(2-halo-1-nitroso) compounds*:

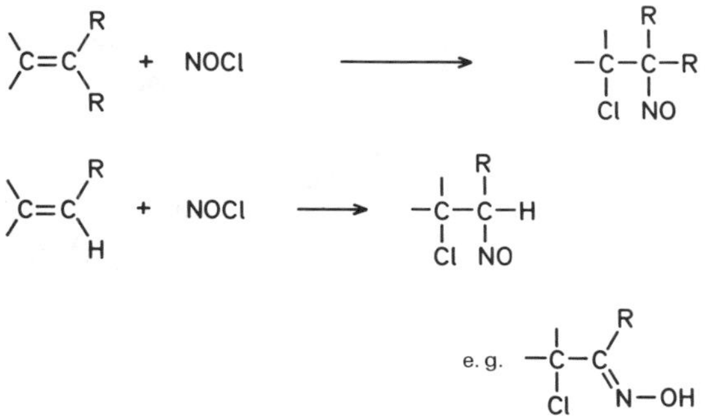

Addition of nitrosyl halides proceeds more readily to olefins of the type[47]

than to

$$R-CH=CH-R \quad \text{or} \quad R-CH=CH_2$$

The yields obtained are strongly dependent on the reaction conditions, especially the temperature and solvent. Thus, at higher temperatures chlorination and oxidation reactions predominate. Sometimes the initially formed nitroso compound can be oxidized to the *nitro compound* by excess nitrosating agent[48]:

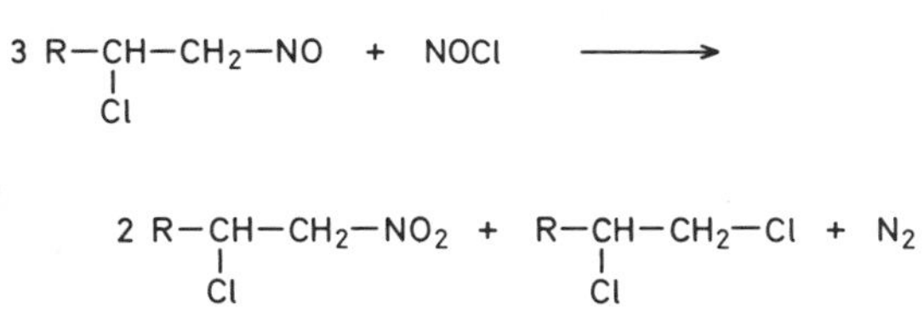

Disproportionations, too, can afford nitro compounds; this is observed especially during addition of nitrosyl chloride to fluoroalkenes. From chlorofluoronitrosoalkanes nitrogen is eliminated under these circumstances and fluoronitroalkanes and chlorofluoroalkanes are formed[49].

Diolefins can add nitrosyl halides on one or both double bonds[50]. The triolefin 1,5,9-cyclododecatriene adds one molecule nitrosyl chloride to form *bis(2-chloro-1-nitroso-5,9-cyclododecadiene)*. Here, the addition takes place preferentially on a *trans* double bond[51].

Enol ethers[52] and *ketene acetals*[53] add one molecule nitrosyl chloride at low temperatures. In correspondence with the polarity of their double bonds the nitrosyl group adds to the carbon atom in the β-position. From 2-methoxypropene the colorless dimeric *2-chloro-2-methoxy-1-nitrosopropane* is formed *via* the stage of the blue monomeric compound[50]:

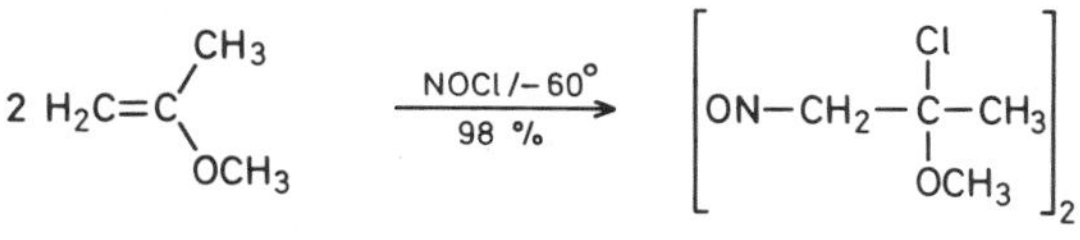

[40] Note: some acyl nitrites **explode** at high temperatures.
[41] W. Prandte, K. Sennewald, Ber. 62, 1754 (1929); W. Prandte, W. Dollfus, Ber. 65, 756 (1932); H. Sutcliffe, J. Org. Chem. 30, 3221 (1965).
[42] J. Schmidt, H. Dieterle, Justus Liebigs Ann. Chem. 377, 54 (1910); J. Schmidt, K.T. Widmann, Ber. 42, 1891 (1909); J. Schmidt, A. Haid, Justus Liebigs Ann. Chem. 377, 25 (1910); J. Schmidt, K.T. Widmann, Ber. 42, 498 (1909).
[43] J. Schmidt, H. Dieterle, Justus Liebigs Ann. Chem. 377, 33, 34 (1910).
[44] K.A. Ogloblin, M.A. Samarcev, Z. Org. Khim. 1, 860 (1965).
[45] A.J. Titow, Dokl. Akad. SSSR 149, 612 (1963); C.A. 59, 7361 (1963).
[46] J. Meinwald, Y.C. Meinwald, T.N. Baker, J. Amer. Chem. Soc. 86, 4074 (1964); J. Meinwald, Y.C. Meinwald, T.N. Baker, J. Amer. Chem. Soc. 85, 2513 (1963).
[47] V.N. Ipatieff, J. Russ. Phys. Chem. Soc. 31, 426 (1899); Chem. Zentr. II, 178 (1899); V.N. Ipatieff, J. Chem. Soc. (London) 78, I, 14 (1900); V.N. Ipatieff, J. Prakt. Chem. [2] 61, 114 (1900); V.N. Ipatieff, A.A. Solonina, J. Russ. Phys.-Chem. Soc. 33, 496 (1901); Chem. Zentr. II, 1202 (1901); V.N. Ipatieff, A.A. Solonina, J. Chem. Soc. (London) 82, I, 1 (1902); W.A. Tilden, M.O. Forster, J. Chem. Soc. (London) 65, 334 (1894); M. Tuot, Compt. Rend. 204, 697 (1937); USP. 2417675 (1943), Solvay Process Comp., Inv.: L.J. Beckman; C.A. 41, 3478 (1947); USP. 2394430 (1943), Solvay Process Comp., Inv.: J.A. Crowder, M.A. Kise, G.A. Nesty; C.A. 40, 6093 (1946).
[48] K.A. Ogloblin, Zh. Obshch. Khim. 28, 3245 (1958); C.A. 53, 12158 (1959); K.A. Ogloblin, V.N. Kalichebic, A.A. Potechin, V.P. Semenov, Zh. Obshch. Khim. 34, 170 (1964); engl.: 168.
[49] D.E. O'Connor, P. Tarrant, J. Org. Chem. 29, 1793 (1964).

α-Branched, α,β-unsaturated aldehydes or ketones form dimeric *chloronitroso* compounds in ether solution at −50 to 0°[54].

As may be anticipated, the reaction follows an *anti*-Markovnikov orientation. In the presence of alcohols cleavage into α-*chlorooximes* and *carboxylic acid esters* takes place:

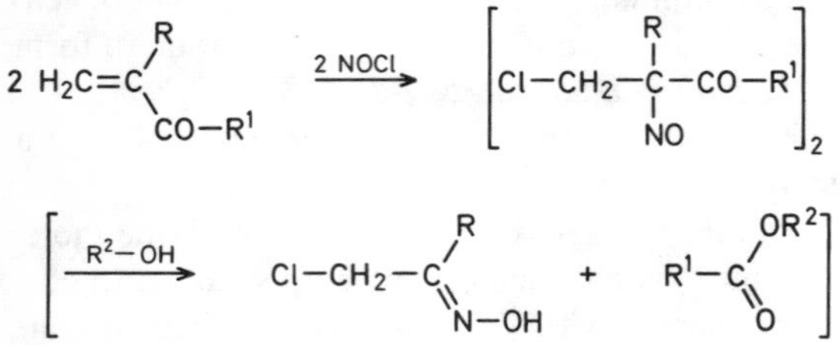

Many unsaturated carboxylic acids, their esters, dialkylamides, and nitriles likewise react with nitrosyl halides to form *halonitroso* compounds[55]. The presence of a catalyst may determine whether a β-*chloro*-α-*nitrosocarboxylic ester* is formed. Thus, methyl methacrylate forms *methyl 3-chloro-2-methyl-2-nitrosopropionate* only in the presence of aluminum chloride, while without the catalyst a mixture of *methyl 2-chloro-3-nitropropionate* and *2,3-dichloro-2-methylpropionate* is obtained[56].

Nitrosylsulfuric acid[57] and nitrosyl nitrate[58] add to olefins analogously to nitrosyl halides:

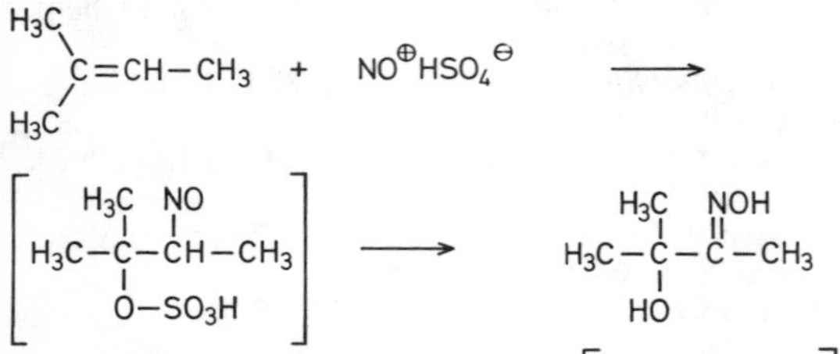

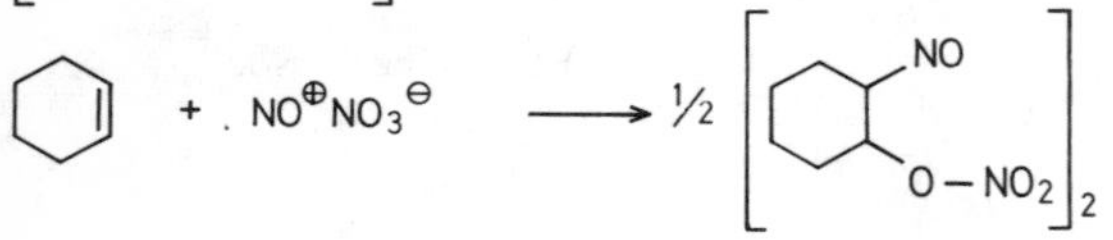

The following *methods of working* have proven successful for preparing vicinal halonitroso compounds:

1 With reactive olefins the calculated amount of nitrosyl halide is introduced into the liquid or (chloroform) solution of the olefin at −10 to 0° (or at 25° under pressure[59]). The addition to less reactive olefins is catalyzed with aluminum[60], or iron(III) chloride[60], nickel(II)[61], copper(II) salts[61], or light[62].
2 Nitrosyl halide is used *in situ* by allowing halogen hydride acid to act on a mixture of 1 mol olefin and 1.5 mol alkyl nitrite at −15°. Addition of glacial acetic acid or ethanol is recommended[63].
3 1.5 mol thionyl chloride are added to 1 mol olefin at −5° and dinitrogen trioxide is passed in.
4 A concentrated aqueous solution of sodium nitrite is added drop by drop to the olefin dissolved in excess alcoholic hydrochloric acid.

The following Table 2 surveys the 2-halo-1-nitroso compounds prepared by these techniques.

Nitrosations of olefins with *nitrogen oxides* generally suffer from the drawback that they proceed very nonuniformly. However, if appropriate reaction conditions are employed an up to 90% yield of *tetrafluoro-1-nitro-2-nitrosoethane* can be obtained from tetrafluoroethylene, for example[71].

When *dinitrogen trioxide* adds to olefinic double bonds *1-nitro-2-nitroso compounds* (as a rule as dimers) are formed[72]. In general the nitro group adds to the carbon atom with more hydrogen atoms, the nitroso group to the carbon atom with fewer hydrogens. The action of pure *dinitrogen tetraoxide* on olefins leads to vicinal *nitronitrosooxy compounds* and, by oxidation (or hydrolysis of the nitrites), further to *nitroalkanes*, vicinal *nitronitrooxy compounds*, and *nitroalcohols*[73].

Addition of dinitrogen tetraoxide to cycloalkanes in the presence of nitric acid leads to the corre-

Table 2. 2-Halo-1-nitrosoalkanes from alkenes

Alkene	Method	Reaction product	M. p. [°C]	Yield [% of theory]	Ref.
Tetrafluoroethylene	1	*2-Chlorotetrafluoro-1-nitrosoethane*	($b_{760} = -2°$)	68	[64]
2-Butene	1	*Bis-[3-chloro-2-nitrosobutane]*	106	29	[65]
Cyclododecene	1	*Bis-[2-chloro-1-nitrosocyclododecane]*	135	60	[66]
2,3-Dimethyl-2-butene	2	*3-Bromo-2,3-dimethyl-2-nitrosobutane*	—	~ 100	[67]
	4	*3-Chloro-2,3-dimethyl-2-nitrosobutane*	123	~ 100	[68]
Styrene	2	*Bis-[2-chloro-1-nitroso-1-phenylethane]*	103	30	[69]
2,5-Dimethyl-2-hexene	3	*2-Chloro-2,5-dimethyl-3-nitrosohexane*	123	—	[70]

[50] *E.V. Lynn, F.A. Lee*, J. Amer. Pharm. Ass., Sci. Ed. *15*, 174 (1926);
G. Kraemer, A. Spilker, Ber. *29*, 558 (1896).

[51] DAS 1 094 741 (1959), Studiengesellschaft Kohle, Erf.: *G. Wilke, E.W. Müller, J. Stedefelder*; C.A. *55*, 25808 (1961).

sponding dimeric *1-nitroso-2-nitrooxycyloal-kanes*[74]:

$$n = 3,4,5$$

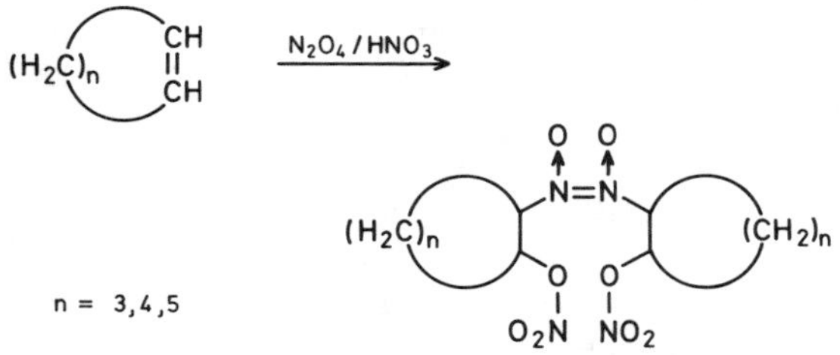

In addition to the $C{=}C$ double bond of the olefins it is the $C{=}N$ double bond of the oximes, above all, that is suitable for synthesizing *nitroso compounds:*

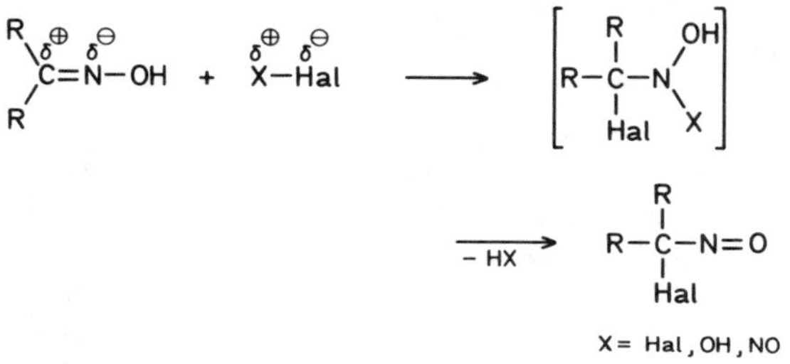

$$X = Hal, OH, NO$$

Hydroxamoyl chlorides[75], esters of hydroxamic acids[76], and nitrolic acids[77] can be employed as well as a number of aldoximes and ketoximes.

The following *techniques* lend themselves to the preparation of *geminal halonitroso compounds* from oximes:

1 Chlorine or bromine is passed into a solution of the oxime in aqueous mineral acid, aqueous base, or in an organic solvent (in acid medium the halogen itself reacts, in alkali the hypohalite).
2 Nitrosyl chloride in excess is allowed to act on an ether solution or suspension of the oxime.

Table 3 illustrates the use of these two techniques.

Treatment of dialkyl ketoximes with dinitrogen tetraoxide often affords *pseudonitroles*[83]:

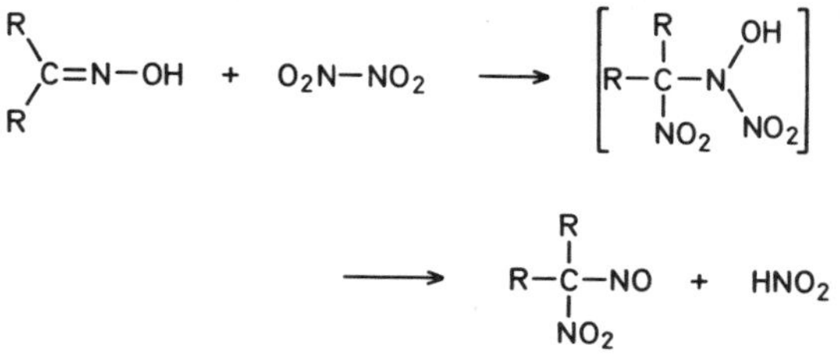

Table 3. Geminal halonitroso compounds from oximes

Oxime	Method	Reaction product		M.p. [°C]	Yield [% of theory]	Ref.
Acetaldoxime	2	*Bis-[1-chloro-1-nitrosoethane]*		65	7	78
Acetone oxime	1	*2-Bromo-2-nitrosopropane*		—	39	79
	2	*2-Chloro-2-nitrosopropane*		$(b_{760}=70)$	—	80
Cyclohexanone oxime	1	*1-Chloro-1-nitrosocyclohexane*		$(b_{12}=51)$	83–89	81
	2			$(b_{12}=52)$	60	80
1,4-Cyclohexanedione dioxime	1	*1,4-Dichloro-1,4-dinitrosocyclohexane*	*trans*	109	65	82
			cis	160–165	10	82
	2	*1,4-Dichloro-1,4-dinitrosocyclohexane*	*trans*	109	51	80
			cis	—	10	80

[52] *K.A. Ogloblin, V.P. Semenov,* Dokl. Akad. Nauk SSSR *155,* 145 (1964);
B.L. Djatkin, R.A. Bekker, J.L. Knunjanc, Izv. Akad. Nauk SSSR 1121 (1965);
K.A. Ogloblin, D.M. Kunovskaja, Zh. Org. Khim. *6,* 40 (1970).
[53] *K.A. Ogloblin, D.M. Kunovskaja,* Zh. Org. Khim. *1,* 1713 (1965).
[54] *K.A. Ogloblin, A.A. Potechin,* Dokl. Akad. Nauk SSSR *159,* 853 (1964);
K.A. Ogloblin, A.A. Potechin, Zh. Org. Khim. *1,* 408, 865, 1352 (1965).
[55] USP. 2336387 (1941), Solvay Process Co., Inv.: *L.J. Beckham;* C.A. *38,* 30563 (1944);
V.N. Ipatieff, J. Russ. Phys. Chem. Soc. *31,* 426 (1899);
V.N. Ipatieff, J. Prakt. Chem. [2] *61,* 114 (1900).
[56] *K.A. Ogloblin, V.P. Semenov,* Zh. Org. Khim. *1,* 27 (1965).
[57] *D.G. Boller, G.H. Whitfield,* J. Chem. Soc. (London) 2773 (1964).
[58] *J.E. Franz, J.F. Herber, W.S. Knowles,* J. Org. Chem. *30,* 1488 (1965).
[59] USP. 2394430 (1943), Solvay Process Co., Inv.: *J.A. Crowder, M.A. Kise, G.A. Nesty;* C.A. *40,* 6093 (1946).
[60] *A.J. Titow,* Dokl. Akad. Nauk SSSR *149,* 612 (1963); C.A. *59,* 7361 (1963).
[61] USP. 2417675 (1943), Solvay Process Comp., Inv.: *L.J. Beckham;* C.A. *41,* 3478 (1947).
[62] *J.D. Park, A.P. Stefani, J.R. Lacher,* J. Org. Chem. *26,* 4017 (1961);
USP. 3072705 (1963). Minnesota Mining and Manufg. Co., Inv.: *J.D. Park, A.P. Stefani;* C.A. *58,* 10077 (1963).

These techniques for preparing *pseudonitroles* are used especially where the corresponding secondary nitro compounds are inaccessible or accessible only with difficulty. Instead of introducing dinitrogen tetraoxide into the ether solution of the ketoxime as a gas or liquid, the action of sodium nitrite and ~30% sulfuric acid also sometimes produces success[84]. While this procedure furnishes poorer yields, it displays the advantage that the easily accessible sodium nitrite can be made the starting point, quite apart from the shorter reaction time.

On reacting simple α-hydroxyiminocarboxylic acid esters with a mixture of nitric acid and ammonium nitrate *esters* of *α-nitro-α-nitrosocarboxylic acids* were obtained[85].

The action of lead(IV) acylates on aliphatic and cycloaliphatic ketoximes allows the preparation of stable geminal *acyloxynitroso compounds*[86]:

$$R_2C=N-OH \ + \ (R-COO)_4Pb \ \longrightarrow$$

$$R_2C \overset{\displaystyle NO}{\underset{\displaystyle O-CO-R}{\Big\backslash}} \ + \ (R-COO)_2Pb \ + \ R-COOH$$

2.2.1.4 Oxidation of aliphatic nitrogen compounds

Nitroso compounds may be formed when neutralized Caros acid or organic peracids are allowed to act on amines under suitable reaction conditions:

$$R-NH_2 \ + \ 2\,[O] \ \longrightarrow \ R-NO \ + \ H_2O$$

Thus, the corresponding *tertiary dinitroso compounds* were obtained by reacting certain primary aliphatic amines having their amino group linked to a tertiary carbon atom with Caro's acid[87]. *Hydroxylamines* and *nitro compounds* are side-products.

When the amino group is positioned on a primary or secondary carbon atom the oxidation leads to *primary* or *secondary dinitroso compounds*, or to the isomeric *oximes*[88]. *Bis(nitrosocyclohexane)* is obtained from cyclohexylamine with Caro's acid in weakly acid medium but mainly *cyclohexanone oxime* in neutral, weakly alkaline, or strongly acid medium[89]. *Bis(nitrosocyclohexane)* is claimed to be formed in up to 90% yield[90] if the oxidation of the cyclohexylamine is performed with organic peracids such as peroxyacetic acid or peroxybenzoic acid in organic solvents like benzene at 0 to 20°.

Secondary amines, too, can be oxidized to nitroso compounds with organic peroxy acids while losing an alkyl group as carbonyl compound. *Bis(nitrosocyclohexane)* is obtained from the dicyclohexylamine with 3 molecules 78% peroxyacetic acid according to the following equation[91]:

$$2 \ \text{⬡}-NH-\text{⬡} \ + \ 6\,[O] \ \longrightarrow$$

$$\Big[\text{⬡}-NO\Big]_2 \ + \ 2\,\text{⬡}{=}O \ + \ 2H_2O$$

[63] *O. Wallach*, Justus Liebigs Ann. Chem. *245*, 245, 251, 252, 261, (1888); *252*, 109 (1889).

[64] *R.N. Haszeldine*, J. Chem. Soc. (London) 2079 (1953).

[65] *K.A. Ogloblin, M.A. Samartseev*, Zh. Obshch. Khim. *33*, 3257 (1963); engl.: 3184; *K.A. Ogloblin, M.A. Samartseev*, Zh. Obshch. Khim. *34*, 1525 (1964); engl.: 1533.

[66] Fr. P. 1 442 416 (1965), Organico S.A.; C.A. *67*, 108258 (1967).

[67] *E. Schmidt, F. Leipprandt*, Ber. *37*, 546 (1904).

[68] *E. Thiele*. Ber. *27*, 455 (1894); *G.L. Gloss, S.E. Brois*, J. Amer. Chem. Soc. *82*, 6068 (1960).

[69] *R. Perrot*, Compt. Rend. *203*, 330 (1936); *N. Thorne*, J. Chem. Soc. (London) 4275 (1956).

[70] *M. Tuot*, Compt. Rend. *204*, 698 (1937).

[71] *J.D. Park*, J. Org. Chem. *26*, 3316 (1961); *G.H. Grawford*, J. Polym. Sci. *45*, 259 (1960); C.A. *55*, 5327 (1960).

[72] *H.B. Baldock, N. Levy, C.W. Scaife*, J. Chem. Soc. (London) 2628 (1949); *M.L. Scheinbaum*, J. Org. Chem. *35*, 2785 (1970); *A. Michael, C.H. Carlson*, J. Org. Chem. *4*, 169 (1939); *5*, 14 (1940).

[73] *N. Levy, C.W. Scaife*, J. Chem. Soc. (London) 1093, 1100 (1946).

[74] USP. 3 106 575 (1961), Monsanto Chemical Co., Inv.: *J.F. Herber*.

[75] *O. Piloty, H. Steinbock*, Ber. *35*, 3115 (1902).

[76] *J. Houben, E. Schmidt*, Ber. *46*, 3620 (1913).

[77] *O. Graul, A. Hantzsch*, Ber. *31*, 2857 (1898).

[78] *H. Rheinboldt, M. Dewald*, Justus Liebigs Ann. Chem. *451*, 276 (1927).

[79] *O. Piloty*, Ber. *31*, 454 (1898).

[80] *H. Rheinboldt, M. Dewald*, Justus Liebigs Ann. Chem. *455*, 300 (1927).

[81] *Eu. Müller, H. Metzger, D. Fries*, Ber. *87*, 1454 (1954); *O. Wichterle, M. Hudlicky*, Collect. Czech. Chem. Commun. *12*, 661 (1947).

[82] *O. Piloty, H. Steinbock*, Ber. *35*, 3115 (1902).

[83] *R. Scholl*, Ber. *21*, 506 (1888); *J.L. Riebsomer*, Chem. Rev. *36*, 183 (1945).

[84] *A. Schöfer*, Ber. *34*, 1911 (1901).

[85] *H.E. Ungnade, L.W. Kissinger*, J. Org. Chem. *24*, 666 (1959).

[86] *D.C. Iffland, G.X. Griner*, Chem. & Ind. (London) 176 (1956); *E.H. White, W.J. Considine*, J. Amer. Chem. Soc. *80*, 628, 630 (1958).

The oxidation of *N*-substituted *hydroxylamines* is one of the most important methods for preparing *nitroso compounds,* especially as aliphatic hydroxylamines are relatively easily accessible by a variety of routes. Compounds carrying the hydroxyamino group on a tertiary carbon atom are invariably oxidized to *monomeric* or *dimeric nitroso compounds.*

By contrast, the *isomeric oximes* may be formed in addition to the *dinitroso compounds* during oxidation of compounds containing the hydroxyamino group bound to primary or secondary carbon atoms.

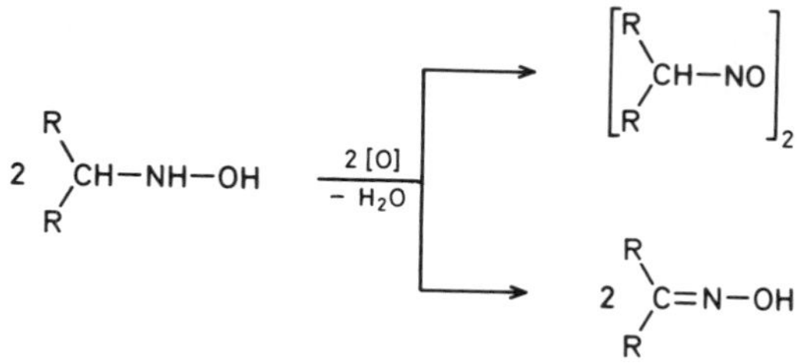

$$2\ \underset{R}{\overset{R}{\diagdown}}CH-NH-OH \xrightarrow[- H_2O]{2\ [O]} \begin{cases} \left[\underset{R}{\overset{R}{\diagdown}}CH-NO \right]_2 \\[2em] 2\ \underset{R}{\overset{R}{\diagdown}}C=N-OH \end{cases}$$

The following techniques have proven successful for the oxidation of *N*-alkylhydroxylamines:

1 Aqueous bromine, chlorine, or hypochlorite is allowed to act on a weakly hydrochloric acid solution of the hydroxylamine.
2 The hydroxylamine dissolved in sulfuric or acetic acid is treated with potassium dichromate solution at 0°.
3 The hydroxylamine dissolved in chloroform, ether, or methanol is treated with freshly precipitated yellow mercury oxide (at boiling point).

In addition to these three techniques there are a number of oxidation procedures of minor importance:

Bis(nitrosocyclohexane) is obtained in 85–90% yield when *N*-cyclohexylhydroxylamine is oxidized in sulfuric acid solution at 0° with potassium permanganate or in aqueous phase at 50° with iron(III) chloride[92]. *N*-Methylhydroxylamine hydrochloride is oxidized to *cis-bis(nitrosomethane)* by periodic acid in 80% yield[93].

In isolated cases atmospheric oxygen may be used as oxidizing agent. Thus, ethyl 2-hydroxyamino-2-methylpropionate is very readily oxidized to *diethyl bis(2-methyl-2-nitrosopropionate)* by atmospheric oxygen[94]. This atmospheric oxidation of hydroxylamines can be catalyzed by mangane-

Table 4. Nitroso compounds by oxidation of *N*-alkylhydroxylamines

Hydroxylamine	Method	Reaction product	M. p. [°C]	Yield [% of theory]	Ref.
N-4-Heptylhydroxylamine	1	*Bis-[4-nitrosoheptane]*	50	80	98
2-Hydroxyamino-2-methyl-propionamide	1	*Bis-[2-methyl-2-nitrosopropionamide]*	158	~ 100	99
cis-N-Decahydro-2-naphthyl-hydroxylamine	1	*cis-Bis-[decahydro-2-nitrosonaphthalene*	133	85	100
1,1,4-Trimethylpentyl-hydroxylamine	2	*Bis-[2,5-dimethyl-2-nitrosohexane]*	54	~ 60	101
2-[Hydroxyamino]-2-methyl-1-propanol acetate	2	*Bis-[2-methyl-2-nitroso-1-propanol acetate]*	68	—	102
N-1-Adamantylhydroxylamine	2	*Bis-[1-nitrosoadamantane]*	—	25	103
4-Hydroxyamino-4-methyl-2-pentanone	3	*Bis-[4-methyl-4-nitroso-2-pentanone]*	76	60	104

[87] *E. Bamberger, R. Seligmann,* Ber. *36,* 685 (1903); *E. Bamberger, R. Seligmann,* Ber. *36,* 695 (1903).
[88] DBP. 953 069 (1952), Farbf. Bayer, Erf.: *H. Krimm, K. Hamann;* C.A. *53,* 5159 (1959); *W.D. Emmons,* J. Amer. Chem. Soc. *79,* 6523 (1957); *E. Bamberger, R. Seligmann,* Ber. *36,* 686 (1903); *I. Okamura, R. Sakurai, T. Tanabe,* Chem. High Polymers (Tokyo) *9,* 279, 284 (1952); C.A. *48,* 4225 (1954); *I. Okamura, R. Sakurai,* Chem. High Polymers (Tokyo) *9,* 434 (1952); C.A. *48,* 9070 (1954); DBP. 948 417 (1954), Farbf. Bayer, Erf.: *H. Krimm;* C.A. *52,* 18315 (1958); Jap. AS 62/11142 (1962), Fujisawa Pharmaceutical Co.; Inv.: *M. Ohara;* C.A. *59,* 10067 (1963).

[89] *I. Okamura, R. Sakurai,* Chem. High Polymers (Tokyo) *8,* 296 (1951); C.A. *47,* 2992 (1953); *I. Okamura, R. Sakurai,* Chem. High Polymers (Tokyo) *9,* 10, 230 (1952); C.A. *48,* 9933, 11794 (1954).
[90] DBP. 953 069 (1952), Farbf. Bayer, Erf.: *H. Krimm, K. Hamann;* C.A. *53,* 5159 (1959); Chem. Zentr. 5706 (1957).
[91] DBP. 948 417 (1954), Farbf. Bayer, Erf.: *H. Krimm;* C.A. *52,* 18315 (1958).
[92] Fr. P. 1 123 977 (1955), Synthese Chemie GmbH; Chem. Zentr. 3304 (1959).
[93] *T. Emery, J.B. Neilands,* J. Amer. Chem. Soc. *82,* 4903 (1960).
[94] *O. Piloty, B. Graf Schwerin,* Ber. *34,* 1867 (1901).

se(IV)[95], cobalt(III)[95], and copper(II)[96] salts. Di-ethyl azodicarboxylate is a suitable oxidizing agent for preparing the sensitive *bis(nitrosocyclopropane)* from *in situ* generated *N*-cyclopropylhydrox-ylamine[97]:

$$\triangleright\!\!-NO_2 \xrightarrow{Zn/NH_4Cl} \left[\triangleright\!\!-NH\!-\!OH\right]$$

$$\xrightarrow[- H_5C_2OOC-NH-NH-COOC_2H_5]{H_5C_2OOC-N=N-COOC_2H_5} \tfrac{1}{2}\left[\triangleright\!\!-NO\right]_2$$

Table 4 surveys the methods **1–3.**

In addition to amines and hydroxylamines *azome-thines* **8**[105], *nitrones* **9**[106], and *oxaziridines* **10**[107] afford nitroso compounds on oxidation. Perox-yacetic acid is the chief oxidizing agent used:

$$2\ R^1\!-\!N\!=\!C\!\!\begin{smallmatrix}R\\R\end{smallmatrix} \xrightarrow{H_3C-CO-OOH} 2\ R^1\!-\!N\!\!\overset{R}{\underset{O}{\diagup}}\!\!R$$

8   **10**

$$\xrightarrow{H_3C-CO-OOH} [R^1\!-\!NO]_2 + 2\ \begin{smallmatrix}R\\\\R\end{smallmatrix}\!\!C\!=\!O$$

$$\Big\uparrow H_3C-CO-OOH$$

$$2\ \overset{R^1}{\underset{O}{N}}\!\!=\!C\!\!\begin{smallmatrix}R\\R\end{smallmatrix}$$

9

2.2.1.5 Reduction of aliphatic nitro compounds

Trapping of the nitroso compound stage during reduction of aliphatic nitro compounds prepara-tively has to date succeeded with few examples only[108].

Thus, cathodic reduction of a solution of trichloro-nitromethane in low alcohols at 20–22° in the presence of 35–40% sulfuric acid yields *trichloro-nitrosomethane*[109] (in up to 70% of theory). If less sulfuric acid is added *N-methylhydroxylamine* and *methylamine* are obtained.

With 21% hydrochloric acid at 100° chlorofluo-ronitroacetic acid is converted into dichlorofluoro-nitrosomethane accompanied by decarboxylation (65% of theory)[110]:

$$O_2N\!-\!\overset{F}{\underset{Cl}{C}}\!-\!COOH\ +\ HCl$$

$$\longrightarrow ON\!-\!\overset{F}{\underset{Cl}{C}}\!-\!Cl\ +\ CO_2\ +\ H_2O$$

α-Nitroolefins with the nitro group on a secondary carbon atom form *1, 2-dichloronitroso* compounds when hydrogen chloride acts on them in ether[111], *e.g.:*

$$\begin{smallmatrix}H_3C\\O_2N\end{smallmatrix}\!\!C\!=\!CH_2\ +\ 2\ HCl\ \xrightarrow[-H_2O]{}\ H_3C\!-\!\overset{Cl}{\underset{NO}{C}}\!-\!CH_2\!-\!Cl$$

35–45%

*1,2-Dichloro-2-
nitrosopropane*

2.2.1.6 Decomposition of alkyl nitrites

Alkoxy radical formation is the first reaction step during photolysis or thermolysis of *alkyl nitrites* with loss of nitrogen monoxide[112]:

$$R\!-\!\overset{R}{\underset{R}{C}}\!-\!O\!-\!N\!=\!\bar{O} \xrightleftharpoons{\nabla\ \text{or}\ h\nu} R\!-\!\overset{R}{\underset{R}{C}}\!-\!\bar{O}\!\cdot\ +\ \dot{N}\!=\!\bar{O}$$

[95] Fr. P. 1 123 977 (1955); DAS. 1 015 799 (1954), Synthese Chemie GmbH; Chem. Zentr. 3304 (1959);
USP. 2 949 490 (1958), Hercules Powder, Inv.: *J.H. Pieper, J.E.V.N. Stauch; C.A. 55*, 4392 (1961).
[96] *N.B. Shitova, K.I. Matveev, M.M. Danilova,* Kineti-ka i Kataliz 7, 995 (1966); C.A. *66*, 54790 (1967).
[97] *R. Stammer, J.B.F.N. Engberts, T.J. de Boer,* Rec. Trav. Chim. Pays-Bas *89*, 169 (1970).

[98] *Eu. Müller, H. Metzger,* Ber. *88*, 176 (1955).
[99] *O. Piloty, B. Graf Schwerin,* Ber. *34*, 1864 (1901); *S. Okano, Y. Fukagawa,* J. Pharm. Soc. Jap. *74*, 1059 (1954); C.A. *49*, 11 629a (1955).
[100] *Eu. Müller, U. Heuschkel,* Ber. *92*, 70 (1959).
[101] *O. Piloty, O. Ruff,* Ber. *31*, 457 (1898).
[102] *E.R. Schwartz,* J. Amer. Chem. Soc. *79*, 4354 (1957).
[103] *E.W. Hartgerink, E.B.F.N. Jungberts, A.E.W. Wajer, T.E. De Boer,* Rec. Trav. Chim. Pays-Bas *88*, 481 (1969).
[104] *C. Harries, L. Eablonsky,* Ber. *31*, 549, 1379 (1898).
[105] *W.D. Emmons,* Rec. Trav. Chim. Pays-Bas *79*, 6523 (1957).
[106] *H. Krimm,* Chem. Ber. *91*, 1063 (1958); *H. Krimm,* Chem. Ber. *91*, 1068 (1958).
[107] *W.D. Emmons,* J. Amer. Chem. Soc. *79*, 6523 (1959); *H. Krimm,* Chem. Ber. *91*, 1067 (1958).
[108] *J. Kovacs,* Acta Universitatis Szegediensis, Physica et Chemica 2, 56 (1948); C.A. *44*, 6384 (1950); *A.M. Unrau,* Can. J. Chem. *42*, 1741 (1967); C.A. *61*, 5500 (1964).

These become stabilized by disproportionation to alcohols and carbonyl compounds, by rupture into alkyl radicals and carbonyl compounds,

$$R_3C-\underline{\overline{O}}\cdot \xrightarrow[-R_2C=O]{} R\cdot \xrightarrow{NO} R-NO$$

or by isomerization with displacement of a δ-hydrogen to oxygen (Barton reaction)[113]:

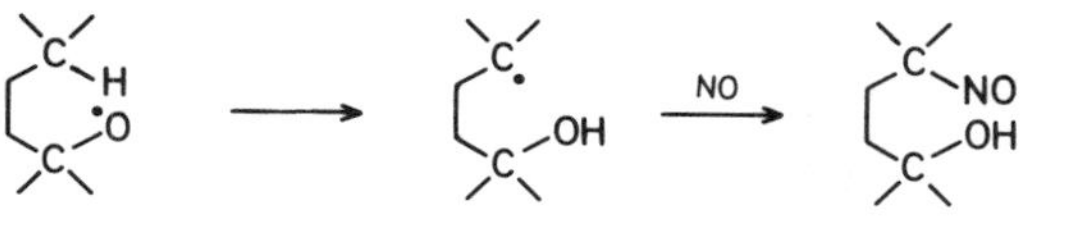

The alkyl radicals formed can react further to *C-nitroso compounds* with the nitrogen monoxide formed or any that is introduced additionally.

Which of the named possible stabilizations of the intermediately arising alkoxy radicals predominates is a question of the reaction kinetics. During *gas-phase pyrolyses* the free-radical decomposition mechanism preponderates. *cis-Bis(tri-deutero-nitrosomethane)* and *perdeuteroacetaldehyde* were obtained from heptadeuteroisopropyl nitrite[114].

Heptafluoronitrosooxycyclobutane forms *hexafluoro-4-nitrosobutyrylfluoride*[115] and *hexafluoro-4-nitrobutyryl fluoride* with rupturing of the ring. Butyl nitrite is converted into *cis*-bis(1-nitrosopropane)[116].

The *pyrolysis conditions* can be varied over wide limits. Thus, from cyclohexylmethyl nitrite and α-alkylated derivatives[117] at 200–1000° and 1–0.001 atmosphere pressure, using contact times between 0.1 and 200 seconds, *bis(nitrosocyclohexane)* is obtained.

Photolysis of alkyl nitrites as a method of preparing *C-nitroso compounds* has been investigated in even greater detail than thermolysis. During gas-phase reactions the same free-radical decomposition mechanism is in evidence as during pyrolysis. From *tert*-butyl nitrite *bis(nitrosomethane)* is formed[116]:

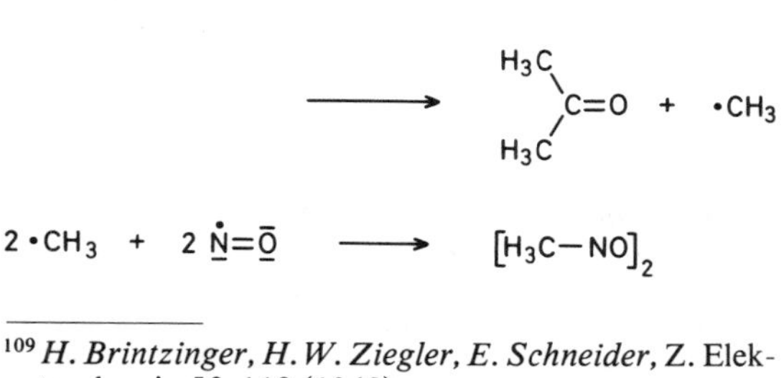

$$2\cdot CH_3 + 2\,\underline{\dot{N}}=\underline{\overline{O}} \longrightarrow [H_3C-NO]_2$$

Primary and secondary alkyl nitrites, too, furnish *C-nitroso compounds* during photolysis in the gas phase[118]. Quite fundamentally, the light used for irradiation must be from the wavelength region below 3300 Å. By contrast, when working in a solvent is employed (heptane, benzene, toluene), it is precisely UV light with a wavelength of >3300 Å that is successful. The *alkoxy radicals* that are formed here as well display a preferential *isomerization* involving displacement of a hydrogen in the δ-position to the oxygen in addition to the above decomposition mechanism. Barton[113] first discovered this behavior during photolysis of 5α-cholestane-3β,6β-diol 3-acetate nitrite to 5α-cholestane-19-nitroso-3β,6β-diol 3-acetate dimer. On account of the general significance of this reaction type it is called the *Barton reaction:*

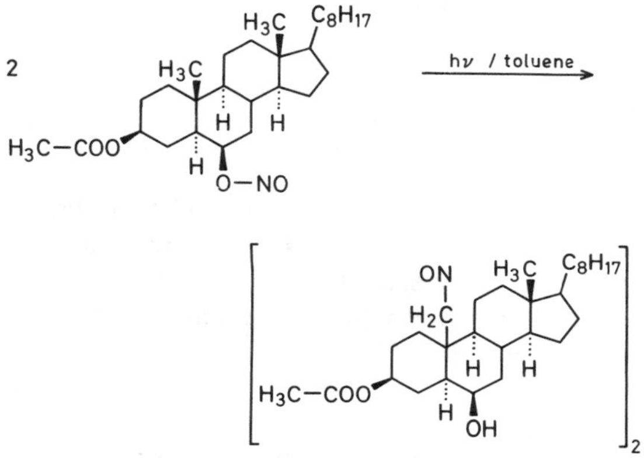

An essential requisite for the Barton reaction is the possibility of intermediate formation of a six-membered ring transition state[119], *i.e.*, the nitrosooxy group needs to be linked to an at least four-membered carbon chain. Accordingly, irradiating methyl, ethyl, and propyl nitrite in solution affords no *C-nitroso compounds*[120].

[109] *H. Brintzinger, H.W. Ziegler, E. Schneider*, Z. Elektrochemie *53*, 110 (1949).

[110] *I.V. Martynov, Y.L. Kruglyak*, Zh. Obshch. Khim. *35*, 248 (1965); engl.· 250.

[111] *R.L. Heath, J.D. Rose*, J. Chem. Soc. (London) 1485 (1947).

[112] *P. Kabasakalian, E.R. Townley*, J. Amer. Chem. Soc. *84*, 2711 (1962); *L. Batt*, J. Chem. Soc. (London) 37 (1962).

[113] *D.H.R. Barton*, J. Amer. Chem. Soc. *82*, 2640 (1960).

[114] *L. Batt*, J. Chem. Soc. (London) 37 (1962).

[115] *S. Andreades*, J. Org. Chem. *27*, 4157 (1962).

[116] *H.W. Thompson, F.S. Dainton*, Trans. Faraday Soc. *33*, 1546 (1967); *C.S. Coe, T.F. Doumani*, J. Amer. Chem. Soc. *70*, 1516 (1948).

[117] Niederl. P. 6500271 (1964), Teijin Co.; C.A. *64*, 1981 (1966).

Butyl nitrite readily passes over into the dimeric 4-nitrosobutanol in a Barton reaction[120], but the yield is only 4% because the detaching of a hydrogen atom in the δ-position must take place from a methyl group. This is more difficult than the abstraction from methylene groups.

The higher alkyl nitrites afford *bis(nitrosoalcohols)* in yields of up to 40%[119]. Formation of nitroso compounds by the alkoxy radical decomposition mechanism is a side-reaction especially with secondary and tertiary alkyl nitrites[120].

2.2.1.7 Special techniques

N-Nitrosohydroxylamine derivatives can decompose into *C-nitroso compounds* in the presence of nitric acid or nitrous acid[121]. In this way *bis(diphenylnitrosomethane)* is obtained from *N*-(diphenylmethyl)-*N*-nitrosohydroxylamine[122]:

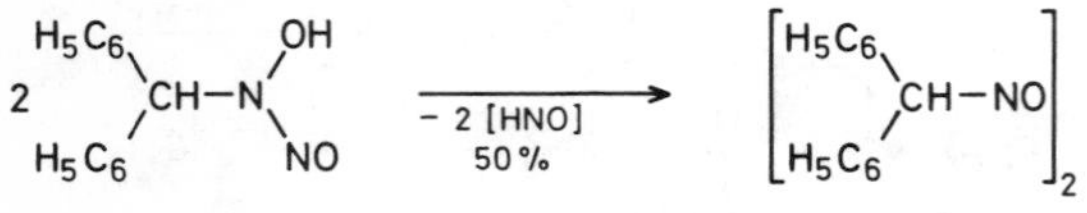

While most metal and ammonium salts of the *N*-nitrosohydroxylamine derivatives are generally stable, the silver salts decompose at a greater or lesser rate with formation of metallic *silver* and *C-nitroso compounds*[123].

Aliphatic nitro compounds and the alkali metal salts of the corresponding nitronic acids, too, can sometimes be decomposed into *nitroso compounds*. Boiling difluoronitroacetic acid or its esters in concentrated hydrochloric acid initially gives rise to the *aci*-form of the difluoronitromethane by decarboxylation; the difluoronitromethane at once loses water or alcohol and, under the action of the hydrochloric acid, forms the blue gas *chlorodifluoronitrosomethane* (67%)[124]:

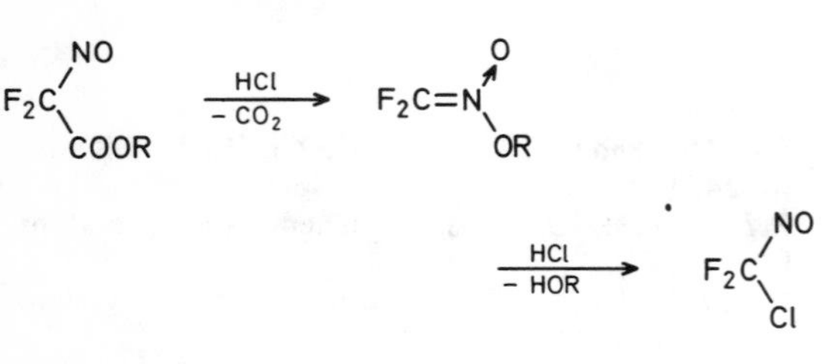

In ether solution containing hydrogen chloride the sodium salt of 2-nitropropane is converted into *2-chloro-2-nitrosopropane*[124] (90%). Analogously *1-Chloro-1-nitrosocyclohexane* (93%) forms from the sodium salt of nitrocyclohexane. Using a solution of excess hydrogen fluoride in ethyl acetate in place of the ethereal hydrogen chloride solution affords the dimeric *1-nitrosocyclohexene* in up to 25% yield[124]:

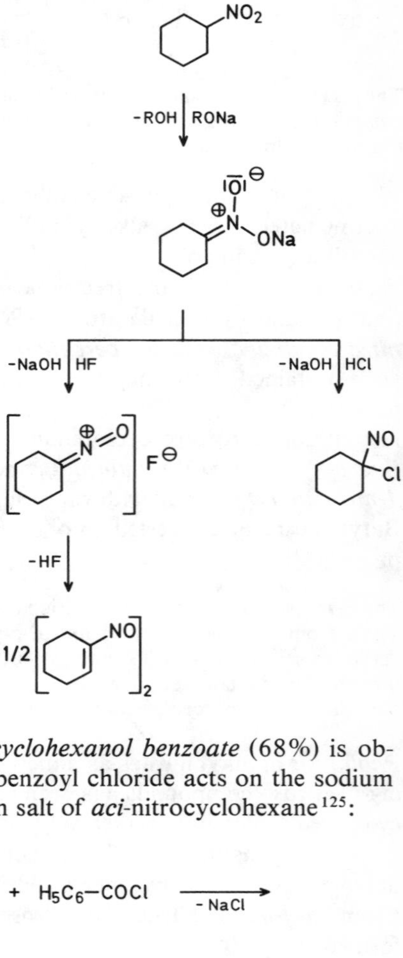

1-Nitroso-1-cyclohexanol benzoate (68%) is obtained when benzoyl chloride acts on the sodium or ammonium salt of *aci*-nitrocyclohexane[125]:

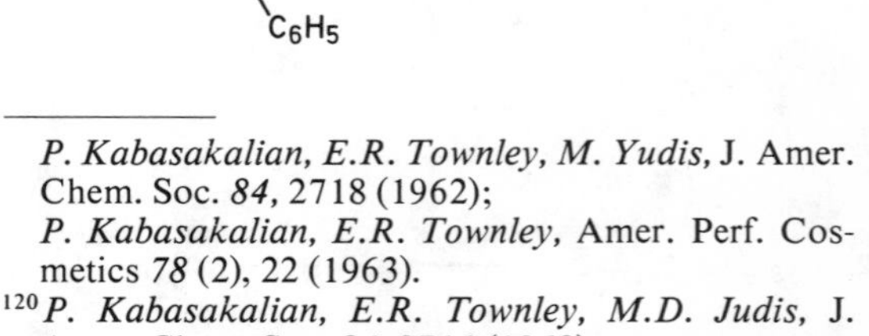

[118] *B.G. Gowenlock, J. Trotman*, J. Chem. Soc. (London) 4190 (1955); 1670 (1956);
P. Tarte, Bull. Soc. Roy. Sci. Liège *22*, 226 (1953); C.A. *48*, 2480 (1954).
[119] *P. Kabasakalian, E.R. Townley, M.D. Judis*, J. Amer. Chem. Soc. *84*, 2716 (1962);
P. Kabasakalian, E.R. Townley, J. Amer. Chem. Soc. *84*, 2711 (1962);

P. Kabasakalian, E.R. Townley, M. Yudis, J. Amer. Chem. Soc. *84*, 2718 (1962);
P. Kabasakalian, E.R. Townley, Amer. Perf. Cosmetics 78 (2), 22 (1963).
[120] *P. Kabasakalian, E.R. Townley, M.D. Judis*, J. Amer. Chem. Soc. *84*, 2716 (1962);
P. Kabasakalian, E.R. Townley, J. Amer. Chem. Soc. *84*, 2724 (1962).

N-Hydroxyamidoximes **11** can be disproportionated into *1-nitrosoaldoximes (nitrosolic acids)* **12** and *amidoximes* **13** *via* an unstable azo compound by brief heating with dilute alkali[126]:

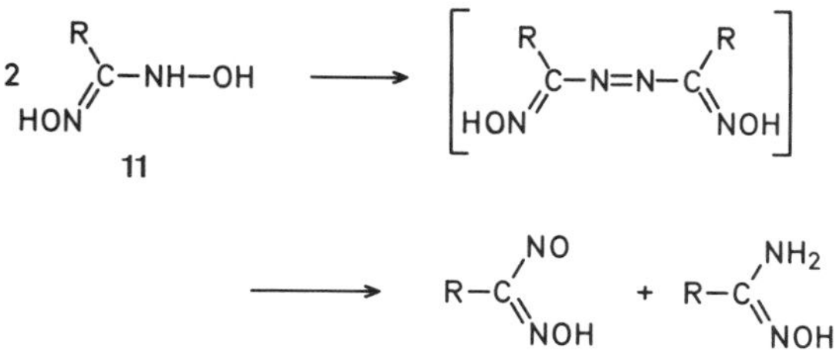

As the free nitrosolic acids easily decompose into fulminic and hyponitrous acids, they are generally isolated as their (silver) salts.

2.2.2 Transformation reactions and analytical hints

Under suitable reaction conditions not involving an attack on the nitroso group itself *aliphatic nitroso compounds* can be prepared by transformation from other nitroso compounds.

Thus, a synthetic pathway leads to *1-nitrosoolefins* from olefins *via* the additon of nitrosyl chloride and elimination of hydrogen chloride[127], *e.g.*:

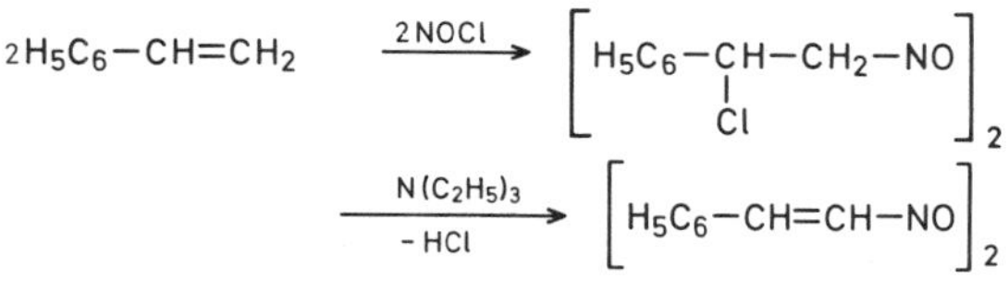

Bis-[2-nitroso-1-phenylethylene]; 81%

In geminal halonitroso compounds the halogen can be replaced by the nitro group with silver nitrate to give *pseudonitroles*[128]:

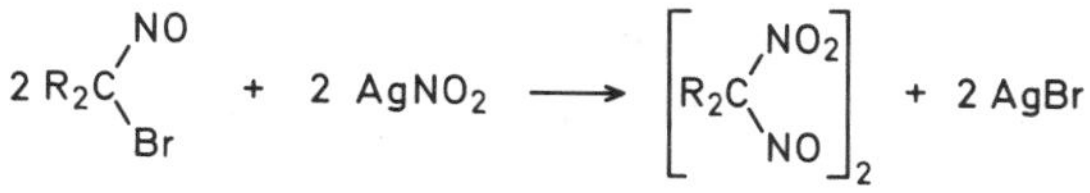

2.2.2.1 Isomerization to oximes

Primary and secondary aliphatic nitroso compounds can isomerize to *oximes* (cf. p. 34):

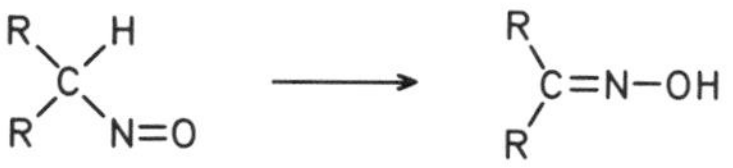

The rearrangement may be accelerated, or is brought about by warming and by the action of acids, bases, polar agents, or nitrogen monoxide[129]. In tertiary nitroso compounds which intrinsically cannot isomerize, the tendency to form the hydroxyimino group may be so highly developed sometimes that rearrangement to the oxime ensues with loss of a substituent group or with ring opening. This fragmentation is observed especially in β-substituted nitroso compounds[130]:

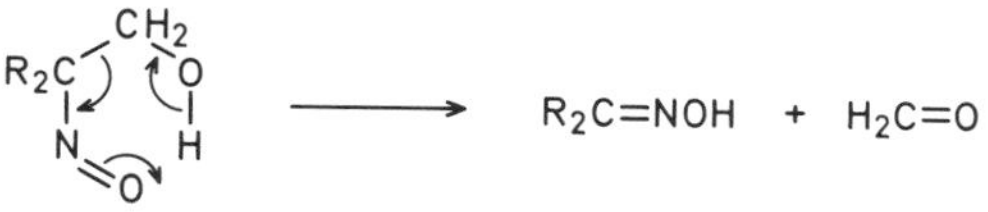

2.2.2.2 Oxidation

Oxidation with manganese(VII), chromium(VI), and lead(IV) oxides, oxygen, hydrogen peroxide, nitrosyl chloride, trifluoroperoxyacetic acid, or nitric acid converts many nitroso compounds into the corresponding *nitro compounds*:

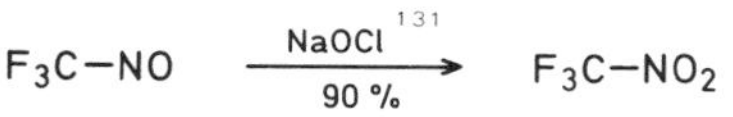

Trifluoronitromethane

[121] *Eu. Müller, H. Metzger,* Chem. Ber. *89*, 398 (1956);
R. Behrend, E. König, Justus Liebigs Ann. Chem. *263*, 221 (1891); *263*, 344 (1891);
R. Behrend, W. Platner, Justus Liebigs Ann. Chem. *278*, 367 (1894);
R. Behrend, D. Nissen, Justus Liebigs Ann. Chem. *269*, 398 (1892);
Eu. Müller, H. Metzger, Chem. Ber. *89*, 404 (1956).
[122] *R. Behrend, W. Platner,* Justus Liebigs Ann. Chem. *278*, 367 (1894).
[123] *A. Angeli, V. Castellana, R. Ferrero,* Atti. Accad. Naz. Linzei, Classe Sci. Fis., Mat. Natur., Rend. *18*, II, 38 (1909); Chem. Zentr. II 975 (1909).
[124] *I.L. Knunyants, A.V. Fokin, V.S. Blagovescenskij,* Doklady Akad. Nauk SSSR *146*, 1088 (1962); C.A. *58*, 7840 (1963).
[125] *E.H. White, W.J. Considine,* J. Amer. Chem. Soc. *80*, 628 (1958).

[126] *H. Wieland, H. Hess,* Ber. *42*, 4176, 4179 (1909);
H. Wieland, Justus Liebigs Ann. Chem. *353*, 91 (1907);
H. Wieland, H. Bauer, Ber. *39*, 1483 (1906);
H. Wieland, Ber. *38*, 1456 (1905).
[127] *W. Pritzkow, H. Schaefer, P. Pabst, A. Ebenroth, J. Beger,* J. Prakt. Chem. 29, 123 (1965);
R.D. Shrivastava, S.S. Deshapande, Proc. Natl. Inst. Sci. India Pt. A. *31*, 180 (1965); C.A. *64*, 6696 9770 (1966);
W.R. Miller, E.H. Pryde, J.C. Cowan, H.M. Teeter, J. Amer. Oil Chemists Soc. *42*, 713 (1965).

2.2.2.3 Reduction

α-Nitroso ketones can be reduced to *azoxy* and *hydrazo* compounds with sodium or aluminum amalgam[132]. By contrast, *α-nitrosocarboxylic acid esters* afford α-amino *carboxylic acid esters*[133]. 2-Aryl-2-halo-1-nitrosoalkanes can be reduced to *primary amines* catalytically on Raney nickel[134] or with lithium tetrahydroborate[135] with reductive splitting off of halogen, *e.g.*:

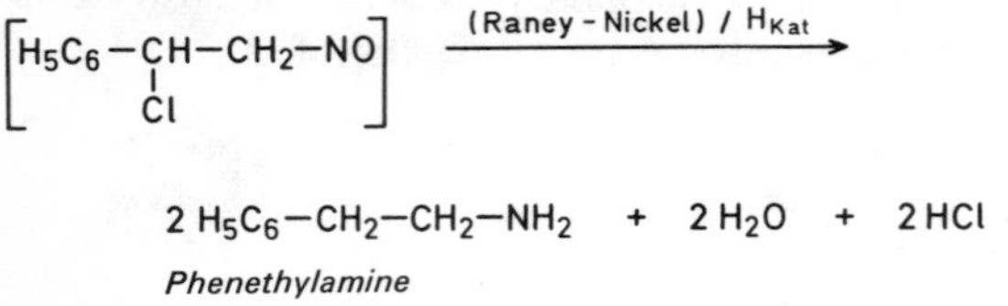
(Raney - Nickel) / H_{Kat}
$H_5C_6-CH-CH_2-NO$
Cl
$2\ H_5C_6-CH_2-CH_2-NH_2$ + $2\ H_2O$ + 2 HCl
Phenethylamine

3-Chloro-2,3-dimethyl-2-nitrosobutane is reduced to *N-(3-chloro-2,3-dimethyl-2-butyl)hydroxylamine*[136] by lithium tetrahydroaluminate, but to *2-chloro-1,2,2-trimethylpropylamine* by tin(II) chloride in concentrated hydrochloric acid[137].
4a-Chloro-1,2,3,4,5,6,7,8-octahydro-8a-nitrosonaphthalene is an example displaying particular variability[138]:

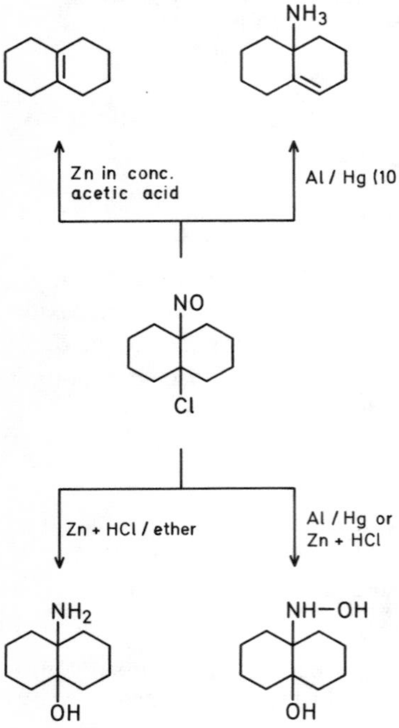
NH_3
Zn in conc. acetic acid
Al / Hg (10 days)
NO
Cl
Zn + HCl / ether
Al / Hg or Zn + HCl
NH_2
OH
NH—OH
OH

2.2.2.4 Addition and condensation

As well as a series of 'linear' additions[139] and the familiar condensation reactions with primary amines and hydroxylamines

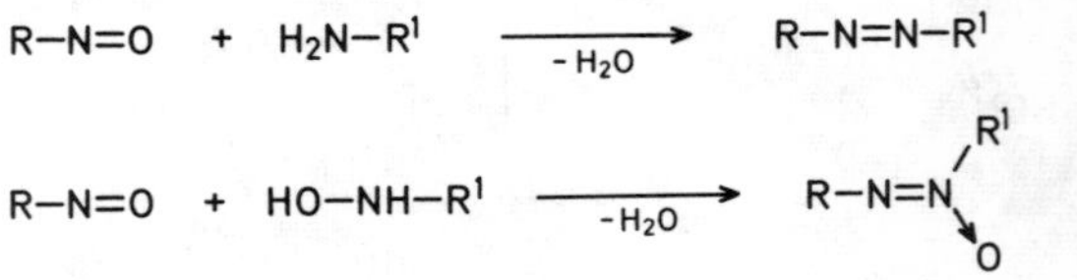
R—N=O + H_2N-R^1
$-H_2O$
$R-N=N-R^1$
R—N=O + $HO-NH-R^1$
$-H_2O$
R—N=N
R^1
O

nitroso compounds under certain circumstances display *cycloadditions* with olefins, dienes, acetylenes, ketenes, azomethines, isocyanides, and azo compounds. Here, the synthetic significance of this reaction type in respect of *N,O-heterocycles* may be illustrated just for the four following general examples:

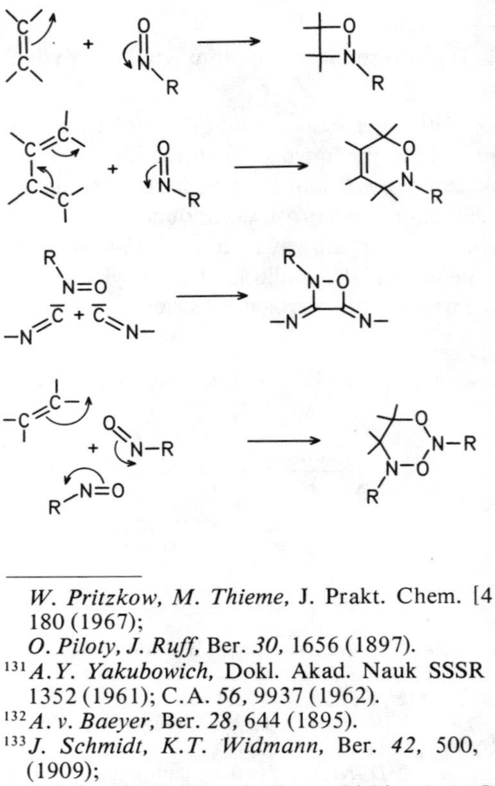

---

[128] *O. Piloty, A. Stock,* Ber. *35,* 3093 (1902);
M.F. Hawthorne, J. Amer. Chem. Soc. *79,* 2515 (1957).

[129] *M.H. Palmer, E.R.R. Russell,* Chem. & Ind. (London) 157 (1966).

[130] *W. Pritzkow, W. Rösler,* Justus Liebigs Ann. Chem. *703,* 66 (1967);
W. Pritzkow, M. Thieme, J. Prakt. Chem. [4] *36,* 180 (1967);
O. Piloty, J. Ruff, Ber. *30,* 1656 (1897).

[131] *A.Y. Yakubowich,* Dokl. Akad. Nauk SSSR *140,* 1352 (1961); C.A. *56,* 9937 (1962).

[132] *A. v. Baeyer,* Ber. *28,* 644 (1895).

[133] *J. Schmidt, K.T. Widmann,* Ber. *42,* 500, 1894 (1909);
J. Schmidt, H. Dieterle, Justus Liebigs Ann. Chem. *377,* 38 (1910).

[134] *R. Perrot, R. Holbein,* Bull. Soc. Chim. France 578, 690 (1949);
R. Perrot, R. Holbein, Compt. Rend. *234,* 2617 (1952);
R. Perrot, P. Wodey, Bull. Soc. Chim. France 797 (1951);
R. Perrot, P. Wodey, Compt. Rend. *240,* 101 (1955).

[135] *A. Burger,* 136th Meeting of the American Chemical Society (1959), Abstracts of Papers, 38 – 0.

[136] *A. Dornow, K.J. Fust,* Chem. Ber. *90,* 1773 (1957).

Certain *polymeric nitrosovinyl compounds* have gained technical importance[140]. Fluorinated nitrosoalkanes such as trifluoronitrosomethane can be copolymerized with fluoroalkenes[140]:

$$F_3C-N=O \ + \ F_2C=CF_2 \ \longrightarrow$$

$$\left[\begin{array}{c} -N-O(CF_2-CF_2)_n- \\ | \\ CF_3 \end{array} \right]_m$$

2.2.2.5 Replacement of the nitroso group

As is shown by the following survey, the nitroso function in aliphatic nitroso compounds is able to undergo a number of substitution reactions, especially if activated groups are present in the adjacent position:

$$(F_3C)_3C-NO \xrightarrow[-NO-Hal]{Hal} (F_3C)_3C-Hal \qquad [141]$$
$$33-58\%$$

$$R_2C\begin{array}{c} NO \\ \diagdown \\ NO_2 \end{array} \xrightarrow[\substack{-H_2O \\ -2N_2}]{2 \ HN_3} R_2C\begin{array}{c} N_3 \\ \diagdown \\ NO_2 \end{array} \qquad [142]$$

$$\left[H_5C_6-CH-CH-CH_2-COOC_2H_5 \atop \hspace{0.8cm} | \hspace{0.6cm} | \atop \hspace{0.8cm} NO \hspace{0.3cm} NO_2 \right]_2 \xrightarrow[\substack{-N_2O \\ -H_2O}]{2 \ H_5C_6-NH_2} \qquad [143]$$

$$2 \ H_5C_6-CH-CH-CH_2-COOC_2H_5 \atop \hspace{0.8cm} | \hspace{0.6cm} | \atop \hspace{0.8cm} NH \hspace{0.3cm} NO_2 \atop \hspace{0.8cm} | \atop \hspace{0.8cm} H_5C_6$$

Ethyl 4-anilino-3-nitro-4-phenylbutyrate

$$\left[H_5C_6-CH-CH_2 \atop \hspace{0.6cm} | \hspace{0.5cm} | \atop \hspace{0.6cm} NO \hspace{0.3cm} NO_2 \right]_2 \xrightarrow[-N_2O]{H_5C_2-ONa} \qquad [144]$$

$$2 \ H_5C_6-CH-CH=N\begin{array}{c} O \\ \diagup \\ \diagdown ONa \end{array} \atop \hspace{0.8cm} | \atop \hspace{0.8cm} OC_2H_5$$

Sodium salt of 2-ethoxy-2-phenyl-1-aci-nitroethane;

$$\left[H_5C_6-CH-CH-CH_2-COOCH_3 \atop \hspace{0.9cm} | \hspace{0.6cm} | \atop \hspace{0.9cm} NO \hspace{0.3cm} NO_2 \right]_2 \xrightarrow[H_3C-CO-O-CO-CH_3]{H_2SO_4 \ /} \qquad [145]$$

$$H_5C_6-CH-CH-CH_2-COOCH_3 \atop \hspace{0.3cm} | \hspace{1.2cm} | \atop H_3C-CO-O \hspace{0.5cm} NO_2$$

Methyl 4-hydroxy-3-nitro-4-phenylbutyrate acetate

$$R-NO \xrightarrow[-N_2]{2 \ NO} R-O-NO_2 \qquad [146]$$

[137] *G. Closs, S.J. Brois*, J. Amer. Chem. Soc. *82*, 6068 (1960).

[138] *W. Hückel, M. Blohm*, Justus Liebigs Ann. Chem. *502*, 114 (1933).

[139] *A.A. Gevorkyan, B.L. Dyatkin, I.L. Knunyants*, Izv. Akad. Nauk SSSR 1599 (1965); C.A. *64*, 1944 (1966).

[140] USP. 3058961 (1960), Inv.: *R.N. Haszeldine*; C.A. *58*, 3585 (1963).

[141] *B.L. Dyatkin, A.A. Gevorkyan, I.L. Knunyants*, Izv. Akad. Nauk SSSR 1879 (1965).

[142] *S. Maffei, G.F. Bettinetti*, Ann. Chim. *47*, 1286 (1957).

[143] *H. Wieland*, Ber. *36*, 2564 (1903);
H. Wieland, Justus Liebigs Ann. Chem. *329*, 237 (1903);
K. Ssidorenko, J. Russ. Phys.-Chem. Soc. *45*, 1590 (1913); Chem. Zentr. I, 1068 (1914);
A. Angeli, Gazz. Chim. Ital. *23*, II, 124 (1893);
P. Toennies, Ber. *20*, 2986 (1887);
E.A. Sommer, Ber. *29*, 360 (1896).

[144] *H. Wieland*, Ber. *36*, 2565 (1903);
H. Wieland, Justus Liebigs Ann. Chem. *329*, 236 (1903).

[145] *H. Wieland*, Justus Liebigs Ann. Chem. *329*, 253 (1903).

[146] *R. Sartorius*, Dissertation, Universität Giessen, 1939;
E. Weitz, L. Müller, K. Dinges, Chem. Ber. *85*, 879 (1952);
H. Metzger, Dissertation, Universität Tübingen, 1954;
L.G. Donaruma, D.J. Carmody, J. Org. Chem. 22, 635 (1957).

[147] *O. Piloty, A. Stock*, Ber. *35*, 3098 (1902).

[148] *M.F. Hawthrone, R.D. Strahm*, J. Amer. Chem. Soc. *79*, 2516 (1957).

[149] *O. Wallach*, Justus Liebigs Ann. Chem. *252*, 132 (1889); *258*, 344 (1890); *346*, 220 (1906);
W.A. Tilden, J. Chem. Soc. (London) *85*, 763 (1904);
F.H. Turber, R.C. Thielke, J. Amer. Chem. Soc. *53*, 1031 (1931);
E.V. Lynn, J. Amer. Chem. Soc. *41*, 361 (1919);
J. Thiele, Ber. *27*, 456 (1894);
J. Schmidt, F. Leipprand, Ber. *37*, 546, 547 (1904);
W. Hückel, Justus Liebigs Ann. Chem. *474*, 132 (1929);
H. Wieland, Ber. *36*, 2564 (1903).

[150] *J.G. Aston, D.F. Menard, M.G. Mayberry*, J. Amer. Chem. Soc. *54*, 1537 (1932).

[151] *E. Strom, A. Bluhm*, Chem. Commun. 115 (1966);
W. Theilacker, A. Knap, H. Uffmann, Angew. Chem. Intern. Ed. Engl. *4*, 688 (1965).

[152] *A. Mackor, Th. Wajer, Th. de Boer, J. van Voorst*, Tetrahedron Lett. 2115 (1966).

[153] *R.N. Haszeldine, B. Mattinson*, J. Chem. Soc. 1741 (1957);
J. Mason, J. Chem. Soc. 4531 (1963).

[154] *K. Anderson, C. Crumpler, D. Hammick*, J. Chem. Soc. 1679 (1935).

2.2.2.6 Elimination of the nitroso group

Many nitroso compounds, in particular geminal halonitroso compounds, pseudonitroles, and pseudonitrosites, but also nitroso ketones and nitrosohydrocarbons, suffer more or less spontaneous decompositions at room temperature or on warming. Often the molecule decomposes into small fragments while liberating nitrogen and oxygen containing gases. Acids and bases also can frequently bring about such a decomposition.

Thus, for example, 2-bromo-2-nitrosopropane affords *acetone* in aqueous pyridine[147]:

$$2\ \begin{array}{c} H_3C \\ H_3C \end{array}\!\!C\!\!\begin{array}{c} Br \\ NO \end{array} + H_2O \xrightarrow{Pyridine} \begin{array}{c} H_3C \\ H_3C \end{array}\!\!C=O$$

$$+\ N_2O\ +\ 2\,HBr$$

With silver perchlorate, too, geminal chloronitroso compounds react to *ketones* practically quantitatively[148]:

$$R_2C\!\!\begin{array}{c} Cl \\ NO \end{array} \xrightarrow[-AgCl]{H_2O\,/\,Ag^{\oplus}/\,H^{\oplus}} \left[R_2C\!\!\begin{array}{c} OH \\ NO \end{array}\right] \xrightarrow{-NOH}$$

$$R_2C=O$$

Elimination of the nitroso group from 2-halo-1-nitro-2-nitroso compounds often leads to formation of *olefins*. Certain pseudonitrosites, however, can suffer more extensive transformation into *aldehydes* or *ketones* and *nitroalkanes*[149]:

$$R-\overset{\textstyle |}{\underset{\textstyle |}{C}}-\overset{\textstyle |}{\underset{\textstyle |}{C}}-R^1 \longrightarrow$$
(ON, NO2)

In nitroso ketones the nitroso group can be frequently split off reductively[150]:

$$\left[(H_3C)_2\underset{NO}{\overset{|}{C}}-CH_2-CO-CH_3\right]_2 \xrightarrow[-2\,NH_3]{SnCl_2\,/\,HCl}$$

$$2\ (H_3C)_2C=CH-CO-CH_3$$

4-Methyl-3-penten-2-one; 30%

2.2.2.7 Photochemistry

In contrast to aromatic nitroso compounds aliphatic nitroso compounds decompose also on receiving radiation at their longest-wave absorption maximum at ~ 7000 Å (n→π^* transition)[151]:

$$R-N=O \xrightarrow[-\dot{N}O]{h\nu} R\cdot \xrightarrow{R-N=O} R_2\dot{N}O$$

---

[155] *W. Lüttke*, Journal de Physique et le Radium *15*, 635 (1954);
W. Lüttke, Angew. Chem. *66*, 159 (1954);
Eu. Müller, H. Metzger, Chem. Ber. *87*, 1284 (1954);
Eu. Müller, H. Metzger, D. Fries, Chem. Ber. *87*, 1457 (1954); *88*, 1896 (1955);
J. Jander, R.N. Haszeldine, J. Chem. Soc. (London) 915 (1954);
E.H. White, W.J. Considine, J. Amer. Chem. Soc. *80*, 627 (1958);
J. Goubeau, I. Fromme, Z. Anorg. Allgem. Chemie *258*, 22 (1949);
W. Lüttke, Z. Elektrochemie *61*, 312 (1957);
P. Tarte, Bull. Soc. Chim. Belg. *63*, 528 (1954).

[156] *Eu. Müller, H. Metzger*, Chem. Ber. *88*, 168 (1955);
H. Metzger, Dissertation, Universität Tübingen, 1954;
Eu. Müller, D. Fries, H. Metzger, Chem. Ber. *88*, 1899 (1955);
P. Tarte, Bull. Soc. Chim. Belg. *63*, 534 (1954);
J. Jander, R.N. Haszeldine, J. Chem. Soc. (London) 916 (1954);
J.F. Brown, J. Amer. Chem. Soc. *77*, 6348 (1955);
B.G. Gowenlock, J. Trotman, J. Chem. Soc. (London) 4193 (1955); 1673 (1956);
B.G. Gowenlock, J. Chem. Soc. (London) 3929 (1957);
W. Lüttke, Angew. Chem. *68*, 417 (1956);
W. Lüttke, Z. Elektrochemie *61*, 307 (1957);
L.G. Donaruma, D.J. Carmody, J. Org. Chem. *22*, 636 (1957);
W. Lüttke, Z. Elektrochemie *61*, 976 (1957).
[157] *P. Tarte*, Bull. Soc. Chim. Belg. *63*, 531 (1954);
J. Mason, J. Chem. Soc. (London) 3904 (1957);
K.A. Anderson, C.J. Crumpler, D.L. Hammick, J. Chem. Soc. (London) 1680 (1935);
D.L. Hammick, M.W. Lister, J. Chem. Soc. (London) 491 (1937);
J. Jander, R.N. Haszeldine, J. Chem. Soc. (London) 913 (1954);
P. Tarte, Bull. Soc. Chim. Belg. *63*, 536 (1954);
J.R. Schwartz, J. Amer. Chem. Soc. *79*, 4354 (1957).

Araliphatic *nitroxides*, too, can be prepared in this way[152]:

$$(H_3C)_3C-N=O \ + \ H_5C_6-N=O \ \xrightarrow{h\nu}$$

$$(H_3C)_3C-\overset{|}{\underset{C_6H_5}{N}}-\overline{O}\cdot \ + \ (H_3C)_3C-\overset{|}{\underset{C(CH_3)_3}{N}}-\overline{O}\cdot$$

Irradiating trifluoronitrosomethane in the gas phase (with red light) affords a dimer having the nitritoamine structure[153]:

$$2\,F_3C-N=O \ \xrightarrow{h\nu} \ \overset{F_3C}{\underset{F_3C}{>}}N-O-N=O$$

*Bis-(trifluoromethyl)-N-
nitritoamine; 96%*

A further interesting photoreaction is the β-elimination of hyponitrous acid from nitroso ketones[154], *e.g.:*

$$(H_3C)_2\overset{|}{\underset{NO}{C}}-CH_2-CO-CH_3 \ \xrightarrow[-HNO]{h\nu}$$

$$(H_3C)_2C=CH-CO-CH_3$$

*4-Methyl-3-penten-
2-one*

2.2.2.8 Analysis

The following spectroscopic data can serve for the rapid identification of aliphatic nitroso compounds:

IR:	No valence vibration		Ref.
	of the monomers	1540–1621 cm^{-1}	155
	of the *trans* dimers	1160–1300 cm^{-1}	156
	of the *cis* dimers	1323–1350 cm^{-1}	156
		1387–1426 cm^{-1}	
UV:	$n_N \to \pi^*$-transition	625–790 mμ (ε: ~ 50)	157
	$n_O \to \pi^*$-transition	270–290 mμ (ε: ~ 80)	158
	$\pi \to \pi^*$-transition	~ 220 mμ (ε: ~ 5000)	158
	Absorption by the dimer	275–300 mμ (ε: 5000–12000)	159

---

[158] *J. Mason*, J. Chem. Soc. (London) 3904 (1957).
[159] *E.C.C. Baly, C.H. Desch*, J. Chem. Soc. (London) *93*, 1754 (1908);
H.T.J. Chilton, B.G. Gowenlock, J. Chem. Soc. (London) 3235 (1953);
Eu. Müller, H. Metzger, Chem. Ber. *88*, 1893 (1955);
H. Metzger, Eu. Müller, Chem. Ber. *90*, 1186 (1957);
B.G. Gowenlock, J. Trotman, J. Chem. Soc. (London) 1673 (1956);
J. Jander, R.N. Haszeldine, J. Chem. Soc. (London) 914 (1954).

2.3 Bibliography

Aromatic nitroso compounds

B.G. Gowenlock, W. Lüttke, Structure and Properties of C-Nitroso-Compounds, Quart. Rev. Chem. Soc. *12*, 321 (1958).
T.A. Turney, G.A. Wright, Nitrous Acid and Nitrosation, Chem. Rev. *59*, 497 (1959).
A.T. Austin, Nitrosation in Organic Chemistry, Sci. Progr. *49*, 619 (1961).
P.A.S. Smith, The Chemistry of Open-Chain Organic Nitrogen Compounds, Vol. II, p. 355, W.A. Benjamin Inc., New York, Amsterdam 1966.
N.V. Sidgwick in *I.T. Millar, H.D. Springall*, The Organic Chemistry of Nitrogen, 3. Ed., p. 339, new ed., Clarendon Press, Oxford 1966.
S. Nishigaki, F. Yoneda, Recent Topics on the Reaction of Nitroso Compounds, Kagaku No Ryoiki *22*, 753 (1968); C.A. *70*, 3862 (1968);
Reactions of Aromatic Nitroso Compounds I. Kagaku No Ryoiki *22*, 660 (1968); C.A. *69*, 85813 (1968).
H. Feuer, The Chemistry of the Nitro and Nitroso Groups, Part 1 (1969), Part 2 (1970). Interscience Publishers, John Wiley & Sons, New York, London, Sydney, Toronto 1970.
W. Seidenfaden in *Houben-Weyl*, Methoden der organischen Chemie, Aromatische Nitroso-Verbindungen, Bd. X/1, p. 1017, Georg Thieme Verlag, Stuttgart 1971.

Aliphatic nitroso compounds

J.L. Riebsommer, The Reactions of Nitrogen Tetroxide with Organic Compounds, Chem. Rev. *36*, 183, 196 (1945).
E.F. Degering, An Outline of Organic Nitrogen Compounds, p. 107, 150, University Lithoprinters, Ypsilanti, Mich., 1950.
L.J. Beckham, W.A. Fessler, M.A. Kise, Nitrosyl Chloride, Chem. Rev. *48*, 369 (1951).
O. Touster, The Nitrosation of Aliphatic Carbon Atoms, Org. Reactions *7*, 327 (1953).
B.G. Gowenlock, W. Lüttke, Structure and Properties of C-Nitroso Compounds, Quart. Rev. *12*, 321 (1958).
T.A. Turney, G.A. Wright, Nitrous Acid and Nitrosation, Chem. Rev. *59*, 497 (1959).
A.T. Austin, Nitrosation in Organic Chemistry, Sci. Progr. *49*, 619 (1961).
P.A.S. Smith, The Chemistry of Open Chain Organic Nitrogen Compounds, Vol. II, p. 355, W.A. Benjamin Inc., New York, Amsterdam 1966.
N.V. Sidgwick, The Organic Chemistry of Nitrogen, 3. Ed., new ed. by *I.T. Millar, H.D. Springall*, p. 339, Clarendon Press, Oxford 1966.
H. Feuer, The Chemistry of the Nitro and Nitroso Groups, Part 1 (1969), Part 2 (1970), Interscience Publishers, John Wiley & Sons, New York.
H. Metzger, H. Meier in *Houben-Weyl*, Methoden der organischen Chemie, Aliphatische Nitroso-Verbindungen, Bd. X, p. 1893, Georg Thieme Verlag, Stuttgart 1971.

3 Hydroxylamines

Contributed by

J. Keck
Dr. Karl Thomae GmbH,
Biberach an der Riß

This chapter describes compounds in which the hydrogen atoms of hydroxylamine

$$\text{H}_2\text{N}-\text{O}-\text{H}$$

have been replaced by one or more alkyl or aryl groups. The products are substances of the general formulae

$$R^1\text{-N(H)}-\text{OH} \qquad R^2\text{-N}(R^1)-\text{OH} \qquad \text{H-N(H)}-\text{OR}^3$$

$$R^1\text{-N(H)}-\text{OR}^3 \qquad R^2\text{-N}(R^1)-\text{OR}^3$$

R^1 , R^2 , R^3 = Alkyl , Aryl

The nitrogen atom of the hydroxylamine, alone or together with the oxygen atom, may belong to a heterocyclic ring to give compounds types such as

N-Hydroxy-pyrrolidines	N-Hydroxy-morpholines	N-Hydroxy-9-acridanones

or

Oxaziridines	1,2-Oxazetidines	Isoxazolidines	Tetrahydro-1,2-oxazines

R = H, Alkyl, Aryl

which can be considered to be cyclic hydroxyl-amine derivatives.

The *nomenclature* of the open-chain compounds is not uniform. Hydroxylamines substituted on the nitrogen are named as *N-alkyl-* and *N-arylhydroxyl-amines* (since 1972 the Chemical Abstracts list them as *N*-hydroxyalkyl- and *N*-hydroxyaryl-amines). Analogously, the compounds substituted on the oxygen are termed *O-alkyl-* and *O-arylhy-droxylamines* but often also alkoxy- and aryloxy-amines (as in C.A. before 1972) or hydroxylamine-*O*-alkyl and -aryl ethers. Alternatively again, the prefixes α and β are used to denote substitution on the oxygen and nitrogen respectively (*e.g.* in the older C.A.).

Most hydroxylamines are *not very stable*, especially in the presence of atmospheric oxygen. However, they do form stable *salts* with mineral acids which serve for isolating and storing the compounds. An exception are the *O-(phenethyl)hy-droxylamines (phenethyloxyamines)*, which form quite stable bases, while their hydrochlorides decompose rapidly[1].

Differences in *basicity* are encountered. Thus, the *O*-substituted hydroxylamines are somewhat less basic than the corresponding *N*-substituted compounds[2]. However, hydroxylamines with a free hydroxy group also possess weakly *acid* properties. *Mono-* and *disodium salts* of *N*-phenylhydrox-ylamine have been described[3].

Hydroxylamines substituted with low alkyl and aryl groups are liquids at room temperature or low-melting solids. Expectedly, the melting points and boiling points of the *O*-hydroxylamine bases are substantially lower than those of the corresponding *N*-hydroxylamines. Table 1 lists the physical data of the methyl-substituted compounds and includes the melting points of the hydrochlo-rides and the pK_a values[2].

Table 1. Physical data of the methyl-substituted hydroxylamines

	Mp °C Base	Bp$_{760}$ °C Base	Mp °C Hydrochloride	pK_a
N-Methylhydroxylamine	38.5	115	87–89	5.96
N,N-Dimethylhydroxylamine	17.6	100.6	103–106	5.2
O-Methylhydroxylamine (Methoxyamine)	−86.4	48.1	151	4.6
O,N-Dimethyl-hydroxylamine	−97	42.3	115–116	4.75
Trimethylhydroxylamine	−97.2	30	122.5–123.5	3.65

N- and *O*-substituted hydroxylamines possess *reducing* properties; they are less marked in the *O*-substituted compounds[4]. For example, *N-methyl-hydroxylamine* reduces Fehling's solution and ammoniacal silver nitrate solution, *O-methylhydrox-ylamine* only ammoniacal silver nitrate solution[5]. Fehling's solution is not reduced by *O*-alkylhy-droxylamines even on heating[6]. *O*- and *N*-substi-tuted hydroxylamines can act also as *oxidizing agents,* for example, toward iron(II) hydroxide, which is oxidized to iron(III) hydroxide in aqueous suspension[7].

[1] *B.J.R. Nicolaus, G. Pagani, E. Testa,* Helv. Chim. Acta *45,* 1384 (1962).

[2] *T.C. Bissot, R.W. Parry, D.H. Campbell,* J. Amer. Chem. Soc. *79,* 796 (1957).

[3] *E. Bamberger, F. Brady,* Ber. *33,* 271 (1900); *J. Schmidt,* Ber. *32,* 2911 (1899); *G.F. White, K.H. Knight,* J. Amer. Chem. Soc. *45,* 1780 (1923).

3.1 Preparation of organic hydroxylamine compounds

Proven methods for preparing organic hydroxylamines can be divided into the following four groups:

1 *Substitution* of hydrogen atoms of hydroxylamine by organic substituent groups
2 Preparation of organic hydroxylamines by *addition* to multiple bonds
3 Preparation of the hydroxyamino group from nitrogen compounds of other *oxidation* stages
4 *Cleavage* and *rearrangement* reactions.

Replacing the hydroxylamine hydrogen by other groups can be effected in directed manner either on the nitrogen or oxygen according to the choice of the method. In individual cases a simultaneous N,O-substitution is possible. Table 2 lists the important and usually generally applicable methods.

Table 2. Methods for preparing variously substituted hydroxylamines

Hydroxylamines	Method of preparation		Page
Alkyl−NH−OH	Reduction of aliphatic nitro and nitroso compounds	$R-NO_2 \longrightarrow R-NH-OH$ $R-NO \longrightarrow R-NH-OH$	65
	Cleaving of nitrones		67
	Reduction of oximes		66
	Addition of H_2N-OH to activated $C{=}C$ double bonds	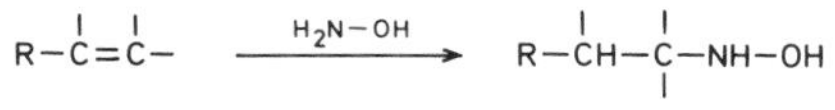 $R = CO, R^1OOC, O_2N, SO_2$	59
Aryl−NH−OH	Reduction of aromatic nitro and nitroso compounds	$R-NO_2 \longrightarrow R-NH-OH$ $R-NO \longrightarrow R-NH-OH$	65
	Replacement of activated groups in aromatic and heteroaromatic compounds by H_2N-OH	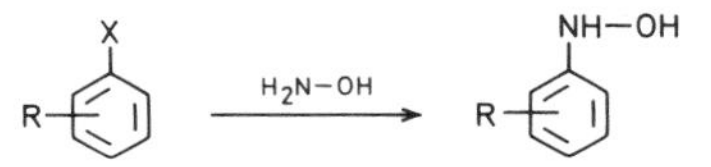 $X = OH, OR, Halogen, NO_2, NH_2, SCH_3, SO_3H$	58 59
H_2N-O-Alkyl	Cleavage of O-alkyl oximes	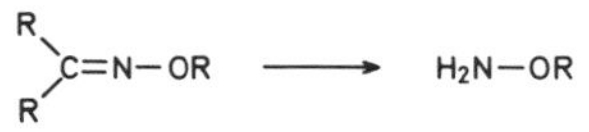	68
	Cleavage of N-alkoxyphthalimide	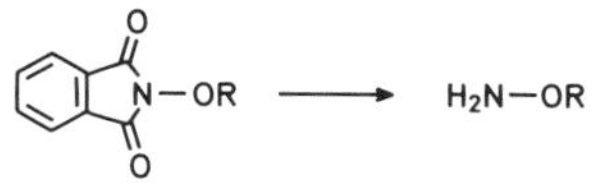	69
	Cleavage of hydroxamic acid esters	$R-CO-NH-OR \longrightarrow H_2N-OR$	70
	Cleavage of N-alkoxyurethans	$H_5C_2O-CO-NH-OR \longrightarrow H_2N-OR$	70
	Reaction of hydroxylamine-O-sulfonic acid or chloramine with alcoholates	$H_2N-O-SO_3Na + NaOR \longrightarrow H_2N-OR$ $H_2N-Cl + NaOR \longrightarrow H_2N-OR$	59 71

[4] *P.J. Baker jr.*, Kirk-Othmer Encycl. Chem. Technol., 2nd Ed. Vol. I, p. 495, 502, 503, John Wiley & Sons, New York 1966.
[5] *W. Traube, H. Ohlendorf, H. Zander*, Ber. *53*, 1485 (1920).
[6] *W. Theilacker, K. Ebke*, Angew. Chem. *68*, 303 (1956).
[7] *P. Baumgarten, H. Erbe*, Ber. *71*, 2603 (1938).

Table 2. (continued)

Hydroxylamines	Method of preparation		Page
Alkyl–N(Alkyl)–OH	Alkylation of H_2N–OH or Alkyl–NH–OH	H_2N–OH + 2 RX ⟶ R–N(R)–OH R^1–N(H)–OH + RX ⟶ R^1–N(R)–OH	57
	Oxidation of secondary amines	R–N(R)–H $\xrightarrow{H_2O_2}$ R–N(R)–OH	63
	Thermal cleavage of trialkyl aminooxides	R–N(R)(→O)(CH₂–CH₂–R¹) ⟶ R–N(R)–OH	64
	Reduction of nitrones	R–C(R)=N(→O)–R ⟶ R–CH(R)–N(R)–OH	66
Alkyl–N(Aryl)–OH	Alkylation of *N*-aryl hydroxylamines	Aryl–N(H)–OH + RX ⟶ Aryl–N(R)–OH	58
Aryl–N(Aryl)–OH	Reaction of nitroso compounds with Grignard compounds	Aryl–NO $\xrightarrow[\text{2. }H_2O]{\text{1. Aryl–Mg–Br}}$ Aryl–N(Aryl)–OH	62
H–N(Alkyl)–O–Alkyl	Cleavage of *N*-alkyl-*N*-alkoxyurea	R^3NH–CO–N(R^1)–OR² ⟶ H–N(R^1)–OR²	71
	Reduction of *O*-alkyl oximes	R–C(R)=N–OR¹ ⟶ R–CH(R)–N(H)–OR¹	69
H–N(Aryl)–O–Alkyl	Replacement of activated groups in aromatic and heteroaromatic compounds by H_2N-*O*-Alkyl	R–C₆H₄–X $\xrightarrow{H_2N–O–Alkyl}$ R–C₆H₄–NH–O–Alkyl X = OH, OR, Halogen, NO_2, NH_2, SCH_3, SO_3H	58 59
Alkyl–N(Alkyl)–O–Alkyl	Alkylation of H_2N-*O*-Alkyl or Alkyl-NH-*O*-Alkyl	H_2N–O–Alkyl + 2 RX ⟶ R–N(R)–O–Alkyl Alkyl–NH–O–Alkyl + RX ⟶ R–N(Alkyl)–O–Alkyl	58 58
	Rearrangement of *N*-oxides of tertiary amines	R–N(R¹)(→O)–R $\xrightarrow{\triangle}$ R–N(R)–OR¹ R¹ = Benzyl, Allyl	63
Alkyl–N(Aryl)–O–Alkyl	Alkylation of Aryl-NH-*O*-Alkyl	H–N(Aryl)–O–Alkyl + RX ⟶ R–N(Aryl)–O–Alkyl	58

Table 2. (continued)

Hydroxylamines	Method of preparation		Page
	Rearrangement of *N*-oxides of tertiary amines	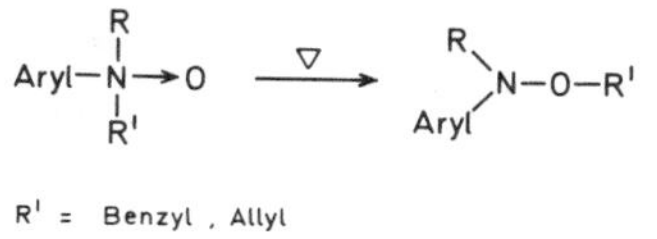 R[^I] = Benzyl , Allyl	63
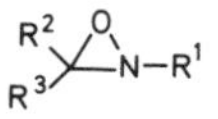	Oxidation of Schiff's bases with peracids		71
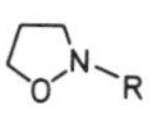	1,3-Dipolar cycloaddition of nitrones to alkenes	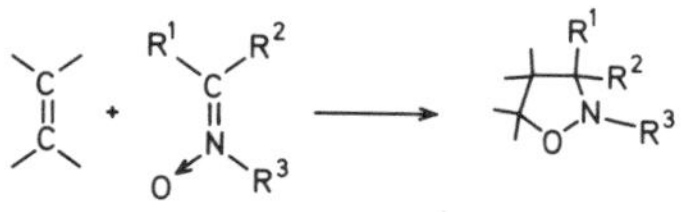	62
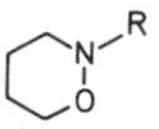	Reaction of dienes with nitroso compounds and catalytic hydrogenation	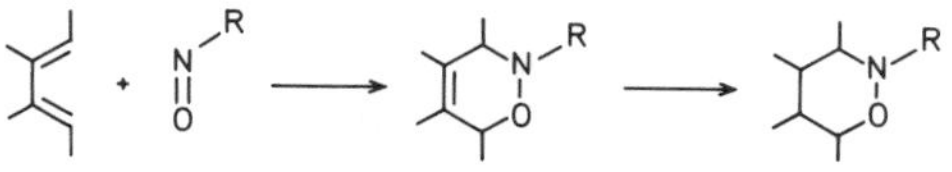	63

3.1.1 Replacement of reactive groups by the hydroxyamino group

The reaction can be formulated as follows by way of example of a monosubstitution:

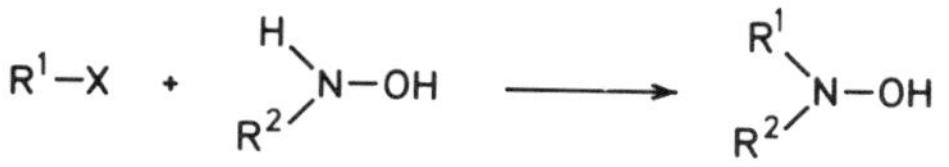

$$R^1\!-\!X \;+\; \underset{R^2}{\overset{H}{>}}N\!-\!OH \;\longrightarrow\; \underset{R^2}{\overset{R^1}{>}}N\!-\!OH \quad (1)$$

$$R^1\!-\!X \;+\; \underset{R^3}{\overset{R^2}{>}}N\!-\!OH \;\longrightarrow\; \underset{R^3}{\overset{R^2}{>}}N\!-\!OR^1 \quad (2)$$

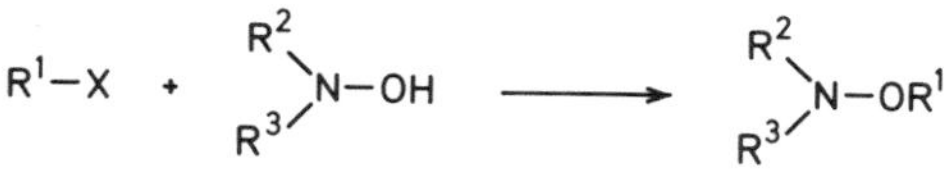

The reactive group *X* may be, *e.g.*:

Aliphatically or arylaliphatically bound halogen
An alkylsulfonic acid group
Activated hydroxy or alkoxy
Nitro-group activated aromatic halogen
Activated nitro, amino, methylthio, and sulfonic acid groups in aromatic and heteroaromatic compounds.
Amino groups in tertiary Mannich bases

N-Unsubstituted or *N*-monosubstituted hydroxylamines react almost exclusively according to Equation 1 to give *N*-substitution. The *O*-substitution according to 2 succeeds only in certain special cases (see p. 58) or by blocking the amine function with suitable protective groups (see p. 68).

3.1.1.1 Alkylation of hydroxylamines with alkyl halides and other alkylating agents

Direct alkylation of hydroxylamine leads almost exclusively to *N*-disubstituted derivatives. *N*-monoalkylated hydroxylamines are accessible in

this way only in isolated instances and can be obtained more advantageously by partial reduction of nitro compounds (see p. 21) or by cleavage of nitrones (see p. 349, 350). The method is well suited, however, for preparing *N*,*N*-dialkylhydroxylamines. Further alkylation leads to trialkylhydroxy ammonium compounds, especially when methyl iodide is used as alkylating agent:

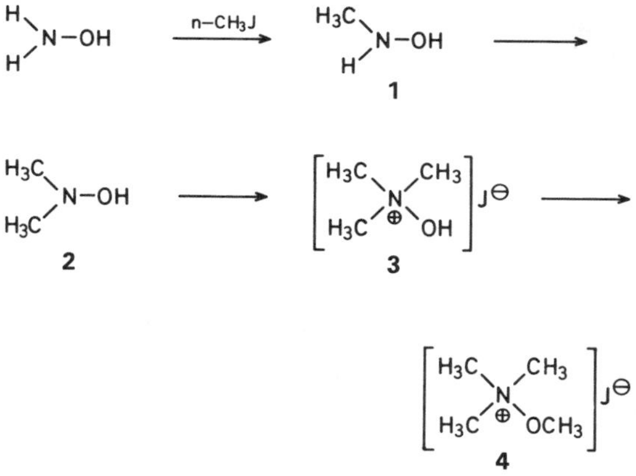

The first isolatable reaction product is the quaternary salt 3[8], which is identical with the hydriodide of trimethylamine oxide[9]. On reacting 3 further with methyl iodide *O*-alkylation occurs and tetramethylhydroxy ammonium iodide 4 is formed[10], which is obtained also from *O*,*N*,*N*-trimethyl-

---

[8] *A. Hantzsch, W. Hilland,* Ber. *31,* 2058 (1898).
[9] *W.R. Dunstan, E. Goulding,* J. Chem. Soc. *75,* 792 (1899).

hydroxylamine (N-methoxydimethylamine) and methyl iodide[11].

An example for the preparation of N-monoalkylated hydroxylamines is the synthesis of *N-[(4-chlorophenyl)diphenylmethyl]hydroxylamine* from the corresponding substituted trityl chloride and hydroxylamine. Here the monosubstituted compound can be isolated if hydroxylamine is used in excess (in 78% yield)[12]:

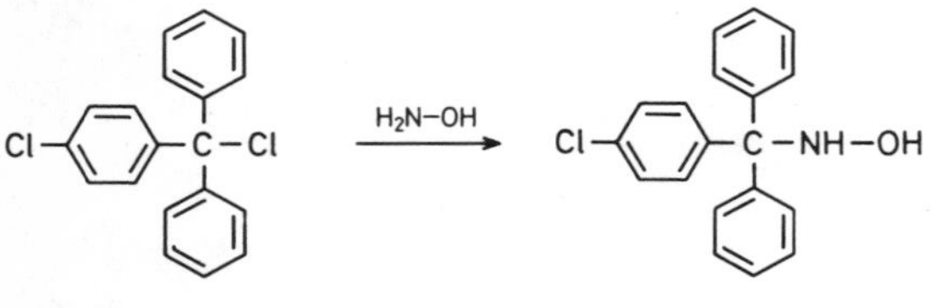

With less bulky substituents direct alkylation of hydroxylamine furnishes N,N-dialkylhydroxylamines[9] such as, for example, *N,N-dibenzylhydroxylamine* from hydroxylamine and benzyl chloride[13]:

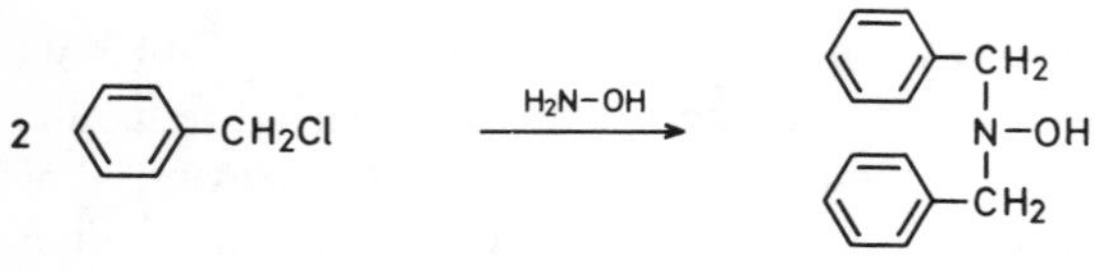

Compounds with *different* substituents on the nitrogen can be obtained by alkylating monosubstituted hydroxylamines. *N-Ethyl-N-(4-tolyl)hydroxylamine* and other N-alkyl-N-arylhydroxylamines can be synthesized in this way, for example[14]:

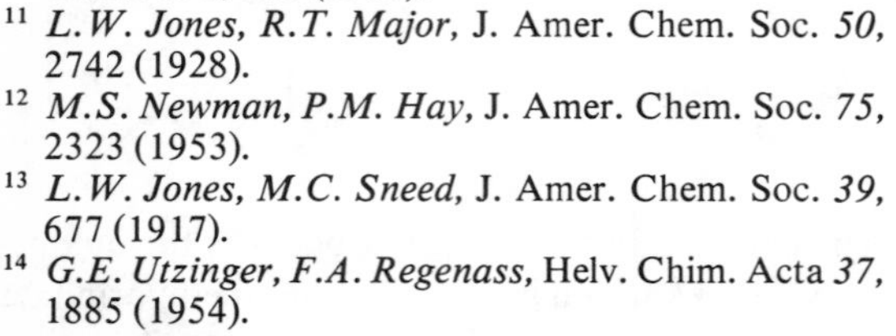

O,N,N-trisubstituted hydroxylamines are accessible by alkylation or arylation of O-alkylhydroxylamines. The reaction can be carried out *in steps*, as is revealed by the example of the synthesis of *O-allyl-N-(2,4-dinitrophenyl)-N-methylhydroxylamine*[15]:

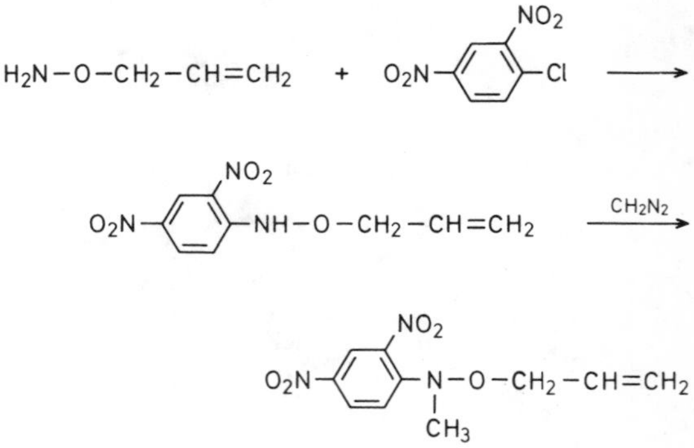

Here the second N-alkylation is performed with diazomethane. In other cases dialkyl sulfates[16] and epoxides can be employed successfully. Thus, *2-(ethoxyethylamino)ethanol [O,N-diethyl-N-(2-hydroxyethyl)hydroxylamine]* is obtained from O,N-diethylhydroxylamine and ethylene oxide[17]:

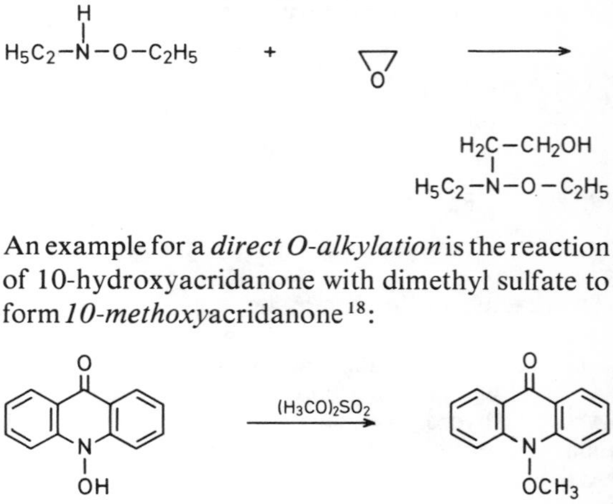

An example for a *direct O-alkylation* is the reaction of 10-hydroxyacridanone with dimethyl sulfate to form *10-methoxy*acridanone[18]:

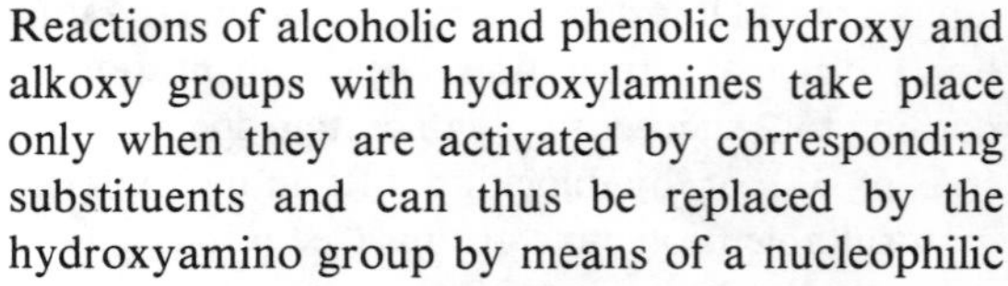

3.1.1.2 Exchange of hydroxy and alkoxy groups against the hydroxyamino group

Reactions of alcoholic and phenolic hydroxy and alkoxy groups with hydroxylamines take place only when they are activated by corresponding substituents and can thus be replaced by the hydroxyamino group by means of a nucleophilic mechanism. Thus, diphenylmethanol does not react with N-phenylhydroxylamine[19], while from

[10] *J. Meisenheimer, K. Bratring*, Justus Liebigs Ann. Chem. *397*, 288 (1913).

[11] *L.W. Jones, R.T. Major*, J. Amer. Chem. Soc. *50*, 2742 (1928).

[12] *M.S. Newman, P.M. Hay*, J. Amer. Chem. Soc. *75*, 2323 (1953).

[13] *L.W. Jones, M.C. Sneed*, J. Amer. Chem. Soc. *39*, 677 (1917).

[14] *G.E. Utzinger, F.A. Regenass*, Helv. Chim. Acta *37*, 1885 (1954).

[15] *R.F. Kleinschmidt, A.C. Cope*, J. Amer. Chem. Soc. *66*, 1931 (1944).

[16] *A.C. Cope, P.H. Towle*, J. Amer. Chem. Soc. *71*, 3427 (1949).

[17] *R.T. Major, H.J. Hess, C.A. Stone*, J. Med. Pharm. Chem. *1*, 382 (1959).

[18] *A. Kliegl, A. Fehrle*, Ber. *47*, 1635 (1914); *R.M. Acheson, B. Adcock, G.M. Glover, L.E. Sutton*, J. Chem. Soc. 3367 (1960).

[19] *H. Rupe, R. Wittwer*, Helv. Chim. Acta *5*, 217 (1922).

[20] *A. Mothwurf*, Ber. *37*, 3150 (1904).

triphenylmethanol and hydroxylamine *N,N-bis-(triphenylmethyl)hydroxylamine* is formed[20].

$$2\,(H_5C_6)_3C-OH \xrightarrow{H_2N-OH} (H_5C_6)_3C-\overset{\overset{\displaystyle OH}{|}}{N}-C(C_6H_5)_3$$

The same compound is accessible by allowing excess trityl chloride to act on hydroxylamine. With excess hydroxylamine the monosubstituted *N-triphenylmethylhydroxylamine*[20] is formed. 2,4-Dinitroanisole forms *N-(2,4-dinitrophenyl)hydroxylamine* with hydroxylamine while losing methanol[21].

$$O_2N-\underset{OCH_3}{\overset{NO_2}{\bigcirc}} \xrightarrow[-H_3C-OH]{H_2N-OH} O_2N-\underset{NH-OH}{\overset{NO_2}{\bigcirc}}$$

In aldehydes present as hydrates or hemiacetals the hydroxy group can also react with hydroxylamines by undergoing *O-* or *N-*substitution. Thus, from *N*-hydroxypiperidine and chloral hydrate *1-piperidinooxy-2,2,2-trichloroethanol* is formed[22].

$$\underset{}{\bigcirc}N-OH \;+\; \underset{HO}{\overset{HO}{\diagup}}CH-CCl_3 \longrightarrow$$

$$\underset{}{\bigcirc}N-O-\overset{\overset{\displaystyle OH}{|}}{C}H-CCl_3$$

The hemiacetal of 5-hydroxyvaleraldehyde reacts with *N*-methylhydroxylamine to form *N-methyl-N-tetrahydropyran-2-ylhydroxylamine*[23]:

$$\underset{O}{\bigcirc}OH \xrightarrow{H_3C-NH-OH} \underset{O}{\bigcirc}\underset{\overset{|}{OH}}{N-CH_3}$$

An exchange of $-ONa^{\ominus}$ takes place formally during the reaction between hydroxylamine-*O*-sulfonic acid and sodium alcoholates and leads to *O*-alkylhydroxylamines[24]:

$$H_2N-O-SO_3Na \xrightarrow[-Na_2SO_4]{RONa} H_2N-O-R$$

O-Phenylhydroxylamine (phenoxyamine), which is difficult to prepare, is formed by an analogous reaction[25].

---

[21] *O. Neunhoeffer, W. Ruske,* Justus Liebigs Ann. Chem. *610,* 152 (1957).

[22] *G. Zinner, W. Ritter, W. Kliegel,* Pharmazie *20,* 294 (1965).

[23] *H. Ulrich, A.A.R. Sayigh,* Angew. Chem. *74,* 468 (1962).

[24] *G. Bargigia,* Atti Accad. Nazl. Lincei, Classe Sci. Fis., Mat. Nat., Rend. *39,* 83 (1965); C.A. *64,* 17405 (1966).

[25] *C.L. Bumgardner, R.L. Lilly,* Chem. & Ind. (London) 559 (1962).

3.1.1.3 Replacement of further reactive groups by the hydroxyamino group

If sufficiently activated by further substituents in the molecule, aromatically or heteroaromatically bound halogen can be exchanged against the hydroxyamino group. In addition to halogen, this holds for activated nitro, amino, methylthio, and sulfonic acid groups in aromatic and heteroaromatic compounds, and for amino groups in tertiary Mannich bases or their quaternary salts. Some of the desired hydroxylamines are obtained in very good yield (see Table 3).

Instead of using quaternary salts of Mannich bases to *N*-alkylate hydroxylamines (last example in Table 3), it is simpler to react *C–H* acid components (*e.g.*, cyclohexanone), formaldehyde, and *N*-monosubstituted hydroxylamine directly under the conditions of a Mannich reaction. Thus, *N,N-disubstituted* hydroxylamines can generally be prepared in good yield according to the general equation,

$$R^1-H \;+\; CH_2O \;+\; H-\underset{R^2}{\overset{OH}{N}} \xrightarrow{-H_2O}$$

$$R^1-CH_2-\underset{R^2}{\overset{OH}{N}}$$

and especially if R^2 is an alkyl group[32].

3.1.2 Hydroxylamines by addition to multiple bonds

3.1.2.1 Addition to *C=C* double bonds

Hydroxylamines can be added on only to those *C=C* double bonds which are *activated* by adjacent electron-attracting groups. The hydroxylamine group adds in the *β*-position with respect to the activating group X, which may be a carbonyl group or carboxy group (or one of its functional derivatives), or a nitro or sulfone group.

$$X-\overset{|}{C}=\overset{|}{C}- \;+\; \underset{H}{\overset{R^1}{N}}-O-R^2 \longrightarrow$$

$$X-\overset{|}{C}H-\overset{\overset{\displaystyle R^1}{|}}{\underset{|}{C}}-N-O-R^2 \qquad \textbf{3}$$

$$X-\overset{|}{C}=\overset{|}{C}- \;+\; \underset{R^1}{\overset{R^1}{N}}-OH \longrightarrow$$

$$X-\overset{|}{C}H-\overset{|}{\underset{|}{C}}-O-N\overset{R^1}{\underset{R^1}{}} \qquad \textbf{4}$$

Table 3. Hydroxylamines by substitution of functional groups on aromatic or heteroaromatic compounds

Starting compound	Hydroxylamine	End product	Yield [% of theory]	Ref.
	$H_2N-O-CH_2-C_6H_5$	*2-Amino-4-[benzyloxyamino]-6-methyl-5-nitropyrimidine*	78	[26]
	H_2N-OH	*N-[4,6-dinitro-3-tolyl]-hydroxylamine*		[27]
	$2\ H_2N-OH$	*5,6-Dihydro-4,6-bis(hydroxy-amino)-2-pyrimidinol*	95	[28]
	H_2N-OH	*6-(Hydroxyamino)purin-2-ol*	85	[29]
	H_2N-OH	*6-(Hydroxyamino)purine 3-oxide*	59	[30]
		N-[2-Oxocyclohexylmethyl]-N-phenylhydroxylamine	90	[31]

Normally alkylation on the nitrogen occurs according to Equation **3** even when R^2 is hydrogen. For example, carvone (*p*-mentha-6,8-dien-2-one) adds on hydroxylamine on the $C=C$ double bond conjugate to the carbonyl group[33]:

6-Hydroxyamino-p-menth-8-en-2-one

[26] *E.C. Taylor, J.W. Barton,* J. Org. Chem. *24,* 127 (1959).
[27] *M. Giua,* Gazz. Chim. Ital. *53,* 659 (1923).
[28] *D.M. Brown, P. Schell,* J. Chem. Soc. 208 (1965).
[29] *A. Giner-Sorolla, S. O'Bryant, J.H. Burchenal, A. Bendich,* Biochemistry *5,* 3057 (1966).
[30] *A. Giner-Sorolla,* J. Med. Chem. *12,* 718 (1969).
[31] *J. Thesing, A. Müller, G. Michel,* Chem. Ber. *88,* 1033 (1955).
[32] *J. Thesing, H. Uhrig, A. Müller,* Angew. Chem. *67,* 31 (1955).

[33] *G. Baddeley, K. Brocklehurst,* Proc. Chem. Soc. 145 (1962).

N,N-Disubstituted hydroxylamines react according to Equation **4** giving *O*-alkylation.

Hydroxylamine is able to react also with two molecules of an α,β-*unsaturated ketone*. Thus, *3,3'-(hydroxyimino)dipropiophenone* is formed from acrylophenone[34]:

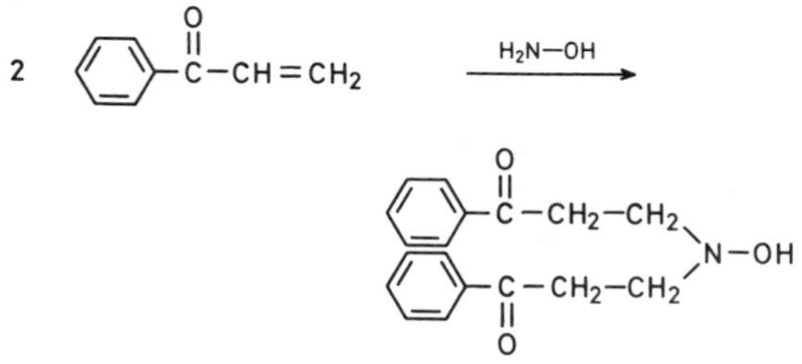

Addition of hydroxylamine to α,β-unsaturated carboxylic acids leads to formation of *β-hydroxyamino carboxylic acids*[35]:

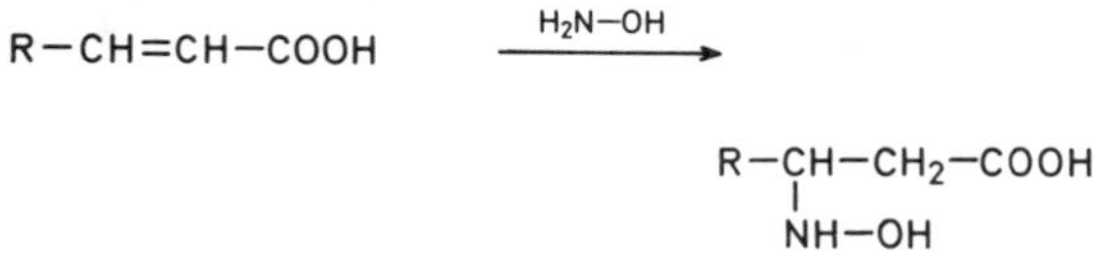

With α,β-unsaturated *esters* (*e.g.* cinnamic esters) an additional reaction takes place on the ester grouping[36], but if 2 molecules ester are reacted with 1 molecule hydroxylamine, then *diethyl 3,3'-(hydroxyimino)dipropionate* is obtained smoothly from ethyl acrylate[37]:

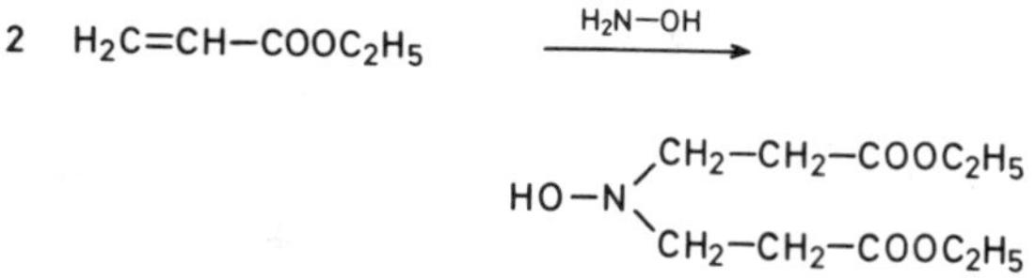

From 2-nitro-1-phenylethylene and hydroxylamine a good yield of *N-(2-nitro-1-phenylethyl)-hydroxylamine* is obtained[38]:

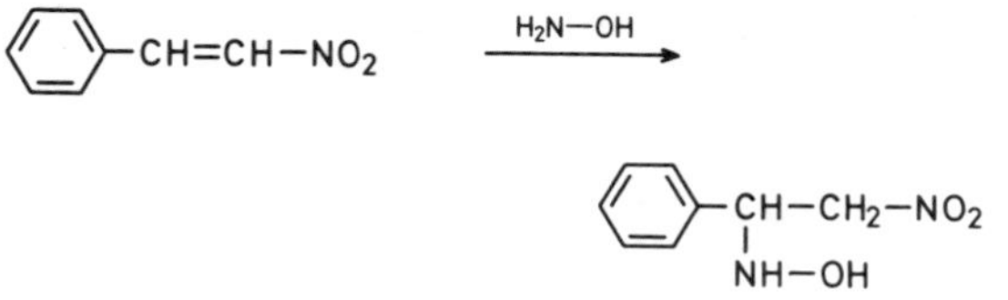

Divinylsulfone reacts with two molecules *N*-methylhydroxylamine with formation of *N,N'-*

(sulfonyldiethylene)bis[N-methyl]hydroxylamine, while with one molecule hydroxylamine *4-hydroxythiomorpholine 1,1-dioxide* is formed[39]:

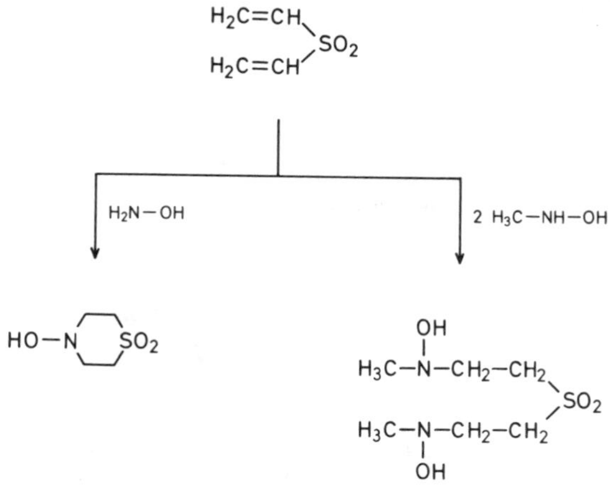

From *N*-hydroxypiperidine and *N,N*-diethylhydroxylamine addition to activated double bonds has led to the preparation of a series of *O*-substituted hydroxylamines[40]:

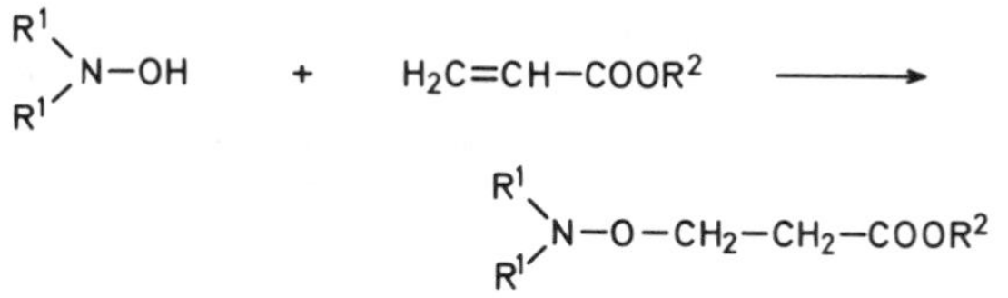

3.1.2.2 Addition to C=N double bonds

A series of reactive compounds can add on to the C=N double bond of *nitrones*, for example, Grignard compounds, nitroalkanes, and hydrogen cyanide. Here 1,3-addition leads to *N,N-disubstituted* hydroxylamines. Alkenes, too, add on with formation of *isoxazolidines*.

Thus, for example, *N-(α-benzylcinnamyl)-N-phenyl-hydroxylamine* is formed from cinnamaldehyde *N*-phenylnitrone (*N*-phenyl-α-styrylnitrone) and benzylmagnesium chloride[41]:

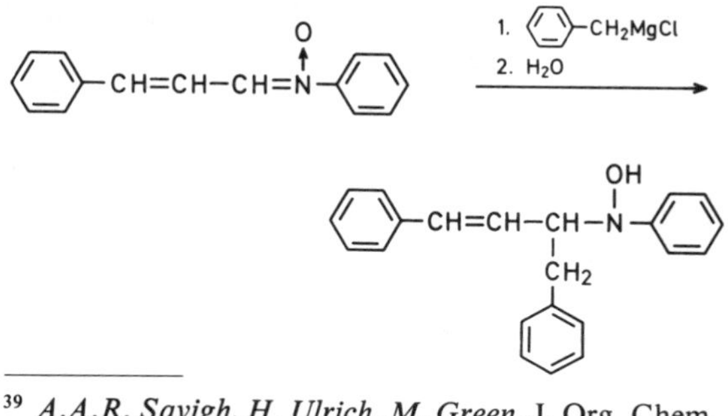

[34] *D.J. Casey, C.S. Marvel*, J. Org. Chem. **24**, 1023 (1959).

[35] *T. Posner*, Ber. **39**, 3515 (1906); *T. Posner, G. Schreiber*, Ber. **57**, 1127 (1924).

[36] *T. Posner*, Ber. **40**, 218 (1907).

[37] *F. Becke, G. Mutz*, Chem. Ber. **98**, 1322 (1965).

[38] *C.D. Hurd, J. Patterson*, J. Amer. Chem. Soc. **75**, 285 (1953).

[39] *A.A.R. Sayigh, H. Ulrich, M. Green*, J. Org. Chem. **29**, 2042 (1964).

[40] *G. Zinner*, Angew. Chem. **71**, 311 (1959).

[41] *G.E. Utzinger, F.A. Regenass*, Helv. Chim. Acta **37**, 1890 (1954).

The addition of nitroethane to ethyl 5-methyl-4, 5-dihydro-3*H*-pyrrole-5-carboxylate 1-oxide affords the *ethyl ester* of *1-hydroxy-2-methyl-5-(1-nitroethyl)proline*[42]:

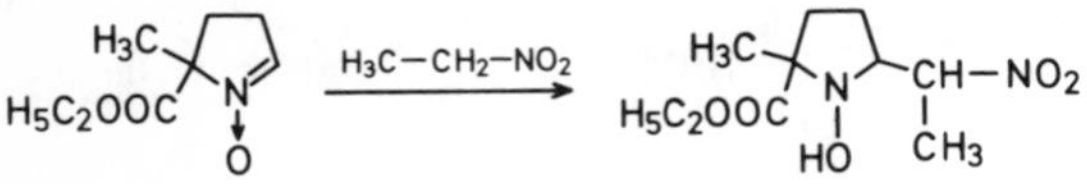

Addition of hydrogen cyanide to 5, 5-dimethyl-4, 5-dihydro-3H-pyrrole 1-oxide gives *1-hydroxy-5, 5-dimethyl-2-pyrrolidinecarbonitrile*[43]:

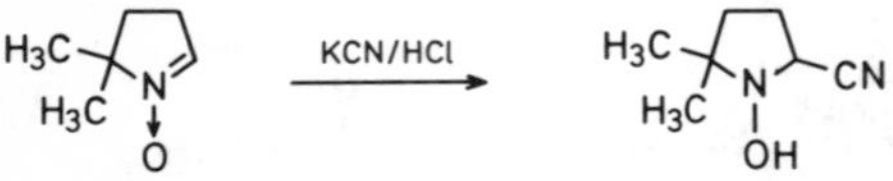

1, 3-Dipolar cycloaddition of nitrones to alkenes leads to *isoxazolidines* according to the following scheme[44]:

Thus, α-methylstyrene and benzaldehyde phenyl-nitrone (*N, α-diphenylnitrone*), for example, afford two diastereomeric *5-methyl-2, 3, 5-triphenylisoxazolidines*[45]:

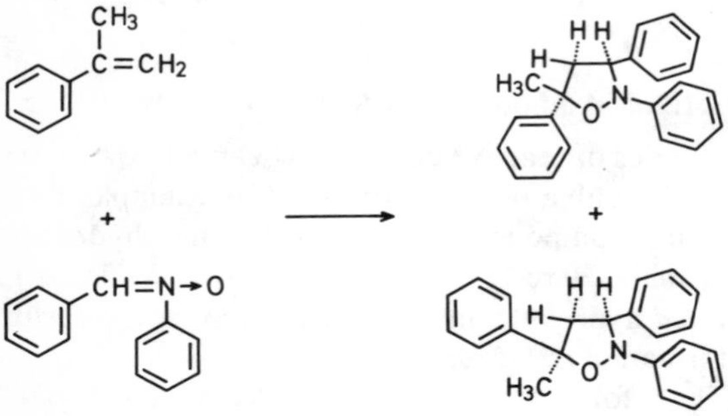

Addition of hydrogen cyanide to *oximes* of aliphatic aldehydes or ketones makes α-*hydroxyamino carboxylic acid nitriles* **5** accessible. It is not essential to isolate the oxime **6**. The reaction fails with aromatic aldehydes and ketones[46]:

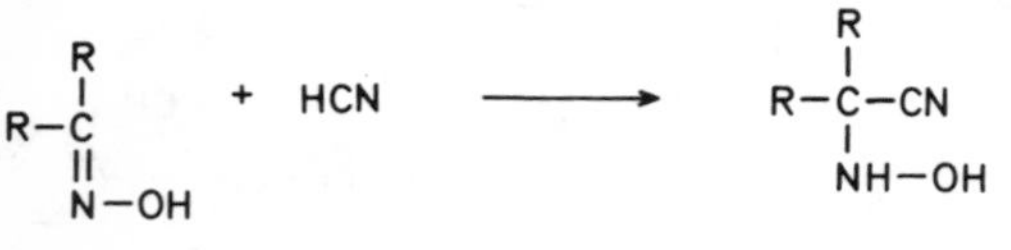

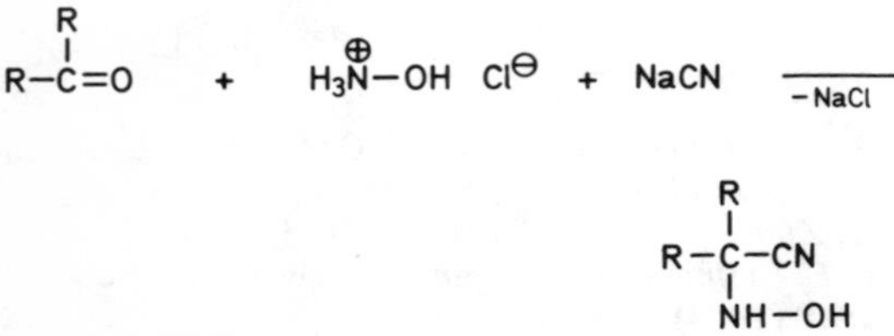

3.1.2.3 Addition to *N=O* double bonds

Hydroxylamines are formed from aromatic and aliphatic nitro compounds by addition of Grignard compounds, olefins, or dienes and also by dimerizing aromatic nitroso compounds.

Nitroso compounds react with *Grignard* reagents to form *N, N-disubstituted* hydroxylamines. The reaction is fundamental for preparing *N, N-dialkyl* hydroxylamines; it first enabled labile *N, N-diphenylhydroxylamine* to be prepared from nitroso benzene and phenylmagnesium bromide[47].

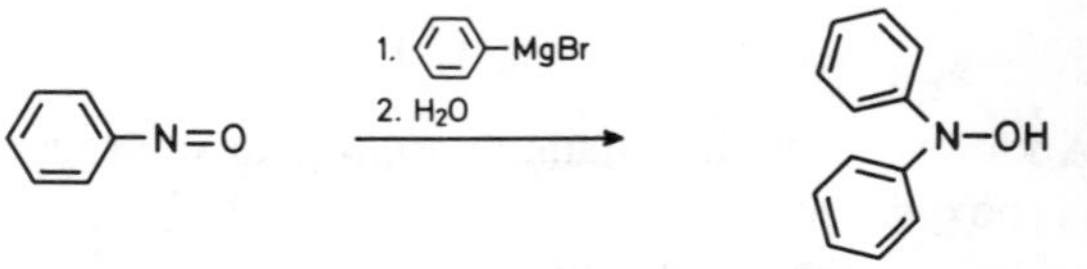

Biphenyl and diphenylamine are side-products of this reaction[48].

1, 2-Addition of nitroso compounds to *olefins* in accord with the general equation[49]

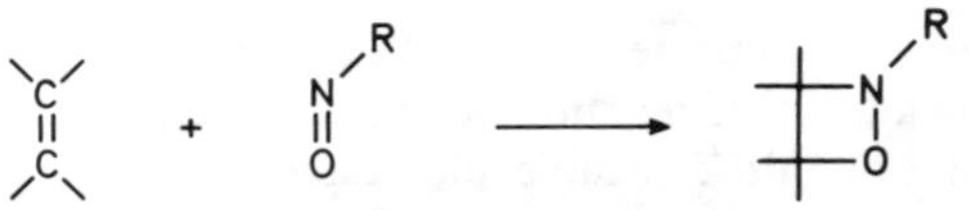

has so far gained importance only among aliphatic perfluoronitroso compounds. Thus, from pentafluoronitrosoethane and tetrafluoroethylene *2-pentafluoroethyl-3, 3, 4, 4-tetrafluoro-1, 2-oxazetidine* was obtained[50]:

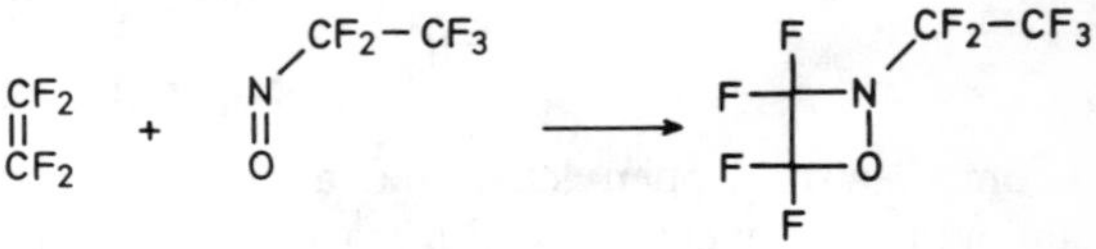

[42] W.D.S. Bowering, V.M. Clark, R.S. Thakur, Lord A. Todd, Justus Liebigs Ann. Chem. *669*, 110 (1963).

[43] R. Bonnett, R.F.C. Brown, V.M. Clark, I.O. Sutherland, A. Todd, J. Chem. Soc. 2100 (1959).

[44] R. Huisgen, Angew. Chem. *75*, 627 (1963);
R. Huisgen, H. Seidl, I. Brüning, Chem. Ber. *102*, 1102 (1969);
W. Oppolzer, K. Keller, Tetrahedron Lett. 1117 (1970).

[45] R. Huisgen, R. Grashey, H. Seidl, H. Hauck, Chem. Ber. *101*, 2559 (1968).

[46] L. Neelakantan, W.H. Hartung, J. Org. Chem. *23*, 964 (1958).

[47] H. Wieland, A. Roseeu, Ber. *45*, 494 (1912).

[48] H. Gilman, R. McCracken, J. Amer. Chem. Soc. *49*, 1052 (1927).

[49] L.L. Muller, J. Hamer, 1, 2-Cycloaddition Reactions, Chap. Oxazetidine, p. 257, Interscience Publishers, John Wiley & Sons, New York 1967.

[50] D.A. Barr, R.N. Haszeldine, J. Chem. Soc. 1153 (1960).

The Diels-Alder reaction of nitroso compounds as dienophiles with 1,3-butadiene and its derivatives leading *to 3,6-dihydro-2H-1,2-oxazines* has found wide application[51].

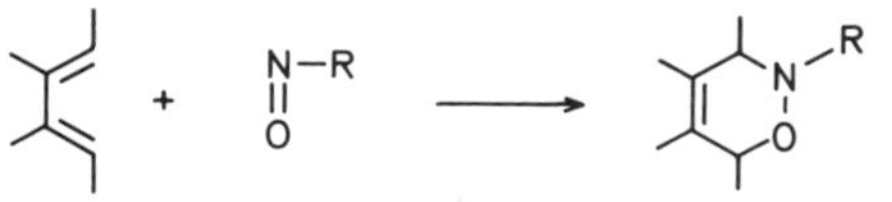

An example of such a reaction is the preparation of *1-(3,6-dihydro-2H-1,2-oxazin-2-yl)cyclohexanecarbonitrile* from 1-nitrosocyclohexanecarbonitrile and 1,3-butadiene. The dihydro compound can then be converted into *1-(tetrahydro-2H-1,2-oxazin-2-yl)-cyclohexanecarbonitrile* by catalytic hydrogenation[52]:

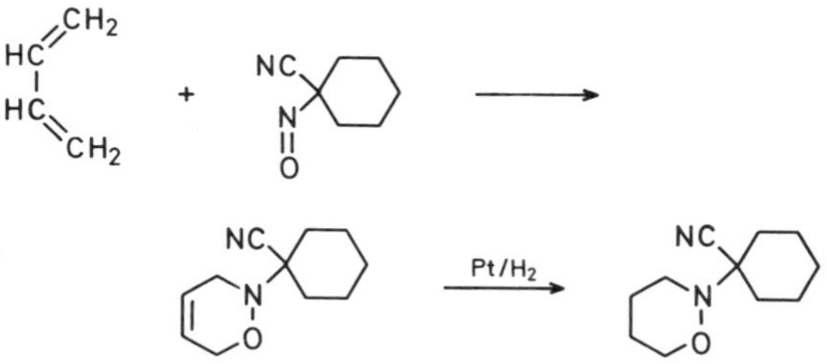

The analogous reaction with 1-fluoronitrosoalkanes generally proceeds very smoothly and in good yield as a result of the activation of the NO group by the fluorine atoms[53].

Nitrosobenzene or substituted nitrosobenzenes can *dimerize* in the presence of sulfuric acid to form *N,N-diaryl* hydroxylamines. In general, the *p*-position in at least one molecule must be unoccupied in this reaction[54]:

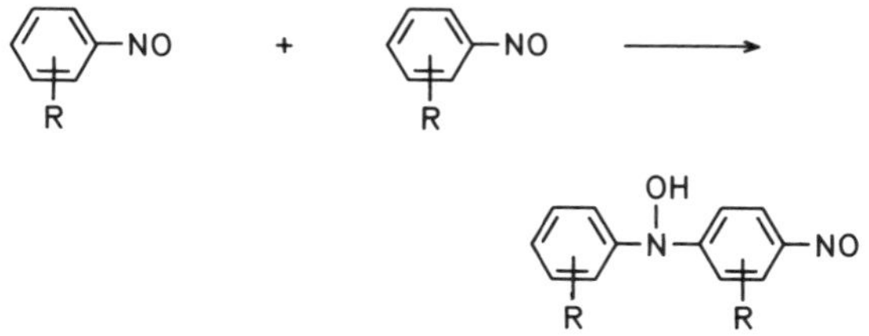

3.1.3 Hydroxylamines by oxidation of amines

On oxidation of *primary* but especially *secondary aliphatic* amines with hydrogen peroxide hydrox-ylamines are obtained in accord with the following equation

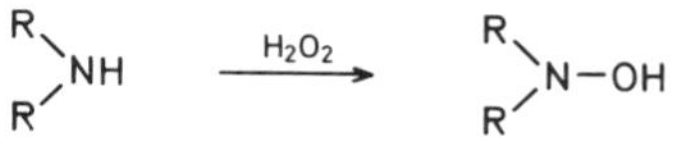

The yield is generally very low, for *N-hydroxymorpholine*, for instance, it is 20%[55]. It is claimed that it can be increased by working in the presence of *catalysts* such as sodium tungstate or formic acid esters. *N-Hydroxypiperidine*, for example, has been prepared in this way from piperidine and hydrogen peroxide in 78% yield[56]:

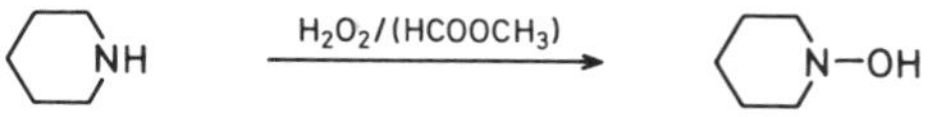

The reaction could be transferred to *aromatic* amines such as aniline only by oxidizing the magnesium salt of aniline with hydrogen peroxide at $-25°$[57]. However, the technique is not generally applicable.

Oxidation of *tertiary* amines with hydrogen peroxide or with peracids proceeds in a different manner. *N*-Oxides are formed initially (see p. 406) and can be rearranged thermally to *trisubstituted* hydroxylamines if benzyl or alkyl groups are present which migrate from the nitrogen to the oxygen (*Meisenheimer rearrangement*)[58]:

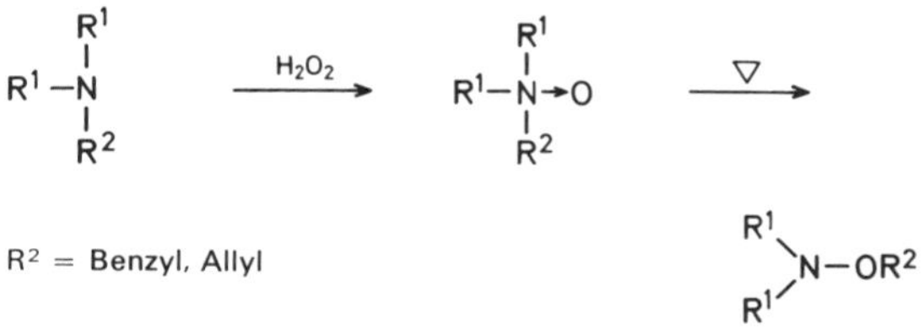

R^2 = Benzyl, Allyl

O-Benzyl-N-methyl-N-phenylhydroxylamine, for instance, has been prepared in this way[59]:

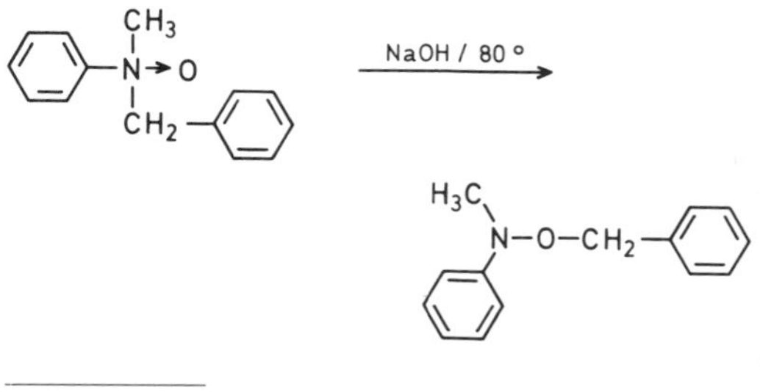

[51] *G. Kresze J. Firl,* Fortschr. Chem. Forsch. *11,* 245 (1969) (Kap. Diensynthesen mit Nitroso-Verbindungen).

[52] *O. Wichterle, V. Gregor,* Collect. Czech. Chem. Commun. *24,* 1158 (1959).

[53] *R.E. Banks, M.G. Barlow, R.N. Haszeldine,* J. Chem. Soc. 4714 (1965).

[54] DOS. 2 020 043 (29. 10. 1970), I.C.I., Erf.: *D. Dodman, K.W. Pearson, J.M. Woolley.*

[55] *G. Zinner, W. Kliegel,* Arch. Pharm. (Weinheim, Ger.) *299,* 166 (1966).

[56] DBP 1 004 191 (1957), BASF, Erf.: *W. Ruppert;* C.A. *54,* 584 (1960).

[57] *E. Bamberger, F. Tschirner,* Ber. *32,* 1675 (1899); *J.F. Durand, R. Naves,* Compt. Rend. *180,* 521 (1925).

[58] *J. Meisenheimer, H. Greeske, A. Willmersdorf,* Ber. *55,* 513 (1922).

[59] *U. Schöllkopf, U. Ludwig,* Chem. Ber. *101,* 2224 (1968).

Fundamentally, the technique lends itself well for preparing *N,N*-dialkyl-*O*-allyl- and *N,N*-dialkyl-*O*-benzylhydroxylamines, because *N*-oxides of tertiary amines are easily accessible and the rearrangement gives good yields.

*Trialkyl*amino oxides without benzyl or allyl groups with hydrogen on the carbon in the *β*-position to the nitrogen are cleaved thermally to *N,N*-dialkyl hydroxylamines and olefins (*Cope-Mamlock-Wolffenstein* elimination)[60]:

$$R^1 - \overset{\underset{\displaystyle CH_2}{|}}{\underset{\displaystyle CH_2-R^2}{N}} \rightarrow O \longrightarrow \overset{\underset{\displaystyle R^1}{|}}{\underset{\displaystyle R^1}{N}} - OH \;+\; H_2C = CH - R^2$$

Thus, *N,N*-dipentylhydroxylamine is obtained from tripentylamine with loss of 1-pentene[61]:

$$H_3C-(CH_2)_4-\overset{\underset{\displaystyle (CH_2)_4-CH_3}{|}}{\overset{\displaystyle (CH_2)_4-CH_3}{N}}\rightarrow O \xrightarrow{\triangledown} \begin{array}{c} H_3C-(CH_2)_4 \\ \\ H_3C-(CH_2)_4 \end{array}\!\!\!N-OH$$

$$+ \; H_2C=CH-CH_2-CH_2-CH_3$$

Many of the *N,N*-dialkylhydroxylamines obtained in this way are accessible more conveniently in alternative manner. However, the application of this hydroxylamine synthesis to *N*-oxides of complex natural substances is of interest.

Table 4. N-Hydroxylamines by reduction of nitro compounds

Starting compound	Reducing agent	End product	Yield [% of theory]	Ref.
(structure) *o-nitro CH=C(NH-CO-Ph)COOC$_2$H$_5$*	Zinc dust/NH$_4$Cl	(structure) Ethyl α-benzamido-2-hydroxyaminocinnamate	75	63
(structure) (CH$_3$)$_3$C-NO$_2$	Zinc dust/NH$_4$Cl	(structure) H$_3$C-C(CH$_3$)$_2$-NH-OH N-tert-Butylhydroxylamine	64	64
Cl-⬡-NO$_2$	NaHS/CaCl$_2$	Cl-⬡-NH-OH N-[4-Chlorophenyl]hydroxylamine	67	65
⬡-NO$_2$	Zn/Hg/Al$_2$(SO$_4$)$_3$	⬡-NH-OH N-Phenylhydroxylamine	85	66
(structure) decalin with NO$_2$ groups	Al/Hg	(structure) trans-N,N'-[decahydro-4a,8a-naphthylene]dihydroxylamine	65–75	67
(structure) NO, Cl decalin	Al/Hg	(structure) N-[8a-Chlorodecahydro-4a-naphthyl]hydroxylamine	66	68
O$_2$N-⬡-NO$_2$	Ascorbic acid	O$_2$N-⬡-NH-OH N-[4-Nitrophenyl]hydroxylamine	65	69
(structure) H$_3$C, NO, NO$_2$, CH$_3$ benzene	Ascorbic acid	(structure) H$_3$C, NH-OH, NO$_2$, CH$_3$ benzene N-[3,5-Dimethyl-6-nitrophenyl]-hydroxylamine	18	70

[60] *L. Mamlock, R. Wolffenstein*, Ber. *33*, 159 (1900); *A.C. Cope, Th.T. Foster, Ph.H. Towle*, J. Amer. Chem. Soc. *71*, 3929 (1949).

[61] *A.C. Cope, H.H. Lee*, J. Amer. Chem. Soc. *79*, 965 (1957).

3.1.4 *N*-Hydroxylamines by reduction of *C–N*-compounds

3.1.4.1 Reduction of nitro and nitroso compounds

Complete reduction of aromatic and aliphatic nitro and nitroso compounds leads to primary amines (see p. 494)[62]:

$$R-NO_2 \longrightarrow R-NO \longrightarrow R-NH-OH$$
$$\longrightarrow R-NH_2$$

The corresponding *N-monosubstituted* hydroxylamines may be formed as an intermediate stage of this reaction. They are accessible in many cases, and sometimes in good yield, by maintaining certain reaction conditions. The reduction can be performed either with chemical reducing agents, catalytically, or electrochemically. A photochemical reduction, too, has been described. Where the molecule contains several nitro or nitroso groups a *partial reduction* often succeeds. Among numerous chemical reducing agents that have been used for preparing *N*-monosubstituted hydroxylamines the following merit particular mention:

Zinc dust in neutral or weakly acid solution
Sulfides such as sodium hydrogen sulfide in the presence of calcium chloride
Amalgamated zinc and *aluminum amalgam*
Ascorbic acid
Diborane

Table 4 contains some examples

Alkali metal salts of primary and secondary nitro compounds can be reduced with *diborane* to give *N-monosubstituted* hydroxylamines. The reaction fails if the free nitro compounds are used in place of the salts. *N-Benzylhydroxylamine*, for example, has been obtained from the potassium salt of α-nitrotoluene by this technique[71]:

N-Hydroxylamines are accessible also by *catalytic* reaction of aromatic and aliphatic nitro compounds. Palladium or platinum are generally employed as catalysts (see Table 5).

Table 5. N-Hydroxylamines by catalytic reduction of nitro compounds

Starting compound	Reducing agent	End product	Yield [% of theory]	Ref.
	Pd/C, H_2N-NH_2	*N-[2,5-Dichlorophenyl]-hydroxylamine*	Almost quantitative	72
	PtO_2/H_2	*N,N'-[Sulfonyldi-o-phenylene]bishydroxylamine*	95	73
$H_3C-C-CH-CH_3$ HON $\ $ NO_2	Pd/C, H_2	$H_3C-C-CH-CH_3$ HON $\ $ NH-OH *3-Hydroxyamino-2-butanone-oxime*	90	74
	Pd/C, H_2	*N-[3-Nitrophenyl]hydroxylamine*		75

[62] *N.V. Sidgwick, I.T. Millar, H.D. Springall,* The Organic Chemistry of Nitrogen, p. 387, Clarendon Press, Oxford 1966.
[63] *A.P. Martinez, W.A. Skinner, W.W. Lee, L. Goodman, B.R. Baker,* J. Org. Chem. *26,* 863 (1961).
[64] *O. Exner, B. Kakáč,* Collect. Czech. Chem. Commun. *28,* 1658 (1963).
[65] *R.D. Haworth, A. Lapworth,* J. Chem. Soc. *119,* 770 (1921).
[66] *E. Bamberger, M. Knecht,* Ber. *29,* 863 (1896).
[67] *E. Müller, U. Heuschkel,* Z. Naturforsch. B *19,* 1024 (1964).
[68] *W. Hückel, M. Blohm,* Justus Liebigs Ann. Chem. *502,* 126 (1933).

Electrochemical reduction of aromatic and aliphatic nitro compounds to the corresponding *N*-hydroxylamines proceeds in high yield when a constant electrode potential is used. Thus, at between 10° and 25° numerous substituted nitrobenzenes have been converted into the hydroxylamines. The reduction was performed on a mercury cathode at −250 to −450 mV in dilute sulfuric acid[76]. Lower aliphatic hydroxylamines were obtained by reduction of the associated nitro compounds on a stirred mercury electrode under controlled conditions in 3N hydrochloric acid[77].

The *photochemical* reduction of 4-nitropyridine 1-oxide and its methyl-substituted derivatives, too, has been described. For example, *4-hydroxyaminopyridine 1-oxide* (~ 100% of theory) was prepared by irradiation of the nitro compound with a 450 W high-pressure mercury lamp in absolute ethanol[78].

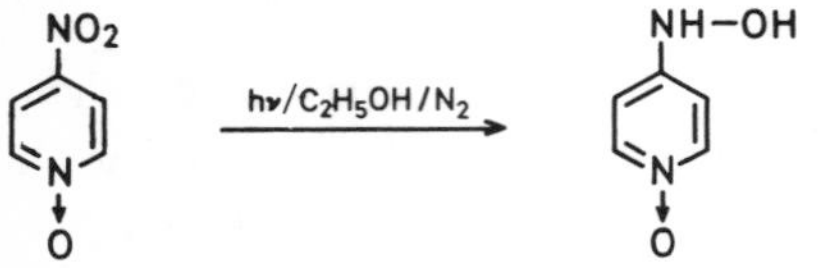

3.1.4.2 Reduction of oximes

Reduction of oximes (see p. 433) normally yields amines. Hydroxylamines[79] and imines[80] are intermediate stages in this reaction:

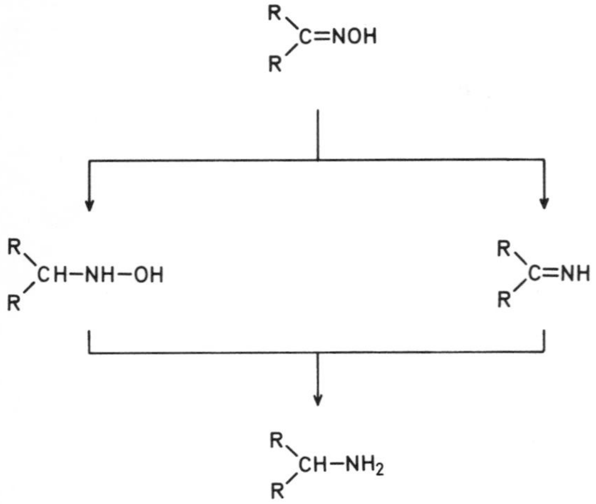

Under certain conditions the hydroxylamines are preparatively accessible, especially if the reduction is carried out with hydrogen and platinum as catalyst or with diborane.

For example, 4-heptanone oxime forms *N-(1-propylbutyl)hydroxylamine* by catalytic hydrogeneration with platinum[79]:

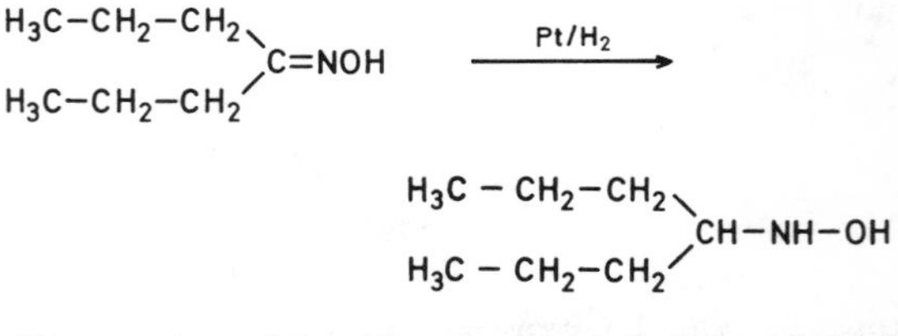

The reaction with hydrogen proceeds in good yield even without isolation of the oxime if it is conducted in a water-methanol suspension of ketone and hydroxylamine hydrochloride with platinum as catalyst. In this way *trans-N-(decahydro-1-naphthyl)hydroxylamine* has been prepared from *trans*-1-decahydronaphthalenone[81]:

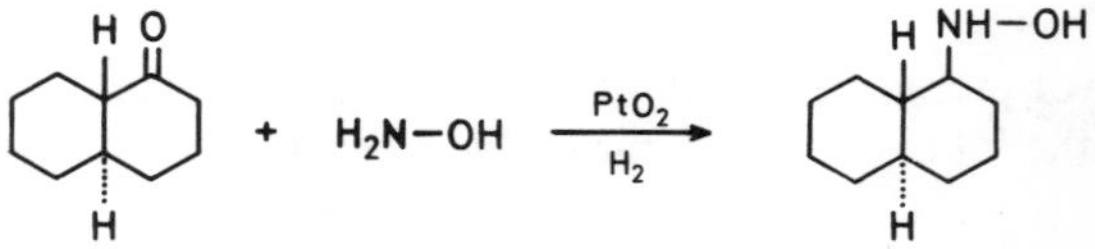

After hydrolysis the reaction of aliphatic and aromatic aldoximes and ketoximes with diborane leads to *N*-monosubstituted hydroxylamines. From polymeric formaldehyde oxime *N-methylhydroxylamine* is formed, from benzaldehyde oxime *N-benzylhydroxylamine*[82]:

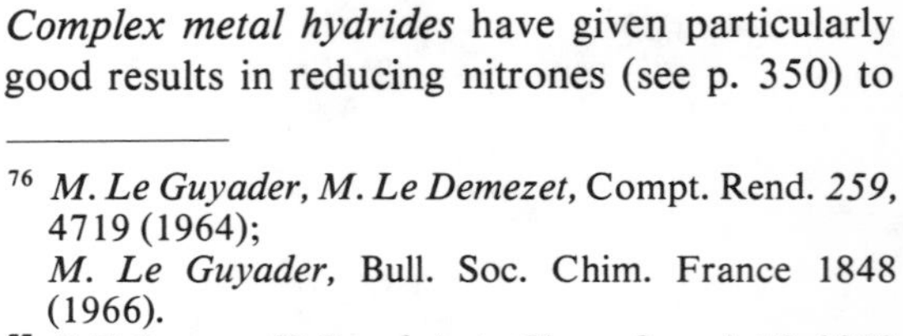

3.1.4.3 Reduction of nitrones

Complex metal hydrides have given particularly good results in reducing nitrones (see p. 350) to

[69] *R. Kuhn, F. Weygand*, Ber. *69*, 1973 (1936).
[70] *R. Kuhn, H. Vetter, P. Desnuelle*, Ber. *70*, 1317 (1937).
[71] *H. Feuer, R.S. Bartlett, B.F. Vincent jr., R.S. Anderson*, J. Org. Chem. *30*, 2880 (1965).
[72] *P.M.G. Bavin*, Can. J. Chem. *36*, 239 (1958).
[73] *K. Michel, M. Matter*, Helv. Chim. Acta *44*, 2206 (1961).
[74] *M.L. Scheinbaum*, J. Org. Chem. *35*, 2790 (1970).
[75] *K. Brand, J. Steiner*, Ber. *55*, 881 (1922).

[76] *M. Le Guyader, M. Le Demezet*, Compt. Rend. *259*, 4719 (1964);
M. Le Guyader, Bull. Soc. Chim. France 1848 (1966).
[77] *P.E. Iversen, H. Lund*, Acta Chem. Scand. *19*, 2303 (1965).
[78] *C. Kaneko, S. Yamada, I. Yokoe, N. Hata, Y. Ubukata*, Tetrahedron Lett. 4729 (1966).
[79] *G. Vavon et Krajcinovic*, Bull. Soc. Chim. France [*4*] *43*, 231 (1928).
[80] *G. Mignonac*, Compt. Rend. *170*, 936 (1920).
[81] *E. Müller, U. Heuschkel*, Chem. Ber. *92*, 69 (1959).
[82] *H. Feuer, B.F. Vincent jr., R.S. Bartlett*, J. Org. Chem. *30*, 2877 (1965).

the corresponding *N,N-disubstituted* hydroxyl-amines. Lithium aluminum hydride is frequently employed, but in some instances sodium or potassium tetrahydroborate can be used.

Catalytic hydrogenation here proceeds nonuniformly and success is only seldom achieved in preparing hydroxylamines from nitrones, for instance, with platinum as catalyst. Generally, secondary amines or Schiff's bases are obtained[83]. One example of reduction with lithium aluminum hydride is the preparation of *N, N-dibenzylhydroxylamine* (85% of theory) from *N-benzyl-α-phenylnitrone (benzaldehyde benzylnitrone)*[84]:

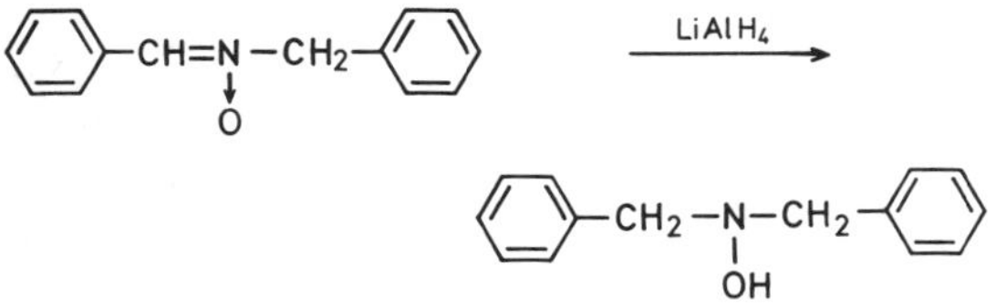

Alkali metal tetrahydroborates have been employed mainly during reduction of cyclic nitrones. Thus, *1-hydroxy-2,4,4-trimethylpyrrolidine* is formed from 2,4,4-trimethyl-4,5-dihydro-3*H*-pyrrole 1-oxide by the action of potassium tetrahydroborate[85]:

In the 1,4-benzodiazepine series, too, these reductions to cyclic hydroxylamines succeed both with lithium aluminum hydride[86] and with sodium tetrahydroborate[87].

3.1.5 Other methods for preparing *N*-hydroxylamines

Nitrones can be cleaved hydrolytically into *N*-monosubstituted hydroxylamines and carbonyl compounds in accord with the equation

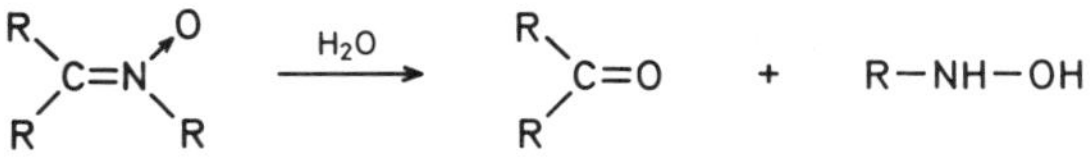

The reaction is sometimes used for preparing hydroxylamines, more often for obtaining the carbonyl compounds formed.

Acid hydrolysis of *N*-benzyl-α-phenylnitrone furnishes *N-benzylhydroxylamine* and *benzaldehyde*[88]:

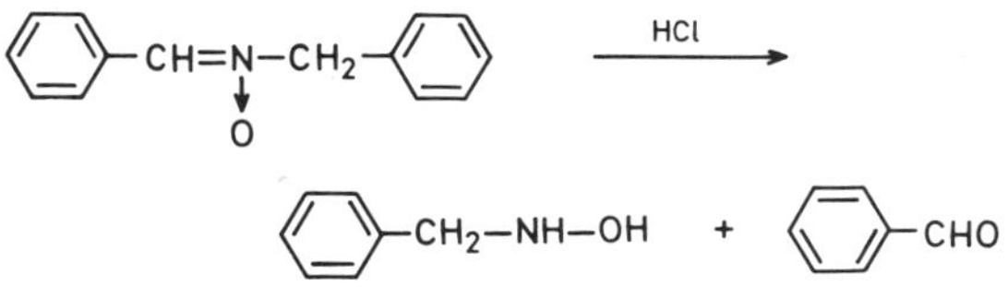

Isolation of the nitrones is not necessary in every case. One way in which they can be obtained is by alkylating oximes, and the reaction product can be hydrolyzed directly without further purification. In this way, *3-hydroxyaminopropionic acid*, for example, has been prepared from the sodium salt of *anti*-benzaldoxime and ethyl 3-bromopropionate[89]:

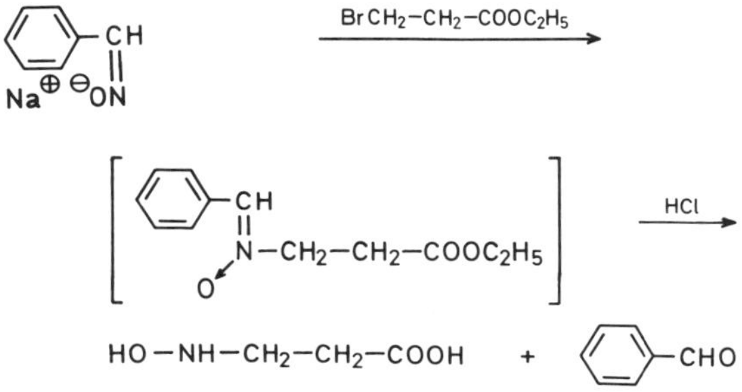

In addition to acid hydrolysis hydrazine has been found to be successful for bringing about the nitrone cleavage[90].

It should be noted that the desired substituted hydroxylamines do not form in every case during hydrolysis of nitrones. For example, *N*-diphenyl-methyl-α,α-diphenylnitrone (benzophenone diphenylmethylnitrone) affords benzophenone and hydroxylamine hydrochloride on treatment with concentrated hydrochloric acid[91].

Oxaziridines are converted into *N*-monosubstituted hydroxylamines by acid hydrolysis, especially the 2-alkyl-3-aryl substituted compounds. Dilute sulfuric acid is employed for the hydrolysis be-

[83] *R.F.C. Brown, V.M. Clark, Sir A. Todd*, J. Chem. Soc. 2107 (1959);
L.H. Sternbach, E. Reeder, J. Org. Chem. *26*, 1114 (1961).
[84] *O. Exner*, Collect. Czech. Chem. Commun. *20*, 205 (1955).
[85] *R. Bonnett, R.F.C. Brown, V.M. Clark, I.O. Sutherland, Sir A. Todd*, J. Chem. Soc. 2098 (1959).
[86] *L.H. Sternbach, E. Reeder*, J. Org. Chem. *26*, 1115 (1961).
[87] *G.F. Field, W.J. Zally, L.H. Sternbach*, Tetrahedron Lett. 2610 (1966).

[88] *L.W. Jones, M.C. Sneed*, J. Amer. Chem. Soc. *39*, 677 (1917).
[89] *E. Buehler, G.B. Brown*, J. Org. Chem. *32*, 265 (1967).
[90] *E. Bellasio, F. Parravicini, A. Vigevani, E. Testa*, Gazz. Chim. Ital. *98*, 1014 (1968).
[91] *A.C. Cope, A.C. Haven jr.*, J. Amer. Chem. Soc. *72*, 4897 (1950).
[92] *L. Horner, E. Jürgens*, Chem. Ber. *90*, 2184 (1957).
[93] *W.D. Emmons*, J. Amer. Chem. Soc. *79*, 5750 (1957).
[94] *G. Zinner, W. Kliegel*, Chem. Ber. *99*, 2686 (1966).

cause oxaziridines oxidize hydrochloric acid to chlorine[92]. An example of such a cleavage is the preparation of *N-tert-butylhydroxylamine* from 2-*tert*-butyl-3-phenyloxaziridine[93].

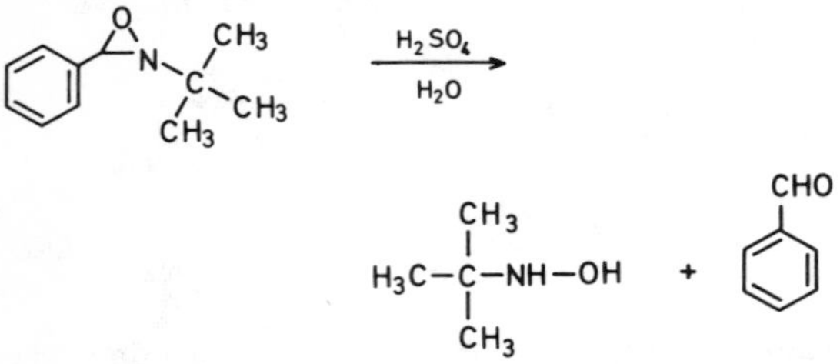

Bis(hydroxyamino)methanes are accessible by reacting hydroxylamines with formaldehyde. Thus, for example, *N,N'-dihydroxy-N,N'diisopropylmethanediamine* (89%) is obtained from *N*-isopropylhydroxylamine and half the molar quantity formaldehyde[94]:

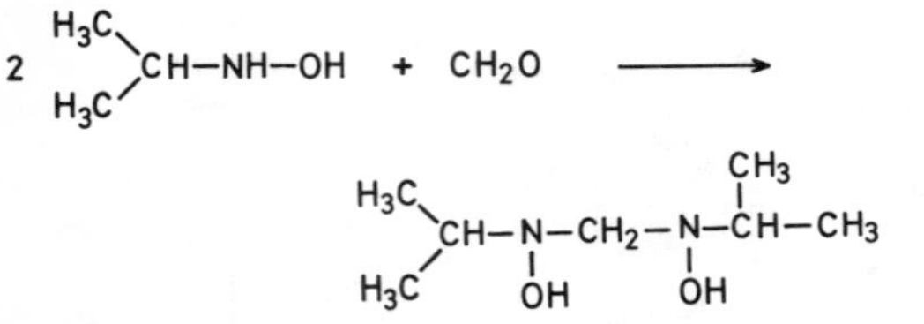

N-arylhydroxylamines react analogously. *N*-phenylhydroxylamine and formaldehyde form *N,N'-dihydroxy-N,N'-diphenylmethanediamine*[95].

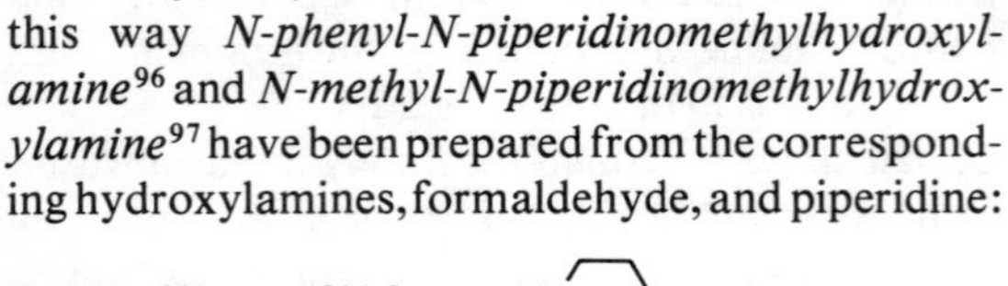

Mixed substituted hydroxylamines can be prepared by reacting formaldehyde simultaneously with a hydroxylamine and a secondary amine. In this way *N-phenyl-N-piperidinomethylhydroxylamine*[96] and *N-methyl-N-piperidinomethylhydroxylamine*[97] have been prepared from the corresponding hydroxylamines, formaldehyde, and piperidine:

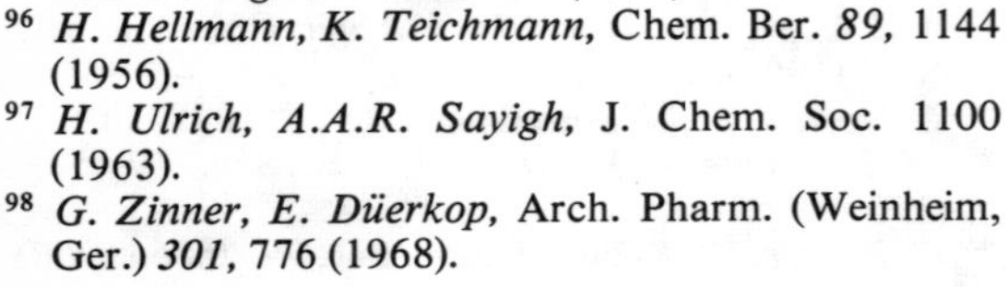

$R=C_6H_5, CH_3$

N-Hydroxy dicarboxylic acid imides can be reduced to *N,N-dialkylhydroxylamines* with lithium aluminum hydride, for example, *N*-hydroxy-1,2-cyclohexanedicarboximide to *8-azabicyclo[4.3.0]nonan-8-ol*[98]:

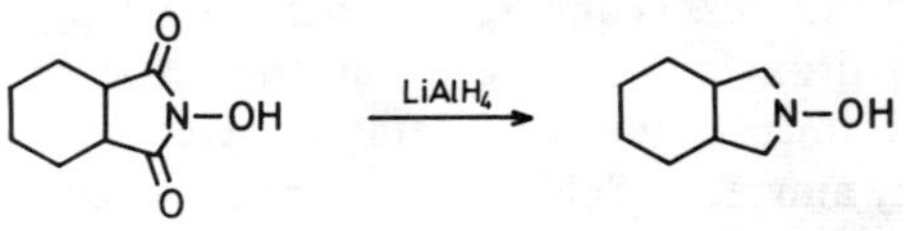

3.1.6 Other methods for preparing *O*-hydroxylamines

Oximes of aldehydes and ketones can be used for preparing *O*-hydroxylamines. Alkylation allows them to be converted into *O*-alkyl oximes, from which *N*-unsubstituted *O-alkylhydroxylamines* are obtained by heating with acids or cleaving with hydrazine derivatives. Catalytic hydrogenation enables *O,N-disubstituted* hydroxylamines to be prepared:

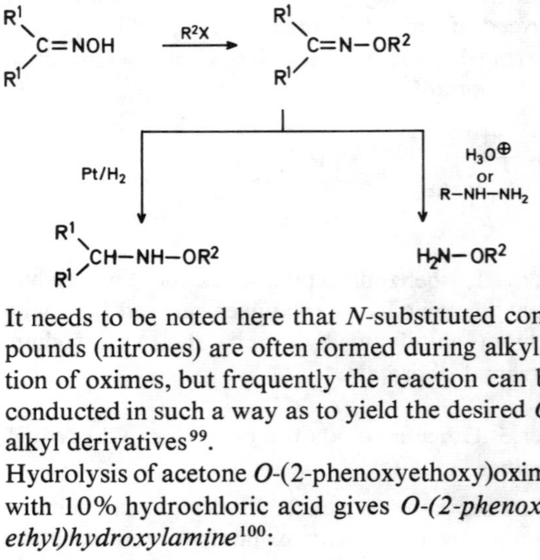

It needs to be noted here that *N*-substituted compounds (nitrones) are often formed during alkylation of oximes, but frequently the reaction can be conducted in such a way as to yield the desired *O*-alkyl derivatives[99].

Hydrolysis of acetone *O*-(2-phenoxyethoxy)oxime with 10% hydrochloric acid gives *O-(2-phenoxyethyl)hydroxylamine*[100]:

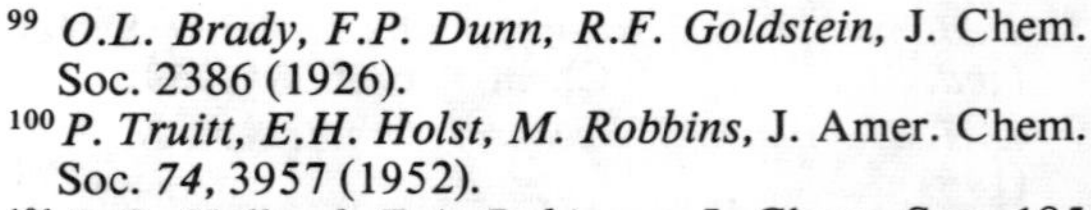

O-(2-Diethylaminoethyl)hydroxylamine is obtained analogously[101]. Acetaldehyde *O*-allyloxime has been cleaved into *O-allylhydroxylamine (alloxy-*

[95] E. Bamberger, Ber. *33*, 947 (1900).

[96] H. Hellmann, K. Teichmann, Chem. Ber. *89*, 1144 (1956).

[97] H. Ulrich, A.A.R. Sayigh, J. Chem. Soc. 1100 (1963).

[98] G. Zinner, E. Düerkop, Arch. Pharm. (Weinheim, Ger.) *301*, 776 (1968).

[99] O.L. Brady, F.P. Dunn, R.F. Goldstein, J. Chem. Soc. 2386 (1926).

[100] P. Truitt, E.H. Holst, M. Robbins, J. Amer. Chem. Soc. *74*, 3957 (1952).

[101] D.O. Holland, F.A. Robinson, J. Chem. Soc. 185 (1948).

amine) and the corresponding hydrazone by trapping the free carbonyl compound with 2,4-dinitrophenylhydrazine[102]:

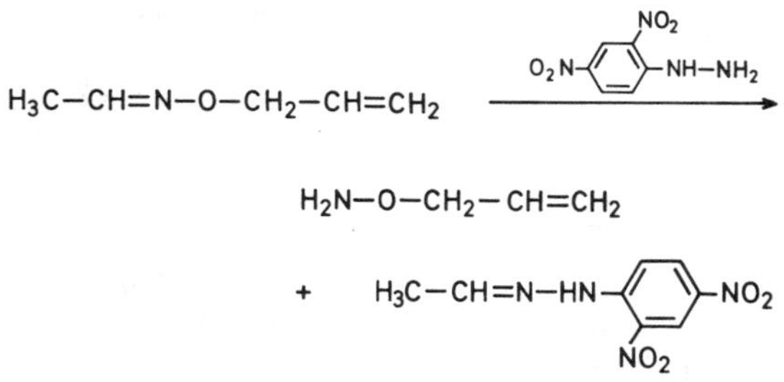

$$H_3C-CH=N-O-CH_2-CH=CH_2 \quad \xrightarrow{\quad O_2N\text{—}\begin{array}{c}NO_2\end{array}\text{—}NH-NH_2 \quad}$$

$$H_2N-O-CH_2-CH=CH_2$$

$$+ \quad H_3C-CH=N-HN-\!\!\!\begin{array}{c}NO_2\\ \end{array}\!\!\!-NO_2$$

An example of the preparation of *O,N-disubstituted* hydroxylamines by catalytic hydrogenation of *O*-alkyl oximes is the synthesis of *O-ethyl-N-(1-phenyl-2-propyl)hydroxylamine*[103]:

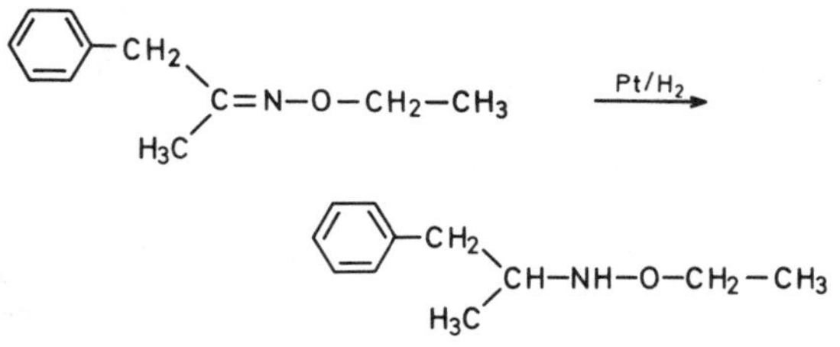

$$\xrightarrow{\quad Pt/H_2 \quad}$$

N-Hydroxyphthalimide can be alkylated similarly to the oximes to furnish *O-alkylhydroxylamines* by either hydrolysis or hydrazinolysis:

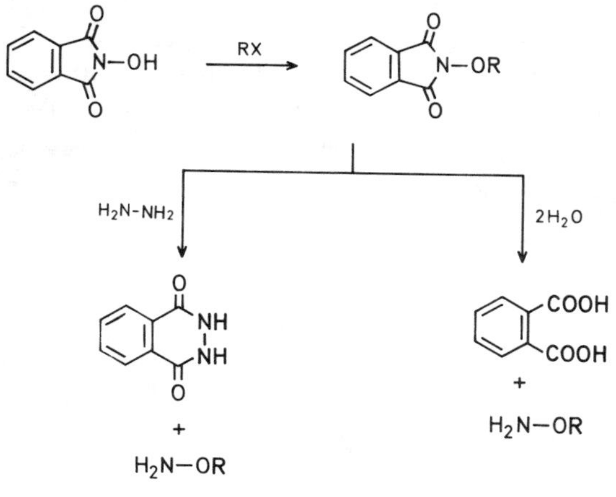

Thus, *O,O'-ethylenebishydroxylamine dihydrobromide*, for example, is formed from *N,N-ethylenedioxyphthalimide* by hydrolysis with 48% hydrobromic acid[104]:

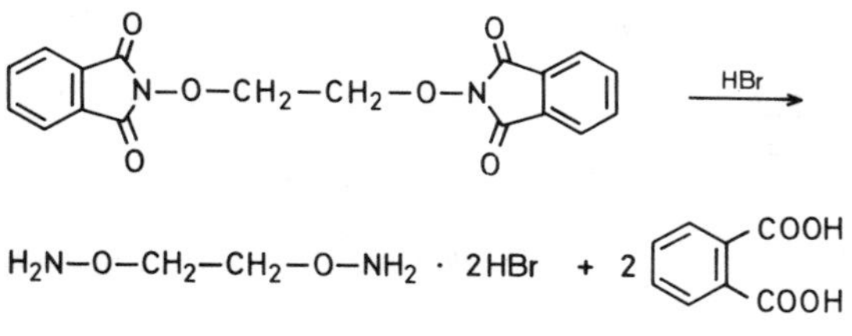

$$H_2N-O-CH_2-CH_2-O-NH_2 \cdot 2HBr \ + \ 2\!\begin{array}{c}COOH\\COOH\end{array}$$

while *O-ethylhydroxylamine hydrochloride* is formed from *N*-ethoxyphthalimide by cleaving with hydrazine and hydrochloric acid[105]:

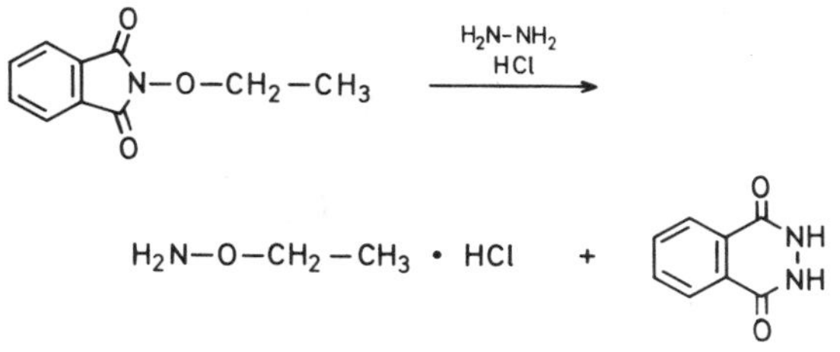

$$H_2N-O-CH_2-CH_3 \cdot HCl \ + \ \begin{array}{c}O\\NH\\NH\\O\end{array}$$

The synthesis of *O*-arylhydroxylamines by this technique has not been successful so far[106].

Hydroxylamine-*N,N*-disulfonic acid can likewise be alkylated and converted into *O-alkyl*hydroxylamines by hydrolysis with dilute sulfuric acid. *O-Methylhydroxylamine (methoxyamine)* is one compound that can be obtained in this way[107, 108]:

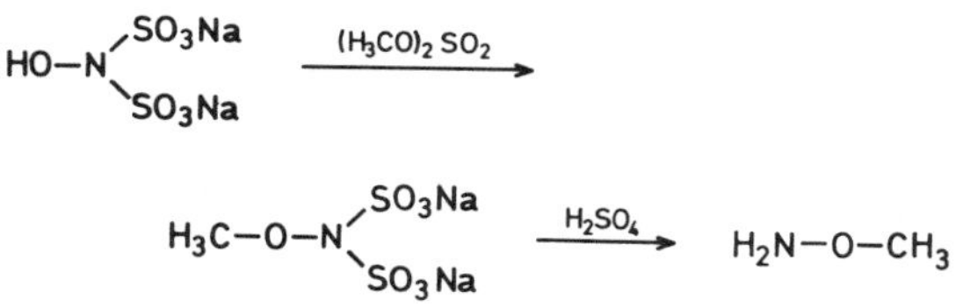

$$HO-N\!\!\begin{array}{c}SO_3Na\\SO_3Na\end{array} \quad \xrightarrow{\quad (H_3CO)_2SO_2 \quad}$$

$$H_3C-O-N\!\!\begin{array}{c}SO_3Na\\SO_3Na\end{array} \quad \xrightarrow{\quad H_2SO_4 \quad} \quad H_2N-O-CH_3$$

The procedures for preparing *O*-alkylhydroxylamines in this section display the common feature that both hydrogen atoms of the nitrogen are replaced prior to the *O*-alkylation. In what follows techniques are described which enable a selective alkylation on the oxygen to be performed by means of an *N*-monoacylation; subsequent hydrolysis also leads to *O*-alkylhydroxylamines **7**. An *O,N*-dialkylation in accord with **8** leads *O,N-dialkyl*hydroxylamines:

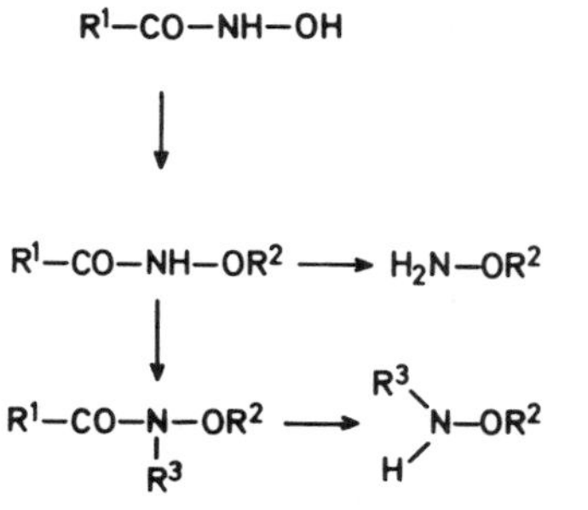

$$R^1-CO-NH-OH$$

$$\downarrow$$

$$R^1-CO-NH-OR^2 \ \longrightarrow \ H_2N-OR^2 \quad \textbf{7}$$

$$\downarrow$$

$$R^1-CO-N\!\!\begin{array}{c}OR^2\\R^3\end{array} \ \longrightarrow \ \begin{array}{c}R^3\\H\end{array}\!\!N-OR^2 \quad \textbf{8}$$

[102] *A.N. Nesmeyanov, A.K. Kochetkov, R.K. Freidlina,* Izv. Akad. Nauk SSSR, Otdel. Khim. Nauk 512 (1951); C.A. *46*, 7071 (1952).

[103] *R.T. Major, K.W. Ohly,* J. Med. Pharm. Chem. *4*, 58 (1961).

[104] *L. Bauer, K.S. Suresh,* J. Org. Chem. *28*, 1604 (1963).

[105] *J.R. Nicolaus, L. Mariani, E. Testa,* Ann. Chim. (Rom) *53*, 288 (1963).

[106] *A.O. Ilvespää, A. Marxer,* Chimia *18*, 9 (1964).

Compound classes such as hydroxamic acid, *N*-hydroxyurethans, and *N*-hydroxyureas belong in this category.

For example, benzohydroxamic acid yields allyl benzohydroxamate with allyl bromide from which *O-allylhydroxylamine hydrochloride* is obtained by hydrolysis with ethanolic hydrochloric acid[109]:

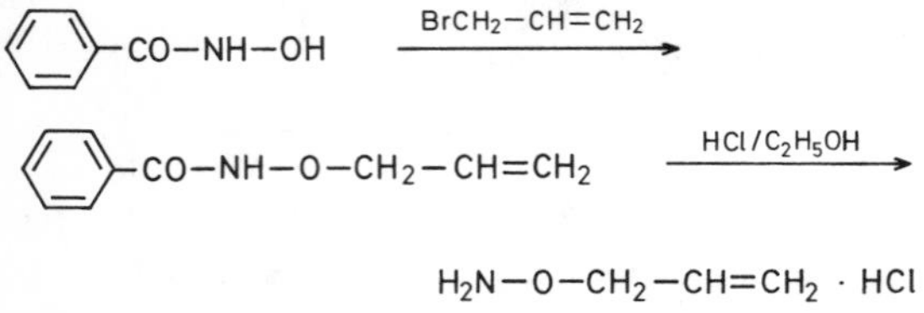

A series of ring-substituted *O-aralkyl*hydroxylamines, too, has been prepared by using this method. For example, *O-(methoxybenzyl)hydroxylamine* was obtained from (4-methoxybenzyl) benzohydroxamate by cleaving with ethanolic hydrochloric acid[110]

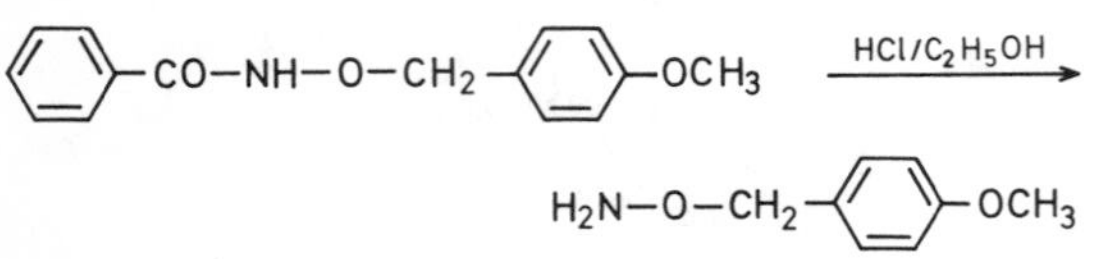

α-Aminooxy carboxylic acids[111] and *α-amino-ω-aminooxy carboxylic acids*[112] also are accessible in this way.

N-Acylation and *O*-alkylation succeed in a *one-step reaction* on reacting hydroxylamine with methyl 2-amino-3-chloropropionate in aqueous triethylamine. The antibiotic *cycloserine* is the product obtained. As the reaction proceeds very rapidly no intermediate products haven been isolated to date[113]:

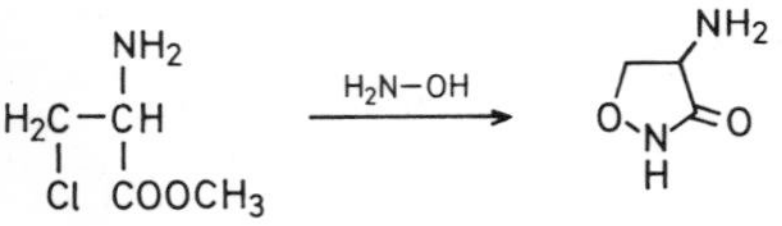

In the frequently employed urethan method for preparing *O*-hydroxylamines use is generally made of *N*-hydroxyurethan obtainable from hydroxylamine and ethyl chloroformate. This compound is a viscous oil which may decompose **explosively** on distillation and, therefore, should not be purified if possible. For further reactions the crude product is employed[114].

Numerous *O-alkyl*- and *O-aralkyl*hydroxylamines are accessible from this starting material, for instance, *O-phenethylhydroxylamine* by alkylation of *N*-hydroxyurethan with phenethyl bromide and subsequent alkaline hydrolysis[115]:

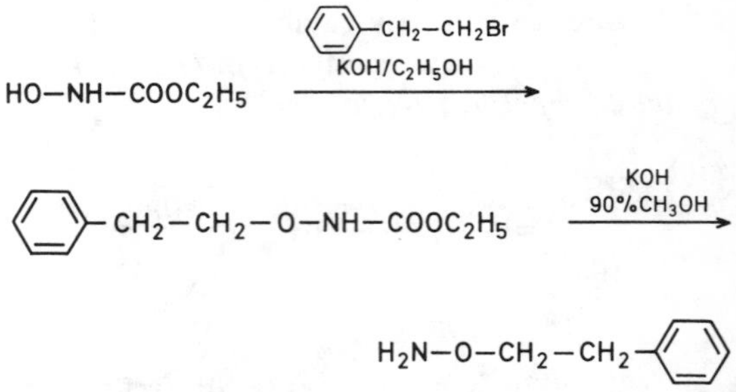

*O,N-dialkyl*hydroxylamines with the same alkyl substituent groups are obtained by alkylating *N*-hydroxyurethan with, for example, dialkyl sulfate and submitting the *O,N*-dialkyl-*N*-hydroxyurethan to alkaline hydrolysis[116].

α-Aminooxy carboxylic acids are formed from the corresponding urethans by acid hydrolysis **9**. If the nitrogen is alkylated additionally at the urethan stage, then *α-alkylaminooxy carboxylic acids* **10** are obtained after the hydrolysis[117]:

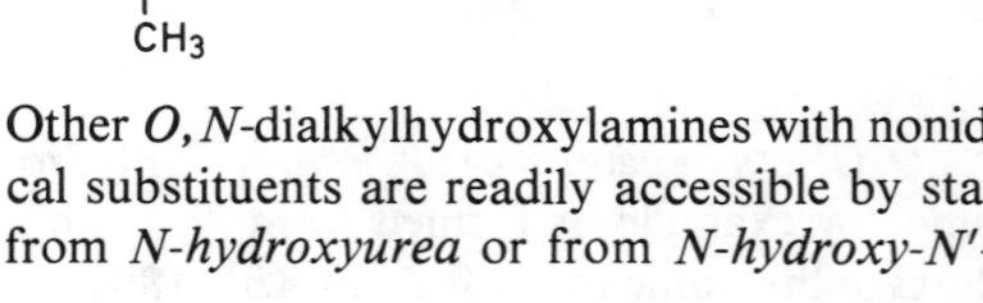

Other *O,N*-dialkylhydroxylamines with nonidentical substituents are readily accessible by starting from *N-hydroxyurea* or from *N-hydroxy-N'-phe-*

[107] *O. Scherer, G. Hörlein, K. Härtel*, Angew. Chem. *75*, 853 (1963);
DBP 1112082 (1959), Farbw. Hoechst, Erf.: *O. Scherer, G. Hörlein, R. Hübner, G. Schneider;* C.A. *56*, 8561 (1962).

[108] *H. Hjeds*, Acta Chem. Scand. *19*, 1764 (1965).

[109] *O.L. Brady, F.H. Peakin*, J. Chem. Soc. 228 (1930).

[110] *P. Mamalis, J. Green, D.J. Outred, M. Rix*, J. Chem. Soc. 3924 (1962).

[111] *D. McHale, J. Green, P. Mamalis*, J. Chem. Soc. 225 (1960).

[112] *Y. Knobler, M. Frankel*, J. Chem. Soc. 1632 (1958).

[113] *P.A. Plattner, A. Boller, H. Frick, A. Fürst, B. Hegedüs, H. Kirchensteiner, St. Majnoni, R. Schläpfer, H. Spiegelberg*, Helv. Chim. Acta *40*, 1531 (1957).

[114] *B.J.R. Nicolaus, G. Pagani, E. Testa*, Helv. Chim. Acta *45*, 359 (1962).

[115] *B.J.R. Nicolaus, G. Pagani, E. Testa*, Helv. Chim. Acta *45*, 1387 (1962).

[116] *R.T. Major, E.E. Fleck*, J. Amer. Chem. Soc. *50*, 1479 (1928).

[117] *E. Testa, B.J.R. Nicolaus, L. Mariani, G. Pagani*, Helv. Chim. Acta *46*, 766 (1963).

nylurea. Both compounds can be alkylated in *stages* on the *hydroxyamino* group, with substitution on the oxygen taking place first followed by reaction on the nitrogen. According to Equation **11** cleaving is performed with caustic soda, according to Equation **12** with aniline; in either case *O,N*-dialkylhydroxylamines are formed[118].

Reacting *N*-chloroalkylamines or hydroxylamine-*O*-sulfonic acids in alkaline solution with aldehydes or ketones gives *oxaziridines*[121]:

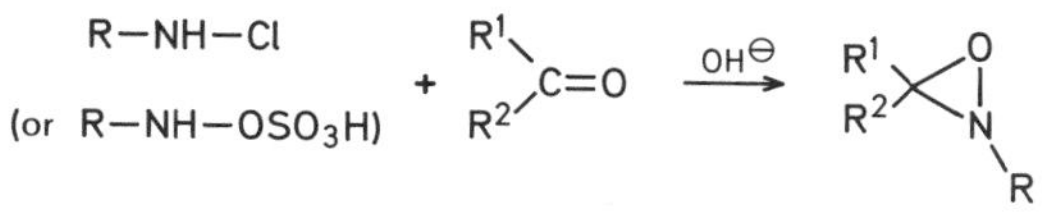

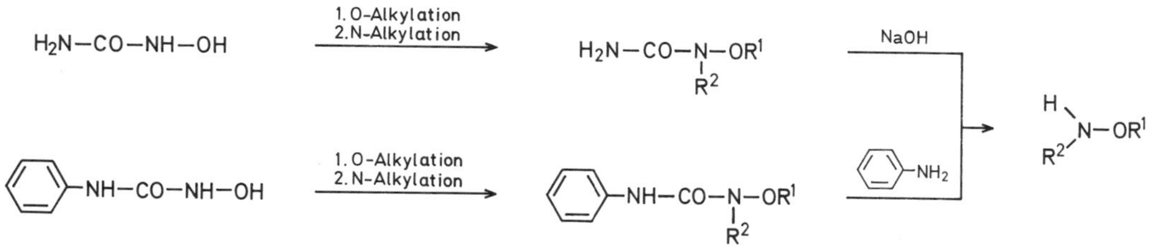

$$R-NH-Cl$$
$$\text{(or } R-NH-OSO_3H)$$
$$+ \quad \begin{matrix} R^1 \\ R^2 \end{matrix}C=O \quad \xrightarrow{OH^\ominus} \quad \begin{matrix} R^1 \\ R^2 \end{matrix}\begin{matrix} O \\ N \\ R \end{matrix}$$

11

12

Reacting *N*-hydroxyurethan with 1,3-dibromopropane or with 1,4-dibromobutane leads to cyclization with formation of *ethyl oxazolidine-2-carboxylate* **5** and *ethyl tetrahydro-2 H-1,2-oxazine-α-carboxylate* **6**. From these hydrolysis with 16% hydrochloric acid affords *isoxazolidine* **7** and *tetrahydro-1,2-oxazine* **8** respectively[119]:

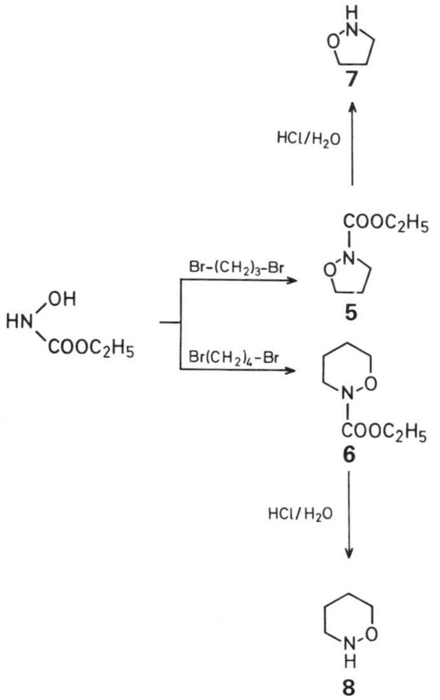

A more generally applicable method for preparing oxaziridines is oxidation of Schiff's bases with peracids[122]:

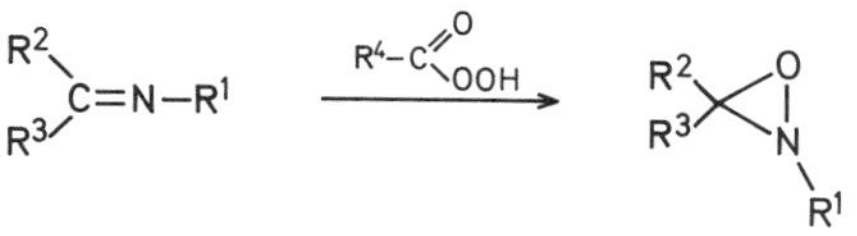

$$\begin{matrix} R^2 \\ R^3 \end{matrix}C=N-R^1 \quad \xrightarrow{R^4-C\begin{smallmatrix}O\\OOH\end{smallmatrix}} \quad \begin{matrix} R^2 \\ R^3 \end{matrix}\begin{matrix}O\\N\\R^1\end{matrix}$$

3.2 Transformation reactions

Acylation of *N*-alkylhydroxylamines with acyl chlorides, carboxylic acid anhydrides, or esters in general leads to *N-acylated* compounds, which are *hydroxamic acid* derivatives **13**. Further acylation leads to *O,N-diacylhydroxylamines* **14**. *N,N-Disubstituted O-acylhydroxylamines* are accessible by acylating *N,N*-dialkylhydroxylamines **15**:

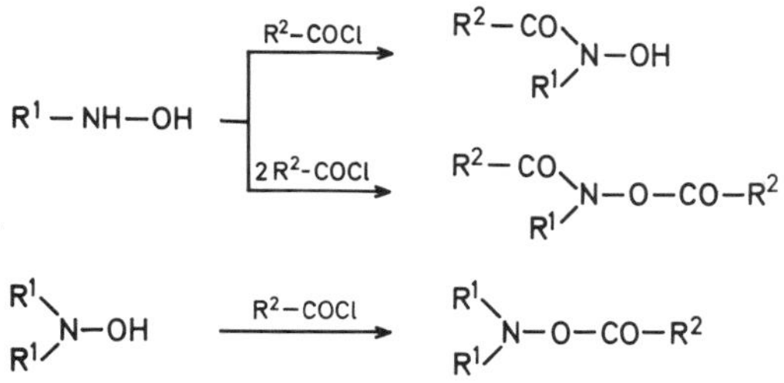

*O-Alkyl*hydroxylamines can be prepared also by reacting chloramine with alcoholates in accord with the equation[120]:

$$H_2N-Cl + NaOR \longrightarrow H_2N-OR + NaCl$$

O-Hydroxylamines *add on* isocyanates, isothiocyanates, and cyanamide with formation of urea derivatives **16**, thiourea derivatives **17**, and guanidine derivatives **18**:

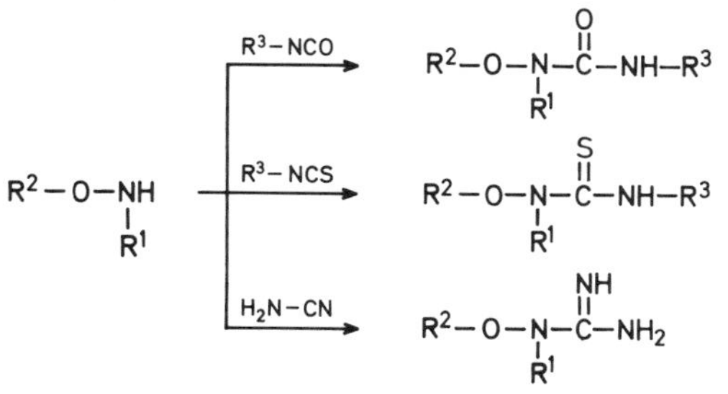

[118] *N. Kreutzkamp, P. Messinger,* Chem. Ber. *100,* 3463 (1967).
[119] *H. King,* J. Chem. Soc. 432 (1942).

During *oxidation* of N-arylhydroxylamines nitroso compounds are formed initially with condense with not yet oxidized hydroxylamine to form the azoxy compound (see p. 137):

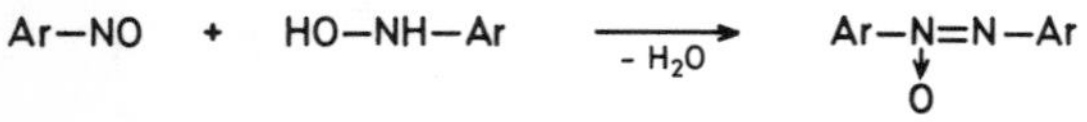

$$Ar-NO \quad + \quad HO-NH-Ar \quad \xrightarrow[-H_2O]{} \quad Ar-N=N-Ar$$
$$\downarrow$$
$$O$$

N,N-disubstituted hydroxylamines can be dehydrogenated to *nitrones* (see p. 345):

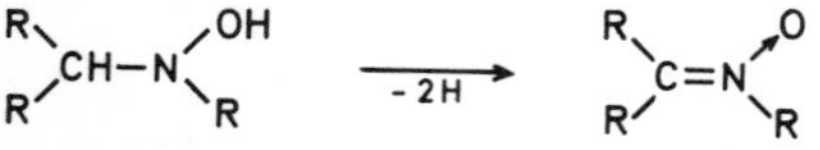

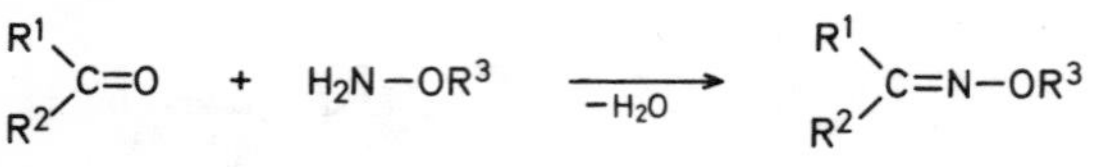

N-Monosubstituted and O,N-disubstituted hydroxylamines are nitrosated on the nitrogen with nitrous acid or its esters with formation of *N-nitrosohydroxylamines:*

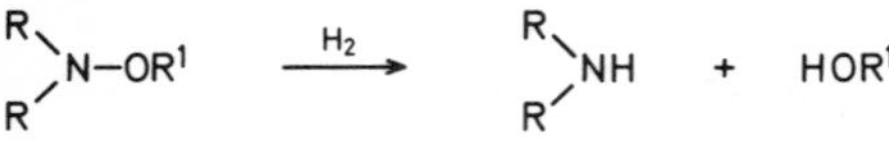

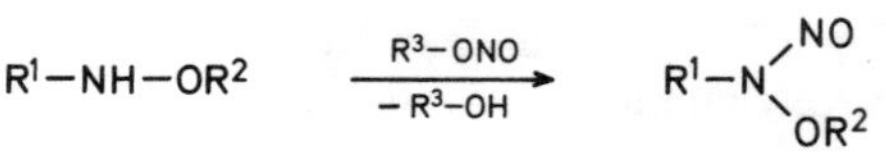

Reduction of N-hydroxylamines leads to the corresponding amines **19**. O-Hydroxylamines can be cleaved into amine or ammonia and alcohol reductively **20**:

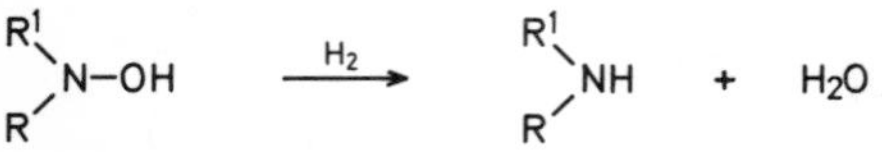

On heating with dilute sulfuric acid N-arylhydroxylamines rearrange to form 4-aminophenols (Bamberger rearrangement) (see p. 569):

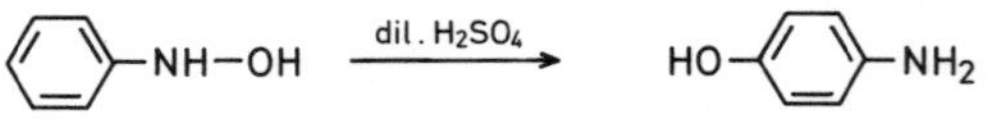

Under the action of phosphorus(V) chloride N-triarylmethylhydroxylamines can be rearranged with formation of benzophenone imines (Stieglitz rearrangement):

N-Hydroxylamines can be *condensed* to nitrones with carbonyl compounds (see p. 346):

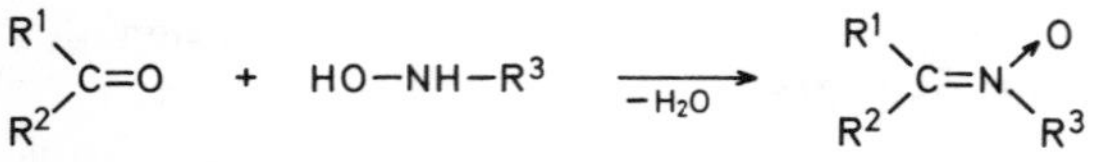

N-Unsubstituted O-alkylhydroxylamines react with carbonyl compounds with formation of O-alkyl oximes (see p. 434):

3.3 Bibliography

Houben-Weyl, Methoden der organischen Chemie, Bd. X/1, Abschnitt Hydroxylamine, p. 1091, Georg Thieme Verlag, Stuttgart 1971.

P.A.S. Smith, The Chemistry of Open-Chain Organic Nitrogen Compounds, Vol. II, W.A. Benjamin, New York 1966.

N.V. Sidgwick, I.T. Millar, H.D. Springall, The Organic Chemistry of Nitrogen, Clarendon Press, Oxford 1966.

A.O. Ilvespää, A. Marxer, Chimia 18, 1 (1964).

L.L. Muller, J. Hamer, 1,2-Cycloaddition Reactions, Chap. Oxaziridine, p. 103, Chap. Oxazetidine, p. 257, Interscience Publishers, John Wiley & Sons, New York 1967.

J. Hamer, M. Ahmad in *J. Hamer,* 1,4-Cycloaddition Reactions, Chap. Nitroso Compounds as Dienophiles, p. 419, Academic Press, New York 1967.

---

[120] *W. Theilacker, K. Ebke,* Angew. Chem. *68,* 303 (1956).

[121] *E. Schmitz, R. Ohme, D. Murawski,* Chem. Ber. *98,* 2516 1965).

[122] *H. Krimm,* Chem. Ber. *91,* 1057 (1958).

4 Hydrazines and Hydrazones

Contributed by

E. V. Brown
Department of Chemistry,
University of Kentucky,
Lexington, Kentucky

L. Caglioti
Cattedrà di Chimica Organica,
Facoltà di Farmacìa, Università di Roma,
Rome

G. Paolucci
G. Rosini
Istituto di Chimica Industriale dell'Università di Bologna,
Bologna

W. Sucrow
Organisch-chemisches Institut der Technischen
Universität Berlin,
Berlin

On replacing hydrogen atoms of hydrazine by alkyl or aryl groups organic hydrazines are formed. While the strong basicity of hydrazine is weakened as substitution increases — aromatic substituents produce a greater effect than aliphatic ones — the great majority of organic derivatives of hydrazine are, nonetheless, relatively strong, monoacidic bases which form stable salts. Monoacylation leaves the basicity intact, *N,N'*-diacylation leads to acid compounds. Provided one of the two nitrogen atoms remains linked to hydrogen the *N,N'*-diacyl-hydrazides can be converted into metal salts with, for example, alcoholates; in aqueous solution these salts undergo partial hydrolysis.

Most hydrazines are sensitive to oxidation. Phenylhydrazine rapidly turns brown in the air; others, depending on their constitution, reduce Fehling's solution, ammoniacal silver salt solutions, *etc.* On the other hand, the reductive cleavage of the *N,N*-bond represents a possible synthetic pathway to amines. On heating in toluene tetra-phenylhydrazine is partially split into free radicals homolytically. Some hydrazines represent valuable therapeutic agents (hydrazinophthalazines, INH, etc.), others, notably arylhydrazines (and also aryl hydrazides and aryl hydrazones), are *blood* and *protoplasm poisons*. External contact can cause severe skin damage, and chronic absorption of small quantities, too, is hazardous. Phenylhydrazine when finely divided (e.g., on cotton fabric or asbestos) will *ignite spontaneously*. Handling novel representatives of this substance class demands care in every case.

4.1 Arylhydrazines and Arylhydrazones

The usual nomenclature for the arylhydrazines is the name of the aryl group followed by hydrazine, as, phenylhydrazine, 4-pyridylhydrazine, etc. Occasionally this is reversed as, 2-hydrazinoquinoline. The replacement of the hydrogens of hydrazine (H_2N-NH_2) by aryl groups can lead to five different types of arylhydrazines:

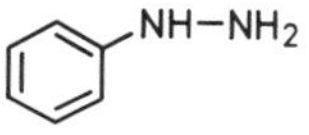

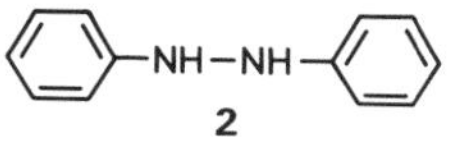

Compound 2 is a hydrazo compound and this type has been discussed separately because of its importance (p. 153). Among the remaining four, 1 and 3 are of interest because they react with aldehydes and ketones to Schiff's bases. Compounds 4 and 5 are named 1, 1, 2-triphenylhydrazine and 1, 1, 2, 2-tetraphenylhydrazine, or N, N, N'-triphenylhydrazine and N, N, N', N'-tetraphenylhydrazine.

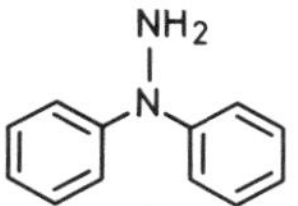

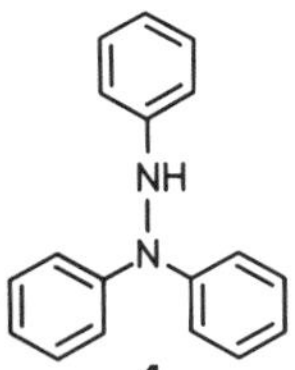

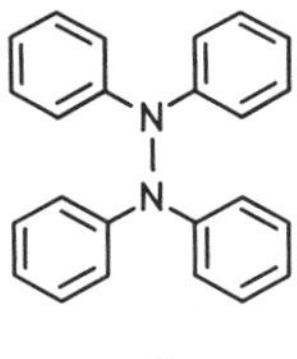

4.1.1 Preparation of arylhydrazines

4.1.1.1 By reduction

4.1.1.1.1 Reduction of diazonium compounds

Historically the reduction of diazonium salts by *bisulfites* led to the first preparation of arylhydra-

zines and it is still an important preparatory reaction[1]. The course of this reaction[2, 3, 4] is as follows:

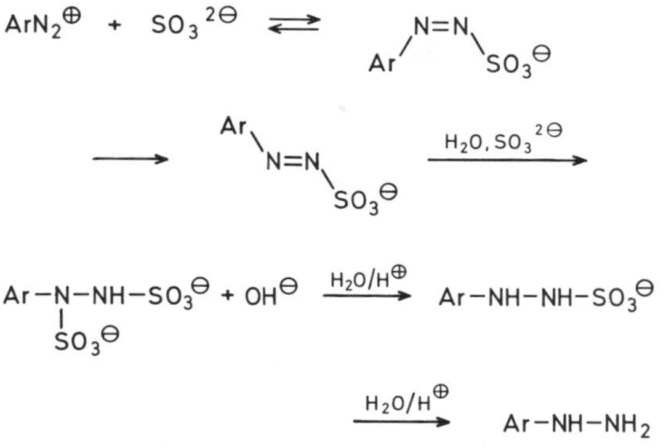

The details may vary somewhat with the substitution on the phenyl ring[5]. In addition to sodium sulfite[6], potassium[7] and ammonium sulfites may also be used to some advantage. By the use of either of these three reagents many substituted aryldiazonium salts have been converted to the arylhydrazines[8]. Aqueous sulfur dioxide solutions have been used for this reduction[9-11].

Sodium dithionite ($Na_2S_2O_4$) may be used to reduce aryldiazonium salts[12] or to convert arylhydrazine-β-sulfonic acids[13] to the corresponding arylhydrazines.

Stannous chloride[14, 15] in acid solution will reduce aryldiazonium salts in excellent yields[16]. When other reducible groups are present precautions must be taken.

Zinc powder[17, 18] in acetic acid and *sodium amalgam*[19] have been used to reduce diazonium salts, diazotates or arylhydrazine-β-sulfonic acids to the corresponding hydrazines, but seem to have little advantage over the previously mentioned reagents. A rather recently employed reagent for this reduction is *triphenylphosphine*[20] which has given good yields of aromatic hydrazines[21].

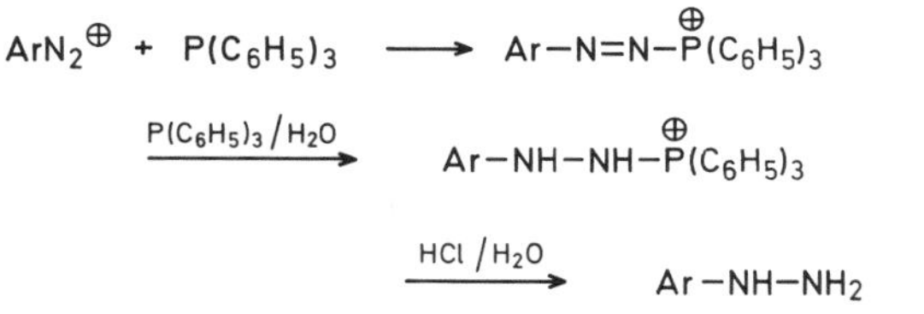

[1] E. Fischer, Ber. 8, 590 (1875).
[2] W. Davies, J. Chem. Soc. 121, 715 (1922).
[3] E.S. Lewis, H. Suhr, Chem. Ber. 92, 3031 (1959).
[4] R. Huisgen, R. Lux, Chem. Ber. 93, 540 (1960).
[5] Houben Weyl, Methoden der organischen Chemie, Bd. X/2, p. 180, Georg Thieme Verlag, Stuttgart 1967.
[6] A.J. Vogel, A Textbook of Practical Organic Chemistry, p. 607, Longmans, Green and Co., London 1948.

Another unusual set of reducing agents that have been applied to arylhydrazine syntheses is ascorbic acid, isoascorbic acid[22] and pyridoine[23] and related *ene-diol types*. This reaction proceeds *via* a split product of the ene-diol which can then be hydrolyzed to the hydrazine.

Electrochemical reduction has been very successful for the preparation of arylhydrazines although probably not as generally useful in the laboratory as the chemical methods[24-26].

4.1.1.1.2 Reduction of *N*-nitroso and *N*-nitro compounds

N-Nitroso secondary amines may be reduced to substituted hydrazines by a number of reagents. One such reagent is zinc and acetic acid[27] and a large number of nitrosamines have been successfully reduced in this manner[28]. Lithium aluminum hydride gives good yields of hydrazines from *N*-nitrosamines[29].

A number of other agents have been used occasionally. These are sodium dithionite[30], aluminum amalgam[31], ammonium sulfide[32], cuprous halides[33], platinum plus hydrogen[34], palladium and hydrogen[35] or cathodic reduction[36].

A number of *heterocyclic hydrazines* have also been made by the reduction of nitrosamines or nitramines, among them those of benzothiazolines[37],

1,3,4-oxadiazoles[38], 1,2,4-triazoles[39], 1,3,4-thiadiazoles[38], pyridines[40], and pyrimidines[41].

4.1.1.2 Exchange reactions

A considerable number of different groups can be replaced by the hydrazino group by either heating with hydrazine itself or by treatment with sodium hydrazide.

4.1.1.2.1 Replacement of halogen

Fluorine in many polyfluoroaromatics may be converted to the aromatic hydrazine by heating with hydrazine or hydrazine hydrate. By this procedure 1,2,3,5,6-pentafluorobenzene gives *2,3,5,6-tetrafluorophenylhydrazine*[42,43]. Aromatic chlorine and bromine, particularly when activated by nitro groups, can be readily replaced by the hydrazino group[44,45]. In some cases the hydrazine group condenses with the activating group to give a heterocycle as, for instance, 2-nitrophenylhydrazine to *1-hydroxybenzotriazole*[46] or 2-cyano-4-nitrophenylhydrazine giving *3-amino-5-nitroindazole*[47].

Typical of active halogen replacement by hydrazine in the heterocyclic field are

2-Chloropyridine ⟶ *2-Hydrazinopyridine*[48]

2-Chloro-4,6-dimethylpyridine ⟶ *2-Hydrazino-4,6-dimethylpyridine*[49]

2-Chlorobenzoxazole ⟶ *2-Hydrazinobenzoxazole*[50]

2-Chlorobenzothiazole ⟶ *2-Hydrazinobenzothiazole*[51]

2-Chloroquinoline ⟶ *2-Hydrazinoquinoline*[52,53]

[7] *E. Bamberger, E. Kraus,* Ber. *29,* 1834 (1896).

[8] *Houben-Weyl,* Methoden der organischen Chemie, Bd. X/2, p. 171, Georg Thieme Verlag, Stuttgart 1967.

[9] *H. Brintzinger, K. Pfannstiel, J. Janecke,* Angew. Chem. *56,* 233 (1943).

[10] *E. Stephenson,* Org. Syntheses, Coll. Vol. III, 475 (1955).

[11] *Houben-Weyl,* Methoden der organischen Chemie, Bd. X/2, p. 197, Georg Thieme Verlag, Stuttgart 1967.

[12] *L. Thompson,* J. Soc. Dyers Colour. *37,* 8 (1921).

[13] *G.M. Robinson, R. Robinson,* J. Chem. Soc. *125,* 833 (1924).

[14] *V. Meyer, M.T. Lecco,* Ber. *16,* 2976 (1883).

[15] *I.M. Hunsberger,* J. Org. Chem. *21,* 394 (1956).

[16] *Houben-Weyl,* Methoden der organischen Chemie, Bd. X/2, p. 209, Georg Thieme Verlag, Stuttgart 1967.

[17] *F.J. Stevens, T.D. Griffin, T.L. Fields,* J. Amer. Chem. Soc. *77,* 43 (1955).

[18] *Houben-Weyl,* Methoden der organischen Chemie, Bd. X/2, p. 216, Georg Thieme Verlag, Stuttgart 1967.

[19] *E. Bamberger,* Ber. *29,* 499, 473 (1896).

[20] *L. Horner, H. Stöhr,* Chem. Ber. *86,* 1076 (1953).

[21] *L. Horner,* Fortschr. Chem. Forsch. *7,* 48 (1966).

[22] USP. 2,706,732 Hoffmann-La Roche Inc., Inv.: *R. Duschinsky, M. Schmall;* C.A. *50,* 2666h (1956).

[23] *B. Eistert, H. Munder,* Chem. Ber. *88,* 230 (1955).

[24] *P. Rüetschi, G. Trümpler,* Helv. Chim. Acta *36,* 1649 (1953).

[25] *E.W. Cook, W.G. France,* J. Amer. Chem. Soc. *56,* 2225 (1934).

[26] *Houben-Weyl,* Methoden der organischen Chemie, Bd. X/2, p. 222, Georg Thieme Verlag, Stuttgart 1967.

[27] *W.W. Hartmann, L.J. Roll,* Org. Syntheses Coll. Vol. II, 418, 460 (1943).

[28] *Houben-Weyl,* Methoden der organischen Chemie, Bd. X/2, p. 227, Georg Thieme Verlag, Stuttgart 1967.

An alternate method involves the replacement of halogen using arylsulfonic acid hydrazides. An example of this reaction is the formation of *5-nitro-2-pyridyltosylhydrazide*[54].

When sodium hydrazide is used to replace a halogen, most often this strong base reacts *via* an aryne intermediate[55].

4- and 3-Tolylhydrazine

4.1.1.2.2 Replacement of nitro groups

Occasionally a nitro group in a di- or polynitroaromatic can be displaced by hydrazine[56]. In fact in some cases the nitro group is replaced ahead of a halogen[57]. Nitro groups in heterocyclic compounds also have undergone hydrazinolysis[58].

4.1.1.2.3 Replacement of ether groups

Alkyl and aryl ether groups activated by *ortho* and *para* nitro groups can be replaced by hydrazine[59] and it often happens that an *ortho* nitro group reacts with the hydrazine thus formed to give a benzotriazole[60].

2,4,6-Trinitrophenylhydrazine

4,6-Dinitro-1,2,3-benzotriazole

This replacement also takes place in nitrogen heterocycles where the ether group is adjacent to the nitrogen[61].

4.1.1.2.4 Replacement of sulfur-containing groups

Aromatic *sulfonic acids* and their derivatives can be replaced by treatment with hydrazine[62, 63] or sodium hydrazide[64]. *Mercapto* groups[65] and *alkylthio* ethers[66] can be converted to hydrazines by hydrazine.

4.1.1.2.5 Replacement of aromatic amino and hydroxy groups

The replacement of the primary *amino* group by hydrazine is in some cases possible[67] but in general substituted amino groups replace more readily[68]. A much better reaction is the replacement of the aryl *hydroxy* group and this is done using *Bucherer conditions*[69]; *e.g.*

1,3-Phenylenedihydrazine

[29] *H. Zimmer, L.F. Audrieth, M.Z. Zimmer, R.A. Rowe,* J. Amer. Chem. Soc. *77,* 790 (1955).
[30] *C.G. Overberger, J.G. Lombardino, R.G. Hiskey,* J. Amer. Chem. Soc. *80,* 3009 (1958).
[31] *R. Huisgen,* Justus Liebigs Ann. Chem. *559,* 182 (1948).
[32] *A. Hempel,* J. Prakt. Chem. [2] *41,* 168 (1890).
[33] *T. Yoshikawa,* J. Pharm. Soc. Jap. *76,* 776 (1956); C.A. *51,* 1198a (1957).
[34] *P.W. Neber, G. Knöller, K. Herbst, A. Trissler,* Justus Liebigs Ann. Chem. *471,* 113 (1929).
[35] *C. Paal, W.-N. Yao,* Ber. *63,* 57 (1930).
[36] *P.E. Iversen,* Acta Chem. Scand. *25,* 2337 (1971).
[37] *H.J. Backer,* Rec. Trav. Chim. Pays-Bas *32,* 44 (1913).
[38] *R. Stollé, K. Fehrenbach,* J. Prakt. Chem. [2] *122,* 315 (1929).
[39] *H. Gehlen, J. Dost,* Justus Liebigs Ann. Chem. *665,* 148 (1963).
[40] *K. Arvarson,* Ark. Kemi *7,* 266 (1955); C.A. *50,* 334i (1956).
[41] *S. Gabriel, J. Coleman,* Ber. *34,* 1240 (1901).
[42] *G.M. Brooke, J. Burdon, M. Stacey, J.C. Tatlow,* J. Chem. Soc. (London) 1768 (1960).
[43] *Houben-Weyl,* Methoden der organischen Chemie, Bd. X/2, p. 236, Georg Thieme Verlag, Stuttgart 1967.
[44] *C.F.H. Allen,* Org. Syntheses, Coll. Vol. II, 228, (1943).
[45] *E. Königs, A. Wylezich,* J. Prakt. Chem. [2] *132,* 24 (1932).
[46] *O.L. Brady, C.V. Reynolds,* J. Chem. Soc. (London) 193 (1928).
[47] *E.W. Parnell,* J. Chem. Soc. (London) 2363 (1959).

This reaction may actually proceed through a ketone type intermediate[70] and therefore involve hydrazone formation.

4.1.1.3 Condensation of aromatic amines with various nitrogen compounds

Chloramine condenses with aniline a 50% yield of *phenylhydrazine*[71]. Somewhat better is the reaction between aniline and hydroxylamine-*O*-sulfonic acid which gives 50% yield of phenylhydrazine[72]. A modification of this method uses aniline-*N*-sulfonic acid with hydroxylamine-*O*-sulfonic acid to prepare phenylhydrazine-α-sulfonic acid[73]. The heating of aromatic amines with azides can lead to the formation of aromatic hydrazine derivatives. An example is the preparation of *ethyl phenylhydrazine-β-carboxylate* from aniline and ethyl azidocarbonate[74, 75]. Another method leading to a hydrazine derivative is the reaction of *N*-chloroacetanilides with sodium hydrazide leading to α-*acetylphenylhydrazines*[76].

4.1.1.4 Heteroaromatic and pseudoaromatic hydrazines by cyclization reactions

Ethyl 5-hydrazino-3-methyl-4-pyrazolecarboxylate (30%) can be synthesized by reacting ethyl 2-chloroacetoacetate with thiocarbonylhydrazide in ethanolic hydrochloric acid[77].

An alternative synthetic approach to this heterocyclic compound is diazotization and sulfite reduction of ethyl 5-amino-3-methyl-4-pyrazolecarboxylate.

Reacting 2-benzylidenehydrazino-4,6-dimethylpyrimidine [prepared from (benzylideneamino)guanidine] and 2,4-pentanedione with hydrazine hydrate forms *2-hydrazino-4,6-dimethylpyrimidine*[78].

Benzylideneaminoguanidine is used wherever formation of pyrazoles must be avoided. *2-Hydrazino-4,6-dimethylpyrimidine* is obtained also on acid hydrolysis of the *N*-benzoyl derivative[79].

Hydrazinophthalazines are prepared by reacting suitable derivatives of benzoic or phthalic acid with hydrazines.

1-Hydrazinophthalazine is formed by the reaction between 2-dihydroxymethylbenzonitrile diacetate and hydrazine[80], *1,4-dihydrazinophthalazine* (65%) by that between phthalonitrile and hydrazine hydrate in 1,4-dioxane/acetic acid[81].

[48] *R.G. Fargher, H.R. Furness*, J. Chem. Soc. (London) *107*, 688 (1915).
[49] *M.P.V. Boarland, J.F.W. McOmie, R.N. Timms*, J. Chem. Soc. (London) 4691 (1956).
[50] *L. Katz*, J. Amer. Chem. Soc. *75*, 714 (1953).
[51] *L. Katz*, J. Amer. Chem. Soc. *73*, 4009 (1951).
[52] *W.H. Perkin, R. Robinson*, J. Chem. Soc. (London) *103*, 1978 (1913).
[53] *Houben-Weyl*, Methoden der organischen Chemie, Bd. X/2, p. 260, Georg Thieme Verlag, Stuttgart 1967.
[54] *A. Deavin, C.W. Rees*, J. Chem. Soc. (London) 4970 (1961).
[55] *T. Kauffmann, H. Henkler*, Chem. Ber. *96*, 3159 (1963).
[56] *O.L. Brady, J.H. Bowman*, J. Chem. Soc. (London) *119*, 894 (1921).
[57] *M. Giua*, Gazz. Chim. Ital. *52* I, 346 (1922).
[58] *E.C. Taylor, J.S. Driscoll*, J. Amer. Chem. Soc. *82*, 3141 (1960).
[59] *W. Borsche*, Ber. *56*, 1488 (1923).
[60] *W. Borsche, W. Trautner*, Justus Liebigs Ann. Chem. *447*, 1 (1926).

[61] *J. Chesterfield, J.F.W. McOmie, E.R. Sayer*, J. Chem. Soc. (London) 3478 (1955).
[62] *H. Bradburg, F.J. Smith*, J. Chem. Soc. (London) 2943 (1952).
[63] *Houben-Weyl*, Methoden der organischen Chemie, Bd. X/2, p. 277, Georg Thieme Verlag, Stuttgart 1967.
[64] *W. Burkhardt, T. Kauffmann*, Angew. Chem. *79*, 57 (1967).
[65] *J. Druey, K. Meier, K. Eichberger*, Helv. Chim. Acta *37*, 131 (1954).
[66] *E.J. Birr, W. Walther*, Chem. Ber. *86*, 1401 (1953).
[67] *T. Curtius, G.M. Dedichen*, J. Prakt. Chem. [2] *50*, 247 (1894).
[68] *L. Spiegel*, Ber. *41*, 886 (1908).
[69] *L. Hoffmann*, Ber. *31*, 2909 (1898).
[70] *H. Seeboth, D. Bärwolf, B. Becker*, Justus Liebigs Ann. Chem. *683*, 85 (1965).
[71] *H.H. Sisler, H.S. Ahuja, N.L. Smith*, J. Org. Chem. *26*, 1819 (1961).
[72] *R. Gösl, A. Meuwsen*, Chem. Ber. *92*, 2521 (1959).
[73] *A. Meuwsen, M. Wilhelm*, Z. Anorg. Chem. *302*, 219 (1959).
[74] *K. Hafner, D. Zinser, K.L. Moritz*, Tetrahedron Lett. 1733 (1964).
[75] *T.J. Prosser, A.F. Marcantonio, D.S. Breslow, C.A. Genge*, Tetrahedron Lett. 2482 (1964).
[76] *W.F. Short*, J. Chem. Soc. (London) *119*, 1446 (1921).
[77] *H. Beyer, G. Wolter, H. Lemke*, Chem. Ber. *89*, 2550 (1956).
[78] *K. Sirakawa*, J. Pharm. Soc. Jap. *79*, 1477 (1959).
[79] *R. Giuliano, G. Leonardi*, Farmaco Ed. Sci. *11*, 389 (1956); C.A. *53*, 12756 (1957).

1,4-Dihydrazino-6-(trifluoromethyl)phthalazine has been prepared from 5-(trifluoromethyl)phthalic anhydride and hydrazine hydrate by the following sequence of reactions[82]:

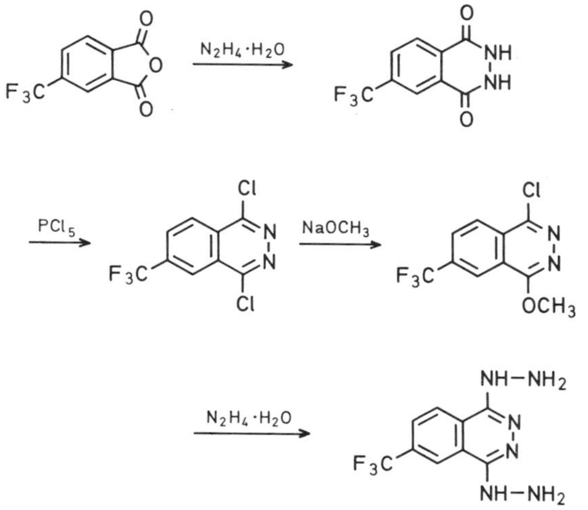

Derivatives of *2-hydrazino-1,3-thiazole* have been synthesized by reacting thiosemicarbazones or 1-acetylthiosemicarbazide with α-halo ketones[85], *e.g.*:

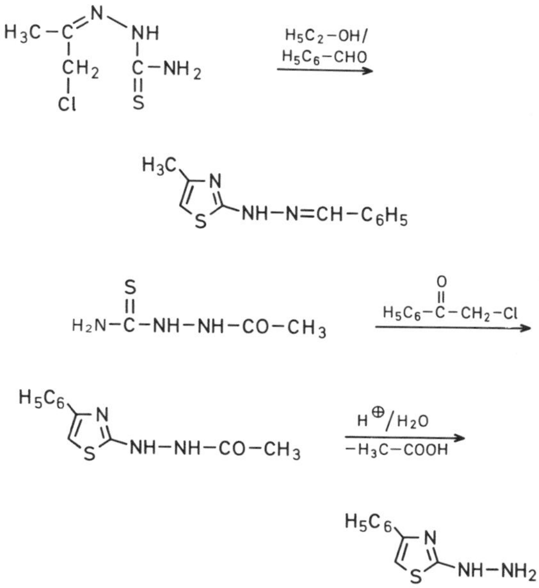

The synthesis of *2-hydrazino-1,3-thiazoles* succeeds by reacting thiosemicarbazide hydrochloride with 1,2-dichloroethyl ethyl ether[83]:

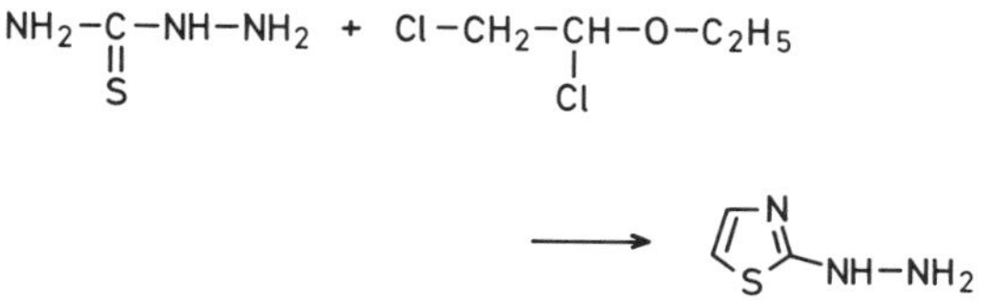

$$NH_2-C(=S)-NH-NH_2 \ + \ Cl-CH_2-CH(Cl)-O-C_2H_5 \longrightarrow$$

Analogously, the reaction of ethyl 2-chloroacetoacetate with thiosemicarbazide in neutral ethanolic solution affords *ethyl 2-hydrazino-4-methyl-1,3-thiazole-5-carboxylate*[84]:

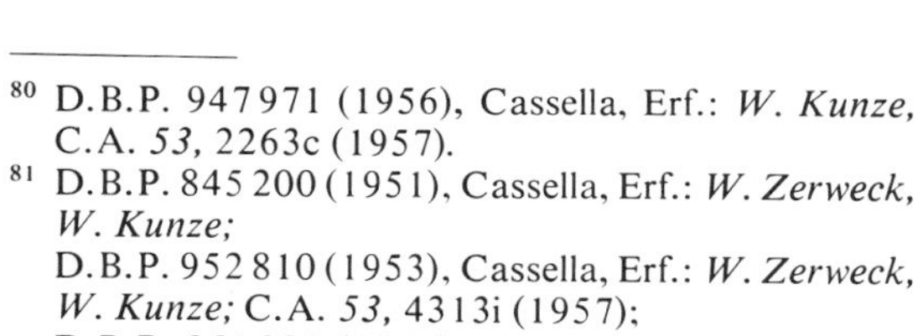

$$H_5C_2OOC-CH(CO-CH_3)-Cl \ + \ H_2N-C(=S)-NH-NH_2 \longrightarrow$$

In acid solution the reaction follows a different course.

Condensation of 4-(chloroacetyl)antipyrine with 4-methoxybenzaldehyde thiosemicarbazone affords *4-anisaldehyde (4-antipyrinyl-1,3-thiazol-2-yl)-hydrazone*[86]:

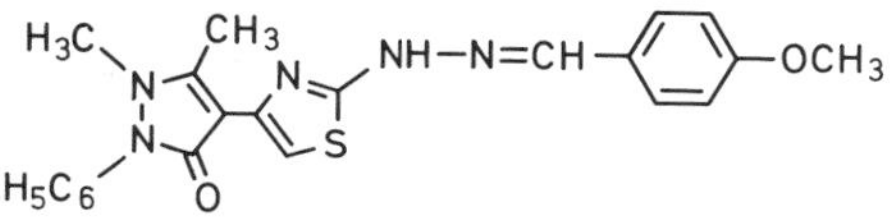

The reaction of 1,4-dicarbonyl compounds with hydrazine leads to the formation of *N-aminopyrroles*:

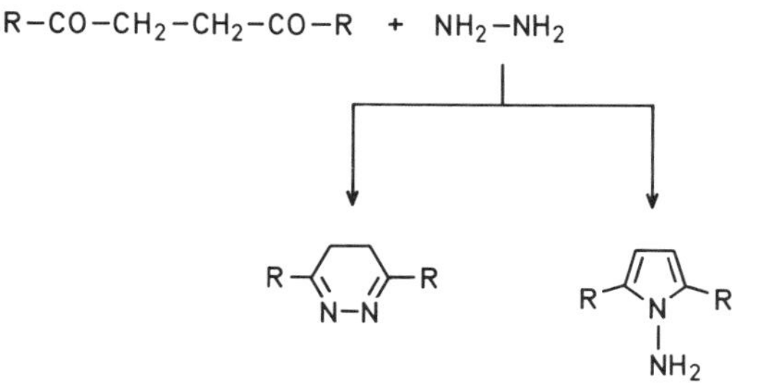

$$R-CO-CH_2-CH_2-CO-R \ + \ NH_2-NH_2$$

[80] D.B.P. 947971 (1956), Cassella, Erf.: *W. Kunze*, C.A. *53*, 2263c (1957).
[81] D.B.P. 845200 (1951), Cassella, Erf.: *W. Zerweck, W. Kunze;* D.B.P. 952810 (1953), Cassella, Erf.: *W. Zerweck, W. Kunze;* C.A. *53*, 4313i (1957); D.B.P. 951995 (1954), Cassella, Erf.: *W. Kunze;* C.A. *53*, 4313f (1957).
[82] *B. Cavalieri, E. Bellasio, A. Sardi,* J. Med. Pharm. Chem. *13*, 148 (1970).
[83] *H. Beyer, H. Höhn, W. Lässig,* Chem. Ber. *85*, 1126 (1952).

[84] *H. Beyer, W. Lässig, E. Bulka,* Chem. Ber. *87*, 1387 (1954); *H. Beyer, G. Wother,* Chem. Ber. *89*, 1652, 1655 (1956).
[85] *J. McLean, J.F. Wilson,* J. Chem. Soc. (London) 556 (1937); *G. Fodor, G. Wilheim,* Acta Chim. Acad. Sci. (Budapest) *2*, 189, 193 (1952); C.A. *48*, 3346a (1954);

In order to avoid the simultaneous formation of 4,5-dihydropyrazines, 1,4-diketones are condensed with *N*-aminophthalimide in hot acetic acid[87] and subsequently treated with hydrazine to form l-aminopyrroles, *e.g.*:

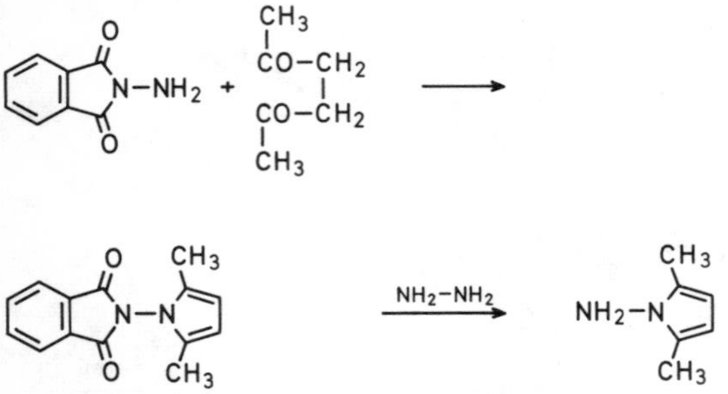

tert-Butyl carbazate was successfully employed in the synthesis of *1-amino-2,5-diphenylpyrrole* from 1,4-diphenyl-1,4-butanedione; the initially formed protected derivative was cleaved with gaseous hydrogen chloride[88].

1-Amino-2-pyridones are prepared by treating the sodium salts of a 2-pyridone with chloramine[89] or by the action of hydrazine hydrate on the corresponding *2H*-pyranones[90]. Derivatives of *1-amino-2-oxo-nicotinonitriles(1-amino-3-cyanopyridones)* are obtained by the base-catalyzed condensation of cyanoacetylhydrazide with 1,3-diketones or β-keto carboxylic acid esters[91].

1-Amino-2,3-indolediones (1-aminoisatins) are obtained by the rapid distillation of 2-hydrazino-phenylglyoxylic acids[92]. Oxalic acid 1-phenyl-2-benzylidenehydrazide hydrochloride cyclizes to *1-(benzylideneamino)-2,3-indolediones* at 150° in the presence of aluminum chloride.

H. Berger, H. Höhn, W. Lässig, Chem. Ber. *85*, 1122, 1125, 1127, 1129 (1952);
H. Berger, H. Höhn, W. Lässig, Chem. Ber. *87*, 1391, 1397 (1954);
H. Taniyama, Y. Tanaka, H. Uchida, U.Uchida, J. Pharm. Soc. Jap. *74*, 370 (1954); C.A. *49*, 5442e (1955);
G. Henseke, W. Krüger, Chem. Ber. *88*, 1642 (1955);
G. Henseke, M.Winter, Chem. Ber. *89*, 961 (1956).

[86] *N. Ergenc*, Istanbul Univ. Eczacilik, Fak. Mecm. *6*, 1 (1970); C.A. *73*, 130929c (1970).

[87] *R. Epton*, Chem. & Ind. (London) 425 (1965).

[88] *L.A. Carpino*, J. Org. Chem. *30*, 736 (1965).

[89] *K. Hegerle, H. Erlenmeyer*, Helv. Chim. Acta *39*, 1203 (1956).

[90] *El-Sayed, El-Kholy, F.K. Rafla, G. Soliman*, J. Chem. Soc. (London) 4490 (1961).

[91] *W. Ried, A. Meyer*, Chem. Ber. *90*, 2841, 2845 (1957).

[92] *P.W. Weber*, Ber. *55*, 826 (1922);
P.W. Neber, H. Keppler, Ber. *57*, 778 (1924).

Treatment with dilute hydrochloric acid does not afford free 1-aminoisatin but *indazole-3-carboxylic acid,* while dilute sodium hydroxide affords *3-phenyl-4-cinnoline-carboxylic acid*[93]:

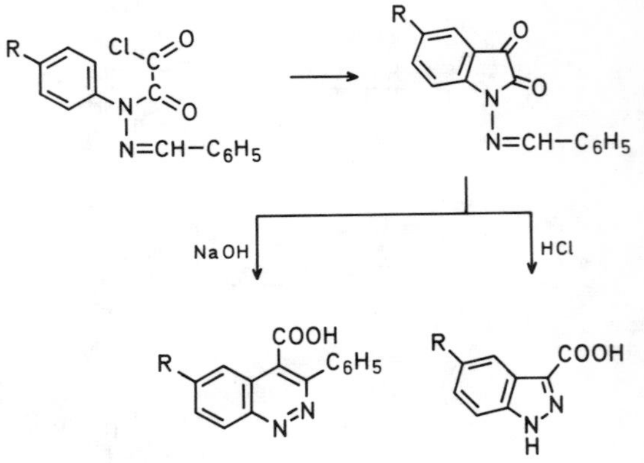

R = H, CH₃

N'-Acyl-*N*-(2-aminophenyl)hydrazines (readily obtained by catalytic reduction of the corresponding nitro compound) cyclize under acid conditions to *N*-aminobenzimidazoles by acyl migration[94].

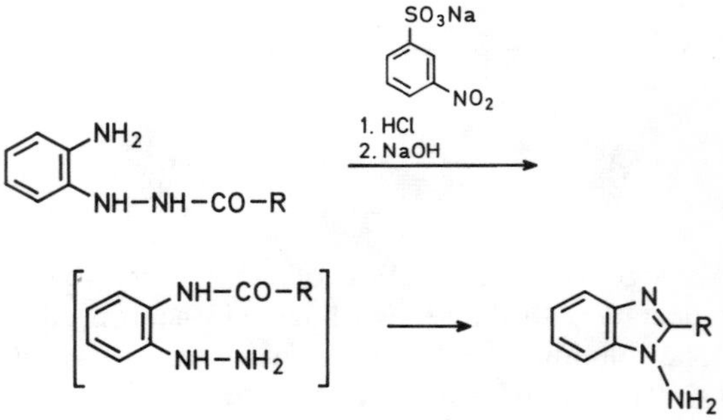

R=H   ; 1-Amino-
R=CH₃ ; 1-Amino-2-methyl-   } -benzimidazole   55 %
R=C₆H₅; 1-Amino-2-phenyl-
R=NH₂ ; 1,2-Diamino-

Analogously, derivatives of *1-(formylamino)benz-imidazole* are formed from *N*-(2-aminophenyl)-*N'*-formylhydrazine with organic acids[95]:

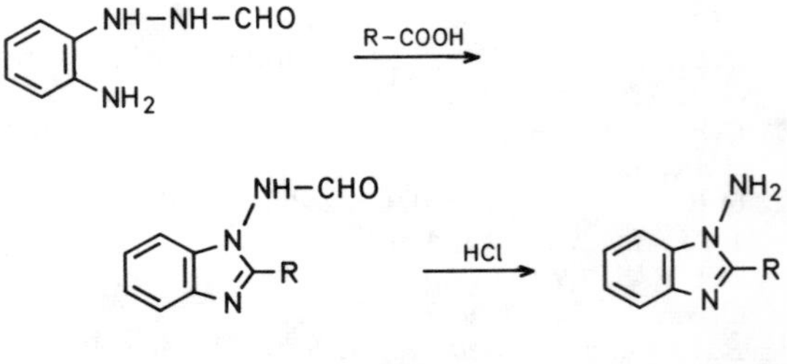

R = H;   1-Amino-   89.5%
R = C₂H₅;   1-Amino-2-ethyl-   } -benzimidazole 94.6%
R = CH₂—C₆H₅;   1-Amino-2-benzyl-   24 %

1-Amino-1,2,3-triazoles can be obtained from dibenzoylhydrazones of 1,2-diketones by oxidation with potassium hexacyanoferrate(III) or potassium permanganate[96]:

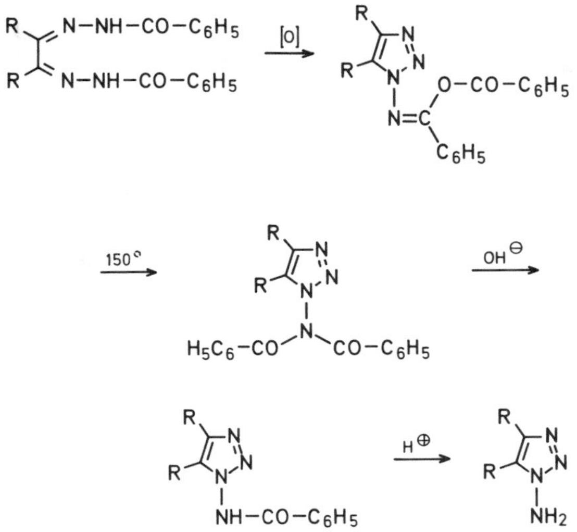

1-Amino-5-methyl-1,2,3-triazole-4-carboxylic acid was obtained by the reaction between ethyl 5-methyl-1,2,3-oxadiazole-4-carboxylate and semicarbazide[97].

3-Methyl-*3H*-diazirine-3-carboxaldehyde reacts with phenylhydrazine to give *4-methyl-1-phenyl-amino-1,2,3-triazole*[98].

Tetraazido-1,4-benzoquinone on treatment with triphenylphosphine in dichloromethane solution furnishes *2,6-bis(triphenylphosphoranylidene)-aminobenzo[1,2-d;4,5-d']bistriazole-4,8(2H, 6H)-dione* which, reacted with hydrochloric acid

in glacial acetic acid, furnishes *2,6-diaminobenzo[1,2-d;4,5-d']bistriazole-4,8(2H,6H)-dione*[99]:

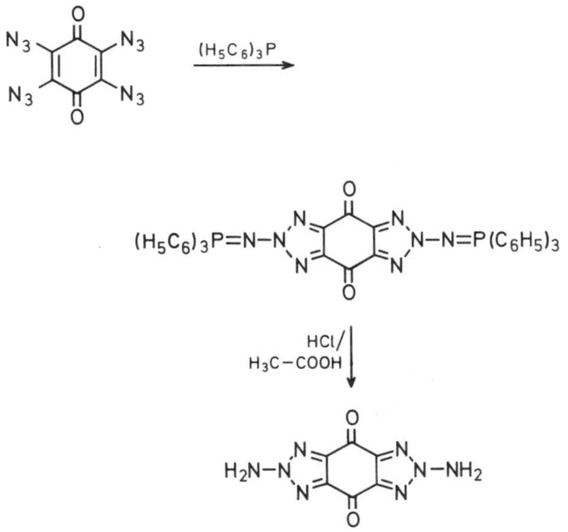

2-Amino-2H-naphtho[2,3-d]triazole-4,9-dione is obtained analogously[100].

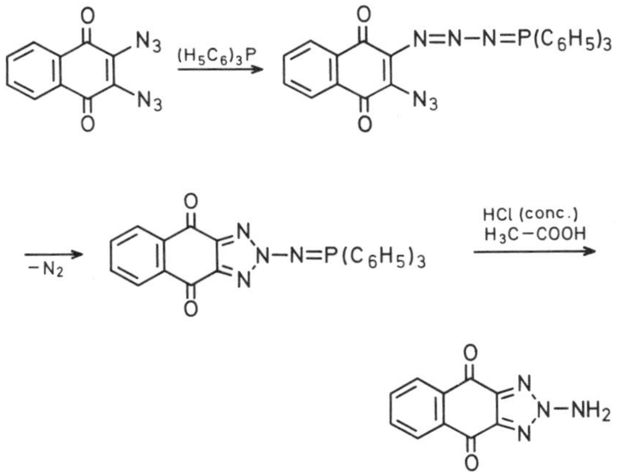

On treatment with acids or on heating 1,2-dihydro-1,2,4,5-tetrazines isomerize to *4-amino-1,2,4-triazoles*[101]:

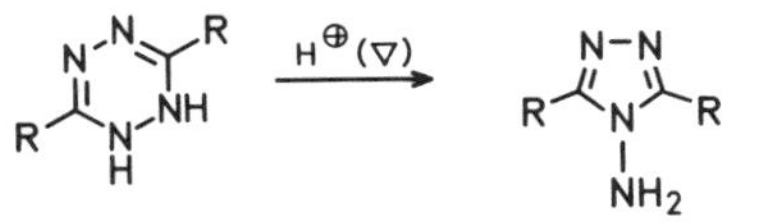

[93] R. Stollé, W. Becker, Ber. 57, 1123 (1924);
H.E. Baumgarten, J.L. Furmas, J. Org. Chem. 26, 1536 (1961);
H.E. Baumgarten, W.F. Wittman, G.I. Lehmann, J. Heterocycl. Chem. 6, 333 (1969).
[94] R.A. Abramovitch, K. Schofield, J. Chem. Soc. (London) 2326, 2328, 2332 (1955).
[95] M.N. Sheng, A.R. Day, J. Org. Chem. 28, 736 (1963).
[96] W. Bauer, Chemiker-Ztg., I, 267 (1901);
R. Stollé, Ber. 59, 1742 (1926);
H.W. Pechmann, W. Bauer, Ber. 42, 659, 669 (1909);
D.Y. Curtin, N.E. Alexandrov, Tetrahedron 19, 1697 (1963);
D.Y. Curtin, L.L. Miller, Tetrahedron Lett. 1869 (1965).
[97] L. Wolff, Justus Liebigs Ann. Chem. 325, 160 (1902);
L. Wolff, A.A. Hall, Ber. 36, 3613 (1903).
[98] E. Schmitz, C. Hoerig, C. Gründemann, Chem. Ber. 100, 2093 (1967);
E. Schmitz, C. Hoerig, Chem. Ber. 100, 2101 (1967).

[99] W.L. Mosby, M.L. Silva, J. Chem. Soc. (London) 1003 (1965).
[100] F. Dallacker, Monatsh. Chem. 91, 294 (1960).
[101] T. Curtius, A. Darapsky, F. Müller, Ber. 40, 822, 837 (1907);
T. Curtius, A. Darapsky, F. Müller, Ber. 48, 1622 (1915);
K. Macura, T. Lieser, Justus Liebigs Ann. Chem. 564, 66 (1949).

The course of the isomerization under acid conditions is dependent largely on the substituents in positions 3 and 6; thus, electronegative groups stabilize the dihydrotetrazine system, electropositive groups promote the formation of 4-amino-1,2,4-triazoles.

4-Amino-1,2,4-triazoles result also from reacting carboxylic acids or esters, amides, and nitriles with hydrazine:

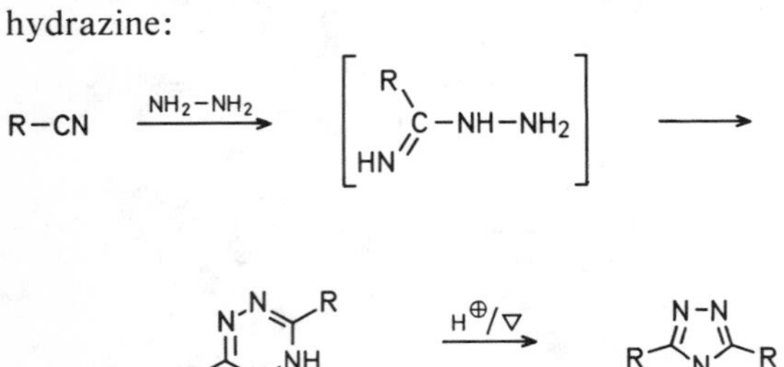

4-Amino-1,2,4-triazole itself is accessible in this way[102].

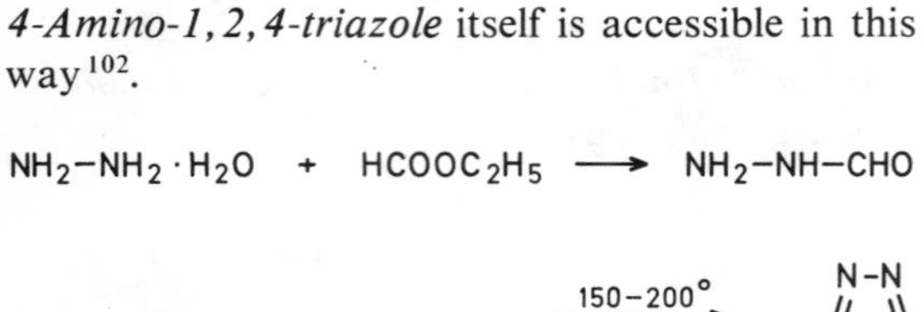

$$NH_2-NH_2 \cdot H_2O \ + \ HCOOC_2H_5 \longrightarrow NH_2-NH-CHO$$

Mercapto derivatives of 4-amino-1,2,4-triazoles have been synthesized by reacting thiocarbohydrazides with carbon disulfide in pyridine, potassium ethyl dithiocarbonate[103], formic anhydride, formamide, or triethyl orthoformate[104].

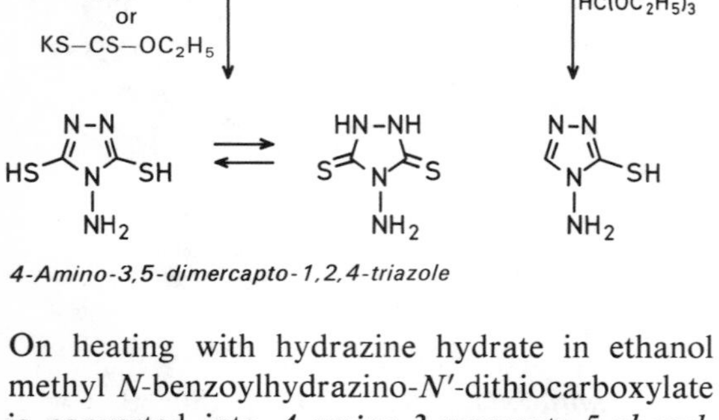

4-Amino-3,5-dimercapto-1,2,4-triazole

On heating with hydrazine hydrate in ethanol methyl *N*-benzoylhydrazino-*N'*-dithiocarboxylate is converted into *4-amino-3-mercapto-5-phenyl-1,2,4-triazole* (50%)[105].

Treatment of 1,5-bis(phenylthiocarbamoyl)-3-thiocarbohydrazide with pyridine furnishes both *4-amino-3,5-dimercapto-1,2,4-triazole* (44%) and *4-amino-3-anilino-5-mercapto-1,2,4-triazole* (33%)[106].

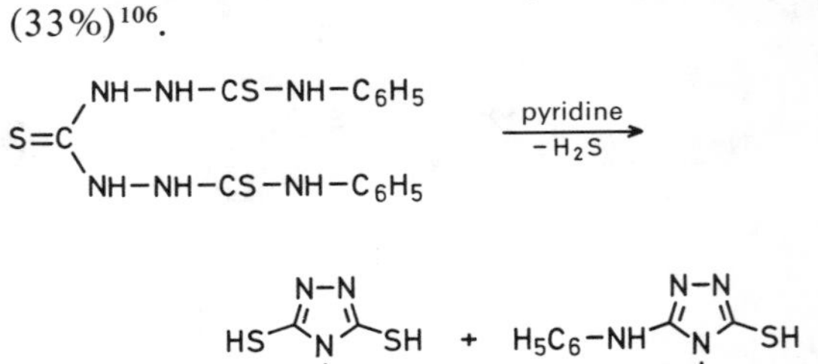

Alkaline or acid hydrolysis of ethyl 1*H*-benzotriazol-1-ylamidinoformate leads to *1-aminobenzotriazole*[107]:

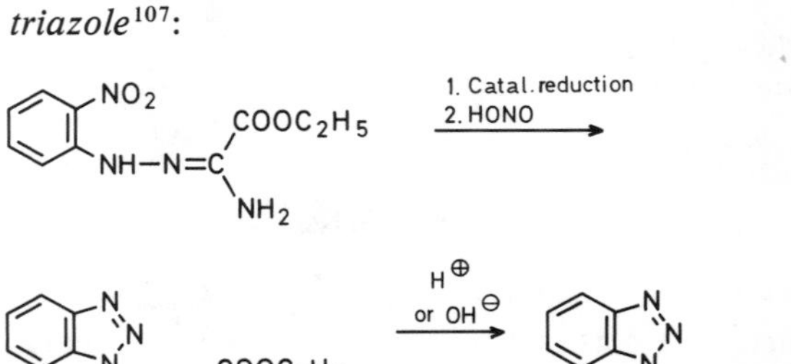

1-Amino-1H-benzotriazole and *2-amino-2H-benzotriazole* can be prepared from 2-nitroaniline *via* the following steps[108]:

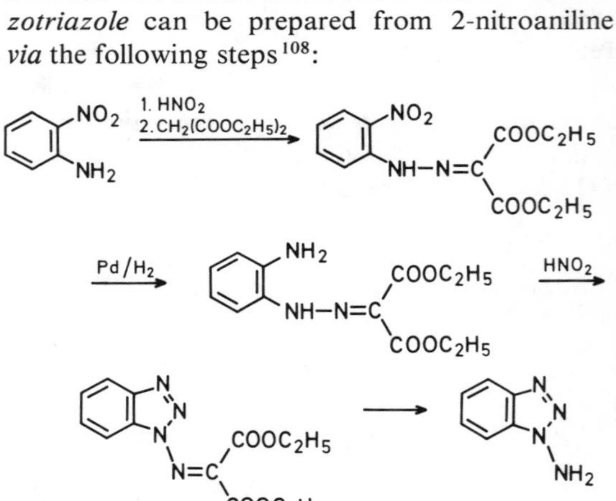

[102] *C.H.F. Allen, A. Bell*, Org. Syntheses, Coll. Vol. III, 96 (1955).
[103] *P.C. Gulia, M.K. De*, J. Chem. Soc. (London) *125*, 1215 (1924);
P.C. Gulia, M.K. De, J. Indian Chem. Soc. *1*, 144 (1924);
N. Petri, Z. Naturforsch. *16B*, 767 (1961);
J. Sandström, Acta Chem. Scand. *15*, 1295 (1961);
A.W. Lutz, J. Org. Chem. *29*, 1174 (1964).

1-Benzylideneamino-1,2,3,4-tetrazole reacts with dilute hydrochloric acid to form *1-aminotetrazole*[109].

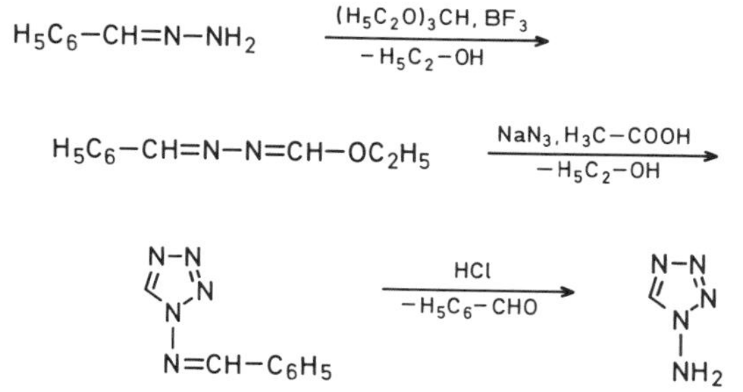

$$H_5C_6-CH=N-NH_2 \xrightarrow[-H_5C_2-OH]{(H_5C_2O)_3CH,\ BF_3}$$

$$H_5C_6-CH=N-N=CH-OC_2H_5 \xrightarrow[-H_5C_2-OH]{NaN_3,\ H_3C-COOH}$$

Thiosemicarbazide, 4-phenylthiosemicarbazide, and thiocarbohydrazide on treatment with lead(II) oxide and sodium azide in warm ethanol afford respectively *1,5-diaminotetrazole, 1-amino-5-anilinotetrazole*, and *1-amino-5-hydrazinotetrazole*[110].

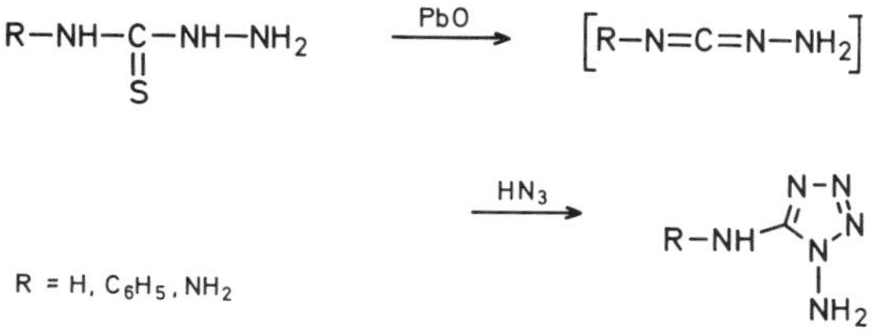

R = H, C₆H₅, NH₂

4.1.2 Transformation reactions of arylhydrazines

4.1.2.1 Reaction with carbonyl compounds

Arylhydrazines react with carbonyl compounds such as aldehydes and ketones to form the corresponding arylhydrazones (see p. 87), which can be used to purify and characterize these compounds[111]:

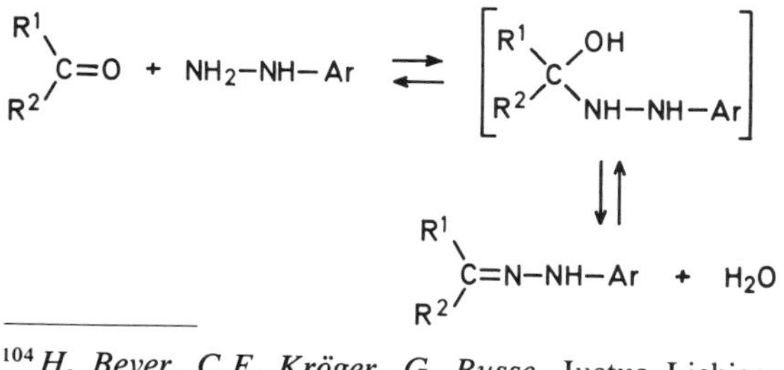

Greater details of the preparation of hydrazones are described on pp. 87, 118. Re the formation of osazones see p. 144 and ref.[183].

In the case of α-negatively substituted carbonyl compounds the reaction with arylhydrazines leads to α-substituted arylhydrazones, which are converted into *1-arylazoalkenes*[112] by 1:4-elimination (see p.145):

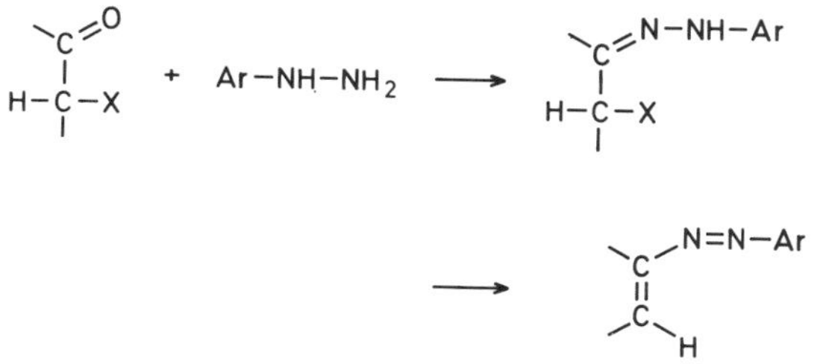

Areneazo-alkenes (and alkyl or acyl azoalkenes generally) can tautomerize to the arylhydrazone of the corresponding α,β-unsaturated ketones[113] (see p. 143)

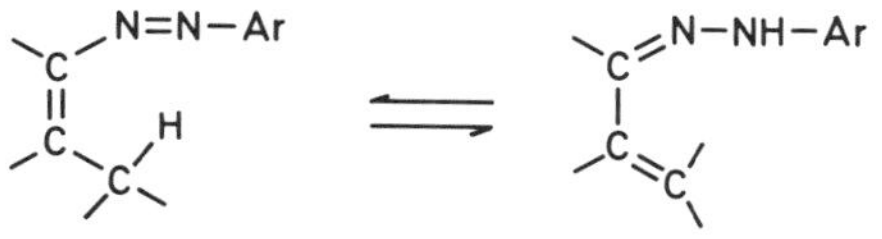

or undergo 1:4-addition with nucleophiles such as thiophenols, sulfinic acids, and hydrazines[114]. The reaction of an aryl azoalkene with an arylhydrazine

[104] *H. Beyer, C.F. Kröger, G. Busse,* Justus Liebigs Ann. Chem. *637*, 135 (1960);
C.F. Kröger, E. Tenor, H. Beyer, Justus Liebigs Ann. Chem. *643*, 121 (1961).

[105] *E. Hoggarth,* J. Chem. Soc. (London) 4811 (1952).

[106] *A. Dornow, H. Pancksch,* Chem. Ber. *99*, 81 (1966).

[107] *R. Trave, G. Bianchetti,* Atti Acad. Naz. Lincei Classe Sci. Fis., Mat. Natur., Rend. *28*, 652 (1960).

[108] *C.D. Campbell, C.W. Ress,* J. Chem. Soc. C 742 (1969).

[109] *I. Hagedorn, H.D. Winkelmann,* Chem. Ber. *99*, 850 (1966).

[110] *R. Stollé, E. Gaertner,* J. Prakt. Chem. *132*, 209 (1931).

[111] *E. Fisher, J. Hirschberger,* Ber. *21*, 1805 (1888);
H. Rupe, A. Gassmann, Helv. Chim. Acta *19*, 569 (1936);
J.C. Sowden, H.O.L. Fischer, J. Amer. Chem. Soc. *69*, 1963 (1947);
A.E. Arbuzov, Y.P. Kitaev, Kirov's Inst. Chem. Technol. Kazan *23*, 60 (1957); C.A. *52*, 9980a (1958).

[112] *L. Caglioti, P. Grasselli, G. Rosini,* Tetrahedron Lett. 4545 (1965);
L. Caglioti, G. Rosini, F. Rossi, J. Amer. Chem. Soc. *88*, 3865 (1966);
J. Buckingham, R.D. Guthrie, Chem. Commun. 781 (1966);
A. Hassner, P. Catsoulacos, Chem. Commun. 121, (1967);
H. Simon, G. Heubach, H. Wacker, Chem. Ber. *100*, 3101 (1967).

[113] *A. Dondoni, G. Rosini, G. Mossa, L. Caglioti,* J. Chem. Soc. B 1404 (1968);
H. Simon, S. Brodka, Tetrahedron Lett. 4991 (1969).

[114] *L. Caglioti, G. Rosini, F. Rossi,* J. Amer. Chem. Soc. *88*, 3865 (1966);
L. Caglioti, A. Dondoni, G. Rosini, Chim. Ind. (Milan) *50*, 122 (1968);
H. Simon, W. Moldenbauer, Ber. *101*, 2124 (1968);
L. Caglioti, G. Rosini, Chem. & Ind. (London), 1093 (1969).

leads to an arylhydrazinoarylhydrazone (see p. 145) which is also an intermediate compound in the formation of osazones (see above).

Arylsulfonylhydrazines react with carbonyl compounds having a leaving group X in the α-position to give α-substituted tosylhydrazones (p. 121), which yield tosyl azoalkenes by 1:4-elimination of $H-X$ (see p. 123). Arylhydrazones of ketones or aldehydes with a reactive group in the β-position in general react further to heterocycles (see p. 91).

Arylhydrazines can be *acylated* by reacting with free carboxylic acids, the unsubstituted nitrogen being attacked first. Many acids react at once, in other instances the arylhydrazide can be obtained better by reacting the arylhydrazine with a suitable acyl derivative such as an acid chloride or an ester. For polyfunctional compounds, which require mild conditions in order to avoid undesired changes in structure, acyl derivatives of the type RCOOX=Y[115] are recommended. These compounds, which can be prepared quickly from carboxylic acids by reacting with suitable reagents such as carbonyldiimidazole, dicyclohexylcarbodiimide, *etc.*, form the corresponding hydrazides in high yield by nucleophilic attack on the hydrazino group.

2-Phenylhydrazides readily break down on oxidation in aqueous medium to form the carboxylic acid and benzene. This reaction has been applied in peptide chemistry as a conservative method for eliminating protective groups[116].

$$R-CO-NH-NH-C_6H_5 \xrightarrow[\text{MnO}_2]{\text{CuSO}_4 \text{ or}} R-CO-N=N-C_6H_5$$

$$\xrightarrow{\text{H}_2\text{O}} R-COOH + \left[H_5C_6-N=NH\right] \longrightarrow C_6H_6 + N_2$$

4.1.2.2 Splitting off of the hydrazino group

Exchanging the hydrazino group ($-NH-NH_2$) of an arylhydrazine against hydrogen can be effected by oxidizing agents *via* an aryldiazene (aryldiimide) first reaction stage. Aryldiazenes are extremely unstable compounds that decompose rapidly into nitrogen and the corresponding arene (see Vol. IV of this handbook).

$$Ar-NH-NH_2 \xrightarrow{[O]} \left[Ar-N=NH\right] \longrightarrow N_2 + ArH$$

---

[115] *J. Rudinger*, Pure Appl. Chem. *7*, 335 (1963).
[116] *R.M. Hann, C.S. Hudson*, J. Amer. Chem. Soc. *56*, 957 (1934);
R.B. Kelley, G. Umbreit, W.F. Ligget, J. Org. Chem. *29*, 1273 (1964).

There is a difference of four electrons (or four hydrogen atoms) between the oxidation levels of hydrazine and nitrogen. Not surprisingly, the oxidation can occur in different stages and the products obtained vary with both the amount and nature of the oxidizing agent. Monosubstituted hydrazines usually furnish products resulting from two-electron or four-electron oxidations. A two-electron oxidation gives rise to diazenes or their tautomers, $R-N=NH$ or $R^{\oplus}-NH=N^{\ominus}$, which decompose rapidly into RH and nitrogen, while a four-electron oxidation leads to diazonium salts or their transformation products (see p. 177).

A variety of reagents including copper(II) sulfate, periodate, iron(III) chloride, potassium hexacyanoferrate(III), mercury(II) oxide, and oxygen, under suitable conditions, convert arylhydrazines into aromatic compounds and nitrogen.

In acid solution and using excess oxidizing agent four-electron oxidation takes place with formation of amines, azides, or tetrazenes[117]:

$$H_5C_6-NH-NH_2\cdot H_2SO_4 \xrightarrow{\text{HgO}} \left[H_5C_6-\overset{\oplus}{N}\equiv N\right]\cdot HSO_4^{\ominus}$$

$$+ \left[H_5C_6-N=N-NH-NH-C_6H_5\right]$$

$$\longrightarrow H_5C_6-N_3 + H_5C_6-NH_2$$

Replacement of the arylsulfonylhydrazino group by hydrogen is based on the fact that the arylsulfonyl derivatives of hydrazines (and of hydrazones) decompose on heating or on treatment with alkali to give rise to a redox process in which the original sulfonate group is reduced to the corresponding sulfinate ion (two-electron oxidation):

$$Ar-NH-NH_2 \xrightarrow{\text{ArSO}_2\text{Cl}} Ar-NH-NH-SO_2-Ar$$

$$\longrightarrow \left[Ar-N=NH\right] + ArSO_2H \longrightarrow ArH + N_2$$

Decomposition of benzenesulfonic acid hydrazides occurs under mild treatment with dilute sodium hydroxide, for example, the formation of *pyrimidine* from 2,4-bis[2-(phenylsulfonyl)hydrazino]pyrimidine[118]:

In relation to the above-described redox process the $-NH-NH-SO_2-AR$ moiety of a molecule is equivalent to a hydrogen atom in that it is easily decomposed into hydride hydrogen by treatment with alkaline reagents. An extension of this concept is the well-known McFayden-Stevens reaction. Here the benezenesulfonic acid 2-acylhydrazides are decomposed with alkali to form the corresponding *aldehydes*[119]:

$$R-CO-NH-NH-SO_2-Ar \xrightarrow{\text{OH}^\ominus}$$

$$R-CHO \ + \ N_2 \ + \ ArSO_2{}^\ominus$$

Oxidation of phenylhydrazines with copper(II) sulfate in concentrated hydrochloric acid or hydrobromic acid or with iodine-potassium iodide affords the corresponding *haloaromatic compounds* (see Vol. VII). Treatment of arylhydrazines with mercury(II) acetate in the presence of catalytic amounts of copper(II) salts leads to *arylmercury(II) compounds* (see Vol. VIII, p. 54), which on a repeated reaction with arylhydrazine are converted into *diarylmercury, e.g.*[120]:

$$3 \ H_5C_6-NH-NH_2 \ \longrightarrow$$

$$3 \ H_5C_6-NH-NH-Hg-O-CO-CH_3$$

$$\longrightarrow \ 3 \ H_5C_6-N=N-Hg-O-CO-CH_3$$

$$\xrightarrow{-3\,N_2} \ 3 \ H_5C_6-Hg-O-CO-CH_3$$

$$\xrightarrow[-Hg,\,-3\,H_3C-COOH,\,-N_2]{H_5C_6-NH-NH_2} 2 H_5C_6-Hg-C_6H_5$$

Organoarsenic compounds (see Vol. VII) can be obtained from phenylhydrazine and arsenic acid.

Oxidation of arylhydrazines in the presence of acrylonitrile with copper(II) chloride leads to *3-aryl-2-chloropropionitriles, e.g.*[121]:

$$H_5C_6-NH-NH_2 \xrightarrow{[o]} C_6H_5 \cdot \xrightarrow{H_2C=CH-CN}$$

$$H_5C_6-CH_2-\overset{\bullet}{C}H-CN \xrightarrow{CuCl_2} H_5C_6-CH_2-\overset{\overset{\text{Cl}}{|}}{C}H-CN$$

4.1.2.3 Reduction of *N*-substituted arylhydrazines to amines

The reduction of hydrazines by Raney nickel or with metals in strong acids results in a reductive

cleavage of the *N–N* bond to produce amines. Details are discussed on pp. 510–520.

Treatment of arylsulfonic acid hydrazones with suitable reducing agents leads to alkanes, nitrogen and arylsulfinic acid.

This reaction may be considered as an extension of the exchange of arylsulfonylhydrazino groups against hydrogen.

$$Ar-SO_2-NH-N=C\overset{\displaystyle R^1}{\underset{\displaystyle R^2}{\big<}} \longrightarrow Ar-SO_2-NH-NH-\overset{\overset{\displaystyle R^1}{|}}{\underset{\underset{\displaystyle H}{|}}{C}}-R^2$$

$$\longrightarrow \ Ar-SO_2H \ + \ N_2 \ + \ \overset{\displaystyle H}{\underset{\displaystyle H}{\big>}}C\overset{\displaystyle R^1}{\underset{\displaystyle R^2}{\big<}}$$

In arylsulfonic acid 2-acylhydrazides, too, the carbonyl group is reduced to alkane by lithium tetrahydroaluminate (see Vol. IV).

$$R-CO-NH-NH-Ts \xrightarrow{LiAlH_4} R-CH_3 \ + \ N_2 \ + \ TsH$$

Two hydrogen atoms of the methyl group originate from the lithium tetrahydroaluminate and the third from water added during working-up of the reaction mixture. It is possible, therefore, to start from a carboxylic acid and to obtain a singly, doubly, or triply deuterated methyl group in dependence on the nature of the deuterated reagent employed[122].

[117] *F.D. Chattaway*, J. Chem. Soc. (London) *91*, 1323 (1907);
F.D. Chattaway, J. Chem. Soc. (London) *93*, 852 (1908);
L. Maaskant, Rec. Trav. Chim. Pays-Bas *56*, 211 (1937);
B. Vis, Rec. Trav. Chim. Pays-Bas *58*, 387 (1939);
R.H. Poirier, F. Beenington, J. Org. Chem. *19*, 1187 (1954);
D.J. Cram, J.S. Bradshaw, J. Amer. Chem. Soc. *85*, 1108 (1963).
[118] *M.P.V. Boarland, J.F.W. McOmie, R.M. Timms*, J. Chem. Soc. (London) 4693 (1952).
[119] *J.S. McFadyen, T.S. Stevens*, J. Chem. Soc. (London) 584 (1936);
E. Moseting, Org. Reactions 8, 232 (1954);
M.S. Newman, E.G. Caflisch, J. Amer. Chem. Soc. *80*, 862 (1958);
M. Sprecher, M. Feldikimel, M. Wilchek, J. Org. Chem. *26*, 3664 (1961).
[120] *O.A. Seide, S.M. Scherlin, G.J. Bros*, J. Prakt. Chem. (2), *138*, 59, 66 (1933).
[121] *F. Minisci, R. Galli, M. Cecere*, Gazz. Chim. Ital. *95*, 751 (1965).
[122] *L. Caglioti, P. Grasselli, G. Zubiani*, Chim. Ind. (Milan) *47*, 62 (1965);
L. Caglioti, Tetrahedron *22*, 487 (1966).

4.1.2.4 Oxidation of N-substituted arylhydrazines

The oxidation level of hydrazine is intermediate between that of ammonia and nitrogen. In general, all types of substituted hydrazines can be oxidized even under mild conditions; the specific reactivity depends on the class of compound and the oxidizing agent employed.

There is a difference of four electrons between the oxidation levels of hydrazine and nitrogen and, consequently, either two-electron or four-electron oxidations occur (see p. 84).

Symmetrical diaryl-substituted hydrazines are easily oxidized to azo compounds by a two-electron oxidation (cf. p. 132) using the usual oxidizing agents[123–128]

$$Ar-NH-NH-Ar \xrightarrow{[o]} Ar-N=N-Ar$$

or to azoxy compounds (cf. p. 138) by peroxyacetic acid[129]:

$$Ar-NH-NH-Ar \xrightarrow{H_3C-COOOH} Ar-N=N-Ar \atop O$$

Oxidation of N'-acyl-N-arylhydrazines leads to *arylazoalkanes* (p. 138) which rapidly isomerize to the corresponding *arylhydrazones* (p. 142).

$$Ar-NH-NH-CH_2-CH_3 \xrightarrow{[o]} Ar-N=N-CH_2-CH_3$$

$$\longrightarrow Ar-NH-N=CH-CH_3$$

The behavior of N'-acyl-N-arylhydrazines toward oxidation depends markedly on the oxidizing agent employed. Treatment of N'-acetyl-N-phenylhydrazine with copper(II) acetate in boiling ethanol results in N'-*acetyl-N,N-diphenylhydrazine*[130]:

$$H_5C_6-NH-NH-CO-CH_3 \xrightarrow{Cu^{2\oplus}} (H_5C_6)_2N-NH-CO-CH_3$$

An improved yield of N'-acyl-N,N-diarylhydrazines is obtained if an independent source of phenyl radicals, for instance as phenylhydrazine or benzenediazonium salt, is added.

The reaction has been extended to other acylarylhydrazines[131]. Employing alternative oxidizing agents such as mercury(II) oxide[132], iron(III) chloride[133], potassium permanganate[134], chromium-(VI) oxide[135], lead(IV) acetate[136], N-bromosuccinimide[137], leads to acyl azoaromatic compounds as well[138] (see p. 144):

$$R-CO-NH-NH-Ar \xrightarrow{[o]} R-CO-N=N-Ar$$

Arylazosulfonates are formed by oxidizing salts of N-arylhydrazine-N'-sulfonic acids with hydrogen peroxide[139], mercury(II) oxide[140], potassium chromate[140], or ammoniacal silver(II) oxide[141]:

$$Ar-NH-NH-SO_3^{\ominus} \xrightarrow{[o]} Ar-N=N-SO_3^{\ominus}$$

Aryldiazolsulfones are obtained by oxidation of arylsulfonic acid hydrazides with mercury(II) oxide[142], lead(IV) oxide[143], sodium hypobromite[144], or bromine in acetic anhydride[144].

[131] S. Goldschmidt, J. Bader, Justus Liebigs Ann. Chem. 473, 150 (1929);
S.G. Cohen, J. Nicholson, J. Amer. Chem. Soc. 86, 3892 (1964).
[132] G. Heller, Justus Liebigs Ann. Chem. 263, 279 (1891);
W. Mackwald, P. Wolff, Ber. 25, 3116 (1892);
O. Widman, Ber. 28, 1925 (1895);
M. Bush, H. Holzmann, Ber. 34, 320 (1901);
M. Bush, H. Brandt, Ber. 39, 1395 (1906).
[133] G. Young, J. Chem. Soc. (London) 67, 1067 (1895);
M. Bush, R. Frey, Ber. 36, 1362 (1903);
E. Jolles, B. Bini, Gazz. Chim. Ital. 68, 510 (1938);
Chem. Zentr. I, 1750 (1939).
[134] J. Thiele, Ber. 28, 2599 (1895);
O. Widman, Ber. 28, 1925 (1895);
E. Bamberger, Ber. 35, 1429 (1902).
[135] O. Widman, Ber. 28, 1925 (1895);
E. Jolles, Gazz. Chim. Ital. 68, 496 (1938); Chem. Zentr. I, 1749 (1939).
[136] W. Borsche, Justus Liebigs Ann. Chem. 334, 143 (1904).
[137] L.A. Carpino, P.H. Terry, P.J. Crowley, J. Org. Chem. 26, 4336 (1961).
[138] E. Fahr, H. Lind, Angew. Chem. Intern Ed. Engl. 5, 372 (1966).
[139] E. Bomberger, A. Meyenberg, Ber. 30, 374 (1897).
[140] E. Fischer, Justus Liebigs Ann. Chem. 190, 98 (1878).
[141] J. Tröger, W. Hille, P. Vasterling, J. Prakt. Chem. (2), 72, 511 (1905);
J. Tröger, G. Puttkammer, Ber. 40, 206 (1907).

[123] T. Curtius, K. Heidenreich, Ber. 27, 773 (1894).
[124] N. Rabjohn, Org. Syntheses, Coll. Vol. III, 375 (1955).
[125] J. Thiele, Justus Liebigs Ann. Chem. 271, 127 (1892).
[126] B.W. Konings, Ber. 10, 1532 (1877).
[127] H. Bock, G. Rudolph, E. Baltin, Chem. Ber. 98, 2054 (1965).
[128] C.W. Rees, R.C. Stozz, J. Chem. Soc. C 1474 (1969).
[129] B. Newbold, J. Org. Chem. 27, 3919 (1962).
[130] J. Tafel, Ber. 25, 413 (1892);
L. Gatterman, E.S. Johnson, R. Holzle, Ber. 25, 1075 (1892).

2-Aryl-1-phosphoryl hydrazides afford the corresponding azo derivatives on oxidation with mercury(II) oxide[145].

$$Ar-NH-NH-PO(OR)_2 \xrightarrow{HgO} Ar-N=N-PO(OR)_2$$

2-Aryl-1-trialkylsilylazo compounds are obtained by oxidation of N'-aryl-N-(trialkylsilyl)hydrazines with di-*tert*-butyl peroxide at elevated temperature[146]:

$$\begin{array}{c} R \\ R^1-Si-NH-NH-C_6H_5 \\ R^1 \end{array} \xrightarrow{(H_3C)_3C-O-O-C(CH_3)_3} $$

$$\begin{array}{c} R \\ R^1-Si-N=N-C_6H_5 \\ R^1 \end{array}$$

R = Ph, R¹ = CH₃
R = R¹ = C₂H₅

Asymmetrically disubstituted hydrazines on treatment with oxidizing agents such as iron(III) chloride or potassium permanganate yield secondary amines and nitrogen[147]:

$$2\ R_2N-NH_2 \xrightarrow{[O]} 2\ R_2NH + N_2$$

Electrochemical oxidation of N,N-diarylhydrazines in organic medium leads to diazenium cations under acid conditions and to tetraphenyltetrazene under basic conditions[148], e.g.:

$$(H_5C_6)_2N-\overset{\oplus}{N}H_3 \rightleftharpoons 2\,e^\ominus + 2\,H^\oplus + \left[(H_5C_6)_2N=NH\right]^\oplus$$

1,1-Diphenyl-diazenium cation

$$(H_5C_6)_2N-NH_2 + 2\ \overset{}{\underset{N}{\bigcirc}} \longrightarrow 2\,e^\ominus$$

$$+\ 2\ \overset{\oplus}{\underset{\underset{H}{N}}{\bigcirc}} + (H_5C_6)_2\overset{\oplus}{N}=\overset{\ominus}{N}$$

$$2\ (H_5C_6)_2\overset{\oplus}{N}=\overset{\ominus}{N} \longrightarrow (H_5C_6)_2N-N=N-N(C_6H_5)_2$$

1,1,4,4-Tetraphenyltetrazene

Tetrazenes are obtained also by treating N,N-diarylhydrazines and N-alkyl-N-arylhydrazines with conventional oxidizing agents[149], e.g.:

$$2\ \begin{array}{c} H_5C_6-N-NH_2 \\ CH_3 \end{array} \xrightarrow{[O]} \begin{array}{c} H_5C_6-N-N=N-N-C_6H_5 \\ CH_3 \qquad CH_3 \end{array}$$

1,4-Dimethyl-1,4-diphenyltetrazene

Triaryl-substituted hydrazines are less easily oxidized than the mono- and disubstituted compounds, as only one $N-H$ bond can be oxidized with formation of a free radical, which dimerizes to the tetrazene[150]:

$$2\ \begin{array}{c} Ar-N-NH \\ R\ \ R \end{array} \xrightarrow{[O]} 2\ \begin{array}{c} Ar-N-N\cdot \\ R\ \ R \end{array}$$

$$\longrightarrow \begin{array}{c} Ar-N-N-N-N-Ar \\ R\ \ R\ \ R\ \ R \end{array}$$

Oxidation of tetraarylhydrazines with strong oxidizing agents such as bromine and lead(IV) oxide produces a colored radical cation, which on reduction with potassium iodide regenerates the initial hydrazine[151].

$$\begin{array}{c} Ar \quad\quad Ar \\ \diagdown \ddot{N}-\ddot{N} \diagup \\ Ar \diagup \quad \diagdown Ar \end{array} \underset{KJ}{\overset{PbO_2, H^\oplus}{\rightleftarrows}} \begin{array}{c} Ar \quad\quad Ar \\ \diagdown \ddot{N}-\overset{\oplus}{N}\cdot \diagup \\ Ar \diagup \quad \diagdown Ar \end{array}$$

4.1.3 Preparation of arylhydrazones

Arylhydrazones are derived from either aldehydes **1** or ketones **2**

$$\begin{array}{c} R^2 \\ \diagdown C=N \\ H \diagup \end{array}\begin{array}{c} NH-Ar \end{array}$$

1a

$$\begin{array}{c} R^2 \\ \diagdown C=N \\ R^1 \diagup \end{array}\begin{array}{c} NH-Ar \end{array}$$

2a

$$\begin{array}{c} R^2 \\ \diagdown C=N \\ H \diagup \quad\quad NH-Ar \end{array}$$

1b

$$\begin{array}{c} R^2 \\ \diagdown C=N \\ R^1 \diagup \quad\quad NH-Ar \end{array}$$

2b

[142] *W. Königs*, Ber. *10*, 1533 (1877).

[143] *M. Claasz*, Ber. *44*, 1415 (1911).

[144] *R. Rane*, Dissertation, Universität Marburg 1952.

[145] *F. Suckfüll, H. Hanbrich*, Angew. Chem. *70*, 238 (1958).

[146] *H. Watanabe, K. Inone, Y. Nagai*, Bull. Chem. Soc. Japan, *43*, 2660 (1970).

[147] *R.H. Poirer, F. Bennington*, J. Org. Chem. *19*, 1157 (1954).

[148] *G. Cauquis, M. Genies*, Tetrahedron Lett. 2903 (1970).

[149] *R.L. Hinnaw, K.L. Hannu*, J. Amer. Chem. Soc. *81*, 3294 (1949);
W.R. McBride, E.M. Bens, J. Amer. Chem. Soc. *81*, 5546 (1959);
L.A. Carpino, A.A. Santelli, R.W. Murray, J. Amer. Chem. Soc. *82*, 2728 (1960);

Except with derivatives of symmetrical ketones there is also the possibility of geometrical isomers (**1a–1b** and **2a–2b**). These *syn* and *anti* isomers can in many cases be distinguished by their *NMR* spectra [152–156]. Many phenylhydrazones are phototropic in turning colored on exposure to light and reverting to the colorless form when placed in the dark [157].

The reaction of aldehydes, ketones, and quinones with arylhydrazines is acid-catalyzed and occurs in steps [158]. It is not always necessary to use the carbonyl compound itself; several precursors of carbonyl compounds will react with arylhydrazines to give the hydrazones. For example, α,α-dichlorotoluene affords *benzaldehyde phenylhydrazone* with phenylhydrazine [159, 160], 9,9-dichlorofluorene gives *fluorenone phenylhydrazone* [161]. In addition, acetals and acetates of aldehydes often can be converted into the phenylhydrazones [162, 163]. A number of aldehydes and ketone derivatives too can form phenylhydrazones, among them thioketones [164], imines [165], anils [166], oximes [167], and even

other hydrazones [168, 169]. This latter case may be considered as a *trans*-hydrazonization.

Acid-sensitive heterocyclic compounds may open during treatment with phenylhydrazine and acid to give the phenylhydrazones of open-chain aldehydes and ketones, *e.g.*:

5,6-Dihydro-4*H*-pyran ⟶
         5-Hydroxyvaleraldehyde phenylhydrazone [170]

and

2,1-Benzisoxazole ⟶
         2-Aminobenzaldehyde phenylhydrazone [171, 172]

1-(2,4-Dinitrophenyl)-pyridinium chloride ⟶   *5-(2,4-Dinitro-anilino)-2,4-pentadienal phenylhydrazone* [173, 174]

4,5-Dihydro-2-methylfuran ⟶
         5-Hydroxy-2-pentanone phenylhydrazone [175]

5-(2-Chlorophenyl)-4,5-dihydro-2,4-diphenyl-1,3,4-oxadiazole ⟶ *2-Chlorobenzaldehyde phenylhydrazone* [176]

Simultaneously acting on naphthols with arylhydrazine and sodium hydrogen sulfite reduces them to *tetrahydronaphthalene hydrazones* [177], *e.g.*:

Sodium 1,2,3,4-tetrahydro-3-oxo-1-naphthalene-sulfonate diphenyl-hydrazone

C.G. Overberger, N.P. Mariullo, R.G. Hiskey, J. Amer. Chem. Soc. *83*, 1374 (1961);
W.R. McBride, L.P. Herin, J. Org. Chem. *27*, 417, 2423 (1962);
G.S. Hammond, B. Seidel, R.E. Pincock, J. Org. Chem. *28*, 3277 (1963).

[150] *A. Michaelis, C. Claessen*, Ber. *22*, 2238 (1889).
[151] *R.H. Poirer, F. Benington*, J. Org. Chem. *19*, 1157, 1847 (1954);
A.E. Arbuzov, F.G. Valitova, A.V. Ilyosov, B.M. Kozyrev, Yu.B. Yablokov, Dokl. Akad. Nauk SSR, *147*, 99 (1962); Engl. 949.
[152] *R.M. Silverstein, J.W. Schoolery*, J. Org. Chem. *25*, 1355 (1960).
[153] *G.J. Karabatsos, J.D. Graham, F.M. Vane*, J. Amer. Chem. Soc. *84*, 753 (1962).
[154] *G.J. Karabatsos, B.L. Shapiro, F.M. Vane, J.S. Fleming, J.S. Ratka*, J. Amer. Chem. Soc. *85*, 2784 (1963).
[155] *G.J. Karabatsos, R.A. Taller*, J. Amer. Chem. Soc. *85*, 3624 (1963).
[156] *Houben-Weyl*, Methoden der organischen Chemie, Bd. X/2, p. 479, Georg Thieme Verlag, Stuttgart 1967.
[157] *G.H. Brown, W.G. Shaw jr.*, J. Org. Chem. *24*, 132 (1959).
[158] *J.D. Roberts, M.C. Caserio*, Basic Principles of Organic Chemistry, p. 452, W.A. Benjamin, New York 1965.
[159] *S. Bodforss*, Ber. *69*, 668 (1926).
[160] *L.A. Jones, C.K. Hancock, R.B. Seligman*, J. Org. Chem. *26*, 230 (1961).
[161] *J. Schmidt, H. Wagner*, Ber. *43*, 1801 (1910).
[162] *K. Hess, C. Uibring*, Ber. *50*, 367 (1917).
[163] *A. Schönberg, K. Praefke*, Chem. Ber. *99*, 196 (1966).
[164] *R. Mayer, J. Morgenstern, J. Fabian*, Angew. Chem. *76*, 166 (1964).

[165] *R. Cantarel, J. Guenzet*, Bull. Soc. Chim. France 1549 (1960).
[166] *G. Reddelien*, Ber. *46*, 2717 (1913).
[167] *W. Meister*, Ber. *40*, 3443 (1907).
[168] *J. Wolf*, Chem. Ber. *86*, 840 (1953).
[169] *H. El Khadem*, J. Org. Chem. *29*, 2073 (1964).
[170] *C. Glacet*, Compt. Rend. *234*, 635 (1952).
[171] *G. Heller*, Ber. *36*, 4184 (1903).
[172] *S. Gabriel, E. Leupold*, Ber. *31*, 2186 (1898).
[173] *Th. Zincke, Fr. Krollpfeiffer*, Justus Liebigs Ann. Chem. *408*, 285 (1915).
[174] *H. Beyer, E. Thieme*, J. Prakt. Chem. [4] *31*, 293 (1966).
[175] *F. Boberg, A. Marei, H. Stegemeyer*, Justus Liebigs Ann. Chem. *655*, 110 (1962).
[176] *R. Huisgen, R. Grashey, M. Seidel, H. Knupper, R. Schmidt*, Justus Liebigs Ann. Chem. *658*, 169 (1962).
[177] *H. Seeboth, D. Bärwoulf, B. Becker*, Justus Liebigs Ann. Chem. *683*, 87 (1965).

Acetylenes add phenylhydrazines with formation of hydrazones of the corresponding methylene ketones[178, 179], *e.g.*:

$$H_3COOC-C\equiv C-COOCH_3 \xrightarrow{H_5C_6-NH-NH_2}$$

$$\underset{\underset{N-NH-C_6H_5}{|}}{H_3COOC-\overset{\displaystyle ||}{C}-CH_2-COOCH_3}$$

Dimethyl oxosuccinate phenylhydrazone; 85%

Some acid derivatives can be reduced to the corresponding *aldehyde hydrazones* with arylhydrazines. For example, imidic acid esters furnish the aldehyde hydrazone with arylhydrazines in the presence of sodium amalgam[180]:

$$\underset{\underset{NH\cdot HCl}{|}}{R-\overset{\displaystyle ||}{C}-OC_2H_5} + H_5C_6-NH-NH_2$$

$$\xrightarrow{Na-Hg_x} R-CH=N-NH-C_6H_5$$

Nitriles, too, can be used for preparing hydrazones. For example, phenylacetonitrile is hydrogenated catalytically to *phenylacetaldehyde phenylhydrazone* in the presence of excess phenylhydrazine with Raney nickel and hydrogen[181]. α,α,α-Trichlorotoluene reacts with phenylhydrazine and copper powder to form *benzaldehyde phenylhydrazone*[182].

With many α-hydroxy and α-amino aldehydes and ketones the reaction with phenylhydrazone proceeds beyond the hydrazone stage to *osazones*[183]. Diarylhydrazones have been reacted with naphthol under oxidizing conditions to give *naphthoquinone diarylhydrazones*[184].

Occasionally, indirect methods are employed for synthesizing arylhydrazones. Examples are the reactions between hydrazine derivatives such as the sodium N^2-sulfonates[185, 186] or various acylhydrazines[187] and aldehydes or ketones. Diazo-

alkanes furnish aromatic hydrazones with aryl Grignard reagents[188], *e.g.*:

$$(H_5C_6)_2CN_2 \quad + \quad 2\,H_5C_6-MgBr$$

$$\xrightarrow{H_2O} \quad (H_5C_6)_2C=N-NH-C_6H_5$$

Benzophenone phenylhydrazone

Similar reactions take place between Grignard reagents and *N*-nitrosoamines[189], *e.g.*:

$$(H_5C_6)_2N-NO \quad + \quad H_5C_2-Mg\,I$$

$$\xrightarrow{H_2O} \quad (H_5C_6)_2N-N=CH-CH_3$$

Acetaldehyde diphenylhydrazone

4.1.4 Transformation reactions of arylhydrazones

4.1.4.1 The Fischer indole synthesis[190, 191]

Arylhydrazones, and especially phenylhydrazones, form suitable and readily available starting materials for synthesizing indole derivatives, into which they can be converted by cyclization in acid solution. This cyclization reaction thus ensues at a faster rate than the hydrolysis of the hydrazones back to the corresponding hydrazines and carbonyl compounds.

The structure and yield of the products obtained are very dependent on the reaction conditions and the proton or Lewis acid used as catalyst.

[178] *E. Buchner*, Ber. 22, 2930 (1889).
[179] *R. Heilmann, R. Glénat*, Compt. Rend. 234, 1557 (1952).
[180] *F. Henle*, Ber. 38, 1362 (1905).
[181] *A. Gaiffe*, Chim. Ind. (Milan) 93, 259, 268 (1965).
[182] *S. Bodforss*, Ber. 59, 666 (1926).
[183] *Houben-Weyl*, Methoden der organischen Chemie, Bd. X/2, p. 434, Georg Thieme Verlag, Stuttgart 1967.
[184] *S. Hünig, F. Brühne*, Justus Liebigs Ann. Chem. 667, 86 (1963).
[185] *J. Tröger, G. Puttkammer*, Ber. 40, 207 (1907).
[186] *S. Hünig, T. Utermann*, Chem. Ber. 88, 423 (1955).
[187] *C. U. Rogers, B. B. Corson*, J. Amer. Chem. Soc. 69, 2910 (1947).

[188] *G. H. Coleman, H. Gilman, C. E. Adams, P. E. Pratt*, J. Org. Chem. 3, 99 (1939).
[189] *H. Wieland, H. Fressel*, Ber. 44, 901 (1911).
[190] *P. L. Julian, E. W. Meyer, H. C. Printy* in: *R. C. Elderfield*, Heterocyclic Compounds, Vol. III: The Chemistry of Indoles, John Wiley & Sons, New York 1961;
B. Robinson, The Fischer Indole Synthesis, Chem. Rev. 63, 373 (1963);
Houben-Weyl, Methoden der organischen Chemie, Bd. X/2, Georg Thieme Verlag, Stuttgart 1967;
B. Robinson, Recent Studies on the Fischer Indole Synthesis, Chem. Rev. 69, 227 (1969).
[191] *E. Fischer, F. Jourdan*, Ber. 16, 2241 (1883).
[192] *R. B. Carlin*, J. Amer. Chem. Soc. 74, 1077 (1952).
[193] *P. Grammaticakis*, Bull. Soc. Chim. France 14, 438 (1947).
[194] *G. N. Robinson, R. Robinson*, J. Chem. Soc. (London) 113, 639 (1918).
[195] *N. N. Sovorov, N. P. Sorokina, J. N. Sheinker*, J. Gen. Chem. USSR 28, 1058 (1958).

Although the reaction has been known for a long time its mechanism has not yet been completely elucidated. A persuasive theory which at present finds no objections[192], cf. [193, 194] is the following:

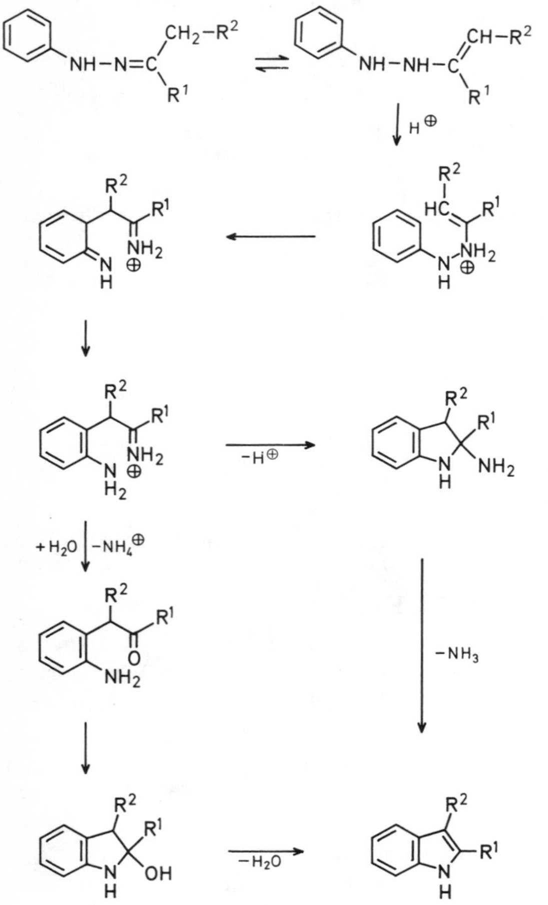

The enehydrazine intermediate can be trapped as the diacetyl derivative **1** when 2-butanone phenyl-hydrazone is reacted with *p*-toluenesulfonic acid in the presence of acetic anhydride. On treatment with acid or by distillation with zinc dust[195] this diacetyl derivative is converted into 2,3-dimethyl-indole **2**:

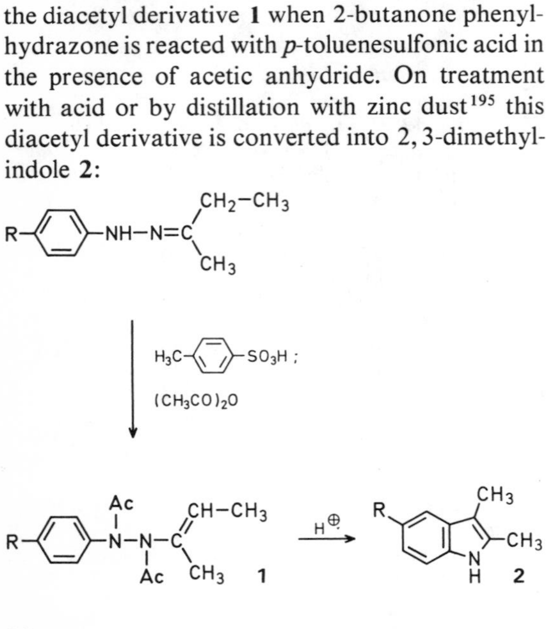

Ketone acyl phenylhydrazones, too, are transformed into *1-acylindoles* in warm glacial acetic acid[196].

This result suggests that 1-acylindoles themselves may be intermediates during the transformation of **1** and **2** and that the primary deacetylation on the nitrogen is not absolutely essential.
In respect of which of the two nitrogen atoms of the phenylhydrazone is eliminated, experiments with labeled compounds have shown clearly that it is the imine nitrogen which is involved[197].

Arylhydrazones of unsymmetrical acyclic ketones can furnish two different products on indolization. Normally, exclusively *3 H*-indoles are formed from the phenylhydrazones **3**; by contrast, the hydrazones **4** afford both *1H*- and *3 H*-indoles, and the hydrazones **5** *3-alkyl-2-methylindole*[198]:

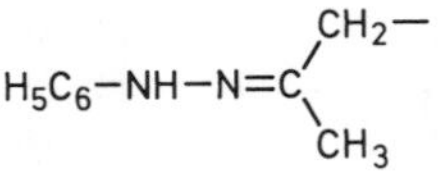

From arylhydrazones of cyclic ketones *carbazoles* are obtained. Applying the Fischer reaction to arylhydrazones of 2-alkylcyclohexanones furnishes mixed[199] *1-alkyl-1, 2, 3, 4-tetrahydrocarbazole* **6** and *11-alkyl-2, 3, 4, 9b-tetrahydro-1 H-carbazole* **7**.

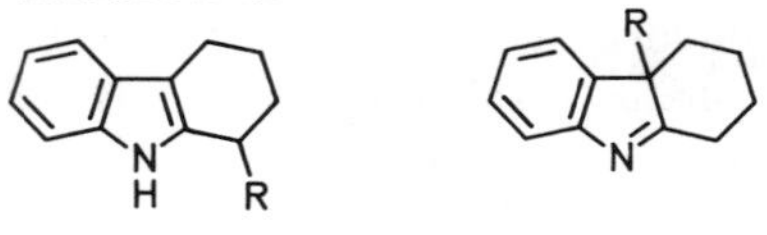

With *m*-substituted arylhydrazones the Fischer reaction leads to two isomers in relative proportions dependent on the nature of R[200].

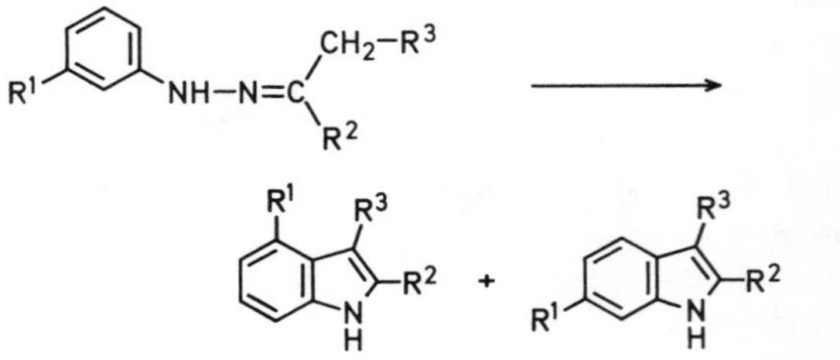

[196] *H. Yamamoto*, Bull. Chem. Soc. Japan *40*, 425 (1967):
   H. Yamamoto, J. Org. Chem. *32*, 3693 (1967).
[197] *C.H.F. Allen, C.V. Wilson*, J. Amer. Chem. Soc. *65*, 611 (1943).

Cyclization of the phenylhydrazones of 1,4-diacetylbenzene, and 1,3,5-triacetylbenzene, and 2,6-diacetylpyridine with zinc chloride or polyphosphoric acid gave nitrogen analogs of polyaryl compounds with indole structures in 50% yield[201]:

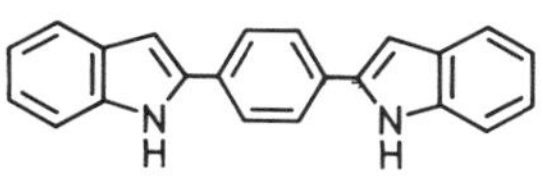

2,2'-p-phenylenediindole

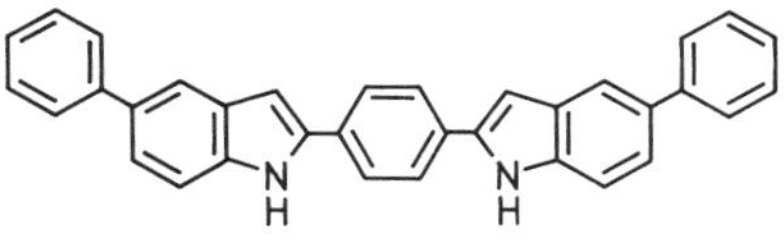

2,2'-p-phenylene-bis[5-phenylindole]

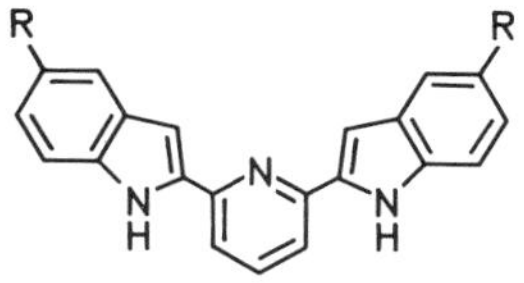

R = H;   2,2'-(2,6-pyridinediyl)diindole
R = CH₃; 2,2'-(2,6-pyridinediyl)bis[5-methylindole]

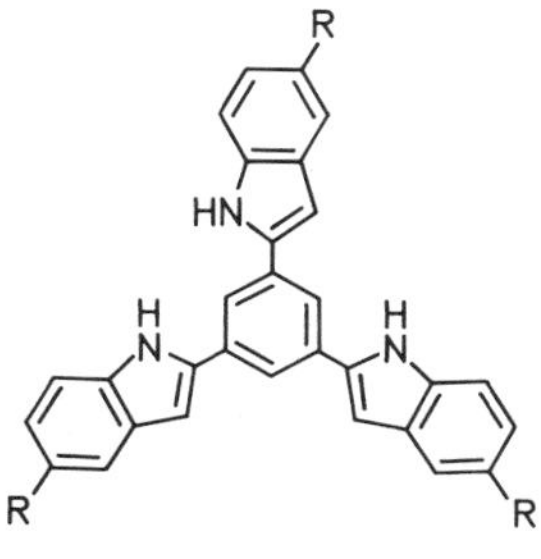

R = H;   2,2',2''-1,3,5-phenenyltriindole
R = CH₃; 2,2',2''-1,3,5-phenenyltris[5-methylindole]

4.1.4.2 Other arylhydrazone cyclization reactions

Arylhydrazines of aldehydes or ketones with a reactive group in the α or β position (acyl-, alkoxycarbonyl group or double bond) display an anomalous behavior under the conditions of the Fischer indolization reaction because they cyclize to other ring systems without loss of nitrogen.

For example, reacting 1,3-dicarbonyl compounds with arylhydrazines under Fischer reaction conditions leads to 1-arylpyrazoles via the monoarylhydrazone intermediate stage:

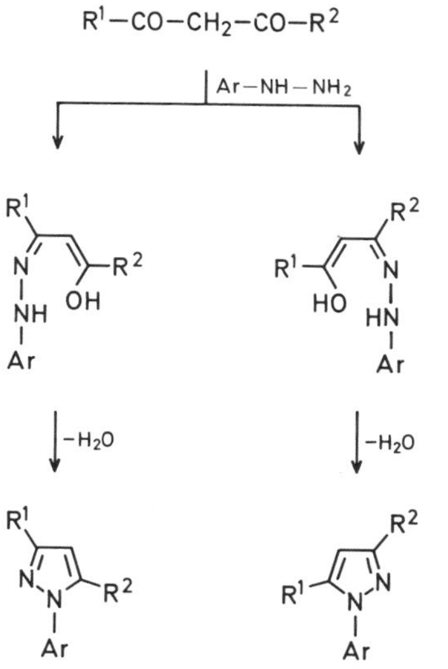

α,β-Unsaturated arylhydrazones cyclize to 1-aryl-4,5-dihydropyrazoles in the presence of acid reagents.

Arylhydrazones of β-keto carboxylic acid esters are converted into 1-arylpyrazolones under the influence of acids or bases:

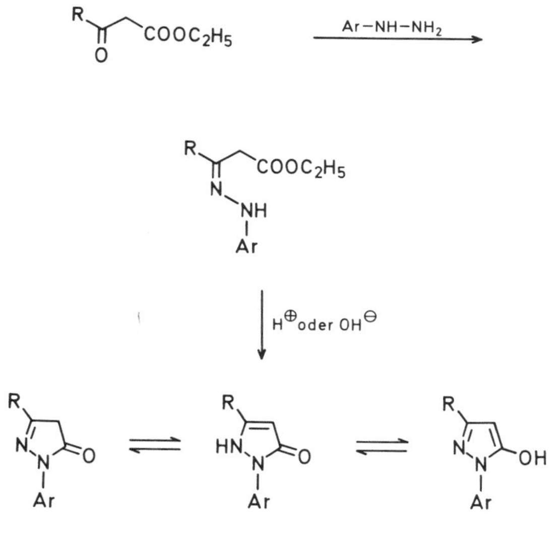

4.2 Aliphatic hydrazines, hydrazo compounds, hydrazones, and azines

4.2.1 Preparation of aliphatic hydrazines and hydrazo compounds

Caution! Alkylhydrazines are **toxic** and **carcinogenic** and need to be handled with great care.

[198] G. Plaucher, A. Bonavia, Gazz. Chim. Ital. 32, 414 (1902).
[199] K.H. Pausacker, C.J. Schubert, J. Chem. Soc. (London) 1384 (1949).
[200] D.W. Ockenden, K. Schofield, J. Chem. Soc. (London) 3175 (1957).
[201] N.P. Buu-Hoï, F. Perin, P. Jacquignon, J. Heterocycl. Chem. 2, 7 (1965).

4.2.1.1 Alkylation of hydrazines and their derivatives

4.2.1.1.1 With alkyl halides and dialkyl sulfates

4.2.1.1.1.1 Alkylation to monoalkylhydrazines

Alkylation of hydrazine leads to *unsymmetrical N,N*-dialkylated hydrazines *via* a monoalkylhydrazine stage, and further to the *N,N,N-trialkylhydrazinium cation*:

$$H_2N-NH_2 + RX \longrightarrow R-NH-NH_2 \xrightarrow{RX} R_2N-NH_2$$
$$\downarrow RX$$
$$\overset{\oplus}{R_3N}-NH_2 \ X^{\ominus}$$

The reason is that the positive charge on the nitrogen which arises during the transitional state is stabilized by already present alkyl groups, and that the nucleophilic properties of the substituted *N* are of similar magnitude as those of the unsubstituted hydrazine. Consequently, especially the lower alkyl halides tend to produce overalkylation, but this event can be avoided by adding the halide slowly or, better, its alcoholic solution[202] to a large excess (1:5–1:10) of hydrazine hydrate or hydrazine in isopropyl alcohol[203]. The products are extracted with ether and distilled over barium or potassium oxide. In this way, monoalkylhydrazines, beginning with *ethylhydrazine,* are obtained in yields varying from 40–70%. Monoalkylation with *long-chain* alkyl halides containing six and more C atoms is carried out at elevated temperature[203–205].

For steric reasons there is less risk of overalkylation with secondary halides such as isopropyl bromide[203] or cyclopentyl bromide[205]. Halogen hydride is easily split from tertiary halides; where this is not possible the monoalkylhydrazines are formed, *e.g., 1-hydrazinoadamantane*[206] from 1-bromoadamantane:

Allylhydrazine plus a little disubstitution product is obtained best from allyl bromide and excess hydrazine hydrate[207]; the same applies to *2-propynylhydrazine*[208] and *benzylhydrazines*[209, 210]. For benzylhydrazine itself reacting the 1-acetyl-2-isopropylidenehydrazine anion with benzyl chloride is recommended[211, 212]:

Phenethyl halides, too, furnish high yields of phenethylhydrazines by the use of the above techniques[213, 214].

1,1'-Hexamethylenedihydrazine   (1,6-Dihydrazinohexane) can be obtained from 1,6-dichlorohexane but shorter 1,ω-dihaloalkanes also give rise to ring closure[215, 216].
Alkylation of hydrazine with dimethyl or diethyl sulfate affords an only small yield of the monoalkylation product[217]; better results are recorded by methylating benzaldehyde azine[218] or acetone azine[219] and decomposing the *N*-methylimmonium ion formed hydrolytically.

4.2.1.1.1.2 Alkylation to *N,N*-dialkylhydrazines

Unlike bis-hydroxyalkylation (see p. 95) and dialkylation with electron deficient olefins (see p. 96), preparation of *N,N*-dialkylhydrazines from hydrazine with two equivalents alkyl halide is

[202] *H.-H. Stroh, H.G. Scharnow,* Chem. Ber. *98,* 1588 (1965).
[203] *A.N. Kost, R.S. Sagitullin,* Zh. Obshch. Khim. *33,* 867 (1963); Engl.: 855.
[204] *O. Westphal,* Ber. *74B,* 759 (1941).
[205] *E. Müller* in: *Houben-Weyl,* Methoden der organischen Chemie, Bd. X/2, p. 7, Georg Thieme Verlag, Stuttgart 1967.
[206] F.P. 1 561 947 (1969), Farbenfabriken Bayer A.G., C.A. *72,* 42964j (1970).

[207] *B.V. Ioffe, Z.I. Sergeeva, A.P. Kochetov,* Zh. Org. Khim. *3,* 983 (1967); C.A. *67,* 99552z (1967).
[208] A.P. 3 083 229 (1963), Dow Chemical Co., Inv.: *P.D. Oja, W. Creek;* C.A. *59,* 8593c (1963).
[209] *H. König, R. Huisgen,* Chem. Ber. *92,* 429 (1959).
[210] *J.H. Biel und Mitarbb.,* J. Amer. Chem. Soc. *81,* 2805 (1959).
[211] *A.N. Kost, R.S. Sagitullin,* Zh. Obshch. Khim. *27,* 3338 (1957); C.A. *52,* 9071c (1958).
[212] *H. Dorn, H. Dilcher,* Justus Liebigs Ann. Chem. *717,* 104 (1968).
[213] USP. 3 334 017, 3 359 316 (1967), Colgate-Palmolive Co., Inv.: *J.H. Biel;* C.A. *68,* 12676s (1968), *69,* 10210g (1968).
[214] *A.N. Kost, R.S. Sagitullin, S. Yü-shan,* Zh. Obshch. Khim. *30,* 3280 (1960); C.A. *55,* 19906f (1961).
[215] *E. Müller* in: *Houben-Weyl,* Methoden der organischen Chemie, Bd. X/2, p. 7, Georg Thieme Verlag, Stuttgart 1967.
[216] *A.N. Kost, R.S. Sagitullin,* Usp. Khim. *33,* 361 (1964); Engl.: 159.
[217] *R.D. Brown, R.A. Klearly,* J. Amer. Chem. Soc. *72,* 2762 (1950).
[218] *H.H. Hatt,* Org. Syntheses, Coll. Vol. II, 395 (1943).
[219] USP. 2 926 532 (1960), Aerojet-General Corp., Inv.: *A.F. Graefe;* C.A. *55,* 9281c (1961).

of only minor importance. *N,N-Dimethylhydrazine* can be obtained in moderate yield from dimethyl sulfate and hydrazine[219]. As overalkylation and underalkylation go in parallel, this method can be applied successfully only for introducing larger alkyl groups[204, 205].

However, clear-cut results are obtained during alkylation of monoacylhydrazines. The hydrazinecarboxylic acid esters obtainable from carbonic acid esters and hydrazine can be dialkylated with allyl and benzyl bromides[220]. Cleaving the carbazates with hydrazine, hydrogen bromide, or potassium hydroxide then furnishes the N,N-dialkylhydrazines[221], *e.g.*

$$H_2N-NH-COOC_2H_5 \quad + \quad 2\ H_2C=CH-CH_2Br$$

$$\longrightarrow \quad (H_2C=CH-CH_2)_2N-NH-COOC_2H_5$$

$$\xrightarrow{N_2H_4} \quad (H_2C=CH-CH_2)_2N-NH_2$$

N,N-Diallylhydrazine

The above-mentioned ring closure reactions include also the reaction of hydrazine with 1,4-dibromobutane to *N-aminopiperidine*[222] and with 1-Phenyl-1,2-ethanediol dimethanesulfonate to form *N-aminoaziridines*[223-225]:

$$Br(CH_2)_4Br \ + \ N_2H_4 \ \longrightarrow \ \langle N-NH_2 \rangle$$

N-Amino-2-phenylaziridine

2-Hydrazinoethanol (2-hydroxyethylhydrazine) can undergo ring closure with acids to form *N-aminoaziridine*.

4.2.1.1.1.3 Alkylation to N,N'-dialkylhydrazines

As direct alkylation of hydrazine *via* the N,N-dialkylhydrazines leads to formation of N,N,N-trialkylhydrazinium ions, it is necessary to prepare N,N'-dialkylhydrazines by alkylating the easily prepared N,N'-diacylhydrazines. To obtain N,N'-dimethylhydrazine, for example, one can methy-

late N,N'-dibenzoylhydrazine[226], N,N'-diformylhydrazine[227], or diethyl dicarbamate (N,N'-hydrazinedicarboxylate)[228] with dimethyl sulfate; the acyl group is split off with hydrochloric acid in each case:

$$H_5C_6-CO-NH-NH-CO-C_6H_5 \ + \ (H_3CO)_2SO_2 \longrightarrow$$

$$H_5C_6-CO-\underset{\underset{H_3C}{|}}{N}-\underset{\underset{CH_3}{|}}{N}-CO-C_6H_5 \xrightarrow[H_2O]{HCl} H_3C-NH-NH-CH_3$$

$$+ \quad H_5C_6-COOH$$

Diethyl sulfate affords inferior yields of N,N'-diethylhydrazine[228], allyl bromides give satisfactory results[229]:

$$OCH-NH-NH-CHO \ + \ 2\ BrCH_2-CH=CH_2$$

$$\longrightarrow \quad H_2C=CH-CH_2-\underset{\underset{OHC}{|}}{N}-\underset{\underset{CHO}{|}}{N}-CH_2-CH=CH_2$$

$$\xrightarrow{HCl} \quad H_2C=CH-CH_2-NH-NH-CH_2-CH=CH_2$$

N,N'-Diallylhydrazine

By comparison, allylation *via* the N anions offers only a small advantage[230]:

$$H_5C_2OOC-NH-NH-COOC_2H_5 \xrightarrow{NaH} H_5C_2OOC-\overset{\ominus}{N}-NH-COOC_2H_5$$

While direct alkylation of hydrazine with 1,3-dibromopropane yields only little *pyrazolidine* admixed with other compounds, N,N'-diisobutyrylhydrazine is particularly suitable for preparing cyclic N,N'-hydrazines[231], *e.g.*:

[220] *F. Eloy, C. Moussebois*, Bull. Soc. Chim. Belg. *68*, 409 (1959); C.A. *54*, 3179g (1960).

[221] *O. Diels*, Ber. *56*, 1933 (1923); *E. Müller* in: *Houben-Weyl*, Methoden der organischen Chemie, Bd. X/2, p. 11, Georg Thieme Verlag, Stuttgart 1967.

[222] *M. Rink, M. Mehta, R. Lux*, Arch. Pharm. (Weinheim, Ger.) *294*, 640 (1961).

[223] *D. Felix, J. Schreiber, K. Pries, U. Horn, A. Eschenmoser*, Helv. Chim. Acta *51*, 1461 (1968).

[224] *R.K. Müller, D. Felix, J. Schreiber, A. Eschenmoser*, Helv. Chim. Acta *53*, 1479 (1970).

[225] *H. Paulsen, D. Stoye*, Angew. Chem. *80*, 120 (1968); *H. Paulsen, D. Stoye*, Angew. Chem. Intern. Ed. Engl. *7*, 134 (1968).

[226] *H.H. Hatt*, Org. Syntheses, Coll. Vol. II, 208 (1943).

[227] *J. Thiele*, Ber. *42*, 2576 (1909); *E. Müller* in: *Houben-Weyl*, Methoden der organischen Chemie, Bd. X/2, p. 16, Georg Thieme Verlag, Stuttgart 1967.

[228] *C.D. Hurd, F.F. Cesark*, J. Amer. Chem. Soc. *89*, 1417 (1967).

[229] *F. Hrábak*, Collect. Czech. Chem. Commun. *34*, 4010 (1969); Synthesis 307 (1970).

[230] *B.H. Al-Sader, R.J. Crawford*, Can. J. Chem. *48*, 2745 (1970).

[231] *H. Stetter, H. Spangenberger*, Chem. Ber. *91*, 1982 (1958).

1,2,3,4-Tetrahydrophthalazine[232], which is a reagent for formaldehyde, is obtained analogously from phthalic acid hydrazide and 1,2-bis(chloromethyl)benzene:

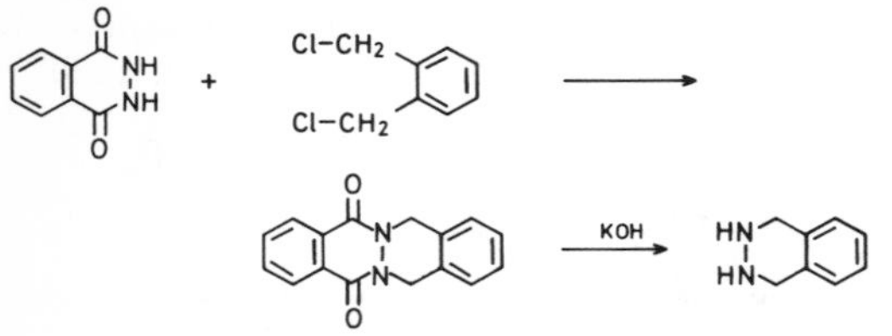

4.2.1.1.1.4 Alkylation to tri- and tetraalkylhydrazines

During alkylation of hydrazine with three equivalents of alkyl halide quarternary compound formation is a major feature but can be suppressed by working with *longer* alkyl chlorides. Thus the series of compounds *tripropyl-* to *trioctylhydrazines* was obtained in yields varying from 25 to 40%[204, 233]. Somewhat better results are achieved by forming the anion from N,N'-dialkylhydrazines with sodium hydride in 1,2-dimethoxyethane and alkylating[234]. Special circumstances obtain where steric hindrance makes quaternization difficult. Thus *tribenzylhydrazine* can be prepared directly from hydrazine and benzyl chloride[235] and alkylated further to *tetrabenzylhydrazine*[236]:

$$3\ H_5C_6-CH_2Cl\ +\ N_2H_4$$

$$\longrightarrow\ (H_5C_6-CH_2)_2N-NH-CH_2-C_6H_5$$

$$\xrightarrow{H_5C_6-CH_2Cl}\ (H_5C_6-CH_2)_2N-N(CH_2-C_6H_5)_2$$

2-Propynyl bromide affords *tri-2-propynylhydrazine*[237] in the presence of bis(triphenylphosphine)-platinum dichloride. While *triisopropylhydrazine* can be obtained with isopropyl bromide[238], this compound cannot be alkylated further although alkylation of N,N'-dimethyl-N-isopropylhydrazine does succeed[238]. In other cases tetraalkylhydrazines are prepared from trialkylhydrazines with alkyl bromide in the presence of magnesium oxide[204, 239], *e.g.*:

$$(H_7C_3)_2N-NHC_3H_7\ +\ BrC_3H_7\ \longrightarrow\ (H_7C_3)_2N-N(C_3H_7)_2$$

Tetrapropylhydrazine

Di- and trialkylhydrazines can be methylated to tetraalkylhydrazines under *Eschweiler-Clarke* conditions[240].

1,2-Diazetidines are formed from N,N'-dialkylhydrazine and 1,2-dibromoethane[241, 242] or 1,2-dibromopropane (*3-methyl-1,2-diazetidine*[243]). N-Allyl-N,N-dialkylhydrazinium salts and corresponding N-benzyl-N,N-dialkyl compounds or their acyl derivatives form N'-*allyl-* or N'-*benzyl-N,N-dialkylhydrazines* by a base-catalyzed rearrangement[244], *e.g.*:

$$\overset{\oplus}{(H_3C)_2N}-NH_2 \atop |\ \ CH_2-CH=CH_2} \xrightarrow{NaOH} (H_3C)_2N-NH-CH_2-CH=CH_2$$

N-Allyl-N',N'-dimethylhydrazine

4.2.1.1.1.5 Alkylation with special alkyl halides

Because of their pharmaceutical importance innumerable *(X-aminoalkyl)hydrazines* have been prepared[245]. Halides which carry their halogen and amino group on adjacent C atoms are particularly suitable for alkylating hydrazine directly, but N-dimethyl-3-chloropropylamine, too, has been used successfully[246]. In the simplest case *(2-dimethylaminoethyl)hydrazine* is obtained[247, 248],

$$(H_3C)_2N-CH_2-CH_2Cl\ +\ N_2H_4$$

$$\longrightarrow\ (H_3C)_2N-CH_2-CH_2-NH-NH_2$$

[232] R. Ohme, E. Schmitz, Z. Anal. Chem. 220, 105 (1966).

[233] E. Müller in: Houben-Weyl, Methoden der organischen Chemie, Bd. X/2, p. 9, Georg Thieme Verlag, Stuttgart 1967.

[234] D.B.P. 1102170 (1961), U.S. Rubber Co., Erf.: H.W. Stewart; C.A. 56, 4618b (1962).

[235] H. Franzen, H. Kraft, J. Prakt. Chem. [2] 84, 137 (1911).

[236] H. Wieland, E. Schamberg, Ber. 53, 1329 (1920).

[237] J.N. Nelson, H.B. Jonassen, D.M. Roundhill, Inorg. Chem. 8, 2591 (1969).

[238] F. Klages und Mitarb., Justus Liebigs Ann. Chem. 547, 1 (1941).

[239] E. Müller in: Houben-Weyl, Methoden der organischen Chemie, Bd. X/2, p. 10, Georg Thieme Verlag, Stuttgart 1967.

[240] USP. 3013076 (1956), Nat. Dist. and Chemical Corp., Inv.: D. Horvitz; C.A. 56, 8562c (1962).

[241] USP. 3129215 (1964), FMC Corp., Inv.: D. Horvitz; C.A. 60, 15874f (1964).

[242] E. Müller in: Houben-Weyl, Methoden der organischen Chemie, Bd. X/2, p. 14, Georg Thieme Verlag, Stuttgart 1967.

[243] C. Dittli, J. Elguero, R. Jacquier, Bull. Soc. Chim. France 696 (1970).

[244] S. Wawzonek, E. Yeakey, J. Amer. Chem. Soc. 82, 5718 (1960);
M.G. Indzhikyan, Z.G. Gegelyan, A.T. Babayan, Arm. Khim. Zh. 19, 674 (1966); C.A. 66, 104531c (1967);
K.-H. König, B. Zeeh, Chem. Ber. 103, 2052 (1970).

[245] E. Jucker, Angew. Chem. 71, 321 (1959);
E. Jucker, Pure Appl. Chem. 6, 409 (1963).

[246] G. Leclerc, P. Mélounoun, C.G. Wermouth, Bull. Soc. Chim. France 1099 (1967).

[247] J.H. Biel, W.K. Hoya, H.A. Leiser, J. Amer. Chem. Soc. 81, 2527 (1959).

[248] D.B.P. 1095841 (1960), Farbenfabriken Bayer. Erf.: H. Klös, H.A. Offe; C.A. 57, 11019h (1962).

but versatilely substituted derivatives can be prepared[248-250]. If an excess of hydrazine is not used then *1,2-bis[2-dialkylamino(ethyl)]hydrazines* result[249]. Acyl hydrazines give particularly clean alkylations[251], e.g.[252]:

$$(H_5C_2)_2N-CH-CH_2Cl \quad + \quad H_2N-NH-COCH_3$$
$$\underset{CH_2-CH(CH_3)_2}{|}$$

$$\longrightarrow \quad (H_5C_2)_2N-CH-CH_2-NH-NH-COCH_3$$
$$\underset{CH_2-CH(CH_3)_2}{|}$$

*Acetic acid
(2-diethylamino-4-
methylpentyl)hydrazine*

The acyl group can be eliminated subsequently by acid hydrolysis to give *(2-diethylamino-4-methylpentyl)hydrazine.*

Numerous other alkylation products with additional functional groups have been described, among them *β-hydrazino acetals* from 3-chloropropionaldehyde diethylacetal[253] and *α-hydrazinecarboxylic acids* from α-halo carboxylic acids and hydrazine or methylhydrazine[254]. Depending on the experimental conditions, either *hydrazinoacetic acid* or *hydrazonodiacetic acid* is obtained from chloroacetic acid[255]:

$$Cl-CH_2-COOH + N_2H_4 \longrightarrow H_2N-NH-CH_2-COOH$$

$$\xrightarrow{Cl-CH_2-COOH} H_2N-N(CH_2-COOH)_2$$

4.2.1.1.2 Alkylation with oxiranes, imines, and sultones

Ethylene oxide (oxirane) reacts smoothly with hydrazines to form *(2-hydroxyethyl)hydrazines;* whether *2-hydrazinoethanol* or *2,2'-hydrazo-diethanol* arises is again governed by the proportion of the two reactants employed[256]:

$$N_2H_4 \quad + \quad \triangledown \longrightarrow HO-CH_2-CH_2-NH-NH_2$$

$$\xrightarrow{\triangledown} (HO-CH_2-CH_2)_2N-NH_2$$

Substituted oxiranes such as propylene oxide[256], 2,3-epoxybutane[257] 1,2-epoxybutane[257], 7-oxabicyclo[4.1.0]heptane (1,2-epoxycyclohexane)[258], etc.[256], may be used in place of ethylene oxide, and mono- to trialkylhydrazines instead of simple hydrazine itself[259]. *2-(1-Methylhydrazino)ethanol* is formed from methylhydrazine

$$H_3C-NH-NH_2 \quad + \quad \triangledown \longrightarrow$$

$$\underset{HO-CH_2-CH_2}{\overset{H_3C}{\diagdown}}N-NH_2$$

and *2-(2,2-dimethylhydrazino)ethanol* from *N,N*-dimethylhydrazine:

$$(H_3C)_2N-NH_2 \quad + \quad \triangledown$$

$$\longrightarrow (H_3C)_2N-NH-CH_2-CH_2-OH$$

By contrast, under mild conditions a *N-(2-hydroxypropyl)-N,N-dimethylhydrazinium betain* is obtained from propylene oxide and *N,N*-dimethylhydrazine[260]:

$$\underset{H_3C}{\overset{H_3C}{\diagdown}}N-NH_2 \quad + \quad \overset{H_3C}{\underset{}{\diagdown}}\triangledown \longrightarrow$$

$$\underset{H_3C}{\overset{H_3C}{\diagdown}}\overset{\oplus}{N}-\overset{\ominus}{NH}$$
$$\underset{CH_2-CH-CH_3}{|}$$
$$\underset{OH}{|}$$

Numerous combinations of substituted oxiranes and substituted hydrazines, too, have been studied[259, 261]. Diaziridines form *2-hydrazinoethanols* and *2,2'-hydrazodiethanols* with oxiranes[262].

Epoxides with other functional groups often undergo *ring closure* reactions with hydrazine. Chloromethyloxirane (epichlorohydrin) forms *pyrazole*

$$Cl-CH_2 \triangledown \quad + \quad N_2H_4 \longrightarrow \underset{H}{\overset{HO}{\diagdown}}\text{[ring]}NH \xrightarrow[\text{Oxidation}]{ZnCl_2/} \underset{H}{\text{[ring]}}N$$

4-Pyrazolidinol

[249] USP. 3 272 807 (1966), Colgate-Palmolive Co., Inv.: *J.H. Biel, W.K. Hoya; C.A. 66,* 2489v (1967).

[250] *J.H. Biel und Mitarb.,* J. Org. Chem. *26,* 3338 (1961).

[251] *A. Ebnöther und Mitarb.,* Helv. Chim. Acta *42,* 533 (1959).

[252] *E. Müller* in: *Houben-Weyl,* Methoden der organischen Chemie, Bd. X/2, p. 8, Georg Thieme Verlag, Stuttgart 1967.

[253] *A. Wohl,* Ber. *64,* 1381 (1931).

[254] *A. Carmi, G. Pollack, H. Yellin,* J. Org. Chem. *25,* 44 (1960).

[255] *E. Müller* in: *Houben-Weyl,* Methoden der organischen Chemie, Bd. X/2, p. 13, Georg Thieme Verlag, Stuttgart 1967.

[256] *E. Müller* in: *Houben-Weyl,* Methoden der organischen Chemie, Bd. X/2, p. 17, Georg Thieme Verlag, Stuttgart 1967.

[257] *G. Gever,* J. Amer. Chem. Soc. *76,* 1283 (1954).

[258] Jap. P. 5458 ('66), Sankyo Ltd., Inv.: *T. Taguchi; C.A. 65,* 5381a (1966).

[259] *G. Benoit,* Bull. Soc. Chim. France 708 (1939); *G. Benoit,* Bull. Soc. Chim. France 242 (1947).

[260] D. Off. 1914032 (1969), Ashland Oil & Refining Co., Erf.: *R.C. Slagel; C.A. 72,* 54724u (1970).

[261] *E. Müller* in: *Houben-Weyl,* Methoden der organischen Chemie, Bd. X/2, p. 18, 19, Georg Thieme Verlag, Stuttgart 1967.

and 4,5-epoxy-4-methyl-pentyne forms *1-amino-2,4-dimethylpyrrole*:

$$CH_3-C(O)(CH_3)-C\equiv C-CH_3 \xrightarrow{N_2H_4}$$

$$H_2N-NH-CH_2-C(CH_3)(OH)-C\equiv C-CH_3 \longrightarrow$$

Hydrazine reacts with *ethyleneimine(aziridine)* in analogous manner to form *(2-aminoethyl)hydrazine* and bis(2-aminoethyl)hydrazine[263], while methylhydrazine gives *1-(1-methylhydrazino)-2-propanethiol* with 1,2-epithiopropane (propylene sulfide) in benzene[264]

$$H_3C-\underset{S}{\triangle} + H_3C-NHNH_2 \longrightarrow H_3C-CH-CH_2-\underset{NH_2}{\overset{CH_3}{N}}$$
$$\overset{|}{SH}$$

Sultones also are attacked by hydrazines with ring opening and formation of a *C–N* bond. For example, propanesultone forms *3-hydrazino-1-propanesulfonic acid*[265]:

$$\underset{O}{SO_2} \xrightarrow{N_2H_4} H_2N-NH-(CH_2)_3-SO_3H$$

$$\xrightarrow{} \text{NH-NH}_2 \quad H_2N-N(CH_2)_3-SO_3H$$

N-Cyclohexyl-N-(3-sulfopropyl)hydrazine

and the correspondingly substituted products with alkylhydrazines[266].

4.2.1.1.3 Alkylation with active olefins

Electron deficient olefins add on hydrazine with formation of *C–N* bonds. Here, too, the *N,N-dialkyl*hydrazine is formed subsequently to the *alkylhydrazine* if the olefin concentration is high enough. On adding the calculated amount of acrylonitrile to hydrazine hydrate an excellent yield of *2-hydrazinopropionitrile* is obtained[267], while on adding hydrazine hydrate to excess acrylonitrile *3,3'-hydrazonodipropionitrile* is formed[268]:

$$H_2C=CH-CN + N_2H_4 \longrightarrow H_2N-NH-CH_2-CH_2-CN$$

$$\xrightarrow{H_2C=CH-CN} H_2N-N(CH_2-CH_2-CN)_2$$

Alkylhydrazine can straightforwardly take the place of hydrazine with formation of *2-(N-alkylhydrazino)propionitriles*[266, 269], e.g.:

$$\text{cyclohexyl}-NH-NH_2 + H_2C=CH-CN$$

$$\longrightarrow H_2N-N(\text{cyclohexyl})-CH_2-CH_2-CN$$

3-(N-Cyclohexylhydrazino)-propionitrile

Diaziridines also react in this way with acrylonitrile[270]. Acyl hydrazines add to the free amino group of acrylonitrile[251, 271]. e.g.:

$$H_5C_6-CO-NH-NH_2 + H_2C=CH-CN$$

$$\longrightarrow H_5C_6-CO-NH-NH-CH_2-CH_2-CN$$

N-(N'-Benzoylhydrazino)propionitrile

The addition products of hydrazine and monoalkylhydrazines with α,β-unsaturated carboxylic acid derivatives generally suffer spontaneous ring closure with formation of *pyrazolidones*. By contrast, the *3-(N,N'-dialkylhydrazino)propionates* can be isolated from *N,N'-*dialkylhydrazines and esters of acrylic acid and cyclize only when sodium methoxide is allowed to act on them[272]:

$$H_3C-NH-NH-CH_3 + H_2C=CH-COOC_2H_5 \longrightarrow$$

$$H_3C-NH-\underset{CH_3}{N}-CH_2-CH_2-COOC_2H_5 \longrightarrow$$

Ethyl (N,N'-dimethylhydrazino)-N-propionate   *1,2-Dimethyl-3-pyrazolidinone*

From two molecules ethyl acrylate and N,N-dimethylhydrazine the following noteworthy product is obtained which can be cyclized to *1-(dimethylamino)-4-piperidone* by a Dieckmann condensation[273]:

[262] E.P. 1085794 (1967), Whiffen & Sons Ltd., Inv.: *M.D. Hinchliffe, J. Miller;* C.A. *68,* 29243w (1968).

[263] *E. Müller* in: *Houben-Weyl*, Methoden der organischen Chemie, Bd. X/2, p. 20, Georg Thieme Verlag, Stuttgart 1967.

[264] *D.L. Trepanier, P.E. Krieger,* J. Heterocycl. Chem. *4,* 254 (1967).

[265] *W. Schindler,* Veröffentl. Wiss. Photo-Lab. Wolfen *10,* 277 (1965); C.A. *66,* 46385v (1967).

[266] *H. Dorn, K. Walter,* Z. Chem. *7,* 151 (1967).

[267] D.R.P. 598185 (1931), I.G. Farb., Erf.: *U. Hoffmann, B. Jacobi;* C.A. *28,* 5473[9] (1934);
E. Müller in: *Houben-Weyl*, Methoden der organischen Chemie, Bd. X/2, p. 21, Georg Thieme Verlag, Stuttgart 1967.

[268] *M.A. Iorio,* Gazz. Chim. Ital. *94,* 1391 (1964).

[269] *H. Dorn, A. Zubeck, G. Hilgetag,* Chem. Ber. *98,* 3377 (1965).

[270] E.P. 1081292 (1967), Whiffen & Sons Ltd., Inv.: *J. Miller;* C.A. *68,* 114071h (1968).

[271] *S.I. Suminov, A.N. Kost,* Zh. Obshch. Khim. *33,* 2208 (1963); Engl.: 2152.

[272] *M.J. Kornet, S.I. Tan,* J. Heterocycl. Chem. *5,* 397 (1968).

[273] *A.H. Beckett, J.V. Greenhill,* J. Med. Chem. *4,* 423 (1961).

$(H_3C)_2N-NH_2 \quad + \quad 2\,H_2C=CH-COOC_2H_5 \quad \longrightarrow$

$(H_3C)_2N-N\begin{cases} CH_2-CH_2-COOC_2H_5 \\ CH_2-CH_2-COOC_2H_5 \end{cases} \longrightarrow$ [piperidinone ring with N(CH₃)₂ substituent]

Diethyl
(N,N-dimethylhydrazino)-N',N'-dipropionate

Ethylenesulfonic acid derivatives, too, add on hydrazine[274, 275], while addition of aliphatic hydrazines to α,β-unsaturated ketones and aldehydes leads to *pyrazolines* via the hydrazones and is discussed there (see p. 116).

The nucleophilicity of hydrazine itself is not sufficient to afford addition to the double bond of styrenes and dienes. However, with sodium hydrazide formation of *phenethylhydrazine* and *N,N'-diphenethylhydrazine*[276], as well as of numerous other phenethylhydrazines is accomplished successfully[277]:

$H_3C-O-\langle\text{C}_6\text{H}_4\rangle-CH=CH-CH_3 \quad \xrightarrow{\text{NaNH-NH}_2}$

$H_3C-O-\langle\text{C}_6\text{H}_4\rangle-CH_2-\underset{\overset{|}{CH_3}}{CH}-NH-NH_2$

N-(4-methoxy-α-methylphenethyl)-hydrazine

N-Sodium-*N*-methylhydrazide also can be added[278]. A drawback of the technique is the risk of an **explosion** during working with sodium hydrazide, which needs to be prepared from hydrazine and sodium amide in pure nitrogen at 0° while following certain precautionary measures very strictly[279]. Recently, it has been proposed to replace the dangerous sodium hydrazide by the less hazardous barium hydrazide[280].

4.2.1.1.4 Alkylation with cyanhydrins

Cyanhydrins of aldehydes or ketones alkylate hydrazine smoothly to *2,2'-hydrazo-bis[alkanenitriles]* [*N,N'-bis(α-cyanoalkyl)hydrazines*]. The reaction proceeds very readily even in aqueous solution and even the crude hydrazine obtained from the urea process affords a good yield of *2,2'-hydrazo-bis[2-methylpropionitrile]* with acetone cyanhydrin[281]. It is unnecessary to preform the cyanhydrin, one can equally react a mixture of acetone and sodium cyanide with hydrazine sulfate[282]:

$2\,H_3C-CO-CH_3 \quad + \quad N_2H_4 \cdot H_2SO_4 \quad + \quad 2\,NaCN$

$\longrightarrow \quad H_3C-\underset{\overset{|}{NC}}{\overset{\overset{\displaystyle H_3C}{|}}{C}}-NH-NH-\underset{\overset{|}{CN}}{\overset{\overset{\displaystyle CH_3}{|}}{C}}-CH_3$

Other open-chain ketones and cyclohexanone, and also acetophenones can be used; so can acetaldehyde[282]. Monoalkyl-[283] and *N,N*-dialkylhydrazines[284] may be employed in a modification of the principle. Often the reaction takes on more the character of a Mannich condensation, *e.g.*:

$H_5C_6-SO_2-O-CH_2-CN \quad + \quad H_2N-NH-CH_2-CH_2-CN$

$\longrightarrow \quad H_2N-N\begin{cases} CH_2-CN \\ CH_2-CH_2-CN \end{cases}$

3-[1-(Cyanomethyl)-hydrazino]propionitrile

$HCN \quad + \quad CH_2O \quad + \quad H_2N-N(CH_3)_2$

$\longrightarrow \quad (H_3C)_2N-NH-CH_2-CN$

(2,2-Dimethylhydrazino)-acetonitrile

4.2.1.2 Alkylhydrazines from amines with *N*-functionalized ammonia or amine

By comparison to the above described techniques, preparation of alkylhydrazines by forming the

[274] D.R.P. 696 776 (1938), I.G. Farb., Erf.: *H. Zischler, G. Wilmanns*; C.A. *35*, 5913⁴ (1941); *E. Müller* in: *Houben-Weyl*, Methoden der organischen Chemie, Bd. X/2, p. 22, Georg Thieme Verlag, Stuttgart 1967.

[275] *H. Dorn, K. Walter*, Justus Liebigs Ann. Chem. *720*, 98 (1968).

[276] *T. Kauffmann, C. Kossel, D. Wolf*, Chem. Ber. *95*, 1540 (1962); *E. Müller* in: *Houben-Weyl*, Methoden der organischen Chemie, Bd. X/2, p. 24, Georg Thieme Verlag, Stuttgart 1967.

[277] *E. Müller* in: *Houben-Weyl*, Methoden der organischen Chemie, Bd. X/2, p. 25, Georg Thieme Verlag, Stuttgart 1967.

[278] *T. Kauffmann, K. Lötzsch, D. Wolf*, Chem. Ber. *99*, 3148 (1966).

[279] *E. Enders* in: *Houben-Weyl*, Methoden der organischen Chemie, Bd. X/2, p. 293, Georg Thieme Verlag, Stuttgart 1967.

[280] *K.-H. Linke, R. Taubert, K. Bister, W. Bornatsch, B.J. Liem*, Z. Naturforsch. *26b*, 296 (1971).

[281] Tschech. P. 122 204 (1967), Erf.: *S. Kratky, E. Pavlacka, M. Marko*; C.A. *68*, 49129s (1968).

[282] *E. Müller* in: *Houben-Weyl*, Methoden der organischen Chemie, Bd. X/2, p. 27, Georg Thieme Verlag, Stuttgart 1967.

[283] *S. Grudzinski*, Acta Pol. Pharm. *24*, 145 (1967); C.A. *67*, 90509h (1967).

[284] *M. Götz, K. Zeile*, Tetrahedron *26*, 3185 (1970).

N–N bond (see also p. 105) displays the fundamental advantage that overalkylation is avoided.

4.2.1.2.1 Alkylhydrazines from alkylamine and chloramine

This method corresponds to the Raschig hydrazine synthesis and allows *monoalkylhydrazines* to be prepared from primary amines and chloramine[285]:

$$R-NH_2 \ + \ NH_2Cl \longrightarrow R-NH-NH_2 \ HCl$$

and *1,1-dialkylhydrazines* from secondary amines and chloramine[286]:

$$R_2NH \ + \ NH_2Cl \longrightarrow R_2N-NH_2 \ \cdot \ HCl$$

A further parallel with the Raschig synthesis is the need to add gelatin to bind traces of heavy metals and to maintain alkaline conditions. On the other hand, the amine excess required to prevent an attack by chloramine on already formed hydrazine may be reduced (amine:chloramine = 5:1) by comparison to that of ammonia during Raschig synthesis (ammonia : cloramine = 20 : 1). The more nucleophilic nature of the substituted amines enables the reaction to proceed at between 0° and room temperature, while the Raschig procedure requires 60°.

Details of the preparation of chloramine and the general preparation of mono- and 1,1-dialkylhydrazines are readily available[287]. Tables 1 and 2 survey the products obtained by this method.

Table 1. Alkylhydrazines[288] by reacting primary alkylamines with chloramine

R–NH–NH$_2$	Yield [% of theory] (on NH$_2$Cl)	Hydrazine isolated as
Methylhydrazine	64	Sulfate
Ethylhydrazine	67	Sulfate
Propylhydrazine	62	Sulfate
Isopropylhydrazine	55	Sulfate
Butylhydrazine	68	Sulfate
Isobutylhydrazine	59	Sulfate
tert-Butylhydrazine	71	Hydrochloride
Hexylhydrazine	57	Oxalate
Cyclohexylhydrazine	60	Sulfate
Allylhydrazine	52	Hydrochloride
2-Hydrazinoethanol	58	Oxalate
(2-Aminoethyl)hydrazine	75	Oxalate

Table 2. N,N-Dialkylhydrazines[288] by reacting secondary alkylamines with chloramine

R$_2$N–NH$_2$	Yield [% of theory] (on NH$_2$Cl)	Hydrazine isolated as
N,N-Dimethylhydrazine	53	Oxalate
N,N-Diethylhydrazine	41	Oxalate
N,N-Dipropylhydrazine	40	Oxalate
N,N-Dibutylhydrazine	42	Oxalate
N-Aminomorpholine	—	Hydrochloride
N-Aminopiperidine	51	—
1-Amino-2(1H)-pyridone	28	Free base

The yield of *methylhydrazine*, which is prepared technically in this way[289], can be increased to 75% by adding triethylamine[290]. When reacting secondary amines the yield is improved by using an excess of ammonia[286].

Alternatively, the reaction between chloramine and *primary* and *secondary* amines can be conducted in anhydrous medium[291] by introducing a gaseous mixture of chloramine and ammonia, produced from ammonia and chlorine, into the liquid primary[292] or secondary[293] amines. Here the use of gelatin and sodium hydroxide are unnecessary.

N-chloroalkylamines cannot be used for this reaction and so, as a rule, it is not possible to prepare *N,N'*-dialkylhydrazines in this way. However 1,3-propanediamine, and 1,4-butanediamine (giving *hexahydropyridazine*) can be cyclized *via* their *N*-chloro derivatives[294]:

Pyrazolidine

[285] *L.F. Audrieth, L.H. Diamond,* J. Amer. Chem. Soc. *76,* 4869 (1954); *77,* 3131 (1955).

[286] *R.A. Rowe, L.F. Audrieth,* J. Amer. Chem. Soc. *78,* 563 (1956);
E.P. 800248 (1956), Olin Mathieson Chem. Corp., Inv.: *C.C. Clark, R.E. Morningstar;* C.A. *53,* 5130f (1959).

[287] *E. Müller* in: *Houben-Weyl,* Methoden der organischen Chemie, Bd. X/2, p. 31, Georg Thieme Verlag, Stuttgart 1967.

[288] *W. Teilacker, E. Wegner,* Angew. Chem. *72,* 129 (1960);
E. Müller in: *Houben-Weyl,* Methoden der organischen Chemie, Bd. X/2, p. 32, Georg Thieme Verlag, Stuttgart 1967.

[289] *R. Ohme, A. Zubeck,* Z. Chem. *8,* 41 (1968).

[290] USP. 2 901 511 (1959), W.R. Grace & Co., Inv.: *F.R. Hurley;* C.A. *54,* 5464d (1960).

[291] *H.H. Sisler, F.T. Neth, R.S. Drago, D. Yaney,* J. Amer. Chem. Soc. *76,* 3906 (1954).

[292] *K. Hoegerle, H. Erlenmeyer,* Helv. Chim. Acta *39,* 1203 (1956).

[293] *G.M. Omietanski, A.D. Kelmers, R.W. Shellman, H.H. Sisler,* J. Amer. Chem. Soc. *78,* 3874 (1956).

[294] *A. Lüttringhaus, J. Jander, R. Schneider,* Chem. Ber. *92,* 1756 (1959);
E. Müller in: *Houben-Weyl,* Methoden der organischen Chemie, Bd. X/2, p. 33, Georg Thieme Verlag, Stuttgart 1967.

Tertiary amines furnish hydrazinium salts with the chloramine reagent[295], *e.g.*:

$$(H_3C)_3N \quad + \quad Cl-NH_2 \quad \longrightarrow \quad (H_3C)_3\overset{\oplus}{N}-NH_2 \; Cl^{\ominus}$$

N,N,N-Trimethylhydrazinium chloride

Many examples of this reaction have been performed, including that using 1,4-diazobicyclo[2.2.2]octane[296].

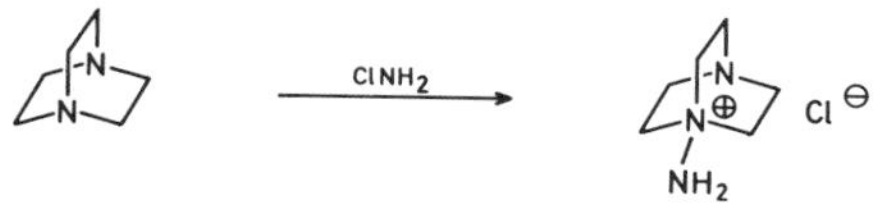

1-Amino-4-aza-1-azoniabicyclo[2.2.2.]octane chloride

4.2.1.2.2 Alkylhydrazines from alkylamine and hydroxylamine-*O*-sulfonic acid

For laboratory use hydroxylamine-*O*-sulfonic acid can replace chloramine in many instances[297, 298]. While the yield of hydrazine is as a rule somewhat less, there is the advantage of greater stability, better dosability, and the freedom from smell of the reagent.

Primary or *secondary* amines are heated briefly with one equivalent potassium hydroxide and half an equivalent hydroxylamine-*O*-sulfonic acid in aqueous solution[299]. Secondary amines react somewhat more smoothly than primary ones, but experimental data relating to the preparation of *N,N*-dialkylhydrazines by this technique are still limited. By contrast, numerous *monoalkylhydrazines*, notably the C$_{2-5}$ compounds, have been prepared in yields of around 50%[299, 300]:

$$R-NH_2 \quad + \quad H_2N-O-SO_3H \quad \longrightarrow \quad R-NH-NH_2 \quad + \quad H_2SO_4$$

The yield can be increased by using a three- to four-fold excess of amine in place of potassium hydroxide[300].

Tertiary amines give a very good yield of quaternary hydrazinium compounds[299, 301], but this technique, too, is bettered slightly by the chloramine procedure. Certain *heterocycles, e.g.,* pyridine, also furnish hydrazinium compounds[299]. With this method also the attempt to employ *N*-alkylhydroxylamine-*O*-sulfonic acid produces only moderately good results. It forms *N,N'-dimethyl-N-phenylhydrazine* with *N*-methylaniline[302] and the *N,N'-dialkylated* or trialkylated hydrazines[303] with alkylamines or dialkylamines. *N*-Acylhydroxylamine-*O*-sulfonic acid derivatives can be used and play an important role during the preparation of *1-aminoaziridines*[223], *e.g.*:

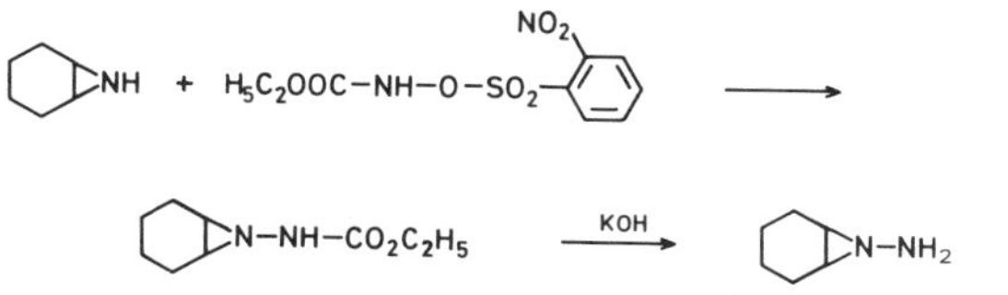

7-Amino-7-azabicyclo[4.1.0]heptane

because *N*-nitrosoaziridines are unstable at room temperature.

4.2.1.2.3 Alkylhydrazines from alkylurea and hypochlorite

Hydrazines can be obtained from 1-chloroureas in way similar to that using chloramine from 1-chloroureas; the two nitrogen atoms to be linked are held together by a carbonyl group:

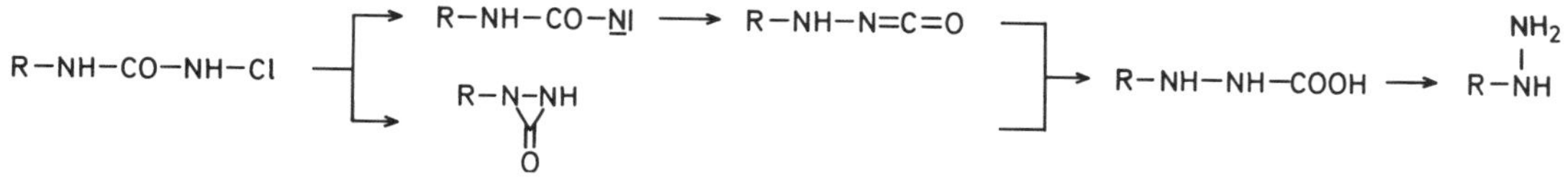

The reaction can proceed as a special case of Hofmann carboxamide degradation (upper pathway) or *via* a diaziridinone intermediate stage (lower pathway)[304]. In some cases, *e.g.*, during prepara-

[295] *G.M. Omietanski, H.H. Sisler*, J. Amer. Chem. Soc. 78, 1211 (1956).

[296] *H.H. Sisler, H.S. Ahuja, N.L. Smith*, J. Org. Chem. 26, 1819 (1961).

[297] *F. Sommer, O.F. Schultz, M. Nassau*, Z. Anorg. Chem. 147, 143 (1925).

[298] *H.J. Matsuguma, L.F. Audrieth*, Inorg. Synth. 5, 122 (1957);
E. Enders in: *Houben-Weyl*, Methoden der organischen Chemie, Bd. X/2, p. 293, Georg Thieme Verlag, Stuttgart 1967.

[299] *R. Gösl, A. Meuwsen*, Chem. Ber. 92, 2521 (1959);
E. Müller in: *Houben-Weyl*, Methoden der organischen Chemie, Bd. X/2, p. 37, Georg Thieme Verlag, Stuttgart 1967.

[300] *G. Gever, K. Hayes*, J. Org. Chem. 14, 813 (1949).

[301] *H.H. Sisler, R.A. Bafford, G.M. Omietanski, B. Rudner, R.J. Drago*, J. Org. Chem. 24, 859 (1959).

[302] *O. Westphal, O. de Burlet*, Angew. Chem. 58, 77 (1945).

[303] *N.V. Khromov-Borisov, T.N. Kononova*, Probl. Poluch. Poluprod. Prom. Org. Sin., Akad. Nauk SSSR, Otd. Obsh. Tekh. Khim. 10 (1967); C.A. 68, 48947v (1968).

tion of *N,N-dialkyl*hydrazines from 1,1-dialkyl-ureas only the upper route is feasible.

Monoalkyl- and *N,N*-dialkylhydrazines, particularly, can be prepared by this method. The monoalkyl- or 1,1-dialkylurea is reacted with the equivalent amount of sodium hypochlorite solution in aqueous caustic soda and the mixture is subsequently heated[304, 305]. *Methylhydrazine*[305], *N,N-dimethylhydrazine, N,N-diethylhydrazine, 1-aminopiperidine,* and *1-aminomorpholine* have all been prepared in high yield in this way[304]. As the products are obtained in solution, the latter is suitably evaporated following making acid and the bases are then liberated with caustic soda.

To carry out the reaction in organic solvents the chlorine is introduced in the form of *tert*-butyl hypochlorite, *e.g.,* in methanol[306].

The reaction becomes particularly versatile when the 1-alkyl-3-chloroureas are isolated, because the subsequent rearrangement can then be performed with different nucleophiles to form the corresponding carbamates. Thus, the 1-alkyl-3-chloroureas obtained by chlorination of alkylureas can be converted into *methyl 3-alkylcarbazates (1-alkylhydrazine-2-carboxylates)* with sodium methoxide and into alkylhydrazins with caustic soda[307]:

$$(H_3C)_3C-NH-CO-NH_2 \xrightarrow{Cl_2} (H_3C)_3C-NH-CO-NH-Cl$$

$$\xrightarrow{NaOCH_3} (H_3C)_3C-NH-NH-COOCH_3$$

Methyl 3-tert-butylcarbazate

$$\xrightarrow{NaOH} \left((H_3C)_3C-NH-NH-COOH\right)$$

$$\longrightarrow (H_3C)_3C-NH-NH_2$$

tert.-Butylhydrazine

The 1-acyl-1-chloroureas obtained by chlorinating 1-acylureas behave correspondingly. Alkoxycarbonylureas thus yield the symmetrical or unsymmetrical *dicarbamic acid (N,N'-hydrazinedicarboxylic acid)* derivatives, *e.g.:*

$$H_3COOC-NH-CO-NH_2 \xrightarrow{Cl_2} H_3COOC-\overset{Cl}{\underset{|}{N}}-CO-NH_2$$

$$\xrightarrow{H_5C_2-ONa} H_3COOC-NH-NH-COOC_2H_5$$

Ethylmethyl dicarbamate

$$\xrightarrow{H_9C_4-NH_2} H_3COOC-NH-NH-CO-NH-C_4H_9$$

Little is known to date about obtaining *N,N'*-dialkylhydrazines from 1,3-dialkylureas, but the reaction is feasible, as is shown by the example of 1,3-di-*tert*-butylurea[308]:

$$(H_3C)_3C-NH-CO-NH-C(CH_3)_3 \xrightarrow{(H_3C)_3C-O-Cl}$$

$$(H_3C)_3C-\underset{\underset{Cl}{|}}{N}-CO-NH-C(CH_3)_3$$

$$\xrightarrow{(H_3C)_3COK}$$

(H₃C)₃C—N—N—C(CH₃)₃ (diaziridinone with C=O)

$$\xrightarrow[\text{cleavage}]{+ Y} (H_3C)_3C-NH-\overset{R}{\underset{|}{N}}-C(CH_3)_3$$

Y = HCl/H₂O;   *N,N'-Di-tert-butylhydrazine* (R = H)

Y = KOC(CH₃)₃; *tert-Butyl 2,3-di-tert-butylcarbazate* (R = COOC(CH₃)₃)

Y = HCOOH;   *2,3-Di-tert-butylcarbazic acid anhydride with formic acid* (R = CO—O—CHO)

Y = HCl;   *2,3-Di-tert-butylcarbazoyl chloride* (R = COCl)

However, *diaziridinones* are accessible also in alternative manner[309], *e.g.:*

$$(H_3C)_3C-NO + \overset{\ominus}{C}\equiv\overset{\oplus}{N}-C(CH_3)_3 \longrightarrow$$

(H₃C)₃C—N—N—C(CH₃)₃ (diaziridinone with C=O)

1,2-Di-tert-butyldiaziridinone

In place of haloureas the less accessible 1-alkyl-3-hydroxyurea-*O*-sulfonic acids can be used for the

[305] USP. 2 917 545 (1959), Nat. Dist. and Chemical Corp., Inv.: *D.W. Lum, I.L. Mador;* C.A. *54,* 6549e (1960); *E. Müller* in: *Houben-Weyl,* Methoden der organischen Chemie, Bd. X/2, p. 35, Georg Thieme Verlag, Stuttgart 1967.

[306] USP. 3 442 612 (1969), Pennsalt Chemicals Corp., Inv.: *L.K. Huber, L.R. Ocone;* C.A. *71,* 30044a (1969).

[307] D.Off. 1 810 164, 1 810 165 (1969), Wallace and Tiernan Inc., Erf.: *C.S. Sheppard, L.E. Korczykowski;* C.A. *72,* 54725v (1970), *71,* 80756e (1969).

[308] *F.D. Greene, J.C. Stowell, W.R. Bergmark,* J. Org. Chem. *34,* 2254 (1969).

[309] *F.D. Greene, J.F. Pazos,* J. Org. Chem. *34,* 2269 (1969);

E. Schmitz, Dreiringe mit zwei Heteroatomen, Oxaziridine, Diaziridine, cyclische Diazoverbindungen, p. 112, Springer Verlag Berlin, Heidelberg 1967.

[304] *R. Ohme, H. Preuschhof,* J. Prakt. Chem. *312,* 349 (1970).

rearrangement reaction[304]. The alkali-catalyzed conversion of the alkylcarbamoyl oxaziridines obtainable from isocyanates and oxaziridines proceeds similarly[310], *e.g.*:

$$H_9C_4-NCO \; + \; HN-O \longrightarrow H_9C_4-NH-CO-N-O$$

$$\xrightarrow{NaOH} (H_9C_4-NH-NH-COOH) \longrightarrow \underset{\textit{Butylhydrazine}}{H_9C_4-NH-NH_2}$$

4.2.1.2.4 Alkylhydrazines from alkylsulfonamides and hypochlorite

In analogy with the method starting from alkylureas, it is possible also to prepare alkylhydrazines from alkylsulfamides in alkaline solution with sodium hypochlorite at elevated temperature[311]. Here, the symmetrical sulfuric acid bisalkylamides are especially important. They are simple to prepare and yield symmetrical *N,N'-dialkylhydrazines*:

$$2R-NH_2 + SO_2Cl_2 \rightarrow R-NH-SO_2-NH-R$$

$$\xrightarrow{NaOCl} \left(\begin{array}{c} \underset{R-N-SO_2-NH-R}{\overset{Cl}{|}} \longrightarrow \underset{R}{\overset{O\diagdown S\diagup O}{N-N}}\diagup R \\[2ex] \rightarrow \underset{\underset{SO_3Na}{|}}{R-NH-N-R} \end{array} \right) \longrightarrow R-NH-NH-R$$

Excess sodium hypochlorite should be avoided because it oxidizes the hydrazines to azo compounds. *N,N'-Dimethyl-, N,N'-diethyl-, N,N'-dipropyl-,* and *N,N'-dibutylhydrazines* are obtained in good yield in this way.

N,N-Dialkylhydrazines and monoalkylhydrazines can, however, be prepared in the same way. The starting materials are obtained by transamidation of sulfamide with secondary or primary amines. High yields of *N,N*-dimethyl- and *N,N*-diethylhydrazine, *1-aminopiperidine, 1-aminomorpholine,* and of *butyl-* and *cyclohexylhydrazine* are obtained[311].

4.2.1.2.5 Preparation and hydrolysis of diaziridines

The concept of preparing hydrazines *via* diaziridines has been outstandingly fruitful[312]:

$$\underset{R}{\overset{R}{>}}\hspace{-0.3em}\diagup\hspace{-0.3em}\underset{N-R^I}{\overset{N-R^I}{}} \xrightarrow{H_3O^\oplus} R-CO-R \; + \; R^I-NH-NH-R^I$$

Monoalkyl- and *N,N'-dialkylhydrazines* are the products obtained. Since the diaziridines are prepared from the same carbonyl compound *R–CO–R* that is reliberated on hydrolysis, the overall procedure is equivalent to a Raschig synthesis but without some of the latter's drawbacks. Thus, as diaziridines are stable towards chloramine and hydroxylamine-*O*-sulfonic acid, it is, unlike with Raschig, unnecessary to employ excess amine; conversely, unlike with the urea method, an excess of these reagents is not harmful. In addition, preparation of the diaziridines succeeds surprisingly also with *N*-chloroalkylamines, which cannot be used for the Raschig method.

4.2.1.2.5.1 Diaziridines from Schiff's bases and *N*-functional amines

During reaction of Schiff's bases with chloramine stoichiometric amounts of the two reactants add together in the presence of an equivalent of the amine, forming the constituent of the Schiff's base, in ether at room temperature[313]:

$$H_3C-CH_2-CH=N-\bigcirc \; + \; NH_2Cl$$

$$\longrightarrow \underset{\underset{H}{\overset{}{}}}{H_3C-CH_2-\diagup\hspace{-0.3em}\overset{N}{\underset{N}{}}}\hspace{-0.3em}\diagdown\bigcirc$$

1-Cyclohexyl-3-ethyldiaziridine

$$\bigcirc=N-\bigcirc \; + \; NH_2Cl$$

$$\longrightarrow$$

7-Cyclohexyl-7,8-diazaspiro[5.2]octan

The amine serves to trap the released hydrogen chloride; the course of the reaction may be gauged from the precipitation of the hydrochloride. No restrictions apply to the choice of the primary

[310] *E. Schmitz, R. Ohme, S. Schramm,* Chem. Ber. *100,* 2600 (1967).

[311] *R. Ohme, H. Preuschhof,* Justus Liebigs Ann. Chem. *713,* 74 (1968).

[312] *E. Schmitz,* Dreiringe mit zwei Heteroatomen, Oxaziridine, Diaziridine, cyclische Diazoverbindungen, p. 67, Springer Verlag Berlin, Heidelberg 1967.

[313] *E. Schmitz, D. Habisch,* Chem. Ber. *95,* 680 (1962); *E. Müller* in: *Houben-Weyl,* Methoden der organischen Chemie, Bd. X/2, p. 76, Georg Thieme Verlag, Stuttgart 1967.

Table 3. 1-Alkyldiaziridines and alkylhydrazines from Schiff's bases and chloramine[313]

Schiff's base from Oxo compound	Amine	Diaziridine	Yield* [% of theory]	Alkylhydrazine	Yield** [% of theory]
Acetaldehyde	Ethylamine	*1-Ethyl-3-methyldiaziridine*	18	*Ethylhydrazine*	48
	Cyclohexylamine	*1-Cyclohexyl-3-methyldiaziridine*	46	*Cyclohexylhydrazine*	79
Propionaldehyde	Cyclohexylamine	*1-Cyclohexyl-3-ethyldiaziridine*	55	*Cyclohexylhydrazine*	88
Butyraldehyde	Benzylamine	*1-Benzyl-3-propyldiaziridine*	26	*Benzylhydrazine*	87
Heptanal	Butylamine	*1-Butyl-3-hexyldiaziridine*	53	*Butylhydrazine*	77
	Cyclohexylamine	*1-Cyclohexyl-3-hexyldiaziridine*	54	*Cyclohexylhydrazine*	88
Acetone	Isopropylamine	*1-Isopropyl-3,3-dimethyldiaziridine*	40	*Isopropylhydrazine*	83
	Cyclohexylamine	*1-Cyclohexyl-3,3-dimethyldiaziridine*	64	*Cyclohexylhydrazine*	81
Cyclohexanone	Cyclohexylamine	*1-Cyclohexyl-1,2-diazaspiro[2.5]octane*	71	*Cyclohexylhydrazine*	87

* calculated on titrimetric values of 80–100% pure products
** as oxalates

amine, and both aldimines and ketimines can be employed. The diaziridines listed in Table 3 can be distilled unchanged in vacuum, and all except the first representative crystallize.

Hydrolysis of diaziridines proceeds at varying rates. Diaziridines from aldehydes hydrolyze so slowly that stable *oxalates* can be obtained, those from ketones decompose more rapidly. In every case hot aqueous oxalic acid or 2 N mineral acid liberates the *alkylhydrazine* in good yield.

The identical technique is used during reacting of Schiff's bases with *N*-chloroalkylamines following preparation of the reagent from primary amine and sodium hypochlorite[314]. However, here the size of the alkyl substituent in the *N*-chloroalkylamine occasionally interferes, unlike that in the Schiff's base, in that amine exchange or transchlorination occurs. It is therefore advisable to employ the larger alkyl group in the Schiff's base when unlike amines are concerned.

N,N'-Dialkylhydrazines are obtained by hydrolysis with warm dilute hydrochloric acid[314].

4.2.1.2.5.2 Diaziridines from oxo compounds, amines, and chlorine

There is no need to prepare the Schiff's base separately and react it with the *N*-chloroamine in order to prepare *symmetrical* diaziridines. Instead, the oxo compound, alkylamine, and chlorine can be reacted directly in the gas phase to form a diaziridine. *Tetramethyldiaziridine* is obtained in this way from acetone, methylamine, and chlorine; hydrolysis with sulfuric acid converts it into *1,2-dimethylhydrazinium sulfate*[315]:

$$\begin{array}{c} H_3C \\ H_3C \end{array} C{=}O \ + \ 2 \ H_3C{-}NH_2 \ + \ Cl_2 \ \longrightarrow$$

$$\xrightarrow{H_3O^{\oplus}} \ H_3C{-}NH{-}NH{-}CH_3 \ + \ H_3C{-}CO{-}CH_3$$

Table 4. 1,2-Dialkyldiaziridines and 1,2-dialkylhydrazines from Schiff's bases and *N*-chloroalkylamines[314]

Schiff's base from Oxo compound	Amine R¹–NH₂ R¹ =	N-Chloro-alkylamine R²–NHCl R² =	Diaziridine	Yield [% of theory]	1,2-Dialkylhydrazine	Yield* [% of theory]
Acetaldehyde	Butyl	Butyl	*1,2-Dibutyl-3-methyldiaziridine*	64		
Propionaldehyde	Cyclohexyl	Methyl	*1-Cyclohexyl-3-ethyl-2-methyl-diaziridine*	57	*1-Cyclohexyl-2-methyl-hydrazine*	63
Butyraldehyde	Butyl	Methyl	*1-Butyl-2-methyl-3-propyl-diaziridine*	42	*1-Butyl-2-methylhydrazine*	83
	Butyl	Ethyl	*1-Butyl-2-ethyl-3-propyl-diaziridine*	55	*1-Butyl-2-ethylhydrazine*	73
	Butyl	Propyl	*1-Butyl-2,3-dipropyldiaziridine*	50	*1-Butyl-2-propylhydrazine*	7
	Butyl	Butyl	*1,2-Dibutyl-3-propyldiaziridine*	71	*1,2-Dibutylhydrazine*	85
Heptanal	Methyl	Methyl	*3-Hexyl-1,2-dimethyldiaziridine*	68		
	Methyl	Butyl	*1-Butyl-3-hexyl-2-methyl-diaziridine*	19		
			1,2-Dibutyl-3-hexyldiaziridine	11		
			3-Hexyl-1,2-dimethyldiaziridine	1		
	Butyl	Methyl	*1-Butyl-3-hexyl-2-methyl-diaziridine*	63		
	Butyl	Butyl	*1,2-Dibutyl-3-hexyldiaziridine*	53		

* as hydrochloride

In place of acetone other ketones can be used[316].

Aldehydes need to be reacted at low temperature (−40°) with ammonia and chloramine in order to avoid amide formation and to obtain diaziridines. However, these react with further aldehyde to form *1,3,5-triazabicyclo-[3.1.0]hexanes*[317]:

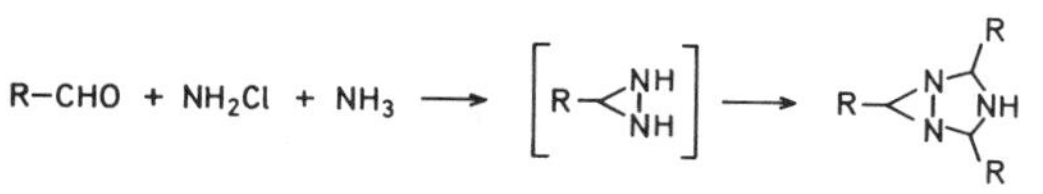

whose acid hydrolysis furnishes unsubstituted hydrazine. Diaziridines of formaldehyde are suitably prepared from aminals and sodium hypochlorite[318]:

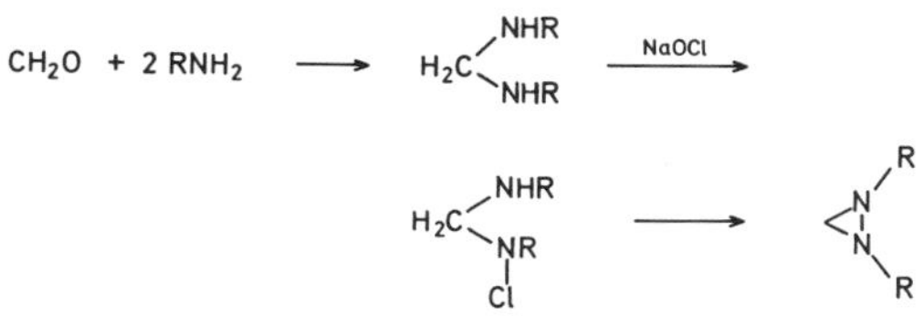

but a patent[319] claims a process for their direct preparation, *e.g.*:

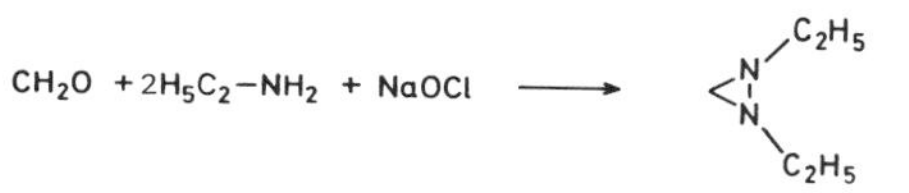

1,2-Diethyldiaziridine

4.2.1.2.5.3 Alkylhydrazines *via* diaziridines from oxo compounds, amines, and hydroxylamine-*O*-sulfonic acid

During direct conversion of ketones into diaziridines, too, hydroxylamine-*O*-sulfonic acid offers the above advantages by comparison to chlor-amine. It is more reactive and thus even iner-ter ketones which do not react with chloramine may be used. The preferred active ketones such as cyclohexanone or acetone react even in aqueous solution, others in methanol.

Cyclohexanone, ammonia, and hydroxylamine-*O*-sulfonic acid afford *1,2-diazaspiro[2.5]octane* smoothly[320, 321]. Using alkylamines in place of ammonia leads to *N*-alkyldiaziridines[320, 322]:

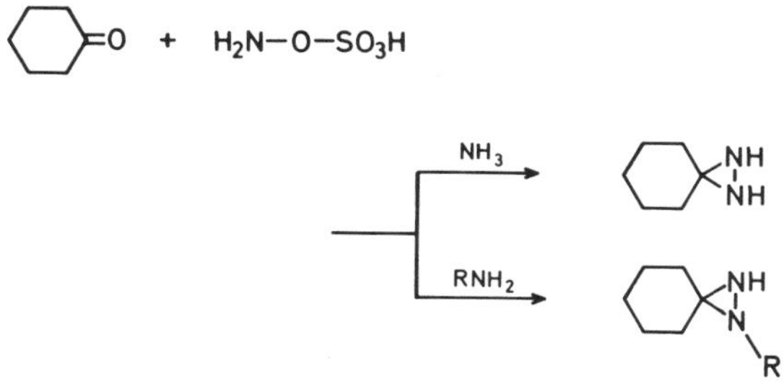

The yield of *1-methyl-1,2-diazaspiro[2.5]octane* by this method excels that from the alternative combination cyclohexanone, ammonia, and *N*-methylhydroxylamine-*O*-sulfonic acid[322].

2-Hydroxyketones also form diaziridines, but ketones with other functional groups such as an α-carbonyl or α-amino group or an α,β double bond do not[323].

The technique becomes especially valuable for obtaining alkylhydrazines because the diaziridines need not be isolated. Excess hydroxylamine-*O*-sulfonic acid is added to the cold aqueous solution of the equivalent amounts of cyclohexanone and ethylamine, the diaziridine is extracted with toluene, excess amine is washed out with oxalic acid, and warm hydrolysis with oxalic acid is then performed[324].

Table 5. Alkylhydrazines from primary amine, cyclohexanone, and hydroxylamine-*O*-sulfonic acid[324]

Primary amine	Alkylhydrazine	Yield [% of theory] as oxalate
Methylamine	*Methylhydrazine*	60
Ethylamine	*Ethylhydrazine*	67
Propylamine	*Propylhydrazine*	65
Butylamine	*Butylhydrazine*	53
tert-Butylamine[325]	*tert-butylhydrazine*	59 (as hydrochloride)
Benzylamine	*Benzylhydrazine*	70
Cyclohexylamine	*Cyclohexylhydrazine*	61

[314] E. Schmitz, K. Schinkowski, Chem. Ber. 97, 49 (1964);
E. Müller in: Houben-Weyl, Methoden der organischen Chemie, Bd. X/2, p. 77, Georg Thieme Verlag, Stuttgart 1967.
[315] D.A.S. 1127907 (1959), Farbf. Bayer, Erf.: H.J. Abendroth; C.A. 57, 9664i (1962);
E. Müller in: Houben-Weyl, Methoden der organischen Chemie, Bd. X/2, p. 79, Georg Thieme Verlag, Stuttgart 1967.
[316] D.A.S. 1123330 (1959), Bergwerksverband G.m.b.H., Erf.: S. Paulsen, G. Huck; C.A. 57, 7275g (1962).
[317] E. Schmitz, Chem. Ber. 95, 688 (1962);
E. Müller in: Houben-Weyl, Methoden der organischen Chemie, Bd. X/2, p. 79, Georg Thieme Verlag, Stuttgart 1967.
[318] R. Ohme, E. Schmitz, P. Dolge, Chem. Ber. 99, 2104 (1966).
[319] Belg. P. 715868 (1968), Solvay & Cie.; C.A. 71, 81327c (1969).
[320] H.J. Abendroth, Angew. Chem. 73, 67 (1961).
[321] E. Schmitz, R. Ohme, Org. Synth. 45, 83 (1965).
[322] E. Schmitz, R. Ohme, R.-D. Schmidt, Chem. Ber. 95, 2714 (1962);
E. Müller in: Houben-Weyl, Methoden der organischen Chemie, Bd. X/2, p. 81, Georg Thieme Verlag, Stuttgart 1967.

4.2.1.2.5.4 Diaziridines from diazirines

Diaziridines[326] react with Grignard compounds to form *N*-alkyldiaziridines[327, 318], *e.g.:*

1-tert-Butyl-1,2-
diazaspiro[2.5]octane

However, as diaziridines are not readily accessible and are frequently obtained best *via* diaziridines[326], this technique is unlikely to be often usable for preparing *alkyl*hydrazines.

4.2.1.2.5.5 Transformation of the diaziridines

In connection with alkylhydrazine preparation the greatest interest pertains to the *hydrolysis* of the *diaziridines*. The weakly basic properties of the diaziridines enables them to be extracted from organic solvents in the cold with dilute mineral acid. If rapid working is employed, they can be reliberated with alkali, toward which they are stable[313]. On heating with dilute aqueous mineral acid or aqueous oxalic acid they are split into carbonyl compound and hydrazine[328]. While the nature of the substituents on the nitrogen is not of major importance, hydrolysis of diaziridines derived from ketones proceeds more rapidly than those based on aldehydes. The yield of *alkylhydrazine* is around 80%. Diaziridines unsubstituted in the 3-position (from formaldehyde) react with acids mainly with *N,N*-cleavage.

At *pH* 2.5 *N,N*-unsubstituted diaziridines yield unsymmetrical azines with ketones in aqueous solution where applicable[329]. Acylation of diaziridines

with a free *NH* group is possible. 1,2-Dibenzoyl-aziridines can be resaponified under alkaline conditions without decomposing the three-membered ring[330].

Diaziridines are strong *oxidizing* agents which liberate two equivalents iodine from acid iodide solution, and this reaction serves for their quantitative determination.

Raney nickel or lithium tetrahydroaluminate cleaves the *N-N* bond with formation of *primary* or *secondary amines* and, where applicable, ammonia.

Dehydrogenating nitrogen-unsubstituted diaziridines leads to diazirines. Silver oxide or iodine are particularly appropriate reagents[331] but other dehydrogenating agents have also been used successfully[326, 327].

A monograph describes the reactions of the diaziridines in detail[312].

4.2.1.2.6 Other *N—N* bond formation reactions

1,1'-Biaziridine is prepared by *N—N* bond formation in a type of Wurtz synthesis from 1-chloroaziridine and 1-lithiumaziridine[332]:

Bi-piperide-
1-yl

Piperidine and morpholine (giving *4,4'-bimorpholine*) can be linked oxidatively in the form of the *N*-copper(I) salts[333].

4.2.1.3 Alkylhydrazines and hydrazo compounds by reduction

4.2.1.3.1 Reduction of *N*-nitroso compounds

Reducing *N*-nitroso compounds is the classic method for preparing *N,N*-dialkylhydrazines.

[323] *E. Schmitz, C. Hörig, C. Gründemann*, Chem. Ber. *100*, 2093 (1967).

[324] *R. Ohme, E. Schmitz, L. Sterk*, J. Prakt. Chem. [4] *37*, 257 (1968).

[325] *G. Büttner, S. Hünig*, Chem. Ber. *104*, 1088 (1971).

[326] *M. Bauer, E. Müller* in: *Houben-Weyl*, Methoden der organischen Chemie, Bd. X/4, p. 894, Georg Thieme Verlag, Stuttgart 1968.

[327] *E. Schmitz, R. Ohme*, Chem. Ber. *94*, 2166 (1961); *E. Müller* in: *Houben-Weyl*, Methoden der organischen Chemie, Bd. X/2, p. 82, Georg Thieme Verlag, Stuttgart 1967.

[328] *E. Müller* in: *Houben-Weyl*, Methoden der organischen Chemie, Bd. X/2, p. 51, 52, Georg Thieme Verlag, Stuttgart 1967.

[329] *A. Jankowski*, Angew. Chem. *77*, 1026 (1965); *A. Jankowski*, Angew. Chem. Intern. Ed. Engl. *4*, 978 (1965).

[330] *E. Schmitz, D. Habisch, C. Gründemann*, Chem. Ber. *100*, 142 (1967).

[331] *C.F.R. Church, M.J. Weiss*, J. Org. Chem. *35*, 2465 (1970).

[332] *A.F. Graefe, R.E. Meyer*, J. Amer. Chem. Soc. *80*, 3939 (1958); *E. Müller* in: *Houben-Weyl*, Methoden der organischen Chemie, Bd. X/2, p. 35, Georg Thieme Verlag, Stuttgart 1967.

[333] *T. Kauffmann, J. Albrecht, D. Berger, J. Legler*, Angew. Chem. *79*, 620 (1967); *T. Kauffmann*, Angew. Chem. Intern. Ed. Engl. *6*, 633 (1967).

Many different reducing agents are in use for this purpose[334,335].

They must under no circumstances split the *N–N* bond. One of many publications that gives generally applicable details about the preparation of a *N*-nitrosoamine is ref.[336].

Caution: *N*-Nitroso compounds are **carcinogenic**.

4.2.1.3.1.1 Reduction of *N*-nitroso compounds with metals

The classic laboratory method is reduction with zinc in aqueous acetic acid, which in general affords high yields[336]. Even sensitive *benzylhydrazines* can be obtained in satisfactory yield using a particularly mild variant of this reaction[337], while other reducing agents fail. For technical purposes the expensive acetic acid is advantageously replaced by mineral acid; in this way a more than 90% yield of the important rocket propellant, *N,N-dimethylhydrazine*, is obtained[338]:

$$(H_3C)_2N-NO \xrightarrow{Zn/HCl} (H_3C)_2N-NH_2$$

However, in some instances aqueous hydrochloric acid also gives higher yields than acetic acid or other acids[339]. Zinc in ammoniacal ethanol is an alternative successful reagent for araliphatic *N*-nitrosoamines, including *N*-benzylnitrosoamine[340]. *Aluminum* in dilute caustic soda may be used in place of zinc[341] or else activated aluminum in organic solvents[342]. Thus, *N*-Allyl-*N*-benzylhydrazine, for example, can be prepared with aluminum amalgam in ether[343]:

$$\begin{array}{c} H_2C=CH-CH_2 \\ H_5C_6-CH_2 \end{array} N-NO \xrightarrow{Al/Hg} \begin{array}{c} H_2C=CH-CH_2 \\ H_5C_6-CH_2 \end{array} N-NH_2$$

Sodium in ethanol or better in liquid ammonia also is used as the reducing agent[344].

Reducing the *N*-nitrosoamines serves practically exclusively for the preparation of *N,N-dialkylhydrazines*. Methylhydrazine alone has been prepared from 1-methyl-1-nitrosourea by reduction with zinc in aqueous acetic acid[345].

4.2.1.3.1.2 Catalytic reduction of *N*-nitroso compounds

The increasing demand for *N,N-dimethylhydrazine* as rocket propellant has led to numerous investigations on the catalytic hydrogenation of aliphatic *N*-nitrosoamines in water or aqueous ethanol. Platinum, palladium, and nickel are the principal catalysts used, but the tendency for overhydrogenation to occur increases in the order given and with the size of the alkyl groups, *i.e.*, the *N,N* bond is cleaved hydrogenolytically.

While platinum on alumina under pressure[346] gives excellent yields:

1-Methyl-1-nitrosourea $\longrightarrow$ *Methylhydrazine*[347]; 75%

1,3-Dimethyl-1-nitrosourea $\longrightarrow$ *1,2-Dimethylhydrazine*[347]

as does *palladium* on carbon at normal pressure[348] without special additives, these latter are often admixed to palladium catalysts in order to suppress the *N–N* cleavage. Iron(II) salts[349] especially have proven successful here, and calcium chloride, lithium chloride, ammonium salts[350], and urea[351],

[334] *A.L. Fridman, F.M. Mukhametshin, S.S. Novikov*, Usp. Khim. *40*, 64 (1971); Engl.: 34.

[335] *E. Müller* in: *Houben-Weyl*, Methoden der organischen Chemie, Bd. X/2, p. 38, Georg Thieme Verlag, Stuttgart 1967.

[336] *H.H. Hatt*, Org. Syntheses, Coll. Vol. II, 211 (1943).

[337] *A.N. Kost, M.A. Yurowskaya*, Zh. Obshch. Khim. *39*, 2723 (1969); Engl.: 2662.

[338] A.P. 3317607 (1967), FMC Corp., Inv.: *H.K. Latourette, J.A. Pianfetti*; C.A. *67*, 53647f (1967).

[339] *C.G. Overberger, M. Valentine, J.-P. Anselme*, J. Amer. Chem. Soc. *91*, 687 (1969).

[340] *B.T. Hayes, T.S. Stevens*, J. Chem. Soc. C 1088 (1970).

[341] USP. 2802031 (1957), Metalectro Corp., Inv.: *D. Horvitz*; C.A. *52*, 1201f (1958); *E. Müller* in: *Houben-Weyl*, Methoden der organischen Chemie, Bd. X/2, p. 39, Georg Thieme Verlag, Stuttgart 1967.

[342] D.B.P. 1115138(1963), Arzneimittelwerk Dresden, Erf.: *H.G. Kazmirowski, H. Goldhahn, E. Castens*; C.A. *60*, 2826c (1964).

[343] *E. Müller* in: *Houben-Weyl*, Methoden der organischen Chemie, Bd. X/2, p. 39, Georg Thieme Verlag, Stuttgart 1967.

[344] *H. Zimmer, L.F. Audrieth, M. Zimmer, R.A. Rowe*, J. Amer. Chem. Soc. 77, 790 (1955); *E. Müller* in: *Houben-Weyl*, Methoden der organischen Chemie, Bd. X/2, p. 39, Georg Thieme Verlag, Stuttgart 1967.

[345] *G. v. Brüning*, Justus Liebigs Ann. Chem. *253*, 7 (1889); *E. Müller* in: *Houben-Weyl*, Methoden der organischen Chemie, Bd. X/2, p. 38, Georg Thieme Verlag, Stuttgart 1967.

[346] USP. 3271454 (1966), Allied Chemical Corp., Inv.: *D. Pickens*; C.A. *65*, 18416h (1966).

[347] A.P. 3387030 (1968), Allied Chemical Corp., Inv.: *C.R. Walter jr.*; C.A. *69*, 28968v (1968).

[348] *K. Klager, E.M. Wilson, G.H. Helmkamp*, Ind. Eng. Chem. *52*, 119 (1960); A.P. 3214474 (1965), Aerojet-General Corp., Inv.: *K. Klager*; C.A. *64*, 1957h (1966).

[349] USP. 2979505 (1961), Food Machinery and Chemical Corp., Inv.: *W.B. Tuemler, H. Winkler*; C.A. *55*, 15349d (1961); *E. Müller* in: *Houben-Weyl*, Methoden der organischen Chemie, Bd. X/2, p. 40, 41, Georg Thieme Verlag, Stuttgart 1967.

too, have been employed. A tabular survey lists the most important results obtained with iron chloride[349].

In a simplified process dialkylamine, sodium nitrate, iron(II) chloride, hydroxyl ions, and palladium on carbon are used on which the intermediately formed nitrosoamine is hydrogenated[352].

The cheaper *Raney nickel,* too, may be employed for the reduction[353, 347]. Stirring the hydrogenation composition initially for one hour at normal pressure under hydrogen before increasing the pressure to about 40 atmospheres avoids the *N–N* cleavage[353].

4.2.1.3.1.3 Reduction of *N*-nitroso compounds with lithium tetrahydroaluminate

Lithium tetrahydroaluminate reduction of *N*-nitrosoamines[354] is almost certainly the most convenient laboratory method but entails the risk of an *N–N* cleavage if aromatic groups are present on the nitrogen[355]. This can be countered by using stoichiometric quantities of the reagent and by slow introduction of the latter into the solution of the *N*-nitrosoamine. It is remarkable that an *N–N* cleavage is much more liable to happen with fresh than with old lithium tetrahydroaluminate[356].

Equally, high to very high yields of *N, N-dialkylhydrazines* and *N, N-alkylphenylhydrazines* have been obtained by introducing *N*-nitrosoamine into excess tetrahydroaluminate in ether[344, 357–359] or tetrahydrofuran[355], among them *N-aminopyrazolidine* and *N-aminopiperidine* from cyclic *N*-nitrosoamines[355, 358].

4.2.1.3.1.4 Electrolytic reduction of *N*-nitroso compounds

Very good yields of *N, N-dialkylhydrazines* can be obtained by reduction of *N*-nitrosoamines in sulfuric acid solution on lead or cadmium cathodes[360, 361, 573]. Particularly good results are obtained in an internally amalgamated copper tube constructed as a flow cell[362].

Preparation of *1-deoxy-1-(1-methylhydrazino)-mannitol* and *isomers* from the corresponding *N*-nitrosoamines on a mercury cathode of large area is noteworthy[363, 573].

4.2.1.3.1.5 Sydnone method

Using the methods described on pp. 104, 105, monoalkylhydrazines can be prepared only by the roundabout route *via* the not easily accessible *1-alkyl-1-nitrosoureas.* An essentially similar method consists in the hydrolytic cleaving of *N*-alkylsydnones obtained from 1-alkyl-1-nitrosoglycines by ring closure.

Two pathways lead to these starting materials: alkylation of a primary amine with α-haloacetic acid esters followed by nitrosation[364, 365]

or a Strecker synthesis with primary amine, prussic acid, and formaldehyde followed by saponifying the nitrile and nitrosation[366]:

[350] *G. W. Smith, D. N. Thatcher,* Ind. Eng. Chem., Prod. Res. Development *1,* 117 (1962).

[351] USP. 3 153 095 (1964), Commercial Solvents Corp., Erf.: *J. B. Tindall;* C. A. *62,* 5192f (1965).

[352] USP. 3 154 538 (1964), FMC Corp., Inv.: *D. A. Lima;* C. A. *62,* 3937a (1965).

[353] D.B.P. 1 064 954 (1956), 1 138 789 (1959), Olin Mathieson Chem. Corp., Erf.: *D. J. Jaszaka;* C. A. *53,* 9058h (1959), *56,* 9962b (1962); *E. Müller* in: *Houben-Weyl,* Methoden der organischen Chemie, Bd. X/2, p. 40, Georg Thieme Verlag, Stuttgart 1967. F.P. 1 364 573 (1964), Société d'Electro, Inv.: *P. Besson, A. Nallet, G. Luiset;* C. A. *61,* 11892b (1964).

[354] *A. Hajós,* Komplexe Hydride, p. 202, Deutscher Verlag der Wissenschaften, Berlin 1966.

[355] *C. Hanna, F. W. Schueler,* J. Amer Chem. Soc. *74,* 3693 (1952).

[356] *H. Smith und Mitarbb.,* J. Heterocycl. Chem. *5,* 757 (1968).

[357] *F. W. Schueler, C. Hanna,* J. Amer. Chem. Soc. *73,* 4996 (1951); *E. Müller* in: *Houben-Weyl,* Methoden der organischen Chemie, Bd. X/2, p. 41, Georg Thieme Verlag, Stuttgart 1967.

[358] *C. G. Overberger, L. P. Herlin,* J. Org. Chem. *27,* 417 (1962).

[359] *G. Neurath, B. Pirmann, M. Dünger,* Chem. Ber. *97,* 1631 (1964).

[360] F.P. 1 186 902 (1959), Frz. Staat, Inv.: *Desseigne, Cohen;* C. A. *56,* 320d (1962).

[361] USP. 2 916 426 (1959), Nat. Dist. and Chemical Corp., Inv.: *D. Horvitz, E. Cerwonka;* C. A. *54,* 6370c (1960).

[362] D.B.P. 1 078 134 (1960), 1 085 535 (1958), Chemische Fabrik Kalk, Erf.: *H. J. Schmidt, H. Nees;* C. A. *55,* 14309d (1961), *56,* 8562a (1962); *E. Müller* in: *Houben-Weyl,* Methoden der organischen Chemie, Bd. X/2, p. 42, Georg Thieme Verlag, Stuttgart 1967.

[363] *H. Dorn, H. Dilcher, K. H. Schwarz,* Chem. Ber. *99,* 2620 (1966).

[364] *E. Müller* in: *Houben-Weyl,* Methoden der organischen Chemie, Bd. X/2, p. 51, Georg Thieme Verlag, Stuttgart 1967.

[365] *J. Deles,* Roczniki Chem. *39,* 317 (1965); C.A. *63,* 1719f (1965).

[366] *H. U. Daeniker,* Helv. Chim. Acta *50,* 2008 (1967).

$$R-NH_2 + CH_2O + HCN \longrightarrow R-NH-CH_2-CN \xrightarrow{HCl}$$

$$R-NH-CH_2-CO_2H \xrightarrow{HNO_2} R-N-CH_2-CO_2H \xrightarrow{(CF_3CO)_2O}$$
$$\underset{NO}{}$$

$$R-N\overset{\oplus}{\underset{N-O}{\bigcirc}} \xrightarrow{HCl} R-NH-NH_2$$

*1-Hydrazino-
adamantane*

R = 1-Adamantyl

Using this technique other *monoalkylhydra-zines*[367] and *1,ω-alkanedihydrazines*[364, 368], too, have been prepared.

4.2.1.3.2 Reduction of alkyl and acyl hydrazones

For the preparation of aliphatic hydrazones see p. 118. Reduction of the *C=N* double bond is an important method for preparing alkylhydrazines. Unsubstituted keto hydrazones afford branched monoalkylhydrazines, monoalkyl hydrazones give *N,N'-dialkyl*hydrazines, and 1,1-dialkylhydrazones *trialkylhydrazines*.

$$\underset{R^2}{\overset{R^1}{>}}C=N-N\underset{R^4}{\overset{R^3}{<}} \longrightarrow \underset{R^2}{\overset{R^1}{>}}CH-NH-N\underset{R^4}{\overset{R^3}{<}}$$

R^1, R^2 = Alkyl, R^3, R^4 = H or Alkyl

Hydrogenation of aldehyde and ketone acyl hydrazones has proven to be particularly successful. *N'-Acyl-N-alkylhydrazines* are obtained whose hydrolysis leads to *monoalkyl*hydrazines:

$$\underset{R^2}{\overset{R^1}{>}}C=N-NH-COR^3 \longrightarrow \underset{R^2}{\overset{R^1}{>}}CH-NH-NH-COR^3$$

$$\longrightarrow \underset{R^2}{\overset{R^1}{>}}CH-NH-NH_2$$

This is at the same time the *best* route for preparing *N'-acyl-N-alkylhydrazines*, because acylation of alkylhydrazines seldom furnishes a uniform product[369].

4.2.1.3.2.1 Catalytic reduction of alkyl and acyl hydrazones

Mainly on account of the tendency for the *N—N* bond to undergo hydrogenolysis, catalytic hydrogenation of the *C=N* double bond in nonacy-

lated hydrazones is of only limited importance. Symmetrical *N,N'-dialkylhydrazines* are prepared from azines (see p. 109). Hydrogenation of monoalkyl hydrazones with platinum to give acceptable yields of *unsymmetrical N,N'-dialkylhydrazines* is as a rule favored by starting with keto hydrazones[370, 371]. Aldehyde hydrazones suffer *N—N* cleavage more readily[371]. *Trialkylhydrazines* are generally more advantageously prepared by reduction with complex hydrides from dialkyl hydrazones (see below), but a platinum-catalyzed hydrogenation often affords satisfactory results[372].

Preparation of *monoalkylhydrazines* from unsubstituted hydrazones is made harder by the difficulties encountered in preparing the hydrazones in the pure state, *e.g.*, free from azine. Nonetheless, many articles describe the hydrogenation of such hydrazones with platinum catalysts[210, 370, 373]. A considerable simplification consists in hydrogenating the mixture of equivalent amounts ketone, hydrazine, and hydrochloric acid with platinum directly without isolating the hydrazone[371, 374, 375]:

$$\bigcirc\!\!=\!\!O + N_2H_4 + HCl \xrightarrow{Pt/H_2} \bigcirc\!\!-NH-NH_2 \cdot HCl$$

Cyclopentylhydrazine

Working with acyl hydrazones offers several advantages. Thus, they are formed as uniform products from acyl hydrazine and the carbonyl compound, the risk of an *N—N* cleavage is very much reduced, so that even Raney nickel can be used for the hydrogenation[251, 376]. Aldehyde acylhydrazones give results which are as successful as ketone acylhydrazones and, consequently, unbranched products as well can be obtained. Alkoxycarbonyl hydrazones of aldehydes or ketones can be hydrogenated with a platinum-on-

[367] W. Baker, W.D. Ollis, V.D. Poole, J. Chem. Soc. (London) 307 (1949).

[368] N.P. Zapevalova und Mitarbb., Izv. Akad. Nauk SSSR 1369 (1969); Engl.: 1266.

[369] H. Dorn, A. Zubeck, K. Walter, Justus Liebigs Ann. Chem. 707, 100 (1967).

[370] USP. 3 000 903 (1959), Lakeside Labs., Inv.: J.H. Biel; C.A. 56, 1393e (1962).

[371] H. Dorn, K. Walter, Justus Liebigs Ann. Chem. 720, 98 (1968).

[372] USP. 3 517 064 (1970), Tenneco Chemicals Inc., Inv.: H. Sidi; C.A. 73, 120065d (1970).

[373] USP. 3 095 448 (1963), Lakeside Labs., Inv.: J.H. Biel; C.A. 59, 12661a (1963).

[374] D.A.S. 1 082 258 (1958), Ciba, Erf.: J. Druey und Mitarbb.; C.A. 55, 23384b (1961);
E. Müller in: Houben-Weyl, Methoden der organischen Chemie, Bd. X/2, p. 44, Georg Thieme Verlag, Stuttgart 1967.

[375] D.A.S. 1 106 335, 1 107 236, 1 107 237 (1958), Ciba, Erf.: J. Druey und Mitarbb.; Chem. Zentr. 16197 (1961), 4256 (1962).

charcoal catalyst under normal conditions to form *1-alkylhydrazine-2-carboxylic acid esters* whose alkaline splitting yields the *monoalkylhydrazine*[377], *e.g.:*

$$H_5C_2OOC-NH-NH_2 \xrightarrow{H_3C-CO-CH_3}$$

$$H_5C_2OOC-NH-N=C\begin{array}{c}CH_3\\CH_3\end{array} \xrightarrow{Pt/C/H_2}$$

$$H_5C_2OOC-NH-NH-CH(CH_3)_2 \xrightarrow{NaOH}$$

Ethyl 3-isopropylcarbazate

$$H_2N-NH-CH(CH_3)_2 \; + \; CO_2 \; + \; C_2H_5OH$$

Isopropylhydrazine

Hydrogenation of benzyloxycarbonyl hydrazones furnishes the hydrazine directly by hydrogenolysis of the ester group[378].
Acid hydrolysis of *N'-acyl-N-alkylhydrazines* obtainable from formyl hydrazones[369], acetyl hydrazones[251, 379, 380], propionyl hydrazones[380], or benzoyl hydrazones[381] of ketones or aldehydes by hydrogenation with platinum catalysts proceeds especially smoothly, and this method, too, can be employed for a whole series of *monoalkylhydrazines* in the form of a *single-vessel technique* with Raney nickel[376] *e.g.:*

$$\langle\rangle-CHO \; + \; H_2N-NH-CO-CH_3 \longrightarrow$$

$$\langle\rangle-CH=N-NH-CO-CH_3 \xrightarrow{Ni/H_2}$$

$$\langle\rangle-CH_2-NH-NH-CO-CH_3$$

$$\xrightarrow{HCl} \langle\rangle-CH_2-NH-NH_2 \cdot HCl$$

(Cyclohexylmethyl)hydrazine

It is advisable to acylate before carrying out the hydrogenation of monoalkylhydrazones[251], *e.g.:*

$$(H_3C)_2CH-NH-N=\langle\rangle N-CH_3 \xrightarrow{H_3C-COCl}$$

$$(H_3C)_2CH-N-N=\langle\rangle N-CH_3$$
$$\qquad\qquad \underset{CO-CH_3}{|}$$

$$\xrightarrow{Pt/H_2} (H_3C)_2CH-N-NH-\langle\rangle N-CH_3$$
$$\qquad\qquad\qquad \underset{CO-CH_3}{|}$$

*N-Acetyl-N-isopropyl-N'-
(1'-methyl-4-piperidylidene)-
hydrazine*

$$\xrightarrow{HCl} (H_3C)_2CH-NH-NH-\langle\rangle N-CH_3$$

*N-Isopropyl-N'-(1-methyl-4-
piperidylidene)hydrazine*

Finally, a reminder is given of the possibility of reducing hydrazones with sodium amalgam[382, 383].

4.2.1.3.2.2 Reduction with complex hydrides or diborane

A considerable number of hydrazones have been reduced to hydrazines with lithium tetrahydroaluminate. Phenylhydrazones of aliphatic aldehydes give *unbranched N-alkyl-N'-phenylhydrazines* in these circumstances[384], those of ketones *branched* ones[251].
Trialkylhydrazines have been obtained predominantly from ketone dialkylhydrazones[385], in the simplest case *trimethylhydrazine* from *N,N*-dimethyl-*N'*-methylenehydrazine[386]:

$$(H_3C)_2N-N=CH_2 \xrightarrow{LiAlH_4} (H_3C)_2N-NH-CH_3$$

The cleanest procedure is probably reduction of benzaldehyde alkylhydrazones with diborane to

[376] D.B.P. 1003215 (1954), Farbw. Hoechst, Erf.: *G. Ehrhart, W. Krohs*; C.A. *54*, 7736c (1960);
E. Müller in: *Houben-Weyl*, Methoden der organischen Chemie, Bd. X/2, p. 47, Georg Thieme Verlag, Stuttgart 1967.
[377] Schweiz. P. 307629, 309770 (1952), I.R. Geigy, Erf.: *C. Simon*; C.A. *51*, 5113f.g(1957);
E. Müller in: *Houben-Weyl*, Methoden der organischen Chemie, Bd. X/2, p. 46, Georg Thieme Verlag, Stuttgart 1967.
[378] *C.G. Overberger, L.C. Palmer, B.S. Marks, N.R. Byrd*, J. Amer. Chem. Soc. *77*, 4100 (1955).
[379] *E. Bellasio, A. Ripamonti, E. Testa*, Gazz. Chim. Ital. *98*, 3 (1968).
[380] *H. Röhnert*, Z. Chem. *5*, 302 (1965).
[381] *M. Freifelder, W.B. Martin, G.S. Stone, E.L. Coffin*, J. Org. Chem. *26*, 383 (1961).

[382] *H. Franzen, F. Krafft*, J. Prakt. Chem. [2] *84*, 137 (1911).
[383] *B. Gisin, M. Brenner*, Helv. Chim. Acta *53*, 1030 (1970).
[384] *K. Kratzl, K.P. Berger*, Monatsh. Chem. *89*, 83 (1958).
[385] *G. Zinner, H. Böhlke, W. Kliegel*, Arch. Pharm. (Weinheim, Ger.) *299*, 245 (1966).
[386] *J.B. Class, J.G. Aston, T.S. Oakwood*, J. Amer. Chem. Soc. *75*, 2937 (1953);
E. Müller in: *Houben-Weyl*, Methoden der organischen Chemie, Bd. X/2, p. 48, Georg Thieme Verlag, Stuttgart 1967.

give *N-alkyl-N'-benzylhydrazines* in very good yield without the need to isolate the hydrazones[387], *e.g.:*

$$H_5C_2-NH-NH_2 \ + \ OCH-C_6H_5 \longrightarrow$$

$$(H_5C_2-NH-N=CH-C_6H_5) \xrightarrow{B_2H_6} H_5C_2-NH-NH-CH_2-C_6H_5$$

N-Benzyl-N'-ethylhydrazine

It is in many instances possible to employ sodium tetrahydroborate instead of lithium tetrahydroaluminate. The reduction with this reagent succeeds best if 1,1-dialkyl hydrazones are used[388]. Bis(2-chloroethyl) hydrazones are reduced efficiently only if they are derived from ketones[389], *e.g.:*

$$(Cl-CH_2-CH_2)_2N-N=C\begin{matrix}CH_3\\CH_3\end{matrix}$$

$$\xrightarrow{NaBH_4} (Cl-CH_2-CH_2)_2N-NH-CH(CH_3)_2$$

N,N-Bis-(2-chloroethyl)-N'-
isopropylhydrazine

The reaction has been transferred to keto steroids[390]. Tetrahydroborate reduction of acyl hydrazones in boiling ethanol gives *N-acyl-N'-alkylhydrazines*[391–393,574], *e.g.:*

$$H_2N-\underset{Cl}{\overset{Cl}{\bigcirc}}-CO-NH-N=CH-C_6H_5$$

$$\xrightarrow{NaBH_4} H_2N-\underset{Cl}{\overset{Cl}{\bigcirc}}-CO-NH-NH-CH_2-C_6H_5$$

N-(4-Amino-3,5-dichloro)-
benzoyl-N'-benzylhydrazine

but the catalytic hydrogenation is of more general significance.

During reduction of acyl hydrazones with lithium tetrahydroaluminate the aminocarbonyl group is reduced along with the *C=N* double bond. Mixtures of *N'-acyl-N-alkylhydrazines* and *N,N'-di-*

alkylhydrazines are obtained[394]; the latter can be isolated in good yields only if keto hydrazones are the starting materials[395], *e.g.:*

$$H_3C-CO-NH-N=C\begin{matrix}CH_3\\CH_3\end{matrix} \xrightarrow{LiAlH_4}$$

$$H_3C-CH_2-NH-NH-CH(CH_3)_2$$

N-Ethyl-N'-isopropylhydrazine

Aldehyde acyl hydrazones gave inferior results.

4.2.1.3.3 Preparation and reduction of azines

4.2.1.3.3.1 Preparation of azines

Symmetrical aldazines and *ketazines* are often formed spontaneously from two equivalents of the carbonyl compound and one equivalent hydrazine. The hydrazone stage which the reaction passes through is not noticeable at least with aldazines. Aliphatic aldazines are prepared by adding hydrazine hydrate drop by drop to the ether solution of the aldehyde in the cold[396–398], *e.g.:*

$$2 \ H_3C-CHO \ + \ N_2H_4 \longrightarrow H_3C-CH=N-N=CH-CH_3$$

Acetaldazine

As the azines are resistant to bases, working up and drying of the products ensues in the presence of potassium carbonate. *Isobutyraldehyde azine*[399] is prepared in the presence of sodium carbonate. *Benzaldehyde azine* is obtained from benzaldehyde and hydrazine in aqueous ammonia[400]. *Formaldehyde azine* behaves anomalously; when prepared from formaldehyde and hydrazine it is present as *octahydro-1,2,4,5-tetrazino[1,2-a]-1,2,4,5-tetrazine*, but pyrolysis furnishes the unstable monomeric *formaldehyde azine*[401]:

$$3 \ N_2H_4 \ + \ 4 \ CH_2O \longrightarrow$$

$$\underset{HN}{\overset{HN}{\diagup}}\underset{N}{\diagdown}\underset{N}{\overset{N}{\diagup}}\underset{NH}{\overset{NH}{\diagdown}} \xrightarrow{\triangledown} H_2C=N-N=CH_2$$

[387] *J.A. Blair, R.J. Gardner,* J. Chem. Soc. C. 1714 (1970).

[388] *G.N. Walker, M.A. Moore, B.N. Weaver,* J. Org. Chem. *26*, 2740 (1961).

[389] *W. Schulze, G. Letsch, H. Fritzsche,* J. Prakt. Chem. [4] *33*, 96 (1966).

[390] *I.B. Dlikman, D.S. Bidnaya,* Zh. Obshch. Khim. *39*, 1642 (1969); Engl.: 1610.

[391] *R.S. Varma,* J. Prakt. Chem. [4] *38*, 260 (1968).

[392] E.P. 921322 (1963), C.E. Frosst & Co., Inv.: *C.H. Yates, L.M. Thompson;* C.A. *60*, 4064f (1964).

[393] F. Med. P. 5439 (1967), S.I.F.A., Inv.: *J.R. Boissier, R. Ratouis;* C.A. *71*, 38449g (1969).

[394] *R.L. Hinman,* J. Amer. Chem. Soc. *79*, 414 (1957).

[395] *L. Spialter und Mitarbb.,* J. Org. Chem. *30*, 3278 (1965).

[396] *D. Kolbach, D. Korunčev* in: *Houben-Weyl,* Methoden der organischen Chemie, Bd. X/2, p. 91, Georg Thieme Verlag, Stuttgart 1967.

[397] *R. Renaud, L.C. Leitch,* Can. J. Chem. *32*, 545 (1954).

[398] *J. Elguero, R. Jacquier, C. Marzin,* Bull. Soc. Chim. France 713 (1968).

[399] *A. Franke,* Monatsh. Chem. *20*, 847 (1899); E. Müller in *Houben-Weyl,* Methoden der organischen Chemie, Bd. X/2, p. 92, Georg Thieme Verlag, Stuttgart 1967.

Simple aliphatic ketazines such as *acetone azine* are often prepared without solvent by heating[398] similar to the aldazines[402]. With branched aliphatic and especially araliphatic or *benzophenone azines* it is necessary to ensure that the hydrazone stage is reacted completely to azine. Heating the components in ethanol[398], if necessary with added hydrochloric acid[403], or boiling in benzene using a water separator[398] is then necessary. Benzophenone azine is prepared best from the hydrazone[404] in ethanol with sulfuric acid:

$$2 \ (H_5C_6)_2C=N-NH_2 \longrightarrow$$

$$(H_5C_6)_2C=N-N=C(C_6H_5)_2 \quad + \quad N_2H_4$$

In general, hydrazones, acyl hydrazones, and semicarbazones undergo disproportionation to azines on heating with or without acid.

From 1, 5-diketones and hydrazine *cyclic azines* are obtained[405]. *Sterically hindered* azines are formed from the carbonyl compounds with salts of hydrazine in polyphosphoric acid[406].

Unsymmetrical azines can be prepared by heating hydrazones with the desired carbonyl compound[398, 407], but disproportionation into two symmetrical azines easily takes place. This result can be avoided by employing the carbonyl component as a Schiff's base or the hydrazine as a phosphazine[408].

Symmetrical and unsymmetrical azines can be obtained from diaziridines and carbonyl compounds with accompanying acid catalysis[329]:

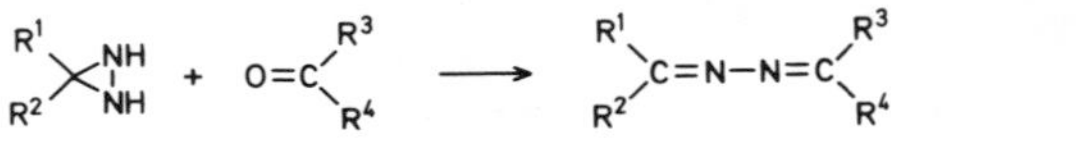

Several patent specifications claim the preparation of symmetrical azines by this method[409–411].
The configuration of azines and their *NMR* spectra have been investigated in detail[398, 412].

4.2.1.3.3.2 Catalytic hydrogenation of azines

Azines can be reduced to the symmetrical N,N'-dialkylhydrazines in acid solution with platinum or palladium catalysts[413, 414]. N,N'-Diisopropylhydrazine is obtained by hydrogenating acetone azine with platinum in glacial acetic acid[413],

$$(H_3C)_2C=N-N=C(CH_3)_2$$

$$\xrightarrow{Pt/H_2} \quad (H_3C)_2CH-NH-NH-CH(CH_3)_2$$

but it is not necessary to prepare the azine separately. The mixture of hydrazine and acetone can be hydrogenated in ethanolic hydrochloric acid[415]. N,N'-*Dicycloalkylhydrazines*, too, are obtained by the platinum-catalyzed hydrogenation of cycloalkanone and hydrazine in ethanol-glacial acetic acid[416], *e.g.*:

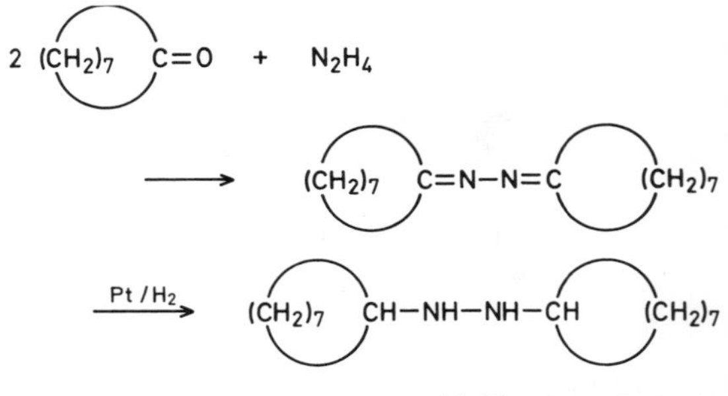

1,2-Dicyclooctylhydrazine

[400] *H.H. Hatt*, Org. Syntheses, Coll. Vol. II, 395 (1943).
[401] *N.P. Neureiter*, J. Amer. Chem. Soc. *81*, 2910 (1959).
[402] USP. 2 962 532 (1957), Aerojet-General Corp., Inv.: *A.F. Graefe*; C.A. *55*, 9281b (1961); *E. Müller* in: *Houben-Weyl*, Methoden der organischen Chemie, Bd. X/2, p. 95, Georg Thieme Verlag, Stuttgart 1967.
[403] *D. Kolbach, D. Korunčev* in: *Houben-Weyl*, Methoden der organischen Chemie, Bd. X/2, p. 95, Georg Thieme Verlag, Stuttgart 1967.
[404] *H.H. Szmant, C. McGinnis*, J. Amer. Chem. Soc. *72*, 2890 (1950).
[405] *C.G. Overberger, J.J. Monagle*, J. Amer. Chem. Soc. *78*, 4470 (1956); *E. Müller* in: *Houben-Weyl*, Methoden der organischen Chemie, Bd. X/2, p. 96, Georg Thieme Verlag, Stuttgart 1967.
[406] *D.B. Mobbs, H. Suschitzky*, J. Chem. Soc. C 175 (1971).
[407] *E. Müller* in: *Houben-Weyl*, Methoden der organischen Chemie, Bd. X/2, p. 104, Georg Thieme Verlag, Stuttgart 1967.
[408] *E. Müller* in: *Houben-Weyl*, Methoden der organischen Chemie, Bd. X/2, p. 105, 106, 110, Georg Thieme Verlag, Stuttgart 1967.

[409] Niederl. Anw. 6 411 384 (1965), Bergwerksverband G.m.b.H.; C.A. *64*, 3358d (1966).
[410] Niederl. Anw. 6 516 176, 6 605 269 (1966), Whiffen & Sons Ltd.; C.A. *65*, 14904g (1966), *66* 75662p (1967).
[411] E.P. 1 164 435 (1969), Fisons Ind. Chem. Ltd., Inv.: *B.J. Needham, M.A. Smith*; C.A. *71*, 112906x (1969).
[412] *J. Elguero, R. Jacquier, C. Marzin*, Bull. Soc. Chim. France 1375 (1969).
[413] *K.A. Taipale*, Ber. *56*, 954 (1923).
[414] *H. Röhnert*, Arch. Pharm. (Weinheim, Ger.) *296*, 296 (1963).
[415] *H.L. Lochte, W.A. Noyes, J.R. Bailey*, J. Amer. Chem. Soc. *44*, 2564 (1922).
[416] *A.C. Cope, J.E. Engelhart*, J. Amer. Chem. Soc. *90*, 7092 (1968).

Benzaldehyde azine is hydrogenated to *N,N'-dibenzylhydrazine* with platinum in ethanolic hydrochloric acid[417], benzophenone azine requires forced hydrogenation conditions in order to form *N,N'-bis(diphenylmethyl)hydrazine*[418]. Certain cyclic azines have been hydrogenated to the cyclic hydrazines with palladium on charcoal in tetrahydrofuran[419].

A special variant for forming *branched N,N'-dialkyl-hydrazines* consists in initially reacting the azine with one equivalent phenylmagnesium bromide or phenyllithium and then hydrogenating the hydrazone obtained[420], *e.g.*:

$$H_3C-CH=N-N=CH-CH_3$$

$$\xrightarrow{H_5C_6-MgBr}$$

$$H_5C_6-\underset{\underset{CH_3}{|}}{CH}-NH-N=CH-CH_3$$

$$\xrightarrow{Pt/H_2}$$

$$H_5C_6-\underset{\underset{CH_3}{|}}{CH}-NH-NH-CH_2-CH_3$$

N-Ethyl-N'-(1-phenylethyl)-hydrazine

4.2.1.3.3.3 Reduction of the azines with lithium tetrahydroaluminate

Reduction of the azines with lithium tetrahydroaluminate leads to *N,N'-dialkylhydrazines* in good yield. This technique is particularly suitable with the lower aliphatic azines[421], *e.g.*:

$$H_3C-CH=N-N=CH-CH_3$$

$$\xrightarrow{LiAlH_4}   H_3C-CH_2-NH-NH-CH_2-CH_3$$

N,N'-Diethylhydrazine

Substituted azines also are a possibility[422].
Only one single example of the reduction of azines with diborane has become known to date[387].

4.2.1.3.4 Reduction of acylhydrazines

It has already been emphasized in the section describing the lithium tetrahydroaluminate reduction of the acyl hydrazones (see p. 108) that reduction of amides to amines is generally difficult. As was shown there, acylhydrazines with a free *NH* group are reduced very slowly and often incompletely[423], while *N*-alkylated hydrazides react more smoothly. Thus *tetramethylhydrazine* is obtained from *N,N'-diformyl-N,N'-dimethylhydrazine*[424, 575], or, in better yield, from trimethylformylhydrazine[425]. Other *N,N'-diacyl-N,N'-dialkylhydrazines*, too, are reduced smoothly[423], and the same is true for example, of benzoyltrimethylhydrazine (giving *benzyltrimethylhydrazine*)[426].

$$H_3C-\underset{\underset{}{|}}{\overset{\overset{OHC}{|}}{N}}-\underset{\underset{}{|}}{\overset{\overset{CHO}{|}}{N}}-CH_3$$

$$(H_3C)_2N-\underset{\underset{CH_3}{\diagdown}}{N}^{\diagup CHO}$$

$$\longrightarrow   (H_3C)_2N-N(CH_3)_2$$

Of derivatives with an *NH* group diethyl bicarbamate (hydrazine-1,2-dicarboxylate) gives a good yield of *N,N'-dimethylhydrazine*[423] on prolonged reaction:

$$H_5C_2-OOC-NH-NH-COO-C_2H_5$$

$$\xrightarrow{LiAlH_4}   H_3C-NH-NH-CH_3$$

Certain other results also are not in accord with the rule in respect of yield. Thus, *N,N*-dimethyl-*N'*-formylhydrazine is converted in good yield to *trimethylhydrazine*[425], and *N*-acyl-*N'*-phenylhydrazines were invariably reduced smoothly[384]. Pivalic acid hydrazide affords a good yield of *neopentylhydrazine*[325] if aluminum chloride is present during the reaction. Likewise, diborane is a better reducing agent for hydrazides than lithium tetrahydroaluminate. *N,N'*-Diacylhydrazines are reduced in high yield at 130° in diglyme[427], al-

[417] *H.H. Fox, J.T. Gibas*, J. Org. Chem. *20*, 60 (1955).
[418] *S.G. Cohen, C.H. Wang*, J. Amer. Chem. Soc. *77*, 2457 (1955).
[419] *C.G. Overberger, I. Tashlick*, J. Amer. Chem. Soc. *81*, 217 (1959);
E. Müller in: *Houben-Weyl*, Methoden der organischen Chemie, Bd. X/2, p. 45, Georg Thieme Verlag, Stuttgart 1967.
[420] *C.G. Overberger, A.V. DiGiulio*, J. Amer. Chem. Soc. *80*, 6562 (1958).
[421] *R. Renaud, L.C. Leitch*, Can. J. Chem. *32*, 545 (1954);
E. Müller in: *Houben-Weyl*, Methoden der organischen Chemie, Bd. X/2, p. 43, Georg Thieme Verlag, Stuttgart 1967.
[422] *J.H. Biel, A.F. Drukker, T.F. Mitchell jr.*, J. Amer. Chem. Soc. *82*, 2204 (1960).

[423] *R.L. Hinman*, J. Amer. Chem. Soc. *78*, 1645, 2463 (1956).
[424] *J.B. Class, J.G. Aston, T.S. Oakwood*, J. Amer. Chem. Soc. *75*, 2937 (1953);
E. Müller in: *Houben-Weyl*, Methoden der organischen Chemie, Bd. X/2, p. 50, Georg Thieme Verlag, Stuttgart 1967.
[425] *R.T. Beltrami, E.R. Bissell*, J. Amer. Chem. Soc. *78*, 2467 (1956).
[426] *D.L. Griffith, J.D. Roberts*, J. Amer. Chem. Soc. *87*, 4089 (1965).
[427] *H. Feuer, F. Brown jr.*, J. Org. Chem. *35*, 1468 (1970).

though here, too, the *N*-alkylated acylhydrazines react more smoothly [427, 428].

$$H_5C_2 - CO - NH - NH - CO - C_2H_5$$

$$\xrightarrow{B_2H_6 \ / \ THF} \quad H_7C_3 - NH - NH - C_3H_7$$

N,N'-Dipropylhydrazine

4.2.1.4 Addition to azodicarboxylic acid diesters

Azodiformic acid diesters advantageously [429] undergo *diene synthesis* on irradiation to form *1,2, 3,6-tetrahydro-1,2-pyridazine dicarboxylic acid diesters*, which are hydrogenated, hydrolyzed, and decarboxylated to cyclic hydrazines. *2,3-Di-azabicyclo[2.2.1]heptane* [430]

is obtained from cyclopentadiene, *2,3-diazabicy-clo[2.2.2]octane* from cyclohexadiene [429, 430], and *hexahydropyridazine* [431] from butadiene. A survey describes these and other additions to azodiformic acid [432].

4.2.2 Transformation of aliphatic and araliphatic hydrazines and analysis

4.2.2.1 Properties and protonation

Aliphatic and araliphatic hydrazines are colorless, strongly basic compounds which are generally liquid or crystallize with difficulty. Surprisingly, the degree of basicity relative to hydrazine itself decreases as the number of alkyl groups increases. The toxic and carcinogenic properties have already been referred to.

While there is unanimity that alkylation (see p. 92) and nitrosation of monoalkylhydrazines ensues on the substituted *N*, the site of *protonation* is not fixed unambiguously. In mono- and *N,N*-dial-kylhydrazines the correlation between basicity and Taft constant [433] and also *IR* data [434] point to protonation of the substituted *N*, while phenylhydrazines from *IR* data [434], and *p*-substituted benzylhydrazines also from their ρ value [435] should be protonated on the free $-NH_2$ group. Excess acid converts them into the *divalent salts* from which the free hydrazines can be reliberated with concentrated alkali. This reaction can be appended to all methods of preparation leading to hydrazinium salts.

4.2.2.2 Acylation

With isocyanates and isothiocyanates monoalkyl-hydrazines react on the substituted *N* to form *2-al-kylsemicarbazides* [216, 324].

$$R-NH-NH_2 \xrightarrow{KOCN} \overset{\overset{\displaystyle R}{|}}{H_2N-CO-N-NH_2}$$

The results of acylation with anhydrides and esters were confusing until a systematic study on methylhydrazine showed that with anhydrides it was predominantly the substituted *N*, and with esters mainly the free NH_2 that was acylated [436, 576]. Preferment for the free NH_2 is the greater the more demanding the hydrazine substituents and the esters are in respect of space. While methylhydrazine still reacts with a variety of esters [369], *N,N*-dimethylhydrazine can be acylated solely with esters of formic acid. These latter quite generally acylate mor cleanly than others.

Methylhydrazine, butylhydrazine, benzylhydrazine, and 2-hydrazinopropionitrile [(2-cyanoethyl)hydrazine] are converted uniformly and in good yield into *N-formyl-N-methylhydrazine* and the corresponding *butyl, benzyl,* and *2-cyanoethyl* compounds with methyl formate, while by contrast cyclohexylhydrazine yields uniformly *N-cyclohexyl-N'-formylhydrazine* because of the bulk of the cyclohexyl group [369]:

[428] *M.J. Kornet, P.A. Thio, S.L. Tan,* J. Org. Chem. *33* 3637 (1968).

[429] *R. Askani,* Chem. Ber. *98,* 2551 (1965).

[430] *S.G. Cohen, R. Zand,* J. Amer. Chem. Soc. *84,* 586 (1962); *E. Müller* in: *Houben-Weyl,* Methoden der organischen Chemie, Bd. X/2, p. 54, Georg Thieme Verlag, Stuttgart 1967.

[431] *K. Alder, H. Nicklas,* Justus Liebigs Ann. Chem. *585,* 81 (1954).

[432] *E. Fahr, H. Lind,* Angew. Chem. *78,* 376 (1966); *E. Fahr, H. Lind,* Angew. Chem. Intern. Ed. Engl. *5,* 372 (1966).

[433] *F.E. Condon,* J. Amer. Chem. Soc. *87,* 4491 (1965).

[434] *R.F. Evans, W. Kynaston,* J. Chem. Soc. 3151 (1963).

[435] *J. Deles,* Roczniki Chem. *43,* 1165 (1969); C.A. *72,* 2912j (1970).

[436] *R.L. Hinman, D. Fulton,* J. Amer. Chem. Soc. *80,* 1895 (1958).

[437] *F. Höfler, U. Wannagat,* Monatsh. Chem. *97,* 1598 (1966).

$$H_9C_4-NH-NH_2 \quad + \quad HCOOCH_3$$

$$\longrightarrow \quad \underset{\underset{CHO}{|}}{H_9C_4-N-NH_2}$$

Catalytic hydrogenation of the acyl hydrazones is the best method for cleanly preparing *N-acyl-N'-alkylhydrazines* (see p. 107).

Reacting monoalkylhydrazines with excess anhydride leads to *1,2-diacyl-alkylhydrazines*.

4.2.2.3 Silazanes from alkylhydrazines

An extensive literature describes the silylation of alkylhydrazines. The following example is characteristic. On reacting methylhydrazine with trimethylchlorosilane first *N-methyl-N'-trimethylsilylhydrazine* and then *N-methyl-N,N'-bis(trimethylsilyl)hydrazine* is obtained; the latter can be silylated further to *methyltris(trimethylsilyl)hydrazine* only indirectly *via* the anion[437]:

$$H_3C-NH-NH_2 \xrightarrow{(H_3C)_3SiCl} H_3C-NH-NH-Si(CH_3)_3$$

$$\xrightarrow{(H_3C)_3SiCl} \underset{\underset{Si(CH_3)_3}{|}}{H_3C-N-NH-Si(CH_3)_3} \xrightarrow{n-H_9C_4-Li}$$

$$\left[\underset{\underset{\overset{\ominus}{\ominus}}{H_3C-N-\bar{N}-Si(CH_3)_3}}{\overset{Si(CH_3)_3}{|}} \right] \xrightarrow{(H_3C)_3SiCl}$$

$$\underset{\underset{Si(CH_3)_3}{|}}{H_3C-N-N[Si(CH_3)_3]_2}$$

The reactions with *N,N'-* and *N,N*-dimethylhydrazine proceed analogously[438]:

$$H_3C-NH-NH-CH_3 \xrightarrow{(H_3C)_3SiCl} \underset{\underset{Si(CH_3)_3}{|}}{H_3C-N-NH-CH_3}$$

N,N'-Dimethyl-N-trimethylsilylhydrazine

$$\xrightarrow{n-H_9C_4-Li} \left[\underset{\underset{\overset{\ominus}{\ominus}}{H_3C-N-\bar{N}-CH_3}}{\overset{Si(CH_3)_3}{|}} \right] \xrightarrow{(H_3C)_3SiCl}$$

$$\underset{\underset{H_3C-N-N-CH_3}{}}{(H_3C)_3Si \quad Si(CH_3)_3}$$

N,N'-Dimethyl-N,N'-bis-(trimethylsilyl)hydrazine

---

[438] *U. Wannagat, F. Höfler*, Monatsh. Chem. 97, 976 (1966).

$$(H_3C)_2N-NH_2 \xrightarrow{(H_3C)_3SiCl} (H_3C)_2N-NH-Si(CH_3)_3$$

N,N-Dimethyl-N'-trimethylsilylhydrazine

$$\xrightarrow{n-H_9C_4-Li} [(H_3C)_2N-\overset{\ominus}{\bar{N}}-Si(CH_3)_3]$$

$$\xrightarrow{(H_3C)_3SiCl} (H_3C)_2N-N[Si(CH_3)_3]_2$$

N,N-Dimethyl-N',N'-bis-(trimethylsilyl)hydrazine

All products and especially the *N*-anions tend to undergo rearrangements[439] which can easily lead to erroneous conclusions.

4.2.2.4 Enehydrazines from alkyl- and aralkylhydrazines

Aliphatic hydrazines that are unable to form hydrazones can form enehydrazines or their transformation products with enolizable carbonyl compounds. With butyraldehyde trimethylhydrazine gives *(1-buten-1-yl)trimethylhydrazine*[440]:

$$(H_3C)_2N-NH-CH_3 \quad + \quad OCH-C_3H_7$$

$$\longrightarrow \quad \underset{\underset{CH_3}{|}}{(H_3C)_2N-N-CH=CH-CH_2-CH_3}$$

N,N'-Dialkyl-*N*-phenylhydrazines form enehydrazines which readily undergo Fischer ring closure with formation of indoles[441, 442]. A remarkable example of this reaction is the preparation of *9-(3-aminopropyl)-1,2,3,4-tetrahydrocarbazole* from 1-phenylpyrazolidine[443]:

2-(1-Cyclohexen-1-yl)-1-phenylpyrazolidine

---

[439] *R. West, M. Ishikawa, R.E. Bailey*, J. Amer. Chem. Soc. 89, 4068, 4072 (1967).

[440] *G. Zinner und Mitarbb.*, Chem. Ber. 99, 1678 (1966).

[441] *P. Schieß, A. Grieder*, Tetrahedron Lett. 2097 (1969);

A. Grieder, P. Schieß, Chimia 24, 25 (1970).

N,N'-Disubstituted hydrazines are able to undergo similar reactions. Thus N-methyl-N'-phenyl-hydrazines with cyclohexanone among other reactions undergo Fischer ring closure and form *1,2,3,4-tetrahydrocarbazole* via the enehydrazine [N-(1-cyclohexen-1-yl)-N-methyl-N'-phenyl-hydrazine][444], while N,N'-dimethylhydrazine and cyclohexanone react together under acid catalysis by similar ring closure to give *1,2,3,4,5,6,7,8-octahydro-9-methylcarbazole*[445−447]:

N,N'-Bis(1-cyclohexen-1-yl)-N,N'-dimethylhydrazine

The same method is suitable for preparing *symmetrical N-methylpyrroles* from open ketones and aldehydes[446−448], e.g.:

3,4-Diphenyl-1-methylpyrrole

Occasionally, the bis-enehydrazines can be isolated. As an alternative the reaction can be performed with monomethylhydrazine[448]. Carbonyl-stabilized, isolatable enehydrazines are obtained from 1,3-diketones and N,N'-dimethylhydra-

zine[449], and they, too, can still undergo Fischer ring closure under drastic conditions, e.g.:

3-(N,N'-Dimethylhydrazino)-5,5-dimethyl-2-cyclohexen-1-one

3,4,6,7,8,9-Hexahydro-3,3,9-trimethyl-carbazol-1(2H)-one

4.2.2.5 Hexahydrotetrazines

The hydrazones expected from monalkylhydrazines and aldehydes are unstable if α-unbranched aldehydes are used. They dimerize to form hexahydro-1,2,4,5-tetrazines.

Methyl-, benzyl-, and phenylhydrazine[450−452] give the 1,4-disubstituted hexahydrotetrazines with formaldehyde under mild conditions:

R = CH₃	*1,4-Dimethyl-*	
R = C₆H₅	*1,4-Diphenyl-*	*hexahydro-1,2,4,5-*
R = H₅C₆−CH₂	*1,4-Dibenzyl-*	*tetrazine* 75%

[442] *I.I. Grandberg, N.M. Przhevalskij*, Khim. Geterot-sikl. Soedin. 943 (1969); C.A. *72*, 111211a (1970).

[443] *A.N. Kost, P.A. Sviridova, L.A. Golubeva, Y.N. Portnov*, Khim. Geterotsikl. Soedin. 371 (1970); C.A. *73*, 25205r (1970).

[444] *J.P. Chapelle, J. Elguero, R. Jacquier, G. Tarrago*, Bull. Soc. Chim. France 240 (1970).

[445] *E. Schmitz, H. Fechner-Simon*, Org. Prep. Proced. *1*, 253 (1969).

[446] *W. Sucrow, G. Chondromatidis*, Chem. Ber. *103*, 1759 (1970).

[447] *J.P. Chapelle, J. Elguero, R. Jacquier, G. Tarrago*, Bull. Soc. Chim. France 3147 (1970).

[448] *H. Fritz, P. Uhrhan*, Justus Liebigs Ann. Chem. *744*, 81 (1971).

[449] *W. Sucrow, G. Wiese*, Chem. Ber. *103*, 1767 (1970).

[450] *E. Schmitz, R. Ohme*, Monatsber. Deutsch. Akad. Wiss. Berlin *6*, 425 (1964); C.A. *62*, 9136b (1965).

[451] *H. Dorn, H. Dilcher*, Justus Liebigs Ann. Chem. *717*, 104 (1968).

[452] *E. Schmitz, R. Ohme*, Justus Liebigs Ann. Chem. *635*, 82 (1960).

[453] *W. Skorianetz, E. Kováts*, Helv. Chim. Acta *53*, 251 (1970).

[454] *E. Schmitz*, Justus Liebigs Ann. Chem. *635*, 73 (1960).

With acetaldehyde *hexahydro-1,3,4,6-tetramethyl-1,2,4,5-tetrazine* is obtained[453]:

$$2\ H_3C-NH-NH_2\ +\ 2\ H_3C-CHO$$

With formaldehyde *N,N'-dimethylhydrazine* affords *hexahydro-1,2,4,5-tetramethyl-1,2,4,5-tetrazine*[454]. *N*-Alkylsulfuryl-*N'*-methylhydrazines also undergo the reaction[275], and 1,2,3,4-tetrahydrophthalazine serves as a reagent for formaldehyde[232]:

5,9,14,18-Tetrahydro-7H,16H-1,2,4,5-
tetrazino[1,2-b; 4,5-b']diphthalazine

Finally, *N,N'*-dialkylhydrazines can be reacted widely with aromatic[455] and aliphatic[239, 252, 456] aldehydes:

$$2\ H_5C_2-NH-NH-C_2H_5\ +\ 2\ Cl-\!\!\!\big\langle\ \big\rangle\!\!\!-CHO$$

3,6-Bis(4-chlorophenyl)hexahydro-
1,2,4,5-tetramethyl-1,2,4,5-tetrazine

but, depending on the reaction conditions, *2,3,4,5-tetraalkyl-1,3,4-oxadiazolidines* may arise as main products[456-458], *e.g.*:

$$H_3C-NH-NH-CH_3\ +\ 2\ H_5C_2-CHO$$

2,5-Diethyl-3,4-dimethyl-
1,3,4-oxadiazolidine

4.2.2.6 Pyrazoles, pyrazolines, pyrazolidines

Alkylhydrazines undergo ring closure with 1,3-difunctional compounds, albeit often less smoothly than the arylhydrazines[459-463].

4.2.2.6.1 Pyrazoles, pyrazolones

2,4-Pentanedione forms 1,3,5-trimethylpyrazole with methylhydrazine, while unsymmetrical 1,3-dicarbonyl compounds almost invariably afford two isomeric products with alkylhydrazines[459-461, 464], *e.g.*:

$$H_5C_6-CO-CH_2-CO-CH_3$$

| 1,3-Dimethyl-5-phenylpyrazole | 1,5-Dimethyl-3-phenylpyrazole |

The same holds for the preparation of pyrazoles from ethynyl ketones[464]. By contrast, the reaction between alkylhydrazines and β-keto carboxylic acid esters to pyrazolones proceeds more cleanly but not always smoothly[462, 465], *e.g.*:

$$H_3C-CO-CH_2-COOR\ +\ H_3C-NH-NH_2$$

1,3-Dimethyl-2-
pyrazolin-5-one

[455] *R. Grashey, R. Huisgen, K.K. Sun*, J. Org. Chem. *30*, 74 (1965).

[456] *W. Sucrow, H. Bethke, G. Chondromatidis*, Tetrahedron Lett. 1481 (1971).

[457] *B. Zwanenburg, W.E. Weening, J. Strating*, Rec. Trav. Chim. Pays-Bas *83*, 877 (1964).

[458] *L. Eberson, K. Persson*, Acta Chem. Scand. *18*, 721 (1964).

[459] *G. Coispeau, J. Elguero*, Bull. Soc. Chim. France 2717 (1970).

[460] *R. Fusco*, Pyrazoles, Pyrazolines, Pyrazolidines, Indazoles and Condensed Rings, p. 3, R.H. Wiley, Interscience Publ., New York, London 1967.

[461] *C.H. Jarboe*, Pyrazoles, Pyrazolines, Pyrazolidines, Indazoles and Condensed Rings, p. 177, R.H. Wiley, Interscience Publ., New York, London 1967.

[462] *R.H.P. Wiley*, Pyrazolones, Pyrazolidones and Derivatives, Interscience Publ., New York, London 1964.

[463] *A.N. Kost, I.I. Grandberg*, Advances in Heterocyclic Chemistry, Vol. VI, p. 347, Academic Press New York 1966.

[464] *G. Coispeau, J. Elguero, R. Jacquier*, Bull. Soc. Chim. France 689 (1970).

[465] *J. Elguero, R. Jacquier, G. Tarrago*, Bull. Soc. Chim. France 3780 (1967).

but here, too, isomer mixtures sometimes result. In this respect all results of older work should be regarded with some reserve[459].

The similarly proceeding reaction with α-chloro-β-chlorocarbonylenamines[466] leads to mixtures with methylhydrazine:

$$H_3C-CH_2-\underset{\underset{COCl}{|}}{\overset{\overset{Cl}{|}}{C}}=C-N(C_2H_5)_2$$

$\xrightarrow{H_3C-NH-NH_2}$

3-(Diethylamino)-4-ethyl-1-methyl-3-pyrazolin-5-one

+

3-(Diethylamino)-4-ethyl-2-methyl-3-pyrazolin-5-one

4.2.2.6.2 Pyrazolines

α,β-Unsaturated aldehydes and ketones give 2-pyrazolines with alkylhydrazines[461]. Since the attack of the NH_2 group presumably ensues on the carbonyl group, the first step of this reaction is hydrazone formation followed directly by ring closure. Here, too, isomer mixtures are often formed, but ketones more readily lead to uniform products than aldehydes[467], e.g.:

$$H_3C-CH=\underset{\underset{CH_3}{|}}{C}-CHO$$

$\xrightarrow{H_3C-NH-NH_2}$

1,3,4-Trimethyl-2-pyrazoline

+

1,4,5-Trimethyl-2-pyrazoline

$$H_2C=\underset{\underset{CH_3}{|}}{C}-CO-CH_3 \xrightarrow{H_3C-NH-NH_2}$$

N,N'-Dialkylhydrazines afford 3-pyrazolines with α,β-unsaturated ketones[468, 469], e.g.:

$$H_5C_2-NH-NH-C_2H_5 \quad + \quad CH_2O$$

$$+ \quad H_3C-CO-C_6H_5 \longrightarrow$$

1,2-Diethyl-5-phenyl-3-pyrazoline

4.2.2.6.3 Pyrazolidines

The addition of monoalkylhydrazine to α,β-unsaturated carboxylic acid derivatives leading to pyrazolidones is of little practical significance[459]. Symmetrical N,N'-dialkylhydrazines yield open adducts which can be cyclized with sodium methoxide[470]:

$$H_3C-NH-NH-CH_3 \quad + \quad \underset{\underset{CH_3}{|}}{H_2C=C}-COOC_2H_5$$

$$\longrightarrow H_3C-NH-\underset{\overset{\overset{CH_3}{|}}{N}}{\;}-CH_2-\underset{\underset{CH_3}{|}}{CH}-COOC_2H_5$$

Ethyl 3-(1,2-dimethylhydrazino)isobutyrate

$\longrightarrow$

1,2,4-Trimethyl-pyrazolidin-3-one

In addition, malonic diesters cyclize with N,N'-dialkylhydrazines to form pyrazolidinediones[471] and 3-bromopropanoyl chloride with 1-acyl-2-alkylhydrazines to form pyrazolidinones[472], e.g.:

$$(H_7C_3)_2C\overset{\overset{COCl}{}}{\underset{\underset{CH_2-Br}{}}{}} \quad + \quad \overset{HN-CH_2-CH_3}{\underset{HN-CO-CH_3}{}}$$

$\longrightarrow$

1-Acetyl-2-ethyl-4,4-dipropylpyrazolidin-3-one

All pyrazolidinones can be reduced smoothly to pyrazolidines with lithium tetrahydroaluminate.

[466] R. Buyle, G. Viehe, Tetrahedron 25, 3453 (1969).
[467] B.V. Ioffe et al., Khim. Geterotsikl. Soedin. 932 (1966); 1061 (1969); 1249 (1970); C.A. 66, 115641q (1967), 72, 132606y (1970), 75, 5794v (1971).
[468] R.L. Hinman, R.D. Ellefson, R.F. Campbell, J. Amer. Chem. Soc. 82, 3988 (1960).
[469] J.P. Chapelle, J. Elguero, R. Jacquier, G. Tarrago, Bull. Soc. Chim. France 240 (1970).
[470] M.J. Kornet, S.I. Tan, J. Heterocycl. Chem. 5, 397 (1968).
[471] E. Jucker, Angew. Chem. 71, 321 (1959).
[472] E. Bellasio, A. Ripamonti, E. Testa, Gazz. Chim. Ital. 98, 3 (1968).

N-Alkyl-*N*-(2-cyanoethyl)hydrazines   obtained by cyanoethylation cyclize in alkaline medium[473]:

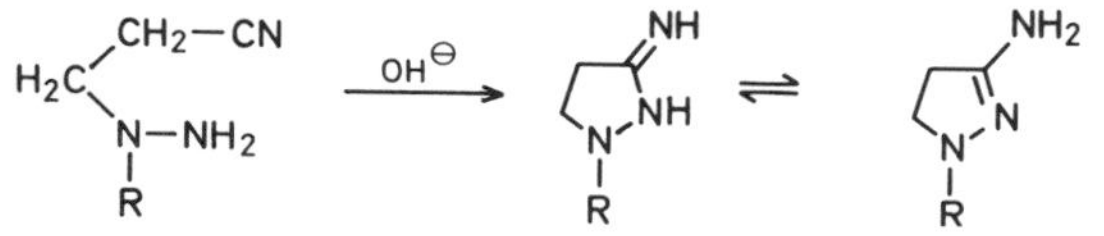

A manifestly very versatile procedure is suggested by the reaction[474] between *N*-acyl-*N'*-alkylhydrazines with formaldehyde to form *azomethine imines,* which are trapped as pyrazolidines with olefins, *e.g.:*

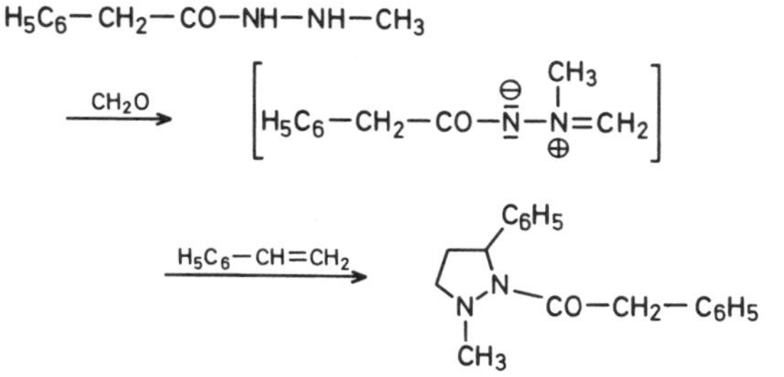

*1-Methyl-3-phenyl-2-
(phenylacetyl)pyrazolidine*

4.2.2.7 Oxidation of alkylhydrazines

Like the parent substance, alkylhydrazines are reducing agents which are oxidized by ammoniacal silver salt solutions and potassium hexacyanoferrate(III). Hydrocarbons, nitrogen, and water are the products formed[475]. Atmospheric oxygen, too, attacks hydrazines[476] and, for this reason, distillations should always be conducted under nitrogen.

Oxidation of monoalkylhydrazines with copper(II) or iron(III) chloride leads to *alkyl radicals,* which can be trapped by chlorine ions or active double bonds just as in the Sandmeyer or Meerwein reaction of aromatic diazonium compounds[477].

N,N'-Dialkylhydrazines are dehydrogenated to the yellow aliphatic *azo* compounds with mercury oxide[395, 396, 419], sodium hypochlorite[311], or atmospheric oxygen[419]. (Re the dehydrogenation of di-

aziridines see (p. 104). Oxidation of *N,N*-dialkyl-hydrazines, in particular with mercury oxide[478], lead(IV) acetate[479], or pyrolusite[480] leads to *nitrenes,* which are quenched as such[481] or can dimerize to *tetrazenes* or, while giving off nitrogen, to *hydrocarbons:*

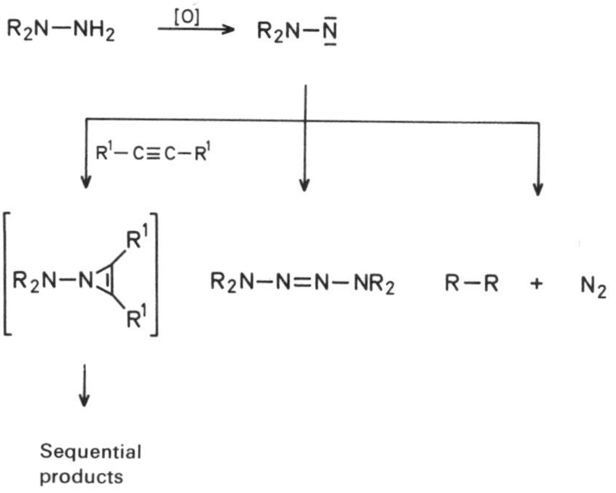

Sequential products

4.2.2.8 *N–N* cleavage

Although the risk of an *N–N* cleavage during hydrogenation of *N*-nitrosoamines, hydrazones, and azines is considerable, this *N–N* hydrogenolysis is seldom employed preparatively[482, 483]. The splitting of certain *acyl hydrazines*[484] with Raney nickel in boiling ethanol is mainly of *analytical* importance:

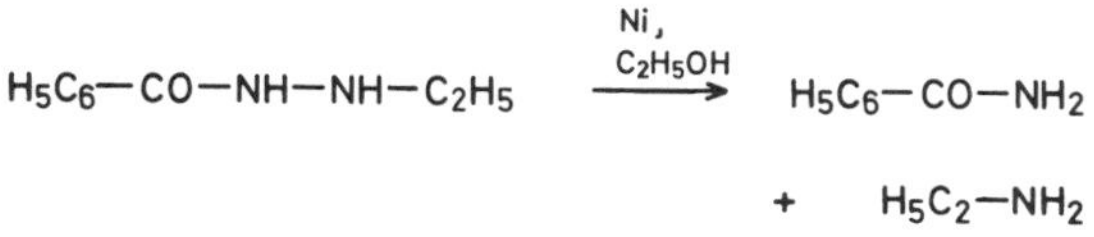

N–N cleavage of alkylhydrazines with Raney nickel which may be preparatively useful, has been described, such as during formation of *diazacycloalkanes* of medium ring size[231], *e.g.:*

[473] *H. Dorn, A. Zubeck, G. Hilgetag,* Chem. Ber. *98,* 3377 (1965).

[474] *W. Oppolzer,* Tetrahedron Lett. 2199 (1970).

[475] *E. Müller* in: *Houben-Weyl,* Methoden der organischen Chemie, Bd. X/2, p. 64, Georg Thieme Verlag, Stuttgart 1967.

[476] *E. Höft, H. Schultze,* Z. Chem. *7,* 137 (1967).

[477] *F. Minisci, R. Galli, M. Cecere,* Gazz. Chim. Ital. *95,* 751 (1965).

[478] *R.A. Abramovitch, B.A. Davis,* Chem. Rev. *64,* 149, 174 (1964).

[479] *C. Koga, J.-P. Anselme,* J. Amer. Chem. Soc. *91,* 4323 (1969).

[480] *I. Bhatnagar, M.V. George,* J. Org. Chem. *33,* 2407 (1968).

[481] *D.J. Anderson, T.L. Gilchrist, C.W. Rees,* Chem. Commun. 147 (1969).

[482] *R. Schröter* in: *Houben-Weyl,* Methoden der organischen Chemie, Bd. XI/1, p. 531, Georg Thieme Verlag, Stuttgart 1957.

[483] *F. Zymalkowski,* Katalytische Hydrierung im Organisch-Chemischen Laboratorium, p. 307, Ferdinand Enke Verlag, Stuttgart 1965.

[484] *R.L. Hinman,* J. Org. Chem. *22,* 148 (1957).

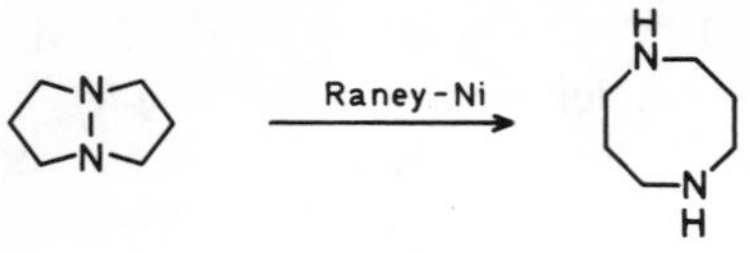

1,5-Diazacyclooctane

or for preparing certain *open-chain amines* with rhodium on alumina[378] or with Raney nickel[269, 322], *e.g.*[266]:

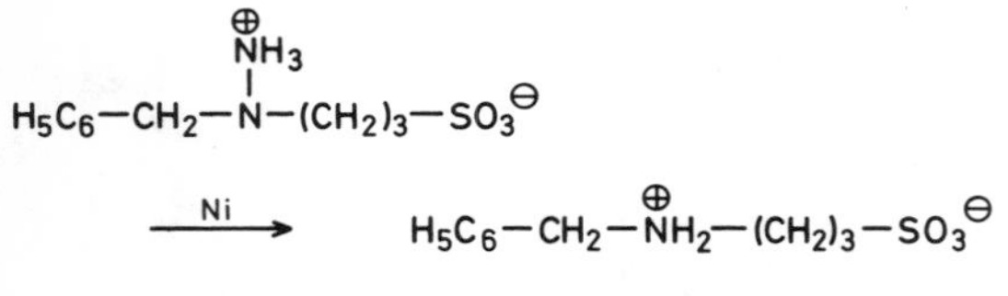

3-(Benzylamino)propanesulfonic acid betaine

4.2.2.9 Analytical hints

The crystalline *benzoyl* derivatives[485] serve as a general method of identifying hydrazines. More specific reagents for *N,N-dialkylhydrazines* are the *N,N-dialkyl-N'-(2-hydroxy-4-nitrobenzylidene)hydrazines*[359, 486]:

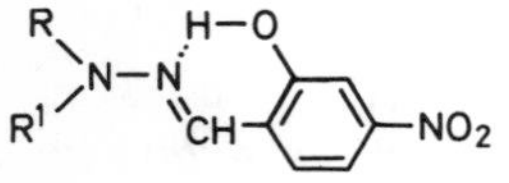

differing in their melting point, *UV* spectrum, and gas-chromatographic retention time[487]. With mercury sulfate *N,N*-dialkylhydrazines are oxidized to aldehydes, which are determined photometrically[488].

It is best to convert the monoalkylhydrazines into the 4-nitrobenzylidene compounds not directly but in the form of *semicarbazides*[324].

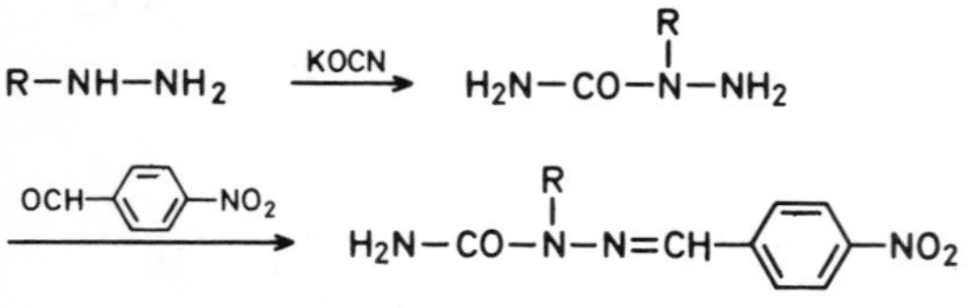

The hydrocarbon formed with hydroxylamine-*O,N*-disulfonic acid in accordance with

$$R-NH-NH_2 \quad + \quad HO_3S-NH-O-SO_3H$$

$$\longrightarrow \quad RH \ + \ N_2 \ + \ H_2N-SO_3H \ + \ H_2SO_4$$

too, can be determined by gas chromatography[324]. *4-Dimethylaminobenzylidene* derivatives[489] serve for the colorimetric determination of methylhydrazine.

Basic stationary phases are used preferably for separating alkylhydrazines by *gas chromatography*[490, 491]. There are literature data about the *paper chromatography* of acylated hydrazines[436, 492]. Numerous studies on the *IR* spectra of free alkylhydrazines (included deuterated representatives) and their *Raman spectra*[493] as well as of their hydrochlorides[434, 494], and an article dealing with the *NMR spectra* of salts of some alkylhydrazines in deuterated dimethyl sulfoxide[495] are available.

4.2.3 Preparation of aliphatic hydrazones

Hydrazones of aliphatic hydrazines, acyl, or sulfonyl hydrazines with desired aldehydes or ketones are formed readily from the components. Sometimes the reaction proceeds spontaneously, at other times warming or mild acid catalysis is required.

4.2.3.1 Monoalkylhydrazones

Monalkylhydrazones of aliphatic aldehydes and ketones, and of aromatic aldehydes have often been prepared by mixing and heating the components in the absence of solvent[496] or in ethanol[497]. The reaction with aliphatic aldehydes and

[485] *E. Müller* in: *Houben-Weyl*, Methoden der organischen Chemie, Bd. X/2, p. 127, Georg Thieme Verlag, Stuttgart 1967.
[486] *G. Neurath, M. Dünger*, Chem. Ber. *97*, 2713 (1964).
[487] *G. Neurath, W. Lüttich*, J. Chromatogr. *34*, 257 (1968).
[488] *R. Preussmann, H. Hengy, A.v. Modenberg*, Anal. Chim. Acta *42*, 95 (1968).

[489] *H. McKennis jr., A.S. Yard*, Anal. Chem. *26*, 1960 (1954).
[490] *C. Bighi, A. Betti, G. Saglietto*, Bull. Soc. Chim. France 4637 (1967).
[491] *G.D. Lakata*, J. Gaschromatogr. *5*, 41 (1967).
[492] *R.L. Hinman*, Anal. Chim. Acta *15*, 125 (1956).
[493] *D. Hadži, J. Jan, A. Ozvirk*, Spectrochim. Acta *25A*, 97 (1969);
  U. Anthoni, C. Larsen, P.H. Nielsen, Acta Chem. Scand. *22*, 1025 (1968);
  J.R. Durig, W.C. Harris, J. Chem. Phys. *51*, 4457 (1969).
[494] *J.A. Blair, R.J. Gardner*, J. Chem. Soc. C. 2707 (1970).
[495] *J.-L. Aubagnac, J. Elguero, R. Jacquier*, C.R. Acad. Sci., Ser. C *263*, 739 (1966).
[496] *R.H. Wiley, G. Irick*, J. Org. Chem. *24*, 1925 (1959).
[497] *D. Todd*, J. Amer. Chem. Soc. *71*, 1353 (1949).

ketones in ether or without solvent proceeds even in the cold[498, 499]; having barium oxide present may be useful[500], e.g.:

$$H_3C-NH-NH_2 \quad + \quad OCH-CH_3$$

$$\longrightarrow \quad H_3C-NH-N=CH-CH_3$$

Acetaldehyde methylhydrazone[500]

$$(H_3C)_2CH-NH-NH_2 \quad + \quad O=C\begin{smallmatrix}CH_3\\C_2H_5\end{smallmatrix}$$

$$\longrightarrow \quad (H_3C)_2CH-NH-N=C\begin{smallmatrix}CH_3\\C_2H_5\end{smallmatrix}$$

2-Butanone isopropylhydrazone[498]

Especially with the lower representatives mild conditions need to be maintained[577], because these easily dimerize to 1,2,4,5-tetrahydrotetrazines (see p. 114). For example, formaldehyde phenylhydrazone is obtained as monomer only in the absence of acid or base, and the dimerization tendency is the less the more carefully the product is purified[501, 578]:

$$H_5C_6-NH-NH_2 \quad + \quad CH_2O$$

$$\longrightarrow \quad H_5C_6-NH-N=CH_2 \quad \longrightarrow$$

Hexahydro-1,4-diphenyl-1,2,4,5-tetrazine

In addition, it is appropriate, by adding the aldehyde component to the hydrazine drop by drop, to avoid an excess of aldehyde giving rise to the formation of pyrazoline or hydrazone aminal, which represent intermediate products during hydrazone formation in the reverse reaction[502, 503]:

$$3\ H_{13}C_6-CHO \quad + \quad 2\ H_2N-NH-CH_3$$

$$\longrightarrow$$

Heptanal heptylidene-bis-(methyl)hydrazone

$$\xrightarrow{H_3C-NH-NH_2} \quad 3\ H_{13}C_6-CH=N-NH\underset{|}{\overset{}{}}CH_3$$

Heptanal methylhydrazone

α,β-Unsaturated carbonyl compounds afford pyrazolines readily (see p. 116), methylhydrazones only in exceptional cases[499]. N-Allylhydrazones of aliphatic aldehydes and acetone are also known[504].
In rare cases methylhydrazones are obtained by allowing diazomethane to act on activated methylene groups[505, 506].
A patent describes the preparation of butanone methylhydrazone from ketone, methylamine, and chloramine[507].

4.2.3.2 Dialkylhydrazones

Dialkylhydrazones are more stable than the monoalkyl derivatives. On this account they can be prepared safely by heating the components without solvent[496, 508-510], in ethanol[511, 512, 497, 579], or in benzene with a water separator[513, 514] even with p-toluenesulfonic acid present[515] and distilling over potassium hydroxide. Some dimethylhydrazones have been obtained in the presence of barium oxide[516], e.g.:

$$\begin{smallmatrix}H_3C\\H_3C\end{smallmatrix}CH-CHO \quad + \quad H_2N-N(CH_3)_2$$

$$\longrightarrow \quad \begin{smallmatrix}H_3C\\H_3C\end{smallmatrix}CH-CH=N-N(CH_3)_2$$

2-Methylpropionaldehyde dimethylhydrazone

$$O_2N-\!\!\!\!\bigcirc\!\!\!\!-CHO \quad + \quad H_2N-N(CH_2-C_6H_5)_2$$

$$\longrightarrow \quad O_2N-\!\!\!\!\bigcirc\!\!\!\!-CH=N-N(CH_2-C_6H_5)_2$$

4-Nitrobenzaldehyde dibenzylhydrazone

[499] B.T. Gillis, M.P. LaMontagne, J. Org. Chem. 33, 762 (1968).

[500] G.J. Karabatsos, R.A. Taller, Tetrahedron 24, 3557 (1968).

[501] B.V. Ioffe, V.S. Stopskij, Dokl. Akad. Nauk SSSR 175, 1064 (1967); Engl.: 712.

[502] N. Rabjohn, K.B. Sloan, J. Heterocycl. Chem. 6, 187 (1969).

[503] C. Harries, T. Haga, Ber. 31, 56 (1898).

[504] B.V. Ioffe, A.P. Kochetov, Zh. Org. Khim. 6, 36 (1970); C.A. 72, 78305j (1970).

[505] R. Schmiechen, Tetrahedron Lett. 4995 (1969).

[506] W. Uhde, K. Hartke, Chem. Ber. 103, 2675 (1970).

[507] F.P. 1539669 (1968), Fisons Ind. Chemicals Ltd.; C.A. 71, 101318f (1969).

[508] R.H. Wiley, S.C. Slaymaker, H. Kraus, J. Org. Chem. 22, 204 (1957).

[509] P.A.S. Smith, E.E. Most jr., J. Org. Chem. 22, 358 (1957).

[510] H. Röhnert, Arch. Pharm. (Weinheim, Ger.) 296, 296 (1963).

[511] G.R. Newkome, D.L. Fishel, J. Org. Chem. 31, 677 (1966).

[512] M.A. Iorio, Gazz. Chim. Ital. 94, 1391 (1964).

[513] R.F. Smith, L.E. Walker, J. Org. Chem. 27, 4372 (1962).

[498] B.V. Ioffe, V.S. Stopskij, Z.I. Sergeeva, Zh. Org. Khim. 4, 986 (1968); C.A. 69, 43356n (1969).

Here, too, the reaction often proceeds in the cold, for instance, that between N,N-dialkylhydrazines and benzaldehyde[517] or with the more sensitive N,N-bis(2-chloroethyl)hydrazine. The latter reacts as the hydrochloride in boiling ethanol to form the hydrazone hydrochlorides[518, 389, 519] but affords the hydrazone at room temperature[389] if used as the free base in water, e.g.:

Acetone bis[2-chloroethyl]hydrazone

Particular care is indicated during formation of *formaldehyde dialkylhydrazones*[386, 508, 385]. In the same way hydrazones of 1-aminoaziridines are prepared by carefully introducing the aldehyde into the ether solution of the aminoaziridine whilst cooling[520]. Similar considerations apply to the preparation of the thermolabile hydrazones from 1-aminoaziridines and α,β-epoxy ketones[223] or phenylglyoxal[224].

By contrast, drastic conditions are required for reacting acetophenone[509, 511] or benzophenone[511]. Acid catalysis is additionally necessary in the latter case, but it was used also during the reaction of steroid ketones with N,N-dimethylhydrazine[521] and of phenylglyoxals with N,N-dialkylhydrazines[522].

During the reaction of α,β-unsaturated carbonyl compounds with N,N-dialkylhydrazines an attack in the β-position occurs in addition to that on the carbonyl group[523].

4.2.3.3 Acylhydrazones

Acylhydrazones[524] (the C.A. names them as acyl-alkylidene hydrazines, etc.) are formed smoothly from acyl hydrazines and carbonyl compounds in warm alcohol; occasionally the addition of acetic acid is necessary. In the extended sense the important *semicarbazones* and *thiosemicarbazones*[524], too, are acyl hydrazones, e.g.:

Camphor semicarbazone

Acyl hydrazines are obtained by heating aliphatic or aromatic carboxylic acid esters with hydrazine hydrate in alcoholic solution[525]. From these, acyl hydrazones are normally formed by boiling with the carbonyl component in ethanol[380, 251, 526, 527], optionally with added acetic acid[528], or in methanol[529], or isopropyl alcohol[380]. Liquid carbonyl components can at the same time act as solvent[526, 381], e.g.:

Acetic acid
(3-cyclohexen-1-
ylmethylene)hydrazide[527]

N-Formyl-N'-methylenehydrazine[394]

N-Isonicotinoyl-N'-
isopropylidenehydrazine[526]

---

[514] P. Foley jr., E. Anderson, F. Dewey, J. Chem. Eng. Data 14, 272 (1969).

[515] F.P. 1 455 835 (1966), Shell Internat. Res. Maatschapij N.V.; C.A. 67, 11327w (1967).

[516] G.J. Karabatsos, R.A. Taller, Tetrahedron 24, 3923 (1968).

[517] O.L. Brady, G.P. McHugh, J. Chem. Soc. (London) 121, 1648 (1922).

[518] W. Schulze, G. Letsch, J. Prakt. Chem. [4] 14, 11 (1961).

[519] L.N. Volovelskij, G.V. Knorozova, A.B. Simkina, Zh. Obshch. Khim. 39, 2585 (1969); Engl.: 2523.

[520] S. Hillers, A.V. Eremeev, M. Lidaks, Khim. Geterotsikl. Soedin. 1 (1970); C.A. 72, 100390g (1970).

[521] R.H. Wiley, S.H. Chang, J. Med. Chem. 6, 610 (1963).

[522] E. Massarani, R. Pozzi, L. Degen, J. Med. Chem. 13, 157 (1970).

[523] B.V. Ioffe, K.N. Zelenin, Zh. Org. Khim. 5, 183 (1969); C.A. 70, 86929q (1969).

[524] O. Bayer in: Houben-Weyl, Methoden der organischen Chemie, Bd. VII/1, p. 467, Georg Thieme Verlag, Stuttgart 1954.

[525] H. Henecka, P. Kurtz in: Houben-Weyl, Methoden der organischen Chemie, Bd. III, p. 676, Georg Thieme Verlag, Stuttgart 1952.

[526] H.L. Yale et al., J. Amer. Chem. Soc. 75, 1933 (1953).

In many cases, however, hydrazone formation ensues even in the cold[526], and this is important in respect of the more sensitive representatives[530], *e.g.*:

$$Cl-CH_2-CO-NH-NH_2$$

$$\xrightarrow{H_3C-CO-CH_3} \quad Cl-CH_2-CO-NH-N=C\begin{smallmatrix}CH_3\\CH_3\end{smallmatrix}$$

1-(Chloroacetyl)-2-isopropylidinehydrazine

Girard reagents[531], oxalic acid dihydrazides and other carboxylic acid dihydrazides represent special acyl hydrazides[528]. Numerous acyl hydrazones have been prepared from steroid ketones[532].

The *1,2-dibenzoyldiaziridines* obtainable from diaziridines by benzoylation easily isomerize thermally to form *dibenzoylhydrazones*[330], *e.g.*:

1,2-Dibenzoyl-3,3-dimethyl-diaziridine

1,1-Dibenzoyl-2-isopropylidene-hydrazine (Acetone dibenzoyl-hydrazone)

4.2.3.4 Sulfonylhydrazones

The most important representatives of this group are the *p-toluenesulfonyl hydrazones (alkylene p-tolylsulfonylhydrazines)* obtained from *p*-toluenesulfonic acid hydrazide[533] and carbonyl compounds in cold[534] or boiling ethanol[535], occasionally with added hydrochloric acid, *e.g.*:

N-Cyclohexylidene-N'-p-tolylsulfonyl hydrazine

For preparing *p*-toluenesulfonylhydrazones of unsaturated carbonyl compounds reacting at 40° in methanol or benzene is recommended[536], while *p*-toluenesulfonylhydrazones of keto steroids[537] are formed in the cold[538].

Like toluenesulfonic acid hydrazide, methane- and 2,4-dinitrophenylsulfonic acid hydrazides, too, can be prepared from the sulfonyl chlorides and converted into hydrazones[539], *e.g.*:

$$H_3C-SO_2Cl \xrightarrow{N_2H_4} H_3C-SO_2-NH-NH_2$$

$$\xrightarrow{H_3C-CO-CH_3} H_3C-SO_2-NH-N=C\begin{smallmatrix}CH_3\\CH_3\end{smallmatrix}$$

N-Isopropylidine-N'-(methyl-sulfonyl)hydrazine

(*p*-Phenylazobenzene) sulfonic acid hydrazide (azobenzenesulfonic acid hydrazide) is recommended as a reagent for aromatic aldehydes and ketones[540].

4.2.4 Transformation reactions of aliphatic hydrazones and analysis

4.2.4.1 Reactions on the amine nitrogen

Aliphatic hydrazones[541] are basic and are *protonated* predominantly on the *amine nitrogen*[542]. *Acylation* and *nitrosation*, too, take place at this point. *Methylation* of aldehyde[514, 543] and ketone

[527] *J.R. Boissier* und *Mitarb.*, Chimie Thérapeutique *1*, 320 (1966).

[528] *D.M. Wiles, T. Suprumchuk*, Can. J. Chem. *46*, 701 (1968).

[529] *O. Isler et. al.*, Helv. Chim. Acta *38*, 1046 (1955).

[530] *S. Ito, T. Narusawa*, Bull. Chem. Soc. Japan *43*, 2257 (1970); C.A. *73*, 87122v (1970).

[531] *O. Bayer* in: *Houben-Weyl*, Methoden der organischen Chemie, Bd. VII/1, p. 478, Georg Thieme Verlag, Stuttgart 1954.

[532] *L.N. Volovelskij, A.B. Simkina*, Zh. Obshch. Khim. *37*, 1571 (1967); Engl.: 1491; USP. 3 264 331 (1966), Schering Corp., Inv.: *C.H. Robinson, L.E. Finckenor;* C.A. *65*, 20187d (1966).

[533] *L. Friedman, R.L. Litle, W.R. Reichle*, Org. Synth. *40*, 93 (1960).

[534] *K. Geibel*, Chem. Ber. *103*, 1637 (1970).

[535] *W.R. Bamford, T.S. Stevens*, J. Chem. Soc. (London) 4735 (1952).

[536] *G.L. Closs, L.E. Closs, W.A. Böll*, J. Amer. Chem. Soc. *85*, 3796 (1963).

[537] *J.E. Herz, E. González, B. Mandel*, Australian J. Chem. *23*, 857 (1970).

[538] *M. Fischer, Z. Pelah, D.H. Williams, C. Djerassi*, Chem. Ber. *98*, 3236 (1965).

[539] *J.W. Powell, M.C. Whiting*, Tetrahedron 7, 305 (1959).

[540] *R.J.W. Cremlyn*, J. Chem. Soc. (London) 6235 (1964); *R.J.W. Cremlyn, D.N. Waters*, J. Chem. Soc. (London) 6243 (1964).

[541] *Y.P. Kitaev, B.I. Buzykin, T.V. Troepolskaja*, Usp. Khim. *39*, 961 (1970); Engl.: 441.

[542] *J. Elguero, R. Jacquier, C. Marzin*, Tetrahedron Lett. 3099 (1970).

[543] *B.V. Ioffe, N.L. Zelenina*, Zh. Org. Khim. *4*, 1558 (1968); C.A. *69*, 105848k (1968).

dimethylhydrazones[509, 544] with methyl iodide or methyl *p*-toluenesulfonate to the *trimethylhydrazinium salts* proceeds very smoothly. While *trimethylhydrazonium salts* derived from aldehydes decompose into nitriles under the action of bases[514, 543, 545], *e.g.*:

$$H_3CO-\langle\rangle-CH=N-N(CH_3)_2$$

$$\xrightarrow{H_3C-\langle\rangle-SO_3-CH_3} H_3CO-\langle\rangle-CH=N-\overset{\oplus}{N}(CH_3)_3$$

$$\xrightarrow{KOH} H_3CO-\langle\rangle-CN \;+\; (CH_3)_3N$$

those derived from ketones form *azirines* with alkoxide in a Neber reaction[546, 547], or *pyridines* on pyrolysis[544].

Careful distillation of the monoalkylhydrazones over potassium hydroxide effects a rearrangement to the thermodynamically less stable but lower boiling, isomeric aliphatic *azo* compounds[548, 549], *e.g.*:

$$(H_3C)_2CH-NH-N=CH_2 \qquad (H_3C)_2C=N-NH-CH_3$$

$$\downarrow$$

$$(H_3C)_2CH-N=N-CH_3$$

Methyl-(isopropyl)diazene
(1-Methylethaneazomethane)

Monomethylhydrazones furnish *pyrazolines*[550] with aldehydes, and *N-methylpyrroles*[551] with suitable ketones under acid catalysis *via* the enehydrazine stage.

4.2.4.2 Reactions on the methine carbon

The action of hydrazine[580] on numerous dimethylhydrazones represents an elegant method for obtaining azinefree, unsubstituted hydrazones[511, 579], *e.g.*:

$$\langle\rangle=N-N(CH_3)_2 \xrightarrow{N_2H_4} \langle\rangle=N-NH_2$$

Cyclohexanone hydrazone

$$+ \quad H_2N-N(CH_3)_2$$

but excess *N,N*-diethylhydrazine, too, can be used for the exchange[552]. Prussic acid adds on to the C=N double bond[553]; the adduct can be hydrolyzed to α-hydrazinocarboxylic acids[554]:

$$\begin{matrix}R\\\end{matrix}C=N-NH-CO-NH_2 \xrightarrow{HCN} R-\underset{CN}{\overset{R}{C}}-NH-NH-CO-NH_2$$

$$\xrightarrow{HCl} \underset{R}{\overset{R}{C}}\underset{COOH}{\overset{NH-NH_2}{}}$$

Monoalkylhydrazones of ketones are relatively smoothly oxidized to α-acetoxyazo compounds with lead(IV) acetate or peroxyacetic acid,

$$\langle\rangle=N-NH-CH_3$$

$$\xrightarrow{Pb(O-CO-CH_3)_4} \langle\rangle\overset{N=N-CH_3}{\underset{O-CO-CH_3}{}}$$

1-(Methaneazo)cyclo-hexanol acetate

while aldehyde and dialkyl hydrazones react less uniformly[555].

4.2.4.3 Fragmentation of hydrazones of 1-aminoaziridines

Hydrazones derived from 1-aminoaziridines and α,β-epoxy ketones (acyloxiranes) easily decom-

[544] *G.R. Newkome, R.L. Fishel*, J. Heterocycl. Chem. *4*, 427 (1967).

[545] *R.F. Smith, A.C. Bates*, J. Chem. Educ. *46*, 174 (1969).

[546] *R.F. Parcel*, Chem. & Ind. (London) 1396 (1963).

[547] *S. Sato*, Bull. Chem. Soc. Japan *41*, 1440 (1968); C.A. *69*, 96354s (1968).

[548] *B.V. Ioffe, Z.I. Sergeeva, V.S. Stopskij*, Dokl. Akad. Nauk SSSR *167*, 831 (1966); Engl.: 393.

[549] USP. 3 350 385 (1967), Inv.: *L. Spialter, G.L. Untereiner*; C.A. *68*, 59093q (1968).

[550] *B.V. Ioffe, V.S. Stopskij, N.B. Burmanova*, Khim. Geterotsikl. Soedin. 1066 (1969); C.A. *72*, 132607z (1970).

[551] *J.-P. Chapelle, J. Elguero, R. Jacquier, G. Tarrago*, Bull. Soc. Chim. France 4464 (1969).

[552] *G.S. Goldin, S.N. Tsiomo, G.S. Shor*, Zh. Org. Khim. *6*, 745 (1970); C.A. *73*, 14045n (1970).

[553] D. Off. 1921878 (1969), Erf.: *R.E. McLeay, S. Sheppard*; C.A. *72*, 89843j (1970).

[554] *W. Knobloch, G. Subert*, J. Prakt. Chem. [4] *36*, 29 (1967).

[555] *J. Warkentin*, Synthesis 279, 293 (1970).

[556] *W. Kirmse*, Carbene, Carbenoide und Carbenanaloge, p. 137, Verlag Chemie, Weinheim/Bergstr. 1969.

[557] *L. Friedman, H. Shechter*, J. Amer. Chem. Soc. *81*, 5512 (1959).

[558] *J.H. Bayless, L. Friedman, F.B. Cook, H. Shechter*, J. Amer. Chem. Soc. *90*, 531 (1968).

[559] *W. Kirmse, B.-G. v. Bülow, H. Schepp*, Justus Liebigs Ann. Chem. *691*, 41 (1966).

pose thermally into *acetylenic aldehyde, olefin,* and nitrogen[223, 581], *e.g.:*

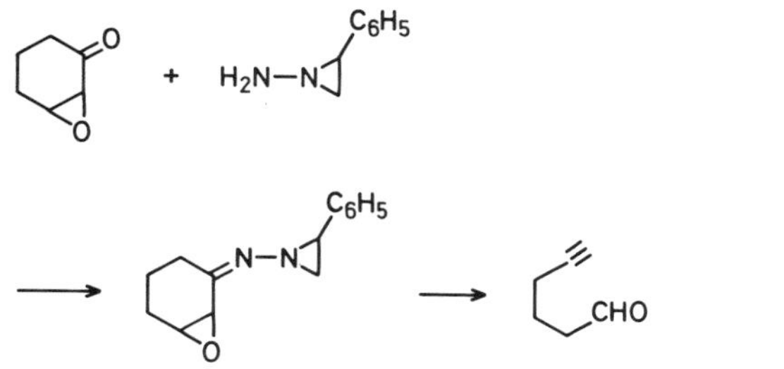

1-(7-Oxabicyclo[4.1.0]hept-2-ylideneamino)-2-phenyl-aziridine

5-Hexynal.

+ N2 + H5C6—CH=CH2

Styrene

2-Diazoacetophenone and *propenylbenzene* are obtained from phenylglyoxal[224, 581]:

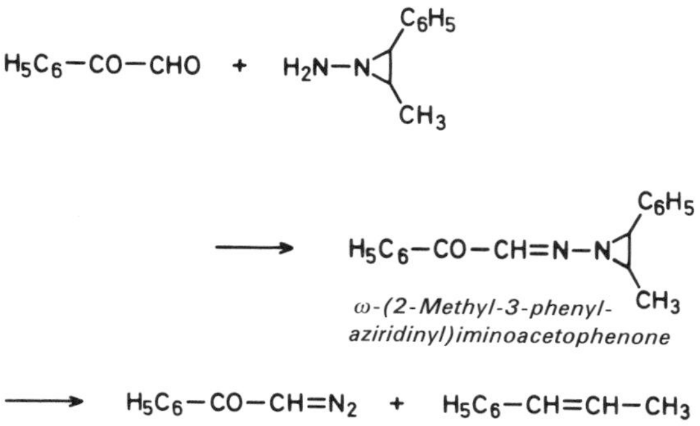

H5C6—CO—CHO + H2N—N

ω-(2-Methyl-3-phenyl-aziridinyl)iminoacetophenone

⟶ H5C6—CO—CH=N2 + H5C6—CH=CH—CH3

4.2.4.4 Reactions of *p*-tolylsulfonylhydrazones

4.2.4.4.1 Bamford-Stevens reaction

Alkali salts of alkylene-*p*-tolylsulfonylhydrazines (*p*-toluenesulfonyl hydrazones) decompose readily into *diazo compounds* and *toluenesulfinic acid*. In dependence on the reaction conditions either carbenes or carbonium ions form from the diazo compounds and finally lead to mainly olefins[556]. In the typical Bamford-Stevens reaction *p*-toluenesulfonyl hydrazones are heated with the sodium salt of the glycol to 180° in glycol, *e.g.:*

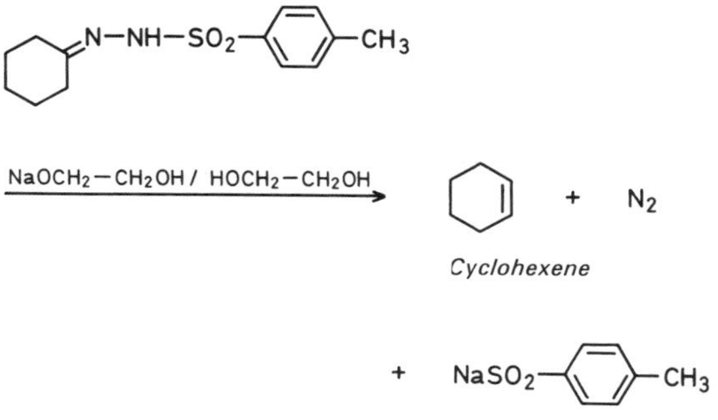

Cyclohexene

+ NaSO2—⟨⟩—CH3

In this reaction, which proceeds *via* the carbonium ion, rearrangements are observed with branched molecules[557, 558]. When aprotic solvents are used the carbenoid pathway leads to formation of cyclopropane side-products[536]. Strong bases such as sodium hydride or sodium amide[559], and, in particular, alkyllithium additionally detach the proton from the α-position and, in an exceptionally smooth reaction, lead to optionally terminal olefins without a rearrangement:

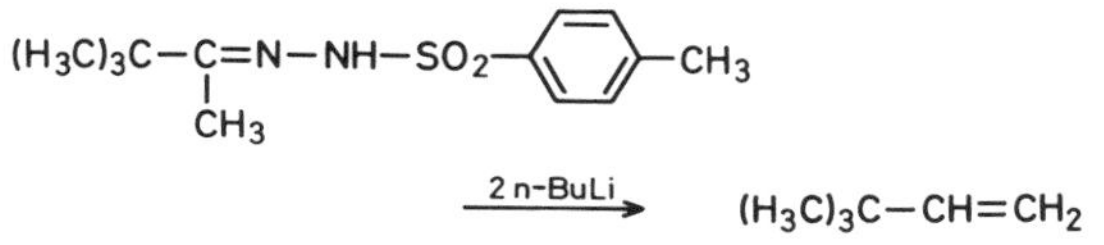

3,3-Dimethyl-1-butene[560]

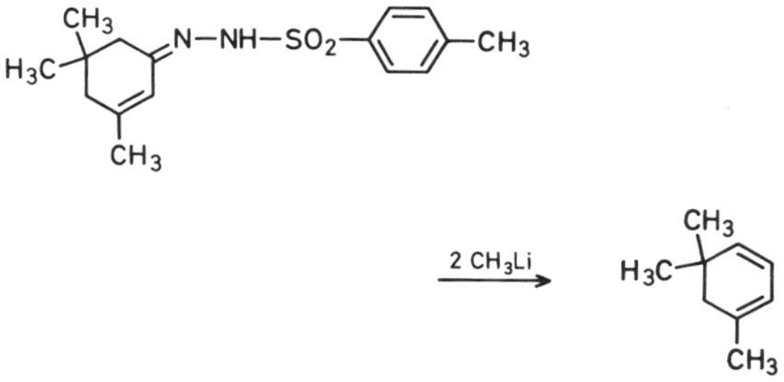

1,5,5-Trimethyl-1,3-cyclohexadiene[561]

Even camphor *p*-toluenesulfonylhydrazone yields only *norbornene*[562].

[560] *G. Kaufman, F. Cook, H. Shechter, J. Bayless, L. Friedman,* J. Amer. Chem. Soc. **89**, 5736 (1967).

[561] *W.G. Dauben et. al.,* J. Amer. Chem. Soc. **90**, 4762 (1968).

[562] *R.H. Shapiro, M.J. Heath,* J. Amer. Chem. Soc. **89**, 5734 (1964).

[563] *A. Eschenmoser, D. Felix, G. Ohloff,* Helv. Chim. Acta **50**, 708 (1967).

[564] *M. Tanabe et. al.,* Tetrahedron Lett. 3739, 3943 (1967).

[565] *L. Caglioti, M. Magi,* Tetrahedron **19**, 1127 (1963).

[566] *L. Caglioti, P. Grasselli,* Chem. & Ind. (London) 152 (1964).

[567] *L. Caglioti,* Tetrahedron **22**, 487 (1966).

[568] *B.V. Ioffe, A.G. Vitenberg, V.N. Borisov,* Zh. Org. Khim. **5**, 1706 (1969); C.A. **72**, 2974f (1970).

[569] *G.J. Karabatsos, J.D. Graham, F.M. Vane,* J. Amer. Chem. Soc. **84**, 753 (1962).

[570] *B.V. Ioffe, V.S. Stopskij,* Zh. Org. Khim. **4**, 1312 (1968); C.A. **69**, 81911x (1968).

[571] *D. Hadži, J. Jan,* Spectrochim. Acta **23A**, 1571 (1967).

4.2.4.4.2 *p*-Toluenesulfonylhydrazones of α,β-epoxy ketones (acyloxiranes)

p-Toluenesulfonyl hydrazones of α,β-epoxy ketones fragment remarkably easily thermally or under the action of sodium alkoxide to form *alkynones*[563, 564, 581], *e.g.:*

4,4-Dimethyl-6-heptyn-2-one

4.2.4.4.3 Reduction with complex hydrides

As a milder variant of the Wolf-Kishner reduction toluenesulfonyl hydrazones of aldehydes and ketones can be reduced to the corresponding methyl and methylene compounds with lithium tetrahydroaluminate[565, 538]. In boiling methanol or, better, *p*-dioxane as medium sodium tetrahydroborate, too, gives a successful reaction[566, 567], *e.g.:*

Cyclohexane

4.2.4.5 Analytical hints for aliphatic hydrazones

Few analytical procedures have been described. As the simple hydrazones are liquid, methylation will produce solid derivatives. Rather than hydrolyzing to form the components, the aldehyde and ketone components can be embraced analytically by transhydrazonization[568], *e.g.* with 2,4-dinitrophenylhydrazine.

Under these circumstances spectral methods are particularly important. *NMR* data for methylhydrazones[500] and dimethylhydrazones[516], and for semicarbazones[569] are available, and *IR* spectra[508, 496, 570] and *UV* spectra[496], too, have been described, in particular also *IR* spectra of acyl hydrazones[571] and acyl semicarbazones[572].

---

[572] *W.H.T. Davison, P.E. Christie*, J. Chem. Soc. (London) 3389 (1955).

[573] *P.E. Iversen*, Acta Chem. Scand. *25*, 2337 (1971); *P.E. Iversen*, Chem. Ber. *104*, 2195 (1971).

[574] *F.E. Condon*, J. Org. Chem. *37*, 3615 (1972).

4.3   Bibliography

Arylhydrazines and Arylhydrazones

P.L. Julian, E.W. Meyer, H.C. Printy in *R.C. Elderfield*, Heterocyclic Compounds, Vol. III: The Chemistry of Indoles, John Wiley & Sons, New York 1952–1961.

W. Freudenberg in *R.C. Elderfield*, Heterocyclic Compounds, Vol. III: The Chemistry of Carbazole, John Wiley & Sons, New York 1952–1961.

T.L. Jacobs in *R.C. Elderfield*, Heterocyclic Compounds, Vol. V: Pyrazoles and Related Compounds, Vol. VI: Pyridazines and Related Compounds, John Wiley & Sons, New York 1952–1961.

R.C. Elderfield, S.L. Wyte in *R.C. Elderfield*, Heterocyclic Compounds, Vol. VI: Phthalazine and its Derivatives, John Wiley & Sons, New York 1952–1961.

J.H. Boyer in *R.C. Elderfield*, Heterocyclic Compounds, Vol. VIII: Monocyclic Triazoles and Benzotriazoles, John Wiley & Sons, New York 1952–1961.

A. Weissberger, E.C. Taylor, The Chemistry of Heterocyclic Compounds, J. Wiley, Intersci. Publ., New York 1953–1970.

J.C.E. Simpson, Condensed Pyridazine and Pyrazine Rings, Part II, 1953.

J.G. Erickson, P.F. Wiley, V.P. Wistrach, The 1,2,3- and 1,2,4-Triazines, Tetrazines and Pentazines, 1956.

L.C. Behr, R. Fusco, C.H. Jarboe, Pyrazoles, Pyrazolines, Pyrazolidines, Indazoles and Condensed Rings, 1967.

D.J. Brown, The Pyrimidines, Supl. I, 1970.

Houben-Weyl, Methoden der organischen Chemie, Bd. XI/1, Georg Thieme Verlag, Stuttgart 1957.

E.H. Rodd, Chemistry of Carbon Compounds, IV A, 42.

B. Robinson, The Fischer Indole Synthesis, Chem. Rev. *63*, 373 (1963).

R.H. Wiley, P. Wiley, Pyrazolones, Pyrazolidones and Derivatives, Intersci. Publ., New York, London 1964.

Houben-Weyl, Methoden der organischen Chemie, Bd. X/3, Georg Thieme Verlag, Stuttgart 1965.

---

[575] *K.-H. Linke, R. Turley, E. Flaskamp*, Chem. Ber. *106*, 1052 (1973).

[576] *F.E. Condon*, J. Org. Chem. *37*, 3608 (1972).

[577] *D.M. Lemal, F. Menger, E. Coats*, J. Amer. Chem. Soc. *86*, 2395 (1964).

[578] *C.H. Schmidt*, Chem. Ber. *103*, 986 (1970); *B.V. Ioffe, V.S. Stopskii*, Chem. Ber. *104*, 343 (1971).

[579] *G.R. Newkome, D.L. Fishel*, Org. Synth. *50*, 102 (1970).

[580] *E. Nachbaur, G. Leiseder*, Monatsh. Chem. *102*, 1718 (1971).

[581] *D. Felix, J. Schreiber, G. Ohloff, A. Eschenmoser*, Helv. Chim. Acta *54*, 2896 (1971); *D. Felix et. al.*, Helv. Chim. Acta *55*, 1276 (1972).

C.G. Overberger, J.P. Anselme, J.G. Lombardino, Organic Compounds with Nitrogen-Nitrogen Bonds, Ronald Press Co., New York 1966.

P.A.S. Smith, The Chemistry of Open-Chain Organic Nitrogen Compounds, Vol. II, W.A. Benjamin, New York 1966.

Houben-Weyl, Methoden der organischen Chemie, Bd. X/2, Georg Thieme Verlag, Stuttgart 1967.

B. Robinson, Recent Studies on the Fischer Indole Synthesis, Chem. Rev. *69,* 227 (1969).

J. Buckingham, The Chemistry of Arylhydrazones — Quart. Rev. *23,* 37 (1969).

Aliphatic hydrazines, hydrazo compounds, hydrazones, and azines

E. Müller in: *Houben-Weyl,* Methoden der organischen Chemie, Bd. X/2, p. 5, Georg Thieme Verlag, Stuttgart 1967.

A.N. Kost, R.S. Sagitullin, Monoalkylhydrazines, Usp. Khim. *33,* 361 (1964); Engl.: 159.

R. Ohme, A. Zubeck, Synthesen des Hydrazins und seiner Alkylderivate, Z. Chem. *8,* 41 (1968).

E. Schmitz, Dreiringe mit zwei Heteroatomen, Oxaziridine, Diaziridine, cyclische Diazoverbindungen, Springer Verlag, Berlin, Heidelberg, New York 1967.

D. Kolbach, D. Korunčev in: *Houben-Weyl,* Methoden der organischen Chemie, Bd. X/2, p. 85–122, Georg Thieme Verlag, Stuttgart 1967.

Y.P. Kitaev, B.I. Buzykin, T.V. Troepolskaja, The Structure of Hydrazones, Usp. Khim. *39,* 961 (1970); Engl.: 441.

W. Kirmse, Carbene, Carbenoide und Carbenanaloge, p. 137, Verlag Chemie, Weinheim/Bergstr. 1969.

5 Azo and Azoxy Compounds

Contributed by

E.V. Brown
Department of Chemistry,
University of Kentucky,
Lexington, Kentucky

H.G. Padeken
Stuttgart

5.1 Aromatic azo compounds

5.1.1 Introduction

This section discusses compounds of the structure R–N=N–R[1] where the two *R* groups represent aromatic rings, either carbocyclic or heterocyclic. In general, such compounds show a high degree of thermal *stability* as well as maintaining the azo linkage through many types of chemical reactions. The usual electrophilic aromatic *substitutions* can be carried out on aromatic azo compounds, *i.e.*, halogenation[1,2,3], nitration[1,4], and sulfonation[5,6]. Normally the rules of substitution hold for these reactions with the azo group acting as an *ortho-para* director. In some cases, vigorous conditions lead to a reverse coupling reaction[7].

Many *transformations* can be performed on various groups which are *substituted* on the aromatic rings of azo compounds. Alkylations of both hydroxy and amino groups to ethers[8], tertiary amines[9], and quaternary salts[10] are typical. Both carboxylic[11] and sulfonic[12] acids have been converted to their acid chlorides. The Ullmann[13] and other halogen replacements[14] work well with the azo group acting as a mild activating group for the halogen.

Stable *transition metal complexes* of aromatic azo compounds are numerous. Among some that might be mentioned are those of nickel[15], palladium[16], platinum[16], manganese[17], rhenium[17], and cobalt[17]. Reactions of *azobenzene (1,2-diphenyl-diazene)* with various metal carbonyls has afforded stable metal complexes[18]. Some of the general aspects of this chemistry have been covered in a review paper[19]. One should also consult Vol. VIII of this work. Azobenzene[20] and other aromatic azo compounds can exist in *cis* and *trans* modifications around the –N=N– group. Ordinarily the *trans* configuration is the more stable[21]. The *trans* to *cis* shift is usually light-catalyzed

cis   *trans*

while the reverse shift is thermally promoted. Generally, activation energies are low (8–24 *kcal/mole*)[22] and *cis-trans* conversions are rapid even at room temperature. In certain cases flash spectroscopic techniques[23] have been used because of the rapidity of the conversion. However, in other cases, *cis* and *trans*-isomers have been separated by chromatography on activated alumina[24]. The mechanism of this conversion is still highly controversial and the matter is the subject of a review[25] and several reports[26-28].

In addition, some hydroxy substituted azo compounds exhibit *hydroxyazo*→quinone-hydrazone equilibria[29] and the rate and mechanism of this reaction has been discussed[30,31].

The *protonation* of the azo group in strong acids[32] and the structure of the conjugate acids in the *cis* and *trans* series have been studied extensively[33] although the matter is still highly controversial[34-36]. This point arises also in connection with the theory on the *carcinogenicity*[37] of *N,N*-dimethyl-4-(phenylazo)aniline and its analogs[38,39]. Of course, infrared[40,41] and notably electron absorption[33-36] characteristics of the aromatic azo compounds have been observed extensively.

[1] *J. Burns, H. McCombie, H.A. Scarborough*, J. Chem. Soc. 2928 (1928).

[2] *L. Pentimalli, A. Resaliti*, Ann. Chim. (Rome) *46*, 1037 (1956); C.A. *51*, 8744a (1957).

[3] *D.R. Fahey*, J. Organometal. Chem. *27*, 283 (1971).

[4] *G.M. Badger, G.E. Lewis*, J. Chem. Soc. 2150 (1953).

[5] *P. Ruggli, M. Stäuble*, Helv. Chim. Acta *24*, 1080 (1941).

[6] *J.A. Pearl*, J. Org. Chem. *10*, 205 (1945).

[7] *M.P. Schmidt*, J. Prakt. Chem. [2] *85*, 235 (1912).

[8] *H.E. Fierz-David, L. Blangey, H. Streiff*, Helv. Chim. Acta *29*, 1718 (1946).

[9] *L. Horner, H. Müller*, Chem. Ber. *89*, 2756 (1956).

[10] *J. Voltz*, Chimia *15*, 168 (1961).

[11] *G.H. Coleman, G. Nichols, C.M. McCloskey, H.D. Anspen*, Org. Synthese, Coll. Vol. III, 712 (1955).

5.1.2 Preparation

5.1.2.1 Coupling with diazonium compounds

Coupling partners for diazonium salts may be divided into three categories, namely, aromatic hydrocarbons, aromatic amines, and phenols. When sufficient methyl groups in the proper positions are present it is possible to couple the very powerfully reactive 2,4,6-trinitrobenzenediazonium salts with mesitylene[42], 1,2,3,5-tetramethylbenzene, and pentamethylbenzene, but not with 1,2,4,5-tetramethylbenzene, to afford respectively *1,3,5-trimethyl-2',4',6'-trinitro-,1,2,3,5-tetramethyl-2',4',6'-trinitro-, 1,2,3,4,5-pentamethyl-2',4',6'-trinitroazobenzene.* m-Xylene couples with diazonium salts of thiadiazoles[43], so do other hydrocarbons with electron-rich positions[44]. *Pyrrole*[45,46] and *indole*[47,48], which are electron-rich heterocycles, couple readily with many diazonium salts even when electron-withdrawing groups like the carboxy group are present. *Imidazole* couples in the 2-position and the nature of this coupling has been discussed[49].

Coupling of diazonium salts with amines takes place in neutral to weakly acid solution (*pH* 4—7). When using primary or secondary aromatic amines this reaction often leads to the formation of

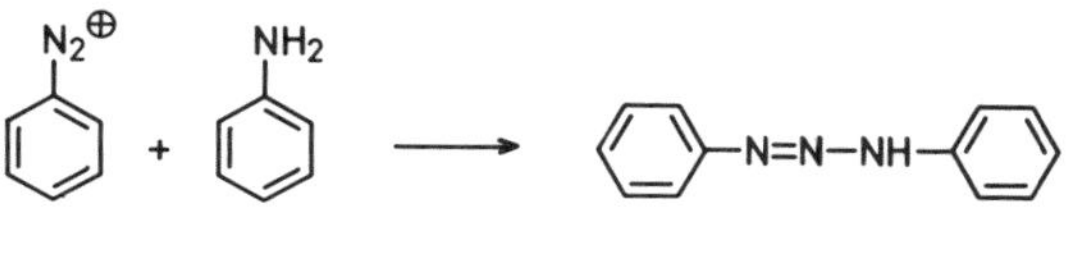

diazoamino compounds which must then be rearranged to the *aminoazo* compounds (cf. p. 131). However, if powerfully coupling diazonium salts[50] are used or very active amines[51], such as those containing additional amino or hydroxy groups, then *direct* coupling to aminoazo compounds can be realized. A popular way to avoid formation of diazoamino compounds with primary and secondary amines is to first form the aminomethanesulfonic acid[52] by the use of formaldehyde and

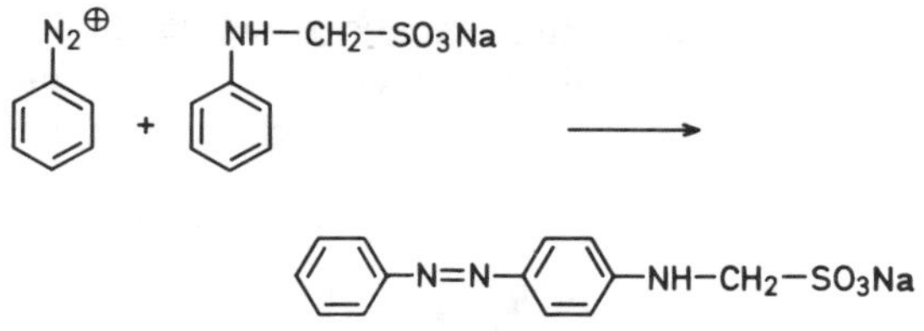

sodium hydrogen sulfite. Subsequent hydrolysis of the methanesulfonic acid group leads to the desired aminoazo compound. Of course, the foregoing complications are not found with *tertiary amines* which usually give the *para* substituted aminoazo compound in good yield.

A typical example is the formation of *Methyl orange*[53] *[4-([4-(dimethylamino)phenyl]azo)benzenesulfonic acid]* and of *Methyl red [2-([4-(dimethylamino)phenyl]azo)benzoic acid]*[54]:

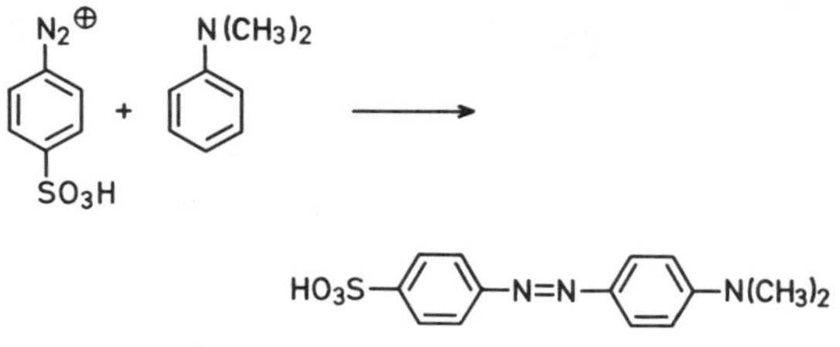

The ability of *N,N*-dialkylanilines to couple can be diminished by substituents and may be almost reduced to zero by two ortho substituents which

[12] M. Schmid, R. Mory, Helv. Chim. Acta 38, 1329 (1955).

[13] DRP 627138 (1936), I.G. Farben; Erf.: D. Delfs; C.A. 30, 4332 (1936).

[14] B.I. Stepanov, M.A. Andreeva, Zh. Obshch. Khim. 30, 2748 (1960); C.A. 53, 17066g (1959).

[15] J.P. Kleiman, M. Dubeck, J. Amer. Chem. Soc. 85, 1544 (1963).

[16] A.C. Cope, R.W. Siekman, J. Amer. Chem. Soc. 87, 3272 (1965).

[17] R.F. Heck, J. Amer. Chem. Soc. 90, 313 (1968).

[18] M.I. Brice, M.Z. Iqbal, F.G.A. Stone, J. Chem. Soc. 3204 (A 1970).

[19] G.W. Parshall, Accounts Chem. Res. 3, 139 (1970).

[20] J.M. Robertson, M. Prasad, I. Woodward, Proc. Roy. Soc. Ser. A 154, 187 (1936).

[21] E. Bergmann, L. Engel, S. Sándor, Ber. 63, 2572 (1930).

[22] R.J.W. LeFevre, J.O. Northcott, J. Chem. Soc. 867 (1953).

[23] P.D. Wildes, J.G. Pacifici, G. Irick jr., D.G. Whitten, J. Amer. Chem. Soc. 93, 2004 (1971).

[24] A.H. Cook, J. Chem. Soc. 876 (1938); A.H. Cook, J. Chem. Soc. 1309, 1315 (1939); L. Zechmeister, O. Frehden, P.F. Jörgensen, Naturwissenschaften 26, 495 (1938).

[25] H.H. Jaffé, Chem. Rev. 53, 191 (1953).

[26] D. Schulte-Frohlinde, Justus Liebigs Ann. Chem. 612, 131, 138 (1958).

[27] E.R. Talaty, J.C. Fargo, Chem. Commun. 65 (1967).

[28] O. Gegian, K.A. Muszkat, E. Fischer, J. Amer. Chem. Soc. 90, 3907 (1968).

[29] G.E.K. Branch, M. Calvin, The Theory of Organic Chemistry, p. 299, Prentice-Hall, New York 1941.

[30] G. Wettermark, M.E. Langmuir, D.G. Anderson, J. Amer. Chem. Soc. 87, 476 (1965).

[31] R.L. Reeves, R.S. Kaiser, J. Org. Chem. 35, 3670 (1970).

[32] I. Gränacher, H. Suhr, A. Zenhäusern, H. Zollinger, Helv. Chim. Acta 44, 313 (1961).

[33] J.H. Collins, H.H. Jaffé, J. Amer. Chem. Soc. 84, 4708 (1962).

[34] G.E. Lewis, Tetrahedron 10, 129 (1960).

tend to twist the dimethylamino group out of the plane of the benzene ring[55]. Generally, *acetylation* of the amino group reduces its electron donating capacity so that coupling does not take place. Exceptions to this take place with acetaminofurans where the electron donation of the ring makes up for the withdrawal by the acyl group[56].

Several cases of coupling with *sulfonamides* of aromatic amines have been reported[57]. Many aminoheterocycles couple to give normal aminoazo compounds, for example, 4,6-diaminopyrimidine[58] which couples in the 5-position and 2,6-diaminopyridine[59] which couples in the 3-position to give 'pyridium'.

Diazonium salts couple with *aromatic hydroxy* compounds in mildly basic solutions (*pH* 7—9). Coupling normally takes place in the para position to the hydroxy group but small amounts of the ortho product can usually be found. When a group is

4-(Phenylazo)phenol

present in the *para* position then *ortho* coupling takes place readily. If both *ortho* and *para* positions are occupied, coupling will not take place unless an *ortho* or *para* group can be displaced in the process. Pyridine has been indicated as a beneficial

solvent for phenol coupling in some cases[60, 61]. A typical phenolic coupling is exemplified by the coupling of phenol with diazotized aniline which yields 4-(phenylazo)phenol with one molecule diazonium salt, a 2,4-bisazo compound with two molecules and, in strong alkaline solution, with a third molecule a 2,4,6-trisazo product[62, 63] [*2,4-bis(phenylazo)phenol* and *2,4,6-tris(phenolazo)phenol* respectively].

Other typical examples are:

Benzenediazonium salt + Resorcinol ⟶
      4-(Phenylazo)resorcinol[64]

4-Nitrobenzenediazonium salt + Phenol ⟶
      4-[(4-Nitrophenyl)azo]phenol[65]

      + Salicylic acid ⟶
      2-Hydroxy-5-[(4-nitrophenyl)azo]benzoic acid[64]

4-Sulfobenzenediazonium salt + 2-Chlorophenol ⟶
      4-[(3-Chloro-4-hydroxyphenyl)azo]benzenesulfonic acid[66]

Many hydroxy-containing heterocycles couple in the same manner as the corresponding phenols and naphthols as, for example, 5-pyrazolol[67], 2-pyrazolin-5-one, 2-carbazolol, 3-dibenzofuranol[64] and 8-quinolinol[68].

When there are both hydroxy and amino groups present, as in 8-amino-1-naphthol, it is possible to obtain coupling in the 5-position under acid conditions and in the 4-position under alkaline conditions. The same or different diazonium salts may be used to synthesize a number of bisazo compounds, *e.g.*:

8-Amino-5-phenylazo-1-naphthol

8-Amino-4-(4-nitrophenyl)azo-5-phenylazo-1-naphthol

[35] F. Gerson, E. Heilbronner, Helv. Chim. Acta 45, 42 (1962).
[36] E. Haselbach, Helv. Chim. Acta 53, 1526 (1970).
[37] B. Pullman, Rev. Franc. Etudes Clin. Biol. 2, 327 (1957).
[38] E.V. Brown, W.H. Kipp, Cancer Res. 29, 1341 (1969).
[39] E.V. Brown, W.H. Kipp, Cancer Res. 30, 2089 (1970).
[40] D. Hadzi, J. Chem. Soc. 2143 (1956).
[41] K. Ueno, J. Amer. Chem. Soc. 79, 3066 (1957).
[42] K.H. Meyer, H. Tochtermann, Ber. 54, 2283 (1921).
[43] L.I. Smith, J.H. Paden, J. Amer. Chem. Soc. 56, 2169 (1934).
[44] L.F. Fieser, W.P. Campbell, J. Amer. Chem. Soc. 60, 1142 (1938).
[45] O. Fischer, E. Hepp, Ber. 19, 2251 (1886).
[46] H. Fischer, F. Rothweiler, Ber. 56, 512 (1923).
[47] W. Madelung, O. Wilhelmi, Ber. 57B, 234 (1924).

[48] J.H. Binks, J.H. Ridd, J. Chem. Soc. 2398 (1957).
[49] R.D. Brown, H.C. Duffin, J.C. Maynard, J.H. Ridd, J. Chem. Soc. 3937 (1953).
[50] Z.J. Allan, J. Podstata, Collect. Czech. Chem. Commun. 29, 2264 (1964); C.A. 61, 14813c (1964).
[51] A.I. Vogel, Practical Organic Chemistry, 3rd Ed., p. 623, Longmans, Green and Co., London 1956.
[52] USP 2 139 325 (1937); C.A. 33, 2151 (1939).
[53] L.F. Fieser, Organic Experiments, p. 234, D.C. Heath and Co., Boston, Mass. 1964.
[54] E. Rupp, R. Loose, Ber. 41, 3905 (1908).
[55] C.D. Nenitzescu, V. Väntu, Ber. 77, 705 (1944).
[56] A.P. Dunlop, F.N. Peters, The Furans, p. 187, Reinhold Publishing Co., New York 1953.
[57] O.N. Witt, G. Schmitt, Ber. 27, 2370 (1894).
[58] B. Lythgoe, A.R. Todd, A. Topham, J. Chem. Soc. 315 (1944).
[59] A.E. Chichibabin, O.A. Zeide, J. Russ. Phys. Chem. Soc. 50, 522 (1920); C.A. 18, 1496 (1924).

Since many important substituted aminonaphthols are commercial azo dye components, there are considerable industrial applications for this idea[69].

Aromatic *hydrazines* and several of their acylated derivatives as well as their corresponding hydrazones have been caused to couple with diazonium salts, although this reaction is much less common than the coupling reactions mentioned above[70].

5.1.2.2 Rearrangement of diazoamino compounds (triazenes)

As mentioned in the previous section, primary and secondary amines often react with diazonium salts to give triazenes.

Typically these are rearranged by heating under acid conditions[71]. If the 4-position is open, the aromatic azo group migrates to this point during the rearrangement.

4-(Phenylazo)aniline

If only a 2-position is free rearrangement to this position will take place[72]. Since this is an intermolecular rearrangement, it is possible to start with a particular triazene and obtain a different aminoazo compound[73]. This situation is further complicated by the fact that unsymmetrical triazenes exist as a tautomeric mixture[74].

4-(4-Tolylazo)aniline

5.1.2.3 Reduction reactions

In a number of cases compounds of higher oxidation states have been reduced to azo compounds.

A general procedure is to combine a *nitro* compound and an *amine* by the use of alkali. This has been accomplished in aqueous solution with some success[75], but most generally by using powdered sodium hydroxide or sodium metal (Martynoff reaction). Nitrobenzene[76], halogenated nitrobenzenes, and nitroaniline[77, 78] are reacted with amines of the benzene and naphthalene series, *e.g.*:

Azobenzene

A modification of the Martynoff method has been used to prepare 2- and 4-(dimethylaminophenylazo)pyridines[79]. A less well known reaction is that of *N*-sulfinyl amines with arylhydroxylamines[80, 81].

For the preparation of symmetrical azo compounds a number of reductions of aryl *nitro* compounds have been performed using various reducing agents. Probably the best of these is *lithium tetrahydroaluminate* because it will reduce both

[60] *G. Heller,* J. Prakt. Chem. [2] *77*, 189 (1908).
[61] *J.W. Haworth, I.M. Heilbron, D.H. Hey,* J. Chem. Soc. 349 (1940).
[62] *E. Grandmougin, H. Freimann,* Ber. *40*, 2662 (1907).
[63] *G. Heller, O. Nötzel,* J. Prakt. Chem. [2] *76*, 58 (1907).
[64] *H. Gilman, M.W. Van Ess,* J. Amer. Chem. Soc. *61*, 3146 (1939).
[65] *K.H. Engel,* J. Amer. Chem. Soc. *51*, 2986 (1929).
[66] *R.C. Farmer, A. Hantzsch,* Ber. *32*, 3098 (1899).
[67] *K. Venkataraman,* Synthetic Dyes, p. 607, Academic Press, New York 1952.
[68] *G.M. Badger, R.G. Buttery,* J. Chem. Soc. 614 (1956).
[69] *Houben-Weyl,* Methoden der organischen Chemie, Bd. X/3, p. 287, Georg Thieme Verlag, Stuttgart 1965.
[70] *Houben-Weyl,* Methoden der organischen Chemie, Bd. X/3, p. 260, Georg Thieme Verlag, Stuttgart 1965.
[71] *E.V. Brown, J.J. Duffy,* J. Org. Chem. *32*, 1124 (1967).
[72] *P. Ruggli, A. Courtin,* Helv. Chim. Acta *15*, 75 (1932).
[73] *R. Nietzki,* Ber. *10*, 662 (1877).
[74] *K.H. Beyer,* J. Amer. Chem. Soc. *64*, 1318 (1942).
[75] *Houben-Weyl,* Methoden der organischen Chemie, Bd. X/3, p. 339, Georg Thieme Verlag, Stuttgart 1965.

aryl nitro compounds and azoxy compounds to azo products in good yields and only seldom will it reduce the azo group further[82, 83]. *Sodium tetrahydroborate* will reduce azoxy and nitro compounds to azo compounds under the proper experimental conditions[84] usually with some reduction down to the amines. Catalytic reduction using hydrazine has been reasonably successful for azo compound preparation[85] although other reduction products are also produced.

Glucose[86], and other *aldehydes* and *ketones*[87] have been used to reduce aromatic nitro compounds, and in some cases have given good yields of azo compounds, but the methods generally lead to mixtures of products. The use of *metals*, such as iron, silicon[88] and zinc[89], is not recommended for azo compound preparation. These metals occasionally lead to fair yields of product but many times mixtures of reduction products occur.

Electrochemical reduction of nitro compounds proceeds to various reduction products and one can often develop the method for the preparation of azo compounds. However, considering the problems with apparatus and conditions[89-91], this technique probably is not practical for the preparation of individual azo compounds.

5.1.2.4 Oxidation reactions

Quite a number of oxidizing agents are able to convert aromatic *amines* to azo compounds. Some of these give acceptable yields while others furnish large amounts of azoxy compounds or other by-products.

Perborates and borate/hydrogen peroxide mixtures have given modest amounts of azo compounds (20–58%)[92, 93]. *Hydrogen peroxide* alone or in acetic acid will often lead to azoxy compounds which might then be reduced to azo compounds[94, 95]. *Persulfates* generally produce mixtures of azo and azoxy compounds[96, 97].

Oxygen or air in the presence of a pyridine copper(I) chloride catalyst has given respectable yields of azo products (25–95%)[98] and also works

$$2 \; \langle\!\!\!\bigcirc\!\!\!\rangle\text{--}NH_2 + O_2 \xrightarrow{\;Cu^{\oplus}\;} \langle\!\!\!\bigcirc\!\!\!\rangle\text{--}N{=}N\text{--}\langle\!\!\!\bigcirc\!\!\!\rangle$$

Azobenzene

well in the presence of *tert*-butyl alcohol, its alcoholate[99, 100] or alkali[101]. *Hypochlorite* is effective as an oxidizing agent for both benzene[102] and pyridine[103] amines although occasionally halogenation occurs concurrently. Chromic acid[104] and permanganate[105, 106] have given success in special cases.

Aromatic *hydrazo* compounds (diarylhydrazines) can be oxidized to azo compounds by many oxidizing agents but this method is of value only where the hydrazo compound is more readily prepared than the azo compound itself. Some of the reagents[107] that have been used are air, peroxides, hypochlorite, nitrous acid, nitric acid, ferric chloride, and chromic acid.

[76] *M. Martynoff*, C.R. Acad. Sci., Paris *223*, 747 (1946).
[77] *M. Martynoff*, C.R. Acad. Sci., Paris *225*, 1332 (1947).
[78] *M. Martynoff*, C.R. Acad. Sci., Paris *227*, 1371 (1948).
[79] *R.W. Faessinger, E.V. Brown*, J. Amer. Chem. Soc. *73*, 4606 (1951).
[80] *H.E. Fierz-David, L. Blangey, E. Merian*, Helv. Chim. Acta *34*, 846 (1951).
[81] *O.M. Friedmann, R.M. Gofstein, A.M. Seligmann*, J. Amer. Chem. Soc. *71*, 3010 (1949).
[82] *E. Wiberg, A. Jahn*, Z. Naturforsch. B *7*, 580 (1952).
[83] *W. Reid, F. Müller*, Chem. Ber. *85*, 470 (1952).
[84] *R.O. Hutchins, D.W. Lamson, L. Rua, C. Milewski, B. Maryanoff*, J. Org. Chem. *36*, 803 (1971).
[85] *A. Furst, R.C. Berlo, S. Houton*, Chem. Rev. *65*, 51 (1965).
[86] DRP 1098 644 (1957); Erf.: *K.H. Schündehütten, K.H. Schmidt, F. Suckfill, H. Nickel*; C.A. *55*, 27901c (1961).
[87] USP 2 804 453 (1954); Inv.: *L.C. Anderson, C.E. Smith jr.*; C.A. *52*, 2065 (1958).

[88] *R. Meier, F. Böhler*, Chem. Ber. *89*, 2303 (1956).
[89] *P. Ruggli, M. Hinovker*, Helv. Chim. Acta *17*, 396 (1934).
[90] *K. Elbs, W. Kirsch*, J. Prakt. Chem. *67*, 265 (1903).
[91] *C.K. Mann, K.K. Barnes*, Electrochemical Reactions in Nonaqueous Systems, Marcel Dekker, New York 1970.
[92] *P. Santurni, F. Robbins, R. Stubbings*, Org. Synth. *40*, 18 (1960).
[93] *S.M. Mehta, M.V. Vakilwala*, J. Amer. Chem. Soc. *74*, 563 (1952).
[94] *R. Ruggli, C. Petitjean*, Helv. Chim. Acta *21*, 711 (1938).
[95] *E. Pfeil, K.H. Schmidt*, Justus Liebigs Ann. Chem. *675*, 36 (1964).
[96] *O.N. Witt, E. Kopetschni*, Ber. *45*, 1134 (1912).
[97] *M.K. Seikel*, J. Amer. Chem. Soc. *62*, 1214 (1940).
[98] *A.P. Terentév, J.D. Magilyanskii*, Zh. Obshch. Khim. *28*, 1959 (1958); C.A. *53*, 1327b (1959).
[99] *G.A. Russel, E.G. Jansen, H.D. Becker, F.J. Smentowski*, J. Amer. Chem. Soc. *84*, 2652 (1962).
[100] *L. Horner, J. Dehnert*, Chem. Ber. *96*, 786 (1963).
[101] *E. Bamberger, S. Wildi*, Ber. *39*, 4276 (1906).
[102] *A.G. Green, F.M. Rowe*, J. Chem. Soc. *101*, 2443 (1912).
[103] *A. Kirpal*, Ber. *67*, 70 (1934).
[104] *P. Ruggli, B. Hegedüs*, Helv. Chim. Acta *24*, 703 (1941).
[105] *C. Laar*, Ber. *14*, 1928 (1881).
[106] *G. Canquis, G. Fauvelot*, Bull. Soc. Chim. France 2014 (1964).

5.1.2.5 Condensation reactions

The condensation of aromatic nitroso compounds with *arylamines* proceeds either under acid[108, 109] or alkaline conditions[110], *e.g.:*

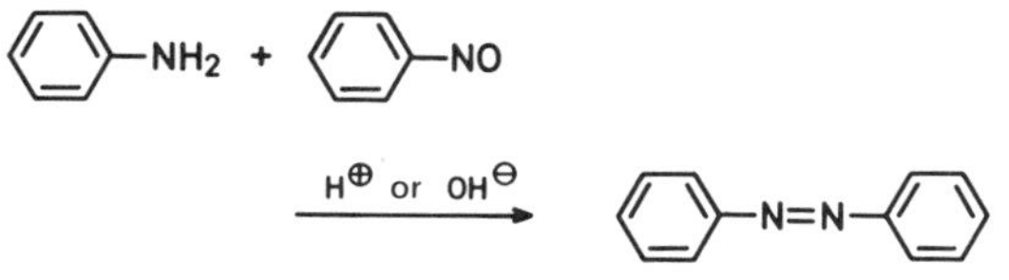

Diphenyldiazene
(Azobenzene)

Under acid conditions the reaction is normally conducted by mixing the reactants in glacial acetic acid and, if necessary, warming; the nitrogen atom of the activated nitroso group attacks the aniline nitrogen atom[111].

In practise, this means that electron-donating groups in the aniline and electron-abstracting groups in the nitrosobenzene promote the reaction. Conversely in basic media electron-donating groups in nitrosobenzene and electron-withdrawing groups in aniline aid the reaction[112]. Consequently, the basic condensation is eminently suitable for a number of aminoheterocycles, especially those with an amino group adjacent a nitrogen atom[113-114]. Compounds such as N,N-dimethyl-4-nitrosoaniline, by contrast, unfortunately afford only poor yields.

A condensation reaction of less general utility is that between an *arylhydrazine* and a *quinone* in acid medium. *1,4-Benzoquinone*[114-116], 1,4-naphthoquinone[117], phenanthraquinone[118], and several other quinones have been used successfully, *e.g.:*

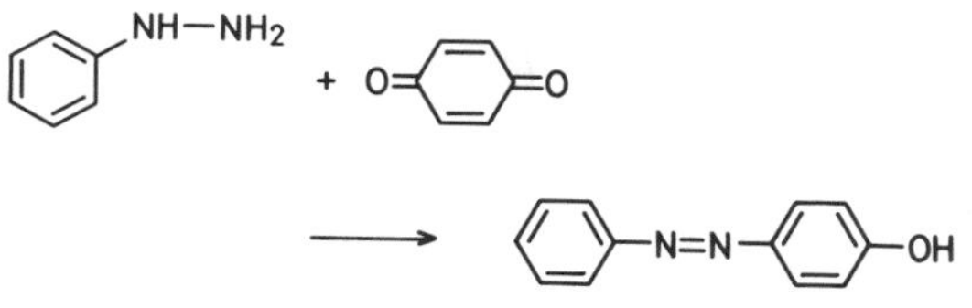

4-(Phenylazo)phenol

Aryl isocyanates and *nitrosobenzenes* can be condensed together to azo compounds[119], *e.g.:*

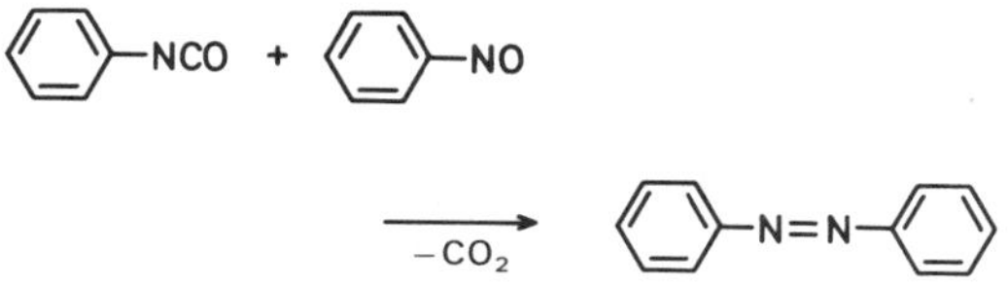

Azobenzene

Self-condensation of *diazonium salts* catalyzed by copper(I) salts leading to symmetrical azo compounds[120] is somewhat akin to a coupling reaction, *e.g.:*

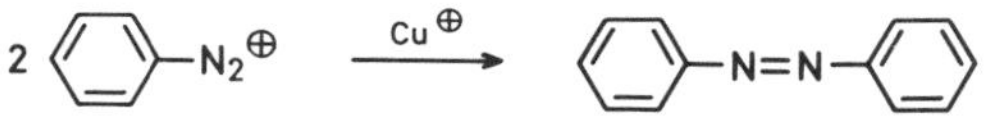

N-Heteroaromatic compounds with an amino group in the 2- or 4-position to the nitrogen can be diazotized but only seldom couple to form azo compounds, because their diazonium salts are too unstable. In such cases oxidative coupling of the corresponding hydrazones with phenols or aromatic amines brings success[121], *e.g.:*

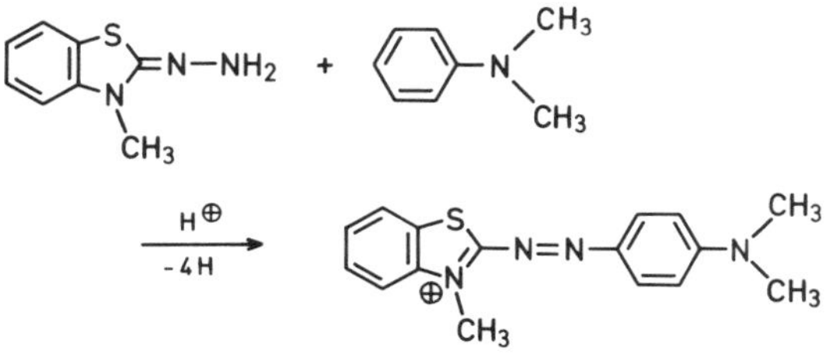

2-[(4-Dimethylaminophenyl)azo]-3-methylbenzothiazolinium salt

[107] *Houben-Weyl*, Methoden der organischen Chemie, Bd. X/3, p. 377, Georg Thieme Verlag, Stuttgart 1965.

[108] *E. Bamberger*, Ber. *29*, 102 (1896).

[109] *C. Mills*, J. Chem. Soc. 925 (1895).

[110] *N. Campbell, A.W. Henderson, D. Taylor*, J. Chem. Soc. 1281 (1953).

[111] *Y. Ogata, Y. Takagi*, J. Amer. Chem. Soc. *60*, 3591 (1958).

[112] *E.V. Brown, W.H. Kipp*, J. Org. Chem. *36*, 170 (1971).

[113] *E.V. Brown*, J. Heterocycl. Chem. *6*, 571 (1969).

[114] *P. Tomasik*, Roczniki Chem. 44 (3), 509 (1970); C.A. *74*, 3477z (1971).

[115] *W.M. Lauer, S.E. Miller*, J. Amer. Chem. Soc. *57*, 520 (1935).

[116] *S.S. Joshi, D.S. Deorha, P.C. Joshi*, J. Indian Chem. Soc. *38*, 395 (1961); C.A. *56*, 5875 (1962).

[117] *T. Zincke, H. Bindewald*, Ber. *17*, 3026 (1884).

[118] *M. Kamel, R. Wizinger*, Helv. Chim. Acta *42*, 129 (1959).

[119] *H. Staudinger, R. Endle*, Ber. *50*, 1042 (1917).

[120] *D.C. Freeman, C.E. White*, J. Org. Chem. *21*, 379 (1956).

[121] *S. Hünig*, Angew. Chem. *70*, 215 (1958); *S. Hünig*, Angew. Chem. *74*, 818 (1962); *S. Hünig*, Chimia *15*, 133 (1961).

[122] *R. Pfister, F. Häflinger*, Helv. Chim. Acta *40*, 399 (1957).

[123] *R.A. Cartwright, J.C. Tatlow*, J. Chem. Soc. 1997 (1953).

[124] *T. Kametani, Y. Nomura*, J. Pharm. Soc. Jap. *74*, 1037 (1954); C.A. *49*, 11593g (1955).

5.1.3 Transformations and analytical reactions

5.1.3.1 Reduction

One of the commoner reactions of azo compounds is *reduction* to the *hydrazo* compound or to the amine (see pp. 154, 509).

Reduction to the arylhydrazo compound may be carried out using a large number of *reducing agents,* among them zinc powder[122], sodium amalgam[123], lithium amalgam[124], sodium[125], lithium[126], potassium[127], sulfides[128], formaldehyde[129], certain ketones[130], aminoiminomethanesulfinic acid (thiourea dioxide)[131], sodium dithionate[132], stannous chloride[133], hydrogen bromide[134], phenylhydrazine[135], light plus isopropyl alcohol[136], and thioacetic acid[137].

The difficulty with many of these reagents is that they hydrogenate the hydrazo compound further to the amine

In acid solution there is also the tendency for the hydrazo compound to undergo benzidine rearrangements[132]. Probably the most reliable reagents for the transformation of azo compounds to hydrazo compounds in the benzene and naphthalene series are zinc[138] and sulfides[139], while hydrogen chloride and bromide work well with nitrogen heterocycles[140]. Catalytic hydrogenation is often quite successful since there is usually an appreciable difference in the conditions producing reduction to hydrazo compounds compared to those giving complete reduction to amine.

Some of the catalysts that have been used are palladium both on carbon[141] and on calcium carbonate[142], platinum oxide[143] and Raney nickel[144]. As substitutes for gaseous hydrogen one has used hydrazine[145] or, more recently, diazene[146].

Besides the addition of hydrogen to the azo linkage, a number of reagents add across the nitrogen double bond nitrogen linkage to give substituted hydrazo compounds. In a number of instances in the pyridine and quinoline series, phenylazo compounds have added hydrogen chloride or bromide across the whole phenylazo system giving *4-halophenylhydrazo* compounds[147]. Arylsulfinic acids can be added to the azo linkage to give the

4-[2-(4-Chlorophenyl)hydrazino]-pyridine 1-oxide

arylsulfonylhydrazo[148] compound which in some cases can be rearranged to a 4-arylsulfonylazo compound[149]. Ketene[150] and substituted ketenes[151]

4-(Phenylazo)phenyl phenyl sulfone

[125] *G.F. White, K.H. Knight,* J. Amer. Chem. Soc. *45,* 1780 (1923).

[126] *H. Normant, M. Larchevêque,* C.R. Acad. Sci., Paris *260,* 5062 (1965).

[127] *J.W.B. Reesor, G.F. Wright,* J. Org. Chem. *22,* 375 (1957).

[128] *M.P. Cava, J.F. Stucker,* J. Amer. Chem. Soc. *79,* 1708 (1957).

[129] USP 2 794 046 (1957); Inv.: *A.W. Sogu;* C.A. *52,* 428e (1958).

[130] *A.A. Sayigh,* J. Org. Chem. *25,* 1709 (1960).

[131] *H. Beyer, H.I. Hasse, W. Wildgrube,* Chem. Ber. *91,* 252 (1958).

[132] *M.K. Seikel,* J. Amer. Chem. Soc. *62,* 1214 (1940).

[133] *J. Thiele, W. Manchot,* Justus Liebigs Ann. Chem. *303,* 49 (1898).

[134] *M. Colonna, L. Pentimalli,* Ann. Chim. (Rome) *48,* 1403 (1958); C.A. *53,* 20061b (1959).

[135] *R. Walther,* J. Prakt. Chem. [2], *53,* 463 (1896).

[136] *B.E. Blaisdell,* J. Soc. Dyers Colour *65,* 618 (1949); C.A. *44,* 2755f (1950).

[137] *B.M. Mikhailov, I.S. Saveléva,* Izv. Akad. Nauk SSR, Otdel. Khim. Nauk 1304 (1959); C.A. *54,* 1372e (1960).

[138] *Houben-Weyl,* Methoden der organischen Chemie, Bd. X/3, p. 708, Georg Thieme Verlag, Stuttgart 1965.

[139] *Houben-Weyl,* Methoden der organischen Chemie, Bd. X/3, p. 714, Georg Thieme Verlag, Stuttgart 1965.

[140] *Houben-Weyl,* Methoden der organischen Chemie, Bd. X/3, p. 719, Georg Thieme Verlag, Stuttgart 1965.

can add across the azo linkage and in some instances have been hydrolyzed to the corresponding *N-carboxymethylhydrazo* compound[152]. Diazomethane[153] also has been added across the azo bond and in one case the product was then air-oxidized to an *N-formylhydrazo* compound[154], *e.g.*:

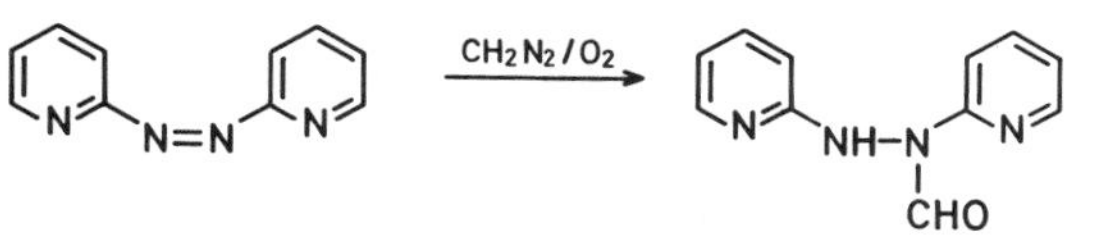

Alkali metals can be added to the azo group and subsequently replaced by alkyl groups[127], *e.g.*:

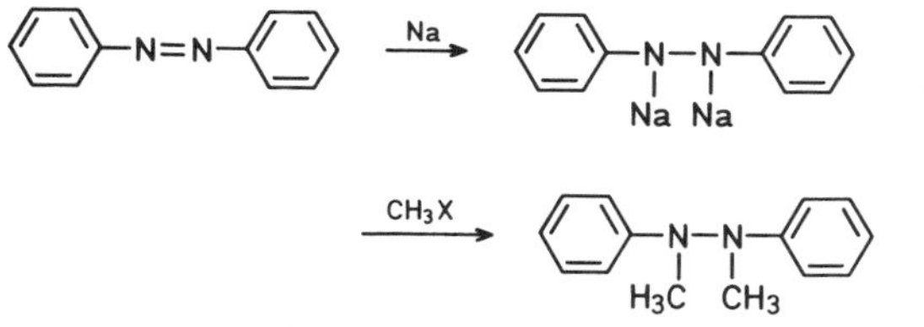

Two molecules of a Grignard compound[155] (RMgX) likewise add to form *N,N'*-bis(halo-magnesium)hydrazines, which can then react further.

Complete reduction of the azo compounds to amines has practical significance for at least two reasons:

1 It affords amines which are otherwise difficult to obtain[58].
2 It serves for the qualitative and quantitative analysis of azo compounds.

In the case of symmetrical azo compounds it is not difficult to determine the nature and amount of the amine formed. By contrast, with unsymmetrical azo compounds it is necessary to separate the two amines after the reaction and determine them individually. The nature of the amines obtained decides the procedure to be followed.

A number of *reducing agents* have been used including sodium dithionite[156, 157], sulfides[158], sodium bisulfite[159], stannous chlorides[160], zinc[161], iron[162], hydrogen iodide[163], and catalytic hydrogenation[164]. Sodium dithionate or catalytic hydrogen is recommended for this splitting of the azo linkage for the purpose of analysis. The advantage of dithionate is its generality and the fact that it is unnecessary to employ highly purified azo compounds. On the other hand, catalytic hydrogenation leaves a very pure solution ready for further reactions after filtration of the catalyst[165].

5.1.3.2 Oxidations

Aromatic azo compounds are readily oxidized to azoxy compounds (see p. 132).

The chief *oxidizing agents* are organic peracids[166] or a mixture of hydrogen peroxide[167] and an organic acid such as acetic acid.

With unsymmetrical azo compounds it is possible sometimes to isolate both isomeric azoxy compounds, in other cases only one of the two is obtained[168]. When oxidizable groups are present in addition to the azo linkage, these can be attacked as well and lead to complications. *N,N'*-Dimethylamino groups[169] or other tertiary amino groups such as those in the pyridine or quinoline ring[170] are particularly prone to this effect, *e.g.*:

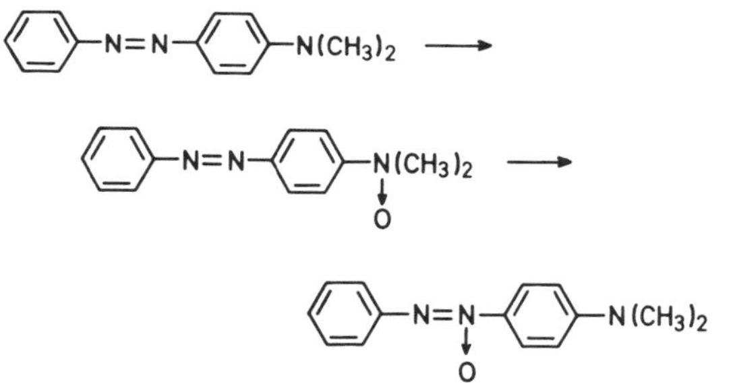

[141] *R. Pfister, F. Häflinger*, Helv. Chim. Acta *40*, 399 (1957).
[142] *M. Busch, K. Schulz*, Ber. *62*, 1458 (1929).
[143] *C.M. McCloskey*, J. Amer. Chem. Soc. *74*, 5922 (1952).
[144] *L. Horner, U. Schwenk*, Justus Liebigs Ann. Chem. *579*, 207 (1953).
[145] *E.J. Corey, W.L. Mock, D.J. Pasto*, Tetrahedron Lett. 347 (1961).
[146] *C.E. Müller*, J. Chem. Educ. *42*, 254 (1965).
[147] *A. Risaliti, L. Pentimalli*, Ann. Chim. (Rome) *46*, 1046, 1057 (1956); C.A. *51*, 8742i (1957).

[148] *W. Bradley, J.D. Hannon*, J. Chem. Soc. 2713 (1962).
[149] *W. Bradley, J.D. Hannon*, Chem. & Ind. (London) 540 (1959).
[150] *E. Fahr, W. Fischer, A. Jung, L. Sauer*, Tetrahedron Lett. 161 (1967).
[151] *L. Horner, L. Spielschka*, Chem. Ber. *89*, 2765 (1956).
[152] *G.O. Schenck, N. Engelhard*, Angew. Chem. *68*, 71 (1956).
[153] *B.M. Colonna, A. Risaliti*, Gazz. Chim. Ital. *89*, 2493 (1953); C.A. *55*, 5494a (1961).
[154] *A.R. Katritzky, S. Musierowicz*, J. Chem. Soc. 78c (1966).
[155] *H. Gilman, J.C. Bailie*, J. Org. Chem. *2*, 84 (1938).
[156] *E. Grandmougin*, Ber. *39*, 2494, 2561 (1906).

5.1.3.3 Ring closure reactions

A typical ring closure of suitably substituted is that of *2H-1,2,3-triazole* formation. For example, 2-(phenylazo)aniline cyclizes to *2-phenyl-2H-benzotriazole* on dehydrogenation[171]:

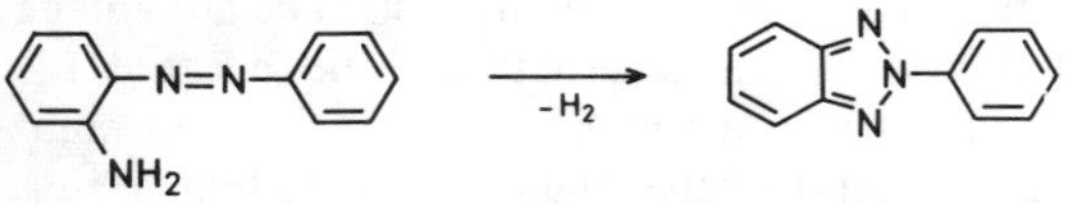

The same type of reaction can be accomplished by reductive cyclization of *o*-nitroarylazo compounds[172]. *Ortho*-azido or sulfamylamino groups can be ring-closed in this same manner by heating[173-174].

A different type of ring closure takes place when 2-aminoazo compounds are reacted with phosgene; *1,2,4-triazines*[175] are obtained, *e.g.*:

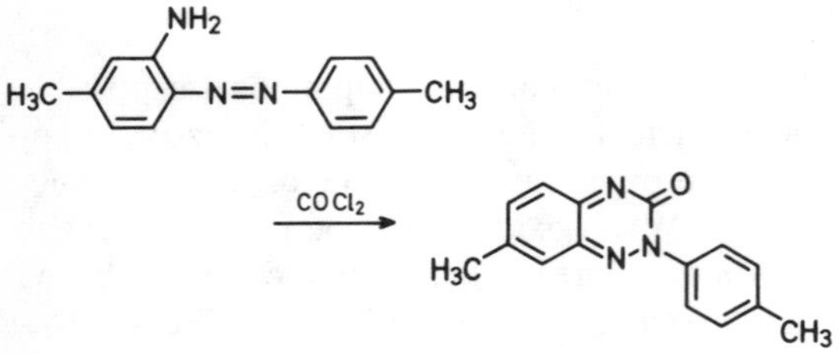

Phenazines are obtained from 2-phenylazobenzenes by acid treatment[176] with loss of a fragment containing one of the nitrogen atoms. Substituted aminobenzimidazoles[177] are formed, too, without loss of nitrogen.

Azobenzene[178] and its substitution products[179] are cyclized to *benzo[c]cinnolines* in acceptable yields, *e.g.*:

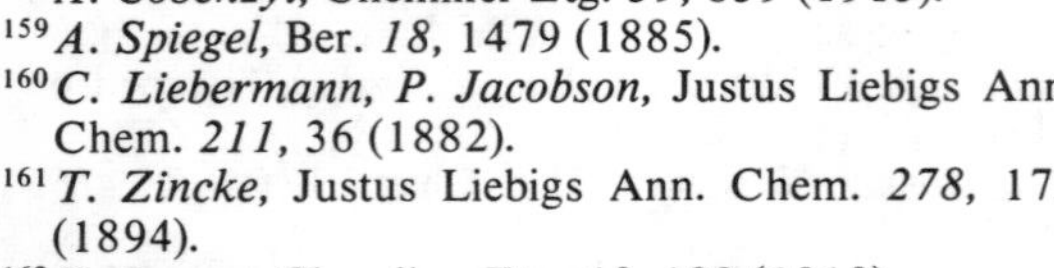

5.2 Aromatic azoxy compounds[180]

5.2.1 General

The compounds to be discussed in this section have the structure

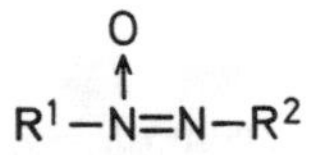

$$R^1-\overset{\overset{\textstyle O}{\uparrow}}{N}=N-R^2$$

with R^1 and R^2 denoting aromatic rings. Unsymmetrical azoxybenzenes may be named in one of the following ways:

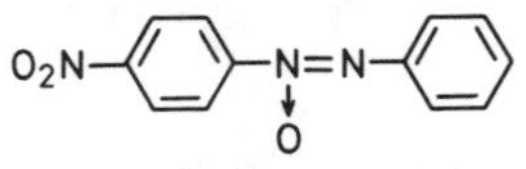

4'-Nitroazoxybenzene (β-4-)
or
(4-Nitrophenyl)phenyl-diazene 1-oxide

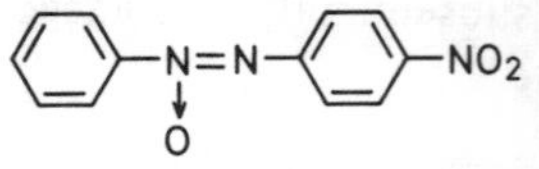

4-Nitroazoxybenzene (α-4-)
or
(4-Nitrophenyl)phenyl-diazene 2-oxide

In addition to position isomerism *cis-trans* isomerism has been observed and, with appropriately substituted compounds, even optical isomerism[181] (differentiation by their dipole moments[182-183], *NMR* spectrometry[184], *IR*-spectrometry[185])

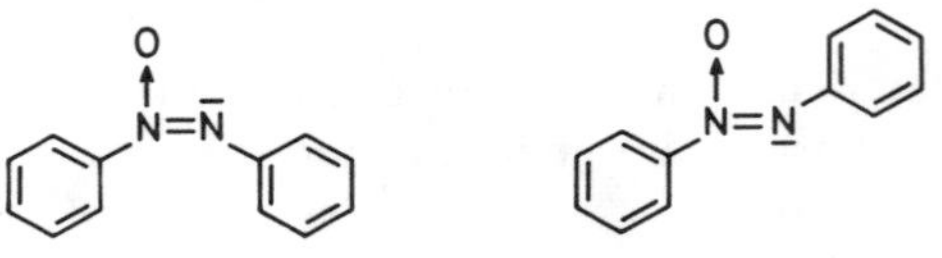

In several cases these isomers have been obtained by the oxidation of known *cis* and *trans* azobenzenes[186]. *Trans-cis* photoisomerizations proceed

[157] *W.J. Close, B.D. Tiffany, M.A. Spielman*, J. Amer. Chem. Soc. *71*, 1265 (1949).

[158] *A. Cobenzyl*, Chemiker Ztg. *39*, 859 (1915).

[159] *A. Spiegel*, Ber. *18*, 1479 (1885).

[160] *C. Liebermann, P. Jacobson*, Justus Liebigs Ann. Chem. *211*, 36 (1882).

[161] *T. Zincke*, Justus Liebigs Ann. Chem. *278*, 173 (1894).

[162] *R. Jansen*, Chemiker Ztg. *12*, 109 (1913).

[163] *R. Meyer*, Ber. *53*, 1265 (1920).

[164] *M.S. Raasch*, J. Amer. Chem. Soc. *75*, 2956 (1953).

[165] *K. Venkataraman*, Synthetic Dyes, p. 448, Academic Press, New York 1952.

[166] *G.M. Badger, G.E. Lewis*, J. Chem. Soc. 2147 (1953).

[167] *B.T. Newbold*, J. Org. Chem. *27*, 3919 (1962).

[168] *G. Leandri, A. Risaliti*, Ann. Chim. (Rome) *44*, 1036 (1954); C.A. *49*, 8849h (1955).

[169] *A.F. Douglas, P.H. Gore, J.W. Hooper*, J. Chem. Soc. 674c (1967).

[170] *M. Colonna, A. Risaliti, L. Pentimalli*, Gazz. Chim. Ital. *86*, 1067 (1956).

[171] *M.P. Schmidt, A. Hagenböcher*, Ber. *54*, 2191, 2201 (1921).

[172] USP 2 362 988 (1944); Geigy, Inv.: *A. Conzetti, O. Schmid*, C.A. *39*, 2886 (1945).

[173] USP 2 904 544 (1959); DuPont, Inv.: *R.A. Carboni*, C.A. *54*, 11062d (1960).

[174] *A. Michaelis, G. Erdmann*, Ber. *28*, 2192 (1895).

[175] *M. Busch*, Ber. *32*, 2959 (1899).

[176] *O.N. Witt*, Ber. *20*, 571 (1887).

[177] *G.B. Crippa*, Gazz. Chim. Ital. *63*, 251 (1933).

[178] *G.M. Badger, R.J. Drewer, G.E. Lewis*, Australian J. Chem. *16*, 1042 (1963).

[179] *G.M. Badger, C.P. Joshua, G.E. Lewis*, Australian J. Chem. *18*, 1639 (1965).

[180] *H.E. Bigelow*, Chem. Rev. *9*, 117 (1931).

rapidly[184], the reverse *cis-trans* shift very rapidly on careful heating[187]. (Re the obtaining of the *cis* and *trans* azoxy compounds from the corresponding *cis* and *trans* azo compounds, see refs. [188-190].) Azoxy compounds of the type

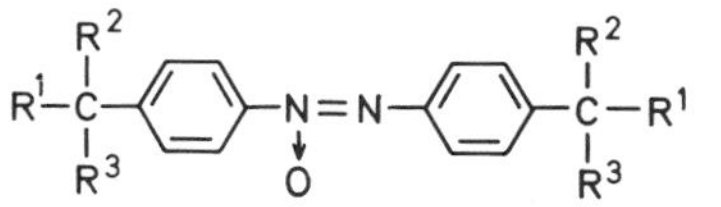

are optically active[191].

The *UV* and *visible* spectra of many azoxy compounds have been studied[192, 193]. Some studies have been reported of *L.C.A.O.* methods of *M.O.* calculations of *electron density* around the heteroatoms of the azoxy group[194].

Raman spectra of substituted azoxybenzenes have been determined[195]. An extensive *mass spectra* investigation of aromatic azoxy compounds has been reported[196]; some investigations have been carried out on the inorganic coordination compounds[197-198].

5.2.2 Preparation

5.2.2.1 Condensation reactions

The condensation of nitrosobenzene with *N*-phenylhydroxylamine proceeds under mild conditions to give *azoxybenzene (diphenyldiazene N-oxide)*[199]:

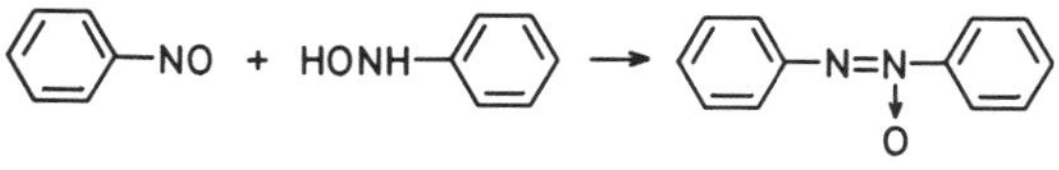

However, this reaction is of little value for preparing unsymmetrical azoxybenzenes since the major products are the two possible symmetrical compounds and only a very minor amount is the unsymmetrical one[182]. The rates and mechanism of the reaction have been examined and the abnormal course of the reaction is believed to be caused by a rapid equilibrium between the nitroso compound and the hydroxylamine[200, 201]:

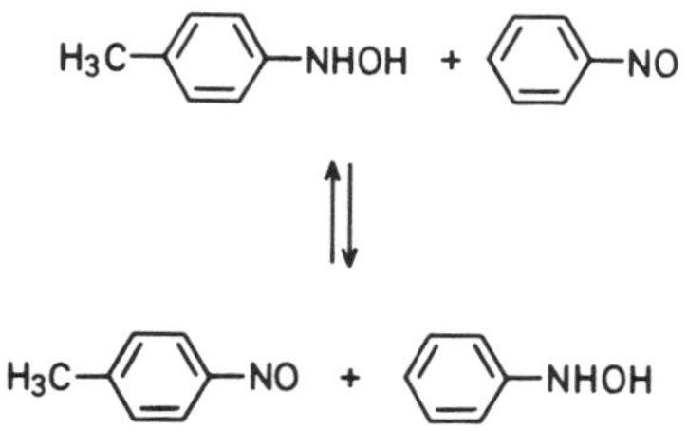

In the special case of the reaction of *N*-phenylhydroxylamine with *N, N-dimethyl-4-nitrosoaniline* both *N, N-dimethyl-4-(phenyl-NNO-azoxy)aniline* and *N, N-dimethyl-4-(phenyl-ONN-azoxy)aniline* are obtained[202].

A 38% yield of *azoxybenzene* was formed in the unusual condensation of nitrosobenzene with benzylamine[203]. 1-(4-Nitrobenzyl)pyridinium salts condense with *N*-phenylhydroxylamine to give *4-(Phenylazoxy)benzaldehyde phenylnitrone*[204]:

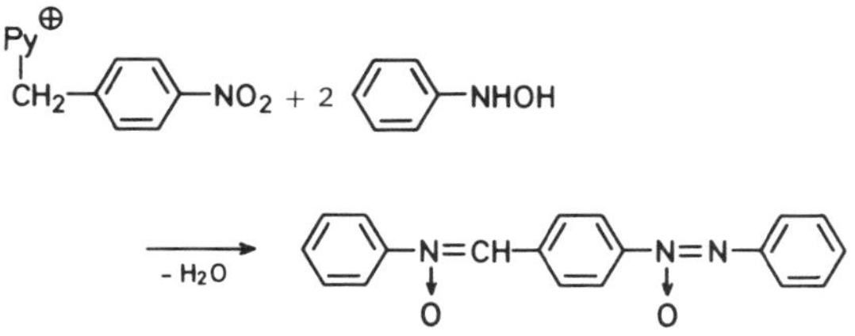

Aromatic nitrosohydroxylamine tosylates can be condensed with aryl Grignard reagents to give azoxy compounds in good yields[205]:

[181] *A. Hantzsch, A. Werner,* Ber. *23,* 22 (1890).

[182] *E. Müller,* Justus Liebigs Ann. Chem. *495,* 132 (1932).

[183] *K.A. Gehrckens, E. Müller,* Justus Liebigs Ann. Chem. *500,* 296 (1933).

[184] *D.L. Webb, H.H. Jaffé,* J. Amer. Chem. Soc. *86,* 2419 (1964).

[185] *W. Luettke,* Justus Liebigs Ann. Chem. *668,* 184 (1963).

[186] *G.M. Badger, R.G. Buttery, G.E. Lewis,* J. Chem. Soc. 2143 (1953).

[187] *D. Webb, H.H. Jaffé,* Tetrahedron Lett. 1875 (1964).

[188] *G.M. Badger, G.E. Lewis,* J. Chem. Soc. 2147 (1953).

[189] *G.M. Badger, G.E. Lewis,* J. Chem. Soc. 2151 (1953).

[190] *G.M. Badger, R.G. Buttery,* J. Chem. Soc. 2156 (1953).

[191] *T.T. Chu, C.S. Marvel,* J. Amer. Chem. Soc. *55,* 2841 (1933).

[192] *J. Weinstein, J.A. Sousa, A.L. Blahm,* J. Org. Chem. *29,* 1586 (1964).

[193] *C. Tosi,* Corsi Semin. Chem. 50 (1968); C.A. *71,* 75768f (1969).

[194] *M.F. Shostakovskii, Y.L. Frolov, C.G. Skvortsova, G.R. Kontarev,* Radiospektrosk. Kvantovokhim Metody Strukt. Issled. 62 (1967); C.A. *70,* 10959d (1969).

[195] *A.S. L'vova, V.A. Chirkov, V.A. Zubov, M.M. Sushchinskii,* Opt. Spektrosk. *23,* 168 (1967); C.A. *67,* 103784w (1967).

[196] *J.H. Bowie, R.G. Cooks, G.E. Lewis,* Australian J. Chem. *20,* 1601 (1967).

[197] *A.L. Blach, D. Petridis,* Inorg. Chem. *8,* 2247 (1969).

[198] *E. Wieteska, J. Minczewski,* Chem. Anal. (Warsaw) *12,* 1119 (1967); C.A. *68,* 73617v (1968).

[199] *E. Bamberger, R. Rising,* Justus Liebigs Ann. Chem. *316,* 357 (1901).

4'-Chloroazoxybenzene

5.2.2.2 Reduction reactions

One of the most useful methods for the preparation of *symmetrical* azoxy compounds is the reduction of aromatic *nitro* compounds by various reagents. Although a great variety of reducing agents has been used, probably the three most practical are alcohols, aldehydes, and certain complex metal hydrides. Heating an aromatic nitro compound in alcoholic sodium hydroxide solution or in sodium methoxide-methanol solution usually produces excellent yields of the symmetrical azoxy compound[206, 207].

The kinetics and the mechanism of this reaction have been studied[208]. Among the aldehydes used as reducing agents, *glucose* has been rather well investigated and gives an 83% yield of *azoxybenzene*[209] and fairly good yields with a number of substituted nitrobenzenes[210].

Other aldehyde and ketone reducing agents that have been used include lactose, maltose, galactose, fructose, and formaldehyde[211, 212].

Among the *complex hydrides* the best reducing agent for the preparation of azoxybenzenes seems to be sodium[213] or potassium tetrahydroborate[214].

The sodium borohydride-cobaltous chloride system has also been used to give good yields of azoxybenzenes and the conclusion has been drawn that the relative yields are related to the Hammett constants of the nitro compound. The more electron-attracting it is, the higher is the yield[215].

Lithium tetrahydroaluminate performs the reduction also but tends to carry the molecule to the azo compound[216].

Besides the complex metal hydrides a number of *inorganic reducing agents* have been used to convert aromatic nitro compounds to azoxy compounds. Sodium arsenite has been useful in some cases[217]. Yields of about 70% have been obtained with stannous chloride and sodium hydroxide[218]. In a group of aromatic nitro compounds the yields of azoxy products varied from 30–90% using magnesium metal and methanol with and without ammonium chloride[219].

Silicon and sodium hydroxide were unsatisfactory for nitrobenzene and many of its substituted compounds but did show promise where a carboxy group was attached to the ring[220]. Zinc and sodium hydroxide have given some azoxy compounds but generally afford the aromatic hydroxylamines[221]. Recently, thallium in ethanol has been successful as a reducing agent[222] with yields varying from 64–93% with various nitrobenzenes. Catalytic hydrogenation has been used to obtain azoxy compounds from nitro compounds[223] but conditions must be very carefully controlled to avoid further reduction[224]. Aromatic nitroso compounds have been reduced in good yield to azoxy compounds by heating with trialkylammonium formates[225].

5.2.2.3 Oxidation reactions

Where the proper azo compound is available, a very satisfactory synthesis of azoxy compounds is oxidation of the azo compounds.

Peracids (*e.g.* peroxybenzoic acid[176, 188–190]) seem to be generally applicable but other oxidizing agents, too, have been used[226]. In many cases both position isomers of the azoxybenzenes are obtained from unsymmetrical azo compounds, while in certain others only one isomer is found.

[200] *Y. Ogata, M. Tsuchida, Y. Tagaki,* J. Amer. Chem. Soc. *79*, 3397 (1957).

[201] *S. Oae, T. Fukumoto, M. Yamagami,* Bull. Chem. Soc. Japan *36*, 728 (1963); C.A. *59*, 9849 (1963).

[202] *W. Anderson,* J. Chem. Soc. 1722 (1952).

[203] *K. Suzuki, E.K. Weisburger,* Tetrahedron Lett. 5409 (1966).

[204] *H. Rembges, F. Krohnke, I. Vogt,* Chem. Ber. *103*, 3427 (1970).

[205] *T.E. Stevens,* J. Org. Chem. *29*, 311 (1964).

[206] *P. Starke,* J. Prakt. Chem. *59*, 204 (1899).

[207] *B.T. Newbold,* J. Chem. Soc. 4260 (1961).

[208] *Y. Ogata, J. Mibae,* J. Org. Chem. *27*, 2048 (1962).

[209] *H.E. Bigelow, A. Palmer,* Org. Syntheses, Coll. Vol. II, 58 (1943).

[210] *H.W. Galbraith, E.F. Degering, E.F. Hitch,* J. Amer. Chem. Soc. *73*, 1323 (1951).

[211] *B.T. Newbold, R.P. LeBlanc,* J. Org. Chem. *27*, 312 (1962).

[212] *P.E. Gagnon, K.F. Keirstead, B.T. Newbold,* Can. J. Chem. *35*, 1304 (1957).

[213] *C.E. Weill, G.S. Pauson,* J. Org. Chem. *21*, 803 (1956).

Spectrophotometric analysis has been used to *determine* the percentage of each isomer formed and the results were interpreted in terms of conjugative and inductive effects[227]. Nuclear magnetic resonance has also been used in distinguishing between the position isomers formed[184].

Probably the commonest oxidizing agent for this reaction is peroxyacetic acid or its equivalent, namely, 30% hydrogen peroxide in acetic acid[228]. Another excellent method for synthesizing unsymmetrical azoxy compounds is the Reissert *2H-indazole 1-oxide* oxidation. The method proceeds *via* nitriles which are cyclized to 3-cyano-2-aryl-*2H*-indazole 1-oxides; subsequent oxidation affords azoxy compounds with an easily split off carboxy group.

Oxidation of aniline to azoxybenzene succeeds with peracetic acid. This reaction probably proceeds *via* the intermediate nitrosobenzene stage, condensation with a second molecule aniline to azobenzene, and further oxidation[231]. Several references describe investigations aimed at determining the optimum experimental conditions[232, 233].

5.2.2.4 From other azoxy compounds by substitution

Electrophilic reaction partners can ring-substitute aromatic compounds but only in the ring remote from the *N*-oxide group[234].

In this method the *N-O* of the azoxy group is always adjacent the benzene ring which had the original 2-nitrobenzaldehyde structure. This synthesis of a known position isomer of substituted azoxybenzene is often used[229, 230].

The inductive effect of the *N*-oxide group reduces the electron density of the adjacent ring so much that it is not substituted, *e.g.*:

Main product:
4-Nitroazoxybenzene

Side product:
2-Nitroazoxybenzene

In addition to nitration it has been possible to halogenate[235] and mercurate[236] azoxy compounds. Generally, when the *ortho* and *para* positions in the ring farthest from the *N—O* linkage are filled, it is not easy to substitute the other ring[180]. Success or failure of an electrophilic substitution under standard conditions therefore sometimes distinguishes structural isomers (*e.g.*, 4- and 4'-bromoazoxybenzenes).

[214] *H.J. Shine, H.E. Mallory*, J. Org. Chem. *27*, 2390 (1962).

[215] *T. Satoh, S. Suzuki, T. Kikuchi, T. Okada*, Chem. & Ind. (London) 1626 (1970).

[216] *A. Rassat, J.R. Revet*, Bull. Soc. Chim. France 3679 (1968).

[217] *H.J. den Hertog, C.H. Henkens, J.H. van Roon*, Rec. Trav. Chim. Pays-Bas *71*, 1145 (1952).

[218] *G. Lock, E. Bayer*, Ber. *69*, 2666 (1936).

[219] *L. Zechmeister, P. Rom*, Justus Liebigs Ann. Chem. *468*, 117 (1929).

[220] *R. Meier, F. Böhler*, Chem. Ber. *89*, 2301 (1956).

[221] *L. Wacker*, Justus Liebigs Ann. Chem. *317*, 375 (1901).

[222] *A. McKillop, R.A. Raphael, E.C. Taylor*, J. Org. Chem. *35*, 1670 (1970).

[223] *K. Brand, J. Steiner*, Ber. *55*, 875 (1922).

[224] *M. Busch, K. Schultz*, Ber. *62*, 1458 (1929).

[225] *M. Sekiya, S. Takayama*, Chem. Pharm. Bull. (Tokyo) *18*, 2146 (1970).

[226] *A. Werner, E. Strasny*, Ber. *32*, 3256 (1899).

[227] *A. Risaliti, A. Monti*, Gazz. Chim. Ital. *91*, 299 (1961).

[228] *Houben-Weyl*, Methoden der organischen Chemie, Bd. X/3, p. 762, Georg Thieme Verlag, Stuttgart 1965.

[229] *L.C. Behr*, J. Amer. Chem. Soc. *76*, 3672 (1954).

[230] *L.C. Behr, E.G. Alley, O. Levard*, J. Org. Chem. *27*, 65 (1962).

[231] *A.L. Bluhm, J. Weinstein, J.A. Sousa*, J. Org. Chem. *28*, 1989 (1963).

5.2.3 Transformations and analytical reactions

5.2.3.1 The Wallach rearrangement

The Wallach[237] rearrangement of azoxybenzenes occurs in concentrated sulfuric acid solutions and brings about the formation of *hydroxyazobenzenes*

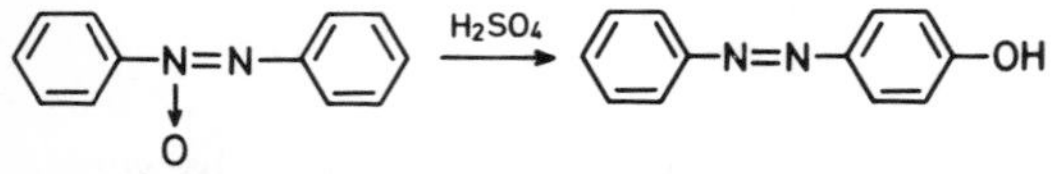

It may be accomplished also by chlorosulfonic acid, in which case the chlorosulfonic ester of the hydroxyazobenzene is the product[238]. Generally, the product is a 4-hydroxyazobenzene when a *para* position is available. The case of 4-bromoazoxybenzene and 4'-bromoazoxybenzene [4-(bromophenyl)phenyldiazene 1- and 2-oxides] which both afford *[(4-bromophenyl)azo]phenol*[239] illustrates this point.

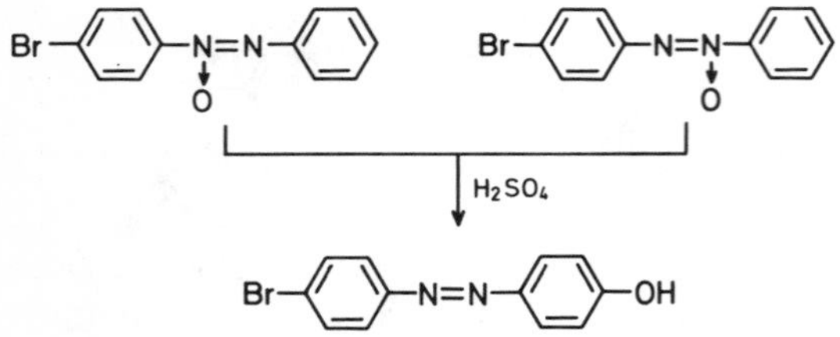

One case in which the *ortho*-hydroxy compound is the major product is the Wallach rearrangement of the two forms of 2-phenylazoxynaphthalene[240]:

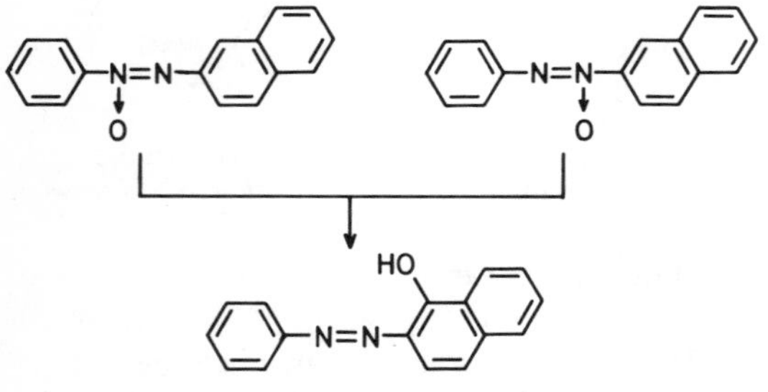

2-Phenylazo-1-naphthol

There may be other cases where *ortho* products have been obtained[241], but in general they represent minor products.

Much effort has been devoted to elucidating the mechanism of the Wallach rearrangement including isotopic labeling. The situation with ^{15}N in one of the azoxy nitrogen, with ^{18}O in the azoxy group and alternatively in the sulfuric acid have been examined[242]. *Azoxybenzene-d$_{10}$* was also synthesized and rearranged while looking for a primary kinetic isotope effect[243]. There seems little doubt that azoxybenzene is monoprotonated in the acid at least[244, 245], but much controversy has arisen about the possibility of a stable dication. Much kinetic and equilibria data have been amassed to bolster the various mechanisms[246-250]. The whole matter has been discussed in an excellent review[251] which includes the application of molecular orbital calculations to the problem.

The *photoisomerization*[252, 253] of azoxybenzene to hydroxyazobenzenes is often discussed along with the Wallach rearrangement but by mechanism and products is an entirely different reaction. While the Wallach rearrangement is predominantly intermolecular and gives the *para*-hydroxy compound whenever possible, the photoisomerization is intramolecular and affords *ortho*-hydroxyazo products.

The irradiation ($\lambda \sim 300$ mμ) of an alcoholic solution of azoxybenzene leads to *trans-cis* equilibrium and rearrangement to 2-(phenylazo)phenol after longer reaction times[187, 254-256]. The most probable intramolecular mechanism is the following:

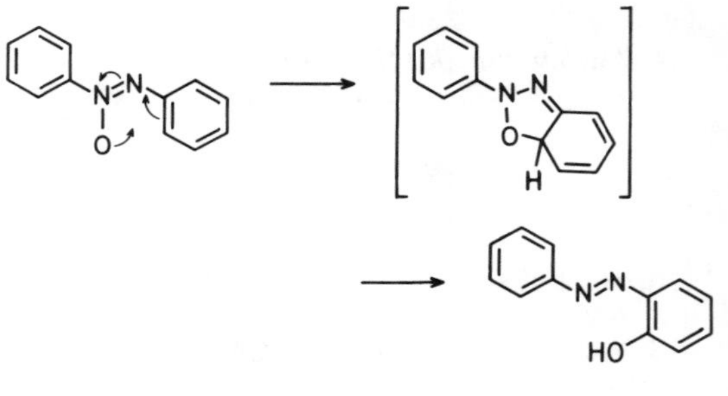

Two survey articles review details[251, 257]. A similar mechanism has been proposed for the conversion of azoxybenzene to *ortho*-hydroxyazobenzene by heating in acetic anhydride[115, 251].

[232] *F.P. Greenspan*, Ind. Eng. Chem. *39*, 847 (1947).

[233] *G.B. Payne, P.H. Deming, P.H. Williams*, J. Org. Chem. *26*, 659 (1961).

[234] *N. Zinin*, Justus Liebigs Ann. Chem. *114*, 218 (1860).

[235] *L.C. Behr*, J. Amer. Chem. Soc. *76*, 3672 (1954).

[236] *T. Vkai, Y. Ito*, J. Pharm. Soc. Jap. *73*, 821 (1953); C.A. *48*, 9945i (1954).

[237] *O. Wallach, L. Belli*, Ber. *13*, 525 (1880).

[238] *V.O. Lukashevich, T.N. Kurdynmova*, Zh. Obshch. Khim. *18*, 1963 (1948); C.A. *43*, 38008 (1949).

[239] *C.S. Hahn, H.H. Jaffé*, J. Amer. Chem. Soc. *84*, 946 (1962).

[240] *E. Buncel, A. Dolenko*, Tetrahedron Lett. 113 (1971).

[241] *M.M. Shemyakin, Ts.E. Agadzhanyan, V.I. Maimind, R.V. Kudryavtsev*, Isz. Akad. Nauk SSSR, Ser. Khim., 1339 (1963).

The ketone-*sensitized* photoreduction of azoxybenzene furnishes azobenzene (1, 2-diphenyldiazene)[258, 259], with hydrogen abstraction by the sensitizer (usually benzophenone)[260] from the solvent (usually alcohol) possibly playing a part.

5.2.3.2 Isomerization

Conversion of *4'-nitro-* into *4-nitroazoxybenzene* succeeds with chromic acid[261] or sulfuric acid[262]. *4-Phenylazoxy-ONN-benzoic acid*[263] and *4-(phenylazoxy)azoxybenzene*, too, can be isomerized[264]. Since these are the only cases reported, it seems that the reaction is little capable of general use.

5.2.3.3 Oxidation reactions

Generally azoxy compounds are stable to oxidation as is exemplified by the oxidation of 4-methylazoxybenzene to *4-(phenyl-ONN-azoxy)benzoic acid*[265]:

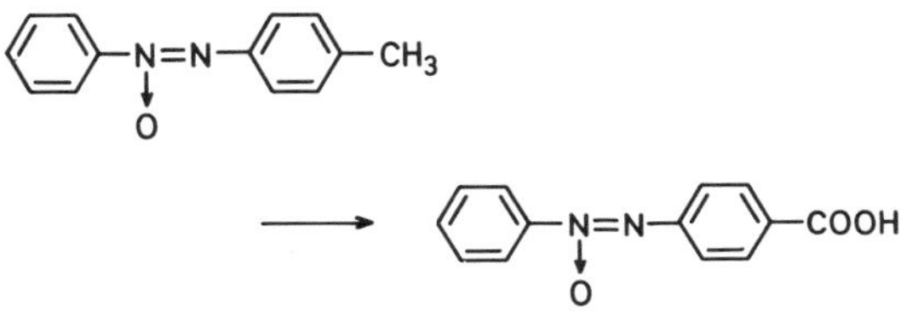

Exceptions are hydroxy-substituted azoxybenzenes, where often one isomer is more easily oxidized than the other. With potassium permanganate the product is the *trans*-diazotate[266, 267].

5.2.3.4 Reduction reactions

The reduction of azoxy compounds to azo compounds is discussed in greater detail on p. 131.

A comparison of the various reagents showed that the following furnished good yields: Raney nickel and hydrogen[268], zinc and sodium hydroxide[269] or pyridine[268], lithium tetrahydroaluminate[270, 186], potassium tetrahydroborate[271], trialkylphosphines[272], or phosphorus trichloride[273].

Vigorous reduction methods such as using metals and acid could be expected to reduce azoxybenzene to *anilines*. Photoreduction to aniline is also possible in the presence of disodium 9, 10-dihydro-9, 10-dioxo-2, 6-anthracenedisulfonate and isopropyl alcohol[274]. *Analytical* procedures for azoxy compounds have been developed using polarography with stannous and titanium chlorides[275].

[253] *H.M. Knipscheer*, Rec. Trav. Chim. Pays-Bas *22*, 1 (1903).

[254] *H. Mouser, G. Gauglitz, F. Stier*, Justus Liebigs Ann. Chem. *739*, 84 (1970).

[255] *R. Tanikaga*, Bull. Chem. Soc. Japan *41*, 2151 (1968).

[256] *G.M. Badger, R.G. Buttery*, J. Chem. Soc. 2243 (1954).

[257] *G.G. Spence, E.C. Taylor, O. Buchardt*, Chem. Rev. *70*, 231 (1970).

[258] *R. Tanikaga, K. Maruyama, R. Goto, A. Kaji*, Tetrahedron Lett. 5925 (1966).

[259] *R. Tanikaga*, Bull. Chem. Soc. Japan *41*, 1664 (1968).

[260] *B.M. Monroe, C.C. Wamser*, Mol. Photochem. *2*, 213 (1970).

[261] *A. Angeli*, Gazz. Chim. Ital. *46*, II., 67 (1916).

[262] *P.H. Gore*, Chem. & Ind. (London) 191 (1959).

[263] *A. Angeli, B. Valori*, Atti. Acad. Lincei I., *22*, 132 (1913); C.A. *7*, 2223 (1913).

[264] *A. Angeli*, Atti. Acad. Lincei I., *22*, 844 (1913); C.A. *8*, 76 (1914).

[265] *D. Bigiavi, V. Saratelli*, Gazz. Chim. Ital. *57*, 555 (1927).

[266] *A. Angeli*, Gazz. Chim. Ital. I., *51*, 35 (1921).

[267] *D. Bigiavi, R. Poggi*, Gazz. Chim. Ital. *54*, 114 (1924).

[268] *P. Ruggli, G. Bartusch*, Helv. Chim. Acta *27*, 1371 (1944).

[269] *R. Medola, L.J. Andrews*, J. Chem. Soc. *69*, 7 (1896).

[270] *R.F. Nystrom, W.G. Brown*, J. Amer. Chem. Soc. *70*, 3738 (1948).

[271] *H.E. Brearley, H. Gott, H.A.O. Hill, M. O'Reordan, J.M. Pratt, R.J.P. Williams*, J. Chem. Soc. 612 (A 1971).

[272] *L. Horner, H. Hoffman*, Angew. Chem. *68*, 480 (1956).

[273] *J.F. Vozza*, J. Org. Chem. *34*, 3219 (1969).

[274] *S. Hashimoto, J. Sunamoto, H. Fuji, K. Kano*, Bull. Chem. Soc. Japan *41*, 1249 (1968).

[275] *Y.A. Gawargious, S.W. Bishara*, Fresenius. Z. Anal. Chem. *245*, 366 (1969); C.A. *71*, 77095h (1969).

[242] *M.M. Shemyakin, V.I. Maimind* in *T.S. Gore*, Recent Progress in the Chemistry of Natural and Synthetic Colouring Matters and Related Fields, p. 441, Academic Press, New York 1962.

[243] *E.C. Hendley, D. Duffey*, J. Org. Chem. *35*, 3579 (1970).

[244] *D. Duffey, E.C. Hendley*, J. Org. Chem. *33*, 1918 (1968).

[245] *J. Singh, P. Singh, J.L. Boivin, P.E. Gagnon*, Can. J. Chem. *41*, 499 (1963).

[246] *E. Buncel, B.T. Lawton*, Can. J. Chem. *43*, 862 (1965).

[247] *E. Buncel, B.T. Lawton*, Chem. & Ind. (London) 1835 (1963).

[248] *P.H. Gore*, Chem. & Ind. (London) 191 (1959).

[249] *C.S. Hahn, K.W. Lee, H.H. Jaffé*, J. Amer. Chem. Soc. *89*, 4975 (1967).

[250] *E. Buncel, W.M.J. Strachan, R.J. Gillespie, R. Kapoor*, Chem. Commun. 765 (1969).

[251] *E. Buncel* in *B.S. Thyagarajan*, Mechanisms of Molecular Migrations, Vol. I, p. 61, Interscience Publishers, New York 1968.

[252] *L. Wacker*, Justus Liebigs Ann. Chem. *317*, 375 (1901).

5.3 Aliphatic and mixed aliphatic-aromatic azo compounds

5.3.1 Dialkyl and alkylaryl diazenes

Diazenes are derived from diazene (diimide) $HN=NH$. The azo nomenclature adopted for aromatic compounds is normally applied only to symmetrical aliphatic derivatives. In this section (like in the new Chemical Abstracts listing) alkyl and alkylaryl azo and azoxy compounds will therefore be named as diazenes and diazene oxides respectively. The method has the additional advantage of unambiguously characterizing the diazene oxides.

This contribution discusses dialkyl and alkylaryl diazenes and their *N*-oxides; the following are not dealt with: arylazo compounds of the type

$$Aryl-N=N-X$$

X = OR, SR, O-Acyl, SO_2–R, SO_2–OR, COOH
  (and their derivatives), CN, Acyl and their derivatives
  (*e.g.*, formazans, p.723)

diacyl diazenes of carboxylic acids or carbonic acid, heteroaromatic compounds containing an azo grouping, *e.g.*

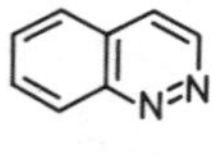

The chapter does include the corresponding vinyl derivatives unless they are part of an aromatic system, *e.g.*

$$R-CH=CH-N=N-R^1$$

The *N*-oxides are dealt with analogously.
Diazenes are obtained either in the *cis* or the *trans* form. Details are given in the section on aromatic azo compounds (p. 128). Re spectroscopic details see p. 149.

5.3.1.1 Preparation

In general, both dialkyl and alkylaryl diazenes are best prepared from hydrazine derivatives. The following methods are available:

1 Oxidation of *N,N-dialkyl* or *N*-alkyl-*N'*-arylhydrazines
2 Cleavage of *N,N'*-dinitroso-*N,N'*-dialkylhydrazines
3 Preparation from hydrazones or azines
4 Preparation from arylhydrazines and α-halocarbonyl compounds

An additional useful procedure for preparing dialkyl diazenes is *direct N–N interlinking* and also the reaction of 2-diazo 1,3-diketones or tosyl azides with methylene compounds. Preparation of alkylaryl diazenes by means of the coupling reaction of aryl diazonium salts with *CH*-acid compounds is very important especially in connection with the Japp-Klingemann reaction.

5.3.1.1.1 From hydrazine derivatives

The ready oxidizability of the *N,N'*-dialkyl- and *N*-alkyl-*N'*-arylhydrazines quite generally affords a good to very good yield of the corresponding diazenes

$$R-NH-NH-R^1 \xrightarrow{-H_2} R-N=N-R^1$$

Mercury oxide is the *oxidizing agent* used predominantly for medium-size and higher *N,N'*-dialkyl- and *N*-alkyl-*N'*-arylhydrazines, while copper(II) salts are employed for preparing volatile diazenes (see Table 1, p. 143). Halogens, atmospheric oxygen, hydrogen peroxide[276, 277], and other oxidizing agents[278–281], too, have been used. The oxidation is generally carried out in neutral aqueous solution (in acid and alkaline media isomerization to hydrazones takes place).

[276] *S.G. Cohen, C.H. Wang*, J. Amer. Chem. Soc. *77*, 2460 (1955).
[277] *J. Thiele*, Justus Liebigs Ann. Chem. *376*, 267 (1910).
[278] *J. Thiele*, Ber. *42*, 2578 (1909) (potassium chromate).
[279] *J.E. Leffler, L.M. Barbato*, J. Amer. Chem. Soc. *77*, 1690 (1955) (lead(IV)-oxide).
[280] *H. Wieland, E. Popper, H. Seefried*, Ber. *55*, 1823 (1922) (Brom).
[281] *S. Goldschmidt, S. Nathan*, Justus Liebigs Ann. Chem. *437*, 224 (1924).
[282] *R. Renaud, L.C. Leitch*, Can. J. Chem. *32*, 549 (1954).
[283] *C.G. Overberger, I. Tashlick*, J. Amer. Chem. Soc. *81*, 217 (1959).
[284] *S.G. Cohen, R. Zand*, J. Amer. Chem. Soc. *84*, 586 (1962).
[285] *S.G. Cohen*, J. Amer. Chem. Soc. *79*, 2661 (1957).
[286] *R. Askani*, Chem. Ber. *102*, 3304 (1969).
[287] *O. Diels, W. Koll*, Justus Liebigs Ann. Chem. *443*, 270 (1925).
[288] *M. Bögemann, A. von Friedrich*, Privatmitteilung, Leverkusen; vgl. *Houben-Weyl*, Bd. X/2, p. 764.
[289] *C.G. Overberger, P.T. Huang, M.B. Berenbaum*, Org. Syntheses, Coll. Vol. IV, 66 (1944).
[290] *F.P. Jahn*, J. Amer. Chem. Soc. *59*, 1761 (1937).
[291] *E. Bamberger, W. Pemsel*, Ber. *36*, 56 (1903); *E. Fischer*, Ber. *29*, 797 (1896).
[292] *J.N. Brough, B. Lythgoe, P. Waterhouse*, J. Chem. Soc. (London) 4076 (1954).
[293] USP 2 556 876 (1947), DuPont, Inv.: *J.W. Hill*, Chem. Zentr. 8056 (1952).

Table 1. Dialkyl- and Alkylaryldiazenes by oxidation of corresponding hydrazines

R—NH—NH—R¹ R	R¹	Oxidizing agent	Diazene	Yield [%]	Ref.
C_2H_5	C_2H_5	HgO	Diethyldiazene	97	[282]
$H_5C_6-CH-(CH_2)_4-CH-C_6H_5$		HgO	3,4,5,6,7,8-Hexahydro-3,8-diphenyl-1,2-diazocine	50	[283]
(bicyclic structure NH/N—H)		O_2	1,2-Diazabicyclo[2.2.1]hept-2-ene	23	[284]
H_5C_6-CH- CH_3	$-CH-C_6H_5$ CH_3	O_2	Bis-(1-phenylethyl)diazene	72	[285]
(bicyclic structure NH/N—H)		HgO	7,8-Diazatricyclo[4.2.2.0²,⁵]deca-3,7,9-triene	80	[286]
(polycyclic structure N/N—H)		HgO	9,10-Diazapentacyclo[4.4.0.0²,⁵.0³,¹⁰.0⁴,⁷]dec-8-ene	70	[286]
$(H_3C)_2CH-$	$(H_3C)_2CH-$	Cu^II	Bis-(1-methylethyl)diazene	95	[287]
$(H_3C)_2C-$ CN	$(H_3C)_2C-$ CN	Cl_2	2,2'-Diazenebis(methylpropanonitrile)	88	[288]
(cyclohexane-CN)	(cyclohexane-CN)	Cl_2	Diazenebis(1-cyclohexanecarbonitrile)	90	[289]
CH_3	CH_3	Cu^II	Dimethyldiazene	95	[290]
C_2H_5	C_6H_5	HgO	Ethylphenyldiazene		[291]
C_6H_5	$Cl-\langle\rangle-CH_2-$	Fe^III	(4-Chlorophenylmethyl)phenyldiazene	70	[292]

Re the preparation of *polyazonitriles* see ref. [293].

The N,N'-dialkyl-N,N'-dinitrosohydrazines accessible by treating N,N'-dialkylhydrazines with nitrous acid are often converted into the corresponding diazenes even at room or slightly elevated temperature while losing nitrogen monoxide[294]. This roundabout path offers advantages where the yield of diazenes from N,N'-dialkylhydrazines by the direct route is poor:

$$H_5C_6-CH_2-\underset{\underset{NO}{|}}{\overset{\overset{NO}{|}}{N}}-N-CH_2-C_6H_5 \longrightarrow$$

$$H_5C_6-CH_2-N=N-CH_2-C_6H_5 \;+\; 2\,NO$$

Dibenzyldiazene; 72%

Hydrazones and azines are alternative starting products. At elevated temperature aldehyde alkylhydrazones isomerize to diazenes with caustic potash; *e.g.:*

$$H_3C-NH-N=CH-R \xrightarrow{KOH,\,40°} H_3C-N=N-CH_2-R$$

R = C_2H_5; *Methylpropyldiazene*[295]; 60%
R = C_3H_7; *Butylmethyldiazene*[295]; 64%

Oxidation of ketone alkyl- and arylhydrazones with lead(IV) acetate affords *α-acetoxydiazenes* in good to very good yields[296]:

$$\underset{R}{\overset{R^1}{>}}C=N-NH-R^2 \xrightarrow{Pb(O-CO-CH_3)_4} R-\underset{\underset{O-CO-CH_3}{|}}{\overset{\overset{R^1}{|}}{C}}-N=N-R^2$$

R = R¹ = CH_3; R² = $-\overset{\overset{CH_3}{|}}{CH}-CH_2-C_6H_5$

2-(1-Benzylethylazo)-2-propanol acetate[297]; 77%

R = R¹ = CH_3; R² = $2,4-(NO_2)_2-C_6H_3$

2-(2,4-Dinitrophenylazo)-2-propanol acetate[298]; 89%

In the presence of an α-hydrogen atom N,N'-dialkylhydrazones are converted into diazenes with splitting off of a ketone[296]:

$$\underset{R}{\overset{R^1}{>}}C=N-\underset{\underset{\underset{R^4}{|}}{CH-R^3}}{\overset{R^2}{N}} \xrightarrow{Pb(O-CO-CH_3)_4}$$

$$R-\underset{\underset{O-CO-CH_3}{|}}{\overset{\overset{R^1}{|}}{C}}-N=N-R^2 \;+\; O=C\underset{R^4}{\overset{R^3}{<}}$$

As already described above, oxidation of arylhydrazones leads to diazenes; where the arylhydra-

[294] J. Thiele, Justus Liebigs Ann. Chem. 376, 242, 265 (1910).
[295] USP 3350385 (1967), DuPont, Inv.: L. Spiralter, G.L. Untereiner.
[296] D.C. Iffland, L. Salisbury, W.R. Schafer, Angew. Chem. 72, 501 (1960); J. Amer. Chem. Soc. 83, 747 (1961).
[297] G.T. Whyburn, J.R. Bailey, J. Amer. Chem. Soc. 50, 911 (1928).
[298] W. Theilacker, H.J. Tomuschat, Chem. Ber. 88, 1086 (1955).
[299] M. Busch, G. Friedenberger, W. Tischbein, Ber. 57, 1784 (1924).

zone contains an α-hydrogen atom *arylvinyl-diazenes* are obtained, *e.g.*:

N-(β-Phenylazostyryl)-aniline[299]; 80–90%

Mercury oxide and oxygen are the most frequently used oxidizing agents. The reaction can be applied to osazones[300], *e.g.*:

1,2-Ethenediylbis(phenyldiazene)
[1,2-Bis(phenylazo)ethylene]

Autoxidation of benzaldehyde phenylhydrazone affords α-*benzeneazobenzyl hydroperoxide*[301–303], atmospheric oxidation[302] in alkaline solution, by contrast, gives *benzil bis(phenylhydrazone)*[302].

While oxidation of the arylhydrazones with peroxy acids (*e.g.*, peroxyacetic acid) leads to diazene *N*-oxides, treatment with benzoyl peroxide[304] and with lead(IV) acetate[305] affords (α-*acyloxyalkyl*)-*aryldiazenes*, *e.g.*:

2-[(4-Bromobenzene)azo]-2-propanol acetate; 48%

---

[300] H. von Pechmann, Ber. *21*, 2751 (1888);
P. Grammaticakis, Compt. Rend. *224*, 1509 (1947).
[301] K.H. Pausacker, J. Chem. Soc. (London) 3478 (1950).
[302] H. Biltz, F. Sieden, Justus Liebigs Ann. Chem. *324*, 310 (1902).
[303] R. Criegee, G. Lohaus, Chem. Ber. *84*, 219, 221 (1951).
[304] J.T. Edward, S.A. Samad, Can. J. Chem. *41*, 1638 (1963).

Using potassium permanganate as oxidizing agent as a rule leads to formation of additional *C–N* links, *e.g.*:

Alkylvinyldiazenes are obtained on reacting α-halocarbonyl compounds with, for example, two molecules methylhydrazine

Methyl-(2-methylpropenyl)-diazene[306]; 35%

(1-Cyclohexenyl)-methyldiazene[306]; 61%

Arylhydrazines react in exactly the same way with α-halocarbonyl compounds in alcoholic solution. When one molecule arylhydrazine is used *arylvinyldiazenes* are obtained, while with excess arylhydrazine either hydrazones or hydrazinohydrazones are formed, *e.g.*:

Phenylhydrazine + 2-Bromoacetophenone ⟶

1,3,5,6-Tetrahydro-6-(phenylazo)triphenylpyridazine[307, 308]; 60%

Phenylhydrazine + Ethyl α-chloroacetate ⟶

Ethyl 3-Phenylazo-2-butenoate[309]; 50%

2,4,6-Trichlorophenylhydrazine + Chloral ⟶

(Trichloroethenyl)-(2,4,6-trichlorophenyl)diazene[310]

Reducing α-oxoarylhydrazones with sodium hydrogen sulfite in ammonia/ethanol likewise affords arylvinyldiazenes[311], e.g.:

*(1-Cyclohexyl)(2-nitrophenyl)-
diazene*

In a general reaction symmetrical and unsymmetrical ketazines add chlorine at around −60° to form bis-[α-chloralkyl]diazenes, e.g.:

*Bis-(1-chlorocyclo-
hexyl)diazene*[37]; 78%

The *(α-chloroalkyl)(α-cyanoalkyl)diazenes* and *bis(α-acyloxyalkyl)diazenes*[314] accessible by reaction with hydrogen cyanide[312] or sodium salts of carboxylic acids and chlorine[313] respectively are obtained in good yield:

*1-(Chlorodiphenylmethyl)azo-
cyclohexanecarbonitrile*[314]; 92%

Treatment of bis(α-chloroalkyl)diazenes with silver salts of carboxylic acids, too, affords *bis(α-acyl-oxyalkyl)diazenes*. However, the technique usually fails at the stage of isolating the chlorine derivatives.

Catalytic hydrogenation of ketazines, too, furnishes the corresponding diazenes, e.g.:

Dicyclodecyldiazene[315]; 77%

3,4,5,6,7,8-Hexahydro-3,8-diphenyl-1,2-diazocine[316]; 98%

5.3.1.1.2 By *N–N* bond formation

Preparation of dialkyldiazenes involving fresh *N–N* bond formation possesses importance above all with polyfluoro derivatives and α-substituted dialkyldiazenes. The starting compounds may be of quite varied nature, but not all the methods can be recommended, e.g.:

1 Reaction of isocyanates with hydrogen peroxide

$$R-N=C=O \; + \; H_2O_2 \; + \; O=C=N-R \longrightarrow$$

$$R-NH-CO-O-O-CO-NH-R \xrightarrow{-2\,CO_2}$$

$$R-NH-NH-R \xrightarrow{H_2O_2} R-N=N-R$$

up to 40%

As *N,N'*-dialkylhydrazine is an intermediate product stage, it is more advantageous to use it directly[317].

2 Reaction of ketones with hydroxylamine-*O*-sulfonic acid[318], e.g.:

*1,1'-Azodicyclo-
hexanol*; 44%

3 Reaction of *N,N'*-dichloroamines and sodium or potassium hydroxide at −10 to +10°, e.g.[319]:

*4,4'-Azobis(cyclohexane-
carbonitrile)*; 85%

[305] *D.C. Iffland, E. Cerda*, J. Org. Chem. *28*, 2769 (1963);
R.W. Hofmann, Chem. Ber. *97*, 2765 (1964).
[306] *B.T. Gillis, J.T. Hagarty*, J. Amer. Chem. Soc. *87*, 4576 (1965).

[307] *D.Y. Curtin, W.E. Tristram*, J. Amer. Chem. Soc. *72*, 5238 (1950).
[308] 4-Nitrophenylhydrazine therefore forms Phenylgly-oxal-bis[4-nitrophenylhydrazone].
[309] *J. van Alphen*, Rec. Trav. Chim. Pays-Bas *64*, 109, 112, 305 (1945).
[310] *F.D. Chattaway, F.G. Daldy*, J. Chem. Soc. (London) 2756, 2760 (1928).
[311] *F. Sparatore*, Gazz. Chim. Ital. *85*, 1098 (1955).
[312] *S. Goldschmidt, B. Acksteiner*, Justus Liebigs Ann. Chem. *618*, 173 (1958).
[313] *E. Benzing*, Chem. Ber. *72*, 709 (1960).

Amines and α-halonitriles are more important starting compounds. Thus *tert*-alkylamines are interlinked to diazenes with hypochlorite[320, 321], cf. refs.[322, 323] (with ozone azoxy compounds are obtained, see p. 150), *e.g.*:

$$2\ \text{H}_3\text{C}-\underset{\underset{\text{CN}}{|}}{\overset{\overset{\text{CH}_3}{|}}{\text{C}}}-\text{NH}_2 \xrightarrow[\substack{-\ 2\ \text{H}_2\text{O} \\ -\ 2\ \text{NaCl}}]{2\ \text{NaOCl}} \text{H}_3\text{C}-\underset{\underset{\text{CN}}{|}}{\overset{\overset{\text{CH}_3}{|}}{\text{C}}}-\text{N}=\text{N}-\underset{\underset{\text{CN}}{|}}{\overset{\overset{\text{CH}_3}{|}}{\text{C}}}-\text{CH}_3$$

2,2'-Azobis[methyl-
propanonitrile] (Azobis-
isobutyronitrile); 86%

Reacting alkylamines with nitrosoalkanes is used above all for preparing unsymmetrical dialkyldiazenes, especially trifluoromethyl derivatives[324, 325]:

$$\text{H}_3\text{C}-\text{NH}_2 + \text{ON}-\text{CF}_3 \xrightarrow[-\ \text{H}_2\text{O}]{} \text{H}_3\text{C}-\text{N}=\text{N}-\text{CF}_3$$

Methyl(trifluoro-
methyl)diazene; 57%

Fluorinated dialkyldiazenes are obtained from α-halonitriles and metal fluorides. Thus *bis(trifluoromethyl)diazene* (40%) is obtained on passing cyanogen chloride over silver fluoride whilst cooling; sodium fluoride/chlorine may be used in place of silver fluoride (31%)[326]. Fluoro- or perhalonitriles can be reacted to diazenes with chlorine or bromine at 50–190° in the presence of excess silver fluoride; thus the following are obtained[326]:

Bis(pentafluoroethyl)diazene;	60%
Bis(nonafluorobutyl)diazene;	88%
Bis(2,2,2-trichloro-1,1-difluoroethyl)diazene;	65%

5.3.1.1.3 From *CH*-acid or organometallic compounds

Active methylene compounds are converted into diazenes in partly very good yield with 2-diazo 1,3-diketones[327] and tosyl azides[328] in ethanol/potassium ethoxide:

$$\underset{\underset{\text{O}}{\|}}{\overset{\overset{\text{O}}{\|}}{\underset{R^2-C}{R^1-C}}}\!\!>\!\!C=N_2 + H_2\overset{\overset{O}{\|}}{\underset{\underset{R^4}{}}{C}}\!\!-R^3 \xrightarrow[]{\substack{H_5C_2-OH \\ KOC_2H_5}}$$

$$\underset{\underset{\text{O}}{\|}}{\overset{\overset{\text{O}}{\|}}{\underset{R^2-C}{R^1-C}}}\!\!>\!\!CH-N=N-CH-R^4 \quad CO-R^3$$

50 - 95 %

$$2\ \underset{\underset{\text{O}}{\|}}{\overset{\overset{\text{O}}{\|}}{\underset{R^2-C}{R^1-C}}}\!\!>\!\!CH_2 + \text{Tosyl}-N_3 \xrightarrow[-\ \text{Tosyl}-NH_2]{\substack{H_5C_2-OH \\ KOC_2H_5}}$$

$$\underset{\underset{\text{O}}{\|}}{\overset{\overset{\text{O}}{\|}}{\underset{R^2-C}{R^1-C}}}\!\!>\!\!CH-N=N-CH-CO-R^2 \quad CO-R^1$$

50 - 90 %

Probably the best method for preparing alkylaryldiazenes is to couple *CH*-acid or organometallic compounds with aryldiazonium salts (cf. Japp-Klingemann reaction, p. 190 ff.). As a rule the reaction is carried out with tertiary *CH*-acid compounds (secondary ones are oxidized to hydrazones) at as low a temperature and in as neutral a solution as possible, *e.g.*:

$$\text{H}_3\text{C}-\text{CO}-\underset{\underset{\text{CH}_3}{|}}{\text{CH}}-\text{COOC}_2\text{H}_5 \xrightarrow[\text{weakly acid}]{H_5C_6-\overset{\oplus}{N_2}\ \overset{\ominus}{Cl}}$$

$$\text{H}_5\text{C}_6-\text{N}=\text{N}-\underset{\underset{\text{CH}_3}{|}}{\overset{\overset{\text{CO}-\text{CH}_3}{|}}{\text{C}}}-\text{COOC}_2\text{H}_5$$

Ethyl 2-phenylazo-2-methyl-
3-oxobutanoate[329]; 78%

[314] E. Benzing, Justus Liebigs Ann. Chem. *631*, 19 (1960);
S. Goldschmidt, B. Acksteiner, Justus Liebigs Ann. Chem. *618*, 182 (1958).

[315] H. Meister, Justus Liebigs Ann. Chem. *679*, 92 (1964).

[316] J. Kossanyi, Compt. Rend. *257*, 929 (1963).

[317] H. Esser, K. Rastädter, G. Reuter, Chem. Ber. *89*, 685 (1956).

[318] E. Schmitz, R. Ohme, S. Schramm, Angew. Chem. *75*, 208 (1963).

[319] USP 2 346 554 (1967), DuPont, Inv.: *J.J. Fuchs*.

[320] DBP 849 109 (1949), Rohm & Haas Co., Erf.: *P. Laroche DeBenneville, J.S. Strong*; Chem. Zentr. 416 (1954);

[321] USP 2 711 405 (1949), DuPont, Inv.: *A.W. Anderson*; C.A. *50*, 5725 (1956).

[322] USP 3 207 714 (1963), Rohm & Haas Co., Inv.: *P. Laroche DeBenneville*.

[323] R. Ohme, E. Schmitz, Angew. Chem. *77*, 429 (1965).

[324] A.H. Dinwoodie, R.N. Hazeldine, J. Chem. Soc. (London) 2266 (1965).

[325] V.A. Ginsburg, A.N. Medvedev, Zh. Obshch. Khim. *37*, 611 (1967); engl.: 572.

[326] W.J. Chambers, C.W. Tullock, D.D. Coffman, J. Amer. Chem. Soc. *84*, 2337 (1962).

The *CH*-acid compounds can vary. Thus the diazenes obtained often serve as initial products for preparing heteroaromatic compounds[330], *e.g.*:

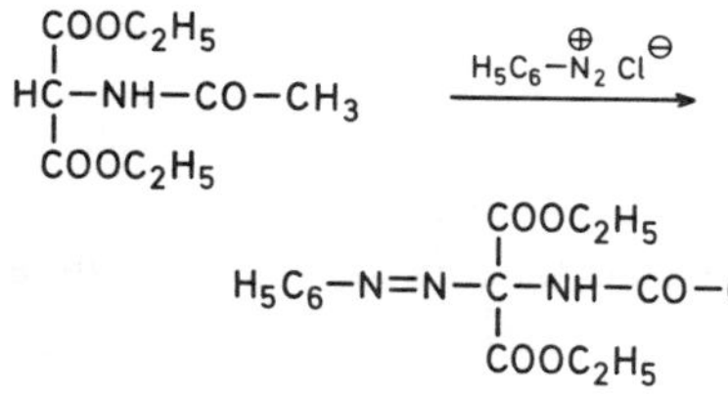

Diethyl acetamido
(phenylazomalonate; 90%

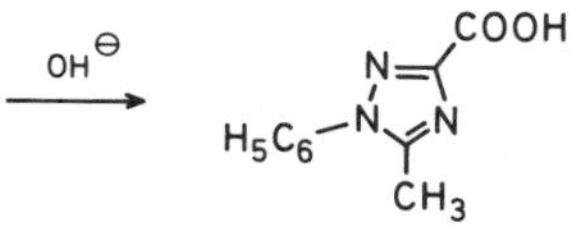

5-Methyl-1-phenyl-
1,2,4-triazole-3-
carboxylic acid; 95%

If the reaction conditions are not maintained accurately acyl groups are lost and hydrazones are formed (see pp. 191, 196, 197).
Conjugated dienes[331, 332] and thus also polymethine dyes[333] and even 1, 1-diphenylethylene[332] couple with aryldiazonium salts:

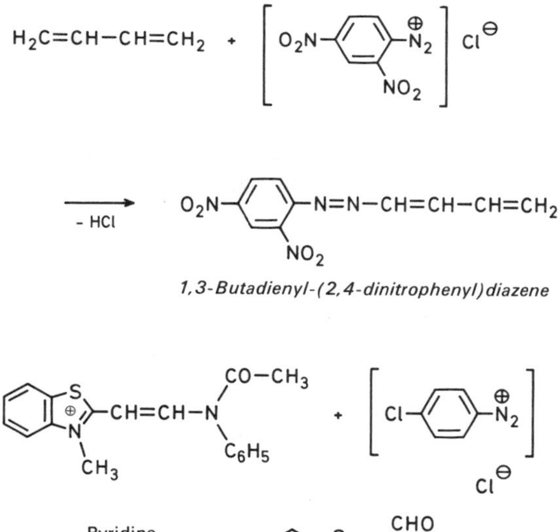

1,3-Butadienyl-(2,4-dinitrophenyl)diazene

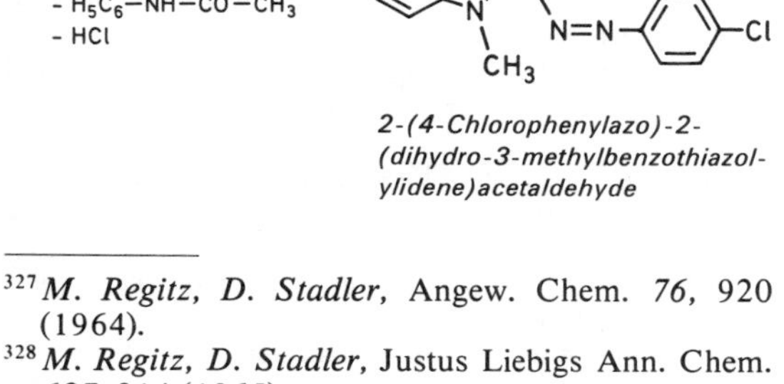

2-(4-Chlorophenylazo)-2-
(dihydro-3-methylbenzothiazol-
ylidene)acetaldehyde

A detailed survey is given in *Houben-Weyl*, Vol. X/3, p. 480.
As a rule, reaction with organometallic compounds is limited to alkylzinc halides, *e.g.*:

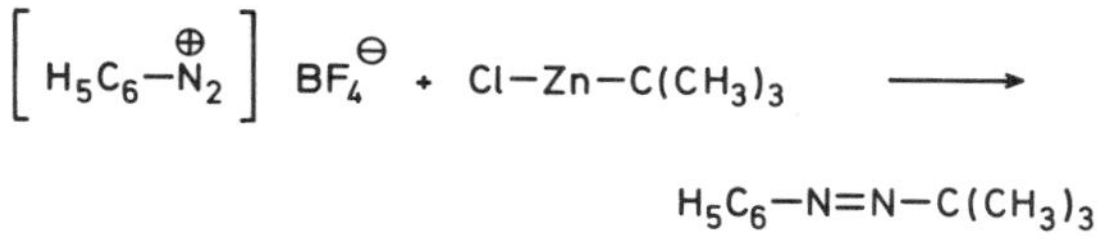

$H_5C_6-N=N-C(CH_3)_3$

(1,1-Dimethylethyl)-
phenyldiazene[334], cf. [335];
40%

5.3.2 Transformation reactions

At elevated temperature[336] and on *UV* irradiation[337] dialkyldiazenes lose nitrogen and form free radicals which combine to form a mixture of alkanes, alkenes, *etc.* (**Caution:** an **explosive** decomposition may occur on overheating, cf. refs. [336–339]):

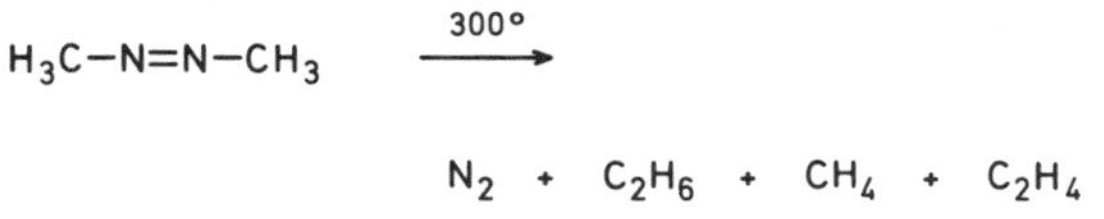

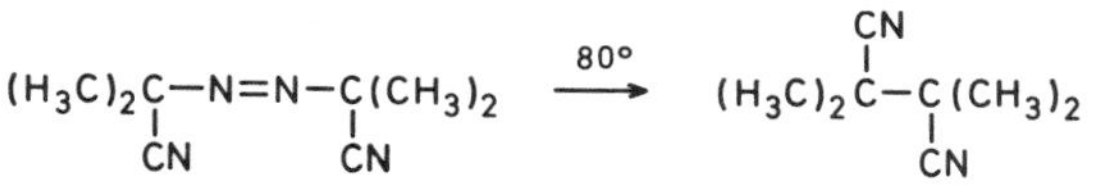

$$N_2 + C_2H_6 + CH_4 + C_2H_4$$

This decomposition into free radicals is utilized in the α-cyanodialkyldiazene and α-alkoxycarbonyl-dialkyldiazene series for initiating olefin polymer-

[329] *F.R. Japp, F. Klingemann*, Ber. 20, 2942, 3284, 3384 (1887);
F.R. Japp, F. Klingemann, Justus Liebigs Ann. Chem. 247, 190 (1888).
[330] *M. Regitz, B. Eistert*, Chem. Ber. 96, 3120, 3126 (1963).
[331] *A.P. Terentiev*, Chem. Zentr. I, 4043 (1936); II, 1628, 1629 (1937); I, 4315 (1938); I, 640; II, 3066 (1939);
H. Marxmeier, E. Pfeil, Chem. Ber. 97, 820 (1964).
[332] *C.F.H. Allen, C.G. Elliot, A. Bell*, Can. J. Res. 17 [B], 85 (1939).
[333] *H. Wahl*, Compt. Rend. 250, 2908 (1960);
A. Treibs, R. Zimmer-Galler, Justus Liebigs Ann. Chem. 627, 166, 179 (1959).
[334] *D.Y. Curtin, J.A. Ursprung*, J. Org. Chem. 21, 1221 (1956);
R. O'Connor, J. Org. Chem. 26, 4380 (1961).
[335] *R. O'Connor, W. Rosenbrook*, J. Org. Chem. 26, 5208 (1961).
[336] *H.C. Ramsperger*, J. Amer. Chem. Soc. 49, 912 (1927);
A.O. Allen, O.K. Rice, J. Amer. Chem. Soc. 57, 311 (1935).
[337] *M. Burton, T.W. Davis, H.A. Taylor*, J. Amer. Chem. Soc. 59, 1038 (1937).

[327] *M. Regitz, D. Stadler*, Angew. Chem. 76, 920 (1964).
[328] *M. Regitz, D. Stadler*, Justus Liebigs Ann. Chem. 687, 214 (1965).

ization[340, 341], cf. [326], and for initiating free-radical chains (chlorination, sulfochlorination)[342], cf. [343]. In the presence of easily split compounds the thermolysis leads to mixtures, *e.g.*[344]:

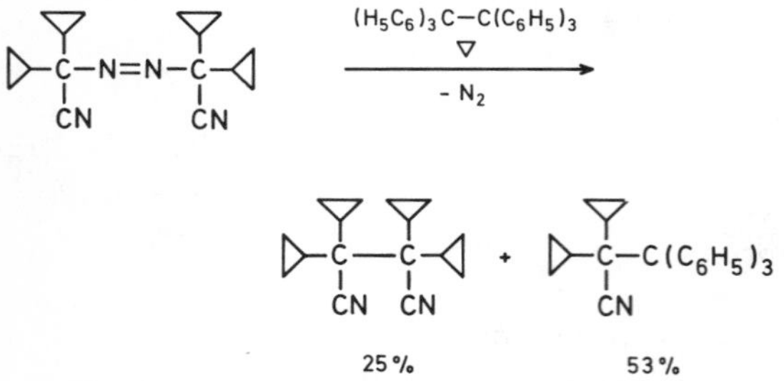

25 %   53 %

This photolysis or thermolysis of cyclic diazenes is important above all for preparing cyclic compounds, *e.g.*:

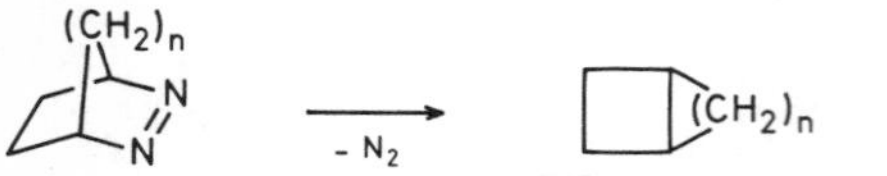

n = 1   345
n = 2   346

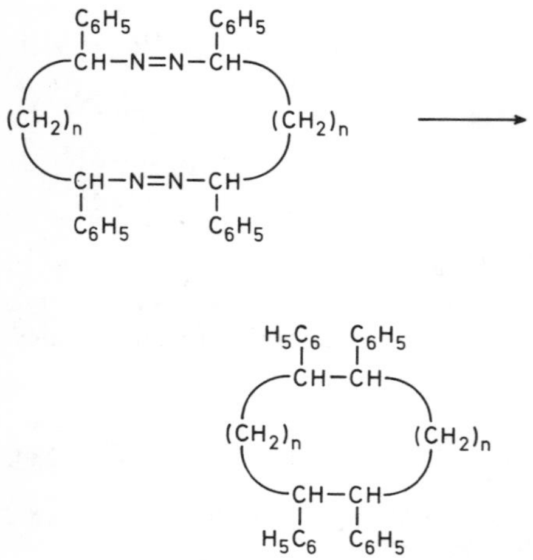

347

n = 6, 8 ...

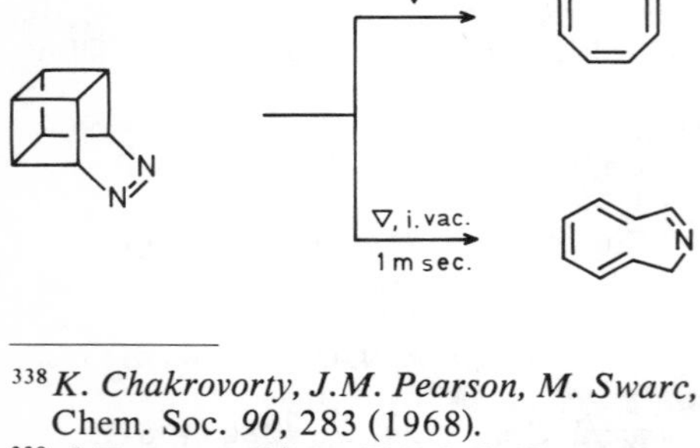

348

As the decomposition of the diazenes involves formation of gas, dialkyldiazenes are also employed as blowing agents in the plastics industry[349].

Alcoholic acid fragments dialkyldiazenes into carbonyl compounds and hydrazine[350]. The hydrazones formed as a result of rearrangement are seldom obtained, but they can be isolated on reaction with alkali metal hydroxides and alkoxides[350]. Alkali metals themselves add to the $N=N$ double bond forming N,N'-dialkali-N,N'-dialkylhydrazines[351].

Re the Diels-Alder reactions of vinyldiazenes see ref.[352]. Alkylaryldiazenes possess some importance as intermediate products in the Japp-Klingemann reaction (see p. 196, 197) and as starting derivatives in the synthesis of heteroaromatic compounds.

5.4 Dialkyl and alkylaryl diazene *N*-oxides

Dialkyldiazene *N*-oxides form *cis* and *trans* derivatives which are in photochemical and pyrolytic equilibrium[353], cf.[354, 355]:

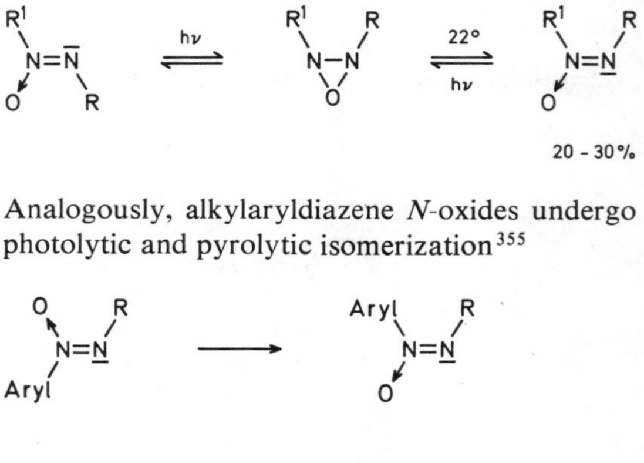

20 – 30 %

Analogously, alkylaryldiazene *N*-oxides undergo photolytic and pyrolytic isomerization[355]

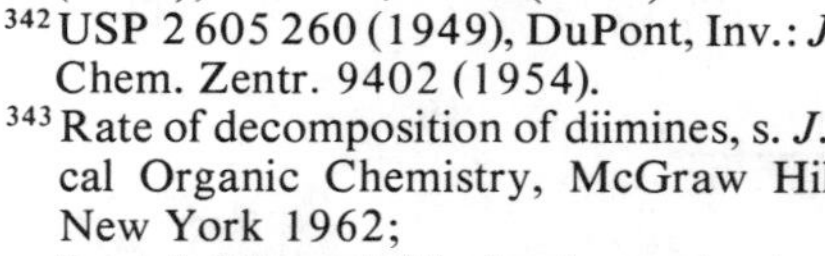

[341] *V.A. Ginsburg,* Dokl. Akad. Nauk SSR *149,* 97 (1963); C.A. *59,* 5008 (1963).

[342] USP 2 605 260 (1949), DuPont, Inv.: *J.R. Johnson;* Chem. Zentr. 9402 (1954).

[343] Rate of decomposition of diimines, s. *J. Hine,* Physical Organic Chemistry, McGraw Hill Book Co., New York 1962;
Reaktivität und Mechanismus in der organischen Chemie, 2. Aufl., Georg Thieme Verlag, Stuttgart 1966;
Activation energy and decomposition of azonitriles; *D. Lim,* Collect. Czech. Chem. Commun. *33,* 1122 (1968).

[344] *J.C. Martin, J.E. Schultz, J.W. Timberlake,* Tetrahedron Lett. 4269 (1967).

[345] *R. Criegee, A. Rimmelin,* Chem. Ber. *90,* 414 (1957).

[346] *S.G. Cohen,* J. Amer. Chem. Soc. *86,* 679 (1964).

[338] *K. Chakrovorty, J.M. Pearson, M. Swarc,* J. Amer. Chem. Soc. *90,* 283 (1968).

[339] cf. Decomposition of Methyltrifluoromethyldiimine, *O. Dobis, J.M. Pearson, M. Swarc,* J. Amer. Chem. Soc. *90,* 178 (1968).

[340] *K. Ziegler,* Kunststoffchemie *39,* 45 (1949); *K. Ziegler,* Brennstoffchem. *30,* 181 (1949); *C.G. Overberger,* J. Amer. Chem. Soc. *71,* 2661 (1949) (vgl. Übersicht im Houben-Weyl, Bd. X/2, p. 799; XIV/1, p. 219).

In addition, the *N—O* group, too, suffers rearrangement to the extent of ~ 15%, *e.g.*:

1-Ethyl-2-phenyldiazene
2-oxide

cis-(5%) trans-(10%)
1-Ethyl-2-phenyldiazene 1-oxide

As a result of its remarkable effect on the *NMR, IR,* and *UV* spectra the position of the *NO* group can be readily established[356]. See ref.[358] for further spectroscopic investigations[353, 357] including circular dichroism of corresponding cyclohexyl derivatives. Merely a reference can be made to the separation of alkali and alkaline earth metals with the aid of diazene *N*-oxides[359] and to the polarographic determination of azoxy groups[360].

5.4.1 Preparation

Dialkyldiazene *N*-oxides are obtained by oxidation of corresponding diazenes with peroxybenzoic acid[361] or peroxyacetic acid[362] (in part even with atmospheric oxygen[363]) in weakly acid medium, *e.g.*:

Dipropyldiazene N-oxide 75%
Dicyclodecyldiazene N-oxide 93%

Hydrazones accessible by acid catalysis of diazenes, too, are oxidized to diazene *N*-oxides by peroxybenzoic acid.

As a rule the *NO* group is on the bulky alkyl group in unsymmetrical dialkyldiazenes and on the vinyl group in alkylvinyl derivatives, *e.g.*[363]:

1-(1-Cyclohexenyl)-2-methyl-
diazene 1-oxide; 60%

Oxidation of arylhydrazones with peroxy acids[364] leads to formation of diazene *N*-oxides with the *NO* group adjacent the aryl group:

R = H; 2-Benzyl-1-phenyldiazene 1-oxide; 54%

R = NO₂; 2-(4-Nitrobenzyl)-1-phenyldiazene 1-oxide; 85%

N-Alkylhydroxylamines, too, are oxidized with oxygen in the presence of heavy metal salts of carboxylic acids to form new *N—N* links. Thus 81% *dicyclohexyldiazene N-oxide* is obtained from *N*-cyclohexylhydroxylamine[365]. The reaction is strongly dependent on the reaction conditions and, for example, can be directed to give exclusively oximes.

On mild reaction with zinc chloride/hydrochloric acid bis-(α-nitroso) ketones, carboxylic acid esters, and nitriles are converted into the corresponding diazene *N*-oxides in good yield. The following are obtained, *e.g.*:

3,3'-Azobis[3-methyl-2-butanone] N-oxide[366]; 60%
2,2'-Azobis[methylpropionitrile] N-oxide[367]; 56%

Catalytic reduction with palladium/calcium carbonate, too, is successful, *e.g.*:

Dicyclohexyldiazene N-oxide; 72%[368]; 83%[369]

1-Aryl-2-arylsulfonyloxydiazene 1-oxides are alkylated to 2-alkyl-1-aryldiazene 1-oxides by alkylmagnesium halides[370], *e.g.*:

2-Butyl-1-phenyldiazene 1-oxide; 62%
2-Ethyl-1-phenyldiazene 1-oxide; 44%

[347] *C.G. Overberger, J.P. Anselme, J.R. Rall*, J. Amer. Chem. Soc. *85*, 2752 (1963);
C.G. Overberger, J. Amer. Chem. Soc. *80*, 6556 (1958).
[348] *R. Askani*, Chem. Ber. *102*, 3305 (1969).
[349] *F. Lober*, Angew. Chem. *64*, 65 (1952);
H. Hartmann, Rec. Trav. Chim. Pays-Bas *46*, 150 (1927).
[350] *H.L. Lochte, W.A. Noyes, J.R. Bailey*, J. Amer. Chem. Soc. *44*, 2556 (1922).
[351] *W. Schlenk, E. Bergmann*, Justus Liebigs Ann. Chem. *463*, 315 (1928).
[352] *K.N. Zelenin, Z.M. Matveeva*, Zh. Org. Khim. *4*, 532 (1968); C.A. *68*, 105, 135 (1968).
[353] *J. Swigert, G.K. Taylot*, J. Amer. Chem. Soc. *93*, 7337 (1971).
[354] *S.H. Stephan, D.G. Frederick*, J. Amer. Chem. Soc. *89*, 6761 (1967).
[355] *G.G. Hecht, T.D. Greene*, J. Amer. Chem. Soc. *89*, 6761 (1967).
[356] *K.G. Taylor, T. Riehl*, J. Amer. Chem. Soc. *94*, 250 (1972).
[357] *T.G. Bernard, D.H. Jack*, J. Org. Chem. *32*, 95 (1967);

R. Biela, R. Hoernig, W. Pritzkow, J. Prakt. Chem. *36*, 197 (1967).
[358] *W.J. Gahren, M.P. Kunstmann*, J. Org. Chem. *37*, 902 (1972).
[359] *K. Kodama, T. Asai*, Bunseki Kagaku *21*, 584 (1972); C.A. 77, 42612 (1972).
[360] *Y.A. Gawargins, S.W. Bishara*, Fresenius Z. Anal. Chem. *245*, 366 (1969).

The value of condensing nitrosoalkanes with *N*-al-kylhydroxylamines[368, 371] resides in the directed synthesis of asymmetric dialkyldiazene *N*-oxides (the *NO* group is introduced by the nitroso compound), *e.g.*[372]:

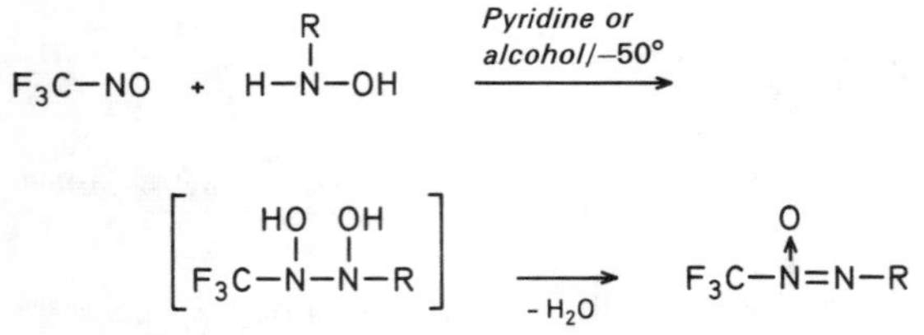

R = Alkyl, Aryl

During preparation of symmetrical dialkyldiazene *N*-oxides the yields are generally lower than in the above-described methods.

Directed synthesis of asymmetric *N*-oxides is achieved also on cleaving *N*-nitrosourethanes with potassium *tert*-butoxide and subsequent treatment with triethyloxonium tetrafluoroborate, *e.g.*[373]:

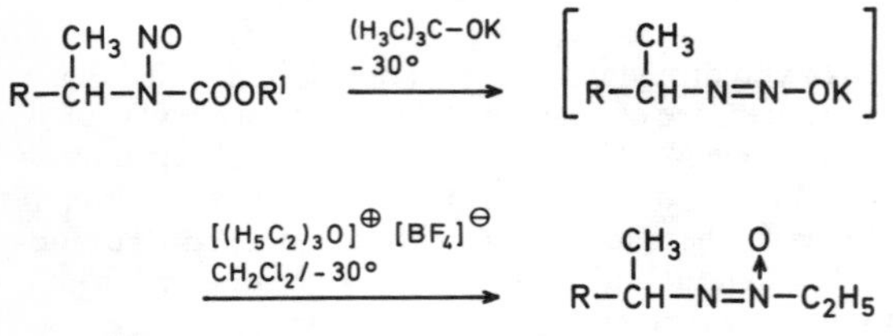

e.g., R = C₆H₁₁; *1-Ethyl-2-(1-cyclohexylethyl)diazene 1-oxide; 30% and 40%*[373]

Disproportionation of nitrosotrifluoromethane with alkali metal hydroxide affords bis(trifluoromethyl)-diazene *N*-oxide side by side with trifluoronitromethane[374]:

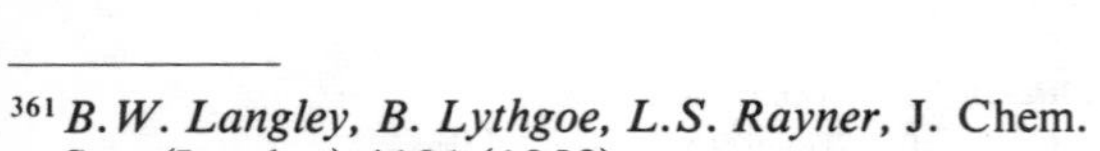

$$3\ F_3C-NO \xrightarrow{OH^\ominus} F_3C-\overset{O}{\overset{\uparrow}{N}}=N-CF_3 + F_3C-NO_2$$

In the presence of reducing agents the yield of *N*-oxide increases markedly.

Ozonolysis of *tert*-butylamine furnishes *di-tert-butyldiazene N-oxide* side by side with 2-methyl-2-nitropropane[375].

5.4.2 Transformation reactions

Dialkyldiazene *N*-oxides decompose when hot hydrochloric acid acts on them. The compounds formed are determined by the structure of the *N*-oxides[361]:

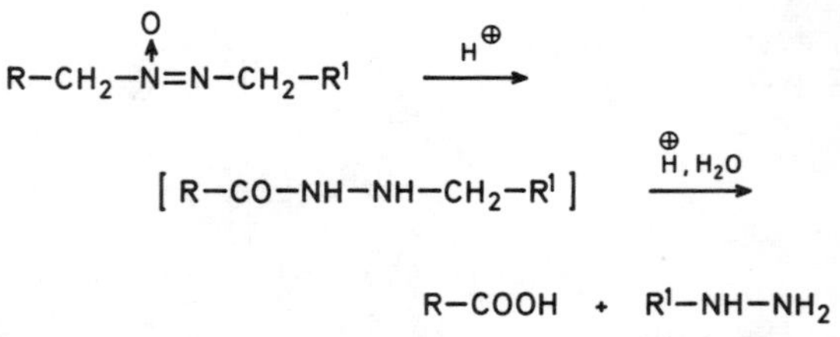

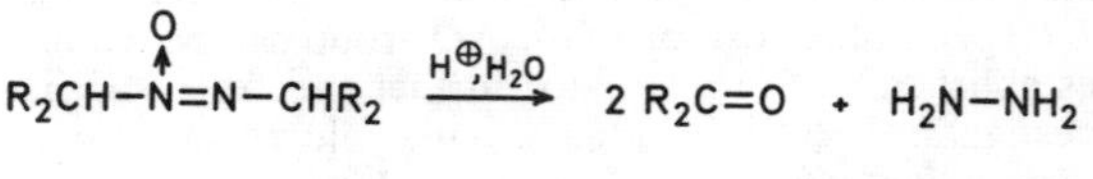

By contrast, reaction with hydrochloric acid at room temperature affords carboxylic acid hydrazides in the first case and azines in the second[369]:

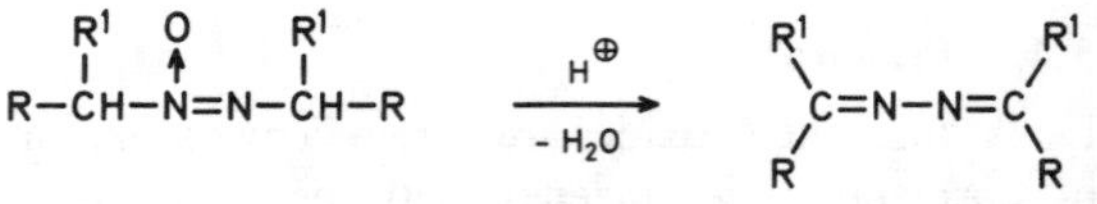

Reduction furnishes the corresponding *N*,*N*′-disubstituted hydrazines. Re the photolysis and thermolysis of the *N*-oxides see p. 399.

Iron pentacarbonyl reduces bis(trifluoromethyl)-diazene *N*-oxide to *bis(trifluoromethyl)diazene* among other products (50%)[376].

[361] *B.W. Langley, B. Lythgoe, L.S. Rayner*, J. Chem. Soc. (London) 4191 (1952).

[362] *H. Meister*, Justus Liebigs Ann. Chem. *679*, 92 (1964).

[363] *E.H. Rodd*, Chemistry of Carbon Compounds, 1. Aufl., p. 410, Elsevier Publ. Co., New York 1950; *J.P. Snyder, D.G. Farum*, J. Amer. Chem. Soc. *93*, 3815 (1971).

[364] *J.N. Brough, B. Lythgoe, P. Waterhouse*, J. Chem. Soc. 4069, 4076 (1954); *B.T. Gillis, K.F. Schimmel*, J. Org. Chem. *27*, 413 (1962).

[365] *H. Meister*, Justus Liebigs Ann. Chem. *679*, 90 (1964).

[366] *J.G. Aston, D.F. Menard, M.G. Mayberry*, J. Amer. Chem. Soc. *54*, 1535 (1932).

[367] *J.G. Aston, G.T. Parker*, J. Amer. Chem. Soc. *56*, 1387 (1934).

[368] *H. Meister*, Justus Liebigs Ann. Chem. *679*, 83, 89 (1964).

[369] *R. Biela, R. Hoernig, W. Pritzkow*, J. Prakt. Chem. *36*, 197 (1967).

[370] *T.E. Stevens*, J. Org. Chem. *29*, 311 (1964).

[371] *J.G. Aston, D.E. Ailman*, J. Amer. Chem. Soc. *60*, 1930 (1926).

[372] *V.A. Ginsburg, L.L. Martynova, N.F. Prinezentseva, Z.A. Buchek*, Zh. Obshch. Khim. *38*, 2505 (1968); engl.: 2422.

[373] *R.A. Moss, M.J. Landon*, Tetrahedron Lett. 3897 (1969).

[374] *V.A. Ginsburg, L.L. Martynova, M.N. Vasileva*, Zh. Org. Khim. 7, 2074 (1971); C.A. 76, 13717 (1972).

[375] *P.S. Bailey, J.E. Keller*, J. Org. Chem. *33*, 2680 (1968).

[376] *A.S. Filatov, M.A. Englin*, Zh. Obshch. Khim. *39*, 783 (1969); engl.: 743.

5.5 Bibliography

Aromatic azo and azoxy compounds

H.E. Bigelow, Chem. Rev. *9,* 117 (1931).

E. Buncel, in *B.S. Thyagarajan,* Azoxy Rearrangements in Mechanisms of Molecular Migrations, Vol. I, Interscience Publishers, New York 1968.

T.S. Gore, B.S. Joshi, S.V. Sunthankor, B.D. Tilak, Recent Progress in the Chemistry of Natural and Synthetic Coloring Matters and Related Fields, Academic Press, New York 1962.

G.G. Spence, E.C. Taylor, O. Buchardt, Chem. Rev. *70,* 231 (1970).

K. Venkataram, The Chemistry of Synthetic Dyes, Vol. I, II, Academic Press, New York 1952.

H. Zollinger, Azo and Diazo Chemistry, Interscience New York 1961.

Aliphatic and mixed aliphatic-aromatic azo compounds

H. Zollinger, Azo and Diazo Chemistry, Interscience Publishers, New York 1961.

P.A.S. Smith, The Chemistry of Open Chain Organic Nitrogen Compounds, Vol. I, W.A. Benjamin, New York, Amsterdam 1965.

C.G. Overberger, J.P. Anselme, J.G. Lombardino, Organic Compounds with Nitrogen-Nitrogen Bonds, Ronald Press Co., New York 1966.

V.T. Bandurco, Dissertation Abstracts, Int. B *32,* 2062 (1971).

6 Aromatic Hydrazo Compounds

Contributed by

E.V. Brown
Department of Chemistry,
University of Kentucky,
Lexington, Kentucky

6.1 General

The aromatic hydrazo compounds are *N, N*-diaryl-hydrazines (cf. p. 75) but they are generally discussed separately because of their relation to the aromatic azo compounds, to which they can be oxidized and from which they can be obtained by reduction. They also undergo disproportionation on heating and the benzidine rearrangement with acids which distinguish them from the other hydrazines.

6.2 Preparation

6.2.1 Reduction methods

Reduction of aromatic nitro, azo and azoxy compounds may lead to the formation of hydrazo products. For the preparation of *symmetrical hydrazo compounds* reduction of the corresponding nitro aromatic is usually convenient. A variety of reagents may be used for this reaction.

According to Haber[1] the reduction of nitrobenzene in alkaline media proceeds to nitrosobenzene and *N*-phenylhydroxylamine which combine to give azoxybenzene, which is reduced to azobenzene and finally *hydrazobenzene*.

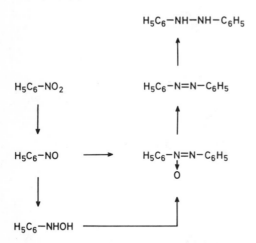

One of the most general *reducing agents* is zinc dust and sodium hydroxide in aqueous ethanol as solvent[2,3]. The yields usually vary from 40 to 75%[4].

Other finely divided metals have been used including iron[5] silicon[6] and aluminum[7]. This latter agent

seems to offer promise as the method of choice because of availability and good yields[8]. Sodium can be used in the form of sodium amalgam in aqueous ethanol[9]. Various carbonyl compounds have been used as reducing agents, namely, 9-fluorenone[10], formaldehyde[11] and glucose[12], although these reagents are somewhat less attractive for laboratory preparations. Catalytic hydrogenation can also convert nitrobenzenes to hydrazo compounds. Palladium on carbon[13], Raney nickel[14], and ruthenium on carbon[15] have been used. Hydrazine hydrate may be the source of the hydrogen[15, 16]. Electrochemical reduction is probably of more importance from the industrial standpoint than as a laboratory preparation[17–21].

Unsymmetrical hydrazo compounds are prepared from the corresponding azo and azoxy compounds by reduction. Of course symmetrical compounds can be prepared in this way as well but the method of choice would be from aromatic nitro compounds as indicated above. Powdered zinc in alkaline media[22] or in mildly acidic media[23] is an excellent reducing agent for this purpose. Pyridine is often used as a solvent for this reduction[24]; under various conditions yields of 40–95% are obtained[25].

[3] *R.B. Carlin*, J. Amer. Chem. Soc. *67*, 930 (1945).
[4] *Houben-Weyl*, Methoden der organischen Chemie, Bd. X/2, p. 699, Georg Thieme Verlag, Stuttgart 1967.
[5] Brit. P. *381*, 774 (1933), Inv.: *H. Dreyfus*; C.A. *27*, 5081 (1933).
[6] *R. Meier, F. Böhler*, Chem. Ber. *89*, 2305 (1956).
[7] USP. 2 570 866 (1948), General Aniline and Film Corp., Inv.: *D.E. Sargent, G.W. Pedlow jr.*; C.A. *46*, 4568b (1952).
[8] *K. Oie, J. Kosugi, H. Takahashi, Y. Takata*, Hokkaido Daigaku Kogakubu Kenkyu Hokoku *60*, 75 (1971); C.A. *75*, 76302u (1971).
[9] USP. 2 486 358 (1949), Directe van de Statsmijnen in Limburg, Inv.: *G. Hallie*; C.A. *44*, 2023f (1950).
[10] *A.A.R. Sayigh*, J. Org. Chem. *25*, 1707 (1960).
[11] USP. 2 794 046 (1954), Allied Chemical and Dye Corp., Inv.: *A.W. Sogn;* C.A. *52*, 428e (1958).
[12] USP. 2 794 047 (1954), Allied Chemical and Dye Corp., Inv.: *A.W. Sogn;* C.A. *52*, 428h (1958).
[13] *K. Brand, J. Steiner*, Ber. *55*, 886 (1922).
[14] *H.H. Szmant, R.L. Lapinski*, J. Amer. Chem. Soc. *78*, 458 (1956).
[15] *S. Pietra, M. Res*, Ann. Chim. (Rome) *48*, 299 (1958).
[16] *A. Furst, R.C. Berlo, S. Hooton*, Chem. Rev. *65*, 51 (1965).
[17] *K. Sugino, T. Sekine*, J. Electrochem. Soc. *104*, 500 (1947).
[18] *J.S. Clovis, G.S. Hammond*, J. Org. Chem. *28*, 3291 (1963).
[19] *Houben-Weyl*, Methoden der organischen Chemie, Bd. X/2, p. 494, Georg Thieme Verlag, Stuttgart 1967.

[1] *F. Haber, C. Schmidt*, Z. Physik. Chem. *32*, 271 (1900).
[2] *R. Adams, J.R. Johnson, C.F. Wilcox*, Laboratory Experiments in Organic Chemistry, 5th Ed., p. 332, MacMillan Co., New York 1963.

Several other metals including lithium[26], sodium[27], and potassium[26] may be used and various metal amalgams are also suitable, i.e., of sodium[28], lithium[29] and aluminum[30]. Occasionally the use of certain other metal and metal derivatives has been successful for this reduction, such as trialkyl and triarylstannanes[31], sodium pentoxide[32], phenyllithium[33], magnesium iodide and magnesium[34], magnesium metal and ammonium chloride[35], lithium tetrahydroaluminate[36], stannous chloride[37] and iron with acid[38].

Several of these must be used cautiously to avoid overreduction and with selected compounds to avoid benzidine rearrangement. A number of azo and azoxy compounds have been reduced to the hydrazobenzenes with sulfides, particularly ammonium sulfide[39].

In addition to the foregoing generally useful methods a variety of other reducing agents have been tried. Formaldehyde with catalytic amounts of a quinone reduces azo and azoxybenzenes to the hydrazo compounds[10]. Aminoiminosulfinic acid has also brought about this reduction[40]. Heating with phenylhydrazine has reduced azobenzene to hydrazobenzene[41]. Stannous chloride[37], sodium tetrahydroborate[42] and sodium dithionite[43] have been used. Somewhat more unusual is the use of hydrogen bromide[44] or irradiation in solvents such as isopropyl alcohol[45]. Catalytic hydrogenation has been applied but there is always the danger of reducing on to the amine. As catalysts, palladium[46-48], platinum[46], and Raney nickel[49] have been most popular. In place of hydrogen one can use hydrazine[50], various hydrazine derivatives[51, 52] or diimide[53] with catalysts to accomplish this reduction. As with reduction of nitrobenzenes to hydrazobenzenes it is also possible to use electrolytic reduction of azo and azoxybenzenes[20, 54].

In several cases the reducing agent causes substitution in one of the benzene rings as well as reduction to the hydrazo compound. This is particularily true of the hydrogen halides. Hydrogen chloride in methanol converts 4-(phenylazo)pyridine to *4-[(4-chlorophenyl)hydrazino]pyridine* probably according to the following scheme[54, 55].

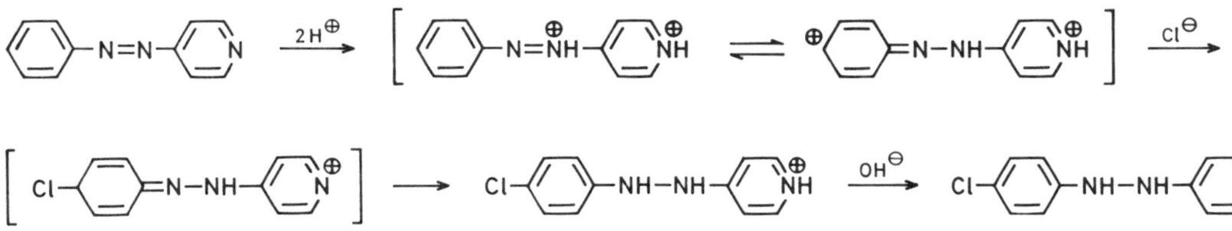

20 *Houben-Weyl*, Methoden der organischen Chemie, Bd. IV/2, p. 494, Georg Thieme Verlag, Stuttgart 1955.
21 *R.H. McKee, B.G. Gerapostolan*, Trans. Electrochem. Soc. 68, 329 (1935).
22 *R. Pfister, F. Häflinger*, Helv. Chim. Acta, 40, 399 (1957).
23 *R.B. Carlin, W.O. Forshey*, J. Amer. Chem. Soc. 72, 798 (1950).
24 *P. Ruggli, E. Iselin*, Helv. Chim. Acta 27, 1715 (1944).
25 *Houben-Weyl*, Methoden der organischen Chemie, Bd. X/2, p. 708, Georg Thieme Verlag, Stuttgart 1967.
26 *H. Normant, M. Larchenêque*, C.R. Acad. Sci. (Paris) 260, 5062 (1965).
27 *G.F. White, K.H. Knight*, J. Amer. Chem. Soc. 45, 1780 (1923).
28 *R.A. Cartwright, J.C. Tatlow*, J. Chem. Soc. 1997 (1953).
29 *T. Kametani, Y. Nomina*, J. Pharm. Soc. Jap. 74, 1037 (1964); C.A. 49, 11593g (1955).
30 *K. Auwers, M. Eckardt*, Justus Liebigs Ann. Chem. 359, 370 (1908).
31 *J.G. Noltes*, Rec. Trav. Chim. Pays-Bas. 83, 515 (1964).
32 *O. Diels, R. Rhodius*, Ber. 42, 1075 (1909).
33 *P.F. Holt, B.P. Hughes*, J. Chem. Soc. 764 (1954).
34 *W.E. Bachmann*, J. Amer. Chem. Soc. 53, 1524 (1931).
35 *L. Zechmeister, P. Rom*, Ber. 59, 873 (1926).
36 *F. Bohlmann*, Chem. Ber. 85, 394 (1952).
37 *A. Kirpal, E. Reiter*, Ber. 60, 665 (1927).
38 *J.A. Pearl*, J. Org. Chem. 10, 205 (1945).
39 *Houben-Weyl*, Methoden der organischen Chemie, Bd. X/2, p. 713, Georg Thieme Verlag, Stuttgart 1967.
40 *H. Breyer, H.J. Haase, W. Wildgrube*, Chem. Ber. 91, 247 (1958).
41 *R. Walther*, J. Prakt. Chem. 2, 53, 463 (1896).
42 *T. Nielson, H.C.S. Wood, A.G. Wylie*, J. Chem. Soc. 371 (1962).
43 *M.K. Seikel*, J. Amer. Chem. Soc. 62, 1214 (1940).
44 *M. Colonna, L. Pentimalli*, Ann. Chim. (Rome) 48, 1403 (1958).
45 *B.E. Blaisdell*, J. Soc. Dyers Colour 65, 618 (1949); C.A. 44, 2755f (1950).
46 *M. Busch, K. Schulz*, Ber. 62, 1458 (1929).
47 *R. Pfister, F. Häflinger*, Helv. Chim. Acta 40, 399 (1957).
48 *A. Étienne, R. Piat*, Bull. Soc. Chim. France 550 (1962).
49 *L. Hormer, U. Schwenk*, Justus Liebigs Ann. Chem. 579, 207 (1953).
50 *E.J. Corey, W.L. Mock, D.J. Pasto*, Tetrahedron Lett. 347 (1961).
51 *E.J. Corey, W.L. Mock*, J. Amer. Chem. Soc. 84, 685 (1962).
52 *S. Hünig, H.R. Müller, W. Thier*, Tetrahedron Lett. 353 (1961).
53 *F. Aylward, M.H. Sawistowska*, J. Chem. Soc. 1435 (1964).
54 *M. Colonna, A. Risaliti*, Gazz. Chim. Ital. 86, 288, 699 (1956); C.A. 52, 377c, 2006d (1958).

In similar manner

4-[(4-Chlorophenyl)azo]pyridine $\xrightarrow{HCl}$
  4-[N-(2,4-Dichlorophenyl)hydrazino]pyridine[56]

4-Phenylazopyridin $\xrightarrow{HBr}$
  4-[N'-(4-bromphenyl)-hydrazino]pyridin[56]

4-(Phenylazo)pyridine $\xrightarrow{HBr}$
  2-[N'-(4-Chlorophenyl)hydrazino]thiazole[57]

Azobenzene $\xrightarrow{HCl}$ 4-Chlorohydrazobenzene[58]

The action of ultraviolet light and acetyl chloride on azobenzene leads to *4-chloro-N, N'-diacetylhydrazobenzene* presumably by the following scheme[59].

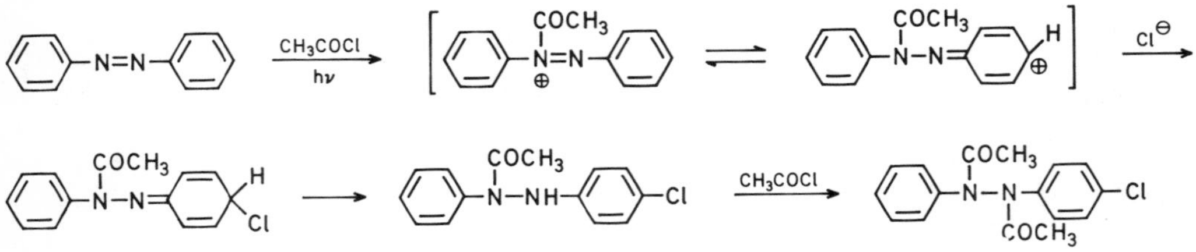

Other reagents which add to the —N=N— linkage of aromatic azo compounds to give hydrazo compounds are benzenesulfinic acids[60], diazomethanes[61], and ketones[62] (see p. 134).

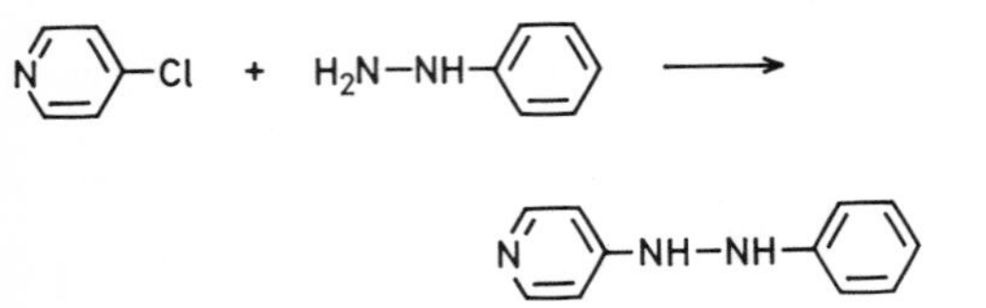

Nitro groups have occasionally been replaced in this way but the reaction is of little usefulness.

A number of specialized methods of only limited utility have been used for the preparation of hydrazo compounds. In one example, *1,1'-hydrazoisoquinoline* has been prepared from isoquinoline and the sodium derivative of hydrazine[66]. Condensing *N*-chloroacetanilide with the sodium derivative of acetanilide eliminates sodium chloride and affords *N,N'*-diacetyl-*N,N'*-diphenylhydrazine[67, 68]. Oxidation has converted some secondary amines to hydrazo compounds as, for example, 1,2,3,4-tetrahydroquinoline[69] to *1,2,3,4,1',2',3',4'-octahydro-1,1'-biquinoline* while phenylcarbamoyl chloride has been oxidized to *hydrazobenzene*[70].

6.2.2 Exchange and miscellaneous reactions

A number of hydrazo compounds have been prepared by the replacement of an active halogen on either a benzene or heterocyclic ring using phenylhydrazine or a substituted phenylhydrazine[63]. Typical is the preparation of *2,4,6-trinitrohydrazobenzene*[64]

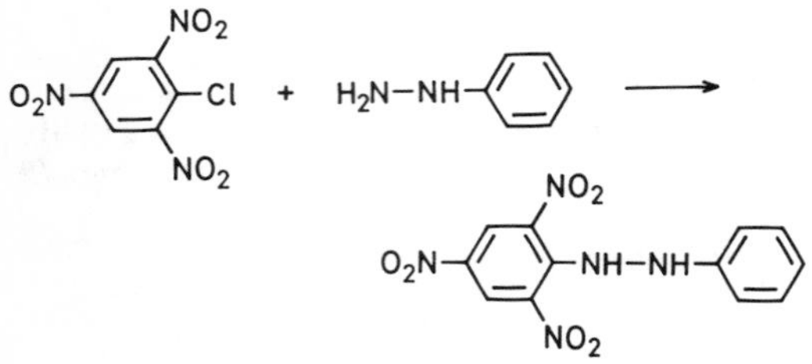

and of *4-(N'-phenylhydrazino)pyridine*[65].

55 *Houben-Weyl*, Methoden der organischen Chemie, Bd. X/2, p. 717, Georg Thieme Verlag, Stuttgart 1967.
56 *E. Koenigs, W. Freigang, G. Lobmayer, A. Zscharn,* Ber. *59,* 321 (1926).
57 *L. Pentimalli, L. Lolli,* Ann. Chim. (Rome) *47,* 36 (1957).
58 *P. Jacobson,* Justus Liebigs Ann. Chem. *367,* 304, 308 (1909).
59 *G.E. Lewis, R.J. Mayfield,* Australian J. Chem. *19,* 1445 (1966).
60 *W. Bradley, J.D. Hannon,* J. Chem. Soc. 2713 (1962).
61 *A.H. Cook, D.G. Jones,* J. Chem. Soc. 185 (1941).
62 *A.R. Katritsky, S. Musierowicz,* J. Chem. Soc. 78 (1966).
63 *Houben-Weyl*, Methoden der organischen Chemie, Bd. X/2, p. 724, Georg Thieme Verlag, Stuttgart 1967.
64 *E. Fischer,* Justus Liebigs Ann. Chem. *253,* 2 (1889).
65 *E. Koenigs, W. Freigang, G. Lobmayer, A. Zscharn,* Ber. *59,* 324 (1926).
66 *T. Kauffmann, H. Hacker, C. Kosel,* Z. Naturforsch. *146,* 602 (1959).
67 *H. Schmidt, G. Schultz,* Justus Liebigs Ann. Chem. *207,* 327 (1881).
68 *W.F. Short,* J. Chem. Soc. *119,* 1447 (1921).
69 *H. Willard, E. Haas,* Ber. *53,* 1336 (1920).
70 DBP 1072627 (1957), Phoenix Gummiwerke AG; Erf.: *K. Rastädter; C.A. 55,* 10386b (1961).
71 *P.K. Bose,* J. Indian Chem. Soc. *4,* 334 (1927).
72 *E. Bulka, H. Beyer,* Chem. Ber. *92,* 1447 (1959).
73 *B. Rassow, K. Berger,* J. Prakt. Chem. 2, *84,* 260 (1911).
74 *J.N. Ashley, S.S. Berg,* J. Chem. Soc. 3091 (1957).
75 *J.W.B. Reesor, G.F. Wright,* J. Org. Chem. *22,* 377 (1957).
76 *G. Wittig, J.E. Grolig,* Chem. Ber. *94,* 2148 (1961).

6.2.3 Ring syntheses

Numerous hydrazo compounds with heterocyclic rings have been prepared by ring closure reactions. Starting materials are semicarbazide derivatives which are reacted with halogenated ketones or aldehydes. Thus, N-phenylthiosemicarbazide and 2-bromoacetophenone give *4-phenyl-2(phenylhydrazino)thiazole*[71], while hydrazobis(thiocarboxamide) and chloro-2-propanone give *2,2'-hydrazobis[4-methylthiazole]*[72]:

$$H_2N-\underset{\underset{S}{\|}}{C}-NH-NH-\underset{\underset{S}{\|}}{C}-NH_2 \;+\; 2\; H_3C-CO-CH_2-Cl$$

+ 2 HCl

6.3 Reactions and analysis

6.3.1 Substitutions

Because of the great tendency for hydrazo compounds to undergo rearrangements with acids (see p. 158), few substitution reactions of the electrophilic aromatic type are possible. However, alkylation or replacement of the hydrogens at the hydrazino group is possible and with either methyl iodide or dimethyl sulfate one can prepare *methyl-* and *dimethylhydrazo* compounds[73, 74]. A particularly useful technique for dialkylation is the prior formation of a dialkali metal derivative of the hydrazo compound followed by treatment with the alkyl halide[75, 76], *e.g.*:

$$H_5C_6-NH-NH-C_6H_5 \;+\; 2\; LiCH_3 \longrightarrow$$

(Diphenylhydrazo)-
dilithium

N,N'-Dimethyl-
hydrazobenzene

Formaldehyde can condense two hydrazobenzenes *via* a methylene group[77] and higher aldehydes can react in a similar manner[78]. Acetylenedicarboxylic acid esters can add to give an enamine which can be used in the synthesis of various heterocycles[79]. *N-Acyl* and *N,N-diacyl* derivatives of hydrazo compounds can be prepared from acid anhydrides[80] or acid chlorides. In the latter case it is best to use a tertiary amine for removing the acid as formed. *Acetyl*[81], *chloroacetyl*[82], *benzoyl*[83], *hexanoyl*[84], and *phthaloyl*[85] derivatives have been prepared. Using chlorosulfonic acid the pyridine and potassium *N,N-diphenylhydrazine-N-sulfonates* were isolated[86]. *N-Phenylsilyl* derivatives have also been made from hydrazobenzene[87, 88]. Nitrous acid converts hydrazobenzene to the unstable *N,N'-dinitrosohydrazobenzene*[89].

6.3.2 Oxidations and reductions

One of the outstanding properties of aromatic hydrazo compounds is their relative ease of *oxidation* to *azo* compounds (see p. 132).

The air oxidation of normally colorless hydrazo compounds is the cause of their rapid discoloration especially while still moist. In addition to air[90], various common oxidizing agents have been effective, such as hypohalite[91], nitric acid[92], chromic acid[93], and nitrous acid[94].

Reduction of hydrazobenzene gives *aniline*[95], and it is best to avoid acid conditions because of the possibility of benzidine rearrangement.

[77] B. Rassow, M. Lummerzheim, J. Prakt. Chem. 2, 64, 138 (1901).
[78] B. Rassow, F. Burmeister, J. Prakt. Chem. 2, 84, 249 (1911).
[79] E.H. Huntress, J. Borstein, W.M. Hearon, J. Amer. Chem. Soc. 78, 2225 (1956).
[80] J.J. Ritter, J.O. Ritter, J. Amer. Chem. Soc. 52, 2817 (1930).
[81] A.G. Green, E.H. Bearder, J. Chem. Soc. 99, 1968 (1911).
[82] S. Goldschmidt, Justus Liebigs Ann. Chem. 437, 213 (1924).
[83] M.P. Freundler, Bull. Soc. Chim. France 3, 29, 822 (1903).
[84] G. Hallmann, J. Ringhardtz, N. Fischer, Chem. Ber. 90, 540 (1957).
[85] H.P. Kaufmann, Angew. Chem. 40, 78 (1927).
[86] V.N. Ufimtsev, Zh. Obshch. Khim. 10, 1757 (1940); C.A. 35, 3819⁵ (1941).
[87] D. Wittenberg, M.V. George, T.C. Wu, D.H. Miles, H. Gilman, J. Amer. Chem. Soc. 80, 4532 (1958).
[88] M.V. George, D. Wittenberg, H. Gilman, J. Amer. Chem. Soc. 81, 361 (1959).
[89] A. Baeyer, Ber. 2, 683 (1869).
[90] D.A. Blackadder, C. Hinshelwood, J. Chem. Soc. 2898 (1957).
[91] DBP 352, 354 (1914), Kalle and Co., AG, Chem. Zentr. 93, IV, 164 (1922).
[92] H. Beyer, G. Henseke, Chem. Ber. 82, 148 (1949).
[93] C. Willgerodt, F. Herzog, J. Prakt. Chem. 2, 71, 385 (1905).
[94] USP. 2893816, American Cyanamide Co., Inv.: S.M. Toang, C.E. Lewis, A.P. Paul; C.A. 54, 1868b (1960); L. Pentimalli, Gazz. Chim. Ital. 93, 404 (1963); C.A. 59, 5127 (1963).
[95] M. Tanaka, Chemiker Ztg. 48, 25 (1924).

6.3.3 Rearrangement and disproportionations

Hydrazo compounds are very sensitive to aqueous acids and under their influence suffer the *benzidine rearrangement*, which is discussed in detail on p. 558.

A reaction which occurs concurrently with the acid-catalyzed benzidine rearrangement is in many cases a likewise acid-catalyzed *disproportionation*[96]. It involves oxidation of one molecule hydrazobenzene to *azobenzene* and simultaneous reduction of a second molecule to two molecules *aniline*. Re the mechanism see refs. [97-99].

Finally, in addition to the acid-catalyzed benzidine rearrangement, there is a thermal or noncatalyzed rearrangement[100]. Disproportionation often proceeds concurrently with this thermal reaction[101].

[96] *R.B. Carlin, G.S. Wich*, J. Amer. Chem. Soc. *80*, 4023 (1958).

[97] *H.J. Shine*, in Mechanism of Molecular Migrations, John Wiley & Sons, New York 1969.

[98] *H.J. Shine, J.P. Stanley*, J. Org. Chem. *32*, 905 (1967).

[99] *H.J. Shine, J.T. Chamness*, J. Org. Chem. *32*, 901 (1967).

[100] *J. Meisenheimer, K. Witte*, Ber. *36*, 4153 (1903).

[101] *D.V. Banthrope, E.D. Hughes*, J. Chem. Soc. 2849 (1964).

7 Aromatic Diazonium Salts

Contributed by

K. Schank
Institut für Organische Chemie,
Universität des Saarlandes,
Saarbrücken

7.1 Preparation and general features of diazonium salts[1]

Diazotization of primary amines with $NO^{\oplus}$ donors may be regarded as the most usual method of preparing compounds containing the diazonium group:

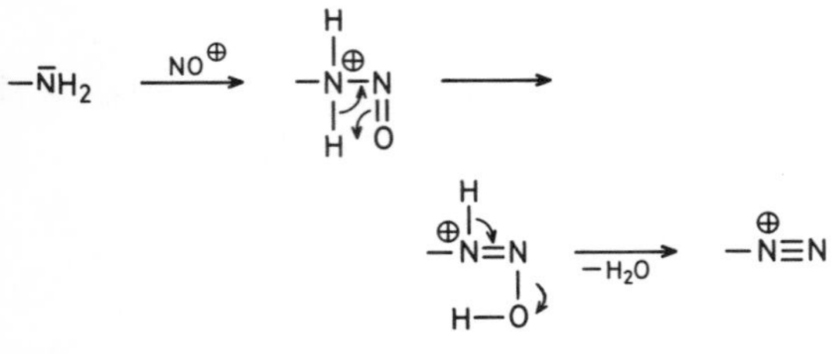

Since the electronic configuration of molecular nitrogen is substantially preformed in the diazonium group, elimination of N_2 from diazonium salts occurs particularly readily:

Mesomerism can fundamentally *stabilize* the diazonium stage, as is the case, for example, in aromatic diazonium compounds[1-4].

Purely *aliphatic* diazonium ions do no behave in this way and therefore decompose spontaneously losing nitrogen and forming carbenium ions, and are frequently[5] used as a source of carbenium ions. The bonding carbon atom passes from the nonplanar sp^3 state to a state of planar sp^2 hybridization accompanied by shortening of the remaining three bonds. Preventing this

transhybridization by making the bonding C atom the bridgehead of a bicyclic system and thus imposing a *nonplanar* arrangement upon it allows aliphatic diazonium ions, too, to be detected by azo coupling reactions, provided certain conditions are fulfilled[6], *e.g.*:

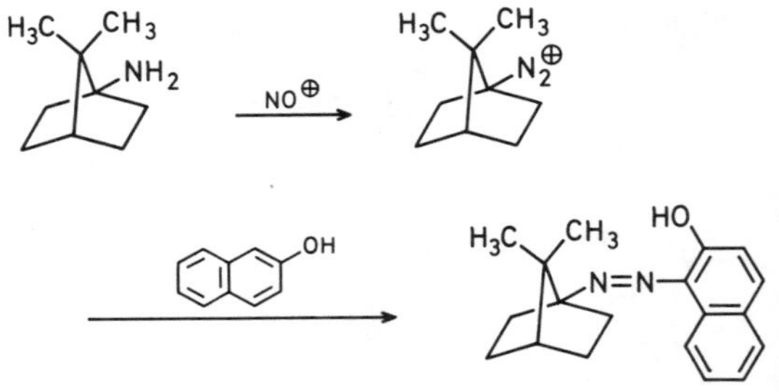

Table 1. Important methods for preparing aromatic diazonium salts

Method		Page
Diazotization of primary arylamines with nitrosyl derivatives	$Ar-NH_2 + [NO^{\oplus}] \longrightarrow$ $Ar-N_2^{\oplus} + H_2O$	164 168
Substitutions on aryldiazonium salts	$Ar-N_2^{\oplus} \longrightarrow Ar^1-N_2^{\oplus}$	171
Various selected reactions:		
a) Transdiazotization	$Ar^1-N_2^{\oplus} + Ar^2-NH_2 \longrightarrow$ $Ar^1-NH_2 + Ar^2-N_2^{\oplus}$	173
b) Nitrosation + fission of acylated arylamines	$Ar-NH-X \longrightarrow Ar-N-X$ $\underset{NO}{} \longrightarrow Ar-N_2^{\oplus}$	175
c) Nitrosation + fission of arylimine derivatives	$Ar-N=Y \xrightarrow{NOX}$ $Ar-N_2^{\oplus} X^{\ominus} + Y=O$	175
d) Direct diazotization of electron-rich aromatic compounds with HNO_2	$Ar-H \xrightarrow{HNO_2} Ar-N_2^{\oplus} NO_3^{\ominus}$	176

Zwitterionic combination of the diazonium nitrogen atom with a carbanionic carbon atom ultimately leads to the relatively stable aliphatic *α-diazo carbonyl* com-

The Author wishes to express his thanks to Dr. R. Pütter for valuable advice
The literature cited has been taken from the Chemical Abstracts (to c. 1969) and the Chemische Informationsdienst up to and including 1970.

[1] *R. Pütter* in *Houben-Weyl*, Methoden der organischen Chemie, 4. Aufl., Bd X/3, Georg Thieme Verlag, Stuttgart; Über Reaktivität und Stabilität von Arendiazoniumionen, *H. Zollinger*, Accounts Chem. Res. *6*, 335 (1973).

[2] *A.V. Pankratov, N.I. Savenkova*, Zh. Neorg. Khim. *13*, 2610 (1968); C.A. *70*, 16733x (1969);
H.W. Roesky, D. Bormann, O. Glemser, Kurznachr. Akad. Wiss. Göttingen, Sammelh. *2*, 51 (1966); C.A. *67*, 60460t (1967);
H.W. Roesky, O. Glemser, D. Bormann, Chem. Ber. *99*, 1589 (1966);
J.K. Ruff, Inorg. Chem. *5*, 1791 (1966); C.A. *65*, 16482f (1966);
D. Moy, A.R. Young, J. Amer. Chem. Soc. *87*, 1889 (1965).

[3] *K. Bott*, Angew. Chem. *77*, 683 (1965); C.A. *63*, 14900c (1965);
A. Schmidt, Chem. Ber. *99*, 2976 (1966);
G.F. Terescenko, G.I. Koldobskij, L.I. Bagal, Zh. Org. Khim. *6*, 1132 (1970);
R. Kreher, Angew. Chem. *83*, 915 (1971).

[4] *K. Bott*, Angew. Chem. *76*, 992 (1964); C.A. *62*, 7624c (1965);
K. Bott, Tetrahedron *22*, 1251 (1966);
K. Bott, Angew. Chem. *82*, 931. 953 (1970);
Diazonium salt of a β-diketone-metal chelate: *J.P. Collmann, M. Yamada*, J. Org. Chem. *28*, 3017 (1963);
C. Wentrup, H. Dahn, Helv. Chim. Acta *53*, 1637 (1970).

[5] *J.H. Bayless, A.T. Jurevicz, L. Friedman*, J. Amer. Chem. Soc. *90*, 4466 (1968);
D.E. Pearson, C.V. Breder, J.C. Craig, J. Amer. Chem. Soc. *86*, 5054 (1964);
K.H. Scheit, W. Kampe, Angew. Chem. *77*, 811 (1965); C.A. *63*, 17951d (1965);
G.A. Olah, N.A. Overchuk, J.C. Lapierre, J. Amer. Chem. Soc. *87*, 5785 (1965);
W. Kirmse, F. Scheidt, Chem. Ber. *103*, 3711 (1970).

[6] *D.Y. Curtin, B.H. Klandermann, D.F. Tavares*, J. Org. Chem. *27*, 2709 (1962).

pounds, which are dealt with on p. 208. Their cyclic vinylogs are the *quinone diazides;* when these are written in their mesomeric, diazonium phenolate form, they can be seen to be intermediate in structure between diazo compounds and aromatic diazonium compounds:

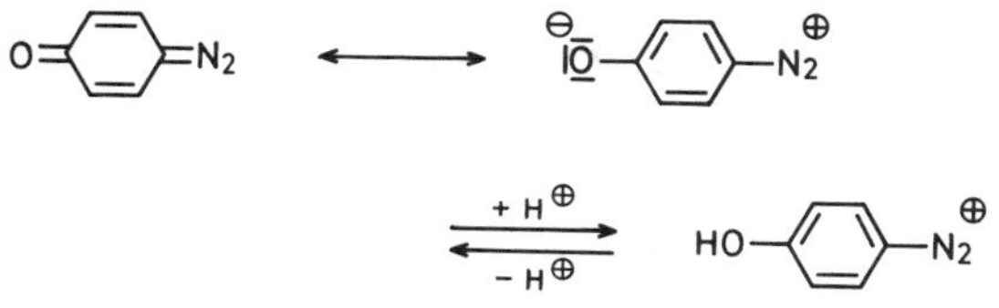

On protonation in concentrated acids they yield hydroxybenzenediazonium salts[7], which are strong acids because of the large substituent effect exerted by the diazonium group, when their formation is reversed. Their acidity also depends upon the steric effects of other substituent groups[8] as well as on the associated anionic center (imide[9], carbanion[10], etc.).

The *reaction characteristics* of diazonium salts are due to the *ambifunctional* character of the nitrogen of the diazonium group:

1 the very powerful substituent group of Type 2[11] causes the attached aromatic moiety to become so strongly positively charged that *electrophilic* substitution reactions do not occur unless substituent groups of Type 1 are also present;

2 because of the powerful activating mesomeric effects at the *o*- and *p*-positions of the aromatic nucleus, substituent groups at these positions can undergo *nucleophilic* replacement, provided that they form satisfactory anionic leaving groups[12];

3 one of the most important properties of the diazonium group is its ability to *couple*, at the β-nitrogen atom, with anions or nucleophiles; this reaction is generally termed *azo-coupling* (see p. 191);

4 as mentioned already, molecular nitrogen is an almost ideal leaving group; it can therefore be readily *eliminated* also from aromatic diazonium salts, by substitution, thermolysis, or photolysis. Numerous *arylation* reactions are based on this property (see p. 178). The dangerously **explosive** nature of many diazonium salts in the isolated form[13] is related to this feature however.

In *acid* aqueous solution diazonium salts decompose readily; the primary reaction is hydrolysis to a phenol followed by *secondary azo-coupling*[14], e.g.:

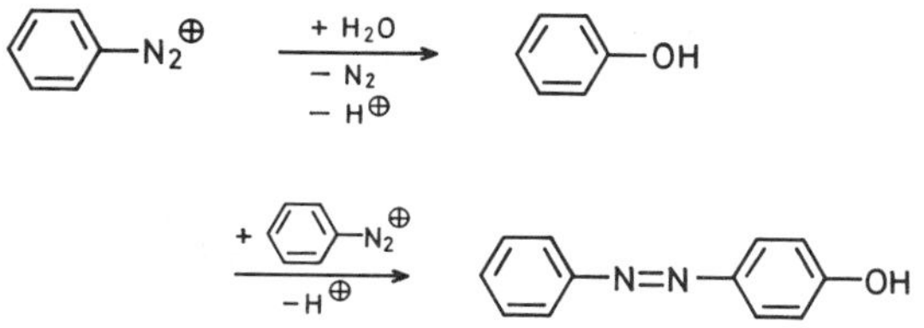

4-(Phenylazo)phenol

In *neutral* media, the **highly explosive** *diazo-anhydrides* are formed; these exist in equilibrium with their dissociation products, the *diazonium diazotates*[15]:

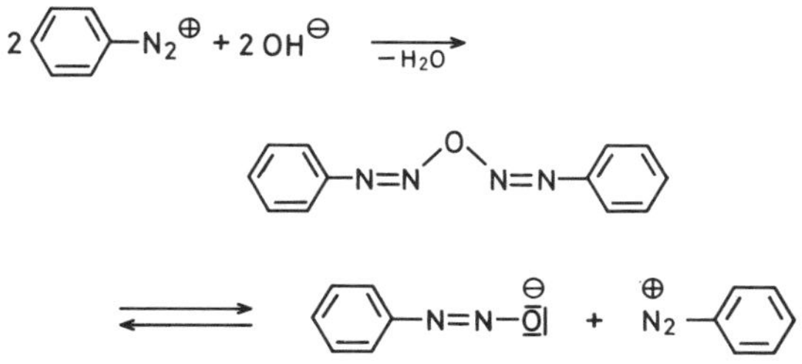

In *weakly basic* media (sodium bicarbonate or borate buffer), the products of a *partial hydrolysis* at the diazonium nitrogen atoms[16], proceeding in the opposite direction to diazotization are found:

$$2\ \text{Ar}-\text{N}_2^{\oplus}\ \text{X}^{\ominus}\ +\ 2\ \text{H}_2\text{O}\ \longrightarrow$$

$$\text{Ar}-\text{NH}-\text{N}=\text{N}-\text{Ar}\ +\ \text{HNO}_2\ +\ 2\ \text{HX}$$

[7] *R. Pütter* in *Houben-Weyl*, Methoden der organischen Chemie, 4. Aufl., Bd. X/3, p. 8, 51, Georg Thieme Verlag, Stuttgart 1965;
L.A. Kazitsyna, B.S. Kikot, L.E. Vinogradova, O.A. Reutov, Dokl. Akad. Nauk SSSR *158*, 1369 (1964); C.A. *62*, 2731f (1965).

[8] *V.V. Ershov, G.A. Nikiforov*, Dokl. Akad. Nauk SSSR *158*,1362 (1964); C.A. *62*, 2694g (1965).

[9] *L.A. Kazitsyna, A.V. Upadysheva, O.A. Reutov*, Dokl. Akad. Nauk SSSR *164*, 110 (1965); C.A. *63*, 16177d (1965).

[10] *R. Pütter* in *Houben-Weyl*, Methoden der organischen Chemie, 4. Aufl., Bd. X/3, p. 8, Georg Thieme Verlag, Stuttgart 1965;
Th. Severin, J. Dahlstroem, Angew. Chem. *76*, 954 (1964); C.A. *62*, 6595 (1965).

[11] Hammett substituent constant for the diazonium group; *S.C. Gardner, E.C. Lupton jr.*, J. Amer. Chem. Soc. *90*, 4328 (1968).

[12] *R. Pütter* in *Houben-Weyl*, Methoden der oganischen Chemie, 4. Aufl., Bd. X/3, p. 94, Georg Thieme Verlag, Stuttgart 1965.

[13] *R. Pütter* in *Houben-Weyl*, Methoden der organischen Chemie, 4. Aufl. Bd. X/3, p. 32, Georg Thieme Verlag, Stuttgart 1965.

[14] *M. Matrka, Z. Sagner, V. Chmatal, V. Sterba, M. Vesely*, Collect. Czech. Chem. Commun. *32*, 1462 (1967); C.A. *67*, 2620j (1967).

[15] *R. Pütter* in *Houben-Weyl*, Methoden der organischen Chemie, 4. Aufl., Bd. X/3, p. 11, 553, Georg Thieme Verlag, Stuttgart 1965.

[16] *M. Matrka, Z. Sagner, V. Chmatal, V. Sterba*, Collect. Czech. Chem. Commun. *32*, 2679 (1967); C.A. *67*, 63540e (1967).

A similar reversal of diazotization had been observed previously with 2,6-dinitrobenzenediazonium tetrafluoroborate[17].

In *strongly alkaline* medium, diazonium salts first undergo conversion into the *syn-diazotates,* which themselves rearrange[18] to form the stable *anti-diazotates:*

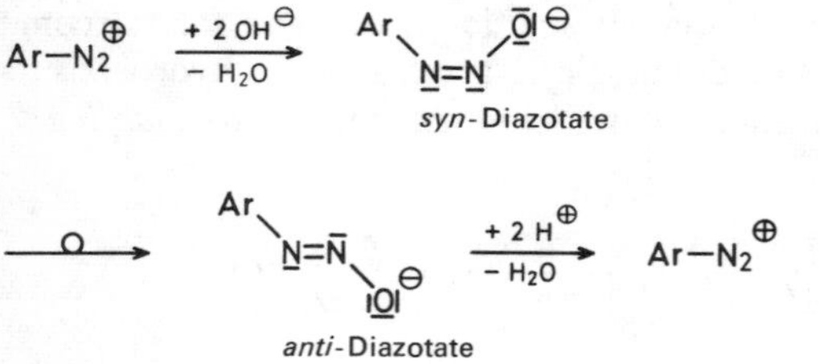

*syn-*Diazotate

*anti-*Diazotate

The diazotates regenerate the initial diazonium ion on acidification[19]. On the basis of spectroscopic evidence, it has been concluded that 4-(nitrosoamino)benzophenone is formed as a primary product when potassium benzophenonediazotate is acidified with acetic acid[20]:

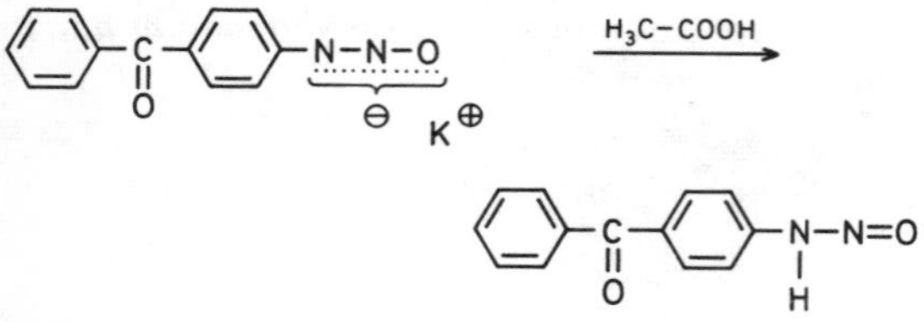

The corresponding *N*-alkyl-*N*-nitrosoanilines were prepared by nitrosation of the anilines and diazo ethers from silver isodiazotates and alkyl iodides, and were used for spectral comparison:

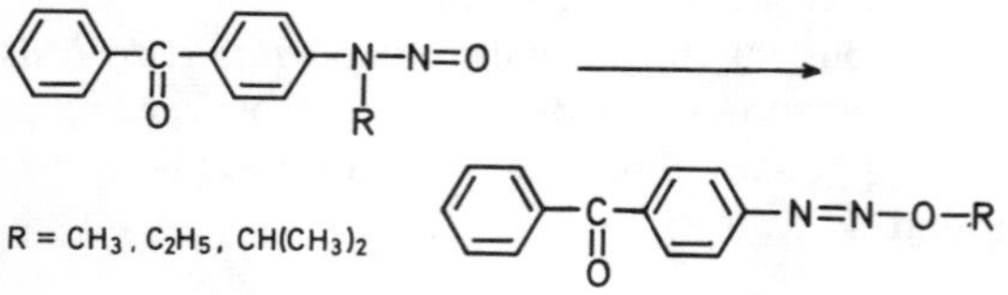

R = CH₃, C₂H₅, CH(CH₃)₂

Arylnitrosoamines are obtained by the action[21] of nitrosyl chloride on aniline and its nuclear-substituted derivatives in ether at −78°.

A reaction which is of interest as regards the interaction between diazonium and diazotate ions, and whose mechanism has been investigated, is the Suckfüll-Dittmer synthesis[22]:

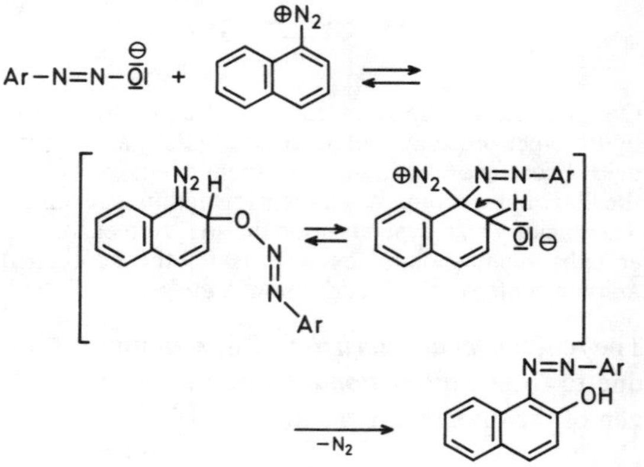

In this reaction, a naphthalene-1-diazonium salt is oxidized by an aryldiazotate, forming the corresponding arylazo dye derived from 2-naphthol.

A relative of this synthetic method is the behavior toward aryldiazosulfonates, but in this case no hydroxy group is introduced into the naphthalene nucleus:

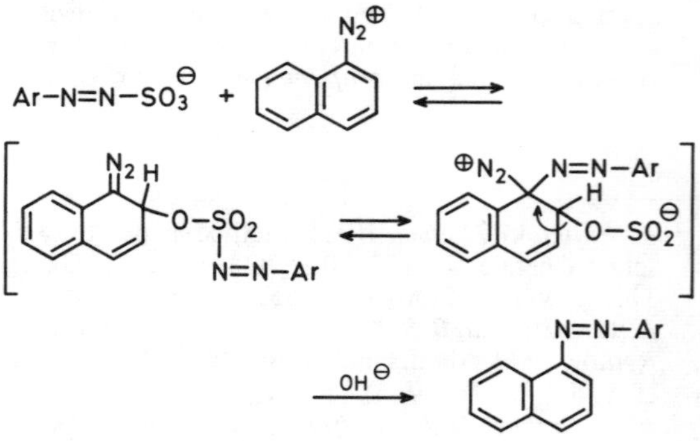

[17] *R. Pütter* in *Houben-Weyl,* Methoden der organischen Chemie, 4. Aufl., Bd. X/3, p. 12, Fußnote 6, Georg Thieme Verlag, Stuttgart 1965.

[18] *R. Pütter* in *Houben-Weyl,* Methoden der organischen Chemie, 4. Aufl., Bd. X/3, p. 552, Georg Thieme Verlag, Stuttgart 1965.

[19] *R. Pütter* in *Houben-Weyl,* Methoden der organischen Chemie, 4. Aufl., Bd. X/3, p. 559, Georg Thieme Verlag, Stuttgart 1965;
E.S. Lewis, M.P. Hanson, J. Amer. Chem. Soc. *89,* 6268 (1967);
B.A. Porai-Koshits, V.V. Shaburov, Zh. Org. Khim. *2,* 510 (1966); C.A. *65,* 8733d (1966);
B.A. Porai-Koshits, Zh. Org. Khim. *2,* 1125 (1966); C.A. *65,* 16887c (1966).

[20] *B.A. Porai-Koshits, V.I. Zaionts, T.A. Eiges,* Zh. Org. Khim. *3,* 1262 (1967); C.A. *67,* 99769a (1967);
T.A. Eiges, B.A. Porai-Koshits, V.I. Minkin, Zh. Org. Khim. *4,* 322 (1968); C.A. *68,* 104459b (1968).

[21] *E. Müller, H. Haiss,* Chem. Ber. *96,* 570 (1963).

[22] *M. Christen, L. Funderburk, E.A. Halevi, G.E. Lewis, H. Zollinger,* Helv. Chim. Acta *49,* 1376 (1966);
K.H. Schündehütte in *Houben-Weyl,* Methoden der organischen Chemie, 4. Aufl., Bd. X/3, p. 324, 327, Georg Thieme Verlag, Stuttgart 1965.

7.1.1 Mechanism of the diazotization reaction

The action of nitrous acid on primary *aromatic* amines in the presence of strong acids in aqueous solution leads to diazonium salts. Diazotization (doubling of the nitrogen) involves primary addition of a nitrosyl cation to the free electron pair of the amine nitrogen atom[23], this being the rate-determining step of the reaction:

$$① \quad Ar\!-\!\bar{N}H_2 + NO^{\oplus} \xrightarrow{slow} Ar\!-\!\overset{\overset{\displaystyle H}{|_{\oplus}}}{\underset{\displaystyle H}{N}}\!-\!NO$$

$$② \quad Ar\!-\!\overset{\overset{\displaystyle H}{|_{\oplus}}}{\underset{\displaystyle H}{N}}\!-\!NO \xrightarrow{fast} Ar\!-\!\overset{}{\underset{\displaystyle H}{\bar{N}}}\!-\!NO + H^{\oplus}$$

$$③ \quad Ar\!-\!\overset{}{\underset{\displaystyle H}{\bar{N}}}\!-\!NO + H^{\oplus} \xrightarrow{fast} Ar\!-\!\overset{\oplus}{N}\!\equiv\!N + H_2O$$

Sodium nitrite is the usual source of nitrous acid[24]. Nitrous acid itself does not act as a nitrosating agent since $OH^{\ominus}$ is not a suitable leaving group. However, on protonation the neutral water molecule can be eliminated, as it is an efficient leaving group; consequently, $NO^{\oplus}$ is formed in concentrated sulfuric acid:

$$HNO_2 + H^{\oplus} \rightleftarrows H_2NO_2^{\oplus} \xrightarrow[-H_2O]{\underset{conc.}{H_2SO_4}} NO^{\oplus}$$

$$\Big\downarrow {\scriptstyle +X^{\ominus}} \atop {\scriptstyle -H_2O}$$

$$X\!-\!NO$$

$$X = Cl, Br, NO_2, SCN, etc.$$

In dilute acids, either the protonated nitrous acid can react directly with the amine, nucleophilic replacement of water occurring, or, depending on the nature of $X^{\ominus}$, other intermediates suitable for the nitrosation process may be involved. In the protic solvent water compounds of the type X–NO have electrophilic reactivities in the following order:

$$NO^{\oplus} > H_2O^{\oplus}\!-\!NO > N \equiv C\!-\!S\!-\!NO > Br\!-\!NO$$
$$> Cl\!-\!NO > O_2N\!-\!NO$$

In recent years, many papers have appeared[25] which deal with the *reaction kinetics* of diazotization in terms of *pH*, nucleophile $X^{\ominus}$, solvent, *etc.* Furthermore, a continuous process for use on the industrial scale has been worked out[26] for determining the end-point in the diazotization of aromatic amines with nitrous acid.

Diazotization *techniques* in general

Normal diazotizations of aromatic amines with nitrous acid are carried out in dilute mineral acids by either direct or indirect methods (see p. 164). Weakly basic amines are often diazotized in concentrated acids, partly because of solubility considerations, and partly in order to minimize side-reactions such as hydrolysis or formation of diazoamino compounds. However, this procedure generally furnishes only diazonium salt solutions. To obtain diazonium salts in the *undiluted* form (caution: **danger of explosion!**), organic solvents are mainly used in which some of the amines have high solubilities; examples are alcohols, acetic acid, formamide, and dimethylformamide. The preferred *diazotization reagents* in such cases are the esters of nitrous acid, and also nitrosyl hydrogen sulfate or nitrosyl tetrafluoroborate. Nowadays, hardly any use is made of nitrous gases as in the classical Griess method[27], or of dinitrogen trioxide and dinitrogen tetroxide. A new diazotization method employing pure nitrogen monoxide has been described[28].

[23] *R. Pütter* in *Houben-Weyl*, Methoden der organischen Chemie, 4. Aufl., Bd. X/3, p. 12, Georg Thieme Verlag, Stuttgart 1965.

[24] *D.H.Wilcox jr.*, Amer. Dyest. Rept. 55, 891 (1966); C.A. 66, 16441t (1967);
Diazotization with N_2O_3: *L.A. Kazitsyna, N.D. Klyueva, A.V. Upadysheva, E.E. Chaikina*, Zh. Org. Khim. 4, 1620 (1968); C.A. 70, 3401n (1969);
J. Rignady, J. Barcelo, Bull. Soc. Chim. France 3783 (1968); C.A. 70, 37525w (1969);
W. Wojtkiewicz, J. Szadowski, Chem. Stosow., Ser. A8, 77 (1964); C.A. 62, 6593f (1962).

[25] *V. Sterba, Z. Sagner, M. Matrka*, Collect. Czech. Chem. Commun. 30, 3333 (1965); C.A. 63, 14668b (1965);
V. Sterba, Z. Sagner, M. Matrka, Chem. Listy 59, 1361 (1965); C.A. 63, 6429g (1965);
J.H. Ridd, J. Soc. Dyers Colour. 81, 355 (1965); C.A. 63, 11736g (1965).
A. Aboul-Seoud, Bull. Soc. Chim. Belg. 75, 249 (1966); C.A. 65, 13498d (1966);
H. Schmid, Chemiker-Ztg. 90, 351 (1966); C.A. 65, 13584b (1966);
E. Kalatzis, J. Chem. Soc. B 277 (1967);
A. Aboul-Seoud, Bull. Soc. Chim. Belg. 76, 221 (1967); C.A. 67, 26139h (1967);
A. Aboul-Seoud, Anales Real Soc. Espan. Fis. Quim. (Madrid), Ser. B 63, 661 (1967); C.A. 67, 107858w (1967);
A. Aboul-Seoud, A. Kasem, Z. Physik. Chem. (Leipzig) 235, 325 (1967); C.A. 68, 48712q (1968);
Z.A. Schelly, J. Phys. Chem. 74, 4062 (1970).

[26] Fr. P. 1419092 (C1 C o7c), 26. 11. 65. Ger. Appl. 8. 11. 63; BASF, Erf.: *H. Kindler, D. Schuler;* C.A. 65, 17092f (1966);
H. Kindler, D. Schuler, Chem. Ing. Techn. 37, 402 (1965).

[27] *R. Pütter* in *Houben-Weyl*, Methoden der organischen Chemie, 4. Aufl., Bd. X/3, p. 29, Georg Thieme Verlag, Stuttgart 1965.

7.1.1.1 Diazotization in aqueous medium

In the usual general *direct method* for preparing aqueous solutions of diazonium salts, the solution or suspension in dilute hydrochloric acid of the amine concerned ($\sim$2.5 *mols* acid per *mol* amine) is treated with the molar amount of concentrated sodium nitrite solution at 0–10°. The reactions mostly proceed rapidly and almost quantitatively[29]. The end-point can be determined by using potassium iodide-starch paper (blue coloration when excess nitrite is present), but numerous other methods have also been used[30]. Excess of nitrous acid can easily be removed with urea or sulfamic acid:

$$HNO_2 \; + \; H_2N{-}CO{-}NH_2 \longrightarrow \quad N_2 \; + \; NH_4HCO_3$$

$$HNO_2 \; + \; H_2N{-}SO_3H \longrightarrow \quad N_2 + H_2SO_4 + H_2O$$

It is advisable to replace hydrochloric by sulfuric or phosphoric acid[31] if chloride ions are likely to initiate undesired side-reactions during subsequent reaction stages (*e.g.*, on heating to form the corresponding phenol). The diazotization reaction then proceeds somewhat more slowly, and a disadvantage in this case is sometimes the appreciably lower solubility of the diazonium hydrogen sulfate as compared with the readily soluble chlorides.

Aromatic amines which contain strongly acidic substituent groups often exist in acid solution as sparingly soluble betaines and cannot be diazotized efficiently by the method outlined above. In such cases, it is preferable to carry out diazotization by the *indirect method*, in which advantage is taken of the solubility of the betaines in dilute alkalis, and a solution of the amine and sodium nitrite in dilute alkali is slowly added, dropwise and with vigorous stirring, to the acid.

Both procedures can be used for the diazotization of primary aromatic amines of basicities similar to that of aniline. Weakly basic amines such as 2,4-dinitroaniline, and other amines with Type 2 substituent groups on the aromatic nucleus need to be diazotized in acids of the highest possible concentration. Under these conditions, arylamines of normal basicity exist in the form of their ammonium salts, and can no longer be diazotized.

7.1.1.2 Diazotization in concentrated acids[32]

Diazotized weakly basic arylamines have aromatic nuclei with extremely strong positive charges, and consequently are very easily attacked by the nucleophile water. Hydrolysis results; either nitrogen is eliminated and a phenol formed[14], or else nitrite is eliminated and the amine is regenerated[17]:

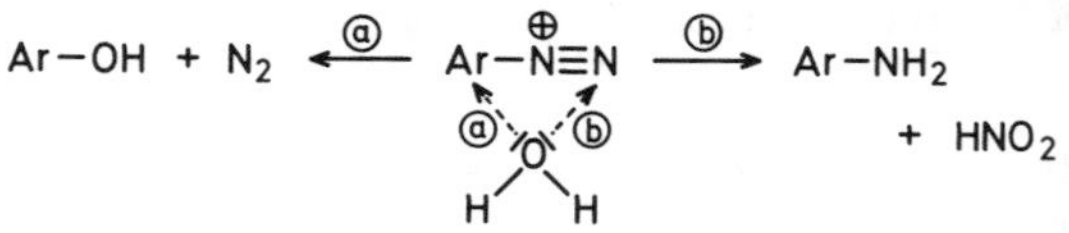

It is consequently preferable in the case of such amines to use concentrated acids, these being *solvents* of lower nucleophilism, and to employ nitrosylsulfuric acid as the diazotizing agent. Concentrated sulfuric acid has given very good results; by adding the sodium nitrite solution dropwise and with cooling, the diazotization reagent can be prepared directly *in situ*[33]. Thus, tetrachloroanthranilic acid was readily diazotized in concentrated sulfuric acid[34], whereas acetic acid proved to be an unsuitable medium:

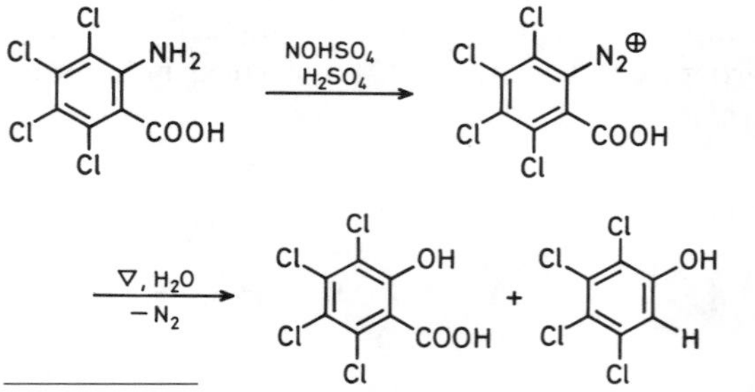

---

[28] Neth. Appl. 6 411 182 (Cl C o7c), 20. 4. 65, Roussel-UCLAF; Fr. Appl. 15. 10. 63; C.A. *63*, 16277f (1965);
Jap. P. 4331 ('67) (Cl 16C 3), 22. 2. 1964, Appl. 11. 5. 64; Sankyo Chemical Industries Co., Inv.: *S. Miyadokoro;* C.A. *67*, 21733q (1967);
J. Rigaudy, J.C. Vernieres, C.R. Acad. Sci., Paris *261*, 5516 (1965); C.A. *65*, 8797e (1966).

[29] *R. Pütter* in *Houben-Weyl*, Methoden der organischen Chemie, 4. Aufl., Bd. X/3, p. 16, Georg Thieme Verlag, Stuttgart 1965.

[30] *R. Pütter* in *Houben-Weyl*, Methoden der organischen Chemie, 4. Aufl., Bd. X/3, p. 17, Georg Thieme Verlag, Stuttgart 1965;
T. Takeuchi, T. Miwa, Kogyo Kagaku Zasshi *69*, 398 (1966); C.A. *65*, 16051e (1966);
Czech. P. 120 137 (Cl. G o1n), 15. 10. 66, Appl. 23. 5. 64; Erf.: *M. Vesely, V. Kmonicek, Z. Bohac;* C.A. *67*, 55 152q (1967).

[31] In aqueous metaphosphoric acid: *V.V. Kozlov, T.D. Silaeva, S.K. Eremin*, Zh. Org. Khim. *4*, 2145 (1968); C.A. *70*, 67784x (1969);
In aqueous orthophosphoric acid: *V.V. Kozlov, T.D. Silaeva*, Zh. Org. Khim. *1*, 1663 (1965); C.A. *64*, 1988g (1966).

[32] *R. Pütter* in *Houben-Weyl*, Methoden der organischen Chemie, 4. Aufl., Bd. X/3, p. 22, Georg Thieme Verlag, Stuttgart 1965.

[33] *H.E. Fierz-David, L. Blangey*, Grundlegende Operationen der Farbenchemie, 8. Aufl., p. 244, Springer Verlag, Wien 1952;
R. Pütter in *Houben-Weyl*, Methoden der organischen Chemie, 4. Aufl., Bd. X/3, p. 23, Georg Thieme Verlag, Stuttgart 1965.

[34] *R. Howe*, J. Chem. Soc. C 478 (1966).

Boiling with water caused replacement of the nitrogen atoms of the diazonium group by the hydroxy group together with partial decarboxylation.

In many cases, however, a *mixture* of glacial acetic and sulfuric acids has given good results, since it can be used for the diazotization of aromatic amines of basicity similar to that of aniline as well as of weakly basic amines; tetraazotization of diamines is even possible. The acetic acid can easily be prevented from freezing by adding propionic acid, methanol, or acetone[35].

Diazotization of weakly basic amines likewise proceeds more rapidly in a mixture of sulfuric and phosphoric acids than in concentrated sulfuric acid alone; the $NO^{\oplus}$ donor in such cases is nitrosylsulfuric acid.

Concentrated nitric acid is also occasionally used as the medium for diazotization of weakly basic amines. The advantage of the higher reaction rate than with sulfuric acid is, however, counterbalanced by its usually undesired nitration and oxidation properties and by the enormously high danger of **explosion** of the *diazonium nitrates* formed. For example, 2, 4, 5-trichloroaniline is oxidatively diazotized by a mixture of concentrated and fuming nitric acid[36]:

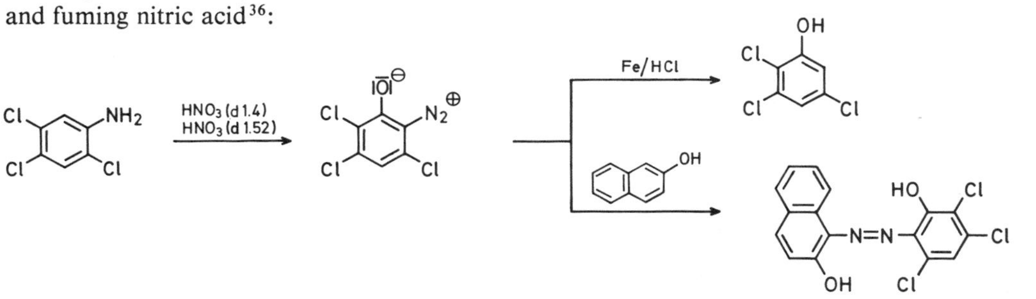

Structural proof was provided by reduction to the corresponding phenol, and by azo coupling with 2-naphthol.

Diazotization of aromatic amines in *concentrated fluoroboric acid* is achieving increasing popularity. This medium can be used for the diazotization of strongly basic aminophenols[37]:

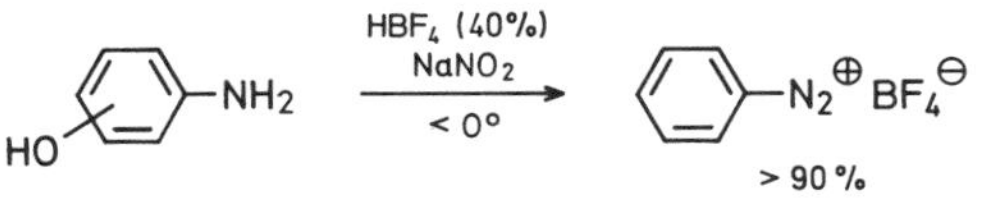

and diamines[38]:

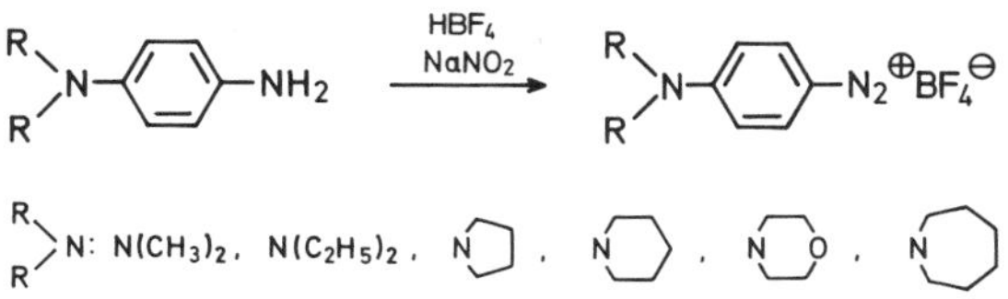

as well as of weakly basic amines[39]:

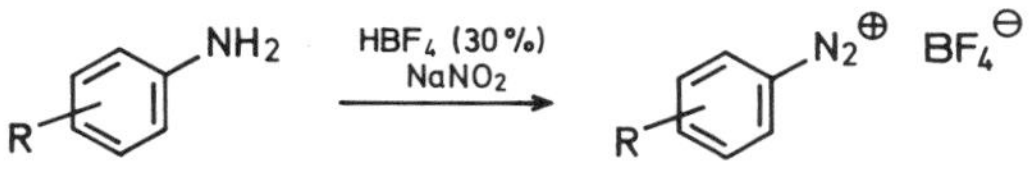

R = 2- and 4−NO$_2$
2−COOH
4−SO$_3$H

4-Chloroaniline has been diazotized in a mixture of boric acid and 35−46% hydrofluoric acid[40].

Diazotizations in highly concentrated or anhydrous hydrofluoric acid[41] are principally of importance for the preparation of organofluorine compounds, *e.g.*, of perfluorobenzene derivatives[42]:

[35] USSRP.187034(Cl.Co7c),11.10.66,Appl.3.1.66, Institute of Chemistry of the High-Molecular-Weight Compounds, Academy of Sciences, Ukrainian S.S.R., Inv.: *V.T. Burmistrov, A.L. Kravchenko, N.A. Lipatnikov;* C.A. *67,* 73335f (1967).

[36] *H. Koopmann,* Tetrahedron Lett. 513 (1965).

[37] *O. Danek, D. Snobl, J. Knizek, S. Nouzova,* Collect. Czech. Chem. Commun. *32,* 1642 (1967); C.A. *66,* 116676s (1967).

[38] *C. Sellers, H. Suschitzky,* J. Chem. Soc. (London) 6186 (1965).

[39] *A.M. Lukin, N.A. Bolotina, G.B. Zavarikhina,* Tr. Vses, Nauk Issled. Inst. Khim. Reaktivov Osobo Chist. Khim. Veshchestv. No. *28,* 198 (1966); C.A. *67,* 2835h (1967);
K. Sturm, W. Siedel, R. Weyer, H. Ruschig, Chem. Ber. *99,* 328 (1966).

[40] *G.P. Schiemenz,* Chem. Ber. *99,* 504 (1966).

[41] *R. Pütter* in *Houben-Weyl,* Methoden der organischen Chemie, 4. Aufl., Bd. X/3, p. 22, Fußnote *4,* Georg Thieme Verlag, Stuttgart 1965.

[42] *G.M. Brooke, E.J. Forbes, R.D. Richardson, M. Stacey, J.C. Tatlow,* J. Chem. Soc. (London) 2088 (1965).

7.1.1.3 Diazotization in organic solvents[43]

The usual aim of diazotization reactions conducted with aromatic amines in organic solvents is to prepare *solid* diazonium salts. The diazotization reagents employed are esters of nitrous acid (preferably pentyl nitrite). With glacial acetic acid as solvent, nitrosylsulfuric acid has proved to be the most suitable reagent; nitrogen oxides are only seldom used nowadays. *Glacial acetic acid* has proved a convenient solvent for the production of crystalline diazonium chlorides, whereas methanol and ethanol are useful for the preparation of the corresponding sulfates. Because of their excellent solvent power, *formamide* and *dimethylformamide* are especially useful for the diazotization of higher molecular weight amines. Other aprotic solvents such as acetone, ethyl acetate, ether, chloroform, or even hydrocarbons permit solid diazonium salts to be isolated directly after diazotization with alkyl nitrite, but there is the difficulty that the amine salts have to be used in suspension, *i.e.*, the reaction is necessarily heterogeneous.

The properties of isolated diazonium salts depend primarily on the nature of the *anion*. Only the diazonium salts of strong acids are stable, particularly acid salts. It must be emphasized most strongly that crystalline diazonium salts must be handled with **extreme care** because of the **danger of explosion,** and that it is imperative to avoid elevated temperatures or impact. The commercial stabilized diazonium *chlorides* are mostly zinc chloride *double salts*, but many other metal chlorides are also used. These double salts, and the diazonium sulfates, are preferred for use in dyeing practise.

Acid diazonium salts of organic sulfonic acids are likewise stable, and are occasionally used in diazotype copying processes.

Diazonium *tetrafluoroborates* are very popular for laboratory use because of their relatively high stability and storage life. In addition they readily precipitate from solution. The rising interest in aromatic fluorine compounds, which are readily available through the Schiemann reaction (see p. 185) also ensures an extensive use for these salts.

In contrast, diazonium salts of the fluorine acids of other elements such as tin, titanium, silicon, arsenic, antimony, or zinc have achieved little importance, although the thermolysis of diazonium hexafluoroantimonates gave the best fluorination yields in the Schiemann reaction[44]. Other diazonium salts with more or less complex anions[45] have not attained any special importance. Aryldiazonium salts of 2-nitro-1,3-indandione are stable and light-sensitive[46].

7.1.1.4 Diazotization of aromatic diamines
7.1.1.4.1 Monocyclic aromatic diamines

a) *o*-Diamines

As mentioned at the outset, a characteristic of aromatic diazonium ions is the high electrophilism of their β-nitrogen atoms. If an aromatic amine is diazotized which has a powerfully nucleophilic

[43] *R. Pütter* in *Houben-Weyl*, Methoden der organischen Chemie, 4. Aufl., Bd. X/3, p. 28, Georg Thieme Verlag, Stuttgart 1965.

[44] *C. Sellers, H. Suschitzky*, J. Chem. Soc. C 2317 (1968).

[45] *R. Pütter* in *Houben-Weyl*, Methoden der organischen Chemie, 4. Aufl., Bd. X/3, p. 38, Georg Thieme Verlag, Stuttgart 1965;
V.P. Sagalovich, S.A. Gavrilova, V.V. Kozlov, Y.M. Kulikov, Z.F. Shakhova, Dokl. Akad. Nauk SSSR *168*, 832 (1966); C.A. *65*, 5385g (1966);
Silicon- and cerium-molybdenum heteropoly acids: *V.V. Kozlov, V.P. Sagalovich, L.D. Ashkinadze*, Zh. Obshch. Khim. *36*, 1304 (1966); C.A. *65*, 18512a (1966);
R. Korewa, F. Karczynski, Roczniki Chem. *38*, 705 (1964); C.A. *62*, 1582e (1965) (Tetrachloroferrates)
T.N. Polynova, N.G. Bokii, M.A. Porai-Koshits, Zh. Strukt. Khim. *6*, 878 (1965); C.A. *64*, 11978e (1966);
A.N. Nesmeyanov, L.G. Makarova, I.V. Polovyanyuk, Zh. Obshch. Khim. *35*, 681 (1965); C.A. *63*, 4320c (1965) (Chloromercurates);
M.F. Hawthorne, F.P. Olsen, J. Amer. Chem. Soc. *86*, 4219 (1964) (Polyhydrodecaborates);
Brit. P. 985 707 (Cl. C o7d), 10.3.65, Appl. 5.9.63, E.I. du Pont de Nemours & Co., Inv.: *W.H. Knoth*; C.A. *63*, 11359c (1965);
G.R. Chalkley, D.J. Snodin, G. Stevens, M.C. Whiting, J. Chem. Soc. C 682 (1970) (Hexafluorophosphate).

[46] Brit.P.1124948(Cl.Co7c),21.8.1968,Appl.14,12. 1964,Algraphy Ltd.,Inv.:*F.E.Smith;*C.A.*70*,5234j (1969).

group in the immediate vicinity of the amino group, the diazonium group first formed may at once undergo cyclic azo-coupling[47].

A prerequisite for this to happen is generally the formation of a resonance-stabilized heterocyclic system of low energy; *o-quinone diazides* exist in the noncyclized diazonium phenolate form, which is more favored energetically[48].

In the case of 1,2-phenylenediamine and its homologs diazotization in dilute acid results in the formation of *1,2,3-benzotriazoles* of low energy by spontaneous coupling of the diazotized amino group with the nondiazotized one[49]. If, however, the nucleophilism of the nondiazotized amino group is considerably depressed by protonation, then preparation of *1,2-bis-diazonium salts* also becomes possible by the use of the phosphoric acid, sulfuric acid, or sulfuric/acetic acid techniques (see pp. 164–165):

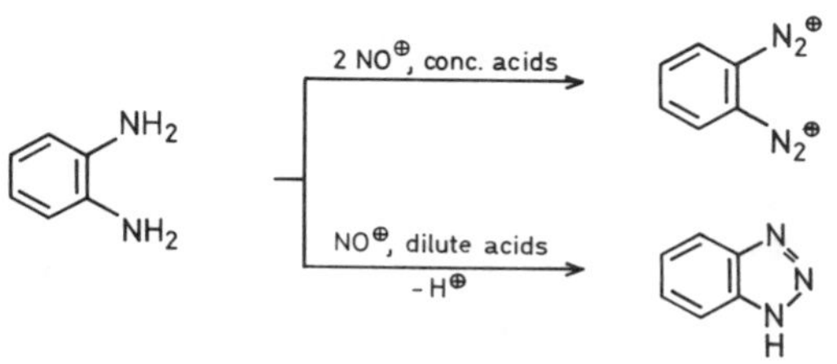

Other easily removable groups besides protons can be eliminated, *e.g.*, acyl or sulfonyl groups.

N-Monosubstituted diamines form *N*-monosubstituted triazoles.

b) *m*-Diamines

Properties corresponding to those of the *o*-isomers are also observed with *m*-diamines, except that in this case ring-formation involving *intra*molecular azo-coupling is no longer possible. Diazotization of 1,3-phenylenediamine in dilute hydrochloric acid solution leads to formation of the azo dye Bismarck Brown as the result of *inter*molecular azo-coupling. In contrast, diazotization in concentrated acids yields *1,3-bis-diazonium salts*[50].

c) *p*-Diamines

The amino groups in 1,4-phenylenediamine are diazotized at different rates, since the diazonium group which is first formed acts as a substituent group of Type 1 and greatly depresses the nucleophilism of the second amino group because of mesomeric and inductive effects. Diazotization in dilute hydrochloric acid leads to the formation of both *4-aminobenzenediazonium chloride* and *benzene-1,4-bis-diazonium chloride*[51]; partial hydrolysis and oxidation are invariably observed during diazotizations in dilute acid solutions[52]. Pure bisdiazonium salts can be prepared in concentrated acids, as with the *o*- and *m*-isomers. The pure monodiazonium salt can be readily prepared by an indirect method[53], in which 4-acetamidoaniline is diazotized in the normal way, and the acetyl group is then removed in *2N* hydrochloric acid at 70°.

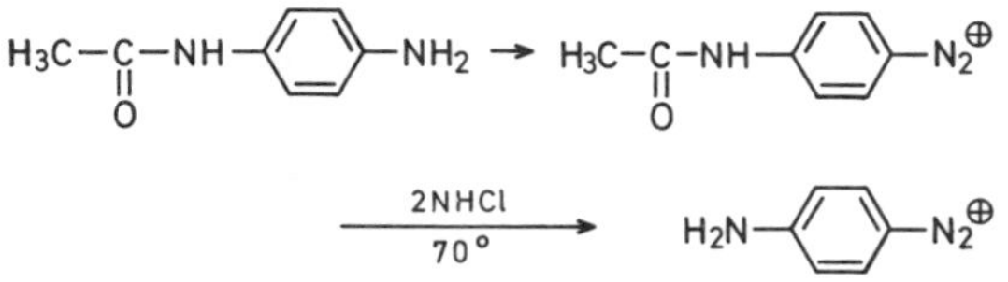

1,4-Phenylenediamine derivatives substituted on the nucleus by halogen, nitro, sulfonyl, carboxy or arsinic acid groups are only monodiazotized even with excess diazotization reagent, and preferentially on the *o*-unsubstituted amino group.

The *4-(alkylamino)anilines* are also of interest in this connection; they are diazotized in the normal way with 1 molar equivalent nitrous acid in dilute mineral acid solution. However, since the coupling rate of the diazonium salts formed is now slow because of their low electrophilism, treatment with a further molar equivalent of nitrous acid forms the 4-(N-alkyl-N-nitrosoamino)benzenediazonium salts, which are able to couple efficiently; the nitroso group can be removed easily from their azo-coupling products by hydrolysis[54]:

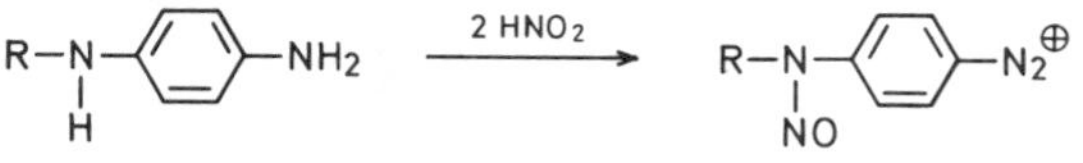

[47] R. Pütter in Houben-Weyl, Methoden der organischen Chemie, 4. Aufl., Bd. X/3, p. 47, Georg Thieme Verlag, Stuttgart 1965.

[48] L.C. Anderson, M.J. Roedel, J. Amer. Chem. Soc. 67, 955 (1945).

[49] R. Pütter in Houben-Weyl, Methoden der organischen Chemie, 4. Aufl., Bd. X/3, p. 39, Georg Thieme Verlag, Stuttgart 1965.

[50] R. Pütter in Houben-Weyl, Methoden der organischen Chemie, 4. Aufl., Bd. X/3, p. 40, 41, Georg Thieme Verlag, Stuttgart 1965.

[51] P. Griess, Ber. 17, 603 (1884); P. Griess, Ber. 19, 317 (1886).

[52] R. Pütter in Houben-Weyl, Methoden der organischen Chemie, 4. Aufl., Bd. X/3, p. 42, Georg Thieme Verlag, Stuttgart 1965.

[53] DRP 205037 (1907), BASF; Fortschritte der Teerfabrikation und verwandter Industriezweige 9, 300 (1911).

[54] DRP 154336 (1903), BASF; Fortschritte der Teerfabrikation und verwandter Industriezweige 7, 367 (1905).

However, hydrolysis and oxidation to *p*-quinones also occur readily here. The secondary amino group of *4-anilinobenzenediazonium salts* is so acid that the orange-colored, **explosive** *N-(4-diazo-2,5-cyclohexadien-1-yliden)aniline* precipitates out of alkaline solution[55]:

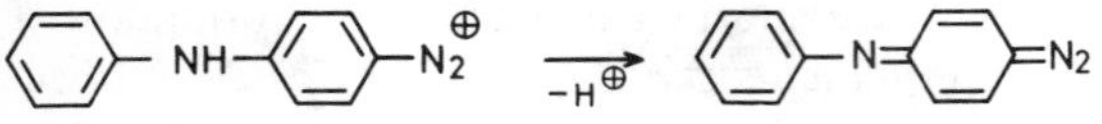

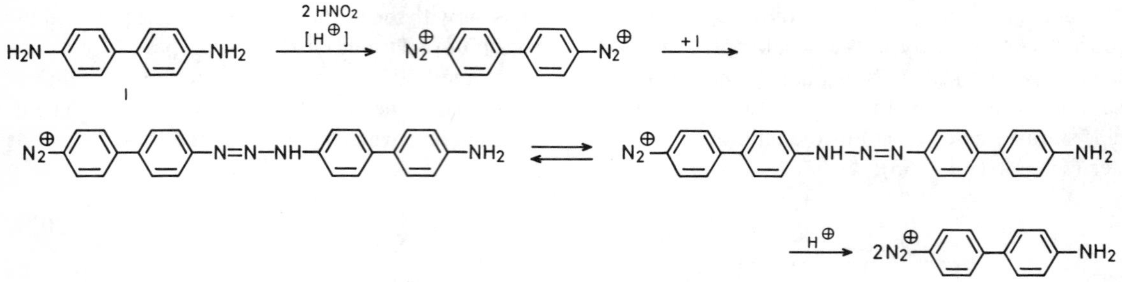

7.1.1.4.2 Bicyclic aromatic diamines[56]

Of this class of compounds only the 1,5-, 2,6-, 2,7-, and 1,8-naphthalenediamines and benzidine have gained interest. While the first three naphthalenediamines can be diazotized normally — suitably in concentrated acids in order to eliminate undesired azo coupling side-reactions — the 1,8-naphthalenediamine cyclizes with nitrous acid to form *1H*-naphtho[1,8-d,e]-1,2,3-triazine[57]:

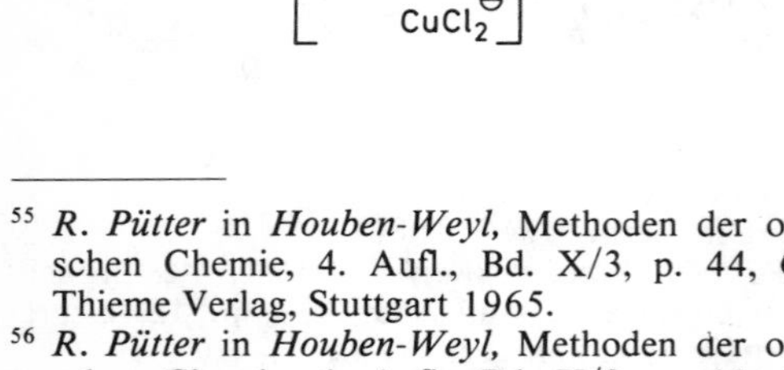

Unlike 1,2,3-benzotriazole, however, 1H-naphtho[1,8-d,e]-1,2,3-triazine does not display true aromatic compound properties but, under the conditions of the Sandmeyer reaction, behaves like a normal diazonium salt[58]. Consequently double diazotization is possible when strong acids are used.

In benzidine the linked-together arylamine units behave like aniline on diazotization. While the *bisdiazonium* salts can be prepared directly, a trans-diazotization technique (see p. 173) is employed for preparing the monodiazonium salts:

The bisdiazonium salt can be used as a reagent for olefins and conjugated dienes[59].

7.1.2 Diazotization of heteroaromatic amines[60]

7.1.2.1 Diazotization of monoamines

A particularly successful method for hydrolyzing *carboxamides* involves diazotization of the amides with nitrous acid in concentrated sulfuric or phosphoric acid[61]:

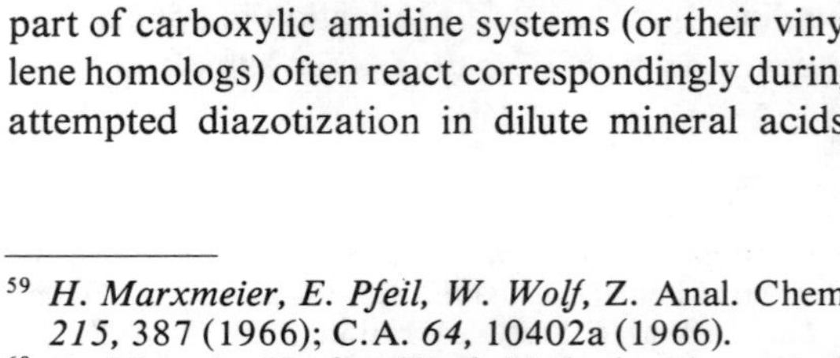

Heteroaromatic amines whose amino groups form part of carboxylic amidine systems (or their vinylene homologs) often react correspondingly during attempted diazotization in dilute mineral acids:

[55] *R. Pütter* in *Houben-Weyl*, Methoden der organischen Chemie, 4. Aufl., Bd. X/3, p. 44, Georg Thieme Verlag, Stuttgart 1965.

[56] *R. Pütter* in *Houben-Weyl*, Methoden der organischen Chemie, 4. Aufl., Bd. X/3, p. 45, Georg Thieme Verlag, Stuttgart 1965.

[57] *A. De Aguiar*, Ber. *7*, 315 (1874).

[58] DRP 147852 (1903), BASF; Fortschritte der Teerfabrikation und verwandter Industriezweige *7*, 131 (1905).

[59] *H. Marxmeier, E. Pfeil, W. Wolf,* Z. Anal. Chem. *215*, 387 (1966); C.A. *64*, 10402a (1966).

[60] *R. Pütter* in *Houben-Weyl*, Methoden der organischen Chemie, 4. Aufl., Bd. X/3, p. 53, Georg Thieme Verlag, Stuttgart 1965.

[61] *L. Bouveault*, Bull. Soc. Chim. France (3) *9*, 368 (1893);
H. Henecka in *Houben-Weyl*, Methoden der organischen Chemie, 4. Aufl., Bd. VIII, p. 432, Georg Thieme Verlag, Stuttgart 1952.

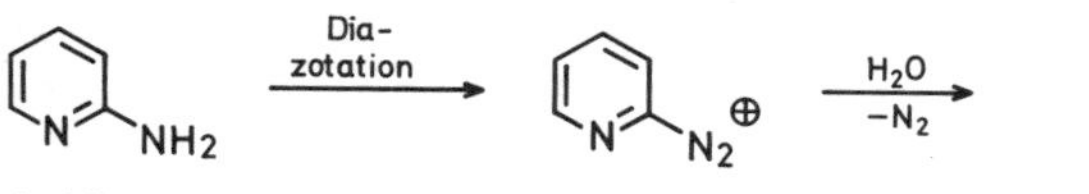

Amidine

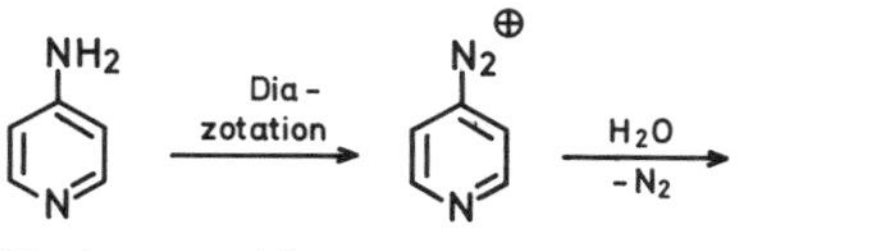

Vinylogous amidine

If the electron-withdrawing power of the ring-nitrogen atoms in 2- and 4-aminopyridines[62] is taken away by oxidation to the *N*-oxide, then the corresponding diazonium salts are easily prepared[63]. The *2-pyridinediazonium N*-oxide *salt* is here assigned the structure of a *1,2,3,5-oxatriazolium ion*, but in the solution equilibrium this structure must be in equilibrium with the open form since normal azo coupling is observed with 2-naphthol[64]:

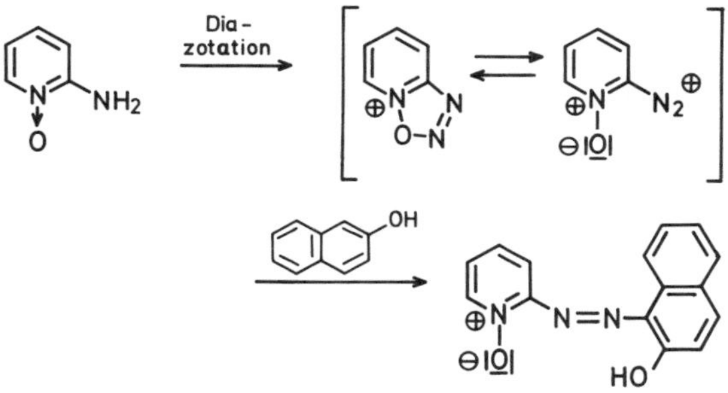

By employing the technique of diazotization in concentrated acid (see p. 164), however, the heteroaromatic carboxylic acid amidines (or vinylogs) can mostly be diazotized satisfactorily and, because their coupling reactivity is high, can be directly coupled in strongly acid media[65].

If the position of the amino group on the heteroaromatic nucleus is such that no mesomeric effects operate, the amines concerned can be diazotized without difficulty in dilute mineral acid;

examples are 3-aminopyridine[66] (forming *3-pyridine-diazonium salts*), 3-amino-2-chloropyridine[67], and 5-aminouracil[68] (forming *uracil-5-diazonium salts*).

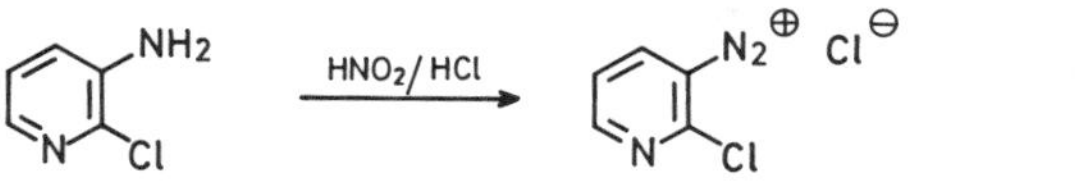

2-Chloro-3-pyridinediazonium chloride

Details have been published concerning azo coupling or substitution reactions of diazonium salts of derivatives of pyrrole[69], pyrazole[70], imidazole[71], 1,2,4-triazole[72], 1,3-thiazole[73], and 1,3,4-thiadiazole[74].

On diazotization amino groups attached to heterocyclic nitrogen atoms are replaced by hydrogen[75]:

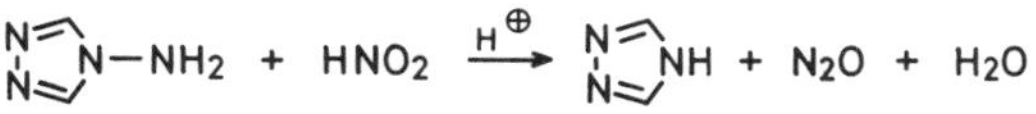

In certain cases, diazotization reactions of heteroaromatic amines may proceed differently to those of the analogous benzene derivatives. As the number of electronegative atoms in the heterocyclic ring increases, the diazonium group has an increasing tendency to add a hydroxyl ion in dilute mineral acid solution, forming an *N*-nitrosoamine[76] (*cf.* keto-enol tautomerism in this connection):

$$R-\overset{\oplus}{N}\equiv N + OH^{\ominus} \rightleftharpoons \left[R-N=N-OH\right] \rightleftharpoons R-NH-NO$$

[62] *E. Kalatzis*, J. Chem. Soc. B 273, 277 (1967).

[63] *E. Ochiai*, J. Org. Chem. *18*, 534 (1953).

[64] *H.G.O. Becker, H. Böttcher, H. Haufe*, J. Prakt. Chem. *312*, 433 (1970).

[65] *E. Koenigs, G. Kinne, W. Weiss*, Ber. *57*, 1172 (1924).

[66] *R. Pütter* in *Houben-Weyl*, Methoden der organischen Chemie, 4. Aufl., Bd. X/3, p. 55, Georg Thieme Verlag, Stuttgart 1965;
P. Tomasik, E. Kucharzewska-Rusek, A. Thomas, Roczniki Chem. *44*, 1131 (1970).

[67] *M.M. Kochhar*, J. Heterocycl. Chem. *6*, 977 (1969).

[68] *S.H. Chang, I.K. Kim, D.S. Park, B.-S. Hahn*, Daehan. Hwahak Hwoejee *9*, 29 (1965); C.A. *64*, 15876f (1966).

[69] *A. Mineo*, Corriere Farm. *21*, 318 (1966); C.A. *67*, 43617g (1967).

[70] *H. Reimlinger, G.S.D. King, M.A. Peiren*, Chem. Ber. *103*, 2821 (1970);
H. Reimlinger, R. Merényi, Chem. Ber. *103*, 3284 (1970).

[71] *K. Komatsu, K. Konishi, N. Kuroki, M. Kaimori*, Kogyo Kagaku Zasshi *73*, 1694 (1970).

[72] Roum. P. 50786 (C1 C o9b), 6.4.68, Appl. 9.7.65, Romania, Ministry of the Chemical Industry, Inv.: *A. Spiliadis, D. Bretcanu, C. Eftimescu, R.T. Schip*; C.A. *70*, 4114h (1969);
A.N. Frolov, M.S. Pevzner, I.N. Sochor, A.G. Gal'kovskaja, L.I. Bagal, Khim. Geterotsikl. Soedin. 705 (1970) (Diazonium nitrates, fluoroborates, and betaines).

[73] *A. Spiliadis, M. Hilsenrath, V. Cornea, E. Balta, G. Molau, E. Teodorescu, L. Predescu*, Rev. Chim. (Bucharest) *17*, 271 (1966); C.A. *66*, 11843r (1967);
C. Grünert, K. Wiechert, Z. Chem. *10*, 396 (1970).

[74] *G. Asato, G. Berkelhammer, E.L. Moon*, J. Med. Chem. *13*, 1015 (1970).

[75] *Th. Curtius, A. Darapsky, E. Müller*, Ber. *40*, 836 (1907).

This reaction is normally reversible in concentrated acids. However, if the heteroaromatic ring also bears a positive charge as, for example, in 2,3-diphenyl-5-nitrosoaminotetrazolium chloride[77], then no diazonium salt is formed even in concentrated hydrochloric acid.

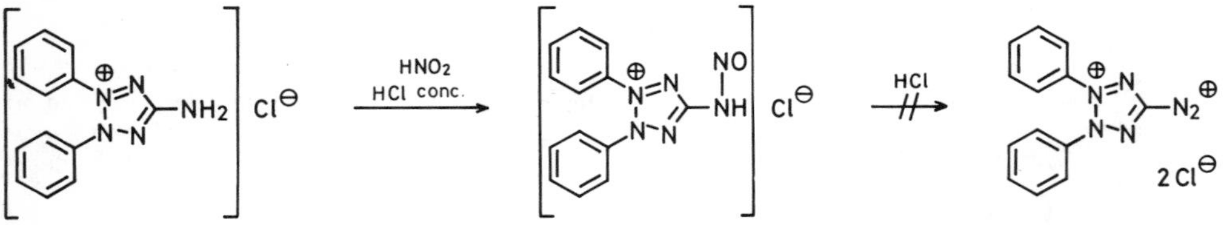

both amino groups in 2,4-thiophenediamine can be diazotized in dilute sulfuric acid, forming *2,4-thiophene-bis-diazonium salts,* other heteroaromatic diamino compounds are usually either unreactive or react only incompletely with 2 molar equivalents of nitrous acid[81].

The formation of *diazonium betaines*[78] is occasionally observed; formally, these are comparable to the quinone diazides in the aromatic series. In addition, examples are known in which, instead of diazotization at the nitrogen atom, nuclear nitrosation or some other reaction differing from the usual occurs[79].

7.1.2.2 Diazotization of diamines

Relative little experimental information is at present available about the diazotization of heteroaromatic diamines. In the reaction of 4,5-diamino-2-phenyl-*2H*-1,2,3-triazole with nitrous acid in dilute hydrochloric acid simple diazotization first occurs, and is followed (after buffering) by slow *triazole formation*[80]:

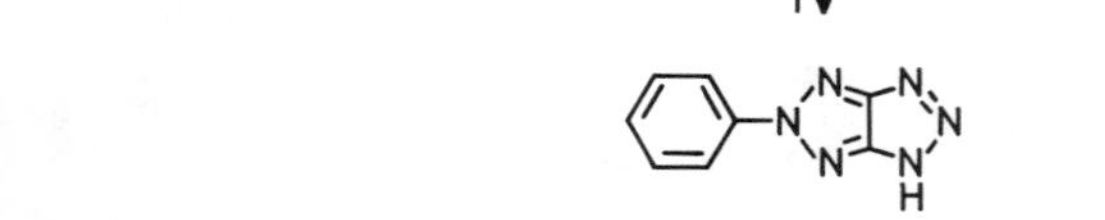

4-Amino-2-phenyl-2H-1,2,3-triazole-5-diazonium chloride

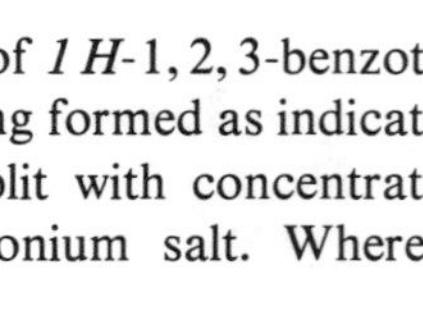

1,5-Dihydro-5-phenyl-1,2,3-triazolo[4,5-d]-1,2,3-triazole

In contrast to the behavior of *1H*-1,2,3-benzotriazole, the second triazole ring formed as indicated above can be reversibly split with concentrated acids, regenerating a diazonium salt. Whereas

On diazotization, 3,5-diamino-1,2,4-triazole undergoes self-coupling, forming a *polymeric triazolotriazene*[82]:

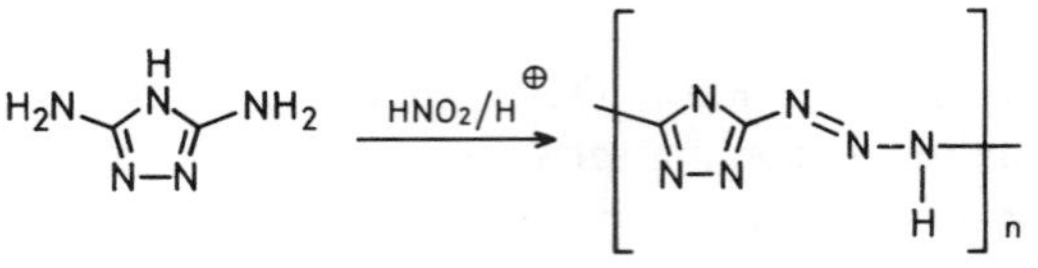

A tabular summary of diazotizations of heterocyclic mono- and diamines is given in ref.[1].

7.1.3 Preparation of aromatic diazonium salts by reactions of other diazonium salts

7.1.3.1 By rearrangement

The most surprising rearrangement which has been observed with aromatic diazonium salts is the ability of the two nitrogen atoms of the diazonium group to undergo mutual exchange, as evidenced by isotope labeling[83]:

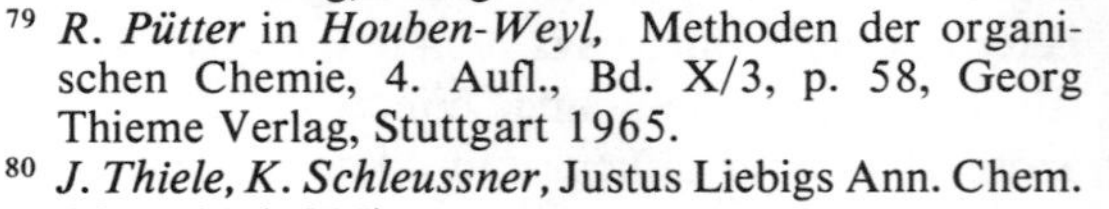

[76] *R. Pütter* in *Houben-Weyl,* Methoden der organischen Chemie, 4. Aufl., Bd. X/3, p. 55, Georg Thieme Verlag, Stuttgart 1965.

[77] *E. Bamberger, R. Padova, E. Ormerod,* Justus Liebigs Ann. Chem. *446,* 266 (1926).

[78] *R. Pütter* in *Houben-Weyl,* Methoden der organischen Chemie, 4. Aufl., Bd. X/3, p. 57, Georg Thieme Verlag, Stuttgart 1965.

[79] *R. Pütter* in *Houben-Weyl,* Methoden der organischen Chemie, 4. Aufl., Bd. X/3, p. 58, Georg Thieme Verlag, Stuttgart 1965.

[80] *J. Thiele, K. Schleussner,* Justus Liebigs Ann. Chem. *295,* 129 (1897).

[81] *R. Pütter* in *Houben-Weyl,* Methoden der organischen Chemie, 4. Aufl., Bd. X/3, p. 59, Georg Thieme Verlag, Stuttgart 1965.

[82] *M. Hauser,* J. Org. Chem. *29,* 3449 (1964).

[83] *A.K. Bose, I. Kugajewsky,* J. Amer. Chem. Soc. *88,* 2325 (1966);
E.S. Lewis, R.E. Holliday, J. Amer. Chem. Soc. *88,* 5043 (1966); *E.S. Lewis, R.E. Holliday,* J. Amer. Chem. Soc. *91,* 426, 430 (1969);
G.W. Van Dine, R. Hoffmann, J. Amer. Chem. Soc. *90,* 3227 (1968).

It was found that diazonium salts containing a *fluctuating 1, 2, 3-thiadiazole ring* are formed on di-azotization of 7-amino-6-substituted 1, 2, 3-benzo-thiadiazoles[84]:

X = Cl, CN
R¹ = F, Cl, OCH₃, NO₂

In contrast, diazonium salts containing substituent groups at the 4-position instead of the 6-position did not rearrange.

A spontaneous quantitative *hydrolytic diazotiza-tion* has been reported[85] to occur with 2-amino-phenol 2, 6-dimethylbenzoate:

2-Hydroxybenzene-diazonium chloride

However, it is noteworthy that the corresponding 2, 6-dichlorobenzoate ester undergoes normal diazotiza-tion.

7.1.3.2 By substitution

As already indicated, the diazonium ion (acting as a Type 2 substituent group) retards electrophilic substitution on aromatic nuclei — especially at the *o*- and *p*-positions of the same nucleus — much more

even than the nitro group. In the case of quinone diazides this effect is partially annulled by the ox-ide-oxygen atom, which behaves as a powerful Type 1 substituent[86]:

2,4-Dibromo-6-diazo-2,5-cyclohexadien-1-one betain

2,6-Dibromo-4-diazo-2,5-cyclohexadien-1-one betain

A tabular summary of halogenation, nitration, and sulfonation reactions of quinone diazides is given in ref. [1].

4-Anilinobenzenediazonium salts react with form-aldehyde to yield polycondensed diazonium salts, which are deprotonated under neutral conditions, forming diazoimino compounds[87]:

[84] E. Haddock, P. Kirby, A. W. Johnson, J. Chem. Soc. C 2514 (1970).

[85] D.J. Triggle, S. Vickers, Chem. Commun. 544 (1965);
S. Vickers, D.J. Triggle, D.R. Garrison, J. Chem. Soc. C 632 (1968).

[86] E. Bamberger, E. Kraus, J. Prakt. Chem. [2] 105, 264 (1922).

[87] V.V. Kozlov, S.K. Eremin, Teor. Eksp. Khim. 6, 795 (1970);
R. Pütter in Houben-Weyl, Methoden der organi-schen Chemie, 4. Aufl., Bd. X/3, p. 89, Georg Thieme Verlag, Stuttgart 1965.

[88] E. Bamberger, Ber. 55, 3383 (1922);
J. Sauer, R. Huisgen, Angew. Chem. 72, 303 (1960) (Mechanism).

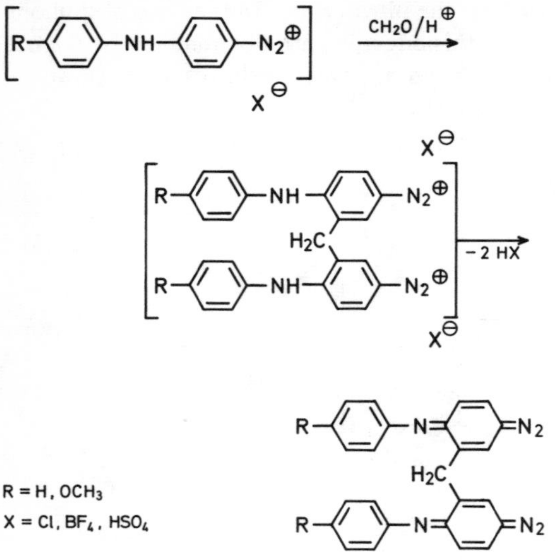

R = H, OCH₃

$R = H, OCH_3$

$X = Cl, BF_4, HSO_4$

and higher polycondensation products

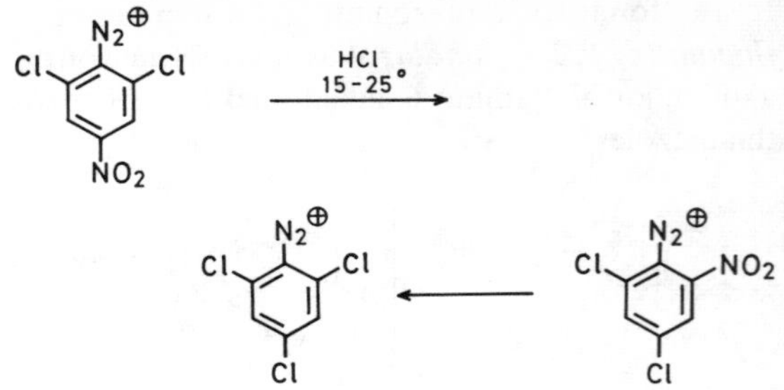

A direct oxidation of diazonium ions in alkaline potassium hexacyanoferrate(III) solution has been reported[88] to yield *1-diazo-2(1H)naphthalenone:*

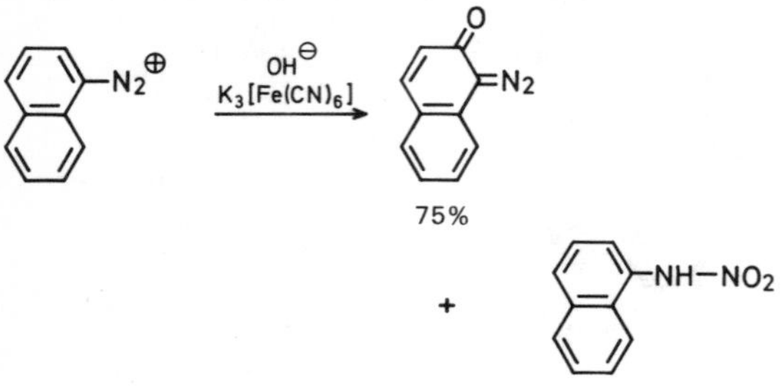

75%

+

Substituent groups *o-* or *p-* to the diazonium group on the aromatic nucleus are more prone than hydrogen to undergo *nucleophilic* substitution, *o-* situated groups being easier to replace than *p-* groups. The ease of replacement of substituent groups decreases in the order:

$$NO_2 > Br > Cl > OCH_3 > SO_3H$$

The factors governing this order in aqueous medium are the electronegativity of the various substituent groups and their ability to act as leaving groups. If the aromatic nucleus is rendered sufficiently positive[90], both *o-*[89] and *p-*situated nitro groups can be replaced, *e.g.:*

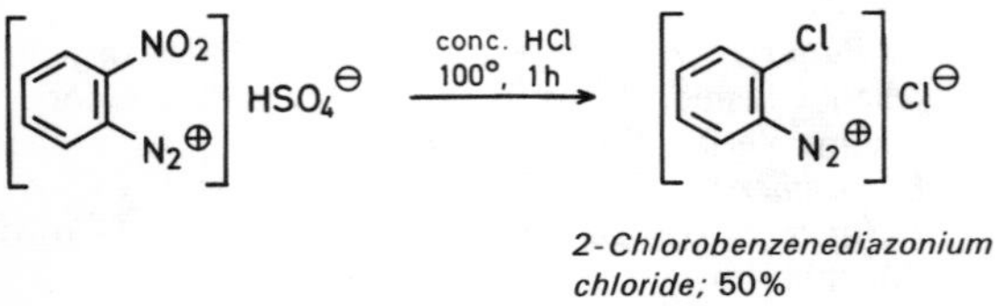

2-Chlorobenzenediazonium chloride; 50%

2,4,6-Trichloroben-zenediazonium salt

When heated in a mixture of acetic and hydrochloric acids, *o-* and *p-*nitro-substituted arylamines are often converted into the corresponding chloro-substituted *diazonium salts.* The first step in this reaction is nucleophilic replacement of the nitro group by chlorine, and the amine is then diazotized by the liberated nitrite ion in acid solution (see the tabular summary containing many examples in ref.[1]).

Halogen-exchange of bromide and chloride (or of labeled chloride, and unlabeled chloride) has likewise been observed, *e.g.:*

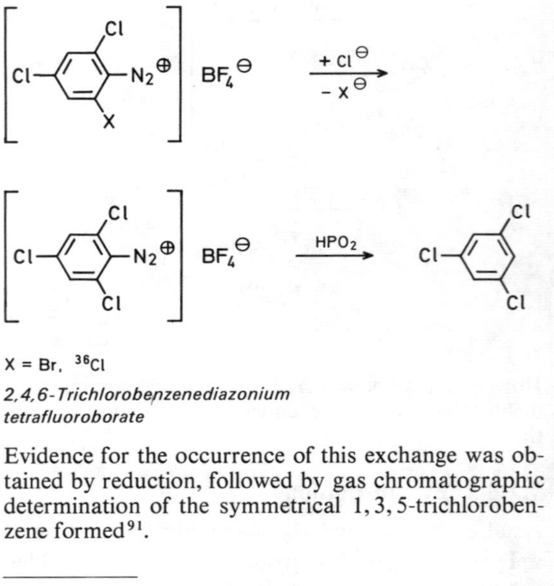

$X = Br, {}^{36}Cl$

2,4,6-Trichlorobenzenediazonium tetrafluoroborate

Evidence for the occurrence of this exchange was obtained by reduction, followed by gas chromatographic determination of the symmetrical 1,3,5-trichlorobenzene formed[91].

[89] Z. J. Allan, J. Podstata, Collect. Czech. Commun. *31*, 3418 (1966); C.A. *66*, 47285z (1967).
[90] B. Andersson, B. Lamm, Acta Chem. Scand. *23*, 2983 (1969);
R. Pütter in *Houben-Weyl*, Methoden der organischen Chemie, 4. Aufl., Bd. X/3, p. 94, Georg Thieme Verlag, Stuttgart 1965.
[91] B. Lamm, B. Andersson, Ark. Kemi *25*, 367 (1966); B. Lamm, B. Liedholm, Acta Chem. Scand. *21*, 2679 (1967);
R. Pütter in *Houben-Weyl*, Methoden der organischen Chemie, 4. Aufl., Bd. X/3, p. 99, Georg Thieme Verlag, Stuttgart 1965.

The solvent has a decisive influence on the rate at which this halogen-exchange reaction occurs. Less solvation of the attacking chloride imparts it enhanced nucleophilic properties. The rate of bromide-chloride exchange increases markedly in the series:

$$H_2O < CH_3OH < C_5H_5OH$$

Nucleophilic replacement by the *hydroxy* group of halogen, nitro, sulfonyl, or alkoxy groups positioned *o*- or *p*- to the diazonium group is of particular importance. The tables in ref. [1] provide a comprehensive survey of these reactions.

Elimination of an acyl group during diazotization of a 2-acyloxyaniline has already been mentioned[85]. Conversely, however, a quinone diazide has been acylated to an acyloxydiazonium salt capable of coupling powerfully[92], e.g.:

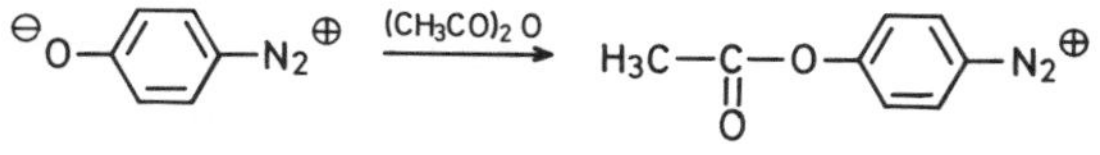

4-Acetoxybenzenediazonium salt

7.1.4 Special methods

7.1.4.1 Transdiazotization

Preparation of monodiazotized benzidine by transdiazotization was described on p. 168[93]. It proceeds particularly readily (sometimes even quantitatively) if the diazonium salt of a weakly basic amine acts on a more strongly basic amine in acid medium[94], e.g.:

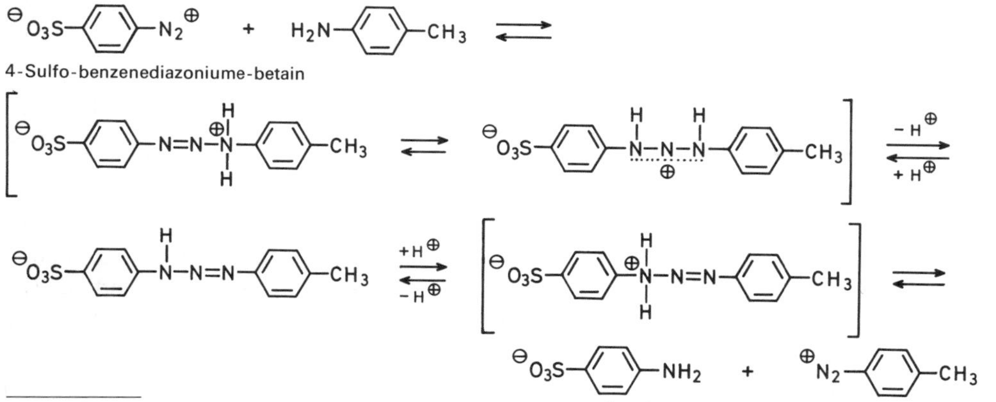

4-Sulfo-benzenediazoniume-betain

4-amino-benzene-sulfonate   *4-Toluenediazonium*

This transdiazotization representing the *aminolysis* of the diazonium ion of a weakly basic amine may be regarded as related directly to the *hydrolysis* of a diazonium ion[17]:

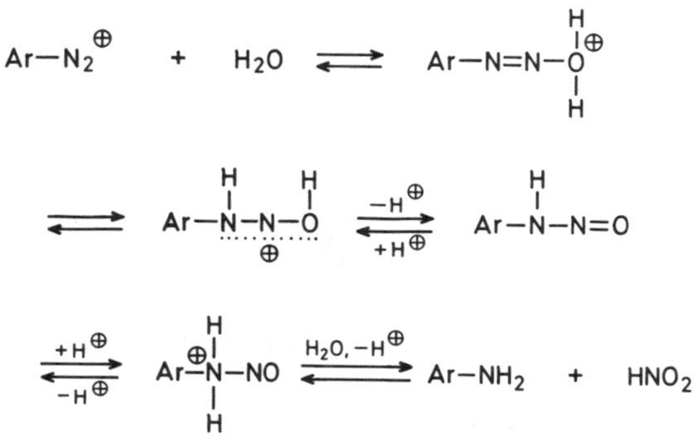

The *N*-azo coupling products, which are called triazenes or diazoamino compounds, can be readily isolated (see p. 200).

This reaction is suitable also for preparing aliphatic diazo compounds from diazonium salts. Thus *ethyl diazoacetate* is obtained by thermal triazene fission of the diazoamino compound from 4-nitrobenzenediazonium salt and ethyl aminoacetate[95].

While only one nitrogen atom is transferred during transdiazotization, diazo group transfer from ben-

[92] DRP 206 455 (1907), Farbf. Bayer; Fortschritte der Teerfabrikation und verwandter Industriezweige *9*, 300 (1911).

[93] *R. Pütter* in *Houben-Weyl*, Methoden der organischen Chemie, 4. Aufl., Bd. X/3, p. 46, Georg Thieme Verlag, Stuttgart 1965;
C. Süling in *Houben-Weyl*, Methoden der organischen Chemie, 4. Aufl., Bd. X/3, p. 728, Georg Thieme Verlag, Stuttgart 1965.

[94] *C. Schraube, M. Fritsch*, Ber. *29*, 287 (1896);
A. Hantzsch, F.M. Perkin, Ber. *30*, 1412 (1897);
H. Mehner, J. Prakt. Chem. [2] *63*, 266 (1901).

[95] *R.J. Baumgarten*, J. Org. Chem. *32*, 484 (1967).

[96] *J.M. Tedder, B. Webster*, J. Chem. Soc. (London) 4417 (1960);
R. Pütter in *Houben-Weyl*, Methoden der organischen Chemie, 4. Aufl., Bd. X/3, p. 88, Georg Thieme Verlag, Stuttgart 1965.

[97] *W.R. Bamford, T.S. Stevens*, J. Chem. Soc. (London) 4735 (1952);
R. Pütter in *Houben-Weyl*, Methoden der organischen Chemie, 4. Aufl., Bd. X/3, p. 84, 85, Georg Thieme Verlag, Stuttgart 1965.

zenesulfonyl azide or from 4-toluenesulfonyl *azide* to phenolates[96] involves the migration of two nitrogen atoms:

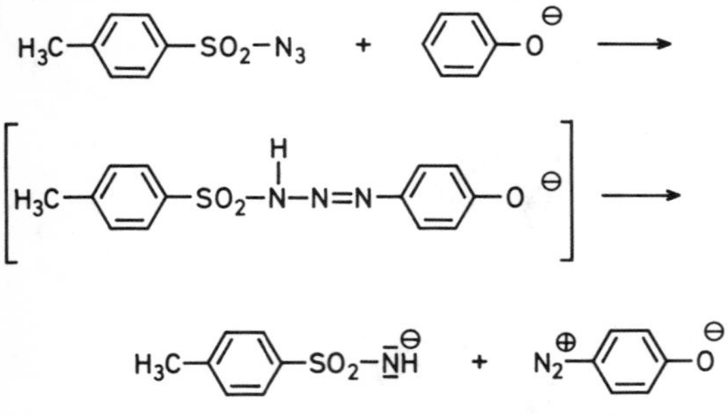

4-Diazo-2,5-cyclo-
hexadien-1-one betain

Because the electrophilism of the sulfonyl azides is slight, however, this reaction is possible only with very nucleophilic aromatic compounds.

As a variant of the Bamford-Stevens reaction, the reaction between *o*- and *p*-quinones with corresponding sulfonic acid *hydrazides* is an alternative to diazo group transfer with sulfonyl azides[97]:

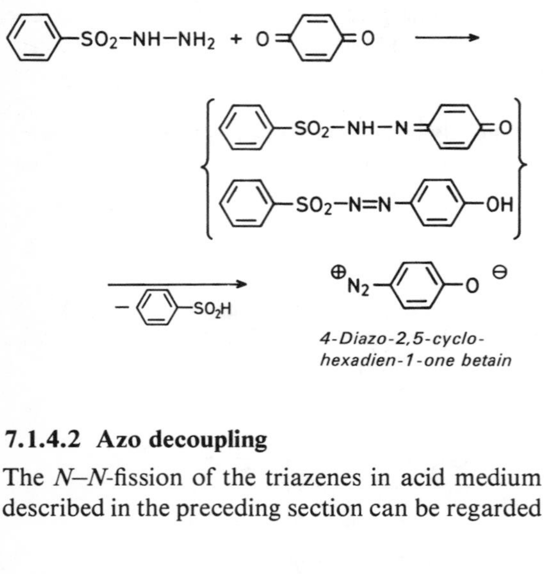

4-Diazo-2,5-cyclo-
hexadien-1-one betain

7.1.4.2 Azo decoupling

The *N–N*-fission of the triazenes in acid medium described in the preceding section can be regarded

as the reversal of an azo coupling at the nitrogen. Such retro-reactions in acid medium have been observed also with *C*-azo coupling products[98], and are called azo decoupling[99], *e.g.*:

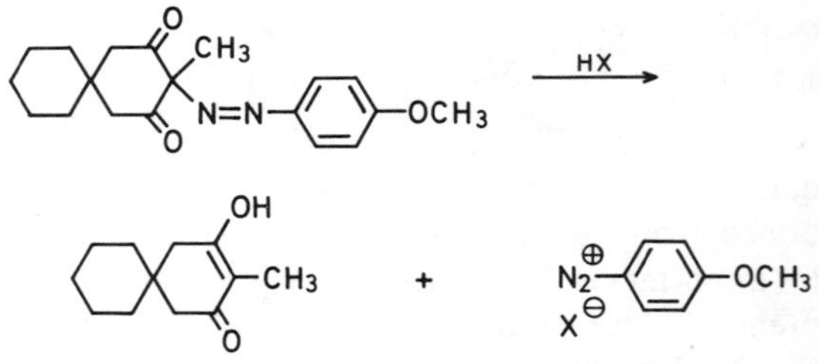

4-Anisolediazonium salt

When hydrogen chloride is passed into a benzene solution of the yellow azo dye decolorization occurs and the colorless diazonium salt precipitates[100].

Recently, similar azo decoupling reactions have been carried out by treating azo dyes based on monoalkylated 1,3-dicarbonyl compounds also with boron trifluoride[101], *e.g.*:

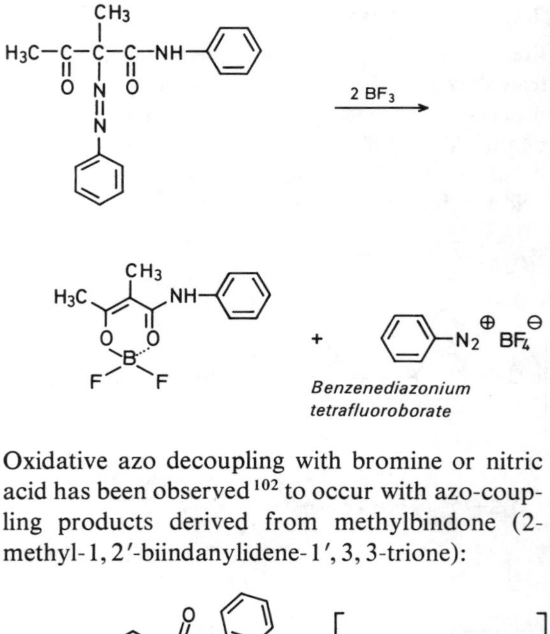

Benzenediazonium
tetrafluoroborate

Oxidative azo decoupling with bromine or nitric acid has been observed[102] to occur with azo-coupling products derived from methylbindone (2-methyl-1,2'-biindanylidene-1',3,3-trione):

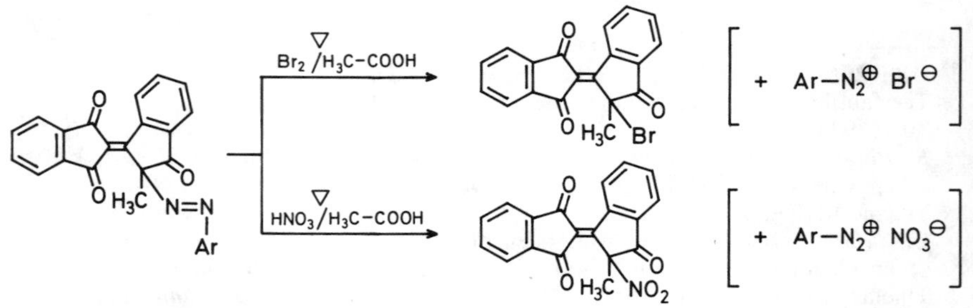

[98] *E.H. Rodd,* Chemistry of Carbon Compounds, Vol. IVb, p. 1280, Elsevier Publ. Co., London 1959; *O.A. Stamm, H. Zollinger,* Helv. Chim. Acta *40,* 1955 (1957).

[99] *B. Eistert, K. Schank,* Chem. Ber. *96,* 2304 (1963); *K. Schank,* Dissertation, Universität Saarbrücken 1962.

7.1.4.3 Nitrosation of acylated amines

Acylated aromatic amines (Ar—NH—X; X = SO₃H, NO₂ or COOR) can be readily diazotized by using NO⊕-donors[103]:

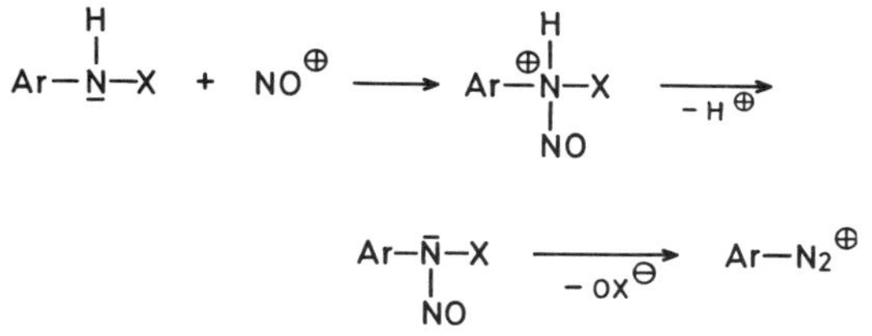

With sulfamic acids and *N*-nitroamines the intermediate nitroso compounds cannot be isolated and the diazonium salts are obtained directly. In contrast, however, the corresponding *N*-nitroso-carboxamides can be isolated easily when a suitable practical procedure is employed (dinitrogen trioxide in acetic acid + acetic anhydride or nitrosyl chloride in acetic anhydride + potassium acetate). These intermediates also are not particularly stable, and in solution they undergo an intramolecular rearrangement by an uncoiling mechanism[104], forming the corresponding diazo esters, which exist in equilibrium with their isomeric *diazonium carboxylates*[105]:

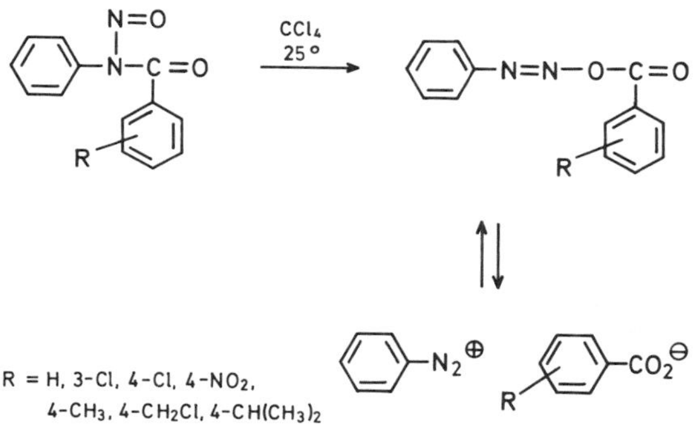

This rearrangement can also be catalyzed by protons, but denitrosation occurs as a competitive reaction[106]. Removal of the acyl group with bases

is a more successful technique[107]. Isatins (indole-2,3-diones) can be diazotized with nitrosylsulfuric acid in 99% acetic acid, forming *2-(carboxycarbonyl)benzenediazonium hydrogen sulfate*[108]:

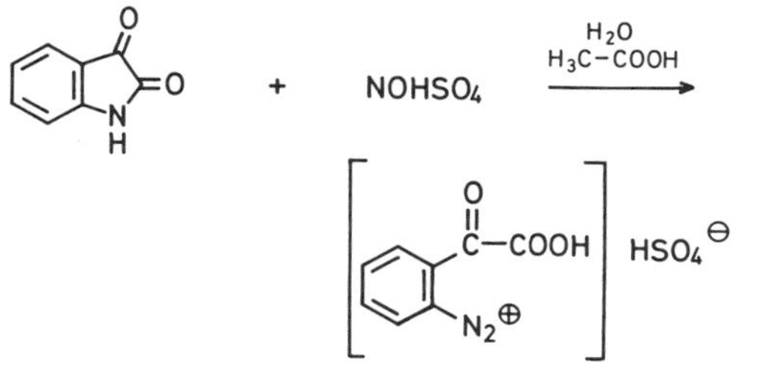

2-(Carboxycarbonyl)benzene-diazonium hydrogen sulfate[108]

7.1.4.4 Nitrosation of other functional arylamine derivatives

Treatment of *azomethines* in inert solvents such as benzene with NOX (X = Cl[109] or NO₃[110]) results in the formation of diazonium salts, together with the aldehyde component, *e.g.*:

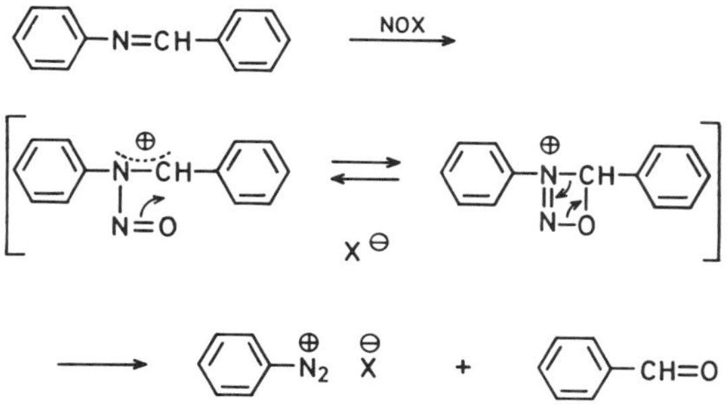

Correspondingly, phenyl isocyanate can be converted into a diazonium salt by treatment with dinitrogen tetraoxide in chloroform, carbon dioxide being eliminated[111]:

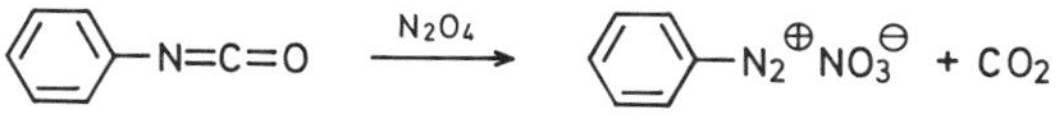

Sulfinylamines react in a similar manner with nitrosyl salts in 1,2-dichloroethane, sulfur dioxide is eliminated[112]:

[100] *K. Schank*, unpubl.
[101] *G.C. Barrett, M.M. El-Abadelah, M.K. Hargreaves*, Chem. & Ind. (London) 2130 (1966); C.A. *66*, 64877m (1967);
G.C. Barrett, M.M. El-Abadelah, M.K. Hargreaves, J. Chem. Soc. C 1986 (1970).
[102] *R. Zagats, M. Buka, G. Vanags*, Zh. Org. Khim. *1*, 535 (1965); C.A. *63*, 2936e (1965).
[103] *R. Pütter* in *Houben-Weyl*, Methoden der organischen Chemie, 4. Aufl., Bd. X/3, p. 66, Georg Thieme Verlag, Stuttgart 1965.
[104] *R. Pütter* in *Houben-Weyl*, Methoden der organischen Chemie, 4. Aufl., Bd. X/3, p. 68, Georg Thieme Verlag, Stuttgart 1965.
[105] *Ch. Rüchardt, Chuan Cheng Tan*, Chem. Ber. *103*, 1774 (1970).
[106] *R. Huisgen, H. Reimlinger*, Justus Liebigs Ann. Chem. *599*, 161 (1956).
[107] *R. Huisgen*, Justus Liebigs Ann. Chem. *573*, 163 (1951).
[108] Ger. P. 1212098 (Cl. C o7c), 10.3.66, Appl. 25.2.64, BASF, Erf.: *M. Seefelder, H. Armbrust*; C.A. *64*, 17494f (1966).
[109] *J. Turcan*, Bull. Soc. Chim. France 627 (1935).
[110] *R.M. Scribner*, J. Org. Chem. *29*, 3429 (1964).
[111] *G.B. Bachmann, W. Michalowicz*, J. Org. Chem. *23*, 1800 (1958).

$$Ar-N=S=O \;+\; NO^{\oplus}X^{\ominus} \;\xrightarrow[-10° \rightarrow 0°]{(CH_2Cl)_2}\;$$

$$X = ClO_4, SbCl_6 \qquad\qquad Ar-N_2^{\oplus}X^{\ominus} \;+\; SO_2$$

By treatment with nitrous acid, *N*-phenylhydroxyl-amine can be nitrosated to form *N*-nitroso-*N*-phe-nylhydroxylamine. If an excess of nitrous acid is used, *benzenediazonium nitrate* is formed by a redox process:

$$\text{Ph}-NHOH \;+\; HNO_2 \;\xrightarrow[-H_2O]{(H^{\oplus})}\;$$

$$\text{Ph}-N\!\!\begin{array}{c}O-H\\|\\N=O\end{array} \;\xrightarrow[-H_2O]{+HNO_2}\; \text{Ph}-N_2^{\oplus}NO_3^{\ominus}$$

Conversely, however, the *N*-nitroso-*N*-phenylhydroxyl-amine can be regenerated from the diazonium salt by treatment with a nitric oxide-saturated solution of ferrous sulfate in the presence of cuprous chloride; this is a variant of the Sandmeyer reaction[113]:

$$\text{Ph}-N_2^{\oplus} \;\xrightarrow[-N_2]{FeSO_4\cdot NO/CuCl}\; \text{Ph}-N\!\!\begin{array}{c}O-H\\|\\N=O\end{array}$$

The ammonium salt is used in analytical chemistry[114] as a precipitant for heavy metal ions under the name Cupferron®.

Formation of the diazonium compound proceeds *via* reduction of the *N*-nitrosohydroxylamine; the reducing agent involved may be either excess nitrous acid, nitrate being formed, or zinc dust in acetic acid. The isomeric *N*-nitroamines, too, have been reduced to diazonium ions[115].

Aromatic *nitroso* compounds, which represent the next higher oxidation stage of aromatic amines above hydroxylamines, are likewise *redox-diazotized* by excess nitrous acid[116]:

$$Ar-N=O \;+\; 3\,HNO_2 \;\xrightarrow[-HNO_3]{-H_2O}\; Ar-N_2^{\oplus}NO_3^{\ominus}$$

Since the ring of electron-rich aromatic compounds such as phenols and *N,N*-dialkylanilines is easily nitrosated by nitrous acid, these compounds can be converted directly into diazonium salts in a *combined reaction in one flask*[38]:

$$R_2N-\text{C}_6H_5 \;+\; 4\,HNO_2 \;\xrightarrow[-HNO_3]{-2\,H_2O}\;$$

R = CH₃, C₂H₅
R,R = (CH₂)₂O(CH₂)₂–,
– (CH₂)ₙ–
where n = 4,5,6

$$R_2N-\text{C}_6H_4-N_2^{\oplus}NO_3^{\ominus}$$

More active nitrosating agents such as nitrosylsul-furic acid are necessary with aromatic compounds such as anisole and mesitylene whose electron density is lower[117].

Simultaneous chlorination and diazotization has been observed[118] to occur during the reaction of aromatic nitroso compounds with nitrosyl chloride:

$$\text{(1-nitrosonaphthalene)} \;\xrightarrow{NOCl/HCl}\; \left[\text{(naphthalene with }N_2^{\oplus}\text{ and Cl)}\right] Cl^{\ominus}$$

4-Chloro-1-naphthalene-diazonium chloride

The diazonium group can be introduced in one step even into aromatic compounds of low reactivity by using metal catalysis ($Hg^{2\oplus}$!)[119].

Hydroxylamine can be used for *inverse diazotization*[120] instead of nitrous acid (or its derivatives), *e.g.*:

$$\text{Ph}-N=O \;+\; H_2NOH \;\xrightarrow[-H_2O]{[Na_2CO_3]}\;$$

$$\text{Ph}-N_2^{\oplus}OH^{\ominus} \;\xrightarrow{\text{(2-naphthol)}}\; \text{(1-phenylazo-2-naphthol)}$$

Benzenediazonium hydroxide

Nitrobenzene, the next higher oxidation stage, reacts correspondingly with the next lower oxida-

[112] *K. Bott*, Angew. Chem. *77*, 132 (1965).

[113] *F. Minisci, R. Galli*, Chim. Ind. (Milan) *46*, 423 (1964); C.A. *64*, 12578f (1966).

[114] *R. Pütter* in *Houben-Weyl*, Methoden der organi-schen Chemie, 4. Aufl., Bd. X/3, p. 71, Georg Thieme Verlag, Stuttgart 1965.

[115] *R. Pütter* in *Houben-Weyl*, Methoden der organi-schen Chemie, 4. Aufl., Bd. X/3, p. 83, Georg Thieme Verlag, Stuttgart 1965.

[116] *R. Pütter* in *Houben-Weyl*, Methoden der organi-schen Chemie, 4. Aufl., Bd. X/3, p. 73, Georg Thieme Verlag, Stuttgart 1965.

[117] *J.M. Tedder*, Tetrahedron *1*, 270 (1957); *J.M. Tedder*, J. Chem. Soc. (London) 4003 (1957).

[118] Jap. P. 10 343 ('66) (Cl. 23 D o1), 4.6.63, Appl. 27.12.63, Toyo Rayon Co., Ltd., Inv.: *S. Torimitsu, M. Ohno*; C.A. *65*, 15548f (1966).

[119] *R. Pütter* in *Houben-Weyl*, Methoden der organi-schen Chemie, 4. Aufl., Bd. X/3, p. 77, Georg Thieme Verlag, Stuttgart 1965.

tion stage, namely, ammonia (as sodium amide); a diazonium compound is again formed[121]:

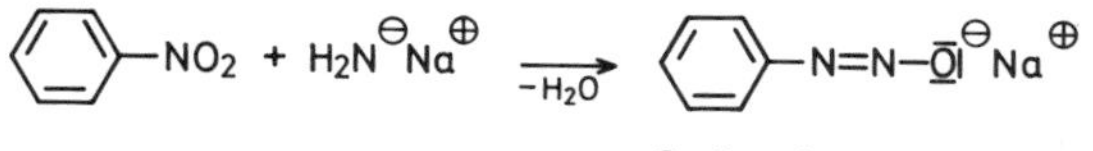

Sodium benzenediazotate

The intermediately formed diazonium ion can be identified by azo coupling.

7.1.4.5 Diazonium salts from arylhydrazines and their derivatives

Like N-phenylhydroxylamine, phenylhydrazine can be reacted with nitrous acid. In this case, however, nitrosation can occur at either the α- or the β-nitrogen atom depending on the experimental conditions, and different reaction products are formed[122]:

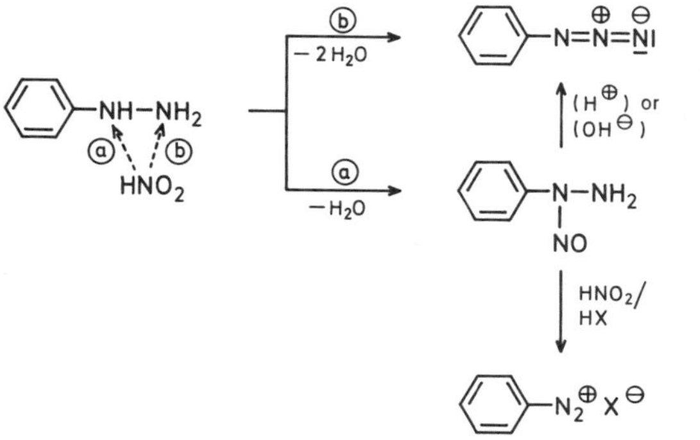

In mineral acid solution the main product is the azide[123] if the calculated amount of nitrous acid is used and the hydrazine is always in excess. If nitrous acid is constantly in excess, the diazonium salt predominates[124]. N-Nitroso-N-phenylhydrazine forms preferentially in weakly acid solution[125]; it can be converted into phenyl azide[126] by treatment with acids or bases, while treatment with nitrous acid gives the diazonium salt[127].

Direct oxidation of arylhydrazines with various metal derivatives[128] has not achieved any preparative importance. The reaction of arylhydrazine and its derivatives with bromine is of some interest; in dilute acid solution oxidation to diazonium salts only occurs, in concentrated acid medium some ring bromination takes place as well[129], *e.g.:*

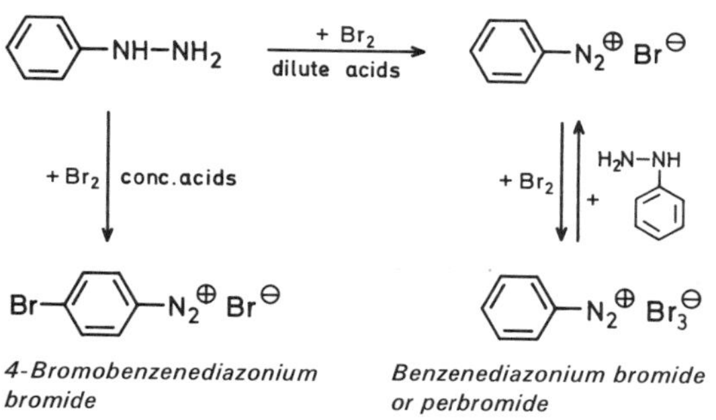

4-Bromobenzenediazonium bromide

Benzenediazonium bromide or perbromide

With an excess of bromine aryldiazonium bromides form[130] so-called perbromides, which in turn can oxidize more arylhydrazine to diazonium bromide.

Thionylphenylhydrazine can be correspondingly oxidized to the diazonium salt without nuclear bromination occurring[131].

On treatment with bromine, both acetone phenylhydrazone[132] and N'-acetyl-N-phenylhydrazine[133] first undergo nuclear o-, p-dibromination; subsequent oxidation to *2, 4-dibromobenzenediazonium bromide* then occurs.

Mention should be made in this connection of the *oxidative coupling* of heteroaromatic hydrazines[134]; this represents a valuable complement to the azo coupling of aromatic and heteroaromatic diazonium salts.

7.1.4.6 Miscellaneous

Tritium-labeled diazonium salts[135] and *polymeric* diazonium salts[136] derived from 3-vinylaniline have been described in the literature. Benzenediazosul-

[120] *R. Pütter* in *Houben-Weyl*, Methoden der organischen Chemie, 4. Aufl., Bd. X/3, p. 86, Georg Thieme Verlag, Stuttgart 1965.

[121] *F.W. Bergstrom, J.S. Buehler*, J. Amer. Chem. Soc. *64*, 19 (1942).

[122] *K. Clusius, K. Schwarzenbach*, Helv. Chim. Acta *42*, 739 (1959).

[123] *R.O. Lindsay, C.F.H. Allen*, Org. Syntheses Coll. Vol. III, 710 (1955).

[124] *J. Thiele*, Ber. *41*, 2807 (1908).

[125] *J. Altschul*, J. Prakt. Chem. [2] *54*, 496 (1896).

[126] *K. Clusius, H.R. Weisser*, Helv. Chim. Acta *35*, 1557 (1952).

[127] *L. Rügheimer*, Ber. *33*, 1718 (1900).

[128] *R. Pütter* in *Houben-Weyl*, Methoden der organischen Chemie, 4. Aufl., Bd. X/3, p. 78, Georg Thieme Verlag, Stuttgart 1965.

[129] *F.D. Chattaway*, J. Chem. Soc. (London) *93*, 852 (1908);
F.D. Chattaway, J. Chem. Soc. (London) *95*, 862 (1909).

[130] *M. Aroney, R.J.W. Le Fèvre*, J. Chem. Soc. (London) 1630 (1955).

[131] *J. Klieeisen*, Ber. *27*, 2549 (1894).

[132] *E. Votucek, P. Piru*, Bull. Soc. Chim. France [4] *33*, 918 (1923).

[133] *L. Michaelis*, Ber. *26*, 2190 (1893).

[134] *R. Pütter* in *Houben-Weyl*, Methoden der organischen Chemie, 4. Aufl., Bd. X/3, p. 82, Georg Thieme Verlag, Stuttgart 1965.

[135] *P.S. Traylor, S.J. Singer*, Biochemistry *6*, 881 (1967).

[136] *C.L. Arcus, R.H. Still*, J. Chem. Soc. (London) 4340 (1964).

fonic acid undergoes photolytic fission as indicated in the scheme below[137]:

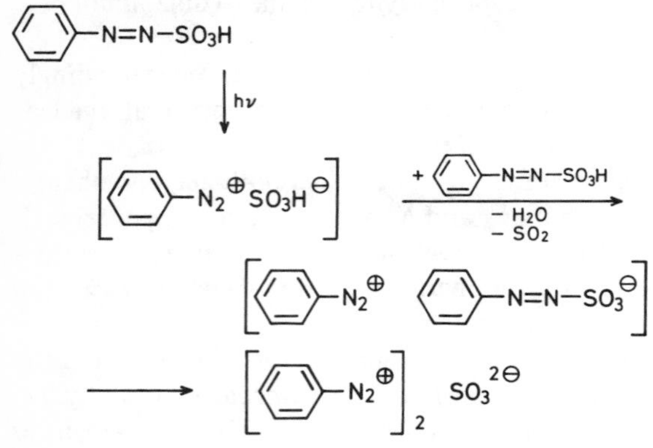

Bis[benzenediazonium sulfite]

Benzenesulfonylhydroxamic acid reacts with phenylhydroxylamine in alkaline alcoholic solution, forming *phenyl 4-phenylazo sulfone*[138]:

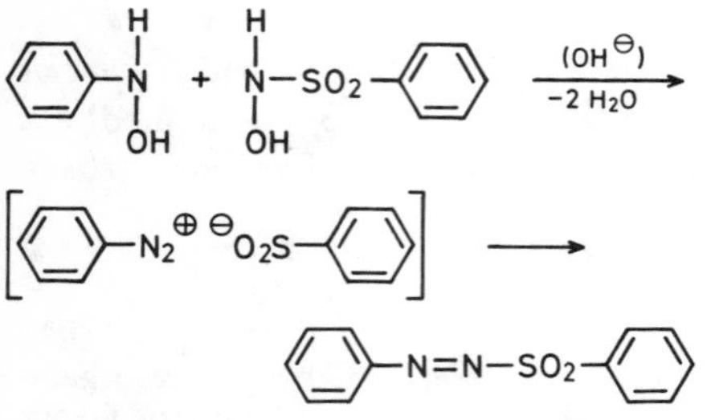

7.2 Transformations of aromatic diazonium salts

The two main classes of reactions of diazonium salts are those in which the diazonium nitrogen atoms are lost, and those in which they are retained in the molecule[138].

7.2.1 Reactions involving exchange of the diazonium nitrogen

Aromatic diazonium salts readily lose their nitrogen forming aryl *carbenium* ions; if electron donors are present aryl *free radicals* may be formed. These can undergo versatile addition or substitution reactions to yield products in which the original diazonium group has been replaced by other groups. The table in ref.[139] provides a survey.

7.2.1.1 Replacement of the diazonium group by hydrogen

Various techniques can be used to replace the diazonium group by hydrogen. Aromatic amines can be deaminated *via* intermediate diazonium compounds; this behavior is made use of both for analytical (constitution determinations) and preparative purposes[140]:

$$Ar-NH_2 \longrightarrow Ar-N_2^{\oplus} \xrightarrow[-N_2]{\text{Reduction}} Ar-H$$

In the classical method originated by P. Griess[141], which is still employed, diazonium salts are heated in anhydrous or aqueous ethanol. Here, the diazonium ions behave as oxidizing agents and are themselves reduced to the corresponding aromatic hydrocarbons while nitrogen is lost and the ethanol is oxidized to acetaldehyde:

$$Ar-N_2^{\oplus} + H_3C-CH_2-OH \xrightarrow{-N_2}$$

$$Ar-H + H_3C-CH=O + H^{\oplus} (+ Ar-OC_2H_5)$$

Only the hydrogen of the CH_2 group is transferred as has been shown by isotope-labeling experiments[142]. Aryl ethers may also be formed as

Table 2. Important methods for replacing the nitrogen in aromatic diazonium salts

Reactions involving loss of the diazonium nitrogen	Nucleophilic reactants	Page
Replacement by 1) Hydrogen $Ar-N_2^{\oplus} + H + e \longrightarrow Ar-H + N_2$	pot. hydride donor	178
2) Organic substituent groups $Ar-N_2^{\oplus} + Y=Z-H \xrightarrow{-H^{\oplus}} Ar-Z=Y + N_2$	π-electron systems	181
3) Halogens $Ar-N_2^{\oplus} + [X^{\ominus}] \longrightarrow Ar-X + N_2$	(complexed) halides	185
4) Oxygen- and sulfur-containing groups $Ar-N_2^{\oplus} + OR^{\ominus} \longrightarrow Ar-OR + N_2$ $Ar-N_2^{\oplus} + SR^{\ominus} \longrightarrow Ar-SR + N_2$	nucleophilic chalcogen groups	187
5) Nitrogen- and phosphorus- containing groups $Ar-N_2^{\oplus} + NR_3 \longrightarrow Ar-\overset{\oplus}{N}R_3 + N_2$ $Ar-N_2^{\oplus} + PR_3 \longrightarrow Ar-\overset{\oplus}{P}R_3 + N_2$	nucleophilic groups of trivalent N and P	188
Aryne formation followed by trapping reactions		189

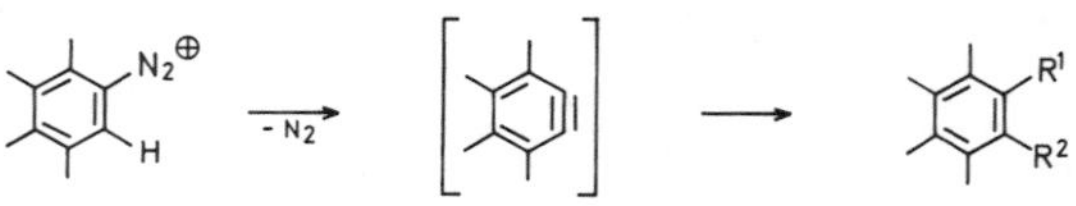

[137] *J. Reichel, R. Vidac*, Rev. Roumaine Chim. *15*, 1227 (1970).

[138] *B.S. Kikot, L.A. Kazitsyna, O.A. Reutov*, Izv. Akad. Nauk SSSR, Ser. Khim. 966 (1966); C.A. *65*, 18460a (1966).

[139] *R. Pütter* in *Houben-Weyl*, Methoden der organischen Chemie, 4. Aufl., Bd. X/3, p. 113, 114, Georg Thieme Verlag, Stuttgart 1965.

[140] *R. Pütter* in *Houben-Weyl*, Methoden der organischen Chemie, 4. Aufl., Bd. X/3, p. 115, Georg Thieme Verlag, Stuttgart 1965.

[141] *P. Griess*, Justus Liebigs Ann. Chem. *113*, 210 (1860); *P. Griess*, Justus Liebigs Ann. Chem. *117*, 63 (1861).

[142] *L. Melander*, Ark. Kemi *3*, 525 (1952).

by-products in very varying amounts, but no simple relationship is found between the nature of the

substituent groups on the aromatic nucleus and the course of the reaction[143]. In many reactions copper (in various oxidation states) has proven to be catalytically active[144], as have zinc dust and ammonium pentachloromolybdate[145].

A purely ionic mechanism is ascribed to the formation of ethers from diazonium salts and alcohols, whereas reductive deamination appears to proceed by a free-radical mechanism, at least under certain conditions. During decomposition of diazonium salts in methanol-D_1 a characteristic isotope effect which depended on the concentration of the sodium methoxide used was found[146]:

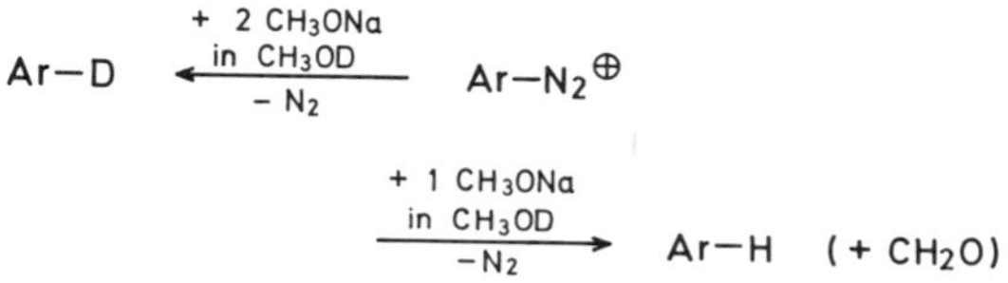

$$Ar-D \xleftarrow[\; -N_2 \;]{\substack{+\,2\,CH_3ONa \\ in\ CH_3OD}} Ar-N_2^{\oplus}$$

$$\xrightarrow[\; -N_2 \;]{\substack{+\,1\,CH_3ONa \\ in\ CH_3OD}} Ar-H \quad (+\,CH_2O)$$

Since (almost) no deuterium appears in the ring when molar amounts of sodium methoxide were added, it was concluded that a free-radical reaction occurs, which according to findings to date suggests an attack on the C–H bond:

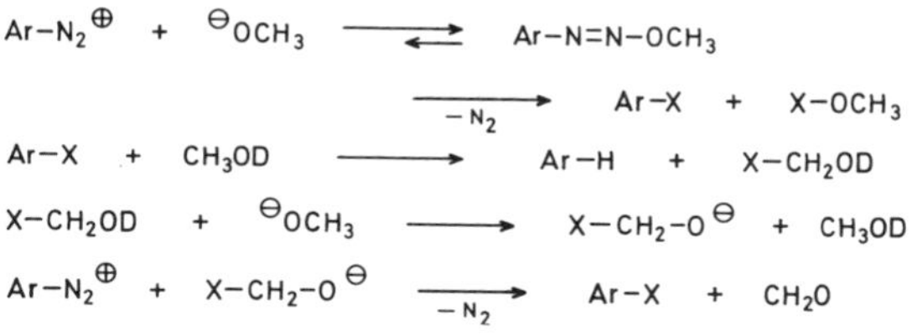

$$Ar-N_2^{\oplus} + {}^{\ominus}OCH_3 \rightleftharpoons Ar-N=N-OCH_3$$
$$\xrightarrow[-N_2]{} Ar-X + X-OCH_3$$
$$Ar-X + CH_3OD \longrightarrow Ar-H + X-CH_2OD$$
$$X-CH_2OD + {}^{\ominus}OCH_3 \longrightarrow X-CH_2-O^{\ominus} + CH_3OD$$
$$Ar-N_2^{\oplus} + X-CH_2-O^{\ominus} \xrightarrow[-N_2]{} Ar-X + CH_2O$$

The radical anion formulated should thus be regarded as the essential reaction intermediate, but it has not yet been identified.
Bunnett interprets the deuteration of the aromatic nucleus when excess methoxide was used as evidence for the involvement of an aryl anion. Two alternatives

are postulated for the formation of this anion, route **a** being regarded as the more likely:

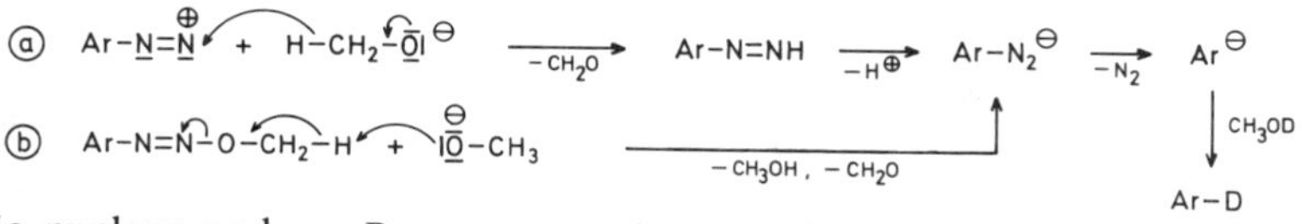

Route **a** can be regarded as a hydride transfer process[147], **b** as a fragmentation reaction. If the reaction is carried out in *tert*-butanol with *tert*-butoxide as base practically no reduction product is formed.

An important step is to suppress the formation of phenol ethers during the decomposition of diazonium salts in alcohols by adding alkali[148].
Other solvents which readily lose a hydrogen atom by a free-radical attack are also suitable for reductive deamination of diazonium ions. Water-soluble ethers such as tetrahydrofuran, 1,4-dioxane, *etc.*, which can simultaneously be oxidized to hemiacetals, are especially suitable[149]:

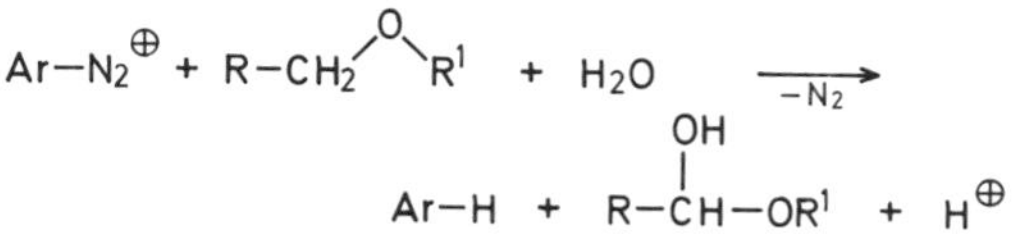

Tertiary amines react correspondingly[149].
Intramolecular hydrogen transfer has been observed during the decomposition of diazotized *N,N*-disubstituted anthranilamides[150]; other reaction products are also formed:

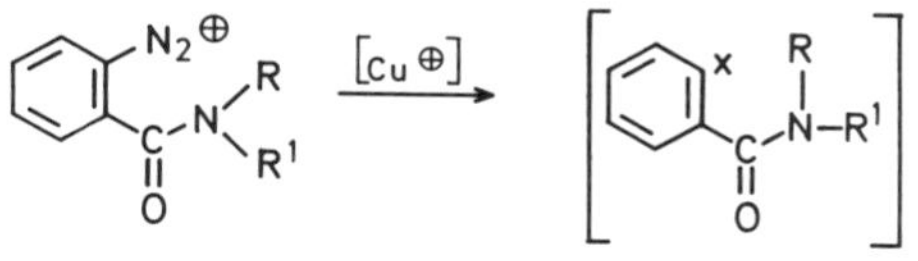

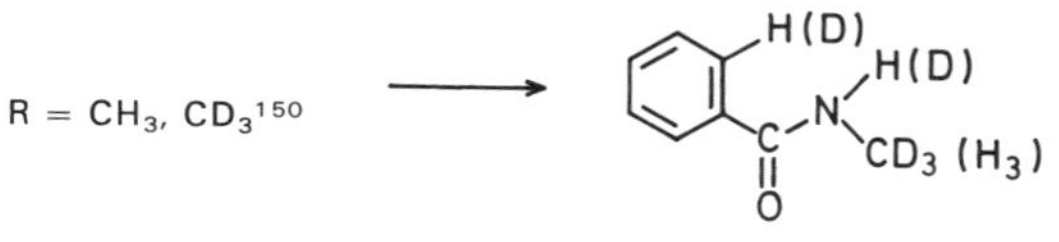

$R = CH_3, CD_3$[150]

With 2,3,5,6-tetrachlorobenzenediazonium sulfate it was possible to effect a reductive deamina-

[143] R. Pütter in Houben-Weyl, Methoden der organischen Chemie, 4. Aufl., Bd. X/3, p. 124, Georg Thieme Verlag, Stuttgart 1965.
[144] R. Pütter in Houben-Weyl, Methoden der organischen Chemie, 4. Aufl., Bd. X/3, p. 119, Georg Thieme Verlag, Stuttgart 1965;
A. Mineo, Corriere Farm. 21, 318 (1966); C.A. 67, 43617g (1967).
[145] R. Pütter in Houben-Weyl, Methoden der organischen Chemie, 4. Aufl., Bd. X/3, p. 122, Georg Thieme Verlag, Stuttgart 1965; s. a. Lit. [175];
R. Korewa, F. Karczynski, Roczniki Chem. 38, 705 (1964); C.A. 62, 1582e (1965).

[146] J.F. Bunnett, H. Takayama, Chem. Commun. 367 (1966);
J.F. Bunnett, D.A.R. Happer, J. Org. Chem. 32, 2701 (1967);
J.F. Bunnett, H. Takayama, J. Org. Chem. 33, 1924 (1968);
J.F. Bunnett, H. Takayama, J. Amer. Chem. Soc. 90, 5173 (1968).
[147] M. Bloch, H. Musso, U. Zahorsky, Angew. Chem. 81, 392 (1969);
M. Bloch, H. Musso, U. Zahorsky, Angew. Chem. Intern. Ed. Engl. 8, 370 (1969) (Hydride transfer with NaBH₄); A. Rieker, P. Niederer, D. Leibfritz, Tetrahedron Lett. 4287 (1969).

tion even by using a mixture of acetic and sulfuric acids; *2,3,5,6-tetrachlorophenylacetic acid* is formed as well as *1,2,4,5-tetrachlorobenzene*[151].

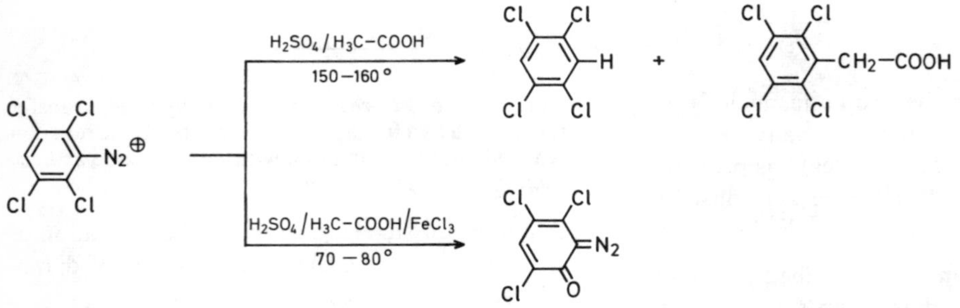

The addition of ferric chloride as oxidizing agent affords *2,4,5-trichloro-6-diazo-2,5-cyclohexadien-1-one.*

Although a large number of other reducing agents[152] have been used for deamination of diazonium salts, only hypophosphorous acid[153] and its sodium salt[154] have achieved real importance, and are generally superior to the alcohols for reduction; *e.g.:*

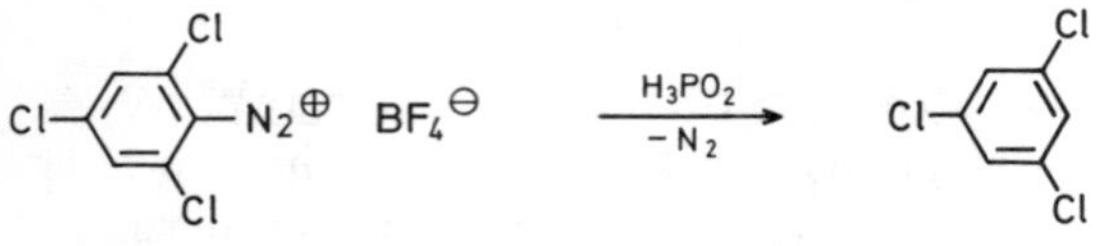

1,3,5-Trichlorobenzene

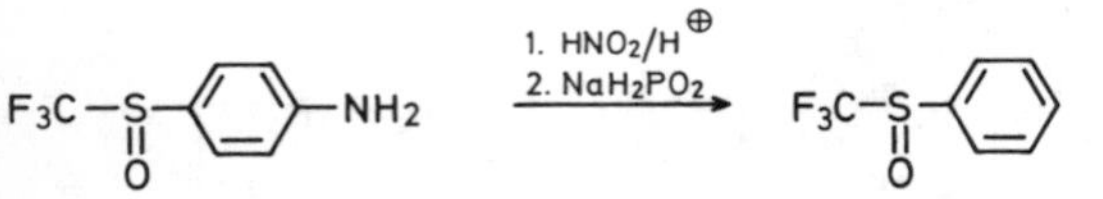

Phenyl trifluoromethyl sulfoxide

It remains to be seen how much use will be made of new reduction methods involving silyl and stannyl hydrides[155]:

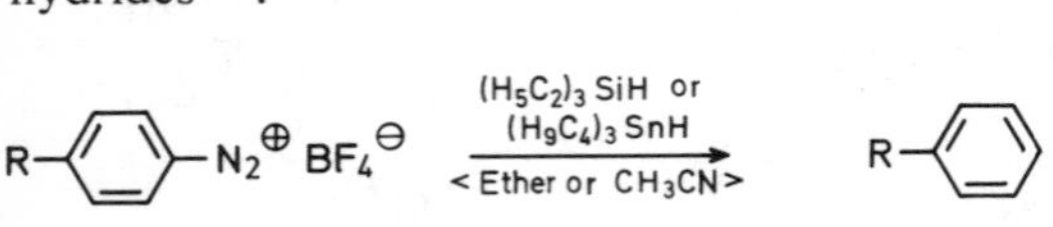

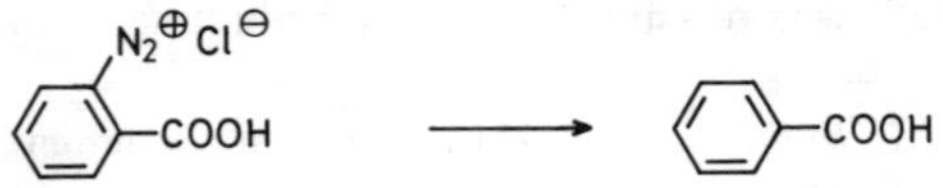

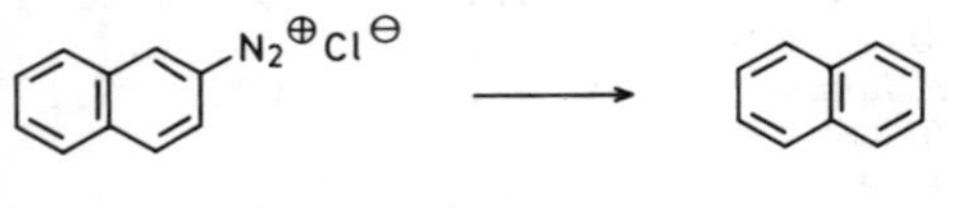

7.2.1.2 Replacement of the diazonium group by organic groups

Just as the diazonium group can be replaced by hydrogen, its replacement by organic groups, involving *C—C* linking, is also possible. Particularly

of interest is the *Gomberg-Bachmann reaction, i.e.,* the *arylation* of aromatic hydrocarbons (and their derivatives[156]).

7.2.1.2.1 Arylation of aromatic and heteroaromatic compounds

In the case of aliphatic compounds, alkyl halides, sulfates or sulfonates are generally used as alkylating agents; corresponding reactions of aromatic derivatives do not occur because of the stability of the bonds between the aromatic nuclei and the leaving groups. Because the nitrogen atoms of the diazonium moiety constitute an exceptionally efficient leaving group, diazonium salts provide a

[148] *R. Pütter* in *Houben-Weyl*, Methoden der organischen Chemie, 4. Aufl., Bd. X/3, p. 127, 128, Georg Thieme Verlag, Stuttgart 1965.

[149] *H. Meerwein, H. Allendörfer, P. Beckmann, F. Kunert, H. Morschel, F. Pawellek, K. Wunderlich,* Angew. Chem. **70**, 211 (1958);
W. Rüdiger, Chr. Rüchardt, Tetrahedron Lett. 2407 (1969).

[150] *A.H. Lewin, A.H. Dinwoodie, Th. Cohen,* Tetrahedron **22**, 1527 (1966);
Th. Cohen, C.H. McMullen, K.Smith, J. Amer. Chem. Soc. **90**, 6866 (1968);
E.J. Forbes, C.J. Gray, Tetrahedron **24**, 6223 (1968);
H. Suschitzky, C.F. Sellers, Tetrahedron Lett. 1105 (1969).

[151] *R. Nishiyama,* Tetrahedron Lett. 1403 (1967);
R. Nishiyama, Yuki Gosei Kagaku Kyokai Shi **25**, 681 (1967); C.A. **67**, 90504c (1967).

[152] *R. Pütter* in *Houben-Weyl*, Methoden der organischen Chemie, 4. Aufl., Bd. X/3, p. 135, Georg Thieme Verlag, Stuttgart 1965.

[153] *B. Lamm, B. Andersson,* Ark. Kemi **25**, 367 (1966);
R. Pütter in *Houben-Weyl*, Methoden der organischen Chemie, 4. Aufl., Bd. X/3, p. 131, Georg Thieme Verlag, Stuttgart 1965;
N. Kornblum, Org. Reactions **2**, 262 (1947).

[154] *L.M. Yagupol'skii, M.S. Marenets, N.V. Kondratenko,* Zh. Obshch. Khim. **35**, 377 (1965); C.A. **62**, 14551c (1965);
P. Ruggli, A. Staub, Helv. Chim. Acta **20**, 52 (1937).

[155] *J. Nakayama, M. Yoshida, O. Simamura,* Tetrahedron **26**, 4609 (1970).

[156] *W.E. Bachmann, R.A. Hoffman,* Org. Reactions **2**, 224 (1947).

valuable alternative. Formally, an ionic mechanism should be expected for the reaction:

$$Ar-N_2^{\oplus} \; + \; Ar'-H \longrightarrow Ar-Ar' \; + \; N_2 \; + \; H^{\oplus}$$

Free aryl cations, which are regarded as *hard acids*[157] according to Pearson's concept, tend to prefer to react with *hard bases* (such as $OH^{\ominus}$, H_2O, ROH), however, and are not very suitable for arylating aromatic compounds (or olefins). However, if reaction conditions are chosen under

which aryl radicals (behaving as *soft acids*) are formed rather than aryl cations, then added aromatic compounds can be readily alkylated. For this purpose, the diazonium salt solution is adjusted to a weakly alkaline *pH* and at the same time is stirred with the aromatic compound to be arylated[158]. The diazo anhydride (which is insoluble in the aqueous phase) is formed[159], and is immediately dissolved by the organic phase in which it decomposes to form nitrogen and a radical:

$$ArX \; + \; Ar'H \longrightarrow Ar-Ar' \; + \; HX$$

The aryl free-radical generated arylates the aromatic coreactant Ar'H concerned, preferably in

the *o*-position, irrespective of its nature as a Type 1 or 2 substituent[160]. Pyridine can be arylated analogously[161]. Carbocyclic aromatic compounds are, however, more reactive, as is indicated by the use of a mixture of pyridine and a diazonium fluoroborate to arylate aromatic compounds in homogeneous medium (added solvent: acetonitrile or sulfolane)[162]

The reaction of an aniline with pentyl nitrite in an aromatic hydrocarbon and the Kobayashi technique[164] are further variants for arylation in homogeneous solution[163, 164]. It seems that in some cases an ionic substitution mechanism cannot entirely be excluded[165]. A tabular survey of reactions involving arylation of aromatic compounds with aromatic diazonium salts and other potential diazonium group derivatives is given in ref.[2] (p. 148 ff.).

Aromatic diazonium salts can be dimerized reductively to biphenyl derivatives by reducing agents, among them cuprous salts[166], *e.g.*:

2,2'-Biphenyl-dicarboxylic acid

[157] *R.G. Pearson*, J. Amer. Chem. Soc. *85*, 3533 (1963);
R.G. Pearson, Science *151*, 172 (1966);
R.G. Pearson, Chem. Eng. News *43*, Nr. 22, p. 90 (1965);
B. Saville, Angew. Chem. *79*, 966 (1967).
[158] *R. Pütter* in *Houben-Weyl*, Methoden der organischen Chemie, 4. Aufl., Bd. X/3, p. 145, Georg Thieme Verlag, Stuttgart 1965.
[159] *Chr. Rüchardt, E. Merz*, Tetrahedron Lett. 2431 (1964).
[160] *D.F. De Tar, H.J. Scheifele*, J. Amer. Chem. Soc. *73*, 1442 (1951).
[161] *J.W. Haworth, J.M. Heilbron, D.H. Hey*, J. Chem. Soc. (London) 349, 1284 (1940).
[162] *R.A. Abramovitch, J.G. Saha*, Can. J. Chem. *43*, 3269 (1965);
R.A. Abramovitch, J.G. Saha, Tetrahedron *21*, 3297 (1965);
R.A. Abramovitch, F.F. Gadallah, J. Chem. Soc. B 497 (1968);
R.A. Abramovitch, O.A. Koleoso, J. Chem. Soc. B 1292 (1968).

[163] *J.I.G. Cadogan*, J. Chem. Soc. (London) 4257 (1962);
L. Freedman, J.F. Chlebowski, J. Org. Chem. *33*, 1633 (1968);
C.M. Camaggi, R. Leardini, M. Tiecco, A. Tundo, J. Chem. Soc. B 1683 (1970).
[164] *M. Kobayashi, H. Minato, E. Yamada, N. Kobori*, Bull. Chem. Soc. Japan *43*, 215 (1970);
M. Kobayashi, H. Minato, N. Watanabe, N. Kobori, Bull. Chem. Soc. Japan *43*, 258 (1970).
[165] *R. Pütter* in *Houben-Weyl*, Methoden der organischen Chemie, 4. Aufl., Bd. X/3, p. 148, Georg Thieme Verlag, Stuttgart 1965;
B.L. Kaul, H. Zollinger, Helv. Chim. Acta *51*, 2132 (1968);
A.J. Sisti, R.L. Cohen, Can. J. Chem. *44*, 2580 (1966) (1, 3-rearrangement!).
[166] *R. Pütter* in *Houben-Weyl*, Methoden der organischen Chemie, 4. Aufl., Bd. X/3, p. 160, Georg Thieme Verlag, Stuttgart 1965;
E.R. Atkinson, H.J. Lawler, Org. Syntheses Coll. Vol. I, 222 (1941) (Diphenic acid).

In connection with this reaction, an interesting mechanistic study has been carried out on the reaction of 4-methylbenzenediazonium tetrafluoroborate in sulfolane with cuprous benzoate in diglyme[167]:

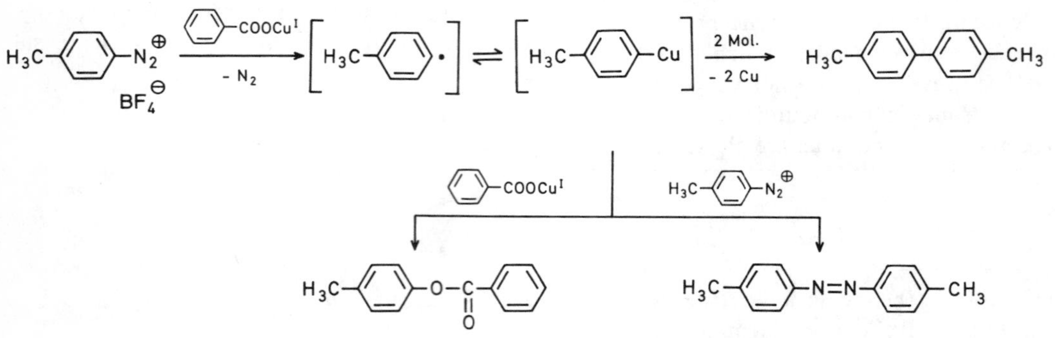

If the redox catalysts used are cupric salts, unsaturated compounds can be arylated in particularly good yields. When diazonium chlorides and cupric

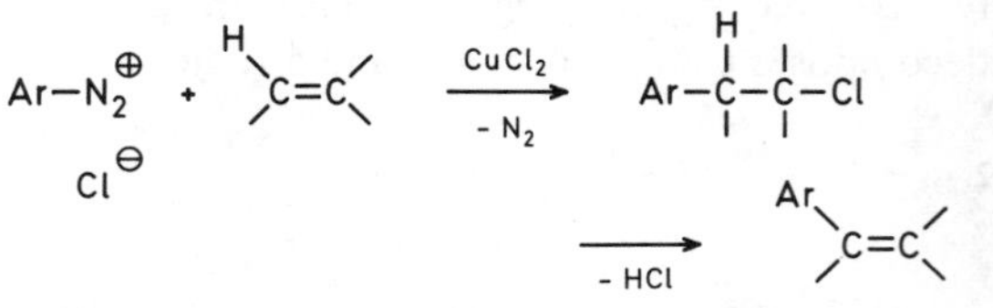

7.2.1.2.2 Arylations of olefins

In the presence of a redox catalyst, olefins (especially α,β-unsaturated carbonyl compounds) can readily be arylated with diazonium salts. These arylations proceed especially easily with *quinones* and their derivatives[168]; quinhydrones which are present in traces catalyze the reactions:

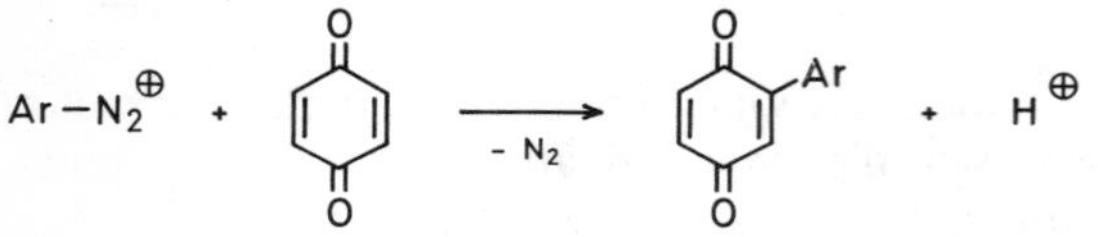

chloride are used, *chloroarylation (Meerwein reaction)* occurs, and in many cases is followed by spontaneous elimination of hydrogen halide to form *arylolefins*[169]:

Since cuprous ions and chloride or bromide ions are indispensable for this reaction, the following mechanism has been assumed[170]:

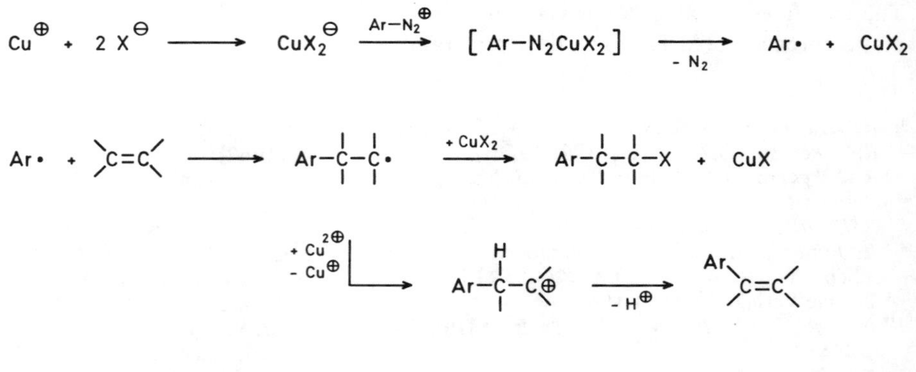

[167] *Th. Cohen, A.H. Lewin*, J. Amer. Chem. Soc. *88*, 4521 (1966).
[168] *R. Pütter* in *Houben-Weyl,* Methoden der organischen Chemie, 4. Aufl., Bd. X/3, p. 167, Georg Thieme Verlag, Stuttgart 1965;
J. Asselin, P. Brassard, P.L'Ecuyer, Can. J. Chem. *44*, 2563 (1966).
[169] *R. Pütter* in *Houben-Weyl,* Methoden der organischen Chemie, 4. Aufl., Bd. X/3, p. 171, Georg Thieme Verlag, Stuttgart 1965 (Tabellen p. 178–183);

In the Sandmeyer reaction[171] (see p. 186), the aryl radical interacts directly with the cupric salt, whereas in the Meerwein reaction it behaves as a *soft acid* in adding preferentially to the *soft base*, the olefin. Since electron-deficient olefins and conjugated diolefins and polyolefins are *softer bases* than ethylene, propene, *etc.,* they undergo arylation especially well.

7.2.1.2.3 Arylation of azomethines

Like $C=C$ double bonds, $C=N$ double bonds can also act as *soft bases* in being arylated by the mechanism described above; indeed, phenylhydrazones react even in the absence of catalyst, while semicarbazones and oximes react under Meerwein reaction conditions[172]:

$$Ar-N_2^{\oplus} \;+\; \underset{H}{\overset{R}{\underset{\parallel}{\overset{N}{}}}}\!C\!-\!R^1 \xrightarrow{-N_2} \underset{Ar}{\overset{R}{\underset{\parallel}{\overset{N}{}}}}\!C\!-\!R^1 \;+\; H^{\oplus}$$

$R = NH-C_6H_5 , OH , NH-CO-NH_2$

$R^1 = H , Alkyl , Aryl , Acyl$

The synthesis of *aryloxysulfonium salts* from sulfones and diazonium salts can be regarded as involving arylation of $S=O$ double bonds[173]:

$$R^1-\!\!\!\!\bigcirc\!\!\!\!-N_2^{\oplus} X^{\ominus} + \underset{R^3}{\overset{R^2}{\underset{}{S}}}\!\!\overset{O}{\underset{O}{\rlap{$=$}}} \xrightarrow[-N_2]{\triangledown} \underset{R^3}{\overset{R^2}{\underset{\oplus}{S}}}\!\!\overset{O}{\underset{}{}}O-\!\!\!\!\bigcirc\!\!\!\!-R^1$$

$$(+ R^1-\!\!\!\!\bigcirc\!\!\!\!-F)$$

$$X = BF_4$$

7.2.1.2.4 Internuclear cyclization of polynuclear diazonium salts

Intramolecular arylations of aryldiazonium salts can occur in cases where a second aryl group is joined to the position *ortho* to the diazonium group *via* a bridge containing one or two members. Depending on the nature of the link, the reaction can form 5- or 6-membered *carbocycles* or *heterocycles*[174]. Many such cyclizations (especially those forming 5-membered rings) appear to proceed by ionic mechanisms, presumably on account of the close proximity of the coupling positions, as is indicated by good results in mineral acid media and poor results if the reaction is carried out under alkaline conditions. The most important types of cyclization are the following:

G.H. Cleland, J. Org. Chem. *34*, 744 (1969) (α,β-unsaturated carboxylic acids and esters);
R.K. Freidlina, B.V. Kopylova, L.V. Yashkina, Dokl. Akad. Nauk SSSR *183*, 1113 (1968); C.A. *70*, 67934p (1969) (α,β-unsaturated nitrils);
C. Nakashima, S. Tanimoto, R. Oda, Kogyo Kagaku Zasshi *67*, 1705 (1964); C.A. *62*, 10357f (1965);
C. Nakashima, S. Tanimoto, R. Oda, Nippon Kagaku Zasshi *86*, 442 (1965); C.A. *63*, 8239g (1965) (α,β-unsaturated sulfo-compounds);
Y. Wada, R. Oda, Kogyo Kagaku Zasshi *67*, 2093 (1964); C.A. *62*, 13177e (1965) (Vinyl phosphonate);
V.M. Naidan, Nauk Zap., Chernivets'k Derzh. Univ., Ser. Prirodn. Nauk *51*, 40 (1961); C.A. *62*, 10353g (1965) (Vinylidene halogenides);
G. Rockstroh, F. Wolf, G. Schwachula, Z. Chem. *11*, 61 (1971);
V.M. Naidan, A.V. Dombrovskii, Zh. Org. Khim. *2*, 888 (1966); C.A. *65*, 13580b (1966);
N.I. Ganushak, N.F. Stadmichuk, A.V. Dombrovskii, Zh. Org. Khim. *5*, 691 (1969); C.A. *71*, 21764h (1969) (Vinyl chlorides);
Chiu-Hsien Wang, Chi Chou Ch'en, Chung-Chie Li, K'o Hsueh T'ung Pao *17*, 419 (1966); C.A. *66*, 37707z (1967) (Furans);
R.Frimm, J.Kovao, A. Jurasek, Sb. Pr. Chem. Fak. Slov. Vyss. Sk. Tech. 35 (1967); C.A. *69*, 106365n (1968);
S. Yoshina, A. Tanaka, G.-N. Yang, H.-C. Hsiu, Yakugaku Zasshi *90*, 1150 (1970);
H.D. Hoffmann, W. Müller, Biochim. Biophys. Acta *123*, 421 (1966) (Guanosine);
N.I. Ganushchak, K.G. Zolotukhina, A.V. Dombrovskii, Zh. Org. Khim. 2, 1066 (1966); C.A. *65*, 18510f (1966) (conjugated dienes);
N.I. Ganushchak, V.A. Vengrzhanovskii, A.V. Dombrovskii, Zh. Org. Khim. *5*, 113 (1969); C.A. *70*, 87181b (1969);
H.S. Mehra, J. Indian Chem. Soc. *45*, 178 (1968) (Acrylic acid ester);
N.O. Pastushak, A.V. Dombrovskii, A.N. Mukhova, Zh. Org. Khim. *1*, 572 (1965); C.A. *63*, 1727c (1965);

A.V. Dombrovskii, B.S. Fedorov, Z.V. Belitskaya, Probl. Poluch. Poluprod. Prom. Org. Sin., Akad. Nauk SSSR, Otd. Obshch. Tekh. Khim. 173 (1967) C.A. *68*, 77903h (1968) (Acrylonitrile).
[170] *J.K. Kochi*, J. Amer. Chem. Soc. *78*, 1228 (1956); *J.K. Kochi*, J. Amer. Chem. Soc. *79*, 2944 (1957).
[171] *E.Pfeil*, Angew.Chem.*65*, 155(1953)(Mechanisms); *S.C. Dickerman, D.J. De Souza, N. Jacobson*, J. Org. Chem. *34*, 710 (1969).
[172] *R. Pütter* in *Houben-Weyl*, Methoden der organischen Chemie, 4. Aufl., Bd. X/3, p. 183, Georg Thieme Verlag, Stuttgart 1965.
[173] *G.R. Chalkley, D.J. Snodin, G. Stevens, M.C. Whiting*, J. Chem. Soc. C 682 (1970).
[174] *D.F. De Tar*, Org. Reactions 9, 409 (1957); *R. Pütter* in *Houben-Weyl*, Methoden der organischen Chemie, 4. Aufl., Bd. X/3, p. 188, Georg Thieme Verlag, Stuttgart 1965.
[175] *A.H. Lewin, Th. Cohen*, J. Org. Chem. *32*, 3844 (1967); *R. Pütter* in *Houben-Weyl*, Methoden der organischen Chemie, 4. Aufl., Bd. X/3, p. 195, Georg Thieme Verlag, Stuttgart 1965.

7.2.1.2.4.1 Formation of carbocyclic five-membered rings, fluorene derivatives[175]

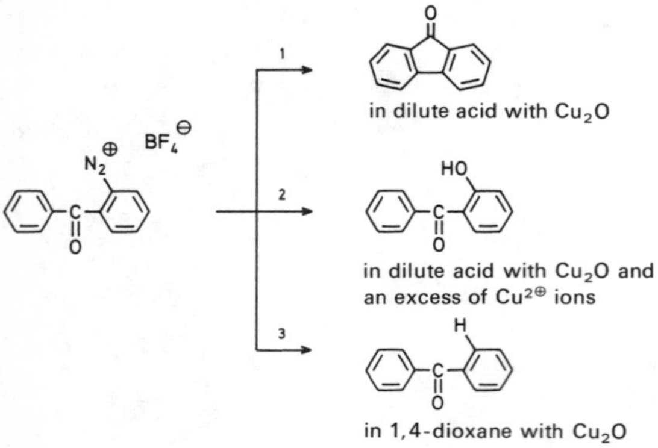

in dilute acid with Cu_2O

in dilute acid with Cu_2O and an excess of $Cu^{2\oplus}$ ions

in 1,4-dioxane with Cu_2O

A reaction following path **1** has also been achieved by using a mixture of acetone and pyridine as reaction medium[176].

On thermolysis in mineral acid medium, 2-alkyl-

$R = C_2H_5$, $R^1 = 4'$-OCH_3

$R = C_2H_5$, $R^1 = 3'$-OCH_3

$R = CH_3$, $R^1 = 2'$-OCH_3

benzenediazonium sulfonates yield not only phenols, but also *indan* derivatives[177]:

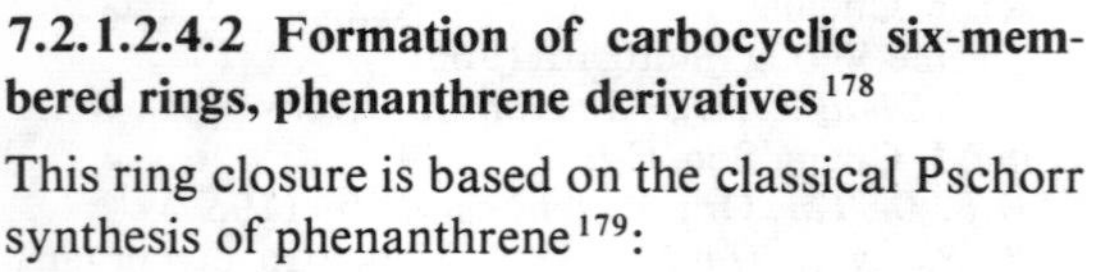

$R = R^1 = H$; *2-Propylphenol*; 87% + *Indan*; 13%

$R = R^1 = CH_3$; *2-(2,2-Dimethylpropyl)phenol*
            + *2,2-Dimethylindan*; 88%

7.2.1.2.4.2 Formation of carbocyclic six-membered rings, phenanthrene derivatives[178]

This ring closure is based on the classical Pschorr synthesis of phenanthrene[179]:

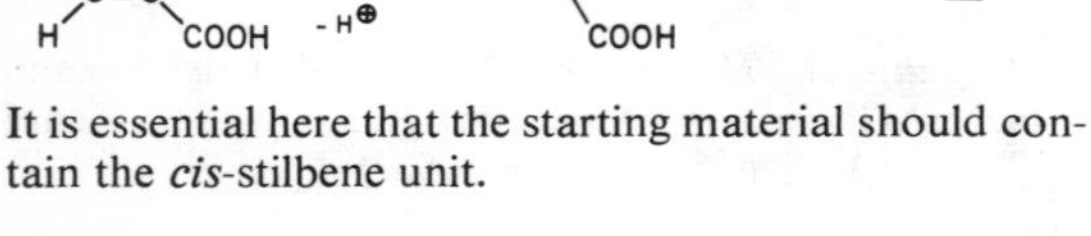

It is essential here that the starting material should contain the *cis*-stilbene unit.

7.2.1.2.4.3 Formation of heterocyclic five-membered rings[180]

A classical example of this reaction type is the synthesis of *carbazole*[181]:

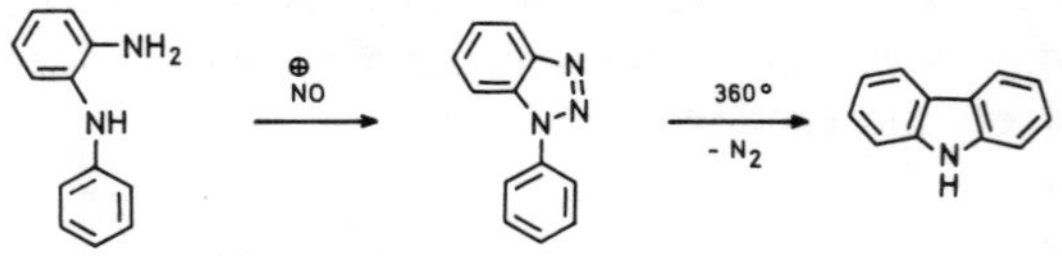

The *1-phenyl-1H-1,2,3-benzotriazole* which is formed as the primary product on diazotization loses nitrogen at an elevated temperature and forms carbazole.

Diazotized anthranilanilides react with sodium iodide in acetone to form spiro-cyclohexadienyl radicals, which can undergo various further reactions[182]:

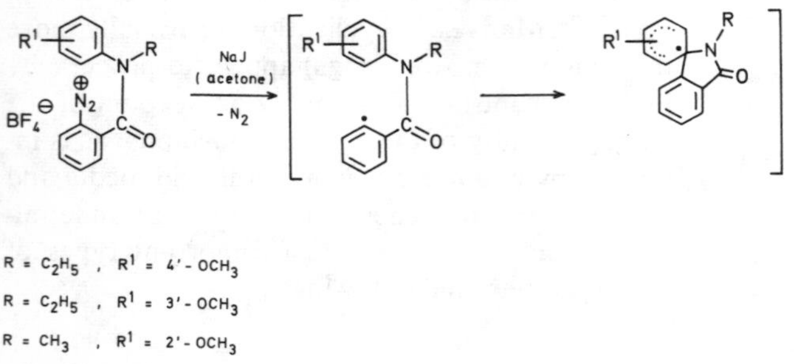

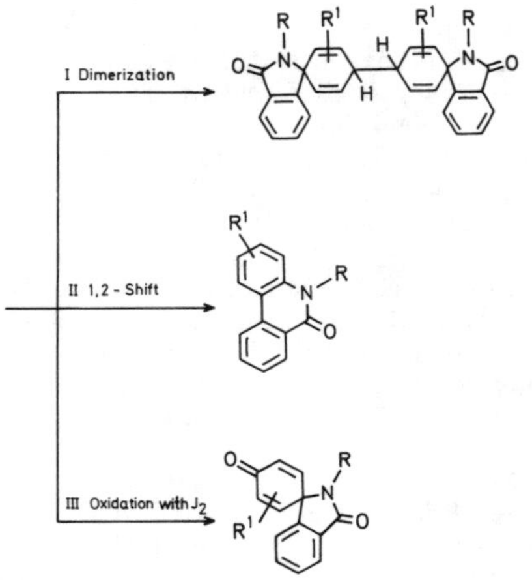

I Dimerization

II 1,2 - Shift

III Oxidation with J_2

[176] *R.A. Abramovitch, A. Robson,* J. Chem. Soc. C 1101 (1967).

[177] *P. Martinson,* Acta Chem. Scand. *22,* 1357 (1968).

[178] *R. Pütter* in *Houben-Weyl,* Methoden der organischen Chemie, 4. Aufl., Bd. X/3, p. 198, Georg Thieme Verlag, Stuttgart 1965;
P.H. Leake, Chem. Rev. *56,* 27 (1956).

[179] *R. Pschorr,* Angew. Chem. *43,* 245 (1930).

[180] *R. Pütter* in *Houben-Weyl,* Methoden der organischen Chemie, 4. Aufl., Bd. X/3, p. 196, Georg Thieme Verlag, Stuttgart 1965.

[181] *C. Graebe, F. Ullmann,* Justus Liebigs Ann. Chem. *291,* 16 (1896).

[182] *D.H. Hey, G.H. Jones, M.J. Perkins,* Chem. Commun. 1438 (1970);
D.H. Hey, J.A. Leonard, C.W. Rees, A.R. Todd, J. Chem. Soc. C 1513 (1967);
D.H. Hey, C.W. Rees, A.R. Todd, J. Chem. Soc. C 1518 (1967);
D.M. Collington, J. Chem. Soc. C 1017, 1021 (1968) (Spirocyclohexadienes).

Corresponding hydrazides of diazotized anthranilic acids yield *3-indazolin-3-ones*[183].

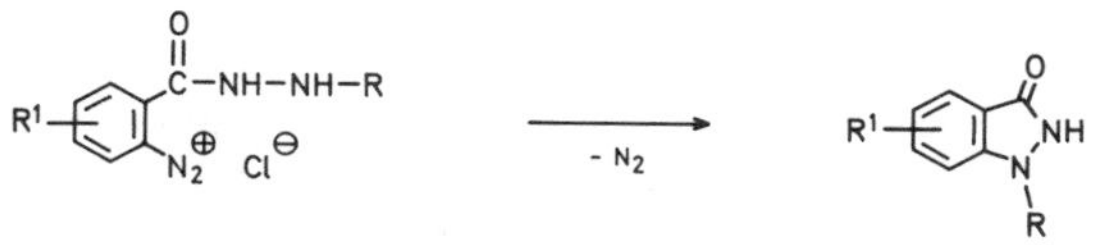

For the synthesis of heterocycles, too, the variant in pyridine[176] has been applied successfully.

7.2.1.2.4.4 Six- and seven-membered heterocyclic rings[184]

It is sometimes found that an intermediate azo coupling product is formed first, which can then undergo rearrangement in the presence of a copper catalyst[185], *e.g.*:

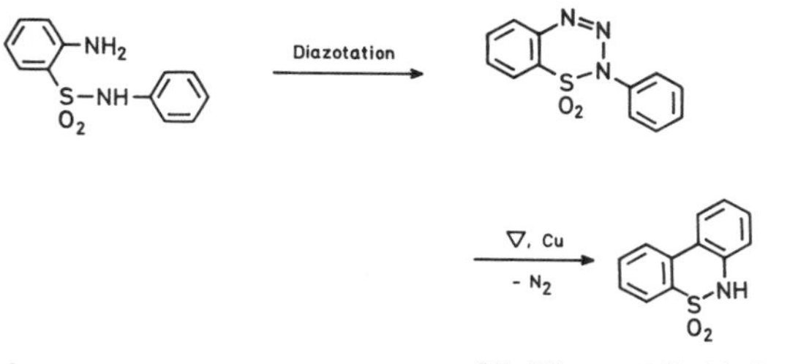

6H-Dibenzo-1,2-thiazine 5,5-dioxide

8H-Benzo[c]naphtho[1,8a, 8e,f]-1,2-thiazepine 7,7-dioxide (which is a sultam containing seven-membered ring) can be prepared analogously from 1-anilinosulfonylnaphthalene-8-diazonium chloride[186].

---

[183] *T. Kametani, K. Sota, M. Shio,* J. Heterocycl. Chem. *7*, 815 (1970).

[184] *R. Pütter* in *Houben-Weyl*, Methoden der organischen Chemie, 4. Aufl., Bd. X/3, p. 202, Georg Thieme Verlag, Stuttgart 1965.

[185] *F. Ullmann, C. Gross,* Ber. *43*, 2694 (1910).

[186] *R.E. Steiger,* Bull. Soc. Chim. France [4] *53*, 1254 (1933).

[187] *G. Balz, G. Schiemann,* Ber. *60*, 1186 (1927).

[188] *E. Forche* in *Houben-Weyl*, Methoden der organischen Chemie, 4. Aufl., Bd. V/3, p. 213, 232, Georg Thieme Verlag, Stuttgart 1962;
T. Nozaki, Kagaku No Ryoiki *19*, 764 (1965); C.A. *69*, 26370v (1968);
L.M. Yagupol'skii, M.S. Marenets, N.V. Kondratenko, Zh. Obshch. Khim. *35*, 377 (1965); C.A. *62*, 14551c (1965);
V.M. Ivanova, Z.N. Seina, S.A. Nemleva, S.S. Gitis, Zh. Obshch. Khim. *36*, 127 (1966); C.A. *64*, 14125d (1966);
J.I. De Graw, M. Cory, W.A. Skinner, J. Chem. Eng. Data *13*, 587 (1968); C.A. *69*, 106150p (1968);
N. Ishikawa, S. Sugawara, T. Watanage, Kogyo Kagaku Zasshi *71*, 519 (1968); C.A. *69*, 106168a (1968).

[189] *T. Nozaki, Y. Tanaka,* Intern. J. Appl. Radiation Isotopes *18*, 111 (1967); C.A. *67*, 11283d (1967).

[189a] *S. Yasuda, K. Yamanaka, T. Kurihara, N. Saito,* Yakugaku Zasshi *88*, 209 (1968); C.A. *69*, 67022b (1968).

7.2.1.3 Replacement of the diazonium group by groups of other elements

7.2.1.3.1 Replacement by halogen

Because of the recent increase in interest in organofluorine compounds, replacement of the diazonium group in aromatic compounds *(Balz-Schiemann reaction*[187]*)* has become a procedure of growing importance. The Sandmeyer reaction, proven for introducing chloro-, bromo- or cyanogroups, forms phenols when fluoride is used rather than aromatic fluoro compounds. On the technical scale it is better to prepare aromatic fluoro compounds by diazotization of aromatic amines in *anhydrous hydrofluoric acid*, followed by thermolysis of the resulting diazonium fluorides. For laboratory work, however, the Balz-Schiemann reaction, in which the dry diazonium *tetrafluoroborate* is thermolyzed (sometimes in the presence of sodium fluoride), has achieved greater importance[188]:

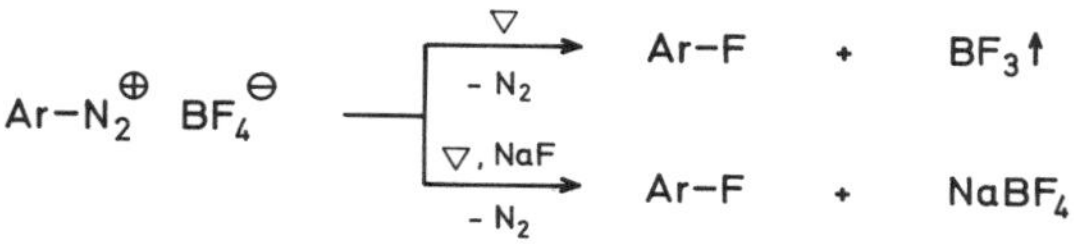

This method provides an especially simple means of preparing [18]F-*labeled* aryl fluorides[189].

An interesting rearrangement has been found to occur on thermolysis of 2-nitrobenzenediazonium tetrafluoroborate in xylene[189a]:

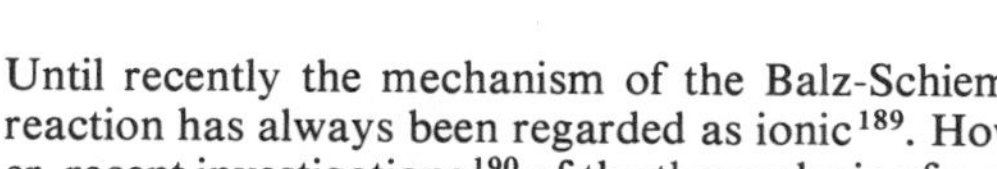

The sole product obtained was *4-fluoro-1-nitrobenzene*, which is naturally also obtainable from 4-nitrobenzenediazonium tetrafluoroborate, and not the expected *o*-isomer.

Until recently the mechanism of the Balz-Schiemann reaction has always been regarded as ionic[189]. However, recent investigations[190] of the thermolysis of a number of aryldiazonium tetrafluoroborates and *hexafluorophosphates* directly in the resonator of an ESR apparatus provided evidence for the occurrence of a *radical* type of process.

---

[190] *T.K. Al'sing, A.G. Boldyrev,* Zh. Org. Khim. *3*, 2249 (1967); C.A. *68*, 68264y (1968).

[191] *L.G. Makarova, I.V. Polovyanyuk,* Izv. Akad. Nauk SSSR, Ser. Khim. 2750 (1967); C.A. *69*, 18733a (1968).

Besides the hexafluorophosphates mentioned, *hexafluorogermanates*[191], *hexafluoroarsenates*, and *hexafluoroantimonates* can be used for introducing fluorine; the antimony compounds provide better yields of aromatic fluoro compounds than the other complex ions[192]. Organoarsenic compounds are also formed, in a side-reaction, when hexafluoroarsenates are employed:

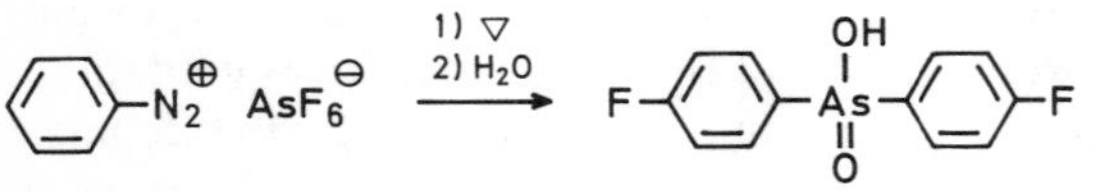

Bis-(4-fluorophenyl)hydroxy-arsine oxide

During decomposition of 4-nitrobenzenediazonium tetrafluoroborate in the dipolar aprotic dimethyl sulfoxide, it has been found that a characteristic halide-catalysis occurs; the yields of aryl halide obtained increase as the nucleophilism of the halide ion increases[193]:

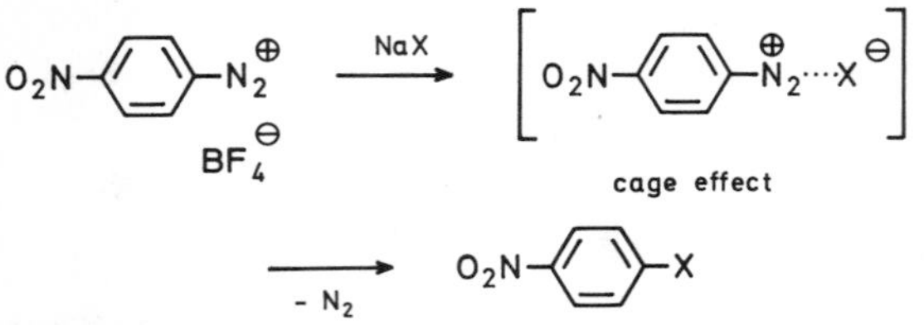

cage effect

X = F;	4-Fluoro-		0%
X = Cl;	4-Chloro-	1-nitrobenzene	$24 \pm 5\%$
X = Br;	4-Bromo-		$70 \pm 5\%$
X = I;	4-Iodo-		$85 \pm 5\%$

Mechanisms of the replacement of the diazonium groups by numerous nucleophiles have been discussed in the literature[194].

The available chemical evidence indicates that the *softer* the entering group is as a base, the more easily does it effect uncatalyzed replacement of the diazonium group. In the presence of their anions, the halogens bromine and iodine tend to form perhalides $Br_3^{\ominus}$[195] or $I_3^{\ominus}$[196], which behave as particularly *soft bases* capable of replacing the diazonium group directly. *Aryl iodides* can consequently be prepared, using pentyl nitrite/I_2, without isolating the intermediate diazonium compound[197]. The nucleophilism of iodine is so intense that even aryl iodides can be arylated with diazonium salts, iodonium salts being formed[198]:

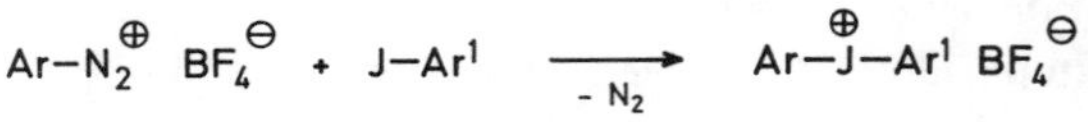

The free *bromide ion* is not suitable for replacement of the diazonium group for preparative purposes, but can be activated with mercuric bromide, as well as by complexing with molecular bromine[199]. On the other hand, decomposition of the corresponding diazonium trichloromercurates in the presence of copper powder leads to *arylmercury chlorides* (the Nesmeyanov reaction)[200]:

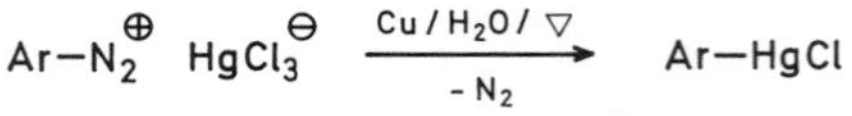

The classical Sandmeyer reaction is still the most important method for preparing *aryl chlorides* and *bromides* from diazonium salts[201]; the catalytic activity of the added copper salts arises from their complex-forming and charge-transfer capacity. In its mechanism this reaction corresponds to the Meerwein arylation (see p. 182):

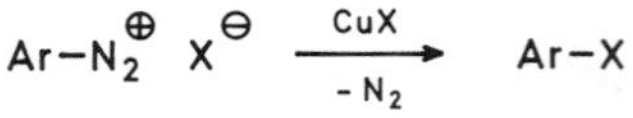

The Sandmeyer reaction can, however, be used to introduce groups other than chloride and bromide onto aromatic nuclei see ref. [139] (Tab. 13, p. 113). 2-Substituted thiazoles can be prepared from thiazole-2-diazonium tetrafluoroborates[202].

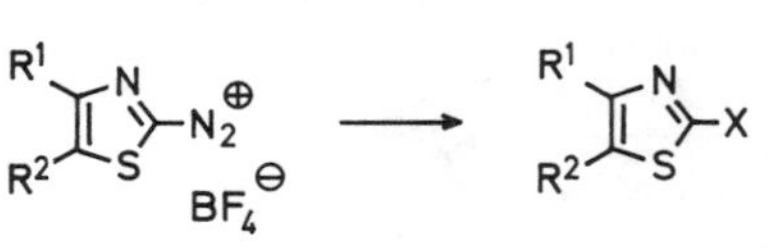

$X = Cl, Br, J, SCN, SH, (N_3)$

[192] *C. Sellers, H. Suschitzky*, J. Chem. Soc. C 2317 (1968).

[193] *M.D. Johnson*, J. Chem. Soc. (London) 805 (1965).

[194] *E.S. Lewis, J.E. Cooper*, J. Amer. Chem. Soc. *84*, 3847 (1962);
E.S. Lewis, L.D. Hartung, B.M. McKay, J. Amer. Chem. Soc. *91*, 419 (1969).

[195] *A. Roedig* in *Houben-Weyl*, Methoden der organischen Chemie, 4. Aufl., Bd. V/4, p. 448, Georg Thieme Verlag, Stuttgart 1960.

[196] *A. Roedig* in *Houben-Weyl*, Methoden der organischen Chemie, 4. Aufl., Bd. V/4, p. 639, Georg Thieme Verlag, Stuttgart 1960.

[197] *L. Freedman, J.F. Chlebowski*, J. Org. Chem. *33*, 1636 (1968).

[198] *A.N. Nesmeyanov, T.P. Tolstaya, I.N. Lisichkina*, Izv. Akad. Nauk SSSR, Ser. Khim. 194 (1968); C.A. *69*, 18736d (1968).

[199] *A. Roedig* in *Houben-Weyl*, Methoden der organischen Chemie, 4. Aufl., Bd. V/4, p. 447, Georg Thieme Verlag, Stuttgart 1960.

[200] *A.N. Nesmeyanov, L.G. Makarova, I.V. Polovyanyuk*, Zh. Obshch. Khim. *35*, 681 (1965); C.A. *63*, 4320c (1965);
O.A. Reutov, Angew. Chem. *72*, 198 (1960) (Mechanism).

If *azide ions* are present, a secondary cyclization reaction occurs, so that *1,3-thiazolo[2,3-d]tetrazoles* are formed:

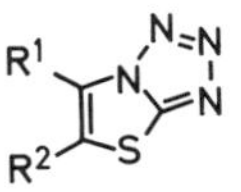

It has been found that using ferric chloride as catalyst induces a considerable substituent effect[203]:

7.2.1.3.2 Replacement by oxygen- and sulfur-containing groups

The so-called *boiling* of aromatic diazonium salts is a procedure of general applicability for converting primary aromatic amines into phenols. A usual technique is to warm diazonium sulfates to hydrolyze them[204]. If acid-sensitive materials are involved, the decomposition can be carried out in alkaline medium[205]:

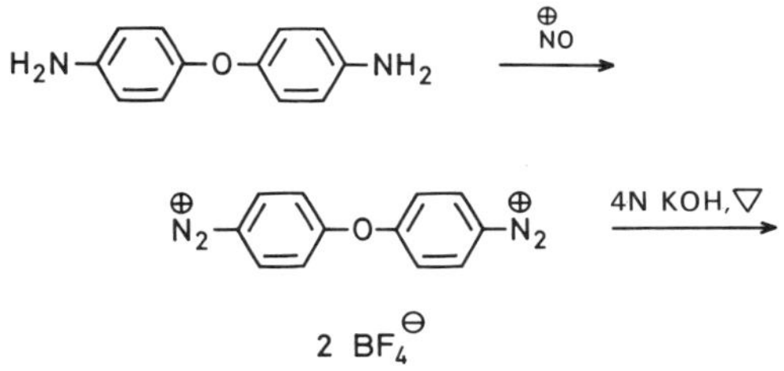

4,4'-Oxydi(benzenediazonium)-bis[tetrafluoroborate]; 98%

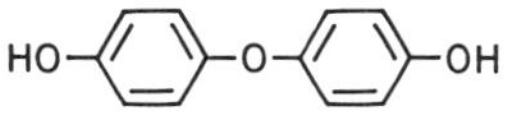

4,4'-Oxydiphenol; 87%

Reactions analogous to the nucleophilic replacement of the diazonium group by the hydroxy group in aqueous medium are known. Use of carboxylic acids as reaction medium allows acyloxy groups to be introduced[205], in alcoholic media alkoxy groups enter (see p. 178).

Thiophenols can be prepared from aromatic diazonium salts by forming intermediate xanthates, which are then hydrolyzed by alkali[206].

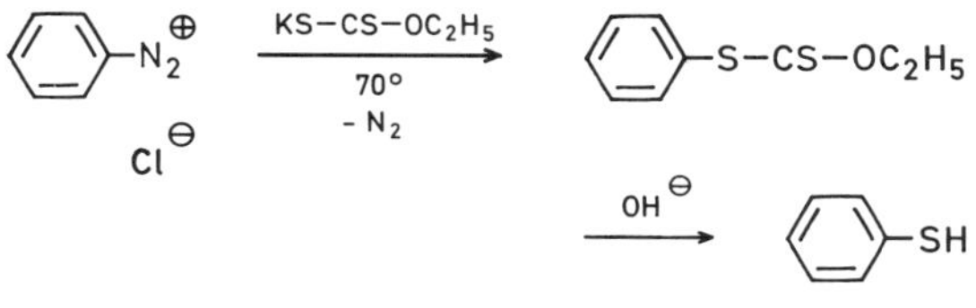

Other *S*-nucleophilic thio compounds can be used in place of potassium *O*-ethyl xanthate[207]; the *thiocarboxamide derivatives* or *thiocyanates* obtained can be decomposed to form *thiophenols*[208].

Analogously to the preparation of phenolic ethers, aromatic *sulfides* also can be prepared from diazonium salts and mercaptans in alkaline solution; diazo-sulfides (some of which are very **explosive**) may be formed as intermediates[209]. This danger can, however, be avoided by employing a suitable reaction procedure:

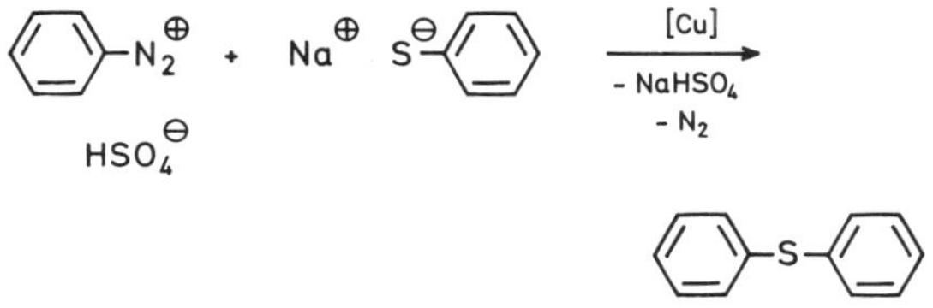

Diphenyl sulfide

[201] *T. Sandmeyer*, Ber. *17*, 1633, 2650 (1884);
A. Roedig in *Houben-Weyl*, Methoden der organischen Chemie, 4. Aufl., Bd. V/4, p. 438, Georg Thieme Verlag, Stuttgart 1960;
E. Forche in *Houben-Weyl*, Methoden der organischen Chemie, 4. Aufl., Bd. V/3, p. 846, Georg Thieme Verlag, Stuttgart 1962.

[202] *C. Grünert, K. Wiechert*, Z. Chem. *10*, 396 (1970).

[203] *Y. Nakatani*, Tetrahedron Lett. 4455 (1970).

[204] *H.E. Ungnade, E.F. Orwoll*, Org. Synth. *23*, 11 (1943);
E. White, D.J. Woodcock in *S. Patai*, The Chemistry of the Amino Group, p. 407, Interscience Publishers, London 1968; s. a. Lit. [175].

[205] *G. Koga, M. Yasaka, Y. Nakano*, Org. Prep. Proced. *1*, 205 (1969); C.A. *71*, 70229w (1969).

[206] *A. Schöberl, A. Wagner* in *Houben-Weyl*, Methoden der organischen Chemie, 4. Aufl., Bd. IX, p. 12, Georg Thieme Verlag, Stuttgart 1955;
M. Bögemann, S. Petersen, O.-E.Schultz, H. Söll in *Houben-Weyl*, Methoden der organischen Chemie, 4. Aufl., Bd. IX, p. 817, 840, Georg Thieme Verlag, Stuttgart 1955.

[207] *E.P. Nesynov, P.S. Pel'kis*, Zh. Org. Khim. *4*, 1425 (1968); C.A. *69*, 86509g (1968) (Thiocarbonamides).
R.K. Freidlina, B.V. Kopylova, Dokl. Akad. Nauk SSSR *173*, 839 (1967); C.A. *67*, 73273j (1967);
E.P. Nesynov, M.M. Besprozvannaja, P.S. Pel'kis, Zh. Org. Khim. *6*, 540 (1970);
M. Bögemann, S. Petersen, O.-E. Schultz, H. Söll in *Houben-Weyl*, Methoden der organischen Chemie, 4. Aufl., Bd. IX, p. 863, Georg Thieme Verlag, Stuttgart 1955 (Rhodanides).

[208] *A. Schöberl, A. Wagner* in *Houben-Weyl*, Methoden der organischen Chemie, 4. Aufl., Bd. IX, p. 16, Georg Thieme Verlag, Stuttgart 1955.

[209] *A. Schöberl, A. Wagner* in *Houben-Weyl*, Methoden der organischen Chemie, 4. Aufl., Bd. IX, p. 116, Georg Thieme Verlag, Stuttgart 1955;
G.E. Hilbert, T.B. Johnson, J. Amer. Chem. Soc. *51*, 1526 (1929).

Disulfides are obtained from sodium disulfide in a corresponding manner[210]. *Sulfinic acids*[211], the next higher stable oxidation state above thiophenols, can likewise be prepared directly from aromatic diazonium salts:

$$(Ar-N_2^{\oplus})_2 SO_4^{2\ominus} + 2 SO_2 + H_2SO_4 + 2 Cu$$
$$\longrightarrow 2 Ar-SO_2H + 2 N_2 + 2 CuSO_4$$

In this case the copper behaves as a reducing agent. If alcohol or sodium hydrogen sulfite is present, the sulfur dioxide itself can cause reduction, so that the reaction can be accomplished using a smaller amount of copper:

$$Ar-N_2^{\oplus} \; Cl^{\ominus} + 2 SO_2 + 2 H_2O$$
$$\xrightarrow[- N_2]{[Cu] \, , \, C_2H_5OH} Ar-SO_2H + HCl + H_2SO_4$$

A method which is of preparative interest for the preparation of *sulfonyl chlorides,* and consequently also of sulfonic acids[212], involves the reaction of aromatic diazonium chlorides with sulfur dioxide in a nonaqueous medium (glacial acetic acid, formic acid, tetrahydrofuran, or alcohol) in the presence of cupric chloride:

$$Ar-N_2^{\oplus} \; Cl^{\ominus} + SO_2 \xrightarrow[- N_2]{CuCl_2} Ar-SO_2Cl$$

A reaction which is analogous to the formation of aromatic thioethers from diazonium salts and mercaptans is the formation of *sulfones*[213] from sulfinic acids *via* intermediate diazosulfinates at elevated temperatures:

$$Ar-N_2^{\oplus} \; X^{\ominus} + R-SO_2H \xrightarrow[- HX]{}$$

$$\underset{O_2}{Ar-N=N-S-R} \xrightarrow[- N_2]{\triangledown} \underset{O_2}{Ar-S-R}$$

7.2.1.3.3 Replacement by nitrogen and phosphorus derivatives

Reactions involving replacement of the diazonium group by nitrogen-containing groups are only of little preparative importance, since the reaction products concerned can mostly be prepared from aromatic amines without diazotization. Diazonium salts react with potassium cyanate to form *isocyanates,* but these are readily hydrolyzed in the aqueous medium required for the initial reaction[214]. The more stable *aryldicyandiamides* can be obtained by treatment of diazonium salts with dicyandiamide (cyanoguanidine)[215]:

$$Ar-N_2^{\oplus} \; Cl^{\ominus} + \underset{NH}{H_2N-C-NH-CN} \xrightarrow[- HCl]{}$$

$$\underset{NH}{Ar-N=N-NH-C-NH-CN} \xrightarrow[- N_2]{[H^{\oplus}]}$$

$$\underset{NH}{Ar-NH-C-NH-CN}$$

The diazonium group can be replaced easily by the nitro group by using the so-called *nitro-Sandmeyer reaction;* copper catalysts are generally used (in favorable cases none)[216]. This method is especially suitable for the *conservative* introduction of the nitro group into aromatic or heteroaromatic compounds[217]; *e.g.:*

5-Nitro-1,3,4-thiadiazole-2-carboxaldehyde

[210] *A. Schöberl, A. Wagner* in *Houben-Weyl,* Methoden der organischen Chemie, 4. Aufl., Bd. IX, p. 67, Georg Thieme Verlag, Stuttgart 1955.

[211] *F. Muth* in *Houben-Weyl,* Methoden der organischen Chemie, 4. Aufl., Bd. IX, p. 320, Georg Thieme Verlag, Stuttgart 1955.

[212] *F. Muth* in *Houben-Weyl,* Methoden der organischen Chemie, 4. Aufl., Bd. IX, p. 579, Georg Thieme Verlag, Stuttgart 1955;
USSR P. 205013 (Cl C o7c), 13.11.67, Appl. 31.5.65, All Union Scientific-Research Institute of Monocrystalls, Inv.: *N.N. Dykhanov, A.B. Dzhizhelava, T.S. Ryzhova;* C.A. *69,* 27036c (1968).

[213] *F. Muth* in *Houben-Weyl,* Methoden der organischen Chemie, 4. Aufl., Bd. IX, p. 334, Georg Thieme Verlag, Stuttgart 1955.

[214] *S. Petersen* in *Houben-Weyl,* Methoden der organischen Chemie, 4. Aufl., Bd. VIII, p. 127, Georg Thieme Verlag, Stuttgart 1952.

[215] *H.-F. Piepenbrink* in *Houben-Weyl,* Methoden der organischen Chemie, 4. Aufl., Bd. VIII, p. 210, 213, Georg Thieme Verlag, Stuttgart 1952.

[216] *W. Seidenfaden, D. Pawellek* in *Houben-Weyl,* Methoden der organischen Chemie, 4. Aufl., Bd. X/1, p. 836, Georg Thieme Verlag, Stuttgart 1971;
B.V. Tronov, I.M. Yakovleva, Org. Komplek. Soedin. (Tomsk) 175 (1965); C.A. *66,* 65307n (1967).

[217] *G. Asato, G. Berkelhammer, E.L. Moon,* J. Med. Chem. *13,* 1015 (1970).

Copper catalysts are not required for the preparation of aromatic *azides* from diazonium salts by substitution, *i.e.*, no Sandmeyer reaction is involved. Experiments with ^{15}N-labeled compounds have shown that there is more than one reaction path: aryl azide and nitrogen may be formed directly from the arylazoazide, but in addition a side-reaction occurs in which ring-closure takes place to form an arylpentazole, which itself decomposes to aryl azide and nitrogen[218]:

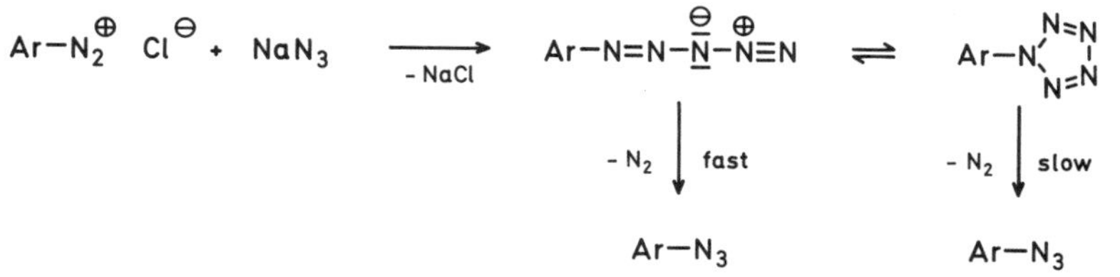

1,2-Diazidobenzene can be prepared in a corresponding manner from the 2-azidobenzenediazonium ion; thermolysis of the diazide yields *2,4-hexadienonitrile*[219]:

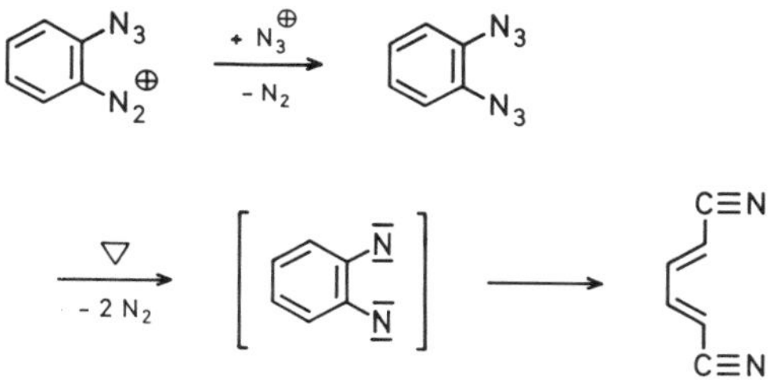

Organophosphorus compounds also can be prepared from aromatic diazonium salts by means of the Sandmeyer reaction[220]. Diazonium fluoroborates react with phosphorus trichloride in the presence of cuprous chloride or bromide, forming *aryltrichlorophosphonium salts*[221], which can be hydrolyzed to phosphonic acids, reduced to dichlorophosphines by magnesium[222], or converted into *thiophosphonic dichlorides (phosphonothioic dichlorides)* by treatment with hydrogen sulfide[223]:

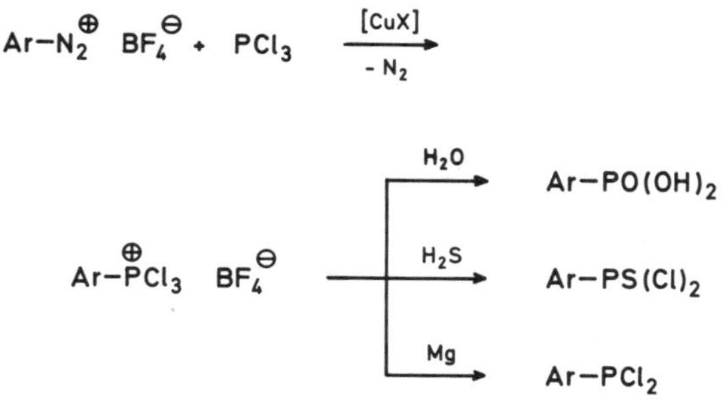

Dichlorophosphines themselves react with diazonium salts analogously yielding *diaryldichloro-*

phosphonium salts, which form *phosphinic acids* on hydrolysis[224]:

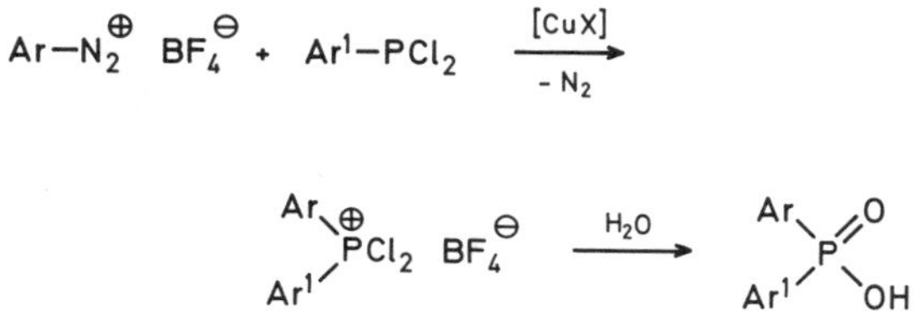

7.2.1.3.4 Special derivatives

Selenium and tellurium compounds can participate in substitution reactions with diazonium compounds analogous to those involving sulfur compounds[225]. Organomercury[226] and organotin[227] compounds can be prepared from aromatic compounds by means of the Nesmeyanov reaction[200] mentioned previously; organic derivatives of arsenic[228], antimony[229], and bismuth[230] can be prepared similarly.

[218] *C. Grundmann* in *Houben-Weyl*, Methoden der organischen Chemie, 4. Aufl., Bd. X/3, p. 800, Georg Thieme Verlag, Stuttgart 1965.

[219] *J.H. Hall*, J. Amer. Chem. Soc. *87*, 1147 (1965).

[220] *K. Sasse* in *Houben-Weyl*, Methoden der organischen Chemie, 4. Aufl., Bd. XII/1, p. 231, 366, Georg Thieme Verlag, Stuttgart 1963.

[221] *G.O. Doak, L.D. Freedman*, J. Amer. Chem. Soc. *73*, 5658 (1951).

[222] *I.G. Malakhova, E.N. Tsvetkov, I. Kabachnik*, Izv. Akad. Nauk SSSR, Ser Khim. 2602 (1970).

[223] *Z.E. Golubsk, B. Glowiak*, Bull. Acad. Polon Sci., Ser. Sci. Chim. *12*, 471 (1964); C.A. *62*, 1311d (1965).

[224] *L.D. Freedman, G.O. Doak*, J. Amer. Chem. Soc. *74*, 2884 (1952);
J.B. Levy, L.D. Freedman, G.O. Doak, J. Org. Chem. *33*, 474 (1968).

[225] *H. Rheinboldt* in *Houben-Weyl*, Methoden der organischen Chemie, 4. Aufl., Bd. IX, p. 942, 987, 995, 1089, Georg Thieme Verlag, Stuttgart 1955.

[226] *C. Weygand, G. Hilgetag*, Organisch-chemische Experimentierkunst, 3. Aufl., p. 759, J.A. Barth Verlag, Leipzig 1964.

[227] *C. Weygand, G. Hilgetag*, Organisch-chemische Experimentierkunst, 3. Aufl., p. 773, J.A. Barth Verlag, Leipzig 1964.

[228] *C. Weygand, G. Hilgetag*, Organisch-chemische Experimentierkunst, 3. Aufl., p. 777, J.A. Barth Verlag, Leipzig 1964.

[229] *C. Weygand, G. Hilgetag*, Organisch-chemische Experimentierkunst, 3. Aufl., p. 779, J.A. Barth Verlag, Leipzig 1964.

7.2.2 Aryne formation on elimination of diazonium nitrogen[231]

Thermolysis of benzenediazonium-2-carboxylate[232] or of 1, 2, 3-benzothiadiazole 1, 1-dioxide is a convenient means of preparing *dehydrobenzene*. By trapping the appropriate intermediates, it was shown[233] that decarboxylation and elimination of nitrogen occur independently of each other:

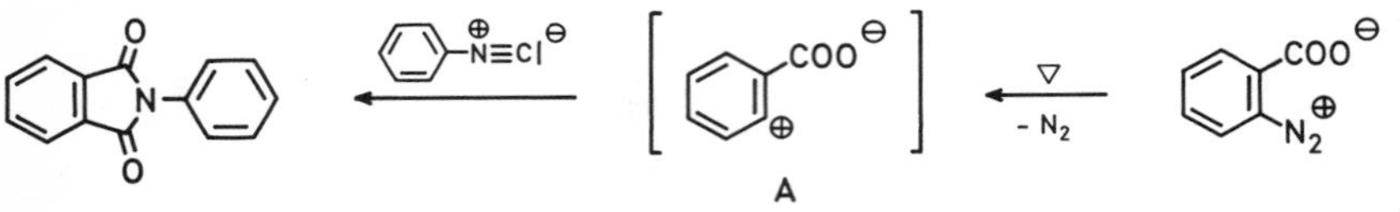

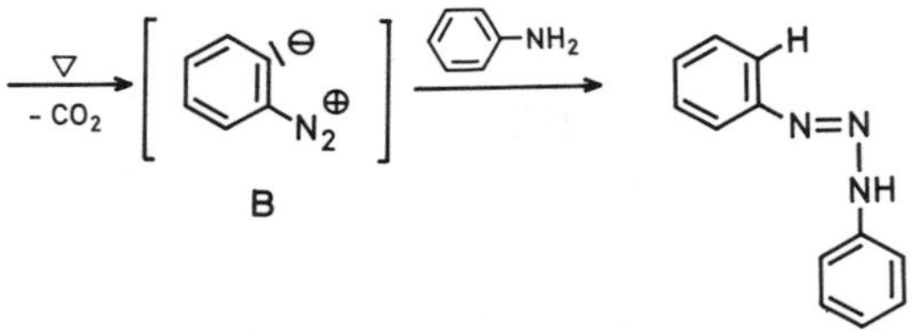

Evidence for the intermediate *A* has also been obtained with other cycloaddition reactions[234].

The existence of the aryne stage has been demonstrated not only by the usual trapping methods involving dienes[231], but also by the formation of products containing four-membered rings. *Dibenzo[b, h]biphenylene(cyclobuta[1, 2-b;   3, 4-b']dinaphthalene)* has been formed in small amounts by heating   naphthalene-2-diazonium-3-carboxylate[235]:

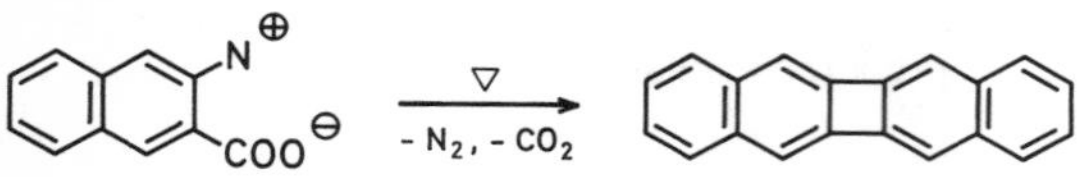

Benzenediazonium 2-carboxylate reacts with ethyl acrylate and acrylonitrile to form *bicyclo[4.2.0]octa-1, 3, 5-triene-7-carbonitrile* and *ethyl bicyclo[4.2.0]octa-1, 3, 5-triene-7-carboxylate* respectively[236]:

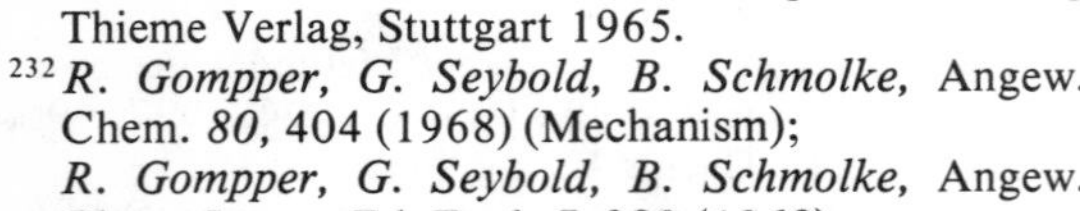

Similarly, normal addition of bromine to form *1, 2-dibromobenzene*[237], or of methanol to 4-chlorodehydrobenzene[238] to yield *3-* and *4-chloro-1-methoxybenzenes* are possible. These reactions involve stepwise additions.

Trapping reactions have demonstrated the probability of direct formation of arynes (under acetate catalysis)[239]. The intermediate occurrence of *1, 3-dehydrobenzene* on flash photolysis of benzenediazonium-3-carboxylate has also been demonstrated[240]. Formation of *1, 3-* or *1, 4-arynes* from 3- or 4-carboxybenzenediazonium salts is discussed in ref.[241].

Investigations of the mechanisms of photolysis of diazonium compounds in aqueous solution, or of their thermolysis in the presence or absence of nucleophiles, have been many and varied but have not yet yielded a clear picture. Thus, the anion of the diazonium salt may determine whether a reaction is ionic or proceeds at least partially by a free-radical mechanism. In the case of photolysis, whether radicals are involved or not depends on the wavelength used for irradiation. Substituent groups on the aromatic ring affect the relative rate

[230] *H. Gilman, H.L. Yale,* Chem. Rev. *30,* 281 (1942).

[231] *R. Pütter* in *Houben-Weyl,* Methoden der organischen Chemie, 4. Aufl., Bd. X/3, p. 207, Georg Thieme Verlag, Stuttgart 1965.

[232] *R. Gompper, G. Seybold, B. Schmolke,* Angew. Chem. *80,* 404 (1968) (Mechanism);
R. Gompper, G. Seybold, B. Schmolke, Angew. Chem. Intern. Ed. Engl. *7,* 389 (1968).

[233] *S. Yaroslavsky,* Chem. & Ind. (London) 765 (1965).

[234] *D.C. Dittmer, E.S. Whitman,* J. Org. Chem. *34,* 2004 (1969);
R.R. Schmidt, W. Schneider, Tetrahedron Lett. 5095 (1970);
S.F. Dyke, A.J. Floyd, S.E. Ward, Tetrahedron *26,* 4005 (1970).

[235] *C.F. Wilcox jr., S.S. Talwar,* J. Chem. Soc. C 2162 (1970).

[236] *T. Matsuda, T. Mitsuyasu,* Bull. Chem. Soc. Japan *39,* 1342 (1966).

[237] *L. Friedman, F.M. Logullo,* Angew. Chem. *77,* 217 (1965).

[238] *J.F. Bunnett, D.A.R. Happer, M. Patsch, C. Pyun, H. Takayama,* J. Amer. Chem. Soc. *88,* 5250 (1966).

[239] *R.W. Frank, K. Yanagi,* J. Amer. Chem. Soc. *90,* 5814 (1968);
J.I.G. Cadogan, J.R. Mitchell, J.T. Sharp, Chem. Commun. 1 (1971);
Ch. Rüchardt, C.C. Tan, Angew. Chem. *82,* 547 (1970);
S. Yaroslavsky, Tetrahedron Lett. 1503 (1965).

[240] *R.-S. Berry, J. Clardy, M.E. Schafer,* Tetrahedron Lett. 1011 (1965).

[241] *R. Hoyos de Rossi, H.E. Bertorello, R.A. Rossi,* J. Org. Chem. *35,* 3328, 3332 (1970).

[242] *R. Pütter* in *Houben-Weyl,* Methoden der organischen Chemie, 4. Aufl., Bd. X/3, p. 123, Georg Thieme Verlag, Stuttgart 1965.

[243] *L. Horner, H. Stöhr,* Chem. Ber. *85,* 993 (1952).

[244] *T.W.M. Herbertz, H.J. Wetzchewald,* Monatsh. Chem. *98,* 1364 (1967);
S. Kikuchi, K. Honda, M. Sukigara, Seisan-Kenkyn *19,* 151 (1966); C.A. *69,* 43231t (1968);
T. Takahiro, Y. Tsuguo, Nippon Shashin Gakkai Kaishi *29,* 27 (1966); C.A. *68,* 50952m (1968).

of photolysis[242-249]. Information has been published[250] concerning the effect of the structures of diazonium compounds on their reactivity.

7.2.3 Reactions with retention of the diazonium nitrogen; coupling reactions with nucleophiles

Acyldiazonium ions are electrophilic and consequently react readily with *nucleophiles*. Since most of the colored azo coupling products from compounds containing *CH*-acid groups are crystalline, *azo coupling* has become the most popular reaction for the detection of such compounds. Recently the azo coupling has been established to be a free-radical coupling reaction[250a]. Coupling with anions can be formulated as follows in general:

$$Y|^{\ominus} + \overset{\oplus}{N_2}-Ar \rightleftharpoons Y-N=N-Ar$$

The stability of the *Y–N* bond in solution, and hence the position of the equilibrium, depend on the coupling partners. A more acidic compound *Y–H*, i.e., one with a less energetic $Y^{\ominus}$ anion and a more completely compensated positive charge on the nitrogen by virtue of appropriate, mesomerism-inducing nuclear substitution (quinone diazides!), makes coupling more difficult and facilitates the azo-decoupling backreaction.

Table 3. Reactions of aromatic diazonium salts in which the nitrogen is retained

Azo coupling at	Coupling partners	Reaction products	Page
Carbon	1) Nucleophilic olefins	Arylhydrazones	192
	2) Nucleophilic aromatic compounds	Azo dyes	
	3) Nucleophilic hetero-aromatic compounds	Azo dyes	
	4) Carbanions	Azo dyes and arylhydrazones	
	5) Arylhydrazones (as heterologous enamines)	Formazans	
Nitrogen	1) Amine derivatives	Triazenes, polyazenes	200
	2) Cyanide	Diazoisocyanides	
Oxygen Phosphorus Sulfur	Groups which contain these elements as nucleophiles	See examples cited	201
Reductions to arylhydrazines			72

[245] *M. Sukigara, S. Kikuchi,* Bull. Chem. Soc. Japan *40*, 461 (1967).

[246] *M. Breitenbach, K.-H. Heckner, D. Jäkel,* Z. Physik. Chem. (Leipzig) *244*, 377 (1970);
D. Schulte-Frohlinde, H. Blume, Z. Physik. Chem. (Frankfurt/Main) *59*, 282 (1968); C.A. *69*, 105635p (1968).

Thus, for example, the anion of *trinitromethane* initially forms true salts with aryldiazonium ions, which are then converted slowly into azo coupling products containing *C–N* bonds[251]:

$$\overset{\oplus}{Ar-N_2}\ |\overset{\ominus}{C}(NO_2)_3 \xrightarrow{\ 2-24h\ } Ar-N=N-C(NO_2)_3$$

2-Nitro-1,3-indandione is acid enough for its adducts with diazonium ions to exist as true salts[252]. A similar situation holds for the adducts from diazonium ions and arylsulfinates; these normally exist as covalent diazosulfinates[253] but as *diazonium sulfinates* in the case of the strongly acid 2,4-dinitrobenzenesulfinic acid[254].

If the nucleophilic center still bears a hydrogen atom the azo coupling product may isomerize to an arylhydrazone:

$$\begin{array}{c} Y-N=N-Ar \\ |\quad\quad\;\;: \\ H\cdot\cdot X-H \end{array} \xrightarrow{\ -HX\ } \begin{array}{c} Y=N-N-Ar \\ |\\ H \end{array}$$

$$\text{e.g.: } HX = H_2O$$

Other groups on *Y* that are easily displaced by nucleophilic attack also make arylhydrazone formation possible (*e.g.*, the acyl groups: *Japp-Klingemann* reaction):

$$\begin{array}{c} Y-N=N-Ar \\ |\quad\quad\;\;: \\ O=C\cdot\cdot X-H \\ | \\ R \end{array} \xrightarrow{\ -R-COX\ } Y=N-NH-Ar$$

$$\uparrow -CO_2$$

$$\begin{array}{c} Y-N=N-Ar \\ |\quad\quad\;\;: \\ \underset{\|}{C}-O-H \\ O \end{array}$$

[247] *D. Schulte-Frohlinde, H. Blume,* Z. Physik. Chem. (Frankfurt/Main) *59*, 299 (1968); C.A. *69*, 95719c (1968).

[248] *E.S. Lewis, L.D. Hartung, B.M. McKay,* J. Amer. Chem. Soc. *91*, 419 (1969).

[249] *M.Y. Turkina, A.F. Levit, B.V. Chizhov, J.P. Gragerov,* Zh. Org. Khim. *5*, 1238 (1969); C.A. *71*, 101068z (1969);
V.V. Kozlov, V.P. Sagalovich, Y.M. Kulikov, Zh. Org. Khim. *7*, 133 (1971).

[250] *B.A. Porai-Koshits,* Usp. Khim. *39*, 608 (1970);
Z. Simon, I. Badilescu, Rev. Roumaine Chim. *8*, 207 (1963); *12*, 243 (1967); C.A. *67*, 116423 (1967);
G. Thirot, Ingenieurs E.P.C.I. *44*, 11 (1965); *45*, 11 (1965); C.A. *64*, 19382d (1966).

[250a] *N.N. Bubnov, K.A. Bilevitch, L.A. Poljakova, O.Y. Okhlobystin,* J. Chem. Soc. C 1058 (1972).

[251] *M.S. Pevzner, A.K. Pan'kov, A.G. Gal'kovskaja, I.N. Sochor, L.I. Bagal,* Zh. Org. Khim. *6*, 1944 (1970);
R. Pütter in *Houben-Weyl,* Methoden der organischen Chemie, 4. Aufl., Bd. X/3, p. 611, Georg Thieme Verlag, Stuttgart 1965.

The nucleophilic center Y can belong to a carbon, nitrogen, or oxygen atoms, *etc.*

Azo coupling[255] follows second-order reaction kinetics. The effect of various factors on the kinetics and mechanism have been studied[256, 257]. Substituent groups in the *p*- and *m*-positions of the diazonium ion affect the rate of azo coupling in accord with the Hammett relationship[258] and the same is true of substituent effects in phenolic components involved as coreactants in the coupling[259]. Phenols and anilines have been found to couple in the *o*- and *p*-position, or in both. Within certain limits the nature of the attack can be directed by introducing substituent groups into the phenol, varying the electrophilism of the diazonium component involved, changing the *pH,* and adding hydroxy-group containing substances such as glycerol, sugars, *etc.*[260].

7.2.3.1 Azo coupling at the carbon

When diazonium ions of especially high reactivity or electron-rich olefins are employed, *olefins* can couple as nucleophiles with aromatic diazonium salts. Highly reactive diazonium salts which are used for the detection and cleaving (depending on the structures concerned) of monoenes and dienes[263] include bis-diazotized 1,4-phenylenediamine[261] and diazotized 2,4-dinitroaniline[262]. In such reactions the properties of the diazonium group in its role as substituent correspond approximately to the effect of two nitro groups[264] ($\sigma_p = 0.78$ for NO_2, $\sigma_p = 1.91$ for $N_2^{\oplus}$), as indicated by the values of the Hammett substituent constants (calculated on the dissociation constants of *p*-substituted benzoic acids). The attack is generally at the more electron-rich *C* atom of the double bond[262]; *3-methyl-3-buten-2-one, 2,4-dinitrophenylhydrazone* is formed from *2-methyl-2-butene:*

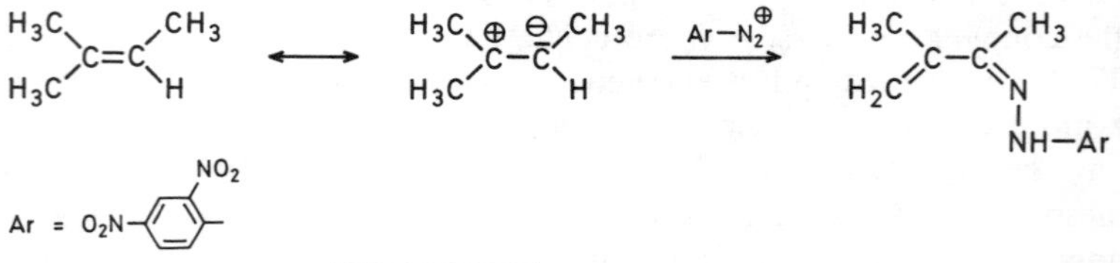

[252] Brit. P. 1 124 948 (Cl. C o7c), 21. 8. 1968, Appl. 14. 12. 1964, Algraphy Ltd., Inv.: *F.E. Smith;* C.A. *70,* 5234j (1969).

[253] *R. Pütter* in *Houben-Weyl,* Methoden der organischen Chemie, 4. Aufl., Bd. X/3, p. 578, Georg Thieme Verlag, Stuttgart 1965.

[254] *H. Meerwein, G. Dittmar, G. Kaufmann, R. Raue,* Chem. Ber. *90,* 853 (1957).

[255] *N.A. Frigerio,* J. Chem. Educ. *43,* 142 (1966); C.A. *64,* 13360c (1966);
K.H. Schündehütte in *Houben-Weyl,* Methoden der organischen Chemie, 4. Aufl., Bd. X/3, p. 226, Georg Thieme Verlag, Stuttgart 1965.

[256] *J.R. Penton, H. Zollinger,* Helv. Chim. Acta *54,* 573 (1971);
O. Sziman, A. Messmer, Tetrahedron Lett. 1625 (1968).

[257] *J. Kavalek, J. Panchartek, V. Sterba,* Collect. Czech. Chem. Commun. *35,* 3470 (1970);
M. Matrka, V. Chmatal, Chem. Listy *60,* 1530 (1966); C.A. *66,* 18547f (1967).

[258] *W. Wojtkiewicz, J. Szadowski,* Chem. Stosow., Ser. A *8,* 233 (1964); C.A. *62,* 13275f (1965);
L.N. Golubkin, A.A. Spryskov, Izv. Vyssh. Ucheb. Zaved., Khim. Khim. Tekhnol. *11,* 171 (1968); C.A. *69,* 76136n (1968);
L.M. Rozhdestvenskaya, I.L. Bagal, B.A. Porai-Koshits, Reacts. Sposobnost. Org. Soedin. *6,* 114 (1969); C.A. *71,* 90677n (1969);
Y. Hashida, K. Nakajima, S. Sekiguchi, K. Matsui, Kogyo Kagaku Zasshi *72,* 1132 (1969), C.A. *71,* 92619a (1969); *73,* 180, 184 (1970).

[259] *I. Dobas, V. Sterba, M. Vecera,* Chem. & Ind. (London) 1814 (1968); C.A. *70,* 46565m (1969);
L.N. Golubkin, A.A. Spryskov, Izv. Vyssh. Ucheb. Zaved., Khim. Khim. Tekhnol. *9,* 901 (1966); C.A. *67,* 10915z (1967).

[260] *L.N. Ogoleva, B.I. Stepanov,* Zh. Org. Khim. *1,* 2083 (1965); C.A. *64,* 12844h (1966); *2,* 108 (1966); C.A. *64,* 14142f (1966);
B.I. Stepanov, Z.F. Sergeeva, Probl. Poluch. Poluprod. Org. Sin., Akad. Nauk SSSR, Otd. Obshch. Tekhnol. Khim. 101 (1967); C.A. *68,* 12154u (1968);
Z.F. Sergeeva, B.I. Stepanov, Tr. Mosk. Khim.-Tekhnol. Inst. *52,* 134, 138 (1967); C.A. *68,* 95574f, 95575g (1968);
B.I. Stepanov, L.N. Ogoleva, Zh. Org. Khim. *3,* 371 (1967); C.A. *67,* 10912w (1967);
B.I. Stepanov, L.N. Ogoleva, Z.F. Sergeeva, Zh. Org. Khim. *3,* 902, 906 (1967); C.A. *67,* 63967z, 43077f (1967);
Z.F. Sergeeva, B.I. Stepanov, Zh. Org. Khim. *4,* 638 (1968); C.A. *69,* 2735s (1968);
S. Kishimoto, T. Hirashima, O. Manabe, H. Hiyama, Kogyo Kagaku Zasshi *71,* 1195 (1968); C.A. *70,* 28133q (1969).

[261] *H. Marxmeier, E. Pfeil, W. Wolf,* Z. Anal. Chem. *215,* 387 (1966);
J. Goerdeler, H. Haubrich, Chem. Ber. *93,* 397 (1960).

[262] *H. Marxmeier, E. Pfeil,* Justus Liebigs Ann. Chem. *678,* 28 (1964).

[263] *H. Marxmeier, E. Pfeil,* Chem. Ber. *97,* 815 (1964).

[264] *E.S. Lewis, M.D. Johnson,* J. Amer. Chem. Soc. *81,* 2070 (1959).

Anethole (1-methoxy-4-propenylbenzene)[265] splits into *acetaldehyde* and *4-methoxybenzaldehyde arylhydrazones* (re the mechanism see ref.[263]):

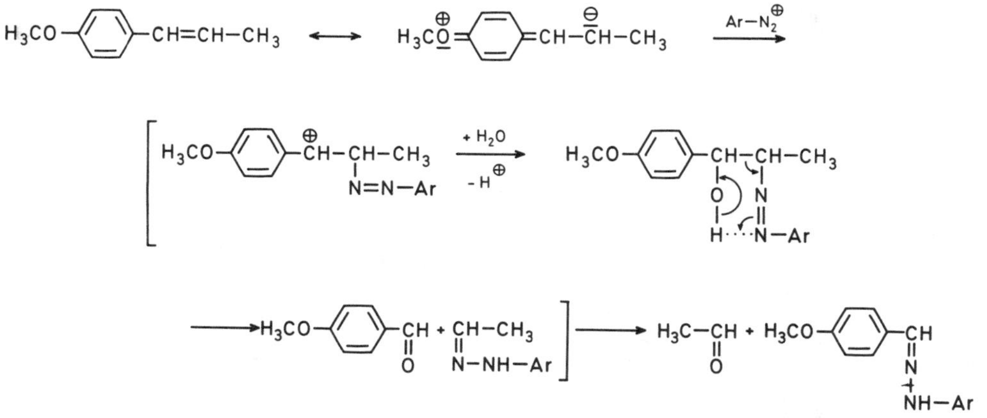

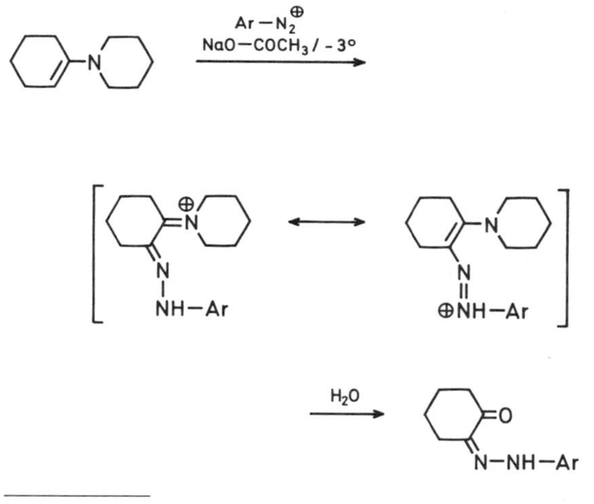

After azo coupling β,β-disubstituted enamines cannot be stabilized by formation of a mesomeric cation, and they suffer a type of hydrolytic fission which corresponds to a retroaldol addition (possi-

Enol ethers such as ethyl vinyl ether undergo azo coupling at the β-carbon atom with simultaneous hydrolysis to aldehydes[266]. Enamines with at least one hydrogen atom at the β-carbon atom react in corresponding manner[267], *e.g.*:

bly involving a sigmatropic 1,5-hydrogen migration). Thus, 1-isopropylidenepiperidine is cleaved on reaction with 4-nitrobenzenediazonium chloride to form *acetone 4-nitrophenylhydrazone* and *1-formylpiperidine*[268]:

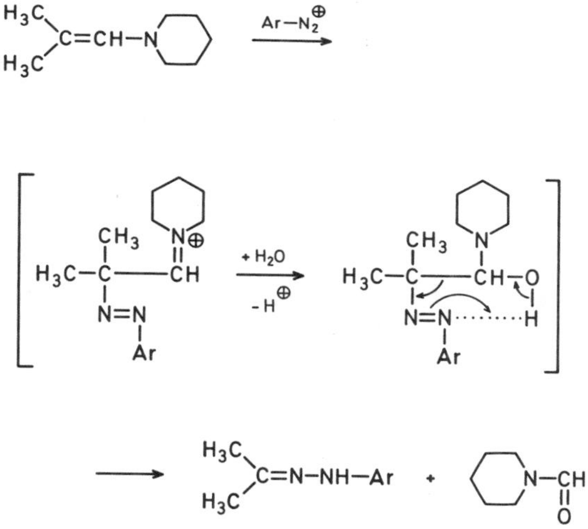

Enamines with an acyl group on the β-carbon atoms (vinylogous carboxamides) can likewise couple readily to form *azo dyes*[269]. Conjugated

[265] A. Quilico, M. Freri, Gazz. Chim. Ital. *58*, 380 (1928).

[266] M. Seefelder, H. Eilingsfeld, Angew. Chem. *75*, 724 (1963);
E. Enders in Houben-Weyl, Methoden der organischen Chemie, 4. Aufl., Bd. X/3, p. 543, Georg Thieme Verlag, Stuttgart 1965.

[267] V.I. Shvedov, L.B. Altukhova, A.N. Grinev, Zh. Org. Khim. *1*, 879 (1965); C.A. *63*, 6853h (1965); M.E. Kuehne, J. Amer. Chem. Soc. *84*, 841, 846 (1962).

[268] E. Enders in Houben-Weyl, Methoden der organischen Chemie, 4. Aufl., Bd. X/3, p. 540, Georg Thieme Verlag, Stuttgart 1965.

[269] A. Singh, B. Hirsch, Indian J. Chem. *8*, 756 (1970); A.N. Grinev, N.N. Arkhangel'skaja, G.Y. Uretskaya, Zh. Org. Khim. *5*, 1472 (1969); C.A. *72*, 112678z (1969).

[270] U.K. Pandit, M.J.M. Pollmann, H.O. Huisman, J. Chem. Soc. D 527 (1969);
M.J.M. Pollmann, H.R. Reus, U.K. Pandit, H.O. Huisman, Rec. Trav. Chim. Pays-Bas *89*, 929 (1970).

[271] K.H. Schündehütte in Houben-Weyl, Methoden der organischen Chemie, 4. Aufl., Bd. X/3, p. 230–263, Georg Thieme Verlag, Stuttgart 1965;
Ch.-H. Wang, Bull. Inst. Chem. Acad. Sinica *11*, 72 (1965); C.A. *67*, 2557u (1967).

dienamines can undergo azo coupling at both the β- and the δ-carbon atom[270]; solvent effects are of decisive importance:

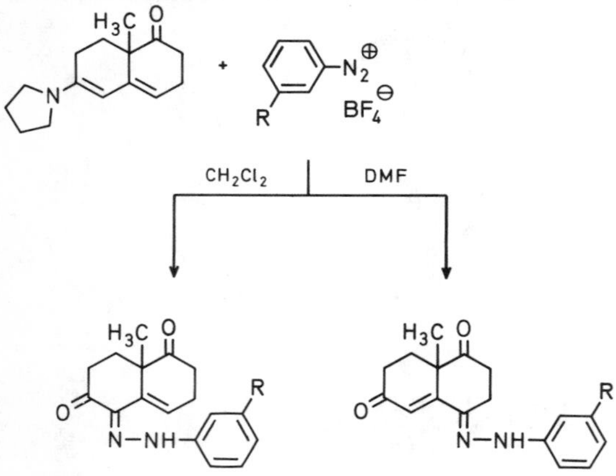

R = H; OCH₃; *4,4a,6,7-Tetrahydro-4a-methyl-1,2,5(3H)-naphthalenetrione 1-(phenylhydrazone) and 1-[(3-methoxyphenyl)hydrazone]*

R = H; OCH₃; *2,3,8,8a-Tetrahydro-8a-methyl-1,4,6(7H)-naphthalenetrione 1-(phenylhydrazone) and 1-[(3-methoxyphenyl)hydrazone]*

Available experimental information suggests that the azo coupling of olefinic components proceeds as represented in the following scheme:

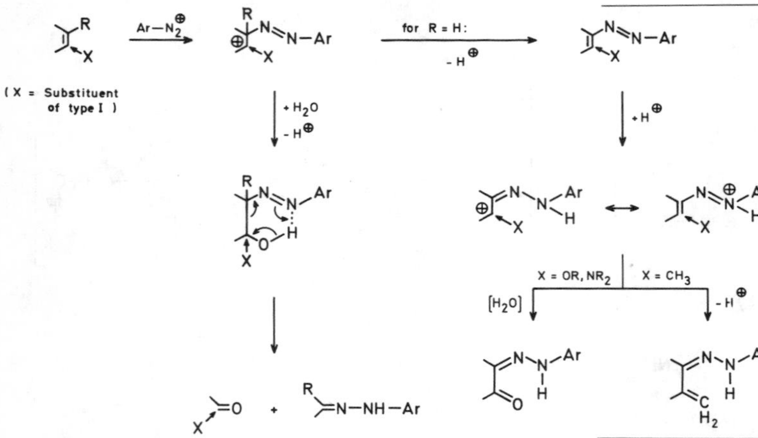

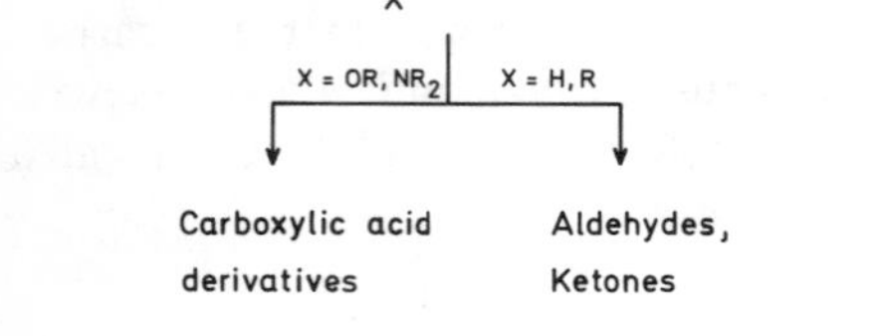

Carboxylic acid derivatives   Aldehydes, Ketones

The laws observed with olefins are followed also by aromatic compounds, *i.e.*, azo coupling reactions at aromatic nuclei occur most readily with electron-rich aromatic compounds. Invariably, the nuclear positions of highest electron density are the sites where coupling occurs, *i.e.*, the positions *o*- and *p*- to Type 1 substituent groups; unsubstituted positions[271] are attacked preferentially, *e.g.*:

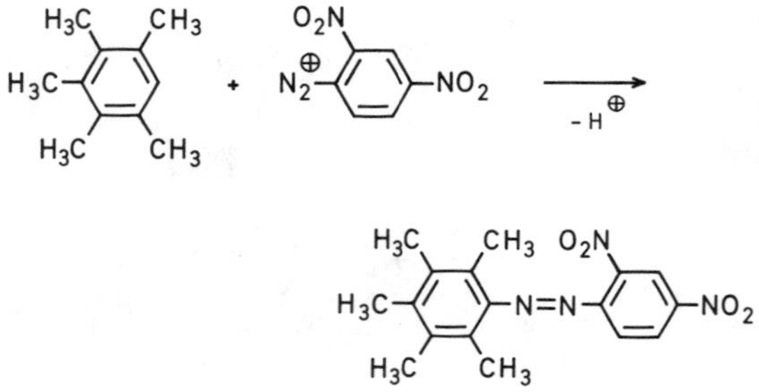

1-Pentamethylphenyl-2-(2,4-dinitrophenyl)diazene

It should be borned in mind that fission of ether linkages may occur during azo coupling of phenolic ethers; consequently, such coupling reactions should be carried out in acetate-buffered solution or in anhydrous acetic acid[272]:

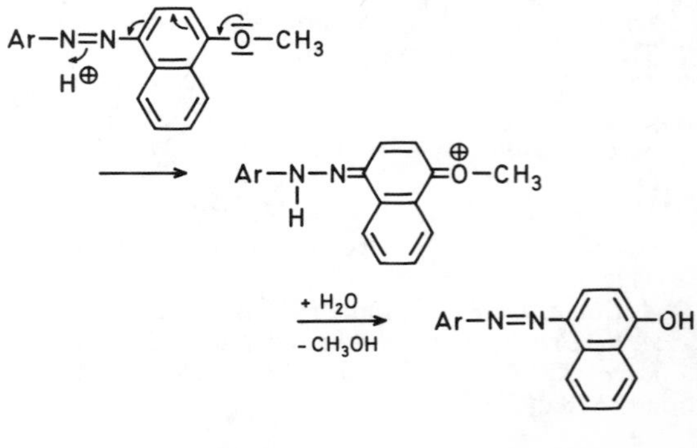

[272] *J.F. Bunnett, G.B. Hoey*, J. Amer. Chem. Soc. *80*, 3142 (1958).

[273] *J.B. Hendrickson, D.J. Cram, G.S. Hammond* in Organic Chemistry, 3. Ed., p. 949, McGraw-Hill Book Co., New York 1970.

Depending on the substituent group present, on the diazonium component, and the reaction conditions, primary and secondary aromatic amines can undergo not only *C*-azo coupling but also *N*-azo coupling to form *triazenes* (see p. 253). Provided the electron density of the aromatic amine moiety is not greatly reduced by additional Type 2 substituents, the triazenes can readily undergo acid-catalyzed rearrangement to form *aminoazo* compounds (ref. [139], p. 238), *e.g.*:

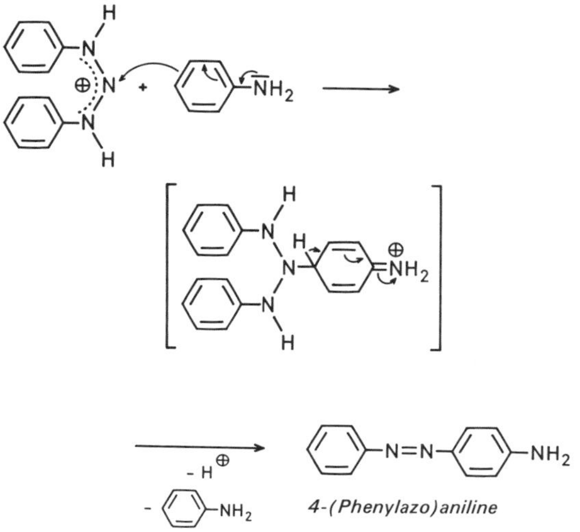
4-(Phenylazo)aniline

Like electron-rich aromatic compounds, *heteroaromatic* compounds, too, can be submitted to electrophilic azo coupling as long as they are sufficiently nucleophilic. A recent classification divides the simple heteroaromatic compounds into two groups[273] according to their formal derivation from benzene:

a Those in which one *CH* group of the benzene molecule has been replaced, for example, by a nitrogen atom. Heteroaromatic compounds containing six-membered rings result which are formally derivatives of carbonyl compounds (cyclic azomethines) and which do not undergo substitution reactions unless powerful Type 1 substituent groups are also present.

b Those in which formally two *CH* groups linked by a double bond in the benzene molecule are replaced by one heteroatom (principally *N, O, S*). Heteroaromatic compounds containing five-membered rings result which can be regarded as derivatives of enols or dienols and which consequently undergo electrophilic substitution readily.

c A combination of both features yields 5- and 6-membered heteroaromatic compounds whose properties correspond formally to those of groups **a** and **b.**

For example, 4,5-disubstituted *imidazoles* couple at their 2-position with aryldiazonium salts to form dyestuffs useful for dyeing polypropylene fibers[274]:

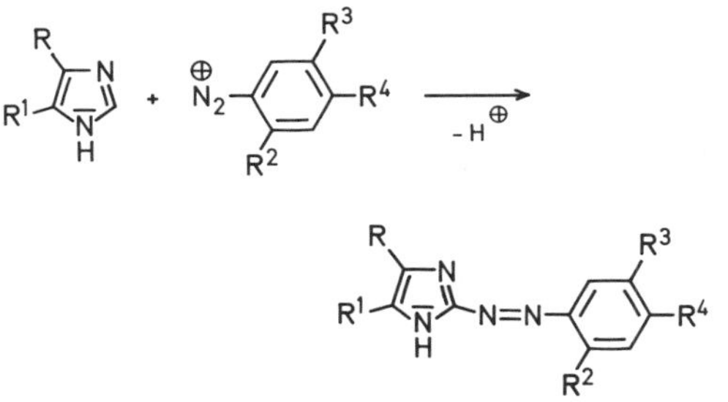

As heteroaromatic compounds containing a six-membered ring, *pyridines* must be activated by at least one powerful Type 1 substituent group such as the hydroxy group if azo coupling is to occur; in such cases coupling occurs *ortho* or *para* to the hydroxy group regardless of the heteroatom[275]:

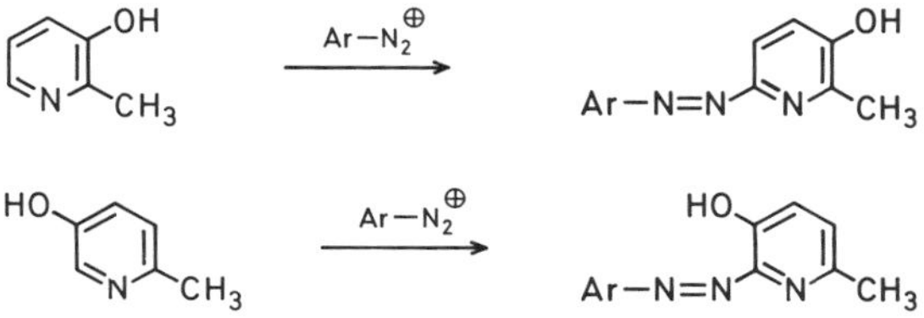

Table 4 below gives a tabular summary[276] of azo coupling reactions of heterocyclic compounds.

Table 4. Position of azo coupling reactions with heterocyclic compounds

	Ref.		Ref.		Ref.
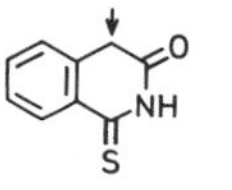	276	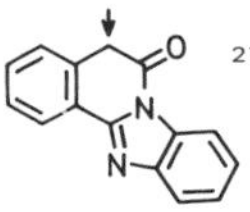	277	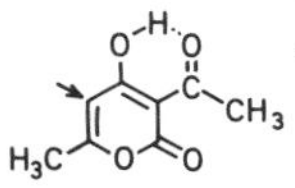	278
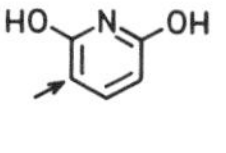	279	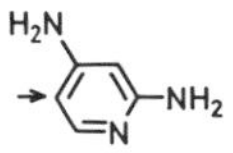	280	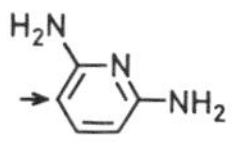	280
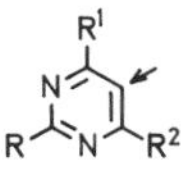	281	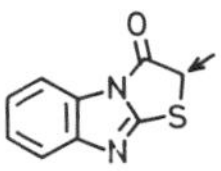	282	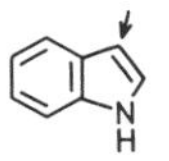	283
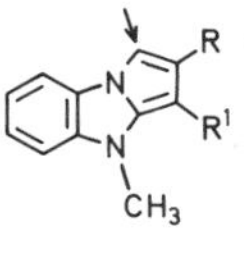	284	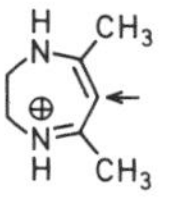	285	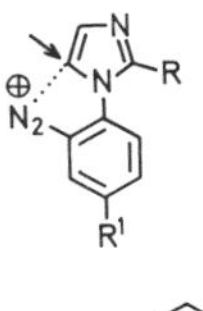	286
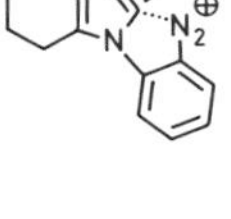	286	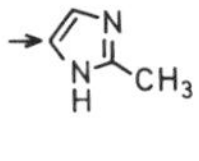	287	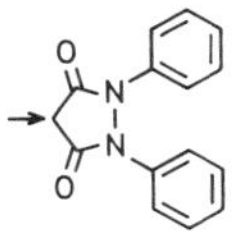	288

[274] *K. Komatsu, N. Kuroki,* Kogyo Kagaku Zasshi *73,* 2190 (1970).

[275] *K.N. Dyumaev, R.N. Lokhov, B.E. Zaitsev, L.D. Smirnov,* Izv. Akad. Nauk. SSSR, Ser. Khim. 2601 (1970).

Phenols and stable *enols* or their anions are particularly reactive coupling partners for aromatic diazonium salts. Phenolate anions can also be coupled with diazonium salts of low electrophilism (quinone diazides).

The *hydroxyazo compounds* obtained, together with the derivative metal complexes, form the largest class of synthetic colorants ranging from water-soluble textile dyes to pigments that are practically insoluble in most solvents[289].

The *p-azo* coupling product from phenol and a 4-butylbenzenediazonium salt is used as the starting material for the preparation of nematic phases (*O*-methyl and *O*-ethyl ethers)[290].

$$H_3C(CH_2)_3 - \bigcirc - N_2^{\oplus} \ + \ \bigcirc - OH \ \xrightarrow{-H^{\oplus}}$$

$$H_3C(CH_2)_3 - \bigcirc - N=N - \bigcirc - OH$$

4-[(4-Butylphenyl)azo]phenol

Coupling can occur either *o-* or *p-* (or both) to the hydroxy group[260].

Coupling reactions entirely analogous to those involving phenols occur with enols or enolates in

which conjugation with a carbonyl group is possible instead of with an aromatic nucleus and thus represent derivatives of vinylogous carboxylic acids[291]:

$$\begin{array}{c} R-C \overset{OH}{\underset{}{\Vert}} \\ R^1 - C \\ \Vert \\ O \end{array} - R^2 \ + \ N_2^{\oplus} - Ar \ \xrightarrow{-H^{\oplus}}$$

$$\begin{array}{cc} \text{A} & \text{B} \end{array}$$

Unsubstituted 1,3-dicarbonyl compounds spontaneously yield stable, strongly chelated 2-arylhydrazones **H** of 1,2,3-tricarbonyl compounds; those which are monosubstituted at the 2-position afford readily decomposing azo dyes **A**[292]. In such cases the tendency to form α-oxoarylhydrazones (*i.e.*, heterovinylogous carboxamides of especially low energy) is so marked that an acyl group is eliminated from the azo dyes **A** by solvolysis (Japp-Klingemann fission)[293]:

[276] *A. Moustafa, M.I. Ali, A.A. El-Sayed,* Justus Liebigs Ann. Chem. *739*, 63 (1970).

[277] *A. Moustafa, M.I. Ali, A.A. El-Sayed,* Justus Liebigs Ann. Chem. *739*, 68 (1970).

[278] *D.R. Gupta, R.S. Gupta,* J. Indian Chem. Soc. *43*, 377 (1966); C.A. *65*, 10518a (1966).

[279] *H.J. Knackmus,* J. Heterocycl. Chem. 7, 733 (1970).

[280] *P. Tomasik, E. Kucharzewska-Rusek, A. Thomas,* Roczniki Chem. *44*, 1131 (1970).

[281] *D. Sen, S.D. Chaudhuri,* J. Indian Chem. Soc. *47*, 369 (1970).

[282] *J.M. Singh,* J. Med. Chem. *13*, 1018 (1970).

[283] *V.G. Avramenko, V.D. Nazina, N.N. Suvorov,* Khim. Geterotsikl. Soedin. 1071 (1970).

[284] *F.S. Babicev, G.P. Kutrov, M.J. Kornilov,* Ukr. Khim. Zh. *36*, 909 (1970).

[285] *E.M. Grant, D. Lloyd, D.R. Marshall,* Chem. Commun. 1320 (1970).

[286] *A.M. Simonov, L.M. Sitkina, A.F. Pozharskii,* Chem. & Ind. (London) 1454 (1967); C.A. *68*, 2881 (1968).

[287] *K. Komatsu, K. Nagasawa, K. Konishi, N. Kuroki,* Kogyo Kagaku Zasshi *73*, 989 (1970);
K. Komatsu, M. Kaimori, K. Konishi, N. Kuroki, Kogyo Kagaku Zasshi *73*, 991 (1970);
K. Komatsu, T. Ando, K. Konishi, N. Kuroki, Kogyo Kagaku Zasshi *73*, 995 (1970).

[288] *V.G. Yakutovich, B.L. Moldaver, Y.P. Kitaev, Z.S. Titova,* Izv. Akad. Nauk SSSR, Ser. Khim. 877 (1968); C.A. *70*, 3928q (1969).

[289] *K.H. Schündehütte* in *Houben-Weyl*, Methoden der organischen Chemie, 4. Aufl., Bd. X/3, p. 263, Georg Thieme Verlag, Stuttgart 1965;
C.V. Stead, J. Chem. Soc. C 693 (1970);
Z.J. Allan, J. Podstata, Collect. Czech. Chem. Commun. *35*, 2444 (1970).

[290] *H. Kelker, B. Scheurle, R. Hatz, W. Bartsch,* Angew. Chem. *82*, 984 (1970).

[291] *E. Enders* in *Houben-Weyl*, Methoden der organischen Chemie, 4. Aufl., Bd. X/3, p. 477, 480, 498, Georg Thieme Verlag, Stuttgart 1965;
J. Elguero, R. Jacquier, G. Tarrago, Bull. Soc. Chim. France 2981 (1966);
A. Balog, L. Beu, E. Hamburg, Rev. Chim. (Bucharest) *21*, 523 (1970).

[292] *K. Schank,* Dissertation, Saarbrücken 1962, p. 23.

The more electrophilic acyl group is naturally split off preferentially; analogous *N* or *S* acyl groups behave correspondingly. Even hydrolytic elimination of vinylogous acyl groups in alkaline medium has been observed[294], *e.g.*:

Corresponding results have also been obtained in other cases[296]. Rupture of *C,C* bonds evidently occurs particularly easily when the substituent *R* can be displaced as a resonance-stabilized carbenium ion, or when a sigmatropic 1,5-hydrogen migration proceeding *via* a cyclic transition state (see p. 193) is possible[296]:

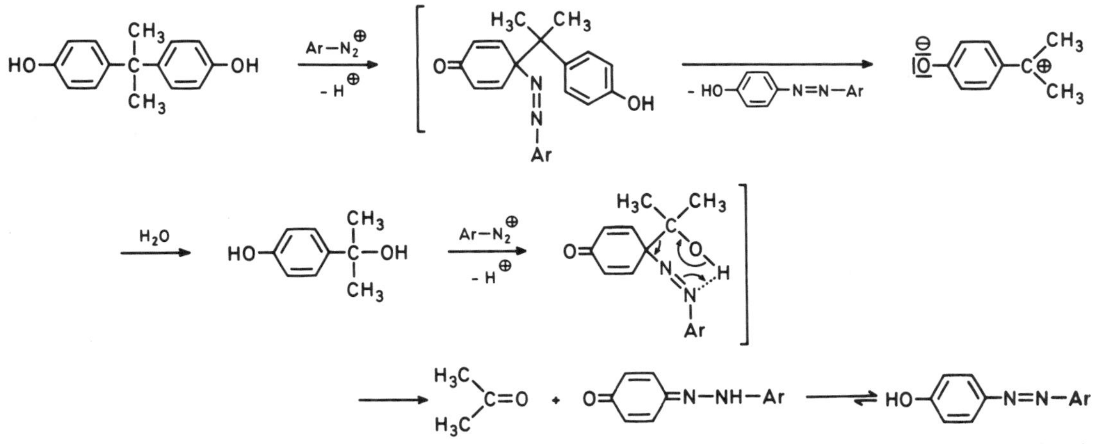

4-[(4-Nitrophenyl)azo]phenol

$Ar = $ —⟨⟩—NO$_2$

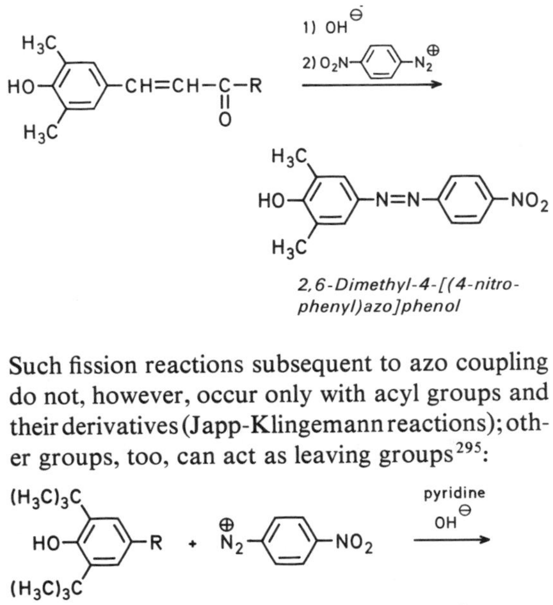

2,6-Dimethyl-4-[(4-nitro-phenyl)azo]phenol

In reactions analogous to those of enamines, aldehyde arylhydrazones or acyl*hydrazones* couple with aromatic diazonium salts to give *formazans*[297]; the bisazoalkanes formed as intermediates can sometimes be isolated, and isomerized to *formazan*[298]:

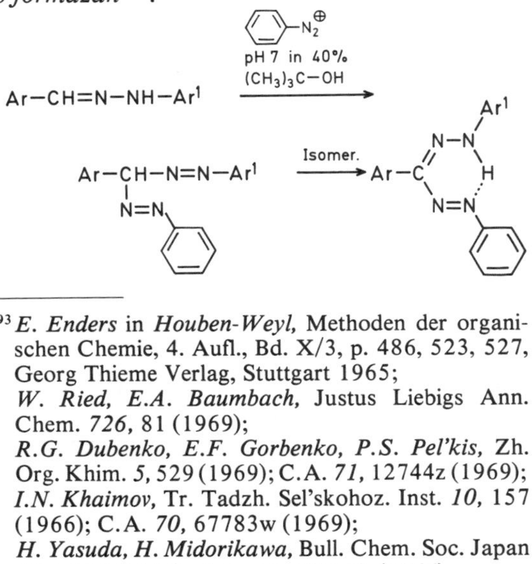

Such fission reactions subsequent to azo coupling do not, however, occur only with acyl groups and their derivatives (Japp-Klingemann reactions); other groups, too, can act as leaving groups[295]:

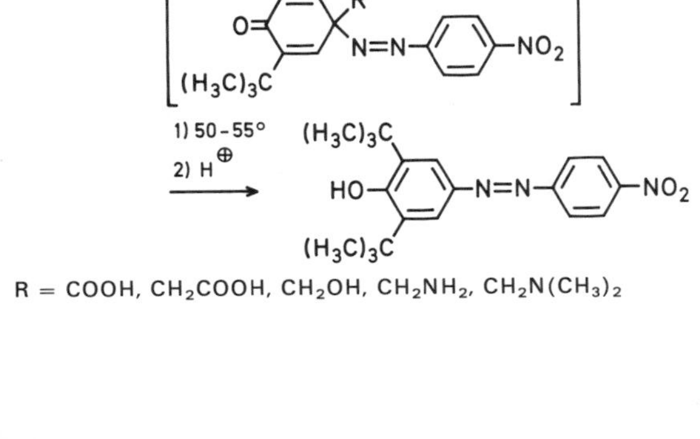

R = COOH, CH$_2$COOH, CH$_2$OH, CH$_2$NH$_2$, CH$_2$N(CH$_3$)$_2$

[293] *E. Enders* in *Houben-Weyl,* Methoden der organischen Chemie, 4. Aufl., Bd. X/3, p. 486, 523, 527, Georg Thieme Verlag, Stuttgart 1965;
W. Ried, E.A. Baumbach, Justus Liebigs Ann. Chem. *726*, 81 (1969);
R.G. Dubenko, E.F. Gorbenko, P.S. Pel'kis, Zh. Org. Khim. *5*, 529 (1969); C.A. *71*, 12744z (1969);
I.N. Khaimov, Tr. Tadzh. Sel'skohoz. Inst. *10*, 157 (1966); C.A. *70*, 67783w (1969);
H. Yasuda, H. Midorikawa, Bull. Chem. Soc. Japan *39*, 1596 (1966); C.A. *65*, 12191h (1966);
M.M. Kochhar, J. Heterocycl. Chem. *6*, 977 (1969);
H. Ishii, Y. Murakami, S. Tani, K. Abe, N. Ikeda, Yakugaku Zasshi *90*, 724 (1970).
[294] *H. Wittmann, H. Uragg,* Monatsh. Chem. *88*, 404 (1967); C.A. *67*, 32159h (1967).

The substituent effects of groups on Ar[1] on the kinetics of formazan formation are about three times those of groups on Ar!

The intense color of the arylformazans can be attributed to chelate formation and resonance stabilization in the anions and cations which are readily obtained, since the *formazan [3-(phenylazo)-1H-indazole]* formed by diazotization and self-coupling of 2-aminobenzaldehyde phenylhydrazone is only yellow[299]:

3-Benzeneazo-1H-
⟨benzo[c]pyrazole⟩

Compounds containing active methyl groups can also be coupled to form *formazans* directly[300], *e.g.:*

Phenylglyoxal 1-(4-nitro-phenyl)hydrazone

3-Benzoyl-1,5-bis(4-nitrophenyl)formazan

Moreover, ketone arylhydrazones can undergo azo coupling to produce formazans provided that they are substituted at the α-position by readily eliminated acyl groups[301], *e.g.:*

3-Methyl-1,5-diphenylformazan

The course of this reaction corresponds entirely to that of the Japp-Klingemann reaction.

1,5-Diarylformazans in which the central carbon atom is substituted couple further to form *arylazo-1,5-formazans* if the substituent is an acyl group[302]

[295] *L.S. Geidysh, G.A. Nikiforov, V.V. Ershov,* Izv. Akad. Nauk SSSR, Ser. Khim. 2386 (1968); C.A. *70,* 19694w (1969).

[296] *H. Wittmann, U. Hehenberger,* Monatsh. Chem. *100,* 455 (1969); C.A. *71,* 21773k (1969); *D. Nardi, A. Tajana, E. Massarani,* Chem. Ber. *101,* 2285 (1968).

[297] *R. Pütter* in *Houben-Weyl,* Methoden der organischen Chemie, 4. Aufl., Bd. X/3, p. 627, Georg Thieme Verlag, Stuttgart 1965; *G.N. Tjurenkova, N.P. Bednjagina,* Khim. Geterotsikl. Soedin. 1198 (1970); *J.A. Rybakova, L.F. Lipatova, R.O. Matevosjan,* Zh. Org. Khim. *6,* 182 (1970).

[298] *A.F. Hegarty, F.L. Scott,* J. Org. Chem. *32,* 1957 (1967); *A.F. Hegarty, F.L. Scott,* J. Chem. Soc. C 2507 (1967); *F.A. Neugebauer,* Tetrahedron *26,* 4843 (1970).

[299] *H. Auterhoff, R.D. Aye,* Arch. Pharm. (Weinheim, Ger.) *300,* 800 (1967); C.A. *68,* 59502x (1968); a lightening of color on acylation; *R. Pütter* in *Houben-Weyl,* Methoden der organischen Chemie, 4. Aufl., Bd. X/3, p. 682, Georg Thieme Verlag, Stuttgart 1965.

[300] *R. Pütter* in *Houben-Weyl,* Methoden der organischen Chemie, 4. Aufl., Bd. X/3, p. 644, Georg Thieme Verlag, Stuttgart 1965; *V.V. Razumovskii, E.F. Rychkina,* Zh. Org. Khim. *5,* 1255 (1969); C.A. *71,* 101470z (1969); *O.M. Staskevic, G.T. Piljugin, V.V. Staskevic,* Khim. Geterotsikl. Soedin. 1104 (1970).

[301] *R. Pütter* in *Houben-Weyl,* Methoden der organischen Chemie, 4. Aufl., Bd. X/3, p, 644, 653, Georg Thieme Verlag, Stuttgart 1965.

[302] *R. Pütter* in *Houben-Weyl,* Methoden der organischen Chemie, 4. Aufl., Bd. X/3, p. 645, Georg Thieme Verlag, Stuttgart 1965; *E. Bamberger, J. Lorenzen,* Ber. *25,* 3539 (1892); *E. Bamberger, J. Müller,* J. Prakt. Chem. [2] *64,* 199 (1901).

or a similarly easily eliminated substituent such as an alkylol group[303], e.g.:

1,5-Diphenyl-3-phenylazoformazan

$$Ar = 4-CH_3, Cl, Br, NO_2-C_6H_4 \text{ or } C_6H_5$$

Besides the azo coupling reactions of electron-rich olefins, aromatic and heteroaromatic compounds, enol derivatives, *etc.*, described here, all carbanions and their potential precursors can fundamentally couple at the carbon atom. The *CH* bond is activated by Type 2 substituent groups in the following order of decreasing effectiveness[304]:

$$NO_2 > CHO > COCH_3 > CN > CO_2C_2H_5 > CONH_2 > CO_2H >$$
$$SO_2CH_3 > SOCH_3 > C_6H_5$$

Table 5 lists some azo coupling reactions at nucleophilic carbon atoms.

Table 5. Position of azo coupling with *CH*-acid compounds and some nucleophiles

	Reference	
$Ar-CO-CH-SO_2-Ar^1$ $\quad\quad\quad OCH_3$	305	
$Ar-CO-CH-SO_2-Ar^1$ $\quad\quad\quad OCOCH_3$	306	
$X-CH_2-Y$	307	x, y = *acidifying* groups
$O_2N-CH_2-S-CCl_3$	308	
$O_2N-CH_2-(CH_2)_2-CH_2-NO_2$	309	
$CO-NR_2$	310	

[303] V.V. Kozlov, Y.M. Kulikov, Dokl. Akad. Nauk. SSSR, Ser. Khim. *196*, 103 (1971).

[304] S. Hünig, O. Boes, Justus Liebigs Ann. Chem. *579*, 28, 33, 37 (1953).

[305] K. Schank, Justus Liebigs Ann. Chem. *716*, 87 (1968).

[306] K. Schank, Chem. Ber. *99*, 48 (1966); K. Schank, Chem. Ber. *103*, 3087 (1970).

[307] W. Ried, M. Butz, Justus Liebigs Ann. Chem. *716*, 190 (1968).

[308] USSR P. 188 983 (Cl. C o7c), 17.11.66, Appl. 23.12.65; Firma Institute of Organic Chemistry, Academy of Sciences Ukrainian SSR, Inv.: L.S. Pupko, A.I. Dychenko, P.S. Pel'kis; C.A. *68*, 12677t (1968).

[309] G. Rembarz, B. Ernst, Z. Chem. 7, 309 (1967); C.A. *67*, 99795f.

Table 5 (continued)

	Reference	
$Ar-SO_2-CH_2-SO_2-Ar$	311	
C₆H₅—C≡C—Ag	312	
(pyrazolidinedione structure with H_5C_6-N, H_5C_6-N ring, R^1, $C-CH_2$, R)	313	
(thiazolidinone structure with HN, S, R^1, $C-CH_2$, R)	314	
$HO_3S-CH_2-CO-CH_2-SO_3H$	315	
$Ar-SO_2-CH_2-COOC_2H_5$	316	
$H_3C-C=CH_2-COOC_2H_5$, $N-NH-CS-NH_2$	317	
$H_3C-CH=CH-CHO$	318	
$CH_2=C=O$	319	
(cyclopentadienylidene=P(C₆H₅)₃)	320	
$(H_3C)_3C-ZnCl$	321	
$H_2C=N_2$	322	Tetrazole formation occurs
C₆H₅—CH—COOC₂H₅, HgBr	323	

7.2.3.2 Azo coupling at the nitrogen

Nucleophilic nitrogen can react with aromatic diazonium salts like nucleophilic carbon to produce coupling. *Triazenes*[324] are formed from primary and secondary amines:

$$Ar-N_2^{\oplus} + H-N\genfrac{}{}{0pt}{}{R^1}{R^2} \xrightarrow{-H^{\oplus}} Ar-N=N-N\genfrac{}{}{0pt}{}{R^1}{R^2}$$

With primary amines, double azo coupling occurs and affords *1,4-pentaazadienes*[325].

$$2Ar-N_2^{\oplus} + H_2N-R \xrightarrow{-2H^{\oplus}} $$

$$Ar-N=N-\underset{\underset{R}{|}}{N}-N=N-Ar$$

Analogously, hydrazines and its derivatives yield *tetrazenes*[326] and higher azo homologs (see Table 6).

Table 6. Azo coupling reactions of nucleophilic amines

	Reference	
$HN(CH_2-CH_2-X)_2$; $X = F, Cl$	327	
HNR_2; $R = CH_3, C_2H_5, C_3H_7$	328	Where $R=C_6H_5$ N and C azo coupling occur[329]
$H_3C-NH-CH_2-COOC_2H_5$	330	$SOCl_2$/pyridine form: (triazolinone/oxadiazolium structure with $H_3C-N^{\oplus}$, $O^{\ominus}$, Ar)
C₆H₅—CH₂—NH₂	331	Use of excess NaOH can lead to pentaazadiene formation
$F-NH-COOC(CH_3)_3$	332	α-Elimination from a fluorotriazene forming an azide
$H-NF_2$	332	S_N substitution by $F^{\ominus}$ occurs at the phenyl ring; o- and p-F—C₆H₄—N₃ are formed

[310] *J. Rakotondraibe, C. Bertrand, P. Bedos,* C.R. Acad. Sci., Paris, Ser. C *267,* 576 (1968); C.A. *70,* 29078n (1969);
C. Bertrand, C.R. Acad. Sci., Paris *261,* 1012 (1965); C.A. *63,* 14914a (1965).

[311] *R.G. Dubenko, V.M. Neplyuev, P.S. Pel'kis,* Zh. Org. Khim. *3,* 33 (1967); C.A. *66,* 94767t (1967).

[312] *A.M. Sladkov, L.Y. Ukhin,* Izv. Akad. Nauk SSSR, Ser. Khim. 1552 (1964); C.A. *64,* 14116a (1966).

[313] *V.G. Jakutovic, B.L. Moldaver, J.P. Kitaev, Z.S. Titova,* Izv. Akad. Nauk SSSR, Ser. Khim. 2086 (1970).

[314] *T.V. Perova, S.N. Baranov,* Khim. Geterotsikl. Soedin. 232 (1969); C.A. *71,* 30389s (1969).

[315] *W.G. Grot,* J. Org. Chem. *30,* 515 (1965).

[316] *R.G. Dubenko, Y.N. Usenko, P.S. Pel'kis,* Zh. Org. Khim. *1,* 570 (1965); C.A. *63,* 1723g (1965).

[317] *H.G. Garg, R.A. Sharma,* J. Med. Chem. *13,* 991 (1970).

[318] *J. Podstata, Z.J. Allan,* Collect. Czech. Chem. Commun. *31,* 3829 (1966); C.A. *65,* 20253d (1966).

[319] *W. Ried, P. Junker,* Justus Liebigs Ann. Chem. *709,* 85 (1967).

[320] *S.C. Sehgal, A.V. Patwardhan,* Indian J. Chem. *8,* 900 (1970).

[321] *R. O'Connor,* J. Org. Chem. *26,* 4380 (1961).

[322] *H. Reimlinger, G.S.D. King, M.A. Peiren,* Chem. Ber. *103,* 2821 (1970).

[323] *I.P. Beletskaya, K.P. Butin, O.A. Reutov,* Izv. Akad. Nauk SSSR, Ser. Khim. 1382 (1966); C.A. *66,* 64783c (1967).

[324] *C. Süling* in *Houben-Weyl,* Methoden der organischen Chemie, 4. Aufl., Bd. X/3, p. 695, Georg Thieme Verlag, Stuttgart 1965.

[325] *C. Süling* in *Houben-Weyl,* Methoden der organischen Chemie, 4. Aufl., Bd. X/3, p. 736, Georg Thieme Verlag, Stuttgart 1965.

[326] *C. Süling* in *Houben-Weyl,* Methoden der organischen Chemie, 4. Aufl., Bd. X/3, p. 733, Georg Thieme Verlag, Stuttgart 1965.

Table 6 (continued)

	Reference

183

333

Adenosine oxide

334 Coupling with diazotized 5-, 6- and 8-aminoquinolines gives triazenes. In contrast, reaction of these amines with benzenediazonium chloride leads only to C azo coupling

R—NH—OH 335

$H_2C(CH=NOH)_2$ 336 Triazene formation occurs by cyclization:

H_2N—CHO 337 Reaction with $SOCl_2$/pyridine forms diazoisocyanides

---

N-Aminopyridinium chloride undergoes azo coupling with 4-nitrobenzenediazonium chloride, the reaction leading spontaneously to *4-azido-1-nitrobenzene*[338]:

---

[327] *Y.F. Shealy, C.A. O'Dell,* J. Pharm. Sci. *59,* 1358 (1970).

[328] *V.V. Ershov et al.,* Izv. Akad Nauk SSSR, Ser. Khim. 146 (1971).

[329] *Z.J. Allan,* Tetrahedron Lett. 2425 (1970).

[330] *K.T. Potts, S. Husain,* J. Org. Chem. *35,* 3451 (1970);
J.H. Phillips jr., S.A. Robrish, C. Bates, J. Biol. Chem. *240,* 699 (1965); C.A. *62,* 5472c (1965).

[331] *H. Moll, R. Vuille,* Helv. Chim. Acta *52,* 665 (1969); C.A. *70,* 114499v (1969).

[332] *K. Baum,* J. Org. Chem. *33,* 4333 (1968).

Analogous α-eliminations are known to occur with the azo coupling products of primary sulfonamides, azides being produced[339]. The coupling reaction[340] between aryldiazonium ions and cyanide at −15° is particularly interesting:

R	A Mp	B Mp
NO_2	49–51°	86°
Br	48°	132°
OCH_3	48–50°	120–121°

In this reaction *diazo isocyanides*[337] are formed first, and these rearrange to the stable cyanides (*i.e., cis-trans* isomerization is not involved)[341].

7.2.3.3 Coupling at the oxygen

Coupling products representing aryldiazotate derivatives of varying stability[346] and obtained from aromatic diazonium salts and materials containing nucleophilic oxygen, include *diazoalkyl ethers* and *diazoaryl ethers*[342], *diazo anhydrides*[343], *diazo carboxylates*[344], and *diazosulfonates*[345]. In addition, the formation of biphenyl by reaction of *N,N*-diphenylhydroxylamine

---

[333] *H. Koessel, S. Doehring,* Biochim. Biophys. Acta *95,* 663 (1965); C.A. *63,* 5716h (1965).

[334] *E.V. Brown, J.J. Duffy,* J. Org. Chem. *32,* 1124 (1967).

[335] *S.M. Dugar, N.C. Sogani,* J. Indian Chem. Soc. *43,* 289 (1966); C.A. *65,* 5385e (1966).

[336] *H.-J. Sturm, H. Armbrust,* Justus Liebigs Ann. Chem. *729,* 139 (1969).

[337] *J. Susko, T. Ignasiah,* Bull. Acad. Polon Sci., Sér. Sci. Chim. *18,* 657, 663, 669, 672 (1970).

[338] *T. Okamoto, S. Hayashi,* Yakugaku Zasshi *86,* 766 (1966); C.A. *65,* 20116h (1966).

[339] *C. Grundmann* in *Houben-Weyl,* Methoden der organischen Chemie, 4. Aufl., Bd. X/3, p. 808, Georg Thieme Verlag, Stuttgart 1965; cf.a.:
R.C. Kerber, J. Org. Chem. *33,* 4442 (1968).

[340] *L.A. Kazitsyna, E.S. Kozlov, O.A. Reutov,* Dokl. Akad. Nauk SSSR *160,* 600 (1965); C.A. *62,* 13068c (1965).

[341] *R. Pütter* in *Houben-Weyl,* Methoden der organischen Chemie, 4. Aufl., Bd. X/3, p. 593, Georg Thieme Verlag, Stuttgart 1965.

[342] *R. Pütter* in *Houben-Weyl,* Methoden der organischen Chemie, 4. Aufl., Bd. X/3, p. 560, 561, Georg Thieme Verlag, Stuttgart 1965.

[343] *R. Pütter* in *Houben-Weyl,* Methoden der organischen Chemie, 4. Aufl., Bd. X/3, p. 145, 553, Georg Thieme Verlag, Stuttgart 1965.

with benzenediazonium tetrafluoroborate[347] probably involves a coupling at the oxygen as the first stage:

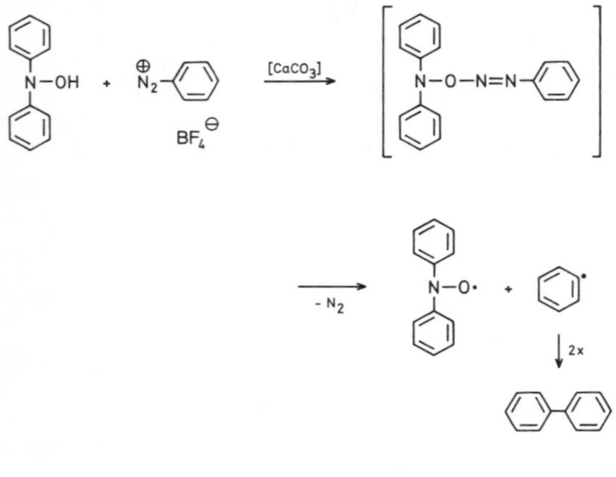

7.2.3.4 Coupling at the phosphorus

Various coupling reactions of aromatic diazonium salts with nucleophilic phosphorus(III) compounds have been surveyed in a review, see ref.[348].

7.2.3.5 Coupling at the sulfur

As the basically *soft* position in sulfides[349], sulfinates[350], sulfur dioxide, and sulfites[351], sulfur is very reactive towards aromatic diazonium salts. 4-Aminothiophenol reacts quite differently to the

analogous oxygen compound in that on diazotization it forms a **highly explosive** *polymer*[352]:

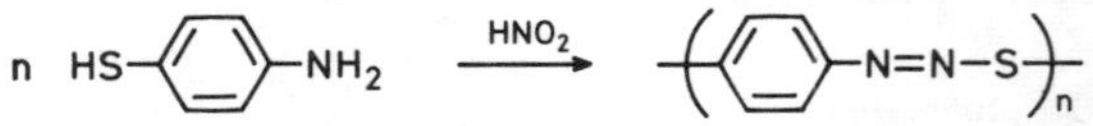

instead of a quinone diazide.

7.3 Analysis

7.3.1 Spectral data

The *IR spectra* of aromatic diazonium salts contain a characteristic band (N_2 valence vibration) within the 2000–3000 cm^{-1} region, with the exact position depending to some extent on the associated anion and on the substituent groups on the aromatic nucleus[353]. The effect exercised by substituents on the aromatic ring both on the N_2 valence vibration band and on the *Ar–N* valence vibration over the 1300–1400 cm^{-1} region is more significant. It has been established[354] that a linear relationship exists between the observed frequencies and the Hammett σ_p values. *IR spectroscopy*

[344] *R. Pütter* in *Houben-Weyl*, Methoden der organischen Chemie, 4. Aufl., Bd. X/3, p. 149, Georg Thieme Verlag, Stuttgart 1965.

[345] *L. Benati, M. Tiecco, A. Tundo*, Boll. Sci. Fac. Chim. Ind. Bologna *24*, 219 (1966); C.A. *67*, 2847p (1967).

[346] *R. Pütter* in *Houben-Weyl*, Methoden der organischen Chemie, 4. Aufl., Bd. X/3, p. 551, Georg Thieme Verlag, Stuttgart 1965.

[347] *R.M. Cooper, M.J. Perkins*, Tetrahedron Lett. 2477 (1969).

[348] *R. Pütter* in *Houben-Weyl*, Methoden der organischen Chemie, 4. Aufl., Bd. X/3, p. 587, Georg Thieme Verlag, Stuttgart 1965.

[349] *R. Pütter* in *Houben-Weyl*, Methoden der organischen Chemie, 4. Aufl., Bd. X/3, p. 564, Georg Thieme Verlag, Stuttgart 1965;
H. van Zwet, E.C. Kooyman, Chem. Commun. 313 (1965); C.A. 63, 9849e (1965);
H. van Zwet, E.C. Kooyman, Rec. Trav. Chim. Pays-Bas *86*, 993, 1143 (1967); C.A. *68*, 38807k (1968);
T. Yamada, Bull. Chem. Soc. Japan *43*, 1506 (1970).

[350] *R. Pütter* in *Houben-Weyl*, Methoden der organischen Chemie, 4. Aufl., Bd. X/3, p. 578, Georg Thieme Verlag, Stuttgart 1965;
W.V. Farrar, Chem. & Ind. (London) 1985 (1964);
K. Garves, J. Org. Chem. *35*, 3273 (1970).

[351] *R. Pütter* in *Houben-Weyl*, Methoden der organischen Chemie, 4. Aufl., Bd. X/3, p. 570 Georg Thieme Verlag, Stuttgart 1965.

[352] *W. Tagaki, M. Nakagawa, S. Oae*, Mem. Fac. Eng., Osaka City Univ. *6*, 217 (1964); C.A. *63*, 17949a (1965).

[353] *V.P. Sagalovich, L.D. Ashkinadze, V.V. Kozlov, L.A. Kazitsyna*, Zh. Obshch. Khim. *36*, 1934 (1966); C.A. *66*, 94564z (1967) (Heteropolyacids);
L.A. Kazitsyna, N.B. Dzegilenko, A.A. Kibkalo, V.V. Kolzlov, B.I. Belov, Zh. Org. Khim. *2*, 1822 (1966); C.A. *66*, 50503z (1967);
L.A. Kazitsyna, N.B. Dzegilenko, Zh. Org. Khim. *4*, 2153 (1968); C.A. *70*, 67400n (1969) (Salts of organic acids);
V.V. Kozlov, T.D. Silaeva, Dokl. Akad. Nauk SSSR *117*, 116 (1967); C.A. *69*, 35609g (1968) (Orthophosphates).

[354] *R.J. Cox, J. Kumamoto*, J. Org. Chem. *30*, 4254 (1965);
K. Tabei, C. Ito, Bull. Chem. Soc. Japan *41*, 514 (1968); C.A. *68*, 91391w (1968);
L.A. Kazitsyna, L.D. Ashkinadze, O.A. Reutov, Izv. Akad. Nauk SSSR, Ser. Khim. 702 (1967); C.A. *67*, 77611r (1967);
L.A. Kazitsyna, L.D. Ashkinadze, A.V. Upadysheva, L.E. Vinogradova, O.A. Reutov, Dokl. Akad. Nauk SSSR *167*, 835 (1966); C.A. *65*, 1600h (1966);
D.A. Bochvar, N.P. Gamgaryan, V.V. Mischenko, L.A. Kazitsyna, Dokl. Akad. Nauk SSSR *175*, 829 (1967); C.A. *68*, 77604m (1968) (Quinone diazides);
B.A. Porai-Koshits, I.L. Bagal, Latv. PSR Zinat. Akad. Vestis, Khim. Ser. 569 (1965); C.A. *64*, 9570e (1966);
I.L. Bagal, B.A. Porai-Koshits, Reakts. Sposobnost. Org. Soedin., Tartu Gos. Univ. *3*, 89, 102 (1966); C.A. *68*, 82729k, 82731e (1968);
P. Schuster, O.E. Polansky, Monatsh. Chem. *96*, 396 (1965); C.A. *63*, 9226f (1965).

has been used also to investigate the *pH*-dependence of the acid-base equilibria of *p*-aminosubstituted diazonium salts[355]. The electronic configuration of ring-substituted diazonium salts has been studied using (*UV* and visible) electron-excitation spectroscopy[357].

7.3.2 Analytical methods

Crystal structure investigations[357–359]
Thin-layer chromatography
    Salt separation[36]
    Amine determination[361]
    Phenol determination[362]
Quantitative determinations
    Phenols, amines[363–366]
Oxidation potential, potentiometry[367]
Half-wave potential, polarography[368]
Ultraviolet spectrometry[369]
Colorimetry[370]

[355] *L.A. Kazitsyna, A.V. Upadysheva, A.A. Stepanova, O.A. Korytina, O.A. Reutov*, Izv. Akad. Nauk SSSR, Ser. Khim. 1917 (1967); C.A. *68*, 64120a (1968);
L.A. Kazitsyna, N.B. Dzegilenko, A.V. Upadysheva, V.V. Mishchenko, O.A. Reutov, Izv. Akad. Nauk SSSR, Ser. Khim. 1925 (1967); C.A. *68*, 68244s (1968).
[356] *M. Sukigara, S. Kikuchi*, Bull. Chem. Soc. Japan *40*, 1077, 1082 (1967); C.A. *68*, 73707z, 73708a (1968).
[357] *C. Roemming, K. Eaerstad*, Chem. Commun. 299 (1965); C,A. *63*, 9160c (1965).
[358] *C. Roemming, T. Tjoernhom*, Acta Chem. Scand. *22*, 2934 (1968); C.A. *70*, 62147g (1969).
[359] *B. Greenberg, Y. Okaya*, Acta Crystallogr., Sect. B. *25*, 2101 (1969); C.A. *71*, 129686y (1969).
[360] *R.J. Gritter*, J. Chromatogr. *20*, 416 (1965); C.A. *64*, 5728d (1966).
[361] *V.C. Quesnel*, J. Chromatogr. *24*, 268 (1966); C.A. *65*, 20489 (1966).
[362] *A. Fono, A.M. Sapse, T.S. Ma*, Mikrochim. Ichnoanalyt. Acta 1098 (1965); C.A. *64*, 16019a (1966); *T. Kondo, I. Kawashiro*, Shokuhin Eiseigaku Zasshi *6*, 433 (1965); C.A. *64*, 16633c (1966).
[363] *M. Matrka, A. Spevak, E. Verisova*, Chem. Listy *61*, 532 (1967); C.A. *67*, 17622k (1967); *M. Matrka, E. Verisova*, Chem. Průmysl. *16*, 660 (1966); C.A. *66*, 34638s (1967).
[364] *B.A. Porai-Koshits, A.B. Tomchin*, Zh. Org. Khim. *2*, 2238 (1966); C.A. *66*, 85551r (1967); *T. Jasinski, R. Korewa, H. Smagowski*, Chem. Anal. (Warsaw) *9*, 655 (1964); C.A. *62*, 2231h (1965).
[365] *L. Legradi*, Magy. Kem. Folyoirut *72*, 336 (1966); C.A. *65*, 14432c (1966); *L. Legradi*, Mikrochim. Ichnoanal. Acta 865, 870 (1965); C.A. *65*, 14a (1966).
[366] *L. Nebbia, V. Bellotti*, Chim. Ind. (Milan) *52*, 791 (1970).
[367] *M. Matrka, Z. Sagner, V. Chmatal, V. Sterba*, Collect. Czech. Chem. Commun. *32*, 2532 (1967); C.A. *67*, 53513j (1967).
[368] *R.M. Elofson, F.F. Gadallah*, J. Org. Chem. *34*, 854 (1969).
[369] *I. Balakrishnan, M.P. Reddy*, Indian J. Chem. *6*, 257 (1968); C.A. *70*, 20958s (1969).
[370] *V.T. Kaplin, Y.Y. Lur'e, S.E. Pachenko, N.G. Fesenko*, Gidrokhim. Mater. *41*, 84 (1966); C.A. *66*, 61637x (1967).

8 Aliphatic Diazo Compounds

Contributed by

M. Regitz
Fachbereich Chemie der Universität
Trier/Kaiserslautern,
Kaiserslautern

I.K. Korobizina
L.L. Rodina
Department of Chemistry,
University of Leningrad,
Leningrad

8.1 Preparation of aliphatic diazo compounds

Ever since their discovery by Curtius in 1883, aliphatic diazo compounds have played an outstanding role in synthetic organic chemistry. Today their photolytic decomposition occupies the centre of interest because it affords a reliable means of entry into carbene chemistry. Their *identification* can be performed by means of the *IR* diazo vibrational band which occurs over the 2000 to 2250 cm^{-1} region in dependence of the substituents. As a rule substituents with electron-acceptor properties produce a shift of the diazo absorption to higher wave numbers. A confusion with acetylenes, nitriles, cumulenes, or heterocumulenes is feasible; these can absorb over the same region.

A quantitative determination is indicated above all in those diazo compounds which cannot be isolated in substance. It is performed by reacting with acid and is described on p. 226.

A special reference to the liability of certain diazo compounds to **decompose** or **explode** is appropriate, although this is often overestimated. Particular care is necessary during purification of diazo compounds by *distillation* (vent, protective screen, goggles); large quantities are avoided when this operation is carried out. Handling *diazomethane* itself and its homologs, which are particularly sensitive to decomposition induced by traces of catalysts, is made easier by employing it only in dilute solution. In addition, they are strong **poisons** whose breathing and touching must be strictly eschewed. Potential dangers are pointed out in individual cases. However, overall it can be said that the great majority of the diazo compounds described in the present chapter are quite stable and can be handled without difficulty.

Synthesis of aliphatic diazo compounds is effected according to four fundamental principles.

1 The starting materials used are compounds which already contain a functional group with a N atom on the subsequent diazo C atom; this functional group is converted into the N_2 group by condensation (*amine diazotization, Forster reaction*).
2 Compounds containing a functional group with two N atoms on the relevant carbon are converted into diazo derivatives (*dehydrogenation of hydrazones, Bamford-Stevens reaction, acyl cleavage of N-acyl-N-nitrosoamines*).
3 Introduction of the entire N_2 group to an acceptor by a donor (*diazo group transfer*)
4 Substitution reactions on *ready* existing diazo compounds[1].

Table 1. a-Diazo ketones by Forster reaction (see pp. 211, 212)

1,2-diketone monooxime	Reaction partner Reaction conditions	a-Diazo ketone	Yield [% of theory]	M. p. [°]	Ref.
	Aq. chloramine, caustic soda; 1 hr. 0–3°, 3 hrs. at ~ 20°	5-(Diazoacetyl)uracil	15	> 300	49
	Aq. chloramine, caustic soda; 1 hr. at 2°, 5 hrs. at ~ 20°	7-Chloro-2-diazo-4-methyl-1-indanone	63	173–180 decomp.	50
	Aq. chloramine, caustic soda; 5–6 hrs. at 0°	2-Diazo-1H-benz[e]inden-3-one	65	102 decomp.	51
	Aq. chloramine, caustic soda; 6 hrs. at 0°	2-Diazo-5,6-dihydro-1H-cyclobuta[f]inden-1-one	61	129–130	52
	Aq. chloramine, caustic soda; 5–6 hrs. at 0°	2-Diazo-3,4-dihydro-1(2H)-naphthalenone	57	52–53	51
	Aq. chloramine, caustic soda; 30 min at 0°	3-Diazo-2-pinanone (3-Diazo-7,7-dimethylbicyclo[3.1.1] heptan-2-one)	85	oil	53
	Aq. chloramine, caustic soda; 1 hr. in melting ice-bath	5-Diazo-2-carenone (5-Diazo-3,7,7-trimethylbicyclo[4.1.0] hept-3-en-2-one)	67	40 decomp.	54
	Aq. chloramine, caustic soda, methanol; 7 hrs. at ~ 20°	16-Diazo-3 β-hydroxy-androst-5-en-17-one	77	220	55

Table A surveys the preparation of the various types of aliphatic diazo compounds (p. 244).

certain constitutional requisites in the amine component.

8.1.1 Diazotization of amines

Diazotization of *primary* aliphatic amines to form diazo compounds (1→3), which was used initially for preparing *ethyl diazoacetate*[2], is bound up with

[1] *Review: B. Eistert, M. Regitz, G. Heck, H. Schwall* in *Houben-Weyl*, Methoden der organischen Chemie, 4. Aufl., Bd. X/4, p. 557, Georg Thieme Verlag, Stuttgart 1968.

[2] *T. Curtius*, Ber. *16*, 2230 (1883).

Assuming the alkyldiazonium salt **2** to be an intermediate stage, then the same can form the diazoalkane **3** while losing a proton (or *HX*). However, the detaching of $H^\oplus$ takes place only where it is facilitated by dipoles in the α-position having a proton-loosening effect such as the halomethyl, cyano, or acyl group. This condition defines the *limits* of the method; where it is not met **2** loses nitrogen to form the carbenium ion **4**, which can react in various ways that are of no interest here.

8.1.1.1 Diazoalkanes and fluorodiazoalkanes

The conversion of methylamine into *diazomethane* with nitrosyl chloride is something of an anomaly, because proton-activating substituents are absent[3, 4]. It becomes comprehensible when it is remembered that, unlike diazotisation with nitrite, it takes place over the alkaline region:

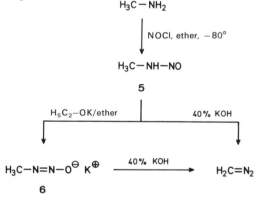

$$H_3C-NH_2$$

NOCl, ether, $-80°$

$$H_3C-NH-NO$$

5

H_5C_2-OK/ether   40% KOH

$$H_3C-N=N-O^\ominus \ K^\oplus \xrightarrow{\ 40\% \ KOH\ } H_2C=N_2$$

6

The unstable methylnitrosamine **5** can be identified as an intermediate stage by *UV*-optical evidence. With potassium ethoxide in ether the reaction solution affords potassium methyl diazotate **6**, whose alkaline hydrolysis results in 48–54% *diazomethane* (on nitrosyl chloride). The diazotate **6**, which is stable for quite some time if moisture is excluded, is also known as *stabilised diazomethane* (cf. p. 223). Direct alkaline hydrolysis of the nitrosoamine solution yields the same result (74–78% diazomethane). *1-Diazooctane* ($\sim$40%) is accessible in the same way[5].

Acid diazotization of alkylammonium salts with trifluoromethyl or other highly fluorinated alkyl groups on the amine carbon generally proceeds as

expected. Thus 2,2,2-trifluoroethylammonium chloride and sodium nitrite afford up to 70% *1-diazo-2,2,2-trifluoroethane*[6, 7]:

$$F_3C-CH_2-\overset{\oplus}{N}H_3 \ Cl^\ominus \xrightarrow[-NaCl, \ -2H_2O]{NaNO_2} F_3C-CH=N_2$$

The following are obtained analogously

2-Diazo-1,1,1-trifluoropropane[8]

2-Diazo-1,1,1,3,3,3-hexafluoropropane[9]; 48%

1-Diazo-2,2,3,3,4,4,4-heptafluorobutane[10]

1-Diazo-2,2,3,3-tetrafluoropropane[11]; 45%

1-Diazo-2,2,2-trifluorophenylethane[12]; 17%

2-Diazo-1,1,1-trichloroethane cannot be obtained in this way[13]. Care is indicated during distillation of fluorine-substituted diazo compounds, an additional reason being that the diazo compounds prepared by this route still contain traces of acid which catalyze the decomposition.

8.1.1.2 α-Diazocarbonyl compounds

Ethyl diazoacetate is prepared by diazotization of amines[2, 14]:

$$H_2C-COOC_2H_5 \atop \underset{Cl^\ominus}{\overset{\oplus}{N}H_3} \xrightarrow[CH_2Cl_2, <0°]{H_2SO_4/NaNO_2} HC-COOC_2H_5 \atop N_2$$

The ester group can be varied widely[15], *e.g.: tert-butyl diazoacetate*[16] and D- and L-*diazoacetylserine (azaserine)*[17, 18]:

$$H_2C-CO-O-CH_2-CH-COOH \atop \underset{Cl^\ominus}{\overset{\oplus}{N}H_3} \qquad NH_2$$

AgNO$_2$ (L – Form)
NaNO$_2$ (D – Form)

$$HC-CO-O-CH_2-CH-COOH \atop N_2 \qquad\qquad NH_2$$

3 *E. Müller, W. Rundel,* Chem. Ber. *91*, 466 (1958).
4 *E. Müller, H. Haiss, W. Rundel,* Chem. Ber. *93*, 1541 (1960).
5 *J. Bakke,* Acta Chem. Scand. *22*, 1833 (1968).

6 *H. Gilman, R.G. Jones,* J. Amer. Chem. Soc. *65*, 1458 (1943).
7 *B.L. Dyatkin, E.P. Mochalina,* Izv. Akad. Nauk SSSR, Ser. Khim. 1225 (1964); C.A. *61*, 11881f (1964).
8 *R.A. Shepard, P.L. Sciaraffa,* J. Org. Chem. *31*, 964 (1966).
9 *E.P. Mochalina, B.L. Dyatkin,* Izv. Akad. Nauk SSSR, Ser. Khim. 926 (1965); C.A. *63*, 5515c (1965).
10 *L.C. Krogh, T.S. Reid, H.A. Brown,* J. Org. Chem. *19*, 1124 (1954).
11 *J.H. Atherton, R. Fields, R.N. Haszeldine,* J. Chem. Soc. C 366 (1971).
12 *R.A. Shepard, S.E. Wentworth,* J. Org. Chem. *32*, 3197 (1967).
13 *A. Roedig, K. Grohe,* Tetrahedron *21*, 2375 (1965).
14 *N.E. Searle,* Org. Syntheses Coll. Vol. IV, 424 (1963).

Diazotization of glycine hydrazide affords *diazoacetic acid azide* (Caution: **explosive**)[19]:

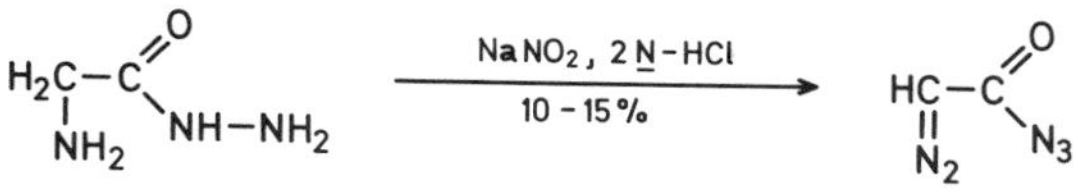

While *long-chain 2-diazoalkane acid esters* are accessible by amine diazotization (*ethyl 2-diazopropionate* <10%[20], *ethyl 2-diazohexanoate*, 30%[21]), they are obtained in higher yield by deformylating diazo group transfer (see p. 235).

Amine diazotization is not particularly suitable for the synthesis of diazo ketones because it is often accompanied by acid decomposition of the diazo compound. Thus, from 2-aminocyclohexanone it is not the diazo derivative that is formed but cyclopentanecarboxylic acid, probably in a sequential reaction (involving a carbenium ion mechanism) accompanied by ring contraction[22]:

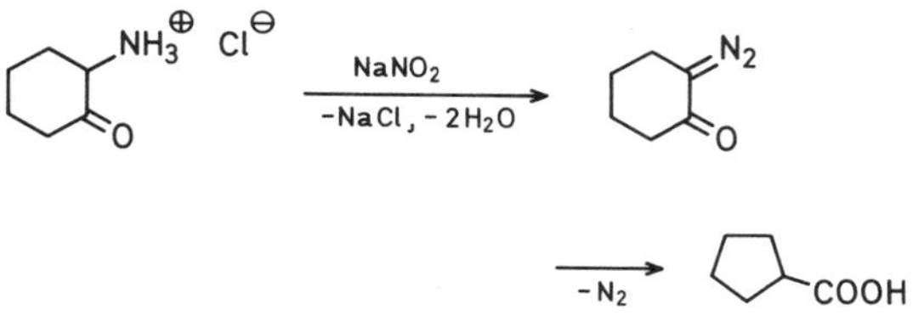

In other cases, for instance, *3-diazocamphor (3-diazo-2-bornanone)*[23] and *2-diazoacetophenone*[24], it can be employed. Analogous behavior applies to *o-quinone diazides*, which are structurally closely related to the α-diazo ketones[25], *e.g.*:

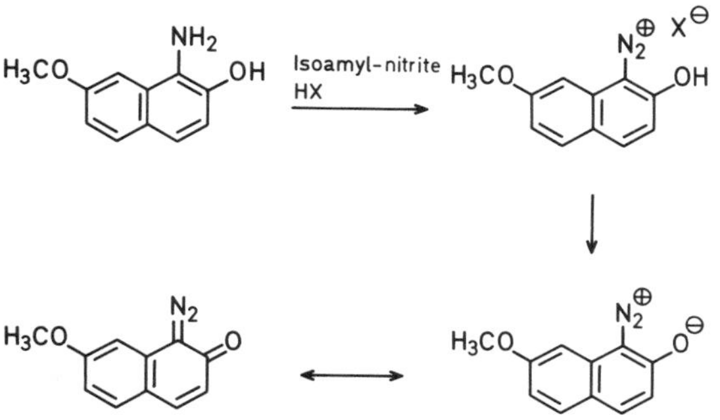

1-Diazo-7-methoxy-2(1H)-naphthalenone

In the case of 3,6-diamino-1,2,4,5-benzenetetrol diazotization is additionally accompanied by dehydrogenation, so that *3,6-bis(diazo)-1,2,4,5-cyclohexanetetrone* is formed as the final stage. Its quinone diazide character is expressed in its diazonium enolate limiting formula[26]:

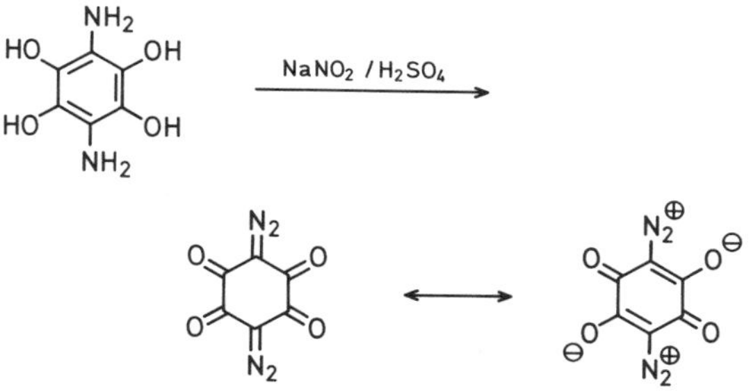

2-Diazo-1,3-dioxo compounds are generally accessible without complications, because the **2→3** transition (p. 207) is guaranteed by two proton-activating groups and the exceptional acid stability prevents decomposition.

The stability is due to an extensive delocalization of the electrons on the diazo carbon hindering the *C*-protonation:

The following compounds can be prepared analogously:

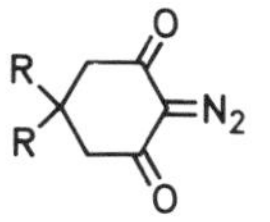

R = H; *2-Diazo-1,3-cyclo-hexanedione*[28]

R = CH₃; *2-Diazo-5,5-dimethyl-1,3-cyclo-hexanedione*[27]

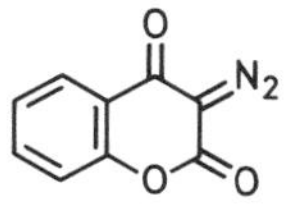

3-Diazo-2,4-chromandione[29]

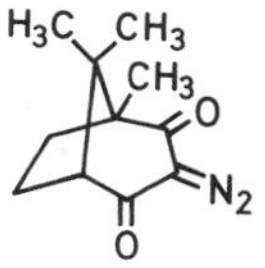

3-Diazo-1,8,8-trimethylbicyclo-[3.2.1]octane-2,4-dione[30]

[15] USP. 2 490 714 (1949), DuPont, Inv.: *N.E. Searle*; C.A. *44*, 3519d (1950).
[16] *E. Müller, H. Huber-Emden*, Justus Liebigs Ann. Chem. *660*, 54 (1962).
[17] *J.A. Moore, J.R. Dice, E.D. Nicolaides, R.D. Westland, E.L. Wittle*, J. Amer. Chem. Soc. *76*, 2884 (1954).
[18] *E.D. Nicolaides, R.D. Westland, E.D. Wittle*, J. Amer. Chem. Soc. *76*, 2887 (1954).
[19] *H. Neunhoeffer, G. Cuny, W.K. Franke*, Justus Liebigs Ann. Chem. *713*, 96 (1968).

[20] *T. Curtius, E. Müller*, Ber. *37*, 1261 (1904).
[21] *C.S. Marvel, W.A. Noyes*, J. Amer. Chem. Soc. *42*, 2259 (1920).
[22] *O.E. Edwards, M. Lesage*, J. Org. Chem. *24*, 2071 (1959).
[23] *R. Schiff*, Ber. *14*, 1375 (1881);
A. Angeli, Gazz. Chim. Ital. *23*, II, 351 (1893);
A. Angeli, Gazz. Chim. Ital. *24*, II, 318 (1894).
[24] *A. Angeli*, Ber. *26*, 1715 (1893).
[25] *O. Süs*, Justus Liebigs Ann. Chem. *579*, 133 (1953);
O. Süs, K. Möller, Justus Liebigs Ann. Chem. *593*, 91 (1955);
O. Süs, M. Glas, K. Möller, H.-D. Eberhardt, Justus Liebigs Ann. Chem. *583*, 150 (1953).

Synthesis of the diazo compound involves the steps of nitrating the 1, 3-diketone, forming the oxime or performing an azo coupling, reducing to amine, and then diazotizing.

As β-diketones can be converted into diazo derivatives by diazo group transfer in one step (see p. 231), amine diazotization has lost much of its importance.

8.1.1.3 α-Diazophosphono and α-diazophosphinyl compounds

Dimethyl and *diethyl esters* of *diazomethanephosphonic acid* ($R^1 = OCH_3$, OC_2H_5; $R^2 = H$)[31, 32] are formed by amine diazotization in acetic acid solution. The same is true of *diazodiphenylphosphinylmethane* ($R^1 = C_6H_5$; $R^2 = H$[32, 33]), *diazodibenzylphosphinylmethane* ($R^1 = C_6H_5 - CH_2$; $R^2 = H$)[33], and *diazodiphenylphosphinylphenylmethane* ($R^1 = R^2 = C_6H_5$)[34] which display a reactivity matching that of the α-diazo ketones[35]. Admittedly, in the last-named case partial decomposition of the diazo compound is observed, possibly as a consequence of a reaction of the nitrite in the mineral acid medium[36].

The extraordinary thermal stability of *PO-substituted* diazo compounds is noteworthy; it far exceeds that of the α-diazocarbonyl compounds[37].

8.1.1.4 Heteroaromatic diazo compounds

Diazotization of heteroaromatic amines initially affords diazonium salts in analogy with aromatic amines (see p. 168), which, if suitably constituted, can be deprotonated to diazo compounds with weak bases. Thus, 3-aminopyrazole furnishes the diazonium salt in methanolic hydrochloric acid from which hydrogen chloride can be split off with triethylamine or sodium bicarbonate to form *3-diazopyrazole*[38].

Substituted 3- and 4-diazopyrazoles are accessible in the same way[39].

4-Diazo-5-imidazolecarboxamide[40] and the *8-diazopurines*[41] belonging to the same basic system are formed by diazotization of the corresponding amines. Both these diazo compounds cyclize under acid and base catalysis[40]:

X = CH; *7H-Imidazolo[4,5-d]-1,2,3-triazin-4-one;* 85%

X = N;   *7H-1,2,3-Triazolo[4,5-d]-1,2,3-triazin-4-one;* 51%

Diazotetrazoles[42] and 3-diazopyrazoles[43] are prepared in analogous manner.

[26] *F. Henle*, Justus Liebigs Ann. Chem. *350*, 344 (1906);
G.B. Ansell, P.R. Hammond, S.V. Hering, P. Corradini, Tetrahedron *25*, 2549 (1969).
[27] *P. Haas*, J. Chem. Soc. (London) *91*, 1433 (1907).
[28] *H. Stetter, K. Kiehs*, Chem. Ber. *98*, 1184 (1965).
[29] *F. Arndt, L. Loewe, R. Ün, E. Ayaça*, Chem. Ber. *84*, 319 (1951);
C.F. Huebner, K.P. Link, J. Amer. Chem. Soc. *67*, 99 (1945).
[30] *B. Eistert, D. Greiber, I. Caspari*, Justus Liebigs Ann. Chem. *659*, 64 (1962).
[31] *D. Seyferth, R.S. Marmor*, Tetrahedron Lett. 2493 (1970).
[32] *M. Regitz, A. Liedhegener, U. Eckstein, M. Martin, W. Anschütz*, Justus Liebigs Ann. Chem. *748*, 207 (1971).
[33] *N. Kreutzkamp, E. Schmidt-Samoa, K. Herberg*, Angew. Chem. 77, 1138 (1965); *N. Kreutzkamp, E. Schmidt-Samoa, K. Herberg*, Angew. Chem. Intern. Ed. Engl. *4*, 1078 (1965).
[34] *M. Regitz, H. Eckes*, unpubl. results, Saarbrücken 1971.
[35] *M. Regitz, A. Liedhegener, W. Anschütz, H. Eckes*, Chem. Ber. *104*, 2177 (1971).
[36] *L. Horner, H. Hoffmann, H. Ertel, G. Klahre*, Tetrahedron Lett. *9* (1961).

[37] *M. Regitz, W. Bartz*, Chem. Ber. *103*, 1477 (1970).
[38] *H. Reimlinger, A. van Overstraeten, H.G. Viehe*, Chem. Ber. *94*, 1036 (1961).
[39] *D.G. Farnum, P. Yates*, J. Amer. Chem. Soc. *84*, 1399 (1962).
[40] *Y.F. Shealy, R.F. Struck, L.B. Holum, J.A. Montgomery*, J. Org. Chem. *26*, 2396 (1961);
Y.F. Shealy, C.A. Krauth, J.A. Montgomery, J. Org. Chem. *27*, 2150 (1962).
[41] *J.W. Jones, R.K. Robins*, J. Amer. Chem. Soc. *82*, 3773 (1960).
[42] *J. Thiele*, Justus Liebigs Ann. Chem. *270*, 59 (1892); *J. Thiele, J.T. Marais*, Justus Liebigs Ann. Chem. *273*, 147 (1893);
J. Thiele, H. Ingle, Justus Liebigs Ann. Chem. *287*, 235 (1895).
[43] *F. Angelico*, Atti Acad. Naz. Lincei, Classe Sci. Fis., Mat. Natur., Rend. II 5, *14*, II, 169 (1905).

8.1.2 Forster reaction

Condensation of oximes with chloramine *(Forster reaction)* represents a simple method for synthesizing the diazo group[45]:

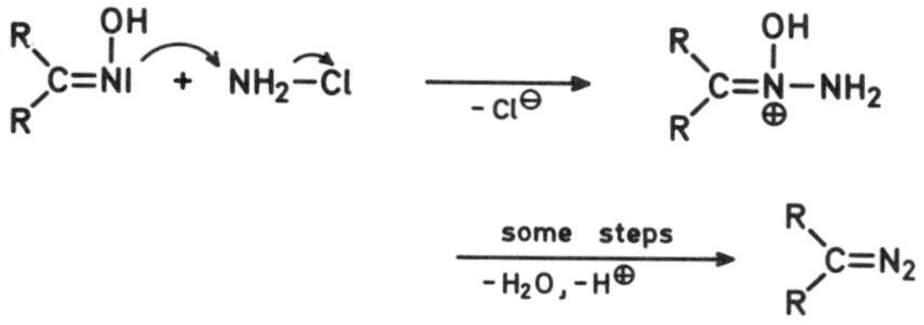

The Forster reaction[44] is used predominantly for preparing[45] α-*diazocarbonyl* compounds.

8.1.2.1 Diazomethane and aryldiazomethanes

Diazomethane can be prepared in 70–75% yield using the Forster reaction[46]; by contrast, applying the procedure to *C*-alkylated oximes is not very successful[46]:

$$H_2C=NONa \xrightarrow[-NaCl, -H_2O]{NH_2Cl, Ether/Methanol} H_2C=N_2$$

During preparation of aryl-substituted diazoalkanes by the Forster reaction *(9-diazofluorene, diazodiphenylmethane, etc.)* the chloramine is generated in the reaction mixture from ammonia and sodium hypochlorite[45]. Replacing chloramine by hydroxylamine-*O*-sulfonic acid affords no preparative advantage here[45] but does give success during synthesis of 1-(2-diazoethylidene)indene[47]:

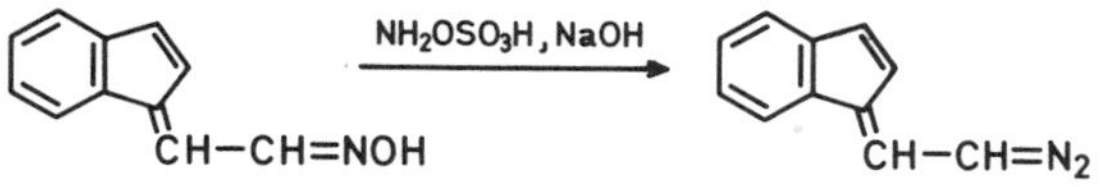

8.1.2.2 α-Diazo ketones

Using the Forster reaction for preparing diazo ketones (significance see p. 212) enables α-methylene ketones to be converted into diazo derivatives in two steps. The chloramine condensation is preceded merely by an oximation (the same transformation can be effected by deformylating diazo group transfer, see p. 235):

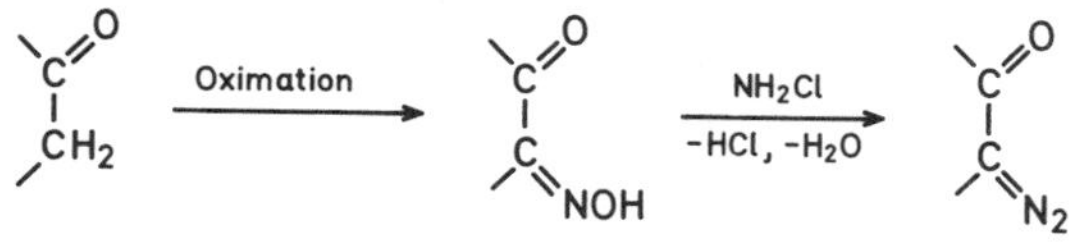

The Forster reaction proceeds particularly successfuly in the *indan series*. Thus, 3,3-diphenyl-1,2-indandione 2-oxime affords *2-diazo-3,3-diphenyl-1-indanone* (65%)[48]. The reaction is interesting in that the last-named compound cannot be obtained from 3,3-diphenyl-1,2-indandione and *p*-toluenesulfonic acid hydrazide (Bamford-Stevens reaction, see p. 215); the positional isomer *1-diazo-3,3-diphenyl-2-indanone* is formed exclusively[48].

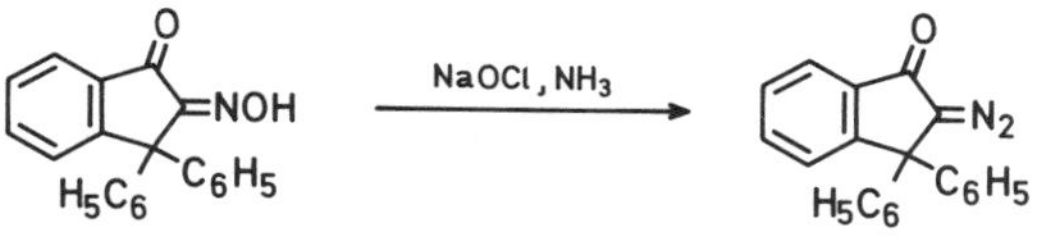

In the *steroid* series, too, numerous applications are known. Table 1 (see p. 207) portrays the potential variability of the Forster reaction.

α,α'-*Bisdiazo ketones*, whose carbene reactivity merits attention[56, 57], are accessible by double Forster reaction of the corresponding 1,2,3-trione 1,3-dioximes

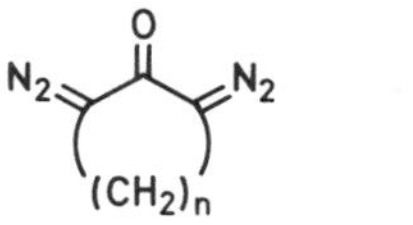

n = 2; *2,5-Bis(diazo)-cyclopentanone*
n = 3; *2,6-Bis(diazo)-cyclohexanone*

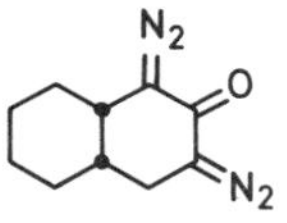

cis-1,3-Bis(diazo)-2-decahydronaphthalone

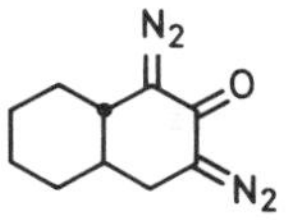

trans-1,3-Bis(diazo)-2-decahydronaphthalone

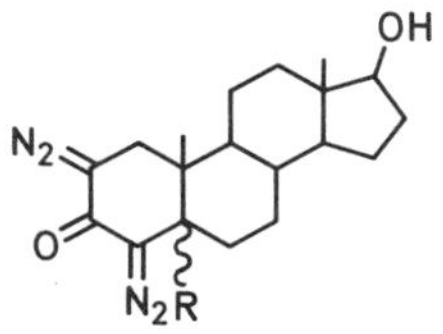

R = αH; βH; *2,4-Bis(diazo)-17β-hydroxy-5α-(or 5β)-androstan-3-one*

[44] *M.O. Forster*, J. Chem. Soc. (London) *107*, 260 (1915).

[45] *J. Meinwald, P.G. Gassman, E.G. Miller*, J. Amer. Chem. Soc. *81*, 4751 (1959);
K. Muth, Dissertation, Universität Mainz, 1958;
M. Schmid, Diplomarbeit, Universität Mainz, 1958.

[46] *W. Rundel*, Angew. Chem. *74*, 469 (1962).

[47] *T. Severin, H. Krämer, P. Adhikary*, Chem. Ber. *104*, 972 (1971).

[48] *M.P. Cava, R.L. Litle, D.R. Napier*, J. Amer. Chem. Soc. *80*, 2257 (1958).

[49] *L.O. Ross, E.M. Acton, W.A. Skinner, L. Goodman, B.R. Baker*, J. Org. Chem. *26*, 3395 (1961).

[50] *M.P. Cava, R.L. Litle, D.R. Napier*, J. Amer. Chem. Soc. *80*, 2257 (1958).

[51] *L. Horner, W. Kirmse, K. Muth*, Chem. Ber. *91*, 430 (1958).

[52] *L. Horner, K. Muth, H.G. Schmelzer*, Chem. Ber. *92*, 2953 (1959).

In particular cases chloramine can be replaced by phenylhydrazine as condensation component, as is demonstrated by the reaction of 3-hydroxyimino-2,4-chromandione to *3-diazo-2,4-chromandione*[58]:

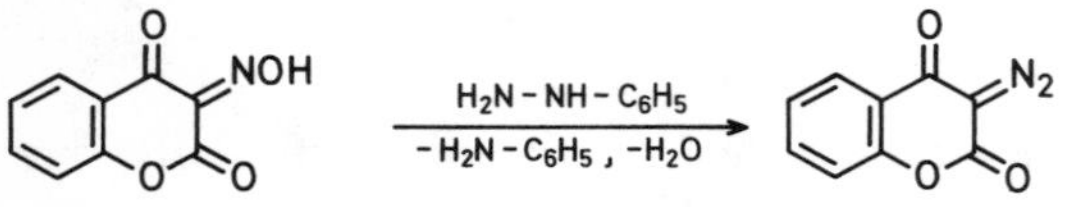

No assessment will be made here whether all variants of the Forster reaction proceed in mechanistically uniform manner (see p. 211).

8.1.3 Dehydrogenation of hydrazones

Dehydrogenation of hydrazones is one of the oldest[59] and most frequently used methods of diazo synthesis:

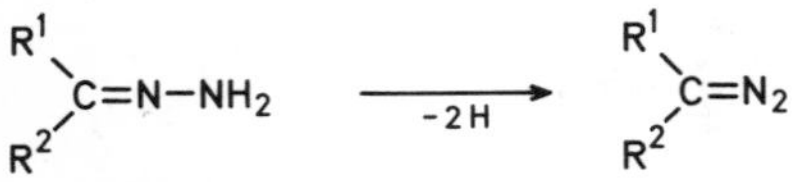

In addition to mercury(II) oxide, silver(I) oxide, manganese(IV) oxide, and lead(IV) acetate are often employed as *dehydrogenating agents*. Sodium and calcium hypochlorite, mercury(II) trifluoroacetate and trifluoroacetamide, nickel peroxide, hydrogen peroxide, iodine, even atmospheric oxygen have all been used, but less frequently.

aldehydes and ketones, and with di- and triketones in which one carbonyl group reacts preferentially such as, for example, in isatin[61]:

By contrast, if two or more adjacent carbonyl groups possess comparable reactivity isomer formation must be anticipated, as during the reaction of phenyl-2-pyridylglyoxal with hydrazine[62], *e.g.:*

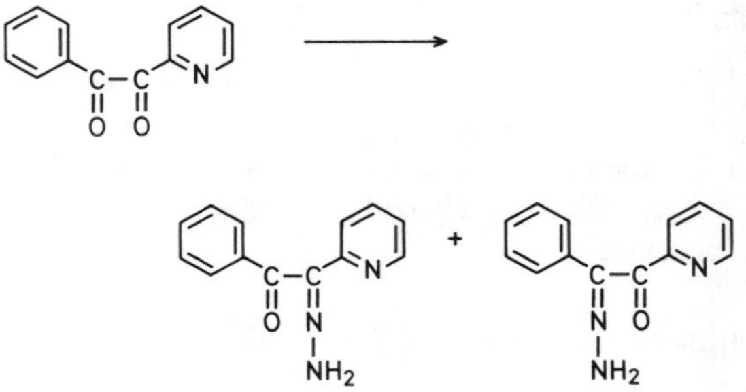

An important variant for synthesizing α-diazo ketones avoids this problem by the steps of bromination of an α-methylene ketone and reacting with hydrazine to form hydrazones[63, 64]:

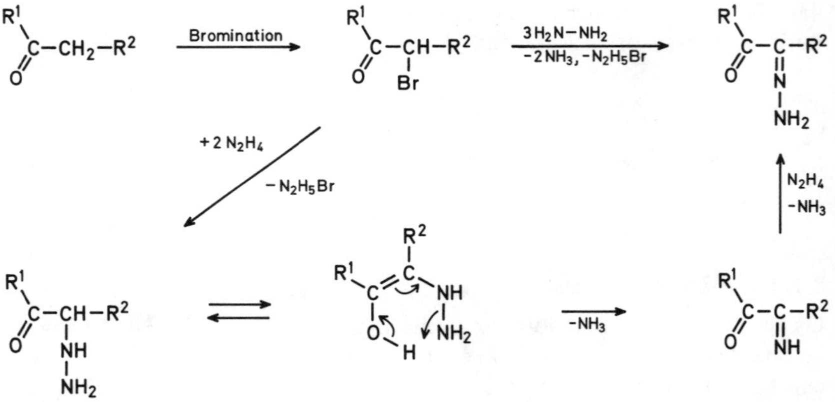

Azines are occasionally obtained as *side-products* of the dehydrogenation but because of their sparing solubility can generally be easily separated. They are formed exclusively when hydrazones are dehydrogenated with *N*-bromosuccinimide[60]. The success of the diazo synthesis is often decided at the hydrazone preparation stage from carbonyl compounds and hydrazine; it, too, may be accompanied by azine formation. No problems arise with

[53] *J. Meinwald, P. G. Gassman,* J. Amer. Chem. Soc. *82*, 2857 (1960).
[54] *J. W. J. Still, D. T. Wang,* Can. J. Chem. *46*, 1583 (1968).
[55] *G. Muller, C. Huynh, J. Mathieu,* Bull. Soc. Chim. France 296 (1962);
M. P. Cava, E. Moroz, J. Amer. Chem. Soc. *84*, 115 (1962);
J. Meinwald, G. G. Curtius, P. G. Gassman, J. Amer. Chem. Soc. *84*, 116 (1962).
[56] *W. Kirmse,* Angew. Chem. *71*, 537 (1959).

The following sections discuss the advantages and drawbacks of the individual dehydrogenating agents briefly. Table 2 lists numerous individual examples (see p. 216); as far as is possible individual dehydrogenation variants have been juxtaposed there for comparison.

8.1.3.1 Mercury(II) oxide

Dehydrogenation reactions with mercury(II) oxide are generally performed in benzene, toluene, petroleum ether, ether, or in chloroform at room temperature; occasionally cooling is required. Addition of anhydrous sodium sulfate is recommended for binding the water formed. Ethanolic caustic potash catalyzes the dehydrogenation of, for example, benzophenone hydrazone, appreciably[65]. Adding both these aids cause the conversion of 9-hydrazonofluorene into *9-diazofluorene* to proceed almost quantitatively in a short time in ether[66]. Numerous *aryldiazomethanes* of varying degree of substitution and *α-diazocarbonyl* compounds, too, have been synthesized (see Table 2). In particular instances mercury(II) acetamide (giving *3-diazocamphor*)[67] and mercury(II) trifluoroacetate in acetonitrile/triethylamine (giving *4-diazo-2,2,5,5-tetramethyl-3-hexanone*)[68] have proven themselves. In addition to azine formation (see p. 212) dehydrogenation of α-oxo aldehyde hydrazones can also lead to *mercury bis-diazo* compounds. These are reaction products of the primarily formed α-diazo ketones with unused mercury oxide[69] (see p. 241), *e.g.*:

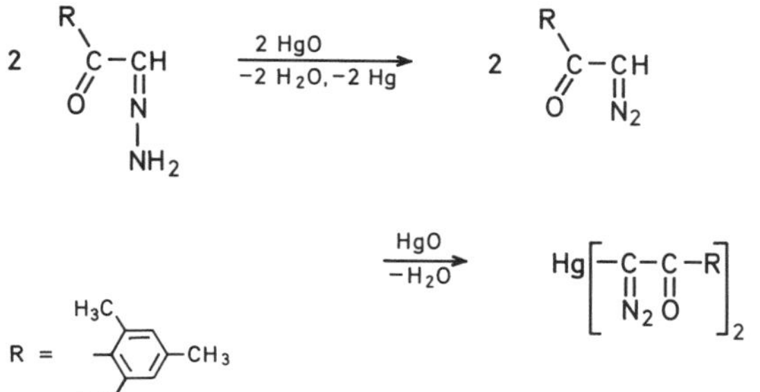

$$R = \text{mesityl}$$

Bis(1-diazo-2-mesityl-2-oxoethyl)mercury

Attempts to prepare 1,2-bis(diazo) compounds from 1,2-bis(hydrazono) compounds by the above technique have so far remained unsuccessful; evidently they change into acetylenes spontaneously while losing nitrogen twice over:

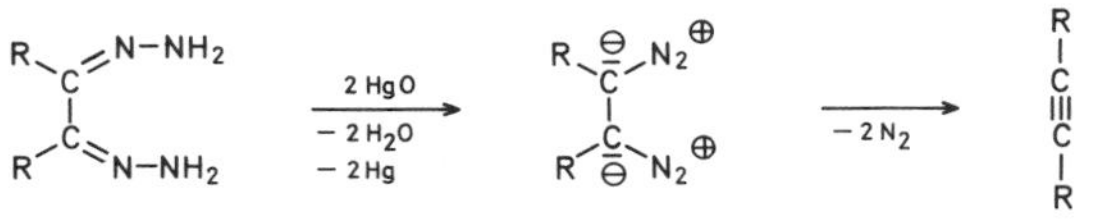

This sequence of reactions is well known in the case of *diphenylacetylene*[70]; it has gained fresh importance during preparation of cycloalkynes such as *cycloheptyne*[71], *cyclooctyne*[71,72], *cyclononyne*[73], *etc.*[74].

8.1.3.2 Silver(I) oxide

Dehydrogenations with silver(I) oxide are generally carried out in ether, tetrahydrofuran, pentane, petroleum ether, or aromatic hydrocarbons. Water-binding additives and base catalysis offer the benefits described on p. 212. The reactions generally proceed more rapidly than with mercury(II) oxide, and their superiority is thus most clearly marked during preparation of *unstable* diazo compounds. Thus mercury oxide dehydrogenation of cyclohexanone hydrazone affords merely an azine despite variations in the reaction conditions[75],

[57] *R. Tasovac, M. Stefanović, A. Stojiljković*, Tetrahedron Lett. 2729 (1967).

[58] *G. Casini, F. Gualtieri, M.L. Stein*, Gazz. Chim. Ital. *95*, 983 (1965).

[59] *T. Curtius*, Ber. *22*, 2161 (1889).

[60] *M.Z. Barakat, M.F.A. El-Wahab, M.M. El-Sadr*, J. Amer. Chem. Soc. *77*, 1670 (1955).

[61] *T. Curtius, K. Thun*, J. Prakt. Chem. [2] *44*, 188 (1891).

[62] *B. Eistert, E. Endres*, Justus Liebigs Ann. Chem. *734*, 56 (1970).

[63] *S. Hauptmann, M. Kluge, K.D. Seidig, H. Wilde*, Angew. Chem. *77*, 678 (1965);
S. Hauptmann, M. Kluge, K.D. Seidig, H. Wilde, Angew. Chem. Intern. Ed. Engl. *4*, 688 (1965);
M.P. Cava, E.J. Glamkowsky, P.M. Weintraub, J. Org. Chem. *31*, 2755 (1966).

[64] *S. Hauptmann, H. Wilde*, J. Prakt. Chem. *311*, 604 (1969).

[65] *J.B. Miller*, J. Org. Chem. *24*, 560 (1959), here is also discussed the mechanism of KOH-katalysis.

[66] *A. Schönberg, W.I. Awad, N. Latif*, J. Chem. Soc. (London) 1368 (1951).

[67] *M.O. Forster, A. Zimmerli*, J. Chem. Soc. (London) *97*, 2156 (1910).

[68] *M.S. Newman, A. Arkell*, J. Org. Chem. *24*, 385 (1959).

[69] *P. Yates, F.X. Garneau*, Tetrahedron Lett. *71*, (1967).

[70] *T. Curtius*, Ber. *22*, 2161 (1889).

[71] *G. Wittig, A. Krebs*, Chem. Ber. *94*, 3260 (1961).

[72] *A.T. Blomquist, L.H. Liu*, J. Amer. Chem. Soc. *75*, 2153 (1953).

[73] *A.T. Blomquist, R.E. Burge, L.H. Liu, J.C. Bohrer, A.C. Sucsy, J. Kleis*, J. Amer. Chem. Soc. *73*, 5510 (1951);
V. Prelog, K. Schenker, W. Küng, Helv. Chim. Acta *36*, 471 (1953).

[74] *R.W. Hoffmann*, Dehydrobenzene and Cycloalkynes, 1. Aufl., p. 332, Verlag Chemie, Weinheim/Bergstr. 1967.

[75] *K. Heyns, A. Heins*, Justus Liebigs Ann. Chem. *604*, 133 (1957).

while dehydrogenation with silver oxide at any rate furnishes 27–28% *diazocyclohexane*[75]. Silver oxide also serves for preparing the quite unstable *2-diazopropane* (−60°, 20–30%)[76], mercury oxide fails entirely here[76]. In other respects the range of application corresponds to that of mercury oxide dehydrogenation although to date far fewer examples have become known (cf. Table 2).

8.1.3.3 Manganese(IV) oxide

Manganese(IV) oxide as dehydrogenating agent possesses the advantage of being very active and cheap. Advantageously *activated* manganese(IV) oxide is employed; it is prepared by alkaline conproportionation of potassium permanganate and manganese(II) sulfate tetrahydrate before use[77]. It

8.1.3.4 Lead(IV) acetate

Lead(IV) acetate has so far proven to be a successful dehydrating reagent in *special cases* only. It is employed either in acetonitrile, benzonitrile, or dichloromethane. *Diazomalononitrile*[80], *diazobis-(trifluoromethyl)methane*[81], and *methyl diazophenylacetate*[82] are successful examples of its application. The *limits* of this reactions are often set by the sensitivity of many diazo compounds toward lead(IV) acetate itself or to the acetic acid formed during the dehydrogenation. Thus, while acetophenone hydrazone still affords *α-diazoethylbenzene,* this compound then reacts further to *1-phenyl-1,1-ethanediol diacetate* and *phenylethylacetate* and *acetophenone azine* respectively[83]:

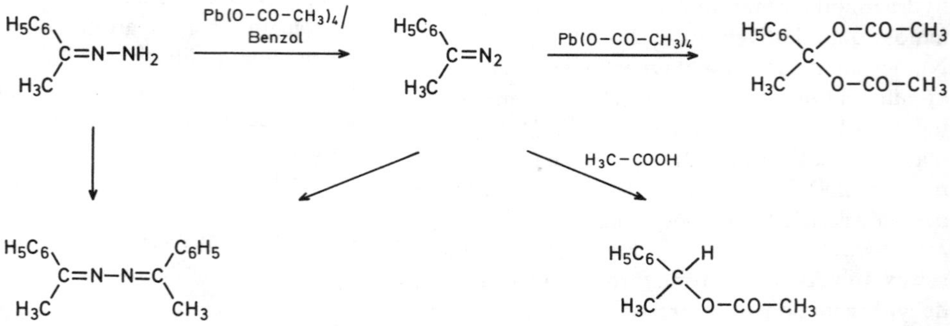

enables one to reduce the necessary excess of dehydrogenating agent drastically, as was shown during the preparation of (4-chlorophenyl)diazophenylmethane[78]. The preferred solvent is chloroform, but the solvents mentioned in the two preceeding sections, too, can be employed. This variant has proved successful for *aryl, heteroaryl,* and *sulfonyl diazomethanes;* for *α-diazocarbonyl* compounds it can hardly be bettered[64, 79] (see Table 2, p. 216).

3-Hydrazono-*5α*-androstan-17*β*-ol acetate reacts with lead(IV) acetate to form *5α-androstane-3α, 17β-diol* and *5α-androstane-3β,17β-diol diacetates* (65:35) as well as *2-androsten-17β-ol acetate* with loss of all its nitrogen[84]:

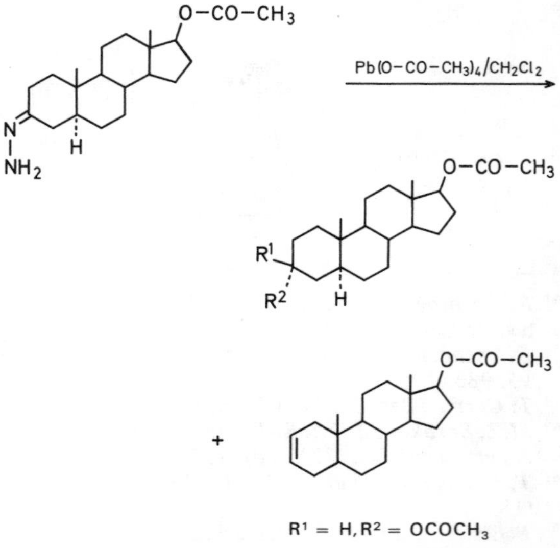

R¹ = H, R² = OCOCH₃
R¹ = OCOCH₃, R² = H

[76] *D.E. Applequist, H. Babad,* J. Org. Chem. *27,* 288 (1962).
[77] *J. Attenburrow, A.F.B. Cameron, J.H. Chapman, R.M. Evans, B.A. Hems, A.B.A. Jansen, T. Walker,* J. Chem. Soc. (London) 1094 (1952).
[78] *W. Schroeder, L. Katz,* J. Org. Chem. *19,* 718 (1954).
[79] *H. Morrison, S. Danishefsky, P. Yates,* J. Org. Chem. *26,* 2617 (1961).
[80] *E. Ciganek,* J. Org. Chem. *30,* 4198 (1965).
[81] *D.M. Gale, W.J. Middleton, C.G. Krespan,* J. Amer. Chem. Soc. 87, 675 (1965).
[82] *E. Ciganek,* J. Org. Chem. *35,* 862 (1970).

8.1.3.5 Miscellaneous dehydrogenation reagents

Dehydrogenation of 9-hydrazonofluorene to *9-diazofluorene* with molecular oxygen takes place in ethanol/sodium ethoxide[85]; oddly, it fails with the pure hydrazone[86]. A hydrazone anion intermediate stage occurs; that hydrogen peroxide is formed can be interpreted mechanistically[87]. The following were obtained analogously using methyllithium as base[87]:

Diazodiphenylmethane
2-Diazo-2-phenylacetophenone
α-Diazoethylbenzene

Nickel(III) oxide in ether can be employed very successfully for dehydrogenating benzophenone hydrazone[88], while alkaline *hydrogen peroxide* solution converts 5-hydrazonobarbituric acid into *5-diazobarbituric acid* (85–88%)[89]. However, the range of application of the two reagents cannot as yet be surveyed.

Certain 1,2-diketone monohydrazones can be dehydrogenated with *calcium hypochlorite* in methanol (see Table 2)[90], without revealing special advantages. *Sodium hypochlorite* was used to prepare *diazotetrachlorocyclopentadiene* analogously[91].

The last-named reaction can be performed, alternatively, by using *iodine* in ether/triethylamine[92]. *Diazodiphenylmethane* is accessible by the same technique, while if the base is left out exclusively benzophenone azine is formed[93]. The importance of the base and the possible sequential reactions are shown by the following formula scheme[93]:

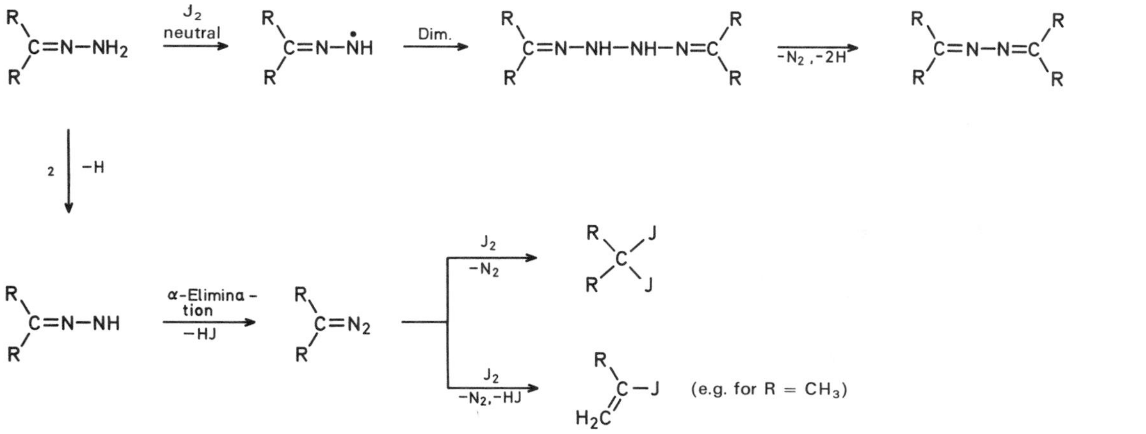

Formation of 1,1-diiodo compounds and of iodoalkenes appear to limit the preparative significance of the basic iodine dehydrogenation severely[93].

8.1.4 Bamford-Stevens reaction

In addition to hydrazone dehydrogenation the Bamford-Stevens reaction[115] offers an alternative means for converting carbonyl groups into diazo groups. It consists in base cleavage of arylsulfonylhydrazones, which are synthesized from carbonyl compounds and arylsulfonylhydrazine in the usual way.

Salts of sulfinic acids are obtained side by side with the diazo compounds:

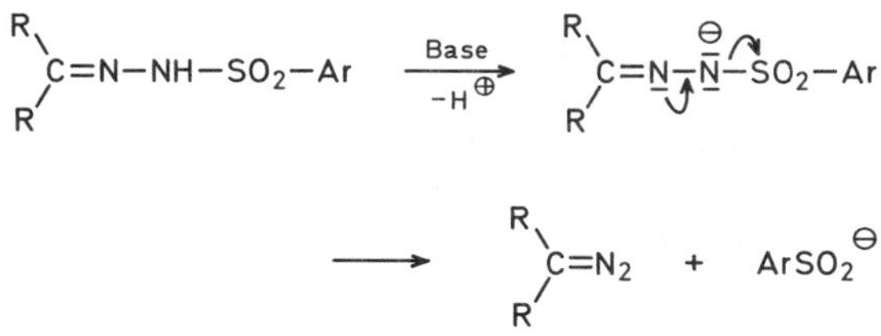

$$R_2C=N-NH-SO_2-Ar \xrightarrow[-H^{\oplus}]{Base} R_2C=N-\overset{\ominus}{N}-SO_2-Ar$$

$$\longrightarrow R_2C=N_2 + ArSO_2^{\ominus}$$

Here, too, the preparation of the hydrazone is often decisive for the success of the synthesis; the remarks made on p. 212 apply analogously. Other *limits* are imposed by numerous side, by-pass, and sequential reactions, but these can be utilized in directed manner and have achieved exceptional importance.

Primarily, formation of *carbenes* may be mentioned. These are obtained by decomposition of initially formed diazo compounds in aprotic media (ethers) at higher

[83] *A. Stojiljković, N. Orbović, S. Sredojević, M.L. Mihailović,* Tetrahedron 26, 1101 (1970).
[84] *M. Debono, R.M. Molloy,* J. Org. Chem. 34, 1454 (1969);
D.H.R. Barton, P.L. Blatten, J.F. McGhie, Chem. Commun. 450 (1969).

[85] *H. Staudinger, A. Gaule,* Ber. 49, 1951 (1916).
[86] *H. Staudinger, K. Miescher,* Helv. Chim. Acta 2, 554, 578 (1919).
[87] *W. Fischer, J.-P. Anselme,* J. Amer. Chem. Soc. 89, 5312 (1967).

Table 2. Diazo compounds by hydrazone dehydrogenation

Hydrazone	Dehydrogenating agent	Solvent	Diazo compound	Yield [% of theory]	M. p. [°C]	Ref.
Benzaldehyde hydrazone	HgO	Petroleum ether	*α-Diazotoluene*	45–56	−29	94
		Pentane		38	—	95
		Ether		⩾ 54	—	96
Benzophenone hydrazone	HgO	Petroleum ether	*Diazodiphenylmethane*	95–98	29–30	97, 98
Fluorenone hydrazone	HgO	Ether	*9-Diazofluorene*	almost 100	94–95	99, 100
1-Indanone hydrazone	Ag$_2$O	Ether	*1-Diazoindan*	93	—	101
2,2,2-Trifluoroacetophenone hydrazone	HgO	Ether	*(1-Diazo-2,2,2-trifluoroethyl)-benzene*	84	—	102
9-Anthracenecarboxaldehyde hydrazone	HgO	Ether	*9-Diazomethylanthracene*	65	63–64 decomp.	103
3-Pyridinecarboxaldehyde hydrazone	HgO	Ether	*3-Diazomethylpyridine*	68	—	104
1,5-Diphenyl-1,2,3-triazole-4-carboxaldehyde hydrazone	MnO$_2$	Ether	*4-Diazomethyl-1,5-diphenyl-1,2,3-triazole*	48	110 decomp.	105
Bis(2-thienyl) ketone hydrazone	Ag$_2$O	Hexane	*Diazobis(2-thienyl)methane*	—	—	106
Phenyl 4-pyridyl ketone hydrazone	MnO$_2$	Chloroform	*4-(α-Diazobenzyl)pyridine*	80	64.5–65	106
Benzil monohydrazone	HgO	Ether	*2-Diazo-2-phenylacetophenone*	87–94	79	107
	MnO$_2$	Chloroform		90–100	—	108
	Ca(OCl)$_2$	Methanol		75–85	—	108
Camphor 3-hydrazone	HgO	Benzene	*3-Diazocamphor*	81	74–75	109
	MnO$_2$	Chloroform	*(3-Diazo-2-bornanone)*	90–100	—	108
	Ca(OCl)$_2$	Methanol		75–85	—	108
1-Phenylglyoxal 2-hydrazone	MnO$_2$	Chloroform	*2-Diazoacetophenone*	85	49	110
1-Naphth-1-yl-glyoxal 2-hydrazone	MnO$_2$	Chloroform	*ω-Diazo-1'-acetonaphthone*	55	90	110
9-Hydrazonoanthrone	HgO	Tetrahydrofuran	*9-Diazoanthrone*	96	—	111
Diethyl mesoxalate hydrazone	Ag$_2$O	Tetrahydrofuran	*Diethyl diazomalonate*	89	—	112
[Bis(phenylsulfonyl)methylene]-hydrazine	MnO$_2$	Tetrahydrofuran	*Diazobis(phenylsulfonyl)methane*	52	99–100 decomp.	113
[Bis(4-nitrophenylsulfonyl)-methylene]hydrazine	MnO$_2$	Chloroform	*Diazobis(4-nitrophenylsulfonyl)-methane*	—	205–209 decomp.	114

[88] K. Nakagawa, H. Onoue, K. Minami, Chem. Commun. 730 (1966).

[89] E. Fahr, Justus Liebigs Ann. Chem. 627, 213 (1959).

[90] R.C. Fuson, L.J. Armstrong, W.J. Schenk, J. Amer. Chem. Soc. 66, 964 (1944).

[91] H. Disselnkötter, Angew. Chem. 76, 431 (1964); H. Disselnkötter, Angew. Chem. Intern. Ed. Engl. 3, 379 (1964).

[92] USP. 3 422 158 (1969); Hooker Chemical Corp., Inv.: D. Knutson; C.A. 70, 67877e (1969).

[93] D.H. Barton, R.E. O'Brien, S. Sternhell, J. Chem. Soc. (London) 470 (1962).

[94] H. Staudinger, A. Gaule, Ber. 49, 1897 (1916).

[95] W.M. Jones, W.T. Tai, J. Org. Chem. 27, 1324 (1962).

[96] J.-P. Anselme, Organic Preparations and Procedures 1, 73 (1969).

[97] H. Staudinger, E. Anthes, F. Pfenninger, Ber. 49, 1928 (1916).

[98] L.I. Smith, K.L. Howard, Org. Syntheses Coll. Vol. III, 351 (1955).

[99] H. Staudinger, O. Kupfer, Ber. 44, 2197 (1911).

[100] A. Schönberg, W.I. Awad, N. Latif, J. Chem. Soc. (London) 1368 (1951).

[101] R.A. Moss, J.D. Funk, J. Chem. Soc. C 2026 (1967).

[102] R.A. Shepard, S.E. Wentworth, J. Org. Chem. 32, 3197 (1967).

[103] T. Nakaya, T. Tomomoto, M. Imoto, Bull. Chem. Soc. Japan 40, 691 (1967).

[104] B. Eistert, W. Kurze, G.W. Müller, Justus Liebigs Ann. Chem. 732, 1 (1970).

[105] P.A.S. Smith, J.G. Wirth, J. Org. Chem. 33, 1145 (1968).

[106] H. Reimlinger, Chem. Ber. 97, 3493 (1964).

[107] C.D. Nenitzescu, E. Solomonica, Org. Syntheses Coll. Vol. II, 496 (1947).

temperature[116]; by contrast, in protic media (alcohols) *carbenium* ion formation *via* loss of nitrogen must be anticipated[117].

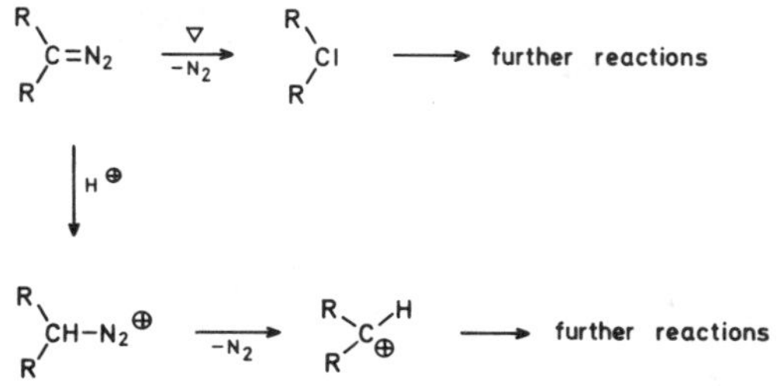

Carbenium ions can arise also where sulfonylhydrazones are only incompletely converted into their salts; in this case the hydrazone functions as proton donor leading to formation of *N-alkylated hydrazones*[118]:

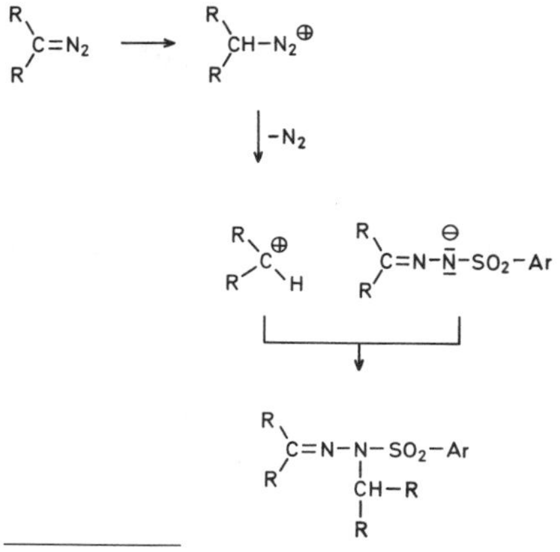

Bases such as sodium hydride or sodium amide (excess)[119] and alkyllithium (2 *mols*)[120] can also deprotonate α-*C* atoms in sulfonylhydrazones in addition to the nitrogen if the constitutional requisites are met. This leads to formation of alkenes without the necessity of a diazo compound intermediate stage[121] (mechanism[119–121]), *e.g.*:

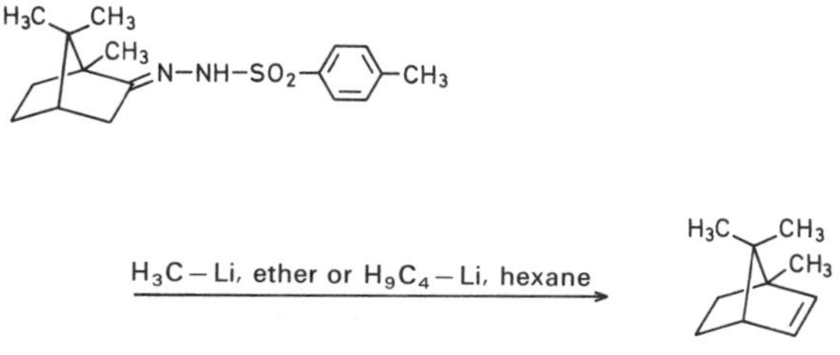

Bornene; ~ 100%

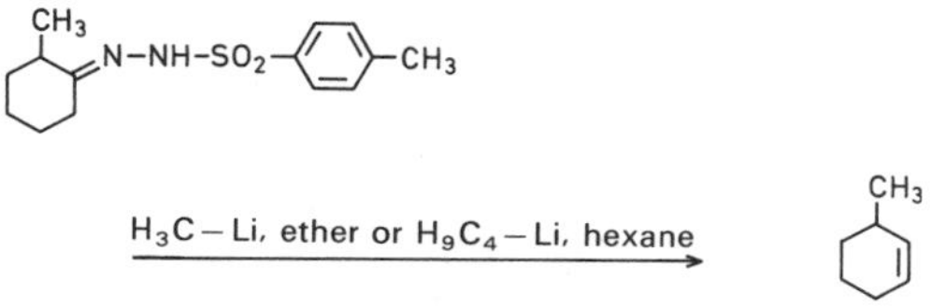

3-Methylcyclohexene; 98%

During alkali splitting of the bis(*p*-toluenesulfonyl)-hydrazones of 1,2-diketones no more success in isolating *1,2-bis(diazo)* compounds is achieved than during hydrazone dehydrogenation (see p. 213). In addition to acetylene formation that of *1,2,3-triazoles* is observed; it evidently comes about *via* a 1,5-cyclization of *3-diazo-2-butanone 4-tolylsulfonylhydrazone*[115]:

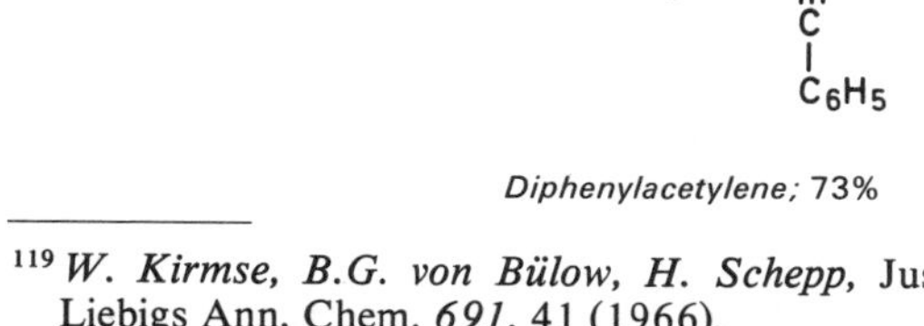

Diphenylacetylene; 73%

[108] *H. Morrison, S. Danishefsky, P. Yates,* J. Org. Chem. *26*, 2617 (1961).

[109] *J. Bredt, W. Holz,* J. Prakt. Chem. [2] *95*, 133 (1917) [Dehydrogenation with mercury(II)-acetamide s. p. 213].

[110] *S. Hauptmann, H. Wilde,* J. Prakt. Chem. *311*, 604 (1969) (Preparation of substituted Phenylbenzoyldiazomethane).

[111] *J.C. Fleming, H. Shechter,* J. Org. Chem. *34*, 3962 (1969).

[112] *E. Ciganek,* J. Org. Chem. *30*, 4366 (1965).

[113] *J. Diekmann,* J. Org. Chem. *30*, 2272 (1965).

[114] USP. 3 332 936 (1967), DuPont, Inv.: *J. Diekmann;* C.A. *68*, 59304j (1968).

[115] *W.R. Bamford, T.S. Stevens,* J. Chem. Soc. (London) 4735 (1952).

[116] *W. Kirmse,* Carbene, Carbenoide und Carbenanaloge, 1. Aufl., p. 137, Verlag Chemie Weinheim/Bergstr. 1969.

[117] *J.W. Powel, M.C. Whiting,* Tetrahedron *7*, 305 (1959);
J.W. Powel, M.C. Whiting, Tetrahedron *12*, 168 (1961);
L. Friedman, H. Shechter, J. Amer. Chem. Soc. *81*, 5512 (1959);
C.H. De Puy, D.H. Froemsdorf, J. Amer. Chem. Soc. *82*, 634 (1960).

[118] *D.M. Lemal, A.J. Fry,* J. Org. Chem. *29*, 1673 (1964).

[119] *W. Kirmse, B.G. von Bülow, H. Schepp,* Justus Liebigs Ann. Chem. *691*, 41 (1966).

[120] *G. Kaufman, F. Cook, H. Shechter, J. Bayless, L. Friedman,* J. Amer. Chem. Soc. *89*, 536 (1967).

[121] *R.H. Shapiro, M.J. Heath,* J. Amer. Chem. Soc. *89*, 5734 (1967).

[122] *G.L. Closs, R.A. Moss,* J. Amer. Chem. Soc. *86*, 4042 (1964).

[123] *H.W. Davies, M. Schwarz,* J. Org. Chem. *30*, 1242 (1965).

Other sequential reactions are discussed on p. 219. The Bamford-Stevens reaction is limited almost entirely to *p*-toluenesulfonylhydrazones, cleaving other sulfonylhydrazones brings no advantages.

8.1.4.1 Diazoalkanes

The Bamford-Stevens reaction proceeds smoothly with *arylated diazomethanes*. Cleaving the *p*-toluenesulfonyl hydrazones is carried out in, for instance, ethanol/sodium ethoxide[115], triethylene glycol/sodium methoxide[122], dimethylformamide/diethyl amine[123], or pyridine/sodium methoxide[124], generally with gentle warming. The last-named solvent is superior to ethanol, 1,2-dimethoxyethane, dimethylformamide, and dimethyl sulfoxide for preparing α-*diazotoluene*[124]. Table 3 provides some idea of the range of application of the reaction.

N-(4,5-Dimethyl-1H-1,2,3-triazol-1-yl)-p-toluenesulfonamide

Table 3. Aryldiazomethanes by Bamford-Stevens reaction

p-Toluenesulfonylhydrazone of	Solvent, Base; Reaction conditions	Diazo compound	Yield [% of theory]	M. p. [°C]	Ref.
Benzaldehyde	Ethanol, sodium ethoxide; 6 hrs. at 50°	α-*Diazotoluene*	62	—	125
	Triethylene glycol, sodium methoxide; 45 min at 60°		55	—	126
	Pyridine, sodium methoxide; 30 min at 55–60°		65–70	—	127
4-Nitrobenzaldehyde	Dimethylformamide/diethylamine; 1 hr. at 20°	α-*Diazo-4-nitrotoluene*	59	80 decomp.	128
Benzophenone	Ethanol, sodium ethoxide; 5 hrs. at 65–70°	*Diazodiphenylmethane*	58	—	125
Acetophenone	Ethanol, sodium ethoxide; 7 hrs. at 65–70°	α-*Diazoethylbenzene*	27	—	125
	Pyridine, sodium methoxide; 40 min at 75–80°		32	—	127
	Methanol, sodium methoxide; 2 hrs. at 65°	*11-Diazo-11H-benzo[a] fluorene*	90	140 decomp.	129
2-Allyloxybenzaldehyde	1,2-Dimethoxyethane, sodium methoxide; 5 min at 50°	*2-Allyloxy-α-diazotoluene*	30–35	—	130
	Pyridine/sodium methoxide; 90 min at 70°		76	71 decomp.	131
		5-Diazo-5H-dibenzo[a, d] cycloheptene			
3-Pyridinecarboxaldehyde	Ethanol, sodium ethoxide; some minutes at 60°	*3-Diazomethylpyridine*	75	—	132
	Pyridine, sodium ethoxide; 1 hr. at 60–65°		80	—	
	Dimethylformamide, diethylamine; briefly at 50° then 45 min at 20°		60	—	

[124] *D.G. Farnum*, J. Org. Chem. *28*, 870 (1963).

[125] *W.R. Bamford, T.S. Stevens*, J. Chem. Soc. (London) 4735 (1952).

Heterocyclic diazomethanes such as *3-* and *4-diazomethylpyridines* are accessible *via* the Bamford-Stevens reaction [133] (cf. Table 3). By contrast, *2-(α-diazobenzyl)pyridine*, which may be expected to form during alkali cleavage of phenyl 2-pyridyl ketone *p*-tolylsulfonylhydrazone, at once cyclizes to *3-phenyltriazolo[4,5-c]pyridine*[134].

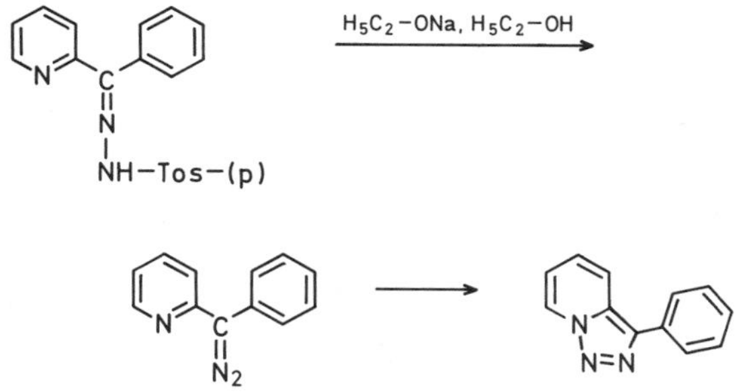

For *alkyldiazomethanes* the above decomposition conditions are little suitable because they are extremely sensitive to protons; these can be furnished even by alcohols. Even if the Bamford-Stevens reaction medium is protonfree (*e.g.*, pyridine) *methanol* is formed during salt formation of the *p*-tolylsulfonylhydrazone, for example, with methoxide, and can initiate decomposition of the carbenium ions. As, on the other hand, diazomethanes are thermally quite stable, *vacuum pyrolysis* of the *p*-tolylsulfonyl hydrazone salts is of assistance[135]. Lithium compounds obtained from hydrazones with butyllithium are employed preferentially (see Table 4, p. 220):

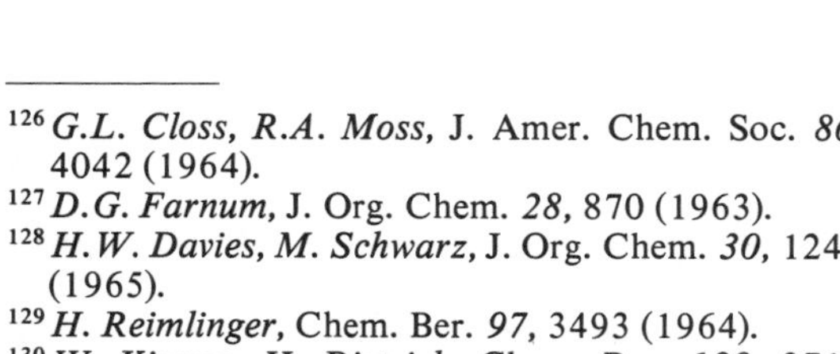

As ketone *p*-toluenesulfonylhydrazones in general require higher decomposition temperatures than the corresponding aldehyde derivatives, it is not surprising that *2-diazopropane, diazocyclopentane,* and *diazocyclohexane* are formed in yields of <5%; they do not withstand the thermal stress. Others, such as, for example, *1-diazo-2,2,4,4-tetramethylcyclobutane* or *cyclopropyldiazophenylmethane* (cf. Table 4) survive the vacuum pyrolysis without decomposition. For aryldiazomethanes this variant furnishes no advantage[135].

8.1.4.2 α,β-Unsaturated diazo compounds

Preparation of α,β-unsaturated diazo compounds by Bamford-Stevens reaction is fairly complicated because of two sequential reactions: in addition to *3H-pyrazoles* **10**, which arise from the α-diazoolefin **8** by 1,5-cyclization, *cyclopropenes* **9** can form. While at high temperatures cyclopropenes are formed[141-143], at low temperatures *diazo* compounds[142, 144] or *3H-pyrazoles* **10**[145-147] often arise. Reactions 7→8 and 8→9 can be performed either thermally or photolytically[142, 148], converting 3H-pyrazoles into cyclopropenes is possible only by photolytic means[145, 146]:

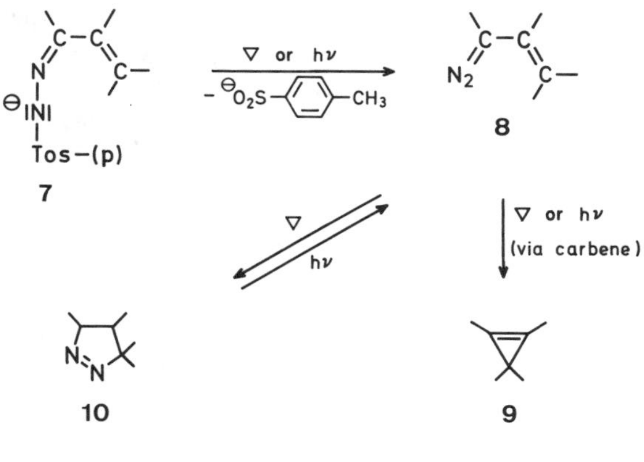

[126] *G.L. Closs, R.A. Moss,* J. Amer. Chem. Soc. *86,* 4042 (1964).
[127] *D.G. Farnum,* J. Org. Chem. *28,* 870 (1963).
[128] *H.W. Davies, M. Schwarz,* J. Org. Chem. *30,* 1242 (1965).
[129] *H. Reimlinger,* Chem. Ber. *97,* 3493 (1964).
[130] *W. Kirmse, H. Dietrich,* Chem. Ber. *100,* 2710 (1967).
[131] *J. Moritani, S.I. Murahashi, K. Yoshinaga, H. Ashitaka,* Bull. Chem. Soc. Japan *40,* 1506 (1967).
[132] *B. Eistert, W. Kurze, G.W. Müller,* Justus Liebigs Ann. Chem. *732,* 1 (1970).
[133] *B. Eistert, W. Kurze, G.W. Müller,* Justus Liebigs Ann. Chem. *732,* 1 (1970).

[134] *H. Reimlinger,* Chem. Ber. *97,* 3493 (1964) [Pyridyl(2)-diazomethane (through hydrazone dehydration) reacts analogously];
J.H. Boyer, R. Borgers, L.T. Wolford, J. Amer. Chem. Soc. *79,* 678 (1957);
J.D. Bower, G.R. Ramage, J. Chem. Soc. (London) 4506 (1957).
[135] *G.M. Kaufman, J.A. Smith, G.G. Van der Strouw, H. Shechter,* J. Amer. Chem. Soc. *87,* 935 (1965).
[136] *G.M. Kaufman, J.A. Smith, G.G. Van der Strouw, H. Shechter,* J. Amer. Chem. Soc. *87,* 935 (1965).
[137] *J.H. Atherton, R. Fields, R.N. Haszeldine,* J. Chem. Soc. C 366 (1971).
[138] *Z. Majerski, S.H. Liggero, P. von R. Schleyer,* J. Chem. Soc. D 949 (1970).
[139] *M. Rey, R. Begrich, W. Kirmse, A.S. Dreiding,* Helv. Chim. Acta *51,* 1001 (1968).

Table 4. Diazoalkanes by vacuum pyrolysis of *p*-toluenesulfonylhydrazone salts

p-Toluenesulfonylhydrazone of	Reaction conditions	Diazo compound	Yield [% of theory]	Ref.
Pivalaldehyde	Lithium salt at 80–135°/0.3 torr, 45 min	*1-Diazo-2,2-dimethylpropane*	84–96	[136]
Propionaldehyde	Lithium salt at 80–135°/0.3 torr, 45 min	*1-Diazopropane*	46–56	[136]
Isobutyraldehyde	Lithium salt at 80–135°/0.3 torr, 45 min	*1-Diazo-2-methylbutane*	63–65	[136]
2,2,4,4-Tetramethylcyclobutanone	Sodium salt at 80–135°/0.3 torr, 45 min	*1-Diazo-2,2,4,4-tetramethyl-cyclobutane*	72	[136]
3,3,4,4,5,5,5-Heptafluoro-2-pentanone	Sodium salt, heat slowly to 150°	*2-Diazo-3,3,4,4,5,5,5-heptafluoropentane*	49	[137]
	Sodium salt at 170–180°/0.1 torr, 1–2 hrs.	*2-Diazohomoadamantane*		[138]
3-Cyclohexenecarboxaldehyde	Sodium salt in silicone oil at 80–90°/0.1 torr, 4 hrs.	*4-Diazomethylcyclohexene*	80	[139]
1-Methyl-2-cyclopentenecarboxal-dehyde	Sodium salt in triglyme at 80°, vacuum	*3-Diazomethyl-3-methylcyclo-pentene*	—	[140]

This problem explains why to date only few α,β-unsaturated diazo compounds have been synthesized by Bamford-Stevens reaction. For example, from the *p*-tolylsulfonylhydrazones of 2,3-dimethyl-2-butenol **11** and 4-methyl-3-penten-2-one **12** mild pyrolysis ($T_{decomp} \leqq 100°$) affords *1-diazo-2,3-dimethyl-2-butene* **13** and *2-diazo-4-methyl-3-pentene* **14**, while pyrolysis at 160° furnishes *1,3,3-trimethylcyclopropene* **15**. The latter is accessible also from the diazo compounds[142]:

[140] *G.L. Closs, R.B. Larrabee*, Tetrahedron Lett. 287 (1965).

[141] *G.L. Closs, L.E. Closs*, J. Amer. Chem. Soc. *83*, 2015 (1961).

[142] *G.L. Closs, L.E. Closs, W.A. Böll*, J. Amer. Chem. Soc. *85*, 3796 (1963).

[143] *H.H. Stechl*, Chem. Ber. *97*, 2681 (1964).

It is worth mentioning also the *6-diazo-3,3-dimethyl-1,4-cyclohexadiene* obtained by hydrazone pyrolysis (sodium hydride in tetraglyme, 70–80°/0.2 torr). For constitutional reasons it cannot cyclize but on standing in air passes over into the corresponding azine[149]. Decomposition of nitrosourethans has given more frequent success in synthesizing the relatively unstable α,β-unsaturated diazo compounds than the Bamford-Stevens reaction (see p. 225).

8.1.4.3 α-Diazocarbonyl compounds

During cleavage of *p*-toluenesulfonylhydrazones of 1,2-diketones organometallic bases and high decomposition temperatures such as are necessary with diazoalkanes can be avoided. It generally proceeds with aqueous alkali hydroxide at room temperature to give satisfactory yields[48]:

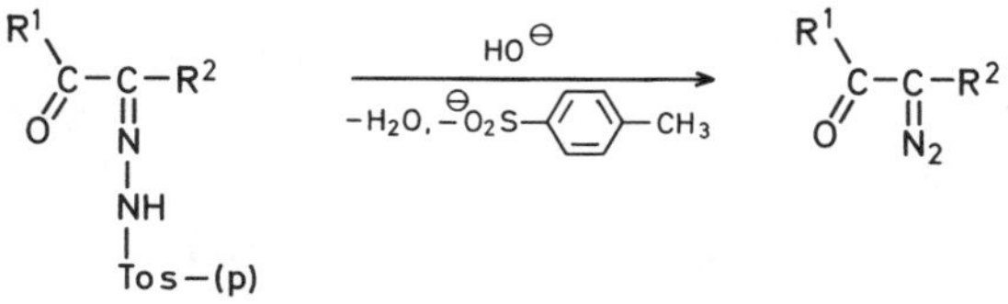

The cleavage can be carried out also with basic alumina in dichloromethane or ethyl acetate[150]. These are conditions under which aldehyde and ketone *p*-toluenesulfonylhydrazones are not amenable to the Bamford-Stevens reaction. *o*-Quinones can be converted into *quinone diazides* especially smoothly. In the case of phenanthrene quinone

reaction with *p*-toluenesulfonyl hydrazide affords *phenanthrene quinone diazide* directly without a possible isolation of the hydrazone[48, 151]. By contrast, with mono- and dinuclear systems the hydrazone intermediates can be isolated[152].

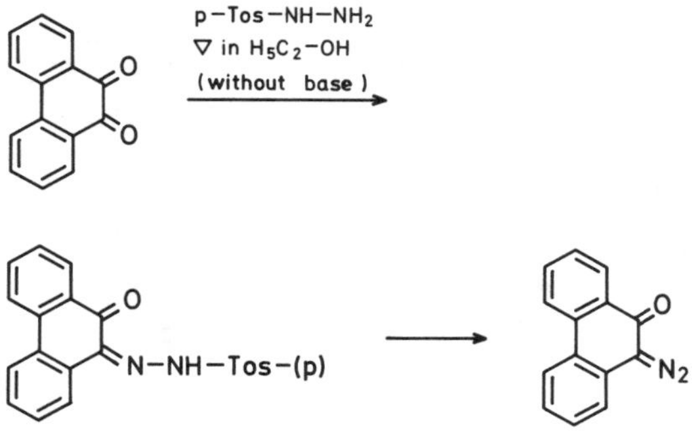

Table 5 lists some representative examples of the very many α-diazocarbonyl compounds that have been prepared by the Bamford-Stevens reaction. The problems encountered during preparation of α-diketone *p*-tolylsulfonylhydrazones are pointed out on p. 217.

A typical example of a Bamford-Stevens reaction with a 1,2-diketone of comparable carbonyl reactivity is the reaction between phenyl-2-pyridylglyoxal and *p*-toluenesulfonic acid hydrazide[62]. On the one hand, it leads to *3-benzoyltriazolo[1,2-c]pyridine* without isolatable intermediates, on the other, to the isolatable *phenyl-2-pyridylglyoxal 1-(p-tolylsulfonyl)hydrazone*, which undergoes normal Bamford-Stevens cleavage to α-*diazobenzyl 2-pyridyl ketone*:

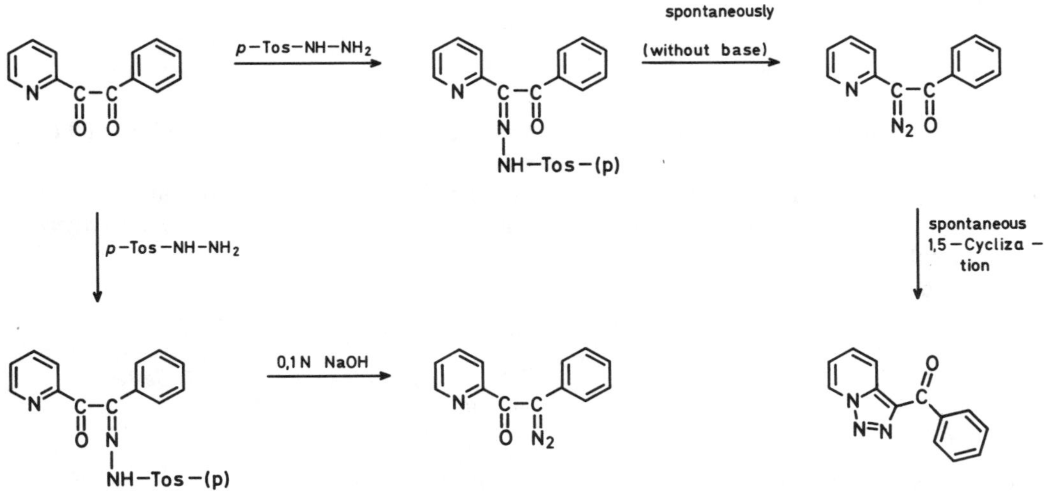

[144] *T. Sato, S. Watanabe*, Bull. Chem. Soc. Japan *41*, 3017 (1968).
[145] *G.L. Closs, W.A. Böll*, Angew. Chem. *75*, 640 (1963);

G.L. Closs, W.A. Böll, Angew. Chem. Intern. Ed. Engl. *2*, 399 (1963).
[146] *G.L. Closs, W.A. Böll, H. Heyn, V. Der*, J. Amer. Chem. Soc. *90*, 173 (1968).

Table 5. α-Diazocarbonyl compounds by Bamford-Stevens reaction

1,3-Diketone p-toluenesulfonyl hydrazone	Solvent, base; Reaction conditions	α-Diazo ketone	Yield [% of theory]	M.p. [°C]	Ref.
2,3-Butanedione p-tolylsulfonylhydrazone	Dichloromethane, basic alumina; 16 hrs. at ~ 20°	3-Diazo-2-butanone	86	oil	153
Benzil p-tolylsulfonylhydrazone	Dichloromethane, basic alumina; 6 hrs. at ~ 20°	2-Diazo-2-phenylacetophenone	84	73–76	153
1,2-Cyclohexanedione p-tolylsulfonylhydrazone	Dichloromethane, sodium hydroxide in water, 1 hr. at ~ 20°	2-Diazocyclohexanone	75	bp$_{1.5}$ =69–70°	154
(structure: N–NH–Tos–(p))	Sodium hydroxide in water	trans-5-Diazobicyclo[6.1.0]-nonan-4-one	—	—	157,155
(structure: H Cl ... N–NH–Tos–(p))	Pentane, sodium hydroxide in water; 2 hrs. at 0°	Syn-7-Chloro-3-diazo-2-bornanone	94	68	156
Camphor 3-p-toluene sulfonylhydrazone	Petroleum ether, sodium hydroxide in water; 2 hrs. at ~ 20°	3-Diazocamphor (3-Diazo-2-bornanone)	76	75	157
1,2-Indandione 2-p-toluene-sulfonylhydrazone	Dichloromethane, sodium hydroxide in water; 2.5 hrs. at ~ 20°	2-Diazo-1-indanone	58	86–88	157
(structure: N–NH–Tos–(p))	Sodium hydroxide in water	(structure)	high	149–151 decomp.	158
9,10 Dihydro-9,10-ethano-anthracene-11,12-dione 12-p-tolylsulfonylhydrazone		12-Diazo-9,10-dihydro-9,10-ethanoanthracen-11-one			
2-Indoline-2,3-dione 3-p-tolylsulfonylhydrazone	Sodium hydroxide in water; ~ 15 hrs. at ~ 20°	3-Diazo-2-indolinone	94	168 decomp.	157

Diazoacetic acid esters with *unsaturated alcohols* are accessible in an elegant synthetic procedure in which the nitrogen group is built up by hydrazone cleavage[159]. From 2-buten-1-ol and p-toluene-sulfonic acid (1-chloroethylidene)hydrazide the hydrazones *syn* and *anti* to the *C=N* double bond are synthesized (with a *trans* configuration on the *C=C* double bond) and cleaved to *2-buten-1-ol* diazoacetate. As the reaction conditions of the two individual steps correspond, the overall synthesis can, alternatively, be carried out in a single step using two molecules of base. *4,5,6,7-tetrahydro-5-*

[147] R.H. Findlay, J.T. Sharp, J. Chem. Soc. D 909 (1970).
[148] H. Dürr, Chem. Ber. 103, 369 (1970).
[149] M. Jones, A.M. Harrison, K.R. Rettig, J. Amer. Chem. Soc. 91, 7462 (1969).
[150] J.M. Muchowski, Tetrahedron Lett. 1773 (1966).
[151] O. Süs, H. Steppan, R. Dietrich, Justus Liebigs Ann. Chem. 617, 20 (1958).
[152] W. Ried, R. Dietrich, Chem. Ber. 94, 387 (1961).

[153] J.M. Muchowski, Tetrahedron Lett. 1773 (1966).
[154] H. Stetter, K. Kiehs, Chem. Ber. 98, 1181 (1965).
[155] P.G. Gassman, F.J. Williams, J. Seeter, J. Amer. Chem. Soc. 90, 6893 (1968);
A.J. Ashe, Tetrahedron Lett. 523 (1969).
[156] J. Meinwald, C. Blomquist Jensen, A. Lewis, C. Swithenbank, J. Org. Chem. 29, 3469 (1964).
[157] M.P. Cava, R.L. Litle, D.R. Napier, J. Amer. Chem. Soc. 80, 2257 (1958).
[158] J. Meinwald, E.G. Miller, Tetrahedron Lett. 253 (1961).
[159] H.O. House, C.J. Blankley, J. Org. Chem. 33, 53 (1968).

indanol and *1,2,3,4,5,6,7,8-octahydro-2-naph-thol diazoacetates*, too, are obtained by a combined reaction in one vessel[159]:

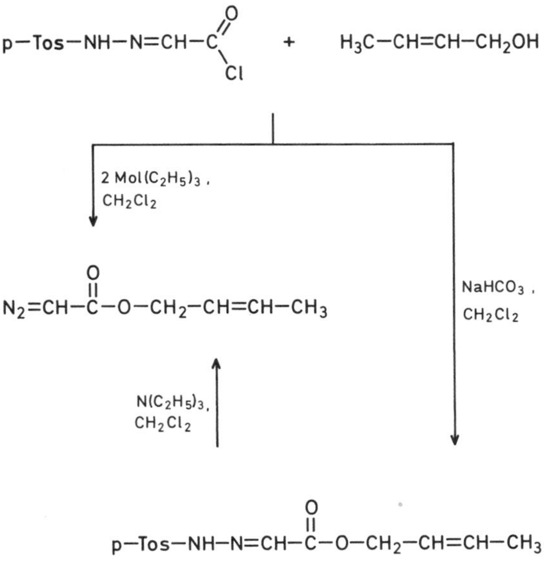

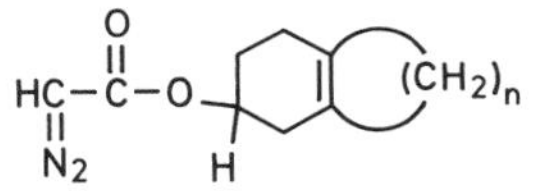

$$p-Tos-NH-N=CH-C\overset{O}{\underset{Cl}{<}} \quad + \quad H_3C-CH=CH-CH_2OH$$

$$2\ Mol(C_2H_5)_3, CH_2Cl_2 \downarrow$$

$$N_2=CH-\overset{O}{\overset{\|}{C}}-O-CH_2-CH=CH-CH_3 \qquad \begin{array}{c} NaHCO_3, \\ CH_2Cl_2 \end{array}$$

$$\uparrow \begin{array}{c} N(C_2H_5)_3, \\ CH_2Cl_2 \end{array}$$

$$p-Tos-NH-N=CH-\overset{O}{\overset{\|}{C}}-O-CH_2-CH=CH-CH_3$$

$$HC-\overset{O}{\overset{\|}{C}}-O-\ldots(CH_2)_n \qquad n = 3,4$$

The Bamford-Stevens reaction is suitable for preparing *2-diazo-1,3-dicarbonyl* compounds from 1,2,3-tricarbonyl compounds only where the two outer carbonyl groups are inactivated toward the condensation with *p*-toluenesulfonic acid hydrazide; the reaction proceeds to the diazo compound stage without addition of base. *5-Diazobarbituric acid* **16**[160], *diethyl diazomalonate* **17**[160], and *2-diazo-1,3(2H)-phenalenedione* **18**[160] are examples of successful syntheses. By contrast, preparation of *2-diazo-1,3-indandione* **19** from ninhydrin is devalued by the simultaneous formation of isomeric bis-hydrazones and of the tris-hydrazone[161]:

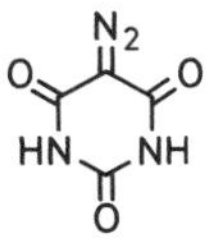

16

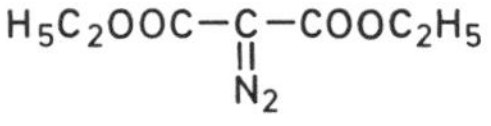

$$H_5C_2OOC-\overset{N_2}{\overset{\|}{C}}-COOC_2H_5$$

17

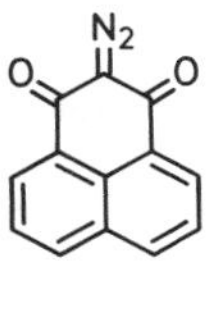

18

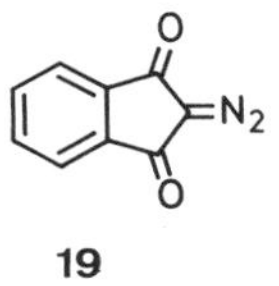

19

8.1.4.4 Phosphorus, silicon, and germanium substituted diazomethanes

Alkali cleavage of *p*-toluenesulfonylhydrazones of acylphosphonic acid diesters creates a widely based route to diazomethanephosphonic acid diesters of widely varying substitution. The carbonyl components employed for the hydrazone condensation are accessible in one stage from acyl halides and trialkyl phosphites by the Michaelis-Arbusov method[162–164]:

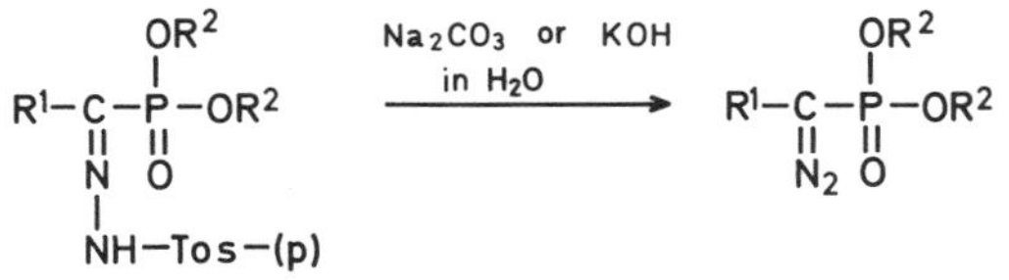

$$R^1-\overset{OR^2}{\underset{\underset{NH-Tos-(p)}{\overset{\|}{N}}}{\overset{\|}{C}}-\overset{}{\overset{\|}{P}}}-OR^2 \xrightarrow[\text{in } H_2O]{Na_2CO_3 \text{ or } KOH} R^1-\overset{OR^2}{\underset{\overset{\|}{N_2}}{\overset{\|}{C}}}-\overset{}{\overset{\|}{P}}-OR^2$$

Diazotriphenylsilylmethanes with aliphatic and aromatic second substituents are accessible by the same route[165]:

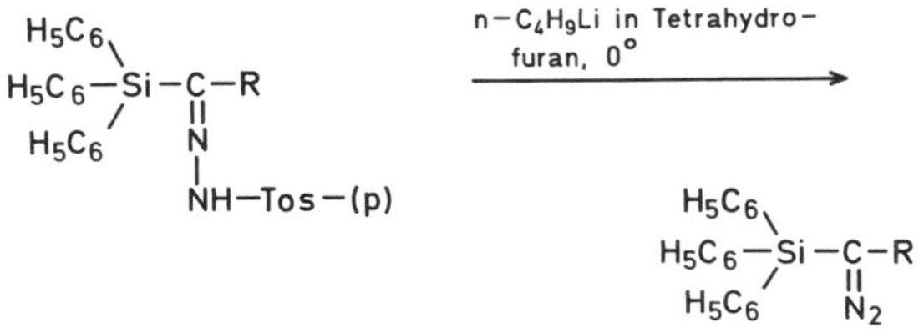

$$\begin{array}{c} H_5C_6 \\ H_5C_6-Si-\overset{}{\overset{\|}{C}}-R \\ H_5C_6 \quad \overset{}{\underset{NH-Tos-(p)}{N}} \end{array} \xrightarrow{\begin{array}{c} n-C_4H_9Li \text{ in Tetrahydro-} \\ \text{furan, } 0° \end{array}} \begin{array}{c} H_5C_6 \\ H_5C_6-Si-\overset{}{\overset{\|}{C}}-R \\ H_5C_6 \quad N_2 \end{array}$$

Use of the organometallic base for cleaving the hydrazone is reminiscent of the preparation of diazoalkanes by the same technique (see p. 219). The cleavage remains incomplete with sodium methoxide in pyridine. *(α-Diazobenzyl)triphenylgermane* too, has been obtained by Bamford-Stevens reaction[165].

8.1.5 Acyl cleavage of *N*-nitrosoalkyl carboxamides

Alkaline cleavage of *N*-nitrosoalkyl carboxamides preferentially serves for synthesizing *diazome-*

[160] *G. Heck*, unpubl. results, University Saarbrücken, 1963;
B. Eistert, M. Regitz, G. Heck, H. Schwall in *Houben-Weyl*, Methoden der organischen Chemie, 4. Aufl., Bd. X/4, p. 565, Georg Thieme Verlag, Stuttgart 1968.

[161] *M. Regitz, G. Heck*, Chem. Ber. *97*, 1482 (1964).

[162] *D. Seyferth, P. Hilbert, R.S. Marmor*, J. Amer. Chem. Soc. *89*, 4811 (1967);
R.S. Marmor, D. Seyferth, J. Org. Chem. *36*, 128 (1971).

[163] *M. Regitz, W. Anschütz, A. Liedhegener*, Chem. Ber. *101*, 3734 (1968).

[164] *W. Jugelt, D. Schmidt*, Tetrahedron *25*, 5569 (1969).

[165] *A.G. Brook, P.F. Jones*, Can. J. Chem. *47*, 4353 (1969).

thane and its homologs; in the case of *N*-methyl-*N*-nitrosourethane it led to the discovery of diazomethane[166]. In addition to urethans (**20**, R^1=OR), ureas (**20**, R^1=NR_2), and carboxamides (**20**, R^1=*Alkyl* or *Aryl*) *N*-alkyl-*N*-nitrosoguanidines and *N*-alkyl-*N*-nitrososulfonamides too, are accessible to acyl cleavage. The advantages and disadvantages of the individual variants are discussed in what follows. Formation of diazotates **21** as a precursor of the diazoalkanes **23** is now accepted as certain; they could be isolated during decomposition of *N*-alkyl-*N*-nitrosourethans[167–169] and of a corresponding urea[170] and benzamide derivative[171]. Nucleophilic incorporation of the base (generally alkoxide) into the carbonyl group of **20** precedes their formation and initiates the cleavage **20**→**21**. The esters that arise, too, can be isolated in many instances[168, 172-175]. During acyl cleavage the diazotates **21** are in equilibrium with diazonium ions **24** *via* diazohydroxides **22**; these ions are converted into diazoalkanes **23**, or carbenium ions **25**, or their sequential products:

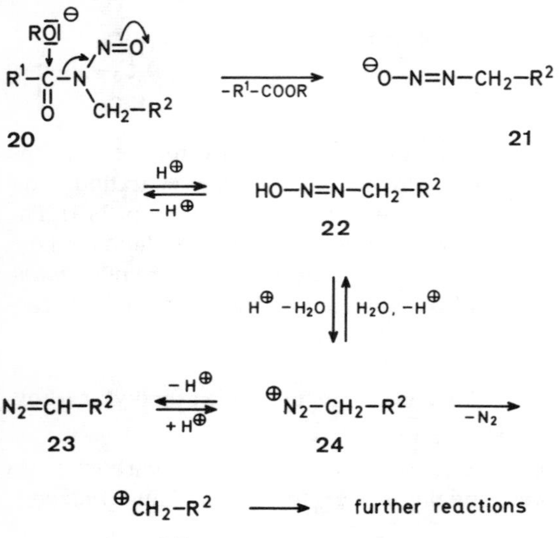

A specific mechanism has been proposed for the formation of diazotate from *N*-alkyl-*N*-nitrosoureas and numerous arguments have been adduced in its favor[175]:

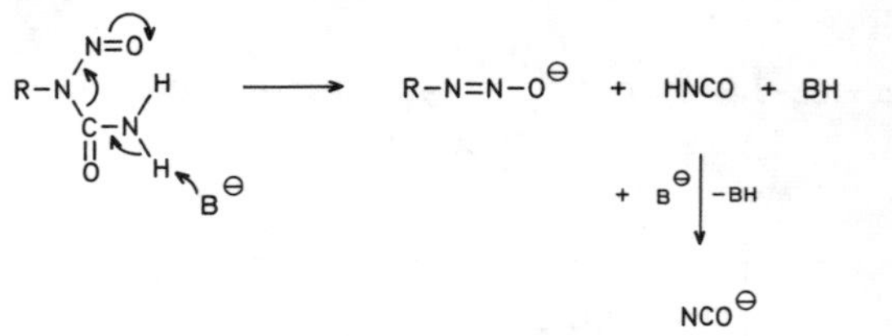

Another opinion interprets the formation of cyanate as due to base attack on the *NO* group[174, 176].

8.1.5.1 *N*-Alkyl-*N*-nitrosourethans

N-Methyl-*N*-nitrosourethan is readily available from ethyl chloroformate and methylamine and then nitrosation[177]. It keeps well but displays strong skin-irritant properties to make its use difficult. Decomposition of the urethane takes place even in ethanol with catalytic amounts of alkali metal carbonate and is suitable above all for *in situ* (nascent state) reactions with *diazomethane*[178]:

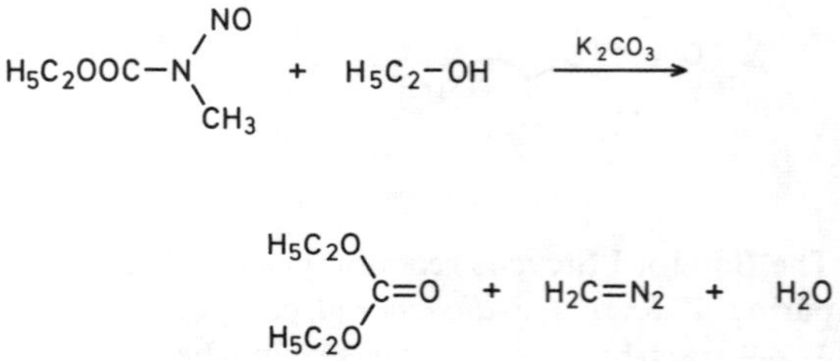

Homologization of cyclohexanone to *2-phenylcycloheptanone* with analogously produced *α-diazotoluene* (41–47%) is a smoothly proceeding application example[179]:

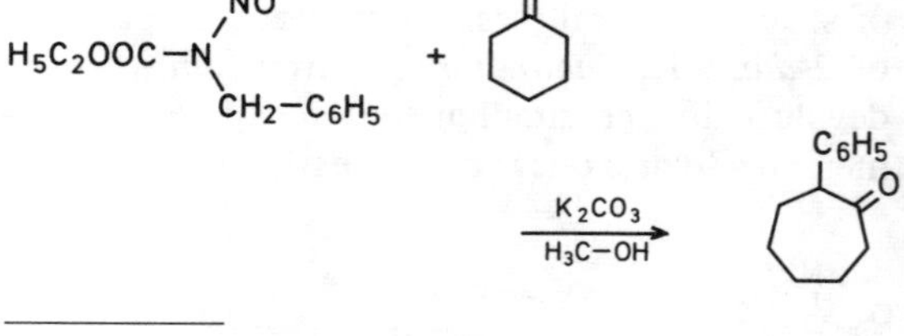

[166] *H. v. Pechmann*, Ber. **27**, 1888 (1894).
[167] *A. Hantzsch, M. Lehmann*, Ber. **35**, 897 (1902).
[168] *R.A. Moss*, J. Org. Chem. **31**, 1082 (1966).
[169] *R.A. Moss, F.C. Shulman*, Tetrahedron **24**, 2881 (1968).
[170] *T.K. Tandy, W.M. Jones*, J. Org. Chem. **30**, 4257 (1965).
[171] *H. Hart, J.L. Brewbaker*, J. Amer. Chem. Soc. **91**, 706 (1969).
[172] *R. Huisgen*, Justus Liebigs Ann. Chem. **573**, 163 (1951);
R. Huisgen, J. Reinertshofer, Justus Liebigs Ann. Chem. **575**, 174 (1952).
[173] *C.D. Gutsche, H.E. Johnson*, J. Amer. Chem. Soc. **77**, 109 (1955).
[174] *W.M. Jones, D.L. Muck*, J. Amer. Chem. Soc. **88**, 3798 (1966).
[175] *W. Kirmse, G. Wächtershäuser*, Justus Liebigs Ann. Chem. **707**, 44 (1967).
[176] *W.M. Jones, D.L. Muck, T.K. Tandy*, J. Amer. Chem. Soc. **88**, 68 (1966).
[177] *W.W. Hartman, M.R. Brethen*, Org. Syntheses Coll. Vol. II, 278 (1943);
[178] DRP 579 309 (1933), Schering, Erf.: *H. Meerwein*; C.A. **27**, 4546 (1933).
[179] *C.D. Gutsche*, J. Amer. Chem. Soc. **71**, 3513 (1949);
C.D. Gutsche, H.E. Johnson, Org. Syntheses Coll. Vol. IV, 780 (1963).

When a *solution* of *diazomethane* in an organic solvent (*e.g.*, ether) is needed for synthetic purposes, alkali cleavage of *N*-methyl-*N*-nitrosourethane with methanolic potassium hydroxide in ether is performed and the diazomethane is distilled into a cooled ether receiver[180]. Admittedly the solution still contains methanol. This drawback is avoided by performing the urethane cleavage with sodium glycolate in glycol, a nonvolatile alcohol, and conducting the diazomethane into the desired solvent in a current of nitrogen[181, 182]. In this way *cyclopropyldiazomethane* was prepared in reasonable yield[183]. For some further examples of the preparation of homologous and substituted diazomethanes using this technique, see Table 6. It cannot be employed for the synthesis of 2-diazo-1, 1-difluoroethane[11]:

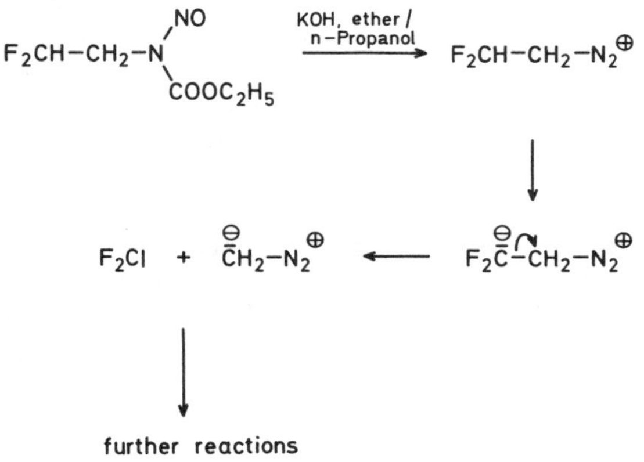

α,ω-bisdiazoalkanes are accessible from the corresponding dinitroso compounds:

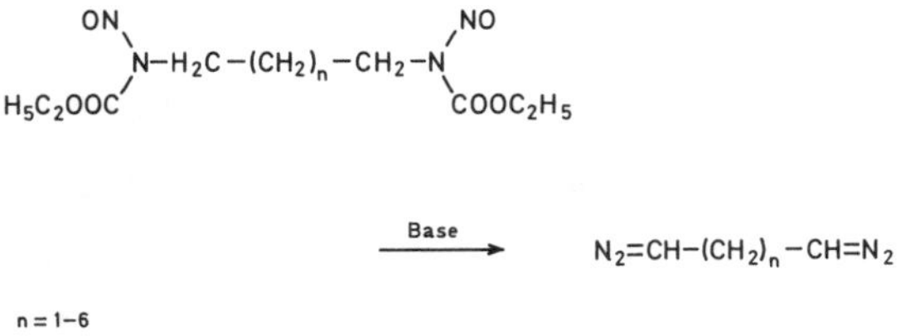

n = 1–6

The first member of this series, *bis(diazo)methane*, has not been isolated but has been identified by various reactions[184]. The same holds for *1,2-bis(diazo)-ethane*[185, 187]. Other homologs are tolerably stable in solution or substance[185–187].

Numerous *α,β*-unsaturated diazoalkanes have been synthesized *via* nitrosourethanes; they are only moderately stable because, in general, they isomerize to pyrazoles by 1,5-cyclization. Thus, potassium hydroxide cleavage of *N*-allyl-*N*-nitrosourethane affords 3-diazopropene (22–25%)[188, 189], which slowly passes over into pyrazole[189, 190] (for further examples, see Table 6).

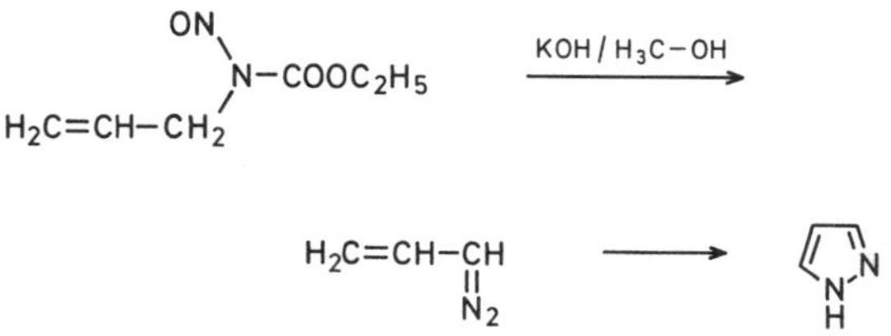

Preparation of *α*-diazocarbonyl compounds *via* nitrosourethanes has so far attracted little interest. Thermal cleavage of *N*-ethoxycarbonylmethyl- and *N*-1-ethoxycarbonylethyl-*N*-nitrosourethan in the presence of catalytic amounts of potassium carbonate is the only instance that can be used preparatively[191]:

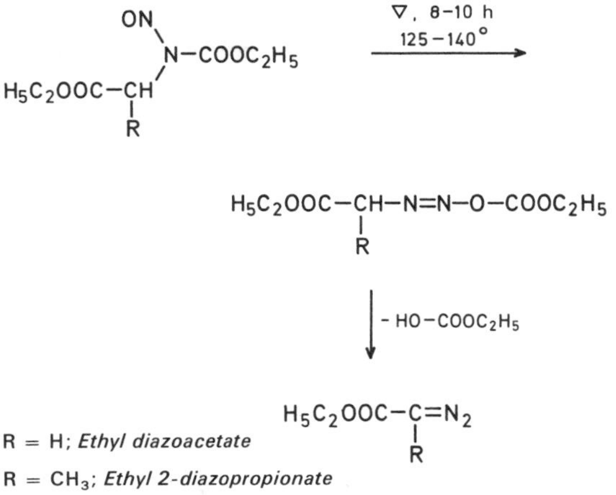

R = H; *Ethyl diazoacetate*

R = CH$_3$; *Ethyl 2-diazopropionate*

[180] *B. Eistert* in Neuere Methoden der präparativen organischen Chemie, 3. Aufl., p. 359, 398, Verlag Chemie, Weinheim/Bergstr. 1949.

[181] *H. Meerwein, W. Burneleit*, Ber. *61*, 1840 (1928).

[182] During distilling of diazomethane in N$_2$ or with organic solvents **care** is indicated (goggles, screen, vent). Sharp glass edges and ground joints should be avoided.

[183] *R.A. Moss, F.C. Shulman*, Tetrahedron *24*, 2881 (1968) (with triethyleneglycol).

[184] *H. Holter, H. Bretschneider*, Monatsh. Chem. *53/54*, 963 (1929).

[185] *E. Müller, S. Petersen*, Angew. Chem. *63*, 18 (1951).

[186] *T. Lieser, G. Beck*, Chem. Ber. *83*, 137 (1950).

[187] *C.M. Samour, J.P. Mason*, J. Amer. Chem. Soc. *76*, 441 (1954).

[188] *S. Nirdlinger, S.F. Acree*, Amer. Chem. J. *43*, 381 (1910).

[189] *C.D. Hurd, S.C. Lui*, J. Amer. Chem. Soc. *57*, 2656 (1935).

[190] *D.W. Adamson, J. Kenner*, J. Chem. Soc. (London) 286 (1935).

[191] *E.H. White, R.J. Baumgarten*, J. Org. Chem. *29*, 2070 (1964).

By contrast, alkylsulfonyl- and arylsulfonyldia-zomethanes only became available in this way[192, 193]. It is noteworthy that the splitting of the urethans already takes place during chromatography on alumina[193].

8.1.5.2 N-Alkyl-N-nitrosoureas

N-Methyl-*N*-nitrosourea is one of the most used starting materials for preparing *diazomethane;* it is readily accessible from urea and methylamine hydrochloride on subsequent nitrosation[194–197]. If necessary, for instance, on prolonged storage, it may be recrystallized from methanol.

Prolonged storage at 20–30° leads to **spontaneous decomposition** and formation of *methyl isocyanate*[194, 195, 198, 199]; at <20° the same decomposition ensues very slowly and is accompanied by trimerization of the methyl isocyanate[199]. Storing *N*-methyl-*N*-nitrosourea at ≦0° makes it suitable for preparing diazomethane even after substantial periods of time:

$$H_2N-\underset{\underset{CH_3}{|}}{\overset{\overset{NO}{|}}{\underset{\underset{O}{\|}}{C}}-N} \xrightarrow{20-30°} N_2 + H_2O + H_3C-N=C=O$$

$$\xrightarrow{\text{Trimerization}} \text{(triazine structure)}$$

N-Methyl-*N*-nitrosourea is cleaved in a two-phase system (40–70% aqueous caustic potash/ether) and affords ether solutions of *diazomethane* (~70%)[198, 200–202]. Other organic solvents, (*e.g.*, benzene, acetone[200]), too, are suitable for dissolving the diazomethane.

$$H_2N-\underset{\underset{CH_3}{|}}{\overset{\overset{NO}{|}}{\underset{\underset{O}{\|}}{C}}-N} + KOH \longrightarrow$$

$$H_2C=N_2 + KOCN + 2H_2O$$

Ether solutions of diazomethane can be distilled for *purification* purposes[200–202] (cf. p. 224); they can be dried over potassium hydroxide, optionally with squeezing in of sodium wire[203]. Traces of methylamine can be removed with water-soluble starch[203]. Diazomethane *gas*, free from solvent, is obtained by introducing an aqueous suspension of the *N*-methyl-*N*-nitrosourea in caustic potash; the diazomethane is removed from the reaction mixture in a current of nitrogen[204]. To *determine* the *content* of diazomethane (or diazoalkane) solutions these are reacted with excess benzoic acid and back-titrated with alkali hydroxide. Alternatively, the ester can be determined gravimetrically with 4-nitrobenzoic acid[205].

Numerous homologs and substitution products of diazomethane have been prepared by alkaline-cleavage of *N*-alkyl-*N*-nitrosoureas. *Diazoethane*[198], *1-diazopropane*[206], *1-diazo-2-methylpropane*[207], *1-diazobutane*[198, 208], *1-diazododecane*[209], *α-diazotoluene*[174, 198], *diazotrimethylsilylmethane*[210] are examples of successful syntheses; some further compounds are listed in Table 6.

Cleavage of *N*-cycloalkyl-*N*-nitrosoureas is carried out at low temperatures (down to −50°) with alkoxides. *Diazocyclobutane*[211] and *diazocyclohexane*[212], which are accessible in this way, are evidently less stable than the open-chain diazoalkanes. This is true, especially, of the angularly strained *1-diazo-2,2-diphenylcyclopropane*, which quickly changes into *1,1-diphenylallene via* the carbene route[213]:

$$H_2N-\underset{\underset{O}{\|}}{C}-N\underset{\overset{|}{H}}{\overset{\overset{NO}{|}}{}} \overset{C_6H_5}{\underset{C_6H_5}{}} \xrightarrow{LiOC_2H_5} N_2 \overset{C_6H_5}{\underset{C_6H_5}{}}$$

$$\longrightarrow H_2C=C=C\overset{C_6H_5}{\underset{C_6H_5}{}}$$

[192] *J. Strating, A.M. van Leusen*, Rec. Trav. Chim. Pays-Bas *81*, 966 (1962).

[193] *A.M. van Leusen, J. Strating*, Rec. Trav. Chim. Pays-Bas *84*, 151 (1965).

[194] *F. Arndt, H. Scholz*, Angew. Chem. *46*, 47 (1933).

[195] *F. Arndt, L. Loewe, S. Avan*, Ber. *73*, 606 (1940).

[196] *F. Arndt*, Org. Syntheses Coll. Vol. II, 461 (1943).

[197] *B. Eistert* in Neuere Methoden der präparativen organischen Chemie, 3. Aufl., p. 395, Verlag Chemie, Weinheim/Bergstr. 1949.

[198] *E.A. Werner*, J. Chem. Soc. (London) 1093 (1919).

[199] *K. Clusius, F. Endtinger*, Helv. Chim. Acta *43*, 2063 (1960).

[200] *F. Arndt, J. Amende*, Angew. Chem. *43*, 444 (1930).

[201] *F. Arndt*, Org. Syntheses Coll. Vol. II, 165 (1943).

[202] *A. Roedig, K. Grohe*, Tetrahedron *21*, 397 (1965).

[203] *G. Wittig, K. Schwarzenbach*, Justus Liebigs Ann. Chem. *650*, 12 (1961).

[204] *W.H. Urry, N. Bilow*, J. Amer. Chem. Soc. *86*, 1815 (1964).

[205] *E.K. Marshall, S.F. Acree*, Ber. *43*, 2323 (1910).

[206] *J.R. Dyer, R.B. Randall, H.M. Deutsch*, J. Org. Chem. *29*, 3423 (1964).

[207] *B. Eistert, M. Regitz, G. Heck, H. Schwall* in Houben-Weyl, Methoden der organischen Chemie, 4. Aufl., Bd. X/4, p. 540, Georg Thieme Verlag, Stuttgart 1968.

[208] *W. Kirmse, H.A. Rinkler*, Justus Liebigs Ann. Chem. *707*, 57 (1967).

[209] *G.D. Buckley, N.H. Ray*, J. Chem. Soc. (London) 3701 (1952).

[210] *D. Seyferth, A.W. Dow, H. Menzel, T.C. Flood*, J. Amer. Chem. Soc. *90*, 1080 (1968).

[211] *D.E. Applequist, D.E. McGreer*, J. Amer. Chem. Soc. *82*, 1965 (1960).

[212] *K. Heyns, A. Heins*, Justus Liebigs Ann. Chem. *604*, 133 (1957).

α,ω-Bis(diazo)alkanes[186, 187, 214], too, are accessible in this way, albeit no advantages over the urethan cleavage are revealed (see p. 225). In conclusion, it is worth mentioning the preparation of *diazoacetaldehyde acetals* by urea cleavage[215]; it is made possible by the alkali resistance of the acetal groups:

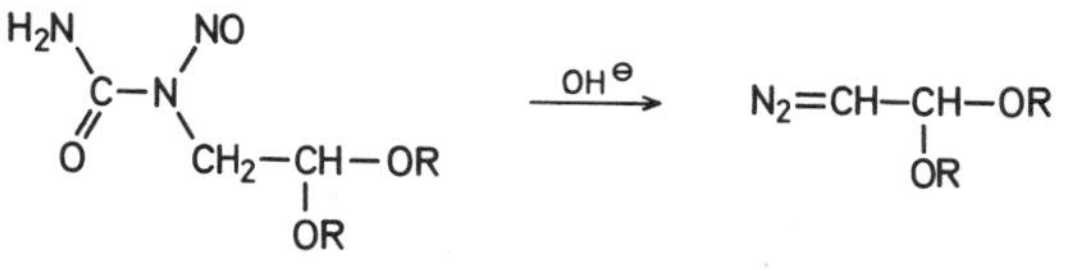

R = C_2H_5; *Diazoacetaldehyde diethyl acetal*

R = $-CH_2-CH_2-$; *2-(Diazomethyl)-1,3-dioxolane*

8.1.5.3 *N*-Alkyl-*N'*-nitro-*N*-nitrosoguanidines

Base cleavage of *N*-alkyl-*N'*-nitro-*N*-nitrosoguanidines has revealed itself to be a useful method for preparing diazoalkanes. These compounds display the advantage over nitrosoureas of greater thermal stability (store in dark bottles) but, like nitrosourethanes, are skin irritants. The potassium salt of nitrocyanamide formed during cleavage of nitrosoguanidines with potassium hydroxide can be isolated[216]:

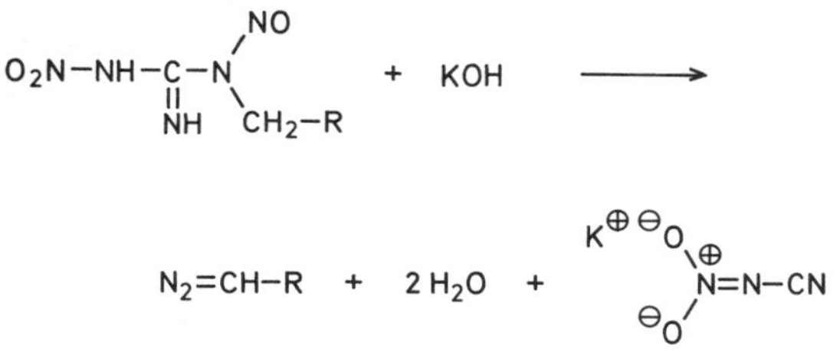

Diazomethane is accessible in 83–93% yield[216, 218] from *N*-methyl-*N'*-nitro-*N*-nitrosoguanidine[217] analogously to urea cleavage[201]. Homologs, too (*diazoethane, 1-diazopropane, 1-diazo-*

butane, 1-diazopentane), can be obtained satisfactorily by this method (55–65%, with distillation the yields are somewhat lower)[216]; synthesis of α-diazotoluene[216, 219] proceeds particularly smoothly. By contrast, preparation of 2-diazopropane and diazocyclohexane fails already at the nitrosation stage of the relevant guanidines[216].

8.1.5.4 *N*-Alkyl-*N*-nitrosocarboxamides

Although *N*-alkyl-*N*-nitrosobenzamide was recognized to be suitable for the preparation of *diazomethane* at a very early date[166], useful laboratory techniques for diazomethane and its homologs from carboxamides were not developed until recently. *N*-Methyl-*N*-nitrosoacetamide[220] is cleaved smoothly with methanolic potassium hydroxide but affords methanol-containing diazomethane solutions[221, 222]. Organometallic bases, too, detach acyl by attacking on the carbonyl carbon, as is shown by the formation of tertiary lithium alkoxides[223]. Lithium methyldiazotate is isolated at the same time; its hydrolysis leads to diazomethane[223]:

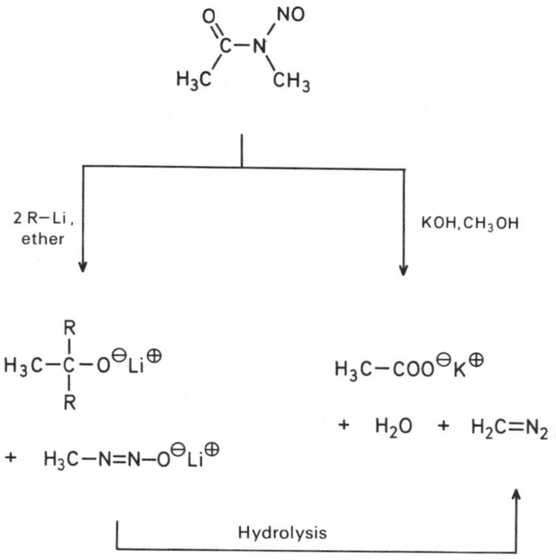

Base cleavage of *N,N'*-dialkyl-*N,N'*-dinitrosooxalamide is probably more interesting preparatively. Carried out with methylamine, for instance, it affords 65% *diazomethane* in addition to 82%

[213] *W.M. Jones, D.L. Muck, T.K. Tandy,* J. Amer. Chem. Soc. *88*, 68 (1966).

[214] *H. Lettré, M. Brose,* Naturwissenschaften *36*, 57 (1949).

[215] *W. Kirmse, M. Buschoff,* Angew. Chem. *77*, 681 (1965); *W. Kirmse, M. Buschoff,* Angew. Chem. Intern. Ed. Engl. *4*, 692 (1965).

[216] *A.F. McKay, W.L. Ott, G.W. Taylor, M.N. Buchanan, J.F. Crooker,* Can. J. Chem. *28B*, 683 (1950).

[217] *A.F. McKay, G.F. Wright,* J. Amer. Chem. Soc. *69*, 3028 (1947).

[218] *A.F. McKay,* J. Amer. Chem. Soc. *70*, 1974 (1948).

[219] *B. Eistert, M. Regitz, G. Heck, H. Schwall* in *Houben-Weyl,* Methoden der organischen Chemie, 4. Aufl., Bd. X/4, p. 543, Georg Thieme Verlag, Stuttgart 1968;

[220] *G.F. D'Alelio, E.E. Reid,* J. Amer. Chem. Soc. *59*, 109 (1937).

[221] *K. Heyns, O.F. Woyrsch,* Chem. Ber. *86*, 76 (1953).

[222] *K. Heyns, W. v. Bebenburg,* Chem. Ber. *86*, 278 (1953).

[223] *H. Reimlinger,* Angew. Chem. *73*, 221 (1961).

N,N'-dimethyloxamide[224]. As the latter is once more available for reaction following nitrosation, methylamine can be converted into diazomethane in a cyclic process in this way (cf. p. 207):

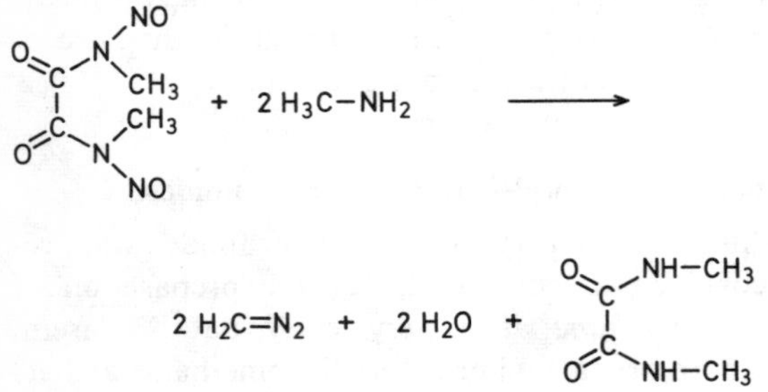

By employing different bases the above dinitroso compound affords also *gaseous* diazomethane (potassium hydroxide in water/cellosolve) and *in situ* diazomethane (potassium carbonate in methanol) as well as distilled diazomethane (sodium butoxide in butanol/ether is an alternative to methylamine)[225]. In addition *N,N'*-dimethyl-*N,N'*-dinitrososuccinamide[226] and -terephthalamide[227] have been used for preparing diazomethane. Homologs of diazomethane, too, are accessible by this route (for examples see Table 6), and *α,ω-bis(diazo)alkanes*. One characteristic example is acyl detachment from nitrosated *nylon 66*, which affords *1,6-bis(diazo)hexane*[228]:

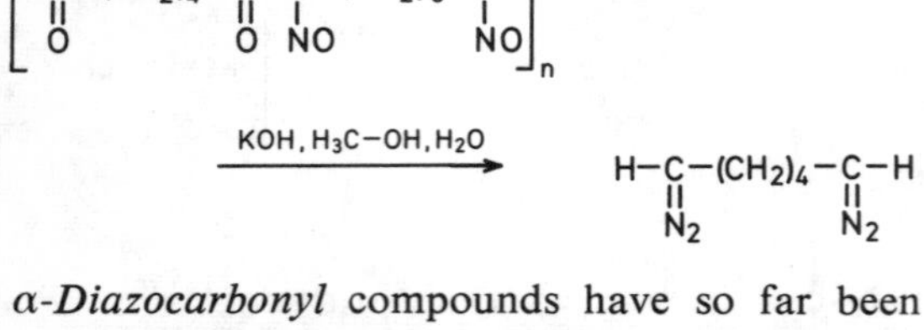

α-Diazocarbonyl compounds have so far been synthesized from nitrosated carboxamides in modest measure only. In addition to certain *α-diazo ketones*[229] interest has been concentrated on *ethyl diazoacetate*[191, 230]. It is formed in 84% yield during cleavage of *N*-ethoxycarbonylmethyl-*N*-nitrosoacetamide with barium oxide/barium hydroxide in methanol[230], while the same reaction

using ammonia as base affords *diazomethane via* a mechanistically different path[231].

Acyl detachment in *cyclic* nitrosated carboxamides involves ring opening:

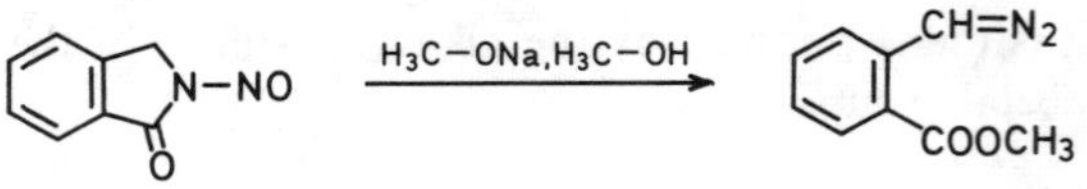

Methyl 2-diazomethyl-benzoate[232]

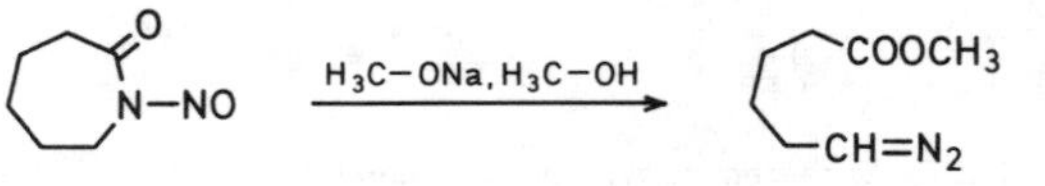

Methyl 6-diazohexanoate[233]

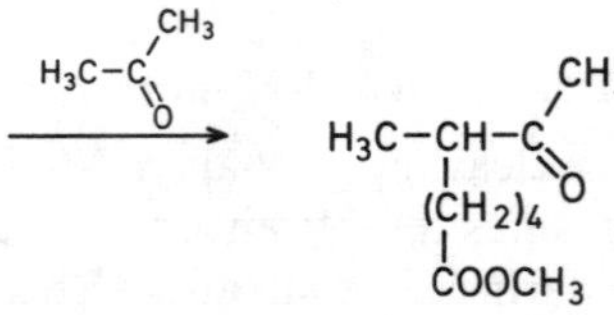

Methyl 6-methyl-7-oxo-octanoate

In the last example the unstable diazo derivative is trapped with acetone.

8.1.5.5 *N*-Alkyl-*N*-nitroso-*p*-toluenesulfonamides

Temperatures of 50–75° are necessary for cleaving *N*-alkyl-*N*-nitroso-*p*-toluenesulfonamides with potassium hydroxide[234], but these do not govern the yield of *diazomethane*. The procedure used enables either gaseous diazomethane (potassium hydroxide in carbitol/water)[235] or an ether solution (analogously but with added ether) to be prepared[235, 236]:

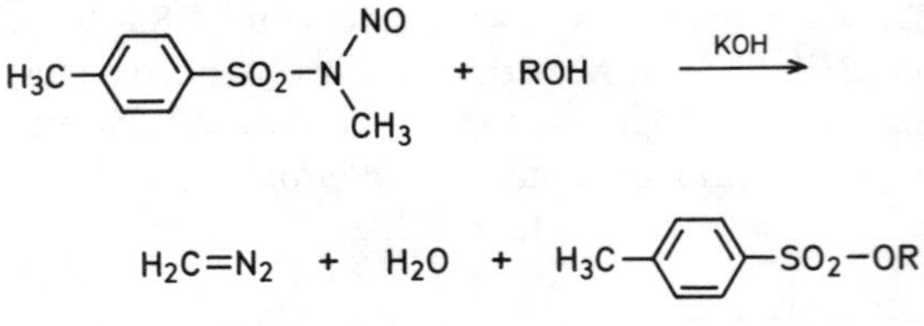

The undenied *advantage* of the method is the storability of the nitrosoamide at room temperature (dark bottle) and the fact that it can be obtained

[224] USP. 2 675 378 (1954), DuPont, Inv.: *F.S. Fawcett;* C.A. *49,* 1777b (1955).
[225] *H. Reimlinger,* Chem. Ber. *94,* 2547 (1961).
[226] DBP. 841 747 (1952), BASF, Erf.: *H. Stummeyer, G. Hummel;* C.A. *52,* 10162c (1958).
[227] *J.A. Moore, D.E. Reed,* Org. Synth. *41,* 16 (1962).
[228] *E. Pfeil, O. Weissel,* Chem. Ber. *91,* 1170 (1958).
[229] *V. Franzen,* Justus Liebigs Ann. Chem. *602,* 199 (1957).
[230] *H. Reimlinger, L. Skattebøl,* Chem. Ber. *93,* 2162 (1960).

[231] *H. Reimlinger, L. Skattebøl,* Chem. Ber. *94,* 2429 (1961).
[232] *A. Oppé,* Ber. *46,* 1095 (1913).
[233] *W. Pritzkow, P. Dietrich,* Justus Liebigs Ann. Chem. *665,* 88 (1963).
[234] *T.J. De Boer, H.J. Backer,* Org. Syntheses Coll. Vol. IV, 943 (1963).
[235] *T.J. De Boer, H.J. Backer,* Rec. Trav. Chim. Pays-Bas *73,* 229 (1954).
[236] *T.J. De Boer, H.J. Backer,* Org. Syntheses Coll. Vol. IV, 250 (1963).

commercially. In addition it exhibits no skin-irritant properties. Evidently the full range of application has not yet been made use of. In addition to α-*diazotoluene*[237] and *diazocyclohexane*[212] only a few *alkoxy*-substituted diazoalkanes[238] have been synthesized in this way (cf. Table 6).

8.1.6 Alkoxide cleavage of 4-(N-alkyl-N-nitrosoamino)-4-methyl-2-pentanones

A *reliable* method for preparing *diazoalkanes* consists in the alkoxide cleavage of 4-(N-alkyl-N-nitrosoamino)-4-methyl-2-pentanones.

The quite stable nitroso compounds are accessible by means of two simple synthetic steps, namely, addition of alkylamine to 4-methyl-3-penten-2-one and subsequent nitrosation[239, 240]. During their cleavage, which evidently ensues according to a β-elimination mechanism with alkoxide attack on the hydrogen adjacent the *CO, 4-methyl-3-penten-2-one* and an *alkyl diazotate* changing over into a diazoalkane are formed (see pp. 208, 224):

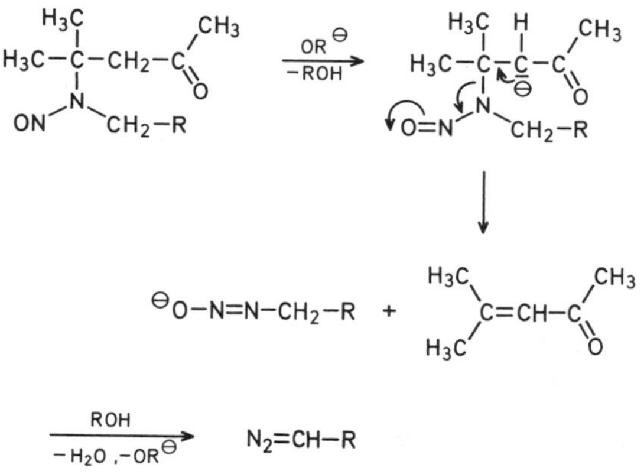

In this method for preparing diazomethane, sodium cyclohexyloxide in cyclohexanol/ether[240–242] has proven successful above all in addition to the use of sodium isopropoxide[239, 240]; diazomethane yields of up to 85% are achieved. It is necessary to allow for the fact that diazomethane can to a modest degree form a *methyl 4,4-dimethyl-1-pyrazolin-3-yl ketone* with the 4-methyl-3-penten-2-one formed during the cleavage. This is

Table 6. Aliphatic diazo compounds from *N*-alkyl-*N*-nitrosocarboxamides and 4-(*N*-alkyl-*N*-nitrosoamino)-4-methyl-2-pentanones

Diazo compound	Method*	Yield [% of theory]	Obtained as	Ref.
Diazoethane	A	75	Ether solution	245
	C	51	Ether solution	246
	E	50	Ether solution	247
1-Diazobutane	C	25–29	Ether solution	246, 248
1-Diazohexane	E	23	Ether solution	247
1-Diazo-2,2,2-triphenylethane	A	100	Crystals; mp 78–80° (decomp.)	249
1-Diazo-2-ethoxyethane	C	—	Ether solution	250
	D	47–49	Ether solution	251
2-Butyloxy-1-diazoethane	D	44–46	Ether solution	251
1-Diazo-3-methoxypropane	B	82	Benzene solution	252
	D	5–8	Ether solution	251
3-Chloro-1-diazopropane	B	47	Benzene solution	252
Methyl 4-diazobutyrate	C	—	Ether solution	253
α-*Diazotoluene*	C	40	Ether solution	248
(2-Allyloxyphenyl)-diazomethane	B	21	*n*-Pentane/glycol dimethyl ether solution	254
cis-3-Diazomethylbicyclo-[3.1.0]hexane	B	85–90	*n*-Pentane solution	255
trans-3-Diazomethylbicyclo-[3.1.0]hexane	B	87	*n*-Pentane solution	255
3-Diazopropene	B	65–70	Tetrahydrofuran solution	256
	E	41	Ether solution	247
	A	65	Cyclohexene solution	257
trans-1-Diazo-3-(3-nitrophenyl)-propene	A	41	Cyclohexene solution	257
Diazopropyne	C	—	Crystals, mp ∼ −65°	258
Benzenesulfonyl-diazomethane	A	81	Oil	259
Diazomethyl-α,α,α-trifluoro-3-tolyl sulfone	A	51	Crystals, mp < 20°	260

* A: Cleavage of *N*-alkyl-*N*-nitrosourethanes
B: Cleavage of *N*-alkyl-*N*-nitrosoureas
C: Cleavage of *N*-alkyl-*N*-nitroso carboxamides
D: Cleavage of *N*-alkyl-*N*-nitroso-*p*-toluenesulfonamides
E: Cleavage of 4-(*N*-alkyl-*N*-nitrosoamino)-4-methyl-2-pentanone

[237] *C.G. Overberger, J.-P. Anselme*, J. Org. Chem. *28*, 592 (1963).
[238] *E. Pfeil, E. Weinrich, O. Weissel*, Justus Liebigs Ann. Chem. *679*, 42 (1964).
[239] *E.C.S. Jones, J. Kenner*, J. Chem. Soc. (London) 363 (1933).
[240] *C.E. Redemann, F.O. Rice, R. Roberts, H.P. Ward*, Org. Syntheses Coll. Vol. III, 244 (1955).

[241] *D.W. Adamson, J. Kenner*, J. Chem. Soc. (London) 1551 (1937).

feasible both in the reaction space and in the distillate because some ketone is entrained during the diazomethane distillation[241]:

$$H_3C \atop H_3C \! > \!\! C\!=\!CH\!-\!CO\!-\!CH_3 \quad + \quad CH_2N_2$$

$$\longrightarrow \quad H_3C \atop H_3C \! > \!\! \underset{N \diagdown N}{\overset{H_3C \; CO-CH_3}{\underset{\diagup}{\bigtriangleup}}} H$$

In addition to *cyclopropyldiazomethane*[243] numerous diazomethane homologs[244] provide evidence of the efficacy of this synthetic principle (for examples see Table 6).

8.1.7 Diazo syntheses with azides

A further possible method of preparing aliphatic diazo compounds consists in the *direct* introduction of both *N* atoms, *i.e.*, the entire azo group, in

a single reaction step. It is transferred by a donor (azides) to an acceptor (*CH*-acid compounds, activated ethylenes, and acetylenes), a process that is called diazo group transfer[261]. The suitability of azides for this reaction becomes clear when one considers that they possess an N_2 group linked to nitrogen and that they are isoelectronic with diazo compounds.

8.1.7.1 Diazo group transfer to methylene compounds with *p*-toluenesulfonic acid azide

So far the readily accessible and stable *p*-toluenesulfonic acid azide has been used almost exclusively as the diazo group transfer agent[262, 263]; the *p*-toluenesulfonamide formed during the reaction is separated by virtue of its acid properties:

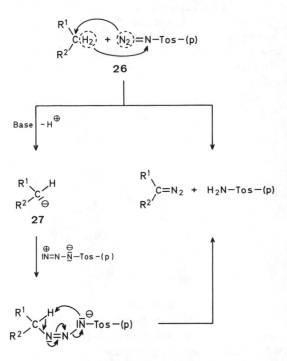

The carbanion **27** of the *CH*-acid compound **26** and not the latter itself is the reaction partner of the azide. Thus, the diazo group transfer is performed in the presence of a base whose strength is

[242] *M. Berenbom, W.S. Fones*, J. Amer. Chem. Soc. *71*, 1629 (1949).

[243] *P.D. Shevlin, A.P. Wolf*, J. Amer. Chem. Soc. *88*, 4735 (1966).

[244] *D.W. Adamson, J. Kenner*, J. Chem. Soc. (London) 286 (1935).

[245] *A.L. Wilds, A.L. Meader*, J. Org. Chem. *13*, 763 (1948).

[246] *H. Reimlinger*, Chem. Ber. *94*, 2547 (1961).

[247] *D.W. Adamson, J. Kenner*, J. Chem. Soc. (London) 286 (1935).

[248] *K. Heyns, W. v. Bebenburg*, Chem. Ber. *86*, 278 (1953).

[249] *L. Hellerman, R.L. Garner*, J. Amer. Chem. Soc. *57*, 139 (1935).

[250] *E. Pfeil, O. Weissel*, Chem. Ber. *91*, 1170 (1958).

[251] *C. Groth, E. Pfeil, E. Weinrich, O. Weissel*, Justus Liebigs Ann. Chem. *679*, 42 (1964).

[252] *W. Kirmse, H.J. Schladetsch, H.W. Bücking*, Chem. Ber. *99*, 2579 (1966).

[253] *H. Reimlinger*, Chem. Ber. *97*, 331 (1964).

[254] *W. Kirmse, H. Dietrich*, Chem. Ber. *100*, 2710 (1967).

[255] *W. Kirmse, K. Pöhlmann*, Chem. Ber. *100*, 3564 (1970).

[256] *J. Hooz, H. Kono*, Organic Preparations and Procedures International *3*, 47 (1971).

[257] *J.L. Brewbaker, H. Hart*, J. Amer. Chem. Soc. *91*, 711 (1969).

[258] *H. Reimlinger*, Angew. Chem. *74*, 252 (1962); *H. Reimlinger*, Angew. Chem. Intern. Ed. Engl. *1*, 216 (1962); *H. Reimlinger*, Justus Liebigs Ann. Chem. *713*, 113 (1968).

[259] *A.M. van Leusen, J. Strating*, Rec. Trav. Chim. Pays-Bas *84*, 151 (1965).

[260] *J.B.F.N. Engberts, G. Zuidema, B. Zwanenburg, J. Strating*, Rec. Trav. Chim. Pays-Bas *88*, 641 (1969).

[261] *M. Regitz*, Angew. Chem. *79*, 786 (1967); *M. Regitz*, Angew. Chem. Intern. Ed. Engl. *6*, 733 (1967); *M. Regitz* in Neuere Methoden der präparativen organischen Chemie, Bd VI, p. 76, Verlag Chemie, Weinheim/Bergstr. 1970; *M. Regitz*, Synthesis 351 (1972).

[262] *O. Dimroth*, Justus Liebigs Ann. Chem. *373*, 356 (1910).

[263] *T. Curtius, W. Klavehn*, J. Prakt. Chem. [2] *112*, 76 (1926).

governed by the proton-loosening effect of the substituents R^1 and R^2. In general a triazene is an intermediate stage; following the proton transfer to the amide nitrogen it decomposes into a diazoalkene and the *p*-toluenesulfonamide anion.

Preparation of *diazocyclopentadiene* was the start[264] of a rapid development. By varying the reaction conditions (acetonitrile/diethylamine[265] or diethylamine without solvent)[266] the synthesis can be made considerably simpler:

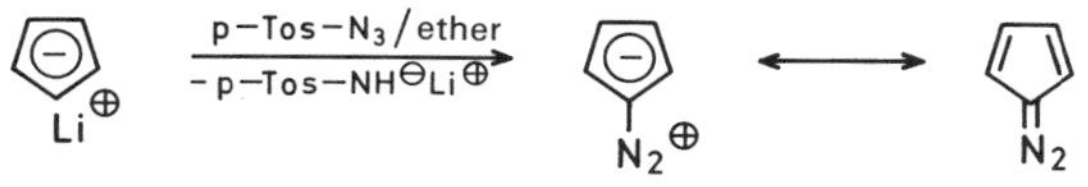

2,5-Diphenylcyclopentadiene gives *5-diazo-1,4-* and *5-diazo-1,3-diphenylcyclopentadiene* side by side; they are formed *via* the mesomeric carbanion, which contains two nucleophilic centers for the attack by azide[265]:

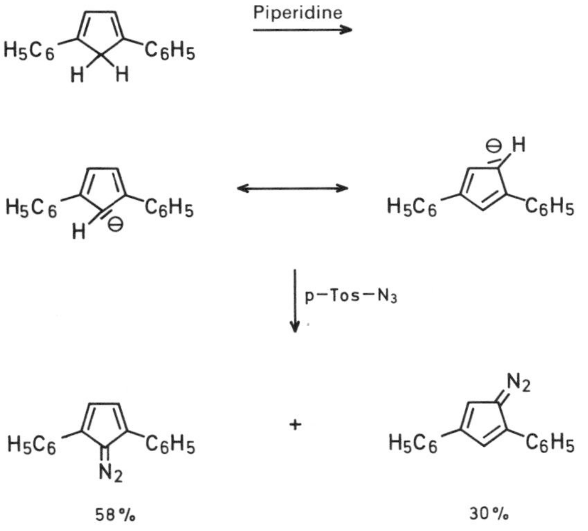

The results of the diazo group transfer to 2-cyclopentadienecarboxylic acid esters, likewise leading to positional isomer diazocyclopentadienes, can be interpreted analogously[267].

For further examples of synthesis of *diazocyclopentadienes* and *diazocyclohexadienes* see Table 7.

Diazo group transfers to 1,3-dicarbonyl compounds ensue as a result of the high proton activation by carbonyl groups in the presence of weak bases (organic amines) and furnish high yields of *2-diazo-1,3-dioxo* compounds[268–270]:

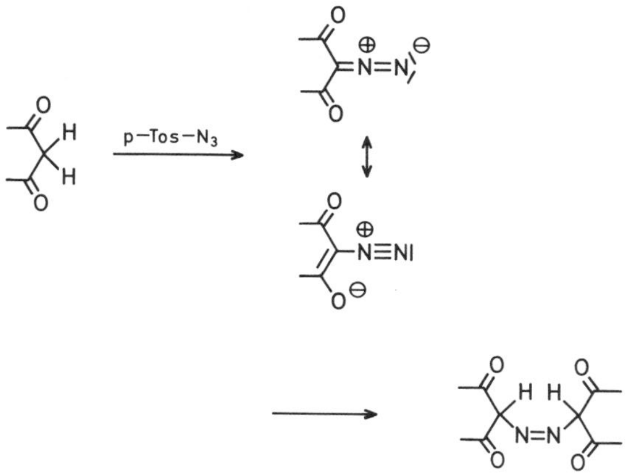

Because of their ready handling properties they have entirely displaced the formerly usual amine diazotization (see p. 210). Table 7 lists individual examples.

As 2-diazo-1,3-dicarbonyl compounds possess an extremely electrophilic diazo group (see the limiting formulae)[271], an azo coupling with the 1,3-diketone sometimes cannot be entirely suppressed[268].

The azo coupling arises preferentially in cyclic 1,3-dicarbonyl compounds and it may occur exclusively if the diazo group transfer agent is used in half the molar quantity[269]. With phenols and naphthols this azo group transfer entirely dominates the quinone diazide synthesis[272].

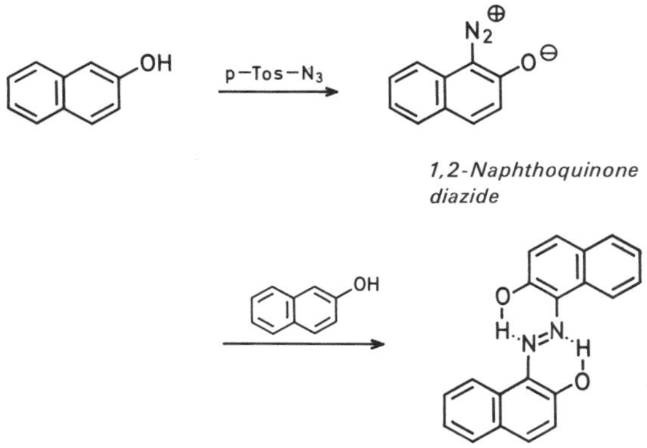

1,2-Naphthoquinone diazide

1,1'-Azo-di-2,2'-naphthol

[264] *W. von E. Doering, C.H. De Puy*, J. Amer. Chem. Soc. 75, 5955 (1953).

[265] *M. Regitz, A. Liedhegener*, Tetrahedron 23, 2701 (1967).

[266] *T. Weil, M. Cais*, J. Org. Chem. 28, 2472 (1963).

[267] *J.C. Martin, D.R. Bloch*, J. Amer. Chem. Soc. 93, 451 (1971).

[268] *M. Regitz*, Justus Liebigs Ann. Chem. 676, 101 (1964).

[269] *M. Regitz, D. Stadler*, Justus Liebigs Ann. Chem. 687, 214 (1965).

[270] *M. Regitz, A. Liedhegener*, Chem. Ber. 99, 3128 (1966).

[271] *M. Regitz, A. Liedhegener, D. Stadler*, Justus Liebigs Ann. Chem. 713, 101 (1968).

[272] *J.M. Tedder, B. Webster*, J. Chem. Soc. (London) 4417 (1960).

An intramolecular azo coupling is observed during diazo group transfer to 1,3,5-triketones[273]; it gives *3,5-diacyl-4H-pyrazol-4-ones*. Ring closure can be substantially suppressed in favor of the 2,4-bis(diazo)-1,3,5-triketo compounds by skilful choice of the reaction conditions (addition of the 1,3,5-triketone to the diazo group transfer agent)[274]:

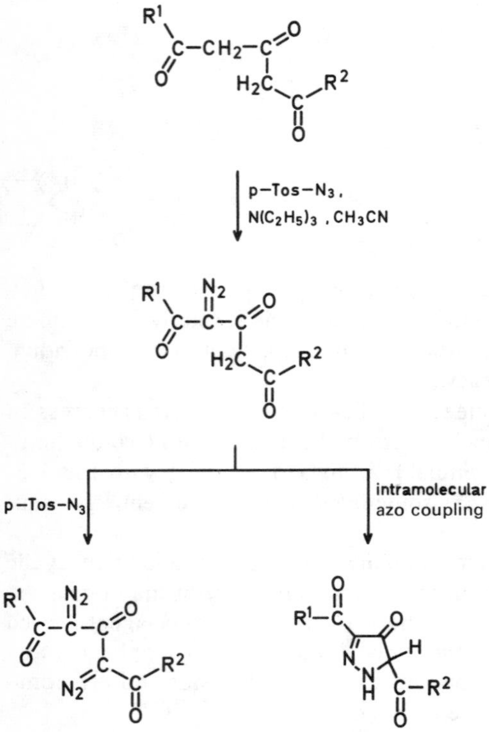

$R^1 = R^2 = C_6H_5$; *2,4-Bis(diazo)-1,5-diphenyl-1,3,5-pentanetrione; 64%*

$R^1 = R^2 = 4-CH_3O-C_6H_4$; *2,4-Bis(diazo)-1,5-bis(4-methoxyphenyl)-1,3,5-pentanetrione; 81%*

$R^1 = C_6H_5$; $R^2 = 4-Cl-C_6H_4$; *2,4-Bis(diazo)-1-phenyl-5-(4-chlorophenyl)-1,3,5-pentanetrione; 62%*

$R^1 = C_6H_5$; $R^2 = CH_3$; *2,4-Bis(diazo)-1-phenyl-1,3,5-hexanetrione; 80%*

Diazo group transfers to β-imino and β-thioxo ketones can also involve ring closure[275-277]. Thus, from 3-arylimino-1-indanones either *2-diazo-3-arylimino-1-indanones (X=H, Cl, NO$_2$)* or, by virtue of a subsequent 1,5-cyclization, *3-arylindeno[1,2-d]-1,2,3(H)-triazol-8-ones* are formed *(X=CH$_3$, I OCH$_3$)* depending on the substituent X[276]. The latter compounds can be subsequently isomerized to the diazo derivatives by opening the ring with mineral acids to form *3-arylimmonium-2-diazo-1-indanone* salts and deprotonating these[276]:

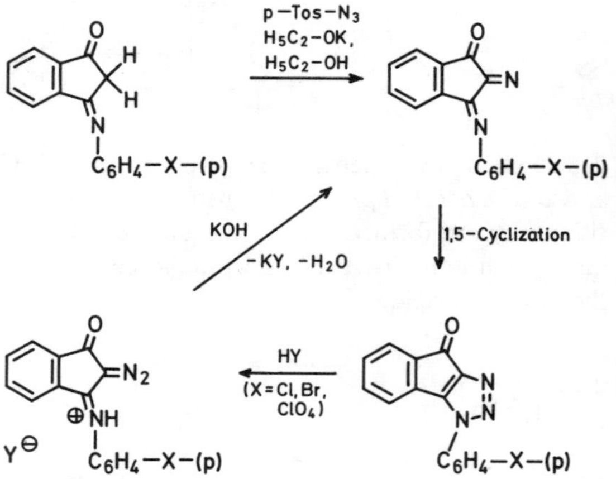

Diazo group transfer to diethyl malonates in the presence of an organic base ensues in the desired sense[270] (cf. Table 7). By contrast, the same reaction using sodium ethoxide as base affords almost solely *ethyl 2-diazo-N-(p-tolylsulfonyl)malonamate*[268, 270] with formal displacement of an ethoxy radical. Since a subsequent conversion of the diazo ester into diazoamide can be ruled out[270], it is logical to postulate a change in mechanism from a linear to a cyclic course with a 4,5-dihydro-1,2,3-triazole intermediate stage[270]:

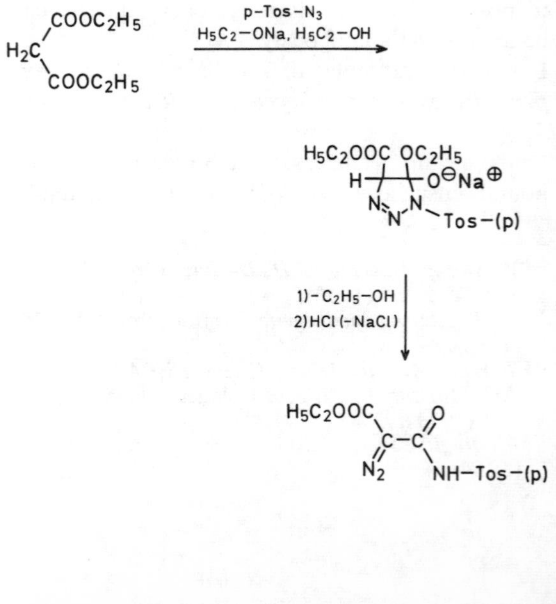

[273] *M. Regitz, H.J. Geelhaar*, Chem. Ber. *101*, 1473 (1968).

[274] *M. Regitz, H.J. Geelhaar, J. Hocker*, Chem. Ber. *102*, 1743 (1969).

[275] *M. Regitz*, Chem. Ber. *99*, 2918 (1966).

[276] *M. Regitz, H. Schwall*, Justus Liebigs Ann. Chem. *728*, 99 (1969).

[277] *M. Regitz, A. Liedhegener*, Justus Liebigs Ann. Chem. *710*, 118 (1967).

Analogously the following conversions can be effected:

Ethyl acetoaceate ⟶ 2-Diazo-N-p-tolylsulfonyl-acetoacetamide[268, 272]

Ethyl (4-nitrophenyl)acetate ⟶ Diazo-N-p-tolylsulfonyl-(4-nitrophenyl)acetamide[278]

The synthesis of α-*diazocarbonyl* compounds by diazo group transfer is limited by the proton activity of the *CH$_2$* components. Compounds of the type *R^1–CO–CN$_2$–R^2* are accessible only if *R^2* itself is still proton-activating such as, for instance, phenyl[278] or substituted phenyl[279].

Preparation of the *diazosulfonylmethanes*[280] (cf. p. 226) is considerably enriched by diazo group transfer where the sulfonyl methanes *(R^1–SO$_2$–CH$_2$–R^2)* are additionally activated by *R^2*. The reaction conditions are largely the same as those for diazo group transfer to *CO*-activated methanes; some of the side-reactions and sequential reactions mentioned there also arise. Thus, from ethyl 4-tolylsulfonylacetate either the *N-sodium derivative* of *2-diazo-N,2-bis(4-tolylsulfonyl)acetamide* (70%) or *ethyl diazo-4-tolylsulfonylacetate* (84%) is formed according to the base used[281]:

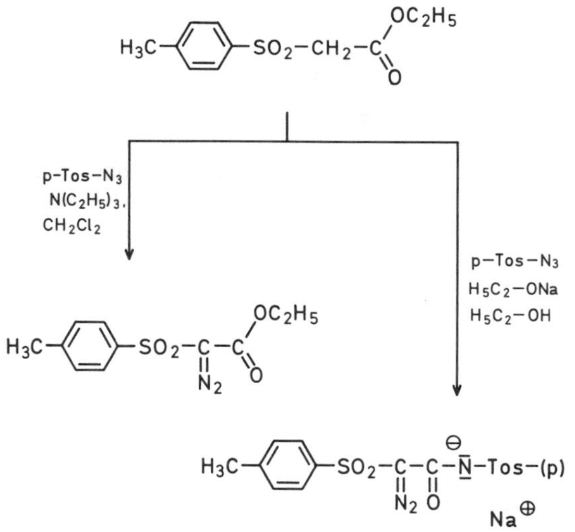

Azo group transfer occurs on reacting p-toluenesulfonic acid azide with 2H-benzo[b]thiophene-3-one 1,1-dioxide. The initially formed *tautomeric hydrazone* undergoes azo decoupling on heating

in glycol, so that *2-diazo-2H-benzo[b]thiophen-3-one 1,1-dioxide* is still accessible albeit by a detour[282]:

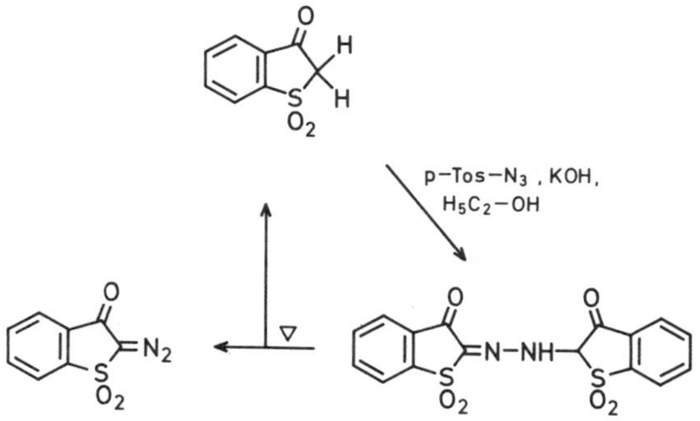

Table 7 lists some smoothly proceeding syntheses of diazosulfonylmethanes.

PO-substituted diazomethanes, whose preparation by amine diazotization and Bamford-Stevens reaction (see p. 210) has already been dealt with, are obtainable by diazo group transfer if they possess a second proton-activating group on the diazo carbon. The transfer proceeds predominantly in the presence of organometallic bases, because the *PO* group is a weaker proton activator than the carbonyl group (Table 7 contains several examples):

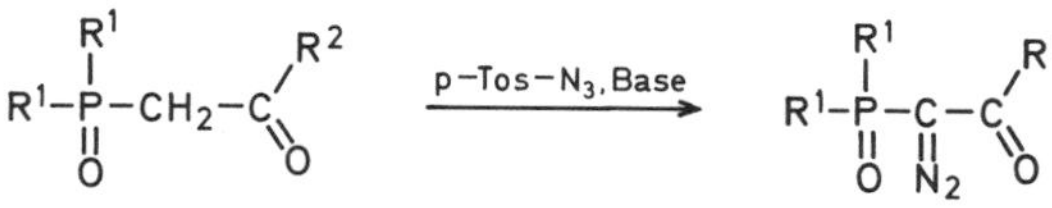

R^1 = C$_6$H$_5$; *1-Diazo-1-diphenylphosphinyl-2-alkanones*[283]

R^1 = OC$_2$H$_5$; *Diethyl 1-diazo-2-oxoalkanephosphonate*[163]

8.1.7.2 Diazo group transfer to methylene compounds with azidinium salts

Diazo group transfer with azidinium salts[293], which can be interpreted as *N-diazonium salts*[294], represents a substantial enrichment of possible syntheses. Their *N-diazonium* character explains their excellent *N$_2$-transfer* properties:

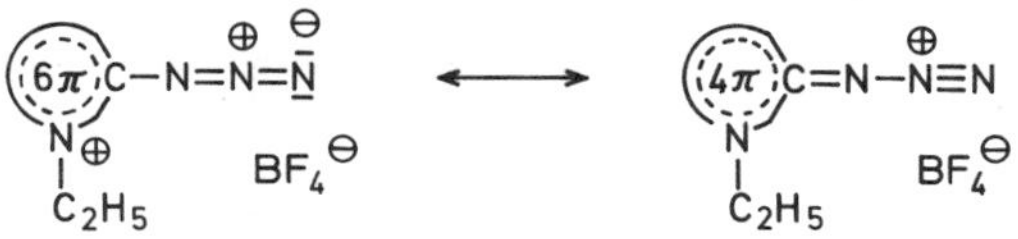

[278] *M. Regitz*, Chem. Ber. *98*, 1210 (1965).
[279] *W. Jugelt*, Z. Chem. *5*, 456 (1965).
[280] *A.M. van Leusen, J. Strating*, Quarterly Reports on Sulfur Chemistry *5*, 67 (1967).
[281] *A.M. van Leusen, P.M. Smid, J. Strating*, Tetrahedron Lett. 337 (1965).

[282] *M. Regitz*, Chem. Ber. *98*, 36 (1965).
[283] *M. Regitz, W. Anschütz*, Chem. Ber. *102*, 2216 (1969).
[284] *M. Regitz, A. Liedhegener*, Tetrahedron *23*, 2701 (1967).

Table 7. Diazo compounds by diazo group transfer

Diazo compound	Solvent/base (for the diazo group transfer)	Yield [% of theory]	Physical data	Ref.
Diazotetraphenylcyclopentadiene	Acetonitrile/piperidine	95	mp 148° (decomp.)	284
10-Diazoanthrone	Ethanol/piperidine	94	mp > 280° (decomp.)	285
2-Diazo-1,3-diphenyl-1,3-propanedione	Dichloromethane/piperidine	70	mp 107°	286
2-Diazo-1,3-indandione	Ethanol/triethylamine	80–85	mp 149°	287
2-Diazo-5,5-dimethyl-1,3-cyclohexanedione	Ethanol/triethylamine	73	mp 108°	288
Ethyl 2-diazo-3-oxobutyrate	Acetonitrile/triethylamine	84	liquid (cannot be distilled)	286
Diethyl diazomalonate	Acetonitrile/triethylamine	95	liquid (cannot be distilled)	286
ω-Benzenesulfonyl-ω-diazoacetophenone	Ethanol/triethylamine	63	mp 124–125° (decomp.)	289
Bis(benzenesulfonyl)diazomethane	Ethanol-water/sodium hydroxide	70	mp 99° (decomp.)	290
ω-Diazo-ω-diphenylphosphinylacetophenone	Dichloromethane/piperidine	100	mp 155° (decomp.)	291
Bis(diethoxyphosphono)diazomethane	Benzene/potassium tert-butoxide	35	$bp_{0.4}$ 131–133°	292

The overall transfer reaction in the case of 2-azido-ethyl-2,3-benzothiazolium tetrafluoroborate proceeds according to the following scheme[295]:

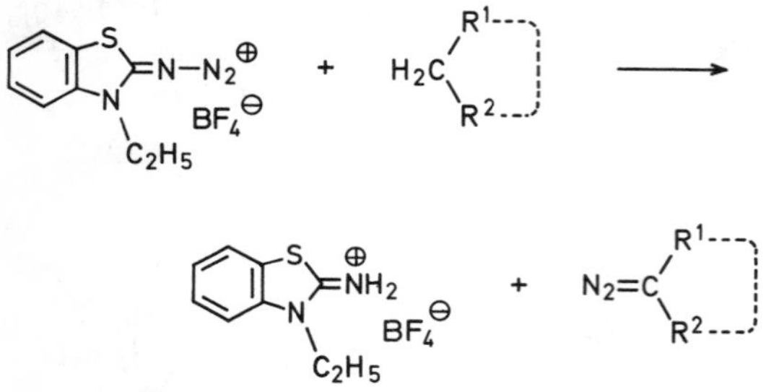

The mesomerism-stabilized amidinium salt is probably a cofactor in providing for a generally smooth reaction. As a rule, the reaction is carried out in aqueous alcoholic solution or in suspension at temperatures of up to 80° and pH values of 0–8, i.e., over a range of pH where sulfonic acid azides no longer display reactivity. It has revealed itself to be superior to diazo group transfer with p-toluenesulfonic acid azide where sequential reactions such as, for instance, azo coupling need to be suppressed. Thus, 2H-benzo[b]thiophen-3-one 1,1-dioxide reacts smoothly to the 2-diazo derivative because the latter does not couple in the acid medium[296] (cf. p. 233). Diazonitromethanes and diazoacetonitriles, whose preparation by diazo group transfer with p-toluenesulfonic acid azide had previously failed[261], are accessible by azidinium salt transfer[297, 298] (Table 8 provides information about individual examples).

Table 8. Diazo group transfers with 2-azido-3-ethylbenzothiazolium tetrafluoroborate

Methylene compound	Reaction conditions	Diazo compound	Yield [of of theory]	M.p. [°C]	Ref.
1,3-Indandione	None given	2-Diazo-1,3-indandione	84	—	299
Phloroglucinol (1,3,5-benzenetriol)	None given	2,4,6-Tris(diazo)-1,3,5-cyclo-hexanetrione	95	—	299
Barbituric acid	None given	5-Diazobarbituric acid	84	—	299
5-Methyl-1-phenyl-4H-pyrazol-3-one	Dimethylformamide/water 4:1, pH 3–4	4-Diazo-5-methyl-1-phenyl-4H-pyrazol-3-one	97	96	300
Methyl nitroacetate	Methanol/water 9:1, sodium acetate	Methyl diazonitroacetate	18	59–60	301
2-Nitroacetophenone	Methanol/water 9:1, sodium acetate	2-Diazo-2-nitroacetophenone	23	86–88	301
Benzenesulfonylacetonitrile	Methanol/water, pH 4–5	Benzenesulfonyldiazoacetonitrile	55	103–108 (decomp.)	302
Ethyl cyanoacetate	Methanol/water, pH 6.9	Ethyl cyanodiazoacetate	73	liquid (cannot be distilled)	302

The restricting reaction conditions, namely, acid to neutral, at the same time define the limit of the diazo group transfer with azidinium salts. It is reached when the CH_2 component no longer forms anions under the above conditions or the diazo compound is unstable to acids.

8.1.7.3 Diazo group transfer to formyl compounds

If one CH_2 component is only monoactivated, then the diazo group transfer generally fails because of deficient proton activity. This synthetic gap can be closed by Claisen introduction of a formyl group as a second proton-activating group and removing it again during the course of the transfer reaction

29   **28**

To carry out this deformylating diazo group transfer[303-305], either the Claisen condensation product **29** ($Na^\oplus$ is the counterion) is subjected to the azide reaction *(Variant A)* or the tautomer mixture **28** is isolated and the deformylating diazo group transfer is performed in the presence of an organic base *(Variant B)*.

As Table 9 shows, varying the substituents R^1 and R^2 makes available the entire spectrum of α-*diazocarbonyl* compounds; the preparation of α,β-*unsaturated diazo ketones* may be emphasized here[306]. *Diethyl diazomethanephosphonates,* too, are accessible to a modest degree[307].

Table 9. α-Diazocarbonyl compounds by deformylating diazo group transfer

α-Diazocarbonyl compound	Variant	Yield [% of theory]	Physical data	Ref.
1-Diazo-3,3-dimethyl-2-butanone	A	83	bp$_{15}$ 69°	308
2-Diazoacetophenone	A	73	mp 48–49°	308
Diazomethylferrocene	A	84	mp 68°	308
3-Diazo-2-butanone	B	60	bp$_{25}$ 56–59°	308
3-Diazo-4-heptanone	B	65	bp$_{22}$ 81–82°	308
4-Diazo-6-methyl-5-hepten-3-one	B	90	Oil (cannot be distilled)	309
2-Diazopropionaldehyde	B	64	bp$_{12}$ 35°	308
2-Diazovaleraldehyde	B	68	bp$_{14}$ 53–54°	308
2-Diazohydrocinnamaldehyde	B	94	Decomp. on distillation	308
Ethyl diazoacetate	A	69	bp$_{12}$ 45°	308
Ethyl diazophenylacetate	B	75	Oil (cannot be distilled)	308
Ethyl 2-diazobutyrate	B	43	bp$_{11}$ 63–65°	308
1-Diazo-4-methyl-3-penten-2-one	A	69	bp$_{0.03}$ 39–40°	310
1-Diazo-4-(4-methoxy-phenyl)-3-buten-2-one	A	54	mp 123° (decomp)	310
1-Diazo-6-phenyl-3,5-hexadien-2-one	A	73	mp 123–124 (decomp)	310

[285] M. Regitz, Chem. Ber. *97*, 2742 (1964).
[286] M. Regitz, A. Liedhegener, Chem. Ber. *99*, 3128 (1966).
[287] M. Regitz, H. Schwall, G. Heck, B. Eistert, G. Bock, Justus Liebigs Ann. Chem. *690*, 125 (1965).
[288] M. Regitz, D. Stadler, Justus Liebigs Ann. Chem. *687*, 214 (1965).
[289] M. Regitz, W. Bartz, Chem. Ber. *103*, 1477 (1970).
[290] F. Klages, K. Bott, Chem. Ber. *97*, 735 (1964).
[291] M. Regitz, W. Anschütz, Chem. Ber. *102*, 2216 (1969).
[292] M. Regitz, W. Anschütz, A. Liedhegener, Chem. Ber. *101*, 3734 (1968).
[293] H. Balli, F. Kersting, Justus Liebigs Ann. Chem. *647*, 1 (1961).
[294] H. Balli, Justus Liebigs Ann. Chem. *647*, 11 (1961).
[295] H. Balli, V. Müller, Angew. Chem. *76*, 573 (1964); H. Balli, V. Müller, Angew. Chem. Intern. Ed. Engl. *3*, 644 (1964).

[296] H. Balli, University Basel, private Communication (9. Juli 1964).
[297] H. Balli, R. Löw, Tetrahedron Lett. 5821 (1966).
[298] H. Balli, H. Rempfler, unpubl. results, Basel 1970.
[299] H. Balli, V. Müller, Angew. Chem. *76*, 573 (1964); H. Balli, V. Müller, Angew. Chem. Intern. Ed. Engl. *3*, 644 (1964).
[300] H. Balli, R. Gipp, Justus Liebigs Ann. Chem. *699*, 133 (1966).
[301] H. Balli, R. Löw, Tetrahedron Lett. 5821 (1966).
[302] H. Balli, H. Rempfler, unpubl. results, Basel 1970.
[303] M. Regitz, F. Menz, J. Rüter, Tetrahedron Lett. 739 (1967).

Cyclic α-*diazo ketones* are accessible by an analogous two-stage reaction. The homologous series *2-diazo-1-cyclopentanone* to *2-diazo-1-cyclododecanone* can be synthesized gaplessly in high yield[305]; a competing reaction branching off at the triazoline stage and involving loss of nitrogen is observed strongly only in the case of 2-diazo-1-cyclododecanone[305]:

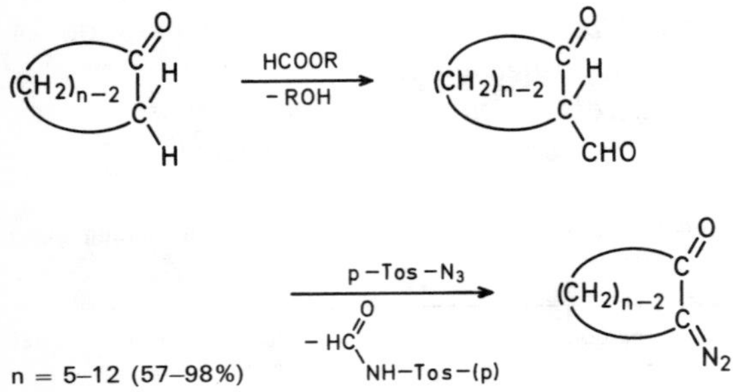

n = 5–12 (57–98%)

This technique can be applied with success to *bicyclic* ketones as well[311–313].

8.1.7.4 Diazo group transfer to activated olefins

Diazo group transfer to *enamines* of suitable structure proceeds entirely independently of proton-activating substituents; naturally, it ensues without bases. The sulfonic acid azide adds in a *specific direction* to form a 4,5-dihydro-1,2,3-triazole, whose spontaneous decomposition furnishes the *diazoalkane* and an *amidine*. Side-reactions commencing at the 4,5-dihydro-1,2,3-triazole stage and generally proceeding with loss of nitrogen can severely diminish the diazoalkane yield[314]:

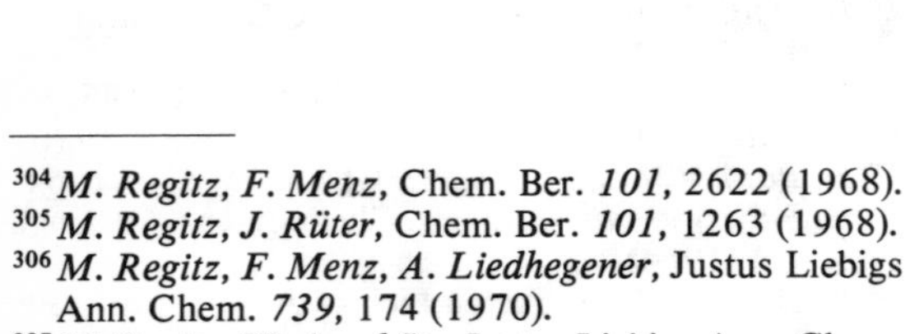

In individual cases the arising of *diazomethane* and α-*diazo ketones* during reaction of enamines with *p*-toluenesulfonic acid azide[314–316] or of *enol ethers* with *PO*-substituted azides in the sense of the above scheme is reported[316]. Systematic investigations have shown that from the point of view of diazomethane yield and the accessibility of the enamine a diazo group transfer to the commercially available Fischer base (2-methylene-1,3,3-trimethyl-2-indoline) can hardly be matched. In addition to the amidines *diazomethane* is taken from the reaction medium in a current of nitrogen in 81–87% yield[317].

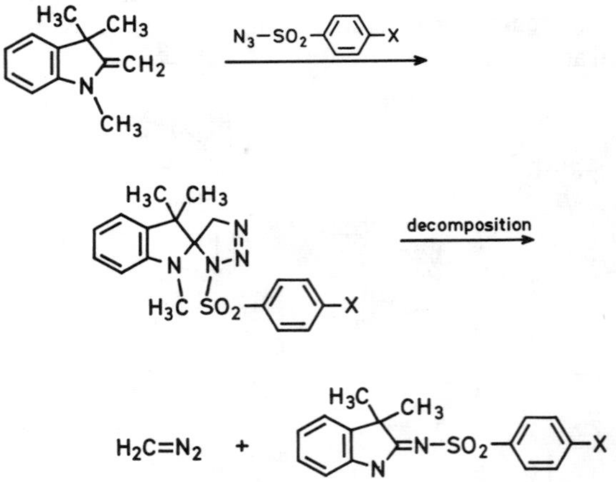

X = CH₃ , OCH₃ , NO₂

Mention should also be made of a *combined technique,* in which *enamine condensation* of acetaldehyde and piperidine is allowed to proceed first of all and *p*-toluenesulfonic acid azide is then added; the yield of *diazomethane* is 20–37%[317]:

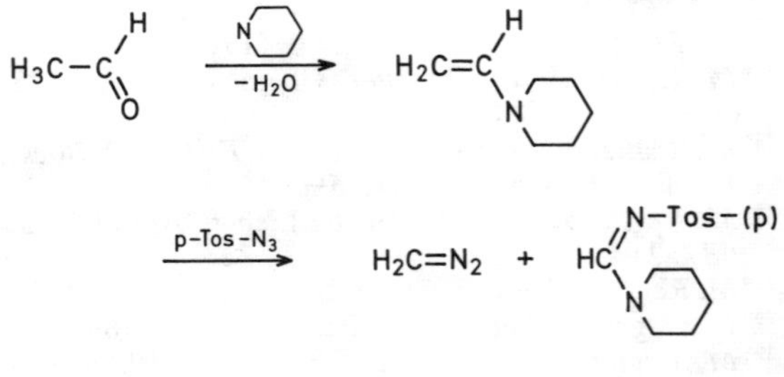

[304] M. Regitz, F. Menz, Chem. Ber. *101*, 2622 (1968).

[305] M. Regitz, J. Rüter, Chem. Ber. *101*, 1263 (1968).

[306] M. Regitz, F. Menz, A. Liedhegener, Justus Liebigs Ann. Chem. *739*, 174 (1970).

[307] M. Regitz, W. Anschütz, Justus Liebigs Ann. Chem. *730*, 194 (1969).

[308] M. Regitz, F. Menz, Chem. Ber. *101*, 2622 (1968).

[309] S. Julia, G. Linstrumelle, Bull. Soc. Chim. France 3490 (1966).

[310] M. Regitz, F. Menz, A. Liedhegener, Justus Liebigs Ann. Chem. *739*, 174 (1970).

[311] M. Rosenberger, P. Yates, J.B. Hendrickson, W. Wolf, Tetrahedron Lett. 2285 (1964).

[312] T. Gibson, W.F. Erman, J. Org. Chem. *31*, 3028 (1966).

[313] R.T. Brook, B.V. Brophy, Tetrahedron Lett. 4187 (1969).

α-*Diazotoluene* and *3-diazo-1-propene* were like-wise synthesized by the enamine route[317, 318]. This is true also of α-*diazocarbonyl* compounds; but here the techniques described on p. 235 are superior. α-*Diazo aldehydes* are an exception; they were discovered in this manner[319, 320]. Starting from 2-amino-1-alkenals the entire homologous series as far as *2-diazoheptanal* is accessible[321]. *Diazoacetaldehyde* alone requires a variation of the substituent on the amino nitrogen (*NC–CH₂–CH₂* in place of *CH₃*):

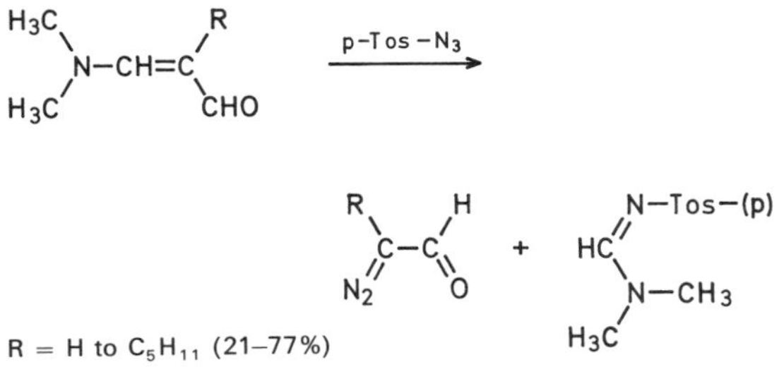

R = H to C₅H₁₁ (21–77%)

In conclusion, reference is made to the cycloaddition of azides to the low-electron double bond of α,β-unsaturated carbonyl compounds which affords open-ring diazo isomers *via* 4,5-dihydro-1,2,3-triazoles[322]. Thus, methyl acrylate and arylazides form 1-aryl-4-methoxycarbonyl-4,5-di-hydro-1,2,3-triazoles, which are converted into *methyl 3-arylamino-2-diazopropionates* by base catalysis. In other instances the ring opens even without addition of base.

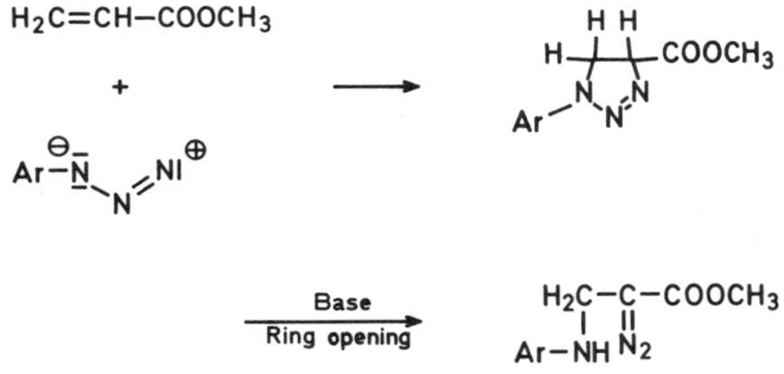

[314] R. Fusco, G. Bianchetti, D. Pocar, R. Ugo, Chem. Ber. 96, 802 (1963).
[315] G. Bianchetti, D. Pocar, P. Dalla Croce, A. Vigevani, Chem. Ber. 98, 2715 (1965).
[316] G. Bianchetti, D. Pocar, R. Stradi, Gazz. Chim. Ital. 100, 726 (1970).
[317] M. Regitz, G. Himbert, Justus Liebigs Ann. Chem. 734, 70 (1970).
[318] D. Pocar, G. Bianchetti, P. Dalla Croce, Gazz. Chim. Ital. 95, 1220 (1965).
[319] J. Kučera, Z. Arnold, Tetrahedron Lett. 1109 (1966).
[320] Z. Arnold, Chem. Commun. 229 (1967).
[321] J. Kučera, Z. Janoušek, Z. Arnold, Collect. Czech. Chem. Commun. 35, 3618 (1970).
[322] R. Huisgen, G. Szeimies, L. Möbius, Chem. Ber. 99, 475 (1966).

8.1.7.5 Diazo group transfer to acetylenes

Alkoxy- and aminoacetylenes with specific orientation add sulfonic acid azides to form 1,2,3-triazoles; the products can either be isolated, change into the open-ring diazo isomers, or are in equilibrium with these in solution. Thus, ethoxyacetylene affords *ethyl diazoacetate sulfonylimides* (*R¹ = H; R² = CH₃, C₆H₅, 4–Br–C₆H₄*) directly without detection of the isomeric 1,2,3-triazoles[323]. Starting from 1-ethoxy-1-propyne either *1-arylsulfonyl-5-ethoxy-4-methyl-1,2,3-triazoles [R¹ = CH₃; R² = 4-(CH₃)₂N–C₆H₄, 2,4,6-(CH₃)₃–C₆H₂]* or *ethyl 2-diazo-N-(4-methyl [or methoxy]benzenesulfonyl)propionimidate (R¹ = CH₃; R² = 4-CH₃–C₆H₄, 4-CH₃O–C₆H₄)* are obtained in dependence on the substituent; in deuterochloroform they exist side by side in equilibrium[324]:

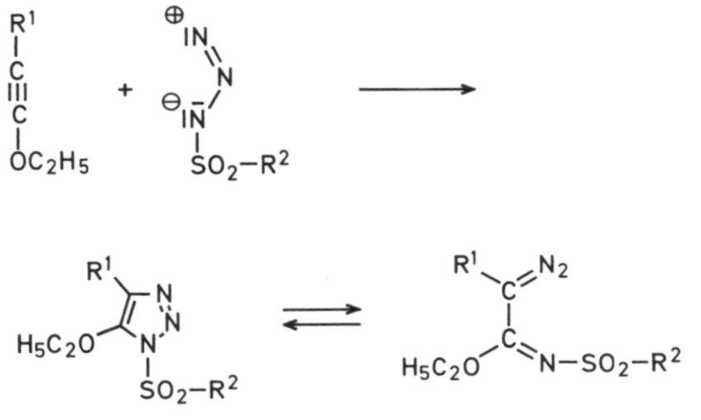

Similar relationships are encountered with *ynamines;* alkyl and arylamino acetylenes and also acylated compounds of the same type react with sulfonic acid azides exclusively to form α-*diazoamidines* **30**[324, 325] and **31**[324] respectively, while the analogously obtained 1,2,3-triazole **32**[324] exhibits no tendency to diazo isomerization even when the sulfonyl group is nitro-substituted (*cf.* the ring opening effect to the nitro group in the following example).

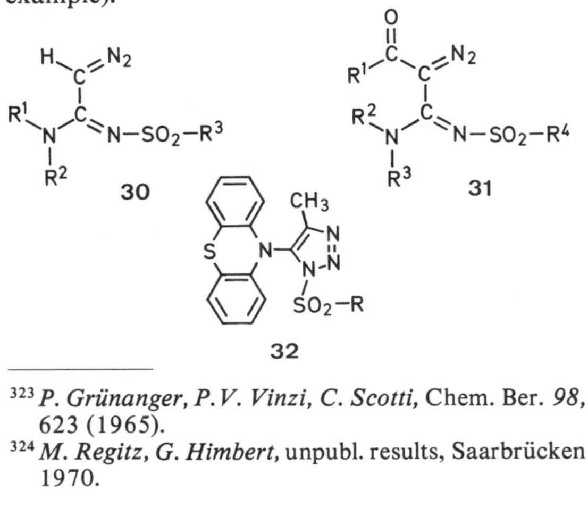

[323] P. Grünanger, P.V. Vinzi, C. Scotti, Chem. Ber. 98, 623 (1965).
[324] M. Regitz, G. Himbert, unpubl. results, Saarbrücken 1970.

The sulfonic acid azide component is seen to exert a strong effect on the 5-amino-1,2,3-triazole/diazoamidine *equilibrium* during its reaction with *N*,*N*-diethylpropynylamine[325, 326]. Although in the crystalline state one isomer is strongly predominant or occurs exclusively, in deuterochloroform *NMR*-determinable equilibria once more occur. It is revealed clearly that electron-attracting substituents in *R* favor the diazoamidine:

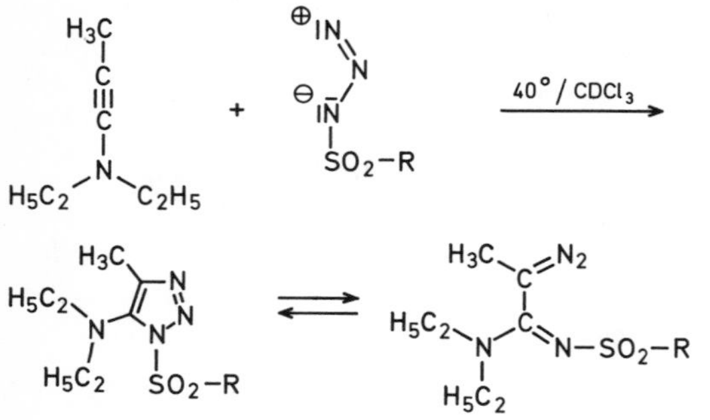

R	Tri-azole %	Amid-ine %	
C$_6$H$_4$—OCH$_3$—(p)	84	16	N'-4-Methoxybenzene-sulfonyl-2-dda
CH$_3$	63	37	N'-Methanesulfonyl-2-dda
C$_6$H$_4$—Cl—(p)	30	70	N'-(4-Chlorobenzene-sulfonyl)-2-dda
C$_6$H$_4$—NO$_2$—(o, m, p)	0	100	N'-(2-,3-,4-Nitrobenzene-sulfonyl)-2-dda

2-dda = 2-diazo-N,N-diethylacetamidine

Cyanogen azide reacts even with unsubstituted acetylene to form an isomer mixture of *1,2,3-triazole-1-carbonitrile* and *(2-diazoethylidene)cyanamide*[327].

8.1.8 Diazo compounds from other diazo compounds

Transformations of diazo compounds (substitution, addition, cleavage) which leave the diazo group intact have achieved topical importance. They proceed predominantly on the diazo carbon atom and are thus limited to diazomethyl compounds or disubstituted diazoalkanes with easily exchangeable groups. The diazo group often displays astonishing stability here even under quite strange conditions.

8.1.8.1 Nitration, halogenation

Ethyl diazoacetate can be nitrated smoothly to *ethyl diazonitroacetate* with dinitrogen pentoxide[328, 329], albeit half the diazo ester is lost with formation of ethyl nitratoacetate, as is shown in the reaction scheme. Attempts to replace the second ethyl diazoacetate molecule by an auxiliary base (pyridine, triethylamine) remained fruitless[329]:

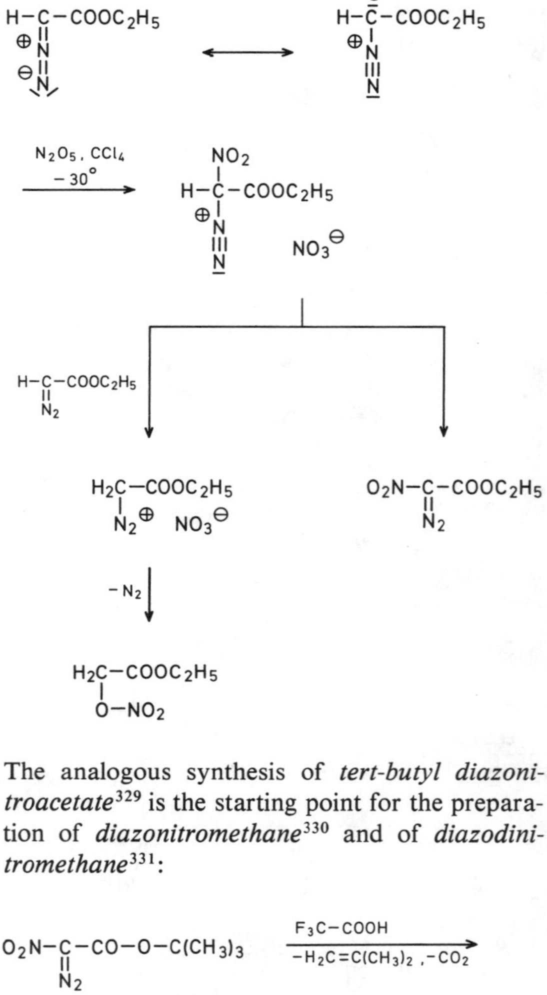

The analogous synthesis of *tert-butyl diazonitroacetate*[329] is the starting point for the preparation of *diazonitromethane*[330] and of *diazodinitromethane*[331]:

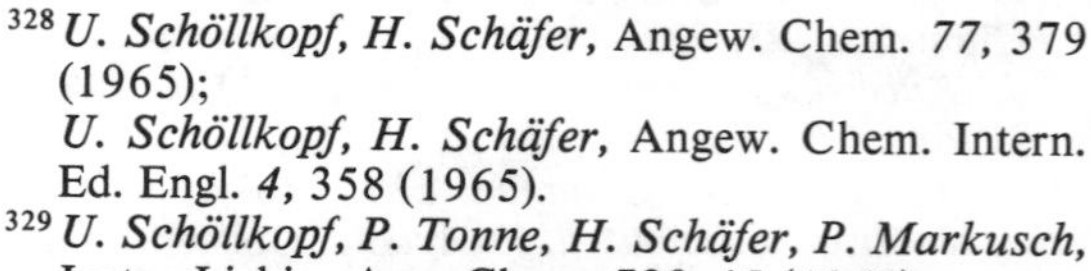

[325] *M. Regitz, G. Himbert*, Tetrahedron Lett. 2823 (1970).

[326] *R.E. Harman, F. Stanley, S.K. Gupta, J. Johnson*, J. Org. Chem. *35*, 3444 (1970).

[327] *M.E. Hermes, F.D. Marsh*, J. Amer. Chem. Soc. *89*, 4760 (1967).

[328] *U. Schöllkopf, H. Schäfer*, Angew. Chem. *77*, 379 (1965);
U. Schöllkopf, H. Schäfer, Angew. Chem. Intern. Ed. Engl. *4*, 358 (1965).

[329] *U. Schöllkopf, P. Tonne, H. Schäfer, P. Markusch*, Justus Liebigs Ann. Chem. *722*, 45 (1969).

[330] *U. Schöllkopf, P. Markusch*, Tetrahedron Lett. 6199 (1966).

Diazocyclopentadiene with its strongly nucleophilic 6π-system is substituted to form *diazonitrocyclopentadienes* by isomerization during nitration in the 2- and 3-positions[332]:

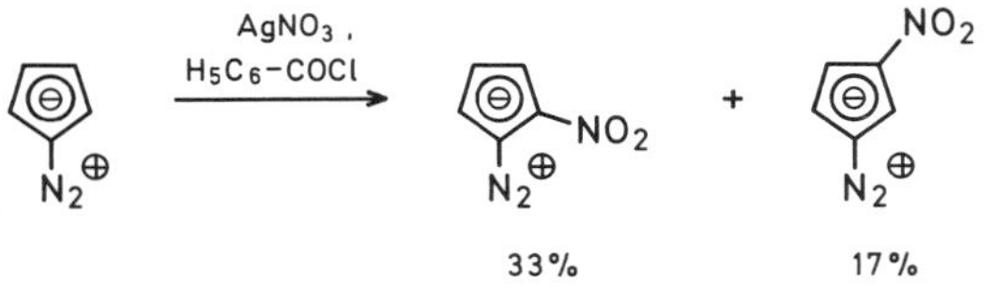

Direct *halogenation* of aliphatic diazo compounds is of only moderate importance and is inferior to metal/halogen exchange (see p. 242). The *chlorodiazomethane* accessible with *tert*-butyl hypochlorite at $-100°$ decomposes at above $-40°$ with loss of nitrogen[333]. Diazocyclopentadiene reacts with *N*-bromosuccinimide to form *diazotetrabromocyclopentadiene*[332].

8.1.8.2 Acylation

Acylation reactions of aliphatic diazo compounds[334] occupy an outstanding place in *diazo ketone* synthesis[335]. The most frequently carried out procedure has been reaction of *acid chlorides* with diazomethane. The hydrogen chloride formed can react further with the resulting diazo ketone with loss of nitrogen. To prevent this happening the acid chloride is introduced into an at least 100% excess of a solution of diazomethane in ether (not *vice versa*); the more nucleophilic diazomethane now traps the hydrogen chloride with formation of methyl chloride and nitrogen. The *base function* of the second molecule diazomethane can sometimes be taken on by tertiary organic amines[342, 336].

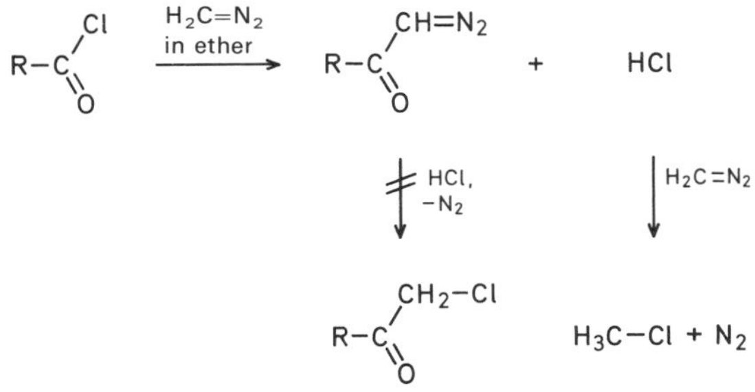

Steric influences can prevent diazo ketone formation from acid chlorides. Thus, neither 2,4,6-trimethylbenzoyl chloride[337] nor 2-chloro-6-methoxybenzoyl chloride[338] react with diazomethane. Double diazo ketone synthesis is observed in terephthaloyl chloride[339], oxalyl chloride[340], and alkane diacid dichlorides with 3–8 CH_2 members[441]. The extraordinary range of application of the diazomethane acylation is demonstrated by Table 10.

Table 10. α-Diazo ketones by Arndt-Eistert diazomethane acylation

α-Diazo ketone	Yield [% of theory]	Physical data	Ref.
ω-Diazoacetophenone	72	mp 49°	342
2-(Diazoacetyl)benzonitrile	81	mp 139–141° (decomp)	343
ω-Diazo-ω-(4-toluenesulfonyl-amino)acetophenone	73	mp 126–127° (decomp)	344
1-Diazoacetonaphthalene	92	mp 54–55°	345
4-Diazo-2-methyl-1-phenyl-1-butene	67	mp 88–89°	346
1-Diazo-3-phthalimido-2-propanone	88	mp 169° (decomp)	347
3-Diazo-1-(3-indolyl)-1,2-propanedione	86	mp 149° (decomp)	348
Diazoacetaldehyde	—	Liquid	349
Phenyl diazoacetate	—	bp$_{0'01}$ 35–40°	350
3-(Adamant-1-yl)-1-diazo-2-propanone	77	mp 34–35°	351
3,3-Diallyl-1,5-bisdiazo-2,4-pentanedione	—	mp 81–82°	352

[331] *U. Schöllkopf, P. Markusch*, Angew. Chem. *81*, 577 (1969);
U. Schöllkopf, P. Markusch, Angew. Chem. Intern. Ed. Engl. *8*, 612 (1969).

[332] *D.J. Cram, R.P. Partos*, J. Amer. Chem. Soc. *85*, 1273 (1963).

[333] *E.L. Closs, J.J. Coyle*, J. Amer. Chem. Soc. *87*, 4270 (1965).

[334] *F. Arndt, B. Eistert, W. Partale*, Chem. Ber. *60*, 1364 (1927);
F. Arndt, J. Amende, Ber. *61*, 1122 (1928).

[335] *B. Eistert, M. Regitz, G. Heck, H. Schwall* in *Houben-Weyl*, Methoden der organischen Chemie, 4. Aufl., Bd. X/4, p. 589, Georg Thieme Verlag, Stuttgart 1968;
B. Eistert in Neuere Methoden der präparativen organischen Chemie, 3. Aufl., p. 378, Verlag Chemie, Weinheim/Bergstr. 1949.
W.E. Bachmann, W.S. Struve, Org. Reactions *1*, 38 (1942).

[336] *M.S. Newman, P. Beal*, J. Amer. Chem. Soc. *71*, 1506 (1949).

[337] *W.E. Bachmann, W.S. Struve*, Org. Reactions *1*, 38, 45 (1942).

[338] *T.P.C. Mulholland, R.J.W. Honeywood, H.D. Preston, D.T. Rosevear*, J. Chem. Soc. (London) 4939 (1965).

[339] *W.C.J. Ross*, J. Chem. Soc. (London) 752 (1950).

[340] *M. Frankel, M. Harnik*, J. Amer. Chem. Soc. *74*, 2120 (1952).

[341] *E. Fahr*, Justus Liebigs Ann. Chem. *638*, 1 (1960).

[342] *F. Arndt, J. Amende*, Ber. *61*, 1122 (1928).

In place of acid chlorides *acid anhydrides* can act as acylating reagents; here, too, 2 molecules diazomethane are consumed[353, 354]:

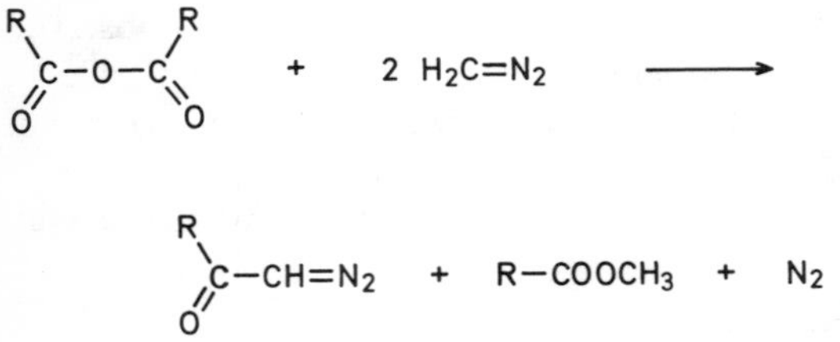

One variant starts directly from carboxylic acids, which in a combined reaction technique are first submitted to anhydride formation with carbohexylcarbodiimide and then reacted with diazomethane[355]:

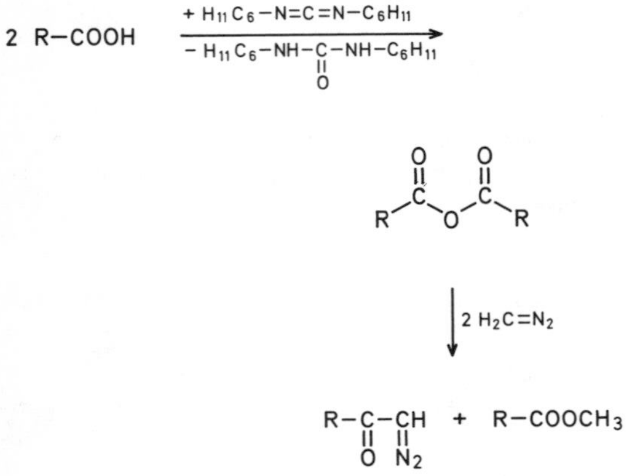

Diazomethane acylations using α,β-*unsaturated acid chlorides* are usually accompanied by sequential reactions. Thus, the several times investigated reaction of cinnamoyl chloride with excess diazomethane leads to *diazomethyl 4-phenyl-2-*

pyrazolin-3-yl ketone **33** and *diazomethyl 1-cinnamoyl-4-phenyl-2-pyrazolin-3-yl ketone* **34** respectively according to the reaction time[356]:

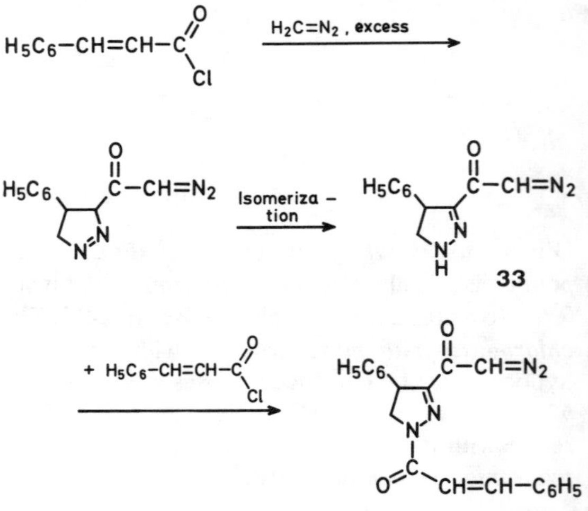

As the cycloaddition to the activated double bond ensues quite rapidly, the 1:2 reaction corresponding to stoichiometric acylation, too, affords no 1-diazo-4-phenyl-3-buten-2-one but, instead, **33**, which is acylated to **34** by unchanged cinnamoyl chloride[306, 357].

With certain homologs of diazomethane (diazoethane, 1-diazopropane, 1-diazobutane) successful acylation reactions have also been carried out[229, 358, 359]. Diazoacetic acid esters were reacted analogously[341, 360–362].

8.1.8.3 Metallation

Following the discovery of the metallability of aliphatic diazo compounds by the successful mercuration of ethyl diazoacetate[363], the question of the

[343] G. Holt, D.K. Wall, J. Chem. Soc. (London) 1428 (1965).

[344] W. Hampel, J. Prakt. Chem. *311*, 78 (1969).

[345] F. Arndt, B. Eistert, Ber. *68*, 201 (1935).

[346] J.A. Moore, J. Org. Chem. *20*, 1607 (1955).

[347] W. Jugelt, P. Falck, J. Prakt. Chem. *310*, 88 (1968).

[348] N. Erdmann, Z. Chem. *9*, 269 (1969).

[349] F. Kaplan, G.K. Meloy, J. Amer. Chem. Soc. *88*, 950 (1960).

[350] H. Chaimovich, R.J. Vaughan, F.H. Westheimer, J. Amer. Chem. Soc. *90*, 4088 (1968).

[351] J.K. Chakrabarti, S.S. Szinai, A. Todd, J. Chem. Soc. C 1303 (1970).

[352] S. Bien, D. Ovadia, J. Org. Chem. *35*, 1028 (1970).

[353] W. Bradley, R. Robinson, J. Chem. Soc. (London) 1310 (1928).

[354] D.S. Tarbell, J.A. Price, J. Org. Chem. *22*, 245 (1957).

[355] D. Hodson, G. Holt, D.K. Wall, J. Chem. Soc. C 971 (1970).

[356] M. Regitz, F. Menz, A. Liedhegener, Justus Liebigs Ann. Chem. *739*, 174 (1970).

[357] A. Nabeya, J.A. Moore, J. Org. Chem. *35*, 2022 (1970).

[358] A.C. Wilds, A.L. Meader, J. Org. Chem. *13*, 763 (1948).

[359] P. Yates, D.G. Farnum, D.W. Wiley, Chem. Ind. (London) 69 (1958).

[360] H. Staudinger, J. Becker, H. Hirzel, Ber. *49*, 1978 (1916).

[361] J.H. Locker, C.H. Hayes, J. Org. Chem. *28*, 1342 (1963).

[362] A. Roedig, H. Aman, E. Fahr, Justus Liebigs Ann. Chem. *675*, 47 (1964).

[363] E. Buchner, Ber. *28*, 215 (1895).

[364] E. Müller, P. Kästner, R. Beutler, W. Rundel, H. Suhr, B. Zeeh, Justus Liebigs Ann. Chem. *713*, 87 (1968);
E. Müller, R. Beutler, B. Zeeh, Justus Liebigs Ann. Chem. *719*, 72 (1968).

evidence of isodiazoacetate[364] and the problems of carbene chemistry produced a dramatic development in respect of further metallations. Hydrogen/lithium exchange on diazomethane proceeds smoothly with methyllithium in ether[365]; the same metal derivative is formed when dinitrogen monooxide is reacted with methyllithium[366]:

$$H_2C=N_2 \qquad\qquad N_2O$$

$$\xrightarrow[\text{H}_3\text{C–Li, ether}]{\quad} \Big| \xleftarrow[\text{H}_3\text{C–Li, ether}]{\quad}$$

$$LiCH=N_2$$

Diazomethyllithium

Triphenylmethylsodium was employed to introduce sodium into diazomethane (giving *diazomethylsodium*)[367], (Polymeric) *zinc, cadmium,* and *mercury* substituted *diazomethanes* are formed by loss of amine from silylated metal amides with diazomethane[368]:

$$M\left[N\left(Si\begin{smallmatrix}CH_3\\[-2pt]\diagup\\[-2pt]—CH_3\\[-2pt]\diagdown\\[-2pt]CH_3\end{smallmatrix}\Big/_2\right)\right]_2 + H_2C=N_2 \longrightarrow$$

$$M_2C=N_2 + 2\,HN\left(Si\begin{smallmatrix}CH_3\\[-2pt]\diagup\\[-2pt]—CH_3\\[-2pt]\diagdown\\[-2pt]CH_3\end{smallmatrix}\right)$$

Ethyl diazoacetate[369] and its phosphorus analog diethyl diazomethanephosphonate[370] can be smoothly metallated with *n*-butyllithium in ether (giving *ethyl lithiumdiazoacetate* and *diethyl lithiumdiazophosphonate*). Mercuration of acylated diazomethane takes place exclusively with mercury oxide in aprotic solvents. In this way diazoacetic acid esters[363,371,372], diazomethyl ketones[69,371,374], and *PO*-substituted diazomethanes[32] afford stable mercury derivatives that can undergo further substitution reactions at the diazo carbon (see p. 242):

$$2\,R{-}\underset{N_2}{\overset{\parallel}{C}}H + HgO \longrightarrow \underset{N_2}{\overset{R}{\underset{\parallel}{\diagup}}}C{-}Hg{-}C\underset{N_2}{\overset{R}{\underset{\parallel}{\diagdown}}} + H_2O$$

R = COOR¹, COR¹, PO(OR¹)₂, PO(R¹)₂

Silver diazo compounds, likewise accessible by direct metallation of acylated diazoalkanes, are considerably less stable than mercury derivatives, at least in the carbonyl series. They are prepared at $\leq 0°$ in order to avoid a Wolff rearrangement and are not isolated but submitted to further substitution reactions directly[373] (see p. 242). By contrast, silver derivatives of *PO*-substituted diazomethanes can be isolated without difficulty[32].

$$2\,R{-}\underset{N_2}{\overset{\parallel}{C}}H + Ag_2O \longrightarrow 2\,R{-}\underset{N_2}{\overset{\parallel}{C}}{-}Ag + H_2O$$

R = COOR', COR', PO(OR')₂, PO(R')₂

An elegant technique for introducing substituent groups containing germanium, tin, and lead twice into diazomethane consists in splitting off amine from trisubstituted metal amide and the diazo compound; the metal derivatives are quite stable thermally[368,374]:

$$2\,R_3M{-}N\underset{R^1}{\overset{R^1}{\diagup\diagdown}} + H_2C=N_2 \xrightarrow{\text{ether}}$$

$$(R_3M)_2C=N_2 + 2\,HN\underset{R^1}{\overset{R^1}{\diagup\diagdown}}$$

R	CH₃	CH₃	C₂H₅	C₆H₅	CH₃
M	Ge	Sn	Sn	Sn	Pb
Yield % of theory	58	95	82	90	100

Using the same procedure, groups containing tin[374], germanium[368], and lead[368] can be introduced into ethyl diazoacetates to form the corresponding metallated ethyl diazoacylates.

[365] *E. Müller, D. Ludsteck,* Chem. Ber. *87,* 1887 (1954).

[366] *E. Müller, W. Rundel,* Chem. Ber. *90,* 1302 (1957).

[367] *E. Müller, H. Disselhoff,* Justus Liebigs Ann. Chem. *512,* 250 (1934).

[368] *J. Lorberth,* J. Organometal. Chem. *27,* 303 (1971).

[369] *U. Schöllkopf, H. Frasnelli,* Angew. Chem. *82,* 291 (1970);
U. Schöllkopf, H. Frasnelli, Angew. Chem. Intern. Ed. Engl. *9,* 301 (1970).

[370] *M. Regitz, B. Weber,* unpubl. results, Saarbrücken 1970.

[371] *T. Dominh, O.P. Strausz, H.E. Gunning,* Tetrahedron Lett. 5237 (1968).

[372] *M. Regitz, U. Eckstein,* unpubl. results, Saarbrücken 1970;
M. Regitz, Synthesis 351 (1972).

[373] *U. Schöllkopf, N. Rieber,* Chem. Ber. *102,* 488 (1969).

[374] *M.F. Lappert, J. Lorberth, J.S. Poland,* J. Chem. Soc. A 2954 (1970).

8.1.8.4 Substitution *via* metal derivatives

Digermanyl- and distannyldiazomethanes are accessible also from trisubstituted metal chlorides and diazomethyllithium in addition to *via* the above amine elimination. Evidently, the second molecule diazomethyllithium acts as the metallating agent for intermediately formed monogermanyl and monostannyl diazomethanes. Silylation is the only case where the secondary substitution does not succeed[374]:

$$2\ R_3MCl\ +\ 2\ LiCH=N_2 \xrightarrow{\text{ether}}$$

$$(R_3M)_2C=N_2\ +\ 2\ LiCl\ +\ CH_2N_2$$

M = Ge; R = C₆H₅; *(Diazomethylene)bis[trimethylgermane];* 44%

M = Sn; R = CH₃; *(Diazomethylene)bis[trimethylstannane];* 80%

R = C₆H₅; *(Diazomethylene)bis[triphenylstannane];* 56%

Ethyl lithiumdiazoacetate is the starting point of numerous substitution reactions[369], *e.g.*:

Ethyl 2-diazo-propionate; 35%

Ethyl benzoyl-diazoacetate; 45%

Ethyl diazotrimethyl-silylacetate; 47%

Ethyl diazotributyl-stannylacetate; 61%

Ethyl silverdiazoacetate[373, 375], silver diazo ketones[373], and silver dialkoxyphosphonodiazomethanes[370] above all are characterized by reactivity toward S_N1-active alkyl halides (allyl iodide, 1-bromo-2-butene, benzyl bromide, *etc.*); *C*-alkylated α-diazocarbonyl and phosphono compounds respectively result. Using silver diazo compounds also for preparing *diazohaloalkanes* is noteworthy. In addition to conversion of *ethyl silverdiazoacetate* into *ethyl diazoiodoacetate* with iodine in ether[373], an analogous procedure for preparing dimethyl bromodiazomethane

phosphonates and *diazoiodomethanephosphonates* is known[370].

Diethyl mercury-bis(diazoacetate) opens up a wide range of possible syntheses.

Ethyl bromo-, chloro-, iododiazoacetate[376, 377]

Ethyl 2-diazo-3,3,3-triphenylpropionate[373]

Ethyl diazotrimethyl-silylacetate[378, 379]

Ethyl diazotrimethyl-stannylacetate[378, 380]

8.1.8.5 Addition reactions

Reactions of this type are made possible by the proton mobility of the diazomethyl hydrogen that, ultimately, also explains the occurrence of direct substitution.

[375] *U. Schöllkopf, N. Rieber*, Angew. Chem. *79*, 238 (1967);
U. Schöllkopf, N. Rieber, Angew. Chem. Intern. Ed. Engl. *6*, 261 (1967).

[376] *F. Gerhardt, U. Schöllkopf, H. Schumacher*, Angew. Chem. *79*, 50 (1967);
F. Gerhardt, U. Schöllkopf, H. Schumacher, Angew. Chem. Intern. Ed. Engl. *6*, 74 (1967).

[377] *U. Schöllkopf, F. Gerhardt, M. Reetz, H. Frasnelli, H. Schumacher*, Justus Liebigs Ann. Chem. *716*, 204 (1968).

[378] *U. Schöllkopf, N. Rieber*, Angew. Chem. *79*, 906 (1967);
U. Schöllkopf, N. Rieber, Angew. Chem. Intern. Ed. Engl. *6*, 884 (1967).

[379] *U. Schöllkopf, D. Hoppe, N. Rieber, V. Jacobi*, Justus Liebigs Ann. Chem. *730*, 1 (1969).

[380] *B. Banhidai*, Diplomarbeit Univ. Göttingen 1969.

[381] *H. Biltz, E. Krämer*, Justus Liebigs Ann. Chem. *436*, 154 (1924).

Aldollike addition of diazomethyl compounds to *ketones* was first observed during reaction of ethyl diazoacetate with alloxan tetrahydrate[381]:

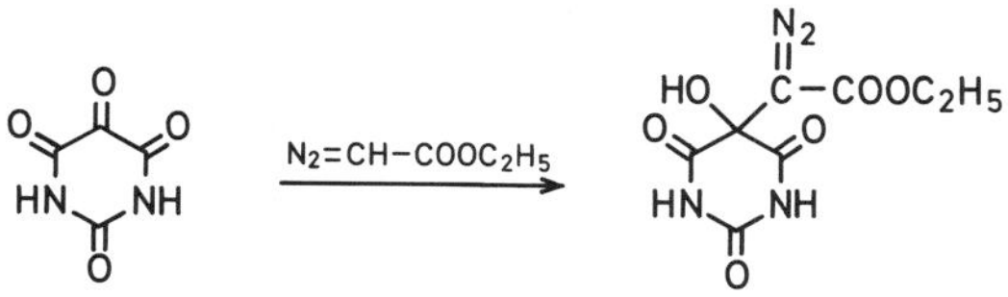

Ethyl α-diazohexahydro-5-hydroxy-2,4,6-trioxo-5-pyrimidineacetate

It naturally proceeds particularly smoothly in di- and tricarbonyl compounds and can be catalyzed with bases such as diethylamine[382]. Its variability, which extends to *PO*-substituted diazo compounds is apparent from the following examples.

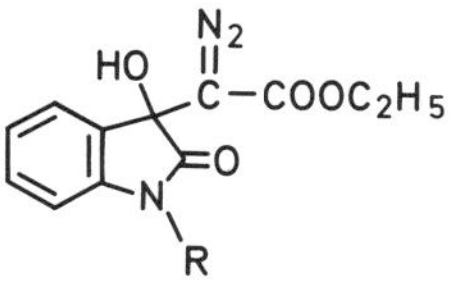

R = H;   *Ethyl α-diazo-3-hydroxy-2-oxo-2-indoline-acetate*[382, 383]

R = OCOCH₃;   *Ethyl 1-acetoxy-α-diazo-2-oxo-2-indoline-acetate*[382, 383]

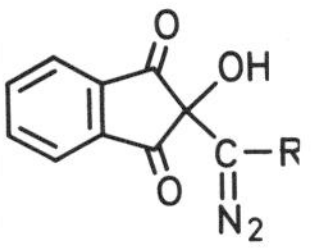

R = COOC₂H₅;   *Ethyl α-diazo-2-hydroxy-1,3-dioxo-2-indanacetate*

R = PO(OCH₃)₂;   *2-(Diazodimethoxyphosphonomethyl)-2-hydroxy-1,3-indandione*[384, 370]

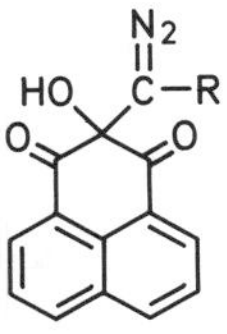

R = COOC₂H₅;   *Ethyl α-diazo-2-hydroxy-1,3-dioxo-2-phenaleneacetate*

R = PO(C₆H₅)₂;   *2-(Diazodiphenylphosphinylmethyl)-2-hydroxy-1,3-phenalenedione*[385, 372]

Direct aldol addition of ethyl diazoacetate to nonactivated ketones is unknown; *hydroxy* deri-

vatives can nonetheless be obtained on subsequent hydrolysis *via* the more nucleophilic ethyl lithiumdiazoacetate[369]:

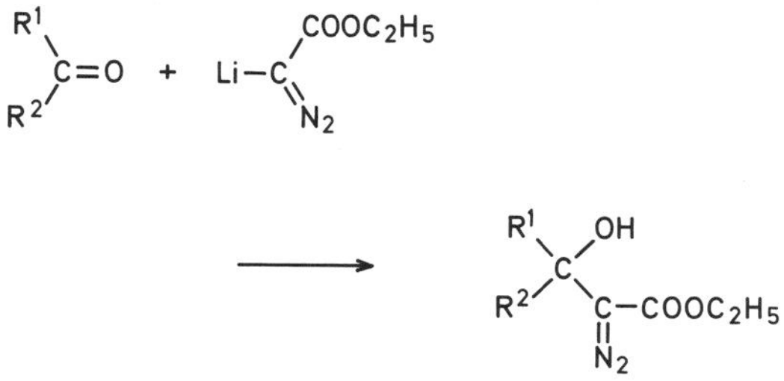

R¹ = C₆H₅; R² = H;   *Ethyl 2-diazo-3-hydroxyhydrocinnamate; 41%*

R² = CH₃;   *Ethyl 2-diazo-3-hydroxybutyrate; 50%*

Isolation of *3-(diazomethyl)-3-hydroxy-1-phenyl-2,4(1H, 3 H)quinolinedione* also succeeds[386]:

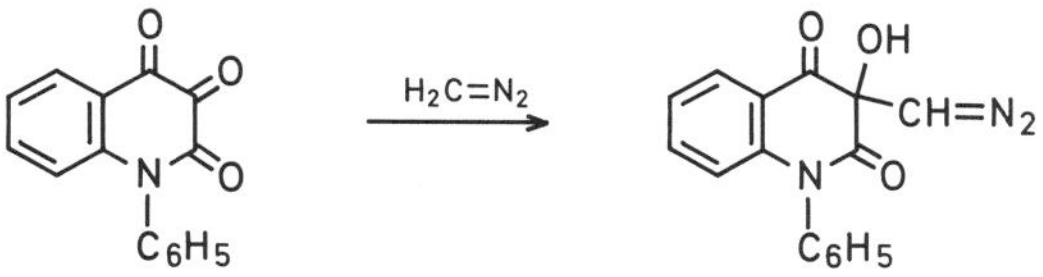

While analogous adducts with 1,2-diketones and chloral cannot be isolated, they have been trapped by 1,3-dipolar cycloaddition or by 1,3-dioxole formation[387, 388]. Allowing the aldol addition to proceed intramolecularly leads to the formation of cyclic α-hydroxydiazo compounds[389] in a new ring closure reaction, *e.g.*:

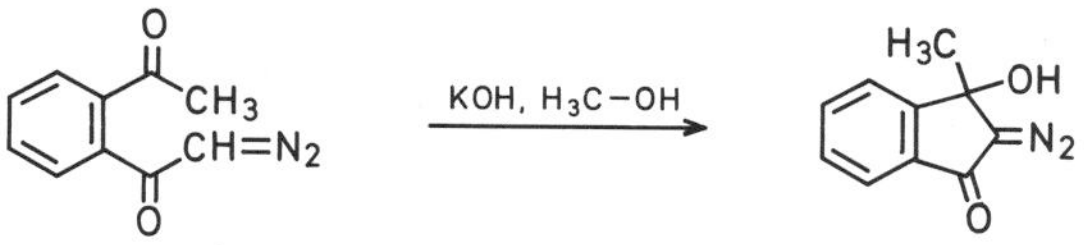

2-Diazo-3-hydroxy-3-methyl-1-indanone

Electronrich olefins of the type of 1,1′,3,3′-tetraphenyl-2,2′-bisimidazolidinylidene react with *2 molecules* diazomethyl compound with cleavage of the olefin and formation of *α-diazo aminals*[390]. As the reaction is initiated by protonation of the olefin, its success, too, requires an adequate proton mobility of the diazomethyl hydrogen[390].

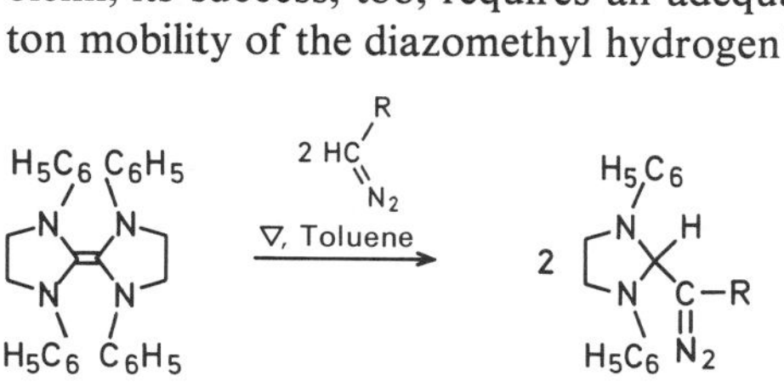

R = COR′, COOR′, CONR′₂, PO(OR′)₂, PO(R′)₂

[382] *B. Eistert, G. Borggrefe*, Justus Liebigs Ann. Chem. *718*, 142 (1968).

[383] *B. Eistert, G. Borggrefe, H. Selzer*, Justus Liebigs Ann. Chem. *725*, 37 (1969).

[384] *B. Eistert, P. Donath*, unpubl. results, Saarbrücken 1970.

[385] *B. Eistert, W. Eifler, O. Ganster*, Chem. Ber. *102*, 1988 (1969).

[386] *B. Eistert, P. Donath*, Chem. Ber. *103*, 993 (1970).

[387] *B. Eistert, O. Ganster*, Chem. Ber. *104*, 78 (1971).

Starting from *enamines* a high yield of *2-amino-1-sulfonyldiazoalkanes* is obtained with diazomethyl sulfones[391]. The reaction proceeds *via* a proton bridge complex between diazomethyl hydrogen and the β-carbon of the enamine:

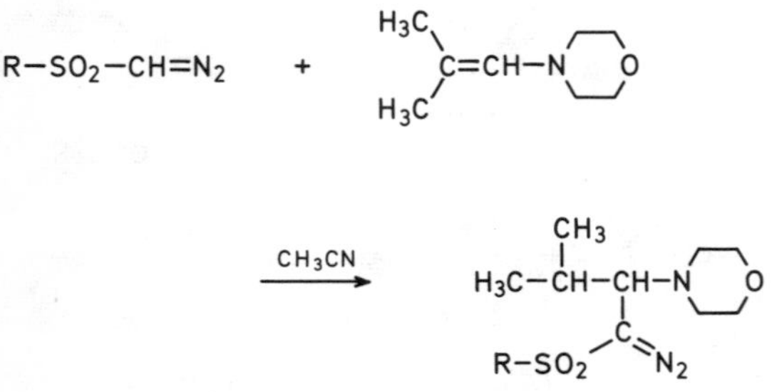

R—SO₂—CH=N₂   +   [structure: H₃C—C(=CH—N(morpholine)) with two H₃C groups]

$\xrightarrow{\text{CH}_3\text{CN}}$ [structure: H₃C—CH—CH—N(morpholine) with CH₃ and R—SO₂—C=N₂ substituents]

R = 4–CH₃–C₆H₄; *4-[α-[Diazo(p-tolylsulfonyl)methyl]-2-methylpropyl]morpholine; 89%*

R = CH₃–C₆H₅;   *4-[α-[Diazo(phenylsulfonyl)methyl]-2-methylpropyl]morpholine; 82%*

8.1.8.6 Cleavage reactions

Leaving aside the sodium hydroxide cleavage of the 2-diazo-2-phenylacetophenone to *α-diazotoluene* and sodium benzoate[392], acyl cleavages from diazocarbonyl compounds, although known for a long time[360, 393], have only recently attracted some attention. This is true expecially of diazocarbonyl compounds that are readily accessible by diazo group transfer (see p. 231). Thus, a high yielding synthesis of *tert-butyl diazoacetate* starts from *tert*-butyl 2-diazoacetoacetate, which is ruptured smoothly with sodium methoxide in methanol[394]:

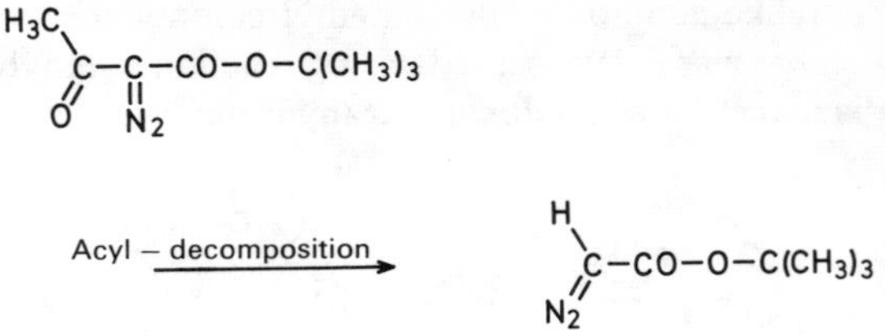

[structure: H₃C—C(=O)—C(N₂)—CO—O—C(CH₃)₃]

$\xrightarrow{\text{Acyl — decomposition}}$ [structure: H—C(N₂)—CO—O—C(CH₃)₃]

In the same way further diazoacetic acid esters and *diazoacetamides* can be prepared[395]. 2-Diazo 1,3-diketones, too, are amenable to acyl cleavage. Thus, *1-diazo-2-propanone* is readily prepared from 3-diazo-2,4-pentanedione[396], while 2-diazo-5,5-dimethyl-1,3-cyclohexanedione gives *6-diazo-3,3-dimethyl-5-oxohexanoic acid*[396] with simultaneous ring opening:

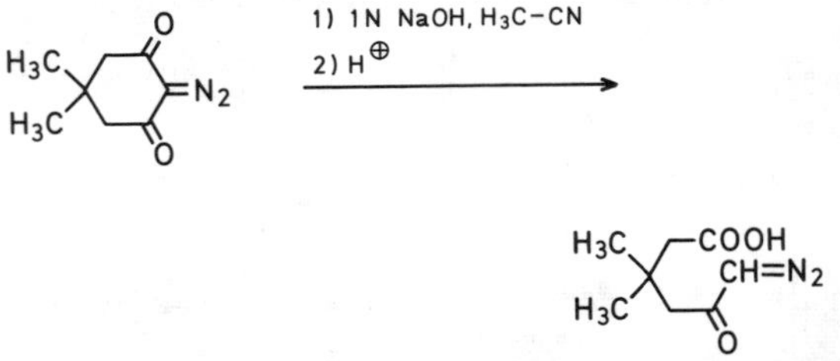

[structure: 5,5-dimethyl-2-diazo-1,3-cyclohexanedione] $\xrightarrow[\text{2) H}^{\oplus}]{\text{1) 1N NaOH, H}_3\text{C—CN}}$ [structure: H₃C, H₃C—C with CH₂COOH, CH=N₂, and C=O]

Finally, acetyl cleavage from 1-arylsulfonyl-1-diazo-2-propanones affords a reasonable yield of *diazo-p-tolylsulfonylmethane* and *diazo-p-nitrophenylsulfonylmethane*[397]:

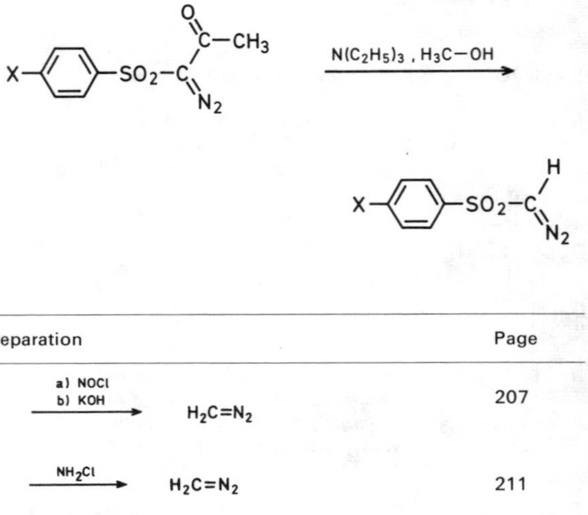

[structure: X—C₆H₄—SO₂—C(N₂)—C(=O)—CH₃] $\xrightarrow{\text{N(C}_2\text{H}_5)_3, \text{ H}_3\text{C—OH}}$ [structure: X—C₆H₄—SO₂—C(N₂)—H]

Table A. Methods for preparing aliphatic diazo compounds

Diazo compound	Method of preparation		Page
H₂C = N₂ *Diazomethane*	Diazotization of methylamine	H₃C—NH₂ $\xrightarrow[\text{b) KOH}]{\text{a) NOCl}}$ H₂C=N₂	207
	Forster reaction	H₂C=NONa $\xrightarrow{\text{NH}_2\text{Cl}}$ H₂C=N₂	211

[388] *B. Eistert, H. Juraszyk*, Chem. Ber. *103*, 2707 (1970).

[389] *T.L. Burkoth*, Tetrahedron Lett. 5049 (1969).

[390] *J. Hocker, M. Regitz, A. Liedhegener*, Chem. Ber. *103*, 1486 (1970).

[391] *A.M. van Leusen, B.A. Reith, R.J. Mulder, J. Strating*, Angew. Chem. *83*, 290 (1971);
A.M. van Leusen, B.A. Reith, R.J. Mulder, J. Strating, Angew. Chem. Intern. Ed. Engl. *10*, 271 (1971).

[392] *P. Yates, B.L. Shapiro*, J. Org. Chem. *23*, 759 (1958).

[393] *L. Wolff*, Justus Liebigs Ann. Chem. *394*, 23 (1912).

[394] *M. Regitz, J. Hocker, A. Liedhegener*, Org. Synth. *48*, 36 (1968).

[395] *M. Regitz, J. Hocker, A. Liedhegener*, Organic Preparations and Procedures *1*, 99 (1969).

[396] *J.B. Hendrickson, W.A. Wolf*, J. Org. Chem. *33*, 3610 (1968).

[397] *D. Hodson, G. Holt, D.K. Wall*, J. Chem. Soc. C 2201 (1968).

Table A. Methods for preparing aliphatic diazo compounds

Diazo compound		Method of preparation	Page
	Acyl cleavage of N-nitroso compounds	$H_3C-N \begin{smallmatrix} X \\ NO \end{smallmatrix}$ $\xrightarrow{\text{Base}}$ $H_2C=N_2$	223
		$X = COOC_2H_5$ $X = CONH_2$ $X = C \begin{smallmatrix} NH \\ NH-NO_2 \end{smallmatrix}$ $X = CO-CO-N \begin{smallmatrix} CH_3 \\ NO \end{smallmatrix}$ or $CO-CH_3$ $X = SO_2-C_6H_4-CH_3-(p)$	
	Cleavage of 4-methyl-4-(N-methylnitrosoamino)-2-pentanone	$H_3C-\underset{\underset{ON}{\overset{H_3C}{\mid}}{\overset{\mid}{C}}-CH_2-CO-CH_3$ with $N-CH_3$ $\xrightarrow{RO^{\ominus}}$ $H_2C=N_2$	229
	Diazo group transfer to enamines	indole structure $=CH_2$, CH_3, CH_3, N-CH_3 $\xrightarrow{p-Tos-N_3}$ $H_2C=N_2$	236
$R-CH=N_2$	Hydrazone dehydrogenation	$R-CH=N-NH_2$ $\xrightarrow{\text{Dehydrogenation}}$ $R-CH=N_2$ $R = Aryl, Heteroaryl$	(Table 2, p. 216)
	Bamford-Stevens reaction	$R-CH=N-NH-SO_2-C_6H_4-CH_3-(p)$ $\xrightarrow{\text{Base}}$ $R-CH=N_2$ $R = Alkyl, Aryl, Heteroaryl$	(Table 4, p. 220)
	Acyl cleavage of N-nitroso compounds	$R-CH_2-N \begin{smallmatrix} X \\ NO \end{smallmatrix}$ $\xrightarrow{\text{Base}}$ $R-CH=N_2$ $R = Alkyl, Aryl$ $X = COOC_2H_5$ $X = CONH_2$ $X = C \begin{smallmatrix} NH \\ NH-NO_2 \end{smallmatrix}$ $X = CO-R^1$ $X = SO_2-C_6H_4-CH_3-(p)$	(Table 6, p. 229)
	Cleavage of 4-methyl-4-(N-alkyl-N-nitrosoamino)-2-pentanone	$H_3C-\underset{\underset{ON}{\overset{H_3C}{\mid}}{\overset{\mid}{C}}-CH_2-CO-CH_3$ with $N-CH_2-R$ $\xrightarrow{RO^{\ominus}}$ $R-CH=N_2$ $R = Alkyl$	(Table 6, p. 229)
$\underset{N_2}{\overset{H}{C}}-(CH_2)_n-\underset{N_2}{\overset{H}{C}}$	Acyl cleavage of bis-N-(nitroso) compounds	$\underset{ON}{\overset{X}{N}}-(CH_2)_{n+2}-\underset{NO}{\overset{X}{N}}$ $\xrightarrow{\text{Base}}$ $\underset{N_2}{\overset{H}{C}}-(CH_2)_n-\underset{N_2}{\overset{H}{C}}$ $X = COOC_2H_5$ $X = CONH_2$ $X = COR$	224 226 227
Perfluoroalkyl$-\underset{N_2}{\overset{R}{C}}$	Amine diazotization	Perfluoroalkyl$-\underset{NH_2}{\overset{}{CH}}-R$ $\xrightarrow{HX, NaNO_2}$ **Perfluoroalkyl**$-\underset{N_2}{\overset{R}{C}}$ $R = H, CH_3, CF_3, C_6H_5$	208

8.1 Preparation of aliphatic diazo compounds

Table A. Methods for preparing aliphatic diazo compounds

Diazo compound	Method of preparation		Page
R–C–R^1 (N$_2$)	Forster reaction	R–C–R^1 (NOH) $\xrightarrow{NH_2Cl}$ R–C–R^1 (N$_2$) R, R^1 = Aryl	211
	Hydrazone dehydrogenation	R–C–R^1 (N–NH$_2$) $\xrightarrow{Dehydrogenation}$ R–C–R^1 (N$_2$) R = R^1 = CN, CF$_3$, Aryl, Heteroaryl	(Table 2, p. 216)
	Bamford-Stevens reaction	R–C–R^1 (N–NH–SO$_2$–C$_6$H$_4$–CH$_3$–(p)) $\xrightarrow{Base}$ R–C–R^1 (N$_2$) R, R^1 = Alkyl, Aryl	(Tables 3, 4, p. 218, 220)
	Acyl cleavage of *N*-nitrosoureas	R/R^1 CH–N(NO)–CONH$_2$ $\xrightarrow{Base}$ R/R^1 C=N$_2$ R, R^1 = Alkyl	226
C=C–C (N$_2$)	Bamford-Stevens reaction	C=C–C–N–NH–SO$_2$–C$_6$H$_4$–CH$_3$–(p) $\xrightarrow{Base}$ C=C–C (N$_2$)	219
	Acyl cleavage of *N*-nitroso compounds	C=C–CH–N(X)(NO) $\xrightarrow{Base}$ C=C–C (N$_2$) X = COOC$_2$H$_5$ X = CONH$_2$	(Table 6, p. 229)
	Cleavage of 4-(*N*-alkyl-*N*-nitrosoamino)-4-methyl-2-pentanones	H$_3$C–C(CH$_3$)–CH$_2$–C(O)–CH$_3$, N(ON)–CH–C=C $\xrightarrow{RO^\ominus}$ C=C–C (N$_2$)	(Table 6, p. 229)
(N$_2$)	Diazo group transfer to cyclopentadienes	$\xrightarrow{Base, p-Tos-N_3}$	(Table 7, p. 234)
H–C(COOR)(N$_2$)	Amine diazotization	H$_2$C(COOR)(NH$_2$) $\xrightarrow{HX, NaNO_2}$ H–C(COOR)(N$_2$) R = Alkyl	208
	Bamford-Stevens reaction	H–C(COOR)–N–NH–SO$_2$–C$_6$H$_4$–CH$_3$–(p) $\xrightarrow{Base}$ H–C(COOR)(N$_2$) R = Alkyl	222

Table A. Methods for preparing aliphatic diazo compounds

Diazo compound		Method of preparation	Page
	Acyl cleavage of *N*-nitroso compounds	X—N(NO)—CH$_2$—COOR $\xrightarrow{\text{Base}}$ H—C(COOR)=N$_2$ R = Alkyl; X = COOC$_2$H$_5$; X = COCH$_3$	226 228
	Diazo group transfer to malonaldehydic acid ester	H—C(CHO)—CH$_2$—COOR $\xrightarrow{\text{Base, p–Tos–N}_3}$ H—C(COOR)=N$_2$ R = Alkyl	(Table 9, p. 235)
	Diazomethane acylation	H—C(H)=N$_2$ $\xrightarrow{\text{Cl–COOR}}$ H—C(COOR)=N$_2$ R = Aryl	(Table 10, p. 239)
	Acyl cleavage of α-diazo-β-oxocarboxylic acid esters	R^1—C(O)—C(COOR)=N$_2$ $\xrightarrow{\text{Base}}$ H—C(COOR)=N$_2$ R = Alkyl	245
R—C(COOR^1)=N$_2$	Amine diazotization	R—CH(NH$_2$)—COOR^1 $\xrightarrow{\text{HX, NaNO}_2}$ R—C(COOR^1)=N$_2$ R,R^1 = Alkyl	208
	Hydrazone dehydrogenation	R—C(COOR^1)=N—NH$_2$ $\xrightarrow{\text{Dehydrogenation}}$ R—C(COOR)=N$_2$ R = Aryl, R^1 = Alkyl	213
	Diazo group transfer with azidinium salts	R—CH$_2$—COOR^1 $\xrightarrow{\text{Azidinium salt}}$ R—C(COOR^1)=N$_2$ R = CN, R^1 = Alkyl	(Table 8, p. 234)
	Diazo group transfer to malonaldehydic acid ester	R—CH(CHO)—COOR^1 $\xrightarrow{\text{Base, p–Tos–N}_3}$ R—C(COOR^1)=N$_2$ R = Alkyl, Aryl; R^1 = Alkyl	(Table 9, p. 235)

Table A. Methods for preparing aliphatic diazo compounds

Diazo compound		Method of preparation		Page
	Substitution of metallized diazoacetates	$M-C(COOR^1)=N_2 \xrightarrow{RCl} R-C(COOR^1)=N_2$		
		$M = Li; R = Alkyl, (CH_3)_3Si, (C_4H_9)_3Sn$		
		$R' = Alkyl$		
		$M = Ag, R = S_N1$-active halide group		
		$R' = Alkyl$		242
		$Hg[C(COOR^1)=N_2]_2 \xrightarrow{R_2S} R-C(COOR^1)=N_2$		243
				243
		$R = (H_3C)_3Si, (H_3C)_3Sn;\ R^1 = Alkyl$		
$O_2N-C(COOR)=N_2$	Diazo group transfer with azidinium salts	$O_2N-CH_2-COOR \xrightarrow{\text{Azidinium salt}} O_2N-C(COOR)=N_2$		(Table 8, p. 234)
		$R = Alkyl$		
	Nitration of diazoacetates	$H-C(COOR)=N_2 \xrightarrow{N_2O_5} O_2N-C(COOR)=N_2$		238
		$R = Alkyl$		
$Hal-C(COOR)=N_2$	Substitution of mercury bisdiazoacetates	$Hg[C(COOR)=N_2]_2 \xrightarrow{SOCl_2,\ Br_2,\ J_2} Hal-C(COOR)=N_2$		243
		$R = Alkyl$		
$R-C(=O)-C(H)=N_2$	Hydrazone dehydrogenation	$R-C(=O)-C(H)=N-NH_2 \xrightarrow{\text{Dehydrogenation}} R-C(=O)-C(H)=N_2$		(Table 2, p. 216)
		$R = Aryl$		
	Diazo group transfer to 3-oxoaldehydes	$R-C(=O)-CH_2-CHO \xrightarrow{\text{Base, } p-Tos-N_3} R-C(=O)-C(H)=N_2$		(Table 9, p. 235)
		$R = Alkyl, Aryl$		
	Diazomethane acylation	$R-C(=O)-Cl \xrightarrow{H_2C=N_2} R-C(=O)-C(H)=N_2$		(Table 10)
		$R = Alkyl, Aryl$		
	Acyl cleavage of 2-diazo-1,3-diketones	$R-C(=O)-C(N_2)-C(R)=O \xrightarrow{\text{Base}} R-C(=O)-C(H)=N_2$		245
		$R = Alkyl$		

Table A. Methods for preparing aliphatic diazo compounds

Diazo compound		Method of preparation	Page
$R-\underset{\underset{O}{\|\|}}{C}-\underset{\underset{N_2}{\|\|}}{C}-R^1$	Forster reaction	$R-\underset{\underset{O}{\|\|}}{C}-\underset{\underset{NOH}{\|\|}}{C}-R^1 \xrightarrow{NH_2Cl} R-\underset{\underset{O}{\|\|}}{C}-\underset{\underset{N_2}{\|\|}}{C}-R^1$ R,R^1 = Alkyl, Aryl	(Table 1, p. 207)
	Hydrazone dehydrogenation	$R-\underset{\underset{O}{\|\|}}{C}-\underset{\underset{N-NH_2}{\|\|}}{C}-R^1 \xrightarrow{Dehydrogenation} R-\underset{\underset{O}{\|\|}}{C}-\underset{\underset{N_2}{\|\|}}{C}-R^1$ R,R^1 = Alkyl, Aryl	(Table 2, p. 216)
	Bamford-Stevens reaction	$R-\underset{\underset{O}{\|\|}}{C}-\underset{\underset{N-NH-SO_2-C_6H_4-CH_3-(p)}{\|\|}}{C}-R^1 \xrightarrow{Base} R-\underset{\underset{O}{\|\|}}{C}-\underset{\underset{N_2}{\|\|}}{C}-R^1$ R,R^1 = Alkyl, Aryl	(Table 5, p. 222)
	Diazo group transfer	$\underset{O}{\overset{R}{\diagdown}}C-CH_2-R^1 \xrightarrow{Base,\ p-Tos-N_3} \underset{O}{\overset{R}{\diagdown}}C-\underset{N_2}{\overset{R^1}{\diagup}}C$ R = Alkyl, Aryl; R^1 = Aryl	232
	Diazo group transfer to 2-formyl ketones	$R-\underset{\underset{O}{\|\|}}{C}-\underset{\underset{CHO}{\|}}{CH}-R^1 \xrightarrow{Base,\ p-Tos-N_3} R-\underset{\underset{O}{\|\|}}{C}-\underset{\underset{N_2}{\|\|}}{C}-R^1$ R,R^1 = Alkyl	(Table 9, p. 235)
	Diazoalkane acylation	$\underset{O}{\overset{R}{\diagdown}}C-Cl \xrightarrow{\underset{N_2}{\overset{HC-R^1}{\|\|}}} \underset{O}{\overset{R}{\diagdown}}C-\underset{N_2}{\overset{R^1}{\diagup}}C$ R = Alkyl, Aryl; R^1 = Alkyl	210
	Reaction of silver diazo ketones with S_N1-active halides	$\underset{O}{\overset{R}{\diagdown}}C-\underset{N_2}{\overset{Ag}{\diagup}}C \xrightarrow{R^1-Hal} \underset{O}{\overset{R}{\diagdown}}C-\underset{N_2}{\overset{R^1}{\diagup}}C$ R = Alkyl, Aryl, Heteroaryl; R' = S_N1-active halide group	242
$\overset{}{\diagup}C=\overset{\|}{C}-\underset{\underset{O}{\|\|}}{C}-\underset{N_2}{\overset{}{C}\diagdown}$	Diazo group transfer to 3-oxo-4,5-unsaturated aldehydes	$\overset{}{\diagup}C=\overset{\|}{C}-\underset{\underset{O}{\|\|}}{C}-\underset{\underset{CHO}{\|}}{CH}- \xrightarrow{Base,\ p-Tos-N_3} \overset{}{\diagup}C=\overset{\|}{C}-\underset{\underset{O}{\|\|}}{C}-\underset{N_2}{\overset{}{C}\diagdown}$	(Table 9, p. 235)
(p-Quinone diazide structure with O, N$_2$)	Amine diazotization	(phenol-amine ring) $\xrightarrow[b)\ Base]{a)\ HX,\ NaNO_2}$ (quinone diazide)	208

p-Quinone diazides behave analogously

Table A. Methods for preparing aliphatic diazo compounds

Diazo compound	Method of preparation		Page
	Bamford-Stevens reaction	[structure] $=N-NH-SO_2-C_6H_4-CH_3-(p)$ $\xrightarrow{\text{or Base}}$ [structure with N_2]	221
[structure: R, H, N_2, $C-C$, O]	Diazo group transfer to alkylmalonaldehydes	$R-CH-CHO$ with CHO $\xrightarrow{\text{Base, } p-Tos-N_3}$ [product] $R = $ Alkyl	(Table 9, p. 235)
	Diazo group transfer to 3-aminoacroleins	$R-C$ with CHO and $CH-N(CH_3)_2$ $\xrightarrow{p-Tos-N_3}$ [product] $R = $ Alkyl	236
[structure: O_2N, Ar, N_2]	Diazo group transfer with azidinium salts	O_2N-CH_2-C with Ar, O $\xrightarrow{\text{azidinium salt}}$ [product]	(Table 8, p. 234)
[structure: N_2C, O, CN_2]	Forster reaction	HON [structure] NOH $\xrightarrow{NH_2Cl}$ N_2C [structure] CN_2	211
[structure: $R-C=O$, $R^1-C=O$, $C=N_2$]	Amine diazotization	$R-C=O$, $R^1-C=O$, $CH-NH_2$ $\xrightarrow{HX, NaNO_2}$ [product] $R, R^1 = $ Alkyl, Aryl	210
	Bamford-Stevens reaction	$R-C=O$, $R^1-C=O$, $C=O$ $\xrightarrow[\triangledown]{H_2N-NH-SO_2-C_6H_4-CH_3-(p)}$ [product] usable only if R and R′ deactivate the 1,3-oxo groups	222
	Diazo group transfer with *p*-toluenesulfonic acid azide or azidinium salts	$R-C=O$, $R^1-C=O$, CH_2 $\xrightarrow[\text{or azidinium salt}]{\text{Base, } p-Tos-N_3}$ [product] $R, R^1 = $ Alkyl, Aryl, Alkoxy	(Table 7, p. 234) (Table 8, p. 234)
	Diazoacetate acylation	$R-C=O$, Cl $\xrightarrow{H-C(COOR^1)(N_2)}$ [product] $R-C=O$, $R^1O-C=O$, $C=N_2$ $R = $ Alkyl, Aryl; $R^1 = $ Alkyl	240

Table A. Methods for preparing aliphatic diazo compounds

Diazo compound		Method of preparation		Page
$R-SO_2-C\!\!<^H_{N_2}$	Acyl cleavage of *N*-nitrosourethanes	$R-SO_2-CH_2-N\!\!<^{COOC_2H_5}_{NO}$ $\xrightarrow{Al_2O_3}$ $R-SO_2-C\!\!<^H_{N_2}$ R = Alkyl, Aryl		(Table 6, p. 229)
	Acyl cleavage of acetylarene-sulfonyldiazomethanes	$R-SO_2-C\!\!<^{COCH_3}_{N_2}$ $\xrightarrow{N(C_2H_5)_3/H_3C-OH}$ $R-SO_2-C\!\!<^H_{N_2}$ R = Aryl		244
$R-SO_2-C\!\!<^{SO_2-R^1}_{N_2}$	Hydrazone dehydrogenation	$R-SO_2-C\!\!<^{SO_2-R^1}_{N-NH_2}$ $\xrightarrow{Dehydrogenation}$ $R-SO_2-C\!\!<^{SO_2-R^1}_{N_2}$ R,R^1 = Aryl		(Table 2, p. 216)
	Diazo group transfer	$R-SO_2-CH_2-SO_2-R^1$ $\xrightarrow[\text{}]{Base, p-Tos-N_3}$ $R-SO_2-C\!\!<^{SO_2-R^1}_{N_2}$ R,R^1 = Alkyl, Aryl		(Table 7, p. 234)
$R-SO_2-C\!\!<^{\overset{R^1}{C=O}}_{N_2}$	Diazo group transfer with *p*-toluenesulfonic acid azide or azidinium salts	$R-SO_2-CH_2-C\!\!<^{R^1}_{O}$ $\xrightarrow[\text{or azidinium salt}]{Base, p-Tos-N_3}$ $R-SO_2-C\!\!<^{\overset{R^1}{C=O}}_{N_2}$ R,R^1 = Alkyl, Aryl		(Table 7, p. 234)
$R-SO_2-\underset{N_2}{\overset{-CH-}{C}}-C-N\!\!<$	Alkylation of diazosulfonyl-methanes with enamines	$R-SO_2-C\!\!<^H_{N_2}$ $\xrightarrow{>C=C-N<}$ $R-SO_2-\underset{N_2}{C}-C-N\!\!<$ R = Alkyl, Aryl		244
$R-\underset{O}{\overset{R}{P}}-C\!\!<^H_{N_2}$	Amine diazotization	$R-\underset{O}{\overset{R}{P}}-\underset{NH_2}{CH_2}$ $\xrightarrow{HX, NaNO_2}$ $R-\underset{O}{\overset{R}{P}}-C\!\!<^H_{N_2}$ R = Alkyl, Aryl, Alkoxy		209
$RO-\underset{O}{\overset{RO}{P}}-C\!\!<^{R^1}_{N_2}$	Bamford-Stevens reaction	$RO-\underset{O}{\overset{RO}{P}}-C\!\!<^{R^1}_{N-NH-SO_2-C_6H_4-CH_3-(p)}$ $\xrightarrow{Alkali\ hydroxide}$ $RO-\underset{O}{\overset{RO}{P}}-C\!\!<^{R^1}_{N_2}$ R = Alkyl, R^1 = Alkyl, Aryl		223
$R-\underset{O}{\overset{R}{P}}-C\!\!<^{\overset{R^1}{C=O}}_{N_2}$	Diazo group transfer	$R-\underset{O}{\overset{R}{P}}-CH_2-\underset{O}{C}-R^1$ $\xrightarrow{Base, p-Tos-N_3}$ $R-\underset{O}{\overset{R}{P}}-C\!\!<^{\overset{R^1}{C=O}}_{N_2}$ R = Aryl, Alkoxy; R^1 = Alkyl, Aryl, Alkoxy		233 (Table 7, p. 234)

Table A. Methods for preparing aliphatic diazo compounds

Diazo compound	Method of preparation		Page
  (OR)	Diazo group transfer to ynamines or yne ethers	 R can be varied widely	237
  1,2-dicarbonyl compounds behave analogously	Aldol addition of acyl-diazomethanes to tricarbonyl compounds	 R = Acyl	242, 249
	Reacting 1,1′,3,3′-tetraphenyl bis(2-imidazoylidene) with acyldiazomethanes	 R = Acyl	243
	Metallizing diazomethyl compounds with organic lithium compounds or phenyloxirane	 M = Li, R = H, Acyl M = Ag, R = Acyl	241
$M_2C=N_2$	Metallizing diazomethane with metal amides	 M = Zn, Cd, Hg	241
	Metallizing diazomethyl compounds with mercury oxide	 R = Acyl or other proton-activating groups	241
	Metallizing diazomethane with metal amides	 M = Ge, Sn, Pb ; R = Alkyl, Aryl	241
	Substituting diazomethyllithium with metal halides	 M = Ge, Sn ; R = Alkyl, Aryl	241

8.2 Transformations of aliphatic diazo compounds

8.2.1 Decomposition reactions

8.2.1.1 Thermal, catalytic and photolytic decomposition

8.2.1.1.1 General

Aliphatic diazo compounds are one of the most reactive classes of organic substances. Depending on the reaction conditions, they can behave as carbene sources, as 1,3-dipoles, as nucleophiles, as electrophiles, and as acids or bases. Although not everything is yet known about the fine structure of aliphatic diazo compounds, and many of their reaction mechanisms are still the subject of debate, they are used very widely in organic synthesis. Because of their high reactivity, the diazo compounds are not highly selective, and this makes working with them more difficult.

One of the most important characteristics of the diazoalkanes is that they easily split off nitrogen, often spontaneously and accompanied by vigorous **explosions.** Consequently, when working with diazoalkanes, all appropriate precautions must be taken. Increase in temperature, exposure to radiation, rough surfaces, and catalysts promote the decomposition of diazoalkanes[400].

In most cases dilute *solutions* (up to 10%) of the diazoalkanes in inert solvents are used.

The *ease* with which aliphatic diazo compounds split off nitrogen depends on the character of the substituent groups present in the molecule. Often, electron-attracting substituent groups which are attached to the carbon atom bearing the diazo group increase the thermal and photochemical stability of diazoalkanes, and their resistance toward catalytic decomposition; electron donors have the converse effect[401–403].

Although exact kinetic investigations of the relative stabilities of different types of aliphatic diazo compounds are still at an early stage[404–411], numerous experimental data[401, 412–414] indicate the following order of relative stability:

$(RS)_2CN_2 < ClCHN_2 < (CH_3)_2CN_2$, diazocycloalkanes $< CH_2N_2 < H_5C_6CHN_2 < (C_6H_5)_2CN_2 <$ 9-diazofluorene, diazocyclopentadiene

The relatively high stability of *9-diazofluorene*[412], *diazocyclopentadiene*[415], *diazocyclononatetraene*[416], and of five-membered heterocyclic compounds such as *diazopyrrole* and *diazoindole*[417] can be attributed to the mesomerism of these compounds:

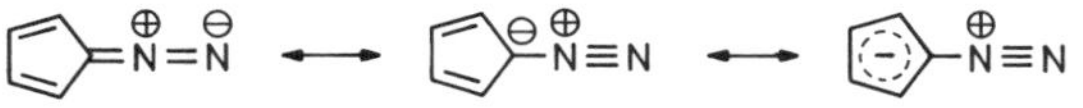

The α-diazocarbonyl compounds (especially 2-diazo-1,3-dicarbonyl compounds) are more stable because of the greater contribution of the diazonium or diazoanhydride structures to the mesomeric ground state[403, 418].

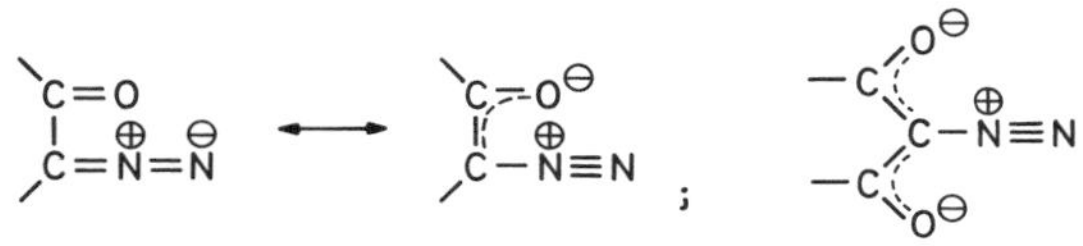

Other strongly electron-withdrawing substituent groups also enhance the stability of aliphatic diazo compounds.

[398] *J.B. Hendrickson, W.A. Wolf*, J. Org. Chem. *33*, 3610 (1968).

[399] *D. Hodson, G. Holt, D.K. Wall*, J. Chem. Soc. C 2201 (1968).

[400] *B. Eistert, M. Regitz, G. Heck, H. Schwall*, in Houben-Weyl, Methoden der organischen Chemie, Bd. X/4, p. 482, Georg Thieme Verlag, Stuttgart 1968.

[401] *I.A. D'yakonov*, Aliphatic Diazocompounds, their Structure, Properties and Reactions (russ.) Leningrad. Gosudarst. Univ. 1958; C.A. *53*, 2088e (1959).

[402] *R. Huisgen*, Angew. Chem. *67*, 439 (1955).

[403] *O.P. Studzinskii, I.K. Korobizina*, Usp. Khim. *39*, 1754 (1970) (russ.).

[404] *E. Fahr, K.-H. Keil, H. Lind, F. Scheckenbach*, Z. Naturforsch. B *20*, 526 (1965).

[405] *M. Regitz, W. Bartz*, Chem. Ber. *103*, 1477 (1970).

[406] *C.E. Bawn, T.B. Rhodes*, Trans. Fraday. Soc. *50*, 934 (1954).

[407] *J. Feltzin, A. Restaino, R. Mesrobian*, J. Amer. Chem. Soc. 77, 206 (1955).

[408] *A.G. Nasini, L. Trossarelly, G. Saini*, Makromol. Chem. *44–46*, 550 (1960).

[409] *I.A. D'yakonov, A.G. Vitenberg, M.I. Komendantov*, Zh. Organ. Khim. *1*, 1183 (1965); C.A. *63*, 13023f (1965).

[410] *A.G. Vitenberg, I.A. D'yakonov, A. Sindel*, Zh. Organ. Khim. *2*, 1532 (1966); C.A. *66*, 64881h (1967).

[411] *A.G. Vitenberg, I.A. D'yakonov*, Zh. Org. Khim. *5*, 1036 (1969); C.A. *71*, 69923m (1969).

[412] *H. Staudinger, A. Gaule*, Ber. *49*, 1897 (1916).

[413] *U. Schöllkopf, E. Wiskott*, Justus Liebigs Ann. Chem. *694*, 44 (1966).

[414] *G. Gloss, J. Coyle*, J. Amer. Chem. Soc. *84*, 4350 (1962).

[415] *W.E. Doering, C.H. De Puy*, J. Amer. Chem. Soc. *75*, 5955 (1956).

[416] *D. Lloyd, N.W. Preston*, Chem. & Ind. (London) 1039 (1966).

[417] *J.M. Tedder*, Advan. Heterocycl. Chem. *8*, 1 (1967).

[418] *E. Fahr*, Justus Liebigs Ann. Chem. *617*, 11 (1958); *E. Fahr*, Justus Liebigs Ann. Chem. *638*, 1 (1960).

Thus, for example, *diazonitromethane*[419, 420], *2-diazo-1,1,1-trifluoropropane*[421], *1-diazo-1-phenyl-2,2,2-trifluoroethane*[422], *diazosulfonylmethane*[423], *bis-(sulfonyl)-diazoalkanes*[424], *α-diazophosphine oxides*[425-428], *diazoperfluoroalkanes*[429, 431] and *diazomalononitrile*[431] are relatively stable. The *o-* and *p-*quinone diazides[432] *occupy an intermediate position between the aliphatic diazo compounds and the diazonium salts.*

Many diazo compounds form *carbenes* on photolysis, thermolysis, and on catalytic decomposition (especially when copper or its salts are used):

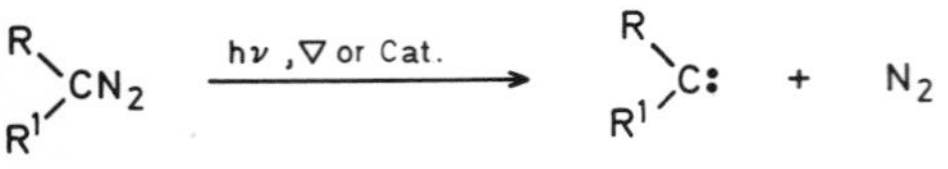

Since these fission reactions proceed easily in most cases, diazoalkanes are the most important sources of carbenes[433, 434].

Evidence for the formation of carbenes during thermolysis and photolysis of diazoalkanes has been obtained not only from numerous trapping reactions, but also with the aid of physical methods[433-436]. Since the chemistry of carbenes is a large field of modern organic chemistry in its own right (see, *e.g.*, refs. [401, 433-435, 437, 438]), it will be dealt with here only so far as is necessary to understand the reactions of aliphatic diazo compounds.

8.2.1.1.2 Decomposition to ethylene derivatives (olefins) and azines

The decomposition of aliphatic diazo compounds in the absence of reaction partners can lead to *azines* 1 and to *alkenes* 2:

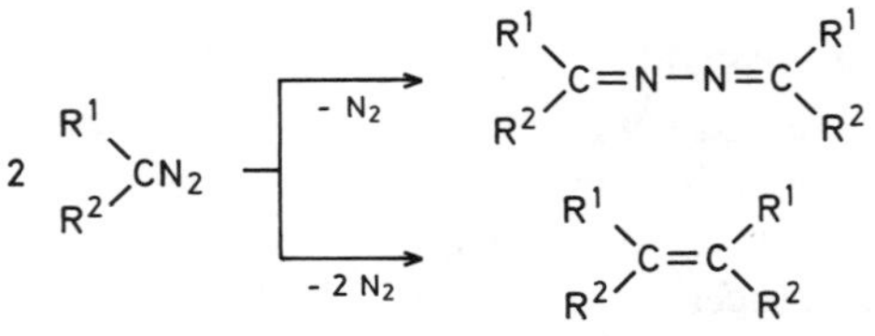

It is usually assumed that carbenes are formed as intermediates in these reactions[436-442], but this route is not invariably the only possible one[441].

Formation of azines and olefins is a characteristic[400] of the relatively stable diazo compounds, *e.g.*:

Diazodiphenylmethane	*Tetraphenylethylene*
9-Diazofluorene	*9,9'-Bifluorenylidene*
1-Diazoethanol acetate	*2-Butene-2,3-diol diacetate*
α-Diazo ketones	*1,2-Diacylethylenes*

Diazomethane decomposes instead to *polymethylene*[435].

Azines are often formed by heating diazo compounds in the *absence* of *catalysts*, particularly on the preparative scale.

By using a mixture of two different diazo compounds, it is possible to prepare *unsymmetrical azines;* if one of the components is diazodiphenyl-

[419] U. Schöllkopf, D. Hoppe, N. Rieber, V. Jacobi, Justus Liebigs Ann. Chem. 730, I (1969).

[420] U. Schöllkopf, P. Markusch, Tetrahedron Lett. 6199, (1966).

[421] R. Shepard, P. Sciaraffa, J. Org. Chem. 31, 964 (1966).

[422] R. Shepard, S. Wentworth, J. Org. Chem. 32, 3197 (1967).

[423] A.M. van Leusen, J. Strating, Rec. Trav. Chim. Pays-Bas 84, 151 (1965).

[424] J. Diekmann, J. Org. Chem. 28, 2933 (1963).

[425] L. Horner, H. Hoffmann, H. Ertel, G. Klahre, Tetrahedron Lett. 9 (1961).

[426] N. Kreutzkamp, E. Schmidt-Samoa, K. Herbert, Angew. Chem. 77, 1138 (1965).

[427] D. Seyferth, P. Hilbert, R. Marmor, J. Amer. Chem. Soc. 89, 4810 (1967).

[428] M. Regitz, W. Anschütz, W. Bartz, A. Liedhegener, Tetrahedron Lett. 3171 (1968).

[429] B.L. Dyatkin, E.P. Mochalina, Isv. Akad. Nauk SSSR, Ser. Khim. 1035 (1965); C.A. 63, 8185d (1965);
E.P. Mochalina, B.L. Dyatkin, Isv. Akad. Nauk SSSR, Ser. Khim. 926 (1965); C.A. 63, 5515b (1965).

[430] D. Gale, W. Middleton, C. Krespan, J. Amer. Chem. Soc. 87, 657 (1965); 88, 3617 (1966).

[431] E. Ciganek, J. Org. Chem. 30, 4198 (1965).

[432] L.A. Kazitsyna, B.S. Kikot, A.V. Upadysheva, Usp. Khim. 35, (5), 881 (1966); C.A. 65, 7087 a (1966).

[433] W. Kirmse, Carbene, Carbenoide und Carbenanaloge, Verlag Chemie, Weinheim/Bergstr. 1969.

[434] D. Bethell, Structure and Mechanism in Carbene Chemistry, in V. Gold, Advances in Physical Organic Chemistry, Vol. VII, p. 153, Academic Press, New York 1969.

[435] A.M. Trozzolo, Accounts Chem. Res. 1, 329 (1968).

[436] J.E. Dove, J. Riddick, Can. J. Chem. 48, 3623 (1970).

[437] L.P. Danilkina, M.I. Komendantov, R.R. Kostikov, T.V. Mandel'Shtam, V.V. Razin, E.M. Kharicheva, Vestn. Leningrad. Univ., Fiz. Khim. 1970, (I), 123; C.A. 73, 34367x (1970).

[438] B.J. Herold, P.P. Gaspar, Fortschr. Chem. Forsch. 5, 89 (1965).

[439] T. Sasaki, Sh. Eguchi, A. Kojima, Bull. Chem. Soc. Japan 41, 1658 (1968).

[440] J.W.J. Still, D.T. Wang, Can. J. Chem. 46, 1583 (1968).

[441] G. Cowell, A. Ledwith, Quart. Rev. Chem. Soc. 24, 119 (1970).

[442] D. Schumann, E. Frese, A. Schönberg, Chem. Ber. 102, 3192 (1969).

methane yields can reach 70–90%[443, 444]. Unsymmetrical azines (20–30%) are also formed from α-diazocarbonyl compounds and diazoalkanes[445]. Heating aliphatic diazo compounds in inert solvents with added *copper powder* or *copper salts* promotes the formation of olefins. Thus, from 9-diazofluorene one can obtain 9, 9'-bifluorenylidene (91%)[446], while diazodiphenylmethane gives *tetraphenylethylene* in 86% yield[446, 447]. Intramolecular cyclizations are also known, for example, *2-cycloalkene-1,4-diones* are formed from bis-(α-diazo ketones)[448]:

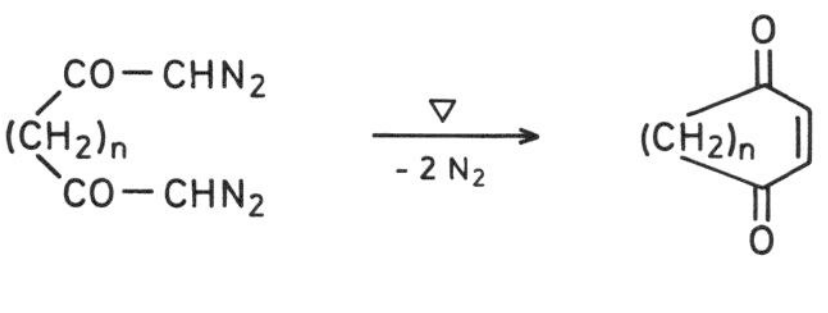

Naturally, the *parallel* formation of azines and open-chain alkenes is often observed.

The catalyzed thermal decomposition of ethyl diazoacetate gives *triethyl 4,5-dihydropyrazole-3,4,5-tricarboxylate*[400] together with a mixture of *diethyl fumarate* and *diethyl maleate*.

Heating α-diazoketones in the presence of copper and its salts likewise causes elimination of nitrogen and the formation of unsaturated compounds. In some cases, this reaction can be used for the synthesis of symmetrical (and even unsymmetrical) *γ-diketones*[448–453]. *Ethylene derivatives* can also be formed from diaryldiazomethanes and diazomethane itself by interaction with carbenes which do not originate from aliphatic diazo compounds but which are generated in some other way. The reaction with dichlorocarbene[454, 455] is, for example:

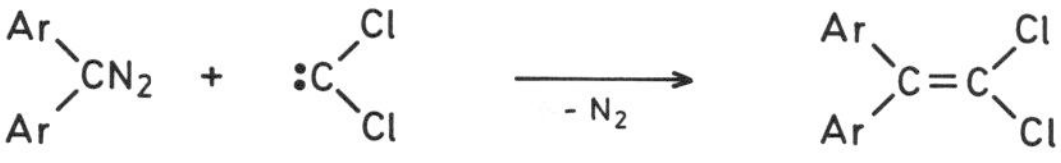

Photolysis of diazoalkanes is used much less often than catalytic thermal decomposition for the preparation of *ethylene derivatives* and *azines*, since it proceeds less cleanly[456–458]. However, photolytic and thermal decompositions of diazodiphenylmethane in benzene give similar yields (80–90%) of *azine*[459].

8.2.1.1.3 Decomposition accompanied by intramolecular rearrangement

Carbenes which are formed by the decomposition of aliphatic diazo compounds can also become stabilized by isomerization; this is particularly characteristic of branched diazoalkanes and diazocycloalkanes. Formation of *cyclopropanes* 3 has been observed, involving insertion of the carbon atom of the carbene in a *CH* bond which is in a γ-position; *olefins* 4 are also formed. Since formation of olefins requires a hydride shift, it can only occur when a hydrogen atom is present in the α-position relative to the diazo group. Isomerization processes which proceed exclusively *via* carbene intermediates can only take place under strictly *aprotic* conditions[400, 460]; otherwise, isomers resulting from protonation products are also formed, *e.g.* 5 and the ether 6:

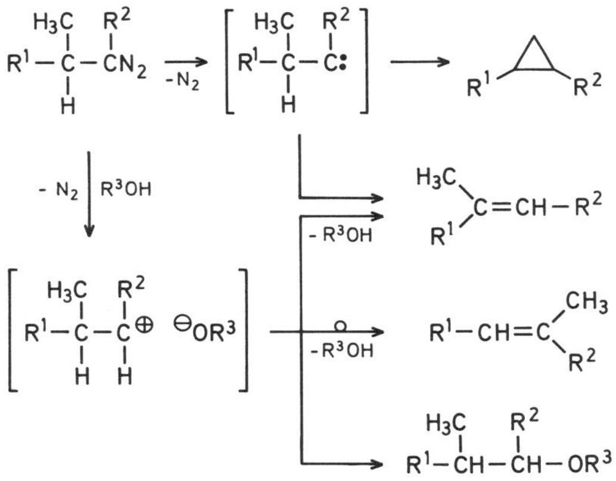

[443] *R. Huisgen, R. Fleischmann,* Justus Liebigs Ann. Chem. *623,* 47, 65 (1959).

[444] *A. Schönberg, K. Junghans,* Chem. Ber. *98,* 820 (1965).

[445] *P. Yates, D.G. Farnum, D.W. Wiley,* Tetrahedron *18,* 881 (1962).

[446] *H. Nozaki, S. Moriuti, M. Yamabe, R. Noyori,* Tetrahedron Lett. 59 (1966).

[447] *H. Nozaki, H. Takaya, S. Moriuti, R. Noyori,* Tetrahedron *24,* 3655 (1968).

[448] *J. Font, F. Serratosa, J. Valls,* Chem. Commun. 721 (1970).

[449] *P.K. Banerjee, D. Mukhopadhyay, D.N. Chaudhury,* J. Indian. Chem. Soc. *42,* 115 (1965).

[450] *J. Ernest,* Chem. Listy *45,* 262 (1951); *J. Ernest,* Collcet. Czech. Chem. Commun. *19,* 1179 (1954).

[451] *V. Hnevsowa, V. Smely, J. Ernest,* Collect. Czech. Chem. Commun. *219,* 1459 (1956).

[452] *J. Ernest,* Collect. Czech. Commun. *24,* 530, 1022, 2072, 3341 (1959).

[453] *Jr.A. Zhdanov, V.I. Kornilov, G.V. Bogdanova,* Carbohyd. Res. *3,* 139 (1966).

[454] *H. Reimlinger,* Chem. & Ind. (London) 1306 (1969).

[455] *S.P. Mantus, J.T. Carroll, Ch. Dodson,* J. Org. Chem. *33,* 4272 (1968).

[456] *I.A. D'yakonov, A.G. Vitenberg,* Zh. Org. Khim. *5,* 1032 (1969).

[457] *W. Kirmse, L. Horner, H. Hoffmann,* Justus Liebigs Ann. Chem. *614,* 19 (1958).

Consequently, a firm distinction between the *carbenoid* and *carbocationoid* mechanisms cannot always be made on the basis of the structure of the reaction products[461-463].

Open-chain diazoalkanes mostly form cyclopropanes whose relative yields increase with increasing branching of the alkyl group[400, 464]. The formation of *olefins* is characteristic of 1-aryl-2-diazoethanes[465, 466]. Cyclopropyldiazomethanes undergo ring-enlargement to *cyclobutenes,* which in some cases are formed in satisfactory yields[467].

The decomposition of *higher* diazocycloalkanones usually leads to a complex reaction mixture[468-476], although the exclusive formation of the corresponding *cycloalkenes* can be observed, for example, in the decomposition of bicyclo[3.1.0]-hexan-3-one *p*-tosylhydrazone with sodium ethoxide[472]:*

Bicyclo[3.1.0]hex-2-ene; 91%

Investigations of the copper-catalyzed decomposition of a large series of diazomethanephosphonate diesters have shown that the nature of the endproduct depends on the structure of the substituent groups in the diazo compound. For example yields of α,β-unsaturated compounds can be up to 80–87% if an α-hydrogen atom is present[467]. In the case of the decomposition of unstable diazoalkanes and diazo ketones, a reaction between the diazo group and the $C=C$ double bond is often observed.

8.2.1.2 Decomposition by mineral acids

Mineral acids in general cause the decomposition of aliphatic diazo compounds, with elimination of nitrogen; the main products are *hydroxy* or *halo derivatives* (when hydrogen halides are used)[400]. This reaction is in fact the first qualitative means of detecting the presence of the diazo group.

α-Diazocarbonyl compounds form *hydroxy* or *halo ketones*** or *enediols* analogously[477-482]. Formation of enediols is promoted by catalytic amounts of copper. In this reaction, too, the reactivity of the diazo compounds depends on the nature of the substituent groups in the molecule[401, 402].

Substituent groups at the α-carbon atom (*e.g.,* phenyl, carbonyl, or carboxy) able to resonate with the diazo group increase the stability of aliphatic diazo compounds toward mineral acids, and at the same time reduce their nucleophilism[441].

---

* The Bamford-Stevens method is being used increasingly for synthesizing *diazoalkanes* (see p. 215). The unstable diazoalkanes are not isolated as such, but are produced under conditions which simultaneously cause decomposition[468-471].

** α-Thiocyanato ketones are formed by the action of mineral acids and ammonium thiocyanate[481].

[458] *R. Moss, J.D. Funk,* J. Chem. Soc. C 2026 (1967).

[459] *D.R. Dalton, S.A. Liebmann,* Tetrahedron *25,* 3321 (1969).

[460] *J.H. Atherton, R. Fields, R.N. Haszeldine,* J. Chem. Soc. C 366 (1971).

[461] *R.H. Shapiro, J.H. Duncan, J.C. Clopton,* J. Amer. Chem. Soc. *89,* 471, 1442 (1967).

[462] *H.M. Frey, J.D.R. Stevens,* J. Chem. Soc. 1700 (1965).

[463] *K.B. Wiberg, J.M. Lavonish,* J. Amer. Chem. Soc. *88,* 5272 (1966).

[464] *B. Singh,* J. Org. Chem. *31,* 181 (1966).

[465] *P.B. Sargeant, H. Shechter,* Tetrahedron Lett. 3957 (1964).

[466] *W.J. Baron, M. Jones, P.P. Caspar,* J. Amer. Chem. Soc. *92,* 4739 (1970).

[467] *R.S. Marmor, D. Seyferth,* J. Org. Chem. *36,* 128 (1971).

[468] *W. Kirmse, H.D. von Scholz, H. Arnold,* Justus Liebigs Ann. Chem. *711,* 22 (1968).

[469] *W. Kirmse, H.W. Bücking,* Justus Liebigs Ann. Chem. *711,* 31 (1968).

[470] *G. Maier, M. Strasser,* Tetrahedron Lett. 6453 (1966).

[471] *D.H. Paskovich, W.N. Kwok,* Tetrahedron Lett. 2227 (1967).

[472] *P.K. Freeman, D.G. Kuper,* J. Org. Chem. *30,* 1047 (1965).

[473] *L. Friedman, H. Schechter,* J. Amer. Chem. Soc. *83,* 3159 (1961).

[474] *F.T. Bond, D.E. Bradway,* J. Amer. Chem. Soc. *87,* 4977 (1965).

[475] *P.K. Freeman, R.C. Johnson,* J. Org. Chem. *34,* 1751 (1969).

[476] *M. Tanabe, D.F. Crowe, R.L. Dehn,* Tetrahedron Lett. 3943 (1967).

[477] *G. Massaroli, Del Corona, L.G. Signorelli,* Bull. Chim. Farm. 107 (II), 679 (1968).

[478] *W. Jugelt, L. Berseck,* Tetrahedron Lett. 2659, 2665 (1968).

[479] *A. Benkö, P. Teszler,* Chem. Ber. *100,* 2184 (1967).

[480] *W. Hampel, T. Török,* Z. Chem. *8,* 226 (1968).

[481] *W. Hampel, M. Kapp,* Z. Chem. *11,* 12 (1971).

[482] *F. Haupter, A. Pucek,* Chem. Ber. *93,* 249 (1960).

[483] *F. Klages, K. Bott,* Chem. Ber. *97,* 735 (1964).

[484] *D.G. Cram, R.D. Partos,* J. Amer. Chem. Soc. *85,* 1273 (1963).

[485] *O.W. Webster,* J. Amer. Chem. Soc. *88,* 4055 (1966).

2-Diazo-1,3-dicarbonyl compounds often do not react even with concentrated mineral acids; also very *resistant* are diazocyclopentadiene[483–485] and diazo compounds which contain a fluoro[430] or nitro group[486] or certain other electron-withdrawing substituents[479].

The effect of the electron-withdrawing substituent groups is due to the fact that they stabilize the diazonium ions which are formed as intermediates during acid hydrolysis.

Diazo compounds which are not decomposed by the action of mineral acids yield fairly stable *diazonium salts* whose structures are similar to those of the aromatic diazonium salts[487–489].

For example, *ethylenediazonium hexachloroantimonate* is known[490]; other Lewis acids can also be used in place of antimony(V) pentachloride.

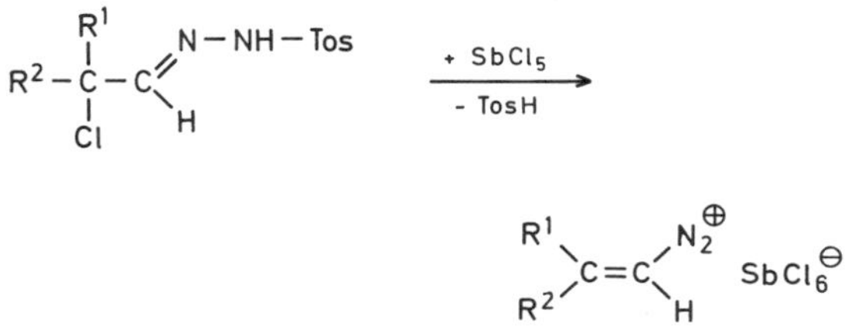

2-Diazo-1,3-dicarbonyl compounds[400, 491, 492] also give stable *complexes.* An interesting example is the *ω-(diazoalkyl)diazonium ion* which exists in equilibrium with a bis(diazoalkane) and an alkane-bis-(diazonium)ion[493].

The *mechanism* of the hydrolysis of aliphatic diazo compounds by acids depends upon the structure of the initial reactants and upon the reaction conditions. Although there are large differences in the kinetics of the reaction, the same intermediate stages always occur. Protonation first takes place reversibly or irreversibly, to form a diazonium ion, which either immediately reacts with a nucleophile (*e.g.*, with water) in a bimolecular process or else forms a carbonium ion in a monomolecular process.

Hydrolysis of diazoalkanes, diazo-(4-nitrophenyl)methane, 2-diazo-1,1,1-trifluoropropane[478], and sec-α-diazocarbonyl compounds *(R^1–CO–CN_2–R^2, R^1O–CO–CN_2–R^2)* follows an A–S_E2 mechanism[478, 494–500]:

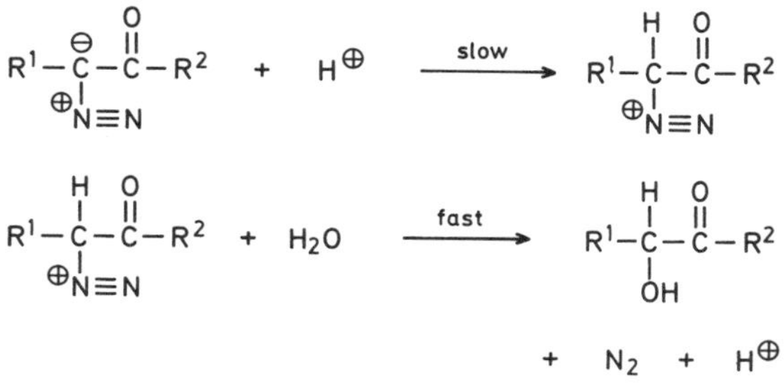

An *A–2* mechanism is more likely for primary α-diazo ketones *(R–CO–CHN_2),* ethyl diazoacetate, 2-diazo-1,1,1-trifluoroethane, and α-diazosulfones *(R–SO_2–CHN_2),* especially if powerfully nucleophilic reagents are present[498, 500–506]:

$$R-CO-\overset{\ominus}{C}H-N_2^{\oplus} + H^{\oplus} \rightleftharpoons R-CO-CH_2-N_2^{\oplus}$$

$$R-CO-CH_2-N_2^{\oplus} + H_2O \xrightarrow{\text{slow}} R-CO-CH_2-OH + N_2 + H^{\oplus}$$

In some cases this hydrolysis may also proceed by an *A–1* mechanism[507, 508]:

$$R-CO-CH=N_2 + H^{\oplus} \rightleftharpoons R-CO-CH_2-N_2^{\oplus}$$

$$R-CO-CH_2-N_2^{\oplus} \xrightarrow{\text{slow}} R-CO-\overset{\oplus}{C}H_2 + N_2$$

$$R-CO-\overset{\oplus}{C}H_2 + H_2O \xrightarrow{\text{fast}} R-CO-CH_2-OH + H^{\oplus}$$

---

[486] *U. Schöllkopf, P. Tonne, H. Schäfer, P. Markusch,* Justus Liebigs Ann. Chem. *722,* 45 (1969).
[487] *K. Bott, Angew. Chem. 76,* 992 (1964).
[488] *H. Reimlinger,* Angew. Chem. *75,* 788 (1963).
[489] *J.R. Mohrig, A. Keegstra,* J. Amer. Chem. Soc. *89,* 5492 (1967).
[490] *K. Bott,* Angew. Chem. *82,* 953 (1970).
[491] *J.P. Collman, M. Jamada,* J. Org. Chem. *28,* 3017 (1963).
[492] *K. Bott,* Tetrahedron *22,* 1251 (1966).
[493] *W. Kirmse, B. Brinkmann,* Chem. Ber. *103,* 925 (1970).
[494] *W. Jugelt, L. Berseck,* Z. Chem. *6,* 420 (1966).
[495] *J.B.F.N. Engberts, N.F. Borsch, B. Zwanenburg,* Rec. Trav. Chim. Pays-Bas *85,* 1068 (1966).
[496] *J.B.F.N. Engberts, B. Zwanenburg,* Tetrahedron Lett. 831 (1967).
[497] *W. Jugelt, D. Schmidt,* Tetrahedron Lett. 985 (1967).
[498] *H. Dahn, H. Gold, M. Ballenegger, J. Lenoir, G. Diderich, R. Malherbe,* Helv. Chim. Acta *51* (8), 2065 (1968).
[499] *W. Jugelt, D. Schmidt,* Tetrahedron *24,* 59 (1968).
[500] *K.D. Warren, J.R. Yandle,* J. Chem. Soc. 4221 (1965).
[501] *C.W. Thomas, L.L. Levenson,* Chem. & Ind. 1264 (1966).
[502] *L.L. Levenson, C.W. Thomas,* J. Chem. Soc. B 680 (1967).
[503] *S. Aziz,* J. Chem. Soc. B 1302 (1968).
[504] *S. Aziz, J.G. Tillett,* Tetrahedron Lett. 2321 (1968).
[505] *B. Zwanenburg, J.B.F.N. Engberts,* Rec. Trav. Chim. Pays-Bas *84,* 165 (1965).

In contrast to diazoalkanes, α-diazocarbonyl compounds can be protonated both at the α-carbon atom and at the oxygen atom of the carbonyl group; in the latter case, *enediols* are the final hydrolysis products[482]:

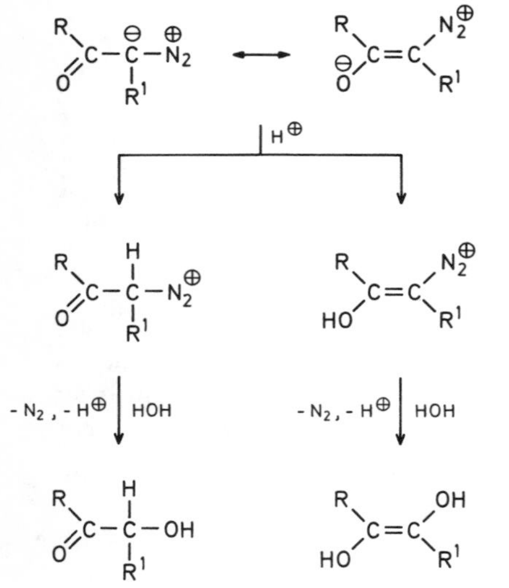

Acid hydrolysis of the aliphatic α-diazo ketones makes it possible to synthesize α-*ketols* and their derivatives in preparative yields[400].

When working with acid-resistant diazo compounds *copper* catalysts often bring on the reaction[509].

The formation of *enediols* from the hydrolysis of α-diazo ketones with acids has only been observed rarely[482]. In contrast, *trans*-fixed α-diazo-β-dicarbonyl compounds form the corresponding enediols, so-called *reductones*, relatively easily[510-512]. Dilute and concentrated *sulfuric acids* react differently with α-diazo ketones. *Ketols* are formed in the reaction with dilute acid, whereas the concentrated acid gives α-*ketol sulfates*[513]. Similar reaction products are also formed with other acids, e.g., *phosphoric acid*[514]. The reaction between

diazomethane and hydrogen halides in the *gas phase* is a convenient method for the preparation of *hot methyl halides*[441]:

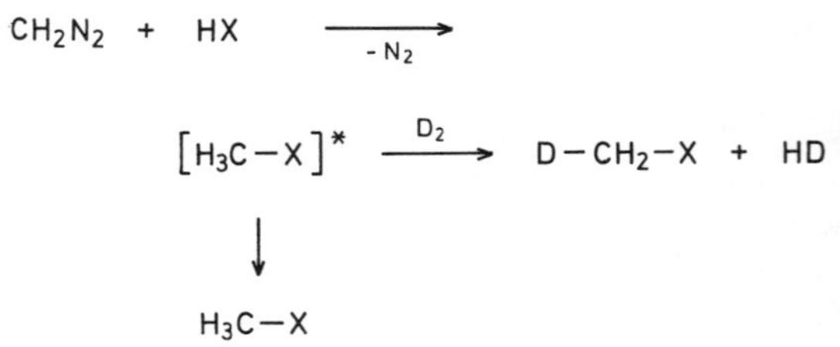

$$CH_2N_2 \;+\; HX \xrightarrow{\;-N_2\;}$$

$$[H_3C-X]^* \xrightarrow{D_2} D-CH_2-X \;+\; HD$$

$$\downarrow$$

$$H_3C-X$$

8.2.1.3 Decomposition by bases

At low temperatures aliphatic diazo compounds are quite stable toward aqueous alkalis. Signs of decomposition are observed on heating, or on treatment with *concentrated* alkalis, or with *nonaqueous* alkali solutions.

The action of concentrated alkali on α-diazoacetophenone and 2-diazo-2-phenylacetophenone leads to a complex reaction mixture[515-518]. Formation of the reaction products can be explained in terms of the simultaneous occurrence of fission, solvolysis, reduction, and condensation of the starting materials*.

The alkali-induced fission of α-diazocarbonyl and α-diazo-1,3-dicarbonyl compounds[519], of α-diazocarboxylic esters[520]. and of α-diazo-1,3-ketosulfones[521] is of practical interest. This reaction can be used for the preparation of *diazoalkanes, α-diazocarbonyl compounds*, and *α-diazosulfones* (see p. 245).

The action of *excess* alcoholic potassium hydroxide solution on 4-diazo-2,2,5,5-tetraalkyltetrahydro-3-furanones gives preparative yields of *tetraalkyl-substituted diglycolic acids*[522]:

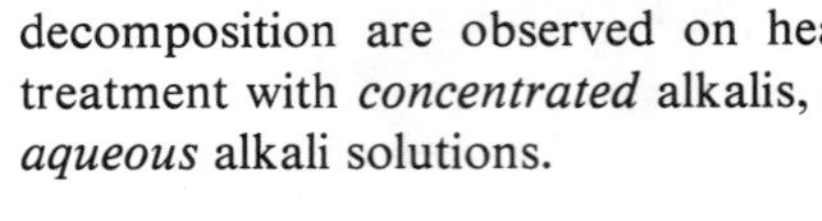

---

* Potassium *tert*-butoxide causes mainly dimerization of diazo ketones to the corresponding tetrazoles in relatively high yields[518].

[506] B. Zwanenburg, J.B.F.N. Engberts, Tetrahedron 24, 1737 (1968).
[507] L.L. Levenson, J. Chem. Soc. B 1051 (1969).
[508] H. Dahn, Chem. & Ind. 37 (1963).
[509] F. Weygand, H.J. Bestmann, Angew. Chem. 72, 535 (1960).
[510] F. Arndt, Chem. Ber. 84, 319 (1951).
[511] B. Eistert, D. Greiber, J. Caspari, Justus Liebigs Ann. Chem. 659, 64 (1962).
[512] B. Eistert, G. Heck, Justus Liebigs Ann. Chem. 681, 123 (1965).
[513] M.S. Newman, P.F. Beal, J. Amer. Chem. Soc. 72, 5161 (1950).
[514] B.A. Arbuzov, A.O. Vizel, Isv. Akad. Nauk SSSR, Otd. Khim. Nauk 749 (1963); C.A. 59, 7362g (1963).

[515] P. Yates, B.L. Shapiro, J. Amer. Chem. Soc. 81, 212 (1959).
[516] P. Yates, D.G. Farnum, Tetrahedron Lett. 22 (1960).
[517] P. Yates, D.G. Farnum, J. Amer. Chem. Soc. 85, 2967 (1963).
[518] P. Yates, C. Meresz, H. Morrison, Tetrahedron Lett. 77, 1576 (1967).
[519] M. Regitz, J. Hocker, A. Liedhegener, Organic Preparations and Procedures I, 99 (1969).
[520] V.B. Hendrickson, W.A. Wolf, J. Org. Chem. 33, 3610 (1968).
[521] D. Hodson, G. Hold, D.K. Wall, J. Chem. Soc. C 2201 (1968).

The *mechanism* of this reaction is not clear, although the occurrence of free radicals as intermediates has been demonstrated with the aid of *ESR* techniques.

In all the reactions mentioned alkali is observed to cause fission of the *C–C* bond between the carbonyl group and the carbon atom bearing the diazo group.

It is evident that the first stage of the reaction[523] in which *2-benzopyran-1,4(3H)-dione* (86%) is formed from 2-diazo-1,3-indandione is analogous in character:

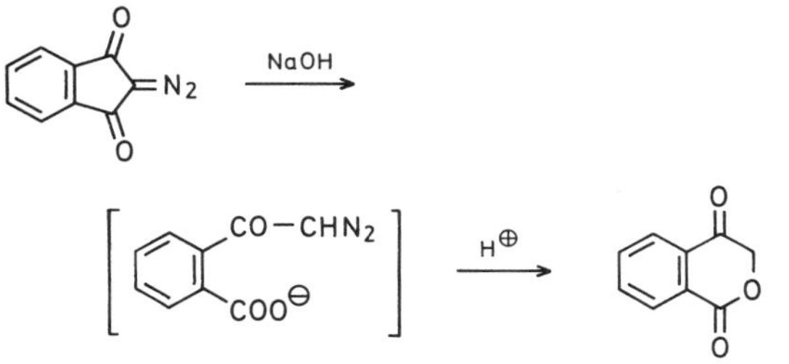

A unique variation which has been observed, and which resembles an aldollike rearrangement, is the reaction of 2-diazo-1-phenyl-1,3-butanedione with potassium hydroxide solution[524]:

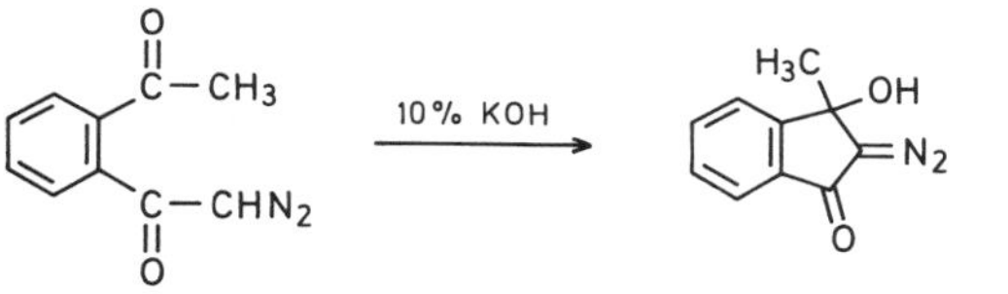

2-Diazo-3-hydroxy-3-methyl-1-indanone; 60%

8.2.2 Reactions with multiple bond systems

8.2.2.1 Reactions with *C=C* double bond systems

8.2.2.1.1 Reactions with *C=C* double bond systems leading to formation of pyrazolines

The reaction between aliphatic diazo compounds and substances containing *C–C* multiple bonds is suitable for the preparation of a very wide variety of cyclic systems such as *pyrazolines, pyrazoles, diazepines, indazoles, cyclopropanes, cyclopropenes,* and their derivatives. This reaction, which can proceed both intermolecularly and intramolecularly[400, 525–529], is often regarded as *cycloaddition* of the diazo compound or of the carbene to the multiple bond[525]. This branch of the chemistry of aliphatic diazo compounds is being investigated intensively, partly because the reactions involved are of considerable importance for preparative use, and also because the study of these reactions can solve many interesting theoretical problems. The theory of the conservation of orbital symmetry has provided new stimuli for the development of this field[530].

The addition of diazo compounds to olefins can be regarded as a 1,3-dipolar cycloaddition[525]:

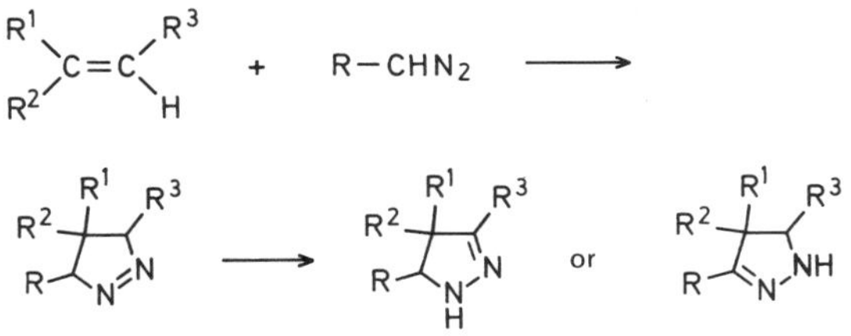

The primary reaction products are *4,5-dihydro-3H-pyrazoles (1-pyrazolines).* If these possess a hydrogen atom at the 3- or 5-positions, they usually rearrange irreversibly to *4,5-dihydro-1H-pyrazoles* (2-pyrazolines) either spontaneously or under certain reaction conditions (mild heating, addition of bases or acids)*.

The relative reactivity of diazo compounds as 1,3-dipoles in diminishing order[419, 533–537] is:

---

* In some cases *2-pyrazolines* have been isolated at low temperatures[531]. Some are also stable at room temperature[532].

[522] *L.L. Rodina, I.K. Korobizina,* Zh. Org. Khim. *4,* 2212 (1968); C.A. *70,* 68019p (1969).
[523] *G. Holt, D.K. Wall,* J. Chem. Soc. C, 857 (1966).
[524] *T.L. Burkoth,* Tetrahedron Lett. 5049 (1969).
[525] *R. Huisgen, R. Grashey, J. Sauer,* in *S. Patai,* The Chemistry of Alkenes, Chap. X, Interscience Publishers, London New York 1964.
[526] *A.N. Kost, V.V. Ershov,* Usp. Khim. *27,* 431, (1958); C.A. *52,* 16357h (1958).
[527] *J.L. Brewhaker, H. Hart,* J. Amer. Chem. Soc. *91,* 711 (1969).
[528] *R.H. Findlay, J.T. Sharf,* Chem. Commun. 909 (1970).
[529] *J.T. Sharp,* Chem. Commun. 1197 (1970).

[530] *R.B. Woodward, R. Hoffmann,* Angew. Chem. *21,* 797 (1969).
[531] *H. Kisch, O.E. Polansky, P. Schuster,* Tetrahedron Lett. 805 (1969).
[532] *C.G. Overberger,* J. Amer. Chem. Soc. *84,* 869 (1962);
C.G. Overberger, J. Amer. Chem. Soc. *86,* 658 (1964).
[533] *R. Huisgen,* Angew. Chem. *75,* 604 (1963).
[534] *B.L. Dyatkin, E.P. Mochalina,* Isv. Akad. Nauk SSSR, Ser. Khim. *7,* 1225 (1964).
[535] *H. Paul, J. Lange, A. Kausmann,* Chem. Ber. *98,* 1789 (1965).
[536] *D. Seyferth, A.W. Dow, H. Menzel, Th. C. Flood,* J. Amer. Chem. Soc. *90,* 1080 (1968).
[537] *A.G. Brook, P.F. Jones,* Can. J. Chem. *49,* 1841 (1971).

1-diazoalkanes and diazoperfluoroalkanes > diazomethane > diazodiphenylmethane > ethyl diazoacetate > 2-acetoxy-1-diazo-1-trimethylsilylethane > 1-alkoxy-1-diazoethane > α-diazo ketones > α-diazo-1,3-dicarbonyl compounds.

$C=C$ double bonds in strained rings and bicyclic systems [535, 538–542], and double bonds activated by electron-withdrawing substituent groups [543–548] are the most powerful dipolarophiles. While *nonactivated* olefins such as 2,3-dimethyl-1-butene, 2,3-dimethyl-2-butene, and 1-hexene do not react with diazomethane [549, 550], stable *cycloaddition products* are formed from diazomethane and styrene, 1,1-diphenylethylene, and 9-methylfluorene. Electron-withdrawing groups can be arranged in the following order of (increasing) *activating power* [400, 543, 549]:

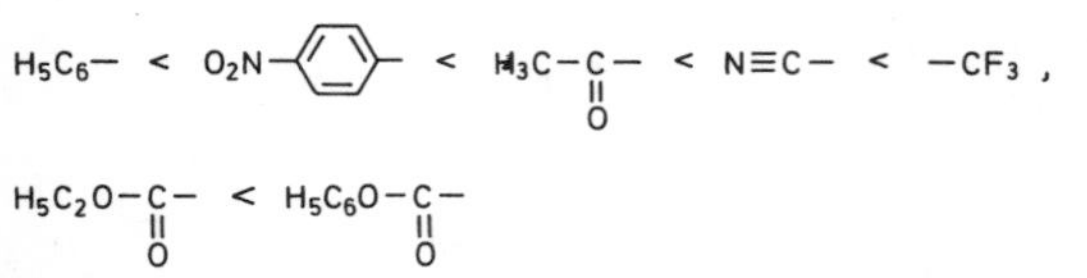

$$H_5C_6- \ < \ O_2N-\!\!\langle \ \rangle\!\!- \ < \ \underset{\overset{\|}{O}}{H_3C-C-} \ < \ N\equiv C- \ < \ -CF_3 \ ,$$

$$H_5C_2O-\underset{\overset{\|}{O}}{C-} \ < \ H_5C_6O-\underset{\overset{\|}{O}}{C-}$$

The following *reaction conditions* are generally used for the preparation of *pyrazolines*:

The mixture of diazo compound and unsaturated component is heated. An alternative technique is to keep the reaction mixture in a sealed tube for a considerable time (several days to several months)*. In some cases addition of small amounts of water or alcohol to accelerate the reaction is recommended [400].

α,β-Unsaturated carbonyl compounds, their acetals, α,β-unsaturated sulfones, nitro compounds, and nitriles [400, 551–556] form 1:1 adducts with diazoalkanes and diazocarbonyl compounds.

The structure of the pyrazoline formed is generally (but not invariably [557–559]) in accordance with the *Auwers rule*:

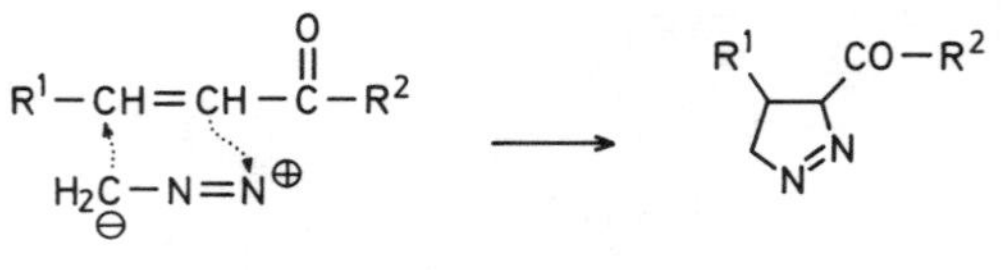

Addition of the diazoalkanes to α,β-unsaturated carbonyl compounds proceeds *stereoselectively* [560–562]. Reaction of β-chlorovinyl ketones with diazomethane is a convenient method for preparing *3-acylpyrazoles* (65–90%) [563]:

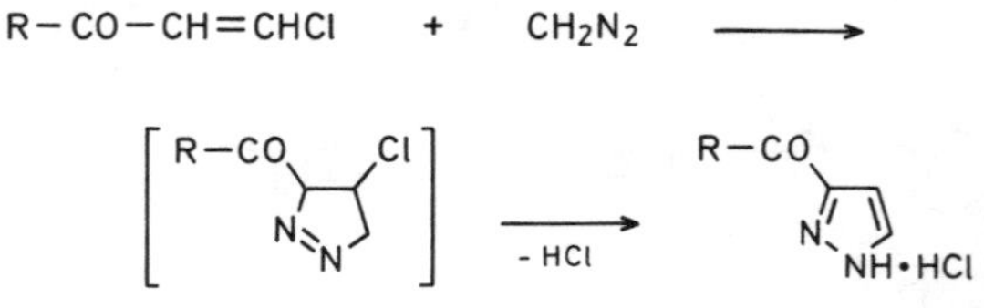

* According to the Woodward-Hoffmann rule [530], the reactions between diazoalkanes and olefins are classified as $[\pi^4 + \pi^2]$ cycloadditions; these take place, under thermal control, as symmetry-allowed $[4_s + 2_s]$ or $[4_a + 2_a]$ processes.

[538] *N. Filipescu, J.R. De Member*, Tetrahedron *24*, 5181 (1968).
[539] *M. Franck-Neumann*, Tetrahedron Lett. 2979 (1968).
[540] *P.G. Gassman, K.T. Mansfield*, J. Org. Chem. *32*, 915 (1967).
[541] *T.H Kinstle*, J. Amer. Chem. Soc. *89*, 3660 (1967).
[542] *H. Prinzbach*, Chimia *23*, 37 (1969).
[543] *J.H. Atherton, R. Fields*, J. Chem. Soc C 1507 (1968).
[544] *S. Hauptmann, K. Hirschberg*, J. Prakt. Chem. *308*, 73 (1967).
[545] *J.F. Keana, Ch.U. Kim*, J. Org. Chem. *35*, 1093 (1970).
[546] *A. Nabeya, J.A. Moore*, J. Org. Chem. *35*, 2022 (1970).
[547] *A. Nabeya, F.B. Culp, J.A. Moore*, J. Org. Chem. 2015 (1970).
[548] *Ju. A. Kotijan, P.V. Petrovskii, B.L. Djatkin, I.L. Knunjantz*, Zh. Org. Khim. *7*, 1363 (1971) (russ.).
[549] *R. Huisgen, H. Stangl, H.J. Sturm, H. Wagenhofer*, Angew. Chem. *73*, 170 (1961).
[550] *A.P. Meshcheryakov, I.E. Dolgii*, Isv. Akad. Nauk SSSR, Otdel. Khim. Nauk 931 (1960); C.A. *54*, 24436d (1960).
[551] *J. Bus, H. Steinberg, Th.J. de Boer*, Monatsh. Chem. *98*, 1155 (1967).
[552] *J. Hamelin, R. Carrie*, Bull. Soc. Chim. 2513 (1968).
[553] *R. Danion-Bougot, R. Carrie*, Bull. Soc. Chim. 313 (1969).
[554] *A. Ledwith, D. Parry*, J. Chem. Soc. C *16*, 1408 (1966).
[555] *J.M. Stewart, C. Carlisle, K. Kom, G. Lee*, J. Org. Chem. *35*, 2040 (1970).
[556] *M. Regitz, W. Anschütz, A. Liedhegener*, Chem. Ber. *101*, 3734 (1968).
[557] *W.E. Parham, C. Serres, P.R.O. Connor*, J. Amer. Chem. Soc. *80*, 588 (1958).
[558] *W.E. Parham, H.G. Braxton, P.R. O'Connor*, J. Org. Chem. *26*, 1805 (1961).
[559] *R. Hüttel, J. Riedl, H. Martin, K. Franke*, Chem. Ber. *93*, 1425 (1960).
[560] *T.V. Van Auken, K.L. Rinehart*, J. Amer. Chem. Soc. *84*, 3736 (1962).
[561] *J. Tabushi, K. Tagaki, R. Oda*, Tetrahedron Lett. 2075 (1964); *J. Tabushi, K. Tagaki, R. Oda*, Tetrahedron *213*, 2621 (1967).
[562] *D.T. Witiak, B.K. Sinha*, J. Org. Chem. *35*, 501 (1970).
[563] *A.N. Nesmeyanov, N.K. Kochetkov*, Izv. Akad. Nauk SSSR, Otdel. Khim. Nauk 686 (1951); C.A. *46*, 7565i (1952).

3,4-Dihydro-2-(methoxymethylene)-1*(2H)*-naphthalenone[564, 565] and tetrahydro-8-(methoxymethylene)-5*(4H)*-indanone react with diazomethane and diazoethane to form the corresponding *pyrazolines* (87–95%), but do not react with 2-diazopropane or α-diazotoluene:

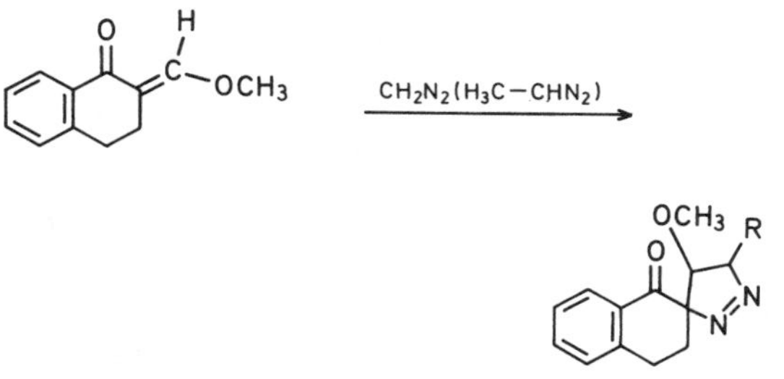

$CH_2N_2(H_3C-CHN_2)$

R = H; *3,4-Dihydro-4'-methoxyspiro[naphthalene-2(1H), 3'-[1]pyrazolin]-1-one; 92%*
R = CH₃; *3,4-Dihydro-4'-methoxy-5'-methylspiro[naphthalene-2(1H),3'-[1]pyrazolin]-1-one; 95%*

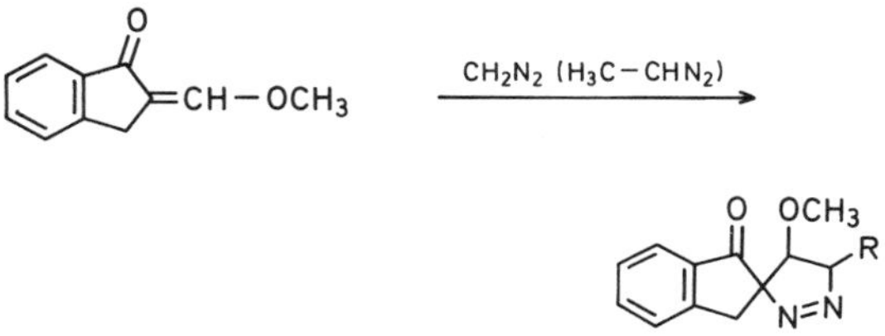

$CH_2N_2 (H_3C-CHN_2)$

R = H; *4'-Methoxyspiro[indan-2,3'-[1]pyrazolin]-1-one; 97%*
R = CH₃; *4'-Methoxy-5'-methylspiro[indan-2,3'-[1]-pyrazolin]-1-one; 87%*

A recent example of a 1,3-dipolar cycloaddition is the addition of diazomethane to 6,6-dimethylfulvene **7**[566].

The primary adducts **8** and **9** *(7,7-dimethyl-1,4-dihydro-1H-* and *5,7a-dihydro-4H-cyclopenta[c]pyridazine)* rearrange further on thermolysis. This cycloaddition is a $[\pi_s{}^6 + \pi_s{}^4]$ process, which is allowed in the ground state:

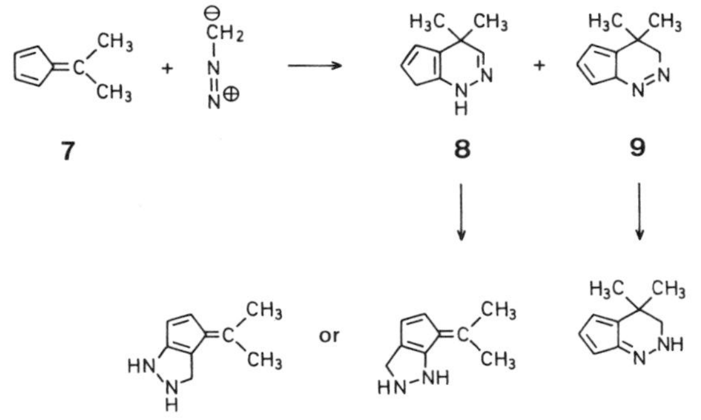

Cycloalkenes react more easily with simple diazoalkanes the greater the strain is in their ring; the products are *bicyclic pyrazolines*[535]:

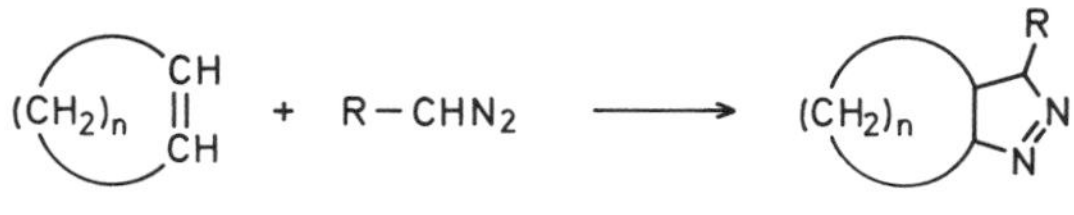

Diazomethane can add as a 1,3-dipole to both the strained double bond in trans-cyclooctene[567] and the activated cyclobutene system[539–542]:

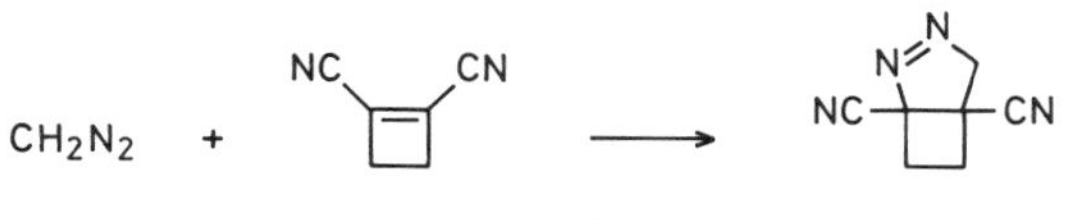

2,3-Diazabicyclo[3.2.0]-hept-2-ene-1,5-dicarbonitrile

The very reactive 2-diazopropane adds to *cis*-3,4-dichlorocyclobutene **10** to give a quantitative yield of the isomeric pyrazolines **11** and **12** in a 1:2.5 ratio[539, 568]:

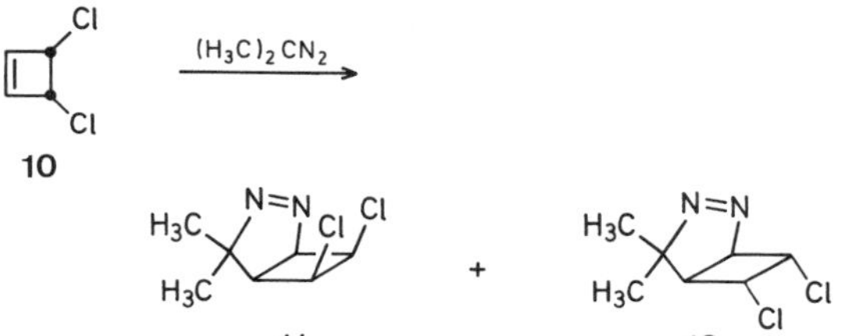

6,7-Dichloro-4,4-dimethyl-2,3-diazabicyclo[3.2.0]hept-2-ene

Diazoalkanes and α-diazocarboxylic acid esters can also add to the double bonds of unsaturated four-membered cyclic sulfones; isomeric pyrazolines are formed[569]. When diazodiphenylmethane

[566] *K.N. Houk, L.J. Luskus,* Tetrahedron Lett. 4029 (1970).
[567] *E.J. Corey, J.I. Shulman,* Tetrahedron Lett. 3655 (1968).
[568] *M. Franck-Neumann,* Angew. Chem. *81,* 189 (1969).
[569] *D.C. Dittmer, H. Glassman,* J. Org. Chem. *35,* 999 (1970).
[570] *K.B. Wiberg,* J. Amer. Chem. Soc. *82,* 6375 (1960).
[571] *J. Ciabattoni, G.A. Berchtold,* J. Org. Chem. *31,* 1336 (1966).
[572] *M. Franck-Neumann, C. Buchecker,* Tetrahedron Lett. 2659 (1969).
[573] *M.I. Komendantov,* Zh. Org. Khim. *7,* 423 (1971) (russ.).
[574] *L.A. Carpino,* J. Amer. Chem. Soc. *93,* 476 (1971).
[575] *G. Manecke, W. Hübner,* Tetrahedron Lett. 2443 (1971).
[576] *F.M. Dean, L.E. Houghton,* J. Chem. Soc. C 2060 (1968).
[577] *H. Brockmann, A. Zeeck,* Chem. Ber. *100,* 2885 (1967).

[564] *D. Nasipuri, K.K. Biswas,* J. Indian. Chem. Soc. *44,* 620 (1967);
D. Nasipuri, K.K. Biswas, Tetrahedron Lett. 2963 (1966).
[565] *D. Nasipuri, K.K. Biswas,* J. Indian. Chem. Soc. *45,* 146 (1968).

and (2-diazoethyl)benzene are used, these give off nitrogen on heating and on *UV*-irradiation to form 2-thiabicyclo[2.1.0]pentane 2, 2-dioxides. Bicyclic pyrazolines derived from α-diazotoluene and 4-di-azoanisole lose sulfur dioxide to form the corresponding *pyrazoles*:

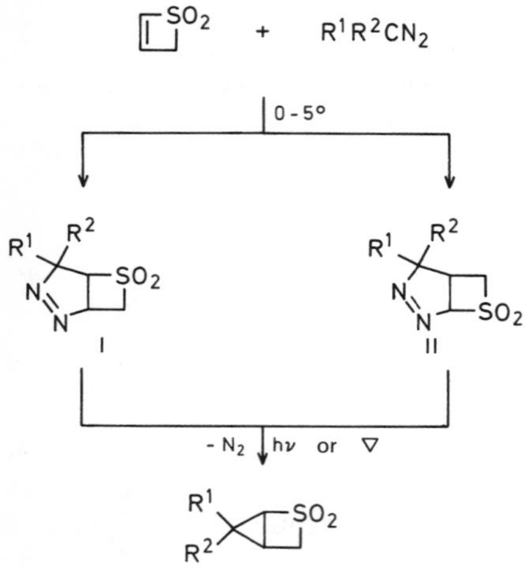

R¹ = R² = C₆H₅;   4,4-Diphenyl- ⎫ 2-thia-bicyclo[2.1.0]-
R¹ = CH₃; R² = C₆H₅; 4-Methyl-   ⎬ pentane-2,2-dioxide
                       4-Phenyl-   ⎭

I; R¹ = R² = C₆H₅;   4,4-Diphenyl- ⎫ 6-thia-2,3-diaza-
   R¹ = C₆H₅; R² = CH₃; 4-Methyl-   ⎬ bicyclo[3.2.0]hept-
                          4-phenyl-   ⎭ 2-ene 6,6-dioxide

II; R¹ = R² = C₆H₅;   4,4-Diphenyl- ⎫ 7-thia-2,3-diaza-
    R¹ = C₆H₅; R² = CH₃; 4-Methyl-   ⎬ bicyclo[3.2.0]hept-
                          4-phenyl-   ⎭ 2-ene 7,7-dioxide

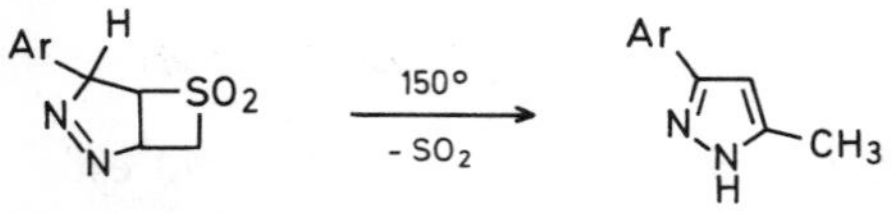

Ar = C₆H₅; 3-Methyl-5-phenylpyrazole

Ar = 4–CH₃–O–C₆H₄; 3-(4-Methoxyphenyl)-5-methylpyrazole

The reaction of diazoalkanes with *cyclopropene* derivatives[542, 570–573] and episulfones[574] also leads to *1,1a,4,4a-tetrahydrocyclopropa[c]pyrazoles*:

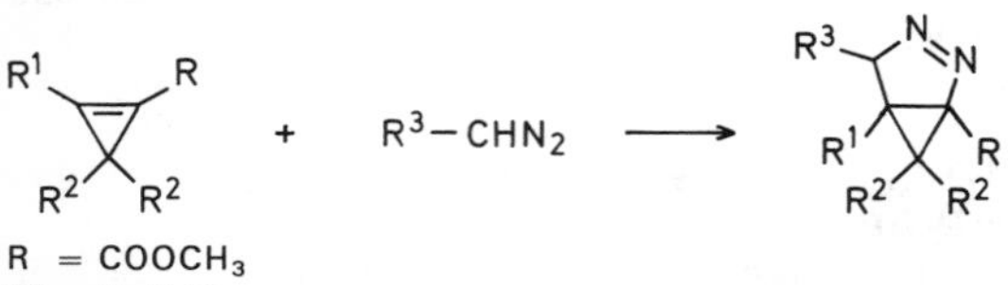

R = COOCH₃
R¹ = H, C₆H₅
R² = CH₃, C₆H₅
R³ = H, CH₃

During reaction of diazoalkanes with *p-quinones* the carbonyl group generally does not participate

(unless it is highly activated); instead, addition to the double bond produces the corresponding *pyrazolines* and *bipyrazolines*[409, 575–579].

The reaction between *tetracyclones*[580] or *inda-cyclones*[581] and diazoalkanes proceeds analogously.

Because of their strongly unsaturated character, *acecyclones* can add two molecules diazomethane or *α-diazotoluene*[582]:

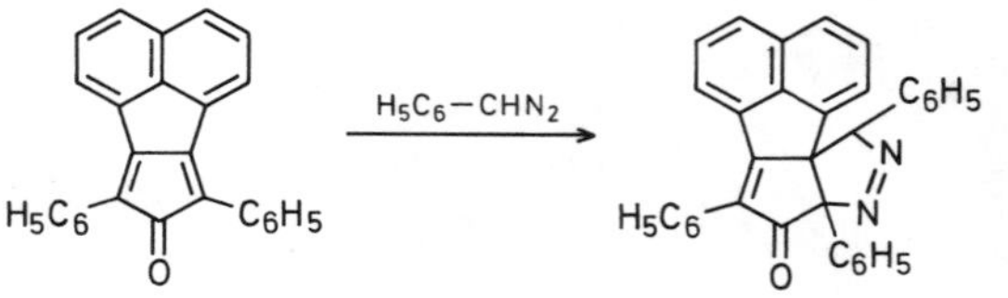

3a,4-Dihydro-1,3a,5-
triphenyl-1H-naphtho-
[1',8':4,5,6]pentaleno-
[1,6-c]pyrazol-4-one

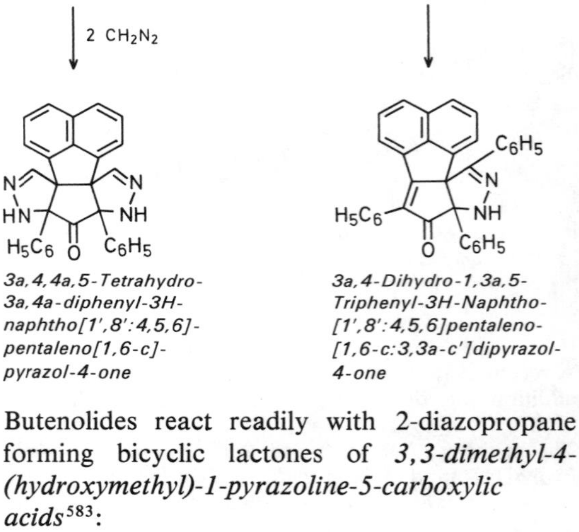

3a,4,4a,5-Tetrahydro-   3a,4-Dihydro-1,3a,5-
3a,4a-diphenyl-3H-   Triphenyl-3H-Naphtho-
naphtho[1',8':4,5,6]-   [1',8':4,5,6]pentaleno-
pentaleno[1,6-c]-   [1,6-c:3,3a-c']dipyrazol-
pyrazol-4-one   4-one

Butenolides react readily with 2-diazopropane forming bicyclic lactones of *3,3-dimethyl-4-(hydroxymethyl)-1-pyrazoline-5-carboxylic acids*[583]:

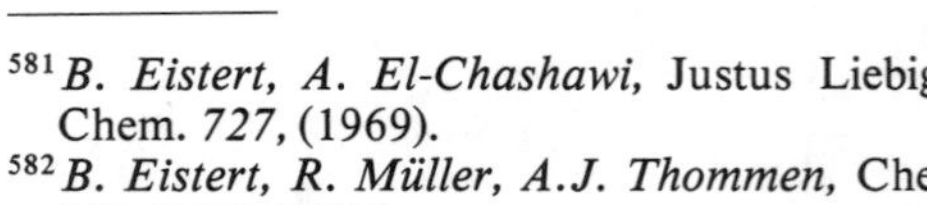

Allyldiazo compounds often cyclize spontaneously at room temperature in a unimolecular reaction and quantitative yields of *pyrazoles* are formed[400, 527], *e.g.*:

[578] H. Brockmann, H. Greve, A. Zeeck, Tedrahedron Lett. 1929 (1971).

[579] B. Eistert, H. Fink, J. Riedinger, H.-G. Hahn, H. Dürr, Chem. Ber. 102, 3111 (1969).

[580] B. Eistert, A. Langbein, Justus Liebigs Ann. Chem. 678, 78 (1964).

[581] B. Eistert, A. El-Chashawi, Justus Liebigs Ann. Chem. 727, (1969).

[582] B. Eistert, R. Müller, A.J. Thommen, Chem. Ber. 101, 3138 (1968).

[583] M. Franck-Neumann, Angew. Chem. 80, 42 (1968).

[584] C.G. Overberger, R.E. Zangaro, R.E.K. Winter, J.P. Anselme, J. Org. Chem. 36, 975 (1971).

[585] I.A. D'yakonov, J. Gen. Chem. (USSR) 15, 473 (1945); C.A. 40, 4718 (1946).

[586] S.D. Andrews, A.C. Day, Chem. Commun. 17, 902, (1967).

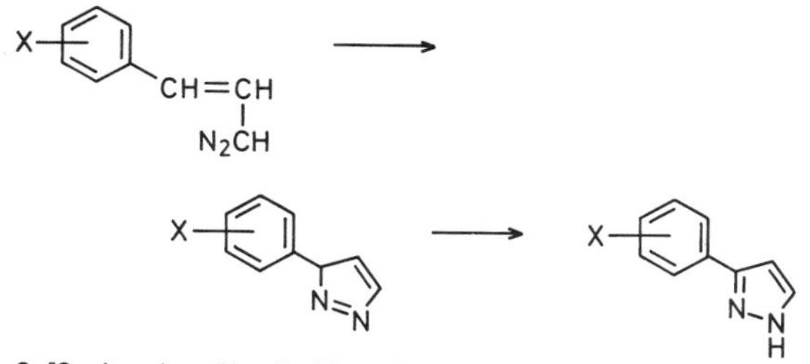

$$H_5C_2-AlCl_2 + CH_2N_2 \longrightarrow H_5C_2-Al(CH_2Cl)Cl + N_2$$

3-[3-nitrophenyl(or 4-chlorophenyl or 4-tolyl)]-1H-(or 3H-)pyrazole

The corresponding tosylhydrazones can be used as starting materials for this interamolecular cyclization, and the reaction may be performed in such a way that the diazo compound reacts further without isolation [528, 529].

Aliphatic and alicyclic *dienes* can likewise react with diazoalkanes to form pyrazolines. If an excess of the diazo compound is used, both double bonds in the diene react [400, 538, 584].

Pyrazolines are formed in good yield from *allene* and diazomethane [585, 586], and from alkoxyallene and diazomethane or diazophenylmethane. In the last case the reaction takes several days and is carried out in the absence of light [587]. Cyanoallene can react with diazomethane and with ethyl diazoacetate, forming the corresponding *cyanopyrazoles* [400]. Aliphatic, aliphatic-aromatic, and heterocyclic α-diazo ketones also react readily with equimolar amounts of cyanoallene, whereas 2-diazo-2-phenylacetophenone and 3-diazo-2,4-pentanedione cannot react with cyanoallene.

8.2.2.1.2 Reactions with C=C double bond systems producing cyclopropanes

When the reaction of diazo compounds with olefins is carried out photochemically or in the presence of catalysts, nitrogen is eliminated and the corresponding cyclopropanes are formed:

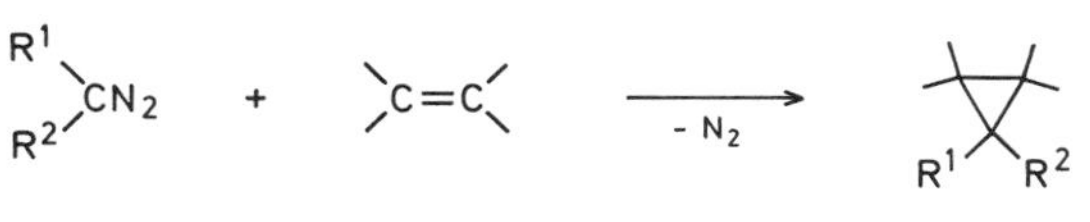

The *mechanism* of the reaction depends on the reaction conditions. There is evidence for the intermediate formation of organometallic compounds if halides of zinc or alkylaluminum are used as catalysts [400, 441, 588]. For example, the reaction between cyclohexene and diazomethane (in pentane at 50°) to form *norcarane (bicyclo[4.1.0]heptane)* proceeds in the presence of ethylaluminum chloride as shown below [588]:

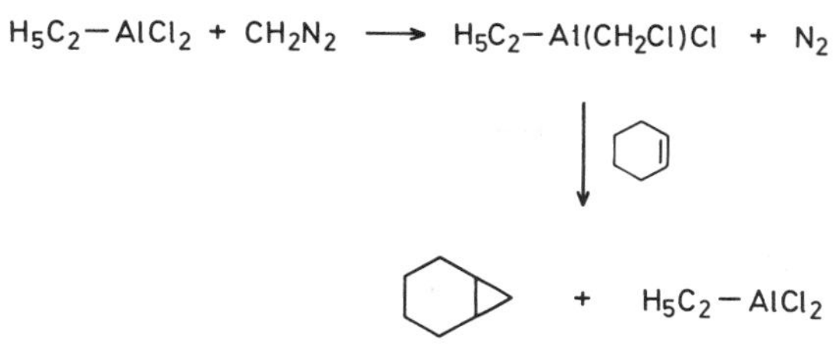

It is assumed that a carbenoid mechanism operates when the reaction with unsaturated compounds is carried out photolytically or in the presence of copper or copper salts. Evidently, the rate-determining step is the formation of the carbene, whereas the stereochemistry of the carbene addition [433, 434, 437, 589–592] depends upon the character (*i.e.*, singlet or triplet) of the carbene formed. The *reactivity* of the carbene depends upon the method used for decomposition of the diazo compound; the most reactive carbenes are those which are produced photolytically [593, 594]. The carbenes which are formed by catalytic decomposition form complexes with the catalysts, and are consequently more *selective* with a lower tendency to involvement in side-reactions [433, 434].

In these reactions, the carbene is the electrophilic component, and the unsaturated compound is the nucleophile [401, 433].

The electrophilic nature of the carbene has also been evidenced by the results of many competitive reactions with olefins of differing structures: the yields of cyclopropanes are largest when olefins are used which contain electron-donating substituent groups [433, 437, 595].

Diazomethane reacts the most easily with C=C double bonds [433].

Diazo compounds which contain O, P, S, Se, and Si likewise react with unsaturated materials, forming cyclopropanes (in 40–70% yield) [438, 596–598]. For example, diazo-(4-methoxybenzenesulfonyl)-methane reacts with isobutene to form *cyclopropyl 4-methoxyphenyl sulfone* [596]:

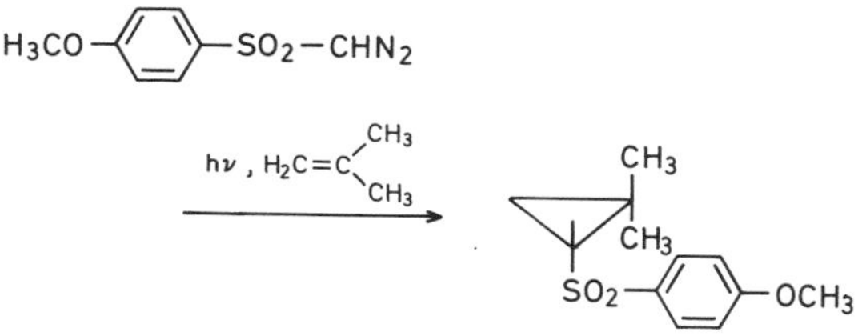

[587] P. Battioni, M. Yon Vo-Quang, Compt. Rend. C **266**, 1310 (1968).

[588] U. Burger, R. Huisgen, Tetrahedron Lett. 3057 (1970).

[589] M. Jones, Jr. A.M. Harrison, K.B. Rettig, J. Amer. Chem. Soc. **91**, 7462 (1969).

[590] M. Takebayashi, T. Ibata, H. Kohara, K. Ueda, Bull. Chem. Soc. Japan **42**, 2938 (1969).

[591] S.H. Goh, J. Chem. Soc. C 2275 (1971).

[592] J.E. McMurry, Th.E. Gloss, Tetrahedron Lett. 2575 (1971).

[593] P.S. Skell, R.M. Etter, Proc. Chem. Soc. 443 (1961).

[594] F.D. Mango, J. Dvoretzky, J. Amer. Chem. Soc. **88**, 1654 (1966).

[595] J. Novak, J. Ratusky, V. Sneberk, F. Sorm, Collect. Czech. Chem. Commun. **22**, 1836 (1957).

The addition of carbene *(R—C—R¹)* to an unsymmetrically substituted olefin can lead to the formation of two isomeric cyclopropanes, the *syn-* and *anti-*forms. The *ratio* of the stereoisomeric cyclopropane derivatives depends upon the nature of the carbene formed as intermediate[433, 437], and upon the substituent groups at the double bond. Thus, the reaction of ethyl diazoacetate (in the presence of copper) with allyl alcohol, allyl chloride, 3-buten-2-one and 4-penten-2-one[599-602], with styrene[603, 604], or with 1-alkenes[593] leads mainly to the *anti*-isomer of the cyclopropane derivative; the yield of this isomer increases with increasing size of the substituent groups at the double bond[437].

The *1 + 2 cycloaddition* to *C=C* double bonds of carbenes generated from diazo compounds has become the most important method for the synthesis of *cyclopropane derivatives,* and has made possible the preparation of some very interesting cyclic and *polycyclic* systems. For example, ethoxycarbonylcarbene (generated by heterogeneous catalytic decomposition of ethyl diazoacetate) has been successfully added to dialkyl esters of dialkyl- and diarylcyclopropenedicarboxylic acids, yielding *dialkyl bicyclobutane-2,4-dicarboxylates*[605-608]:

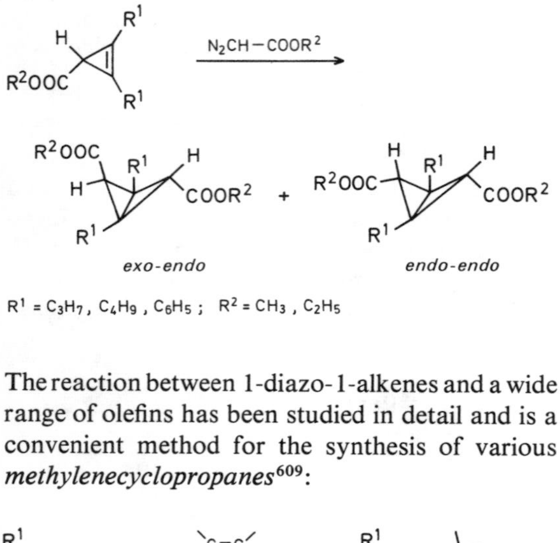

exo-endo   endo-endo

$R^1 = C_3H_7, C_4H_9, C_6H_5$; $R^2 = CH_3, C_2H_5$

The reaction between 1-diazo-1-alkenes and a wide range of olefins has been studied in detail and is a convenient method for the synthesis of various *methylenecyclopropanes*[609]:

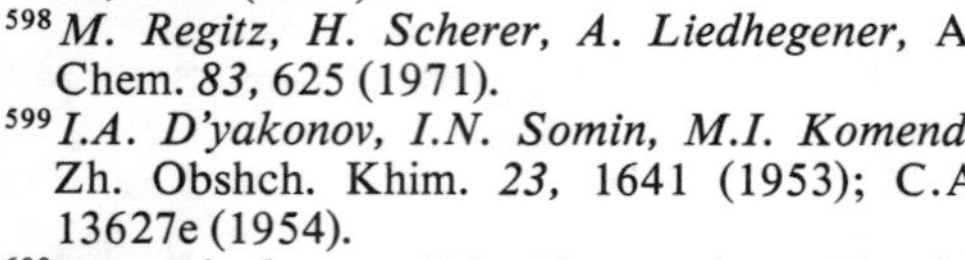

The many *intramolecular* addition reactions which have been observed with numerous unsaturated diazo compounds[610-615] are of particular importance.

For example, the cyclopropane ring can be formed successfully by intramolecular addition of the *C* atom of a carbene (formed by photolysis of the alkali metal salt of a tosylhydrazone of an α,β-unsaturated carbonyl compound) to a double bond in the same molecule[610, 612]:

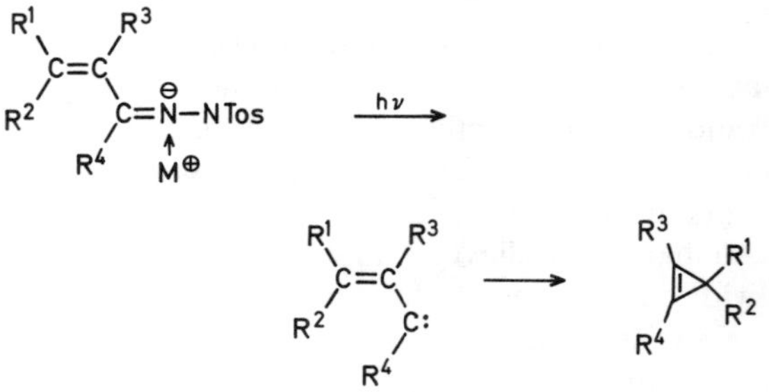

Photolysis of 4-diazo-1-butene (in heptane at −78°) leads to a mixture of *butadiene* and *bicyclobutane* (5:1)[616]. Intramolecular additions have not yet been successful where the diazo group and the *C=C* double bond are widely separated.

[596] *A.M. van Leusen, R.J. Mulder, J. Strating,* Tetrahedron Lett. 543 (1964).
[597] *D. Seyferth, R.S. Mormor, P. Hilbert,* J. Org. Chem. *36,* 1379 (1971).
[598] *M. Regitz, H. Scherer, A. Liedhegener,* Angew. Chem. *83,* 625 (1971).
[599] *I.A. D'yakonov, I.N. Somin, M.I. Komendantov,* Zh. Obshch. Khim. *23,* 1641 (1953); C.A. *48,* 13627e (1954).
[600] *I.A. D'yakonov, N.B. Vinogradova,* Zh. Obshch. Khim. *22,* 1349 (1952); C.A. *47,* 4293c (1953).
[601] *I.A. D'yakonov, N.D. Pirogova,* Zh. Obshch. Khim. *21,* 1971 (1951); C.A. *46,* 6590i (1952).
[602] *I.A. D'yakonov, O.V. Guseva,* Zh. Obshch. Khim. *22,* 1355, (1952); C.A. *47,* 4293e (1953).
[603] *A. Burger, W.L. Yost,* J. Amer. Chem. Soc. *70,* 2198 (1948).
[604] *C.H. de Puy, G.M. Dappen, K.L. Eilers, R.A. Klein,* J. Org. Chem. *29,* 2813 (1964).
[605] *I.A. D'yakonov, M.I. Komendantov, V.V. Razin,* Zh. Obshch. Khim. *33,* (7), 2420 (1963); C.A. *59,* 13839e (1963).
[606] *I.A. D'yakonov, V.V. Razin, M.I. Komendantov,* Tetrahedron Lett. 1135, 1127 (1966).
[607] *I.A. D'yakonov, V.V. Razin, M.I. Komendantov,* Dokl. Akad. Nauk SSSR, *177,* 354 (1967); C.A. *68,* 49163y (1968).
[608] *V.V. Razin, M.I. Mostova, I.A. D'yakonov,* Zh. Org. Khim. *4* (3), 535 (1968); C.A. *68,* 104582m (1968).
I.A. D'yakonov, V.V. Razin, M.I. Komendantov, Zh. Org. Khim. 1969, *5,* (2), 368; C.A. *70,* 10667w.
[609] *M.S. Newman, T.B. Patrick,* J. Amer. Chem. Soc. *91,* 6461 (1969).
[610] *H. Dürr,* Chem. Ber. *103,* 369 (1970).
[611] *H. Dürr,* Justus Liebgis Ann. Chem. *723,* 102 (1969).
[612] *H. Tsuruta,* Tetrahedron Lett. 3775 (1967).
[613] *O. House, C.J. Blankley,* J. Org. Chem. *33,* 53 (1968).
[614] *V. Ion, M. Popovici, C.D. Nenitzecu,* Tetrahedron Lett. 3383 (1965).
[615] *T. Severin, H. Krämer, P. Adhikary,* Chem. Ber. *104,* 972 (1971).
[616] *D.M. Lemal, F. Menger, C.W. Clark,* J. Amer. Chem. Soc. *85,* 2529 (1963).

The extension of this intramolecular addition reaction to cycloalkenylcarbenes generated from the corresponding cycloalkenyldiazoalkanes has made many *tricyclic* systems available[433].

Best results with intramolecular cycloadditions are often obtained by using copper or a copper salt as *catalyst*[617-619].

Copper catalysts *modify* carbenes so much that both inter- and intramolecular addition reactions can take place where the carbene sources are α-diazo ketones which have a pronounced tendency to undergo Wolff rearrangement[400, 433]. Intramolecular additions occur more smoothly with diazo ketones (in which hydride shifts cannot occur) than with diazoalkanes.

The *yields* of the intramolecular addition products increase as the distance between the diazo group and the double bond decreases[620]:

$$R-CH=CH-(CH_2)_n-CO-CHN_2$$

n = 2; R = H;	Bicyclo[3.1.0]hexan-2-one; 59%	
R = C_6H_5;	6-Phenylbicyclo[3.1.0]hexan-2-one; 59%	
n = 3; R = H;	Bicyclo[4.1.0]heptan-2-one; 37%	
R = C_6H_5;	7-Phenylbicyclo[4.1.0]heptan-2-one; 30%	
n = 4; R = H;	Bicyclo[5.1.0]octan-2-one; 3%	
R = C_6H_5;	traces	

Interesting results are obtained with the intramolecular additions of diazocycloalkenyl ketones; these reactions provide a simple method for the synthesis of *tricyclic ketones* which cannot be prepared in other ways[433, 620-624].

Such reactions can be carried out either photochemically[625] or in the presence of Lewis acids[623]. For example, irradiation of diazomethyl 2,2,3-trimethyl-3-cyclopenten-1-yl ketone in heptane gives *6,7,7-trimethyltricyclo[3.2.1.0³,⁶]octan-4-one* (79%)[625]:

Intramolecular cyclization has also been noted with unsaturated bis-diazo ketones[626].

Dienes having *isolated* double bonds readily undergo *double* addition, even when the olefinic component is in excess. The double bond which has the largest number of adjacent alkyl or other electron-donating substituent groups reacts first[437, 627].

Spiro compounds are formed by irradiation of solutions of some diazocyclopentadienes in norbornadiene; only one double bond reacts[628].

The reaction of norbornadiene with an excess of methyl diazoacetate forms isomeric diadducts *(tetracyclo[3.3.1.0²,⁴0⁶,⁸]nonane derivatives)* in good yields[629].

As a rule, only one of the two double bonds reacts in the case of conjugated dienes[437, 630]. However, some examples are known of the formation of 1,4-addition products[631, 633].

It should be emphasized again that *conjugated* C=C double bonds and C=C double bonds in strained rings exhibit the highest reactivity in the cyclopropane-formation reaction. Thus, the carbenes generated by copper-catalyzed decomposition of ethyl or butyl diazoacetate react readily with cyclopentadiene, forming *mono-* and *di-adducts* in yields of 40% and 20% respectively[634].

[617] W. Kirmse, H. Dietrich, Chem. Ber. 98, 4027 (1965).

[618] O. Tsuge, I. Shinkai, M. Koga, J. Org. Chem. 36, 745 (1971).

[619] P.M. McCurry jr., Tetrahedron Lett. 1845 (1971).

[620] M. Fawsi, D. Gutsche, J. Org. Chem. 31, 1390 (1966).

[621] W. von E. Doering, E.T. Fossel, R.L. Kaye, Tetrahedron 21, 25 (1965).

[622] M. Mongrain, J. Lafontaine, A. Belanger, P. Deslongchamps, Can. J. Chem. 48, 3273 (1970).

[623] W.F. Erman, L.C. Stone, J. Amer. Chem. Soc. 93, 2821 (1971).

[624] A. Costantino, G. Linstrumelli, S. Julia, Bull. Soc. Chim. France 907 (1970).

[625] P. Yates, A.G. Fallis, Tetrahedron Lett. 2493 (1968).

[626] S. Bien, D. Ovadia, J. Org. Chem. 35, 1028 (1970).

[627] S.D. Koch, D.V. Lopiekes, R.M. Kliss, R.J. Wineman, J. Org. Chem. 26, 3122 (1961).

[628] H. Dürr, G. Scheppers, L. Schrader, Chem. Commun. 257 (1969).

[629] S.C. Clarke, B.L. Johnson, Tetrahedron 24, 5067 (1968).

[630] V.A. Kalinina, Ju.I. Kheruse, A.A. Petrov, Zh. Org. Khim. 3, 637 (1967); C.A. 67, 43527c (1967); V.A. Kalinina, Ju.I. Kheruse, A.A. Petrov, Zh. Org. Khim. 5, 1903 (1969); C.A. 72, 54830a (1970); V.A. Kalinina, Ju.I. Kheruse, Zh. Org. Khim. 4, 963 (1968); C.A. 69, 4353f (1968); V.A. Kalinina, Ju.I. Kheruse, Zh. Org. Khim. 4, 1347 (1968); C.A. 69, 86427d (1968).

[631] V. Franzen, Chem. Ber. 95, 571 (1962).

[632] I.A. D'yakonov, T.V. Domareva-Mandel'shtam, K.K. Preobrazhenskii, Zh. Org. Khim. 5, 1114, (1969); C.A. 71, 70368r (1969).

As already mentioned (see p. 266), *pyrazolines* are generally the first products formed from the reaction of α,β-unsaturated carbonyl compounds with diazoalkanes. However, the formation of *cyclopropanes* even under very mild reaction conditions[400, 635–638] is known; for example, 1,3-dioxoindanylidenemalononitrile reacts with diazodiphenylmethane to form *1',3'-dioxo-3,3-diphenylspiro[cyclopropane-1,2'-indan]2,2-dicarbonitrile* (67%)[639]:

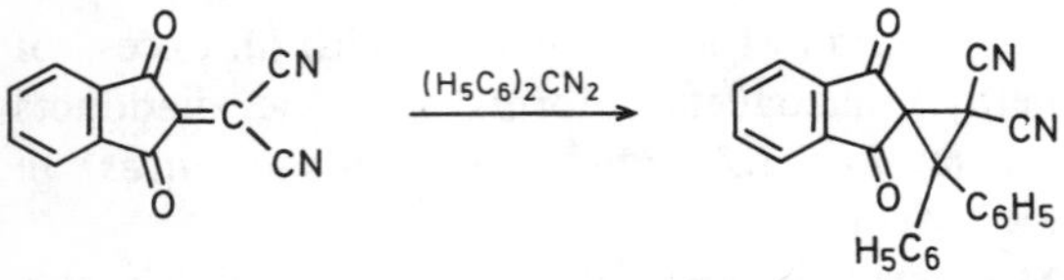

The reaction of diazo compounds with *allenes* can lead both to *cyclopropane derivatives* and to *pyrazolines*[640, 641].

Reacting diazoalkanes or diazocarbonyl compounds with *ketenes* and their derivatives is of considerable importance for synthesis.

The simplest ketenes can react with diazomethane even at low temperatures ($-78°$) forming *cyclopropanones;* if an excess of diazomethane is present, *cyclobutanones* are formed[642–644]:

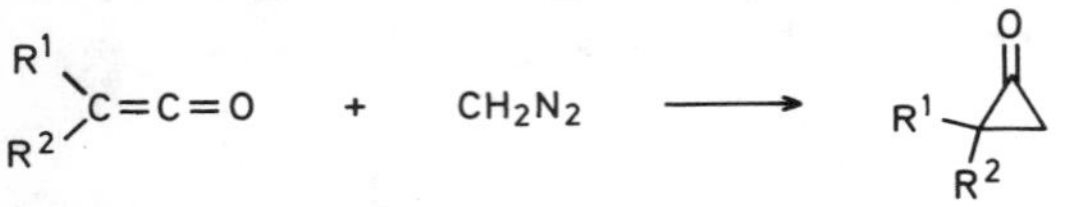

$R^1 = R^2 = H$;   *Cyclopropanone; 90%*

$R^1 = R^2 = CH_3$; *2,2-Dimethylcyclopropanone; 93%*

Esters of α-diazo carboxylic acids do not furnish uniform products with ketenes[400].

3-Butenoic acid γ-lactones (γ-but-3-enolides) are formed fairly easily by reaction of ketenes with α-diazo ketones which contain aliphatic, aromatic, or heterocyclic substituent groups[400]. Alkyldiazo ketones are especially reactive. *Ketenes,* which are formed as intermediates by the decomposition of diazocarbonyl compounds, can participate in this reaction[400]. For example, thermolysis of 2-diazoacetophenones yields 4-hydroxy-*2,4-diphenyl-3-butenoic acid γ-lactone*[645]:

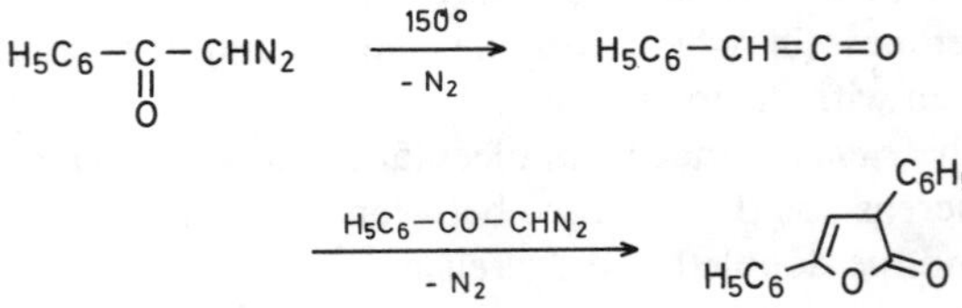

The examples mentioned demonstrate that the main course of the photolytic or copper-catalyzed reactions of diazo compounds with olefins is addition to $C=C$ bonds to form cyclopropane derivatives. This reaction does not always proceed smoothly, however. *Pyrazolines* can be formed as *by-products* by 1,3-dipolar cycloaddition (see p. 258).

When carbenes are formed as intermediates, insertion of the carbene into C–C, C–H, C–O and other bonds[400, 617, 646–648] may be observed, as may interaction with the solvent[400, 649] (see p. 279), hydride or alkyl migration (see p. 255), or Wolff rearrangement (in the case of keto carbenes (see p. 284). Oxidation of the carbene formed has been established, especially with sensitized photolysis; this is typical of triplet-carbenes[433, 650].

8.2.2.2 Reactions with aromatic and heteroaromatic compounds

Reactions of diazo compounds with aromatic compounds open up the possibility of synthesizing *nonbenzenoid* aromatic systems such as *azu-*

[633] *J.A. Berson, E.S. Hand,* J. Amer. Chem. Soc. *86,* 1978 (1964).

[634] *B. Föhlisch,* Chem. Ber. *97,* 88 (1964).

[635] *G. Swoboda, A. Eitel, J. Swoboda, F. Wessely,* Monatsh. Chem. *95,* 1355 (1964).

[636] *G. Billek, O. Saiko, A. Stephen, F. Wessely,* Monatsh. Chem. *97,* 633 (1966).

[637] *F. Nierlich, P. Schuster, O.E. Polansky,* Monatsh. Chem. *102,* 428 (1971).

[638] *H. Kisch, F. Mark, O.E. Polansky,* Monatsh. Chem. *102,* 448 (1971).

[639] *A. Schönberg, E. Singer,* Chem. Ber. *103,* 3871 (1970).

[640] *A.T. Blomquist, D.J. Connolly,* Chem. & Ind. (London) 310 (1962).

[641] *I.A. D'yakonov,* Zh. Obshch. Khim. *15,* 473 (1945); C.A. *40,* 4718 (4) (1946).

[642] *W. Hammond, N. Turro,* J. Amer. Chem. Soc. *88,* 2880 (1966);
N. Turro, W. Hammond, J. Amer. Chem. Soc. *88,* 3673 (1966).

[643] *S.E. Schaafsma, H. Steinberg, Th.J.De Boer,* Rec. Trav. Chim. Pays-Bas *86,* 651 (1967).

[644] *N. Turro, R. Gagosian,* J. Amer. Chem. Soc. *92,* 2036 (1970).

[645] *R. Huisgen, H. König, G. Binsch, H.J. Sturm,* Angew. Chem. *73,* 368 (1961).

[646] *G. Cauquis, G. Reverdy,* Tetrahedron Lett. 1085 (1968).

[647] *M. Kapps, W. Kirmse,* Angew. Chem. *81,* 86 (1969).

[648] *M. Regitz, H. Scherer, W. Anschütz,* Tetrahedron Lett. 753 (1970).

[649] *U. Simon, O. Süs, L. Horner,* Justus Liebigs Ann. Chem. *697,* 17 (1966).

[650] *J. Moritani, Y. Yamomoto, Shun-Ichi Murahashi,* Tetrahedron Lett. 5697 (1968).

lenes, tropones, tropolones, tropylium salts, and their *homologs*[433, 651].

Ring-enlargement is observed in the addition of diazoalkanes to benzene and its homologs. If the reaction is carried out photolytically or in the presence of copper catalysts, the formation of carbenes as intermediates is usually assumed. *Cycloheptatriene* is formed in the reaction between diazomethane and benzene[652]:

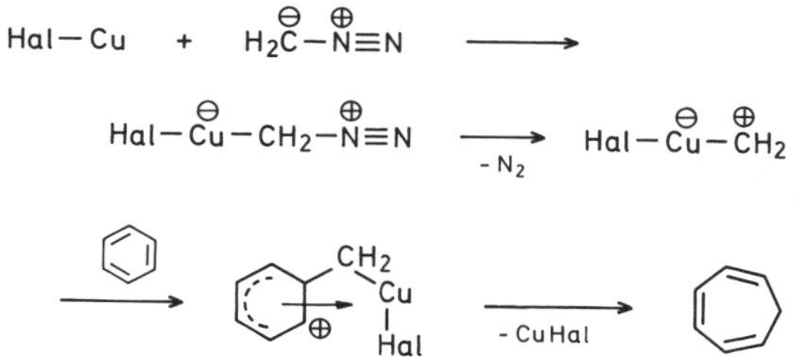

At the same time, the intermediate formation in some cases of *norcaradiene* derivatives, which then isomerize to cycloheptatrienes, cannot be excluded[651, 653–655], *e.g.:*

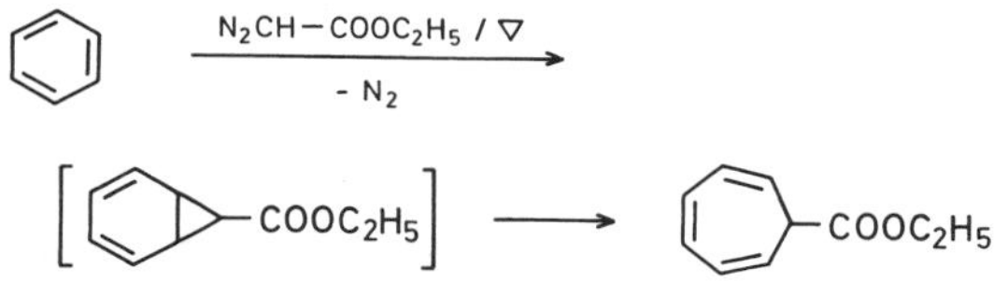

Ethyl 7-cycloheptatriene-carboxylate

Other aromatic compounds besides benzene have been reacted with alkoxycarbonylcarbenes. Many authors have assigned norcaradiene-type structures to the reaction products. If, however, one bears in mind the ease with which norcaradienes containing no stabilizing substituent groups undergo the Cope rearrangement, it would seem more correct to assume cycloheptatriene-type structures[433, 651, 656].

The best yields (50–85%) of cycloheptatrienes are obtained from the reactions of diazo compounds with aromatic compounds by using cuprous salts as catalysts[651, 652, 657, 658].

[651] *T.V. Domareva-Mandel'shtam, I.A. D'yakonov,* Sovrem. Probl. Org. Khim. 151 (1969), (russ.) Izd. Leningrad. Univ., Leningrad USSR; C.A. *73,* 143 19e (1970).

[652] *E. Müller,* Justus Liebigs Ann. Chem. *661,* 38 (1963).

[653] *E. Ciganek,* J. Amer. Chem. Soc. *87,* 652 (1965); *E. Ciganek,* J. Amer. Chem. Soc. *89,* 1454 (1967).

[654] *E. Vogel, W. Wiedemann, H. Kiefer, W.F. Harrison,* Tetrahedron Lett. 673 (1963).

[655] *H. Dürr, H. Kober,* Angew. Chem. *83,* 362 (1971).

[656] *G.G. Linstrumelle,* Tetrahedron Lett. 85 (1970).

Carbenes generated from diazo compounds react with aromatic *CH* bonds less readily than with *C=C* double bonds. Since methylene reacts with the *C=C* bond only 3.5 times faster than with the *CH* bond on photolysis (or thermolysis)[659], *toluene* is formed as well as *cycloheptatriene* on reaction with benzene. Photolytic or catalytic decompositions of diazomethane in halobenzenes or phenolic ethers give a somewhat similar picture[651].

Benzocycloheptatriene derivatives are formed by photolysis of 1,4- or 1,2,3,4-substituted 5-diazo-cyclopentadienes in benzene. 5-Diazocyclopentadiene itself, which is unsubstituted at the 4-position, gives *3aH-cyclopentacyclooctene* in 70% yield under the same conditions[660]:

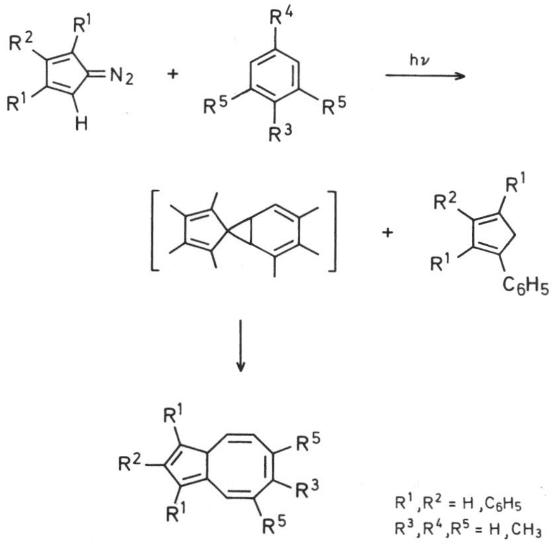

The sole reaction product from thermolysis or photolysis of diazomalononitrile in benzene is *7,7-norcaradienedinitrile* (82%)[653].

If this reaction is carried out in toluene, the three isomeric *4-tolylmalononitriles* are formed as well as the three norcaradiene derivatives[661]. α-Diazo ketones do not react with benzene[595]. If the diazo ketone group is present in the same molecule as an aromatic nucleus, however, *intramolecular* cyclization may occur, accompanied by a *CH* insertion reaction[662, 663], *e.g.:*

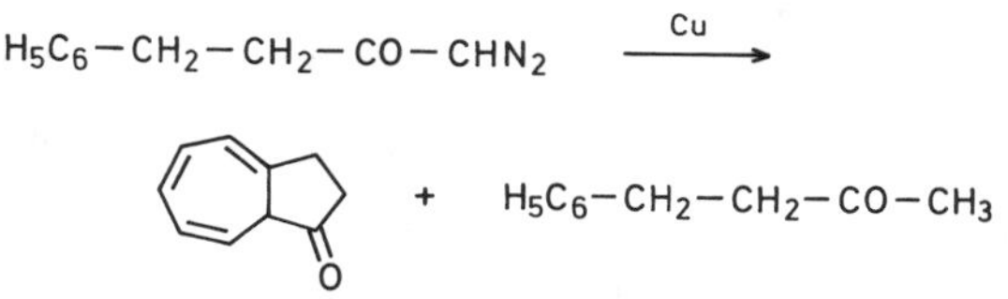

$$H_5C_6-CH_2-CH_2-CO-CHN_2 \xrightarrow{Cu}$$

3,8a-Dihydro-1(2H)azulenone

[657] *E. Müller, H. Kessler, B. Zeeh,* Fortschr. Chem. Forsch. *7,* 128 (1966).

Carbenes generated by photolysis of diethyl diazo-methanephosphonate add to benzene[664] to form the stable *diethyl (7-phenyl-2,4-norcaradien-7-yl)phosphonate* **13** plus the bis-adduct *tetraethyl (tricyclo[5.1.0.0²,⁴]oct-5-ene-3,8-di-yl)diphosphonate* **14**:

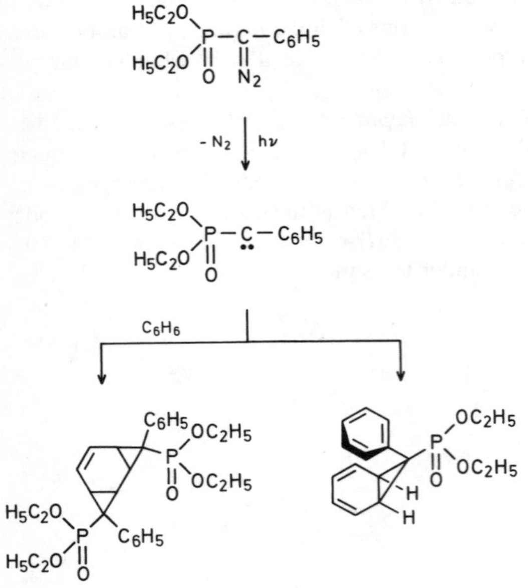

14   **13**

On reaction with diazomethane or diazofluorene, indan, fluorene, and tetralin undergo ring-enlargement; in some cases, the reaction products can be used for the preparation of *azulenes*[400, 651].

Reactions of diazoalkanes and α-diazo esters with *polynuclear* aromatic compounds lead to *cyclopropane derivatives;* here too, the catalytic reaction procedure has advantages.

Cyclopropane derivatives are similarly formed (22–50%) during the reaction of diazoalkanes and α-diazo esters with *aromatic heterocyclic compounds (furan* and *thiophene),* using cuprous salt catalysts or photolysis[400, 433]. By contrast, *pyrrole*

and *indole* react with ethyl diazoacetate and di-azomethyl ketones to form *CH* insertion products at the 2- or 3-positions[400, 433].

Analogous reactions occur on photolysis of di-azomethane in *pyridine*[665]; the yield of *2-methyl-pyridine (picoline)* is 83%. Insertion into the *NH* bond can also occur (see p. 287).

With *benzofuran* addition of ethoxycarbonylcar-bene occurs at the furan ring[666], whereas with *benzo[b]thiophene* the addition takes place at the benzene ring[667].

The reaction of diazomethane with certain stable organic *cations* can lead to the formation of homologous compounds; for example, xanthy-lium perchlorate is converted into *dibenzo[b,f]ox-epine* (60%)[668].

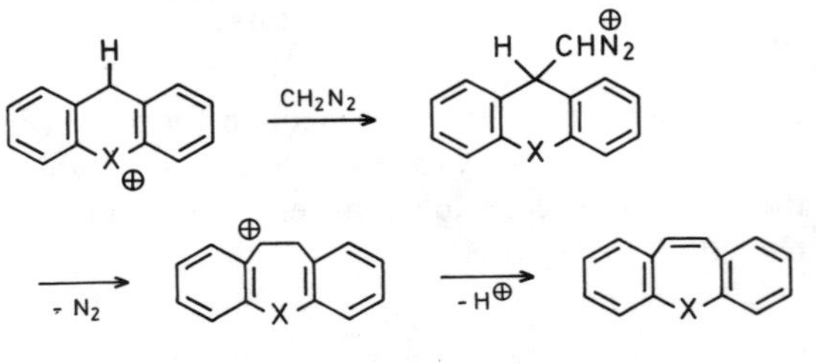

X = O, S

8.2.2.3 Reactions with $C \equiv C$ triple bond systems

8.2.2.3.1 Reactions with $C \equiv C$ triple bond systems producing pyrazoles and 3H-pyrazoles

The $C \equiv C$ triple bond of acetylenes, cycloalkynes, and even arynes can also behave as dipolarophile in 1,3-cycloaddition reactions with diazo compounds[400]. The reactions involving *acetylenes* and cycloalkynes yield *5 H-pyrazoles,* which mostly isomerize to pyrazoles*.

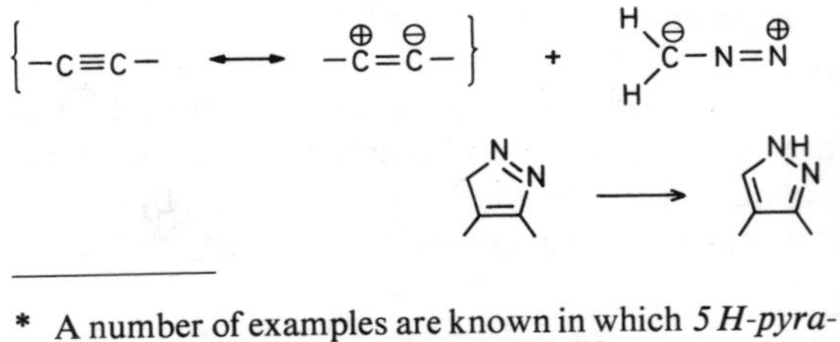

* A number of examples are known in which *5 H-pyra-zole* itself has been isolated[400, 669, 670].

[658] *E. Müller, H. Fricke, H. Kessler,* Tetrahedron Lett. 1501 (1963).

[659] *R.M. Lemmon, W. Strohmeier,* J. Amer. Chem. Soc. *81,* 106 (1959).

[660] *H. Dürr, G. Scheppers,* Tetrahedron Lett. 6059 (1968).

[661] *P.W. Grubb, R.A. Clark, D.R. Hartter, M.R. Wilcott,* J. Amer. Chem. Soc. *89,* 4076 (1967).

[662] *H. Ledon, G. Cannie, G. Linstrumelle, S. Jullia,* Tetrahedron Lett. 3971 (1970).

[663] *S. Jullia, A. Constantino, G. Linstrumelle,* C.R. Acad. Sci. C *264,* 407 (1967).

[664] *M. Regitz,* Angew. Chem. *82,* 224 (1970).

[665] *R. Daniels, O. Salerni,* Proc. Chem. Soc. 286 (1960).

[666] *G.M. Badger, B.J. Christie, H.H. Rodda, J.M. Pirkle,* J. Chem. Soc. 1179 (1958).

[667] *D. Sullivan, R. Pettit,* Tetrahedron Lett. 401 (1963); *R.G. Turnbo, D.L. Sullivan, R. Pettit,* J. Amer. Chem. Soc. *86,* 5630 (1964).

[668] *H.W. Whitlock,* Tetrahedron Lett. 593 (1961); *H.W. Whitlock, M.R. Pesce,* Tetrahedron Lett. 743 (1964).

Isomerization is generally observed when mono-substituted acetylenes are employed, or when diazo compounds with a hydrogen atom attached to the carbon atom bearing the diazo group are used [559, 671, 672].

Compounds with electron-withdrawing substituent groups have the highest reactivity [543, 549]. Thus, the reactions of 2-diazo-1,1,1-trifluoroethane with alkynes in the absence of light yield *1,3-dipolar cycloadducts*. The reactivities of acetylenic compounds decrease in the following order [543]:

$$F_3C-C\equiv C-CF_3 \approx H_3COOC-C\equiv CH > F_3C-C\equiv CH >$$
$$> Br-C\equiv CH > HC\equiv CH$$

Unsymmetrically substituted acetylenes form two *isomeric pyrazoles*, whose relative yields depend upon the polarizing effect of the substituent groups. The *Auwers rule* is only obeyed when the substituents at the triple bond are alkyl, nitrile, acyl, or alkoxycarbonyl groups [400, 669, 673].

Alkynes react not only with diazomethane and 1-diazoalkanes, but also with bis(diazoalkanes) [400, 674], aryldiazoalkanes, 9-diazofluorene [400], diazocyclopentadiene [675], 3-diazopropyne [676], esters of α-diazo carboxylic acids, and α-diazo ketones [400]. Organometallic diazomethane derivatives $[(CH_3)_3M-CHN_2; M=SN_2Si]$ can also be used as 1,3-dipoles [677].

Diazodiphenylmethane adds to cyclooctyne, forming *9,9-diphenylbicyclo[6.1.0]nonane* [678].
Cycloaddition of diazocyclopentadiene to dimethyl acetylenedicarboxylate gives *dimethyl 2H-cyclopenta[d]pyridazine-3,4-dicarboxylate* [679]:

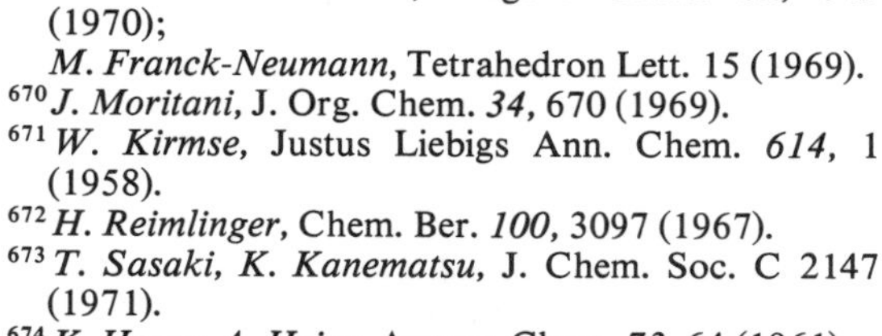

This reaction can be regarded as a thermally allowed $[\pi_s^8 + \pi_s^2]$ process.
Pyrazoles are formed also during cycloaddition with dimethyl acetylenedicarboxylates, e.g. [680, 681]:

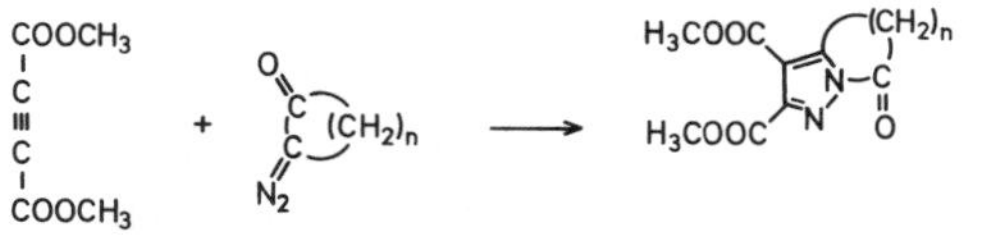

n = 4; *Dimethyl 5,6,7,8-tetrahydro-8-oxo-4H-pyrazolo-[1,5-a]azepine-2,3-dicarboxylate;* 78%

n = 3; *Dimethyl 4,5,6,7-tetrahydro-7-oxo-pyrazolo-[1,5-a]-pyridine-2,3-dicarboxylate;* 81%

α-Diazo carboxylic acid esters and α-diazo ketones trap arynes effectively [682].

8.2.2.3.2 Reactions with $C\equiv C$ triple bond systems producing cyclopropenes

The reaction of diazo compounds with *acetylenes* leads to *cyclopropene derivatives* if the reaction conditions allow elimination of nitrogen from the diazo compound and formation of a carbene. It was first performed between ethyl diazoacetate and (2-propynyl)benzene in the presence of copper catalysts [683, 684]:

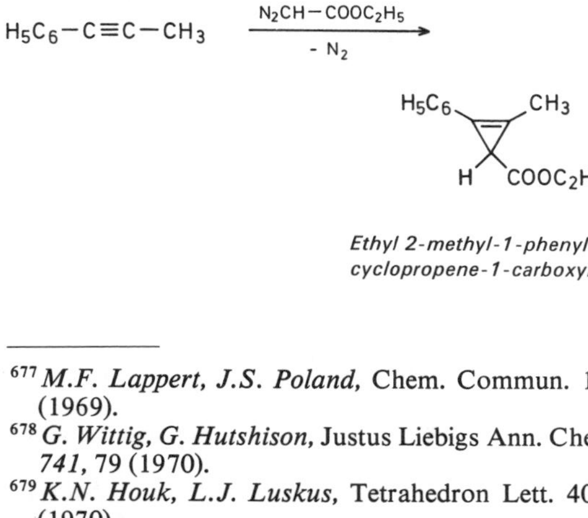

Ethyl 2-methyl-1-phenyl-3-cyclopropene-1-carboxylate

[669] M. Franck-Neumann, Angew. Chem. 82, 549 (1970); M. Franck-Neumann, Tetrahedron Lett. 15 (1969).
[670] J. Moritani, J. Org. Chem. 34, 670 (1969).
[671] W. Kirmse, Justus Liebigs Ann. Chem. 614, 1 (1958).
[672] H. Reimlinger, Chem. Ber. 100, 3097 (1967).
[673] T. Sasaki, K. Kanematsu, J. Chem. Soc. C 2147 (1971).
[674] K. Heyns, A. Heins, Angew. Chem. 73, 64 (1961).
[675] H. Dürr, L. Schrader, Angew. Chem. 81, 426 (1969).
[676] H. Reimlinger, Justus Liebigs Ann. Chem. 713, 113 (1968).
[677] M.F. Lappert, J.S. Poland, Chem. Commun. 156 (1969).
[678] G. Wittig, G. Hutshison, Justus Liebigs Ann. Chem. 741, 79 (1970).
[679] K.N. Houk, L.J. Luskus, Tetrahedron Lett. 4029 (1970).
[680] A.S. Katner, J. Org. Chem. 38, 825 (1972); M. Franck-Neumann, Tetrahedron Lett. 937 (1972).
[681] L.L. Rodian, V.V. Buluscheva, I.K. Korobizina, J. Org. Chem. in print (1973).
[682] W. Ried, Z. Chem. 6, 356 (1966).
[683] I.A. D'yakanov, M.I. Komendantov, Vestnik Leningrad-Univ. 11, N 22, Ser. Fiz. i. Khim. N4, 166 (1956); C.A. 52, 2762i (1958). I.A. D'yakonov, M.I. Komendantov, S.P. Korshunov, Zh. Obshch. Khim. 32, 923 (1962); C.A. 58, 2375e (1963).

Ethoxyfurans are sometimes formed during the thermally induced isomerization of ethyl 3-cyclopropene-carboxylates; these reactions are catalyzed by cuprous salts produced from cupric salts during decomposition of the diazo compounds[685, 686].

α-Diazo carboxylic esters react with both mono- and disubstituted acetylenes forming *cyclopropenes*. High yields of *cyclopropene derivatives* have also been obtained by photolysis of diazodiphenylmethane in various acetylenic compounds[687].

Addition of two molecules ethyl diazoacetate to acetylenedicarboxylic esters leads to *bicyclobutane derivatives*[605, 606].

α-Diazo ketones react with disubstituted acetylenes in an analogous way[688]:

$$R^l-C\equiv C-R^l \quad + \quad N_2CH-\overset{R^2}{\underset{O}{\overset{\displaystyle |}{C}}} \quad \xrightarrow{-N_2} \quad$$

The yields of cyclopropene derivatives are higher when α-diazo carboxylic esters are used than with α-diazo ketones[688−690].

The reaction between diazomethane and acetylenes has been carried out photolytically in the gas phase[691, 692]. Although it was not possible to isolate *cyclopropene*, the formation of *allene* and *propyne* can be taken as evidence for the intermediate formation of cyclopropene:

$$HC\equiv CH \quad + \quad N_2CH_2 \quad \xrightarrow[-N_2]{h\nu}$$

$$[\triangledown] \longrightarrow H_2C=C=CH_2 \quad + \quad H_3C-C\equiv CH$$

In the reactions under consideration diazomethane is less selective than ethyl diazoacetate[433]. Thus, for example, the rate of addition of diazomethane to the triple bond in 4-octyne is only eight times the rate of the *CH* insertion reaction (the yield[693] of *1,2-dipropylcyclopropene* is 23%). With the same coreactant ethyl diazoacetate gave a 61% yield of *ethyl 1,2-dipropyl-3-cyclopropene-carboxylate*, the rate of addition to the triple bond being 22 times that of the *CH* insertion reaction[693].

Satisfactory yields of *cyclopropene-3,3-dicarbonitriles* can also be obtained by thermolysis of diazomalononitrile in the presence of substituted acetylenes[694].

The use of *competitive reactions* has allowed a *comparison* to be made of the relative reactivities of triple and double bonds; the results showed that double bonds are considerably more reactive toward photolytically produced ethoxycarbonylcarbene than are triple bonds[695]. Ethoxycarbonylcarbene produced thermally from ethyl diazoacetate (using copper-bronze or copper sulfate as catalyst) adds three times as fast to triple bonds than to double bonds[695].

When the double and triple bonds are both present in the same molecule, their relative reactivities depend on their respective positions. The reaction of 2-methyl-1-hexen-4-yne with ethyl diazoacetate (in the presence of cupric salts) leads to adducts both at the double bond and at the triple bond (3:1)[696, 697].

In contrast, ethoxycarbonylcarbene adds only at the double bonds of conjugated enynes such as 2-methyl-1-penten-3-yne, forming *ethyl 2-propynyl-1-cyclopropane-carboxylate*[437].

[684] *R. Breslow, M. Battiste*, Chem. & Ind. (London) 1143 (1958).
R. Breslow, R. Winter, M. Battiste, J. Org. Chem. *24*, 415 (1959);
R. Breslow, D. Chipman, Chem. & Ind. (London) 1105 (1960).
[685] *M.I. Komendantov, I.A. D'yakonov, T.S. Smirnova*, Zh. Org. Khim. *2*, 559 (1966); C.A. *65*, 7124c (1966).
[686] *I.A. D'yakonov, M.I. Komendantov, T.S. Smirnova*, Zh. Org. Khim. 1742 (1969); C.A. *72*, 12455b (1970).
[687] *M. Hendrick, W. Baron, M. Jones*, J. Amer. Chem. Soc. *93*, 1558 (1971).
[688] *N. Obata, I. Moritani*, Tetrahedron Lett. 1503 (1966);
N. Obata, I. Moritani, Bull. Chem. Soc. Japan *39*, 1975, 2250 (1966).
[689] *M. Vidal, E. Chollet, P. Arnold*, Tetrahedron Lett. 1073 (1967).

[690] *M.I. Komendantov, T.S. Smirnova, I.A. D'yakonov*, Zh. Org. Khim. *3*, (10), 1903 (1967); C.A. *68*, 12145t (1968).
[691] *H.M. Frey*, Chem. & Ind. (London) 1266 (1960).
[692] *T. Terao, N. Sakai, S. Shida*, J. Amer. Chem. Soc. *85*, 3919 (1963).
[693] *H. Lind, A.J. Deutschmann jr.*, J. Org. Chem. *32*, 326 (1967).
[694] *E. Ciganek*, J. Amer. Chem. Soc. *88*, 1979 (1966).
[695] *I.A. D'yakonov, M.I. Komendantov, L.P. Danilkina, R.N. Gmyzina, T.S. Smirnova, A.G. Vitenberg*, Zh. Org. Khim. 383 (1969); C.A. *70*, 105633g (1969).
[696] *L.P. Danilkina, I.A. D'yakonov*, Zh. Org. Khim. *2*, (1), 3 (1966); C.A. *64*, 14101b (1966).
[697] *I.A. D'yakonov, R.N. Gmyzina, L.P. Danilkina*, Zh. Org. Khim. *2*, 2079 (1966); C.A. *66*, 75710c (1967).
[698] *B. Eistert*, Synthesen mit Diazomethan, in Neuere Methoden der präparativen organischen Chemie, Bd. I, p. 359, Verlag Chemie, Weinheim/Bergstr. 1949.
[699] *C.D. Gutsche, D. Redmore*, Carbocyclic Ring Expansion Reactions, p. 81, Academic Press, New York, London 1968.

8.2.2.4 Reactions with carbonyl compounds

8.2.2.4.1 Reactions with monocarbonyl compounds

8.2.2.4.1.1 Reactions with open-chain aldehydes and ketones

The reaction of diazo compounds and especially of diazomethane with carbonyl compounds is one of the most fully investigated and widely used reactions of this class of compounds; it leads mainly to homologization (formation of homologous ketones and aldehydes) and formation of oxiranes (epoxides)[400, 441, 657, 698, 699]. Although many normal features of this reaction are known, it is not always possible to predict the nature and amount of the reaction products.

The *course of the reaction* depends not only upon the structure of the starting materials but also on the catalyst used, the methods employed for preparation and purification of the diazoalkane concerned, the order of mixing the reactants, and other factors which are difficult to control[700-702].

The following general scheme can be regarded as valid for most reactions of carbonyl compounds with diazo compounds[703]. The diazoalkane, reacting in the carbanionic form 15, adds to the electrophilic carbon atom of the carbonyl group 16 forming a *zwitterionic* diazonium alkoxide 17*.

Elimination of nitrogen leads to the *zwitterion* 18, which can be stabilized either by cyclization to the epoxide 19 or by migration of the groups R^3 and R^4 (1,2-nucleophilic shift) to form homologous carbonyl compounds 20 and 21:

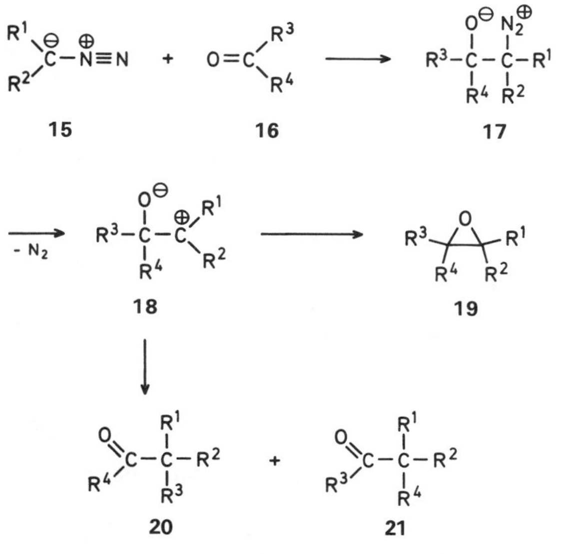

In this reaction mechanism the carbonyl component is the electrophile and the diazoalkane is the nucleophilic reagent. This fits in well with known experimental facts: aldehydes react faster than ketones, aliphatic carbonyl compounds react faster than aromatic ones. Electron-withdrawing substituent groups (*e.g.*, chloro, nitro, or carbonyl groups) at the α-position of the carbonyl compound increase the rate of reaction with diazoalkanes. The reactivity of diazo compounds is higher with diazomethane and its homologs than with α-diazotoluene, ethyl diazoacetate, or α-diazo ketones, corresponding to decreasing nucleophilism. The *direction* in which the intermediate *zwitterion* 18 becomes stabilized depends on electronic and steric factors, as well as upon the catalyst used[705]. Electron-attracting and bulky substituent groups *alpha* or *ortho* to the carbonyl group promote the formation of *epoxide (oxirane)*[400, 705, 706].

The tendency to form epoxides is less with the higher diazoalkanes[705]. The relative *ratio* of the carbonyl compounds 20 and 21 formed depends upon the rates of migration of the groups R^3 and R^4. Groups with sp^2-hybridized carbon atoms (vinyl, aryl, acyl) generally migrate more easily than those with sp^3-hybridized carbon atoms[707-709].

The reactions between carbonyl compounds and diazoalkanes are *catalyzed* by water and alcohols (especially methanol). Addition of methanol accelerates the whole reaction and raises the yield of epoxides. Lewis acids[657], particularly boron trifluoride, aluminum chloride and trialkyloxonium salts, are more powerful catalysts[710].

15	16	17

18	19

20	21

---

* A number of such addition products have been isolated as such, after isomerization to the *aldol adducts*, from 15 and 16 (R and $R^1 = H$)[704].

[701] *L. Capuano, F. Jamaigne*, Chem. Ber. *96*, 798 (1963).

[702] *B. Eistert, Mustafa A. El-Chahawi*, Monatsh. Chem. *98*, 941 (1967).

[703] *F. Arndt, B. Eistert*, Chem. Ber. *68*, 196 (1935).

[704] *B. Eistert, O. Ganster*, Chem. Ber. *104*, 78 (1971).

[705] *C.D. Gutsche*, The Reaction of Diazomethane and its Derivatives with Aldehydes and Ketones, Org. Reactions *8*, 364 (1954).

[706] *H.W. Moore, M.W. Grayston*, J. Org. Chem. *35*, 2832 (1970).

[707] *H.O. House, E.J. Grubbs, W.F. Gannon*, J. Amer. Chem. Soc. *82*, 4099 (1960).

[708] *M. Hanack, H.M. Ensslin*, Justus Liebigs Ann. Chem. *697*, 100 (1966).

[709] *R.G. Carlson, N.S. Behn*, J. Org. Chem. *33*, 2069 (1968).

[710] *W.L. Mock, M.E. Hartmann*, J. Amer. Chem. Soc. *92*, 5767 (1970).

[700] *L. Capuano*, Chem. Ber. *98*, 3187 (1965).

Using Lewis acids as catalysts admittedly precludes the formation of epoxides and consequently increases the yields of *homologs* and reduces the reaction time[657]. Addition of Lewis acids permits reactions to be carried out successfully between carbonyl compounds and inert diazo compounds (such as diazodiphenylmethane and ethyl diazoacetate) which, when mild or no catalysts are employed, either do not react at all or not in the way indicated. Thus, for example, α,β-unsaturated ketones form pyrazolines in the absence of catalysts, but react at the carbonyl group when Lewis acids are present[707, 710, 712].

The scheme set out on p. 271 is of wide validity but does not explain every experimental observation[713-716]; other mechanisms have therefore also been suggested[558, 713, 716, 717].

Aldehydes and ketones of any desired structure can enter into reaction with diazoalkanes. Reactions of carbonyl compounds with diazoalkanes can be carried out by both *ex situ* and *in situ* procedures.

The following *side-reactions* can occur. The carbonyl compound **20**, **21** (p. 271) formed can subsequently react with a second molecule of the diazo compound forming further *aldehyde, ketone,* or *epoxide*. This may become the main reaction if the carbonyl compound initially produced is more reactive than the original carbonyl component or if a large excess of diazoalkane is used. An example is the formation of *3-benzo[b]thien-3-yl ethyl ketone* from benzo[b]thiophene-3-carboxaldehyde[700].

In the reactions of diazomethane with aldehydes, especially when the latter is in excess, stabilization of the intermediate *zwitterions* can lead to *glycidols* (oxiranemethanols), β-*hydroxy ketones*, and *methyl ketones*[400, 700] (**22, 23, 24, 25**):

$$R-\overset{\overset{H}{|}}{\underset{\underset{O^{\ominus}}{|}}{C}}-CH_2\overset{\oplus}{N_2} \longrightarrow$$

22

$$R-\underset{\underset{OH}{|}}{CH}-CHN_2 \xrightarrow{R-CHO} \underset{\underset{OH}{|}}{\overset{R-CH}{|}}\overset{R}{\underset{O}{\diagup}}$$

23

$$\downarrow$$

$$R-\underset{\underset{OH}{|}}{CH}-CH_2-\overset{R}{\underset{O}{C}} \longrightarrow H_3C-\overset{R}{\underset{O}{C}} + R-CHO$$

24   **25**

8.2.2.4.1.2 Reactions with cyclic ketones

Reactions of diazo compounds with saturated cyclic ketones, likewise proceed by the mechanism illustrated on p. 271, and give *homologous cyclic ketones* and *spirooxiranes*:

$$(CH_2)_n\!\!-\!\!C\!\!=\!\!O \xrightarrow{R-CHN_2} (CH_2)_n\overset{C=O}{\underset{CH-R}{|}}$$

$$+ \quad (CH_2)_n \overset{R}{\underset{O}{\diagup\!\!\!\diagdown}}$$

These reactions and their general behavior have been studied in detail, and are often used as a means of ring-expansion[699, 705].

Cyclopropanones react with diazomethane and diazoethane even at $-78°$ forming *cyclobutanone* and *2-methylcyclobutanone* derivatives[718].

A 58:17:25 mixture of cyclopentanone, cycloheptanone, and cyclooctanone is formed during the reaction of cyclobutanone with diazomethane[719].

Use of ring enlargement from cyclopentanones to *cyclohexanones* is fairly rare since the six-membered ketones formed are more reactive than the starting materials, and consequently the main products obtained are those from double or triple homologization[699]. The five-membered cyclic ketone tetrahydro-2,2,5,5-tetramethyl-3-furanone, however, is converted into the corresponding *2,2,6,6-tetramethyltetrahydropyranone* by reaction with diazomethane or ethyl diazoacetate[720, 721].

Reactions of cyclohexanones with diazoalkanes generally proceed smoothly. The main products are the seven-membered cyclic ketones, which are

[711] *Eu. Müller, M. Bauer, W. Rundel,* Justus Liebigs Ann. Chem. *654*, 92 (1962).

[712] *W.S. Johnson, M. Neeman, S.P. Birkeland, N.A. Fedoruk,* J. Amer. Chem. Soc. *84*, 989 (1962).

[713] *A. Schönberg, K. Junghans,* Chem. Ber. *96*, 3328 (1963).

[714] *I.K. Korobizina, O.P. Studzinskii,* Zh. Org. Khim. *4*, 524 (1968); C.A. *68*, 104956m (1968).

[715] *I.K. Korobizina, O.P. Studzinskii, Z.I. Ugorets,* Zh. Org. Khim. *5*, 1109 (1969) (russ.); C.A. *71*, 70218s (1969).

[716] *C.D. Gutsche, J. Bowers,* J. Org. Chem. *32*, 1203 (1967).

[717] *R.S. Bly, F.B. Culp jr., R.K. Bly,* J. Org. Chem. *35*, 2235 (1970).

[718] *N.J. Turro, R.B. Gagosian,* Chem. Commun. 949 (1969).

[719] *J. Jaz, I. Davreux,* Bull. Soc. Chim. Belg. *74*, 370 (1965).

[720] *I.K. Korobizina, K.K. Pivnitskii,* Zh. Obshch. Khim. *30*, 4008 (1960); C.A. *55*, 22277d (1961).

[721] *I.K. Korobizina, O.P. Studzinskii,* Khim. Geterotsikl. Soedin 848 (1966); C.A. *67*, 21753w (1967).

obtained in 40–80% yield. Since side-reactions only occur to a minor extent, this reaction provides a convenient and widely applicable method for converting symmetrically substituted six-membered into seven-membered cyclic ketones[699]. Unsymmetrically substituted cyclohexanones yield mixtures of isomeric reaction products, which are often difficult to separate. Steric factors also have a considerable influence on the course of the reaction (see, *e.g.*, refs. [699, 722]).

An elegant intramolecular variation of this type of reaction has been worked out for synthesizing *bicyclo[4.3.0]alkanones*. For example, treating 2-[(N-acetyl-N-nitrosoamino)propyl]cyclo-alkanones with sodium ethoxide yields 80–98% *bicycloalkanones*[710, 723]:

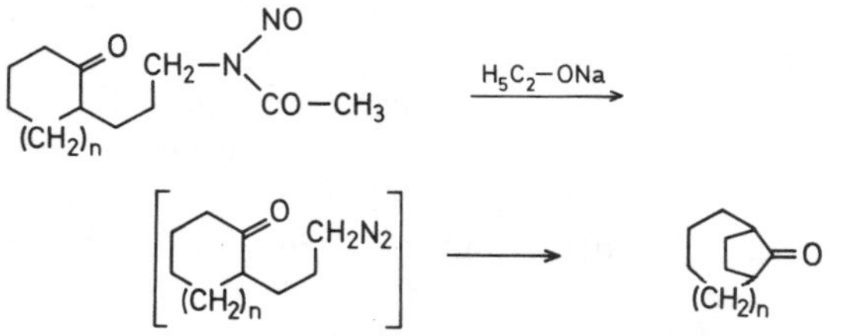

n = o; *Bicyclo[3.2.1]octan-8-one; 88%*

n = 1; *Bicyclo[4.2.1]nonan-9-one; 81%*

n = 2; *Bicyclo[5.2.1]decan-10-one; 60%*

In contrast, 2-[(*N*-acetyl-*N*-nitrosoamino)propyl]-1,6,6-trimethylcyclohexanone forms almost only epoxide and no larger rings[716]. Diazoalkanes have been used also for expanding the *A* ring in steroids to form *3-oxo-A-homosteroids*[657, 699].

Reacting diazoalkanes with the six-membered *heterocycles* tetrahydro-4*H*-pyran-4-one and tetrahydro-4*H*-thiopyran-4-one gives both epoxides and the corresponding *seven-membered cyclic ketones*[400].

Lewis acids have proven to be very effective catalysts for preparing homologous compounds containing medium and *large rings* (C_8-C_{15})[657]. The relative reactivities of C_8-C_{15} cycloalkanones have been calculated from kinetic and statistical models and agree well with experimental data[724]. 2-Oxo-[2.2.n]cyclophane is particularly reactive toward diazomethane in methanolic solution[725]. The reactivity of various *bicyclo[n.2.1]alkanones* toward diazomethane has been studied[726].

2-Cycloalk-1-en-1-ones such as substituted cyclopentadienones and *p*-quinones react with diazoalkanes mainly by attacking the *C=C* double bond (see p. 258). An interesting example is the reaction between 2-diazopropane and 2,4,6-cyclohepta-trien-1-one (tropone)[727]:

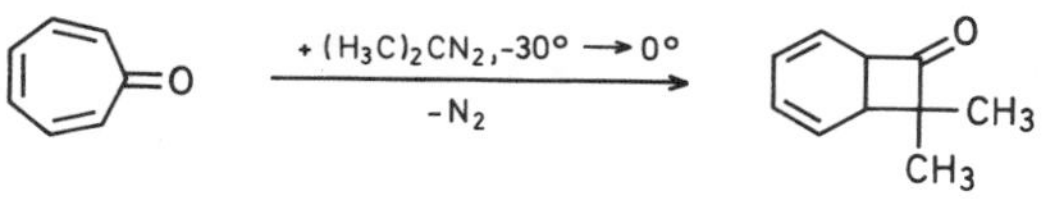

7,7-Dimethylbicyclo-[4.2.0]octa-2,4-dien-8-one

The carbonyl group participates in the reaction only if the reactivity of the double bond is greatly reduced because of steric or electronic substituent effects[728, 729].

8.2.2.4.2 Reactions with 1,2-dicarbonyl compounds

As a rule, 1,2-diketones are more reactive toward diazoalkanes than are monocarbonyl compounds. Depending on the structure of the 1,2-dicarbonyl compound concerned, reactions with diazoalkanes can proceed in four different ways, in parallel or sequentially[730].

1,2,3-Tricarbonyl compounds react similarly[731]. Formation of *enol ethers* is observed whenever enolized diketones are treated with a sufficiently nucleophilic diazoalkane:

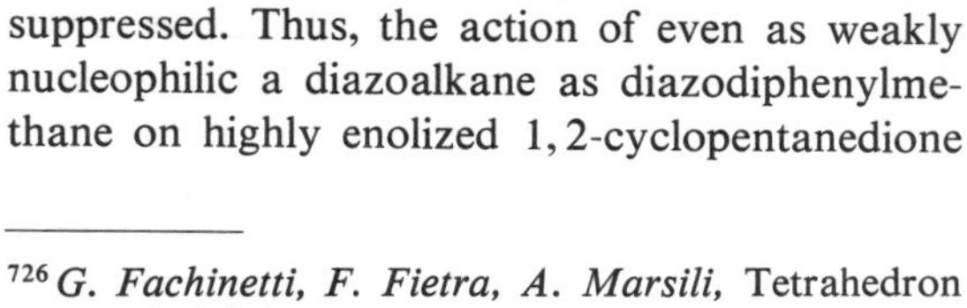

The more acid the enol is, the higher are the yields of enol ethers and the more parallel reactions are suppressed. Thus, the action of even as weakly nucleophilic a diazoalkane as diazodiphenylmethane on highly enolized 1,2-cyclopentanedione

[722] *T.D. Inch, G.J. Lewis, R.P. Peel, N. Williams*, Chem. Commun. 1549 (1970).

[723] *C.D. Gutsche, D.M. Bailey*, J. Org. Chem. *28*, 607, 610 (1963).

[724] *J. Heiss, M. Bauer, E. Müller*, Chem. Ber. *103*, 463 (1970).

[725] *D.J. Cram, R.C. Helgeson*, J. Amer. Chem. Soc. *88*, 3515 (1966).

[726] *G. Fachinetti, F. Fietra, A. Marsili*, Tetrahedron Lett. 393 (1971).

[727] *M. Franck-Neumann*, Tetrahedron Lett. 2143 (1970).

[728] *B. Eistert*, Justus Liebigs Ann. Chem. *735*, 145 (1970).

[729] *W. Rundel, P. Kästner*, Justus Liebigs Ann. Chem. *737*, 87 (1970).

[730] *O.P. Studzinskii*, in *I.A. D'yakonov*, Sovrem. Probl. Org. Khim. p. 224, Izd. Leningrad, Univ., Leningrad, USSR 1969.

[731] *B. Eistert, J. Grammel*, Chem. Ber. *104*, 1942 (1971).

produces *2-(diphenylmethoxy)-2-cyclopenten-1-one*[715]. Formation of *1,3-dioxolanes* is often seen in reactions of diazoalkanes with 1,2-diketones with an *o*-quinonoid structure:

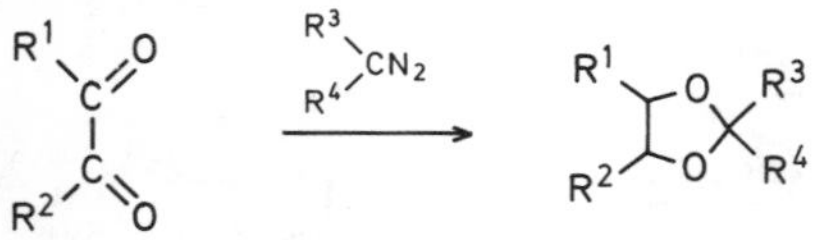

It cannot be excluded that such a reaction course is related to the formation of stable aromatic structures (see, *e.g.*, refs. [730, 732]).

Since the same substances can be formed by the hydrolysis of 1,3-dioxolanes as from isomeric α-ketoepoxides, the structures of the 1,3-dioxolanes must be proven spectroscopically, not merely by chemical means. Cases of incorrect structural assignment are known[707, 733, 734]. The mechanism of formation of the 1,3-dioxolanes has not been completely elucidated[735]; in the present authors' view a scheme in which intermediate *epoxides* are formed and then isomerize to *1,3-dioxolanes* appears to be most probable[730].

Formation of *oxiranes* and *bis-oxiranes* (*mono-* and *diepoxides*) is typical of reactions of 1,2-dicarbonyl compounds with diazoalkanes[698, 730, 734]:

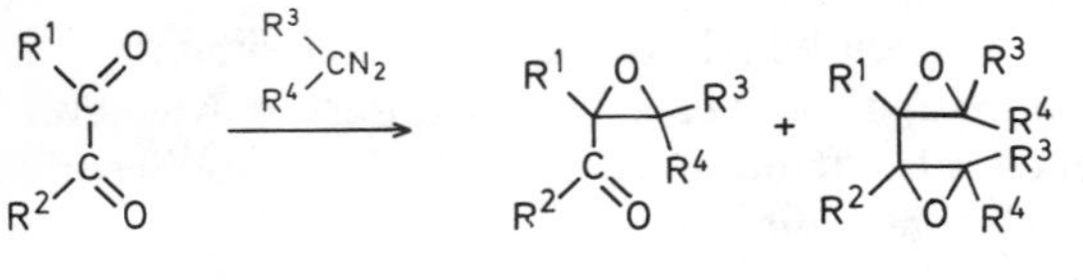

Oxiranes (epoxides) are formed especially easily if the dicarbonyl component used is a diaryl diketone or an *o*-quinone[730], and the diazo compound is diazomethane, diazoethane, or diazodiphenylmethane[713–715]. The formation of epoxides from reactions with ethyl diazoacetate is seldom observed[736–738].

Formation of homologous α- and *β-diketones* takes place in accordance with the scheme shown on p. 271:

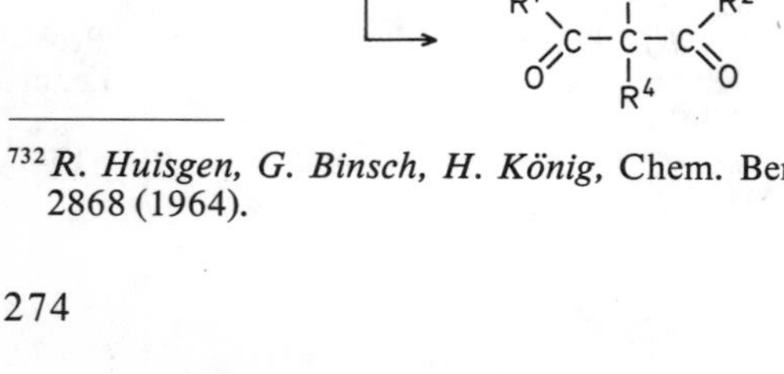

α-Diketones **26** and **27** result when insertion of the diazoalkane group occurs between a carbonyl group and a hydrocarbon moiety; *β*-diketones **28** are formed when insertion is between the carbonyl groups[698, 730]. Aliphatic diazo compounds of a wide variety of structures can participate in this reaction, but not diazodiphenylmethane. Catalysts are often used to accelerate the reaction; Lewis acids are most commonly employed. Formation of homologs proceeds especially well when cyclic α-dicarbonyl compounds are used.

α-*Diketones* are formed *predominantly* when the dicarbonyl group is attached to an aromatic unit[735] or is condensed with one, as, for instance, in phenanthraquinone[736, 737], isatin[739, 740], *N*-hydroxyisatin[741], 1,2-acenaphthylenedione[400]. Formation of *β*-diketones is especially characteristic of 2,3-bornanedione(camphorquinone)[742, 743], 2,2,5,5-tetraalkyltetrahydro-3,4-furandiones[721, 744–746], 4,5-dihydroxy-4-cyclopentene-1,2,3-trione (croconic acid)[748], and 1,2,3-indantrione[740]. In these cases the reactions with diazomethane do not yield the free *β*-diketones but their enol ethers. Thus, reacting 2,2,5,5-tetraalkyltetrahydro-3,4-furandiones with excess ethereal diazomethane solution and added methanol gives good yields of *2,2,6,6-tetraalkyl-5-methoxy-2H-pyran-3(6H)-one*[744].

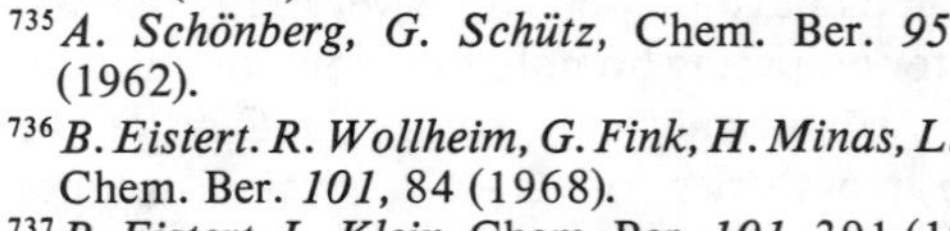

[733] *H. Biltz, H. Petzold*, Justus Liebigs Ann. Chem. *433*, 71 (1923).

[734] *B. Eistert, G. Fink, R. Wollheim*, Chem. Ber. *91*, 2710 (1958).

[735] *A. Schönberg, G. Schütz*, Chem. Ber. *95*, 2386 (1962).

[736] *B. Eistert. R. Wollheim, G. Fink, H. Minas, L. Klein*, Chem. Ber. *101*, 84 (1968).

[737] *B. Eistert, L. Klein*, Chem. Ber. *101*, 391 (1968).

[738] *B. Eistert, W. Eifler, O. Ganster*, Chem. Ber. *102*, 1988 (1969).

[739] *B. Eistert, G. Borggrefe*, Justus Liebigs Ann. Chem. *718*, 142 (1968).

[740] *B. Eistert, W. Kurze, G. Müller*, Justus Liebigs Ann. Chem. *732*, 1 (1970).

[741] *B. Eistert, G. Borggrefe, H. Selzer*, Justus Liebigs Ann. Chem. *725*, 37 (1969).

[742] *H. Rupe, C. Frey*, Helv. Chim. Acta *27*, 627 (1944).

[743] *H. Rupe, F. Hafflinger*, Helv. Chim. Acta *23*, 139 (1940).

[744] *I.K. Korobizina, K.K. Pivnitskii*, Zh. Obshch. Khim. *30*, 4016 (1960); C.A. *55*, 22302f (1961).

[745] *I.K. Korobizina, K.K. Pivnitzkii*, Dokl. Akad. Nauk SSSR *132*, 127 (1960); C.A. *54*, 21076c (1960).

[732] *R. Huisgen, G. Binsch, H. König*, Chem. Ber. *97*, 2868 (1964).

Table 1. Reaction of 1,2-dicarbonyl and 1,2,3-tricarbonyl compounds with diazo compounds

Starting substances		Reaction products	Ref.
Ketone	Diazo compound		

The table contains the following chemical structures and text:

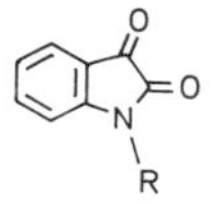

Diazo compound: $N_2CH-CO-R^1$, $R^1 = OC_2H_5, C_6H_5, CH_3$; Ref. 739

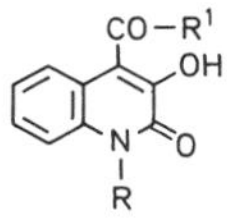

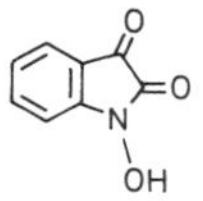

Diazo compound: $R-CHN_2$, $R = H, CH_3$; Ref. 741

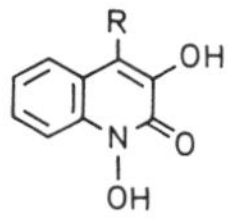

1,3-Dihydroxy-2(1H)-quinolinone; 1,3-Dihydroxy-4-methyl-2(1H)-quinolinone

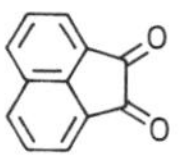

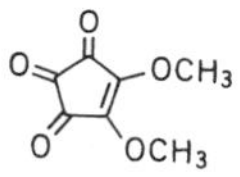

Diazo compound: $R-CHN_2$; Ref. 400

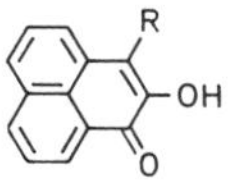

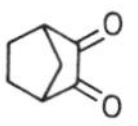

2-Hydroxy-1-phenalenone

Diazo compound: CH_2N_2; Ref. 748

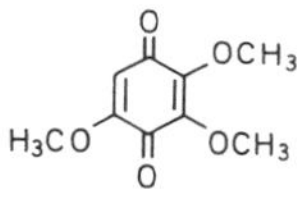

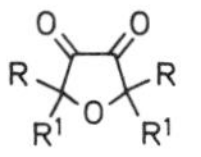

Trimethoxy-1,4-benzoquinone

Diazo compound: CH_2N_2; Ref. 747

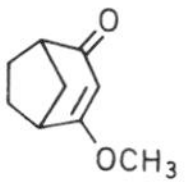

2-Methoxybicyclo[3.2.1]oct-2-en-3-one

Diazo compound: CH_2N_2, $R = R^1 = CH_3$; $R = CH_3, R^1 = C_2H_5$; Ref. 744

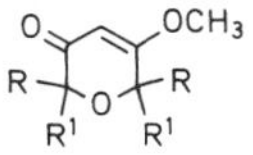

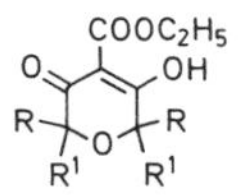

5-Methoxy-2,2,6,6-tetramethyl-2H-pyran-3(6H)-one; 2,6-Diethyl-5-methoxy-2,6-dimethyl-2H-pyran-3(6H)-one

Diazo compound: $N_2CH-COOC_2H_5$; Ref. 746

Ethyl 3,6-dihydro-5-hydroxy-2,2,6,6-tetramethyl-2H-pyran-4-carboxylate; Ethyl 2,6-diethyl-3,6-dihydro-5-hydroxy-2,6-dimethyl-2H-pyran-4-carboxylate

Cases are also known in which the reaction proceeds simultaneously in both directions[749].

8.2.2.4.3 Reactions with carboxylic acid derivatives

Carbonyl groups of carboxylic acid derivatives are generally inert toward diazoalkanes.

Carboxylic acid *esters* do not react with diazoalkanes[715]. The reaction of acid *chlorides* or *anhydrides* with diazomethane is an important method *(Arndt-Eistert procedure)* for the preparation of α-diazocarbonyl compounds, and is discussed in detail on p. 240. Carbox*amides* are generally stable to diazoalkanes, but fission of the *NH–CO* link may be observed if small amounts of water or methanol are added[400].

The reaction of lactams with diazoalkanes is discussed on p. 286.

8.2.2.5 Reactions with compounds containing *C=S* groups

In many cases, *thiocarbonyl* compounds react more readily with aliphatic diazo compounds than do the corresponding carbonyl compounds, because the thiocarbonyl group is more polarized than the carbonyl group. Reaction can proceed in two directions:

either 1,3-dithiolanes **29** or thiiranes **30** are formed[750]:

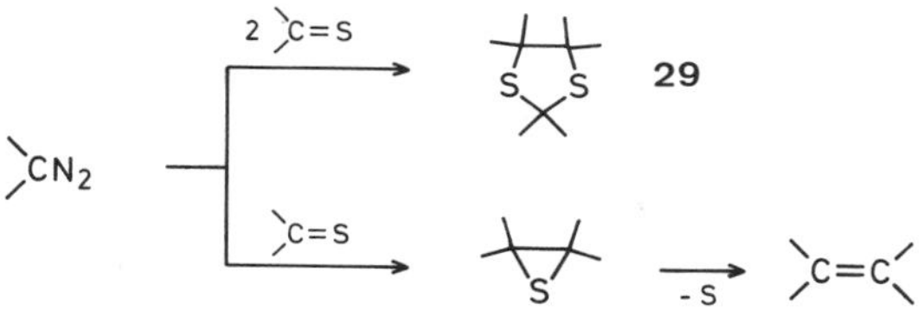

Formation of *1,3-dithiolanes* **29** is observed with reactions involving diazomethane, whereas *thiiranes* **30** are formed from disubstituted diazoalkanes; both can be obtained from monosubstituted diazoalkanes. The course of the reaction is determined mainly by steric factors[750].

[746] *I.K. Korobizina, O.P. Studzinskii,* Zh. Org. Khim. *5,* 1276 (1969) (russ.); C.A. *71,* 112 739v (1969).

[747] *I.G. Durocher, H. Favre,* Can. J. Chem. *42,* 264 (1964).

[748] *H. Iamada, I. Hirata,* Bull. Chem. Soc. Japan *31,* 550 (1958).

[749] *B. Eistert, R. Müller, I. Mussler, H. Selzer,* Chem. Ber. *102,* 2429 (1969).

[750] *A. Schönberg, B. König, E. Singer,* Chem. Ber. *100,* 767 (1967).

Since thiiranes eliminate sulfur on heating (especially if copper is present), forming *olefins* **31**, the reaction under discussion is an important method for the preparation of *tetrasubstituted ethylene* derivatives, both symmetrical and unsymmetrical, particularly the latter[751-755]. A general method has been developed from trithiocarbonic acid triesters and diaryldiazomethanes[756].

Sulfenes readily add diazomethane forming high yields of *thiirane 1,1-dioxides,* which can form olefins by elimination of sulfur dioxide.

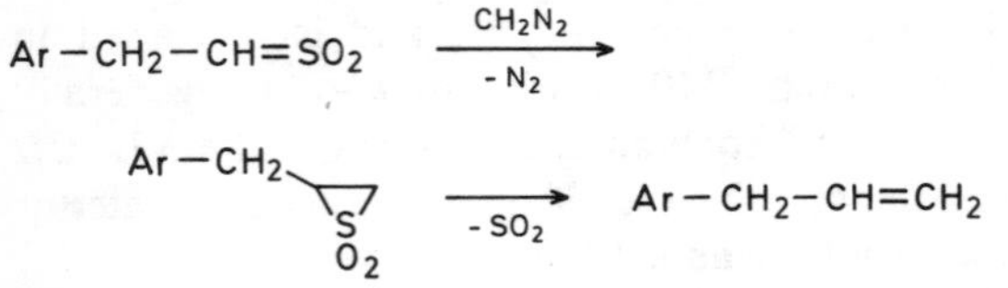

$$Ar-CH_2-CH=SO_2 \xrightarrow[-N_2]{CH_2N_2}$$

$$\xrightarrow{-SO_2} Ar-CH_2-CH=CH_2$$

This reaction, too, is used for the synthesis of olefins[757, 758].

The reaction between *N-sulfinylaniline* and diazodiphenylmethane proceeds similarly, but involves addition to the $N=S$ double bond; the product is *benzylideneaniline*[759]:

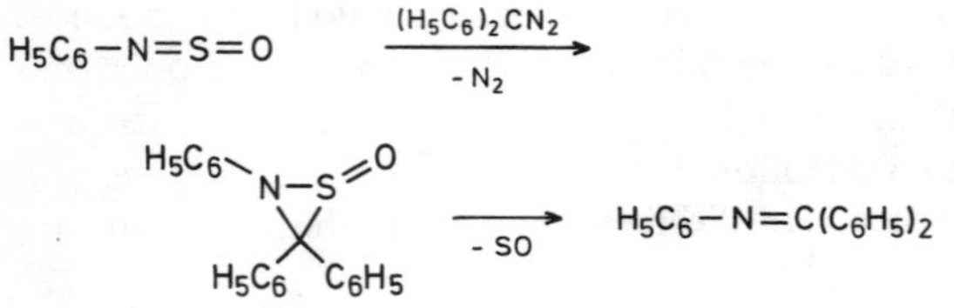

$$H_5C_6-N=S=O \xrightarrow[-N_2]{(H_5C_6)_2CN_2}$$

$$\xrightarrow{-SO} H_5C_6-N=C(C_6H_5)_2$$

Heterocyclic systems can be formed by reaction of diazo compounds with thiourea and thiosemicarbazide, nitrogen being eliminated. Thus, *5-aryl-2-arylamino-6H-1,3,4-thiadiazines* are formed from aryldiazo ketones and arylthiosemicarbazides (50–70%)[760]:

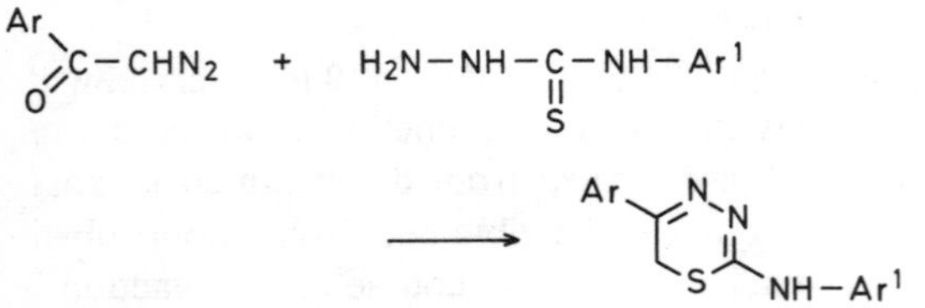

Diazomethyl aryl ketones react with thiourea and its N-substituted derivatives forming *4-aryl-2-alkylaminothiazoles* (or *2-arylamino-4-arylthiazoles)* (50–90%)[761].

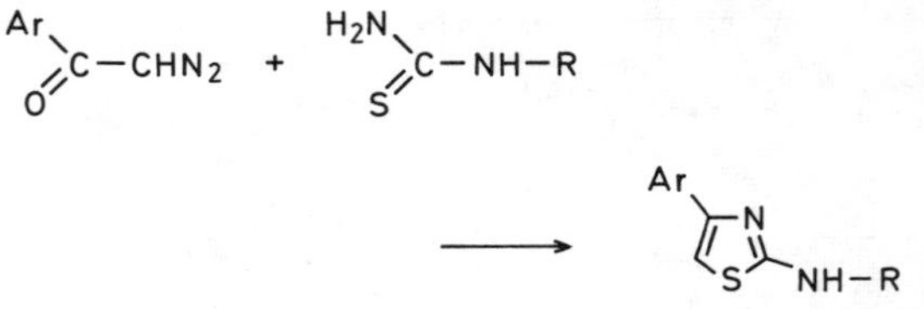

Instances are also known of the *1,3-dipolar cycloaddition* of diazoalkanes (without elimination of nitrogen) to $C=S$ double bonds. For example, diazoalkanes add to thioisocyanates, forming substituted *1,2,3-thiadiazoles* (60%)[762, 763]:

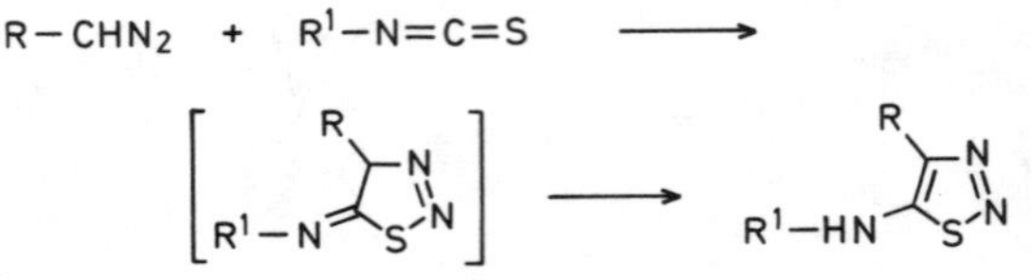

$$R-CHN_2 + R^1-N=C=S \longrightarrow$$

With phenyl and benzoyl isothiocyanate α-diazo carboxylic acid esters and α-diazo ketones form the analogous *1,2,3-thiadiazoles* (50–60%)[764]. When substituted 2-diazoacetophenones[765] or ethyl diazotate[766] are employed the relatively high yields of 60–80% are obtained.

8.2.2.6 Reactions with compounds containing $C=N$ and $C\equiv N$ groups

Addition of aliphatic diazo compounds to the $C=N$ double bond, like that to $C=C$ double bonds, can take place either with or without loss of nitrogen.

[751] *M. Sander,* Chem. Rev. *66*, 297 (1966).
[752] *A. Schönberg, K. Junghans,* Chem. Ber. *99*, 1241 (1966).
[753] *I.K. Korobizina, O.P. Studzinskii,* Zh. Org. Khim. *5*, 1493 (1969) (russ.); C.A. *71*, 112708j (1969).
[754] *Y. Poirier, N. Lozach,* Bull. Soc. Chim. France 2090 (1967).
[755] *A. Schönberg, E. Frese,* Chem. Ber. *96*, 2420 (1963);
A. Schönberg, E. Frese, Chem. Ber. *101*, 694 (1968).
[756] *A. Schönberg,* Tetrahedron Lett. 2487 (1968).
[757] *G. Opitz, K. Fischer,* Angew. Chem. *77*, 41 (1965).

[758] *N. Fischer,* Synthesis 393 (1970).
[759] *J.G. Steffer, H.R. Musser,* Chem. Commun. 481 (1970).
[760] *W. Hampel,* Z. Chem. *9*, 61 (1969).
[761] *W. Hampel, J. Müller,* J. Prakt. Chem. *310*, 320 (1968).
[762] *D. Martin, W. Mucke,* Justus Liebigs Ann. Chem. *682*, 90 (1965).
[763] *M. Regitz,* Angew. Chem. *79*, 786 (1967).
[764] *S. Hauptmann, K. Hirschberg,* J. Prakt. Chem. *308*, 82 (1967).
[765] *W. Ried, B.M. Beck,* Justus Liebigs Ann. Chem. *673*, 124, 128 (1964).
[766] *J. Goerdeler, G. Gnad,* Tetrahedron Lett. 795 (1964).

To azomethines (Schiff's bases) diazomethane adds in the manner of a 1,3-dipolar cycloaddition and leads to substituted *4,5-dihydro-1,2,3-tri-azoles* (60–80%)[400, 767]:

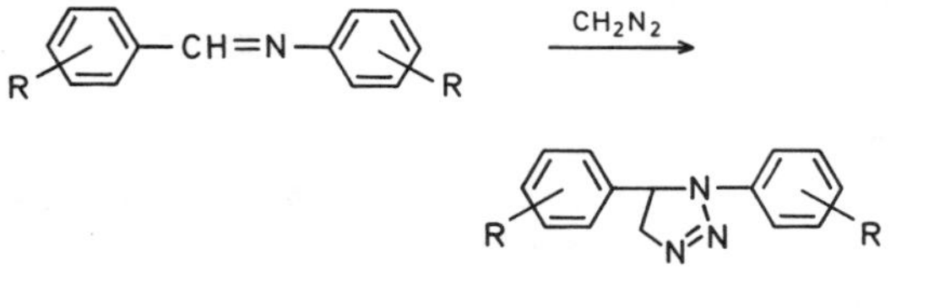

Addition to *ketenimines* furnishes 1,2,3-tri-azoles[768].

4,5-Dihydro-1,2,3-triazoles are formed when per-fluoroketo fluoroimines react with diazomethane at −50°[769], *e.g.*:

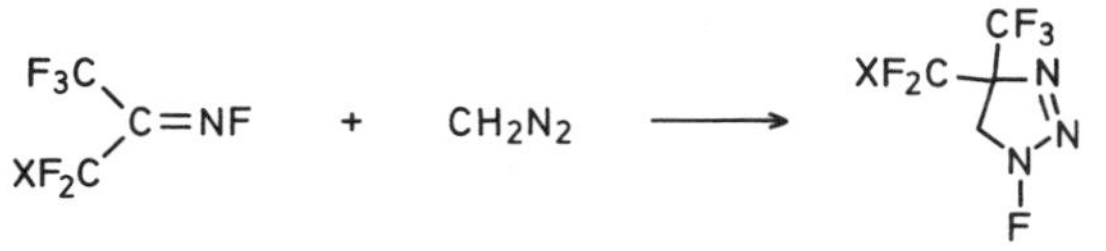

X = F;   *1-Fluoro-4,5-dihydro-4,4-bis(trifluoromethyl)-1,2,3-triazole*

X = CF₃; *1-Fluoro-4,5-dihydro-4-pentafluoroethyl-4-trifluoromethyl-1,2,3-triazole*

Where the C=N double bond stands *alpha* to the diazo group an intramolecular 1,5-cyclization is possible[770, 771]; thus, for example, 3-arylimino-2-diazo-1-indanones are converted into *3-arylindeno[1,2-d]triazol-8(3H)-ones*[771]:

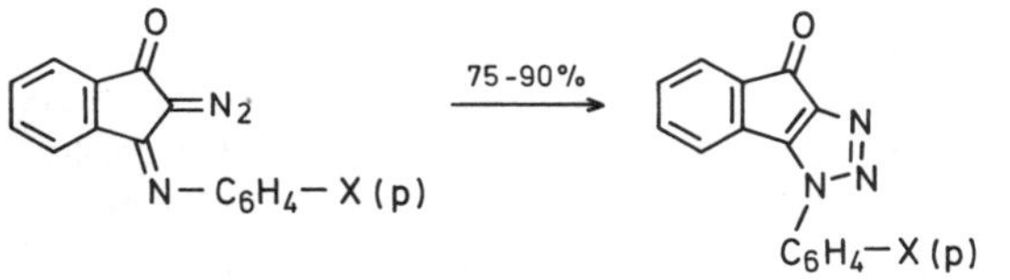

α-Iminodiazoalkanes and 5-amino-1,2,3-tri-azoles are in a mutual equilibrium whose position is determined by the nature of the substituents R^1 and R^2 and by the solvent used[772]:

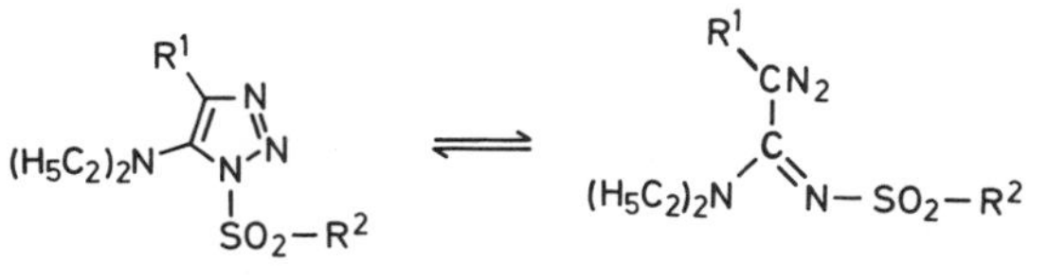

1,2,3-Triazoles are formed from diazo compounds and carbodiimides; both diazomethane[400] and ω-diazo carboxylic acid esters[764] can add to the C=N double bond.

If nitrogen is split off during the addition of di-azoalkanes to the C=N double bond, aziri-dines[769, 764] or their transformation products[773] result.

(Trifluoromethyl)imidocarbonyl fluoride adds diazomethane in xylene at 0° to form *2,2-di-fluoro-1-trifluormethylaziridine*[774]:

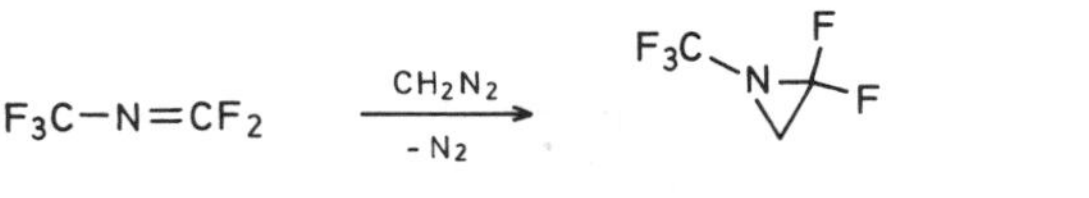

Immonium salts (above all the perchlorates) react smoothly with diazoalkanes to form *aziridinium salts* (50–90%)[400, 775]. Thus, 1-cyclohexylidenepi-peridinium perchlorate affords *6-azoniadispi-ro[5.0.5.1]tridecane perchlorate* on reaction with diazomethane[776]:

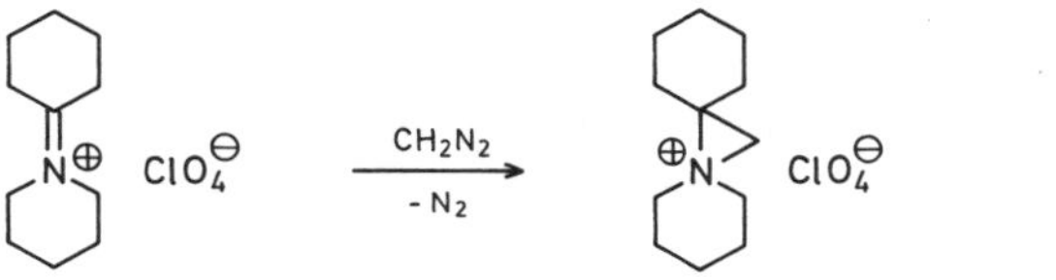

Unlike thioisocyanates (see p. 276) isocyanates furnish no uniform reaction product with diazo compounds[400]; it is apparently difficult to estab-lish any generally applicable laws governing this reaction.

With *isonitriles* diazo compounds form *keten-imines* (50%)[777, 778]. The reaction may be initiated both photolytically[777] and thermally[778]. It can be interpreted as the addition of two formally bivalent reagents; it is electrophilic relative to the carbon atom of the isocyanide group[777]:

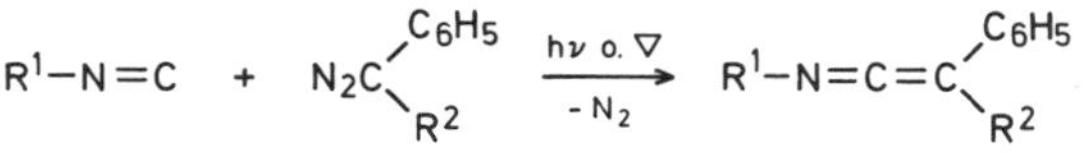

R^1 = tert.−C_4H_9 R^2 = C_6H_5, $COOCH_3$

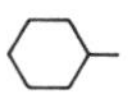

[767] *P. Kadaba*, Tetrahedron 22, 2453 (1966); *P. Kadaba*, Angew. Chem. 79, 281 (1967).

[768] *G.R. Krow*, Angew. Chem. 83, 455 (1971).

[769] *B.L. Dyatkin, K.N. Makarov, I.L. Knunyants*, Te-trahedron 27, 51 (1971).

[770] *M. Regitz, W. Anschütz*, Chem. Ber. 102, 2216 (1969).

[771] *M. Regitz, H. Schwall*, Justus Liebigs Ann. Chem. 728, 99 (1969).

[772] *M. Regitz, G. Himbert*, Tetrahedron Lett. 2823 (1970).

[773] *A.J. Speziale, C.C. Tung, K.W. Ratts, A. Yao*, J. Amer. Chem. Soc. 87, 3460 (1965).

[774] *A.L. Logothetis*, J. Org. Chem. 29, 3049 (1964).

[775] *H.O. Bernhard, V. Snieckus*, Tetrahedron 27, 2091 (1971).

[776] *N.J. Leonard, K. Jann*, J. Amer. Chem. Soc. 84, 4806 (1962).

[777] *J.H. Boyer, W. Beverung*, Chem. Commun. 1377 (1969).

[778] *E. Ciganek*, J. Org. Chem. 35, 862 (1970).

Of the reactions between diazo compounds and substances containing $C\equiv N$ triple bonds, two are of particular interest.

The first is addition of diazomethane, diazoethane, or ethyl diazoacetate to activated *nitriles* (dicyanogen, cyanogen halides, phenyl cyanate), which leads to the formation of *2H-1,2,3-triazoles* or their *2-methyl* derivatives[400]:

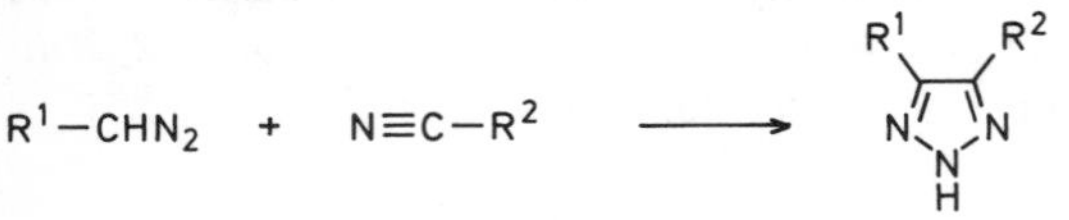

Examples are also known of *intramolecular* additions of diazo ketones to $C\equiv N$ triple bonds, *e.g.*, 2-diazoacetylbenzonitrile rearranges to *3H-indeno[1,2-d]triazol-8-one* in 84% yield[779]:

It is also noteworthy that *1-nitroso-3-phenyl-2-pyrazoline* is formed as the main product from the addition of diazomethane to *benzonitrile N-oxide*[780, 781].

The second reaction, which is more widely applicable and has been extensively studied, is the 1,3-dipolar cycloaddition of acylcarbenes and of ethoxycarbonylcarbenes (generated by the copper-catalyzed decomposition of diazo compounds) to nitriles, forming *oxazoles*. For instance, the decomposition of α-diazoacetophenone in the presence of benzonitrile and of copper cyanide gives *2,5-diphenyloxazole*[782]:

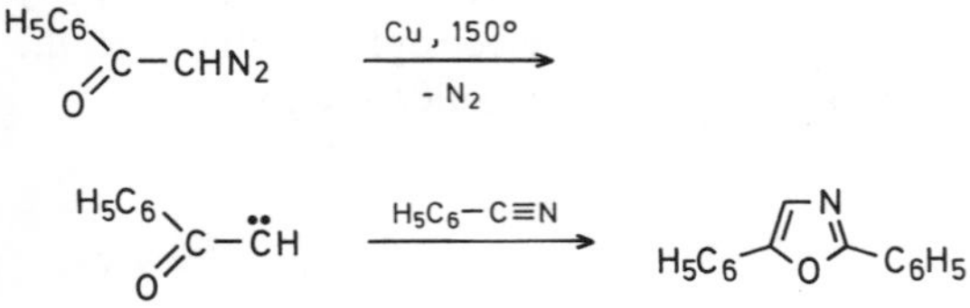

8.2.2.7 Reactions with compounds containing $N=N$ or $N\equiv N^{\oplus}$ groups

Aliphatic diazo compounds only react with $N=N$ double bonds when the azo group is activated by electron-withdrawing groups. No general rules concerning these processes can be formulated,

however[400]. Diazomethane adds to hexafluoroazomethane, without nitrogen being eliminated, forming *2,5-dihydro-1,2-bis(trifluoromethyl)-tetrazole*[783]. The reaction of diazoalkanes and of ethyl diazoacetate with dibenzoyldiazene or azodicarboxylic acid diesters[784-786] can proceed as a 1,3-dipolar addition, the carbenes formed under the reaction conditions reacting so as to produce *1,2,4-oxadiazoles*. Substituted *hydrazones* are formed in some cases, especially at elevated temperatures[784, 787-790]:

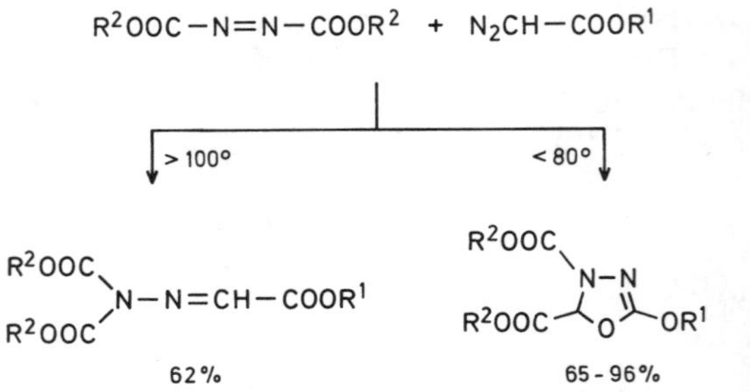

Thus, the appropriate *oxadiazole derivatives* are formed in the reactions of dibenzoyldiazene with diazodiphenylmethane or α-diazo ketones, or of azodicarboxylic diesters with alkyl diazoacetates or 9-diazofluorene at temperatures below 80°. Substituted hydrazones are formed at higher temperatures[764, 768].

Aliphatic α-diazo ketones, which are less nucleophilic, only react with azodicarboxylic diesters at temperatures above 100°, and the sole products are the bis(alkoxycarbonyl)-substituted monohydrazones of the corresponding 1,2-dicarbonyl compounds[400, 788, 789]. Since *1,2-dicarbonyl compounds* can be isolated easily by hydrolysis of these monohydrazones, this reaction is of some interest for preparative use[400].

2,2'-azodipyridine and 2,2'-azodiquinoline react with diazoalkanes, forming *4,5-dihydro-1,2,4-triazoles*[790]:

[779] G. Holt, D.K. Wall, J. Chem. Soc. 1428 (1965).
[780] G. Lo. Vecchino, M. Crisafulli, M. Aversa, Tetrahedron Lett. 1909 (1966).
[781] K. Nagarajan, P. Rajagopalan, Tetrahedron Lett. 5525 (1966).
[782] R. Huisgen, Chem. Ber. 97, 2628, 2864 (1964).

[783] V.A. Ginsburg, A.Ja. Yakubovich, Dokl. Akad. Nauk SSSR 142, 354 (1962); C.A. 57, 4518d (1962).
[784] R. Breslov, C. Yaroslavsky, S. Yaroslavsky, Chem. & Ind. (London) 1961 (1961).
[785] E. Fahr, H. Lind, Angew. Chem. 78, 381 (1966).
[786] E. Fahr, K. Königsdorfer, Tetrahedron Lett. 1873 (1966).
[787] E. Fahr, K. Döppert, K. Königsdorfer, F. Scheckenbach, Tetrahedron, 24, 1011 (1967).
[788] E. Fahr, K. Königsdorfer, F. Scheckenbach, Justus Liebigs Ann. Chem. 690, 138 (1965).
[789] E. Fahr, F. Scheckenbach, Justus Liebigs Ann. Chem. 655, 86 (1962).
[790] J. Market, E. Fahr, Tetrahedron Lett. 4337 (1967).

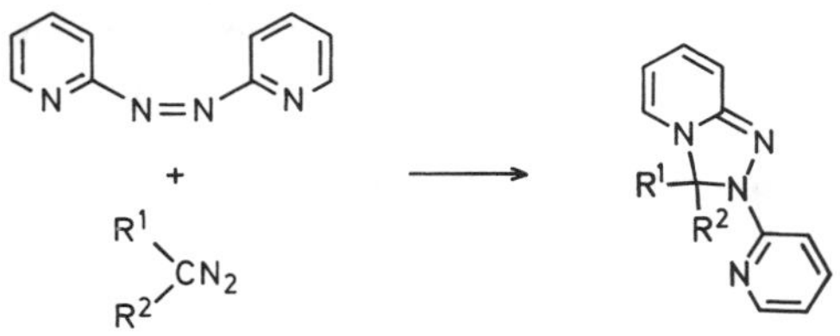

R^1 = H, R^2 = C$_6$H$_5$; *2,3-Dihydro-3-phenyl-2-(2-pyridyl)-1,2,4-triazolo[4,3-a]pyridine; 79%*

R^1 = R^2 = C$_6$H$_5$; *2,3-Dihydro-3,3-diphenyl-2-(2-pyridyl)-1,2,4-triazolo[4,3-a]pyridine; 71%*

R^1 = R^2 = Biphenyl-2,2'-diyl; *Fluorene-9-spiro-3-2-(2-pyridyl)-1,2,4-triazolo[4,3-a]pyridine; 20%*

Diazomethane is an exception in that on reacting with 2,2'-azodipyridine *formic acid 1,2-di-2-pyridylhydrazide* is formed[791].

Aryl-(α-chloroalkylidene)hydrazines result from reaction of aliphatic diazo compounds (diazomethane, diazoethane, or ethyl diazoacetate) with *diazonium salts,* i.e., with compounds which formally contain an $N \equiv N^{\oplus}$ triple bond[400,792]:

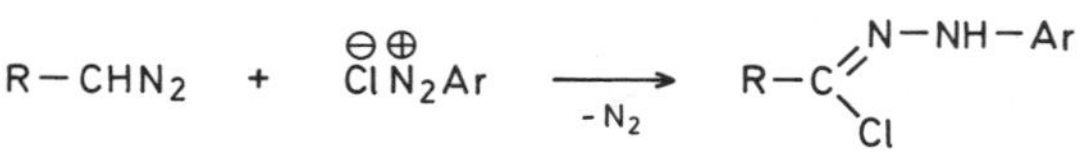

The reaction of diazonium salts with diazomethane can also proceed without nitrogen elimination, *tetrazoles*[793] being formed, *e.g.:*

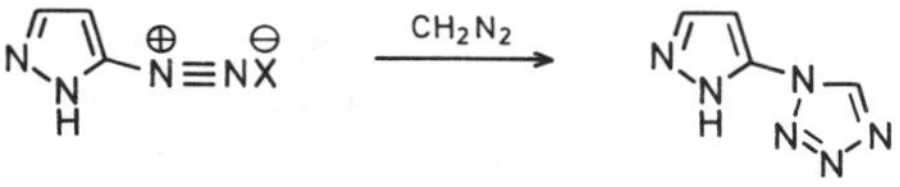

1-Pyrazol-5-yltetrazole

8.2.3 Insertion reactions

8.2.3.1 Reactions with *CH* bonds

Insertion of a diazoalkane group into a *CH* bond is often observed during the decomposition of aliphatic diazo compounds, particularly under photolytic conditions, but also when thermally induced (in the presence or absence of copper or cupric salts as catalysts).

The reactions mainly involve the carbenes formed by elimination of nitrogen from the diazo compounds, but this is not the only possibility in every case.

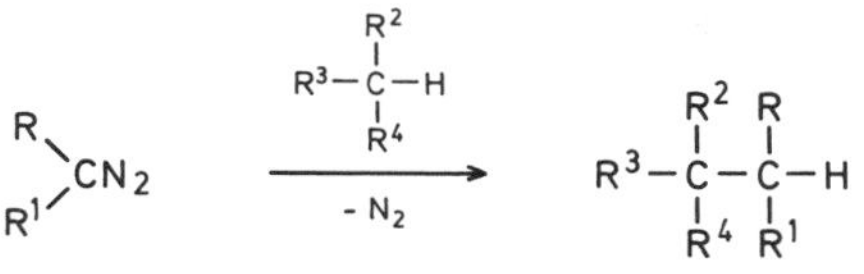

Methylene (which can be obtained by photolysis in both the gas and liquid phases) can be inserted into the *CH* bond of the solvent. Comparison of relative reaction rates shows that methylene has little[795] or no[794] selectivity toward *CH* bonds of various types. All other carbenes, generated by photolysis or thermolysis of diazoalkanes in solution, are much more selective as regards reactivity toward different types of *CH* bonds[433]:

tertiary > secondary > primary.

These insertion reactions often give preparative yields. For example, photolysis of ethyl diazoacetate in cyclohexane gives a good yield of *ethyl cyclohexylacetate*[456]; photolysis of diazocyclopentadiene in the same solvent produces *5-cyclohexylcyclopentadiene* (57%)[457].

Thermolysis of 2-diazohexafluoropropane in cyclohexane (in the presence of hydroquinone) gives a 60% yield of *1,1,1,3,3,3-hexafluoro-2-cyclohexylpropane* 32. If the reaction is carried out in the presence of *tert*-butyl hydroperoxide, nitrogen is not eliminated and two different products result, namely, *hexafluoro-2-propanone cyclohexylhydrazone* 33 and *(2-cyclohexylazo)-1,1,1,3,3,3-hexafluoropropane* 34[796].

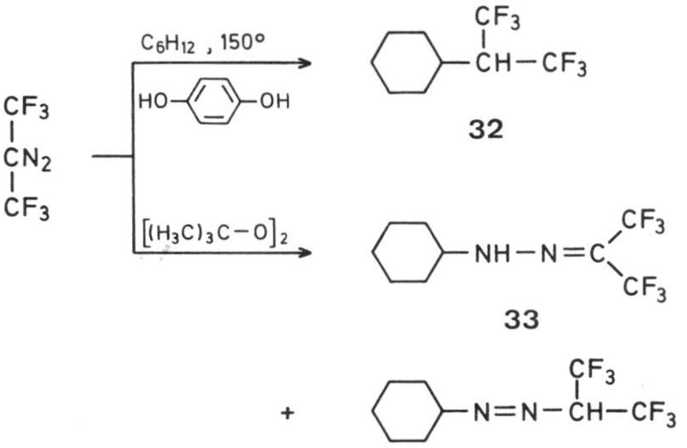

[791]A. Katritzky, S. Musierowicz, J. Chem. Soc. 78 (1966).

[792]R. Huisgen, H.J. Koch, Naturwissenschaften *41*, 16 (1954);
R. Huisgen, H.J. Koch, Justus Liebigs Ann. Chem. *591*, 200 (1955).

[793]H. Reimlinger, G.S.D. King, M.A. Peiren, Chem. Ber. *103*, 2821 (1970).

[794]W. von E. Doering, R.G. Buttery, R.G. Langhlin, N. Chandhuri, J. Amer. Chem. Soc. 78, 3224 (1956).

[795]B.M. Herzog, R.W. Carr, J. Phys. Chem. *71*, 2688 (1967).

The thermolysis of 5-(diazomethyl)-1,4-diphenyl-1,2,3-triazole in cyclohexane affords *5-(cyclohexylmethyl)-1,4-diphenyl-1,2,3-triazole* (67%)[797]:

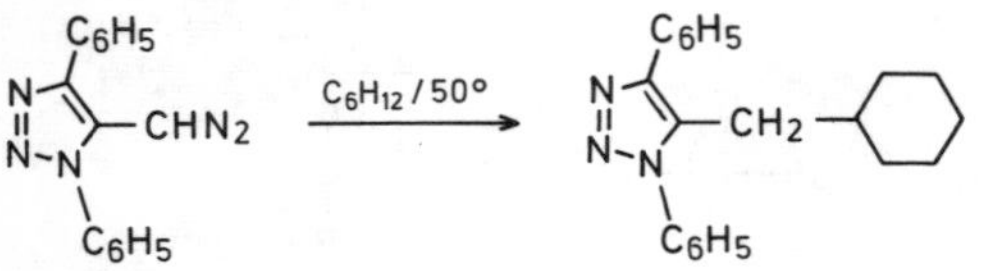

A peculiar example of an intramolecular reaction involving insertion in a *CH* bond is the photolysis of diethyl diazomalonate (in the presence of thiobenzophenone) to 4-hydroxybutyric acid γ-lactone (γ-butyrolactone) (80%)[798].

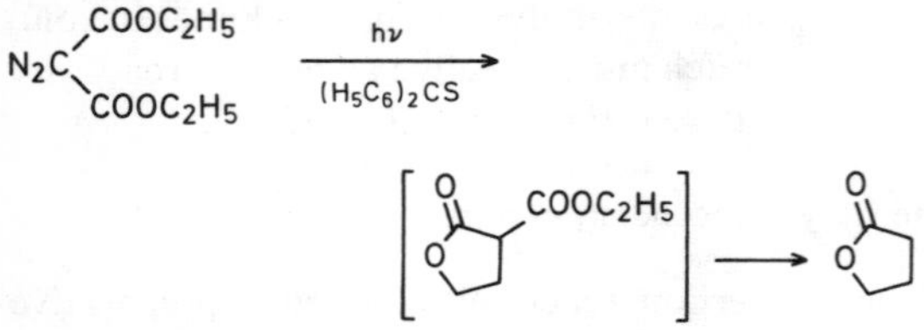

Photolysis of diazomethane in the presence of unsaturated hydrocarbons, in the gas phase, shows that addition of methylene to double bonds is about ten times faster than insertion into the allylic *CH* bond which, in turn, is faster than insertion in the vinyl position[433]. This is not a general rule, however.

Insertion products are formed also in the reactions of diazoalkanes with *aromatic CH* bonds. For example, thermolysis (at 20°) of 2-diazohexafluoropropane in benzene yields *β,β,β,β',β',β'-hexafluorocumene* as well as cycloheptatriene derivatives[430]. Pyrolysis of ethyl diazoacetate in molten hexamethylbenzene gives *ethyl 3-(pentamethylphenyl)propionate* as well as ring-fission products[799].

Photolysis of allyl diazoacetate in benzene affords only insertion products, namely, *allyl phenylacetate* and *allyl cycloheptatrienecarboxylate*[617]:

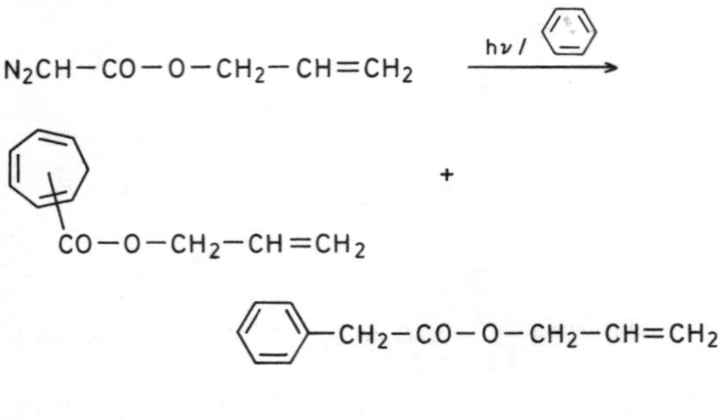

In certain cases insertion into *CH* bonds of aromatic and heteroaromatic compounds may become the main reaction[400,433].

The reaction of diazoalkanes with azulenes under photolytic conditions or in the presence of Lewis acids is a convenient method for the preparation of *1-alkylazulenes*[800]:

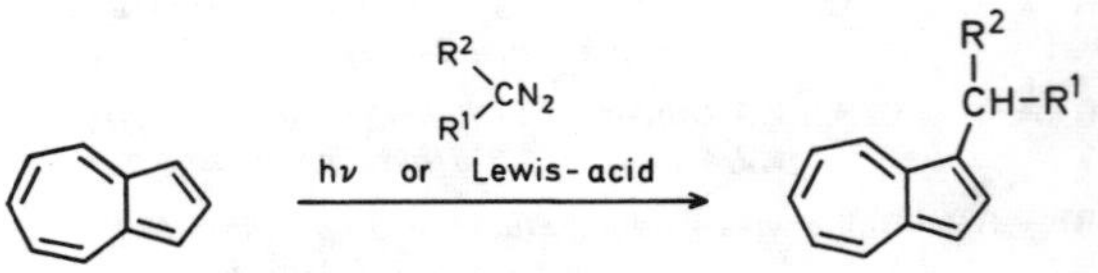

Intramolecular *C*-alkylation with the aid of diazo ketones has been used for the synthesis of *tetracyclic diterpenoids;* trifluoroacetic acid or fluoroboric acid are used as catalyst[801,802].

The photolysis of diazomethane in ethers (*e.g.,* methyl propyl ether[803] or tetrahydrofuran[858]) likewise yields *CH* insertion products, resulting mainly from attack at the *CH* bond adjacent to the oxygen atom[433,804]:

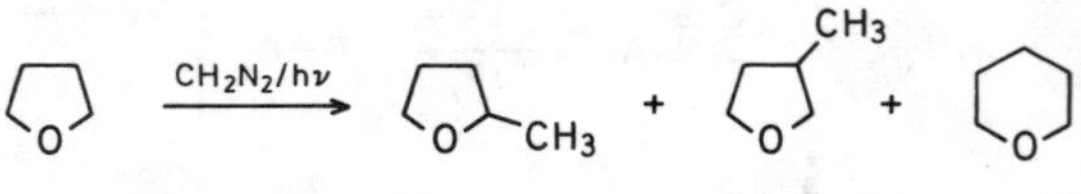

| Tetrahydro-2-methylfuran | Tetrahydro-3-methylfuran | Tetrahydropyran |

Electron-withdrawing substituent groups (*CO*, *SO₂*, *CF₃*, etc.) activate neighboring *CH* bonds in sertion reactions with diazoalkanes. An example is the conversion of 1,3,5-trithiane 1,1,3,3,5,5-hexaoxide (cyclic trimethyl sulfone) into its hexamethyl derivative, *2,2,4,4,6,6-hexamethyl-1,3,5-trithiane hexaoxide* (~100%)[805]:

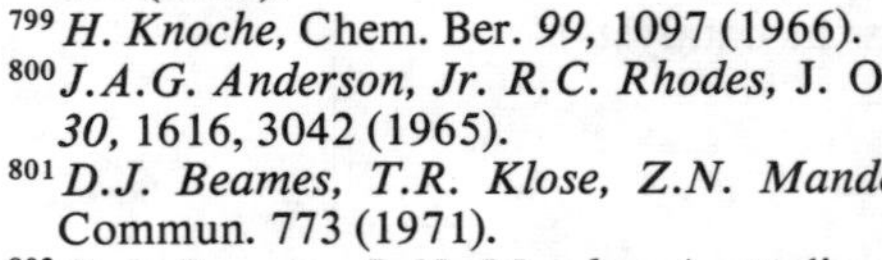

[797] *P.A.S. Smith, J.G. Wirth,* J. Org. Chem. *33,* 1145 (1968).
[798] *J.A. Kaufman, S.J. Weininger,* Chem. Commun. 593 (1969).
[799] *H. Knoche,* Chem. Ber. *99,* 1097 (1966).
[800] *J.A.G. Anderson, Jr. R.C. Rhodes,* J. Org. Chem. *30,* 1616, 3042 (1965).
[801] *D.J. Beames, T.R. Klose, Z.N. Mander,* Chem. Commun. 773 (1971).
[802] *D.J. Beames, L.N. Mander,* Australian J. Chem. 343 (1971).
[803] *H.M. Frey,* Rec. Trav. Chim. Pays-Bas *83,* 117 (1964).
[804] *H.M. Frey, M.A. Voisey,* Chem. Commun. 454 (1966).
[805] *F. Arndt, C. Martins,* Justus Liebigs Ann. Chem. *499,* 246 (1932).

[796] *W.J. Middleton, D.M. Gale, C.G. Krespan,* J. Amer. Chem. Soc. *90,* 6813 (1968).

Reacting diazomethane with 1,1,1,3,3,3-hexafluoro-2-propanol gives *hexafluoro-2-methoxy-2-methylpropane* (73%)[806].

Small amounts of *CH* insertion products are also obtained from reactions of ketones with ethyl diazoacetate in the presence of silver oxide[807].

In summary, it can be stated that the formation of *CH* insertion products must always be regarded as a possibility in reactions of diazoalkanes at high temperatures in the presence of copper, particularly during photolysis.

8.2.3.2 Reactions with carbon-halogen bonds

Aliphatic diazo compounds can be inserted into *C-Hal* bonds, nitrogen being eliminated. Results of the photochemical reaction of diazomethane with *tert*-butyl chloride have shown that in this case insertion into the *C-Hal* bond is easier than into a *CH* bond[433]:

$$(H_3C)_3CCl \xrightarrow{\ CH_2N_2\ /\ h\nu\ }$$

$$(H_3C)_3C-CH_2Cl \quad + \quad H_3C-CH_2-\underset{\underset{Cl}{|}}{C}(CH_3)_2$$

3-Chloro-2,2-dimethylpropane; 60%   *2-Chloro-2-methylbutane*; 40%

Ethyl 2-halo-4-pentenoates can be obtained in good yields by treatment of allyl halides with ethyl diazoacetate in the presence of copper or its salts[600,808], cf.[433,809].

$$H_2C=CH-CH_2-X \xrightarrow[-N_2]{\ N_2CH-COOC_2H_5\ }$$

$$H_2C=CH-CH_2-\underset{\underset{X}{|}}{CH}-COOC_2H_5$$

With ethoxycarbonylcarbene, 2,3-dichloropropene and 2,3-dibromopropene afford the products of insertion into the *CH* bond and also cyclopropane derivatives by addition to the double bond[810].

The reactions of ethyl diazoacetate with diaryl- and triarylhalomethanes, and with 8-bromo-9-phenylfluorene and 8-bromo-9-(4-tolyl)fluorene, are complex[600,811,812], *e.g.*:

$$(H_5C_6)_3C-Br \xrightarrow[-N_2]{\ N_2CH-COOC_2H_5\ /\ CuSO_4\ }$$

$$(H_5C_6)_2C=C\underset{COOC_2H_5}{\overset{C_6H_5}{<}} \quad + \quad Br-CH_2-COOC_2H_5$$

Ethyl triphenylacrylate   *Ethyl bromoacetate*

The high quantum yield of the photochemical reaction of diazomethane with *polyhalomethanes* clearly shows that this is a radical-chain reaction. The main product from carbon tetrachloride is *1,3-dichloro-2,2-bis(chloromethyl)-propane* (60%)[813,814]. A detailed study has been made of the reaction mechanism and the intermediates involved[433].

The reaction of *acyl halides* with excess diazomethane is an important method for the preparation of *α-diazo ketones* (see p. 239).

However, if no excess of diazomethane is present, or if diazoalkanes are used which possess no hydrogen atom attached to the carbon atom bearing the diazo group, insertion into the *C-Hal* bond may occur[400].

8.2.3.3 Reactions with *CO*, *CS* and *CN* bonds

Ethers are often used as solvents when working with diazoalkanes since they do not react mutually in the dark and in the absence of catalysts.

In the gas phase, and under *UV*-irradiation, diazomethane and ethers mainly form *CH* insertion products; it has been observed that small amounts of *CO* insertion products are formed on photolysis of diazomethane in tetrahydrofuran (see p. 280).

A high temperatures or on photolysis ethyl diazoacetate causes ethers (including phenolic ethers) to undergo an unusual *ether interchange*, 2-hydroxyalkanecarboxylic or α-aryl-α-hydroxyacetic acid esters being formed in yields of up to 40%[400].

[806] *H.J. Kötzsch*, Chem. Ber. *99*, 1143 (1966).

[807] *A. Wasfi*, J. Indian Chem. Soc. *47*, 341 (1970).

[808] *I.A. D'yakonov, N.B. Vinogradova*, Zh. Obshch. Khim. *21*, 851 (1951); C.A. *46*, 440d (1952); C.A. *23*, 244 (1953); C.A. *48*, 3318i (1954).

[809] *W. Kirmse, M. Buschhoff*, Chem. Ber. *102*, 1087 (1969).

[810] *I.A. D'yakonov, T.V. Domareva*, Zh. Obshch. Khim. *25*, 934 (1955); C.A. *50*, 3221i (1956); C.A. *25*, 1486 (1955); C.A. *50*, 4795h (1956).

[811] *I.A. D'yakonov, T.V. Domareva*, Zh. Obshch. Khim. *29*, 3098 (1959); C.A. *54*, 14211f (1960).

[812] *L.P. Danilkina, T.V. Domareva, I.A. D'yakonov*, Vestnik Leningrad. Univ. *12, N16, Ser. Fiz. Khim. N3, 131 (1957)*: C.A. *52*, 6907g (1958).

[813] *W.H. Urry, J.R. Eiszner*, J. Amer. Chem. Soc. *73*, 2977 (1951);
W.H. Urry, J.R. Eiszner, J. Amer. Chem. Soc. *74*, 5822 (1952).

[814] *W.H. Urry, N. Bilow*, J. Amer. Chem. Soc. *86*, 1815 (1964).

Under the influence of mineral or Lewis acids α-diazo ketones[400, 815–817] and α-diazosulfones[818] which contain an alkoxy or mercapto group at the 4- or 5-position can undergo intramolecular cyclization. Thus, 4-alkoxy and 2-(alkylthio)-1-diazo-2-butanones rearrange respectively to *tetrahydro-3-furanones* and *tetrahydro-3-thiophenones* on treatment with dilute hydrochloric acid, for example[816]:

$$RX-CH_2-CH_2-CO-CHN_2 \xrightarrow{-N_2}$$

X = O, S

Cyclic ethers such as 2-phenyloxirane [(epoxyethyl)benzene] and 2-phenyloxetane [(1,3-epoxypropyl)benzene] afford a mixture of *cis* and *trans 3-phenyloxetane* and *ethyl tetrahydro-3-phenyl-2-furoate* on reaction with ethyl diazoacetate (photolysis or thermocatalytic decomposition) – the reaction is more selective in the second case[433, 819].

$$(CH_2)_n \underset{O}{\diagdown} C_6H_5 \xrightarrow{N_2CH-COOC_2H_5 / h\nu \text{ or } \triangledown}$$

$$(CH_2)_n \underset{O}{\diagdown} \overset{C_6H_5}{\diagup} COOC_2H_5$$

n = 1, 2

During the copper-catalyzed decomposition of diazomethane in vinyloxirane (3,4-epoxy-1-butene) the insertion of methylene into the *CO* bond was observed; considerable amounts of *3-vinyloxetane (3,5-epoxy-1-pentene)* (29%) were obtained in addition to *cyclopropyloxirane [(epoxyethyl)cyclopropane]* (64%)[647]:

$$\xrightarrow{CH_2N_2/CuJ \text{ or } CuCl_2}$$

64%   +   29%   +   7%

1,4-Benzodioxane and *2-phenyl-1,3-dioxolane* likewise afford insertion products with ethyl diazoacetate[400].

Formation of *CO* insertion products has been demonstrated during photolysis of diazomethane in the gas phase in acetals or orthocarboxylic acid triesters[809].

A preparatively important insertion reaction of aliphatic diazo compounds into *CO* and *CS* bonds[820, 821] is the reaction of ethyl diazoacetate with acetals and thioacetals in the presence of boron trifluoride etherate to form *2,3-dialkoxy* and *2,3-dialkylthiocarboxylic acids*. When α-diazo ketones are employed, *2,3-dialkoxy* and *2,3-dialkylthio ketones* **35** respectively are formed:

$$\underset{R}{\overset{H}{>}}C\underset{XR^1}{\overset{XR^1}{<}} \; + \; N_2CH-\underset{O}{\overset{\parallel}{C}}-R^2$$

$$\xrightarrow[-N_2]{BF_3}$$

$$R-\underset{}{\overset{R^1-X}{\underset{}{CH}}}-\underset{}{\overset{X-R^1}{\underset{}{CH}}}-CO-R^2$$

35

X = O, S ; R = H, CH$_3$, C$_6$H$_5$

R^1 = CH$_3$, C$_2$H$_5$; R^2 = OC$_2$H$_5$, CH$_3$, o-Cl-C$_6$H$_4$, p-CH$_3$O-C$_6$H$_4$
 p-NO$_2$-C$_6$H$_4$

Trialkyl orthocarboxylates and *orthothiocarboxylates* react analogously with ethyl diazoacetate or α-diazo ketones:

$$\underset{O}{\overset{R}{>}}C-CHN_2 \; + \; R^2-C(XR^1)_3 \xrightarrow[-N_2]{BF_3 - etherate}$$

$$\underset{O}{\overset{R}{>}}C-\underset{XR^1}{\overset{XR^1}{\underset{|}{CH}}}-\underset{XR^1}{\overset{XR^1}{\underset{|}{C}}}-R^2 \xrightarrow{-HXR^1} \underset{O}{\overset{R}{>}}C-\underset{XR^1}{\overset{XR^1}{C}}=C\underset{XR^1}{\overset{R^2}{<}}$$

36   **37**

X = O, S
R = CH$_3$, OC$_2$H$_5$, Ar
R^1 = CH$_3$
R^2 = H, CH$_3$

Elimination of alcohol or mercaptan from the ketones **36** enables esters of *2,3-dialkoxyacrylic*

---

[815] *W. Hampel, J. Friedrich*, Z. Chem. Soc. *86*, 343 (1970).

[816] *J.H. Sperna-Weiland*, Rec. Trav. Chim. Pays-Bas *83*, 81 (1964).

[817] *W. Hampel*, J. Prakt. Chem. *311*, 78 (1969).

[818] *A.M. van Leusen, P. Richters, J. Strating*, Rec. Trav. Chim. Pays Bas *85*, (3), 323 (1966).
 H. Nozaki, H. Takaya, R. Noyori, Tetrahedron Lett. 2563 (1965).

[819] *H. Nozaki, H. Takaya, R. Noyori*, Tetrahedron Lett. 5239 (1966);
 H. Nozaki, H. Takaya, R. Noyori, Tetrahedron *22*, 3393 (1966).

[820] *A. Schönberg, K. Praefcke*, Tetrahedron Lett. 2043 (1964).

[821] *A. Schönberg, K. Praefcke*, Chem. Ber. *100*, 778 (1967).

and *2,3-(dialkylthio)acrylic acids* and of correspondingly substituted α,β-*unsaturated ketones* **37** to be prepared.

The *limits* of the *reaction* are defined by spatial factors. Thus, no reaction takes place when the central carbon atom of the acetals or orthocarboxylic acid triesters (or their sulfur analogs) carries space-filling substituents[821].

A more complex interaction takes place between 2-diazo-2'-methoxyacetophenone and orthoformic and orthotrithioformic acid triesters (again with added boron trifluoride etherate). Here, the result of the *CO* and *CS* insertion reactions and subsequent ring closure is the formation of *2,3-dihydrobenzofuranone* derivatives[822]:

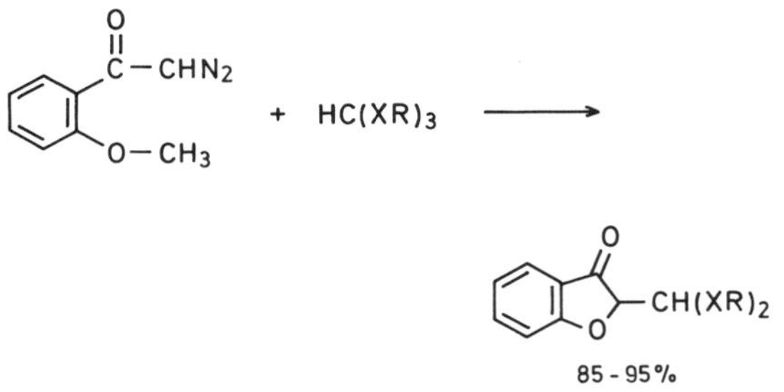

85 - 95%

During the *sensitized* photolysis (in the presence of benzophenone) of dimethyl diazomalonate in alkyl allyl sulfides addition products **39** as well as formation of the *CS*-insertion products **38** were observed[823-825], *e.g.*:

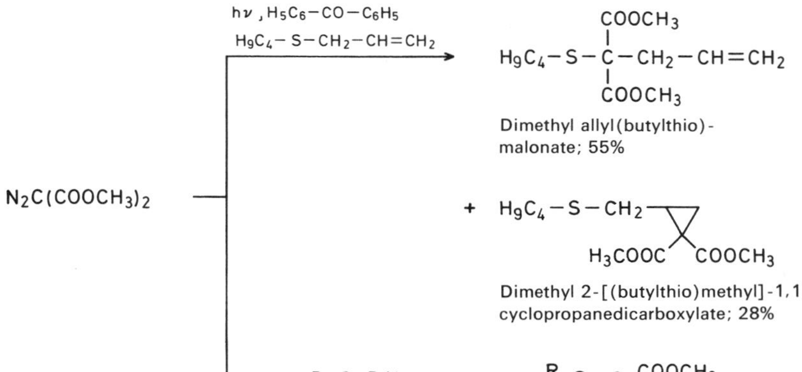

Direct photolysis or the copper(II) sulfate catalyzed decomposition of diethyl diazomalonate in dialkyl sulfides or sulfoxides leads to the formation of stable *ylides* (50–90%)[825, 826].

8.2.3.4 Reactions with *OH* bonds

8.2.3.4.1 Reaction with carboxylic acids

Reacting diazoalkanes with carboxylic acids represents one of the most important methods for the preparation of *carboxylic acid esters*, particularly methyl esters.

In contrast to the reactions with aqueous mineral acids (see p. 256) no free alkyldiazonium ions are formed during reaction between carboxylic acids and diazoalkanes in nonpolar solvents. Instead the diazonium carboxylate **41** loses nitrogen and rearranges into an ester **42** directly. A second reaction path consists in the formation of a secondary ion pair **43** which transforms into the carboxylic acid ester **42** or (if the reaction is conducted in alcohol) dissociates into a carbonium ion **44** and carboxylate anion **45**, from which ethers **46** and carboxylic acids **47** are formed with alcohol[827-830]:

[822] *A. Schönberg, K. Praefcke, J. Kohtz*, Chem. Ber. *99*, 3076 (1966).
[823] *W. Ando, K. Nakayama, K. Ichibori, T. Migita*, J. Amer. Chem. Soc. *91*, 5164 (1969).
[824] *W. Ando, T. Yagihara, S. Kondo, K. Nakayama, H. Yamato, S. Nakaido, T. Migita*, J. Org. Chem. *36*, 1732 (1971).

[825] *W. Ando, T. Yagihara, S. Tozune, S. Nakaido, Migita*, Tetrahedron Lett. 1979 (1969);
W. Ando, T. Yagihara, S. Tozune, S. Nakaido, T. Migita, J. Amer. Chem. Soc. *91*, 2786 (1969).
[826] *F. Dost, I. Gosselck*, Tetrahedron Lett. 5091 (1970).

$$\underset{R^2}{\overset{R^1}{>}}CN_2 \;+\; R-COOH \;\xrightarrow{\text{slow}}\; \left[\underset{R^2}{\overset{R^1}{>}}CHN_2^{\oplus}\right]\left[\overset{\ominus}{O}\overset{O}{\overset{\|}{-}}C-R\right]$$

41

$$-N_2 \downarrow$$

$$\underset{R^2}{\overset{R^1}{>}}\overset{\oplus}{C}H \;+\; \overset{\ominus}{O}OC-R \;\longleftarrow\; \left[\underset{R^2}{\overset{R^1}{>}}\overset{\oplus}{C}H\right]\left[\overset{\ominus}{O}\overset{O}{\overset{\|}{-}}C-R\right] \;\underset{\longleftarrow}{\longrightarrow}\; \underset{R^2}{\overset{R^1}{>}}CH-O-\overset{O}{\overset{\|}{C}}\!\!\overset{}{\underset{R}{}}$$

44   **45**   **43**   **42**

$$\downarrow {\scriptstyle H_5C_2-OH}$$

$$\underset{R^2}{\overset{R^1}{>}}CH-OC_2H_5 \;+\; R-COOH$$

46   **47**

The studies of the rate of decomposition of di-azoalkanes induced by carboxylic acids led to the working out of a method for *determining* the *acidity* tf the carboxylic acids[400].

Complications can arise during the reactions between the carboxylic acids and diazo compounds[827, 828, 831]. By reacting carboxylic acids with aliphatic diazo compounds whose protonation products tend to undergo rearrangements, the isomers of the anticipated esters and olefins, too, can be obtained[400, 832].

From α-diazo ketones and carboxylic acids the corresponding α-acetoxy ketones can be obtained in good yield. *Trans*-fixed α-diazo-β-dicarbonyl compounds react with carboxylic acids only on adding copper catalysts[400]. Despite the great reactivity of diazo compounds during reaction with carboxylic acids, the *diazoacetic acid* was isolated in substance[833].

8.2.3.4.2 Reaction with alcohols and phenols

Alcohols without a polar substituent are not acid enough and hence are inert toward diazoalkanes. With the aid of *catalysts* (boron trifluoride, alumi-num chloride, and borofluoric acid are the most effective) it is possible to alkylate alcohols to ethers with diazoalkanes (40–90%)[657, 834–836]:

$$R^1-OH \;+\; N_2C\underset{R^3}{\overset{R^2}{<}} \;\xrightarrow[-N_2]{\text{Cat.}}\; R^1-O-\overset{R^2}{\underset{|}{CH}}-R^3$$

Numerous investigations show that the reactivity of the alcohols toward diazoalkanes falls in the order primary > secondary > tertiary[835, 836].

Sterically hindered alcohols and those which readily dissociate at the *CO* bond (*e.g.*, triphenylmethanol) cannot be alkylated under these conditions[657]. There is no concordant opinion concerning the mechanism of this reaction, cf. ref.[657].

If the alcohol contains activating groups (*Cl, CN, COOH, CF₃*), then the methylation with diazomethane can be carried out directly without the addition of catalysts.

Alkylation of *sugars* and *cellulose* with diazomethane[400] and higher diazoalkanes[837] has been described in a number of papers.

[827] *R.A. More, O. Ferrall, W.K. Kwok, S.I. Miller,* J. Amer. Chem. Soc. *86,* 5553 (1964).

[828] *R.A. Moss, F.C. Shulman,* Tetrahedron, *24,* 2881 (1968).

[829] *W. Jungelt, F. Pragst,* Tetrahedron *24,* 5123 (1968).

[830] *N.B. Chapman, M.R.J. Dack, J. Shorter,* J. Amer. Chem. Soc. B, 834 (1971).

[831] *N.B. Chapman, A.Ehsan, J. Shorter, K.J. Toyne,* Tetrahedron Lett. 1049 (1968).

[832] *W. Kirmse, K. Horn,* Tetrahedron Lett. 1827 (1967).

[833] *K. Schank,* Justus Liebigs Ann. Chem. *723,* 505 (1969).

[834] *E. Müller, W. Rundel,* Angew. Chem. *70,* 105 (1958).

[835] *M.C. Caserio, J.D. Roberts, M. Neeman, W.S. Johnson,* J. Amer. Chem. Soc. *80,* 2584 (1958); *M.C. Caserio, J.D. Roberts, M. Neeman, W.S. Johnson,* Tetrahedron *6,* 36 (1959).

[836] *E. Müller, M. Bauer, W. Rundel,* Justus Liebigs Ann. Chem. *677,* 55 (1964).

[837] *M. Parmale, E.A. Plisko, S.N. Danilov,* Zh. Prikl. Khim. (Leningrad) *39,* 2310 (1966); C.A. *66,* 20068 (1967).

In general, *phenols* are sufficiently acid to be able to react with diazoalkanes in the absence of catalysts. The rate of etherification is diminished if the hydroxy group is chelated, for instance, by an carbonyl group in the *ortho* position. Sterically hindered 2,4,6-tri-*tert*-butylphenol could not be etherified with diazomethane[400].

α-Diazo ketones, 2-diazo 1,3-diketones, and α-diazo carboxylic esters do not react with alcohols; their degree of stability is so high that the solid representatives of this class of compounds can be recrystallized from alcohols. Replacement of the diazo group by the alkoxy group in these compounds may be effected in the presence of boron trifluoride etherate[400] or sulfuric acid[449].

Photolysis and *thermocatalytic decomposition* (in the presence of copper or its salts) of the aliphatic diazo compounds in alcohols, too, can lead to ether formation:

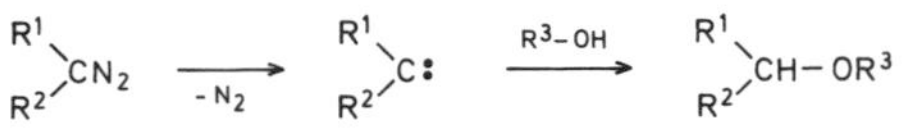

$$R^1R^2CN_2 \xrightarrow{-N_2} R^1R^2C: \xrightarrow{R^3-OH} R^1R^2CH-OR^3$$

However, the yield of ether is seldom satisfying[424, 457, 838-840] because the corresponding carbenes can undergo a variety of side-reactions (see p. 265).

One possible side-reaction is reduction of the diazo group[649, 840]. Thus, for example, during photolysis of diazo-1,3-sulfones in different alcohols the following reaction product ratio was observed[840]:

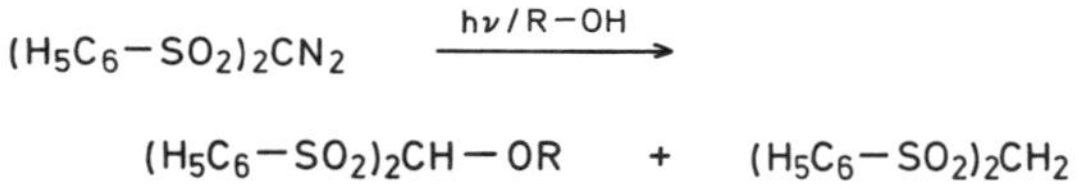

$$(H_5C_6-SO_2)_2CN_2 \xrightarrow{h\nu / R-OH}$$

$$(H_5C_6-SO_2)_2CH-OR \quad + \quad (H_5C_6-SO_2)_2CH_2$$

R—OH:		
H_3C-OH	82	18
H_5C_2-OH	66	34
$(H_3C)_2CH-OH$	15	85
$(H_3C)_3C-OH$	100	0

An interesting example of a reaction in which the replacement of the diazo group by the alkoxy group and the transetherification reaction are observed to proceed simultaneously is photolysis of the esters of diazoacetic acid in various alcohols[841]. The main products are compounds of Type **48**:

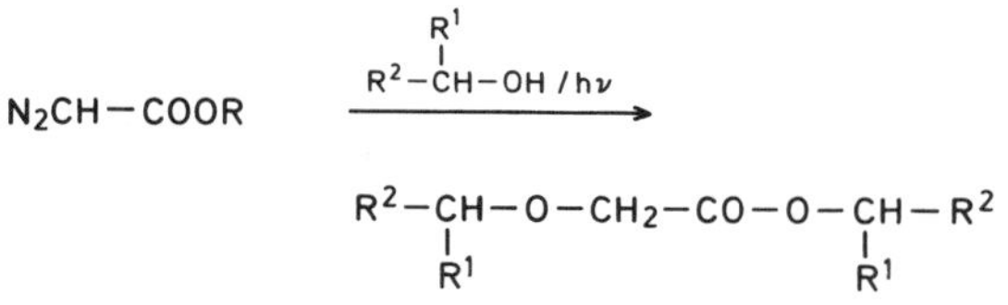

$$N_2CH-COOR \xrightarrow{R^2-CH(R^1)-OH / h\nu}$$

$$R^2-CH(R^1)-O-CH_2-CO-O-CH(R^1)-R^2$$

48

65 - 68%

8.2.3.4.3 Reactions with tautomeric systems

In tautomeric and potentially tautomeric systems it is mainly the acid hydrogen atom that reacts with diazoalkanes, especially diazomethane, and two series of derivatives can form. The reaction proceeds the more readily the more acid the character of the hydrogen atom is.

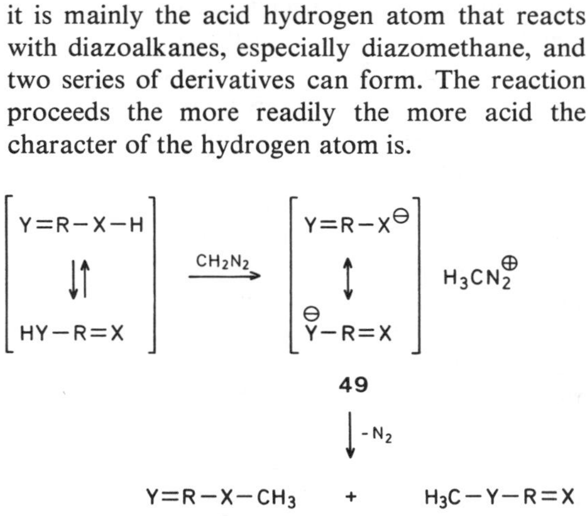

$$\begin{bmatrix} Y=R-X-H \\ \updownarrow \\ HY-R=X \end{bmatrix} \xrightarrow{CH_2N_2} \begin{bmatrix} Y=R-X^\ominus \\ \updownarrow \\ {}^\ominus Y-R=X \end{bmatrix} H_3C N_2^\oplus$$

49

$$\downarrow -N_2$$

$$Y=R-X-CH_3 \quad + \quad H_3C-Y-R=X$$

It is postulated that the reaction proceeds *via* the stage of a pair of methyldiazonium ions **49**[441], and this explains why during methylation with diazomethane the reaction products obtained are often not the same as when other methylating agents are employed.

Enolizable α-dicarbonyl compounds

$$\left(-CH_2-\underset{O}{\overset{||}{C}}-\underset{O}{\overset{||}{C}}- \quad \rightleftharpoons \quad -CH=\underset{HO}{\overset{|}{C}}-\underset{O}{\overset{||}{C}}- \right)$$

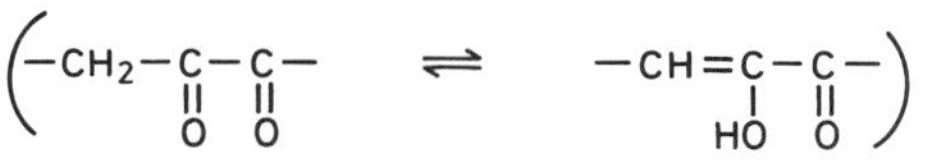

afford mainly *O*-derivatives with diazoalkanes (see p. 273), but in some cases formation of small quantities of *C*-methylation products was observed as well on reaction with diazomethane[400].

For the reaction of the enolizable β-dicarbonyl compounds

$$\left(-\underset{O}{\overset{||}{C}}-CH_2-\underset{O}{\overset{||}{C}}- \quad \rightleftharpoons \quad -\underset{OH}{\overset{|}{C}}=CH-\underset{O}{\overset{||}{C}}- \right)$$

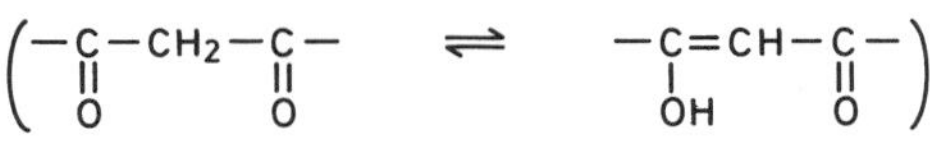

formation of the *enol ethers,* too, is a typical manifestation. The conversion of the *trans*-fixed enols which are entirely or approximately coplanar (vinylogous carboxylic acids) proceeds unambiguously and rapidly. Examples are 1,3-cyclobutanediones, 1,3-cyclopentanediones, and 1,3-cyclohexanediones[400] and 2,2,6,6-tetraalkylte-

[838] D. Bethell, G. Stevens, P. Tickle, Chem. Commun. 793 (1970).

[839] P. Yates, J. Amer. Chem. Soc. 74, 5376 (1952).

[840] J. Dieckmann, J. Org. Chem. 30, 2272 (1965).

[841] T. DoMinh, O.P. Strausz, H.E. Gunning, J. Amer. Chem. Soc. 91, 1261 (1969).

trahydro-3, 5-pyrandiones[744], which furnish *O-methyl ethers* quantitatively during the reactions. No indications of the simultaneous formation of identifiable amounts of *C*-methyl derivatives have been obtained to date[842]. Unsymmetrically substituted 1, 3-cycloalkanediones generally give mixtures of the two structurally isomeric enol ethers (see, *e.g.*, the reaction between *trans*-decahydro-1, 3-naphthalenedione with diazomethane to *4a, 5, 6, 7, 8, 8a-hexahydro-3-methoxy-1(4H)-naphthalenone* and *4a, 5, 6, 7, 8, 8a-hexahydro-4-methoxy-2(1H)-naphthalenone*[843].

Open-chain 1, 3-dicarbonyl compounds, for instance, 2, 4-pentanedione, in which the acidity of the enolic hydroxy group is sharply diminished by chelate formation, therefore react with diazoalkanes more slowly. The addition of methanol has been found to promote the reaction. The primary product are the *cis-syn-enol ethers* **51** which rearrange further[400]. That small amounts of *3-methyl-2, 4-pentanedione* **52** are formed is considered to be a further proof of the existence of the ion pair **50**[844]:

2-Methoxy-pent-2-en-4-one

In general, unsymmetrical acyclic β-diketones and 2-acylcycloalkanones furnish two isomeric enol ethers each of which may be present in different stereoisomeric forms[400].

β-Formyl carbonyl compounds afford mainly *O*-derivatives of hydroxymethylene form[567, 845, 846]:

3-Oxocarboxylic acid esters (*e.g.*, ethyl acetoacetate) require the presence of methanol for a reaction with ethereal diazomethane solution to take place. The main product is *ethyl 3-methoxy-2-butenoate* side by side with small amounts of *C*-methyl derivatives and epoxides[400]. Cyclic 3-oxocarboxylic esters such as ethyl tetrahydro-4-oxo-3-thiophenecarboxylate smoothly afford enol ethers with ethereal diazomethane solutions[847], for example, *ethyl 2, 5-dihydro-4-methoxy-3-thiophenecarboxylate*.

β-Oxolactones (five- and six-membered ones), which cannot form intramolecular hydrogen bonds, and whose hydrogen atom is therefore sufficiently acid, react rapidly and quantitatively with diazomethane to form enol ethers[512], *e.g.*:

1-Hydroxy-β-methoxycyclo-hexanecrotonic acid δ-lactone

Lactams able to react in the tautomeric lactim-lactam form yield *N*- and *O*-derivatives on reacting with diazoalkanes. The numerous experimental data available, which were obtained with a varying degree of precision, do not permit soundly based generalizations about the relationship between the course of the reaction and the structure of the reagents[400, 848].

Aci-nitro compounds

in general do not react with diazoalkanes[849], but if the nitro compound is activated by strongly electron-withdrawing substituents *methyl esters of nitronic acids* are formed[400]. For accelerating the

[842] *R. Gomper*, Chem. Ber. *93*, 189, 197 (1960).

[843] *H. Mühle, C. Tamm*, Helv. Chim. Acta *45*, 1475 (1962).

[844] *G.S. Hammond, R.M. Williamo*, J. Org. Chem. *27*, 3775 (1962).

[845] *H. Henecka*, Chemie der β-Dicarbonylverbindungen, Springer Verlag, Berlin (1950).

[846] *I.K. Korobizina, I.I. Popova, N.N. Gaidamovich, Yu.K. Yur'ev*, Zh. Obshch. Khim. *31*, 2542 (1961); C.A. *56*, 8662c (1961).

[847] *B.A. Arbutzov, O.A. Erastov, A.B. Remizov*, Isv. Akad. Nauk SSSR, Ser. Khim. (1), 180 (1969); C.A. *70*, 106285g (1969).

[848] *J.L. Wong, D.S. Fuchs*, J. Org. Chem. *36*, 848 (1971).

[849] *N.R. Ghosh, C.R. Ghoshal, S. Shah*, Chem. & Ind. (London) *15*, 493 (1969).

process it is recommended to employ freshly prepared *aci*-nitro compounds in the reaction[850]. The reaction between trinitromethane and diazomethane leads to mixed *dinitroformaldehyde methoxynitrone, 1,1,1-trinitroethane*, and *3-nitro-2-isoxazoline N-oxide*. How much of each reaction product is formed depends on the character of the solvent; thus, in ether the nitronic acid ester is obtained in 90% yield[851, 852].

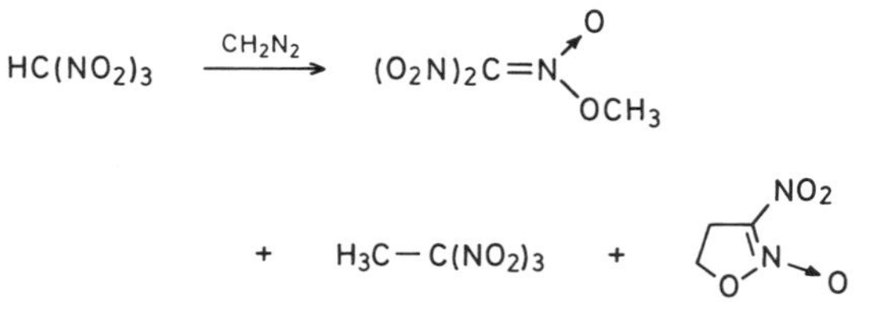

$$HC(NO_2)_3 \xrightarrow{CH_2N_2} (O_2N)_2C=N\underset{OCH_3}{\overset{O}{\diagup}}$$

$$+ \quad H_3C-C(NO_2)_3 \quad +$$

During the reaction of the diazoalkanes with oximes both *O*-alkylation (leading to *O*-alkyloximes) and *N*-alkylation (leading to nitrones) can take place[853, 854]:

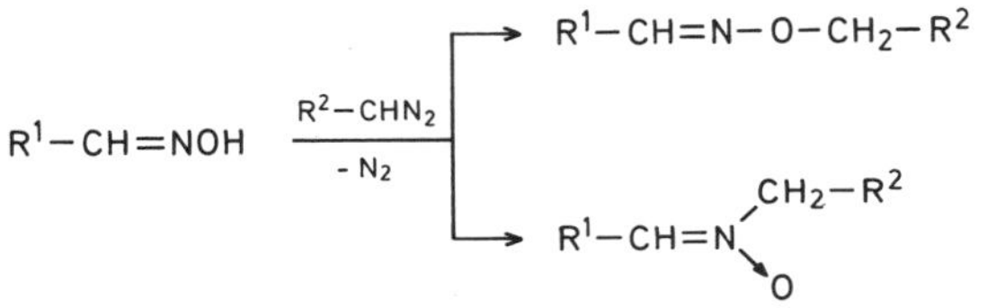

$$R^1-CH=NOH \xrightarrow[-N_2]{R^2-CHN_2} \begin{cases} R^1-CH=N-O-CH_2-R^2 \\ \\ R^1-CH=N\underset{O}{\overset{CH_2-R^2}{\diagup}} \end{cases}$$

Nitrone formation ensues smoothly when monooximes of various cyclic 1,2-diketones[855, 856] are employed, *e.g.*:

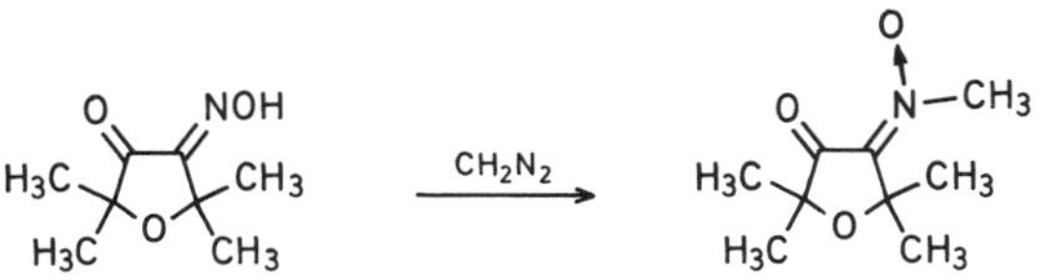

2,2,5,5-Tetramethyl-
3,4(2H, 5H)-furandione
monomethylnitrone

Monooximes of some higher condensed *o*-quinones such as phenanthraquinone do not afford nitrones on reaction with ethereal diazomethane solution but suffer cyclization to oxazoles[855].

O,O-Diethylphosphorothioate forms *O,O*-diethyl-*S*-methyl and *O,O*-diethyl-*O*-methyl phosphorothioate[857]:

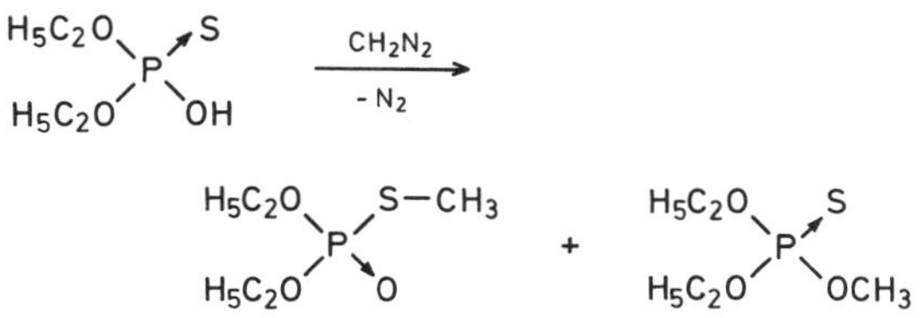

8.2.3.5 Reactions with *NH* and *SH* bonds

In aprotic solvents diazoalkanes do not react with *ammonia* and *amines; N-methylation* is observed only in the presence of strongly electron-withdrawing substituents adjacent to the amino group[858].

Salts of amines, for instance, those with boron trifluoride, are methylated in high enough yield by diazomethane or its ether solution[657]:

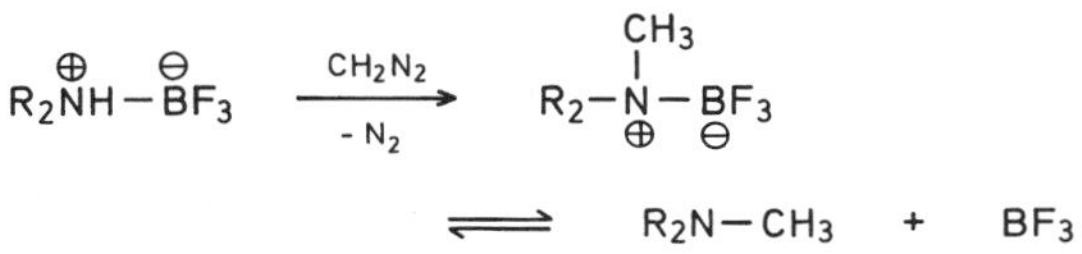

$$R_2\overset{\oplus}{N}H-\overset{\ominus}{B}F_3 \xrightarrow[-N_2]{CH_2N_2} R_2-\overset{CH_3}{\underset{\oplus}{N}}-\overset{\ominus}{B}F_3$$

$$\rightleftharpoons R_2N-CH_3 + BF_3$$

Analogously, the pyridinium, quinolinium, and isoquinolinium perchlorates can be converted into perchlorates of the quaternary bases[859]:

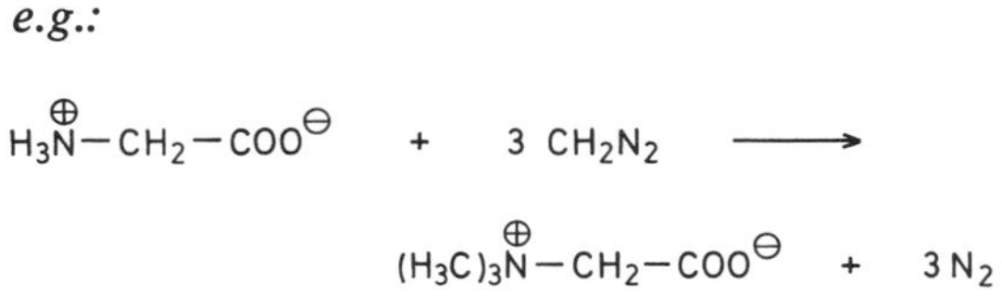

$$Ar-CH_2-\text{[pyridinium]}\overset{\oplus}{N}ClO_4^{\ominus} \xrightarrow{R-CHN_2}$$

$$Ar-CH_2-\text{[pyridinium]}\overset{\oplus}{N}CH_2-R \quad ClO_4^{\ominus}$$

α-Amino carboxylic acids and *α-sulfo* carboxylic acids, too, are methylated by diazomethane[400], *e.g.*:

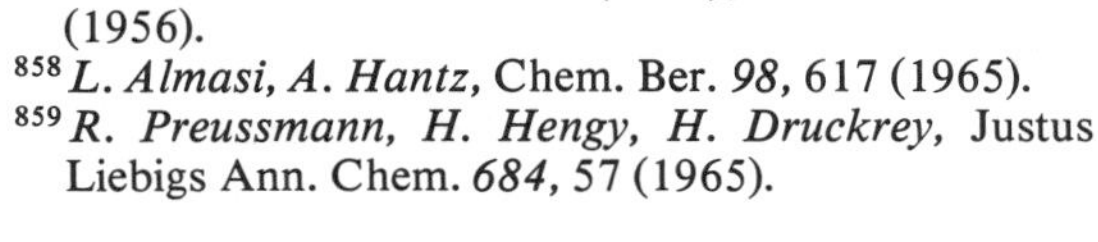

$$H_3\overset{\oplus}{N}-CH_2-COO^{\ominus} + 3\ CH_2N_2 \longrightarrow$$

$$(H_3C)_3\overset{\oplus}{N}-CH_2-COO^{\ominus} + 3\ N_2$$

Trimethylglycine betain

[850] F. Arndt, J.D. Rose, J. Chem. Soc. I, 1, (1935).
[851] A.A. Onichshenko, I.E. Chlenov, L.M. Makarenkova, V.A. Tartakowskii, Izv. Akad. Nauk SSSR, Ser. Khim, 1560 (1971), (russ.).
[852] A.A. Onichshenko, V.A. Tartakowskii, Izv. Akad, Nauk SSSR, Ser. Khim. 948 (1970) (russ.).
[853] H. Metzger, Herstellung und Umwandlung von Oximen, in Houben-Weyl, Methoden der organischen Chemie Bd. X/4, p. 315, Georg Thieme Verlag, Stuttgart 1968.
[854] W. Rundel, Methoden zur Herstellung und Umwandlung von Nitronen, in Houben-Weyl, Methoden der organischen Chemie, Bd. X/4, p. 315, Georg Thieme Verlag, Stuttgart 1968.
[855] B. Eistert, R. Müller, H. Selzer, E.A. Hackmann, Chem. Ber. 97, 2469 (1964).
[856] L.L. Rodina, L.V. Koroljeva, I.K. Korobizina, Zh. Org. Khim. 6, 2336 (1970) (russ.).

[857] M.I. Kabachnik, S.T. Joffe, T.A. Mastryukova, Zh. Obshch. Khim. 25, 684 (1955); C.A. 50, 3850i (1956).
[858] L. Almasi, A. Hantz, Chem. Ber. 98, 617 (1965).
[859] R. Preussmann, H. Hengy, H. Druckrey, Justus Liebigs Ann. Chem. 684, 57 (1965).

Alkylation of amines with diazomethane, diazo-acetic acid and α-diazo ketones can be effected also by adding copper(I) salts (25–70% yields)[860].

Intramolecular insertions of the diazo ketones (with loss of nitrogen) into the *NH* bond are known as well[547].

Thiols and *thiophenols* are more acidic in character than the alcohols and phenols and, hence, react more easily with diazoalkanes[861].

8.2.4 Wolff rearrangements of α-diazocarbonyl compounds

The rearrangement of diazo ketones into ketenes and their further transformation products, namely, carboxylic acids and their derivatives, discovered by L. Wolff[862] and G. Schroeter[863] has been

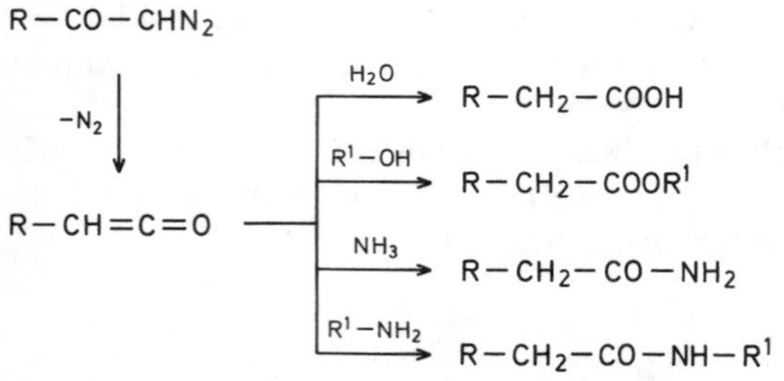

R—CO—CHN$_2$

$$R-CH=C=O \xrightarrow{\begin{array}{l}H_2O\\R^1-OH\\NH_3\\R^1-NH_2\end{array}}$$

R—CH$_2$—COOH
R—CH$_2$—COOR^1
R—CH$_2$—CO—NH$_2$
R—CH$_2$—CO—NH—R^1

(−N$_2$)

investigated in detail[402, 509, 864–868]. Either thermolysis, photolysis, or addition of catalysts achieves the rearrangement of the diazo ketones. In the catalytic variant of the rearrangement the diazo compounds are warmed in the appropriate sol-

vent[513, 703] in the presence of silver oxide or silver salts. Platinum[862] and copper powder[839] are used only seldom.

Copper and copper compounds do not represent typical catalysts of the Wolff rearrangement (see p. 255), but copper iodide[869, 870] and copper oxide[871, 872] are exceptions and are sometimes used.

Thermolysis at more elevated temperatures is used less often than the catalytic decomposition for carrying out the rearrangement. Nonetheless, this reaction variant enables *carboxanilides* and some esters to be synthesized in high yields[864, 867]. In general, the *photolytic* rearrangement of diazocarbonyl compounds displays the greatest importance[873].

Numerous studies demonstrate that the splitting off of the nitrogen from the α-diazo ketones **53** is the first stage of the reaction. The octet deficiency in the intermediate keto carbene can be filled in the following manner:

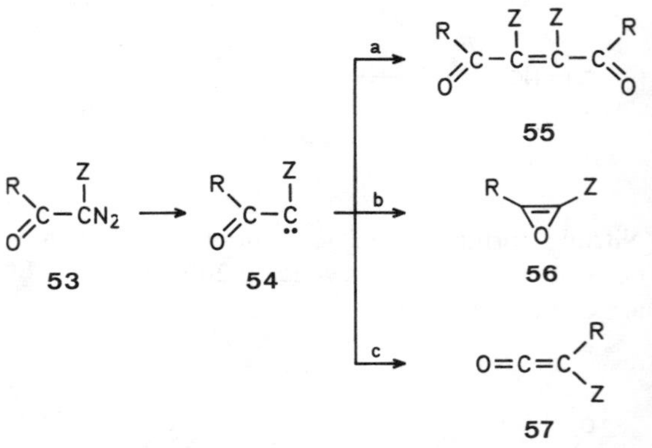

In diazo ketones the bimolecular reaction **a** occurs only seldom (see p. 255). Formation of oxirene as intermediate stage **b** was for a long time held to be improbable[874], but evidence for it was obtained during the photolytic rearrangement of α-diazo ketones and diazoacetic acid esters in the

[860] T. Saegusa, Y. Izo, Sh. Kobayashi, K. Hirota, Tetrahedron Lett. 6131 (1966).
[861] A. Schönberg, A. Wagner, Methoden zur Herstellung und Umwandlung von Sulfiden (Thioäthern), in Houben-Weyl, Methoden der organischen Chemie, Bd. IX, p. 116, Georg Thieme Verlag, Stuttgart 1955.
[862] L. Wolff, Justus Liebigs Ann. Chem. 325, 129 (1902); L. Wolff, Justus Liebigs Ann. Chem. 394, 23 (1912).
[863] G. Schröter, Ber. 42, 2346 (1909); G. Schröter, Ber. 49, 2704 (1916).
[864] W.E. Bachmann, W.S. Struve, in Organic Reactions, Vol. I, p. 38, John Wiley & Sons, New York 1942.
[865] V. Franzen, Chemiker-Ztg. 81, 359 (1957).
[866] P.A.S. Smith, Rearrangements Involving Migration to an Electron-Deficient Nitrogen or Oxygen in Pe. de Mayo, Molecular Rearrangements, p. 558, John Wiley & Sons, New York, London 1963.
[867] L.L. Rodina, I.K. Korobizina, Usp. Khim. 34, 611 (1967) (russ.); C.A. 67, 116250u (1967).
[868] R.C. Denney, Named Organic Reactions, Butterworths & Co. London 1969.

[869] P. Yates, J. Fugger, Chem. & Ind. (London) 1511 (1957).
[870] K.S. Bhandari, V. Snieckus, Can. J. Chem. 49, 2354 (1971).
[871] H. Erlenmeyer, M. Äberli, Helv. Chim. Acta, 31, 28 (1948).
[872] R. Casanova, T. Reichstein, Helv. Chim. Acta 33, 417 (1950).
[873] A. Schönberg, Präparative organische Photo-Chemie, Springer Verlag, Berlin, Göttingen, Heidelberg 1958.
[874] V. Franzen, Justus Liebigs Ann. Chem. 614, 31 (1958).
[875] I.G. Csizmadai, J. Font, O.P. Strausz, J. Amer. Chem. Soc. 90, 7360 (1968).

gas phase and in solution[875–877]. During reaction **c** the electron gap in the keto carbene is filled by addition of the group in the β-position which migrates together with its pair of electrons. As a result the Wolff rearrangement may be regarded as a nucleophilic *1,2-rearrangement*. The ketene which forms can transform into *carboxylic acid* or its *esters* or *amides* in dependence on the reaction conditions. Sometimes stable ketenes can be isolated in substance[863, 867], but generally the products of the further reaction of ketenes and not these themselves are formed, namely, polymers and adducts with the starting diazo ketone or keto carbenes[433, 867, 878, 879]. (Re the formation of *4-hydroxy-3-butenoic acid γ-lactones* from diazo ketones and ketones, see p. 266). Ketenes may be trapped by means of cycloadditions with azomethines[880], azo compounds[881], and *o*-quinones[882]. The conversion of the carbonyl group of α-diazo ketones into the carboxy group during this rearrangement has been unambiguously demonstrated by the use of radioactive carbon[867].

The *kinetic* investigation of the thermal Wolff rearrangement has confirmed the two-stage mechanism (involving keto carbene and ketene formation)[883–886].

Alternative mechanisms (ionic[887–889], free-radical[513] ones *via zwitterions*[890]) have been proposed without always having adequate backing.

The results of the photolytic rearrangement of diazo esters in alcohols show that this reaction follows not just the carbene mechanism[841]. Dewar ascribes the classic form of a π-complex to the intermediate compound, and this is in accord with the intramolecular character of the Wolff rearrangement.

Using optically active diazo ketones (with the center of chirality in the α-position with respect to the carbonyl group) and *cis-trans* isomers, it was determined that this rearrangement proceeds with *configuration retention*[867, 891].

Spectroscopic investigations show that the *cis*-conformation of the diazo and carbonyl group is the predominant one in α-diazo ketones. The alkyl group is in the *trans*-position to the leaving nitrogen atom to lead to promotion of the synchronous mechanism[892]. In 4-diazo-2,2,5,5-tetramethyl-3-hexanone the *cis*-conformation is hindered by the steric interaction with the *tert*-butyl groups; in fact, this diazo ketone reacts mainly with migration of the methyl group and furnishes the rearrangement products in only small yield[893, 894]:

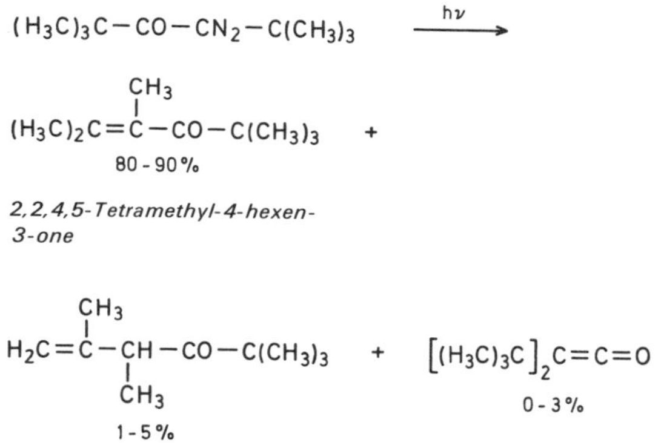

$$(H_3C)_3C-CO-CN_2-C(CH_3)_3 \xrightarrow{h\nu}$$

$$(H_3C)_2C=\overset{\overset{\textstyle CH_3}{|}}{C}-CO-C(CH_3)_3 \quad +$$
80-90%

2,2,4,5-Tetramethyl-4-hexen-3-one

$$H_2C=\overset{\overset{\textstyle CH_3}{|}}{\underset{\underset{\textstyle CH_3}{|}}{C}}-CH-CO-C(CH_3)_3 \quad + \quad [(H_3C)_3C]_2C=C=O$$
1-5%　　　　　　　　　　　　　　　0-3%

The sharp dependence of the yields of rearrangement product on temperature is caused also by the equilibrium between the *cis* and the *trans* forms of the α-diazo ketones[885, 895]. Side-reactions during the Wolff rearrangement characteristic of the carbenes which can take place are the following:

[876] *D. Thoraton, R. Gosavi, O.P. Strausz*, J. Amer. Chem. Soc. *92*, 1768 (1970).
[877] *S.A. Matlin, P.G. Sammes*, Chem. Commun. 1, 11 (1971).
[878] *M. Takebayashi, T. Ibata*, Bull. Chem. Soc. Japan *41*, 1700 (1968).
[879] *W. Ried, H. Mengler*, Fortschr. Chem. Forsch. *5*, 1 (1965).
[880] *W. Kirmse, L. Horner*, Chem. Ber. *89*, 2759 (1956).
[881] *W. Fischer, E. Fahr*, Tetrahedron Lett. 5245 (1966).
[882] *W. Ried, W. Radt*, Justus Liebigs Ann. Chem. *676*, 110 (1964);
W. Ried, W. Radt, Justus Liebigs Ann. Chem. *688*, 170 (1965).
[883] *A. Melzer, E.F. Jenny*, Tetrahedron Lett. 4503 (1968).
[884] *W. Jugelt, D. Schmidt*, Tetrahedron *25*, 969 (1969).
[885] *W. Jugelt, D. Schmidt, R. Bernhardt, B. Stolze*, Z. Chem. *II*, 146 (1971).
[886] *W. Bartz, M. Regitz*, Chem. Ber. *103*, 1463 (1970).
[887] *Y. Yukawa, T. Ibata*, Bull. Chem. Soc. Japan *42*, 802 (1969).
[888] *U. Schöllkopf, N. Rieber*, Chem. Ber. *102*, 488 (1969).
[889] *E. Wenkert, B.L. Mylari, L.L. Davis*, J. Amer. Chem. Soc. *90*, 3870 (1968).
[890] *A.L. Wilds, N.F. Woolsey, J. van Den Berghe, C.H. Winestock*, Tetrahedron Lett. 4841 (1965).

[891] *J.F. Lane, E.S. Wallis*, J. Org. Chem. *6*, 443 (1941);
J.F. Lane, E.S. Wallis, J. Amer. Chem. Soc. *63*, 1674 (1941).
[892] *F. Kaplan, G.K. Meloy*, J. Amer. Chem. Soc. *88*, 950 (1966).
[893] *M.S. Newman, A. Arkell*, J. Org. Chem. *24*, 385 (1959).
[894] *H.N.A. Al-Jallo, B.J. Arnold, B.J. Rothwell, E.S. Waight*, J. Chem. Soc. B 846 (1971).
[895] *D.M. Jordan*, Dissertion Abstr. *26*, 3633 (1966); C.A. *64*, 12499 (1966).

Insertion and interaction with solvents[896], alkyl migration[893, 894], and hydride migration leading to formation of unsaturated ketones[897].

With α-acetoxydiazo ketones[898] and certain α-halodiazo ketones[899, 900] secondary elimination is observed simultaneously with the rearrangement, *e.g.*:

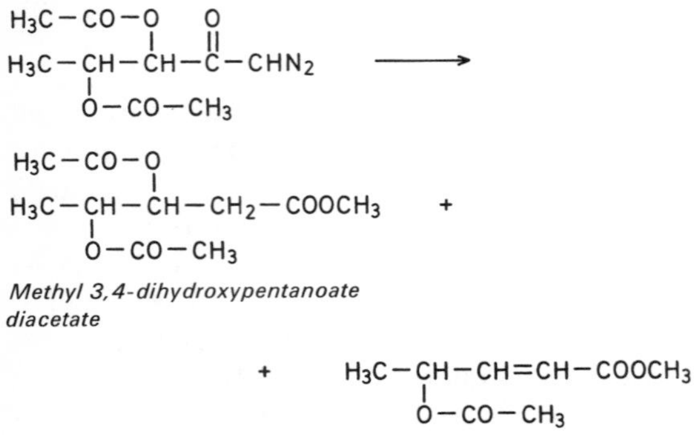

$$H_3C-CH(\!-O-CO-CH_3)-CH(\!-O-CO-CH_3)-\overset{O}{\overset{\|}{C}}-CHN_2 \longrightarrow$$

$$H_3C-CH(\!-O-CO-CH_3)-CH(\!-O-CO-CH_3)-CH_2-COOCH_3 \quad +$$

Methyl 3,4-dihydroxypentanoate diacetate

$$+ \quad H_3C-CH(\!-O-CO-CH_3)-CH=CH-COOCH_3$$

Methyl 4-hydroxy-2-pentenoate acetate

Sometimes the intermediate keto carbene is oxidized[867, 901, 902].

Steric factors can interfere with the normal course of the reaction, for instance in diazomethyl 9-phenyl-9-fluorenyl ketone[903]:

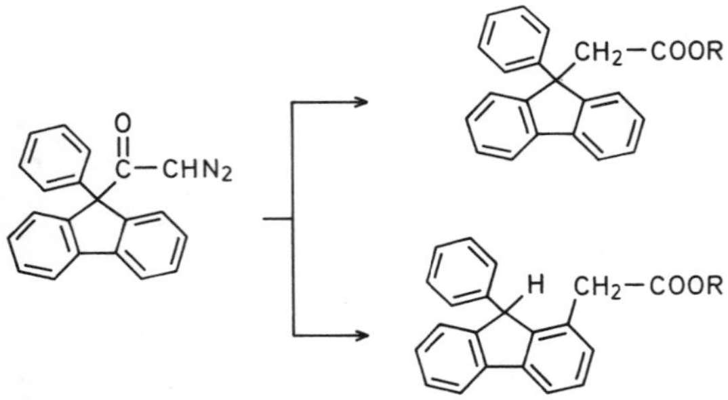

An anomalous reaction of often observed in α-diazo ketones containing a cycloalkenyl group[867, 904, 905] (cf. p. 264).

Despite these possible side-reactions the Wolff rearrangement is of quite substantial preparative importance – it is employed in the aliphatic, aliphatic-aromatic, aromatic, and alicyclic/heterocyclic series for preparing carboxylic acids and their derivatives[698, 867, 906–911].

The rearrangement of most α-diazo ketones of the type $X_3CCO–CHN$ $(X=Alk, Cl, etc.)$ proceeds smoothly[433, 867].

Thus, this reaction was accomplished with 1-diazo-3,3,3-trifluoro-2-propanone both photolytically and with adding silver oxide in alcohols, ammonia, and amines[912]:

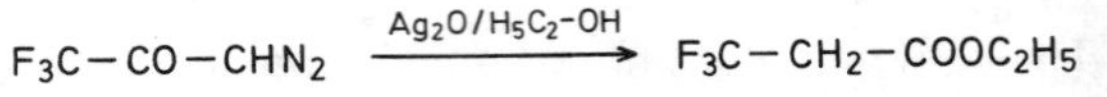

$$F_3C-CO-CHN_2 \xrightarrow{\text{Ag}_2\text{O}/\text{H}_5\text{C}_2\text{-OH}} F_3C-CH_2-COOC_2H_5$$

Ethyl 3,3,3-trifluoro-propionate

When the hydrogen atom is replaced by electron-attracting groups, the tendency to undergo rearrangement is reduced; trichloro- and trifluoroacetyldiazoacetate afford a carbene on photolysis, while losing nitrogen, which does not suffer rearrangement but undergoes an insertion reaction[429, 913]. Ethyl diazoacetate in alcohol furnishes a rearrangement product in 29% yield (side by side with insertion products, exchange products, etc.)[914, 915]. *Polymeric* products are formed during the photolysis of methyl diazoacetate in benzene;

[896] *N.R. Ghosh, C.R. Ghosal, S. Shah*, Chem. Commun. 151 (1969).

[897] *V. Franzen*, Justus Liebigs Ann. Chem. *602*, 199 (1957).

[898] *F.W. Bachelor, G.A. Miana*, Tetrahedron Lett. 4733 (1967).

[899] *J.P. Freeman*, Chem. & Ind. 1254 (1959).

[900] *A. Roedig, H. Lunk*, Chem. Ber. *87*, 971 (1954).

[901] *L. Horner, W. Kirmse, K. Muth*, Chem. Ber. *91*, 430 (1958).

[902] *I.K. Korobizina, L.L. Rodina, L.M. Stashkova*, Zh. Obshch. Khim *33*, 3109 (1963) (russ.); C.A. *60*, 1673h (1964);
I.K. Korobizina, L.L. Rodina, L.M. Stashkova, Khim. Geterotsikl. Soedin. 843 (1966) (russ.); C.A. *67*, 21752v (1967).

[903] *A.L. Wilds, R.L. von Trebra, N.F. Woolsey*, J. Org. Chem. *34*, 2401 (1969).

[904] *A.S. Monahan, Sh. Tang*, J. Org. Chem. *33*, 1445 (1968).

[905] *S. Masamune, K. Fukumoto*, Tetrahedron Lett. 4647 (1965).

[906] *J.N. Chatterjea, K.D. Banerji, N.M. Sahay*, J. Indian Chem. Soc. *45*, 171 (1968).

[907] *W. Jugelt*, Z. Chem. *11*, 174 (1971).

[908] *F.A. Trofimov, T.I. Mukhanova, A.N. Grinev, K.S. Shadurskii*, Khim. Farm. Zh. *2*, 9 (1968); C.A. *70*, 68029s (1969).

[909] *Ph. E. Eaton, K. Nyi*, J. Amer. Chem. Soc. *93*, 2786 (1971).

[910] *K.-P. Zeller, H. Meier, E. Müller*, Justus Liebigs Ann. Chem. *749*, 178 (1971).

[911] *W.C. Agosta, A.B. Smith*, Tetrahedron Lett. 4517 (1969).

[912] *F. Brown, W.K.R. Musgrave*, J. Chem. Soc. 2087 (1953).

[913] *F. Weygand, K. Koch*, Angew. Chem. *73*, 531 (1961).

[914] *O.P. Strausz, Th. Do Minh, H.E. Gunning*, J. Amer. Chem. Soc. *90*, 1660 (1968).

[915] *H. Chaimovich, R.J. Vaughan, F.H. Westheimer*, J. Amer. Chem. Soc. *90*, 4088 (1968).

their formation can be explained by the reaction between ketene and the intermediate carbene (or the initial diazo compound)[916].

The rearrangement of α-diazo ketones furnished active peptides containing *optically active β-aminocarboxylic acid* groups[917, 918], amides which are employed for the synthesis of alkaloids[849], and derivatives of deoxy acids[919]. *Phosphinic acid esters* are formed in good yield (60–77%) during the photolysis of the corresponding phosphorus-containing diazo ketones[920]:

$$H_5C_6 - \overset{\overset{\displaystyle C_6H_5}{|}}{\underset{\underset{\displaystyle O}{\|}}{P}} - CN_2 - R$$

$$-N_2 \quad \Big\downarrow \quad h\nu$$

$$H_5C_6 - \overset{\overset{\displaystyle C_6H_5}{|}}{\underset{\underset{\displaystyle O}{\|}}{P}} - \ddot{C} - R$$

$$\boxed{H_3C-OH}$$

$$O=\overset{\overset{\displaystyle H_5C_6 \quad C_6H_5}{|\quad\quad|}}{\underset{\underset{\displaystyle OCH_3}{|}}{P}} - CH - R \qquad\qquad H_5C_6 - \overset{\overset{\displaystyle C_6H_5}{|}}{\underset{\underset{\displaystyle O \quad OCH_3}{\|\quad\quad|}}{P}} - CH - R$$

R = H:	*Methyl benzyl-phenyl-phosphinate;* 61%	R = H;	*(Methoxymethyl)-diphenyl-phosphine oxide; 39%*
R = COC₆H₅:	*3-(Diphenyl-phosphino)-3-phenyl-propiophenone; 77%*	R = COC₆H₅;	*3-(Diphenyl-phosphinyl)-3-phenyl-propiophenone; 23%*

However, often this reaction was accompanied by parallel insertion processes and by reduction and migration of the methyl group[921–923].

Publications describe the rearrangement of α-diazo β-keto sulfones; in this case the yield of rearrangement product is only 10–20%. The reaction proceeds photolytically in alcohol and leads preferentially to an interaction with the solvent[924, 925].

The rearrangement succeeds also with *unsaturated* diazo ketones; the best results are obtained if the transformation takes place under homogeneous conditions – in the presence of silver benzoate in triethylamine[867].

Great importance attaches to the use of the Wolff rearrangement for reducing the ring size of cyclic α-diazo ketones. Particular mention should be made of the photolytic method, by which it is possible to prepare *strained cyclic systems*[433, 867]. For example, diazo ketones containing bicyclo[2, 2, 2]octane and bicyclo[2, 2, 2]octene systems rearrange on photolysis, forming *norbornane*[926] and *norbornene*[927] derivatives. The same method has been applied to the homocubane system, the six-membered ring rearranging to form a five-membered ring[928]. Pentalene derivatives have been prepared[929], and ring-contraction of α-diazo-cycloalkanones has been achieved. For example, 2-diazocyclohexanone and 2-diazocyclopentanone formed *cyclopentanecarboxylic acid* and *cyclobutanecarboxylic acid* (80–90%) on irradiation involving ring contraction[930]:

$$(H_2C)_n \overset{\displaystyle O}{\underset{\displaystyle N_2}{\diagup}} \quad \xrightarrow{h\nu / H_2O} \quad (H_2C)_n - CO_2H$$

$$n = 2,1$$

It is interesting that rearrangement of 2-diazocyclopentanone does not occur when silver oxide is added; under analogous conditions 2-diazocyclohexanone forms *adipic acid*, which probably results from the 1,2-cyclohexanedione formed as intermediate[931].

[916] G.O. Schenck, A. Ritter, Tetrahedron Lett. 3189 (1968).

[917] B. Penke, J. Czombos, L. Baláspiri, J. Petres, K. Kovacs, Helv. Chim. Acta 53, 1057 (1970).

[918] W. Jugelt, P. Falck, J. Prakt. Chem. 38, 88 (1968).

[919] Yu. A. Zhdanov, Luu Dinh Chung, V.I. Kornilov, Zh. Obshch. Khim 38, 1411 (1968); C.A. 69, 67680g (1968).

[920] M. Regitz, Angew. Chem. 82, 224 (1970).

[921] M. Regitz, H. Scherer, W. Anschütz, Tetrahedron Lett. 753 (1970).

[922] M. Regitz, A. Liedhegener, W. Anschütz, H. Eckes, Chem. Ber. 104, 2177 (1971).

[923] L.I. Zakharkin, V.N. Kalinin, V.V. Gedymin, Tetrahedron 27, 1317 (1971).

[924] R.J. Mulder, A.M. van Leusen, J. Strating, Tetrahedron Lett. 3057 (1967).

[925] A. Leusen, P. Smid, J. Strating, Tetrahedron Lett. 1165 (1967).

[926] K. Wiberg, A. Hess, J. Org. Chem. 31, 2250 (1966).

[927] L. Horner, D.W. Baston, Chem. Ber. 98, 1252 (1965).

[928] W.G. Dauben, D.L. Whalen, Tetrahedron Lett. 3743 (1966).

[929] L. Horner, H.-G. Schmelzer, H.-U. von Eltz, K. Habig, Justus Liebigs Ann. Chem. 661, 44 (1963).

[930] I.K. Korobizina, L.L. Rodina, T.P. Sushko, Zh. Org. Khim. 4, 175 (1968); C.A. 68, 86612g (1968).

[931] M. Regitz, J. Rüter, Chem. Ber. 102, 3877 (1969).

High yields of *cycloalkanecarboxanilides* are formed by thermolysis in aniline[932] of cyclic diazo ketones containing from six to twelve ring members. In the case of 3-diazobicyclo[3.1.0]hexan-2-ones a better yield of the rearranged product is, in fact, obtained by thermolysis in *N*-methylaniline than by *UV*-irradiation in methanol[933]. Photolysis is the best method for preparing *benzocyclobutene* derivatives[867, 934, 935].

Heterocyclic diazo ketones also rearrange in high yield on irradiation. This method has been used successfully not only for the synthesis of *pyrrole-2-carboxylic acid*,

and of *indolecarboxylic* and *azoindolecarboxylic acids*[936], but has been employed, too, for converting 4-diazotetrahydro-3-furanones into *oxetane-carboxylic acids*[902, 937, 938]:

R = R¹ = CH₃; 98%

R = CH₃, R¹ = C₂H₅; 92%

R = R¹ = C₂H₅; 72%

R = R¹ = C₆H₅; 58%

5-Diazo-3,3,6,6-tetramethyl-4-thiepanone rearranges very easily[939], as the diazo and carbonyl groups on the seven-membered ring are *cis* to each other[892].

On Wolff rearrangement 2-diazo-1,3-diketones furnish β-keto acids or their derivatives. However, during thermal or photolytic decomposition in alcohols or other protic solvents the yields are admittedly only small (20–40%) because side-reactions step into the foreground[940–944].

UV-irradiation of aliphatic, carbocyclic, and heterocyclic 2-diazo-1,3-diketones in aprotic solvents which contain the necessary water can serve as a convenient method for preparing the corresponding *β-keto carboxylic acids*[945]:

R = R¹ = C₂H₅;   2-Ethyl-3-oxopentanoic acid; 70%

R = R¹ = (H₃C)₂CH;   2-Isopropyl-4-methyl-3-oxopentanoic acid; 85%

R = (H₃C)₃C, R¹ = CH₃;   2,4,4-Trimethyl-3-oxopentanoic acid; 80–85%

$\begin{smallmatrix}R\\R^1\end{smallmatrix}$ –CH₂–C(CH₃)₂–CH₂–;   4,4-Dimethyl-2-oxocyclopentanoic acid; 95%

$\begin{smallmatrix}R\\R^1\end{smallmatrix}$ –C(CH₃)₂–O–C(CH₃)₂–;   Tetrahydro-2,2,5,5-tetramethyl-4-oxo-3-furoic acid; 95%

The relative migrational capacity of two different substituent groups is dependent not only their nature but also on the way the reaction is conducted[946].

Examples of the rearrangement of 1,3-bisdiazo ketones, too, are known[947–949].

---

[932] *M. Regitz, J. Rüter,* Chem. Ber. *101*, 1263 (1968).

[933] *P.R. Brook, B.V. Braphy,* Tetrahedron Lett. 4187 (1969).

[934] *L. Horner, K. Muth, H. Schmelzer,* Chem. Ber. *92*, 2953 (1959);
L. Horner, K. Muth, H. Schmelzer, Chem. Ber. *92*, 188 (1959).

[935] *J.L. Mateos, O. Chao, H. Flores,* Tetrahedron *19*, 1051 (1963).

[936] *O. Süs, K. Möller,* Justus Liebigs Ann. Chem. *593*, 91 (1955);
O. Süs, K. Möller, Justus Liebigs Ann. Chem. *599*, 233 (1956);
O. Süs, K. Möller, Justus Liebigs Ann. Chem. *612*, 153 (1958).

[937] *I.K. Korobizina, L.L. Rodina,* Zh. Org. Khim. *1*, 932 (1965); C.A. *63*, 6939a (1965).

[938] *I.K. Korobizina, L.L. Rodina, V.G. Shuvalova,* Zh. Org. Khim. *4*, 2016 (1968) (russ.); C.A. *70*, 28 725j (1969).

[939] *Ae. de. Groot, J.A. Boerma, J. de. Valk, H. Wynberg,* J. Org. Chem. *33*, 4025 (1968).

[940] *H. Schwall, V. Schmidt, B. Eistert,* Chem. Ber. *102*, 1731 (1969).

[941] *J. Hayasi, T. Okada, M. Kawanisi,* Bull. Chem. Soc. Japan *43*, 2506 (1970).

[942] *H. Veschambre, D. Vocelle,* Can. J. Chem. *47*, 1981 (1969).

[943] *W.D. Barker,* Can. J. Chem. *47*, 2853 (1969).

[944] *M.P. Cava, R.J. Spangler,* J. Amer. Chem. Soc. *89*, 4550 (1967).

[945] *I.K. Korobizina, V.A. Nikolaev,* Zh. Org. Khim. *7*, 413 (1971) (russ.).

[946] *V.A. Nikolaev, S.D. Kotok, J.K. Korobizina,* Zh. Org. Khim. *10* (in print) (1974) (russ.).

8.2.5 Aliphatic azo coupling

Azo coupling is not as typical a reaction for aliphatic diazo compounds as it is for aromatic derivatives and procedurally significant only with 2-diazo-1,3-carbonyl compounds. The most exhaustive studies conducted have been on the azo coupling of 5,5-dimethyl-1,3-cyclohexanedione, which reacts with phloroglucinol[950] and with compounds (e.g., β-diketones) containing active hydrogen[951]:

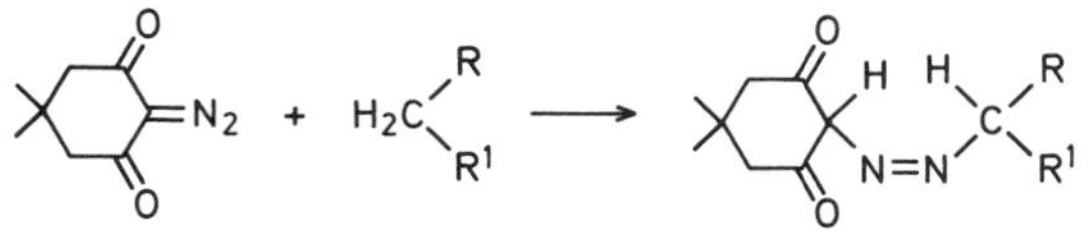

Symmetrical aliphatic azo compounds can be prepared without prior isolation of the diazo component: the 1,3-dicarbonyl compound is merely treated with half the theoretical amount of tosyl azide[951-956]:

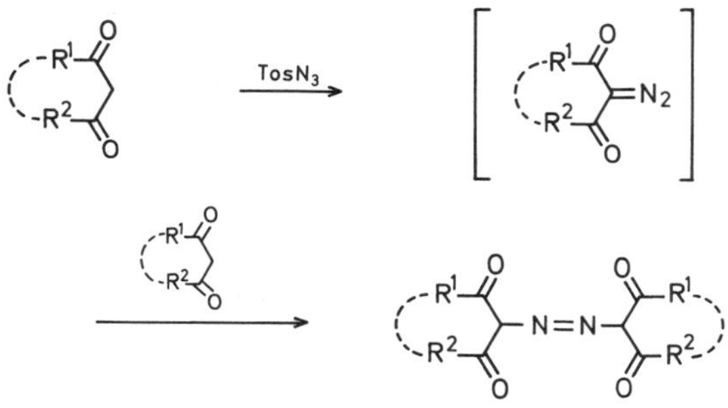

Diazomethane as the azo component gives both methylhydrazones[957-959] and unsymmetrical azo compounds[960]:

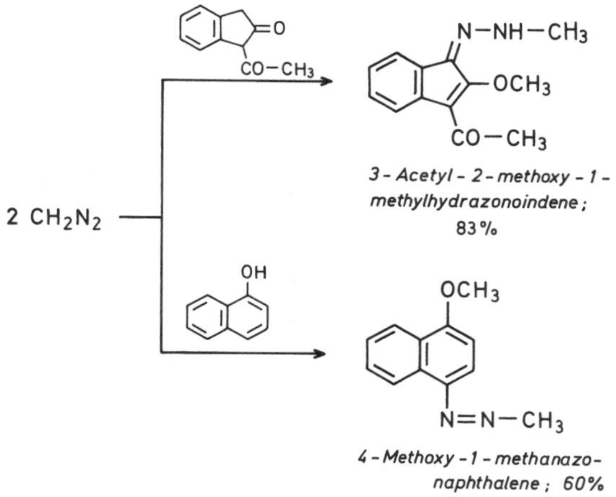

3-Acetyl-2-methoxy-1-
methylhydrazonoindene;
83%

4-Methoxy-1-methanazo-
naphthalene; 60%

8.2.6 Oxidation and reduction

Diazoalkanes are slowly oxidized by oxygen. It has been demonstrated that formaldehyde is formed on passing air through irradiated ethereal solutions of diazomethane; under the same conditions, diazodiphenylmethane forms benzophenone[400]. Oxidation of diazo ketones can occur as a side-reaction during rearrangements[902].

Ozone reacts vigorously with diazoalkanes. Ozonization of solutions of diazodiphenylmethane, 9-diazofluorene, or 2-diazo-2-phenylacetophenone (azibenzil) forms the corresponding carbonyl compounds in almost quantitative yields[961]. On ozonization of monosubstituted diazo compounds (diazophenylmethane, ethyl diazoacetate, 2-diazoacetophenone, 4'-chloro-2-diazoacetophenone), the CH bond adjacent to the diazo group may be attacked as well as the C=N bond[962].

Treatment with tert-butyl hypochlorite is a convenient method for the conversion of 2-diazo-1,3-dicarbonyl compounds into 1,2,3-tricarbonyl compounds[963, 964]. For example, reacting 2-diazo-1,3-indandione with tert-butyl hypochlorite in ethanol yields 2,2-diethoxy-1,3-indandione, which is converted into ninhydrin by treatment with aqueous sulfuric acid[964]:

[947] R. Tasovac, M. Stafonovic, A. Stojiljkovic, Tetrahedron Lett. 2792 (1967).

[948] R.F. Borch, D.L. Filds, J. Org. Chem. 34, 1480 (1969).

[949] P. Whitman, B. Trost, J. Amer. Chem. Soc. 91, 7534 (1969).

[950] Th. Severin, Angew. Chem. 70, 745 (1958).

[951] M. Regitz, D. Stadler, Angew. Chem. 76, 920 (1964).

[952] M. Regitz, D. Stadler, Justus Liebigs Ann. Chem. 687, 214 (1965).

[953] M. Regitz, A. Liedhegener, D. Stadler, Justus Liebigs Ann. Chem. 713, 101 (1968).

[954] W. Ried, E. Kahr, Justus Liebigs Ann. Chem. 725, 228 (1969).

[955] W. Ried, B. Peters, Justus Liebigs Ann. Chem. 729, 119 (1969).

[956] M. Colonna, L. Greci, Gazz. Chim. Ital. 293 (1971).

[957] R. Schmiechen, Tetrahedron Lett. 4995 (1969).

[958] K. Hartke, W. Uhde, Tetrahedron Lett. 1697 (1969).

[959] W. Uhde, K. Hartke, Chem. Ber. 103, 2675 (1970).

[960] J. Pyrek, O. Achmatowicz, Tetrahedron Lett. 2651 (1970).

[961] A.M. Reader, P.S. Bayley, H.M. White, J. Org. Chem. 30, 784 (1965).

[962] P.S. Bailey, A.M. Reader, P. Kolsaker, H.M. White, J.C. Barborak, J. Org. Chem. 30, 3042 (1965).

[963] M. Regitz, H.G. Adolph, Justus Liebigs Ann. Chem. 723, 47 (1969).

[964] M. Regitz, H.G. Adolph, Chem. Ber. 101, 3604 (1968).

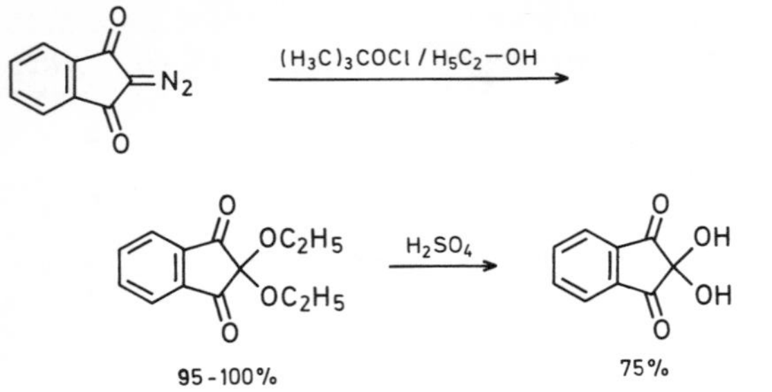

95 - 100%   75%

Oxidation of 9-diazofluorene with *peracids* or with *dibenzoyl peroxide* mainly gives *fluorenone,* but the occurrence of insertion and dimerization reactions has also been noted[965]. The relative ratios of the products formed depend on the peroxy compound used for oxidation. High yields of *substituted benzophenones* are obtained by oxidation of the corresponding diaryldiazomethanes with perbenzoic acid[966].

Oxidation of 9-diazofluorene with *lead tetraacetate* gives *dimethyl fluorene-9,9-diacetate* in high yield[967].

Electrochemical oxidation of diazoalkanes, a reaction which leads to good yields of *olefins,* proceeds *via diazonium radical-cation* intermediates[968].

Reduction of the catalyst by diazoalkanes has been observed to occur in many reactions involving heterogenous catalysis[409, 410, 969].

Reduction of aliphatic diazo compounds may be accompanied by elimination of nitrogen, or one or both of the nitrogen atoms may remain in the molecule, depending on the reducing agent and on the reaction conditions. If nitrogen is eliminated, the diazo group is replaced by two hydrogen atoms; if nitrogen is retained, the products are *amines, hydrazines,* or *hydrazones*[400].

8.2.7 Reactions with elements or with inorganic or organoelemental compounds

8.2.7.1 Reactions with halogens

Reactions of aliphatic diazo compounds with *bromine* or *iodine* generally proceed smoothly, the nitrogen being replaced by *two* halogen atoms[400].

Examples are the reaction of diazo-(4-tolyl)methane with iodine, in which *4-(diiodomethyl)-1-toluene* is formed[970], and that of diazophenyltriphenylsilylmethane with bromine, where *dibromophenyltriphenylsilylmethane* is produced[971]:

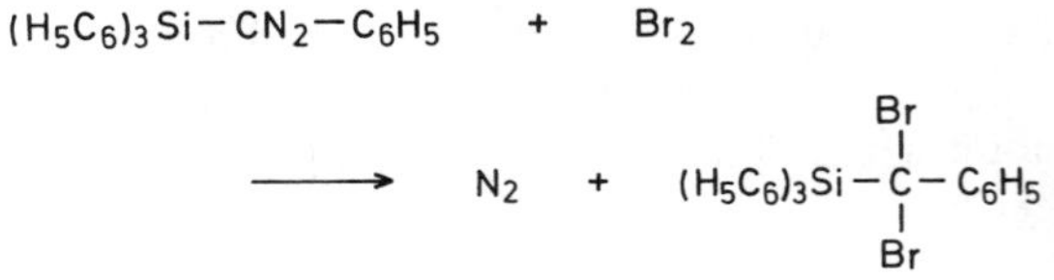

$$(H_5C_6)_3Si-CN_2-C_6H_5 \quad + \quad Br_2$$

These reactions can be regarded as electrophilic additions of the halogen to the nucleophilic centers of the diazoalkanes, they do not take place with weakly nucleophilic diazoalkanes[402].

Replacement of nitrogen is also observed on treatment with *chlorine,* but considerable resinification often occurs. Consequently, it is better to use phenyliodosochloride instead of free chlorine[400], *e.g.:*

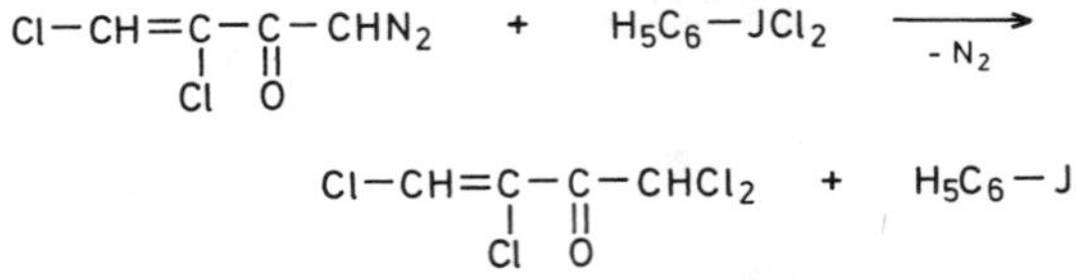

1,1,3,4-Tetrachloro-4-buten-2-one

8.2.7.2 Reactions with sulfur and organosulfur compounds

Aliphatic diazo compounds react with sulfur, even at room temperature; *thiiranes (ethylene sulfides)* are formed in high yield, the reaction can be accelerated by irradiation[972, 973]:

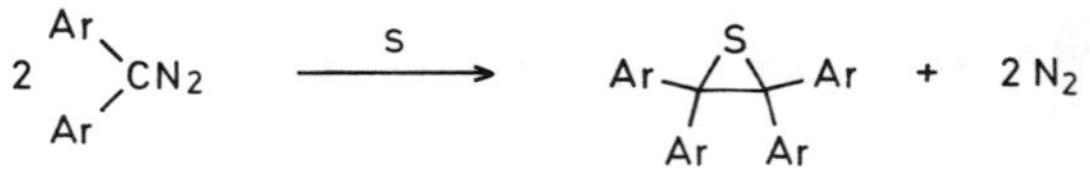

Reaction with *hydrogen sulfide* can give various products depending on the structure of the aliphatic diazo compounds concerned; diazomethane and diazodiphenylmethane form thiols, while 2-diazoacetophenone and ethyl diazoacetate are reduced to *phenylglyoxal 1-hydrazone* and

[965] *H. Lind, E. Fahr,* Tetrahedron Lett. 4505 (1966).
[966] *R. Curci, F. Furia, F. Marcuzz,* J. Org. Chem. *36,* 3774 (1971).
[967] *H.R. Hensel,* Chem. Ber. *88,* 527 (1955).
[968] *W. Jugelt, F. Pragst,* Tetrahedron, *24,* 5123 (1968); *W. Jugelt, F. Pragst,* Elektrochim. Acta *15,* 1543 (1970).
[969] *S. Trahanovsky, D. Kobbins, D. Smick,* J. Amer. Chem. Soc. *93,* 2086 (1971).

[970] *R.A. Moos,* J. Org. Chem. *30,* 3261 (1965).
[971] *K.D. Kaufmann, B. Auräth, P. Träger, K. Rühlmann,* Tetrahedron Lett. 4973 (1968).
[972] *N. Latif, J. Fathy,* J. Org. Chem. *27,* 1633 (1962).
[973] *A. Schönberg, E. Frese,* Chem. Ber. *95,* 2810 (1962).

ethyl glyoxalate hydrazone. 2-Diazo-1,3-dicarbonyl compounds and 2-diazo-2-phenylacetophenone (azibenzil) cyclize to form *1,2,3-thiadiazoles* on reaction with hydrogen sulfide[400, 402], *e.g.:*

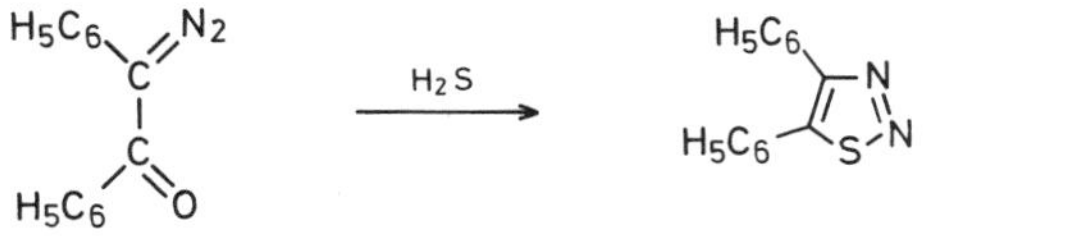

4,5-Diphenyl-1,2,3-
thiadiazole

On reaction with *sulfur dioxide* diazo compounds eliminate nitrogen and form *episulfones*[400].
It has been found that addition of liquid or gaseous sulfur dioxide to α-diazotoluene proceeds *stereospecifically* giving good yields of *cis-dibenzyl sulfone*[974]. Reactions of diazomethane, diazophenylmethane, or ethyl diazoacetate with sulfur dioxide in alcoholic media yield mixed *sulfite esters,* while diazodiphenylmethane forms *diphenylmethanesulfonate esters* under the same reaction conditions[400]:

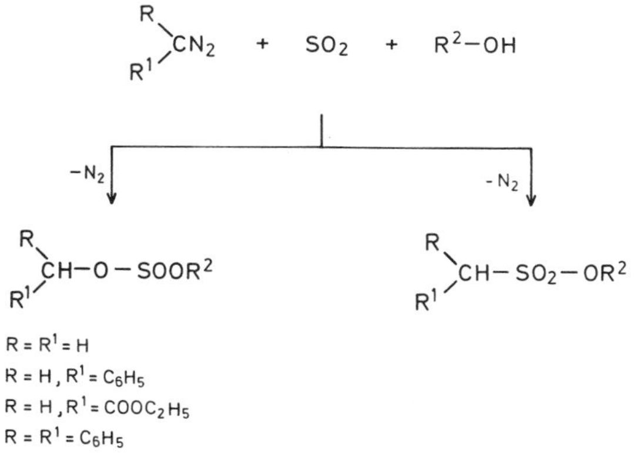

R = R¹ = H
R = H, R¹ = C₆H₅
R = H, R¹ = COOC₂H₅
R = R¹ = C₆H₅

3-Diazopropene reacts with sulfur dioxide to form *4,5-dihydrothiepine 1,1-dioxide*[975]:

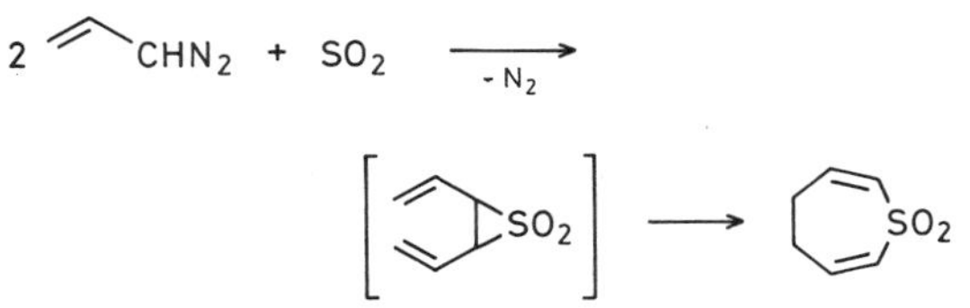

The reaction between 2-diazo-2-phenylacetophenone (azibenzil) and sulfur dioxide (in benzene at elevated temperature) leads to the (4 + 2)-cycloaddition product[976]:

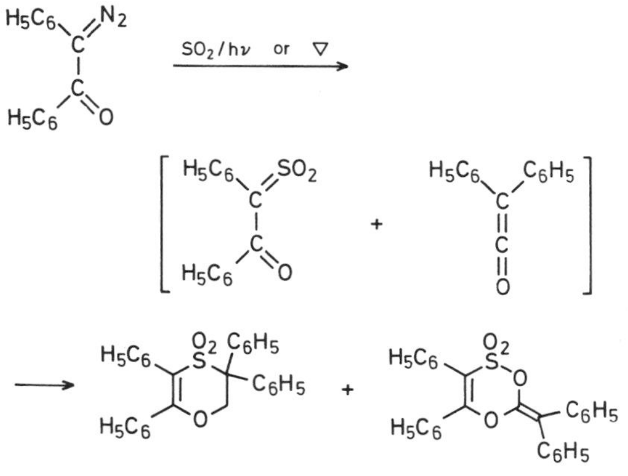

2,3-Dihydro-3,3,5,6-
tetraphenyl-1,4-
oxathiin 4,4-dioxide

2-(diphenylmethylene)-
5,6-diphenyl-1,3,4-
dioxathiin 4,4-dioxide

It is noteworthy that whereas diazo compounds do not react with carbon dioxide, they are able to react with *carbon disulfide* at room temperature; for example, 9-diazoxanthene is converted into *3,6-dixanthenylidene-1,2,4,5-tetrathiane* in 88% yield[400].

The structures of the reaction products from carbon disulfide and 3-diazo-2-butanone[977] or 2-diazo-2-phenylacetophenone[978] have also been elucidated.

Reactions of substituted 2-diazoacetophenones with *thiophosgene* lead to 1,3-cycloaddition products, namely, *1,2,4-thiadiazoles*[879, 979] (1,2,3-thiadiazoles are not formed)[765]:

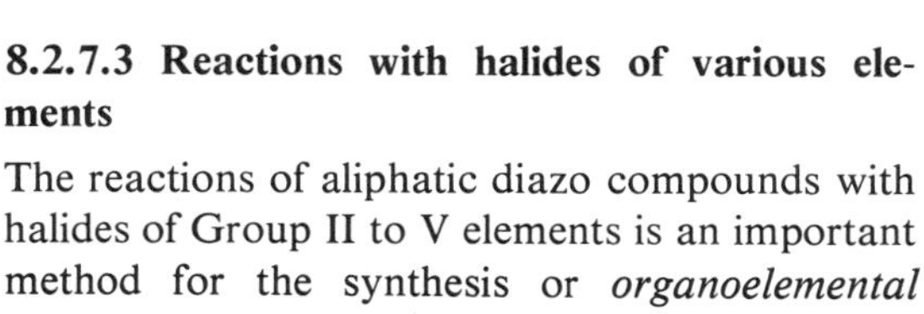

e.g., R = H; 5-Chloro-1,2,4-thiadiazolinyl phenyl ketone

8.2.7.3 Reactions with halides of various elements

The reactions of aliphatic diazo compounds with halides of Group II to V elements is an important method for the synthesis or *organoelemental*

[974] *N. Tokura, T. Nagai, Sh. Matsumura,* J. Org. Chem. *31,* 349 (1966).
[975] *W. Mock,* Chem. Commun. 1254 (1970).
[976] *A.J. Stothers, L. Danks, J. King,* Tetrahedron Lett. 2551 (1971).

[977] *J.A. Kapecki, J.E. Boldwin, J.C. Paul,* J. Amer. Chem. Soc. *90,* 5800 (1968).
[978] *P. Yates, L. Williams,* Tetrahedron Lett. 1205 (1968).
[979] *T. Bacchetti, A. Alemagna, B. Danieli,* Tetrahedron Lett. 3569 (1964).

compounds[400, 402, 441, 980, 981] (the diazo technique for synthesizing aliphatic organometallic compounds):

$$E^nX_n + mR_2CN_2 \longrightarrow (XCR_2)_mE^nX_{n-m} + mN_2$$

This reaction gives preparative yields when a covalent elementcarbon bond is formed. Since all elements which react with diazoalkanes possess a vacant *p*-orbital, it is assumed that the first stage in these processes is the addition of the nucleophilic carbon atom of the diazo compound at the vacant *p*-orbital of the element (polar mechanism)[441, 980]. The detailed study of the reaction of diazodiphenylmethane with mercuric chloride in tetrahydrofuran has confirmed the correctness of this view:

$$HgCl_2 + (H_5C_6)_2\overset{\ominus}{C}-\overset{\oplus}{N}\equiv N \rightleftharpoons$$

$$\begin{bmatrix} (H_5C_6)_2\overset{\oplus}{C}N_2 \\ \overset{|}{Cl}-\overset{\ominus}{Hg}-Cl \end{bmatrix} \xrightarrow{-N_2} \begin{bmatrix} (H_5C_6)_2\overset{\oplus}{C} \\ \overset{|}{Cl}-\overset{\ominus}{Hg}-Cl \end{bmatrix}$$

(H₅C₆)₂CN₂

$$\begin{bmatrix} (H_5C_6)_2C-C(C_6H_5)_2\overset{\oplus}{N_2} \\ \overset{|}{Cl}-\overset{\ominus}{Hg}-Cl \end{bmatrix} \qquad \begin{matrix} (H_5C_6)_2C-HgCl \\ | \\ Cl \end{matrix}$$

58

$-N_2 \downarrow$

$$\begin{bmatrix} (H_5C_6)_2C-\overset{\oplus}{C}(C_6H_5)_2 \\ \overset{|}{Cl}-\overset{\ominus}{Hg}-Cl \end{bmatrix} \longrightarrow \begin{matrix} H_5C_6 \quad C_6H_5 \\ | \qquad | \\ ClHg-C-C-Cl \\ | \qquad | \\ H_5C_6 \quad C_6H_5 \end{matrix}$$

59

Depending on the molar ratios of the reactants, the two different reaction products *(chlorodiphenylmethyl)mercury chloride* and *(2-chlorotetraphenylethyl)mercury chloride)* **58**, **59** are obtained in quantitative yields[982].

The most fully investigated of the halides of Group II metals is mercuric chloride, whose reactions with various aliphatic diazo compounds mainly yield mono- and dialkylation products*[400, 980, 983, 984], *e.g.:*

$$HgCl_2 \begin{cases} \xrightarrow[-N_2]{+ CH_2N_2} & Cl-CH_2-HgCl \\ & \text{(Chloromethyl)mercury chloride} \\ \xrightarrow[-2N_2]{+ 2CH_2N_2} & Hg(CH_2-Cl)_2 \\ & \text{Bis(chloromethyl)mercury} \end{cases}$$

Iodomethylmagnesium iodide is formed from magnesium iodide and diazomethane, and has been successfully separated at low temperatures[986]:

$$MgJ_2 + CH_2N_2 \xrightarrow{-N_2} JMg-CH_2-J$$

Iodomethylzinc iodide is the primary product from treatment of zinc iodide with ethereal diazomethane solution; it is of importance in synthesis since it can convert olefins into cyclopropane derivatives[400, 441], *e.g.:*

$$ZnJ_2 + CH_2N_2 \xrightarrow{-N_2} JZn-CH_2-J$$

$$H_3C-CH=CH-CH_3 \longrightarrow \begin{matrix} H_3C \quad\quad CH_3 \\ \triangle \end{matrix}$$

cis-1,2-Dimethylcyclopropane

Halides of Group III elements react likewise with diazoalkanes[400, 980, 981]. Boron and aluminum halides are widely used as catalysts for many reactions involving diazoalkanes; if no other reactants are present, aluminum chloride (or bromide) and ethereal diazomethane solution react at −50° to −70° to form the *tris(halomethyl)aluminum*[987]. Reaction of boron trifluoride with diazomethane in the gas phase gives *fluoromethylboron difluoride*[988]**.

---

* Treating α-diazo ketones with mercuric oxide affords their mercury derivatives in good yield[985]:

$$\begin{matrix} R \\ \diagdown \\ \quad C-CHN_2 \\ \diagup \\ O \end{matrix} \xrightarrow{HgO} Hg(CN_2-CO-R)_2$$

**The use of aluminum and boron compounds for polymerization of diazoalkanes is described in ref.[441].

[980] *D. Seyferth*, Chem. Rev. *55*, 1155 (1955).
[981] *A. Ya. Yakubovich, V.A. Ginsburg*, Usp. Khim. *20*, 734 (1951) (russ.).
[982] *A. Ledwith, L. Phillips*, J. Chem. Soc. 5969 (1965).

[983] *L. Hellerman, M.D. Newmann*, J. Amer. Chem. Soc. *54*, 2859 (1932).
[984] *A.N. Nesmeyanov, G.S. Povch*, Ber. *67B*, 971 (1934).
[985] *P. Yates, F. Garneau*, Tetrahedron Lett. 71 (1967).
[986] *G. Wittig, F. Wingler*, Chem. Ber. *97*, 2139 (1964).
[987] *L. Al'Mashi, I. Fel'meri, A. Gants*, Dokl. Akad. Nauk SSSR *118*, 1121 (1958); C.A. *52*, 12753f (1958).
[988] *J. Goubeau, K. Rohwedder*, Justus Liebigs Ann. Chem. *604*, 168 (1957).

The halides of Group IV elements are used more often for the synthesis of organoelemental compounds than are those of Group II and Group III elements.

Unlike carbon tetrachloride (see p. 284), the tetrahalides of silicon[981, 989], germanium[990], tin[981, 991] and lead[992, 993] react vigorously with diazomethane in ethereal solution, even at low temperatures, and stepwise methylation occurs, e.g.:

$$SiCl_4 + n\,CH_2N_2 \longrightarrow$$

Cl₃Si—CH₂—Cl

(Chloromethyl)silicon trichloride

Cl₂Si(CH₂—Cl)₂

Bis(chloromethyl)-silicon dichloride

Cl—Si(CH₂—Cl)₃

Tris(chloromethyl)-silicon chloride

The number of methyl groups introduced can sometimes be controlled by the amount of diazomethane employed, the temperature, and by adding cupric sulfate.

Reactions of diazomethane with Group V metal halides proceed analogously[981, 994]. For example, phosphorus pentachloride reacts with diazomethane to form *tris(chloromethyl)-phosphine dichloride*, which can be partially hydrolyzed to *tris(chloromethyl)phosphine oxide*:

$$PCl_5 + 3\,CH_2N_2 \xrightarrow[-3N_2]{} (Cl—CH_2)_3PCl_2$$

$$\xrightarrow{H_2O} (Cl—CH_2)_3PO$$

8.2.7.4 Reactions with metals and organometallic compounds

The reactions of diazomethane with alkali metals, with methyllithium or phenyllithium, or with tri-

phenylmethylsodium, form *metal derivatives* of diazomethane, *i.e.*, in these reactions diazomethane behaves as a pseudoacid[995, 996]:

$$CH_2N_2 + LiCH_3 \longrightarrow CH_4 + [HCN_2]^{\ominus}Li^{\oplus}$$

The conditions of hydrolysis can be so chosen that either diazomethane is regenerated ($pH > 8$), or *isodiazomethane* ($pH < 8$) is formed:

$$\left[\overset{\ominus}{C}\equiv\overset{\oplus}{N}-\overset{\ominus}{N}-H \longleftrightarrow H-\overset{\ominus}{C}=\overset{\oplus}{N}=\overset{\ominus}{N}\right]Me^{\oplus}$$

pH < 8 | H₂O   pH > 8

$$\overset{\ominus}{C}\equiv\overset{\oplus}{N}-NH_2 \longleftrightarrow C=N-NH_2 \quad CH_2N_2$$

Many investigations have produced evidence that isodiazomethane has the structure of *fulminic acid amide (N-isocyanamine)*[997, 998].

Its *properties* differ from those of diazomethane. It is much more **explosive,** but more stable to acids. Caustic alkalis cause rapid and quantitative rearrangement of isodiazomethane to diazomethane.

Isodiazomethane gives many of the reactions characteristic of isonitriles[997-1000].

It has been shown that radical-anions are formed by the action of sodium on 9-diazofluorene and diazodiphenylmethane[1001].

Treatment of diazomethane with finely dispersed cesium in an inert solvent at −60° yields a mixture of acetylene, tetrahedrane, benzene, cubane, *etc.*; the formation of all of these reaction products can be explained in terms of the intermediate occurrence of *methine*[1002]:

$$CH_2N_2 \xrightarrow{Cs} \left[\begin{array}{c}H-\overset{\ominus}{C}-\overset{\oplus}{N}\equiv N \\ \updownarrow \overset{\ominus}{} \\ H-\overset{\ominus}{C}=\overset{\oplus}{N}=\overset{\ominus}{N}\end{array}\right] + Cs^{\oplus} + \tfrac{1}{2} H_2$$

$$\downarrow O_2$$

$$\left[\begin{array}{c}H-\overset{\ominus}{C}-\overset{\oplus}{N}\equiv N \\ \updownarrow \\ H-\overset{\oplus}{C}-\overset{\ominus}{N}=\overset{\ominus}{N}\end{array}\right] \xrightarrow{-N_2} H-\overset{\cdot\cdot}{\overset{\cdot}{C}}$$

[989] A. Ya. Yakubobich, V.A. Ginsburg, Zh. Obshch. Khim. 22, 1783 (1952); C.A. 47, 9256e (1953).

[990] D. Seyferth, E. Rochow, J. Amer. Chem. Soc. 77, 907 (1955).

[991] A. Ya. Yakubobich, S.P. Makarov, G.I. Gavrilov, Zh. Obshch. Khim 22, 1788 (1952); C.A. 47, 9257b (1953).

[992] A. Ya. Yakubobich, E.N. Merkulova, S.P. Kakarov, G.I. Gavrilov, Zh. Obshch. Khim. 22, 2060 (1952); C.A. 47, 9257h (1953).

[993] A. Ya. Yakubobich, S.P. Makarov, Zh. Obshch. Khim. 22, 1528 (1952); C.A. 47, 8010c (1953).

[994] A. Ya. Yakubobich, V.A. Ginsburg, Zh. Obshch. Khim. 22, 1534 (1952); C.A. 47, 9254g (1953).

[995] E. Müller, H. Disselhoff, Justus Liebigs Ann. Chem. 512, 250 (1934).

[996] E. Müller, W. Kreutzmann, Justus Liebigs Ann. Chem. 512, 264 (1934).

The reactions between diazoalkanes and organo-magnesium compounds or phenyllithium form the basis of a convenient method for the preparation of *hydrazones:*

$$(H_5C_6)_2CN_2 \xrightarrow[\text{2. } H_2O]{\text{1. } H_5C_6-MgBr} (H_5C_6)_2C=N-NH-C_6H_5$$

The reaction mechanism is not yet completely clear[402, 980].

Phenylcopper reacts differently with relatively stable diazo compounds (diazodiphenylmethane, ethyl diazoacetate, α-diazo ketones): nitrogen is eliminated, and benzene derivatives are formed in yields of 30–85%[1003]:

$$\underset{R^1}{\overset{R}{\diagup}}CN_2 \xrightarrow[\text{2. } H_2O]{\text{1. ArCu}} Ar-\underset{R^1}{\overset{R}{CH}}$$

Diazomethane reacts with trialkylaluminum; nitrogen is eliminated, and *polymethylenation* occurs[1004]:

$$Al\overset{R}{\underset{R}{\diagdown R}} \xrightarrow[-3n N_2]{+3n CH_2N_2} Al\overset{(CH_2)_{n_1}-R}{\underset{(CH_2)_{n_3}-R}{- (CH_2)_{n_2}-R}}$$

R = Alk

$n_1 + n_2 + n_3 = n$

This last reaction proceeds by the mechanism shown on p. 296.

Organoboron compounds, principally *alkylboranes* and *arylboranes*, react with diazoalkanes in such a way that the diazoalkane group is inserted into the *C–B* bond[441]. Subsequent treatment with hydrogen peroxide makes possible the preparation of *alcohols, ketones, diketones, aldehydes, etc.*[1005–1011], *e.g.:*

$$R_3B + N_2CH-CO-CH_3 \longrightarrow$$

$$R_3\overset{\ominus}{B}-\underset{\overset{|}{\overset{\oplus}{N_2}}}{CH}-CO-CH_3 \longrightarrow R_2B-\underset{\overset{|}{R}}{CH}-CO-CH_3 + N_2$$

$$\downarrow H_2O_2$$

$$R-CH_2-CO-CH_3$$

It is probable that this reaction will acquire considerable importance for synthesis.

8.2.7.5 Reactions with organophosphorus compounds

Aliphatic diazo compounds react with derivatives of trivalent phosphorus without loss of nitrogen, *phosphazines* being formed:

$$\underset{R^2}{\overset{R^1}{\diagup}}CN_2 + PR_3 \rightleftharpoons \underset{R^2}{\overset{R^1}{\diagup}}C=N-N=PR_3$$

$$\longleftrightarrow \underset{R^2}{\overset{R^1}{\diagup}}\overset{\ominus}{C}-N=\overset{\oplus}{N}=PR_3$$

These reactions involve electrophilic addition of the diazo compound to the nucleophilic phosphine; they are quite general, and can be carried out with aliphatic diazo compounds of various structures[400, 467, 1012–1014].

[998] *E. Müller, R. Beutler, B. Zeeh*, Justus Liebigs Ann. Chem. *719*, 72 (1968).

[999] *R. Beutler, B. Zeeh, E. Müller*, Chem. Ber. *102*, 2636 (1969).

[1000] *E. Müller, V. Nespital, R. Beutler*, Tetrahedron Lett. 525 (1971).

[1001] *Th. Kauffmann, S.M. Hage*, Angew. Chem. *75*, 248 (1963).

[1002] *A. Troischose*, Angew. Chem. *82*, N7, (253) (1970).

[1003] *T. Sato, S. Watanabe*, Chem. Commun. 515 (1969).

[1004] *H. Hoberg*, Justus Liebigs Ann. Chem. *656*, I (1962); *H. Hoberg*, Justus Liebigs Ann. Chem. *695*, I (1966).

[1005] *J. Hooz, S. Linke*, J. Amer. Chem. Soc. *90*, 5936, 6891 (1968).

[1006] *J. Hooz, D.M. Gunn*, Chem. Commun. 139 (1969).

[1007] *J. Hooz, D.M. Gunn*, J. Amer. Chem. Soc. *91*, 6195 (1969).

[1008] *J. Hooz, G.F. Morrison*, Can. J. Chem. *48*, 868 (1970).

[1009] *J. Hooz, D.M. Gunn*, Tetrahedron Lett. 3455 (1969).

[1010] *J. Hooz, D.M. Gunn, H. Kono*, Can. J. Chem. *49*, 2371 (1971).

[1011] *D.J. Pasto, P.W. Wojtkowski*, J. Org. Chem. *36*, 1790 (1971).

[1012] *K. Sasse*, Methoden zur Herstellung und Umwandlung von organischen Phosphorverbindungen in *Houben-Weyl*, Methoden der organischen Chemie, Bd. XII/1, p. 180, Georg Thieme Verlag, Stuttgart 1963.

[1013] *A.V. Kirsanov*, Phosphazo Compounds, Chap. 8, Naukova Dumka, Kiev 1965 (russ.).

[1014] *J. Strating, J. Hures, M.M. van Leusen*, Rec. Trav. Chim. Pays-Bas. *85*, 1061 (1966).

[997] *E. Müller, P. Kästner, R. Beutler, W. Rundel, H. Suhr, B. Zeeh*, Justus Liebigs Ann. Chem. *713*, 87 (1968).

Phosphazines are generally well-defined crystalline materials with sharp melting points, and have proven to be very convenient derivatives for the *characterization* of aliphatic diazo compounds. Triphenylphosphine, which is readily available, is mostly used; it reacts at room temperature with diazo compounds dissolved in ether, benzene, or ethyl acetate.

In certain cases heating or the addition of Lewis acids is necessary[1015]. Phosphazine formation is an equilibrium reaction[1016].

The *stability of the phosphazines* is dependent on the nature of the substituent groups at the α-carbon atom of the diazo component and at the phosphorus. Heating also accelerates the decomposition of phosphazines. Triphenylphosphazines derived from alkyl- or aryldiazomethanes decompose only with difficulty[1016], whereas α-ketophosphazines *(R—CO—CH=N—N=P(C$_6$H$_5$)$_3$)*, in which the α-carbon atom is linked to an electron-withdrawing group decompose under fairly mild conditions (*e.g.*, in chloroform solution[1013] at 20°) to form triphenylphosphine and α-diazo ketones. Some 2-diazo-1,3-diketones do not react with triphenylphosphine, although they can react with the more nucleophilic tris(morpholino)phosphine[1017–1019].

α-Ketophosphazines react readily with methyl iodide forming *methyltriphenylphosphonium iodide* and liberating the α-*diazo ketone*[1020]:

$$R-CO-CH=N-N=P(C_6H_5)_3 \rightleftharpoons$$

$$R-CO-CHN_2 \;+\; P(C_6H_5)_3$$

$$\xrightarrow{H_3C-J} \left[(H_5C_6)_3P-CH_3\right]^{\oplus} J^{\ominus}$$

In this way it is *possible* to purify the α-diazo ketones *via* their triphenylphosphazines[1020, 1021].

[1015] *W. Ried, H. Appel*, Z. Naturforsch. B. *15*, 684 (1960).

[1016] *H.J. Bestmann, L. Gotlich*, Justus Liebigs Ann. Chem. *655*, I (1963).

[1017] *M. Regitz, A. Liedhegener, O. Stadler*, Justus Liebigs Ann. Chem. *713*, 101 (1968).

[1018] *W. Ried, H. Appel*, Justus Liebigs Ann. Chem. *646*, 82 (1961);
W. Ried, H. Appel, Justus Liebigs Ann. Chem. *679*, 56 (1964).

[1019] *I.K Korobizina, V.A. Nikolaev*, Zh. Org. Khim. *8* (in print) (1972) (russ.).

[1020] *H.J. Bestmann, H. Buckschweski, H. Leube*, Chem. Ber. *92*, 1345 (1959).

[1021] *M. Regitz, A. Liedhegener*, Chem. Ber. *99*, 3143 (1966).

8.3 Bibliography

B. Eistert, Synthesen mit Diazomethan, Angew. Chem. *54*, 99, 124 (1941);

B. Eistert, in Neuere Methoden der präparativen organischen Chemie, Bd. I, p. 359, Verlag Chemie, Weinheim/Bergstr. 1949.

A.Ya. Yakubovich, V.A. Ginsburg, The Diazo Method for the Synthesis of Organo-Element Compounds in the Aliphatic Series, Usp. Khim. *20*, 734 (1951) (russ.).

C.D. Gutsche, The Reaction of Diazomethan and its Derivatives with Aldehydes and Ketones, Org. Reactions 8, 364 (1954).

R. Huisgen, Altes und Neues über aliphatische Diazoverbindungen, Angew. Chem. *67*, 439 (1955).

D. Seyferth, Preparation of Organometallic and Organometalloidal Compounds with Diazoalkane Method, Chem. Rev. *55*, 1155 (1955).

I.A. D'yakonov, Aliphatic Diazo Compounds, Their Structure, Preparations and Reactions (russ.), Gosudarst, Univ. Leningrad, 1958; C.A. *53*, 2088e (1959).

F. Weygand, R. Bestmann, Synthesen unter Verwendung von Diazoketonen, Angew. Chem. *72*, 535 (1960).

H. Zollinger, Azo and Diazo Chemistry, Interscience Publishers, New York, London 1961.

W. Ried, H. Mengler, Zur präparativen Chemie der Diazocarbonylverbindungen, Fortschr. Chem. Forsch. 5, 1 (1965).

Eu. Müller, H. Kessler, B. Zeeh, Katalysierte Diazoalkan-Reaktionen, Fortschr. Chem. Forsch. 7, 128 (1966).

P.A.S. Smith, The Chemistry of Open-Chain Organic Nitrogen Compounds, Vol. II, Chap. Azides and Diazo-Compounds, p. 211, W.A. Benjamin, New York, Amsterdam 1966.

N.V. Sidgwick, J.T. Millar, H.D. Springall, The Organic Chemistry of Nitrogen, 3. Ed. Chap. Aliphatic Diazo Compounds and Derivatives of Hydrogen Azide, p. 472, Clarendon Press, Oxford 1966.

L.L. Rodina, I.K. Korobizina, Wolff Rearrangement, Usp. Khim. *36*, 611 (1967) (russ.); C.A. *67*, 116250u (1967).

B. Eistert, M. Regitz, G. Heck, H. Schwall, Methoden zur Herstellung und Umwandlung von aliphatischen Diazoverbindungen in *Houben-Weyl*, Methoden der organischen Chemie Bd. X/4, p. 482, Georg Thieme Verlag, Stuttgart (1968).

C.D. Gutsche, D. Redmore, Carbocyclic Ring Expansion Reactions. Chap. IV. Diazoalkane Ring Expansions of Cycloalkanones, Academic Press, New York, London 1968.

O.P. Studzinskii in *I.A. D'yakonov*, Sovrem. Probl. Org. Khim. p. 224, Reactions of α-Dicarbonyl Compounds with Diazo Compounds (russ.), Izd. Leningrad, Univ. Leningrad, USSR 1969.

O.P. Studzinskii, I.K. Korobizina, Concerning the Structure of Aliphatic Diazo Compounds and their Isomers, Usp. Khim. *39*, 1754 (1970) (russ.).

G. Cowell, A. Ledwith, Developments in the Chemistry of Diazo-alkanes, Quart. Rev. *24*, 119 (1970).

9 Azido Compounds

Contributed by

H. Scherer
M. Regitz
Fachbereich Chemie der Universität
Trier/Kaiserslautern,
Kaiserslautern

9.1 Preparation of azido compounds

9.1.1 of organic azido compounds

Azides are useful intermediates for the synthesis of numerous heterocyclic compounds, and in addition provide a good means of generating nitrenes by elimination of nitrogen.

The *UV* absorption band at 285 mμ can be used for their *identification*[1]. In addition, azides have a characteristic *IR* band between 2080 and 2170 cm^{-1}, which is attributed to the asymmetrical valence vibration. The symmetrical valence vibration at between 1200 and 1270 cm^{-1} is considerably weaker; since its position is more substituent dependent and is in a region where many other bands occur, it is not suitable for the identification of azides[2].

With inorganic azido compounds, the N_3-group is sufficiently similar to the halides for it to be regarded as a pseudohalogen. In contrast, this similarity is much less marked with the organic derivatives because organic azido compounds are sensitive to heat and to acids, and their reactions often involve *decomposition,* which frequently proceeds **explosively.** Extreme care is essential when working with azides. Explosions can be caused thermally (hence, care is required with distillation), by shock, and also by the action of chemical reagents (*e.g.,* concentrated sulfuric acid) on the undiluted materials. The lower the molecular weight, the greater the tendency to decompose explosively. Thus, for example, a particularly treacherous explosion of *methyl azide* has been reported[3]. Unexpected explosions of organic and organometallic azides which had been regarded as stable have also been described[4].

Azido compounds are usually *synthesized* by addition of hydrazoic acid, azide ions, or halogen azides to unsaturated compounds, or by replacement of functional groups by the azide group. A third possibility is formation of the N_3-group in a stepwise manner, starting with compounds which already contain one or two nitrogen atoms on suitable carbon atoms. Chemical rearrangements of

the carbon skeletons of organic azides have achieved some importance[5]. For a survey describing the preparation of organic azides up to 1965, see ref.[5].

9.1.1.1 Addition of inorganic azides to multiple bonds

9.1.1.1.1 Addition of hydrazoic acid and azide ions

Addition of hydrazoic acid or of azide ions to nonactivated *olefins* is difficult. Following earlier unsuccessful experiments[6, 7], the first successful addition of hydrazoic acid to alkenes, cycloalkenes, and arylalkenes was described[8] in 1951. For example, *2-azidopropane* was prepared from propene at 150° in a steel autoclave:

$$H_3C-CH=CH_2 \xrightarrow{HN_3} H_3C-\underset{\underset{N_3}{|}}{CH}-CH_3$$

Addition occurs more easily if the $C=C$ double bond is conjugated with a nitro, cyano, azomethine, or carbonyl group; in such cases, the azide group adds exclusively at the β-carbon atom. Examples of this type of reaction are shown in Tab. 1

Only in exceptional cases do *acetylenes* react with hydrazoic acid to form vinyl azides[12, 13]; the usual products are 1, 2, 3-triazoles[7] and isoxazoles[12]. The formation of *dimethyl azidofumarate* from dimethyl acetylenedicarboxylate[12] is worthy of mention:

$$H_3COOC-C\equiv C-COOCH_3$$

$$\xrightarrow{HN_3} \underset{H_3COOC}{\overset{H}{\diagdown}}C=C\underset{N_3}{\overset{COOCH_3}{\diagup}}$$

[1] *F.R. Benson, L.W. Hartzel, E.A. Otten,* J. Amer. Chem. Soc. *76,* 1858 (1954).
[2] *E. Lieber, C.N.R. Rao, A.E. Thomas, E. Oftedahl, R. Minnis, C.V.N. Namburg,* Spectrochim. Acta *19,* 1135 (1963).
[3] *C. Grundmann, H. Haldenwanger,* Angew. Chem. *62,* 410 (1950).
[4] Nachrichten aus Chemie und Technik *18,* 26 (1970).

[5] *J.H. Boyer, F. Canter,* Chem. Rev. *54,* 1 (1954); *C. Grundmann* in *Houben-Weyl,* Methoden der organischen Chemie, 4. Aufl., Bd. X/3, p. 777, Georg Thieme Verlag, Stuttgart 1965.
[6] *E. Oliveri-Mandala, G. Caronna,* Gazz. Chim. Ital. *71,* 182 (1941); C.A. *36,* 2804 (1942).
[7] *O. Dimroth, G. Fester,* Ber. *43,* 2219 (1910).
[8] USP. 2 557 924 (1951), Universal Oil Products Co., Inv.: *R.E. Schaad;* C.A. *46,* 1028 (1952).
[9] *J.H. Boyer,* J. Amer. Chem. Soc. *73,* 5248 (1951).
[10] *W.J. Awad, S.M.A.R. Omran, F. Nagieb,* Tetrahedron *19,* 1591 (1963).
[11] *R. Adams, W. Joje,* J. Amer. Chem. Soc. *74,* 5560 (1952).
[12] *U. Türck, H. Behringer,* Chem. Ber. *98,* 3020 (1965).

Table 1. Azido compounds by addition of hydrazoic acid to olefins

Olefin	Reaction conditions	Azido compound	Physical constants	Yield [% of th.]	Ref.
Acrolein	Glacial acetic acid/sodium azide, ice/common salt bath	*3-Azidopropionaldehyde*		71	9
N-Phenylmaleimide	Glacial acetic acid/sodium azide, 45 min at 70–80°	*Azido-N-phenylsuccinimide*	m 85–86°	80	10
Acrylonitrile	Glacial acetic acid/sodium azide, 1–3 days at 20°	*3-Azidopropionitrile*	b₁ 64°	17	9
4-Methyl-2-penten-2-one (Mesityl oxide)	Glacial acetic acid/sodium azide, 24 h at 100°	*4-Azido-4-methyl-2-pentanone*	b₃ 57–58°	38	9
2-Vinylpyridine	Glacial acetic acid/sodium azide, 24 h at 100°	*2-(2-azidoethyl)pyridine*	b₁ 65°	50	9
1,4-Naphthoquinone bis-(benzenesulfonyl-imide)	Glacial acetic acid/sodium azide, elevated temperature, spontaneous reaction	*2-Azido-N,N'-bis-(benzenesulfonyl)-1,4-naphthalenediamine*	m 185–186°	94	11

Similarly, *2-azido-1,1,1,4,4,4-hexafluoro-2-butene* is obtained from hexafluoro-2-butyne [14]. In the case of ethoxyacetylene, two molecules of hydrazoic acid add, giving the completely saturated product *1,1-diazido-1-ethoxyethane* [15]:

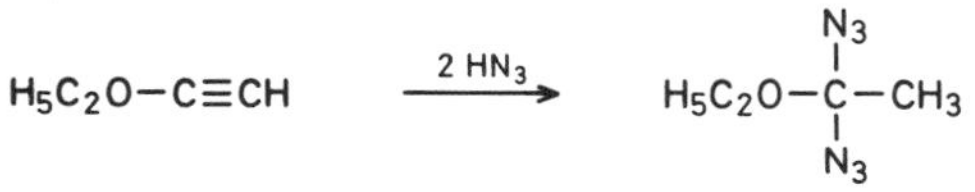

$$H_5C_2O-C{\equiv}CH \xrightarrow{2\ HN_3} H_5C_2O-\underset{N_3}{\overset{N_3}{C}}-CH_3$$

The addition of azide ions to alkenes only proceeds smoothly with perfluorinated olefins. The reaction of perfluoropropene with sodium azide in ethanol, for example, yields *1-azido-1,1,2,3,3,3-hexafluoropropane* [16]:

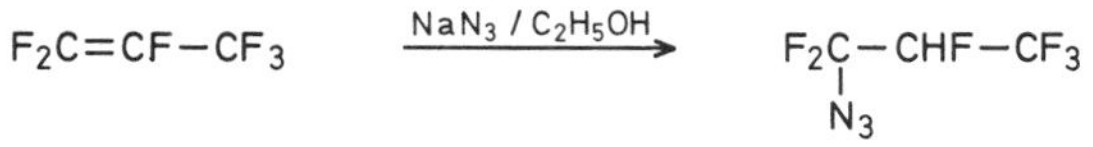

$$F_2C{=}CF-CF_3 \xrightarrow{NaN_3\ /\ C_2H_5OH} F_2C-\underset{N_3}{CHF}-CF_3$$

The addition of azide to *cumulated* double bonds can be carried out successfully in some cases. *Ethyl 3-azido-2-butenoate* is obtained from ethyl butadienoate [17]:

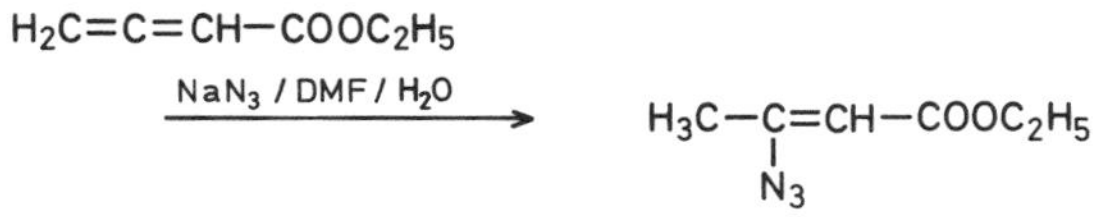

$$H_2C{=}C{=}CH-COOC_2H_5$$
$$\xrightarrow{NaN_3\ /\ DMF\ /\ H_2O} H_3C-\underset{N_3}{C}{=}CH-COOC_2H_5$$

The reaction fails with terminally disubstituted butadienoic acid esters and amides, however [17]. The use of *mercury-containing catalysts* for the addition of azides to olefins has proved to have advantages. Reaction of an olefin with mercuric azide (which is prepared *in situ* from mercuric chloride and sodium azide) first forms the intermediate **1** which gives the α-*azidoalkylmercury* compound **2**; reduction with sodium borohydride then provides [18] the *alkyl azide* **3**.

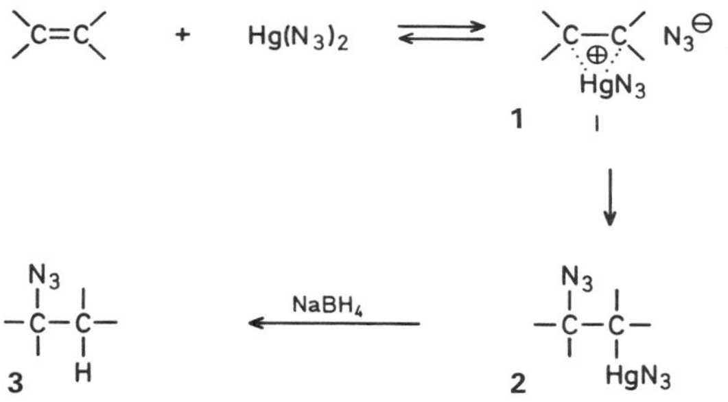

$$\underset{}{\overset{}{C}}{=}\overset{}{C} + Hg(N_3)_2 \rightleftharpoons \underset{\underset{\mathbf{1}}{HgN_3}}{C{-}C}\ N_3^{\ominus}$$

$$\underset{\mathbf{3}}{\overset{N_3}{\underset{H}{-C-C-}}} \xleftarrow{NaBH_4} \underset{\mathbf{2}}{\overset{N_3}{\underset{HgN_3}{-C-C-}}}$$

These variations are very suitable for the synthesis of *secondary* and *tertiary* alkyl azides by azide addition to olefins with terminal double bonds. Additions to compounds such as norbornene, in which the double bond is in a strained ring system, also proceed smoothly. The reaction fails, however, with alkenes having nonterminal double bonds [18]. Tab. 2 shows the scope of the procedure.

Numerous aliphatic azido compounds can be prepared in good yields [24, 25] by reaction of olefins with the *lead(IV)acetate/trimethylsilyl azide* system. The reaction proceeds as shown in the following scheme, and yields various azide group-containing products, depending on the substituent groups. In

[13] *G. L'abbé, A. Hassner,* Angew. Chem. *83,* 103 (1971); Angew. Chem. Intern. Ed. Engl. *10,* 98 (1971).

[14] *R.E. Banks, M.J. McGlinchey,* J. Chem. Soc. (C) 2172 (1970).

[15] *Y.A. Sinnema, J.F. Arens,* Rec. Trav. Chim. Pays-Bas *74,* 901 (1955).

[16] *I.L. Knunyantz, E.G. Bykhovskaya,* Dokl. Akad. Nauk SSSR *131,* 1338 (1960); C.A. *54,* 20840d (1960).

[17] *G.R. Harvey, K.W. Ratts,* J. Org. Chem. *31,* 3907 (1966).

[18] *C.H. Heathcock,* Angew. Chem. *81,* 148 (1969); Angew. Chem. Intern. Ed. Engl. *8,* 134 (1969).

Table 2. Azides by azide addition to $C=C$ double bonds

Olefin	Reaction conditions	Azido compound	Physical constants	Yield [% of th.]	Ref.
1-Octene	Mercury(II) acetate/sodium azide in tetrahydrofuran/water, 24 h at 30°	2-Azidooctane	b_9 68°	55	19
Methylene-cyclohexane	Mercury(II) acetate/sodium azide in tetrahydrofuran/water, 68 h at 90°	1-Azido-1-methylcyclohexane		60	19
trans-Stilbene	Dichloromethane, lead(IV) acetate, trimethylsilyl azide, −20°	a,a'-Diazidobibenzyl (±): meso = 70:30		42	20
		2-Azido-1,2-diphenylethanol acetate erythro : threo = 75:25		42	
Styrene	Dichloromethane, lead(IV) acetate, trimethylsilyl azide, −20°	(1,2-Diazidoethyl)benzene	$b_{0.00005}$ 60°	39	20
		a-(Azidomethyl)benzyl acetate	$b_{0.0005}$ 63°	31	
	Dichloromethane, lead(IV) acetate, trimethylsilyl azide, 12 h at 0°	2,9-Diazido-1,2,3,4-tetrahydro-1,4-methano-naphthalene	m 37–39°	9	21, 22
		9-Azido-1,2,3,4-tetrahydro-1,4-methanonaphthalen-2-ol acetate	m 45–47°	83	
	Dichloromethane, lead(IV) acetate, trimethylsilyl azide, 1–2 h at −20°	3a-Azidocholestan-2-one	m 102–105°	50–60	22, 23

addition, azidefree derivatives are sometimes formed (path D), especially in the case of conformationally rigid olefins.

$$Pb(OCOCH_3)_4 \; + \; n(H_3C)_3SiN_3$$

$$\downarrow \; -n(H_3C)_3SiOCOCH_3$$

$$Pb(OCOCH_3)_{4-n}(N_3)_n$$

19 C.H. Heathcock, Angew. Chem. 81, 148 (1969); Angew. Chem. Intern. Ed. 8, 134 (1969).
20 E. Zbiral, K. Kischa, Tetrahedron Lett. 1167 (1969).
21 E. Zbiral, A. Stütz, Tetrahedron 27, 4953 (1971).
22 E. Zbiral, A. Stütz, Tetrahedron 27, 4953 (1971).
23 E. Zbiral, G. Nestler, Tetrahedron 27, 2293 (1971).
24 E. Zbiral, K. Kischa, Tetrahedron Lett. 1167 (1969).
25 E. Zbiral, Synthesis 285 (1972).
26 G. Nestler, E. Zbiral, Monatsh. Chem. 102, 115 (1971).
27 E. Zbiral, G. Nestler, Tetrahedron 27, 2293 (1971).
28 O.E. Edwards, K.K. Purushothaman, Can. J. Chem. 42, 712 (1964).

With strainfree olefins, the reaction proceeds along paths B and C; paths A and D are only followed to a minor extent[24, 26]. However, when the double bond forms part of the relatively rigid molecular framework of steroids, paths A and D are mainly followed. Reaction route A is followed with doubly branched *steroid* olefins; the azido group is *axially* positioned[27]. This reaction is of some importance, since α-*azido steroid ketones* are very difficult to prepare by bromide/azide exchange[28, 29]. Triply branched steroidal olefins (*e.g.*, $\Delta^{5, 6}$-steroids) react as indicated by path D, however[30]. Examples of the use of this reaction are given in Tab. 2.

These reactions are generally carried out at low temperature (−20°). The complex product distribution can be regarded as the result of a transfer-reaction, in which the positive azide formed in a polar reaction is transferred to the substrate[25]. Rapid decomposition of the reagent may even occur at room temperature, resulting in radical substitution at the allylic position[31]. Under these conditions, cholest-5-en-3β-ol acetate is converted into *7α-azidocholest-5-en-3β-ol acetate*[31]:

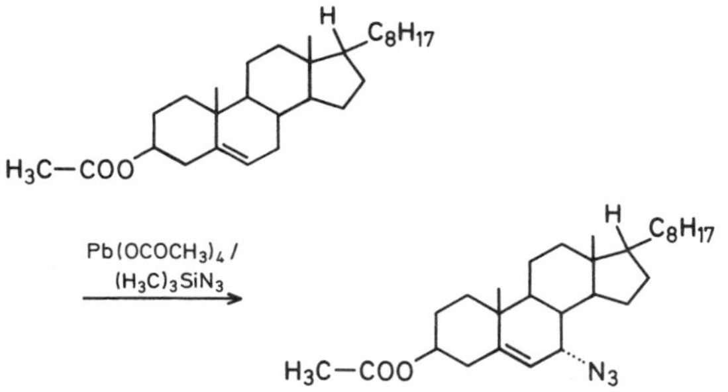

The $C \equiv C$ triple bond can also be involved in such azide-transfers[25]. Because of the **explosive character** of the lead(II) azide formed, these reactions should only be carried out on the small scale. This disadvantage can, however, be circumvented by using *phenyliodosodiacetate* instead of lead tetraacetate, especially when reactions following pathways A or D are anticipated[32]. The phenyliodoso diacetate/acetyl chloride/trimethylsilyl azide system can also be employed for azide-transfers to olefins[33]; phenyliodoso chloride-azide is formed as intermediate, and the reaction products are α-*azidohalogen* compounds.

Oxidative, dye-sensitized azide addition to indene forms a mixture of different hydroperoxides, from which *2-azido-1-indanol* can be isolated after reduction[34]:

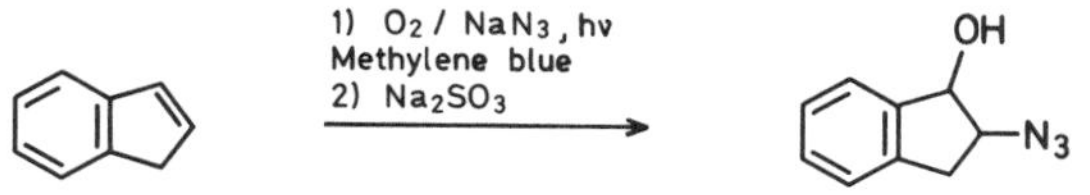

trans-2-Azido-1, 2-dimethylcyclohexanol can be obtained from 1, 2-dimethylcyclohexene in similar manner[34].

A further variation of oxidative azide addition is *electrolysis* of azide ions in acetic acid/olefin mixtures; this technique can be used for the preparation of *1, 2-* and *1, 4-diazidoalkanes* in a single-stage reaction[35].

Azide free-radicals are first formed by anodic oxidation, and react with the double bond to give radical-type primary products, whose subsequent fate depends very much on the structure of the olefin[36], as the following examples show. Thus, the radical formed from styrene and the azide radical dimerizes, giving *1, 4-diazido-2, 3-diphenylbutane*[36]:

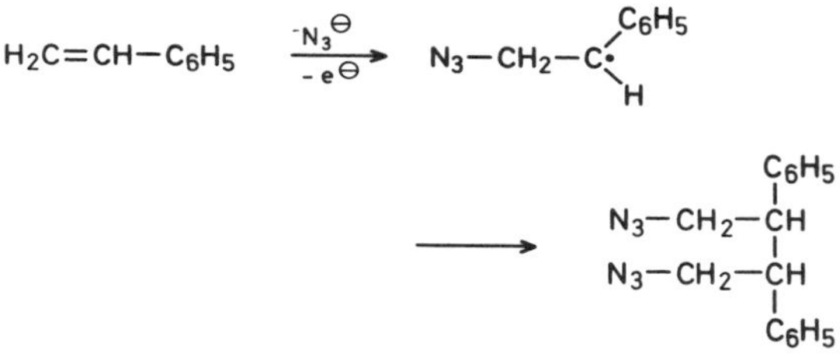

With 1, 1-diphenylethylene, the radical first formed is oxidized to the corresponding cation, which reacts further with excess azide and acetate ions to give *1, 2-diazido-1, 1-diphenylethane* and *2-azido-1, 1-diphenylethanol acetate*

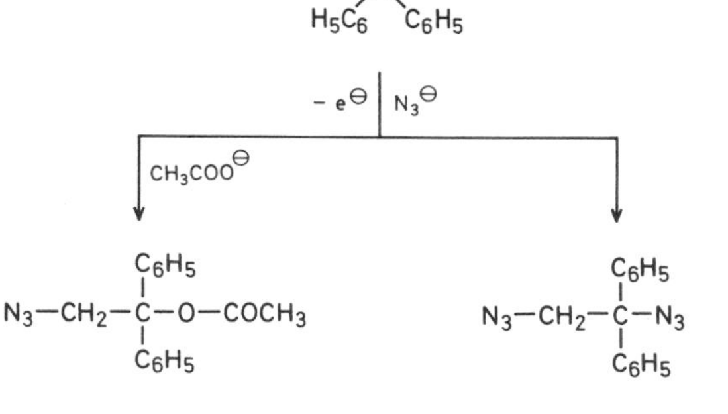

[29] *J.G.L. Jones, B.A. Marples*, J. Chem. Soc. (C) 1188 (1970).
[30] *E. Zbiral, G. Nestler, K. Kischa*, Tetrahedron **26**, 1427 (1970).
[31] *K. Kischa, E. Zbiral*, Tetrahedron **26**, 1417 (1970).
[32] *E. Zbiral, G. Nestler*, Tetrahedron **26**, 2945 (1970).

In the case of purely aliphatic olefins such as cyclohexene, the radical-intermediate may combine with azide radicals to form *1,2-diazidocyclohexane*. A further possibility is, however, hydrogen-abstraction from cyclohexene, leading to *azidocyclohexane* and the cyclohexenyl radical, which reacts with a further azide radical to form *3-azidocyclohexene*:

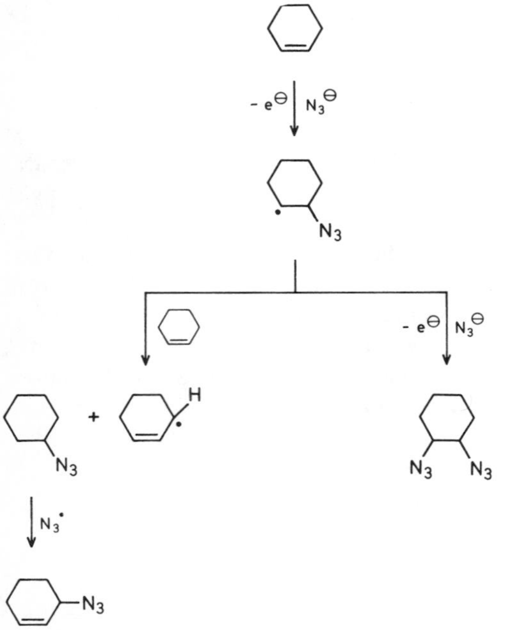

9.1.1.1.2 Addition of halogen azides

Addition of halogen azides to olefins is a good method for the *stereospecific* and *regiospecific*[37] introduction of the azide function into these starting materials[38]. For example, 4,4-dimethyl-1-pentene and *iodine azide* yield *4-azido-5-iodo-2,2-dimethylpentane,* not the regioisomeric 5-azido-4-iodo-2,2-dimethylpentane[39]:

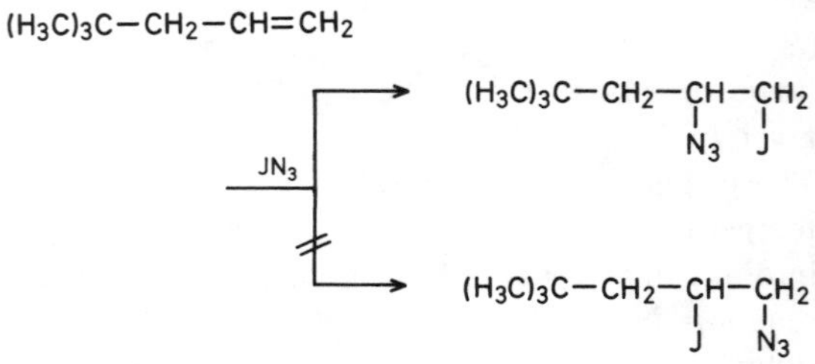

The iodine azide necessary for these reactions can be prepared safely as shown in the following equation[40, 41]:

$$JCl \; + \; NaN_3 \xrightarrow{\text{polar solvents}} JN_3 \; + \; NaCl$$

Chlorine azide and bromine azide can be prepared in a two-phase system from chlorine (or bromine) and sodium azide in the presence of acid, in an analogous manner[42].

Table 3. α-Azido halides by addition of halogen azide to olefins

Olefin	Reaction conditions	Azido compound	Physical constants	Yield [% of th.]	Ref.
trans-Stilbene	Iodine azide/acetonitrile, 8–20 h ~ 20°	*erythro-α-Azido-α'-iodobibenzyl*	m 131–133°	73	44
Isobutene	Iodine azide/acetonitrile, 8–20 h ~ 20°	*2-Azido-1-iodo-2-methylpropane*	b$_7$ 68–69°	60	44, 45
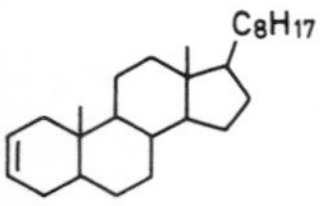	Iodine azide/acetonitrile, 8–20 h ~ 20°	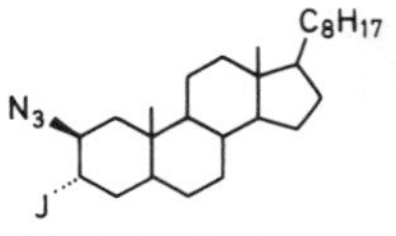 *2β-Azido-3α-iodocholestane*	m 101–102°	53	44
Chalcone	Iodine azide/acetonitrile, 8–20 h ~ 20°	*3-Azido-2-iodo-1,3-diphenylpropane*	m 104°	100	44, 46
Cyclooctadiene	Iodine azide/acetonitrile, 8–20 h ~ 20°	*3-Azido-4-iodo-1-cyclooctene*		72	47
Cyclopentene	Iodine azide/acetonitrile	*trans-1-Azido-2-iodocyclopentane*		89	47
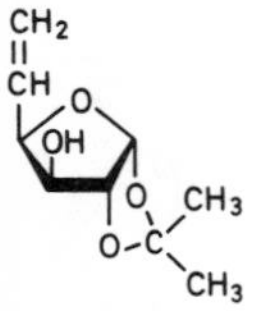	Iodine azide/acetonitrile, 17 h ~ 20°	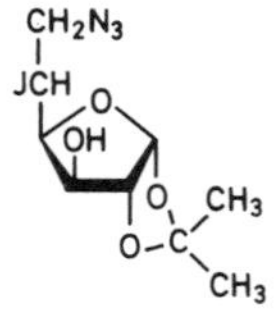 *β-L-6-Azido-5,6-dideoxy-5-iodo-1,2-O-isopropylidene-idofuranose*	m 81–83°	70	48

The reaction of iodine azide to olefins is a *trans*-addition; consequently, *cis*-2-butene yields *threo-2-azido-3-iodobutane,* whereas the *trans*-olefin gives the *erythro*-product[39, 40, 43]. Analogously, the reaction of cyclohexene with iodine azide yields *trans-2-azido-1-iodocyclohexane*[40]. Tab. 3 contains further examples.

A cyclic halonium ion has been proposed as intermediate[40], consistent with the high steric and regional specificities; electronic factors determine its ring-opening to the α-azidohalogen compound:

Steric control of the reaction is only observed when the double bond is flanked by bulky groups. The reaction of 3,3-dimethyl-1-butene with iodine azide to form *1-azido-2-iodo-3,3-dimethylbutane* illustrates this[41]. Although the reaction proceeds by an ionic mechanism, rearrangements are only observed in cases where the activation energy is particularly low. Thus, *1-azido-3-iodo-1,1,2-triphenylpropane* is obtained from 1,1,1-triphenylpropene[41]:

The formation of *2-(azidomethyl)-5-iodotricyclo-[2.2.1.0^{3,5}]heptane* from 5-methylene-2-norbor-

nene or of *exo-anti-2-azido-1,2,3,4-tetrahydro-9-iodo-1,4-methanonaphthalene* from 1,2,3,4-tetrahydro-1,4-methanonaphthalene are also almost quantitative rearrangements which occur when iodine azide adds to olefines[41]:

Triple bonds also can add iodine azide; for example, 1-phenyl-1-propyne yields *2-azido-1-iodo-1-phenyl-1-propyne*[49]:

The ease of homolytic splitting of halogen azides increases in the order

$$IN_3 < BrN_3 < ClN_3$$

Whereas iodine azide reacts almost exclusively by an ionic mechanism, with *bromine azide* the reaction conditions can be chosen so that the process is either ionic or radical in nature. An example is the reaction of styrene with bromine azide; in nitromethane *(2-azido-1-bromoethyl)benzene* is obtained, while in pentane the regioisomeric *(1-azido-2-bromoethyl)benzene* results[42, 50]:

---

[33] E. Zbiral, J. Ehrenfreund, Tetrahedron *27*, 4125 (1971).

[34] W. Fenical, D.R. Kearns, P. Radlick, J. Amer. Chem. Soc. *91*, 7771 (1969).

[35] H. Schäfer, Angew. Chem. *82*, 134 (1970); Angew. Chem. Intern. Ed. Engl. *9*, 134 (1970).

[36] H. Schäfer, Chem.-Ing.-Tech. *42*, 164 (1970).

[37] A. Hassner, J. Org. Chem. *33*, 2684 (1968).

[38] A. Hassner, Accounts Chem. Res. *4*, 9 (1971).

[39] A. Hassner, F.W. Fowler, J. Org. Chem. *33*, 2686 (1968).

[40] A. Hassner, F.W. Fowler, L.A. Levy, J. Amer. Chem. Soc. *89*, 2077 (1967).

[41] A. Hassner, J.S. Teeter, J. Org. Chem. *35*, 3397 (1970).

[42] A. Hassner, F.P. Boerwinkle, A.B. Levy, J. Amer. Chem. Soc. *92*, 4879 (1970).

[43] A. Hassner, L.A. Levy, J. Amer. Chem. Soc. *87*, 4203 (1965).

[44] F.W. Fowler, A. Hassner, L.A. Levy, J. Amer. Chem. Soc. *89*, 2077 (1967).

[45] Destillation is not recommended as the compounds decompose.

Chlorine azide tends to undergo radical-type additions[42, 50, 51]. Ionic addition only occurs in very strongly polar media such as fuming sulfuric acid.

9.1.1.2 Azido compounds by exchange

9.1.1.2.1 Replacement of halogen

The halogen/azide exchange method has proved very useful for the preparation of *azidoalkanes*. Only the most simple organic azides, such as methyl and ethyl azide, have not yet been prepared by this means. The reaction can be carried out with commercial sodium azide, but the use of activated or freshly prepared sodium azide[52] can be advantageous[53]. Chlorides and bromides react smoothly; only where particularly mild reaction conditions are essential is the reaction between alkyl iodides and silver azide recommended. Glycol monoalkyl ethers have proved to be efficient *solvents*, since they allow reactions to be carried out in a homogeneous phase and at elevated temperatures[54]. Dimethylformamide can also be used as solvent[55]. Tab. 4 surveys the scope of the reaction.

It is possible to replace *several* halogen atoms attached to the same carbon atom; for example, treatment of carbon tetrachloride with tetrachloroantimony azide yields *triazidocarbenium hexachloroantimonate*[64].

$$3/2\,(SbCl_4N_3)_2 \quad + \quad CCl_4$$

$$\xrightarrow[-\,2\,SbCl_5]{} \quad [C(N_3)_3]^{\oplus} [SbCl_6]^{\ominus}$$

Nucleophilic substitutions proceed extraordinarily slowly in the case of vinyl halides; consequently, *vinyl azides* can only be obtained by halogen/azide exchange when the vinylic halogen atom is activated by electron-withdrawing substituent groups. For example, hexafluoropropene can be converted at 0° into unstable *trans-1-azido-1,2,3,3,3-penta-*

Table 4. Azido compounds by halogen/azide exchange

Halogen compound	Reaction conditions	Azido compound	Physical constants	Yield [% of th.]	Ref.
1-Bromohexane	Sodium azide, methanol/water, 6 h at 90°	*1-Azidohexane*	b_{758} 156–157°	50	56
Bromocyclopentane	Sodium azide, ethanol/water, 12 h reflux	*Azidocyclopentane*	b_{20} 51–52°	good	57
3-Chloro-1-heptene	Sodium azide, methanol/water, 1 week ~ 20°	*D-3-Azido-1-heptene*	b_{32} 78–81°		58
Chlorodiphenyl-phosphinylmethane	Sodium azide, dimethyl-formamide, 2 h at 110°	*(Azidomethyl)diphenyl-phosphine oxide*	m 110–111°	51	59
Dichlorodiphenyl-methane	Silver azide, ether, several hours ~ 20°	*Diazidodiphenylmethane*	m 40–42°	88	60
3-Chlorotropone	Sodium azide, dimethyl sulfoxide, 18 h at 300°	*3-Azidotropone*	m 61–63°	90	61
4-Iodo-1-nitrobenzene	Sodium azide, water, 17 days ~ 20°	*4-Azido-1-nitrobenzene*	m 71°	50	62
3-Bromodihydro-2(3H)-furanone	Sodium azide, ethanol, 40 h at 80°	*3-Azidodihydro-2(3H)-furanone*	$b_{0.2}$ 72°	80	63

[46] *G. L'abbé, A. Hassner*, J. Org. Chem. *36*, 258 (1971).

[47] *A. Hassner, F.W. Fowler*, J. Org. Chem. *33*, 2686 (1968).

[48] *J.S. Brimacombe, J.C.H. Bryan, T.A. Hamor*, J. Chem. Soc. (B) 514 (1970).

[49] *A. Hassner, A. Friederang, R.J. Isbister*, Tetrahedron Lett. 2939 (1969).

[50] *A. Hassner, F. Boerwinkle*, J. Amer. Chem. Soc. *90*, 216 (1968);
A. Hassner, F. Boerwinkle, Tetrahedron Lett. 3309 (1969).

[51] *F. Minisci, R. Galli, M. Cecere*, Chim. Ind. (Milan) *48*, 347 (1966);
F. Minisci, Chim. Ind. (Milan) *49*, 705 (1967).

[52] *J. Thiele*, Ber. *41*, 2681 (1908).

[53] *P.A.S. Smith*, Org. Reactions *3*, 382 (1946).

[54] *E. Lieber, T.S. Chao, C.N.R. Rao*, J. Org. Chem. *22*, 238 (1957).

[55] *J.A. Durden, H.A. Standburg, W.H. Catlette*, J. Chem. Eng. Data *9*, 228 (1964);
H. Boyer, R. Moriarty, B. Darwent, P.A.S. Smith, Chem. Eng. News *3*, 56 (1964).

[56] *K. Henkel, F. Weygand*, Ber. *76*, 812 (1943).

fluoropropene, which at room temperature rearranges to *2,3-difluoro-2-(fluoromethyl)-2H-azirine*[65]:

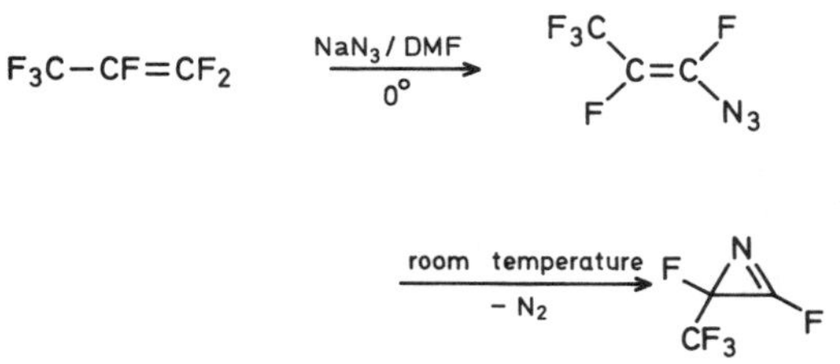

Further similar examples are the syntheses of *1-azido-2,3,3,3-tetrafluoropropene*[66] and *1-azido-3,3,3-trifluoro-1-methoxy-2-(trifluoromethyl)propene*[67]. Halogen/azide exchange is also possible with aryl β-chlorovinyl sulfones[68] and with β-halovinyl ketones[69].

Diethyl (azidomethylene)malonate can be prepared as follows[70]:

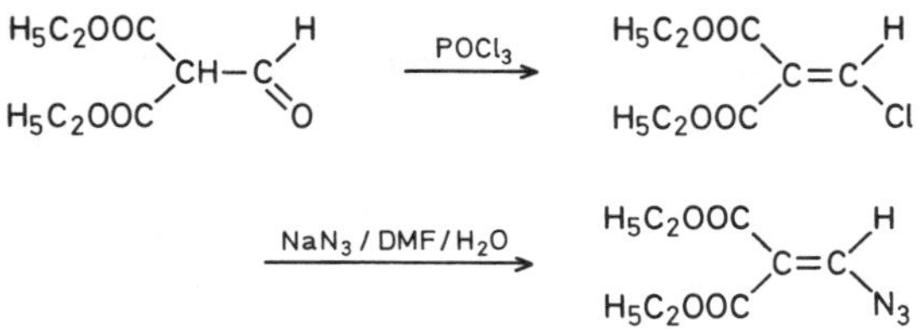

The same reaction can also be carried out with the formyl derivatives of ethyl cyanoacetate and malononitrile[71].

Aromatically bound halogen can only be replaced by the azide group when it is highly activated by other substituent groups. For example, picryl chloride yields *1-azido-2,4,6-trinitrobenzene*[72].

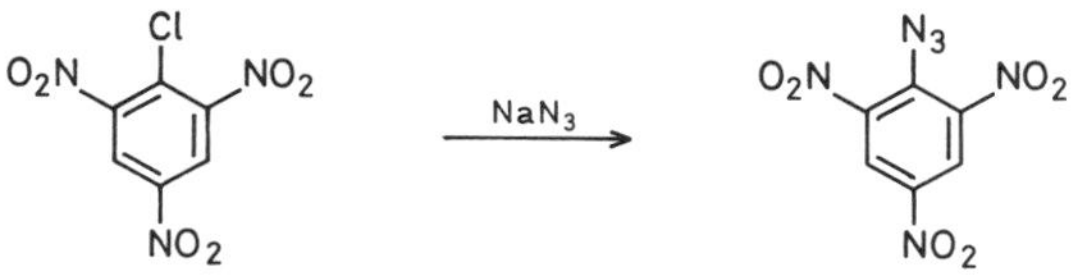

Halogen/azide exchange can also be carried out with highly fluorinated aromatic compounds[73], e.g.:

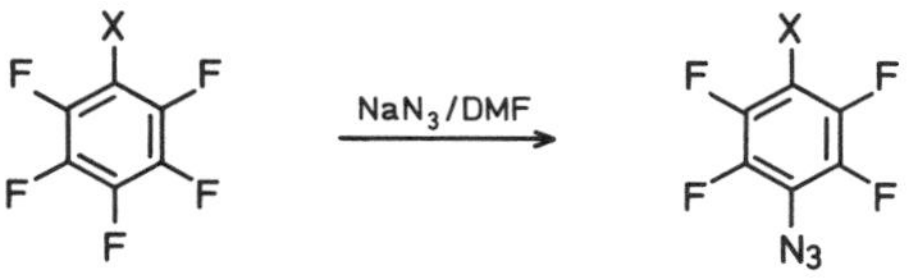

X = Cl; *6-Azido-3-chloro-1,2,4,5-tetrafluorobenzene*; 63%

X = CF₃; *4-Azido-2,3,5,6-tetrafluoro-1-(trifluoromethyl)-benzene*; 83%

X = F; *Azidopentafluorobenzene*; 60%

X = N₃; *3,6-Diazidotetrafluorobenzene*

This kind of reaction is also of importance for the preparation of *cyclic imide azides* from imidoyl chlorides. Often, however, one obtains not the azido-compounds, which are doubtless first formed, but isomeric tetrazoles formed by 1,5-cyclization[74]. Examples are the conversion of 2-chloropyrimidine into *tetrazolo[1,5-a]pyrimidine*[75] and of 7-benzyl-6-chloropurine into *6-benzyltetrazolo[1,5-f]purine*[76]:

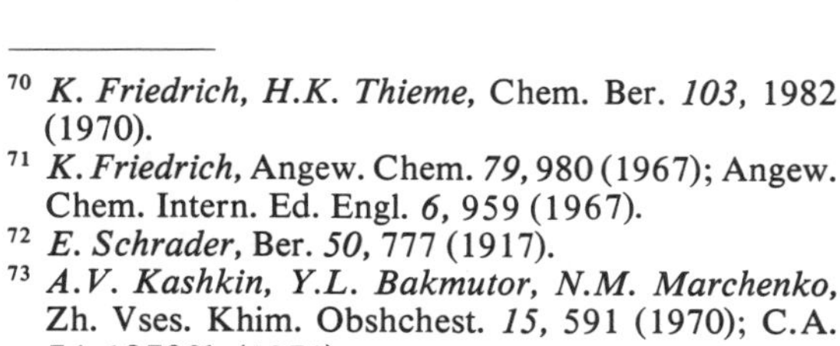

[57] *T.F. Fagley, H.W. Myers*, J. Amer. Chem. Soc. *76*, 6001 (1954).

[58] *P.A. Levene, A. Rothen, M. Kuna*, J. Biol. Chem. *120*, 777 (1937).

[59] *D. Seyferth, P. Hilbert*, Organic Preparations and Procedures International *3*, 51 (1971).

[60] *S. Götzky*, Ber. *64*, 1555 (1931).

[61] *J.D. Hobbson, J.R. Malpass*, J. Chem. Soc. (C) 1499 (1969).

[62] *D. Vorländer*, Ber. *70*, 146 (1937).

[63] *U. Kraatz, W. Hasenbrink, H. Wamhoff, F. Korte*, Chem. Ber. *104*, 2458 (1971).

[64] *U. Müller, K. Dehnicke*, Angew. Chem. *78*, 825 (1966); Angew. Chem. Intern. Ed. Engl. *5*, 841 (1966).

[65] *R.E. Banks, G.J. Moore*, J. Chem. Soc. (C) 2304 (1966).

[66] *R.E. Banks, M.J. McGlinchey*, J. Chem. Soc. (C) 2172, 1970. In attempts to react 1,1,3,3,3-pentafluoropropene with triethylammonium azide violent **explosions** occured.

[67] *C.G. Krespan*, J. Org. Chem. *34*, 1278 (1969).

[68] *G. Modena, P.E. Todesco*, Gazz. Chim. Ital. *89*, 866 (1959).

[69] *M.J. Rybinskaya, A.N. Nesmeyanov, N.K. Kochetkov*, Usp. Khim. 961 (1969); Russian Chem. Rev. *38*, 433 (1969); C.A. *71*, 48884 (1969).

[70] *K. Friedrich, H.K. Thieme*, Chem. Ber. *103*, 1982 (1970).

[71] *K. Friedrich*, Angew. Chem. *79*, 980 (1967); Angew. Chem. Intern. Ed. Engl. *6*, 959 (1967).

[72] *E. Schrader*, Ber. *50*, 777 (1917).

[73] *A.V. Kashkin, Y.L. Bakmutor, N.M. Marchenko*, Zh. Vses. Khim. Obshchest. *15*, 591 (1970); C.A. *74*, 12729k (1971).

[74] *H. Reimlinger*, Chem. Ber. *103*, 1900 (1970).

[75] *C. Wentrup*, Tetrahedron *26*, 4969 (1970).

[76] *C. Temple, C.L. Kussner, J.A. Montgomery*, J. Org. Chem. *31*, 2210 (1966).

One sometimes observes *equilibria* between the azido and tetrazolo forms.

These are dependent upon temperature[77, 78], solvent[75, 76, 78, 79], and substituent groups[75, 80]. A protic medium always favors the azido-form[78]. If the nitrogen atom in the α-position to the azido group has ammonium character, as, for instance, in quaternary heterocyclic bases, ring closure is not possible. Accordingly, heterocyclic α-haloammonium salts react smoothly with sodium azide to so-called *azidinium salts*[81]:

Acyclic members of this class of materials can be prepared in the same way[82]:

Y = H, Cl, CH$_2$Cl, CHCl$_2$, CCl$_3$

This reaction fails with substituent groups having a + I effect.

9.1.1.2.2 Replacement of the diazonium group

The diazonium group of aromatic and heteroaromatic compounds can be replaced by the azido group by treatment with sodium azide. For example, 1-naphthylamine can be converted, after diazotisation, into *1-azidonaphthalene*[83]:

Substituent groups only cause difficulties if they interfere with the diazotization.

This method, which for long was considered to be a variation of the Sandmeyer reaction[84], probably proceeds *via* an aryl azoazide and an arylpentazole[85].

9.1.1.2.3 Replacement of the sulfate or sulfonate groups

The sulfate group is readily displaced by the more strongly nucleophilic azide ion. Thus, *methyl azide* and *ethyl azide* are best prepared by reaction of sodium azide with dimethyl sulfate[86] or diethyl sulfate[87]. Other examples of successful syntheses involving this reaction principle are the preparation of *4-azidomethyl-2,2-diphenyl-1,3-dioxolane*[88] and of *2-acetamido-3-azido-4,6-O-(benzylidenemethyl)-2,3-dideoxy-α-D-glucopyranoside*[89]:

[77] *R. Fusco, S. Rossi, S. Maiorana*, Tetrahedron Lett. 1965 (1965).

[78] *C. Temple, J.A. Montgomery*, J. Amer. Chem. Soc. *86*, 2946 (1964).

[79] *B. Stanovnik, M. Tisler*, Tetrahedron *23*, 387 (1967).

[80] *S. Carboni, A. Da Settimo, P.L. Ferrarini*, Gazz. Chim. Ital *97*, 1061 (1967);
T. Sasaki, K. Kanematsu, M. Murata, J. Org. Chem. *36*, 446 (1971)

[81] *H. Balli, F. Kersting*, Justus Liebigs Ann. Chem. *647*, 1 (1961);
H. Balli, Justus Liebigs Ann. Chem. *647*, 11 (1961);
H. Balli, D. Schelz, Helv. Chim. Acta *53*, 1903 (1970);
H. Balli, D. Schelz, Helv. Chim. Acta *53*, 1913 (1970).

[82] *A. Schmidt*, Chem. Ber. *100*, 3319 (1967);
A. Schmidt, Chem. Ber. *100*, 3725 (1967).

[83] *M.O. Forster, H.E. Fierz*, J. Chem. Soc. (London) 1942 (1907).

[84] *A. Hantsch*, Ber. *36*, 2056 (1903);
G.T. Morgan, E.G. Couzens, J. Chem. Soc. (London) 1691 (1910).

[85] *R. Huisgen, J. Ugi*, Angew. Chem. *68*, 705 (1956);
R. Huisgen, J. Ugi, Chem. Ber. *90*, 2914 (1957);
J. Ugi, H. Perlinger, L. Behringer, Chem. Ber. *91*, 2324 (1958).

[86] *O. Dimroth, W. Wislicenus*, Ber. *38*, 1573 (1905).

[87] *H. Staudinger, E. Hauser*, Helv. Chim. Acta *4*, 872 (1921).

cis-1,2-Bis[p-tosyl)ethylene yields *cis-* and *trans-azidovinyl p-tolyl sulfones*, which in the presence of azide ions or other proton acceptors slowly cyclize to *4-(4-tolylsulfonyl)-1H-1,2,3-triazole*[90]:

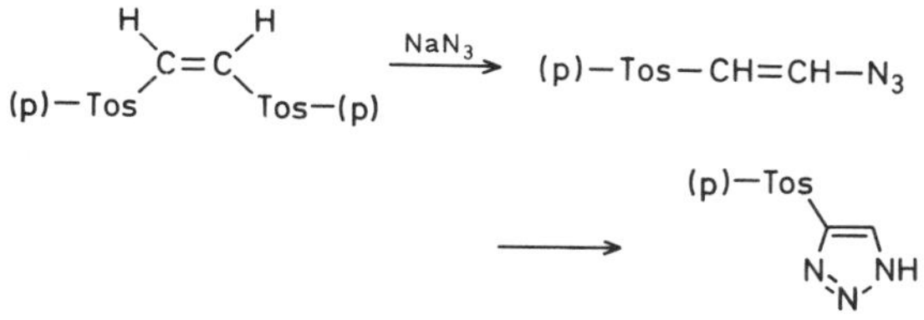

9.1.1.2.4 Replacement of other functional groups

The nitroxy group can be replaced by the azido group. 3-Methyl-3-nitroxy-2-butanone oxime yields *3-azido-3-methyl-2-butanone oxime*[91]. Formation[91] of *2-azido-5-isopropenyl-2-methylcyclohexanone oxime* from the corresponding 1-nitroxy compound, and the analogous preparation of *2-azido-2-methylpropane*[92] are further examples of the use of this method:

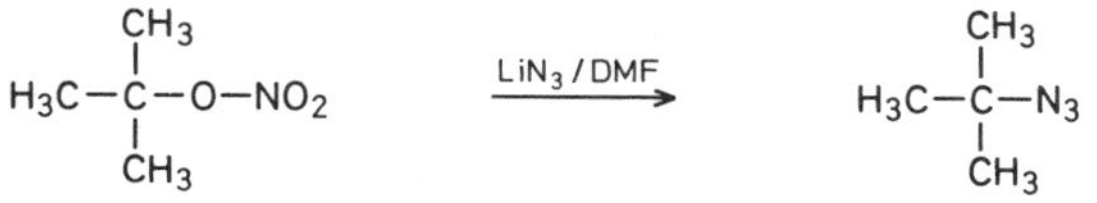

The well-known *ring-opening* of *oxiranes* with hydrogen halides can be transferred to hydrazoic acid, giving azido compounds[93,94]. For example, epoxycyclohexane forms *trans-2-azido-cyclohexanol*[93] by the usual *trans-*opening method:

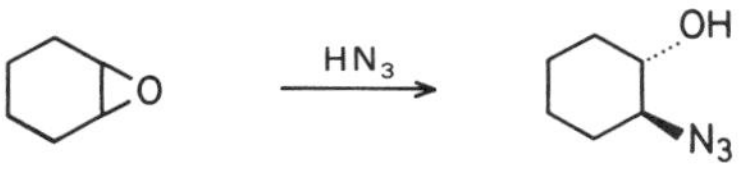

This kind of ring-opening can also be carried out with *aziridines*. Thus, *2-amino-3-azido-4,6-O-(benzylidenemethyl)-2,3-dideoxy-α-D-altroside* is formed from the sugar derivative **1**[95]:

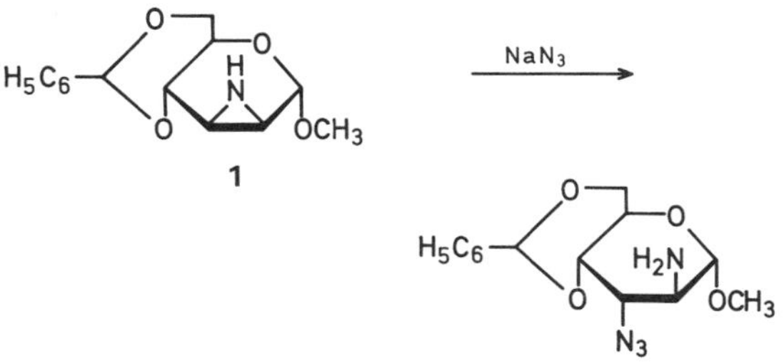

9.1.1.3 Synthesis of the azido group from nitrogen-containing compounds

9.1.1.3.1 Azido compounds from amines

Primary aliphatic and aromatic amines can be converted into amines by diazo group transfer (cf. p. 232). Because of their low acidity, they are treated with *p*-toluenesulfonyl azide in the presence of organometallic bases. In this way, *phenyl azide* is obtained from aniline, *cyclohexyl azide* from cyclohexylamine, and *butyl azide* from butylamine[96-98]. Diazo group transfer to the anion of bis(trimethylsilyl)amine gives *trimethylsilyl azide* in quantitative yield[99]:

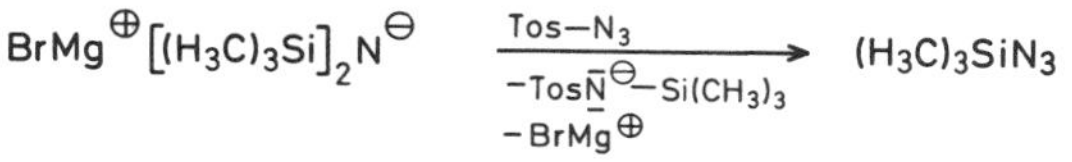

The reaction of 1,1-diphenylmethylenimine with *p*-toluenesulfonyl azide to form *1-azido-N-(diphenyl-methylene)-1,1-diphenylmethylamine* probably also proceeds in the sense of a transfer process[100]:

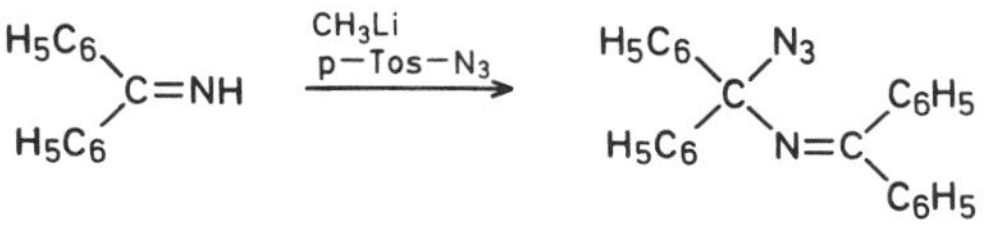

9.1.1.3.2 Azido compounds from hydrazines

Monosubstituted hydrazines react with nitrous acid to form unstable *N*-nitroso compounds, which

[88] *P.M. Carabateas, A.R. Surrey*, J. Heterocycl. Chem. **6**, 945 (1969).

[89] *Y. Ali, A.C. Richardson*, J. Chem. Soc. (C) 1764 (1968).

[90] *J.S. Meek, J.S. Fowler*, J. Amer. Chem. Soc. **89**, 1967 (1967).

[91] *M.O. Forster, F.M. Van Gelderen*, J. Chem. Soc. (London) **239**, 2059 (1911).

[92] *P. Margaretha, S. Solar, O.E. Polansky*, Angew. Chem. **83**, 410 (1971); Angew. Chem. Intern. Ed. Engl. **10**, 412 (1971).

[93] *J.P. Ingham, W.L. Petty, P.L. Nichols*, J. Org. Chem. **21**, 373 (1956).

[94] *M. Hasegawa, H.Z. Sable*, Tetrahedron **25**, 3567 (1969).

[95] *R.D. Guthrie, P. Murphy*, J. Chem. Soc. (London) 3828 (1965).

[96] *J.P. Anselme, W. Fischer*, Tetrahedron **25**, 855 (1969);
J.P. Anselme, W. Fischer, J. Amer. Chem. Soc. **89**, 5284 (1967).

[97] *J.B. Hendrickson, W.A. Wolf*, J. Org. Chem. **33**, 3610 (1968).

[98] *T.R. Steinheimer, D.S. Wulfman, L.N. McCullagh*, Synthesis 325 (1971).

[99] *N. Wiberg, W. Ch. Joo, M. Veith*, Inorg. Nucl. Chem. Lett. **4**, 223 (1968).

[100] *J.P. Anselme, W. Fischer, N. Koga*, Tetrahedron **25**, 89 (1969).

yield the corresponding azides in the presence of alkalis or acids[101]:

$$R-NH-NH_2 \xrightarrow[-H_2O]{HNO_2} \underset{\underset{NO}{|}}{R-N-NH_2} \xrightarrow{-H_2O} R-N_3$$

This nitrosation reaction is a convenient method for the preparation of *aromatic* and *heteroaromatic azides, e.g.*:

3-Azido-4-(benzyl-idineamino)-5-methyl-4H-1,2,4-triazole; 47%[102]

6-Azidotetrazolo-[1,5-b]pyridazine; 64%[103]

6-Azido-7,8-dihydrobenzo[h]-tetrazolo[5,1-a]-phthalazine[104]

9.1.1.3.3 Azido compounds from diazonium salts

Diazonium salts can be converted into azido compounds by treatment of their perbromides with aqueous ammonia (do not confuse with the exchange reaction, p. 310). This reaction is of historical significance, since it led to the discovery of the azides[105]. It is generally applicable to other perhalides, albeit individual cases of failure have been reported[106]. The reactions of diazonium salts with *N*-haloimines[107], hydroxylamine[108] and its derivatives[109], aldoximes[110], and hydrazine[111] may be regarded as variants of this procedure. Reaction of aryldiazonium salts with alkylhydrazines leads to the unstable *1,4-diaryl-3,4-dihydrotetrazenes*, which decompose mainly into the aromatic *azide*[112]:

$$Aryl-N_2^{\oplus}Cl^{\ominus} \xrightarrow[-HCl]{H_2N-NH-Alkyl}$$

$$Aryl-N=N-NH-NH-Alkyl$$

side-reaction   main-reaction

Alkyl—N₃ + Aryl—NH₂   Aryl—N₃ + Alkyl—NH₂

Diazonium salts react with the alkali metal salts of aliphatic and aromatic sulfonamides, forming the corresponding triazene compounds, which spontaneously decompose into *aryl azides* and *sulfinates*[113]:

$$Aryl-N_2^{\oplus}Cl^{\ominus} \xrightarrow[-HCl]{Na^{\oplus}\overset{\ominus}{H}N-SO_2-R}$$

$$Aryl-N=N-\overset{\ominus}{\underset{}{N}}-SO_2-R \xrightarrow[-R-SO_2^{\ominus}Na^{\oplus}]{} Aryl-N_3$$

9.1.1.3.4 Azido compounds from other nitrogen-containing compounds

Aromatic sulfonyl azides add arylmagnesium halides, forming arylsulfonyl aryltriazenes, which can be split thermally[114, 115] or hydrolytically[116] into *aryl azide* and arylsulfinate:

$$Ar-MgX + Ar^1-SO_2-N_3$$

$$\xrightarrow{} Ar-N=N-\overset{\ominus}{\underset{}{N}}\overset{MgX^{\oplus}}{}-SO_2-Ar^1$$

$$\xrightarrow[-XMg-SO_2-Ar^1]{} ArN_3$$

Phenyl azide has been prepared in high yield in this manner[116]. In contrast, the oxidation of triazenes to azides is of practical importance only in exceptional cases[117], since in general triazenes are prepared by the reduction of azides (see p. 322).

[101] *E. Fischer*, Justus Liebigs Ann. Chem. *190*, 89 (1878).
[102] *H.H. Takimoto, G.C. Denault, S. Hotta*, J. Org. Chem. *30*, 711 (1965).
[103] *A. Kovačić, B. Stanovnik, M. Tišler*, J. Heterocycl. Chem. *5*, 351 (1968).
[104] *B. Stanovnik, M. Tišler*, Chimia *22*, 141 (1968).
[105] *P. Griess*, Justus Liebigs Ann. Chem. *137*, 65 (1866).

[106] *J.H. Boyer, F. Canter,* Chem. Rev. *54*, 1 (1954); *C. Grundmann* in *Houben-Weyl*, Methoden der organischen Chemie, 4. Aufl., Bd. X/3, p. 777, Georg Thieme Verlag, Stuttgart 1965.
[107] D.R.P. 456857 (1925), Rheinische Kampferfabrik K.G.mbH., Erf.: *S. Skraup, K. Steinruck;* Chem. Zentr. I, 3111 (1928).
[108] *M.O. Forster, H.E. Fierz*, J. Chem. Soc. (London) 1350 (1907).
[109] *H. Rupe, K. Majewski*, Ber. *33*, 3401 (1900).
[110] *J. Mai*, Ber. *24*, 3418 (1891); *J. Mai*, Ber. *25*, 372, 1685 (1892).
[111] *E. Noelting, O. Michel*, Ber. *26*, 88 (1893).
[112] *T. Curtius*, Ber. *26*, 1263 (1893).
[113] *P.K. Dutt, H.R. Whitehead, A. Wormall*, J. Chem. Soc. (London) 2088 (1921); 1463 (1924); *A. Key, P.K. Dutt*, J. Chem. Soc. (London) 2035
[114] *S. Ito, T. Hirabayashi, K. Matsumoto*, Bull. Chem. Soc. Japan *43*, 2254 (1970).
[115] *S. Ito*, Bull. Chem. Soc. Japan *39*, 635 (1966).
[116] *P.A.S. Smith, C.P. Rowe, L.B. Bruner*, J. Org. Chem. *34*, 3430 (1969).
[117] *A. Dorapski*, Ber. *40*, 3033 (1907).

In this connection, reference should also be made to the *α-azidoimine/tetrazole equilibrium* (see p. 310), which can be displaced fully in favor of the azide form if this is removed by reaction. Thus, *2-azido-4,6-pyrimidinediol* is formed from 5-aminotetrazole by condensation with diethyl malonate[74].

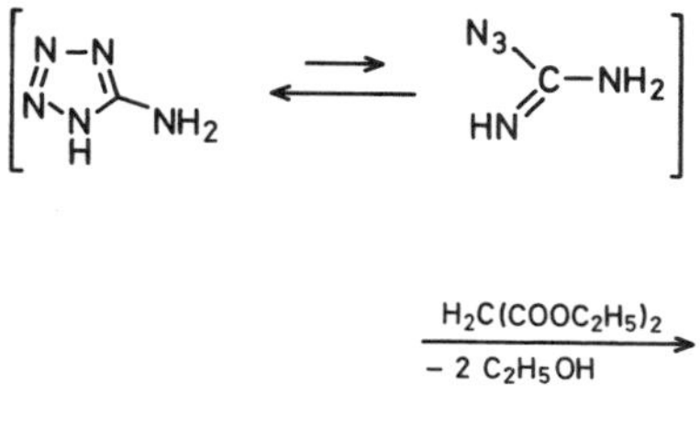

The conversion of tetrazole[1,5-b]pyridazine into *3-azidopyridazine 1-oxide* by oxidation with hydrogen peroxide may be mentioned as a further example of ring-opening[118]:

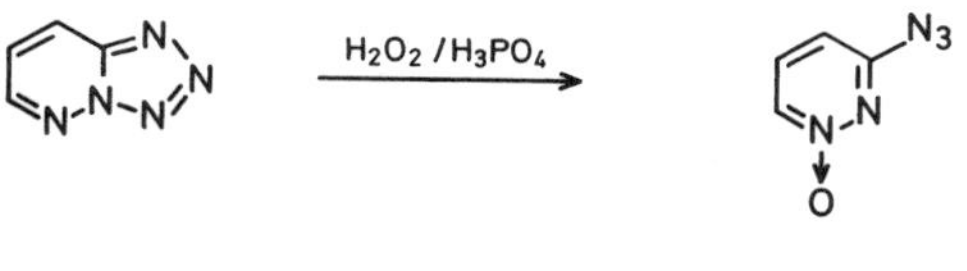

2*H*-Azirines, too, can be converted ito azides. Reaction of 3-methyl-2-phenyl-2*H*-azirine with diazomethane forms *3-azido-2-phenyl-1-butene* together with the *cis-* and *trans-1-azido-2-phenyl-2-butenes*[119]:

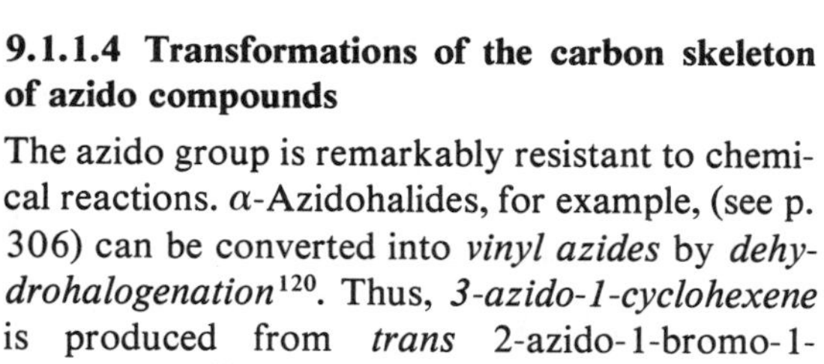

9.1.1.4 Transformations of the carbon skeleton of azido compounds

The azido group is remarkably resistant to chemical reactions. α-Azidohalides, for example, (see p. 306) can be converted into *vinyl azides* by *dehydrohalogenation*[120]. Thus, *3-azido-1-cyclohexene* is produced from *trans* 2-azido-1-bromo-1-cyclohexene[39].

Condensation reactions at the carbon skeleton can also be carried out; for example, *ethyl α-azidocinnamates* are obtained from ethyl azidoacetate and substituted benzaldehydes[121]:

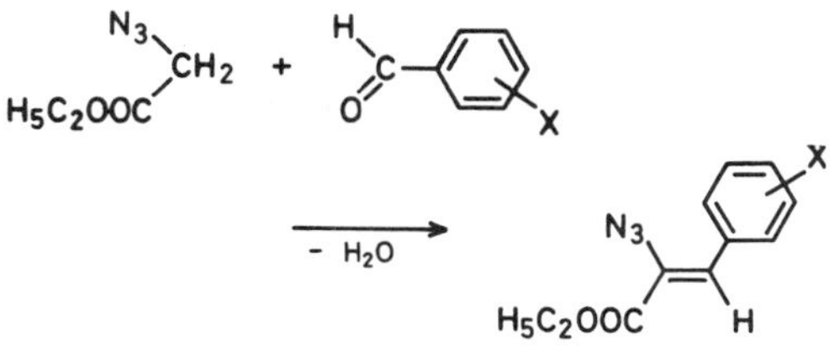

If acetophenone is used as the reaction partner, analogous *α-azidochalcones* are obtainable[122]. Substitution, addition, and oxidation reactions can also be carried out with retention of the N_3-group[106].

The usual carbonyl reactions can be carried out with α-azidocarbonyl compounds[123]. Thionyl chloride converts α-azidocarboxylic acids into reactive *α-azido acid chlorides*[123,124], which can be subjected to other exchange reactions[125]. Dehydrohalogenation to *azidoketenes (in situ)* which may be trapped as azetidinones by reaction with Schiff bases is particularly noteworthy[126]:

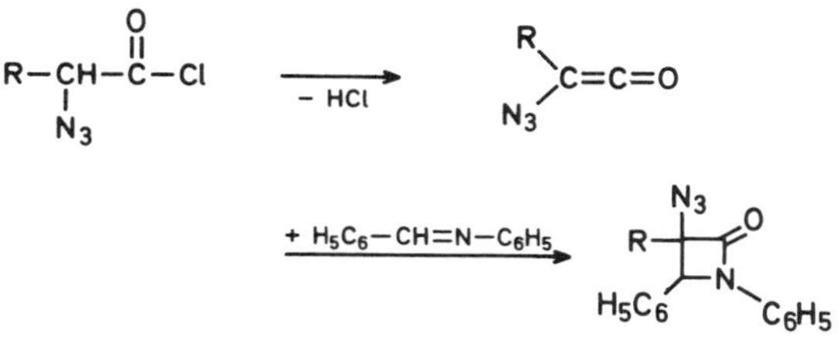

9.1.2 Preparation of organoelemental azido compounds

The halogen/azide exchange method is suitable also for preparing azido compounds whose N_3 group is attached to a heteroatom. Tab. 5 illustrates the scope of this method.

[118] *B. Stanovnik, M. Tisler, M. Ceglar, V. Bah*, J. Org. Chem. *35*, 1138 (1970).

[119] *V. Nair*, J. Org. Chem. *33*, 2121 (1968).

[120] *G. L'abbé, A. Hassner*, Angew. Chem. *83*, 103 (1971); Angew. Chem. Intern. Ed. Engl. *10*, 98 (1971).

[121] *H. Hemetsberger, D. Knittel, H. Weidmann*, Monatsh. Chem. *100*, 1599 (1969).

[122] *D. Knittel, H. Hemetsberger, H. Weidmann*, Monatsh. Chem. *101*, 157 (1970).

[123] *M.O. Forster, H.E. Fierz*, J. Chem. Soc. (London) 72 a. 669 (1908);
M.O. Forster, R. Müller, J. Chem. Soc. (London) 126 (1910).

[124] *K. Freudenberg, W. Kuhn, I. Bumann*, Ber. 63, 2380 (1930);
K. Freudenberg, G. Piazolo, C. Knoevenagel, Justus Liebigs Ann. Chem. *537*, 197 (1939).

[125] *J.M. Muchowski*, Can. J. Chem. *48*, 1946 (1970).

Table 5. Preparation of organometallic azides by halogen/azide exchange

Halogen compound	Reaction conditions	Azido compound	Physical constants	Yield [% of th.]	Ref.
p-Toluenesulfonyl chloride	Sodium azide in ethanol/water, 2.5 h ~ 20°	p-Toluenesulfonic acid hydrazide		81–86	[127]
Benzenesulfinyl chloride	Sodium azide in acetonitrile, 3 h −35°	Benzenesulfinyl azide			[128]
$(H_5C_2)_2N$–P(O)–Cl, $(H_5C_2)_2N$	Sodium azide in acetone, 20 h reflux	$(H_5C_2)_2N$–P(O)–N_3, $(H_5C_2)_2N$ Tetraethylphosphoro-diamidic azide	$b_{0.8}$ 104° $n_D^{20°}$ = 1.4678	97	[129]
H_5C_6–P(O)–Cl, H_5C_6	Lithium azide in acetonitrile, 16 h ~ 20°	H_5C_6–P(O)–N_3, H_5C_6 Azidodiphenyl phosphine oxide	$b_{0.05}$ 137–144	68	[130]
H_3C, H_3C–As⊕–Cl, H_3C $SbCl_6^{\ominus}$	Sodium azide in nitrobenzene, 36 h reflux	H_3C, H_3C–As⊕–N_3, H_3C $SbCl_6^{\ominus}$ Azidotrimethylarsonium hexachloroantimonate	m 185°	89	[131]
$(p)-H_3C-H_4C_6$, $(p)-H_3C-H_4C_6$–B–Cl	Lithium azide in benzene, 15 h ~ 20°	$(p)-H_3C-H_4C_6$, $(p)-H_3C-H_4C_6$–B–N_3 Azido-di-4-tolylborane	$b_{0.0001}$ 96° m 39–40°	90	[132]
$(H_3C)_2N$, $(H_3C)_2N$–B–Cl	Lithium azide in benzene, 12 h reflux	$(H_3C)_2N$, $(H_3C)_2N$–B–N_3 Azidobis(dimethylamino)-borane	b_{11} 40–41°	69	[133]
$(H_5C_6)_3$Si–Cl	Lithium azide in tetrahydrofuran, 24 h ~ 20°	$(H_5C_6)_3$Si–N_3 Azidotriphenylsilane	m 83°	70	[134]

Other methods are of limited importance. Besides halogen, amino or hydroxy groups can be replaced by the azido group [135–138]. It is sometimes advantageous to use silicon azide as the reagent for introducing the azido group instead of lithium azide or sodium azide [139, 140]. The reaction of appropriate trialkyl compounds with chlorine azide has proven useful for the preparation of azides of some Group 3A elements [141]:

$$(H_5C_2)_3M \xrightarrow[-\; C_2H_5Cl]{ClN_3} (H_5C_2)_2M-N_3$$

M = Al , Ga , In , Tl

Organocadmium and organomercury azides are prepared in the same way [142]. Arylsulfonyl azides and, also, O,O-diethyl phosphonothioate azide can be obtained by treatment of the corresponding hydrazides with sodium nitrite, as well as by halogen/azide exchange [143, 144].

[126] A.K. Bose, B. Anjaneyulu, S.K. Bhattacharga, M.S. Manhas, Tetrahedron 23, 4769 (1967); A. Hassner, R.J. Isbister, R.B. Greenwald, J.T. Klug, E.C. Taylor, Tetrahedron 25, 1637 (1969); M.S. Manhas, J.S. Chib, Y.H. Chiang, A.K. Bose, Tetrahedron 25, 4421 (1969).

[127] M. Regitz, J. Hocker, A. Liedhegener, Org. Synth. 48, 36 (1968).

[128] T.J. Maricich, J. Amer. Chem. Soc. 90, 7179 (1968).

[129] F.L. Scott, R. Riordan, P.D. Morton, J. Org. Chem. 27, 4255 (1962).

[130] K.L. Paciorek, Inorg. Chem. 3, 96 (1964).

[131] A. Schmidt, Chem. Ber. 101, 4015 (1968).

[132] P.J. Paetzold, P.P. Haberer, R. Müllbauer, J. Organometal. Chem. 7, 45 (1967).

[133] P.J. Paetzold, G. Maier, Chem. Ber. 103, 281 (1970).

[134] N. Wiberg, B. Neruda, Chem. Ber. 99, 740 (1966).

[135] N. Wiberg, B. Neruda, Chem. Ber. 99, 740 (1966).

[136] K. Rühlmann, A. Reiche, M. Becker, Chem. Ber. 98, 1814 (1965).

[137] E. Lieber, C.N. Rao, F.M. Keane, J. Inorg. Nucl. Chem. 25, 631 (1963).

9.2 Transformations of azido compounds

The reactivity of organic azido compounds is determined by two main features: their tendency to *eliminate nitrogen*, and their *1,3-dipolar character*. In agreement with dipole moment measurements[145] and the results of electron diffraction studies on *methyl azide*[146], their *structure* can be described by the mesomeric formulae **1** and **2**[147]:

From these, it can be seen that the bond order of the $N-N_2$ link is about 1.5. The partial double bond character is reduced by carbonyl and sulfone groups, as indicated in formulae **3** and **4**.

Azides of the types indicated above are consequently less stable to heat[1] than are alkyl and aryl azides.

9.2.1 Transformations of azido compounds depending on their dipolar character

9.2.1.1 1,3-Dipolar cycloadditions

The addition of a 1,3-dipole to a dipolarophile proceeds by a onestep *synchronous* mechanism[148, 149], in which the new σ-bonds are formed simultaneously, though not necessarily at the same speed[150]:

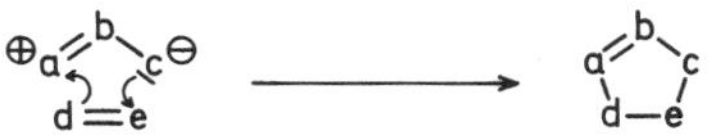

Such a reaction involves *stereospecific cis*-addition. For example, dimethyl fumarate adds 4-methoxyphenyl azide to form *dimethyl trans-4,5-dihydro-1-(4-methoxyphenyl)-1H-1,2,3-triazole-4,5-dicarboxylate*[151]. Phenyl azide and (−)-*trans*-cyclooctene   (*3aS,9aS)-trans-(+)-3a,4,5,6,7,8,9,9a-octahydro-1-phenyl-1H-cyclooctatriazole*[152]:

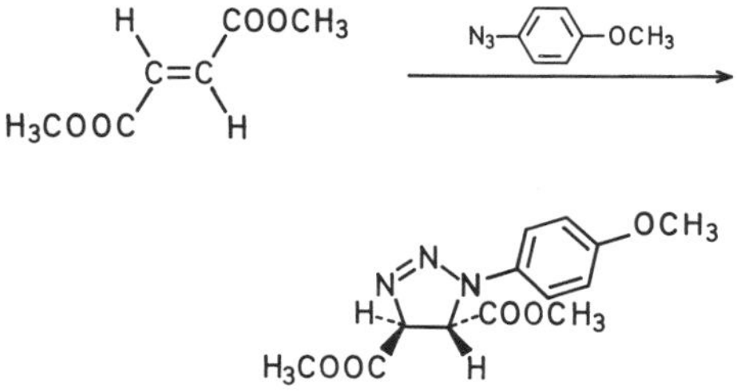

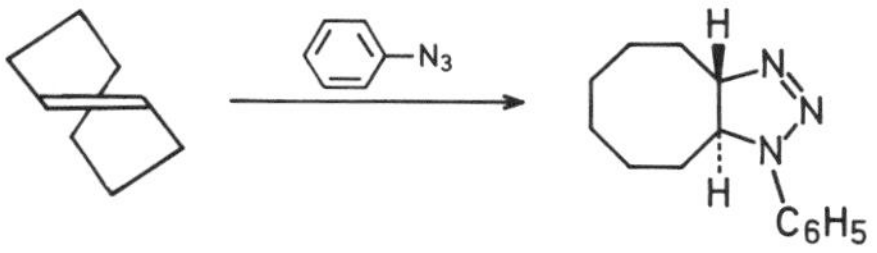

Large negative entropies of activation, low enthalpies of activation, and reaction rates which are mainly independent of the solvent are characteristics of azide addition. In the absence of steric factors the direction of addition is determined by electronic effects[145].

9.2.1.1.1 Addition to alkenes

1,3-Dipolar cycloaddition of azides to *olefins* is followed by eliminating nitrogen from the thermally labile *4,5-dihydro-1,2,3-triazoles*. The latter reaction, which may occur more or less readily, has the higher temperature coefficient, and consequently it is advisable to prepare the 4,5-dihydro-1,2,3-triazoles at as low a temperature as possible, using long reaction times when necessary[148]. For example, reaction times of up to 5 months have

[138] *J. Lorberth, H. Krapf, H. Nöth*, Chem. Ber. *100*, 3511 (1967).

[139] *W. Sundermeyer*, Chem. Ber. *96*, 1293 (1963).

[140] *N. Wiberg, W.Ch. Joo, H. Henke*, Inorg. Nucl. Chem. Lett. *3*, 267 (1967).

[141] *J. Müller, K. Dehnike*, J. Organometal. Chem. *12*, 37 (1968).

[142] *K. Dehnike, J. Strähle, D. Seybold, J. Müller*, J. Organometal. Chem. *6*, 298 (1966).

[143] *T. Curtius, G. Kraemer*, J. Prakt. Chem. [2] *125*, 326 (1930).

[144] *G. Schrader* in *Houben-Weyl*, Methoden der organischen Chemie, 4. Aufl., Bd. XII/2, p. 808, Georg Thieme Verlag, Stuttgart 1964.

[145] *G. L'abbé*, Chem. Rev. *69*, 345 (1969).

[146] *R.L. Livingstone, C.N.R. Rao*, J. Phys. Chem. *60*, 756 (1960).

[147] *L. Pauling*, Die Natur der chemischen Bindung, 3. Aufl., p. 255, Verlag Chemie, Weinheim/Bergstr. 1968.

[148] *R. Huisgen*, Angew. Chem. *75*, 604 (1963); Angew. Chem. Intern. Ed. Engl. *2*, 565 (1963).

[149] *R. Huisgen*, Angew. Chem. *75*, 742 (1963); Angew. Chem. Intern. Ed. Engl. *2*, 633 (1963).

[150] *P. Schreiner, J.H. Schomaker, S. Deming, W.J. Libbey, G.P. Nowack*, J. Amer. Chem. Soc. *87*, 306 (1965).

[151] *R. Huisgen, G. Szeimies, L. Möbius*, Chem. Ber. *99*, 475 (1966).

[152] *T. Aratani, Y. Nakanisi, H. Nozaki*, Tetrahedron *26*, 4339 (1970).

[153] *P. Schreiner*, Tetrahedron *24*, 349 (1968).

Table 6. Addition of azido compounds to strained olefins

Reaction partner	Reaction conditions	Reaction product	Physical constants	Yield [% of th.]	Ref.
Diethyl phosphoroazidate	No solvent 40 h at 45°	Diethyl (3a,4,5,6,7,7a-hexahydro-4,7-methano-1H-benzotriazol-1-yl)phosphonate		92	[157]
Cyanogen azide	Ethyl acetate 10 h at ~ 20°	6,6a-Dihydro-3,3a,4,5,6a-pentamethyl-6-methylene-1(3aH)-cyclopentapyrazole-carbamonitrile	m 149–150°	30	[158]
Phenyl azide	Benzene, 15 h at 38°	3a,4,5,6,7,7a-Hexahydro-7a-morpholino-1-phenyl-4,7-methano-1H-benzotriazole	m 136–138°	85	[159]
	Carbon tetrachloride, 5 weeks at 0° (in the dark)	2-(2-Norbornen-7-ylamino)-2,4,6-cycloheptatrien-1-one	m 129–132°	57	[160]
		3-Azatricyclo[3.2.1.0²,⁴]octan-3-yl-2,4,6-cycloheptatrien-1-one	m 108–110°	6	[160]
Phenyl azide	Petroleum ether, several days ~ 20°	1,3a,4,7,8,8a-hexahydro-1-phenyl-4,8-methanocyclohepta-1H-triazole	m 68–69°[161]	85[161]	[162]
		1,3a,4,5,8,8a-hexahydro-1-phenyl-4,8-methanocyclohepta-1H-triazole			

been recorded for the reaction of 1-hexene with 4-chlorophenyl azide to form *5-butyl-1-(4-chlorophenyl)-4,5-dihydro-1,2,3-triazole*[153]:

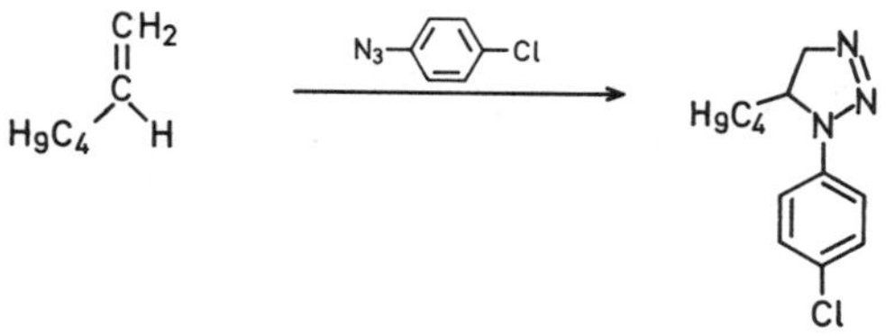

Azide additions to *allenes* and angularly strained olefins proceed particularly readily. *4,5-Dihydro-4-isopropylidene-5,5-dimethyl-1-(4-nitrophenyl)-1,2,3-triazole* is obtained from 2,4-dimethyl-2,3-pentadiene and 4-nitrophenyl azide[154]:

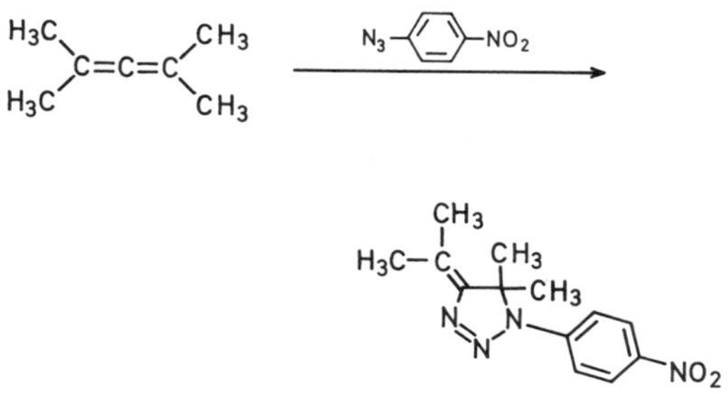

In the reaction of *dimethyl 3,3-dimethyl-1-cyclopropene-1,2-dicarboxylate* with ethyl azide, the final product is dimethyl 2-diazo-4-(ethylimino)-3,3-dimethylglutarate, undoubtedly *via* 4,5-dihydro-1,2,3-triazole as intermediate[155]:

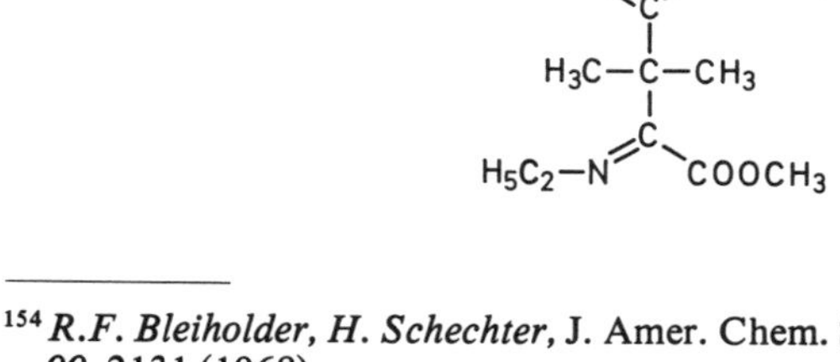

Many additions of azides to the double bonds of bicyclic systems are known. For example, norbornene and phenyl azide yield *3a,4,5,6,7,7a-hexahydro-1-phenyl-4,7-methano-1H-benzotriazole*[156].

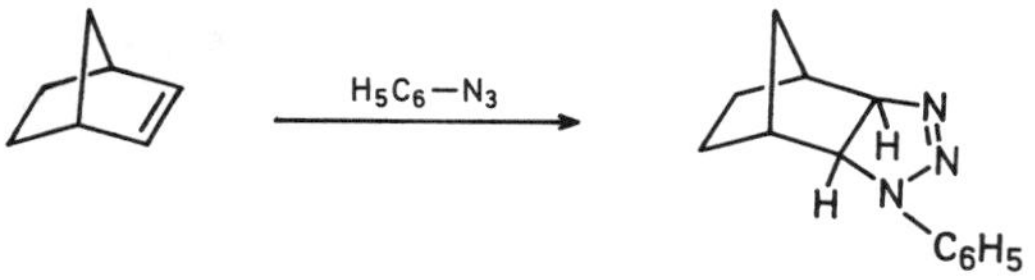

Further examples of the addition of azido compounds to strained olefins are given in Tab. 6.

Adducts from norbornene derivatives are relatively stable to heat; nitrogen is only lost at about 150° yielding mixtures of phenylimines and aziridines[163]. These last are the main products from the photolytic decomposition of 4,5-dihydro-1,2,3-triazoles[164-166]. If the electron density of the lone pair on the trigonal nitrogen atom is reduced by an adjacent electron-withdrawing group, the 4,5-dihydro-1,2,3-triazole derivative is considerably less stable. Thus, for example, 1-benzoyl-3a,4,5,6,7,7a-hexahydro-4,7-methano-*1H*-benzotriazole decomposes at temperatures even as low as 40° forming *3-benzoyl-exo-3-azatricyclo-[3.2.1.0²,⁴]-octane*[164]:

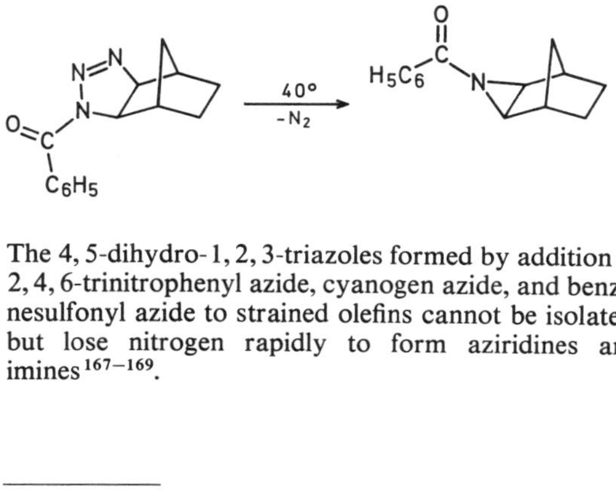

The 4,5-dihydro-1,2,3-triazoles formed by addition of 2,4,6-trinitrophenyl azide, cyanogen azide, and benzenesulfonyl azide to strained olefins cannot be isolated, but lose nitrogen rapidly to form aziridines and imines[167-169].

[154] *R.F. Bleiholder, H. Schechter*, J. Amer. Chem. Soc. *90*, 2131 (1968).

[155] *M. Franck-Neumann, C. Buchecker*, Tetrahedron Lett. 2659 (1969).

[156] *R.S. McDaniel, A.C. Oehlschlager*, Tetrahedron *25*, 1381 (1969).

[157] *K.D. Berlin, L.A. Wilson, L.M. Raff*, Tetrahedron *23*, 965 (1967).

[158] *A.G. Anastassiou, S.W. Eachus*, Chem. Commun. 429 (1970).

[159] *J.F. Stephen, E. Marcus*, J. Heterocycl. Chem. *6*, 969 (1969).

[160] *J.D. Hobson, J.R. Malpass*, J. Chem. Soc (C) 1935 (1970).

[161] refering to mixture of the two structural isomers.

[162] *R.S. McDaniel, A.C. Oehlschlager*, Can. J. Chem. *48*, 345 (1970).

[163] *K. Alder, G. Stein, W. Friedrichsen*, Justus Liebigs Ann. Chem. *501*, 1 (1933).

[164] *R. Huisgen, L. Möbius, G. Müller, H. Stangl, G. Szeimies, J.M. Vernon*, Chem. Ber. *98*, 3992 (1965).

[165] *P. Schreiner*, J. Amer. Chem. Soc. *90*, 988 (1968).

[166] *P. Schreiner*, J. Org. Chem. *30*, 7 (1965).

[167] *A.S. Bailey, J.J. Wedgwood*, J. Chem. Soc. (C) 682 (1968).

[168] *J.E. Franz, C. Osuch, M.W. Dietrich*, J. Org. Chem. *29*, 2922 (1964).

[169] *F.D. Marsh, M.E. Hermes*, J. Amer. Chem. Soc. *86*, 4506 (1964).

9.2.1.1.2 Addition to α,β-unsaturated carbonyl compounds and nitriles

Azido compounds add to α,β-unsaturated carboxylic acid esters, ketones, and nitriles; the electron-withdrawing substituent group appears at the 4-position of the 4,5-dihydro-1,2,3-triazole formed[151]. The preparation of *methyl 4,5-dihydro-1-phenyl-1,2,3-triazole-4-carboxylate* from methyl acrylate and phenyl azide can be cited as an example[151]:

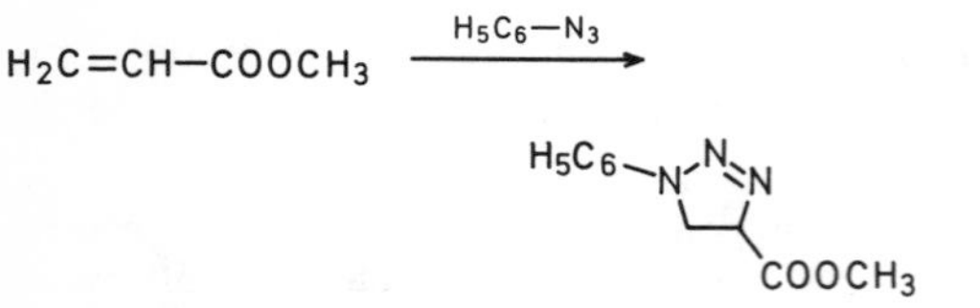

Analogously, *4,5-dihydro-1-(4-nitrophenyl)-1,2,3-triazole-4-carbonitrile*[151] is obtained from acrylonitrile and 4-nitrophenyl azide. On treatment with bases, 4,5-dihydro-1,2,3-triazoles can isomerize (with ring-opening) to form diazo compounds (see also p. 237).

The 4,5-dihydro-1,2,3-triazoles formed by addition of azides to ethylenic compounds having geminal activating substituent groups can seldom be isolated; they usually eliminate nitrogen and add further olefin, to form *pyrrolidines*.

Benzylidenemalononitrile and 4-methoxyphenyl azide thus give *3,5-diphenyl-1-(4-methoxyphenyl)-2,2,4,4-tetracyanopyrrolidine*; the corresponding 4,5-dihydro-1,2,3-triazole and aziridine derivatives cannot be isolated[170]:

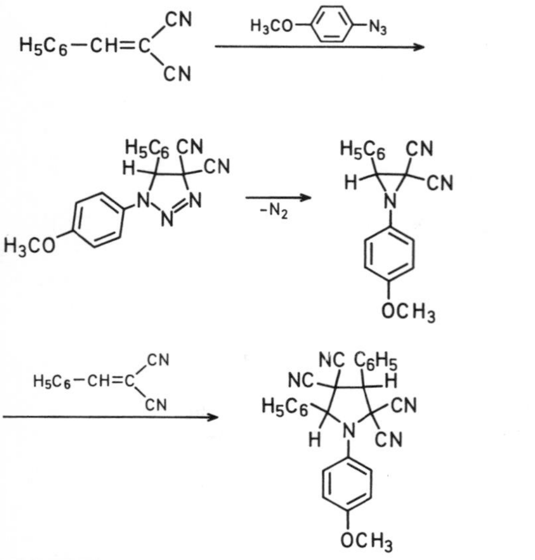

9.2.1.1.3 Addition to enamines and enol ethers

Enamines possess electron-rich double bonds and with electronic control add azides smoothly to form *5-amino-4,5-dihydro-1,2,3-triazoles*[171, 172]. Thus, 4-nitrophenyl azide and 1-butenyl-1-morpholine together afford *4-ethyl-4,5-dihydro-5-morpholino-1-(4-nitrophenyl)-2H-1,2,3-triazole*[173]:

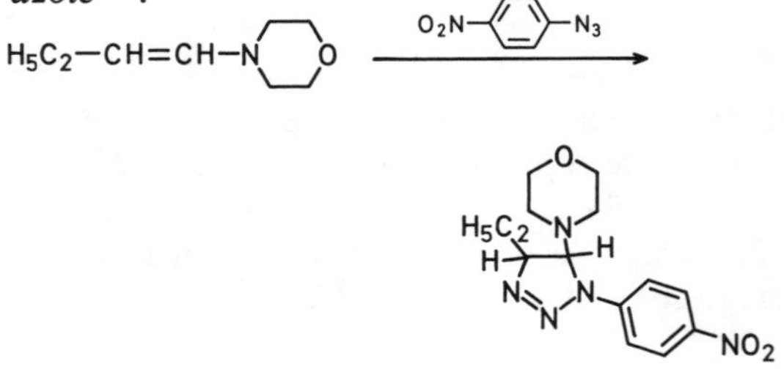

If the enamine concerned has a carbonyl, sulfonyl, or nitro group *beta* to its nitrogen atom, then the 4,5-dihydro-1,2,3-triazole first formed cannot be isolated; it spontaneously eliminates a molecule of secondary amine and forms a *1,2,3-triazole*[174, 175]. This is illustrated by the synthesis of *5-methyl-1-(4-nitrophenyl)-1H-1,2,3-triazol-4-yl phenyl ketone*[174].

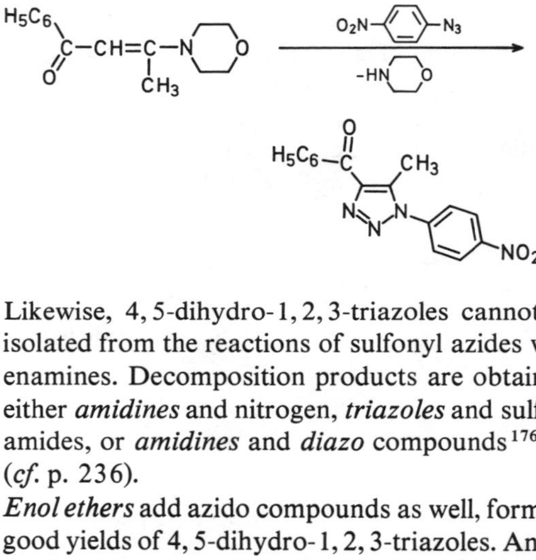

Likewise, 4,5-dihydro-1,2,3-triazoles cannot be isolated from the reactions of sulfonyl azides with enamines. Decomposition products are obtained, either *amidines* and nitrogen, *triazoles* and sulfonamides, or *amidines* and *diazo* compounds[176-180] (*cf.* p. 236).

Enol ethers add azido compounds as well, forming good yields of 4,5-dihydro-1,2,3-triazoles. An example is the preparation of *4,5-dihydro-5,5-dimethoxy-4,4-dimethyl-1-phenyl-1,2,3-triazole* from dimethylketene dimethylacetal and phenyl azide[181]:

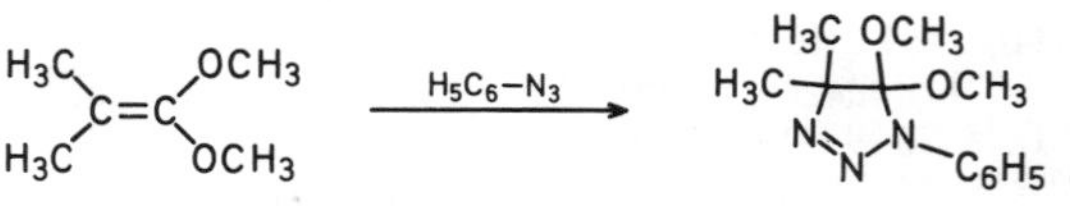

[170] *F. Fexier, R. Carrie*, Tetrahedron Lett. 823 (1969).
[171] *J.F. Stephen, E. Marcus*, J. Heterocycl. Chem. *6*, 969 (1969).

[172] *M.E. Munk, Y.K. Kim*, J. Amer. Chem. Soc. *86*, 2213 (1964).

Table 7. 1,3-Dipolar cycloadditions of azido compounds to alkynes

Reactants	Reaction conditions	Reaction product	Physical constants	Yield [% of th.]	Ref.
$H_3C-S-CH_2-N_3$ Azidodimethyl sulfide Phenylacetylene	Toluene, 12 h reflux	H_5C_6 — N—N$-CH_2-S-CH_3$ *1-[(Methylthio)methyl]-4-phenyl-1H-1,2,3-triazole*	m 56–57°	26	189
		H_5C_6 $H_3C-S-CH_2-N$—N *1[(Methylthio)methyl]-5-phenyl-1,2,3-triazol*	m 114–115	22	189
H_5C_6 O P H_5C_6 N_3 *N,N*-Diethyl-1-propynamine	Benzene, 30 min at 60–65°, then 3 h reflux	H_3C $N(C_2H_5)_2$ C_6H_5 N—N—P—C_6H_5 O *[5-(Diethylamino)-4-methyl-1H-1,2,3-triazol-1-yl]diphenyl-phosphine oxide*	m 121–122	43	190
CH_3 N H_3C N N_3 Dimethyl acetylenedicarboxylate	No solvent, 70 h at 80°	H_3COOC COOCH$_3$ CH_3 N—N—N N CH_3 *Dimethyl 1-(4,6-dimethyl-pyrimid-2-yl)-1,2,3-triazole-4,5-dicarboxylate*	m 146–147°	83	191
cis-2-Azido-2-butene Diethyl acetylenedicarboxylate	No solvent, ~20°	H_5C_2OOC COOC$_2H_5$ H N—N—C=C H_3C CH_3 *Dimethyl 1-(2-buten-2-yl)-1,2,3-triazole-4,5-dicarboxylate*	m 58–60°	80	192
4-Nitrophenyl azide Phenylethynyllithium	Ether, 48 h ~20° (subsequent hydrolysis)	H_5C_6 NH—N=N—C_6H_4—NO$_2$—(p) (p)—$O_2N-H_4C_6$—N—N *1-(4-Nitrophenyl)-4-[3-(4-nitrophenyl)-2-triazeno]-5-phenyl-1H-1,2,3-triazole*	m 189–190°	98	193

The *4,5-dihydro-1,2,3-triazoles* derived from aziridines and enol ethers are *thermally unstable* above 100°, and can decompose in two ways [182]. 4,5-Dihydro-1,2,3-triazoles derived from open-chain enol ethers eliminate alcohols, and form *1,2,3-triazoles* (*cf.* the analogous behavior of enamine adducts mentioned above), whereas those derived from cyclic enol ethers expel nitrogen and form *arylimines* [183]. These decomposition reactions occur at a considerably lower temperature in the case of sulfonyl azide adducts [182, 184].

9.2.1.1.4 Addition to alkynes

Azides add to $C\equiv C$ triple bonds forming *1,2,3-triazoles* [145, 185]:

$$R^1-C\equiv C-R^2 \xrightarrow{N_3-R^3}$$

R^1 R^2 N—N—R^3 + R^1 R^2 R^3—N—N

With unsymmetrical *acetylenes* both possible isomers are formed, as a rule [186, 187]. *Arynes* can also readily be trapped with azides. Dehydrobenzene generated from benzenediazonium *o*-carbox-

[173] R. Fusco, G. Bianchetti, D. Pocar, Gazz. Chim. Ital. **91**, 933 (1961).

[174] R. Fusco, G. Bianchetti, D. Pocar, R. Ugo, Gazz. Chim. Ital. **92**, 1040 (1962).

[175] S. Maiorana, D. Pocar, P.D. Croce, Tetrahedron Lett. 6043 (1966).

[176] G. Bianchetti, P.D. Croce, D. Pocar, Tetrahedron Lett. 2043 (1965).

ylate reacts with benzoyl azide to form *1,2,3-benzotriazol-1-yl phenyl ketone*[188]:

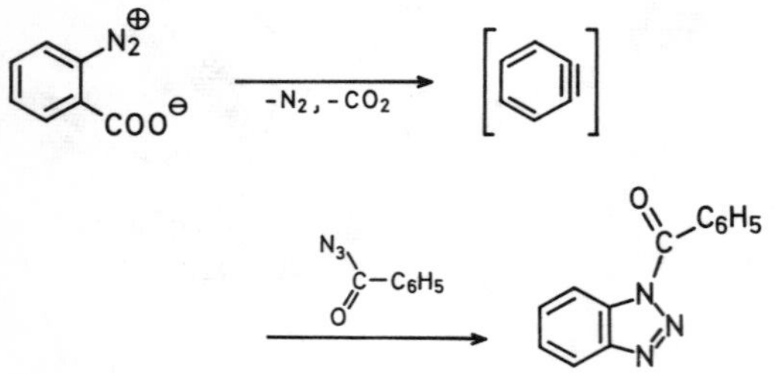

Further examples of 1,3-dipolar cycloaddition to alkynes are listed in Tab. 7.

Ynamines react with azides to form *5-amino-1,2,3-triazoles*, the orientation of these products being the same as in the enamine reactions (see p. 318)[194]. Depending on the nature of the substituent groups, the aminotriazoles may exist in equilibrium with open-chain α-*diazoamidines* (see p. 236):

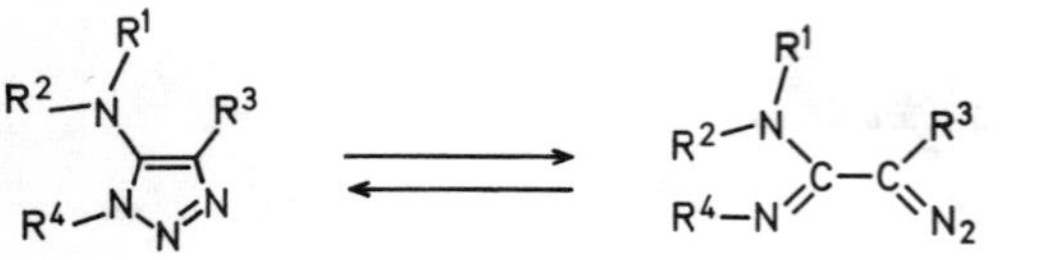

In the reaction of acetylenes with trimethylsilyl azide, cycloaddition is followed by migration of the silyl residue from the 1- to the 2-position. 2-Butyne and trimethylsilyl azide thus yield *4,5-dimethyl-2-trimethylsilyl-1,2,3-triazole*[195]:

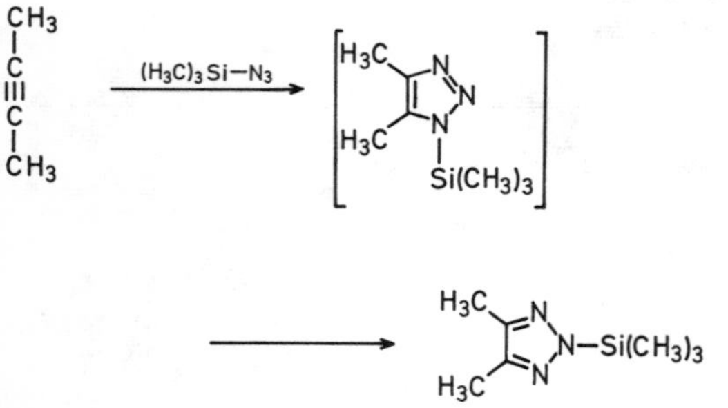

9.2.1.1.5 Addition to heteromultiple bonds

1,3-Dipolar cycloadditions proceed more easily, the more the loss of the π-bonding energy is compensated for by the gain in energy due to the formation of the new σ-bond[149]. Since the bond energy decreases in the order:

$$C-N > N-N > O-N$$

it is clear that azides do not add to aldehydes and ketones. *Imines* react preferentially in their tautomeric form, provided that their constitution permits it; *5-amino-4,5-dihydro-1,2,3-triazoles* (or products formed from them) are isolated[179, 196, 197]. Thus, cyclohexylideneisopropylamine (or the enamine tautomer) and benzoyl azide eliminate nitrogen and form *N-(N-isopropylcyclopentanecarboximidoyl)benzamide*[198]:

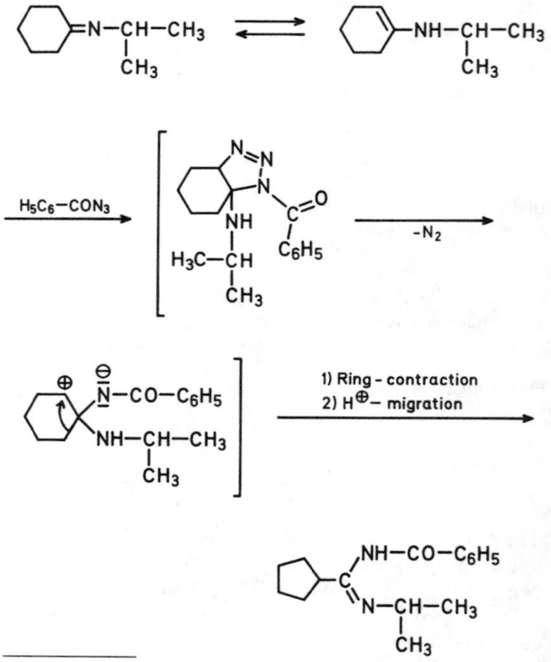

[184] *D.L. Rector, R.E. Harmon*, J. Org. Chem. *31*, 2837 (1966).

[185] *F.R. Benson, W. Savell*, Chem. Rev. *46*, 1 (1950).

[186] *W. Kirmse, L. Horner*, Justus Liebigs Ann. Chem. *614*, 1 (1958).

[187] *J.C. Sheeham, C.A. Robinson*, J. Amer. Chem. Soc. *73*, 1207 (1951).

[188] *W. Ried, M. Schön*, Chem. Ber. *98*, 3142 (1965).

[189] *G. Coreia-Munoz, R. Madronero, M. Rico, M.C. Saldana*, J. Heterocycl. Chem. *6*, 921 (1969).

[190] *K.D. Berlin, S. Rengaraju, T.E. Snider, N. Mandava*, J. Org. Chem. *35*, 2027 (1970).

[191] *R. Huisgen, K.v. Fraunberg, H.J. Sturm*, Tetrahedron Lett. 2589 (1969).

[192] *G. L'abbé, J.E. Galle, A. Hassner*, Tetrahedron Lett. 303 (1970).

[177] *R. Fusco, G. Bianchetti, D. Pocar, R. Ugo*, Chem. Ber. *96*, 802 (1963).

[178] *J. Kucera, Z. Arnold*, Tetrahedron Lett. 1109 (1966).

[179] *A.C. Ritchie, M. Rosenberger*, J. Chem. Soc. (C) 227 (1968).

[180] *A.S. Bailey, M.C. Chum, J.J. Wedgwood*, Tetrahedron Lett. 5953 (1968).

[181] *R. Scarpati, M.L. Graziano, R.A. Nicolaus*, Gazz. Chim. Ital. *100*, 665 (1970).

[182] *R. Huisgen, L. Möbius, G. Szeimies*, Chem. Ber. *98*, 1138 (1965).

[183] *P. Scheiner*, J. Org. Chem. *32*, 2022 (1967).

With heterocumulenes, cycloaddition occurs at one of the double bonds. For example, reaction of triphenyltin azide with phenyl isothiocyanate yields *1-phenyl-4-(triphenylstannyl)-2-tetrazoline-5-thione*[199]:

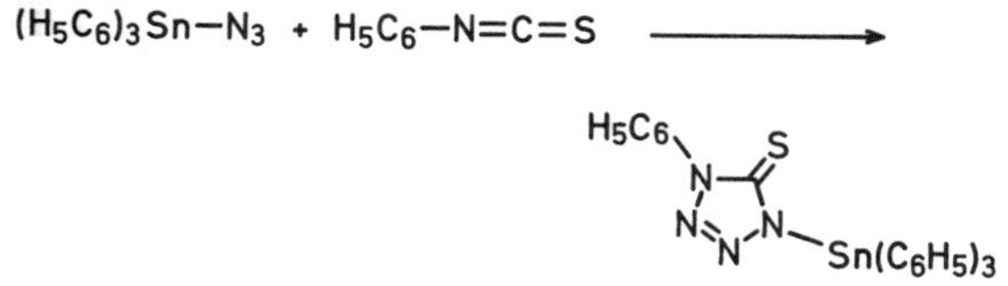

$$(H_5C_6)_3Sn-N_3 + H_5C_6-N=C=S \longrightarrow$$

The nitrile group participates in cycloadditions with azides only if it is activated by electron-withdrawing groups[200], or if the addition can proceed intramolecularly[201].

Both *1,2,3-triazoles* and *diazo* compounds can be obtained by reacting *1-(acylalkylidene)phosphoranes* with azides[202-205]:

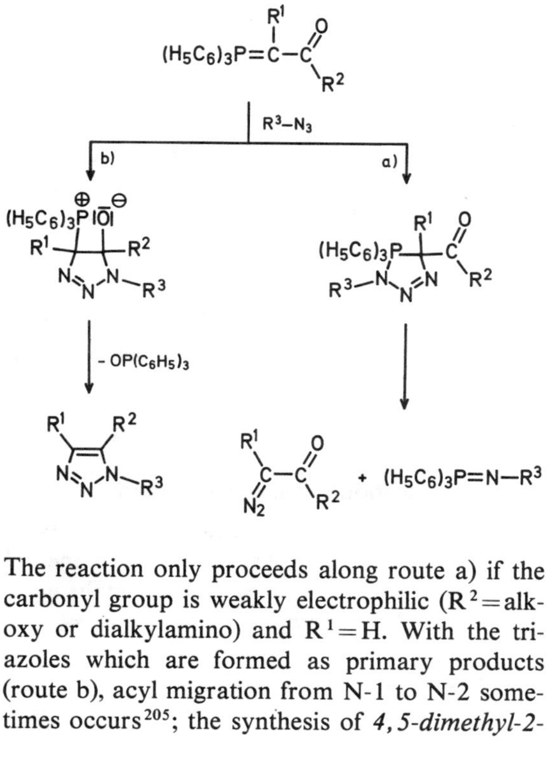

The reaction only proceeds along route a) if the carbonyl group is weakly electrophilic (R^2 = alkoxy or dialkylamino) and R^1 = H. With the triazoles which are formed as primary products (route b), acyl migration from N-1 to N-2 sometimes occurs[205]; the synthesis of *4,5-dimethyl-2-*

(4-nitrobenzoyl)-2H-1,2,3-triazole[205] is an example of this behavior:

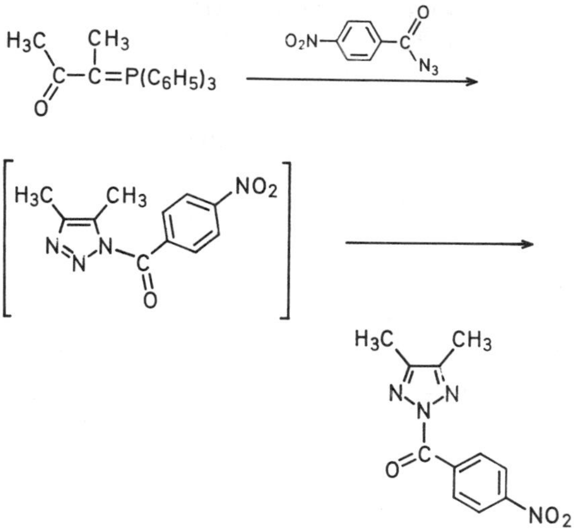

9.2.1.2 Base-catalyzed addition to compounds with activated methylene groups

Since its first observation[206], the base-catalyzed addition of azides to compounds with activated methylene groups (Dimroth reaction) has achieved increasing importance for the synthesis of *1,2,3-triazoles*. The reactions **1** to **3** shown below can be distinguished, according to whether a nitrile, a carbonyl, or an ester group participates in the reaction:

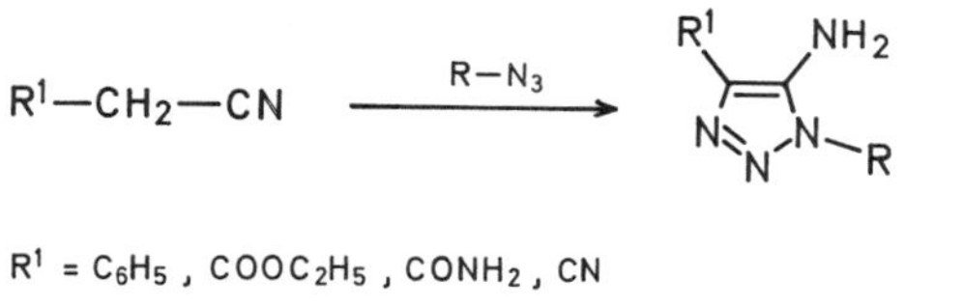

$$R^1-CH_2-CN \xrightarrow{R-N_3} \qquad ①$$

R^1 = C_6H_5, $COOC_2H_5$, $CONH_2$, CN

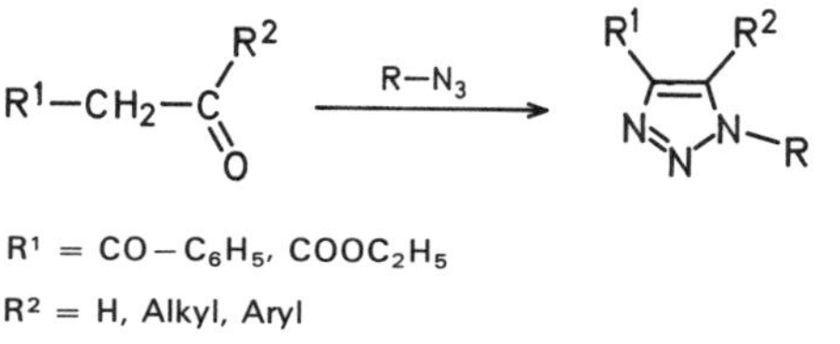

$$R^1-CH_2-C\!\!<^{R^2}_{O} \xrightarrow{R-N_3} \qquad ②$$

R^1 = $CO-C_6H_5$, $COOC_2H_5$
R^2 = H, Alkyl, Aryl

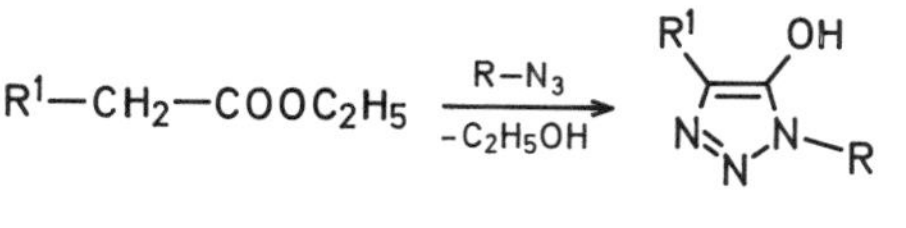

$$R^1-CH_2-COOC_2H_5 \xrightarrow[-C_2H_5OH]{R-N_3} \qquad ③$$

R^1 = H, CH_3, C_6H_5, $COOC_2H_5$

[193] G.S. Akimova, V.N. Cistokletov, A.A. Petrov, Zh. Org. Khim. *4*, 389 (1968); C.A. *68*, 105, 100 (1968).

[194] R. Fuks, R. Boijle, H.G. Viehe, Angew. Chem. *78*, 594 (1966); Angew. Chem. Intern. Ed. Engl. *5*, 585 (1966).

[195] L. Birkofer, P. Wegner, Chem. Ber. *99*, 2512 (1966).

[196] K. Alder, G. Stein, W. Friedrichsen, K.A. Hornung, Justus Liebigs Ann. Chem. *515*, 165 (1935).

[197] G. Bianchetti, D. Pocar, P.D. Croce, A. Vigevani, Gazz. Chim. Ital. *97*, 289 (1967).

[198] R.D. Burpitt, V.W. Goodlett, J. Org. Chem. *30*, 4308 (1965).

[199] P. Dunn, D. Oldefield, Australian J. Chem. *24*, 645 (1971).

[200] W.R. Carpenter, J. Org. Chem. *27*, 2085 (1962).

The reaction succeeds with alkyl, aryl, and vinyl azides. It fails with acyl and sulfonyl azides, however; with the former, nucleophilic displacement of the azido group by the carbanion from the methylene compound occurs, whereas with the latter, diazo group transfer (see p. 230) takes place under the conditions of the Dimroth reaction. Details and examples of the Dimroth reaction are given in the literature[207].

9.2.1.3 Reactions with tertiary phosphines

Tertiary phosphines add very readily to azides[208]; the adducts which are formed first are in general not particularly stable, and spontaneously eliminate nitrogen to form *phosphinimines (Staudinger reaction)*. Thus, triphenylphosphine and phenyl azide yield *N-phenyl-P,P,P-triphenylphosphine imide*[209]:

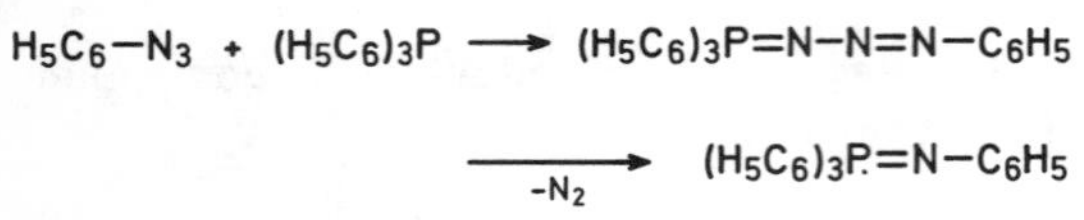

$$H_5C_6-N_3 + (H_5C_6)_3P \longrightarrow (H_5C_6)_3P=N-N=N-C_6H_5$$

$$\xrightarrow{-N_2} (H_5C_6)_3P=N-C_6H_5$$

Other compounds containing trivalent phosphorus, such as *phosphinous* esters and amides, and *phosphonous* diesters and diamides, undergo the same reaction[210].

The phosphinimines which are initially formed in the reaction of α-azido ketones (41) with triphenylphosphine cannot be isolated; *pyrazines* are obtained as the final products[211] *via* 3,6-dihydropyrazines as intermediates. If an acyl halide is present,

acylation occurs at the nitrogen atom of the phosphinimine, forming a *phosphonium salt* which then eliminates triphenylphosphine oxide; the resulting intermediate then yields an oxazole[212]:

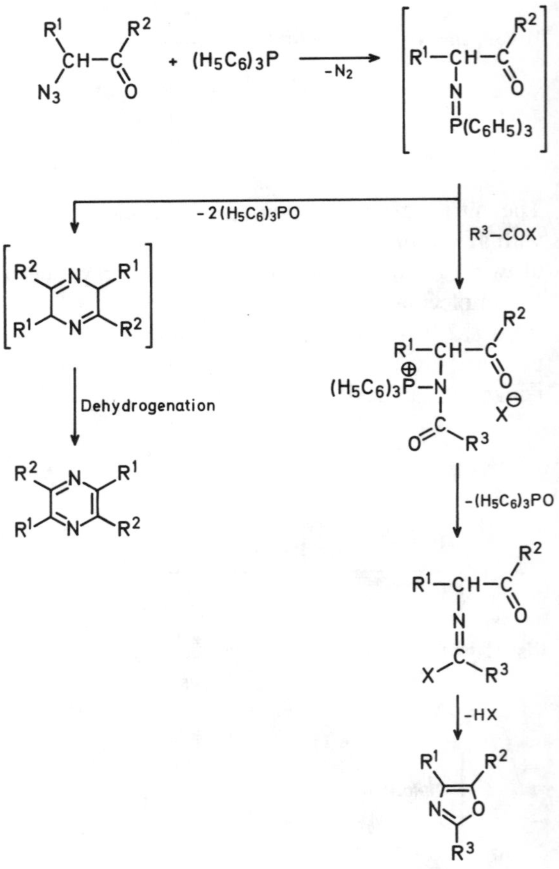

9.2.1.4 Reduction to triazenes and amines

Grignard reagents add to azido compounds to form *1,3-disubstituted triazenes*. Thus, methyl azide and methylmagnesium iodide yield the **highly explosive** *1,3-dimethyltriazene*[213]:

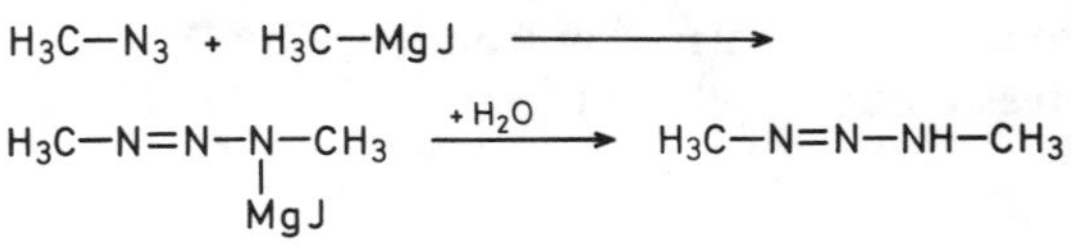

$$H_3C-N_3 + H_3C-MgJ \longrightarrow$$

$$H_3C-N=N-\underset{\underset{MgJ}{|}}{N}-CH_3 \xrightarrow{+H_2O} H_3C-N=N-NH-CH_3$$

This reaction can be carried out successfully with aliphatic, aromatic, heteroaromatic[214-216], and bifunctional[217] azido compounds. The reaction of

[201] *P.A.S. Smith, J.M. Clegg, J.H. Hall*, J. Org. Chem. *23*, 524 (1958).

[202] *G.R. Harvey*, J. Org. Chem. *31*, 1587 (1966).

[203] *G. L'abbé, H.J. Bestmann*, Tetrahedron Lett. 63 (1969).

[204] *G. L'abbé, P. Ykman, G. Smets*, Tetrahedron *25*, 5421 (1969).

[205] *P. Ykman, G. L'abbé, G. Smets*, Tetrahedron Lett. 5225 (1970).

[206] *O. Dimroth*, Ber. *35*, 1029 (1902); Ber. *35*, 4041 (1902).

[207] *G. L'abbé*, Ind. Chim. Belge *36*, 3 (1971).

[208] *C. Grundmann* in *Houben-Weyl*, Methoden der organischen Chemie, 4. Aufl., Bd. X/3, p. 777, Georg Thieme Verlag, Stuttgart 1965.

[209] *H. Staudinger, J. Meyer*, Helv. Chim. Acta *2*, 635 (1919).

[210] *V.A. Shokol, V.A. Molyavko, L.J. Mikhailyuchenko, N.K. a. L.J. Derkach*, Zh. Obshch. Khim. *41*, 318 (1971); C.A. *75*, 20513w (1971);
V.A. Sholkol, V.J. Molyavko, L.J. Derkach, Zh. Obshch. Khim. *40*, 998 (1970); C.A. *73*, 77321 (1970);
G.K. Genkina, B.A. Korolev, V.A. Gilyarov, M.J. Kabachnik, Zh. Obshch. Khim. *41*, 80 (1971); C.A. *75*, 129210 (1971).

[211] *E. Zbiral, J. Stroh*, Justus Liebigs Ann. Chem. *727*, 231 (1969).

[212] *E. Zbiral, J. Stroh, E. Bauer*, Monatsh. Chem. *102*, 168 (1971).

[213] *O. Dimroth*, Ber. *39*, 3905 (1906).

triphenylsilyl azide with phenylmagnesium bromide follows a different course: the magnesium salt of *phenyltriphenylsilylamine* is obtained, together with *tetraphenylsilane*[218]:

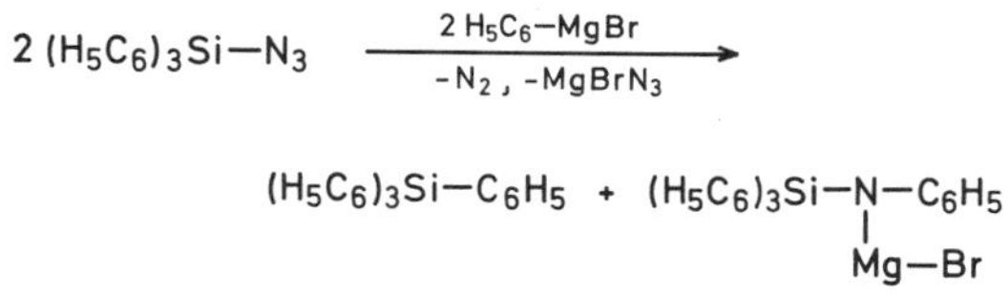

$$2\,(H_5C_6)_3Si-N_3 \xrightarrow[-N_2,\,-MgBrN_3]{2\,H_5C_6-MgBr}$$

$$(H_5C_6)_3Si-C_6H_5 \;+\; (H_5C_6)_3Si-\underset{\underset{Mg-Br}{|}}{N}-C_6H_5$$

Azides can be *reduced* to *monosubstituted triazenes* by mild reducing agents[209]. An example is the reduction of phenyl azide to *1-phenyltriazene* by stannous chloride in ether[219]:

$$H_5C_6-N_3 \xrightarrow{SnCl_2\,/\,ether} H_5C_6-N=N-NH_2$$

In contrast, powerful reducing agents such as zinc in acetic acid, aluminum amalgam in moist ether, titanous chloride, sodium borohydride, lithium aluminium hydride, *etc.*, reduce azides directly to *amines*[209]. Catalytic hydrogenation with a palladium or platinum catalyst[209] is generally to be preferred, however, because the reaction conditions are milder.

9.2.2 Decomposition of azido compounds to nitrenes

Thermal or *photolytic* elimination of nitrogen from organic azides without involvement of other reactants leads to nitrenes[220]:

$$R-N_3 \xrightarrow[-N_2]{\triangledown\ or\ h\nu} R-\underline{\bar{N}}$$

[214] *L.J. Skripnik, V.Ya. Pocinok,* Khim. Geterotsikl. Soedin. 474 (1968); C.A. *69,* 96600u (1968).

[215] *L.J. Skripnik, V.Ya. Pocinok, T.F. Prikhod'ko,* Khim. Geterotsikl. Soedin. 201 (1971); C.A. *75,* 48111n (1971).

[216] *G.S. Akimova, J.G. Kolokol'ceva, V.N. Cistokletov, A.A. Petrov,* Zh. Org. Khim *4,* 954 (1968); C.A. *69,* 43551x (1968).

[217] *O. Dimroth,* Ber. *38,* 670 (1905).

[218] *N. Wiberg, W.C. Joo,* J. Organometal. Chem. *22,* 333 a. 349 (1970).

[219] *O. Dimroth,* Ber. *40,* 2376 (1907); *O. Dimroth, K. Pfister,* Ber. *43,* 2757 (1910).

[220] *H. Kwart, A.A. Khan,* J. Amer. Chem. Soc. *89,* 1950 (1967); *H. Kwart, A.A. Khan,* J. Amer. Chem. Soc. *89,* 1951 (1967); *K.v. Frauenberg, R. Huisgen,* Tetrahedron Lett. 2599 (1969) (the decomposition temperature can be reduced by copper catalysis).

[221] *W. Lwowski,* Nitrenes, p. 3, Interscience Publishers, New York 1970.

The intermediate occurrence of nitrenes cannot always be deduced from the nature of the reaction products; special investigations may be necessary in individual cases. This question has been discussed thoroughly in the literature[221]. The following properties can be regarded as characteristic of nitrenes[145]:

a They have a triplet ground state, following Hund's rule; this has been demonstrated by ESR measurements at low temperature.

b They can react in both the triplet and singlet states.

c Their lifetime is short (only a few microseconds[222]), and they become stabilized in various ways.

9.2.2.1 1,2-Hydrogen migration and rearrangement

The main product from the thermal and photochemical decomposition of alkyl azides are *aldimines,* which are formed by H-migration. Thus, the gas-phase pyrolysis of ethyl azide at 380–400° mainly gives *polymeric ethylidenimine* together with *aziridine* and *methylene(methylimine).*

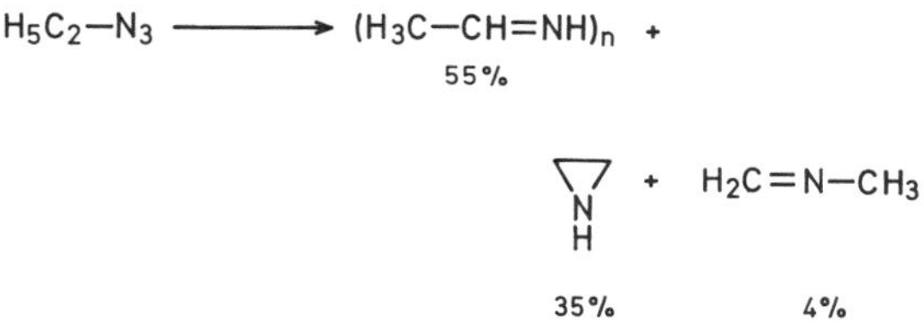

$$H_5C_2-N_3 \longrightarrow \underset{55\%}{(H_3C-CH=NH)_n} +$$

$$\underset{35\%}{\underset{H}{\overset{\triangledown}{N}}} + \underset{4\%}{H_2C=N-CH_3}$$

Polymeric ethylidenimine is also the main product (68%) from photolysis in methanol[223]. In analogous reactions, monomeric imines (*butylidenimine* and *octylidenimine*) are obtained in good yield from butyl and octyl azides[224]. In addition, 9-azidofluorene is converted smoothly into *fluoren-9-imine*[225].

By contrast, formation of pyrrolidines from butyl azide and other long-chain azido compounds[226] could not be reproduced[224].

Thermolysis of α-azido ketones yields *imidazoles*[227, 228] *via* the corresponding aldimines:

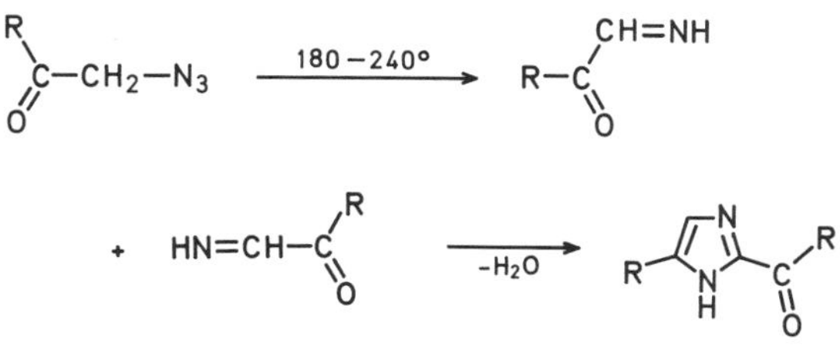

$$\underset{O}{\overset{R}{\underset{\|}{C}}}-CH_2-N_3 \xrightarrow{180-240°} R-\underset{\underset{O}{\|}}{C}\overset{CH=NH}{}$$

$$+ \; HN=CH-\underset{\underset{O}{\|}}{C}\overset{R}{} \xrightarrow{-H_2O} R-\underset{\overset{|}{H}}{\overset{N}{\diagup}}\underset{}{C}-\underset{\underset{O}{\|}}{C}\overset{R}{}$$

[222] *D.W. Cornell, R.S. Berry, W. Lwowski,* J. Amer. Chem. Soc. *87,* 3626 (1965).

[223] *E. Koch,* Tetrahedron *23,* 1747 (1967).

If no α-hydrogen atoms are present, nitrenes can be stabilized also by migration of other groups. Probably, the best known example is the *Curtius degradation* of acyl azides (see p. 564). Azido compounds of boron, silicon, or phosphorus undergo similar rearrangements[229−231] but sulfonyl azides and azidoformic acid derivatives do not[232].

Compounds in which the azido group is attached to a tertiary carbon atom undergo alkyl or aryl migration to form *azomethines* or products derived from these[233]. Table 8 provides some indication of the preparative versatility of the nitrene rearrangement.

Table 8. Rearrangements of azido compounds or nitrenes

Azido compound	Reaction conditions	Reaction product	Physical constants	Yield [% of th.]	Ref.
	Photolysis in benzene	4-Azido-2,5-di-tert-butyl-1,3-dioxo-2-cyclopentenecarbonitrile	m 75–77°	41	234
	Photolysis in benzene	1,5-Diphenyltetrazole	m 146°	14	235
		2-Phenylbenzimidazole	m 291°	52	
		$(H_5C_6-N=C=N-C_6H_5)_3$ Trimeric diphenylcarbodiimide	m 168–170°	10	
	Thermolysis in toluene	2-Benzenediglyoxylonitrile	m 103°	52	236
	Thermolysis in toluene	Tetrazolo[1,5-b]isoquinoline-5,10-dione	m 216°	95	237
	Trichloroacetic acid, 30 min at 65°	3-Amino-5-oxo-(2H,5H)-furanylideneacetonitrile	m 201–204°	44	238
	Photolysis in methanol	Methyl P-(1,1,2,3-tetramethyl-3-butenylphosphoamidate	m 107–115°	16	239
		2-Methoxy-3,3,4,5,5-pentamethyl-1,2-azophospholidine 2-oxide	m 137–138°	60	240

Table 9. Intramolecular *CH*-insertion reactions of nitrenes

Azido compound	Reaction conditions	Reaction product	Physical constants	Yield [% of th.]	Ref.
2-Azidobiphenyl	Photolysis in benzene	Carbazole	m 247°	71	[245]
		2,2'-Azobiphenyl	m 144.5°	12	
(biphenyl-SO₂N₃)	Thermolysis at 150°, no solvent	6H-Dibenzo-1,2-thiazine 5,5-dioxide	m 200–202°	70	[246]
(naphthyl S-phenyl N₃, CH₃CO)	Thermolysis at 140–160° in decalin	12H-Benzo[a]phenothiazin-10-yl methyl ketone	m 209°	70	[247]
Hexanoyl azide	Photolysis in cyclohexane	6-Methyl-2-piperidinone	m 86–88°	13	[248]
		5-Ethyl-2-pyrrolidinone	b₁₅ 133–134°	8	
		N-Cyclohexylhexanamide	m 72–73°	3	
(adamantyl-CH₂-CO-N₃)	Photolysis in acetonitrile	Octahydro-3a,7,5,9-dimethano-3aH-cycloocta[b]pyrrol-2(3H)-one (Adamantano-[2,1-b]pyrrol-2(3H)-one)	m 153–155°	35	[249]

9.2.2.2 Intramolecular *CH*-insertion and dimerization

Intramolecular *CH*-insertion opens up an elegant route to the synthesis of numerous *heterocyclic compounds*. A number of examples of the use of this principle are summarized in Tab. 9. Those reactions are included which proceed only formally *via CH-insertion*. Formation of azo compounds in various yields from decomposition of aryl azides has been observed[241–244].

[224] *D.H.R. Barton, A.N. Starratt*, J. Chem. Soc. (London) 2444 (1965).

[225] *J.P. Anselme*, Organic Preparations and Procedures *1*, 201 (1969).

[226] *D.H.R. Barton, L.R. Morgan*, J. Chem. Soc. (London) 622 (1962).

[227] *J.H. Boyer, D. Straw*, J. Amer. Chem. Soc. 75, 1642 u. 2683 (1953).

[228] *R.Y. Ning, L.H. Sternbach*, J. Med. Chem. *13*, 1251 (1970).

[229] *W.T. Reichele*, Inorg. Chem. *3*, 402 (1964).

[230] *P.J. Paetzold, P.P. Habereder, R. Müllbauer*, J. Organometal. Chem. 7, 61 (1967).

[231] *M.P. Harger*, Chem. Commun. 442 (1971).

[232] *W. Lwowski, R. De Mauriac, Th. W. Mattingly, E. Scheiffele*, Tetrahedron Lett. 3285 (1964).

[233] *R.A. Abramovitch, E.P. Kyba*, J. Amer. Chem. Soc. *93*, 1537 (1971).

9.2.2.3 Intermolecular *CH*-insertion

Intermolecular *CH*-insertion, which leads to *secondary amines,* is of less importance for preparative purposes. The formation of primary amines, in varying yields, is observed as a side-reaction, as with almost all nitrene reactions. Thus, the thermolysis of phenyl azide in cyclohexane yields both *N-cyclohexylaniline* and *aniline*[250].

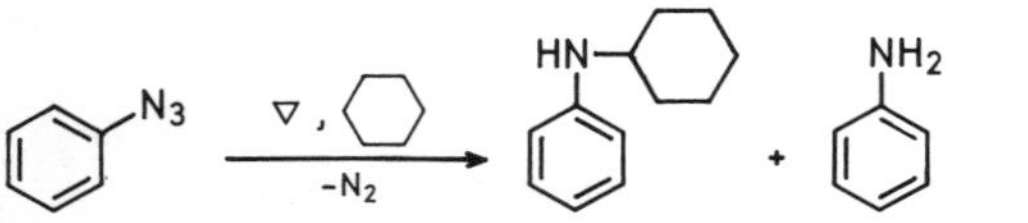

Analogously, the thermolysis of phenyl azide in triethylsilane yields *N-triethylsilylaniline* as well as a small amount of aniline[251]. A further example is the formation of *N–1*- and *N–2-adamantylglycine ethyl esters* by thermolysis of ethyl azidoformate in the presence of adamantane[252].

9.2.2.4 Reactions with nucleophilic partners

Nitrenes have a sextet of electrons, and add readily to nucleophilic partners. Thermal or photolytic decomposition of vinyl azides is consequently a good method for the preparation of *azirines* (intramolecular nucleophilic saturation), which has achieved preparative importance since its discovery[253, 254]. An example is the synthesis of *3-methyl-2-phenyl-2 H-azirine* from 2-azido-1-phenyl-1-propene[255]. Many further examples are found in the literature[145].

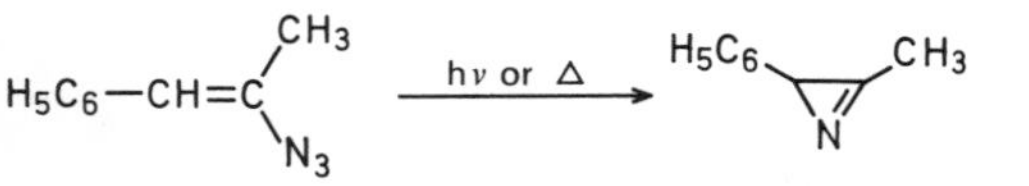

Azirines are also involved as intermediates during the photolysis of aryl azides in the presence of amines[256]. As the photolysis of the phenyl azide/diethylamine system shows, these reactions proceed *via* ring-enlargement and incorporation of the amine into *2-amino-3 H-azepins*[257], e.g.:

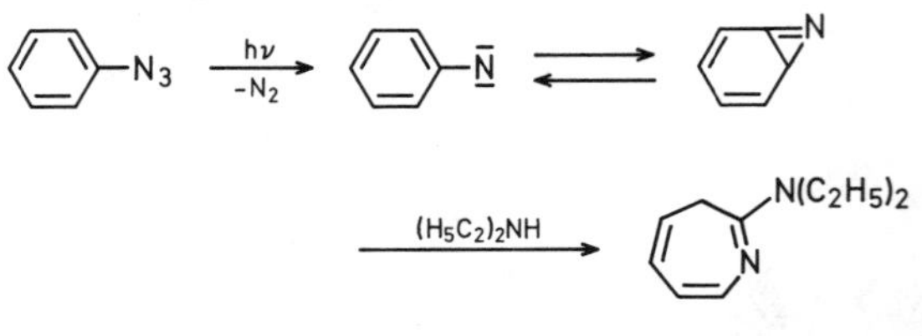

2-Diethylamino-3H-azepine

Intermolecular nitrene addition to double bonds is also possible, and offers a good method for the synthesis of *aziridines*[221]. Thus, the photolytic decomposition of ethyl azidoformate in cyclohexene yields mainly *ethyl 7-azabicyclo[4.1.0]heptane-7-carboxylate* plus the three positionally isomeric *ethyl cyclohexenylcarbamates*[258]:

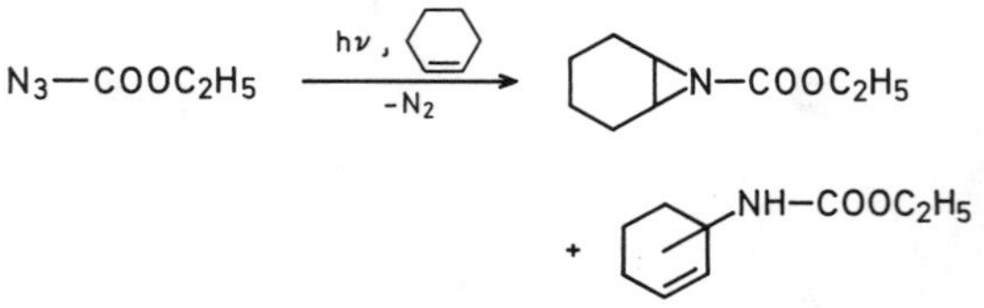

If the decomposition of ethyl azidoformate, cyanogen azide, or azidoformyl azide is carried out in benzene, azepines (*1-azidocarbonyl, 1-ethoxycarbonyl, 1-cyano* derivatives) are obtained instead of aziridines[259]. Evidently, the *7-azanorcaradienes* formed as intermediates are unstable and isomerize to azepines spontaneously:

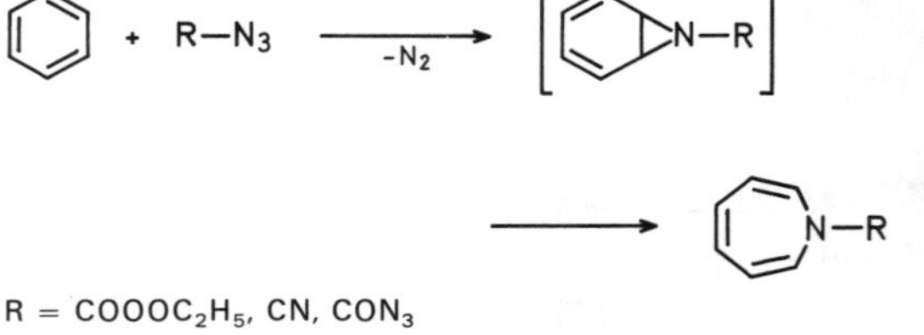

R = COOOC$_2$H$_5$, CN, CON$_3$

[234] *H.W. Moore, W. Weyler,* J. Amer. Chem. Soc. *93,* 2812 (1971).

[235] *R.M. Moriarty, J.M. Kliegman,* J. Amer. Chem. Soc. *89,* 5959 (1967).

[236] *J.A. van Allen, W.J. Priest, A.S. Marshall, G.A. Reynolds,* J. Org. Chem. *33,* 1100 (1968).

[237] *H.W. Moore, D.S. Pearce,* Tetrahedron Lett. 1621 (1971).

[238] *H.W. Moore, H.R. Shelden, P.F. Shellhammer,* J. Org. Chem. *34,* 1999 (1969).

[239] *M.J.P. Harger,* Chem. Commun. 442 (1971).

[240] (mixture of the two structural isomers in ratio of 1:3).

[241] *J.S. Swenton,* Tetrahedron Lett. 3421 (1968).

[242] *L. Horner, A. Christmann, A. Gross,* Chem. Ber. *96,* 399 (1963).

[243] *B. Singh, J.S. Brinen,* J. Amer. Chem. Soc. *93,* 540 (1971).

[244] *D.M. Lemal, T.W. Rave, S.D. McGregor,* J. Amer. Chem. Soc. *85,* 1944 (1963).

[245] *J.S. Swenton, J.T. Ikeler, B.H. Williams,* J. Amer. Chem. Soc. *92,* 3103 (1970).

[246] *R.A. Abramovitch, C.J. Azogu, J.T. McMaster,* J. Amer. Chem. Soc. *91,* 1219 (1969).

[247] *M. Messer, D. Farge,* Bull. Soc. Chim. France 4395 (1969).

[248] *T. Brown, O.E. Edwards,* Can. J. Chem. *45,* 2599 (1967).

[249] *J.K. Chakrabarti, S.S. Szinai, A. Todd,* J. Chem. Soc. (C) 1303 (1970).

[250] *J.H. Hall, J.W. Hill, J.M. Fargher,* J. Amer. Chem. Soc. *90,* 5313 (1968).

Table 10. Reaction of nitrenes with weakly nucleophilic partners

Reactants	Reaction conditions	Reaction products	Physical constants	Yield [% of th.]	Ref.
Azidocyclohexane tert-butylisonitrile	24 h at 90° with iron pentacarbonyl present	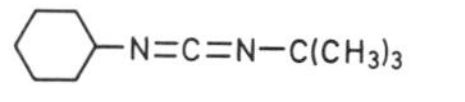 *tert-Butylcyclohexylcarbodiimide*		60	262
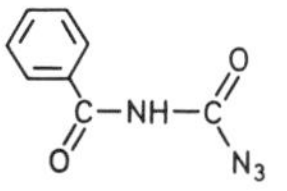	Photolysis in benzene	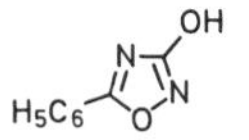 *5-Phenyl-1,2,4-oxadiazol-3-ol*	m 200–201°	77	263
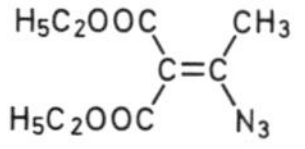	2 h reflux in toluene	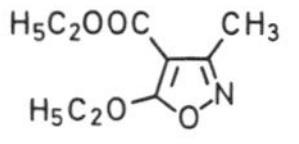 *Ethyl 5-ethoxy-3-methyl-4-isoxazolecarboxylate*	m 80°	70	264
Phenyl azide Carbon monoxide	In 1,2,2-trichloro-1,2,2-trifluoroethane, autoclaving at 180°	*Phenyl isocyanate*		100	265
p-Toluenesulfonic acid azide Dimethyl sulfoxide	Photolysis in dimethyl sulfoxide	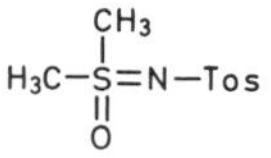 *Dimethyl-N-(4-tolylsulfonyl)-sulfoximine*	m 168°	32	266

Ethoxycarbonylnitrene can also add as a 1,3-dipole to multiple bonds[260, 261]. Thus, for example, *2-ethoxy-5-methyl-1,3,4-oxadiazole* can be obtained by trapping with acetonitrile[260]:

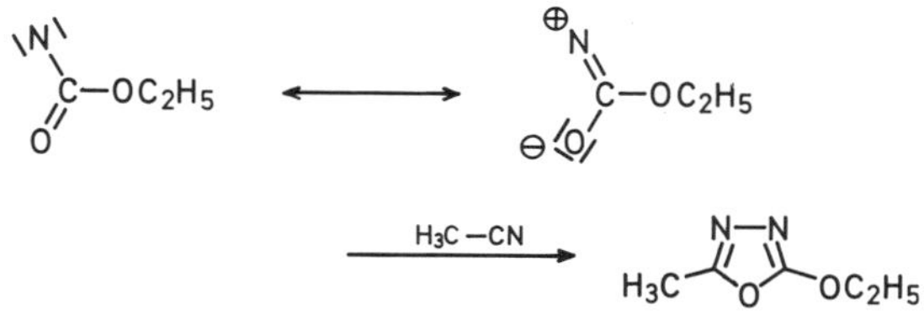

Further examples of the reaction of nitrenes with weakly nucleophilic partners are shown in Tab. 10.

9.2.3 Decomposition by acids and bases

Decomposition of organic azides by acids has been conducted principally with mechanistic aspects in mind[145, 267]. Exceptions apart,[268, 269] these reactions are of little importance for preparative purposes because of the wide product spectrum. Both nitrogen and the azide ion can be split off under the influence of acids[270, 271]. Reaction products that have been isolated include imines, amines, azomethines, alcohols, and ketones[209, 267]. o- and p-Unsubstituted aryl azides are converted into 2- and 4-haloanilines on treatment with halogen hydrides[272], while dilute sulfuric acid yields 2- and 4-aminophenols[273].

In general, organic azido compounds are quite stable toward alkali. α-Azidocarbonyl compounds alone react, forming α-imino ketones provided at least one hydrogen atom is attached to the carbon atom bearing the azide group. An example is the synthesis of *α-iminocamphor (3-imino-2-bornanone)*[274]:

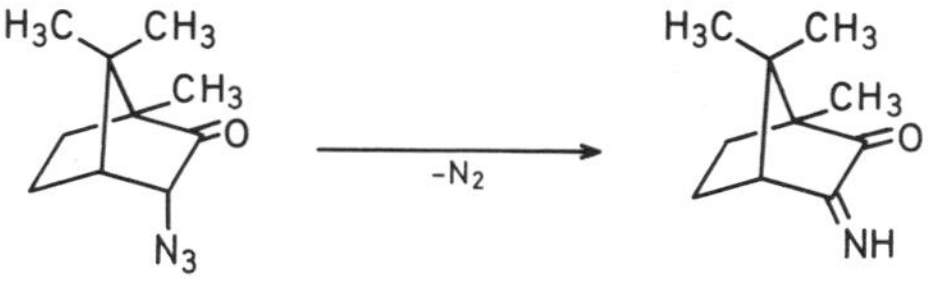

Nitro groups in o- and p-positions activate the azide group in phenyl azides to such an extent that they can be replaced readily by the hydroxy group with aqueous alkali solution[275] (giving 2- and 4-azidophenol). Organoelemental azido compounds are often hydrolyzed even by atmospheric moisture; the hydrazoic acid thus liberated can cause **explosions.**

[251] *F.A. Carey, C.W. Hsu,* Tetrahedron Lett. 3885 (1970).

[252] *D.S. Breslow, E.J. Edwards, R. Leone, P.v.R. Schleyer,* J. Amer. Chem. Soc. *90,* 7097 (1968).

[253] *G. Smolinsky,* J. Amer. Chem. Soc. *83,* 4483 (1961).

[254] *G. Smolinsky,* J. Org. Chem. *27,* 3557 (1962).

[255] *K. Isomura, S. Kobayashi, H. Taniguchi,* Tetrahedron Lett. 3499 (1968).

[256] *W. Lwowski,* Nitrenes, p. 3, Interscience Publishers, New York 1970.

[257] *W. von E. Doring, R.A. Odum,* Tetrahedron *22,* 81 (1966).

[258] *W. Lwowski, T.J. Maricich,* J. Amer. Chem. Soc. *87,* 3630 (1965).

[259] *W. Lwowski,* Nitrenes, p. 197, 211, 274, 339, Interscience Publishers, New York 1970.

[260] *R. Huisgen, H. Blaschke,* Justus Liebigs Ann. Chem. *686,* 145 (1965).

[261] *R. Huisgen, H. Blaschke,* Chem. Ber. *98,* 2985 (1968).

[262] *T. Salgusa, Y. Ito, T. Shimizu,* J. Org. Chem. *35,* 3995 (1970).

[263] *R. Neidlein, H. Krüll,* Justus Liebigs Ann. Chem. *716,* 156 (1968).

[264] *K. Friedrich, H.K. Thieme,* Chem. Ber. *103,* 1982 (1970).

[265] *R.P. Bennett, W.B. Hardy,* J. Amer. Chem. Soc. *90,* 3295 (1968).

[266] *L. Horner, A. Christmann,* Chem. Ber. *96,* 388 (1963).

[267] *A. Hassner, E.S. Ferdinandi, R.J. Isbister,* J. Amer. Chem. Soc. *92,* 1672 (1970).

[268] *H.W. Moore, H.R. Schelden, W. Weyler,* Tetrahedron Lett. 1243 (1969).

[269] *J.H. Boyer, F.C. Canter, J. Hamer, D.K. Putney,* J. Amer. Chem. Soc. *78,* 325 (1956).

[270] *R. Kreher, G. Jäger,* Z. Naturforsch. *19b,* 657 (1964).

[271] *R. Kreher, G. Jäger,* Angew. Chem. *77,* 730 (1965); Angew. Chem. Intern. Ed. Engl. *4,* 706 (1965).

[272] *P.A.S. Smith, B.B. Brown,* J. Amer. Chem. Soc. *73,* 2438 (1951).

[273] *E. Bamberger,* Justus Liebigs Ann. Chem. *424,* 233 (1921).

[274] *M.O. Forster, H.E. Fierz,* J. Chem. Soc. (London) 826 (1905) a. 867 (1907).

[275] *E. Noelting, E. Grandmougin, O. Michel,* Ber. *25,* 3328 (1892).

10 Organic *N*-Oxides

Contributed by

H. Stamm
Pharmazeutisch-Chemisches Institut der Universität
(TH) Karlsruhe/Heidelberg,
Karlsruhe/Heidelberg

10.1 Nitrile oxides

In the majority of cases nitrile oxides arise only as nonisolatable and unstable intermediate products. Most nitrile oxides survive for long periods only below −15° [1]; aromatic oxides are more stable than aliphatic ones (for an exception see ref. [2]). In general they dimerize spontaneously to form *furoxans* (furazan N-oxides) (see p. 391), in acid medium to *1,2,4-oxadiazole 4-oxides* (see p. 392) or, alternatively yet, to the *isomeric isocyanates* by rearranging [3]. Sterically hindered nitrile oxides alone (generally 2,6-substituted benzonitrile oxides but cf. ref. [2]) are stable in the normally accepted sense of this term. The methods discussed here are therefore rather techniques for preparing *furoxans, 1,2,4-oxadiazole 4-oxides, isocyanates*, other *cycloadducts* (p. 333), and *acyclic addition products* (p. 332), with the *nitrile oxide* being set free *in situ* from stable precursors or being prepared immediately prior to their conversion. Several

detailed surveys cover the overall field of the nitrile oxides [1, 4, 5, 6].

10.1.1 Preparation

A novel thermolysis of oxime dehydro dimers serves as a source of nitrile oxide under neutral conditions [632].

10.1.1.1 By dehydrogenation of aldoximes

Aldoximes can be dehydrogenated in various different ways but the only preparatively useful tech-

[1] *Ch. Grundmann* in *Houben-Weyl*, Methoden der organischen Chemie, 4. Ausgabe, Bd. X/3, p. 838, Georg Thieme Verlag, Stuttgart 1965.

[2] *S. Ranganathan, B.B. Singh, C.S. Panda*, Tetrahedron Lett. 1225, (1970).

[3] *Ch. Grundmann, P. Kochs*, Angew. Chem. 82, 637 (1970).

[4] *Ch. Grundmann*, Fortschr. Chem. Forsch. 7, 62 (1966).

[5] *Ch. Grundmann* in *Z. Rapoport*, The Chemistry of the Cyano Group, Interscience Publishers, London 1970.

nique is probably dehydrogenation of aromatic aldoximes with sodium hypobromite solution at 0°.

$$Ar-CH=NOH \xrightarrow{\text{BrO}^{\ominus}/0°} Ar-C\equiv N\to O$$

Lead(IV) acetate in dichloromethane, too, can be
used with success[6,7].

10.1.1.2 Nitrile oxides from hydroxamic acid compounds

Nitrile oxides are obtained by eliminating *HX*
from the hydroxamic acid compounds **1**:

$$\underset{\substack{| \\ X}}{\overset{\substack{R \\ |}}{C}}=NOH \xrightarrow[-HX]{\text{Base}} R-C\equiv N\to O$$

1

$$R-CH=NOH \xrightarrow[-HCl]{\text{Hal}_2} \underset{\substack{| \\ R}}{\overset{\substack{Cl \\ |}}{C}}=N-OH$$

$$R-CH_2-NO_2 \xrightarrow[-H_2O]{\text{HNO}_2} R-\underset{\substack{| \\ NO_2}}{\overset{\substack{N-OH \\ \|}}{C}}$$

The technique is preparatively important for
X = Hal (**1** = hydroxamoyl halide) and X = NO$_2$
(**1** = nitrolic acid). Eliminating hydrogen chloride
from hydroxamoyl chlorides (X = Cl) is by far the
most widely preferred method of preparing nitrile
oxides. Using aqueous sodium bicarbonate or
sodium carbonate solutions affords the nitrile oxides in generally good yield at 0°. Caustic soda,
too, may be used, and triethylamine also accom-
plishes the elimination when used in ether at −20°
(triethylamine hydrochloride is precipitated), but
sometimes the nitrile oxide formed reacts with the
amine. The hydroxamoyl chloride is obtained
from the aldoxime by halogenation with elementary halogen or with nitrosyl chloride[7].
Alternatively, converting the oxime to the
nitrile oxide can be effected in one step with *N*-bro-
mosuccinimide in diethylformamide in the presence of sodium ethoxide or triethylamine[6].
With the nitrolic acids obtainable from primary
nitroalkanes and nitrous acid, water is sufficient to
form the nitrile oxides. An acid nitrous acid
medium also yields acyl nitrile oxides (*R = acyl*)
from primary α-diazocarbonyl compounds *via* **1**
(X = N$_2^{\oplus}$)[629].

[6] *Ch. Grundmann*, Synthesis 344, (1970).
[7] *B.J. Wakefield, D.J. Wright*, J. Chem. Soc. (C)
1165 (1970).

10.1.1.3 Nitrile oxides by dehydration of primary nitroalkanes

$$R-CH_2-NO_2 \xrightarrow[-H_2O]{} R-C\equiv N\to O$$

2

Dehydrating 1-nitroalkanes **2** is performed best
with triethylamine and phenyl isocyanate. This
method can be applied only as an *in situ* technique.
Re the dehydrogenation with sulfuric acid cf. ref. [8],
with acetic anhydride/boron trifluoride *cf.*
ref. [9].

10.1.2 Transformations of nitrile oxides

Thermal rearrangement of nitrile oxides to *isocyanates*[2] has already been referred to (see p. 331).
Re *dimerizations cf.* p. 331. Retrogressive cleavage of furoxans to nitrile oxides ensues only at
temperatures at which the nitrile oxides at once
rearrange to isocyanates (but *cf.* ref. [633]). Re formation of trimers and polymers cf. refs. [1,4].
Dimerization with excess boron trifluoride in hex-
ane affords *1,4,2,5-dioxadiazines*[6].
Nucleophilic *additions* to nitrile oxides proceed
readily[1]:

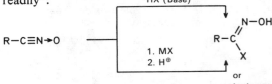

HX = HCN, HOR, HSR, HHal, H$_2$O, H$_2$S, HN(R)$_2$, HO−CO−R[6],
HO−P(O)YZ[10], HC≡C−R (as side-reaction)[11,12],
R−MgX, HSCN (product rearranges)

Tertiary amines sometimes furnish the *zwitterions*
3:

$$R-\underset{\substack{| \\ N(R)_3 \\ \oplus}}{\overset{\substack{N-O^{\ominus} \\ \|}}{C}}$$

3

Deoxygenation of nitrile oxides can be effected
with zinc/acetic acid or with phosphines.

[8] *J.T. Edward, P.H. Tremaine*, Can. J. Chem. *49*,
3483, 3489, 3493 (1971).
[9] *K. Hirai, H. Matsuda, Y. Kishida*, Chem. Pharm.
Bull. (Tokyo) *20*, 97 (1971).
[10] *J.I.G. Cadogan, J.A. Challis, D.T. Eastlick*, J.
Chem. Soc. (B) 1988 (1971).
[11] *A. Battaglia, A. Dondini, A. Mangini*, J. Chem. Soc.
(B) 554 (1971).
[12] *S. Morrocchi, A. Ricca, A. Zanarotti*, Tetrahedron
Lett. 3329 (1969).
A. Battaglia, A. Dondoni, Tetrahedron Lett. 1221
(1970).
P. Beltrame, P. Sartirana, C. Vintani, J. Chem. Soc.
(B) 814 (1971).

The most important property of the nitrile oxides is their propensity to undergo cycloadditions with multiple bond systems serving for synthesizing heterocycles (isoxazole derivatives **4** in the widest sense[1, 4, 6, 13]); suitably the nitrile oxides are generated *in situ*:

$$R-C\equiv N\rightarrow O$$
$$+$$
$$X=Y$$

4

$X = Y =$ Alkenes[13], Enol ethers, Alkynes[14], Alkinyl ethers, Carbonyl compounds[15], Azomethines[16], Nitriles, Thiocarbonyl compounds[17], Sulfinylamines, Furan[6], Pyrrole[6], Thiophene[6], Fulvenes[18], Boron imides[6], Phosphonium ylides[6], Sulfur dioxide[19].

In the following cases the cycloaddition to **4** is succeeded by elimination of a respective substituent of X and Y, leading to the subsequent formation of a double bond between X and Y[6]:

$X = Y =$ Vinyl chloride, Enol acetate, Acetoacetates, Enamine[20], Ketene dithioacetals.

Nitroso compounds (p. 392), azodicarboxylic acid diesters[4, 6], diazomethane[4], and azirines[21] are added to form other types of products.

10.2 Nitrones

10.2.0 Nomenclature, structure, properties, and scope

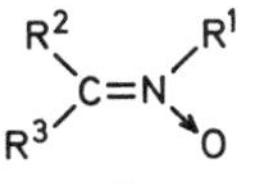

5

Nitrones[664] **5** are *N*-oxides of Schiff's bases and thus carbonyllike compounds. Aldo nitrones *(R^2 or $R^3 = H$)* and keto nitrones *(R^2, $R^3 = alkyl$ or aryl)* are distinguished. Naming the nitrones can be performed in various different ways. Thus, to take an example,

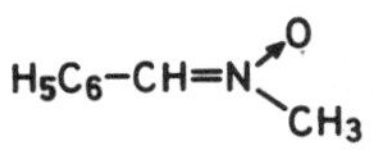

$$H_5C_6-CH=N\begin{smallmatrix}O\\CH_3\end{smallmatrix}$$

can be named

— as a derivative of the basic carbonyl compound: *benzaldehyde methylnitrone;*
— by specifying the substituents on the nitrone group: *C-phenyl-N-methylnitrone,* or *N-methyl-α-phenylnitrone*;*
— by using the prefix *nitrono,* for instance, *methylnitronobenzene* (this method is employed in complicated cases only where the others fail or are too cumbersome);
— as the *N*-oxide of a Schiff's base: *N-benzylidenemethylamine N-oxide;*
— as the *N*-methyl ether of *benzaldoxime* or *N-methylbenzaldoxime* — often found in the older literature;
— as *benzaldehyde N-methyloxime,* as has been proposed in *tentative rules* of the *IUPAC* (not a logical or a happy proposal).

The names of compounds **6** with an endocyclic nitrone function

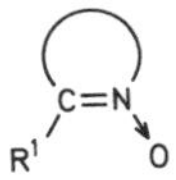

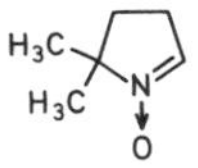

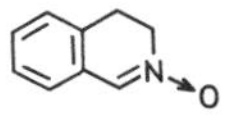

are derived from the parent heterocycles, for instance, *5,5-dimethyl-1-pyrroline N-oxide* or *1-oxide* **7** and *3,4-dihydroisoquinoline N-oxide* or *2-oxide* **8**.

Compounds with the endocyclic nitrone function forming part of an aromatic system are dealt with separately as *N*-oxides of azaaromatic compounds (see p. 355). Isatogens (3 *H*-indol-3-one *N*-oxides) and *N*-oxides of 4*H*-pyrazol-4-ones, too, are discussed separately on p. 351.
Compounds with the structure **5**, in which the first atom of R^1, R^2, or R^3 is a heteroatom (neither C nor H) are not embraced (*e.g.*, ethers of *aci*-nitroalkanes and aldazine *N*-oxides).
Geometric isomerism is possible in acyclic nitrones:

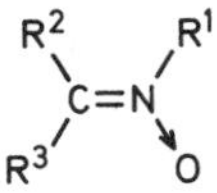

5-Z 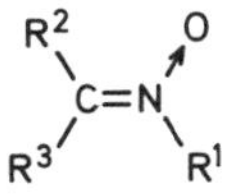   **5-E**

---

* As used in the Chemical Abstracts until 1971. Can confuse because α-substitution in carbonyl and carbonyllike compounds has a different meaning.

[13] *R. Huisgen,* Angew. Chem. *75,* 604 (1963).
[14] *A. Battaglia, A. Dondini, F. Taddei,* J. Heterocycl. Chem. *7,* 721 (1970).
[15] *W.I. Awad, M. Sobhy,* Can. J. Chem. *47,* 1473 (1969).
[16] *K.-H. Magosch, R. Feinauer,* Angew. Chem. *83,* 882 (1971).
[17] *A. Battaglia, A. Dondini, G. Maccagnani, G. Mazzanti,* J. Chem. Soc. (B) 2096 (1971).

Using the E–Z notation[22] is suitable for characterizing the isomers. In some cases $(R^2=Ar, R^3=CN)$ it has been possible to isolate both isomers and to assign the structures with the aid of the dipole moments.

Aldo nitrones probably exist almost exclusively in the Z form $(5$–$Z, R^2 = H)$ unless ring closure necessarily brings about the E form $(7, 8)$[23-29]. Rapid E–Z isomerization appears to take place at least at elevated temperature[29]. The *rotation barrier* was determined to be 24.7 *kcal/mole*[30] and 23.2 *kcal/ mole*[29] respectively for two nitrones.

Formula **5** reproduces the bond and charge relationships only incompletely; a number of classical limiting formulae need to be adduced to describe them[26, 31, 32]:

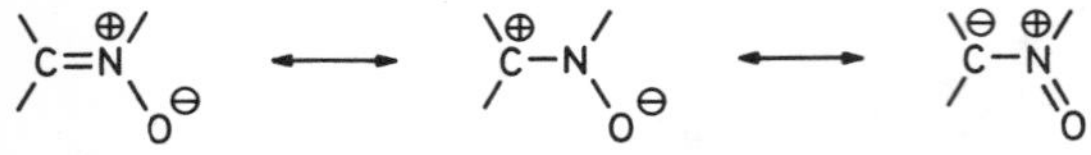

The dipole moment[24, 25] of the nitrones (3.49 D for *N-benzaldehyde methylnitrone* and 4.31 D for *benzophenone methylnitrone* and the $C=N$ bond length $(1.309\,\text{Å}$[28, 33]$)$ that is greater than in oximes, as well as the shorter N, O-bond length $(1.284\,\text{Å}$[28, 33]$)$ seem to be compatible. From the bond length, the π-bond order of the $C=N$ bond is approximately 0.65[32, 33].

The *UV spectra* of the α-aryl nitrones show the intense (log $\epsilon \sim 4.1$) $\pi \rightarrow \pi^*$ band at 290–350 nm[23, 24, 26, 34, 35] as the longest-wave band. This band is weaker in nonconjugated nitrones (log $\epsilon \sim 3.8$–3.95) and of shorter wavelength (~ 230–240 nm)[26, 36-39]. The position of the band is solvent-dependent in that polar solvents stabilize the ground state and produce a hypsochromic shift. During *UV* measurements the photochemical instability of the nitrones must be allowed for (see p. 348). The *IR spectra* of simple nitrones display the $C=N$ band at 1500–1655 cm^{-1}[40-42, 37] and the strong N–O band between 1150 and 1280 cm^{-1}[43, 44] or near 1070 cm^{-1}[35].

In the ^1H–*NMR spectra*[26, 27, 45, 46] of *N-methyl-C-phenyl aldonitrones* the chemical shift for the aldehyde H is found to be 7.19–8.04 ppm depending on the substitution in the benzene ring (in CDCl$_3$ downfield from TMS internally). 6.97 ppm was found in an aliphatic aldo nitrone[47]. The *N*-methyl signal is at 3.68–3.98 ppm. In trifluoroacetic acid these signals are displaced strongly downfield as a result of protonation on the oxygen, *e.g.*, to 8.35 and 4.25 ppm respectively in *4-chlorobenzaldehyde methylnitrone*. Where a single *ortho* group on the phenyl ring has forcibly produced a conformation in such a nitrone, in which the *ortho* proton still present is held close to the nitrone oxygen, combined anisotropy and field effects produce a strong paramagnetic shift to ~ 9.30 ppm. Without an *ortho* substitution a rotation-induced average of about 8.2 ppm is found for the two ortho protons. This paramagnetic shift of the *ortho* protons is intensified by electron-donating *para* groups.

Mass spectra of substituted *C-phenyl-N-phenyl aldo*

nitrones[48] display the molecular ion $M^{\cdot\oplus}$ and the respective 2-arylbenzisoxalinium ion $(M-1)^{\oplus}$ as well as the fragments X–C_6H_4–$C\equiv O^{\oplus}$ and Y–C_6H_4–$N^{\cdot\oplus}$ (as basis peak), which are probably formed *via* the nitrone isomer oxaziridine. $M^{\cdot\oplus}$ and $(M-1)^{\oplus}$ are formed also in the corresponding *N*-methyl nitrones[49]. The $(M-O)^{\cdot\oplus}$ ion generally encountered in the *C*-aryl nitrones[48, 49] may be absent in the purely aliphatic nitrones[50, 51]. An $(M-OH)^{\oplus}$ ion is apparently formed by all aldo nitrones. Re the mass spectra of *fluorenone phenyl nitrone* see ref.[52].

Nitrones generally crystallize readily, and from anhydrous medium also as adducts with various metal halides[53] (important for p. 339).

Nitrones dissolve best in polar solvents. They are often hygroscopic and in general sufficiently basic to form salts[53] with strong acids (including picric acid) and *adducts*[54, 55] with Lewis acids. The structure of the cations formed

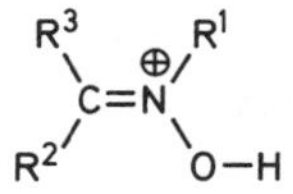

[18] *P. Caramella, P. Frattini, P. Grünanger*, Tetrahedron Lett. 3817 (1971).

[19] *E.H. Burk, D.D. Carlos*, J. Heterocycl. Chem. *7*, 177 (1970).

[20] *P. Caramella, E. Cereda*, Synthesis 433 (1971).

[21] *V. Nair*, Tetrahedron Lett. 4831 (1971).

[22] *J.E. Blackwood, C.L. Gladys, K.L. Loening, A.E. Petrarca, J.E. Rush*, J. Amer. Chem. Soc. *90*, 509 (1968).

[23] *J. Thesing, W. Sirrenberg*, Chem. Ber. *91*, 1978 (1958).

[24] *T. Kubota, M. Yamakawa, Y. Mori*, Bull. Chem. Soc. Japan *36*, 1552 (1963) .

[25] *E.J. Grubbs, J.A. Villareal*, Tetrahedron Lett. 1841 (1969).

[26] *K. Kojano, H. Suzuki*, Tetrahedron Lett. 1859 (1968).

[27] *J. Hoenicke*, Dissertation, Universität Marburg/Lahn 1971.

[28] *K. Folting, W.N. Lipscomb, B. Jerslev*, Acta Cryst. *17*, 1263.

[29] *L.W. Boyle, M.J. Peagram, G.H. Whitham*, J. Chem. Soc. (B) 1728 (1971).

[30] *K. Kojano, I. Tanaka*, J. Phys. Chem. *69*, 2545 (1965).

[31] *J. Hamer, A. Macaluso*, Chem. Rev. *64*, 473 (1964).

[32] *R. Huisgen*, Angew. Chem. *75*, 604 (1963).

[33] *J.E. Bayer, J. Häfelinger*, Chem. Ber. *99*, 1689 (1966).

[34] *W. Rundel* in *Houben-Weyl*, Methoden der organischen Chemie, Band X/4 p. 310, Georg Thieme Verlag, Stuttgart 1968.

[35] *S. Tamagaki, S. Kozuka, S. Oae*, Tetrahedron *26*, 1795 (1970).

[36] *L.S. Kaminsky, M. Lamchen*, J. Chem. Soc. (B) 1085 (1968).

[37] *J.F. Elsworth, M. Lamchen*, J. Chem. Soc. (C) 2423 (1968).

[38] *L.S. Kaminsky, M. Lamchen*, J. Chem. Soc. (B) 715 (1968).

reveals the genetic relationship between oximes and nitrones; its preparative use is discussed on p. 340. Electron-attracting groups in R^2 and R^3 can reduce the basicity enough to stop salt formation in alcoholic medium; in aqueous solution the nitrone is likely to cleave hydrolytically.

Spectroscopic methods and determination of dipole moments allow a reliable *identification* of the nitrone nature of a compound; chemical methods such as deoxygenation to the Schiff's base and hydrolysis gives revealing hydrolysis products (see p. 349). The frequent need for determining the structure precisely is due to the tendency of many nitrones to isomerize and polymerize, and because some syntheses can give closely related products side by side with or even instead of nitrones. Among these changes of the nitrones are the following:

1 *Behrend rearrangement*[56] with strong bases (*via* an anion with nitratelike mesomerism):

2 *Photochemical isomerization* to oxaziridines and other products (see p. 348).

3 *Rearrangement* to oxime ethers[57] (*e.g.* if R^1 is a low energy free-radical (diphenylmethyl[58, 59]):

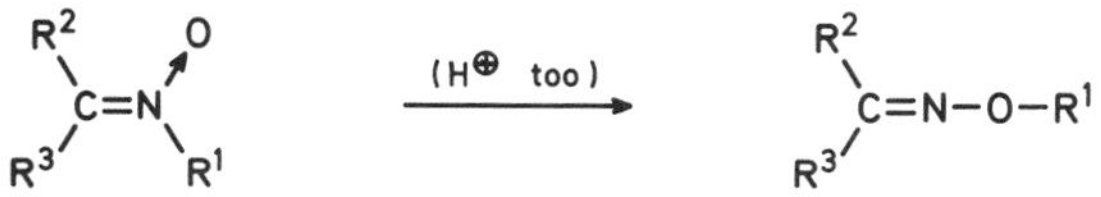

4 Rearrangement to carboxamides by the agency of acetic anhydride, acids, Lewis acids, or bases[60], or photochemically or thermally (see p. 348); the migrating group R^3 need not be just hydrogen:

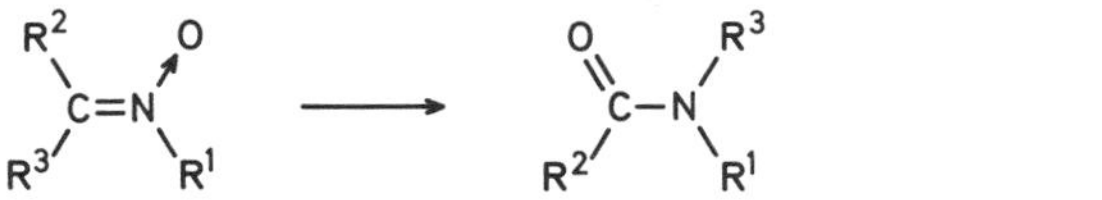

5 *Wallach-like rearrangement*[60] to imines:

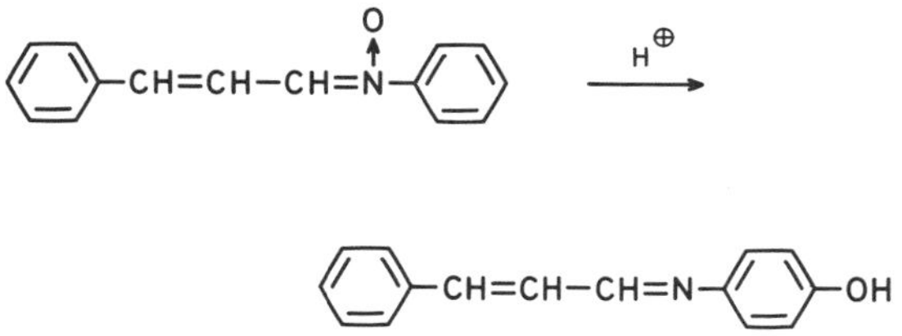

6 *Rearrangement to α-hydroxyimines*[61] or their derivatives:

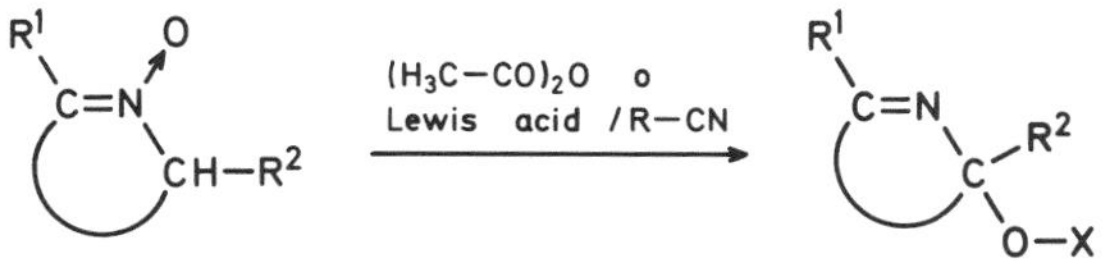

X = H , CH$_3$—CO—

7 *Nitrone-N-hydroxyenamine tautomerism*[37, 62]:

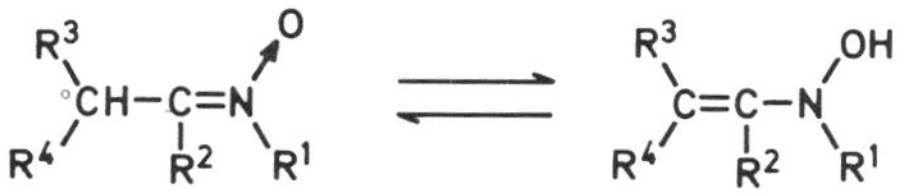

or, in α,β-unsaturated nitrones, tautomeric equilibrium with the corresponding *N*-hydroxydienamine[63].

8 *Oligomerization* or *polymerization*.

[39] *R. Bonnet, R.F.C. Brown, V.M. Clark, I.O. Sutherland, A. Todd,* J. Chem. Soc. (London) 2094 (1959).

[40] *P.A.S. Smith,* The Chemistry of Open-Chain Organic Nitrogen Compounds, Vol. II, Benjamin, New York 1966.

[41] *R.F.C. Brown, V.M. Clark, M. Lamchen, A. Todd,* J. Chem. Soc. 2116 (1959).

[42] *G.R. Hansen, R.L. Boyd,* J. Heterocyclic Chem. 7, 911 (1970).

[43] *W. Rundel* in *Houben-Weyl,* Methoden der organischen Chemie, Bd. X/4, p. 316, Georg Thieme Verlag, Stuttgart 1968.

[44] *H.G. Aurich,* Chem. Ber. 98, 3917 (1965).

[45] *K. Kojano, H. Suzuki,* Bull. Chem. Soc. Japan 42, 3306 (1969).

[46] *K. Tori, M. Ohtsuru, T. Kubota,* Bull. Chem. Soc. Japan 39, 1089 (1966).

[47] *H.O. House, R.W. Magin,* J. Org. Chem. 28, 647 (1963).

[48] *B.S. Larsen, G. Schroll, S.-O. Lawesson, J.H. Bowie, R.G. Cooks,* Tetrahedron 24, 5193 (1968).

[49] *H. Stamm, J. Hoenicke, H. Steudle,* Arch. Pharm. (Weinheim, Ger.) 305, 619 (1972).

[50] *M. Masui, Ch. Yijima,* Chem. Pharm. Bull. (Tokyo) 17, 1517 (1969).

[51] *H. Gnichtel,* Chem. Ber. 103, 2411, 3442 (1970).

[52] *D.R. Eckroth, T.H. Kinstle, D.O. DeLaCruz, J.K. Sparacino,* J. Org. Chem. 36, 3619 (1971).

[53] *W. Rundel* in *Houben-Weyl,* Methoden der organischen Chemie, Bd. X/4, p. 418, Georg Thieme Verlag, Stuttgart 1968.

[54] *B. Zeeh,* Synthesis 1, 37 (1970).

[55] *W.D. Emmons,* J. Amer. Chem. Soc. 79, 5739 (1957).

[56] *W. Rundel* in *Houben-Weyl,* Methoden der organischen Chemie, Bd. X/4, p. 414, Georg Thieme Verlag, Stuttgart 1968.

[57] *W. Rundel* in *Houben-Weyl,* Methoden der organischen Chemie, Bd. X/4, p. 430, Georg Thieme Verlag, Stuttgart 1968.

[58] *E.J. Grubbs, J.D. McCullough jr., B.H. Weber, J.R. Maley,* J. Org. Chem. 31, 1098 (1966).

[59] *J.S. Vincent, E.J. Grubbs,* J. Amer. Chem. Soc. 91, 2022 (1969).

[60] *W. Rundel* in *Houben-Weyl,* Methoden der organischen Chemie, Bd. X/4, p. 428, Georg Thieme Verlag, Stuttgart 1968.

[61] *L.H. Schlager,* Tetrahedron Lett. 4519 (1970).

[62] *M. Mousseron-Canet, J.-P. Boca,* Bull. Soc. Chim. France 1296 (1967).

[63] *P. Schiess, P. Ringele, H.L. Chia,* Chimia 24, 24 (1970).

C-Alkyl nitrones which still carry an H atom on the first C atom of the alkyl group probably exist as dimers, trimers, or polymers in the majority of cases[36, 39, 64, 65]. The exact structure is unknown or the published structure may require revision. It is probable that oligomerization equilibria are a general feature[64]; they may allow reactions of the monomeric nitrone even where the latter cannot be isolated[37]. C-Aryl nitrones and 1-pyrroline 1-oxides[65] exist as true monomeric nitrones, while 2,3,4,5-tetrahydropyridine 1-oxide[64] and 3,6-dihydro-2H-1,4-oxazine 4-oxide[37] form polymers or lower oligomers; the homologous seven-membered ring compound 4,5,6,7-tetrahydro-3H-azepin 1-oxide polymerizes and dimerizes with opening of the ring[66, 67].

Originally an aldol structure 9 was postulated for the dimeric nitrones. Recently, however, the structure 9 has been amended in favor of the structure 10 in at least two cases ($R' = C_6H_5$, $R^2 = R^4 = H$, $R^3 = C_2H_5$[68, 69] and $R^1 = C_6H_5$, $R^2 = CH_3$, $R^3 = R^4 = H$[70]). A critical review is thus indicated when old oligomer structures are accepted. Two pathways are feasible for the formation of the cyclic dimers 10:

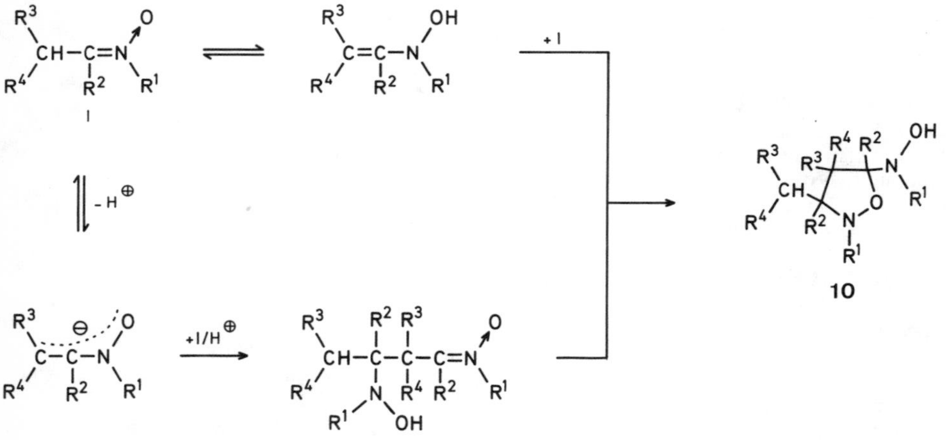

10.2.1 Preparation of nitrones

10.2.1.1 Dehydrogenation of hydroxylamines

Preparation of nitrones from hydroxylamines by dehydrogenation is not always clear in respect of the direction of the formation of a $C=N$ double bond:

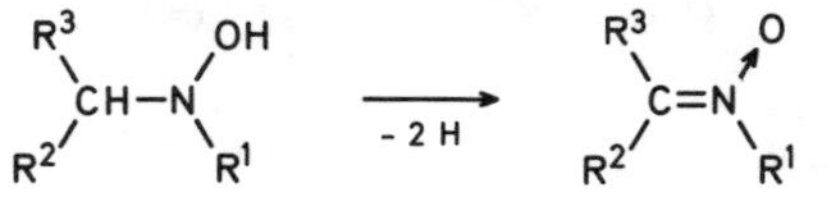

$$\text{(1)}$$

Where R' carries an sp^3-bound H atom in the α-position the nitrone function can be formed also toward R^1, either directly or subsequently by Behrend rearrangement (cf. p. 334). In these cases mixtures are obtained, but one isomer may be formed preferentially[64]. Under defined oxidation conditions a once-formed nitrone displays no secondary changes.

An oxidative dimerization is found in formaldehyde nitrones, which are converted spontaneously into glyoxal dinitrones (see p. 338). This change is prevented or slowed down only in formaldehyde nitrones with bulky N-substituents (see p. 347).

10.2.1.1.1 Air oxidation and disproportionation[37, 71]

Some hydroxylamines 11 are converted into the corresponding nitrones 12 either as a result of atmospheric oxidation (Equation 1) or dispropor-

[64] J. Thesing, H. Mayer, Justus Liebigs Ann. Chem. 609, 46 (1957); J. Thesing, H. Mayer, Chem. Ber. 89, 2160 (1956).
[65] R.F. Brown, V.M. Clark, A. Todd, Proc. Chem. Soc. 97 (1957).
[66] M.A.T. Rogers, Nature 177, 128 (1956).

[67] E.J. Alford, J.A. Hall, M.A.T. Rogers, J. Chem. Soc. (C) 1103 (1966).
[68] A.D. Baker, J.E. Baldwin, D.P. Kelly, J. DeBernardis, Chem. Commun. 344 (1969).
[69] W. Kliegel, Tetrahedron Lett. 2627 (1969).
[70] R. Foster, J. Iball, R. Nash, Chem. Commun. 1414 (1968).

tionation (Equation **2**) either spontaneously or on heating:

$$2 \quad \underset{R^2}{\overset{R^3}{>}} CH-N \underset{R^1}{\overset{OH}{<}} \longrightarrow \qquad (2)$$

11

$$\underset{R^2}{\overset{R^3}{>}}C=N \overset{O}{\underset{R^1}{\nearrow}} \quad + \quad \underset{R^2}{\overset{R^3}{>}}CH-NH-R^1 \quad + \quad H_2O$$

12

As is revealed by the formation of *glyoxal bis-N-phenyl nitrone* from 1,2-dibromoethane and *N*-phenylhydroxylamine[72], the transition to the nitrone can ensue during the preparation of the disubstituted hydroxylamine **11**. It is not possible to give a verdict about the scope of the method[71].

Investigations of the mechanism (*via* the corresponding nitroxide) are available[73].

The air oxidation is catalyzed by ammoniacal copper(II) solution (the reaction is complete when the blue color of the copper(II) tetrammine persists) (cf. p. 338).

10.2.1.1.2 Dehydrogenation with peroxy compounds

Dehydrogenation with hydrogen peroxide/glacial acetic acid is probably applicable in exceptional cases because of the sensitivity to hydrolysis of most nitrones[74, 75]. The use of anhydrous peroxy acid has not been described so far. By contrast, dehydrogenation with *tert*-butyl hydroperoxide may be more generally applicable[76]:

$$\underset{R^2}{\overset{R^3}{>}} CH-N \underset{R^1}{\overset{OH}{<}} \quad + \quad (H_3C)_3C-OOH \longrightarrow$$

11

$$\underset{R^2}{\overset{R^3}{>}}C=N \overset{O}{\underset{R^1}{\nearrow}} \quad + \quad H_2O \quad + \quad (H_3C)_3C-OH$$

12

[71] W. Rundel in *Houben-Weyl*, Methoden der organischen Chemie, Bd. X/4, p. 317, Georg Thieme Verlag, Stuttgart 1968.

[72] G.E. Utzinger, Justus Liebigs Ann. Chem. 566, 63 (1944).

[73] D.J. Cowley, W.A. Waters, J. Chem. Soc. (B) 96 (1970).

[74] W. Rundel in *Houben-Weyl*, Methoden der organischen Chemie, Bd. X/4, p. 319, Georg Thieme Verlag, Stuttgart 1968.

10.2.1.1.3 Dehydrogenation with other oxidizing agents

Aqueous, hydroxide or bicarbonate containing potassium hexacyanoferrate(III) solution[37, 77] likewise dehydrogenates hydroxylamines to nitrones (see Table 1, p. 337). In practise a two-phase procedure is employed and the nitrone formed is taken up in ether or chloroform. Yellow mercury oxide[37, 78], too, for example, in acetone or chloroform can be used as oxidizing agent (see Table 1).

Table 1. Nitrones by dehydrogenation of hydroxylamines with respectively potassium hexacyanoferrate(III) **1** and mercury(II) oxide **2**

Method	R^1	R^2	R^3
1	Ar–CH$_2$	Ar	H
	Ar	Alkyl	H
	Ar	R–CO–CH$_2$–	H
	–(CH$_2$)$_3$–		H 3,4-Dihydro-2H-pyrrole 1-oxide
	–(CH$_2$)$_4$–		H 2,3,4,5-Tetrahydropyridine 1-oxide
2	–(CH$_2$)$_3$–		H 3,4-Dihydro-2H-pyrrole 1-oxide
	–(CH$_2$)$_4$–		H 2,3,4,5-Tetrahydropyridine 1-oxide
	–CH$_2$–O–(CH$_2$)$_2$–		H 5,6-Dihydro-3H-1,4-oxazine 4-oxide
	Ar–CH– $|$ R	Ar	H Ar
	–CH$_2$–CH$_2$–		H 3,4-Dihydroquinoline 1-oxide
	–CH$_2$–CH–NH– $|$ Y		Ar

In individual cases the following oxidizing agents have been used[79]:

Lead(IV) oxide (from lead(IV) acetate)[79]
Copper(II) acetate[79]
Silver nitrate/ammonia[79]
Potassium permanganate/acetone[79]
Potassium dichromate/acetic acid[79]
Sodium periodate[79]
Nitrosobenzene[80]
p-Benzoquinone[80]
Chloranil[81]
2,4,6-Tri-tert-butylphenoxyl
Dimethyl sulfoxide/dicyclohexylcarbodiimide/acid[82]

10.2.1.2 Nitrones from carbonyl compounds and N-monosubstituted hydroxylamines

Condensation of carbonyl compounds with N-monosubstituted hydroxylamines is a rapidly balancing equilibrium reaction:

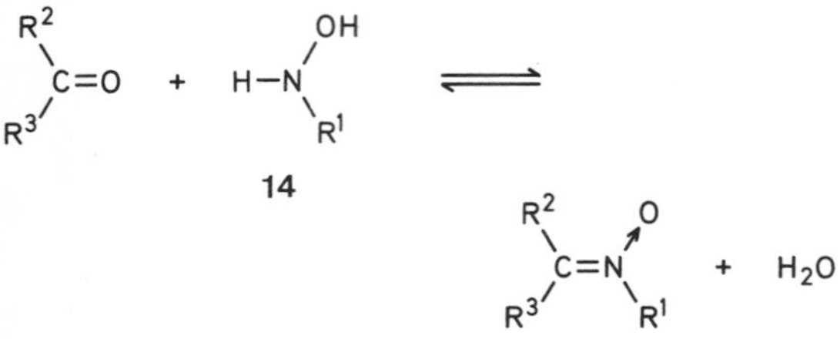

For the reaction between 4-chlorobenzaldehyde and N-methylhydroxylamine in acid solution the equilibrium constant is 1.54[83]. The usefulness of this general method is limited primarily by steric factors; thus, when N-tritylhydroxylamine is employed the temperatures required are so high that the nitrones formed rearrange[58]. Ortho-substituted arylhydroxylamines do not react or give poor yields. 2-Chloro- and 2-nitrobenzaldehydes, too, react poorly, 2-hydroxybenzaldehyde relatively smoothly by comparison[27].

10.2.1.2.1 Nitrones from aldehydes or aldehyde derivatives and N-monosubstituted hydroxylamines[34, 86, 87]

The reaction of aldehydes or aldehyde derivatives and N-monosubstituted hydroxylamines (Equation 3; $R^3 = H$) generally gives the corresponding nitrones smoothly and in good yield. Either room temperature (e.g., water or ethanol as solvent) or elevated temperature, e.g., in benzene solution are employed[84]. In general, the hydroxylamines react more rapidly with carbonyl compounds than anilines[83–85, 630]; this method is thus the synthesis of choice provided the hydroxylamine is not too inaccessible. N-Alkyl- and N-arylhydroxylamines 14 are reacted as salts in the presence of an acid scavenger (bicarbonate, acetate, alkoxide). (The scavenger is unnecessary if the nitrone formed is only weakly basic.)

In general formaldehyde nitrones $(R^2, R^3 = H)$ cannot be isolated and glyoxal dinitrones (cf. p. 336) are obtained in their place:

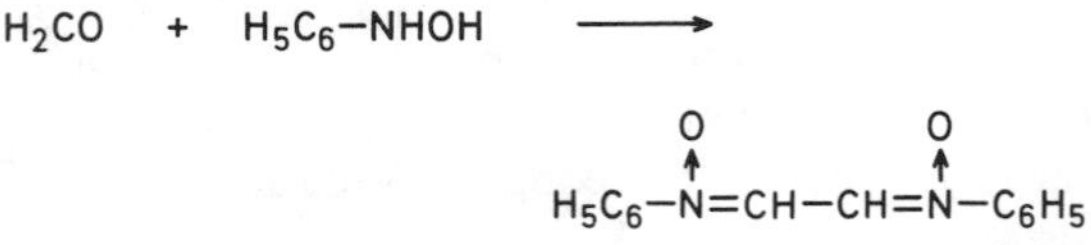

Glyoxal bis(phenylnitrone)

The structure of a formaldehyde nitrone[88] claimed to be monomeric requires to be examined by modern means.

Enolized aldehydes (enols or enolates of β-keto aldehydes) react like simple aldehydes; the problem of tautomerism (oxo nitrone, enol nitrone, or oxo-ene hydroxylamine) has evidently not yet been solved[85, 89].

75 L.H. Sternbach, E. Reeder, J. Org. Chem. 26, 4963 (1962).
76 H.E. DeLaMare, G.M. Coppinger, J. Org. Chem. 28, 1068 (1963).
77 W. Rundel in Houben-Weyl, Methoden der organischen Chemie, Bd. X/4, p. 320, Georg Thieme Verlag, Stuttgart 1968.
78 W. Rundel in Houben-Weyl, Methoden der organischen Chemie, Bd. X/4, p. 322, Georg Thieme Verlag, Stuttgart 1968.
79 W. Rundel in Houben-Weyl, Methoden der organischen Chemie, Bd. X/4, p. 325, Georg Thieme Verlag, Stuttgart 1968.
80 W. Rundel in Houben-Weyl, Methoden der organischen Chemie, Bd. X/4, p. 327, Georg Thieme Verlag, Stuttgart 1968.
81 J.B. Bapat, D.St. C. Black, Australian J. Chem. 21, 2483 (1968).
82 A.H. Fenselau, E.H. Hamamura, J.G. Moffatt, J. Org. Chem. 35, 3546 (1970).
83 J.E. Reimann, W.P. Jencks, J. Amer. Chem. Soc. 88, 3973 (1966).
84 G. Klopman, K. Tsuda, J.B. Louis, R.E. Davis, Tetrahedron 26, 4549 (1970).
85 G.E. Utzinger, F.A. Regenass, Helv. Chim. Acta 37, 1892, 1898 (1954).
86 H.K. Kim, R.E. Bambury, J. Med. Pharm. Chem. 12, 719 (1969);
H.K. Kim, R.E. Bambury, J. Med. Pharm. Chem. 14, 366 (1971).
87 H.K. Kim, H.K. Yaktin, R.E. Bambury, J. Med. Pharm. Chem. 13, 238 (1970);
H.K. Kim, R.E. Bambury, H.K. Yaktin, J. Med. Pharm. Chem. 14, 301 (1971).
88 C. Runti, F. Collino, Ann. Chim. (Rome) 49, 1472 (1959).
89 J. Thesing, A. Müller, G. Michel, Chem. Ber. 88, 1027 (1955).

Imine derivatives (not hydrazones) and bisulfite derivatives of aldehydes react just as smoothly as the free aldehydes:

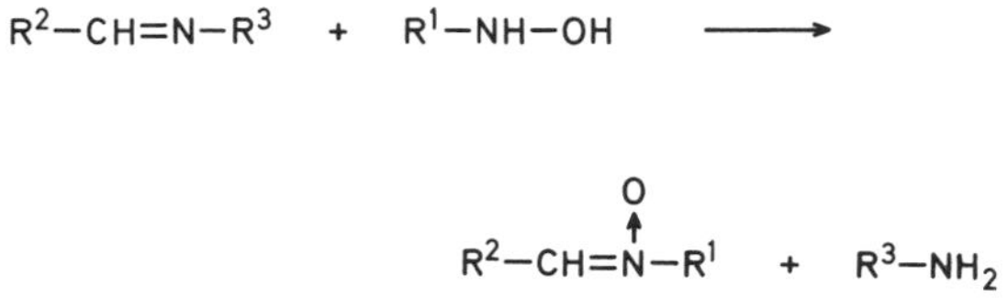

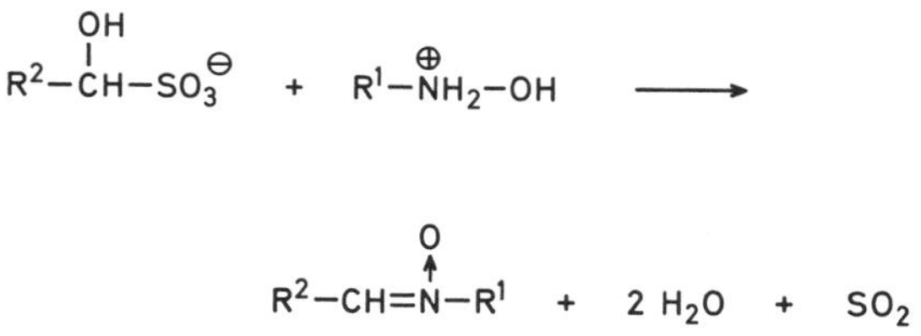

$$R^2{-}CH{=}N{-}R^3 \ + \ R^1{-}NH{-}OH \ \longrightarrow$$

$$\overset{O}{\overset{\uparrow}{R^2{-}CH{=}N}}{-}R^1 \ + \ R^3{-}NH_2$$

$$\underset{\underset{\textstyle OH}{|}}{R^2{-}CH{-}SO_3^{\ominus}} \ + \ R^1{-}\overset{\oplus}{N}H_2{-}OH \ \longrightarrow$$

$$\overset{O}{\overset{\uparrow}{R^2{-}CH{=}N}}{-}R^1 \ + \ 2\,H_2O \ + \ SO_2$$

Dinitrones of malonaldehyde are obtained from its tetraalkyl acetals.

10.2.1.2.2 Nitrones from ketones or ketone derivatives and *N*-monosubstituted hydroxylamines[90]

The effect of steric hindrance, which increases in the sequence formaldehyde < aldehydes < ketones, is so strong during nitrone preparation that keto nitrones are only seldom obtained by several hours' heating in, for instance, boiling ethanol according to Equation **3** (reaction with 2- and 4-hydroxyacetyl aromatic compounds). In practise, the reaction is once more performed with the free hydroxylamine base **14** or its salts with acid scavengers either present or absent:

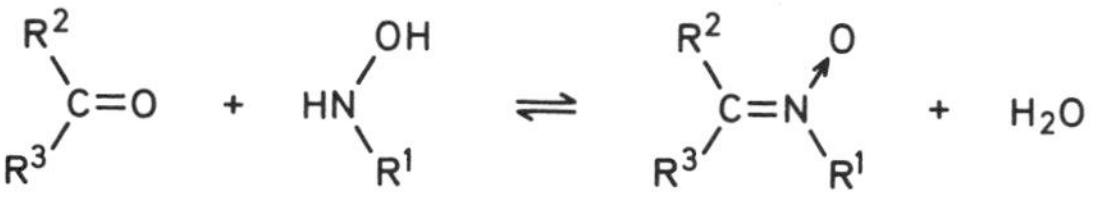

$$\underset{R^3}{\overset{R^2}{}}{>}C{=}O \ + \ HN\underset{R^1}{\overset{OH}{<}} \ \rightleftharpoons \ \underset{R^3}{\overset{R^2}{}}{>}C{=}\overset{\uparrow}{N}\underset{R^1}{\overset{O}{<}} \ + \ H_2O$$

$R^1 = CH_3$, Aryl
$R^2 = $ Alkyl, $-CO{-}NH{-}Ar$
$R^3 = $ Aryl (including with OH substituents), Alkyl, $-CO{-}NH{-}Ar$

α-Diketones react only singly with *N*-methyl-hydroxylammonium chloride[91], while with *N,N'*-dihydroxy-2,3-dimethyl-2,3-butanediamine a cyclic dinitrone is formed, namely a *2,3-dihydropyrazine 1,4-dioxide*[92].

The ketimines **15** $(X{=}H)$[58,91] or their *N*-nitroso derivatives **15** $(X{=}NO)$[93] react better than the free ketones:

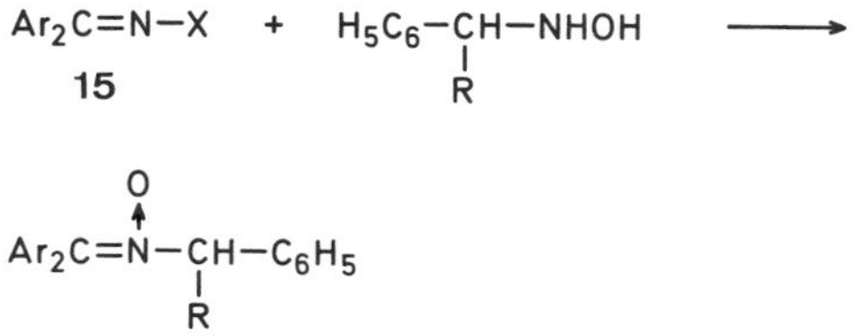

$$\underset{\mathbf{15}}{Ar_2C{=}N{-}X} \ + \ \underset{\underset{\textstyle R}{|}}{H_5C_6{-}CH{-}NHOH} \ \longrightarrow$$

$$\overset{O}{\overset{\uparrow}{Ar_2C{=}N}}{-}\underset{\underset{\textstyle R}{|}}{CH}{-}C_6H_5$$

X = H, NO
R = H, C_6H_5

Ketone acetals, too, can be reacted to nitrone salts with hydroxylamine salts[91]:

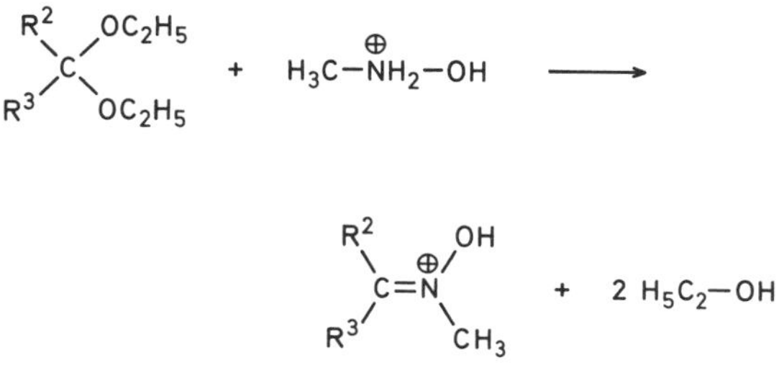

$$\underset{R^3}{\overset{R^2}{}}{>}\underset{\underset{\textstyle OC_2H_5}{|}}{\overset{\overset{\textstyle OC_2H_5}{|}}{C}} \ + \ H_3C{-}\overset{\oplus}{N}H_2{-}OH \ \longrightarrow$$

$$\underset{R^3}{\overset{R^2}{}}{>}C{=}\overset{\oplus}{\underset{CH_3}{N}}{-}OH \ \ + \ 2\,H_5C_2{-}OH$$

10.2.1.2.3 Nitrones *via in situ* generated hydroxylamines (cyclic nitrones from ω-nitrocarbonyl compounds)[94]

Reduction of nitro ketones affords *1-pyrroline 1-oxides*:

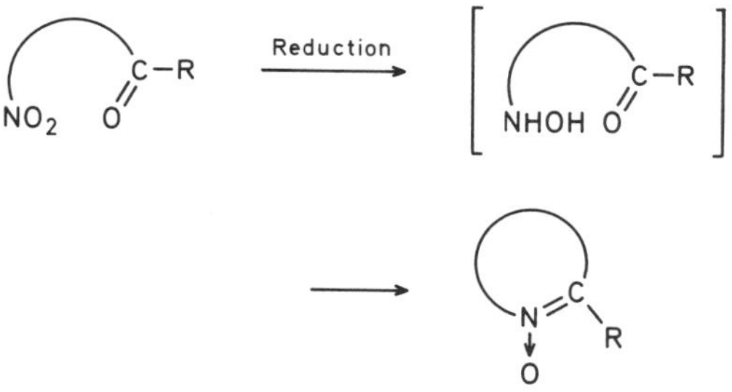

Where a nitro group is in the δ-position, formation of a *2,3,4,5-tetrahydropyridine 1-oxide* is an alternative possibility[41]. In analogous γ- or δ-nitroaldehydes the aldehyde group is suitably pretreated by acetalizing during the reduction and reliberated subsequently by acid hydrolysis. γ- or δ-nitro nitriles afford *amino nitrones* $(R{=}NH_2)$.

[90] *W. Rundel* in *Houben-Weyl*, Methoden der organischen Chemie Bd. X/4, p. 341, Georg Thieme Verlag, Stuttgart 1968.
[91] *W. Rundel* in *Houben-Weyl*, Methoden der organischen Chemie Bd. X/4, p. 343, Georg Thieme Verlag, Stuttgart 1968.
[92] *M. Lamchen, T.W. Mittag*, J. Chem. Soc. (C) 2300 (1966).
[93] *D.G. Morris*, Chem. Commun. 221 (1971).

[94] *W. Rundel* in *Houben-Weyl*, Methoden der organischen Chemie Bd X/4, p. 345, Georg Thieme Verlag, Stuttgart 1968.
[95] *R.F.C. Brown, V.M. Clark, A. Todd*, J. Chem. Soc. (London) 2105 (1959).
[96] *E.J. Grubbs, R.J. Milligan, M.H. Goodrow*, J. Org. Chem. *36*, 1780 (1971).

The yields are often good. Zinc/ammonium chloride in cold aqueous or alcoholic solution is the most frequently used reducing agent. The hotter the solution is, the more entirely reduced products without nitrone structure are formed[81]; the quality of the zinc also plays a part[81]. These difficulties have led to the use of alternative reducing agents:

iron in weakly acid solution[81], sodium tetrahydroborate together with palladium/carbon[81], Raney nickel/hydrazine[94], and catalytic hydrogenation[81, 94].

The technique can be used also for preparing *acyclic* nitrones by reducing a nitro compound in the presence of an aldehyde.

Zinc reduction of a γ-nitrocarboxylic acid ester leads to cyclic hydroxamic acid, a 1-hydroxy-2-pyrrolidone, which can be reduced to *1-pyrroline 1-oxide* with lithium tetrahydroaluminate[94].

An alternative type of generation of the required hydroxylamine *in situ* is based on the following isomerization[95]:

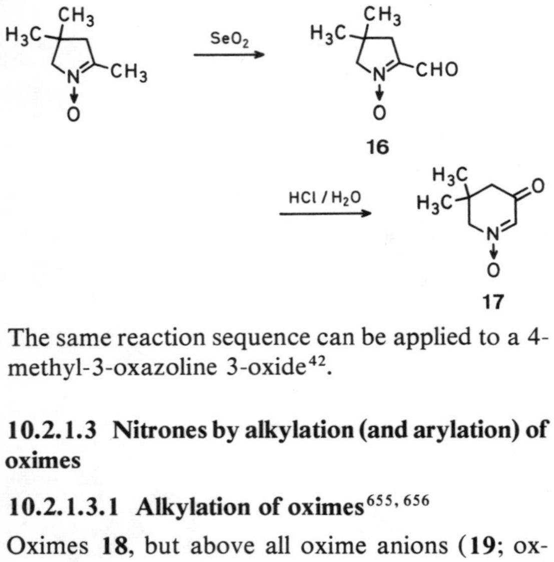

16

17

The same reaction sequence can be applied to a 4-methyl-3-oxazoline 3-oxide[42].

10.2.1.3 Nitrones by alkylation (and arylation) of oximes

10.2.1.3.1 Alkylation of oximes[655, 656]

Oximes 18, but above all oxime anions (19; oximates), can react ambidently toward nucleophiles; as a result, generally two products, oxime *O*-ethers 20 and nitrones 21 form:

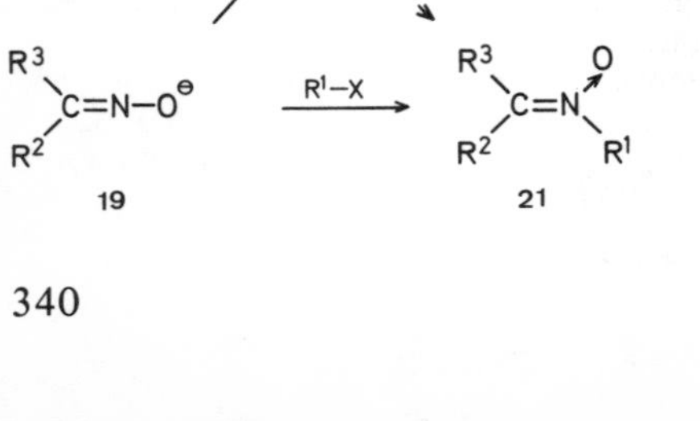

R^1–X =
Benzyl halides
Benzhydryl halides
Allyl halides
α-Halocarboxylic acid derivatives[655, 656, 102, 105, 657, 658]

α-Halo oximes[659]
Alkyl halides
Dialkyl sulfates
Chlorodimethyl sulfide[82]
R^1–X ≠ $(C_6H_5)_3CCl$[25]

Literature errors relating to uniformity, yield ratios, or structure of oxime alkylation products are probably fairly frequent occurrences and are not excluded even in the most recent literature[25].

Admittedly, the nitrones 21 can often be separated readily from the oxime *O*-ethers due to their greater polarity and tendency to crystallize, or by salt formation with acids (for spectroscopic *UV*-differentiation see ref. [96]). The yield ratio depends both on the reaction pathway and the reaction conditions[97]:

1 The free, neutral oximes afford preferentially nitrones during slow reaction with dimethyl sulfate (or methyl bromide)
2 Sterically demanding radicals R^1 in the alkylating agent R^1–X favor oxime *O*-ether formation.
3 Sodium salts of ketoximes afford preferentially oxime *O*-esthers.
4 Sodium salts of the *E*-aldoximes *(syn, α)* afford almost exclusively oxime ethers
5 Sodium salts of the z-aldoximes *(anti, β)* form oxime *O*-ethers and nitrones side by side
6 Silver oximates yield practically only oxime *O*-ethers
7 An electron-withdrawing substituent Y in

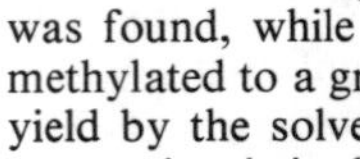

favors oxime *O*-ether formation[98].

In part these findings can be explained by steric or electronic effects (partly adducing the general behavior of ambident nucleophiles[99]). During concentration studies[100, 101] relating to the methylation of a sodium oximate a more rapid *O*-methylation of the free oximate was found, while the unseparated ion pairs were *N*-methylated to a greater extent; a control of the nitrone yield by the solvent may be anticipated[624, 99]. In this connection it is doubtful whether the reaction conditions employed hitherto are already the optimum ones. Suppression of the dissociation of the ion pair by adding an extraneous, highly dissociated sodium salt (sodium tetraphenylborate) increases the yield of nitrone[101]. The nature (hard or soft) of the group X to

97 E. *Bühler*, J. Org. Chem. *32*, 261 (1967).
98 P.A.S. *Smith*, J.E. *Robertson*, J. Amer. Chem. Soc. *84*, 1197 (1962).
99 R. *Gompper*, Angew. Chem. *76*, 412 (1964).
100 St.G. *Smith*, D.V. *Milligan*, J. Amer. Chem. Soc. *90*, 2393 (1968).

be displaced by the oxime anion, too, influences the product ratio. Methyl 1,4-methylbenzenesulfonate ($X=CH_3C_6H_5SO_3$, *hard*) affords less nitrone and more oxime *O*-ether than methyl iodide *(X=I soft)*[101].

During intramolecular oxime alkylation[102] the geometrical isomer of the oxime used determines which product is formed. Where the halogen atom to be displaced is located *trans* with respect to the oxime hydroxyl the relevant cyclic nitrone is formed, otherwise the cyclic oxime *O*-ether is the product[75, 103, 104]. In place of a halogen atom a trimethylammonium group can be displaced[103]. Probably, the rearrangement[105–107] of 2-chlorome-thylquinazoline 3-oxide brought about by hard and not too large nucleophiles (primary aliphatic amines, hydroxide ion, methoxide ion, hydrazine), too, is probably based on an intramolecular oxime alkylation.

Simple oximes either do not react at all with diazomethane[108, 660] or they afford predominantly oxime *O*-ethers. By contrast, the corresponding nitrones 23 are formed with monooximes 22 of 1,2-dicarbonyl compounds, often in excellent yield:

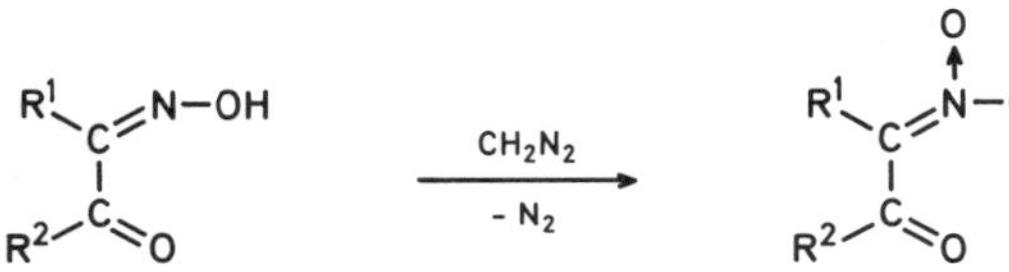

To bring the reaction about, the solution or suspension of a monooxime 21 in ether is treated with an ether solution of diazomethane. If necessary the reaction (evolution of nitrogen) is speeded up by adding methanol. Diazoethane reacts analogously[108].

Aromatic oximes (R^1, R^2=aryl) 24 form *N-(methylthio)methyl nitrones* 25[82, 661] plus a little *O*-(methylthio)methyl oxime 26 and the correspond-

ing benzanilide 27 (by Beckmann rearrangement) under Moffatt conditions, *i.e.*, with dimethyl sulfoxide (DMSO), dicyclohexylcarbodiimide (DCC), and phosphoric acid or trifluoroacetic acid:

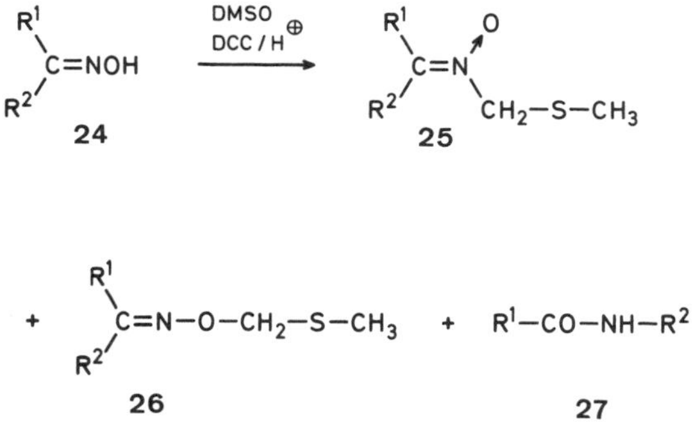

R^1 = Aryl, (Alkyl), R^2 = Aryl

On prolonged reaction the Moffatt reagents effect an (essentially) intramolecular rearrangement of the nitrone 25 to the oxime ether 26. Good nitrone yields are obtained with excess dicyclohexylcarbodiimide and nonpolar diluents (benzene, ether). Electron-donating substitution (4-methoxy) in R^1 or R^2 promotes formation of 27 at the expense of nitrone formation.

Under Moffatt conditions aliphatic ketoximes generally suffer only small conversions; Beckmann rearrangements occur preferentially and complex mixtures result. *E*- and *Z*-benzaldehyde oximes afford the corresponding nitrone in yields of only up to 39% and give mainly arylnitriles.

Z-Benzaldoxime 28 forms the nitrones 30 with acrylic acid derivatives 29 by Michael addition[109]

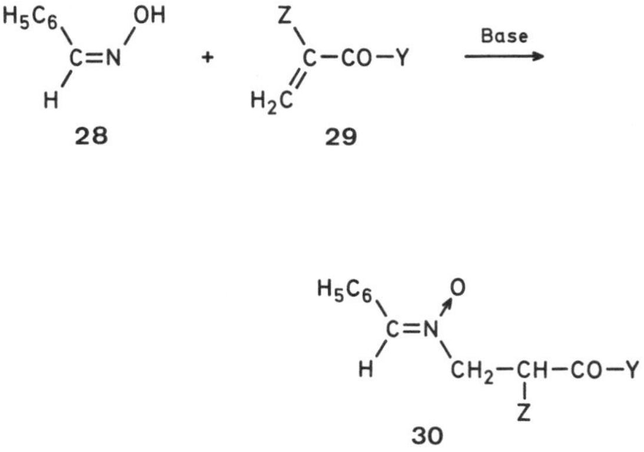

The addition ensues on letting the methanolic or ethanolic solution stand (5 days to 4 weeks) in the presence of triethylamine. Methyl α-chloroacrylate and *N,N*-diethyl-α-phenylacrylamide are

[101] *S.G. Smith, M.P. Hanson*, J. Org. Chem. *14*, 1931 (1971).
[102] *W. Rundel* in *Houben-Weyl*, Methoden der organischen Chemie, Bd. X/4, p. 410, Georg Thieme Verlag, Stuttgart 1968.
[103] *E.E. Garcia, J.G. Riley, R. Fryer*, J. Org. Chem. *33*, 1360 (1968).
[104] *H.A. Lutz*, J. Pharm. Sci. *58*, 1460 (1969).
[105] *A. Stempel, I. Douvan, L.H. Sternbach*, J. Org. Chem. *33*, 2963 (1968).
[106] *M.E. Derieg, R.I. Fryer, L.H. Sternbach*, J. Chem. Soc. (C) 1103 (1968).
[107] *K.-H. Wünsch, I. Krumpholz, J. Perez-Zayas, R. Tapanes-Peraza, G. Schulze*, Z. Chemie *10*, 113 (1970).

reacted in 1,4-dioxane in the presence of sodium ethoxide. From the *E*-oxime the oxime *O*-ether is formed.

Table 2. Nitrones **28** formed from benzaldehyde oxime **30** by Michael addition

Z	Y	Benzaldehyde	Yield [%]
H	OCH_3	*(2-methoxycarbonyl)ethyl nitrone*	84
C_6H_5	OC_2H_5	*(2-ethoxycarbonyl)phenethyl nitrone*	72
$NHCOCH_3$	OCH_3	*2-acetamino-2-methoxy-carbonylethyl nitrone*	67
Cl	OCH_3	*2-chloro-2-methoxy-carbonylethyl nitrone*	43
H	$N(C_2H_5)_2$	*2-(diethylaminoethyl) nitrone*	17
C_6H_5	$N(C_2H_5)_2$	*2-(diethylaminophenethyl) nitrone*	58

10.2.1.3.2 Cyclizing alkylation with carbonyl compounds [111, 112, 625–628]

The *E*-isomers **31** and **33** of α-amino and α-amino oximes *(X=H)* react with aldehydes and acetone to give a good yield of the corresponding *3-imidazoline 1-oxides* **32** *(X=H)* and *1,2-dihydroquinazoline 3-oxides* **34**

The *3-imidazolin-3-ol 1-oxides* **32** *(X=OH)* are formed analogously from the *E*-isomers **31** of hydroxyamino oximes *(X=OH)*. Aldehydes *(R⁴=H)* are reacted in ethanol at room or elevated temperature, acetone either at 140–150° in a bomb tube (with α-amino oximes **31**) or at the boiling point with addition of a little acetic acid (α-amino oxime **33**). The *Z*-α-amino oximes afford *1,2,5-oxadiazines*. *Z*-(2-aminophenyl) ketoximes can be converted into the reactive *E*-isomers **33** during the reaction by adding copper(II) sulfate and subsequently reacting to *1,2-dihydroquinazoline 3-oxide* **34** [110].

The action of alcoholic hydrochloric acid converts the 3-imidazolin-3-ol 1-oxides **32** *(X=OH; R)* into the *2H-imidazole 1-oxides* **35** [111]. By contrast, from the 3-imidazolin-3-ol acetate 1-oxides the isomeric *N*-oxides **36** are formed with acid [111]. From **31** *(X=H)* 3-imidazolin-2-one and 3-imidazolin-2-thione 1-oxides **32** *(X=H; R⁴=R⁵=O, S)* are obtained with phosgene and thiophosgene respectively [112]. For recent work see ref. [634].

[108] B. Eistert, H.-K. Witzmann, Justus Liebigs Ann. Chem. **744**, 105 (1971).

[109] E. Bellasio, F. Parravincini, A. Vigevani, E. Testa, Gazz. Chim. Ital. **98**, 1014 (1968).

During acid cleavage of the acetal **37** in glyme solution *5,6-dihydro-3,6-diphenyl-2,5-pyrazine-diol 1,4-dioxide* **39** is formed *via* phenylglyoxal 2-oxime **38**[113]. In dimethyl sulfoxide **39** decomposes slowly into the oxime **35**:

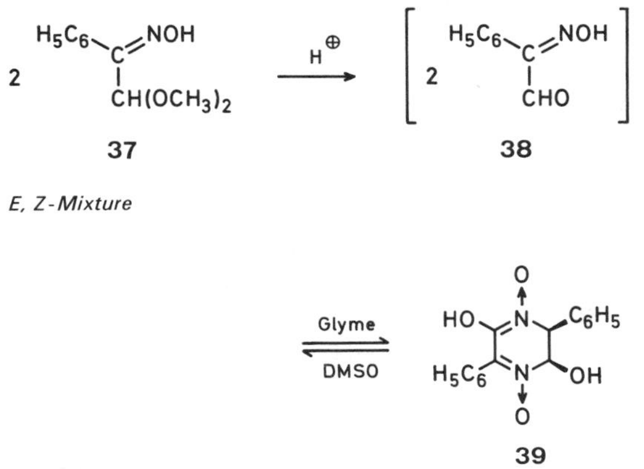

isomerization *via* the oximate ion formed during the reaction):

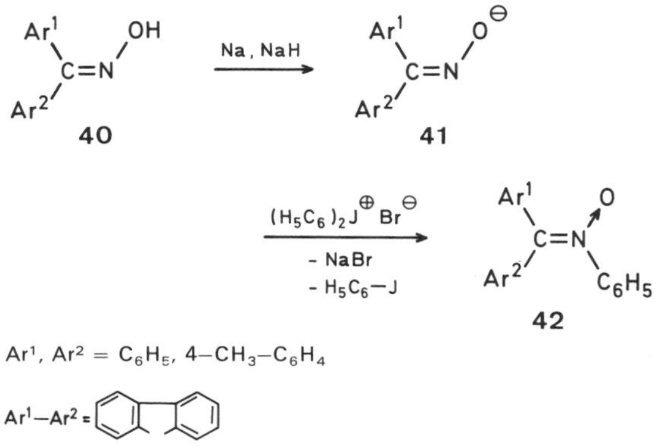

$Ar^1, Ar^2 = C_6H_5, 4-CH_3-C_6H_4$

$Ar^1-Ar^2 = $

Arylation of oximes with haloaromatic compounds activated by strongly electron-attracting substituents affords almost exclusively oxime *O*-ethers[96, 114].

E, Z-Mixture

10.2.1.4 Nitrones from acid methyl or methylene compounds and aromatic nitroso compounds

As a carbonyl-analogous group, the nitroso group undergoes aldollike condensations with carb-anions. The intermediate hydroxylamine derivative **45** is partly dehydrogenated to the nitrone **46** under the reaction conditions and partly (sometimes predominantly) dehydrated to the imine **47**. In general, the nitroso compound used **44** serves as hydrogen acceptor and is reduced to *N*-arylhydroxylamine in the process, which, in turn, condenses with **44** to the azoxy compound. (Re the mechanism of the dehydrogenation, see refs. [115, 116].)

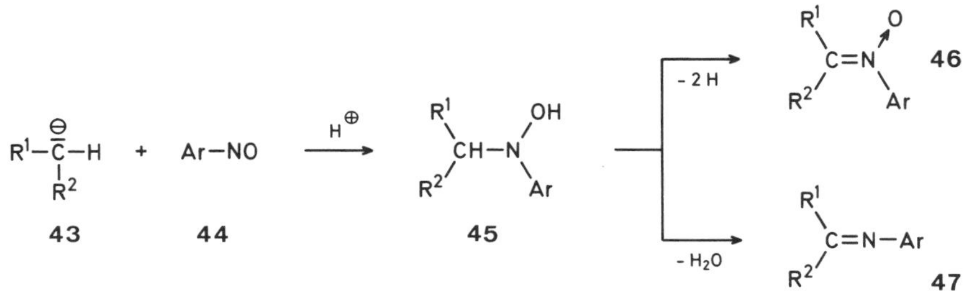

10.2.1.3.3 Arylation of oximes

The sodium oximate **41** (from **40** with sodium in tetrahydrofuran or with sodium hydride in dimethylformamide) reacts with diphenyliodonium bromide at room temperature in tetrahydrofuran/dimethyl sulfoxide or in dimethylformamide to a mixture of *N*-phenyl nitrone **42** (8–29%) and the isomeric *O*-phenyl oxime (44–79%)[96]. The configuration (*E,Z*) of the oxime ($Ar^1 \neq Ar^2$) remains intact in the nitrone. A mixture of the *E*- and *Z*-nitrones is obtained only on reacting in tetrahydrofuran-dimethyl sulfoxide (probably by

[110] *F. Field, W.J. Zally, L.H. Sternbach,* J. Org. Chem. **30**, 3957 (1965).

[111] *L.B. Volodarskii, A.N. Lysak, V.A. Koptyut,* Khim. Geterosikl. Soedin. 766 (1966); C.A. **66**, 115644t (1967).

[112] *H. Gnichtel,* Angew. Chem. **83**, 904 (1971).

[113] *T.P. Karpetsky, E.H. White,* J. Org. Chem. **37**, 339 (1972).

[114] *C.J. Cattanach, R.G. Rees,* J. Chem. Soc. (C) 53 (1971).

[115] *G.T. Knight, M.J.R. Loadman,* J. Chem. Soc. (B) 2107 (1971).

[116] *I. Tanasescu, I. Nanu,* Ber. **75**, 650 (1942).

[117] *W. Rundel* in *Houben-Weyl,* Methoden der organischen Chemie Bd. X/4, p. 364, Georg Thieme Verlag, Stuttgart 1968.

In order to obtain *nitrones* preferentially it is necessary to work with a corresponding excess of nitroso compound or in the presence of other hydrogen acceptors (*e.g.*, methylene blue). Small quantities of a base are required for the catalysis (carbonate, piperidine, sometimes also hydroxide or alkoxide ions).

In general, the reaction is carried out in alcoholic solution by allowing to stand at room temperature or brief heating. Overlong *heating* reduces the yield of nitrone and leads to imines and carboxamides instead. Separating the reaction products has presented difficulties; modern chromatographic techniques are claimed to be especially effective and valuable. In older work the nitrone was possibly misinterpreted or missed as reaction product. Where electron-attracting substituents (**43**, R and R^2) stabilize the intermediate carbanion the reaction is called an *Ehrlich-Sachs reaction*[117, 118] in the narrow sense.

With $R^1 = H$ the reaction succeeds only if R^2 is at the same time an electron-attracting aryl group (π-deficiency aromatic compound).

In some cases the reaction can be catalyzed with *acid* as well, namely, where a small aromatization energy permits tautomerism to the analogous phenyl enamine **48**, *e.g.*:

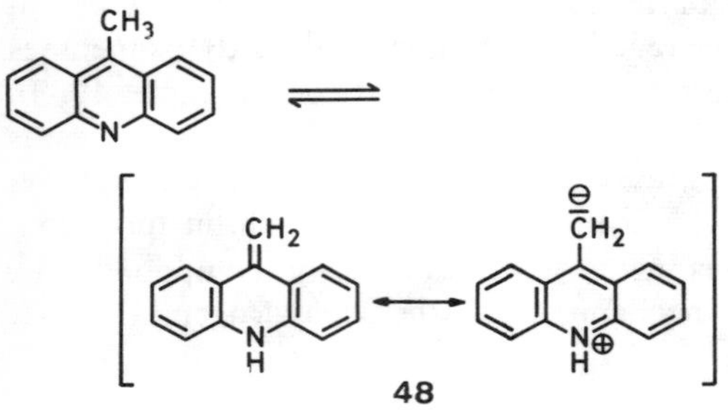

Catalysis by *UV* irradiation, too, has been described (in 9-methylacridine)[119].

Preparing *nitrones* from nitroso compounds and *carbanions* requires no dehydrogenation of the intermediate hydroxylamine compound if the substituent X introduced by the carbanion can be eliminated together with the pair of bonding electrons:

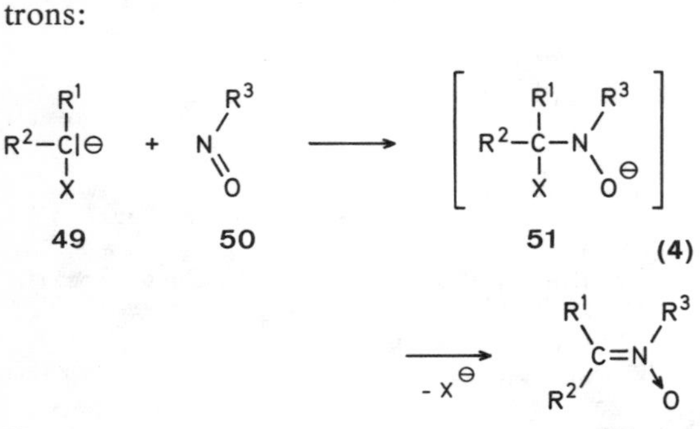

R^3 = Aryl, tert-Alkyl

Only aromatic nitroso compounds (monomeric and dimeric) and monomeric nitrosoalkanes (dimeric ones are too sluggish) can be used as the nitroso compounds. In place of the carbanion **49** a similarly structured free radical can be reacted to a nitrone by elimination of the substituent radical X[120].

Preparing *nitrones* from aromatic nitroso compounds and benzyl halides[121] is limited primarily by the slight acidity of the benzyl halides.

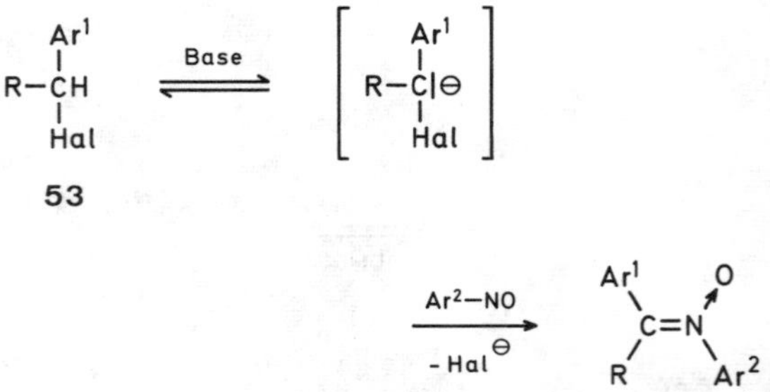

Relatively strong bases are therefore required for the reaction (hydroxide, alkoxide ions); in addition, the benzyl hydrogen must be activated sufficiently by stabilization of the carbanion. Suitably the condensation should be performed rapidly and at as low a temperature as possible in order to suppress the base-catalyzed rearrangement of the nitrone into the isomeric anilide. Alcohols are used as solvent. Preparing nitrones from aromatic nitroso and nitro compounds and α-substituted arylacetonitriles[37, 122, 123] proceeds in accordance with Equation:

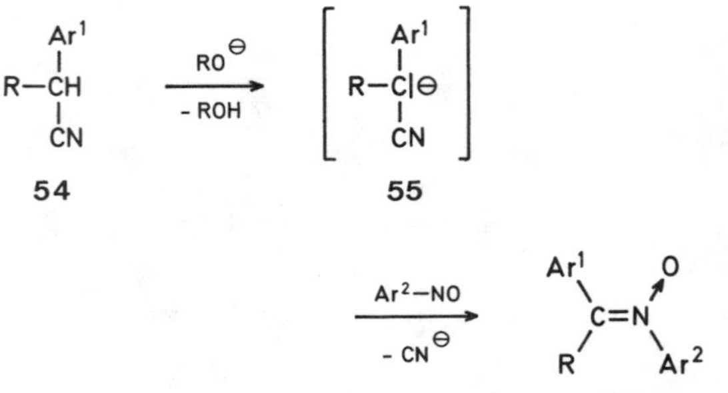

R = Aryl, —NH—Aryl, Benzyl

[118] *H. Krauch, W. Kunz*, Reaktionen der organischen Chemie, 3. Aufl., p. 193, Dr. Alfred Hüthig Verlag, Heidelberg 1966.

[119] *W. Rundel* in *Houben-Weyl*, Methoden der organischen Chemie Bd. X/4, p. 365, Georg Thieme Verlag, Stuttgart 1968.

[120] *T. Hokosogai, N. Inamoto, R. Okazaki*, J. Chem. Soc. (C) 3399 (1971).

Evidently α-substitution of the arylacetonitriles **54** is necessary in order to avoid dehydration of the aldollike adduct with the nitroso compound to imine. From 2-aryl-3-phenylpropanonitrile (**54**; $R = -CH_2-C_6H_5$) *benzil dinitrone* is obtained with excess nitrosobenzene. The reactions are conducted in alcoholic solution and afford very good yields.

Reacting the nitrile **54** with nitrosobenzene, 2-chloronitrosobenzene, and 2-bromonitrosobenzene (**57**) in a 2:1 molar ratio can likewise lead to *nitrones*[123] (*cf.* p. 375):

R = Ar¹ = C₆H₅
X = H, Cl, Br

57   **58**   **59**

85 – 100 %

The reaction proceeds with concentrated methanolic potassium hydroxide or sodium methoxide solution at 30–60°.

A synthetic pathway preferred for preparing the *N*-aryl nitrones **62** (*C*-aryl, *C*-heteroaryl, *C*-acyl nitrones) starts from the pyridinium salts **60** (*cf.* Equation **4**, p. 344),
X = pyridinium, X⊖ = pyridine; Kröhnke reaction: *cf.* Table p. 346)[124]:

R = Ar¹ = C₆H₅
X = H, Cl, Br

60   **61**   **62**

Base
– HY

Ar–NO
– Pyridine

In general the ylides **61** are liberated *in situ* from the corresponding *CH*-acid pyridinium salts **60** with bases. One factor determining the reaction conditions is the acidity of the pyridinium salts **60** and the stability of the ylides **61**.

Hydroxide ion, carbonate, or hydrogen carbonate in aqueous alcohol, aqueous alkali, and pyridine or alkoxides or piperidine in alcohol, or piperidine in acetone-pyridine are used as *bases*. In some instances the pyridine salts **60** are acid enough to react without addition of base. Only the eliminated pyridine is then available for binding the liberated hydrogen halide *HY*.

In other cases the preformed ylide **61** is used directly and without addition of base.

Low temperatures (≤20°) are advisable for the reaction in order to avoid rearrangements of the product taking place. The end of the reaction is indicated by the change in color. A test for the reactivity of the pyrimidinium compound **60** rests on the color change with *N,N*-dimethyl-4-nitrosoaniline/acetone in aqueous alcoholic medium from green to red or brown.

The Kröhnke technique generally affords the *nitrones* in excellent yield because side-reactions are substantially suppressed. It is particularly suitable for preparing *C-aryl-N-aryl nitrones* (starting from **60**, $R^1 = acyl$) and can serve also for synthesizing *di-* and *trinitrones* from corresponding di- and trifunctional starting materials.

The *C-acyl nitrones* obtained are stable for only days or weeks and are readily cleaved to α-keto aldehydes by acids.

The *pyridinium salts* needed are obtained either from pyridine and the corresponding arylmethyl-, heteroarylmethyl-, and phenacyl halides or by the Ortoleva-King method from active methyl compounds (heteroaromatic *C*-methyl compounds, acetophenone derivatives) with iodine and pyridine.

In the ylide precursor **60** the pyridinium group can be replaced by homologous and analogous groups (methylpyridinium, quinolinium, *etc.*) without advantage, and in individual cases by a trialkylammonium group.

[121] *W. Rundel* in *Houben-Weyl*, Methoden der organischen Chemie Bd. X/4, p. 372, Georg Thieme Verlag, Stuttgart 1968.
[122] *W. Rundel* in *Houben-Weyl*, Methoden der organischen Chemie Bd. X/4, p. 374, Georg Thieme Verlag, Stuttgart 1968.
[123] *M. Jawdosiuk, B. Ostrowska*, Chem. Commun. 548 (1971).
[124] *W. Rundel* in *Houben-Weyl*, Methoden der organischen Chemie Bd. X/4, p. 375, Georg Thieme Verlag, Stuttgart 1968.
[125] *H. Rembges, F. Kröhnke, I. Vogt*, Chem. Ber. *103*, 3427 (1970).
[126] *W. Rundel* in *Houben-Weyl*, Methoden der organischen Chemie Bd. X/4, p. 404, Georg Thieme Verlag, Stuttgart 1968.
[127] *A.W. Johnson, R.T. Amel*, J. Org. Chem. *34*, 1240 (1969).
[128] *B.H. Freeman, D. Lloyd, M.I.C. Singer*, Tetrahedron *28*, 343 (1972).

Table 3. Nitrones which can be prepared by the Kröhnke method

$$\left[\begin{array}{c} R^1 \\ \overset{|}{C}H - N^{\oplus} \\ R^2 \end{array} \right] X^{\ominus}$$

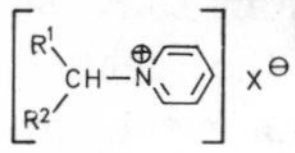	R^1	R^2	Yield [%]
Phenacyl type	—CO—Ar (Ar = also heteroaryl)	H, C_6H_5	44–100 (sometimes not stated)
	—COOC₂H₅	2-Pyridyl	41
	—CO—N=	H, C_6H_5, CH_3	67–100
	—CS—N=	H	47
	—CS—SCH₃	H	66–73
	—CN	H	85
	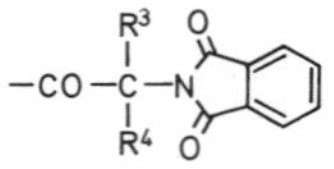 (structure with R^3, R^4, phthalimide) —CO—C—N R^3, R^4 = H, Alkyl	H	50–98
Benzyl type	Aryl (if at all possible without an amino, hydroxy, or alkoxy group)	H, Aryl, —CN —CO—C_6H_5 —COOC₂H₅	60–100
	Tropolonyl	H	good
	Heteroaryl [108, 59, 60]	H	37–100 mainly good
	Heteroaryl with 'onium' group	H	70–100
Allyl type	—CHCH—R	H	not stated

If R^1 or R^2 in the ylide precursor stage **60** is a heteroaromatic compound with onium structure, as, for instance, in the 1-methyl-2-(pyridino-methyl)quinolinium salt **63**, then the Kröhnke reaction may fail to take place because charge equalization stabilizes the ylide **61** (the aromatization energy of the heteroaromatic compound to be furnished by the charge equalization is evidently a codetermining factor):

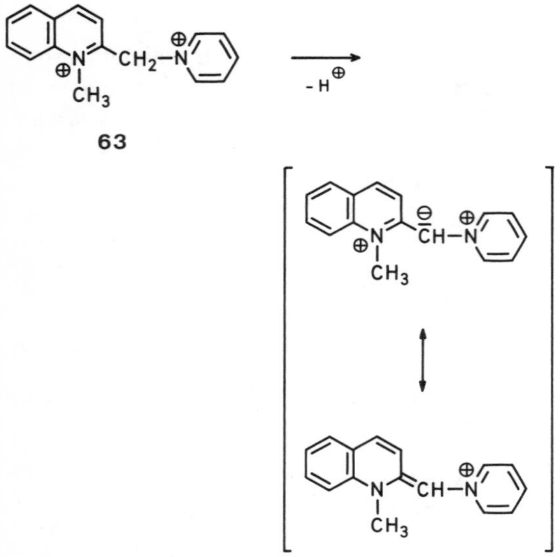

63

N-Arylhydroxylamines can be used for the Kröhnke reaction if the hydroxylamine is dehydrogenated under the reaction conditions, for instance, by the nitro group of a nitro-substituted pyridinium ion. Because of sequential reactions *N*-phenylhydroxylamine and *N*-(4-nitrobenzyl)pyridinium salt afford a mixture of *C*-(4-azoxyphenyl) nitrones [125].

Sulfonium ylides and aromatic nitroso compounds together form nitrones **62** in boiling dichloromethane or ether [126, 127]:

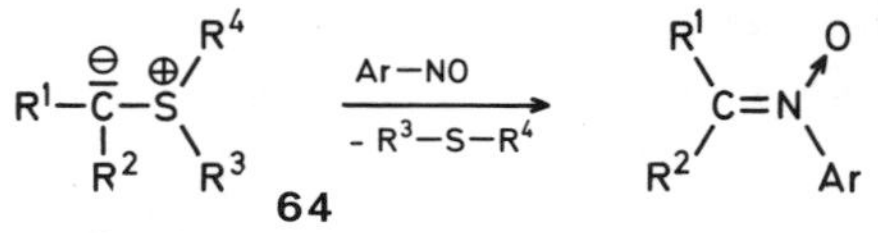

64

R^1, R^2 = Aryl
R^1 = H, R^2 = CO—C_6H_5

In the few investigated cases the yields vary from good to very good. The range of application is probably limited by the availability of the ylides. The corresponding reaction with phosphonium ylides gives the corresponding imines in place of the nitrones.

Analogously, sulfonium, selenonium, arsonium, and stibonium ylides afford nitrones with nitrosobenzene in boiling benzene after some hours (selenonium and stibonium ylides afford the best yields, 80% [128, 129]):

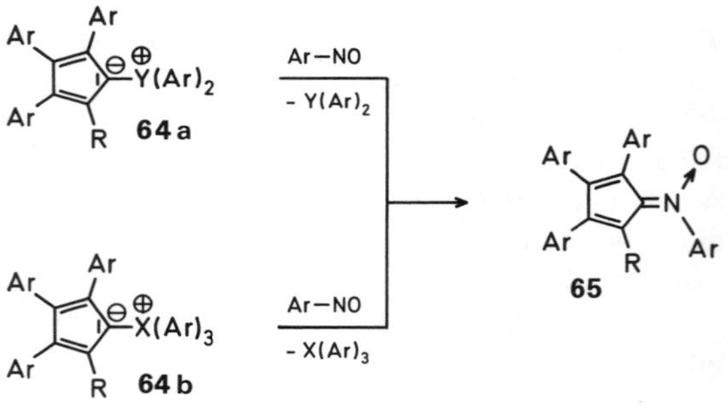

64a

64b

65

Y = S, Se
X = As, Sb
Ar = C_6H_5
R = H, C_6H_5

Reacting diazoalkanes with nitroso compounds, too, gives nitrones [662, 130]:

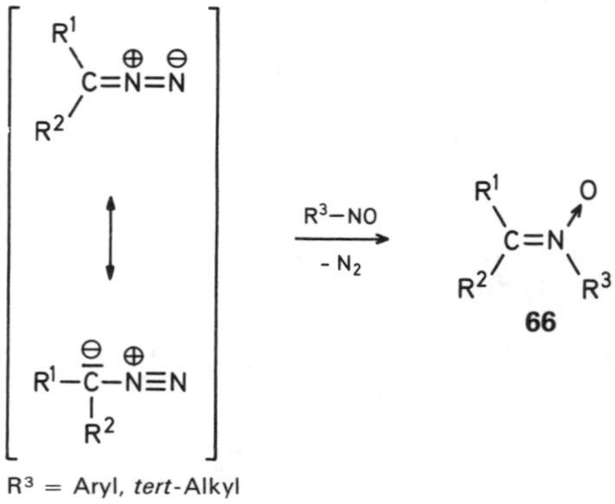

66

R^3 = Aryl, *tert*-Alkyl

If the nitrosoalkanes are bulky enough even monomeric formaldehyde nitrones (**66**, $R^1 = R^2 = H$) can be obtained[29, 130].

The mixture of the reactants — generally in ether — is allowed to stand until evolution of nitrogen ceases. Direct titration of the nitroso compounds with diazomethane is possible[130], α-carbonyldiazo compounds react more slowly or not at all. Otherwise the yields are generally good to very good.

Synthesis of nitrones from nitrosobenzene and phenylhydrazones of aromatic aldehydes or ketones[131] has not yet been studied in detail in respect of mechanism and scope.

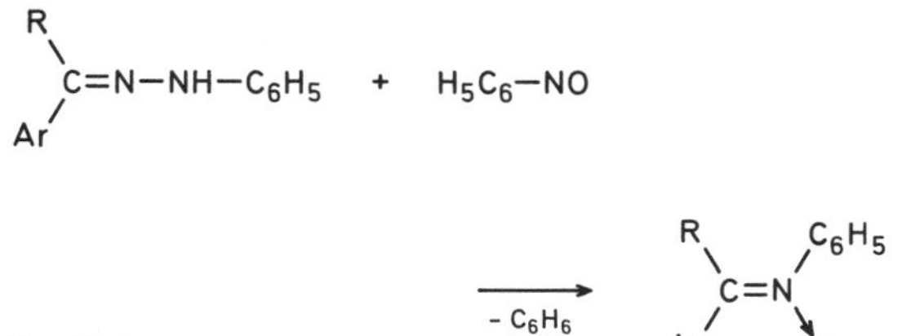

$$R = H, Ar$$

The reactants are admixed at room temperature with or without solvent, best of all under nitrogen in order to prevent air oxidation of the phenylhydrazone.

10.2.1.5 Addition of aromatic nitroso compounds to multiple bonds[132]

A number of interesting synthetic methods, admittedly of apparently very restricted procedural value, are formulated in Equations 5—7:

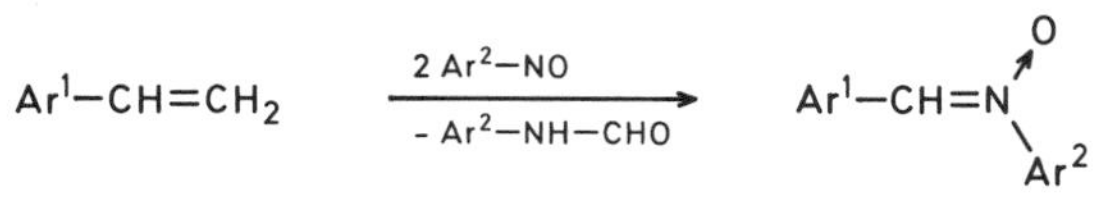

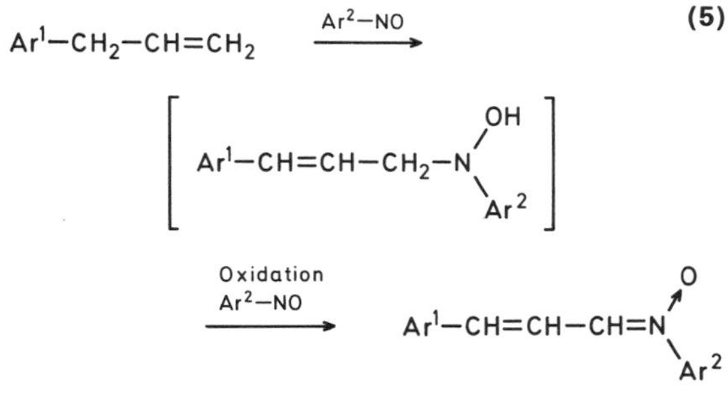

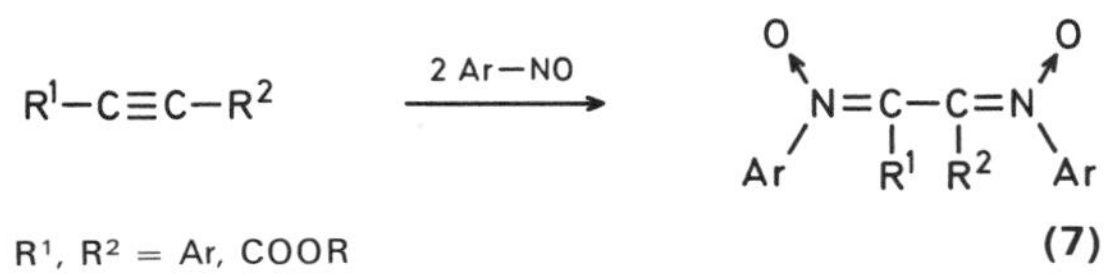

They furnish moderate yields and in part (Equation 7) require reaction times of weeks or months. For the mechanism of Reaction 6, see refs.[115, 133].

From quinones *cyclic dinitrones* or *hydroxy nitrones* are obtained by a similar reaction, sometimes in better yield[132].

Reaction **6** appears to be of more general nature, but the nitrone formed can be isolated only if it is sufficiently stabilized[134] (*cf.* ref.[135]). For example, a small yield of *1,4-benzoquinone 2-methyl-3-penten-1-yl nitrone* is obtained from 4-nitrosophenol and 2-methyl-2-pentene[134]:

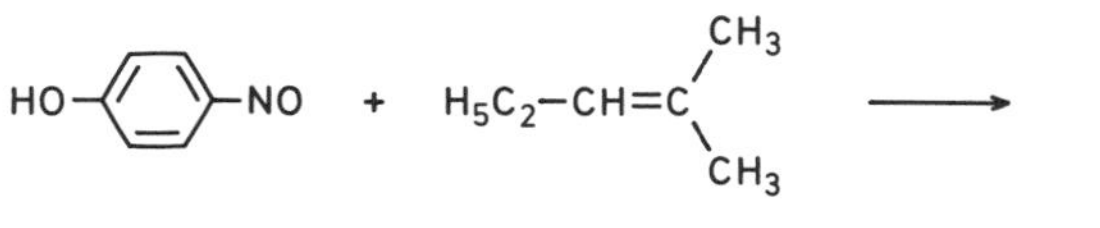

The benzofuroxan, which is in equilibrium with 2-dinitrosobenzene, adds to *N,N*-dimethylisobutylamine to form *3,3-dimethyl-2-dimethylamino-2,3-dihydroquinoxaline 1,4-dioxide*[136] (*cf.* p. 378):

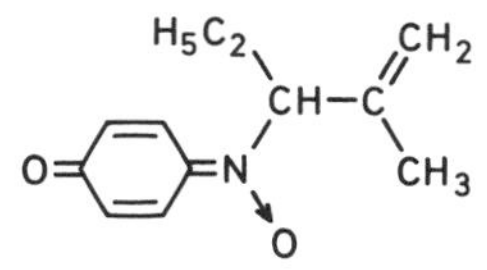

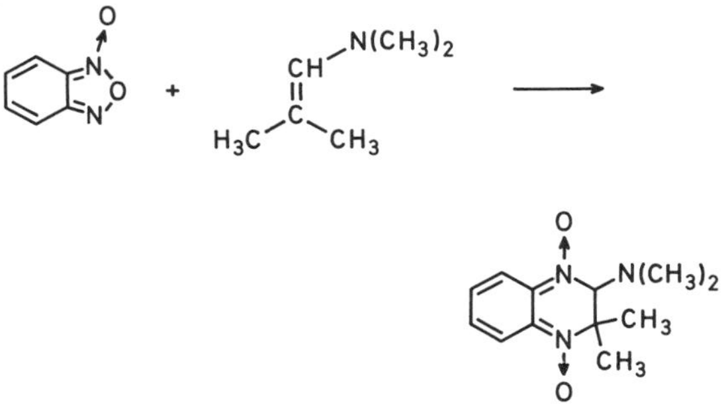

[129] *D. Lloyd, M.I.C. Singer*, Tetrahedron *28*, 353 (1972).

[130] *J.E. Baldwin, A.K. Qureshi, B. Sklarz*, Chem. Commun. 373 (1968);
J.E. Baldwin, A.K. Qureshi, B. Sklarz, J. Chem. Soc. (C) 1073 (1969).

[131] *D.W. Berry, R.W. Bryant, J.K. Smith, R.G. Landdt*, J. Org. Chem. *35*, 845 (1970).

[132] *W. Rundel* in *Houben-Weyl*, Methoden der organischen Chemie Bd. X/4, p. 405, Georg Thieme Verlag, Stuttgart 1968.

[133] *G.T. Knight*, Chem. Commun. 1016 (1970).

[134] *G.T. Knight, B. Pepper*, Chem. Commun. 1507 (1971).

[135] *P. Minisci, R. Galli, A. Quilico*, Tetrahedron Lett. 788 (1963).

[136] *J.W. McFarland*, J. Org. Chem. *36*, 1842 (1971).

Oxidation of the 4-*sec*-alkylamino-5-nitrosouracil leads directly to a *xanthine 7-oxide*[137] *via* an imine by intramolecular addition of the nitroso group to the azomethine group:

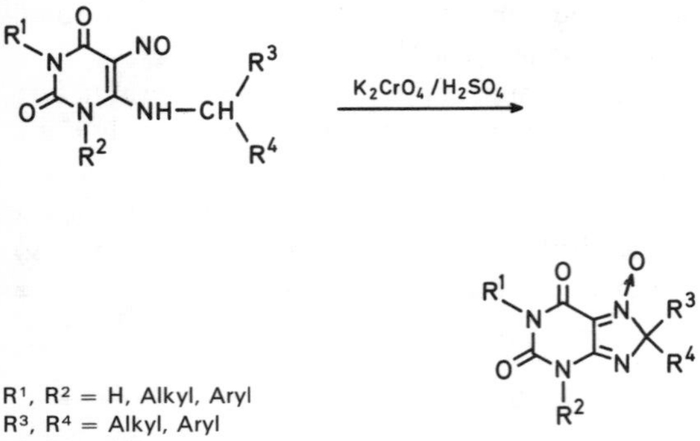

R^1, R^2 = H, Alkyl, Aryl
R^3, R^4 = Alkyl, Aryl
R^3–R^4 = (CH$_2$)$_5$

10.2.1.6 Isomerization of oxaziridines and oxidation of Schiff bases[138–148]

Nitrones **68** and the isomeric oxaziridines **67** can be interconverted *via* an electrocyclic reaction[149] (and including when R^1 and R^3 form a ring):

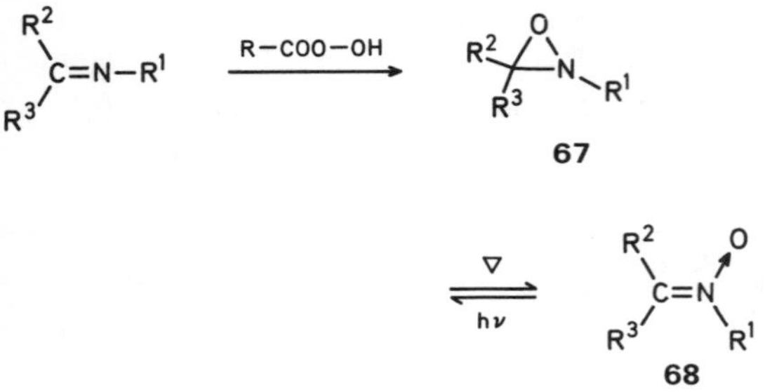

In general the equilibrium lies entirely on the side of the *nitrones*, which form spontaneously even at low temperatures or require heating in dependence on the structure from the oxaziridines. However, steric and electronic factors can lead to a reversal of the stability relationships[150] (cf. ref. [151]). Photochemical conversion of nitrone to oxaziridine proceeds stereoselectively, the thermal back-reaction does not[149], but in practical terms a subsequent direct *Z–E* isomerization of the nitrone formed must be anticipated. Sometimes the oxaziridines can be rearranged by acids to nitrones[55, 152, 153] (but see, *e.g.*, ref. [154] also [631]).

The conversion possesses importance as a method of preparing nitrones if the oxaziridine is readily and cheaply available (by oxidation of Schiff bases with peroxy acids).

Whether the oxidation of the imines occasionally leads to nitrones directly requires to be checked[138, 152, 42] because normally oxaziridines are formed[153]. In a few cases only nitrone formation has been reliably established, and it is unresolved whether they arise by the agency of a subsequent transformation of primarily formed oxaziridines. This subsequent transformation has been excluded during formation of a *cyclic nitrone*[152] (cf. ref. [155]).

Thermolysis of the oxaziridines can give rise to *carboxamides* side by side with nitrones[156, 157], and sometimes the former even predominate[145, 148]. Deoxygenation can afford *imines*[158], while migration of the N-substituent leads to *oximes*[55]. The nitrones formed are themselves decomposed into a variety of products (*cf.* p. 349) when heated to sufficiently high temperatures[159]. The photochemical[153, 160] back-reaction may not take place[145]. Sometimes it is diminished by side-reactions and sequential reactions[161, 162], namely

1 Rearrangement to a carboxamide[163, 164]
2 Rearrangement to an imino ester[164]
3 Conversion into an N-acyl nitrate or a nitroso compound and aldehyde[156]
4 Deoxygenation to the Schiff base by shortwave light[160]

10.2.1.7 Special and potential synthetic pathways

In individual cases novel routes to nitrones or nitrone derivatives, or novel reactions proceeding *via* intermediate nitrone stage have been described. For the time being it is difficult to assess whether these pathways can be expanded preparatively and they will therefore be sketched here briefly only.

[137] *H. Goldner, G. Dietz, E. Carstens*, J. Prakt. Chem. *692*, 134 (1966).
[138] *W. Rundel* in *Houben-Weyl*, Methoden der organischen Chemie Bd. X/4, p. 408, Georg Thieme Verlag, Stuttgart 1968.
[139] *W. Rundel* in *Houben-Weyl*, Methoden der organischen Chemie, Bd. X/4, p. 451, Georg Thieme Verlag, Stuttgart 1968.
[140] *E. Schmitz* in *A.R. Katritzky*, Advances in Heterocyclic Chemistry, Vol. II, p. 83, Academic Press, New York, London 1963;
E. Schmitz in Dreiringe mit zwei Heteroatomen, p. 6, Springer Verlag, Berlin 1967.
[141] *G.G. Spence, E.C. Taylor, O. Burchardt*, Chem. Rev. *70*, 231 (1970).
[142] *J. Keck*, in this Vol., Chapter 3, «Hydroxylamines».
[143] *D.R. Boyd, W.B. Jennings, R. Spratt, D.M. Jerina*, Chem. Commun. 745 (1970).
[144] *H. Krimm*, Chem. Ber. 91, 1057 (1958).
[145] *R. Bonnet, V.M. Clark, A. Todd*, J. Chem. Soc. (London) 2102 (1959).
[146] *M. Lamchen, T.W. Mittag*, J. Chem. Soc. (C) 1917 (1968).
[147] *J.S. Splitter, M. Calvin*, J. Org. Chem. *23*, 651 (1958).
[148] *J.S. Splitter, M. Calvin*, J. Org. Chem. *30*, 3427 (1965).
[149] *J.S. Splitter, T.-M. Su, H. Ono, M. Calvin*, J. Amer. Chem. Soc. *93*, 4075 (1971).

1 Phenylnitromethane reacts with the ynamines to form mixtures of substances from which in two cases the nitrones were isolated in moderate yield[165]; the main products are the *oxime ethers:*

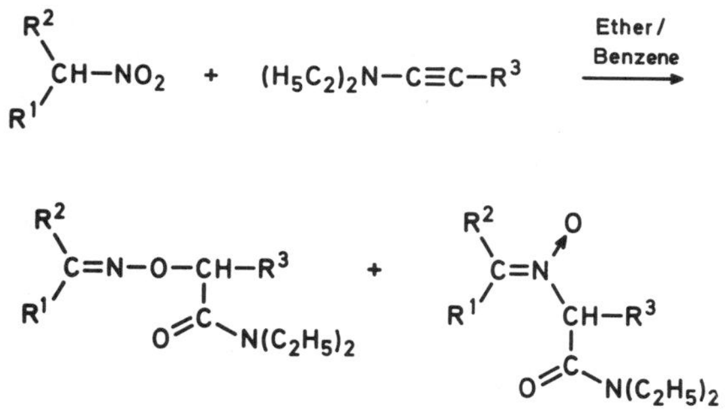

R^1 = H; R^2 = C$_6$H$_5$; R^3 = CH$_3$; *2-(Benzylidenaminoxy)-N,N-diethylpropionamide*
+
Benzaldehyde 1-(diethyl-aminocarbonylethyl)nitrone

R^3 = C$_6$H$_5$; *2-(Benzylidenaminoxy)-N,N-diethyl-2-phenylacetamide*
+
Benzaldehyde α-(diethyl-aminocarbonylbenzyl)nitrone

Nitromethane and nitroethane afford exclusively the oxime *O*-ether.

2 Cyclic hydroxamic acids are reduced to the cyclic nitrones by lithium tetrahydroaluminate[47, 81] (but *cf.* ref.[92]):

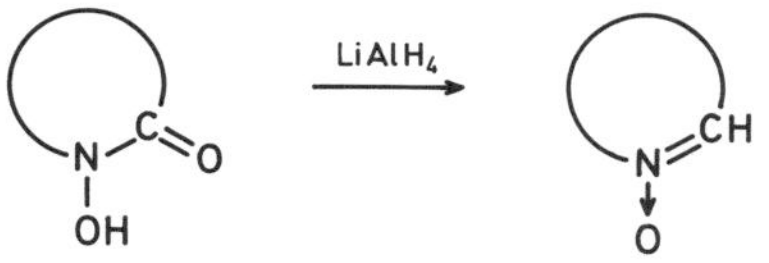

3 During reaction of hydroxylamine with α,β-unsaturated ketones the saturated β-hydroxy-*N*-methyl nitrones are formed as side-products in some instances[663]:

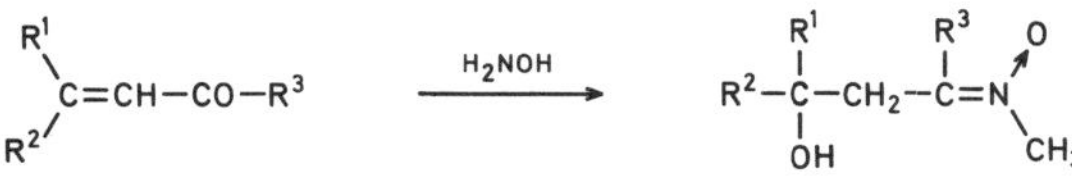

4 Photolysis of 2-*tert*-butyl-1-nitrobenzenes leads to a reaction mixture from which by chromatography *3,3-dimethyl-3H-indole 1-oxide* can be obtained as one product[166] (*cf.* pp. 370, 383).

5 A cyclopentyl nitrite that is fused into a larger, nonaromatic ring system was isomerized photochemically to the corresponding 2,3,4,5-tetrahydro-2-pyridinol 1-oxide in good yield[167, 168].

[150] *H.O. Larson, K.Y.W. Ing, D.L. Adams,* J. Heterocyc. Chem. 1227 (1970).

[151] *F. Montanari, I. Moretti, G. Torre,* Chem. Commun. 1086 (1969).

[152] *B. Singh,* J. Amer. Chem. Soc. *90,* 3893 (1968).

[153] *J.B. Bapat, D.St.C. Black,* Australian J. Chem. *21,* 2507 (1968).

[154] *P. Milliet, X. Lusinchi,* Tetrahedron Lett. 3763 (1971).

6 *Tert*-alkyl-nitroso compounds can be alkylated on the nitrogen with triethyloxonium tetrafluoroborate to form nitrone salts *via* nitrosonium compounds[169]:

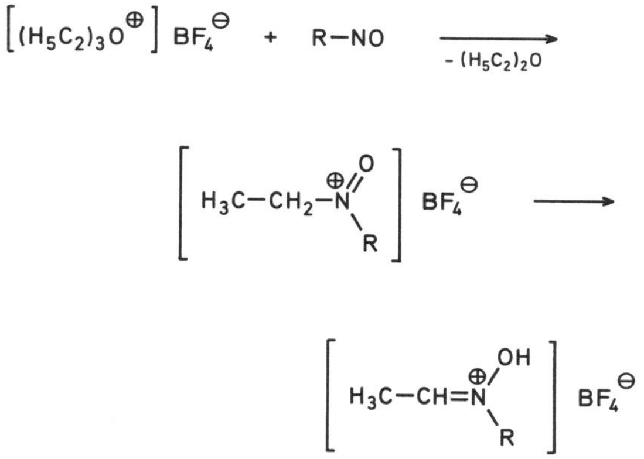

It remains open to what extent this method can be used for preparing nitrones.

7 *N-(α-Hydroxyaryl) nitrones* presumably arise as intermediate products during addition of azomethine ylides to 1-nitroso-2-naphthol leading to *2H*-naphtho[1,2-d]oxazoles[170].

8 Intermediate formation of a nitrone is postulated also during photochemical reaction of nitroethane and cyclohexane[171].

10.2.2 Transformation reactions of the nitrones

Undoubtedly the greatest importance of the nitrones as starting points for syntheses is their use in *1,3-dipolar cycloadditions*[32]:

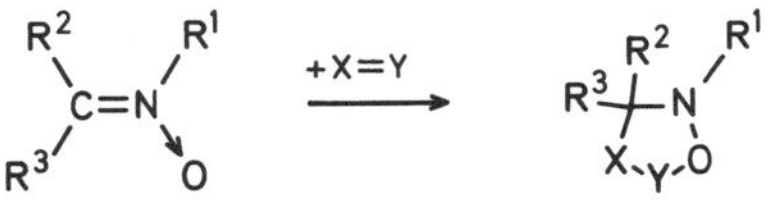

X=Y =	Alkenes[158, 172–176]	Ketenimines[183]
	Alkynes[173, 177, 178]	Isocyanates[173, 184]
	Allenes[179]	Sulfenes[173, 185]
	Enol ethers[180]	Sulfinylamines[173, 186]
	Ketenes[181]	Methylenephosphoranes[187]
	Thioketenes[182]	

Nitrones display many typical *carbonyl reactions* but the end-products often differ from those of the actual carbonyl reaction:

1 Piperidine addition affords a poor yield of *carboximide piperidide* **70**; *X* = piperidino)[173].

2 Addition of hydrogen cyanide seldom leads to type **69** derivatives but generally to α-iminonitriles **70** (*X* = *CN*, Bellavita reaction) or sequential products[173].

3 Nitroalkanes furnish hydroxylamines **60**[112] (*X* = −CR^{4}R^5−NO$_2$)[173].

[155] *G. Roblot, G. Lukacs, X. Lusinchi,* Tetrahedron Lett. 505 (1972).

[156] *A.L. Bluhm, J. Weinstein,* J. Amer. Chem. Soc. *92,* 1444 (1970).

[157] *H. Izawa, P. De Mayo, T. Tabata,* Can. J. Chem. *47,* 51 (1969).

4 With *N*-alkyl aldonitrones, esters and amides of acetoacetic, cyanoacetic, malonic, and phenylacetic acids give *5-isoxazolidinones* **71**[189, 190], but cf. ref. [188].

5 *C*-Methyl and *C*-methylene nitrones add to form *γ*-hydroxyamino nitrones **9** (Type **69**)[173].

6 Grignard compounds likewise form hydroxylamines **69** with nitrones *(X = R⁴)* in good yield[173, 92].

7 Reformatzky reagent (α-bromocarboxylic ester/ zinc) forms *5-isoxazolidinone* **71** with *N*-alkyl nitrones[191].

8 Acid hydrolysis leads to *carbonyl compounds* **72** (see Vol. V) *(Y = O)*[192].

9 Hydrazines and hydroxylamine give *hydrazones* and *oximes (**72**, Y = N–N<, Y = N–OH)*[192].

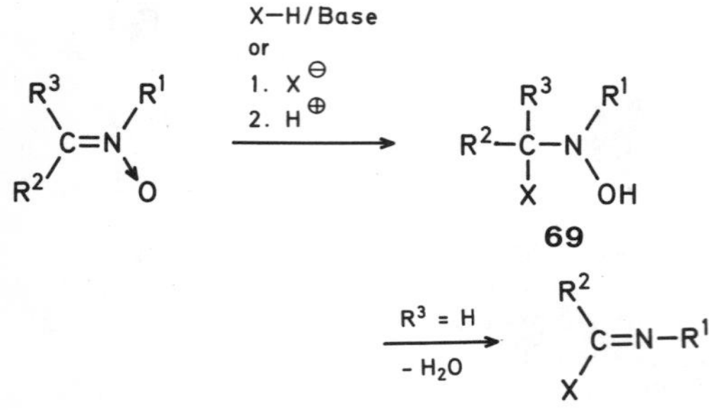

An acid-catalyzed aldol addition, too, has been described[188].

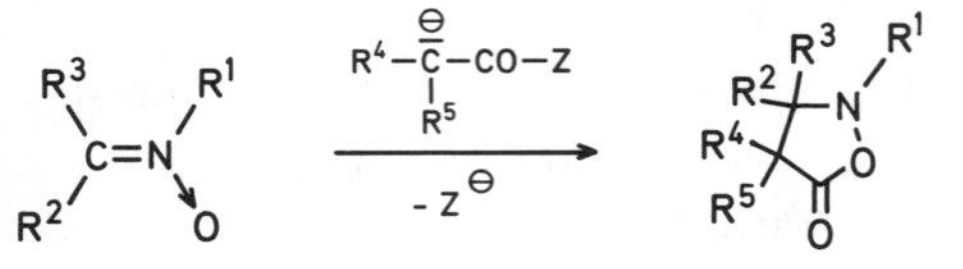

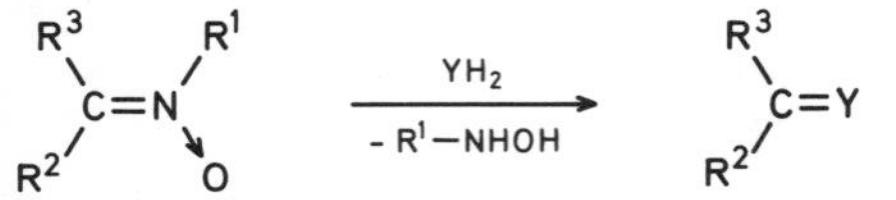

C-Methyl and *C*-methylene nitrones can undergo base-catalyzed aldol additions; they serve as the methylene component, while a different nitrone[173] or an aldehyde[39, 42, 193] functions as *carbonyl component*. With aldehydes condensation to the *α,β-unsaturated nitrone* ensues.

N-Methyl nitrones can be *alkylated* on the oxygen with dialkyl sulfates or alkyl halides to form *N*-alkoxy-*N*-methylimmonium salts[173]. With quaternary aziridinium ions nitrones form *quaternary tetrahydro-1,2,4-oxadiazinium ions*[194].

Nitrones can add free radicals to the nitrone carbon atom with formation of long-lived *nitroxyl radicals*. This reaction is recommended for trapping and identifying short-lived radicals[195–198].

The products that arise most frequently during *reduction* of nitrones are *hydroxylamines* **73**, imines **74**, and secondary amines **75**:

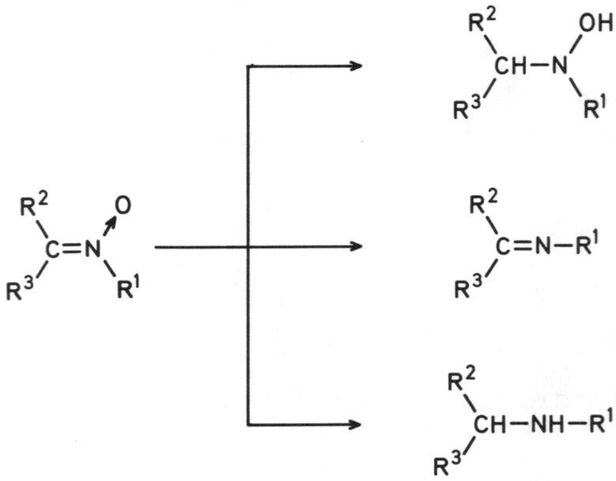

In dependence on the working conditions catalytic hydrogenation affords hydroxylamines, imines, or secondary amines[37, 105, 199]. Reduction with complex hydrides normally leads to hydroxylamines[37, 199], with sodium in alcohol generally to secondary amines[199]. Zinc/acetic, zinc/hydrochloric acid, or zinc/ammonium chloride reduction (*cf.* p. 339) in water generally produces imines[199], sometimes also secondary amines[37].

[158] *J.B. Bapat, D.St.C. Black*, Australian J. Chem. *21*, 2521 (1968).

[159] *W. Rundel* in *Houben-Weyl*, Methoden der organischen Chemie Bd. X/4, p. 443, Georg Thieme Verlag, Stuttgart 1968.

[160] *J.S. Splitter, M. Calvin*, Tetrahedron Lett. 3995 (1970).

[161] *J.S. Splitter, M. Calvin*, Tetrahedron Lett. 1445 (1968).

[162] *E. Meyer, G.W. Griffin*, Angew. Chem. *79*, 648 (1967).

[163] *L.S. Kaminsky, M. Lamchen*, J. Chem. Soc. (C) 2295 (1966).

[164] *G.F. Field, L.H. Sternbach*, J. Org. Chem. *33*, 4438 (1968).

[165] *J. Ficini, A. Bonenfant, C. Barbara*, Tetrahedron Lett. 41 (1972).

[166] *D. Döpp, K.-H. Sailer, E. Brugger*, Angew. Chem. *83*, 898 (1971).

[167] *H. Suginome, N. Sato, T. Masamune*, Tetrahedron Lett. 3353 (1969); *H. Suginome, N. Sato, T. Masamune*, Tetrahedron 27, 4863 (1971).

[168] *H. Suginome, T. Mizuguchi, T. Masamune*, Tetrahedron Lett. 4723 (1971); *H. Suginome, T. Mizuguchi, T. Masamune*, Chem. Commun. 376 (1972).

[169] *J.E. Baldwin, R.G. Pudussery, B. Sklarz, M.K. Sultan*, Chem. Commun. 1361 (1968).

[170] *J.W. Lown, J.P. Moser*, Chem. Commun. 247 (1970).

[171] *S.T. Reid, J.N. Tucker*, Chem. Commun. 1286 (1970).

The last method, like reduction with sodium, can afford *1,2-diamines* from 1-pyrroline 1-oxides by reductive dimerization corresponding to the formation of pinacols from ketones[199, 200].

A specific deoxygenation to imines can be achieved with triphenylphosphine and compounds with similar affinity for oxygen[158, 199].

Nitrones are cleaved by two equivalents ozone into nitro and carbonyl compounds (ketones and aldehydes)[201]. Iron(III) chloride[37, 201] or manganese(IV) oxide[202] convert cyclic aldo nitrones into hydroxamic acids, while sodium periodate cleaves the aldo nitrone function into a nitroso function and a carbonyl group[37, 201]. With neutral potassium permanganate solution hydroxamic acid formation, nitrosocarboxy, and nitrocarboxy cleavage are observed side by side[202]. Lead(IV) acetate oxidizes aldo nitrones to *O*-acetylhydroxamic acids[202, 203, 204].

C-Methyl and *C*-methylene nitrones are oxidized on the methyl and methylene group by selenium-(IV) oxide to form aldehyde and ketone respectively (*cf.* p. 339)[42, 92, 205].

10.3 Isatogens (3*H*-indol-3-one 1-oxides) and related compounds

Isatogens **77** differ from the ordinary α-keto nitrones (*C*-acyl nitrones) in their methods of preparation and certain reactions. The same is true of the related *4 H-pyrazol-4-one N*-oxides. Both compound classes are deeply colored (orange-red) and in some respects behave like quinones.

For data relating to the polarographic reduction of isatogens[206], their *UV* spectra[207, 208], and their mass spectra[209], see the literature.

Isatogens take the indole numbering; the Chemical Abstracts list them as 3 *H*-indol-3-one 1-oxides.

10.3.1 Preparation
10.3.1.1 From 2-nitrostyrenes or (2-nitrophenyl)acetylenes

2-Nitrophenylacetylenes **76** can be isomerized to the isatogens **77** in various different ways[210, 211] (apparently treatment with cold concentrated sulfuric acid is now hardly used). Isomerization is effected by illuminating (sunny daylight for several days) a solution of (2-nitrophenyl)acetylene **76** in pyridine or quinoline[212]. Pyridine or quinoline are absolutely essential; see ref.[213] for the probable mechanism. Alternatively, isatogens **77** can be obtained directly starting from the acetylene precursors **80**[214] or **79**[207] by irradiation in pyridine. This technique may be advantageous if preparation of the acetylene **76** is the most difficult step in the sequence[214].

R = COOAlkyl, (subst.) C_6H_5, naphthyl, Pyridyl, Thienyl,   **80**
Anthrachinonyl
R ≠ H

[172] *R. Huisgen, R. Grashey, H. Hauck, H. Seidl*, Chem. Ber. *101*, 2548, 2559, 2568 (1968);
R. Huisgen, H. Hauck, R. Grashey, H. Seidl, M. Burger, Chem. Ber. *102*, 736, 1117 (1969).
[173] *W. Rundel* in *Houben-Weyl*, Methoden der organischen Chemie Bd. X/4, p. 418, Georg Thieme Verlag, Stuttgart 1968.
[174] *R.R. Fraser, Y.S. Lin*, Can. J. Chem. *46*, 801 (1968).
[175] *N. Singh, S. Mohan*, Chem. Commun. 787 (1968).
[176] *W. Oppolzer, K. Keller*, Tetrahedron Lett. 1121, 4313 (1970).
[177] *E. Winterfeldt, W. Krahn, H.-V. Stracke*, Chem. Ber. *102*, 2346 (1969).
[178] *R. Huisgen, H. Seidl, R. Knorr*, Chem. Ber. *102*, 904 (1969).
[179] *N.A. LeBel, E. Banucci*, J. Amer. Chem. Soc. *92*, 5278 (1970).

[180] *R. Paul, S. Tchelitcheff*, Bull. Soc. Chim. France 4179 (1967).
[181] *R.N. Pratt, D.P. Stokes, G.A. Taylor, P.C. Brookes*, J. Chem. Soc. (C) 2086 (1968).
[182] *M.S. Raasch*, J. Org. Chem. *35*, 3470 (1970).
[183] *M.W. Barker, J.H. Gardner*, J. Heterocycl. Chem. *5*, 881 (1968).
[184] *R. Huisgen, H. Seidl, R. Grashey*, Chem. Ber. *102*, 926 (1969).
[185] *W.E. Truce, J.W. Fieldhouse, D.J. Vrencur, J.R. Nozell, R.W. Campbell, D.G. Brady*, J. Org. Chem. *34*, 3097 (1969).
[186] *O. Tsuge, M. Tashiro, S. Mataka*, Tetrahedron Lett. 3877 (1968).
[187] *R. Huisgen, J. Wulff*, Chem. Ber. *102*, 746 (1969).
[188] *R.F.C. Brown, W.D. Crow, L. Subrahmanyan, C.S. Barnes*, Australian J. Chem. *20*, 2485 (1967).
[189] *H. Stamm, J. Hoenicke*, Synthesis 145 (1971);
H. Stamm, J. Hoenicke, Justus Liebigs Ann. Chem. *748*, 143 (1971).
[190] *L.S. Kaminsky, M. Lamchen*, J. Chem. Soc. (C) 1683 (1967).
[191] *H. Stamm, J. Hoenicke*, Justus Liebigs Ann. Chem. *749*, 146 (1971).

Often heating in dry pyridine[212] is sufficient to make the acetylene **76** react. Thus, if the synthesis of the $C \equiv C$ triple bond (**80**→**76**) in the tolan (diphenylacetylene) series *(R = alkyl)* is by-passed by the tolan synthesis from copper(I) 2-nitrophenyl acetylide (**76**, *R = Cu*) and an aromatic iodine compound *(ArI)*, then heating these reactants in dry pyridine (8 hours or longer) affords the *isatogens* (**77**, *R = Ar*) directly[207]. The reaction stops at the tolan stage to a greater or lesser degree only if the aryl iodide is *ortho-substituted*[207]. Furthermore, the 1-(2-nitrophenyl)-1-alkynes **76** are converted into isatogen by nitrosobenzene in chloroform[207, 212, 215]. Probably the nitrosobenzene first adds to the acetylene derivative to form a *C*-(2-nitrosobenzyl)-*N*-phenylnitrone and is regenerated when the latter cyclizes.

The reaction mixture is stood for some weeks or heated for some days. *2-(2-Pyridyl)-6-azaisatogen [2-(2-pyridyl)-3 H-pyrrolo[2,3-b]pyridin-3-one 1-oxide]*, too, can be prepared by this technique[215].

In all these methods the yields are very variable and, depending on the isatogen, may be very small or very good. Where *R = 2-nitroaryl* two *different* isatogens can form in theory with unsymmetrical 1-(2-nitrophenyl)-1-alkynes according to which nitro group effects the oxygen transfer to the acetylene group. However, in practise and in conformity with the proposed mechanism[213], ring closure to the isatogen always ensues on the acyl group exerting the strongest electron pull on the acetylene group[212], and 2-(3, 4-dinitrophenyl)-isatogen cannot be obtained in this way.

Isatogens with the 2-position unsubstituted (**77**, *R=H*) are unknown. On reacting (2-nitrophenyl)-acetylenes (**76**, *R=X=Y=Z=H*) with nitrosobenzene *2, 2′-bis-3 H-indole-3, 3′-dione 1, 1′-dioxide (2, 2′-biisatogenyl)* is obtained side by side with other products, but it is probably better to prepare it from 1, 4-bis(2-nitrophenyl)butadiyne[211].

1, 4-Phenylene-bis-2-(3 H-indol-3-one) N, N′-dioxide [1, 4-bis-(2-isatogenyl)benzene], too, has been prepared. By contrast, no success was obtained in condensing two *3 H-pyrrol-3-one N-oxide* structures onto a single benzene ring. Stilbenes **76** (R = 4-substituted phenyl) can furnish isatogens **77** *(R = 4-substituted phenyl)* in moderate yield on irradiation of their benzene solution, and the better the more likely the potential electron displacement from R to the *2-nitrophenyl group* occurs[207]. Accordingly, the yields decrease if the phenyl substituent is changed in the following manner

$$N(CH_3)_2 > OH > OCH_3 > O-CO-CH_3$$

The intrinsically small yields (at most 44%) apparently can be increased sharply by adding a little acetic acid or a free-radical (for dehydrogenating an intermediate stage). Brief irradiation is sufficient to perform the reaction, the further reaction sequence proceeds on its own within a few days. Using this method *4-aza-* and *6-azaisatogen (3 H-pyrrolo[3, 2-b]pyridin-3-one* and *3 H-pyrrolo[2, 3-b]pyridin-3-one 1-oxides [R = (dimethylamino)phenyl]*, too, were obtained, and also *2-vinylisatogens (R=−CH=CH−Ar, −CH−CH−CH=CH−Ar)*[216]. *2-Phenylbenzo-[d, e]quinolin-3-one 1-oxide* (homoisatogen) was prepared from 1-nitro-2-phenylethynylnaphthalene by photochemical means[216]:

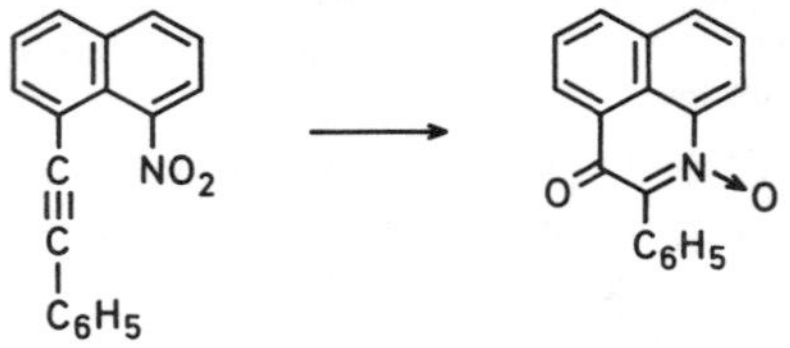

10.3.1.2 Isatogens from 2-nitrobenzaldehydes and *N*-phenylmethylpyridinium bromides or from 2-nitrobenzoyl compounds

The pyridinium salts **83** formed from 2-nitrobenzaldehydes **81** and 1-benzylpyridinium salts **82** with the aid of bases are converted into the isatogens **84** on *UV* irradiation in partly very good yield[208]. *2-(2-Pyridyl)isatogen,* which is accessible also from the corresponding azastilbene was obtained similarly (in moderate yield in either case)[217].

---

[192] *J. Hamer, A. Macaluso,* Chem. Rev. *64,* 461 (1964).

[193] *W. Rundel* in *Houben-Weyl,* Methoden der organischen Chemie Bd. X/4, p. 413, Georg Thieme Verlag, Stuttgart 1968.

[194] *N.J. Leonard, D.A. Durand, F. Uchimaru,* J. Org. Chem. *32,* 3607 (1967).

[195] *E.G. Janzen, B.J. Blackburn,* J. Amer. Chem. Soc. *91,* 4481 (1969).

[196] *E.G. Janzen, J.L. Gerlock,* J. Amer. Chem. Soc. *91,* 3108 (1969).

[197] *H. Iwamura, M. Iwamura,* Tetrahedron Lett. 3723 (1970).

[198] *M. Iwamura, N. Inamoto,* Bull. Chem. Soc. Japan *43,* 856, 860 (1970).

[199] *W. Rundel* in *Houben-Weyl,* Methoden der organischen Chemie Bd. X/4, p. 432, Georg Thieme Verlag, Stuttgart 1968.

[200] *J.B. Bapat, D.St.C. Black,* Australian J. Chem. *21,* 2497 (1968).

[201] *W. Rundel* in *Houben-Weyl,* Methoden der organischen Chemie Bd. X/4, p. 440, Georg Thieme Verlag, Stuttgart 1968.

[202] *N.J.A. Gutteridge, F.J. McGillan,* J. Chem. Soc. (C) 641 (1970).

The mechanism of this reaction still seems uncertain, but a reaction sequence *via* stilbene or tolan compounds is feasible.

The α-pyridinostilbenes formed by dehydration of the pyridinium salts **83** can be converted into the same isatogens by bases, either by heating in pyridine with addition of a little diethylamine or without heating in aqueous alcoholic solution on adding sodium carbonate solution[218].

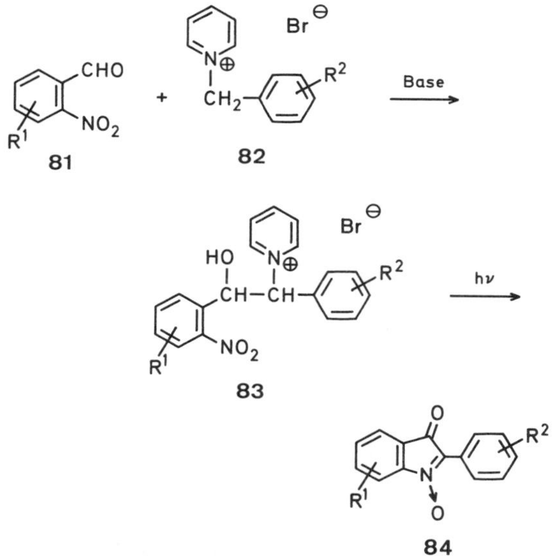

81   82   83   84

R¹ = H, 5-OCH₃, 6-OCH₃, 5-Cl
R² = H, 3-NO₂, 4-NO₂

R^1 = H, 5-OCH$_3$, 6-OCH$_3$, 5-Cl
R^2 = H, 3-NO$_2$, 4-NO$_2$

Stirring methyl or ethyl 2-nitrobenzoylacetate or 1-(2-nitrophenyl)-1,3-butanedione with aqueous sodium bicarbonate solution for some days affords a purple-colored mixture from which, following filtration, orange-colored isatogens are obtained on making acid[219]:

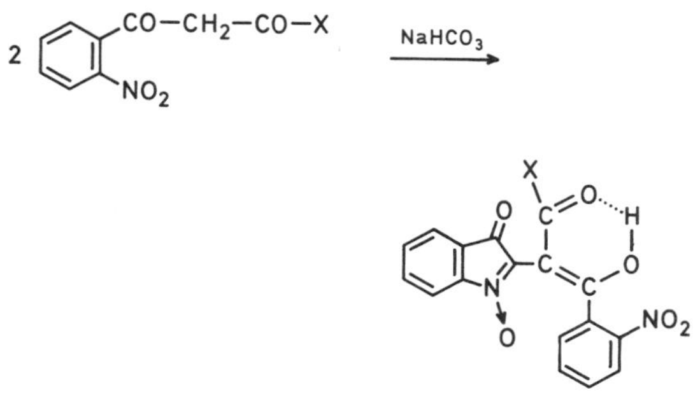

X = OCH$_3$; *Methyl α-(2-nitrobenzoyl)-3-oxo-3H-indole-2-acetate 1-oxide*
X = OC$_2$H$_5$; *Ethyl α-(2-nitrobenzoyl)-3-oxo-3H-indole-2-acetate 1-oxide*
X = CH$_3$; *1-(2-Nitrophenyl)-2-(3-oxo-3H-indol-2-yl)-1,3-butanedione N-oxide*

10.3.1.3 Isatogen imines (3-iminoindole 1-oxides)

1-Hydroxy-2-phenylindole reacts with aromatic nitroso compounds in ethanol in the presence of an equivalent sodium ethoxide to give a very good yield of *isatogen imines*[220]. *3-(2-Pyridylimino)indole 1-oxides* (**85**, R = 2-pyridyl) are obtained by thermal decomposition of an adduct of 1-hydroxy-2-phenylindole and 2,2'-azopyridine[220]. Using tosyl azide in place of an aromatic nitroso compound leads to formation of *2-phenyl-3H-indol-3-one azine 1-oxide* **86**[221].

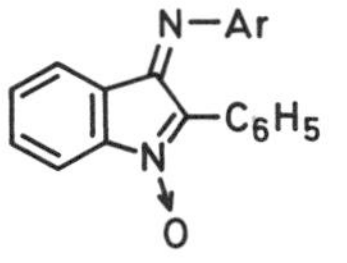

85   86

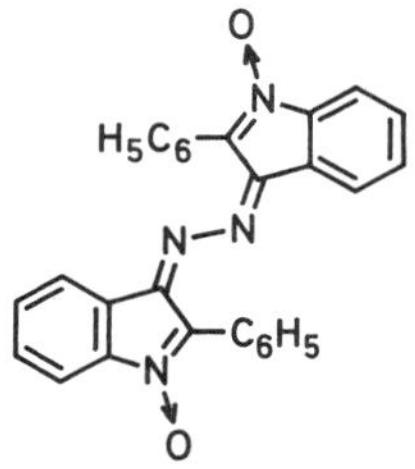

[203] *L.A. Nejmann, S.V. Zukova, L.B. Senjavina, M.M. Semjakin*, Zh. Obshch. Khim. *38*, 1480 (1968).

[204] *S. Tamagaki, S. Oae*, Bull. Chem. Soc. Japan *43*, 1573 (1970).

[205] *W. Rundel* in *Houben-Weyl*, Methoden der organischen Chemie Bd X/4, p. 417, Georg Thieme Verlag, Stuttgart 1968.

[206] *J.E. Bunney, M. Hooper*, J. Chem. Soc. (B) 1239 (1970).

[207] *J.S. Splitter, M. Calvin*, J. Org. Chem. *20*, 1086 (1955).

[208] *R. Danieli, G. Maccagnani*, Boll. Sci. Fac. Chim. Ind. Bologna *23*, 347, 353, 405 (1965); C.A. *64*, 17523 1752317523b, 15702 (1966).

[209] *D.R. Eckroth*, Chem. Commun 465 (1970).

[210] *W.C. Sumpter, F.M. Miller*, The Chemistry of Heterocyclic Compounds, Vol. VIII, p. 154, Interscience Publishers, New York, London 1954.

[211] *J.D. Loudon, G. Tennant*, Quart. Rev. Chem. Soc. *18*, 389 (1964).

[212] *C.C. Bond, M. Hooper*, J. Chem. Soc. (C) 2453 (1969).

[213] *R. Huisgen*, Angew. Chem. *75*, 604 (1963).

[214] *P. Ruggli, A. Disler*, Helv. Chim. Acta *10*, 938 (1927).

[215] *M. Hooper, D.A. Patterson, D.G. Wibberley*, J. Pharm. Pharmacol. *17*, 734 (1965).

[216] *C.C. Leznoff, R.J. Hayward*, Can. J. Chem. *49*, 3596 (1971).

[217] *D.A. Patterson, D.G. Wibberley*, J. Chem. Soc. (London) 1706 (1965).

[218] *F. Kröhnke, M. Meyer-Delius*, Chem. Ber. *84*, 932 (1951).

[219] *R.T. Coutts, M. Hooper, D.G. Wibberley*, J. Chem. Soc. (London) 5205 (1961);
M. Hooper, D.G. Wibberley, J. Chem. Soc. (C) 1596 (1966).

10.3.1.4 4H-Pyrazol-4-one 1,2-dioxides and 1-oxides

From α,β-unsaturated oximes **87** half-concentrated nitric acid and a few small crystals sodium nitrite afford a small quantity of light-red *4H-pyrazol-4-one 1,2-dioxides* **88**[222, 223]:

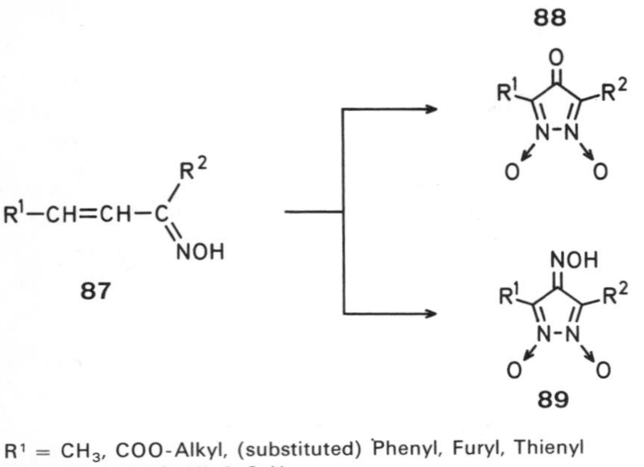

88

87

89

R^1 = CH$_3$, COO-Alkyl, (substituted) Phenyl, Furyl, Thienyl
R^2 = Alkyl, COO-Alkyl, C$_6$H$_5$

Reacting the same oxides in acetic acid during some days with adequate quantities of sodium nitrite likewise leads to formation of sparingly water and solvent soluble light-red *oximes* **89** side by side with the *4-oxo derivatives* **88** as main product[222, 223].

Ensuring an oxygenfree medium during the reaction affords the oximes in good yield, while, conversely, passing through oxygen improves the yield of ketone (up to 40%)[223]:

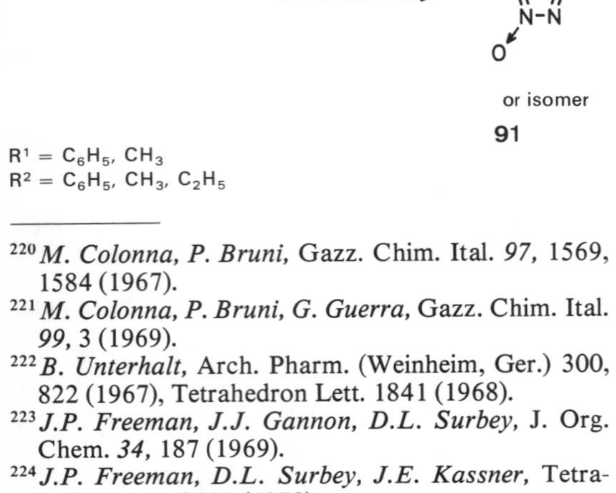

88

90

or isomer

Fremy's salt

91

or isomer

R^1 = C$_6$H$_5$, CH$_3$
R^2 = C$_6$H$_5$, CH$_3$, C$_2$H$_5$

4H-Pyrazol-4-one 1-oxides **91** (which can form dimers) are obtained in excellent yield by oxidizing the 1-hydroxypyrazol-4-ols **90** with Fremy's salt [ON(SO$_3$K$_2$][223]. Oxidation with peroxy acids affords different products[224] (see p. 380).

10.3.2 Transformation reactions of isatogen and related compounds

4H-Pyrazol-4-one 1,2-dioxides **88** reduce to either *pyrazol-4-ols*[224] or *1-hydroxypyrazol-4-ols* **90** (p. 353) in dependence on conditions. The nitrone groupings of the 4H-pyrazol-4-one 1,2-dioxides **88** and 1-oxides **91** undergo 1,3-dipolar cycloadditions[225], with **88** adding olefinic dipolarophiles on only one nitrone grouping, while two equivalents acetylenedicarboxylic acid diester produce addition with elimination of dinitrogen monooxide.

Isatogens are converted readily into free radicals thermally if they can abstract hydrogen or hydroxyl from the solution[226]. Reducing isatogens with phenylhydrazine can lead to the corresponding indol-3-ol (indoxyl)[217]. With phosphorus(III) chloride deoxygenation takes place[227]. Oxidation with alkaline hexacyanoferrate(III) solution affords *2H-3,1-benzoxazine-2,4(1H)-dione*[219].

Rearrangement of 2-phenylisatogen with methanolic sulfuric acid leads to *2,1-benzisoxazol-3-yl phenyl ketone*[228]. With hydroxylamine, 2-phenylisatogen forms two oximes, the normal ketoxime and a rearranged product, *2,1-benzisoxazol-3-yl phenyl ketone oxime*[229] (formerly regarded as the *N-oxime*). Ring expansion of isatogens occurs with ammonia (to a *2-substituted 4H-1,3-benzoxazin-4-one*)[230] (see p. 378)[231], with nitriles such as ethenetetracarbonitrile or trichloroacetonitrile[232] (to a *4(3H)-quinazolinone*), or with acetylenes (to *4-quinolinols*)[212, 232]. Isatogens add olefins to form 1,3-cycloadducts **92**, which can be cleaved to 2-disubstituted 3-indolinones **93**[233]. Under piperidine catalysis ethyl cyanoacetate is added on with formation of *ethyl 2-amino-3a,4-dihydro-4-oxoisoxazolo[2,3-a]indole-3-carboxylates*[234] **94**, while nucleophilic *XH* compounds otherwise add to the corresponding 2-disubstituted *1-hydroxy-3-indolinones* **95**. Like 2,2-dihydroxy-1H-

[220] *M. Colonna, P. Bruni*, Gazz. Chim. Ital. *97*, 1569, 1584 (1967).

[221] *M. Colonna, P. Bruni, G. Guerra*, Gazz. Chim. Ital. *99*, 3 (1969).

[222] *B. Unterhalt*, Arch. Pharm. (Weinheim, Ger.) 300, 822 (1967), Tetrahedron Lett. 1841 (1968).

[223] *J.P. Freeman, J.J. Gannon, D.L. Surbey*, J. Org. Chem. *34*, 187 (1969).

[224] *J.P. Freeman, D.L. Surbey, J.E. Kassner*, Tetrahedron Lett. 3797 (1970).

[225] *J.P. Freeman, M.L. Hoare*, J. Org. Chem. *36*, 19 (1971).

[226] *L. Lunazzi, G.F. Pedulli, G. Maccagnani, A. Mangini*, Tetrahedron Lett. 5807 (1966);
L. Lunazzi, G.F. Pedulli, G. Maccagnani, A. Mangini, J. Chem. Soc. (B) 1072 (1967).

[227] *R.J. Richman, A. Hassner*, J. Org. Chem. *33*, 2548 (1968).

indene-1,3-(2*H*)-dione (ninhydrin) and 1*H*-indole-2,3-dione (isatin), isatogens can deaminate and decarboxylate α-amino acids oxidatively, and are themselves converted into *3-indolinones* in the process[235].

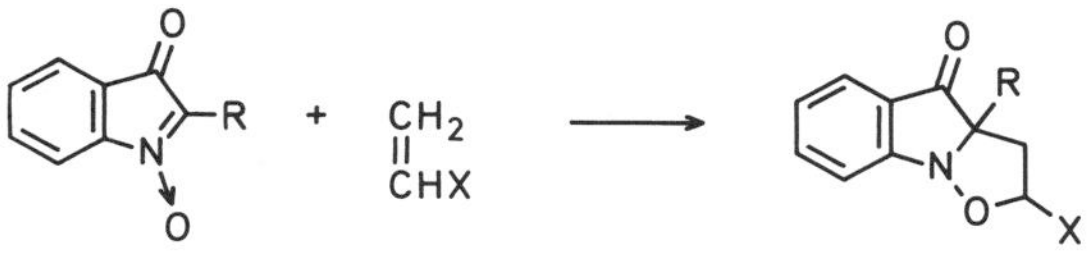

92

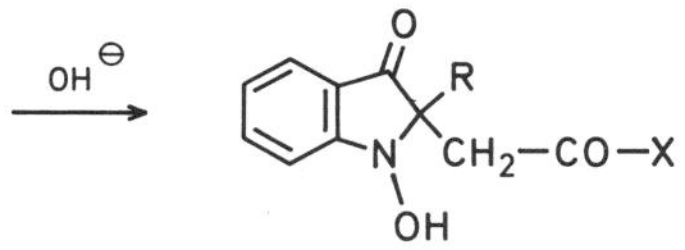

93

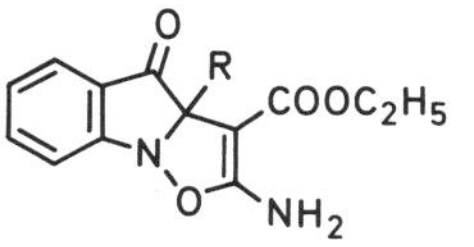

94 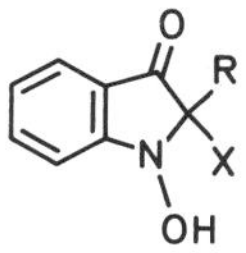   **95**

10.4 *N*-Oxides of azaaromatic compounds

10.4.0   Structure, properties, tautomerism, scope

N-Oxides of azaaromatic compounds can be polarized both ways from the *N–O* bond[236] in correspondence with the canonical limiting formulae for pyridine *N*-oxide:

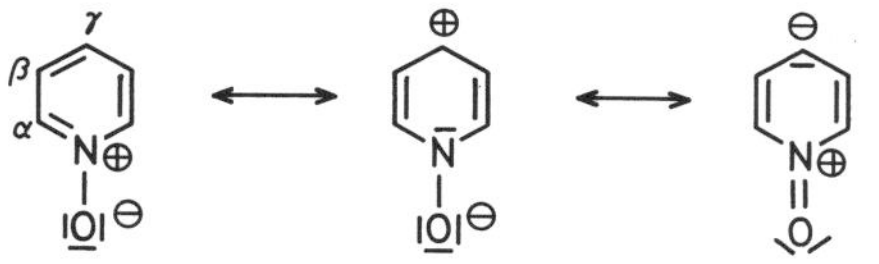

1a   **1b**   **1c**

The α-positions are analogous

[228] *J.L. Pinkus, T. Cohen, M. Sundaralingam, G.A. Jeffrey*, Proc. Chem. Soc. *70*, (1960).

[229] *M. Sax, J. Pletcher, D. Scholtz, R.M. Gerkin, J.L. Pinkus*, J. Chem. Soc. (B) 560 (1971).

[230] *D.R. Eckroth, R.H. Squire*, Chem. Commun. 312 (1969); *D.R. Eckroth, R.H. Squire*, J. Org. Chem. *36*, 224 (1971).

[231] *W.E. Noland, D.A. Jones*, J. Org. Chem. *27*, 341 (1962).

[232] *W.E. Noland, R.F. Modler*, J. Amer. Chem. Soc. *86*, 2086 (1964).

[233] *W.E. Noland, D.A. Jones*, Chem. Ind. 363 (1962).

[234] *J.E. Bunney, M. Hooper*, Tetrahedron Lett. 3857 (1966).

In the following the position markings α,β,γ are used purely formally, including for bridgehead atoms in condensed azaaromatic compounds.

The following general statements refer fundamentally to azines (6-membered azaaromatic rings), but probably apply analogously in substantial measure to azoles (5-membered azaaromatic rings).

Almost invariably, the *nomenclature* is based on the parent compound; the older literature contains many erroneous formulae and, correspondingly, misnomers (*e.g.*, furoxans = glyoxime peroxides).

The change from pyridine to its *N*-oxide increases the *dipole moment* from *2.20* to *4.24* D. This increase is much less than in aliphatic amines; an explanation put forward is that an appreciable portion of 1c contributes to the ground state[237]. Nonetheless, the tendency to form hydrogen bonds on the oxygen and hence to form hydrates is clearly marked[238, 239]. In *N*-monooxides of diazines the dipole moment often enables isomeric structures to be identified[240, 241]. The *N–O* bond is stronger than that in *N*-oxides of tertiary amines, so that in many cases it withstands a vacuum distillation of the compounds. Polarographic reduction of the *N*-oxide group ensues at a much more negative potential than in aliphatic *N*-oxides[242, 243]. The length of the *N–O* bond is *1.24–1.34* $\mathring{A}$[244, 245, 239], and possibly more in hydrates[246]. Some of these properties are less marked in condensed azaaromatic compounds.

The pK_A values of *N*-oxides of azaaromatic compounds lie at around 1 or less[247]. Thus, they are appreciably weaker bases than the parent compounds (the azaaromatic compounds themselves) or *N*-oxides of tertiary amines.

Like in nitrones (p. 333), it is often necessary to confirm or establish the *structure* of prepared *N*-oxides of azaaromatic compounds, because products without an *N*-oxide group can arise (*e.g.*, pyridone structures) under the conditions of the synthesis.

One of the most important tests is deoxygenation with PX_3 compounds (X = hal, *O*-alkyl, phenyl)[248], hydrogen-Raney nickel[249], sulfurous acid[250], iron (or zinc) in acetic acid[251], or by other methods[249]. The oxidizing property of the *N*-oxides gives rise to a (not strictly specific) color test, in which a blue dye of the crystal violet type[252] is formed from *N,N*-dimethylaniline in the presence of hot hydrochloric acid.

[235] *M. Hooper, J.W. Robertson*, Tetrahedron Lett. 2139 (1971).

[236] *A.R. Katritzky*, Quart. Rev. Chem. Soc. *10*, 395 (1956).

[237] *E. Ochiai*, Aromatic Amine Oxides, p. 78, Elsevier Publ., Co., Amsterdam, London, New York 1967.

Physicochemical identifications tests are primarily the *IR* and *UV* spectra.

The most characteristic feature of the *IR* spectrum is the strong $N-O$ band between 1200 and 1360 cm^{-1}, whose position is affected by substituents (for the relationship between ν and Hammett's σ constant of the α- and γ-substituents, cf. refs. [253-256]) and by hydrogen bonds[253, 255, 257-259]. In *pyridazine 1, 2-dioxides* and derivatives[243], and in *quinoxaline 1, 4-dioxides*[260], the $N-O$ bands are strongly displaced, sometimes to 1400 cm^{-1}. *Furoxans* absorb at 1150–1190 cm^{-1} [261].

The *UV absorption* maxima λ_{NO} ($\pi-\pi^*$) of the *N*-oxides[254, 256, 262] are at longer wavelengths and display a stronger absorbance than the maxima λ_N of the respective parent compound. However, as the solvent increases in polarity (and above all with hydrogen bonds), λ_{NO} approaches λ_N[243, 262]. In the protonated form (*e.g.*, in 1 *N* HCl) or in the *O*-alkylated form the *UV* spectra hardly differ from those of the protonated parent heterocycles; λ_{NO} is then only slightly longer than λ_N.

The ^1H-*NMR* spectra[254, 263-265] of the *N*-oxides display solvent-dependent signals of the ring protons that are displaced to a greater or lesser extent relative to the parent heterocycles (order of magnitude to about 1 ppm). In aprotic solvents an appreciable difference is found in that the order of the increasing-field signals in pyridine is Hα–Hγ–Hβ, in pyridine *N*-oxide it is Hα–Hβ–Hγ. This relationship together with the coupling constants can serve to locate the $N-O$ group in pyridazine 1-oxides definitively.

Eliminating the $^{14}N-C-H$ coupling in the *N*-oxide, which leads to an intensification of the α-*H* signal by comparison to that in the parent aromatic compound, likewise can have diagnostic value in monooxides of polyazaaromatic compounds[266].

One feature of the *mass spectra* is the ion $(M$-16$)^\oplus$ formed by deoxygenation in addition to the molecular ion $M^\oplus$ [267-270].

A *quantitative* determination has been carried out with perchloric acid in acetic anhydride[271]. Re the electromeric reduction, see ref. [272].

X = O, S, NR

Problems of tautomerism (Equations 1–3) in *N*-oxides of *NH*-, *OH*-, or *SH*-substituted azaaromatic compounds, or in *N*-oxides of diazoles have been elucidated spectroscopically, more particularly by comparing the *UV* spectra with those of *O*-methylated or *O*-protonated *N*-oxides[238, 273-276, 669]. The hydroxamic acid compounds 2 *(X = O)* are stained red (or possibly blue or violet) with iron-(III) chloride[240]. As the *position* of the tautomeric equilibrium is not always known and is often dependent on the solvent, this chapter does not distinguish between true *N*-oxides and compounds which require a tautomerization to convert them into true *N*-oxides.

[238] H. Lettau, Z. Chem. *10*, 211 (1970).

[239] R. Desiderato, J.C. Terry, J. Heterocycl. Chem. *8*, 617 (1971).

[240] M. Tisler, B. Stanovnik in A.R. Katritzky, A.J. Boulton, Advances in Heterocyclic Chemistry, Vol. IX, p. 285, Academic Press, New York, London 1968.

[241] E. Ochiai, Aromatic Amine Oxides, p. 48, 78, Elsevier, Publ., Co., Amsterdam, London, New York 1967.

[242] E. Ochiai, Aromatic Amine Oxides, p. 91, Elsevier, Publ., Co., Amsterdam, London, New York 1967.

[243] I. Suzuki, M. Nakadate, S. Sueyoshi, Tetrahedron Lett. 1855 (1968).

[244] E. Ochiai, Aromatic Amine Oxides, p. 89, Elsevier, Publ., Co., Amsterdam, London, New York 1967.

[245] R. Desiderato, J.C. Terry, Tetrahedron Lett. 3203 (1970).

[246] W.E. Oberhänsli, Helv. Chim. Acta *53*, 1787 (1970).

[247] E. Ochiai, Aromatic Amine Oxides, p. 97, 139, Elsevier Publ., Co., Amsterdam, London, New York 1967.

[248] E. Ochiai, Aromatic Amine Oxides, p. 190, 202, Elsevier Publ., Co., Amsterdam, London, New York 1967.

[249] E. Ochiai, Aromatic Amine Oxides, p. 184, Eslevier Publ., Co., Amsterdam, London, New York 1967.

[250] E. Ochiai, Aromatic Amine Oxides, p. 184, 201, Elsevier Publ., Co., Amsterdam, London, New York 1967.

[251] E. Ochiai, Aromatic Amine Oxides, p. 196, Elsevier Publ., Co., Amsterdam, London, New York 1967.

[252] N.A. Coats, A.R. Katritzky, J. Org. Chem. *24*, 1836 (1959).

[253] E. Ochiai, Aromatic Amine Oxides, p. 114, Elsevier Publ., Co., Amsterdam, London, New York 1967.

[254] S. Okada, A. Kosasayama, T. Konno, F. Uchimaru, Chem. Pharm. Bull. (Tokyo) *19*, 1337, 1344 (1971).

[255] S. Ghersetti, G. Maccagnani, A. Mangini, F. Montanari, J. Heterocycl. Chem. *6*, 859 (1969).

[256] A.R. Katritzky, Quart. Rev. Chem. Soc. *13*, 353 (1959).

A special type of tautomerism (activation threshold ~ 14 *kcal/mole*) is possible with benzofuroxans[454, 455]. Unsymmetrically substituted benzofuroxans are, consequently, obtained in only one isomeric form in general, independently of the synthetic pathway. In solution both isomers can be identified unless one of the positions in the benzene portion immediately adjacent the furoxan ring is substituted while the other is not.

Re the question of the existence of *2,1-benzisoxazole 1-oxides* see ref.[277, 649].

For the extensive literature dealing with the biological action of azaaromatic *N*-oxides, see refs.[278-284].

10.4.1 Preparation of *N*-oxides of azaaromatic compounds

10.4.1.1 By direct oxidation of the azaaromatic compounds

10.4.1.1.1 Reaction conditions

A very extensive literature describes the direct oxidation of the azaaromatic compounds. The selected references are intended merely to demarcate the applications:

Pyridine, its analogous benzo and aza derivatives and their substitution products, azines in general, can generally be oxidized to *N-oxides* very smoothly with peroxy acids[285]:

$$\text{Pyridine} + X{-}OOH \xrightarrow{\sim 0-100°} \text{Pyridine-}N\text{-oxide} + X{-}OH \quad (4)$$

X = Acyl or inorganic acid substituent

Other oxidizing agents (*e.g.*, diacyl peroxides)[286] hardly play a part.

Of the available methods peroxy acid oxidation is the most widely applicable one for preparing *N*-oximes of azines. Direct oxidation cannot be used or is inferior to other techniques if the relevant azines are either difficult to obtain, sensitive to oxidation, or possess pyridonelike structures.

In general, azoles cannot be converted into their *N*-oxides directly[238, 285]. Exceptions are thiazoles **3**[285], pyrazoles **4**[287], 2,2'-biimidazole and 2,2'-bibenzimidazole[238], and also 2-acetamidobenzothiazole[288]. *2,2-Bibenzimidazole 1,1'-dioxide* **5** exists partly as a diradical[238]. Oxidizing guanine leads to *2-amino-6-hydroxypurine 3-oxide*[290] and not, as postulated originally[289], to guanine 7-oxide. Re the oxidation of 1,2,3-triazoles, *cf.* ref.[291]. On account of the conversion of Eq. 4 into an equilibrium reaction 1,2,3-thiadiazole 2-oxides are formed in moderate yield only[652].

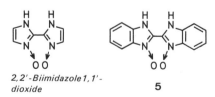

R = CH$_3$; *2,4-Dimethylthiazole 3-oxide;* 60%
R = H;   *4-Methylthiazole 3-oxide;* 14% (mainly oxidative ring opening)

X = Y = H;   *1-Methylpyrazole 2-oxide*
X = NH$_2$; Y = NO$_2$; *1-Methyl-5-nitropyrazole 2-oxide;* 10%

2,2'-Biimidazole 1,1'-dioxide

5

2,2'-Bibenzimidazole 1,1'-dioxide

That *N*-oxides of azoles are so seldom accessible by peroxy acid oxidation is due partly to

[257] *A.R. Katritzky, A.P. Ambler,* Physical Methods in Heterocyclic Chemistry, Vol. II, Academic Press, New York 1963.
[258] *H.H. Jaffe, H.L. Jones* in *A.R. Katritzky, A.J. Boulton,* Advances in Heterocyclic Chemistry, Vol. III, p. 232, Academic Press, New York, London 1964;
M.A. Weiner, J. Organometallic Chem. (Lausanne) *23,* (1970). 2, C 20 − C 22;
J.H. Nelson, R.G. Garvey, R.O. Ragsdale, J. Heterocyclic Chem. *4,* 591 (1967).
[259] *W. Dürckheimer,* Justus Liebigs Ann. Chem. *756,* 145 (1972).
[260] *M.J. Haddadin, C.H. Issidorides,* Tetrahedron Lett. 3253 (1965).
[261] *A. Dornow, K.J. Fürst, H.D. Jordan,* Chem. Ber. *90,* 2124 (1957).
[262] *E. Ochiai,* Aromatic Amine Oxides, p. 126, 174, Elsevier Publ., Co., Amsterdam, London, New York 1967.

[263] *E. Ochiai,* Aromatic Amine Oxides, p. 101, Elsevier Publ., Co., Amsterdam, London, New York 1967.
[264] *K. Tori, M. Ogata, H. Kano,* Chem. Pharm. Bull (Tokyo) *11,* 681 (1963).
[265] *P. Hamm, W.v. Philipsborn,* Helv. Chim. Acta *54,* 2363 (1971).
[266] *A. Pollak, B. Stanovnik, M. Tisler,* J. Org. Chem. *35,* 2478 (1970).

supervening side-reactions (*e.g.*, with imidazoles[292]) and partly to resistance toward peroxy acids (*e.g.*, with isoxazoles[293]).

The oxidizing agents are generally peroxy carboxylic acids, rarely inorganic peracids, but in recent times hydrogen peroxide/tungstic acid has been used more often[294, 295]. Hydrogen peroxide (Equation 4, *X* = *H*) alone is inadequate, but will serve when admixed with a carboxylic acid (or an inorganic oxygen acid) as a result of the peroxy acid equilibrium:

$$R-C\overset{O}{\underset{OH}{\diagup}} + H_2O_2 \rightleftharpoons R-C\overset{O}{\underset{OOH}{\diagup}} + H_2O$$

N-Oxidation can therefore often be effected by heating in hydrogen peroxide admixed with glacial acetic acid, benzoic acid, or other carboxylic acids. When highly concentrated hydrogen peroxide is used, this reaction can be performed also in water-immiscible solvents. Preformed peroxy acid achieves the same object. The very stable 3-chloroperbenzoic acid is much used[296]. On the whole, the oxidation proceeds the more readily the more basic the amine is and the stronger the acid is; the choice of the peroxy acid may thus be decisive. Somewhat the following order applies in respect of the efficacy of the peroxy acids:

peroxyacetic acid < peroxybenzoic acid < monoperoxyphthalic acid < monoperoxymaleic acid < trifluoroperoxyacetic acid.

Very recently two combinations have proven to be even more efficacious in two individual cases. Azole azines with bridgehead nitrogen were oxidized to *N*-oxides in at least small yields by a mixture of 85% hydrogen peroxide and polyphosphoric acid (with formic, acetic, or trifluoroacetic acid present, no oxidation takes place)[266]:

X = CH, N

4a   R = Cl;   *6-Chloro-3-nitropyridazine 1-oxide*
     R = OCH₃; *6-Methoxy-3-nitropyridazine 1-oxide*

4b   X = CH;   *6-Nitroimidazolo[1,2-b]pyridazine 5-oxide*
     X = N;   *6-Nitro-1,2,3-triazolo[4,3-b]pyridazine 5-oxide*

4c   X = CH;   *Imidazolo[1,2-b]pyridazine 5-oxide*
     X = N;   *1,2,3-Triazolo[4,3-b]pyridazine 5-oxide*

The addition of concentrated sulfuric acid to a mixture of 90% hydrogen peroxide and trifluoroacetic acid or acetic acid can dramatically increase the yield from resistant starting substances (polyhalopyridines and polyhalodiazines)[297].

The action of these inorganic acids is attributable presumably less to their water binding power than to the protonation of the peroxy acid.
Using 2-sulfoperoxybenzoic acid, too, should therefore be of interest[298]. Sulfuric acid (conc.) and hydrogen peroxide (60%) is another efficacious mixture[299].

Where an aqueous acid medium (H₂O₂/H₂O/R—COOH) is not employed, chloroform or ether (when using monoperoxyphthalic acid) are generally used as solvent. Less polar solvents often enable other sensitive groups in the molecules to be left unaffected. The reaction is then performed in combination with preformed peroxy acids, generally at low temperatures. Employing hydrogen peroxide/glacial acetic acid is the most convenient way of carrying out the reaction, although prolonged heating is often required and, as a result, can easily produce changes in groupings sensitive to oxidation.

[267] *T.A. Bryce, J.R. Maxwell*, Chem. Commun. *11*, 206 (1965);
R. Grigg, B.G. Odell, J. Chem. Soc. (B) 218 (1966);
A. Tatematsu, MM Yoshizumi, Tetrahedron Lett. 2985 (1967);
N. Bild, M. Hesse, Helv. Chim. Acta *50*, 1885 (1967);
O. Buchardt, A.M. Duffield, R.H. Shapiro, Tetrahedron *24*, 3139 (1968);
A. Kubo, Chem. Pharm. Bull. (Tokyo) *15*, 1079 (1967); 16, 1533 (1968).
[268] *I. Beger*, J. Prakt. Chem. *311*, 746 (1969).
[269] *H.G.O. Becker, D. Beyer, H.-J. Timpe*, J. Prakt. Chem. *312*, 869 (1970).
[270] *F. Uchimaru, S. Okada, A. Kosayama, T. Konno, E. Brown, R.J. Moser; W.W. Paudler, The-Kuei Chen, H. Ogura, Sh. Sugimoto*, J. Heterocycl. Chem. 8, 99, 189, 317, 391 (1971).
[271] *C.W. Muth, R.S. Darlak, W.H. English, A.T. Hammer*, Justus Liebigs Ann. Chem. *34*, 1163 (1962).

10.4.1.1.2 Side-reactions[285]

Alcohol and *amino* groups are suitably protected by *acylating* before carrying out the peroxy acid oxidation and are regenerated subsequently by hydrolysis. A primary amino group may be oxidized as far as nitro under certain circumstances (see p. 357). Sulfonamides can be oxidized to *N*-hydroxysulfonamides. Tertiary amino groups are generally converted into the *N*-oxides before the ring nitrogen in the azaaromatic compound is attacked. A marked electron displacement from the tertiary amino group to the azine nitrogen alone can enable the oxidation of the ring nitrogen to proceed preferentially, *e.g.*, in 4-dimethylamino-pyridine[300]. While (2-dimethylamino)pyridine is oxidized on the exocyclic *N* due to steric effects[300], α-arylaminopyridines afford the corresponding pyridine *N*-oxides with hydrogen peroxide/glacial acetic acid[301].

Azo groups are oxidized to azoxy groups by peroxy acids. *Olefinic* bonds are attacked more slowly than the nitrogen, and so quinine forms first quinuclidine *N*-monooxide and then the dioxide. The vinyl group of quinine remains intact[236]. As a mixed dioxide the dioxide of quinine can be deoxygenated selectively to the quinoline *N*-monooxide of the quinine with aqueous sulfurous acid[285, 312].

Thioether groups, but not the sulfur of attached thiophene rings[302], are in every case oxidized to sulfoxides or sulfones before the *N*-oxide is formed. *Formyl, carbinol,* and active *methylene* groups are often converted into carboxylic acids (sometimes also to keto groups)[303]:

Picolinic acid 1-oxide

Formyl groups can be protected by acetalization (and regenerated subsequently) if the following-on peroxy oxidation is conducted in nonaqueous medium[304]. Easily nucleophilically substituted azaaromatic compounds such as 1,3- and 1,4-diazines (pyrimidines, quinazolines, quinoxalines) or compounds whose azine ring has diminished aromatic character (5-phenanthridine, acridine) are sometimes hydroxylated in the α-position or γ-position[285, 306] (rarely in the β-position[305]); alternatively, they are cleaved oxidatively[285, 305, 307, 308] (the former view concerning the oxidation products of acridine needs to be revised[308]):

4-Isopropylquinazoline 1-oxide

Acridine 9-oxide

[272] R.W. Janssen, C.A. Discher, J. Pharm. Sci. 60, 798 (1971).

[273] A.R. Katritzky, J.M. Lagowski in A.R. Katritzky, A.J. Boulton, Advances in Heterocyclic Chemistry, Vol. I, p. 359, Academic Press, New York, London 1963.

[274] G. Zvilichovsky, Israel J. Chem. 6, 123 (1968).

[275] H. Goldner, G. Dietz, E. Carstens, Justus Liebigs Ann. Chem. 693, 233 (1966).

[276] J.C. Parham, T.G. Winn, G.B. Brown, J. Org. Chem. 36, 2639 (1971).

[277] K.-H. Wünsch, A.J. Bouldon in A.R. Katritzky, A.J. Boulton, Advances in Heterocyclic Chemistry, Vol. VIII, p. 277, Academic Press, New York, London 1967.

[278] E. Ochiai, Aromatic Amine Oxides, Elsevier Publ., Co., Amsterdam, London, New York 1967.

[279] P.H. Gund, G. Berkelhammer, J. Med. Chem. 14, 992 (1971).

Oxidative ring cleavages can arise not only in the azine ring being oxidized but also in a fused ring[385].

Hydrolytic exchange of ring substituents ensues readily if the ring nitrogen atom is part of an imidoyl chloride or imidic ester structure[385, 309]. (For *hydrolyses* on substituents [ester, acetal, *etc.*] when the reaction mixture contains water, see p. 359.)

1-Phenoxyphthalazine
3-oxide

R = CH₃, C₆H₅, OH

R = CH_3, C_6H_5, OH

R¹; R²≠H, or R¹≠C₆H₅

R^1; $R^2 \neq H$, or $R^1 \neq C_6H_5$

Replaced halide may have a rehalogenating action in the oxidizing medium[310, 311].

10.4.1.1.3 Influence of structure and substituents on oxidizability[285]

During peroxy acid oxidation the azaaromatic compound is the nucleophilic reaction partner, and it therefore reacts more slowly the less basic it is. The electrophilic partner in the reaction, the hydroxy oxygen of the peroxy acid, is spatially more demanding than a proton. As a result, steric hindrance can supervene on the influence of the basicity of the azaaromatic compound. If only one of the two α-positions is occupied this effect is generally not yet apparent. For example, 2-methylpyridines with a vacant 6-position are readily oxidized[312], because alkyl groups in the α-position facilitate the oxidation by virtue of enhancing the basicity. A certain degree of steric hindrance is produced only when both α-positions are alkylated[285, 313], or if a single alkyl group is particularly bulky (*tert*-butyl[313], *sec*-alkyl[285]). Under these circumstances it is necessary to employ stronger peroxy acids or make do with a lower yield, for instance, during oxidation of 2,6-diphenylpyridine (to *2,6-diphenylpyridine 1-oxide*) and 2-phenylquinoline (to *2-phenylquinoline 1-oxide*).

An approximate indication of the relative reactivity of the substituted azaaromatic compound — leaving aside steric effects — can be obtained from the basicity series of the substituted pyridines. The principal basicity-reducing ring substituents, approximately in decreasing order of their effect, are: phenyl and alkoxy (neither in the 4-position), iodo, bromo, acyl, chloro, cyano, fluoro, nitro (varies somewhat with the site of substitution; trifluoromethyl and sulfonyl are not allowed for).

The extent to which a particular substituent reduces the basicity is in general markedly dependent on the site of substitution and increases in the order 4- < 3- < 2-.

[280] *K. Fukui, A. Imamura, Ch. Nagata*, Chem. Pharm. Bull. (Tokyo) *33*, 122 (1960).

[281] *A.C. Fürst, C. Klausner, W.C. Cutting*, Nature *18*, 908 (1959).

[282] *A. Myler, G.B. Brown*, J. Biol. Chem. *244*, 4072 (1969).

[283] *G. Stöhrer, G.B. Brown*, Angew. Chem. *83*, 934 (1971).

[284] *P.B. Ghosh, B. Ternai, M.W. Whitehouse*, J. Med. Chem. *15*, 255 (1972).

[285] *E. Ochiai*, Aromatic Amine Oxides, p. 19, Elsevier Publ., Co., Amsterdam, London, New York 1967.

[286] *D.H. Hey, K.S.Y. Liang, M.J. Perkins*, J. Chem. Soc. (C) 7679 (1967).

[287] *E.W. Parnell*, Tetrahedron Lett. 3941 (1970).

[288] *T.A. Lisi*, Chem. Ind. 368 (1964).

[289] *T.J. Delia, G.B. Brown*, J. Org. Chem. *31*, 178 (1966).

[290] *V. Wölcke, G.B. Brown*, J. Org. Chem. *34*, 978 (1969).

[291] *A.J. Hubert*, Bull. Soc. Chim. Belges *79*, 195 (1970).

[292] *E.S. Schipper, A.R. Day* in *R.C. Elderfield*, Heterocyclic Compounds, Vol. V, p. 194, John Wiley & Sons, London 1957.

[293] *N.K. Kochetkov, S.D. Sokolov* in *A.R. Katritzky, A.J. Boulton*, Advances in Heterocyclic Chemistry, Vol. II, p. 421, Academic Press, New York, London 1963.

[294] *W.W. Paudler, D.J. Pokorny*, J. Org. Chem. *36*, 1720 (1971).

[295] U.S.P. 3047579 (1962), Shell Oil Co., Inv.: *R.C. Whitman*; C.A. *58*, 7916c (1963).

It is thus understandable that the strongest inhibition of peroxy acid oxidation is produced by α-substituents. A halogen or an alkoxy group in the α-position is quite strongly deactivating, the latter presumably by virtue of a combined reduction in basicity and steric hindrance. α-Cyanoazaaromatic compounds, which are N-oxidized with astonishing ease, display a very strong and inexplicable deviation from the correlation between basicity and reactivity toward peroxy acids. By contrast, β- and γ-cyano aromatic compounds behave in the expected manner. Singly trifluoromethylated pyridines can be oxidized relatively easily, while compounds with two trifluoromethyl groups react with greater difficulty (trifluoroacetic acid/hydrogen peroxide is required)[312, 314, 315]. Pentahalopyridines demand particularly active oxidizing agents[297], while at the same time prolonged heating (50°) of the mixture reduces the yield by deoxygenation[316]. Quinolines with an electron-withdrawing substituent in the 8-position are oxidized to the corresponding oxindoles via 3-quinolinols[317].

10.4.1.1.4 Oxidation of polyazaaromatic compounds

During oxidation of polyazaaromatic compounds, N-dioxides and isomeric N-monooxides can be formed, sometimes together, in addition to the side-reactions and deviation reactions mentioned on p. 359. Where the nitrogen atoms belong to different noncondensed rings (polyazabiphenyls and others), the oxidation is subject to the usual electronic and steric effects (p. 370), and N-monooxides and N-dioxides can also be obtained[318].

Among diazines N-dioxides have been obtained side by side with or from the monooxides in the case of pyridazines (yields <10%)[243], cinnolines[243, 305], quinoxalines[313, 319, 320], phenazines[320], quinoxalino[2,3-b]quinoxaline[321], 1,2,4-benzotriazine[322], and pyrazines[254, 299].

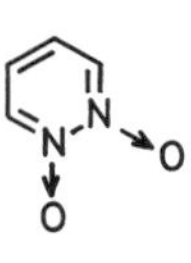

Pyridazine 1,2-dioxide 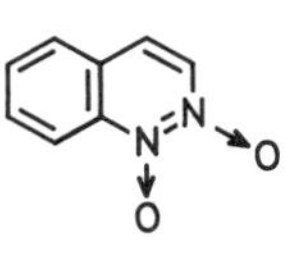   Cinnoline 1,2-dioxide 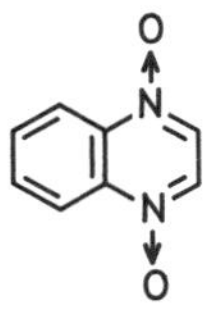   Quinoxaline 1,4-dioxide

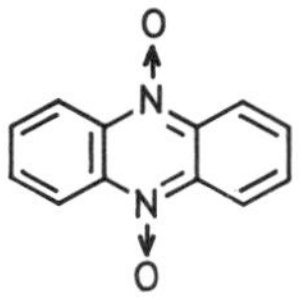

Phenazine 9,10-dioxide 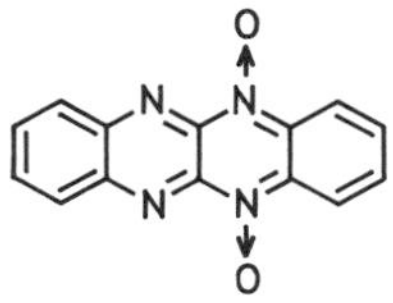   Quinoxalino[2,3-b]quinoxaline 5,12-dioxide

N-Dioxides containing one respective N-oxide function in two neighboring rings of a condensed system have also been prepared:

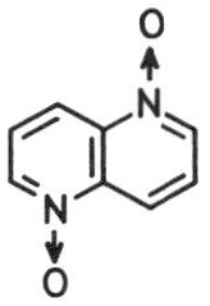

1,5-Naphthyridine 1,5-dioxide[320] 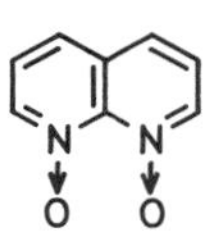   1,8-Naphthyridine 1,8-dioxide[323] 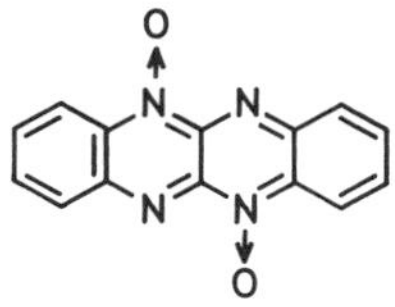   Quinoxalino[2,3-b] quinoxaline 5,11-dioxide[321]

With naphthyridines (1,5- and 1,6-) N-monooxides are also obtained[294].

Where the nitrogen atoms are located in farther spaced apart rings of a condensed system a double oxidation should not be difficult[302], e.g.:

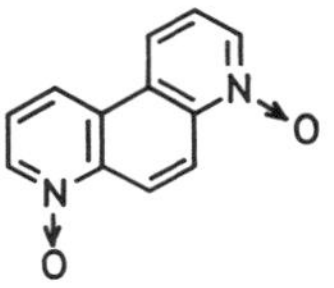

4,7-Phenanthroline 4,7-dioxide[320]

Which monooxide of a polyazaaromatic compound is formed exclusively or preferentially is dependent substantially on the previously related effects of basicity and steric hindrance (p. 360); interactions between two substituents can produce apparent anomalies[324]. From substituted pyridines, pyridazines[240], and cinnolines[305] both isomers are frequently formed, while from phthalazines and quinoxalines one only arises[320]. Quinazolines generally afford the 3-oxides unless a deactivating 4-substituent favors formation of the 1-oxide[325]:

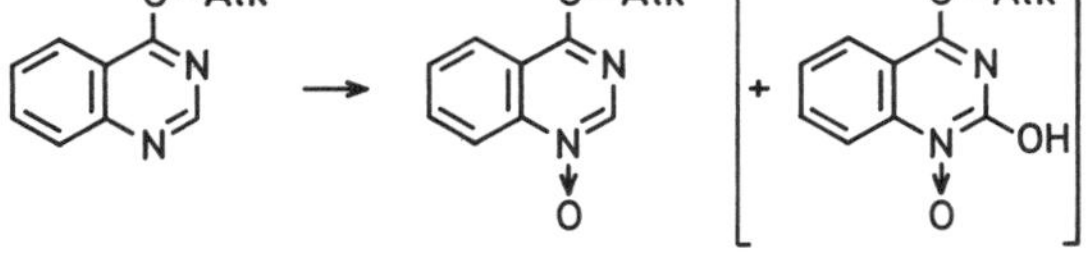

or tautomer

[296] L.F. Fieser, M. Fieser, Reagents for Organic Synthesis, John Wiley & Sons, New York, London, Sidney 1967.

[297] G.E. Chivers, H. Suschitzky, Chem. Commun. 28 (1971);
G.E. Chivers, H. Suschitzky, J. Chem. Soc. (C) 2867 (1971).

1,2,4-Triazines are oxidized in the 1-position; by contrast, *3-amino-1,2,4-triazines* give rise to the *2-oxide* and *1-oxide* side by side[326] (new ref.[635]).

Purines are evidently substantially destroyed by hydrogen peroxide/carboxylic acid[327] unless a methyl or amino group in the pyridine portion of the molecule facilitates the *N*-oxidation of the adjacent nitrogen[290, 327–329]. 3-Chloroperoxybenzoic acid in glacial acetic acid at 65° oxidizes cytosine arabinoside to *cytosine arabinoside 3-oxide*[330].

3-Phenylbenzo[f]quinoxaline is oxidized in the 4-position to give *3-phenylbenzo[f]quinoxaline 4-oxide*[331].

During peroxy acid oxidation of 3-phenyl-1,2,4-benzotriazine **5** in the cold 3-phenyl-1,2,4-benzotriazine 1-oxide **6** is formed, at 50° the 2-oxide **7**[332]:

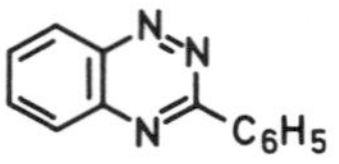

5

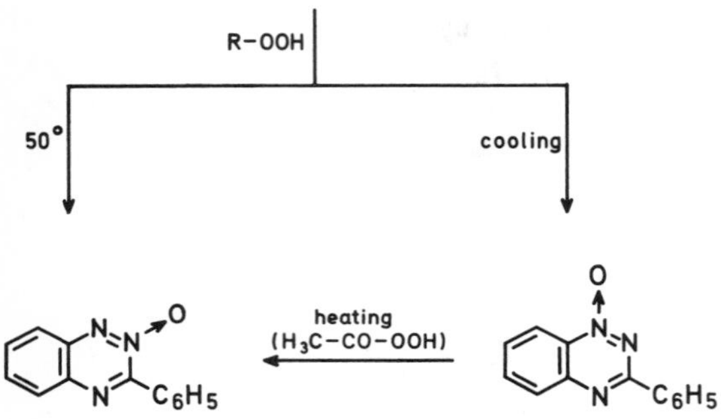

7   **6**

10.4.1.2 Transformation of *N,N*-dioxides into *N*-monooxides

10.4.1.2.1 Partial deoxygenation of *N,N*-dioxides

The partial deoxygenation of the *N*-dioxides of polyazaromatic compounds is generally of no synthetic significance. It serves a useful purpose only where the alternative preparation of the isomeric *N*-oxides concerned is impossible or difficult.

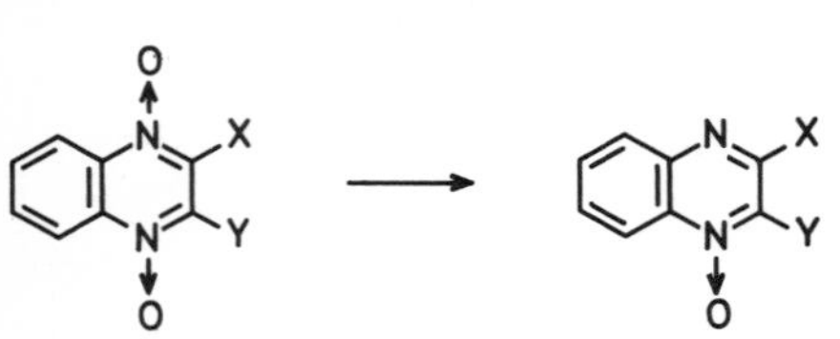

or tautomer

Reagent	
H$_2$SO$_3$[333]	X = H, Y = C$_6$H$_5$; *2-Phenylquinoxaline 1-oxide*
H$_2$-Raney nickel[334]	X = H, Y = CH$_3$; *2-Methylquinoxaline 1-oxide*; 50% X = OH, Y = NH$_2$; *3-Amino-2-quinoxalinol 1-oxide* (accompanied by complete deoxygenation)
PCl$_3$[335]	X = H, Y = −CO−N(CH$_3$)C$_6$H$_5$; *2-(N-Methyl-N-phenyl quinoxaline-carboxamide 1-oxide)*

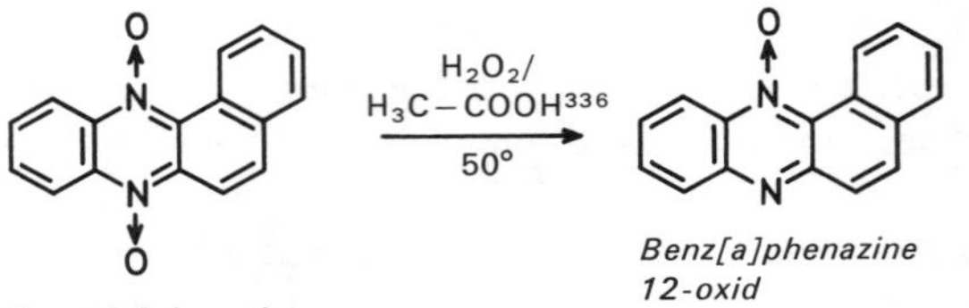

Benzo[a]phenazine 7,12-dioxide   *Benz[a]phenazine 12-oxid*

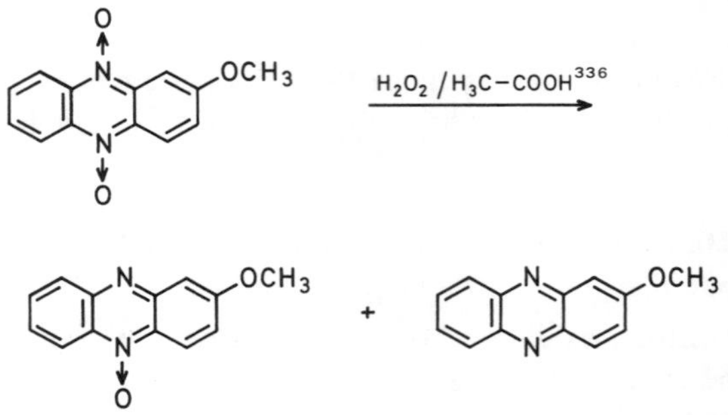

2-Methoxyphenazine 9-oxide   *2-Methoxyphenazine*

1-Hydroxypyrazole 2-oxides (**6, 7a**, p. 383) reduce to the corresponding *1-hydroxypyrazoles* **8** by reaction with sodium dithionite[337]. The same reducing agent furnishes 1-hydroxy-pyrazol-4-ols (**8**, $R^2=OH$) from 4*H*-pyrazol-4-one 1,2-dioxides(**9**)[338], which suffer a further reduction step in addition to the deoxygenation:

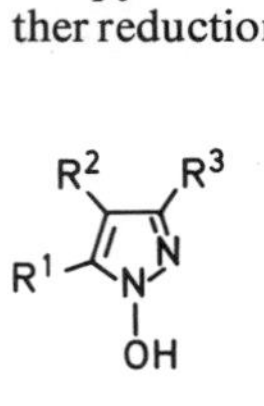

or tautomer   **9**

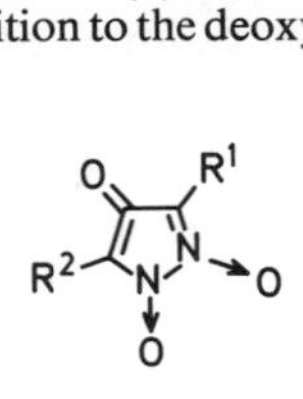

8

[298] *J.M. Bachbawat, N.K. Mathur*, Tetrahedron Lett. 691 (1971).

[299] *D. Kyriacou*, J. Heterocycl. Chem. **8**, 697 (1971).

[300] *R. Frampton, C.D. Johnson, A.R. Katritzky*, Justus Liebigs Ann. Chem. 749, 12 (1971).

[301] *D.M. Bailey, R.E. Johnson, J.D. Conolly, R.A. Ferrari*, J. Med. Chem. **14**, 439 (1971).

[302] *L.H. Klemm, S.B. Mathur, R. Zell, R.E. Merrill*, J. Heterocycl. Chem. **8**, 931 (1971).

[303] *E. Ochiai*, Aromatic Amine Oxdies, p. 34, Elsevier Publ., Co., Amsterdam, London, New York 1967.

[304] *W. Krüger, G. Krüger*, Z. Chemie **10**, 184 (1970).

[305] *M.H. Palmer, E.R.R. Russel*, J. Chem. Soc. (C) 2621 (1968).

[306] *E. Ochiai*, Aromatic Amine Oxides, p. 331, Elsevier Publ., Co., Amsterdam, London, New York 1967.

10.4.1.2.2 Partial *N*-oxide-pyridone rearrangement in *N*-dioxides

The transformations of the *N*-oxides into the corresponding chloro[254] or hydroxy azaaromatic compounds described on p. 354 and p. 395 occur on only one *N*-oxide function in some *N*-dioxides, *e.g.* in quinoxaline *N*-dioxides[339, 340].

10.4.1.3 Azine *N*-oxides by cyclization of oximes and hydroxylamine compounds or their precursors and hydroxylamine

In simple pyridine derivatives direct oxidation of the parent aromatic compounds is generally the most advantageous route to the *N*-oxides. The cyclization reactions described in what follows are appropriate where difficulties arise during direct oxidation or where the aromatic starting compounds are less easily accessible. Also, these cyclization reactions may give particular and specifically substituted *N*-oxides, more particularly clearly defined and anticipated isomers of *N*-monooxides of polyazaaromatic compounds. On the whole cyclization reactions are less suitable for obtaining monocyclic compounds than for preparing condensed systems.

The following scheme reproduces the main variants of the ring closure principle involved in the methods described here; some tautomerism-linked modifications have been embraced:

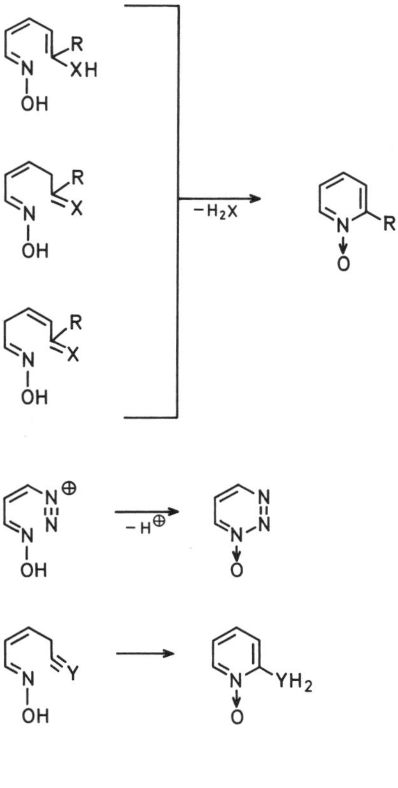

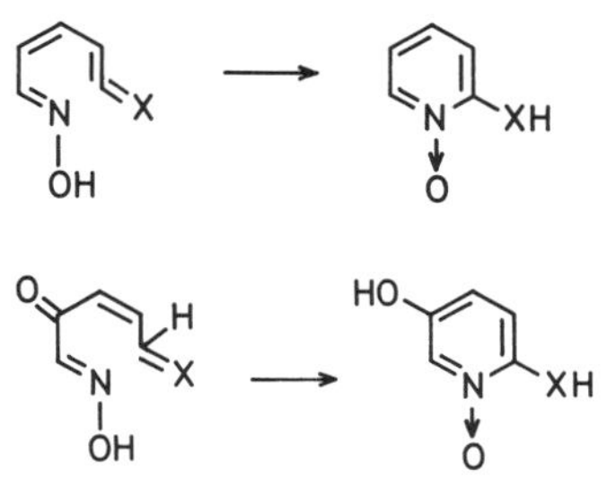

The acyclic oximes required are generally generated *in situ* from other precursors and often cannot be isolated.

10.4.1.3.1 Azine *N*-oxides *via* oximes of unsaturated 1,5-dicarbonyl compounds ($5+1$ or $6_{acycl.} \rightarrow 6_{cycl.}$) (*cf.* p. 367)

Glutaconaldehyde (2-pentenedial) and its derivatives can form *pyridine N-oxides* with hydroxylamine[341, 342]. α-Formyl-2-tolualdehyde (homophthalaldehyde) forms *isoquinoline N-oxide* with hydroxylamine[341]. This nonoxidative route is in itself of no preparative interest, but could achieve importance as a result of a recently investigated variant for special cases where easily oxidized groups in an azaaromatic compound forbid the peroxy acid oxidation.

According to this variant, which has been studied on pyridine, methylated pyridines, and isoquinoline, pyridine is, for instance, reacted with 4-chloro-1,3-dinitrobenzene to give *N*-(2,4-dinitrophenyl)pyridinium chloride. The latter is cleaved to the oxime in a Zincke-cleavage with 2 equivalents hydroxylammonium chloride and 3 equivalents triethylamine in methanolic solution, and the oxime is heated in suitable solvents (possibly with added acid) to afford the pyridine *N*-oxide (overall yield 45–88%)[343]:

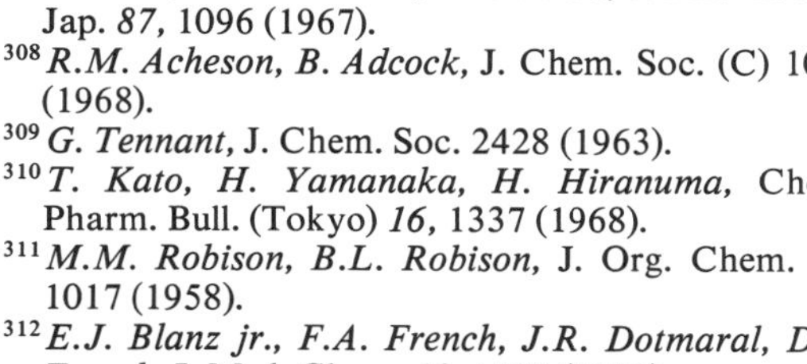

[307] *T. Kato, H. Yamanaka, T. Shibata*, J. Pharm. Soc. Jap. *87*, 1096 (1967).

[308] *R.M. Acheson, B. Adcock*, J. Chem. Soc. (C) 1045 (1968).

[309] *G. Tennant*, J. Chem. Soc. 2428 (1963).

[310] *T. Kato, H. Yamanaka, H. Hiranuma*, Chem. Pharm. Bull. (Tokyo) *16*, 1337 (1968).

[311] *M.M. Robison, B.L. Robison*, J. Org. Chem. *23*, 1017 (1958).

[312] *E.J. Blanz jr., F.A. French, J.R. Dotmaral, D.A. French*, J. Med. Chem. *13*, 1124 (1970).

[313] *E. Hayashi, Y. Miura*, J. Pharm. Soc. Jap. *87*, 643 (1967).

On letting the barium salt of 2,4,6-heptanetrione (diacetylacetone; prepared from 2,6-dimethyl-4*H*-pyran-4-one and barium hydroxide) stand with hydroxylammonium chloride in ethanol *2,6-dimethyl-4-hydroxyaminopyridine 1-oxide* (R=CH₃) is formed[341, 344]:

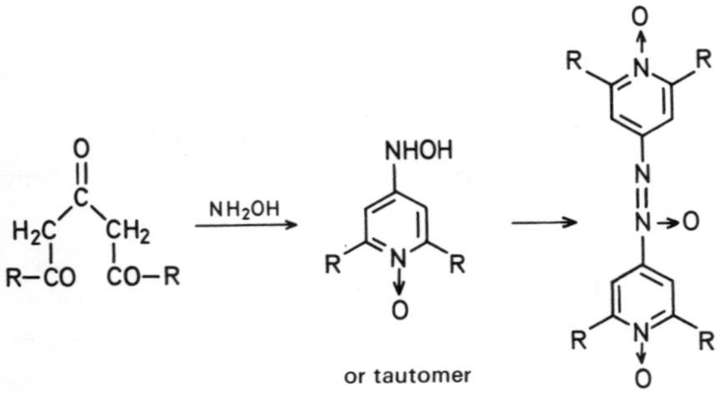

or tautomer

Holding 2,4,6-heptanetrione with 3 equivalents hydroxylamine for five days at 65° affords *azoxy dimers* (70%), as recent investigations have shown[345, 346].

The pyridinium salts **10** obtained from 2-pyridyl ketone oximes by quaternization with phenacyl bromide or bromoacetaldehyde can be cyclocondensed to the *pyrido[1,2-a]pyrazinium 2-oxide bromides* **11**[347–349]:

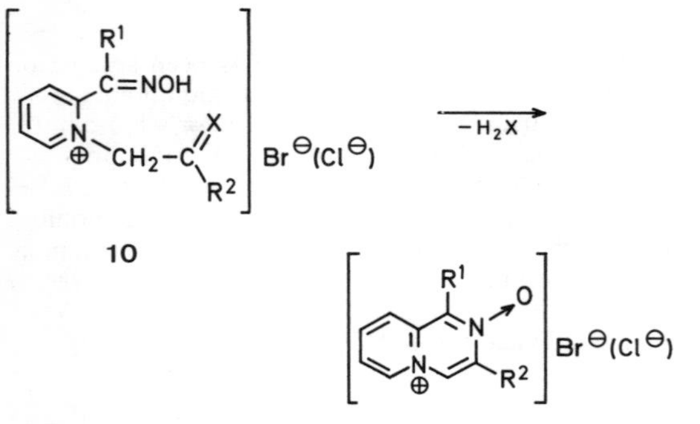

10

11

X = O, NOH (for R¹ = −C(CH₃)₃)
R¹ = H, CH₃, C₂H₅, −CH(CH₃)₂, −C(CH₃)₃, C₆H₅
R² = H, C₆H₅, CH₃
(R¹ and R² must not both be aryl together)

To bring the reaction about, heating with concentrated acids, generally with addition of a suitable solvent, is necessary. In the case of the *tert*-butyl ketone **10** (R¹ = *tert*-butyl, R² = H) the dioxime (X = NOH) is used instead of the monooxime. The yields are moderate to very good. In aromatically substituted derivatives **10** (R¹, R² = aryl) the reaction deviates to become a Beckmann rearrangement[349]. Starting from Type **10** isoquinolinium compounds it is possible also to obtain *isoquinolino[1,2-b]pyrazinium 2-oxide* salts[348].

2-Acylaminobenzophenone oximes (**12**; R¹ = Aryl) and 2-acylaminobenzaldehyde oximes (**12**; R¹ = H) cyclize to *quinazoline 3-oxides* (**13**) in the presence of acids, generally in good yields[350–352]. The solution in glacial acetic acid saturated with hydrogen chloride is allowed to stand one or two days. In the case of the aldoxime (**12**; R¹ = H) alone boron trifluoride is preferred as condensation agent[352]. Good yields of the 3-oxides **13** are obtained also from the 2-amino oximes **14** (letting stand in glacial acetic acid with excess acyl chloride[350, 351] or reacting with formic acid)[353]. 2-Acylaminoacetophenone oximes, too (**12**; R¹ = CH₃) afford the corresponding *4-methylquinazoline 3-oxides* (**13**; R¹ = CH₃), either already at the oxime preparation stage or during the subsequent treatment with acids under certain conditions[354]:

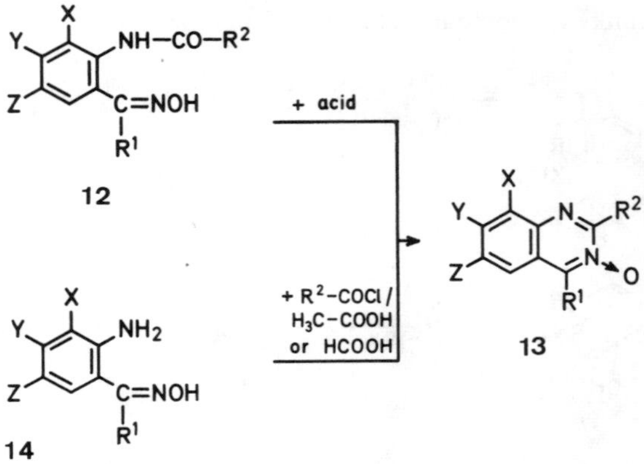

12

13

14

[314] *Y. Kobayashi, I. Kumadaki, S. Taguchi*, Chem. Pharm. Bull. (Tokyo) *17*, 2335 (1969).

[315] *Y. Kobayashi, I. Kumadaki*, Chem. Pharm. Bull. (Tokyo) *17*, 510 (1969).

[316] *S.M. Roberts, H. Suschitzky*, Chem. Commun. 893 (1967);
S.M. Roberts, H. Suschitzky, J. Chem. Soc. (C) 1537 (1968).

[317] *T. Nakashima, L. Suzuki*, Chem. Pharm. Bull. (Tokyo) *17*, 2293 (1969).

[318] *H. Igeta, T. Tsuchiya, C. Okuda, H. Yokogawa*, Chem. Pharm. Bull. (Tokyo) *18*, 1340 (1970).

[319] *G.W.H. Cheeseman* in *A.R. Katritzky, A.J. Boulton*, Advances in Heterocyclic Chemistry, Vol. II, p. 234, Academic Press, New York, London 1963.

[320] *E. Ochiai*, Aromatic Amine Oxides, p. 30, Elsevier Publ., Co., Amsterdam, London, New York 1967.

[321] *R. Kuhn, P. Skrabal*, Chem. Ber. *101*, 3913 (1968).

[322] *J.P. Horwitz, R.C. Elderfield*, Heterocyclic Compounds, Vol. VII, p. 745, John Wiley & Sons, London 1961.

[323] *W.W. Paudler, T.H. Kress* in *A.R. Katritzky, A.J. Boulton*, Advances in Heterocyclic Chemistry, Vol. XI, p. 139, Academic Press, New York, London 1969.

[324] *E. Ochiai*, Aromatic Amine Oxides, p. 47, Elsevier Publ., Co., Amsterdam, London, New York 1967.

[325] *W.L.F. Armarego* in *A.R. Katritzky, A.J. Boulton*, Advances in Heterocyclic Chemistry, Vol. I, p. 276, Academic Press, New York, London 1963.

Oxidation of the corresponding quinazolines with hydrogen peroxide/glacial acetic acid affords *quinazoline 1-oxides*.

2-(Formylmethyl)benzoic acid oxime is converted in 75% extent to *1-isoquinolinol 2-oxide* or its tautomeric form in boiling chlorobenzene[355]:

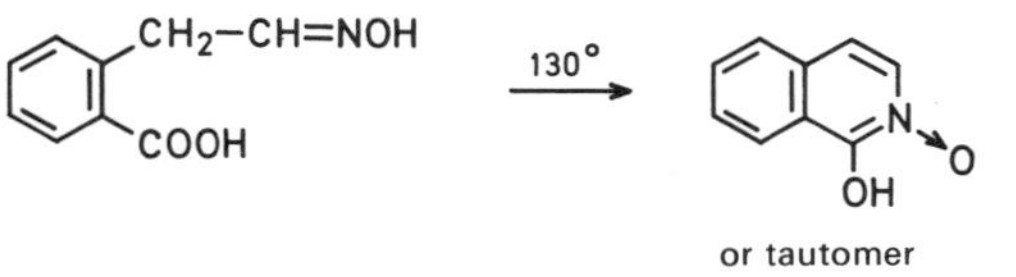

Benzil oxime guanylhydrazone cyclizes to *4-amino-5,6-diphenyl-1,2,4-triazine 4-oxide* in acetic acid with loss of ammonia[356]:

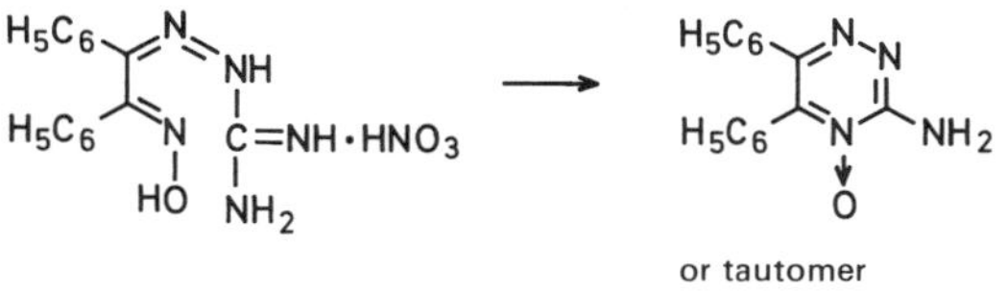

The reaction evidently possesses more general character, as is demonstrated by the following reaction[357]:

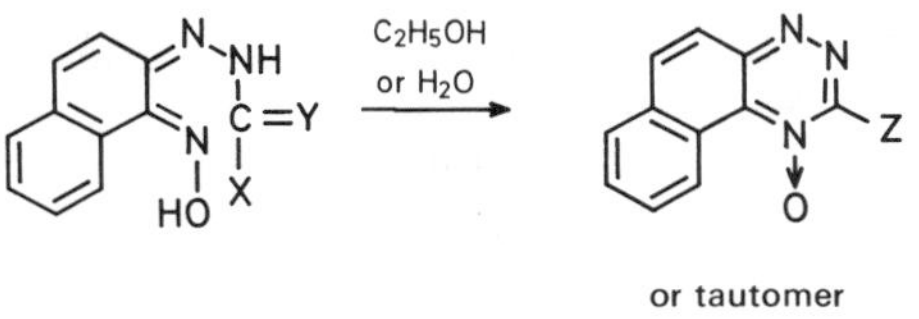

X = NH$_2$, NH–C$_6$H$_5$, OC$_2$H$_5$, C$_6$H$_5$
Y = NH, N–C$_6$H$_5$, O, S
Z = NH$_2$; *3-Aminonaphtho[2,1-e]-1,2,4-triazine 1-oxide*
Z = NH–C$_6$H$_5$; *3-Anilinonaphtho[2,1-e]-1,2,4-triazine 1-oxide*
Z = OH; *Naphtho[2,1-e]-1,2,4-triazin-3-ol 1-oxide*
Z = C$_6$H$_5$; *3-Phenylnaphtho[2,1-e]-1,2,4-triazine 1-oxide*

Depending on the reaction conditions, the yields of the *naphtho[2,1-e]-1,2,4-triazine 4-oxides* are sometimes very good.

10.4.1.3.2 1,3,5-Triazine *N*-oxides from *N,N'*-dicyanoamidines and hydroxylamine (5 + 1)

4-Substituted 2,6-diamino-1,3,5-triazine 1-oxides are formed from alkali metal salts of *N,N'*-dicyanoamidines and hydroxylammonium chloride in 2-ethoxyethanol at room temperature[358, 359]:

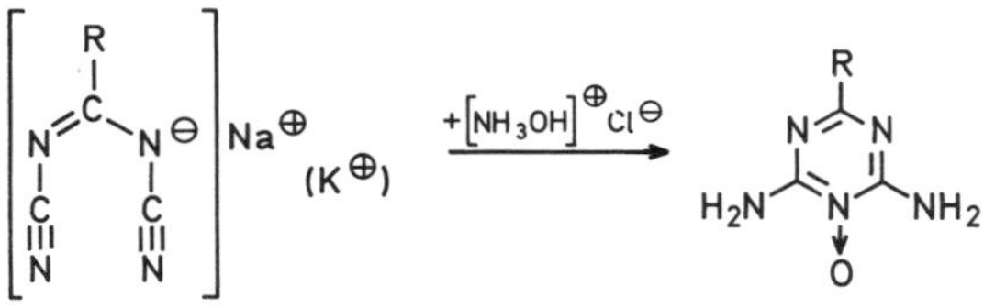

R = NH$_2$, CH$_3$, C$_2$H$_5$, C$_3$H$_7$, –CH$_2$–CH(CH$_3$)$_2$, C$_{11}$H$_{23}$, C$_6$H$_5$,
4–Cl–C$_6$H$_5$

Yields are invariably higher (24–69%) than during direct oxidation of the 2,4-diamino-1,3,5-triazines.

10.4.1.3.3 Pyridine 1-oxides and quinazoline 3-oxides from heterocycles and hydroxylamine (6$_{cycl.}$ + 1–1)

4H-Pyran-4-ones form *4-pyridinol 1-oxides* with hydroxylamine on heating[341, 360]

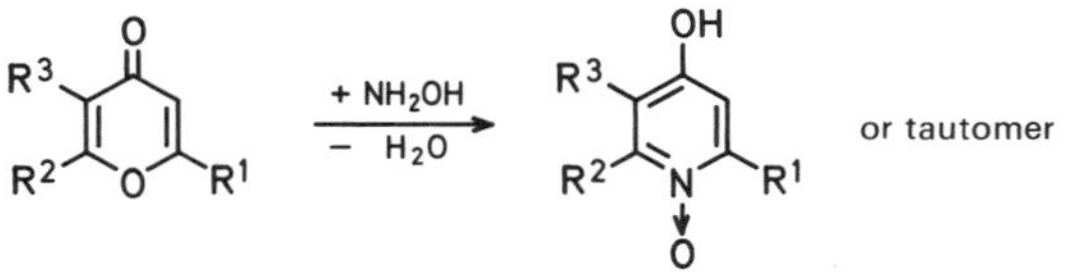

R^1 = COOH, R^2 = R^3 = H: *4-Hydroxypicolinic acid 1-oxide*
R^1 = COOH, R^2 = H, R^3 = O-Alkyl: *5-Alkoxy-4-hydroxypicolinic acid 1-oxide*
R^1 = R^2 = C$_6$H$_5$; R^3 = H: *2,6-Diphenyl-4-pyridinol 1-oxide*

By contrast, after several days at room temperature[341, 361], or with excess hydroxylammonium

[326] *W.W. Paudler, Teh-Kuei Chen,* J. Org. Chem. *36*, 787 (1971).
[327] *M.A. Stevens, A. Giner-Sorolla, H. Smith, G.B. Brown,* J. Org. Chem. *27*, 567 (1962).
[328] *A. Giner-Sorolla, C. Gryte, M.L. Cox, J.C. Parham,* J. Org. Chem. *36*, 1228 (1971).
[329] *N.J.M. Birdsall, T.-Ch. Lee, T.J. Delia, J.C. Parham,* J. Org. Chem. *36*, 2635 (1971).
[330] *R.P. Panzica, R.K. Robins, L.B. Townsend,* J. Med. Chem. *14*, 259 (1971).
[331] *J.W. Lown, M.H. Akhtar,* Can. J. Chem. *49*, 1610 (1971).
[332] *J.P. Horwitz* in *R.C. Elderfield,* Heterocyclic Compounds, Vol. VII, p. 720, John Wiley & Sons, London 1961.
[333] *E. Hayashi, Ch. Iijima,* Yakugaku Zasshi *82*, 1093 (1962); C.A. *58*, 4551 (1963).
[334] *E. Ochiai,* Aromatic Amine Oxides, p. 193, Elsevier Publ., Co., Amsterdam, London, New York 1967.

[335] *E. Ochiai,* Aromatic Amine Oxides, p. 198, Elsevier Publ., Co., Amsterdam, London, New York 1967.
[336] *E. Ochiai,* Aromatic Amine Oxides, p. 205, Elsevier Publ., Co., Amsterdam, London, New York 1967.
[337] *J.P Freeman jr., I.J. Gannon,* J. Org. Chem. *34*, 194 (1969).
[338] *J.P. Freeman, J.J. Gannon, D.L. Surbey,* J. Org. Chem. *34*, 187 (1969).
[339] *G.H.W. Cheeseman,* J. Chem. Soc. 1246 (1961).
[340] *J.K. Landquist,* J. Chem. Soc. 2830 (1953).
[341] *E. Ochiai,* Aromatic Amine Oxides, p. 51, Elsvier Publ., Co., Amsterdam, London, New York 1967.
[342] *J.B. Bapat, D.St.C. Black, R.F.C. Brown* in *A.R. Katritzky, A.J. Boulton,* Advances in Heterocyclic Chemistry, Vol. X, p. 206–215, Academic Press, New York, London 1969.

chloride and pyridine in boiling ethanol[346], *4-hydroxyaminopyridine N-oxides* are obtained in good yield (possibly *via* a subsequent oximation of the tautomeric form of the primarily formed pyridinol 1-oxide)[345]. These latter are converted rapidly into the red *4,4'-azodipyridine 1,1'-dioxides* with 10% caustic soda[346, 344, 361]:

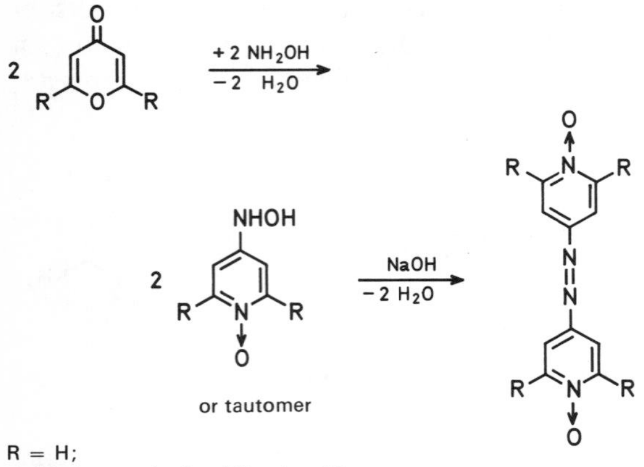

R = H;
4-(Hydroxyamino)pyridine 1-oxide ⟶
    4,4'-Azodipyridine 1,1'-dioxide

R = CH₃;
4-Hydroxyamino)-2,6-dimethylpyridine 1-oxide ⟶
    2,2',6,6'-Tetramethyl-4,4'-azodipyridine 1,1'-dioxide

R = C₂H₅;
2,6-Diethyl-4-(hydroxyamino)pyridine 1-oxide ⟶
    2,2',6,6'-Tetraethyl-4,4'-azodipyridine 1,1'-dioxide

Condensed 4*H*-pyran-4-ones, for instance, flavones[362], apparently do not form azine *N*-oxides with hydroxylamine but, instead *(2-hydroxyphenyl)isoxazoles*. Sulfur in a thiapyranonelike bond can be replaced by hydroxylamine[342] (recent literature, see ref. [636]).

Pyrylium salts afford pyridine *N*-oxides with ethanolic hydroxylamine solution on shaking briefly[363, 364]:

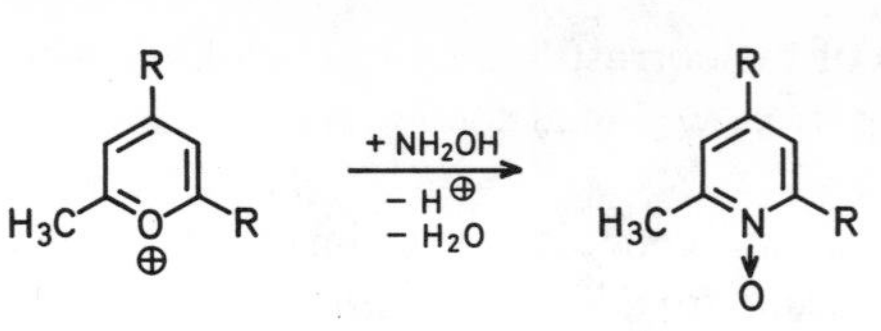

R = CH₃; *2,4,6-Trimethylpyridine 1-oxide*
R = C₆H₅; *4,6-Diethyl-2-methylpyridine 1-oxide*

A reaction sequence starting from *2-furyl methyl ketone* leads to *2-methyl-3,6-pyridinediol N-oxide*[365]:

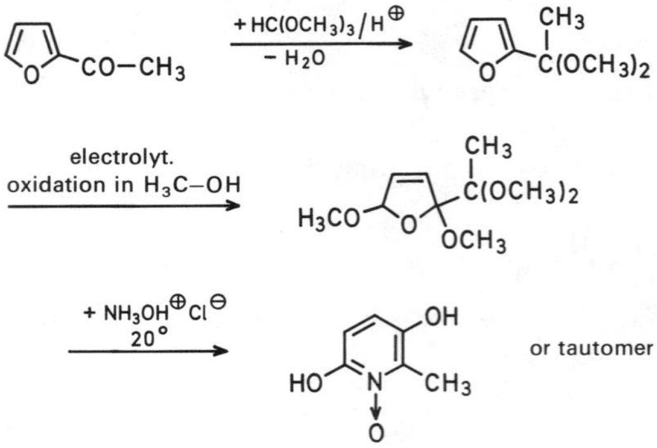

This technique can be varied in terms of the acyl group. *Quinazoline 3-oxides* **13** are obtained from quinazolines and hydroxylammonium chloride in cold caustic soda[341]. The *2-hydroxyiminomethylaminobenzaldehyde oxime* or corresponding *acetophenone oxime* coformed to the extent of one half is likewise converted into *quinazoline 3-oxide* **13a** by transoximation in hot acetone. At elevated temperatures the same components form 2-aminoquinazoline 3-oxide **15** and 2-aminobenzaldehyde oxime (cf. refs. [350, 351]):

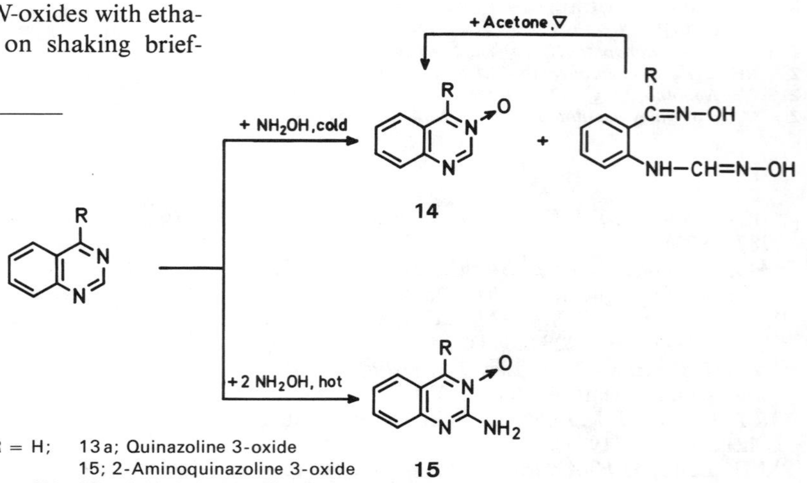

R = H;   **13a**; Quinazoline 3-oxide
          15; 2-Aminoquinazoline 3-oxide
R = CH₃; **13a**; 4-Methylquinazoline 3-oxide
          15; 2-Amino-4-methylquinazoline 3-oxide

[343] *Y. Tamura, N. Tsujinoto, M. Mano*, Chem. Pharm. Bull. (Tokyo) *19*, 130 (1971);
Y. Tamura, N. Tsujimoto, M. Uchimura, Chem. Pharm. Bull. (Tokyo) *19*, 143 (1971);
Y. Tamura, N. Tsujimoto, Chem. Ind. 926 (1970).

[344] *F. Parisi, P. Bovina, A. Quilico*, Gazz. Chim. Ital. *92*, 1138 (1962).
[345] *L.G. Wideman*, Chem. Commun. 1309 (1970).
[346] *P. Yates, M.J. Jorgenson, S.K. Roy*, Can. J. Chem. *40*, 2146 (1962).

In 4-methoxy- and 4-phenoxyquinazoline in boiling methanol a type of Dimroth rearrangement of 4-aminoquinazoline 3-oxide occurs after substitution:

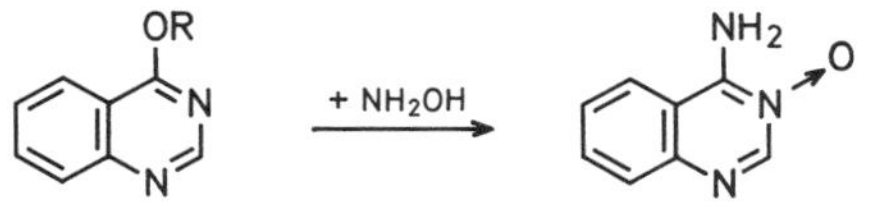

These quinazoline rearrangement techniques are very useful because the direct oxidation of quinazolines often affords either *4-quinazolinols* or, if the 4-position is occupied, *quinazoline 1-oxides*.

Re the displacement of the substituted *N*-atom of pyridinium compounds by hydroxylamine, see p. 363.

10.4.1.3.4 Diazine and triazine *N*-oxides from α-hydrazono oximes, aromatic 2-amino oximes, or 2-amino hydroxamic acids by condensation or diazotization (5 + 1 or $6_{open} \rightarrow 6_{cycl}$) (cf. p. 363)

α-(2-Aminoaryl) oximes **16**[341], α-(2-aminoaryl) hydroxamic acids[366, 637] **17**, or α-(2-aminoaryl) amidoximes **18** react with orthoformic acid triesters to condensed pyrimidine *N*-oxides, sometimes in excellent yield **19**:

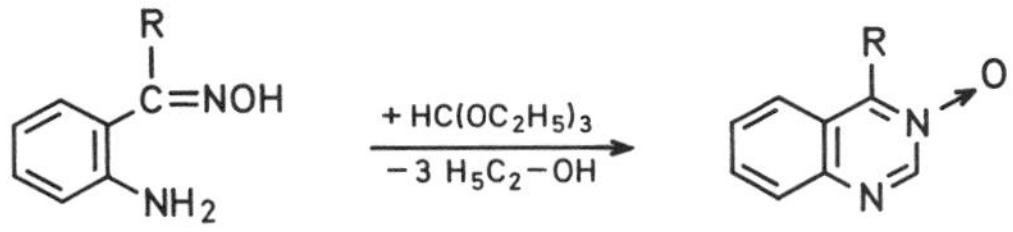

16

R = H;   *Quinazoline 3-oxide*
R = CH$_3$; *4-Methylquinazoline 3-oxide*

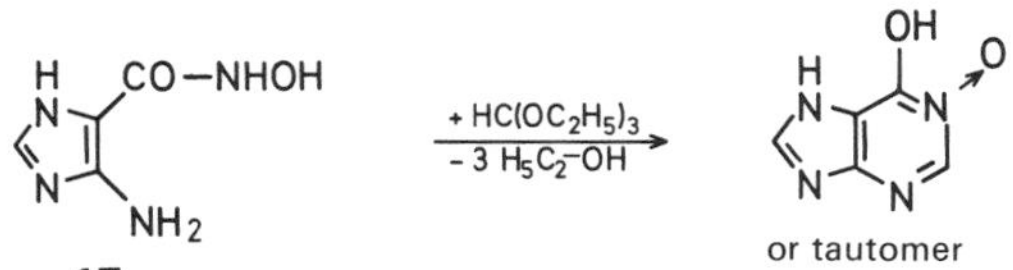

17

Purine 1-oxide

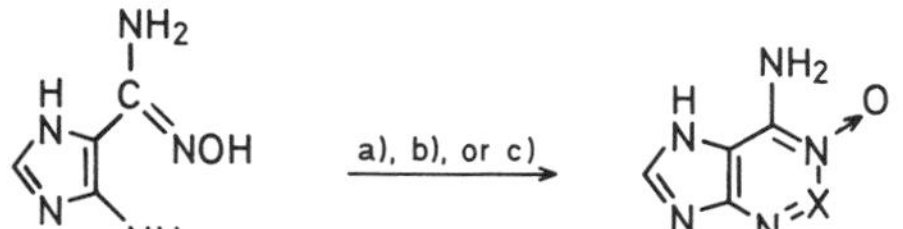

18

or tautomer

a) + HC(OC$_2$H$_5$)$_3$; X = CH $\longrightarrow$ **19** *6-Aminopurine 1-oxide*
b) + CS$_2$;   X = C—SH $\longrightarrow$ **20** *2-Mercaptopurine 1-oxide*[367]
c) + HNO$_2$;   X = N $\longrightarrow$ **21** *4-Amino-5H-imidazo[4,5-d]-1,2,3-triazine 3-oxide*[368]

Similar syntheses of quinoxaline 3-oxides and *pteridine 1-oxides* succeed also with acid anhydrides[342].

Preparing the quinazoline 3-oxides from **16** is probably even more advantageous than obtaining them from heterocycles and hydroxylamine (see p. 366).

With benzeneboronic acid in boiling xylene 2-aminobenzaldehyde oxime (**16**; *R = H*) condenses to the hydrolysis-sensitive *1,2-dihydro-2-phenyl-1,3,2-benzodiazoborine 3-oxide*[369].

With carbon disulfide **18** furnishes *6-amino-2-mercaptopurine 1-oxide* **20**[367]. Diazotization of the primary amino group in **18** or **22** leads to the *1,2,3-triazine 3-oxides* **21** and **23** respectively[370, 371, 341] provided that the oximes **22** are used in the *E*-form (the *Z*-isomers yield instead *1,2-benzisoxazoles* while losing nitrogen):

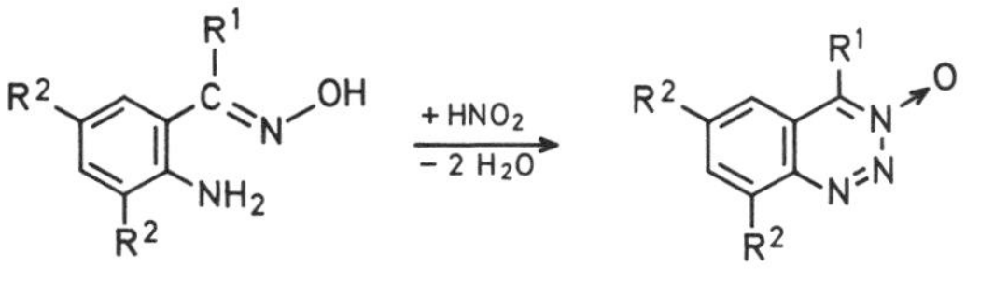

or tautomer (R^1 = OH)
22

or tautomer (R^1 = OH)
23

R^1 = CH$_3$, C$_6$H$_5$, NH$_2$, OH[342]
R^2 = H, CH$_3$

[348] *C.K. Bradsher, S.A. Telang*, J. Org. Chem. *31* (1966).

[349] *J. Adamson, E.E. Glover*, J. Chem. Soc. (C) 861 (1971).

[350] *L.H. Sternbach, S. Kaiser, E. Reeder*, J. Amer. Chem. Soc. *82*, 475 (1960).

[351] *L.H. Sternbach, E. Reeder, O. Keller, W. Metlesies*, J. Org. Chem. *26*, 4488 (1961).

[352] *A. Stempel, I. Douvan, L.H. Sternbach*, J. Org. Chem. *33*, 2963 (1968).

[353] *L.H. Sternbach, E. Reeder*, J. Org. Chem. *26*, 4936 (1961).

[354] *A. Kovendi, M. Kirez*, Chem. Ber. *98*, 1049 (1965).

[355] *M.M. Robison, B.L. Robison*, J. Amer. Chem. Soc. *80*, 3443 (1958).

[356] *J.P. Horwitz* in *R.C. Elderfield*, Heterocyclic Compounds, Vol. VII, p. 743, John Wiley & Sons, London 1961.

[357] *R. Fusco, G. Bianchetti*, Gazz. Chim. Ital. *87*, 446 (1957).

[358] *F.J. Lalor, F.L. Scott*, J. Chem. Soc. (C) 1034 (1969).

[359] *T.J. Shaw*, J. Org. Chem. *27*, 3890 (1962).

[360] *G. Soliman, I. El-Sayed El-Khaly*, J. Chem. Soc. (C) 1755 (1954).

[361] *F. Parisi, P. Bovina, A. Quilico*, Gazz. Chim. Ital. *90*, 903 (1960).

[362] *W. Baker, J.B. Harborne, M.D. Ollis*, J. Chem. Soc. (C) 1303 (1952).

[363] *E. Schmitz*, Chem. Ber. *91*, 1488 (1958).

[364] *C.L. Pedersen, N. Harrit, O. Buchardt*, Acta Chem. Scand. *24*, 3435 (1970).

[347] *E.E. Glover, M.J.R. Loadman*, J. Chem. Soc. (C) 2391 (1967).

Evidently an aminooxime **22** can form *in situ* from the activated methyl of 3-amino-4-methylquinolines **24** provided diazotization is slowed by profound protonation of the amino group. In concentrated HCl the *1,2,3-triazino[4,5-c]quinoline 2-oxides* form almost quantitatively with 2 *mols* sodium nitrite[372], while in dilute H$_2$SO$_4$ the more rapid diazotization expectedly gives the indazole derivative **26**:

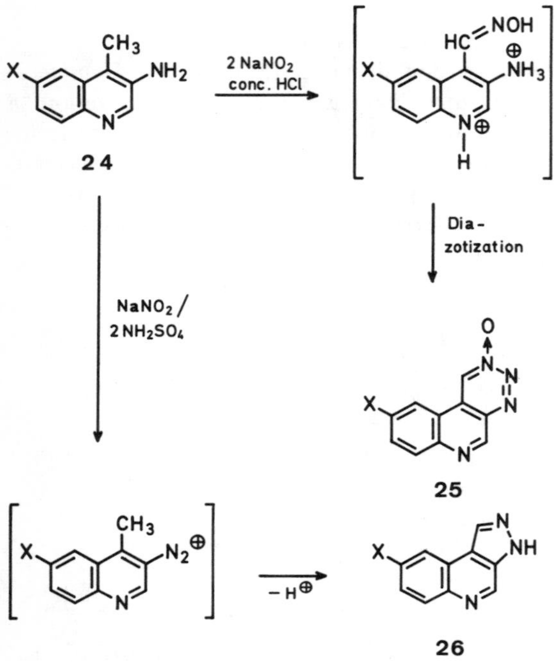

25; X = H;   *1,2,3-Triazino[4,5-c]quinoline 2-oxide*
   X = Cl; *9-Chloro-1,2,3-triazino[4,5-c]quinoline 2-oxide*

Heating α-hydrazono oximes with orthocarbonic acid triesters in the presence of *p*-toluenesulfonic acid often leads to *1,2,4-triazine 4-oxides*[373]:

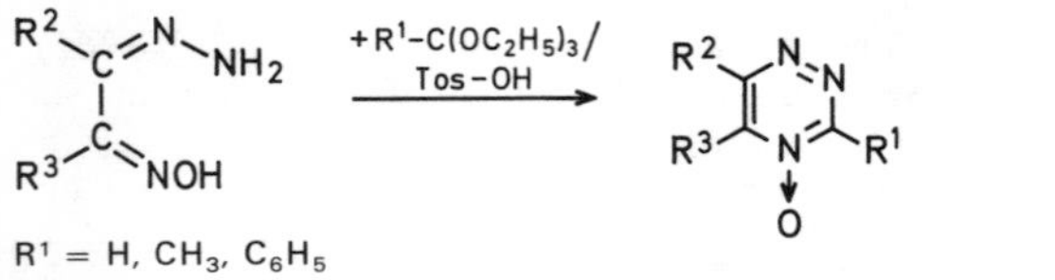

R^1 = H, CH$_3$, C$_6$H$_5$
R^2 = CH$_3$, C$_6$H$_5$
R^3 = H, CH$_3$

[365] *J.T. Nielsen, N. Elming, N. Clauson-Kaas*, Acta Chem. Scand. *9*, 14, 30 (1955).
[366] *E.C. Taylor, C.C. Cheng, O. Vogl*, J. Org. Chem. *24*, 2019 (1959).
[367] *R.M. Cresswell, G.B. Brown*, J. Org. Chem. *28*, 2560 (1963).
[368] *M.A. Stevens, H.W. Smith, G.B. Brown*, J. Amer. Chem. Soc. *82*, 3189 (1960).
[369] *H.L. Yale*, J. Heterocycl. Chem. *8*, 205 (1971).
[370] *I. Meisenheimer, O. Senn, P. Zimmermann*, Ber. *60*, 1736 (1927).

10.4.1.3.5 4-Aminopyrimidine 1-oxide synthesis from unsaturated β-hydroxyaminonitriles and amidines (4 + 2)

A reaction which presumably can serve generally for synthesizing 4-aminopyrimidine 1-oxides is the synthesis of *6-amino-7-methylpurine 3-oxide*[374] from 4-hydroxyamino-1-methyl-5-imidazolecarbonitrile and formamidine in boiling ethanol:

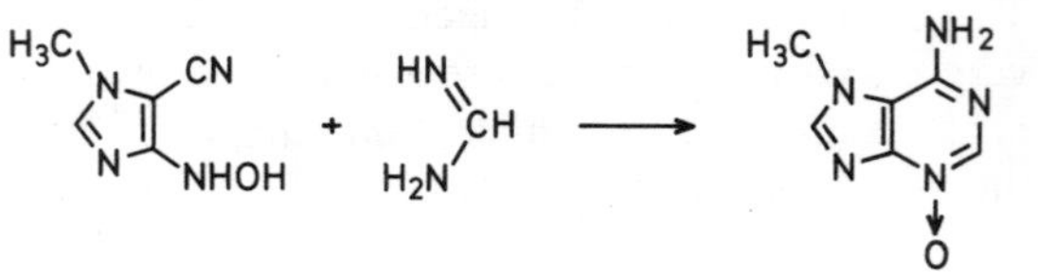

10.4.1.3.6 Quinoxaline 1,3-dioxides from 1,2-quinone dioximes (or benzofuroxans) and 1,2-dicarbonyl or α-hydroxycarbonyl compounds[375] (4 + 2)

1,2-benzoquinone dioxime **27** reacts with α-keto aldehydes on brief (5-minute) heating in aqueous solution on the steam bath to give *2-quinoxalinol 1,3-dioxides* **28** (50–65%).

The technique is claimed to be superior to other methods of preparation.

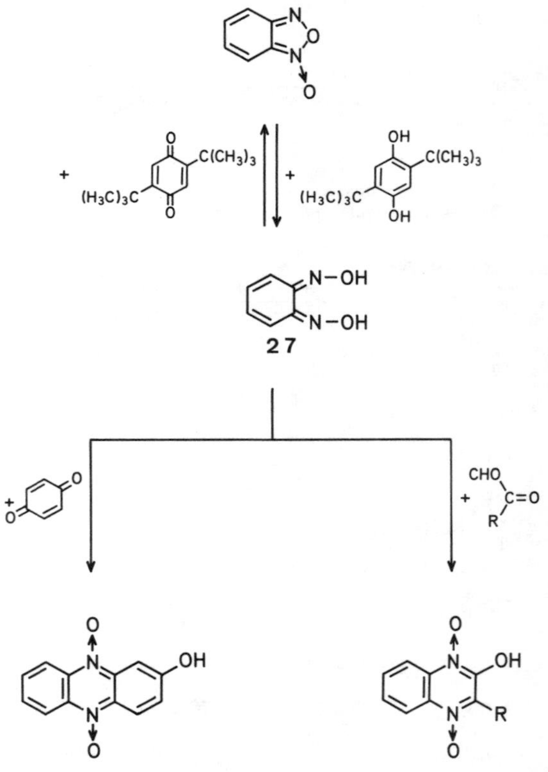

28; R = H;   *2-Quinoxalinol 1,4-dioxide*
   R = CH$_3$; *2-Methyl-3-quinoxalinol 1,4-dioxide*

29;   *2-Phenazinol 9,10-dioxide*[375, 376]

The reaction of 1, 3-diketones, α-hydroxy aldehydes, and α-hydroxy ketones with 1, 2-benzoquinone dioxime (affording up to 20% quinoxaline 1, 3-dioxides) is less important preparatively[375]. Likewise, synthesizing *2-phenazinol 9, 10-dioxide* **29** from **27** and *p*-benzoquinone[375, 376] cannot complete with method 10.4.1.6.2.

10.4.1.3.7 *N*-Oxides of pyrazines and 1, 2, 4-triazines from diketones and hydroxylamine compounds with an amino group in the *β*-position (4 + 2)

At −60° in methanol 2-aminopropionohydroxamic acid furnishes pyrazine *N*-oxides with 1, 2-diketones[377, 342]. Like α-keto aldehydes, 2-bromocinnamaldehyde undergoes a base-catalyzed reaction to 6-benzyl-3-methyl-2-pyrazinol 1-oxide $(R^1 = H; R^2 = -CH_2-C_6H_5)$[342]:

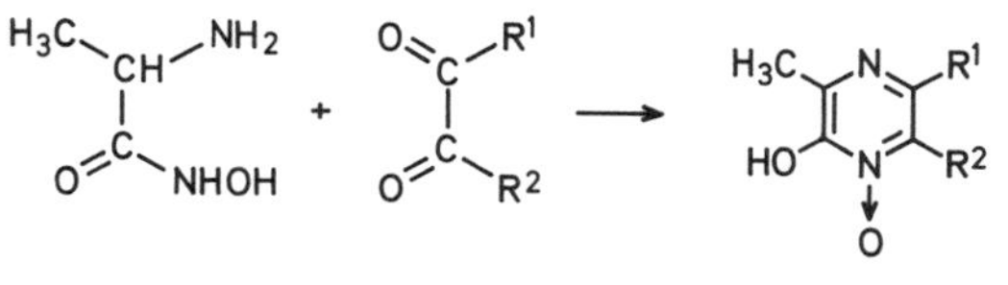

or tautomer

e.g., $R^1 = CH_3$; $R^2 = H$; *3,6-Dimethyl-2-pyrazinol 1-oxide;*
$R^1 = R^2 = H$;   *3-Methyl-2-pyrazinol 1-oxide;*
$R^1 = R^2 = C_6H_5$; *3-Methyl-5,6-diphenyl-2-pyrazinol*
    1-oxide; (poor yield)
$R^1 = R^2 = CH_3$; *3,5,6-Trimethyl-2-pyrazinol 1-oxide*

Using the same synthetic principle, benzoic acid hydrazide oxime can be reacted with 1, 2-diketones[373]:

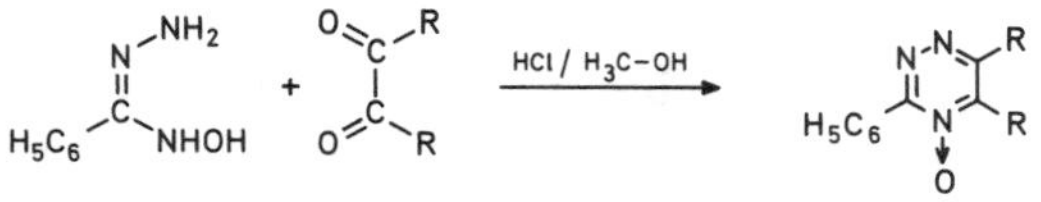

R = CH_3; *5,6-Dimethyl-3-phenyl-1,2,4-triazine 4-oxide;*
R = C_6H_5; *3,5,6-Triphenyl-1,2,4-triazine 4-oxide;* (impure)

[371] *J.P. Horwitz* in *R.C. Elderfield,* Heterocyclic Compounds, Vol. VII, p. 779, John Wiley & Sons, London 1961.

[372] *D.W. Ockenden, K. Schofield,* J. Chem. Soc. (London) 1915 (1953).

[373] *H. Neunhoeffer, F. Weischedel, V. Bömisch,* Justus Liebigs Ann. Chem. 750, 12 (1971).

[374] *E.C. Taylor, P.K. Löffler,* J. Org. Chem. *24,* 2035 (1959).

[375] *E. Abushanab,* J. Org. Chem. *35,* 4279 (1970).

[376] *M.J. Abu El-Haj, B.W. Dominy, J.D. Johnston,* J. Org. Chem. *37,* 589 (1972).

[377] *G. Dunn, J.A. Elvidge, G.T. Newbold, D.W.C. Ramsay, F.S. Spring, W. Sweeney,* J. Chem. Soc. (London) 2707 (1949).

[378] *W. Sharp, F.S. Spring,* J. Chem. Soc. (London) 932 (1951).

10.4.1.3.8 *N*-Oxides of pyrazines and 1, 2, 4-triazines from α-hydroxyimino ketones (1, 2-ketone monooximes) and α-aminonitriles or amidrazones (3 + 3)

On heating with α-aminonitriles in chloroform, α-hydroxyimino ketones condense to *2-aminopyrazine 1-oxides*[378]; addition of titanium(IV) chloride increases the intrinsically low yields considerably[379].

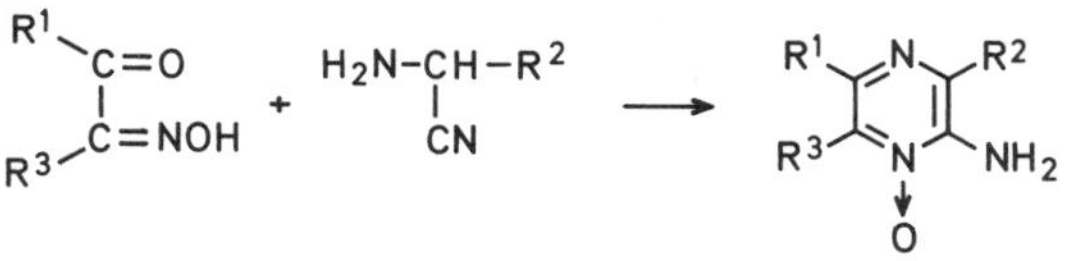

or tautomer

$R^1 = CH_3, C_2H_5, C_6H_5, 3$-Indolyl
$R^2 = C_2H_5, C_6H_5, CH_3, 3$-Phthalimidopropyl
$R^3 = H, CH_3, C_6H_5$

A high yield of *3-methyl-6-phenyl-1, 2, 4-triazine 4-oxide* is obtained from phenylglyoxal monooxime and acetamidrazone[373]:

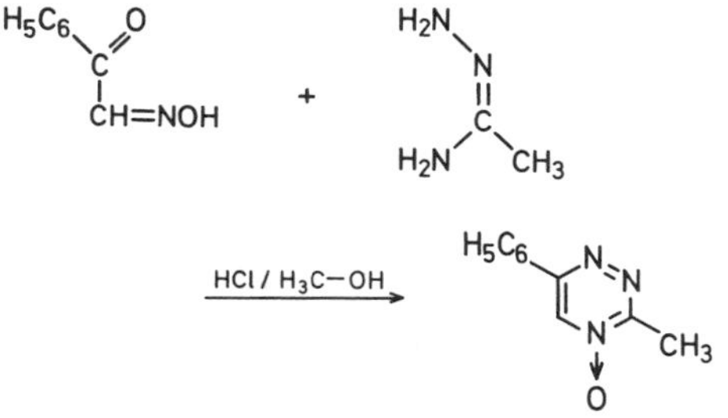

10.4.1.3.9 2-Pyrimidinol 1-oxides from hydroxyurea compounds and 1, 3-dicarbonyl compounds or their analogs (3 + 3)

Alcoholate-catalyzed condensation of *β*-keto carboxylic acid esters with hydroxyurea in ethanol leads to *3-hydroxyuracils*[380], the tautomers of the *2, 6-pyrimidinediol 1-oxides*[380, 381]:

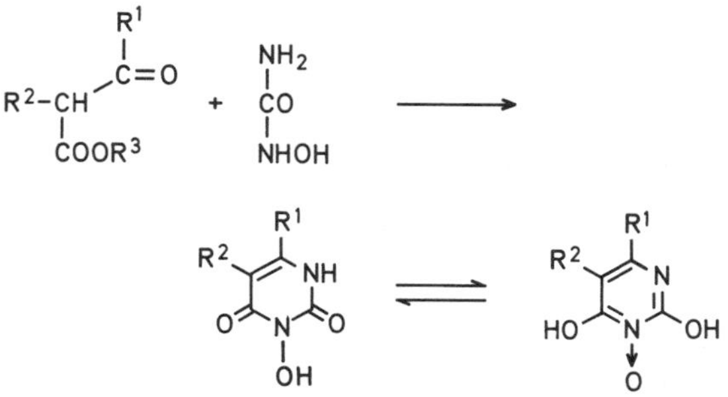

$R^1 = CH_3, CH_2-COOC_2H_5$
$R^2 = H$
$R^1 - R^2 = -(CH_2)_4-$
$R^3 = CH_3; C_2H_5$

[379] *T.P. Karpetzky, E.H. White,* J. Amer. Chem. Soc. *93,* 2333 (1971).

For the analogous preparation of the 2-pyrimidinol 1-oxide the diacetal of malonaldehyde is condensed with the methyl ether or benzyl ether of hydroxyurea by means of hydrogen chloride in ethanol at 20°, initially to 1-alkoxy-2-pyrimidinone. Cleavage of the ether by boiling in 27% hydrogen bromide/glacial acetic acid for 10 minutes affords the *N*-oxide in good yield[274]:

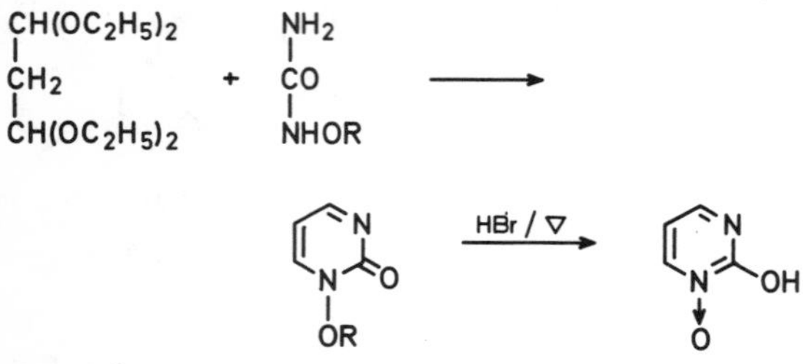

R = CH$_2$–C$_6$H$_5$, CH$_3$ or tautomer

The direct route of condensation with hydroxyurea instead of its ethers is not recommended because of poor yields. Analogously, 1,3-diketones furnish *4,6-disubstituted* 2-pyrimidinol 1-oxides[342].

The *2,4-pyrimidinediol 1-oxides* **30** (Y = O), **30a**, and **30b**, and the *4-amino-2-pyrimidinol 1-oxide* (**30**, R = R^1 = R^2 = H; Y = NH) are likewise obtained from benzyloxyurea with the aid of alkoxide by condensation with the acetals **31**, ethoxymethylene malonic acid diester **32**, or with cyanoacetic acid ester **33**. The initially formed 1-benzyloxy-2-pyrimidinones **34** need to be debenzylated solvolytically (hydrogen bromide/glacial acetic acid) or hydrogenolytically (H$_2$/Pd-C)[382] (direct condensation of hydroxyurea affords isoxazoles):

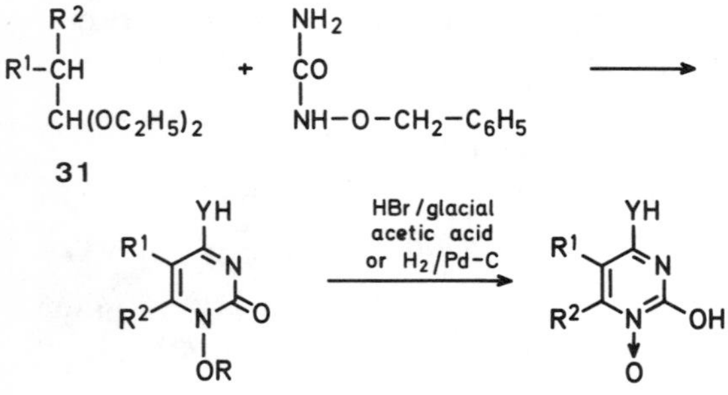

31

X^1 = H; X^2 = COOC$_2$H$_5$;   Y = O; R^1 = R^2 = H
X^1 = H; X^2 = CN;   Y = NH; R^1 = R^2 = H

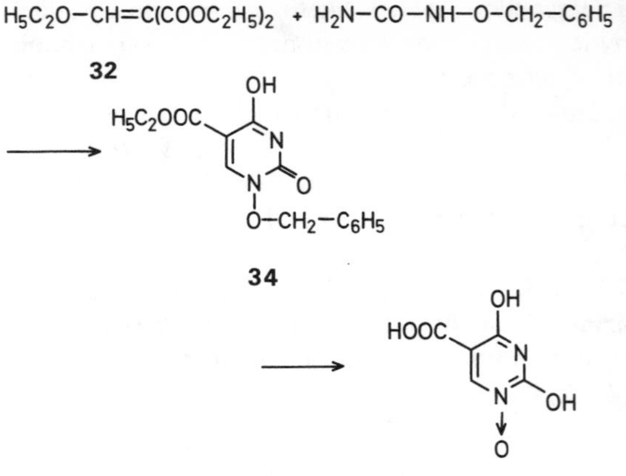

32

34

30a

2,4-Dihydroxypyrimidine-5-carboxylic acid 1-oxide

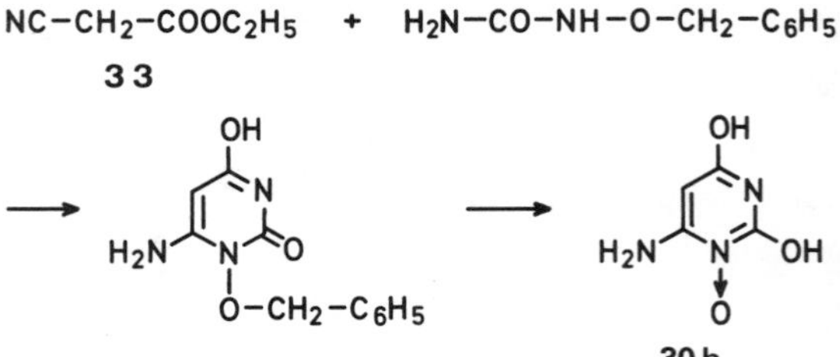

NC–CH$_2$–COOC$_2$H$_5$ +   H$_2$N–CO–NH–O–CH$_2$–C$_6$H$_5$

33

34

30b

6-Amino-2,4-pyrimidinediol 1-oxide

10.4.1.3.10 Three-component reaction of hydroxylamine with two carbonyl compounds (3 + 1 + 2)

The synthesis principle of cyclizing derivatives of hydroxylamine can be realized also in a three-component reaction. Thus, 2,4-pentanedione (acetylacetone), cyanoacetic acid ester, and hydroxylammonium chloride together form *2-hydroxy-4,6-dimethyl-3-pyrimidinecarbonitrile 1-oxide* (20%) with alcoholic caustic potash and piperidine on heating[383]:

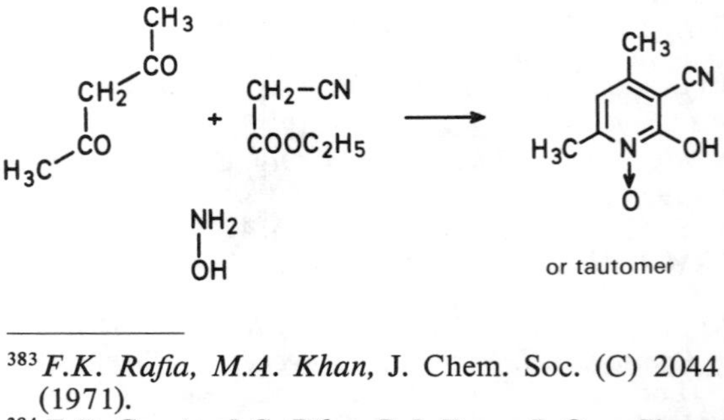

or tautomer

[380] *E. Ajello, T. Ajello, V. Sprio*, Ric. Sci. Rend. Sec. B *4*, 105 (1964).
[381] *A.L. Cossey, J.N. Phillips*, Chem. Ind. *58*, (1970).
[382] *W. Klötzer*, Monatsh. Chem. *95*, 265, 1729 (1964); *W. Klötzer*, Monatsh. Chem. *96*, 169 (1965).

[383] *F.K. Rafia, M.A. Khan*, J. Chem. Soc. (C) 2044 (1971).
[384] *E.E, Garcia, J.G. Riley, R.I. Fryer*, J. Org. Chem. *33*, 1359 (1968).

10.4.1.3.11 Oxidative cyclization of unsaturated oximes ($6_{acycl.} \rightarrow 6_{cycl.}$)

Two examples of the following reaction have been described[384]:

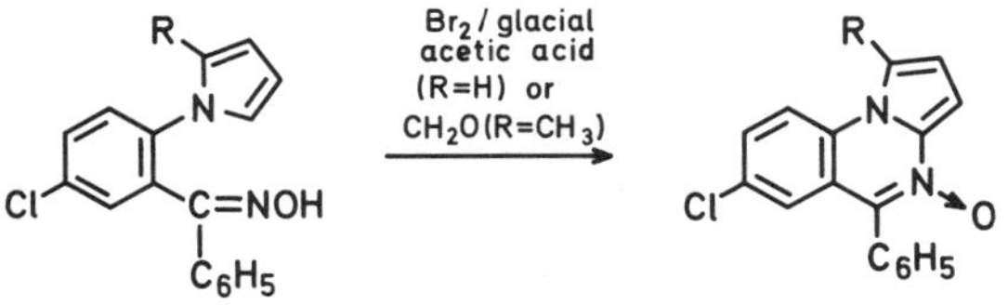

R = H;   *7-Chloro-5-phenylpyrrolo[1,2-a]quinazoline 4-oxide*

R = CH₃; *7-Chloro-1-methyl-5-phenylpyrrolo[1,2-a]-quinazoline 4-oxide*

10.4.1.4 *N*-Oxides of benzo azines from 2-nitrophenyl compounds

Using the following techniques enables aromatic nitro compounds to be cyclized to benzazine *N*-oxides directly or indirectly *via* a reduction.

10.4.1.4.1 Cyclodehydration of 2-methylene-1-(2-nitrophenyl) and 2-amino-1-(2-nitrophenyl) compounds (Arndt reaction) ($6_{acycl.} \rightarrow 6_{cycl.}$)

2-Nitrobiphenyls carrying an acid methylene group in the 2'-position afford 5-phenanthridine 5-oxides on brief heating in methanolic caustic soda[385, 386, 638]:

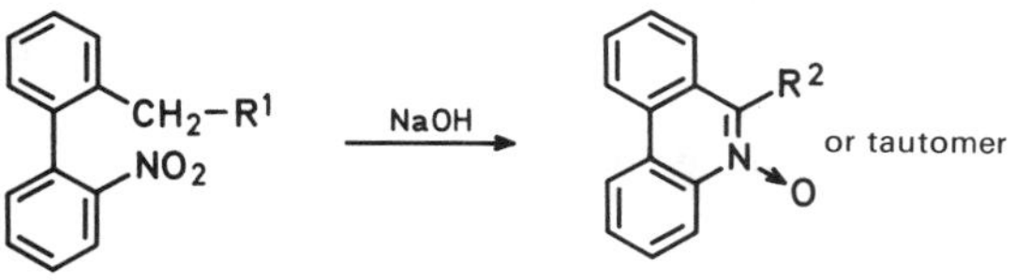

R¹ = R² = COOCH₃, CONH₂, CN; *Methyl 6-phenanthridine-carboxylate 5-oxide*
6-Phenanthridinecarbox-amide 5-oxide
*6-Phenanthridinecarbo-nitrile 5-oxide**

R¹ = COC₆H₅, R² = H;   *Phenanthridine 5-oxide*

R¹ = SO₂C₆H₅, R² = OH;   *6-Phenanthridinol 5-oxide*

R¹ = COOCH₃; R² = COOH   *6-Phenanthridinecarboxylic*
(with more than 1 equivalent   *acid 5-oxide*
NaOH);

* From the 2'-bromomethyl derivative in aqueous-ethanolic sodium cyanide solution

---

The acidifying group R¹ can be removed solvolytically or be modified (R¹ ≠ R²). Good to very good yields are obtained.

Sodium methoxide can be used in place of sodium hydroxide; acidifying the methylene group by means of a phenyl group is not adequate[386].

Other aromatic systems of corresponding constitution can likewise be condensed; thus, for instance, ring closure to *6-arylquinolino[3,4-b]quinoxaline 5-oxide* occurs even in 3-benzyl-2-(2-nitrophenyl)quinoxalines[387]:

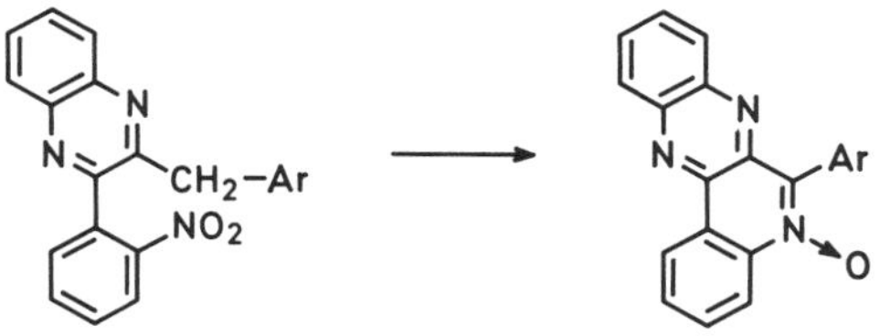

Re a similar synthesis *cf.* ref.[639].

An amino group reacts similarly to an activated methylene group, albeit somewhat slower. *Benzo-[c]cinnoline 5-oxides* are obtained from 2-(2-nitrophenyl)anilines[385, 388]:

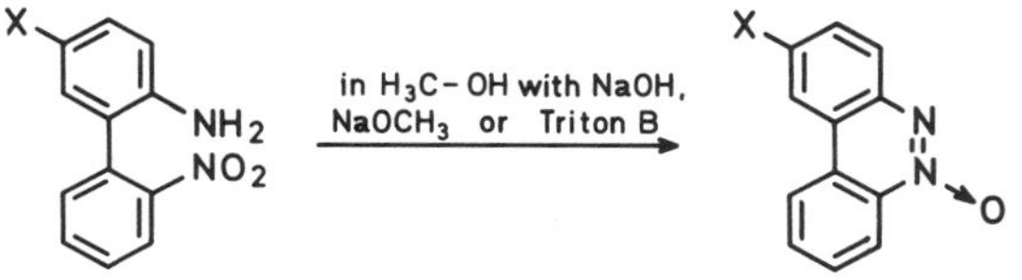

5-Amino-1-(2-nitroaryl)pyrazoles react analogously to *pyrazolo[5,1-c]benzo-1,2,4-triazine 5-oxides*[389].

This principle of ring closure is not limited to biaryls with their favorable molecule geometry, as is demonstrated by the preparation of *3-quinolinol 1-oxides* and *3-cinnolinol 1-oxides* substituted in the 4-position[386]:

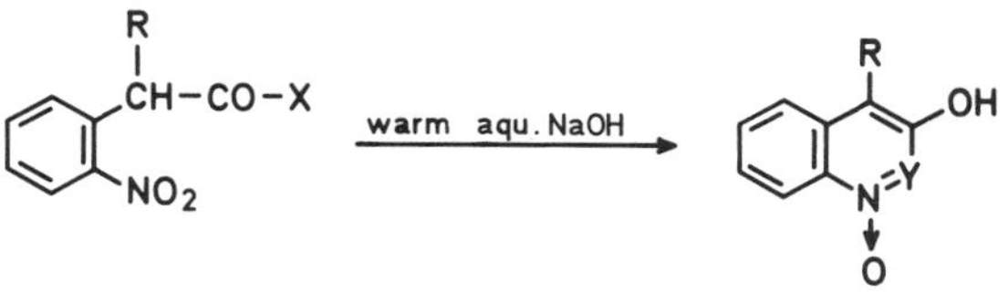

X = CH₂C₆H₅, Y = C–C₆H₅
X = NH₂,   Y = N

---

[385] *C.W. Muth, J.C. Ellers, O.F. Folmer,* J. Amer. Chem. Soc. **79**, 6500 (1957);
C.W. Muth, N. Abraham, M.L. Linfield, R.B. Wotring, E.A. Pacofsky, J. Org. Chem. **25**, 736 (1960).

[386] *J.D. Loudon, G. Tennant,* Quart. Rev. Chem. Soc. **18**, 389 (1964).

[387] *R.P. Barnes, J.H. Graham, M.A.S. Aureshi,* J. Org. Chem. **28**, 2890 (1963).

---

[388] *J.W. Barton, J.F. Thomas,* J. Chem. Soc. (London) 1265 (1964).

[389] *Y. Ahmad, P.A.S. Smith,* J. Org. Chem. **36**, 2972 (1971).

[390] *G. Tennant,* J. Chem. Soc. (London) 2666 (1964).

The 2-nitroanilides **35** of the cyano- or acylacetic acids, or of malonic acid monoester likewise suffer an intramolecular Knoevenagel condensation under the influence of bases; here the nitro group takes the place of the carbonyl function[309, 390, 391, 392]:

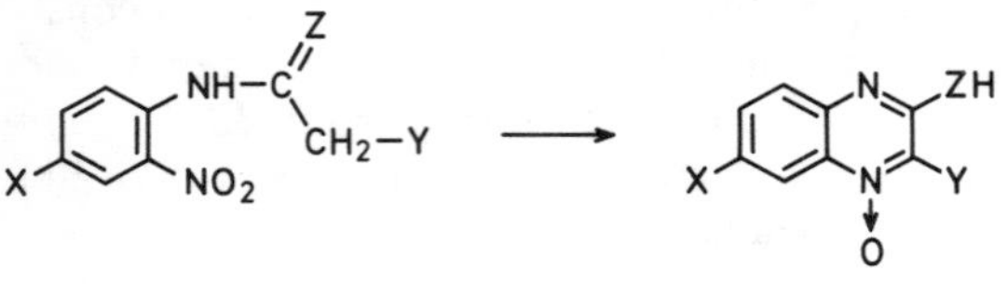

35

36
or tautomer

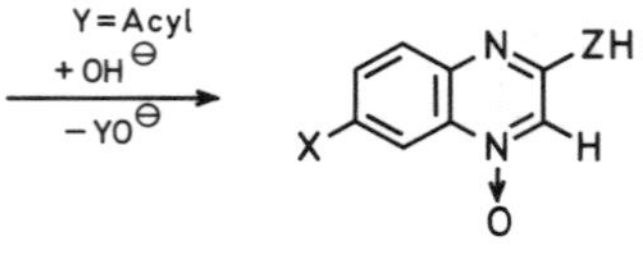

37
or tautomer

X = H, CH_3, OCH_3, OC_2H_5, Cl
Y = CN, CO-Alkyl, CO—C_6H_5, COOC_2H_5, 2—NO_2—C_6H_4,
    2,4—$(NO_2)_2$—C_6H_3
Z = O, S[391]

Yield and structure of the products evidently are strongly dependent on the reaction conditions and can exceed 50%[391]. The initially formed 3-quinoxalinol 1-oxides can be modified at the *N*-oxide preparation stage or by a subsequent hydrolysis in dependence on the reaction conditions and the nature of the substituent *Y*:

1 With Y = Acyl or alkoxycarbonyl to 3-quinoxalinol *1-oxides* **37**.
2 With 2-nitroanilides of α-methyl-β-oxo-carboxylic acids to *2-methyl-3-quinoxalinol 1-oxide* (**36**, *Y = CH_3*) in every case[390, 391].
3 With 2-nitroalinides of α-methyl-β-oxo *2,3-quinoxalinediol 1-oxides* (**36**, *Y = OH*)[390, 392].
4 With 2-nitroanilides of 2-oxo-1-cycloalkanecarboxylic acids (cyclic β-keto acids) to the *2-(ω-carboxyalkyl)-3-hydroxyquinoxaline 1-oxides(Z = O)*[391]:

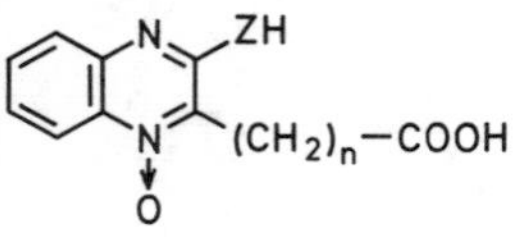

n = 3,4,5,6
Z = O,S

38

3-Quinoxalinethiol 1-oxides *(Z = S)* and **38***[Z = S, n = 4; 2-(3-mercaptoquinoxaline)butyric acid 1-oxide]*, too, have been prepared starting from the corresponding thioanilides[391].

In practise the condensation of the 2-nitroanilides is performed as follows:

1 With dilute aqueous barium hydroxide solution at 60–70° (the products are precipitated as barium salts)
2 With 1 *N* caustic soda in pyridine at room temperature[390, 391]
3 With aqueous potassium cyanide solution (**35**; *Y = CN*, *Cl* to **36** with *Y = CN*)[390]
4 With sodium amide in boiling benzene
5 With sodium ethoxide in ethanol *(Y = acyl)*.

2-Aryl-3-quinoxalinol 1-oxides (**36**; *Y = Ar*) can be prepared analogously from 2-nitroanilides of arylacetic acids. The yields fall very markedly if the arylacetyl group is inadequately activated[390, 391] or the nitrogen of the anilide group is not *N*-methylated[390]. With *N*-methylanilides *2-aryl-4-methyl-3(4H)-quinoxalinone 1-oxides* are formed.

Re the preparation of the 2-nitroanilides see refs.[309, 390, 391].

The pyridine-2-nitroacetanilide hydrochloride obtained from *N*-chloroacetyl-2-nitroaniline with pyridine can also by cyclized to *2-amino-3-quinoxalinol 1-oxide* (50%) with sodium hydroxide in ethanol[391] or, better, with piperidine in warm methanol[309]; Zincke cleavage of the pyridine ring occurs simultaneously:

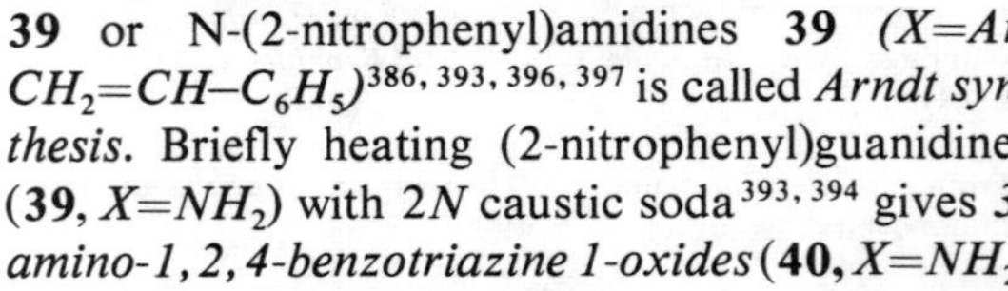

or tautomer

Preparation of *1,2,4-benzotriazine 1-oxides* **40** from (2-nitrophenyl)urea derivatives[322, 386, 393–395] **39** or N-(2-nitrophenyl)amidines **39** *(X=Ar, CH_2=CH—C_6H_5)*[386, 393, 396, 397] is called *Arndt synthesis*. Briefly heating (2-nitrophenyl)guanidines (**39**, *X=NH_2*) with 2*N* caustic soda[393, 394] gives *3-amino-1,2,4-benzotriazine 1-oxides* (**40**, *X=NH_2*) in sometimes quantitatively yield. The same is true of the preparation of 1,2,4-benzotriazin-3-ol 1-oxides, **40**, *X=OH;* from (2-nitrophenyl)ureas, the

---

[391] *R. Fusco, S. Rossi*, Gazz. Chim. Ital. **94**, 1, 3 (1964); C.A. 61, 4352b (1964).
[392] *Y. Ahmad, M.S. Habib, Ziauddin*, Tetrahedron **20**, 1107 (1964).
[393] *F. Arndt*, Ber. **46**, 3522 (1913); *F. Arndt, B. Rosenau*, Ber. **50**, 1248 (1917).

[394] *F.J. Wolf, K. Pfister, R.M. Wilson jr., C.A. Robinson*, J. Amer. Chem. Soc. **76**, 3551 (1954); U.S.P. 2 489 352 (1949), Inv.: *F.J. Wolf, K. Pfister*, C.A. *44*, 3536 (1950).

corresponding 3-aryl derivatives (**40**, X=Ar) from *N*-(2-nitrophenyl)amidines, and of the 3-mercapto derivatives (**40**, X=SH) and other condensed triazine *N*-oxides, for instance *3-aminopyridino[b]-1,2,4-triazine 1-oxide*[386, 322].

or tautomer

39

or tautomer

40

X = NH$_2$, NH—Ar, OH, SH, Ar, CH$_2$=CH—C$_6$H$_5$

Variations in R^1, R^2, and R^3 apply exclusively for X = NH$_2$:
R^1 = H, Cl, CH$_3$
R^2 = H, Cl
R^3 = H, Cl, Br, I, CH$_3$, OCH$_3$, OC$_2$H$_5$, COCH$_3$

The yield of *3-aryl-1,2,4-benzotriazine 1-oxides* can be improved considerably by using sodium alcoholate in place of caustic soda[397]. Overlong heating with alkali during preparation of the 3-amino and 3-hydroxy derivatives may produce ring contraction to 1,2,3-benzotriazole. Cyclization of the (2-nitrophenyl)ureas (**39**, *X=OH*, or tautomeric form) is suitably carried out with 30% sodium hydroxide, because lower alkali concentrations favor cleavage of the urea to *(2-nitrophenyl) isocyanate*[395]. In *N*-(2-nitrophenyl)thioureas desulfurization to the *cyanamide derivative* may arise as a competing reaction.

The *main application* of the Arndt ring closure is in relation to the (2-nitrophenyl)guanidines easily accessible from 2-nitroanilines and cyanamide which, prepared *in situ*, cyclize after boiling for a few minutes with caustic soda. Yields of 20—60% (on the 2-nitroaniline used) are obtained. An initial reaction in glacial acetic acid[394] is recommended for substantial batches and in general improves the yield. Where these procedures fail, heating the 2-nitroaniline to 180° with cyanamide sometimes affords success[394], but 2,4-dinitroaniline does not react even then[394]. The Arndt synthesis is so convenient to carry out that after subsequent deoxygenation the *1,2,4-benzotriazines* themselves are prepared in this way. Peroxide treatment of 3-amino-1,2,4-benzotriazine affords a different *N*-monooxide[322]. Alkali treatment of (2-nitrobenzenesulfonyl)gua-

nidine gives *N'-(2-nitrophenyl)guanidine-N²-sulfinic acid via* a type of Smiles rearrangement; the product is likewise cyclized to *3-amino-1,2,4-benzotriazine 1-oxide* after hydrolysis[386]. The Arndt synthesis principle forms the basis also of the synthesis of *3-phenyl-* and *7-methyl-3-phenyl-1,2,3-triazolo[5,1-c]benzo-1,2,4-triazine oxide*[398, 399]:

X = H, CH$_3$

10.4.1.4.2 Azine *N*-oxides by reductive cyclization of 2-substituted nitrobenzenes ($6_{acycl.} \rightarrow 6_{cycl.}$)

R = H, COOC$_2$H$_5$, CONH$_2$, COOH, COCH$_3$
X = COOC$_2$H$_5$; Z = OH
X = CN;   Z = NH$_2$
X = CO—CH$_3$; Z = CH$_3$

41 (R—X = *o*-C$_6$H$_4$—N—CO—) —— ⟶
        |
        CO—CH$_3$

43 (Z = OH; R = *o*-C$_6$H$_4$—NHCOCH$_3$)

Quinoline N-oxides **43** are obtained by reduction of α-(2-nitrobenzylidene)carboxylic acid esters, nitriles, or α-(2-nitrobenzylidene) ketones **41** with zincacetic acid[400], ammonium sulfide[401], sodium tetrahydroborate/palladium/charcoal[342], hydrogen/palladium/kieselguhr[402], and electrolytically[403].

[395] *F.J. Wolf, R.M. Wilson jr., K. Pfister, M. Rishler*, J. Amer. Chem. Soc. 76, 4611 (1954).
[396] *R.F. Robbins, K. Schofield*, J. Amer. Chem. Soc. 3186 (1957); C.A. 52, 4663a (1958).
[397] *R. Fusco, G. Bianchetti*, C.A. 53, 9243 (1959).
[398] *E. Lieber, T.S. Chao, C.N.R. Rao*, J. Org. Chem. 22, 654 (1957).
[399] *G. Tennant*, J. Chem. Soc. (C) 1279 (1967).

[400] *E.C. Taylor, N.W. Kalenda*, J. Org. Chem. 18, 1755 (1953).
[401] *P. Friedländer*, Ber. 47, 3369 (1914).
[402] *K.H. Bauer*, Ber. 71, 2226 (1938).
[403] *H. Lund, S. Kwee*, Acta Chem. Scand. 23, 2711 (1969);
   H. Lund, L.G. Feoktistov, Acta Chem. Scand. 23, 3482 (1969).
[404] *J. Meisenheimer, E. Stotz*, Ber. 58, 2334 (1925).
[405] *J.D. Loudon, I. Wellings*, J. Chem. Soc. (London) 3462 (1960).
[406] *R.T. Coutts, D.G. Wibberley*, J. Chem. Soc. (London) 2518 (1962).

The preparatively unimportant reduction of the compounds of the 2-nitrophenyl derivatives **42** accompanied by *HY* elimination is carried out with zinc/acetic acid[404] or ammonium chloride[405].

By contrast, reducing the α-(2-nitrobenzoyl)carboxylic esters and α-(2-nitrobenzoyl) ketones of analogous structure affords mixtures of *4-quinolinol N-oxides*[236, 406] and *4-quinolinols*:

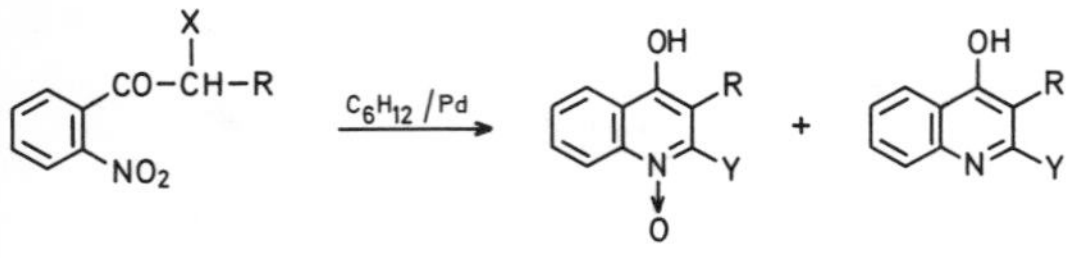

or tautomer

R	X	Y	
H	COCH$_3$	CH$_3$	*2-Methyl-4-quinolinol 1-oxide;* 61%
COOC$_2$H$_5$	COCH$_3$	CH$_3$	*Ethyl 4-hydroxy-2-methyl-3-quinolinecarboxylate 1-oxide;* 14%
COOC$_2$H$_5$	CN	NH$_2$	*Ethyl 2-amino-4-hydroxy-3-quinolinecarboxylate; 1-oxide;* 21%*
COOCH$_3$	CN	NH$_2$	*Methyl 2-amino-4-hydroxy-3-quinolinecarboxylate 1-oxide;* 25%*

* as salt of the starting ester

The amides **44a**, **44b** accessible from 2-nitrophenyl isocyanate and 2-nitrobenzoyl chloride can be hydrogenated in good yield to *2-amino-3-quinoxalinol 1-oxide* **45** and *2-amino-4-quinazolinol 1-oxide* **46** respectively on Adams catalysts[407]:

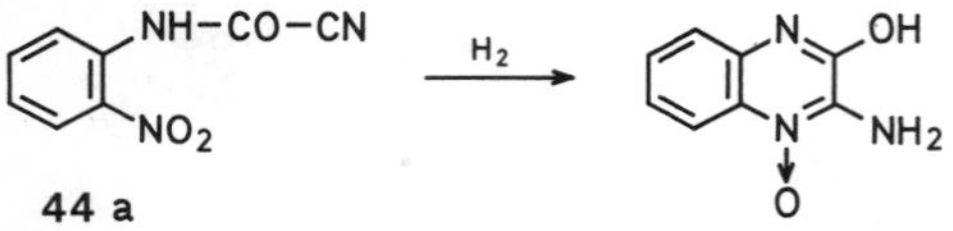

44 a

or tautomer

45

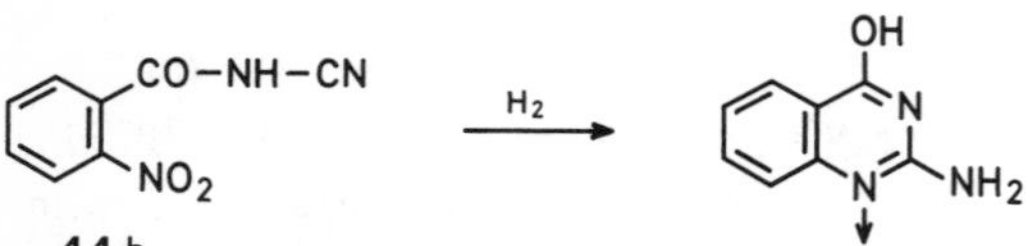

44 b

or tautomer

46

Reduction of 2,2'-dinitrobiphenyls with sodium sulfide in ethanol affords *benzo[c]cinnoline 5-oxides* in good yield[408] (new ref. [640]):

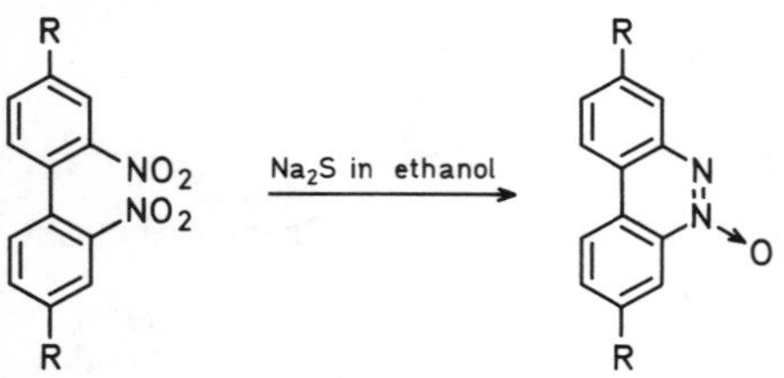

R = CH$_3$; *3,8-Dimethylbrenzo[c]cinnoline 5-oxide*
R = N(CH$_3$)$_2$; *3,8-Bis(dimethylamino)benzo[c]cinnoline 5-oxide*

Analogously, *benzo[c]-2,7-naphthyridine 6-oxides* **48** *(Z=N)* form from 4-(2-nitrophenyl)pyridines **47** *(Z=N)* during reduction with zinc and ammonium chloride in aqueous ethanol, or with tin(II) chloride and boiling hydrochloric acid, generally in moderate yield[409] (only with X=R^2=COOC$_2$H$_5$ is it 75%):

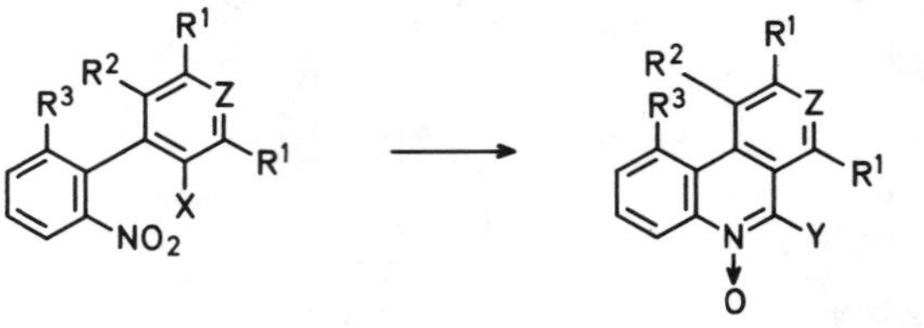

47

48

or tautomer

1 Z = N; R^1 = CH$_3$, R^3 = H, R^2= COOC$_2$H$_5$, CN, COCH$_3$, H;
   X = COOC$_2$H$_5$ Y = OH
        Ethyl 5-hydroxy-2,4-dimethylbenzo[c]-2,7-naphthyridine-1-carboxylate 6-oxide; 75%
   X = COCH$_3$ Y = CH$_3$
   X = CN   Y = NH$_2$

2 Z = CH; R^1 = R^2 = R^3 = H
   X = COCH$_3$, COOH, CONH$_2$; Y = OH
        6-Phenanthridinol 5-oxide
   X = CN; Y = NH$_2$
        6-Aminophenanthridine 5-oxide

Catalytic hydrogenation (Pt) of 2-(2-nitrophenyl)benzoic acid derivatives **47** *(Z = CH)* in the presence of mineral acid leads to good yields of phenanthridine 5-oxides **48** *(Z = CH)*, which can be easily separated from conformed phenanthridines by virtue of their alkali solubility[410].

Where the reductive cyclization can occur twice (R^2 = NO$_2$, R^3 = X) a quantitative yield of *pyrido[2,3,4,5-lmn]phenanthridine 4-oxide* is obtained **49**[410].

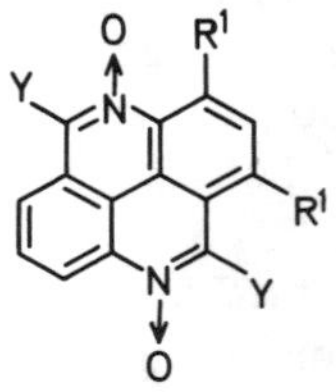

49

Synthesis of *4-aminopyrrolo[1,2-a]quinoxaline 5-oxide* demonstrates that this method can be applied generally[411]:

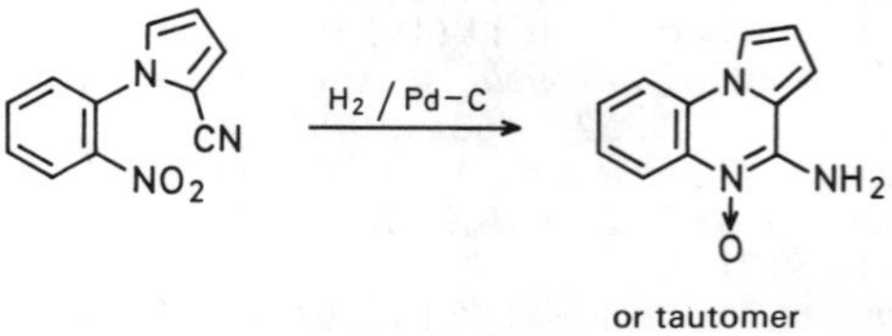

or tautomer

10.4.1.4.3 4-Quinolinol *N*-oxides from 2-nitrobenzylidene compounds and halogen hydride (4 + 2 or $6_{acycl.} \rightarrow 6_{cycl.}$)

2-(2-Nitrobenzylidene)-1,3-dioxo compounds* **50** are, on saturating their ether solutions with hydrogen chloride or hydrogen bromide, converted into 4-quinolinol 1-oxides* (70–90%)[386, 412, 413], which precipitate as unstable salts on standing:

R = OC$_2$H$_5$, CH$_3$, C$_6$H$_5$[109]
R^1 = CH$_3$, CH$_2$–COOR
X = H, Cl, Br
Hal = Br, Cl
Y = H, Cl

or tautomer
52

The conversion probably proceeds according to an N–C oxygen transfer[213] *via* an intermediate nitroso stage **51**. When hydrogen chloride is used the nitroso derivative reduces to the *N*-(4-chlorophenyl)hydroxylamine derivative (for *X = H*) or the *N-(2-chlorophenyl)hydroxylamine derivative* (for *X = Cl, Br*) by addition of hydrogen chloride. The end-products are the *4-quinolinol 1-oxides* **52**, which contain one more chlorine atom than the starting substances. This ring substitution does not take place when hydroquinone is added, nor when hydrogen bromide is used in place of hydrogen chloride.

---

* If these are produced *in situ* the yield of the *N*-oxides is reduced appreciably.

[407] *E.C. Taylor, C.W. Jefford*, Chem. & Ind. 1559 (1963).
[408] *E. Ullmann, P. Dieterle*, Ber. *37*, 24 (1904).
[409] *S.B. Hansen, V. Petrow*, J. Chem. Soc. (London) 350 (1953).
[410] *C.W. Muth, J.R. Elkins, M.L. DeMatte, S.T. Chiang*, J. Org. Chem. *32*, 1106 (1967).
[411] *G.W.H. Cheeseman, M. Rafig*, J. Chem. Soc. (C) 2732 (1971).
[412] *J.D. Loudon, I. Wellings*, J. Chem. Soc. (London) 3470 (1960).
[413] *J.D. Loudon, G. Tennant*, J. Chem. Soc. (London) 3092 (1962).
[414] *I.P. Sword*, J. Chem. Soc. (C) 820 (1971); *T.W.M. Spence, G. Tennant*, J. Chem. Soc. (C) 3712 (1971).
[415] *T.W.M. Spence, G. Tennant*, Chem. Commun. 1100 (1970).

The epoxides **53** related to the benzylidene derivatives **50** are likewise converted into the otherwise inaccessible *3,4-quinolinediol 1-oxides* **54** in ether by saturating with hydrogen chloride[415, 414]; once again, on adding hydroquinone the chlorinefree compound *(Y = H)* is obtained. Very good yields are obtained when the *cis*-3-acyl-2-(2-nitrophenyl)epoxides *(R^2 = H)* are made the starting point, medium yields when the corresponding *trans* derivatives *(R^2 = H)* are employed. When 3,3-diacyl-2-(2-nitrophenyl)epoxides *(R^2 = acyl)* are used an acyl group is split off:

R^2 = H, COCH$_3$, COC$_6$H$_5$
R^1 = CH$_3$, C$_6$H$_5$
Y = H, Cl

With manganese(IV) oxide the 3,4-quinolinediol 1-oxides can be dehydrogenated to the corresponding 1,2-quinones, which condense with 1,2-phenylenediamine to the *quinolino[3,4-b]quinoxaline 5-oxides* **55**[415].

10.4.1.4.4 4-Quinolinecarbonitrile 1-oxides from 2-nitrobenzylidene compounds and potassium cyanide ($6_{acycl.} \rightarrow 6_{cycl.}$)

4-Quinolinecarbonitrile 1-oxides **57** are obtained in moderate yield from the benzylidene compounds

---

[416] *J.D. Loudon, G. Tennant*, J. Chem. Soc. (London) 3466 (1960).
[417] *A. Wohl, W. Aue*, Ber. *34*, 2442 (1901).
[418] *B. Cross, P.J. Williams, R.E. Woodall*, J. Chem. Soc. (C) 2085 (1971).
[419] *E. Ochiai*, Aromatic Amine Oxides, p. 62, Elsevier Publ., Co., Amsterdam, London, New York 1967.
[420] *I.J. Pachter, M.C. Kloetzel*, J. Amer. Chem. Soc. *73*, 4958 (1951); *I.J. Pachter, M.C. Kloetzel*, J. Amer. Chem. Soc. *74*, 971 (1952).
[421] *Y. Maki, M. Sako, E.C. Taylor*, Tetrahedron Lett. 4271 (1971).

56 with potassium cyanide in boiling ethanol[405, 416]:

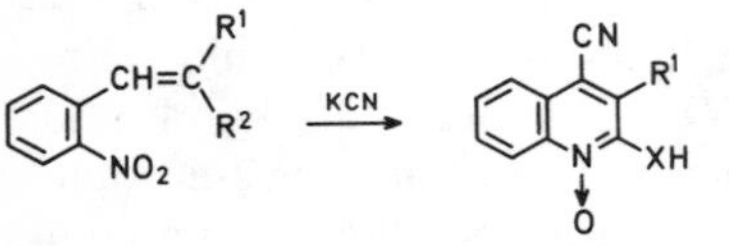

56   **57**

or tautomer

$R^1 = R^2 = COOC_2H_5$; $X = O$;   *Ethyl 4-cyano-2-hydroxy-3-quinolinecarboxylate 1-oxide*

$R^1 = C_6H_5$; $R^2 = CN$; $X = NH$; *2-Amino-3-phenyl-4-quinolinecarbonitrile 1-oxide*

The quinoline 1-oxide is initially precipitated as the potassium salt and is obtained in the free state on making acid. Possibly, the reaction is basically an $N \rightarrow C$ oxygen transfer of the kind involved in the procedure described on p. 375 with a subsequent reduction of the intermediate nitroso stage by cyanide.

10.4.1.4.5 Phenazine *N*-oxides from aromatic amines and aromatic nitro compounds (Wohl-Aue reaction) (3 + 3)

Nitrobenzene can be condensed to a mixture of *phenazine 9-oxide* and *phenazine* with primary aromatic amines by strong bases (powdered potassium hydroxide, sodium amide) with or without solvents (toluene, xylene, chlorobenzene, *etc.*) above 100° [417-420]:

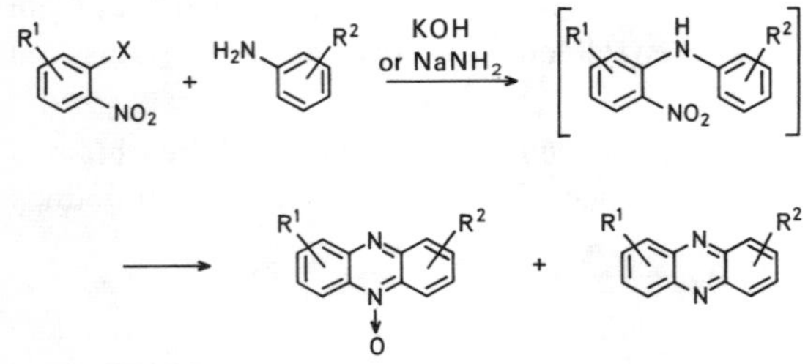

X = H, Cl, OCH₃

Overhigh temperatures (>160°) favor phenazine formation at the expense of the *N*-oxide yield. Yields are invariably poor, and the phenazines must be separated. The advantage is that a single reaction affords the *N*-oxides directly from cheap and simple starting compounds. This route is, however, of practical importance only where the direct oxidation of the phenazines fails or furnishes only the other isomeric *N*-oxide. Thus, for example, the Wohl-Aue technique affords *benzo[a]phenazine 12-oxide* (57%) from nitrobenzene and 2-naphthylamine, while peroxy acid oxidation of benzo[a]phenazine gives the *benzo[a]phenazine 7-oxide*[420]:

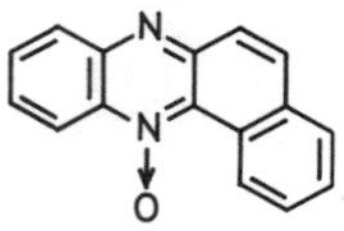

57

It is possible, alternatively, to commence directly from a 2-nitrodiphenylamine, but the ring closure is performed best with fuming sulfuric acid at room temperature[418] (to 80%, especially if a 4-substituent in the nonnitrated phenyl ring prevents the competing 4-sulfonation). New ref. [641].

The Wohl-Aue reaction can serve also for preparing *pyrimido[4,5-b]pteridine 10-oxides* (cf. p. 377)[421].

10.4.1.5 Condensed pyrazine monooxides from (2-nitrosoaryl)amines or 2-nitrosoenamines

10.4.1.5.1 Condensation of (2-nitrosoaryl)-amines with pyridinium ylides (4 + 2)

The ylides **59** formed from 1-(2-oxo-2-phenethyl)- and 1-(2-oxopropyl)pyridinium salts **58** react with primary (2-nitrosoaryl)amines in boiling methanol or ethanol to condensed *pyrazine N-monoxides* **60**[422, 423]:

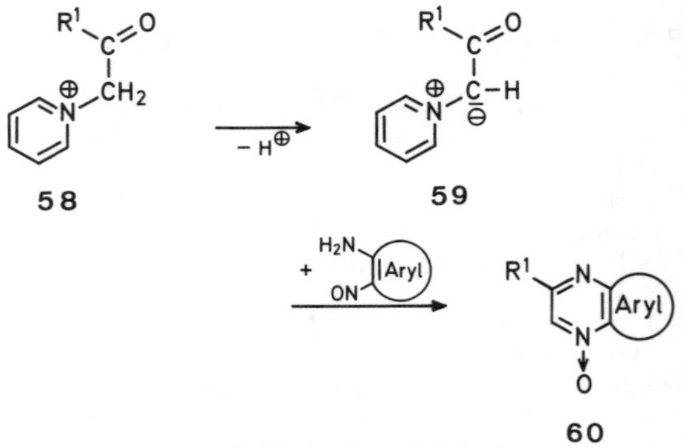

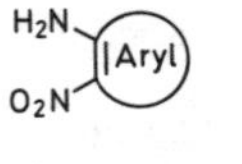

58   **59**

60

1-Nitroso-2-naphthylamine
2-Nitroso-1-naphthylamine
4,6-Diamino-5-nitrosopyrimidine
4,6-Diamino-5-nitroso-2-phenylpyrimidine

$R^1 = C_6H_5$, $4-Br-C_6H_4$, CH_3

[422] *I.J. Pachter, P.E. Nemeth, A.J. Villiani*, J. Org. Chem. *28*, 1197 (1963).

[423] *K. Gerlach, F. Kröhnke*, Chem. Ber. *95*, 1124 (1962).

[424] *H. Goldner, G. Dietz, E. Carstens*, Z. Chem. *4*, 454 (1964);
    H. Goldner, G. Dietz, E. Carstens, Justus Liebigs Ann. Chem. *691*, 142 (1966).

The precursors of the pyridinium salts **58**, too, can be used directly if sodium acetate is added to trap the acid[422]. Good yields result. This technique displays the advantage that the position of the *N*-oxide function in phenazine derivatives can be determined by selecting the starting substances. Thus, the *7-oxides* and *12-oxides* respectively of *benzo[a]phenazines* were obtained selectively[423]. By using 4,6-diamino-5-nitrosopyrimidines as the (2-nitrosoaryl)amine component *pteridine 5-oxides* were prepared for the first time[422].

In the same way (using potassium cyanide as base) the α-cyanobenzylpyridinium salts can be reacted to *4,7-diamino-6-phenylpteridine 5-oxides*[422]:

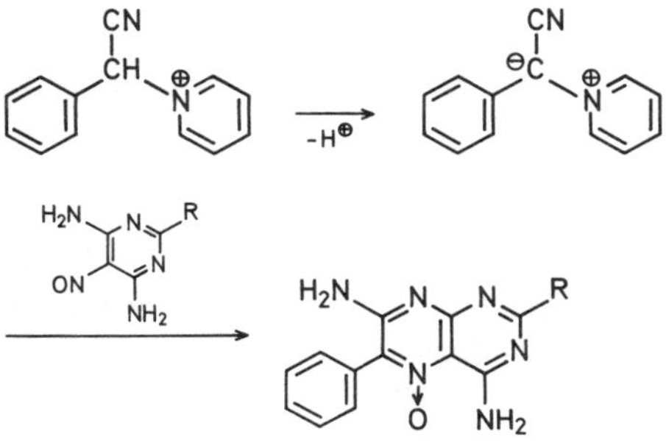

R = C₆H₅, SCH₃

10.4.1.5.2 Pyrimido[5,4-g]pteridine 5-oxides (4 + 2)

6-Chloro-1,3-dimethyl-5-nitrouracil **61** can be reacted to the *N*-oxides **64** and **65** in boiling dimethylformamide with the aminonitroso compounds **62** and **63** respectively[421] (the yields are good, see p. 376):

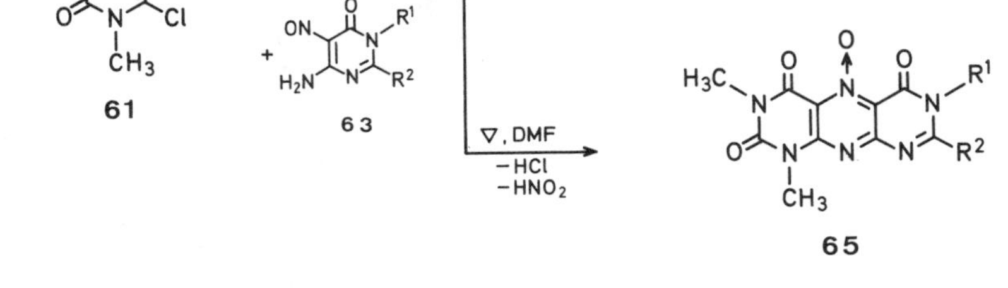

64; *1,3,7,9-Tetramethyl-2,4,6,8(1H,3H,7H,9H)-pyrimido-[5,4-g]pteridinetetrone 5-oxide*

65; R¹ = H, CH₃
R² = C₆H₅, SCH₃

10.4.1.5.3 Alloxazine 5-oxides from 6-anilino-5-nitrosouracils ($6_{acycl.} \rightarrow 6_{cycl.}$)

The 6-arylamino-5-nitrosouracils prepared by nitrosation of the corresponding 6-arylaminouracils cyclize either spontaneously during their preparation ($R^2 = CH_3$, $R^3 = H$), or on heating in the presence of isopentyl nitrite, to a mixture of the *benzo[g]pteridines* ($Z = N$) and their *5-oxides* ($Z = N \rightarrow O$)[424, 425]:

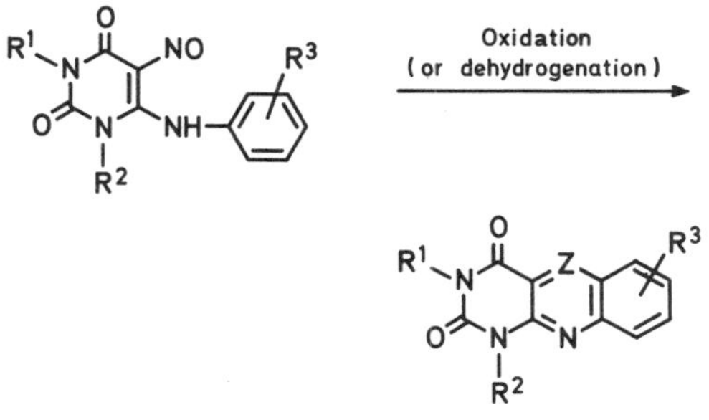

Z = N→O, N

R¹ = H, CH₃
R² = H, CH₃
R³ = H, CH₃, OC₂H₅, Cl, N(CH₃)₂

For a similar synthesis *cf.* refs. [639, 642].

10.4.1.6 Azine *N*-oxides by ring expansion

10.4.1.6.1 Ring expansion of 2,1-benzisoxazole by acid methylene compounds ($5_{cycl.} + 2$)

With *CH*-acid compounds of the malonic acid series, 2,1-benzisoxazole forms *2-aminoquinoline 1-oxides* ($Z = NH_2$) and *2-quinolinol 1-oxides* ($Z = OH$) in ethanol in the presence of catalytic

64

65

[425] H.-G. Kazmirowski, H. Goldner, E. Carstens, J. Prakt. Chem. *32*, 43 (1966).

[426] E.C. Taylor, J. Bartulin, Tetrahedron Lett. 2337 (1967).

amounts of piperidine[426]. The yields are almost quantitative. Where piperidine is not an adequate base the sodium or potassium salts of the *CH*-acid compounds may be employed.

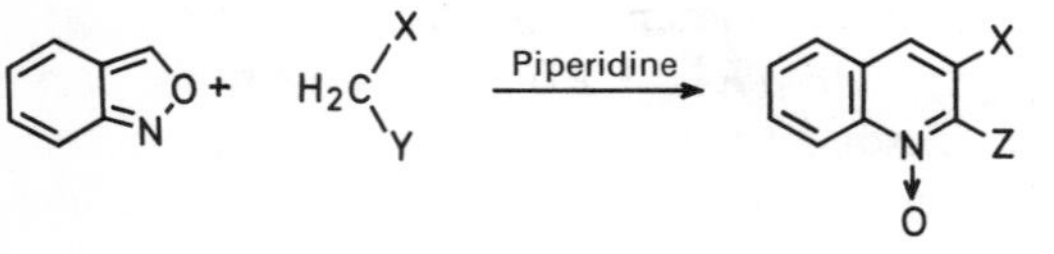

or tautomer

X	Y	Z	1-Oxide
CN	CN	NH_2	*2-Amino-3-quinoline-carbonitrile 1-oxide*
CN	$COOC_2H_5$	OH	*2-Hydroxy-3-quinoline-carbonitrile 1-oxide*
$CO-NH_2$	CN	NH_2	*2-Amino-3-quinolin-carboxamide 1-oxide*
$COOCH_3$	$COOCH_3$	OH	*Methyl 2-hydroxy-3-quinolinecarboxylate 1-oxide*
$SO_2-C_6H_5$	CN	NH_2	*2-Amino-3-quinoline-benzenesulfonic acid 1-oxide*

10.4.1.6.2 Quinoxaline and phenazine *N*-dioxides from benzofuroxans (benzofurazan *N*-oxides) and carbonyl compounds, enamines, ynamines, or phenols ($5_{cycl.} + 2$)

An evidently very productive synthesis principle consists in condensing benzofuroxans with enol and enolate groupings (new ref.[643]). In this way *quinoxaline 1,3-dioxides*[427, 428] are formed from benzofuroxans and carbonyl compounds with ammonia or primary amines in methanolic solution at 20–55° (for reacting aldehydes primary amines are preferable to ammonia):

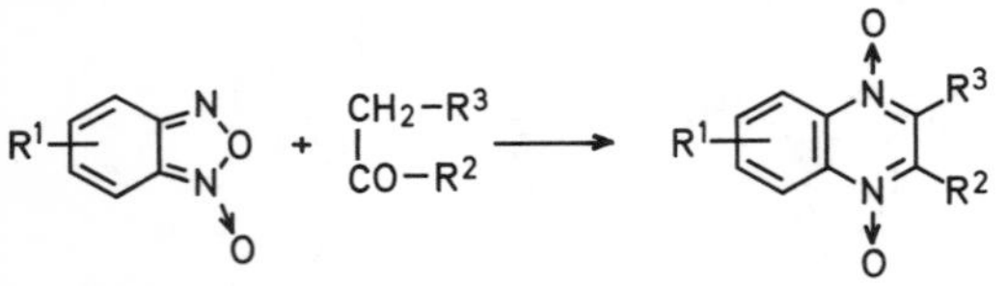

R^1 = H, Alkyl, OCH_3, OC_2H_5, Halogen, COOR
R^2 = CH_3, H, C_6H_5
R^3 = H, Alkyl, Aryl, $COCH_3$, COC_6H_5, COOR, $CONH_2$,
    $CO-NH-R$, $CO-N(R)_2$, $-CH_2-COOR$,
    $-CH_2-CO-NH-R$, $-CH_2-CO-N(R)_2$,
    $-C=NOH$, Halogen, CN
        |
        R

R^2-R^3 = cyclic

In an alternative embodiment benzofuroxan is allowed to stand with 1,3-dicarbonyl compounds for a day in triethylamine or diethylamine follow-

ing dissolution of the substances in these by warming[429, 430].

From 5,6-methylenedioxybenzofurazan *N*-oxide and β-keto carboxylic acid ethyl esters, for example, *3-hydroxy-6,7-methylenedioxy-2-quinoxalinyl ketone 1,4-dioxides* are formed[259].

On condensing 1,3-diketones with caustic soda the acyl group of the quinoxaline 1,4-dioxide is cleaved solvolytically[429]; alternatively the cleaving can be performed subsequently[430]:

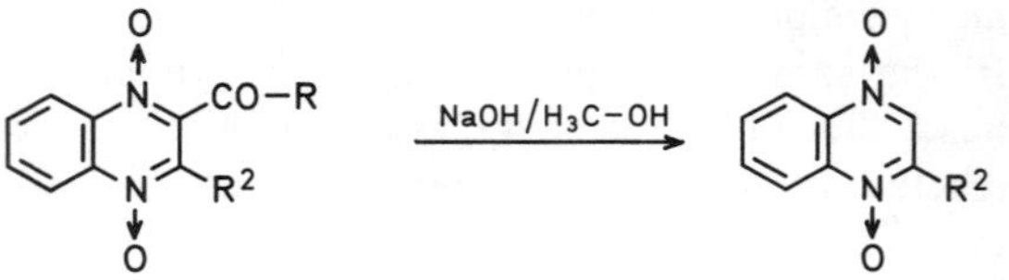

During the hydroxide-catalyzed reaction of benzofuroxans with 2-pentanone a benzimidazole derivative is obtained[259].

Ammonia in methanolic solution enables ester groups to be aminolyzed and halogen (*e.g.*, when chloroacetaldehyde is employed) to be replaced by primary amine. The yields are high[427] (up to 91%[427]; during condensations with 1,3-diketones in triethylamine up to 78%[429]).

Reacting benzofuroxans with unsymmetrically substituted 1,3-diaryl-1,3-propanediones[430] in general affords *aroylarylquinoxaline 1,3-dioxide mixtures* **66** and **67** which cannot be separated by column chromatography:

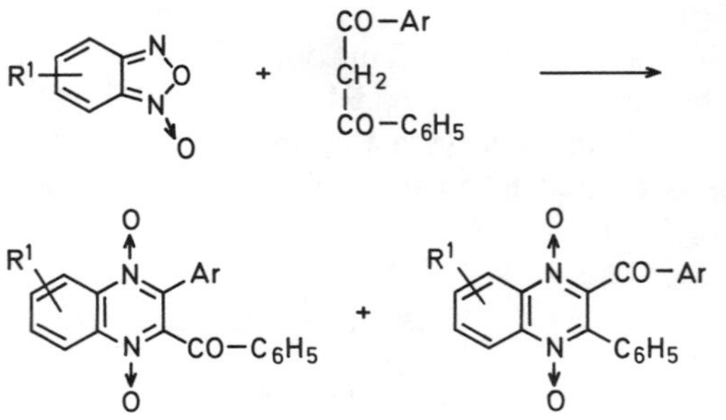

66   **67**

R^1 = H; Ar = 2–CH_3–C_6H_4 } only 67
                  2–OCH_3–C_6H_4
            Ar = 2–NO_2–C_6H_4 ⟶ 66 and 67
            Ar = 4–X–C_6H_4
                  X = Br, OCH_3, CH_3, NO_2
R^1 = CH_3, OCH_3, Cl ⟶ mixture always[430]

[427] *K. Ley, F. Seng, U. Eholzer, R. Nast, R. Schubart*, Angew. Chem. *81*, 569 (1969).
[428] *J.C. Mason, G. Tennant*, Chem. Commun. 586 (1971).
[429] *C.H. Issidorides, M.J. Haddadin*, J. Org. Chem. *31*, 4067 (1966).
[430] *M.J. Haddadin, G. Agopian, C.H. Issidorides*, J. Org. Chem. *36*, 514 (1971).
[431] *W.E. Noland, D.A. Jones*, J. Org. Chem. *27*, 341 (1962).
[432] *J.P. Freeman, D.L. Surbey, J.E. Kassner*, Tetrahedron Lett. 3797 (1970).
[433] *F. Eloy*, J. Org. Chem. *26*, 952 (1961).

5-Monosubstituted benzofuroxans form exclusively *7-R¹-3-phenyl-2-quinolinecarbonitrile 1,4-dioxides*[428], with 3-oxo-3-phenylpropionitrile in ethanolic ammonia; the products are cleaved to *7-R¹-2-quinoxalinol 1,4-dioxides* by ethanolic sodium ethoxide solution[428].

In analogy with the ketones, malononitrile affords the deep red *3-amino-2-quinoxalinecarbonitrile 1,4-dioxides* with benzofuroxans[427] (the strongly exothermic reaction in dimethylformamide is kept at between 20 and 30° by adding ammonia or amines gradually):

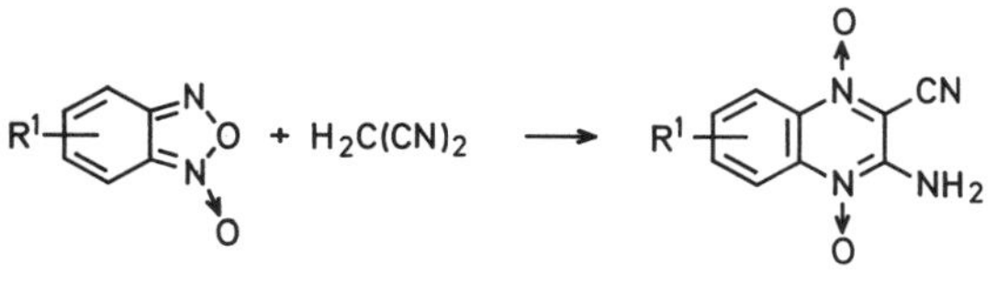

or tautomer

R¹ = H; *3-Amino-2-quinoxalinecarbonitrile;*
1,4-dioxide; 83%

With hydroquinones in aqueous alkaline suspension or other basic media benzofuroxans form *2-phenazinol 5,10-dioxides* at 20–30°[427]:

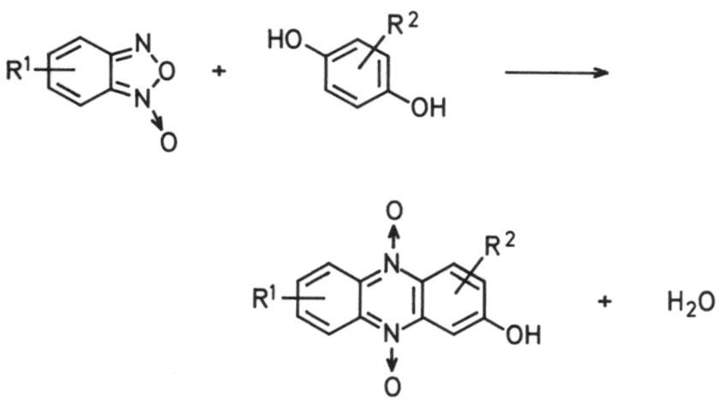

R² = H, Alkyl, Aryl, Halogen, NO_2, SO_2–R, COOH,
    –CH_2–CO–N(R)$_2$, $COOCH_3$
R¹ = CH_3, OCH_3, Cl

After stirring several hours the phenazine 5,10-dioxides are precipitated by making acid; practically quantitative yields are obtained in some cases and 97% with hydroquinone.

1,4-Naphthalenediols, 2-naphthols, 4-substituted 1-naphthols, 8-quinolinols, 2,3-pyridinediols, 1-phenazinol, and indole can be reacted correspond-ingly in alcohol or carbon tetrachloride by allowing ammonia to act at 20–40°. With phenols the reaction is performed at above 60°[427, 376].

Enamines, too, can be used in the reaction[260, 644]:

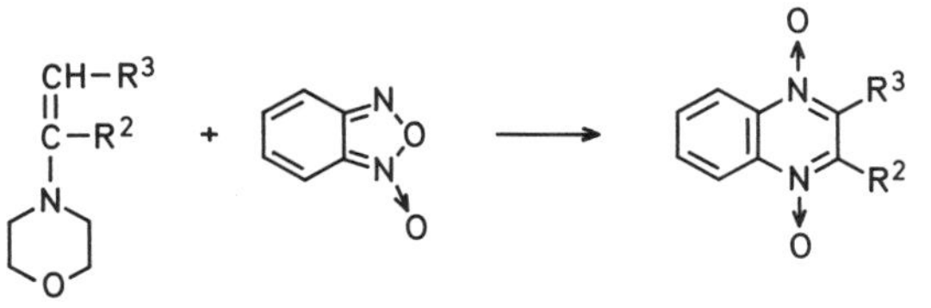

R² = C_6H_5; R³ = H
R²–R³ = –$(CH_2)_3$–

R³–CH=C–R² = Cholest-2-en-3-yl;
with R²–R³ = –$(CH_2)_4$–; *Phenazine 5,10-dioxide;* 48%

Reacting ynamines with benzofuroxans in tetra-hydrofuran at 10–30° affords, for example, *3-ethyl-2-(ethylmethylamino)quinoxaline N-dioxides* and *3-ethyl-2-(diethylamino)quinoxaline N-dioxides*[427]:

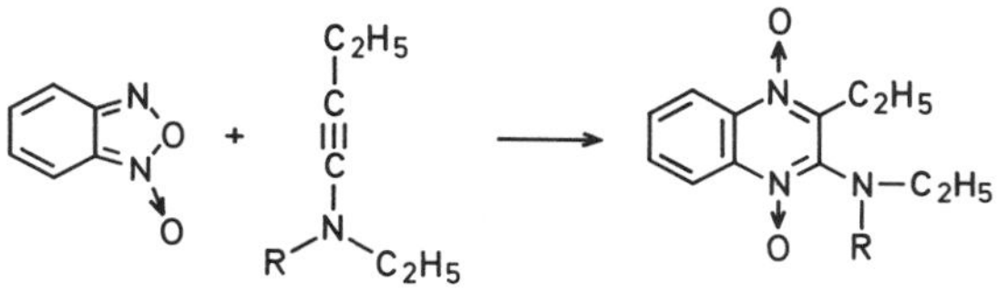

R = CH_3, C_2H_5

α-Halocarbonyl compounds, too, give *2-amino-quinoxaline 1,4-dioxides* with benzofuroxans by condensation with amines.

10.4.1.6.3 Ring expansion of isatogens and azole *N*-oxides ($5_{cycl.} + 1$)

Heating 2-phenyl-3*H*-indol-3-one 1-oxide (2-phenylisatogen) with ethanolic ammonia affords *3-phenyl-4-cinnolinol 1-oxide* (26%)[431]:

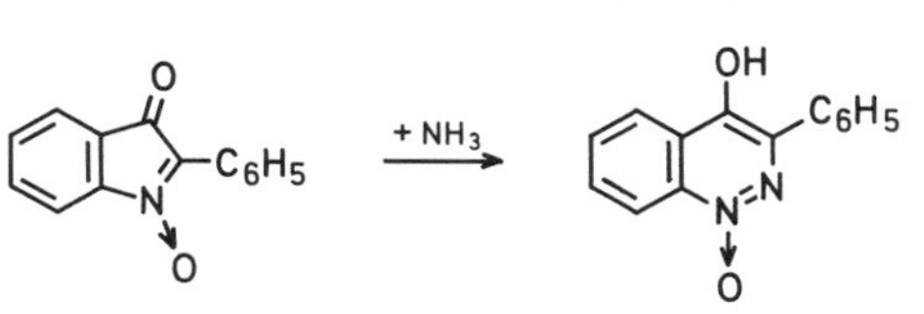

[434] O. Diels, F. Riley, Ber. *48*, 897 (1915).
[435] W. Dilthey, J. Friedrichson, J. Prakt. Chem. *127*, 292 (1930).
[436] Ch.M. Selwitz, A.I. Kosak, J. Amer. Chem. Soc. 77, 5370 (1955).
[437] K. Bodendorf, H. Towliati, Arch. Pharm. (Weinheim, Ger.) *298*, 293 (1965).

[438] A.W. Allan, B.H. Walter, J. Chem. Soc. (C) 1397 (1968).
[439] Y. Goto, M. Yamazaki, M. Hamana, Chem. Pharm. Bull. (Tokyo) *19*, 2050 (1971).
[440] P.M. Weintraub, J. Med. Chem. *15*, 419 (1972).
[441] H. Lettau, Z. Chem. *10*, 431 (1970).
[442] H. Lettau, Z. Chem. *10*, 338 (1970).
[443] H. Lettau, Z. Chem. *10*, 462 (1970).
[444] K. Volkamer, H. Baumgärtel, H. Zimmermann, Angew. Chem. *79*, 941 (1967).

Oxidation with peroxy acids converts 1-hydroxy-pyrazol-4-ols into *6H-1,3,4-oxadiazin-6-one 4-oxides via* the stage of a *pyrazol-4-one 1-oxide* in the manner of a Baeyer-Villiger oxidation[432]:

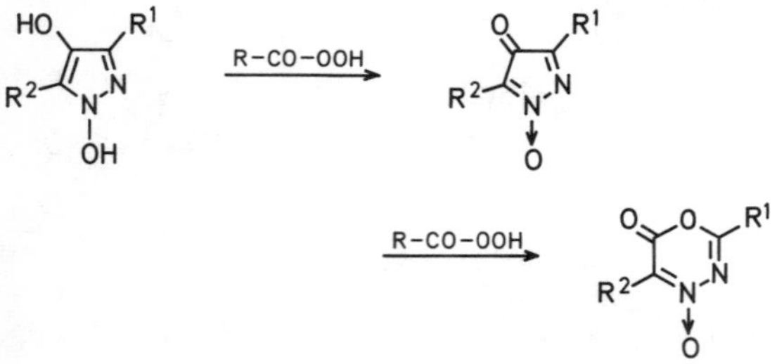

$R^1 = R^2 = C_6H_5$;   *2,5-Diphenyl-6H-1,3,4-oxadiazin-6-one 4-oxide;*
$R^1 = C_6H_5$; $R^2 = CH_3$; *5-Methyl-2-phenyl-6H-1,3,4-oxadiazin-6-one 4-oxide;*
$R^1 = C_6H_5$; $R^2 = C_2H_5$; *5-Ethyl-2-phenyl-6H-1,3,4-oxadiazin-6-one 4-oxide;*

10.4.1.7 Cyclization of oximes and hydroxyl-amines to azole *N*-oxides

A variety of condensation and oxidation reactions of oximes lead to azole *N*-oxides. By contrast, a previously postulated isomerization of azide oximes $R–C(N_3)=NOH$ to 1-tetrazolols does not take place[433]. For syntheses which proceed *via* nonisolated hydroxylamine intermediate stages.

10.4.1.7.1 Oxazole and imidazole *N*-oxides from mono- or dioximes of 1,2-diketones or 2-nitrosoenamines (4 + 1 or 3 + 2)

To prepare *oxazole 3-oxides* from 1,2-diketone monooximes and aldehydes the glacial acetic solution of the components is saturated with hydrogen chloride; hydrochlorides of the *N*-oxides either precipitate spontaneously or on adding ether:

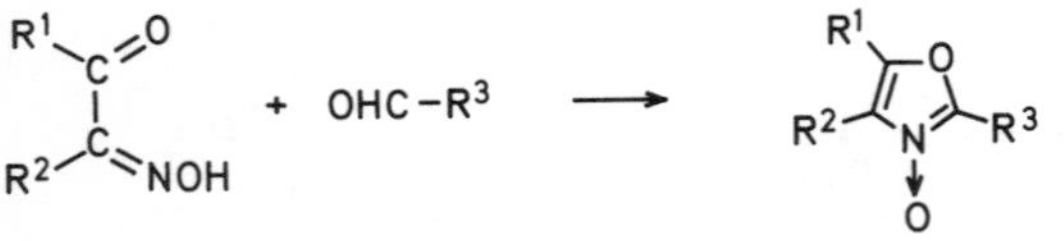

$R^1 = CH_3, C_6H_5, C_2H_5$
$R^2 = H, COCH_3, CH_3, C_6H_5$
$R^3 = C_6H_5$, subst. C_6H_5, $-CH=CH-C_6H_5$, 2-Furyl, 2-Thienyl, 1-Oxido-4-pyridyl
$R^1 - R^2 = -(CH_2)_3-C-$
                      $\|$
                      NOH

The yields are between 55 and 92% but are less for purely aliphatic components[434-440]. 5-Phenyl-2,3-thiophenedione 2-oxime forms no *N*-oxide[436]. Formaldehyde yields other products with 1,2-diketone monooximes (α-hydroxyimino ketones) in alcoholic hydrochloric acid[437], but in hydrochloric acid/glacial acetic acid *4,5-diphenyloxazole 3-oxide* is obtained[440].

If the reaction is performed in the presence of ammonia or primary amines, or if the aldehydes are employed in the form of imines, then the light-sensitive *imidazole 3-oxides*[238, 441–445, 268, 669] are formed (this reaction displays wide applicability $R^1 = H$, alkyl, aryl):

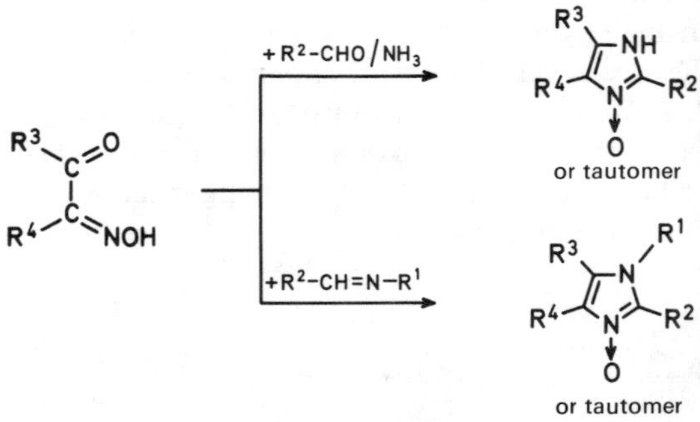

1,2-Diketone monooximes, aldehyde, and ammonia or ammonium acetate are reacted in alcohol, and the initially precipitated monohydrates are dehydrated, either by recrystallizing from anhydrous solvents or by heating above the melting point. Purely aliphatic 1,2-diketone monooximes alone form the *imidazole 3-oxides* without supplying heat.

Alkanals *(R² = alkyl)* react only with the most reactive 1,2-diketone monooximes. Terephthalaldehyde forms *2,2'-p-phenylenediimidazole 3,3'-dioxides* smoothly, while *p*-phenylenediamine and 4,4'-biphenyldiamine give *p-phenylenediimidazoline 3,3'-dioxide* and the *4,4'*-derivative of *biphenyl* respectively. When the aldehyde alkyl or aryl imines are employed the reaction is conducted in hot acetic acid[441].

1-Hydroxyimidazole 3-oxides form from 1,2-diketone monooximes and aldoximes[268, 437, 444–447] (35–94%) by condensing in ethanol by saturating with hydrogen chloride or with concentrated acid. In glyoxal dioxime only one group reacts[447].

Carrying out the reaction unsuitably may give an *1,3,4-oxadiazine N-oxide* as side-product[445]. The fairly hydrolysis-resistant 1-hydroxyimidazole 3-oxides are acid and can be titrated with phenolphthalein; they also form salts with mineral acids. *2-Aryl-1-hydroxy-4,5-dimethyl-* and *1-hydroxy-4,5-dimethylimidazole-2-carboxaldehyde oxime 3-oxides* are formed from 1,2-dioximes and aldehydes under these conditions[238, 444, 448, 449]. Re syntheses of condensed imidazole 3-oxides from enamines see[645].

10.4.1.7.2 2-Imidazolol and 2-imidazolethiol 3-oxides from α-amino ketoximes (4 + 1)

Synthesis of one *1-hydroxy-4-methyl-5-phenyl-imid-azol-2-ol 3-oxide* and one *-2-thiol 3-oxide* by the following scheme has been described[450]:

───────────
[445] *K. Volkamer, H.W. Zimmermann,* Chem. Ber. *102,* 4177 (1969).

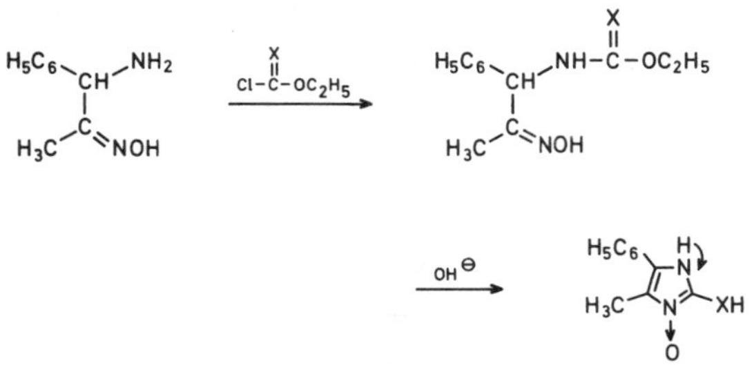

X = O, S

or tautomer

Similar syntheses of *imidazole 3-oxides* *(XH = CH₃)* are described in ref. [646].

10.4.1.7.3 Azole *N*-oxides by oxidation of 1,2-dioximes and 1,2-oxime hydrazones ($5_{acycl.} \rightarrow 5_{cycl.}$)

Furoxans[451–453, 633, 654], *benzofuroxans*[451, 454, 455] and a *2 H-1,2,3-triazine 1-oxide* can be made by dehydrating 1,2-dioximes or intermediate 1,2-oxime hydrazones, especially where the 1,2-dioximes are readily to hand. Alkaline hexacyanoferrate(III), chlorine in ethanol or benzene, alkali hypochlorite, dinitrogen trioxide or tetraoxide is used for oxidation[456]. *E.g.*, 1,2-cyclohexanedione dioxime is dissolved in 10% caustic soda and poured into 7% NaClO; the furoxan (66%) at once precipitates[452].

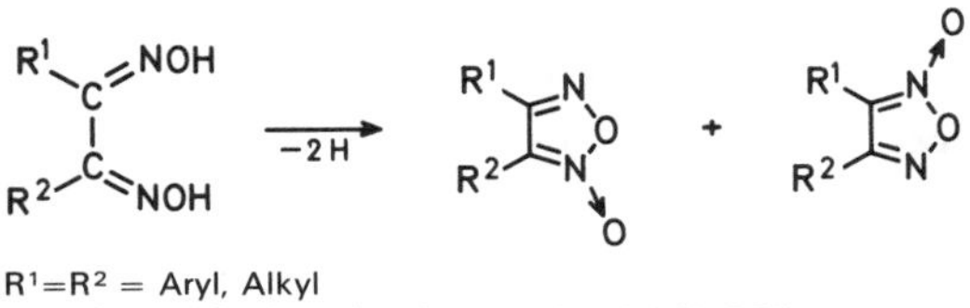

R¹=R² = Aryl, Alkyl
R¹–R² = –(CH₂)₄– (oxidation only with NaOCl)
    4,5,6,7-tetrahydrobenzofuroxan
R¹–R² = (CH₂)₃!

[446] *J.B. Wright*, J. Org. Chem. *29*, 1620 (1964).
[447] *H. Towliati*, Chem. Ber. *103*, 3952 (1970).
[448] *G. La Parola*, Gazz. Chim. Ital. *75*, 216 (1945); C.A. *41*, 4146 (1947).
[449] *P. Franchetti, M. Grifantim*, J. Heterocycl. Chem. 7, 1295 (1970).
[450] *A. Dornow, H.H. Marquardt*, Chem. Ber. *97*, 2169 (1964);
H.G.O. Becker, G. Görmar, H.-J. Timpe, J. Prakt. Chem. *312*, 610 (1970);
W.T. Flowers, D.R. Taylor, A.E. Tipping, C.N. Wright, J. Chem. Soc. (C) 1986 (1971).
[451] *J.H. Boyer* in *R.C. Elderfield*, Heterocyclic Compounds, Vol. VII, p. 469, John Wiley & Sons, London 1961.
[452] *A. Dornow, K.J. Fust, H.D. Jordan*, Chem. Ber. *90*, 124 (1957).

2,3-Bornanedione dioxime (camphorquinone dioxime) analogously affords *4,5,6,7-tetrahydro-8,8-dimethyl-4,7-methanobenzofurazan N-oxide*, while oxidation with bromine in sulfuric acid is necessary with 1-amino-3-phenylpropanetrione 1,2-dioxime in order to obtain *4-amino-3-furazanyl phenyl ketone N-oxide:*

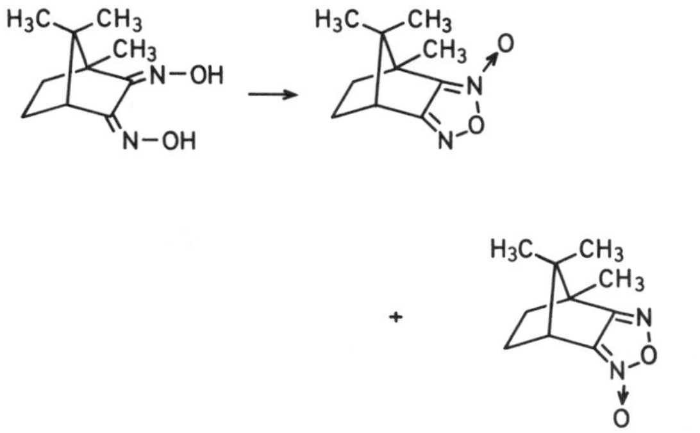

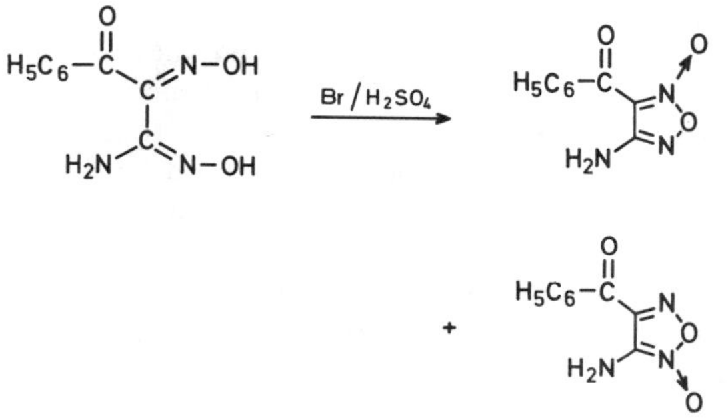

Oxidation of the three *cis-trans* isomeric phenylglyoximes with dinitrogen tetraoxide in ether invariably affords one and the same *4-phenylfuroxan*[456]. Otherwise, preparation of the furoxans during oxidation of unsymmetrical glyoximes may be dependent on the isomer used. At least in one case that furoxan is formed from the two *Z,E*-forms of the glyoxime whose structure is anticipated by the glyoxime geometry, while the *E,E* and *Z,Z* forms afford mixtures of the two furoxan isomers[451]. The energy threshold for the occurrence of tautomerization of the two furoxan isomers is thus high enough to maintain each form intact. For recent work see refs. [633, 650, 654].

Carrying out the oxidation of an aryl glyoxime (*amphi* form!) in the presence of a secondary amine or ammonia furnishes useful yields of 3-

[453] *D.J. Chadwick, W.R.T. Cottrell, G.D. Meakins*, J. Chem. Soc. Perkin I, 655 (1972).
[454] *A.J. Boulton, P.B. Gosh* in *A.R. Katritzky, A.J. Boulton*, Advances in Heterocyclic Chemistry, Vol. X, p. 1, Academic Press, New York, London 1969.
[455] *A.J. Boulton, A.R. Katritzky, M.J. Sewell, B. Wallis*, J. Chem. Soc. (B) 914 (1967).

amino-4-arylfuroxans, which on heating rearrange to the isomeric *3-aryl-4-aminofuroxans*[457].

R¹ = R² = H, X = H, 3−CF₃, CH₃, 2−Cl, 4−Cl
R¹ = R² = CH₃, X = H
R¹−R² = CH₂−CH₂−NH−CH₂−CH₂, X = H
R¹ = CH₂−CH₂Cl, R² = H, X = H

The reaction, which proceeds *via* the relevant arylfuroxan and ultimately *via* the amidoxime, is carried out either in one stage with four equivalents hexacyanoferrate(III), or the variously accessible amidoxime is oxidized subsequently (generally with bromine).

Transformation of the silver salt of *4-hydroxyimino-3-methyl-5-isoxazolinone* with nitric acid may be regarded as the oxidation of a masked 1,2-dione[458]:

4-Methyl-3-furazancarboxylic acid 2-oxide

Oxidation of 1,2-quinone dioximes to *benzofuroxans* can be carried out substantially with the above oxidizing agents. A drawback is the often difficult access to the 1,2-quinone dioximes, an advantage is that only one benzofuroxan is invariably formed independently of the geometry of the starting material (cf. p. 357), e.g.:

Phenanthro-[9,10-c][1,2,5]-oxadiazole, 1-oxide, 10-Nitrophenanthro-[9,10-c][1,2,5]-oxadiazole 1-oxide and isomers

The classical structural formulae reveal that the aromatic character in the benzene portion of the benzofuroxans is reduced. This finds expression in the ready bromine addition to form a *4,5,6,7-tetrabromo-4,5,6,7-tetrahydrobenzofuroxan*[451], and may possibly explain why, for instance, *naphtho[2,3]furoxan* is unknown while *6,7-diphenylquinoxalino[6,7]furoxan*[454] has been prepared.

Evidently, the last step of the conversion of 3,5-dimethyl-4-nitroisoxazole with benzenediazonium salts is similar to the oxidation of the 1,2-dioximes[459, 460]:

5-Methyl-2-phenyl-4-(phenylazo)-1,2,3-triazole 1-oxide

10.4.1.7.4 Furoxans from alkenes and nitrogen oxides[461, 462] (2 + 2 + 1 or 5$_{acycl.}$ → 5$_{cycl.}$)

α-Nitro oximes (pseudonitrosites) are formed from alkenes with higher nitrogen oxides; they are dehydrated to *furoxans* from their solutions in dilute aqueous alkali on acidifying with cold acid:

or isomer

[456] *J.V. Burakevich, A.M. Lore, G.P. Volpp,* J. Org. Chem. *36*, 5 (1971).

10.4.1.7.5 1,2,4-Triazole 4-oxides by dehydration of hydroxamic acid acylhydrazones ($5_{acycl.} \rightarrow 5_{cycl.}$) or from hydroxylamine and 1,4-dichloro-2,3-diazabutadienes (4 + 1)

Boiling N^3-hydroxy-N^1-acylamidrazones (or the tautomeric hydroxamic acid acylhydrazones) for two hours in molar amounts with 3% sodium hydroxide solution furnishes *1,2,4-triazole 4-oxides* (43–78%)[620]:

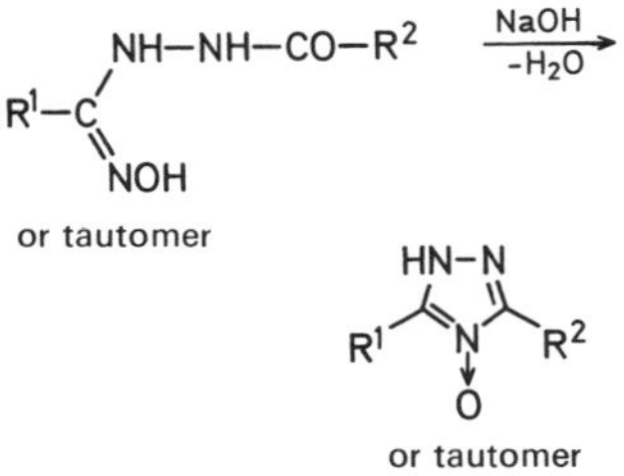

R^1 = H, CH$_3$, Br, C$_6$H$_5$, CH$_2$−NH$_2$, CH$_2$−NH−CO−C$_6$H$_5$
R^2 = Aryl, CH$_2$−C$_6$H$_5$

3,5-Diphenyl-1,2,4-triazole 4-oxide ($R^1 = R^2 = C_6H_5$) is obtained also from hydroxylamine and 1,4-dichloro-1,4-diphenyl-2,3-diazabutadiene[621].
1-Alkenes yield no furoxans but α-nitro ketones.

10.4.1.7.6 1-Hydroxypyrazole 2-oxides by nitrosation of α,β-unsaturated ketoximes (4 + 1)

α,β-Unsaturated oximes, which must not carry an H atom in the α-position *($R^2 = H$)*, are converted into *1-hydroxypyrazole 2-oxides* (~ 90%) in cooled glacial acetic acid on adding aqueous sodium nitrite solution drop by drop[337]:

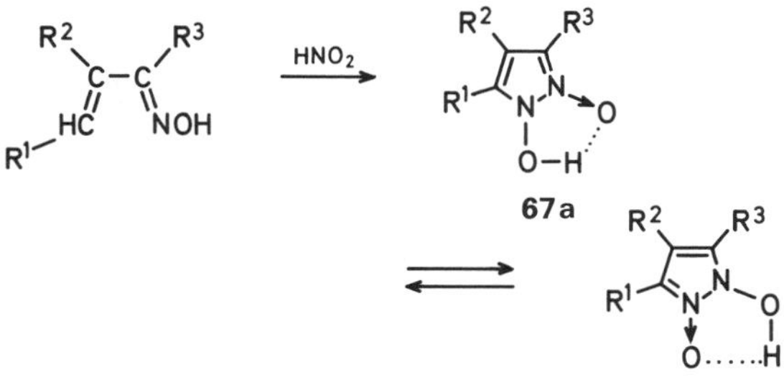

R^1, R^2, R^3 = CH$_3$; *1-Hydroxy-3,4,5-trimethylpyrazole 2-oxide*
C$_6$H$_5$; *1-Hydroxy-3,4,5-triphenylpyrazole 2-oxide*

The light-sensitive, high-melting *N*-oxides are precipitated during the reaction. They are little soluble in most solvents but dissolve in dilute caustic soda, from which they can be recovered unchanged or precipitated as the chelate complex by adding transition metal ions.

10.4.1.7.7 3-Hydroxysydnone imines from α-hydroxyaminonitriles ($5_{acycl.} \rightarrow 5_{cycl.}$)

The *N*-nitroso compounds prepared *in situ* from α-hydroxyaminonitriles by nitrosation are converted into the corresponding *3-hydroxysydnone imines* on treatment with acid in ice-cold methanol (sometimes in poor yield)[463]:

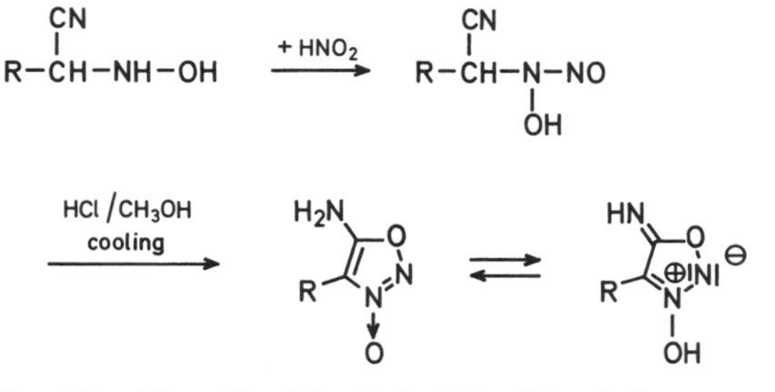

R = CH$_2$−CH$_2$−CH$_3$, CH$_2$−C$_6$H$_5$, CH$_2$−CH$_2$−C$_6$H$_5$

10.4.1.7.8 2-Aminothiazole-3-oxides from α-chloro oximes or α-oxo thiocyanates (3 + 2 or 4 + 1)

Reacting α-aryl-α-chloro oximes **68** with barium thiocyanate[464] or α-oxo thiocyanates **69** with hydroxylammonium chloride[465] in boiling ethanol in some instances affords good to very good yields of *2-aminothiazole 3-oxides* **70**:

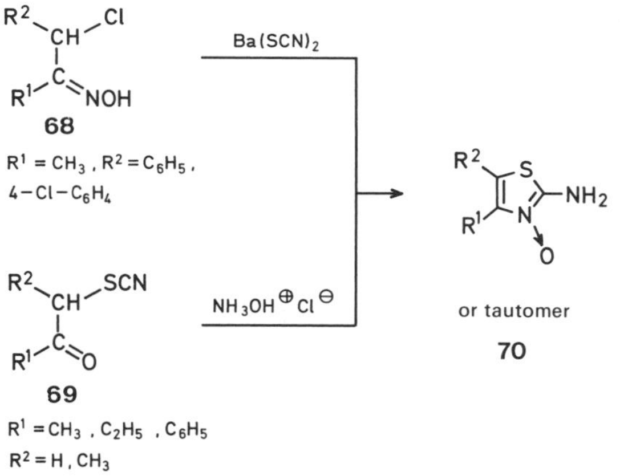

R^1 = CH$_3$, R^2 = C$_6$H$_5$,
4−Cl−C$_6$H$_4$

R^1 = CH$_3$, C$_2$H$_5$, C$_6$H$_5$
R^2 = H, CH$_3$

Conversion of 2-oxo-1,2-diphenylethanol thiocyanate (**69**, $R^1 = R^2 = C_6H_5$) to the *N-oxide* does not succeed.

[457] *A.R. Gagneux, R. Meick*, Helv. Chim. Acta **53**, 1883 (1970).

[458] *A. Quilico* in The Chemistry of Heterocyclic Compounds, Vol. XVII, p. 139, Interscience Publishers, New York, London 1962.

[459] *A. Quilico* in The Chemistry of Heterocyclic Compounds, Vol. XVII, p. 68, Interscience Publishers, New York, London 1962.

[460] *A. Quilico*, Gazz. Chim. Ital. **72**, 399 (1942); C.A. **38**, 4598 (1944).

10.4.1.7.9 Tetrasubstituted imidazole 3-oxides from nitrile, olefin, and nitrosyl hydrogen sulfate (2 + 2 + 1)

Two *mols* nitrosyl hydrogen sulfate react with 1 mol of a nitrile and two mols of an olefin to give *imidazole N-oxides*[268]:

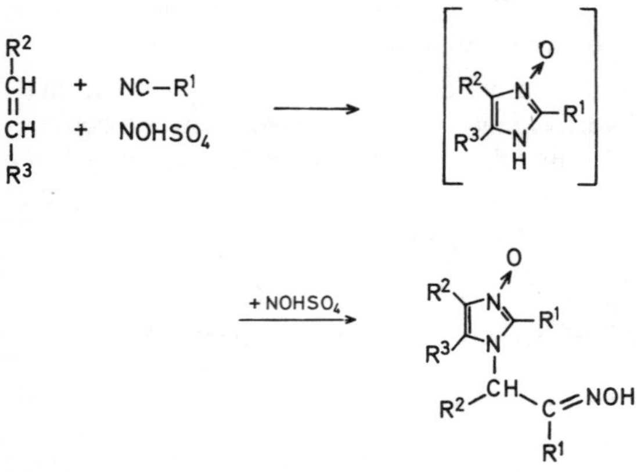

The olefin is added drop by drop to the suspension of nitrosyl hydrogen sulfate in the nitrile at below 0°. On pouring the reaction mixture into aqueous ammonia the *imidazole 3-oxides* (<30%) crystallize.

10.4.1.8 *N*-Oxides of benzo azoles from 2-nitroanilines or from 2-nitrophenyl thiocyanates

10.4.1.8.1 Benzimidazole 3-oxides by dehydration of *N*-substituted 2-nitroanilines (5$_{acycl.}$ → 5$_{cycl.}$)

N-(2-nitrophenyl)benzylamines cyclize in boiling methanolic caustic soda in sometimes good yield to form 2-arylbenzimidazole 3-oxides[466, 467] (re similar syntheses *cf.* ref.[647]).

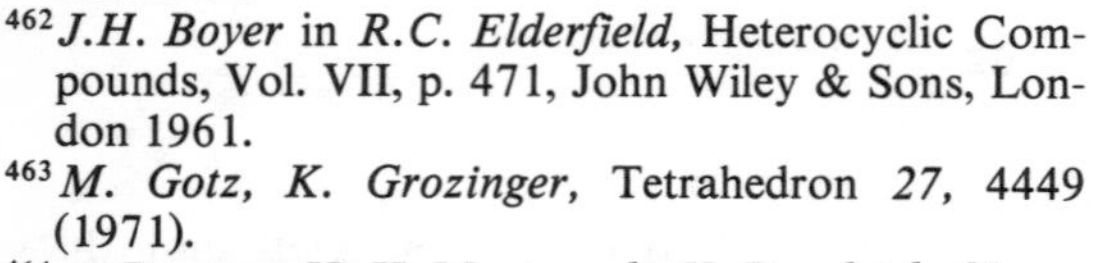

X = H, Cl, NO$_2$

N-Benzylidene-2-nitroaniline affords *2-phenyl-benzimidazole 3-oxide* (79%[666]) in the presence of cyanide ions (a similar synthesis of benzothiazole *N*-oxides see ref.[667]).

N,N-Dialkyl-2-nitroanilines cyclize to *1-alkyl-benzimidazole 3-oxides* under acid conditions:

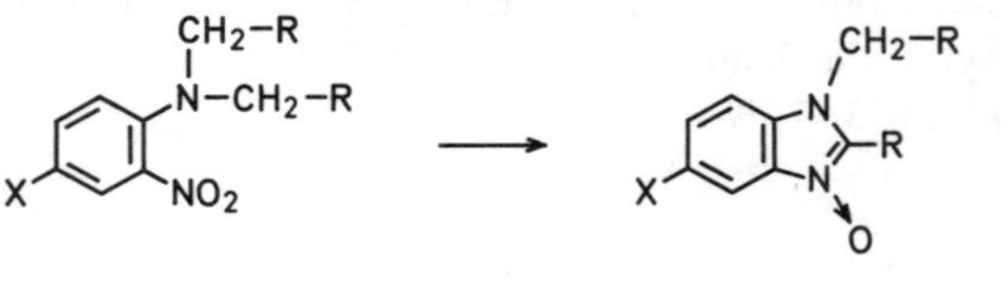

1) With hydrochloric acid at 110–160°
   X = H, COOH, CF$_3$, NO$_2$
   R = CH$_3$ or R—R = —(CH$_2$)$_n$—, —CH$_2$—O—CH$_2$—;

2) Photochemically:
   X = H;   R—R = —(CH$_2$)$_2$—;   2,3-Dihydro-1H-pyrrolo[1,2-a]-benzimidazole 4-oxide
              R—R ≠ —(CH$_2$)$_3$—, —(CH$_2$)$_4$—;
   X = Cl;   R—R = —(CH$_2$)$_2$—;   6-Chloro-2,3-dihydro-1H-pyrrolo-[1,2-a]benzimidazole 4-oxide
              R—R = —(CH$_2$)$_4$—;   3-Chloro-7,8,9,10-tetrahydro-1H-azepino[1,2-a]-benzimidazole 5-oxide
   X = NO$_2$; R—R = —(CH$_2$—O—CH$_2$)—; 3,4-Dihydro-8-nitro-1H-1,4-oxazino-[4,3-a]benzimidazole 10-oxide

The reaction in aqueous hydrochloric acid at 110–160° affords medium yields[468], photolysis in dilute aqueous methanolic hydrochloric acid either good yields of *N*-oxides[469] or, instead, depending on the structure, the corresponding benzimidazoles themselves. It has not yet been elucidated when benzimidazoles form and when their *N*-oxides[469] (steric and electronic effects appear to play a part). For recent work see ref.[653].

Evidently, the photochemical condensation can be transferred also to analogs of the 2-nitroanilines, as is shown by the preparation of *7,8-dihydro-6H-pyrrolo[2',1' : 2,3]imidazolo[4,5-b]pyridine 5-oxide* (78%)[469] from 3-nitro-2-pyrrolidino-pyridine[469].

Photolysis[470] of α-(2,4-dinitroanilino)carboxylic

[461] *J.H. Boyer* in *R.C. Elderfield*, Heterocyclic Compounds, Vol. VII, p. 477, John Wiley & Sons, London 1961.

[462] *J.H. Boyer* in *R.C. Elderfield*, Heterocyclic Compounds, Vol. VII, p. 471, John Wiley & Sons, London 1961.

[463] *M. Gotz, K. Grozinger*, Tetrahedron *27*, 4449 (1971).

[464] *A. Dornow, H.-H. Marquardt, H. Paucksch*, Chem. Ber. *97*, 2165 (1964).

[465] *H. Beyer, G. Ruhlig*, Chem. Ber. *89*, 107 (1956).

[466] *G. De Stevens, A.B. Brown, D. Rose, H.I. Chernov, A.J. Plummer*, J. Med. Chem. *10*, 211 (1967).

[467] *G.W. Stacy, B.V. Ettling, A.J. Popa*, J. Org. Chem. *29*, 1537 (1964).

[468] *R. Fielden, O. Meth-Cohn, D. Price, H. Suschitzky*, Chem. Commun. 772 (1969).

[469] *R. Fielden, O. Meth-Cohn, H. Suschitzky*, Tetrahedron Lett. 1229 (1970).

[470] *P.H. MacFarlane, D.W. Russell*, Tetrahedron Lett. 725 (1971).

[471] *R.J. Pollitt*, Chem. Commun. 262 (1965).

[472] *D.J. Neadle, R.H. Pollitt*, J. Chem. Soc. (C) 1764 (1967).

[473] *O. Meth-Cohn*, Tetrahedron Lett. 1235 (1970).

acids in acid medium gives *5-nitrobenzimidazole 3-oxides*[471, 472, 473]:

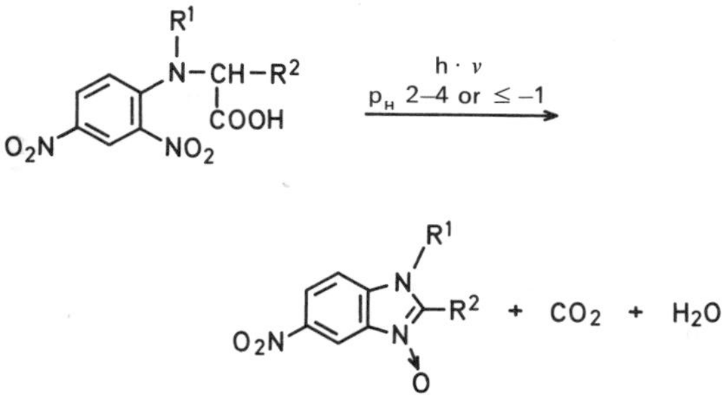

or tautomer

R^1 = H; R^2 = H, CH_3, $CH(CH_3)_2$, $-CH(CH_3)-C_2H_5$,
        CH_2OH, C_6H_5
R^1 = H, R^2 = $CHOH-CH_3$ affords two compounds R^2 = H
        and R^2 = $CHOH-CH_3$
R^1 = CH_3, R^2 = H; R^1-R^2 = $(CH_2)_3$

With secondary amines the yields are 70–80%, from tertiary amines considerably less[472]. If $R^1 = H$ and $R^2 = C_6H_5$ pH 10 is more favorable. *N*-(2, 4-Dinitrophenyl)glycine and peptides derived from it are not decarboxylated during ring closure to benzimidazole 3-oxides[474]. Re syntheses of *2-benzimidazolyl ketone 3-oxides* cf. refs. [475, 476] and re *pyridino[b-5, 4]imidazole N-oxides* cf. ref. [648].

10.4.1.8.2 Benzotriazole *N*-oxides from 2-nitrophenylhydrazines ($3 + 2$ or $5_{acycl.} \rightarrow 5_{cycl.}$)

2-Nitrophenylhydrazines cyclize to *1H*-benzotriazole 3-oxides on heating accompanied by base catalysis:

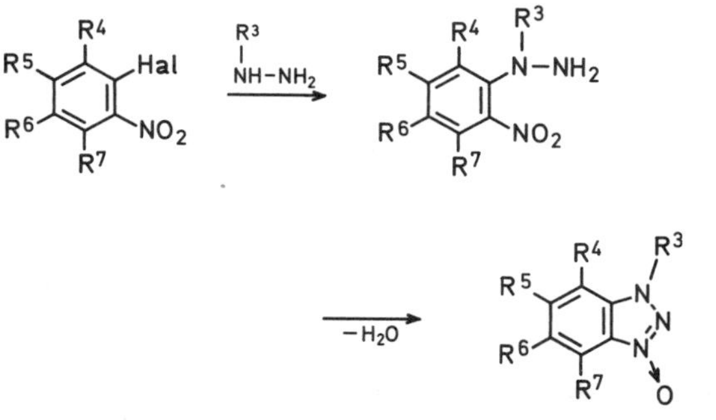

Hal mostly = Cl   or tautomer
R^3 = H
R^4 = H, CH_3, Cl, Br
R^5 = H, CH_3, $CH(CH_3)_2$, OCH_3, CN, Cl, Br, I
R^6 = H, CH_3, CF_3, $CO-NH_2$, OCH_3, $SO_2-N(C_2H_5)_2$, Cl, Br, NO_2
R^7 = H, Cl

[474] *L.A. Ljublinskaya, V.M. Stepanov*, Tetrahedron Lett. 4511 (1971).
[475] *A.E. Luetzow, J.R. Vercellotti*, J. Chem. Soc. (C) 1750 (1967).
[476] *A.E. Luetzow, N.E. Hofmann, J.R. Vercellotti*, Chem. Commun. 301 (1966).

The general procedure is to react the precursors of the 1-nitrophenylhydrazines (2-halo-1-nitrobenzenes) directly to the *benzotriazole 3-oxides* with hydrazine hydrate under appropriate conditions[477–479], *e.g.*, by heating in ethanol with or without triethylamine at the boiling point or at 110° in autoclaves[479]. The yields are dependent on the substituents but are generally good (> 60%) or very good. Analogously, 1-chloro-2, 4-dinitronaphthalene affords a small yield of *5-nitro-1H-naphtho[1, 2-d]triazole 3-oxide*[480].
4, 6-Dinitro-1H-benzotriazole 3-oxide formed from 2, 4, 6-trinitrophenylhydrazine with sodium acetate in boiling 0.8 *M* acetic acid[481]. For further syntheses see ref. [668].
2-Substituted 2H-benzotriazole 1-oxides are formed in over 50% yield when 2-halo-1-nitrobenzenes are reacted with monosubstituted hydrazines in boiling glacial acetic acid or acetic anhydride[482–484] (in boiling ethanol benzotriazoles are obtained):

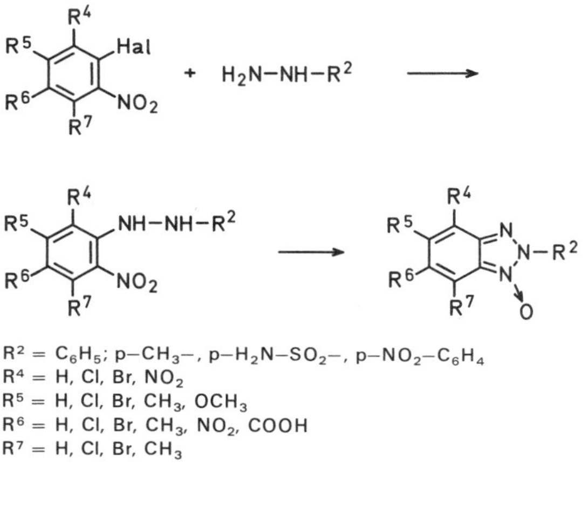

R^2 = C_6H_5; p–CH_3–, p–H_2N-SO_2–, p–$NO_2-C_6H_4$
R^4 = H, Cl, Br, NO_2
R^5 = H, Cl, Br, CH_3, OCH_3
R^6 = H, Cl, Br, CH_3, NO_2, COOH
R^7 = H, Cl, Br, CH_3

10.4.1.8.3 Benzimidazole 3-oxides and purine *N*-oxides from *o*-nitrosoanilines or β-nitrosoenamines *via* Schiff bases ($4 + 1$ or $5_{acycl.} \rightarrow 5_{cycl.}$)

4-Nitro-2-nitrosoaniline can be condensed with aromatic aldehydes in boiling acetic acid with addition of *p*-toluenesulfonic acid to form *2-arylbenzimidazole 3-oxides* ($\sim$ 60%)[238, 485]:

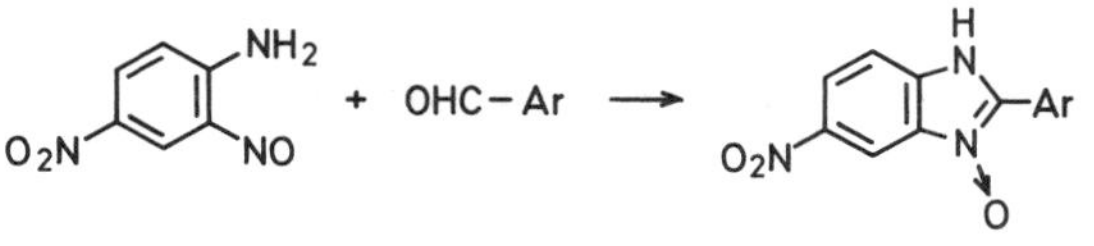

or tautomer

Ar = 2-naphthyl, 4-pyridyl
Ar = C_6H_4X (X = p–Cl, o–CH_3O, m–CH_3O, p–CH_3O,
        3, 4–$(CH_3O)_2$, p–CN)

Analogously, *3-methyl-6-nitrobenzimidazole 1-oxide* (35%) is formed from *N*-methyl-4-nitro-2-nitrosoaniline with formaldehyde[486].

Evidently the reduction of a 1-nitro group to the nitroso group can be accomplished also *in situ* with excess aldehyde; thus 2-phenylbenzimidazole 1-oxide is obtained from 2-nitroaniline with two equivalents benzaldehyde on prolonged boiling[467].

4-Amino-1,3-dimethyl-5-nitrosouracil **71** condenses to *8-phenyltheophylline 7-oxide* **75** with benzaldehyde in dimethylformamide[487] or, better, with *N*-benzylideneaniline in glacial acetic acid (65%)[424]; the product is reduced to the purine derivative **74** with dimethylformamide[487]:

dimethylnitrosouracil **71** when R = phenyl, so that in this case the condensation technique (starting from **71**) needs to be used. Employing nitric acid or permanganate in boiling butanol to carry out the oxidation of the 4-alkylamino-5-nitrosouracils **72** affords theophylline 7-oxide (**75**, R = H) with loss of the alkyl group[275].

10.4.1.8.4 Benzimidazole 3-oxides and benzothiazole 3-oxides by reduction of 2-nitroanilides and 2-nitroaryl thiocyanates ($5_{acycl.} \rightarrow 5_{cycl.}$)

Reduction of 2-nitroanilides of carboxylic acids with alkaline hyposulfite solution, alcoholic solution of a sulfide, zinc/ammonium chloride solution,

71

$+ H_5C_6-CHO$
or $H_5C_6-CH=N-C_6H_5$

72   Oxidation   **73**

74   HCON(CH$_3$)$_2$ **75**

or tautomer

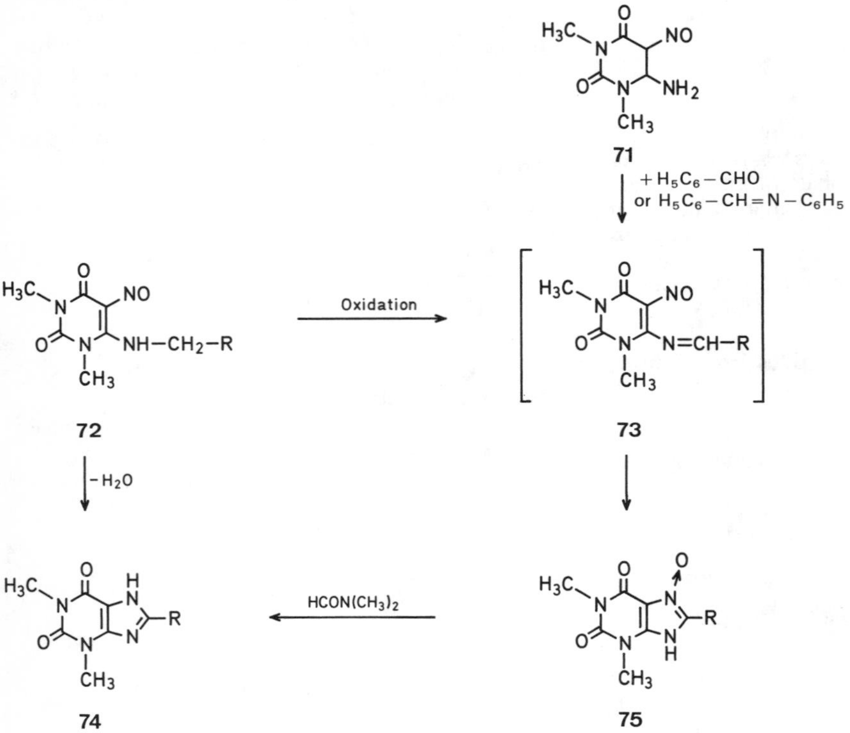

R = H, CH$_3$, C$_2$H$_5$, C$_3$H$_7$, CH(CH$_3$)$_2$, CH$_2$–C$_6$H$_5$

N-Alkylated 4-amino-5-nitrosouracils **72** too, can be dehydrogenated to the purine 7-oxide **75** (~ 50–60°) on heating[424]. The *8-alkyl-1,2,3,6-tetrahydropurines* **74** formed simultaneously by dehydration can be separated easily by virtue of their slight acidity. Small amounts of the 7-oxides **75** are formed already during preparation of the 4-alkylamino-5-nitrosouracils **72** with excess nitrosating agent [nitrous acid, isopentyl nitrite]. The nonisolatable imine **73** is hydrolyzed to 4-amino-1,3-

or catalytically (Pd) in general leads to *benzimidazole 3-oxides* in good yields[341, 466, 472, 488–490]:

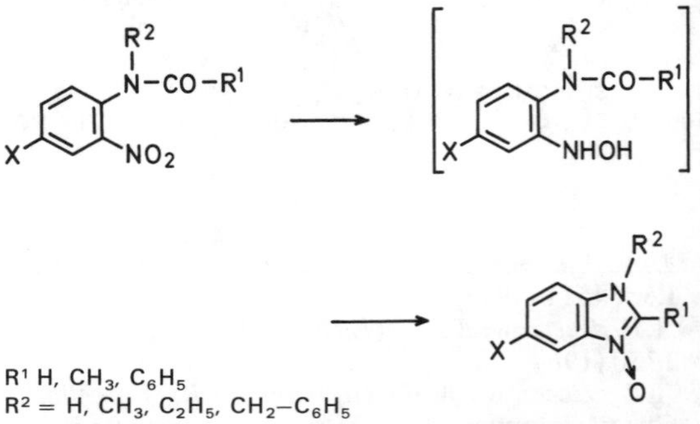

R^1 H, CH$_3$, C$_6$H$_5$
R^2 = H, CH$_3$, C$_2$H$_5$, CH$_2$–C$_6$H$_5$
X H, OCH$_3$, NO$_2$

or tautomer

[477] *H. Singh, R.S. Kapil*, J. Org. Chem. **25**, 657 (1960).
[478] *N.L. Leonard, K. Golankiewiez*, J. Org. Chem. **34**, 359 (1969).
[479] *W. König, R. Geiger*, Chem. Ber. **103**, 788 (1970).
[480] *E. Müller, K. Weisbrod*, J. Prakt. Chem. **111**, 307 (1925).

A second nitro group $(X=NO_2,\ R^1=R^2=H)$ remains intact if the reduction with sulfide is performed at 0–20° [488, 472] (giving *5-nitrobenzimidazole 3-oxide*).

For $R^1=H$, *2-benzimidazolinone* is the main product, for $R^1=pyrrolidinomethyl$ the technique fails.

2-Nitroaryl thiocyanates are reduced to *2-aminobenzothiazole 3-oxides* in alcoholic solution with hydrogen (3 atm) and Raney nickel [288]:

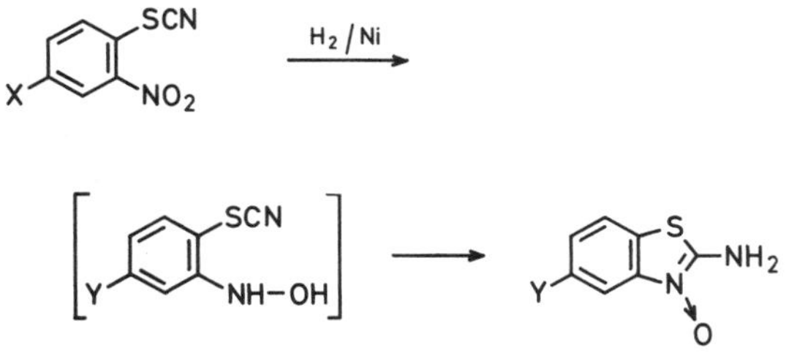

or tautomer

X = Y = H;   *2-Aminobenzothiazole 3-oxide;* ~ 80 %
X = NO₂, Y = NH₂; *2,5-Diaminobenzothiazole 3-oxide*

Re the formulated intermediate stages with hydroxylamine structure, see p. 380.

10.4.1.8.5 2*H*-Benzotriazole 1-oxides and 2*H*-indazole 1-oxides by oxidation of 2-aminoazo compounds or by reduction of 2-nitroazo compounds or benzylidene-2-nitroanilines ($5_{acycl.} \to 5_{cycl.}$).

Reduction with ammonium sulfide [491-494] or sodium sulfide [495] in boiling ethanol (possibly also hydrazine) converts 2-nitro azoaromatic compounds into *2-aryl-2H-benzotriazole 1-oxides*, sometimes in very good yield:

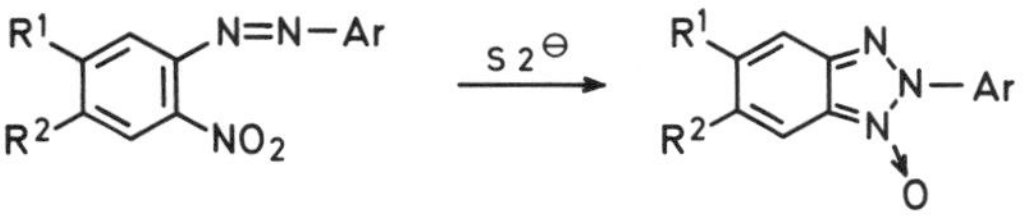

R¹ = H, Cl
R² = NO₂, H, Cl, CH₃
Ar = 4–X–C₆H₄; X = H, Br, Cl, N(CH₃)₂, NHC₆H₅, N(CH₂–CH₂–Cl)₂

[481] *R. Huisgen, V. Weberndörfer*, Chem. Ber. *100*, 71 (1967);
[482] *H. Goldstein, A. Jaquet*, Helv. Chim. Acta *24*, 30 (1941).
[483] *A. Mangini*, Gazz. Chim. Ital. *65*, 1191 (1935); C.A. *30*, 5521⁵ (2936);
A. Mangini, Atti. Accad. Naz. Lincei, Classe Sci. Fis. Mat. Natur. Rend. *21*, 759 (1935); C.A. *30*, 3416⁶ (1936).

Oxidizing 2-aminoazo compounds, too, leads to benzotriazole 1-oxide (the yields are small and the isomerism relationship is unclear [496]).

Ethyl(2-nitrophenyl)acetate forms *ethyl 2-(4-dimethylaminophenyl)-2H-indazole-3-carboxylate 1-oxide* in boiling ethanol in the presence of bases (sodium carbonate, trisodium phosphate, piperidine, pyridine) with *N,N*-diethyl-4-nitrosoaniline [497]:

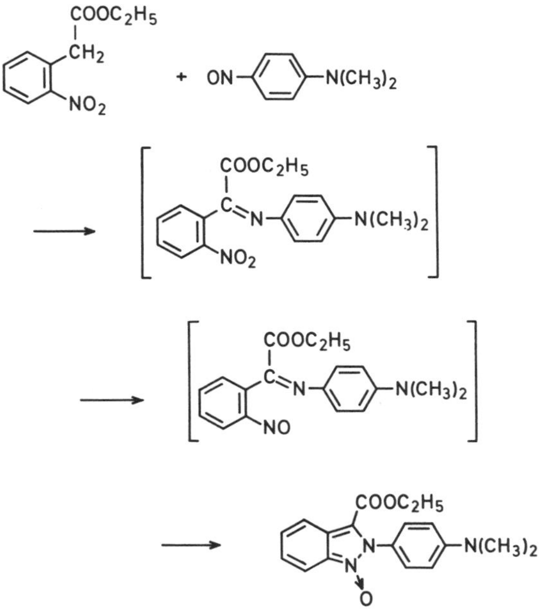

It is difficult to convert the 2-aryl-2*H*-indazole-3-carbonitrile 1-oxides described on p. 390 into the corresponding carboxylic acids and, for this reason, the above reaction sequence could assume interest as an alternative (cf. p. 384).

10.4.1.8.6 Benzofuroxans by oxidation of 2-nitroanilines ($5_{acycl.} \to 5_{cycl.}$)

Oxidation of primary 2-nitroarylamines with alkaline hypochlorite solution furnishes *benzofuroxans* in almost quantitative yield [454, 455, 498-500]:

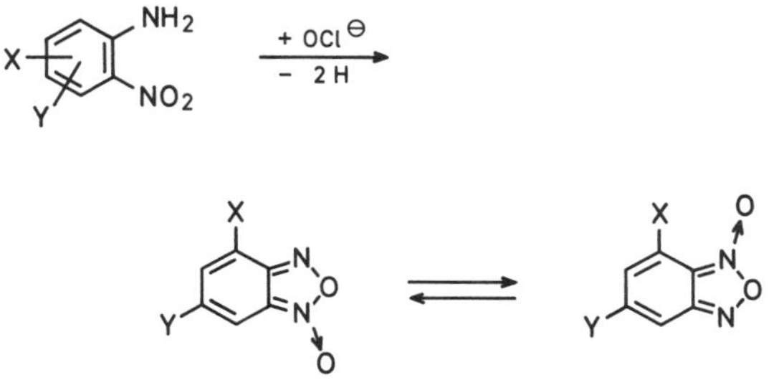

X = H; Y = H, Cl, CH₃, OCH₃
X = Y = CH₃
X = Y ≠ NH₂, R–CO–NH, SO₃H, NO₂ [501]

To carry out the oxidation, sodium hypochlorite solution is added to an aqueous alcohol suspension of the 2-nitroaniline with cooling until decolorized.
Phenyl iodosoacetate in benzene solution (not in acetic acid) is used as oxidizing agent[502].
In neutral solution the oxidation leads to 2,2'-dinitroazo compounds.

2,4-Dinitroaniline is simultaneously methoxidized and chlorinated during the oxidation in methanolic solution to give *6-chloro-5-methoxybenzofurazan N-oxide:*

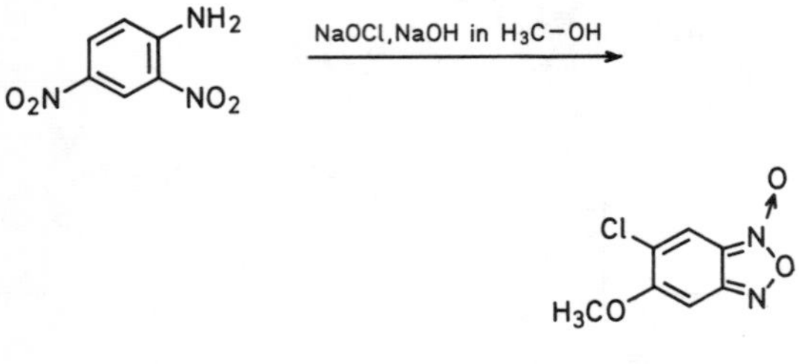

NaOCl, NaOH in H₃C—OH

or isomer

Re the position of the tautomeric equilibrium in unsymmetrically substituted benzofuroxans, *cf.* refs.[454, 455].
Higher condensed furoxans and bis(benzofuroxans), too, can be obtained in this way, cf. ref.[504] *e.g.:*

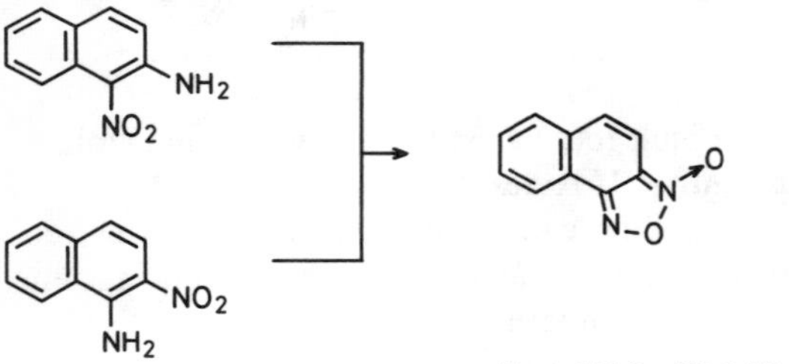

Naphth[1,2-c][1,2,5]-oxadiazole N-oxide (Naphtho[1,2]furoxan)

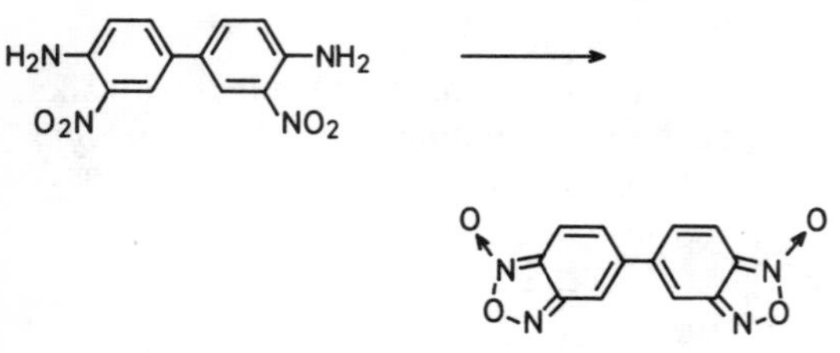

5,5'-Bibenzofurazan 3,3'-dioxide

10.4.1.8.7 Benzofuroxans from 2-nitroaryl azides (7-2) and furoxans from 1,2-dinitro-1-alkenes

Thermal or photochemical decomposition of the 2-nitroaryl azides is the most reliable method for preparing *benzofuroxans* in good yield and for wide application[284, 454, 455, 498, 494].

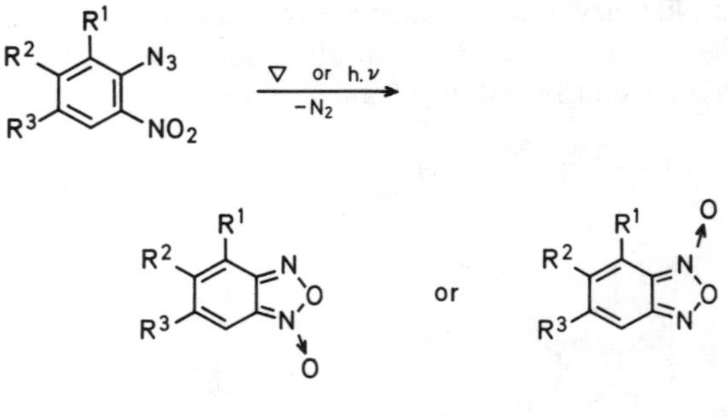

∇ or h.ν
−N₂

R¹ = H, NO₂, C₆H₅
R² = H, NO₂, N₃, OCH₃, O—COCH₃, NH—COCH₃, N(CH₃)₂, Cl, COOH, COOC₂H₅
R³ = H, NO₂, Cl

The reaction is carried out without solvent or in benzene, toluene, acetic acid, *etc.*[503]. Deoxygenation to benzofurazan can ensue if the temperature is too high[504].
5-Aminobenzofurazan and *5-benzofurazanol N-oxides (5-amino-, 5-hydroxybenzofuroxan)* are not accessible directly. The unstable amino compound can be obtained as the impure hydrochloride by careful hydrolysis of the *acetamido*benzofurazan N-oxide obtainable by decomposition of azides or from 5-benzofurazancarboxylic acid N-oxide *via* a series of intermediate stages[501]. The likewise unstable *5-benzofurazanol N-oxide* is obtained similarly by hydrolysis of the 5-Acetoxybenzofurazanol N-oxide prepared by thermolysis of 2-nitrophenyl-5-azidol acetate[501].

One way of preparing the azide required for the thermolysis is from *N,N*-dimethyl-2,4-dinitroaniline with sodium azide *in situ* in hot dimethyl sulfoxide and then decomposing it directly to 5-dimethylaminobenzofurazan N-oxide[501].

[484] *S.S. Joshi, S.P. Gupta*, J. Indian Chem. Soc. *35*, 681 (1958); C.A. *53*, 14092f (1959);
R.S. Kapil, S.S. Joshi, J. Indian Chem. Soc. *36*, 417 (1959); C.A. *54*, 9949 (1960);
A. Prakash, I.R. Gambhir, J. Indian Chem. Soc. *41*, 845 (1964); C.A. *62*, 11804c (1965).

[485] *D.W. Russel*, J. Med. Chem. *10*, 984 (1967); *D.W. Russel*, Chem. Commun. 498 (1965).
[486] *D.J. Needle*, J. Chem. Soc. (C) 2127 (1969).
[487] *E.C. Taylor, E.E. Garcia*, J. Amer. Chem. Soc. *86*, 4721 (1964).
[488] *S. Takahashi, H. Kana*, Chem. Pharm. Bull. (Tokyo) *11*, 1375 (1963);
S. Takahashi, H. Kana, Chem. Pharm. Bull. (Tokyo) *12*, 282 (1964).
[489] *F. Minisci, R. Galli, A. Quilico*, Tetrahedron Lett. 785 (1963).
[490] *K. Fries, H. Reity*, Justus Liebigs Ann. Chem. *527*, 38 (1937).

Higher condensed benzofuroxans, too, are accessible in this way, even those with a fused-on heterocycle. Thus, the imidazo[4,5]benzofuroxans **78** are formed in yields of between 45 and 80% by thermolysis of the 5-amino-6-nitrobenzimidazoles **76**[504] (starting from azide **77** affords lower yields except for $R^1 = R^2 = CH_3$):

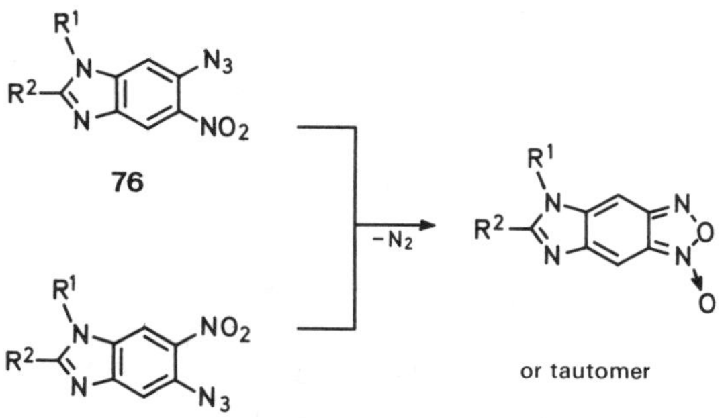

76

77   **78**

$R^1 = R^2 = CH_3$; *5,6-Dimethylimidazo[4,5-f]benzofuroxan*
$R^1-R^2 = -(CH_2)_3-, -(CH_2)_4-, -(CH_2)_5-,$
$-CH_2-CH_2-O-CH_2-$

If no nitrogen is given off in boiling glacial acetic acid (with R^1, $R^2 \neq CH_3$), then the thermolysis is performed in boiling 2-ethoxyethanol ($\sim 135°$) (photolysis affords lower yields).

Benzo[1,2-c; 4,5-c']bis[1,2,5]oxadiazole 3,6-dioxide (benzo[a,d]difuroxan) is unobtainable; the double *o*-quinonoid structure in the 1,2- and 4,5-positions of the benzene ring is evidently impossible. For this reason, 1,3-diazido-4,6-dinitrobenzene splits off nitrogen only one-sidedly to form *5-azido-6-nitrobenzofurazan 3-oxide,* which explodes when heated vigorously[500]. By contrast, an unsymmetric *benzo[1,2-c;3,4-c']bis[1,2,5]oxadiazole 3,6-dioxide (benzo[a,c]difuroxan)* can be obtained by double azide decomposition[500], and also *benzotris[1,2,5]oxadiazole 3,6,9-trioxide (benzotrifuroxan)*[503] by triple azide decomposition.

[1,2,5]-Thiadiazolo[e]benzofuroxan is accessible under similar conditions[505].

Attempts to prepare pyridofuroxans by this procedure generally lead to polymerization[506]. Particular members of this class only can be obtained in this way[506]; the azide required during their preparation may under certain circum-

stances be generated from a nitrazinotetrazole compound[507-509]:

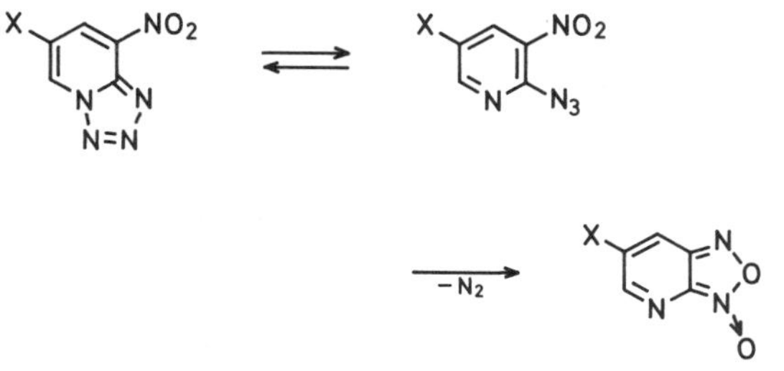

X = H;   *Pyrido[2,3]furoxan*
X = NO₂; *6-Nitrofurazano[b]pyridine*

In analogy with the decomposition of 2-nitroaryl azides rapid conversion of 1,2-dinitro-1-alkenes with sodium azide leads to furoxans[498]:

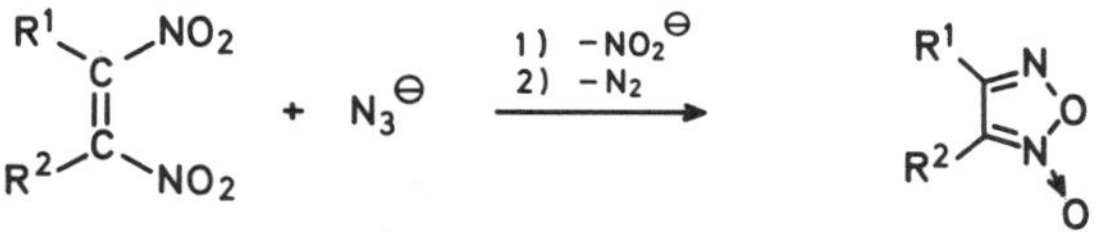

(*cis* or *trans*)   or isomer

$R^1 = R^2 = CH_3$;   *4,5-Dimethylfuroxan*
          C_2H_5;   *4,5-Diethylfuroxan*
          C_6H_5;   *4,5-Diphenylfuroxan*
$R^1 = CH_3, R^2 = C_3H_7$; *5-Methyl-4-propylfuroxan*

10.4.1.8.8 2,1-Benzisoxazole 1-oxides (anthranil oxides) from 2-nitroaryl compounds ($5_{acycl.} \rightarrow 5_{cycl.}$)

The nitrilimines generated *in situ* during hydrogen bromide elimination from 2-nitrobenzoyl bromide hydrazones with the aid of bases (sodium acetate, ammonia solution, pyridine) or by heating[510] cyclize to yellow, explosive *3-arylazo-2,1-benzisoxazole oxide*[277, 386, 511-513]:

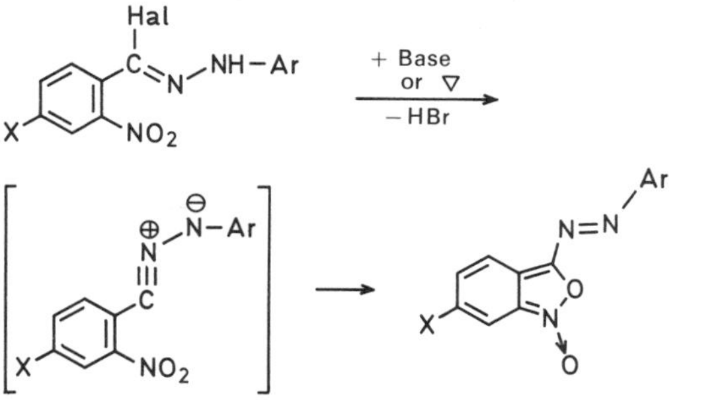

X = H, NO₂

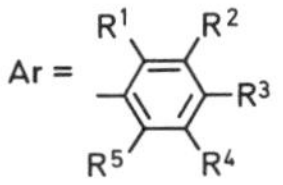

Ar =

$R^1 = R^2 = H, Br, Cl, CH_3$
$R^3 = Br, Cl, CH_3, NO_2$
$R^4 = H, Br, Cl$
$R^5 = H, Br, Cl, CH_3, NO_2$

[491] A. Werner, E. Stiasny, Ber. 32, 3256 (1899).
[492] K. Elbs, J. Prakt. Chem. 108, 209 (1924).
[493] W.C.J. Ross, G.P. Warwick, J. Chem. Soc. (London) 1724 (1956).
[494] F.B. Mallory, C.S. Wood, J. Org. Chem. 27, 4109 (1962).
[495] E. Bamberger, R. Hübner, Ber. 36, 3822 (1903).

During preparation of the starting substances halogenation of *Ar* must be anticipated.

Recently, a mesoionic 1, 2, 3-benzotriazine structure has been ascribed to these anthranil *N*-oxides[649].

The recently discovered route *via* the 2-nitrobenzoic acid hydrazides and phosphorus(V) chloride, too, should be feasible[514]. 2-Nitrobenzonitrilimines can be formed also by lead(IV) acetate oxidation of 2-nitrobenzaldehyde arylhydrazones; in this way halogenfree *N*-oxides are obtained, e.g.[510]:

X = H; Ar = C₆H₅;   *3-(Phenylazo)-2,1-benzisoxazole 1-oxide;* 76%

Ar = 4−NO₂−C₆H₄; *3-[4-Nitrophenyl)azo]-2,1-benzisoxazole 1-oxide;* 54%

Adding concentrated sulfuric acid to molten 2′, 3′, 4′, 5′, 6′-pentafluoro-2-nitrobenzhydrol affords a good yield of *3-pentafluorophenyl-2, 1-benzisoxazole 1-oxide*[515].

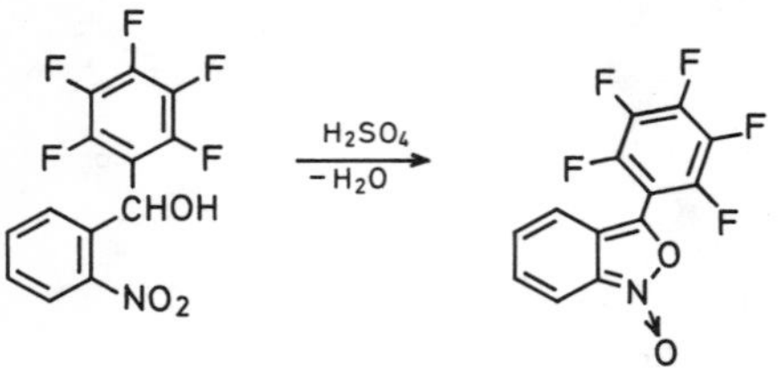

10.4.1.8.9 2*H*-Benzotriazole 1-oxides from 2-nitrophenylhydrazines ($5_{acycl.} \rightarrow 5_{cycl.}$)

The following azomethine 2-nitrophenylimines can be converted into *2-[2-(2 H-benzotriazol-2-yl)-ethyl]benzaldehyde 1-oxides* in very good yield by oxygen transfer[516]:

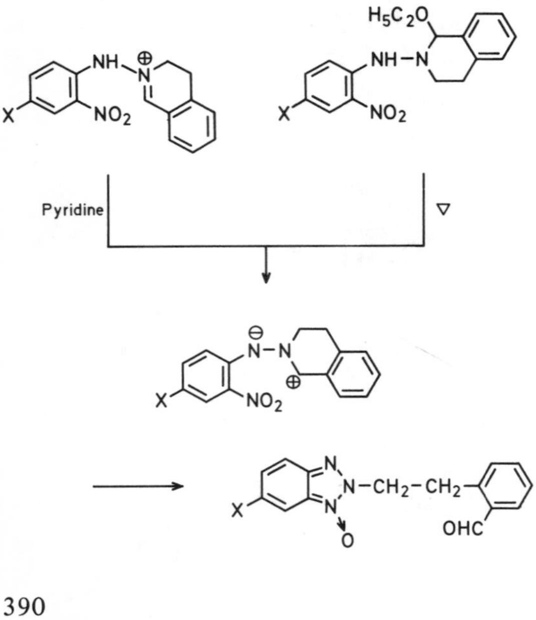

10.4.1.8.10 2-Aryl-2*H*-indazole 1-oxides from *N*-(2-nitrobenzylidene)anilines or their precursors ($5_{acycl.} \rightarrow 5_{cycl.}$)

N-(2-Nitrobenzylidene)anilines react to form *2-aryl-2H-3-indazolol 1-oxides* in boiling ethanol in the presence of sodium carbonate[517]:

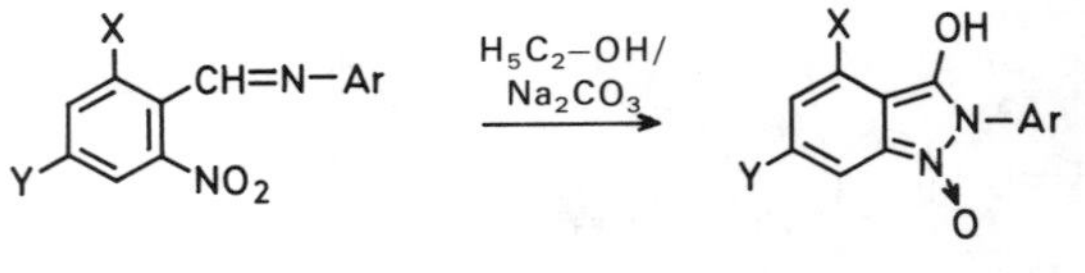

or tautomer

X = H, NO₂
Y = H, NO₂
Ar = 4−C₆H₄−R, R = H, CH₃, OC₂H₅, N(CH₃)₂

The yields obtained so far vary and are sometimes poor, but may be capable of improvement. For X = H the *N*-oxide is formed only as side-product. Preparing the arylimines and ring closure to the *N*-oxide can be performed in a closed vessel when 2, 4, 6-trinitrotoluene and *N*, *N*-diethyl-4-nitroso-aniline are employed.

While in the above method the ring closure entails the transfer of oxygen[213], 2-aryl-2 *H*-indazole-3-carbonitrile 1-oxides are formed from 2-nitrobenzaldehyde, primary arylamine, and prussic acid by dehydration (mostly in ethanol):

[496] *G. Charrier, G.B. Crippa*, Gazz. Chim. Ital. *53*, 462 (1923); C.A. *265* (1924);
G. Charrier, G.B. Crippa, Gazz. Chim. Ital. *58*, 11 (1925); C.A. *2495* (1925).

[497] *I. Tanasescu, E. Tanasescu*, Bull. Soc. Chim. France *55*, 1016 (1935).

[498] *J.H. Boyer* in *R.C. Elderfield*, Heterocyclic Compounds, Vol. VII, p. 479, John Wiley & Sons, London 1961.

[499] *F.B. Mallory*, Org. Synthesis, Coll. Vol. IV, 74 (1963).

[500] *R.J. Gaughran, J.P. Picard, J.V.R. Kaufmann*, J. Amer. Chem. Soc. *76*, 1233 (1954).

[501] *A.J. Boulton, P.B. Ghosh, A.R. Katritzky*, J. Chem. Soc. (C) 971 (1966).

[502] *L.K. Dyall, K.H. Pausacker*, Australian J. Chem. *11*, 491 (1958).

[503] *A.S. Bailey, J.R. Case*, Tetrahedron *3*, 113 (1958).

[504] *R.C. Perera, R.K. Smalley, L.G. Rogerson*, J. Chem. Soc. (C) 1348 (1971).

[505] *P.B. Ghosh*, Tetrahedron Lett. 2999 (1971).

[506] *A.S. Bailey, M.W. Heaton, J.I. Murphy*, J. Chem. Soc. (C) 1211 (1971).

[507] *J.H. Boyer, E.J. Miller*, J. Amer. Chem. Soc. *81*, 4671 (1959);
J.H. Boyer, D.J. McCane, W.J. McCarville, A.T. Tweedie, J. Amer. Chem. Soc. *75*, 5298 (1953).

[508] *J.H. Boyer, H.W. Hyde*, J. Org. Chem. *25*, 458 (1960).

[509] *B. Stanovnik, M. Tisler*, Chimia *25*, 272 (1971).

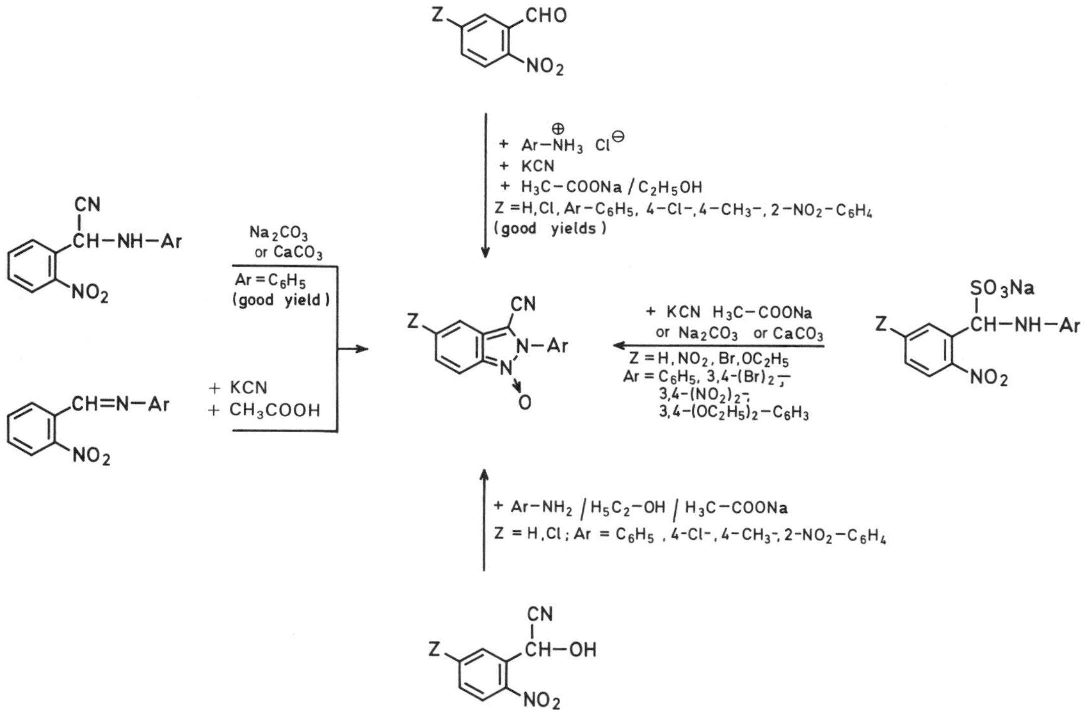

Z———CHO
———NO₂

+ Ar–NH₃⁺ Cl⁻
+ KCN
+ H₃C–COONa / C₂H₅OH
Z = H, Cl, Ar–C₆H₅, 4-Cl-, 4-CH₃-, 2-NO₂–C₆H₄
(good yields)

CN
CH–NH–Ar
NO₂

Na₂CO₃
or CaCO₃
Ar = C₆H₅
(good yield)

+ KCN
+ CH₃COOH

CH=N–Ar
NO₂

Z———[indazole N-oxide]
CN
N–Ar
N
O

+ KCN H₃C–COONa
or Na₂CO₃ or CaCO₃
Z = H, NO₂, Br, OC₂H₅
Ar = C₆H₅, 3,4-(Br)₂–
3,4-(NO₂)₂–
3,4-(OC₂H₅)₂–C₆H₃

SO₃Na
Z———CH–NH–Ar
NO₂

+ Ar–NH₂ / H₅C₂–OH / H₃C–COONa
Z = H, Cl; Ar = C₆H₅, 4-Cl-, 4-CH₃-, 2-NO₂–C₆H₄

CN
Z———CH–OH
NO₂

10.4.1.9 Azole *N*-oxides from nitrile oxides

10.4.1.9.1 Furoxans by dimerization of nitrile oxides[451, 521] (3 + 2)

Except for a few members with a particularly sterically demanding group *R*, nitrile oxides are unstable and dimerize spontaneously even at room temperature at a greater or lesser rate (seconds to days or weeks) to furoxans[522]:

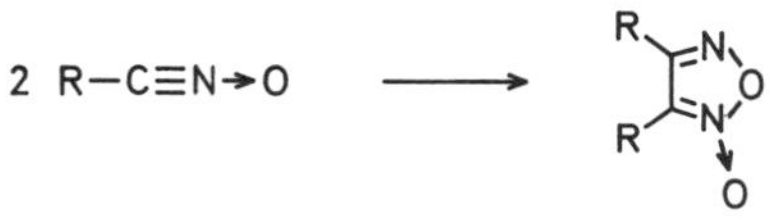

2 R–C≡N→O ⟶
[furoxan structure]
R
N
R
N
O

R = Aryl, Alkyl, Aryl–C   Halogen, COOC₂H₅
‖
NOH

If necessary the reaction can be speeded up by heating. Rearrangement of the nitrile oxide to isocyanate (*R–N=C=O*) is a possible side-reaction and becomes the main reaction in nitrile oxides having particularly bulky groups *R*, for instance, in 2,6-disubstituted benzonitrile oxides with substituents such as CH_3, CH_3O, CH_3S, *Br*, *I*, NO_2, or SO_2R[523]. 2,6-Dichloro- and pentafluorobenzonitrile oxides dimerize to *bis(pentafluorophenyl)-furoxan*, but the pentachloro compound does not[524]. In bis-nitrile oxides 1,3-dipolar cycloaddi-

tion followed by dimerization can lead to heterocyclically substituted furoxans[525, 526], *e.g.:*

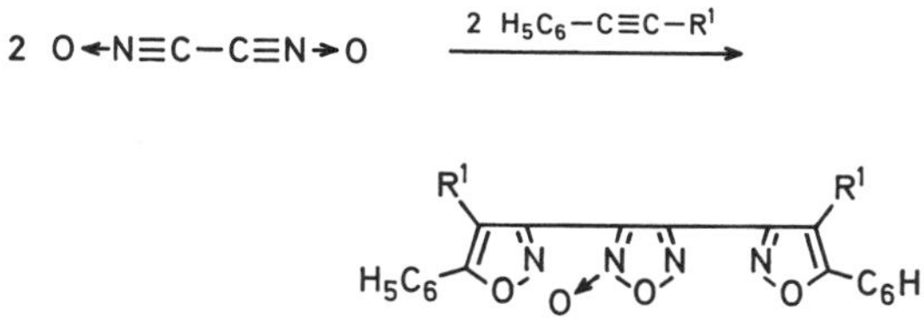

2 O←N≡C–C≡N→O   2 H₅C₆–C≡C–R¹ ⟶

R¹
H₅C₆   N   N   N   R¹
O   O   O   O
C₆H₅

R¹ = H;   *Bis(5-phenyl-3-isoxalolyl)furoxan*
R¹ = C₆H₅; *Bis(4,5-diphenyl-3-isoxalolyl)furoxan*

[510] *W.A.F. Gladstone, J.B. Aylward, R.O.C. Norman*, J. Chem. Soc. (C) 2587 (1969).

[511] *J.E. Erickson* in *A. Weissberger*, The Chemistry of Heterocyclic Compounds, Vol. X, p. 27, Interscience Publishers, John Wiley & Sons, New York, London, Sydney 1956.

[512] *M.S. Gibson*, Tetrahedron *18*, 1377 (1962).

[513] *A.F. Hegarty, M. Cashman, J.B. Aylward, F.L. Scott*, J. Chem. Soc. (B) 1879 (1971).

[514] *R. Huisgen*, Angew. Chem. *75*, 604 (1963).

[515] *P.L. Coe, A.E. Jukes, J.C. Tatlow*, J. Chem. Soc. (C) 2020 (1966).

[516] *R. Grashey*, Angew. Chem. *74*, 155 (1962).

[517] *S. Secareanu, I. Lupas*, Bull. Soc. Chim. France *53*, 1436 (1933).

[518] *G. Heller, G. Spidmeyer*, Ber. *58*, 835 (1925).

[519] *L.A. Reissert, F. Lemmer*, Ber. *3*, 351 (1926).

[520] *L.C. Behr, E.G. Alley, O. Levand*, J. Org. Chem. *27*, 65 (1962);
L.C. Behr, J. Amer. Chem. Soc. *76*, 3672 (1954).

The dimerization of the nitrile oxides is reversible, but the back-reaction generally ensues only at temperatures at which the nitrile oxides formed rearrange to isocyanates (but cf. ref. [650]). It is perfectly possible to convert monosubstituted furoxans to symmetrically disubstituted furoxans *via* thermal decomposition[523].

In practise, the technique of preparing symmetrically disubstituted furoxans discussed here consists in preparing the nitrile oxides (cf. p. 331). Either aldoximes (-2H) form the start or nitrolic acids are used as intermediate stages. The second pathway is probably the one by which symmetrical diacylfuroxans are formed from methyl ketones or methylene ketones with concentrated or anhydrous nitric acid; nitrosation initially leads to formation of the α-*hydroxyimino ketone*[451, 527]:

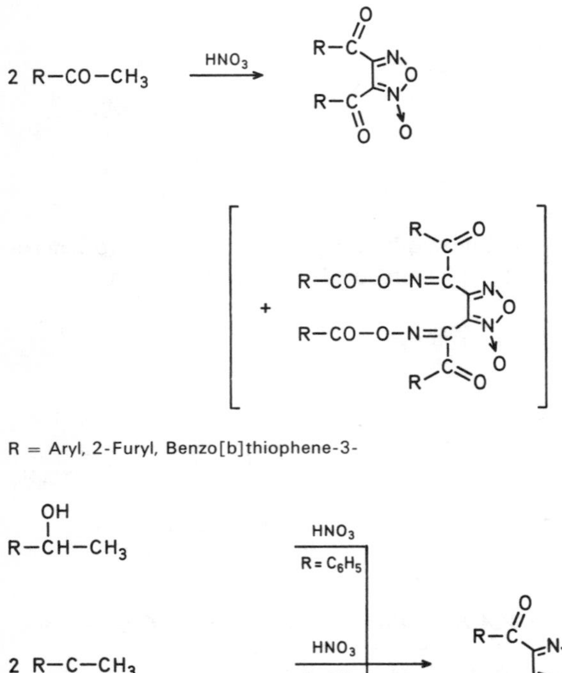

Re a new synthesis of diacylfuroxans, *cf*. ref. [629].

In connection with transformation of nitrile oxides to azole *N*-oxides, it may be pointed out that treating nitrolic acids with weak aqueous alkali does not lead to 1,3,5-triazole 1,3,5-trioxides as was originally assumed[523].

10.4.1.9.2 1,2,4-Oxadiazole 4-oxides from nitrile oxides (3 + 2)

Nitrile oxides (possibly as precursors) dimerize in the presence of catalytic amounts of hydrogen chloride or boron trifluoride[622] to form *1,2,4-oxadiazole 4-oxides via* a 1,3-dipolar cycloaddition of intermediately formed hydroximoyl chloride[525, 527, 528, 451]:

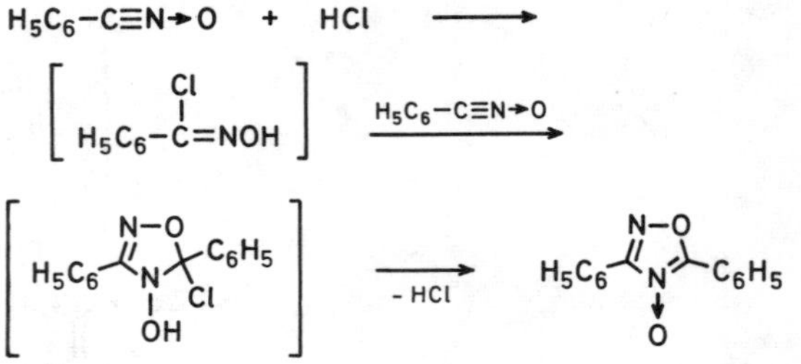

3,5-Diphenyl-
1,2,4-oxadiazole
4-oxide

Concerning a recent synthesis, see ref. [632].

10.4.1.9.3 1-Hydroxybenzimidazole 3-oxides from nitrile oxides and aromatic nitroso compounds (2 + 3)

In ether solution benzonitrile oxide adds to aro-

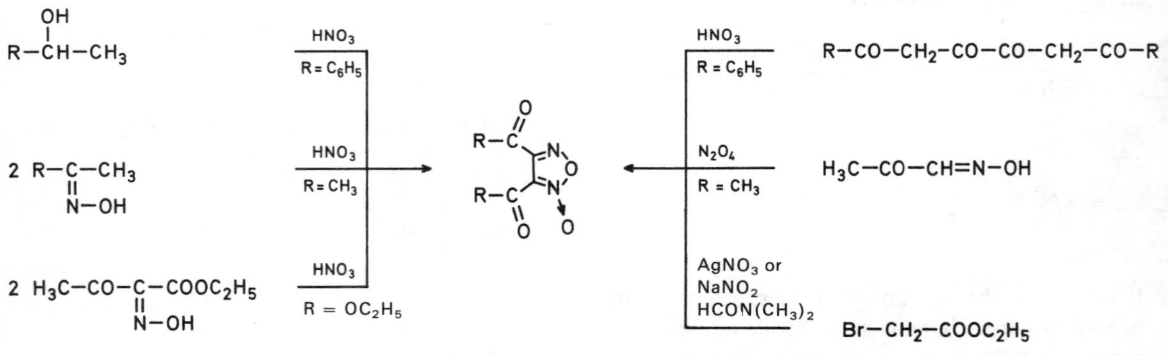

R = Aryl, 2-Furyl, Benzo[b]thiophene-3-

R = CH₃;   *Diacetyl-*
R = C₆H₅;   *Dibenzoyl-*
R = OC₂H₅; *Bis[ethoxycarbonyl]furoxan*

[521] *Ch. Grundmann* in *Houben-Weyl*, Methoden der organischen Chemie Band X/3, Stickstoffverbindungen I, Teil 3, p. 841, Georg Thieme Verlag, Stuttgart 1965.
[522] *K. Hirai, H. Matsuda, Y. Kishida*, Chem. Pharm. Bull. (Tokyo) *20*, 97 (1972).
[523] *Ch. Grundmann*, Fortschritte der chemischen Forschung, Bd. VII, p. 83, Springer Verlag, Berlin, Heidelberg, New York, 1966—1967.
[524] *B.J. Wakefield, D.J. Wright*, J. Chem. Soc. (C) 1165 (1970).

matic nitroso compounds to form *1-hydroxybenzimidazole 3-oxides via* an α-nitroso nitrone[489, 466]:

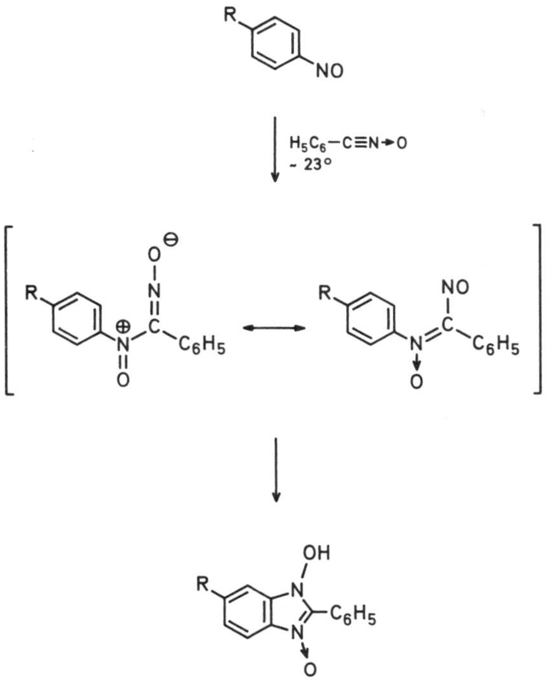

R = H;   *1-Hydroxy-2-phenylbenzimidazole 3-oxide*
R = N(CH₃)₂;   *6-Dimethylamino-1-hydroxy-2-phenylbenzimidazole 3-oxide*
R = OH;   *1,6-Dihydroxy-2-phenyl-6-benzimidazole 3-oxide*

With nitrosobenzene (*R = H*) the intermediate product can be isolated at −20°, with *N,N*-dimethyl-4-nitrosoaniline [*R = N(CH₃)₂*] it cannot. 4-Nitrosophenol initially reacts only to the intermediate product, which is stabilized as the *p*-quinone imine structure by tautomerism; formation of the *N*-oxide requires heating in an inert solvent to 60—70°.

The reaction with 1-nitroso-2-naphthol fundamentally stops at the intermediate stage (no free *ortho* position).

10.4.1.10 *N*-Oxides of azaaromatic compounds from other *N*-oxides

On treatment with alkaline hydrogen peroxide solution quinoxaline *N*-oxides suffer ring contraction to benzimidazole *N*-oxides[333, 529]:

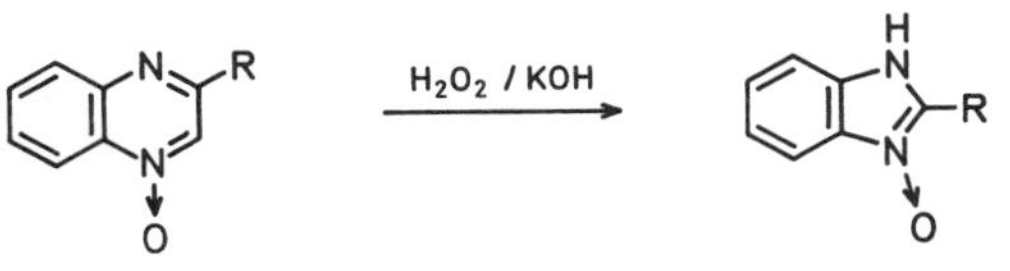

R = Alkyl [including C(CH₃)₃], OCH₃, OC₂H₅
R = H

Condensation of benzofuroxan with ethyl acetoacetate in ethanolic caustic potash at 50° leads to *ethyl benzimidazole-2-carboxylate 3-oxide* (R = COOC₂H₅ > 60%)[259, 623]. For further examples see ref.[651].

Acid ether cleavage of the 2-alkoxypyridine 1-oxides obtainable by direct oxidation affords *2-pyridinol 1-oxides,* which are not accessible directly from the α-pyridinones by peroxy acid oxidation[342].

Ring substitutions and conversions of substituents which preserve the *N*-oxide group are numerous (cf. p. 394). In what follows some examples will be adduced to show how to condense further rings onto an *N*-oxidized azaaromatic compound without loss of the *N*-oxide function.

The hydrazones accessible in good yield from 3-hydrazinopyridazine *N*-oxide with aldehydes are oxidized with lead(IV) acetate in acetic acid at room temperature to *triazolo-[4,3-b]pyridazine 5-oxides* (about 40%)[530].

[525] *Ch. Grundmann,* Fortschritte der chemischen Forschung, Bd. VII, p. 110, Springer Verlag, Berlin, Heidelberg, New York, 1966—1967.
[526] *Ch. Grundmann, V. Mini, J.M. Dean, H.-D. Frommeld,* Justus Liebigs Ann. Chem. *687,* 191 (1965).
[527] *L.C. Behr,* in The Chemistry of Heterocyclic Compounds, Vol. XVII, p. 245, Interscience Publishers, New York, London 1962;
A. Quilico, The Chemistry of Heterocyclic Compounds, Vol. XVII, p. 27, Interscience Publishers, New York, London 1962.
[528] *J.H. Boyer* in *R.C. Elderfield,* Heterocyclic Compounds, Vol. VII, p. 514, John Wiley & Sons, New York, London 1961.
[529] *E. Hayashi, J. Miura,* J. Pharm. Soc. Japan *87,* 648 (1967).
[530] *A. Pollack, B. Stanovnik, M. Tisler,* J. Heterocycl. Chem. *5,* 513 (1968).
[531] *J.H. Boyer* in *R.C. Elderfield,* Heterocyclic Compounds, Vol. VII, p. 481, John Wiley & Sons, New York, London 1961.
[532] *W. Herz, D.R.K, Murty,* J. Org. Chem. *26,* 418 (1961).
[533] *E. Ochiai,* Aromatic Amine Oxides, p. 399, Elsevier Publ., Co., Amsterdam, London, New York 1967.
[534] *J.G. Murray, C.R. Hauser,* J. Org. Chem. *19,* 2008 (1954).
[535] *A.R. Katritzky, J.M. Lagowski,* Chemistry of the Heterocyclic N-Oxides, Academic Press, London, New York 1971.
[536] *A.R. Katritzky,* Quart. Rev. Chem. Soc. *10,* 395 (1956).
[537] *D.V. Ioffe, L.S. Efros,* Russian Chem. Reviews (Uspekhi Khimii) *30,* 569 (1961).
[538] *A.R. Katritzky, A.J. Boulton,* Advances in Heterocyclic Chemistry, Vol. I—XII, Academic Press, New York, London 1963—1970.
[539] *A. Weissberger,* in The Chemistry of Heterocyclic Compounds, Vol. I—XXIV, 1. Ed., Interscience Publishers, New York, London 1951—1971.

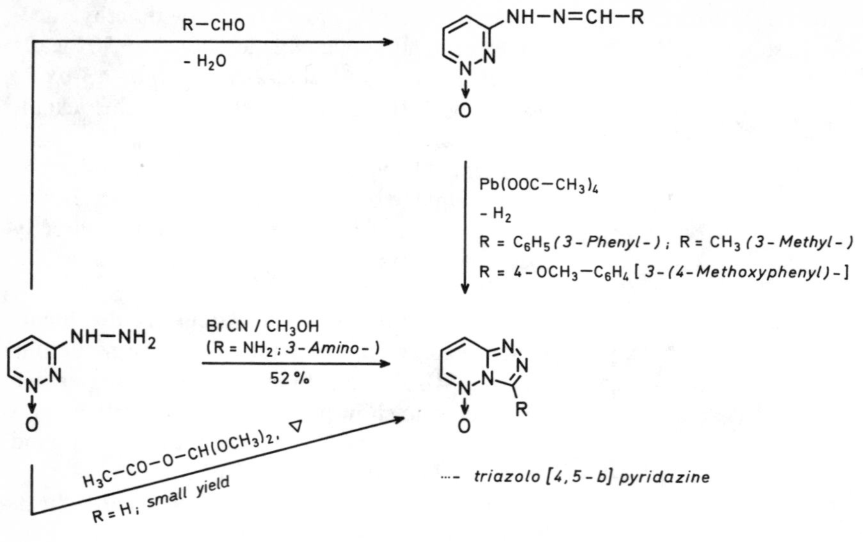

···– *triazolo [4,5-b] pyridazine*

Re further conversions see the following equations[530]:

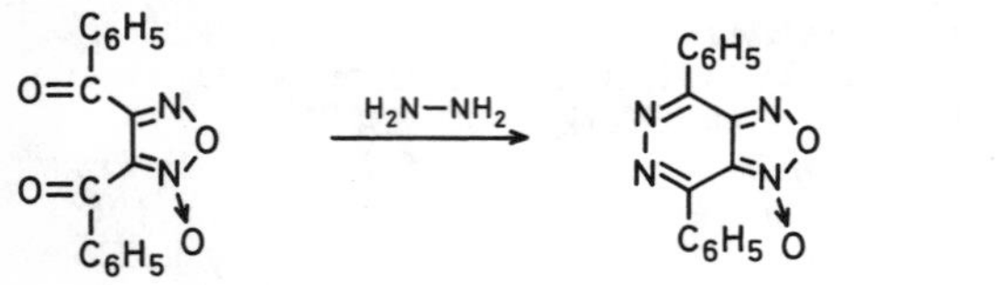

4,7-Diphenylpyridazino-[4,5-c]furoxan[531]

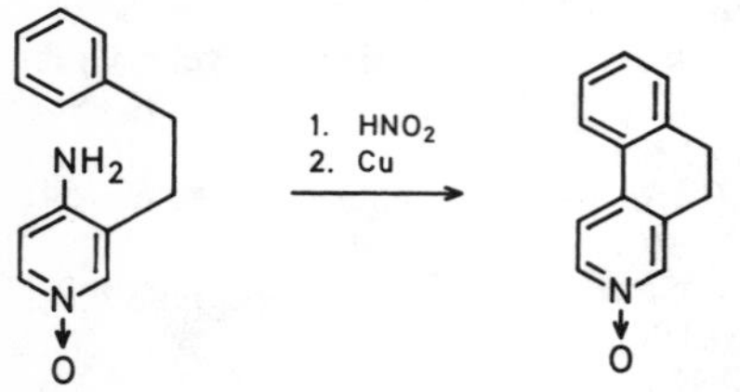

5,6-Dihydrobenzo[f]-isoquinoline 3-oxide[532]

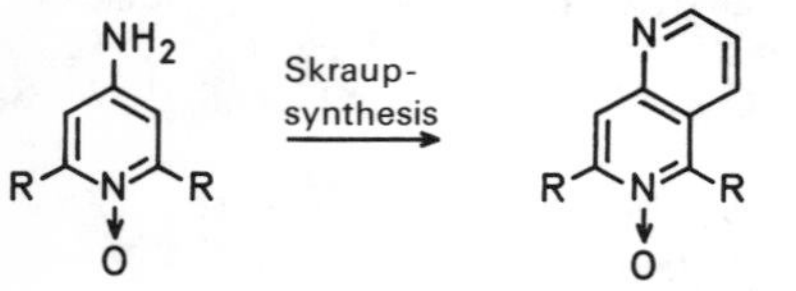

R = H;   *1,6-Naphthyridine 6-oxide*[533]
R = CH₃; *5,7-Dimethyl-1,6-naphthyridine 6-oxide*[533]

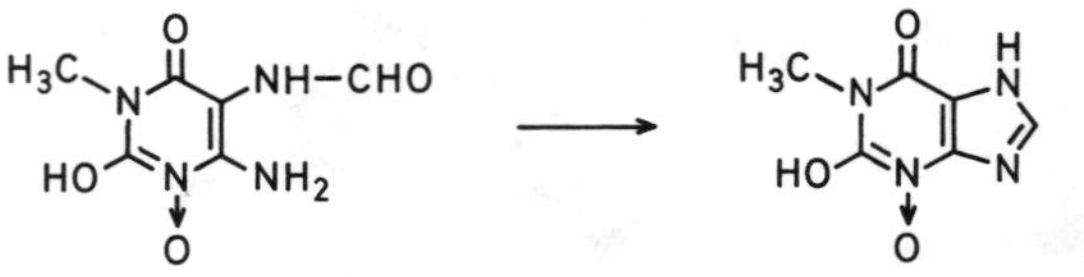

2-Hydroxy-1-methyl-6-purinone 3-oxide[329]

In the above pyridine derivatives the corresponding ring closure does not take place in the oxygen-free parent compound, but it does in the following reaction[534] (cf. p. 375):

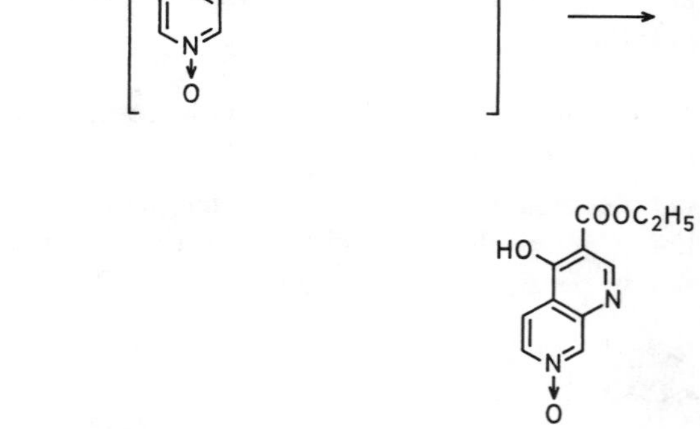

Ethyl 4-hydroxy-1,7-naphthyridine-3-carboxylate 7-oxide

[540] *R.C. Elderfield,* Heterocyclic Compounds, Vol. I–IX, John Wiley & Sons, New York, London 1950–1967.

10.4.2 Transformation reactions of *N*-oxides of azaaromatic compounds

The reactivity and orientating effect of the aromatic rings is modified strongly by the *N*-oxide function. Thus, many interesting preparative possibilities result and often the detour *via* an *N*-oxide is recommended or is the only possible route for obtaining particular azaaromatic compounds without an *N*-oxide function, *cf.* refs. [31, 238, 535–540]. Only a few brief points of the wealth of material are stated here.

The following reactions can ensue on the nitrogen without participation of the aromatic ring:

1 *Deoxygenation:* sometimes is possible thermally, otherwise see pp. 355–357, 362, 389
2 *Salt formation:* see pp. 355–357
3 *Metal complex formation*[541, 542]
4 *O-Alkylations* are performed with the usual alkylating agents to give isolatable products which can, however, rearrange by alkyl migration[543, 544].

Versatile reactions on the ring are feasible. Nucleophilic substitution in the α- or γ-position (cf. limiting formula 1b, p. 356) and sometimes also in the β-position occur readily. If a halogen atom or a nitro group is displaced, the *N*-oxide functions remains intact (for example see p. 388) and ref.[545]); if a H atom is displaced the *N*-oxide oxygen needs to be split off to restore the aromatic character:

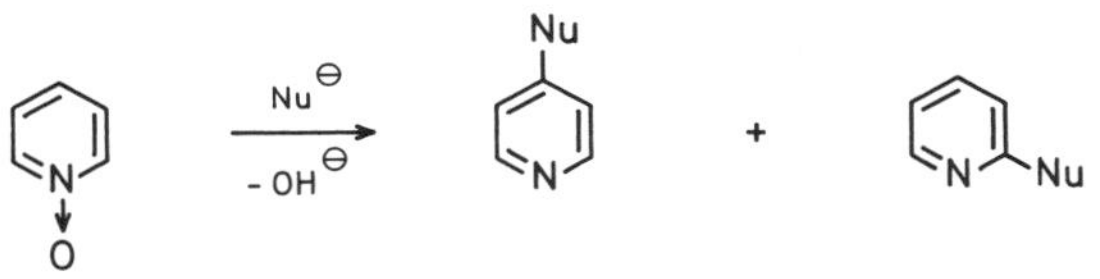

Nucleophilic reactions are facilitated by simultaneously acylating, sulfonating, or phosphorylating the oxygen to be eliminated[254, 439]. Thus, a nitrile group (Nu = CN) can be introduced with potassium cyanide-benzoyl chloride, and a chlorine atom with sulfuryl chloride or phosphoryl chloride. The following can act as nucleophiles in dependence on the *N*-oxide and the reaction conditions:

H_2O or $OH^\ominus$, cf. p. 363	CH-acid compounds
$HSO_3^\ominus$	Enamines[547]
$CN^\ominus$	Pyridine
$PO(OR)_2^\ominus$	$R-S^{\ominus}$[548, 549]
$RSO_3^\ominus$	$R-SH$[548, 549]
Amines	$(H_3C-CO)_2S$[550]
$R-MgX$[546]	Ketones
$R-C\equiv C^\ominus$	Nitroalkanes
	Halides

On treatment with tosyl chloride both chlorine and the tosyloxy group (with entrainment of the *N*-oxide oxygen in analogy with the acyloxy rearrangement) can enter the ring. Reactions with Grignard reagent and other carbanions often proceed with partial maintenance of the *N*-oxide function; thus quinoline *N*-oxide and phenylmagnesium bromide in hot tetrahydrofuran afford 30% *2-phenylquinoline* and 60% *2-phenylquinoline 1-oxide*.

In the α-position oxidative hydroxylation is possible[551] (see p. 359).

Where the compound to be substituted has a methyl group in the α-position, a nucleophilic (and also electrophilic) substitution can sometimes ensue here as well; this is true, for instance, for the reaction with phosphoryl chloride (to the α-chloromethyl derivative of the *O*-free compound) and especially also for the acyloxy rearrangement. *O*-Acylations (primarily with acetic anhydride) proceed readily, but the products at once rearrange with migration of the acyloxy group from the nitrogen to the ring, or, if one is present, also to the α-methyl group[337, 439, 552, 553].

Thus acetylation of 2-collidine *N*-oxide affords mixed *2-pyridinemethanol acetate,* and *2-methyl-3-* and *2-methyl-5-pyridinol acetates*. Alternatively, the acyloxy migration can ensue to an attached ring. During reaction of 2,4-diphenylquinoline *N*-oxide with potassium cyanide/benzoyl chloride a mixture of *8-, 3-,* and *6-benzyloxy-2,4-diphenylquinolines* are formed.

One of the most important properties of the *N*-oxides of azaaromatic compounds is their enhanced readiness to undergo electrophilic ring substitution (*cf.* limiting formula 1c, p. 356), preferably in the α- and γ-positions. By contrast, electrophilic substitutions in the parent

[541] *W.E. Hatfield, J.A. Barnes, D.Y. Jeter, R. Whyman, E.R. Jones Jr.,* J. Amer. Chem. Soc. *92*, 4982 (1970);
S.A. Cotton, J.F. Gibson, J. Chem. Soc. (A) *12*, 2105 (1970);
A.D. Mighell, C.W. Reimann, A. Santoro, Chem. Commun. 204 (1970).
[542] *M. Sundaralingam, C.D. Stout, S.M. Hecht,* Chem. Commun. 240 (1971).
[543] *T. Fujii, T. Itaya,* Chem. Pharm. Bull. (Tokyo) *19*, 1611 (1971).

[544] *T. Fujii, C.C. Wu, T. Itaya,* Chem. Pharm. Bull. (Tokyo) *19*, 1368 (1971);
R.A. Abramovitch, S. Kato, G.M. Singer, J. Amer. Chem. Soc. *93*, 3074 (1971);
W.N. Marmer, D. Swern, J. Amer. Chem. Soc. *93*, 2719 (1971).
[545] *A. Giner-Sorolla,* J. Heterocyclic Chem. *8*, 651 (1971).

aromatic compounds are difficult to carry out and occur in the β-position if at all. The *N*-oxide gives the γ-nitro *N*-oxide with nitrating acid at elevated temperature, and the β-nitro *N*-oxide with acyl nitrate.

Hydroxy groups on the ring (or equally strong directing groups) can outweigh the directing effect of the *N*-oxide function.

Mercurations in the α- and β-positions are feasible by heating to elevated temperature with mercury(II) acetate. *Bromination* of the *N*-oxides proceeds less smoothly, both in respect of yield and directing influence. In acetic anhydride the bromine is β-directed[554]. The *N*-oxides display no activation toward *sulfonation;* a reaction takes place only at high temperatures and, like in the parent aromatic compounds, to β-sulfonated products. *Mannich condensation* has been described for the α-position[555]. Substitutions with reactive aryl halides may be accompanied by an oxygen transfer to the entering aryl group. Thus, *1-(2-quinolyl)-2(1H)pyridone* is formed from quinoline *N*-oxide and 2-bromopyridine[556]. An electrophilic substitution is possible *via* an α-metallization. With butyllithium and elementary sulfur pyridine 1-oxide affords *1-hydroxy-2H-thiopyranone*[557].

Any α- or γ-methyl groups present readily undergo condensations.

A large number of photochemical reactions have been described. Oxygen displacement (to the ring), ring opening, ring expansion, or ring contraction occur preferentially[558–561, 670].

1,3-Dipolar cycloadditions with the *N*-oxide as dipole occur at elevated temperature. The primary products are unstable and rearrange with hydrogen migration, so that ultimately α-substitution involving deoxygenation takes place[562].

10.5 *N*-Oxides of tertiary amines

Unlike in *N*-oxides with a multiple *N–C* bond cf. Chapters 10.1, 10.2, 10.3, 10.4, in *N*-oxides of tertiary amines the polarity of the *N–O* group cannot be modified by resonance. The dipole moments of the *N*-oxides of tertiary amines are therefore large (around 5 D, bonding moment 4.38 D[563]) and considerably greater than those of the parent amines. Correspondingly, they possess relatively high melting points and their distillation or sublimation presents difficulties, to which their thermal lability also contributes. They are hygroscopic, readily soluble in water and sparingly soluble in ethers or hydrocarbons. The hydrate

form is more stable than the anhydrous amine oxide, which easily suffers Cope elimination or Meisenheimer rearrangement. The lenght of the *N–O* bond is ~1.36 Å and its bonding energy has been determined to be *66 kcal/mole*. In the *IR* spectra the *N–O* bond is revealed by its stretching vibration at 950–970 cm^{-1}.

The pK$_A$ values of normal trialkylamine oxides are at around 5 (but *cf.* ref.[564]). Re the polarography, *cf.* ref.[565, 566].

Assays[563, 567] have been performed colorimetrically following cleavage with sulfur dioxide by determining the aldehyde formed[568], by reduction with titanium(III) chloride, and by potentiometric titration with alcoholic hydrochloric acid. In the latter case titration in the presence of methyl iodide can distinguish from tertiary amine present[569, 567]. An iodometric procedure is described below. Re thin-layer chromatography, see refs.[570, 571] and re gas chromatography, see ref.[567].

The amine oxides and their salts possess oxidizing properties. Very likely, the oxidizing action toward iodide in acid solution is due in many cases at least partly to the bound peroxide. However, peroxidefree amine oxides, too, liberate iodine, provided the solution under investigation is not so strongly acid that the amine oxide exists mainly in the protonated form[572]. In the presence of strong mineral acids the amine oxides do not oxidize iodide and any iodine that is liberated points to the presence of peroxides. By contrast, in glacial acetic acid the peroxidefree amine oxides do liberate iodine[572] on heating and, where applicable, buffering with sodium acetate even quantitatively[572]. Hydriodic acid, too, enables the qualitative determination to be performed[572].

Peroxides, but not amine oxide, oxidize thionaphthol to disulfide in basic medium[573].

[546] *T. Kametani, T. Suzuki*, Chem. Pharm. Bull. (Tokyo) *19*, 1424 (1971);
T. Kametani, T. Suzuki, J. Org. Chem. *36*, 1291 (1971);
T.J. van Bergen, R.M. Kellogg, J. Org. Chem. *36*, 1705 (1971);
H. Igeta, T. Tsuchiya, T. Nakai, Tetrahedron Lett. 3117 (1971).

[547] *M. Hamana, I. Kumadaki*, Chem. Pharm. Bull. (Tokyo) *19*, 1669 (1971).

[548] *F.M. Hershenson, I. Bauer*, J. Org. Chem. *34*, 655 (1969).

[549] *B.A. Mikrut, F.M. Hershenson, K.F. King, L. Bauer*, J. Org. Chem. *36*, 3749 (1971).

[550] *J.H. Markgraf, Myong-Ku Ahn, Ch.G. Carson III, G.A. Lee*, J. Org. Chem. *35*, 3983 (1970).

[551] *M. Hamana, M. Yamazaki*, Chem. Pharm. Bull. (Tokyo) *10*, 51 (1962); C.,A. *58*, 504b (1963).

[552] *S. Oae, S. Tamagaki, T. Negoro, K. Ogino, S. Kozuka*, Tetrahedron Lett. 917 (1968);
R. Bodalski, A.R. Katritzky, J. Chem. Soc. (B) 831 (1968).

[553] *N.J.M. Birdsall, U. Wölcke, T.C. Lee, G.B. Brown*, Tetrahedron *27*, 5969 (1971);
T. Cohen, G.L. Deets, J. Amer. Chem. Soc. *94*, 932 (1972).

[554] *M. Yamazaki, Y. Chono, K. Noda, M. Hamana*, Yakugaku Zasshi *85*, 62 (1965); C.A. *62*, 10409d (1965).

Re the *conformation* and *configuration* of cyclic amine oxides see refs.[574-576]. The absolute configuration of chiral aryldialkylamine oxides can be determined by nuclear resonance spectroscopy in optically active 2,2,2-trifluoroethanol[577].

Physiologically, amine oxides can arise as metabolites of tertiary amines and probably also as intermediates during biological degradation and restructuring of amines[578-583]. *Technically,* above all long-chain amine oxides[563,584] are used principally on account of their wetting, foam-generating, and bactericidal effect. These long-chain amine oxides can be isolated from aqueous solution as urea adducts[585].

10.5.1 Preparation

10.5.1.1 Oxidation of tertiary amines with hydrogen peroxide

The most widely used method of preparing the *amine oxides* is oxidation of tertiary amines with excess (rather more than one equivalent to an approximately tenfold amount) dilute or concentrated hydrogen peroxide solution[563,567,586,587] at ~20–75° (sometimes also with cooling as far as to −40°). With water-insoluble amines, lower alcohols are suitably added. Too low a water content in the reaction mixture is a disadvantage[563,567]

and, as a result, too high a concentration of hydrogen peroxide (even 70% solutions) has an adverse effect[563,567]. For technical purposes the reaction may be catalyzed by tungstic or molybdic acid[588] or by sodium pyrophosphate/sodium bicarbonate[589].

In general, the oxidation proceeds smoothly as long as at most one aryl group is linked to the nitrogen. The oxidation of the amine becomes the more difficult the less basic it is. With β-(1,2,3,4-tetrahydroisoquinolino)acrylates and nitrites hydrolysis to the free carboxylic acid, *3,4-dihydro-2(1H)-isoquinolineacrylic acid N-oxide,* takes place simultaneously with the oxidation[590].

Normally the amine oxides bind hydrogen peroxide in a manner similar to hydrates[573,591-593], and even after extraction in chloroform[591] or using equivalent amounts amine and peroxide (the hydrogen peroxide hydrogen bonds are stronger than those of water)[573]. It requires several hours' standing in sodium carbonate solution or caustic soda before the amine oxide is obtained free from peroxide[591]; alternatively, this is achieved *via* the picrate[573], which serves also for isolating the amine oxide in difficult cases. Isolation is possible additionally by evaporation of the reaction mixture in the desiccator[594,567].

[555] *G. Okusa, S. Kamiya,* Chem. Pharm. Bull. (Tokyo) *16,* 142 (1968).

[556] *E.A. Mailey, L.R. Ocone,* J. Org. Chem. *33,* 3343 (1968).

[557] *R.A. Abramovitch, E.E. Knaus,* J. Heterocycl. Chem. *6,* 989 (1969).

[558] *A. Kubo, S. Sakai, S. Yamada, I. Yokoe, Ch. Kaneko, A. Tatematsu, H. Yoshizumi, E. Hayashi, H. Nakata,* Chem. Pharm. Bull. (Tokyo) *15,* 1079 (1967);
A. Kubo, S. Sakai, S. Yamada, I. Yokoe, Ch. Kaneko, Chem. Pharm. Bull. (Tokyo) *16,* 1533 (1968).

[559] *F. Bellamy, L.G.R. Barragan, J. Streith,* Chem. Commun. 456 (1971);
H.G.O. Becker, H.-J. Timpe, J. Prakt. Chem. *312,* 586 (1970);
K.H. Wünsch, H. Bojdala, Z. Chem. *4,* 144 (1970);
M. Ogata, H. Matsumoto, S. Tokahashi, H. Kano, Chem. Pharm. Bull. (Tokyo) *18,* 964 (1970);
O. Buchardt, Tetrahedron Lett. 1911 (1968);
T. Tsuchiya, H. Arai, H. Igeta, Tetrahedron Lett. 2579 (1971);
Ch. Lohse, Tetrahedron Lett. 5625 (1968);
Ch. Kaneko, S. Yamada, I. Yokoe, Tetrahedron Lett. 2333 (1970);
G. Buchardt, K.B. Tomer, V. Madsen, Tetrahedron Lett. 1311 (1971);
J. Streith, P. Martz, Tetrahedron Lett. 4899 (1969);
C. Leibovici, J. Streith, Tetrahedron Lett. 387 (1971).

[560] *J. Streith, P. Martz,* Tetrahedron Lett. 4899 (1969);
P.L. Kumler, O. Buchardt, Chem. Commun. 1321 (1968).

[561] *G.F. Field, L.H. Sternbach,* J. Org. Chem. *33,* 4438 (1968);
E.C. Taylor, G.G. Spence, Chem. Commun. 1037 (1968);
J. Streith, B. Danner, Ch. Sigwalt, Chem. Commun. 979 (1967);
A. Alkyitis, M. Calvin, Chem. Commun. 292 (1968).

[562] *R. Huisgen, H. Seidl, J. Wulff,* Chem. Ber. *102,* 915 (1969);
R. Huisgen, H. Seidl, R. Grashey, Chem. Ber. *102,* 926 (1969).

[563] *D.B. Lake, G.L.K. Hoh,* J. Amer. Oil Chemists Soc. *40,* 628 (1963).

[564] *P. Baranowsky, J. Skolik, M. Wiewiorowski,* Tetrahedron *20,* 2383 (1964).

[565] *Y. Date, S. Tadano,* Nippon Kagaku Zasshi *85,* 525 (1964); C.A. *62,* 8983c (1965).

[566] *J. Krupicka, J. Zavada,* Collect. Czech. Chem. Commun. *32,* 2797 (1967); C.A. *67,* 96271w (1967).

[567] *G.L.K. Hoh, D.O. Barlow, A.F. Chadwick, D.B. Lake, S.R. Sheeran,* J. Amer. Oil Chemists Soc. *40,* 268 (1963).

[568] *C.H. Mitchell, D.M. Ziegler,* Anal. Biochem. 261 (1969); C.A. *70,* 93863s (1969).

[569] *L.D. Metcalfe,* Anal. Chem. *34,* 1849 (1962).

10.5.1.2 Oxidation of tertiary amines with organic hydroperoxides

Very recently, technically easily accessible organic hydroperoxides $R–O–O–H$, too, have been used for the *N*-oxidation of tertiary amines. Without catalyst *(R = tert-butyl)* titrimetric yields of between 44 and 79% were recorded[595]. Group V B and VI B transition metals catalyze the reaction, above all vanadium[596–599], titanium[596], and molybdenum[596, 597, 599] compounds. Under these conditions the yields are about 60–80% on isolated amine oxide[596, 598] or ~80–100% by titrating the reaction mixture (in this case on reacted hydroperoxide because excess amine is used)[597]. The yields are strongly dependent on the reaction conditions. Vanadium(III) compounds[597] are the best catalysts, but titanium(IV) isobutoxide is superior during oxidation of alkyl bis-(2-hydroxyethyl)amines[596]. Cumene and pentylene hydroperoxides are more reactive than *tert*-butyl hydroperoxide[597]. Lower reaction temperatures are possible with the more reactive compounds, but are also necessary, because competing reactions at higher temperatures cause more rapid decomposition[596, 597]. Benzene[596], ethyl acetate[596], tetrahydrofuran[597], and diethyl ether[597] are used as *solvent;* alcohols are less suitable[597]. On the other hand, the favorable effect of water (in the absence of catalysts) has been reported[600]. According to the amine used and the temperature (~20–80°) the reaction is complete in less than one hour to several hours.

A direct comparison with the hydrogen peroxide method (p. 397) is not yet available; it could offer benefits during isolation and prevent formation of perhydrates.

10.5.1.3 Oxidation of tertiary amines with peroxy acids

Peroxy acids (pure peroxy carboxylic acids, mixtures of hydrogen peroxide and carboxylic acids or acid anhydrides, sulfomonoperoxy acids), too, have been used for the *N*-oxidation of tertiary amines[586, 593] (cf. p. 359). However, in general, this method becomes necessary only with weakly basic amines, with which the hydroperoxide method fails or affords poor yields, for instance, tertiary arylamines, and above all with electron-withdrawing substituents or with tertiary diarylamines. Sulfomonoperoxy acid fails with the sterically screened nitrogen in 2, 6-disubstituted *N, N*-dimethylanilines[586]. In 2-haloethylamines, which rapidly undergo change as free bases, the rapidly proceeding peroxy acid oxidation in nonaqueous medium (ether, benzene) is apparently often advantageous[586]. Normal trialkylamines have been oxidized with peroxy acid in acetic acid[567, 601], with peroxyformic acid in 1, 4-dioxane[567], or, giving an almost quantitative yield of pure *N*-oxide, with one equivalent of 3-chloroperoxybenzoic acid in chloroform and subsequent column chromatography[602, 578].

[570] *J.R. Pelka, L.D. Metcalfe,* Anal. Chem. *37*, 603 (1965).

[571] *J.H. Ross,* Anal. Chem. *42*, 564 (1970).

[572] *E. Höft, H. Schultze,* J. Prakt. Chem. *4*, 19, 260 (1963).

[573] *A.A. Oswald, D.L. Guertin,* J. Org. Chem. *28*, 651 (1963).

[574] *Y. Shvo, E.D. Kaufman,* Tetrahedron *28*, 573 (1972);
G. Fodor, Chimia *25*, 282 (1971);
M.J. Cook, A.R. Katritzky, M. Moreno Manas, J. Chem. Soc. (B) 1330 (1971).

[575] *K. Bachmann, W. von Philippsborn,* Helv. Chim. Acta *55*, 637 (1972).

[576] *R.L. Muntz, W.H. Pirkle, I.C. Paul,* J. Chem. Soc. Perkin II, 483 (1972).

[577] *W.H. Pirkle, R.L. Muntz, I.C. Paul,* J. Amer. Chem. Soc. *93*, 2817 (1971).

[578] *N. Castagnoli jr., J.C. Craig, A.P. Melikian, S.K. Roy,* Tetrahedron *26*, 4319 (1970);
N. Castagnoli jr., J.C. Craig, A.P. Melikian, S.K. Roy, Chem. Commun. 1327 (1970).

[579] *P.A. Bather, J.R. Lindsay Smith, R.O.C. Norman, J.S. Sadd,* Chem. Commun. 1116 (1969).

[580] *A. Ahond, A. Cave, C. Kan-Fan, Y. Langlois, P. Potier,* Chem. Commun. 517 (1970).

[581] *J.R. Lindsay Smith, R.O.C. Norman, A.G. Rowley,* Chem. Commun. 1238 (1970).

[582] *P.A. Bather, J.R. Lindsay Smith, R.O.C. Norman,* J. Chem. Soc. (C) 3060 (1971).

[583] *J.T. Edward, J. Whiting,* Can. J. Chem. *49*, 3502 (1971).

[584] *T.P. Matson,* J. Amer. Oil Chemists Soc. *40*, 640 (1963).

[585] U.S.P. 3 316 236 (1967), Continental Oil Co., Inv.: *C.M. Starks, K.E. Harwell;* C.A. *67*, 45110k (1967).

[586] *H. Freytag* in *Houben-Weyl,* Methoden der organischen Chemie, 4. Ausgabe, Band XI/2, p. 190, Georg Thieme Verlag, Stuttgart 1958.

[587] *A.C. Cope, E.R. Trumbull,* Organic Reactions, Vol. XI, p. 317, John Wiley & Sons, New York 1960.

[588] U.S.P. 3 047 579 (1962), Shell Oil Co., Inv.: *R.C. Whitman;* C.A. *58*, 7916c (1963.

[589] Neth. Appl. 6 506 312 (1965), Pennsalt Chemicals Corp., C.A. *64*, 12548b (1966).

[590] *J. Hoenicke,* Dissertation Universität Marburg/Lahn, 1971;
H. Stamm, J. Hoenicke, Arch. Pharm. *307* (1974, in print).

[591] *K. Bodendorf, B. Binder,* Arch. Pharm. 287, 326 (1954).

[592] U.S.P. 3 252 979 (1966), Esso Research and Engineering Co., Inv.: *A.A. Oswald, F. Noel;* C.A. *65*, 7057c (1966).

10.5.1.4 Oxidation of tertiary amines with ozone

Trialkylamines can be oxidized to *N*-oxides with ozone (ozonized oxygen containing 0.01–15% ozone), best of all in chloroform [603, 586] and other chlorinated solvents [604] at −25 to −80° or in alcohols at 1–40° [605]. Contrary to the claim in an older patent specification, no *N*-oxide is formed in aqueous sulfuric acid [603], while during oxidation in alcohols mineral acids are stated to act as catalysts [605]. Other sources recommend the addition of carbonate for avoiding salt formation [604]. However, two primary important requirements are evidently a minimum reaction temperature and a suitable solvent, both in order to avoid degradation of the amine backbone and other undesired oxidation processes as far as possible. Even under favorable conditions *benzylamines* furnish small amounts benzaldehyde as side-product [603]. *Arylamines* and *pyridine bases* react so extensively on C–C bonds that the amine oxide is no longer formed or cannot be isolated [603]. However, certain *N*-substituted piperidines and morpholines, too, gave ill-defined products [603]. In the remaining investigated cases the yields vary from good to very good. On account of the yields the ozone oxidation is claimed to be superior to the hydrogen peroxide method (p. 397) and, because of the cost, to the peroxy acid oxidation (p. 398) [605]. In addition, the amine oxides can be isolated either as the anhydrous compound (p. 398), as base, or as salt in this way.

10.5.2 Transformation reactions of the amine oxides

Amine oxides derived from an allylamine or benzylamine rearrange to the *N*, *N*-disubstituted allyloxy- and benzyloxyamines on heating (up to 170° but partly even during the amine oxide preparation [606] or through photochemical causes [607, 586, 593]:

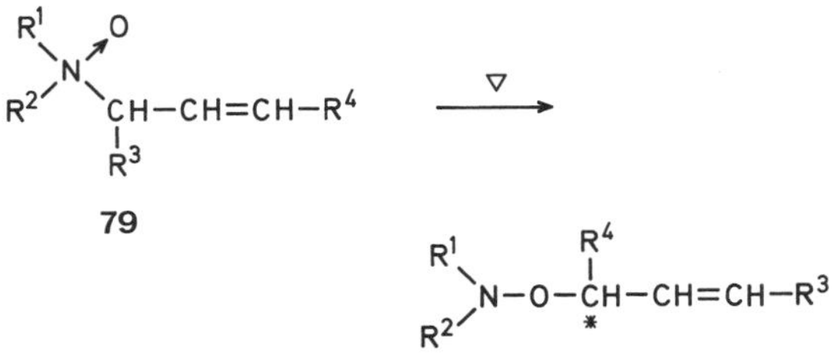

With allyl groups an allyl shift [79, 80] occurs simultaneously ($79 \rightarrow 80$). Contradicting older postulates, this *Meisenheimer rearrangement* proceeds to the hydroxylamine derivative 82 *via* a pair of free-radicals with homolysis of the C–N bond [606–608] and partly by intramolecular and partly intermolecular recombination [578, 593, 608]. On the other hand, if an alkyl *N*-oxide 79 with a chiral nitrogen atom ($R^1 \neq R^2$) is rearranged and $R^4 \neq H$, then the chirality is at least partly transferred from the nitrogen to the carbon (C^* in 80); this may be considered to be a pointer to a sigmatropic *no-mechanism mechanism* [609]. It is therefore possible that the mechanisms in the benzyl and allyl series differ from one another.

[593] *P.A.S. Smith*, The Chemistry of Open-Chain Organic Nitrogen Compounds, Vol. II, p. 21, Benjamin, New York, Amsterdam 1966.

[594] Fr. P. M2782 (1964), Laboratoires Amido, C.A. *62*, 5228 (1965).

[595] U.S.P. 3410903 (1968), Philipps Petroleum Co., Inv.: *P.W. Solomon;* C.A. *70*, 87012x (1969).

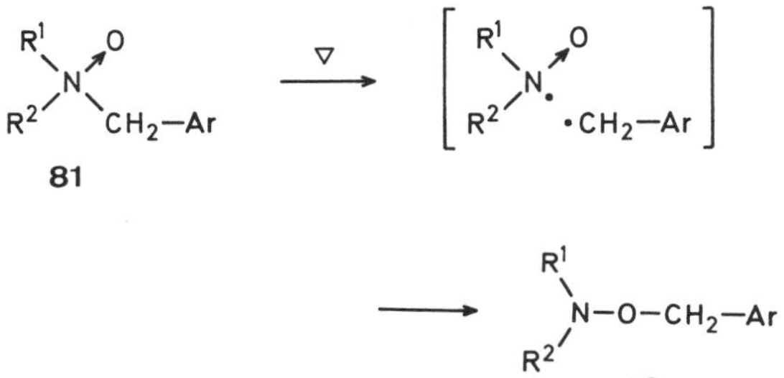

[596] *L. Kuhnen*, Chem. Ber. *99*, 3384 (1966).

[597] *M.N. Sheng, J.G. Zajacek*, J. Org. Chem. *33*, 588 (1968).

[598] Fr. 1471967 1471967 (1967), Chemische Werke Hüls A.G.; C.A. *68*, 86837s (1968).

[599] U.S.P. 3390182 (1968), Halcon International Inc., Inv.: *J. Kollar, R.S. Barker;* C.A. *69*, 96175j (1968).

[600] U.S.P. 3494924 (1970), Atlantcic Richfield Co., Inv.: *G.A. Bonetti, R. Rosenthal, R.L. Shubkin;* C.A. *72*, 100006m (1970).

[601] *A. Lespagnol, M. Debaert, M. Devergnies*, Bull. Soc. Chim. France 2744 (1963).

[602] *J.C. Craig, K.K. Purushothaman*, J. Org. Chem. *35*, 1721 (1970).

[603] *L. Horner, H. Schaefer, W. Ludwig*, Chem. Ber. *91*, 75 (1958).

[604] U.S.P. 3336388 (1967), Archer-Daniels-Midland Co., Inv.: *G.P. Shulman;* C.A. *67*, 108192m (1967).

[605] U.S.P. 3332999 (1967), Ethyl Corp., Inv.: *L.C. Mitchell, T.H. Coffield;* C.A. *67*, 73134q (1967).

[606] *I. Tabushi, J. Hamuro, R. Oda*, Nippon Kagaku Zasshi *90*, 197 (1969); C.A. *70*, 86807y (1969).

[607] *U. Schöllkopf, M. Patsch, H. Schäfer*, Tetrahedron Lett. 2515 (1964);
U. Schöllkopf, G. Ostermann, Justus Liebigs Ann. Chem. *737*, 170 (1970).

[608] *A.R. Lepley, P.M. Cook, G.F. Willard*, J. Amer. Chem. Soc. *92*, 1101 (1970).

[609] *M. Moriwaki, S. Sawada, Y. Inouye*, Chem. Commun. 419 (1970).

[610] *M.R.V. Sahyun, D.J. Cram*, J. Amer. Chem. Soc. *85*, 1263 (1963).

[611] *C. Pinazzi, J.C. Brosse, J. Brossas, A. Pleurdeau*, C.R. Acad. Sci., Paris, Ser. C 266, 443 (1968).

Deoxygenation and loss of the benzyl group as aldehyde are *side-reactions*[578]. Where a Meisenheimer rearrangement is not possible, heating the *N*-oxides **83** affords a Cope elimination in high yield (under certain conditions even at room temperature[610] if free from water)[586, 593, 611]:

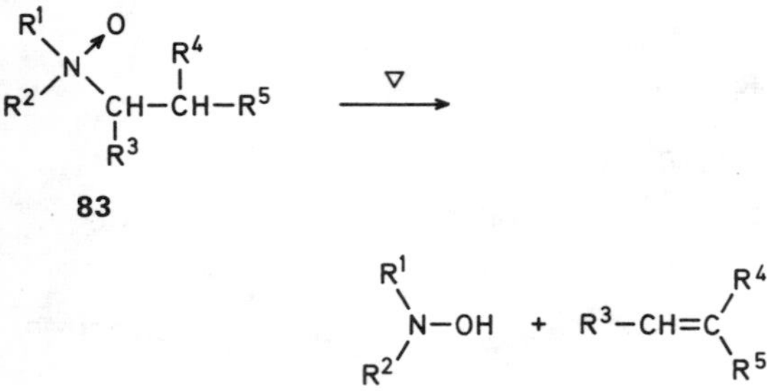

83

Evidently a five-membered cyclic transition state *(O–N–C–C–H* in one plane) needs to be passed through. Where steric factors or salt formation prevent this state the elimination does not occur or requires higher temperatures accompanied by inferior yields. Even hydrogen bonds to the oxygen cause inhibition; this explains why the elimination in protic solvents proceeds very slowly[610]. Ring size in cyclic amine oxides[593, 612, 613], branches in the grouping being eliminated (R^4, $R^5 \neq H$)[593], and additional activation of the migrating hydrogen atom (*e.g.*, $R^4 = C_6H_5$)[593] influence the speed and direction of the elimination, while steric factors govern the *cis-trans* ratio in the olefin formed (R^3, $R^4 \neq H$)[614]. In amine oxides **83** with $R^4 = CN$, COC_6H_5, $COOR$, $COOH$ elimination can be catalyzed by bases (*retrogressive Michael addition*)[590, 615]. Methylamine *N*-oxides, whose structure prevents them undergoing either a Cope or a Meisenheimer reaction, may be cleaved into formaldehyde and secondary amine on very strong heating.

While protonations with acids and alkylations on the oxygen furnish the corresponding hydroxylamine derivatives, the *O*-acyl hydroxylammonium compounds formed by acylation cleave to aldehydes and secondary amine or its acyl derivative spontaneously (*Polonovski reaction*[582, 593]; it is catalyzed by bases; re the mechanism cf. ref. [582, 616, 617]):

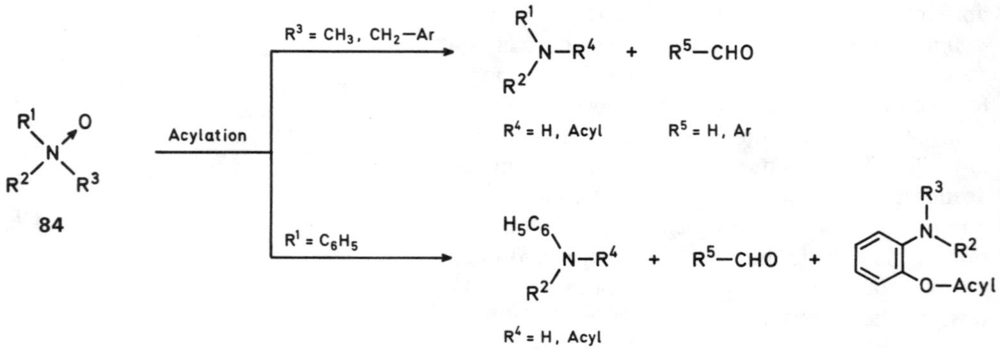

84

In aniline *N*-oxides ($R^1 = C_6H_5$) formation of 2-acyloxyanilines predominates when bases are absent. Where no methyl group or benzyl group is present, Polonovski reaction conditions convert an amine oxide either into an *enamine*[617] or produce a different sort of fragmentation provided a further stabilized cation can be formed as intermediate product side by side with a carbimmonium ion[616, 580]:

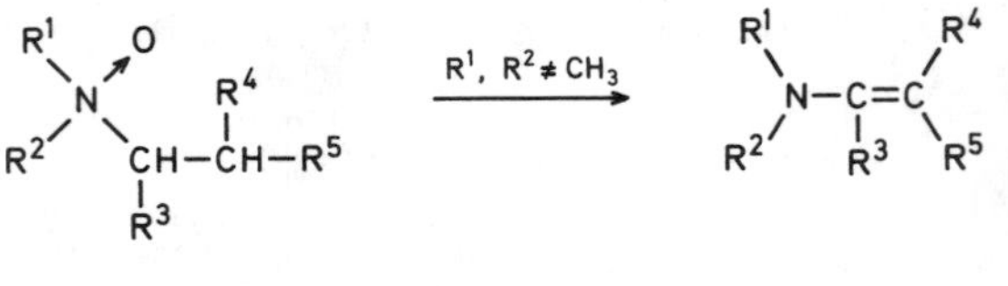

[612] *A.C. Cope, E. Ciganek, J. Lazar*, J. Amer. Chem. Soc. **84**, 2591 (1962).

[613] *J. Sicher, J. Zavada, J. Krupicka*, Tetrahedron Lett. 1619 (1966).

[614] *H. Möhrle, R. Kilian*, Pharmazie **25**, 296 (1970).

[615] *T. Rogers*, J. Chem. Soc. (London) 769 (1955).

[616] *R. Huisgen, W. Kolbeck*, Tetrahedron Lett. 783 (1965).

[617] *R.T. LaLonde, E. Auer, C.F. Wong, V.P. Muralidharan*, J. Amer. Chem. Soc. **93**, 2501 (1971).

[618] *H. Alper, J.T. Edward*, Can. J. Chem. **48**, 1543 (1970).

[619] *A. Hussain Khuthier*, Tetrahedron Lett. 4627 (1970).

[620] *H.G.O. Becker, G. Görmar, H.-J. Timpe*, J. Prakt. Chem. **312**, 610 (1970);
H.G.O. Becker, G. Görmar, H. Haufe, H.-J. Timpe, J. Prakt. Chem. **314**, 101 (1972).

[621] *W.T. Flowers, D.R. Taylor, A.E. Tipping, C.N. Wright*, J. Chem. Soc. (C) 1986 (1971).

[622] *S. Morrocchi, A. Ricca, A. Selva, A. Zanarotti*, Gazz. Chim. Ital. **99**, 165 (1969).

[623] *W. Dürckheimer*, Justus Liebigs Ann. Chem. **756**, 145 (1972).

Iron(III) ions[581, 593] or sulfur dioxide[582, 583] can bring about the Polonovski cleavage in place of acylating agents, but competing reactions take place[586], *e.g.*, deoxygenation (p. 359) and formation of an adduct $R^1R^2R^3N^{\oplus}–O–SO_2^{\ominus}$, which rearranges thermally to the *zwitterion* $R^1R^2R^3N^{\oplus}–SO_3^{\ominus}$[582, 583].

Deoxygenation of an amine oxide can be achieved in very widely different ways[586, 593]:

Sn/HCl   P^{III} derivatives
(NH₄)₂S   Carbenes
H₂/Pd   R–MgX
Pb/Fe^{II}   Fe(CO)₅[618]
NaBr/1,4-dioxane (containing peroxide)[619], *etc.*

[624] *N. Kornblum, R. Seltzer, P. Haberfield*, J. Amer. Chem. Soc. *85*, 1148 (1963).

[625] *L.B. Volodarskii, G.A. Kutikova, R.Z. Sagdeev, Yu.N. Molin*, Tetrahedron Lett. 1065 (1968).

[626] *H. Lettau*, Z. Chemie *10*, 211, 338, 431, 462 (1970).

[627] *W. Rundel* in *Houben-Weyl*, Methoden der organischen Chemie, Bd. X/4, p. 363, Georg Thieme Verlag, Stuttgart 1966.

[628] *S. Petersen, H. Heitzer*, Justus Liebigs Ann. Chem. *740*, 180 (1970).

[629] *H. Dahn, B. Favre, J.-P. Leresche*, Helv. Chim. Acta *56*, 457 (1973).

[630] *M. Masui, Ch. Yijima*, J. Chem. Soc. (B) 56 (1966).

[631] *Y. Ogata, Y. Sawaki*, J. Amer. Chem. Soc. *95*, 4687, 4692 (1973).

[632] *Ch. Grundmann, G.F. Kite*, Synthesis 156 (1973).

[633] *J. Ackrell, M. Altaf-ur-Rahman, A.J. Boulton, R.C. Brown*, J. Chem. Soc. Perkin I, 1587 (1972).

[634] *H. Gnichtel, R. Walentowski, K.-E. Schuster*, Chem. Ber. *105*, 1701 (1972).

[635] *H. Neunhoeffer, H.-W. Frühauf*, Justus Liebigs Ann. Chem. *758*, 111 (1972).

[636] *P. Crabbé, J. Haro, C. Rius, E. Santos*, J. Heterocycl. Chem. *9*, 1189 (1972).

[637] *C.B. Schapira, S. Lamdan*, J. Heterocycl. Chem. *9*, 569 (1972).

[638] *C.W. Muth, R.S. Darlak, J.C. Patton*, J. Heterocycl. Chem. *9*, 1003 (1972).

[639] *F. Yoneda, Y. Sakuma*, Chem. Pharm. Bull. (Tokyo) *21*, 448 (1973).

[640] *E. Laviron, D. Bernard, G. Tainturier*, Tetrahedron Lett. 3643 (1972).

[641] *Y. Maki, T. Hosokami, M. Suzuki*, Chem. Commun. 693 (1972).

[642] *F. Yoneda, S. Nishigaki, K. Shinomura*, Chem. Pharm. Bull. (Tokyo) *19*, 2647 (1971).

[643] *M.J. Haddadin, J.J. Zamet, C.H. Issidorides*, Tetrahedron Lett. 3653 (1972).

[644] *N.A. Mufarrij, M.J. Haddadin, C.H. Issidorides, J.W. McFarland, J.D. Johnston*, J. Chem. Soc. Perkin I, 965 (1972).

[645] *G. Zvilichovsky, G.B. Brown*, J. Org. Chem. *37*, 1871 (1972).

[646] *H. Gnichtel, W. Griebenow, W. Löwe*, Chem. Ber. *105*, 1865 (1972).

[647] *D.B. Livingstone, G. Tennant*, Chem. Commun. 96 (1973).

[648] *G.G. Aloisi, E. Bordignon, A. Signor*, J. Chem. Soc. Perkin II, 2218 (1972).

[649] *R.C. Kerber*, J. Org. Chem. *37*, 1587 (1972).

[650] *A. Gasco, V. Mortarini, G. Ruà, G.M. Nano, E. Menziani*, J. Heterocycl. Chem. *9*, 577, 837 (1972).

[651] *F. Seng, K. Ley*, Angew. Chem. *84*, 1060 (1972); *F. Seng, K. Ley*, Synthesis 606 (1973); *D.W.S. Latham, O. Meth-Cohn, H. Suschitzky*, Chem. Commun. 1040 (1972); *M.J. Abu El-Haj*, J. Org. Chem. *37*, 2519 (1972); *D.P. Claypool, A.R. Sidani, K.J. Flanagan*, J. Org. Chem. *37*, 2372 (1972).

[652] *P. Braun, K.-P. Zeller, E. Müller*, Tetrahedron *28*, 5655 (1972).

[653] *R. Fielden, O. Meth-Cohn, H. Suschitzky*, J. Chem. Soc. Perkin I, 696, 705 (1973).

[654] *J. Ackrell, A.J. Boulton*, J. Chem. Soc. Perkin I, 351 (1973).

[655] *W. Rundel* in *Houben-Weyl*, Methoden der organischen Chemie, Bd. X/4, p. 349, Georg Thieme Verlag, Stuttgart 1968.

[656] *W. Rundel* in *Houben-Weyl*, Methoden der organischen Chemie, Bd. X/4, p. 362, Georg Thieme Verlag, Stuttgart 1968.

[657] *E. Falco, G.B. Brown*, J. Med. Pharm. Chem. *11*, 142 (1968).

[658] *E. Buehler, G.B. Brown*, J. Org. Chem. *32*, 268 (1967).

[659] *L.B. Volodarskii, A.Y. Tikhonov*, Zh. Org. Khim. *6*, 307 (1970), Eng. Edit. p. 294; *L.B. Volodarskii, A.Y. Tikhonov*, Synthesis 210 (1971).

[660] *W. Rundel* in *Houben-Weyl*, Methoden der organischen Chemie, Bd. X/4, p. 360, Georg Thieme Verlag, Stuttgart 1968.

[661] *D.A. Kerr, D.A. Wilson*, Tetrahedron Lett. 2885 (1968); *D.A. Kerr, D.A. Wilson*, J. Chem. Soc. (C) 1718 (1970).

[662] *W. Rundel* in *Houben-Weyl*, Methoden der organischen Chemie, Bd. X/4, p. 402, Georg Thieme Verlag, Stuttgart 1968.

[663] *I.M. Chau, C. Beaute, S. Cornuel, N. Thoai*, Tetrahedron Lett. 4313 (1971).

[664] *M. Lamchen, G.R. Delpierre*, Quart. Rev. *19*, 329 (1965).

[665] *D. Hadzi*, J. Chem. Soc. 5128 (1962).

[666] *R. Marshall, D.M. Smith*, J. Chem. Soc. (C) 3510 (1971).

[667] *K. Wagner, H. Heitzer, L. Oehlmann*, Chem. Ber. *106*, 640 (1973).

[668] *H.J. Shine, L.-T. Fang, H.E. Mallory, N.F. Chamberlain, F. Stehling*, J. Org. Chem. *28*, 2326 (1963); *H.H. Shine, J.Y.-F. Tsai*, J. Org. Chem. *29*, 443 (1964).

[669] *S.O. Chua, M.J. Cook, A.R. Katritzky*, J. Chem. Soc. (B) 2350 (1971).

[670] *G.G. Spence, E.C. Taylor, O. Buchardt*, Chem. Rev. *70*, 231 (1970); *Ch. Lohse*, J. Chem. Soc. Perkin II 229 (1972); *F. Bellamy, L.G.R. Barragan, J. Streith*, Chem. Commun. 456 (1971); *S. Yamada, M. Ishikawa, C. Kaneko*, Tetrahedron Lett. 971, 977 (1972); *Ch. Lohse*, Angew. Chem. *84*, 220 (1972).

[671] *M.J. Haddadin, H.N. Alkaysi, S.E. Saheb*, Tetrahedron *26*, 1115 (1970).

11 Oximes

Contributed by

B. Unterhalt
Institut für Pharmazeutische Chemie
und Lebensmittelchemie der Universität Marburg,
Marburg

11.1 Preparation[1]

11.1.1 Replacement of carbonyl oxygen by the hydroxyimino group

Aldehydes and ketones react with hydroxylamine as nucleophile in neutral or weakly acid solution according to the general scheme

The reaction passes through a maximum velocity at *pH* 4.5–5.5 which is set as a result of the change of the rate-determining step on passing from a neutral to a strongly acid medium. In neutral solution the proton-catalyzed dehydration of the adduct is the slowest step of the reaction, in acid medium it is the attack of the base on the carbonyl group[2-5].

In some cases, especially with sterically demanding ketones, it is advisable to perform the conversion in the presence of excess alkali metal hydroxide or alcoholate. It is assumed that, here, a nucleophilic attack by the base $NHOH^{\ominus}$ or the detaching of the proton linked to the nitrogen atom of the adduct is the initial step[6-9].

With different groups R^1 and R^2 *isomer mixtures are normally formed*. They can be separated by fractional crystallization or, better, in many cases by preparative layer chromatography. The isomers may be designated as *(E)-* and *(Z)-*forms respectively using the Cahn-Ingold-Prelog nomenclature.

The *(E)*-isomer is the form in which the hydroxyl group and the group of greater priority, *e.g.*, R^1, is in the *trans* position, while in *(Z)*-isomers the hydroxyl group and the group of greater priority are in the *cis* position to each other[10].

11.1.1.1 In aromatic and saturated aliphatic aldehydes and ketones

Conversion of aldehydes and ketones with hydroxylamine often requires milder conditions than other amine condensations, and a clearly defined reaction ensues leading to the desired products. In general these are mixtures of isomers, as is shown by thin-layer chromatographic and proton resonance spectroscopy investigations. Throughout, aldoximes form more rapidly than ketoximes. There are also differences in the reactivity of individual aldehydes and ketones[11-13].

[1] *H. Metzger* in *Houben-Weyl*, Methoden der organischen Chemie, Bd. X/4, 4. Aufl., p. 1, Georg Thieme Verlag, Stuttgart 1968.

[2] *E. Barrett, A. Lapworth*, J. Chem. Soc. Abstracts 93, 85 (1908).

[3] *W. Hückel, M. Sachs*, Justus Liebigs Ann. Chem. 498, 176 (1932).

[4] *W.P. Jencks*, J. Amer. Chem. Soc. 81, 475 (1959).

[5] *T.J. Mikkelson, J.R. Robinson*, J. Pharm. Sci. 57, 1180 (1968).

[6] *D.E. Pearson, O.D. Keaton*, J. Org. Chem. 28, 1557 (1963).

[7] *A. Williams, M.L. Bender*, J. Amer. Chem. Soc. 88, 2508 (1966).

[8] *W.P. Jencks*, Progr. Phys. Org. Chem. 2, 63 (1964), Mechanism and Catalysis of Simple Carbonyl Group Reactions.

[9] *Y. Ogata, A. Kawasaki* in *S. Patai*, The Chemistry of the Carbonyl Group, Vol. II, Chap. I, Interscience Publishers, London, New York, Sydney, Toronto 1970.

[10] *J.E. Blackwood, C.L. Gladys, K.L. Loening, A.E. Petrarca, J.E. Rush*, J. Amer. Chem. Soc. 90, 509 (1968).

[11] *M. Pasteka*, Chem. Zvesti 20, 855 (1966); C.A. 67, 64650c (1967).

[12] *H. Sihtola, L. Neimo*, Tappi 46, 730 (1963); C.A. 60, 10911f (1964).

Because of its liability to decompose, *hydroxyl-amine* is employed in the form of a salt and is liberated from the same by adding bases. The easiest medium to work in is an ethanolic or aqueous ethanolic solution containing a small excess of hydroxylamine hydrochloride, sulfate, or acetate in the presence of pyridine, sodium acetate, or alkali metal and alkaline earth metal carbonates (Table 1).

Dimethyl sulfoxide seems to offer considerable advantages as solvent[14].

Where difficulties arise during the oximation in neutral or weakly acid medium, *e.g.,* due to steric effects, the conversion should be carried out in the *strongly alkaline* region (Table 1) and using elevated pressure[15].

Reference is made to the use of 1,2-epoxycyclohexane as hydrogen chloride trap during preparation of *camphor oxime* and *mannose oxime* and to the conversion of sodium hydroxylaminomonosulfonate with carbonyl compounds which yield 67–95% *oxime*[16, 17].

In some cases it is worthwhile to use derivatives of the carbonyl compounds instead of the compounds themselves in the oximation reaction. *Geminal chlorine* and *bromine* atoms are exchanged readily against the hydroxyimino group but geminal *fluorine* atoms are not. Dichloroacetaldehyde reacts with hydroxylammonium chloride and sodium carbonate by double oximation to form *glyoxime*[28]. 3,3-Dibromo-1,1,1-trifluoro-2-propanone reacts in the presence of sodium acetate at 100° to form *1,1,1-trifluoro-2,3-propanedione dioxime*[29]. 2,2-Dichloro-3,3-dimethylbutane reacts smoothly with hydroxylammonium chloride and sodium carbonate in ethanol to give *3,3-dimethyl-2-butanone oxime*[30]. The hydrogen sulfite addition compounds stand out from carbonyl derivatives such as hydrates, aldehyde-ammonia compounds, hydrogen sulfite adducts, and acetals which are formed by simple addition to the carbonyl group, because under normal conditions they often afford astonishingly high yields of

Table 1. Selected methods for preparing oximes from carbonyl compounds and hydroxylamine hydrochloride

Carbonyl compound	Soluent-base	Time [hrs]	Temperature [°C]	Oxime	M. p. [°C]	Yield [% of theory]	Ref.
Acetaldehyde	Water/Na$_2$CO$_3$	15	~20	Acetaldehyde oxime	bp 112–114	80	18
Geranylacetone	Water/NaHCO$_3$	24	~20	6,10-Dimethyl-5,9-undecadien-2-one oxime	bp$_{0.05}$ 107–108	83	19
Fluorenone	Ethanol/BaCO$_3$	5	Reflux	Fluorenone oxime	193–194	high	20
4-Tetrahydrothio-pyranone	Ethanol/water/NaOAc	3	Reflux	4-Tetrahydrothio-pyranone oxime	84–85	90	21
trans-2-Decahydro-naphthalenone	Methanol/water/NaOAc	12	~20	trans-2-Decahydro-naphthalenone oxime	62–76	~100	22
1-Phenyl-2-propanone	Pyridine or water/NaOH			1-Phenyl-2-propanone oxime	69–71(E)		23
4-Formyl-1-phenylpyrazole	Ethanol/pyridine Ethanol/sodium ethoxide	2 60	Reflux ~20	1-Phenyl-4-pyrazole-carboxaldehyde oxime	173(Z) 135(E)	94 33	24
3,3-Dimethyl-2-butanone	Ethanol/water/NaOH	1	~20	3,3-Dimethyl-2-butanone oxime	77–78	85	25
Benzaldehyde	Water/NaOH	1–2	~20	Benzaldehyde oxime	35 (bp$_{12}$ 122–124°)	50	26
Camphor	Ethanol/water/NaOH	1	100	Camphor oxime	115	75	27

[13] P.G. Kletzke, J. Org. Chem. 29, 1363 (1964).
[14] H.-L.Pan, T.L.Fletcher, Chem. & Ind. (London) 240 (1969).
[15] W.H.Jones, E.W.Tristram, W.F.Benning, J. Amer. Chem. Soc. 81, 2151 (1959); US. P. 3 256 331 (1963), Merck and Co., Inv.: W.H. Jones, E.W. Tristram; C.A. 65, 13797h (1966).
[16] M. Miyazaki, T. Takahashi, J. Jap. Forest. Soc. 34, 313 (1952); C.A. 47, 12226i (1953).
[17] W.L. Semon, V.R. Damerell, J. Amer. Chem. Soc. 46, 1290 (1924).
[18] H. Wieland, Ber. 40, 1677 (1907).
[19] G. Pala, A. Mantegani, G. Coppi, J. Med. Chem. 10, 980 (1967).
[20] J. Schmidt, J. Söll, Ber. 40, 4258 (1907).
[21] C. Barkenbus, J.F. Diel, G.R. Vogel, J. Org. Chem. 20, 873 (1955).
[22] W. Hückel, Justus Liebigs Ann. Chem. 451, 128 (1926).
[23] A.C. Huitric, D.B. Roll, J.R. DeBoer, J. Org. Chem. 32, 1661 (1967).

oxime[31-33]. *Thioketones* and *ketimines*, too, can be converted into oximes. Thus, thioxanthone and hydroxylamine hydrochloride in pyridine or sodium acetate-ethanol form *xanthone oxime*[34], 2,4,6-trimethylbenzophenonimine the associated *2,4,6-trimethylbenzophenone oxime*[7, 35, 36]:

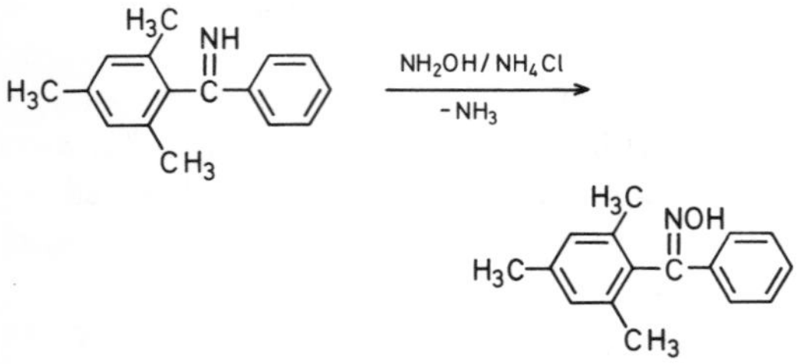

11.1.1.2 Replacement of carbonyl oxygen by the hydroxyimino group in α,β-unsaturated aldehydes and ketones

α,β-Unsaturated aldehydes and ketones react to form the expected oximes by the normal reaction, as is shown, for example, by reactions with substituted acroleins, 3-chloroacroleins, and benzylidenealkanones[37-39]. Often *mixtures* of the geometrical isomers are formed whose separation no longer poses a problem[40]. The reaction proceeds in several different directions if performed with chalcone. In acid solution it leads to formation of the *(E)-* or *syn-chalcone oxime*, in alkaline medium *3,5-diphenyl-2-isoxazoline* is formed side by side with other products[41-45]. With suitable substituted chalcones a partial ring closure occurs even in acid solution with formation of *2-isoxazoline*[43]:

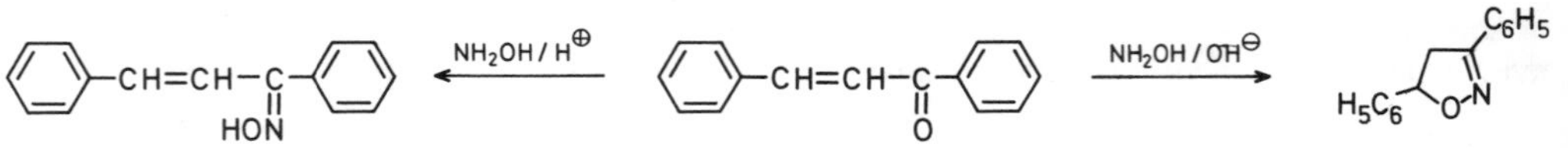

Oximes with $C \equiv C$ triple bonds are of interest. Thus, *propiolaldehyde oxime* was prepared in the normal way[46], while unstable *higher* alkynal oximes were prepared from nitrile oxides and alkynes[47], *e.g.*:

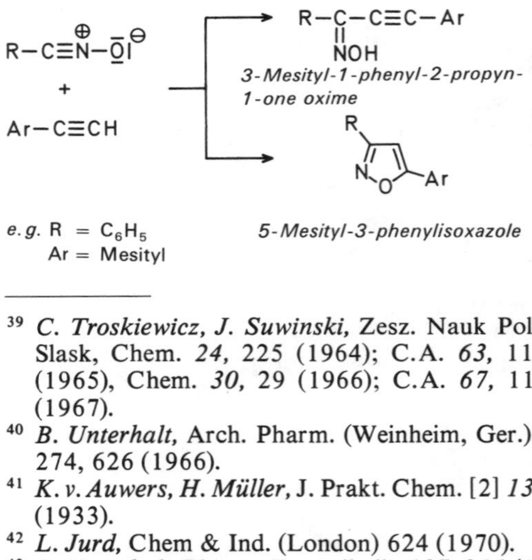

[24] *I.L. Finar, G.H. Lord,* J. Chem. Soc. (London) 3314 (1957);
I.L. Finar, H.E. Saunders, J. Chem. Soc. (C) 1495 (1969).

[25] *O. Piloty, A. Stock,* Ber. *35,* 3097 (1907).

[26] *E. Beckmann,* Ber. *23,* 1684 (1890);
A.I. Vogel, A Textbook of Practical Org. Chem., 3. Ed., p. 719, Longmans, Green and Co., London 1964.

[27] *K. v. Auwers,* Ber. *22,* 605 (1889).

[28] *E. Reisse,* Justus Liebigs Ann. Chem. *257,* 332 (1890);
N. Wittorf, J. Russ. Phys. Chem. Soc. *32,* 88 (1900); Chem. Zentr. II, 30 (1900).

[29] *R. Belcher, A. Sykes, J.C. Tatlow,* J. Chem. Soc. (London) 2393 (1957).

[30] *F. Couturier,* Ann. Chim. (Paris) (6) *26,* 452 (1892); Chem. Zentr. II, 510 (1892).

[31] *I.L. Knunyants, B.P. Fabrichnyi,* Dokl. Akad. Nauk SSSR *68,* 523 (1949); C.A. *44,* 1469d (1950).

[32] *P. Grammaticakis,* Compt. Rend. *224,* 1568 (1947); C.A. *41,* 6548d (1947).

[33] *A. Ahmad, I. Eelnurme, I.D. Spenser,* Can. J. Chem. *38,* 2523 (1960).

[34] *C. Graebe, P. Röder,* Ber. *32,* 1690 (1899).

[35] *C.R. Hauser, D.S. Hoffenberg,* J. Amer. Chem. Soc. *77,* 4885 (1955).

[36] *E.H. Cordes, W.P. Jencks,* J. Amer. Chem. Soc. *84,* 826 (1962) (Amine catalysis for the oximation).

[37] *B. Unterhalt,* Arch. Pharm. (Weinheim, Ger.) *303,* 661 (1970).

[38] *W.R. Benson, A.E. Pohland,* J. Org. Chem. *30,* 1126 (1965).

[39] *C. Troskiewicz, J. Suwinski,* Zesz. Nauk Politech. Slask, Chem. *24,* 225 (1964); C.A. *63,* 11292b (1965), Chem. *30,* 29 (1966); C.A. *67,* 11121z (1967).

[40] *B. Unterhalt,* Arch. Pharm. (Weinheim, Ger.) *299,* 274, 626 (1966).

[41] *K. v. Auwers, H. Müller,* J. Prakt. Chem. [2] *137,* 65 (1933).

[42] *L. Jurd,* Chem & Ind. (London) 624 (1970).

[43] *B. Unterhalt,* Pharm. Zentralhalle *107,* 356 (1968).

[44] *R.P. Barnes, G.E. Pinkney, G. McK. Phillips,* J. Amer. Chem. Soc. *76,* 276 (1954).

[45] *A.T. Balaban, I. Zugravescu, S. Avramovici, W. Silhan,* Monatsh. Chem. *101,* 704 (1970).

[46] US.P. 3006948, Inv.: *J. Happel, C.J. Marzel, A.A. Reidlinger;* Chem. Zentr. 36–2237 (1964).

[47] *S. Morrocchi, A. Ricca, A. Zanarotti, G. Bianchi, R. Gandolfi, P. Grünanger,* Tetrahedron Lett. 3329 (1969).

The following method, giving purer products in better yield, appears to be generally applicable[48]:

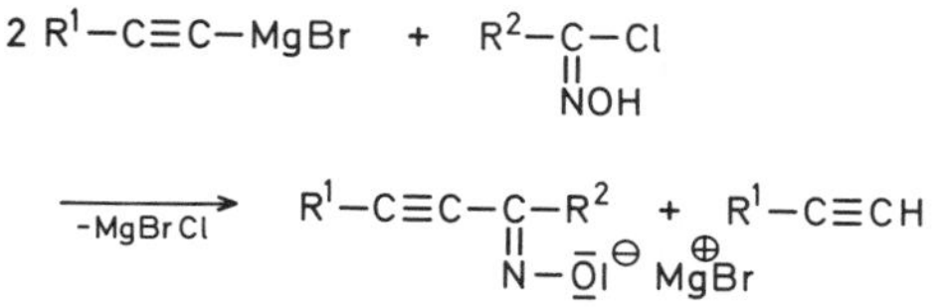

$$2\ R^1-C\equiv C-MgBr\ +\ R^2-\underset{\underset{NOH}{\|}}{C}-Cl$$

$$\xrightarrow[-MgBrCl]{}\ R^1-C\equiv C-\underset{\underset{N-\underset{\ominus}{\overline{\underline{O}}}|^{\ominus}\ \overset{\oplus}{MgBr}}{\|}}{C}-R^2\ +\ R^1-C\equiv CH$$

11.1.1.3 Replacement of carbonyl oxygen by the hydroxyimino group in aldehydes and ketones with a further functional group

In addition to the $C=C$ double bond (see p. 406) other functional groups can influence the oximation of aldehydes and ketones, for instance, a second carbonyl function, halogen, or amine.

11.1.1.3.1 Replacement of carbonyl oxygen by the hydroxyimino group in aldehydes and ketones with a second carbonyl group

1,2-Diketones react to form *dioximes* with excess hydroxylamine *via* the isolatable monooxime stage[49]. Thus 1,2-cyclohexanedione forms *1,2-cyclohexanedione dioxime* at 40—60° and *pH* 7 with aqueous hydroxylammonium chloride solution; the same compound can be obtained in 90% yield[50—52] from 2-morpholino-2-cyclohexen-1-one oxime

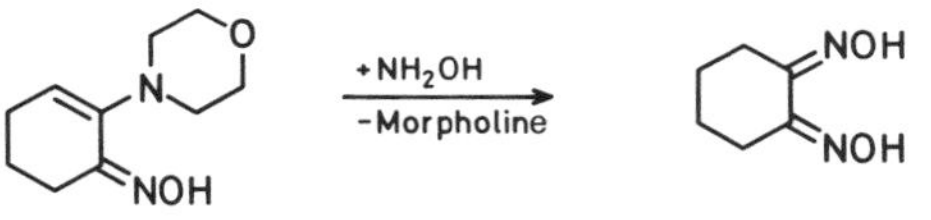

α-Furil dioxime, which normally is formed in moderate yield only, can be obtained in 55% yield by soxhlet extraction of furil with excess methanolic hydroxylammonium chloride solution[53].

1,3-Diketones such as 1-phenyl-1,3-butanedione, furnish either *mono-* or *dioximes* with hydroxylamines in alkaline solution at room temperature in dependence on the working conditions[54]. 3-Phenyl-2,4-pentanedione leads to *3-phenyl-2,4-pentanedione dioxime* and *3-phenyl-2,4-pentanedione 2-oxime* together with *3,5-dimethyl-4-phenyl-isoxazole* on 30 minutes heating in 70% ethanol with hydroxylammonium chloride and sodium acetate[55]:

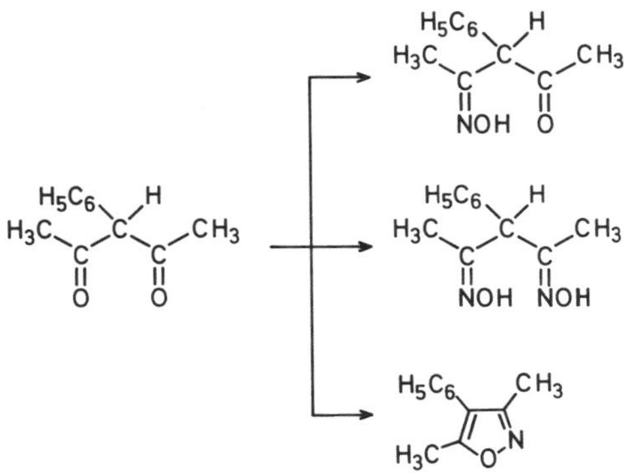

With hydroxylamine in absolute ethanol only the monooxime is formed under these conditions. Boiling in absolute pyridine/ethanol leads to mixed *monooxime* and *dioxime*. β-Oxocarboxylic acid esters and hydroxymethylene ketones react with hydroxylamine preferentially with ring closure[56—59]. Finally, 1-phenyl-1,3-propanedione and also 1-phenyl-1,3-butanedione furnish monooximes and undergo ring-chain tautomerism[60]:

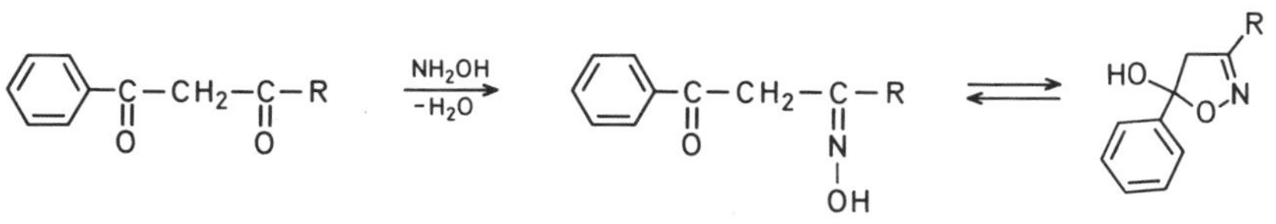

R = H;   1-Phenyl-1,3-   5-Phenyl-2-
         propanedione 3-oxime isoxazolin-5-ol
R = CH₃; 1-Phenyl-1,3-   3-Methyl-5-phenyl-
         butanedione 3-oxime   2-isoxazolin-5-ol

[48] *Z. Hamlet, M. Rampersad, D.J. Shearing,* Tetrahedron Lett. 2101 (1970).
[49] *J.C. Danilewicz,* J. Chem. Soc. (C) 1049 (1970) (Configuration of α-hydroxyiminoketones or 1,2-dioximes); *J.V. Burakevich, A.M. Lore, G.P. Volpp,* J. Org. Chem. *36,* 1 (1971).
[50] Fr. P. 1 339 224 (1961), DuPont, Inv.: *W.J. Arthur;* C.A. *60,* 5358h (1964).
[51] Jap. P. 7009 531 (1967), Kobagashi Perf., Inv.: *R. Sudo, M. Takahashi;* C.A. *73,* 14296v (1970).
[52] *G. Drefahl, G. Heublein, G. Tetzlaff,* J. Prakt. Chem. *311,* 162 (1969).
[53] *S.A. Reed, C.V. Banks, H. Diehl,* J. Org. Chem. *12,* 792 (1947).

[54] *K. v. Auwers, H. Müller,* J. Prakt. Chem. [2] *137,* 86, 90 (1933).
[55] *B. Bobranski, R. Wojtowski,* Roczniki Chem. *38,* 1327 (1964); Chem. Zentr. 22—0911 (1966).
[56] *A.R. Katritzky, S. Oksne,* Proc. Chem. Soc. 387 (1961).
[57] *R. Jacquier, C. Petrus, F. Petrus, J. Verducci,* Bull. Soc. Chim. France 2685, 2690 (1970); C.A. *73,* 119907y, 120541f (1970).
[58] *C.H. Eugster, L. Leichner, E. Jenny,* Helv. Chim. Acta *46,* 543 (1963).

11.1.1.3.2 Replacement of carbonyl oxygen by the hydroxyimino group in aldehydes and ketones with a halogen function

On reacting hydroxylamine with α-halocarbonyl compounds *1,2-dione 1-oximes* or *1,2-dioximes* may form in a redox reaction. Two or three mols hydroxylamine are required. Thus, *1,2-propanedione oxime* is obtained from chloro-2-propanone with three mols hydroxylammonium chloride and two mols sodium carbonate in aqueous solution[61]. *1,2-Cyclopentanedione* and *1,2-cyclohexanedione dioximes* are obtained from 2-bromocyclopentanone and 2-bromocyclohexanone in an aqueous or aqueous ethanolic solution of hydroxylamine hydrochloride solution buffered with sodium acetate[62, 63]. 2-Chlorocyclohexanone does not react in this way; instead, with hydroxylamine sulfate and potassium carbonate in water at 0° *2-chloro-1-cyclohexanone oxime*[64] is formed which in methanol can be dehalogenated with hydroxylamine hydrochloride-palladium chloride at room temperature to give *cyclohexanone oxime*[65].

Synthesis of *2-hydroxyamino oximes* succeeds by starting from (dimeric) nitrosochlorides. Alkanedione dioximes are invariably formed as side-products unless the halogen atom is located on a tertiary carbon atom[66–68].

If the halogen atom is attached to a *C=C* double bond, ring closure to *isoxazole* can occur in the alkaline region with splitting off of hydrogen halides[69].

β-Halogenocarbonyl compounds react with hydroxylamine to form *2-isoxazolines*, but often also to open-chain hydroxylamine derivatives[70].

11.1.1.3.3 Replacement of carbonyl oxygen by the hydroxyimino group in aldehydes and ketones with an amine function

Certain 2-aminocarbonyl compounds can be converted into *alkanedione dioximes* with hydroxylamine[71], *e.g.*:

1,2-Heptanedione dioxime

Others again react conventionally and form *2-amino oximes*, the isomer mixtures can be separated[72–75]. The mode of reaction of enamines of corresponding structure is described on p. 407.

11.1.2 Replacement of hydrogen by the hydroxyimino group

Activated hydrogen atoms can be replaced by nitrosyl cations by using an ionic reaction. Oximes form by isomerization *via* nitroso compounds. By contrast, *nonactivated* hydrogen atoms at most suffer homolytic cleavage and lead to the desired products with nitrosyl radicals (nitrogen monoxide)[76–78].

11.1.2.1 Replacement of hydrogen by the hydroxyimino group in saturated hydrocarbons

Nitrosyl chloride, which can split into nitric oxide and chlorine, or a mixture of nitrosyl chloride, nitric oxide, and hydrogen chloride react with sat-

[59] *N.K. Kochetkov, E.D. Khomutova,* Zh. Obshch. Khim. *30*, 954 (1960); C.A. *55*, 512f (1961).
[60] *J. Castells, A. Colombo,* Chem. Commun. 1062 (1969).
[61] *A. Hantzsch, W. Wild,* Justus Liebigs Ann. Chem. *289*, 292 (1896).
[62] *N. Tokura, I. Shirai, T. Sugahara,* Bull. Chem. Soc. Japan *35*, 722 (1962); C.A. *57*, 8450i (1962).
[63] *R. Belcher, W. Hoyle, T.S. West,* J. Chem. Soc. (London) 2744 (1958).
[64] *A.F. Childs, L.J. Goldsworthy, G.F. Harding, S.G.P. Plant, G.A. Weeks,* J. Chem. Soc. (London) 2321 (1948).
[65] Jap. P. 6 816 124 (1965), Toyo Rayon, Inv.: *S. Wakamatsu, K. Yamamoto, M. Nishimura, M. Ohno*; C.A. *70*, 57291v (1969).
[66] *L.B. Volodarskii, V.A. Koptyug, A.N. Lysak,* Zh. Org. Khim. *2*, 114 (1966); C.A. *64*, 14139h (1966).
[67] *L.B. Volodarskii, Y.G. Putsykin,* Zh. Org. Khim. *3*, 1686 (1967); C.A. *68*, 39169r (1968).
[68] *Y.G. Putsykin, L.B. Volodarskii,* Izv. Sibirsk, Otd. Akad. Nauk SSSR, Ser. Khim. Nauk 101 (1968); C.A. *70*, 46764a (1969).

[69] *K. v.-Auwers,* Ber. *62*, 1321 (1929).
[70] *E. Profft, F. Runge, A. Jumar,* J. Prakt. Chem. *273*, 74 (1955).
[71] *A. Kirrmann, P. Duhamel,* Compt. Rend. *249*, 424 (1959); C.A. *54*, 18508a (1960).
[72] *B. Unterhalt,* Arch. Pharm. (Weinheim, Ger.) *300*, 38 (1967).
[73] *Y.L. Chow, C.J. Colon,* J. Org. Chem. *33*, 2598 (1968).
[74] *Z. Tegyey, B. Matkovics,* Magy. Kem. Folyoirut *74*, 116 (1968); C.A. *69*, 2780c (1968).
[75] *H. Möhrle, B. Gusowski, R. Feil,* Tetrahedron *27*, 221 (1971).
[76] DDRP. 71 763 (1967), Erf.: *W. Kiessling, W. Pritzkow, P. Rudolff*; C.A. *73*, 98395n (1970); *C. Förster,* J. Prakt. Chem. *311*, 370 (1969).
[77] *O. Touster,* Org. Reactions VII, 6, 327 (1953) (The Nitrosation of Aliphatic Carbon Atoms).
[78] *C.V. Banks,* Record Chem. Progr. *25*, 85 (1964).

urated hydrocarbons, while being irradiated, by the chlorine taking up a hydrogen atom, while the carbon radical yields the oxime with nitric oxide *via* the nitroso compound[79]. The oxime separates as the hydrogen chloride adduct and is liberated by adjusting to *pH* 5–6[80]. A four-center mechanism is proposed to explain the photooximation of cyclohexane to *cyclohexanone oxime*[81]:

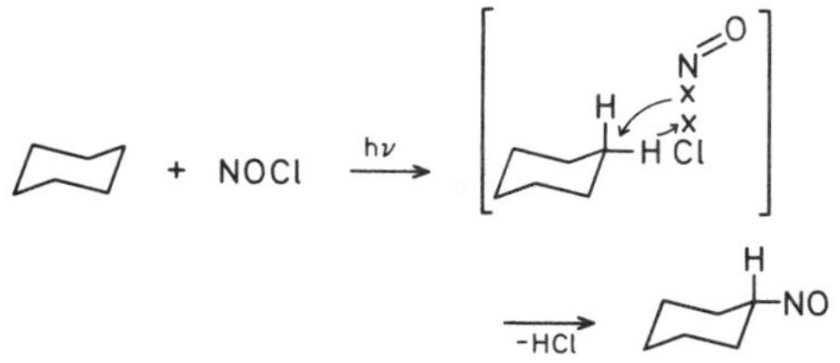

Nitrosyl chloride may be replaced by nitrosylsulfuric acid in the presence of a small quantity of chlorosulfuric acid or hydrogen chloride, by nitric acid and hydrochloric acid, dinitrogen tetroxide or dinitrogen trioxide and hydrogen chloride, alkyl nitrite and alkali nitrite, and hydrogen chloride.

Photonitrosation leads to oximes only when hydrogen atoms are replaced on primary and secondary carbon atoms. Secondary carbon atoms are attacked before primary ones; mixtures may be formed.

The photooximation of 2,2,3,3-tetramethylbutane and 2,3-dimethylbutane furnishes *2,2,3,3-tetramethylbutyraldehyde* and *2,3-dimethylbutyraldehyde oximes*[82, 83]:

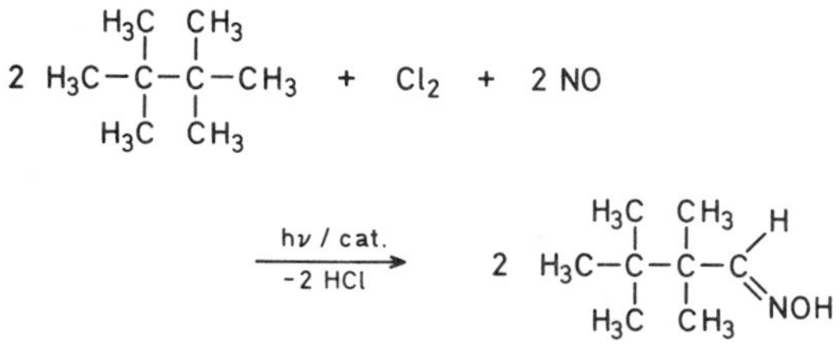

Nitric oxide reacts with cyclohexane to form *cyclohexanone oxime via* nitrosocyclohexane also when acted on by ionizing radiation[84]; without irradiation *cycloal-*

kanone oximes are formed when cycloalkanes are reacted with nitrosyl chloride or nitric oxide-chlorine at 200–450° in the presence of hydrogen chloride[85].

11.1.2.2 Replacement of hydrogen by the hydroxyimino group in compounds with activated C–H bonds

Activation of the *C–H* bond is obtained best by electronegative groups on the adjacent carbon atom, but its effect varies, as is revealed by positive results with ketones and generally negative ones with simple esters. Nonetheless, a 78% yield of *methyl 2-hydroxyiminobutyrate* was obtained from methyl butyrate and methyl nitrite in the presence of sodium methoxide/dimethylformamide[76].

In general, nitrous acid or derivatives such as the nitrosyl chlorides, anhydride and esters were employed, the esters in the presence of hydrogen chloride, alkali alcoholate, sodium amide, or metallic sodium.

Excess nitrosating agent should be avoided because of cleavage with reformation of the carbonyl compound. Oximation by photochemically excited nitrite anions, too, has been described[86].

11.1.2.2.1 Replacement of hydrogen by hydroxyimino in ketones and 1,3-diketones

Ketones and 1,3-diketones afford the anticipated hydroxyimino compounds with the stated nitrosating agents at low temperature (0–50° or less)[87–90]. Furthermore, transnitrosations with substituted diphenylnitrosamines, too, lead to oxime formation, *e.g.*, with camphor and 2-phenylacetophenone[91, 92] to *2,3-bornanedione 3-oxime (camphorquinone oxime)* and *benzil monooxime*. In some cases *dimeric nitroso compounds* form in dependence on the nitrosating agent and solvent, *e.g.*, invariably with tertiary carbon atoms in the α-position. With ketones containing a tertiary and

[79] E. Müller, H.-G. Padeken, Chem. Ber. *99*, 2971 (1966).
[80] M. Pape, Fortschr. Chem. Forsch. *7*, 559 (1966/67) (Photooximation of saturated hydrocarbons).
[81] E. Müller, G. Fiedler, J. Heiß, Chem. Ber. *101*, 765 (1968).
[82] E. Müller, A.E. Böttcher, Tetrahedron Lett. 3083 (1970).
[83] M.W. Mosher, N.J. Bunce, Can. J. Chem. *49*, 28 (1971).
[84] US.P. 3 062 812 (1959), Amer. Chem. Corp., Inv.: R.P. Taylor; C.A. *58*, 8913c (1963).

[85] Holl. P. 6 407 452 (1963), Soc. Edison; C.A. *63*, 2908f (1965).
[86] K. Pfoertner, Helv. Chim. Acta *53*, 922 (1970).
[87] J.J. Norman, R.M. Heggie, J.B. Larose, Can. J. Chem. *40*, 1547 (1962).
[88] Brit. P. 1 140 754 (1966), Pfizer, Inv.: J.C. Danilewicz, M. Szelke; C.A. *71*, 38566f (1969).
[89] W.H. Hartung, J.C. Munch, J. Amer. Chem. Soc. *51*, 2264 (1929).
[90] T. Shono, H. Nii, K. Shinra, Kogyo Kagaku Zasshi *72*, 1669 (1969); C.A. *72*, 21437z (1970).
[91] D.B. Parihar, S.P. Sharma, Chem. & Ind. (London) 1227 (1966).
[92] C.H. Schmidt, Angew. Chem. *75*, 169 (1963).

a primary or secondary α-carbon atom frequently merely *nitroso ketones* or mixtures of *diketone monooximes* and *nitroso ketones* are formed. 2-Methyl-3-hexanone and camphor, *e.g.*, give exclusively *diketone monooximes*[93, 94].

With alkyl methyl ketones the attack proceeds preferentially on the secondary carbon atom, as is demonstrated by the preparation of *2,3-butane-dione 3-oxime*[95], while with unsubstituted dialkyl ketones *isomeric* diketone monooximes may be produced in dependence on the chain length. Oximation of coumaranones[96], chromanones[97], steroid ketones[98], and indanones[99] presents no difficulties, *e.g.*:

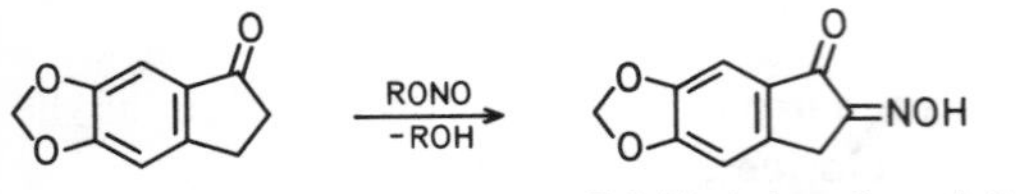

5,6-Methylenedioxy-1,2-indanedione 2-oxime

1,2-Dione 1-oximes (2-oxo aldoximes) are obtained advantageously *via* the β-oxocarboxylic ester (see below). *Cyclic* ketones in particular are able to react in both α-positions. The structure of the nitrosating agent determines the course of the reaction. With secondary and 2-branched alkyl nitrites and alcoholate, cyclohexanone, *e.g.*, gives 1,2-cyclohexanedione 2-oxime; with the other nitrosating agents 2,6-cyclohexanonedione dioximes are obtained throughout[100]. From cyclopentanone and 4-methylcyclohexanone and methyl nitrite and hydrogen chloride in ether *1,2,3-cyclopentanetrione 1,3-dioxime* and *5-methyl-1,2,3-cyclohexanetrione 1,3-dioxime* are obtained[101].

11.1.2.2.2 Replacement of hydrogen by the hydroxyimino group in carboxylic acid derivatives

Normally, esters, nitriles, and free carboxylic acids cannot be converted into oximes without a

further activating group[76]. A link between the vulnerable methylene group and an aromatic substituent is sufficient for activation, as is demonstrated by the oximation of phenylacetonitrile to *hydroxyiminophenylacetonitrile* in 50% yield and of ethyl phenylacetate into *ethyl hydroxyimino-phenylacetate* in over 79% yield[102, 103]. In other cases it is necessary to start with β-oxocarboxylic acid esters, malonic acid, malononitrile, malonamides, malonic acid diesters, cyanoacetates, and from α-oxonitriles and α-oxocarboxylic acid esters.

Sometimes *sequential reactions* take place in dependence on the medium. Thus, methyl 2,4-dinitrophenylacetate reacts with sodium methoxide and isopentyl nitrite to form *methyl 6-nitro-1,2-benzisoxazole-3-carboxylate* while losing nitrous acid intramolecularly[104]:

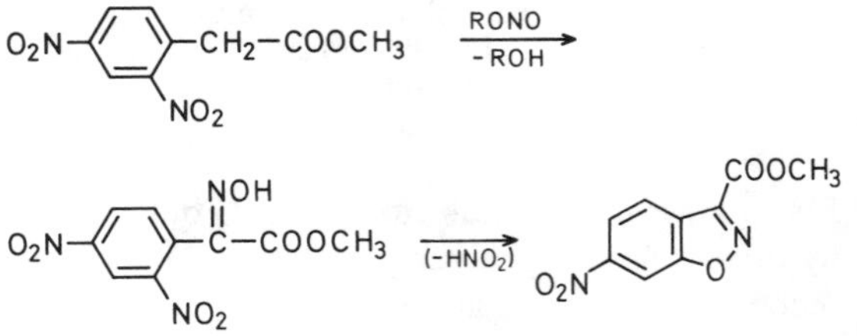

1,3-Dicarboxylic compounds without 2-substituents of above structure and their derivatives in general yield the *1,3-dioxo-2-hydroxyimino* compounds[105]. Glacial acetic acid and concentrated aqueous sodium nitrite solution as well as nitrosyl chloride or alkyl nitrite and sodium ethoxide are employed. However, if the ester function is split hydrolytically during the ester nitrosation, then *1,2-diketone monooximes* form by *ketone cleavage* accompanied by decarboxylation. *1,2-Propanedione 1-oxime* is formed from acetoacetic acid ester in this way.

By contrast, nitrosation of unsubstituted acetoacetic acid esters with nitrosylsulfuric acid in concentrated sulfuric acid leads to a 75% yield of *hydroxyiminoacetic acid esters* with retention of

[93] J.G. Aston, M.G. Mayberry, J. Amer. Chem. Soc. 57, 1888 (1935).

[94] L. Claisen, O. Manasse, Justus Liebigs Ann. Chem. 274, 73 (1893).

[95] W.L. Semon, V.R. Damerell, Org. Syntheses, Coll. Vol. II, 204 (1943).

[96] L.J. Smith, R.R. Holmes, J. Amer. Chem. Soc. 73, 4294 (1951).

[97] P. Pfeiffer, H. Oberlin, E. Konermann, Ber. 58, 1947 (1925).

[98] M.P. Cava, P.M. Weintraub, Steroids 4, 41 (1964); C.A. 61, 10738d (1964).

[99] W.H. Perkin jr., R. Robinson, J. Chem. Soc. 91, 1076 (1907).

[100] A. Treibs, A. Kuhn, Chem. Ber. 90, 1691 (1957).

[101] A.F. Ferris, F.E. Gould, G.S. Johnson, H. Stange, J. Org. Chem. 26, 2602 (1961).

[102] M.R. Zimmermann, J. Prakt. Chem. [2] 66, 359 (1902).

[103] W. Wislicenus, R. Grützner, Ber. 42, 1934 (1909).

[104] W. Borsche, Justus Liebigs Ann. Chem. 390, 1 (1912).

[105] D.S. Jackson, J. Chem. Soc. (B) 785 (1969) (Hydrogen bonds in hydroxyiminomalonic and cyanoacetic esters).

[106] L. Bouveault, A. Wahl, Bull. Soc. Chim. France 31, 675 (1904).

[107] R. Locquin, Bull. Soc. Chim. France 31, 1068 (1904).

the ester function and splitting off of the acetyl group (acid cleavage)[106-108].

α-Monosubstituted β-oxocarboxylic acids, their esters, and malonic acid derivatives suffer cleavage of their molecule during nitrosation. For example, 2-cyclohexanecarboxylic acid under the above conditions gives a 90% yield of *1,2-cyclohexanedione 1-oxime*[109] (see p. 409).

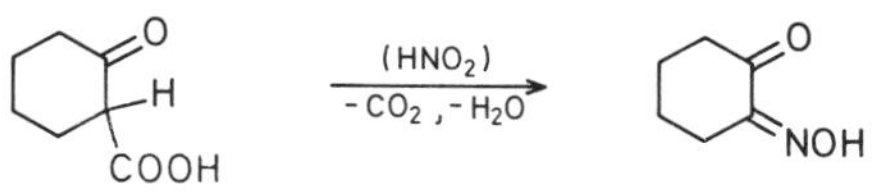

With ethyl nitrite and sodium ethoxide ethyl cyclopentanone-2-carboxylate undergoes 60% ring fission to form *diethyl 2-hydroxyiminoadipate*[110]:

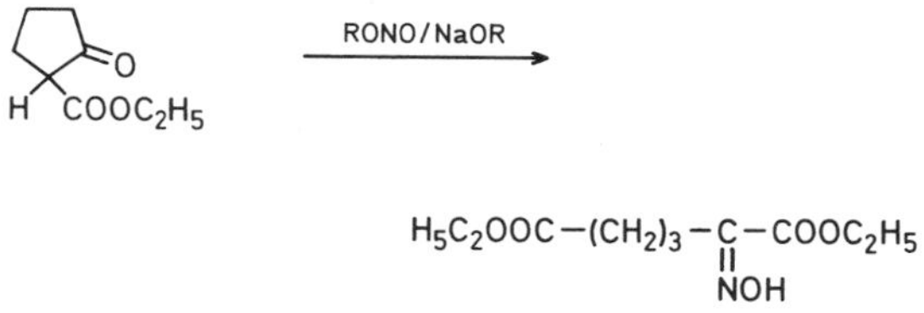

$$H_5C_2OOC-(CH_2)_3-\underset{\underset{NOH}{\|}}{C}-COOC_2H_5$$

Monosubstituted malonic acids and their esters all split off a hydroxycarbonyl or alkoxycarbonyl group on nitrosation with either alkyl nitrite and hydrogen chloride or alkyl nitrite and sodium ethoxide to give good yields of α-hydroxyimino carboxylic acids and their esters[111,112]. It is possible to extend this reaction to α-substituted cyanoacetic acid esters but not to the associated acids[113]. Nitrosation of diethyl acetylmalonate appears to be of interest: with nitrous acid, by splitting off an alkyloxycarbonyl group, *ethyl α-hydroxyiminoacetoacetate* is formed, *i.e.*, the acetyl group remains attached to the molecule[114]:

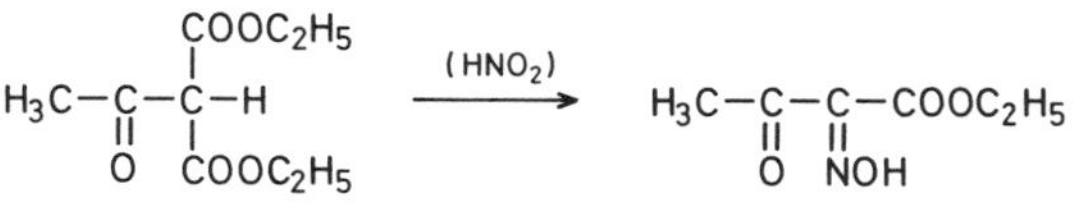

11.1.2.2.3 In aliphatic nitro compounds

Primary nitro compounds yield *nitrolic acids (α-hydroxyiminonitro compounds* 1, secondary ones give *pseudonitroles* 2:

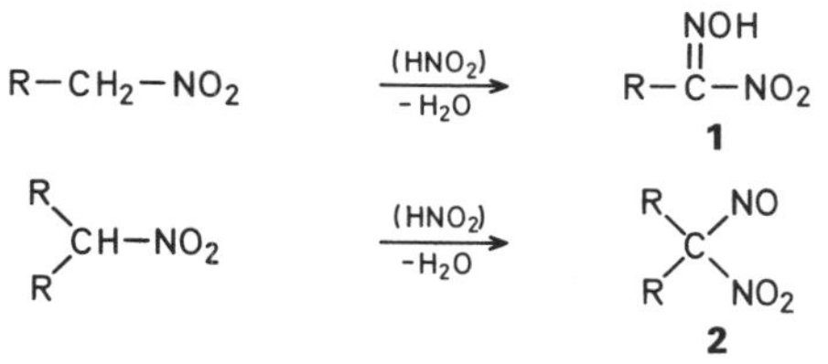

Here, only the nitrolic acids are of interest. They are synthesized by treating an alkaline solution of the nitro compound with alkali metal nitrite and carefully acidifying with dilute sulfuric acid whilst cooling. In individual cases dinitrogen trioxide or dinitrogen tetroxide may be used instead of nitrous acid, and oxalic acid, for example, in place of sulfuric acid[115].

Some pseudonitriles of particular structure can form nitrolic acids 3 by breaking a C–C-bond or give α-hydroxyimino carboxylic acid esters 4 with formal leaving of a nitrate ion[116-117].

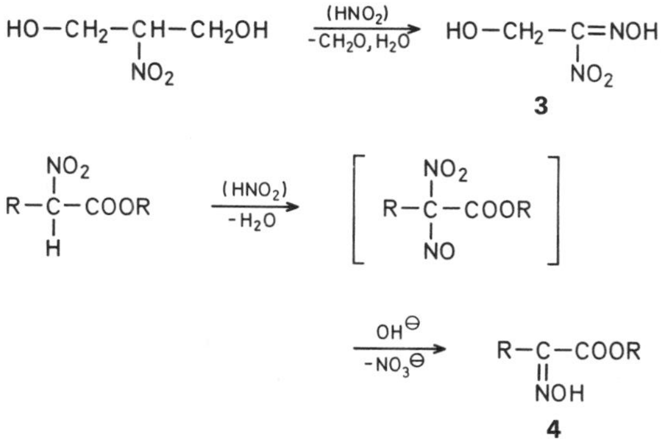

11.1.2.2.4 Replacement of hydrogen by the hydroxyimino group in methylene and methyl groups activated by double-bond systems

Methyl or methylene groups linked to aromatic structures with electron-attracting *ortho* and *para* substituents, and thus able to form a quinonoid

[108] R.H. Barry, W.H. Hartung, J. Org. Chem. 12, 460 (1947) (Use of butyl nitrite and applications of this method for monosubstituted esters).

[109] T.A. Geissman, M.J. Schlatter, J. Org. Chem. 11, 771 (1946).

[110] W. Dieckmann, Ber. 33, 579 (1900).

[111] R.H. Barry, A.M. Mattocks, W.H. Hartung, J. Amer. Chem. Soc. 70, 693 (1948).

[112] J.C. Shivers, C.R. Hauser, J. Amer. Chem. Soc. 69, 1264 (1947).

[113] T.K. Walker, J. Chem. Soc. (London) 125, 1622 (1924).

[114] E. Lang, Ber. 20, 1327 (1880).

[115] L.I. Khmel'nitskii, S.S. Novikov, O.V. Lebedev, Izv. Akad. Nauk SSSR, Otdel. Khim. Nauk 2019 (1960); C.A. 55, 19833b (1961).

[116] J.C. Earl, F.C. Ellsworth, E.C.S. Jones, J. Kemer, J. Chem. Soc. (London) 2702 (1928).

[117] N. Kornblum, J.H. Eicher, J. Amer. Chem. Soc. 78, 1494 (1956).

[118] H. Bredereck, G. Simchen, P. Speh, Justus Liebigs Ann. Chem. 737, 39 (1970).

system in the presence of bases, react with sodium alcoholate and alkyl nitrite to form the oximes[118], *e.g.:*

4-Nitrobenzaldehyde oxime

The aromatic compound must not be carrying halogen or nitro in the *o*-position because otherwise a ring closure side-reaction occurs to form the *1,2-benzisoxazole* derivative[119], *e.g.:*

2-Chloro-4-nitroacetophenone oxime

3-Methyl-6-nitro-1,2-benzisoxazole

On account of the inductive effect, *azoaromatic* compounds also facilitate substitution at alkyl groups in 2- and 4-positions relative to nitrogen, as is illustrated by conversion of alkylpyridines and other alkylheterocycles with sodium amide and butyl nitrite in liquid ammonia[120, 121]. No nitrosation occurs in acid medium. By contrast, 4-methylpyrimidine gives only 40% of the *oxime salt* with potassium *tert*-butoxide and isopentyl

nitrite but an 80% yield of 4-pyrimidinecarboxaldehyde oxime hydrochloride with hydrochloric acid and isopentyl nitrite. While in alkaline medium the reaction ensues *via* resonance-stabilized carbanions, the initial step under acid conditions is protonation of one of the two nitrogens. Detaching of a proton of the methyl group leads to a quasi-quinonoid system of low energy content on which the electrophilic attack of the nitrosyl cation can ensue. 4, 6-dimethylpyrimidine correspondingly furnishes the *4,6-pyrimidinedicarboxaldehyde dioxime*.

Nitrosation of 2-methylpyrimidine with potassium *tert*-butoxide and pentyl nitrite gives 20% 4-pyrimidinecarboxaldehyde oxime, in the presence of 2 molecules acid no nitrosation takes place because a quasi-quinonoid structure cannot form[118]

11.1.3 Oximes by addition reactions

A predominant place among additive oxime formation reactions is held by the action of nitrosyl chloride on C=C double bonds. Dimeric 1-chloro-1-nitroso compounds are formed or, as a result of the latter's dissociation and isomerization, *2-chlorooximes*, which are accessible to further reactions (p. 413)[122].

[119] *A. Kövendi, M. Kircz*, Chem. Ber. *97*, 1902 (1964).
[120] *S.E. Forman*, J. Org. Chem. *29*, 3323 (1964);
US. P. 3 150 135 (1962), F.M.C. Corp., Inv.: *S.E. Forman; C.A. 62*, 2765h (1965).

[121] *T. Kato, H. Yamanaka, H. Hiranuma*, Yakagaku Zasshi *90*, 877 (1970); C.A. *73*, 77180y (1970).
[122] *W. Pritzkow, H. Schaefer, P. Pabst, A. Ebenroth, J. Beger*, J. Prakt. Chem. *301*, 123 (1965).

Nitrosyl chloride too adds on to $C=N$ double bonds, but this reaction requires oximes as starting material and will not be discussed further here.

11.1.3.1 Oximes by addition reactions at $C=C$ double bonds

11.1.3.1.1 Oximes by addition reactions at $C=C$ double bonds with nitrosyl halides and nitrosylsulfuric acid

During addition of nitrosyl compounds such as nitrosyl halides or nitrosylsulfuric acid to olefins the *ionic* reaction proceeds by addition of the nitrosyl cation on the electron-rich end in accord with Markovnikov's rule and the remainder on the less electronegative end:

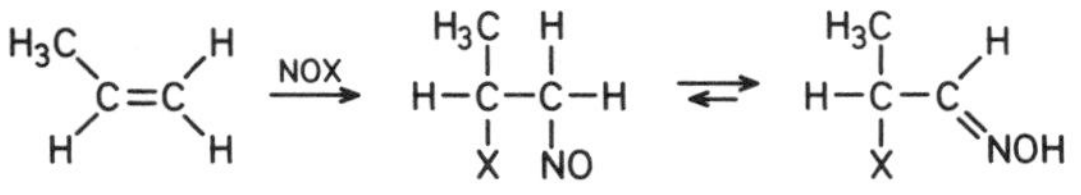

X = Hal, OSO₃H

Thus, butenone and acrylonitrile yield *1-chloro-2,3-butanedione 2-oxime* and *3-chloro-2-hydroxyiminopropionitrile*[123–124]. The initially formed dimeric primary and secondary nitroso compounds isomerize to the oxime, but with excess nitrosyl chloride *1,2-dihalo-1-nitroso* compounds or *hydroxamoyl chlorides* can form[125]. Finally, oxidation to *2-halo-1-nitro* compounds is a possibility[126, 127].

To *prevent* side-reactions, excess hydrochloric acid is employed. As a result, *oxime hydrochloride* is removed from the medium, especially in the case of secondary nitrosoalkanes and bis(chloronitroso) compounds, and oxime formation is almost quantitative[128, 130]. A slow rate of addition of hydrogen chloride to the $C=C$ double bond is a

requisite. The oxime hydrochlorides, especially those of the *2-halo oximes*, can be converted into the free oximes under mild conditions[131].

2-Chlorooximes react with a very wide range of nucleophilic reagents to form 2-substituted oximes[132, 136]. Often, α-cyano and thiocyanatooximes cannot be prepared in this way because ring closure to *5-aminoisoxazoles*[137] and *2-aminothiazole N-oxides*[138] takes place:

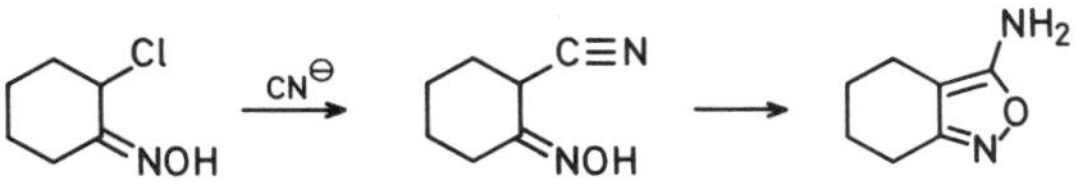

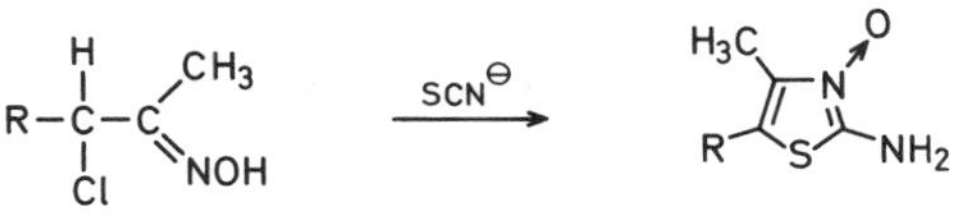

Preparation of *2-methylene oximes* with dimethylsulfonium methylide in dimethylsulfoxide/tetrahydrofuran rates a mention[139].

11.1.3.1.2 Oximes by addition reactions at $C=C$ double bonds with oxides of nitrogen

Addition of dinitrogen trioxide in a free-radical reaction leads to 2-nitroso-1-nitro compounds

[123] *K.A. Ogloblin, A.A. Potekhin*, Zh. Obshch. Khim. *34*, 2688 (1964); C.A. *61*, 14519f (1964).

[124] *K.A. Ogloblin, W.P. Ssemenow*, Zh. Org. Khim. *1*, 1361 (1965); Chem. Zentr. 42–0860 (1966).

[125] *K.A. Ogloblin, V.N. Kalikhevich, A.A. Potekhin, W.P. Ssemenow*, Zh. Obshch. Khim. *34*, 1227 (1964); C.A. *61*, 2967b (1964).

[126] *K.A. Ogloblin, V.N. Kalikhevich, A.A. Potekhin, W.P. Ssemenow*, Zh. Obshch. Khim. *34*, 170 (1964); C.A. *61*, 10525b (1964).

[127] *A. Dornow, H.D. Jordan, A. Müller*, Chem. Ber. *94*, 67 (1961) (Preparation of 2-chlorooximes from nitroalkenes).

[128] *G. Collin, R. Höhn, H.G. Hauthal, H. Hübner, W. Pritzkow, W. Rolle, H. Schaefer, M. Wahren*, Justus Liebigs Ann. Chem. *702*, 55 (1967).

[129] *E. Müller, D. Fries, H. Metzger*, Chem. Ber. *88*, 1898 (1955).

[130] DBP. 1 082 253 (1960), BASF, Erf.: *O.v. Schickh, H. Metzger*; C.A. *55*, 17547b (1961).

[131] Ital. P. 712 470 (1964), Soc. Edison, Inv.: *C. Brichta, G. Ribaldone, G. Borsotti, A. Nenz*; C.A. *68*, 68533k (1968).

[132] *M. Ohno, N. Naruse, S. Torimitsu, M. Okamoto*, Bull. Chem. Soc. Japan *39*, 1119 (1966); C.A. *65*, 13563h (1966).

[133] Jap. P. 6814, 932 (1965), Toyo Rayon, Inv.: *M. Ohno*; C.A. *70*, 57293x (1969);
Jap. P. 6907, 696 (1964), Toyo Rayon, Inv.: *M. Ohno, S. Torimitsu*; C.A. *71*, 38439d (1969);
Jap. P. 7019, 902 (1967), Toray Ind., Inv.: *M. Ohno, N. Naruse, I. Terasawa, T. Matsuzaki*; C.A. *73*, 98469q (1970);
Jap. P. 7009, 533 (1967), Toyo Rayon, Inv.: *M. Ohno, I. Terasawa*; C.A. *73*, 14294f (1970).

[134] *A. Dornow, H.D. Jordan, A. Müller*, Chem. Ber. *94*, 73 (1961);
W. Pritzkow, H. Schaefer, P. Pabst, A. Ebenroth, J. Beger, J. Prakt. Chem. [4] *29*, 123 (1965).

[135] *M. Ohno, S. Torimitsu, N. Naruse, M. Okamoto, I. Sakai*, Bull. Chem. Soc. Japan *39*, 1129 (1966); C.A. *65*, 13564d (1966).

[136] Jap. P. 10, 500 (1965), Toyo Rayon, Inv.: *M. Ohno, N. Naruse*; C.A. *67*, 90462n (1967).

[137] *M. Ohno, N. Naruse*, Bull. Chem. Soc. Japan *39*, 1125 (1966); C.A. *65*, 13678g (1966);
M. Ohno, N. Naruse, Tetrahedron Lett. 2151 (1964).

[138] *A. Dornow, H.-H. Marquardt, H. Pauksch*, Chem. Ber. *97*, 2165 (1964);
W. Walter, E. Schaumann, Synthesis 117 (1971).

[139] *P. Bravo, G. Gaudiano, C. Ticozzi, A. Umani-Ronchi*, Chem. Commun. 1311 (1968).

which can isomerize to *2-nitrooximes* in the usual way:

$$2 \; \text{C}=\text{C} \overset{R}{\underset{H}{}} \xrightarrow{N_2O_3} \left[\underset{O_2N}{-}\text{C}-\underset{NO}{}\text{C}-H \overset{R}{} \right]_2 \xrightarrow{cal} -\underset{NO_2}{}\text{C}-\text{C} \overset{R}{\underset{NOH}{}}$$

With palladium-animal charcoal these can be reduced to *2-hydroxyamino oximes*[140].

Addition of dinitrogen tetroxide furnishes either 1,2-dinitro compounds or 2-nitroalkyl nitrites and, by oxidation, 2-nitroalkyl nitrates[141].

11.1.3.2 Oximes by addition reactions with nitrile oxides

Grignard compounds and nitrile oxides react together to *ketoximes* by 1,3-addition. Thus *acetophenone oxime* is obtained from benzonitrile oxide and methyl magnesium iodide side by side with acetophenone[142]:

$$\text{C}\equiv\text{N}\rightarrow\text{O} + CH_3MgJ \longrightarrow$$

$$\underset{CH_3}{}\text{C}=\text{NOMgJ} \xrightarrow{H_2O} \underset{CH_3}{}\text{C}=\text{NOH}$$

11.1.4 Oximes from aliphatic nitroso compounds

11.1.4.1 Oximes from aliphatic nitroso compounds by isomerization of primary and secondary nitroso compounds

Aliphatic nitroso compounds with the nitroso group on the primary or secondary carbon atom are usually present as *dimers*. Using thermal energy and surface catalysts and with acid and base catalysis they depolymerize reversibly to monomer which, by virtue of conversion into oxime, is removed from the equilibrium mixture[143].

11.1.4.1.1 Oximes from thermal aliphatic nitroso compounds by isomerization

Unless the dimer-monomer equilibrium is far in favor of the dimer, spontaneous isomerization to

oxime is often observed both in bulk and in solution[144]. This is true especially for still impure compounds. *Addition* of alumina, silica gel, or activated charcoal causes certain dinitroso compounds to rearrange into oximes at room temperature[145], with others dissociation and isomerization demand drastic conditions. Thus bis(nitrosocyclohexane) requires 10 minutes heating to 120° to give an 86% yield of *cyclohexanone oxime*[146], while bis(2-nitro-3-nitrosobutane) yields in dimethylformamide 98% of slowly decomposing *3-nitro-2-butanone oxime* under the same conditions[147].

11.1.4.1.2 Oximes from aliphatic nitroso compounds by isomerization with acids

Conversion of *secondary* nitroso compounds into oximes *via* proton catalysis has already been described on p. 409. It can be accelerated further by *UV* light and other energy-rich rays[148, 149]. The dissociation of the secondary dinitroso compounds appears to be the rate-determining step of the isomerization. *Primary* dinitroso compounds should react analogously but, in fact, behave entirely differently. With hydrogen chloride they rearrange with disproportionation to *aldehyde, ester,* and *hydrazine dihydrochloride*[150]:

$$R-CH_2-\underset{\underset{O}{\downarrow}}{\overset{O}{N}}=N-CH_2-R \xrightarrow{(HCl)} R-CH=N-NH-\underset{\underset{O}{\parallel}}{C}-R$$

$$\xrightarrow{CH_3OH/HCl} R-CH(OCH_3)_2 + R-COOCH_3 + N_2H_4 \cdot 2HCl$$

The rearrangement proceeds *via* acylhydrazones, which can be isolated as the hydrochlorides[128].

11.1.4.1.3 Oximes from aliphatic nitroso compounds by isomerization with bases

The action of bases on primary and secondary dinitroso compounds leads to the formation of the mesomeric oxime anion *via* deprotonation of the nitroso group carbon atom. For steric reasons,

[140] M.L. Scheinbaum, J. Org. Chem. *35*, 2785, 2790 (1970).
[141] H. Shechter, J.J. Gardikes, T.S. Cantrell, G.V.D. Tiers, J. Amer. Chem. Soc. *89*, 3005 (1967).
[142] H. Wieland, Ber. *40*, 1669 (1907).
[143] B.G. Gowenlock, W. Lüttke, Quart. Rev. Chem. Soc. 12, 321 (1958) (Structure and Properties of C-Nitroso-Compounds).
[144] B.G. Gowenlock, J. Trotman, J. Chem. Soc. (London) 4190 (1955).
[145] E. Müller, D. Fries, H. Metzger, Chem. Ber. *88*, 1891 (1955).
[146] E. Müller, H. Metzger, Chem. Ber. *88*, 176 (1955).
[147] DBP. 1238461 (1963), Esso, Erf.: W.W. Koser; C.A. *67*, 73238b (1967).
[148] E. Müller, U. Heuschkel, Chem. Ber. *92*, 68 (1959).
[149] DBP. 1092911 (1960), Erf.: E. Müller, G. Schmidt; C.A. *55*, 25806b (1961).
[150] R. Höhn, H. Schaefer, H. Hübner, M. Wahren, W. Pritzkow, G. Lauterbach, P. Fulde, P. Herrmann, Tetrahedron Lett. 2581 (1965).

Table 2 Conversion of tertiary nitroso compounds into oximes

Nitroso compound	Cleaving agent	Products	Yield [% of theory]	Ref.
(cyclohexanone with CH₃, O, ON–CH(CH₃)₂ substituents)	Hydrochloric acid	*3,7-Dimethyl-6-hydroxyimino-octanoic acid*	60	[159]
(cyclohexanone with ON, CH₃, O, CH(CH₃)₂ substituents)	Hydrogen chloride	*3-Isopropyl-6-hydroxyimino-heptanoic acid*	little	[160]
$H_3C-\overset{O}{\overset{\|}{C}}-\overset{NO}{\underset{CH_3}{\overset{\|}{C}}}-CH_3$	Hydrochloric acid	*Acetone oxime* + acetic acid*		[161]
(cyclopentanone ring with H₃C, COOC₂H₅, NO, O)	Sodium ethoxide or HCl-ethanol	*Diethyl 2-hydroxyimino-4-methylhexanoate*	50	[162]
(cyclohexanone ring with O, CH₃, NO, O)	Potassium hydroxide 1,4-dioxane-water	*6-Hydroxyimino-5-oxo-heptanoic acid*	23	[163]
(cyclohexane ring with NO, COOH)	Sulfuric acid	*Cyclohexanone oxime* + carbon dioxide*		[164]
$H_3C-\overset{NO}{\underset{NO_2}{\overset{\|}{C}}}-CH_3$	Sodium hydroxide/Methanol	*Acetone oxime + sodium nitrate*	little	[165]
$Cl-CH_2-\overset{NO}{\underset{CH_3}{\overset{\|}{C}}}-CHO$	Methanol	*1-Chloro-2-propanone oxime + methyl formate*		[166]
$H_3C-\overset{Cl}{\overset{\|}{C}}H-\overset{NO}{\underset{CH_3}{\overset{\|}{C}}}-CHO$	Methanol	*3-Chloro-2-butanone oxime + methyl formate*		[166]
$H_3C-\overset{O}{\overset{\|}{C}}-\overset{NO}{\underset{CH_3}{\overset{\|}{C}}}-CH_2-Cl$	Methanol	*1-Chloro-2-propanone oxime + methyl acetate*		[166]
(cyclohexanone ring with Cl, NO, CH₃, O)	Methanol	*Methyl 5-chloro-6-hydroxy-iminoheptanoate*		[166]
$H_3C-CH_2-\overset{Cl}{\underset{NO}{\overset{\|}{C}}}-CH_3$	hν (650 nm)-methanol	*Butanone oxime + 2,3-Butanedione 2-oxime, etc.*	18	[167]
$H_3C-\overset{Cl}{\underset{NO}{\overset{\|}{C}}}-CH_2-CH_2-COOH$	hν (610 nm)-methanol	*Methyl 4-hydroxyiminovalerate*		[168]

* Intermediate products

[151] *R. Behrend, W. Platner,* Justus Liebigs Ann. Chem. *278*, 369 (1894).

[152] *C. Kjellin, K.G. Kuylenstjerna,* Ber. *30*, 1899 (1897).

[153] *N. Thorne,* J. Chem. Soc. (London) 2587 (1956).

[154] US. P. 2 852 561 (1958), Hercules Powder Co., Inv.: *J.H.F. Pieper, J.E.V.N. Stauch;* C.A. *53*, 6115h (1959).

[155] *J. Schmidt, H. Dieterle,* Justus Liebigs Ann. Chem. *377*, 41 (1910).

the rate of dissociation is greater in secondary dinitroso compounds than in primary ones[128]. Often, alkoxides are used for the isomerization[151-153], aqueous alkali hydroxide is less advantageous[152, 154], alkali hydrogen carbonate may be employed[155], and ammonia and amines, too, can be utilized[156-158].

Transferring this reaction to bis(2-chloro-1-nitroso) compounds leads to an elimination-addition reaction proceeding in parallel with the dissociation, as kinetic measurements have shown to be likely (see p. 412; ref.[122]).

11.1.4.2 Oximes from aliphatic nitroso compounds by isomerization of tertiary nitroso compounds

Tertiary nitroso compounds carrying a second electronegative group on the carbon atom can split off this group under certain conditions and isomerize to oximes. Ring systems are cleaved (Table 2).

$$\left[\begin{array}{c} R^1 \\ R^2 \end{array} \begin{array}{c} X \\ C \\ N=\underline{O}| \end{array} \right]_2 \longrightarrow 2 \begin{array}{c} R^1 \\ R^2 \end{array} C=N-\underline{O}|^{\ominus} + \left[X^{\oplus} \right]$$

X = Acyl, NO₂, poss. Hal.

11.1.5 Oximes by oxidation

11.1.5.1 Oximes by oxidation of amines

Primary amines with amino groups linked to primary or secondary carbon atoms furnish aldoximes and ketoximes on oxidation *via* the hydrox-

ylamine stage. Where the amino group is on the tertiary carbon atom nitroso compounds are formed analogously.

A high yield of the desired oximes is obtained if the oxidation is performed with 30% excess *hydrogen peroxide* at room temperature in methanol-ethanol solution — primary alcohols prevent the formation of stoichiometric adducts of oxime and amine and hence promote complete oxidation[169-171]. Below 0° hydroxylamines can be isolated[172, 173].

An even better oxidation is obtained if sodium or amine salts of *molybdic* or *tungstic* acid are added[169, 174-177]. The reaction can be extended to amines with additional nonsensitive functional groups[178]. For the synthesis of *cyclohexanone oxime,* which is the most important one technically, *Trilon B®* is, in addition, used as hydrogen peroxide stabilizer[169-175].

Neutral, aqueous *peroxymonosulfate* serves as a further oxidizing agent, but may cause oxidation to the nitro compound stage[179-181]. Preparation of oximes therefore requires the maintenance of a definite temperature and *pH*. For example, at 30—35° in neutral or weakly alkaline solution up to 85% *cyclohexanone oxime* is obtained from cyclohexylamine and mixtures of hydrogen peroxide and sulfuric acid, while, by contrast, in weakly acid medium bis(nitrosocyclohexane)[182] is formed, and the same compound is obtained when

[156] H. Wieland, S. Bloch, Ber. 37, 1531 (1904).
[157] A. DiGiacomo, J. Org. Chem. 30, 2614 (1965).
[158] US. P. 2 805 253 (1957), Hercules Powder Co., Inv.: J.H.F. Pieper, J.E.V.N. Stauch; C.A. 52, 6401b (1958).
[159] A. v. Bayer, O. Manasse, Ber. 27, 1912 (1894).
[160] A. v. Bayer, E. Oehler, Ber. 29, 33 (1896).
[161] J.G. Aston, D.F. Menard, M.G. Mayberry, J. Amer. Chem. Soc. 54, 1530 (1932);
V. Meyer, Ber. 21, 1293 (1888).
[162] W. Dieckmann, A. Groeneveld, Ber. 33, 605 (1900).
[163] S.I. Zav'yalov, V.M. Medvedeva, Izv. Akad. Nauk SSSR 2165 (1959); C.A. 54, 10929c (1960).
[164] H. Metzger, L. Beer, Z. Naturforsch. 18b, 986 (1963).
[165] A. Schöfer, Ber. 34, 1911 (1901).
[166] K.A. Ogloblin, A.A. Potekhin, Dokl. Akad. Nauk SSSR 159, 853 (1964); C.A. 62, 8996g (1965).
[167] S. Mitchell, J. Cameron, J. Chem. Soc. (London) 1967 (1938).
[168] S. Mitchell, K. Schwarzwald, G.K. Simpson, J. Chem. Soc. (London) 604 (1941);
E. Müller, H. Metzger, D. Fries, Chem. Ber. 87, 1456 (1954).

[169] K. Kahr, Angew. Chem. 72, 135 (1960).
[170] DBP. 951 211 (1956), BASF, Erf.: W. Ruppert; C.A. 53, 260f (1959).
[171] DBP. 951 720 (1956), Synthese-Chemie, Erf.: J.H.F. Pieper; C.A. 53, 260i (1959).
[172] US. P. 2 795 611 (1957), Hercules Powder Co., Inv.: H. List; C.A. 52, 429a (1958).
[173] DBP. 951 933 (1956), Synthese-Chemie; C.A. 53, 3060b (1959).
[174] Schweiz. P. 288 168 (1953), Inventa AG; C.A. 49, 2484d (1955).
[175] US. P. 2 718 528 (1954), Hercules Powder Co., Inv.: J.H.F. Pieper; C.A. 50, 4207e (1956).
[176] Holl. P. 6 503 548 (1964), Halcon Intern. (Reaction with, for example, cumene peroxide); C.A. 64, 6527d (1966).
[177] P. Burckard, J.P. Fleury, F. Weiss, Bull. Soc. Chim. France 2730 (1965); C.A. 64, 1988e (1966).
[178] K. Kahr, C. Berther, Chem. Ber. 93, 132 (1960).
[179] E. Bamberger, T. Scheutz, Ber. 34, 2262 (1901).
[180] E. Bamberger, Ber. 35, 4293 (1902).
[181] E. Bamberger, R. Seligmann, Ber. 35, 4299 (1902).
[182] I. Okamura, R. Sakurai, Chem. High Polymers Japan 8, 296 (1951); C.A. 47, 2992d (1953);
I. Okamura, R. Sakurai, Chem. High Polymers Japan 9, 10 (1952); C.A. 48, 9933c (1954);
I. Okamura, R. Sakurai, Chem. High Polymers Japan 9, 230 (1952); C.A. 48, 11794d (1954).

organic peracids are employed [183, 184]. *Secondary amines* also can be converted into oximes with three times the molar quantity of peracid or hydrogen peroxide in the presence of molybdenum and tungsten compounds [185]:

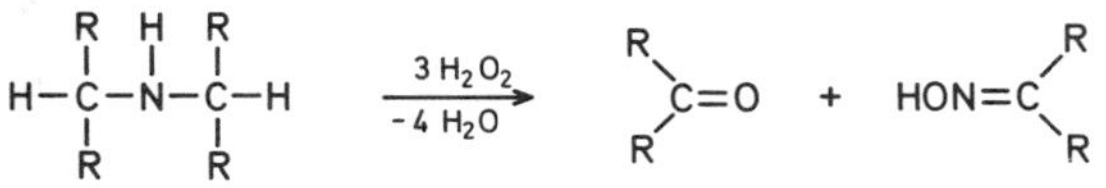

$$\underset{\substack{R\;H\;R\\|\;\;|\;\;|}}{H-C-N-C-H} \;\;\xrightarrow[-4\,H_2O]{3\,H_2O_2}\;\; \underset{R}{\overset{R}{>}}C=O \;+\; HON=C\underset{R}{\overset{R}{<}}$$

Working at room temperature in addition to carbonyl gives the dimeric nitroso compounds, which can be rearranged to oximes.

11.1.5.2 Oximes by oxidation of hydroxylamines

Hydroxylamines are formed intermediately during the oxidation of amines to oximes. If they are themselves used in oxidation reactions, then either tertiary *nitroso compounds* or *oximes* and *dinitroso* compounds together are obtained.
Numerous patent specifications describe the oxidation of *N*-cyclohexylhydroxylamine to cyclohexanone oxime.

Dehydrogenation is achieved, for example, with oxygen or air in the presence of cobalt-lead-manganese naphthenate or copper-cobalt oleate. The reaction is performed under pressure at at most 60° in cyclohexane, benzene, or methanol as solvent. *Cyclopentanone oxime* can be obtained similarly [186].

Other processes for obtaining cyclohexanone oxime by oxidation proceed *via* the reaction of cyclohexylhydroxylamine with ethyl nitrate and sodium ethoxide [187] and *via* a spontaneous disproportionation to cyclohexylamine and cyclohexanone oxime at 170° [188].

Other noteworthy conversions include autoxidation of (α-hydroxyaminobenzyl)acetophenone oxime to 1,3-diphenyl-1,3-propanedione dioxime in the presence of some ammonia [189] and the dehydrogenation of α-hy-

droxyaminonitriles, which can be prepared readily from hydrogen cyanide and aldoximes, to α-hydroxyiminonitriles with *p*-benzoquinone [190] (see p. 410).

11.1.5.3 Oximes by oxidation of oxaziridines

Oxaziridines of suitable structure react with *peracids* to form *ketones* and *bis-nitroso* compounds or the isomeric oximes in the following way:

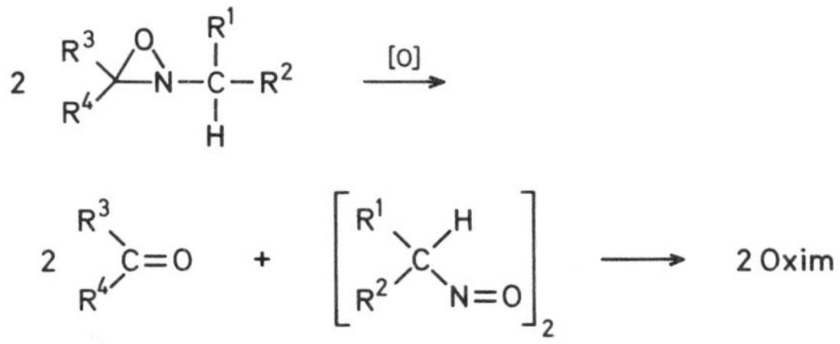

$$2\;\;\underset{R^4}{\overset{R^3}{>}}\underset{\substack{\backslash\\N-\underset{H}{\overset{O}{C}}-R^2\\|\;\;R^1}}{}\;\;\xrightarrow{[O]}$$

$$2\;\;\underset{R^4}{\overset{R^3}{>}}C=O \;+\; \left[\underset{R^2}{\overset{R^1}{>}}C\underset{N=O}{\overset{H}{<}}\right]_2 \;\longrightarrow\; 2\;Oxim$$

Thus, 2-benzyl-3,3-diethyloxaziridine ($R^1=C_6H_5$, $R^2=H$, $R^3=R^4=C_2H_5$) with peracetic acid gives a substance in 60% yield which, according to its *IR* spectrum, is mainly *benzaldoxime* side by side with 37% crude α-nitrosotoluene dimer [184], and from 2-cyclohexyl-3-methyloxaziridine ($R^1=R^2=-(CH_2)_5-$, $R^3=CH_3$, $R^4=H$) *nitrosocyclohexane dimer* [191]. On the other hand, when *2-tert*-butyl-3-phenyloxaziridine is heated to 200°, then unlike in the above reaction mainly *benzaldoxime* besides *benzaldehyde tert-butylnitrone* are formed with elimination of isobutene [192]. For preparing specific oximes this technique is likely to be suitable in *exceptional* cases only.

11.1.6 Oximes by reduction

11.1.6.1 Oximes by reduction of primary and secondary nitro compounds

Nitro compounds are readily reduced to amines and, under certain conditions, to hydroxylamines. In addition, there are procedures which lead directly to oximes *via* the rearrangement of the intermediately formed nitroso compound. The starting compound may be a salt of the nitro compound, *i.e.*, the *aci-nitro* compound, and either inorganic or organic reducing agents are employed.

11.1.6.1.1 Oximes by reduction of primary and secondary nitro compounds with metal salts

A practical laboratory method for preparing oximes from nitro compounds at room temperature uses *tin(II) chloride* in excess concentrated hydrochloric acid. The aqueous solution of the alkali metal salt of the nitro compound is added

[183] DBP. 953 069 (1956), Bayer, Erf.: *H. Krimm, K. Hamann*; C.A. *53*, 5159c (1959).
[184] *W.D. Emmons*, J. Amer. Chem. Soc. *79*, 6523 (1957).
[185] DBP. 948 417 (1959), Bayer, Erf.: *H. Krimm*; C.A. *52*, 18315e (1958).
[186] Schweiz. P. 324 434 (1957), Inventa AG, Erf.: *K. Kahr, A. Flam*; C.A. *52*, 17138h (1958).
[187] *P.A.S. Smith, G.E. Hein*, J. Amer. Chem. Soc. *82*, 5734 (1960).
[188] DBP. 939 810 (1956), Bayer, Erf.: *G. v. Schuckmann*; C.A. *52*, 14673a (1958).
[189] *K. v. Auwers, H. Müller*, J. Prakt. Chem. [2] *137*, 77 (1933).
[190] *L.W. Kissinger, H.E. Ungnade*, J. Org. Chem. *25*, 1471 (1960).
[191] DBP. 1 073 468 (1960), Bayer, Erf.: *H. Krimm, H. Schnell*; C.A. *55*, 17544f (1961).
[192] *W.D. Emmons*, J. Amer. Chem. Soc. *79*, 5753 (1957).

drop by drop to a solution of 1.3–3 equivalents tin(II) chloride[193, 194]. To save using tin(II) chloride, the reduction may be performed in the presence of zinc, aluminum, or magnesium[195]. *Iron-(II) chloride* and *sulfate* also reduce nitro compounds to oximes[196]. A concentrated aqueous solution of the alkali metal salt of the nitro compounds is added continuously to concentrated mineral acid together with the solution of the iron-(II) salt; excess iron salt must always be present.

In alkaline medium nitro compounds can be reduced with *arsenic(III), antimony(III),* and *tin(II)* oxide. A 2:1 to 1:1 ratio of nitroalkane to reducing oxide and ~110° are employed[197].

11.1.6.1.2 Oximes by reduction of primary and secondary nitro compounds with other inorganic compounds

Sodium salts of primary and secondary nitro compounds are reduced to oximes by *hydrogen sulfide* at *pH* 3[198]. Nitroethane gives 86% *acetaldoxime*, nitrocyclohexane 78% *cyclohexanone oxime* side by side with cyclohexanone, cyclohexanol, and cyclohexylamine. A better yield is obtained when 5–15 molar % of the nitro compound of hydroxylamine hydrochloride is added, because the carbonyl compounds formed also are then converted into oximes. For example, 92% *acetaldoxime* is obtained from nitroethane at *pH* 4, but only 80% when hydroxylamine hydrochloride is omitted[199].

The reduction of nitrocyclohexane with *alkali metal* or *ammonium sulfide* or *polysulfide* takes place in water or alcohols, *e.g.,* with sodium hydrogen sulfide in butanol or with ammonium hydrogen sulfide in ethanol[200]; alternatively hydrogen sulfide is passed into at least one equivalent of an amine such as piperidine, 2,2′,2″-nitrilotriethanol (triethanolamine), or cyclohexylamine[201].

The reaction is not applicable to aromatic nitro compounds, which are reduced to amines.

Heating of a mixture of carbon disulfide, aqueous ammonia, and a primary or secondary nitro compound to 70–100° under pressure also leads to oximes, probably *via* intermediately formed ammonium hydrogen sulfide[202, 203].

Similarly a number of additional inorganic sulfur compounds can be used. Sulfur dioxide[204, 205], sodium thiosulfate[206–209], and sodium dithionite and its formaldehyde adduct[210] are the most important ones. Thus *acetone oxime* is obtained in respectively 90 and 52% yield[1] from the sodium salt of 2-nitropropane and sodium thiosulfate or sodium dithionite by introducing the mixture into 2*N* sulfuric acid.

Primary nitro compounds furnish reduction products of not yet certain constitution[211]. By contrast, ω-nitroacetophenone reacts in the expected way with sodium dithionite to form *2-phenylquinoxaline*[212] in the presence of *o*-phenylenediamine *via* the oxime stage:

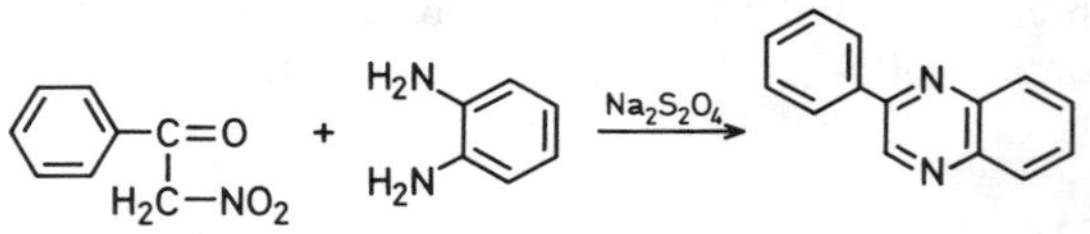

Other reducing agents used advantageously for synthesizing *cyclohexanone oxime* are hydrogen iodide and hydroxylamine[213]. However, here the *aci*-nitro compound is assumed to have been pre-

[193] *J. v. Braun, O. Kruber,* Ber. *45*, 396 (1912).
[194] *J. v. Braun, W. Sobecki,* Ber. *44*, 2533 (1911).
[195] Brit. P. 710 142 (1954), Bayer; C.A. *49*, 11686e (1955).
[196] US. P. 2 861 142 (1957), DuPont, Inv.: *L.G. Donaruma;* C.A. *52*, 6400i (1958).
[197] US. P. 2 820 826 (1958), Ind. Rayon Co., Inv.: *S.C. Temin, M. Levine;* C.A. *52*, 9202f (1958).
[198] DBP. 825 544 (1951), Bayer, Erf.: *H. Welz;* C.A. *50*, 398g (1956).
[199] DBP. 855 254 (1952), Bayer, Erf.: *H. Welz;* C.A. *52*, 15576b (1958).
[200] Holl. P. 69 568 (1952), Dir. van de Staatsmijnen; C.A. *47*, 1734c (1953).
[201] Holl. P. 72 867 (1953), Dir. van de Staatsmijnen; C.A. *48*, 3384c (1954).

[202] US. P. 2 763 686 (1956), DuPont, Inv.: *L.G. Donaruma;* C.A. *51*, 3660b (1957).
[203] Brit. P. 1 166 577 (1967), Olin Mathieson Co. (Reduction of nitrocyclohexane to the oxime and ketone with carbon disulfide and potassium hydroxide or sodium methoxide); C.A. *72*, 12213w (1970).
[204] DBP. 1 140 567 (1961), Bayer, Erf.: *W. Mueller, H. Schnell;* C.A. *58*, 11240h (1963).
[205] US. P. 3 136 756 (1961), Foster, Grant Co., Inv.: *H. Hopff, M. Kawara;* Chem. Zentr. 21–2650 (1966).
[206] Brit. P. 796 726 (1958), ICI, Inv.: *K.B. Wilson;* C.A. *53*, 261b (1959).
[207] Brit. P. 738 888 (1955), BASF; C.A. *51*, 2857d (1957).
[208] *A.A. Artem'ev, E.V. Genkina, A.B. Malimonova, V.P. Trofil'kina, M.A. Isaenkova,* Zh. Vses. Khim. Obshchest. *10*, 588 (1965); C.A. *64*, 1975h (1966).
[209] Brit. P. 784 608 (1953), Hoechst, Inv.: *M. Rieber;* Chem. Zentr. 16481 (1959).
[210] US. P. 2 800 508 (1957), Hoechst, Inv.: *M. Rieber, L. Orthner;* C.A. *51*, 17987d (1957).
[211] DBP. 936 629 (1955), BASF, Erf.: *O. v. Schickh, H.J. Riedl;* C.A. *52*, 19951f (1958).
[212] *A. Dornow, W. Sassenberg,* Justus Liebigs Ann. Chem. *594*, 185 (1955).
[213] *B.L. Moldavskii, I.I. Ivanova,* Zh. Analit. Khim. *12*, 274 (1957); C.A. *52*, 173b (1958).

viously *disproportionated* into ketone and nitrous oxide and to form the oxime subsequently, *i.e.*, no true reduction takes place[1]:

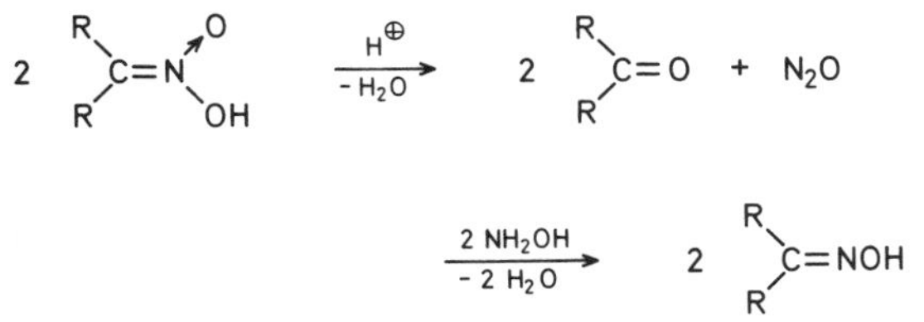

The *pH* is significant during the action of *nascent hydrogen* on primary and secondary nitrogen compounds. In a strongly acid medium reduction to amine takes place, while under weakly acid, neutral, and weakly alkaline conditions oximes and *N*-substituted hydroxylamines are sometimes formed. *Carbon monoxide,* too, is used for reducing secondary nitro compounds in 96% sulfuric acid at 20–60°. As an example, the yield of *cyclohexanone oxime* is 34%[214]. Table 3 illustrates the further possible reductions.

11.1.6.1.3 Oximes by reduction of primary and secondary nitro compounds by catalytic hydrogenation

Catalytic hydrogenation of nitro compounds to oximes requires the aid of special catalyst systems. Thus, *cyclohexanone oxime* is obtained in up to 70% yield by employing a silver oxide-zinc oxide-chromium oxide contact at 100–105° and 150 atmospheres. Its life can be lengthened with calcium, barium, or strontium oxide; formation of amine is largely suppressed[215].

In other procedures primary and secondary nitro compounds are transformed into oximes with ammonia or amines in the presence of copper powder at 80–160°[222], and with platinum(IV) oxide[223] or Raney nickel[224]. This technique can be improved by supplying hydrogen and using 300 atmospheres to give a yield of more than 90% *cyclohexanone oxime*[225].

11.1.6.1.4 Oximes by reduction of primary and secondary nitro compounds with alkylating agents

Primary and secondary nitro compounds react with typical alkylating agents such as alkyl halide,

Table 3. Reduction of nitro compounds to oximes

Nitro compound	Reducing agent	Oxime	Yield [% of theory]	Ref.
Nitrophenylmethane	Na amalgam; Zn dust/ alkali metal hydroxide	*Benzaldoxime*		216
Nitrophenylacetonitrile	Zn dust/alkali metal hydroxide	*Hydroxyiminophenyl- acetonitrile*	70	217
3,5-Dimethyl-*a*,2-dinitrotoluene	Na amalgam/ethanol	*2-Amino-3,5-dimethyl- benzaldoxime*	74	218
Nitrocyclohexane	Zn dust/dilute methanol/ hydrochloric acid	*Cyclohexanone oxime*	70	219
	Zn/HCl Fe/HCl		75 74	219
Nitroethane	Mg/H_2SO_4	*Acetaldoxime*	70	220
1-Nitropropane	Zn dust-glacial/acetic acid	*Propanal oxime*	43 (aldehyde)	221

[214] US. P. 3 341 590 (1965), Commerc. Solv. Co., Inv.: *L.R. Jones;* C.A. *67,* 108319h (1967).

[215] *C. Grundmann,* Angew. Chem. *62,* 558 (1950); Brit. P. 921 944 (1959), Commerc. Solv. Co.; C.A. *59,* 8617b (1963).

[216] *A. Hantzsch, O.W. Schultze,* Ber. *29,* 2252 (1896).

[217] *W. Wislicenus, A. Endres,* Ber. *35,* 1759 (1902); *G. Ponzio, L. Avogadro,* Gazz. Chim. Ital. *57,* 124 (1927).

[218] *E. Bamberger, M. Weiler,* J. Prakt. Chem. [2] *58,* 339 (1898).

[219] DBP. 910 647 (1954), Bayer, Erf.: *H. Welz;* C.A. *51,* 4420h (1957).

[220] Brit. P. 722 745 (1955), Bayer; C.A. *50,* 8710c (1955);

O. Convert, J. Armand, Compt. Rend. (C) *262,* 1013 (1966); C.A. *64,* 19397e (1966).

[221] *K. Johnson, E.F. Degering,* J. Amer. Chem. Soc. *61,* 3194 (1939).

[222] DBP. 855 555 (1953), BASF, Erf.: *G. Wiest;* Chem. Zentr. 1591 (1954).

[223] DBP. 916 948 (1954), Bayer, Erf.: *J. Weise, H. Welz, G. v. Schuckmann, H. Danziger;* C.A. *52,* 14673b (1958).

[224] US. P. 2 638 482 (1953), Mathieson Chem. Corp. Inv.: *C. Grundmann;* C.A. *48,* 5214d (1954).

[225] DBP. 1 188 585 (1963), BASF, Erf.: *K. Adam, W. Arend;* C.A. *62,* 16085f (1965).

[226] *J.T. Thurston, R.L. Shriner,* J. Amer. Chem. Soc. *57,* 2164 (1935).

dimethyl sulfate, triethyloxonium tetrafluoroborate, and diazomethane to give *oximes* and *aldehydes via* nitronic acid esters by a disproportionation reaction:

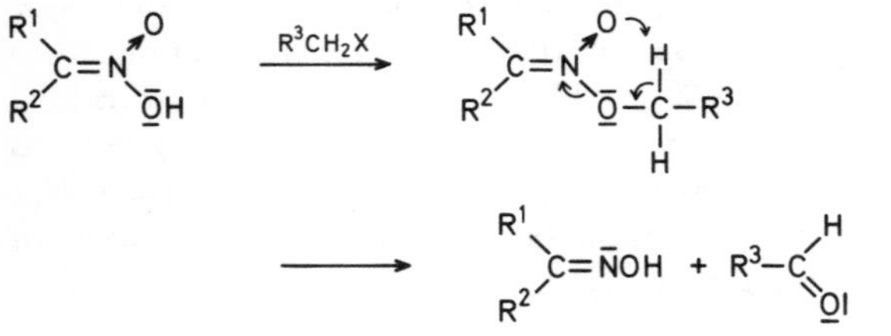

The nitronic acid esters have been isolated in some instances and on decomposition also yielded oximes or their reaction products[226–230]. Table 4 gives further information.

phenone by way of an intramolecular Cannizzaro reaction. Lithium ethoxide is advantageously used as basic condensation agent[242].

11.1.6.1.5 Oximes by reduction of primary and secondary nitro compounds with alkylating agents

Oximes may be formed when Grignard reagents are allowed to act on *primary* nitro compounds. For example, ethyl nitroacetate and methyl magnesium iodide yield *3-hydroxy-3-methyl-2-butanone oxime* as a result of the simultaneous reaction of the ester function[243]:

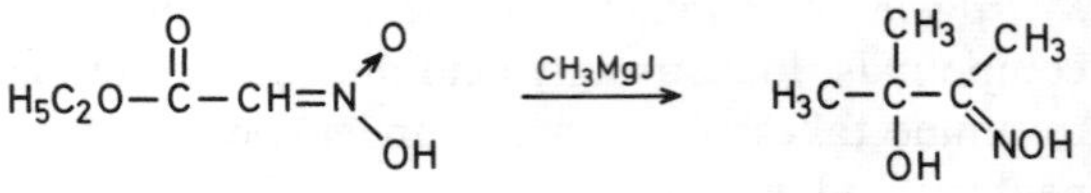

Table 4. Preparation of oximes by alkylating nitro compounds

Nitro compound	Alkyl halide	Conditions	Product	Yield [% of theory]	Ref.
2-Nitropropane	4-Methylbenzyl bromide	Na ethoxide	*Acetone oxime* + *4-Tolualdehyde*		232
Nitrophenylmethane	Benzyl chloride	Na ethoxide	*Benzaldoxime* + benzaldehyde	80	233
Nitrocyclohexane	1,4-Bis-(chloro methyl)-benzene	Na ethoxide	*Cyclohexanone oxime* + Terephthalaldehyde		234
9-Nitrofluorene	Benzyl chloride	K salt-ethanol	*Fluorenone oxime* + Benzaldehyde	87	235
Nitrocyclohexane	Benzyl chloride	Na ethoxide	*Cyclohexanone oxime* + Benzaldehyde	80	236
	Dimethyl sulfate	Na methoxide	*Cyclohexanone oxime*	76	237
	Diethyl sulfate	KOH-H$_2$O	*Cyclohexanone oxime* (+ *O*-ether)	51	238
	Triethyloxonium-BF$_4$	NaOH-H$_2$O	*Cyclohexanone oxime* (+ *O*-ether)	93	239
2-Nitromethyl-quinoxaline	Diazomethane-ether		*2-Quinoxalinecarboxaldehyde oxime*		240

Oxiranes, too, form oximes with *aci*-nitro compounds *via* the nitronic acid esters[231]. Thus, nitrocyclohexane gives 70% *cyclohexanone oxime*[241] with ethylene oxide (oxirane) and basic catalysts, and the reaction of 2-nitrobutane with phenyloxirane gives *butanone oxime* and *butanone O-(2-hydroxy-2-phenylethyl)oxime* together with mandelaldehyde, which rearranges to ω-hydroxyaceto-

[227] *F. Ratz*, Monatsh. Chem. *26*, 1499 (1905).
[228] *K. v. Auwers, B. Ottens*, Ber. *57*, 456 (1924).
[229] *T. Severin, H. Kullmer*, Chem. Ber. *104*, 440 (1971).
[230] *T. Severin, H. Krämer, P. Adhikary*, Chem. Ber. *104*, 972 (1971).
[231] *M.J. Astle, F.J. Donat*, J. Org. Chem. *25*, 507 (1960).

[232] *H.B. Hass, M.L. Bender*, J. Amer. Chem. Soc. *71*, 1767 (1949);
H.B. Hass, M.L. Bender, Org. Synth. *30*, 99 (1950).
[233] *L. Weisler, R.W. Helm*, J. Amer. Chem. Soc. *67*, 1167 (1945).
[234] Jap. P. 19 280 (1961), Toyo Rayon, Inv.: *A. Miyake, N. Yola;* C.A. *60*, 2847d (1964).
[235] *C.D. Nenitzescu, D.A. Isacescu*, Ber. *63*, 2494 (1930).
[236] DBP. 825 547 (1951), Bayer, Erf.: *K. Hamann, K. Bauer;* C.A. *47*, 2204f (1953).
[237] *R.E. McCoy, R.S. Gohlke*, J. Org. Chem. *22*, 286 (1957).
[238] US. P. 2791 611 (1954), DuPont, Inv.: *L.G. Donaruma;* C.A. *51*, 15561c (1957).
[239] *L.G. Donaruma*, J. Org. Chem. *22*, 1024 (1957).
[240] *P.E. Fanta, R.M.W. Rickett, D.S. James*, J. Org. Chem. *26*, 938 (1961).
[241] DBP. 877 303 (1953), BASF, Erf.: *H. Ufer;* C.A. *52*, 8192g (1958).

Correspondingly, ω-nitroacetophenone reacts with isopropyl magnesium chloride to give mainly *4-hydroxy-4-phenyl-2,5-dimethyl-3-hexanone oxime*. Nitrophenylmethane and phenyl magnesium bromide furnish a small quantity of *benzophenone oxime*, ethyl magnesium bromide is said to give *benzaldoxime* and *N-ethyl-N-benzylhydroxylamine* plus a little *N*-ethylbenzylamine[243].

11.1.6.1.6 Oximes by reduction of primary and secondary nitro compounds with other organic compounds

Certain oximes are accessible by reduction of nitro compounds with *aldehydes*[244], *alcohols*, and *phenols*[245]. At the optimum *pH* of 3 the yield is around 70%, aldehydes furnish carboxylic acids, while alcohols afford aldehydes or carboxylic acids.

11.1.6.2 Oximes by reduction of nitroalkenes

With the normal reducing agents nitroalkenes are converted into saturated oximes. The reaction could begin with 1,4-addition followed by reduction of the *aci*-nitro compound formed to give the oxime. *Catalytically* excited hydrogen, nascent hydrogen, tin(II) chloride, or Grignard compounds are all suitable for this purpose[246-247]:

$$\mathrm{C=C-N{\nearrow}^O_{\searrow O|}} \xrightarrow{[H]} \mathrm{-\overset{H}{\underset{|}{C}}-\overset{|}{\underset{|}{C}}=N{\nearrow}^O_{\searrow OH}} \xrightarrow[-H_2O]{[H]} \mathrm{-\overset{H}{\underset{|}{C}}-\overset{|}{\underset{|}{C}}=NOH}$$

ω-Nitroalkenes or primary nitroalkenes give aldoximes with aluminium amalgam or zinc dust and acetic acid, *e.g.*, convert 4-(nitrovinyl)anisole into 4-methoxyphenylacetaldoxime[248]. Secondary nitroalkenes react with iron and hydrochloric acid to form ketoximes[249], 2-nitro-1-phenyl-1-propene forms *1-phenyl-2-propanone oxime*[250] with iron and water.

Adding aqueous methanolic tin(II) chloride solution and hydrochloric acid to the methanolic solution or suspension of a nitroalkene leads to a good yield of *2-methoxy oximes*, and, in ethanolic solution, to *2-ethoxy oximes* (p. 413).

Using the inverse reaction at -10 to $-15°$, ω-nitrostyrene, *e.g.* gives *phenylacetaldoxime* and *2-nitro-1-phenylethane*[246].

2-Chloro oximes are obtained in medium yield if the reduction with tin(II) chloride is carried out in ether in the presence of hydrogen chloride at -10 to $0°$[251].

$$\mathrm{R^1-CH=C-R^2} \xrightarrow{HCl} \mathrm{R^1-\overset{H}{\underset{|}{C}}-\overset{|}{\underset{|}{C}}-R^2} \xrightarrow{SnCl_2} \mathrm{R^1-\overset{H}{\underset{|}{C}}-\overset{|}{\underset{|}{C}}-R^2}$$

The *catalytic hydrogenation* is performed advantageously in acid alcoholic solution. Thus, *cyclooctanone oxime* is formed in 78% yield from 1-nitro-1-cyclooctene over palladium-carbon in a weakly hydrochloric acid solution of methanol[252]. 1,4-Dioxane[253] and pyridine[252] are alternative solvents. In pyridine the hydrogenation can be carried out at 50° with palladium/carbon without using pressure.

11.1.7 Oximes by ring cleavage

11.1.7.1 Oximes by ring cleavage of pyrrole derivatives

Pyrrole and substituted pyrroles suffer nucleophilic attack by hydroxylamine and hydroxylammonium chloride, open their ring and form *1,4-dioximes* with loss of ammonia[254]:

$$\mathrm{pyrrole} + 2\,NH_2OH \xrightarrow{-NH_3} \mathrm{R^4-\underset{NOH}{\overset{R^3}{\underset{\|}{C}}}-CH-CH-\underset{NOH}{\overset{R^2}{\underset{\|}{C}}}-R^1}$$

If 2,5-disubstituted 3*H*-pyrrol-3-one oximes are used for this reaction, then an analogous ring

[242] *G.B. Bachman, T. Hokama*, J. Amer. Chem. Soc. *81*, 4223 (1959);
US. P. 3 123 639 (1959), Purdue Research Found., Inv.: *G.B. Bachman, T. Hokama*; C.A. *60*, 14390g (1964).

[243] *A. Dornow, H. Gehrt, F. Ische*, Justus Liebigs Ann. Chem. *585*, 220 (1954).

[244] DBP. 837 691 (1952), Bayer, Erf.: *H. Welz*; C.A. *52*, 9201f (1958).

[245] DBP. 837 692 (1952), Bayer, Erf.: *H. Welz, J. Weise*; C.A. *47*, 1729a (1953).

[246] *A. Dornow, A. Müller*, Chem. Ber. *93*, 32 (1960).

[247] *G.D. Buckley*, J. Chem. Soc. (London) 1495 (1947).

[248] *K.W. Rosenmund*, Ber. *42*, 4781 (1909).

[249] *H.B. Hass, A.G. Susie, R.L. Heider*, J. Org. Chem. *15*, 10 (1950).

[250] US. P. 2 233 823 (1941), Purdue Res. Found., Inv.: *A.G. Susie, H.B. Hass*; C.A. *35*, 3650 (1941).

[251] *A. Dornow, H.D. Jordan, A. Müller*, Chem. Ber. *94*, 67 (1961).

[252] *W.K. Seifert, P.C. Condit*, J. Org. Chem. *28*, 265 (1963);
US. P. 3 156 723 (1961), Calif. Res. Co., Inv.: *W.K. Seifert*; Chem. Zentr. 41–2580 (1966).

[253] *V.J. Traynelis, R.F. Lore*, J. Org. Chem. *29*, 369 (1964).

[254] *S.P. Findley*, J. Org. Chem. *21*, 644 (1956).

opening occurs, but *3-aminotriphenylpyrrole* is reported to be formed in the alkaline region with *3H*-pyrrol-3-one oxime[255, 256].

11.1.7.2 Oximes by ring cleavage of furazans

Treatment of dibenzoylfurazan in ethanolic solution with 20% caustic soda leads primarily to splitting off of a benzoyl group.
Secondarily, stabilization of the carbanion formed occurs with formation of *2-hydroxyimino-3-oxo-3-phenylpropionitrile* accompanied by ring opening[257]:

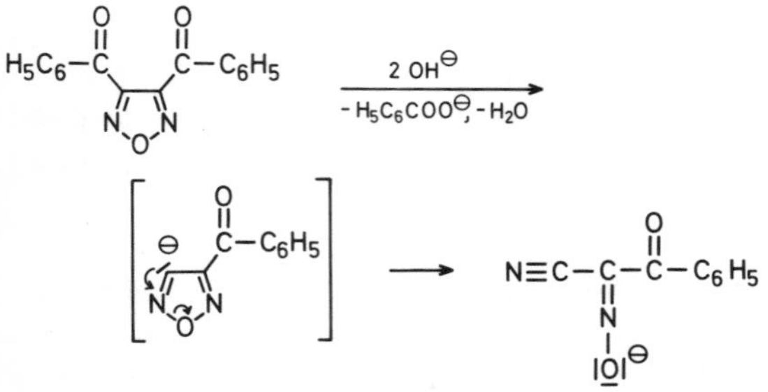

Naturally, this cleavage reaction applies also to all monosubstituted furazans. Accordingly, furazancarboxylic acid yields *(Z)-hydroxyiminoacetonitrile via cyanohydroxyiminoacetic acid* by decarboxylation[258].
On heating in dilute caustic soda, diaroylfurazan dioximes suffer an intramolecular rearrangement. A good yield of *aryl(arylfurazanyl)glyoxime* is obtained together with *5,5'-diaryl-4,4'-bifurazanyls* as side-product[259]:

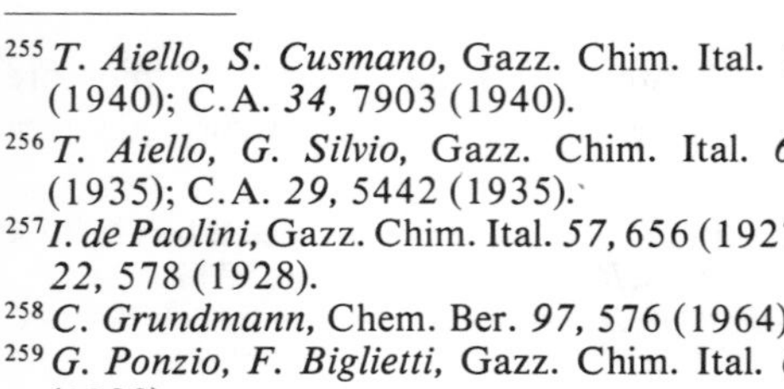

11.1.7.3 Oximes by ring cleavage of furoxans

Zinc and acetic acid[260-263] and also *Raney nickel* and *hydrogen*[264] reduce furoxans to the corresponding *dioximes, e.g.:*

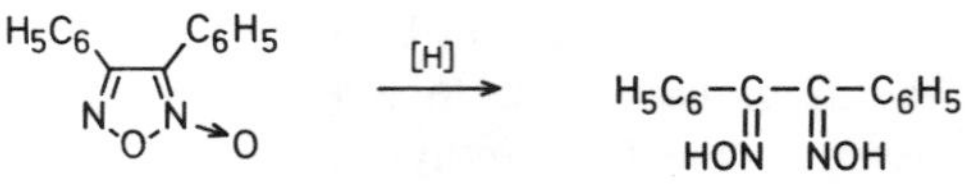

Diphenylglyoxime

Other reducing agents lead to either *furazans* or *diamines*[265].
Under certain conditions the furoxan ring is attacked by bases. With phenylfuroxan, for example, aniline is one reagent which abstracts the proton, and the nonisolatable nitrile oxide undergoes 1,3-addition with a second molecule of aniline to form *anilinophenylglyoxime*[266]:

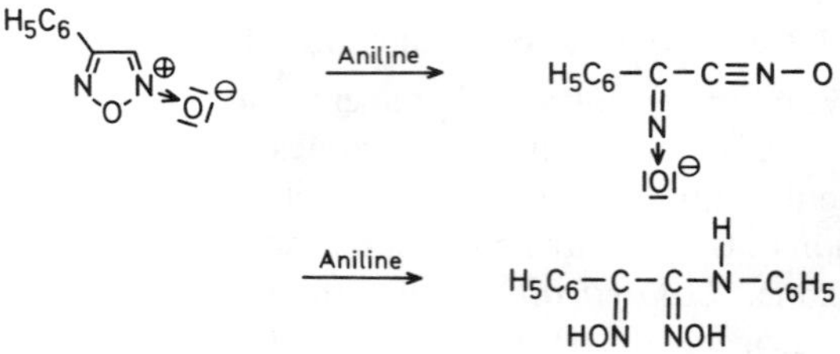

Where R = R' = acyl or alkoxycarbonyl, the base takes up one substituent group and adds on to the intermediately formed nitrile oxide. The aminoglyoxime (hydroxamic acid amide) then forms a *2-isoxazoline* by ring closure in a subsequent reaction.

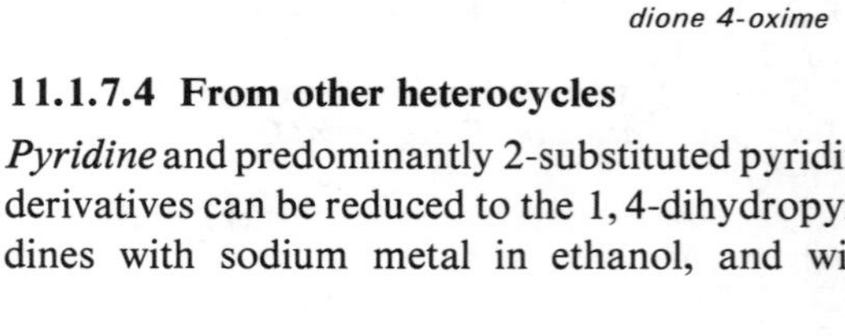

R = R' = COOR

3-Amino-2-isoxazolin-4,5-dione 4-oxime

11.1.7.4 From other heterocycles

Pyridine and predominantly 2-substituted pyridine derivatives can be reduced to the 1,4-dihydropyridines with sodium metal in ethanol, and with

[255] *T. Aiello, S. Cusmano,* Gazz. Chim. Ital. *70,* 127 (1940); C.A. *34,* 7903 (1940).

[256] *T. Aiello, G. Silvio,* Gazz. Chim. Ital. *65,* 176 (1935); C.A. *29,* 5442 (1935).

[257] *I. de Paolini,* Gazz. Chim. Ital. *57,* 656 (1927); C.A. *22,* 578 (1928).

[258] *C. Grundmann,* Chem. Ber. *97,* 576 (1964).

[259] *G. Ponzio, F. Biglietti,* Gazz. Chim. Ital. *63,* 159 (1933).

[260] *A. Angeli,* Ber. *26,* 528 (1893).

[261] *A. Werner, C. Bloch,* Ber. *32,* 1982 (1899).

[262] *L. Avogadro,* Gazz. Chim. Ital. *54,* 545 (1924); C.A. *19,* 262 (1925).

[263] *J. Meisenheimer, H. Lange, W. Lamparter,* Justus Liebigs Ann. Chem. *444,* 111 (1925).

[264] *T. Mukaiyama, T. Hoshino,* J. Amer. Chem. Soc. *82,* 5340 (1960).

[265] *J.V.R. Kaufmann, J.P. Picard,* Chem. Rev. *59,* 429 (1959) (The Furoxans).

[266] *H. Wieland, E. Gmelin,* Justus Liebigs Ann. Chem. *375,* 297 (1910).

hydroxylamine hydrochloride liberate ammonia to form open-chain *1,5-dioximes*[267], *e.g.*:

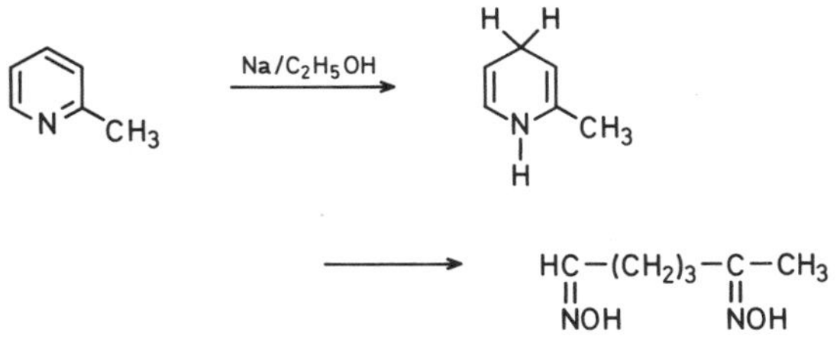

1,5-Hexadione
dioxime

In the reaction of hydroxylamine with 2,8-dimethylchromone and 2,8-dimethyl-4-thiochromone in neutral solution the associated oxime is formed. It, too, is isolated on acidifying an alkaline solution of the product with mineral acid. By contrast, careful addition of acetic acid to the alkaline solution gives the dioxime[268]:

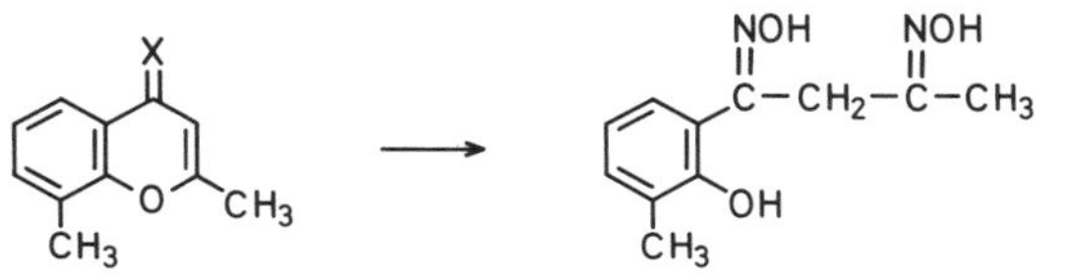

X = O, S, NOH

1-(2-Hydroxy-3-methyl-
phenyl)-1,3-butanedione
dioxime

11.1.8 Preparation of *O*-alkyl and *O*-aryl oximes

Alkylation and arylation reactions with oximes are manifestly largely dependent on the *steric make-up, the reagent,* and the *solvent.* Both *O*- and *N*-alkyl or *O*-alkyl and *N*-aryl products (*E*- and *Z*-forms) can result; they differ in basicity, solubility, and steam volatility:

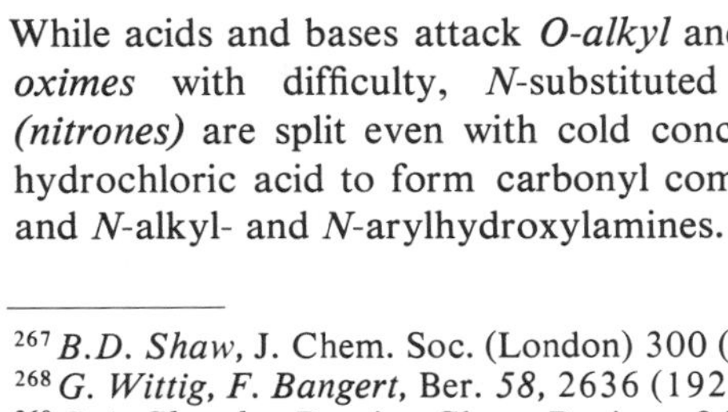

While acids and bases attack *O-alkyl* and *O-aryl oximes* with difficulty, *N*-substituted oximes *(nitrones)* are split even with cold concentrated hydrochloric acid to form carbonyl compounds, and *N*-alkyl- and *N*-arylhydroxylamines.

11.1.8.1 Preparation of *O*-alkyl and *O*-aryl oximes with alkylating and arylating agents

Alkylating agents such as *alkyl halides, diazoalkanes,* and *dialkyl sulfates* attack the ambient oxime anions either at the oxygen atom or at the nitrogen atom[269]. With substituted benzophenone oximes, *e.g.*, the conversion takes place with benzyl halides in ethanolic solution almost independently of the alkali metal cation approximately in a 3:1 *O*- to *N*-alkylation ratio *(benzophenone O-benzyl oxime to benzophenone benzylnitrone).* Using methyl bromide displaces the yield toward the *N*-alkylated product[270].

Alkylation of benzophenone oxime sodium in benzene with chloromethyl methyl sulfide gives 47% *benzophenone O-methylthiomethyl oxime* and 38% *benzophenone methylthiomethyl nitrone*[271, 272].

The action of chlorotriphenylmethane, which produces mainly *O*-alkylation of both isomers, is in conflict with earlier observations, according to which *(E)*-benzaldoxime is alkylated at the oxygen with simple alkyl halides, while *(Z)*-benzaldoxime is alkylated on the nitrogen[273, 274]:

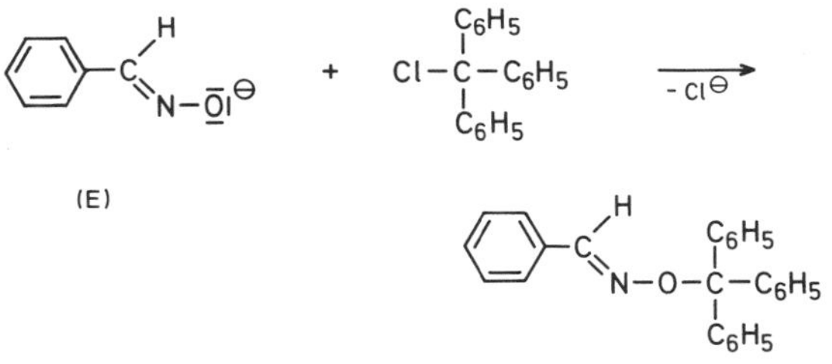

(E)

Benzaldehyde O-triphenyl-
methyl oxime

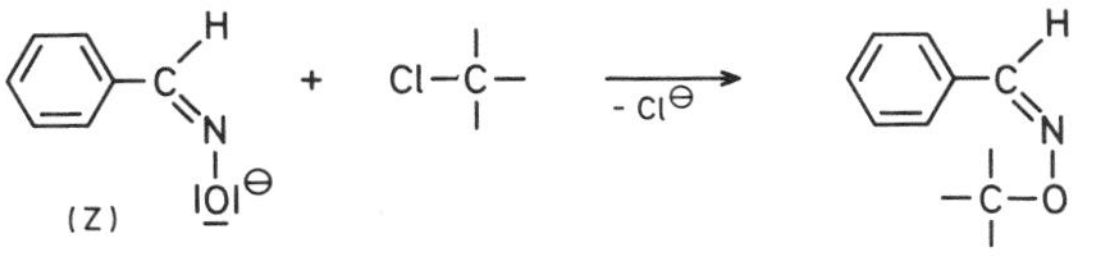

(Z)

In the molten state (200°) isomerization occurs until a 10:1 ratio of *(E)*- to *(Z)*-isomer has been reached! Establishment of equilibrium proceeds

[267] *B.D. Shaw,* J. Chem. Soc. (London) 300 (1937).
[268] *G. Wittig, F. Bangert,* Ber. *58,* 2636 (1925).
[269] *S.A. Shevelev,* Russian Chem. Reviews 844 (1970) (Dual Reactions of Ambient Anions).

[270] *P.A.S. Smith, J.E. Robertson,* J. Amer. Chem. Soc. *84,* 1197 (1962).
[271] *A.H. Fenselau, E.H. Hamamura, J.G. Moffatt,* J. Org. Chem. *35,* 3546 (1970).
[272] *D.A. Kerr, D.A. Wilson,* J. Chem. Soc. (C) 1718 (1970).
[273] *E. Buehler,* J. Org. Chem. *32,* 261 (1967).
[274] *E.J. Grubbs, J.A. Villarreal,* Tetrahedron Lett. 1841 (1969).

via a splitting into iminoxy and triphenylmethyl free radicals.

The action of *diazomethane* on the isomeric nitrobenzaldoximes leads to the *O*-methyl ethers with retention of configuration. Nitrones form side by side; a *cis/trans* isomerism is postulated[275].

In alkaline solution *dimethyl sulfate* normally reacts with isomeric nitrobenzaldoximes to a mixture of *O*-alkyl and *N*-alkyl products. Evidently, with *(E)*-isomers mainly methyloxime is formed and from *(Z)*-isomers mainly methylnitrone. The proportion of methyl oxime is said to be increased by raising the pH[276]. Ketoximes, too, yield mixtures with dialkyl sulfates, *O*-alkyl oximes predominate[277-279].

In general, it is found that when alkali salts of oximes are used which are prepared without a change in configuration with sodium methoxide, ethoxide, hydride, amide, and potassium *tert*-butoxide in alcohols, dimethylformamide, dimethylsulfoxide, benzene, and xylene, then alkyl oximes as well as nitrones can result[280]. Under certain circumstances the nitrones decompose at the working-up stage, and they are therefore often overlooked or else their isolation is undesired[281].

An *exclusive O*-alkylation is achieved when *alkyl iodides* act on the *silver* salts of the oximes or on oximes and silver oxide in ether or ethanol solution[282]. The *(E)*- and *(Z)*-configurations do not necessarily stay intact[273]. *Oxidizing N,O-dialkylhydroxylamines* too, leads to *O*-alkyloximes[283], *e.g.:*

$$H_3C-CH_2-NH-O-C_2H_5 \xrightarrow[-H_2O]{OCl^{\ominus}}$$

$$H_3C-CH=N-O-C_2H_5 \quad + \quad Cl^{\ominus}$$

Acetaldehyde O-ethyloxime

The reaction of both aldehydes and ketones with *O*-alkyl *hydroxylamines*, which unambiguously leads to *O*-alkyloximes, is more elegant and generally applicable[284]:

$$\begin{array}{c} R^1 \\ R^2 \end{array}\!C=O \quad + \quad H_2N-OR \xrightarrow{-H_2O} \begin{array}{c} R^1 \\ R^2 \end{array}\!C=NOR$$

This reaction is often made use of for building up products with a basic side-chain for pharmaceutical use[285], for example, from the group of the *5H-dibenzo[a,d]cycloheptenone oximes*[286], *e.g.:*

5 H-Dibenzo[a,d]cyclohepten-5-one
O-[2-(dialkylamino)ethyl]oxime

Admittedly, in these cases *O*-alkyl oximes are often obtained during the alkali salt alkylation stage, as was shown also with unsaturated oximes[287]:

4-Phenyl-3-buten-2-one
O-[2-(dialkylamino)eth]oxime

Reactions of benzaldoximes with 3-bromo-4-propyne[288], and of aldoximes and ketoximes with halogen carboxylic acids and derivatives[289,290] lead to *O*-substituted oximes.

[275] *A.F. Thompson jr., M. Baer*, J. Amer. Chem. Soc. *62*, 2094 (1940).

[276] *O.L. Brady, R.F. Goldstein*, J. Chem. Soc. (London) 2405 (1926).

[277] *O.L. Brady, N.M. Chokshi*, J. Chem. Soc. (London) 2272 (1929).

[278] *M.M. Kochhar, R.G. Brown, J.N. Delgado*, J. Pharm. Sci. *54*, 394 (1965).

[279] *A.F. Ferris*, J. Org. Chem. *24*, 1726 (1959).

[280] *G. Kamai, A.D. Nikolaeva, V.S. Perekhod'ko*, Zh. Org. Khim. *4*, 567 (1968); C.A. *69*, 43348m (1968).

[281] *J. Hamer, A. Macaluso*, Chem. Rev. *64*, 473 (1964) (Nitrones).

[282] *G. Vermillion, C.R. Hauser*, J. Org. Chem. *6*, 512 (1941).

[283] *A.B. Boese jr., L.W. Jones, R.T. Major*, J. Amer. Chem. Soc. *53*, 3533 (1931).

[284] *J.A. Skorcz, J.T. Suh, C.I. Judd*, J. Med. Chem. *9*, 658 (1966); *M. Chaillet, A. Dargelos, J. Deschamps*, Compt. Rend. *259*, 3288 (1964); C.A. *62*, 3970h (1965).

[285] Holl. P. 6 810 133 (1968), N.V. Philips' Gloeilampenfabrieken; C.A. *72*, 121354p (1970).

[286] *G. Aichinger, O. Behner, F. Hoffmeister, S. Schütz*, Arzneimittel-Forsch. *19*, 838 (1969).

[287] *B. Unterhalt*, Arch. Pharm. (Weinheim, Ger.) *304*, 454 (1971).

[288] S. Afrik. P. 6 802 699 (1967), Fisons Pest. Contr., Inv.: *P.L. Carter, G.T. Newbold, D.T. Saggers;* C.A. *71*, 3143a (1969).

[289] DBP. 1 960 910 (1968), N.V. Philips' Gloeilampenfabrieken, Erf.: *J. v. Dijk, J.M.A. Zwagemakers, V. Claassen;* C.A. *73*, 76833h (1970).

[290] *P. Block jr.*, J. Org. Chem. *30*, 1307 (1965).

However, bromoacetophenone oxim yields 60% nitrone with *(Z)*-benzaldoxime which cleaves with hydrazine to form hydroxyaminoacetophenone oxime[291].

Alkylation of oximes with *oxiranes* proceeds as anticipated; the reaction between 4-pyridinecarboxaldehyde oxime and phenoxymethyloxirane serves as an illustration here[292]:

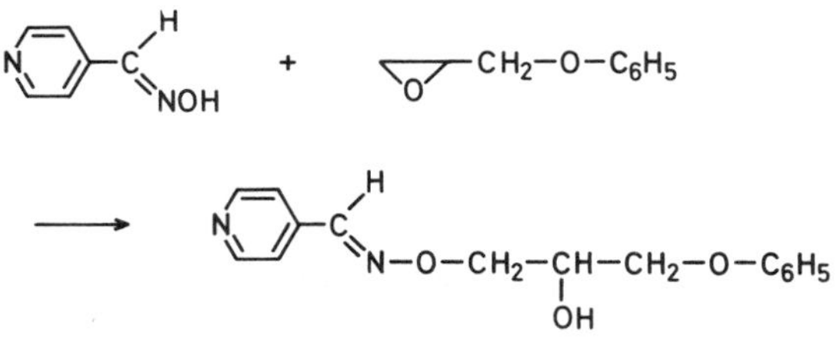

4-Pyridinecarboxaldehyde O-(2-hydroxy-3-phenoxypropyl)oxime

O-(2-hydroxyalkyl) oximes are formed[293]. The associated nitrone has been isolated in only a few cases; presumably, it undergoes ring-chain tautomerism[294]:

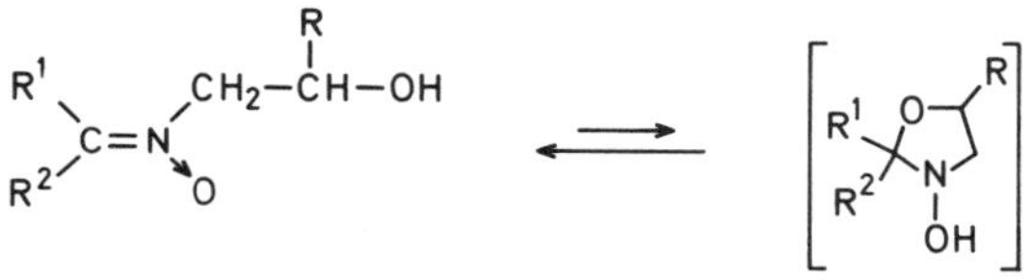

Intramolecular O-alkylation takes place when oximes of certain Mannich bases effect ring closure to form *2-isoxazolines* under the influence of, for example, sodium ethoxide[295, 296], *e.g.*:

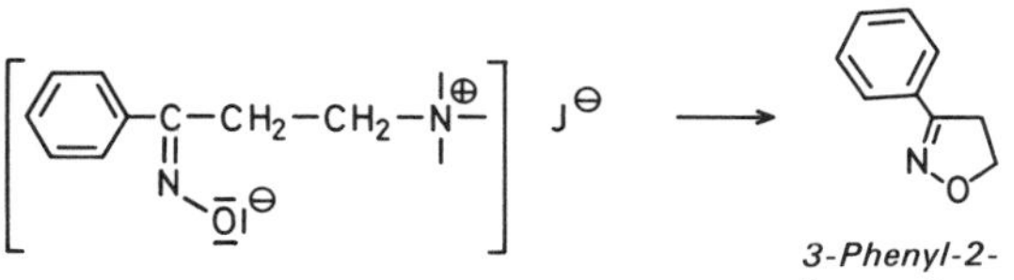

3-Phenyl-2-isoxazoline

2-Isoxazolines are formed also when 2-halo oximes are reacted with dimethylsulfoxonium methylide[297]:

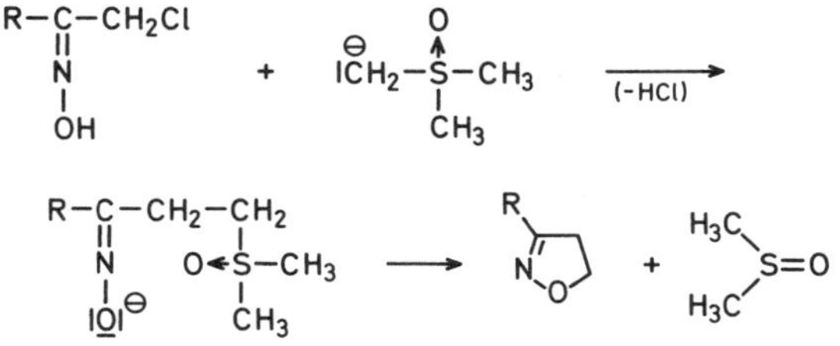

On treating benzil monooxime with dimethylsulfonium methylide the expected 3,4-diphenylisoxazole is not formed; instead, *3,4-diphenyl-2-isoxalin-5-ol* is obtained surprisingly *via* an intermediate oxirane stage[298]:

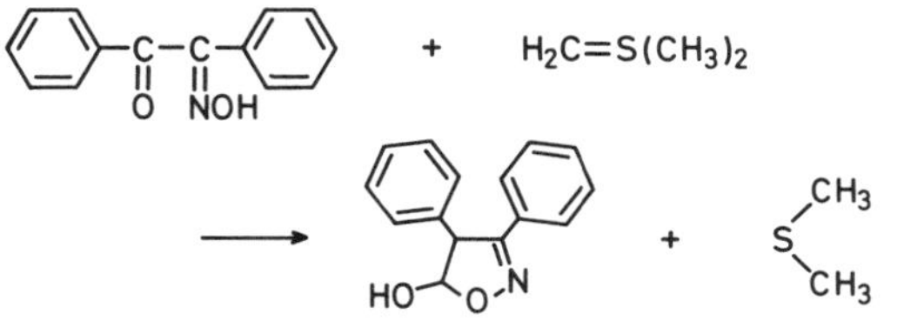

O-Aryl oximes can be synthesized in one of two ways:

1 In 70–95% yield from carbonyl compounds and *O*-arylhydroxylamines[299] — these can be generated also *in situ* from the corresponding carbamates[300, 301].
2 From the alkali metal salts of the oximes in polar solvents such as dimethylformamide and dimethyl sulfoxide using activated aryl halides to give invariably a 60–90% yield.

Possible activated aryl halides are fluorobenzenes and chlorobenzenes having electron-attracting groups in the *o*- or *p*-position[302–305] and also 2- and 4-bromopyridines[306].

[291] *L.B. Volodarskii, A.Y. Tikhonov*, Izv. Akad. Nauk SSSR, Ser. Khim. 2341 (1969); C.A. *72*, 43048 (1970).
[292] Poln. P. 59077 (1967), Instytut Farmaceutyczny, Inv.: *M. Krajewska, R. Palanowski, J. Wolf*; C.A. *73*, 35231k (1970).
[293] *J.A. Skorcz, J.T. Suh, C.I. Judd*, J. Med. Chem. *9*, 658 (1966).
[294] *W. Kliegel*, Justus Liebigs Ann. Chem. *733*, 192 (1970).
[295] *F.L. Scott, R.J. MacConaill*, Tetrahedron Lett. 3685 (1967).
[296] *R.J. MacConaill, F.L. Scott*, Tetrahedron Lett. 2993 (1970);
R.J. MacConaill, F.L. Scott, J. Chem. Soc. (C) 584 (1971).
[297] *P. Bravo, G. Gaudiano, C. Ticozzi, A. Umani-Ronchi*, Chem. Commun. 1311 (1968);

P. Bravo, G. Gaudiano, P.P. Ponti, A. Umani-Ronchi, Tetrahedron *26*, 1315 (1970).
[298] *P. Bravo, G. Gaudiano, C. Ticozzi*, Tetrahedron Lett. 3223 (1970).
[299] *T. Sheradsky*, Tetrahedron Lett. 5225 (1966).
[300] *T. Sheradsky*, J. Heterocycl. Chem. *4*, 413 (1967).
[301] *Z. Rappoport, T. Sheradsky*, J. Chem. Soc. (B) 902 (1967).
[302] *A. Mooradian*, Tetrahedron Lett. 407 (1967).
[303] *A. Mooradian, P.E. Dupont*, Tetrahedron Lett. 2867 (1967).
[304] Holl. P. 6 600 834 (1966), Ciba; C.A. *66*, 28501a (1967).
[305] *R. Huisgen, I. Ugi, M.T. Arsemi, J. Witte*, Justus Liebigs Ann. Chem. *602*, 133 (1957);
G. Vermillion, A.E. Rainsford, C.R. Hauser, J. Org. Chem. *5*, 68 (1940);
H.P. Fischer, C.A. Grob, Helv. Chim. Acta *45*, 2538 (1962);
M.M. Kochhar, B.B. Williams, J.H. Fan, J. Pharm. Sci. *58*, 1382 (1969).

11.1.8.2 Preparation of *O*-alkyl and *O*-aryl oximes by Michael addition

Activated alkenes are suitable for alkylating oximes. Esters of acrylic acid, acrylamide, acrylonitrile, acrolein, *N*-substituted maleinimides[307], and fluorinated alkenes[308, 309] undergo base catalysis to furnish *O*-alkyl oximes:

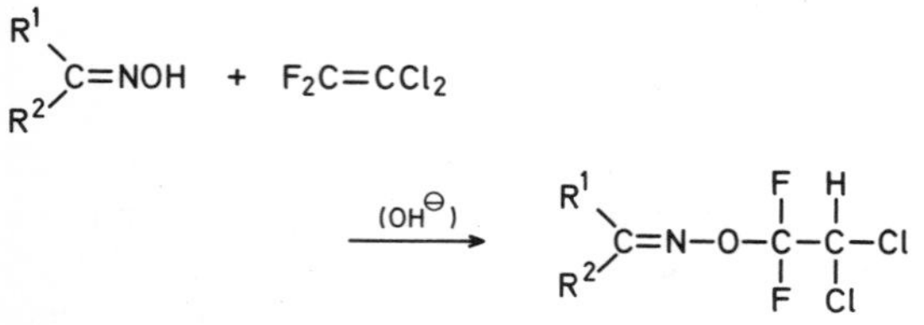

Two successive additions of oximes to ethyl ethynyl ether have been described[310]:

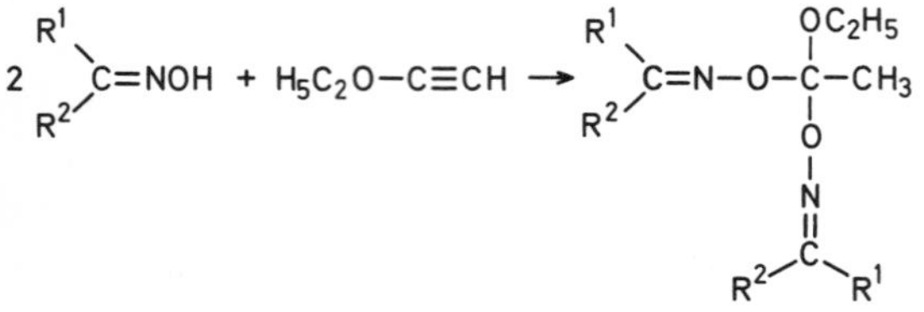

11.1.9 Preparation of *O*-acyl oximes

The reaction between oximes and acyl chlorides leads exclusively to *O*-acyl oximes[311]. When pure isomers are used there is the risk, and especially in the acid region, of configuration reversal, of loss of water to nitrile with aldoximes, and of a possible Beckmann rearrangement or fission with ketoximes. In addition to acyl chlorides[312], anhydrides[313], ketenes, isocyanates, cyanates, and isothiocyanates form *O*-acyl oximes. *O*-Acyl oximes

are used as transacylating agents in peptide synthesis[314].

11.1.9.1 Preparation of *O*-acyl aldoximes

While acetylation of *(E)*-benzaldoximes with acetic anhydride presents no difficulties[315], special steps need to be taken with the *(Z)*-form, because even the slightest trace of acid and certain solvents produce isomerization to the *(E)*-form[316].

Acetic anhydride may be replaced by *ketene*[317]. When sodium carbonate is added, nitriles are usually formed from the *O*-acetyl-*(Z)*-oximes, *(E)*-oximes from the *(E)*-forms. This behavior is utilized for configuration assignment.

Benzoylation, performed in alkaline solution according to Schotten-Baumann or in pyridine, leads to *O*-benzoyl *(E)*-oximes[317−319], *e.g.:*

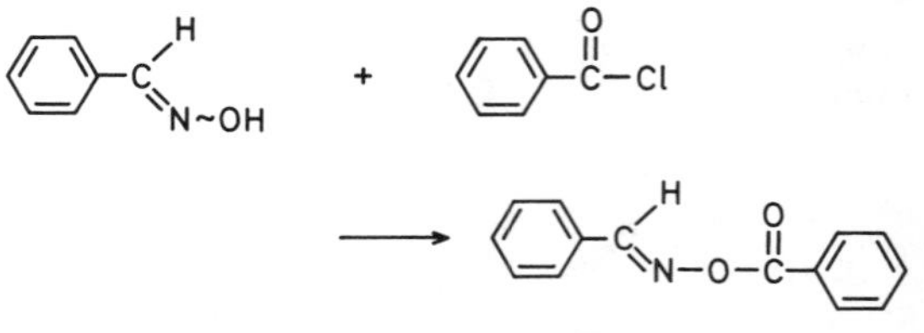

Benzaldehyde O-benzoyl-oxime

Isocyanates and carbamoyl chlorides furnish the corresponding carbamic acid esters[320−325], aryl

---

[306] US. P. 3 218 329 (1963), Upjohn, Inv.: *L.A. Paquette;* C.A. *64,* 3494h (1966).

[307] *H.A. Bruson, T.W. Riener,* J. Amer. Chem. Soc. *65,* 24 (1943);
US. P. 2 352 514 (1944), Resinous Prod., Inv.: *H.A. Bruson, T.W. Riener;* C.A. *38,* 5506 (1944).

[308] *A.P. Stefani, J.R. Lacher, J.D. Park,* J. Org. Chem. *25,* 676 (1960).

[309] *D.C. England, L.R. Melby, M.A. Dietrich, R.V. Lindsey jr.,* J. Amer. Chem. Soc. *82,* 5116 (1960).

[310] *H.D.A. Tigchelaar-Lutjeboer, H. Bootsma, J.F. Arens,* Rec. Trav. Chim. Pays-Bas *79,* 888 (1960); C.A. *55,* 4353i (1961).

[311] *O. Exner,* Chem. Listy *48,* 1634 (1954); C.A. *49,* 14674i (1955);
O. Exner, Chem. Listy *50,* 779 (1956); C.A. *50,* 15477f (1956).

[312] *J.W. Churchill, M. Lapkin, F. Martinez, J.A. Zaslowsky,* J. Amer. Chem. Soc. *81,* 2110 (1959) (Kinetics of oxime acylations).

[313] US. P. 3 503 732 (1967), Gulf Res., Inv.: *R.P. Cahoy;* C.A. *72,* 131365g (1970).

[314] *G. Losse, K.-H. Hoffmann, G. Hetzer,* Justus Liebigs Ann. Chem. *684,* 236 (1965);
R. Buyle, Helv. Chim. Acta *47,* 2444 (1964);
N. Cagnoli Bellavita, A. Colonna, Ann. Chim. (Rome) *59,* 630 (1969); C.A. *71,* 90756u (1969).

[315] *C.R. Hauser, E. Jordan,* J. Amer. Chem. Soc. *57,* 2450 (1935).

[316] *C.R. Hauser, C.T. Sullivan,* J. Amer. Chem. Soc. *55,* 4611 (1933);
B. Unterhalt, Arch. Pharm. (Weinheim, Ger.) *303,* 661 (1970) (Isomerisation of cinnamaldoximes on addition of protons).

[317] *O.L. Brady, G.P. McHugh,* J. Chem. Soc. (London) 2414 (1925);
US. P. 3 483 231 (1966), Union Carbide Co., Inv.: *E. Marcus, J.L. Hughes;* C.A. *72,* 43163j (1970) (Formation of O-acetoacetyloximes with diketene).

[318] *G. Vermillion, E. Jordan, C.R. Hauser,* J. Org. Chem. *5,* 75 (1940).

[319] *G. Vermillion, C.R. Hauser,* J. Amer. Chem. Soc. *62,* 2939 (1940).

[320] *G. Vermillion, A.E. Rainsford, C.R. Hauser,* J. Org. Chem. *5,* 68 (1940).

[321] Fr. P. 1 377 474 (1962), Union Carbide Co., Inv.: *L.K. Payne jr., M.H.J. Weiden;* C.A. *63,* 2900a (1965).

[322] Fr. P. 2 005 668 (1968), Bayer; C.A. *73,* 14521q (1970).

cyanates give *O-aryloxy carbimidoyl aldox-imes*[326], chloroformic acid esters *O-alkoxy carbonyl aldoximes,* and isothiocyanates *thiocarbamic acid esters*[327]. Generally *(E)*-isomers are obtained. Salts of benzaldoximes can be converted into *O*-tosyl-*(E)*-oximes in ether or acetone[328]; with benzylsulfonyl chloride *benzylsulfonates* form *via* phenylsulfene which, like tosylates, are easily cleaved to *nitriles*[329]. The reaction of, *e.g.,* *N*-alkyl-pyridinium aldoxime salts with esters and halides of phosphoric acid and thiophosphoric acid is of *biochemical* interest[330].

11.1.9.2 Preparation of *O*-acyl ketoximes

Ketoximes behave fundamentally analogously to aldoximes, but it is frequently possible to isolate both the geometric isomers. Greater detail is given in the following compilation[331]: reaction with

Diketene[332]	Sulfonyl chloride-pyridine[346]
Dicyclohexylcarbodiimide-phenylacetic acid[333]	Phosphoryl chloride[347]
(Subst.) Benzoyl chloride-base[334-336]	Pseudo-saccharin chloride[348]
Isocyanates[337-341]	Phosgene[349]
Acyl isocyanates[342]	Thiophosgene and *N,N*-dimethylcarbamoyl chloride[350]
Carbamoyl chlorides[334, 343, 344]	
Cyanates[345]	

11.1.9.3 Preparation of *O*-acyl dioximes

Monooximes and dioximes of 1, 2-dioxo compounds can be acetylated, benzoylated, and converted into the carbamic acid esters in the usual way. With toluenesulfonyl chloride and methane-sulfonyl chloride tosylates or mesylates are formed[351]. Phosgene, succinyl chloride, and also diisocyanates yield polymers[352, 353]:

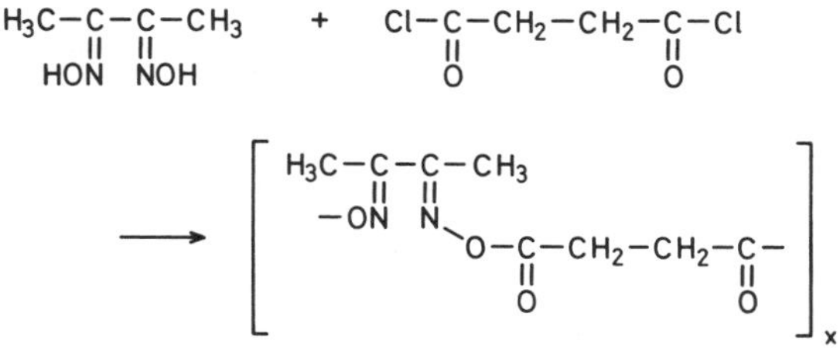

[335] *W.B. Renfrow, J.F. Witte, R.A. Wolf, W.R. Bohl,* J. Org. Chem. *33,* 152 (1968).
[336] *L. Lang, G. Horvath, L. Vargha, G. Ocskay,* Bull. Soc. Chim. France 2724 (1965); C.A. *64,* 559f (1966).
[337] DBP. 2 003 748 (1969), Ciba, Erf.: *O. Rohr;* C.A. *73,* 87691c (1970).
[338] DDRP. 1 232 947 (1963), VEB Fahlberg-List, Erf.: *A. Jumar, P. Held, W. Schulze;* C.A. *66,* 104807x (1967).
[339] *A. Jumar, H. Gruenzel,* Arch. Pflanzenschutz *2,* 163 (1966); C.A. *68,* 21095d (1968); Brit. P. 1 165 220 (1966), Fahlberg-List, Inv.: *A. Jumar, H. Gruenzel;* C.A. *72,* 11573b (1970).
[340] Jap. P. 6 722 938 (1965), Yoshitomi Pharm. Ind., Inv.: *M. Nakanishi, A. Tsuda, S. Saheki;* C.A. *69,* 43630x (1968).
[341] *A. Iovtchev, H. Reinheckel, N. Bonchev, S. Spasov,* Monatsh. Chem. *96,* 1639 (1965); C.A. *64,* 4974b (1966).
[342] USSR. P. 232 245 (1967), Inv.: *K.A. Nuridzha-nyan, L.M. Nesterova, N.E. Mironova;* C.A. *70,* 114665w (1969).
[343] US. P. 3 483 246 (1966), Mobil Oil, Inv.: *H.A. Kaufmann;* C.A. *72,* 66674v (1970).
[344] Fr. P. 1 343 654 (1961), Union Carbide Corp., Inv.: *J.R. Kilsheimer, D.T. Manning;* C.A. *60,* 10568f (1964).
[345] *E. Grigat, R. Pütter,* Chem. Ber. *99,* 2361 (1966); DBP. 1 242 218 (1965), Bayer, Erf.: *E. Grigat, R. Pütter;* C.A. *68,* 21719k (1968).
[346] Jap. P. 6 923 495 (1964), Fujisawa Pharm., Inv.: *S. Unio, K. Kariyone, K. Tanaka;* C.A. *72,* 31447s (1970).
[347] S. Afrik. P. 6 804 420 (1967), Ciba, Inv.: *A. Hubele;* C.A. *71,* 123964r (1969).
[348] *H. Hettler, H. Neygenfind,* Tetrahedron Lett. 6031 (1966); *H. Neygenfind, H. Hettler,* Tetrahedron Lett. 5509 (1968).
[349] *A. Jumar, P. Held, W. Schulze,* Z. Chem. *7,* 344 (1967). Brit. P. 1 153 263 (1967), Bayer, Inv.: *H. Krimm, H. Schnell;* C.A. *71,* 38367d (1969); DBP. 1 809 385 (1968), Bayer, Erf.: *W. Daum, H. Krimm, H. Scheinpflug, P.E. Frohberger, F. Grewe;* C.A. *73,* 66265d (1970).
[350] *B. Cross, R.J.G. Searle, R.E. Woodall,* J. Chem. Soc. (C) 1833 (1971).
[351] US. P. 3 497 554 (1966), Amer. Cyanamid, Inv.: *W.A. Reiners, M.J. Weiss;* C.A. *72,* 100148j (1970).

[323] US. P. 3 522 287 (1968), Shell, Inv.: *C. Donninger, J.H. Davies, H.R. Davis;* C.A. *73,* 76698m (1970).
[324] DBP. 2 015 527 (1969), Amer. Cyanamid, Erf.: *R.W. Addor, D.E. Ailman;* C.A. *73,* 120133z (1970).
[325] S. Afrik. P. 6 903 964 (1968), BASF, Inv.: *K. Kiehs, A. Fischer;* C.A. *73,* 45184m (1970).
[326] *E. Grigat, R. Pütter,* Chem. Ber. *99,* 2364 (1966).
[327] *C.V. Gheorghiu, E. Rucinschi,* Rev. Chim. Acad. rep. popul. Roumaine *2,* 1 (1957); C.A. *52,* 14635h (1958).
[328] *R.J. Crawford, C. Woo,* Can. J. Chem. *43,* 1541 (1965).
[329] *W.E. Truce, A.R. Naik,* Can. J. Chem. *44,* 299 (1966).
[330] *Y. Ashani, S. Cohen,* J. Med. Chem. *13,* 471 (1970).
[331] *V. Sprio, P. Madonia,* Ann. Chim. (Rome) *50,* 1627 (1960) C.A. *56,* 4713f (1962); *V. Sprio, G.C. Vaccaro,* Ann. Chim. (Rome) *49,* 2075 (1959); C.A. *54,* 16443 (1960).
[332] *E. Marcus, J.K. Chan, J.L. Hughes,* J. Chem. Eng. Data *12,* 151 (1967); C.A. *66,* 54976d (1967).
[333] *R. Rigny, S. Samne,* Compt. Rend. (C) *266,* 1303 (1968); C.A. *69,* 43584k (1968).
[334] *M.M. Kochhar, B.B. Williams, J.H. Fan,* J. Pharm. Sci. *58,* 1383 (1969).

11.1.10 Organometallic oximes

Oximes can be transformed into organometallic compounds. Thus, when aldoximes and ketoximes are reacted with chlorotrialkylsilanes in pyridine, hydrolysis-resistant *O-trialkylsilyl oximes* are obtained[354],[355]:

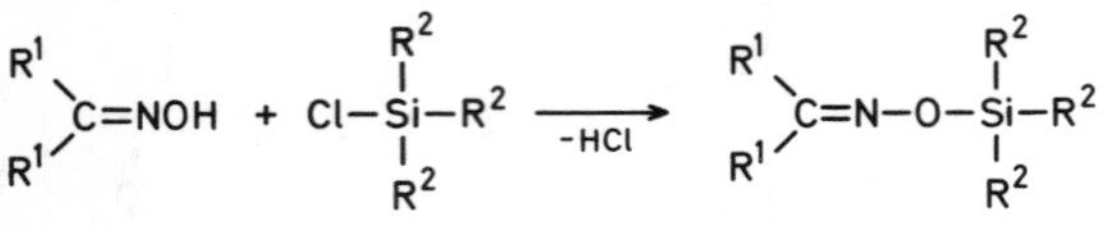

Chlorotrialkyl tin yields herbicidal and fungistatic *trialkyltin oximes*[356]. Alternatively, these compounds are formed from oximes and trialkyltin or triaryltin hydroxide by azeotropic distillation[357] or with bis(trialkyltin) oxide[358]. Similarly, dichloroethylarsine reacts with butanone oxime in triethylamine ether (at 0–5° and then 1-hour heating) to give *ethyl bis(alkylidenaminoxy)arsine*[359]:

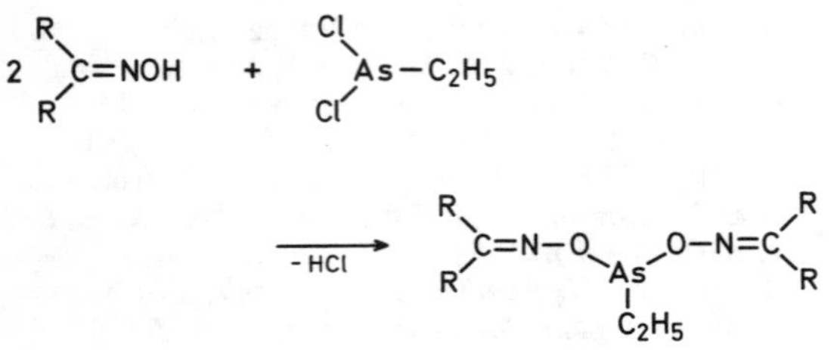

Other reactions lead to the analogous organolead, organogermanium, and organoantimony compounds[360]. Some interest accrues to the prepara-

tion of *2,10-Dioxa-3,9-diaza-1-selena-Se(IV)-tricyclo[6.2.1.0^{4,11}]undecatriene-(1^{11},3,8)* and its *6,6-dimethyl* derivate[361].

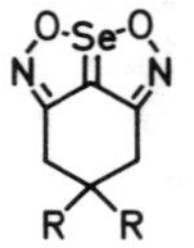

R = H, CH₃

11.2 Transformations

11.2.1 Condensation reactions

Oximes with further reactive functional groups give rise to *heterocycle* formation. Ethyl acetoacetate and benzaldoxime, for example, give *4-benzylidene-3-methyl-2-isoxalin-5-one* after a previous transoximation in the presence of zinc chloride or phosphoric acid[362]:

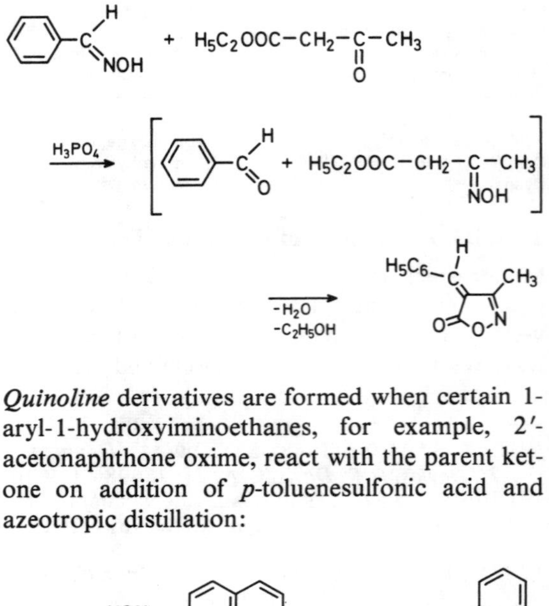

Quinoline derivatives are formed when certain 1-aryl-1-hydroxyiminoethanes, for example, 2′-acetonaphthone oxime, react with the parent ketone on addition of *p*-toluenesulfonic acid and azeotropic distillation:

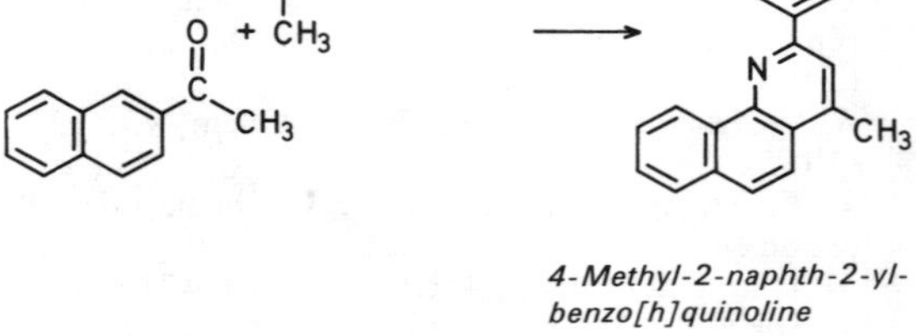

4-Methyl-2-naphth-2-yl-benzo[h]quinoline

[352] US. P. 3 026 303 (1957), DuPont, Inv.: *L.G. Donaruma;* C.A. *57,* 9984i (1962);
L.G. Donaruma, J. Org. Chem. *26,* 577 (1961).

[353] *M.S. Mirkamilova, N.I. Bekasova, V.V. Korshak,* Vysokomolekul Soedin., Ser. A *10,* 771 (1968); C.A. *69,* 19922y (1968).

[354] *B.N. Dolgov, Z.I. Sergeeva, N.A. Zubkova, M.G. Voronkov,* Zh. Obshch. Khim. *30,* 3347 (1960); C.A. *55,* 19765g (1961).

[355] *Z.I. Sergeeva, Z.M. Matveeva, M.G. Voronkov,* Zh. Obshch. Khim. *31,* 2017 (1961); C.A. *55,* 27176g (1961).

[356] US. P. 3 282 672 (1961), Monsanto, Inv.: *G. Weissenberger;* C.A. *66,* 28891c (1967).

[357] Jap. P. 6 724 573 (1964), Yoshitomi Pharm., Inv.: *M. Nakanishi, S. Inamasu;* C.A. *69,* 44022n (1968).

[358] *P.G. Harrison, J.J. Zuckerman,* Inorg. Chem. *9,* 175 (1970); C.A. *72,* 43818b (1970).

[359] *G. Kamai, R.G. Miftakhova, N.G. Gazetdinova,* Zh. Obshch. Khim. *39,* 1798 (1969); C.A. *71,* 124603j (1969).

[360] *P.G. Harrison, J.J. Zuckerman,* Inorg. Nucl. Chem. Lett. *6,* 5 (1970); C.A. *72,* 111579b (1970).

[361] *D. Paquer, M. Perrier, J. Vialle,* Bull. Soc. Chim. France 4517 (1970); C.A. *74,* 87900t (1971).

[362] *J.J. Donleavy, E.E. Gilbert,* J. Amer. Chem. Soc. *59,* 1072 (1937);
W. Ried, A. Czack, Justus Liebigs Ann. Chem. *676,* 130 (1964).

[363] *A. Rosenthal,* J. Org. Chem. *26,* 1638 (1961).

[364] *E. Wenkert, B.F. Barnett,* J. Amer. Chem. Soc. *82,* 4671 (1960).

The reaction cannot be performed with acetophenone and its oxime[363]. Pyrolysis of 2-phenylcyclohexanone oxime leads ultimately to *1,2,3,4-tetrahydrocarbazole* or *carbazole* with ring closure side by side with numerous other products[364]:

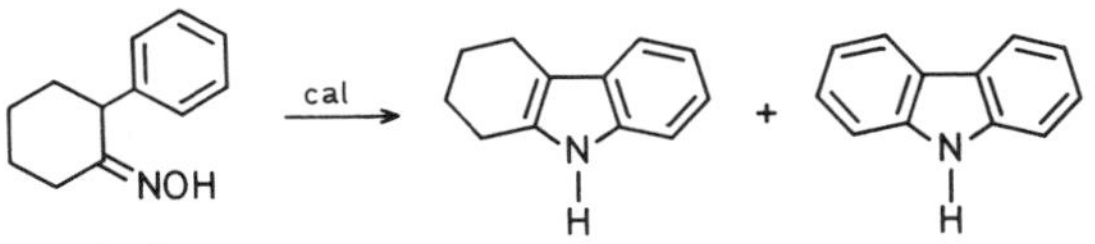

In addition to the reaction of α-oxo oximes with *o*-phenylenediamine to form quinoxalines and the reactions of, in particular, 1-hydroxyimino-2-hydrazones to *1,2,3-triazoles*[365-367] or to *1,2,4-triazine 4-oxides*[368], that of aldehydes and aldoximes deserves a mention.

For example, allowing 2,3-butanone monooxime to stand for some time in the cold with benzaldehyde in hydrogen chloride-saturated acetic acid leads to a 27% yield of *4,5-dimethyl-2-phenyloxazole 3-oxide:*

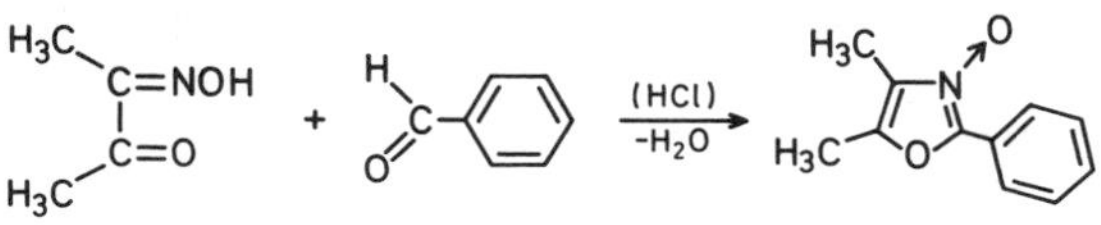

Benzil monooxime reacts analogously to form *2,4,5-triphenyloxazole 3-oxide* in 62% yield[369]. By contrast, working with ammonia in 95% ethanol or in 1,4-dioxane leads to an exothermic reaction producing *1-hydroxyimidazoles*[370, 371], *e.g.:*

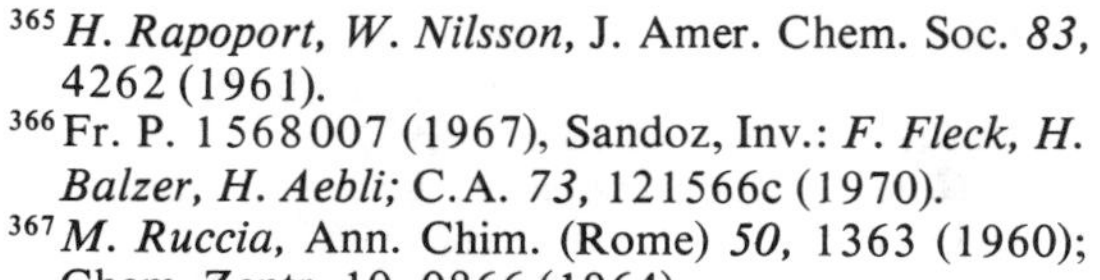

1-Hydroxy-4,5-dimethyl-2-phenylimidazole

With primary amines 1-substituted imidazole 3-oxides are formed which are readily reduced to *imidazoles* with zinc dust in glacial acetic acid in a combined technique in one and the same vessel.

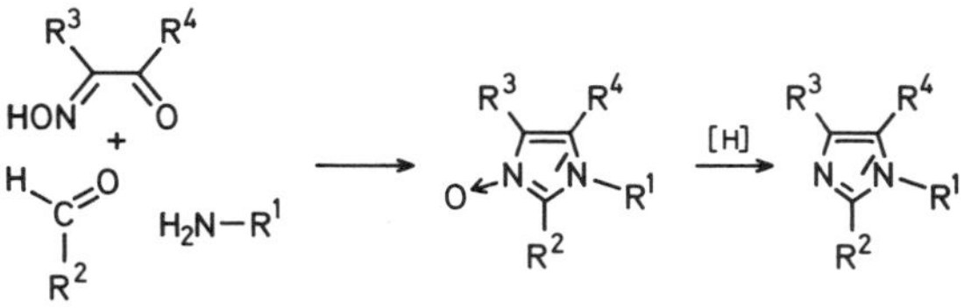

N-Oxides that are sparingly soluble in glacial acetic acid are isolated and reduced in formic acid[372].

1-Hydroxyimidazole 3-oxides are obtained with aldoximes in ethanol containing hydrochloric acid and can be hydrogenated to the corresponding imidazoles under mild conditions with Raney nickel[373-374]:

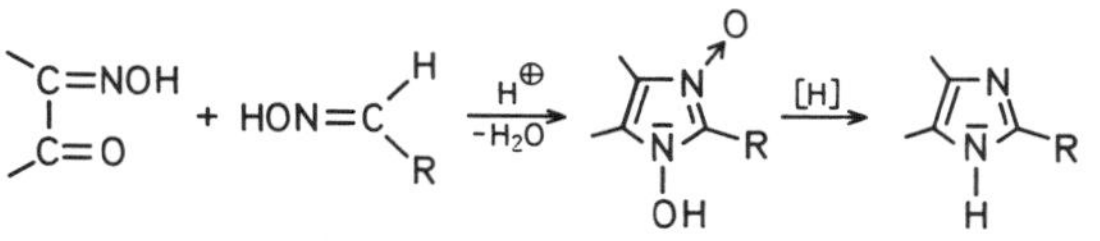

1,2-Dioximes behave similarly towards 2-nitrobenzaldehyde[375]:

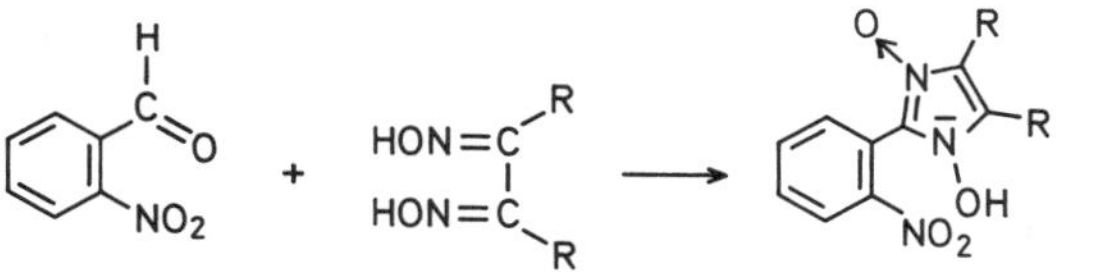

Dehydrogenation with lead(IV) oxide in 1,4-dioxane gives free radicals just as with 1-hydroxyimidazoles[376]. The condensation proceeds also in the absence of solvent and of hydrogen chloride[377].

1,3-Diketone monooximes and oximes of β-oxo carboxylic acids and their derivatives form *isoxazole* or *2-isoxazolin-5-one* compounds in the presence of acids[378, 379], acid chlorides[380], or acid

[365] *H. Rapoport, W. Nilsson*, J. Amer. Chem. Soc. *83*, 4262 (1961).

[366] Fr. P. 1 568 007 (1967), Sandoz, Inv.: *F. Fleck, H. Balzer, H. Aebli;* C.A. *73*, 121566c (1970).

[367] *M. Ruccia*, Ann. Chim. (Rome) *50*, 1363 (1960); Chem. Zentr. 10–0866 (1964).

[368] *F.J. Lalor, F.L. Scott*, J. Chem. Soc. (C) 1034 (1969).

[369] *C.M. Selwitz, A.J. Kosak*, J. Amer. Chem. Soc. *77*, 5372 (1955).

[370] *F.J. Allan, G.G. Allan*, Chem. & Ind. (London) 1837 (1964).

[371] *K. Akagne, F.J. Allan, G.G. Allan, T. Friberg, S.O. Mnircheartaigh, J.B. Thomson*, Bull. Chem. Soc. Japan *42*, 3204 (1969);
K. Akagne, F.J. Allan, G.G. Allan, T. Friberg, S.O. Mnircheartaigh, J.B. Thomson, Synthesis 108, 1226 (1971).

[372] *H. Lettau*, Z. Chem. *11*, 10 (1971).

[373] *K. Bodendorf, H. Towliati*, Arch. Pharm. (Weinheim, Ger.) *298*, 293 (1965).

[374] *H. Towliati*, Chem. Ber. *103*, 3952 (1970).

[375] *P. Franchetti, M. Grifantini*, J. Heterocycl. Chem. *7*, 1295 (1970).

[376] *K. Volkamer, H.W. Zimmermann*, Chem. Ber. *103*, 299 (1970).

[377] *J.B. Wright*, J. Org. Chem. *29*, 1620 (1964).

[378] *O. Diels, K. Schleich*, Ber. *49*, 285 (1916).

[379] *J. Wislicenus*, Justus Liebigs Ann. Chem. *308*, 248 (1899).

[380] *A. Hantzsch*, Ber. *24*, 505 (1891).

anhydrides[381]. A 1,3-diketone monooxime is formed intermediately if an oxime with an α-hydrogen atom is converted into the 1,4-dilithium salt and then reacted with esters of benzoic acid[382, 383]:

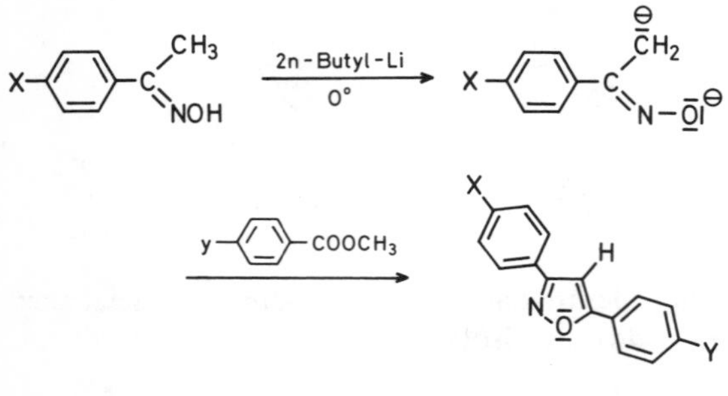

Isoxazoles are invariably formed also when β-oxo-aldehydes or β-oxoaldehyde acetals are used[384]. The following reaction in acetic anhydride seems noteworthy[381, 385]:

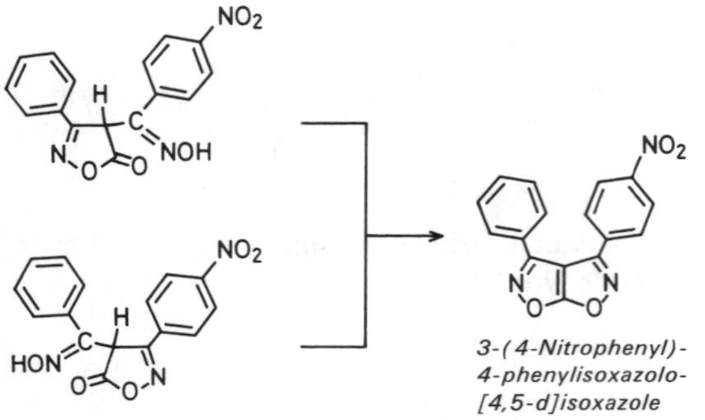

3-(4-Nitrophenyl)-
4-phenylisoxazolo-
[4,5-d]isoxazole

1-Hydroxypyrrole derivatives form on reacting salts of 1,3-dicarbonyl compounds with 2-halo ketoximes[331], e.g.:

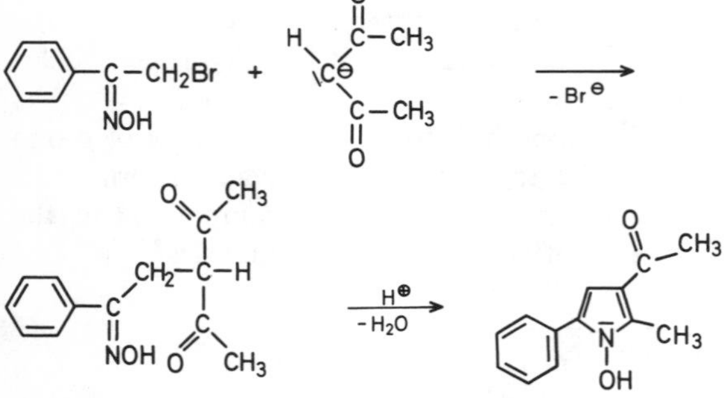

1-Hydroxy-2-
methyl-5-phenyl-
pyrrol-3-yl methyl
ketone

2-(2-Hydroxyiminoethyl)benzoic acid cyclizes at 130° to form *2-hydroxy-1,2-dihydro-1-isoquinoli-none*[386]:

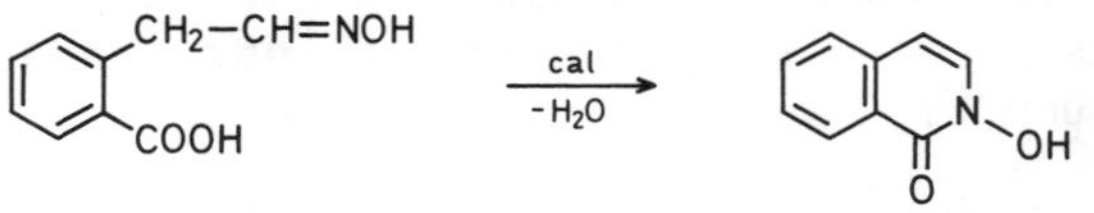

On boiling with 18% hydrochloric acid, ring closure occurs in 1,5-pentanedione dioxime with formation of *pyridine,* but the reverse reaction is more important[387].

Oximes of α,β-unsaturated compounds can be cyclized to *isoxazoles* by acids. 6-Phenyl-3,5-hexadien-2-one oxime is converted into *2-methyl-6-phenylpyridine* on dry distillation[388]:

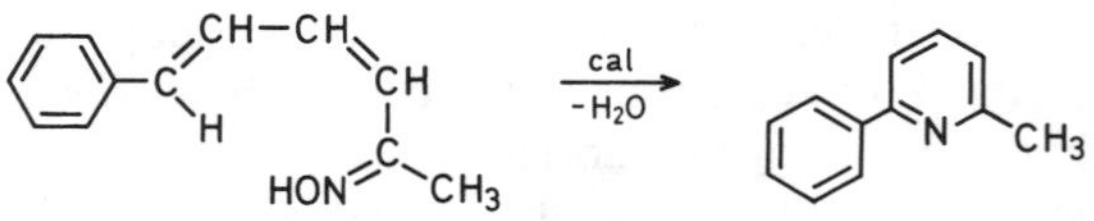

Cinnamaldehyde oxime and other substituted phenylacrolein oximes are reported to give *quino-lines* in acetic anhydride[389], substituted 1-phenyl-1-buten-3-one oxime in the presence of Lewis acids[390] or after *UV* irradiation[391], and *O*-acyl oximes on heating in decalin, nitrobenzene, or formamide[392]. With phosphorus(V) oxide or polyphosphoric acid, *isoquinoline* is formed from cinnamaldehyde oxime, while from 4-phenyl-3-buten-2-one oxime *1-methylisoquinoline* and from 3-methyl-4-phenyl-3-buten-2-one oxime *1,3-dimethylisoquinoline* are obtained in poor yield[393].

[384] *J. Castells, A. Colombo,* Chem. Commun. 1062 (1969).

[385] *G. Lo Vecchio, G. Lamonica, G. Cum,* Gazz. Chim. Ital. *93,* 15 (1963); C.A. *59,* 12779b (1963).

[386] *M.M. Robison, B.L. Robison,* J. Amer. Chem. Soc. *80,* 3444 (1958).

[387] *B.D. Shaw,* J. Chem. Soc. (London) 300 (1937); *J. Colonge, J. Dreux, H. Delplace,* Bull. Soc. Chim. France 447 (1957); C.A. *51,* 17864b (1957).

[388] *M. Scholtz,* Ber. *32,* 1935 (1899).

[389] *J. Glinka, C. Troszkiewicz,* Roczniki Chem. *37,* 1643 (1963); C.A. *61,* 1828e (1964); *C. Troszkiewicz, J. Glinka,* Roczniki Chem. *36,* 1387 (1962); C.A. *59,* 2766e (1963); *J. Glinka,* Roczniki Chem. *39,* 885 (1965); C.A. *64,* 3470f (1966).

[390] *S. Goszczynski,* Zesz. Nauk Politech. Slask Chem. *24,* 231 (1964); C.A. *63,* 11291h (1965); *S. Goszczynski,* Zesz. Nauk Politech. Slask Chem. *25,* 5 (1964); C.A. *63,* 4253d (1965).

[391] *C. Troszkiewicz, S. Goszczynski,* Roczniki Chem. *37,* 919 (1963); C.A. *60,* 500c (1964).

[392] *S. Goszczynski, E. Salwinska,* Zesz. Nauk Politech. Slask Chem. *50,* 83 (1969); C.A. *73,* 44552t (1970).

[381] *A. Quilico, G. Gaudiano, L. Merlini,* Gazz. Chim. Ital. *89,* 571 (1959); C.A. *54,* 12121b (1960).

[382] *C.F. Beam, M.C.D. Dyer, R.A. Schwarz, C.R. Hauser,* J. Org. Chem. *35,* 1806 (1970).

[383] *J.S. Griffiths, C.F. Beam, C.R. Hauser,* J. Chem. Soc. (C) 974 (1971).

Nitro, amino, sulfhydryl, and nitrile groups, and also halogen atoms in the 2- or 3-position are able to undergo condensation reactions with the hydroxyimino group. 2-nitro oximes give furoxans with alkali metal hydroxide or with 98% sulfuric acid[394]:

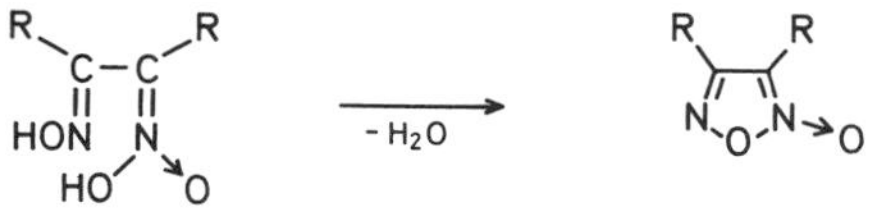

On heating 4-anilino-4-phenyl-2-butanone oxime with dicyclohexylcarbodiimide in acetonitrile cyclization occurs and *1,5-diphenyl-3-methyl-2-pyrazoline* is formed (33% of theory)[395]:

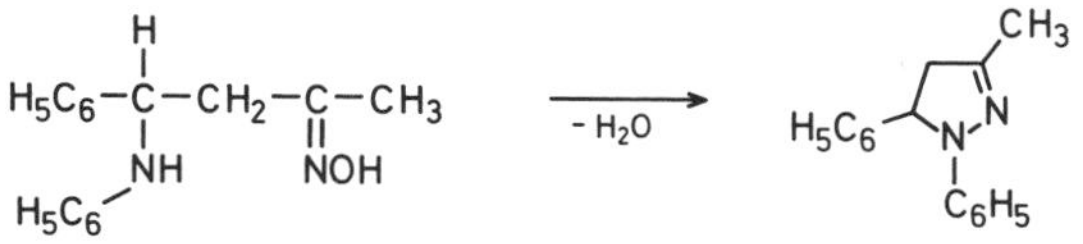

2-Mercaptobenzaldoxime and analogous compounds undergo ring closure with polyphosphoric acid to form benzisothiazoles[396]:

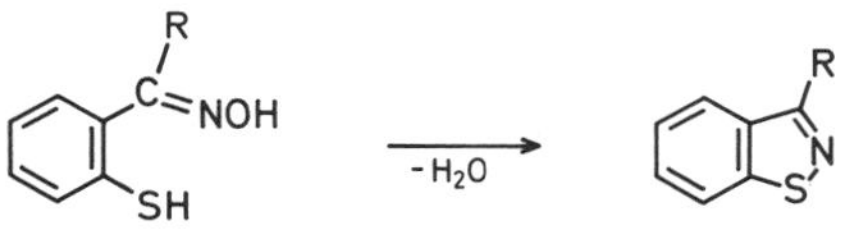

In contrast, 2-halo benzaldoximes are converted into *salicylic acid nitriles* with alkali *via* unstable benzisoxazoles[397]. However, ring closure of, for instance, *salicaldehyde O-acetyloxime* in the presence of silica gel or alumina is claimed in a patent[398]. Oximes of 2-halo benzophenones and 2-halo acetophenones also afford the desired *oxazoles*[399].

In correspondence with their configuration, 2-hydroxyamino oximes react with aldehydes in boiling benzene either to *nitrones* or to *3-hydroxy-3-imidazoline 1-oxides*, e.g.:

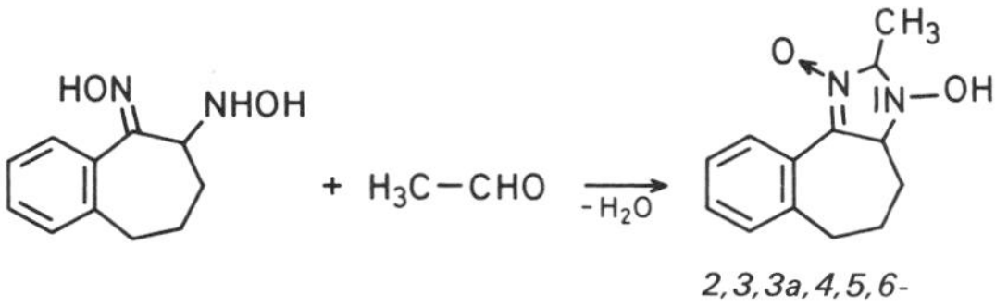

2,3,3a,4,5,6-
Hexahydro-3-
hydroxy-2-
methylbenzo-
[3,4]cyclohept-
[1,2-d]imidazole
1-oxide

The same reaction takes place with ketones[400]. With sulfuric acid autocondensation of the 2-hydroxyamino oximes to *phenazine N,N-dioxides* occurs[401].

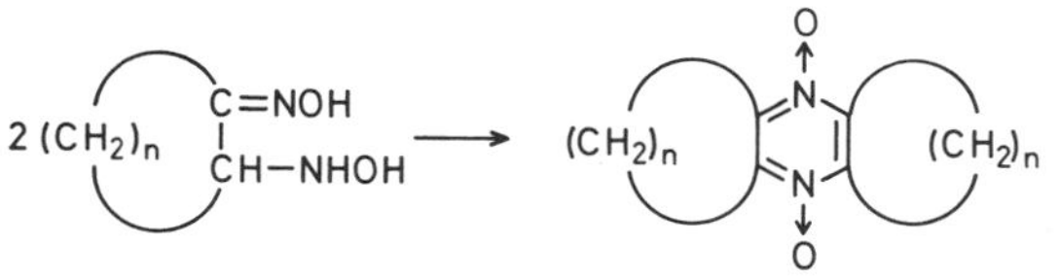

During reaction of 2-aminooximes with aldehydes the oxime configuration is decisive. The *(E)*-isomers form *3-imidazoline N-oxides* in ethanolic solution at room temperature[402], e.g.:

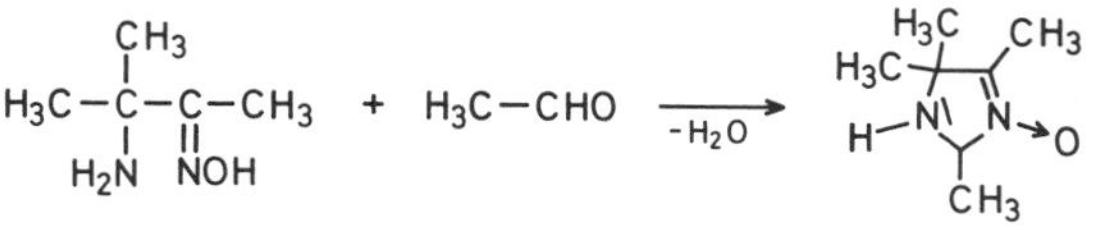

2,4,5,5-Tetra-
methyl-3-imid-
azoline 3-oxide

From *(Z)*-isomers 5,6-dihydro-4H-1,2,5-oxadiazines can be obtained in some instances[403]:

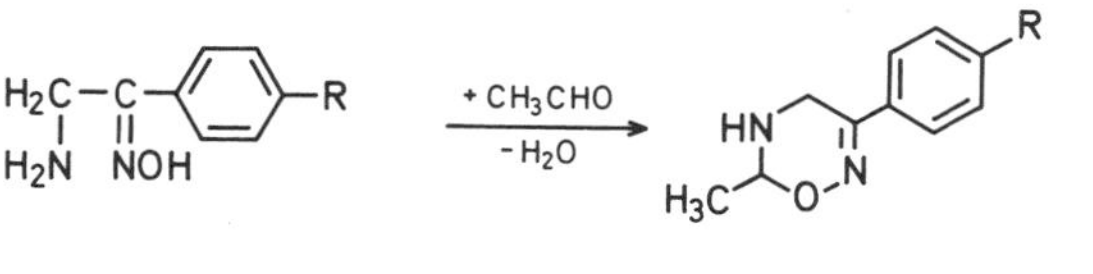

Conversion of the *(E)*-oximes into urethan with ethyl chloroformate followed by alkaline saponification produces ring closure and gives *3-imid-*

[393] *B. Unterhalt*, unpubl.;
S. Goszczynski, Zesz. Nauk Politech. Slask Chem. *25*, 5 (1964); C.A. *63*, 4253d (1965).

[394] *A. Dornow, K.J. Fust, H.D. Jordan*, Chem. Ber. *90*, 2125 (1957);
D. Klamann, W. Koser, P. Weyerstahl, M. Fligge, Chem. Ber. *98*, 1833 (1965);
M.L. Scheinbaum, J. Org. Chem. *35*, 2785 (1970) (Preparation of 2-nitrooximes).

[395] *A. Hassner, M.J. Michelson*, J. Org. Chem. *27*, 300 (1962).

[396] *A. Ricci, A. Martani*, Ann. Chim. (Rome) *53*, 577 (1963); C.A. *59*, 8721d (1963).

[397] *O.L. Brady, A.N. Cosson, A.J. Roper*, J. Chem. Soc. (London) 2427 (1925).

[398] DBP. 1 903 701 (1969), BASF, Erf.: *F. Becke, H. Hagen*; C.A. *73*, 87910a (1970).

[399] *J. Meisenheimer, P. Zimmermann, U. v. Kummer*, Justus Liebigs Ann. Chem. *446*, 205 (1926).

[400] *L.B. Volodarskii, A.N. Lysak, V.A. Koptyug*, Khim. Geterotsikl. Soedin. 766 (1966); C.A. *66*, 115644f (1967);
L.B. Volodarskii, G.A. Kutekova, Tetrahedron Lett. 1065 (1968);
L.B. Volodarskii, A.Y. Tikhonov, Zh. Org. Khim. *6*, 307 (1970); C.A. *72*, 110955j (1970).

[401] *M.L. Scheinbaum*, J. Org. Chem. *35*, 2790 (1970).

[402] *H. Gnichtel*, Chem. Ber. *103*, 2411 (1970).

[403] *H. Gnichtel*, Chem. Ber. *103*, 3442 (1970).

azolin-2-one 3-oxide[404]. With thiophosgene and *O*-ethyl thiocarbonyl chloride the analogous sulfur compounds, probably present in the thione form, are formed[405]:

4-Methyl-5-phenyl-3-imidazolin-
2-one (or 2-thione) 3-oxide

The *(Z)*-oximes furnish *4 H-1,2,5-oxadiazin-6-(5 H)-one* or corresponding *6-thione* with phosgene and thiophosgene[405, 406]:

X = O, S

The reactions with 2-aminobenzophenone oximes have become very important. When ortho carboxylic acid triesters or halocarboxylic acid halides are added, they do not yield benzopyrazoles on heating, but instead lead to *quinazoline 3-oxides* by insertion of a carbon atom, *e.g.*:

6-Chloro-2-chloromethyl-4-
phenylquinazoline 3-oxide

Reacting the above chloromethyl derivative with methylamine, for example, leads to ring expansion, and the pharmaceutically important *7-chloro-2-methylamino-5-phenyl-3 H-1,4-benzodiazepine 4-oxide* is obtained[407, 408].

11.2.2 Removal of the hydroxyimino group

Replacement of the hydroxyimino group by the basic carbonyl function is very important in respect of aldehyde and ketone *purification via* their oximes and for preparing carbonyl compounds *via* a nitrosation reaction.

Hydrolysis of the oximes to aldehydes and ketones is the counterpart of the oxime synthesis (p. 404); by adding protons the equilibrium is displaced to the side of the starting materials[409]. The fissility of the oximes depends on their constitution and increases approximately in the order

aliphatic aldoximes < α,β-unsaturated aldoximes < aromatic aldoximes < ketoximes

The optimum *pH* is said to be 2.3[410]. Isolation of the carbonyl compounds is performed either by extraction with an appropriate solvent or by steam distillation, as was shown with experiments on cycloalkanone oximes[411].

Acid chlorides[412] or salts sometimes serve to regenerate carbonyl compounds[412]. Thus *benzophenone* is formed from benzophenone oxime with nitrosyl chloride, phosgene, ammonium chloride, anhydrous zinc sulfate, or various metal oxides[413].

To enhance the yield of the oxime hydrolysis, the equilibrium constituent hydroxylamine can be removed by adding one of the following oxidizing agents:

[404] *A. Dornow, H.H. Marquardt,* Chem. Ber. *97,* 2169 (1964).

[405] *H. Gnichtel, S. Exner, H. Bierbüße, M. Alterdinger,* Chem. Ber. *104,* 1514 (1971).

[406] *H. Gnichtel, S. Thiele,* Chem. Ber. *104,* 1507 (1971).

[407] *L.H. Sternbach,* Angew. Chem. *83,* 70 (1971).

[408] *A.V. Bogatskii, S.A. Andronati,* Russian Chem. Reviews 1064 (1970).

[409] *B.J. Gregory, R.B. Moodie, K. Schofield,* J. Chem. Soc. (B) 1687 (1970) (Kinetics of hydrolyses of some oximes in sulfuric acid).

[410] *M.I. Winnik, N.G. Sarachani,* J. Physik. Chem. (USSR) *34,* 2671 (1960); Chem. Zentr. 8105 (1963).

[411] Fr. P. 1 247 736 (1960), BASF, Inv.: *O.v. Schickh, H. Metzger;* Chem. Zentr. 52–2408 (1964); DBP. 1 080 102 (1959), BASF, Erf.: *O. v. Schickh, H. Metzger;* Chem. Zentr. 8473 (1961).

[412] *H. Leuchs, H. Rauch,* Ber. *48,* 1531 (1915).

[413] *E. Beckmann, E. Bark,* J. Prakt. Chem. [2] *105,* 342 (1922/23).

[414] *F. Henle, G. Schupp,* Ber. *38,* 1372 (1905); *L. Claisen, O. Manasse,* Ber. *22,* 530 (1889); *T. Wieland, D. Grimm,* Chem. Ber. *96,* 275 (1963).

[415] *C. Mannich, H. Budde,* Arch. Pharm (Weinheim, Ger.) *270,* 285 (1932); *C. Harries,* Ber. *35,* 1184 (1902).

[416] DBP. 1 078 570 (1959), BASF, Erf.: *O. v. Schickh, H. Metzger;* C.A. *55,* 25808e (1961); *S. Gabriel, R. Meyer,* Ber. *14,* 2334 (1881).

Nitrous acid[414] and alkyl nitrites[415], potassium permanganate[416], potassium dichromate[416], iron(III) chloride[417], bromine[418], ozone[419], and cerium ammonium nitrate[420].

Side-reactions, for example, 'pernitroso' compound formation, needs to be guarded against (see p. 435).

Alternatively, the free hydroxylamine can be removed by reduction. Thus, mild hydrolysis can be achieved with *chromium(II) acetate*[421]. The carbonyl compounds are formed in 55–81% yield if the oximes are boiled with equimolar iron pentacarbonyl and a catalytic amount of boron trifluoride in dibutyl ether in an atmosphere of nitrogen[422].

Reaction with *titanium(III)* chloride in sodium acetate-buffered aqueous methanol or 1,4-dioxane at room temperature also yields carbonyl compounds. *Imines* can be isolated from sterically hindered ketoximes[423].

Treatment with a *second,* more reactive carbonyl compound such as formaldehyde[424] or 4-oxopentanoic acid (levulinic acid) can afford the carbonyl compound from the oxime. The levulinic acid can be employed selectively[425]. Steroid ketones can be liberated from their oximes with pyruvic acid in dilute acetic acid[426].

The *stability* of *carbonyl* derivatives increases in the order azomethine < oxime < hydrazone < semicarbazone < thiosemicarbazone.

Thus, normally an oxime can be converted into a derivative on its right which, in addition, is less soluble. As a result, certain oximes, *e.g.,* 4-nitrobenzaldoxime, can be determined gravimetrically as 2,4-dinitrophenylhydrazones[427]. In dependence on their configuration, monooximes of 1,2-diketones react either to form monohydrazone or monohydrazone oxime[428]. 1,4-Dihydro-4-pyridazinone oximes are formed analogously when 3-pyrrolinone oximes are reacted with hydrazine hydrochloride in aqueous solution[429]:

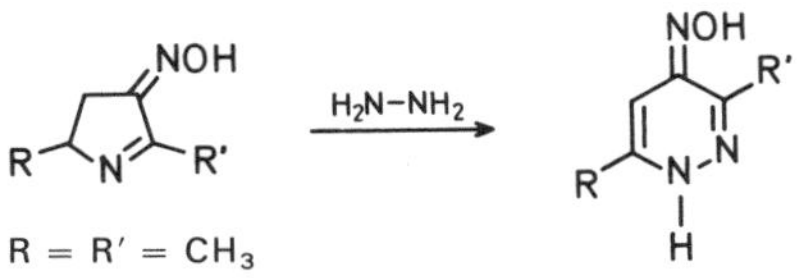

11.2.3 Reduction of oximes

Oximes can be converted reductively into *hydroxylamines* (see p. 66), or amines (see p. 503) and, when certain structural conditions are met, into *aziridines* (see p. 594).

11.2.4 Oxidation of oximes

Under certain conditions ketoximes are oxidized to *nitro* compounds, for example, by using trifluoroperacetic acid[430, 431].

Treatment of aldoximes with peroxymonosulfuric acid or hydrogen peroxide gives *hydroxamic* acids.

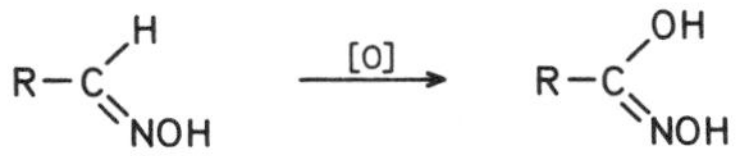

With sodium hypobromite in alkaline solution and with *N*-bromosuccinimide in dimethylformamide-sodium ethoxide aromatic aldoximes form *nitrile oxides*[432]. Simple *(E)*-aldoximes also react to nitrile oxides with lead(IV) acetate in dichloromethane at −78°; as the temperature rises the yield falls sharply[433].

(Z)-Aldoximes vary in their behavior. Among aliphatic compounds dimeric *1-nitrosoalkyl esters* are one product formed *via* free-radical intermediate stages. They are in equilibrium with the monomeric form and readily rearrange to acetyl-

[417] S. Gabriel, Ber. *15*, 2004 (1882).
[418] O. Piloty, Ber. *30*, 3164 (1897).
[419] C. Harries, Justus Liebigs Ann. Chem. *343*, 323 (1905).
[420] J.W. Bird, D.G.M. Diaper, Can. J. Chem. *47*, 145 (1969).
[421] E.J. Corey, J.E. Richman, J. Amer. Chem. Soc. *92*, 5276 (1970).
[422] H. Alper, J.T. Edward, J. Org. Chem. *32*, 2938 (1967).
[423] G.H. Timms, E. Wildsmith, Tetrahedron Lett. 195 (1971);
G.H. Timms, E. Wildsmith, Synthesis 272 (1971).
[424] B. Eistert, H.-K. Witzmann, Justus Liebigs Ann. Chem. *744*, 106 (1971);
R. Fischer, T. Wieland, Chem. Ber. *93*, 1387 (1960);
A. Dornow, A. Müller, Chem. Ber. *93*, 33 (1960).
[425] C.H. De Puy, B.W. Ponder, J. Amer. Chem. Soc. *81*, 4629 (1959).
[426] E.B. Hershberg, J. Org. Chem. *13*, 543 (1948).
[427] O.L. Brady, F.H. Peakin, J. Chem. Soc. (London) 478 (1929).

[428] M.O. Forster, E. Kunz, J. Chem. Soc. (London) *105*, 1718 (1914);
H. Rupe, S. Kessler, Ber. *42*, 4717 (1909).
[429] T. Aiello, M. Miraglia, R. Torcetta, Gazz. Chim. Ital. *77*, 525 (1947); C.A. *42*, 3409h (1948).
[430] R.J. Sundberg, P.A. Bukowick, J. Org. Chem. *33*, 4098 (1968).
[431] R. El Bacha, A. Deluzarche, C. Tanielian, A. Pousse, Compt. Rend. (C) *271*, 648 (1970);
R. El Bacha, A. Deluzarche, C. Tanielian, A. Pousse, Chem. Inform. 50–181 (1970).
[432] C. Grundmann, R. Richter, J. Org. Chem. *33*, 476 (1968).
[433] G. Just, K. Dahl, Tetrahedron *24*, 5251 (1968).

hydroxamic acids. Aromatic *(Z)*-aldoximes furnish *aldoxime anhydride N-oxides via* iminoxy radicals[486]; cf. refs. [433, 434], *e.g.*:

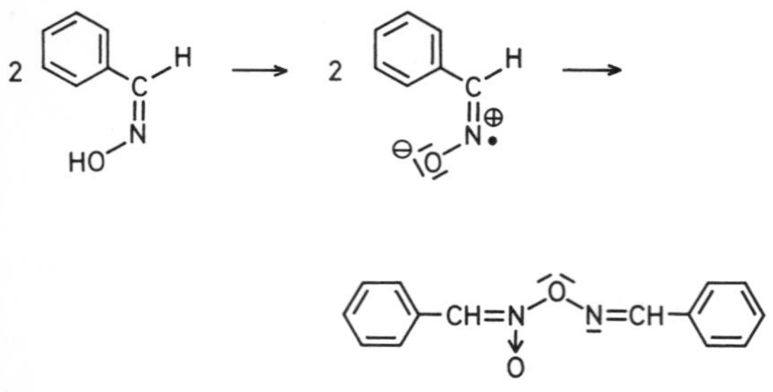

Benzaldehyde oxime anhydride
N-oxide

On oxidation with lead(IV) acetate or benzoate, aliphatic and alicyclic ketoximes are converted into 1-nitrosoalkyl esters. While the *acetyloxy* derivatives dimerize only with difficulty and as a result are not easily purified, the more readily dimerizing *benzoyloxy* compounds are easier to handle. In the presence of other carboxylic acids interchange of the acyloxy groups may take place during the reaction[435, 436].

Aliphatic ketoximes with a branched alkyl group furnish *free radicals* with potassium hexacyanoferrate(III), lead(IV) oxide, and silver oxide in benzene or carbon tetrachloride, as was shown by the electron resonance spectra[437, 438]. Oximes of bicyclic ketones behave likewise when lead(IV) acetate is added. The end-products were not isolated[439–443].

With hindered alicyclic oximes such as, for instance, 2, 2, 6, 6-tetramethylcyclohexanone oxime ring fission occurs and nitrile oxides are formed, which react with the acetic acid to give *acetyl hydroxamic acids:*

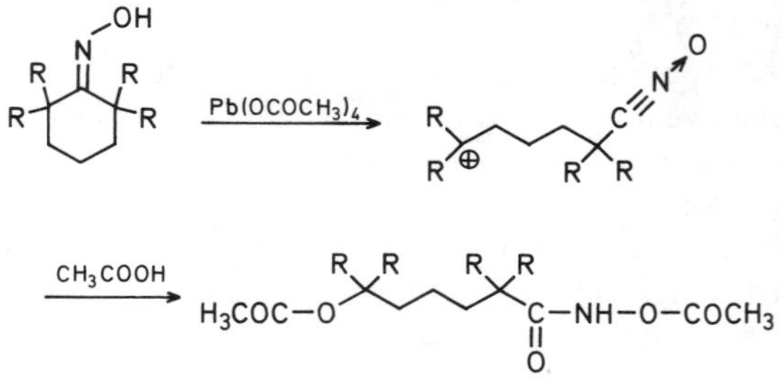

Like aliphatic ketoximes, camphor oxime evidently forms *2-nitrosobornyl acetate,* which subsequently opens its ring[444].

With aromatic ketoximes the oxidation with lead(IV) acetate likewise proceeds *via* free-radical stages[434, 445].

The reaction of oximes with nitrous acid or nitrosyl halide proceeds *via N-nitroso* compounds to

the *initial carbonyl* compounds (**1**)
nitrimines (**2**)
'pernitroso' compounds (**3**)

In simple oximes reactions **1** and **2** proceed in dependence on the substituent in the 2-position. Thus, camphor oxime (2-bornanone oxime), 1, 3, 3-trimethyl-2-norbornanone oxime (fenchone oxime), and 2, 2-dimethylbutanone oxime form nitrimines (*2-nitriminobornane, 2-nitrimino-1,3,3-trimethylnorbornane,* and *2,2-dimethyl-3-nitriminobutane* respectively)

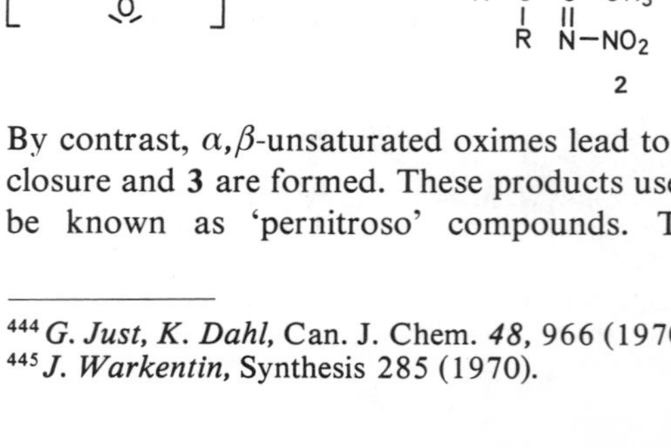

By contrast, α,β-unsaturated oximes lead to ring closure and **3** are formed. These products used to be known as 'pernitroso' compounds. Thus,

[434] *M.M. Frojmovic, G. Just,* Can. J. Chem. *46,* 3719 (1968).
[435] *H. Kropf, R. Lambeck,* Justus Liebigs Ann. Chem. *700,* 1 (1966).
[436] *Y. Yukawa, M. Sakai, S. Suzuki,* Bull. Chem. Soc. Japan *39,* 2266 (1966) [Deoximation with lead (IV) acetate]; C.A. *66,* 10695n (1967).
[437] *M. Fedtke, H. Mitternacht,* Z. Chem. *4,* 389 (1964).
[438] *B.C. Gilbert, R.O.C. Norman,* J. Chem. Soc. (B) 86 (1966).
[439] *A. Caragheorgheopol, M. Hartmann, R. Kühmstedt, V.E. Sahini,* Tetrahedron Lett. 4161 (1967).
[440] *A. Caragheorgheopol, U. Gräfe, M. Hartmann, V.E. Sahini, K. Wermann,* Tetrahedron Lett. 3035 (1970).
[441] *H. Caldararu, N. Barbulescu, L. Ivan, V.E. Sahini,* Tetrahedron Lett. 3039 (1970).
[442] *R.O.C. Norman, B.C. Gilbert,* Adv. Phys. Org. Chem. *5,* 83 (1967), Oximes (radicals).
[443] *E.G. Rozantsev,* Free Nitroxyl Radicals, p. 41, Plenum Press, New York, London 1970 (Org. nitroxyls with localized unpaired electrons).

[444] *G. Just, K. Dahl,* Can. J. Chem. *48,* 966 (1970).
[445] *J. Warkentin,* Synthesis 285 (1970).

3,5,5-trimethyl-5 H-pyrazole, 1,2-dioxide is obtained from 4-methyl-3-penten-2-one oxime with pentyl nitrite in glacial acetic acid[446–448]:

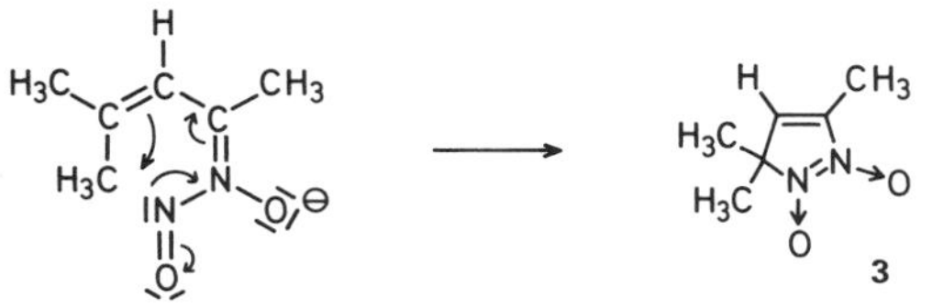

Unbranched α,β-unsaturated oximes such as 4-aryl-3-buten-2-one oximes yield *3-aryl-5-methyl-4 H-pyrazol-4-one 1,2-dioxide oximes* with nitrous acid in glacial acetic acid and the corresponding ketones with nitric acid; 1,3-diarylpropen-3-one oximes largely ketones:

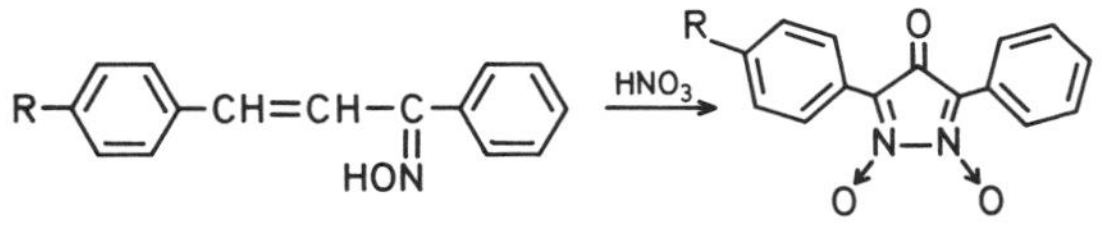

Their structure has been secured by synthesis and by proton resonance spectroscopy[449]. The symmetrical make-up of the *3,5-dimethyl-4 H-pyrazolone 1,2-dioxide*, too, has been confirmed[450,451].

4-Aryl-3-buten-2-one oximes branched in the α-position display ring closure with formation of *2-hydroxypyrazole 1-oxides*[449,452]:

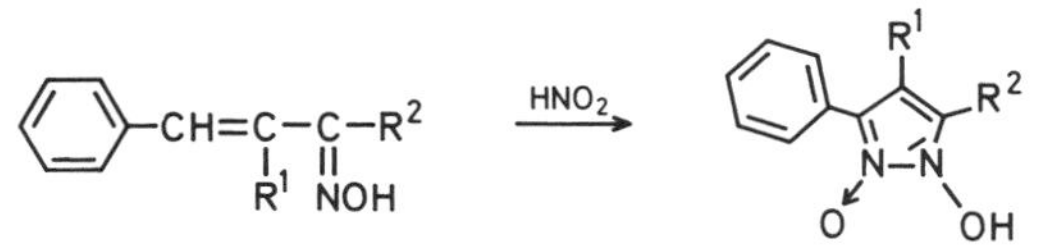

11.2.5 Rearrangements and other reactions of oximes

Under mild conditions, for instance, with phosphorus(V) chloride in ether or benzene, the *Beckmann rearrangement*[453] leads to carboxamides by intramolecular migration of the group

that is *trans* with respect to the hydroxy group[454] (p. 566). *Determining* whether oximes have the *(E)-* or *(Z)*-configuration is made possible in this way. The photorearrangement also has become very important[455–461].

Ketoxime *O*-tosylates with α-hydrogen atoms give a good yield of *2-amino ketones*[462] with bases. *O*-Aryl oximes rearrange to *benzofurans* under the influence of acids in analogy with the Fischer indole synthesis[299,300,302,463], e.g.:

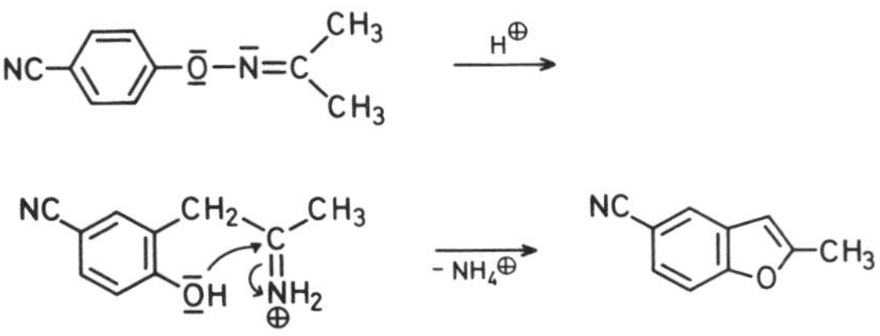

4-Hydroxy-3-(2-iminopropyl)-benzonitrile   2-Methyl-5-benzofurancarbonitrile

In the example cited the intermediate can be isolated[303]. 1-Alkyl-4-piperidone *O*-aryloximes also react in this way in ethanolic hydrochloric acid; the acetal 1 can be separated[464,465]

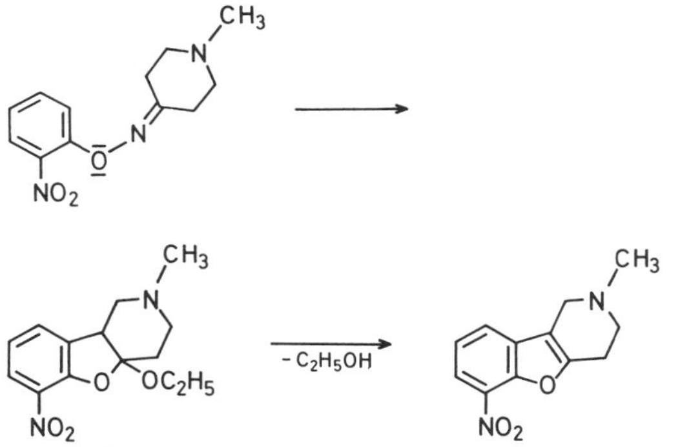

4a-Ethoxy-1,2,3,4,4a,9b-hexahydro-2-methyl-6-nitrobenzofuro[3,2-c]-pyridine   *1,2,3,4-Tetrahydro-2-methyl-6-nitrobenzo-furo[3,2-c]pyridine*

From *O*-vinyl oximes substituted pyrroles and not the expected furans are obtained[466], e.g.:

[446] T. Wieland, D. Grimm, Chem. Ber. *96*, 275 (1963).

[447] J.P. Freeman, J. Org. Chem. *27*, 1309 (1962).

[448] C.Y. Shine, K.P. Park, L.B. Clapp, J. Org. Chem. *35*, 2063 (1970).

[449] B. Unterhalt, Tetrahedron Lett. 1841 (1968);
B. Unterhalt, Arch. Pharm. (Weinheim, Ger.) *300*, 822 (1967).

[450] J.P. Freeman, D.L. Surbey, Tetrahedron Lett. 4917 (1967).

[451] J.P. Freeman, J.J. Gannon, D.L. Surbey, J. Org. Chem. *34*, 187 (1969).

[452] J.P. Freeman, J.J. Gannon, J. Heterocycl. Chem. *3*, 544 (1966);
J.P. Freeman, J.J. Gannon, J. Org. Chem. *34*, 194 (1969).

[453] E. Beckmann, Ber. *19*, 988 (1886).

[454] R.K. Hill, R.T. Conley, O.T. Chortyk, J. Amer. Chem. Soc. *87*, 5646 (1965) (Intermolecular migration);
R.T. Conley, J. Org. Chem. *28*, 278 (1963).

[455] H. Izawa, P. de Mayo, T. Tabata, Can. J. Chem. *47*, 51 (1969).

[456] T. Sato, H. Obase, Tetrahedron Lett. 1633 (1967).

[457] H. Suginome, H. Takahashi, Tetrahedron Lett. 5119 (1970).

[458] B.L. Fox, H.M. Rosenberg, Chem. Commun. 1115 (1969).

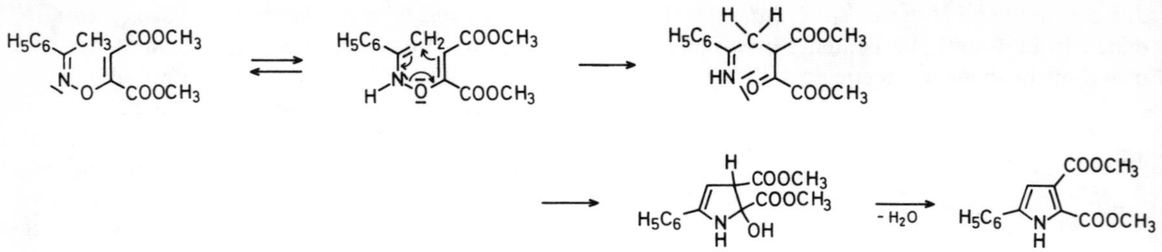

Dimethyl 5-phenyl-
2,3-pyrrole-
dicarboxylate

(Z)-Aldoximes are cleaved particularly readily into nitrile either on alumina or thoria contacts or when one of numerous dehydrating agents is used[467–470]. One important use of this reaction is chain degradation in sugar chemistry.

Isonitriles are an alternative product depending on the reaction conditions, the configuration of the oximes, and the substituents. Thus, with toluenesulfonyl chloride *(E)*-3,5-dimethyl-4-hydroxybenzaldoxime yields 70% *3,5-dimethyl-4-hydroxybenzylisonitrile*, but the *(Z)*-isomer gives the expected *3,5-dimethyl-4-hydroxybenzonitrile*[471,472].

Ketoximes with α-substituents relative to the oxime function often suffer fragmentation[473]:

2-Hydroxy ketoximes[474]
2-Alkoxy ketoximes[475,476]
2-Alkylthio ketoximes[477,478]
2-Amino ketoximes[476,479,480]
Diketone mono-oximes
2-Hydroxyimino carboxylic acids[481]

2-Hydroxyimino carboxylic acid esters[481]
α,β-Unsaturated ketoximes[482,483]
2,2-Diaryl ketoximes[484]
2-Trisubstituted ketoximes[485]

11.3 Bibliography

P. Kurtz in *Houben-Weyl*, Methoden der organischen Chemie, Bd. VIII, p. 252, Georg Thieme Verlag, Stuttgart 1952 (Methoden zur Herstellung und Umwandlung von Nitrilen und Isonitrilen).

O. Touster, Org. Reactions VII, 6, 327 (1953) (The nitrosation of aliphatic Carbon atoms).

B.G. Gowenlock, W. Lüttke, Quart. Rev. Chem. Soc. 12, 321 (1958) (Structure and properties of C-Nitroso-Compounds).

[459] *G. Just, L.S. Ng*, Can. J. Chem. *46*, 3381 (1968).

[460] *T. Sasaki, S. Eguchi, T. Toru*, Chem. Commun. 1239 (1970).

[461] *R. Beugelmans, J.-P. Vermes*, Bull. Soc. Chim. France 342 (1970); *R. Beugelmans, J.-P. Vermes*, Chem. Inform. 26–161 (1970).

[462] *C. O'Brien*, Chem. Rev. *64*, 81 (1964) (The Neber Rearrangement).

[463] *D. Kaminsky, J. Shavel jr., R.I. Meltzer*, Tetrahedron Lett. 859 (1967).

[464] *C.J. Cattanach, R.G. Rees*, J. Chem. Soc. (C) 53 (1971).

[465] *T. Sheradsky, G. Salemnick*, J. Org. Chem. *36*, 1061 (1971) [Rearrangement of O-(2-pyridyl)-oximes].

[466] *T. Sheradsky*, Tetrahedron Lett. 25 (1970).

[467] *I. Hagedorn, W.H. Gündel, K. Schoene*, Arzneimittel-Forsch. *19*, 603 (1969) (Formation of nitriles from phosphorylated oximes under physiological conditions); *L.J. Lohr, R.W. Warren*, J. Chromatogr. *8*, 127 (1962) (Gas chromatographic dehydration of benzaldoxime and salicylic aldehyde oxime); *G. Just, C. Pace-Asciak*, Tetrahedron *22*, 1072 (1966) (Photochemical dehydration).

[468] *R. Appel, R. Kleinstück, K.-D. Ziehn*, Chem. Ber. *104*, 2025 (1971).

[469] *W. Lehnert*, Tetrahedron Lett. 559 (1971).

[470] *J.M. Prokipcak, P.A. Forte*, Can. J. Chem. *49*, 1321 (1971).

[471] *E. Müller, B. Narr*, Z. Naturforsch. *16b*, 845 (1961).

[472] *I. Ugi, U. Fetzer, U. Eholzer, H. Knupfer, K. Offermann*, Angew. Chem. *77*, 502 (1965).

[473] *R.K. Hill*, J. Org. Chem. *27*, 29 (1962).

[474] *C.W. Shoppee, S.K. Roy*, J. Chem. Soc. (London) 3774 (1963).

[475] *M. Ohno, N. Naruse, I. Terasawa*, Org. Synth. *49*, 27 (1969).

[476] *K. Lunkwitz, W. Pritzkow, G. Schmid*, J. Prakt. Chem. *309*, 319 (1968).

[477] *R.L. Autrey, P.W. Scullard*, J. Amer. Chem. Soc. *90*, 4924 (1968).

[478] Jap. P. 7032688 (1967), Toray, Inv.: *M. Ohno, I. Terasawa*; C.A. *74*, 87444x (1971).

[479] *C.A. Grob*, Angew. Chem. *81*, 552 (1969).

[480] DBP. 1955038 (1968), Ajinomoto Co., Erf.: *T. Ito, T. Ichikawa, K. Nagata, T. Kato*; C.A. *73*, 34829z (1970) (With two cyclic 2-aminooximes a Beckmann rearrangement is supposed to occur).

[481] *K. Friedrich, H. Straub*, Chem. Ber. *103*, 3366 (1970).

[482] *B. Unterhalt*, Arch. Pharm. (Weinheim, Ger.) *300*, 748 (1967).

[483] *J. Wiemann, N. Thoai, H. Poksoon*, Bull. Soc. Chim. France 3920 (1967); C.A. *68*, 49004x (1968).

[484] *A. Hassner, E.G. Nash*, Tetrahedron Lett. 525 (1965).

[485] *R.T. Conley, B.E. Nowak*, J. Org. Chem. *27*, 3196 (1962).

[486] *B. Unterhalt, U. Pindur*, Chimia (Zürich) *27*, 210 (1973).

J.V.R. Kaufmann, J.P. Picard, Chem. Rev. *59,* 429 (1959) (The Furoxans).

J.H. Boyer in *R.C. Elderfield,* Heterocycl. Compounds 7, p. 462, John Wiley & Sons, New York 1961 (Oxadiazoles).

L.G. Donaruma, W.Z. Heldt, Org. Reactions XI, 1 (1960) (The Beckmann Rearrangement).

P.A.S. Smith in *P. De Mayo,* Molecular Rearrangements, Vol. I, p. 457, John Wiley & Sons, New York 1963 (Rearrangement involving migration to an electron-deficient *N* or *O*).

W.P. Jencks, Progr. Phys. Org. Chem. *2,* 63 (1964) (Mechanism and Catalysis of Simple Carbonyl Group Reactions).

C.V. Banks, Record Chem. Progr. *25,* 85 (1964).

J. Hamer, A. Macaluso, Chem. Rev. *64,* 473 (1964) (Nitrones).

C. O'Brien, Chem. Rev. *64,* 81 (1964) (The Neber Rearrangement).

G.R. Delpierre, M. Lamchen, Quart. Rev. Chem. Soc. *19,* 329 (1965).

N.V. Sidgwick, The Organic Chemistry of Nitrogen, 3. Aufl., p. 310, Clarendon Press, Oxford 1966 (The Oximes).

P.A.S. Smith, The Chemistry of Open-Chain Org. Nitrogen Compounds, Vol. II, p. 29, W.A. Benjamin, New York, Amsterdam 1966 (Oximes and Derivatives).

R.O.C. Norman, B.C. Gilbert, Adv. Phys. Org. Chem. 5, 83 (1967) (Oximes, Radicals).

H. Metzger in *Houben-Weyl,* Methoden der organischen Chemie, Bd. X/4, 4. Aufl., p. 1, Georg Thieme Verlag, Stuttgart 1968.

A.J. Boulton, P.B. Ghosh, Advan. Heterocycl. Chem. 10, 1 (1969) (Benzofuroxans).

H. Feuer, The Chemistry of the Nitro and Nitroso Groups, Part I (1969); II (1970), Interscience Publishers, London, New York, Sidney, Toronto 1970.

Y. Ogata, A. Kawasaki in *S. Patai,* The Chemistry of the Carbonyl Group, Vol. II, Chap. I, Interscience Publishers, London, New York, Sidney, Toronto 1970.

S.A. Shevelev, Russian Chem. Reviews 844 (1970) (Dual Reactions of Ambident Anions).

S. Patai, The Chemistry of the Carbon-Nitrogen Double Bond, Interscience Publishers, London, New York, Sidney, Toronto 1970.

E.G. Rozantsev, Free Nitroxyl Radicals, p. 41, Plenum Press, New York, London 1970 (Org. Nitroxyls with localized unpaired electrons).

H. Metzger, H. Meier in *Houben-Weyl,* Methoden der organischen Chemie, Bd. X/1, Georg Thieme Verlag, Stuttgart 1971 (Methoden zur Herstellung und Umwandlung von aliphat. Nitroso-Verbindungen).

12 Amines

Contributed by

W. Schneider
J. Hoyer
Pharmazeutisches Institut der Universität,
Freiburg im Breisgau

W. Ehrenstein
Bayer AG
Zentralbereich Patente, Marken und Lizenzen
Leverkusen — Bayerwerk

R. Haller
W. Hänsel
Pharmazeutisches Institut der Universität,
Freiburg im Breisgau

W. Schneider
K. Lehmann
Pharmazeutisches Institut der Universität,
Freiburg im Breisgau

H.J. Roth
Pharmazeutisches Institut der Universität,
Bonn

H. Schönenberger
Pharmazeutisches Institut der Universität,
München

B. Camerino
Monte Edison S.P.A.
Milano

G.F. Cainelli
Università di Bologna, Istituto Chimico,
Bologna

M. Ferles
Faculty of Organic Chemistry,
Vysoká Škola Chemiko-Technologická,
Prague

12.0 Introduction

Amines are organic derivatives of ammonia. In dependence on whether one, two, or three hydrogen atoms of the ammonia molecule have been replaced by hydrocarbon substituents they are classified as primary, secondary, or tertiary amines. All four hydrogen atoms of the ammonium ion can be exchanged against organic substituent groups. Unlike ammonium hydroxide, the hydroxides of such tetrasubstituted ammonium ions cannot decompose into water and anhydro bases. Like the alkali metal hydroxides, they can therefore be isolated in substance and display a similarly strong basicity. Aromatic amines are those in which the amine nitrogen is linked directly with one aromatic or heteroaromatic system. In all other cases the amines are aliphatic amines, even where the open-chain or cyclic aliphatic groups linked to the amine nitrogen contain aromatic substituents.

The substituents exert an important influence on the *basicity*. Since alkyl groups have an electron-repelling effect, primary amines should be more basic than ammonia, secondary amines should be more basic than primary ones, *etc*. Table 1 shows that the experimental pK_b values confirm these expectations.

Tertiary bases alone are weaker bases than secondary or primary ones; steric hindrance of the solvation is adduced as an explanation. The basicity, notably of the tertiary amines, is strongly solvent dependent. In nonaqueous solvents such as 1,4-dioxane or ethyl acetate which are strongly solvating the tertiary amines are once again less basic than secondary amines, like in aqueous solution. However, in chloroform the gradation

tertiary > secondary > primary

that is intrinsically anticipated from the proton affinity is found. Aromatic substituents have the opposite effect. Here, the lone electron pair of the amino group blends with the cloud of the delocalized π-electrons of the aromatic system to cause the basicity to be reduced considerably.

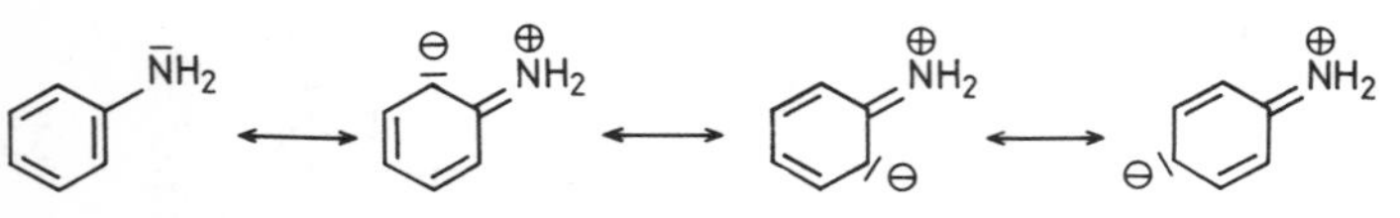

Table 1. Influence of substitution on the basicity of amines

$$K_b = \frac{[\text{Amine} \times \text{H}^\oplus] \times [\text{OH}^\ominus]}{\text{Amine}}; \ pK_b = -\log K_b$$

Amine	pK_b	Amine	pK_b
Ammonia	4.73	Phenethylamine	4.17
Methylamine	3.36	3-Phenylpropyl-	
Dimethylamine	3.29	amine	3.80
Trimethylamine	4.20	Cyclohexylamine	3.39
Ethylamine	3.33	Pyrrolidine	2.9
2-(Hydroxyethyl)-		Piperidine	2.9
amine	4.56	Morpholine	5.6
Bis-(2-hydroxyethyl)-		Aniline	9.42
amine	5.12	*N*-Methylaniline	9.15
Tris-(2-hydroxyethyl)-		*N*,*N*-Dimethylaniline	8.94
amine	6.23	2-Toluidine	9.61
2,2,2-Trifluoroethyl-		2-Chloroaniline	11.23
amine	8.3	3-Chloroaniline	10.48
3,3,3-Trifluoropropyl-		4-Chloroaniline	10.00
amine	5.3	2-Nitroaniline	13.94
Allylamine	4.24	4-Nitroaniline	12.1
N,*N*-Dimethylallyl-		2,4,6-Trinitroaniline	23.3
amine	5.28	*N*,*N*-Dimethyl-2,4,6-	
N,*N*-Dimethyl-2-		trinitroaniline	18.7
propynylamine	6.95	1-Naphthylamine	10.08
N,*N*-Dimethyl-3-		2-Naphthylamine	9.89
butynylamine	5.67	Diphenylamine	13.15
Benzylamine	4.66	Triarylamines	~ 16
Benzylmethylamine	4.42		
N,*N*-Dimethylbenzyl-			
amine	5.07		

an additional proton. Such an amine is no longer basic but acid:

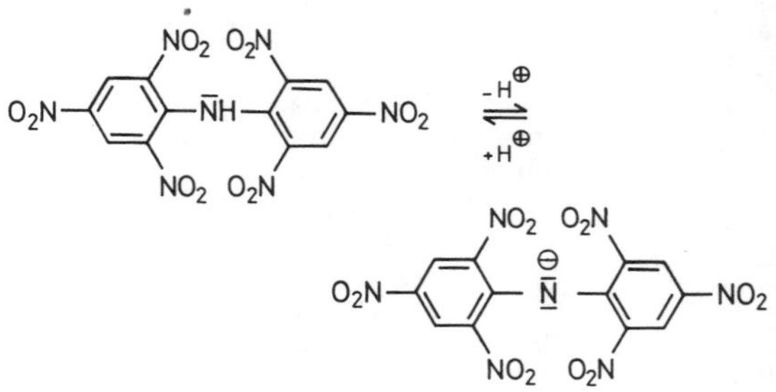

Conversely, the basicity of, for instance, tertiary amines increases markedly where bulky substituents in both *ortho*-positions relative to a tertiary

Substitution with two aromatic groups, especially those carrying electron-attracting substituents, can lead to the *NH* group of such a diarylamine giving up its proton as proton more readily than adding

amino group twist the same out of the plane of the ring and in this way prevent fusion of its lone electron pair with the π-electron cloud of the aromatic system:

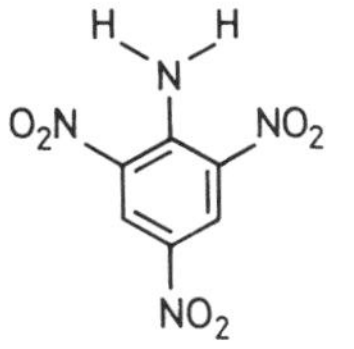

pK$_b$ = 23,3

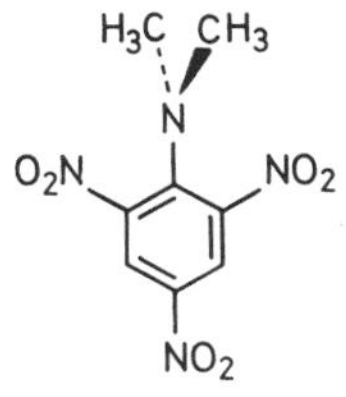

pK$_b$ = 18,7

Three main nomenclature principles are used to *name* the amines. The first emphasises the relationship of the amines with ammonia. Correspondingly, the name begins with the nature and number of the hydrocarbon groups which have replaced the hydrogen atoms and ends with the suffix amine, *e.g.*, butylamine, cyclohexylamine, diphenylamine, *N, N*-dimethylbenzylamine. Using this nomenclature consistently can become clumsy and inappropriate if

1 a simple naming of the hydrocarbon groups is difficult,
2 the molecule contains two or more amino groups,
3 the amino group is only a small constituent of a large complex molecule,
4 the emphasis is to be laid on simultaneously present, further functional groups,
5 trivial names exist for the rest of the molecule.

In all these cases it may be easier to make the rest of the molecule (hydrocarbon, other compound class, trivial name) the basis and to employ *amino* as prefix, *e.g.*, 2-amino-4, 4-diphenylheptane **1**, 1, 4-diaminocyclohexane, 2-aminophenol, 4-dimethylaminobenzaldehyde, 2-aminopropionic acid, 1-aminoadamantane **2**, 3-aminoquinoline.

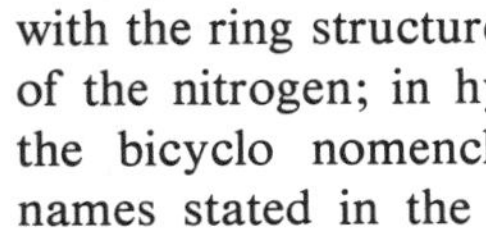

1

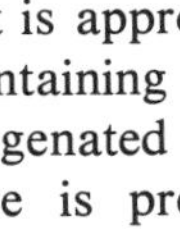

2

Where cyclic amines possess no trivial names (such as, *e.g.*, pyrrolidine, piperidine, quinolizidine, quinacridine, *etc.*), it is appropriate to start with the ring structure containing carbon instead of the nitrogen; in hydrogenated systems either the bicyclo nomenclature is preferred or the names stated in the Ring Index are employed, *e.g.*:

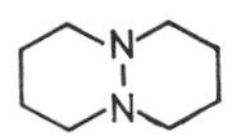

*1,6-Diazabicyclo-
[4.4.0]decane*

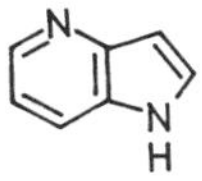

*1 H-Pyrrolo-
[3,2-b]pyridine*

Re the naming of tetraalkylated ammonium compounds see p. 626, of betaines see p. 634.
Much effort has been devoted to investigating whether amines with three different substituents on the nitrogen are optically active or not. Normally the answer is no. Amine molecules form a pyramid with the nitrogen at the apex. The valence angles are practically equal to that of the sp^3-hybridized carbon atom. Not much energy is required to enable the nitrogen to vibrate past the plane of the three substituents. For ammonia the energy required is 5.6 *kcal/mole,* for amines it is 7 *kcal/mole,* so that the isolation or separation of stable optical antipodes at room temperature is not possible. Exceptions are those amines which are so bridged, while having unsymmetrical substituents, that a vibration of the nitrogen through the plane of its substituents is prevented; *e.g.* Tröger's base:

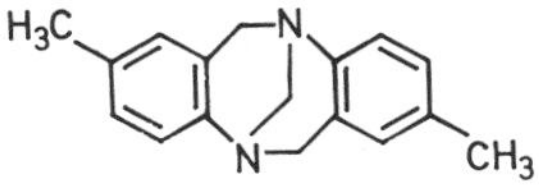

Here, it is actually possible to isolate optical enantiomers. Similar relationships hold for tetraalkylammonium compounds and *N*-oxides in which the nitrogen is surrounded by the four different ligands similarly to a chiral carbon atom. However, if one of the ligands is hydrogen, then evidently the optical activity is prevented *via* the equilibrium with the compound pair tertiary amine plus acid unless the tertiary amine is sterically fixed, like Tröger's base.
Primary and secondary amines are very *reactive* substances. They can be alkylated and acylated in manifold ways and react with carbonyl compounds to form Schiff's bases or enamines. Oxidizing agents, too, attack amines; nitrous acid converts them into primary alcohols, nitrosamines, or diazonium salts. Where amines are to serve for building up complex compounds it is thus necessary to inactivate them with protective groups that can be detached again at the conclusion of the synthesis. This principle is particularly important for peptide syntheses (see p. 600). In other cases it is preferable to introduce the amino group only during the concluding phase of a synthesis by one of the methods described in what follows.
Mixtures of primary, secondary, and tertiary amines are formed during various preparation techniques. Where these cannot be separated by virtue of differences in their physical properties, reaction with *p*-toluenesulfonyl chloride *(Hinsberg method)* furnishes a remedy. Tertiary amines are

not attacked by the reagent. Primary amines form sulfonamides having the character of an acid and are soluble in caustic soda. Secondary amines are converted into sulfonamides which are neither basic or acid in reaction and can thus be separated from the other two types without difficulty. Some trouble is experienced in reliberating primary and secondary amines from the sulfonamides. Hydrolysis requires the use very strong acids (cf. p. 459); a milder procedure is reductive cleavage with sodium in various solvents (cf. p. 459).

The synthesis of numerous *N*-heteroaromatic compounds or pseudoaromatic compounds which are formally amines (pyrrole, indole, acridine, phenothiazine, *etc.*) does not form part of the chapter which follows.

12.1 Preparation of amines by substitution reactions

Direct introduction of the amino group here means replacement of a hydrogen atom linked to carbon by an amino group.

12.1.1 Direct introduction of the amino group with alkali metal amides

Fusion with *sodium amide* successfully introduces the amino group into the 5-position of the unsubstituted nucleus ring in 1- and 2-substituted *naphthalene* derivatives[1]. In the presence of phenol, *1-naphthylamine* and *1,5-naphthylenediamine* are obtained by heating naphthalene with sodium amide; the added phenol is reduced to benzene. *Tertiary* amines can be prepared directly by using *N,N*-disubstituted alkali metal amides. It is advantageous for the aromatic compound to be aminated to carry an electron-attracting substituent. Thus, *1-(2-nitrophenyl)piperidine* is obtained from *nitrobenzene* and lithium piperidide.

Excess nitrobenzene should be employed, because a part of the starting material is destroyed by reduction[2]. With sodium diphenylamide, nitrobenzene gives *4-nitrotriphenylamine*[3]. The same method can be applied to *N-heteroaromatic compounds* (Tschitschibabin reaction)[4]. Pyridine, quinoline, and their derivatives react with metal amides to form amino compounds in 50–100% yield. According to the generally postulated mechanism[5], it is assumed that the initial step in the reaction is addition of the metal amide to the $-CH=N$ group. The resulting product is then converted into the metal derivative either by intramolecular rearrangement or by cleavage into the amino compound and sodium hydride, which react together in their turn. Subsequent hydrolysis yields the amine, *e.g.*:

2-Aminopyridine

As large quantities of ammonia are formed at the outset and throughout the reaction, an intermediate hetaryne stage has been postulated[6], but this mechanism has been disputed[7,8].

Its value as a method of aminating molecules containing $-CH=N$ groups is almost exclusively restricted to heteroaromatic compounds. When certain Schiff's bases were aminated in this way, yields of 20% or less were obtained, and the desired products can be prepared more easily by alternative methods. Of heteroaromatic compounds only pyridine, quinoline, isoquinoline, and their derivatives give satisfactory yields; amino derivatives of pyrazines, pyrimidines, and thiazoles can hardly be prepared by this technique. Ref.[9] presents a survey.

The tendency of substituted bases to undergo amination is influenced by the nature of the substituents. On reacting 2-alkylpyridines with alkali metal amides in liquid ammonia, salt formation of

[1] *F. Sachs*, Ber. *39*, 3006 (1906).
[2] *R. Huisgen, H. Rist*, Justus Liebigs Ann. Chem. *594*, 159 (1955).
[3] *F.W. Bergstrom, I.M. Granara, V. Erickson*, J. Org. Chem. *7*, 98 (1942).
[4] *A.E. Tschitschibabin, O.A. Seide*, J. Russ. Phys.-Chem. Soc. *46*, 1216 (1914).
[5] *C. Deasey*, J. Org. Chem. *10*, 141 (1945).
[6] *L.S. Levitt, B.W. Levitt*, Chem. & Ind. (London) 1621 (1963).
[7] *R.A. Abramowitsch, F. Helmer, J.G. Saha*, Chem. & Ind. (London) 659 (1964).
[8] *G.C. Barrett, K. Schofield*, Chem. & Ind. (London) 1980 (1963).
[9] *M.T. Leffler*, Org. Reactions *1*, 91 (1942).

the enamine form is the sole observed reaction. Formation of the *2-alkyl-6-aminopyridines* requires elevated temperatures and the use of hydrocarbons as solvent, *e.g.*:

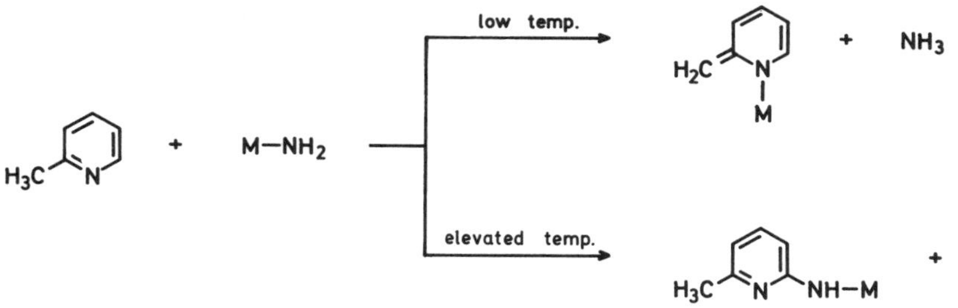

12.1.2 Direct introduction of the amino group with hydroxylamine

Into certain aromatic nitro compounds the amino group can be introduced directly with the aid of

In 2,6-dialkylated *N*-heteroaromatic compounds the amino group is directed into the 4-position. Thus, *4-amino-2,6-dimethylpyridine* is formed from 2,6-dimethylpyridine and sodium amide in boiling xylene. Normally, the amination of pyridine and its derivatives can be controlled in such a way that only one amino group is introduced. *2-aminopyridine* can be obtained in yields of over 75% if the reaction with sodium amide is carried out in *N,N*-dimethylaniline below 120°. Increasing the proportion of metal amide and reacting at $\sim 170°$ either in dimethylaniline or without solvent leads to *2,6-diaminopyridine* as the main product together with small amounts of *4-aminopyridine* but no 2,4-diaminopyridine. Sometimes side-reactions are observed in which the alkali metal salt of the amino heteroaromatic compounds acts as the aminating agent. Thus during the preparation of 2-aminopyridine *2,2'-iminodipyridine* was found as side-product. Coupling is another possible side-reaction, and during amination of pyridines 2,2'- and 4,4'-bipyridines, and also dihydro-4,4'-bipyridines are isolated.

On amination of the quinoline nucleus in liquid ammonia good yields of *aminoquinolines* are generally obtained, but if alkyl groups are in the 2- or 4-position salt formation is the only result[10]. Thus, *2-amino-4-methylquinoline* is obtained only if the reaction is carried out in *N,N*-dimethylaniline at 120°. A carboxy group in the 2- or 4-position increases the reaction rate and improves the yield. *2-aminoquinoline-4-carboxylic acid* (70%) and *4-aminoquinoline-2-carboxylic acid* (81%) are obtained from the corresponding carboxylic acids, potassium amide and potassium nitrate in liquid ammonia; under the same conditions an only 50% yield of *2-aminoquinoline* from quinoline results. An amino group in the 2-position prevents amination, as does a hydroxy group in positions 2 or 8.

hydroxylamine in alkaline solution. The nucleophilic attack of the aminating agent always takes place *ortho* or *para* to an existing nitro group. In the naphthalene series the reaction succeeds even with mononitro compounds, in benzene derivatives at least 2 nitro groups must be present[11], *e.g.*:

2-Nitronaphthalene $\longrightarrow$ *2-Nitro-1-naphthylamine*[11]; 80%

1-Nitronaphthalene $\longrightarrow$ *4-Nitro-1-naphthylamine*[12]; 60%

1,3-Dinitrobenzene $\longrightarrow$ *2,4-Dinitroaniline + 2,4-Dinitro-1,3-phenylenediamine*[13]

2,6-Dinitrotoluene $\longrightarrow$ *2,6-Dinitro-3-toluidine*[11]; 15%

1,3,5-Trinitrobenzene $\longrightarrow$ *2,4,6-Trinitroaniline*[11]; 90%

Nitroquinolines behave towards hydroxylamine like the nitronaphthalenes and yield *aminonitroquinolines*[14]. In a few instances hydroxylamine in alkaline solution successfully introduces the amino group into compounds containing no nitro groups: 1,4-dihydroxy anthraquinone affords *2-amino-1,4-dihydroxyanthraquinone*[15], 8-quinolinol is converted into *5-amino-8-quinolinol*[16]. Heating with hydroxylamine or a hydroxylamine salt in concentrated sulfuric acid in the presence of

[10] *F.W. Bergstrom*, J. Org. Chem. *3*, 233 (1938).

[11] *J. Meisenheimer, E. Patzig*, Ber. *39*, 2533 (1906).

[12] *C.C. Price, S.T. Voong*, Org. Synth. *28*, 80 (1948).

[13] *S.S. Gitis, N.A. Pankova, A.Y. Kaminskii, E.G. Kaminskaya*, Zh. Org. Khim. *5*, 65 (1969); C.A. *70*, 87178 (1969).

[14] *M. Colonna, F. Montanari*, Gazz. Chim. Ital. *81*, 744 (1951).

[15] *C. Marschalk*, Bull. Soc. Chim. France *4*, 629 (1937).

[16] *R. Berg, E. Becker*, Ber. *73*, 172 (1940).

[17] DRP. 287756 (1914), Erf.: *J.F. Turski*; C.A. *10*, 2128 (1916);
Brit. P. 626661 (1949), Inv.: *J.F. Turski*; C.A. *44*, 2761 (1950);
US.P. 2585355 (1952), Inv.: *J.F. Turski*; C.A. *47*, 875 (1953).

iron or vanadium salts allows the direct amination of aromatic compounds such as benzene, naphthalene, anthracene, benzophenone, anthraquinone, benzanthrone, etc.[17]. Introducing the amino group into aromatic compounds with hydroxylamine salts can be accomplished also in the presence of Friedel-Crafts catalysts[18]. With toluene and halobenzenes predominantly o–p orientation obtains accompanied by relatively much m-isomer formation; para isomer is formed the most. The following activity scale has been established for the selectivity factors of the amination of toluene in the presence of aluminum chloride:

$$NH_2OH \cdot HBr \geqslant (NH_2OH)_2 \cdot H_2SO_4 >$$
$$NH_2OH \cdot C_2H_5OSO_3H > NH_2OH \cdot CH_3COOH >$$
$$NH_2OH \cdot HCl > (NH_2OH)_3 \cdot H_3PO_4 = NH_2OH.$$

Aluminum chloride, aluminum bromide, and boron trifluoride are used as catalysts. The yields are dependent on the molar ratio of hydroxylamine salt : catalyst, and for toluene vary from 1–65%, while for halobenzenes they are about 60%. When phenol is used, crude aminophenol (7%), not amenable to purification, is obtained. Various different N-alkylhydroxylamines have been investigated as aminating agents for the toluene-aluminum chloride system[19]. The yield of aromatic amine falls as the alkylation on the nitrogen increases:

$$H_2NOCH_3 : H_2NOH : H_3CHNOCH_3 : (C_2H_5)_2NOH :$$
$$(CH_3)_2NOCH_3 = 42:37:24:2:0 \ (\%)$$

Hydroxylamine-O-sulfonic acid aminates toluene in the presence of aluminum chloride with formation of toluidines in yields of between 36 and 52% and a 51:13:36 ratio of the o:m:p isomers[19a]. Alternatively, the sodium salt of the acid can be employed. The reaction proceeds according to a synchronous exchange mechanism.

Aromatic compounds can also be aminated directly with hydroxylamine-O-sulfonic acid using a redox system[20]. With iron(II) sulfate or iron(II) chloride anisole gives a 38% yield of a mixture of o- and p-anisidines, while toluene gives toluidine (15%) in a 37:21.5:41.5 o–m–p isomer ratio.

Using hydroxylamine hydrochloride/titanium(III) chloride as the aminating system the yield of 2- and 4-anisidines is 18%. With toluene the isomer distribution varies widely (63% o- and 37% p-toluidine in the last-named system; 34% o- and 66% p-toluidine using hydroxylamine-O-sulfonic acid iron(II) salt).

Both acyclic and cyclic olefins can be converted into the corresponding amines with hydroxylamine-O-sulfonic acid in tetrahydrofuran in a one-step reaction via organoboranes[21].

Table 1. Amines from olefins with hydroxylamine-O-sulfonic acid via organoboranes

Olefin	Amine	Yield [% of th.]
1-Octene	Octylamine	64
2-Methyl-1-pentene	2-Methylpentylamine	59
2,4,4-Trimethyl-1-pentene	2,4,4-Trimethyl-1-pentylamine	58
α-Methylstyrene	2-Phenyl-1-propylamine	58
1,1-Diphenylethylene	2,2-Diphenylethylamine	27
Cyclopentene	Cyclopentylamine	59
Cyclohexene	Cyclohexylamine	55
Norbornene	exo-Norbornylamine	52
β-Pinene	cis-Myrtanylamine	55
Ethyl undecenoate	11-Aminoundecanoic acid	30

Amination of both sterically hindered and unhindered olefins is possible also if the borohydride reaction and the subsequent addition of hydroxylamine-O-sulfonic acid takes place in diglyme[22]. Thus, 1-methylcyclopentene, 1-methylcyclohexene, and 1-methylcycloheptene form trans-2-methylcyclopentylamine, -cyclohexylamine, and -cycloheptylamine respectively. Analogously, 1-phenylcyclopentene and 1-phenylcyclohexene are reacted to trans-2-phenylcyclopentylamine and -cyclohexylamine. Starting from norbornene and β-pinene, exo-norbornylamine and cis-Myrtanylamine are obtained as sterically uniform products. This one-step synthesis is therefore especially useful despite the only 40–50% yield. Many functional groups are unaltered when the reaction is carried out in this way.

[18] P. Kovacic, R.P. Bennett, J.L. Foote, J. Amer. Chem. Soc. 84, 759 (1962).
[19] P. Kovacic, J.L. Foote, J. Amer. Chem. Soc. 83, 743 (1961).
[19a] P. Kovacic, R.P. Bennett, J. Amer. Chem. Soc. 83, 221 (1961).
[20] F. Minisci, R. Galli, Tetrahedron Lett. 1679 (1965).
[21] H.C. Brown, W.R. Heydkamp, E. Breuer, W.S. Murphy, J. Amer. Chem. Soc. 86, 3565 (1964).
[22] M.W. Rathke, N. Inoue, K.R. Varma, H.C. Brown, J. Amer. Chem. Soc. 88, 2870 (1966).
[23] F. Minisci, R. Galli, M. Cecere, R. Mondelli, Chim. Ind. (Milan) 47, 994 (1965); C.A. 63, 16234d (1965).

When amination of olefins is performed with *1-hydroxypiperidine* as the *N, N*-dialkylated hydroxylamine and titanium(III) chloride, and in addition oxygen is passed through the reaction solution, *tertiary α-amino ketones* are formed[23]. Styrene yields *2-piperidinoacetophenone*, cyclohexene *2-piperidinocyclohexanone*.

12.1.3 Direct introduction of the amino group with haloamines (including the Hofmann-Löffler-Freytag reaction)

N-Haloamines can be used under various different conditions for aminating aromatic rings directly. In some instances this technique furnishes synthetically useful yields. The reactions proceed by both a polar and a free-radical mechanism.

Chloramine (NH_2Cl) reacts with phenol to give a small yield of *2-aminopyridine* and with quinoline *(4-hydroxyphenyl)-4-aminophenol* and *benzoquinone imine*[24]. *o*-Amination of unsubstituted phenols succeeds by heating with chloramine and sodium hydroxide[25]; by contrast, 2,6-disubstituted phenols rearrange in one step to *dihydroazepinones* with ring expansion but in only moderate yield[26]. Chloramine reacts with pyridine to give a small yield of *2-aminopyridine* and with quinoline to form *2-aminoquinoline* (40%)[27]. *Fluoramine* also has been proposed for introducing the amino group into aromatic compounds[28].

Ring amination with *N-chlorodimethylamine* succeeds by any one of 3 procedures[29]:

1 Heating the components for 3—6 hours in 96% sulfuric acid to 80—100°; addition of metallic salts either increases the yield (Na_2SO_4, CuCl, $NiCl_2$, $FeSO_4 \cdot H_2O$, Hg_2Cl_2) or reduces it ($CoCl_2$).
2 *Irradiating* with visible or ultraviolet light for 8—10 hours in concentrated sulfuric acid.
3 Heating for 1—4 hours with *Lewis acids*, preferably aluminum chloride or iron(III) chloride in nitroalkanes as solvent[30]:

Table 2. Ring amination with *N*-chlorodimethylamine

Substrate	Dimethylamino compound [%]			
	1*	2	3	
Benzene	81	78	90	*N, N-Dimethylaniline*
Toluene	80	61	50	*N, N-Dimethyltoluidine*
tert-Butylbenzene	—	58	43	*tert*-Butyl-*N, N-dimethylaniline***
Naphthalene	21	22	35	*N, N-Dimethylnaphthylamine***
N, N-Dimethylaniline	—	—	21	*N, N, N', N'-tetramethylphenylenediamine***

*   In H_2SO_4 containing 50g/l. Na_2SO_4
**   Precise naming of the end-products is not possible, because mixtures of the *o*-, *m*-, and *p*-products invariably form

The isomer distribution for *toluene* is the same using procedures **1** and **2** (*o:m:p* = 9:53:38). No explanation has been offered for the high proportion of the *m*-compound. During amination of toluene with *N*-chlorodimethylamine using a redox system[31] the same orientation was found. Using procedure **3** the *o:m:p* ratio is 14:27:59.

Suitable reactants are aromatic compounds, optionally with acid-stable first order substituents. With all three methods ring chlorination is occasionally observed. As the alkyl groups lengthen, the ring dialkylamination yield falls in terms of the amine component due to competition with the Hofmann-Löffler reaction.

Aromatic amination can be achieved also with *alkylchloroamines* and *alkyldichloroamines*, but the yields (7—45%) are less than when *N*-chlorodialkylamines are used[32].

N-haloamines can bring about *intramolecular* reactions as well as serve to introduce the amino group directly. The best known method of this type is the Hofmann-Löffler-Freytag reaction[33], which enables *cyclic amines* to be synthesized in an elegant manner.

[24] *F. Raschig*, Angew. Chem. *20*, 2065 (1907).
[25] *W. Theilacker, E. Wegner*, Angew. Chem. *72*, 127 (1960).
[26] *L.A. Paquette*, J. Amer. Chem. Soc. *85*, 3288 (1963);
  L.A. Paquette, W.C. Farley, J. Amer. Chem. Soc. *89*, 3595 (1967).
[27] *M.E. Brooks, B. Rudner*, J. Amer. Chem. Soc. *78*, 2339 (1956).
[28] DRP. 594 900 (1931), Erf.: *O.T. Krefft;* Fortschritte der Teerfabrikation u. verwandter Industriezweige *20*, 440 (1933).
[29] *H. Bock, K. Kompa*, Angew. Chem. *77*, 807 (1965); Angew. Chem. Intern. Ed. Engl. *4*, 783 (1965).
[30] *H. Bock, K. Kompa*, Chem. Ber. *99*, 1347, 1357, 1361 (1966).

[31] *F. Minisci, R. Galli, M. Cecere*, Tetrahedron Lett. 4663 (1965).
[32] *F. Minisci, R. Galli, M. Cecere*, Chim. Ind. (Milan) *48*, 725 (1966);
  V.L. Heasley, P. Kovacic, R.M. Lange, J. Org. Chem. *31*, 3050 (1966).
[33] *K. Löffler, C. Freytag*, Ber. *42*, 3427 (1909);
  K. Löffler, Ber. *43*, 2035 (1910).
[34] *S. Wawzonek, P.J. Thelen*, J. Amer. Chem. Soc. *72*, 2118 (1950).
[35] *G.H. Coleman, G.E. Goheen*, J. Amer. Chem. Soc. *60*, 730 (1938).
[36] US. P. 1 607 605 (1927), Inv.: *E.C. Britton;* C.A. *21*, 249 (1927).
[37] *E.J. Corey, W.R. Hertler*, J. Amer. Chem. Soc. *82*, 1657 (1960).

Acyclic secondary amines form *pyrrolidines* and *piperidines*. Using cyclic secondary amines allows the preparation of bridged and condensed ring systems. The reaction proceeds according to the following mechanism[34]:

X = Halogen

Homolytic cleavage of the N-haloamines dissolved in a strong acid occurs under free-radical forming conditions. The aminium radical abstracts a sterically favorable hydrogen atom intramolecularly to form an alkyl radical which, in its turn, abstracts a halogen atom intermolecularly. In some cases the C-haloamines can be isolated, but normally these intermediates are cyclized directly by alkalizing. N-Bromoamines and N-chloroamines serve as suitable starting materials; any of the following acids may be employed: concentrated sulfuric acid, dilute sulfuric acid[35], phosphoric acid, mixtures of sulfuric and phosphoric acids[36], 5 N sulfuric acid dissolved in acetic acid solution[37], trifluoroacetic acid[38, 39]. Little is known concerning the effect of varying the concentration of the N-haloamines in the sulfuric acid. Generally 0.3 molar solutions are employed, but 1.5 molar solutions are said to give exceptionally good yields[38]. When trifluoroacetic acid is used 10% solutions are adequate[39, 40].

Heating is one way to start the reaction. The temperatures used vary between 60 and 140°, the optimum is dependent on the nature of the starting material and needs to be determined experimentally. A deviation of more than 5° from the optimum can reduce the yield[41].

Using *ultraviolet light* as initiator facilitates the experimental conditions substantially[42]. Advantageously, the reaction is discontinued when iodide solution is no longer positive for N-haloamine[37, 40].

Like other free-radical reactions, the Hofmann-Löffler reaction can be started also with chemical reagents. Hydrogen peroxide[34], potassium persulfate plus iron ammonium sulfate, or iron ammonium sulfate alone may be used[37].

While the principal use lies with N-haloamines, N-haloamides have been occasionally employed. Thus *pyrrolidine* is formed in 50% yield by heating N-butyl-N-chloroacetamide in 95% sulfuric acid for 1 hour to 140° and subsequently treating with alkali[43]. Similar yields are obtained in the reaction of 1, 3-dibutyl-1, 3-dichlorourea under the same conditions[44].

1-, 2-, and 3-substituted pyrrolidines can be prepared using the Hofmann-Löffler reaction. Where alternative reactions are possible the following rules apply[37]. Under thermal conditions secondary hydrogen atoms are abstracted more easily than those from the methyl group. For example, N-butyl-N-chloropentylamine yields exclusively *1-butyl-2-methylpyrrolidine* (35%). By contrast, on irradiation N-chloro-N-methylhexylamine furnishes both *2-ethyl-1-methylpyrrolidine* (17%) and *1, 2-dimethylpiperidine* (2%). Tertiary hydrogen atoms are abstracted preferentially to primary and secondary ones, but the resulting tertiary halides readily undergo solvolysis under the reaction conditions and the expected 2, 2-dialkylpyrrolidines or -piperidines are not obtained.

From 1-bromo-2-(2-methylpropyl)pyrrolidine heating in sulfuric acid and subsequently making alkaline leads to *2-methylpyrrolizidine*[45].

Strangely, analogous treatment of 2-butyl-1-bromopyrrolidine does not give the expected reaction product, but 2-propyl-1-bromopyrrolidine is transformed into *pyrrolizidine* by hot sulfuric acid[46]. An alternative synthesis of this ring system is obtained by the only described use of an N, N-dihaloamine. Irradiation of N, N-dibromoheptylamine gives a 35% yield of *pyrrolizidine*[47]. The

[38] *S. Wawzonek, T.P. Culbertson*, J. Amer. Chem. Soc. *81*, 3367 (1959).

[39] *M.E. Wolff, J.F. Kerwin, F.F. Owings, B.B. Lewis, B. Blank, A. Magnani, V. Georgian*, J. Amer. Chem. Soc. *82*, 4117 (1960).

[40] *J.F. Kerwin, M.E. Wolff, F.F. Owings, B.B. Lewis, B. Blank, A. Magnani, C. Karash, V. Georgian*, J. Org. Chem. *27*, 3628 (1962).

[41] *G.H. Coleman, G. Nichols, T.F. Martens*, Org. Syntheses, Coll. Vol. III, 159 (1951).

[42] *S. Wawzonek, M.F. Nelson jr., P.J. Thelen*, J. Amer. Chem. Soc. *73*. 2806 (1951).

[43] *G.H. Coleman, C.C. Schulze, H.A. Hoppens*, Proc. Iowa Acad. Sci. *47*, 264 (1940).

[44] *G.H. Coleman, G. Alliger*, Proc. Iowa Acad. Sci. *48*, 246 (1941).

[45] *G. Menschikoff*, Ber. *69*, 1802 (1936).

[46] *F. Sorm, J. Brandeis*, Collect. Czech. Chem. Commun. *12*, 444 (1947).

[47] *E. Schmitz, D. Murawski*, Chem. Ber. *93*, 754 (1960).

Hofmann-Löffler reaction has been used to synthesize various basic *alkaloid* structures. Thus from 1-bromo-2-propylpiperidine (*N*-bromoconiine) *octahydroindolizine (δ-coniceine)* is obtained[48]:

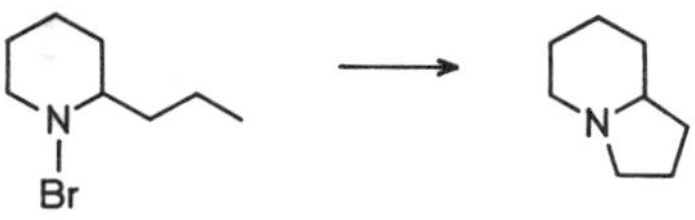

Thermal decomposition of *N*-bromo-*N*-methyl-4-(3-pyridyl)butylamine gives *nicotine*[49]:

Conanine derivatives can be synthesized *via* sterodial intermediates[39, 40, 50], *e.g.*:

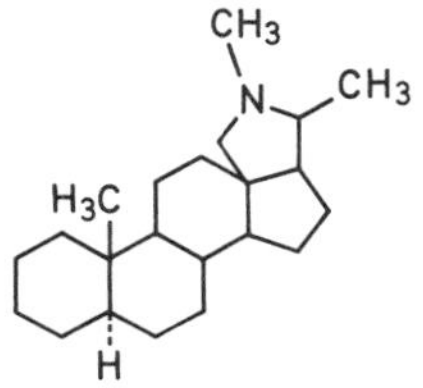

The conanine skeleton is a constituent of many *holarrhena* alkaloids. Subjecting 4-ethylpiperidine to the conditions of the Hofmann-Löffler reaction leads to either *quinuclidine*[42] or *7-methyl-1-azabicyclo[2.2.1]heptane*[51] depending on how the reaction is conducted:

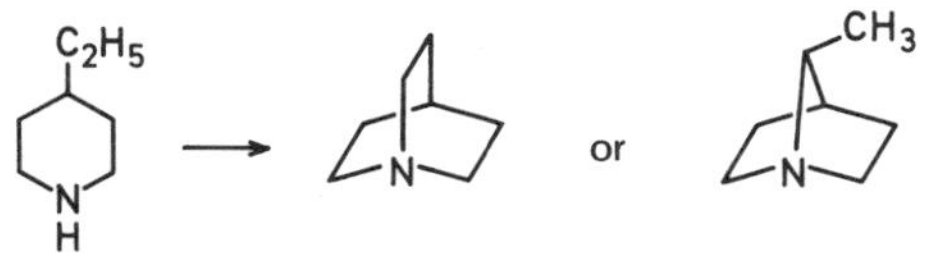

N-chlorocamphidine furnishes *hexahydro-3a,4-dimethyl-1H-2,4-methanocyclopenta[c]pyrrole (cyclocamphidine)* in 67% yield[52]:

Irradiation in sulfuric acid gives 11% *7-methyl-7-azabicyclo[2.2.1]heptane* from *N*-chloro-*N*-cyclohexylmethylamine[37]:

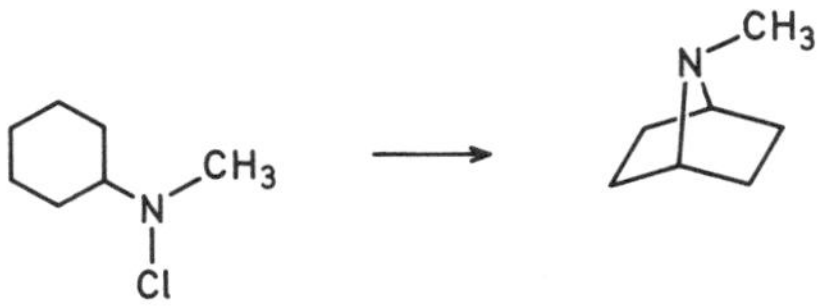

N-bromo-*N*-cycloheptylmethylamine yields 40% *tropane (8-methyl-8-aza-bicyclo[3.2.1]octane)*[53]:

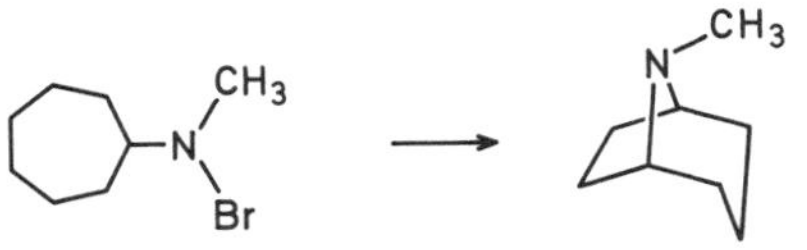

Cyclization of *N*-chloro-*N*-cyclooctylmethylamine leads to 23% *9-methyl-9-azabicyclo[3.3.1]nonane (N-methylgranatanine)* after an only half-hour reaction[34]:

From *N*-ethyl-*N*-chloro-*N*-cyclopentylmethylamine 29% *2-ethyl-2-azabicyclo[2.2.1]heptane* is formed[54], and from *N*-ethyl-*N*-chloro-*N*-cyclohexylmethylamine 10% *6-ethyl-6-azabicyclo[3.2.1]octane*[55]:

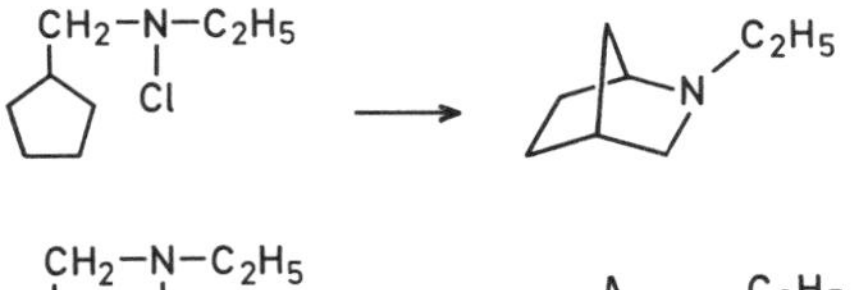

Cyclic and bicyclic *N*-chloroamines will rearrange in the presence of *silver ions*. Solvolysis of 2-chloro-2-azabicyclo[2.2.2]octane with methanol

[48] A.W. *Hofmann*, Ber. *18*, 5, 109 (1885).
[49] K. *Löffler*, S. *Kober*, Ber. *42*, 3431 (1909).
[50] P. *Buchschacher, J. Kalvoda, D. Arigoni, O. Jeger*, J. Amer. Chem. Soc. *80*, 2905 (1958);
E.J. *Corey, W.R. Hertler*, J. Amer. Chem. Soc. *80*, 2903 (1958);
E.J. *Corey, W.R. Hertler*, J. Amer. Chem. Soc. *81*, 5209 (1959).
[51] R. *Lukes, M. Ferles*, Collect. Czech. Chem. Commun. *20*, 1227 (1955).

[52] W.R. *Hertler, E.J. Corey*, J. Org. Chem. *24*, 572 (1959).
[53] G.H. *Coleman, J.J. Carnes*, Proc. Iowa Acad. Sci. *49*, 288 (1942).
[54] P.G. *Gassman, D.C. Heckert*, Tetrahedron *21*, 2725 (1965).
[55] P.G. *Gassman, B.L. Fox*, J. Org. Chem. *32*, 3679 (1967).
[56] P.G. *Gassman, B.L. Fox*, Chem. Commun. 53 (1966);
P.G. *Gassman, B.L. Fox*, J. Amer. Chem. Soc. *89*, 338 (1967).

and silver nitrate leads to *2-methoxy-1-azabicy-clo[3.2.1]octane* (60%)[56] (two alternative rearrangements are possible):

2-chloro-1-azabicyclo[2.2.1]heptane is obtained analogously[57]:

Treatment of *N*-chloroazacyclononane with silver ions gives *indolizidine* (68%)[58]:

The nitrenium ion abstracts a hydride ion transannularly. Subsequently, the carbonium ion attack is on the nitrogen.

N-Chloromethyl-1-phenylcyclobutylamine reacts with silver trifluoroacetate to *1-methyl-2-phenyl-pyrrolidine* with ring expansion *via* a nitrenium ion. With *1-chloro-2-phenylazetidine* and silver trifluoroacetate ring contraction occurs with formation of an unstable *aziridinium salt*, which decomposes further[59].

12.1.4 Direct introduction of the amino group with nitrogen halides

As an introduction, it needs to be pointed out that preparative working with nitrogen halides is hazardous because of their **toxicity**[60] and **explosive nature.** *Nitrogen trichloride* is extremely unstable when pure. While the risk of explosion cannot be eliminated entirely by the use of inert solvents, it is reduced very substantially.

The reaction of nitrogen trichloride with benzene and toluene leads to predominantly chlorinated aromatic compounds in addition to small quantities of chloroanilines[61]. Better yields are obtained by adding *aluminum chloride*[62]. From toluene, *1,3-toluidine* (42%) is formed almost exclusively, and similar results were obtained when other alkylbenzenes were investigated[63].

The unusual *selectivity* of this amination in the case of certain alkyl- and 1,3-dialkylbenzenes is particularly noteworthy[63, 64].

In alkylbenzenes with an alkyl substituent having a tertiary hydrogen atom in the α-position side-chain amination, too, occurs[65].

The relative amination reaction velocities are related linearly to the basicity, chlorination, and bromination[66]. Formation of a σ-complex is therefore assumed as the rate-determining step:

$$Cl_3N + AlCl_3 \rightleftharpoons Cl_3N \text{---} AlCl_3 \rightleftharpoons Cl^{\delta\oplus}(Cl_2NAlCl_3)^{\delta\ominus}$$

Further indications of the σ-substitution mechanism are obtained by the study of the amination of biphenyl and naphthalene[67]. Reacting toluene-aluminum chloride

---

[57] *P.G. Gassman, R.L. Cryberg*, J. Amer. Chem. Soc. *90*, 1355 (1968);
P.G. Gassman, R.L. Cryberg, J. Amer. Chem. Soc. *91*, 2047 (1969).
[58] *O.E. Edwards, D. Vocelle, J.W. Apsimon, F. Haque*, J. Amer. Chem. Soc. *87*, 678 (1965).
[59] *P.G. Gassman, A. Carrasquillo*, Tetrahedron Lett. 109 (1971).
[60] *G.H. Pollock*, J. Appl. Physiol. *1*, 802 (1949); C.A. *44*, 8011 (1950).
[61] *G.H. Coleman, W.A. Noyes*, J. Amer. Chem. Soc. *43*, 2211 (1921).

[62] *P. Kovacic, R.M. Lange, J.L. Foote, C.T. Goralski, J.J. Hiller jr., J.A. Levisky*, J. Amer. Chem. Soc. *86*, 1650 (1964);
P. Kovacic, C.T. Goralski, J.J. Hiller jr., J.A. Levisky, R.M. Lange, J. Amer. Chem. Soc. *87*, 1262 (1965).
[63] *P. Kovacic, J.A. Levisky, C.T. Goralski*, J. Amer. Chem. Soc. *88*, 100 (1966).
[64] *P. Kovacic, K.W. Field, P.D. Roskos, F.V. Scalzi*, J. Org. Chem. *32*, 585 (1967).
[65] *P. Kovacic, R.J. Hopper*, Tetrahedron *23*, 3965, 3977 (1967).
[66] *P. Kovacic, J.A. Levisky*, J. Amer. Chem. Soc. *88*, 1000 (1966).
[67] *P. Kovacic, A.K. Harrison*, J. Org. Chem. *32*, 207 (1967).

with nitrogen trichloride is more successful than with alkylhaloamines and alkyldihalogenoamines[68]. The yields of aromatic amines are small when the two last-named reagents are employed.

The reaction of the NCl_3–$AlCl_3$ amination system with saturated hydrocarbons is of particular importance[69, 70]. A very detailed study has been made of the conditions under which *1-methylcyclohexylamine* is formed in good yield (70–80%) from methylcyclohexane. This latter method comprises adding nitrogen trichloride to a mixture of methylcyclohexane and aluminum chloride in a 1:2:2 molar ratio[71] (for the mechanism see refs.[72–76]).
The applicability of this amination reaction has been investigated in respect of many alicyclic[72], bicyclic[77], and tricyclic[73] hydrocarbons. Adamantane and alkyladamantanes give the corresponding bridgehead amines[73] in good yield (93–95%) on amination with nitrogen trichloride. Various different tricyclic alkanes (adamantane precursors) are converted into adamantylamines in one step with nitrogen trichloride under Friedel-Crafts conditions by rearrangement and amination[73]. In *cis*- and *trans*-decalin and also bicyclo[4.3.0]nonane the attack of the amination reagent occurs analogously on the tertiary position with formation of *cis*-configurated amines[72]. With certain cyclic and bicyclic carbon skeleton rearrangement is the preferred reaction pathway. Thus cycloheptane affords *1-methylcyclohexylamine* (65%)[72]. With cyclohexane the product distribution is temperature dependent; at low temperatures cyclohexylamine (41%) is formed, at higher temperatures *1-methylcyclopentylamine* (49%) predominates[72]. A variety of simultaneous reactions during amination of cyclopentane produces more than one amine, viz., cyclopentylamine (3%), *6-azabicyclo[3.1.0]hexane (36%), 6-cyclopentyl-6-azabicyclo[3.1.0]hexane* (13%), and *dicyclopentylamine* (2%)[72]. Conversion of *exo*-2-chloronorbornane with

NCl_3–$AlCl_3$ yields *2-azabicyclo[3.2.1]octane* (88%) and *3-azabicyclo[3.2.1]octane* (6%) in addition to the expected *2-norbornylamine* (6%)[77] via ring expansion followed by hydrogenation.
Side-chain amination in *p*-cymene succeeds in similar fashion[65]. Adding *tert*-butyl bromide, which prevents ring amination, gives an 80% yield of *p*-cymen-8-amine* calculated on nitrogen trichloride. Evidently, the *tert*-butyl carbonium ion formed acts as a hydride cleaving agent with formation of the *p*-methyl-α,α'-dimethylbenzyl cation, as is supported by the isolation of isobutane. The same method is successful also with *p*-halocumenes and 4-cyclohexyltoluene, and gives *2-(4-halophenyl)-propylamine* and *1-(4-tolyl)cyclohexylamine* respectively[78]. Aminating 3-(4-tolyl)pentane yields *2-(4-tolyl)pentylamine* via a rearrangement; the same amine is obtained from 2-(4-tolyl)pentane. *Tertiary alkylamines* are likewise obtained in good yield (33–90%) by reacting the corresponding *tert*-alkyl halides with NCl_3–$AlCl_3$ at –10° in dichloromethane[79].

12.1.5 Direct introduction of the amino group with azides

Azides may be used to introduce the amino group into aromatic compounds directly under catalytic and thermal conditions. Special precautions are indicated in each case.
Benzene, toluene, *p*-xylene, and *p*-cymene react under pressure with the highly explosive *carbonyl azide* to form the corresponding *aromatic amines* and *pyridine* derivatives[80].
On heating *arylsulfonic acid azides*[81] with aromatic hydrocarbons to 100–140°, arylsulfonic acid arylamides[82] form with evolution of nitrogen. With acids these can be split into *sulfonic acids* and *primary aromatic amines*:

$$R-SO_2-N_3 + H-Ar \xrightarrow{-N_2} R-SO_2-NH-Ar$$

$$\xrightarrow{H_2O} R-SO_3H + H_2N-Ar$$

[68] V.L. Heasley, P. Kovacic, R.M. Lange, J. Org. Chem. 31, 3050 (1966).
[69] P. Kovacic, R.J. Hopper, S.S. Chaudhary, J.A. Levisky, V.A. Liepkalns, Chem. Commun. 232 (1966).
[70] P. Kovacic, S.S. Chaudhary, Tetrahedron 23, 3563 (1967).
[71] P. Kovacic, S.S. Chaudhary, Org. Synth. 48, 4 (1968).
[72] K.W. Field, P. Kovacic, T. Herskovitz, J. Org. Chem. 35, 2146 (1970).
[73] P. Kovacic, P.D. Roskos, J. Amer. Chem. Soc. 91, 6457 (1969).

[74] T.E. Stevens, J. Org. Chem. 33, 2664 (1968).
[75] W.H. Graham, J.P. Freeman, J. Amer. Chem. Soc. 89, 716 (1967).
[76] J.P. Freeman, W.H. Graham, C.O. Parker, J. Amer. Chem. Soc. 90, 121 (1968).
[77] P. Kovacic, M.K. Lowery, P.D. Roskos, Tetrahedron 26, 529 (1970).
[78] P. Kovacic, J.F. Gormish, R.J. Hopper, J.W. Knapczyk, J. Org. Chem. 33, 4515 (1968).
[79] P. Kovacic, M.K. Lowery, J. Org. Chem. 34, 911 (1969).
[80] T. Curtius, A. Bertho, Ber. 59, 565 (1926).
[81] J.F. Tilney-Bassett, J. Chem. Soc. (London) 2517 (1962).
[82] T. Curtius, J. Prakt. Chem. 125, 303 (1930).

Arylsulfonamide with an unsubstituted nitrogen atom is formed as side-product. In other aromatic compounds with *ortho* and *para* directing substituents, too, the arylsulfonylamino group enters the nucleus. Compounds with *meta* directing substituents practically do not react. The orientation suggests an electrophilic replacement, but the velocity constants more nearly fit a free-radical substitution[83, 84].

Direct amination of aromatic compounds succeeds also with *hydrazoic acid* under Friedel-Crafts conditions[85]. To this end a solution of hydrazoic acid is prepared in the compound to be aminated by extracting an aqueous solution of sulfuric acid and sodium azide. To reduce diamination an at least 20:1 molar ratio of aromatic compound to sodium azide is used. A 2:1 molar ratio of catalyst to azide gives the best yield of amination product; the reaction mixture is heated until nitrogen liberation becomes negligible.

$$H_5C_6-X + HN_3 \xrightarrow[\text{or conc. } H_2SO_4]{AlCl_3} X-C_6H_4-NH_2 + N_2$$

X = CH$_3$; *1,4-Toluidine*; 65%
X = Cl; *4-Chloroaniline*; 19%

The yield of *toluidines* using the hydrazoic acid/toluene/aluminum chloride system varies between 24 and 65% calculated on sodium azide (34–88% on actually present hydrazoic acid). Chlorobenzene gives *chloroaniline* (19%). The low figure of 6–15% toluidine obtained in the sulfuric acid-catalyzed reactions is due to conversion of hydrazoic acid into hydrazine. In place of hydrazoic acid *methyl azide* can be used for amination under analogous conditions, but the yield of *N-methyltoluidines* calculated on toluene is only 4–9%.

Solutions of sodium azide in sulfuric acid aminate mesitylene in exceptionally high yield to *2,4,6-trimethylaniline, 2,4,6-trimethyl-1,3-phenylenediamine*, and *3-amino-2,4,6-trimethylbenzenesulfonic acid*[86]. Surprisingly, α-(2-methyl-5-nitrophenyl)-1,4-toluidine is obtained on reacting 4-nitrotoluene analogously.

12.1.6 Direct introduction of the amino group using miscellaneous methods

Amination of acridine and phenanthridine succeeds with sodium *hydrazide* and *sodium N,N-dimethylhydrazide*[87]. *9-Aminoacridine* (65 and 89%) and *9-aminophenanthridine* (2 and 85% respectively) are formed. During reaction with sodium hydrazide the latter must not be present in excess in the reaction mixture; otherwise partial reduction to 9,10-dihydroacridine or 9,10-dihydrophenanthridine occurs. By contrast, during reaction with sodium *N,N*-dimethylhydrazide neither excess hydrazide nor the presence of *N,N*-dimethylhydrazine interfere.

In methanolic solution *p*-benzoquinone reacts with *dimethylamine* (molar ratio 1:6) at 25° in the presence of copper(II) acetate in an atmosphere of oxygen to form *2,5-bis(dimethylamino)-1,4-benzoquinone* (93%). No corresponding amination occurs with methylamine, aniline, and ammonia[88].

Amination of *p*-benzoquinones succeeds also without catalysis. With primary amines (methylamine, ethylamine, propylamine, and butylamine) *bis(alkylamino)-1,4-benzoquinones* are formed, while with secondary amines (dimethylamine, diethylamine, dipropylamine, dibutylamine) exclusively *dialkylamino-1,4-benzoquinones* result. A part of the *p*-benzoquinone used is reduced to *hydroquinone*[89].

12.2 Amines by substitution of functional groups

12.2.1 Amines by substitution of halogens

The long known method for preparing amines by exchanging halogens in organic halides with ammonia, primary, or secondary amines[90] is inherently most suitable for preparing *aliphatic amines*. An analogous exchange in aromatic compounds requires drastic reaction conditions or activating substituents. The reaction is less suitable for preparing just one of the species primary,

[83] O.C. Dermer, M.T. Edmison, J. Amer. Chem. Soc. 77, 70 (1955).

[84] J.F. Heacock, M.T. Edmison, J. Amer. Chem. Soc. 82, 3460 (1960).

[85] P. Kovacic, R.L. Russell, R.P. Bennett, J. Amer. Chem. Soc. 86, 1588 (1964).

[86] G.M. Hoop, J.M. Tedder, J. Chem. Soc. (London) 4685 (1961).

[87] T. Kauffmann, H. Hacker, H. Müller, Chem. Ber. 95, 2485 (1962).

[88] R. Baltzly, E. Lorz, J. Amer. Chem. Soc. 70, 861 (1948).

[89] A. Hikosaka, Bull. Chem. Soc. Japan 43, 3928 (1970); C.A. 74, 64016 (1971).

[90] G. Spielberger in Houben-Weyl, Methoden der organischen Chemie, Bd. XI/1, p. 24, Georg Thieme Verlag, Stuttgart 1957.

[91] J.F. Bunnet, R.F. Zahler, Chem. Rev. 49, 273 (1951); J.F. Bunnet, Quart. Rev. Chem. Soc. 12, 1 (1958).

secondary, or tertiary amines, because alkylation does not halt at a particular step but continues to the quaternary ammonium salt stage. Ultimately, *mixtures* of salts of primary, secondary, and tertiary amines and including quaternary salts are formed (**1–4**):

$$R—Hal + NH_3 \longrightarrow R—NH_3^{\oplus}\ Hal^{\ominus} \quad \mathbf{1}$$

$$R—Hal + R—NH_2 \longrightarrow R_2—NH_2^{\oplus}\ Hal^{\ominus} \quad \mathbf{2}$$

$$R—Hal + R_2—NH \longrightarrow R_3NH^{\oplus} \qquad Hal^{\ominus} \quad \mathbf{3}$$

$$R—Hal + R_3N \longrightarrow R_4N^{\oplus} \qquad Hal^{\ominus} \quad \mathbf{4}$$

The method is predominantly of technical importance. Preparation of primary and secondary amines requires special techniques.

In accord with **1** and **2** primary and secondary amines are obtained only by using varying amounts of excess ammonia. Reactions **2–4** show that mixed secondary and tertiary amines can be obtained from primary amines.

Selective alkylation of an amino group in *diamines* necessitates protecting the other amino group by acylation (p. 585) or benzoylation (p. 671). After carrying out the alkylation the protective group is eliminated.

The reaction rate of the alkyl halides descends in the order

$$I > Br \gg Cl \gg F$$

The reactivity increases from the very sluggish alkyl fluorides to the iodine compounds. The same applies to unsubstituted aryl halides; for *aryl halides* with strongly negative *ortho* and *para* substituents the following reactivity series holds[91]:

$$I < Cl \leq Br \ll F$$

In addition, the reaction velocity is dependent on the nature of the halogen-carbon bond. Primary halogen links are *more reactive* than secondary bonds; branched halides generally react much more slowly because of steric hindrance. From tertiary halides hydrogen halide may be split off with formation of olefins.

Suitable *solvents* alternative to water are ethanol and other alcohols permitting reaction in a homogeneous medium. With aromatic and heteroaromatic halides phenol has given good results, but under certain conditions may itself be alkylated to form phenol ethers.

Halogen exchange is promoted by catalysts (metals and metal salts), for example, sodium or potassium iodide in ethanolic solution. Ammonium salts promote formation of primary amines. Copper and copper(II) salts are advantageous with sluggishly reacting aryl halides[92].

Halogen exchange reactions are important for preparing *polyvalent* and *cyclic* amines. Di- and polyhalogen compounds expectedly yield mixtures of different bases on reacting with ammonia. With diethylamine, 2-cyclohexyl-4, 5-dibromo-2-phenylvaleronitrile yields 62% *bis-(diethylamino)-2-cyclohexyl-2-phenylvaleronitrile* in addition to 5-bromo-2-cyclohexyl-2-phenyl-4-pentenenitrile by elimination[93].

By contrast, dihalides whose halogen atoms are separated by 4, 5, or 6 links are preferentially suited for synthesizing *N-heterocycles:*

Br—$(CH_2)_n$—Br + NH_3 →

n = 4,5,6

1; n = 4, *Pyrrolidine*
n = 5; *Piperidine*
n = 6; *Hexamethylenimine*   **1**

However, the secondary ring amines **1** readily undergo secondary reactions with formation of spirocyclic quaternary salts **2**:

n = 4,5,6

2; n = 4; *1-Azoniaspiro[4.4]nonane bromide*   **2**
n = 5; *1-Azoniaspiro[5.5]undecane bromide*
n = 6; *1-Azoniaspiro[6.6]tridecane bromide*

The manifoldness of the possible reactions is displayed during the synthesis of bridged aza ring systems[94], *e.g.:*

1-Azabicyclo[2.2.1]-heptane

Quinuclidine

[92] *W. Brooks, A.R. Day,* J. Heterocycl. Chem. *6*, 759 (1969).

[93] *E. de Hoffmann, J.P. Schmit, J.J. Charette,* J. Org. Chem. *35*, 4016 (1970).

[94] *W.L. Mosby,* Heterocyclic Systems with Bridgehead Nitrogen Atoms, Part II, Chap. 16–18, p. 1267–1421, Interscience Publishers, New York 1961.

This exchange reaction can generally be postulated to be basically an S_N2 reaction[95, 96], and, as a rule, the more basic amines are more nucleophilic. More weakly basic amines can be made increasingly reactive by converting into alkali metal compounds or Grignard compounds[97].

While simple *aryl halides* are reluctant to undergo nucleophilic substitution, the reactivity is strongly increased by substitution with electron-attracting substituents in the *ortho* or *para* position. The nitro group in particular can product this effect, for instance, in 1-fluoro-2,4-dinitrobenzene[98] or the equally reactive bis(4-fluoro-3-nitrophenyl)-sulfone[99]. A 46–86% yield of *triarylamines* is obtained from suitably substituted diarylamines with fluoronitrobenzenes in the presence of potassium hydroxide in dimethylformamide at 20–110°[100]:

R¹ = H, 4−Br, 4−NO₂, 2,4−(NO₂)₂
R² = H, NO₂

Fluorine is readily exchanged also in bis(4-fluorophenyl)sulfone, with different amines displaying varying reactivities. The second fluorine atom is generally less readily replaced than the first. Thus, it is possible to prepare not just symmetrically substituted *bis(4-aminophenyl)sulfones* but also *4-fluoro-4'*-substituted and unsymmetrically basically substituted *diphenyl sulfones*.

A number of surveys about nucleophilic aromatic substitution are available[91, 101]. Among the many examples the halogen substitution reactions of pyridines, pyridazines, pyrimidines, quinolines, and acridines may be mentioned. The kinetics and mechanism of the substitution reactions of 1-halo-2-naphthols with anilines have been investigated very recently[102]:

R = H, 2−CH₃, 3−CH₃, 4−OCH₃,
4−Cl, 3−Cl, 4−NO₂, 2,6−(CH₃)₂

80–90%

Additional alkylations of a wide variety of amines including ammonia are possible by using alkali metal compounds such as lithium amide, sodium amide, or potassium amide, sodium hydride, and alkali metal compounds of secondary amines. Thus, a 85–95% yield of variously substituted *aminoethoxypyridines* is obtained from bromoethoxypyridines and potassium amide in liquid ammonia with catalytic amounts of iron(III) nitrate at −33° within 10 minutes[103], *e.g.*:

98–99%   1–2%

4-Amino-2-ethoxypyridine

35%

5-Amino-2-ethoxypyridine

65%

4-Amino-2-ethoxypyridine

Quite generally, the elimination-addition reactions proceeding *via* arynes and hetarynes and leading to aromatic or heteroaromatic amines[104] must be included here.

[95] *C.K. Ingold*, Structure and Mechanism in Organic Chemistry, Chap. 7, Bell & Sons, London 1953.
[96] *A. Streitwieser*, Chem. Rev. *56*, 571 (1956).
[97] *F.E. King, T.J. King, I.H.M. Muir*, J. Chem. Soc. (London) 5 (1946).
[98] *F. Sanger*, Biochem. J. *39*, 507 (1945).
[99] *H. Zahn*, Angew. Chem. *67*, 561 (1955).
[100] *G.P. Sharnin, M.I. Shapshin, I.E. Moisak, F.I. Churikov*, Zh. Org. Khim. *7*, 521 (1971).
[101] *S.D. Ross* in *S.G. Cohen, A.Streitwieser, R.W. Taft*, Progress in Physical Organic Chemistry, Vol. I, Chap. 2, Interscience Publishers, New York 1963; *J.F. Bunnet, R.H. Garst*, J. Amer. Chem. Soc. *87*, 3875 (1965).

[102] *M. Bosco, L. Forlani, E.P. Todesco*, J. Chem. Soc. (B) 1742 (1970).
[103] *H.J. ten Hertog, U.J. Pieterse, D.I. Buurman*, Rec. Trav. Chim. Pays-Bas *82*, 1173 (1963).

In substitution reactions *steric* effects play a not insubstantial role in governing the course of the reaction. Both the steric requirements of the amine to be alkylated and the size of the alkyl groups in the halide determine the reaction velocity in the well-established manner, *e.g.*:

$$CH_3I > C_2H_5I > i\text{-}C_3H_7I$$

Alkylation of ammonia and amines with tertiary alkyl halides is made more difficult by the concurrent elimination.

Fundamentally, halides with other functional group react in the same way on amine substitution. Among unsaturated halides, *vinyl halides* differ from *allyl halides* in their reluctance to react, while the latter react to form normal allylamines or rearrange in dependence on the nature of the halogen bond[105, 106].

Rearrangements are observed also in basically substituted halides (cycloammonium rearrangement, alkylation with onium salts)[107, 108].

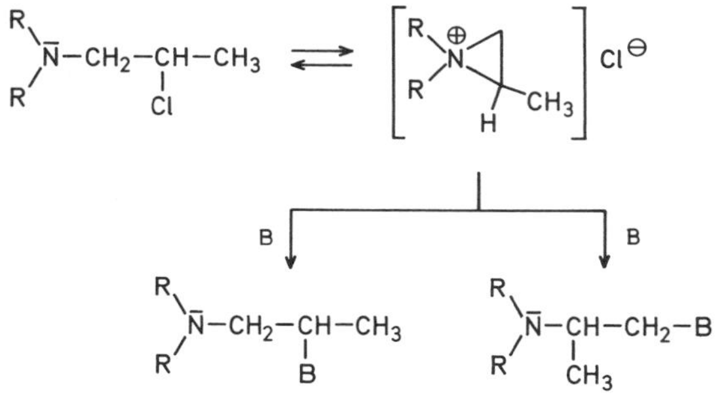

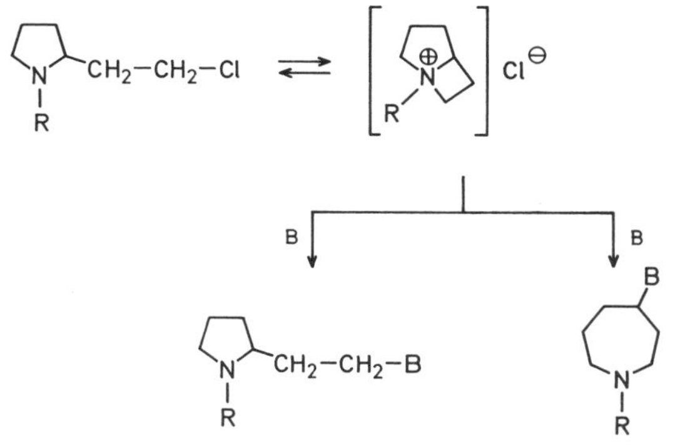

The suitability as alkylating agent is governed decisively by the ring strain in the alkylating bicyclic cation.

As already stated, the alkylation reactions of ammonia and amines described at the outset lead to mixtures (see pp. 452–453). Separating the ionic quaternary salts presents few difficulties, although it needs to be remembered that quaternary ammonium ions are sensitive to nucleophilic substitution reactions[109].

There has therefore been no lack of effort to carry out the alkylation under *controlled* conditions leading to specifically *primary* or secondary amines as end-products:

Primary amines can be obtained from organic halides with *potassium phthalimide* (Gabriel synthesis)[110]:

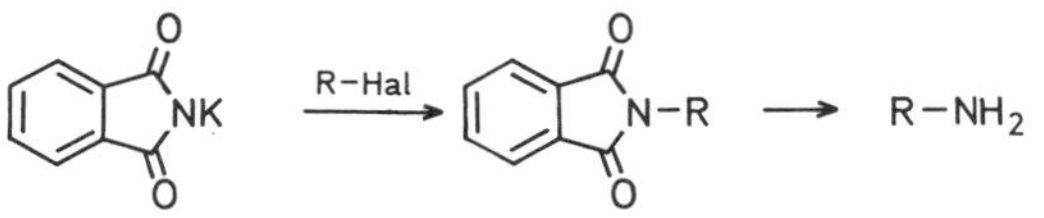

The reaction is not limited to simple alkyl and aryl halides but can be applied to a large number of halides with functional groups. Dihalides can be made to react with either one or both parts. A fruitful alkylation is obtained also by using alkyl toluenesulfonates[111] (p. 461) or epoxides[112]. Hydrolysis of the *N*-substituted phthalimides can be carried out with either concentrated hydrochloric acid or hydrazine[113] under milder conditions. The method is of no importance for preparing primary aromatic amines. For the scope and limitations of the Gabriel synthesis see ref.[114].

B = *prim* or *sec* amines

[104] R. Huisgen, J. Sauer, Angew. Chem. 72, 91 (1960); H. Heaney, Chem. Rev. 62, 81 (1962); T. Kaufmann, Angew. Chem. 77, 557 (1965); T. Kaufmann, R. Wirthwein, Angew. Chem. 83, 21 (1971).
[105] W.G. Young, I.D. Webb, H.L. Goering, J. Amer. Chem. Soc. 73, 1076 (1951).
[106] R.W. De Wolfe, W.G. Young in S. Patai, The Chemistry of Alkenes, Chap. 10, Interscience Publishers, New York 1964.
[107] H. Henecka, U. Hörlein, K.-H. Risse, Angew. Chem. 72, 960 (1960).
[108] A. Ebnöter, E. Jucker, Helv. Chim. Acta 47, 745 (1964).
[109] S. Hünig, W. Baron, Chem. Ber. 90, 395 (1957); D.A. Archer, H. Booth, J. Chem. Soc. (London) 322 (1963); E.D. Hughes, D.J. Whittingham, J. Chem. Soc. (London) 806 (1960).
[110] S. Gabriel, Ber. 20, 2224 (1887).
[111] E.J. Sakellarios, Helv. Chim. Acta 29, 1675 (1946); H. Stetter, W. Böckmann, Chem. Ber. 84, 834 (1951); H. Stetter, W. Böckmann, Angew. Chem. 66, 227 (1954).
[112] S. Gabriel, H. Ohle, Chem. Ber. 50, 804, 819 (1917).
[113] H.R. Ing, R.H.F. Manske, J. Chem. Soc. (London) 2348 (1926).
[114] M.S. Gibson, R.W. Bradshaw, Angew. Chem. 80, 986 (1968).

Saponifiable groups in the same molecule such as cyano, ester, or amide groups, are *disadvantageous* for the preparation of primary amines. Here, the use of *bisbenzenesulfenimide*[115] represents a valuable adjunct:

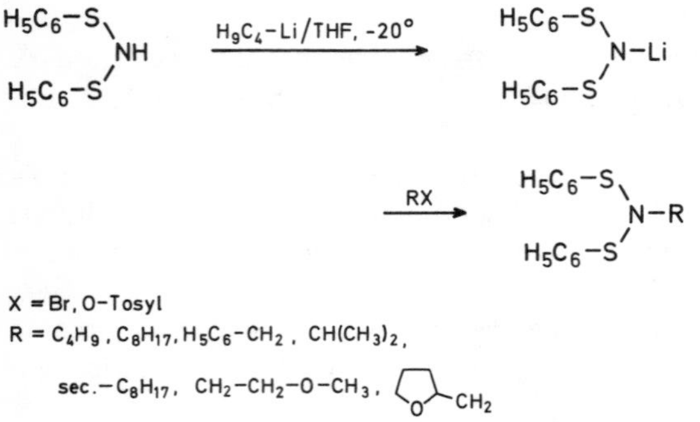

X = Br, O-Tosyl
R = C₄H₉, C₈H₁₇, H₅C₆-CH₂ , CH(CH₃)₂,

$R = C_4H_9, C_8H_{17}, H_5C_6-CH_2, CH(CH_3)_2,$

sec.-C_8H_{17}, $CH_2-CH_2-O-CH_3$,

The *S–N* bond of *N*-alkylated bisbenzenesulfenimides is easily cleaved by hydrochloric acid or thiols to lead to the corresponding *primary amine* and *benzenesulfenyl chloride* or *diphenyl disulfide*:

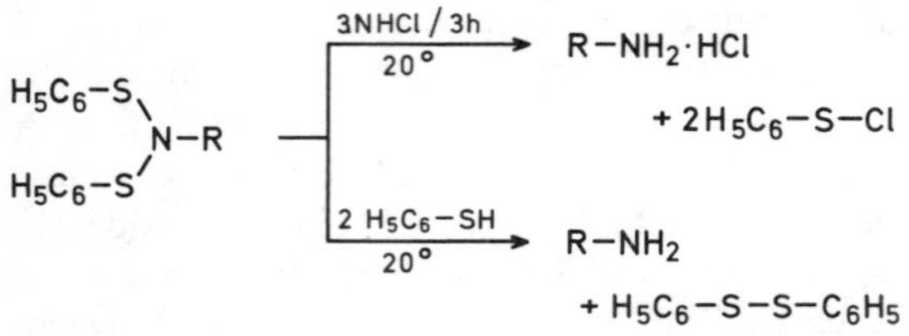

Bisbenzenesulfenimide is equally well suited for preparing primary amines by addition of acrylonitrile or ethyl acrylate (p. 482).

Primary amines can be obtained also from *metallic imines* by alkylation with *alkyl bromides*[116]:

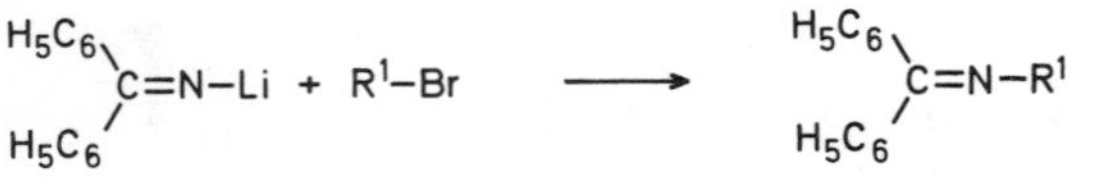

$R^1 = -(CH_2)_n-CH_3$, n = 2, 3, 6

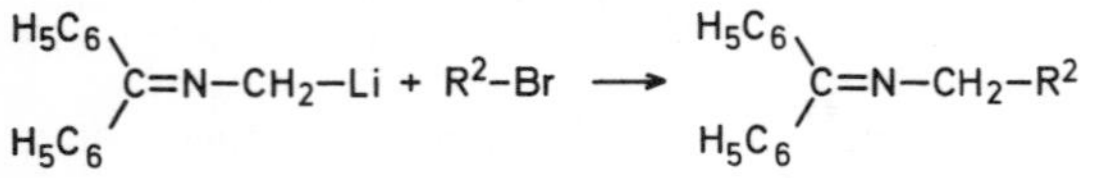

$R^2 = -(CH_3)_2-CH_3$, $-(CH_2)_6-CH_3$, $-CH(CH_3)_2$,
   $-CH(CH_3)-(CH_2)_n-CH_3$ (n = 1, 2, 5)

Acid hydrolysis gives varying yields of the corresponding primary amines.

---

[115] *T. Mukaiyama, T. Taguchi*, Tetrahedron Lett. 3411 (1970).
[116] *T. Cuvigny, P. Hullot*, C.R. Acad. Sci., Paris, Ser. C. *272*, 862 (1971).

Primary amines were prepared in 50–80% yield from *halides* RCH_2-X (X = Cl, Br, I) with *guanidine* and *alkali*[117].

Table 3. Primary amines from alkyl halides and guanidine

Halide RX	h	Temperature [°C]	Mol guanidine : Mol RX	Amine	Yield [% of th.]
$H_7C_3-CH_2-Br$	3	78	2	*Butylamine*	73
$H_9C_3-CH_2-I$	2.5	78	1	*Butylamine*	63
$H_{23}C_{11}-CH_2-Br$	8	78	1	*Dodecylamine*	58
$H_2C=CH-CH_2-Cl$	48	20	2	*Allylamine*	64
$HC≡C-(CH_2)_4-I$	2	78	1	*5-Hexynamine*	34
$H_5C_6-CH_2Cl$	12	20	2	*Benzylamine*	79
$HOOC-CH_2Cl$	4	60	2	*Glycine*	60
$(CH_3)_2CHBr$	12	60	1	*Isopropylamine*	44
$(CH_3)_3CCl$	2	50	1	*tert-Butylamine*	13

Another method of forming *primary amines* is the reaction between organic halides and hexamethylenetetramine (Délépine reaction)[118]:

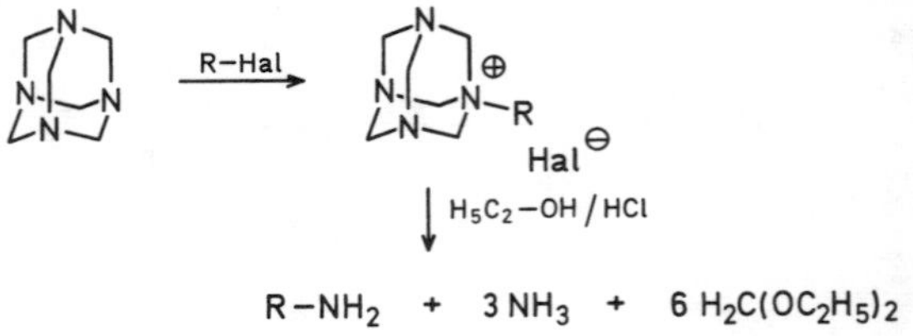

The reaction depends on the quaternary salt formation of hexamethylenetetramine. With sluggishly reacting halides addition of sodium iodide accelerates the reaction satisfactorily. Carbon tetrachloride or chloroform are used as solvent, splitting of the quaternary salts is performed with ethanolic hydrochloric acid or sulfur dioxide[119]. *4-Aminomethyl-1-naphthoic acid* is one compound accessible in this way[120].

---

[117] *P. Hebrard, M. Olomucki*, Bull. Soc. Chim. France 1938 (1970).
[118] *S.J. Angyal*, Org. Reactions, Vol. VIII, Chap. 4, John Wiley & Sons, New York 1954.
[119] *B. Reichert, W. Dornis*, Arch. Pharm. (Weinheim, Ger.) *282*, 100 (1944).
[120] Jap. P. 7 029 179 (1970), Eisai Co., Inv.: *S. Kono, T. Komaki, H. Watanabe*; C.A. *74*, 22608 (1971).
[121] *H. Decker, P. Becker*, Justus Liebigs Ann. Chem. *395*, 362 (1912);
S. Osada, Arch. Pharm. (Weinheim, Ger.) *262*, 501 (1924);
A.L. Morrison, H. Rinderknecht, J. Chem. Soc. (London) 1478 (1950).
J.N. Baxter, J. Cymerman-Craig, J. Chem. Soc. (London) 1940 (1953).

Secondary amines also can be prepared by a variety of substitution reactions.

Alkylation of Schiff's bases with alkyl halides furnishes quaternary ammonium compounds, which on hydrolysis decompose into *secondary amines* and *carbonyl compounds*[121]:

$$R^1-NH_2 + O=C\begin{smallmatrix}R^2\\R^3\end{smallmatrix} \longrightarrow R^1-N=C\begin{smallmatrix}R^2\\R^3\end{smallmatrix} \xrightarrow{R^4-Hal}$$

$$\left[\begin{smallmatrix}R^1\\R^4\end{smallmatrix}\overset{\oplus}{N}=C\begin{smallmatrix}R^2\\R^3\end{smallmatrix}\right]Hal^{\ominus} \longrightarrow \begin{smallmatrix}R^1\\R^4\end{smallmatrix}NH + O=C\begin{smallmatrix}R^2\\R^3\end{smallmatrix}$$

The Schiff's base is best prepared with benzaldehyde. A smooth reaction is obtained with primary aliphatic and arylaliphatic amines (R^1-NH_2), primary aromatic amines often give unsatisfactory results because of side-reactions. The limiting factor of the alkylation reaction is the size of the R^4 alkyl group, and it is not very suitable for introducing large alkyl substituents.

Primary amines and ketones, for instance, benzophenone or acetophenone[122] require high temperatures and specific catalysts (amine-zinc complexes) for conversion into *azomethines, e.g.*:

$$\text{(cyclohexenyl)}-CH_2-NH_2 \xrightarrow[ZnCl_2]{(H_5C_6)_2CO} \text{(azomethine)} \xrightarrow{RX}$$

$$\left[\text{(iminium salt)}\right]X^{\ominus} \longrightarrow \text{(secondary amine)} \quad + (H_5C_6)_2CO$$

This reliable method is supplemented by a *variant*[123] discovered during the monomethylation of amino acids:

$$R-NH_2 \xrightarrow{Cl-(CH_2)_3-COCl} R-NH-C\begin{smallmatrix}O\\Cl\end{smallmatrix} \xrightarrow{AgBF_4}$$

$$R-N=\text{(lactone)} \xrightarrow{H_3C-J} \left[\text{(iminium)}\right]J^{\ominus} \xrightarrow{OH^{\ominus}}$$

$$\begin{smallmatrix}H_3C\\R'\end{smallmatrix}N-C(O)-(CH_2)_3-OH \longrightarrow \begin{smallmatrix}H_3C\\R'\end{smallmatrix}NH + HOOC-(CH_2)_3-OH$$

The reaction is based on acylation of a primary amine with 4-chlorobutyryl chloride in pyridine to the amide, which is cyclized to the *iminolactone* with silver fluoroborate. Alkylation and subsequent hydrolysis gives the *secondary amine*. The yield is about 85%. Here, too, the procedure is *delimited* by the variability of the alkylating agent. A smooth reaction is obtained with methyl iodide, while practically no reaction is obtained with benzyl iodide; steric factors thus play an outstanding part in the halides.

Secondary amines can be obtained similarly *via triphenylphosphinimine*[124]:

$$\left[(H_5C_6)_3\overset{\oplus}{P}-\overset{\ominus}{N}-R \longleftrightarrow (H_5C_6)_3P=N-R\right] + R^1-X \longrightarrow$$

$$\left[(H_5C_6)_3\overset{\oplus}{P}-N\begin{smallmatrix}R\\R^1\end{smallmatrix}\right]X^{\ominus} \xrightarrow{OH^{\ominus}} (H_5C_6)_3P=O + HN\begin{smallmatrix}R\\R^1\end{smallmatrix} + X^{\ominus}$$

R = CH_3, C_2H_5, C_3H_7, i–C_3H_7, CH_2–$CH(CH_3)_2$, $C(CH_3)_3$, 1-adamantyl
R^1 = CH_3, C_2H_5;
X = I, Cl, Br

This synthesis aims especially at *mixed* secondary amines containing a cycloalkyl group. While no difficulties arise during the preparation of triphenylphosphinimines with R=cyclopropyl, cyclopentyl, cyclohexyl, cycloheptyl, and adamantyl, alkylation succeeds only with methyl and ethyl iodides. When a higher alkyl group including cyclopropyl is used, hydrogen halide is eliminated from the alkyl halide with formation of alkylaminotriphenylphosphonium halide and olefin:

$$(H_5C_6)_3P=N-R + R^1-X \longrightarrow \left[(H_5C_6)_3\overset{\oplus}{P}-NH-R\right]X^{\ominus}$$
$$+ \text{ olefine}$$

R^1 = any C_3 and higher alkyl group
X = I

[123] *H. Peter, M. Brugger, J. Schreiber, A. Eschenmoser,* Helv. Chim. Acta **46**, 577 (1963).
[124] *H. Zimmer, M. Jayawant, P. Gutsch,* J. Org. Chem. **35**, 2826 (1970).
[125] *W.J. Hickinbottom,* Reactions of Organic Compounds, Longmans, Green and Co., London 1963.
[126] *R.A. Johnstone, A.W. Robert, D.W. Payling, C. Thomas,* J. Chem. Soc. C 2223 (1969).
[127] *O. Hinsberg,* Ber. **23**, 2962 (1890);
O. Hinsberg, Justus Liebigs Ann. Chem. **265**, 178 (1891);
D. Klamann, G. Hofbauer, Chem. Ber. **86**, 1246 (1953).
[128] *D. Klamann, G. Hofbauer, F. Drahowzal,* Monatsh. Chem. **83**, 870 (1952).
D. Klamann, G. Hofbauer, Chem. Ber. **86**, 1246 (1953).

[122] *R. Grewe, R. Hamann, G. Jacobsen, E. Nolte, K. Riecke,* Justus Liebigs Ann. Chem. **581**, 85 (1953).

Table 4. Secondary amines *via* triphenylphosphinimines (with alkyl iodide)

$$R-NH_2 \longrightarrow R-NH-R^1$$

R	R^1	Yield [% of theory]
▷	CH_3	76
▷	C_2H_5	74
 ◁	CH_3	69
 ◁	C_2H_5	72
 ⬠	CH_3	66
 ⬠	C_2H_5	69
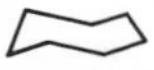 ⬡	CH_3	67
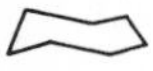 ⬡	C_2H_5	69
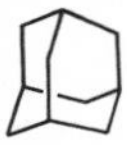 (adamantyl)	CH_3	82

Monomethylation of primary aliphatic and aromatic amines succeeds rapidly and in high yield by treating the readily accessible *trifluoroacetylamides*[125] under basic conditions with alkyl iodide[126]. By modifying the reaction conditions slightly, N,N-dimethylation and N,N-diethylation can be achieved. Special conditions are required for obtaining satisfactory yields with N-propylation reactions:

Table 5. Monoalkylation of primary amines *via* trifluoroacetamides[126]

Amine (R^2NH_2)	R^1	Secondary amine (R^2NHR^1)	Yield [% of th.]
Aniline	CH_3	*N-Methylaniline*	89*
	C_2H_5	*N-Ethylaniline*	83
	C_3H_7	*N-Propylaniline*	18**
4-Nitroaniline	CH_3	*N-Methyl-4-nitroaniline*	86
4-Methoxyaniline	CH_3	*4-Methoxy-N-methylaniline*	98
2,6-Dimethylaniline	CH_3	*2,6,N-Trimethyl-aniline*	77
Cyclohexylamine	CH_3	*N-Methylcyclo-hexylamine*	90
Benzylamine	CH_3	*N-Methylbenzyl-amine*	86

* in aqueous solution 90% N,N-dimethylaniline is obtained
** propyl methanesulfonate instead of propyl iodide gives a 52% yield

The synthesis of secondary amines by the method of Hinsberg[127], like the Gabriel synthesis, is based on the *alkylation* of an anion, in this case a *sulfonamide* ion:

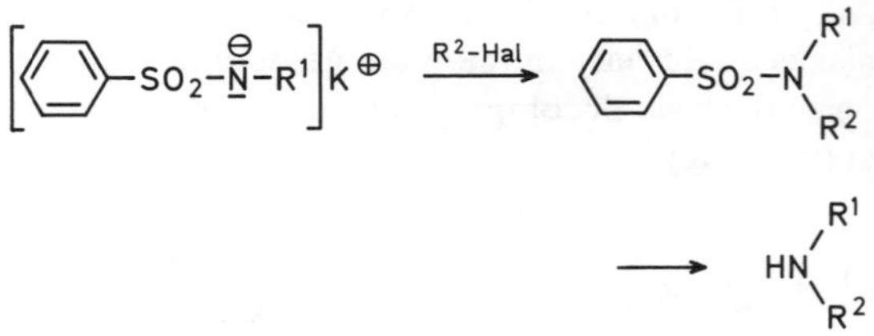

By cleaving with concentrated sulfuric acid or with sodium alcoholate the sulfonamides furnish secondary amines with like groups ($R^1 = R^2$) or mixed secondary amines. Variations of this method are known[128].

The method is of little significance for preparing cyclic secondary amines or alkylarylamines.

Similarly, the reaction of *alkyl halides* or dihalides with *alkali metal salts of saccharin* leads to secondary amines[129]. Repeated alkylation makes mixed secondary amines accessible:

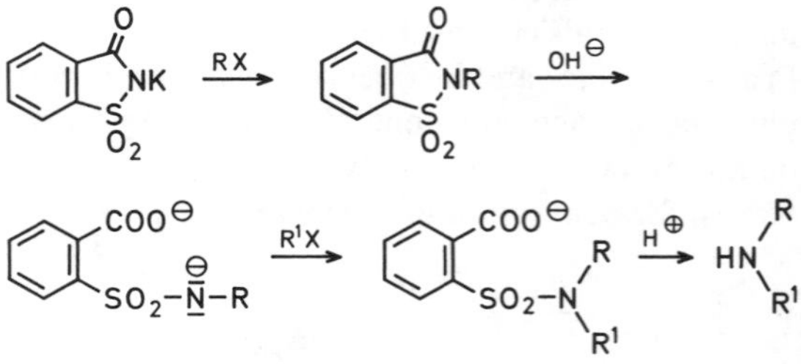

This technique suffers from the *drawback* that splitting of the sulfonamides generally requires drastic acid treatment, even although more conservative methods are available[128]. Milder monoalkylating conditions are desirable, especially with complex and sensitive amines; a suitable reagent for this purpose is a phenacylsulfonamide[130] which may be used both as a protective group for the primary amino group and for monoalkylation:

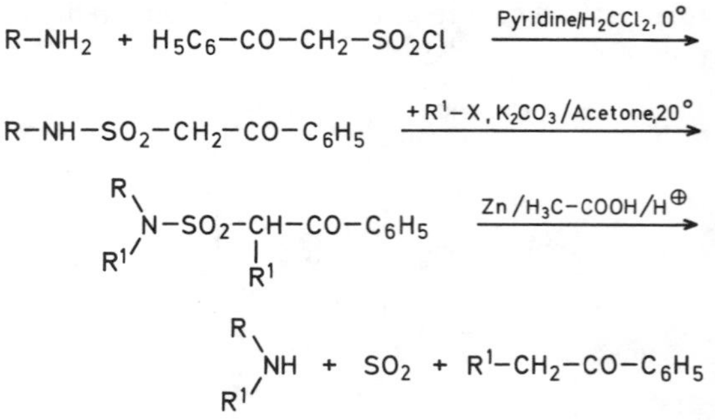

$$R^1-X = CH_3J, \ H_5C_6-CH_2Br$$

[129] *K. Abe,* J. Pharm. Soc. Jap. *75,* 153 (1955).
[130] *J.B. Hendrickson, R. Bergerow,* Tetrahedron Lett. 345 (1970).

Formation of the phenacylsulfonamides takes place readily at 0°. Alkylation likewise proceeds smoothly with dry potassium carbonate in acetone at 20° within 18–24 hours.

The *reductive splitting off* of the protective group ensues with zinc dust in glacial acetic acid plus catalytic amounts of concentrated hydrochloric acid, with the *N*-sulfinate formed decomposing into sulfur dioxide and the secondary amine.

3-Phenyl-3-aza-bicyclo[3.2.2]-nonane; 20%

Unsymmetrically substituted tertiary amines are generally difficult to prepare by normal alkylation reactions, which easily continue to the quaternary

Table 6. Monoalkylation of primary amines with the aid of phenacylsulfonamides[130]

Starting amine	$H_5C_6-CO-CH_2-SO_2-NH-R$		$H_5C_6-CO-CH_2-SO_2-N\begin{smallmatrix}R^1\\R\end{smallmatrix}$		$HN\begin{smallmatrix}R^1 *\\R\end{smallmatrix}$
	[%]	R^1	[%]		[%]
Aniline	94	CH_3	93	N-Methylaniline	78
		$H_5C_6-CH_2$	94	N-Benzylaniline	98
$C_6H_{11}NH_2$	91	CH_3	80	N-Methyl-cyclohexvlamine	—
$H_9C_4-NH_2$	92	CH_3	78	N-Methylbutylamine	75

* isolated as hydrochlorides

Tertiary amines are often difficult to prepare, whether for steric reasons or because of a sluggish reaction. Alkyl bromides and bromobenzene in tetrahydrofuran with naphthalenelithium[131] present alkylate pri and sec amines in good yield to sec and tert bases[132, 133]; bridge ring amines, too, can be arylated in this way[134]:

Table 7. Tertiary amines by alkylation of secondary amines in the presence of naphthalene/lithium[132, 133]

R	R^1	R^2	Amine	Yield [% of th.]
$C_6H_5-CH_2$	$C_6H_5-CH_2$	C_2H_5	N-Ethyl-dibenzylamine	50
		$H_2C=CH-CH_2$	N-Allyl-dibenzylamine	85
		C_6H_5	N-Phenyl-dibenzylamine	70
C_6H_5	C_6H_5	C_4H_9	N-Butyl-diphenylamine	28
	Pyrrolidine	C_4H_9	1-Butyl-pyrrolidine	32

salt stage. *Tertiary amines* with a methyl group substituent can be prepared by alkylating 1, 3, 5-trialkylhexahydro-1, 3, 5-triazines (from formaldehyde and primary amines[135]) with alkyl or aralkyl halides in the presence of water[136]:

Hal = Cl, Br

$R^1 = H_5C_6-CH_2$; $R^2 = CH_3$; N-Dimethylbenzylamine; 27% (no water used)

$R^2 = C_2H_5$; N-Ethyl-N-methylbenzyl-amine; 41% (alkaline decomposition) 65% (decomposition in presence of water)

$R^2 = C_4H_9$; N-Butyl-N-methylbenzyl-amine

$R^1 = C_{12}H_{25}$; $R^2 = CH_3$; N-Dimethyldodecylamine

Similar results are obtained with other alkyl halides.

[131] *U. Schöllkopf* in *Houben-Weyl*, Methoden der organischen Chemie, Bd. XIII/1, p. 87, Georg Thieme Verlag, Stuttgart 1970.

[132] *K. Suga, S. Watanabe, T.P. Pan, T. Fujita,* Chem. & Ind. (London) 78 (1969).

[133] *K. Suga, S. Watanabe, T. Fujita, T.P. Pan,* Bull. Chem. Soc. Japan *42*, 3606 (1969).

[134] *D.-K. Pomorin,* Dissertation, Universität Freiburg i.Br., 1971.

[135] *E.M. Smolin, L. Rapoport,* s-Triazines and Derivatives, p. 477, Interscience Publishers, New York 1959.

[136] *Y. Ohshiro, M. Komatsu, T. Agawa,* Synthesis 89 (1971).

[137] *H. Glaser* in *Houben-Weyl*, Methoden der organischen Chemie, Bd. XI/1, p. 108, Georg Thieme Verlag, Stuttgart 1957.

12.2.2 Amines by substitution of oxygen-containing groups and of enamines

12.2.2.1 Amines by substitution of hydroxy groups

Except for some special cases, replacement of hydroxy by amino[137, 138] cannot be suitably performed in the laboratory. In general, the reaction requires the use of *catalysis* involving either heterogeneous catalysis on dehydrating or hydrogenation contacts or homogeneous catalysis. Its importance lies in its technical application. One of numerous investigations[139] in this field discusses the optimization of alkanol amination with ammonia on an iron contact at 258° and 50 atm for synthesizing primary amines. Amines containing 88–96% primary amino groups are formed in 90–93% yield.

$$H_3C-(CH_2)_n-OH \xrightarrow{NH_3/Cat.} H_3C-(CH_2)_n-NH_2$$

$$n = 5,6,8-12$$

For the readily ensuing replacement of hydroxy groups in cyanohydrins used to prepare α-aminonitriles and α-amino carboxylic acids see p. 604.

Substitution of free *phenolic* hydroxy groups (Bucherer reaction)[140, 141] has preparative importance, but is restricted mainly to the naphthalene series if side-reactions are to be avoided. The reaction converts aromatic hydroxy compounds into primary, secondary, and tertiary aromatic amines relatively simply (and, conversely, amines into hydroxy compounds). Benzene derivatives react in only a few cases, but anthracene, phenanthrene, quinoline, and isoquinoline compounds are converted as readily as naphthalene derivatives. The Bucherer reaction takes place in aqueous phase in the presence of sulfurous acid or sulfites as promoters.

Its mechanism has been elucidated by preparation of the pure intermediates. In contrast to older views, it does not proceed *via* the hydrogen sulfite addition compounds of the keto form of the naphthols or the ketimine form of the naphthylamines but *via* oxotetralin- or iminotetralinsulfonic acids[142], *e.g.*:

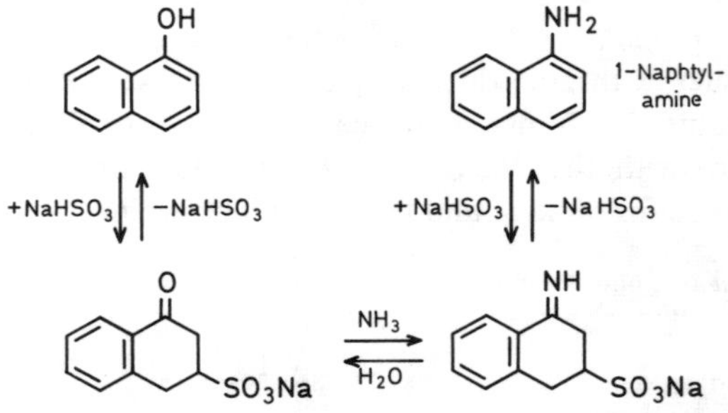

On conversion of naphthols into *naphthylamines*, the oxotetralinsulfonic acid formed from the naphthol and hydrogen sulfit condenses with the amine. Hydrogen sulfite is split off from the iminotetralinsulfonic acid and its enamine form with excess amine, and stabilization as *naphthylamine* occurs. The aminodihydronaphthalenesulfonic acids, which are in equilibrium with their more stable ketimine forms in the case of primary and secondary amines, are unstable.

Direct substitution of free phenolic hydroxy groups is dependent essentially on the structure of the phenol and the related possibility of reacting in the tautomeric keto form. Such promoting effects are produced by condensed aromatic or heteroaromatic rings but also further phenolic hydroxy groups or nitro groups.

12.2.2.2 Amines by substitution of alkoxy groups

Substitution of etherified *alcoholic* hydroxy groups, like that of alcoholic hydroxy groups, requires high temperatures and the use of catalysts. The reactivity is increased by β-carbonyl, β-cyano, and β-carboxy groups. Thus β-amino ketones can be obtained from the corresponding β-methoxy ketones with secondary amines in aqueous solution[143]. *Cyclic ethers* too, which are activated in the β-position by a carbonyl group, react at around 70° with primary aliphatic and

[138] *G.W. Brown* in *S. Patai*, The Chemistry of the Hydroxyl Group, Interscience Publishers, New York 1971.

[139] *G.A. Kliger, A.N. Bashkirov, L.S. Polak, Z.N. Chervochkin, Z.V. Marchevskaya, O.A. Lesik,* Neftekhimiya *10*, 907 (1970); C.A. *74*, 99381 (1971).

[140] *N.L. Drake,* Org. Reactions, Vol. I, Chap. 5, John Wiley & Sons, New York 1942.

[141] *H. Seeboth,* Angew. Chem. *79*, 329 (1967).

[142] *A. Rieche, H. Seeboth,* Justus Liebigs Ann. Chem. *638*, 43, 57, 66, 76, 81, 101 (1960);
A. Rieche, H. Seeboth, Justus Liebigs Ann. Chem. *671*, 77 (1964);
H. Seeboth, D. Bärwolf, B. Becker, Justus Liebigs Ann. Chem. *683*, 85 (1965);
H. Seeboth, H. Neumann, H. Görsch, Justus Liebigs Ann. Chem. *683*, 93 (1965).

[143] *J.N. Nazorov, S.A. Vartangan,* Zh. Obshch. Khim. *22*, 1668, 1794 (1952); C.A. *47*, 9968, 9969 (1953).

aromatic amines and with secondary aliphatic amines by substitution[144]. No wide scope is available for preparing amines by replacing alcoholic alkoxy groups.

Expectedly, this is true, however, for the substitution reaction of *phenolic* ether groups when the aromatic compound has nitro group substituents. Numerous, variously substituted *nitroanilines, dinitroanilines,* and *nitrated naphthylamines* are accessible in this way. Alkoxy groups in nitroheteroaromatic compounds, too, can be substituted by amino groups. *Enol ethers* with activating groups in the β-position again react easily with amines and ammonia (p. 464). By selecting the appropriate enol ether component and amine, normal substitution products or *N*-heterocycles are accessible, either by substitution of oxygen in cyclic ethers with enolic structural elements or by ring closure reactions.

12.2.2.3 Amines by substitution of acyloxy groups

The replacement of esterified hydroxy groups by amine functions is much more important and wide-ranging in preparative chemistry than are substitution reactions of hydroxy and alkoxy groups. *Esters* of *sulfuric acid* and *arylsulfonic acids* are the preferred alkylating agents because they generally react very smoothly and rapidly with ammonia and amines. The mixtures of the different alkylation stages obtained when dialkyl sulfates are used are best separated *via* the p-toluenesulfonamides. Tertiary amines are prepared without difficulty, and especially aromatic and cyclic compounds such as piperidine and like compounds. Primary alcoholic p-toluenesulfonates are probably the most widely used reagents:

$$p-H_3C-C_6H_4-SO_2-O-Alkyl \;+\; RNH_2 \longrightarrow$$

$$p-H_3C-C_6H_4-SO_3H \;+\; R-NH-Alkyl \,(etc).$$

A suitable choice of the molecular relationships, reaction temperature, and solvent allows one to *control* the composition of the alkylation stages obtained. Secondary alcohol arylsulfonates may tend to form olefines during alkylation of amines.

Arylsulfonates of phenols are of little importance for arylating amines because only esters of mono-nitro- and dinitrophenols can be used[145].

The versatile tosyl esters lend themselves also for preparing compounds such as *3-aminosteroids;* configuration inversion occurs[146, 147]. An exceptionally high *stereospecificity* is displayed during the substitution of tosylate, brosylate, and other sulfonate groups by anhydrous ammonia[148]; the axial amines are invariably formed, and its simple, nonreducing nature makes this method an excellent technique for synthesizing *axial amines*.

Amino groups can be introduced into *sugars* and *glycosides* with ammonia *via* the tosyl derivatives. 1,6-Di-*O*-benzenesulfonyl-D-mannitol gives 65% *1,6-dianilinodideoxy-D-mannitol* even without protection of the secondary hydroxy groups with cold aniline[149], and 21% of the corresponding *1,6-diaminodideoxy-D-mannitol* with ammonia. The amino group is introduced in the 6-position in 6-tosyl- and 6-mesyl-α-D-glucosides with ammonia in methanol likewise without masking the secondary hydroxy groups[150]. 9-Alkylation of *adenine* succeeds with 5-(4-bromobenzenesulfonyl)-3-(4-bromobenzoyl)-2-deoxymethyl-α-D-ribofuranoside[151].

The morpholine group can be introduced into 2-(3-tosyloxypropyl)benzotriazole by substitution without the occurrence of the readily proceeding shift of the side-chain to the adjacent nitrogen atom and quaternary salt formation[152]. *2-(3-Morpholinopropyl)benzotriazole* is the product obtained.

Acetoxy groups can be replaced by amino groups in 1-alkoxy-1,2-diacetoxyalkanes with formation of *α-amino-β-hydroxy carboxylic acids* if potassium cyanide is made to react simultaneously with the ammonia or primary amine[153]:

$$
\begin{array}{c}
R-CH-CH-OC_2H_5 \;+\; R^1-NH_2 \;+\; KCN \\
\underset{H_3C-CO-O}{|}\quad \underset{O-CO-CH_3}{|}
\end{array}
$$

$$
\longrightarrow \quad
\begin{array}{c}
R-CH-CH-CN \\
\underset{OH}{|}\quad \underset{NH-R^1}{|}
\end{array}
$$

$$\Big\downarrow HCl/H_2O$$

$$
\begin{array}{c}
R-CH-CH-COOH \\
\underset{OH}{|}\quad \underset{NH-R^1}{|}
\end{array}
$$

[144] *J.N. Nazarov, S.G. Matsoyan, S.A. Vartanyas,* Zh. Obshch. Khim. *23,* 1990 (1953); C.A. *49,* 3002 (1955).

[145] DRP. 194 951 (1906), Erf.: *F. Ullmann;* Fortschritte der Teerfabrikation u. verwandter Industriezweige *9,* 130 (1911); *F. Ullmann, G. Nadai,* Ber. *41,* 1870 (1908).

[146] *L.C. King, M.J. Bigelow,* J. Amer. Chem. Soc. *74,* 3338 (1952).

[147] *J.H. Pierce, H.C. Richards, C.W. Shoppee, R.J. Stevenson, G.H.R. Summers,* J. Chem. Soc. (London) 694 (1955).

[148] *J.L. Pinkus, G. Pinkus, T. Cohen,* J. Org. Chem. *27,* 4356 (1962).

As a modified Strecker synthesis this procedure gives good to very good yields, especially during preparation of serine and its higher homologs (see p. 604).

Esters of carboxylic acids seldom have an alkylating action on amines[154]. Methyl salicylate lends itself to the preparation of both quaternary salts and *tertiary amines*[155, 156]:

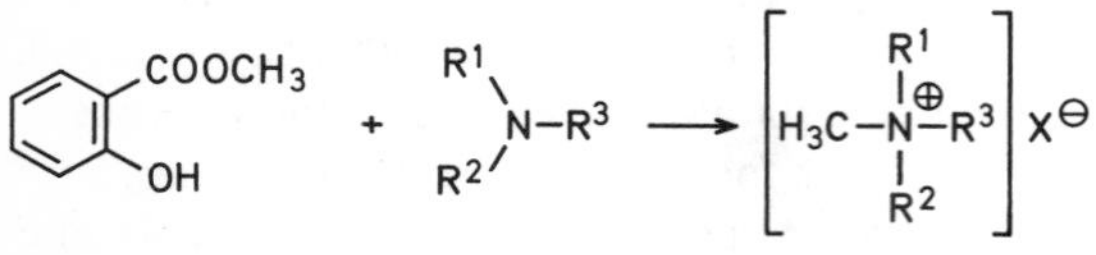

$X^{\ominus}$ = Salicylate, Picrate

R¹ = R² = CH₃; R³ = –CH₂–CH₂–OH: *(2-Hydroxyethyl)- trimethylammonium salicylate* or *picrate*

R¹ = R² = C₂H₅; R³ = –CH₂–CH₂–OH: *Diethyl-(2-hydroxy- ethyl)methyl- ammonium salicylate or picrate*

R¹ = R² = R³ = C₂H₅: *Triethylmethyl- ammonium salicylate or picrate*

From 1,2,3,4-tetrahydroisoquinoline and *N*-methylbenzylamine one obtains respectively *2,2-dimethyl-1,2,3,4-tetrahydroisoquinolinium* salts side by side with *2-methyl-1,2,3,4-tetrahydroisoquinoline* and *benzyltrimethylammonium salts* together with *N,N-dimethylbenzylamine*:

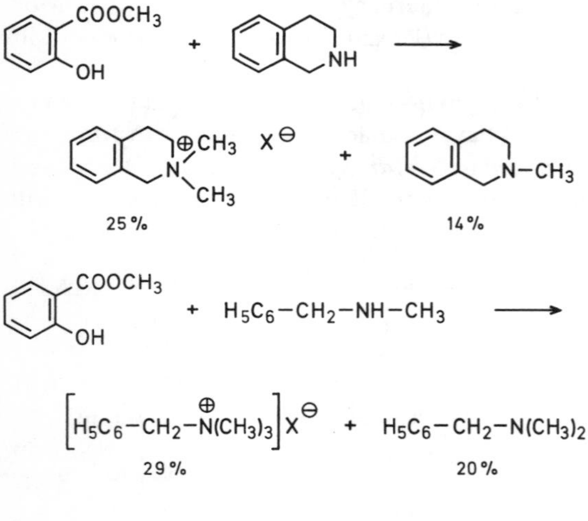

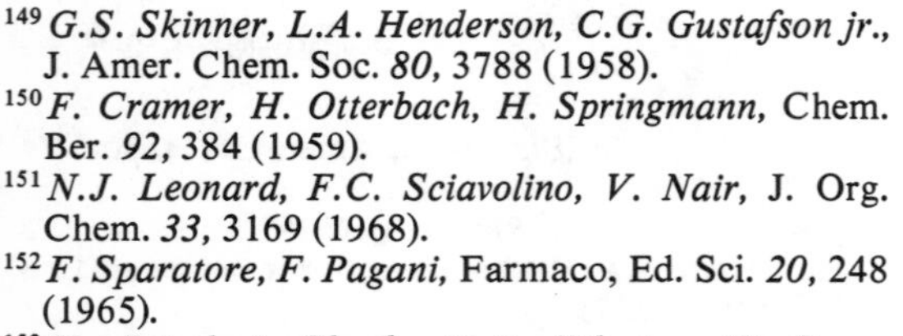

1,2,3,4-Tetrahydroquinoline, *N*-methylaniline or diphenylamine respectively furnish *1,2,3,4-tetrahydro-1-methylquinoline* (75%), *N,N-dimethylaniline* (77%), and *N-methyldiphenylamine* (29%).

As has already been stated in the passage on halogen substitutions (pp. 456–457), those methods are of particular preparative interest that leads to primary and secondary amines. Just like there, derivatives of ammonia or amines are employed whose reactive hydrogen atoms have been protected partially or wholly by blocking.

Primary amines can be prepared not only from hexamethylenetetramine and halides (cf. p. 456); using esters of arylsulfonic acids is fundamentally also possible.

Secondary amines are obtained by alkylating Schiff bases (cf. p. 457), as has been described for halogen exchanges. p-Toluenesulfonic acid esters, too, may be used.

Preparation of primary and secondary amines by alkylation of sulfonamides has also already been described (cf. p. 458). Instead of halides alkyl sulfates and alkyl arylsulfonates, notably alkyl p-toluenesulfonates, are used with advantage. The varying acidity of primary and secondary sulfonamides and their resistance to hydrolysis by aqueous alkalis offers the advantage, during preparative working[157], of, firstly, monoalkylation of unsubstituted sulfonamides in a solution made alkaline with sodium carbonate. Secondly, primary amines can be separated from secondary amines *via* the sulfonamides[158] and tertiary amines, too, may be separated.

Of the carboxamides that can be used for preparing primary and secondary amines by alkylation with esters[111], the already mentioned phthalimide (p. 455)[110] is more important than others such as acetamide or benzamide. Working under anhydrous conditions is generally essential because alkali metal derivatives of carboxamides are hydrolyzed in water.

[149] G.S. Skinner, L.A. Henderson, C.G. Gustafson jr., J. Amer. Chem. Soc. *80*, 3788 (1958).

[150] F. Cramer, H. Otterbach, H. Springmann, Chem. Ber. *92*, 384 (1959).

[151] N.J. Leonard, F.C. Sciavolino, V. Nair, J. Org. Chem. *33*, 3169 (1968).

[152] F. Sparatore, F. Pagani, Farmaco, Ed. Sci. *20*, 248 (1965).

[153] H. Geipel, J. Gloede, K.-P. Hilgetag, H. Gross, Chem. Ber. *98*, 1677 (1965).

[154] C.A. Grob, F. Reber, Helv. Chim. Acta *33*, 1776 (1950); C.A. Grob, W. von Tscharner, Helv. Chim. Acta *33*, 1070 (1950).

[155] T. Kametani, K. Kigasawa, H. Hiiragi, H. Sugahara, T. Hayasaka, H. Iwata, H. Ishimaru, Tetrahedron Lett. 1817 (1965).

[156] T. Kametani, K. Kigasawa, T. Hayasaka, J. Pharm. Soc. Jap. *87*, 265 (1967).

[157] D. Klamann, H. Bertsch, Chem. Ber. *89*, 2007 (1956).

[158] D. Klamann, G. Hofbauer, F. Drahowzal, Monatsh. Chem. *83*, 870 (1952).

Primary *amines* are obtained *via* benzenesulfonamides by reacting the ethyl esters of *N*-(phenylsulfonyl)formidic acid, formed in 90% yield by acid catalysis from triethyl orthoformate and benzenesulfonamide, with 2 mole Grignard compound[159]:

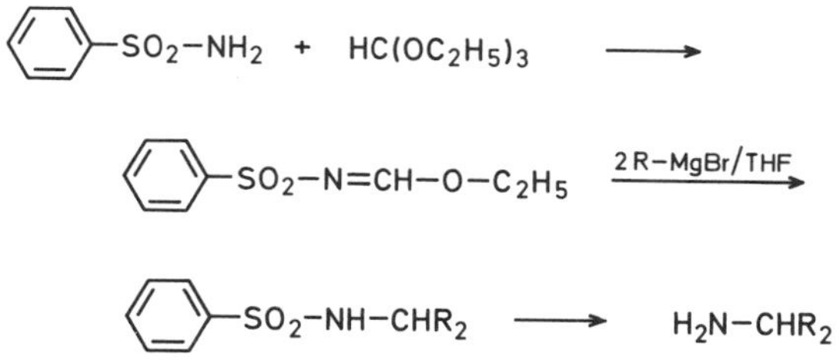

$$\text{(C}_6\text{H}_5)-\text{SO}_2-\text{NH}_2 \ + \ \text{HC(OC}_2\text{H}_5)_3 \longrightarrow$$

$$\text{(C}_6\text{H}_5)-\text{SO}_2-\text{N}=\text{CH}-\text{O}-\text{C}_2\text{H}_5 \ \xrightarrow{2\,\text{R-MgBr/THF}}$$

$$\text{(C}_6\text{H}_5)-\text{SO}_2-\text{NH}-\text{CHR}_2 \longrightarrow \text{H}_2\text{N}-\text{CHR}_2$$

$R = C_2H_5$;	*1-Ethylpropylamine*
$R = C_3H_7$;	*1-Propylbutylamine*
$R = -CH(CH_3)_2$;	*1-Isopropylisobutylamine*
$R = C_6H_5$;	*a-Phenylbenzylamine*

Esters of sulfamic acid, too, can be employed for converting alcohols into tertiary amines[160]:

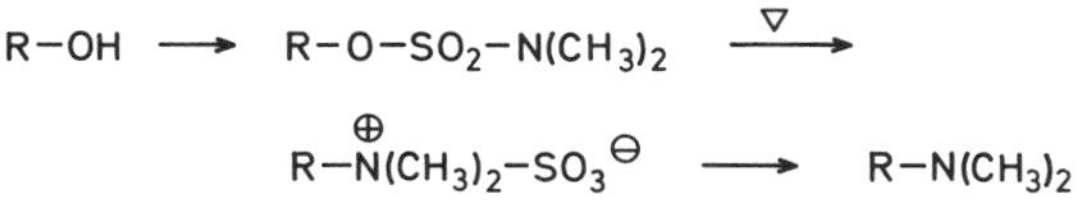

$$R-OH \longrightarrow R-O-SO_2-N(CH_3)_2 \xrightarrow{\triangledown}$$

$$R-\overset{\oplus}{N}(CH_3)_2-SO_3^{\ominus} \longrightarrow R-N(CH_3)_2$$

$R = H_5C_6-\underset{\underset{CH_3}{\vert}}{CH}-$		*N,N,a-Trimethylbenzylamine;* 60%
$= Cl-\underset{\underset{CH_3}{\vert}}{\overset{}{\text{(C}_6\text{H}_4)}}-CH-$		*4-Chloro-N,N,a-trimethylbenzylamine;* 80%
$= H_5C_6-CH=CH-CH_2-$		*N,N-Dimethylcinnamylamine;* 69%
$= (H_5C_6)_2CH-$		*1,1-Diphenyltrimethylamine;* 76%
$= H_5C_6-\underset{\underset{CH=CH_2}{\vert}}{CH}-$		*N,N-Dimethyl-a-vinylbenzylamine;* 25% + *trans-N,N-Dimethylcinnamylamine;* 75%

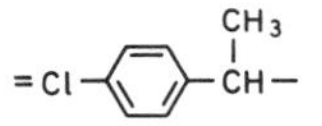

The method is especially suitable for alcohols which tend to form carbonium ions and involves a thermal $S_N i$ rearrangement of the sulfamates to the betaine analogs.

12.2.2.4 Amines by substitution of enamines (cf. 12.3.5)

As already mentioned on p. 460 replacing enolic hydroxy groups by ammonia or primary amines leads to compounds which, as enamines, differ in many ways from the amines because of their reac-

tivity. Splitting off water from carbonyl compounds and the named bases leads to formation of imines or Schiff's bases (azomethines) *via* the generally unstable addition compounds. These azomethines may be in equilibrium with the enamine form:

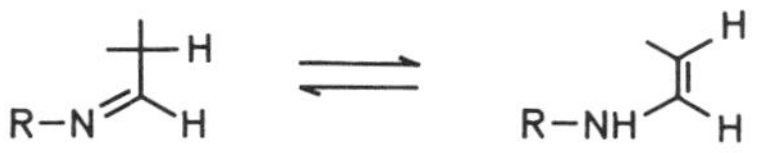

The chemistry of the *tertiary enamines* formed from aldehydes and ketones with secondary amines has made giant strides during recent years. In synthetic chemistry enamines play a very large part as intermediates. For example, they may be used as aids in certain substitution reactions aimed at introducing an amine function by forming a *C–N* bond and for synthesizing heterocycles. As end-products, too, enamines in which mesomerism 1 is hindered for steric reasons[161, 162] are of theoretical and practical importance. The following scheme summarizes the properties and reactivity of tertiary enamines:

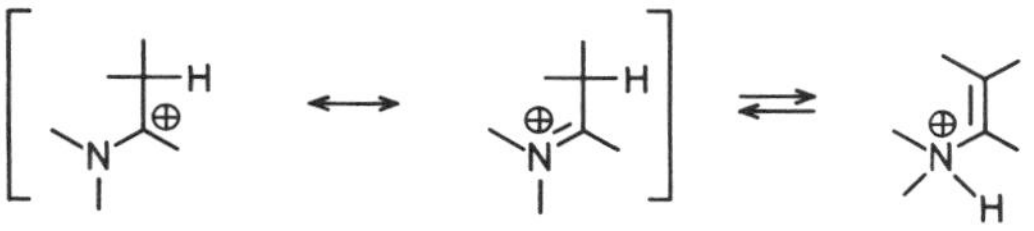

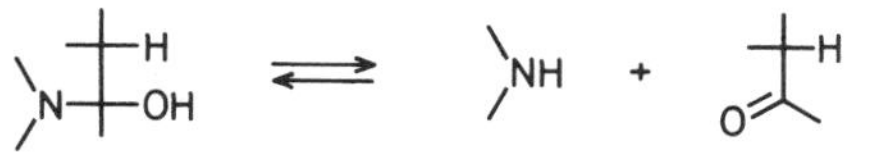

On account of the vinylamine mesomerism 1, enamines possess two basic centers, *viz.*, on the nitrogen and β-carbon atom respectively. Two tautomeric protonated forms 2 are thus possible, the enammonium structure and the mesomeric carbimonium form. The latter explains the reactivity. Normally, hydrolysis produces splitting into amine and carbonyl compound 3 *via* the α-aminoalcohols. As this section is limited to a discus-

[159] *H. Stetter, D. Theisen,* Chem. Ber. *102*, 1641 (1969).

[160] *E.H. White, C.A. Elliger,* J. Amer. Chem. Soc. *87*, 5261 (1965).

[161] *C.A. Grob, A. Kaiser, E. Renk,* Chem. & Ind. (London) 598 (1957);
 C.A. Grob, A. Kaiser, E. Renk, Helv. Chim. Acta *40*, 2170 (1957).

[162] *W. Schneider, F. Schumann,* Tetrahedron Lett. 1583 (1966);
 F. Schumann, Dissertation, Universität (TH) Karlsruhe 1967.

sion of the substitution reaction of enolic hydroxy groups, reference is made to monographs[163-167] for the numerous other methods of preparation. Carbonyl compounds with activating groups which are enolizable or present as hydroxymethylene compounds furnish stable primary and secondary enamines also with ammonia and primary amines. The reaction of ethyl acetate with ammonia, aliphatic amines, and aromatic amines has been investigated in detail as the prototype reaction. During reaction of β-keto carboxylic acid esters with amines, *β-keto carboxamides* may be formed, *e.g., acetoacetarylamides* are easily formed at excessively high temperatures. The reaction conditions and the temperature play a decisive role. In boiling toluene or xylene and with catalytic amounts of *p*-toluenesulfonic acid present, ethyl 2-oxocyclohexanecarboxylate on reacting with secondary amines such as piperidine, morpholine or piperazine yields respectively *2-(1-piperidinylcarbonyl)cyclohexanone,   2-(1-morpholinocarbonyl)cyclohexanone,* or *2-(1-piperazinylcarbonyl)cyclohexanone* plus *2,2'-(1,4-piperazinediyldicarbonyl)dicyclohexanone*[168]:

X = CH₂
  = O
  = NH

2-Oxocyclopentanecarboxylic acid esters react with primary amines at 20° *via* the 2-amino-2-

hydroxycyclopentanecarboxylic acid ester to form methyl and *ethyl 2-(alkylamino)-1-cyclopentene-1-carboxylates*[169]:

R = CH₃ , C₂H₅
R¹ = –CH₂–C₆H₅ , C₃H₇ , –CH₂–CH(CH₃)₂ , –C(CH₃)₃

With dimethylamine[170], piperidine[168], and ethyl 1-piperazinecarboxylate[168] at room temperature the same esters give *methyl and ethyl 2-(dimethylamino)-, 2-piperidino-,* and *2-(4-ethyl carboxylate)piperazinyl-1-cyclopentene-1-carboxylates.* Ethyl 2-oxocyclohexanecarboxylate does not react with secondary amines under these conditions. By contrast, it furnishes the corresponding *enamine carboxylic acid esters* with piperidine and morpholine on heating in benzene with catalytic amounts of *p*-toluenesulfonic acid. While ammonia and primary amines react smoothly to form enamine carboxylic acid esters with cyclic 2-keto carboxylic esters, formation of carboxylic acid enamine esters from cyclic 2-keto carboxylic acid esters and secondary amines is primarily temperature dependent[168], as is revealed by the few examples known to date.

In *1,3-diketones* respectively one carbonyl group reacts, in *1,4-diketones* both carbonyls react with ammonia, and with primary and secondary aliphatic and aromatic amines to readily lead to *pyrroles*[171].

12.2.3 Amines by substitution of carboxy groups
(cf. Amines by rearrangement reactions)
Reactions which lead to an amino group replacing a carboxy group all involve rearrangements and are therefore dealt with on p. 560.

[163] *J. Smuszkoviez*, Advan. Org. Chem. *4*, 1 (1963).
[164] *K. Blaha, O. Cervinka*, Advan. Heterocycl. Chem. *6*, 147 (1966).
[165] *W.J. Taylor*, Indol Alkaloids, An Introduction to the Enamine Chemistry of Natural Products, Pergamon Press, Oxford 1966.
[166] *A.G. Cook*, Enamines: Synthesis, Structure and Reactions, Chap. 2: *Le Roy, W. Haynes*, Methods and Mechanisms of Enamine Formation, Marcel Dekker, New York 1969.
[167] *E.C. Taylor* und *A. McKillop*, The Chemistry of Cyclic Enaminonitriles and *o*-Aminonitriles, Interscience Publishers, New York 1970.
[168] *W. Schneider, E. Schikora*, unpubl.
[169] *F.C. Pennington, W.D. Kehret*, J. Org. Chem. *32*, 2034 (1967).
[170] *J. Schmutz*, Helv. Chim. Acta *38*, 1712 (1955); *E.M. Austin, H.L. Brown, G.L. Buchanan, R.A. Raphael jr.*, Tetrahedron *25*, 5517 (1969).
[171] *N.P. Buu-Hoi*, J. Chem. Soc. (London) 2885 (1949).
[172] *C. Willgerodt, P. Mohr*, J. Prakt. Chem. *34*, 120 (1886).
[173] *H.N. Eisen, S. Belman, M.E. Carsten*, J. Amer. Chem. Soc. *75*, 4583 (1953).
[174] *H.J. Backer, S.K. Wadman*, Rec. Trav. Chim. Pays-Bas *68*, 595 (1949).

12.2.4 Amines by substitution of sulfur-containing groups

12.2.4.1 Amines by substitution of sulfo groups

The action of ammonia or amines on certain *aromatic* or *heteroaromatic* sulfonic acids replaces the sulfo group by an amino group.

In the *benzene* series this amine exchange succeeds only where the sulfo group is activated by at least two nitro groups in the *o-* and *p-*positions. 2,4-Dinitroaniline is formed from 2,4-dinitrobenzenesulfonic acid on heating with ammonia in a tube[172]. 2,4-Dinitrobenzenesulfonic acid reacts with the free amino groups of proteins to form *N-2,4-dinitrophenyl proteins*[173].

In 4-nitrobenzenesulfonic acid replacement of the sulfo group by arylamino groups is possible only indirectly. 4-Nitrobenzenesulfonyl chloride is reacted with cyanamide and sodium hydroxide to form 4-nitrobenzenesulfonylcyanamide, which reacts with primary arylamines to form *N'-aryl-N-(4-nitrobenzenesulfonyl)guanidines*. On heating with 2 N caustic soda, these furnish the *(4-nitrophenyl)arylamines* in quantitative yield[174].

Heating 1-amino-3-naphthalenesulfonic acid with ammonia in the presence of ammonium chloride leads to *1,3-naphthalenediamine*. Where the second ring of the 4-amino-2-naphthalenesulfonic acid contains further sulfo groups, only the sulfo group in the 3-position to amino is exchanged and *1,3-diamino-5-, -6-, -7-* and *-8-naphthalenesulfonic acid, 5,7-* and *6,8-diamino-2-naphthalenesulfonic acid*, and the corresponding *1,3-disulfonic acids* are obtained. On analogously reacting 1-hydroxy-3-naphthalenesulfonic acids the hydroxy group as well as the 3-sulfo group is replaced by amino to furnish *1,3-naphthalenediamine*[175]. If these reactions are performed with primary arylamines in place of ammonia, then the amino group and the sulfo group in the 3-position are replaced simultaneously by arylamino[176]. 8-Amino-1,3,6-naphthalenetrisulfonic acid 1,8-sultam reacts with aniline to form *8-amino-6-anilino-1,3-naphthalenedisulfonic acid 1,8-sultam*[177]:

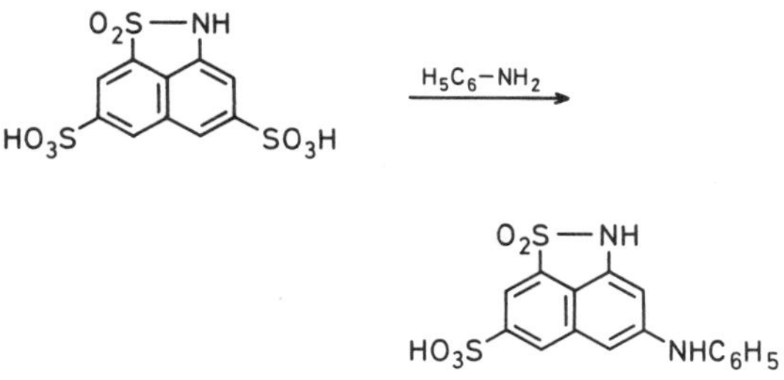

On alkali fusion of 8-cyano-1-naphthalenesulfonic acid intramolecular exchange of a sulfo group against an acylamino group takes place. Ring closure of the intermediately formed 8-sulfo-1-naphthamide leads to *naphthostyril (8-amino-1-naphthoic acid lactam)* which can be hydrolyzed to *8-amino-1-naphthoic acid*[178]:

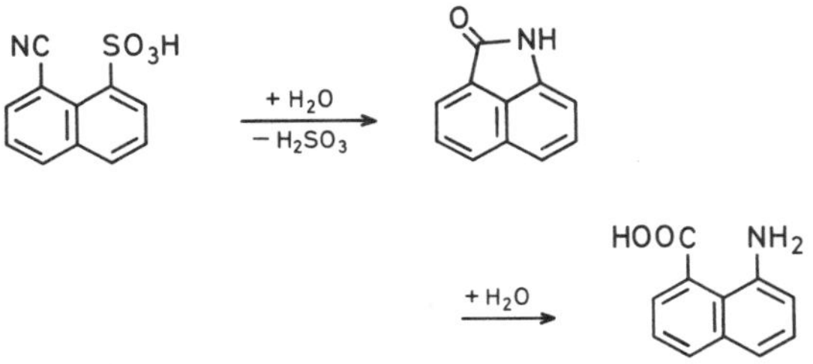

Analogous conversion of 8-cyano-1,5- and 8-cyano-1,6-naphthalenedisulfonic acid gives a good yield of respectively *5-hydroxy-* and *4-hydroxy-2-oxo-1,2-dihydrobenzo[c,d]indoles*[179]. Technical importance accrues to the preparation of amino and diaminoanthraquinones such as *1-amino-, 2-amino-, 1,5-diamino-,* and *2,6-diaminoanthraquinones* from the corresponding mono- and disulfonic acid by reacting with aqueous ammonia under pressure and at elevated temperature[180]. It is necessary to remove the ammonium sulfite formed from the reaction mixture[181]. *(Alkylamino)anthraquinones* can be prepared analogously[182, 183]. 3,4-Dioxo-1-naphthalenesulfonic acid is very reactive toward primary arylamines.

[175] DRP. 89061 (1894), 90905, 90906, 94075 (1895), Kalle & Co.; Fortschritte der Teerfabrikation u. verwandter Industriezweige *4*, 598, 599, 600 (1899).

[176] DRP. 75296 (1893), Farbenfabriken Bayer; Fortschritte der Teerfabrikation u. verwandter Industriezweige *3*, 500 (1896); DRP. 76414 (1893), Farbenfabriken Bayer; Fortschritte der Teerfabrikation u. verwandter Industriezweige *4*, 594 (1899).

[177] DRP. 442610 (1925), I.G. Farben, Erf.: *W. Neelmeier, T. Nocken;* Fortschritte der Teerfabrikation u. verwandter Industriezweige *15*, 322 (1923).

[178] DRP. 441225 (1924), I.G. Farben, Erf.: *R. Herz, F. Schulte;* Fortschritte der Teerfabrikation u. verwandter Industriezweige *15*. 1810 (1923).

[179] *A. Stoll, J. Rutschmann,* Helv. Chim. Acta *34*, 382 (1951).

[180] *H.R.v. Perger,* Ber. *12*, 1567 (1879); *R.E. Schmidt,* Ber. *37*, 69 (1904).

For example, its sodium salt reacts with aniline in very dilute aqueous solution (down to 1 in 300,000) even in the cold to form *1-anilino-3,4-naphthoquinone* and *3-hydroxy-4-(phenylimino)-1(4H)-naphthalenone*.

An analogous reaction occurs with aminoazo compounds, phenylenediamines, aminophenols, aminoarylsulfonic acids, and aminoaryl carboxylic acids[184]. Additional examples from the *N*-heteroaromatic series are listed in Table 8.

Table 8. Heteroaromatic amino compounds by substitution of the sulfonyl group

Sulfonic acid	Amine	Yield [% of th.]	Ref.
R^1 = H, NO_2			185
	e.g., 4-Arylamino-3,5-dibromopyridine		186
	4-Amino-6-methoxy-2-methylquinoline	84	187
	10-Aminobenzo[h]-quinoline	—	188
	2-Aminobenzo-thiazole	—	189
	1-Methyl-4-(methyl-amino)-2(1H)-pyrimidone	84	190

12.2.4.2 Amines by replacement of alkylsulfonyl groups

Substitution reactions of this type have so far been carried out exclusively with methylsulfonyl groups bound to *N*-heterocycles (Table 9).

12.2.4.3 Amines by replacement of arylsulfonyl groups

When ammonia and amines act on *diarylsulfones* in which one nucleus is substituted by one, two, or three nitro groups in the *o*- or *p*-position, the unsubstituted arylsulfonyl group is replaced and sometimes also a nitro group. The course of the reaction is difficult to predict, and the value of the method is thus small[196, 197] (Table 10).

[181] DRP. 267 212 (1912), Farbwerke Hoechst; Fortschritte der Teerfabrikation u. verwandter Industriezweige *11*, 552 (1915);
DRP. 273 810 (1914), Farbwerke Hoechst; Fortschritte der Teerfabrikation u. verwandter Industriezweige *12*, 411 (1917);
DRP. 256 515 (1911), BASF; Fortschritte der Teerfabrikation u. verwandter Industriezweige *11*, 511 (1915);
DRP. 391 073 (1921), Ciba; Fortschritte der Teerfabrikation u. verwandter Industriezweige *14*, 847 (1926).
[182] *C.V. Wilson, J.B. Dickey, C.F.H. Allen*, Org. Synth. *29*, 66 (1949).
[183] *L.F. Fieser, M. Fieser*, J. Amer. Chem. Soc. *57*, 494 (1935).
[184] *M. Böninger*, Ber. *27*, 25 (1894);
F. Sachs, M. Craveri, Ber. *38*, 3685 (1905).
[185] *A. Mangini, M. Colonna*, Gazz. Chim. Ital. *73*, 313 (1943).
[186] *M. Dohrn, P. Diedrich*, Justus Liebigs Ann. Chem. *494*, 288, 301 (1932).
[187] *J. Walker*, J. Chem. Soc. (London) 1552 (1947).
[188] *H. Schenkel-Rudin, M. Schenkel-Rudin*, Helv. Chim. Acta *27*, 1456 (1944).
[189] DRP. 615 526, 617 188 (1933), I.G. Farben, Erf.: *W. Zerweck, H. Salkowski, E. Herdieckerhoff*; Fortschritte der Teerfabrikation u. verwandter Industriezweige *21*, 320; Fortschritte der Teerfabrikation u. verwandter Industriezweige *22*, 297.
[190] *H. Hayatsu, M. Yano*, Tetrahedron Lett. 755 (1969).
[191] *H.S. Forrest, J. Walker*, J. Chem. Soc. (London) 1939 (1948).
[192] *E. Hoggarth*, J. Chem. Soc. (London) 1918 (1949).
[193] *E. Hoggarth*, J. Chem. Soc. (London) 3311 (1949).
[194] *C.W. Noell, R.K. Robins*, J. Amer. Chem. Soc. *81*, 5997 (1959).
[195] *I.G. Nairn, H. Tieckelmann*, J. Org. Chem. *25*, 1127 (1960).
[196] *J.D. Loudon, T.D. Robson*, J. Chem. Soc. (London) 242 (1937).
[197] *A. Livingston, J.D. Loudon*, J. Chem. Soc. (London) 246 (1937).

Table 9. Heteroamino compounds by substitution of methylsulfonyl groups

Sulfone	Amine	Yield [% of theory]	Ref.
O_2N–pyridine–SO_2–CH_3	O_2N–pyridine–NH_2 *2-Amino-5-nitropyridine*	—	191
H_5C_6–oxadiazole–SO_2–CH_3	H_5C_6–oxadiazole–NH–$(CH_2)_2$–$N(C_2H_5)_2$ *5-Phenyl-2-(2-diethylaminoethyl-amino)-1,3,4-oxadiazole*	good	192
benzoxazole–SO_2–CH_3	benzoxazole–NH–R e.g., *2-(Alkylamino)-benzoxazole*	good	193
purine SO_2–CH_3 / H_3C–SO_2	purine $N(CH_3)_2$ / H_3C–SO_2 *6-(Dimethylamino)-2-(methyl-sulfonyl)purine*	90	194
H_3C–CO–O–CH_2, NH_2, pyrimidine–SO_2–CH_3	$HOCH_2$, NH_2, pyrimidine–NH–R e.g., *4-Amino-2-(arylamino)-pyrimidine-5-methanol*	32–73	195

Table 10. Preparation of amino compounds by replacement of arylsulfonyl groups

Sulfone	Amine	Yield [% of theory]	Ref.
phenyl(NO_2)–SO_2–phenyl–CH_3	phenyl(NO_2)–piperidine *1-(2-Nitrophenyl)piperidine*		196
trinitrophenyl–SO_2–trinitrophenyl	trinitrophenyl–NH–R e.g., *N-Alkyl-2,4,6-trinitroaniline*		198
naphthyl($SO_2C_6H_5$)(NO_2)(NO_2)	naphthyl(NH–R)(NO_2)(NO_2) e.g., *N-Alkyl-2,4-dinitro-1-naphthylamine*		199
phenyl–SO_2–phenyl	phenyl–piperidine *1-Phenylpiperidine*		200
	phenyl–hexamethylenimine *1-Phenylhexamethylenimine*	95	201
phenyl(SO_2–C_6H_5)(SO_2–C_6H_5)	phenyl(NR_2)(SO_2–C_6H_5) *N,N-Dialkyl-2-(phenylsulfonyl)aniline*	60–80	202

12.2.5 Substitution of the amino group

The reactions described in this section proceed according to the following general equation:

$$R^1\!-\!N\!\!\begin{array}{c}R^2\\ \\R^3\end{array} + R^4\!-\!N\!\!\begin{array}{c}R^5\\ \\R^6\end{array} \longrightarrow R^1\!-\!N\!\!\begin{array}{c}R^5\\ \\R^6\end{array} + R^4\!-\!N\!\!\begin{array}{c}R^2\\ \\R^3\end{array}$$

Considered formally, substitution of one amino group $(-NR^2R^3)$ by another $(-NR^5R^6)$ or change of place of a substituent R^1 with the substituent R^4 of another amine is involved. The literature calls these reactions variously amine exchange, amine displacement, disproportionation, transamination, or condensation.

12.2.5.1 Substitution of the amino group in the presence of acids, metal salts, or iodine

Diphenylamine can be prepared in the laboratory by heating aniline hydrochloride with excess aniline[203, 204]. On the technical scale aniline is heated with smaller quantities of aniline hydrochloride (10%)[205] or ammonium chloride (0.1–5%, 0.75% is best)[206] to 300°, and the ammonia formed is removed from the reaction mixture continuously.

The catalytic effect of these reagents is enhanced by *adding* zinc chloride, iron(II) chloride[207], or aluminum chloride[208]. 0.3–3% iron(III) chloride or aluminum chloride are also catalytically effective alone[209]. Diphenylamine can be obtained from aniline in particularly good yield in the presence of phosphorus(III) chloride[210].

Heating aniline hydrochloride with toluidine leads to a mixture of *diphenylamine, N-phenyltolylamine* and *ditolylamine*[203]. In other instances such as the reaction of naphthylamines with arylamines in the presence of acids, the unsymmetrical diarylamines can be prepared uniformly and in good yield[211].

Iodine has revealed itself to be an effective catalyst for converting certain primary aromatic amines into secondary amines while ammonia is split off[212]. Thus, heating 2-naphthylamine for 4 hours to 230° in the presence of 0.5–1% iodine gives *di-2-naphthylamine* in almost quantitative yield. The reactions between 1-naphthylamine and arylamines such as aniline, 1, 2-, 1, 3-, and 1, 4-toluidine, 3- and 4-chloroaniline[213], 2- and 4-anisidine[213], and 2, 3 and 3, 4-xylidine[214] giving *N-arylnaphthylamines* also proceed smoothly.

Formation of *N-phenyl-1-naphthylamine* is catalyzed by small amounts of hydrogen iodide or ammonium iodide in the same way as by iodine[215]. The hydrogen iodide formed by the action of iodine on arylamines must therefore be regarded as the true catalyst.

On heating dialkylarylamines with the corresponding primary amines *alkylarylamines* are formed in the presence of hydrogen chloride[216]. With Friedel-Crafts catalysts such as boron trifluoride or aluminum chloride, *N-methylaniline* (47%), *N-methyl-2-toluidine* (68–75%) respectively are formed[217]. Conversion of 3-(diethylamino)propionitrile into *3-(arylamino)propionitriles* succeeds by heating with primary arylamine hydrochlorides, benzenesulfonates, and toluenesulfonates. On heating N,N-dimethylbenzylamine with small amounts of hydrochloric acid or boron trifluoride to 200° disproportionation into N-methyldibenzylamine and *trimethylamine* occurs to a small extent. Better yields (73%) are obtained on adding benzyltrimethylammonium chloride (10 *mol*%). Amines of similar structure in which a

[198] *G. Leandri, A. Tundo,* Ann. Chim. (Rome) *44,* 479 (1954).

[199] *G. Leandri, L. Maioli,* Ann. Chim. (Rome) *45,* 3 (1955).

[200] *W. Bradley,* J. Chem. Soc. (London) 458 (1938).

[201] *H.J. Nitzschke, H. Budka,* Chem. Ber. *88,* 264 (1955).

[202] *G. Köbrich,* Chem. Ber. *92,* 2981 (1959).

[203] *G. De Laire, C. Girard, P. Chapoteaut,* Justus Liebigs Ann. Chem. *140,* 344 (1866).

[204] *H.E. Fierz-David, L. Blangey,* Grundlegende Operationen der Farbenchemie, 8. Aufl., p. 136, Springer-Verlag, Wien 1952.

[205] US. P. 1840576 (1926), DuPont, Inv.: *J. Frei;* Chem. Zentr. I, 3498 (1932).

[206] Brit. P. 432542, 432543 (1933), DuPont, Inv.: *J. Frei;* Chem. Zentr. I, 882 (1936).

[207] US.P. 2120968, 2120969 (1935), DuPont, Inv.: *M.F. Acken;* Chem. Zentr. I, 797 (1939).

[208] US. P. 2645662 (1947), Koppers Co., Inv.: *R.H. Nimmo;* C.A. *47,* 12421h (1953).

[209] Brit. P. 644938 (1948), DuPont; C.A. *45,* 4742h (1951).

[210] BIOS Final Rep. *1157,* 51 (1947).

[211] BIOS Final Rep. *986* II, 363 (1945).

[212] *E. Koenigs, G. Jung,* J. Prakt. Chem. *137,* 141 (1933).

[213] *E. Knoevenagel,* J. Prakt. Chem. *89,* 4, 20 (1914).

[214] *N.P. Buu-Hoi,* J. Chem. Soc. (London) 670 (1949).

[215] *H.H. Hodgson, E. Marsden,* J. Chem. Soc. (London) 1181 (1938);
H.H. Hodgson, E. Marsden, J. Soc. Chem. Ind. *58,* 154T, 290T (1939).

[216] *P.F. Frankland, F. Challenger, N.A. Nichols,* J. Chem. Soc. (London) 198 (1919).

[217] DBP. 865450(1950), BASF, Erf.: *E. Rotter;* Chem. Zentr. 8207 (1953).

Table 11. Preparation of cyclic amines by heating mono- and dihydrochlorides of aliphatic and aromatic diamines

Starting amine	Product	Yield [% of theory]	Ref.
$H_2N-(CH_2)_5-NH_2$	Piperidine		219
$H_2N-(CH_2)_4-NH_2$	Pyrrolidine		220
$H_2N-CH_2-CH-CH_2-CH_2-NH_2$ CH_3	3-Methylpyrrolidine		221
$H_3C-CH-CH_2-CH_2-CH-CH_3$ NH_2 NH_2	2,5-Dimethyl-pyrrolidine		222
$H_2N-(CH_2)_6-NH_2$	2-Ethylpyrrolidine		223
$H_2N-(CH_2)_8-NH_2$	2-Butylpyrrolidine		224
$H_2N-(CH_2)_{10}-NH_2$	2-Hexylpyrrolidine		224
$H_2N-(CH_2)_4-NH-CH_3$ $H_2N-(CH_2)_4-N(CH_3)_2$ $H_3C-NH-(CH_2)_4-NH-CH_3$ $H_3C-NH-(CH_2)_4-N(CH_3)_2$ $(H_3C)_2N-(CH_2)_4-N(CH_3)_2$	1-Methylpyrrolidine		225
$(H_5C_2)_2N-(CH_2)_3-CH-CH_3$ NH_2	2-Methylpyrrolidine	60	226
$CH_2-N(CH_3)_2$ CH_2 $C-CN$ CH_2 $CH_2-N(CH_3)_2$ (phenyl)	1-Methyl-4-phenyl-isonicotinonitrile	79	227
$CH_2-CH_2-NH_2$ NH_2 (benzene)	Indoline	40	228
(biphenyl) H_2N NH_2	Carbazole		229, 230
(tricyclic diamine)	5,10-Dihydropyrido-[3,4-b]quinoxaline		231
(tricyclic diamine)	5,10-Dihydrodipyrido-[3,4-b, 4,3-b]pyrazine		232

[218] *H.R. Snyder, R.E. Carnahan, E.R. Lovejoy*, J. Amer. Chem. Soc. *76*, 1301 (1954).

[219] *A. Ladenburg*, Ber. *18*, 3100 (1885).
[220] *A. Ladenburg*, Ber. *20*, 442 (1887).

furfuryl, thienyl or 1-naphthylmethyl group replaces a benzyl group undergo an analogous disproportionation[218].

Aliphatic or aromatic diamines with mutually favorably spaced amino groups can be converted into *cyclic amines* by heating their mono- and dihydrochlorides.

12.2.5.2 Substitution of the amino group in the presence of dehydration contact catalysts

On leading amine vapors over dehydration contact catalysts migration of alkyl or aryl groups from one nitrogen atom to the other often takes place. These syntheses are particularly of technical interest.

Methylamine yields ammonia (33.2%), *dimethylamine* (31.4%), and *trimethylamine* (3.6%), as well as unchanged starting material (31.8%)[233]. During preparation of *dibutylamine* from butylamine formation of butene as side-product can be avoided by adding ammonia and working under pressure[234]. At 460° aniline vapor is converted into *diphenylamine* (flow-rate-dependent 20–35% yield); side reactions occur only at elevated temperature[235]. Passing hexamethylenediamine over alumina at 350–380° leads to *hexamethylenimine*

in good yield[236]. This result is surprising, because *2-ethylpyrrolidine* is formed during thermolysis of the dihydrochloride.

The opposite reaction too, *viz.*, formation of primary amines from secondary amines and ammonia, is possible. Thus dipentylamine gives *pentylamine* in 75–90% yield at 33–36% conversion with ammonia[237]. Tertiary amines such as trimethylamine or tributylamine can be converted into mixtures of the corresponding primary and secondary amines with ammonia[237, 238]; their composition matches the reaction conditions.

12.2.5.3 Substitution of the amino group in the presence of hydrogenation and dehydrogenation catalysts

Primary aliphatic amines with at least one hydrogen on the α-carbon atom split off ammonia and are converted into the corresponding *secondary amines* in the presence of palladium or nickel and similar *metal catalysts* at temperatures as low as 100–200°. Probably, a part of the primary amine is first dehydrogenated to form the imine, which then adds or condenses with unchanged starting material. The action of the hydrogen split off during the dehydrogenation on these intermediates leads to the secondary amine[239]:

$$R-CH_2-NH_2 \longrightarrow R-CH=NH + H_2$$

$$R-CH=NH + H_2N-CH_2-R$$

$$R-CH-NH_2 \quad\quad\quad R-CH=N-CH_2-R$$
$$|$$
$$NH-CH_2-R$$

$$+H_2 \downarrow \quad\quad\quad\quad\quad \downarrow +H_2$$

$$(R-CH_2)_2NH + NH_3 \quad\quad\quad (R-CH_2)_2NH$$

[221] *H. Oldach*, Ber. **20**, 1644 (1887);
W. Euler, J. Prakt. Chem. **57**, 143 (1898).

[222] *J. Tafel, A. Neugebauer*, Ber. **23**, 1544 (1890).

[223] *A. Müller, E. Feld*, Monatsh. Chem. **58**, 12 (1931).

[224] *E.E. Blaise, L. Houillon*, C.R. Acad. Sci., Paris **142**, 1541 (1906);
E.E. Blaise, L. Houillon, C.R. Acad. Sci., Paris **143**, 361 (1906).

[225] *W. Keil*, HoppeSeyler's Z. Physiol. Chem. **171**, 244 (1927).

[226] *V.J. Stavrovskaja*, Zh. Obshch. Khim. **25**, 148 (1955).

[227] *F.F. Blicke, J.A. Faust, J. Krapcho, E.P. Tsao*, J. Amer. Chem. Soc. **74**, 1844 (1952).

[228] *P. Rugglin, H. Steiger, P. Schobel*, Helv. Chim. Acta **28**, 333 (1945).

[229] *E. Täuber*, Ber. **24**, 200 (1891).

[230] *H. Leditschke*, Chem. Ber. **86**, 522 (1953).

[231] *V. Petrov, J. Saper, B. Sturgeon*, J. Chem. Soc. (London) 2540 (1949).

[232] *E. Koenigs, G. Jung*, J. Prakt. Chem. **137**, 154 (1933).

[233] US.P. 1 926 691 (1931), Commercial Solvents Corp., Inv.: *L.C. Swallen, J. Martin;* Chem. Zentr. I, 125 (1934).

[234] US. P. 2 574 693 (1948), Shell Develop., Inv.: *W.F. Engel, H. Hoog;* C.A. **46**, 6140f (1952).

[235] *H.E. Hoelscher, D.F. Chamberlain*, Ind. Eng. Chem. **42**, 1558 (1950);
H.E. Hoelscher, D.F. Chamberlain, Ind. Eng. Chem. **43**, 1828 (1951).

[236] DRP. 738 448 (1941), I.G. Farben, Erf.: *H. Raab;* Chem. Zentr. II, 2009 (1943).

[237] US.P. 2 192 523 (1937), The Sharpless Solvents Corp., Inv.: *J.F. Olin, T.E. Deger;* Chem. Zentr. I, 1185 (1942).

[238] US. P. 2 112 970 (1931), DuPont, Inv.: *P.E. Millington;* Chem. Zentr. II, 176 (1938);
DRP. 626 923 (1933), I.G. Farben, Erf.: *P. Herold, K. Smeykal;* Fortschritte der Teerfabrikation u. verwandter Industriezweige **22**, 169.

Under certain conditions primary amines are converted into *tertiary* amines. For data relating to the conversion of alkylamines, alkylcyclohexylamines, and aralkylamines, and the required reaction conditions, see ref. [240]. With benzylamine and phenethylamine heating in xylene[241] or ethanol[239], while passing through hydrogen in the presence of palladium catalysts, is sufficient to yield *dibenzylamine* and *diphenethylamine*. The latter compound is formed quantitatively also when nickel in kieselguhr is used[242]. Pentylamine is analogously converted completely into *dipentylamine*[242]. In presence of Raney nickel, C_6–C_{18}-alkyl dialkylamines can be prepared from the corresponding primary amines[243]. *N-cyclohexyl-1-cyclohexen-1-ylamine* is formed in good yield from cyclohexylamine if isolated at once after its formation[244]. Cyclization of diethylenetriamine to *piperazine* over Raney nickel gives best yields (73%) in dipentene or tetrahydrofuran, the optimum reaction temperature is about 150°[245]. During conversion of ethylenediamine to *piperazine* diethylenetriamine is probably formed first. Low-boiling aliphatic primary amines can be converted into secondary amines in the gas phase over metal catalysts (nickel, cobalt, copper, or platinum); small amounts of tertiary amines are also formed[246]. *Mixtures* of primary and secondary aliphatic amines yield tertiary amines in the gas phase over nickel catalysts[247].

Hydrogenation under pressure over Raney nickel converts azomethines into secondary amines with simultaneous amine exchange. For example, *N-(4-anisylidene)-N-phenethylamine* in ethanolic methylamine solution gives *N-methyl-4-anisidine* (89%) and *1,2-Diamino-1-phenylethane*[248].

12.2.5.4 Substitution of the amino group in the presence of sulfurous acid (Bucherer reaction)

Certain *primary* amines of the *naphthalene* series readily exchange their amino groups against alkyl or arylamino groups when heated with primary or secondary amines in the presence of aqueous sodium hydrogen sulfite solution. This method is used chiefly for preparing *N*-substituted *aminohydroxynaphthoic acids*. As this amine exchange is carried out under the same conditions as the replacement of hydroxy by amino, reference is made to p. 460.

12.2.5.5 Substitution of the amino group by combined elimination-addition and substitution mechanism

Tertiary Mannich bases with a *CH*-acid group beta to the amine nitrogen react with primary or secondary amines by amine exchange. Secondary amine is split off from the Mannich base with formation of a reactive unsaturated intermediate to which an amine added in excess adds. From tertiary Mannich bases and primary alkylamines *monosubstituted secondary Mannich bases* are formed which cannot be obtained by the normal Mannich method[249]. Amino acids may be used in place of primary alkylamines (see Table 12).

[239] *K. Kindler*, Justus Liebigs Ann. Chem. *485*, 113 (1931).

[240] *K. Kindler, G. Melamed, D. Matthies*, Justus Liebigs Ann. Chem. *644*, 23 (1961).

[241] *K.W. Rosenmund, G. Jordan*, Ber. *58*, 51 (1925).

[242] *C.F. Winans, H. Adkins*, J. Amer. Chem. Soc. *54*, 307 (1932).

[243] *C.W. Hoerr, H.J. Harwood, A.W. Ralston*, J. Org. Chem. *9*, 201 (1944); *C.W. Hoerr, H.J. Harwood, A.W. Ralston*, J. Org. Chem. *11*, 199 (1946).

[244] *G. Debus*, Bull. Soc. Chim. Belg. *63*, 457 (1954).

[245] *W.B. Martin, A.E. Martell*, J. Amer. Chem. Soc. *70*, 1817 (1948).

[246] DRP. 510439 (1928), I.G. Farben, Erf.: *O. Nicodemus, W. Schmidt*; Fortschritte der Teerfabrikation u. verwandter Industriezweige *17*, 798 (1932).

[247] DRP. 697372 (1937), I.G. Farben, Erf.: *B. Christ*; Chem. Zentr. II, 3744 (1940).

[248] *M. Sekiya, A. Hara, T. Masui*, Chem. Pharm. Bull. (Tokyo) *11*, 277 (1963).

[249] *J.C. Craig, S.R. Johns, M. Moyle*, J. Org. Chem. *28*, 2779 (1963).

[250] *H.R. Snyder, J.H. Brewster*, J. Amer. Chem. Soc. *70*, 4230 (1948).

[251] *H. Hellmann, G. Opitz*, Angew. Chem. *68*, 265 (1956); *H. Hellmann, G. Opitz*, Angew. Chem. *65*, 473, 475 (1953).

[252] *H.R. Snyder, J.H. Brewster*, J. Amer. Chem. Soc. *71*, 1061 (1949).

[253] *E.L. Eliel*, J. Amer. Chem. Soc. *73*, 43 (1951).

[254] *H.R. Snyder, W.E. Hamlin*, J. Amer. Chem. Soc. *72*, 5082 (1950).

[255] *P. Duden, K. Bock, H.J. Reid*, Ber. *38*, 2036 (1905).

[256] *E.E. Howe, A.J. Zambito, H.R. Snyder, M. Tishler*, J. Amer. Chem. Soc. *67*, 38 (1945).

[257] *R.O. Atkinson*, J. Chem. Soc. (London) 1329 (1954).

[258] *A. Butenandt, U. Renner*, Z. Naturforsch. *8* B, 454 (1953).

[259] *H. Hellmann*, Angew. Chem. *65*, 478 (1953).

[260] *H.R. Snyder, E.L. Eliel*, J. Amer. Chem. Soc. *70*, 4233 (1948).

Table 12. Exchange reactions of Mannich bases with primary and secondary amines

Mannich base	Amine	Product	Yield [% of theory]	Ref.
$H_5C_6-\overset{\underset{\|}{O}}{C}-CH_2-CH_2-N(CH_3)_2$	HN O (morpholine)	$H_5C_6-\overset{\underset{\|}{O}}{C}-CH_2-CH_2-N$ O *3-Morpholinopropiophenone*	78	250, 251
$H_5C_6-\overset{\underset{\|}{O}}{C}-CH_2-CH_2-N(C_2H_5)_2$	$H_3C-\underset{\underset{NH_2}{\|}}{CH}-COOH$	$H_5C_6-\overset{\underset{\|}{O}}{C}-CH_2-CH_2-NH$ $H_3C-\overset{\|}{CH}-COOH$ *N-(2-Benzoylethyl)-alanine*	82	249
	$H_5C_6-CH_2-\underset{\underset{NH_2}{\|}}{CH}-COOH$	$H_5C_6-\overset{\underset{\|}{O}}{C}-CH_2-CH_2-NH$ $\underset{\underset{C_6H_5}{\|}}{CH_2-CH}-COOH$ *N-(2-Benzoylethyl)-phenylalanine*	78	249
	$HOOC-\underset{\underset{NH_2}{\|}}{CH}-CH_2-COOH$	$H_5C_6-\overset{\underset{\|}{O}}{C}-CH_2-CH_2-NH$ $HOOC-\overset{\|}{CH}-CH_2-COOH$ *N-(2-Benzoylethyl)-aspartic acid*	72	249
	$HOOC-\underset{\underset{NH_2}{\|}}{CH}-CH_2-CO-NH_2$	$H_5C_6-\overset{\underset{\|}{O}}{C}-CH_2-CH_2-NH$ $HOOC-\overset{\|}{CH}-CH_2-CONH_2$ *N²-(2-Benzoylethyl)-asparagine*	85	249
	H_2N-CH_2-COOH	$\left(H_5C_6-\overset{\underset{\|}{O}}{C}-CH_2-CH_2\right)_2 N-\underset{\underset{COOH}{\|}}{CH_2}$ *N,N-Bis-(2-benzoylethyl)-glycine*	35	249
	$HOOC-\underset{\underset{NH_2}{\|}}{CH}-(CH_2)_2-COOH$	(5-oxo-pyrrolidine ring) $O=$ ring $-COOH$, $N-CH_2-CH_2-\overset{\underset{O}{\|}}{C}-C_6H_5$ *5-Oxo-N-(2-benzoylethyl)-2-pyrrolidinecarboxylic acid*		249
$H_3C-\overset{\underset{\|}{O}}{C}-CH_2-CH_2-N(CH_3)_2$	$H_5C_6-CH_2-\underset{\underset{NH_2}{\|}}{CH}-COOH$	$H_3C-\overset{\underset{O}{\|}}{C}-CH_2-CH_2-NH$ $H_5C_6-CH_2-\overset{\|}{CH}-COOH$ *N-(3-Oxobutyl)-phenylalanine*	81	249
$H_5C_6-\overset{\underset{\|}{O}}{C}-\overset{\overset{CH_3}{\|}}{\underset{\underset{CH_3}{\|}}{C}}-CH_2-N(CH_3)_2$	HN O	Steric hindrance prevents a reaction		252
$\overset{CH_2-N(CH_3)_2}{\underset{OH}{}}$ (naphthol)	HN O	CH_2-N O, OH (naphthol) *1-(Morpholinomethyl)-2-naphthol*	84	250

Table 12. Exchange reactions of Mannich bases with primary and secondary amines   Continuation 1

Mannich base	Amine	Product	Yield [% of theory]	Ref.
	HN (piperidine)	*1-(Piperidinomethyl)-2-naphthol*	92	250
C_6H_5 CH–NH_2 OH (naphthol)	HN O (morpholine)	*1-(1-Morpholino-1-phenyl)-2-naphthol*	88	250
H_3CO OH CH_2–$N(CH_3)_2$	HN (piperidine)	*2-Methoxy-6-(piperidino-methyl)phenol*	93	253
H_3C–CH_2–CH–CH_2–$N(CH_3)_2$ NO_2	HN (piperidine)	H_3C–CH_2–CH–CH_2–N NO_2 *1-(2-Nitrobutyl)piperidine*	23	254
CH_3 H_3C–C–CH_2–$N(CH_3)_2$ NO_2	HN (piperidine)	No reaction		254
$(H_3C)_2N$–CH_2–CH–CH_2–$N(CH_3)_2$ NO_2	H_5C_6–NH_2	H_5C_6–NH–CH_2–CH–CH_2–NH–C_6H_5 NO_2		255
N–CH_2–CH–CH_2–N NO_2	H_5C_6–NH_2	*N,N-Diphenyl-2-nitro-1,3-propanediamine*		255
CH_2–$N(CH_3)_2$ (indole) H	HN (piperidine)	*3-Piperidinomethylindole*	90	256
CH_2–$N(CH_3)_2$ (indole) H	HN (phthalimide)	*3-Phthaliminomethylindole*		257
CH_2–$N(CH_3)_2$ OH (naphthol)	(phthalimide) NH	*1-(Phthalimidomethyl)-2-naphthol*		257

473

Table 12. Exchange reactions of Mannich bases with primary and secondary amines

Mannich base	Amine	Product	Yield [% of theory]	Ref.
		2-Nitro-ω-phthalimido-propiophenone *2,ω-Diaminopropiophenone*		258

Mannich bases without a mobile hydrogen *beta* with respect to the amino group cannot form unsaturated intermediate products and react only as hydrochlorides or quaternary ammonium salts by amine substitution[259]. Mannich bases reacting by the combined elimination-addition mechanism, too, often undergo amine exchange more easily as salts than in the free form.

Table 13. Amine exchange by Mannich bases without a mobile hydrogen atom on the carbon atom beta to the amine nitrogen

Mannich base	Amine	Product	Yield [% of theory]	Ref.
		1-Methyl-3-(piperidino-methyl)indole	60	260
		2-Methoxy-1-(piperidino-methyl)-naphthalene	47	261
	NH_3	*3-(3-Benzoyl-4-hydroxy-4-phenylpiperidino)propiophenone*		262
		3-(4-Chloroanilino)-propiophenone	80–90	263

[261] *H.R. Snyder, J.H. Brewster*, J. Amer. Chem. Soc. *71*, 1058 (1949).

[262] *J.J. Denton, H.P. Schedl, W.B. Meier, M. Brookfield*, J. Amer. Chem. Soc. *72*, 3792 (1950).

12.2.5.6 Substitution of the amino group by other exchange reactions

Table 14. Amines by special exchange reactions

Starting amine	Exchange amine	Product	Yield [% of theory]	Ref.
	R – NH$_2$ (R = Alkyl– or Aryl–)	1-(Alkylamino)- or 1-(Arylamino)-4-aminoanthraquinone		
		1,4-Bis(alkyl)(aryl)amino-anthraquinone		264
	H$_5$C$_6$–CH$_2$–NH$_2$	6-Benzylamino-7H-benz[de]-anthracen-7-one		265
	H$_5$C$_6$–NH$_2$	3-Anilino-1-phenyl-2-pyrazolin-5-one	43	266
	NH$_3$	3,3'-Iminobis(1-phenyl-2-pyrazolin-5-one)	70	267
	HN⟨⟩	1,3-Diphenyl-4-(piperidinomethylene)-2-pyrazolin-5-one	86	268
	H$_5$C$_6$–NH$_2$ (1,3-shift accompanies the exchange)	16 β-Anilino-3 β-hydroxy-17a α-methyl-D-homoandrost-5-en-17-one	45	269
R = (H$_3$C)$_3$C–NH– (H$_5$C$_2$)$_3$C–NH–	R^1–H R^1 = O⟨⟩N– (CH$_3$)$_3$C–NH–	cis-2-Benzylidene-3-tert-butylamino-propiophenone cis-2-Benzylidene-3-morpholino-propiophenone		270

Table 15. Reactions of quaternary salts with ammonia, primary amines, or secondary amines

Quaternary salt	Amine	Product	Yield [% of theory]	Ref.
$H_5C_6-CH_2-\overset{\oplus}{N}(CH_3)_3$ $Br^{\ominus}$	$H_5C_6-CH_2-NH_2$	$H_5C_6-NH-C_6H_5$ *Diphenylamine*	90	271
(pyridine-3-ol-2-yl)$-CH_2-\overset{\oplus}{N}(CH_3)_3$ $Br^{\ominus}$	$H_5C_6-CH_2-NH_2$	(3-pyridinol)$-CH_2-NH-CH_2-C_6H_5$ *2-(Benzylaminomethyl)-3-pyridinol*	71 (as dihydrochloride)	272
(isoindoline-piperidine fused quaternary salt) $Br^{\ominus}$	NH_3	$-CH_2-N$(piperidine), $-CH_2-NH_2$ *1-(2-Aminomethyl)benzyl-piperidine*		273
(isoindoline-piperidine fused quaternary salt) $Br^{\ominus}$	$HN(C_2H_5)_2$	$-CH_2-N$(piperidine), $-CH_2-N(C_2H_5)_2$ *1-(2-Diethylaminomethyl)-benzylpiperidine*		274
(furan-2-yl)$-CH_2-\overset{\oplus}{N}(CH_3)_3$ $J^{\ominus}$	HN(piperidine)	(furan-2-yl)$-CH_2-N$(piperidine) *1-Furfurylpiperidine*	48	275
H_3C-(furan-2-yl)$-CH_2-\overset{\oplus}{N}(CH_3)_3$ $J^{\ominus}$	HN(piperidine)	H_3C-(furan-2-yl)$-CH_2-N$(piperidine) *1-(5-Methylfurfuryl)piperidine*	69	275
(2-methyl quinolizidinium salt) $Br^{\ominus}$	NH_3	(2-methylpiperidine) $N-(CH_2)_5-NH_2$ *1-(5-Aminopentyl)-2-methyl-piperidine*		276
(2-methyl-5-ethyl quinolizidinium salt) $Br^{\ominus}$	NH_3	(5-ethyl-2-methylpiperidine) $N-(CH_2)_5-NH_2$ *1-(5-Aminopentyl)-5-ethyl-2-methylpiperidine*		276
(perhydroquinolizinium salt) $Br^{\ominus}$	NH_3	(perhydroquinoline) $N-(CH_2)_5-NH_2$ *1-(5-Aminopentyl)perhydro-quinoline*		276

[263] *J.C. Craig, M. Moyle*, Chem. & Ind. (London) 690 (1963).

[264] DRP. 172464 (1906), Farbwerke Hoechst; Fortschritte der Teerfabrikation u. verwandter Industriezweige *8*, 316 (1908);
DRP. 205551 (1908), Farbwerke Hoechst; Fortschritte der Teerfabrikation u. verwandter Industriezweige *9*, 728 (1911);
DBP. 911531 (1951), Farbenfabriken Bayer, Erf.: *J. Singer, O. Bayer, H.W. Schwechten;* Chem. Zentr. 11060 (1954).

[265] *W. Bradley*, J. Chem. Soc. (London) 2712 (1949).

[266] *A. Weissberger, H.D. Porter*, J. Amer. Chem. Soc. *64*, 2133 (1942).

[267] *B. Graham, W. Reckhow, A. Weissberger*, J. Amer. Chem. Soc. *76*, 3993 (1954).

[268] *B.A. Porai-Koshits, I.Y. Kvitko, E.A. Shutkova*, Latv. PSR Zinat. Akad. Vestis, Khim. Ser. 587 (1965); C.A. *64*, 8168g (1966).

[269] *D.F. Morrow, M.E. Butler, W.A. Neuklis, R.M. Hofer*, J. Org. Chem. *32*, 86 (1967).

[270] *N.H. Cromwell, K. Matsumoto, A.D. George*, J. Org. Chem. *36*, 272 (1971).

[271] *J.v. Braun, M. Kühn, O. Goll*, Ber. *59*, 2330 (1926).

Table 16. Reactions of quaternary pyridinium salts with ammonia or amines

Pyridinium salt	Amine	Product	Yield [% of theory]	Ref.
O_2N—(ring, NO_2)—pyridinium $Cl^{\ominus}$	$2\ H_5C_6$—NH_2 Ethanol (Zincke cleavage)	O_2N—(ring, NO_2)—NH_2 *2,4-Dinitroaniline*	Glutacondialdehyde + dianil	279
	$2\ H_5C_6$—$\overset{\oplus}{N}H_3\ Cl^{\ominus}$ 200° (Amine exchange)	O_2N—(ring, NO_2)—NH—C_6H_5 *2,4-Dinitrodiphenylamine*		280
O_2N—(ring, NO_2)—pyridinium(R) $Cl^{\ominus}$ $R = CH_3, OH, OCH_3, H_3C\overset{O}{\underset{\parallel}{C}}-NH,$ $(H_3C)_2N$	H_5C_6—NH_2 (Amine exchange)	O_2N—(ring, NO_2)—NH—C_6H_5 *2,4-Dinitrodiphenylamine*	to 85	281
(anthracene-pyridinium, NO_2) $Br^{\ominus}$	H_3C—NH_2 (Zincke cleavage)	(anthracene NH_2, NO_2) *10-Nitro-9-anthramine*	89	282
(dipyridinium) $2\ Cl^{\ominus}$	H_5C_6—NH_2	H_2N—(pyridine) *4-Aminopyridine* (Zincke cleavage)		
		H_5C_6—NH—(pyridine) *4-Anilinopyridine* (Amine exchange)	small	283
(dipyridinium) $2\ Cl^{\ominus}$	NH_3	H_2N—(pyridine) *4-Aminopyridine*	60	284
R^1—CH—CH—pyridinium $Br^{\ominus}$ $\quad\overset{\vert}{R}\quad\overset{\vert}{O}$—$COCH_3$ $R = C_6H_5, 4-Cl-C_6H_4, 3,4-Cl_2-C_6H_3$ $R^1 = 2-Cl-C_6H_4, 3-NO_2-C_6H_4$	HN(piperidine)	R^1—CH—CH—NH_2 $\quad\overset{\vert}{R}$ O—$COCH_3$ (Zincke cleavage) $\downarrow$ HOH $(H^{\oplus})$ R^1—CH—CH—NH_2 $\quad\overset{\vert}{OH}\ \overset{\vert}{R}$	12–95	285 286
H_5C_6—CH=C—pyridinium $Br^{\ominus}$ $\qquad\overset{\vert}{C_6H_5}$	HN(piperidine)	H_5C_6—CH=C—NH_2 $\qquad\overset{\vert}{C_6H_5}$ *1-stilbenamine* (Zincke cleavage) $\downarrow$ $H^{\oplus}$ H_5C_6—CH_2—$\overset{O}{\underset{\parallel}{C}}$—$C_6H_5$ *2-phenylacetophenone*		287

---

[272] *A. Stempel, J.A. Aeschlimann,* J. Amer. Chem. Soc. *74,* 3323 (1952).

[273] *J.v. Braun, F. Zobel,* Justus Liebigs Ann. Chem. *445,* 247 (1925).

Table 16. Reactions of quaternary pyridinium salts with ammonia or amines   Continuation 1

Pyridinium salt	Amine	Product	Yield [% of theory]	Ref.
$H_5C_6-CH=CH-N$ ⊕ ⟩ Br ⊖	HN ⟩X X = CH$_2$,O (Amine exchange)	$H_5C_6-CH=CH-N$ ⟩X *1-Styrylpiperidine* *1-Styrylmorpholine*		288
⟨N⊕⟩ Cl ⊖ ⟨NOH⟩	HN⟨R⟩R^1 R = Alkyl R^1 = H, Alkyl (Amine exchange)	R⟨N⟩R^1 ⟨NOH⟩		289

12.2.5.7 Amine exchange reactions of quaternary ammonium salts

Quaternary ammonium salts and especially those containing benzyl radicals or benzyllike radicals, can react with ammonia, primary, or secondary amines:

$$\begin{bmatrix} R \\ | \\ R^1-N-R \\ | \\ R \end{bmatrix}^{\oplus} X^{\ominus} + HN\begin{array}{c} R^2 \\ \\ R^3 \end{array} \longrightarrow$$

$$R^1-N\begin{array}{c} R^2 \\ \\ R^3 \end{array} + N(R)_3 + HX$$

When methyl iodide reacts with 3-[(dimethylamino)methyl]indole (gramine), the expected quaternary salt is not obtained in the pure form, because the primarily formed 3-(indolylmethyl)trimethylammonium iodide reacts surprisingly rapidly with unchanged gramine to split off trimethylamine and form *bis(3-indolylmethyl)dimethylammonium iodide*. This reversible equilibrium reaction is a special case of a general reaction of quaternary salts of Mannich bases with tertiary amines which proceeds even at room temperature. 3-(Indolylmethyl)trimethylammonium iodide reacts with *N,N*-dimethylbenzylamine and *N*-methylpiperidine to *benzyl(3-indolylmethyl)dimethylammonium iodide* and *1-(3-indolylmethyl)-1-methylpiperidinium iodide* respectively[277]. On alkalizing an aqueous solution of 3-(indolylmethyl)trimethylammonium iodide and *N*-methylaniline acetate, an almost quantitative yield of *3-[(N-methylanilino)methyl]-indole* results[278].

In the reaction with ammonia or amines the quaternary pyridinium salts either exchange the pyridine group against an amino group or else a *Zincke* cleavage takes place in which the 5-carbon pyridine nucleus chain is detached. Thus, a *primary* amine is formed with a nitrogen atom originating from the pyridine ring in addition to a derivative of *glutacondialdehyde (2-pentenedial)*. Which reaction path is taken depends mainly on the constitution of the pyridinium salt, but also on the reaction conditions[279].

The synthesis of *folic acid* (pteroylglutamic acid) also succeeds by amine exchange of a pyridinium salt[290]. In the presence of enzymes analogous reactions proceed even under very mild conditions[291].

[274] *M. Scholtz*, Ber. *31*, 414 (1898).
[275] *E.L. Eliel, P.E. Peckham*, J. Amer. Chem. Soc. *72*, 1209 (1950).
[276] *J.v. Braun, F. Zobel*, Ber. *59*, 1786 (1926).
[277] *C. Schöpf, J. Thesing*, Angew. Chem. *63*, 377 (1951).

[278] *J. Thesing, H. Mayer*, Chem. Ber. *87*, 1084 (1954).
[279] *T. Zincke*, Justus Liebigs Ann. Chem. *330*, 361 (1903);
    T. Zincke, Justus Liebigs Ann. Chem. *333*, 296 (1904).
[280] *F. Kröhnke*, Angew. Chem. *65*, 624 (1953).
[281] *A.F. Vompe, N.F. Turitsyna*, Dokl. Akad. Nauk SSSR *64*, 341 (1949); C.A. *43*, 4671a (1949).
[282] *S. Hünig, K. Requardt*, Angew. Chem. *68*, 152 (1956).
[283] *E. Koenigs, H. Greiner*, Ber. *64*, 1049 (1931).
[284] *I.P. Wibaut, S. Herzberg, J. Schlattmann*, Rec. Trav. Chim. Pays-Bas *73*, 140 (1954).
[285] *F. Kröhnke*, Angew. Chem. *65*, 616 (1953).
[286] *F. Kröhnke, I. Vogt*, Justus Liebigs Ann. Chem. *589*, 45 (1954).
[287] *F. Kröhnke, I. Vogt*, Justus Liebigs Ann. Chem. *589*, 26 (1954).
[288] *F. Kröhnke, I. Vogt*, Justus Liebigs Ann. Chem. *589*, 52 (1954).
[289] *A.J. Birch*, J. Chem. Soc. (London) 314 (1944).
[290] *M.E. Hultquist, E. Kuh, D.B. Cosulich, M.J. Fahrenbach, E.H. Northey, D.R. Seeger, J.P. Sickels, J.M. Smith jr., R.B. Angier, J.H. Boothe, B.J. Hutchings, J.H. Mowat, J.Semb, E.L.R. Stokstad, Y.S. Row, C.W. Waller*, J. Amer. Chem. Soc. *70*, 23 (1948).
[291] *D.W. Wooley*, Nature *171*, 323 (1953).

12.3 Addition of ammonia and amines to C–C multiple bonds, oxiranes, and aziridines

Compounds containing *C–C* multiple bonds can add ammonia and primary and secondary amines; so can oxiranes and aziridines:

X = O, NR

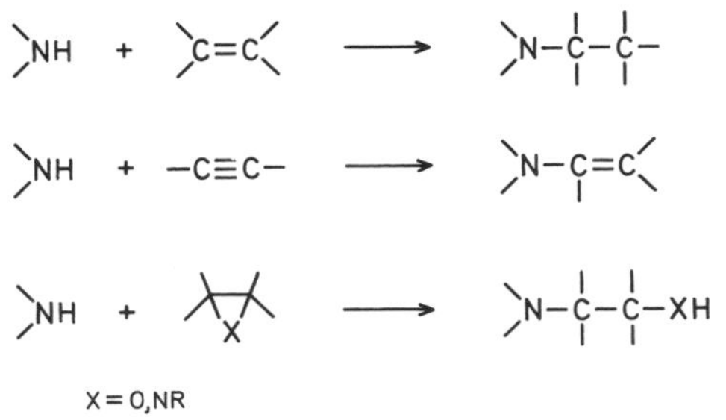

In some cases such addition reaction provide the simplest method of preparing the corresponding amines. The *C–O* bonds of oxirane and its derivatives are sufficiently reactive for the addition reactions usually to proceed under mild conditions. Aziridines can be divided into two classes, activated and basic ones (p. 488). Additions to *C–C* multiple bonds are much more difficult since ethylene, acetylene, and their homologs are nucleophilic in character, whereas the reaction requires the opposite characteristics. Reactions involving the addition of amines are only of interest where substituent groups exerting −*I* and −*M* effects reduce the electron density at the *C–C* multiple bonds so much that nucleophilic addition can occur:

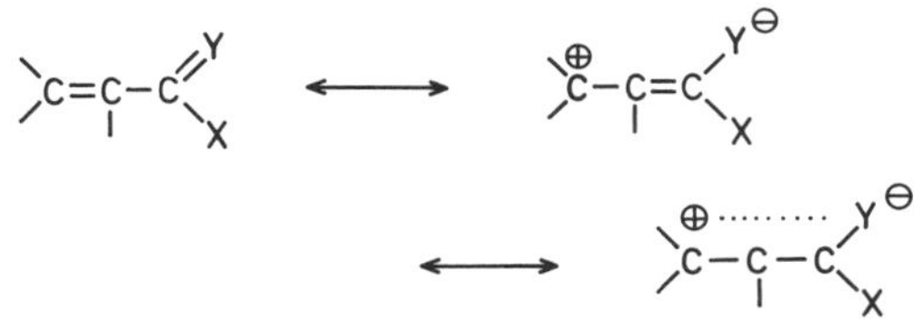

If the electron-attracting power of *Y* is sufficiently high, mesomerism causes the positive charge to be transferred almost undiminished from C_1 to C_3. Consequently, the amine adds at the *β*-position. Styrene yields *phenethylamines*, acrylonitrile gives *3-aminopropionitriles, etc.*

This is also true also of the activation of triple bonds for the formation of *enamines*. Such reactions form the method of choice when the starting materials are readily accessible, the *C–C* multiple bond is sufficiently activated, and the amine used is such that one single reaction path can be expected. Limitations exist when competitive reac-

tions (*e.g.* polymerization) proceed more rapidly, or unexpected secondary reactions are unavoidable, *e.g.*:

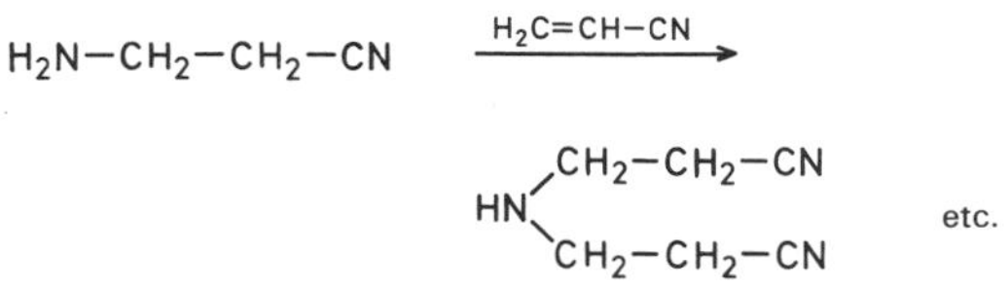

12.3.1 Addition of ammonia and amines to pure alkenes

Addition of ammonia and amines constitutes a nucleophilic addition to the *C–C* multiple bond. Considerable difficulties are encountered, for the reasons mentioned above, especially when the olefin contains +*I* and +*M* substituents; pressure, high temperature, and the presence of a catalyst are therefore necessary. Consequently, many *side-reactions* occur, particularly with aliphatic amines. Ammonia reacts with olefins at 300–350° and 100–200 atmospheres when cobalt catalysts are used[292]. Nitriles and products from fissions and polymerizations are formed as well as the amines. If a catalyst which permits lower reaction temperatures to be used can be found, the yield of amine is increased. For example, secondary amines can be added to olefins containing six carbon atoms in satisfactory yields; alkali metals and their hydrides are used as catalysts. Higher condensation products are only formed in smaller amounts[293].

Alkali metal and alkaline earth metal compounds also promote the addition of olefins to the nitrogen atoms of aromatic amines. The reactivity decreases in the following order:

ethylene > propene > butene > isobutene

Thus, for example, aniline yields *N-ethylaniline* (90%) on reaction with ethylene and *N-isopropylaniline* (45%)[293] with propene. If aluminum is used as catalyst, alkylation of the aromatic ring occurs *ortho* to the amino group. This is attributed to complex formation between aluminum and aniline[294].

---

[292] *F. Möller* in *Houben-Weyl*, Methoden der organischen Chemie, Bd. XI/1, p. 267, Bd. XIV/2, p. 441, Georg Thieme Verlag, Stuttgart 1957.

[293] US. P. 2 501 556 (1950), DuPont, Inv.: *G.M. Whitman*.

[294] *R. Stroh, J. Ebersberger, H. Haberland, W. Hahn*, Angew. Chem. *69*, 124 (1957).

[295] DRP. 496 280 (1930), I.G. Farbenindustrie Ludwigshafen, Erf.: *O. Schmidt, H. Fries*.

12.3.2 Addition of ammonia and amines to dienes, styrene, and vinylpyridines

As might be expected, conjugated dienes react considerably more easily with amines in the presence of alkali metals[295, 296], *e.g.*:

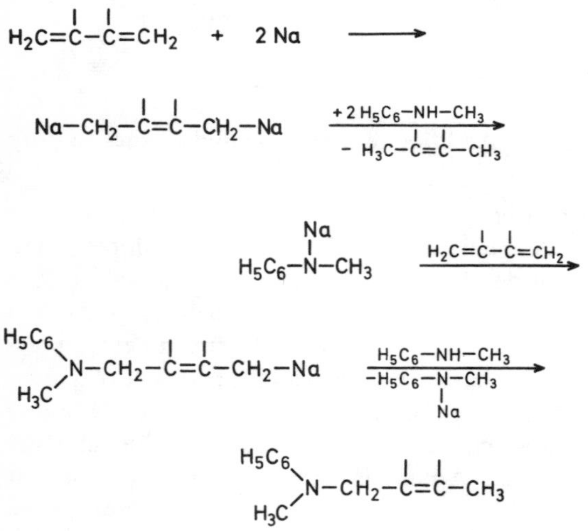

$$H_2C=\overset{|}{C}-\overset{|}{C}=CH_2 \ + \ 2\,Na \ \longrightarrow$$

$$Na-CH_2-\overset{|}{C}=\overset{|}{C}-CH_2-Na \ \xrightarrow[{-\ H_3C-\overset{|}{C}=\overset{|}{C}-CH_3}]{+\,2\,H_5C_6-NH-CH_3}$$

$$H_5C_6-\overset{\overset{\displaystyle Na}{|}}{N}-CH_3 \ \xrightarrow{H_2C=\overset{|}{C}-\overset{|}{C}=CH_2}$$

$$\overset{\displaystyle H_5C_6}{\underset{\displaystyle H_3C}{>}}N-CH_2-\overset{|}{C}=\overset{|}{C}-CH_2-Na \ \xrightarrow[{-H_5C_6-\underset{\underset{\displaystyle Na}{|}}{N}-CH_3}]{H_5C_6-NH-CH_3}$$

$$\overset{\displaystyle H_5C_6}{\underset{\displaystyle H_3C}{>}}N-CH_2-\overset{|}{C}=\overset{|}{C}-CH_3$$

Addition to *styrenes* give good yields of *N*-substituted phenethylamines provided the reactants are not sterically hindered[297-299]. Besides alkali metals, the hydrohalides of the amines to be added can be used as catalysts[300]. Styrene derivatives substituted in the 2- or 4-positions by groups having a −*M* effect require no catalyst for the addition[301, 302]. Primary and secondary amines (except methylamine) react to form 1:1 adducts.

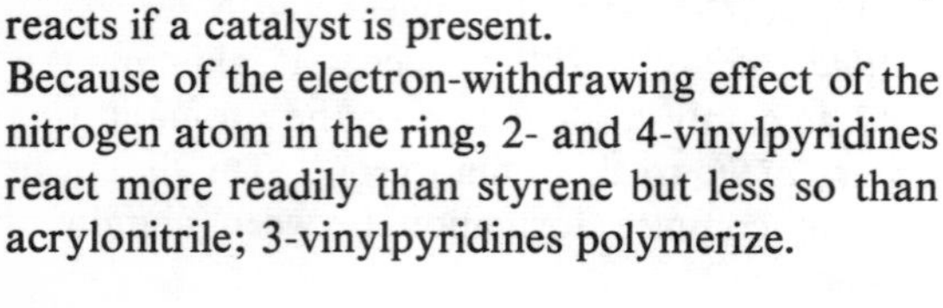

The yields decrease with increasing size of the alkyl group of the alkylamine. 3-Nitrostyrene only reacts if a catalyst is present.

Because of the electron-withdrawing effect of the nitrogen atom in the ring, 2- and 4-vinylpyridines react more readily than styrene but less so than acrylonitrile; 3-vinylpyridines polymerize.

[296] *K. Ziegler*, Angew. Chem. *49*, 499 (1936).
[297] *R. Wegler, G. Pieper*, Chem. Ber. *83*, 1 (1950).
[298] *H. Bestian*, Justus Liebigs Ann. Chem. *566*, 222 (1950).
[299] U.S.P. 2 449 644 (1948), Universal Oil Products Co., Erf.: *D. Danforth;* C.A. *43*, 681 (1949).
[300] *W.J. Hickenboom*, J. Chem. Soc. (London) 319 (1934).
[301] *C.F. Bjork, W.A. Gey, J.H. Robson, R.W. Vandolah*, J. Amer. Chem. Soc. *75*, 1988 (1953).
[302] *W.J. Dale, G. Buell*, J. Org. Chem. *21*, 45 (1956).

Like 2- and 4-nitrostyrenes, 2- and 4-vinylpyridines react with alkylamines or cyclic amines in the absence of catalysts. High yields are obtained from the reaction of aromatic amines if glacial acetic acid is added[303-306], since this protonates the pyridine nitrogen and increases its electron-attracting power.

12.3.3 Addition of ammonia and amines to α,β-unsaturated carbonyl compounds and similar materials

The $C=C$ double bond can participate very readily in electrophilic reactions when it is conjugate to carbonyl or analogous groups. All additions proceed counter to the Markovnikov rule

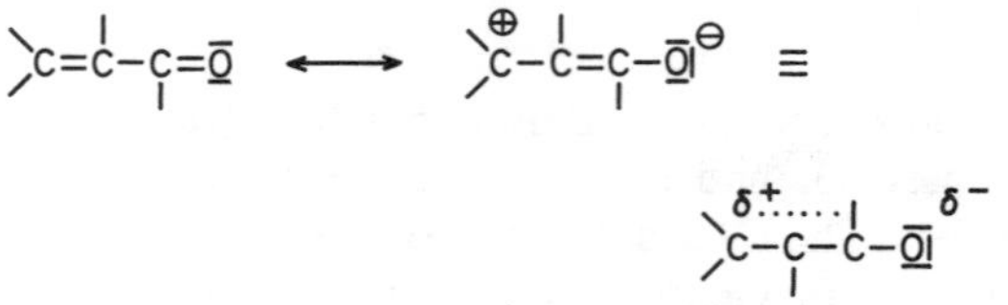

$$\overset{|}{\underset{|}{>}}C=\overset{|}{C}-C=\overline{\underset{\cdot\cdot}{O}} \ \longleftrightarrow \ \overset{\oplus}{\underset{|}{>}}C-\overset{|}{C}=C-\overline{\underset{\cdot\cdot}{O}}|^{\ominus} \ \equiv$$

$$\overset{\overset{\displaystyle \delta\,+}{\cdots}}{>}C-\overset{|}{\underset{|}{C}}-\overset{\delta\,-}{C}-\overline{\underset{\cdot\cdot}{O}}|$$

The ease of reaction follows the magnitude of the −*M* effect, *i.e.*, decreases in the order:

α,β-unsaturated aldehydes >
α,β-unsaturated ketones >
α,β-unsaturated nitriles >
α,β-unsaturated carboxylic acid esters >
α,β-unsaturated carboxamides >
α,β-unsaturated carboxylic acids.

Ammonia and aliphatic amines are sufficiently nucleophilic to be able to react under mild conditions and in the absence of catalysts. Because of the −*I* effect of the aromatic ring, aromatic amines require more vigorous reaction conditions.

12.3.3.1 Addition of ammonia and amines to α,β-unsaturated aldehydes and ketones

α,β-Unsaturated *aldehydes* react with *aliphatic amines* in a 1:2 molar ratio. Good yields are obtained for unsaturated *1,3-diamines*, which can be easily converted into saturated diamines by catalytic hydrogenation:

[303] *H.E. Reich, R. Levine*, J. Amer. Chem. Soc. 77, 4913, 5434 (1955).
[304] *A.J. Matuszko, A. Taurins*, Can. J. Chem. *32*, 538 (1954).
[305] *G. Magnus, R. Levine*, J. Amer. Chem. Soc. 78, 4127 (1956).
[306] *A.R. Phillips*, J. Amer. Chem. Soc. 78, 4441 (1956).

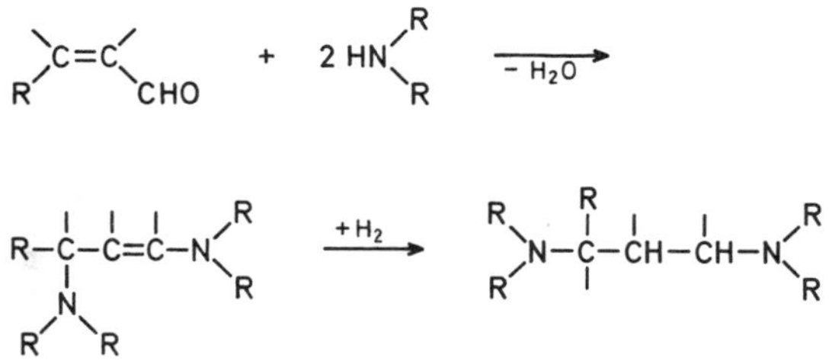

Phthalimide and *succinimide* merely add across the C = C double bond, forming substituted β-amino aldehydes [307].

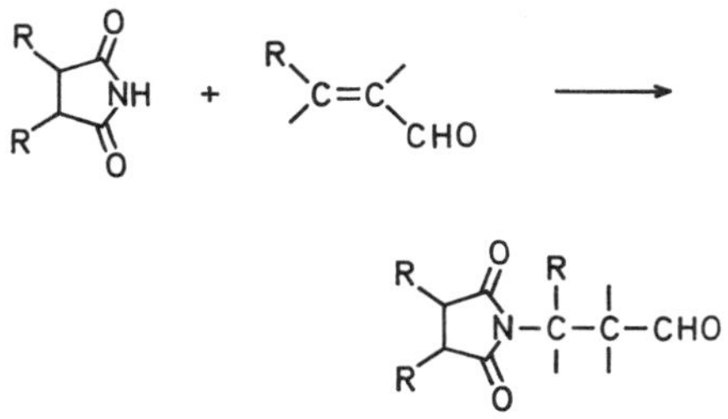

With *aromatic amines* having at least one unsubstituted *o*-position, addition across the C = C double bond occurs first, and is followed by electrophilic substitution of the aromatic ring by ring closure, and finally oxidation to *quinoline* derivatives.

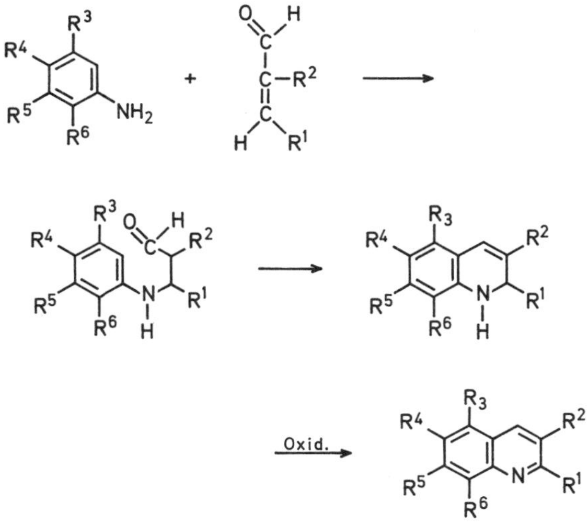

This reaction is of almost universal applicability. Only aromatic amines with labile substituent groups fail to react in the desired way [308-312]. If un-

saturated *ketones* are used instead of unsaturated aldehydes, 4-substituted quinolines result [313].

In a variation of this method, hydrochlorides of aromatic amines are heated with ketones for several days in the presence of aluminum chloride [314, 315]. With acetone, for example, the reaction proceeds as follows:

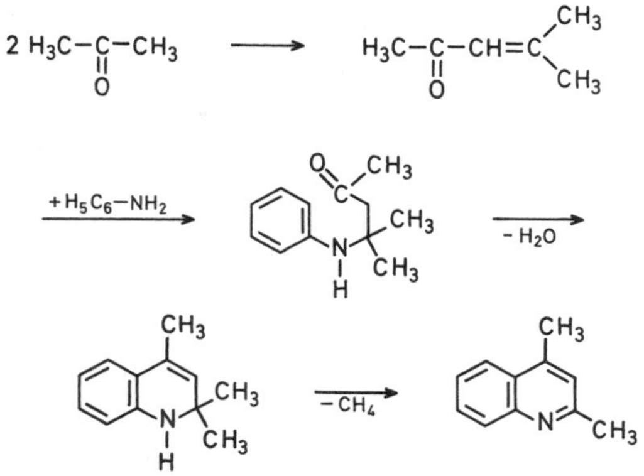

1,2-Dihydro-2,2,4-trimethylquinoline

2,4-Dimethyl-quinoline

Addition of ammonia and amines to α,β-unsaturated ketones does not involve reaction at the carbonyl group [316].

The reaction of 3-bromo-4*H*-chromone with secondary amines to form *N,N-disubstituted 3-amino-4H-chromones* involves several steps [317]:

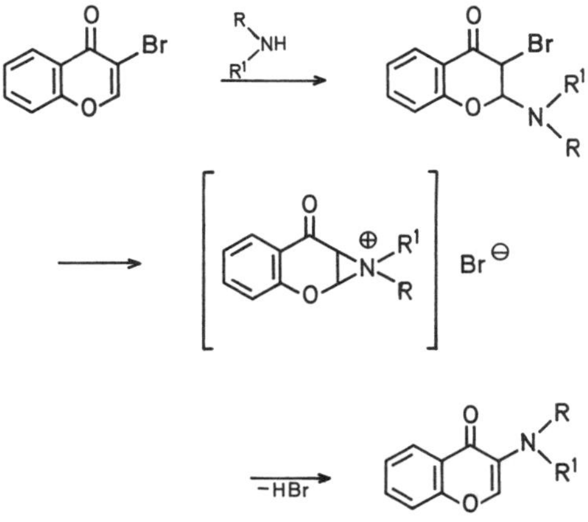

---

[307] *O.A. Moe, D.T. Warner*, J. Amer. Chem. Soc. *71*, 1251 (1949).

[308] *H. Fiedler*, J. Prakt. Chem. [4] *13*, 86 (1961).

[309] *M.J. Weiss, C.R. Hauser* in *R.C. Elderfield*, Heterocyclic Compounds 7, 208 (1961).

[310] *J.L. Finar, R.Y. Hurlock*, J. Chem. Soc. (London) 3259 (1958).

[311] *D.H. Hey, C.W. Rees*, J. Chem. Soc. (London) 905 (1960).

[312] *F.H. Case, H. Idelson*, J. Org. Chem. *27*, 4651 (1962).

[313] *M.J. Weiss, C.R. Hauser* in *R.C. Elderfield*, Heterocyclic Compounds 7, 218 (1961).

[314] *P. Riehm*, Justus Liebigs Ann. Chem. *238*, 9 (1887).

[315] *R.C. Elderfield, J.R. McCarthy*, J. Amer. Chem. Soc. *73*, 975 (1951).

[316] *W.H. Cromwell*, Chem. Rev. *38*, 83 (1946).

[317] *H. Hermann auf dem Keller, F. Zymalkowski*, Arch. Pharm. (Weinheim, Ger.) *304*, 543 (1971).

The course of this reaction has been investigated in detail in order to establish which factors affect the addition[318]. The reaction is determined both by the amine and by the aliphatic or alicyclic unsaturated ketone. For reaction with a given ketone, the steric characteristics of the amine is decisive as well as its basicity. Under standard conditions, dimethylamine and benzylamine add in 40% conversion, cyclic amines such as piperidine, pyrrolidine, N-methylpiperazine, and morpholine add almost quantitatively, but 2-methylpiperidine, 1,2,5-trimethylpiperazine and diethylamine do not react under these conditions.

The following results are obtained on adding piperidine and N-methylpiperidine to various *ketones*. With systems of the type
$H_5C_6-CH=CH-CO-R$
the reaction is independent of R. In 2-benzylidenecycloalkanones the size of the ring is of dominant importance. Five-membered ring ketones hardly react at all (0—15% yield), six-membered ring compounds give yields of up to 90%. Open-chain compounds of the type
$H_5C_6-CH=CR-CO-R$
hardly react.

12.3.3.2 Addition of ammonia and amines to α,β-unsaturated nitriles

Primary and *secondary* amines are the main products of the addition of ammonia to α,β-unsaturated nitriles; only small amounts of tertiary amines are formed[319]:

H₂N... (reaction scheme)

The addition proceeds particularly smoothly in aqueous or alcoholic medium. *3-Aminodipropionitrile* is the main product; ~20% *2-amino propionitrile* is also formed but less than 5% *3-aminotripropionitrile*[320]. Selective preparation of *3-aminopropionic acid* can be achieved readily by thermal fission of the 3-aminodipropionitrile[321].

With methylamine the products are either secondary or tertiary amines, depending on the reaction conditions[320]. Higher primary alkylamines react more selectively; the preparation of 1:1 adducts by variation of the reaction conditions therefore becomes easier as the chain length increases[320, 321—326]. Secondary amines react with acrylonitrile forming tertiary amines; basic catalysts are often used for these reactions[319].

3-(Cyclohexylamino)propionitrile exists in equilibrium with the components from which it was derived[320]. Aliphatic diamines add only one molecule acrylonitrile to each amino group[327]. Hydrazines, hydroxylamines, and (2-hydroxyethyl)amines behave similarly. Aromatic amines or N-heterocycles such as indole or carbazole are considerably less reactive and acid catalysts[320] are generally essential for reaction.

12.3.3.3 Addition of ammonia and amines to α,β-unsaturated carboxylic acids and esters

Additions of aliphatic amines to α,β-unsaturated carboxylic acids or their esters[292] require catalysts such as acetic acid[328], stannic chloride[329], or cupric acetate[330]. In some cases, *e.g.*, the reaction of ammonia with acrylic esters, it is difficult to obtain monoadducts; even with a 5:1 molar ratio mixtures of *alkyl 3-aminopropionates, 3-aminodipropionates,* and *3-aminotripropionates* are obtained[331].

(reaction scheme)

Because of the equilibria between the individual reaction stages, it is, however, possible to trap the bis-addition product as its N-benzoyl derivative by

[318] R. Baltzly, E. Lorz, P.B. Russell, F.M. Smith, J. Amer. Chem. Soc. 77, 624 (1955).
[319] H.A. Bruson, Org. Reactions 5, 82 (1949).
[320] DRP. 598 185 (1931), Fr.P. 742 358 (1932), I.G. Farbenindustrie, Erf.: U. Hoffmann, B. Jakobi.
[321] DRP. 1 480 143 (1942), I.G. Farbenindustrie, Erf.: R. Schröter, L. Knöpfle.
[322] M. Lipp, F. Dallacker, H.G. Rey, Chem. Ber. 91, 2239 (1958).
[323] J.T. Braunholtz, F.G. Mann, J. Chem. Soc. (London) 1817 (1953).
[324] D.W. Adamson, J. Chem. Soc. (London) 144 (1949).
[325] R.W. Holley, A.D. Holley, J. Amer. Chem. Soc. 71, 21276 (1949).
[326] R.C. Fuson, W.E. Parhan, L.J. Read, J. Amer. Chem. Soc. 68, 1239 (1946).
[327] O. Bayer, Angew. Chem. 61, 234 (1949).
[328] H.A. Bruson, Org. Reactions 5, Chap. 2 (1949).

heating the tertiary amine in the presence of ben-zoyl chloride[331]. Addition of secondary amines is hindered progressively with increasing size (*e.g.*, piperidine, morpholine) and branching of the alkyl groups; this is especially the case when branching is close to the nitrogen atom[332–335].

Alkyl substituent groups at the α- or β-positions of the unsaturated carboxylic ester also have some influence; because of their $+I$ effect, the reactivity is depressed somewhat but the selectivity is in-creased. For example[336], methylamine and methyl methacrylate react selectively to form a 1:1 adduct [*methyl 2-methyl-3-(methylamino)butanonate*]. Esters of crotonic acid behave similarly[337]. The addition of anilines which are substituted on the ring by electronegative groups proceeds smoothly if catalytic amounts of glacial acetic acid are pres-ent[338]. By a suitable choice of conditions either *3-aminopropionates* or *bis(2-alkoxycarbonylethyl)-amines* are obtained, as desired. Aniline itself gives only small yields[339].

The reaction of itaconic acid with 4-aminophenols, 4-aminobenzoic acid, *etc.*, in the melt gives[340], for example, *1-(4-hydroxyphenyl)-* and *1-(4-carboxy-phenyl)-5-oxo-3-pyrrolecarboxylic acids* via *2-arylaminomethylsuccinic acid intermediates*[340]:

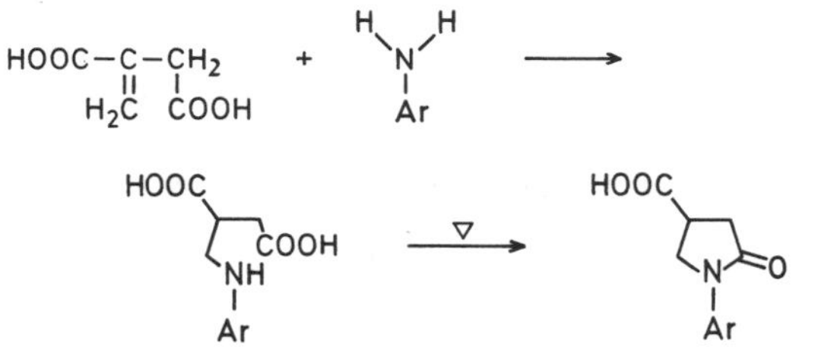

The reaction of substituted ethylenediamines with fumaric diesters, β-benzoylacrylic esters, and β-

acetylacrylic esters proceeds similarly, but under very much milder conditions[341–343]:

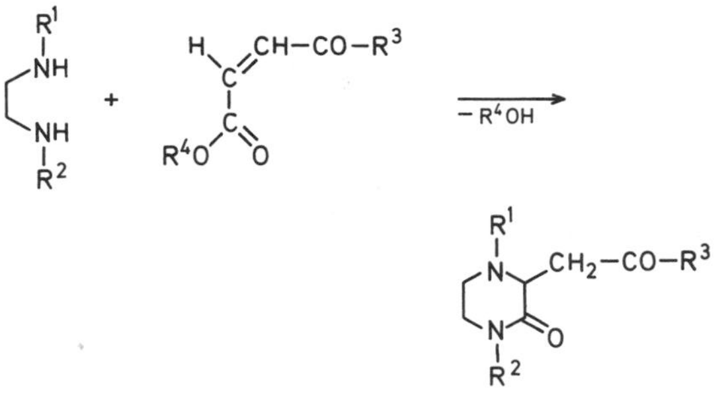

If the reaction mixtures are allowed to stand for several days at room temperature, a wide range of 1,4-disubstituted *3-ethoxycarbonylmethyl-*, *3-(benzoylmethyl)-*, and *3-(2-oxopropyl)-2-piper-azinones* can be obtained. The reaction of 2-amino-pyridine with acrylic esters yields both *3,4-dihydro-2H-pyrido[1,2-a]pyrimidin-2-ones* and *N-2-pyri-dyl-β-alanine esters*[344]:

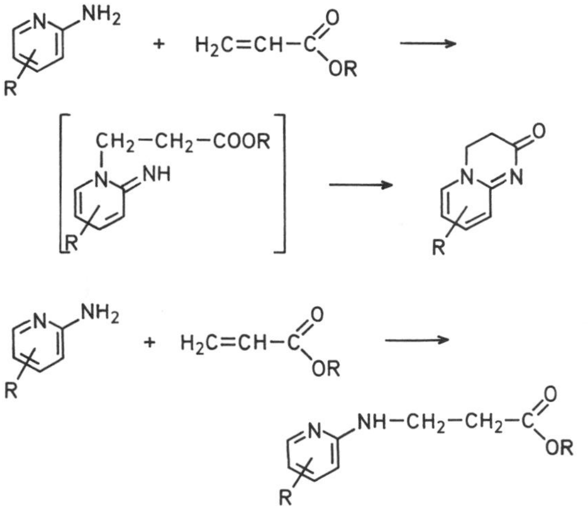

Pyrazoles, too, add in good yields[345].

12.3.3.4 Addition of ammonia and amines to α,β-unsaturated nitro compounds

Most α,β-unsaturated nitro compounds react with aliphatic and aromatic amines, even at room temperature[292]. Although 2-nitroalkylamines prove to be unstable, the corresponding 2-nitro-arylamines are more stable.

[329] *P.L. Southwick, R.T. Crouch*, J. Amer. Chem. Soc. **75**, 3413 (1953).

[330] *J.A. Heininger*, J. Org. Chem. **22**, 1213 (1957).

[331] *S.M. McElvain, G. Stork*, J. Amer. Chem. Soc. **68**, 1069 (1946).

[332] *D.S. Tarbell, N. Shakespeare, C.J. Claus, J.F. Bun-nett*, J. Amer. Chem. Soc. **68**, 1217 (1946).

[333] *D.E. Pearson, W.H. Jones, A.C. Cope*, J. Amer. Chem. Soc. **68**, 1225 (1946).

[334] *R.C. Fuson, W.E. Parhan, L.J. Read*, J. Amer. Chem. Soc. **68**, 1239 (1946).

[335] *O. Hromatha*, Ber. **75**, 131 (1942).

[336] *D.R. Howton*, J. Org. Chem. **10**, 277 (1945).

[337] *J.W. Adamson*, J. Chem. Soc. (London) 885 (1950).

[338] *M. Lipp, F. Dallacker, H.G. Rey*, Chem. Ber. **91**, 2239 (1958).

[339] *N.J. Leonard, R.C. Fox, M. Oki*, J. Amer. Chem. Soc. **76**, 5708 (1954).

[340] *P.L. Paytash, E. Sparrow, J.C. Gathe*, J. Amer. Chem. Soc. **72**, 1415 (1950).

[341] U.S.P. 3 056 786 (1960), Wellcome Foundation, Inv.: *A.P. Philips*.

[342] DBP. 1 135 472 (1962), Wellcome Foundation, Erf.: *A.P. Phillips*.

[343] Brit. P. 901 455 (1962), Burroughs Wellcome & Co.

[344] *G.R. Lappin*, J. Org. Chem. **23**, 1358 (1958).

All nitroamines can be converted into diamines by hydrogenation with Raney nickel catalysts:

$$R-\overset{\underset{\displaystyle |}{N}}{\underset{\underset{\displaystyle R}{|}}{C}}-\overset{|}{\underset{|}{C}}-NO_2 \quad \xrightarrow{Ni/H_2} \quad R-\overset{\underset{\displaystyle |}{N}}{\underset{\underset{\displaystyle H}{|}}{C}}-\overset{|}{\underset{|}{C}}-NH_2$$

12.3.4 Addition of N-halo compounds to C=C double bonds

In the presence of strong acids plus light or iron ions N-chlorodialkylamines add to butadiene[346] to form N,N-dialkyl-4-chloro-2-butenamine:

$$R_2N-Cl \quad + \quad CH_2=CH-CH=CH_2 \quad \longrightarrow$$
$$R_2N-CH_2-CH=CH-CH_2Cl$$

The addition often follows a free-radical mechanism and is complete within 15 minutes[347]. Reactions can also be carried out with monoolefins, allenes, and acetylenes[348]:

$$(H_5C_2)_2N-Cl \quad + \quad H_2C=CH-R \quad \longrightarrow$$
$$(H_5C_2)_2N-CH_2-\underset{\underset{\displaystyle Cl}{|}}{CH}-R$$

$$(H_5C_2)_2N-Cl \quad + \quad H_2C=C=CR_2 \quad \longrightarrow$$
$$(H_5C_2)_2N-CH_2-\underset{\underset{\displaystyle Cl}{|}}{CR_2}$$

Yields from these reactions can be as much as 60%. In some cases addition is followed by a Hoffmann-Loeffler reaction and *pyrrolidines* are obtained. A side-reaction is chlorination of the olefin.

12.3.5 Addition of ammonia and amines to the C≡C bond

Whereas olefins usually react with nucleophilic agents only when they are polarized by an electron-withdrawing group, the ability to react with nucleophiles is an intrinsic characteristic of the C≡C triple bond[349-351].

Enamines are formed by the addition of ammonia and amines to acetylenes:

$$-C≡C- \quad + \quad HNR_2 \quad \longrightarrow \quad \overset{H}{\underset{\displaystyle}{}}C=C-NR_2$$

In addition to the usual method of preparing enamines involving condensation (see p. 463):

$$>NH \quad + \quad O=C< \quad \longrightarrow \quad >N-\overset{|}{C}= \quad + \quad H_2O$$

The addition gains in importance if the reactivity of the acetylene is further increased by mesomeric effects. It is the method of choice for the preparation of certain classes of heterocyclic compounds *via* enamines as intermediates.

Cis-products are usually obtained in the preparation of enamines by addition of amines to acetylene derivatives, and even more with secondary than with primary amines. The steric course depends on a combination of the following factors[352]:

the solvent
the reaction conditions
the constitution of the acetylene
the constitution of the amine.

Detailed investigations of the relationships involved have been made with acetylenedicarboxylic diesters using ice-cold ether as solvent. If the reactants are mixed together rapidly, a mixture of *cis*- and *trans*-adducts is obtained, whereas slow mixing (kinetic control) yields solely 1:1 *cis*-adducts[353-355] (spectroscopic evidence)[353].

Under the above conditions primary alkylamines react rapidly and exothermically primary arylamines more slowly[353]. Ammonia reacts within three hours, cyclohexylamine requires more than 15 hours, aniline several months for completion of the reaction [the products are *diethyl aminomaleates*, *(cyclohexylamino)maleates*, and *anilinomaleates*].

Of the secondary amines, dimethylamine reacts within two hours, diisoproylamine and dicyclohexylamine within 15 hours, and N-methylaniline in 18 hours to give *diethyl (dimethylamino)-maleate*, *(diisopropylamino)maleate*, *(dicyclohexylamino)maleate*, and *N-(methylanilino)maleate*.

[345] R. Reimlinger, J.E.M. Oth, Chem. Ber. 97, 331 (1964).
[346] F. Minisci, R. Galli, Tetrahedron Lett. 167 (1964).
[347] R.S. Neale, R.L. Hinman, J. Amer. Chem. Soc. 85, 2666 (1963).
[348] R.S. Neale, R.L. Hinman, J. Amer. Chem. Soc. 86, 5340 (1964).
[349] F. Bohlmann, Angew. Chem. 69, 82 (1957).
[350] R.S. Mulliken, J. Amer. Chem. Soc. 72, 4493 (1950).
[351] R.A. Raphael, Acetylenic Compounds in Org. Synthesis, Butterworth & Co., London 1955.
[352] E. Winterfeldt, Angew. Chem. 79, 389 (1967).

[353] R. Huisgen, K. Herbig, A. Siegl, H. Huber, Chem. Ber. 99, 2526 (1966).
[354] E.W. Truce, D.G. Brady, J. Org. Chem. 31, 3543 (1966).
[355] K. Herbig, R. Huisgen, H. Huber, Chem. Ber. 99, 2546 (1966).

Reactions with methyl propiolate proceed similarly, and here, too, the inductive effect of substituent groups on the amine clearly influences the reaction rate.

Heteroaromatic amines such as pyrazole[356] also add readily:

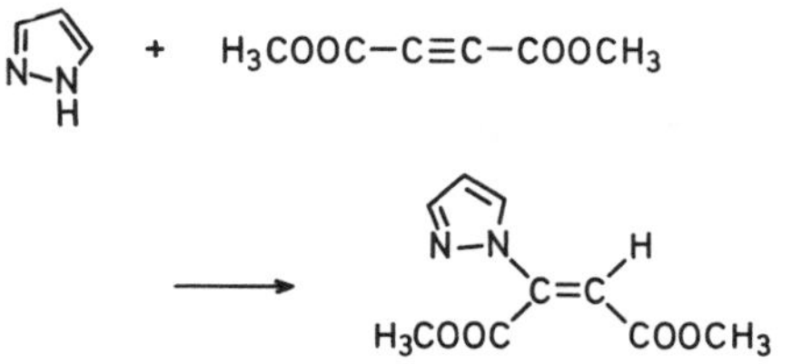

Dimethyl 1-pyrazol-1-ylmaleate

With 2-aminophenol and 1,2-phenylenediamine ring-closure occurs[357]:

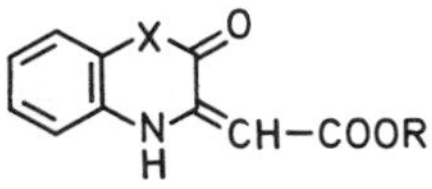

X = O; *Alkyl 2-oxo-2H-1,4-benzoxazin-3(4H)-ylideneacetate*
X = NH; *Alkyl 3,4-Dihydro-3-oxo-[2(1H)-quinoxalin-ylidene]acetate*

This principle can be extended quite generally to the synthesis of heterocycles of the following type[358]:

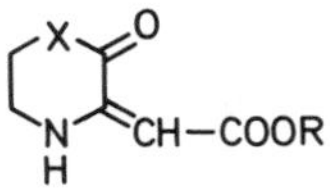

X = O; *Alkyl 5,6-dihydro-2-oxo-(2H-1,4-oxazin-3-ylidene)-acetate*
X = NH; *Alkyl 3-oxo-(2-piperazinyl-idene)acetate*

On using geminal diamines such as guanidine *alkyl 2-imino-5-oxo-4-(imidazolidinylidene)-acetates* are obtained[359]:

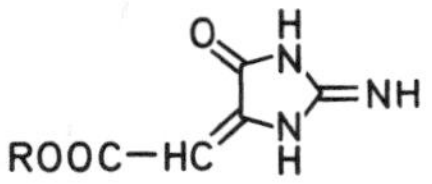

On reacting alkynes of the type $R^1-C\equiv C-COOR^2$
(R^1 = phenyl or *n*-hexyl, R^2 = methyl or ethyl with aniline in the presence of cuprous chloride or oxide, a mixture of *4-hexyl-* or *4-phenyl-*

2(1H)-quinolinone and *2-hexyl-* or *2-phenyl-4-(1H)-quinolinone* respectively is obtained; these products can be separated because their behavior toward alkalis is different[360].

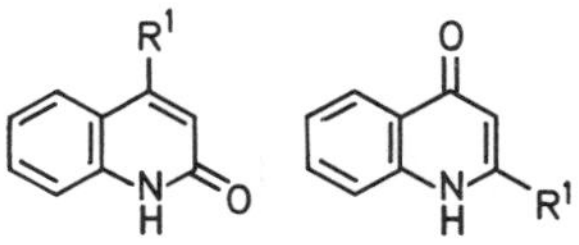

2-Aminopyridine reacts with propiolic acid esters forming a mixture of the *cis-* and *trans-*adduct; the *cis*-adduct reacts further by cyclization[361]:

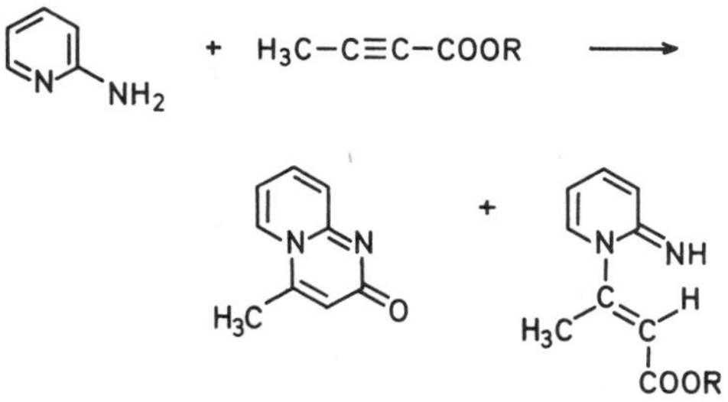

4-Methyl-2H-pyrido-[1,2-a]pyrimidin-2-one   *Alkyl 3-(2-imino-1(2H)-pyridyl)-crotonate*

1-Alkyn-1-ylamines condense with polar multiple bonds and dipolar systems while forming rings[362]. The reaction with cyclic imines leads to ring expansions[363] by respectively two members, *e.g.:*

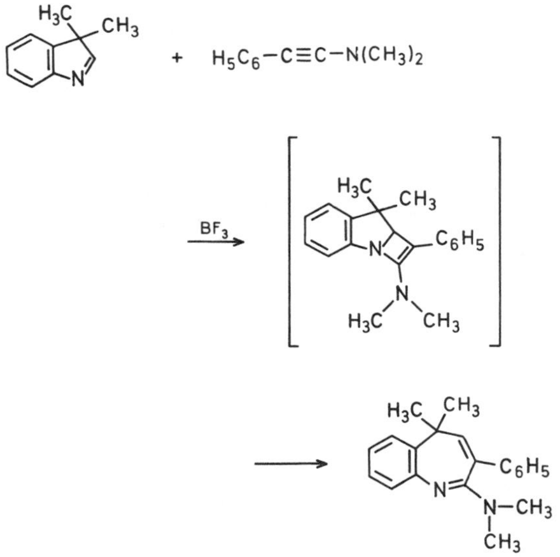

5,5-Dimethyl-2-(dimethylamino)-3-phenyl-5H-benz-[b]azepine

---

[356] *H. Reimlinger, C.H. Moussebois,* Chem. Ber. *98,* 1805 (1965).

[357] *Y. Iwanami,* J. Chem. Soc. Jap., Pure Chem. Sect. *82,* 634 (1964).

[358] *Y. Iwanami,* Bull. Chem. Soc. Japan *37,* 1740, 1745 (1964).

[359] *Y. Iwanami, H. Saaki, H. Sakata,* J. Chem. Soc. Jap., Pure Chem. Sect. (Nippon Kagaku Zasshi) *85,* 704 (1964).

[360] *J. Reisch,* Angew. Chem. *75,* 1203 (1963).

[361] *J.J. Pachter,* J. Org. Chem. *26,* 4157 (1961).

When 3, 4-dihydroisoquinoline is used the intermediately formed azetine can be detected by means of hydrolysis[364], *e.g.*:

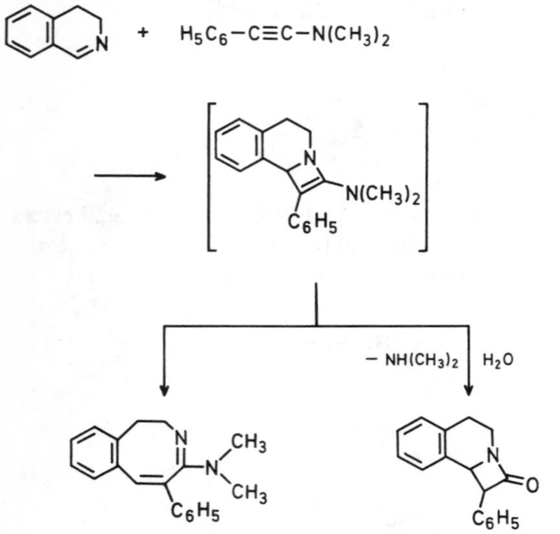

4-(Dimethylamino)-
1,2-dihydro-5-phenyl-
2-benzazocine

1,4,5,9b-Tetra-
hydro-1-phenyl-
azeto[2,1-a]-
isoquinolinon-
2-one

From 1, 2, 4-triazines and *N,N*-diethylpropynyl-amine cycloaddition yields *2-substituted 4-(diethyl-amino)-5-methylpyrimidines*[365]:

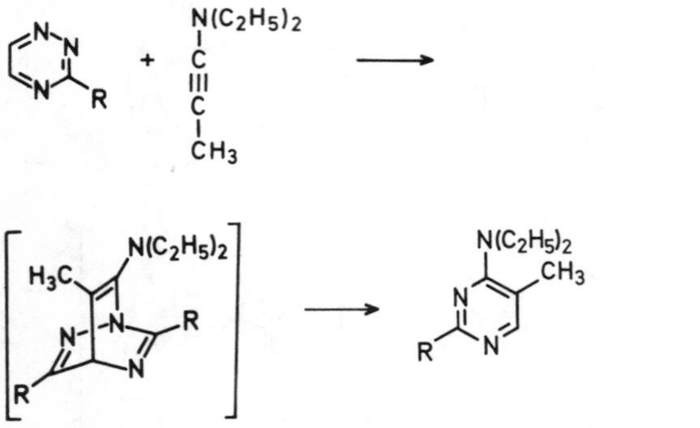

1, 2, 4, 5-tetrazines react to *pyridazine* derivatives[366].

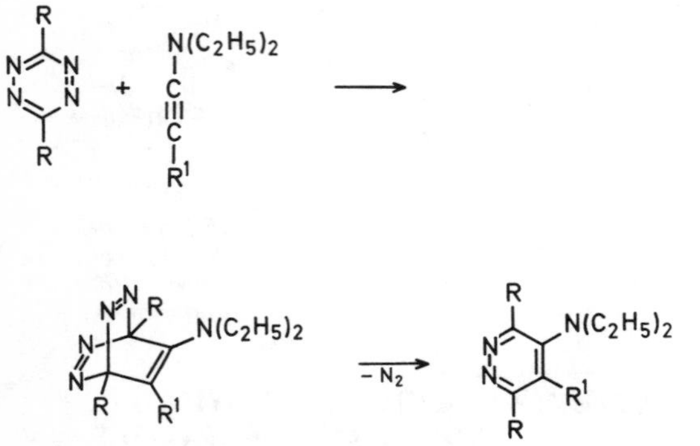

12.3.6 Addition to oxiranes

Ammonia does not react uniformly with *oxirane* (ethylene oxide) because multiple additions generally occur:

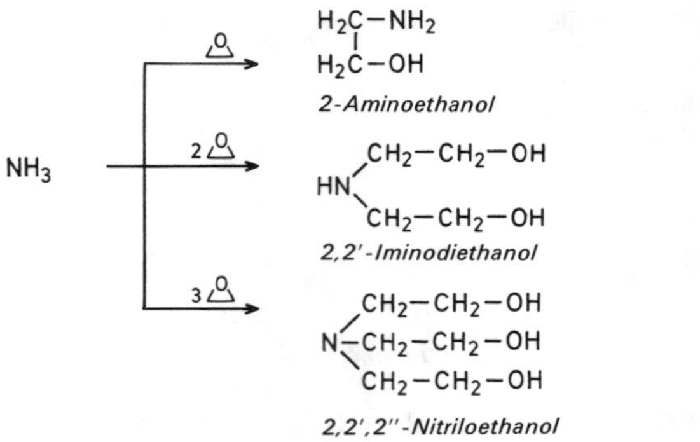

2-Aminoethanol

2,2'-Iminodiethanol

2,2',2''-Nitriloethanol

The sluggish reaction can be accelerated by the addition of water to such an extent that it occurs **explosively.** For reactions in aqueous medium efficient cooling is therefore essential and, moreover, the reaction mixture must never contain excess oxirane. On the technical scale oxirane is introduced into 25–30% aqueous ammonia solution at 30–35° under 2 atmospheres pressure[367, 368].

Primary amines behave similarly to ammonia. Monoadducts are obtained in satisfactory yield only if a large excess of primary amine is used.

In the case of *hydrazine,* monosubstituted hydrazines, and symmetrically and unsymmetrically disubstituted hydrazines, only one amino group reacts with oxiranes. *2-Hydrazinoethanols* or corresponding *N*-alkyl products are obtained[369, 370]. *Hydroxylamine* and its *N*-alkyl derivatives once more react to form multiple addition products.

Alkylated oxiranes react more slowly, because alkyl groups retard the nucleophilic attack by the amine by virtue of their +*I* effect. If the basicity of the amine is too low, addition of a small amount of sodium or an alkali metal hydroxide is necessary.

[362] *H.G. Viehe*, Angew. Chem. 79, 744 (1967).
[363] *R. Fuks, R. Buije, H.G. Viehe*, Angew. Chem. 78, 594 (1966).
[364] *J. Ficini, A. Krief*, Tetrahedron Lett. 2497 (1967).
[365] *H. Neunhoeffer, H.W. Frühauf*, Tetrahedron Lett. 3355 (1970).
[366] *A. Steigel, J. Sauer*, Tetrahedron Lett. 3357 (1970).
[367] BIOS Final Rep. 1059 (1947).
[368] DBP. 535 049 (1953), I.G. Farbenindustrie, Erf.: *H. Ulrich;* Fortschritte der Teerfabrikation u. verwandter Industriezweige 18, 346 (1931).
[369] *G. Benoit*, Bull. Soc. Chim. France 708 (1939).
[370] *G. Benoit*, Bull. Soc. Chim. France 242 (1947).

During addition of ammonia to oxiranecarboxylic acid esters the ring opening ensues on the side of the β-carbon atom with formation of α-hydroxy-β-amino acids[371].

From *trans*-3-substituted 2-oxiranecarboxylic acids *erythro compounds* are formed, from the *cis*-isomers *threo compounds*. The reaction between amines and chloromethyloxirane (epichlorohydrin) has been investigated in detail[373, 374]:

$$\text{oxirane-CH}_2\text{Cl} + R_2NH \longrightarrow$$

$$R_2N-CH_2-\overset{\overset{\displaystyle OH}{|}}{CH}-CH_2Cl$$

Reaction with strong bases leads to renewed oxirane ring formation.

$$R_2N-CH_2-\overset{\overset{\displaystyle OH}{|}}{CH}-CH_2Cl \xrightarrow{-HCl} R_2N-CH_2-\text{oxirane}$$

Further treatment with ammonia, primary, and secondary amines enables unsymmetrically substituted *1,3-diamino-2-propanols* to be prepared:

$$R_2N-CH_2-\text{oxirane} + HN\overset{R^1}{\underset{R^2}{}} \longrightarrow$$

$$R_2N-CH_2-\overset{\overset{\displaystyle OH}{|}}{CH}-CH_2-N\overset{R^1}{\underset{R^2}{}}$$

Morpholine reacts with chloromethyloxirane to form *4,4'-(1,4-dioxane-2,5-diyldimethylene)dimorpholine*. Piperidine and pyrrolidine give similar products[373]:

$$R_2N-CH_2-\text{oxirane} \rightleftharpoons R_2N-CH_2-\overset{\overset{\displaystyle |\overline{\underset{..}{O}}|^{\ominus}}{|}}{\overset{\oplus}{CH}}-CH_2$$

$$\xrightarrow{\text{Cl-CH}_2\text{-oxirane}}$$

1-(5-Chloromethyl-1,4-dioxanemethyl-2-yl-methylene)piperidine and 1-(5-Chloromethyl-1,4-dioxan-2-yl-methylene)-pyrrolidine

$$\xrightarrow{\text{NHR}_2}$$

4,4'-(1,4-dioxane-2,5-diyl-dimethylene)dipiperidine and 4,4'-(1,4-dioxane-2,5-diyl-dimethylene)dipyrrolidine

Cyclohexylamine adds to chloromethyloxirane in 17 hours at room temperature to give *3-chloro-1-(cyclohexylamino)-2-propanol*[375]. Ethylenediamine forms *1,1'-(ethylenediimino)bis[3-chloro-2-propanol]*; only one amino group reacts in each case[376].

Equimolar amounts of aromatic amines and chloromethyloxirane react in alcoholic solution and at room temperature to form monoaddition products[377]. Aminopyridines react as the tautomeric imino forms to give *1-substituted 2-imino-(2H)-pyridines*[378], e.g.:

$$\text{(pyridine)-NH}_2 \rightleftharpoons \text{(pyridine, N-H)}=NH$$

$$+ \text{Cl-CH}_2\text{-oxirane} \longrightarrow$$

$$\text{(pyridyl)}=N\overset{|}{\underset{}{}}\begin{array}{l}CH_2\\CH-OH\\CH_2-Cl\end{array}$$

3-Chloro-1-(2-imino-1(2H)-pyridyl)-2-propanol

If the nitrogen atom is blocked by alkylation the imino group reacts[379]:

$$\text{(pyridine, N-R)}=NH + \text{Cl-CH}_2\text{-oxirane} \longrightarrow$$

$$\text{(pyridine, N-R)}=N-CH_2-\overset{\overset{\displaystyle OH}{|}}{CH}-CH_2Cl$$

[371] F. Möller in Houben-Weyl, Methoden der organischen Chemie, Bd. XI/1, p. 315, Georg Thieme Verlag, Stuttgart 1957.

[372] Epichlorhydrin, Technical, Booklet, 49–35 Shell Chemical Corp. 1949–1953.

[373] D.L. Heywood, B. Phillips, J. Amer. Chem. Soc. 80, 1257 (1958).

[374] J.B. McKellvey, B.G. Webre, R.R. Benerito, J. Org. Chem. 25, 1424 (1960).

[375] J.B. McKelvey, B.G. Webre, E. Klein, J. Org. Chem. 24, 614 (1959).

[376] J.H. Ross, D. Baker, A. Coscia, J. Org. Chem. 29, 824 (1964).

[377] J.R. Merchant, A.S.U. Choughuley, Kum D.D. Vaghani, Current Sci. (India) 4, 142 (1959); C.A. 54, 20827i (1960).

[378] S. Kutkevicius, V. Klusis, Lietuvos TSR Aukstuju Mokyklu Mokslu Darbai, Chem. ir Chem. Technol 6, 51 (1965); C.A. 64, 19607c (1966).

[379] S. Kutkevicius, V. Klusis, Lietuvos TSR Aukstuju Mokyklu Mokslu Darbai, Chem. ir Chem. Technol 5, 33 (1964); C.A. 61, 6985h (1964).

In aqueous or alcoholic medium *tertiary* amines react with oxiranes forming *choline* and its derivatives[380].

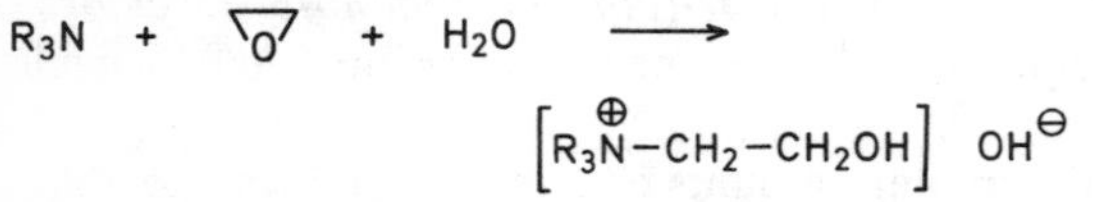

$$R_3N \ + \ \triangle\!\!O \ + \ H_2O \ \longrightarrow$$
$$\left[R_3\overset{\oplus}{N}-CH_2-CH_2OH \right] \ OH^{\ominus}$$

The reaction temperature is between 50 and 100°; water or aqueous alcohol are generally used as solvent. If boron trifluoride is present inner ammonium salts are formed[381,382].

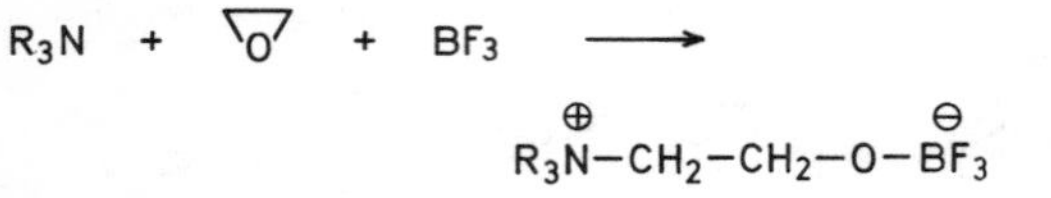

$$R_3N \ + \ \triangle\!\!O \ + \ BF_3 \ \longrightarrow$$
$$R_3\overset{\oplus}{N}-CH_2-CH_2-O-\overset{\ominus}{B}F_3$$

Pure anhydrous tertiary amines do not react with oxiranes. If catalytic amounts of water or alcohol are present, nitrogen-containing *polyaddition* products are formed.

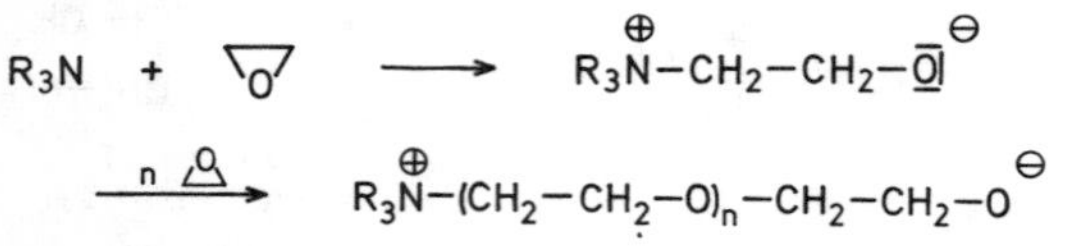

$$R_3N \ + \ \triangle\!\!O \ \longrightarrow \ R_3\overset{\oplus}{N}-CH_2-CH_2-\overset{\ominus}{\underline{O}}$$
$$\xrightarrow{n \ \triangle} \ R_3\overset{\oplus}{N}-(CH_2-CH_2-O)_n-CH_2-CH_2-O^{\ominus}$$

Chain termination ensues by addition of a proton.

If oxiranes are allowed to react with suitable tertiary polyamines of high molecular weight, strongly basic *anion exchangers* are formed.

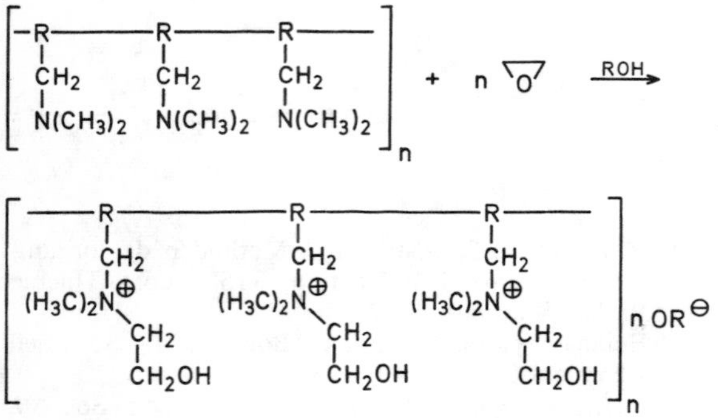

$$\left[\begin{array}{ccc} -R & R & R- \\ | & | & | \\ CH_2 & CH_2 & CH_2 \\ | & | & | \\ N(CH_3)_2 & N(CH_3)_2 & N(CH_3)_2 \end{array} \right]_n \ + \ n \ \triangle\!\!O \ \xrightarrow{ROH}$$

$$\left[\begin{array}{ccc} R & R & R \\ | & | & | \\ CH_2 & CH_2 & CH_2 \\ | & | & | \\ (H_3C)_2\overset{\oplus}{N} & (H_3C)_2\overset{\oplus}{N} & (H_3C)_2\overset{\oplus}{N} \\ | & | & | \\ CH_2 & CH_2 & CH_2 \\ | & | & | \\ CH_2OH & CH_2OH & CH_2OH \end{array} \right]_n \ n \ OR^{\ominus}$$

A summary is given in *Houben-Weyl*, Vol. XIV/2, p. 441.

12.3.7 Addition of ammonia and amines to aziridines

The reactivity of aziridines is mainly influenced by two factors:

the nucleophilism of the ring-nitrogen atom
the strained three-membered ring

In the case of the addition of ammonia and amines to aziridines the three-membered ring is not the predominant factor. Instead, the reactivity of the nitrogen atom in the ring determines the specific properties.
Nitrogen-substituted aziridines can broadly be divided into two groups.

Activated aziridines are compounds in which the substituent groups can enter into conjugation with the unshared pair of electrons on the ring-nitrogen atom and so reduce the latter's basicity.
In contrast, *basic* aziridines behave like secondary or tertiary amines.

The three-membered ring can usually be retained intact if certain precautions are taken; otherwise ring-opening occurs (faster with activated than with basic aziridines). Basic aziridines first add protons, which need to be present in traces only:

$$\underset{R}{\triangle\!\!N} \ + \ H^{\oplus} \ \longrightarrow \ \underset{R}{\triangle\!\!\overset{\oplus}{N}}H$$

Aziridinium cations are very reactive and undergo ring opening. The ring remains intact only if protons are absent.
With activated aziridines ring opening occurs even in the absence of protons after nucleophilic attack of an amine on a carbon atom of the aziridine ring. The ring can be retained intact at least in some products only by working at low temperatures and using a nonpolar solvent.
The ring opening of basic aziridines with amines is of greater preparative importance because it is the most direct method of preparing unsymmetrically *N*-substituted *ethylenediamines* (aminoethylation of amines):

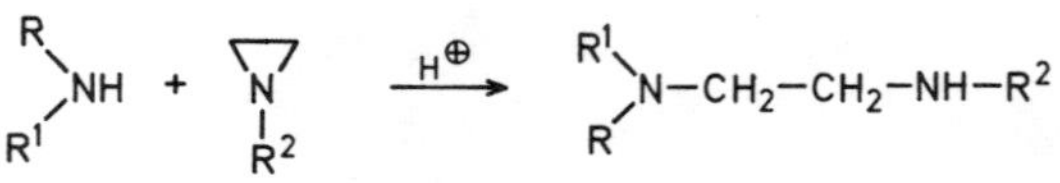

$$\underset{R^1}{\overset{R}{>}}NH \ + \ \underset{R^2}{\triangle\!\!N} \ \xrightarrow{H^{\oplus}} \ \underset{R}{\overset{R^1}{>}}N-CH_2-CH_2-NH-R^2$$

Naturally, other aziridines can function as serve as amines. It is thus not surprising that polymers of the following structure appear as by-products:

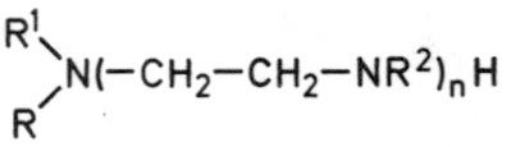

$$\underset{R}{\overset{R^1}{>}}N(-CH_2-CH_2-NR^2)_nH$$

Strongly nucleophilic amines add more smoothly than others; for example, morpholine reacts very much better than aniline. Correspondingly, the

[380] *A.M. Easthan,* Can. J. Chem. *29,* 575, 585 (1951).
[381] *H. Meerwein,* J. Prakt. Chem. *154,* 83 (1940).
[382] *J. Goerdeler* in *Houben-Weyl,* Methoden der organischen Chemie, 4. Aufl., Bd. XI/2, p. 610, Georg Thieme Verlag, Stuttgart 1958.

formation of polymers is less favored with very nucleophilic amines.

Water or alcohol are used as solvent, with or without added acid. As with oxiranes, ring-opening of aziridines proceeds by an S_N2 mechanism.

Reaction with amines in the presence of equimolar aluminum trichloride or boron trifluoride is of interest, since high yields of ethylenediamine derivatives can be obtained without formation of polymeric by-products.

Reaction of 2-piperidinecarboxylic acid esters and of other bifunctional amines yields *heterocycles*, *e.g.*:

Hexahydro-2H-pyrido[1,2-a]-pyrazin-1(6H)-one (1,4-diazabicyclo-[4.4.0]decan-5-one)

Reactions with *hydrazines* proceed like those with amines. *trans*-2-Acylaziridines react with phenylhydrazines *via* Schiff base intermediates; ring expansion to *4-aminopyrazolines* occurs. *cis*-2-Acylaziridines react further, eliminating amine to form *pyrazoles*[386].

12.4 Addition of Schiff bases to *CH*-acid compounds

The reaction of imines with *CH*-acid compounds is formally a component process of the Mannich reaction. In this reaction amine, carbonyl compound and *CH*-acid compound act on each other simultaneously, while imines are actually the reaction products of two of these three components. It is perfectly feasible that during the Mannich reac-

tion the carbonyl compound reacts first with the *CH*-acid compound and only then with the amine. While, accordingly, Schiff bases do not appear as intermediates at all, the end-products are Mannich bases in every case. This does not exclude that the Mannich bases obtained vary in dependence on the reaction conditions or are accessible by only one of the two procedures. A high *CH*-acidity facilitates the addition to imines while less acid compounds require the use of catalysts.

12.4.1 Addition of Schiff bases to dienes

Addition of Schiff bases to dienes proceeds in analogy with a Diels-Alder reaction. The *dienophilic* behaviour of imines has been little investigated to date[384], *e.g.*[385]:

Methyl 1,2,3,6-tetra-hydro-1-phenyl-2-pyridineacetate

As the electron deficiency of the $C=N$ double bond increases the reactivity of the dienophiles grows. Thus, no success is obtained in reacting N-benzylidene-4-chlorobenzenesulfonamide and N-(4-nitrobenzylidene)-p-toluenesulfonamide with 2,3-dimethylbutadiene, cyclopentadiene, and cyclohexadiene, and it requires the strong activation of the $C=N$ double bond in N-(2,2,2-trichloroethylidene)-p-toluenesulfonamide to produce an addition[386]. Boiling for several hours with 2,3-dimethylbutadiene, cyclopentadiene, or 1,3-cyclohexadiene gives respectively *1,2,3,6-tetrahydro-(1-p-tosyl)-6-(trichloromethyl)-3,4-lutidine,2-(p-tosyl)-3-(trichloromethyl)-2-azabicyclo[2.2.1]hept-5-*

[383] *O.C. Dermer, G.E. Ham*, Ethylenimine and other Aziridines, Academic Press, New York 1969.

[384] *M. Lora-Tamayo, R. Madroñero* in *J. Hamer*, 1,4-Cycloaddition Reactions, p. 140, Academic Press, New York 1967.

[385] *K. Alder* in *W. Foerst*, Neuere Methoden der präparativen organischen Chemie, Bd. I, p. 355, Verlag Chemie, Weinheim/Bergstr. 1943.

[386] *G. Kresze, R. Albrecht*, Chem. Ber. *97*, 491 (1964).

[387] *H.R. Snyder, H.A. Kornberg, J.R. Roenig*, J. Amer. Chem. Soc. *61*, 3556 (1939).

ene, or *2-(p-tosyl)-3-(trichloromethyl)-bicyclo-[2.2.2]oct-5-ene:*

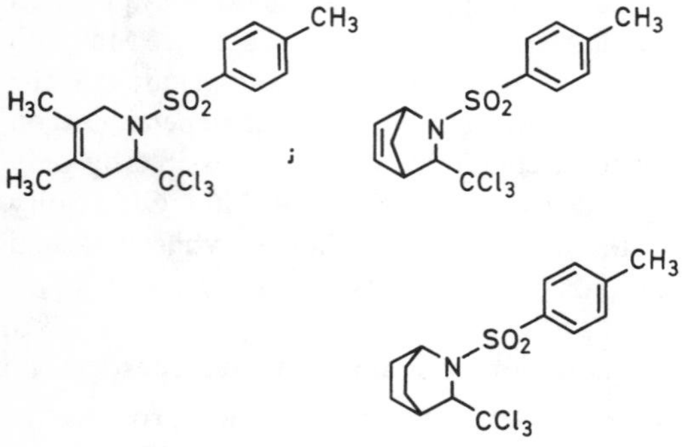

12.4.2 Addition of Schiff bases to ketones

Addition of benzylidenaniline to methyl ketones such as 2-butanone, 4-phenyl-2-butanone, or acetophenone to give respectively 1-anilino-1-phenyl-3-pentanone, 1-anilino-1,5-diphenyl-3-pentanone, and 3-anilino-1,3-diphenyl-1-propanone is a very slow process. It can be speeded up by catalysts. In the presence of boron trifluoride the reaction takes place within a few minutes[387]:

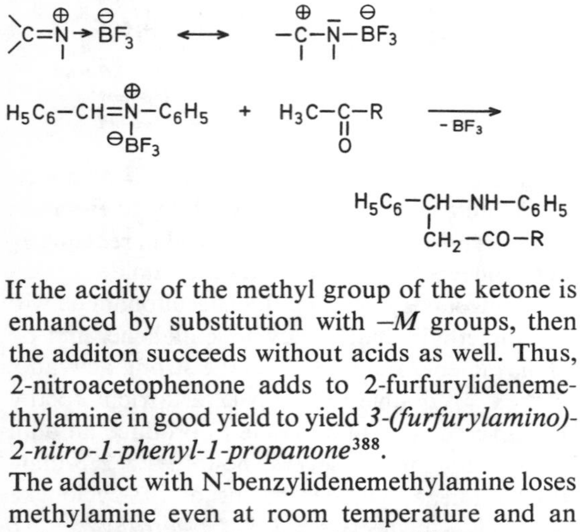

If the acidity of the methyl group of the ketone is enhanced by substitution with $-M$ groups, then the additon succeeds without acids as well. Thus, 2-nitroacetophenone adds to 2-furfurylidenemethylamine in good yield to yield *3-(furfurylamino)-2-nitro-1-phenyl-1-propanone*[388].

The adduct with N-benzylidenemethylamine loses methylamine even at room temperature and an α,β-unsaturated ketone is formed[389]. Evidently, the tendency to form a continuous conjugated system here is greater than the addition of an amine to an activated $C=C$ double bond:

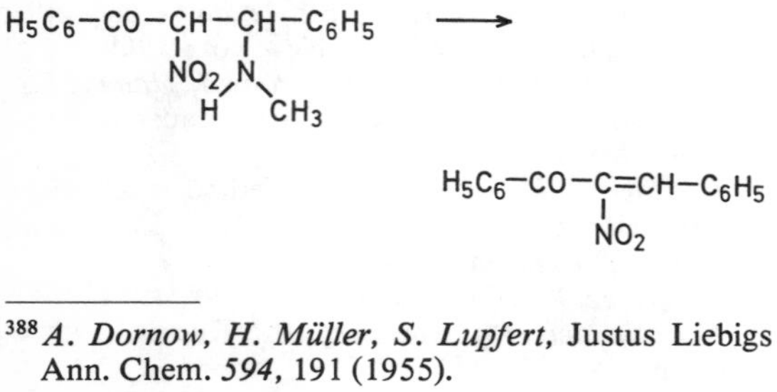

12.4.3 Addition of Schiff bases to carboxylic acid esters

β-Keto carboxylic acid esters such as acetoacetates, esters of benzoylacetic acid, and diesters of malonic acid[390-392], and also, in the presence of sodium amide, esters of 3-oxo-4-phenylbutyric acid[393, 394], add to Schiff bases in good yield:

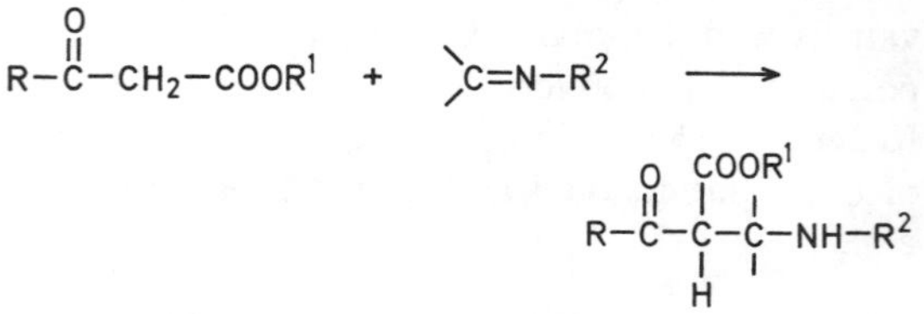

Other acetic acid derivatives with $-M$ substituents, too, undergo the reaction[395].

12.4.4 Addition of Schiff bases to nitro compounds

Addition of nitromethane to benzylidenaniline affords α-(nitromethyl)-N-phenylbenzylamine in good yield:

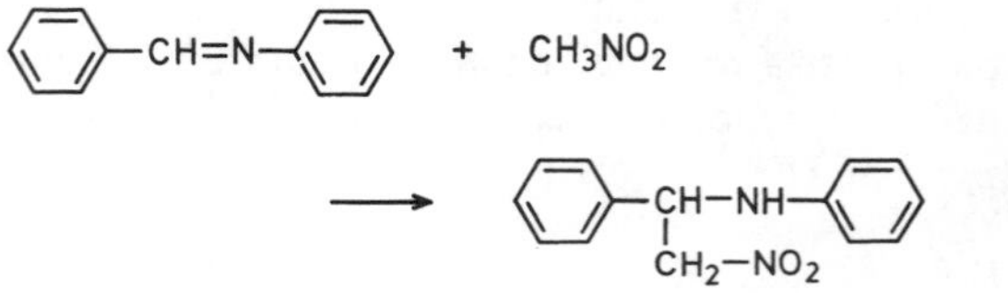

Nitroethane and 2-nitroproane react only in the presence of diethylamine[396, 397]; the products are α-(2-nitroethyl)- and α-(3-nitropropyl)-N-phenylbenzylamine, while nitroacetonitrile gives a substantial yield of *3-anilino-2-nitro-3-phenylpropionitrile*. Schiff bases with $-M$ substituents react especially rapidly; nitroacetanilide reacts instantaneously with 4-nitrobenzylidenaniline to form *3-anilino-2-nitro-3-(4-nitrophenyl)propionitrile*[398].

[388] *A. Dornow, H. Müller, S. Lupfert*, Justus Liebigs Ann. Chem. *594*, 191 (1955).

[389] *N.S. Koslov, J.A. Shur*, Zh. Obshch. Khim. *29*, 2706 (1959); C.A. *54*, 12045 (1960).

[390] *R. Schiff*, Ber. *31*, 607 (1898).

[391] *E.J. Wayne, J.B. Cohen*, J. Chem. Soc. (London) *127*, 450 (1925).

[392] *R. Albrecht, G. Kresze, B. Mlakar*, Chem. Ber. *97*, 483 (1964).

[393] *B.J. Kurtev, N.M. Mollov*, Dokl. Akad. Nauk SSSR *101*, 1069 (1955); C.A. *50*, 3416 (1956).

[394] *Al. Spasov, St. Robev*, Dokl. Akad. Nauk SSSR *95*, 559 (1954); C.A. *49*, 6181 (1955).

[395] *P. Lazzareschi*, Gazz. Chim. Ital. *67*, 371 (1937); C.A. *32*, 1668 (1938).

[396] *M. Katayanagi*, J. Pharm. Soc. Jap. *68*, 232 (1948).

[397] *N.J. Leonard, J.W. Leubner, E.H. Burk*, J. Org. Chem. *15*, 970 (1950).

12.4.5   Addition of Schiff bases to acetylene derivatives

Purely aliphatic azomethines add acetylene in the presence of copper(I) chloride to 1-alkyn-1-yl amines[399].

Addition of N-benzylidene- and N-cinnamylidene-methylamines proceeds *via* a 1,4-dipolar cycloaddition. Benzylidenemethylamine and benzylidenaniline react with two equivalents dialkyl acetylenedicarboxylate to form 1:2 adducts:

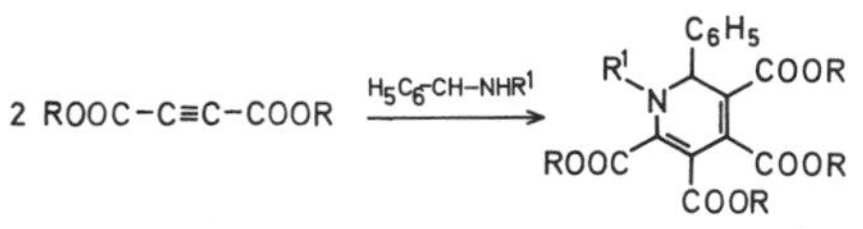

$$2 \; ROOC-C{\equiv}C-COOR \xrightarrow{\;H_5C_6-CH-NHR^1\;}$$

R¹ = CH₃;   *Tetraalkyl 5,6-dihydro-1-methyl-2,3,4,5-pyridinetetracarboxylate*

R¹ = C₆H₅–CH₂–; *Tetraalkyl 1-benzyl-5,6-dihydro-2,3,4,5-pyridinetetracarboxylate*

Such additions have not so far gained procedural value[400].

12.4.6   Addition of Schiff bases to phenols and heteroaromatic compounds

Addition of 8-quinolinol to various different Schiff bases prepared from heterocyclic amines with benzaldehyde requires reaction times of between one hour and eight days[401]:

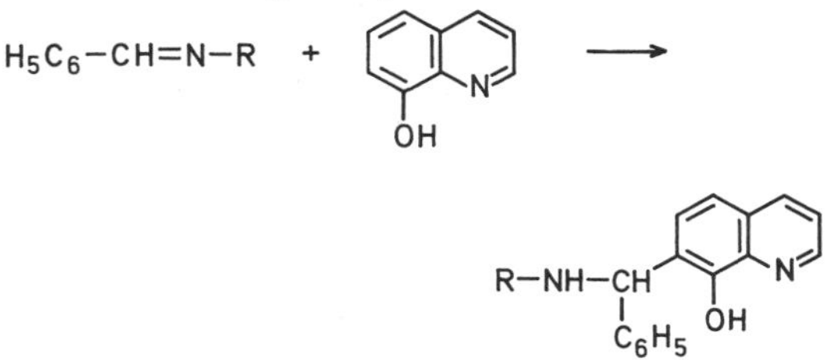

In a little benzene the addition of 2- and 3-methylindole to various Schiff bases, too, proceeds smoothly (80%)[402, 403]:

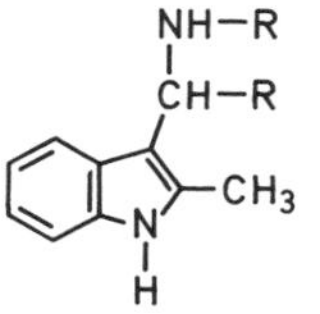

Antipyrine reacts with benzylidenaniline at room temperature during several weeks to form *4-(α-anilinobenzyl)-1,5-dimethyl-2-phenyl-3-pyrazolin-5-one*[404]:

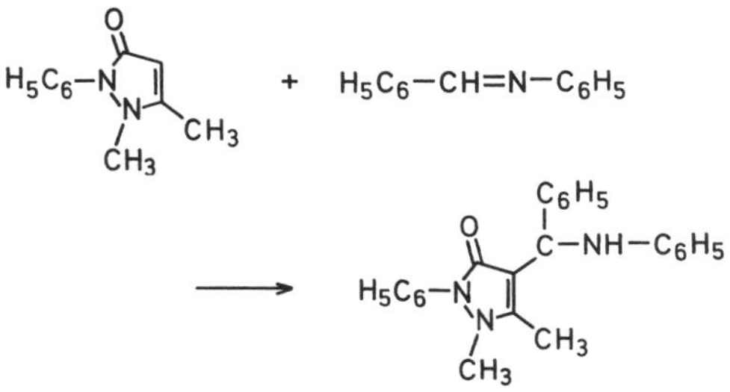

12.5   Amines by reduction

List of methods for synthesizing amines by reduction

Reaction	Method of reduction		Page			Page
Aryl–NO₂	a	H₂/metal catalysts	494	j	Mg–Cu alloy/NH₄Cl/H₂O	498
	b	H₂N–NH₂/metal catalysts	496	k	Na₂S₂O₄	498
Aryl–NH₂	c	Cyclohexene/Pd	497	l	Sulfide	499
Aryl–NO	d	CO/Cu catalysts	497	m	LiAlH₄	499
	e	Isopropyl alcohol/metal catalysts	497	n	NaBH₄/metal catalysts	500
	f	Fe or Fe²⊕/H⊕ donors	497	o	(C₆H₅)₂SnH₂	501
	g	Sn or Sn²⊕/HCl	498	p	B₂H₆	501
	h	Zn/H⊕ donors	498	q	Photochemical reduction	501
	i	Al amalgam/H₂O	498	r	Electrochemical reduction	501

Side reactions: During reduction in strong acid aminophenol formation by virtue of rearrangements (p. 495, 498, 501).

Using *l*, thiophenol formation from aromatic dinitro or halogenonitro compounds is possible (p. 499).

Using *m*, amines are formed only from sterically hindered compounds (p. 499); as a rule azo compounds are formed (p. 499).

[398] *W. Ried, E. Köhler*, Justus Liebigs Ann. Chem. *598*, 145 (1956).

[399] US. P. 2 665 311 (1950), Röhm & Haas Comp., Inv.: *C.H. McKeever, M.F. Fegley*; C.A. *49*, 5516c (1955).

12.5   Amines by reduction

List of methods for synthesizing amines by reduction

Reaction	Method of reduction

Selectivity: Using *p,* only the nitroso group but not the nitro group is reduced (p. 501).
 Partial reduction occurs in polynitro compounds with *l* (p. 499).
 Retention of chlorine and bromine substituents is possible using *a* with Raney nickel or platinum (p. 494),
 using *b* with Raney nickel, platinum, and ruthenium (p. 494), using *d* (p. 497), using *f* (p. 498), and
 using *l* (p. 499); of iodine substituents using *b* with ruthenium (p. 497) and using *f* under alkaline
 conditions (p. 498).
 Dehalogenation occurs frequently with palladium using *a* and *b* (p. 494, 497).
 Retention of *C=C* bonds is possible using *a* with Raney nickel (p. 495), using *b* (p. 497), using *d*
 (p. 497), and using *f* (p. 498).
 Nitrile groups are retained using *a* (p. 495), *f* (p. 498), *g* (p. 498).
 Carbonyl groups are retained using *b* (p. 497), *d* (p. 497), *f* (p. 498), *o* (p. 501) and may be retained
 using *a* (p. 495).

Alkyl−NO₂ ⟶ Alkyl −NH₂

		Page			Page
a	H₂/metal catalysts	495	*d*	Al amalgam/H₂O	498
b	Fe/acid	498	*e*	LiAlH₄	499
c	Zn/acid	498	*f*	NaBH₄/metal catalysts	501

Side reactions: Using *a,* a structure-dependent cleaving of β-hydroxynitro compounds is possible (p. 496).
 Using *e,* cleavage or rearrangements are possible in nitro compounds containing a tertiary *a−C* (p. 500).

Selectivity: In *α, β*-unsaturated nitro compounds: using *e,* the *C=C* bond is coreduced (p. 500), using *a,*
 hydrogenation to oxime or saturated amine occurs in dependence on the conditions (p. 496).
 Using *e,* carbonyl groups are reduced before the nitro group (p. 500).
 Using *a* (p. 496) and *b* (p. 498), the configuration on the *a*-carbon is retained, using *e* racemization
 occurs (p. 500).

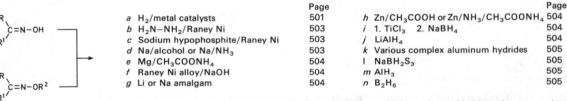

		Page			Page
a	H₂/metal catalysts	501	*h*	Zn/CH₃COOH or Zn/NH₃/CH₃COONH₄	504
b	H₂N−NH₂/Raney Ni	503	*i*	1. TiCl₃ 2. NaBH₄	504
c	Sodium hypophosphite/Raney Ni	503	*j*	LiAlH₄	504
d	Na/alcohol or Na/NH₃	503	*k*	Various complex aluminum hydrides	505
e	Mg/CH₃COONH₄	504	*l*	NaBH₂S₃	505
f	Raney Ni alloy/NaOH	504	*m*	AlH₃	505
g	Li or Na amalgam	504	*n*	B₂H₆	505

Side reactions: Using a secondary amine formation occurs which can be avoided by using appropriate
 conditions (p. 502). Using *j* secondary amines may be formed by rearrangement (p. 506) or aziridines
 may form (p. 507) in dependence on structure and conditions.

Selectivity: Using *j* the carbonyl group is reduced before the oxime group (p. 504, 506); using *a* allows
 reduction to amino ketones (p. 502, 505).
 Using *j C=C* bonds can be preserved (p. 505), using *g* the nitrile group can be preserved (p. 504).
 Using *n* nitro groups remain intact (p. 505).
 Using *n* the reaction can be guided to lead to hydroxylamine (p. 505).
 The stereoselectivity is dependent on the structure and the reducing agent (p. 503, 505).

R¹−N→O (with R and R² substituents) ⟶ R¹−N (with R and R² substituents)

		Page			Page
a	H₂/metal catalysts	508	*e*	Na₂S₂O₄ or SO₂/H₂O	508
b	Fe/CH₃COOH	508	*f*	PCl₃ or P(C₆H₅)₃	508, 509
c	Zn/H⊕ donors	508	*g*	Alkyl halides/NaOH	509
d	TiCl₃	508	*h*	NaBr	509

Selectivity: Aliphatic *N*-oxides are deoxygenated under milder conditions than aromatic ones (p. 509).
 Using *f* and *g* nitro groups remain intact (p. 508), using *a−e* they are coreduced (p. 508).

R−N=N−R¹ ⟶ R−NH₂ + R¹−NH₂

		Page			Page
a	H₂/metal catalysts	509	*c*	Na₂S₂O₄	509
b	H₂N−NH₂/Pd−C	509	*d*	Metals/H⊕ donors	510

		Page			Page
a	H₂/metal catalysts	510	*c*	Zn/CH₃COOH	510
b	H₂N−NH₂/Raney Ni	510			

Selectivity: Using *b N,N*-diacylhydrazines are cleaved to carboxamides (p. 510), with lithium tetrahydroaluminate reduction to primary amine is possible (p. 510).
 Tosylhydrazones are reduced to alkane with lithium tetrahydroaluminate (p. 510).

List of methods for synthesizing amines by reduction

Reaction	Method of reduction

R−C(⊖)(R¹)−N₂⊕ ⟶ R−CH(R¹)−NH₂

	Page		Page
a H₂/metal catalysts	511	c LiAlH₄	511
b Metal/H⊕ donors	511		

Side-reactions: Splitting off of nitrogen is possible (p. 511).

R−N₃ ⟶ R−NH₂

	Page		Page
a H₂/metal catalysts	511	d NaBH₄ in boiling isopropyl alcohol or	511
b H₂N−NH₂/Raney Ni	511	NaBH₄/cobalt catalysts	511
c LiAlH₄	511	e B₂H₆	511
		f 1. P(C₆H₅)₃ 2. H₂O/H⊕	

Remarks: The reaction proceeds with configuration retention on the α-carbon (p. 512).
Aromatic amines can be prepared with Raney Ni alloy *via* sulfonyltriazenes (p. 512).

R−C≡N ⟶ R−CH₂−NH₂

	Page		Page
a H₂/metal catalysts	512	f AlH₃	517
b Na/alcohol	514	g B₂H₆	517
c Raney Ni alloy/NaOH	514	h NaBH₄/Raney Ni	517
d LiAlH₄	514	i NaBH₄/transition metal salts	517
e Various complex aluminum hydrides	517		

Side-reactions: Using *a* secondary or tertiary amine formation is possible (p. 512); methods for avoiding this reaction p. 512.
Using *a* alkane formation is possible with $Pd-C$ and elevated temperature (p. 512).
Methods for avoiding ring closure using *a* on p. 513, cleavage in α-aminonitriles and cyanohydrins using *a* on p. 513–514.
Using *b* reductive hydrogen cyanide cleavage is observed with aromatic nitriles (p. 513–514).
Using *d* a structure-dependent cleavage of α-aminonitriles occurs (p. 516); cleavage is possible in β-alkoxynitriles in dependence on the conditions (p. 515).
Using *d* aziridines are formed from a-chloronitriles (p. 516).

Selectivity: Using *a* (p. 513) and *d* (p. 514) coreduces conjugated $C=C$ bonds, isolated $C=C$ bonds may remain intact (p. 513, 515). Using *f* (p. 517) and *i* (p. 517) $C=C$ bonds may be preserved.
Using *a* may coreduce pyridine rings to piperidine rings (p. 513).
Using *d* coreduces keto groups (p. 515).
Using *g* leaves nitro groups on aromatic compounds intact (p. 516); the same can be achieved using *i* (p. 516). Using *a* leads to coreduction (p. 512).
Using *Pd* in *a* leads to reductive dehalogenation, with other catalysts halogen substituents are preserved (p. 512).

Remarks: The synthesis of secondary amines by reduction with sodium tetrahydroborate succeeds via trialkyloxonium or dialkoxycarbonium fluoroborates (pp. 517, 518).

R−C(=O)−N(R¹)(R²) ⟶ R−CH₂−N(R¹)(R²)

	Page		Page
a H₂/copper chromite catalysts	518	f NaBH₄/transition metal salts	522
b Na/alcohol	518	g 1. (C₂H₅)₃O⊕BF₄⊖ 2. NaBH₄	522
c LiAlH₄	518	h AlH₃	522
d Various complex aluminum hydrides	521	i B₂H₆	523
e NaBH₄ in pyridine	522		

Remarks: Only methods *c–i* are of outstanding importance.
Reductions with hydrogen/Raney nickel are possible following conversion into thioamides (p. 518).
Using *c* for reducing carboxylic acid hydroazides see p. 521, using *i* see p. 523.

Side-reactions: Using *a* (p. 518) many side reactions possible.
Using *b* formation of carboxylic acids and alcohols needs to be allowed for (p. 518).
Using *c* a structure-dependent $C-N$ cleavage is possible (p. 519, 520).
Using *e* (p. 522) and *g* (p. 522) nitriles are formed from N-unsubstituted amides.

Selectivity: Using *c* (p. 521) and *h* (p. 523) the $C=C$ bond can be preserved.
Using *g* (p. 522) and *i* (p. 523) nitro groups on aromatic compounds remain unaltered.
Using *c* (p. 521) a structure- and condition-dependent C-halogen hydrogenolysis is possible.
Especially using *c*, carbonyl groups in esters, ketones, or aldehydes are reduced before the amide group (p. 521); using *i* (p. 523) ester and urethane groups are preserved.
Using *h* (p. 522) successfully reduces sterically hindered amides.
Using *c* partial reductions can be performed in cyclic imides (p. 520) and in lactams (p. 520).

List of methods for synthesizing amines by reduction

Reaction	Method or reduction	
		Page

R−N=C=O ⎤
    ⎬→
R−N=C=S ⎦
    R−NH−CH₃
	a LiAlH₄	523

	Selectivity: Reductive dehalogenation on the *a*-carbon (p. 524). Preservation of *C = C* bonds (p. 524).	

(Reaction scheme: $R-N=C=O$ and $R-N=C=S$ → $R-NH-CH_3$)

12.5.1 Amines from nitro and nitroso compounds

12.5.1.1 Catalytic hydrogenation with molecular hydrogen

Preparation of amines by reduction of the nitro group is a very frequently used technique, and above all with *aromatic* nitro compounds. Accordingly, a number of surveys are available[405-410]. Intermediate nitroso and hydroxylamine stages are involved; thus, firstly, these intermediate stages can lead to side-product formation, for instance, azo compounds, and, secondly, nitroso and hydroxylamine compounds can be reduced to amines by the same fundamental methods used for reducing the nitro group.

Catalytic hydrogenation of nitro compounds may be fundamentally carried out by two methods, firstly, with molecular hydrogen, secondly with *bound* hydrogen, *e.g.*, hydrazine or cyclohexene as hydrogen donor (see p. 496). A large number of *catalysts* have been developed for hydrogenation with molecular hydrogen, notably for continuous technical-scale processes. For discontinuous

working Raney nickel, platinum, and palladium (generally on carriers such as carbon or barium sulfate), and platinum(IV) oxide have proven successful. Hydrogenation in the laboratory is best carried out at normal or slight excess hydrogen pressure (0.5−5 atm), while high-pressure syntheses are restricted to technical processes. The reaction is exothermic and, if heat is allowed to build up, may proceed **explosively.** This risk can be avoided by using low reaction temperatures and low hydrogen pressures and, above all, by using a *solvent.* These measures are also the reaction conditions which substantially prevent side-product formation. Water, dioxane, tetrahydrofuran, or dimethylformamide may be used as solvent, but most reactions proceed smoothly in lower alcohols or glacial acetic acid[406, 409].

Catalytic hydrogenation of *aromatic* nitro compounds generally entails few difficulties. Raney nickel, palladium charcoal, and platinum(IV) oxide are all equally suitable *catalysts*[407, 409]. A survey[409] lists the catalysts and hydrogenation conditions employed for various heterocyclic nitro compounds. Aromatic amines can be prepared also from nitroso compounds using hydrogenation techniques analogous to those for reducing nitro groups[405, 409, 410]. Where the aromatic nitro group contains further reducible groups, the reaction conditions decide whether these are cohydrogenated. As the nitro group is generally more easily reduced catalytically than other groups[407, 409], its *selective* reduction is usually possible by maintaining definite working conditions. Some examples will be named in what follows; the reader is referred also to the detailed literature surveys[405, 409].

Halogen substituents on aromatic compounds are more easily cleaved as the atomic weight increases; fluorine is especially stable towards hydrogenolysis. Reduction over Raney nickel under mild conditions converts aromatic chloro- and bromonitro compounds into the corresponding haloamines[405, 408, 411, 412], but in the presence of alkali metal hydroxides and at elevated temperature hydrogenolysis of the halogen substituent is observed even with Raney nickel[413].

[400] *R. Huisgen, K. Herbig,* Justus Liebigs Ann. Chem. *688*, 105 (1965).

[401] *J.P. Phillips, R.W. Keown, Q. Fernando,* J. Org. Chem. *19*, 907 (1954).

[402] *M. Passerini, T. Bonciani,* Gazz. Chim. Ital. *63*, 138 (1933).

[403] *M. Passerini, F. Albani,* Gazz. Chim. Ital. *65*, 933 (1935).

[404] *M. Passerini, G. Ragni,* Gazz. Chim. Ital. *66*, 684 (1936).

[405] *R. Schröter* in *Houben-Weyl*, Methoden der organischen Chemie, Bd. XI/1, p. 341, Georg Thieme Verlag, Stuttgart 1957.

[406] *F. Zymalkowski*, Katalyt. Hydrierungen im Organisch-chemischen Laboratorium, Ferdinand Enke Verlag, Stuttgart 1965.

[407] *R.L. Augustine*, Catalytic Hydrogenation, Marcel Dekker, New York 1965.

[408] *F. Weygand, G. Hilgetag,* Organisch-chemische Experimentierkunst, J.A. Barth, Leipzig 1970.

[409] *M. Freifelder*, Practical Catalytic Hydrogenation, Wiley Intersci. Publ., New York, London, Sydney, Toronto 1971.

[410] *P.N. Rylander*, Catalytic Hydrogenation over Platinum Metals, Academic Press, New York 1967.

By contrast, untreated palladium contact catalysts have a fairly specific dehalogenating action[405, 414]. Modified palladium catalysts with added copper have, however, been developed which allow the nitro group to be reduced while preserving the halogen substituents[415]. On hydrogenation over platinum contact catalysts under mild reaction conditions a retention of the halogen was observed[406]. Platinum catalysts have been developed which give a selective reduction also under elevated pressure and temperature. These include platinum(IV)-charcoal modified by the addition of morpholine[416] or triethyl phosphite[417] and catalysts based on platinum(IV) sulfide[418, 419]. Detailed investigations about the conditions which avoid dehalogenation are available[409, 416].

Halogen substituents in polynitro compounds, especially if *o*- and *p*- to a nitro group, are particularly sensitive to hydrogenolysis. In alkaline solution such halogens are easily replaced by hydroxy, and in aromatic halomethyl nitro compounds the halogen is easily cleaved hydrogenolytically[405].

During hydrogenation over the usual metal catalysts[406, 409, 410, 420, 421] it is possible to reduce a nitro group before a reductive attack on the *nitrile group* ensues. With nitro and nitrile groups in the *o*-position it was found[422] that *2-aminobenzamide* is formed *via* the 3-aminobenzisoxazole stage:

2-Aminobenzonitriles were obtained with palladium/barium sulfate in dioxane[422].

While *carbonyl groups* in *esters* and *amides* are not reduced under the conditions usual for catalytic reduction of the nitro group, reduction of *acyl* to alkane during hydrogenation of nitroacetophenones[405, 423] and nitrobenzophenones[405] over palladium black and palladized carbon was observed; ethylanilines and benzylanilines respectively were formed. As a rule, however, selective reduction of the nitro group with preserving of the carbonyl function is possible[409]. An example is the synthesis of *aminobenzaldehydes* by hydrogenation over palladium/barium sulfate[424].

In compounds containing a *C=C* bond conjugate with respect to the aromatic part, the nitro group can be reduced to amine over Raney nickel under mild conditions before the *C=C* bond is hydrogenated[405, 409, 425]. Cf. p. 508 concerning the course of the reduction during catalytic hydrogenation of compounds containing an *N-oxide group* in addition to the nitro group. It is not easy to obtain a selective reduction of a nitro group in *polynitro* compounds catalytically[405, 409], and chemical methods have proven themselves for this purpose (p. 499).

A general point to be observed during hydrogenation of aromatic nitro compounds is that in strongly acid solution *4-aminophenols* may be formed by rearrangement of the phenylhydroxylamine stage[406, 426]. Analogously, *4-fluoroaniline* is formed during the reduction of nitrobenzene over platinum(IV) oxide in the presence of hydrogen fluoride[407].

In principle *aliphatic* nitro compounds can be hydrogenated catalytically to primary amines using the same methods as in the reduction of aromatic ones, but the possibility of a number of *side-reactions* occuring needs to be allowed for. Thus, reaction products were identified repeatedly[405, 406, 409] whose formation can be explained by a reaction *via* an intermediate oxime stage, including carbonyl compounds formed by hydrolysis in acid medium and carboxamides formed by Beckmann

[411] *N. Löfgren*, Acta Chem. Scand. *9*, 1079 (1955).

[412] *C. Temple, R.D. Elliot, J.D. Rose, J.A. Montgomery*, J. Heterocycl. Chem. *7*, 451 (1970).

[413] *A.R. Surrey, R.A. Cutler*, J. Amer. Chem. Soc. *73*, 2413 (1951).

[414] *P. Tomasik, E. Plazek*, Roczniki Chem. *39*, 1671 (1965); C.A. *64*, 19549g (1966).

[415] *P.V. Mokrousov, A.I. Naumov*, Katal. Reacts. Zhidk. Faze, Tr. Vses. Konf., 2nd, Alma-Ata, Kaz. SSR 150 (1966) (Publ. 1967); C.A. *69*, 26891 (1968).

[416] *J.R. Kosak*, Ann. N.Y. Acad. Sci. *172*, 175 (1970).

[417] Brit. P. 1 138 567 (1969), GAF Corp., Inv.: *W.C. Craig, G.J. Davis, P.O. Shull*; C.A. *70*, 87253 (1969).

[418] *F.S. Dovell, H. Greenfield*, J. Amer. Chem. Soc. *87*, 2767 (1965).

[419] Neth. Appl. 6 409 250 (1965), United States Rubber Co.; C.A. *63*, 11428b (1965).

[420] *D. Heyl, E. Lutz, S.A. Harris*, J. Amer. Chem. Soc. *78*, 4474 (1956).

[421] *E.C. Taylor, A. McKillop*, Advan. Org. Chem. *7*, 196 (1970).

[422] *H. Musso, H. Schröder*, Chem. Ber. *98*, 1562 (1965); *H. Schröder, U. Schwabe, H. Musso*, Chem. Ber. *98*, 2556 (1965).

[423] *H. Oelschläger, P. Schmersahl*, Arch. Pharm. (Weinheim, Ger.) *296*, 324 (1963).

[424] *W. Borsche, F. Sell*, Chem. Ber. *83*, 78 (1950).

[425] *H. Plieninger, M. Schach von Wittenau*, Chem. Ber. *91*, 1905 (1958).

[426] *P.N. Rylander, I.M. Karpenko, G.R. Pond*, Ann. N.Y. Acad. Sci. *172*, 266 (1970).

rearrangement of the oxime stage. Formation of secondary amines (cf. p. 501) can be suppressed by avoiding elevated temperatures.

During hydrogenation of nitro compounds with a vicinal hydroxy group, as are formed during addition of nitroalkanes to carbonyl compounds cleavage, induced by the base formed during the reaction, was found to be a side-reaction[406],[427]:

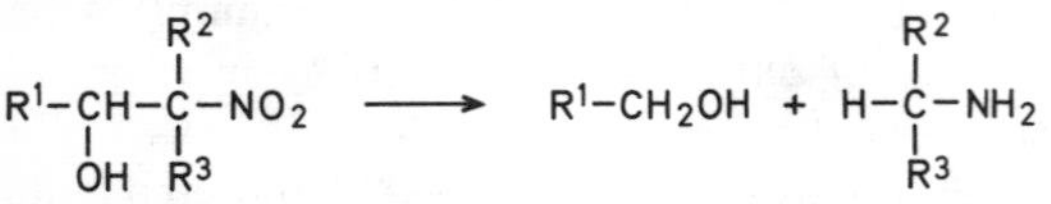

$$R^1-\underset{\underset{OH}{|}}{CH}-\underset{\underset{R^3}{|}}{\overset{\overset{R^2}{|}}{C}}-NO_2 \longrightarrow R^1-CH_2OH + H-\underset{\underset{R^3}{|}}{\overset{\overset{R^2}{|}}{C}}-NH_2$$

This fission can be avoided in simple manner by addition of acid in an amount at least equivalent to amine formed. An example is the synthesis of *1-(aminomethyl)cyclohexanol*, which proceeds smoothly in the presence of glacial acetic acid with hydrogen over Raney nickel[428]; it represents a suitable method of synthesizing *β-hydroxy-amines*:

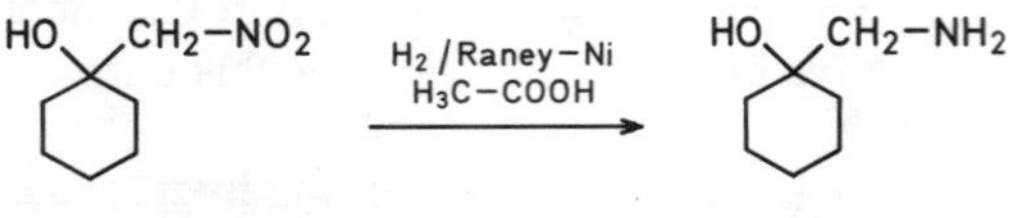

The analogous cleavage can occur during hydrogenation of 1-nitro-2-alkylamines[405],[406]. It was suppressed by using very active catalysts and working under elevated hydrogen pressure while avoiding elevated temperature[406]. A special feature is observed during reduction of *α,β-unsaturated* nitro compounds. Here, under suitable conditions, *e.g.*, when palladium-charcoal is employed in the presence of pyridine, the reaction can be halted at the oxime stage[406],[409],[429],[430], which can be hydrogenated to amine in a separate reaction (p. 501). A number of *2-phenethylamines* have been prepared by this procedure[429],[430]:

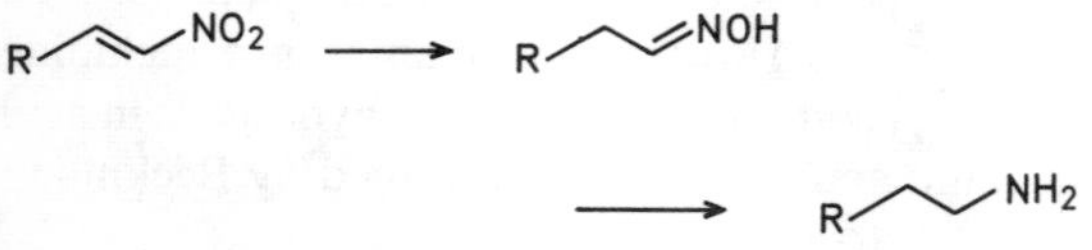

However, hydrogenation of the unsaturated nitro compound to the saturated amine can be per-

formed in one step[405],[409],[431]. Addition of mineral acid to palladium catalysts in glacial acetic acid has been found to be advantageous here[406].

As investigations[432-435] on the stereochemical course of the reduction of aliphatic nitro compounds have shown, hydrogenation to the amine is possible while *preserving* the configuration of the alpha C-atom, provided that conditions are maintained which exclude a preliminary isomerization due to the *CH*-acidity of the nitro compound. Such an isomerization is promoted by bases or elevated temperature. Thus, the configuration remained intact during hydrogenation of *cis*-1-nitro-2-phenylcyclohexane[433] (to *2-phenylcyclohexyl-amine*) over W2 Raney nickel in cold ethanol, of 2-nitrooctane[432] (to *1-methylheptylamine*), and of the *cis* and *trans* isomers of 1,3- and 1,4-dinitro-cyclohexane (to *1,3-* and *1,4-cyclohexanedi-amines*[434]) over platinum(IV) oxide in glacial acetic acid:

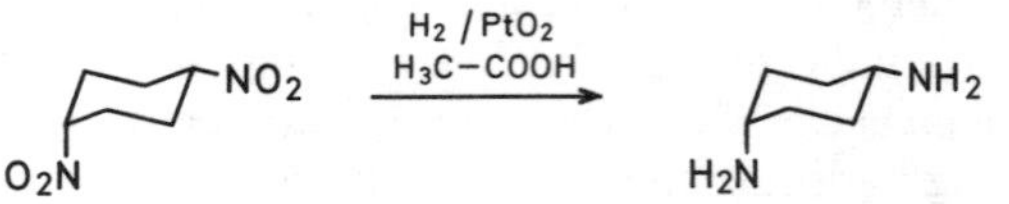

This result speaks against a reduction *via* an oxime stage in these experiments. Hydrogenations of this type proceeding with retention of configuration have gained particular importance for synthesizing *amino sugars* and *aminocyclanols* from nitro compounds[435],[436].

12.5.1.2 Catalytic reduction with miscellaneous hydrogen donors

A number of different techniques have been worked out for catalytically reducing nitro compounds using *bound* hydrogen which in the majority of cases are simpler to carry out than a reduction with molecular hydrogen (see p. 494).

[427] *F. Zymalkowski,* Arch. Pharm. (Weinheim, Ger.) *289,* 52 (1956);
F. Zymalkowski, Arch. Pharm. (Weinheim, Ger.) *291,* 21 (1958).
[428] *H.J. Dauben, H.J. Ringold, R.H. Wade, A.G. Anderson,* J. Amer. Chem. Soc. *73,* 2359 (1951);
H.J. Dauben, Org. Synth. *34,* 19 (1954).
[429] *B. Reichert, W. Koch,* Arch. Pharm. *273,* 265 (1935).
[430] *A. Lindenmann,* Helv. Chim. Acta *32,* 69 (1949).

[431] US.P. 3 458 576 (1969), Great Lakes Carbon Corp., Inv.: *L.A. Bryan;* C.A. *71,* 91049 (1969).
[432] *N. Kornblum, L. Fishbein, R.A. Smiley,* J. Amer. Chem. Soc. *77,* 6261 (1955);
N. Kornblum, L. Fishbein, J. Amer. Chem. Soc. *77,* 6266 (1955).
[433] *F.G. Bordwell, R.L. Arnold,* J. Org. Chem. *27,* 4426 (1962).
[434] *A.T. Nielsen,* J. Org. Chem. *27,* 1998 (1962).
[435] *H.H. Baer,* Advan. Carbohydr. Chem. and Biochem. *24,* 67 (1969).
[436] *F.W. Lichtenthaler* in *W. Foerst,* Neuere Methoden der präparativen organischen Chemie, Bd. IV, p. 140, Verlag Chemie, Weinheim/Bergstr. 1966;
F.W. Lichtenthaler, Fortschr. Chem. Forsch. *14,* 556 (1970).

Hitherto, their importance has been restricted mainly to the reduction of aromatic nitro compounds.

Their widest application consists in the reduction with *hydrazine* as hydrogen donor in the presence of the usual hydrogenation catalysts, *viz.*, Raney nickel, platinum, and palladium (for a review see ref. [437]). The technique is versatile in application and can be performed under mild conditions throughout[437-441]. Generally, the nitro compound and excess hydrizine hydrate are dissolved in methanol or ethanol, and the reaction proceeding with liberation of heat and nitrogen is started by adding the catalyst. Reduction of aromatic nitroso compounds or *N*-arylhydroxylamines likewise proceeds to amine under these conditions[437]. During catalytic reduction with hydrazine *aromatic nitro groups* are generally reduced *selectively* without an attack on carbonyl groups or nonactivated *C*=*C* bonds[406, 437]. Whether a *halogen substituent* on the aromatic compound is split off hydrogenolytically depends on the catalyst in this reaction. Thus, when Raney nickel is employed, fluorine, chlorine, and bromine substituents are observed to remain intact[437, 442], and under special conditions[442] or with ruthenium catalysts[443] even iodine substituents, *e.g.*:

2-Iodoaniline

By contrast, with palladium-carbon a reductive dehalogenation of the corresponding aromatic compounds is often found and can be utilized preparatively[444].

A further possibility for *transfer hydrogenations* of aromatic nitro compounds is offered by making use of *cyclohexene* as the hydrogen donor[445, 446]. Palladium catalysts are superior to platinum and Raney nickel in this method, but in certain compounds[445] the reduction was difficult to bring about even over palladium.

Another method still[447] for selectively reducing nitro groups on aromatic compounds without modifying carbonyl groups, olefinic double bonds, and chlorine substituents uses *complexes* prepared from *copper salts* and *amines* such as ethylenediamine as catalysts. *Carbon monoxide* is the reducing agent, the reaction proceeds under mild conditions in aqueous medium. The same method is apt for the reduction of nitromethane.

A reduction process in the gas phase (at ~260°) with isopropyl alcohol as hydrogen donor, using copper on alumina or vanadium compounds as catalysts, converts nitrobenzene into *aniline* almost quantitatively[448].

12.5.1.3 Reduction with metals and metal salts

There are many long-established methods of synthesizing amines among those which reduce the nitro group by means of chemical reducing agents.

In the laboratory particularly their importance has been diminished somewhat by the development of catalytic hydrogenation procedures and reduction with metal hydrides. For a detailed survey of the numerous techniques see ref. [405]. Here, the discussion will be limited to reactions which are particularly easy to carry out or which are highly selective.

The usual chemical reducing agents can be divided into two groups. *Sulfur* compounds (p. 498) and metals they all act as electron donors. Hydrogenation thus requires the presence of proton donors; as a rule these are the protic solvents. For the mechanism of the reaction see ref. [449].

Iron has remained very important as a reducing agent for aromatic mononitro compounds. Proven methods of carrying out the reaction are warming of the nitro compound with iron filings in acetic acid or in aqueous or aqueous alcoholic hydrochloric acid[405, 450]. Processes which drasti-

[437] *A. Furst, R.C. Berlo, S. Hooton*, Chem. Rev. *65*, 51 (1965).

[438] *P.M.G. Bavin*, Org. Synth. *40*, 5 (1960).

[439] *T.F. Stepanova, O.F. Ginzburg, E.V. Levina*, Zh. Prikl. Khim. (Leningrad) *42*, 2390 (1969); C.A. *72*, 31343 (1970).

[440] *C. Alberti, C. Tironi*, Farmaco, Ed. Sci. *26*, 66 (1971).

[441] *G. Defaye, M. Fétizon*, Bull. Soc. Chim. France 1632 (1969).

[442] *B.E. Leggetter, R.K. Brown*, Can. J. Chem. *38*, 2363 (1960).

[443] *S. Pietra*, Ann. Chim. (Rome) *52*, 727 (1962).

[444] *W.L. Mosby*, J. Org. Chem. *24*, 421 (1959); *W.L. Mosby*, Chem. & Ind. (London) 1348 (1959).

[445] *E.A. Braude, R.P. Linstead, K.H.R. Woolridge*, J. Chem. Soc. (London) 3586 (1954).

[446] *R.A. Clendinning, W.H. Rauscher*, J. Org. Chem. *26*, 2963 (1961).

[447] *H.R. Appell*, J. Org. Chem. *32*, 2021 (1967).

[448] *N.S. Kozlov, M.N. Tovshtein*, Zh. Org. Khim. *3*, 138 (1967); C.A. *66*, 94744 (1967); *N.S. Kozlov, E.S. Osinovik, A.F. Yanchuk*, Vestsi Akad. Nauk Belorus. SSR, Ser. Khim. Nauk 101 (1970); C.A. *72*, 121111 (1970).

[449] *H.O. House*, Modern Synthetic Reactions, p. 73, W.A. Benjamin, New York 1965.

[450] *S.A. Mahood, P.V.L. Schaffner*, Org. Syntheses, Coll. Vol. II, 160 (1943).

cally limit the consumption of acid by cutting it down to catalytic proportions are mainly of technical importance[405]. The reduction with iron can be performed also in neutral solution if electrolytes such as alkali metal and alkaline earth metal chlorides are added. As reduction in alkine medium is possible using iron(II) salts as reducing agents. Generally speaking, reduction of the nitro group with iron displays *high selectivity*. An example is the following conversion[451]:

Methyl 5-amino-2-chloro-3-cyano-6-methylisonicotinate

In general, carbonyl and nitrile groups, olefinic *C=C* bonds, and halogen substituents on aromatic compounds are not altered on reducing the nitro group in acid media. Nitro compounds with acid-sensitive groups are suitably reduced in neutral or alkaline media. Reaction with iron(II) ions in the alkaline region has given particularly good results in the preparation of *aminoarylarsonic acids, aromatic iodoamino* compounds, *aminobenzaldehydes*[452], and *aminocinnamic* acids[405]. *Aminophenylarsonic* acids can be prepared alternatively in simple manner by reducing the nitro compound with iron filings in boiling sodium chloride solution[453].

Aliphatic nitro compounds can be reduced to amines with iron provided acid at least equivalent to the amine is used; the process has often been employed to synthesize *β-aminoalcohols*[405]. Reduction of aliphatic nitro compounds with iron in acetic acid proceeds with configuration retention; this method is thus suitable for *stereoselective* amine syntheses[432, 441, 454].

Tin in strong hydrochloric acid solution has been very frequently used for reducing mainly aromatic and heteroaromatic nitro compounds[405]. Using this technique gives very good yields of, *e.g., perfluoroaminophenols*[455]. By using *tin(II) chloride*, also in acid solution, it is often possible to vary the reaction conditions. A number of side-reactions have been described due partly to the reducing agent itself and partly to the strongly acid conditions[405, 408]; but no generally valid rules can be stated. The possible appearance of chlorinated substitution products[456] and splitting off of a carboxyl group are particularly noteworthy. For example, during the reaction of 4-nitrosalicylic acid with tin and hydrochloric acid *3-aminophenol* is formed, while with tin(II) chloride *4-aminosalicylic acid* can be obtained[457]. The tin(II) chloride reduction of the nitro group has often been used for preparing *aromatic aminonitriles*[421].

Zinc has in many cases been found suitable for preparing aromatic and aliphatic amins by acting as a reducing agent for nitro groups in acidic medium. In particularly strong acidic solution rearrangement to *aminophenols* and ring chlorination has been reported[405]. A mild reduction of aromatic nitro compounds can be carrried out with zinc dust in boiling water or aqueous ethanol in the presence of ammonium chloride or calcium chloride[458]. *β*-Nitroalcohols reduce with zinc in hydrochloric acid (to *β*-amino alcohols) without fission if elevated temperatures are avoided; the same method has been successfully used for reducing nitrodiols (to form *aminodiols*)[459]. Reduction of aromatic and aliphatic nitro compounds with *amalgamated aluminum* and water proceeds selectively except that carbonyl groups can be coreduced. This mild technique is particularly useful for reducing alkali- or acid-labile compounds, *e.g.,* Mannich bases[405]. A *magnesium-copper* alloy has been successfully employed for reducing various different aromatic nitro compounds in water containing ammonium chloride[460].

12.5.1.4 Reduction with sulfur compounds

Sodium dithionite and a variety of sulfides are suitable sulfur compounds particularly for reducing aromatic nitro compounds. Reduction with *sodium dithionite* is generally performed in aqueous, neutral to weakly alkaline medium and occasionally in pyridine or formamide. In the presence of other reducible groups a *selective* reduction of the nitro or nitroso group can be achieved. During

[451] *R.K. Blackwood, G.B. Hess, C.E. Larrabee, F.J. Pilgrim*, J. Amer. Chem. Soc. *80*, 6244 (1958).

[452] *L.I. Smith, J.W. Opie*, Org. Synth. *28*, 11 (1948).

[453] *F.G. Mann, A.J. Wilkinson*, J. Chem. Soc. (London) 3336 (1957).

[454] *E.H. Massey, H.E. Smith, A.W. Gordon*, J. Org. Chem. *31*, 684 (1966).

[455] *J.M. Birchall, R.N. Haszeldine, J. Nikokavouras, E.S. Wilks*, J. Chem. Soc. C 562 (1971).

[456] *C.C. Price, D.D. Guthrie*, J. Amer. Chem. Soc. *68*, 1592 (1946).

[457] *W. Hückel, K. Janecka*, Arch. Pharm. (Weinheim, Ger.) *284*, 341 (1951).

[458] *W.E. Kuhn*, Org. Syntheses, Coll. Vol. II, 447 (1943).

[459] *E.F. Jenny, J. Druey*, Helv. Chim. Acta *42*, 401 (1959).

[460] *G. Faust*, J. Prakt. Chem. [4] *6*, 14 (1958).

reactions in acid solution formation of sulfonic acids was observed as *side-reaction*[405]. A special field of application for sodium dithionite is the reduction of nitrosopyrimidines (to form aminopyrimidines) generally as part of purine syntheses[405, 461, 462]. The reduction is often carried out in the presence of formic acid and the amine is isolated as a *formamide*.

Reduction of the nitro group with *sulfides,* for example, sodium sulfide, sodium disulfide, sodium hydrogen sulfide, and ammonium sulfide, is carried out in weakly alkaline, aqueous, or alcoholic medium. The last-named reagent is generally prepared by introducing hydrogen sulfide into the ammoniacal reaction mixture[408, 463]. Sodium hydrogen sulfide in methanol has been found to be a particularly favorable reducing agent[464]. When sodium sulfide is used, the addition of ammonium chloride[465] can be used to avoid excessively alkaline conditions possibly giving rise to *side-reactions* such as azo compound formation. During reaction of aromatic nitro compounds with sulfides, o- and p-halogens are replaced by the mercapto group with formation of thiophenols; this reaction may be utilized for preparative purposes[466]. A corresponding substitution was observed with aromatic o-dinitro compounds[405]. By contrast, halogens in the *m*-position remain intact[467]. *Aminoazo* compounds can be obtained by selective reduction of *m*- and *p*-nitroazo compounds with sodium sulfide or sodium hydrogen sulfide[405].

Sulfides possess particular significance for the *partial* reduction of nitro groups in aromatic polynitro compounds[405, 408, 414, 464]. *m-Nitroamines* are obtained especially smoothly in this way. During reduction of unsymmetrically substituted aromatic *m*-dinitro compounds a dependence of the course of the reduction on the substituent is observed. Thus, in the presence of amino or hydroxy

substituents, the *o*-nitro group is reduced preferentially[414, 463, 465, 468, 469], *e.g.:*

2-Amino-4-nitrophenol

This course of the reaction is explained by the inductive effect of the substituents[469]; the preferential reduction of the *p*-nitro group observed in presence of alkyl and carboxylate substituents can be interpreted correspondingly[469].

12.5.1.5 Reduction with metal hydrides

The reactions of nitro compounds with the different metal hydrides are described in various surveys[470–475]. While a reduction to amine is possible with complex aluminum hydrides, it generally occurs only with aliphatic nitro compounds[470, 472]. *Azo compounds* are normally obtained on reacting aromatic nitro compounds with *lithium tetrahydroaluminate*[470, 476]; the same is true of other complex aluminum hydrides, for instance, sodium dihydrobis(2-methoxyethoxy)aluminate[477]. However, formation of aromatic amines with lithium tetrahydroaluminate does take place where the nitro group is sterically hindered by *o*-substituents. An example is the reduction of 2-nitromesitylene; here, the proportion of isolated azo compound is small[478]:

2,4,6-Trimethylaniline

[461] *W. Pfleiderer, G. Nübel,* Justus Liebigs Ann. Chem. *631,* 168 (1960).

[462] *G. Levin, A. Kalmus, F. Bergmann,* J. Org. Chem. *25,* 1752 (1960).

[463] *K.P. Griffin, W.D. Peterson,* Org. Syntheses, Coll. Vol. III, 242 (1955).

[464] *J.P. Idoux,* J. Chem. Soc. C 435 (1970).

[465] *W.W. Hartmann, H.L. Silloway,* Org. Syntheses, Coll. Vol. III, 82 (1955).

[466] *H. Wahl, M.T. LeBris,* Bull. Soc. Chim. France 1277 (1954).

[467] *G.A. Timokhin, B.I. Kissin,* Khim. Prom. (Moscow) *42,* 258 (1966); C.A. *64,* 19454e (1966).

[468] *P. Tomasik, Z. Skrowaczewska,* Roczniki Chem. *40,* 637 (1966); C.A. *65,* 3826f (1966).

[469] *H.M. Weiss,* J. Chem. Educ. *43,* 384 (1966).

[470] *V.M. Mićović, M.Lj. Mihailović,* Lithium Aluminum Hydride in Organic Chemistry; Serbian Academy of Sciences, Monographs Vol. 237; Naučna Knjiga, Izdavačko Preduzeće, Beograd 1955.

[471] *N.G. Gaylord,* Reduction with Complex Metal Hydrides; Interscience Publishers, New York 1956.

[472] *M.N. Rerick* in *R.L. Augustine,* Reduction p. 1, Marcel Dekker, New York 1968.

[473] *M. Ferles,* Chem. Listy *64,* 188 (1970).

[474] *H.C. Brown,* Hydroboration, W.A. Benjamin, New York 1962.

[475] *E. Schenker* in *W. Foerst,* Neuere Methoden der präparativen organischen Chemie, Bd. IV, p. 173, Verlag Chemie, Weinheim/Bergstr. 1966.

[476] *H.C. Brown, P.M. Weissman, N.M. Yoon,* J. Amer. Chem. Soc. *88,* 1458 (1966).

[477] *J.F. Corbett,* Chem. Commun. 1257 (1968).

[478] *F.A.L. Anet, J.M. Muchowski,* Can. J. Chem. *38,* 2526 (1960).

Synthesis of *aliphatic* amines from nitro compounds succeeds with various different aluminum alkoxy hydrides[472, 479, 480] as well as with lithium tetrahydroaluminate[472, 476]. If the compound contains groups other than the nitro group reducible by the latter reagent then these are as a rule attacked before the nitro group. Thus, the selectivity here differs materially from that observed during catalytic or other chemical reductions. This feature can be utilized, for example, for preparing *nitroalcohols* from *nitrocarboxylic acid esters*[405, 470, 473]. *Aluminum hydride* is even more suitable for selectively reducing functional groups while preserving the nitro group[481].

During reduction of α,β-unsaturated nitro compounds with lithium tetrahydroaluminate the $C=C$ bond is invariably coreduced[441, 470, 473]. This reaction has become particularly important for synthesizing a large number of pharmacologically interesting *1-phenyl-2-alkylamines, e.g.*[482]:

3-Methoxy-4-methyl-α-methylphenethylamine

On reducing optically active 2-nitrooctane with lithium tetrahydroaluminate to *1-methylheptylamine* complete racemization was observed[432], in contrast methods of reduction capable of being performed in acid medium (pp. 496, 497).

Reduction of *tertiary* aliphatic nitro compounds with lithium tetrahydroaluminate can lead to complications. Thus, during the reduction of vicinal dinitro compounds a reductive splitting of the $C-C$ bond was observed[483], *e.g.*:

Isopropylamine

1,1'-Dinitrobicyclohexyl and *trans*-4a,8a-dinitrodecahydronaphthalene are split by lithium tetra-

hydroaluminate in the same way to give *cyclohexylamine* and *1,6-cyclodecanediamine* respectively[483], but not on reduction with metals in acid (giving instead *[bicyclohexyl]-1,1'-diamine* and *trans-4a,8a-decahydronaphthalenediamine*). A cleavage yielding a similar result is observed during the lithium tetrahydroaluminate reduction of α-nitrocarboxylic acid esters[484, 485]; here an *alkylamine* is formed in a proportion which becomes particularly high during reduction of tertiary nitrocarboxylic acid esters[485] and the corresponding *nitroalcohols*[486]. Using aluminum dialkyl hydrides leads to a similar reductive splitting of the $C-C$ bond[485].

Reduction of tertiary alicyclic nitro compounds, for instance, of 1-methyl-1-nitrocyclohexane[487], with lithium tetrahydroaluminate likewise displays an unexpected course. Secondary amines formed by rearrangement are isolated side by side with the primary amine, in this case *N-methylcyclohexylamine* and *2-methylhexamethylenimine* as the result of a reductive ring expansion. The explanation offered for this finding is a rearrangement starting at the hydroxylamine stage[487]. In the presence of excess aluminum chloride such reactions are observed with secondary as well as primary aliphatic nitro compounds[472].

Complex tetrahydroborates[472, 475] do not reduce nitro to amino groups unless special catalysts are added or special working conditions are employed. *Sodium tetrahydroborate* reduces aromatic nitro compounds in dimethyl sulfoxide at 85° initially to the *azoxy compound* and the latter further to a mixture of *azo compound* and *amine*[488].

[482] *B.T. Ho, W.M. McIsaac, Rong An, L.W. Tansey, K.E. Walker, L.F. Englert, M.B. Noel*, J. Med. Chem. *13*, 26 (1970);
B.T. Ho, L.W. Tansey, R.L. Balster, Rong An, W.M. McIsaac, R.T. Harris, J. Med. Chem. *13*, 134 (1970).
[483] *A. Dornow, K.J. Fust*, Chem. Ber. *90*, 1774 (1957).
[484] *H. Hellmann, D. Starck*, Angew. Chem. *70*, 271 (1958).
[485] *H. Reinheckel, D. Jahnke, G. Tauber*, Monatsh. Chem. *100*, 1881 (1969).
[486] *F.I. Carroll*, J. Org. Chem. *34*, 466 (1969).
[487] *H.J. Barber, E. Lunt*, J. Chem. Soc. (London) 1187 (1960);
G.E. Lee, E. Lunt, W.R. Wragg, H.J. Barber, Chem. & Ind. (London) 417 (1958).
[488] *R.O. Hutchins, D.W. Lamson, L. Rua, C. Milewski, B. Maryanoff*, J. Org. Chem. *36*, 803 (1971).
[489] *T. Neilson, H.C.S. Wood, A.G. Wylie*, J. Chem. Soc. (London) 371 (1962).
[490] *R.L. Edwards, N. Kale*, J. Chem. Soc. (London) 4084 (1964).
[491] *T. Satoh, S. Suzuki, Y. Suzuki, Y. Miyaji, Z. Imai*, Tetrahedron Lett. 4555 (1969).

[479] *G. Hesse, R. Schrödel*, Justus Liebigs Ann. Chem. *607*, 24 (1957).
[480] *H.C. Brown, P.M. Weissman*, J. Amer. Chem. Soc. *87*, 5614 (1965).
[481] *N.M. Yoon, H.C. Brown*, J. Amer. Chem. Soc. *90*, 2927 (1968).

A convenient synthesis for *primary aliphatic* and *aromatic amines* is the reduction of nitro compounds with sodium tetrahydroborate in the presence of metals[489, 490] or metallic salts[491] as catalysts. Reduction in the presence of noble metal catalysts, palladium-carbon being used predominantly[489, 490], can be carried out in aqueous methanol and, during reduction of nitrophenols[489], also in aqueous alkali, *e.g.*:

OH / NO₂ —— NaBH₄ / Pd–C, NaOH, H₂O —→ OH / NH₂

2-Aminophenol

Aromatic nitroso compounds, too, can be converted into amines in this way[489]. For an alternative simple laboratory method for reducing aromatic nitro compounds with sodium tetrahydroborate see ref.[492]. However, the method of conducting the reaction differs fundamentally from the above mentioned technique and substantially resembles the usual catalytic hydrogenation.

In the presence of transition metal salts, *e.g.*, cobalt(II) chloride, aromatic and aliphatic nitro compounds can be reduced to primary amines with sodium tetrahydroborate in solvents such as boiling methanol, ethanol, or dioxane[491].

With *diphenyltin dihydride* nitrobenzaldehydes are successfully converted into *aminobenzaldehydes*[493].

Diborane does not reduce nitro groups[474]. By contrast, aromatic nitroso groups[494] are converted into amines in good yield by B_2H_6 in tetrahydrofuran at 25°; the same holds for phenylhydroxylamine[494].

12.5.1.6 Photochemical reduction

In recent years photochemical processes for preparing *primary aromatic amines* have been utilized also on the preparative scale[495, 496]. In general, primary or secondary alcohols are employed as hydrogen donors. One example is the synthesis of esters of 4-aminobenzoic acid[495]:

O_2N—⟨ ⟩—$COOC_2H_5$

—— C_2H_5OH / hν —→ H_2N—⟨ ⟩—$COOC_2H_5$

Ethyl 4-aminobenzoate

Studies on the relationship between the result of the reduction of aromatic nitro compounds and the solvent and wavelength have been carried out[497].

12.5.1.7 Electrochemical reduction

Aliphatic and especially aromatic nitro compounds can be converted into amines in high yield by cathodic reduction provided certain conditions are maintained[405]. In addition to the electrical variables this includes primarily the *pH* of the electrolyte. In the alkaline region azoxy and azo compounds are formed preferentially from the majority of aromatic nitro compounds, while in *acid solution* amines are obtained. A partial reduction to the arylhydroxylamine stage, followed by a preparatively utilizable rearrangement to *aminophenol*[498], is possible in moderately acid solution. Because of the special techniques required for carrying out electrochemical reductions, it is necessary here to refer to special surveys[499, 500].

12.5.2 Amines from oximes

12.5.2.1 Reduction by catalytic methods

Catalytic hydrogenation of oximes and their *O*-acyl derivatives to *primary* amines can be achieved in principle using the same techniques by which, in particular, aliphatic nitro compounds and also nitriles can be reduced. Hydroxylamines, too, can be reduced to primary amines in this way. Proven methods of reducing oximes are hydrogenation over platinum[409, 410, 501–504] or palla-

[492] *H.C. Brown, K. Sivasankaran*, J. Amer. Chem. Soc. *84*, 2828 (1962).

[493] *H.G. Kuivila, O.F. Beumel*, J. Amer. Chem. Soc. *83*, 1246 (1961).

[494] *H. Feuer, D.M. Braunstein*, J. Org. Chem. *34*, 2024 (1969).

[495] *R.A. Finnegan, D. Knutson*, J. Amer. Chem. Soc. *90*, 1670 (1968).

[496] *S. Hashimoto, H. Fujii, J. Sunamoto*, Kogyo Kagaku Zasshi *70*, 316 (1967); C.A. *67*, 81675 (1967).

[497] *J.A. Barltrop, N.J. Bunce*, J. Chem. Soc. C 1467 (1968);
Y. Kitaura, T. Matsuura, Tetrahedron *27*, 1583 (1971).

[498] *M. LeGuyader, D. Peltier*, C.R. Acad. Sci., Paris *253*, 2544 (1961);
M. Le Guyader, D. Peltier, C.R. Acad. Sci., Paris *254*, 4182 (1962).

[499] *S. Swann* in *A. Weissberger*, Technique in Organic Chemistry, Vol. II, 2nd Ed., p. 385, Interscience Publishers, New York 1956.

[500] *F.D. Popp, H.P. Schultz*, Chem. Rev. *62*, 19 (1962).

[501] *J.G. Murphy*, J. Org. Chem. *26*, 3104 (1961).

[502] *H. Feltkamp*, Arch. Pharm. (Weinheim, Ger.) *295*, 764 (1962).

[503] *W. Hückel, K. Heyder*, Chem. Ber. *96*, 220 (1963);
W. Hückel, G. Stelzer, Chem. Ber. *88*, 984 (1955).

[504] *W.E. Rosen, M.J. Green*, J. Org. Chem. *28*, 2797 (1963).

dium[409, 410, 430, 504–506] catalysts in acid medium, while reacting over Raney nickel[409, 502, 504–507] is usually carried out in alkaline medium. Glacial acetic acid, alcohol acidified with mineral acid, and glacial acetic acid-mineral acid mixtures are suitable in the first case. Ammoniacal alcohol is often used to carry out reductions over the alkaline region; addition of sodium hydroxide[405] or sodium methoxide[504] was some times opportune. In some cases, *e.g.*, during synthesis of *cycloheptylamine*, hydrogenation over rhodium-alumina in methanol has afforded favorable results[508]. Acylating reductions of the oximes over platinum and palladium catalysts are generally carried out in acetic anhydride-glacial acetic acid mixtures[406, 506]; when Raney nickel is employed, acetic anhydride and acetates are used as basic additives[406, 509].

During the catalytic reduction of the oxime group *secondary* amines may be formed as *side-product*:

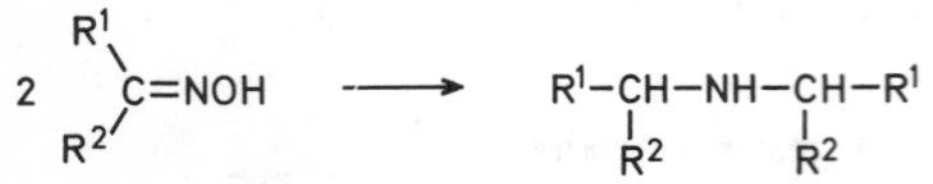

$$2 \begin{matrix} R^1 \\ | \\ R^2 \end{matrix} C=NOH \longrightarrow R^1-\underset{R^2}{CH}-NH-\underset{R^2}{CH}-R^1$$

All the above procedures and the technique of dropwise hydrogenation substantially avoid this side-reaction, which is observed especially over the neutral region[409, 504, 510] and may be utilized preparatively[510]. The cause of secondary amine formation is postulated to be the addition of already formed primary amine to the supposed intermediate imine stage[405, 406, 504, 511]. Other side-reactions such as rearrangements in the presence of Raney nickel[405, 512] have been seldom reported.

Selective hydrogenation of the oxime group in the presence of other reducible groups is less easy[406, 407, 409, 509] than that of, for instance, the

aromatic nitro group. In individual cases the conditions necessary for preserving a nitrile group[409, 513] have been investigated. By contrast, keeping the keto group intact during synthesis of *α-amino ketones* generally presents no difficulties, provided the reduction of the oxime group is performed over palladium catalysts in alcohol acidified with hydrogen chloride[406, 407, 409, 506, 514]. The further reduction to the *β-amino alcohol* is favored by addition of water[406, 506], *e.g.*:

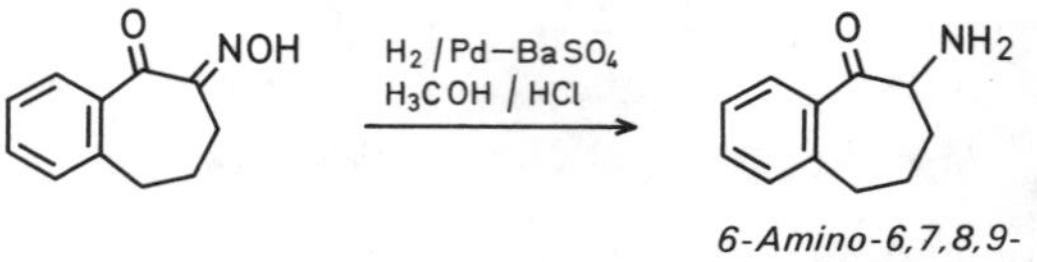

6-Amino-6,7,8,9-tetrahydro-5H-benzocyclohepten-5-one

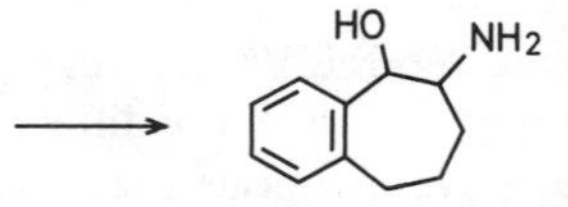

6-Amino-6,7,8,9-tetrahydro-5H-benzocyclohepten-5-ol

Reductions under acylating conditions, too[406, 506], serve for preparing amino ketones. *α-Chloro oximes* which are readily obtained by addition of nitrosyl chloride to alkenes, are invariably dehalogenated on hydrogenation over palladium and platinum catalysts; when the first-named are used the *oxime stage* can be caught[515]:

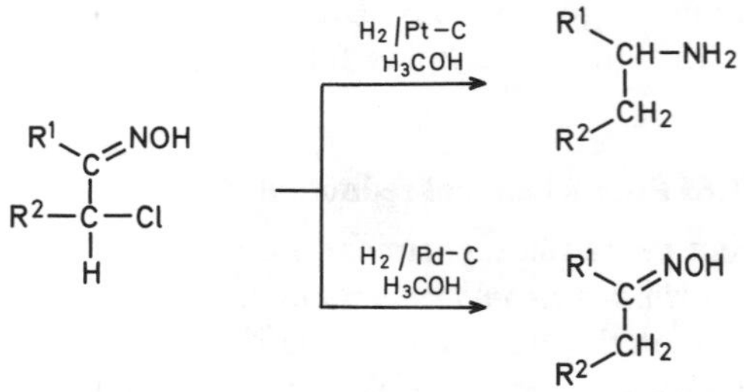

Reduction of the oxime group proceeds with a high degree of *stereoselectivity* under conditions ensuring a kinetically controlled reaction course. These

505 *M.V. Balyakina, E.S. Zhdanovich, N.A. Preobraz-henskii,* Zh. Obshch. Khim. *31,* 2983 (1961); C.A. *56,* 15475b (1962).

506 *T. Yupraphat, H.J. Rimek, F. Zymalkowski,* Justus Liebigs Ann. Chem. *738,* 79 (1970).

507 *D.C. Iffland, Teh-Fu-Yen,* J. Amer. Chem. Soc. *76,* 4180 (1954).

508 *M. Freifelder, W.D. Smart, G.R. Stone,* J. Org. Chem. *27,* 2209 (1962).

509 *A.F. Ferris, G.S. Johnson, F.E. Gould, H. Stange,* J. Org. Chem. *25,* 1302 (1960).

510 *A. Dornow, A. Frese,* Arch. Pharm. (Weinheim, Ger.) *285,* 463 (1952).

511 *V.I. Komarevsky, C.H. Riesz, F.L. Morritz* in *A. Weissberger,* Technique in Organic Chemistry, Vol. II, 2nd Ed., p. 137, Interscience Publishers, New York 1956.

512 *V. Sprio, E. Ajello,* Ann. Chim. (Rome) *56,* 858 (1966).

513 *M. Fields, D.E. Walz, S. Rothschild,* J. Amer. Chem. Soc. *73,* 1000 (1951).

514 *A. Hassner, A.W. Coulter,* Steroids *4,* 281 (1964).

515 *R. Biela, I. Hahnemann, H. Panovsky, W. Pritzkow,* J. Prakt. Chem. *33,* 282 (1966).

516 *E.L. Eliel, N.L. Allinger, S.J. Angyal, G.A. Morrison,* Conformational Analysis, p. 118, Interscience Publishers, New York 1965.

conditions obtain when very active catalysts, particularly those from the platinum and palladium group, are used and the reaction is conducted in an acid medium. The epimeric amine resulting from hydrogen addition to the sterically least hindered side of the molecule is isolated preferentially. On reducing substituted cyclohexanone oximes this means the preferential formation of the epimer with the axial amino group[407, 502, 503, 516–518]:

In this way numerous stereoselective syntheses have been carried out especially in the *steroid field*[519–521].

During synthesis of *amino sugars via* hydrogenation of oximes in the presence of platinum(IV) oxide[522] or palladium-carbon[523] in hydrochloric acid, too, a high degree of stereoselectivity is found.

If the reaction is performed under conditions which allow an epimerization side by side with the reduction, then the proportion of the thermodynamically more stable epimer increases in the isolated product, *i.e.,* the equatorial amines during reduction of substituted cyclohexanone oximes. Thus hydrogenation over Raney nickel in ammoniacal medium often yields equal amounts of the epimers[502, 503, 516]; the epimer ratio is here strongly dependent on the reaction conditions.

When the reduction is carried out under *thermodynamic control,* then the epimer with equatorial amino group is the main product in the conversion of cyclohexanone oxime. An example of such a reaction course is the reduction with sodium in alcohols[502, 503, 516–519, 524, 525] (p. 504).

By applying the stereoselectively proceeding reduction methods, especially with steroids, it has been possible to carry out synthesis of particular epimers[521, 526, 527]. For example, *3α,20β-dihydroxy-5β-pregnan-11β-amine* is obtained on hydrogenating the corresponding oxime over platinum(IV) oxide in glacial acetic acid, while on reduction with sodium/propanol the *11α-amine* is the product[521]. Overall, the results available indicate substantial agreement with the stereoselectivity of the reduction of the carbonyl group.

Initial systematic investigations on the possibility of an asymmetric induction during hydrogenation of oximes over Raney nickel modified with optically active amino acids[528] or palladium on silk fibroin[529] have been communicated. In this connection it could be demonstrated that hydrogenation of *2-acetoxyiminopropionic acid ester* to *2-aminopropionic acid ester* proceeds *via* the acylated hydroxylamine[528].

Catalytic reduction of the oxime group succeeds also with *bound* hydrogen, including, for example, reduction with *sodium hypophosphite* in the presence of Raney nickel in alkaline solution[530]. A detailed study of the reduction of oximes with *hydrazine/Raney nickel* is available[531]. The yield of amine varies widely with the individual compounds, the reaction conditions are much the same as those stated (p. 496) for nitro compounds[531], *e.g.:*

9-Fluorenamine

12.5.2.2 Reduction by miscellaneous chemical methods

Of chemical reducing agents available for converting the oxime group into the amino group, a number of metals have proven themselves as suit-

[517] *W. Hückel, K.D. Thomas,* Justus Liebigs Ann. Chem. *645,* 177 (1961).

[518] *H. Booth, G.C. Gidley, N.C. Franklin,* Tetrahedron *23,* 2421 (1967).

[519] *C.W. Shoppee, R. Cremlyn, D.E. Evans, G.H.R. Summers,* J. Chem. Soc. (London) 4364 (1957); *C.W. Shoppee, D.E. Evans, H.C. Richards, G.H.R. Summers,* J. Chem. Soc. (London) 1649 (1956).

[520] *J. Schmitt, J. Panouse, A. Hallot, P.-J. Cornu, H. Plucket, P. Comoy,* Bull. Soc. Chim. France 1855 (1962).

[521] *R. Rausser, L. Weber, E.B. Hershberg, E.P. Oliveto,* J. Org. Chem. *31,* 1342 (1966); *R. Rausser, C.G. Finckenor, L. Weber, E.B. Hershberg, E.P. Oliveto,* J. Org. Chem. *31,* 1346 (1966).

[522] *A.K. Chatterjee, D. Horton, J.S. Jewell, K.D. Philips,* Carbohyd. Res. *7,* 173 (1968).

[523] *R.U. Lemieux, S.W. Gunner, T.L. Nagabhushan,* Tetrahedron Lett. 2143, 2149 (1965).

[524] *H. Booth, F.E. King,* J. Chem. Soc. (London) 2688 (1958).

[525] *G. Adam, B. Voigt, K. Schreiber,* Tetrahedron *25,* 3783 (1969).

[526] *G. Drefahl, S. Huneck,* Chem. Ber. *93,* 1961 (1960).

[527] *R. Glaser, E.J. Gabbay,* J. Org. Chem. *35,* 2907 (1970).

[528] *R. Imamura, M. Imaida, T. Ninomiya, S. Yajima,* Bull. Chem. Soc. Japan *43,* 1792 (1970).

[529] *S. Akabori, S. Sakurai, Y. Izumi, Y. Fujii,* Nature *178,* 323 (1956).

able in addition to metal hydrides (p. 504). A large number oximes have been reduced with *sodium* in alcohols, *e.g.*, ethanol, propanol, and butanol[502, 503, 519, 521, 524, 525, 532, 533]. For the reduction of cyclohexanone oxime (to *cyclohexylamine*) the dependence of the yield on the reaction conditions and the alcohol used was investigated[534]. High amine yields are obtained by reduction with sodium in liquid ammonia[535, 536]; by modifying the conditions a simultaneous *C*-alkylation can be carried out[537]. The stereochemical course of the reductions with sodium is given in refs. [535, 536] (see p. 503).

Selective reductions of oxime groups with *lithium* or *sodium* amalgam with preservation of nitrile groups has been reported repeatedly[408, 538].

A simple laboratory method for reducing oximes consists in the reaction with *magnesium* in methanol saturated with ammonium acetate[539], *e.g.*:

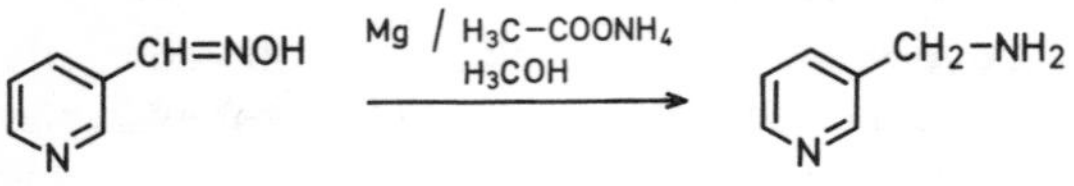

3-(Aminomethyl)pyridine

Nitrile or amide formation is occasionally encountered in dependence on the structure of the oxime[539]. Amine synthesis with *Raney nickel alloy* as reducing agent in aqueous ethanolic sodium hydroxide solution is likewise simple to carry out and versatile[530, 540].

With *zinc*, oximes are generally reduced in acetic acid solution, frequently under acylating conditions. This method has gained particular importance for reducing 1,3-dicarbonyl-2-hydroxyimino

compounds[405, 541], *e.g.*, in the synthesis of *diethyl (acetamino)malonate*[542]:

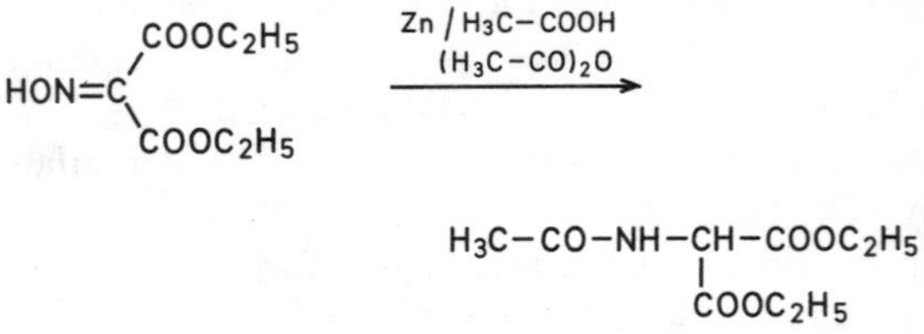

For synthesizing *pyridoxamine*[543], too, reduction of oximes with zinc is appropriate. Reduction of aryl and, particularly, diaryl ketoximes with zinc is conducted best in concentrated ammonia with added ammonium acetate[544].

For the reduction of oximes which are difficult to convert into amine by usual reducing agents because of steric hindrance, the following procedure has given good results: an initial reduction with *titanium(III) chloride* to ketimine is performed, from which the amine is subsequently obtained readily with sodium tetrahydroborate[545].

12.5.2.3 Reduction with metal hydrides

Metalhydrides have found wide application in the synthesis of primary amines from aldoximes and ketoximes. Lithium tetrahydroaluminate is the most important compound used; a number of surveys describe the technique used[470, 471, 473, 546]. The great majority of the investigated oximes could be converted into the corresponding primary amine by heating with lithium tetrahydroaluminate in diethyl ether or tetrahydrofuran without difficulty. Hydroxylamines can be reduced to amines in the same way[546, 547] and likewise *O*-alkyl and *O*-acyl oximes[546].

The oxime group is less easily reduced than carbonyl groups and ester carbonyl groups. Thus, it is possible by careful working to reduce the carbonyl group without altering the oxime grouping[548]; where a reduction of the latter is the aim, *aminoalkanediols*[549] are, for instance, formed from hydroxyiminodicarboxylic acid esters or else *amino alcohols* from α-diketone monooximes[550, 551].

[530] *B. Staskun, T. van Es*, J. Chem. Soc. C 531 (1966).
[531] *D. Lloyd, R.H. McDougall, F.I. Wasson*, J. Chem. Soc. (London) 822 (1965).
[532] *G. Van de Woude, L. van Hove*, Bull. Soc. Chim. Belg. *76*, 566 (1967).
[533] *W.H. Lycan, S.V. Puntambeker, C.S. Marvel*, Org. Syntheses, Coll. Vol. II, 318 (1943).
[534] *J.K. Sugden*, Chem. & Ind. (London) 1224 (1967).
[535] *A.P. Terent'ev, N.I. Gusar*, Zh. Obshch. Khim. *35*, 125 (1965); C.A. *62*, 13068d (1965).
[536] *M. Smith* in *R.L. Augustine*, Reduction, p. 95, Marcel Dekker, New York 1968.
[537] *J.A. Gautier, M. Miocque, C. Fauran, A.-Y. Le Cloarec*, C.R. Acad. Sci., Paris, Ser. C *263*, 1164 (1966).
[538] *J. Hannart, A. Bruylants*, Bull. Soc. Chim. Belg. *72*, 423 (1963).
[539] *J.K. Sugden*, Chem. & Ind. (London) 260 (1969).
[540] *J.L.M.A. Schlatmann, J.G. Korsloot, J. Schut*, Tetrahedron *26*, 949 (1970).
[541] *H. Hellmann, H. Piechota, W. Schwiersch*, Chem. Ber. *94*, 757 (1961).
[542] *A.J. Zambito, E.E. Howe*, Org. Synth. *40*, 21 (1960).

[543] *E. Testa, F. Fava*, Chimia *11*, 310 (1957).
[544] *J.C. Jochims*, Monatsh. Chem. *94*, 677 (1963).
[545] *G.H. Timms, E. Wildsmith*, Tetrahedron Lett. 195 (1971).
[546] *A. Hajós*, Komplexe Hydride, VEB Deutscher Verlag der Wissenschaften, Berlin 1966.
[547] *A. Mustafa, M. Kamel*, J. Amer. Chem. Soc. 76, 124 (1954).
[548] *H. Felkin*, C.R. Acad. Sci., Paris *230*, 304 (1950).
[549] *W. Treibs, H. Reinheckel*, Chem. Ber. *89*, 58 (1956).
[550] *A. Dornow, M. Theidel*, Chem. Ber. 88, 1267 (1955).

However, a selective reduction to amino ketones is possible by the catalytic route. By contrast, reduction of unsaturated oximes with lithium tetrahydroaluminate to the corresponding amines while preserving the $C=C$ bond has been described[552, 553]; the same is true of α,β-unsaturated oximes, as is exemplified by the reduction of 4-phenyl-3-buten-2-one oxime[553]:

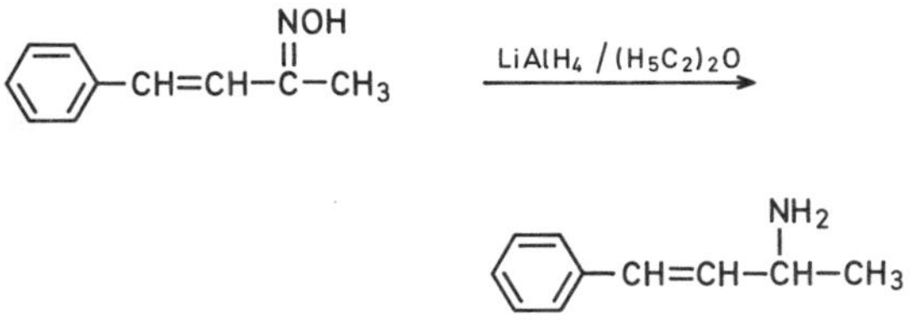

α-Methylcinnamylamine

A series of complex metal hydrides other than lithium tetrahydroaluminate have been tested for their usefulness as reducing agents for the oxime group. It was found that *lithium hydrotri-tert-butoxyaluminate and hydrotrimethoxyaluminate* do not attack the oxime group[476, 480], while with *sodium dihydrobis(2-methoxyethoxy)aluminate*[473, 554] and with *magnesium tetrahydroaluminate*[555] the reaction was successful. *Sodium tetrahydroborate* does not reduce the oxime group[471, 475, 546]; of other complex hydroborates only a sulfurated sodium hydroborate ($NaBH_2S_3$) has so far proven suitable for carrying out reductions to amine[556]. The reaction is conducted in boiling tetrahydrofuran; under special conditions the hydroxylamine can be isolated successfully suggesting a reaction proceeding *via* a hydroxylamine stage[556].

With *diborane* in tetrahydrofuran, oximes are reduced to hydroxylamines in good yield at low temperature[557, 558]. A good yield of amine is obtained also by reacting oximes or hydroxylamines in bis-(2-methoxy-ethyl)ether and tetrahydrofuran with diborane at $105-110°$[559], *e.g.*:

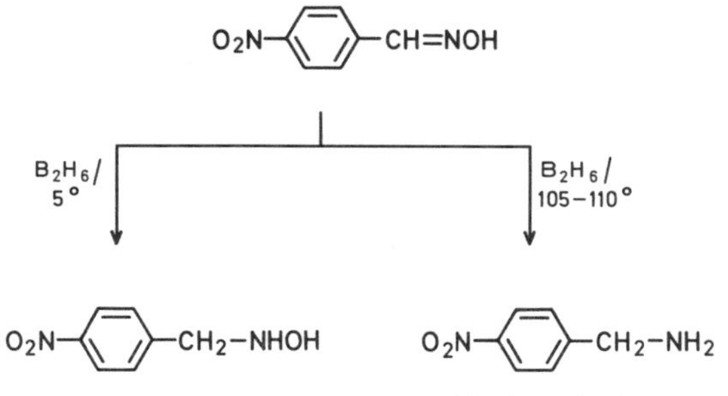

p-Nitrobenzylamine

With *O*-alkyl and *O*-acyl oximes the reduction to amine proceeds even at room temperature[559, 560]:

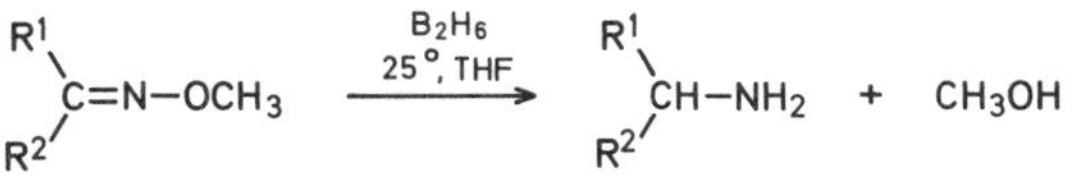

In all reductions with diborane the reaction product is liberated by either acid or alkaline hydrolysis.

Aluminum hydride has revealed itself to be a particularly suitable reducing agent for oximes[481]. The reduction is performed in boiling tetrahydrofuran; it proceeds faster than with lithium tetrahydroaluminate and gives high yields of amine.

The *stereoselectivity* of the reduction of oximes with lithium tetrahydroaluminate has been less extensively investigated than the corresponding reaction of ketones; but the results available point to a substantial similarity in the stereoselectivity of the reduction of the two groups. Thus, on reduction of substituted but sterically unhindered cyclohexanone oximes the epimeric *cyclohexylamine* with the equatorial amino group was formed predominantly[517, 520, 561, 562]. Where addition of the complex metal hydride is distinctly sterically hindered, for example, by *syn*-axial substituents, the epimer with the axial amino group predominates[441, 526, 563-565]. For example, during reduction of friedelin oxime to *aminofriedelane* a 70:30 ratio of epimers with axial and equatorial amino groups respectively is obtained[526]:

[551] *M.E. Wolff, J.F. Oneto*, J. Amer. Chem. Soc. *78*, 2615 (1956).
[552] *D. Arigoni, O. Jeger*, Helv. Chim. Acta *37*, 881 (1954).
[553] *C.R. Walter*, J. Amer. Chem. Soc. *74*, 5185 (1952).
[554] *V. Bažant, M. Čapka, M. Černý, V. Chvalovský, K. Kochloefl, M. Kraus, J. Málek*, Tetrahedron Lett. 3303 (1968).
[555] *B.D. James*, Chem. & Ind. (London) 227 (1971).
[556] *J.M. Lalancette, J.R. Brindle*, Can. J. Chem. *48*, 735 (1970).
[557] *S.C. Ioffe, V.A. Tartakovskii, A.A. Medvedeva, S.S. Novikov*, Izv. Akad. Nauk SSSR, Ser. Khim. 1537 (1964); C.A. *64*, 14114e (1966).
[558] *H. Feuer, B.F. Vincent, R.S. Bartlett*, J. Org. Chem. *30*, 2877 (1965).
[559] *H. Feuer, D.M. Braunstein*, J. Org. Chem. *34*, 1817 (1969).

[560] *P. Catsoulacos*, J. Heterocycl. Chem. *4*, 645 (1967).
[561] *D.R. Smith, M. Maienthal, J. Tipton*, J. Org. Chem. *17*, 294 (1952).
[562] *W. Schneider, K. Lehmann*, Tetrahedron Lett. 4285 (1970).
[563] *C.L. Stevens, R.P. Glinski, K.G. Taylor*, J. Org. Chem. *33*, 1586 (1968).
[564] *C.W. Shoppee, D.E. Evans, G.H.R. Summers*, J. Chem. Soc. (London) 97 (1957).
[565] *D.E. Evans, G.H.R. Summers*, J. Chem. Soc. (London) 906 (1957).

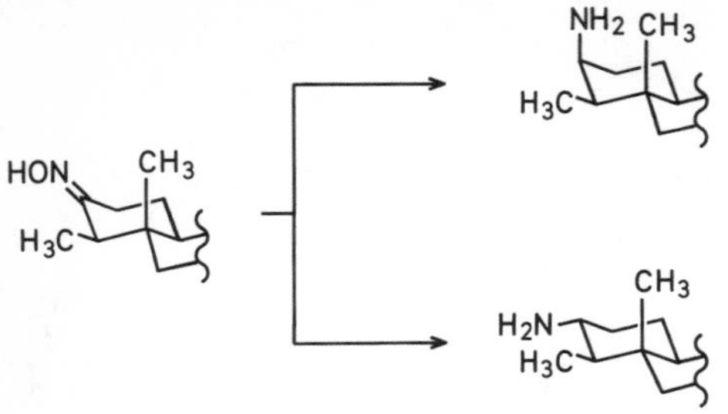

This directed reaction has been made use of for stereoselective syntheses, for example, of *amino sugars*[563] and particularly in the *steroid field*[564, 565]. A discussion of the causes of the observed stereoselectivity for the reduction of ketones is given in the literature[566].

During the reaction of *2,3-bornanedione 3-oxime* with lithium tetrahydroaluminate it was revealed that for reducing both functional groups the addition of the reagent takes place preponderantly from the less hindered *endo*-side; as a result *exo-3-amino-exo-2-bornanol* is isolated as the main product[567–569]:

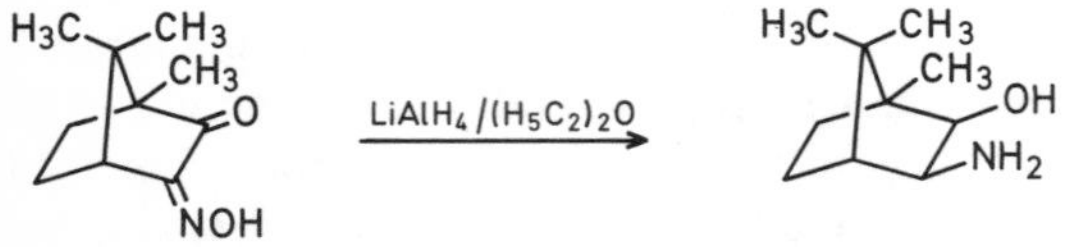

By contrast, on reduction with sodium/ethanol or with zinc the more stable *endo*-amine is formed[569]. The selective reduction of the oxime group by the catalytic route (cf. p. 502) while keeping the carbonyl group intact succeeds also in this example[567, 568].

Formation of *isomeric secondary amines* side by side with primary ones is a side-reaction which has to be considered during reduction of aldoximes or ketoximes with lithium tetrahydroaluminate:

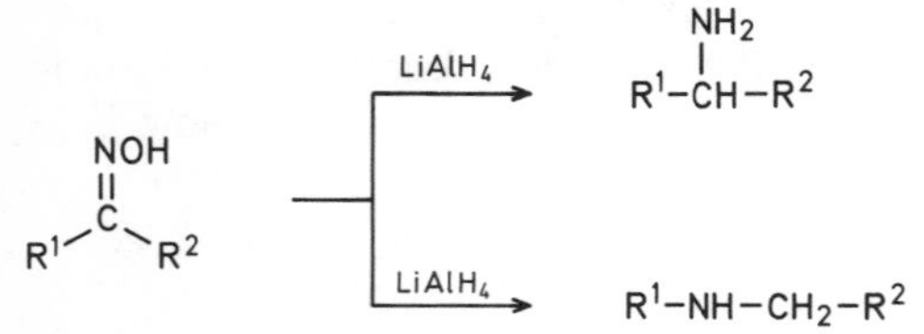

The proportion of secondary amine formed by rearrangement is markedly higher with oximes of acylaromates[561, 570–572] than with aliphatic ones, and, provided they are strainfree, with alicyclic oximes. It was found[570, 571, 573, 574] that it is the aryl group which invariably migrates during the reductive rearrangement of oximes of acylaromates, so that *substituted anilines* are formed. No dependence on the *syn* versus *anti* position of the oxime group could be detected[570, 571, 574].

In addition, it could be shown that reduction of an oxime and of the corresponding hydroxylamine leads to the same composition of the reduction products[571]. These results lead to the conclusion that such a rearrangement does not proceed before the reaction in the sense of a Beckmann rearrangement but, instead, at the hydroxylamine stage[570, 571, 575, 576], as is postulated also for tertiary nitro compounds (p. 500)[487].

The degree of rearrangement is influenced strongly by the nature and position of the substituent on the aromatic ring. Thus, during reduction of substituted acetophenone oximes *para* methoxy or methyl groups promote the formation of secondary amine relative to hydrogen or chlorine as *p*-substituents[570, 573]. With *o*-substituents the reductive rearrangement is suppressed by comparison to the corresponding *p*-substitution[570, 577].

A reductive rearrangement of cyclic ketone oxime leads to *ring expansion*. This is the case especially during reduction of ketoximes of strained rings[575, 578], as is shown by the following example[578]:

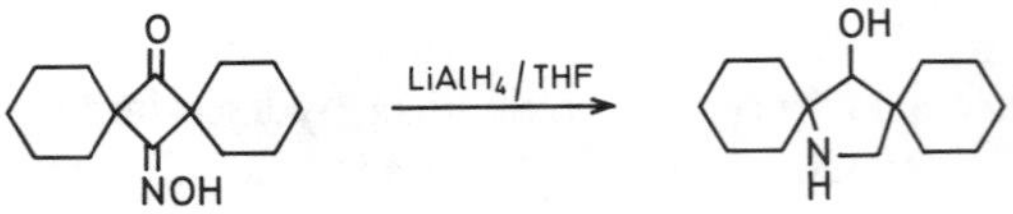

14-Azadispiro-[5.1.5.2]pentadecan-7-ol

[566] *E.L. Eliel, Y. Senda*, Tetrahedron *26*, 2411 (1970).

[567] *A.H. Beckett, N.T. Lan, G.R. McDonough*, Tetrahedron *25*, 5689 (1969).

[568] *A.R. Chittenden, G.H. Cooper*, J. Chem. Soc. C 49 (1970).

[569] *A. Daniel, A.A. Pavia*, Bull. Soc. Chim. France 1060 (1971).

[570] *M.N. Rerick, C.H. Trottier, R.A. Daignault, J.D. DeFoe*, Tetrahedron Lett. 629 (1963).

[571] *S.H. Graham, A.J.S. Williams*, Tetrahedron *21*, 3263 (1965).

[572] *J. Humbert, A. Laurent*, C.R. Acad. Sci., Paris, Ser. C *272*, 1165 (1971).

[573] *R.E. Lyle, H.J. Troscianiec*, J. Org. Chem. *20*, 1757 (1955).

[574] *A.E. Petrarca, E.M. Emery*, Tetrahedron Lett. 635 (1963).

[575] *H.K. Hall*, J. Org. Chem. *29*, 3139 (1964).

[576] *V.A. Zagorevskij, N.V. Dudykina, L.M. Meščerjakova*, Khim. Geterotsikl. Soedin. 302 (1970); Chem. Inform. B 26–322 (1970).

[577] *V.A. Zagorevskij, N.V. Dudykina*, Zh. Obshch. Khim. *34*, 2282 (1964); C.A. *61*, 11960d (1964).

[578] *C. Metzger*, Chem. Ber. *102*, 3235 (1969).

These reductive ring expansions have achieved importance with various heterocycles having condensed aromatic rings for synthesizing seven-membered rings[576, 577, 579–581]. One example is the conversion of a substituted chromanone oxime[581]:

8-Methoxy-3-methyl-
2,3,4,5-tetrahydrobenzo-
[b]-1,4-oxazepine

In such experiments it was found that the ratio of the isomeric primary and secondary amines in the reduction product is dependent also on the substitution in the aliphatic part of the molecule[571, 581]. O-Alkyl oximes behave similarly to oximes during these reductions; with O-tosyloximes the reductive ring expansion is observed in increased measure[576].

Addition of aluminum chloride markedly promotes formation of *secondary amine* during reduction with lithium tetrahydroaluminate[570, 572] and in this way acetophenone oximes can be converted almost completely into *substituted anilines*[570], *e.g.*:

N-Ethylaniline

When aluminum hydride is used as the reducing agent the degree of rearrangement is remarkably small[481].

A further possible side-reaction during reduction of oximes with lithium tetrahydroaluminate is the formation of aziridines[582–584]; it, too, can be utilized preparatively[585], *e.g.*:

*cis-2-Benzyl-3-
phenylaziridine*

A detailed survey about the scope of this reaction is available[582]. The proportion of aziridine formed side by side with amine depends generally on the solvent, possible addition of amine, and above all on the structure of the oxime. Thus, in tetrahydrofuran as solvent substantially more aziridine is formed than in ether[572, 582]. Also, addition of secondary amines, *e.g.*, *N-methylbutylamine*, increases the proportion of aziridine markedly. Concerning structural features of the oxime which promote aziridine formation, see the literature[582]. Thus, oximes with benzyl substituents, *e.g.*, 1,3-diphenyl-2-propanone oxime and also α,β-unsaturated oximes especially tend to form *aziridines* during the reduction. With oximes of five-to seven-membered aliphatic cyclic ketones, aziridine formation was observed only if the relevant ring was part of a bridged system[540, 582, 586], *e.g.*[540]:

2,2a-[piimino-
homoadamantane]

O-alkyloximes and O-acyloximes, too, can undergo such a transformation, as is shown by the reduction of 3-phenyl-2-isoxazolines[582, 587]; here *aziridines* are formed side by side with *amino alcohols*.

Aziridine formation is postulated to take place *via* the azirine stage[582], and, in addition, a dependence of the course of the reaction on the position of the oxime hydroxy group is suggested[540, 582, 584, 586].

[579] *R.J. Fryer, J.V. Earley, E. Evans, J. Schneider, L.H. Sternbach*, J. Org. Chem. *35*, 2455 (1970).

[580] *M. Harfenist, E. Magnien*, J. Amer. Chem. Soc. *80*, 6080 (1958).

[581] *S. Ito*, Bull. Chem. Soc. Japan *43*, 1824 (1970).

[582] *K. Kotera, K. Kitahonoki*, Organic Preparations and Procedures *1*, 305 (1969).

[583] *J. Humbert, A. Laurent*, Bull. Soc. Chim. France 1471 (1967);
G. Alvernhe, A. Laurent, Bull. Soc. Chim. France 3003 (1970).

[584] *J. Fouché*, Bull. Soc. Chim. France 1376 (1970).

[585] *K. Kotera, K. Kitahonoki*, Org. Synth. *48*, 20 (1968).

[586] *K. Kitahonoki, Y. Takano, A. Matsuura, K. Kotera*, Tetrahedron *25*, 335 (1969).

[587] *K. Kotera, Y. Takano, A. Matsuura, K. Kitahonoki*, Tetrahedron *26*, 539 (1970).

That the reduction of oximes with lithium tetrahydroaluminate can lead to surprising results may be illustrated by the following example[584]:

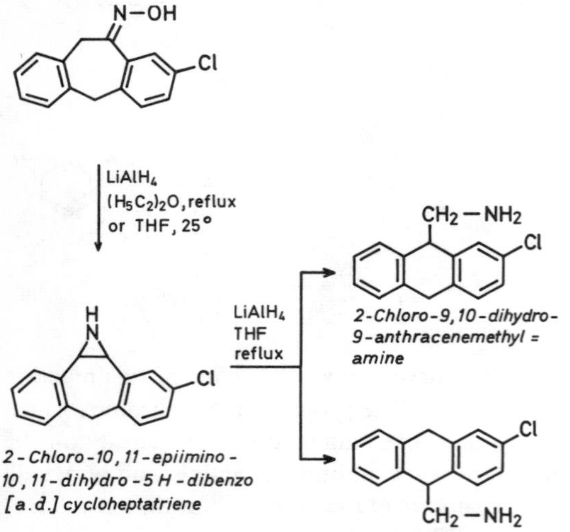

2-Chloro-10,11-epiimino-10,11-dihydro-5H-dibenzo [a.d.] cycloheptatriene

2-Chloro-9,10-dihydro-9-anthracenemethyl = amine

3-Chloro-9,10-dihydro-9-anthracenemethyl = amine

Following reductive aziridine formation a sequence of rearrangements[584] proceeding under the influence of lithium tetrahydroaluminate leads to ring contraction.

In conclusion, the possibility of converting aziridines into amines by catalytic hydrogenation[586, 588] is pointed to. During hydrogenation of optically active aziridines a dependence of retention or inversion of configuration on the catalyst is observed[589].

12.5.3 Amines from amine oxides

Reduction of amine oxides to amines is important especially in *heterocyclic* chemistry. *N*-Oxides display properties during substitution reactions that differ from those of the corresponding bases. As a result it is frequently required to convert the *N*-oxide into the free base following substitution. Surveys[590–594] give information about the versatile methods available for carrying out such reactions.

Fundamentally, the chemical and catalytic techniques customarily used for reducing the nitro groups serve also for deoxygenating *N*-oxides. *Iron* in acetic acid or *zinc* in acid[405, 590, 594] are frequently employed. Of sulfur compounds as reducing agents, *sodium dithionite*[405, 591–594] and *sulfur dioxide* are employed predominantly, and the latter has been used for reducing numerous substituted pyridine *N*-oxides in boiling 1,4-dioxane or water[595]. The rapid and quantitatively proceeding reduction with *titanium(III) chloride* can be applicated in analysis[596]. Of catalytic methods, hydrogenation over Raney nickel in glacial acetic acid/acetic anhydride under normal conditions has gained general acceptance[590, 591, 593, 594]; platinum and palladium catalysts serve a similar purpose[590, 592–594]. A possible hydrogenolysis of halogen substituents needs to be allowed for especially when palladium is employed.

With all the above-named chemical or catalytic procedures nitro groups on the aromatic compound are reduced before deoxygenation of the *N*-oxide takes place. Thus, while it is possible to discover particularly mild conditions which, allow, for instance, nitropyridine *N*-oxides to be *selectively* reduced to *aminopyridine N-oxides, aminopyridines* are invariably obtained if reduction of the *N*-oxide group is the aim[590, 592, 593, 597]. However, selective reduction to nitropyridines succeeds with a group of special deoxygenation reagents, of which phosphorus(III) chloride in chloroform as solvent is in most widespread use[590, 591, 593, 594, 598, 599]. The selectivity of this reagent may be illustrated by the following example[600]:

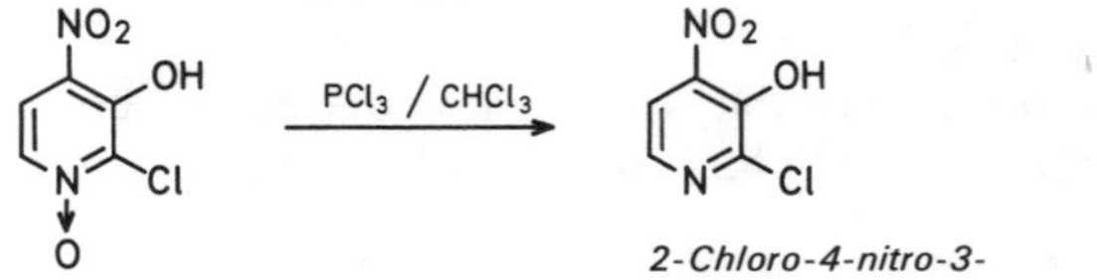

2-Chloro-4-nitro-3-pyridinol

---

[588] *O.C. Dermer, G.E. Ham,* Ethylenimine and other Aziridines, p. 296, Academic Press, New York 1969.

[589] *Y. Sugi, S. Mitsui,* Bull. Chem. Soc. Japan *42,* 2984 (1969).

[590] *E. Ochiai,* J. Org. Chem. *18,* 534 (1953); *E. Ochiai,* Aromatic Amine Oxides, p. 184, Elsevier Publ. Co., Amsterdam 1967.

[591] *K. Thomas, D. Jerchel,* Angew. Chem. *70,* 719 (1958).

[592] *E.N. Shaw* in *E. Klingsberg,* Pyridine and its Derivatives, Part. II, p. 97, Interscience Publishers, New York 1961.

[593] *K. Schofield,* Hetero-Aromatic Nitrogen Compounds, p. 384, Butterworths & Co., London 1967.

[594] *A.R. Katritzky, J.M. Lagowski,* Chemistry of the Heterocyclic N-Oxides, p. 166, Academic Press, London 1971.

[595] *F.A. Daniher, B.E. Hackley,* J. Org. Chem. *31,* 4267 (1966).

[596] *R.T. Brooks, P.D. Sternglanz,* Anal. Chem. *31,* 561 (1959).

[597] *L.H. Klemm, I.T. Barnisch, R. Zell,* J. Heterocycl. Chem. *7,* 81 (1970).

[598] *T.R. Emerson, C.W. Rees,* J. Chem. Soc. (London) 2319 (1964).

Triphenylphosphine is a further reagent of this type. It is used for deoxygenating aromatic *N*-oxides in the absence of solvent and at temperatures allowing distillation of the amine formed[601].

An alternative possibility of selectively deoxygenating aromatic *N*-oxides consists in *O*-alkylation, for instance, with benzyl halides, followed by cleavage with sodium hydroxide[590, 591, 593], *e.g.*:

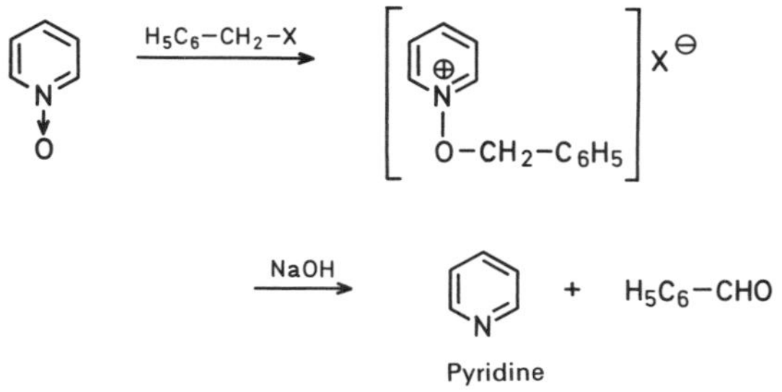

Conversion of 7-chloro-2-methylamino-5-phenyl-3 H-1,4-benzodiazepin-4-oxide with hydrogen-Raney nickel in 1,4-dioxane under normal conditions or with phosphorus(III) chloride[602] to give the deoxy derivative may be adduced as an example[602] of the reduction of an *imine N-oxide* (nitrone).

For the reduction of *aliphatic N*-oxides substantially milder conditions suffice than for aromatic ones. Thus, lower temperatures are adequate for reactions with triphenylphosphine[408, 601]. With sulfur dioxide a reduction of aliphatic *N*-oxides succeeds even in the cold[405]; utilizing this fact allows selective reduction of, *e.g.*, nicotine *N,N'*-dioxide to *nicotine N-oxide* while preserving one of the *N*-oxide groups[603]:

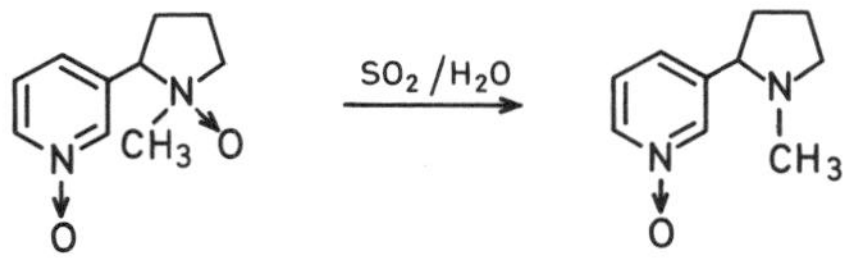

Sodium bromide in boiling 1,4-dioxane, too, has been employed as a mild deoxygenating agent for aliphatic *N*-oxides[604]; the reaction is catalyzed by peroxide.

12.5.4 Amines from azo compounds

Of methods for synthesizing amines which proceed with hydrogenolysis of the *N–N* bond, reductions of azo compounds have gained special importance in that an amino group can be introduced under mild conditions into reactive aromatic compounds, for example, phenols or anilines, *via* azo coupling. In this way, certain amines can be obtained practically free from positional isomers[405, 406, 408]. With corresponding azo couplings involving aliphatic *CH*-acid compounds *arylhydrazones* are generally formed[605].

The same *catalytic* methods[409] used for reducing nitro compounds are available for the reductive splitting of the azo group. A detailed investigation of the reduction of azo dyes in the presence of Raney nickel has been carried out[606]. In neutral to weakly alkaline solution under normal conditions a high yield of amine is obtained. The palladium-carbon catalyzed reaction with hydrazine also is a rapid and successful method for reductively splitting azo groups[607].

Of *chemical* reducing agents sodium dithionite especially has given good results[405, 408, 605, 608, 609]. The reaction is carried out in aqueous or aqueous alcoholic solution with moderate heating. One example is the synthesis of *5-amino-1-benzylindoline*[609]:

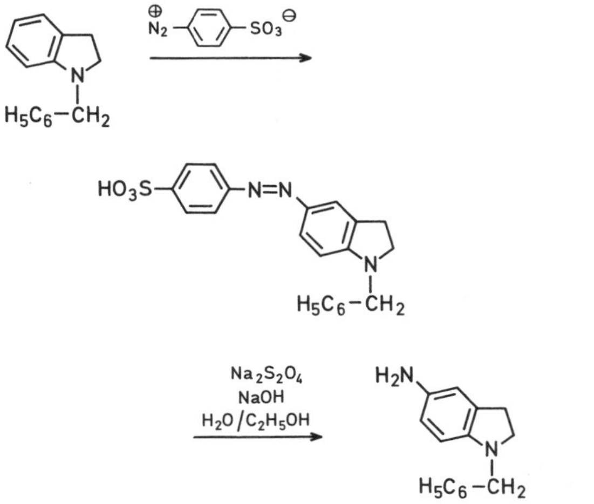

[599] *H.G.O. Becker, G. Görmar, H.-J. Timpe*, J. Prakt. Chem. *312*, 610 (1970).

[600] *K.M. Dyumaev, L.D. Smirnov, R.E. Lokhov, B.E. Zaitsev*, Izv. Akad. Nauk SSSR, Ser. Khim. 2599 (1970); Chem. Inform. B 9–352 (1971).

[601] *E. Howard, N.F. Olszewski*, J. Amer. Chem. Soc. *81*, 1483 (1959).

[602] *L.H. Sternbach, E. Reeder*, J. Org. Chem. *26*, 1111 (1961).

[603] *E.C. Taylor, N.E. Boyer*, J. Org. Chem. *24*, 275 (1959).

[604] *A.H. Khuthier*, Tetrahedron Lett. 4627 (1970).

[605] *Organicum*, 7. Aufl., VEB Deutscher Verlag der Wissenschaften, Berlin 1967.

[606] *W.F. Whitmore, A.J. Revukas*, J. Amer. Chem. Soc. *59*, 1500 (1937); *W.F. Whitmore, A.J. Revukas*, J. Amer. Chem. Soc. *62*, 1687 (1940).

[607] *N. Ichikawa, M.J. Namkung, T.L. Fletcher*, J. Org. Chem. *30*, 3878 (1965).

[608] *S. Ifrim*, Bull. Inst. Politechn. Jaşi [NF] *14*, 265 (1968); C.A. *71*, 61042 (1969).

Cleavage of the azo group with other chemical reducing agents has been reported. These include tin in hydrochloric acid[610], tin(II) chloride, and zinc or iron in acid solution[405]. With lithium aluminum hydride azoxybenzene or azobenzene are reduced only as far as hydrazobenzene[470, 476], but in one case where an azo group formed part of a heterocyclic system reductive fission was accomplished with this reagent[611].

12.5.5 Amines from hydrazones and hydrazines

The pathway of reducing hydrazones or azines for converting a carbonyl into an amine function has been little followed by comparison to oxime reduction (p. 501) or reductive amination (p. 524)[405, 406, 409, 612]. For synthesizing *α-amino ketones* and *α-amino alcohols* reduction of *arylhydrazones* has achieved importance to the extent that these can be prepared *via* azo coupling from aryldiazonium salts and, for example, *β*-dicarbonyl compounds[613, 614]. *Catalytic* hydrogenation of hydrazones has been carried out over palladium-barium sulfate[405] and platinum dioxide in glacial acetic acid[406, 613]. During hydrogenation over Raney nickel, the use of elevated pressure[405, 613, 615] and addition of ammonia[405] are recommended. The reductive cleavage also succeeds well with *chemical* reducing agents, *e.g.*, zinc in acetic acid[612, 614]. Tosylhydrazones are converted into alkanes by lithium tetrahydroaluminate in tetrahydrofuran[616].

With Raney nickel and hydrazine as hydrogen donor, *N,N-diacylhydrazines* are split into carboxamides[617], but reacting acylhydrazines with lithium tetrahydroaluminate leads not only to N–N cleavage but also to reduction to *primary amine*[618] (see p. 521).

Catalytic methods above all have proven themselves for the reductive cleavage of the N–N bond in *alkylhydrazines,* and especially hydrogenation over Raney nickel[406, 409, 615, 619, 620]. For synthesizing *methylamino sugars* in very good yield, conversion of the corresponding tosylates with methylhydrazine and subsequent reductive N–N cleavage is suitable[621], *e.g.:*

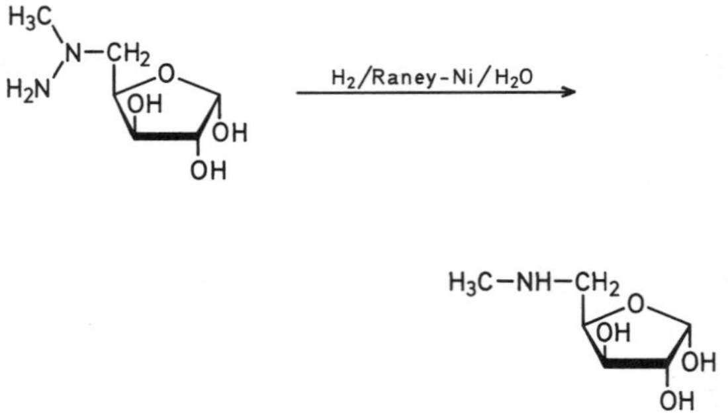

During the reaction of 1-aminoaziridin-2-yl sugars with Raney nickel and hydrazine as hydrogen donor in ethanol, stereospecific reductive opening of the aziridine ring occurs in addition to the N–N cleavage[622]:

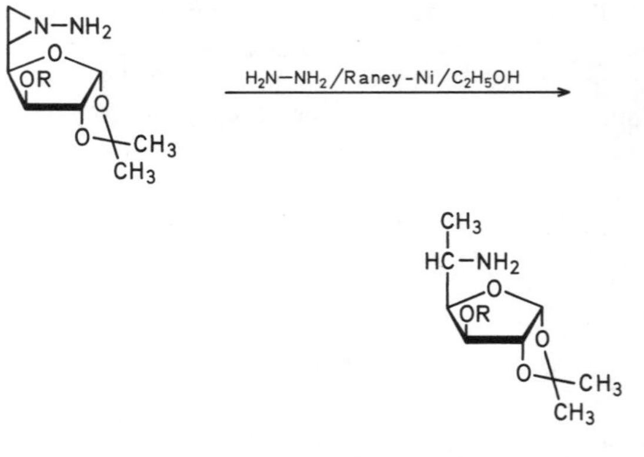

Very high yields were obtained also from the N–N cleavage of hydrazines as part of the asymmetric synthesis of *α-amino acids* by hydrogenation over palladium-carbon under normal conditions in aqueous hydrochloric acid medium[623].

[609] *H.-J. Teuber, G. Schmitt,* Chem. Ber. *102,* 1084 (1969).

[610] *M. Israel, H.K. Protopapa, S. Chatterjee, E.J. Modest,* J. Pharm. Sci. *54,* 1626 (1965).

[611] *N.L. Allinger, G.A. Youngdale,* J. Amer. Chem. Soc. *84,* 1020 (1962).

[612] *D.G. Holland, G.J. Moore, C. Tamborski,* Chem. & Ind. (London) 1376 (1965).

[613] *K. Bodendorf, W. Wössner,* Justus Liebigs Ann. Chem. *623,* 109 (1959).

[614] *D. Shapiro, T. Sheradsky,* J. Org. Chem. *28,* 2157 (1963).

[615] *G. Losse, J. Müller,* J. Prakt. Chem. [4] *12,* 285 (1961).

[616] *L. Caglioti,* Tetrahedron *22,* 487 (1966).

[617] *F.P. Robinson, R.K. Brown,* Can. J. Chem. *39,* 1171 (1961).

[618] *W.F. Armarego, T. Kobayashi,* J. Chem. Soc. C 1597 (1970).

[619] *H. Dorn, A. Zubek, G. Hilgetag,* Chem. Ber. *98,* 3377 (1965).

[620] *K.-D. Hesse,* Justus Liebigs Ann. Chem. *743,* 50 (1971).

[621] *H. Paulsen, G. Steinert,* Chem. Ber. *103,* 475 (1970).

[622] *H. Paulsen, D. Stoye,* Chem. Ber. *102,* 820 (1969).

[623] *E.J. Corey, R.J. McCaully, H.S. Sachdev,* J. Amer. Chem. Soc. *92,* 2476 (1970); *E.J. Corey, H.S. Sachdev, J.Z. Gougoutas, W. Saenger,* J. Amer. Chem. Soc. *92,* 2488 (1970).

12.5.6 Amines from diazo compounds

Reduction of diazo compounds is of little importance for synthesizing amines except in individual cases. It has been carried out successfully with diazo ketones and diazo-β-ketocarboxylic acid esters capable of being built up from diazomethane or diazoacetic acid esters and acyl halides[624], e.g.[625]:

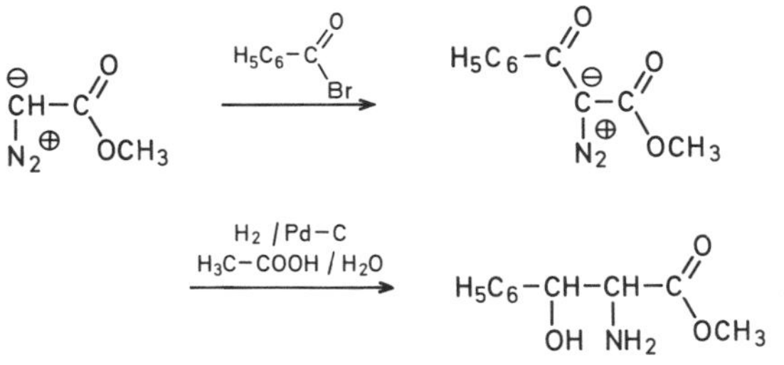

$$CH-C \quad \xrightarrow{H_5C_6-C(Br)=O} \quad H_5C_6-C-C-C \quad \xrightarrow[H_3C-COOH / H_2O]{H_2 / Pd-C} \quad H_5C_6-CH-CH-C$$

3-Phenylserine methyl ester

For such amine syntheses predominantly *catalytic* methods have been used[625, 626]. By varying the reaction conditions it is possible to obtain *amino ketones* or *amino alcohols*[406]; the possible reductive splitting off of nitrogen in this procedure is to be considered. Reduction of diazo compounds to amines can be achieved also with a variety of chemical methods: using metals leads to *amino ketones*, while lithium tetrahydroaluminate gives *amino alcohols*[626].

12.5.7 Amines from azides

Reduction of aromatic and aliphatic *azides* to amines accompanied by nitrogen evolution succeeds readily with numerous reagents[627], just as it has been substantially applied to the reduction of nitro compounds. Catalytic hydrogenations and reductions with complex metal hydrides have proven themselves. *Catalytic* hydrogenation is often carried out over platinum or palladium catalysts[406, 527, 628–631], and generally under normal conditions. Because of the evolution of nitrogen during the reduction, it is advisable to provide for gas exchange. Over Raney nickel hydrogenation succeeds with molecular hydrogen[406, 632] and with hydrazine as hydrogen donor[633].

With *lithium tetrahydroaluminate* conversion of azide into amine succeeds very rapidly and generally in high yield[627, 634–639]. Corresponding reactions with *sodium tetrahydroborate* proceed faster with aromatic azides than with aliphatic ones[640]. The reaction with sodium tetrahydroborate is suitably performed in boiling isopropyl alcohol[641, 642]; this solvent allows the extended action of the reagent required for reducing aliphatic azides to take place. As was shown with various azido steroids, the sodium tetrahydroborate reduction is speeded up markedly by adding tris(2, 2′-bipyridyl)cobalt(II) bromide as catalyst[643]. *Diborane* likewise converts azides into amines[644]. The reaction of *triphenylphosphine* with an azide to form phosphinimine and the latter's subsequent hydrolysis represents a special reduction technique[645]:

[624] F. Weygand, H.J. Bestmann in W. Foerst, Neuere Methoden der präparativen organischen Chemie, Bd. III, p. 280, Verlag Chemie, Weinheim/Bergstr. 1961.
[625] J.H. Looker, D.N. Thatcher, J. Org. Chem. 22, 1233 (1957).
[626] W. Gruber, H. Renner, Monatsh. Chem. 81, 751 (1950).
[627] J.H. Boyer, F.C. Canter, Chem. Rev. 54, 37 (1954).
[628] G. Swift, D. Swern, J. Org. Chem. 32, 511 (1967).
[629] W. Meyer zu Reckendorf, Chem. Ber. 96, 2017 (1963);
W. Meyer zu Reckendorf, Chem. Ber. 102, 4207 (1969);
W. Meyer zu Reckendorf, Chem. Ber. 104, 1976 (1971).
[630] H. Bretschneider, N. Karpitschka, G. Piekarski, Monatsh. Chem. 84, 1084 (1953).
[631] J. Cleophax, A. Goudemer, A.-M. Sepulchre, S.D. Géro, Bull. Soc. Chim. France 4414 (1970).
[632] U.G. Nayak, R.L. Whistler, J. Org. Chem. 33, 3582 (1968).
[633] K. Ponsold, Chem. Ber. 96, 1411, 1855 (1963).
[634] F.-X. Jarreau, Qui Khuong-Huu, R. Goutarel, Bull. Soc. Chim. France 1861 (1963).
[635] A.K. Bose, J.F. Kistner, L. Farber, J. Org. Chem. 27, 2925 (1962).
[636] A. Streitwieser, W.D. Schaeffer, J. Amer. Chem. Soc. 78, 5597 (1956);
A. Streitwieser, C.E. Coverdale, J. Amer. Chem. Soc. 81, 4275 (1959).
[637] P. Pratesi, L. Villa, V. Ferri, E. Grana, O.C. Mastelli, Farmaco, Ed. Sci. 25, 961 (1970).
[638] W.L. Nelson, D.D. Miller, J. Med. Chem. 13, 807 (1970).
[639] N.K. Kochetkov, A.I. Usov, K.S. Adamyants, Tetrahedron 27, 549 (1971).
[640] J.H. Boyer, S.E. Ellzey, J. Org. Chem. 23, 127 (1958).
[641] P.A.S. Smith, J.H. Hall, R.O. Kan, J. Amer. Chem. Soc. 84, 485 (1962).
[642] L. Goodman, J.E. Christensen, J. Org. Chem. 28, 158 (1963);
L. Goodman, J.E. Christensen, J. Org. Chem. 29, 1787 (1964).
[643] K. Ponsold, J. Prakt. Chem. [4] 36, 148 (1967).
[644] J.N. Wells, O.R. Tarwater, J. Pharm. Sci. 60, 156 (1971).
[645] L. Horner, H. Hofmann, Angew. Chem. 68, 473 (1956).

$$R-\underset{\underset{N_3}{|}}{CH}-COOH \xrightarrow[-N_2]{(H_5C_6)_3P} R-\underset{\underset{N=P(C_6H_5)_3}{|}}{CH}-COOH$$

$$\xrightarrow{H_2O\,/\,HBr} R-\underset{\underset{NH_2}{|}}{CH}-COOH \ + \ OP(C_6H_5)_3$$

One technique that lends itself readily to the synthesis of aromatic amines takes the following route. Grignard compounds are prepared from aryl halides and are reacted with p-toluenesulfonyl azide to form the magnesium salt of sulfonyltriazene; the latter is then reduced with Raney nickel alloy in sodium hydroxide to the *amine*[646].

Reduction of azides by catalytic methods or with metal hydrides ensues with *retention of configuration,* as numerous examples have demonstrated. On the other hand, synthesis of aliphatic azides from halides or tosylates, as an S_N2 reaction, is accompanied by in conversion with a very high degree of *stereoselectivity*. This behavior provides a pathway for transforming hydroxy into amino groups under mild reaction conditions accompanied by inversion[454, 527, 628, 629, 631–637, 647]:

$$R^2-\underset{\underset{R^3}{|}}{\overset{\overset{R^1}{|}}{C}}-OTs \longrightarrow N_3-\underset{\underset{R^3}{|}}{\overset{\overset{R^1}{|}}{C}}-R^2$$

$$\longrightarrow H_2N-\underset{\underset{R^3}{|}}{\overset{\overset{R^1}{|}}{C}}-R^2$$

It is utilized especially during syntheses of *amino sugars*[629, 632, 639, 642] and in *steroid* chemistry[527, 633–635]. In cyclic compounds it is thus possible to replace an equatorial hydroxy group by an axial amino group.

12.5.8 Amines from nitriles

12.5.8.1 Reduction using catalytic methods

Platinum and palladium catalysts and also Raney nickel have proven themselves for reducing aliphatic and aromatic nitriles with molecular

hydrogen. In addition, Raney cobalt and rhodium-alumina have gained some importance for carrying out these hydrogenations. Reduction of the nitriles proceeds to *primary amine via* the aldimine stage; reaction of the aldimine with already formed amine is the reason for the appearance of secondary and tertiary amines in the product[406, 409, 648].

A directed course of the reaction enables *primary, secondary,* or *tertiary* amines to be obtained selectively[406, 409, 648, 649].

Primary amines are obtained preponderantly by

a trapping of the primary amine as a salt
b trapping of the primary amine as amide
c addition of ammonia or alkalis.

Salt formation by the amine is achieved by carrying out the hydrogenation in glacial acetic acid or alcohols with added mineral acids such as hydrochloric or sulfuric acid.

Platinum and palladium catalysts are used predominantly for these hydrogenations[405, 406, 409, 410, 540, 648, 650, 651]; as a rule the reduction succeeds at room temperature and ordinary to slightly elevated pressure. Where the molecule contains aromatically bound nitro groups these are also reduced to amine. Halogen substituents remain intact with most hydrogenation procedures[409] but are removed when palladium is employed as catalyst[420, 651], e.g.[420]:

$$\underset{H_3CO-H_2C}{\overset{O_2N}{\diagup}}\overset{\overset{CH_3}{|}}{\underset{N}{\bigcirc}}\overset{CN}{\diagdown}{Cl} \xrightarrow[CH_3OH\,/\,HCl]{H_2\,/\,Pd-C}$$

$$\underset{H_3CO-H_2C}{\overset{H_2N}{\diagup}}\overset{\overset{CH_3}{|}}{\underset{N}{\bigcirc}}\overset{CH_2-NH_2}{\diagdown}$$

3-Amino-5-(aminomethyl)-2-(methoxymethyl)-4-methyl-pyridine

Over palladium-carbon at high reaction temperatures, for example, in boiling p-cymene, the hydrogenation can lead beyond the *amine* stage to *alkane*, especially in the case of aromatic nitriles[648, 652]. Platinum catalysts tend to hydroge-

[646] P.A.S. Smith, C.D. Rowe, L.B. Bruner, J. Org. Chem. 34, 3430 (1969).

[647] D. Arigoni, E.L. Eliel in E.L. Eliel and N.L. Allinger, Topics in Stereochemistry, Vol. IV, p. 226, Wiley Intersci. Publ., New York, London, Sydney, Toronto 1969.

[648] M. Rabinovitz in Z. Rappoport, The Chemistry of the Cyano Group, p. 307, Interscience Publishers, London 1970.

[649] H. Greenfield, Ind. Eng. Chem., Prod. Res. Develop. 6, 142 (1967).

[650] M. Freifelder, R.B. Hasbrouck, J. Amer. Chem. Soc. 82, 696 (1960).

[651] E.M. Godar, R.P. Mariella, J. Org. Chem. 25, 557 (1960).

[652] K. Kindler, K. Luehrs, Justus Liebigs Ann. Chem. 707, 26 (1967).

nate aromatic rings when allowed to act for prolonged periods. If the hydrogenation is performed in strong mineral acid solution over Raney nickel aldehydes are the main product from nitriles; this behavior can be utilized preparatively for synthesizing aldehydes[653].

Where the amine is to be trapped as amide, *acetic anhydride* as solvent is appropriate. Provided basic catalysts such as sodium acetate are used, hydrogenation over Raney nickel gives a very high yield of *acetyl derivatives* of *primary amines* under mild conditions[654], *e.g.*:

$$NC-(CH_2)_4-CN \xrightarrow[\text{(H}_3\text{C}-\text{CO)}_2\text{O / H}_3\text{C}-\text{COONa}]{\text{H}_2\text{/ Raney}-\text{Ni}}$$

N,N'-Hexamethylenebisacetamide

Over noble metal catalysts too, successful hydrogenations can be carried out under acylating conditions[655]. Thus nicotinonitrile forms *N-[(1-Acetyl-3-piperidyl)methyl]acetamide* with a mixed platinum/palladium catalyst[655].

Hydrogenation of nitriles occurs in the presence of *ammonia* and or *alkalis* over Raney nickel[405, 406, 409, 656]. Often this procedure is carried out at elevated temperature and high hydrogen pressure, if necessary with liquid ammonia as solvent[405, 406, 657, 658]. It is possible to carry out hydrogenations under mild conditions also over Raney nickel in alcohol or water saturated with ammonia or in water[659, 660]. Hydrogenations in the presence of ammonia are also possible with noble metal catalysts[661]. With all these catalysts alkali metal hydroxides also promote the formation of *primary amines;* unlike ammonia these additions accelerate the reaction[405, 662]. Rhodium catalysts

in the presence of bases have often proven themselves for these hydrogenations[662, 663], for instance, for the synthesis of *3-(2-aminoethyl)indole (tryptamine)*[663]:

Debenzylation can be substantially prevented with this catalyst[663].

On reducing *α,β-unsaturated nitriles* the *C=C* bond is hydrogenated more easily than the nitrile group, so that in these instances the saturated amine is always obtained[405, 664, 665]. One example is hydrogenation of acrylonitrile[654] to *N-propylacetamide*:

$$H_2C=CH-CN \xrightarrow[\text{(H}_3\text{C}-\text{CO)}_2\text{O, H}_3\text{C}-\text{COONa}]{\text{H}_2\text{/ Raney}-\text{Ni}} H_3C-CH_2-CH_2-NH-CO-CH_3$$

With a nonconjugated position of the *C=C* bond a reduction to amine with retention of the *C=C* bond has been reported in individual cases[406, 409]. During reduction of nitriles containing a pyridine ring, cohydrogenation of the ring to piperidine is often observed in addition to conversion of the nitrile group. A dependence on the structure of the compound and the experimental conditions is noted; a survey describes this behavior[655]. Where the structure is appropriate, ring closure is possible in such cases; such *reductive cyclizations* can be utilized for synthesizing *quinolizidines* and *indolizidines*[666]. It is quite generally necessary to allow for the possibility of ring closure reactions during reduction of nitriles under hydrogenation conditions where further functional groups are present in the molecule permitting formation of five, six or seven membered rings. This complica-

[653] *P. Tinapp*, Chem. Ber. *102*, 2770 (1969).
[654] *F.E. Gould, G.E. Johnson, A.F. Ferris*, J. Org. Chem. *25*, 1658 (1960).
[655] *E. Fischer, G. Rembarz, P. Franke*, Chimia 23, 155 (1969).
[656] *C.D. May, P. Sykes*, J. Chem. Soc. C 649 (1966).
[657] *D. Nowak, W. Jerzykiewicz*, Przem. Chem. *49*, 664 (1970); Chem. Inform. B 6–207 (1971).
[658] *J.C. Robinson, H.R. Snyder*, Org. Syntheses, Coll. Vol. III, 720 (1955).
[659] DDRP. 55 034 (1967), Erf.: *H.G. Kazmirowski, W. Conrad, L. Trampau;* C.A. *68*, 59307 (1968).
[660] *T. Kametani, K. Kigasawa, M. Hiiragi, H. Ishimaru*, J. Heterocycl. Chem. 7, 451 (1970).
[661] *N.I. Shcheglov, D.V. Sokol'skii*, Izv. Akad. Nauk Kaz. SSR, Ser. Khim. 65 (1962); C.A. *59*, 2700f (1963).

[662] *Y. Takagi, S. Nishimura, K. Taya, K. Hirota*, Sci. Pap. Inst. Phys. Chem. Res. Tokyo *61*, 114 (1967); C.A. *68*, 95476 (1968).
[663] *M. Freifelder*, J. Amer. Chem. Soc. *82*, 2386 (1960).
[664] *D.V. Sokol'skii, L.D. Volkova*, Izv. Akad. Nauk. Kaz. SSR, Ser. Khim. *14*, 69 (1964); C.A. *62*, 3933f (1965).
[665] *M. Cariou*, Bull. Soc. Chim. France 205 (1969).
[666] *V. Boekelheide, W.J. Linn, P.O'Grady, M. Lamborg*, J. Amer. Chem. Soc. *75*, 3243 (1953).

tion is most likely to arise during reduction of aminonitriles, dinitriles, or ketonitriles in alkaline medium[406, 409, 667], but as a rule conditions can be found which avoid ring closure reactions. Thus, *amino ketones* can be obtained by hydrogenation of keto nitriles over noble metal catalysts in acid solution[405]; *diamines* are accessible from dinitriles by hydrogenating under acylating conditions[654]. Similar ring closures during reduction of aminonitriles or dinitriles have been successfully avoided by employing high concentrations of ammonia and high pressure[406], and by the use of Raney cobalt[407] or cobalt boride catalysts[668], especially with technical-scale processes.

During reduction of *cyanohydrins* to amine a possible hydrogenolysis of the hydroxy group needs to be allowed for[406, 669]; it occurs particularly readily with cyanohydrins of aromatic aldehydes or ketones and their *O*-alkyl and *O*-acyl derivatives[405]. Cyanohydrins of alicyclic ketones can be successfully hydrogenated over platinum(IV) oxide in an ethanol solution of hydrogen chloride[540] or glacial acetic acid[669] to form *(aminomethyl)cyclohexanol*, e.g.:

1-(Aminomethyl)-
cyclohexanol

With hydrogen/Raney nickel 2-hydroxypropionitrile in water or ethanol[670] and 2-hydroxy-2-methyl-propionitrile under acylating conditions[654] can be converted into respectively *1-amino-2-propanol* and *1-amino-2-methyl-2-propanol*.

During synthesis of *1, 2-diamines* from α-aminonitriles the hydrogenolytic loss of hydrogen cyanide is a competitive reaction which can be substantially eliminated by hydrogenating the *N*-acyl derivatives[405, 407, 409, 671]. However, hydrogenation to diamine in good yield succeeds also by starting from free α-aminonitrile, provided platinum(IV) oxide in ethanolic hydrochloric acid[409, 650] is

selected for *N*-monosubstituted or *N*-unsubstituted compounds, and rhodium-alumina in the presence of ammonia[409, 663] is chosen for *N*-disubstituted compounds, e.g.:

4-(2-Aminoethyl)-1-
methylpiperazine

These techniques allow working at low temperature and moderate pressure. Using rhodium as catalyst[663] is appropriate, too, for reducing 2-cyanoethyl ethers. Conditions allowing such ethers to be hydrogenated to γ-alkoxyalkylamine over Raney nickel while suppressing the fission reactions substantially can be found[665, 672].

12.5.8.2 Reduction with metals

Various chemical methods have been utilized for reducing nitrile to amino groups[405]; only some metals have become important. Sodium in alcohols is paramount[408] here, and has proven itself especially for reducing simple aliphatic nitriles. With aromatic nitriles and α-phenylnitriles substantial hydrogenolytic splitting off of hydrogen cyanide is reported[405].

Various aromatic and aliphatic nitriles can be successfully reduced with Raney nickel alloy and sodium hydroxide in ethanol-water[530]. The action of the same alloy in formic acid on nitriles leads to aldehydes[673]. A good yield of benzylamine from benzonitrile is obtained with nickel precipitated from nickel chloride with zinc dust in boiling water[674].

12.5.8.3 Reduction with metal hydrides

Of metal hydrides available for reducing the nitrile group lithium tetrahydroaluminate has been the one studied by far the most[470-473, 546, 648]. In the majority of cases both aliphatic and aromatic nitriles are reduced to primary amines without complications using this reagent. To avoid side-reactions it is advisable to employ a marked ex-

[667] *H. Mikolajewska, A. Kotelko*, Acta Pol. Pharm. *22*, 219 (1965); C.A. *63*, 17891b (1965).

[668] *B.D. Polkovnikov, L.K. Freidlin, A.A. Balandin*, Izv. Akad. Nauk SSSR, Ser. Khim. 1488 (1959); C.A. *54*, 1264h (1960).

[669] *B. Tchoubar*, Bull. Soc. Chim. France 160 (1949).

[670] *L.D. Volkova, D.V. Sokol'skii*, Izv. Akad. Nauk Kaz. SSR, Ser. Khim. *17*, 60 (1967); C.A. *67*, 53293 (1967).

[671] *A. Kotelko*, Acta Pol. Pharm. *19*, 109 (1962); C.A. *59*, 1482e (1963).

[672] *A. Gaiffe, C. Lannay*, C.R. Acad. Sci., Paris, Ser. C *264*, 1496 (1967).

[673] *T. van Es, B. Staskun*, J. Chem. Soc. (London) 5775 (1965).

[674] *K. Sakai, K. Watanabe*, Bull. Chem. Soc. Japan *40*, 1548 (1967).

cess of at least 1–2 *mols* lithium tetrahydroaluminate per *mol* nitrile. As a rule, the reaction is conducted in ether or tetrahydrofuran and the nitrile is added slowly to the reducing agent. For preparing short-chained aliphatic amines, reduction in highboiling solvents has proven itself; at the end of the reaction the amine can be isolated simply by distillation[675], *e.g.*:

$$H_3C-CH_2-CD_2-CN \xrightarrow{\text{LiAlH}_4 / (H_5C_2O-C_2H_4)_2O}$$

$$H_3C-CH_2-CD_2-CH_2-NH_2$$

Butylamine-2,2-d₂

The reduction can be carried out without epimerization at the α-carbon, as it was shown, *e.g.*, with stereoisomeric cyanoquinolizidines[676] *(aminomethyl-quinolizidines)*. During reduction of nitriles with specially activated α-hydrogen, *e.g.*, phenylacetonitrile or diphenylacetonitrile with lithium tetrahydroaluminate to *phenethylamine* and *2,2-diphenylethylamine* respectively, evolution of hydrogen was frequently observed. The yield of amine was only moderate[677, 678], but can be increased substantially by adding equimolar amounts of aluminum chloride to the reducing agent[678]. While 10,11-dihydro-*5H*-dibenzo[a,d]-cycloheptene-5-carbonitrile is not reduced with lithium tetrahydroaluminate used on its own[679], a good conversion to amine is obtained if aluminum chloride is added[679, 680]:

10,11-Dihydro-5H-dibenzo[a,d]cycloheptene-5-methylamine

If the conditions chosen for the reduction with lithium tetrahydroaluminate prevent the intermediately formed imine stage from being reduced further rapidly, then side-reactions starting from

this imine stage can take place. Thus formation of Schiff's bases, amidines, and aldehydes can be observed especially when a deficiency of lithium tetrahydroaluminate or the inverse method of addition are employed[677, 681]. The occurrence of aldehyde can be utilized for preparative purposes in individual cases[471, 473, 648].

β-Alkoxynitriles with the $RO-CH_2-CH_2-CN$ structure tend to have their molecule split on reaction with lithium tetrahydroaluminate; under these circumstances the *alcohol* (ROH) is obtained[682, 683]. This pathway predominates on reduction in tetrahydrofuran or when the inverse method of addition is employed, but reducing in ether under conditions ensuring a continuous marked excess of lithium tetrahydroaluminate leads to a good yield of the desired *γ-alkoxyamine*[682].

During reduction of substituted cinnamonitriles to amine, coreduction of the *C=C* bond is observed[470, 473, 684]. By performing the reaction carefully, it is even possible to reduce the *C=C* bond without attacking the nitrile group in such compounds[685] and also in 1-cyclohexene-1-carbonitrile. Where the *C=C* bond is more remoter[686], the nitrile group can be reduced selectively[685, 687–690]. α-Ketonitriles are converted into *β-aminoalcohols* in good yield under the normal reaction conditions[550, 691], *e.g.*:

2-Amino-1-(4-chlorophenyl)ethanol

[675] L. Friedman, A.T. Jurewicz, J. Org. Chem. *33*, 1254 (1968).

[676] Y. Yamada, K. Hatano, M. Matsui, Agr. Biol.-Chem. (Tokyo) *34*, 1536 (1970).

[677] L.M. Soffer, M. Katz, J. Amer. Chem. Soc. *78*, 1705 (1956).

[678] R.F. Nystrom, J. Amer. Chem. Soc. *77*, 2544 (1955).

[679] L.G. Humber, M.A. Davis, Can. J. Chem. *44*, 2113 (1966).

[680] L.G. Humber, M.A. Davis, R.A. Thomas, R. Otson, J.R. Watson, J. Heterocycl. Chem. *3*, 247 (1966).

[681] H.U. Sieveking, W. Lüttke, Angew. Chem. *81*, 431 (1969).

[682] L.M. Soffer, M. Katz, E.W. Parrotta, J. Amer. Chem. Soc. *78*, 6120 (1956).

[683] M.F. Shostakovskii, A.S. Atavin, B.A. Trofimov, N.A. Vodbol'skaya, N.K. Gusarova, V.I. Lavrov, Dokl. Akad. Nauk SSSR *166*, 1381 (1966); C.A. *64*, 17409b (1966).

[684] A. Dornow, G. Messwarb, H.H. Frey, Chem. Ber. *83*, 445 (1950).

[685] H. LeMoal, R. Carrié, M. Bargain, C.R. Acad. Sci., Paris *251*, 2541 (1960).

[686] E.A. Braude, O.H. Wheeler, J. Chem. Soc. (London) 327 (1955).

[687] H. Heusser, P.T. Herzig, A. Fürst, Pl.A. Plattner, Helv. Chim. Acta *33*, 1093 (1950).

[688] J.D. Roberts, W.F. Gorham, J. Amer. Chem. Soc. *74*, 2278 (1952).

[689] M. Protiva, J.O. Jilek, V. Hach, E. Adlerová, V. Mychajlyszyn, Chem. Listy *51*, 2109 (1957).

[690] C.J. Cavallito, H.S. Yun, M.L. Edwards, F.F. Foldes, J. Med. Chem. *14*, 130 (1971).

[691] A. Burger, F.D. Hornbaker, J. Amer. Chem. Soc. *74*, 5514 (1952).

The reaction proceeds anomalously with malono-nitriles[692]; even with excess lithium tetrahydroalu-minate *cis-* and *trans-β*-amino acrylonitrile and not diamine is formed[692]:

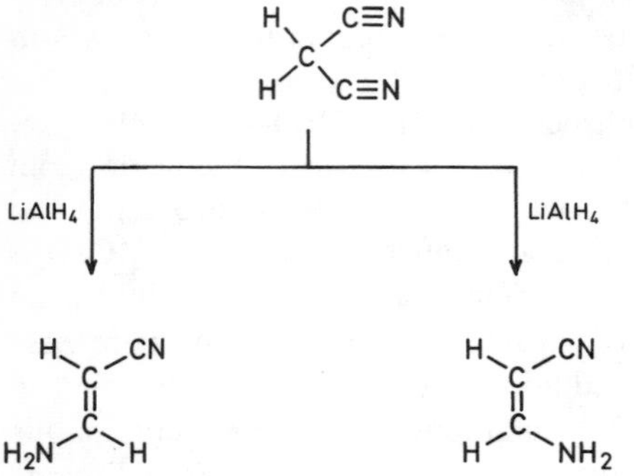

If in dinitriles the two functional groups are linked *via* two carbon atoms so that a strainfree five-membered amidine is able to form, this compound is the predominant product isolated on reduc-tion[681], *e.g.:*

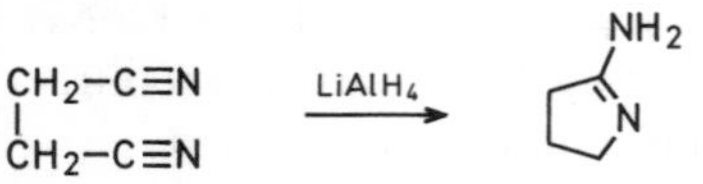

2-Amino-1-pyrroline

This ring closure is explained by the nucleophilic attack of a primarily formed imine group on the second nitrile group provided a suitable spatial orientation is pre-sent[681].

As a rule, cyanohydrins are reduced to *β-aminoal-cohols* with lithium tetrahydroaluminate without difficulty; once more, side-reactions can be avoided by using a marked excess of the reducing agent[470, 693]. Good results have been reported from the reduction of α-acylnitriles[687, 688].

During reaction of α-aminonitriles with lithium tetrahydroaluminate reductive splitting off of hydrogen cyanide is possible in addition to the reduction to *diamine*. Its extent is dependent strongly on the structure of the α-aminonitrile, and in compounds with a completely substituted α-car-bon atom it may predominate[694–697]. Example where this behavior have been observed are the reductions of 2-cyano-1,2-dimethyl piperidine[694], and of the corresponding pyrrolidine[695], and it

takes place, too, with α-aminonitriles whose α-carbon is a constituent of a five-membered ring[696, 698]. A further instance of its occurrence is the formation of substituted *4-aminopiperi-dines*[697], *e.g.:*

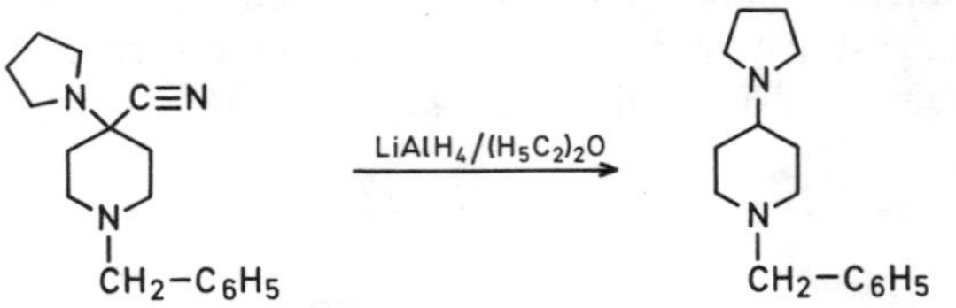

1-Benzyl-4-(1-pyrroli-dinyl)-piperidine

Where it is desired to split the nitrile group off en-tirely, addition of aluminum chloride may be suc-cessful if lithium tetrahydroaluminate alone does not bring about the intended result[696].

α-Aminonitriles with an incompletely substituted α-carbon atom are generally reduced to the di-amine. A special case is observed by compounds with a 3-indolyl substituent; these are converted into *3-aminomethylindoles*[699], *e.g.:*

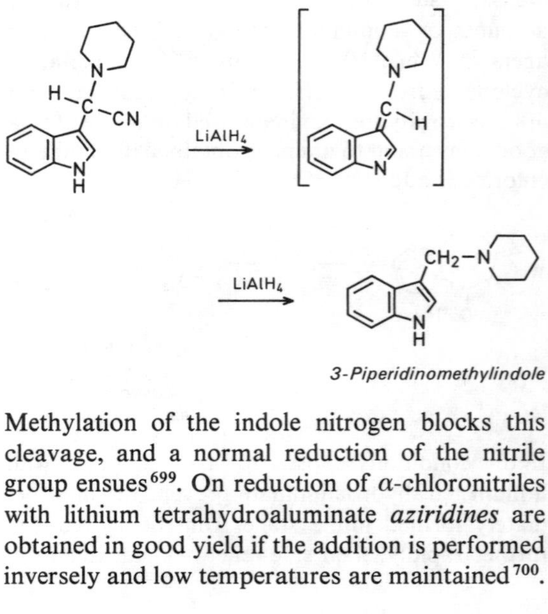

Methylation of the indole nitrogen blocks this cleavage, and a normal reduction of the nitrile group ensues[699]. On reduction of α-chloronitriles with lithium tetrahydroaluminate *aziridines* are obtained in good yield if the addition is performed inversely and low temperatures are maintained[700].

3-Piperidinomethylindole

[692] *H.U. Sieveking, W. Lüttke*, Angew. Chem. *81*, 432 (1969).

[693] *H.R. Nace, B.B. Smith*, J. Amer. Chem. Soc. *74*, 1861 (1952).

[694] *N.J. Leonard, F.P. Hauck*, J. Amer. Chem. Soc. *79*, 5279 (1957).

[695] *E.J. Corey, W.R. Hertler*, J. Amer. Chem. Soc. *82*, 1657 (1960).

[696] *G. Chauvière, B. Tchoubar, Z. Welvart*, Bull. Soc. Chim. France 1428 (1963).

[697] *B. Hermans, P. Van Daele, C. van de Westeringh, C. Van der Eycken, J. Boey, P.A.J. Janssen*, J. Med. Chem. *9*, 49 (1966).

[698] *R. Granger, H. Técher*, Bull. Soc. Chim. France 600 (1960).

[699] *P. Rajagopalan, B.G. Advani*, Tetrahedron Lett. 2197 (1965).

[700] *K. Ichimura, M. Ohta*, Bull. Chem. Soc. Japan *40*, 432 (1967).

An interesting process for synthesizing amines from nitriles consists in Grignard reaction of the nitrile and an immediately following reduction of the concerning adduct with lithium tetrahydroaluminate[701]:

$$R^1-MgX \ + \ R^2-CN \ \longrightarrow \ \underset{R^1-\overset{\|}{C}-R^2}{\overset{N-MgX}{}}$$

$$\xrightarrow{LiAlH_4 \ /(H_5C_2)_2O} \ R^1-\underset{|}{\overset{NH_2}{\underset{CH}{}}}-R^2$$

This pathway also allows *asymmetric* amines to be synthesized. The reduction step is performed with a lithium tetrahydroaluminate reagent modified by optically active alcohols[702]. By comparison to lithium tetrahydroaluminate, the usefulness of other metal hydrides for synthesizing amines from nitriles has been far less investigated. *Lithium hydrotrimethoxyaluminate* has been found to be likewise suitable[480], and for the reduction of aromatic nitriles also *sodium dihydrobis(2-methoxyethoxy)aluminate*[473, 554]. *Sodium tetrahydroaluminate,* too, has been used for reducing nitriles to amines; if the reaction is carried out carefully, aldehydes can be formed with this reagent[473, 546]. The latter reaction is predominant if sodium[479] or *lithium hydrotriethoxyaluminate*[703] are used; these two therefore represent versatile reagents for synthesizing aldehydes from nitriles. By varying the reaction conditions the reduction of benzonitrile with *sodium dihydrodiisobutylaluminate* can be guided to either *benzaldehyde* or *benzylamine;* here, too, amine formation requires the maintenance of a constant excess of reducing agent[704].

Aluminum hydride[481] is a suitable reagent for synthesizing amines from aliphatic and aromatic nitriles[481] and the yields obtained in this way are particularly high. Aluminum hydride serves also for reducing nitriles where lithium tetrahydroaluminate gives unsatisfactory results; one example is the synthesis of *3-butenylamine:*

$$H_2C=CH-CH_2-C\equiv N \ \xrightarrow{AlH_3/THF,25°}$$

$$H_2C=CH-CH_2-CH_2-NH_2$$

Reduction of cinnamonitrile with aluminum hydride yields a mixture of mainly *cinnamylamine* and some *phenylpropylamine*[481].

Both aliphatic and aromatic nitriles react rapidly with diborane to form *primary amines*[705]. The reaction is carried out at room temperature in dimethyl diglycol or tetrahydrofuran and, as far as it has been investigated, gives good yields. The diborane may be introduced as such or formed *in situ* from sodium tetrahydroborate and boron trifluoride etherate. An interesting reaction is the selective reduction with retention of the nitro group[705], *e.g.:*

3-Nitrobenzylamine

Sodium tetrahydroborate does not reduce the nitrile group[475] unless special catalytic additives are present. Compounds with a nitrile carbon atom that is particularly activated towards attack by the tetrahydroborate anion form as special case, *e.g.,* reduction of perfluorinated alkyl nitrile to *perfluoroalkylamine* succeeds in good yield[706]. Additives which make the reduction of aliphatic and aromatic nitriles with sodium tetrahydroborate possible include especially Raney nickel[707] and salts of transition metals[491] as well as aluminum chloride[473, 475].

The reaction of benzonitrile and butyronitrile with sodium tetrahydroborate in the presence of Raney nickel in methanol-water solution made alkaline with sodium hydroxide gives very good yields of *benzylamine* and *butylamine;* its course has been investigated in detail[707].

A reduction technique using a sodium tetrahydroborate-transition metal salt system displays versatile applicability[491]. Cobalt(II) chloride especially has been used as catalyst for this reaction, which is generally conducted in methanol at room temperature; chlorides of nickel, osmium, iridium, and platinum are also suitable. The possibility of carrying out *selective* reductions is noteworthy. Thus, acrylonitrile can be reduced to *allylamine;*

[701] *H. Pohland, H.R. Sullivan,* J. Amer. Chem. Soc. *75,* 5898 (1953).

[702] *O. Červinka, V. Suchan, O. Kotýnek, V. Dudek,* Collect. Czech. Chem. Commun. *30,* 2484 (1965).

[703] *H.C. Brown, C.J. Shoaf,* J. Amer. Chem. Soc. *86,* 1079 (1964);
H.C. Brown, C.P. Garg, J. Amer. Chem. Soc. *86,* 1085 (1964).

[704] *L.I. Zakharkin, V.V. Gavrilenko,* Izv. Akad. Nauk SSSR, Ser. Khim. 2245 (1960); C.A. *55,* 14353a (1961).

[705] *H.C. Brown, B.C. Subba Rao,* J. Amer. Chem. Soc. *82,* 681 (1960).

[706] *S.E. Ellzey, J.S. Wittman, W.J. Connick,* J. Org. Chem. *30,* 3945 (1965).

[707] *R.A. Egli,* Helv. Chim. Acta *53,* 47 (1970).

if low temperatures are maintained a nitro group can be preserved[491], *e.g.*:

$$O_2N-\langle\!\bigcirc\!\rangle-CH_2-C\!\equiv\!N \xrightarrow{\text{NaBH}_4 \,/\, \text{CoCl}_2 \,/\, \text{CH}_3\text{OH}, \, -10°}$$

$$O_2N-\langle\!\bigcirc\!\rangle-CH_2-CH_2-NH_2$$

4-Nitrophenethylamine

Secondary amines can be synthesized elegantly from nitriles by using sodium tetrahydroborate. The nitrile is alkylated at the nitrogen with either trialkyloxonium tetrafluoroborates or dialkoxycarbonium tetrafluoroborates and, after reacting with an alcohol, reduced to amine with sodium tetrahydroborate[708]:

$$H_5C_6-C\!\equiv\!N \xrightarrow{\text{HC}\overset{\oplus}{\underset{OR}{\overset{OR}{|}}} \, BF_4^{\ominus}} \left[H_5C_6-C\!\equiv\!\overset{\oplus}{N}-R \right] BF_4^{\ominus}$$

$$\xrightarrow{\text{R}^1\text{-OH}} H_5C_6-\underset{OR^1}{\overset{OR^1}{\underset{|}{C}}}\!=\!N-R \xrightarrow{\text{NaBH}_4} H_5C_6-CH_2-NH-R$$

12.5.9   Amines from carboxamides

12.5.9.1   Catalytic hydrogenation

Reduction of carboxamides may be used for synthesizing *secondary* or *tertiary amines* as well as for *primary* ones in relation to the substituent on the amide nitrogen. Hydrogenating amides by catalytic techniques is invariably difficult; other reducible groups are always attacked first. Correspondingly, conversion of the amide group into amine requires very *drastic* conditions; often 200–300 atmospheres at around 250° are employed. Copper chromite contact catalysts serve as catalysts and dioxane as solvent[405, 406, 409].

The high reaction temperatures required favor *side-reactions*. Formation of primary alcohols, and, during reduction of *N*-substituted carboxamides, of alkanes are examples; in addition, the amides may be saponified to the carboxylic acid. When alcohols are used as solvent, esterification and subsequent reduction of the ester causes formation of the alcohol derived from the carboxylic acid to be the predominant reaction[406, 409]. Formation of secondary amines during reduction of

amides without substituents at the nitrogen may be substantial; it can be prevented by hydrogenating in the presence of ammonia[709].

The successful reduction of some *cyclic* amides of particular structure, for example, *cytisine*, over platinum in hydrochloric acid solution under normal conditions has been reported[405, 406, 409].

Conversion of carboxamides into amines under *mild* conditions can be achieved following a preliminary conversion into thiocarboxamides. The latter are then reduced and desulfurized with Raney nickel charged with hydrogen[710]. Corresponding reduction of thiobenzamides without a substituent at the nitrogen gave predominantly aldehydes[405]. This pathway to amine *via* a thiolactam has been employed successfully for reducing cyclic amides, *e.g.*, erythrinalactam[711].

12.5.9.2   Reduction with sodium

Reduction of the carboxamide group with sodium in alcohols is no longer important. With noncyclic amides reduction to alcohol and saponification are the predominant results. The method is useful only for reducing *lactams;* and even here the yield of amine is often only moderate because of side-reactions of this type[405]. Reaction with sodium in liquid ammonia enables *aldehydes* to be prepared from amides[712].

12.5.9.3   Reduction with metal hydrides

Reduction of carboxamides with *lithium tetrahydroaluminate* has developed into a *standard method* for synthesizing amines. Using this reagent, amides of very widely different structure and including lactams, imides, and urethans can generally be carried out in good yield and under mild conditions. The many reactions, including in particular the field of *alkaloid* and *heterocyclic* chemistry are described in several surveys[470, 471, 473, 546, 712].

To make the reaction proceed to amine as completely as possible, it is necessary to maintain continuously a marked excess of hydride reagent. For this reason amide is generally added to the lithium tetrahydroaluminate in ether or tetrahydrofuran. Because of the acidity of the amide hydrogen, unsubstituted or monosubstituted amides require a markedly greater excess of lithium tetrahydroaluminate than disubstituted ones; the latter are generally reduced rapidly[476, 713]. Occasionally, the re-

[708] *R.F. Borch*, Chem. Commun. 442 (1969);
    R.F. Borch, J. Org. Chem. *34*, 627 (1969).
[709] *A. Guyer, A. Bieler, G. Gerliczy*, Helv. Chim. Acta *38*, 1649 (1955).

[710] *E.C. Kornfeld*, J. Org. Chem. *16*, 131 (1951).
[711] *A. Mondon, P.R. Seidel*, Chem. Ber. *103*, 1298 (1970).
[712] *B.C. Challis, J.A. Challis* in *J. Zabicky*, The Chemistry of Amides, p. 795, Interscience Publishers, London 1970.

duction has been performed in the presence of aluminum chloride; very good yields of amine were obtained throughout[714, 715].

During reduction of *N-unsubstituted* amides, nitrile formation has been observed in individual cases[679, 716], especially when deficient lithium tetrahydroaluminate is employed[716]. Reacting *N, N-di-substituted* carboxamides under conditions ensuring a constant deficiency of reagent with lithium tetrahydroaluminate at low temperatures results in *C–N cleavage* and formation of aldehyde[471, 712, 713]. This cleavage predominates entirely if the amide is sterically hindered[713], as is the case, for example, with *N*-dimethyl-2, 2-dimethylpropionamide[717] and *N-tert*-butylanisamide[718], or where the amide nitrogen electron pair mesomerism is hindered[717, 719, 720]. This applies, for example, to *N*-methylanilides[721, 722], and diarylcarboxamides, as well as to heterocycles of comparable structure such as *N*-acylphenothiazines[471, 713, 723], *N*-acylaziridines[472, 717] and *N*-acylpyrroles, *N*-acylcarbazoles[713], and *N*-acylimidazoles[719]. In these compounds the mentioned *C–N* cleavage arises also under experimental conditions that normally lead to *amine* formation. Depending on how much hydride is used, either a primary alcohol or the corresponding aldehyde is obtained. This property of the above-named compounds can thus be made use of for synthesizing aldehydes, including unsaturated ones. At the same time it defines the limits of the accessibility of amines by reducing carboxamides with lithium tetrahydroaluminate.

Formation of amines from carboxamides with lithium tetrahydro aluminate can be developed into methods for the reductive alkylation of amines. After formylating or acetylating of the amine reduction takes place with lithium tetrahydroaluminate[470, 471, 473, 546]. The following example, which at the same time demonstrates the *C–N* cleavage on the nitrogen of the aromatic heterocycle, illustrates the principle involved[724]:

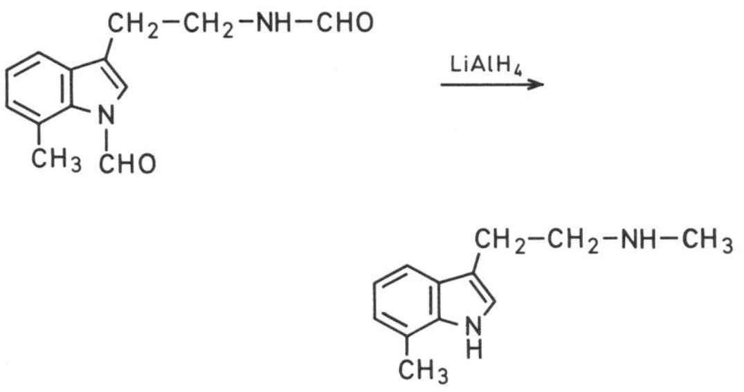

7-Methyl-3-(2-methylamino-ethyl)indole

Alternatively, the reductive alkylation can be performed in such a way as to subject the amine to be alkylated and an ester to reaction with lithium tetrahydroaluminate simultaneously[725, 726]:

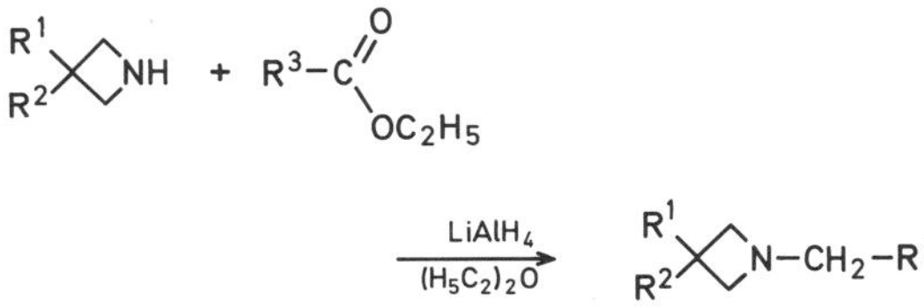

One way to synthesize *N-methylamines* is to reduce urethans with lithium tetrahydroaluminate[471, 473, 727–729]:

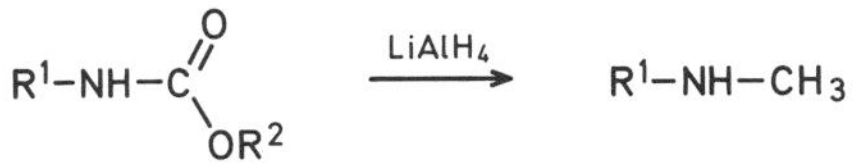

Reduction of lactams to *cyclic amines* with lithium tetrahydroaluminate has found wide application because it makes a large number of heterocyclic

[713] *V.M. Mićović, M.Lj. Mihailović*, J. Org. Chem. *18*, 1190 (1953).

[714] *R.F. Nystrom, C.R.A. Berger*, J. Amer. Chem. Soc. *80*, 2896 (1958).

[715] *M. Shamma, P.D. Rosenstock*, J. Org. Chem. *26*, 718 (1961).

[716] *M.S. Newman, T. Fukunaga*, J. Amer. Chem. Soc. *82*, 693 (1960).

[717] *H.C. Brown, A. Tsukamoto*, J. Amer. Chem. Soc. *86*, 1089 (1964).

[718] *T. Axenrod, L. Loew, P.S. Pregosin*, J. Org. Chem. *33*, 1274 (1968).

[719] *H.A. Staab, H. Bräuling*, Justus Liebigs Ann. Chem. *654*, 119 (1962).

[720] *E.I. Levkoeva, E.S. Nikitskaja, L.N. Jachontov*, Dokl. Akad. Nauk SSSR *192*, 342 (1970); Chem. Inform. B 42–357 (1970).

[721] *C.W. Whitehead, J.J. Traverso, H.R. Sullivan, F.J. Marshall*, J. Org. Chem. *26*, 2814 (1961).

[722] *F. Weygand, H.J. Bestmann*, Chem. Ber. *92*, 528 (1959).

[723] *E. Ziegler, U. Rossmann, F. Litvan, H. Meier*, Monatsh. Chem. *93*, 26 (1962).

[724] *E. Giovannini, T. Lorenz*, Helv. Chim. Acta *40*, 1553 (1957);
E. Giovannini, T. Lorenz, Helv. Chim. Acta *41*, 113 (1958).

[725] *W.B. Wright*, J. Org. Chem. *27*, 1042 (1962).

[726] *E. Testa, L. Fontanella, L. Mariani, G.F. Cristiani*, Justus Liebigs Ann. Chem. *633*, 56 (1960).

[727] *E. Magnien, R. Baltzly*, J. Org. Chem. *23*, 2029 (1958).

[728] *R.L. Dannley, R.G. Taborsky*, J. Org. Chem. *22*, 77 (1957).

[729] *F. Zymalkowski, T. Yupraphat, K. Schmeißer*, Arch. Pharm. (Weinheim, Ger.) *301*, 321 (1968).

compounds accessible[470, 471, 473, 546]. As a rule, the reduction of lactams with five or more members proceeds in high yield without difficulty. Reference is made to several newer heterocycle syntheses of this type[730-737].

During reduction of four-membered ring lactams, the azetidinones and azetidinediones, it is observed that the corresponding *azetidines* are formed provided that the amide nitrogen is unsubstituted[726, 738, 739], *e.g.:*

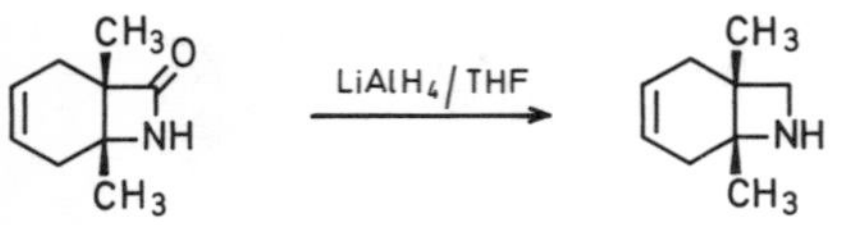

1,6-Dimethyl-7-azabicyclo-
[4.2.0.]oct-3-ene

By contrast, ring fission to *aminoalcohols* is invariably found during the reaction of *N*-substituted azetidinones and azetidinediones with lithium tetrahydroaluminate[471, 726, 738], but *N*-acylazetidines are reduced to *N*-alkylazetidines[726]. During reduction of three-membered lactams no aziridine could be obtained; reacting an aziridinone with lithium tetrahydroaluminate led to the *β-amino alcohol* by reductive ring opening[740].

Reduction of dicarboxylic acid imides, too, has been utilized successfully for synthesizing hetero-

cyclic compounds in many cases, especially for five- to seven-membered rings[471, 741-746]. Examples worth mentioning are conversions of succinimides to *pyrrolidines* and of glutarimides to *piperidines*. The last-named can be obtained also by reducing glutaramides; *trans-3,5-dimethylpiperidine* was obtained in this way, for example[747]. In individual cases a lactam intermediate could be isolated during reduction of imides to amine[471, 546]; *2-pyrrolidinone-5,5-d₂* can be prepared in this way[748]:

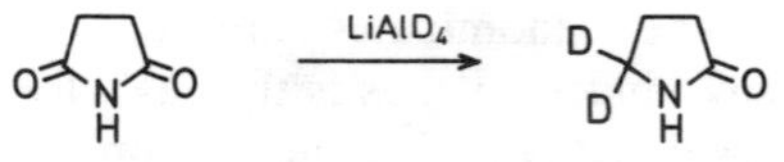

Careful working during reduction of lactams can yield a partial reduction to *cyclic α-hydroxyamines* (cyclic semi-aminals of ω-aminoaldehydes)[470], *e.g.:*

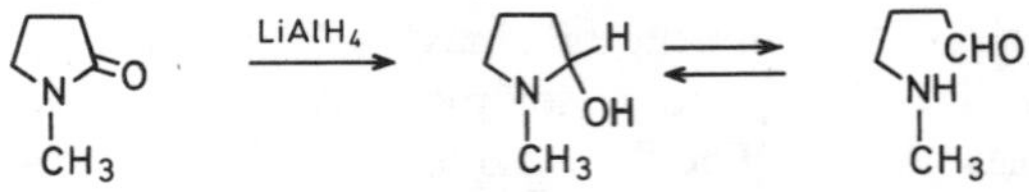

1-Methyl-   4-Methyl-
2-pyrrolidinol   amino-
                    butyral-
                    dehyde

During reduction of imides, in particular, dehydration and hence introduction of a double bond into the ring[745] or a reductive ring opening to the amino alcohol can ensue, starting from such a part-reduction stage. The behavior observed is found to be dependent on the substitution at the nitrogen[742]. Reference is made to further possible structure-dependent cleavage reactions during reduction of amides[749-751].

On reducing lactams with condensed on aromatic rings, *aromatization* of the hetero ring was observed in dependence on the structure and the reaction conditions[471]. Thus, *indoles* can be obtained from isatins[724, 752], *e.g.:*

[730] *W. Schneider, R. Dillmann*, Chem. Ber. *96*, 2377 (1963).
[731] *L.Y. Barkovskaya, M.V. Shostakovskii*, Izv. Akad. Nauk SSSR, Ser. Khim. 2773 (1967); C.A. *69*, 86806 (1968).
[732] *G.N. Walker, D. Alkalay, A.R. Engle, R.J. Kempton*, J. Org. Chem. *36*, 466 (1971); *G.N. Walker, D. Alkalay*, J. Org. Chem. *36*, 491 (1971).
[733] *D.O. Spry, H.S. Aaron*, J. Org. Chem. *34*, 3674 (1969).
[734] *A.A. Santilli, T.S. Osdene*, J. Org. Chem. *31*, 4269 (1966).
[735] *V.L. Narayanan, L. Setescak*, J. Heterocycl. Chem. *7*, 841 (1970).
[736] *H. Fritz, E. Stock*, Tetrahedron *26*, 5821 (1970).
[737] *R. Sarfati, M. Pais, F.-X. Jarreau*, Bull. Soc. Chim. France 255 (1971).
[738] *E. Testa, L. Fontanella, G.F. Cristiani, L. Mariani*, Helv. Chim. Acta *42*, 2370 (1959).
[739] *L.A. Paquette, J.F. Kelly*, J. Org. Chem. *36*, 442 (1971).
[740] *J.L. Sheehan, I. Lengyel*, J. Org. Chem. *31*, 4244 (1966).
[741] *R. Griot*, Helv. Chim. Acta *42*, 67 (1959).
[742] *K.C. Schreiber, V.P. Fernandez*, J. Org. Chem. *26*, 1744 (1961).
[743] *R.T. Major, R.J. Hedrick*, J. Org. Chem. *30*, 1270 (1965).

[744] *G.N. Walker, D. Alkalay*, J. Org. Chem. *36*, 461 (1971).
[745] *E. Tagmann, E. Sury, K. Hoffmann*, Helv. Chim. Acta *37*, 185 (1954).
[746] *E. Dornhege*, Justus Liebigs Ann. Chem. *743*, 42 (1971).
[747] *R. Hiltmann, H. Hulpke*, Synthesis 213 (1971).
[748] *A.M. Duffield, H. Budzikiewicz, C. Djerassi*, J. Amer. Chem. Soc. *86*, 5536 (1964).
[749] *H.J. Bein, A.R. Day*, J. Heterocycl. Chem. *7*, 355 (1970).
[750] *E.E. Garcia, A. Arfaei, R.I. Fryer*, J. Heterocycl. Chem. *7*, 1161 (1970).
[751] *P. Aeberli, W.J. Houlihan*, J. Org. Chem. *34*, 1720 (1969).
[752] *A. Carlsson, A. Corrodi, T. Magnusson*, Helv. Chim. Acta *46*, 1231 (1963).

4,5,6-Trimethoxyindole

Unsaturated amides[753], including in particular α,β-unsaturated ones[473, 754], can be converted into the corresponding *α-alkenylamines* with lithium tetrahydroaluminate while preserving the $C=C$ bonds[754], e.g.:

N,N-Dimethyl-2-phenylcinnamylamine

By contrast, α,β-unsaturated six-membered ring lactams could be reduced in good yield to *piperidines* with lithium tetrahydroaluminate, especially when aluminum chloride was added[715].

Vinylogous carboxamides are reduced to the corresponding unsaturated amines by lithium tetrahydroaluminate[473, 546, 752, 755] (cf. above). The following synthesis of *tryptamine [3-(2-aminoethyl)indole]* is relevant in this connection[756]:

3-(2-Aminoethyl)indole

Ester, keto, or aldehyde carbonyl groups are reduced before the acid amide group is attacked by lithium tetrahydroaluminate; attempts to prepare amines from amides containing these groups therefore leads to *amino alcohols*. With carboxamides containing halogen substituents a *C-halo hydrogenolysis* was repeatedly reported during the

reduction to amine[473, 546, 757, 758]; this is true especially for fluorinated compounds[473, 728, 759]. Addition of aluminum chloride during reduction of *N*-bis(2-chloroethyl)amides suppressed this cleavage[758].

Lithium tetrahydroaluminate reduces *acid hydrazides* similarly to amides; either amines or alkylhydrazines are obtained as products in dependence on their structure. Reduction of simple, *N*-unsubstituted acid hydrazides led to *N—N* cleavage and amine formation[618, 760], reduction of the corresponding *N,N'*-diacylhydrazines gave the corresponding alkylhydrazines[761]. Cyclic hydrazines are formed from cyclic acid hydrazides with lithium tetrahydroaluminate[473, 762, 763]. With pyrazolones the course of the reduction is more complex in that *pyrazolidines* and *pyrazolidones* are isolated in addition to the anticipated *pyrazolines*[764]. *Amines* can be synthesized also from *hydroxamic* acids and their *O*-acyl derivatives with lithium tetrahydroaluminate[765]. In individual cases side-reactions have been reported which lead to isomeric amines as a result of rearrangements[472, 765].

Of the other complex metal hydrides, *magnesium tetrahydroaluminate*[555] and *lithium hydrotrimethoxyaluminate*[480] are suitable reducing agents for converting amides to amines. *Lithium dihydrodiethoxyaluminate* and *lithium hydrotriethoxyaluminate* have become especially important for synthesizing aldehydes from *N,N*-disubstituted amides[472, 717]. Under controlled reaction conditions *sodium tetrahydroaluminate* and its trialkoxy derivatives are suitable for synthesizing aldehydes from such amides[766]. *Sodium dihydrobis(2-methoxyethoxy)aluminate* converts open-chain and cyclic amides into *amines* in good yield[554]; it is noteworthy that this reaction can be performed in nonpolar solvents such as benzene or toluene. From *N*-methylanilides this reagent affords alde-

[753] *L.A. Paquette*, J. Amer. Chem. Soc. *86*, 4092 (1964).

[754] *R. Fuks, H.G. Viehe*, Chem. Ber. *103*, 564 (1970).

[755] *N.A. Nelson, J.E. Ladbury, R.S.P. Hsi*, J. Amer. Chem. Soc. *80*, 6633 (1958).

[756] *F.V. Brutcher, W.D. Vanderwerff*, J. Org. Chem. *23*, 146 (1958).

[757] *A. Halleux, H.G. Viehe*, J. Chem. Soc. C 1726 (1968).

[758] *G.R. Pettit, M.F. Baumann, K.N. Rangammal*, J. Med. Pharm. Chem. *5*, 800 (1962); *G.R. Pettit, D.S. Blonda, E.C. Harrington*, Can. J. Chem. *41*, 2962 (1963).

[759] *Z.B. Papanastassiou, R.J. Bruni*, J. Org. Chem. *29*, 2870 (1964).

[760] *B.I.R. Nicolaus, L. Mariani, G. Gallo, E. Testa*, J. Org. Chem. *26*, 2253 (1961).

[761] *R.L. Hinman*, J. Amer. Chem. Soc. *78*, 1645 (1956).

[762] *E.E. Miklina, N.A. Komarova, L.N. Yachontov*, Khim. Geterotsikl. Soedin. 1369 (1970); Chem. Inform. B 10—347 (1971).

[763] *A. Nakamura, Sh. Kamiya*, Chem. Pharm. Bull. (Tokyo) *18*, 1526 (1970).

[764] *J. Elguero, R. Jacquier, D. Tizané*, Tetrahedron *27*, 133 (1971).

[765] *F. Winternitz, C. Wlotzka*, Bull. Soc. Chim. France 509 (1960).

[766] *L.I. Zakharkin, D.N. Maslin, V.V. Gavrilenko*, Tetrahedron *25*, 5555 (1969).

hydes under similar conditions to those used with lithium tetrahydroaluminate[554]. Reduction of *N*-benzylphthalimide leads to *2-benzylisoindole* involving aromatization side by side with formation of *2-benzylisoindoline;* increasing the temperature and the excess of the reagent favors formation of *2-benzylisoindoline*[767]:

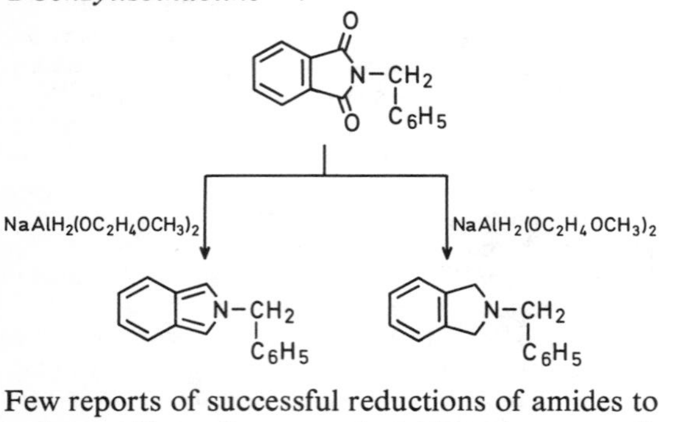

Few reports of successful reductions of amides to amines with *sodium tetrahydroborate* are available[475, 768, 769]. Dicarboxylic acid imides occupy a certain special position here in that under mild conditions reduction does take place but is partial only and, as a rule, stops after one or the two $C=O$ groups has been converted into $CHOH$[743, 770, 771]. However, for *N, N*-disubstituted carboxamides a reduction technique has been discovered which leads to amine and allows sodium tetrahydroborate to be used; the reduction is carried out in boiling pyridine[772, 773]. Reducing α-amino carboxamides to *diamines,* too, can be accomplished in this way[774], *e.g.:*

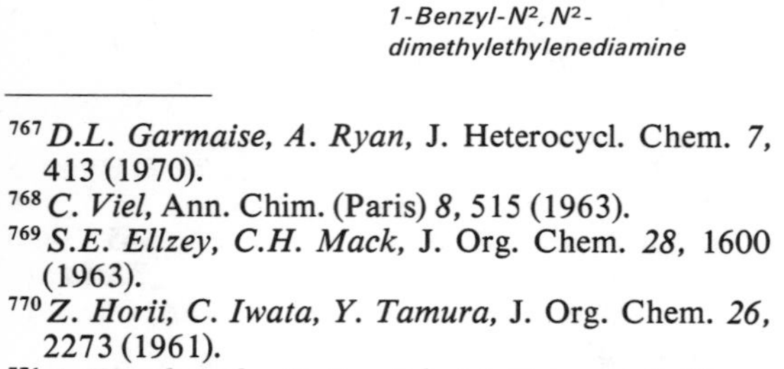

*1-Benzyl-N^2, N^2-
dimethylethylenediamine*

During reactions of *N*-unsubstituted amides with sodium tetrahydroborate nitrile formation was observed[772, 775], but a reduction procedure well suited to such amides utilizes the action of sodium tetrahydroborate in the presence of cobalt(II) chloride in methanol[491]. An interesting and *selective* reduction procedure for *N*-monosubstituted and *N, N*-disubstituted amides, lactams, and urethans consists in converting the amide into an iminoether tetrafluoroborate with triethyloxonium tetrafluoroborate and then treating the product with sodium tetrahydroborate[776]:

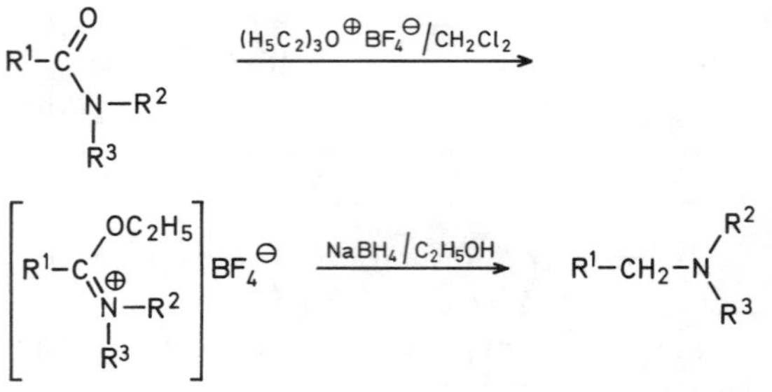

Aromatic nitro groups are not reduced under these conditions. From *N*-unsubstituted amides this technique affords nitriles[776].

Experience available to date[481, 777, 778] shows that reducing amides with *aluminum hydride* gives very good yields of amine under mild conditions; reduction of *N*-disubstituted amides proceeds particularly rapidly[481, 777]. It is noteworthy that no $C–N$ fission leading to aldehyde formation has been observed when aluminum hydride is used, including conditions where this cleavage does take place with lithium tetrahydroaluminate. As a result, even carrying out the reaction with deficient reagent[777] or with sterically hindered carboxamides[481, 778] leads exclusively to amine when aluminum hydride is employed. One example is the reduction in high yield of *N,N,N',N',N",N",N'",N'"-octamethyl-1, 2, 4, 5-tetrakiscarboxamide* to *N,N,N',N',N",N",N'",N'"-octamethyl-1, 2, 4, 5-benzenetetrakis(methylamine)*[778].

[767] *D.L. Garmaise, A. Ryan,* J. Heterocycl. Chem. 7, 413 (1970).

[768] *C. Viel,* Ann. Chim. (Paris) 8, 515 (1963).

[769] *S.E. Ellzey, C.H. Mack,* J. Org. Chem. 28, 1600 (1963).

[770] *Z. Horii, C. Iwata, Y. Tamura,* J. Org. Chem. 26, 2273 (1961).

[771] *A. Warshawsky, D. Ben-Ishai,* J. Heterocycl. Chem. 7, 917 (1970).

[772] *S. Yamada, Y. Kikugawa, S. Ikegami,* Chem. Pharm. Bull. (Tokyo) 13, 394 (1965).

[773] *K. Masuzawa, M. Kitagawa, H. Uchida,* Bull. Chem. Soc. Japan 40, 244 (1967).

[774] *I. Saito, Y. Kikugawa, S. Yamada,* Chem. Pharm. Bull. (Tokyo) 18, 1731 (1970).

[775] *S.E. Ellzey, C.H. Mack, W.J. Connick,* J. Org. Chem. 32, 846 (1967).

[776] *R.F. Borch,* Tetrahedron Lett. 61 (1968).

[777] *H.C. Brown, N.M. Yoon,* J. Amer. Chem. Soc. 88, 1464 (1966).

[778] *H. Schindlbauer,* Monatsh. Chem. 100, 1413 (1969).

[779] *N. Castagnoli, J. Cymerman-Craig, A.P. Melikian, S.K. Roy,* Tetrahedron 26, 4319 (1970).

[780] *H.C. Brown, P. Heim,* J. Amer. Chem. Soc. 86, 3566 (1964);
H.C. Brown, P. Heim, N.M. Yoon, J. Amer. Chem. Soc. 92, 1637 (1970).

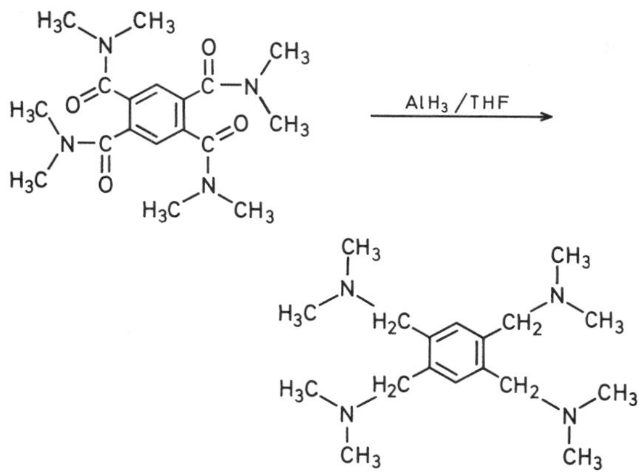

The same technique has given successful results with *o*-phenolic dimethylcarboxamides[778]. Aluminum hydride reduces both the amide and nitro groups in *N,N*-dimethyl-4-nitrobenzamide to give *4-amino-N,N-dimethylbenzylamine*[778]. On reducing *N*-dimethylcinnamamide *N-dimethylcinnamylamine* is formed in very high yield with preservation of the *C=C* bond[481]. *N-Alkoxypiperidines* are formed from *N*-alkoxyglutarimides with aluminum hydride[779].

Diborane represents a useful reagent of good *selectivity* for reducing amide to amine. Reductions using this agent, which are generally conducted in tetrahydrofuran, proceed in high yield and under mild conditions throughout independently of the degree of substitution on the amide nitrogen[759,780,781]. The same is true of vinylogs, *e.g.*, 3-indolyl ketone[782], and the technique has been applied also to lactams[783–785] and to cyclic carboxylic acid hydrazides[785,786] for the synthesis of various *heterocycles*. For example, 1,2-dialkyl- and 1,2-diaryltetrahydropyridazine-3,6-diones can be reduced to the corresponding *1,2-dialkyl-* and *1,2-diaryltetrahydropyridazines*[786]:

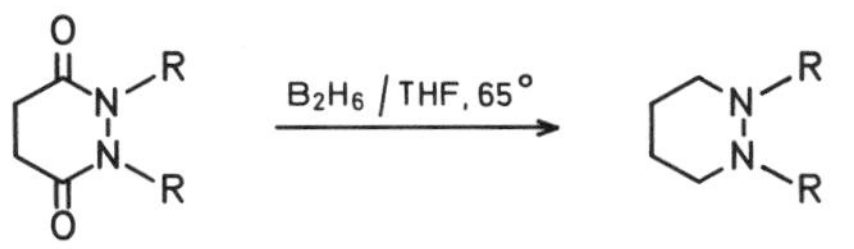

At higher diborane concentrations an additional *N–N* cleavage leading to *1,4-bis(dialkylbutylamines)* and *1,4-bis(diarylbutylamines)* is found with this reaction[786]. Four-membered ring lactams can also be reduced with diborane, as is shown by the conversion of 4-aryl-3-azidoazetidinones into *3-amino-4-arylazetidines*[644].

Using diborane is superior to lithium tetrahydroaluminate for synthesizing *β-fluoroamines* from fluoroacetamides because no dehalogenation is observed[759]. The possibility of a selective reduction of the amide group while preserving a nitro, ester, or urethane group should be emphasized. Thus, *(4-nitrobenzyl)dimethylamine* is formed from the corresponding benzamide[780], while the synthesis of *amino acid esters*[785] and of aminourethans[787] may be illustrated by the following examples:

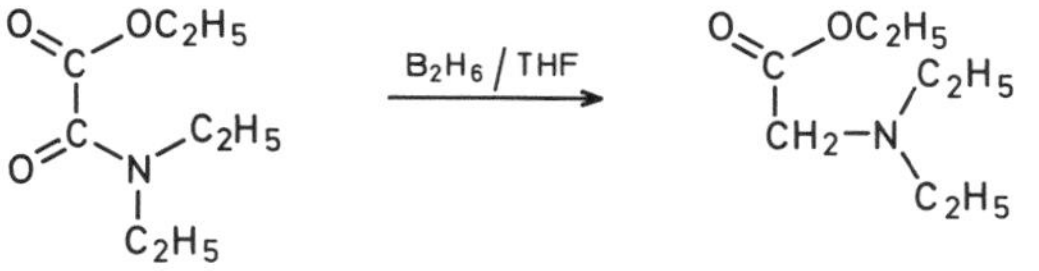

N,N-Diethylglycine ethyl ester

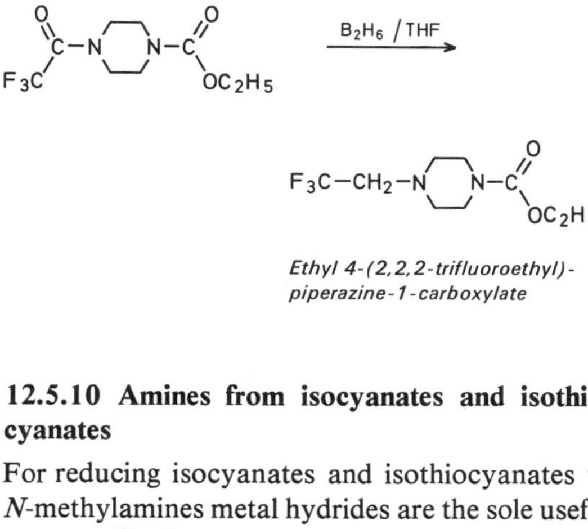

Ethyl 4-(2,2,2-trifluoroethyl)-piperazine-1-carboxylate

12.5.10 Amines from isocyanates and isothiocyanates

For reducing isocyanates and isothiocyanates to *N*-methylamines metal hydrides are the sole useful reagents[405,472]. While attempts to use sodium tetrahydroborate to reduce phenyl isocyanate and phenyl isothiocyanates did lead to *N-methylaniline*, but substantial amounts of side-product were formed[769].

[781] *E.R. Bissel, M. Finger*, J. Org. Chem. *24*, 1256 (1959).

[782] *K.M. Biswas, A.H. Jackson*, Tetrahedron *24*, 1145 (1968).

[783] *D.L. Trepanier, P.E. Krieger*, J. Heterocycl. Chem. *7*, 1231 (1970).

[784] *B. Percherer, F. Humiec, A. Brossi*, Helv. Chim. Acta *54*, 743 (1971).

[785] *M.J. Kornet, A.P. Thio, S.I. Tan*, J. Org. Chem. *33*, 3637 (1968).

[786] *H. Feuer, F. Brown*, J. Org. Chem. *35*, 1468 (1970).

[787] *W.V. Curran, R.B. Angier*, J. Org. Chem. *31*, 3867 (1966).

[788] *R.L. Dannley, R.G. Taborsky, M. Lukin*, J. Org. Chem. *21*, 1318 (1956).

[789] *A.E. Finholt, C.D. Anderson, C.L. Agre*, J. Org. Chem. *18*, 1338 (1953).

By contrast, lithium tetrahydroaluminate gave successful results in a number of instances. Reacting perfluoroalkyl isocyanates with lithium tetrahydroaluminate leads to N-methylamine, although a reductive dehalogenation on the α-carbon atom is observed[788]. A number of aryl and alkyl isocyanates and isothiocyanates can be reduced to the corresponding N-methylamines in high yield, e.g., allyl mustard oil gives *N-methallylamine*[789].

$$H_2C=CH-CH_2-N=C=S \xrightarrow{\;LiAlH_4\,/(H_5C_2)_2O\;}$$

$$H_2C=CH-CH_2-NH-CH_3$$

12.6 Preparation of amines by reductive aminoalkylation

Reacting aldehydes and ketones with ammonia, primary, or secondary amines in the presence of a reducing agent may be termed either reductive aminoalkylation or reductive amination of carbonyl compounds. *Catalytically* activated hydrogen (p. 524), formic acid, and its derivatives (*Leuckart-Wallach reaction*, p. 529), *metals,* or *metal hydrides* (p. 531) are the principal reducing agents used. *Electrolytic* reduction (p. 532) is less important. The initial step during reductive aminoalkylation is addition of ammonia, primary amine, or secondary amine **1** to the carbonyl compound, followed by reduction **2** of either the adduct or the dehydration product. Ammonia and primary amines yield Schiff's bases (azomethines, aldimines, or ketimines) with carbonyl compounds which on reduction give the monoalkylated starting base:

$$\verb|>|C=O + H_2NR \rightleftharpoons \verb|>|C\genfrac{}{}{0pt}{}{OH}{NHR} \rightleftharpoons \verb|>|C=NR + H_2O$$

1

$$\downarrow 2H \qquad 2H$$

2

$$-\underset{H}{\overset{|}{C}}-NHR$$

R = H, Alkyl, Aralkyl, Aryl

Imine formation represents an equilibrium **1** which is displaced to the right by removing the water of reaction. Addition of acid produces a shift to the left by virtue of protonization of the more basic amine, but at the same time accelerates the dehydration of the carbinolamine[790].

Addition of secondary amines to carbonyl compounds leads to α-hydroxyamines (carbinol-

amines) **1**, which are reduced to tertiary amines. *Enamines* **2** may form which on reduction also yield *tertiary amines:*

$$-\underset{H}{\overset{|}{C}}-C=O + HN\verb|<|$$

$$\updownarrow$$

$$-\underset{H}{\overset{|}{C}}-\underset{OH}{\overset{|}{C}}-N\verb|<| \rightleftharpoons \verb|>|C=C-N\verb|<| + H_2O$$

1   **2**

$$\downarrow 2H \qquad\qquad\qquad 2H$$

$$-\overset{|}{C}-\overset{|}{C}-N\verb|<|$$

12.6.1 Reduction with catalytically activated hydrogen

The advantage of this method for preparing amines is that the desired amine can generally be obtained in one step. It is adopted wherever from the first easily accessible carbonyl compounds are concerned or where it is required to introduce amino functions into complex molecules (generally containing a keto group) and to possibly react them further. Care must be taken at the synthesis planning stage that *no* other functional groups are present in the components which are irreversibly altered during the respectively used catalytic process or lead to undesired secondary reactions. The value of the method varies with the structural peculiarities of the carbonyl compounds and amines used; the *requisite* for a successful reaction is a readiness on the part of the carbonyl component to add on to ammonia or amines. In terms of apparatus the procedure is unassuming. Where high pressure autoclaves are not available, hydrogenation can often be performed by a suitable choice of catalyst and other reaction conditions at normal or low pressure.

Reduction with hydrogen may be carried out with almost all hydrogenation catalysts. Raney nickel

[790] *E.H. Cordes, W.P. Jencks,* J. Amer. Chem. Soc. *84,* 826, 832 (1962).
[791] *F. Möller, R. Schröter* in *Houben-Weyl,* Methoden der organischen Chemie, 4. Aufl., Bd. XI/1, p. 602, Georg Thieme Verlag, Stuttgart 1957.
[792] *F. Zymalkowski,* Katalytische Hydrierungen, Ferdinand Enke Verlag, Stuttgart 1965.
[793] *M. Freifelder,* Practical Catalytic Hydrogenation, Wiley Intersci. Publ., New York, London, Sydney, Toronto 1971.

at elevated pressure and temperature, and platinum metals under normal conditions are most often used. Generally, valid recommendations concerning the reaction conditions cannot be given; detailed surveys are found in refs. [791–793].

12.6.1.1 Primary amines from ammonia and carbonyl compounds

Reductive amination of carbonyl compounds with ammonia is a proven technique for preparing *primary amines*. Generally, a large excess of ammonia (to stop secondary and tertiary amine formation) in ethanolic solution and hydrogenation catalysts are used[794].

Aldehydes:

As a rule, higher aliphatic *aldehydes* containing more than five carbon atoms such as heptanal, and aromatic aldehydes[795] yield the corresponding primary amines smoothly and in good yield. By contrast, the more reactive lower aliphatic aldehydes tend to form undesired side-products. Reacting with formaldehyde preferentially yields hexamethylenetetramine, whose hydrogenation leads to a mixture of *mono-*, *di-*, and *trimethylamine*[796]. Reductive amination of acetaldehyde, propionaldehyde, and butyraldehyde leads to some resinous side-products by aldol condensation. Substituted aliphatic aldehydes can be fundamentally reductively aminated just like the simple aldehydes provided the substituent does not upset the hydrogenation. For example, 5-hydroxyvaleraldehyde gives a good yield of *5-amino-1-pentanol*[797]. During conversion of unsaturated aldehydes such as acrolein or crotonaldehyde the reaction conditions determine whether hydrogenation of the double bond[798] or addition of ammonia[799] occur preferentially. *Glycosamines* are easily accessible by reductive amination of aldoses[800].

Ketones:

Reductive amination of simple aliphatic ketones such as acetone[794, 801] or butanone[801] with ammonia entails no difficulties and reaction-kinetic studies of the reaction of acetone with ammonia are available[802]. Higher ketones carrying branched alkyl groups in the α-position to the carbonyl group are reduced to secondary alcohols even with a large excess of ammonia. Adding ammonium chloride or acetic acid may enhance the yield of amine because the acid-catalyzed formation of the Schiff base is accelerated as a result. Using this method, it is possible to carry out hydrogenation at normal pressure with both platinum catalysts[803] and Raney nickel[804]. *Cyclohexylamine* can be obtained in very good yield from cyclohexanone and ammonia[805]. Under certain conditions alkyl aryl ketones such as acetophenone can be smoothly converted into the corresponding primary amines.

Preformed ammonia-carbonyl compound condensation products are seldom hydrogenated. For example, reduction of hexamethylenetetramine yields approximately 70% *trimethylamine*[796]. *Ketimines*, which are accessible either by Grignard reagent treatment of nitriles[806] or by condensation[801], are smoothly hydrogenated to *primary amines*. Primary *3-amino steroids* can be prepared, firstly, by catalytic reduction of the corresponding 3-oxo steroids in the presence of ammonia. A second, very elegant method consists in reductive alkylation with benzylamine. The benzyl group is split off during the hydrogenation and the primary *3-amino steroid* is again formed[807].

[794] *G. Mignonac*, C.R. Acad. Sci., Paris *172*, 223 (1921).
[795] *C.F. Winans*, J. Amer. Chem. Soc. *61*, 3566 (1939).
[796] DBP. 946622 (1953), BASF, Erf.: *O. Sticknoth, L. Wolf, A. Palm.*
[797] *G.F. Woods, H. Sanders*, J. Amer. Chem. Soc. *68*, 2111 (1946);
I. Scriabine, Bull. Soc. Chim. France 454 (1947).
[798] DRP. 527619 (1929), I.G. Farben, Erf.: *K. Baur;* Fortschritte der Teerfabrikation und verwandter Industriezweige *18*, 345.
[799] US.P. 2452602 (1944), Sharples Chemicals Inc., Inv.: *C.N. Robinson jr., J.F. Olin;* C.A. *43*, 2633 (1949).
[800] *F.W. Holly, E.W. Peel, J.J. Cahill, F.R. Komuszy, K. Folkers*, J. Amer. Chem. Soc. *74*, 4047 (1952);
F.W. Holly, E.W. Peel, R. Mozingo, K. Folkers, J. Amer. Chem. Soc. *72*, 5416 (1950).

[801] *D.G. Norton, V.E. Haury, F.C. Davis, L.J. Mitchell, S.A. Ballard*, J. Org. Chem. *19*, 1054 (1954).
[802] *A. LeBris, G. Lefebre, F. Coussemant*, Bull. Soc. Chim. France 1594 (1964).
[803] *E.R. Alexander, A.L. Misegades*, J. Amer. Chem. Soc. *70*, 1315 (1948).
[804] *L. Haskelberg*, J. Amer. Chem. Soc. *70*, 2811 (1948).
[805] *Y. Omote, K. Iwase, I. Nakamara*, Kogyo Kagaku Zasshi *70*, 1355 (1967); C.A. *68*, 29066 (1968).
[806] *P.L. Pickard, D.J. Vaughan*, J. Amer. Chem. Soc. *72*, 876, 5017 (1950);
P.L. Pickard, E.F. Engles, J. Amer. Chem. Soc. *74*, 4607 (1952);
P.L. Pickard, E.F. Engles, J. Amer. Chem. Soc. *75*, 2148 (1953);
P.L. Pickard, S.H. Jenkins jr., J. Amer. Chem. Soc. *75*, 5899 (1953).

12.6.1.2 Secondary amines

12.6.1.2.1 Secondary amines from ammonia and carbonyl compounds

With aliphatic aldehydes and ketones hydrogenation of a mixture of 2 *mols* carbonyl compound and 1 *mol* ammonia generally leads merely to a mixture of mono-, di-, and trialkylamine[808]. *Diethylamine* from acetaldehyde[809] and *dicyclohexylamine* from cyclohexanone[810] alone can be prepared in over 50% yield. By contrast, reaction with aromatic aldehydes proceeds particularly smoothly; benzaldehyde and its substitution products form the corresponding *dibenzylamines* in high yield[795].

12.6.1.2.2 Secondary amines from primary amines and carbonyl compounds

Reductive monoalkylation of primary amines with carbonyl compounds to secondary amines in most cases proceeds *via* the Schiff base as intermediate; above all with aromatic amines the base forms on mere mixing of the components. Where for steric or other reasons formation of the Schiff's base is hindered, substantial quantities of the free carbonyl compound are present in the mixture and the hydrogenation produces alcohols as side-product. In these cases the isolated Schiff's base needs to be used for the reaction. A comparative study of the reduction of a preformed Schiff's base and a mixture of amine and carbonyl compound is available[811].

Tertiary amine formation as a side-reaction, which is always a possibility, can be suppressed substantially by adding the carbonyl compound to the reaction mixture gradually. At the same time aldol reactions are made more difficult.

Often, compounds which are transformed into primary amines on hydrogenation may be used as an alternative for primary amines (nitro, nitroso, azo compounds, oximes, nitriles, *etc.*) and acetals, acylates, or bisulfite adducts in place of the carbonyl compounds.

Aliphatic aldehydes:

Reductive alkylation of primary aliphatic amines with formaldehyde proceeds nonuniformly, because substantial amounts of the *tertiary dimethyl compound* are formed as side-product. By contrast, conversion with higher aldehydes generally gives satisfactory results. Unsaturated aldehydes, too, can be reacted satisfactorily with primary amines to the corresponding saturated *secondary* amines. With primary aliphatic amines glucose yields the corresponding *N-alkylglucosamines*[812].

Primary aromatic amines afford the corresponding *alkylarylamines* in good yield with both formaldehyde and with higher aliphatic aldehydes[813]. Heterocyclic primary amines, too, can be monoalkylated reductively. Thus *5-(ethylamino)-4-methylpyrimidine* and *4-(propylamino)antipyrine* are obtained from 5-amino-4-methylpyrimidine and acetaldehyde[814] and from 4-aminoantipyrine and propionaldehyde[815] respectively.

Aromatic aldehydes:

Fundamentally, both aliphatic and aromatic primary amines can be reductively alkylated with aromatic aldehydes without difficulty, although in many instances the preformed Schiff's base is employed either after isolation or as crude product (see p. 527). Thus, for example, the reactions of benzaldehyde with *m*-toluidine[816] and methylamine[817] to form *N-benzyl-3-toluidine* and *N-methylbenzylamine* respectively have been described. It needs to be remembered that the benzyl group formed is relatively easily cleaved by hydrogenation, especially when palladium catalysts are employed. Platinium catalysts under energetic reaction conditions, can additionally effect hydrogenation of the aromatic compound.

Ketones:

In general, reductive amination of ketones proceeds just as smoothly as with aromatic aldehydes. Where the reaction between amine and ketone is not rapid enough, and alcohols are

[807] *J. Schmitt, J.J. Panouse, P. Comoy, A. Hallot, P.J. Cornu, H. Pluchet*, Bull. Soc. Chim. France 1846, 1855 (1962).
[808] *A. Skita, F. Keil, H. Havemann*, Ber. *66*, 1400 (1933).
[809] *A.M. Grigorovski, A.I. Berkov, G.A. Gorlach, R.S. Margolina, S.B. Levitskaya*, Org. Chem. Ind. (USSR) 7, 671 (1940); C.A. *35*, 5094 (1941).
[810] *A. Skita, F. Keil*, Ber. *61*, 1682 (1928).
[811] *A.L. Klebanskii, M.S. Vilesova*, Zh. Obshch. Khim. *28*, 1767 (1958); C.A. *53*, 1128 (1959).

[812] *E. Mitts, R.M. Hixon*, J. Amer. Chem. Soc. 66, 483 (1944).
[813] *W.S. Emerson, P.M. Walters*, J. Amer. Chem. Soc. *60*, 2023 (1938); *W.S. Emerson, W.D. Robb*, J. Amer. Chem. Soc. *61*, 3145 (1939).
[814] *C.G. Overberger, J.C. Kogon, W.J. Einstman*, J. Amer. Chem. Soc. *76*, 1953 (1954).
[815] *A. Skita, F. Keil, W. Stühmer*, Ber. 75, 1696 (1942).
[816] *C.H.F. Allen, J. Van Allan*, Org. Syntheses, Coll. Vol. III, 827 (1955).

formed as side-product as a result, yields can be improved in these cases, too. Either one uses one component in excess[818] or one adds acid activators such as ammonium chloride, ammonium acetate, or acetic acid. In many cases it is advantageous to hydrogenate the preformed ketimine[801]. During reaction of primary aromatic amines with ketones certain metal-sulfide catalysts display the advantage that ring hydrogenation and catalyst poisoning occur to only a very slight extent[819]. Reductive amination of 3-oxo steroids with primary and secondary amines yields the corresponding *3-mono* or *dialkylamino* steroids. Amino groups in the α-position are formed preferentially if *C-17* carries an α-hydroxy group[820].

Substituted carbonyl compounds:

Reductive alkylation of *aminoalcohols* such as *2-aminoethanol* or *3-aminopropanol* proceeds exceptionally smoothly[821, 822]. Azomethines or heterocyclic compounds, which after isolation can be hydrogenated satisfactorily to the corresponding *N*-substituted amino alcohols, may form intermediately. Thus, *e.g.*, *oxazolidine* derivatives are formed from 1,2-aminoalkanols and *tetrahydro-1,3-oxazines* from 1,3-aminoalkanols[821, 823, 824]. Reductive alkylation of primary diamines can ensue in the same way as is usual with monoamines[825]. For example, hydrazine may be reacted with acetone to give a good yield of *N,N'*-diiso-

propylhydrazine[826]. Reductive amination of ethyl glyoxylate with 3-aminopyridine to form *ethyl 3-pyridylaminoacetate* proceeds equally smoothly[827]. For both preparation of *amino carboxylic acids* from *oxo carboxylic acids* and alkylation of amino carboxylic acids reductive amination and alkylation represents a very valuable technique (p. 600). The course of the reaction during synthesis of optically active α-amino acids has been studied in great detail. One method of preparation is catalytic hydrogenation of the Schiff's base of an α-oxo carboxylic acid and (*S*)(–)- and (*R*)(+)-α-methylbenzylamine followed by hydrogenolysis[828]. The steric course of this reaction has been studied on numerous examples[829–831].

12.6.1.2.3 Secondary amines from Schiff bases

Preparation of *secondary amines* by hydrogenating Schiff's bases is applied especially to primary aromatic amines and aromatic aldehydes[832]. Often, condensation takes place so readily that even working in aqueous solution is possible. The influence of substituents on the rate of hydrogenation has been investigated on *N*-benzylidenaniline and its substitution products by carrying out reaction-kinetic measurements[833].

By contrast, ketones, and particularly aromatic ones, form ketimines only slowly, and as a result alcohols appear as side-products during the hydrogenation. For this reason, elevated temperature and condensing agents such as zinc chloride[834] or phosphorus oxychloride[835] are often employed. Purely *aliphatic ketimines* can be prepared by removing the water of reaction azeotrop-

[817] *R. Wegler, W. Frank*, Ber. 69, 2071 (1936).

[818] *K. Ziegler, H. Wilms*, Justus Liebigs Ann. Chem. 567, 1 (1950).

[819] *F.S. Dovell, H. Greenfield*, J. Org. Chem. 29, 1265 (1964);
H. Greenfield, F.S. Dovell, J. Org. Chem. 31, 3053 (1966).

[820] *J. Schmitt, J.J. Panouse, P. Comoy, A. Hallot, P.J. Cornu, H. Pluchet*, Bull. Soc. Chim. France 455, 463 (1962).

[821] *A.C. Cope, E.M. Hancock*, J. Amer. Chem. Soc. 64, 1503 (1942);
A.C. Cope, E.M. Hancock, J. Amer. Chem. Soc. 66, 1453 (1944);
E.M. Hancock, A.C. Cope, J. Amer. Chem. Soc. 66, 1738 (1944);
E.M. Hancock, E.M. Hardy, D. Heyl, M.E. Wright, A.C. Cope, J. Amer. Chem. Soc. 66, 1747 (1944);
E.M. Hancock, E.M. Hardy, D. Heyl, M.E. Wright, A.C. Cope, Org. Synth. 26, 38 (1946).

[822] *F. Zymalkowski, F. Koppe*, Arch. Pharm. (Weinheim, Ger.) 294, 453 (1951).

[823] *M. Senkus*, J. Amer. Chem. Soc. 67, 1515 (1945).

[824] *E.D. Bergmann*, Chem. Rev. 53, 309 (1953).

[825] *D.E. Pearson, W.H. Jones, A.C. Cope*, J. Amer. Chem. Soc. 68, 1225 (1946).

[826] *Weygand-Hilgetag*, Organisch-chemische Experimentierkunst, p. 537, Johann Ambrosius Barth, Leipzig 1970.

[827] *J.M. Tien, J.M. Hunsberger*, J. Amer. Chem. Soc. 77, 6604 (1955).

[828] *R.G. Hiskey, R.C. Northrop*, J. Amer. Chem. Soc. 83, 4798 (1961).

[829] *K. Harada*, Nature 212, 1571 (1966);
K. Harada, J. Org. Chem. 32, 1790 (1967).

[830] *K. Harada, K. Matsumoto*, J. Org. Chem. 32, 1794 (1967);
K. Harada, K. Matsumoto, J. Org. Chem. 33, 4467 (1968).

[831] *K. Harada, T. Yoshida*, Bull. Chem. Soc. Japan 43, 921 (1970).

[832] *W.S. Emerson*, Org. Reactions 4, 174 (1948).

[833] *A. Roe, J.A. Montgomery*, J. Amer. Chem. Soc. 75, 910 (1953).

[834] *J.H. Billmann, K.M. Tai*, J. Org. Chem. 23, 535 (1958);
G. Reddelien, Ber. 43, 2476 (1910).

[835] *A.W. Weston, R.J. Michaels jr.*, J. Amer. Chem. Soc. 73, 1381 (1951).

ically[801]. By contrast, purely *aliphatic aldimines* are relatively unstable because they tend to undergo trimerization and aldol reaction and their preparation succeeds only under specific conditions[836].

The *steric* course of the reductive amination of cyclic ketones has been studied above all with substituted cyclohexanone derivatives. Condensation of 4-*tert*-butylcyclohexanone with aniline to form the Schiff's base and the latter's catalytic hydrogenation leads to predominantly *cis*-isomer with an axial amino group *(cis-N-phenyl-4-tert-butylcyclohexylamine)*, while reduction with sodium tetrahydroborate affords the *trans* isomer with an equatorial amino group as main product[837].

12.6.1.2.4 Secondary amines from nitro, nitroso, and azo compounds, or from nitriles and carbonyl compounds

Compounds which yield primary amines by reduction can be reductively alkylated also in the presence of carbonyl compounds. Thus, aromatic *nitro* compounds give a good yield of the corresponding *alkylarylamines* with aliphatic aldehydes[838]. An elegant method for preparing *cyclic secondary amines* such as *pyrrolidines* and *piperidines* consists in hydrogenating compounds which contain a carbonyl group and an easily reducible nitrogen function in a suitable position to from each other[791].

Reductive alkylation of *nitroso* compounds and *nitriles* is of little importance. *Azo* compounds generally yield mixtures of secondary and tertiary amines[839].

12.6.1.3 Tertiary amines

In general, preparation of tertiary amines by reductive alkylation using catalysts is unsatisfactory. *Methylation* of secondary amines and *dimethylation* of primary amines with formaldehyde is one reaction which has achieved importance. While trialkylation of ammonia with ketones is not possible, the reaction with, *e.g.*, acetaldehyde, propionaldehyde[840], and also cinnamaldehyde[841] does succeed. *Trimethylamine* can be prepared from ammonium chloride and formaldehyde[842].

12.6.1.3.1 Tertiary amines from primary amines and carbonyl compounds

Generally speaking, primary amines can be alkylated reductively to form *N,N-dimethyl derivatives* in good yield with formaldehyde. A large number of aliphatic primary mono-, di- and polyamines, amino ethers, and amino alcohols have been permethylated[791, 843]. Conversion of a nitro or a nitroso group into the dimethylamino group[844, 845] proceeds equally smoothly. With aliphatic aldehydes, nitromethane and nitrobenzene give the corresponding tertiary amines[846].

12.6.1.3.2 Tertiary amines from secondary amines and carbonyl compounds

Detailed investigations on the effect of the constitution of the secondary amine and of the carbonyl compound on the yield of tertiary amine have been carried out[808, 847]. Dimethylamine and cyclic amines such as pyrrolidine and piperidine can be readily alkylated reductively with both aldehydes and certain ketones. With secondary amines carrying bulky alkyl groups and thus subject to substantial steric hindrance, satisfactory alkylation succeeds only with formaldehyde or after preliminary conversion into an enamine. To prepare a *mixed tertiary amine* by stepwise alkylation it is best to introduce the larger group first and to react the secondary amine obtained further[808].

12.6.1.4 Amines from polycarbonyl compounds

One example of the seldom performed amination of a dialdehyde is the reaction between 2-hydroxyhexanedial and ammonia to form *1,6-diamino-2-hexanol*[848]. Preparation of diamines from diketones succeeds only where the distance between the keto groups prevents interaction. Thus during conversion of 1,2-diketones and primary amines generally only one carbonyl function is reductive-

[836] *K.N. Campbell, A.H. Sommers, B.K. Campbell*, J. Amer. Chem. Soc. *66*, 82 (1944).

[837] *J.R. Bull, D.G. Hey, G.D. Meakins, E.E. Richards*, J. Chem. Soc. C 2077 (1967).

[838] *W.S. Emerson, H.W. Mohrmann*, J. Amer. Chem. Soc. *62*, 69 (1940).

[839] *W.S. Emerson, S.K. Reed, R.R. Merner*, J. Amer. Chem. Soc. *63*, 751 (1941).

[840] *A. Skita, F. Keil*, Ber. *61*, 1452 (1928).

[841] *W. Stühmer, E.A. Elbrächter*, Arch. Pharm. (Weinheim, Ger.) *287*, 139 (1954).

[842] *R. Adams, C.S. Marvel*, Org. Syntheses, Coll. Vol. I, 531 (1941).

[843] *E.H. Woodruff, J.P. Lambooy, W.E. Burt*, J. Amer. Chem. Soc. *62*, 922 (1940).

[844] *W. Krohs*, Chem. Ber. *88*, 866 (1955).

[845] *V.M. Ingram*, J. Chem. Soc. (London) 2247 (1950).

[846] *W.S. Emerson, C.A. Uraneck*, J. Amer. Chem. Soc. *63*, 749 (1941).

[847] *A. Skita, F. Keil*, Ber. *63*, 34 (1930).

[848] *H. Schulz, H. Wagner*, Angew. Chem. *62*, 111, 117 (1950).

[849] *A. Skita, F. Keil*, Ber. *62*, 1142 (1929); *A. Skita, F. Keil, E. Baesler*, Ber. *66*, 858 (1933).

ly aminated[849]. With unsymmetrically substituted diketones the amino group always enters next to the smaller alkyl group, as shown by the preparation of DL-*ephedrine [2-methylamino-1-phenyl-propanol]* from 1-phenyl-1,2-propanedione and methylamine[850]. Reductive aminations which ammonia and methylamine accompanied by ring closure are possible with, for example, 1,4-diketones such as 2,5-hexanedione (giving 2,5-dimethylpyrrolidine and 2,5-dimethylpyrrole)[851] or with compounds such as bicyclo-[3.3.1]nonane-3,7-dione[852]:

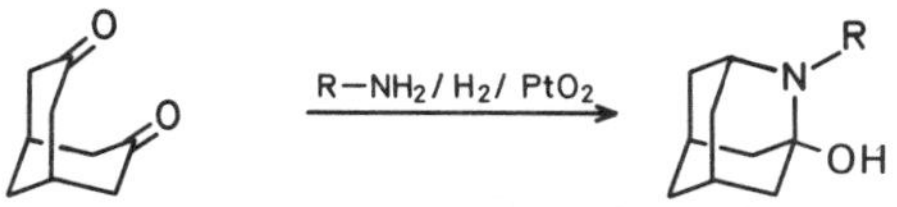

R = H;   *2-Azaadamantan-1-ol; 67%*
R = CH₃; *2-Methyl-2-azaadamantan-1-ol; 49%*

Polyketones can be converted into *polyamines* with ammonia, and primary and secondary amines[853].

12.6.2 Reduction with formic acid (Leuckart-Wallach reaction)

This method is always preferred to other reductive aminations where the reactants carry functional groups that are easily attacked by other reducing agents (see examples in the text) or where compounds are used which poison hydrogenation catalysts. The procedure is very well suited for preparing *tertiary amines* and very economic for obtaining *N,N-dimethylamines*. Aromatic amines react readily, application to low molecular aliphatic aldehydes and ketones is limited. Reductive alkylation of ammonia or amines with carbonyl compounds using formic acid as the reducing agent is known as the *Leuckart-Wallach reaction*[854]. The amine may be available either as the formate, or as the formyl derivative of ammonia or a primary or secondary amine:

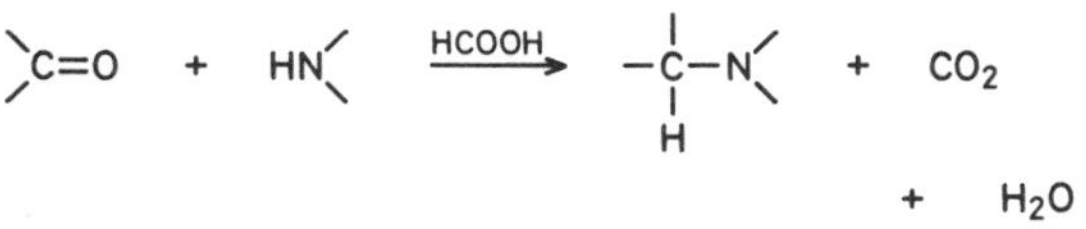

This method of amine preparation, whose scope about equals that of the above-described procedure (p. 524), was first described for the reaction of benzaldehyde and ammonium formate and subsequently also with purely aliphatic carbonyl compounds[855, 856]. *Methylation* of primary and secondary amines with formaldehyde and formic acid became known as the *Eschweiler-Clarke method*[857, 858]. By comparison to catalytic hydrogenation, reduction with formic acid possesses the advantage that carbonyl compounds containing easily reducible groups such as double bonds, nitro, or nitroso groups, too, can be selectively aminated by reduction.

The mechanism of the Leuckart-Wallach reaction has been the subject of numerous investigations[854, 859−863] and has not yet been completely elucidated.

12.6.2.1 Methylation with formaldehyde

Primary and secondary amines can be readily methylated to form *tertiary amines* with formaldehyde and formic acid by the *Eschweiler-Clarke method*. In general, 1−1.25 *mols* formaldehyde and 2−4 *mols* formic acid per *mol* amine are used to introduce a methyl group[858]. In this or a similar manner numerous amines have been permethylat-

[850] *R.H.F. Manske, T.B. Johnson*, J. Amer. Chem. Soc. *51*, 580 (1929);
R.H.F. Manske, T.B. Johnson, J. Amer. Chem. Soc. *54*, 306 (1932);
A. Skita, F. Keil, H. Meiner, Ber. *66*, 974 (1933).
[851] *E.J. Schwoegler, H. Adkins*, J. Amer. Chem. Soc. *61*, 3499 (1939).
[852] *H. Stetter, P. Tacke, J. Gärtner*, Chem. Ber. *97*, 3480 (1964).
[853] *D.D. Coffman, H.H. Hoehn, J.T. Maynard*, J. Amer. Chem. Soc. *76*, 6394 (1954).

[854] *M.L. Moore*, Org. Reactions *5*, 301 (1949).
[855] *R. Leuckart*, Ber. *18*, 2341 (1885);
R. Leuckart, E. Bach, Ber. *19*, 2128 (1886);
R. Leuckart, E. Bach, Ber. *20*, 104 (1887);
R. Leuckart, H. Janssen, Ber. *22*, 1409 (1889);
R. Leuckart, H. Lampe, Ber. *22*, 1851 (1889).
[856] *O. Wallach*, Ber. *24*, 3992 (1891);
O. Wallach, Justus Liebigs Ann. Chem. *269*, 326 (1892);
O. Wallach, Justus Liebigs Ann. Chem. *272*, 99 (1893);
O. Wallach, Justus Liebigs Ann. Chem. *276*, 296 (1893);
O. Wallach, Justus Liebigs Ann. Chem. *289*, 337 (1896);
O. Wallach, Justus Liebigs Ann. Chem. *300*, 278 (1898);
O. Wallach, Justus Liebigs Ann. Chem. *343*, 54 (1905).
[857] *W. Eschweiler*, Ber. *38*, 880 (1905).
[858] *H.T. Clarke, H.B. Gillespie, S.Z. Weisshaus*, J. Amer. Chem. Soc. *55*, 4571 (1933).
[859] *C.B. Pollard, D.C. Young*, J. Org. Chem. *16*, 661 (1951).
[860] *H.W. Gibson*, Chem. Rev. *69*, 673 (1969).
[861] *E. Staple, E.C. Wagner*, J. Org. Chem. *14*, 559 (1949).
[862] *A. Lukasiewicz*, Tetrahedron *19*, 1789 (1963).
[863] *P.L. De Benneville, J.H. Macartney*, J. Amer. Chem. Soc. *72*, 3073 (1950).
[864] *S.H. Pine, B.L. Sanchez*, J. Org. Chem. *36*, 829 (1971).

ed[854, 864]. Substances containing easily reducible groups can be methylated by the Eschweiler-Clarke method without modifying these groups. Examples where this succeeds in good yield are 4-nitrophenethylamine[865] to *N,N-dimethyl-4-nitrophenethylamine* and 1-methyl-3-cyclohexen-1-ylamine to *N,N,1-trimethyl-3-cyclohexen-1-ylamine*[866].

The synthesis of *optically active 2-methylamino carboxylic acids* succeeds in a 3-stage process[867].

First, the optically active amino acid is converted into the Schiff's base with benzaldehyde and the base is reduced to *N*-benzylaminoacides with either sodium tetrahydroborate or catalytically with palladium-carbon without isolation. Following methylation with formaldehyde and formic acid to form *N*-benzyl-*N*-methylaminoacid the benzyl group is split off hydrogenolytically.

On treating certain ethylaryl amines such as mescaline[868] or 3,4-dimethoxyphenethylamine with formaldehyde[869] ring closure occurs with formation of corresponding *6,7,8-trimethoxy-* or *6,7-dimethoxy-1,2,3,4-tetrahydroisoquinolines*. The methoxy groups activate the 6-position in such a way that a *Pictet-Spengler* reaction is made possible[870]. Ring closure also occurs during methylation of 1,3- and 1,2-aminoalcohols[871]. Either smooth alkylation to the tertiary amine takes place or else heterocycles are formed, here, *tetrahydro-1,3-oxazine* or *oxazolidines* according to the configuration, *i.e.*, the reaction can be used for determining the configuration of diastermers. Ring closures of this type are observed also during methylation of bicyclic amino alcohols such as *cis*-3-aminobicyclo [2.2.2]octan-2-ol[872] to form *3-oxa-5-azatricyclo[5.2.2.0^{2,6}]undecane* and with 16α-pregnanamine derivatives[873]. Certain olefinic

amines such as 5-methyl-4-hexenylamine also cyclize under Eschweiler-Clarke conditions[874]. Methylation of an amino group on an asymmetric carbon atom took place without epimerization in the cases that have been investigated[875].

N-Methylation of *aromatic amines* is successful only if the *o*- and *p*-positions of the aromatic compound are substituted because of the tendency of formaldehyde to undergo nuclear condensations. Thus dimethyl derivatives of 2,4,6-trihaloanilines are obtained in good yield[858, 876, 877]. Using a modified method it is, however, possible to permethylate aromatic amines with only one *o*- or *p*-substituent[878].

12.6.2.2 Alkylation with higher aldehydes

Cycloaliphatic aldehydes can be converted into primary amines by reaction with formamide and subsequent hydrolysis of the formyl compound[879]. With ammonium formate aromatic aldehydes yield the corresponding primary amines[880].

Preparing *secondary amines*[881] from aromatic aldehydes and primary amines brings no advantages over other methods (cf. p. 526). By contrast, *tertiary amines* are more easily accessible from aromatic aldehydes and secondary aliphatic amines than is otherwise the case[861]. Likewise, secondary amines such as piperidine or morpholine can be converted smoothly into the corresponding tertiary amines with higher aliphatic aldehydes[863]. In place of the free aldehydes their acetals may be used if a small amount of hydrochloric acid is added to the reaction mixture[882].

[865] *F. Bergel, J.L. Everett, J.J. Roberts, W.C.J. Ross*, J. Chem. Soc. (London) 3835 (1955).

[866] *M. Mousseron, R. Jacquier, R. Zagdoun*, Bull. Soc. Chim. France 974 (1953).

[867] *P. Quitt, J. Hellerbach, K. Vogler*, Helv. Chim. Acta *46*, 327 (1963).

[868] *J.A. Castrillón*, J. Amer. Chem. Soc. *74*, 558 (1952).

[869] *J.S. Buck, R. Baltzly*, J. Amer. Chem. Soc. *64*, 2263 (1942);
R. Baltzly, J. Amer. Chem. Soc. *75*, 6038 (1953).

[870] *W.M. Whaley, T.R. Govindachari*, Org. Reactions *6*, 151 (1951).

[871] *G. Drefahl, H.-H. Hörhold*, Chem. Ber. *94*, 1657 (1961).

[872] *W.L. Nelson*, J. Heterocycl. Chem. *5*, 231 (1968).

[873] *H.P. Husson, P. Poitier, J. Le Men*, Bull. Soc. Chim. France 948 (1966);
H.P. Husson, P. Poitier, J. Le Men, Bull. Soc. Chim. France 1721 (1965).

[874] *A.C. Cope, W.D. Burrows*, J. Org. Chem. *30*, 2163 (1965);
A.C. Cope, W.D. Burrows, J. Org. Chem. *31*, 3099 (1966).

[875] *A.C. Cope, E. Ciganek, L.J. Fleckenstein, M.A.P. Meisinger*, J. Amer. Chem. Soc. *82*, 4651 (1960).

[876] *W.S. Emerson*, J. Amer. Chem. Soc. *63*, 2023 (1941).

[877] *R.B. Sandin, J.R.L. Williams*, J. Amer. Chem. Soc. *69*, 2747 (1947).

[878] *W.L. Borkowski, E.C. Wagner*, J. Org. Chem. *17*, 1128 (1952).

[879] *M. Mousseron, R. Jacquier, R. Zagdoun*, Bull. Soc. Chim. France 197 (1952).

[880] *K.G. Lewis*, J. Chem. Soc. (London) 2249 (1950).

[881] *R. Baltzly, O. Kauder*, J. Org. Chem. *16*, 173 (1951).

[882] *H. Ruschig, K. Schmitt*, Chem. Ber. *88*, 875 (1955).

[883] *A.W. Ingersoll, J.H. Brown, C.K. Kim, W.D. Beauchamp, G. Jennings*, J. Amer. Chem. Soc. *58*, 1808 (1936);
A.W. Ingersoll, Org. Syntheses, Coll. Vol. II, 503 (1943).

12.6.2.3 Alkylation with ketones

Ketones can be smoothly converted into the corresponding *primary* amines with a mixture of ammonium formate and formamide *(Ingersoll reagent)*[883, 884]. Numerous alternative methods are available for converting ketones into primary amines[885, 886]. Amination of acetylpyridines to the corresponding primary amines succeeds with a mixture of formamide and formic acid[887].

Secondary[879, 888, 889] and *tertiary amines*[879, 890, 891] are readily accessible by reacting ketones with amine formates or *N*-alkyl- and *N,N*-dialkylformamides[854].

Reductive amination of cyclic ketones by the Leuckart-Wallach method has been the subject of numerous investigations. C_5–C_{10} cycloalkanones are smoothly converted into the *N*-dimethylcycloalkylamines with dimethylformamide and formic acid[892]. Reductive amination of 4-*tert*-butylcyclohexanone yields *cis/trans* isomer mixtures; smaller alkyl groups such as ethyl form predominantly *cis*-isomers with an axial amino group. With larger alkyl groups such as *tert*-butyl, *trans*-isomers with equatorial amino groups are the main reaction product[893]. For the course and mechanism of the reaction between 2-phenoxycyclohexanone and various secondary amines, see ref. [894].

The Leuckart-Wallach reaction has gained great importance for preparing *amino steroids*. Reacting 5α-cholestan-3-one with primary or secondary amines yields a mixture of *3α-* and *3β-cholestanamine*derivatives[895]. Androstan-17-onederivatives afford preferentially *17β-androstanamines*[896], while androstan-3-ones lead to a mixture of *3α-* and *3β-androstanamines*[897]. In connection with the amination of 3, 20-dioxo steroids[898], 3-amino-20-oxo steroids, too, were prepared by forming the enamines followed by reduction with formic acid[899]. Monoamines can be obtained from dioxo steroids in this way, because enamine formation in the 3-position ensues much more readily.

12.6.3 Reduction with metals or metal hydrides

12.6.3.1 Reduction with metals

Reductive aminations of carbonyl compounds using metals as the reducing agent are relatively seldom carried out and are restricted almost exclusively to the reduction of preformed condensation products, *i.e.*, Schiff bases and imines.

Numerous reductions of this type are today being performed with complex metal hydrides which facilitate to guide the reaction and make it more economic (see p. 532). In general, *N*-benzylamines are obtained readily by reduction with metals. Functional groups in the reactants often behave variably; sometimes they are reduced, at other times they remain unchanged despite the often drastic reducing conditions. Where complex metal hydrides are employed, their selectivity towards reducible groups must be regarded.

Aromatic azomethines can be reduced to the corresponding secondary amines with sodium amalgam in anhydrous ethanol[900]; there is a risk of splitting off the benzyl group. *Aliphatic imines* are reduced smoothly with sodium[901]. In the presence of amalgamated aluminum, aliphatic amines are reductively alkylated with carbonyl compounds without isolating the condensation products[902]. Dimerization can take place during reduction of Schiff's bases formed from aromatic aldehydes and primary aromatic amines[903]; reduction with magnesium in methanol of these compounds in

[884] *F. Nerdel, H. Liebig,* Chem. Ber. *87*, 221 (1954).

[885] *F.S. Crosseley, M.L. Moore,* J. Org. Chem. *9*, 529 (1944).

[886] *V.J. Webers, W.F. Bruce,* J. Amer. Chem. Soc. *70*, 1422 (1948).

[887] *H.E. Smith, V. Rajevsky,* J. Heterocycl. Chem. *5*, 715 (1968).

[888] *A. Novelli,* J. Amer. Chem. Soc. *61*, 520 (1939).

[889] *H. Suter, H. Jutter,* Justus Liebigs Ann. Chem. *576*, 215 (1952).

[890] *P.A.S. Smith, A.J. Macdonald,* J. Amer. Chem. Soc. *72*, 1037 (1950).

[891] *I.F. Bunnett, J.L. Marks, H. Moe,* J. Amer. Chem. Soc. *75*, 985 (1953).

[892] *R.D. Bach,* J. Org. Chem. *33*, 1647 (1968).

[893] *D.G. Hey, G.D. Meakins, T.L. Whateley,* J. Chem. Soc. C 1509 (1967).

[894] *P.F. Coe, B.C. Uff, J.W. Lewis,* J. Chem. Soc. C 2265 (1968).

[895] *R.R. Sauers,* J. Amer. Chem. Soc. *80*, 4721 (1958).

[896] *M. Davis, E.W. Parnell, D. Warburton,* J. Chem. Soc. C 1688 (1966).

[897] *M. Davis, E.W. Parnell, J. Rosenbaum,* J. Chem. Soc. C 1983 (1966).

[898] *J.J. Panouse, J. Schmitt, P.-J. Cornu, A. Hallot, H. Pluchet, P. Comoy,* Bull. Soc. Chim. France 1753 (1963).

[899] *J.J. Panouse, J. Schmitt, P.-J. Cornu, A. Hallot, H. Pluchet, P. Comoy,* Bull. Soc. Chim. France 1761, 1767 (1963).

[900] *O. Fischer,* Justus Liebigs Ann. Chem. *241*, 328 (1887).

[901] *M.R. Tiollais,* Bull. Soc. Chim. France 959 (1947).

[902] *R.A. La Forge, C.R. Whitehead, R.B. Keller, C.E. Hummel,* J. Org. Chem. *17*, 457 (1952).

[903] *W. Stühmer, G. Messwarb,* Arch. Pharm. (Weinheim, Ger.) *286*, 221 (1953).

part gives good yields[904]. Zinc has been successfully as reducing agent in a number of cases, for instance, during reduction of *N*-benzylidene-aniline in acetic acid[905]. Mixtures of ketones and ammonia or primary amines are converted into primary or secondary amines in aqueous solution by the action of powdered zinc[906].

12.6.3.2 Reduction with metal hydrides

Condensation products of ammonia or amines and carbonyl compounds are generally reduced smoothly with complex metal hydrides in good yield. *Lithium tetrahydroaluminate* and *sodium tetrahydroborate* are used above all.

In theory 0.25 *mol* lithium tetrahydroaluminate are required to reduce 1 *mol* azomethine, in practise 0.5—1.0 *mol* are generally employed[907]. An extra consumption of 0.25 *mol* is likely with unsubstituted imines because of the active hydrogen atom[908]. The reduction is generally carried out in diethyl ether[907], 1,4-dioxane[909], or tetrahydrofuran[907]. Aliphatic azomethines are usually reduced smoothly to the corresponding *secondary* amines at room temperature[910]. Sometimes *tertiary* amines are accessible by reduction of methiodides with lithium tetrahydroaluminate[911]. In this way it is possible, for example, to convert primary amines into the corresponding *N,N-dimethyl* compounds[875]. A comparison with the standard method of preparation, the Eschweiler-Clarke method, revealed that epimerization on the C-atom carrying the amino group does not take place in either case[875]. *Monomethylation* of primary aliphatic or secondary alicyclic amines succeeds smoothly after conversion into urethanes and the latter's reduction with lithium tetrahydroaluminate[912]. $H_5C_6-CHR-NH_2$ type *optically active* amines can be prepared by reducing the corresponding imines with optically active lithium hydroalkoxyaluminate[913].

Sodium tetrahydroborate is in many instances superior to lithium tetrahydroaluminate for reducing Schiff bases[914, 915]. Chirality centers and other functional groups are not attacked and the yields are generally higher. In general, a large excess of sodium tetrahydroborate is added to a methanolic solution of the Schiff's base, because the reagent partially decomposes in methanol[915, 916]. The best yields are obtained by the reduction of Schiff's bases from aromatic aldehydes and substituted anilines[914, 915]. Like lithium tetrahydroaluminate, sodium tetrahydroborate possesses the great advantage that *N*-benzyl derivatives can be obtained in high yield. In many cases it is unnecessary to isolate the Schiff base. Reductive aminations can be performed also with the versatilely applicable *lithium cyanohydridoborate*[917]; the reduction proceeds more slowly than with *sodium tetrahydroborate*. What part steric factors play during asymmetric synthesis of α-substituted ethylamines obtained from ketones and optically pure primary amines *via* ketimines, followed by the latter's reduction with lithium tetrahydroaluminate, was investigated by determining the optical purity of the amines obtained after hydrogenolysis[918].

12.6.4 Electrolytic reduction

By comparison to the above methods (p. 523 ff.) electrolytic reduction has not been widely applied[791]. Amination of carbonyl compounds has been investigated with the aid of polarography and macroelectrolysis[919]. Electrochemical reductive amination of ketones to *secondary amines* with methylamine succeeds in yields of 52—73% by first forming the Schiff's base from ketone and methylamine by adding lithium chloride and then

[904] *L. Zechmeister, J. Truka,* Ber. *63,* 2883 (1930).

[905] *O. Anselmino,* Ber. *41,* 621 (1908).

[906] *V. Harlay,* C.R. Acad. Sci., Paris *213,* 304 (1941); *M. Mousseron, P. Froger,* Bull. Soc. Chim. France 843 (1947).

[907] *J.H. Billmann, K.M. Tai,* J. Org. Chem. *23,* 535 (1958).

[908] *M.M. Winternitz, R. Dennilauler,* Bull. Soc. Chim. France 474 (1953).

[909] *E.D. Bergmann, D. Lavie, S. Pinchas,* J. Amer. Chem. Soc. *73,* 5662 (1951).

[910] *A.H. Sommers, S.E. Aaland,* J. Org. Chem. *21,* 484 (1956).

[911] *G.W. Kenner, M.A. Murray,* J. Chem. Soc. (London) 406 (1950).

[912] *F. Zymalkowski, T.Y. Yupraphat, K. Schmeißer,* Arch. Pharm. (Weinheim, Ger.) *301,* 321 (1968).

[913] *O. Červinka, V. Suchan, O. Kotynek, V. Dudek,* Collect. Czech. Chem. Commun. *30,* 2484 (1965); C.A. *63,* 11393 (1965).

[914] *Z. Horii, T. Sakai, T. Inoi,* J. Pharm. Soc. Jap. *75,* 1161 (1955); C.A. *50,* 7756 (1956).

[915] *J.H. Billmann, A.C. Diesing,* J. Org. Chem. *22,* 1068 (1957).

[916] *G.N. Walker, M.A. Moore,* J. Org. Chem. *26,* 432 (1961).

[917] *R.F. Borch, M.D. Bernstein, H.D. Durst,* J. Amer. Chem. Soc. *93,* 2897 (1971); *R.F. Borch, H.D. Durst,* J. Amer. Chem. Soc. *91,* 3996 (1969).

[918] *J.-P. Charles, H. Christel, G. Solladie,* Bull. Soc. Chim. France 4439 (1970).

[919] *H. Muto, E. Ischikawa, K. Odo,* Denki Kagaku Oyoki Kogyo Butsuri Kagaku *36,* 363 (1968); C.A. *70,* 25171 (1969).

electrolyzing the base with platinum electrodes[920]. Small amounts of secondary alcohols are formed from the ketone at the same time. Diisopropyl ketones cannot be aminated by this method; the corresponding carbinol is formed in 83% yield.

12.7 Amines by condensation

12.7.1 Mannich reaction

Compounds with acid, mobile H atoms can be condensed with carbonyl compounds and amines to form α-aminomethyl derivatives. In the narrow sense the term Mannich reaction or Mannich condensation is used for the condensation of CH- and XH-acid compounds with formaldehyde and ammonia or a primary or secondary amine to α-aminomethyl derivatives. In the wider sense it can be taken to include reactions which are not limited to the use of formaldehyde as electrophilic condensation partner between the nucleophilic, H-acid, *active* molecule and the nucleophilic amine. In place of formaldehyde other aliphatic or aromatic aldehydes, ketones, and α-keto acids may be used. A more general term here is α-aminoalkylation.

The Mannich reaction proceeds according to the overall equation:

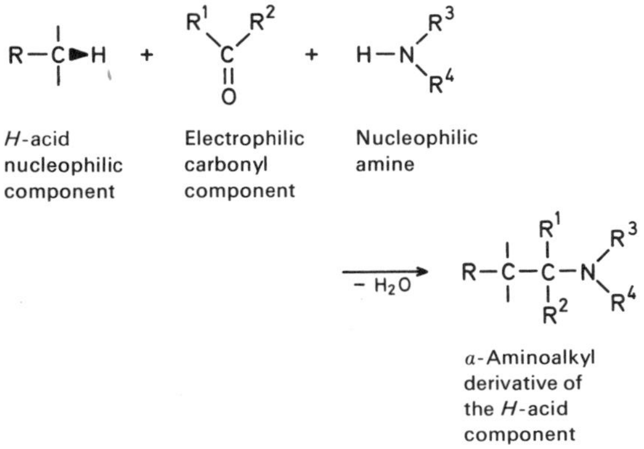

| *H*-acid nucleophilic component | Electrophilic carbonyl component | Nucleophilic amine |

α-Aminoalkyl derivative of the *H*-acid component

It represents an optimum method for synthesizing *N-tertiary β-amino ketones*, *N-tertiary benzylamines* with an *o* or *p hydroxy*, and *aminomethyl or 1-aminoalkyl heterocycles*, and serves as a pathway for preparing the following classes of compounds:

1 *1,3-Amino alcohols* by reduction of the *β*-amino ketones
3 *Methyl-substituted* phenols by hydrogenolysis from phenyl Mannich bases
3 *Chain lengthening* of ketones by one *C* atom
4 *α,β-Unsaturated ketones* by *β*-elimination of the amino group from *β*-amino ketones.

A discussion of the reaction mechanism needs to allow for the fact that a three-component reaction is involved in which the electrophilic carbonyl compound is matched with two nucleophilic partners. The respective nucleophilic potential of the reactants governs whether the carbonyl component reacts first with the amine or with the *H*-acid partner[921].

Where the amine component possesses a higher nucleophilic potential than the acid component, as is generally the case, the carbonyl component first forms a hydroxy compound with the amine which under the action of *H* ions in acid solution is converted into the aminomethylating agent, a mesomerism-stabilized carbonium-immonium ion. In a second electrophilic reaction step this ion combines with the anion of the nucleophilic, *H*-acid partner to form the Mannich base, the aminoalkyl derivative, *e.g.*:

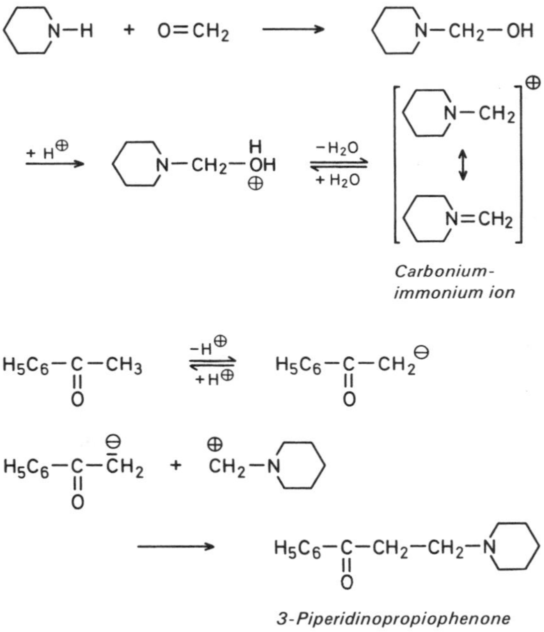

Carbonium-immonium ion

3-Piperidinopropiophenone

With a more highly developed nucleophilic character of the *H*-acid compound the condensation proceeds *via* a corresponding hydroxyalkyl compound of this component. For example, the Mannich reaction of 2,1-benzisoxazolin-3-one with formaldehyde and secondary amines proceeds according to the following scheme:

[920] *R.A. Benkeser, S.J. Mels*, J. Org. Chem. *35*, 261 (1970).
[921] *H. Hellmann, G. Opitz*, Angew. Chem. *68*, 265 (1956).
[922] *C. Mannich*, Arch. Pharm. (Weinheim, Ger.) *255*, 261 (1917).

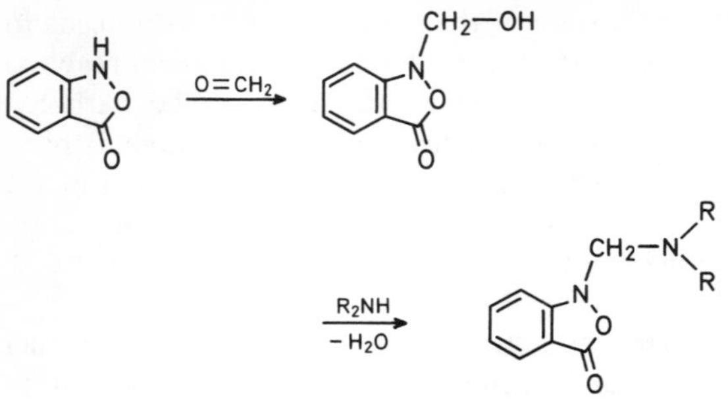

In addition to the above two mechanisms, of which the first generally applies, there are those in which methylenebisamines (*N,N*-acetals) or Schiff's bases appear as intermediate or primary products.

12.7.1.1 Aminoalkylation of *CH*-acid compounds

12.7.1.1.1 Aminoalkylation of carbonyl compounds

Among amino *CH*-acid compounds amenable to aminoalkylation ketones occupy the most important position. In general, they are reacted with the amine hydrochlorides and paraformaldehyde in suitable solvents such as alcohols, benzene, toluene, glacial acetic acid, nitroalkanes, or with aqueous formaldehyde solution using excess ketone to serve as solvent.

12.7.1.1.1.1 Aminoalkylation of aliphatic ketones

When the free amine is used the same reaction product is not always obtained. Thus, heating a mixture of dimethylamine hydrochloride, aqueous formaldehyde solution, and excess acetone leads smoothly to *4-(dimethylamino)-2-butanone*[922], while condensation in alkaline medium affords *4-(dimethylamino)-3-[(dimethylamino)methyl]-2-butanone*[923] as main product side by side with the monoamine:

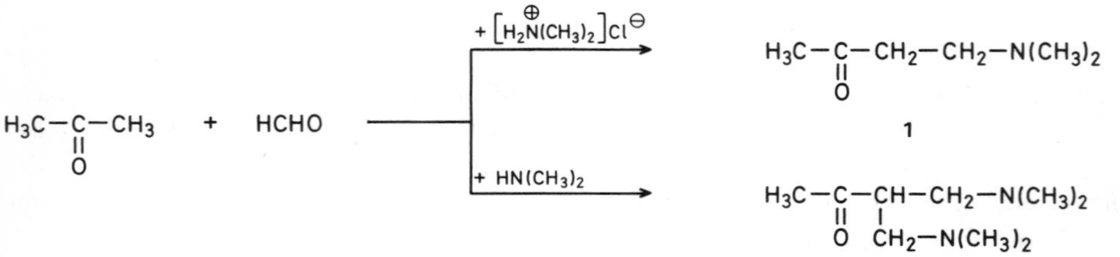

Where the ketone contains different *CH*-acid groups, the empirical rule generally applies that methylene groups react before methyl groups. The *H* atom affording the most stabilized enol by protrotropy is substituted preferentially. Aliphatic ketones are aminomethylated on that *C* atom whose neighboring *C* atom carries the most *H* atoms. In the following examples the boldly printed *H* atoms are substituted:

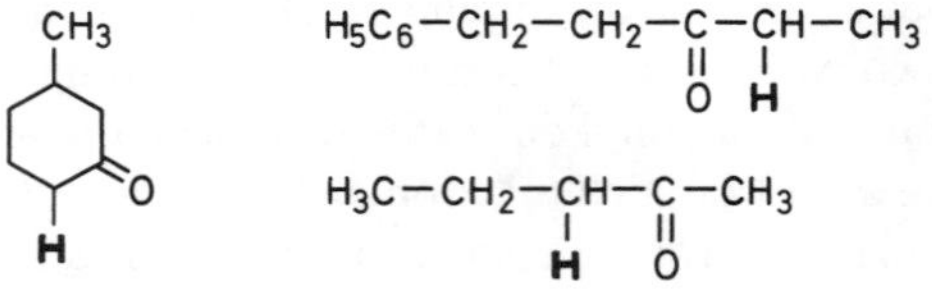

Alternatively, more than one acid H atom may be replaced by aminomethyl groups. In addition to the preceding example **2**, in which the Mannich reaction takes place twice on the same *C* atom, bis-Mannich bases of the general formula **3** may form, which then often appear side by side with Type **1** and Type **2** Mannich bases.

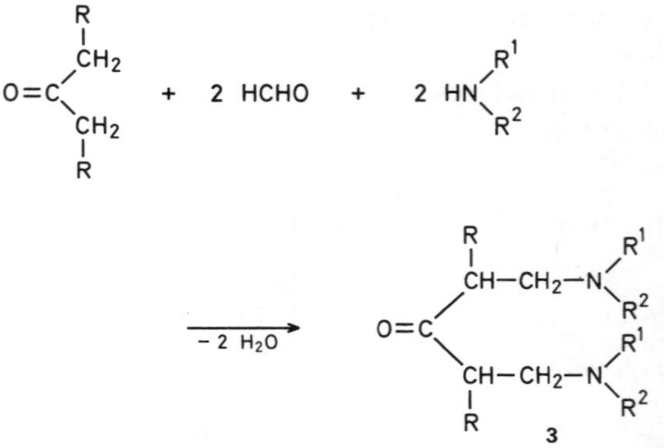

In addition to playing the part of the *H*-acid component, ketones can occasionally act as the carbo-

[923] *C. Mannich, O. Salzmann*, Ber. *72*, 507 (1937).
[924] *T. Götschmann*, Justus Liebigs Ann. Chem. *197*, 27 (1879).
[925] *J.J. Denton, R.J. Turner, W.B. Neier, V.A. Lawson, H.P. Schede*, J. Amer. Chem. Soc. *71*, 2048 (1949).

[926] *C. Mannich, O. Hieronimus*, Ber. *75*, 49 (1942).
[927] *H.J. Roth, Ch. Schwenke*, Arch. Pharm. (Weinheim, Ger.) *297*, 773 (1964).
[928] *H.J. Roth, G. Langer*, Arch. Pharm. (Weinheim, Ger.) *301*, 695 (1968).

nyl component. Thus alkyl methyl ketones can condense with themselves and with diethylamine according to the following scheme[924]:

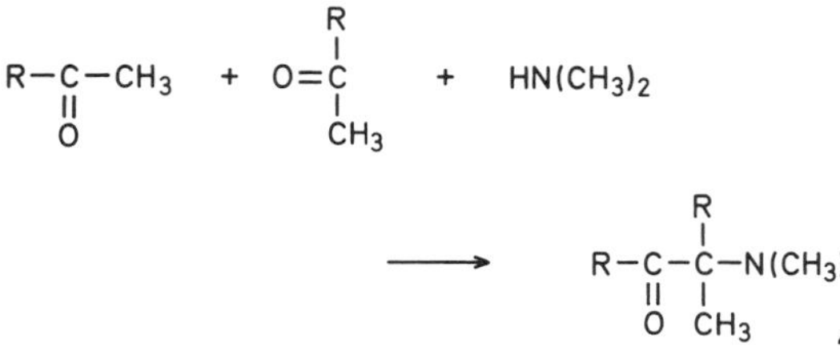

$$R-\underset{\underset{O}{\|}}{C}-CH_3 \;+\; O=\underset{\underset{CH_3}{|}}{\overset{\overset{R}{|}}{C}} \;+\; HN(CH_3)_2$$

$$\longrightarrow \quad R-\underset{\underset{O}{\|}}{C}-\underset{\underset{CH_3}{|}}{\overset{\overset{R}{|}}{C}}-N(CH_3)_2 \qquad 4$$

The Mannich reaction often becomes complicated if the starting and condensation products undergo aldol reactions (see p. 536).

When primary amines or ammonia are employed as the amine component then, as a rule, all *H* atoms on the nitrogen participate in the condensation:

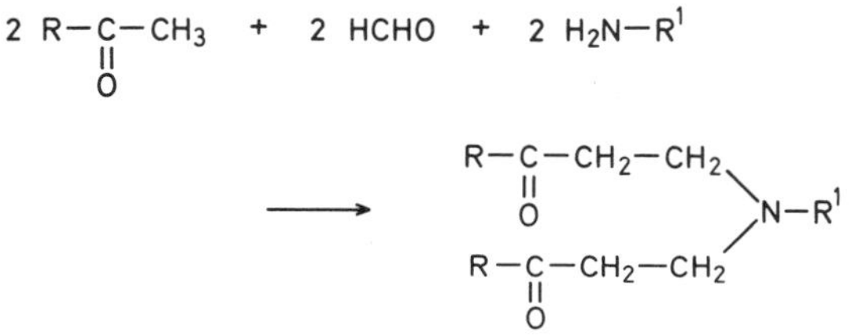

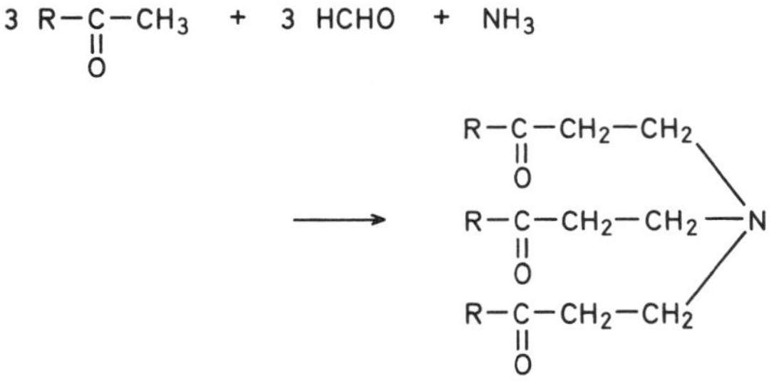

$$2\;R-\underset{\underset{O}{\|}}{C}-CH_3 \;+\; 2\;HCHO \;+\; 2\;H_2N-R^1$$

$$\longrightarrow \quad \underset{R-\underset{\underset{O}{\|}}{C}-CH_2-CH_2}{\overset{R-\underset{\underset{O}{\|}}{C}-CH_2-CH_2}{}}\!\!\!\!\!\!\!\!\!\!\!\! \Big\rangle N-R^1$$

$$3\;R-\underset{\underset{O}{\|}}{C}-CH_3 \;+\; 3\;HCHO \;+\; NH_3$$

$$\longrightarrow \quad \underset{R-\underset{\underset{O}{\|}}{C}-CH_2-CH_2}{\overset{R-\underset{\underset{O}{\|}}{C}-CH_2-CH_2}{}}\!\!\!-N \!\!\!\Big\langle \underset{R-\underset{\underset{O}{\|}}{C}-CH_2-CH_2}{}$$

4-Piperidones are formed by ring closure where two successive aminomethylations of an aliphatic ketone are possible, *viz.*, on two different *C* atoms adjacent to and on either side of the carbonyl group, as well as two successive condensations of a primary amine or of ammonia:

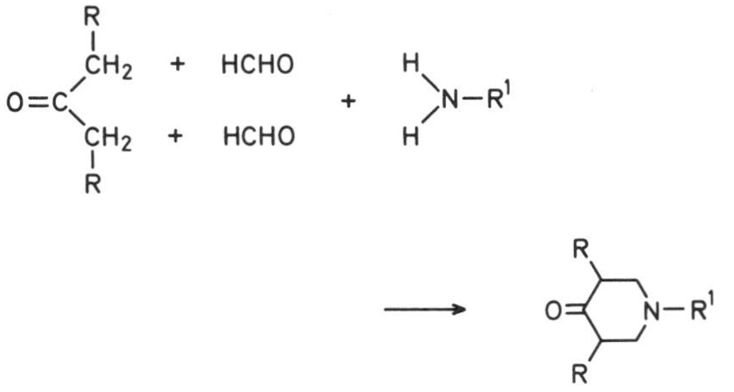

$$O=\underset{\underset{R}{|}}{\overset{\overset{R}{|}}{C}}\!\!\left\langle\begin{array}{l}CH_2 \;+\; HCHO \\ \\ CH_2 \;+\; HCHO\end{array}\right. \;+\; \overset{H}{\underset{H}{}}\!\!\Big\rangle N-R^1$$

$$\longrightarrow \quad O=\!\!\underset{R}{\overset{R}{\bigcirc}}\!\!N-R^1$$

Suitable dialkyl ketones for such a reaction are: acetone, 2-butanone, 2-alkanones, 3-pentanone, and 3- and 4-heptanone.

Finally, it is feasible for suitable ketones to act both as the acid and carbonyl component and to condense together more than once. If ammonia or a primary amine is the amine component, then the bis-Mannich base and a piperidone derivative may be formed in addition to the single Mannich base, *e.g.*:

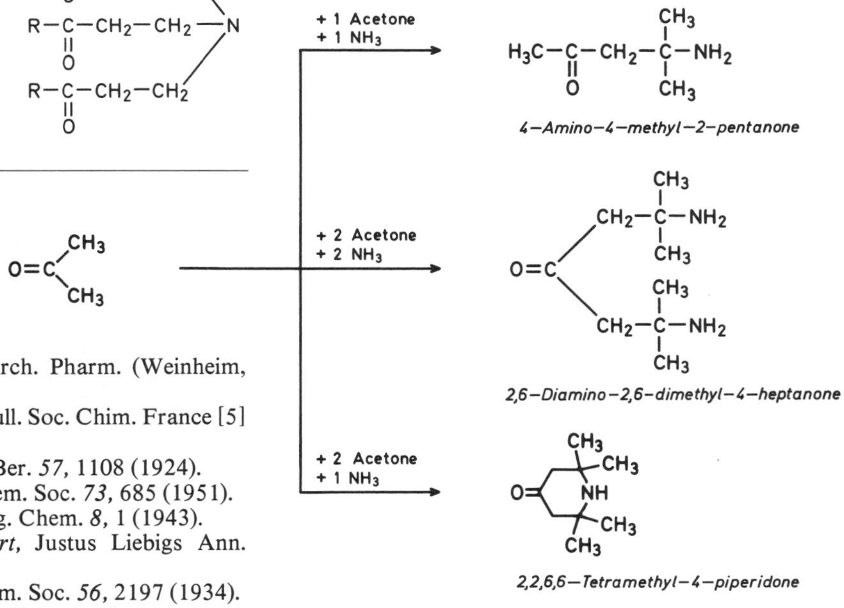

$$O=C\!\!\left\langle\begin{array}{l}CH_3 \\ CH_3\end{array}\right.$$

+ 1 Acetone
+ 1 NH_3 →

$$H_3C-\underset{\underset{O}{\|}}{C}-CH_2-\underset{\underset{CH_3}{|}}{\overset{\overset{CH_3}{|}}{C}}-NH_2$$

4−Amino−4−methyl−2−pentanone

+ 2 Acetone
+ 2 NH_3 →

$$O=C\!\!\left\langle\begin{array}{l}CH_2-\underset{\underset{CH_3}{|}}{\overset{\overset{CH_3}{|}}{C}}-NH_2 \\ \\ CH_2-\underset{\underset{CH_3}{|}}{\overset{\overset{CH_3}{|}}{C}}-NH_2\end{array}\right.$$

2,6−Diamino−2,6−dimethyl−4−heptanone

+ 2 Acetone
+ 1 NH_3 →

$$O=\!\!\overset{CH_3}{\underset{CH_3}{\bigcirc}}\!\!NH$$

2,2,6,6−Tetramethyl−4−piperidone

[929] *H.J. Roth, K. Thaßler*, Arch. Pharm. (Weinheim, Ger.) *304*, 816 (1971).
[930] *C. Sannié, J.J. Panouse*, Bull. Soc. Chim. France [5] *23*, 1541 (1956).
[931] *C. Mannich, M. Bauroth*, Ber. *57*, 1108 (1924).
[932] *L.C. Cheney*, J. Amer. Chem. Soc. *73*, 685 (1951).
[933] *M. Zief, J.P. Moson*, J. Org. Chem. *8*, 1 (1943).
[934] *M. Bockmühl, G. Ehrhart*, Justus Liebigs Ann. Chem. *561*, 52 (1949).
[935] *G.A. Menge*, J. Amer. Chem. Soc. *56*, 2197 (1934).
[936] *H. Böhme, E. Mundlos, G. Keitzer*, Chem. Ber. *91*, 656 (1957).
[937] *R. Oda, M. Nomura, S. Tanimoto, T. Nishimura*, Bull. Inst. Chem. Research, Kyoto Univ. *34*, 224 (1956); Chem. Zentr. 1733 (1959);

R. Oda, S. Tanimoto, M. Nomura, T. Nishimura, K. Kyo, J. Chem. Soc. Jap., Ind. Chem. Sect. *60*, 18 (1957); Chem. Zentr. 5252 (1957).

12.7.1.1.1.2 Aminoalkylation of alkyl aryl ketones

On aminomethylation of aromatic-aliphatic ketones the complications arising as a result of repeated condensations on different C atoms in purely aliphatic ketones are eliminated. When formaldehyde and secondary amines are used the Mannich bases **5** are formed, while with primary amines **6** are formed, and with ammonia compounds **7**:

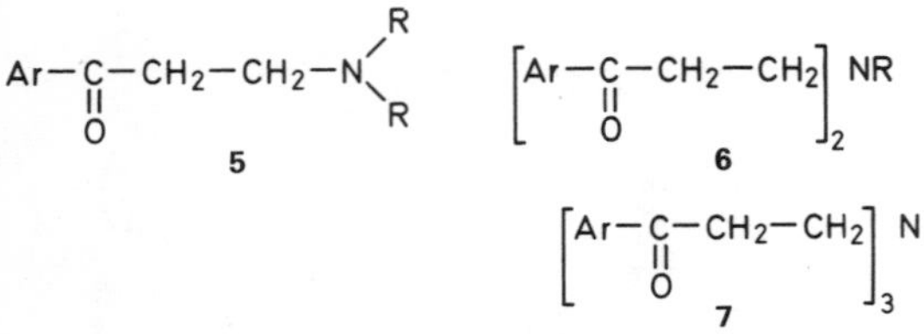

$$Ar-\underset{\underset{O}{\|}}{C}-CH_2-CH_2-N\overset{R}{\underset{R}{\big\langle}}$$

5

$$\left[Ar-\underset{\underset{O}{\|}}{C}-CH_2-CH_2\right]_2 NR$$

6

$$\left[Ar-\underset{\underset{O}{\|}}{C}-CH_2-CH_2\right]_3 N$$

7

Occasionally bis(dialkylaminomethyl) derivatives may be formed, e.g.[925]:

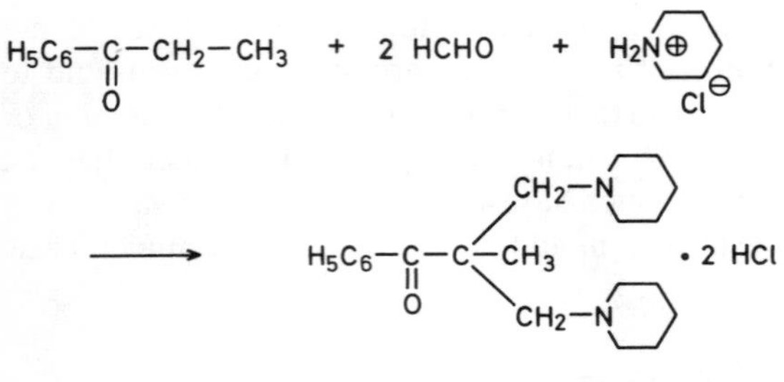

$$H_5C_6-\underset{\underset{O}{\|}}{C}-CH_2-CH_3 \quad + \quad 2\ HCHO \quad + \quad H_2N^{\oplus}\bigcirc\ \ Cl^{\ominus}$$

$$\longrightarrow \quad H_5C_6-\underset{\underset{O}{\|}}{C}-\underset{\underset{CH_2-N\bigcirc}{\overset{CH_2-N\bigcirc}{|}}}{C}-CH_3 \quad \cdot\ 2\ HCl$$

2,2-Bis(piperidinomethyl)-propiophenone; 43%

12.7.1.1.1.3 Aminoalkylation of alicyclic ketones

Cyclic dialkyl ketones are normally aminomethylated on one of the methylene groups adjacent the carbonyl group **8**. Condensation on both α-methylene groups leading to bis-Mannich bases **9** is rarer:

8

9

Using primary amines (or ammonia) as the basic components, the anticipated bis-Mannich bases can react further to form decahydroisoquinolines in aldol-like fashion, particularly where cyclohexanone derivatives are involved[926, 927]:

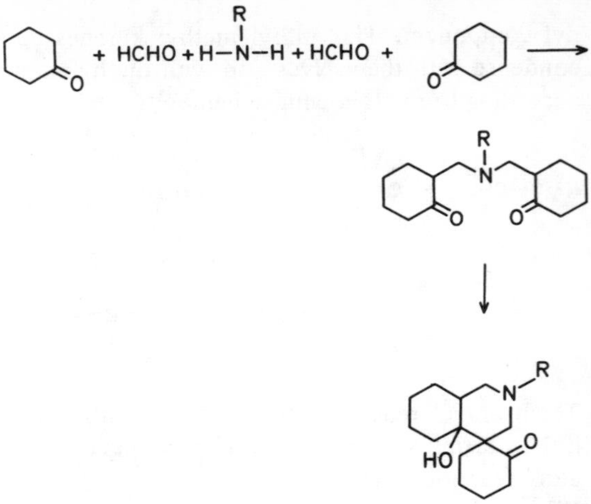

With benzylamine or substituted benzylamines, simple Mannich bases can be prepared which likewise afford the isoquinoline skeleton in a second reaction with formaldehyde and aliphatic ketones; here they act as the amine component[926]:

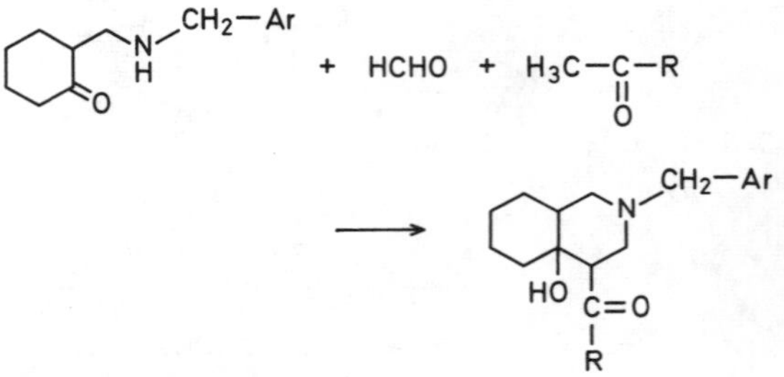

12.7.1.1.1.4 Aminoalkylation of cyclic alkyl aryl ketones

3,4-Dihydro-1(2*H*)-naphthalenones, 1-indanones, 2,2a,3,4-tetrahydro-1(5*H*)acenaphthylenone, and 2,3-dihydro-1(4*H*)- and 2,3-dihydro-4(1*H*)-phenanthrone can be aminomethylated in simple manner on the methylene group adjacent the carbonyl group. Condensing is performed predominantly with paraformaldehyde and the amine hydrochlorides in boiling alcohols. The corresponding amino-

[938] *K. Bredereck, L. Banshaf,* Tetrahedron Lett. 4323 (1970).

[939] *K. Hafner,* Angew. Chem. *70*, 419 (1958).

[940] *C. Mannich, B. Kather,* Arch. Pharm. (Weinheim, Ger.) *257*, 18 (1919).

[941] *W. Logemann, F. Lauria, V. Zamboni,* Chem. Ber. *88*, 1353 (1955).

components are stable as salts, while the free bases tend to split off amine.

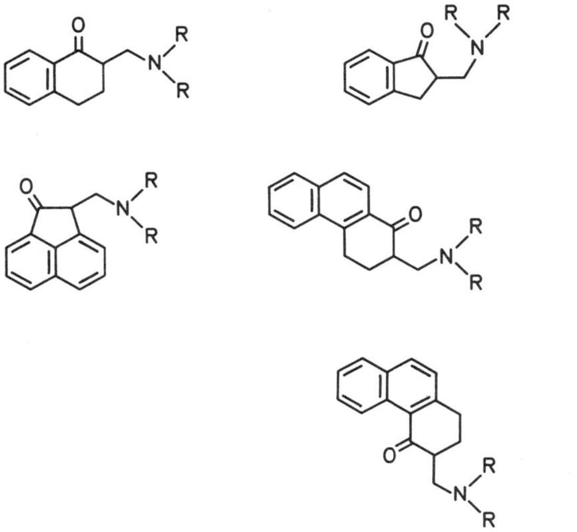

12.7.1.1.1.5 Aminoalkylation of vinylogous ketones

12.7.1.1.1.5.1 Aminoalkylation of vinylogous alkyl aryl ketones

Aminomethylation of α-arylidenedialkyl ketones such as 4-phenyl, 4-(2-furyl)-, and 4-(2-thienyl)-3-buten-2-ones, and of 1-alkyl-3-phenyl-2-buten-1-ones proceeds normally. With primary amines the single aminomethylation products are formed almost exclusively and not the likewise expected bis-Mannich bases:

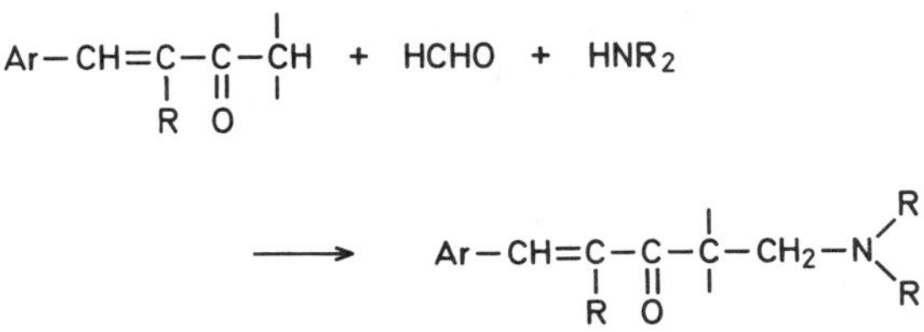

12.7.1.1.1.5.2 Aminoalkylation of vinylogous cycloaliphatic ketones

Vinylogous aminomethylation or γ-aminomethylation of α,β-*unsaturated ketones* appears to succeed only where α-aminomethylation is strongly hindered by steric or energetic factors. Vinylogous aminomethylation in 3,5,5-trimethyl-2-cyclohexen-1-one(isophorone)[928] and in *p*-menth-1-en-3-one (piperitone) has been confirmed[929]:

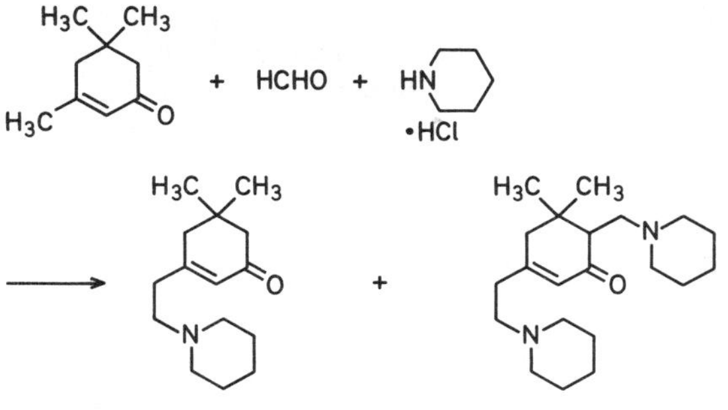

5,5-Dimethyl-3-(2-piperidinoethyl)-2-cyclohexen-1-one; 70–80%

5,5-Dimethyl-3-(2-piperidinoethyl)-(6-piperidinomethyl)-2-cyclohexen-1-one; a little

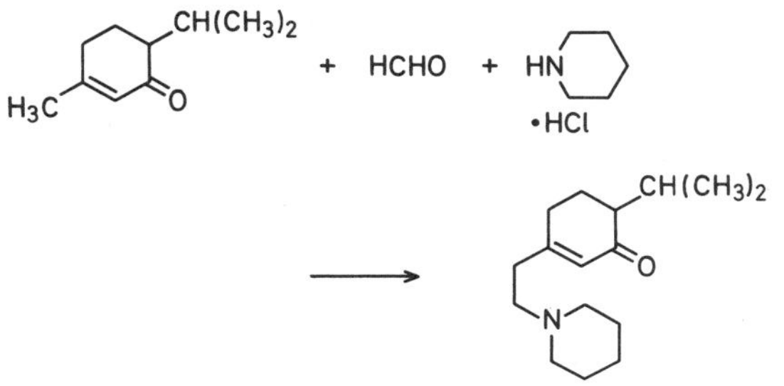

7-(2-Piperidinoethyl)-p-menth-1-en-3-one; 49%

Further, the vinylogous aminomethylation of 5-methyl-3-oxo-1-cyclohexene-1-acetic acid has been described in which introducing the aminomethyl group into the vinylogous position succeeds as a result of additional activation by the carboxy group[930]:

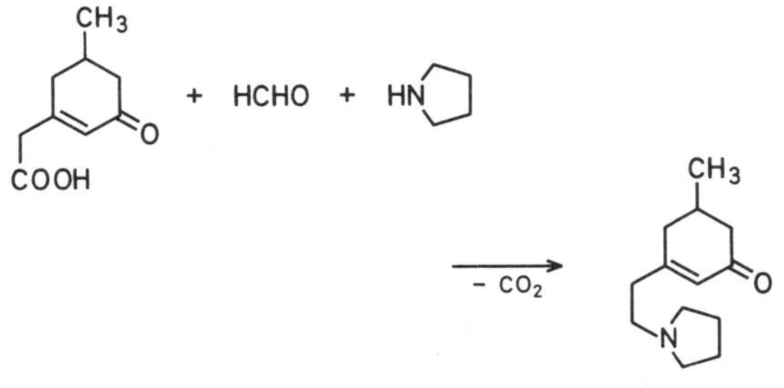

5-Methyl-3-[2-(1-pyrrolidinyl)ethyl]-2-cyclohexen-1-one; 45%

12.7.1.1.1.6 Aminoalkylation of keto carboxylic acids

12.7.1.1.1.6.1 Aminoalkylation of β-keto carboxylic acids

Acetoacetic acid, 3-oxohydrocinnamic acid, and monosubstituted acetoacetic acids can be aminomethylated with formaldehyde and amines even at room temperature, although decarboxylation occurs and the Mannich bases of the corresponding ketones are obtained:

$$R-\underset{\underset{O}{\|}}{C}-\underset{\underset{R}{|}}{C}H-COOH \;+\; HCHO \;+\; HN\diagdown \;\;\xrightarrow[-\;CO_2]{-\;H_2O}\;\; R-\underset{\underset{O}{\|}}{C}-\underset{\underset{R}{|}}{C}H-CH_2-N\diagdown$$

With aldehydes and primary amines or ammonia, acetoacetic esters and diesters of 3-oxoglutaric acid afford the 4-piperidones:

$$H_5C_2O-\underset{\underset{CH_2}{}}{\overset{O}{\underset{|}{C}}}\;\; O=C\;\; \underset{R}{\overset{CH_2}{}} \;+\; 2\;\underset{\underset{H}{|}}{\overset{R^1}{\underset{|}{C}}}=O \;+\; H_2NR^2 \;\longrightarrow\;$$

R= H
R= $-\underset{\underset{O}{\|}}{C}-OC_2H_5$

From 3-oxo-3-phenylpropionic acid, bifunctional aldehydes and primary amines cyclic bis-Mannich bases of the *lobelamine* type are accessible in one step:

$$R-\underset{\underset{O}{\|}}{C}-CH_2 \atop {\underset{COOH}{|}}$$

$$- 2\;H_2O$$
$$- 2\;CO_2$$

Using one *mol* 3-oxoglutaric acid or its diester in place of two *mols* 3-oxo-3-phenylpropionic acid

leads to *heterobicyclic compounds* similar to tropane and pseudopelletrine alkaloids:

$$RO-\underset{\underset{CH_2}{}}{\overset{O}{\underset{\|}{C}}}\;\;O=C\;\;\underset{\underset{RO-C}{}}{\overset{CH_2}{}}\;+\;\left(\underset{\underset{CH}{}}{\overset{CH}{}}\right)\;+\;\underset{\underset{H}{}}{\overset{H}{}}N-R^1\;\longrightarrow$$

12.7.1.1.1.6.2 Aminoalkylation of α-, γ-, δ-keto carboxylic acids

If the oxo group is not in the β-position with respect to the carboxy group, then the keto carboxylic acids react like simple ketones. Levulinic acid and esters and 5-oxohexanoic acid condense exclusively on the methyl group. With aromatic aldehydes and ammonia γ-oxoacid esters form *4-piperidones*:

$$RO-\underset{\underset{O}{\|}}{C}-CH_2\underset{\underset{CH_3}{}}{\overset{CH_2}{\underset{}{C=O}}}\;+\;2\;H\overset{Ar}{\underset{|}{C}}=O\;+\;H_3N\;\longrightarrow$$

Pyruvic acid (2-oxopropionic acid) is exceptional in that aminomethylation, hydroxymethylation, and splitting off of water lead to β-*aminomethyl-α-oxolactones*[931], e.g.:

$$\underset{HO}{\overset{O}{\diagup}}C=C\underset{CH_3}{\overset{O}{\diagdown}}\;+\;2\;HCHO\;+\;HN(CH_3)_2\;\longrightarrow$$

4-(Dimethylaminomethyl)-4-hydroxy-2-oxobutyric acid γ-lactone; 56%

12.7.1.1.1.7 Aminoalkylation of aldehydes

Aldehydes can be aminomethylated like ketones, including on a methine group, but the tendency for side-reactions to occur is much more marked than with ketones. Two successive aminomethylations can take place on methylene and methyl groups:

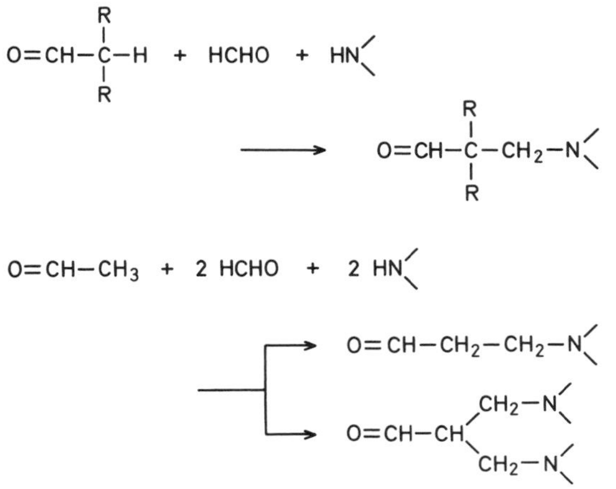

$$O{=}CH-\underset{\underset{R}{|}}{\overset{\overset{R}{|}}{C}}-H \ + \ HCHO \ + \ HN{<}$$

$$\longrightarrow \quad O{=}CH-\underset{\underset{R}{|}}{\overset{\overset{R}{|}}{C}}-CH_2-N{<}$$

$$O{=}CH-CH_3 \ + \ 2\ HCHO \ + \ 2\ HN{<}$$

$$\longrightarrow \quad O{=}CH-CH_2-CH_2-N{<}$$
$$\longrightarrow \quad O{=}CH-CH{\overset{CH_2-N{<}}{\underset{CH_2-N{<}}{}}}$$

$$N{\equiv}CH \ + \ O{=}\overset{|}{\underset{|}{C}} \ + \ HN{\overset{R}{\underset{R}{<}}} \quad \longrightarrow \quad N{\equiv}C-\overset{|}{\underset{|}{C}}-N{\overset{R}{\underset{R}{<}}} \qquad (e.\,g.\ N{\equiv}C-CH_2-NH_2)$$

$$N{\equiv}CH \ + \ O{=}\overset{|}{\underset{|}{C}} \ + \ H_2N-R \quad \longrightarrow \quad N{\equiv}C-\overset{|}{\underset{|}{C}}-NH-R \ + \ (NC-\overset{|}{\underset{|}{C}}-)_2N-R$$

$$N{\equiv}CH \ + \ O{=}\overset{|}{\underset{|}{C}} \ + \ H_3N \quad \longrightarrow \quad N{\equiv}C-\overset{|}{\underset{|}{C}}-NH_2 \ + \ (N{\equiv}C-\overset{|}{\underset{|}{C}}-)_2NH \ + \ (N{\equiv}C-\overset{|}{\underset{|}{C}}-)_3N$$

Under suitable reaction conditions the Mannich reaction may be coupled with a crossed Cannizzaro reaction, the formaldehyde reducing the Mannich bases to the *amino alcohol*[932], e.g.:

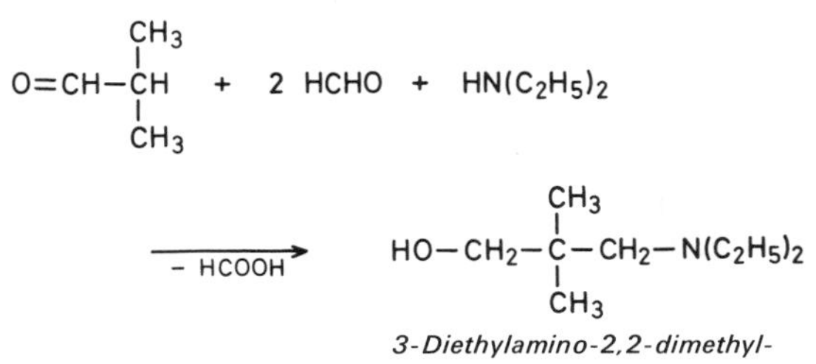

$$O{=}CH-\underset{\underset{CH_3}{|}}{\overset{\overset{CH_3}{|}}{CH}} \ + \ 2\ HCHO \ + \ HN(C_2H_5)_2$$

$$\xrightarrow[-\ HCOOH]{} \quad HO-CH_2-\underset{\underset{CH_3}{|}}{\overset{\overset{CH_3}{|}}{C}}-CH_2-N(C_2H_5)_2$$

3-Diethylamino-2,2-dimethyl-1-propanol

12.7.1.1.1.8 Aminoalkylation of carboxylic acids, nitriles, hydrogen cyanide

The *CH* group of simple *carboxylic* acids and their *esters* is too weakly acid to undergo a Mannich reaction. If the methyl or methylene group adjacent the carboxy group is additionally weakly activated by an electron-attracting function, for-mation of Mannich bases is possible, albeit not by the normal route but by catalytic addition to a Schiff's base.

The activating effect of a nitrile group on a methyl or methylene group also is insufficient to make a Mannich condensation possible. One or two additional *phenyl* groups on the α-C atom facilitate the reaction[933, 934].

Hydrogen cyanide may be regarded as the universally applicable *CH*-acid component. It condenses even with bulky and sterically unfavorable carbonyl compounds. The alkali salts and the amine hydrochlorides are employed for the reaction. With formaldehyde a reaction ensues even at room temperature, with higher aldehydes and ketones heating in water, alcohols, or acetic acid is carried out. In order to obtain mononitriles from primary amines it is necessary to use excess amine. Unsubstituted aminoacetonitrile can be prepared almost quantitatively in liquid ammonia[935]:

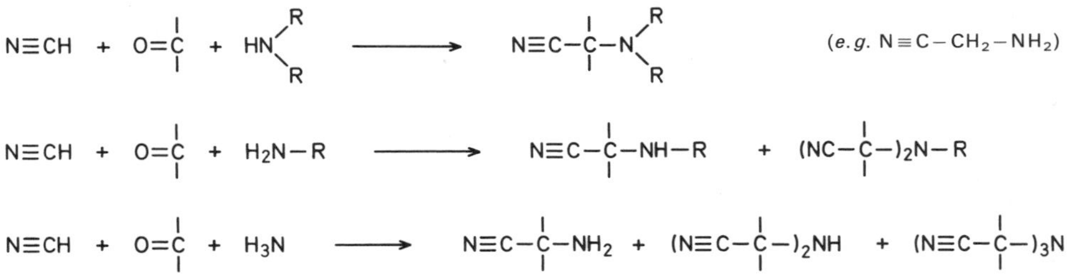

12.7.1.1.1.9 Aminoalkylation of nitro compounds

12.7.1.1.1.9.1 Aminoalkylation of aliphatic nitro compounds

Nitroparaffins are easily aminoalkylated in the α-position with respect to the nitro group; condensation is most readily effected in weakly alkaline medium. The free amines and water are employed, aqueous 1,4-dioxane, or ethanol at room temperature. When formaldehyde is used hydroxymethyl and methylene bis-derivatives are obtained as side-products along with the strongly nucleophilic nitroalkanes. For this reason, the carbonyl component is more appropriately reacted first with the amine component and the nitroparaffin is then added (Route **a**). Conversely, however, it is possible to react the hydroxymethyl compounds of the nitroalkanes specifically with amines to form the Mannich bases. Azeotropic distillation of the water of reaction allows the equilibrium to be shifted in favor of the Mannich base (Route **b**):

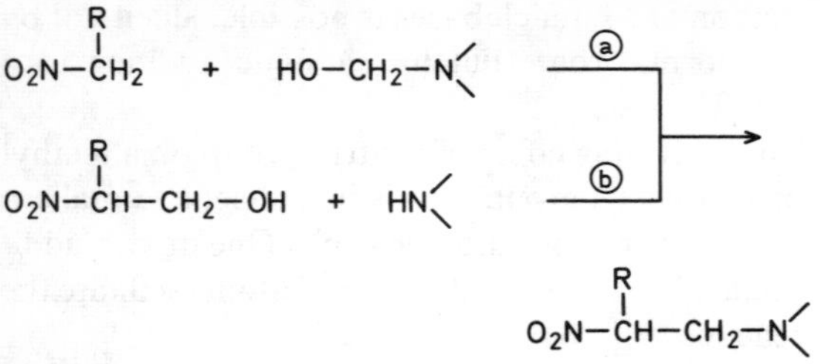

Nitromethane is always aminomethylated twice, the third *H* atom can be reacted further with formaldehyde. To obtain monosubstitution it is necessary to make an ionic reaction between nitromethane sodium and an immonium salt possible[936].

In nitroethane, nitropropane, nitrobutane, *etc.*, nitrocyclohexane, 2-nitropropanol, nitroethanol, and other nitroalkanes the α-methylene groups are aminomethylated either once or twice depending on the experimental conditions.

On reacting with formaldehyde and primary amines polycondensation does not take place; instead, *5-nitro-1,3-tetrahydrooxazines* and *5-nitrohexahydropyrimidines* are formed:

When ammonia is employed as amine component complicated condensations can arise leading to *8-membered* and *bridged* ring systems among others.

12.7.1.1.1.9.2 Aminoalkylation of aromatic nitro compounds

Phenylogous nitroalkanes condense similarly to nitroparaffins if a further *o* or *p* electron-attracting substituent is combined with the activating nitro group. A second nitro group or possibly a carboxy group are appropriate. The condensation ensues with formaldehyde and the free amines, with the *bis-Mannich base* being formed side by side with the *simple base*:

12.7.1.1.1.1.10 Aminoalkylation of alkynes and alkenes

Expectedly, the free hydrogen atom in 1-alkynes is so reactive that it can be exchanged against aminomethyl groups. Aminomethylation of *arylacetylenes* with paraformaldehyde and secondary amines succeeds in boiling 1,4-dioxane without addition of acid. Mannich condensation of *alkylacetylenes* and butadiyne can be catalyzed by copper(I) or iron(III) chloride. A hydroxy group in the acetylene has a notable influence on its ability to condense. While α-hydroxy groups prevent the reaction, β-, γ-, δ-, *etc.*, groups do not. Etherifying or esterifying the α-function restores the possibility of a Mannich reaction.

Acetylene itself can be aminomethylated in autoclaves in the presence of acetic acid and a copper(I) catalyst. Among the resulting Mannich bases monosubstituted products predominate and can be aminomethylated afresh on the other *C* atom to make both *symmetrical* and *unsymmetrical bis(aminoalkyl)acetylenes* accessible.

Olefins carrying at least one *H* atom on the double bond can be aminomethylated only in acid solution because of their weak nucleophilic character. The reaction is performed in acetic acid, optionally with added mineral acid. Often, addition of acetic acid or water to the double bond occurs side by side with the Mannich condensation.

Olefins react with *O,N*-hemiacetals with cooling if boron trifluoride or zinc chloride is added[937]:

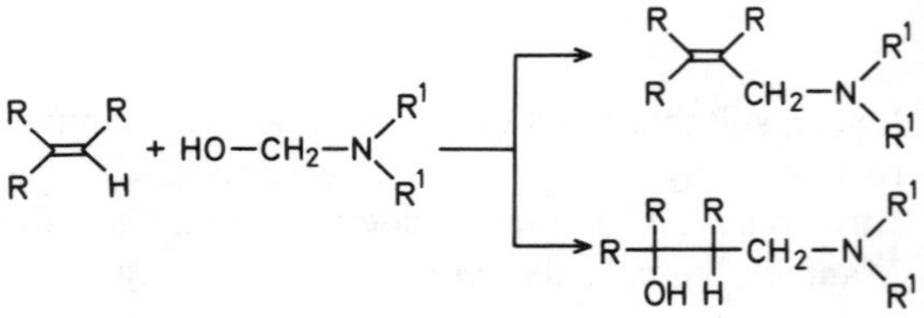

A Mannich base and an addition product are formed side by side. Where the adduct contains a *H* atom on the nitrogen further condensation with a carbonyl component favors the formation of 1,3-tetrahydrooxazines:

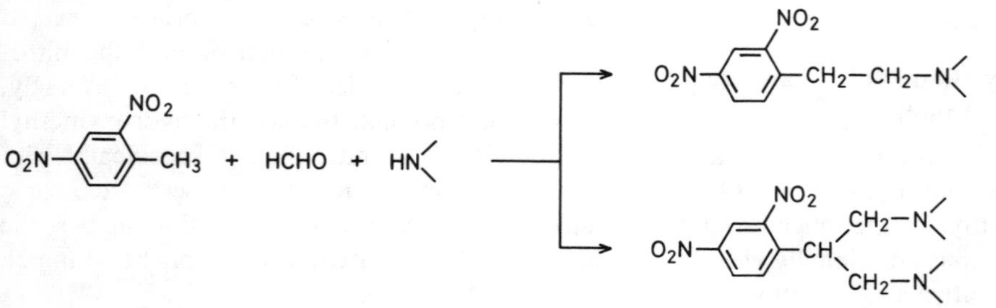

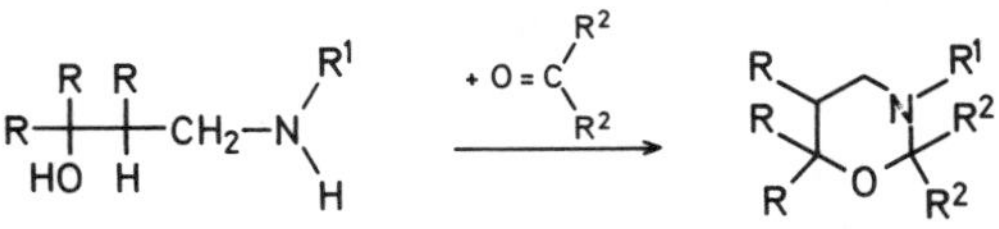

12.7.1.1.1.11 Aminoalkylation of phenols and aromatic amines

Aminomethylation of phenols generally proceeds on the ring and preferentially in the *o*-position with respect to the phenolic hydroxy group. With excess carbonyl and amino components 2- and 3-aminomethyl radicals also can enter the ring. The free amines are employed for the condensation. Substitution occurs in the 2, 4, or 6 position. Amine hydrochlorides should be used only with polyhydric phenols which can form undesired resins with formaldehyde and amines:

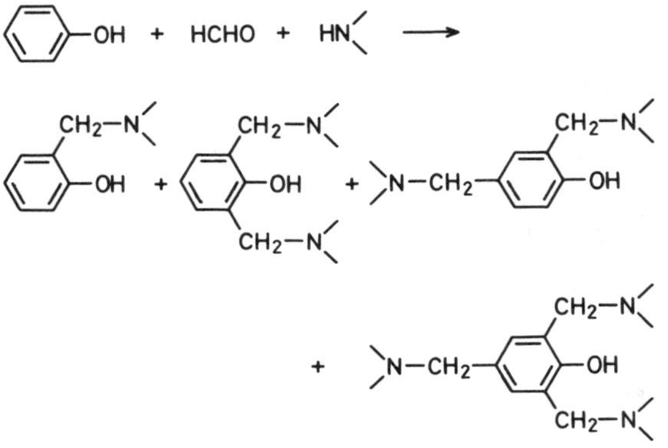

Selective substitution in the *p*-position is *not successful;* only with 2, 5-disubstituted phenols does the aminomethyl enter the 4-position exclusively.

In addition to phenol itself, substituted phenols, naphthols and higher homologs, heterocyclic phenols and tropolone, too, have been aminomethylated.

2 H-benzo[e]-1,3-oxazines are formed preferentially *(Betti reaction)*:

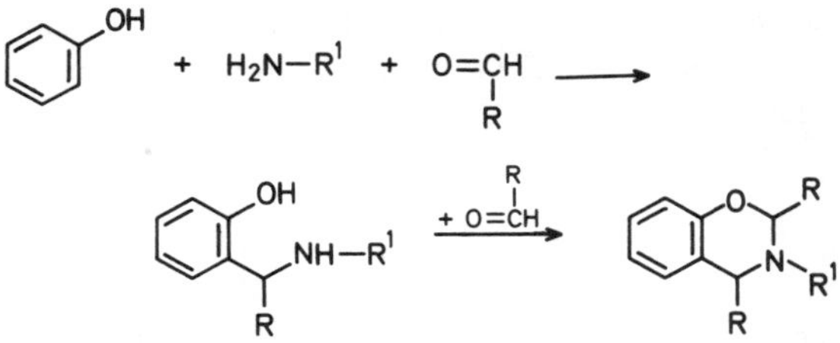

With polyhydric phenols two or three 3, 4-dihydro-*2 H*-1, 3-oxazine rings may be fused on, *e.g.*:

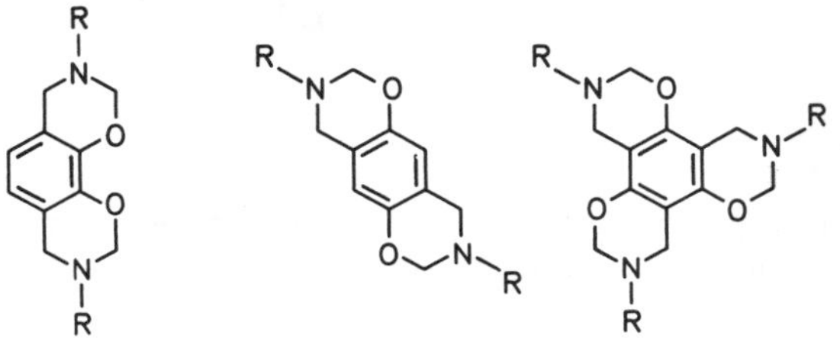

In dependence on the experimental conditions aromatic amines can either function as amine component, be aminomethylated on the nitrogen, or substitute in the *o*- or *p*-position. The latter reaction takes place in the presence of limited quantities of acid, and preferentially in the *p*-position:

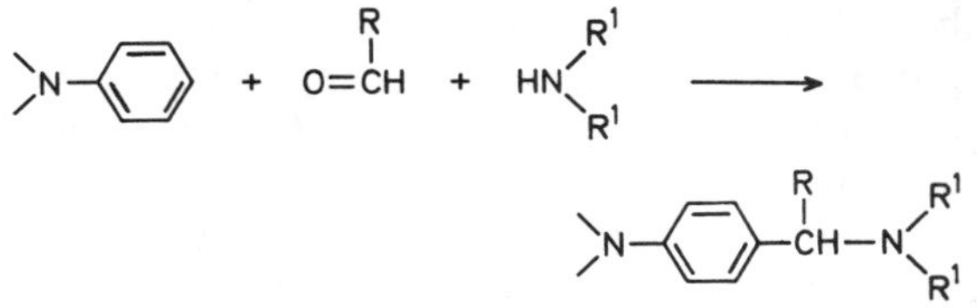

p-Substituted anilines are condensed variously. In addition to methylene-bis derivatives two *mols* aniline derivative and one *mol* formaldehyde can give the Mannich base **1**, while *N, N*-acetalization and renewed aminomethylation can form tetrahydroquinazolines **2** and Tröger bases **3**:

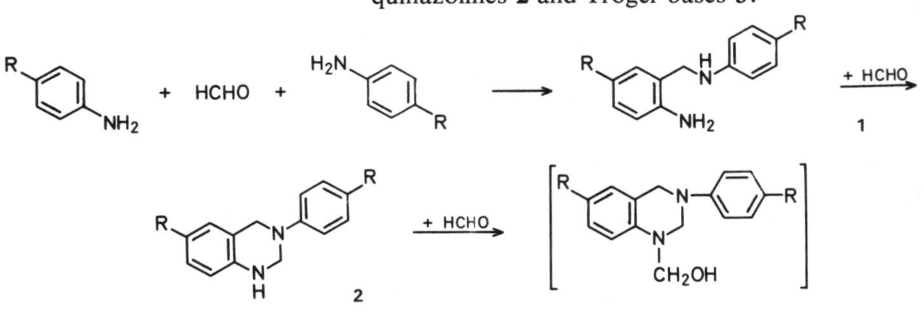

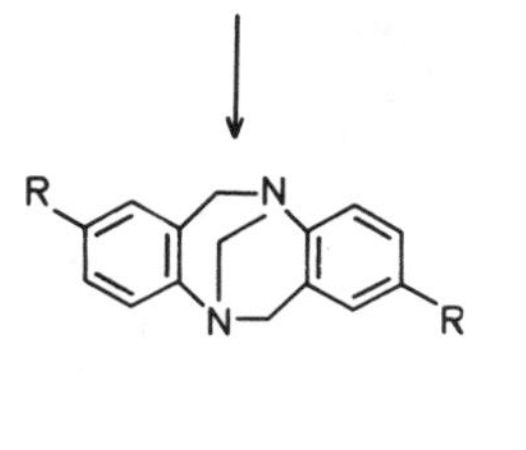

Condensation products form with primary amines and ammonia which, as secondary or primary amines, act the part of the amine component and can react further with the carbonyl compounds and phenols. With excess aldehyde *3,4-dihydro-*

12.7.1.1.1.12 Aminoalkylation of quinones

Certain hydroxyquinones which condense with aldehydes and amines to form Mannich bases in alcoholic solution even at room temperature have been investigated, *e.g.:*

The phenylogous aminomethylation of 1-aminoanthraquinone has been described recently[938].

12.7.1.1.1.13 Aminoalkylation of hydrocarbons

Aliphatic hydrocarbons without activating functions expectedly cannot be aminoalkylated. The electron-attracting effect of three phenyl groups in triphenylmethane likewise is not sufficient. Indene and fluorene do condense in poor yield with cotarnine, which may be regarded as a cyclic *N*-hemiacetal, *i.e.,* as an intramolecular reaction product of an amine and an aldehyde. Condensation with open aminoaldehyde forms or amines and carbonyl compounds does not succeed.

The only exception to date appears to be *azulene,* which can be aminomethylated on the nucleophilic five-membered ring under mild conditions[939]:

R = H;   *N,N-Dimethyl-1-azulenemethylamine*
R = CH₃; *N,N,6-Trimethyl-1-azulenemethylamine*

12.7.1.1.1.14 Aminoalkylation of heterocycles

C-Aminoalkylation is possible in all *heterocycles* displaying *adequate CH*-acidity.

Heterocycles such as 4-tetrahydropyrones and 4-thiapyrone, 4-piperidone, and their benzene homologs (4-chromanone, 4-thiachromanone, and 4-tetrahydroquinolinone) show no anomalies from

simple ketones and, like these, are substituted adjacent the carbonyl group.

Many aminomethylated heterocycles contain the following component structure:

reminiscent of an enolized 1,3-dicarbonyl compound and which can be substituted on the 2-*C*-atom.

Substitution of the 2-*C*-atom is displayed by ring systems with the component structure

Table 1. Aminomethylatable heterocycles
Ⓗ can be substituted by aminomethyl group

	Name	Remarks	Ref.
	Antipyrine		940
	3,5-Pyrazolidine-dione		941
	4*H*-Pyran-4-one	Single and double substitution are possible	942
	Kojic acid	Simultaneous substitution in 3- and 6-positions	943
	4*H*-Chromones	Condense only if positions 2 and 3 are free!	943
	4-Hydroxy-coumarin		944

for instance, 2-indolinone and 3-hydroxy-2-indolinone. By contrast, 4,5-dihydro-3(2*H*)-pyridazinones are substituted in the 4-position.

The possibility of aminomethylating the five-membered heterocycles furan, thiophene, and pyrrole, and also their benzo homologs benzofuran, benzo[b]thiophene, and indole is well known. Here, the Mannich reaction is made difficult by condensation between the heterocycles, by resinification, and by condensation with formaldehyde alone. As addition of acid is necessary to bring about the condensation, but mineral acids have a resinifying effect, it is best to add the solution of the heterocyclic compound drop by drop to the carbonyl compound, amine plus excess acetic acid[945].

Furan itself has afforded only resins on attempted Mannich reaction. 2-Substituted furans can be condensed without difficulty:

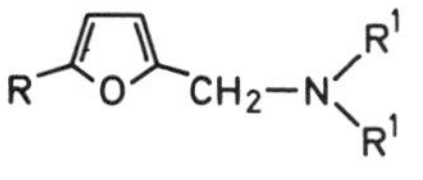

R = H, OH, Cl, Alkyl

Thiophene can be condensed, but the yield of Mannich base is small because drastic conditions also lead to resinification. Several Mannich bases of 2-substituted thiophenes and of 3-methylthiophene are known. The aminomethyl groups always enters the 5-position:

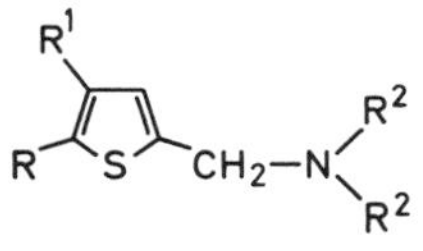

R = H, OH, Cl, Alkyl

Pyrroles are easier to aminoalkylate; substitution in one or both α-positions ensues. If these two positions are already substituted the Mannich condensation occurs in the β-position:

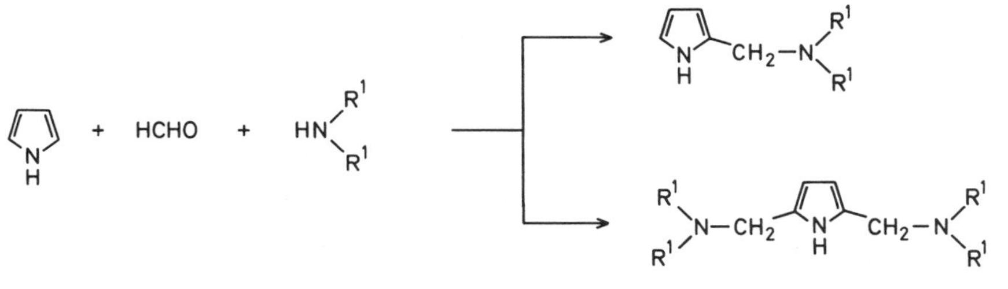

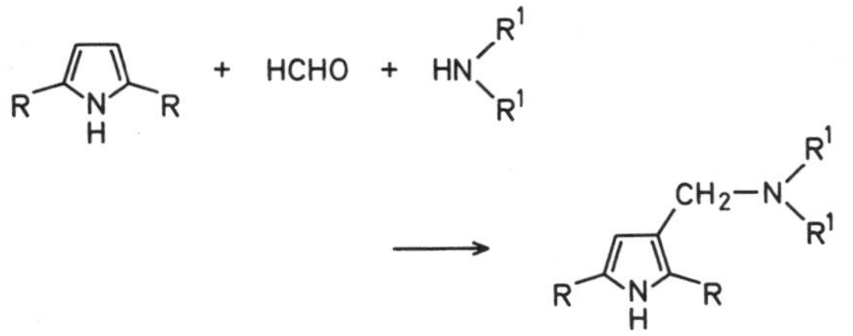

Indoles can be aminomethylated smoothly in the 3-position in acetic acid solution. Where this position is occupied the condensation takes place on the nitrogen. 1,3-Disubstituted indoles condense in position 2:

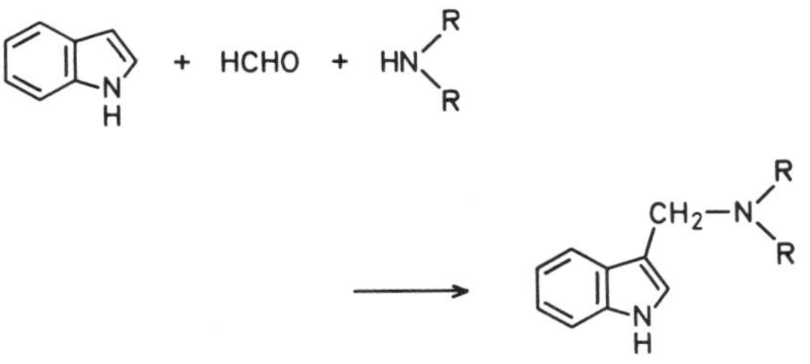

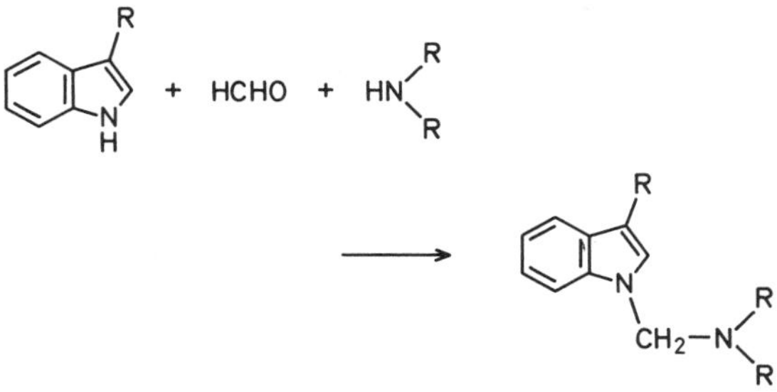

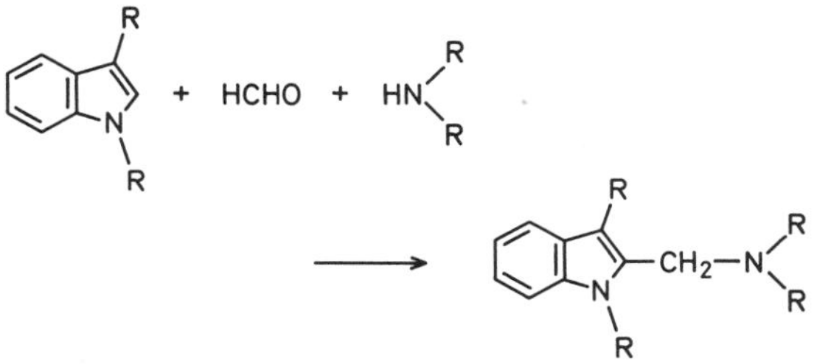

In addition to these heterocycles following a common aspect, *C*-aminomethylation of further individual heterocycles are known: isoindoles (benzo[c]pyrroles), pyrrocolines, thiazoles, imidazoles.

[942] *P.F. Wiley*, J. Amer. Chem. Soc. *74*, 4326 (1952).
[943] *L.L. Woods*, J. Amer. Chem. Soc. *68*, 2744 (1946).
[944] *D.N. Robertson, K.P. Link*, J. Amer. Chem. Soc. *75*, 1883 (1953).
[945] *H. Kühn, O. Stein*, Ber. *70*, 567 (1937).
[946] *H.R. Meyer*, Kunststoffe, Plastics *3*, 160 (1956); C.A. *52*, 11781 (1958).

12.7.1.2 Aminoalkylation of *NH*-acid compounds

Formally speaking, *N, N*-acetalization is a Mannich condensation with the special feature that the *H*-acid and amine components are identical:

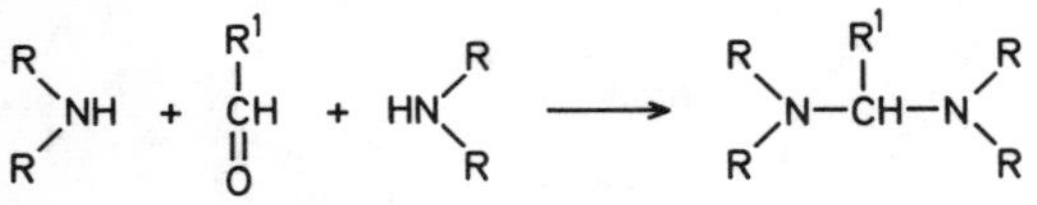

Since it is possible to carry out the condensation in stages with two different amines, *mixed* (unsymmetrical) *N, N*-acetals may be regarded as *N*-Mannich bases. However, the discussion here will be limited to cases where a distinct difference exists between the *NH*-acid compound and the amine components and symmetrical condensations are displaced to the background.

12.7.1.2.1 Aminoalkylation of imides

Carboxylic acid imides such as succinimide or phthalimide are so markedly *NH*-acid that they can be aminomethylated very easily. The condensation is generally performed in boiling ethanol and, especially with *phthalimide*, can be used to characterize amines. Phthalimide Mannich bases crystallize readily throughout.

During condensation of formaldehyde with secondary amines only one Mannich base is expected. Primary aliphatic amines condense with both *H* atoms, primary aromatic amines with one only:

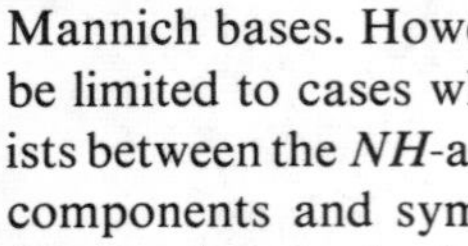

R, R^1 = Alkyl
R = H; R^1 = Aryl
R^2 = Alkyl

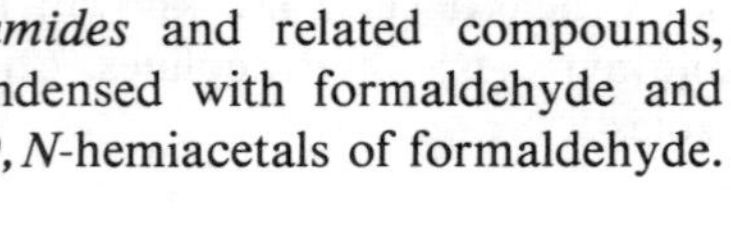

12.7.1.2.2 Aminoalkylation of amides, urethans, carbamates, thiocarbonic acid amides, guanidine, hydroxamic acids, sulfonamides

Simple *carboxamides* and related compounds, too, can be condensed with formaldehyde and amines or the *O, N*-hemiacetals of formaldehyde.

Only one *H* atom per amide group is substituted quite independently of the ratios used and the experimental conditions:

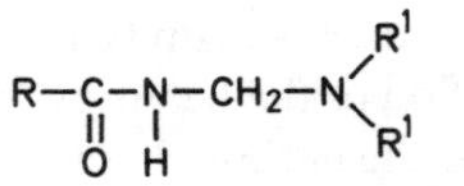

R = CH$_3$, C$_6$H$_5$

Open-chain *N*-substituted carboxamides appear not to condense except where ring closure by an intramolecular Mannich reaction is possible as in *N*-monosubstituted anthranilamides:

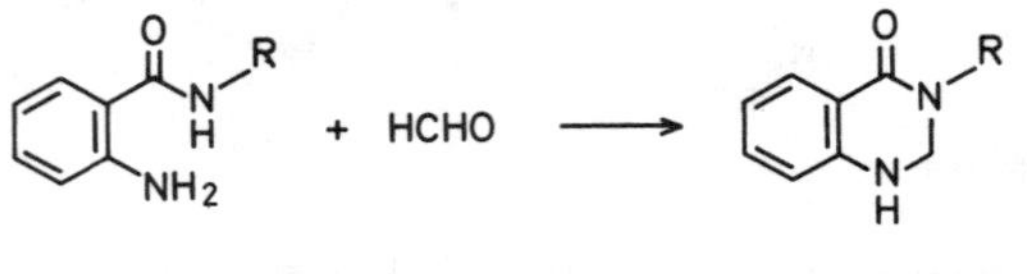

By contrast, *lactams* should be capable of undergoing aminomethylation. Mannich bases of hexahydro-2*H*-azepin-2-one (caprolactam) are known[946]. 2-Pyrrolidinone can be condensed on the nitrogen in moderate yield[947].

Urea and *thiourea* condense with formaldehyde and secondary amines on both amino groups:

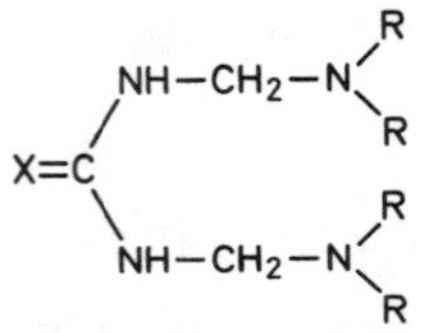

X = O, S

N-monosubstituted ureas or thioureas condense only on the free amino group:

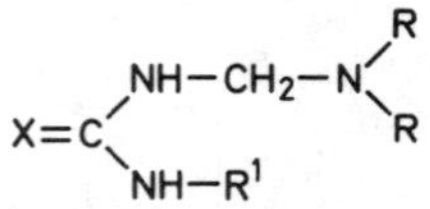

X = O, S

Mannich reaction with aldehydes and primary amines or ammonia leads to *hexahydro-1, 3, 5-triazine* derivatives, and even with *N, N'*-disubstituted ureas and thioureas.

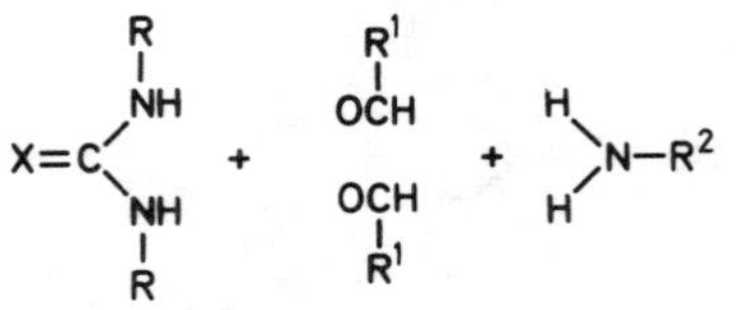

X = O, S
R = R^1 = H, Alkyl, Aryl
R^2 = H, Alkyl

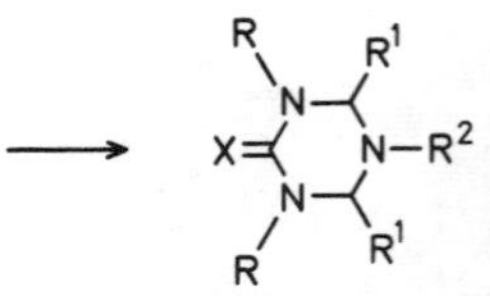

[947] *C.C. Bombardieri, A. Taurius*, Can. J. Chem. *33*, 923 (1955).
[948] *A. Dornow, S. Lüpfert*, Arch. Pharm. (Weinheim, Ger.) *288*, 311 (1955).
[949] *M. Eberhardt*, Dissertation Universität Tübingen 1957 (Ref. *H. Hellmann*).
[950] *H. Hellmann, I. Löschmann*, Chem. Ber. *89*, 594 (1956).

Hydroxamic acid and sulfonamides, too, can be aminomethylated on the nitrogen:

R¹ = H or Allyl

12.7.1.2.3 Aminoalkylation of heterocycles

In many *NH*-acid heterocycles the carboxamide grouping can be recognized and utilized to explain the ability to undergo aminomethylation. In others the quasi-aromatic character plays a part. Of the heterocycles listed in Table 2 *N*-Mannich bases with formaldehyde and amines are known.

12.7.1.3 Aminoalkylation of miscellaneous *XH*-acid compounds

12.7.1.3.1 Aminoalkylation of *OH*-acid compounds

Formation of *O*,*N*-acetals from alcohols, formaldehyde or aromatic aldehydes and amines is a Mannich condensation in which the alcohols play the part of an *OH*-acid compound. Although such condensations proceed even at room temperature, preparation of *O*,*N*-acetals is made difficult by their lability and formation of *N*,*N*-acetals as a competitive reaction.

The following compounds are formed in dependence on the number of the *OH* groups and the use of either primary or secondary amines:

$$R-OH + HCHO + HN{\overset{R^1}{\underset{R^1}{\diagdown}}}$$

$$\longrightarrow R-O-CH_2-N{\overset{R^1}{\underset{R^1}{\diagdown}}}$$

$$2\ R-OH + 2\ HCHO + H_2NR^1$$

$$\longrightarrow {\overset{R-O-CH_2}{\underset{R-O-CH_2}{\diagup\diagdown}}}N-R^1$$

$$\underset{CH_2-OH}{\overset{CH_2-OH}{|}} + 2\ HCHO + 2\ HN{\overset{R^1}{\underset{R^1}{\diagdown}}}$$

$$\longrightarrow \underset{CH_2-O-CH_2-N\diagdown R^1}{\overset{CH_2-O-CH_2-N\diagdown R^1}{|}}$$

2-Aminoethanols can be readily aminomethylated intramolecularly to *1,3-oxazolidines*. In addition

Table 2a. *NH*-acid heterocycles with a component carboxamide structure

	Name	Remarks	Ref.
	Hydantoin	Aminomethylatable on both *NH* groups	947
	2,4-Thiazolidinedione		947
	1,3,4-Oxadiazolin-5-one		948
	2(1*H*)-Pyridones		949
	4(1*H*)-Pyridones		949
	3(2*H*)-Pyridazinones	React only in the carboxamide form, not as lactims	950
	R¹ = OH	Yield only mono-*N*-aminoalkyl derivatives (on the nonlactimized *NH*)	951
	Succinic acid phenylhydrazide		949
	R¹, R² = H; 5-Alkyl	*N*,*C-Bis*-Mannich bases	952
	R¹ = H; 5,5-Disubstituents	Form only mono-aminomethyl derivatives	953
	1,5,5-Trisubstituted barbituric acid derivatives		
	Uracil		947
	Phthalimidine		949
	X = O; 2-Benzoxazolinone		954
	X = S; 2-Benzoxazolinethione		955
	X = O; 2-Benzimidazolinone	Aminomethylatable on both *NH* groups	956
	X = S; 2-Benzimidazolinethione		957, 958

Table 2a (continued)

	Name	Remarks	Ref.
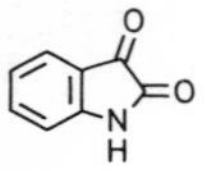	Indole-2,3-dione (Isatin)		959
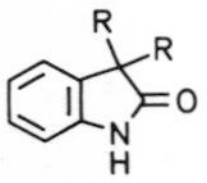	3,3-Dialkylindan		949
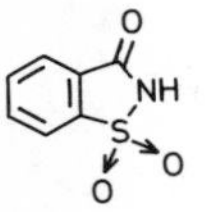	1,2-Benziso-thiazolin-3-one 1,1-dioxide (Saccharin)		960
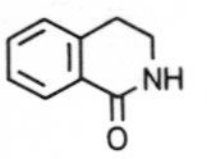	3,4-Dihydro-isocarbostyril	The isomeric 3,4-dihydrocarbostyril yields no Mannich base!	949
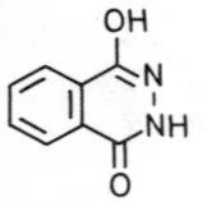	4-Hydroxy-1(2H)-phthalazinone	As one carboxamide group is lactimized condensation occurs on one nitrogen only	961
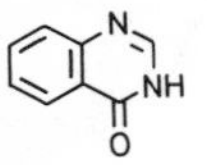	4(3H)-Quinazolinone	Affords very unstable Mannich bases	962
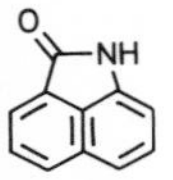	Naphthostyril		949

Table 2b. Quasi-aromatic, *NH*-acid heterocycles

	Name	Remarks	Ref.
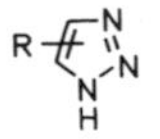	1H-1,2,3-Triazoles		963
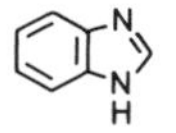	Benzimidazole		964
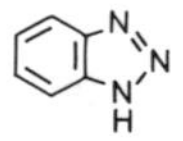	1H-Benzo-1,2,3-triazole		963, 964, 965
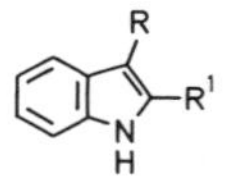	R¹ = H; 3-Substituted indoles	In boiling glacial acetic acid 3-substituted indoles react on the nitrogen; otherwise on the C-2	966, 967, 968
	2,3-Disubstituted indoles		968, 969
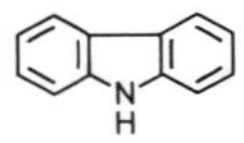	Carbazole		968, 970
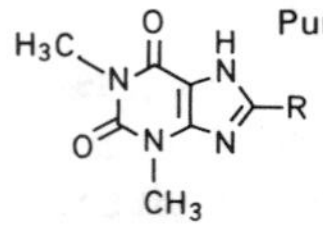	Purines		971, 972

R = H, Alkyl

to formaldehyde and aromatic aldehydes higher aliphatic aldehydes and ketones may be used as carbonyl component.

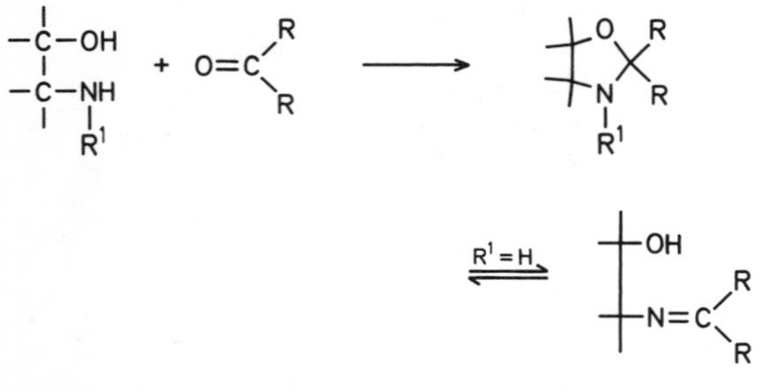

Phenols are substituted on the aromatic nucleus. When primary amines and aldehydes are employed for the condensation, cyclic *O,N*-acetals can form like with hydroxylamines and thus a Mannich condensation ensues on both carbon and oxygen (cf. p. 540).

12.7.1.3.2 Aminoalkylation of *SH*-acid compounds

Aminomethylation of hydrogen sulfide is fundamentally feasible, but in addition to mono- and bis-adducts, which themselves are not very stable, a series of further products are formed:

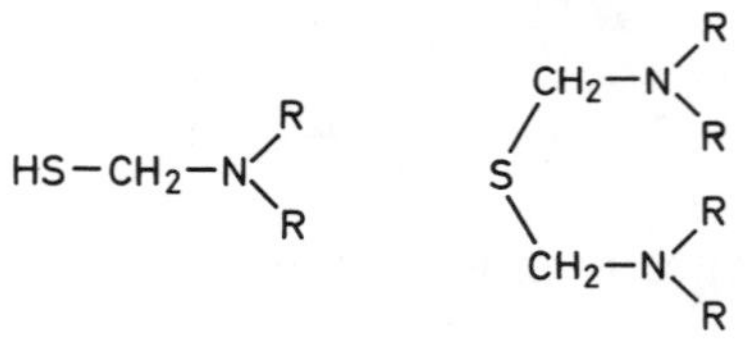

[951] *D.M. Miller, R.W. White,* Can. J. Chem. *34*, 1510 (1956).

[952] *H.J. Roth, R. Brandes,* Arch. Pharm. (Weinheim, Ger.) *299*, 612 (1966).

[953] *H.J. Roth, R. Brandes,* Arch. Pharm. (Weinheim, Ger.) *298*, 888 (1965).

[954] *H. Zinner, H. Herbig, H. Wigert,* Chem. Ber. *89*, 2131 (1956);
H. Zinner, H. Herbig, H. Wigert, Chem. Ber. *90*, 1548 (1957).

[955] *H. Zinner, H. Hübsch, D. Burmeister,* Chem. Ber. *90*, 2246 (1957).

[956] *H. Zinner, B. Spangenberg,* Chem. Ber. *91*, 1432 (1958).

[957] *H. Zinner, O. Schmitt, W. Schritt, G. Rembarz,* Chem. Ber. *90*, 2852 (1957).

[958] DRP. 575 114 (1933), I.G. Farben AG, Erf.: *M. Bögemann, E. Zauker,* Chem. Zentr. II, 3773 (1933).

It is easier to obtain Mannich bases from *thiols*. Mercaptoacetic acid and derivatives can be reacted with the condensates obtained from aldehyde and primary amines to form *4-thiazolidinones*:

α-Oxothiols can react with carbonyl compounds and ammonia to undergo ring closure and form *5-thiazolines*[973]:

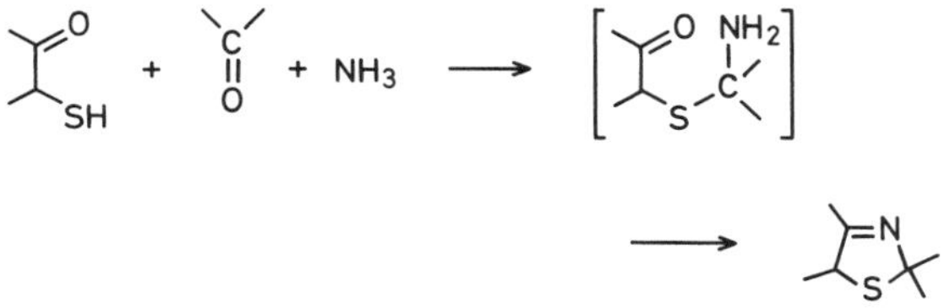

In addition, a number of similar condensations are possible which lead to 5-thiazolines and 1,3-thiazine derivatives. 2-Aminothiols condense with carbonyl compounds to give *1,3-thiazolidines*:

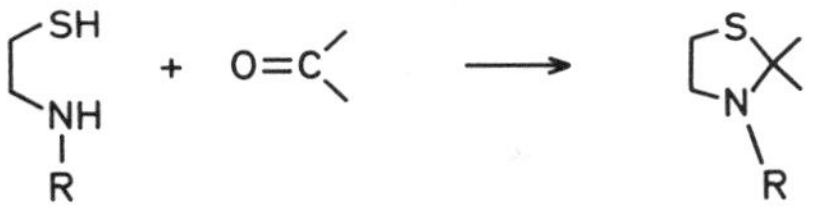

Thiophenols are aminomethylated on the sulfur; ring substitutions of the type that take place with phenols have not been observed so far.

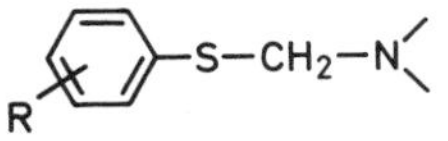

Sulfurous acid and its salts and also sulfinic acid are also aminomethylated:

Further, dithiocarbaminic acids, thioacetic acid, thiourea, and dithiophosphoric acid O,O,S-triesters can be aminoalkylated on the sulfur.

12.7.1.3.3 Aminoalkylation of *PH*- and *SeH*-acid compounds

PH and *SeH* group-containing compounds, too, can undergo condensation on the hetero atoms in the sense of a Mannich reaction. From *hypophosphorous acid*, *dialkyl phosphates*, and *phosphinic acid esters* the following compounds are obtained:

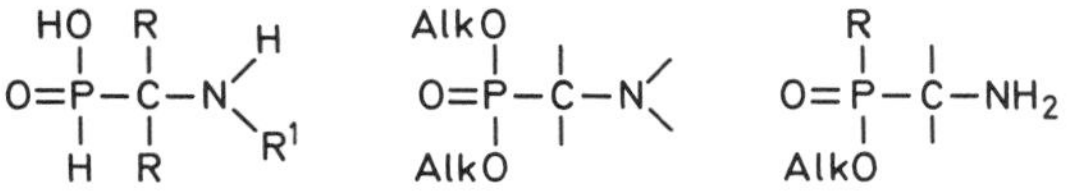

Finally, *hydrogen selenide* also has been aminomethylated:

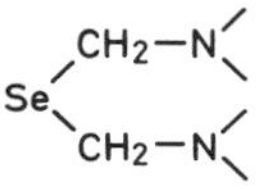

12.7.1.4 Reactions of Mannich bases leading to further amines

12.7.1.4.1 Elimination-addition reactions

The carbon-nitrogen bond in amines is normally so stable that the amino group cannot be exchanged. By contrast, in Mannich bases the bond may be loosened to such an extent that the amino group possesses approximately the reactivity of an aliphatically bound halogen. Of the reactions made possible in this way, exchange of the amino group against another amino group is of interest here. It proceeds in the manner of an *elimination-addition mechanism*[974], which, in tertiary Mannich bases carrying an eliminable hydrogen atom in the β-position with respect to the amino group or in a vinylogous β-position (on the oxygen or a further nitrogen), leads to exchange of the amino group:

[959] *H. Hellmann, I. Löschmann*, Chem. Ber. *87*, 1684 (1954).

[960] *H. Zinner, U. Zelck, G. Rembarz*, J. Prakt. Chem. *280*, 150 (1959).

[961] *H. Hellmann, I. Löschmann*, Angew. Chem. *67*, 110 (1955);
H. Hellmann, I. Löschmann, Chem. Ber. *89*, 594 (1956).

[962] *B.R. Baker, M.v. Querry, A.F. Kadish, J.H. Williams*, J. Org. Chem. *17*, 35 (1952).

[963] *J.J. Licari, L.W. Hartzel, G. Daugherty, F.R. Benson*, J. Amer. Chem. Soc. *77*, 5386 (1955).

[964] *G.B. Bachmann, L.V. Heisey*, J. Amer. Chem. Soc. *68*, 2496 (1946).

[965] *J.H. Burckhalter, V.C. Stephens, L.A.R. Hall*, J. Amer. Chem. Soc. *74*, 3868 (1952).

[966] *J. Thesing, P. Binger*, Chem. Ber. *90*, 1419 (1957).

[967] *S. Swaminathan, S. Ranganathan*, J. Org. Chem. *22*, 70 (1957).

[968] *S. Swaminathan, S. Ranganathan, S. Sulochana*, J. Org. Chem. *23*, 707 (1958).

[969] *F. Troxler, A. Hofmann*, Helv. Chim. Acta *40*, 1706 (1957).

[970] *J.R. Feldmann, E.C. Wagner*, J. Org. Chem. *7*, 31 (1942).

[971] *H.J. Roth, R. Brandes*, Arch. Pharm. (Weinheim, Ger.) *298*, 765 (1965).

[972] *H.J. Roth, R. Brandes*, Arch. Pharmaz. *300*, 1000 (1967).

[973] *F. Asinger, M. Thiel*, Angew. Chem. *70*, 667 (1958).

[974] *H. Hellmann*, Angew. Chem. *65*, 473 (1953).

$$O_2N-CH-CH_2-N(CH_3)_2 \rightleftharpoons \left[O_2N-C=CH_2 \right] \xrightarrow{H_2N-C_6H_5} O_2N-CH-CH_2-\underset{H}{N}-C_6H_5$$
$$\overset{|}{R} \qquad\qquad\qquad \overset{|}{R} \qquad\qquad\qquad\qquad \overset{|}{R}$$

1-(Morpholinomethyl)-2-naphthol; 83.5%

3-(Dialkylaminomethyl)indole

$$H_5C_6-\underset{O}{\overset{\parallel}{C}}-CH_2-CH_2-N(CH_3)_2 \xrightleftharpoons[+\,HN(CH_3)_2]{-\,HN(CH_3)_2} H_5C_6-\underset{O}{\overset{\parallel}{C}}-CH=CH_2$$

ω-Phthalimidopropiophenone; ~ 70%

12.7.1.4.2 Transaminoalkylation

In addition to the elimination-addition reaction, which with amines as reaction partners leads to new Mannich bases **a**, it is worthwhile mentioning transaminoalkylations in which the entire aminoalkyl grouping is transferred to a competing acid compound **b**:

$$X-CH_2-N\overset{R}{\underset{R}{}} \quad + \quad HN\overset{R^1}{\underset{R^1}{}}$$

$$\xrightarrow{\;\;ⓐ\;\;} X-CH_2-N\overset{R^1}{\underset{R^1}{}} \quad + \quad HN\overset{R}{\underset{R}{}}$$

$$X-CH_2-N\overset{R}{\underset{R}{}} \quad + \quad HY \xrightarrow{\;\;ⓑ\;\;} Y-CH_2-N\overset{R}{\underset{R}{}} \quad + \quad HX$$

Transaminoalkylation is a *nucleophilic* displacement of an *H*-acid component **HX** by another **HY**. It is of preparative interest if it leads to Mannich bases which are difficult or impossible to prepare by direct condensation.

Mannich bases *without a mobile hydrogen* atom *on the β-carbon atom and unable to form a resonance-stabilized cation* following elimination of the amino component are able to undergo trans-

aminomethylation. They can exchange their dialkylamino methyl group against a proton of the condensation partner **HY**. Thus, for example, formylaminopiperidinomethylmalonic acid diethyl esters or 1,2-diphenyl-2-hydroxy-3-piperidyl-1-propanone can transfer their aminomethyl[975]:

3-(Piperidinomethyl)indole; ~ 50%

[975] *A. Butenandt, H. Hellmann,* Hoppe Seyler's Z. Physiol. Chem. *284*, 168 (1949).

[976] *H. Hellmann, K. Teichmann,* Chem. Ber. *91*, 2432 (1958).

[977] *W. Ried, G. Keil,* Justus Liebigs Ann. Chem. *616*, 108 (1958).

[978] *H. Hellmann,* Angew. Chem. *65*, 473 (1953).

[979] *H. Hellmann, G. Renz,* Chem. Ber. *84*, 901 (1951).

[980] *A. Einhorn,* Justus Liebigs Ann. Chem. *343*, 207 (1905).

1-Piperidino-2,3-alkanedione
2-phenylhydrazones

Further examples are described in refs.[976–979].

12.7.2 Tscherniac-Einhorn reaction

Under the influence of acid condensation agents, compounds containing mobile hydrogen atoms can be condensed with the reaction products of carboxamides and formaldehyde (*N*-hydroxymethylamides) or with carboxamides and formaldehyde to form *N-substituted carboxamides*. These latter can be subsequently saponified to the aminomethyl derivatives of the acid compounds and acids. Just as the rational term *aminomethylation* is used to describe the Mannich reaction, so the *Tscherniac-Einhorn reaction* can be denoted as an *acylaminomethylation* or *amidomethylation*. The technique lends itself to the introduction of an *N*-unsubstituted aminomethyl group into an acid compound, a result which cannot be achieved with the Mannich reaction because of the multiple condensation that takes place on the ammonia:

Acid compound   Hydroxymethylamide

N-Substituted carboxamide

N-Primary Mannich base

In place of the hydroxymethyl compounds or the individual components (carboxamide and formaldehyde) bis(acylaminomethyl) ethers, ethers of the hydroxymethyl compounds, hydroxymethyl

derivatives of the dicarboxylic acid imides and sulfonic acid imides, and *N*-halomethylcarboxamides and *N*-halomethyl carboxylic acid imides can be used alternatively.

Alcoholic hydrochloric acid, concentrated sulfuric acid, formic acid, 85% phosphoric acid, zinc chloride, and aluminum chloride are used as acid *condensing agents*.

Viewing the Tscherniac-Einhorn reaction as a method for preparing *N*-acylated, primary Mannich bases or *N*-substituted carboxamides, then it possesses a *greater* range of application than the Mannich reaction. On the other hand, its preparative value for synthesizing primary Mannich bases is limited substantially by the need to carry out a hydrolysis, which in labile Mannich bases leads to splitting off of the entire acylated aminomethyl group. The greater scope results from the fact that *hydrogen atoms in aromatic systems* appear to be quite generally sufficiently mobile to be replaced by acylaminomethyl groups[980, 981].

Preparative details, mechanism, and further possibilities for the use of amidomethylation are described in ref.[982].

12.8 Synthesis of amines *via* organometallic compounds

12.8.1 Introduction

Primary, secondary and tertiary amines can be prepared from organometallic compounds by both direct (p. 550) and indirect (p. 550) methods. In the direct method, an organometallic compound reacts with an amine derivative of type $=N-X$, forming a $C-N$ bond (p. 550). With the indirect method, synthesis of an amine from an organometallic compound involves either substitution of aminomethyl derivatives, with formation

[981] *R.O. Cinnéide*, Nature *175*, 47 (1955).
[982] *H. Hellmann*, Aminomethylierungen, in *W. Fœrst*, Neuere Methoden der präparativen organischen Chemie, Band II, Verlag Chemie, Weinheim/Bergstr. 1960.
[983] *R. Kaddatz*, Arzneimittel-Forsch. *7*, 344 (1957).
[984] DBP. 963 424 (1957), Dr. K. Thomae, Erf.: *E. Seeger, A. Kottler*.
[985] *R.B. Moffet, W.M. Hoehn*, J. Amer. Chem. Soc. *69*, 1792 (1947).
[986] *E.C. Dodds*, Proc. Roy. Soc., Ser. B *132*, 119 (1944).
[987] DBP. 1 008 305 (1957), Kali-Chemie, Erf.: *W. Stühmer, S. Funke*.
[988] DBP. 1 008 740 (1957), Dr. K. Thomae, Erf.: *E. Seeger, A. Kottler*.

of a *C–C* bond (p. 550), or addition to compounds having a *C–N* multiple bond (p. 553).

Scheme 1:
Preparation of amines from organometallic compounds

X = Hal, OCH$_3$, OCH$_2$–C$_6$H$_5$;
Y = Cl, OR, CN, SR, NR$_2$, SO$_3$Na

Compound types described:
primary amines by direct methods (p. 550)
    BD I*: R = Alkyl, Aralkyl, Aryl; R^1 = R^2 = H
    BD II: R, R^1 = Alkyl; R^2 = H
    BD III: R, R^1, R^2 = Alkyl

secondary amines from Schiff's bases (p. 553)
    BD II : R = Aryl, R^1 = H; R^2 = Alkyl, Aralkyl, Aryl

tertiary amines from aminomethyl compounds (p. 550)
    BD I : R = Alkyl, Aralkyl, Alkyl; R^1 = R^2 = H
    BD II : R = Alkyl, Aralkyl, Aryl; R^1 = H; R^2 = Alkyl, Aralkyl, Aryl
    BD III : R = Alkyl, Aralkyl, Aryl; R^1 = R^2 = Alkyl

* Branching degree (BD): BD I, II, III = primary, secondary, or tertiary α-*C*-atom, respectively.

The importance of these methods is that they provide ready access to amines which are substituted on the α-carbon atom by several alkyl, aryl or aralkyl groups; such compounds can only be obtained in poor yields by the usual methods for amine preparation. They are of pharmaceutical interest; various compounds of this type are known to be analeptics[983, 984], analgesics[984–988], antiparkinsonian agents[889], cytostatics[990], ganglion blockers[984], circulatory drugs[991, 992], local anesthetics[993], oxytocic substances[994], and antispasmodics[995, 996].

12.8.2 Direct methods

12.8.2.1 Reactions of chloramine and of *O*-methyl- or *O*-benzylhydroxylamine

Whatever the degree of branching of the organic group of the Grignard reagent, the reaction of *chloramine* with organomagnesium chlorides gives good yields (50–90%) of *primary amines*[997–1000], together with *alkyl chlorides*. The results are significantly poorer with organolithium compounds or with organomagnesium bromides or iodides; bromamine also gives inferior results. Nearly quantitative yields are obtained with dialkylmagnesium compounds at low temperatures (–60°)[1000]. Monoalkyl- and dialkylchloramine can only be converted into secondary and tertiary amines in poor yields ($\sim$10%). Direct synthesis of amines can also be carried out using metal amides and alkyl halides (see p. 445). The analogous reactions of *O-methylhydroxylamine* or *O-benzylhydroxylamine* with organomagnesium chlorides and bromides in ether at –10° give 40–90% yields of *primary amines* of the above structural types[997, 1001, 1002]. No information is available about side-reactions and the synthesis of secondary and tertiary amines. Primary, secondary and tertiary organomagnesium chlorides and bromides react equally well, but organomagnesium iodides react poorly.

[989] *H. Thies, H. Schönenberger, M. El-Zanaty*, Arzneimittel-Forsch. *22*, 1138 (1972).
[990] *H. Lettré, H. Fernholz*, Hoppe-Seylers Z. Physiol. Chem. *278*, 175 (1943).
[991] *H. Haury*, Arch. Pharm. (Weinheim, Ger.) *295*, 728 (1962).
[992] *W. Brandt*, Dissertation Universität München 1969.
[993] *H. Thies, H. Schönenberger, K. Borah*, Arch. Pharm. (Weinheim, Ger.) *299*, 1031 (1966).
[994] DBP. 1 116 227 (1961), Dr. K. Thomae, Erf.: *E. Seeger*.
[995] *R. Engelhorn, R. Kaddatz*, Arzneimittel-Forsch. *6*, 454 (1956);
R. Engelhorn, R. Kaddatz, Arzneimittel-Forsch. *12*, 178 (1962).
[996] DBP. 1 011 427 (1954), Dr. K. Thomae, Erf.: *E. Seeger, A. Kottler*.

[997] *R. Schröter* in *Houben-Weyl*, Methoden der organischen Chemie, 4. Aufl., Bd. XI/1, p. 805, Georg Thieme Verlag, Stuttgart 1957.
[998] *G.H. Coleman, C.R. Hauser*, J. Amer. Chem. Soc. *50*, 1193 (1928).
[999] *G.H. Coleman, J.L. Hermanson, H.L. Johnson*, J. Amer. Chem. Soc. *59*, 1896 (1937).
[1000] *M.S. Kharasch, O. Reinmuth*, Grignard Reactions of Nonmetallic Substances, p. 1243, Prentice Hall, New York 1954.
[1001] *M.S. Kharasch, O. Reinmuth*, Grignard Reactions of Nonmetallic Substances, p. 1236 Prentice Hall, New York 1954.
[1002] *N.J. Schewerdina, K.A. Kotscheschkow*, Zh. Obshch. Khim. *8*, 1825 (1938).
[1003] *H. Hellmann, G. Opitz*, α-Aminoalkylierung, p. 230, Verlag Chemie, Weinheim/Bergstr. 1960.

12.8.3   Indirect methods

12.8.3.1   Reactions of aminomethyl compounds

Compounds of formula **1** aminomethylate the carbeniate anion of organometallic compounds **2**, liberating the less nucleophilic component $X^\ominus$. This so-called *transaminomethylation* generally results in excellent yields of *tertiary amines* **3** with various degrees of branching[997, 1003, 1004].

X = Cl, OR, SR, N(R)₂, SO₃Na, CN
M = MgHal, Li

The synthesis of *tertiary* amines (containing primary, secondary or tertiary α-carbon atoms) from *O,N-acetals* and organometallic compounds gives good yields and is of preparative importance[1005]; many examples are available[1003]. This reaction has usually been carried out with Grignard reagents. A side-reaction which has been described is the formation of monomeric reduction products (replacement of *OR* by *H*)[1006]. By applying this procedure to cyclic *O,N-acetals* such as oxazolidines[1007] or 2-dialkylaminotetrahydropyrans[1008], *2-, 3-, 4- or 5-aminoalcohols* can be obtained. Symmetrical *N,N-acetals* react appreciably less well than *O,N-acetals*[1009]. Unsymmetrical *N,N-acetals* of the *N*-dialkylaminomethylcarbazole type give excellent yields (~95%) however[1010].

Little work has been carried out so far on the reaction of chloromethyldialkylamines with organometallic compounds. This method has been used to obtain, for example *N,N-diethyl-2-methoxybenzylamine* from 2-anisyllithium and (chloromethyl)diethylamine[1011]; this product cannot be obtained by direct reaction of anisole with (chloromethyl)diethylamine.

Detailed studies have been made of the reaction of *S,N-acetals* with Grignard reagents[1012, 1013]. *Tertiary* aliphatic amines are formed in 70–90% yields from (dialkylphenylthiomethyl)amines. However, the reaction of *tert*-butyl-magnesium chloride with 1-(phenylthiomethyl)piperidine gives *N-methylpiperidine* (a reduction product), instead of the expected 1-(2,2-dimethylpropyl)piperidine. Reaction of tris(phenylthiomethyl)amine with organomagnesium halides (1:3 molar ratio) forms *tertiary amines* (80–90%). An attempt to obtain primary amines by using equimolar amounts of the reactants met with no success; mixtures of primary, secondary, and tertiary bases resulted. *N-Alkylanilines* and *N,N-dialkylanilines* were obtainable in good yields (80–90%) from *N*-(phenylthiomethyl)aniline and *N,N*-bis(phenylthiomethyl)aniline. Reactions involving S,N-acetals of ketones and of higher aliphatic or aromatic aldehydes have not been described.

Reactions of α-aminoalkylsulfonic acids have only been studied to a limited extent[1014]. In contrast, there is much information on Grignard reactions with α-aminonitriles[1003].

Scheme 2:

by products

R, R¹ = H, Alkyl or Aryl
R² = Alkyl
R³ = Alkyl, Alkenyl, Alkinyl, Aralkyl, Aryl

[1004] *M.S. Kharasch, O. Reinmuth*, Grignard Reactions of Nonmetallic Substances, p. 776–1029, Prentice Hall, New York 1954.
[1005] *G.M. Robinson, R. Robinson*, J. Chem. Soc. (London) *123*, 532 (1923).
[1006] *G. Chauvière, B. Tchoubar, Z. Welvart*, Bull. Soc. Chim. France 1428 (1963).
[1007] *M. Senkus*, J. Amer. Chem. Soc. *67*, 1515 (1954).
[1008] *J. Ficini, H. Normant*, Bull. Soc. Chim. France [5] *24*, 1454 (1957).
[1009] *M. Nomura, K. Yamamoto, R. Oda*, J. Chem. Soc. Jap. Pure Chem. Sect. *57*, 219 (1954).
[1010] *K.G. Mizuch, R.A. Lapina*, J. Gen. Chem. (USSR) *26*, 839 (1956).
[1011] *H. Böhme, U. Bomke*, Arch. Pharm. (Weinheim, Ger.) *303*, 779 (1970).
[1012] *I.E. Pollak, A.D. Trifunac, A.F. Grillot*, J. Org. Chem. *32*, 272 (1967).
[1013] *I.E. Pollak, G.F. Grillot*, J. Org. Chem. *32*, 2892 (1967).

Table 1. Reactions of various α-piperidinonitriles with isomeric butylmagnesium chlorides (1:2 molar ratio)

R¹	R²	H_9C_4-MgCl [% of th.]		$(CH_3)_2CH-CH_2-MgCl$ [% of th.]		$H_5C_2-CH(CH_3)-MgCl$ [% of th.]		$(CH_3)_3C-MgCl$ [% of th.]	
H	H	Adduct formation	64	Adduct formation	39	Adduct formation	18		
		Dimerization	26	Dimerization	48	Dimerization	66	Dimerization	90
H	CH_3	Dimeric reduction	4	Dimeric reduction	11	Dimeric reduction	23	Dimeric reduction	32
		Transaminoalkylation	79	Transaminoalkylation	76	Monomeric reduction	10	Transaminoalkylation	9
						Starting material	44	Monomeric reduction	31
H	C_3H_7	Transaminoalkylation	75	Transaminoalkylation	73	Starting material	63	Transaminoalkylation	7
								Monomeric reduction	69
CH_3	CH_3	Transaminoalkylation	70	Transaminoalkylation	71	Starting material	83	Monomeric reduction	69
CH_3	C_2H_5	Transaminoalkylation	95	Transaminoalkylation	93	Starting material	76	Monomeric reduction	92

With α-amino-α-phenylacetonitriles (**4**, R = aryl) particularly, these reactions give excellent yields of α-*branched amines* **7** (with primary, secondary or tertiary α-carbon atoms), the cyano group being replaced by the carbanionic moiety of the organometallic compound. The aromatic substituent group of the α-amino-α-phenylacetonitriles facilitates the transaminoalkylation reaction by promoting dissociation of the nitrile and formation of a conjugated carbonium-immonium ion **6**. An S_N1 mechanism for this kind of reaction has been discussed[1015]. Reactions of this type between Grignard reagents and α-aminonitriles are widely used for the synthesis of therapeutic agents. The strongly electrophilic nature of α-amino-α-phenylacetonitriles even permits reaction with alkynyl Grignard reagents, which do not react successfully with Schiff bases[1016]. A large number of variously substituted 1-amino-1-phenylalkynes (**7**, R = aryl, $R^3 = -C \equiv C$-alkyl) with antispasmodic properties have been synthesized[989].

In the reaction of Grignard reagents with aliphatic α-aminoacetonitriles (**4**, R=R¹=H; R=alkyl, R¹=H; R and R¹=alkyl), by-products are formed, depending on the structures of the two reactants. Such by-products may result from side-reactions such as adduct formation **8**[1006, 1017–1021], dimerization **9**[1006, 1021–1024], monomeric reduction **10**[1021, 1025], dimeric reduction **11**[1021, 1026–1028], and enamine formation **12**[1015, 1029], and in certain cases these may be the main products formed.

A detailed study of these reaction processes has been carried out[1021], using the reactions of the four isomeric butylmagnesium chlorides with various substituted α-piperidinoacetonitriles as representative examples (see Table 1).

Transaminomethylation products are only formed in good yields from α-aminoacetonitriles with R=alkyl and R¹=H or R and R¹=alkyl, and when the Grignard reagent is not branched at the α-position.
Organometallic derivatives of alkali metals form adducts **8** exclusively[1006].
The preponderance of transaminomethylation products in Grignard reactions of α-alkyl-substituted α-piperidinoacetonitriles can be explained in terms of the +*I* effect of the alkyl group, which facilitates elimination of the cyanide ion and also stabilizes the mesomeric carbonium-immonium ion by hyperconjugation. In this case dimerization to **9** is not possible.
Considerable importance should also be attributed to the Grignard component, which as a Lewis acid interacts with the cyano group and facilitates its removal. Thus, for example, the reaction of a propylmagnesium halide with 2-morpholinopropionitrile gives 57% of the transaminoalkylation product (1-pentylmorpholine), whereas propyllithium gives an addition product in 65% yield[1025]. In addition, replacement of H by Li at the α-carbon atom has been observed. Ketones can be prepared by subsequent treatment with alkyl halides and hydrolysis[1030]. The formation of monomeric and dimeric reduction products **10** and **11** in reactions with *tert*-butylmagnesium chloride can be attributed to steric effects.

[1014] *D. Beke, M. Martos-Bartsai*, J. Gen. Chem. (USSR) *27*, 1836 (1957).

[1015] *D. Cabaret, G. Chauvière, Z. Welvart*, Bull. Soc. Chim. France 4457 (1969).

[1016] *H. Thies, H. Schönenberger, R. Beutel*, Arch. Pharm. (Weinheim, Ger.) *305*, 18 (1972).

[1017] *P. Bruylants*, Bull. Soc. Chim. Belg. *33*, 467 (1924).

[1018] *P. Bruylants*, Bull. Sci. Acad. Roy. Belg. [5] *11*, 261 (1925).

[1019] *Th. S. Stevens, J.M. Cowan, J. Mac. Kinnon*, J. Chem. Soc. (London) 2568 (1931);
Th.S. Stevens, J.M. Cowan, J. Mac Kinnon, J. Chem. Soc. (London) 2607 (1932).

[1020] *C.F. Hammer, R.A. Hines*, J. Amer. Chem. Soc. *77*, 3649 (1955).

[1021] *H. Thies, H. Schönenberger, P.K. Quasba*, Arch. Pharm. (Weinheim, Ger.) *302*, 30 (1969).

Unlike Schiff bases (see p. 554) the reactivity of α-aminonitriles and immonium perchlorates is comparable with that of ketones, as investigations with the compounds **13–15b** have shown[1015]:

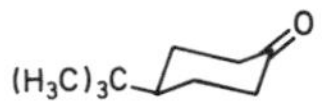

13

14a

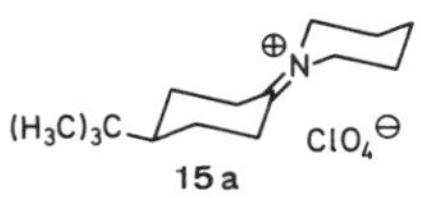

15a

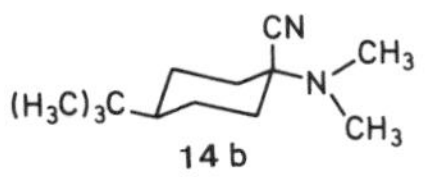

14b

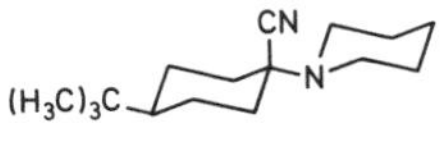

15b

4-tert-Butyl-1-dimethyl-amino-1-cyclohexane-carbonitrile

4-tert-Butyl-1-dimethyl-amino-1-piperidino-1-cyclohexanecarbonitrile

An exception is the reaction with bulky Grignard reagents such as *tert*-butylmagnesium halides which still add to the *C=O* double bond of the ketone **13**, while the derived α-aminonitriles **14a**, **15a** are converted into monomeric reaction products **10** and **12**. Apparently the two alkyl groups on the nitrogen hinder the substitution on the α-C atom. The structure of RMgX was found to govern the transaminoalkylation:reduction product ratio. It decreases in the order

X = Cl > R > Br > I

That this yield ratio remained the same throughout during the reaction of the α-aminonitriles and immonium perchlorates was assessed as proof of the intermediate formation of carbonium-immonium ions (S_N1). The observation that the anionic group ($CN^{\ominus}$ and $ClO_4^{\ominus}$) exerts no appreciable effect on the stereoisomer ratio *(cis : trans)* of the transaminoalkylation reduction product points the same way. Preferential addition of the carbanionic group of the Grignard reagent in the equatorial position with formation of *cis-1,4-tert-butyl-cyclohexylamine* derivative takes place. From *cis-* and *trans-*α-aminonitrile **15b** a uniform product arose by addition of phenylmagnesium bromide in the equatorial position[1031]. These findings led to α-aminonitriles being termed *pseudoimmonium salts*[1015].

Reacting organometallic compounds with the readily accessible aminomethyl derivatives having the formula N–CR_2–X (X = OR, SR, CN) represents a useful method for preparing *tertiary amines* with a secondary or tertiary α-C-atom. (Re the technique see ref.[989].) Even the sluggishly reacting 1-alkyn-1-ylmagnesium halides give this reaction successfully. Compounds of this type can be prepared not at all or in poor yield only by other methods. The limits of the procedure lie with reacting of highly branched Grignard reagents (*e.g., tert*-butylmagnesium halide).

12.8.3.2 Reactions of Schiff bases

Grignard compounds add to the *C=N* bonds of Schiff bases forming *secondary amines* containing secondary α-carbon atoms. The reaction ensues in two stages[1032].

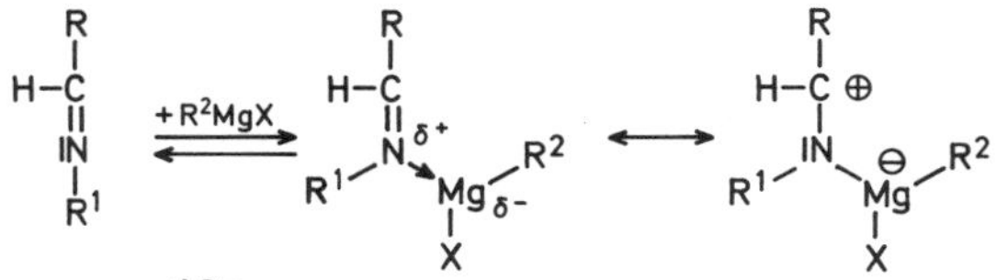

16a

16b

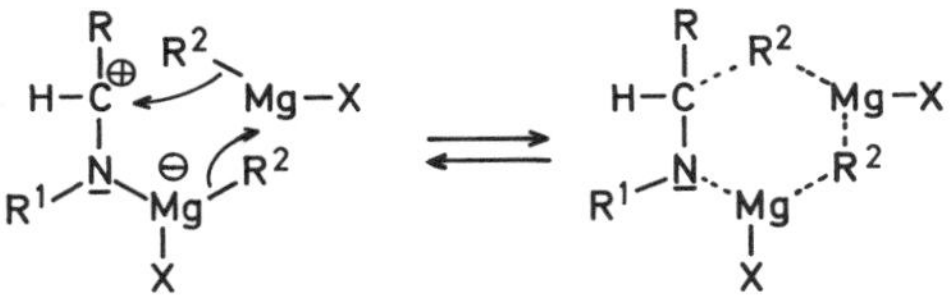

17a

17b

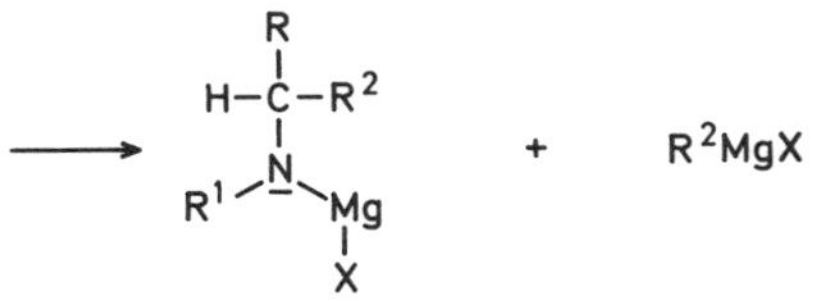

17c

X = Halogen or Alkyl

[1022] *H. Thies, H. Schönenberger, P.K. Quasba,* Angew. Chem. *75*, 91 (1963).

[1023] *H. Thies, H. Schönenberger, P.K. Quasba,* Arch. Pharm. (Weinheim, Ger.) *302*, 161 (1969).

[1024] *H. Thies, H. Schönenberger, P.K. Quasba,* Arch. Pharm. (Weinheim, Ger.) *302*, 168 (1969).

[1025] *H.R. Henze, G.L. Sutherland, G.B. Roberts,* J. Amer. Chem. Soc. *79*, 6230 (1957).

[1026] *P. Bruylants, G. Gevaert,* Bull. Sci. Acad. Roy. Belg. [5] *9*, 27 (1923).

[1027] *M. Velghe,* Bull. Sci. Acad. Roy. Belg. [5] *11*, 301 (1925).

[1028] *A. Christiaen,* Bull. Soc. Chim. Belg. *33*, 483 (1924).

[1029] *J. Sansoulet, Ch. Tackx, Z. Welvart,* Compt. Rend. *250*, 4370 (1960).

[1030] *H.M. Taylor, C.R. Hauser,* J. Amer. Chem. Soc. *82*, 1960 (1960).

On addition of a Grignard reagent to a Schiff base the first reaction stage is rapid addition of the organomagnesium compound (as a Lewis acid) to the nitrogen atom of the $C=N$ double bond with formation of a complex (16a, 16b)[1033]. Only when more than one molecule of Grignard reagent is added does a second, slower stage to the adduct 17c occur *via* a six-membered cyclic transition state (17a, b). The formation of the six-membered ring (17b) depends upon steric factors, and consequently the conversion of Grignard reagents with a bulky structure is slow.

Particularly detailed investigations have been made of the reaction of organomagnesium halides with *N*-benzylidene alkylamines[985, 997, 1034−1039]:

$$H_5C_6-CH=N-R^1 + R^2MgX \longrightarrow H_5C_6-\underset{\underset{R^2}{|}}{C}H-NH-R^1$$

In this reaction the yield of secondary amine depends very much on the chain length and the extent of branching of the alkyl groups in the two reactants[1037, 1038, 1040]. Good yields are obtained mainly with short-chain unbranched Grignard reagents and Schiff bases; for example, 75% α-ethyl-*N*-methylbenzylamine resulted from the use of equimolar amounts of the reactants[1037].

N-Benzylidene alkylamines containing α-branched alkyl groups such as *sec*-butyl and *tert*-butyl (Tab. 2) give especially poor results, as do Grignard reagents with α- and β-branched groups[1040] such as *iso*-butyl and *tert*-butyl.

Compared with alkylmagnesium halides, benzylmagnesium[985], allylmagnesium[1041], and 1-alken-1-ylmagnesium halides[1008] have high reactivities.

Phenylmagnesium halides behave similarly to alkylmagnesium halides; as the chain length of the *N*-benzylidene alkylamine increases the yields of products decrease[1035]. 1-Alkyn-1-ylmagnesium halides do not react[1032, 1037].

The nature of the halogen in the Grignard reagent determines the yield of addition product to a considerable extent. Yields from organomagnesium chlorides and bromides are twice those from the corresponding iodides[1042]. If the molar ratio of ethylmagnesium bromide to benzylidenenbutylamine is altered, then the yield of amine (*N*-butyl-α-ethylbenzylamine*) increases as the proportion of Grignard reagent increases (1:1=31.3%; 2:1=36.4%; 4:1=44.6%)[1032].

An increased yield can be achieved also by using diethylmagnesium[1032] (1:1=29.0%; 2:1=30.0%; 4:1=63.4%).

Attempts to increase the yield of amine by increasing the electrophilic nature of the carbiminic carbon atom with the aid of Lewis acids such as aluminum chloride gave negative results[1032]. No improvement in the yield of adduct resulted from addition of hexamethylphosphoric triamide during the reaction of isopropylmagnesium chloride with benzylidenebutylamine[1040]; the object was to increase the polarity of the *C—Mg* bond by solvation of the magnesium.

The synthesis of *branched secondary amines* (which are difficult or impossible to obtain by Grignard reactions) can be achieved by use of organolithium compounds[1040]; the yields are excellent.

Table 2. Amines by reaction of *N*-benzylidene alkylamines with organomagnesium and organolithium compounds in relation to branching in the *N*-alkyl group.
Experimental conditions: 50 mmole Schiff base/200 mmole Grignard reagent (15 h/35°)
                    100 mmole Schiff base/200 mmole organolithium compound (24 h/20°)

⟨C₆H₅⟩—CH=N—R	Yield [% of theory] from reaction with				
	$(CH_3)_2CH-MgCl$	$(CH_3)_2CH-Li$	H_9C_4-Li	H_9C_4-MgCl	H_4C_4-MgCl
$-CH_2-CH_2-CH_2-CH_3$	*N-Butyl-α-isopropyl-benzylamine* 43.8	67.2	*α,N-Dibutyl-benzylamine* 92.7	91.5	94.5
$-CH(CH_3)_2$	*α,N-Diisopropyl-benzylamine* 43.4	61.0	*α-Butyl-N-isopropylbenzylamine* 86.8	76.6	79.4
$-CH(CH_3)-CH_2-CH_3$	*N-sec-Butyl-α-isopropylbenzylamine* 4.4	80.5	*α-Butyl-N-sec-butylbenzylamine* 94.0	12.3	78.8
$-C(CH_3)_3$	*N-tert-Butyl-α-isopropylbenzylamine* 2.6	57.5	*α-Butyl-N-tert-butylbenzylamine* 98.2	4.5	83.2

[1031] *J.M.Kamenka,* Compt. Rend. *268,* 1620 (1969).
[1032] *R. Beutel,* Dissertation, Universität München 1964.
[1033] *P.M. Maginnity, T.J. Gair,* J. Amer. Chem. Soc. *74,* 4958 (1952).
[1034] *M. Busch,* Ber. *37,* 2691 (1904); *M. Busch,* Ber. *38,* 1761 (1905).

Many *diphenylmethylamines* with a local anesthetic action have been synthesized from benzylidenealkylamines and phenyllithium[993]. 1-Alkyn-1-yllithium and -sodium compounds, like their Grignard derivatives, do not react with benzylidenealkylamines[1032].

Results as good as with organolithium compounds can be obtained by treatment of Schiff bases with Grignard reagents (4 *mols* per *mol* imine) in toluene at 100° for 15 hours. Because Grignard reagents are readily available, this procedure is the method of choice for the synthesis of secondary amines with secondary or tertiary α-carbon atoms; other methods give poor yields.

The low nucleophilicity of the alk-1-yn-1-yl moiety is not sufficient to make an addition to the carbimine C-atom possible. Only the use of more electrophilic reaction partners such as immonium perchlorates[1043] or α-amino nitriles[1032] leads to the desired reaction.

As with α-aminonitriles the reactions of benzylidene alkylamines with Grignard reagents yield monomeric and dimeric reduction products *(alkylbenzylamines* and *N,N'-tetraalkyl-1,2-diphenylethylenediamines)* as well as the α-alkyl-N-alkylbenzylamines[1037–1039, 1041]. Formation of dimeric by-products is promoted to such an extent by high reaction temperatures (especially with branched organomagnesium halides) and by heavy metal ions[1042, 1044–1047] that reaction of benzylidene alkylamines with Grignard reagents can be used as a convenient procedure for preparing *symmetrical N,N'-tetraalkyl-1,2-diphenylethylenediamines.*

Monomeric reduction products are formed only in small amounts compared with analogous reactions of carbonyl compounds. Reactions of aliphatic aldimines with organomagnesium halides yield only small amounts of addition products (up to 17%)[1048]. As a rule, ketimines do not react[1049, 1050]; side-reactions such as dimerization[1051] and enamine formation[1049] are observed. Reacting phenylmagnesium bromide with the imine from benzophenone and 2-naphthylamine results in 1,4-addition giving α-(2-biphenylylbenzyl)-2-naphthylamine[1052].

Some instances of the formation of 1,2-adducts in the react ion of ketimines with organolithium compound and with allylmagnesium bromide have been described[1053].

N-Heteroaromatic compounds such as quinoline and pyridine form 1,2-adducts with organometallic compounds in analogy to the reaction with sodium amide; these adducts then yield *2-substituted quinolines* or *pyridines*[997]:

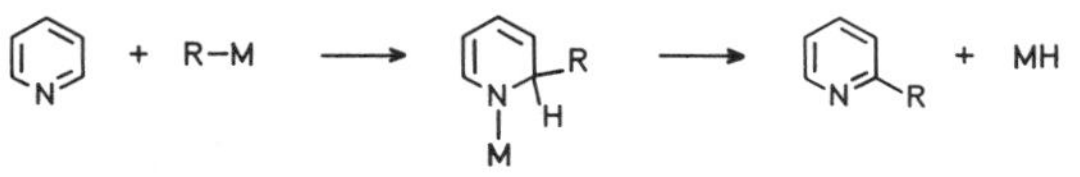

Additions to other unsaturated *C–N* systems are of only limited importance. For example, α-branched primary amines are only formed from nitriles and Grignard reagents (such as alkylmagnesium halides) in exceptional cases[997]. As a rule, the main products from acid amides and Grignard reagents are aldehydes or ketones[997].

Amine formation is most often observed with formamide derivatives. Lactams react with 2 molecules organomagnesium halide forming α-disubstituted cyclic amines; treatment with 1 molecule Grignard reagent yields either *amino ketones* or cyclic enamines, depending on the size of the ring[997].

[1035] *H. Hellmann, A. Opitz,* α-Aminoalkylierung, p. 87, Verlag Chemie, Weinheim/Bergstr. 1960.

[1036] *M.S. Kharasch, O. Reinmuth,* Grignard Reactions of Nonmetallic Substances, p. 1204, Prentice Hall, New York 1954.

[1037] *K.N. Campbell, C.H. Helbing, N.P. Florkowski, B.K. Campbell,* J. Amer. Chem. Soc. *70,* 3868 (1948).

[1038] *H. Thies, H. Schönenberger,* Arch. Pharm. (Weinheim, Ger.) *289,* 408 (1956).

[1039] *H. Thies, H. Schönenberger,* Chem. Ber. *89,* 1918 (1956).

[1040] *D. Stransky,* Dissertation Universität München (in preparation).

[1041] *B.L. Emling, R.J. Horvath, A.J. Saraceno, E.F. Ellermeyer, L. Haile, L.D. Hudac,* J. Org. Chem. *24,* 657 (1959).

[1042] *H. Thies, H. Schönenberger, A. Zeller,* Arch. Pharm. (Weinheim, Ger.) *298,* 26 (1965).

[1043] *B. Karlén, B. Lindeke, S. Lindgren, K.G. Svensson, R. Dahlbohm, D.J. Jenden, J.E. Giering,* J. Med. Chem. *13,* 651 (1970).

[1044] *A. Zeller,* Dissertation, Universität München 1962.

[1045] *H. Thies, H. Schönenberger, K. Borah,* Naturwissenschaften *46,* 378 (1959).

[1046] *H. Schönenberger, H. Thies, A. Zeller, K. Borah,* Naturwissenschaften *48,* 129 (1961).

[1047] *H. Schönenberger, H. Thies, A. Zeller,* Naturwissenschaften *48,* 303 (1961).

[1048] *R. Tiollais,* Bull. Soc. Chim. France 959 (1947).

[1049] *P. Grammaticakis,* Comp. Rend. *223,* 804 (1946).

[1050] *G. Stork, S.R. Dowd,* J. Amer. Chem. Soc. *85,* 2178 (1963).

[1051] *W.F. Short, J.S. Watt,* J. Chem. Soc. (London) 2293 (1930).

[1052] *H. Gilman, J. Morton,* J. Amer. Chem. Soc. *70,* 2514 (1948).

[1053] *H. Gilman, J. Eisch,* J. Amer. Chem. Soc. *79,* 2150 (1957).

[1054] *G. Bender,* Ber. *19,* 2272 (1886); *F.D. Chattaway, K.J.P. Orton,* J. Chem. Soc. (London) *75,* 1046 (1899).

[1055] *E.E. Slosson,* Ber. *28,* 3265 (1895).

[1056] *F.D. Chattaway, K.J.P. Orton,* Ber. *32,* 3573 (1899).

12.9 Amines *via* rearrangements

12.9.1 Rearrangement of *N*-substituted aryl-amines

This section deals with rearrangements involving acid-catalyzed migration into the aromatic ring of a group initially attached to the nitrogen atom of an aromatic amine.

The benzene ring can be substituted further or can be replaced by other aromatic systems including heterocycles. Despite the formally close resemblance, the mechanisms of these processes are not as uniform as might be expected. In some cases the migrating group has electrophilic character, in other cases a nucleophilic one.

Some rearrangements are intramolecular and proceed *via* a cyclic transition state. In *intermolecular* rearrangements the group *X* enters mainly the *para* position of the aromatic ring. Where the same is blocked the *ortho* position is frequently occupied. Sometimes, for instance, in the Fischer-Hepp rearrangement, *X* does not enter the *ortho* position at all (p. 558). Conversely, intramolecular rearrangements furnish mainly *ortho* substitution (*e.g.* the nitramine rearrangement) or an *ortho* and *para* isomer mixture, depending on the structure of the aryl group, as in the benzidine rearrangement. The migrating group never enters the *meta* position.

12.9.1.1 *N*-Haloamides (Orton rearrangement)

A typical example is conversion of *N*-chloro-acetanilide into a mixture of 2- and 4-chloroacetanilide (or the corresponding bromo or iodo compounds) by treatment with hydrochloric acid in protic solvents such as glacial acetic acid, water, or mixtures of the two[1054]:

X = Cl, Br, J

This rearrangement gives mainly the *para* isomer, but often appreciable *ortho* isomer is also isolated. Rearrangements catalyzed by carboxylic acids in

aprotic solvents such as chlorobenzene give the para product almost exclusively.

The reaction has been extended to several *N*-halo-amides[1055-1057].

In addition to the catalyzed rearrangement, isomerization can be brought about by heat[1058], light[1059], or benzoyl peroxide in hot carbon tetrachloride[1060].

N-Haloamides substituted in the ring with halogen atoms react slowly, while *N,N*-dichloroarylamines rearrange easily in ether to give ring-substituted chloroamines[1061]. From a preparative viewpoint the reaction is, in general, inferior to the corresponding direct halogenation of amines and their acyl derivatives.

12.9.1.2 *N*-Arylhydroxylamines (Bamberger rearrangement)

Reaction of *N*-phenylhydroxylamines with dilute aqueous sulfuric acid gives 4-aminophenols as main product[1062].

For preparative purposes it is convenient to carry out the reaction in one step by electrolytic reduction of nitrobenzene under acid conditions[1063].

Alternatively, nitrobenzene can be converted into 4-aminophenol directly (77%) by treatment with aluminum turnings in aqueous sulfuric acid at 90° for 16 hours[1064].

[1057] *F.D. Chattaway, K.J.P. Orton*, Ber. *33*, 2396 (1900).
[1058] *C.W. Porter, P. Wilbur*, J. Amer. Chem. Soc. *49*, 2145 (1927);
A.E. Bradfield, J. Chem. Soc. (London) 351 (1928).
[1059] *J.J. Blanksma*, Rec. Trav. Chim. Pays-Bas *21*, 366 (1902);
H. Matthews, R.V. Williamson, J. Amer. Chem. Soc. *45*, 2574 (1923);
F.W. Hodges, J. Chem. Soc. (London) 240 (1933).
[1060] *K.N. Ayad, C. Beard, R.F. Garwood, W.J. Hickinbottom*, J. Chem. Soc. (London) 2981 (1957).
[1061] *S. Goldschmid, L. Strohmenger*, Ber. *55*, 2450 (1922).
[1062] *E. Bamberger*, Ber. *27*, 1347 (1894).
[1063] *L. Gattermann*, Ber. *26*, 1844 (1893).
[1064] *L. Florn, H. Sanielevici, C. Stoicescu, M. Comananu*, Rev. Chim. (Bucarest) *12*, 649 (1961); C.A. *57*, 14 985i (1962).

12.9.1.3 Arylsulfamic acids

On refluxing in 1,4-dioxane *N*-phenylsulfamic acid was found to rearrange to *sulfanilic acid* in 22% yield[1065]:

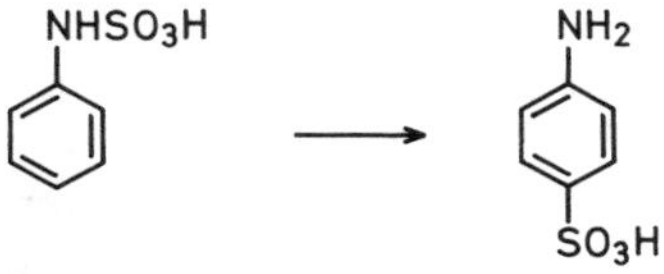

12.9.1.4 Aromatic sulfonamides, sulfinamides, and sulfenamides

On treatment with sulfuric acid (80–98%) at moderate temperatures *N*-substituted arylsulfanilides rearrange to 2- and *4-arylamino diaryl sulfones*[1066]:

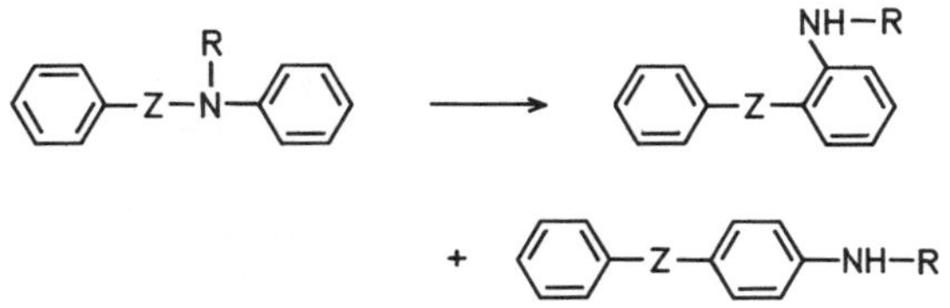

Z = SO₂, SO, S

The rearrangement competes with hydrolysis of the *S–N* bond (when *R=H* the latter is the sole reaction), and also with sulfonation, but under suitable conditions good yields are obtained. The sulfonyl group enters the position *ortho* to nitrogen, indicating that the rearrangement is intramolecular[1067]. Entry of the sulfonyl, sulfoxide, or sulfide group into a free *para* position is observed only under conditions anticipating decompositions[1068].

12.9.1.5 *N*-Nitrosoarylamines (Fischer-Hepp rearrangement)

Treatment of a solution of *N*-nitrosodiphenylamine in dry ether or dry ethanol with a solution of hydrogen chloride in dry ethanol leads to *4-nitrosodiphenylamine* in good yield[1069]:

The nitroso group enters the *para* position of the ring; no rearrangement occurs when this position is blocked.

In the naphthylamine series the nitroso group enters the 4- or 1-position[1070], *e.g.*:

4-Nitroso-N-phenyl-1-naphthylamine; 50%

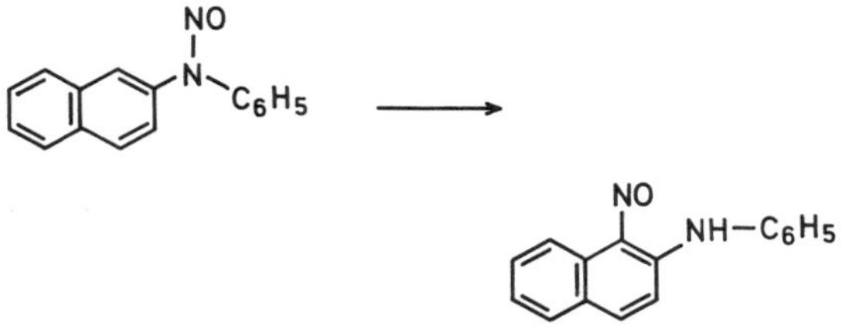

1-Nitroso-N-phenyl-2-naphthylamine

Acids other than hydrogen chloride are not very useful; its specificity suggests that the nitrosating agent is nitrosyl chloride. The reaction fails if the *N*-alkyl substituent is very bulky.

12.9.1.6 Arylnitramines

Arylnitramines undergo rearrangement on treatment with strong aqueous acids or with hydrogen chloride in organic solvents to 2-nitroarylamines. The *para* isomers are at most a side-product[1071].

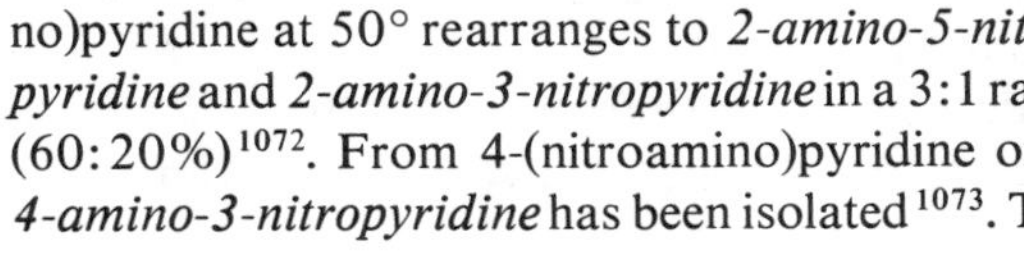

2-Nitroaniline   *4-Nitroaniline*

By contrast, in the pyridine series migration into the *para* position dominates; thus, 2-(nitroamino)pyridine at 50° rearranges to *2-amino-5-nitropyridine* and *2-amino-3-nitropyridine* in a 3:1 ratio (60:20%)[1072]. From 4-(nitroamino)pyridine only *4-amino-3-nitropyridine* has been isolated[1073]. The

[1065] *G. Illuminati*, J. Amer. Chem. Soc. *78*, 2603 (1956).

[1066] *S. Searles, S. Nukina*, Chem. Rev. *59*, 1077 (1959).

[1067] *J. Halberkann*, Ber. *55*, 3074 (1922).

[1068] *A. Mustafa, M.I. Ali*, J. Amer. Chem. Soc. *77*, 4593 (1955).

[1069] *O. Fischer, E. Hepp*, Ber. *19*, 2991 (1886).

[1070] *O. Fischer, E. Hepp*, Ber. *20*, 1247 (1887); *P.W. Neber, H. Rauscher*, Justus Liebigs Ann. Chem. *550*, 182 (1942).

[1071] *E. Bamberger*, Ber. *30*, 1248 (1897).

[1072] *L.N. Pino, W.S. Zehrung*, J. Amer. Chem. Soc. *77*, 3154 (1955).

[1073] *E. Koenigs, M. Mields, H. Gurlt*, Ber. *57*, 1179 (1924).

rearrangement of aromatic nitramines is intramolecular, as has been demonstrated for several substrates over the 0.1–16 M acid range in several solvents [1074].

12.9.1.7 Aromatic diazoamino compounds

Diazoaminobenzene rearranges to *4-aminoazobenzene* by treatment with alcoholic hydrochloric acid or, better, aniline and aluminum chloride (70%) [1075]:

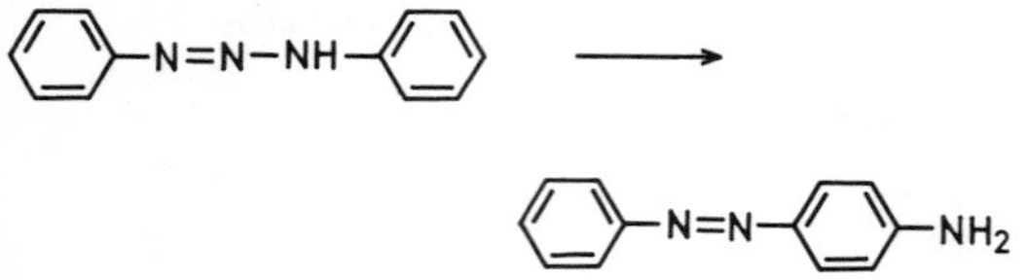

The rearrangement is to the *para* position, but to the *ortho* position if the former is blocked as in 1,3-di-4-tolyltriazene. For a review see ref. [1076].

12.9.1.8 Arylhydrazines

Treatment of phenylhydrazine with concentrated hydrochloric acid at 200° yields *1,4-phenylenediamine* and minor amounts of aniline, ammonium chloride, and nitrogen [1077]:

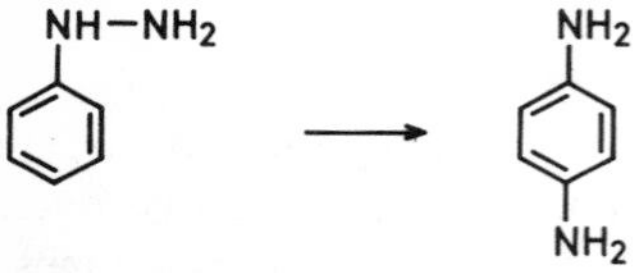

In the same way *N*-methyl-*N*-phenylhydrazine yields *N-methyl-1,4-phenylenediamine*. The reaction fails if the *para* position in the starting material is blocked.

12.9.1.9 Hydrazo compounds (benzidine rearrangement)

The name benzidine rearrangement generally refers to the acid-catalyzed rearrangements of aromatic hydrazo compounds. This rearrangement is usually brought about by treating a solution of the aromatic hydrazo compound with aqueous hydrochloric or sulfuric acid. Often ethanol is used as solvent. Alternatively, the rearrangement can take place in acetic acid or with hydrogen chloride in solvents such as benzene or toluene. A different procedure again is to treat an azo compound suspended or dissolved in ethanol with a solution of stannous chloride in concentrated hydrochloric acid. In this way the azo compound is first reduced to the hydrazo compounds, which then rearranges [1078]. The rearrangement of hydrazobenzene in aqueous ethanol leads to about 70% benzidine 1 and 30% 2,4′-biphenyldiamine 2 [1079]:

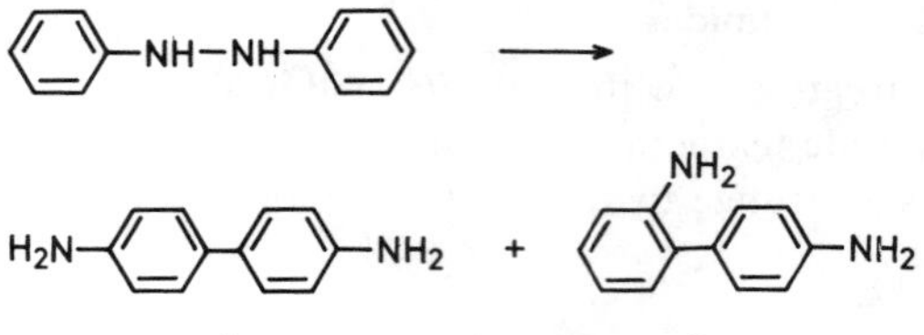

A 2,2′-biphenyldiamine ('*ortho*-benzidine') type product is obtained from dinaphthylhydrazines; thus *N,N′*-di-(2-naphthyl)hydrazine rearranges entirely to *1,1′-bi[2-naphthylamine]* [1080]:

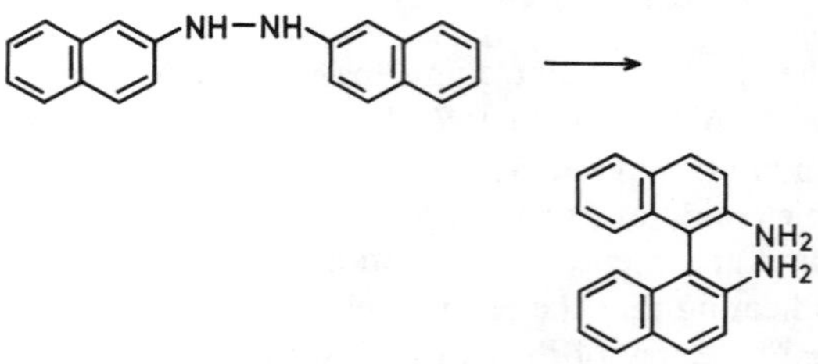

Semidine-type products are obtained from ring-substituted hydrazobenzenes. Unsymmetrical *p,p*-disubstituted hydrazobenzenes may give one or two *o*-semidines in dependence on the inductive properties of the groups.

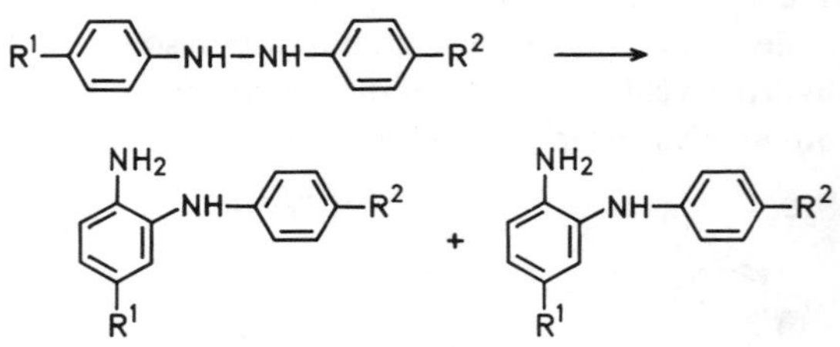

The donor of the two R groups with the greater $+I$ effect is found para to the unsubstituted amino

[1074] *D.V. Banthorpe, E.D. Hughes, D.L. Williams*, J. Chem. Soc. (London) 5349 (1964).

[1075] *R.J. Friswell, A.G. Green*, J. Chem. Soc. (London) 47, 917 (1886).

[1076] *H.J. Shine*, Aromatic Rearrangements, p. 212 Elsevier Publ. Co., Amsterdam, London, New York 1967.

[1077] *J. Thiele, L.H. Wheeler*, Ber. 28, 1538 (1895).

[1078] *P. Jacobsen*, Justus Liebigs Ann. Chem. 428, 76 (1922).

[1079] *D.V. Banthorpe, E.D. Hughes*, J. Chem. Soc. (London) 3308 (1962).

[1080] *D.V. Banthorpe*, J. Chem. Soc. (London) 2407 (1962).

group in the product. From hydrazobenzenes either an ortho-semidine or 2,4′-biphenyldiamine having a single *p*-substituent *R* is formed; the former if *R* is a strong electron donor, a 2,4′-biphenyldiamine if *R* is electron-attracting. In some cases (*R* = halogen) both compounds are obtained. When *R* is carboxy or sulfo the major product is benzidine because the substituent is eliminated. In some instances *para* semidines are obtained as the major rearrangement product, but this can be attributed to the use of Friedel-Crafts catalysts, *e.g.*:

N-(Ethoxyphenyl)-3-ethoxy-1,4-phenylenediamine

With many of the substituted hydrazo compounds disproportionation is observed.

An interesting rearrangement of *N,N′*-substituted cyclic hydrazo compounds has been reported[1081]:

1,2,3,4,5,6,7,8-Octahydro-9,12-etheno-1,8-benzo-diazacyclotetradecine; 77%

Benzidine-like rearrangements are reported on acid treatment of suitable imidazole[1082], pyridine[1083], and thiazole[1084] hydrazo compounds.

The benzidine rearrangement is one of the aromatic rearrangements that may with certainty be

today described to be wholly intramolecular (see the review in ref.[1085]).

Thus, from unsymmetrical hydrazo compounds exclusively unsymmetrical benzidines are obtained[1086]. In the same way rearrangement of a mixture of 2,2′-dimethylhydrazobenzene and 2-^{14}C-methylhydrazobenzene yields only uniform products[1087]:

3,3′-Dimethylbenzidine; 80%

3-^{14}C-Methylbenzidine; 67%

Similar investigations were carried out with mixtures of *N,N′*-di-1- and *N,N′*-di-2-naphthylhydrazines without the discovery of crossed-over products[1088].

Furthermore, the reaction is specifically an acid-catalyzed one and the kinetics indicate that with different substrates the order in respect of the hydrogen ion varies between one and two following the following equation[1089]:

$$\frac{-d\,(Hydrazo)}{dt} = K_2[Hydrazo]\,[H^{\oplus}] + K_3[Hydrazo]\,[H^{\oplus}]^2$$

The rate is unaffected by ring deuteration[1090, 1091] but it does increase with deuterium ions in deuterium oxide over protons in water[1090]. No evidence for free-radical formation during the rearrangement has been found[1092, 1093].

Two main theories for the mechanism thus remain in contention, involving either a polar transition

[1081] *G. Wittig, J.E. Grolig*, Chem. Ber. *94*, 2148 (1961).

[1082] *T. Pyl, H. Labmer, H. Beyer*, Chem. Ber. *94*, 3217 (1961).

[1083] *H. Beyer, H.J. Haase, W. Wildgrube*, Chem. Ber. *91*, 247 (1958).

[1084] *H. Beyer, H.J. Haase*, Chem. Ber. *90*, 66 (1957).

[1085] *H.J. Shine*, Aromatic Rearrangements, p. 126 Elsevier Publ., Co., Amsterdam, London, New York 1967.

[1086] *C.K. Ingold, H.V. Kidd*, J. Chem. Soc. (London) 984 (1933).

[1087] *D.H. Smith, J.R. Schwartz, G.W. Wheland*, J. Amer. Chem. Soc. 74, 2282 (1952).

[1088] *D.V. Banthorpe*, J. Chem. Soc. (London) 2413 (1962).

[1089] *D.A. Blackadder, C. Hinshelwood*, J. Chem. Soc. (London) 2898 (1957).

[1090] *D.V. Banthrope, E.D. Hughes, C.K. Ingold*, J. Chem. Soc. (London) 2864 (1964).

[1091] *G.S. Hammond, W. Grundemeier*, J. Amer. Chem. Soc. 77, 2444 (1955).

[1092] *H.J. Shine, J.P. Stanley*, J. Org. Chem. *32*, 905 (1967).

[1093] *D.V. Banthrope, R. Bramley, J.A. Thomas*, J. Chem. Soc. (London) 2900 (1964).

state[1090] or a π-complex[1094]. For an understanding of these two theories and their pros and cons the reader is referred to several reviews which summarize the present status of the problem[1094–1097].

12.9.1.10 *N*-Alkylanilines (Hofmann-Martius rearrangement)

N-Alkylanilines may be rearranged to *C-alkylated anilines* by heating their hydrochlorides or hydrobromides to 200–300°[1098]. Normally a 4-alkylaniline is the main product; small yields of the 2-alkyl derivative are formed unless the *p*-position is blocked[1099]. This Hofmann-Martius rearrangement succeeds also by heating the amine with a metal halide such as zinc, cobalt(II), nickel(II), cadmium(II) or aluminum chloride[1100]. Certain differences in the distribution of the products obtained by the two respective methods can be observed but are not as marked as was once thought.

In each method polyalkylation can occur. Formation of alkyl halides and olefins, too, may reduce the yield. Products with rearranged alkyl substituents have been obtained. N-Isopentylaniline hydrobromide at 240–270° gives *4-tert-pentylaniline, 2-methyl-2-butene,* and *isopentyl bromide*[1101]:

$$H_2N^{\oplus}-CH_2-CH_2-CH(CH_3)_2$$

Br$^{\ominus}$ ⟶

An example of a Hofmann-Martius rearrangement during Hofmann degradation is the pyrolysis of (α-ethylbenzyl)dimethylphenylammonium hydroxide at 150°[1102]:

α-Ethyl-N,N-dimethyl-α-phenyl-2-toluidine

α-Ethyl-N,N-dimethyl-α-phenyl-4-toluidine

For a review see ref.[1103].

12.9.1.11 Anilides

Conversion of *N,N*-diacetylaniline to *4′-acetamidoacetophenone* ($\sim 15\%$) is brought about by heating with toluenesulfonic acid, phosphoric acid, or methanesulfonic acid. The yield increases to 25% if acetanilide is added. Under similar conditions[1104]

N,N-Diacetyl-2-toluidine ⟶ *4-Acetamido-2-methyl-acetophenone*

N,N-Dipropionylaniline ⟶ *4-Propionamidopropiophenone*

N,N-Dipropionyl-2-toluidine ⟶ *2-Methyl-4-propion-amidopropiophenone*

N-Acylanilines may be converted to amino ketones also by a photochemical rearrangement. Irradiation of a solution of acetanilide in ethanol gives *2-aminoacetophenone* (20%), *4-aminoacetophenone* (25%), and *aniline* (18%), and, similarly, irradiation of propionanilide or butyroanilide gave *4-amino-* and *2-aminopropiophenones* and *ω-amino-* and *2-aminobutyrophenones* respectively[1105].

12.9.2 Rearrangement of isocyanates (amines from carboxylic acids)

Replacement of a carboxy group by an amino group can be achieved by four different but

[1094] *M.J.S. Dewar, A.P. Marchand,* Ann. Rev. Phys. Chem. *16,* 338 (1965).

[1095] *M.J.S. Dewar* in *P. De Mayo,* Molecular Rearrangements, John Wiley & Sons, New York 1963.

[1096] *H.J. Shine* in Mechanism of Molecular Migrations, John Wiley & Sons, New York 1969.

[1097] *C.K. Ingold,* Structure and Mechanism in Organic Chemistry, 2nd Ed., p. 916, Cornell University Press, Ithaca, New York 1969.

[1098] *A.W. Hofmann, C.A. Martius,* Ber. *4,* 742 (1871).

[1099] *W.J. Hickinbottom,* J. Chem. Soc. (London) 1700 (1934).

[1100] *J. Reilly, W.J. Hickinbottom,* J. Chem. Soc. (London) *117,* 103 (1920).

[1101] *W.J. Hickinbottom, S.E.A. Ryder,* J. Chem. Soc. (London) 1281 (1931).

[1102] *D.A. Archer, H. Booth,* Chem. & Ind. (London) 1570 (1962).

[1103] *H.J. Shine,* Aromatic Rearrangements, p. 249, Elsevier, Publ. Co., Amsterdam, London, New York 1967.

[1104] *Saeed Abdalla Abbas, W.J. Hickinbottom,* J. Chem. Soc. C, 1305 (1966).

[1105] *D. Elad,* Tetrahedron Lett. 873 (1963).

mechanistically analogous methods, namely, Hofmann, Curtius, Schmidt, and Lossen rearrangements. Only the first three are of preparative interest.

It has been established that in all these reactions isocyanates are the initial products which usually react further before a substance can be isolated. Isocyanates are readily hydrolyzed to amines by acids or bases and also react with alcohols and amines to yield carbamate esters, ureas, amides, *etc.* All these reactions proceed intramolecularly and with retention of the configuration of the migrating group.

An important preparative advantage of these reactions lies in the formation of primary amines free from contaminating secondary and tertiary amine. The yields vary but in general are satisfactory. When amides are available the Hofmann rearrangement is naturally the most convenient one; with esters the Curtius reaction is favored. For large-scale preparative work the Hofmann reaction is the procedure of choice as it is safe and relatively inexpensive, especially if chlorine is utilized. The Curtius rearrangement occurs under relatively mild conditions and is therefore particularly suitable for sensitive substances. Since hydrazoic acid and many azides are poisonous and violently explosive, it is imprudent to handle large quantities of hydrazoic acid (Schmidt reaction) or to isolate large quantities of azides. Degradation of both or only one of the carboxylic groups of malonic acid and other dicarboxylic acids can be accomplished only by the Curtius reaction. To convert unsaturated acids to amines, the Curtius reaction is once more the method of choice as the Hofmann reaction is likely to halogenate the double bond. Keto acids are degraded best by the Hofmann reaction, while acylated amino acids and peptides are most satisfactorily degraded by the Curtius procedure, and nonacylated amino acids by Hofmann degradation. Aromatic amino acids with substituents such as halogen, amino, and methoxy must normally be degraded by the Curtius reaction in order to avoid halogenation or sulfonation. Only the Hofmann rearrangement has been applied successfully to sugars. For a comparison of the Curtius, Hofmann, and Schmidt reactions see ref. [1106].

12.9.2.1 Hofmann rearrangement

In the Hofmann reaction [1107] an amide is converted to an amine with one less carbon atom by treatment with bromine or chlorine in an alkaline medium.

$$R-CONH_2 \; + \; Br_2 \; + \; n\,OH^{\ominus} \longrightarrow$$

$$R-NH_2 \; + \; CO_3^{2\ominus} \; + \; 2\,Br^{\ominus} \; + \; 2\,H_2O$$

A valuable modification consists in carrying out the reaction in alcoholic solution with subsequent hydrolysis of the carbamic acid ester (urethan) obtained:

$$R-CONH_2 \; + \; Br_2 \; + \; 2\,OH^{\ominus} \; + \; R^1OH$$

$$\longrightarrow \; R-NH-COOR^1 \; + \; 2\,Br^{\ominus} \; + \; 2\,H_2O$$

$$R-NH-COOR^1 \; + \; H_2O \longrightarrow$$

$$R-NH_2 \; + \; CO_2 \; + \; R^1OH$$

The scope of the reaction is wide, ranging as it does from aliphatic and alicyclic to aromatic and heterocyclic amides. Good yields are obtained with amides of fatty acids unless the amide is poorly soluble under the normal conditions of the reaction; in that case the use of an inert solvent, for instance, 1,4-dioxane gives satisfactory results [1108, 1109]. Diamides lead to *diamines* without difficulty. An exception are 1,2-diamides, which rearrange to *uracils* [1110], e.g.:

2-Methylpyrimidino-
[4,5-d]thiazole-5,7-diol

In the same way succinamide is converted to *5,6-dihydropyrimidine-2,4-diol (5,6-dihydrouracil)* [1111].

α-Hydroxy [1112], α-halo [1113], and α,β-unsaturated amides [1113] yield aldehydes (Vol. V) as sequential products of the primarily formed unstable amines or enamines.

X = OH, Halogen

$$\longrightarrow \; R-CHO$$

$$R-CH=CH-CONH_2 \longrightarrow [R-CH=CH-NH_2]$$

$$\longrightarrow \; R-CH_2-CHO$$

[1106] *P.A.S. Smith*, Org. Reactions *3*, 363 (1946).
[1107] *A.W. Hofmann*, Ber. *14*, 2725 (1881).
[1108] *E. Jeffreys*, Ber. *30*, 898 (1897).
[1109] *B.L. Murr, C.T. Lester*, J. Amer. Chem. Soc. *77*, 1684 (1955).

With α,β-triply unsaturated amides the Hofmann degradation furnishes nitriles[1114] (p. 646):

$$R-C\equiv C-CONH_2 \longrightarrow [R-C\equiv C-NH_2]$$
$$\longrightarrow R-CH_2-CN$$

When the amide is substituted in a suitable position by amino or hydroxy, intramolecular reaction of the latter with the intermediate isocyanate can occur with formation of cyclic *ureas* or *urethans*[1115]:

2-(Trifluoromethyl)-8-purinol; 4%

Imides of dicarboxylic acids yield lactams[1116], *e.g.:*

Naphthostyril (8-amino-1-naphthoic acid lactam); 85%

Exceptions are imides of 1,2-dicarboxylic acids, which lead to β-amino acids[1117]:

$$H_2C-COOC_2H_5$$
$$H_2C-NH_2 \cdot HCl$$

β-Alanine ethyl ester hydrochloride

For a review of the Hofmann reaction see ref. [1118].

12.9.2.2 Curtius rearrangement

The thermal decomposition of acid azides to isocyanates and molecular nitrogen is known as the Curtius rearrangement[1119]:

$$R-CO-N_3 \longrightarrow R-N=C=O + N_2$$

Coupled with a subsequent hydrolysis, the Curtius reaction represents a general procedure for replacing carboxy by amino:

$$R-N=C=O \xrightarrow[\text{or } OH^{\ominus}]{H^{\oplus}} R-NH_2$$

Azides can be rearranged under mild conditions merely by warming in a suitable solvent. In benzene and chloroform *isocyanates* are formed, while in alcohols and water the intermediate isocyanates react further to give *urethans* and *ureas* which hydrolyze more or less readily to *amines* or their salts.

The thermal Curtius rearrangement is accelerated in the following order by solvents:

toluene < acetic anhydride < acetic acid < aqueous acetic acid[1120].

Protonic and nonprotonic acids, *e.g.*, sulfuric acid and boron trifluoride, catalyze the thermal reaction. Boron halides and acid azides in toluene at $-60°$ yield 1:1 complexes which between $-20°$ and $+25°$ decompose into the corresponding isocyanate-boron halide complexes. These complexes are easily hydrolyzed to the corresponding urethans on treatment with alcohols[1121].

[1110] *S.J. Childress, P.L. McKee*, J. Amer. Chem. Soc. *73*, 3862 (1951).

[1111] *H. Weidel, E. Roithner*, Monatsh. Chem. *17*, 172 (1896); Chem. Zentr. I, 1266 (1896).

[1112] *R.A. Weerman*, Rec. Trav. Chim. Pays-Bas *37*, 16 (1917); C.A. *12*, 1463 (1918).

[1113] *R.A. Weerman*, Justus Liebigs Ann. Chem. *401*, 1 (1913).

[1114] *I.J. Rinkes*, Rec. Trav. Chim. Pays-Bas *39*, 704 (1920); C.A. *15*, 1510 (1921).

[1115] *J.A. Barone*, J. Med. Chem. *6*, 39 (1963).

[1116] *M.M. Dashevskii*, Izv. Vyssh Ucheb. Zavedenii, Khim. Khim. Tekhnol. *4*, 232 (1961); C.A. *55*, 23490h (1961).

[1117] *A.N. Parshin*, Zh. Obshch. Khim. *20*, 1826 (1950); C.A. *45*, 2407g (1951).

[1118] *E.S. Wallis, J.F. Lane*, Org. Reactions *3*, 267 (1946).

[1119] *T. Curtius*, Ber. *27*, 778 (1894).

[1120] *Y. Yukawa, Y. Tsuno*, J. Amer. Chem. Soc. *81*, 2007 (1959).

[1121] *E. Fahr, L. Neumann*, Angew. Chem. *77*, 591 (1965).

[1122] *L. Horner, G. Bauer, J. Dörges*, Chem. Ber. *98*, 2631 (1965);
R.F.C. Brown, Australian J. Chem. *17*, 47 (1964); C.A. *60*, 7930g (1964);
R. Huisgen, J.P. Anselme, Chem. Ber. *98*, 2998 (1965).

The photoinduced Curtius rearrangement, too, has been studied by many authors[1122]. It proceeds at low temperature and can be conducted in a variety of solvents such as cyclohexane, methylene chloride, benzene, or alcohols.

In addition to the normal rearrangement products, by-products arising from hydrogen abstraction or insertion of an intermediate nitrene species $R—CO\ddot{N}$ into a $C—H$ bond of the solvent are obtained[1123]. A practically constant yield of isocyanate ($\sim40\%$) results under all conditions, while that of the intermolecular nitrene products varies widely.

The thermal Curtius reaction is quite general and embraces aliphatic, alicyclic, aromatic, and heterocyclic acids. Diazides and polyazides give rise to the corresponding *di-* and *polyamines* without difficulty[1124, 1125], monosubstituted malonic acid azides are smoothly converted into the corresponding geminal *diurethans*, which, in turn, are readily hydrolyzed to *aldehydes*[1126] by mineral acid.

$$\begin{array}{c} CO—N_3 \\ | \\ R—CH_2—CH—CO—N_3 \end{array} \longrightarrow$$

$$\begin{array}{c} NH—COOC_2H_5 \\ | \\ R—CH_2—CH—NH—COOC_2H_5 \end{array} \longrightarrow R—CH_2—CHO$$

In hydroxy acids the course of the Curtius reaction is upset severely because the azides or isocyanates can undergo alternative reactions[1127], *e.g.:*

$$\begin{array}{cc} CH_2—CH—NH—CO—O—CH_2—C_6H_5 \\ | \quad\; | \\ OH \quad CO \\ \quad\; | \\ \quad\; N_3 \end{array} \longrightarrow$$

$$\left[\begin{array}{cc} CH_2—CH—NH—CO—O—CH_2—C_6H_5 \\ | \quad\;\; | \\ OH \quad\;\; N \\ \quad\;\; \| \\ \quad\;\; C \\ \quad\;\; \| \\ \quad\;\; O \end{array} \right]$$

$$\longrightarrow \begin{array}{c} CO—O—CH_2—C_6H_5 \\ HN \quad\quad H \\ N \\ O \end{array}$$

Benzyl 2-oxo-4-oxa-zolidinecarbamate; 72%

Acylation or alkylation of the hydroxy groups usually eliminates the difficulty.

α-Hydroxy and α-halo carboxylic acid azides furnish aldehydes or ketones. 1, 5-Dihydroxy-3, 4-isopropylidenedioxy-1-cyclohexanecarboxylic acid azide yields 78% *3-hydroxy-4,5-(isopropylidenedioxy)cyclohexanone,* but its 1-methyl ether forms instead *7,8-(isopropylidenedioxy)-5-methoxy-2-oxa-4-azabicyclo[3.3.1]nonan-3-one* (57%)[1128].

The Curtius reaction is exothermic and care must be taken to maintain control over the potentially explosive system. For a review see ref.[1129].

12.9.2.3 Schmidt rearrangement

The Schmidt rearrangement[1130] may be considered as a modification of the Curtius rearrangement in which the required azide is formed by treatment of a carboxylic acid and a hydrazoic acid with strong mineral acid. Normally, the product obtained after dilution with water is an *amine:*

$$R—COOH \;+\; HN_3 \xrightarrow{\;H_2SO_4\;}$$

$$R—NH_2 \;+\; CO_2 \;+\; N_2$$

Isolation of the intermediate isocyanate has been reported in only a few cases[1131]. Hydrazoic acid is very poisonous and all reactions involving it should be carried out under an efficient hood. *In situ* generation of the hydrazoic acid from sodium azide and concentrated sulfuric acid or 100% sul-

[1123] *S. Linke, G. Y. Tisne, W. Lwowski,* J. Amer. Chem. Soc. *89,* 6308 (1967).
[1124] *P.A.S. Smith,* Org. Synth. *36,* 69 (1956).
[1125] *V. Boekelheide, G.K. Vick,* J. Amer. Chem. Soc. *78,* 653 (1956).
[1126] *T. Curtius, O.E. Mott,* J. Prakt. Chem. *94,* 323 (1916); Chem. Zentr. II, 12 (1917).

[1127] *E.D. Nicolaides,* J. Org. Chem. *32,* 1251 (1967).
[1128] *L.W. Jones, D.H. Powers,* J. Amer. Chem. Soc. *46,* 2518 (1924).
[1129] *P.A.S. Smith,* Org. Reactions *3,* 337 (1946).
[1130] DRP. 500435 (1928), Knoll A.G.; Erf.: *K.F. Schmidt,* C.A. *26,* 2198 (1932).
[1131] *G.K. Rutherford, M.S. Newmann,* J. Amer. Chem. Soc. *79,* 213 (1957).
[1132] *A.J. McNamara, J.B. Stothers,* Can. J. Chem. *42,* 2354 (1964).

furic acid[1132], 20% oleum[1133], or polyphosphoric acid[1133] is possible. The Schmidt degradation of acids to amines displays the advantage of being a one-step reaction which often gives higher yields than either the Hofmann or Curtius rearrangement. It cannot be used with acids which are unstable toward sulfuric acid or with acids containing readily sulfonated aromatic rings. Substituted malonic acids give α-amino acids which do not react further. In this way *homoserine*[1134] and *homolysine*[1135] have been prepared in satisfactory yield from the corresponding triazides and diazides.

$$HOOC-(CH_2)_5-CH(COOH)_2 \longrightarrow$$

$$H_2N-(CH_2)_5-\overset{\overset{\displaystyle NH_2}{|}}{C}H-COOH$$

$$H_3CO-CH_2-CH_2-CH(COOH)_2 \longrightarrow$$

$$H_3CO-CH_2-CH_2-\underset{\underset{\displaystyle NH_2}{|}}{C}H-COOH \xrightarrow{HBr}$$

$$HO-CH_2-CH_2-\underset{\underset{\displaystyle NH_2}{|}}{C}H-COOH$$

The Schmidt reaction proceeds quite smoothly with acids in which the carboxy group is little reactive because it is attached to a tertiary carbon atom, for instance, in podocarpic acid[1136] **1** to give *1,2,3,4,4a,9,10,10a-octahydro-6-hydroxy-1,4a-dimethyl-1-phenanthramine,* and in campholic acid[1137] **2** to give *1,2,2,3-tetramethyl-1-cyclopentanamine* (88%).

Cinnamic acid yields *phenylacetaldehyde* probably *via* intermediate formation of the corresponding vinylamine, tautomerization of the aldimine, and hydrolysis[1138].

$$H_5C_6-CH=CH-COOH \quad + \quad HN_3$$

$$\longrightarrow \quad [H_5C_6-CH=CH-NH_2] \longrightarrow$$

$$[H_5C_6-CH_2-CH=NH] \longrightarrow \quad H_5C_6-CH_2-CHO$$

Nicotinic acid has been degraded in 70% yield to *3-aminopyridine*[1139].

For a review of the Schmidt reaction see ref.[1140].

12.9.2.4 Lossen rearrangement[1141]

Alkali metal salts of *O*-acyl hydroxamic acids rearrange to isocyanates on heating.

The isocyanates react further in aqueous solution to give mainly the corresponding ureas and a little amine. A moderate rate acceleration of the reaction has been reported for benzoic acid and some of its derivatives by using dimethyl sulfoxide as solvent[1142].

In a variant of the Lossen reaction aromatic, but not aliphatic acids can be converted into the corresponding *amines* by treatment with hydroxylamine and polyphosphoric acid.

The seldom used reaction is carried out by heating the mixed reactants at 150–170° for 5–10 minutes. Varying yields are obtained[1143].

12.9.3 Beckmann rearrangement

The acid-catalyzed rearrangement of a ketoxime or an aldoxime to *amides* is called the Beckmann rearrangement[1144].

$$\underset{R'}{\overset{R}{>}}C=NOH \longrightarrow R-NH-CO-R^1$$
$$\text{or} \quad R-CO-NH-R^1$$

Initially, partial heterolysis of the N—O bond forms a positive charge on the nitrogen. Subse-

[1133] *R.F. Stockel, D.M. Hall,* Nature *197,* 787 (1963).
[1134] *K. Hayashi,* Chem. Pharm. Bull. (Tokyo) *7,* 187 (1959).
[1135] *S. Takagi, K. Hayashi,* Chem. Pharm. Bull. (Tokyo) *7,* 183 (1959).
[1136] *L.H. Briggs, G.C. De Ath, S.R. Ellis,* J. Chem. Soc. (London) *61* (1942).
[1137] *J. von Braun,* Justus Liebigs Ann. Chem. *490,* 100 (1931).
[1138] *M. Oesterlin,* Angew. Chem. *45,* 536 (1932).

[1139] *V.L. Zbarskii, G.M. Shutov, V.P. Zhilin, E.Yu. Orlova,* Khim. Geterotsikl. Soedin. 178 (1967); C.A. *67,* 73500 (1967).
[1140] *H. Wolff,* Org. Reactions *3,* 307 (1946).
[1141] *H.L. Yale,* Chem. Rev. *33,* 209 (1943).
[1142] *D.C. Berndt, W.J. Adams,* J. Org. Chem. *31,* 976 (1966).
[1143] *H.R. Snyder, C.T. Elston, D.B. Kellom,* J. Amer. Chem. Soc. *75,* 2014 (1953).
[1144] *E. Beckmann,* Ber. *19,* 988 (1886).

quently, the hydroxy group and its *trans*-positioned substituent exchange their bond partners in a synchronously ensuing intramolecular rearrangement.

More than 99% retention of the configuration of an optically active migrating group obtains[1145]. With most aliphatic ketoximes and with particularly readily isomerizing aryl ketoximes the Beckmann rearrangement gives mixtures of isomeric amides irrespective of whether a pure oxime isomer is used or not. This behavior does not violate the principle of *trans*-migration, as apparently equilibration between oxime isomers in the reaction medium ensues faster than the rearrangement.

The rearrangement can be conducted by treating an oxime with equimolecular quantities of phosphorus(V) chloride in benzene at low temperatures. Other widely used reagents are concentrated sulfuric acid, hydrogen halides, polyphosphoric acid[1146], thionyl chloride[1147], formic acid[1148], and tosyl chloride in pyridine[1149].

Among the more unusual catalysts are copper[1150], alkali, metal, ferric, and aluminum chlorides[1151], trifluoroacetic acid, and Japanese acid earth[1152].

The correct choice of the reaction temperature is vital for obtaining maximum yields.

Many factors determine the optimum value (*e.g.*, solvent, catalyst, nature of oxime, and product) and cannot be predicted accurately. With phosphorus pentachloride, hydrogen fluoride, and boron trifluoride in acetic anhydride optimum results are in general obtained at or below room temperature, with sulfuric or polyphosphoric acid as catalyst the rearrangement usually proceeds best at 100–150°.

Whether isomerization occurs during a Beckmann rearrangement depends to some extent on the conditions as well as on the structure of the oxime. Phosphorus pentachloride in ether or benzene appears to produce least isomerization, while protonic acids such as hydrogen chloride or sulfuric acid in polar solvents (*e.g.*, water or glacial acetic acid) particularly readily cause isomerization.

With fuming sulfuric acid and polyphosphoric acid side-reaction hydrolysis of oximes back to the initial carbonyl compound is minimized, and in general the yield of amide is improved.

The Beckmann reaction is generally applicable since most oximes undergo the rearrangement to yield an amide or a mixture of amides. It has been applied very extensively and successfully to aliphatic, alkylaromatic, aromatic, alicyclic, and heterocyclic ketoximes.

With most aliphatic ketoximes which isomerize particularly readily, Beckmann rearrangement gives a mixture of isomeric amides.

In general araliphatic ketoximes are more stable and the corresponding *anilide* is more often obtained as the main product following their rearrangement:

$$H_3C-CO-NH-C_6H_5$$
Acetanilide; 75–100%

and/or $H_5C_6-CO-NH-CH_3$
N-Methylbenzamide; 0–25%

As aromatic ketones are readily available by a Friedel-Crafts reaction, the Beckmann rearrange-

[1145] *J. Kenyon, D.P. Young,* J. Chem. Soc. (London) 263 (1941).

[1146] *E.R. Ward, T.M. Coulson,* J. Chem. Soc. (London) 4545 (1954).

[1147] *M.S. Ahmad, A.H. Siddiqui,* Indian J. Chem. 403 (1968); C.A. *70,* 88066 (1969).

[1148] *T. Van Es,* J. Chem. Soc. (London) 3882 (1965).

[1149] *H.R. Nace, A.C. Watterson,* J. Org. Chem. *31,* 2108 (1966).

[1150] *S. Yamaguchi,* Mem. Coll. Sci. Kyoto Imp. Univ. *7A,* 281 (1924); C.A. *18,* 2880, (1924).

[1151] *E. Beckmann, E. Bark,* J. Prakt. Chem. *105,* 327 (1923).

[1152] *H. Inoue,* Bull. Chem. Soc. Japan *1,* 177 (1926); C.A. *21,* 892 (1927).

[1153] *D.R. Smith, M. Maienthal, J. Tipton,* J. Org. Chem. *17,* 294 (1952); *R.E. Lyle, H.J. Troscianico,* J. Org. Chem. *20,* 1757 (1955).

ment of their oximes is often the best route to such otherwise not easily accessible *aromatic amines*. Certain aromatic amines seem to owe their formation to a Beckmann rearrangement brought about by lithium tetrahydroaluminate in ether with certain substituted acetophenone oximes[1153] (but cf. p. 504):

Ar—C(CH₃)=N—OH $\xrightarrow{\text{LiAlH}_4}$ [Ar—NH—CO—CH₃]

$\longrightarrow$ Ar—NH—C₂H₅ + Ar—CH(NH₂)—CH₃

Alicyclic ketoximes rearrange to yield lactams under Beckmann reaction conditions. The reaction seems to be quite general for rings of all dimensions. Cleavage of the lactam group leads to *ω-amino acids*.

⬡C=NOH $\longrightarrow$ ⬡(NH)—C=O $\longrightarrow$ ⬡(NH₂)(COOH)

In what follows the limitations of the Beckmann rearrangement as a method for synthesizing amines will be discussed briefly.

Sulfuric acid does not appear to be particularly suitable for bringing about the rearrangement of ketoximes containing aromatic systems, especially if these have electron-donating substituents. Sulfonation occurs as a side-reaction here.

α,β-Unsaturated ketoximes undergo ring closures to *isoxazolines* on treatment with concentrated sulfuric acid, while with phosphorus(V) chloride in ether normal rearrangement products are formed[1154]:

R—CH=CH—C(CH₃)=N—OH

H₂SO₄ ↓   ↓ PCl₅

R-isoxazoline (O—N=C—CH₃)   R—CH=CH—CO—NH—CH₃

[1154] *A.H. Blatt, J.F. Stone*, J. Amer. Chem. Soc. *53*, 4134 (1931);
A.H. Blatt, J. Amer. Chem. Soc. *53*, 1133 (1931).

6-Methyl-5-hepten-2-one oxime when treated with phosphorus pentoxide yields *3,6-dihydro-2,3,3-trimethylpyridine*[1155]:

(H₃C)₂C=CH—CH₂—CH₂—C(=N—OH)—CH₃ $\longrightarrow$

(H₃C)₂C=CH—CH₂—CH₂—NH—CO—CH₃

$\xrightarrow{-\text{H}_2\text{O}}$ (dihydrotrimethylpyridine)

Oximes of *α*-keto acids decarboxylate and dehydrate successively to yield *nitriles*[1156].

R—C(=N—OH)—COOH $\longrightarrow$ R—CN + CO₂ + H₂O

Benzonitriles together with olefins are formed also when arylalkyloximes containing a tertiary *α*-carbon atom are treated with thionyl chloride[1157].

H₃C—C(CH₃)(C₆H₅)—C(=N—OH)—C₆H₅ $\longrightarrow$ H₂C=C(CH₃)—C₆H₅

α-Methylstyrene; 35%

+ H₅C₆—CN

Benzonitrile; 68%

Monooximes derived from *α*-diketones yield one of two possible amides in dependence on the structure and configuration of the oxime[1158].

R—CO—C(R¹)=N—OH $\longrightarrow$ R—CO—NH—CO—R¹

R—CO—C(R¹)=N—OH (HO) $\longrightarrow$ R—CO—CO—NH—R¹

In many cases cleavage to a *nitrile* (see p. 645). and an acid accompanies the rearrangement as side-reaction, or it may be the sole reaction[1159]:

[1155] *O. Wallach*, Justus Liebigs Ann. Chem. *319*, 77 (1901); Chem. Zentr. II, 1348 (1901).
[1156] *W. Dieckmann*, Ber. *33*, 579 (1900).
[1157] *R.E. Lyle, G.G. Lyle*, J. Org. Chem. *18*, 1058 (1953).
[1158] *H.J. Troscianico*, J. Org. Chem. *20*, 1757 (1955);
A. Werner, A. Piquet, Ber. *37*, 4295 (1904);
G. Rule, S.B. Thompson, J. Chem. Soc. (London) 1761 (1937);
L. Francesconi, F. Pirazzoli, Gazz. Chim. Ital. *33*, 36 (1903).

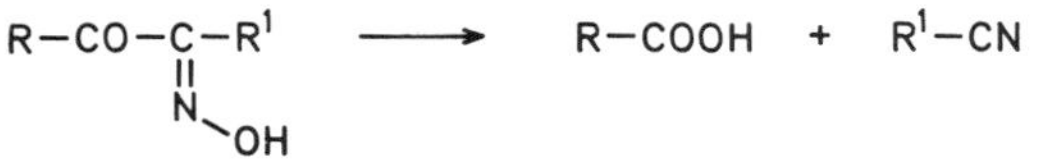

$$R-CO-\underset{\underset{OH}{\overset{\displaystyle N}{\|}}}{C}-R^1 \longrightarrow R-COOH + R^1-CN$$

Such a cleavage to nitriles, sometimes called a second order Beckmann rearrangement, is characteristic of the behavior of other oximes, for example, those derived from α-hydroxy ketones[1160].

Under appropriate conditions aldoximes also undergo the Beckmann rearrangement to yield *amides:*

$$R-CH=NOH \longrightarrow R-CO-NH_2$$
$$+ H-CO-NH-R$$

In general, the N-unsubstituted amide is formed; isolation of a substituted formamide has been reported in only one case. The reaction is thus of no importance for preparing amines; since aldoximes can be dehydrated readily by acidic reagents to form nitriles[1161]

$$R-CH=NOH \longrightarrow R-CN + H_2O$$

these products are often formed from aldoximes under the conditions of the Beckmann rearrangement.

For a survey see ref. [1162].

12.9.4 Schmidt rearrangement of carbonyl compounds

The reaction between equimolar hydrazoic acid and carbonyl compounds in the presence of a strong mineral acid is known as the Schmidt rearrangement[1163]. Aldehydes yield nitriles and formyl derivatives of *amines,* ketones yield *amides:*

$$R-CHO + HN_3 \longrightarrow R-CN +$$
$$R-NH-CHO + H_2$$

$$R-CO-R + HN_3 \longrightarrow R-CO-NH-R$$
$$+ N_2$$

With a large excess of hydrazoic acid ($\geqq 2$ *mols*) aldehydes and ketones yield substituted *tetrazoles*[1163]:

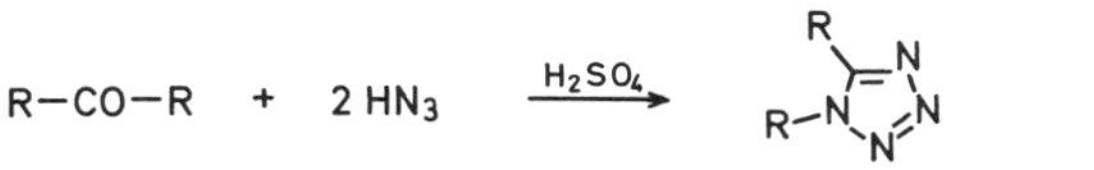

Sometimes *tetrazoles* are obtained as by-products even if equimolar quantities are employed[1164]. Symmetrical ketones yield the corresponding substituted amides. Thus respectively *N-methylacetamide* and *N-benzanilide* are obtained from acetone and benzophenone in quantitative yield[1163]. Unsymmetrical ketones can react in two alternative ways:

$$R-CO-R^1 + HN_3 \longrightarrow R-CO-NH-R^1$$
$$\text{and/or } R-NH-CO-R^1$$

With aryl methyl ketones migration of the phenyl group occurs to give a good yield of the *N*-arylacetamide. Thus, like the Beckmann rearrangement, the Schmidt rearrangement can be used to prepare aromatic amines from aromatic ketones, and in a one-step reaction[1165, 1166], *e.g.:*

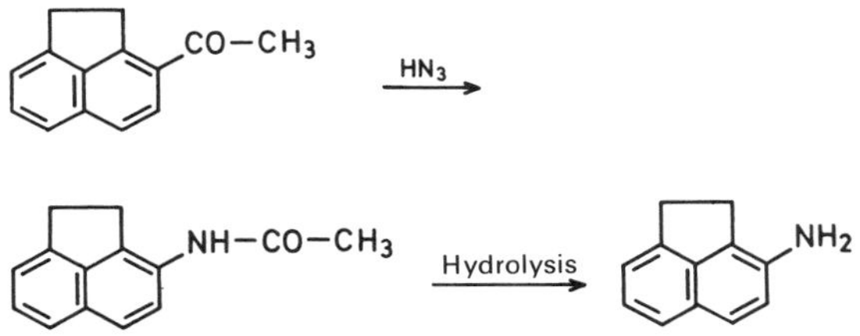

3-Acenaphthylamine; ~ 50%

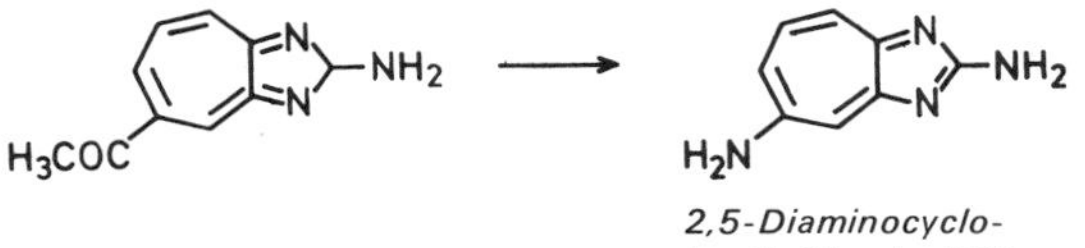

2,5-Diaminocycloheptimidazole; 75%

Ketones are more reactive toward hydrazoic acid than carboxylic acids. It is therefore possible to

[1159] W. Borsche, W. Sander, Ber. *47*, 2815 (1914);
C. Bulow, H. Grotowsky, Ber. *34*, 1479 (1901);
O.L. Brady, G. Bishop, J. Chem. Soc. (London) 810 (1926).
[1160] A. Werner, Th. Detscheff, Ber. *38*, 69 (1905).
[1161] A. Hantzsch, A. Lucas, Ber. *28*, 744 (1895);
J. Meisenheimer, P. Zimmermann, U. von Kummer, Justus Liebigs Ann. Chem. *446*, 205 (1926);
J. Meisenheimer, W. Theilacker, O. Beisswenger, Justus Liebigs Ann. Chem. *495*, 249 (1932);
K. von Auwers, R. Hügel, J. Prakt. Chem. [2] *143*, 179 (1935).

[1162] L.G. Donaruma, W.Z. Heidt, Org. Reactions *11*, 1 (1960).
[1163] K.F. Schmidt, Z. Angew. Chem. *36*, 511 (1923);
K.F. Schmidt, Ber. *57*, 704 (1924);
K.F. Schmidt, Ber. *58*, 2413 (1925).
[1164] L. Ruzicka, M.W. Goldberg, M. Hürbin, H.A. Beeckenoogen, Helv. Chim. Acta *16*, 1323 (1933);
J. v. Braun, A. Heymons, Ber. *63*, 502 (1930).
[1165] W.G.H. Edwards, V. Petrow, J. Chem. Soc. (London) 2853 (1954);
J.R. Dice, P.A.S. Smith, J. Org. Chem. *14*, 179 (1949).

control the reaction of a keto acid or its esters by using equimolecular quantities of hydrazoic acid in such a way that only the keto groups react, *e.g.*[1167]:

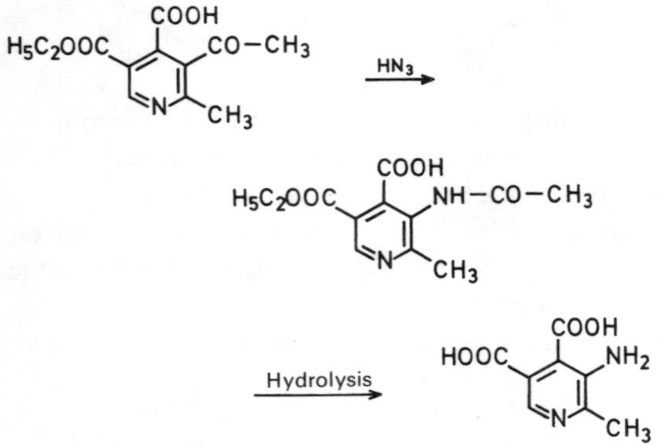

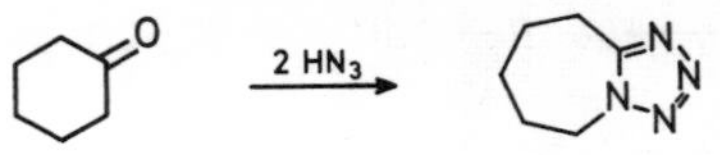

6,7,8,9-Tetrahydro-5H-tetrazolo-
azepine (cardiazol)

12.9.5 Neber rearrangement

The base-catalyzed conversion of *O*-sulfonyl oximes to α-amino ketones *via* azirines is known as the Neber reaction[1169].

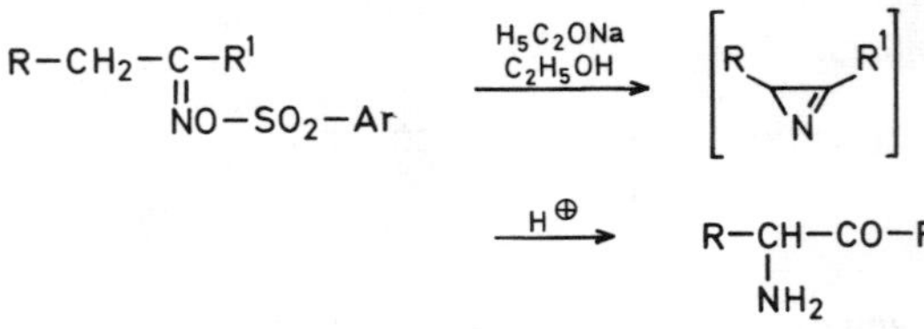

3-Amino-2-methyl-4,5-pyridine-
dicarboxylic acid; 70%

It is applicable to a wide variety of alkylaromatic, heterocyclic, and homocyclic systems. Interestingly, the reaction ranks as the method of choice for preparing intermediates during synthesis of important natural products[1170, 1171]:

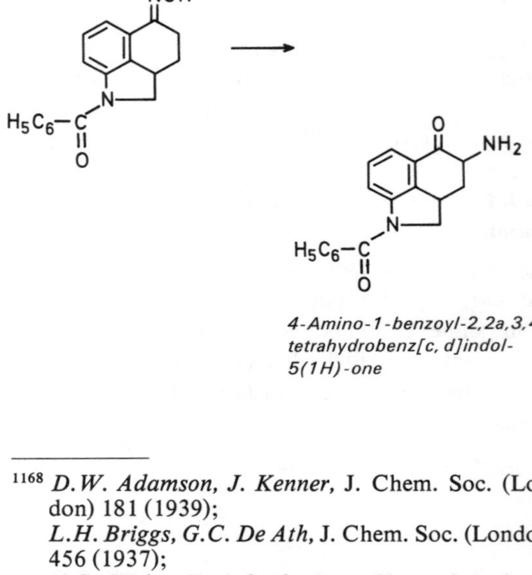

Reacting substituted acetoacetic esters with hydrazoic acid affords a convenient way of synthesizing α-amino carboxylic acids in excellent yield[1163].

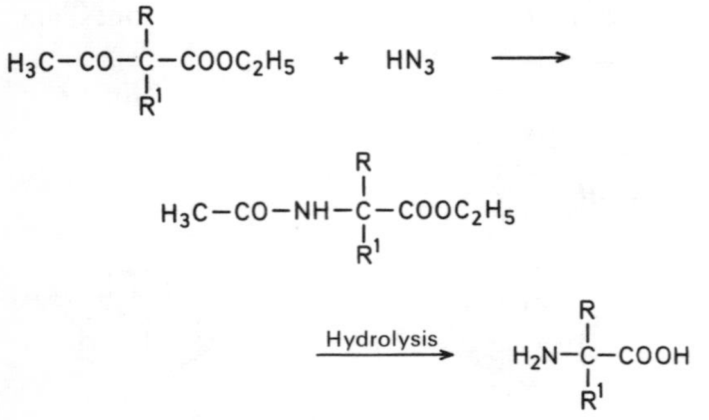

It is particularly useful for preparing α,α-*disubstituted* α-amino acids that are inaccessible by more conventional methods. Hydrazoic acid reacts with cyclic ketones like with open-chain ones to give ring-expanded lactams[1163, 1164, 1168]. The reaction conditions resemble that of a Schmidt reaction (see p. 563; also ref.[1140]).

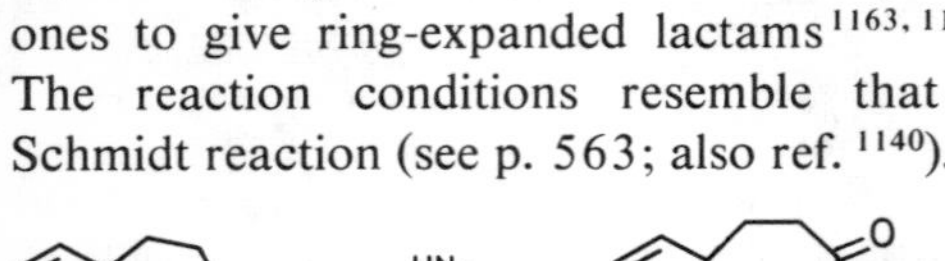

2,3,5,6-Tetrahydro-3-
benzazocin-4(1H)-one

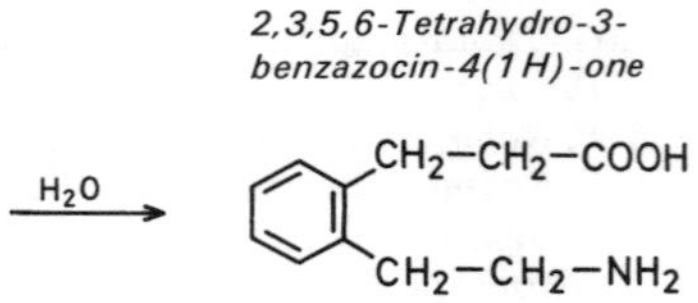

2-(2-Aminoethyl)hydro-
cinnamic acid

4-Amino-1-benzoyl-2,2a,3,4-
tetrahydrobenz[c, d]indol-
5(1H)-one

[1168] *D.W. Adamson, J. Kenner*, J. Chem. Soc. (London) 181 (1939);
L.H. Briggs, G.C. De Ath, J. Chem. Soc. (London) 456 (1937);
N.S. Hjelte, T. Agback, Acta Chem. Scand. *18*, 191 (1964);
K.F. Schmidt, Ber. *57*, 704 (1924);
DRP. 455585 (1925), Knoll AG., Erf.: *K.F. Schmidt.*

[1169] *P.W. Neber, A. Friedolsheim*, Justus Liebigs Ann. Chem. *449*, 109 (1926);
P.W. Neber, A. Uber, Justus Liebigs Ann. Chem. *467*, 52 (1928);
P.W. Neber, A. Burgard, Justus Liebigs Ann. Chem. *493*, 281 (1932).

[1170] *E.C. Kornfeld, E.J. Fornefeld, G.B. Kline, M.J. Mann, D.E. Morrison, R.G. Jones, R.B. Woodward*, J. Amer. Chem. Soc. 78, 3087 (1956).

[1171] *C. O'Brien, E.M. Philbin, S. Ushioda, T.S. Wheeler*, Tetrahedron *19*, 373 (1963).

[1166] *I. Murata*, Bull. Chem. Soc. Japan *34*, 580 (1961).
[1167] *R.G. Jones*, J. Amer. Chem. Soc. *73*, 5244 (1951).

3-Aminoflavanone; 54%

Application of the Neber rearrangement to the synthesis of stereospecific α-amino ketones has been reported in several instances[1172].

Where the oxime tosylate possesses two distinct α-methylene groups, it seems that the rearrangement proceeds so that the amino group is introduced mainly on the more electrophilic α-carbon atom.

On submitting aldoxime tosylates to Neber reaction conditions E_2 elimination of *p*-toluenesulfonic acid occurs with formation of the corresponding *nitriles* or *isonitriles*[1173].

The Neber rearrangement is usually achieved by treating an alcoholic solution or suspension of the oxime tosylate with a sodium or potassium alkoxide followed by acid hydrolysis. Using completely anhydrous conditions in the first stage of the reaction is a prerequisite for good yields.

Two other rearrangements are closely related to the Neber reaction.

Treatment of *N,N*-dichloro-*sec*-alkylamines with sodium methoxide followed with dilute hydrochloric acid gives good yields of the corresponding α-amino ketone hydrochloride[1174].

The same result is in many cases achieved by primary chlorination of a ketimine to an *N*-chloroketimine followed by a base-catalyzed rearrangement[1175]. This modification has been extended to

the synthesis of α-*amino carboxylic acids* from imino ethers[1176].

Another Neber reaction variant is the base-catalyzed rearrangement of dimethylhydrazone methiodides having an α-hydrogen to form α-*amino ketones*. Thus androst-5-ene-3β,17β-diol 17-acetate dimethylhydrazone methiodide is rearranged stereospecifically to an azirine which, on acid hydrolysis, affords *17β-amino-3β-hydroxyandrost-5-en-17α-yl methyl ketone*[1177]:

For a review see ref.[1178].

12.9.6 The Stevens and Sommelet-Hauser rearrangements

Base-catalyzed rearrangements of quaternary ammonium salts can be divided into two categories. In the Stevens rearrangement an alkyl group migrates from the quaternary nitrogen atom to

[1172] *G. Drefahl, D. Martin*, Chem. Ber. *93*, 2497 (1960).
[1173] *M.S. Hatch, D.J. Cram*, J. Amer. Chem. Soc. *75*, 38 (1953).
[1174] *H.E. Baumgarten, F.A. Bower*, J. Amer. Chem. Soc. *76*, 4561 (1954).
[1175] *H.E. Baumgarten, J.M. Petersen*, J. Org. Chem. *28*, 2369 (1963).
[1176] *H.E. Baumgarten, J.E. Dirks, J.M. Petersen, R.L. Zey*, J. Org. Chem. *31*, 3708 (1966).

[1177] *D.F. Morrow, M.E. Butler, E.C.Y. Huang*, J. Org. Chem. *30*, 579 (1965).
[1178] *C. O'Brien*, Chem. Rev. *64*, 81 (1964).

the alpha carbon atom of a second alkyl group. The Sommelet-Hauser rearrangement involves migration to the *ortho* position by a substituent of a benzyl quaternary ammonium salt. In general, rearrangements occur only in those quaternary ammonium salts that contain no beta hydrogen atom and thus cannot undergo Hofmann elimination.

12.9.6.1 Stevens rearrangement

On treatment with 10% sodium hydroxide on the steam-bath for 1 hour benzyldimethylphenacylammonium bromide gives *2-(dimethylamino)-3-phenylpropiophenone* in 90% yield[1179].

The reaction can be extended to other phenacyl[1180–1182] and acetonyl ammonium systems[1183]. It was found that salts without a carbonyl or other electron-withdrawing group, too, rearrange[1184]. Rearrangement decreases in velocity in the order:

phenacyl > propargyl > allyl > benzyl > alkyl

to reflect the decreasing stability of the potential ylide carbanions.

Electron-withdrawing substituents on the migrating group increase the reaction rate[1180, 1185], for instance, in the scheme

the reaction rates decrease in the order

X = NO$_2$ > halogen > CH$_3$ > OCH$_3$

The most important factor during the course of a Stevens reaction is ylide formation. Relatively stable nitrogen ylides such as of the phenacyl series are formed even by sodium hydroxide or sodium alkanolates. However, in the majority of cases stronger bases such as sodium amide or alkyllithium are required.

The Stevens rearrangement is an intramolecularly proceeding, yet highly stereospecific reaction.

Thus, dimethyl-(α-methylbenzyl)phenacylammonium bromide rearranges to *2-(dimethylamino)-1,3-diphenyl-1-butanone* with more than 95% retention[1186, 1187]:

With cyclic compounds either ring expansion or ring contraction can occur[1188, 1189], *e.g.*:

5,7,12,12a-Tetrahydro-isoindolo[2,1-b]iso-quinoline; 41%

N-Cyclobutenyl-N-methyl-aniline; 63%

12.9.6.2 Sommelet-Hauser rearrangement

A related rearrangement which often accompanies a Stevens rearrangement or even suppresses it was described initially by Sommelet[1190] and investigated in greater detail by Hauser[1191]. In this

[1179] *T.S. Stevens, E.M. Creighton, A.B. Gordon, M. MacNicol*, J. Chem. Soc. (London) 3193 (1928).
[1180] *T.S. Stevens*, J. Chem. Soc. (London) 2107 (1930).
[1181] *J.L. Dunn, T.S. Stevens*, J. Chem. Soc. (London) 1926 (1932).
[1182] *J.L. Dunn, T.S. Stevens*, J. Chem. Soc. (London) 279 (1934).
[1183] *T.S. Stevens, W.W. Snedden, E.T. Stiller, T. Thomson*, J. Chem. Soc. (London) 2119 (1930).
[1184] *T. Thomson, T.S. Stevens*, J. Chem. Soc. (London) 1932 (1932).
[1185] *T. Thomson, T.S. Stevens*, J. Chem. Soc. (London) 55 (1932).
[1186] *A. Campbell, A.H.J. Houston, J. Kenyon*, J. Chem. Soc. (London) 93 (1947).
[1187] *J.H. Brewster, M.W. Klyne*, J. Amer. Chem. Soc. 74, 5179 (1952).
[1188] *G. Wittig, H. Tenhaeff, W. Schoch, G. Koenig*, Justus Liebigs Ann. Chem. 572, 1 (1951).
[1189] *G. Wittig, H. Sommer*, Justus Liebigs Ann. Chem. 594, 1 (1955).
Justus Liebigs Ann. Chem. 572, 1 (1951).
[1190] *M. Sommelet*, Compt. Rend. 205, 56 (1937).
[1191] *S.W. Kantor, C.R. Hauser*, J. Amer. Chem. Soc. 73, 4122 (1951).

reaction benzyltrimethylammonium iodide rearranges to the extent of 90–95% to *2,N,N-trimethylbenzylamine* in liquid ammonia under the influence of sodium amide[1192]:

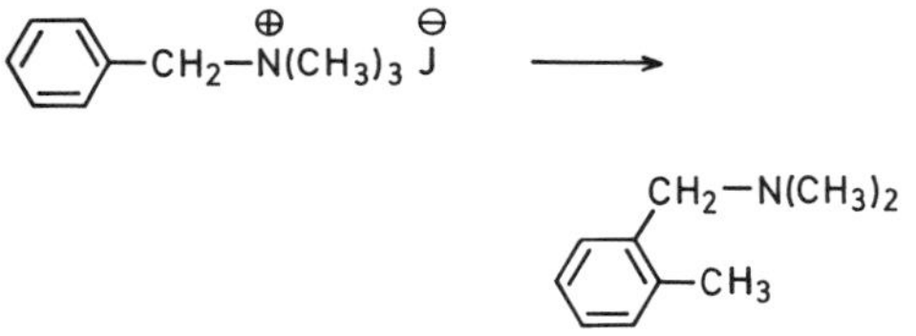

From the same starting compound polymethylbenzylamines are accessible by means of repeated Sommelet-Hauser rearrangement[1191], *e.g.*:

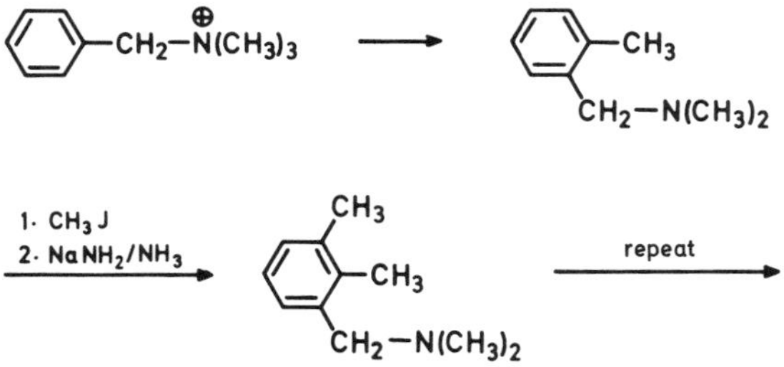

1. CH₃J
2. NaNH₂/NH₃

repeat

2,3,N,N-Tetramethyl-benzylamine; 60%

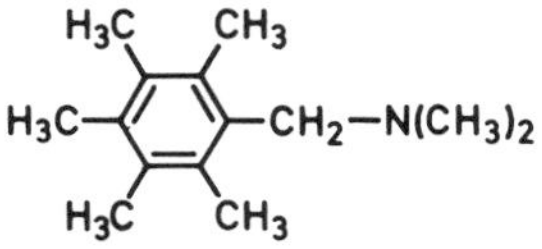

2,3,4,5,6,N,N-Heptamethyl-benzylamine

The Sommelet-Hauser rearrangement succeeds also in the naphthalene series[1193] and with many heteroaromatic compounds[1194], *e.g.*:

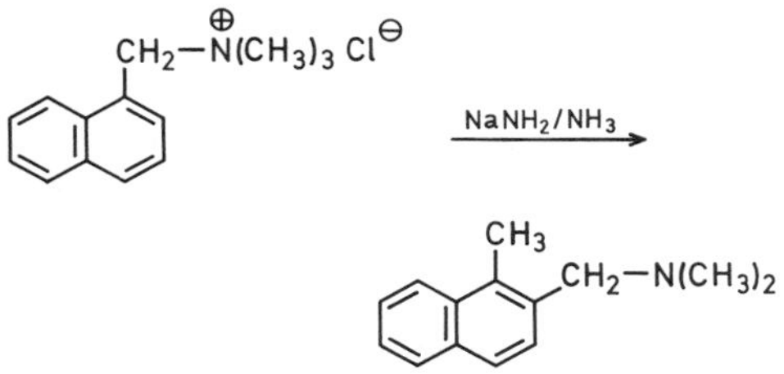

NaNH₂/NH₃

N,N-Dimethyl-(1-methyl-2-naphthyl)methylamine; 75%

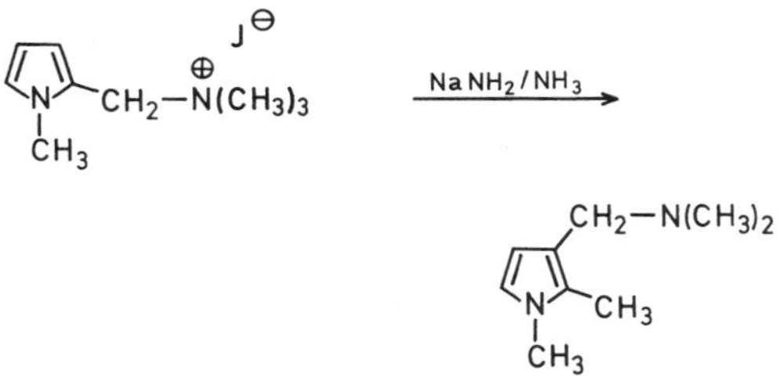

NaNH₂/NH₃

1,2-Dimethyl-3-(dimethyl-amino)methylpyrrole; 79%

In some instances ring expansion has been observed during the rearrangement[1195], *e.g.*:

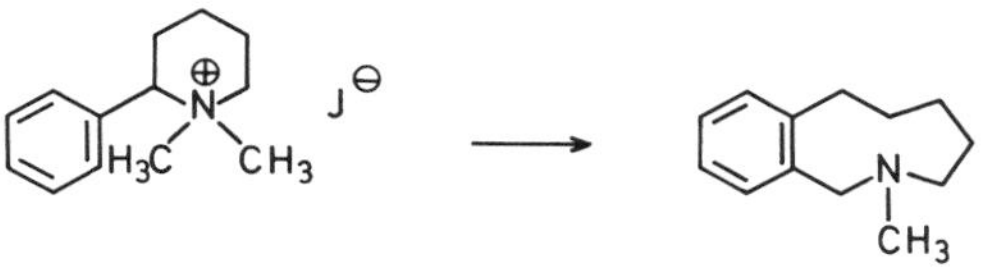

1,3,4,5,6,7-Hexa-hydro-2-methyl-2-benzazonine; 83%

Where the structure allows it Stevens and Sommelet-Hauser rearrangements of quaternary ammonium salts ensue side by side, and thus make their use in synthesis more limited. Sodium amide in liquid ammonia tends to form mainly the Sommelet-Hauser product, while alkyllithium compounds in ether or hydrocarbons favor the Stevens reaction. In general, polar solvents (dimethyl sulfoxide, hexamethylphosphoramide) and ready solubility of the ammonium salts promote the Sommelet-Hauser rearrangement, while elevated temperatures promote the Stevens reaction.

Simple tetraalkylammonium salts in liquid sodium amide-ammonia require to be treated with caution! **Explosions** *have occurred even at low temperatures*[1196].

Excellent surveys describe the Stevens and Sommelet-Hauser rearrangements[1197, 1198].

12.9.7 Chapman rearrangement

Thermal transformation of aromatic *N*-arylbenzimidates into *N*-aroyldiphenylamines is called known as the Chapman rearrangement[1199]:

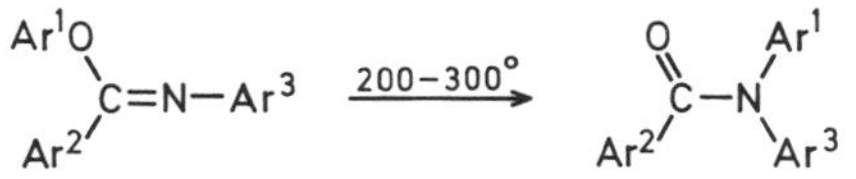

[1192] *W.R. Brasen, C.R. Hauser*, Org. Syntheses, Coll. Vol. IV, 585 (1963).
[1193] *C.R. Hauser, D.N. Van Eenam, P.L. Bayless*, J. Org. Chem. *23*, 354 (1958).
[1194] *R. Paul, S. Tchelitcheff*, Bull. Soc. Chim. France 2134 (1968).
[1195] *G.C. Jones, C.R. Hauser*, J. Org. Chem. *27*, 3572 (1962).
[1196] *W.K. Musker*, J. Org. Chem. *32*, 3189 (1967).
[1197] *S.H. Pine*, Org. Reactions *18*, 403 (1970).
[1198] *G. Wittig*, Bull. Soc. Chim. France 1921 (1971).
[1199] *A.W. Chapman*, J. Chem. Soc. (London) *127*, 1992 (1925).

It can serve as a general method for preparing diarylamines formed by alkaline hydrolysis of the primary reaction products. A large number of substituted diarylamines have been synthesized in this way. In almost every case where the Chapman rearrangement has been applied to imidates of all kinds a successful reaction and high yields have been obtained.

The conversion of *N*-phenylbenzimidates into *N,N-diphenylbenzamides* is considered to be complete after 2 hours at 270–300° and afford no side-products[1199]. For preparing unsymmetrical diarylamines the starting compounds **1** or **2** may be used; the choice is governed by the relative accessibility of the respective amine and the respective phenolic component:

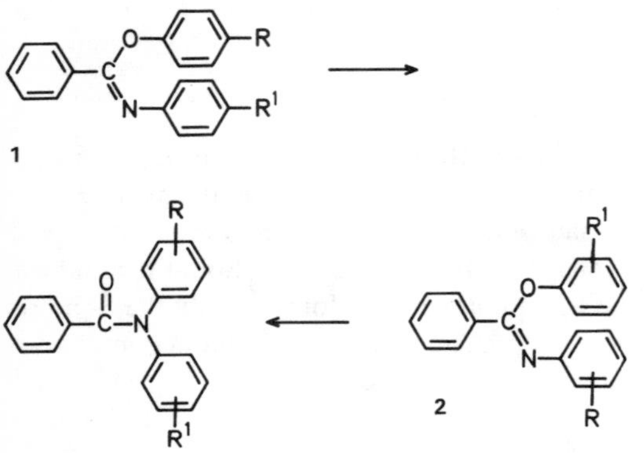

1

2

When both series of reagents are accessible the more acid phenol (*e.g.*, the one with electron-withdrawing substituents) should be chosen, because the corresponding imidate rearranges even at low temperature[1200].

A wide selection of *monohalogenated* and *polyhalogenated*[1201] and also *alkyl*-substituted[1202] diarylamines as well as those with *methoxy*[1203],

nitro[1204], *benzoyl*[1205], *cyano*[1206], and *alkoxycarbonyl* groups[1193] have been prepared with the aid of the Chapman rearrangement.

Imidates containing an aldehyde group cannot be rearranged[1207]. Imidate **3** with an *ortho* acetyl group furnishes *1,2-diphenyl-4(1H)-quinolinone*, with the normal Chapman rearrangement producing an intermediate product of this reaction[1203]:

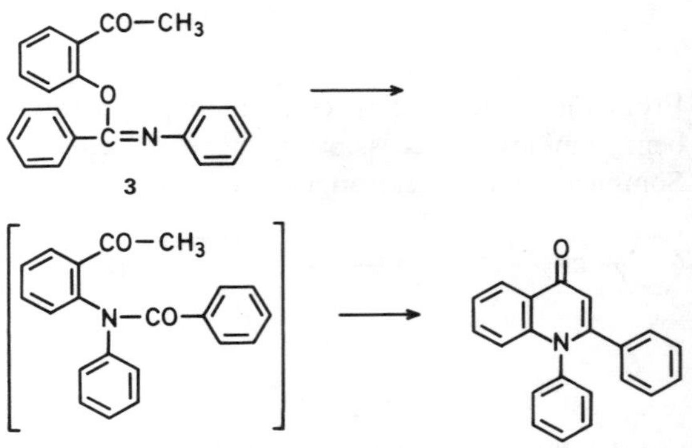

3

Chapman rearrangement of imidates whose phenolic component is derived from salicylic acid serve as starting substances for synthesizing *2-anilinobenzoic acids*[1208]:

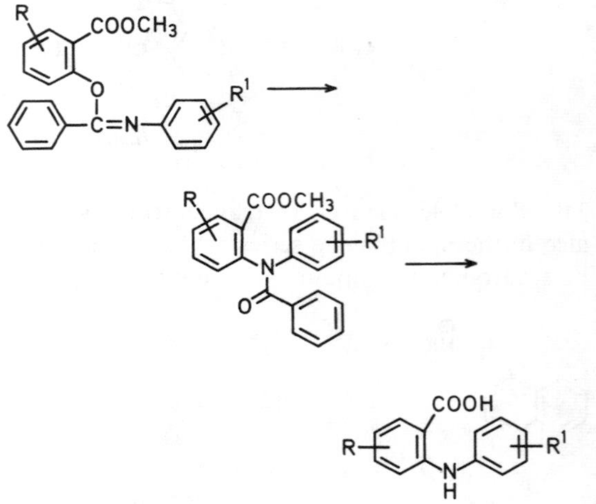

An example of an anomalous Chapman reaction has been reported[1209].

[1200] *A.W. Chapmann*, J. Chem. Soc. (London) 569 (1929).

[1201] *A.W. Chapman*, J. Chem. Soc. (London) 2458 (1930);
L.A. Elson, C.S. Gibson, J. Chem. Soc. (London) 294 (1931);
A.W. Chapman, C.H. Perrot, J. Chem. Soc. (London) 1770 (1932);
M.M. Jamison, E.E. Turner, J. Chem. Soc. (London) 1954 (1937);
P.A.S. Smith, N.W. Kalenda, J. Org. Chem. *23*, 1599 (1958).

[1202] *C.S. Gibson, J.D.A. Johnson*, J. Chem. Soc. (London) 1478, 2743 (1929).

[1203] *A.W. Chapman*, J. Chem. Soc. (London) 1743 (1927).

[1204] *D.M. Hall*, J. Chem. Soc. (London) 1603 (1948).

[1205] *M.P. Lippner, M.L. Tomlinson*, J. Chem. Soc. (London) 4667 (1956).

[1206] *J.N. Ashley, H.J. Barber, A.J. Ewins, G. Newberry, A.D.H. Self*, J. Chem. Soc. (London) 103 (1942).

[1207] *U.M. Brown, P.H. Carter, M. Tomlinson*, J. Chem. Soc. (London) 1843 (1958).

[1208] *D.M.H. Hall, E.E. Turner*, J. Chem. Soc. (London) 694 (1945).

[1209] *J.W. Schulenberg, S. Archer*, J. Amer. Chem. Soc. *82*, 2035 (1960).

Normally, the Chapman rearrangement is effected by heating the imidate without solvent. In most instances temperatures of 250–300° and reaction times of between 1 and 3 hours are employed. Occasionally, solvents (nitrobenzene, diphenyl ether, 1, 2-dichlorobenzene, or biphenyl)[1210] and triglyme and tetraglyme[1211] are used with success.

Chapman rearrangements are intramolecular reactions in which a 1→3 shift of an aryl group from oxygen to nitrogen takes place[1212]. The reaction, which proceeds via a 4-membered transition state, may be considered to be a nucleophilic attack of the nitrogen on the migrating acyl group. Electron-attracting substituents (nitro, chloro) accelerate the intramolecular, nucleophilic substitution of the aryl ring by the nitrogen. For a survey see ref.[1213].

12.9.8 Smiles rearrangement

The Smiles rearrangement[1214] is an intramolecular, base-catalyzed process which proceeds according to the following scheme

A substituent on an aromatic ring possessing a nucleophilic center two or three atoms away from the group X attached to the ring is detached and freshly tied with the nucleophilic center. The carbon atoms which join X and Y together may be either aliphatic or belong to an aromatic system. A very wide scope pertains to the Smiles rearrangement, as is made very clear by the following list (leaving group X, nucleophilic group YH).

YH	X
NHCOR	SO_2, SO, S, O
NHR	SO_2, O
NH_2	SO_2
CONHAr	SO_2, O
$CONH_2$	SO_2, O
SO_2NHR	O
SO_2NH_2	O

and many others

Most Smiles rearrangements require an electronic activation in the migrating aromatic ring. A nitro group in the ortho or para position is particularly effective[1215].

The nucleophilism of the group Y and the firmness of the bond by which the group X is attached are also vital.

Which base is used for the Smiles rearrangement is dependent on the acidity of the YH group. The solvent is determined by the solubility of the compound being rearranged. Normally, the reaction mixtures are heated to 50–100° or at reflux.

Some examples of the Smiles rearrangement are the following[1216–1218]:

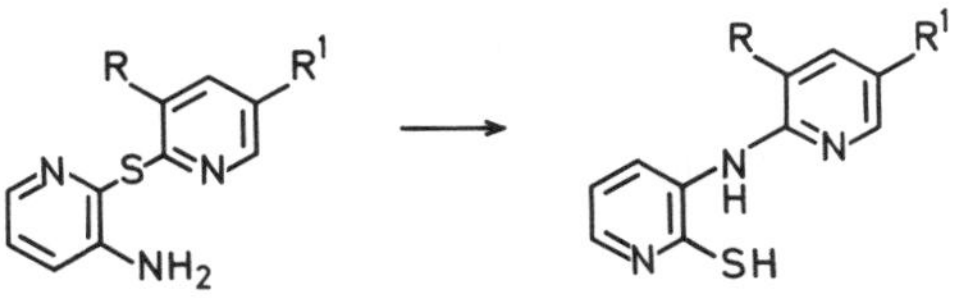

R = NO_2; R^1 = H;   3-[(3-Nitro-2-pyridyl)amino]-2-pyridinethiol; 95%

R = NO_2; R^1 = CH_3;   3-[(5-Methyl-3-nitro-2-pyridyl)-amino]-2-pyridinethiol; 82%

R^1 = NO_2; R = H;   3-[(5-Nitro-2-pyridyl)amino]-2-pyridinethiol; 96%

R^1 = NO_2; R = CH_3;   3-[(3-Methyl-5-nitro-2-pyridyl)amino]-2-pyridinethiol; 92%

2-(2-nitroanilino)benzenesulfinic acid

A reaction of 2-bromoacetanilides with amines leading to a nucleophilic displacement of an amide group by an amine has been reported[1219]:

[1210] R.C. Cookson, J. Chem. Soc. (London) 643 (1953).

[1211] O.H. Wheeler, F. Roman, M.V. Santiago, F. Quiles, Can. J. Chem. 47, 503 (1969).

[1212] O.H. Wheeler, F. Roman, O. Rosado, J. Org. Chem. 34, 966 (1969).

[1213] J.W. Schulenburg, S. Archer, Org. Reactions 18, 403 (1970).

[1214] L.A. Warren, S. Smiles, J. Chem. Soc. (London) 956 (1930).

[1215] L.A. Warren, S. Smiles, J. Chem. Soc. (London) 1040 (1932).

[1216] O.R. Rodig, R.E. Collier, R.K. Schlatzer, J. Org. Chem. 29, 2652 (1964).

[1217] C.F. Wight, S. Smiles, J. Chem. Soc. (London) 340 (1935).

[1218] L.A. Warren, S. Smiles, J. Chem. Soc. (London) 2774 (1932).

[1219] N.W. Gilman, P. Levitan, L.H. Sternbach, Tetrahedron Lett. 4121 (1970).

[1220] T. Harayama, K. Odada, S. Sekiguchi, K. Matsui, J. Heterocycl. Chem. 7, 981 (1970).

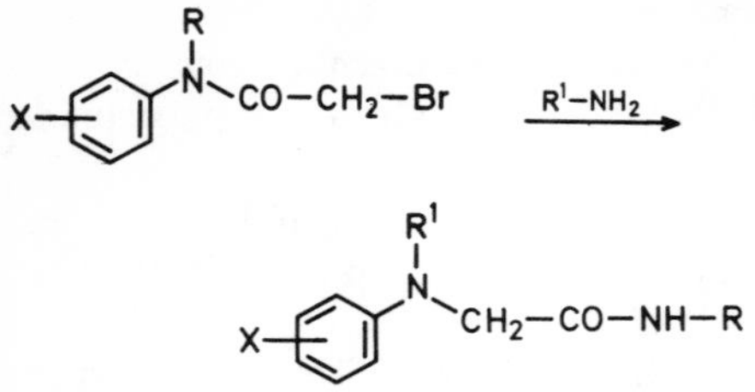

X = Electron-attracting group in *o*- or *p*-position

Ref. [1220] describes a Smiles rearrangement of a 2-nitrophenyl ether during a hydrogenation using Raney nickel; for a survey see ref. [1221].

12.9.9 Cycloammonium rearrangement (β-chloroalkylamines)

Treatment of certain β-chloroalkylamines with aqueous caustic soda readily leads to ring closure and formation of the corresponding aziridine. Treating the aziridine with excess hydrochloric acid (in aqueous ethanol) opens the aziridine ring with formation of an isomeric β-*chloroamine*[1222]:

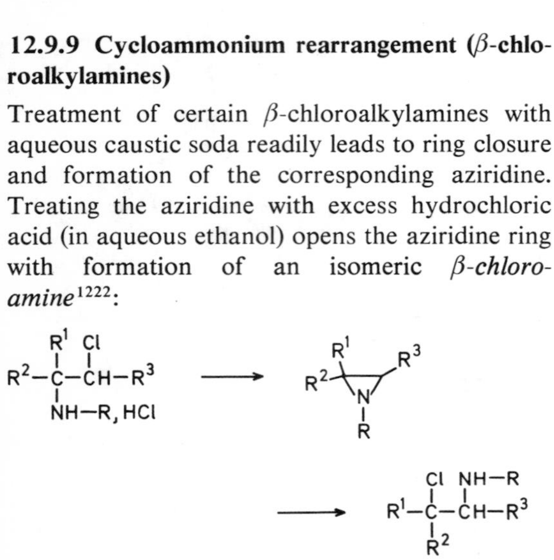

By contrast, the hydrochlorides of β-chloroalkylamines or corresponding aziridines evolve halogen hydride on melting and, in this way, are converted into hydrochlorides of unsaturated amines[1223]:

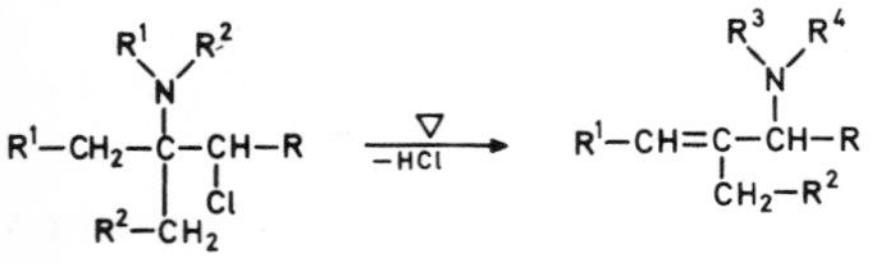

A competing side-reaction, resulting in the formation of ketones and saturated amines, occurs in most, if not all cycloammonium rearrangements and at least in part explains the varying yield of rearrangement products. Thus, 3-methyl-2-buta-

none and pyrrolidine are obtained as side-products during the preparation of *1-(1,2-dimethylallyl)pyrrolidine*:

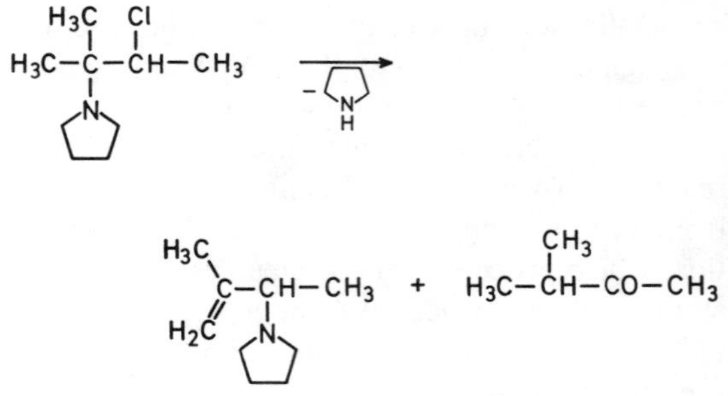

For other examples see ref. [1224].

12.10 Amines by cleavage reactions

The primary products of a whole series of amine syntheses are carboxamides (Gabriel synthesis, Ritter reaction, Leuckart-Wallace reaction, Chapman rearrangement, Beckmann rearrangement, Czerniak-Einhorn reaction, *etc.*); such syntheses therefore require a cleavage of the initially formed amides.

When organic amines are used as synthetic building blocks it is often necessary to protect amino groups to stop them undergoing unintended reactions. The protection can be accomplished by acylation or by an alternative procedure. It needs to be adapted to the reaction conditions so as to leave the protective group unaltered but, in addition, at the end of the synthesis, it must be capable of being eliminated again under conditions causing minimum changes. This aspect is particularly important during *peptide* syntheses.

During certain syntheses mixtures of primary, secondary, and tertiary amines are formed initially and, sometimes, these can be separated most easily *via* derivatives such as those obtained with nitrous acid or toluenesulfonyl chloride. The pure amines are then obtained by decomposing the mutually separated derivatives.

Numerous complex organic amines, for instance, many alkaloids, contain *N*-methyl groups. Demethylation of these groups is often a problem in structure elucidation *via* the *nor* compound or preparation of semisynthetic derivatives with other *N*-alkyl groups.

Overall, it is thus valid to say that obtaining amines by cleavage reactions is a procedure that is frequently in demand, *cf.*, the following summary:

[1221] *W.E. Truce, E.M. Kreider, W.W. Brand*, Org. Reactions *18*, 99 (1970).
[1222] *G.F. Hennion, P.E. Butler*, J. Org. Chem. *27*, 2088 (1962).
[1223] *G.F. Hennion, A.C. Hazy*, J. Org. Chem. *30*, 2650 (1965).
[1224] *Houben-Weyl*, Methoden der organischen Chemie, 4. Aufl., Bd. XI/1, Georg Thieme Verlag, Stuttgart 1957.

Amines by cleavage

Starting substances	Type of cleavage	Reaction scheme	Page
Carboxamides	Hydrolysis	$R-\overset{O}{\underset{\underset{R^2}{|}}{C}}-N-R^1 \longrightarrow R-\overset{O}{C}-OH + HN\overset{R^1}{\underset{R^2}{<}}$	576
$R-\overset{O}{\underset{\underset{R^2}{|}}{C}}-N-R^1$ R = Alkyl-, Aryl-, Alkoxy-	Alcoholysis	$R-\overset{O}{\underset{\underset{R^2}{|}}{C}}-N-R^1 \xrightarrow{H^\oplus, R^3-OH} R-\overset{O}{C}-OR^3 + HN\overset{R^1}{\underset{R^2}{<}}$	579
	Aminolysis (e.g., hydrazinolysis)	$\xrightarrow{H_2N-NH_2}$ NHR ... $\xrightarrow{HCl/H_2O}$ + $[R-\overset{\oplus}{N}H_3]\overset{\ominus}{Cl}$	579
$\langle \rangle CH_2-O-\overset{O}{\underset{\underset{R^2}{|}}{C}}-N-R^1$	Reduction a) with Pd/H$_2$ b) with Na/liq. NH$_3$	$R-NH-\overset{O}{C}-O-CH_2-\langle\rangle \longrightarrow R-NH_2 + CO_2$ $+ H_3C-\langle\rangle$	581
	With H Hal in waterfree medium	$R-NH-\overset{O}{C}-O-CH_2-\langle\rangle \xrightarrow{H Hal} R-NH_2 + CO_2$ $+ Hal-CH_2-\langle\rangle$	
Sulfonamides $R-SO_2-N\overset{R^1}{\underset{R^2}{<}}$	Hydrolysis	$R-SO_2-N\overset{R^1}{\underset{R^2}{<}} \longrightarrow HN\overset{R^1}{\underset{R^2}{<}} + R-SO_3H$	579
CH$_3$-$\langle\rangle$-$SO_2-N\overset{R^1}{\underset{R^2}{<}}$	Reductive cleavage a) with Na/n-BuOH b) with Na/liq. NH$_3$	preferably $\longrightarrow HN\overset{R^1}{\underset{R^2}{<}} +$ reduction products, e.g., toluene, SO$_2$, H$_2$S	581
Amines and quaternary ammonium compounds $[R^2-\overset{R^1}{\underset{R^3}{\overset{|}{\underset{|}{N}}}}-R^4]\,Hal^\ominus$	Dealkylation thermally	$[R^2-\overset{R^1}{\underset{R^3}{\overset{|}{\underset{|}{N}}}}-R^4]Hal^\ominus \xrightarrow{\triangledown} R^2-N\overset{R^1}{\underset{R^3}{<}} + R^4-Hal$	584
	with alkalis	$[R-\overset{CH_2-CH_2-R^1}{\underset{CH_3}{\overset{|}{\underset{|}{\overset{\oplus}{N}}}}}-CH_3]Hal^\ominus \xrightarrow{MeOH} R-N\overset{CH_3}{\underset{CH_3}{<}} + H_2O$ $+ H_2C=CH-R^1 + MeHal$	485
$[R^2-\overset{R^1}{\underset{R^3}{\overset{|}{\underset{|}{N}}}}-R^4]\,OH^\ominus$	thermally (Hofmann cleavage)	$[R^2-\overset{R^1}{\underset{R^3}{\overset{|}{\underset{|}{N}}}}-R^4]OH^\ominus \longrightarrow R^2-N\overset{R^1}{\underset{R^3}{<}} + R^4-OH$	584

Amines by cleavage (continued)

Starting substances	Type of cleavage	Reaction scheme	Page
		$^{\ominus}OH + H-\overset{\displaystyle |}{\underset{\displaystyle |}{C}}-\overset{\displaystyle |}{\underset{\displaystyle |}{C}}-\overset{\oplus}{\underset{R^3}{N}}{-}R^2 \;\longrightarrow\; R^2{-}\overset{R^1}{\underset{R^3}{N}} \;+\; \overset{}{\underset{}{C}}{=}\overset{}{\underset{}{C} }$ $+\;H_2O$	584
$\overset{R^1}{\underset{R^2}{N}}{-}CH_2{-}\bigcirc$	reductively with catalytically excited H_2	$\bigcirc{-}CH_2{-}\overset{R^1}{\underset{R^2}{N}} \;\longrightarrow\; \bigcirc{-}CH_3 \;+\; HN\overset{R^1}{\underset{R^2}{}}$ $R^2 = H$ or Alkyl	584
$\bigcirc{-}CH_2{-}\overset{\oplus}{\underset{|}{N}}{|}$ and $\bigcirc{-}CH{=}CH{-}CH_2{-}\overset{\oplus}{\underset{|}{N}}{|}$	reductively with Na/Hg (Emde degradation)	$\bigcirc{-}CH_2{-}\overset{\oplus}{\underset{|}{N}}{|} \;\longrightarrow\; \bigcirc{-}CH_3 \;+\; {-}N\overset{}{\underset{}{}}$ $\bigcirc{-}CH{=}CH{-}CH_2{-}\overset{\oplus}{\underset{|}{N}}{|} \;\longrightarrow\; \bigcirc{-}CH{=}CH{-}CH_3$ $+\; {-}N\overset{}{\underset{}{}}$	586
$R^1{-}\overset{R^2}{\underset{R^3}{N}}$	with cyanogen bromide (J. v. Braun degradation)	$R^1{-}\overset{R^2}{\underset{R^3}{N}} \;\xrightarrow{BrCN}\; \overset{R^1}{\underset{R^2}{N}}{-}CN \;+\; R^3{-}Br$ $\overset{R^1}{\underset{R^2}{N}}{-}COOH \;\longrightarrow\; \overset{R^1}{\underset{R^2}{N}}H \;+\; CO_2$	587
	by derivatives of carboxylic acids	$R^3{-}\overset{R^1}{\underset{R^2}{N}} \;+\; R{-}\overset{O}{\underset{Cl}{C}} \;\longrightarrow\; \overset{R^1}{\underset{R^2}{N}}{-}\overset{O}{\underset{R}{C}} \;+\; R^3Cl$	587
Particular starting compounds	Hydrolysis	$R{-}N{=}C{=}O \;+\; H_2O \;\longrightarrow\; R{-}NH_2 \;+\; CO_2$	582
$R{-}N{=}C{=}O$ Allyl${-}N{=}C{=}O(S)$		Allyl${-}N{=}C{=}O(S) \;+\; 2\,H_2O \;\longrightarrow$ Allyl${-}NH_2 \;+\; CO_2/H_2S$	
$\overset{R^1}{\underset{R^2}{N}}{-}NO$	Hydrolysis	$\overset{R^1}{\underset{R^2}{N}}{-}NO \;\longrightarrow\; \overset{R^1}{\underset{R^2}{N}}H \;+\; HNO_2$	
$\overset{R^1}{\underset{R^2}{N}}{-}\bigcirc{-}NO$	Hydrolysis	$\overset{R^1}{\underset{R^2}{N}}{-}\bigcirc{-}NO \;\longrightarrow\; \overset{R^1}{\underset{R^2}{N}}H \;+\; HO{-}\bigcirc{-}NO$	

12.10.1 Cleavage of carboxamides

Either *acid* or *basic* catalysis of carboxamide cleavage by hydrolysis may be employed. In many cases it is immaterial which path is chosen; the decisive criterion is which one allows the easier isolation of the amine. In other instances acid hydrolysis displays advantages over the basic technique, or *vice versa*. For example, anticipated formation of amino ketones, haloamines, or other alkali-sensitive compounds as a result of the hydrolysis makes an acid reaction medium mandatory. However, aqueous, acids, too, can initiate unintended sequential reactions which prevent certain amines from being obtained by acid carboxamide hydrolysis. An example is the cleavage of *N-tert*-butyl carboxamides which afford 2-methylpropene instead of *tert*-butylamines on acid splitting. The following rules are important in respect of estimating the rate of hydrolysis:

Table 1. Primary amines by alkaline hydrolysis of some formamides and acetamides

Starting material	Reaction conditions	Product	Yield [%]	Ref.
N-tert-Butylformamide	20% NaOH, 5-h boiling	*tert*-Butylamine	78	[1225]
N-(2-Methyl-3-phenyl-2-propyl)formamide	20% NaOH, 5-h boiling	*1,1-Dimethylphenethylamine*	89	[1225]
N-(2,4,4-Trimethyl-2-pentyl)formamide	20% NaOH, 5-h boiling	*1,1,3,3-Tetramethylbutylamine*	62	[1225]
4-Methoxy-2-nitroacetanilide	KOH + CH$_3$OH, 15 min. on water-bath	*4-Methoxy-2-nitroaniline*	95–97	[1226]
3-Nitro-4-phenylacetanilide	Boiling with dil. KOH	*3-Nitro-4-biphenylamine*	100	[1227]

1 The weaker the base the more rapidly does hydrolysis proceed
2 The following order holds for the rate of hydrolysis of various acid groups:
Formyl and trifluoroacetyl > acetyl > 2- and 4-nitrobenzoyl > benzoyl
3 Steric effects profoundly influence the rate.

12.10.1.1 Alkaline hydrolysis

Table 1 lists a number of examples of the alkaline cleavage of primary products formed in Leuckart-Wallach or Ritter reactions (formamides and acetamides).

The high yield during preparation of *tert*-butylformamide is noteworthy, it would not be obtained in acid medium. Some nitroanilines, too, are formed in high yield. In *N,N'*-(nitro-1,4-phenylene)bisacetamide the alkaline hydrolysis can be guided in such a way by varying the experimental conditions that either *4-amino-3-nitroacetanilide* or *2-nitro-1,4-phenylenediamine* is formed[1228]:

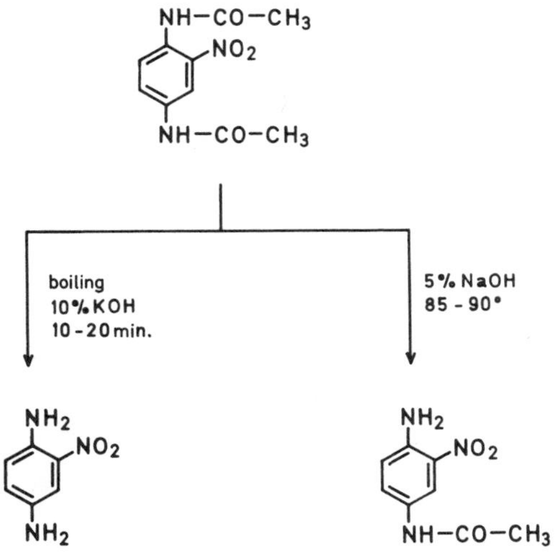

The best method of hydrolyzing 3'-(2-benzoylvinyl)acetamide to *3'-aminochalcone [3-(3-aminophenyl)-1-phenyl-2-propen-1-one]* consists in boiling with an aqueous suspension of magnesium hydroxide[1229]. Cleavage of benzamides to amines succeeds with dilute caustic soda or with a solution of potassium hydroxide in diglyme or ethylene glycol[1230, 1231].

Partial hydrolysis of *N²,N⁵-dibenzoylornithine* (ornithuric acid) with barium hydroxide affords *N²-benzoylornithine*, that with hydrochloric acid, *N⁵-benzoylornithine*[1232]. Barium hydroxide solution is also suitable for hydrolyzing *N-[N-(N-trifluoroacetylglycyl)glycyl]glycine* ethyl ester to *N,N-diglycylglycine*[1233] and L-1-acetamido-1-deoxy-*neo*-inositol to L-*neo*-inosamine[1234].

1,2,3,6-Tetrahydropyridine is formed on heating 1-acetyl-4-piperidone tosylhydrazone with potassium hydroxide in triethylene glycol[1235]. Thus, a Bamford-Stevens reaction occurs here in addition to the hydrolysis:

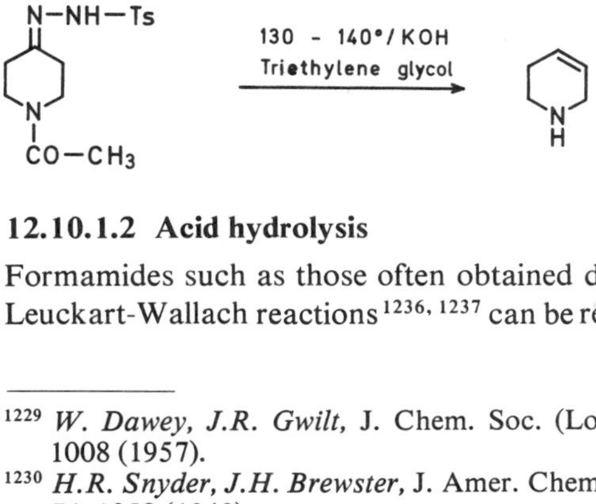

12.10.1.2 Acid hydrolysis

Formamides such as those often obtained during Leuckart-Wallach reactions[1236, 1237] can be readily

[1225] *J.J. Ritter, J. Kalish*, J. Amer. Chem. Soc. *70*, 4048 (1948).

[1226] *P.E. Fanta, D.S. Tarbell*, Org. Syntheses Coll. Vol. III, 661 (1955).

[1227] *N. Campbell, W. Anderson, J. Gilmore*, J. Chem. Soc. (London) 446 (1940).

[1228] *J.B. Polya*, J. Appl. Chem. *1*, 473 (1951).

[1229] *W. Dawey, J.R. Gwilt*, J. Chem. Soc. (London) 1008 (1957).

[1230] *H.R. Snyder, J.H. Brewster*, J. Amer. Chem. Soc. *71*, 1058 (1949).

[1231] *W.M. Lauer, R.G. Lockwood*, J. Amer. Chem. Soc. *76*, 3974 (1954).

[1232] *S.P.L. Sörensen*, Ber. *43*, 643 (1910).

[1233] *F. Weygand, W. Swodenk*, Chem. Ber. *90*, 639 (1957).

[1234] *G.R. Allen*, J. Amer. Chem. Soc. *79*, 1167 (1957).

[1235] *F. Morlacchi, M. Cardellini, F. Liberatore*, Ann. Chim. (Rome) *57*, 1456 (1967); C.A. *69*, 2817 (1968).

[1236] *A. Novelli*, J. Amer. Chem. Soc. *61*, 520 (1939).

hydrolyzed by boiling with dilute hydrochloric acid, e.g.:

N-Ethylformanilide ⟶ N-Ethylaniline; 66% of theory

N-Methyl-3-formotoluidide ⟶ N-Methyl-3-toluidine; 67% of theory

5-Chloro-N-methylformanilide ⟶ 4-Chloro-N-methylaniline; 77% of theory

However, in many cases it is possible to conduct Leuckart-Wallach reactions in such a way that *amines* are obtained *directly*[1238, 1239].

One method for preparing N-alkylanilines consists in reacting aniline with orthoformic acid triesters. N-Alkyl-N-arylformamides[1237] are first formed and these, too, are advantageously cleaved with dilute hydrochloric acid:

$$Ar-NH_2 \ + \ HC(OC_2H_5)_3 \xrightarrow{conc. \ H_2SO_4 \ / \ 115 \ - \ 180°}$$

Hydrolysis of N-(4-chloro-2,6-dimethyl-5-pyrimidinyl)formamide to *5-amino-4-chloro-2,6-dimethylpyrimidine* is accomplished best by allowing concentrated hydrochloric acid to act briefly[1240]. *Arylamine* with *nitro* and *halogen* groups, also, can be readily obtained by acid hydrolysis of the corresponding nitro- and haloacetanilides[1241-1258].

In acetamidosulfonamides the carboxamide structure can be hydrolyzed selectively with dilute sulfuric acid[1259, 1260].

In the same way *sulfanilic acid hydrazide*[1261] is obtained from 4-acetamidobenzenesulfonic acid hydrazide. Table 2 gives information about the reaction conditions and yields during hydrolysis of acetyl and benzoyl derivatives of aliphatic amines with dilute hydrochloric acid[1262-1266].

For hydrolyzing certain benzamidoanthraquinones heating with phosphoric acid is the method of choice[1267]. Pure *cis-decahydroquinoline* is ob-

Table 2. Secondary amines by acid hydrolysis of amides

Starting material	Experimental conditions	Product	Yield [% of theory]	Ref.
DL-N-(1-Acetylpentyl)acetamide	10% HCl, 100°, 2 h	DL-3-Amino-2-heptanone	65	1262
N-Acetyl-β-phenylalanine	Boiling with dil. HCl	β-Phenylalanine	85–86	1263
N-Benzylacetamide	Boiling with conc. HCl	Benzylamine	81	1264
DL-N⁶-Benzoyllysine	Boiling with dil. HCl (10 h)	DL-Lysine	76–85	1265
N,N'-Tetramethylenebisacetanilide	Boiling with dil. HCl	N,N'-Diphenyl-1,4-butanediamine	~90	1266
N,N'-Decamethylenebisacetanilide	Boiling with dil. HCl	N,N'-Diphenyl-1,10-decanediamine	~90	1266

[1237] R.M. Roberts, P.J. Vogt, Org. Synth. 38, 29 (1958).
[1238] A.N. Kost, A.P. Terentěv, G.A. Shvekhgeimer, Izv. Akad. Nauk SSR, Otdel. Khim. Nauk 150 (1951); C.A. 45, 10194 (1951).
[1239] A.W. Ingersoll, J.H. Brown, C.K. Kim, W.D. Beauchamp, G. Jennings, J. Amer. Chem. Soc. 58, 1808 (1936).
[1240] R. Hull, B.J. Lovell, H.T. Openshaw, A.R. Todd, J. Chem. Soc. (London) 41 (1947).
[1241] E. Nölting, A. Collin, Ber. 17, 261 (1884).
[1242] H.H. Hodgson, F. Heyworth, J. Chem. Soc. (London) 1624 (1949).
[1243] J.B. Cohen, H.D. Dakin, J. Amer. Chem. Soc. 79, 1127 (1901).
[1244] J.C. Howard, Org. Synth. 35, 3 (1955).
[1245] A.F. Holleman, J. ter Weel, Rec. Trav. Chim. Pays-Bas 35, 46 (1915).
[1246] Fr. P. 800343 (1936), I.G. Farbenindustrie; Chem. Zentr. II, 2612 (1936).
[1247] J.R. Johnson, L.T. Sandborn, Org. Syntheses, Coll. Vol. I, 111 (1941).
[1248] J. Scott, R. Robinson, J. Chem. Soc. (London) 121, 844 (1922).
[1249] C.W. James, J. Kenner, W.V. Stubbings, J. Chem. Soc. (London) 117, 773 (1920).
[1250] L.H. Welsh, J. Amer. Chem. Soc. 63, 3276 (1941).
[1251] K.H. Pausacker, J.G. Scroggie, J. Chem. Soc. (London) 1897 (1955).
[1252] O.L. Brady, J.N.E. Day, W.J.W. Rolt, J. Chem. Soc. (London) 121, 526 (1922).
[1253] W.L. Mosby, J. Amer. Chem. Soc. 76, 936 (1954).

Table 3. Primary amines by hydrolysis of amides with methanolic or ethanolic hydrogen chloride

Starting materials	Reaction conditions	Product	Yield [% of theory]	Ref.
3'-(Cyanomethyl)acetanilide	CH_3OH + HCl, Boiling 8 h	*Methyl (3-aminophenyl)acetate*	77	[1269]
3'-Cyanoacetanilide	C_2H_5OH + HCl, Boiling 2 h	*3-Aminobenzonitrile*	95	[1270]
4'-Cyanoacetanilide	C_2H_5OH + HCl, Boiling 2.5 h	*4-Aminobenzonitrile*	90	[1270]
Methyl 3-(acetamido)benzoate	CH_3OH + HCl, Boiling 3 h	*Methyl 3-aminobenzoate*	98	[1270]
Methyl 2-acetamido-3-nitrobenzoate	C_2H_5OH + HCl, Boiling 3 h	*Methyl 2-amino-3-nitrobenzoate*	100	[1270]
N-(2-Hydroxy-5-nitrobenzyl)-benzamide	C_2H_5OH + HCl	*2-(Aminomethyl)-4-nitrophenol*	?	[1271]

tained when *cis*-1-benzoyldecahydroquinoline is treated with hydrogen chloride in 1,4-dioxane. Any *trans*-amide impurities in the starting material are left behind unchanged[1268].

12.10.1.3 Alcoholysis and aminolysis

Alcoholytic cleavage, generally performed with alcoholic hydrochloric acid, is often a more conservative method than hydrolysis of carboxamides. Using this procedure, additionally present ester groups, which otherwise are hydrolyzed more rapidly than carboxamides, tend to remain intact; sometimes this holds for cyano groups, too. In particular instances selective cleavage of one of several carboxamide groups succeeds. Acyl migrations by acid catalysis giving aminoalkyl esters from *N*-hydroxyalkylacetamides (acetamido alcohols) are intramolecular alcoholyses of carboxamides. Table 3 gives some examples of reaction conditions and yields[1269-1271].

Carboxamides that are particularly resistant to hydrolysis have been cleaved by heating with a mixture of methanol and boron trifluoride[1272], *e.g.*:

2-Nitroacetanilide ⟶ *2-Nitroaniline*

3-Nitroacetanilide ⟶ *3-Nitroaniline*

2,6-Dimethylacetanilide ⟶ *2,6-Dimethylaniline*

1-(3,4-Dinitronaphthyl)acetamide ⟶ *3,4-Dinitro-1-naphthylamine*

Preparation of *amines* from their acetyl derivatives by heating with alcohols and alcoholates succeeds in some cases where other methods of hydrolysis fail [*e.g.*, with 4'-ethoxy-2',6'-dinitroacetanilide or 1-(4-nitronaphthyl)acetamide][1273]. For cleaving phthalimide groups obtained as primary product in, for instance, the Gabriel synthesis, aminolysis with *hydrazine* is the method of choice[1274].

12.10.2 Cleavage of sulfonamides

12.10.2.1 Acid cleavage

Cleavage of sulfonamides is preparatively important mainly for two reasons:

[1254] *J.H. Kahn, V. Petrow, E.L. Rewald, B. Sturgeon,* J. Chem. Soc. (London) 2128 (1949).

[1255] *V. Veselý, J. Páč,* Collect. Czech. Chem. Commun. *2*, 471 (1930).

[1256] *P. Santurri, F. Robbins, R. Stubbings,* Org. Synth. *40*, 18 (1960).

[1257] *C.M. Atkinson, C.W. Brown, J. McIntyre, J.C.F. Simpson,* J. Chem. Soc. (London) 2023 (1954).

[1258] *H. Gilman, S. Avakian,* J. Amer. Chem. Soc. *68*, 1514 (1946).

[1259] *J.J. Craig, W.E. Cass,* J. Amer. Chem. Soc. *64*, 783 (1942).

[1260] *J.H. Gorvin,* J. Chem. Soc. (London) 736 (1945).

[1261] *T. Curtius, W. Stoll,* J. Prakt. Chem. [2] *112*, 117 (1926).

[1262] *F.E. Lehmann, A. Bretscher, H. Kühne, E. Sorkin, M. Erne, H. Erlenmeyer,* Helv. Chim. Acta *33*, 1217 (1950).

[1263] *R.M. Herbst, D. Shemin,* Org. Syntheses, Coll. Vol. II, 491 (1943).

[1264] *M.A. Phillips,* J. Soc. Chem. Ind. *66*, 325 (1947).

[1265] *J.C. Eck, C.S. Marvel,* Org. Syntheses, Coll. Vol. II, 374 (1943).

[1266] *J.H. Billman, L.R. Caswell,* J. Org. Chem. *16*, 1041 (1951).

[1267] *W. Bradley, J.V. Butcher,* J. Chem. Soc. (London) 2311 (1954).

[1268] *E.A. Mistryukov, V.F. Kucherov,* Izv. Akad. Nauk SSR, Otdel. Khim. Nauk 1345 (1961); C.A. *56*, 2423 (1962).

[1269] *H. Plieninger,* Chem. Ber. *87*, 228 (1954).

[1270] *J.P. van Roon, P.E. Verkade, B.M. Wepster,* Rec. Trav. Chim. Pays-Bas *70*, 1105 (1951).

[1271] *A. Einhorn,* Justus Liebigs Ann. Chem. *343*, 243 (1905).

[1272] *L. Sihlbom,* Acta Chem. Scand. *8*, 529 (1954).

[1273] *P.E. Verkade, P.H. Witjens,* Rec. Trav. Chim. Pays-Bas *62*, 201 (1943).

[1274] *L.I. Smith, O.H. Emerson,* Org. Syntheses, Coll. Vol. III, 151 (1955); *J.C. Sheehan, W.A. Bolhofer,* J. Amer. Chem. Soc. *72*, 2786 (1950).

Table 4. Amines by cleaving sulfonamides with hydrobromic acid and phenol

Amide		Amine	Yield [% of theory]	Ref.
$H_3C-SO_2-NH-C_6H_5$		*Aniline*	83	[1283]
$H_5C_6-SO_2-NH-R$	$R = -CH_2-CH_2-C_6H_5$	*2-Phenylethylamine*	55	[1283]
	$= 2\text{-Naphthyl-}$	*1-Naphthylamine*	84	[1283]
	$= 4-NO_2-C_6H_4$	*4-Nitroaniline*	90	[1283]
	$= 3-NO_2-C_6H_4$	*3-Nitroaniline*	40	[1284]
$4-CH_3-C_6H_4-SO_2-X$	$X = 4-H_3C-C_6H_4-NH-$	*4-Methylaniline*	84	[1284]
	$= H_5C_6-N-$ $\quad\quad\; CH_3$	*N-Methylaniline*	65	[1284] [1286]
	$= (H_3C)_2N-$	*Dimethylamine*	52	[1284]

1 For separating mixtures of primary, secondary, and tertiary amines *via* sulfonamides as intermediate product.

2 For eliminating protective sulfonamide groups from peptides while keeping the carboxamide group intact.

Very severe reaction conditions are required for cleaving sulfonamides hydrolytically. With concentrated sulfuric acid the reaction is usually effected by heating in a sealed tube for several hours[1275-1277]. Boiling with dilute hydrochloric acid is successful in exceptional cases only[1278-1280]; adding acetic acid may enhance the solubility of the sulfonamides in the aqueous acid[1281-1282].

Boiling in a mixture of 48% *hydrobromic acid* and phenol ranks as a convenient method for cleaving arylsulfonamides. The addition of phenol aims not only at solubilizing the sulfonamide but also to trap the bromine liberated during cleavage. For relatively unstable amines, splitting the arylsulfonic acid derivatives with 25–30% hydrogen bromide in glacial acetic acid with added phenol at room temperature is recommended[1283-1286] (Table 4).

Older work describes the hydrolysis of sulfonamides by heating with dilute *sulfuric acid*[1288, 1289]. By contrast, heating with concentrated sulfuric acid or with a mixture of sulfuric and acetic acids is recommended for hydrolyzing *N*-(2-benzoyl-phenyl)toluenesulfonamide to *2'-aminobenzophenone*[1290, 1291].

N-(2, 4-Dinitro-1-naphthyl)toluenesulfonamide is converted quantitatively into *2,4-dinitro-1-naphthylamine* with concentrated sulfuric acid even at 20°[1292].

The importance of the correct choice of the correct reaction conditions is demonstrated by the example of *N*-(4-anisyl)-*N*-methyl-4-toluenesulfonamide. While normal hydrolysis takes place with 60% sulfuric acid at ~20° with formation of *N-methyl-4-anisidine*, reaction with 70% sulfuric acid at 120° gives rearrangement to *2-(4-methoxyphenylsulfonyl)-N-methyl-4-toluidine*[1293].

[1275] *P. Ruggli, G. Geiger*, Helv. Chim. Acta *30*, 2035 (1947).

[1276] *G.E. McCasland, S. Proskow*, J. Amer. Chem. Soc. *76*, 6087 (1954).

[1277] *A. Müller, E. Šrepel, E.F. Fritzsche, F. Dicher*, Monatsh. Chem. *83*, 386 (1952).

[1278] *R.S. Schreiber, R.L. Shriner*, J. Amer. Chem. Soc. *56*, 1618 (1934).

[1279] *D. Klamann, G. Hofbauer*, Monatsh. Chem. *84*, 62 (1953).

[1280] *P. Hemmerich, S. Fallab, H. Erlenmeyer*, Helv. Chim. Acta *39*, 1242 (1956).

[1281] *T.L. Fletcher, M.E. Taylor, A.W. Dahl*, J. Org. Chem. *20*, 1021 (1955).

[1282] *W.S. Johnson, E.L. Woroch, B.G. Buell*, J. Amer. Chem. Soc. *71*, 1901 (1949).

[1283] *H.R. Snyder, R.E. Heckert*, J. Amer. Chem. Soc. *74*, 2006 (1952).

[1284] *D.I. Weisblat, B.J. Magerlein, D.R. Myers*, J. Amer. Chem. Soc. *75*, 3630 (1953).

[1285] *J. Bornstein, S.C. Lashua, A.P. Boisselle*, J. Org. Chem. *22*, 1255 (1957).

[1286] DBP. 830 791 (1950), Upjohn, Erf.: *D.J. Weisblat, B.J. Mayerlein, D.R. Myers;* Chem. Zentr. 7143 (1953).

[1287] *R.C. Fuson, R. Jaunin*, J. Amer. Chem. Soc. *76*, 1171 (1954).

[1288] *W.R. Boon*, J. Chem. Soc. (London) 307 (1947).

[1289] *G.W.H. Cheeseman*, J. Chem. Soc. (London) 3308 (1955).

[1290] *H.J. Scheifele, D.F. De Tar*, Org. Synth. *32*, 8 (1952).

[1291] *F. Ullmann, H. Bleier*, Ber. *35*, 4243 (1902).

The possibility of cleaving *N*-alkyl- and *N*,*N*-dialkyl-*p*-toluenesulfonamides hydrolytically with chlorosulfonic acid at temperatures above 100° is probably not very important procedurally [1294, 1295]. The initial product is an *alkylamino-* or *dialkyl-aminosulfuric acid* which needs to be subsequently hydrolyzed with dilute alkali hydroxide to the free *amine*:

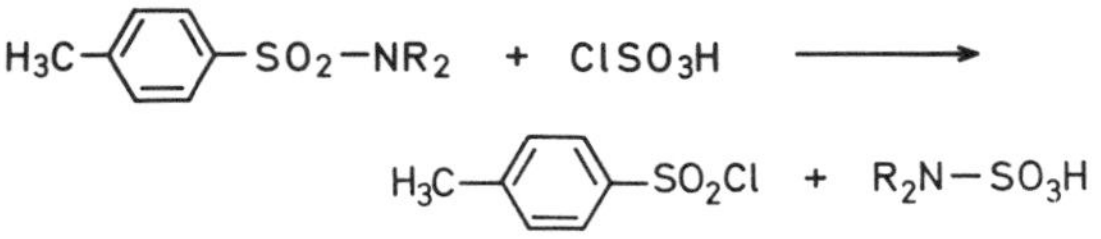

$$H_3C-\langle\rangle-SO_2-NR_2 \ + \ ClSO_3H \longrightarrow$$

$$H_3C-\langle\rangle-SO_2Cl \ + \ R_2N-SO_3H$$

$$R_2N-SO_3H + 2\,NaOH \longrightarrow R_2NH + Na_2SO_4 + H_2O$$

12.10.2.2 Reductive cleavage

Reductive cleavage of sulfonamides is a more conservative technique than their hydrolysis. An example is the reduction of *N*-alkyltoluenesulfonamides by heating with *sodium butoxide* or *isopentyloxide* in corresponding alcohols [1296-1301]. The tosyl groups forms toluene, hydrogen sulfide, and sulfur dioxide (see Table 5).

It is noteworthy that during splitting of *N*-allyl-*N*-methyltoluenesulfonamide the allyl double bond is not reduced [1300].

Table 5. Amines by cleaving sulfonamides with sodium in boiling alcohols
B = Butyl alcohol   A = Isopentyl alcohol

4−H₃C−C₆H₄−SO₂−X	Alcohol	Amine	Yield [% of th.]	Ref.
−N(C₂H₅)(C₆H₅)	A	*N-Ethylaniline*	93	1299
−N(C₆H₁₃)(C₆H₅)	A	*N-Hexylaniline*	80	1299
−N(C₁₂H₂₅)(C₆H₅)	A	*N-Dodecylaniline*	91	1299
−N(CH₃)(CH₂−CH=CH₂)	B	*N-Methylallylamine*	48	1300
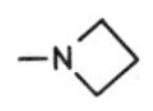 −N⟨ ⟩	A	*Azetidine*		1301

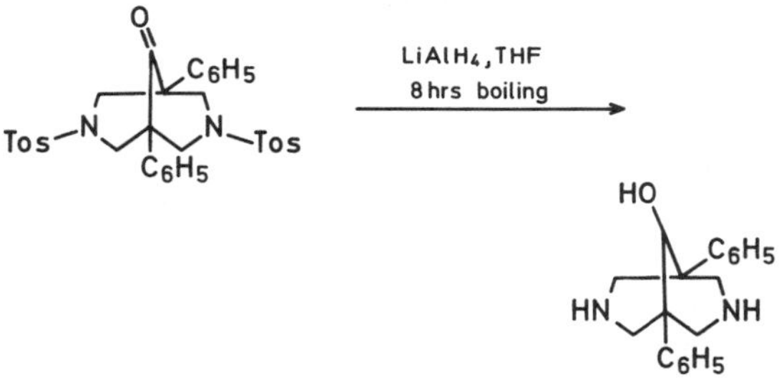

1292 H.H. Hodgson, S. Birtwell, J. Chem. Soc. (London) 433 (1943).
1293 J. Halberkann, Ber. *54*, 1665 (1921).
1294 W. Marckwald, A. Droste-Huelshoff, Ber. *31*, 3261 (1898).
1295 DRP. 634 687 (1936), Erf.: G. Schroeter; C.A. *31*, 710 (1937).

Reductive elimination of sulfonamide groups with *sodium* in *liquid ammonia* is a highly esteemed technique in peptide chemistry because it produces no racemization and existing peptide groups remain intact [1302-1306] (cleavage with hydrogen bromide furnishes inferior results). Removing the protective sulfonamide groups by reduction with *Raney nickel* [1306] is an alternative possibility.

An up-to-date method for splitting sulfonamides consists in allowing *sodium* in naphthalene and *1,2-dimethoxyethane* [1307, 1308] to act. The reaction is performed at room temperature under nitrogen or argon.

In some cases good results are obtained on cleaving sulfonamides with *lithium tetrahydroaluminate*, for example, on detosylating *N*-methyl-*N*-[5-(1-methyl-3-piperidyl)pentyl]-*p*-toluene sulfonamide to *1-methyl-3-(5-methylaminopentyl)-piperidine* [1309]. Similarly, 1,5-diphenyl-3,7-ditosyl-3,7-diazabicyclo [3.3.1]nonan-9-one [1310] is cleaved to *1,5-diphenyl-3,7-diazabicyclo[3.3.1]-nonan-9-ol (1,5-diphenylbispidin-9-ol):*

1296 C.C. Howard, W. Marcwald, Ber. *32*, 2031 (1899).
1297 L.S. Fosdick, O. Fancher, K.F. Urbach, J. Amer. Chem. Soc. *68*, 840 (1946).
1298 H. Stetter, Chem. Ber. *86*, 197 (1953).
1299 D. Klamann, G. Hofbauer, Chem. Ber. *86*, 1246 (1953).
1300 A.W. Weston, A.W. Ruddy, C.M. Suter, J. Amer. Chem. Soc. *65*, 676 (1943).
1301 F.C. Schaefer, J. Amer. Chem. Soc. *77*, 5928 (1955).
1302 P. Karrer, K. Ehrhardt, Helv. Chim. Acta *34*, 2202 (1951).
1303 J. Rudinger, Collect. Czech. Chem. Commun. *19*, 375 (1954).
1304 J.M. Swan, V. du Vigneaud, J. Amer. Chem. Soc. *76*, 3110 (1954).
1305 S. Guttmann, J. Pless, R.A. Boissonnas, Helv. Chim. Acta *45*, 170 (1962).
1306 H.B. Milne, C.H. Peng, J. Amer. Chem. Soc. *79*, 639 (1957).
1307 W.D. Closson, P. Wriede, J. Amer. Chem. Soc. *89*, 5311 (1967).
1308 J. Jacobus, M. Raban, K. Mislow, J. Org. Chem. *33*, 1142 (1968).

Allowing lithium tetrahydroaluminate to act on 1-(p-tolylsulfonyl)-2,3-piperidinedimethanol di-p-toluenesulfonate[1311] and the isomeric 3,4-ester[1312] gives the corresponding *2,3-* and *3,4-dimethyl-piperidines*. It is interesting that 1-(p-tolylsulfo-nyl)-2-piperidinemethanol p-toluenesulfonate under the same conditions loses only the tosyloxymethyl group[1313]. Liberation of *N-ethylaniline* from N-ethyl-N-phenyl-p-toluenesulfonamide is effected by boiling with lithium tetrahydroaluminate in dibutyl ether[1314].

12.10.3 Cleavage of ureas, isothiocyanates, urethans, and cyanamides

Some methods for preparing amines (*e.g.*, Curtius degradation) lead to the initial formation of isocyanates, ureas, or urethans, from which the amines must be liberated subsequently. In individual cases *isothiocyanates* are appropriate starting compounds for synthesizing certain amines. Ureas are generally hydrolyzed with sodium hydroxide in ethylene glycol[1315], with ammonia at 150°[1316], or by heating with phthalic anhydride and hydrazinolysis of the N-alkylphthalimide[1317] formed.

Among the rare cases where amines are obtained by hydrolysis of isothiocyanates (mustard oils) is the preparation of *allylamine* from allyl isothiocyanate by heating with 20% hydrochloric acid[1318], and of *tert-butylamine* from *tert*-butyl isothiocyanate[1319] with 94% formic acid. An alternative possible route is addition of hydrogen sulfide in alkaline solution and subsequent hydrolysis of the dithiocarbonic acid derivative formed in this way[1320].

$$R-N=C=S \xrightarrow{H_2S/NaOH} R-NH-C\underset{S-Na}{\overset{S}{\big\langle}}$$

$$\xrightarrow{2HCl} R-NH_2 \cdot HCl + CS_2 + NaCl$$

Hydrogen chloride[1321, 1322], possibly in the presence of acetic acid[1323] or ethanol[1324], can be employed to convert urethans into amines.

$$R-NH-COOR^1 + 2HX \longrightarrow R-NH_2 \cdot HX$$
$$+ CO_2 + R^1-X$$

$$R-NH-COOR^1 + HX + H_2O \longrightarrow$$

$$R-NH_2 \cdot HX + CO_2 + R^1-OH$$

Splitting urethans with hydrogen bromide[1325-1327] or a solution of hydrogen bromide in either glacial acetic acid[1328-1330] or 1,4-dioxane[1331], or nitromethane[1332] is also very frequently made use of.

Selective decarbobenzoxylation with hydrogen bromide[1325-1331] occupies a special position, *e.g.*:

$$\begin{array}{c} CO-NH_2 \\ | \\ S-CH_2-CH-NH-CO-O-CH_2-C_6H_5 \\ | \\ S-CH_2-CH-NH-CO-O-CH_2-C_6H_5 \\ | \\ CO-NH_2 \end{array} \xrightarrow{HBr/-80°}$$

$$\begin{array}{c} CO-NH_2 \\ | \\ S-CH_2-CH-NH_2 \cdot HBr \\ | \\ S-CH_2-CH-NH_2 \cdot HBr \\ | \\ CO-NH_2 \end{array}$$

Cystinamide; 83%

$$\underset{N}{\overset{CH_2-CH-CO-NH-CH-CO-OC_2H_5}{}} \xrightarrow{HBr}$$

Benzyl 5{2-amino-2{[a-(ethoxycarbonyl)phenethyl]-carbamoyl}ethyl}imidazole-1-carboxylate

[1309] R. Lukeś, J. Kovář, Collect. Czech. Chem. Commun. *19*, 1215 (1954).

[1310] H. Stetter, J. Schäfer, K. Dieminger, Chem. Ber. *91*, 598 (1958).

[1311] A. Šilhánková, D. Doskočilová, M. Ferles, Collect. Czech. Chem. Commun. *34*, 1976 (1969).

[1312] A. Šilhánková, M. Ferles, Collect. Czech. Chem. Commun. *34*, 3186 (1969).

[1313] H. Ripperger, K. Schreiber, Tetrahedron *21*, 1485 (1965).

[1314] D. Klamann, Monatsh. Chem. *84*, 651 (1953).

[1315] D.E. Pearson, J.F. Baxter, K.N. Carter, Org. Syntheses, Coll. Vol. III, 148 (1951).

[1316] D.F. Kupetov, Z.G. Vukolova, Zh. Obshch. Khim. *24*, 698 (1954); C.A. *49*, 5341 (1955).

[1317] L.I. Smith, O.H. Emerson, Org. Syntheses, Coll. Vol. III, 151 (1955).

[1318] M.T. Leffler, Org. Syntheses, Coll. Vol. II, 24 (1943).

[1319] E. Schmidt, W. Striewsky, M. Seefelder, F. Hitzler, Justus Liebigs Ann. Chem. *568*, 192 (1950).

[1320] DBP. 845516 (1940), BASF, Erf.: W. Stade; Chem. Zentr. 1882 (1953).

[1321] L.A. Carpino, D.E. Barr, J. Org. Chem. *31*, 764 (1966).

[1322] P.A.S. Smith, Org. Synth. *36*, 69 (1956).

[1323] W. Siefken, Justus Liebigs Ann. Chem. *562*, 102 (1949).

In individual cases hydrogen iodide[1333], phosphonium iodide[1334, 1335] hydrofluoric acid[1336], and trifluoroacetic acid[1337], and also alcoholic caustic soda[1338], caustic potash[1339, 1340], calcium hydroxide[1341], or barium hydroxide[1342], too, may be used for cleaving urethans.

Palladium catalysts[1343–1346] or sodium in liquid ammonia[1347–1351] cleave benzylurethan groups hydrogenolytically:

$$R-NH-CO-OCH_2-C_6H_5 \ + \ H_2 \longrightarrow$$

$$R-NH_2 \ + \ CO_2 \ + \ C_6H_5-CH_3$$

β-Cyanoalanine is obtained on reacting N-benzyloxycarbonyl-β-cyanoalanine with sodium in liquid ammonia, while sodium in methanol reduces the nitrile group as well[1351] (giving *2,4-diaminobutyric acid*):

3-Cyanoalanine

Hydrogenolytic cleavage of α-piperidyloxycarbonyl amino acids has also been described[1352], *e.g.:*

Phenylalanine; 95%

Cleaving allyl ester groups hydrolytically has been proposed as a method in peptide chemistry[1353].

To avoid undesired additional reactions, it may be useful in individual cases during urethan cleavage to react the latter with phthalic anhydride first and to convert the phthalimides obtained in this way into the free amines by hydrazinolysis[1354], *e.g.:*

1,8-Octanediamine

[1324] *A.E. Barkdoll, W.F. Ross*, J. Amer. Chem. Soc. *66*, 951 (1944).

[1325] *M. Brenner, H.C. Curtius*, Helv. Chim. Acta *46*, 2126 (1963).

[1326] *G.B. Brown, B.R. Baker, S. Bernstein, S.R. Safir*, J. Org. Chem. *12*, 155 (1947).

[1327] *V. Seidlová, M. Protiva*, Collect. Czech. Chem. Commun. *32*, 2826 (1967).

[1328] *D. Ben-Ishai, A. Berger*, J. Org. Chem. *17*, 1564 (1952).

[1329] *R.A. Boissonnas, G. Preitner*, Helv. Chim. Acta *36*, 875 (1953).

[1330] *W.O. Foye, L. Chafetz, E.G. Feldmann*, J. Org. Chem. *22*, 713 (1957).

[1331] *K. Inouye, H. Otsuka*, J. Org. Chem. *26*, 2613 (1961).

[1332] *N.F. Albertson, F.C. McKay*, J. Amer. Chem. Soc. *75*, 5323 (1953).

[1333] *E. Waldschmidt-Leitz, K. Kühn*, Chem. Ber. *84*, 381 (1951).

[1334] *E. Katchalski, I. Grossfeld, M. Frankel*, J. Amer. Chem. Soc. *68*, 879 (1946).

[1335] *W. Lautsch, H.J. Kraege*, Chem. Ber. *89*, 737 (1956).

[1336] *L.A. Carpino*, J. Amer. Chem. Soc. *79*, 98 (1957).

[1337] *F. Weygand, W. Steglich*, Z. Naturforsch. B *14*, 472 (1959).

[1338] *E. Magnien, R. Baltzly*, J. Org. Chem. *23*, 2029 (1958).

[1339] *E. Sawicki, F.E. Ray, V. Glocklin*, J. Org. Chem. *21*, 243 (1956).

[1340] *P. Beranger, J. Levisalles*, Bull. Soc. Chim. France 704 (1957).

[1341] *A. Dornow, O. Hahmann*, Arch. Pharm. (Weinheim, Ger.) *290*, 20 (1957).

[1342] *R. Robinson, W.M. Todd*, J. Chem. Soc. (London) 1743 (1939).

[1343] *M. Bergmann, L. Zervas*, Ber. *65*, 1192 (1932).

[1344] *E.D. Nicolaides, R.D. Westland, E.L. Wittle*, J. Amer. Chem. Soc. *76*, 2887 (1954).

[1345] *E. Wünsch, A. Zwick*, Hoppe-Seylers Z. Physiol. Chem. *333*, 108 (1963).

[1346] *W.J. Schut, H.I.X. Mager, W. Berends*, Rec. Trav. Chim. Pays-Bas *82*, 282 (1963).

[1347] *R.H. Sifferd, V. du Vigneaud*, J. Biol. Chem. *108*, 753 (1935).

[1348] *V. du Vigneaud, G.L. Miller*, Biochem. Prep. *2*, 77 (1952).

[1349] *E. Walton, A.N. Wilson, F.W. Holly, K. Folkers*, J. Amer. Chem. Soc. *76*, 1146 (1954).

[1350] *C. Ressler, V. du Vigneaud*, J. Amer. Chem. Soc. *79*, 4511 (1957).

Alkaline[1355] or acid[1355, 1356] hydrolysis of *dialkyl-cyanamides* to secondary amines is a seldom used method. By contrast, hydrolysis of diallyl cyan-amide to *diallylamine*[1356], which can be performed with 25% sulfuric acid, is of preparative value. Preparation of *4-* and *3-methylaminophenols* succeeds by heating the corresponding cyanamides with water in a closed vessel[1357].

12.10.4 Dealkylation of amines and quaternary ammonium compounds

12.10.4.1 Thermal cleavage of quaternary ammonium hydroxides

Thermal cleavage of quaternary ammonium hydroxides is known as Hofmann elimination. The products are a *tertiary amine* and a fragment with a terminal *C=C* double bond (sometimes an alcohol). During structure elucidation of alkaloids by conventional techniques the Hofmann elimination performs an irreplaceable service. The possibility of utilizing it for preparing compounds with terminal *C=C* double bonds (quinine synthesis) may be a valuable procedure in particular instances. For synthesizing amines this potential technique is significant only where the quaternary bases are saturated *N*-heterocycles[1358–1366].

The normal decomposition temperature of 100–200° can be lowered by working in vacuum; occasionally addition of a little barium hydroxide solution is successful.

Although rules have been established governing the course of a Hofmann elimination[1367], the result of the reaction cannot always be predicted with certainty. For example, during thermal decomposition of 1,1-dimethyl-2-propylpiperidinium hydroxide a mixture of *N,N-dimethyl-1-propyl-4-pentenylamine*, *N,N-dimethyl-5-octenylamine*, and *1-methyl-2-propylpiperidine* are formed[1368]:

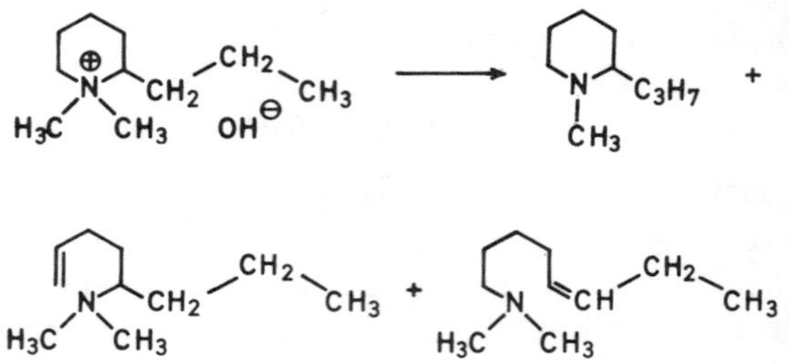

According to the Hofmann rule 1,1,3-trimethylpiperidinium hydroxide affords *N,N,2-trimethyl-4-pentenylamine*, 1,1-dimethyl-3-phenylpiperidinium hydroxide is cleaved to *N,N-dimethyl-4-phenyl-4-pentenylamine*[1369, 1370]:

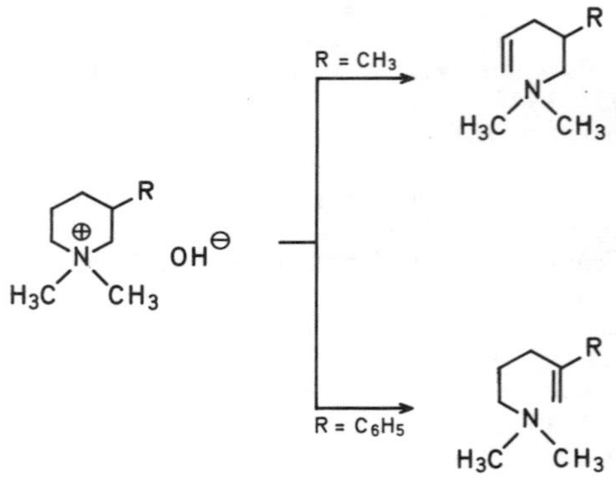

A convenient method for preparing *N,N-dimethyl-4-pentenylamines* consists in catalytic hydrogenation of 1-methylpyridinium methyl sulfate and subsequent decomposition of the initially formed 1,1-dimethylpiperidinium salts[1361].

During Hofmann elimination of 1,2,3,6-tetrahydro-1,1-dimethylpyridinium hydroxides *N,N-dimethyl-2,4-pentadienylamines* form[1371]. Similarly, (8-dimethylaminooctyl)trimethylammonium hydroxide affords *N,N-dimethyl-7-octenylamine*[1364].

[1351] *C. Ressler, H. Ratzkin*, J. Org. Chem. *26*, 3356 (1961).
[1352] *D. Stevenson, G.T. Young*, Chem. Commun. 900 (1967).
[1353] *C.M. Stevens, R. Watanabe*, J. Amer. Chem. Soc. *71*, 725 (1950).
[1354] *R.H.F. Manske*, J. Amer. Chem. Soc. *51*, 1202 (1929).
[1355] *W. Traube, A. Engelhardt*, Ber. *44*, 3149 (1911).
[1356] *E.B. Vliet*, Org. Syntheses, Coll. Vol. I, 201 (1941).
[1357] DRP. 484 906 (1924), Erf.: *W. Traube, E. Hellriegel*, Fortschritte der Teerfabrikation u. verwandter Industriezweige *16*, 393 (1931).
[1358] *A.W. Hofmann*, Ber. *14*, 659 (1881).
[1359] *G. Wittig, T.F. Burger*, Justus Liebigs Ann. Chem. *632*, 85 (1960).
[1360] *G. Merling*, Justus Liebigs Ann. Chem. *264*, 310 (1894).
[1361] *M. Ferles, J. Beran*, Collect. Czech. Chem. Commun. *32*, 2998 (1967).
[1362] *R. Lukeš, J. Hofman*, Chem. Ber. *93*, 2556 (1960).
[1363] *R. Lukeš, M. Ferles*, Collect. Czech. Chem. Commun. *16*, 252 (1951).
[1364] *Z. Polívka, V. Kubelka, N. Holubová, M. Ferles*, Collect. Czech. Chem. Commun. *35*, 1131 (1970).
[1365] *M. Ferles, Z. Polívka*, Collect. Czech. Chem. Commun. *33*, 2121 (1968).
[1366] *J. v. Braun*, Ber. *50*, 45 (1917).
[1367] *C.K. Ingold*, J. Chem. Soc. (London) 997 (1927); *C.K. Ingold*, J. Chem. Soc. (London) 3125 (1928); *C.K. Ingold*, J. Chem. Soc. (London) 68, 69 (1933).
[1368] *C. Glacet, B. Hasiak*, Compt. Rend. *264C*, 1988 (1967).
[1369] *W. Jacobi, G. Merling*, Justus Liebigs Ann. Chem. *278*, 1 (1894).

12.10.4.2 Thermal cleavage of quaternary ammonium salts

The preparatively somewhat tedious liberation of quaternary ammonium bases from their salts with moist silver oxide required for carrying out a Hofmann elimination can sometimes be by-passed by heating the quaternary salts with *alkali hydroxides*[1372-1374]. Thus arylamines afford aryldimethylamines[1375] on heating with sodium hydroxide following a preceding quaternization with dimethyl sulfate.

12.10.4.3 Cleavage of secondary and tertiary benzylamines by reductive debenzylation

Secondary and tertiary benzylamines can be readily debenzylated with the aid of palladium catalysts[1376-1379].

$$R-NH-CH_2-C_6H_5 + H_2 \longrightarrow$$

$$R-NH_2 + H_5C_6-CH_3$$

$$\underset{R^1}{\overset{R}{\diagdown}}N-CH_2-C_6H_5 + H_2 \longrightarrow$$

$$\underset{R^1}{\overset{R}{\diagdown}}NH + H_5C_6-CH_3$$

This possibility possesses preparative value for converting primary amines into *secondary* ones. To carry it out, the Schiff base of the primary amine is first prepared with benzaldehyde and then reduced to the benzylamine stage. Subsequently alkylation to the tertiary amine is performed and the benzyl group is split off hydrogenolytically. In this way *secondary amines* that are reliably free from tertiary amine impurities are obtained[1380].

$$R-NH_2 \xrightarrow{H_5C_6-CH=O} R-N=CH-C_6H_5$$

$$\xrightarrow{H_2/PtO_2} R-NH-CH_2-C_6H_5 \xrightarrow{R^1-X}$$

$$\underset{R^1}{\overset{R}{\diagdown}}N-CH_2-C_6H_5 \xrightarrow{H_2/PtO_2} \underset{R^1}{\overset{R}{\diagdown}}NH +$$

$$H_5C_6-CH_3$$

For monoalkylating *diamines* such as piperazines the monobenzyl derivatives are made the starting point. They are alkylated and subsequently catalytically hydrogenated[1381]. Other amino derivatives can be prepared similarly[1382-1385]. An alternative procedure for reductive carrying out the reductive debenzylation consists in allowing *sodium* in *liquid ammonia* to act[1386, 1387], *e.g.*:

$$\xrightarrow{Na/NH_3}$$

2-Methylimidazole; 92.5%

$$\xrightarrow{Na/NH_3}$$

5-Amino-1,2,3-triazole-4-carboxamide; 66%

12.10.4.4 Emde cleavage

A number of tertiary amines can be prepared by cleaving trialkylbenzylammonium salts with *sodium amalgam*[1388-1390].

[1370] H.W. Bersch, D. Schon, Tetrahedron Lett. 1141 (1966).
[1371] R. Lukeš, Collect. Czech. Chem. Commun. 12, 41 (1947).
[1372] G. Ciamician, P. Magnaghi, Ber. 18, 2079 (1885).
[1373] A. Ladenburg, Justus Liebigs Ann. Chem. 247, 56 (1888).
[1374] R. Grewe, A. Mondon, Chem. Ber. 81, 279 (1948).
[1375] S. Hünig, Chem. Ber. 85, 1056 (1952).
[1376] L. Birkofer, Ber. 75, 429 (1942).
[1377] H. Dahn, U. Solms, Helv. Chim. Acta 35, 1162 (1952).
[1378] H. Dahn, P. Zoller, Helv. Chim. Acta 35, 1348 (1952).
[1379] H. Dahn, P. Zoller, U. Solms, Helv. Chim. Acta 37, 565 (1954).
[1380] J.S. Buck, R. Baltzly, J. Amer. Chem. Soc. 63, 1964 (1941).

[1381] R. Baltzly, J.S. Buck, E. Lorz, W. Schön, J. Amer. Chem. Soc. 66, 263 (1944).
[1382] W. Wenner, J. Org. Chem. 13, 26 (1948).
[1383] M. Frankel, Y. Liwschitz, Y. Amiel, J. Amer. Chem. Soc. 75, 330 (1953).
[1384] Y. Liwschitz, A. Zilkha, J. Amer. Chem. Soc. 76, 3698 (1954).
[1385] S.M. Gadekar, J.L. Frederick, J. Semb, J.R. Vaughan, J. Org. Chem. 26, 468 (1961).
[1386] R.G. Jones, J. Amer. Chem. Soc. 71, 383 (1949).
[1387] J.R.E. Hoover, A.R. Day, J. Amer. Chem. Soc. 78, 5832 (1956).
[1388] H. Emde, Arch. Pharm. (Weinheim, Ger.) 249, 106 (1911).
[1389] H. Emde, Arch. Pharm. (Weinheim, Ger.) 247, 369 (1909).
[1390] H. Emde, H. Schellbach, Arch. Pharm. (Weinheim, Ger.) 249, 111 (1911).

This so-called Emde cleavage, unlike catalytic debenzylation, allows to prepare amines with $C=C$ double bonds, e.g.:

$$H_3C-\overset{\overset{\displaystyle CH_2-C_6H_5}{|}}{\underset{\underset{\displaystyle CH_2-C_6H_5}{|}}{N}}{}^{\oplus}-CH_2-CH=CH_2 \quad J^{\ominus} \xrightarrow{\text{NaHg}}$$

$$\underset{H_3C}{\overset{H_5C_6-CH_2}{>}}N-CH_2-CH=CH_2$$

$$\xrightarrow{H_3C-CH_2-CH_2-J} \quad H_3C-\overset{\overset{\displaystyle CH_2-C_6H_5}{|}}{\underset{\underset{\displaystyle CH_2-CH_2-CH_3}{|}}{N}}{}^{\oplus}-CH_2-CH=CH_2 \quad J^{\ominus}$$

$$\xrightarrow{\text{NaHg}} \quad \underset{H_3C-CH_2-CH_2}{\overset{H_3C}{>}}N-CH_2-CH=CH_2$$

Sodium amalgam can be replaced by sodium amide in liquid ammonia[1391–1393], e.g.:

(tetrahydroisoquinolinium salt) $\xrightarrow{\text{Na/NH}_3\text{ -70}°}$ (2-ethyl product)

2-Ethyl-N,N-dimethylphenethyl-amine

(piperidinium salt) $\xrightarrow{\text{Na/NH}_3}$

4,N,N-Trimethyl-5-phenylpentylamine; 95%

A similar cleavage succeeds with Raney nickel in alkaline solution, for instance, with 1,1-dimethyl-pyrrolidinium salts[1394], e.g.:

(2-benzyl-pyrrolidinium salt) $\xrightarrow{\text{Ra-Ni/NaOH}}$ (product)

R = CH₃;   3-Benzyl-N,N-dimethylpentylamine
R = C₆H₅; 3-Benzyl-N,N-dimethyl-4-phenylbutylamine

12.10.4.5 Cleavage of tertiary amines by oxidation

One way of converting tertiary amines into *secondary* amines by oxidation is with chromium(VI) oxide[1395, 1396], manganese(IV) oxide[1397], mercury(II) acetate[1398], and potassium hexacyanoferrate(III)[1399]. Oxidation with iron(III) salts[1400] proceeds in two stages *via* initially formed *N-oxides*.

Occasionally, tertiary amines can be converted into *N-nitrosoamines* by treatment with tetranitromethane[1401, 1402]. The products can be readily hydrolyzed to *secondary amines*[1403, 1404]. Sometimes *N*-nitrosoamines are cleaved reductively to secondary amines[1405–1408] and these are formed also by treating tertiary amines with 2-nitro-2-propyl hydroperoxide. The initially formed *N*-nitrosoamines are reduced separately. Using this sequence it is possible to degrade *sterically hindered* amines as well[1409].

$$\underset{H_3C}{\overset{H_3C}{>}}CH-NO_2 \xrightarrow{O_2/CuCl} H_3C-\overset{\overset{\displaystyle NO_2}{|}}{\underset{\underset{\displaystyle CH_3}{|}}{C}}-OOH$$

$$\underset{R}{\overset{R}{>}}N-CH_2-R^1 + H_3C-\overset{\overset{\displaystyle NO_2}{|}}{\underset{\underset{\displaystyle CH_3}{|}}{C}}-OOH \longrightarrow$$

$$\underset{R}{\overset{R}{>}}N-N=O + R^1-CH=O + (H_3C)_2CO + H_2O$$

1391 G. Childs, E.J. Forbes, J. Chem. Soc. (London) 2024 (1959).

1392 E. Leete, A.R. Friedman, J. Amer. Chem. Soc. 86, 1224 (1964).

1393 J. Wróbel, A.M. Konowal, Roczniki Chem. 39, 1437 (1965); C.A. 64, 17522 (1966).

1394 S. Sugasawa, S. Ushioda, Tetrahedron 5, 48 (1959).

1395 A. Cavé, C. Kan-Fan, P. Potier, J. Le Men, M. Janot, Tetrahedron 23, 4691 (1967).

1396 I.I. Grandberg, L.I. Gorbacheva, A.N. Kost, D.V. Sibiryakova-Fedotova, Zh. Obshch. Khim. 33, 515 (1963); C.A. 59, 1615 (1963).

1397 H.B. Henbest, A. Thomas, Chem. & Ind. (London) 1097 (1956).

1398 N.J. Leonard, D.F. Morrow, J. Amer. Chem. Soc. 80, 371 (1958).

1399 T.D. Perrine, J. Org. Chem. 16, 1303 (1951).

12.10.4.6 Cleavage of N,N-dialkyl-(4-nitroso-aryl)amines

The traditional Baeyer-Caro cleavage [1410–1413] of N,N-dialkyl-4-nitrosoanilines with alkali metal hydroxide is improved markedly by using heating with sodium hydrogen sulfite [1414, 1415] (see Table 6).

Table 6. Secondary amines by cleaving N,N-dialkyl-4-nitroso-anilines by the methods of Baeyer (A) and von Braun (B)

$ON-\langle\rangle-N\langle^{R^1}_{R^2}$		Meth-od	Amine	Yield [% of th.]	Ref.
CH_3	C_2H_5	A	N-Methyl-ethylamine	75–80	1411
	C_3H_7	A	N-Methyl-propylamine		1412
	C_4H_9	A	N-Methyl-butylamine		1412
$-CH_2-CH_2-N\langle^{CH_3}_{C_6H_4-4-NO}$		B	N,N'-Dimethyl-ethylenediamine	80–88	1414
$-CH_2-CH_2-N(C_4H_9)_2$		B	N,N-Dibutyl-N'-methyl-ethylenediamine	66–70	1415
C_4H_9	C_4H_9	A	Dibutylamine	100	1413

12.10.4.7 Cleavage of tertiary amines with derivatives of carboxylic acids

In some instances cleavage of tertiary amines to form *secondary* ones was performed by heating with benzoyl chloride [1416, 1417]. Initially a benzoyl derivative is formed and is subsequently hydrolyzed:

$$Ar-N(CH_3)_2 \; + \; H_5C_6-CO-Cl \xrightarrow{190°}$$

$$Ar-N\langle^{CH_3}_{CO-C_6H_5} \; + \; CH_3Cl$$

$$Ar-N\langle^{CH_3}_{CO-C_6H_5} \; + \; HCl \; + \; H_2O \longrightarrow$$

$$Ar-NH-CH_3\cdot HCl \; + \; H_5C_6-COOH$$

Degradation of tertiary amines by heating with phosgene or esters of chloroformic acid proceeds similarly. The initially formed urea derivatives [1418], carbamoyl chlorides [1419], and urethans [1420–1422] need to be hydrolyzed subsequently:

$$2\,Ar-N(CH_3)_2 \; + \; COCl_2 \xrightarrow{190°}$$

$$\begin{array}{c} Ar \\ Ar-N-C-N-CH_3 \\ H_3C \;\; O \end{array} \; + \; 2\,CH_3Cl$$

$$2\,R_3N \; + \; COCl_2 \xrightarrow{60-70°/C_6H_5-CH_3}$$

$$R_2N-C\langle^{Cl}_{O} \; + \; R_4N^{\oplus}\,Cl^{\ominus}$$

$$R_3N \; + \; Cl-COOR^1 \longrightarrow R_2N-COOR^1 \; + \; R-Cl$$

12.10.4.8 Cleavage of tertiary amines with cyanogen bromide

This technique for converting tertiary amines into *secondary* ones introduced by v. Braun has been successfully employed for many purposes [1423].

1400 M.S. Fish, C.C. Sweeley, E.C. Horning, Chem. & Ind. (London) R 24 (1956).

1401 E. Schmidt, H. Fischer, Ber. 53, 1537 (1920).

1402 E. Schmidt, R. Schumacher, Ber. 54, 1414 (1921).

1403 J. Reilly, W.J. Hickenbottom, J. Chem. Soc. (London) 115, 175 (1919).

1404 W.G. Macmillan, T.H. Reade, J. Chem. Soc. (London) 585 (1929).

1405 E.C.S. Jones, J. Kenner, J. Chem. Soc. (London) 711 (1932).

1406 W.S. Emerson, J. Amer. Chem. Soc. 63, 2023 (1941).

1407 G.F. Grillot, J. Amer. Chem. Soc. 66, 2124 (1944).

1408 J.S. Buck, C.W. Ferry, Org. Syntheses, Coll. Vol. II, 290 (1943).

1409 B. Frank, J. Conrad, P. Misbach, Angew. Chem. 82, 876 (1970).

1410 A. Baeyer, H. Caro, Ber. 7, 963 (1874).

1411 J. Meisenheimer, Justus Liebigs Ann. Chem. 428, 256 (1922).

1412 R. Stoermer, V.v. Lepel, Ber. 29, 2110 (1896).

1413 J. Reilly, W.J. Hickinbottom, J. Chem. Soc. (London) 113, 107 (1918).

1414 J.v. Braun, K. Heider, E. Müller, Ber. 51, 737 (1918).

1415 R. Munch, G.T. Thannhauser, D.L. Cottle, J. Amer. Chem. Soc. 68, 1297 (1946).

1416 O. Hess, Ber. 18, 685 (1885).

1417 K. Löffler, S. Kober, Ber. 42, 3431 (1909).

1418 A. Wahl, Bull. Soc. Chim. France [5] 1, 244 (1934).

1419 V.A. Rudenko, A.Y. Yakubovich, T.Y. Nikiforova, Zh. Obshch. Khim. 17, 2256 (1947); C.A. 42, 4918 (1948).

1420 DRP. 255942 (1911), Farbenfabriken vorm F. Bayer Elberfeld; Fortschritte der Teerfabrikation u. verwandter Industriezweige 11, 115 (1915).

1421 E.H. Flynn, H.W. Murphy, R.E. McMahon, J. Amer. Chem. Soc. 77, 3104 (1955).

The initially formed disubstituted cyanamides are hydrolyzed and decarboxylated by heating with dilute mineral acid. It is possible that the alkyl bromide formed in the first stage will react with the starting substance, which is then half consumed by a side-reaction:

$$2\,R_3N + BrCN \longrightarrow R_2N{-}CN + R_4N^{\oplus}\,Br^{\ominus}$$

$$R_2N{-}CN + 2HCl + 2H_2O \longrightarrow$$

$$R_2NH\cdot HCl + CO_2 + NH_4Cl$$

From unsymmetrically substituted tertiary amines allyl and benzyl groups are eliminated most easily as bromides followed by methyl, ethyl, and other alkyl groups in the order of their molecular weight, and, finally, aryl substituents[1423–1426]. As a result, in *N*-alkylheterocycles either the alkyl group is split off or *ring-opening* occurs according to the size of the alkyl group:

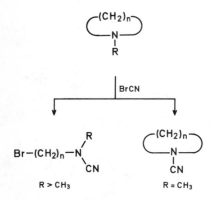

In six-membered[1426–1429] and seven-membered[1430] nitrogen heterocycles the ring remains intact only where the *N*-alkyl group is methyl[1431–1434], but opens in other instances[1431–1434].

[1422] *J. Knabe, H. Gohmert*, Arch. Pharm. (Weinheim, Ger.) *303*, 525 (1970).

[1423] *J.v. Braun*, Ber. *33*, 1438 (1900).

[1424] *J.v. Braun*, Ber. *33*, 2728 (1900).

[1425] *J.v. Braun, K. Moldaenke*, Ber. *56*, 2165 (1923).

[1426] *H.W.J. Cressman*, Org. Synth. *27*, 56 (1947).

[1427] *J.v. Braun*, Ber. *33*, 2734 (1900).

[1428] *R. Lukeš, J. Pliml*, Collect. Czech. Chem. Commun. *19*, 502 (1954).

[1429] *M. Ferles, M. Prystaš*, Collect. Czech. Chem. Commun. *24*, 3326 (1959).

[1430] *B.D. Astill, V. Boekelheide*, J. Amer. Chem. Soc. *77*, 4079 (1955).

[1431] *J.v. Braun*, Ber. *44*, 1252 (1911).

[1432] *J.v. Braun*, Ber. *40*, 3914 (1907).

[1433] *E. Ochiai, K. Tsuda*, Ber. *67*, 1011 (1934).

[1434] *R.C. Elderfield, H.A. Hageman*, J. Org. Chem. *14*, 605 (1949).

12.11 Bibliography

Introduction

H. Kessler, Angew. Chem. *82*, 237 (1970) (Optical antipodes);

F. Montanari, J. Morethi, G. Torre, Chem. Commun. 1694 (1968); 1086 (1969) (Oxaziridines with an optically active *N*);

R. Annunziata, R. Farnasier, F. Montanari, Chem. Commun. 1133 (1972) (Diaziridines with an optically active *N*).

Amines by substitution reactions

F. Möller in *Houben-Weyl*, Methoden der organischen Chemie, 4. Aufl., Bd. XI/1, p. 9, Georg Thieme Verlag, Stuttgart 1957.

M.T. Leffler in Organic Reactions, Vol. I, p. 91, John Wiley & Sons, New York 1942.

P. Kovacic, M.K. Lowery, K.W. Field, Chem. Rev. *70*, 639 (1970).

F. Möller in *Houben-Weyl*, Methoden der organischen Chemie, 4. Aufl., Bd. XI/1, p. 248, Georg Thieme Verlag, Stuttgart 1957.

Amines by reduction

V.M. Mićović, M.Lj. Mihailović, Lithium Aluminium Hydride in Organic Chemistry, Serbian Academy of Sciences, Monographs, Vol. 237, Naučna Knjiga, Izdavačko, Preduzeće, Beograd 1955.

N.G. Gaylord, Reduction with Complex Metal Hydrides, Interscience Publishers, New York 1956.

Houben-Weyl, Methoden der organischen Chemie, 4. Aufl., Bd. XI/1, Georg Thieme Verlag, Stuttgart 1957.

F. Zymalkowski, Katalytische Hydrierungen im Organisch-chemischen Laboratorium, Ferdinand Enke Verlag, Stuttgart 1965.

R.L. Augustine, Catalytic Hydrogenation, Marcel Dekker, New York 1965.

A. Furst, R.C. Berlo, S. Hooton, Chem. Rev. *65*, 51 (1965).

H.O. House, Modern Synthetic Reactions, W.A. Benjamin, New York 1965.

P.A.S. Smith, The Chemistry of Open-Chain Organic Nitrogen Compounds, Vol. I, 1965, Vol. II, 1966, W.A. Benjamin, New York.

A. Hajós, Komplexe Hydride, VEB Deutscher Verlag der Wissenschaften, Berlin 1966.

P.N. Rylander, Catalytic Hydrogenation over Platinum Metals, Academic Press, New York 1967.

R.L. Augustine, Reduction, Marcel Dekker, New York 1968.

S. Patai, The Chemistry of the Amino Group, Interscience Publishers, London 1968.

F. Weygand, G. Hilgetag, Organisch-chemische Experimentierkunst, J.A. Barth, Leipzig 1970.

M. Ferles, Chem. Listy *64*, 188 (1970).

Z. Rappoport, The Chemistry of the Cyano Group, Interscience Publishers, London 1970.

J. Zabicky, The Chemistry of Amides, Interscience Publishers, London 1970.

Amines by reductive aminoalkylation

F. Möller, R. Schröter in *Houben-Weyl*, Methoden der organischen Chemie, 4. Aufl., Bd. XI/1, p. 602, Georg Thieme Verlag, Stuttgart 1957.

W.S. Emerson, Organic Reactions, Vol. IV, p. 174, John Wiley & Sons, New York 1948.

M. Freifelder, Practical Catalytic Hydrogenation, Wiley-Intersci. Publ., New York, London, Sydney, Toronto 1971.

M.L. Moore, Organic Reactions, Vol. V, p. 301, John Wiley & Sons, New York 1949.

H.W. Gibson, The Chemistry of Formic Acid and its Simple Derivatives, Chem. Rev. *69,* 673 (1969).

Amines by condensation

F.F. Blicke, Org. Reactions *1,* 303 (1942).

H. Karbe, Arch. Pharm. (Weinheim, Ger.) *283,* 48 (1950).

K.W. Merz, Pharmazie *11,* 505 (1956).

J.H. Brewster, E.L. Eliel, Org. Reactions 7, 99 (1953).

R. Schröter, « Amine durch Kondensation » in *Houben-Weyl,* Methoden der organischen Chemie, 4. Aufl., Bd. XI/1, p. 731, Georg Thieme Verlag, Stuttgart 1957.

B. Reichert, Die Mannichreaktion, Springer Verlag, Berlin, Göttingen, Heidelberg 1959.

H. Hellmann, G. Opitz, α-Aminoalkylierung, Verlag Chemie, Weinheim/Bergstr. 1960.

Amines *via* organometallic compounds

Houben-Weyl, Methoden der organischen Chemie, Bd. XI/1, p. 805, Georg Thieme Verlag, Stuttgart 1957.

M.S. Kharasch, O. Reinmuth, Grignard Reactions of Nonmetallic Substances, p. 767, 870, 1204, 1235, 1243, Prentice Hall, New York 1954.

S.T. Joffe, A.N. Nesmeyanov, The Organic Compounds of Magnesium, Beryllium, Calcium, Strontium and Barium, p. 356, 392, 547, North-Holland Publishing Co., Amsterdam 1967.

S. Patai, The Chemistry of the Carbon-Nitrogen Double Bond, p. 266, Interscience Publishers, New York 1970.

Amines by rearrangements

H.J. Shine, Aromatic Rearrangements, Elsevier Publishing Co., New York 1967.

P.A.S. Smith, Org. Reactions *3,* 363 (1946).

E.S. Wallis, J.F. Lane, Org. Reactions *3,* 267 (1946).

P.A.S. Smith, Org. Reactions *3,* 337 (1946).

H. Wolff, Org. Reactions *3,* 307 (1946).

H.L. Yale, Chem. Rev. *33,* 209 (1943).

L.G. Donaruma, W.Z. Heidt, Org. Reactions *11,* 1 (1960).

C. O'Brien, Chem. Rev. *64,* 81 (1964).

S.H. Pine, Org. Reactions *18,* 403 (1970).

G. Wittig, Bull. Soc. Chim. France 1921 (1971).

J.W. Schulenberg, S. Archer, Org. Reactions *14,* 1 (1965).

W.E. Truce, E.M. Kreider, W.W. Brand, Org. Reactions *18,* 99 (1970).

Amines by cleavage reactions

Houben-Weyl, Methoden der organischen Chemie, Bd. XI/1, p. 926, Georg Thieme Verlag, Stuttgart 1957.

M.L. Moore, Org. Reactions *5,* 301 (1949) (The Leuckart Reaction).

H.A. Hageman, Org. Reactions 7, 198 (1953) (The von Braun Cyanogen Bromide Reaction).

W.H. Hartung, R. Simonoff, Org. Reactions 7, 263 (1953) (Hydrogenolysis of Benzyl Groups Attached to Oxygen, Nitrogen or Sulfur).

A.C. Cope, E.R. Trumbull, Org. Reactions *11,* 317 (1960) (Olefines from Amines: The Hofmann Elimination Reaction and Amine Oxide Pyrolysis).

13 Aziridines, Azetidines (Alkylenimines)

Contributed by

O. Červinka
Faculty of Organic Chemistry,
Vysoká Škola Chemiko-Technologická,
Prague

13.1 Aziridines (1,2-alkylenimines)

13.1.1 Ring-closure reactions forming aziridine (ethylenimine) and its derivatives

13.1.1.1 Intramolecular alkylation of vicinally substituted amines

The simplest and at the same time most important method for preparing *azleridine (ethylenimine)* and its derivatives is the base-catalyzed intramolecular alkylation of vicinally substituted amines:

The ease of ring-closure depends on a number of factors.

As with similar methods for preparing other small rings, the formation of the aziridine ring depends upon the *number, size,* and *position* of any *alkyl groups* at the carbon atoms involved in the ring-closure[1]. Reaction is faster if the group *L* being displaced is attached to a secondary instead to a primary carbon atom[2]. If the carbon atom is tertiary ring-closure does not occur; instead, *HL* is eliminated, forming an unsaturated amine[3]. Substituent groups on the carbon atom to which the amino-group is attached do not have such a pronounced accelerating effect[4]. The reaction is generally made difficult if space-filling groups, for example, the *tert*-butyl group, are present[5]. The reactivity of *L* decreases in the order[6]:

$$O_3SR > Br > Cl > F > OSO_3^{\ominus} = OPO(OC_2H_5)_2 = OPO(OC_6H_5)_2$$

2-Amino alcohols will not convert directly into aziridines; all attempts at ring-closure merely give unsaturated products such as imines, enamines, *etc.*

The *nucleophilism* of the *nitrogen atom* of the amino group has a considerable influence on the rate of ring-closure. Cyclization is generally faster the larger the number of alkyl groups on the nitrogen atom is, *e.g.:*

$$(CH_3)_2N-CH_2-CH_2-Br > CH_3-NH-CH_2-CH_2-Br > H_2N-CH_2-CH_2-Br$$

Substituent groups with a *−I effect* such as phenyl retard the reaction[7,8].

Since a polar transition state is formed from a nonpolar starting material in the ring-closure reaction, it may be expected that solvents with high dielectric constants will accelerate the reaction. In fact, ring-closure proceeds best in aqueous or aqueous alcoholic media (*pH* 7–8). Higher basicity has no effect on the rate of cyclization, although the aziridinium ion forms aziridine, which can then be isolated, for example, by steam distillation.

Cyclization only proceeds smoothly when the *four centers L–C–C–N* participating in the reaction lie in one plane, *i.e.,* the substituent group *L* and the *amino* group are mutually *trans*-planar. This fact can explain the high *stereospecificity* of the ring closure. For example, *meso*-2, 3-epoxybutane (*meso*-2, 3-dimethyloxirane) and ammonia give *threo*-3-amino-2-butanol, which can be converted into *meso*-2,3-dimethylaziridine via its sulfate monoester.

D-(+)-2,3-dimethyloxirane can be converted into L-(−)-2,3-dimethylaziridine via the (+)-erythro-amino alcohol by the same reaction sequence[9].

---

[1] *B. Capon,* Quart. Rev. Chem. Soc. *18,* 62 (1964).
[2] *B. Hansen,* Acta Chem. Scand. *17,* 1483 (1963).
[3] *R. Adams, T.L. Cairns,* J. Amer. Chem. Soc. *61,* 2464 (1939).
[4] *T.L. Cairns,* J. Amer. Chem. Soc. *63,* 871 (1941).
[5] *C.S. Dewey, R.A. Bafford,* J. Org. Chem. *32,* 3100 (1967).

[6] *C.S. Dewey, R.A. Bafford,* J. Org. Chem. *30,* 495 (1965).
[7] *N.B. Chapman, D.J. Triggle,* J. Chem. Soc. (London) 1385 (1963).

If the substituted carbon atoms form part of a five- or six-membered ring, formation of an aziridine is only possible when the substituent groups involved in the cyclization are able to assume *axial* conformations. Consequently, only *trans*-disubstituted compounds cyclize [10].

In contrast, in the case of the *1,2-epi-iminocyclododecanes*, it has been possible to isolate *cis*- and *trans-aziridines* [11].

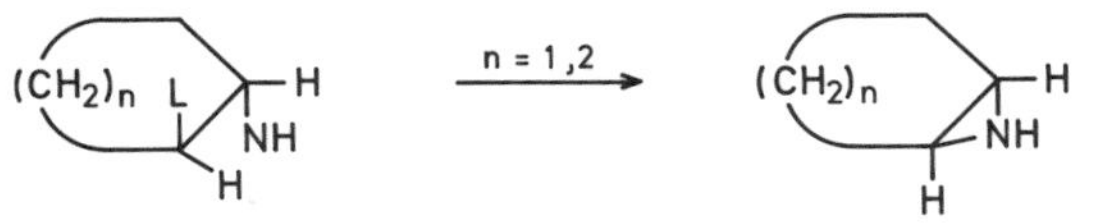

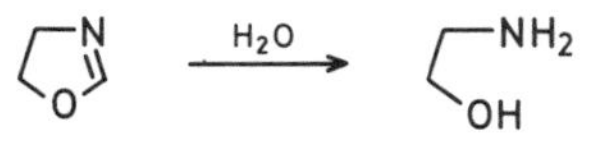

Vicinal amino alcohols (as shown above) can be cyclized by the method of Wenker [12-17].

The α-amino alcohol is converted into the sulfate, which is a stable inner salt obtainable in excellent yield; heating with alkali eliminates sulfuric acid to form *aziridine* [20, 21]. If the reaction is carried out in an autoclave (50–80 atm, 200°), the reaction time may be shortened to a few minutes. This method has the advantage that the sulfate monoester is nonvolatile and, hence, cannot initiate spontaneous polymerization of the corresponding aziridine. In some instances the 2-aminoethyl sulfate esters [18, 19] can be obtained also by a Ritter reaction [14] (cf. p. 701).

In the Gabriel method [22], cf. ref. [23] for preparing *aziridine*, 2-bromoethylamine hydrobromide or 2-chloroethylamine hydrochloride is treated with silver oxide or, preferably, approximately 30% caustic potash. This method displays the disadvantage that even traces of volatile haloethylamines cause polymerization of the aziridine.

α-Iodoallylamines also can be cyclized to aziridines [34].

Reduction of α-chloronitriles with lithium tetrahydroaluminate and subsequent cyclization, too, possesses preparative value [23].

Aziridines can be obtained also by reduction of *2-(azidoalkyl)methanesulfonates* with hydrazine and Raney nickel [24, 25] or with lithium tetrahydroaluminate [26]. These techniques are used in the sugar and steroid series.

13.1.1.2 Intramolecular alkylation of vicinally substituted *N-acylsulfonylamines*, *N-arylsulfonylamines*, and *N-alkylsulfonylamines*

Intramolecular alkylation with amide ions can be used for preparing *N-acylsulfonylamines (R = RCO)*, *N-arylsulfonylaziridines (R = aryl)*, and *N-alkylsulfonylaziridines (R = RSO$_2$)*:

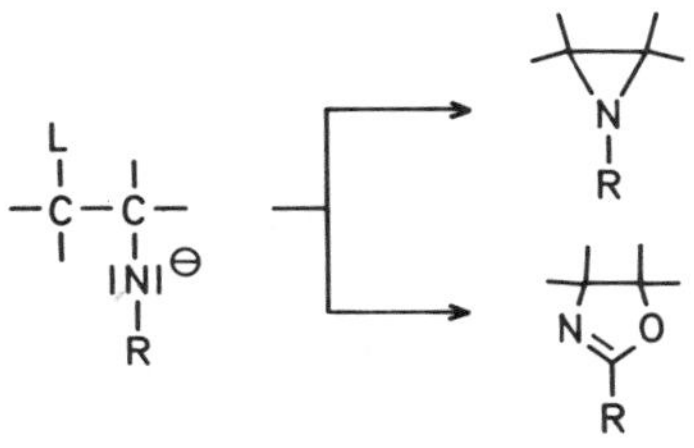

Aziridine derivatives can be obtained readily from substituted sulfonamides, but carboxamides form mainly *2-oxazoline* derivatives; *L* needs to be especially reactive for aziridines to be the main product [27]. Thus, methylpyranosides (but not furanosides) having suitable groups in the 2- and 3-positions can be converted into aziridine deriva-

[8] *P.D. Bartlett, S.D. Ross, C.G. Swain,* J. Amer. Chem. Soc. *71*, 1415 (1949).

[9] *F.H. Dickey, W. Fickett, H.J. Lucas,* J. Amer. Chem. Soc. *74*, 944 (1952).

[10] *L.A. Paquette, W.C. Farley,* J. Amer. Chem. Soc. *89*, 3595 (1967).

[11] *P.E. Fanta, R. Golden, H.J. Hsu,* J. Chem. Eng. Data *9*, 246 (1964).

[12] *A. Rosowaskij,* The Chemistry of Heterocyclic Compounds, Vol. XIX, p. 316, Interscience Publishers, New York 1964.

[13] *Y. Minoura, M. Takebayashi, C.C. Price,* J. Amer. Chem. Soc. *81*, 4689 (1959).

[14] US.P. 3 052 669 (1962), Inv.: *G. Gavlin, K. Hattori.*

[15] *H. Wenker,* J. Amer. Chem. Soc. *57*, 2328 (1935).

[16] *P.A. Leighton, W. Perkins, M.L. Renquist,* J. Amer. Chem. Soc. *69*, 1540 (1947).

[17] DRP. 665 790 (1935), I.G. Farben, Erf.: *H. Ulrych*; Chem. Zentr. II, 4311 (1938).

[18] *S.J. Brois,* J. Org. Chem. *27*, 3532 (1962).

[19] *E. Cherbuliez,* Helv. Chim. Acta *47*, 2106 (1964).

[20] *W.A. Reeves, G.L. Drake jr., C.L. Hoffpauir,* J. Amer. Chem. Soc. *73*, 3522 (1951).

[21] *V.P. Wystrach, D.W. Kaiser, F.C. Schaefer,* J. Amer. Chem. Soc. *77*, 5915 (1955).

[22] *S. Gabriel,* Ber. *21*, 2665 (1888).

[23] *K. Ochimura, M. Ohta,* Bull. Chem. Soc. Japan *40*, 432 (1967).

[24] *R.D. Guthrie, D. Murphy,* J. Chem. Soc. (London) 5288 (1963).

[25] *K. Ponsold,* Chem. Ber. *97*, 3524 (1964).

[26] *K. Ponsold, D. Klemm,* Chem. Ber. *99*, 1502 (1966).

[27] *T. Taguchi, M. Kojima,* J. Amer. Chem. Soc. *81*, 4316 (1959).

tives[28]. With iodocarbamates $(L=I, R=COOR)$ the aziridine ring is formed almost exclusively[29]. This difference can be explained by the stability of the anion derived from the monosubstituted urethan being much less than that of the carboxamide anion, so that intramolecular cyclization moves toward the nitrogen atom. *1-(Alkoxycarbonyl)aziridines* are very easily hydrolyzed and consequently cannot be isolated in some cases.

13.1.1.3 Intramolecular alkylation by vicinally substituted carbanions

The reaction sequence can be illustrated by the following general scheme[30, 31]:

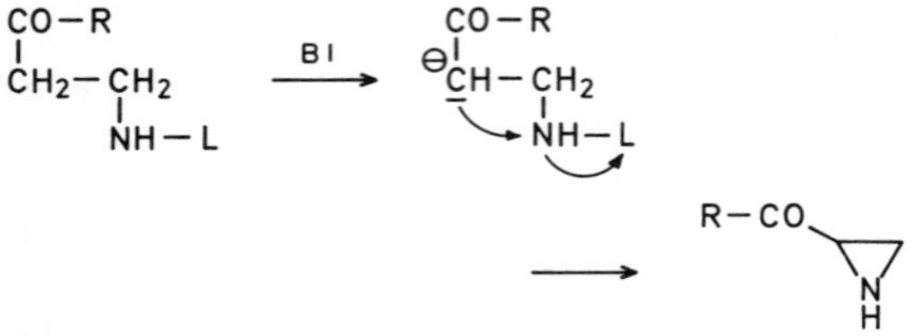

2-Acylaziridines

L is usually halogen or alkoxy. α-Lactams, too, can be prepared by this method[32]:

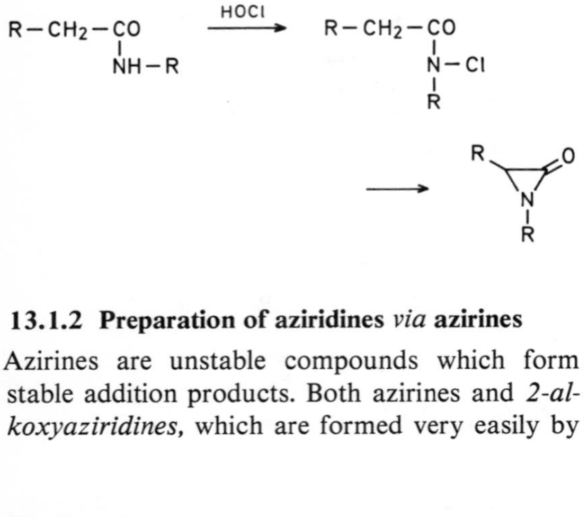

13.1.2 Preparation of aziridines *via* azirines

Azirines are unstable compounds which form stable addition products. Both azirines and *2-alkoxyaziridines*, which are formed very easily by

addition of alcohols, can be reduced to aziridines by lithium tetrahydroaluminate[33-36].

Aziridines are formed also by the action of Grignard reagents[37, 38] $(R^1=alkyl$ or *aryl*) or of lithium tetrahydroaluminate[23] $(R^1=H)$ on α-chloronitriles:

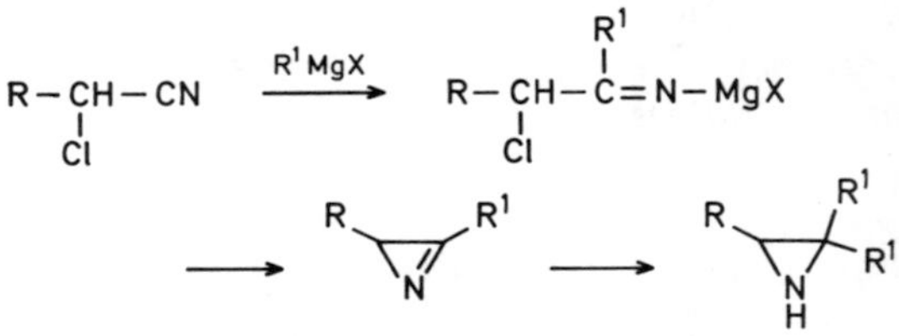

O-Tosyl ketoximes with an active hydrogen atom in the α-position form α-imino ketones[39] under the action of strong bases (Neber rearrangement). In some cases the azirines formed as intermediates can be isolated or trapped as *aziridines* by addition of lithium tetrahydroaluminate.

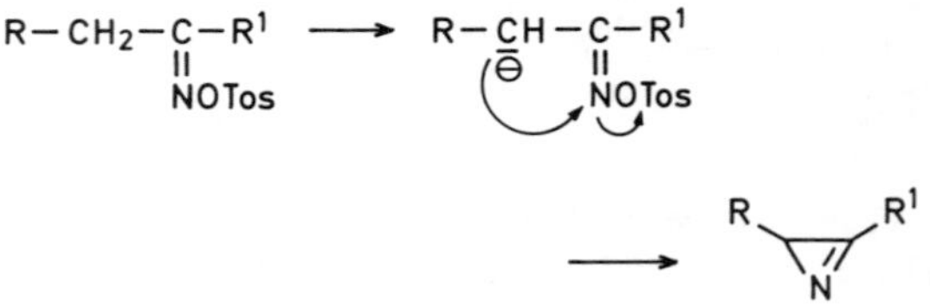

Ring closure of the methiodides of certain nitrogen-substituted ketone hydrazones proceeds similarly[40, 41], *e.g.:*

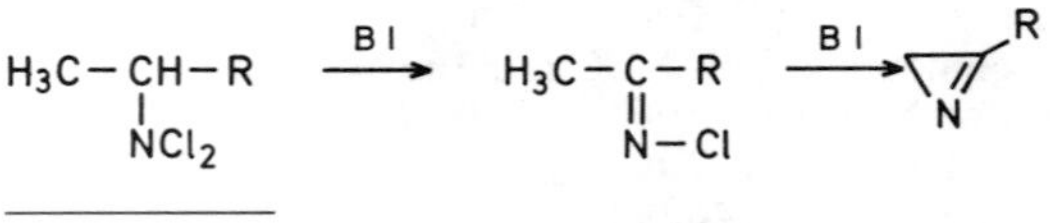

3,3-Dimethyl-2-phenylaziridine

Aziridines can be obtained also from *N,N*-dichloro-sec-alkylamines[42, 43]:

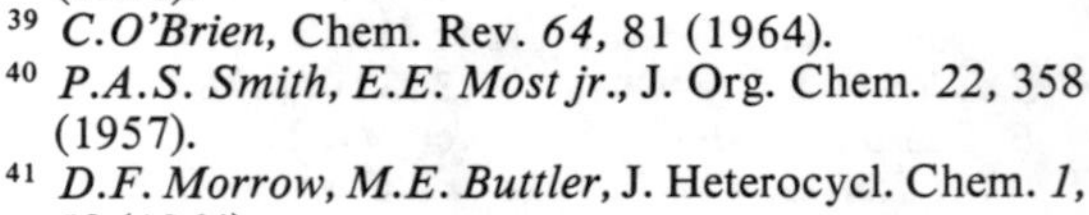

[28] *D.H. Buss, L.D. Hall, L. Hough*, J. Chem. Soc. (London) 1616 (1965).
[29] *C.S. Gebelein, G. Swift, D. Swern*, J. Org. Chem. 32, 3314 (1967).
[30] DBP. 1 054 088 (1959), Erf.: *E. Merck*; C.A. 55, 8439 (1961).
[31] *A.H. Blatt*, J. Amer. Chem. Soc. 61, 349 (1939).
[32] *I. Lengyel, J.C. Sheehan*, Angew. Chem. 80, 27 (1968).
[33] *D.J. Cram, M.J. Hatch*, J. Amer. Chem. Soc. 75, 33 (1953).
[34] *A. Hassner, M.E. Lorber, C. Heathcock*, J. Org. Chem. 32, 540 (1967).
[35] *M.J. Hatch, D.J. Cram*, J. Amer. Chem. Soc. 75, 38 (1953).
[36] *H.E. Baumgarten, J.M. Peterson*, J. Amer. Chem. Soc. 82, 459 (1960).
[37] *O. de Boosoré*, Bull. Soc. Chim. Belg. 32, 26 (1923).
[38] *M. Theunis*, Bull. Sci. Acad. Roy. Belg. 12, 785 (1926).
[39] *C.O'Brien*, Chem. Rev. 64, 81 (1964).
[40] *P.A.S. Smith, E.E. Most jr.*, J. Org. Chem. 22, 358 (1957).
[41] *D.F. Morrow, M.E. Buttler*, J. Heterocycl. Chem. 1, 53 (1964).
[42] *G.H. Alt, W.S. Knowles*, J. Org. Chem. 25, 2047 (1960).
[43] *H.E. Baumgarten*, J. Org. Chem. 31, 3708 (1966).
[44] *S. Eguchi, Y. Ishii*, Bull. Chem. Soc. Japan 36, 1434 (1963).

Grignard reagents or lithium tetrahydroaluminate convert oximes of acyl aromatic compounds into *3-alkyl-2-arylaziridines*[44–46]:

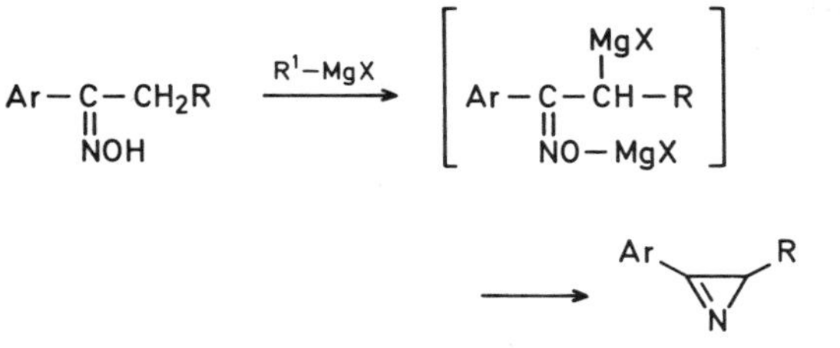

13.1.3 Preparation of aziridines by addition reactions

Aziridines can be formed by addition of alkyl or aryl azides to olefinically unsaturated compounds followed by pyrolysis or photolysis of the resulting *4,5-dihydro-1,2,3-triazoles*. Enamines or imines are sometimes formed as by-products[47]:

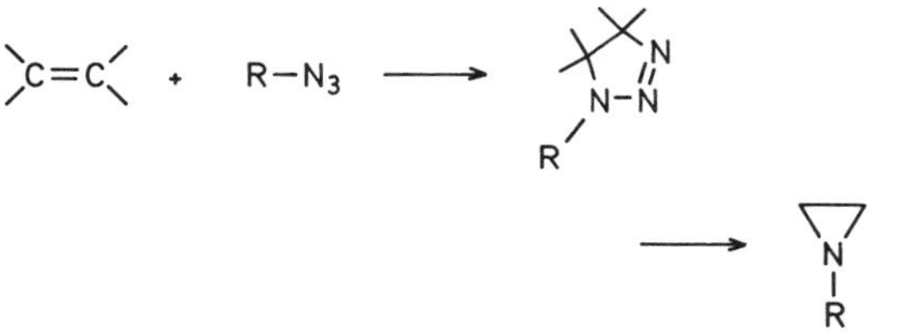

Imidogens (nitrenes) generated pyrolytically or photolytically in the presence of unsaturated compounds, too, can add to the double bonds; for example, *(ethoxycarbonyl)imidogen* arises by photolytic addition of ethyl azidoformate or ethyl *N*-(4-nitrobenzenesulfonyl)carbamate to cyclohexene[48]:

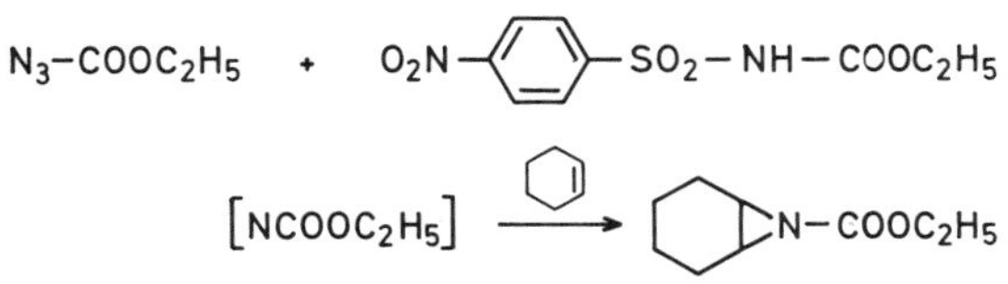

Ethyl 7-azabicyclo-[4.1.0]heptane-7-carboxylate

Sulfonylimidogens[49] and cyanoimidogens react similarly forming *7-azabicyclo[4.1.0]heptane-7-sulfonic acid* and *-7-carbonitrile*[50, 51]. The reaction can be carried out either in solution or in the gaseous phase. Reactions with 1,3-dienes such as 1,3-butadiene[52], isoprene, or cyclohexadiene[53] give only the 1,2-addition products:

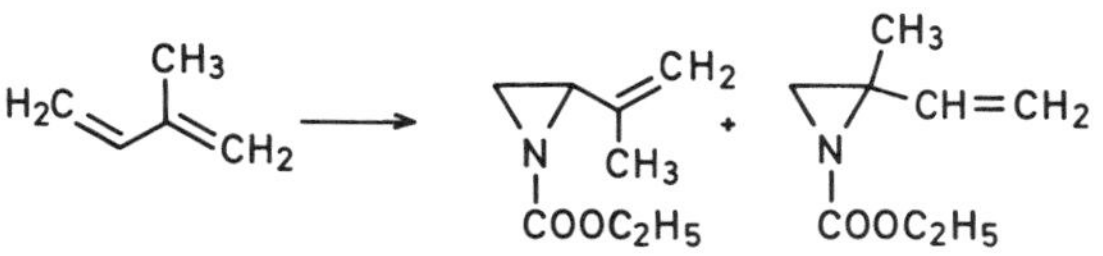

Ethyl 2-iso-propenyl-1-aziridine-carboxylate

Ethyl 2-methyl-2-vinyl-1-aziridine-carboxylate

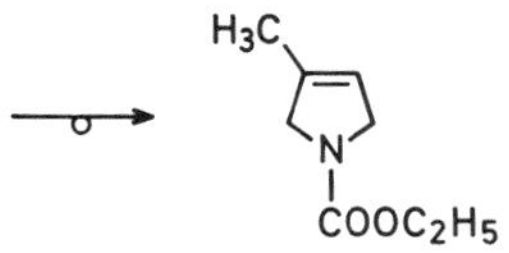

Ethyl 3-methyl-3-pyrroline-1-carboxylate

The primarily formed vinylaziridines can undergo thermal rearrangement to pyrrolines, *i.e.*, 1,4-addition products.

Photolysis of an azidoformate in the presence of *cis*- or *trans*-2-butene followed by hydrolysis of the adduct leads to stereospecific formation of *cis*- or *trans*-2,3-dimethylaziridine.

The *stereospecificity* of imidogen addition reactions can be explained by assuming that under the named conditions the nitrene adds in the singlet state. In solution it changes to the triplet state[52, 54, 55]. The rate of ring closure in the resulting biradical addition product is less than that of the rotation about the *C–C* bond. Hence, in solution the reaction must proceed without stereospecificity and a mixture of *cis* and *trans* isomers are obtained.

[45] *K. Kotera*, Tetrahedron **24**, 6177 (1968).
[46] *K. Kotera, T. Okada, S. Miyazaki*, Tetrahedron **24**, 5677 (1968).
[47] *P. Scheiner*, J. Org. Chem. **30**, 7 (1965).
[48] *W. Lwowski, T.W. Mattinghy jr.*, Tetrahedron Lett. 277 (1962).
[49] *H. Kwart, A.A. Khan*, J. Amer. Chem. Soc. **89**, 1951 (1967).
[50] *F.D. Marsh, M.E. Hermes*, J. Amer. Chem. Soc. **86**, 4506 (1964).

[51] *A.G. Anastassion*, J. Org. Chem. **31**, 1131 (1966).
[52] *K. Hafner, W. Kaiser, R. Puttner*, Tetrahedron Lett. 3953 (1964).
[53] *A. Mishra, S.N. Rice, W. Lwowski*, J. Org. Chem. **33**, 481 (1968).
[54] *P.S. Skell, R.C. Woodworth*, J. Amer. Chem. Soc. **78**, 4496 (1956).
[55] *R. Hoffmann*, J. Amer. Chem. Soc. **90**, 1475 (1968).
[56] *T. Curtius, A. Bertho*, Ber. **59**, 565 (1926).
[57] *A. Bertho*, J. Prakt. Chem. [2] **120**, 89 (1928).
[58] *A.L. Logothetis*, J. Org. Chem. **29**, 3049 (1964).
[59] *I.E. Den Besten, C.R. Wenger*, J. Amer. Chem. Soc. **87**, 5500 (1965).

Aziridines are formed primarily also from the addition of imidogens to aromatic rings[56, 57], of carbenes to Schiff bases[58-60], or of diazomethane to immonium perchlorates[61]. However, these reactions are of more theoretical than practical value. It is interesting also that aziridines have been detected in the products of photolysis[62] or pyrolysis[63] of methylamine and dimethylamine.

13.2   Azetidines (1,3-alkylenimines)

13.2.1   Ring-closure reactions to azetidine and its derivatives

Just as aziridine derivatives can be formed by ring-closure of vicinally substituted amines, it is possible to prepare azetidine derivatives by intramolecular cyclization of γ-substituted alkylamines:

(Y = NH)

Dry distillation of the hydrochloride of 1,3-propanediamine $(X=Y=NH_2)$ gives only traces of *azetidine*. The reaction between arylsulfonamides and 1,3-dibromopropane $(X=Y=Br)$ is synthetically rewarding; the only difficulty is due to hydrolysis of the intermediate 1-sulfonylazetidine since the cyclization product is not sufficiently stable toward acids and bases. Smooth fission to azetidine can be achieved only by treatment of a 1-arylsulfonyl azetidine with sodium ethoxide in boiling ethanol[64]. Cyclization of 3-halopropylamines $(X=Cl$ or Br; $Y=NH_2)$ proceeds less well than that of 2-haloethylamines unless they are substituted on the 2 carbon atom by one or more alkyl groups. Tertiary 3-haloalkylamines form the stable quaternary ammonium salts[65, 66] on heating even in the absence of alkali.

13.2.2   Preparation by 1,2-cycloaddition reactions

The formation of azetidines by 1,2-cycloaddition reactions is limited to starting materials having highly activated double bonds. Ethenetetracarbonitrile adds trifluoromethanesulfenyl chloride; the *2,3,3-tricyano-N-[(trifluoromethyl)-thio]acrylimidoyl* chloride formed adds vinyl ethers and styrenes[67] to the C=N double bond:

[3-Alkyl- or [3-Aryl-2-chloro-1[(trifluoromethyl)thio]-2-azetidinyl]ethenetricarbonitrile

1-(Trifluoromethyl)thioazetidines can be converted into azetidines by warming with alcohols and pyridine. Azetidines are formed also by addition of N-substituted N-halosulfenyl amidines to olefins[68]. The most important method for preparing simple azetidines is *reduction* of the readily accessible β-lactams (2-azetidinones) with lithium tetrahydroaluminate[69, 70]. It may be possible to react azetidine-2-thiones obtainable[71] from β-lactams by heating with phosphorus(V) sulfide to azetidines by treatment with excess Raney nickel.

13.3   Bibliography

H. Bestian in Houben-Weyl, Methoden der organischen Chemie, Bd. XI/2, p. 223, Georg Thieme Verlag, Stuttgart 1958.

[60] F. Weygand, Chem. Ber. 99, 1932 (1966).
[61] N.J. Leonard, K. Jann, J. Amer. Chem. Soc. 82, 6418 (1960);
N.J. Leonard, K. Jann, J. Amer. Chem. Soc. 84, 4806 (1962).
[62] J.V. Michael, W.A. Noyes jr., J. Amer. Chem. Soc. 85, 1228 (1963).
[63] J.R. Anderson, N.J. Clark, J. Catalysis 5, 250 (1966).
[64] G.D. Jones, J. Org. Chem. 9, 484 (1944).
[65] C. Mannich, G. Baumgarten, Ber. 70, 210 (1937).
[66] L. Fowden, Biochem. J. 64, 323 (1956).
[67] H.D. Hartzler, J. Org. Chem. 29, 1194 (1964).

[68] DBP. 1 144 718 (1963), Erf.: R. Graf, D. Guenther, H. Jansen, K. Matterstock.
[69] A. Bonati, G.F. Cristiani, E. Testa, Justus Liebigs Ann. Chem. 647, 813 (1961).
[70] E. Testa, Justus Liebigs Ann. Chem. 626, 114 (1959);
E. Testa, Justus Liebigs Ann. Chem. 633, 56 (1960);
E. Testa, Justus Liebigs Ann. Chem. 635, 119 (1960);
E. Testa, Justus Liebigs Ann. Chem. 647, 92 (1961);
E. Testa, Justus Liebigs Ann. Chem. 639, 157 (1961).
[71] A. Spasov, B. Panaitova, E. Golovinskij, Dokl. Akad. Nauk SSSR 158, 429 (1964).

O.C. Dermer, G.E. Ham, Ethylenimine and Other Aziridines, Academic Press, New York, London 1969.

A. Weissberger, The Chemistry of Heterocyclic Compounds, Vol. XIX, Interscience Publishers, New York 1964.

R.C. Elderfield, Heterocyclic Compounds, Vol. I, John Wiley & Sons, New York 1950.

R.H. Acheson, An. Introduction to the Chemistry of Heterocyclic Compounds, Wiley Intersci. Publishers, New York, London, Sydney 1960.

W.L. Mosby, Heterocyclic Systems with Bridgehead Nitrogen Atoms, Wiley Intersci. Publishers, New York, London, Sydney 1961.

Linda Lee Müller, J. Hamer, 1,2-Cycloaddition Reactions, The Formation of Three and Four Membered Heterocycles, Wiley Intersci. Publishers, New York, London, Sydney 1967.

W. Lwowski, Nitrenes, Wiley Intersci. Publishers, New York, London, Sydney, Toronto 1970.

J.C. Sheehan, E.J. Corey, Organic Reactions, Vol. IX, John Wiley & Sons, New York 1957.

14 Amino Carboxylic Acids

Contributed by

G. Krüger
Dr. Karl Thomae GmbH
Biberach an der Riß

14.0 Introduction

Strictly speaking, amino acids are all those organic acids which contain amino groups in the same molecule. The acid function may be a carboxy, sulfonyl, or phosphonyl group; the simplest aliphatic representatives of such amino acids have all been found in natural organisms:

$$H_2N-CH_2-COOH$$

Glycine
(aminoacetic acid)

$$H_2N-CH_2-CH_2-SO_3H$$

Taurine (2-amino-
ethylsulfonic acid)

$$H_2N-CH_2-CH_2-PO(OH)_2$$

2-Aminoethyl-
phosphonic acid

By comparison to amino carboxylic acids, amino sulfonic and amino phosphonic acids are relatively unimportant and will not be discussed in what follows.

The appearance of the amino and carboxy functions in one and the same molecule is not tied to a particular structure; even compounds as different as *p-aminobenzoic acid, lysergic acid,* or *phenylalanine* are all amino acids:

4-Aminobenzoic acid

Lysergic acid

Phenylalanine

Amino acids in which, like in phenylalanine, amino and the carboxy groups are linked to the same saturated carbon chain possess particular biological significance. Amino acids are distinguished as α-, β-, γ-...ω-amino acids according to the mutual position of the amino and carboxy groups. In iminocarboxylic acids *(carboxyazacycloalkanes)* the nitrogen atom is part of a ring.

$$R-\underset{\underset{NH_2}{|}}{CH}-COOH$$

α-Amino
carboxylic acids

$$R-\underset{\underset{NH_2}{|}}{CH}-CH_2-COOH$$

β-Amino
carboxylic acids

α-imino
carboxylic acids

Many of these amino acids contain further functional groups in the R group, *e.g.*, $-OH$, $-SH$, $-SR'$, $-COOH$, $-NH_2$, heterocycles, *etc.* These impart additional specific properties to them which enable the amino acids to be classified into

Neutral amino acids (without an additional functional group or with hydroxy, thiol or thioalkyl groups

Basic amino acids (with nitrogen substituents)

Acid amino acids (with an additional carboxy group)

α-Amino carboxylic acids are constituents of the naturally widespread peptides and proteins. Many biologically active oligopeptides and polypeptides are being produced synthetically, some on a technical scale. They include kinins, hormones, enzymes, releasing factors, antibiotics, *etc.* (cf. Volume VIII). On account of the special importance of the α-amino acids and because of the large number of different structures, numerous methods have been developed for their preparation; these form the substantive content of the following section. Naturally, a large portion of the techniques named can be employed also for synthesizing amino acids which do not occur in nature.

14.1 Preparation of amino carboxylic acids

Various different possible methods are fundamentally available:

a *Introduction of amino groups* into carboxylic acids by methods used to prepare amines (cf. p. 560)

b *Introduction of carboxy groups* into amines by methods used to prepare carboxylic acids (cf. Volume V, Preparation of carboxylic acids)

c *Introduction of amino and carboxy groups* into compounds containing other functional groups

d *Conversion* of simple, easily accessible amino acids into those of complex structure

e *Special* techniques

As in most amino acids the carbon atom carrying the amino group is chiral, the normal chemical syntheses lead to *racemates*. Where further chirality centers are present *diastereomeric pairs* are obtained whose relative ratios can vary with the nature of the synthetic process. A number of different procedures have been developed for the *separation* into the optical isomers; these will not be discussed here (cf. ref. [1,2]).

The *choice* of a suitable technique for preparing a particular amino acid is determined by various factors which include both the structure and the properties of the desired compound and also the accessibility of a suitable starting material. Synthesizing the latter, alone, can represent a separate problem and the stability of the intermediates under the various reaction conditions, too, has to be considered. In addition, the time needed, the desired quality, the cost, or a special labeling with radioactive isotopes play a part. Finally, the methods available for isolating the synthesized amino acids need to be considered. As they exist as *zwitterions* in neutral solution, their solubility is often little different from that of the inorganic salts in the reaction mixture, and so separation by extraction with organic solvents is often not possible. Frequently, these difficulties can be overcome by making use of ion exchange or gel chromatography.

Tabel 1 surveys important and frequently used methods.

Table 1. Important, and often generally applicable methods for synthesizing amino carboxylic acids

Starting product	Reaction with	Reaction product	Free amino acid by	Page
α-Amino carboxylic acids	$R^1-CH-COOH$ $\quad\quad\mid$ $\quad\quad NH_2$			
$R^1-CH-COOH$ $\quad\quad\mid$ $\quad\quad Hal$	NH_3	$R^1-CH-COOH$ $\quad\quad\mid$ $\quad\quad NH_2$	—	607
	$C_6H_{12}N_4$ Hexamethylenetetramine	$R^1-CH-COOH$ $\quad\quad\mid$ $\overset{\oplus}{N}(C_6H_{12}N_3)\cdot Hal^{\ominus}$	Acid hydrolysis	608
$R^1-CH-COOR^2$ $\quad\quad\mid$ $\quad\quad Hal$	(phthalimide potassium salt) $\text{N}-\text{K}$	$R^1-CH-COOR^2$ $\quad\quad\mid$ (N-phthalimido)	a) Acid hydrolysis b) 1. Ester saponification ($OH^{\ominus}$) 2. N_2H_4, H_2O	607
	NaN_3	$R^1-CH-COOR^2$ $\quad\quad\mid$ $\quad\quad N_3$	1. Ester saponification ($OH^{\ominus}$) 2. Various reduction methods	607
$R^1-C-COOR^2$ $\quad\parallel$ $\quad O$	NH_2OH	$R^1-C-COOR^2$ $\quad\parallel$ $\quad NOH$	1. Ester saponification ($OH^{\ominus}$) 2. H_2/Catalyst	607
	$Ar-NH-NH_2$	$R^1-C-COOR^2$ $\quad\parallel$ $N-NH-Ar$	1. Ester saponification ($OH^{\ominus}$) 2. H_2/Catalyst	607
$R^1-CH-COOR^2$ $\quad\quad\mid$ $\quad\quad COOR^2$	$H_9C_4-NO_2/H_2SO_4$	$R^1-C-COOR^2$ $\quad\parallel$ $\quad NOH$	1. Ester saponification ($OH^{\ominus}$) 2. H_2/Catalyst	607
	$ArN_2^{\oplus}\cdot Hal^{\ominus}$	$R^1-C-COOR^2$ $\quad\parallel$ $N-NH-Ar$	1. Ester saponification ($OH^{\ominus}$) 2. H_2/Catalyst	607
CH_2-COOR^2 $\mid$ NO_2	R^1-Hal	$R^1-CH-COOR^2$ $\quad\quad\mid$ $\quad\quad NO_2$	1. Ester saponification ($OH^{\ominus}$) 2. H_2/Catalyst	609
$R^1-CH-COOH$ $\quad\quad\mid$ $\quad\quad COOH$	HN_3/H_2SO_4	$R^1-CH-COOH$ $\quad\quad\mid$ $\quad\quad NH_2$	—	611
$R^1-CH-COOR^2$ $\quad\quad\mid$ $\quad\quad CO-R^3$	HN_3/H_2SO_4	$R^1-CH-COOR^2$ $\quad\quad\mid$ $NH-CO-R^3$	Acid hydrolysis	611

Table 1 Continuation 1

Starting product	Reaction with	Reaction product	Free amino acid by	Page
furan–C(=O)–R^1	1. NH_2OH 2. Zn/CH_3COOH 3. Acid chloride	furan–CH(R^1)–NH–Acyl	1. $KMnO_4$ 2. Acid hydrolysis	613
R^1–CHO	HCN/NH_3	R^1–CH(NH_2)–CN	Alkaline or acid hydrolysis	604
R^1–CH(CN)–$COOR^2$	1. $N_2H_4 \cdot H_2O$ 2. HNO_2	R^1–CH(CN)–CO–N_3	1. Heating 2. Acid hydrolysis	610
	1. $OH^{\ominus}$ (Ester saponification) 2. Conc. H_2SO_4	R^1–CH(CO–NH_2)–COOH	Br_2, $OH^{\ominus}$	610
Aryl–CHO	H_5C_6–CO–NH–CH_2–COOH	Aryl–CH= oxazolone (C_6H_5)	a) HI/red phosphorus b) 1. H_2/Catalyst 2. Alkaline or acid hydrolysis	605
	hydantoin	Aryl–CH= hydantoin	1. H_2/Catalyst 2. Alkaline or acid hydrolysis	606
Acyl–HN–CH(COOR²)–$COOR^2$	R^1–Hal and other alkylating agents	Acyl–HN–C($COOR^2$)(R^1)–$COOR^2$	Acid hydrolysis	609
	H_2C=CH–Y Y = functional group	Acyl–HN–C($COOR^2$)(CH_2–CH_2–Y)–$COOR^2$	Acid hydrolysis	610
Acyl–HN–CH(CN)–$COOR^2$	R^1–Hal and other alkylating agents	Acyl–HN–C($COOR^2$)(R^1)–CN	Acid hydrolysis	609
N-Substituted α-amino carboxylic acids		R^1–CH(NH–R^2)–COOH		
R^1–CH(Hal)–COOH	R^2–NH_2	R^1–CH(NH–R^2)–COOH	—	606
R^1–CHO	HCN/R^2–NH_2	R^1–CH(NH–R^2)–CN	Alkaline or acid hydrolysis	604
α-Substituted α-amino carboxylic acids		R^1–C(R^2)(NH_2)–COOH		
R^2R^1C=O	HCN/NH_3	R^1–C(R^2)(NH_2)–CN	Alkaline or acid hydrolysis	604

Table 1 Continuation 2

Starting product	Reaction with	Reaction product	Free amino acid by	Page
α-Amino-β-hydroxy carboxylic acids		$R^1-CH-CH-COOH$ $\quad\;\;OH\quad NH_2$		
R^1-CHO	$H_2N\!-\!\!\diagup\!\!\diagdown\!\!O$ (Cu complex, /2)	$R-CH(OH)-$ (Cu glycinate complex)	1. Acid hydrolysis 2. H_2S	615
β-Amino carboxylic acids		$R^1-CH-CH_2-COOH$ $\quad\;\;NH_2$		
$R^1-CH-CH_2-COOR^2$ $\quad\;Hal$	phthalimide N–K	$R^1-CH-CH_2-COOR^2$ (N-phthalimido)	a) Acid hydrolysis b) 1. Ester saponification ($OH^\ominus$) $\quad$ 2. $N_2H_4\cdot H_2O$	608
$R^1-C-CH_2-COOR^2$ $\quad\;\overset{\|}{O}$	NH_2OH	$R^1-C-CH_2-COOR^2$ $\quad\;\overset{\|}{NOH}$	1. Ester saponification ($OH^\ominus$) 2. H_2/Catalyst	608, 609
$R^1-CH\!=\!CH-COOH$	NH_2OH	$R^1-CH-CH_2-COOH$ $\quad\;\;NH_2$	—	
$R^1-CH\!=\!CH-COOR^2$	$C_6H_5-CH_2-NH_2$	$R^1-CH-CH_2-COOR^2$ $\quad\;NH-CH_2-C_6H_5$	1. Acid hydrolysis 2. H_2/Catalyst	616
$R^1-CH-CH_2-COOR^2$ $\quad\;CO-R^3$	HN_3/H_2SO_4	$R^1-CH-CH_2-COOR^2$ $\quad\;NH-CO-R^3$	Acid hydrolysis	612
$R^1-CH-CO-Cl$ $\quad\;NH-Acyl$	CH_2N_2	$R^1-CH-CO-CHN_2$ $\quad\;NH-Acyl$	1. AgO/CH_3OH 2. Acid hydrolysis	612
furan–CHO	1. $R^1-CH_2-NO_2$ 2. $LiAlH_4$ 3. Acid chloride	furan–CH_2-CH-R^1 $\qquad\qquad NH-Acyl$	1. $KMnO_4$ 2. Acid hydrolysis	613
$\gamma,\ \delta\ldots$-Amino carboxylic acids		$R^1-CH-(CH_2)_n-COOH$ $\quad\;\;NH_2$		
$R^1-CH-(CH_2)_n-COOR^2$ $\quad\;Hal$	phthalimide N–K	$R^1-CH-(CH_2)_n-COOR^2$ (N-phthalimido)	a) Acid hydrolysis b) 1. Ester saponification ($OH^\ominus$) $\quad$ 2. $N_2H_4,\ H_2O$	608

Table 1 Continuation 3

Starting product	Reaction with	Reaction product	Free amino acid by	Page
$R^1-\underset{\underset{O}{\parallel}}{C}-(CH_2)_n-COOR^2$	NH_2OH	$R^1-\underset{\underset{NOH}{\parallel}}{C}-(CH_2)_n-COOR^2$	1. Ester saponification ($OH^\ominus$) 2. H_2/Catalyst	608
$R^1-\underset{\underset{CO-R^3}{\mid}}{CH}-(CH_2)_n-COOR^2$	HN_3/H_2SO_4	$R^1-\underset{\underset{NH-CO-R^3}{\mid}}{CH}-(CH_2)_n-COOR^2$	Acid hydrolysis	611
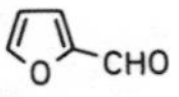 ⟨furan⟩CHO	1. CH_3-CO-R^1 2. NH_3/H_2-Catal. 3. Acid chloride	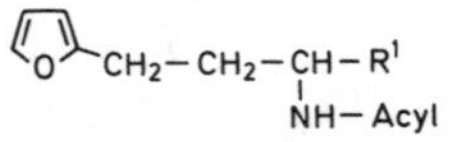 ⟨furan⟩$CH_2-CH_2-\underset{\underset{NH-Acyl}{\mid}}{CH}-R^1$	1. $KMnO_4$ 2. Acid hydrolysis	613

ω-Amino carboxylic acids $H_2N-(CH_2)_n-COOH$ (n > 3)

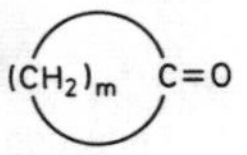 $(CH_2)_m\ C=O$	1. NH_2OH 2. $H^\oplus$ for m ⩾ 4	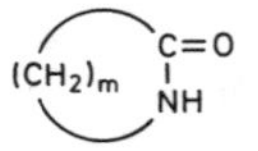 $(CH_2)_m\underset{NH}{\overset{C=O}{\big\langle}}$	Acid or alkaline hydrolysis	612
	HN_3/H_2SO_4 for m ⩾ 3	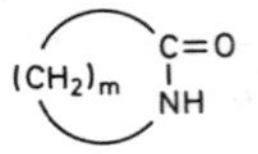 $(CH_2)_m\underset{NH}{\overset{C=O}{\big\langle}}$	Acid or alkaline hydrolysis	612

14.1.1 Preparation from aldehydes and ketones

14.1.1.1 Syntheses with derivatives of hydrogen cyanide and ammonia

In the presence of *ammonia,* aldehydes add on *hydrogen cyanide* to form α-aminonitriles. The subsequent, preferably acid hydrolysis affords α-amino carboxylic acids *(Strecker synthesis[3]),* while from ketones the corresponding α-substituted α-amino acids are obtained[4]. According to the starting material and the order of addition of ammonia and prussic acid either an aldehyde-ammonia adduct or a cyanohydrin (α-hydroxynitrile) are formed intermediately:

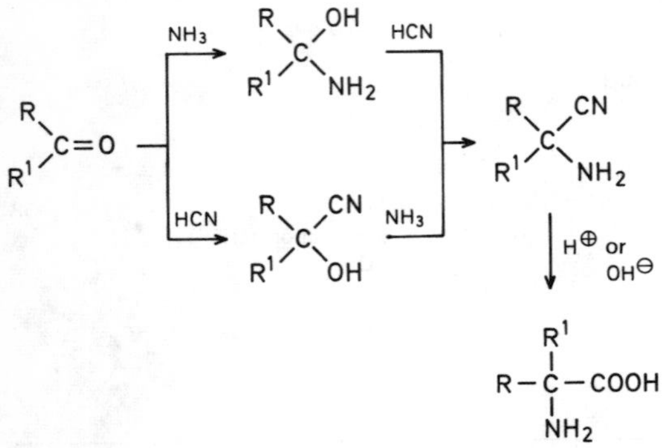

Numerous modifications aiming at both increasing the sometimes only moderate yield and facilitating the handling of the toxic reagents have been described, for instance, the use of aldehyde-bisulfite adducts or of alkali metal cyanide and ammonium chloride[5]. Reacting the cyanohydrins with urea or ammonium carbonate furnishes 5-substituted hydantoins, which generally crystal-

[1] *Th. Wieland, R. Müller, E. Niemann, L. Birkofer, A. Schöberl, A. Wagner, H. Söll,* in *Houben-Weyl,* Methoden der organischen Chemie, 4. Aufl., Bd. XI/2, p. 267, Georg Thieme Verlag, Stuttgart 1958.

[2] *J.P. Greenstein, M. Winitz,* Chemistry of the Amino Acids, Vol. I—III, John Wiley & Sons, New York, London 1961;

E. Schröder, K. Lübke, The Peptides. Methods of Peptide Synthesis, Vol. I, Academic Press, New York, London 1965;

N.V. Sidgwick in *I.T. Millar, H.D. Springall,* The Organic Chemistry of Nitrogen, 3. Ed., p. 195, Clarendon Press, Oxford 1966.

[3] *D.T. Mowry,* Chem. Rev. *42,* 236 (1948).

[4] *D.F. Reinhold, R.A. Firestone, W.A. Gaines, J.M. Chemerda, M. Sletzinger,* J. Org. Chem. *33,* 1209 (1968).

[5] *R. Gaudry,* Can. J. Res. *24B,* 301 (1946);

R. Gaudry, Can. J. Res. *26B,* 773 (1948).

lize well and can be saponified to the *α-amino acids* preferably with alkali *(Bucherer-Bergs modification)*[6]:

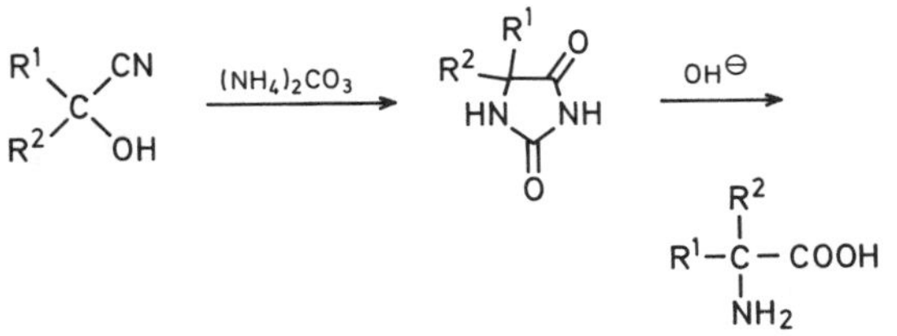

The use of *amines* in place of ammonia yields *N*-substituted α-amino acids[7] the *N-benzyl* derivatives of which can be converted into the free α-amino acids by hydrogenolysis[8]. Analogous asymmetric syntheses with optically active α-methylbenzylamine are feasible[9] (optical yield = 90 to >98%). Some *α-amino-β-hydroxy carboxylic acids* are accessible in up to 85% yield if 1-alkoxy-1,2-diacetoxyalkanes are used in place of the unstable α-hydroxyaldehydes[10].

The *advantages* of the Strecker synthesis lie in the variety of the α-amino acids that can be prepared. In addition, ^{14}C-labeling of the *C–1*[11] or *C–2*[12] is readily accomplished. A *disadvantage* is the high toxicity of prussic acid and its salts.

14.1.1.2 Syntheses with nitrogen-containing heterocycles

Carbonyl compounds condense with the activated methylene group of mostly cyclic glycine derivatives (e.g. 2-oxazolin-5-ones) by linking up of two carbon atoms. Some of these cyclic glycine derivatives react predominantly only with aromatic and heteroaromatic aldehydes, but give well-crystal-

lizing intermediate products, usually in good yield. Others can be reacted also with aliphatic aldehydes and with ketones.

In the original form of this principle *(Erlenmeyer synthesis)*[13] an aromatic aldehyde and an *N*-acylglycine give *4-alkylidene-2-oxazolin-5-ones* (unsaturated azlactone)[14] by heating in acetic anhydride in the presence of sodium acetate. Reduction and hydrolysis afford α-amino acids; whether *4-alkyl-2-oxazolin-5-ones* (saturated azlactone) or the *α,β-unsaturated amino acid* is formed first depends on the sequence of the operation adopted[13, 15]. By boiling with hydriodic acid and red phosphorus the reaction can be performed in one step[16]:

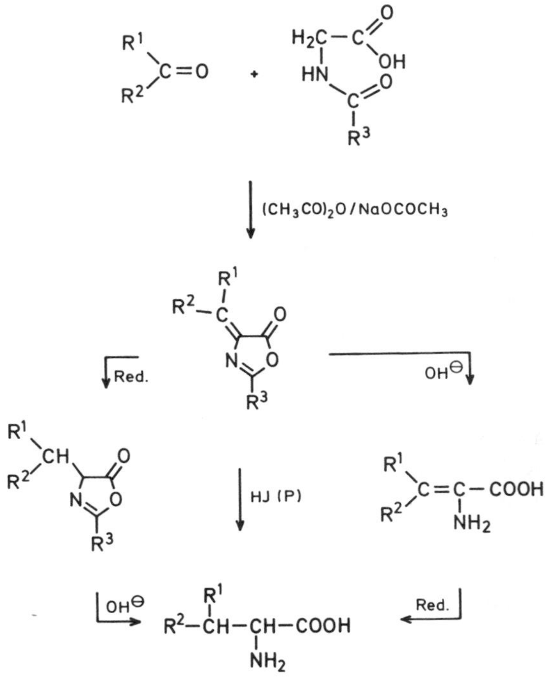

Unsaturated substituted azlactones are obtained also from 4-(chloromethylene)-2-phenyl-2-oxazolin-5-one with aromatic or heteroaromatic organometallic compounds or by Friedel-Crafts alkylation[17]. In this way synthesizing an aldehyde that is difficult to prepare can sometimes be avoided.

The essence of the method lies in the successful preparation of *aromatically* and *heteroaromatically* substituted α-amino acids[18]. Only in the pres-

[6] *E. Ware*, Chem. Rev. *46*, 422 (1950);
M. Viscontini, H. Raschig, Helv. Chim. Acta *42*, 570 (1959).

[7] *W. Staudt*, Z. Physiol. Chem. *146*, 286 (1925);
H. Zahn, H. Wilhelm, Justus Liebigs Ann. Chem. *579*, 1 (1953).

[8] French. P. 1 338 491 (1963), Inv.: *P.M. Theil*; C.A. *60*, 8129b (1964).

[9] *K. Harada*, Nature *200*, 1201 (1963);
M.S. Patel, M. Worsley, Can. J. Chem. *48*, 1881 (1970).

[10] *H. Geipel, J. Gloede, K.P. Hilgetag, H. Gross*, Chem. Ber. *98*, 1677 (1965).

[11] *L. Pichat, J.-P. Guermont, P.N. Liem*, J. Label. Compounds *4*, 251 (1968).

[12] *L. Pichat, P.N. Liem, J.-P. Guermont*, Bull. Soc. Chim. France 837 (1971).

[13] *J. Plöchl*, Ber. *16*, 2815 (1883);
J. Plöchl, Ber. *17*, 1616 (1884).

[14] *E. Baltazzi*, Quart. Rev., Chem. Soc. *9*, 150 (1955).

[15] *E. Erlenmeyer jr.*, Justus Liebigs Ann. Chem. *275*, 1 (1893);
R.M. Herbst, D. Shemin, Org. Syntheses Coll. Vol. II, 491 (1943);
L.H. Briggs, G.C. DeAth, S.R. Ellis, J. Chem. Soc. 61 (1942).

ence of lead(II) acetate as condensing agent it is only sometimes possible to use aliphatic aldehydes[17, 18] and ketones[20, 21] in the Erlenmeyer synthesis.

By means of similar condensations and *via* comparable reaction stages α-amino acids can be prepared also especially from aromatic aldehydes with hydantoin (imidazolidine-2, 4-dione) **1**[22, 23] (cf. p. 617), thiohydantoin (2-thioxo-2, 4-imidazoline-dione) **2**[22, 24], and piperazine-2, 5-dione **3**[25]. With creatinine (2-imino-1-methyl-4-imidazolinone) **4**[26] their *N*-methyl derivatives are formed:

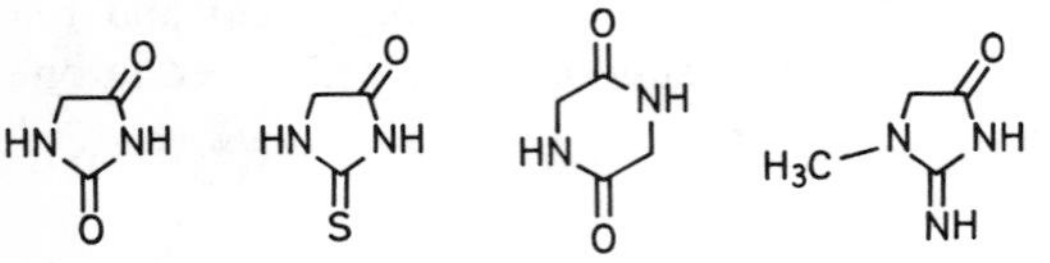

2-Mercapto-2-thiazolin-5-one **5**[27] and 2-benzyl-2-imidazolin-4-one **6**[28] react with both aliphatic aldehydes and with ketones

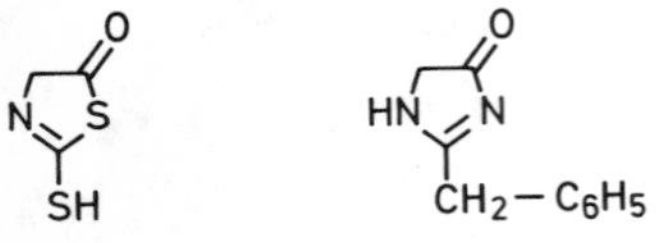

Rhodanine **7**, too, is accessible to condensation with aldehydes. The unsaturated compounds **8** and **9** obtained can be converted into α-amino acids *via* **11** the oximes of α-keto carboxylic acids **10**[29] (cf. p. 608).

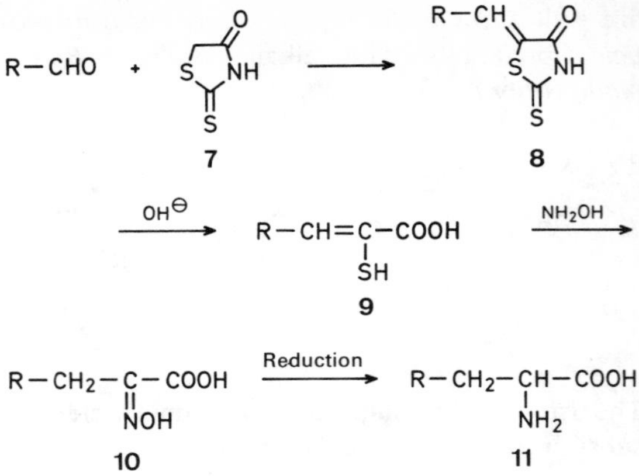

Regarding further condensations with otherwise activated methylene groups, see p. 614.

14.1.2 Preparation from halo carboxylic acids

Amino acids can be obtained from halo carboxylic acids by reacting with ammonia (or amines), potassium phthalimide, hexamethylenetetramine, or sodium azide. With *ammonia*, α-amino acids can be prepared in one step. Primary and secondary *amines* furnish the corresponding *N*-substituted compounds. Side-products cannot always be avoided, other groups (double bonds, esters) may react simultaneously. Introducing an amino group on a carbon atom that is not in the α-position succeeds only rarely in this way (formation of α,β-unsaturated carboxylic acids, lactones, *etc*.). The use of the other reagents leads to a reaction *via* intermediate products but side-reactions do not occur. Other groups in the molecule generally do not react.

By using *potassium phthalimide* the amino group can, in addition, be introduced at any desired place in the molecule. Which one of the methods is

[16] H.B. Gillespie, H.R. Snyder, Org. Syntheses Coll. Vol. II, 489 (1943).

[17] H. Behringer, H. Taul, Chem. Ber. 90, 1398 (1957).

[18] R.K. Griffith, H.J. Harwood, J. Org. Chem. 29, 2658 (1964).

[19] E. Baltazzi, R. Robinson, Chem. & Ind. (London) 191 (1954).

[20] M. Crawford, W.T. Little, J. Chem. Soc. 729 (1959).

[21] E. Galantay, A. Szabo, J. Fried, J. Org. Chem. 28, 98 (1963).

[22] H.L. Wheeler, C. Hoffman, Amer. Chem. J. 45, 368 (1911);
T.B. Johnson, J.S. Bates, J. Amer. Chem. Soc. 38, 1087 (1916).

[23] K. Hiroi, K. Achiwa, S. Yamada, Chem. Pharm. Bull. (Tokyo) 16, 444 (1968).

[24] T.B. Johnson, B.H. Nicolet, J. Amer. Chem. Soc. 33, 1973 (1911);
R. Majima, Ber. 55, 3859 (1922);
H. Behringer, K. Schmidt, Chem. Ber. 90, 2510 (1957).

[25] T. Sasaki, Ber. 54, 163 (1921).

[26] V. Deulofeu, T.J. Guerrero, Org. Syntheses Coll. Vol. III, 586 (1955).

[27] A.H. Cook, I.M. Heilbron, A.L. Levy, J. Chem. Soc. 1594 (1947);
J.D. Billimoria, A.H. Cook, J. Chem. Soc. 2323 (1949);
Brit. P. 666 226 (1952); Beecham Res. Lab. Ltd., Inv.: I.M. Heilbron, A.H. Cook; C.A. 46, 11232h (1952);
K. Balenović, I. Jambrešic, J. Furić, J. Org. Chem. 17, 1459 (1952).

[28] K. Okubo, Y. Izumi, Bull. Chem. Soc. Japan 43, 1541 (1970).

[29] C. Gränacher, Helv. Chim. Acta 5, 610 (1922);
C. Gränacher, M. Gero, A. Ofner, A. Klopfenstein, E. Schlatter, Helv. Chim. Acta 6, 458 (1923);
R. Gaudry, R.A. McIvor, Can. J. Chem. 29, 427 (1951).

chosen is governed by the conditions under which the intermediate products can be converted into amino acids.

14.1.2.1 Syntheses with ammonia and amines

α-Halo carboxylic acids react with aqueous, alcoholic, or liquid *ammonia* to form α-amino carboxylic acids:

$$\underset{\underset{R}{|}}{Hal-CH-COOH} \xrightarrow{\ NH_3\ } \underset{\underset{R}{|}}{H_2N-CH-COOH}$$

$$+ \left[\underset{\underset{R}{|}}{HN(CH-COOH)_2} + \underset{\underset{R}{|}}{N(CH-COOH)_3} \right]$$

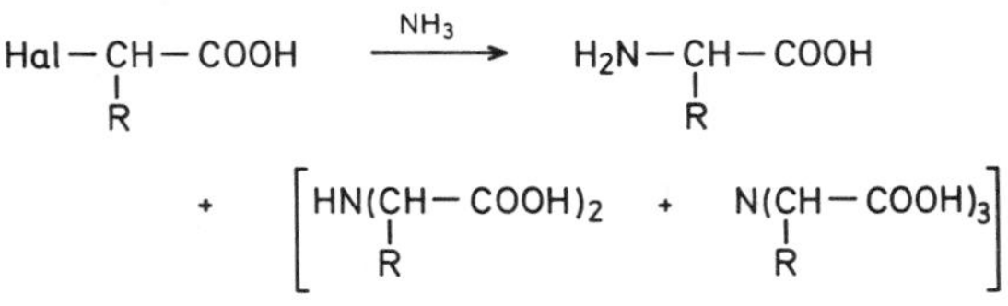

A large excess of ammonia in the presence of ammonium carbonate[30] or formaldehyde[31] substantially suppresses formation of secondary and tertiary amines. With *alkylamines*[32] and *dialkylamines*[33] *N-substituted* and *N,N-disubstituted* amino acids form. When optically active α-bromocarboxylic acids are employed, the reaction generally proceeds with inversion without racemization[34]. β-Substituted α-bromocarboxylic acids, however, are aminated while retaining their configuration[35]. With α-halo carboxylic esters an excess of ammonia or amines generally leads to *α-amino carboxamides*[36], which can be hydrolyzed to the free amino carboxylic acids.

This simplest general process for synthesizing α-amino carboxylic acids is limited solely by the accessibility of the α-halo carboxylic acids; various techniques are known for their preparation[37] (cf. Volume V, Chapter 8). In this connection reference is made also both to synthesis of substituted α-chloro carboxylic acids from $\alpha,\alpha,\alpha,\omega$-tetrachloroalkanes and the related preparation of certain amino acids[38].

14.1.2.2 Syntheses with other nitrogen compounds

α-Halo carboxylic acids or their esters yield α-*phthalimido carboxylic acids* or their esters with potassium phthalimide, especially in dimethylformamide as solvent[39]. Energetic acid hydrolysis[40] or treatment with hydrazine after a preliminary or before a subsequent saponification of the ester group[39, 41] leads to the free amino acid. Methylamine, too, can be used for cleaving the phthaloyl group[42]:

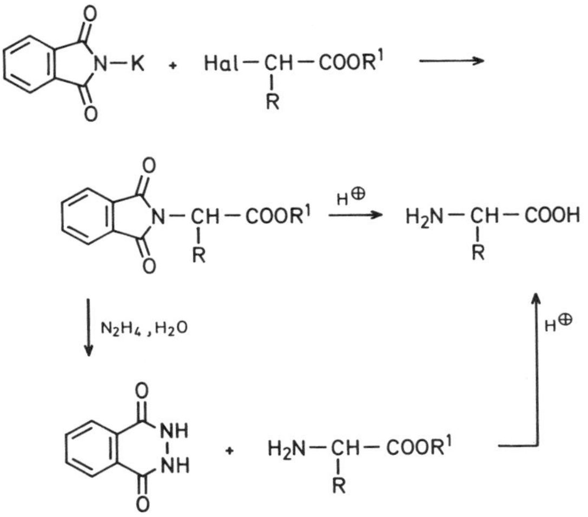

The method is *versatile* in application and generally gives good yields. It can be used also for introducing amino groups that are not in the α-position[43] and is well suited for obtaining ^{15}N-*labeled* amino acids[44].

[30] *N.D. Cheronis, K.H. Spitzmueller*, J. Org. Chem. *6*, 349 (1941).

[31] *G.N. Kulikova, I.N. Strukov*, Khim.-Farm. Zh. *3*, 30 (1969); C.A. *71*, 102 208 (1969).

[32] *D. Billet*, Compt. Rend. *242*, 2159 (1956).

[33] *L. Michaelis, M.P. Schubert*, J. Biol. Chem. *115*, 221 (1936).

[34] *P. Brewster, F. Hiron, E.D. Hughes, C.K. Ingold, P.A.D.S. Rao*, Nature *166*, 178 (1950);
M.B. Watson, G.W. Youngson, J. Chem. Soc. 2145 (1954).

[35] *C.H. Stammer, R.G. Webb*, J. Org. Chem. *34*, 2306 (1969), Fußnote 14.

[36] *J. Volhard*, Justus Liebigs Ann. Chem. *123*, 261 (1862).

[37] *C.S. Marvel*, Org. Syntheses, Coll. Vol. III, 495, 523, 848 (1955);
A.M. Iurkevich, A.V. Dombrowskii, A.P. Terent'ev, J. Gen. Chem. USSR (Eng. Transl.) *28*, 226 (1958);
G.H. Cleland, J. Org. Chem. *34*, 744 (1969).

[38] *A.N. Nesmeyanov, R.Kh. Freidlina, V.N. Kost, T.T. Vassilyeva, B.V. Kopylova*, Tetrahedron *17*, 69 (1962);
K. Saotome, T. Yamazaki, Bull. Chem. Soc. Japan *36*, 1264 (1963).

[39] *J.C. Sheehan, W.A. Bolhofer*, J. Amer. Chem. Soc. *72*, 2786 (1950).

[40] *S. Gabriel, K. Kroseberg*, Ber. *22*, 426 (1889).

[41] *F.E. King, R. Robinson*, J. Chem. Soc. 1433 (1932).

[42] *S. Wolfe, S.K. Hasan*, Can. J. Chem. *48*, 3572 (1970).

[43] *T. Wieland, R. Müller, E. Niemann, L. Birkofer, A. Schöberl, A. Wagner, H. Söll* in *Houben-Weyl*, Methoden der organischen Chemie, 4. Aufl., Bd. XI/2, p. 310, Georg Thieme Verlag, Stuttgart 1958.

[44] *R. Schönheimer, S. Ratner*, J. Biol. Chem. *127*, 301 (1939).

An economic technique in terms of material and effort consists in heating α-halo carboxylic acids with hexamethylenetetramine in anhydrous medium *(Délépine reaction)*[45, 46]. The initially formed quaternary salts are hydrolyzed to α-amino carboxylic acids.

$$H_{12}C_6N_4 \; + \; Hal-\underset{\underset{R}{|}}{CH}-COOH \longrightarrow$$

$$\left[(H_{12}C_6N_3)\overset{\oplus}{N}-\underset{\underset{R}{|}}{CH}-COOH \right] Hal^{\ominus} \xrightarrow{H^{\oplus}} H_2N-\underset{\underset{R}{|}}{CH}-COOH$$

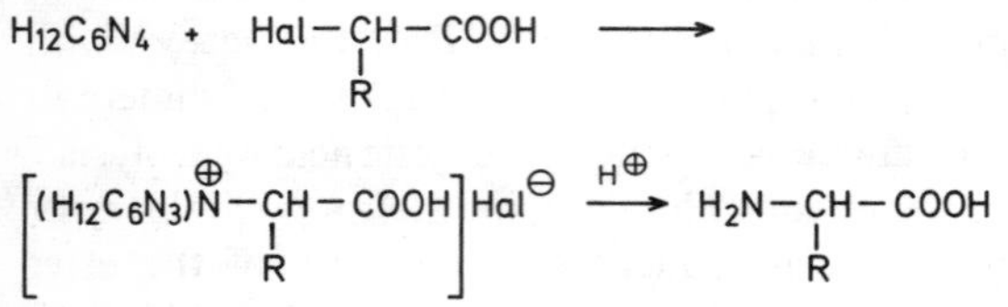

Unlike with the above-named methods, the *α-azido carboxylic acid esters*[47] obtained from α-halo carboxylic acid esters and sodium azide are converted into α-amino carboxylic acids by reduction. After saponification of the ester group[47], this reduction can be performed under widely varying conditions, namely, catalytically[47], with aluminum amalgam[48], with hydrogen bromide in glacial acetic acid[49], or with triphenylphosphine[50].

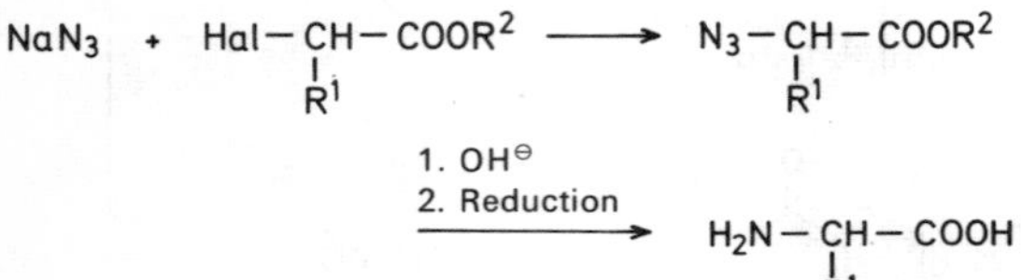

$$NaN_3 \; + \; Hal-\underset{\underset{R^1}{|}}{CH}-COOR^2 \longrightarrow N_3-\underset{\underset{R^1}{|}}{CH}-COOR^2$$

$$\xrightarrow[\text{2. Reduction}]{\text{1. } OH^{\ominus}} H_2N-\underset{\underset{R^1}{|}}{CH}-COOH$$

14.1.3 Preparation from nitrogen-containing carboxylic acid derivatives by reduction

14.1.3.1 Reduction of keto carboxylic acid derivatives

α-Amino carboxylic acid derivatives are obtained from oximes and phenylhydrazones of α-oxo-carboxylic acids or their esters by chemical[51] or catalytic[52, 53] reduction and, in the case of oximes, also by electrolytic[54] reduction. The required starting products can be prepared from the corresponding α-keto acid derivatives (see Volume V, Chapter 8), but above all by direct synthesis from more easily accessible precursors. Thus, *α-hydroxyimino* carboxylic acid esters are obtained by allowing esters of nitrous acid to act on substituted malonic diesters or acetic esters[52, 55], while *α-alkoxyimino* carboxylic acids are obtained from α-alkoxyimino ketones by haloform reaction[56], *α-hydroxyimino* carboxylic acids from alkylidene-rhodanines(5-alkylidene-2-thioxo-1,3-thiazolidin-4-ones)[54] (cf. p. 606). The *phenylhydrazones* are prepared from the same starting products or cyanoacetic acid esters by a Japp-Klingemann[57] reaction (cf. this Volume, p. 87).

Catalytic hydrogenation of α-oxocarboxylic acids in the presence of ammonia, too, leads to α-amino-carboxylic acids[58] and may be used to prepare *15N-labeled* amino acids[59].

The following scheme surveys the above possibilities:

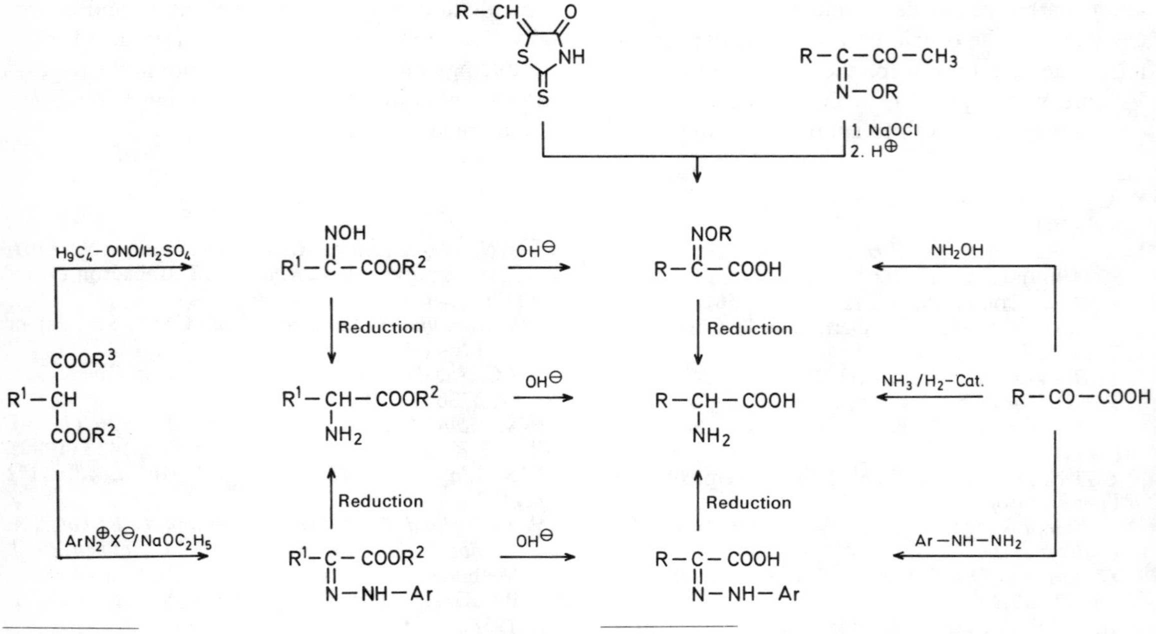

---

[45] *G. Hillmann, A. Hillmann*, Z. Physiol. Chem. *283*, 71 (1948).

[46] French P. 1 237 327 (1960), Soc. Prod. Chim. Ind., Inv.: *H.M. Guinot;* C.A. *55*, 25782i (1961).

From oximes and phenylhydrazones of other keto carboxylic acids, carboxylic acids with a non-α amino group can be prepared[60]. These techniques have lost much of their importance for general synthetic work, but for experiments aimed at an *asymmetric synthesis* of α-amino carboxylic acids they play a significant role. Suitable techniques include the use (a) of hydrogenation catalysts on optically active carrier material[61] and reduction of α-hydroxyimino carboxylic esters with optically active alcohols[62], (b) of azomethines from α-oxo carboxylic acids and optically active benzylamines[63], (c) of α-acylhydrazono carboxylic acids with an optically active acyl[64], (d) of α-hydrazonocarboxylic acids with N-amino derivatives of optically active tertiary bases[65]. The substrate and also the catalyst and solvent effect the optical yield (10–80%)[63].

In an extension of the last-named principle both antipodes of an α-amino carboxylic acid are obtained in up to 100% optical purity by condensing α-oxo carboxylic esters with the 1-amino-2-hydroxymethyl-indoline 12 and stereospecific reduction of the resulting hydrazonolactone 13 to the hydrazinolactone 14[66]. Hydrogenolysis and hydrolysis give the *optically pure* amino carboxylic acid 15 and the optically active indoline 16. The latter can be reused for preparing the starting material 12.

14.1.3.2 Reduction of other carboxylic acid derivatives

The following methods have been used less often hitherto, but if the precursors are readily acces-

[47] A. Bertho, J. Maier, Justus Liebigs Ann. Chem. 498, 50 (1932).

[48] M.O. Forster, H.E. Fierz, J. Chem. Soc. 93, 1859 (1908).

[49] T. Wieland, H. Urbach in Houben-Weyl, Methoden der organischen Chemie, 4. Aufl., Bd. XI/2, p. 354, Georg Thieme Verlag, Stuttgart 1958.

[50] L. Horner, A. Gross, Justus Liebigs Ann. Chem. 591, 117 (1951);
F. Lingens, Z. Naturforsch. B15, 811 (1960).

[51] V.V. Feofilaktov, N.K. Semenova, J. Gen. Chem. USSR (Engl. Transl.) 23, 463, 669, 887 (1953).

[52] K.E. Hamlin, W.H. Hartung, J. Biol. Chem. 145, 349 (1942).

[53] K.L. Walters, W.H. Hartung, J. Org. Chem. 10, 524 (1945).

[54] R. Gaudry, R.A. McIvor, Can. J. Chem. 29, 427 (1951).

[55] W.H. Hartung, A.N. Mattocks, R.I. Ellin, J. Pharm. Sci. 53, 550 (1964).

[56] K. Mori, Nippon Kagaku Zasshi 81, 464 (1960); C.A. 55, 5358a (1961);
A.F. Ferris, J. Org. Chem. 24, 1726 (1959).

[57] R.R. Phillips, Org. Reactions 10, 143 (1959);
G.B. Barlow, A.J. Macleod, J. Chem. Soc. 141 (1964).

[58] F. Knoop, H. Oesterlin, Z. Physiol. Chem. 148, 294 (1925);
F. Knoop, H. Oesterlin, Z. Physiol. Chem. 170, 186 (1927);
M. Murakami, J.-W. Kang, Bull. Chem. Soc. Japan 36, 763 (1963).

[59] T. Wieland, W. Paul, Ber. 77, 34 (1944).

[60] A.P. Terent'ev, N.I. Gusar, J. Gen. Chem. USSR (Engl. Transl.) 35, 125 (1965).

[61] S. Akabori, S. Sakurai, Y. Izumi, F. Fujii, Nature 178, 323 (1956).

[62] cf. H.D. Law, Ann. Rep. Progr. Chem. (Chem. Soc. London) 63, 519 (1966).

[63] R.G. Hiskey, R.C. Northrop, J. Amer. Chem. Soc. 83, 4798 (1961);
R.G. Hiskey, R.C. Northrop, J. Amer. Chem. Soc. 87, 1753 (1965);
cf. T.D. Inch, Synthesis 471 (1970).

[64] S. Akabori, S. Sakurai, Nippon Kagaku Zasshi 78, 1629 (1957); C.A. 53, 21 687b (1959).

[65] A.N. Kost, R.S. Sagitullin, M.A. Yurovskaja, Chem. & Ind. (London) 1496 (1966).

[66] E.J. Corey, R.J. McCaully, H.S. Sachdev, J. Amer. Chem. Soc. 92, 2476 (1970);
E.J. Corey, H.S. Sachdev, J.Z. Gougoutas, W. Saenger, J. Amer. Chem. Soc. 92, 2488 (1970).

[67] USP. 2 570 297 (1951), Upjohn Co., Inv.: D.I. Weisblat, D.A. Lyttle; C.A. 46, 5077g (1952).

[68] S. Zen, S. Kaji, Bull. Chem. Soc. Japan 43, 2277 (1970);
K.K. Babievskii, V.M. Belikov, Yu.N.Belokon, Bull. Acad. Sci. USSR, Div. Chem. Sci. (Engl. Transl.) 1188 (1965).

[69] H. Feuer, H. Hass, K. Warren, J. Amer. Chem. Soc. 71, 3078 (1949).

[70] N. Kornblum, Org. Reactions 12, 101 (1962).

[71] D.I. Weisblat, D.A. Lyttle, J. Amer. Chem. Soc. 71, 3079 (1949);
T. Okuda, Bull. Chem. Soc. Japan 32, 931, 1165 (1959).

sible also represent an advantageous route for synthesizing α-amino carboxylic acids.

Thus, saturated and unsaturated α-*nitro* carboxylic acids and their esters can be converted into α-amino carboxylic acids by reduction and, where applicable, hydrolysis of the ester group[67–74].

$$R^1-CH-COOR^2 \xrightarrow{OH^\ominus} R^1-CH-COOH$$
$$\quad\;\; | \qquad\qquad\qquad\qquad\quad | $$
$$\quad\;\; NO_2 \qquad\qquad\qquad\qquad\; NO_2$$

Reduction   Reduction

$$R^1-CH-COOR^2 \xrightarrow{OH^\ominus} R^1-CH-COOH$$
$$\quad\;\; | \qquad\qquad\qquad\qquad\quad | $$
$$\quad\;\; NH_2 \qquad\qquad\qquad\qquad\; NH_2$$

Reduction   Reduction

$$R^3-CH=C-COOR^2 \xrightarrow{OH^\ominus} R^3-CH=C-COOH$$
$$\qquad\qquad | \qquad\qquad\qquad\qquad\qquad | $$
$$\qquad\qquad NO_2 \qquad\qquad\qquad\qquad\quad NO_2$$

Concerning the reduction of α-azido carboxylic acid esters to α-amino acids see p. 608.

Aziridinecarboxylic acids[75] and *α-acetamido-acrylic acids*[76], too, can be converted into α-amino acids by hydrogenation.

Reduction of *β-hydroxyamino carboxylic acids* by excess hydroxylamine is of some importance for preparing *β*-amino acids from acrylic acids[77].

14.1.4 Preparation from amidomalonic acid diesters and similar compounds

Alkylation of *amidomalonic* acid diesters and of related compounds is at present the most important general method for synthesizing α-amino

carboxylic acids. It is simple to carry out, generally proceeds in high yield, and enables a multitude of different substituents to be introduced:

$$\qquad\qquad COOR^1$$
$$\qquad\qquad\;\; | $$
$$Acyl-HN-CH \; + \; X-R \quad\xrightarrow{Base}$$
$$\qquad\qquad\;\; | $$
$$\qquad\qquad COOR^1$$

$$\qquad\qquad COOR^1$$
$$\qquad\qquad\;\; | $$
$$Acyl-HN-C-R \quad\xrightarrow{H^\oplus} \quad H_2N-CH-COOH$$
$$\qquad\qquad\;\; | \qquad\qquad\qquad\qquad\qquad | $$
$$\qquad\qquad COOR^1 \qquad\qquad\qquad\qquad\; R$$

The original technique using diethyl phthalimidomalonate[78] has been variously modified, frequently in order to adapt it to the requirements in question[79]. Replacement of phthaloyl by benzoyl[80], acetyl[81], and formyl groups[82] is the most significant example. On account of its ready accessibility and reactivity dimethyl formamidomalonate[83] has given good results. *Amidocyanoacetic acid esters* may be used in addition to amidomalonic acid diesters, the phenacetyl group[84] and the acetyl group[85] being particularly suitable acyl substituents:

$$\qquad\qquad COOR^1$$
$$\qquad\qquad\;\; | $$
$$Acyl-HN-CH \; + \; X-R \quad\xrightarrow{Base}$$
$$\qquad\qquad\;\; | $$
$$\qquad\qquad CN$$

$$\qquad\qquad COOR^1$$
$$\qquad\qquad\;\; | $$
$$Acyl-HN-C-R \quad\xrightarrow{H^\oplus} \quad H_2N-CH-COOH$$
$$\qquad\qquad\;\; | \qquad\qquad\qquad\qquad\qquad | $$
$$\qquad\qquad CN \qquad\qquad\qquad\qquad\qquad R$$

Alkylation is performed with a variety of different alkylating agents[86] in the presence of a strong base such as sodium ethoxide in ethanol[81, 82, 84, 85] or sodium hydride in polar, aprotic solvents[87]. Alternatively, it can be carried out *via* a Michael

[72] *M. Stiles, H.L. Finkbeiner*, J. Amer. Chem. Soc. *81*, 505 (1959);
H.L. Finkbeiner, G.W. Wagner, J. Org. Chem. *28*, 215 (1963);
H.L. Finkbeiner, M. Stiles, J. Amer. Chem. Soc. *85*, 616 (1963).
[73] *Y. Mori, S. Kondo, H. Kumagae*, Nippon Kagaku Zasshi *78*, 1174 (1957); C.A. *54*, 5484d (1960).
[74] *K.K. Babievskii, V.M. Belikov, N.A. Tikhonova*, Bull. Acad. Sci. USSR, Div. Chem. Sci. (Engl. Transl.) 76, 733 (1965).
[75] Jap. P. 667 (1961), Osaka University, Inv.: *Y. Yukawa, S. Kimura;* C.A. *56*, 6088i (1962).
[76] Jap. P. 18 357 (1963), Asahi Chemical Industry Co., Ltd., Inv.: *M. Murakawi, S. Senoh, T. Matsusato, H. Itaya, M. Kyo;* C.A. *60*, 3090g (1964).
[77] *R.E. Steiger*, Org. Syntheses, Coll. Vol. III, 91 (1955).

[78] *S.P.L. Sörensen*, Compt. Rend. Trav. Lab. Carlsberg *6*, 1 (1903);
S.P.L. Sörensen, Z. Physiol. Chem. *44*, 448 (1905).
[79] s.a. *H. Gershon, J. Shapira, J.S. Meek, K. Dittmer*, J. Amer. Chem. Soc. *76*, 3484 (1954).
[80] *M.S. Dunn, B.W. Smart, C.E. Redemann, K.E. Brown*, J. Biol. Chem. *94*, 599 (1931).
[81] *H.R. Snyder, C.W. Smith*, J. Amer. Chem. Soc. *66*, 350 (1944);
N.F. Albertson, S. Archer, J. Amer. Chem. Soc. *67*, 308 (1945).
[82] *A. Galat*, J. Amer. Chem. Soc. *69*, 965 (1947).
[83] *J.S. Meek, S. Minkowitz, M.M. Miller*, J. Org. Chem. *24*, 1397 (1959).
[84] *G. Ehrhardt*, Chem. Ber. *82*, 60 (1949).
[85] *N.F. Albertson, B.F. Tullar*, J. Amer. Chem. Soc. *67*, 502 (1945).

reaction with acrylic derivatives to lead to a wide extension in the variability of the technique, especially in respect of α-amino acid derivatives with further functional groups[88]:

$$\text{Acyl}-\text{HN}-\underset{\overset{|}{\text{COOR}^1}}{\overset{\overset{\text{COOR}^1}{|}}{\text{CH}}} \quad + \quad \text{H}_2\text{C}=\text{CH}-\text{Y} \quad \longrightarrow$$

$$\text{Acyl}-\text{HN}-\underset{\overset{|}{\text{COOR}^1}}{\overset{\overset{\text{COOR}^1}{|}}{\text{C}}}-\text{CH}_2-\text{CH}_2-\text{Y} \quad \xrightarrow{\text{H}^\oplus} \quad \text{H}_2\text{N}-\underset{\overset{|}{\text{CH}_2}}{\overset{\overset{\text{COOH}}{|}}{\text{CH}}}-\overset{}{\underset{\text{CH}_2-\text{Y}}{}}$$

The substituted amidomalonic acid diesters and amidocyanoacetic acid esters obtained split off the acyl group and carbon dioxide on heating with mineral acid and afford the desired α-amino carboxylic acids. Alternatively, the phthaliminomalonic acid diesters can be cleaved in stages, first with alkali[89] or hydrazine hydrate[90] (cf. p. 607) and then often under relatively mild conditions with acid.

14.1.5 Preparation from other precursors

14.1.5.1 Synthesis by degradation

Various degradation techniques based on a molecular rearrangement serve for preparing amino carboxylic acids from other precursors (for rearrangements which lead to amino acids without degradation see p. 612).

Starting from the preparatively easily accessible substituted cyanoacetic acid esters, a large number of α-amino carboxylic acids can be prepared either by Curtius degradation of the corresponding carboxylic acid azides[91, 92] or by Hofmann degradation of the malonamic acids[93].

$$\text{R}-\underset{\overset{|}{\text{CN}}}{\text{CH}}-\text{COOR}^1 \quad \xrightarrow{\text{OH}^\ominus} \quad \text{R}-\underset{\overset{|}{\text{CN}}}{\text{CH}}-\text{COOH}$$

left branch:
$$\downarrow \begin{array}{l} 1.\ \text{N}_2\text{H}_4, \text{H}_2\text{O} \\ 2.\ \text{HNO}_2 \end{array}$$

right branch:
$$\downarrow \nabla/\text{conc. H}_2\text{SO}_4$$

$$\text{R}-\underset{\overset{|}{\text{CO}-\text{N}_3}}{\text{CH}}-\text{CN} \qquad\qquad \text{R}-\underset{\overset{|}{\text{CO}-\text{NH}_2}}{\text{CH}}-\text{COOH}$$

left branch:
$$\downarrow \nabla/\text{C}_2\text{H}_5\text{OH}$$

right branch:
$$\downarrow \text{Br}_2, \text{OH}^\ominus$$

$$\text{R}-\underset{\overset{|}{\text{NH}-\text{COOC}_2\text{H}_5}}{\text{CH}}-\text{CN} \quad \xrightarrow{\text{H}^\oplus} \quad \text{R}-\underset{\overset{|}{\text{NH}_2}}{\text{CH}}-\text{COOH}$$

The *choice* between the two methods is determined largely by the group R. For example, Hofmann degradation is not appropriate with substituents sensitive to oxidation. By contrast, it can be applied to substituted acetoacetamides[94], which yield α-amino carboxylic acids in one step:

$$\text{R}-\underset{\overset{|}{\text{CO}-\text{NH}_2}}{\text{CH}}-\text{CO}-\text{CH}_3 \quad \xrightarrow{\text{Br}_2, \text{OH}^\ominus} \quad \text{R}-\underset{\overset{|}{\text{NH}_2}}{\text{CH}}-\text{COOH}$$

The reaction between carbonyl compounds and hydrazoic acid in strong acid solution (*Schmidt degradation*[95]) is related to the above methods. It makes possible a simple synthesis of various amino acids from generally readily accessible precursors. Admittedly, side-reactions are possible with relevant substituents (sensitivity to strong acid, sulfonation of aromatic groups, additional groups able to react with hydrazoic acid). The extreme **toxicity** of hydrazoic acid and its tendency to decompose **explosively,** above all in large batches and at elevated temperature, must always be allowed for.

α-Amino acids are obtained by Schmidt degradation from substituted malonic acids[96] accompanied by shortening of the original chain by one carbon atom:

$$\text{R}-\underset{\overset{|}{\text{COOH}}}{\text{CH}}-\text{COOH} \quad \xrightarrow{\text{HN}_3/\text{H}_2\text{SO}_4} \quad \text{R}-\underset{\overset{|}{\text{NH}_2}}{\text{CH}}-\text{COOH}$$

[86] *K. Shimo*, Kogyo Kagaki Zasshi *64*, 303 (1961); C.A. *57*, 3554h (1962);
H.M. Kissman, B. Witkop, J. Amer. Chem. Soc. *75*, 1967 (1953);
H. Hellmann, F. Lingens, Angew. Chem. *66*, 201 (1954).

[87] *J. Shapira, R. Shapira, K. Dittmer*, J. Amer. Chem. Soc. *75*, 3655 (1953).

[88] *O.A. Moe, D.T. Warner*, J. Amer. Chem. Soc. *70*, 2763, 2765, 3918 (1948).

[89] *R. Kuhn, G. Quadbeck*, Ber. *76*, 529 (1943).

[90] *J.C. Sheehan, D.W. Chapman, R.W. Roth*, J. Amer. Chem. Soc. *74*, 3822 (1952).

[91] *A. Darapsky, H. Decker, E. Steuernagel, O. Schiedrum*, J. Prakt. Chem. *146*, 250 (1936).

[92] *P.E. Gagnon, G. Nadeau, R. Côte*, Can. J. Chem. *30*, 592 (1952);
I. Hori, M. Igarashi, M. Modorikawa, J. Org. Chem. *26*, 4511 (1961).

[93] *Y.-T. Huang, K.-H. Liu, L. Li*, J. Chin. Chem. Soc. (Taipei) *15*, 31, 38, 46 (1947); C.A. *42*, 523b-d, 4941a (1948);
Jap. P. 5265 (1963), Tanabe Seiyaku Co., Inv.: *S. Sugasawa, M. Sasamoto, T. Schichida*; C.A. *59*, 11660d (1963).

[94] *M. Yamoto, M. Fukuyama*, Yakugaku Zasshi *87*, 1431 (1967); C.A. *68*, 87 535 (1968).

[95] *H. Wolff*, Org. Reactions *3*, 307 (1946).

Starting from optionally substituted keto carboxylic acid esters, an amino group can be introduced at any desired site on the molecule[97, 98], with at least two carbon atoms being eliminated:

$$R-\underset{\underset{CO-R^2}{|}}{CH}-(CH_2)_n-COOR^1 \xrightarrow{HN_3/H_2SO_4}$$

$$R-\underset{\underset{NH-CO-R^2}{|}}{CH}-(CH_2)_n-COOR^1 \xrightarrow{H^{\oplus}} R-\underset{\underset{NH_2}{|}}{CH}-(CH_2)_n-COOH$$

α-Amino dicarboxylic acids yield *α,ω-diamino carboxylic acids*, because the carboxy group in the α-position with respect to the amino group is not attacked under the reaction conditions used[99]:

$$HOOC-(CH_2)_n-\underset{\underset{NH_2}{|}}{CH}-COOH \xrightarrow{HN_3/H_2SO_4}$$

$$H_2N-(CH_2)_n-\underset{\underset{NH_2}{|}}{CH}-COOH$$

14.1.5.2 Synthesis by rearrangement

By contrast to the rearrangement-based degradation reaction of the preceding section, *alicyclic ketones* can be converted into *ω-amino acids* with retention of the carbon skeleton. Either a *Beckmann rearrangement* of the ketoximes[100, 101] or a *Schmidt rearrangement* of the ketones with hydrazoic acid[102-104] (see p. 611) to form lactames and the latter's hydrolysis are used:

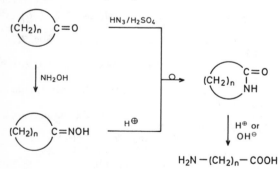

As formulated, the reactions lead to unsubstituted ω-amino carboxylic acids, but substituted derivatives also can be prepared[102, 104]. The separation between the amino and carboxy groups is determined by the size of the ketone ring. For Beckmann rearrangements the lower limit are cyclopentanone derivatives (*e.g.* ref.[100]), while using the Schmidt technique cyclobutanone, too, was successfully rearranged[105].

Acid chlorides of *N*-protected amino acids react with diazomethane to form diazoketones, which can be converted into the homologous amino carboxylic acids under mild conditions in a *Wolff rearrangement*[106]:

$$R-\underset{\underset{NH-Acyl}{|}}{CH}-CO-Cl \xrightarrow{CH_2N_2} R-\underset{\underset{NH-Acyl}{|}}{CH}-CO-CHN_2$$

$$\xrightarrow{AgO/CH_3OH} R-\underset{\underset{NH-Acyl}{|}}{CH}-CH_2-COOH$$

$$\xrightarrow{H^{\oplus}} R-\underset{\underset{NH_2}{|}}{CH}-CH_2-COOH$$

Starting from optically active α-amino carboxylic acids, the reaction, which proceeds without racemization, allows *β-amino carboxylic acids* of known configuration to be built up[107] that are otherwise difficult to prepare.

With higher diazoalkanes branched amino carboxylic acids are formed[108]. In the presence of amines the rearrangement leads to *N-protected amino carboxamides*[109].

[96] *L.H. Briggs, G.C. DeAth, S.R. Ellis*, J. Chem. Soc. 61 (1942);
S. Takagi, K. Hayashi, Chem. Pharm. Bull. (Tokyo) 7, 96 (1959); C.A. *54*, 19516a (1960);
M. Hudlický, V. Jelinek, K. Eisler, J. Rudinger, Collect. Czech. Chem. Commun. *35*, 498 (1970).
[97] *F.P. Boyle, D.O. Holland, W. Marflitt, J.H.C. Nayler, C.M. O'Connor*, J. Chem. Soc. 1719 (1955).
[98] *L. Birkofer, I. Storch*, Chem. Ber. *86*, 749 (1953); *L. Birkofer, I. Storch*, Chem. Ber. *87*, 571 (1954).
[99] *J.V. Rodricks, H. Rapoport*, J. Org. Chem. *36*, 47 (1971).
[100] *S.W. Fox, M.S. Dunn, M.P. Stoddard*, J. Org. Chem. *6*, 410 (1941).

[101] *J.C. Eck, C.S. Marvel*, Org. Syntheses, Coll. Vol. II, 76 (1943);
A. Novotný, Collect. Czech. Chem. Commun. *23*, 1570 (1958).
[102] *D.W. Adamson*, J. Chem. Soc. 1564 (1939).
[103] *R.T. Couley*, J. Org. Chem. *23*, 1330 (1958); *R.T. Couley*, J. Org. Chem. *26*, 692 (1961).
[104] *A. Chimiak*, Bull. Acad. Polon. Sci., Ser. Sci. Chim. *17*, 197 (1969); C.A. *71*, 60617 (1969).
[105] *J. Jaz, J.P. Davreux*, Tetrahedron Lett. 277 (1966).
[106] *F. Weygand, H.J. Bestmann* in Neuere Methoden der präparativen organischen Chemie, Bd. III, p. 287, Verlag Chemie, Weinheim/Bergstr. 1961;
L.L. Rodina, I.K. Korobitsyna, Russian Chem. Reviews (Engl. Transl.) *36*, 260 (1967);
A. Chimiak, Roczniki Chem. *43*, 299 (1969), engl.; C.A. *71*, 3631 (1969);
J. Rudinger, H. Farkašová, Collect. Czech. Chem. Commun. *28*, 2941 (1963).
[107] *K. Balenović, N. Štimac*, Croat. Chem. Acta *29*, 153 (1957).
[108] *K. Balenović, I. Jambrešić, J. Ranogajec*, Croat. Chem. Acta *29*, 87 (1957).
[109] *W. Jugelt, P. Falck*, J. Prakt. Chem. [4] *38*, 88 (1968).

Two mutually similar rearrangement reactions also lead to α-amino carboxylic acids or their derivatives. Whether and where they are preferable to other techniques requires wider investigation. From aliphatic nitriles N-chloro imidic acid esters **17** can be prepared which, under the action of alcoholic bases, suffer a rearrangement akin to the *Neber reaction*[110]. The reaction leads to the α-amino acids **21** *via* azirines **18**, aziridines **19**, and the *triesters of α-amino orthocarboxylic acids* **20** which can be isolated and are normally difficult to prepare[111]:

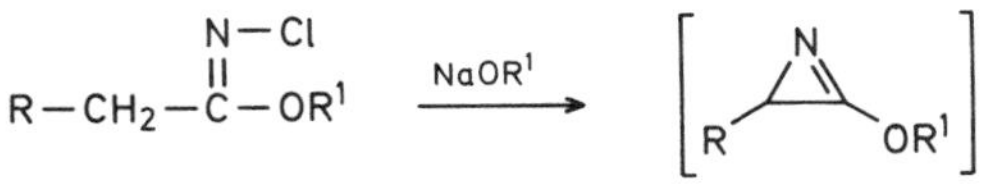

A base-catalyzed rearrangement of secondary α-chlorocarboxamides furnishes *N-substituted* α-amino carboxylic acids[112]. The reaction likewise proceeds *via* an aziridine derivative:

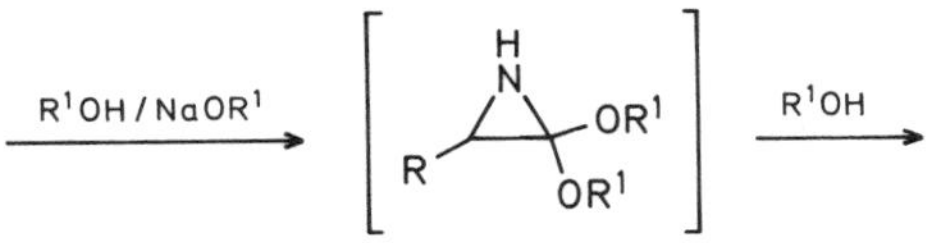

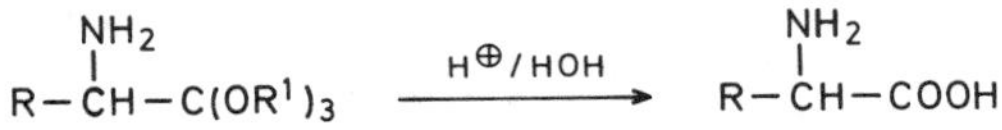

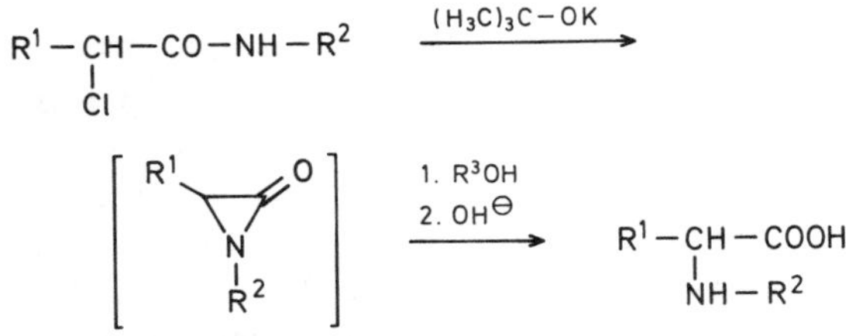

14.1.5.3 Syntheses by oxidation

Amino carboxylic acids can be prepared also from *amines* with oxidizable groups. The oxidation is performed with, for example, potassium permanganate, potassium chromate, iron(III) chloride, *etc.* Further oxidizable groups must be either absent or, like the amino group, capable of being protected appropriately.

During synthesis of α-amino carboxylic acids by oxidation of *amino alcohols* this protection of the amino group is carried out by acylation[113] or ensues by protonation[114]. Alternatively, the oxidation can be performed without protecting the amine by using hot potassium hydroxide solution in the presence of cadmium oxide[115]:

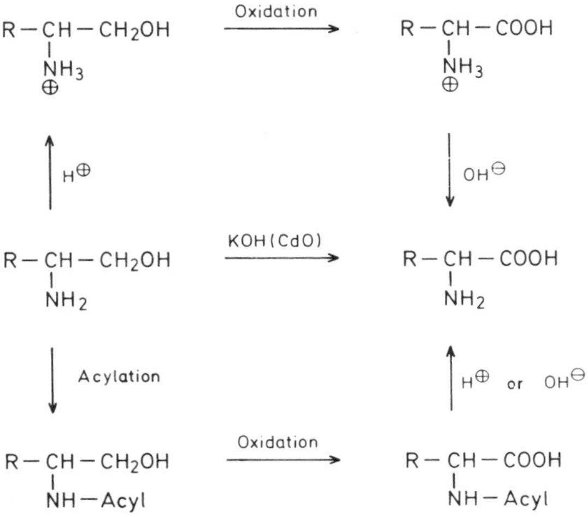

The process is economic to carry out but because of the small number of easily accessible amino alcohols has found only limited application so far.

By contrast, oxidation of *furan derivatives* represents a general method for synthesizing a variety of amino acids from easily accessible precursors. It can be used to prepare α-, β- and γ-*amino carboxylic acids*[116, 117], α-*amino dicarboxylic acids*[117], and *cyclic α-imino carboxylic acids*[118]; failure occurs only with sterically demanding groups[117]:

[110] *H.E. Baumgarten, J.E. Dirks, J.M. Petersen, R.L. Zey*, J. Org. Chem. *31*, 3708 (1966).

[111] *W.H. Graham*, Tetrahedron Lett. 2223 (1969).

[112] *J.C. Sheehan, I. Lengyel*, J. Amer. Chem. Soc. *86*, 1356 (1964).

[113] *H.J. Billmann, E.E. Parker*, J. Amer. Chem. Soc. *65*, 761, 2455 (1943).

[114] *H.J. Billmann, E.E. Parker, W.T. Smith*, J. Biol. Chem. *180*, 29 (1949);
H.N. Christensen, Biochem. Prep. *6*, 49 (1958).

[115] US P. 2 384 816 (1945), Carbide and Carbon Chemicals Corp., Inv.: *G.O. Curme jr., H.C. Chitwood, J.W. Clark*; C.A. *40*, 353 (1946);
US P. 2 384 817 (1945), Inv.: *H.C. Chitwood*; C.A. *40*, 354 (1946).

[116] *A.P. Terent'ev, R.A. Gracheva*, J. Gen. Chem. USSR (Engl. Transl.) *28*, 1225 (1958);
A.P. Terent'ev, R.A. Gracheva, O.P. Shkurko, J. Gen. Chem. USSR (Engl. Transl.) *30*, 3675 (1960);
A.P. Terent'ev, R.A. Gracheva, J. Gen. Chem. USSR (Engl. Transl.) *32*, 2197 (1962);
Â.P. Terent'ev, R.A. Gracheva, L.F. Titova, J. Gen. Chem. USSR (Engl. Transl.) *34*, 516 (1964).

[117] *A.P. Terent'ev, R.A. Gracheva, V.A. Dorokhov*, J. Gen. Chem. USSR (Engl. Transl.) *29*, 3438 (1959).

$$
\text{(furyl)}-\overset{\overset{\displaystyle O}{\|}}{C}-R \qquad\qquad \text{(furyl)}-CHO
$$

$$\downarrow NH_2OH \qquad\qquad \boxed{R-CH_2-NO_2} \qquad\qquad \boxed{H_3C-\overset{\overset{\displaystyle O}{\|}}{C}-R}$$

$$
\text{(furyl)}-\overset{\overset{\displaystyle NOH}{|}}{C}-R \qquad \text{(furyl)}-CH=\overset{\overset{\displaystyle NO_2}{|}}{C}-R \qquad \text{(furyl)}-CH=CH-\overset{\overset{\displaystyle O}{\|}}{C}-R
$$

| 1. Reduction | 1. Reduction | 1. NH_3/H_2 |
| 2. Acylation | 2. Acylation | 2. Acylation |

$$
\text{(furyl)}-\overset{\overset{\displaystyle |}{NH-Acyl}}{CH}-R \qquad \text{(furyl)}-CH_2-\overset{\overset{\displaystyle |}{NH-Acyl}}{CH}-R \qquad \text{(furyl)}-CH_2-CH_2-\overset{\overset{\displaystyle |}{NH-Acyl}}{CH}-R
$$

| 1. Oxidation | 1. Oxidation | 1. Oxidation |
| 2. Hydrolysis | 2. Hydrolysis | 2. Hydrolysis |

$$
R-\overset{\overset{\displaystyle |}{NH_2}}{CH}-COOH \qquad R-\overset{\overset{\displaystyle |}{NH_2}}{CH}-CH_2-COOH \qquad R-\overset{\overset{\displaystyle |}{NH_2}}{CH}-CH_2-CH_2-COOH
$$

Introducing the carboxy function is possible also *via* other oxidizable groups. Thus, in the synthesis of *aminoperfluoro carboxylic acids* starting from acylated aldimines the decisive step is the oxidation of a newly introduced vinyl group (oxidation of *3-alkenamines)* to the carboxylic acid[119]:

$$
R-\overset{\overset{\displaystyle \|}{CH}}{\underset{\underset{\displaystyle N-Acyl}{}}{}} \xrightarrow{\ H_2C=CH-MgHal\ } R-\overset{\overset{\displaystyle |}{NH-Acyl}}{CH}-CH=CH_2
$$

$$
\xrightarrow{\ KMnO_4\ } R-\overset{\overset{\displaystyle |}{NH-Acyl}}{CH}-COOH \xrightarrow{\ H^{\oplus}\ } R-\overset{\overset{\displaystyle |}{NH_2}}{CH}-COOH
$$

14.1.5.4 Syntheses by condensation

Condensation of carbonyl compounds with suitable precursors can be made use of for synthesizing amino acids or their derivatives in some cases (cf. p. 605). The most important of these techniques is synthesis of *α-amino-β-hydroxycarboxylic* acids by aldol condensation of aldehydes and ketones with the activated methylene groups of a glycine derivative.

Thus, *α-amino-β-aryl-β-hydroxy* carboxylic acids are obtained from *N*-arylideneglycine derivatives

and aromatic aldehydes in weakly basic medium after hydrolysis of the primarily formed intermediate *(Erlenmeyer-Früstück reaction)*[120]:

$$
Ar-CHO \ + \ \overset{\overset{\displaystyle }{}}{\underset{\underset{\displaystyle N=CH-Ar}{|}}{H_2C}}-COOH \xrightarrow{\ OH^{\ominus}\ }
$$

$$
Ar-\overset{\overset{\displaystyle |}{OH}}{CH}-\overset{\overset{\displaystyle |}{N=CH-Ar}}{CH}-COOH \xrightarrow{\ H^{\oplus}\ } Ar-\overset{\overset{\displaystyle |}{OH}}{CH}-\overset{\overset{\displaystyle |}{NH_2}}{CH}-COOH
$$

In the *copper(II) complex* of *glycine* (**22**, $R^3 = H$) the activation of the α-methylene group is so strong that aliphatic aldehydes react as well[121]. This method represents the simplest general method of synthesizing *α-amino-β-hydroxy carboxylic* acids[122]. When the copper(II) complex of *N*-pyruvoylglycine **23** is used the reaction proceeds even under mild conditions[123]; with copper(II)

[118] *A.P. Terent'ev, R.A. Gracheva, L.M. Volkova*, J. Gen. Chem. USSR (Engl. Transl.) *31*, 2634 (1961).

[119] *F. Weygand, W. Steglich, W. Oettmeier*, Chem. Ber. *103*, 818 (1970).

[120] *T. Kaneko, K. Harada*, Bull. Chem. Soc. Japan *34*, 1314 (1961);
K. Harada, Bull. Chem. Soc. Japan *42*, 2059 (1969);
M. Frankel, M. Broze, D. Gertner, A. Zilkha, J. Chem. Soc. 249 (1966).

[121] *M. Sato, K. Okawa, S. Akabori*, Bull. Chem. Soc. Japan *30*, 937 (1957).

[122] *Y. Ikutani, T. Okuda, S. Akabori*, Bull. Chem. Soc. Japan *33*, 582 (1960);
T.T. Otani, M. Winitz, Arch. Biochem. Biophys. *102*, 464 (1963).

[123] *T. Ichikawa, S. Maeda, Y. Araki, Y. Ishido*, J. Amer. Chem. Soc. *92*, 5514 (1970).

complexes of other α-amino acids (**22**, $R^3 \neq H$)[124] and with ketones[125, 126] respectively α- and β-branched α-amino carboxylic acids are obtained. From α-keto carboxylic acids β-*branched* β-*hydroxyaspartic acid derivatives* result[127].

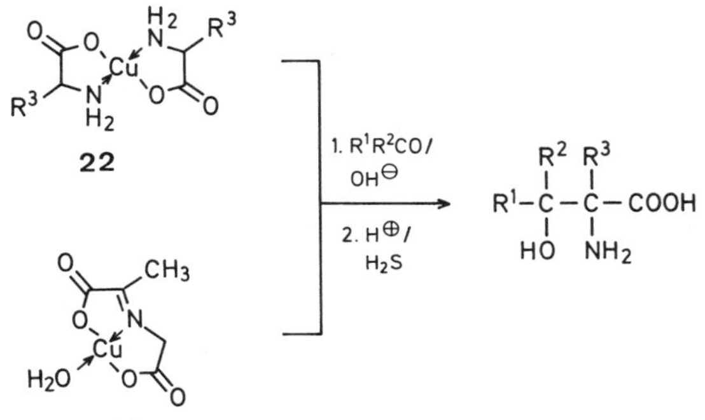

22

23

β-Alkyl- and β-arylsubstituted α-amino-β-hydroxy carboxylic acids are obtained also by the condensation of acetamidomalonic acid and its half esters with aldehydes accompanied by decarboxylation[128]:

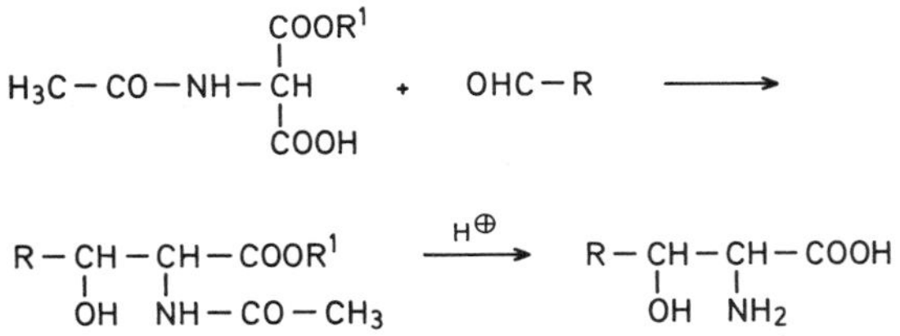

In all the named cases except where formaldehyde is used two chirality centers are newly formed at the same time. As a result, a *mixture* of diastereometric compounds is always obtained (cf. p. 616). Reduction of the α-amino-β-hydroxy carboxylic acids to α-amino carboxylic acids is possible in individual cases[126, 129].

The following methods are less important and not always generally applicable. *N, N-disubstituted* α-amino carboxylic acids are formed by Mannich

condensation of glyoxylic acid with *CH*-activated compounds and secondary amines[130]. Where *N, N-dibenzyl* compounds are formed these can be hydrogenated to the free α-amino carboxylic acids:

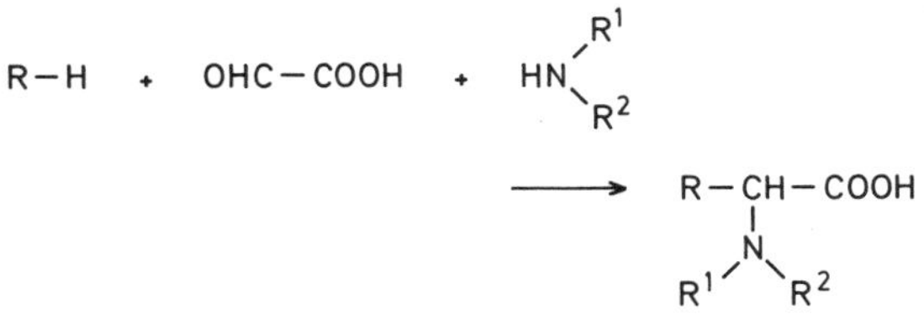

N-substituted glycine derivatives are accessible from amines in a condensation related to the Strecker synthesis with glycolonitrile or by condensation and hydrogenation with glyoxylic acid esters[131]. Using these techniques is appropriate where the amines cannot be reacted with halo carboxylic acid esters:

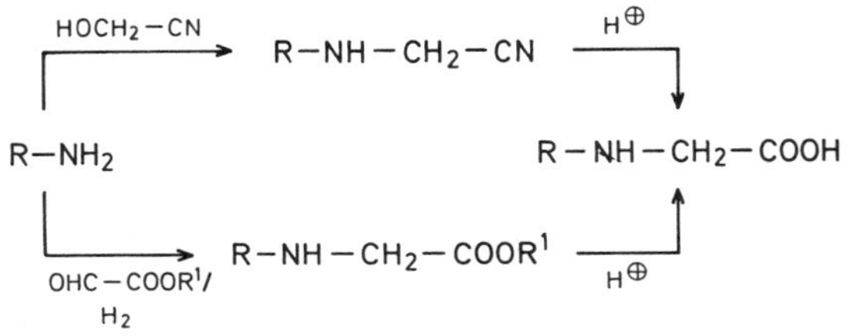

Finally, β-amino-β-arylcarboxylic acids and their *N-alkyl* derivatives are obtained quite generally by condensation of aromatic aldehydes with malonic acid and ammonium acetate or amines[132]:

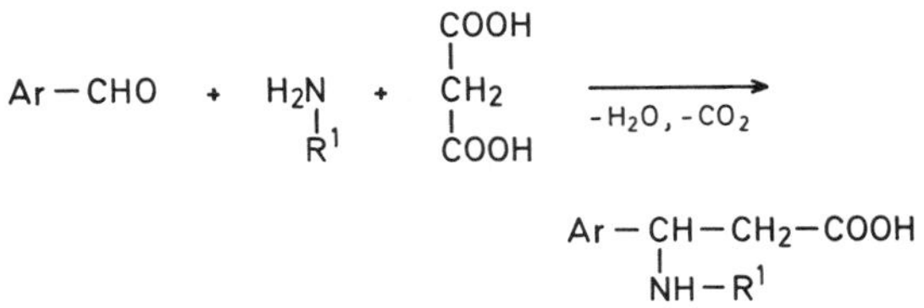

14.1.5.5 Syntheses by addition

Addition procedures have not so far found general application for synthesizing α-amino carboxylic acids but are important in some special cases. Thus, Michael addition of *CH*-acid substances to

[124] *T.T. Otani, M. Winitz*, Arch. Biochem. Biophys. *90*, 254 (1960).

[125] *H. Mix*, Z. Physiol. Chem. *327*, 41 (1962).

[126] *H. Mix*, Z. Physiol. Chem. *337*, 40 (1964).

[127] *L. Benoiton, M. Winitz, R.F. Colman, S.M. Birnbaum, J.P. Greenstein*, J. Amer. Chem. Soc. *81*, 1726 (1959).

[128] *H. Hellmann, H. Piechota*, Z. Physiol. Chem. *318*, 66 (1960);
H. Hellmann, H. Piechota, Justus Liebigs Ann. Chem. *631*, 175 (1960).

[129] Jap. P. 21 319 (1961), Inv.: *M. Sato*; C.A. *57*, 13884i (1962).

[130] *E. Bieckert, T. Funck*, Chem. Ber. *97*, 363 (1964).

[131] *J.M. Tien, I.M. Hunsberger*, J. Amer. Chem. Soc. *77*, 6696 (1955);
J.M. Tien, I.M. Hunsberger, J. Amer. Chem. Soc. *83*, 178 (1961).

[132] *E. Profft, F.-J. Becker*, J. Prakt. Chem. [4] *30*, 18 (1965).

α-amidoacrylic acid esters[133] can introduce substituents into the β-position of α-amino carboxylic acids:

$$R-H \ + \ H_2C=\underset{\underset{NH-Acyl}{|}}{C}-COOR^1 \ \longrightarrow$$

$$R-CH_2-\underset{\underset{NH-Acyl}{|}}{CH}-COOR^1 \ \xrightarrow{H^\oplus} \ R-CH_2-\underset{\underset{NH_2}{|}}{CH}-COOH$$

N-Alkylation of nitriles by carbonium ions from β-hydroxy carboxylic acid esters[134] or unsaturated carboxylic acid esters[135] *(Ritter reaction*[136]*)* is suitable for synthesizing *N-acylated amino carboxylic acids* in some cases.

$$\left.\begin{array}{l} \underset{R^1}{\overset{R^2}{\diagdown}}C=CH-COOR^3 \\[2em] \underset{\underset{OH}{|}}{R^1-\overset{R^2}{\underset{|}{C}}-CH_2-COOR^3} \end{array}\right\} \xrightarrow{R^4-CN} R^1-\underset{\underset{NH-CO-R^4}{|}}{\overset{\overset{R^2}{|}}{C}}-CH_2-COOH$$

Interesting opportunities, above all for *asymmetric syntheses,* are opened up by addition to the double bonds of Schiff bases. Up to 60% optical yields of *N-substituted α-amino carboxylic acids* are obtained in this way by adding Grignard compounds to optically active iminoacetic acid derivatives[137]:

$$R^1-N=CH-COOR^2 \ + \ R-MgHal \ \longrightarrow$$

$$R-\underset{\underset{NH-R^1}{|}}{CH}-COOR^2 \ \xrightarrow{H^\oplus, \ H_2/Cat.} \ R-\underset{\underset{NH_2}{|}}{CH}-COOH$$

A different technique enables the enantiomers of α-amino carboxylic acids to be synthesized in almost quantitative optical yield by addition of prussic acid to optically active azomethines[138]:

$$R^1-CH=N-R^2 \ \xrightarrow{HCN} \ R^1-\underset{\underset{NH-R^2}{|}}{CH}-CN \ \xrightarrow{H^\oplus}$$

$$R^1-\underset{\underset{NH-R^2}{|}}{CH}-COOH \ \xrightarrow{H_2/Cat.} \ R^1-\underset{\underset{NH_2}{|}}{CH}-COOH$$

α-Amino-β-hydroxy carboxylic acids result also from α,β-unsaturated carboxylic acids. From the latter substituted *glycidic acids* are prepared by oxidation, and ammonia or amines are then added on[139]. Unlike the synthesis of α-amino-β-hydroxy carboxylic acids by condensation (cf. p. 614), the reaction proceeds absolutely *stereoselectively.* From *trans-α,β*-unsaturated carboxylic acids pure *erythro* compounds and from *cis*-derivatives pure *threo* compounds are formed by inversion. The technique also enables α-amino-β-hydroxy carboxylic acids not accessible by condensation of ketones and amines to be synthesized.

$$\underset{R^2}{\overset{R^1}{\diagdown}}C=C\underset{H}{\overset{COOH}{\diagup}} \ \xrightarrow{H_2O_2} \ \underset{R^2}{\overset{R^1}{\diagdown}}\overset{O}{\overbrace{C-C}}\underset{H}{\overset{COOH}{\diagup}}$$

$$\xrightarrow{R^3-NH_2} \ R^1-\underset{\underset{R^2}{|}}{\overset{\overset{HO}{|}}{C}}-\underset{\underset{NH-R^3}{|}}{\overset{\overset{COOH}{|}}{C}}-H$$

A general method for synthesizing *β-amino carboxylic acids* and their *N*-substitution products is the addition of ammonia[140] and of amines[141, 142] to α,β-unsaturated carboxylic acids. *N*-Benzyl groups can be removed hydrogenolytically[142]:

$$R^1-CH=CH-COOR^2 \ \xrightarrow{R^3-NH_2}$$

$$R^1-\underset{\underset{NH-R^3}{|}}{CH}-CH_2-COOR^2 \ \xrightarrow[\text{2. } H_2/Cat.]{\text{1. } H^\oplus} \ R^1-\underset{\underset{NH_2}{|}}{CH}-CH_2-COOH$$

[133] *H. Hellmann, K. Teichmann, F. Lingens,* Chem. Ber. *91*, 2427 (1958);
T. Wieland, G. Ohnacker, W. Ziegler, Chem. Ber. *90*, 194 (1957).
[134] *A. Dobrev, C. Ivanov,* Monatsh. Chem. *98*, 2001 (1967);
C. Ivanov, A. Dobrev, Monatsh. Chem. *99*, 1050 (1968).
[135] *G. Jansen, W. Taub,* Acta Chem. Scand. *19*, 1772 (1965);
E.T. Roe, D. Swern, J. Amer. Chem. Soc. *75*, 5479 (1953).
[136] *L.I. Krimen, D.J. Cota,* Org. Reactions *17*, 213 (1969).
[137] *J.C. Fiaud, H.B. Kagan,* Tetrahedron Lett. 1813 (1970).

[138] *M.S. Patel, M. Worsley,* Can. J. Chem. *48*, 1881 (1970).
[139] *J. Liwschitz, J. Robinson, D. Perera,* J. Chem. Soc. 1116 (1962);
T.A. Dobson, L.C. Vining, Can. J. Chem. *46*, 3007 (1968).
[140] *H. Feuer, W.A. Swarts,* J. Amer. Chem. Soc. *77*, 5427 (1955).
[141] *I.L. Honigberg, W.H. Hartung,* J. Org. Chem. *25*, 1822 (1960).
[142] *A. Zilkha, J. Rivlin,* J. Org. Chem. *23*, 94 (1958);
K. Harada, K. Matsumoto, J. Org. Chem. *31*, 2985 (1966).

14.1.5.6 Synthesis by other methods

The use of *hydantoins (imidazolidine-2,4-diones)* for synthesizing α-*amino carboxylic acids* (cf. p. 606) is extended to the preparation of aliphatic amino carboxylic acids and those with additional functional groups in the molecule by carboxylating and alkylating 3-phenylhydantoin 25[143, 144]. Initially, the magnesium chelate 26 is obtained with methoxymagnesium methyl carbonate. This chelate is easily alkylated and, after working up, yields the 5-substituted hydantoin 27, which is hydrolyzed to the corresponding α-amino carboxylic acids:

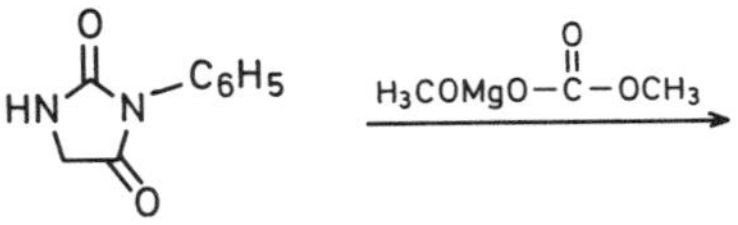

Alkylation with α,ω-*dihaloalkanes* leads to ring closure between the *C–5* and *N–1* atoms and hence to formation of *cyclic α-imino carboxylic acids*. The fact that magnesium chelate 26 can be condensed with aldehydes and substituted by acyl chlorides or acid anhydrides[144] offers further possibilities for synthesizing α-amino carboxylic acid derivatives.

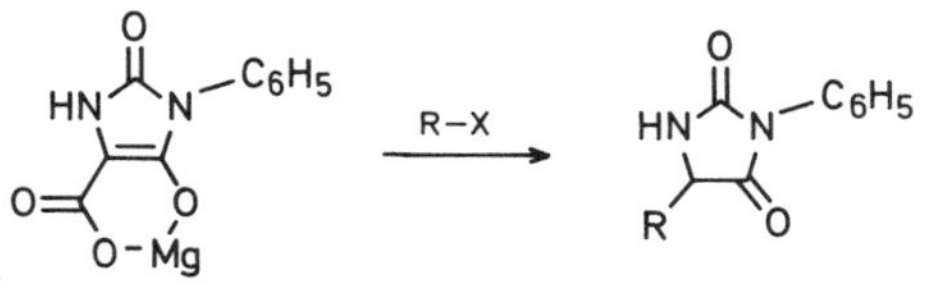

α-*Amino carboxylic acids* are accessible also by alkylation of *N,N-bis(trimethylsilyl)glycine esters*[145]:

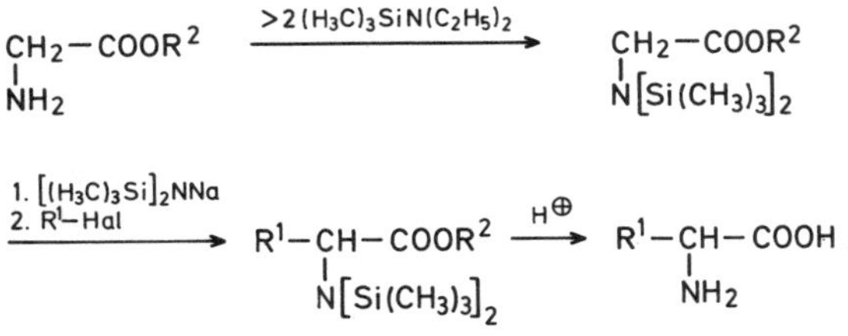

Amino carboxylic acids containing silicon, germanium, or tin in the side-chain can be prepared by this technique[146]. Condensation with aldehydes

ultimately leads to α-*amino-β-hydroxy carboxylic acids*.

From aliphatic α-*isocyanocarboxylic acid esters*[147] (cf. p. 641) it is possible to synthesize α-*amino carboxylic acids*[148, 149], α-*aminoacrylic acids*[150], α-*amino-β-hydroxy carboxylic acids*[151], and *amino dicarboxylic acids*[152] in a variety of different ways. Unlike in the related synthesis using amidomalonic acid esters (cf. p. 609), these techniques allow the preparation of certain α-*alkylated α-amino acids*. For example, *tert*-butyl isocyanoacetate gives *tert*-butyl α-*mono-* and α,α-*dialkyliso-cyanoacetates*[148]. Stepwise acid hydrolysis leads to the unprotected α-amino carboxylic acids *via N*-formamido carboxylic acid esters and *N*-formamido carboxylic acids:

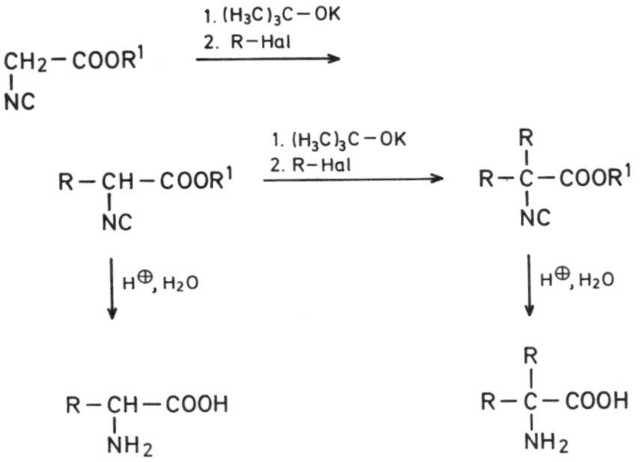

A simple rapid synthesis of α-amino carboxylic acids derivatives from easily accessible precursors which proceeds in high yield is the *four-component-condensation* of amines, carbonyl compounds, isonitriles, and a suitable anion *(Ugi reaction)*[153]. With carboxylate ions, for example, *N-acyl-α-amino carboxamides* are formed, from

[143] *H.L. Finkbeiner*, J. Amer. Chem. Soc. 86, 961 (1964).

[144] *H.L. Finkbeiner*, J. Org. Chem. 30, 3414 (1965).

[145] *K. Rühlmann, G. Kuhrt*, Angew. Chem. 80, 797 (1968).

[146] DDR P. 70097 (1969), Erf.: *K. Rühlmann, G. Kuhrt*; C.A. 73, 15249u (1970).

[147] *I. Ugi, U. Fetzer, U. Eholzer, H. Knupfer, K. Offermann*, Angew. Chem. 77, 492 (1965).

[148] *U. Schöllkopf, D. Hoppe, R. Jentsch*, Angew. Chem. 83, 357 (1971).

[149] *U. Schöllkopf, D. Hoppe*, Angew. Chem. 82, 483 (1970).

[150] *U. Schöllkopf, F. Gerhardt, R. Schröder*, Angew. Chem. 81, 701 (1969).

[151] *D. Hoppe, U. Schöllkopf*, Angew. Chem. 82, 290 (1970).

[152] *U. Schöllkopf, K. Hantke*, Angew. Chem. 82, 932 (1970).

[153] *I. Ugi* in Neuere Methoden der präparativen organischen Chemie, Bd. IV, p. 1, Verlag Chemie, Weinheim/Bergstr. 1966.

which the unsubstituted carboxylic acids can be obtained by hydrolysis. Use of optically active amines leads to asymmetric induction[154]:

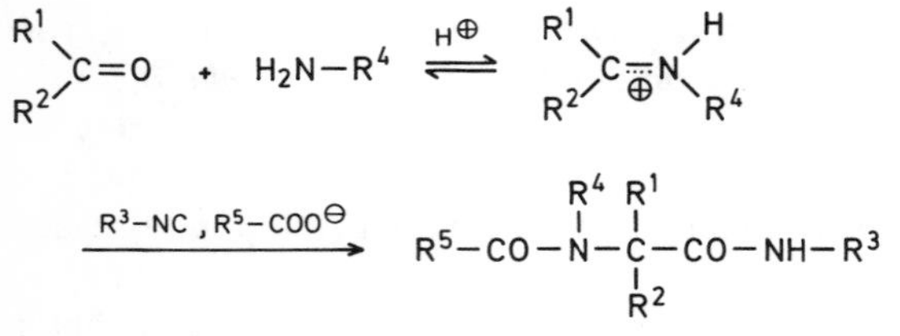

ω-*Cyano carboxylic acids* are converted into the homologous ω-*amino carboxylic acids* by reduction[155]; ^{14}C-labeling in the ω-position is possible by this route[156]:

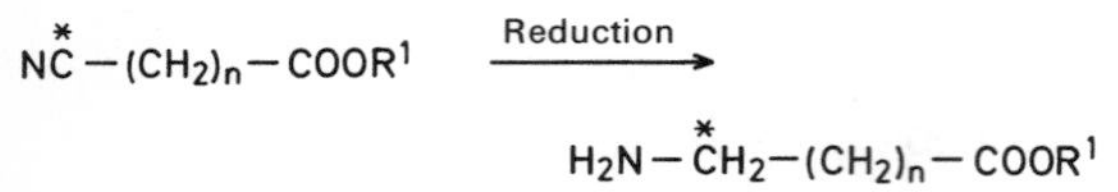

14.1.6 Special synthetic methods

Preparation of amino carboxylic acids with further functional groups in the molecule or of complex overall structure and cyclic imino carboxylic acids often require a special process. For the relevant details reference is made to monographs[1, 2], reviews[157–159], and to the original literature. The following examples are intended merely to indicate the manifold methods that are available for synthesizing such amino acids.

DL-*Aspartic* and DL-*glutamic* acid are obtained by alkylating nitroacetic acid esters with bromocarboxylic acid ester and catalytic reduction of the α-nitrodicarboxylic acid ester and followed by saponification[160]:

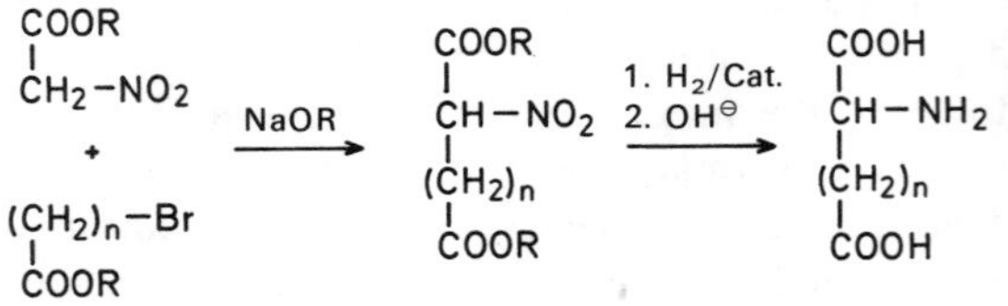

4-*Methyl* L-*aspartates* **32** can be prepared in an asymmetric synthesis from L-*erythro*-2-amino-1, 2-diphenylethanol **28** and dimethyl acetylenedicarboxylate **29** by stereospecific reduction of the exocyclic double bond of the adduct **30** obtained. After addition of acid, hydrogenolysis of **31** gives optically pure **32**[161]:

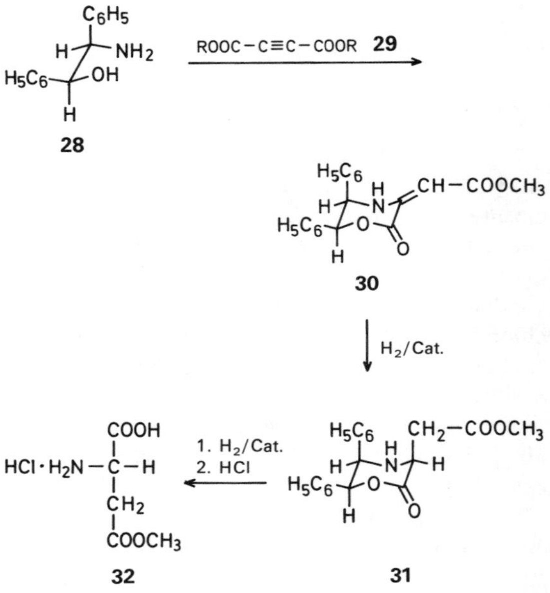

Hexahydropyridazine-3-carboxylic acid **36**, a component of the antibiotic monamycin, is formed by reduction and hydrolysis of the Diels-Alder adduct **35** formed from 1, 4-dihydrophthalazine-1, 4-dione **33** and 2, 4-pentadiene **34**. Hydrogenolysis of the carboxylic acid **36** leads to DL-*ornithine* **37**[162]:

[154] *H. Herlinger, H. Kleinmann, K. Offermann, D. Rucker, I. Ugi*, Justus Liebigs Ann. Chem. *692*, 94 (1966).

[155] *S. Akabori, Y. Izumi, T. Okuda*, J. Chem. Soc. Japan *77*, 490 (1956).

[156] *H.R. Schütte, A. Unverricht*, Z. Chem. *11*, 107 (1971).

[157] *E.N. Safonova, V.M. Belikov*, Russ. Chem. Rev. 375 (1967).

[158] The Chemical Society, Ann. Rep. Progr. Chem. *58*, 300 (1961);
The Chemical Society, Ann. Rep. Progr. Chem. *60*, 448 (1963);
The Chemical Society, Ann. Rep. Progr. Chem. *62*, 389 (1965);
The Chemical Society, Ann. Rep. Progr. Chem. *63*, 517 (1966);
The Chemical Society, Ann. Rep. Progr. Chem. *64B*, 451 (1967);
The Chemical Society, Ann. Rep. Progr. Chem. *65B*, 509 (1968).

[159] Chemical Society Specialist Periodical Report, Aminoacids, Peptides and Proteins, Vol. I, 1969; Vol. II, 1970, The Chemical Society, London; *L.A. Cohen, B. Witkop* in *P. de Mayo*, Molecular Rearrangements, Vol. II, p. 965, Interscience Publishers, New York, London, Sidney 1964.

[160] *S. Zen, E. Kaji*, Bull. Chem. Soc. Japan *43*, 2277 (1970).

[161] *J.P. Vigneron, H. Kagan, A. Horeau*, Tetrahedron Lett. 5681 (1968).

[162] *C.H. Hasall, R.B. Morton, Y. Ogihara, W.A. Thomas*, Chem. Commun. 1079 (1969).

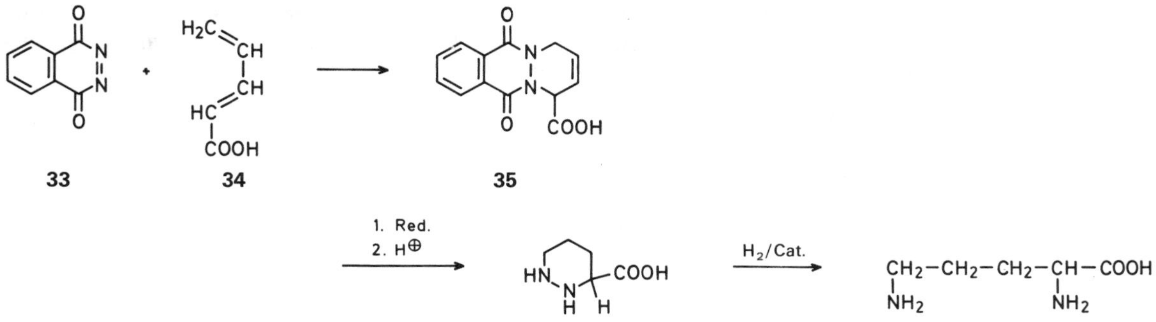

33   **34**   **35**

36   **37**

β,γ-Unsaturated α-amino acids are prepared from diethyl malonates by alkylation, oximation, and selective reduction[163]:

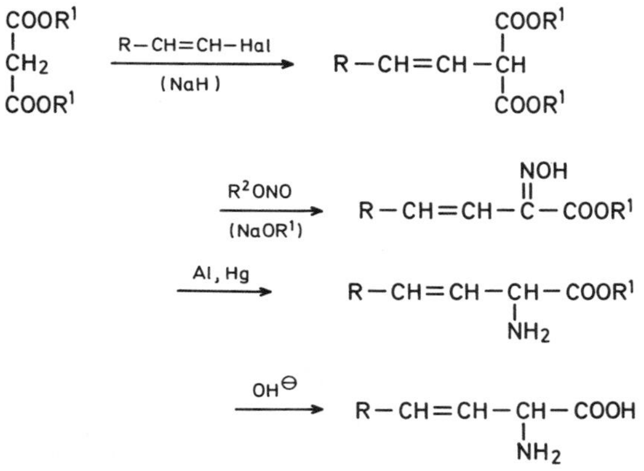

Finally, for synthesizing *2-acetamido-4,4,4-trifluorobutyric, 2-acetamido-4,4,5,5,5-pentafluorovaleric,* and *2-acetamido-4,4,5,5,6,6,6-heptafluorohexanoic acids* (**40**; $R = CF_3$, C_2F_5, C_3F_7) use is made of perfluoroacyldiazoacetic acid esters **38**. Their irridation in acetonitrile leads to 2-methyl-5-perfluoroalkyl-4-oxazole-4-carboxylic acids **40**, which reduction converts into the acetamido carboxylic esters **39**[164]:

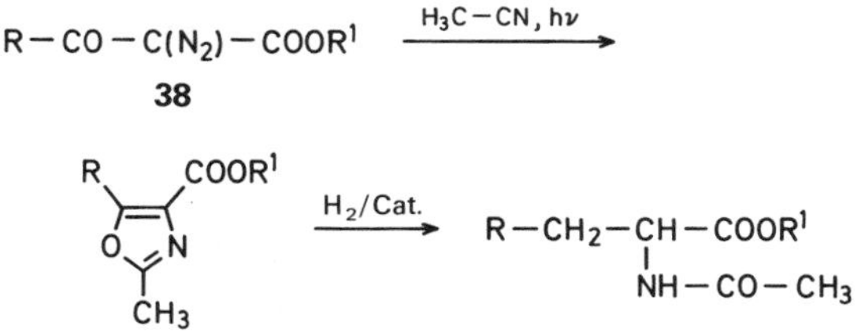

$$R-CO-C(N_2)-COOR^1 \xrightarrow{H_3C-CN,\ h\nu}$$

38

39   **40**

14.2 Transformations

On account of their special structure, transformation of amino carboxylic acids can take place with participation of only one or several of their reactive groups. The following transformations are distinguished

on the amino group
on the carboxy group
on the amino and on the carboxy group
on the α-carbon atom (only with α-amino carboxylic acids)
on additional functional groups
on additional functional groups and on the amino or on the carboxy group

Numerous transformation reactions of aliphatic amines[165], of carboxylic acids[166], and of the substance classes to which the additional functional groups belong can be transferred also to amino carboxylic acids. Their use allows the preparation of certain *N*-amido carboxylic acids, amino carboxylic acid esters, *N*-amido carboxylic esters, and further *amino carboxylic acid derivatives* that are important for peptide syntheses[167–169]. The substituents serve either as selectively cleavable protective groups, or can be employed for coupling with the amino group of other amino carboxylic acids or peptides:

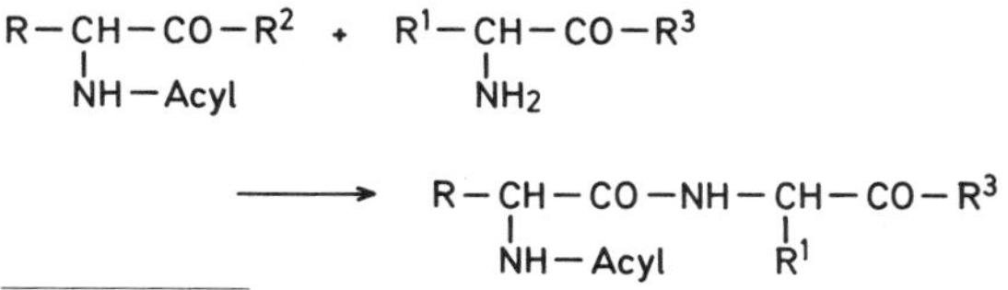

[165] cf. Chapter 12.2, p. 451.

[166] *H. Henecka, E. Ott* in *Houben-Weyl*, Methoden der organischen Chemie, 4. Aufl., Bd. VIII/3, p. 463, Georg Thieme Verlag, Stuttgart 1952.

[167] *J.P. Greenstein, M. Winitz,* Chemistry of the Amino Acids, Vol. II, p. 763, John Wiley & Sons, New York, London 1961.

[168] *E. Schröder, K. Lübke,* The Peptides. Methods of Peptide Synthesis, Vol. I, Academic Press, New York, London 1965.

[169] *E. Schröder, K. Lübke,* Fortschritte der Chemie organischer Naturstoffe *26,* 48 (1968).

[163] *D.J. Drinkwater, P.W.G. Smith,* J. Chem. Soc. (C) 1305 (1971).

[164] *W. Steglich, H.-U. Heiniger, H. Dworschak, F. Weygand,* Angew. Chem. *79,* 822 (1967).

Table 2 shows some important relevant derivatives of α-amino carboxylic acids. Corresponding compounds can be prepared analogously from carboxylic acids with a nonalpha amino group.

Table 2. α-Amino carboxylic acid derivatives important in peptide synthesis

Starting compound	Reaction with	Reaction product	Ref.

R—CH—COOH
    |
    NH$_2$
	HCOOH/(CH$_3$CO)$_2$O	R—CH—COOH
		    |
		    NH—CHO
			170

	(F$_3$C—CO)$_2$O	R—CH—COOH
		    |
		    NH—CO—CF$_3$
			167, 168

	R^1—C$_6$H$_4$—CH$_2$—O—CO—Cl
	R^1 = H, 4-NO$_2$, 4-OCH$_3$
		R—CH—COOH
		    |
		    NH—C(O)—O—CH$_2$—C$_6$H$_4$—R^1
			167–170

	(H$_3$C)$_3$C—O—C(O)—N$_3$	R—CH—COOH
		    |
		    NH—C(O)—O—C(CH$_3$)$_3$
			168–170

	phthalic anhydride	R—CH—COOH, N-phthalimido	167–170

	H$_3$C—C$_6$H$_4$—SO$_2$—Cl	R—CH—COOH
		    |
		    NH—SO$_2$—C$_6$H$_4$—CH$_3$
			167, 168, 170

	2-nitrophenyl—S—Cl	R—CH—COOH
		    |
		    NH—S—(2-O$_2$N—C$_6$H$_4$)
			169, 171

	(H$_5$C$_6$)$_3$C—Cl	R—CH—COOH
		    |
		    NH—C(C$_6$H$_5$)$_3$
			167–170

	CH$_3$OH/SOCl$_2$	R—CH—COOCH$_3$
		    |
		    NH$_2$
			167–169

	R^1—C$_6$H$_4$—CH$_2$OH + H$_3$C—C$_6$H$_4$—SO$_3$H
	R^1 = H, 4-NO$_2$
		R—CH—CO—O—CH$_2$—C$_6$H$_4$—R^1
		    |
		    NH$_2$
			167–169

	H$_2$C=C(CH$_3$)—CH$_3$ + H$_2$SO$_4$	R—CH—CO—O—C(CH$_3$)$_3$
		    |
		    NH$_2$
			168, 169

R—CH—COOCH$_3$
    |
    NH$_2$
	1. Acylation
	2. N$_2$H$_4$, H$_2$O
	3. HNO$_2$
		R—CH—CO—N$_3$
		    |
		    NH—Acyl
			167–169

R—CH—COOH
    |
    NH—Acyl
	R^1—C$_6$H$_4$—OH + cyclohexyl—N=C=N—cyclohexyl
	R^1 = 4-NO$_2$, 2,4,5-(Cl)$_3$
		R—CH—CO—O—C$_6$H$_4$—R^1
		    |
		    NH—Acyl
			167–169

	N-hydroxysuccinimide—OH + cyclohexyl—N=C=N—cyclohexyl	R—CH—CO—O—N(succinimide)
		    |
		    NH—Acyl
			169

Further transformations characteristic of α-amino carboxylic acids are contained in Table 3.

Conversion of one amino carboxylic acid into another is possible in some cases. Some of these con-

Table 3. Transformations of α-amino carboxylic acids

Starting compound	Reaction with	Reaction product	Discussed in Volume	Ref.
$R-\underset{\underset{NH_2}{\vert}}{CH}-COOH$	HNO_2	$R-\underset{\underset{OH}{\vert}}{CH}-COOH$ α-Hydroxy carboxylic acids	V	[172]
	$LiAlH_4$	$R-\underset{\underset{NH_2}{\vert}}{CH}-CH_2OH$ α-Amino alcohols	V	[173]
	1. Excess CH_3I 2. $OH^\ominus$	$R-\underset{\underset{\oplus N(CH_3)_3}{\vert}}{CH}-COO^\ominus$ α-Amino carboxylic acid betaines	VI	[174]
	NaOCl	$R-CHO$ C_{n-1} aldehydes	V	[175]
	$(H_3C)_2N-\!\!\!\bigcirc\!\!\!-CHO$	$R-CH_2-NH_2$ C_{n-1} primary amines	VI	[176]
	1. $(CF_3CO)_2O$ 2. $OH^\ominus$ or $H^\oplus$	$R-\underset{\underset{O}{\Vert}}{C}-COOH$ α-Oxo carboxylic acids	V	[177]
	$(R^1-CO)_2O$/pyridine (Dakin-West reaction)	$R-\underset{\underset{NH-CO-R^1}{\vert}}{CH}-CO-R^1$ α-Acylamino ketones	V	[178]
	$COCl_2/OH^\ominus$	Oxazolidine-2.5-diones	VI	[179]
	1. $R^1-N=C=S$ 2. $H^\oplus$	Substituted thiohydantoins	VI	[180]
	$\bigcirc\!\!\overset{COOCH_3}{\underset{N=C=S}{}}$	1,2,3,4-Tetrahydro-4-oxo-2-thioxo-quinazoline-3-acetic acids	VI	[181]

Table 3 Continuation 1

Starting compound	Reaction with	Reaction product	Discussed in Volume	Ref.
CH_2-COOR^1 $\|$ NH_2 $R^1 = CH_3, C_2H_5, tert-C_4H_9$	HNO_2	CH_2-COOR^1 $\|$ N_2 Diazoacetic esters	VI	182
$R-CH-COOC_2H_5$ $\|$ NH_2	$R^1-CH-COOC_2H_5$ $\|$ NH_2	 2,5-Piperazinedione	IV	183
$R-CH-COOH$ $\|$ $NH-CH_2-C_6H_5$	1. $CH_2O/HCOOH$ 2. $H_2/Catalyst$	$R-CH-COOH$ $\|$ $NH-CH_3$ a-Methylamino carboxylic acids	—	184
$R-CH-COOH$ $\|$ $NH-R^1$ R^1 = alkyl, aryl	1. HNO_2 2. $(CH_3CO)_2O$	 Substituted sydnones	IV	185

[170] *R.A. Boissonas*, Adv. Org. Chem. *3*, 159 (1963).
[171] *L. Zervas, D. Borovas, E. Gazis*, J. Amer. Chem. Soc. *85*, 3660 (1963);
L. Zervas, C. Hamalidis, J. Amer. Chem. Soc. *87*, 99 (1965).
[172] *P. Brewster, F. Hiron, E.D. Hughes, C.K. Ingold, P.A.D.S. Rao*, Nature *166*, 179 (1950);
A.T. Austin, J. Howard, J. Chem. Soc. 3278, 3284 (1961).
[173] *J. Szammer*, Acta Chem. Acad. Sci. Hung. *61*, 417 (1969).
[174] *J.P. Greenstein, M. Winitz*, Chemistry of the Amino Acids, Vol. III, p. 2764, John Wiley & Sons, New York, London 1961.
[175] *R.A. Gray*, Arch. Biochem. Biophys. *81*, 480 (1959).
[176] *K. Dose*, Chem. Ber. *90*, 1251 (1957);
A.F. Al-Sayyab, A. Lawson, J. Chem. Soc. (C) 406 (1968).
[177] *F. Weygand, W. Steglich, H. Tanner*, Justus Liebigs Ann. Chem. *658*, 128 (1962).
[178] *N. Gerenčević, A. Častek, M. Šateva, J. Pluščec, M. Proštenik*, Monatsh. Chem. *97*, 331 (1966);
N.I. Aranova, N.N. Makhova, M.P. Unanyan, G.V. Kondrat'eva, N.A. Rodionova, S.I. Zav'yalov, Bull. Acad. Sci. USSR, Div. Chem. Sci. (Engl. Transl.) 2432 (1968).
[179] *J. Rudinger, Z. Pravda*, Collect. Czech. Chem. Commun. *23*, 1947 (1958);
R. Hirschmann, R.G. Strachan, H. Schwam, E.F. Schoenewaldt, H. Joshua, B. Barkemeyer, D.F. Veber, W.J. Paleveda jr., T.A. Jakob, T.E. Beesley, R.G. Denkewalter, J. Org. Chem. *32*, 3415 (1967).

versions are significant as general techniques or for synthesizing certain types of amino acids which have already been described on p. 605, 612, 614, 617.

Further transformations of individual amino acids into new ones proceed with participation of an additional functional group of the molecule and are mostly of special character. For example, preparation of *proline* from ornithine[186], glutamic acid[187], or glutamine[188], preparation of *arginine*

[180] *E. Ware*, Chem. Rev. *46*, 403 (1950);
J.T. Edward in *N. Kharash, C.Y. Meyers*, The Chemistry of Organic Sulfur Compounds, Vol. II, p. 287, Pergamon Press, London 1966.
[181] *E. Cherbuliez, O. Espejo, B. Willhalm, J. Rabinowitz*, Helv. Chim. Acta *51*, 241 (1968).
[182] *E.B. Womack, A.B. Nelson*, Org. Syntheses Coll. Vol. III, 392 (1955);
N.E. Searle, Org. Syntheses Coll. Vol. IV, 424 (1963);
G.L. Closs, Advan. Alicycl. Chem. *1*, 53 (1966).
[183] *Y.T. Pratt* in *R.C. Elderfield*, Heterocyclic Compounds, Vol. VI, p. 377, John Wiley & Sons, New York 1957;
A.M. Galinsky, J.E. Gearien, E.E. Smissman, J. Amer. Pharm. Assoc. *46*, 391 (1957);
L.C. Vining, Can. J. Chem. *41*, 2903 (1963).
[184] *P. Quitt, J. Hellerbach, K. Vogler*, Helv. Chim. Acta *46*, 327 (1963);
R.K. Olsen, J. Org. Chem. *35*, 1912 (1970).

from ornithine[189], and preparation of *cystine* and *cysteine* from serine[190] are possible.

Methods which enable one to change the *configuration* of amino carboxylic acids by means of chemical reactions and which all proceed with participation of the α-carbon atom are of general importance. One such technique is *racemization* of an optically active α-amino carboxylic acid[191].

cycle of asymmetric transformations[192]; the reader is referred to the original for details. Converting DL-amino carboxylic acids into the L or D enantiomers can also be carried out[193, 194]. The yield can exceed 50% especially with compounds possessing a sterically demanding group[194]. This is not possible with the classical methods of racemate separation, *e.g.*:

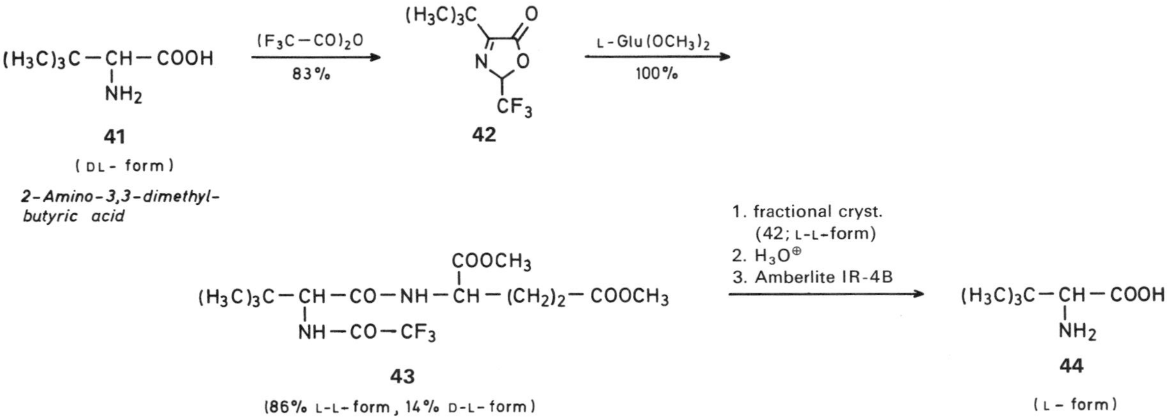

$(H_3C)_3C-CH-COOH$
     |
     NH_2

41
(DL- form)
2-Amino-3,3-dimethyl-butyric acid

$\xrightarrow[83\%]{(F_3C-CO)_2O}$

42

$\xrightarrow[100\%]{L-Glu(OCH_3)_2}$

1. fractional cryst. (42; L-L-form)
2. $H_3O^\oplus$
3. Amberlite IR-4B

$(H_3C)_3C-CH-CO-NH-CH-(CH_2)_2-COOCH_3$
     | |
     $NH-CO-CF_3$ $COOCH_3$

43
(86% L-L-form, 14% D-L-form)

$\longrightarrow$

$(H_3C)_3C-CH-COOH$
     |
     NH_2

44
(L- form)

Controlled conversion of an enantiomer or diastereomer into another is, however, possible under certain circumstances. An example of interconversion of individual diastereomeric amino acids is the preparation of L- and D-*threonine* and of L- and D-*allothreonine* from *trans*-L-5-methyl-2-phenyl-2-oxazoline-4-carboxylic acid by way of a

The decisive criterion is reaction of the 2-trifluoromethyl-3-oxazolin-5-one **42** with dimethyl D- or L-glutamate. It proceeds with strongly asymmetric induction and leads preferentially to the diastereomeric dipeptide **43**, whose asymmetry centers possess the same absolute configuration[195].

[185] *F.H.C. Stewart*, Chem. Rev. *64*, 129 (1964).
[186] *S. Ohshiro, K. Kuroda, T. Fujita*, Yakugaku Zasshi *87*, 1184 (1967); C.A. *68*, 40031 (1968).
[187] *R. Buyle*, Chem. & Ind. (London) 380 (1966).
[188] *T. Itoh*, Bull. Chem. Soc. Japan *36*, 25 (1963).
[189] *A.C. Kurtz*, J. Biol. Chem. *180*, 1253 (1949); Jap. P. 8012 (1967), Kyowa Fermentation Industry Co., Inv.: *K. Fuji, S. Fujisawa, K. Nakano; C.A. 67*, 109006 (1967).
[190] *P. Rambacher*, Chem. Ber. *101*, 2595 (1968).

[191] *A. Neuberger*, Advan. Prot. Chem. *4*, 356 (1948).
[192] *D.F. Elliot*, J. Chem. Soc. 62 (1950).
[193] *W. Steglich, H.-U. Heininger, H. Dworschak, F. Weygand*, Angew. Chem. *79*, 822 (1967).
[194] *W. Steglich, E. Frauendorfer, F. Weygand*, Chem. Ber. *104*, 687 (1971).
[195] *W. Steglich, D. Mayer, X. Barocio de la Lama, H. Tanner, F. Weygand*, Peptides, Proc. VIII. European Peptide Symposium, p. 67, North-Holland Publ. Comp., Amsterdam 1967.

15 Quaternary Ammonium Compounds

Contributed by

J. Goerdeler
Institut für Organische und Bio-Chemie
der Universität Bonn,
Bonn

15.1 Definition, nomenclature, general properties

Quaternary ammonium compounds[1] have the general structure

$$\left[R_4 N^{\oplus} \right] X^{\ominus}$$

The covalently bound organic groups R may be different; in many cases four alkyl groups or one aryl and three alkyl groups are concerned. Ammonium compounds with four directly bound aryl groups are unknown.

The substituent groups are arranged around the central atom as a tetrahedron; if all four are different, separation into optical antipodes is possible[2]. Few restrictions apply to the anion $X^{\ominus}$; more particularly it may be strongly basic *(quaternary ammonium bases)*. This characteristic distinguishes the quaternary compounds from primary to tertiary ammonium compounds. Instead of being separated from the cation, the anion may be joined to it by a covalent link *(zwitterionic compounds, betaines)*.

In an extended sense compounds with only two or three organic radicals bound to the central atom are included; the links here are multiple bonds:

$$\left[R_2 C = \overset{\oplus}{N} R_2 \right] X^{\ominus} \qquad \left[R - C \equiv \overset{\oplus}{N} R \right] X^{\ominus}$$

Immonium
compounds

Nitrilium
compounds

The first-named compounds are often constituents of heterocycles.

As far as possible the *nomenclature* is based on the endings *-ammonium* or *-inium (tetraethylammonium chloride, dimethyl piperidinium iodide)*. Use of the prefix ammonio is occasionally appropriate *(1-triethylammonioacridine bromide)*. Methiodide, methosulfate, iodomethoxide are used to designate compounds which carry (at least) one methyl group on the quaternary nitrogen and whose anion is iodide or methylsulfate. The Chemical Abstracts list the compounds under *Ammonium compounds, substituted*.

Their *saltlike* character makes the compounds generally nicely crystalline and stable at room temperature. They decompose on heating, often during melting[3]. In general, they are readily soluble in polar solvents but sparingly soluble in nonpolar media and thus occasionally difficult to separate from inorganic salts. Some are hygroscopic. Many quaternary ammonium compounds occur in nature, a substantial number are being produced industrially for a variety of purposes[4].

15.2 Preparation of univalent ammonium compounds

15.2.1 With esters of strong acids

The most important method of preparing quaternary ammonium compounds is alkylation of tertiary amines with esters of strong acids[5] *(Menshutkin reaction)*:

$$R_3 N \; + \; RX \quad \longrightarrow \quad \left[R_4 N^{\oplus} \right] X^{\ominus}$$

The speed and the yield of this nucleophilic substitution depend both on the reactants and the conditions.

15.2.1.1 Influence of the amine

Nucleophilic properties and steric relationships govern the reactivity of the amine during quaternization. The rule is that an amine reacts the more smoothly the more basic it is and the less the nitrogen atom is screened. These two effects make the two following velocity series clear:

$(CH_3)_3 N \; > \; H_5 C_6 - N(CH_3)_2 \; > \; (C_5H_6)_2 NCH_3$

$(C_2H_5)_3 N \; > \;$ Pyridine $\; > \;$ 2,6-Dimethylpyridine[6,7]

$H_5 C_6 - N(CH_3)_2 \; > \; H_5 C_6 - N(C_2H_5)_2 \; > \; H_5 C_6 - N(C_3H_7)_2$[8]

[1] *J. Goerdeler* in *Houben-Weyl*, Methoden der organischen Chemie, Bd. XI/2 p. 587, Georg Thieme Verlag, Stuttgart 1958.

[2] *W.J. Pope, S.J. Peachey*, J. Chem. Soc. *75*, 1127 (1899).

[3] *J.E. Gordon*, J. Org. Chem. *30*, 2760 (1965).

[4] *W. Foerst* in: Ullmanns Encyklopädie der technischen Chemie, 3. Aufl., Urban & Schwarzenberg, München, Berlin 1969, Bd. V, p. 763 (Desinfektionsmittel); Bd. IX, p. 302 (Kationenaktive Verbindungen); Bd. X, p. 730 (Haarspülmittel); Bd. XIII, p. 315 (Muskelrelaxantien); Bd. XIII, p. 336 (Blutdrucksenker); Bd. XIV, p. 5 (Fungizide); Bd. XV, p. 157 (Unkrautbekämpfungsmittel); Bd. XV, p. 175 (Materialschutz);
R.F. Kirk, D.F. Othmer, J.A. Cella, Quaternary Ammonium Compounds in: Encyclopedia of Chemical Technology, Vol. XI, p. 375, Interscience Publishers, New York 1953.

[5] *A.W. Hofmann*, Annalen der Chemie und Pharmazie *78*, 253 (1851).

Trimethylamine and quinuclidine[9] react particularly easily, pyridine and dimethylaniline occupy a position in between; alkylation of dialkylanilines with electron-attracting groups[10] and of alkyldiarylamines is difficult[11]. Alkyltriarylammonium salts have not been prepared by this route (see p. 630). Where the amine is part of a saturated ring (azacycloalkanes) the size and conformation of the ring play an important part[12]; medium-sized rings react relatively slowly.

Instead of tertiary amines primary or secondary amines may be used as the *starting product*. In such cases an *auxiliary base* (generally caustic soda or sodium carbonate) must be added, *e.g.*:

$$RNH_2 + 3RHal + 2Base \longrightarrow$$

$$[R_4N^{\oplus}]Hal^{\ominus} + 2Base \cdot HHal$$

This reaction mixture thus contains two different nucleophiles which fundamentally compete for the alkylating agent. The strong base can produce interference at other points of the molecule. To some extent these drawbacks are reduced by employing the reluctantly alkylated 2,6-dimethylpyridine as auxiliary base[13], but only substrates less basic than dimethylpyridine (p$_{Ka}$ 6.8) are feasible, *e.g.*, aniline derivatives.

15.2.1.2 Effect of *RX*

In the majority of cases alkyl halides (chloride, bromide, iodide), dimethyl sulfate, and *p*-toluenesulfonate esters are used for quaternization:

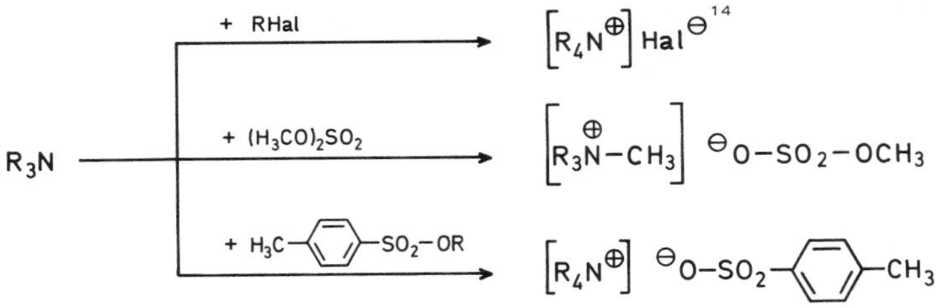

Esters of other acids (*e.g.*, thiocyanates[15], phosphates[16], phosphinates[16], carboxylates[17]) can be used but are practically useful in exceptional cases only, for instance, if the relevant anion is desired directly. With alkyl halides the reactivity follows the sequence

$$RJ > RBr > RCl$$

The higher boiling point of the iodides may be a further advantage. (This aspect is a factor in favor of using dimethyl sulfate or toluenesulfonates). Electron-attracting substituted *sulfonic acid esters* are particularly active alkylating agents, as is shown by the following activity series[18]:

ethyl *p*-toluenesulfonate < ethyl bromide < diethyl sulfate < ethyl iodide ≪ ethyl fluorosulfate

Trifluoromethanesulfonic acid esters deserve a mention in this connection[19].

Within a group the nature of the alkyl, alkenyl, or alkynyl substituent *R* has a substantial effect on the reaction velocity. Secondary halides react more slowly than primary ones. Methyl, allyl, and benzyl compounds react very rapidly. Normal vinyl halides react too sluggishly, but quaternization succeeds if electron-attracting groups are used as activators[20, 21]:

$$R_3N + Cl-CH=CH-CO-Ar(CN)$$

$$\longrightarrow \left[\overset{\oplus}{R_3N}-CH=CH-CO-Ar(CN)\right]Cl^{\ominus}$$

$$R_3N \quad \xrightarrow{+ RHal} \quad [R_4N^{\oplus}]Hal^{\ominus}{}^{14}$$

$$\xrightarrow{+ (H_3CO)_2SO_2} \quad [R_3\overset{\oplus}{N}-CH_3]\ {}^{\ominus}O-SO_2-OCH_3$$

$$\xrightarrow{+ H_3C-\langle\rangle-SO_2-OR} \quad [R_4N^{\oplus}]\ {}^{\ominus}O-SO_2-\langle\rangle-CH_3$$

[6] *K.J. Laidler, C.N. Hinshelwood,* J. Chem. Soc. 858 (1938).

[7] *K. Clarke, K. Rothwell,* J. Chem. Soc. 1885 (1960).

[8] *D.P. Evans,* J. Chem. Soc. 422 (1944).

[9] *H.C. Brown, N.R. Eldred,* J. Amer. Chem. Soc. *71*, 445 (1949).

[10] *A. Zaki, H. Fahim,* J. Chem. Soc. 270 (1942).

[11] *D.A. Archer, H. Booth,* J. Chem. Soc. 322 (1963).

[12] *M. Havel, J. Krupička, M. Svoboda, J. Závada, J. Sicher,* Collect. Czech. Chem. Commun. *33*, 1429 (1968); C.A. *69*, 2308e (1968);
General review on quaternisation of heterocycles: *G.F. Duffin,* Adv. Heterocyclic Chem. *3*, 2 (1964); Re small, charged rings: *D.R. Crist, N.J. Leonard,* Angew. Chem. *81*, 953 (1969).

[13] *H.Z. Sommer, L.L. Jackson,* J. Org. Chem. *35*, 1558 (1970).

Alkynyl halides have been employed occasionally, *e.g.*:

$$R_3N \; + \; Br-C\equiv C-C\equiv C-R \hspace{3cm} [22]$$

$$\longrightarrow \; \left[R_3\overset{\oplus}{N}-C\equiv C-C\equiv C-R \right] Br^{\ominus}$$

Activated *aromatic* and *heteroaromatic* halides can be used for the reaction with limitations, *e.g.*:

$$(H_3C)_3N \; + \; \left[Cl-\!\!\langle\ \rangle\!\!-\overset{\oplus}{N}_2 \right] BF_4^{\ominus} \hspace{1.5cm} [23]$$

$$\longrightarrow \; Cl^{\ominus} \left[(H_3C)_3\overset{\oplus}{N}-\!\!\langle\ \rangle\!\!-\overset{\oplus}{N}_2 \right] BF_4^{\ominus}$$

$$\langle\ \rangle\!\!-N \; + \; \begin{array}{c} Cl-\!\!\langle\ \rangle\!\!-NO_2 \\ O_2N \end{array}$$

$$\longrightarrow \; \left[\begin{array}{c} \langle\ \rangle\!\!-\overset{\oplus}{N}-\!\!\langle\ \rangle\!\!-NO_2 \\ O_2N \end{array} \right] Cl^{\ominus} \hspace{1cm} [24]$$

Chloropyrimidines[25] and chloro-1, 2, 4-thiadiazoles[26] were also reacted analogously, for example. As such ammonium compounds are strong alkylating agents, there is the risk of a splitting off of alkyl during the reaction[24-26].

Further efficient quaternizing agents are *α-halo ketones* and *α-halo esters*, *e.g.*:

$$R_3N \; + \; BrCH_2-CO-C_6H_5 \hspace{2.5cm} [27]$$

$$\longrightarrow \; \left[R_3\overset{\oplus}{N}-CH_2-CO-C_6H_5 \right] Br^{\ominus}$$

The process can be simplified:

$$2\,R_3N \; + \; R-CH_2-CO-Ar \; + \; J_2 \hspace{1.5cm} [28]$$

$$\longrightarrow \; \left[\begin{array}{c} R_3\overset{\oplus}{N}-CH-CO-Ar \\ R \end{array} \right] J^{\ominus} \; + \; \left[R_3\overset{\oplus}{N}H \right] J^{\ominus}$$

The prerequisite is that the salts can be *separated* satisfactorily. Certain methyl groups of activated heterocycles[29] react correspondingly. Because of their *mobile* halogen, α-halo ethers (*e.g.* acetohalo sugars) and α-halo amines can be used even at low temperatures, *e.g.*:

$$R_3N \; + \; \begin{array}{c} Hal-CH-NR_2 \\ R \end{array} \longrightarrow \left[\begin{array}{c} R_3\overset{\oplus}{N}-CH-NR_2 \\ R \end{array} \right] Hal^{\ominus} \hspace{0.5cm} [30]$$

The high water sensitivity of these compounds necessitates taking special precautions[30].

15.2.1.3 The influence of the solvent and of the reaction conditions

Because of the generally ready solubility of the two partners many media are suitable for the Menshutkin reaction. Possible criteria governing their *selection* include the position of the boiling point, the insolubility of the product being formed (see p. 629), and the velocity of the reaction. Much experience and some measurements are available relating to the last-named characteristic. For the reaction of tripropylamine with methyl iodide to form *methyltripropylammonium iodide* at 20°, *e.g.*, the following relative values of the velocity constant k_2 were found[31]:

Cyclohexane	1
Diethyl ether	17
Ethanol	130
Methanol	180
Benzene	250
Ethyl acetate	310
1, 4-Dioxane	520
Acetone	2,100
Dichloromethane	3,900
Acetonitrile	6,600
Dimethylformamide	8,400
Nitromethane	15,400

[14] *A.H. Ford-Moore*, Org. Synth. *30*, 10 (1950); *W.R. Brasen, C.R. Hauser 34*, 61 (1954).

[15] *C.R. McCrosky, F.W. Bergstrom, G. Waitkins*, J. Amer. Chem. Soc. *62*, 2031 (1940).

[16] *N.T. Thuong, M. Lao-Colin, P. Chabrier*, Bull. Soc. Chim. France 932 (1966).

[17] *T. Kametani, K. Kigasawa, T. Hayasaka, M. Hiiragi, H. Ishimaru, S. Asagi*, J. Heterocycl. Chem. *3*, 129 (1966).

[18] *M.G. Ahmed, R.W. Alder, G.H. James, M.L. Sinnot, M.C. Whiting*, Chem. Commun. 1533(1958).

[19] *J. Burdon, V.C.R. McLoughlin*, Tetrahedron *21*, 1 (1965).

[20] *G.W. Fischer, K. Lohs*, Chem. Ber. *103*, 440 (1970).

[21] *F. Scotti, E.J. Frazza*, J. Org. Chem. *29*, 1800 (1964).

[22] *J.L. Dumont, W. Chodkiewicz, P. Cadiot*, Bull. Soc. Chim. France 1197 (1967).

[23] *H. Meerwein, K. Wunderlich, K.F. Zenner*, Angew. Chem. *74*, 807 (1962).

[24] *H. Suhr*, Justus Liebigs Ann. Chem. *701*, 101 (1967).

[25] *W. Klötzer*, Monatsh. Chem. *87*, 131 (1956).

[26] *J. Goerdeler, G. Sperling*, Chem. Ber. *90*, 892 (1957).

[27] *S.D. Ross, M. Finkelstein, R.C. Petersen*, J. Amer. Chem. Soc. *90*, 6411 (1968).

Polar, aprotic solvent thus have an accelerating effect, while the much-used alcohols display a small effect only. It should be emphasized that the order and the relative values are different for other pairings[32, 33].

Normally, the starting components are employed stoichiometrically. With difficult alkylations the use of an excess of one reaction partner is advantageous[34].

In general, side-reactions are avoided more readily by using relatively *low* temperatures for prolonged periods rather than high temperature *briefly*[34]. Reactions under very high pressure may be helpful under difficult steric conditions[35], but necessitate the use of expensive apparatus.

15.2.1.4 Side reactions

It is not necessarily possible to rely on being able to isolate a salt of unambiguous structure or of excellent purity from the quaternization reaction. Whether this does happen depends essentially on two subsequent reactions or side-reactions: transalkylation and elimination.

Transalkylation is based on the fundamental reversibility of the quaternization; it can occur under certain conditions where the groups differ:

$$\left[R_2\overset{\oplus}{\underset{R^2}{N}}{-}R^1 \right] Hal^{\ominus} \rightleftharpoons R{-}Hal + R{-}\underset{R^2}{N}{-}R^1$$

$$\text{and/or} \quad \rightleftharpoons R^1{-}Hal + R_2N{-}R^2 \quad \text{etc.}$$

The *wrong* back-reaction can then lead to new salts. The nature of the halide, temperature, time, and solubility all play a part. Up-to-date examples (without stoichiometry) are:

$$(H_5C_6)_3C{-}N(CH_3)_2 + 2 \ H_3C{-}Br \quad [36]$$

$$\longrightarrow (H_5C_6)_3C{-}Br + \left[(H_3C)_4\overset{\oplus}{N} \right] Br^{\ominus}$$

[cycloheptane ring]$-N(CH_2{-}C_6H_5){-}CH_2{-}C(CH_3)_3 + 2 \ CH_3J \longrightarrow$

$$\left[\text{[cycloheptane ring]}{-}\overset{\oplus}{N}(CH_3)_2(CH_2{-}C_6H_5) \right] J^{\ominus} + (H_3C)_3C{-}CH_2{-}J \quad [37]$$

Avoiding these anomalies may be possible under certain circumstances by selecting a solvent in which the primary salt is precipitated or by prematurely interrupting the reaction. Above all, however, a more active alkylating agent should be considered.

Elimination, a fundamental competitor of nucleophilic substitution, is based on the splitting off of halogen hydride from the alkyl halides if the structure allows it:

$$\overset{}{\underset{}{>}}CH{-}\underset{}{\overset{}{C}}{-}Hal + R_3N$$

$$\longrightarrow \ \overset{}{>}C{=}C\overset{}{<} + \left[R_3\overset{\oplus}{N}H \right] Hal^{\ominus}$$

This side-reaction generally occurs little with primary alkyl halides; with tertiary ones it predominates[38, 39]. Changing the solvent[40] and avoiding high temperatures by selecting an active alkylating agent can bring a remedy here.

Further anomalies relate to allyl and 2-propynyl *rearrangements*, e.g.:

$$Ar{-}\underset{Cl}{\overset{}{C}}H{-}CH{=}CH_2 + (H_5C_2)_3N \longrightarrow$$

$$\left[Ar{-}CH{=}CH{-}CH_2{-}\overset{\oplus}{N}(C_2H_5)_3 \right] Cl^{\ominus} \quad [41]$$

$$R_2\underset{Cl}{\overset{}{C}}{-}C{\equiv}CH + (H_3C)_3N \longrightarrow \left[R_2\overset{}{\underset{\oplus N(CH_3)_3}{C}}{-}C{\equiv}CH \right] Cl^{\ominus}$$

$$+ \left[R_2C{=}C{=}CH{-}\overset{\oplus}{N}(CH_3)_3 \right] Cl^{\ominus} \quad [42]$$

[28] L.C. King, M. McWhirter, R.L. Rowland, J. Amer. Chem. Soc. 70, 239, 242 (1948).
[29] W. Ried, H. Bender, Chem. Ber. 89, 1893 (1956).
[30] H. Böhme, M. Haake, Justus Liebigs Ann. Chem. 705, 147 (1967).
[31] C. Lassau, J.C. Jungers, Bull. Soc. Chim. France 2678 (1968).
[32] N. Menschutkin, Z. Physik. Chem. (Leipzig) 6, 41 (1890).
[33] H. von Halban, Z. Physik. Chem. (Leipzig) 84, 129 (1913).
[34] R.S. Shelton, M.G. vanCampen, C.H. Tilford, H.C. Lang, L. Nisonger, F.J. Bandelin, H.L. Rubenkoenig, J. Amer. Chem. Soc. 68, 753, 755, 757 (1946).
[35] Y. Okamoto, Y. Shimakawa, J. Org. Chem. 35, 3752 (1970) (triphenylmethylpyridiniumchlorid).
[36] E.H. Varga, K. Nador, Arzneimittel. Forsch. 17, 409 (1967); C.A. 67, 90512d (1967).
[37] A.C. Cope, K. Banholzer, F.N. Jones, H. Keller, J. Amer. Chem. Soc. 88, 4700 (1966).

Finally, attention is drawn to the fact that heteroaromatic amines are often not (or not exclusively) alkylated on the amino group but on the ring, *e.g.*:

2-(Dimethylamino)-1-
methylquinolinium iodide

Trimethyl-2-quinolyl-
ammonium iodide[43]

15.2.2 Alkylation with other agents

15.2.2.1 Alkylation with onium compounds

Quaternization with alkyl halides can be accelerated appreciably by the addition of *silver salts*. In the extreme case the alkyl cation is the agent here, *e.g.*[44]:

4-Azoniaspiro[3.5]nonane
perchlorate

Without silver salt the reaction requires much more drastic conditions. *1,1-diethylaziridinium perchlorate*[45] has been prepared analogously.
The technique can probably be made good use of in many difficult cases including with compounds

with alternative types of halogens and with other silver salts. The same is true for the alkylation with tertiary *oxonium salts*[46]. With their aid triphenylamine was for the first time converted into a quaternary salt[47]:

Methyltriphenylammonium
tetrafluoroborate

The conditions (22 days, 75°) demonstrate the degree of *resistance* offered by this amine.
Iodonium[48], and recently also *chloronium* and *bromonium* salts[49] have displayed even greater activity, but have so far been little employed for the reaction under discussion. There is thus much scope for future experimentation.
Reacting salts of amines with *diazoalkanes,* too, proceeds *via* onium stages. Naturally, diazomethane is used predominantly and, evidently, primary and secondary amines, (as salts), too, can be readily quaternized with this agent, *e.g.*[50]:

Dimethyldiphenylammo-
nium tetrafluoroborate

This reaction demonstrates a substantial alkylation activity by the diazo compound. Because of the unpleasant properties of the diazoalkanes, however, this technique is going to remain limited to special cases. An example is the following conversion, which is comparable inprinciple and which for the first time led to the isolation of *aziridinium* salts[51]:

5-Azoniadispiro[4.0.5.1]
dodecane perchlorate

[38] *N. Menschutkin,* Z. Physik. Chem. (Leipzig) *5,* 589 (1890).
[39] *H. Wedekind, R. Oechslen,* Ber. *35,* 3580, 3907 (1902).
[40] *C.D. Hurd, E.H. Ensor,* J. Amer. Chem. Soc. *72,* 5135 (1950).
[41] *G. Cignarella, C. Pasqualucci, E. Ocelli, E. Testa,* Ann. Chimica *58,* 982 (1968).
[42] *G.F. Hennion, C.V. DiGiovanna,* J. Org. Chem. *30,* 3696 (1965).
[43] *D.L. Garmaise, G.Y. Paris,* Chem. & Ind. (London) 1645 (1967).
[44] *N.J. Leonard, D.A. Durand,* J. Org. Chem. *33,* 1322 (1968).
[45] *N.J. Leonard, J.V. Paukstelis,* J. Org. Chem. *30,* 821 (1965).

[46] *H. Meerwein, G. Hinz, P. Hofmann, E. Kroning, E. Pfeil,* J. Prakt. Chem. *147,* 257 (1937).
[47] *S.H. Pine,* J. Org. Chem. *33,* 2554 (1968).
[48] *L.G. Makarowa, A.N. Nesmejanow,* Izv. Akad. Nauk SSSR 617 (1945); C.A. *40,* 46866 (1946).
[49] *G.A. Olah, J.R. DeMember,* J. Amer. Chem. Soc. *92,* 2562 (1970).

15.2.2.2 Alkylation with oxiranes (epoxides)

Tertiary amines react with oxiranes to form quaternary ammonium compounds, e.g.[52]:

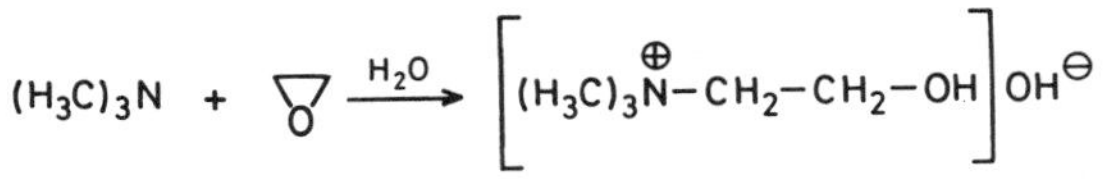

(2-Hydroxyethyl)trimethyl-
ammonium hydroxide

In the presence of acids corresponding salts can form[53]:

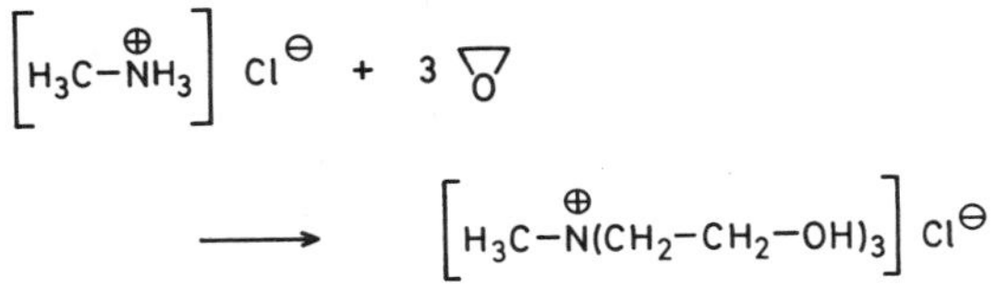

Tris(2-hydroxyethyl)methyl-
ammonium chloride

The reaction with chloromethyloxirane (epichlorohydrin) is relatively sensitive:

in acetonitrile

$\left[(H_3C)_3\overset{\oplus}{N}-CH=CH-CH_2-OH\right]Cl^{\ominus}$ [53]

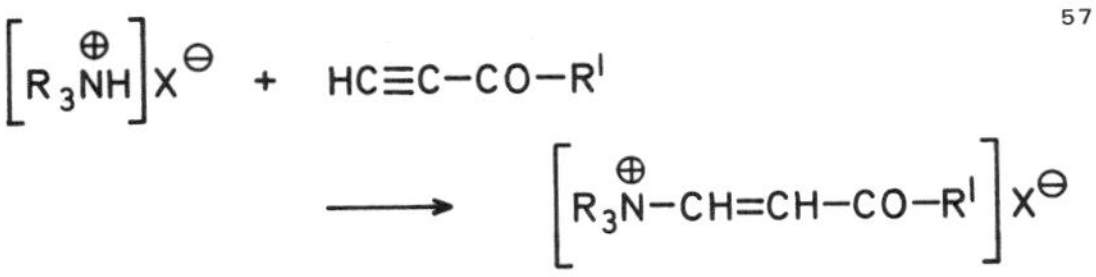

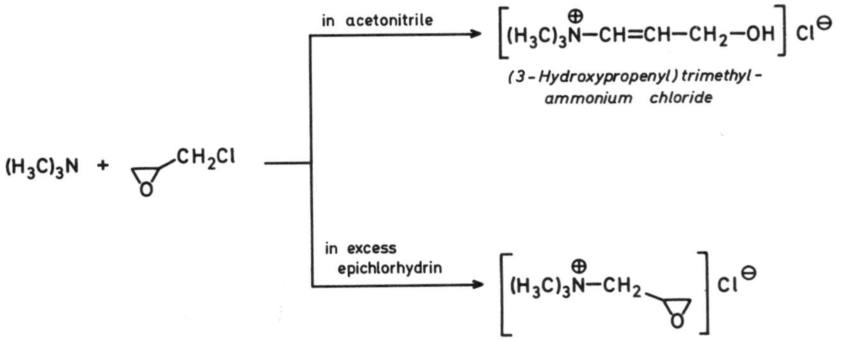

(3-Hydroxypropenyl)trimethyl-
ammonium chloride

in excess
epichlorhydrin

$\left[(H_3C)_3\overset{\oplus}{N}-CH_2-\triangle\right]Cl^{\ominus}$ [54]

(2,3-Epoxypropyl)trimethyl-
ammonium chloride

A reaction leading to the epoxide salt (bottom path) is dependent on the latter's insolubility. Otherwise complex further reactions ensue[54].

15.2.2.3 Reactions with alkynes

Some alkynes react with tertiary amines or their salts to form *vinyl ammonium* compounds, e.g.:

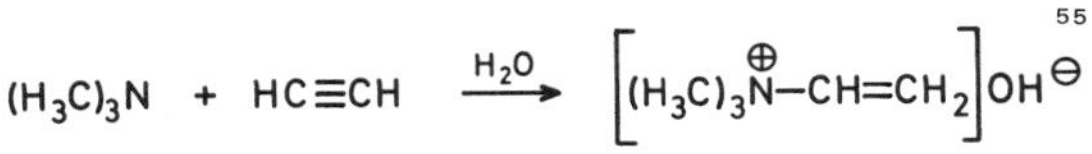

Trimethylvinylammonium
hydroxide

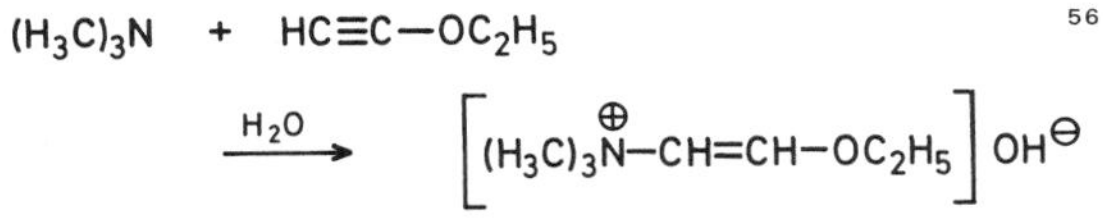

(2-Ethoxyvinyl)trimethylammonium
hydroxide

Some of these reactions require elevated temperatures and pressure. A structural prerequisite of the alkyne appears to be the presence of a methine hydrogen[57].

[50] E. Müller, H. Huber-Emden, W. Rundel, Justus Liebigs Ann. Chem. 623, 34 (1959).

[51] N.J. Leonard, K. Jann, J. Amer. Chem. Soc. 84, 4806 (1962).

[52] K.H. Meyer, H. Hopff, Ber. 54, 2274 (1921).

[53] V.P. Mamaev, G.V. Shishkin, Zh. Org. Khim. 2, 583 (1966); C.A. 65, 8798h (1966).

[54] D.M. Burness, J. Org. Chem. 29, 1862 (1964); J.D. McClure, J. Org. Chem. 35, 2059 (1970).

[55] W. Reppe, Justus Liebigs Ann. Chem. 601, 128 (1956).

[56] J.F. Arens, J.G. Bouman, D.H. Koerts, Rec. Trav. Chim. Pays-Bas 74, 1040 (1955).

[57] G.W. Fischer, K. Lohs, Chem. Ber. 103, 440 (1970).

[58] H. Böhme, N. Kreutzkamp, Sitzungsbericht Ges. Naturw., Marburg 76, 1. Heft, 3 (1953); H. Böhme, W. Lehners, Justus Liebigs Ann. Chem. 595, 169 (1955).

15.3 Preparation of di- and polyvalent ammonium compounds

The two inverse pathways are possible methods for synthesizing *diquaternary* ammonium compounds:

$$R_2N-(CH_2)_n-NR_2 \qquad\qquad Hal-(CH_2)_n-Hal$$
$$+\ 2\ RX \qquad\qquad\qquad\qquad +\ 2\ R_3N$$

$$\left[\overset{\oplus}{R_3N}-(CH_2)_n-\overset{\oplus}{NR_3}\right] 2\ Hal^{\ominus}$$

For $n \geq 3$ the general experience obtained with monovalent compounds holds; for lower values of n limitations generally arise in that only monoquaternary compounds or mixtures are obtained. This behavior may be illustrated by some examples:

1-Piperidino-2-pipecolinium
bromide

Mixed monoquaternary + diquaternary salt

The anomalous results, due to both electronic and steric effects, may be expected especially with longer and branched groups *R*.

Polyammonium salts are prepared either by polymerization of unsaturated univalent compounds (*e.g.*, of allyl derivatives)[62] or by introducing quaternary ammonium groups into preformed polymers. Chloromethylated styrene-divinylbenzene copolymers can be used for this purpose, for example, by reacting with tertiary amines to give the desired products[63]. In the first case water-soluble, in the second case insoluble high-molecular compounds are formed.

The use of alkylamino halides Hal-$(CH_2)n$-NR_2 leads to *linear polyammonium compounds* only if n is substantial. Otherwise, rings with one (or possibly two) quaternary ammonium groups are formed (as for example on p. 630). From a combination of 1,ω-dihaloalkanes and 1,ω-alkyldiamines, too, linear polyammonium compounds can be obtained[64].

15.4 Preparation of acylammonium compounds

These salts play an important role as intermediates in solution, for instance, during Einhorn acylation[65]. The following are recent examples which included isolation and analyses:

$$(H_3C)_3N\ +\ H_5C_6-COCl$$

*Benzoyltrimethylammonium
hexachloroantimonate*

[59] *A.I. Lopushanskii, A.I. Shnarevich, M.I. Shevchuk*, Zh. Org. Khim. *3*, 365 (1967); C.A. *67*, 2851k (1967);
A.P. Gray, T.B. O'Dell, Nature *181*, 634 (1958) (with methyl iodide, the bis-quaternized salt is obtained);
V.P. Denisenko, V.P. Rudi, T.A. Cinaeva, Zh. Obshch. Khim. *38*, 2203 (1968) also with chloracetates; C.A. *70*, 37113k (1969).
[60] *F. Vidal*, J. Org. Chem. *24*, 680 (1959).
[61] *A.P. Gray, T.B. O'Dell*, Nature *181*, 634 (1958).
[62] *G.B. Butler, R.J. Angelo*, J. Amer. Chem. Soc. *79*, 3128 (1957);
G.B. Butler, A. Crawshaw, W.L. Miller, J. Amer. Chem. Soc. *80*, 3615 (1958).
[63] *E.B. Trostyanskaja, I.P. Losev, A.S. Tevlina, S.B. Makarova, G.Z. Nefedova, Hsiang-Jao Lu*, J. Polym. Sci. *59*, 379 (1962); C.A. *58*, 628h (1963).
[64] *W. Kern, E. Brenneisen*, J. Prakt. Chem. *159*, 193 (1941).

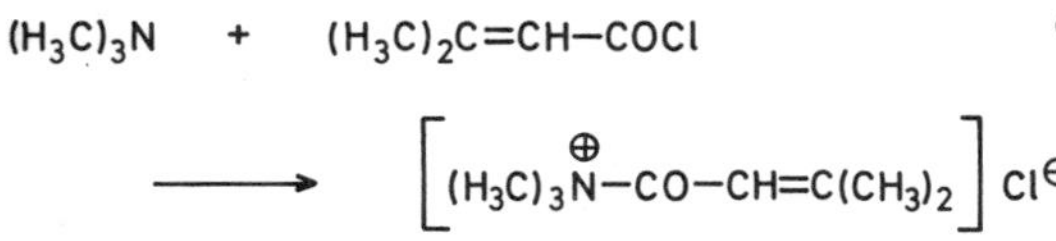

$(H_3C)_3N$ + $(H_3C)_2C=CH-COCl$ [67]

$$\longrightarrow \left[(H_3C)_3\overset{\oplus}{N}-CO-CH=C(CH_3)_2\right]Cl^{\ominus}$$

Trimethyl(3-methyl-2-butenoyl)-ammonium chloride

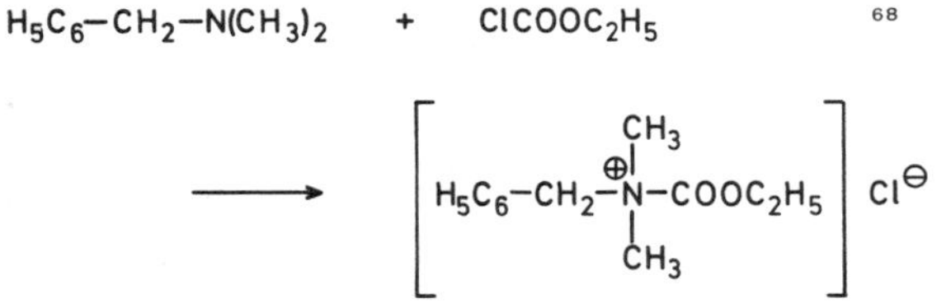

$H_5C_6-CH_2-N(CH_3)_2$ + $ClCOOC_2H_5$ [68]

$$\longrightarrow \left[H_5C_6-CH_2-\overset{\overset{\displaystyle CH_3}{|}}{\underset{\underset{\displaystyle CH_3}{|}}{\overset{\oplus}{N}}}-COOC_2H_5\right]Cl^{\ominus}$$

Benzyl(ethoxycarbonyl)dimethyl-ammonium chloride

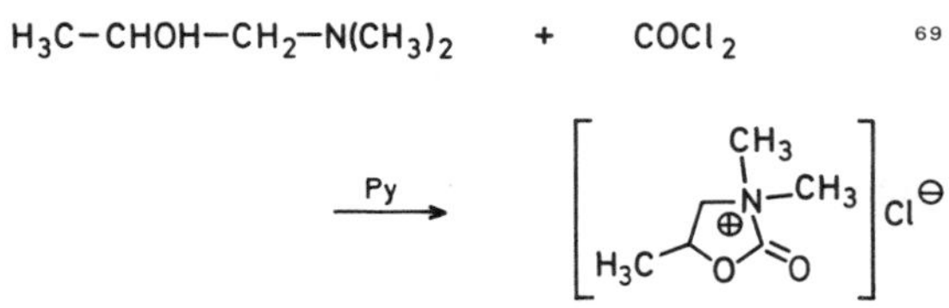

$H_3C-CHOH-CH_2-N(CH_3)_2$ + $COCl_2$ [69]

$$\xrightarrow{Py} \left[\begin{array}{c} \overset{\displaystyle CH_3}{\underset{\displaystyle }{N-CH_3}} \\ H_3C \quad O \end{array}\right]Cl^{\ominus}$$

3,3,5-Trimethyl-2-oxooxazolidinium chloride

Where the structure allows it, *ketene formation* must be anticipated, and especially with strong bases.

On account of their sensitivity to water and ready decomposition, the compounds are difficult to handle (the two last-named compounds split off benzyl chloride and methyl chloride at room temperature respectively). All are strong acylating agents.

15.5 Preparation of immonium and nitrilium compounds

Tertiary *immonium* compounds are formed according to the following scheme:

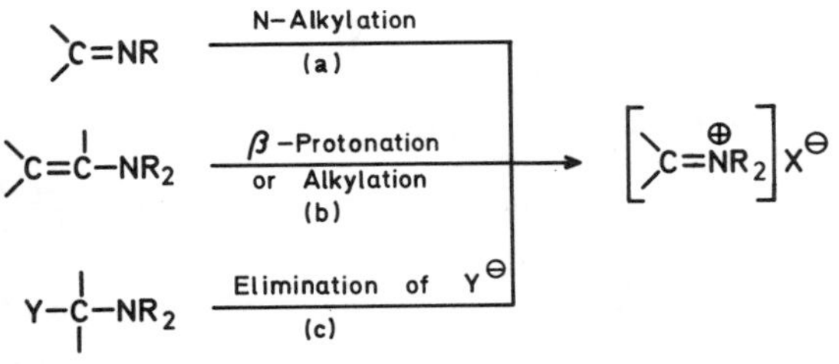

$$\begin{array}{l} \text{\Large$>$}C=NR \xrightarrow[\text{(a)}]{\text{N--Alkylation}} \\[2mm] \text{\Large$>$}C=\overset{|}{C}-NR_2 \xrightarrow[\text{(b)}]{\substack{\beta-\text{Protonation} \\ \text{or Alkylation}}} \\[2mm] Y-\overset{|}{\underset{|}{C}}-NR_2 \xrightarrow[\text{(c)}]{\text{Elimination of } Y^{\ominus}} \end{array} \Biggr\} \left[\text{\Large$>$}C=\overset{\oplus}{N}R_2\right]X^{\ominus}$$

Alkylation **a** has been carried out especially with alkylarylmethylenamines and arylamines[70–72]; it does not always proceed smoothly. *Protonation* **b**, just like alkylation in competition with the reaction on the nitrogen, was detected, for example, during the reaction between 1-cyclohexenylpyrrolidine and perchloric acid[73]. Dienamines can react analogously:

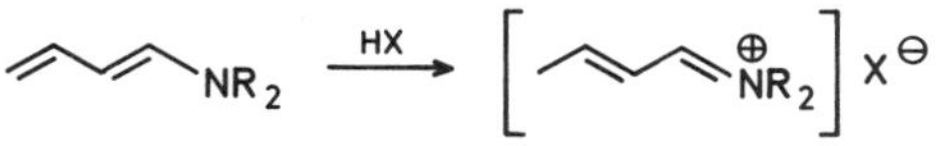

$$\text{/\!\!\backslash\!\!/}NR_2 \xrightarrow{HX} \left[\text{/\!\!\backslash\!\!/}\overset{\oplus}{N}R_2\right]X^{\ominus}$$ [74]

C-alkylation[75] may take place *via* a rearrangement:

$$\text{\Large$>$}C=\overset{|}{C}-\overset{\oplus}{N}R_3 \longrightarrow -\overset{|}{\underset{\underset{\displaystyle R}{|}}{C}}-\overset{|}{C}=\overset{\oplus}{N}R_2$$ [76]

Elimination **c** is the most versatile, because it is possible to abstract amino groups (with protons[77]), cyanide groups (with silver salts[78]), or hydride ions (with carbonium salts[79, 80] or mercury acetate[81]), for example. The shortened procedure consists in reacting carbonyl compounds with secondary ammonium salts:

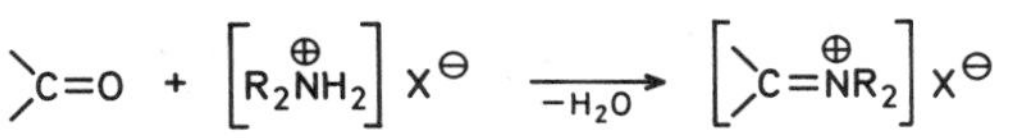

$$\text{\Large$>$}C=O \; + \; \left[R_2\overset{\oplus}{N}H_2\right]X^{\ominus} \xrightarrow{-H_2O} \left[\text{\Large$>$}C=\overset{\oplus}{N}R_2\right]X^{\ominus}$$ [82]

---

[65] *N.O. Sonntag*, Chem. Rev. *52*, 237 (1953).
[66] *F. Klages, E. Zange*, Justus Liebigs Ann. Chem. *607*, 35 (1957).
[67] *G.B. Payne*, J. Org. Chem. *31*, 718 (1966).

---

[68] *H. Böhme, G. Lerche*, Justus Liebigs Ann. Chem. *705*, 154 (1967);
   P. Leduc, P. Chabrier, Bull. Soc. Chim. France 2271 (1963).
[69] *K.C. Murdock*, J. Org. Chem. *33*, 1367 (1968).
[70] *H. Decker, P. Becker*, Justus Liebigs Ann. Chem. *395*, 362 (1913).
[71] *V.I. Minkin, E.A. Medjanceva, J.A. Ostroumov*, Zh. Obshch. Khim. *34*, 1512 (1964); C.A. *61*, 5490e (1964).
[72] *C.R. Hauser, D. Lednicer*, J. Org. Chem. *24*, 46 (1959).
[73] *N.J. Leonard, K. Jann*, J. Amer. Chem. Soc. *84*, 4806 (1962).
[74] *G. Opitz, W. Merz*, Justus Liebigs Ann. Chem. *652*, 139 (1962).
[75] *G. Stork, R. Terrell, J. Szmuszkovicz*, J. Amer. Chem. Soc. *76*, 2029 (1954).
[76] *E. Elkik*, Bull. Soc. Chim. France 903 (1969).
[77] *H. Böhme, E. Mundlos, O.E. Herboth*, Chem. Ber. *90*, 2003 (1957); *H. Böhme, K. Hartke*, Chem. Ber. *93*, 1305 (1960).
[78] *H.G. Reiber, T.D. Stewart*, J. Amer. Chem. Soc. *62*, 3026 (1940).
[79] *H. Meerwein, V. Hederich, H. Morschel, K. Wunderlich*, Justus Liebigs Ann. Chem. *635*, 1 (1960).

Substituted *nitrilium* compounds can be obtained by the following pathways:

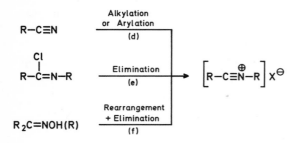

Alkylation **d** is performed with oxonium salts[83] or alkyl chlorides in the presence of antimony(V) chloride[83] or fluoride[84]. Diazonium tetrafluoroborates were used as the *arylating* agent[83]. *Elimination* of the chlorine **e** takes place with complex formers[83, 85] and the same holds for method **f**[86], for which fluorosulfuric acid can be used as an alternative[84].

Both immonium and nitrilium compounds are strong electrophiles. They are important intermediates, for example during α-aminoalkylation and heterocycle syntheses.

15.6 Transformation of quaternary ammonium compounds with retention of the ammonium structure

15.6.1 Anion exchange

Direct synthesis generally leads to the *ammonium halides*. With silver salts they can easily be transformed into compounds with other anions in the usual manner[87]. The *silver salt* need not necessarily be present in the dissolved form[88]. According to a second general method the hydroxides are initially prepared with silver hydroxide or with a strongly basic anion exchanger and are then neutralized with the corresponding base. The relevant ammonium salt can be used as an alternative[89]:

$$\left[R_4N^{\oplus}\right]OH^{\ominus} \; + \; NH_4X \longrightarrow \left[R_4N^{\oplus}\right]X^{\ominus} + NH_3 \; + H_2O$$

This procedure is advantageous with acids which tend to decompose (*e.g.* HSCN, $H_2S_2O_3$). Adsorption of the ammonium ion on a cation exchanger and eluting with the corresponding acid, too, leads to the desired result[90].

Naturally, combinations in which an insoluble salt is precipitated are also suitable, *e.g.*[91]:

$$\left[R_4N^{\oplus}\right]Cl^{\ominus} \; + \; NaOC_2H_5$$

$$\xrightarrow{\text{Ethanol}} \; \left[R_4N^{\oplus}\right]^{\ominus}O-C_2H_5 \; + \; NaCl$$

remains   precip-
in solution   itates

The ammonium salt can be precipitated in the same way, *e.g.*, by adding alkali metal halide to the corresponding methylsulfate in concentrated solution. Typical precipitated forms are the perchlorate, picrates, reineckates, tetraphenylborates, and salts of certain sulfonic acids[92]. Picrates, for example, can be converted into chloride secondarily by treatment with hydrochloric acid and extracting the picric acid with ether. Anion exchange may help to improve the separation and crystallization, avoid hygroscopicity, and aid identification. Anion exchange serving for the *separation* into *optical antipodes* is suitably performed *via* the hydroxide, which is then reacted with an optically active acid to form the diastereomeric salt mixture[93].

15.6.2 Betaines (inner salts), ylides

The principal method for preparing *betaines* is deprotonation of the carboxy group of a quaternary ammonium ion, *e.g.*[91]:

$$\left[\begin{array}{c} R_3\overset{\oplus}{N}-CH-COOH \\ | \\ R \end{array}\right]X^{\ominus} \xrightarrow[-HX]{} \left[\begin{array}{c} R_3\overset{\oplus}{N}-CH-COO^{\ominus} \\ | \\ R \end{array}\right]$$

[80] *S.G. McGeachin*, Can. J. Chem. *47*, 151 (1969); *H. Volz, H.-H. Kiltz*, Justus Liebigs Ann. Chem. *752*, 86 (1971).

[81] *N.J. Leonard, F.P. Hauck*, J. Amer. Chem. Soc. *79*, 5279 (1957).

[82] *N.J. Leonard, J.V. Paukstelis*, J. Org. Chem. *28*, 3021 (1963); thermical preparation of dimethylmethylen-immonium iodide from iodomethyl-trimethyl ammonium iodide: *J. Schreiber, H. Maag, N. Hashimoto, A. Eschenmoser*, Angew. Chem. *83*, 355 (1971).

[83] *H. Meerwein, P. Laasch, R. Mersch, J. Spille*, Chem. Ber. *89*, 209 (1956).

[84] *G.A. Olah, T.E. Kiovsky*, J. Amer. Chem. Soc. *90*, 4666 (1968).

[85] *F. Klages, W. Grill*, Justus Liebigs Ann. Chem. *594*, 21 (1955).

[86] *H. Meerwein*, Angew. Chem. *67*, 374 (1955).

[87] *R.S. Shelton, M.G. Van Campen, C.H. Tilford, H.C. Lang, L. Nisonger, F.J. Bandelin, H.L. Rubenkoenig*, J. Amer. Chem. Soc. *68*, 753, 755, 757 (1946).

[88] *A.H. Ford-Moore*, Org. Synth. *30*, 10 (1950).

[89] *M.M. Markowitz*, J. Org. Chem. *22*, 983 (1957).

[90] *F. Vidal*, J. Org. Chem. *24*, 680 (1959).

All stronger bases, including those in the form of an ion exchanger, are suitable; if $X^{\ominus}$ is a halogen ion the reaction is carried out with silver hydroxide. Alternatively, the *betaine* can be *prepared* in one step by reacting sodium chloroacetate with tertiary amine; separation from sodium chloride succeeds with ethanol. The principle can be applied fundamentally as desired to all those ammonium compounds whose cation contain an acid group, for example, phenol, enol, or sulfonamide groups or an *CH*-acid group (betaines in the extended sense). In addition, there are special techniques such as thermal rearrangement of amino carboxylic or sulfonic acid esters, *e.g.:*

94

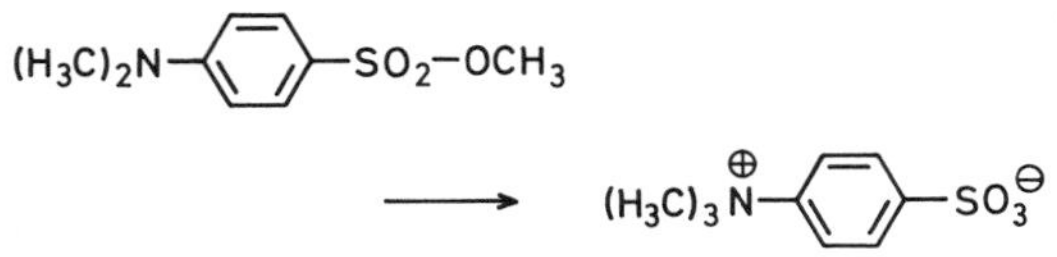

Trimethyl(4-sulfophenyl)-ammonium hydroxide; inner salt

or reaction of tertiary amines with cyclic esters

95

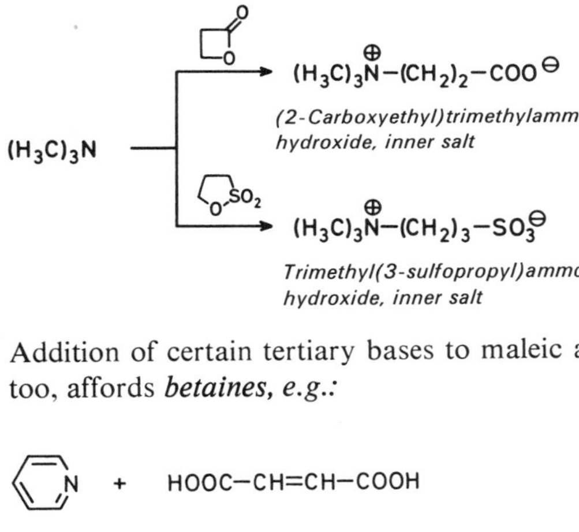

(2-Carboxyethyl)trimethylammonium hydroxide, inner salt

96

Trimethyl(3-sulfopropyl)ammonium hydroxide, inner salt

Addition of certain tertiary bases to maleic acid, too, affords *betaines, e.g.:*

97

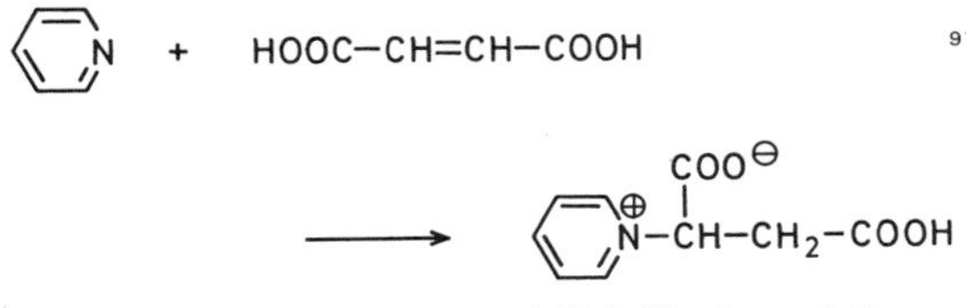

1-(1,2-Dicarboxyethyl)-pyridinium hydroxide, inner salt

Ammonium ylides represent a special type of *zwitterionic* compound

$$R_3\overset{\oplus}{N}-\overset{\ominus}{C}H-R$$

[91] *W.J.C. Croxall, M.F. Fegley, H.J. Schneider*, Org. Synth. *38*, 5 (1958).

The acidity of ordinary methylene groups next to the positively charged N atoms is so slight that only the strongest bases deprotonate:

98

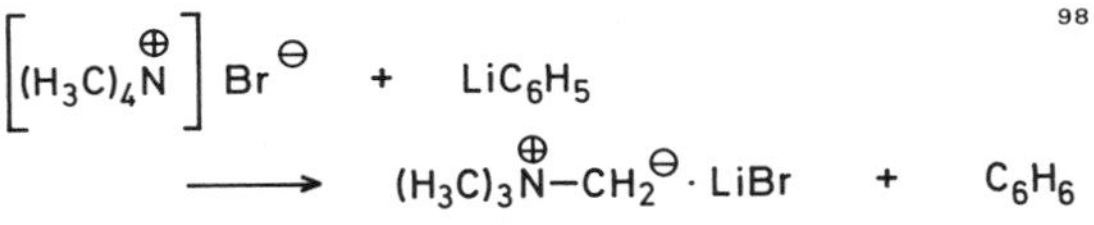

Trimethylammonium-methylide

This simplest representative is stable only in combination with the metal salt; it is often interpreted to be an organometallic compound:

99

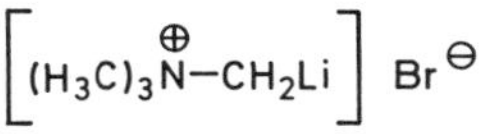

True, relatively stable *zwitterions* are obtained by stronger resonance stabilization, *e.g.:*

100

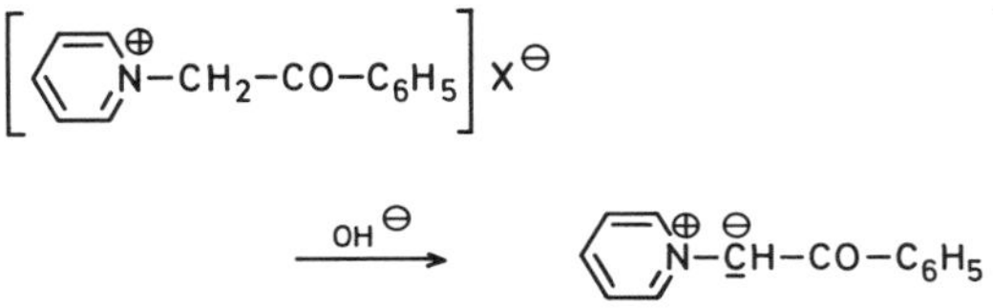

(Benzoylmethyl)-pyridinium hydroxide, inner salt

101

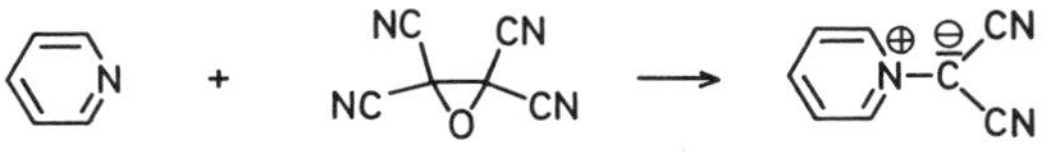

(Dicyanomethyl)-pyridinium hydroxide, inner salt

Further betaine types can be òbtained from ylides, *e.g.:*

102

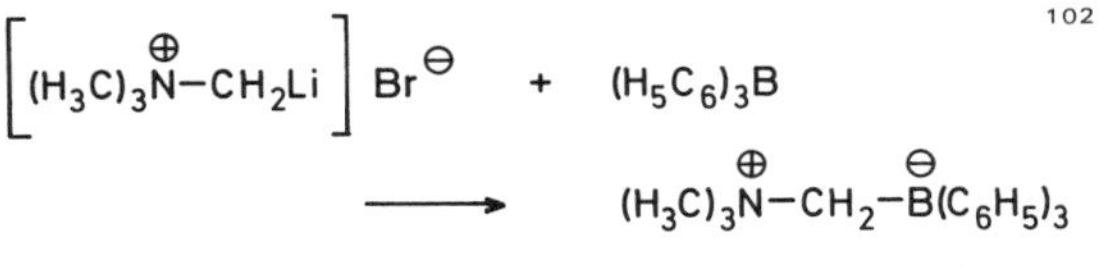

Triphenyl[(trimethyl-ammonio)methyl]borate

[92] *W.G. Leeds, R. Slack*, J. Chem. Soc. 3941 (1956).
[93] *A.C. Cope, W.R. Funke, F.N. Jones*, J. Amer. Chem. Soc. *88*, 4693 (1966).
[94] *R. Kuhn, H.W. Ruelius*, Chem. Ber. *83*, 420 (1950).
[95] *T.L. Gresham, J.E. Jansen, F.W. Shaver, R.A. Bankert, F.T. Fiedorek*, J. Amer. Chem. Soc. *73*, 3168 (1951).
[96] *P. Blumbergs, A.B. Ash, F.A. Daniher, C.L. Stevens, H.O. Michel, B.E. Hackley, J. Epstein*, J. Org. Chem. *34*, 4065 (1969).

Ylides are intermediate stages in important rearrangements leading to tertiary amines (Sommerlet-Hauser, Stevens[103]).

Immonium and *nitrilium ylides* (in solution) are accessible by the agency of 1,3-elimination, *e.g.:*

*Benzylideneammonium-
4-nitrobenzylide*

They form a versatile starting material for 1,3-dipolar cycloadditions[104].

The Author thanks the Bayer AG, Leverkusen and also Messrs. G. Freygang and R. Büchler for help with the literature survey.

[97] *O. Lutz, R. Klein, A. Jirgenson,* Justus Liebigs Ann. Chem. *505,* 307 (1933).

[98] *G. Wittig, M.H. Wetterling,* Justus Liebigs Ann. Chem. *557,* 193 (1945).

[99] *W.K. Musker,* Fortschr. Chem. Forsch. *14,* 295 (1970);
A.W. Johnson, Ylid-Chemistry p. 251, Academic Press, New York, London 1966.

[100] *F. Kröhnke,* Angew. Chem. *65,* 605 (1953).

[101] *W.J. Linn, O.W. Webster, R.E. Benson,* J. Amer. Chem. Soc. *85,* 2032 (1963).

[102] *F. Bickelhaupt, J.W. Barnick,* Rec. Trav. Chim. Pays-Bas *87,* 188 (1968).

[103] *S.H. Pine,* Org. Reactions *18,* 403 (1970).

[104] *R. Huisgen,* Angew. Chem. *65,* 604 (1963).

16 Nitriles

Contributed by

A. Zobáčová
Vysoká Škola Chemicko-Technologická
Prague

16.1 Preparation of nitriles from compounds with the same carbon chain

16.1.1 Nitriles from hydrocarbons and ammonia by catalytic oxidation

The preparation of nitriles from hydrocarbons and ammonia by catalytic oxidation

$$R-CH_3 \xrightarrow[\text{MoO}_3/\text{Al}_2\text{O}_3]{\text{NH}_3/\text{O}_2} R-CN$$

is advantageous particularly for industrial purposes[1-4], but can be used also on the laboratory scale[5]. The vapor of the hydrocarbon is led over the catalyst with excess ammonia and air at ~500°. As a rule, molybdenum(VI) oxide on a suitable carrier is used as catalyst; vanadium, tungsten, and tin oxides are employed less commonly.

Alkanes and *alkenes* are generally transformed during the reaction. Thus, from lower hydrocarbons (with up to five carbon atoms) *C–C* cleavage gives mainly *acetonitrile* (60–95%) side by side with nitriles with 3 and 4 carbon atoms[2]. Higher hydrocarbons are partly cyclodehydrogenated to aromatic nitriles[1]. From heptane a mixture of *acetonitrile, benzonitrile,* and *propionitrile* (1:6:3) is formed[1].

Aromatic hydrocarbons, especially methylaromatic compounds, react more precisely[1, 3, 5]; the following are examples of the reactions that take place:

Toluene	⟶ *Benzonitrile*	
Xylene	⟶ *Tolunitrile*	60–85% [1, 2]
Mesitylene	⟶ *Dimethylbenzonitrile*	
Methylnaphthalene	⟶ *Naphthonitrile*	

Methylheteroaromatic compounds, too, react in clear-cut manner[4], *e.g.:*

Isonicotinonitrile; 70%

Other substituents of the aromatic ring are generally unaltered, *e.g.*[5]:

2,6-Dichlorobenzo-nitrile; 50%

16.1.2 Nitriles from amines and their derivatives

16.1.2.1 Dehydrogenation of *N*-alkyl-*N,N*-dihaloamines

Dehydrohalogenation of *N*-alkyl-*N,N*-dihaloamines, which is a side-reaction during Hofmann degradation of amides[6-8, 231],

$$R-CH_2-NH_2 \longrightarrow [R-CH_2-NX_2]$$
$$\longrightarrow R-CN$$

has been modified suitably to make it into an excellent method of preparative nitrile synthesis.

A first modification consists in using *iodine pentafluoride*[9] as the dehydrating agent. *N*-fluorinated intermediate products are postulated but have not been identified so far. The technique can be applied to both aliphatic and aromatic nitriles:

$$R-CH_2-NH_2 \xrightarrow[\text{IF}_5/\text{C}_5\text{H}_5\text{N}/\text{CH}_2\text{Cl}_2]{15°/2\,\text{h}} R-CN$$

R = C$_6$H$_5$;   *Benzonitrile;* 90%
R = C$_3$H$_7$;   *Butyronitrile;* 45%
R = (CH$_3$)$_2$CH; *Isobutyronitrile;* 40%
R = C$_5$H$_{11}$;   *Hexanenitrile;* 51%

Small amounts of the relevant aldehyde are formed as side-product.

A *two-stage* variant[10, 11], proceeding under very mild conditions and described by the following equation, appears to be even more advantageous:

$$H_9C_4-NH_2 \xrightarrow{\text{Cl}_2/\text{NaHCO}_3/\text{H}_2\text{O}/0°} H_9C_4-NCl_2$$
$$\xrightarrow{\text{CsF/CH}_3\text{CN}/45°/48\,\text{h}} H_7C_3-CN$$

Butyronitrile; 90%

Preparation of the *N,N*-dichloroamine ensues by passing chlorine into an aqueous solution of amine and sodium hydrogen carbonate for several

[1] *A.C. Stevenson,* Ind. Eng. Chem. 41, 1846 (1949).
[2] *W.I. Denton, R.B. Bishop,* Ind. Eng. Chem. *45,* 282 (1953).
[3] *W.I. Denton, R.B. Bishop, H.P. Caldwell, H.D. Chapman,* Ind. Eng. Chem. *42,* 796 (1950).
[4] *B.V. Suvorov, S.R. Rafikov, V.S. Kudinova, B.A. Zhubanova,* Zh. Prikl. Khim. (Leningrad) *32,* 1642 (1959); C.A. 54, 533 (1960).
[5] *H. Koopman,* Rec. Trav. Chim. Pays-Bas *80,* 1075 (1961).

[6] *A.W. Hofmann,* Ber. *17,* 1406 (1884).
[7] *A.W. Hofmann,* Ber. *17,* 1920 (1884).
[8] *E.S. Wallis, J.F. Lane,* Org. Reactions *3,* 267 (1946).
[9] *T.E. Stevens,* J. Org. Chem. *26,* 2531 (1961).
[10] *L.L. Jackson, G.N.R. Smart, G.F. Wright,* J. Amer. Chem. Soc. 69, 1539 (1947).
[11] *C.M. Sharts,* J. Org. Chem. *33,* 1008 (1968).

hours below 12°. With a number of aliphatic amines the yield is approximately 90%[10]. The subsequent dehydrohalogenation is carried out by heating the dichloroamine with cesium fluoride in acetonitrile[11]. Diamines, too, can be reacted in this way. During conversion of 1, 6-hexanediamine into *adiponitrile* the fluorinated derivative *(N, N, N', N'-tetrafluoro-1,6-hexanediamine)* was successfully isolated[11]:

$$H_2N-(CH_2)_6-NH_2 \xrightarrow[NaHCO_3/CCl_3F/0°]{H_2O/F_2}$$

$$F_2N-(CH_2)_6-NF_2 \xrightarrow{CH_3CN/CsF} NC-(CH_2)_4-CN$$

16.1.2.2 Catalytic dehydrogenation of alkyl-amines

The catalytic dehydrogenation of alkylamines to nitriles has been studied in great detail[12-18]:

$$R-CH_2-NH_2 \xrightarrow{-2H_2} R-CN$$

To carry out the reaction the amine vapors are led over metal catalysts (nickel and copper[14], silver[17], or cobalt[18, 19]) at 300–600°. Benzylamines give poorer yields than higher aliphatic amines. From 3-methylbutylamine over 90% *2-methylpropionitrile* is obtained[14]. Side-reaction hydrogenolysis of the amine

$$R-CH_2-NH_2 \begin{cases} \xrightarrow{-2H_2} R-CN \\ \xrightarrow{+H_2} R-CH_3 + NH_3 \end{cases}$$

can be suppressed by adding ammonia[16], easily hydrogenated olefin[18], or by simultaneous oxidation with air[17]

$$H_9C_4-NH_2 + H_3C-CH=CH-CH_2-CH_3$$

$$\xrightarrow{Co/NH_3/180-200°} H_7C_3-CN + C_5H_{12}$$

Butyronitrile; 70%

$$H_2C=CH-CH_2-NH_2 \xrightarrow{Ag/air/450-600°}$$

$$H_2C=CH-CN$$

Acrylonitrile; 90%

Simultaneous air oxidaton is suitable also for preparing unsaturated nitriles. In place of a direct addition of olefin secondary amine, which is cleaved to nitrile and olefin during the reaction, may be added[15], *e.g.:*

$$HN(CH_2-CH_2-CH_3)_2 \xrightarrow{Ni/350-380°}$$

$$H_3C-CH_2-CN + H_2C=CH-CH_3$$

Propionitrile

As alcohols can be converted into amines with ammonia under similar conditions, the nitriles can be obtained alternatively directly from alcohol and ammonia[19], *e.g.:*

$$H_7C_3-CH_2OH + NH_3$$

$$\xrightarrow{Ni/Al_2O_3/300°} H_7C_3-CN$$

Butyronitrile;
81.5%

An elegant laboratory modification for preparing nitriles utilizes nickel peroxide[20] at sharply reduced temperatures as the dehydrating agent, *e.g.:*

$$H_3C-(CH_2)_6-CH_2-NH_2$$

$$\xrightarrow{NiO_2/C_6H_6/1,5\ h\ boiling} H_3C-(CH_2)_6-CN$$

Octanenitrile; 95.8%

16.1.3 Nitriles from aldehydes and their derivatives

16.1.3.1 Dehydration of aldoximes

Acetic anhydride[5, 22-24, 33], thionyl chloride[25], phosphorus(V) oxide[26] (a solution of phosphorus(V) oxide in ethanol[26]), benzenesulfonyl chlo-

[12] *P. Sabatier, G. Gaudion,* Compt. Rend. *165,* 224 (1917).
[13] *A. Mailhe,* Bull. Soc. Chim. France (4) *27,* 229 (1920).
[14] *A. Mailhe, F. deGodon,* Bull. Soc. Chim. France (4) *21,*288(1916);
A. Mailhe, F. deGodon, Compt. Rend. *165,* 557 (1917);
A. Mailhe, F. deGodon, Compt. Rend. *166,* 215 (1918).
[15] *A. Mailhe,* Compt. Rend. *166,*996(1918).
[16] USP 1684634 (1928), Inv.: *M. Luther, K. Pieroh;* C.A. *22,* 4536 (1928);
DRP 730179(1942), Erf.: *K. Baur, K. Vierling, K. Wimmer;* C.A. *38,* 380 (1944).
[17] *L.M. Peters, K.E. Marple, T.W. Evans, S.H. McAllister, R.C. Castner,* Ing. Eng. Chem. *40,* 2046 (1948).
[18] USP 2388218 (1945), Inv.: *J. Olin;* C.A. *40,* 591 (1946).
[19] *M.A. Popov, N.I. Shuikin,* Izv. Akad. Nauk SSSR 713(1958);C.A.*52,*19925(1958).
[20] *K. Nakagawa, T. Tsuji,* Chem. Pharm. Bull. (Tokyo) *11,*296(1963).
[21] *H. Metzger,* in *Houben-Weyl,* Methoden der organischen Chemie, Bd. X/4, 4. Aufl., p. 1, Georg Thieme Verlag, Stuttgart 1968.
[22] *S. Reich,* Bull. Soc. Chim. France(4)*21,*222(1917).
[23] *M.F. Browne, R.L. Shriner,* J. Org. Chem. *22,* 1320 (1957).

ride, and similar compounds, and an ethyl polyphosphate mixture are employed for dehydrating oximes to nitriles [220, 222, 223], *e.g.:*

$$R-CH=NOH \longrightarrow R-C\equiv N$$

$$H_5C_6-CH=NOH$$

$$\xrightarrow{P_2O_5 - C_2H_5OH/C_6H_6 /90°/15\ min} H_5C_6-CN$$

98 %

An equal yield of *benzonitrile* can be obtained with 2,3-dibromopropionitrile as dehydrating agent [27].

For aliphatic nitriles additional mild methods have been developed. In some cases the corresponding aldoxime is isolated [28]:

$$R-CHO + \begin{array}{c} F_3C-CO-NH \\ | \\ F_3C-CO-O \end{array} \xrightarrow{C_5H_5N/C_6H_6/80°/2h}$$

$$R-CN + F_3C-COOH$$

53 - 88 %

Simple boiling of aldehyde, hydroxylamine hydrochloride, and sodium acetate in acetic acid [29] leads to the same result. When dialkyl or diaryl phosphates [30] or 4-chlorophenyl chloroformate are used as dehydrating agent aldoximes are converted into the corresponding nitriles at as low as ~20° [31].

In a modified Wohl sugar degradation [24] the dehydrations succeeds by photochemical means [32]:

81 %

Probably the best method, which has proven itself especially with aliphatic compounds, is based on the action of a titanium(IV) chloride-pyridine complex in tetrahydrofuran on the aldoximes [33]. After about two days at room temperature 80–97% nitrile is obtained.

The purely thermal nitrile preparation from *O-aryl* aldoximes is of preparative importance

although it has been studied mainly in terms of kinetics, mechanism, and catalytic influences [34].

16.1.3.2 From hydrazones and azines by cleavage

Photochemical cleavage

Access of air is the decisive factor for the photochemical preparation of *substituted benzonitriles* (80–95%) [35–37] from the corresponding azines [37]:

$$R-CH=N-N=CH-R \xrightarrow[\substack{air\ excess \\ h\nu/20°/3\ days}]{} 2\ R-CN$$

The same is true of the photolysis of unsymmetrical diphenylhydrazones [38]:

$$R-CH=N-N(C_6H_5)_2 \xrightarrow{h\nu/O_2/CH_3OH/20°/1h}$$

40 - 75 %

Oxidative cleavage

Oxidative cleavage of azines serves primarily for preparing aromatic nitriles

$$R-CH=N-N=CH-R \xrightarrow{Oxid.} 2\ R-CN$$

[27] *T. Mukaiyama, T. Hata*, Bull. Soc. Chem. Japan *33*, 1712 (1960).
[28] *J.H. Pomeroy, C.A. Craig*, J. Amer. Chem. Soc. *81*, 6340 (1959).
[29] *J.H. Hunt*, Chem. & Ind. (London) 1873 (1961).
[30] *P.J. Foley jr.*, J. Org. Chem. *34*, 2805 (1969).
[31] *D.L.J. Clive*, Chem. Commun. 1014 (1970).
[32] *R.W. Binkley*, Tetrahedron Lett. 3439 (1970).
[33] *W. Lehnert*, Tetrahedron Lett. 559 (1971).
[34] *J.H.M. Hill, L.D. Schmookler*, J. Org. Chem. *32*, 4025 (1967).
[35] *J.E. Hodgkins, J.A. King*, J. Amer. Chem. Soc. *85*, 2679 (1963).
[36] *R.W. Binkley*, J. Org. Chem. *34*, 2311 (1969).
[37] *R.W. Binkley*, J. Org. Chem. *34*, 3218 (1969).
[38] *R.W. Binkley*, Tetrahedron Lett. 2085 (1970).

[24] *A. Wohl*, Ber. *24*, 994 (1891);
   A. Wohl, Ber. *26*, 730 (1893).
[25] *F.P. Doyle, W. Ferrier, D.O. Holland, M.D. Mehta, J.H.C. Nayler*, J. Chem. Soc. 2853 (1956).
[26] *T. Mukaiyama, T. Hata*, Bull. Chem. Soc. Japan *34*, 99 (1961).

The oxidation is performed either with *chlorine* in 1,2-dichlorobenzene[39] at 160° or in two stages with trifluoroperacetic acid and *via N-oxides* as intermediates[40]. *9,10-Dioxoanthronitrile* (100%)[39] was obtained by the first technique, *benzonitrile* (83%) and *2-furonitrile* (77%) by the second[40].

Aromatic nitriles are accessible also by oxidizing aldehyde hydrazones with *mercury(II) oxide* when 1,2-diethoxyethane, bis(2-methoxyethyl) ether, tetrahydrofuran, or ethanol are used as solvent[41]. In diethyl ether, 1,4-dioxane, or fura *diazoalkanes* are obtained[41].

Elimination reactions

The preparatively easily conducted conversion of aldehydes to nitriles *via* the hydrazones affords yields of 51–93%[42].

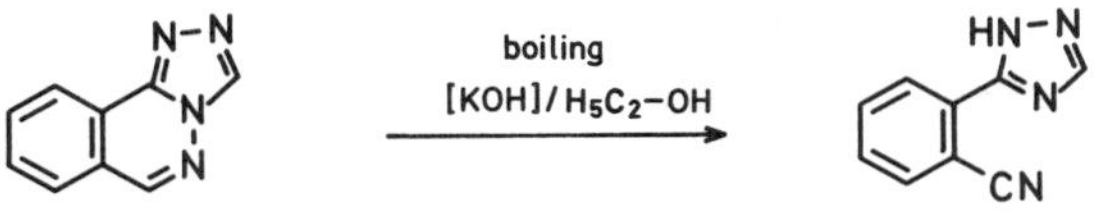

2-(1H-1,2,4-Triazol-5-yl) benzonitrile[47]; 75%

16.1.3.3 By reacting aldehydes with hydrazoic acid

Reacting aldehydes with hydrazoic acid leads to nitriles and *N*-formylamines:

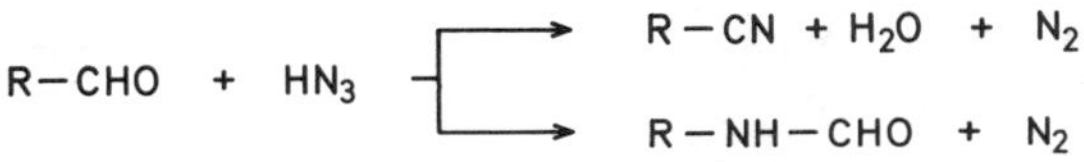

Both competitive reactions can be utilized preparatively. The direction of the reaction is controlled especially by the concentration of the acid[43, 231, 220] used to acidify a mixture of sodium azide and aldehyde in a suitable solvent. ~0° is generally employed for working[43–46]; both aliphatic[44] and aromatic[43, 46] nitriles are accessible in this way.

Often this method of preparation serves merely as the intermediate stage. Boiling the reacted acid mixture briefly leads directly to the carboxylic acid corresponding to the nitrile[44, 46], cf. ref.[224] (see Vol. V), *e.g.*:

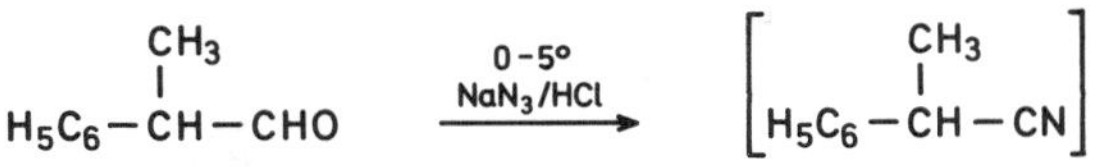

Hydratroponitrile

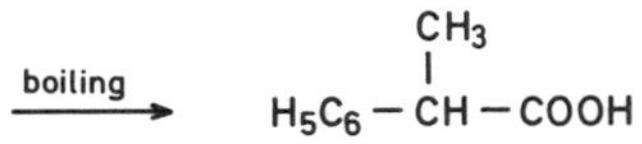

Hydratropic acid[44]; 51%

16.1.3.4 Other methods

Boiling aldehydes with nitroparaffins in acetic acid in the presence of ammonium ions leads to nitriles directly[29, 48], *e.g.*:

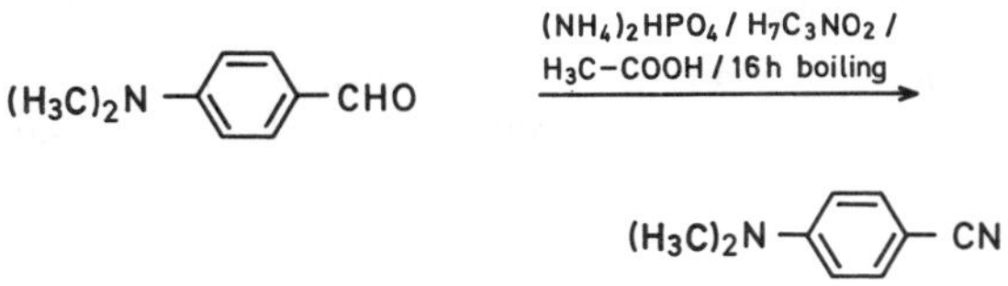

4-(Dimethylamino)-benzonitrile; 77%

Aldoximes[29] probably arise as intermediate products which are dehydrated by acetic acid (cf. p. 639).

The reaction between aldehydes and ammonia in the presence of copper salts and air to form nitriles passes through the aldimine stage[51], *e.g.*:

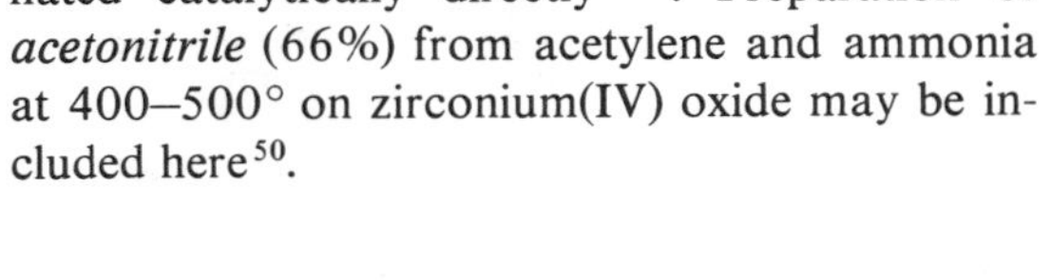

4-Anisonitrile; 90%

In an industrial process the imines are dehydrogenated catalytically directly[1, 49]. Preparation of *acetonitrile* (66%) from acetylene and ammonia at 400–500° on zirconium(IV) oxide may be included here[50].

[39] *E. Klingsberg*, J. Org. Chem. *25*, 572 (1960).

[40] *W. M. Williams, W. R. Dolbier*, J. Org. Chem. *34*, 155 (1969).

[41] *D. B. Mobbs, H. Suschitzky*, Tetrahedron Lett. 361 (1971).

[42] *R. F. Smith, L. E. Walker*, J. Org. Chem. *27*, 4372 (1962).

[43] *W. E. McEwen, W. E. Conrad, C. A. Vander Werf*, J. Amer. Chem. Soc. *74*, 1168 (1952).

[44] *P. A. S. Smith, D. R. Baer, S. N. Ege*, J. Amer. Chem. Soc. *76*, 4564 (1954).

[45] *P. A. S. Smith, Tung-Yin Yu*, J. Org. Chem. *17*, 1281 (1952).

[46] *C. Schuerch*, J. Amer. Chem. Soc. *70*, 2293 (1948).

[47] *K. T. Potts, C. A. Lovelette*, Chem. Commun. 845 (1968).

[48] *H. M. Blatter, H. Lukaszewski, G. de Stevens*, J. Amer. Chem. Soc. *83*, 2203 (1961).

[49] USP 2 525 818, Inv.: *J. E. Mahan*; Chem. Zentr. II, 1661 (1951).

[50] *J. Amiel, G. Nomine*, Compt. Rend. *224*, 483 (1947).

[51] *W. Brackman, P. J. Smit*, Rec. Trav. Chim. Pays-Bas *82*, 757 (1963).

Aromatic nitriles, in particular, are formed by a similar mechanism from aldehydes and chloramine in alkaline medium[52]:

$$R-CHO \; + \; NH_2Cl \; \longrightarrow \; R-CH=NCl$$

$$\longrightarrow \; R-CN$$

or from aldehydes and ammonia in iodine-containing methanolic sodium methoxide solution[53].

In an interesting version of the technique an imine complex consisting of an aldehyde, octacarbonyldicobalt $[Co_2(CO)_8]$, and ammonia is oxidized to nitrile with bromine[54].

16.1.4 From carboxylic acid derivatives

16.1.4.1 From carboxylic acids and ammonia

Reacting carboxylic acids with ammonia serves primarily for preparing aliphatic *nitriles*[1, 55-58], and particularly technically:

$$R-COOH \; + \; NH_3 \; \longrightarrow \; R-CN$$

To carry out the reaction the carboxylic acid vapor and the ammonia are at 350–500° led over silica gel[55] which may be activated with 2% titanium(IV) oxide[58]. Yields in excess of 80% are not exceptional.

16.1.4.2 From carboxylic acids and urea or nitriles

From carboxylic acids and urea[59, 60] (laboratory method)

$$HOOC-(CH_2)_n-COOH \xrightarrow{\;H_2N-CO-NH_2/160-380°\;}$$

$$NC-(CH_2)_n-CN \; + \; NC-(CH_2)_n-COOH$$

n = 7; *Nonane-*   *8-Cyanooctanoic*
   dinitrile; 44%   *acid;* 66%

n = 8; *1,10-Decane-*   *9-Cyanononanoic*
   dinitrile; 48%   *acid;* 33%

or from nitriles[61] carboxylic acid nitriles are obtained

$$R^1-COOH \; + \; R^2-CN \; \longrightarrow \; R^1-CN \; + \; R^2-COOH$$
$$\quad\;\; 1 \qquad\qquad 2 \qquad\qquad\quad 3 \qquad\qquad 4$$

Where the p_K of **4** is greater than that of **1** the yield is excellent. Thus, 4-chlorobenzenzoic acid and excess isophthalonitrile give 93% *4-chlorobenzonitrile*[62].

Aromatic nitriles can be prepared very conveniently and in high yield with the aid of benzenesulfonamide[63]:

$$R-COOH \; + \; H_5C_6-SO_2-NH_2 \xrightarrow{\;220°\;} R-CN$$

16.1.4.3 From carboxamides

Dehydration of carboxamides to nitriles

$$R-CO-NH_2 \xrightarrow{\;-H_2O\;} R-CN$$

is a classic method of preparation. It can be applied to all nitrile types if suitable reaction conditions are chosen; see the surveys in refs.[220, 222, 223]. As amides arise as intermediates in some of the already mentioned methods of preparing nitriles, reference is made also to pp. 639, 640.

The number of dehydrating agents is unusually large; the following are some of those employed: aluminum-(III)-chloride[64], phosphorus(III)[65] and phosphorus-(V)-chloride[66], phosphorus(V)-oxide[67, 68], phosphorus-(V)-oxychloride[25, 69-73], thionyl chloride[74-76], toluenesulfonyl chloride[77], acyl halides[78], phosgene[79], carbox-

[52] C.R. Hauser, A.G. Gillespie, J. Amer. Chem. Soc. *52*, 4517 (1930).

[53] A. Misono, T. Osa, S. Koda, Bull. Chem. Soc. Japan *39*, 854 (1966).

[54] I. Rhee, M. Ryang, S. Tsutsumi, Tetrahedron Lett. 3419 (1970).

[55] J.A. Mitchell, E.E. Reid, J. Amer. Chem. Soc. *53*, 321 (1931).

[56] G.D. Van Epps, E. Reid, J. Amer. Chem. Soc. *38*, 2128 (1916).

[57] A.C. Cope, R.J. Cotter, L.L. Estes, Org. Synth. *34*, 4 (1954).

[58] G. Zemplén, L. Döry, Acta Chem. Acad. Sci. Hung. *14*, 89 (1958).

[59] H. Rapoport, H.D. Baldridge jr., J. Amer. Chem. Soc. *73*, 343 (1951).

[60] B.S. Biggs, W.S. Bishop, Org. Synth. *25*, 95 (1945).

[61] F. Becke, T.F. Burger, Justus Liebigs Ann. Chem. *716*, 78 (1968).

[62] W.G. Toland, L.L. Ferstandig, J. Org. Chem. *23*, 1350 (1958).

[63] P. Oxley, M.W. Partridge, T.D. Robson, W.F. Short, J. Chem. Soc. 763 (1946).

[64] J.F. Norris, A.J. Klemka, J. Amer. Chem. Soc. *62*, 1432 (1940).

[65] F. Salmon-Legagneur, Bull. Soc. Chim. France 580 (1952).

[66] A.W. Titherley, E. Worrall, J. Chem. Soc. *95*, 1147 (1909).

[67] P.C. Teague, W.A. Short, Org. Synth. *33*, 52 (1953).

[68] H.P. Fischer, C.A. Grob, Helv. Chim. Acta *47*, 564 (1964).

[69] E.C. Taylor jr., A.J. Crovetti, Org. Synth. *37*, 12 (1957).

[70] R. Delaby, G. Tsatsas, X. Lusinchi, Compt. Rend. *242*, 2644 (1956);
R. Delaby, G. Tsatsas, X. Lusinchi, Bull. Chim. France 1294 (1956).
R. Delaby, G. Tsatsas, X. Lusinchi, Bull. Soc. Chim. France 409 (1958).

ylic acid anhydrides[65], *N*-dimethyldifluoromethylamine[81], and also Grignard compounds[80]. Splitting off of water with triphenylphosphine and halogen compounds is referred to on p. 644.

The following is limited to the most successful and to recent techniques.

Dry distillation of a mixture of aluminum chloride with primary amide affords aliphatic and aromatic nitriles in up to 97% yield[64]. However, these drastic methods cannot be recommended for more sensitive compounds.

With phosphorus(III)[65] and phosphorus(V)[66] chlorides the reaction succeeds even in boiling benzene from the corresponding amide. Vacuum distillation of a mixture of nicotinamide and phosphorus(V) oxide gives 84% *nicotinonitrile*[67]. In the presence of triethylamine in chloroform, a temperature of 70° is sufficient for obtaining 97% *bicyclo[2.2.2]octane-1-carbonitrile*[68].

Boiling the amides for some minutes with phosphoryl chloride gives, for example, *indole-2-* and *indole-3-carbonitriles* (86 and 94%[25]). Addition of *sodium metabisulfite* was found to be suitable for preparing *phenylacetonitrile* (93% with addition, 34% without)[72]. In pyridine 94% *4-nitrobenzonitrile* is obtained even at room temperature[70]; free hydroxy groups can be exchanged against chlorine at the same time[71], *e.g.:*

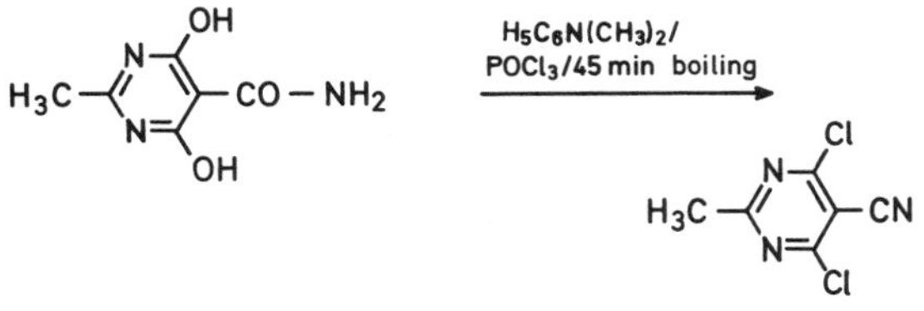

4,6-Dichloro-2-methyl-5-pyrimidine-carbonitrile; 76%

[71] *Z. Buděšinský, J. Kopecký,* Collect. Czech. Chem. Commun. *20,* 52 (1955).
[72] USP 2459128 (1949), Inv.: *M.J. Fahrenbach;* C.A. *43,* 3470 (1949).
[73] *H. Gross, J. Gloede,* Chem. Ber. *96,* 1387 (1963).
[74] *C.R. Hauser, W.R. Brasen,* J. Amer. Chem. Soc. *78,* 494 (1956).
[75] *J.C. Thurman,* Chem. & Ind. (London) 752 (1964).
[76] *E.A. Lawton, D.D. McRitchie,* J. Org. Chem. *24,* 26 (1959).
[77] *C.R. Stephens, E.J. Bianco, F.J. Pilgrim,* J. Amer. Chem. Soc. *77,* 1701 (1955).
[78] *Q.E. Thompson,* J. Amer. Chem. Soc. *73,* 5841 (1951).
[79] *P.M. Brown, D.B. Spiers, M. Whalley,* J. Chem. Soc. 2882 (1957).
[80] *F.C. Whitmore, C.I. Noll, V.C. Meunier,* J. Amer. Chem. Soc. *61,* 683 (1939).
[81] *Z. Arnold,* Collect. Czech. Chem. Commun. *28,* 2047 (1963).

1,2-Phenylene phosphoric acid ester chloride is recommended as an excellent dehydrating agent[73]. In this way, aliphatic and aromatic nitriles are obtained from amides in homogeneous solution in benzene, diethyl ether, or chloroform within one hour in yields of 80–100%[73].

Boiling amides in a mixture of *thionyl chloride and benzene* furnishes *substituted phenylacetonitriles* in 85% yield[74]. Even better results can be obtained with a mixture of thionyl chloride and dimethylformamide[75, 76] (70° must not be exceeded), *e.g.:*

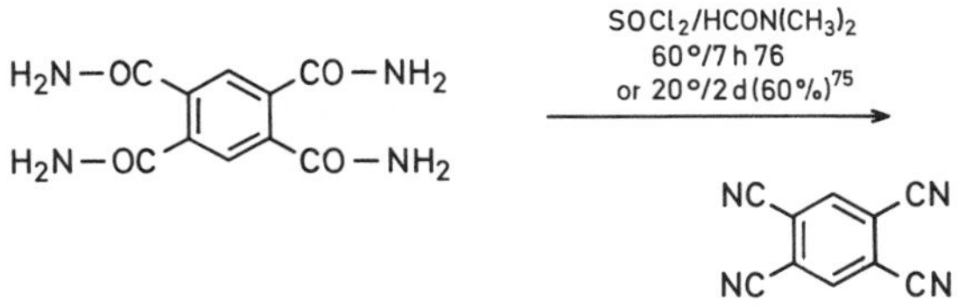

1,2,4,5-Benzenetetra-carbonitrile; 60%

Occasionally, *toluenesulfonyl chloride* in pyridine is used as the dehydrating agent[77]:

$$R-CO-NH_2 \xrightarrow[H_5C_5N/<70°/10\ min]{H_7C_7SO_2Cl/} R-CN$$

The reaction does not succeed so smoothly with *acyl halides*[78]. *Phosgene* gave good results in the synthesis of *citracononitrile* (85%)[79] where alternative methods generally lead to citraconimide.

Branched *Grignard* compounds are able to convert branched carboxamides into nitriles in good yield; thus, from 2,2-dimethylpropionamide and *tert*-butyl magnesium halide up to 72% *pivalonitrile* are obtained[80]. Difluorotrimethylamine is one of the mildest dehydrating agents for preparing nitriles from amides[81], *e.g.:*

$$H_5C_6-CO-NH_2 \xrightarrow[1,4\text{-Dioxan}/-5°/20\ min]{(H_3C)_2NCHF_2 \cdot 3\,HF} H_5C_6-CN$$

Benzonitrile; 81%

Nitriles from secondary and tertiary carboxamides

The *Braun halogenation cleavage*[82–85] originally served mainly for determining constitution in the alkaloid field. However, it has been found that un-

[82] *J. von Braun, C. Müller,* Ber. *39,* 2018 (1906).
[83] *J. von Braun,* Ber. *43,* 2846 (1910).
[84] *J. von Braun, W. Sobecki,* Ber. *44,* 1039 (1911).
[85] *J. von Braun, W. Pinkernelle,* Ber. *67,* 1218 (1934).
[86] *F. Effenberger, R. Gleiter,* Chem. Ber. *97,* 480 (1964).

der certain conditions the method is very suitable for preparing aromatic nitriles (yields much greater than 90%)[86].

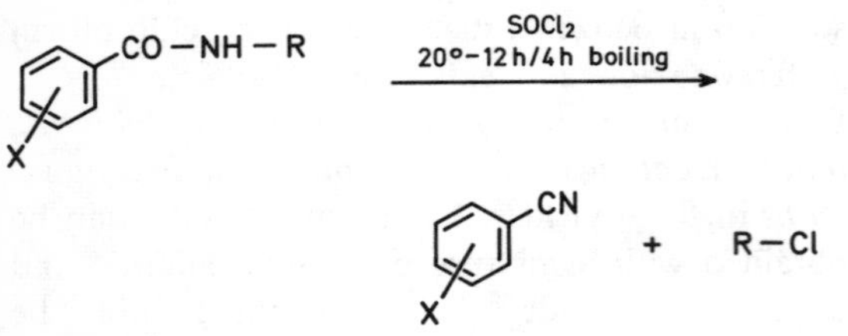

16.1.4.4 From thiocarboxamides

Preparation of nitriles from thiocarboxamides succeeds either thermally[87] or with basic compounds[88, 89]:

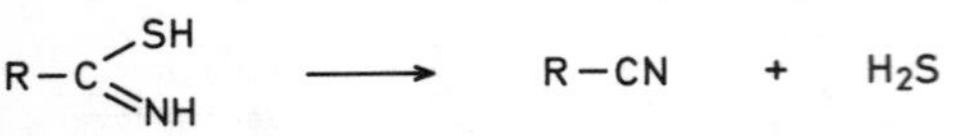

The split-off hydrogen sulfide can be advantageously trapped with mercury(II) chloride[88], *e.g.*:

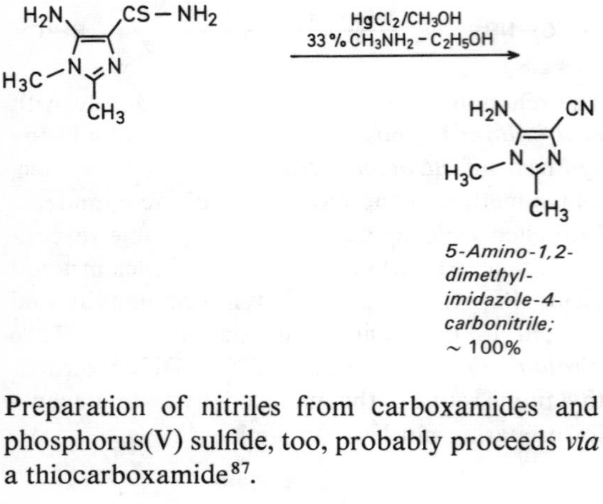

5-Amino-1,2-dimethyl-imidazole-4-carbonitrile; ~ 100%

Preparation of nitriles from carboxamides and phosphorus(V) sulfide, too, probably proceeds *via* a thiocarboxamide[87].

16.1.4.5 From acyl halides

Aliphatic nitriles can be obtained from acyl halides with acetonitrile in the presence of aluminum chloride without amide formation by *transnitrilation*[222]. *Aromatic nitriles* are formed from acyl halides and 2-tolylsulfonylphosphorimidic trichloride[90].

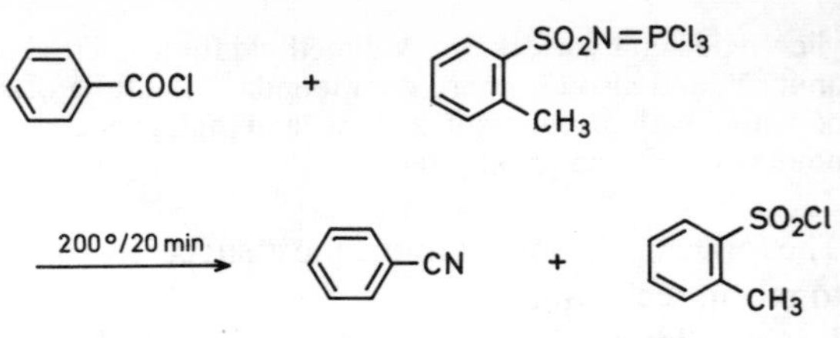

Benzonitrile; 80.5%

16.1.5 Nitriles by cleaving phosphonium salts

Various different methods are characterized by affording nitriles by cleavage of various different phosphorus salts. Often such phosphorus compounds are formed only as likely intermediate products[91-97]. The following examples illustrate some typical possibilities:

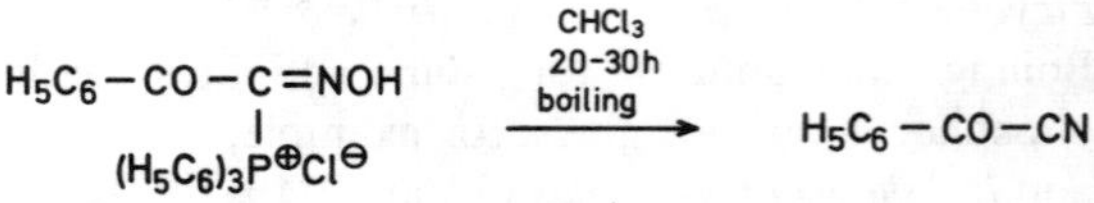

Phenylglyoxylonitrile; 86–98%[91]

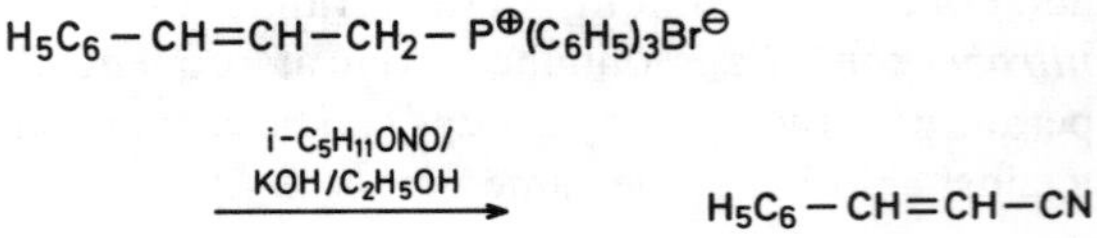

Cinnamonitrile; 54%[92]

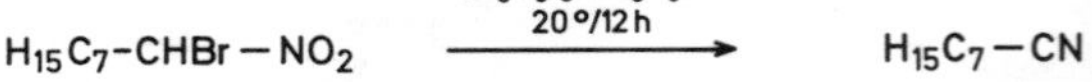

Octanenitrile; 72%[93]

Preparation of nitriles from primary and secondary amides[94, 95] (cf. p. 643) also succeeds in this way:

[87] *L. Henry*, Justus Liebigs Ann. Chem. *152*, 148 (1869).

[88] *G. Shaw, D.N. Butler*, J. Chem. Soc. 4040 (1959).

[89] *B. Prijs, W. Mengisen, S. Fallab, H. Erlenmayer*, Helv. Chim. Acta *35*, 187 (1952).

[90] *A.V. Kirsanov*, Zh. Obshch. Khim. *22*, 274 (1952); C.A. *46*, 11135 (1952).

[91] *K. Akiba, C. Eguchi, N. Inamota*, Bull. Chem. Soc. Japan *40*, 2983 (1967).

[92] *M. Matsui, G. Yabuta*, Agr. Biol. Chem. (Tokyo) *32*, 1044 (1968).

[93] *S. Trippett, D.M. Walker*, J. Chem. Soc. 2976 (1960).

[94] *E. Yamato, S. Sugasawa*, Tetrahedron Lett. 4383 (1970);
R. Appel, R. Kleinstück, K.D. Ziehn, Chem. Ber. *104*, 1030 (1971).

[95] *J. Blum, A. Fisher*, Tetrahedron Lett. 1963 (1970).

[96] *A.M. van Leusen, J.C. Jagt*, Tetrahedron Lett. 967 (1970).

[97] *R.A. Mitsch, E.W. Neuvar*, J. Org. Chem. *33*, 3675 (1968).

$$H_5C_6-CO-NH_2 \quad + \quad P(C_6H_5)_3 \quad + \quad CCl_4$$

$$\xrightarrow{THF/50°}$$

$$H_5C_6-CN \quad + \quad P(C_6H_5)_3 \cdot HCl$$

Benzonitrile; 83.5%

$$R^1-CO-NH-CH_2-R^2 \quad \xrightarrow[\text{}]{+ \; CHCl_3 \; + \; PO(C_6H_5)_3} \quad \xrightarrow{RhCl[P(C_6H_5)_3]_3 \, / 285°}$$

$$R^1-CN \quad + \quad \begin{array}{l} R^2-CH_2OH \quad\quad (R^2 = \text{Alkyl}) \\ \\ R^2-CN \; + \; R^2-CH_3 \quad (R^2 = C_6H_5) \end{array}$$

An excellent method for preparing *polyhalogenated aliphatic nitriles* is reproduced by the following equation[97]:

$$FCl_2C-CF_2-COF \quad \xrightarrow{h\nu/N_2F_4} \quad FCl_2C-CF_2-NF_2$$

$$\xrightarrow{P(C_6H_5)_3/-196° \rightarrow 20°} \quad FCl_2C-CN$$

Dichlorofluoro-acetonitrile; 90%

All the above-mentioned methods of nitrile preparation are characterized by generally leading to the desired result under very mild conditions.

16.2 Nitriles from compounds with an unlike carbon chain by rearrangement optionally accompanied by degradation

16.2.1 Nitriles from isonitriles, *N*-arylformamides, and aryl isothiocyanates

Preparation of nitriles from *N*-arylformamides and *N*-aryl isothiocyanates is based on the isonitrile rearrangement, which has been known for about 100 years[220].

$$\left.\begin{array}{l} R-NH-CHO \\ \\ R-NCS \end{array}\right\} \longrightarrow R-NC \longrightarrow R-CN$$

The starting point is isonitrile directly[98]; alternatively its precursor *N*-formanilide[99] or aromatic isothiocyanates[100] are employed. The actual rear-

rangement proceeds smoothly at temperatures above 200°. *Aromatic* nitriles[99] especially can be prepared by this two-stage synthesis. It allows the direct conversion of aromatic amines into aromatic nitriles, *e.g.*:

6-tert-Butyl-2-tolunitrile; 52%

This very old reaction has been found to be the best procedure for preparing *1,2-dicarbaclosododecaborane-2-carbonitrile*[101]:

$$C_2B_{10}H_{11}-NH-CHO \quad \xrightarrow[\text{H}_5\text{C}_5\text{N}]{POCl_3} \quad C_2B_{10}H_{11}-NC$$

77%

$$\xrightarrow{210°} \quad C_2B_{10}H_{11}-CN$$

71%

A *masked* isonitrile rearrangement could explain the result of the following thermolysis[102]:

Phthalodinitrile; 72%

16.2.2 Nitriles by a second-order Beckmann reaction

A Beckmann reaction of oximes with a keto, carboxy, or hydroxy group, or a quaternary α-carbon atom generally leads directly to nitriles[103-115] (see the surveys in refs.[21, 231, 225]).

[98] *J.U.Nef*, Justus Liebigs Ann. Chem. *280*, 263 (1894).

[99] *P.J.C. Fierens, J. van Rysselberge*, Bull. Soc. Chim. Belg. *61*, 215 (1952).

[100] *H. Schwartz*, Ber. *15*, 2505 (1882).

[101] *L.I. Zakharkin, V.N. Kalinin, V.V. Gedymin*, Synth. Inorg. Chem. *1* (1) 45 (1971).

[102] *M.P. Cava, L. Bravo*, Chem. Commun. 1538 (1968).

[103] *H.R. Snyder, J.K. Williams*, J. Amer. Chem. Soc. *76*, 1298 (1954).

[104] *K.N.F. Shaw, A. McMilan, A.G. Gudmundson, M.D. Armstrong*, J. Org. Chem. *23*, 1171 (1958).

[105] *A. Werner, A. Piguet*, Ber. *37*, 4295 (1904).

[106] *A.F. Ferris*, J. Org. Chem. *24*, 580 (1959); *A.F. Ferris*, J. Org. Chem. *25*, 12 (1960).

[107] *R.T. Conley, F.A. Mikulski*, J. Org. Chem. *24*, 97 (1959).

[108] *A. Stojiljković, R. Tasovac*, Tetrahedron Lett. 1405 (1970).

[109] *C.W. Shoppee, S.K. Roy*, J. Chem. Soc. 3774 (1963).

[110] *S. Wawzonek, J.V. Hallum*, J. Org. Chem. *24*, 364 (1959).

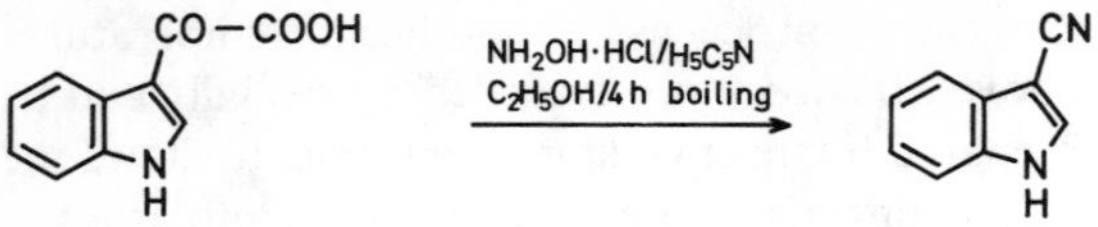

*Indole-3-carbo-
nitrile; 82%*

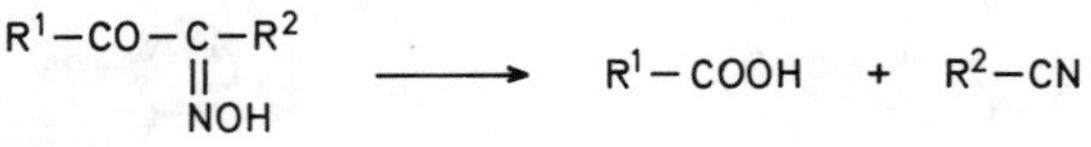

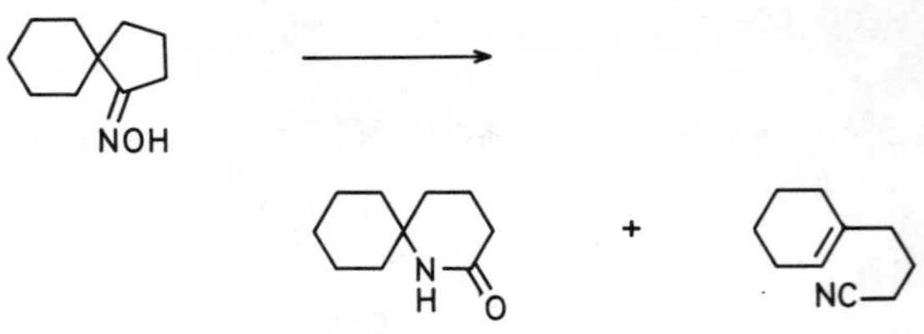

*4-(1-Cyclohexene)-
1-butyronitrile;
96%*

$$R^1-CO-\underset{\underset{NOH}{\|}}{C}-R^2 \longrightarrow R^1-COOH + R^2-CN$$

The yield is very much dependent on the reaction conditions. Benzenesulfonyl chloride in sodium hydroxide gives almost exclusively nitrile, 85% sulfuric acid only amides[106].

Polyphosphoric acid has proven successful in addition to sulfuric acid and benzenesulfonyl chloride[107]; the reaction can also be carried out photochemically[108].

The following example illustrates a second-order Beckmann reaction[109]:

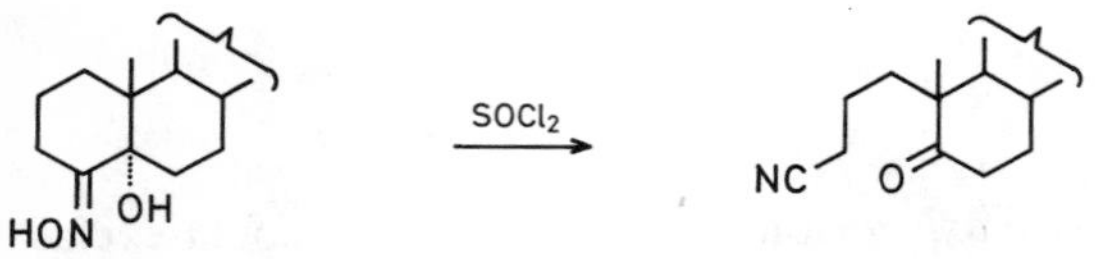

*Dodecahydro-7,9-
dimethyl-6-oxo-7-
benz[e]indenepropio-
nitrile; ~ 100%*

It has been most frequently investigated in ketoximes with quaternary carbon in the *alpha* position[110-115]. Two typical representatives are as follows:

81%

The Schmidt reaction proceeds similarly with compounds of analogous make-up[116], *e.g.*:

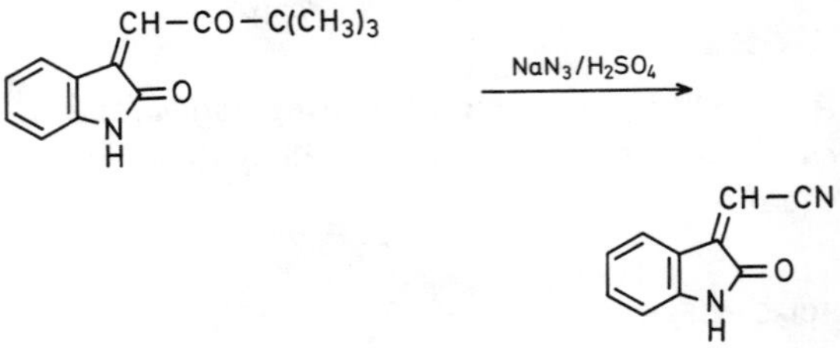

*2-Oxo-3-
indolinylidene-
acetonitrile; 97%*

16.2.3 Nitriles by other reactions

Certain reactions leading directly to nitriles with a shorter carbon chain may be described as a special type of Hofmann amide degradation[117]:

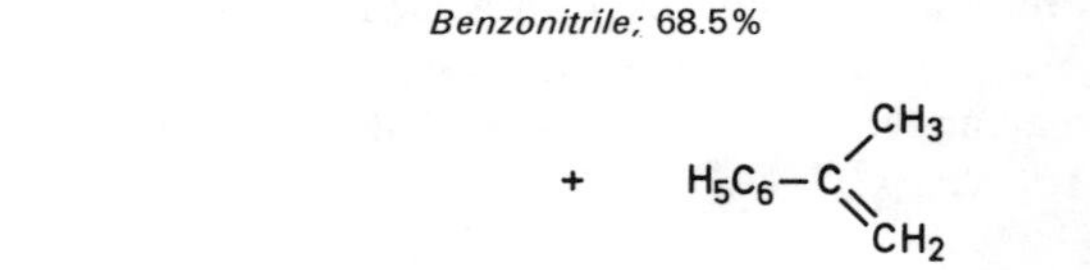

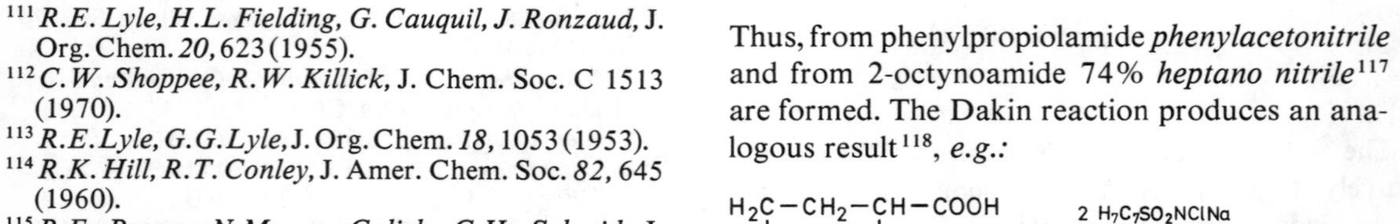

$$R-C\equiv C-CO-NH_2 \xrightarrow{\text{NaOCl}/0^\circ}$$

$$R-C\equiv C-CO-NHCl \xrightarrow{\text{Ba(OH)}_2} R-CH_2-CN$$

Benzonitrile; 68.5%

Thus, from phenylpropiolamide *phenylacetonitrile* and from 2-octynoamide 74% *heptano nitrile*[117] are formed. The Dakin reaction produces an analogous result[118], *e.g.*:

*3-Cyanopropionic
acid*

[111] *R.E. Lyle, H.L. Fielding, G. Cauquil, J. Ronzaud*, J. Org. Chem. *20*, 623 (1955).
[112] *C.W. Shoppee, R.W. Killick*, J. Chem. Soc. C 1513 (1970).
[113] *R.E. Lyle, G.G. Lyle*, J. Org. Chem. *18*, 1053 (1953).
[114] *R.K. Hill, R.T. Conley*, J. Amer. Chem. Soc. *82*, 645 (1960).
[115] *R.F. Brown, N.M. van Gulick, G.H. Schmid*, J. Amer. Chem. Soc. *77*, 1094 (1955).
[116] *G. Tacconi*, Gazz. Chim. Ital. *98*, 344 (1968).
[117] *I.J. Rinkes*, Rec. Trav. Chim. Pays-Bas *46*, 268 (1927).
[118] *H.D. Dakin*, Biochem. J. *11*, 79 (1917).
[119] *H. Bauer, H. Tabor*, Biochem. Prep. *5*, 97 (1957).

Imidazole-4-acetonitrile was formed correspondingly in ~75% yield from histidine with sodium hypochlorite solution[119].

16.3 From compounds of shorter carbon chain

16.3.1 Nitriles by alkylating and arylating metal cyanides

16.3.1.1 From alkyl and aryl halides

Reacting alkyl and aryl halides with metal cyanide is one of the oldest and most versatile nitrile synthesis[220, 222, 223]:

$$RX + M-CN \longrightarrow R-CN + MX \quad (1)$$
$$RX + M-CN \longrightarrow R-NC + MX \quad (2)$$

The reaction lends itself to nitriles, substituted nitriles, and polynitriles of all types[5, 120−130]. Isonitriles are formed as side-products; and more readily the less polar the solvent is and the more covalent the metal cyanide is. With silver cyanide the isonitrile synthesis possesses procedural value:

$$R^1-C=C=CHBr \atop R^2-CH_2 \quad \xrightarrow{CuCN/DMF}$$

$$R^1-C=C=CH-CN \atop R^2-CH_2$$

40−90 %

Reactions in which a reactive alcohol is treated with halide and copper(I) cyanide, too, count as cyanide alkylations. In this way 72% *3-butenenitrile* is obtained from allyl alcohol[131].

16.3.1.2 From esters of sulfonic acids

In place of alkyl halides the corresponding sulfonic acid esters, dialkyl sulfates[132], methane-[133−135], and toluenesulfonyl esters[136−138] can be employed with advantage. The following example illustrates a particularly convenient conversion of alcohols into nitriles[133]:

$$Ar-CH-CH_2OH \atop CH_3 \quad \longrightarrow \quad Ar-CH-CH_2O-SO_2-CH_3 \atop CH_3$$

$$\xrightarrow[40-60°/3h]{NaCN/DMF} \quad Ar-CH-CH_2-CN \atop CH_3$$

85 %

(image of reaction scheme with methoxy-substituted aromatic compounds)

$$H_3CO \quad (CH_2)_3CN \atop H_3CO \quad CH_2OC_2H_5 \atop OCH_3 \quad \xleftarrow{KCN/C_2H_5OH}$$

$$H_3CO \quad (CH_2)_3Cl \atop H_3CO \quad CH_2Cl \atop OCH_3$$

$$\xrightarrow[\substack{1.\ KCN/H_2O \\ 2.\ KCN/[KI]/C_2H_5OH}]{} \quad H_3CO \quad (CH_2)_3CN \atop H_3CO \quad CH_2CN \atop OCH_3$$

4-[(2-Cyanomethyl)-3,4,5-trimethoxyphenyl]-butyronitrile; 86%[123]

Tertiary halogen compounds, too, can be converted into nitriles by making use of this principle[128, 129]:

$$(H_5C_6)_3C-Cl \quad \xrightarrow{CuCN/120-140°} \quad (H_5C_6)_3C-CN$$

Triphenylacetonitrile; 40−90%

[120] E.B. Reid, T.E. Gompe, J. Org. Chem. *18*, 661 (1953).
[121] R. Gaertner, J. Amer. Chem. Soc. *74*, 5319 (1952).

[122] G.A. Page, D.S. Tarbell, J. Amer. Chem. Soc. *75*, 2053 (1953).
[123] H. Rapoport, J.E. Campion, J. Amer. Chem. Soc. *73*, 2239 (1951).
[124] J.H. Burckhalter, P.H. Jackson, J. Sam, H.R. Meyer, J. Amer. Chem. Soc. *76*, 4112 (1954).
[125] R.N. Lewis, P.V. Susi, J. Amer. Chem. Soc. *74*, 840 (1952).
[126] J. Cason, D.D. Phillips, J. Org. Chem. *17*, 298 (1952).
[127] J. Gootjes, A.B.H. Funcke, W.T. Nauta, Arzneimittel Forsch. *19*, 1936 (1969).
[128] P.M. Greaves, S.R. Landor, D.R.J. Laws, J. Chem. Soc. C 291 (1968).
[129] G. Lock, V. Rieger, Chem. Ber. *86*, 74 (1953).
[130] R. Paul, S. Tchelitcheff, Bull. Soc. Chim. France 808 (1952).
[131] C.C. Price, G.A. Cypher, I.V. Krishnamurti, J. Amer. Chem. Soc. *74*, 2987 (1952).
[132] P. Walden, Ber. *40*, 3215 (1907).
[133] M.S. Newman, S. Otsuka, J. Org. Chem. *23*, 797 (1958).
[134] M.S. Newman, R.M. Wise, J. Amer. Chem. Soc. *78*, 450 (1956).
[135] M.S. Newman, D. Lednicer, J. Amer. Chem. Soc. *78*, 4765 (1956).
[136] R. Grewe, H. Pachaly, Chem. Ber. *87*, 46 (1954).
[137] N.L. Allinger, C.L. Neumann, H. Sugiyama, J. Org. Chem. *36*, 1360 (1971).
[138] R. Grewe, E. Nolte, Justus Liebigs Ann. Chem. *575*, 1 (1952).

16.3.1.3 From Mannich bases

Boiling alkali cyanide with free Mannich bases [142] or with their quaternary salts [139-141] in water [139], ethanol [140, 142], or dimethylformamide [141] gives excellent yields of *nitrile, e.g.:*

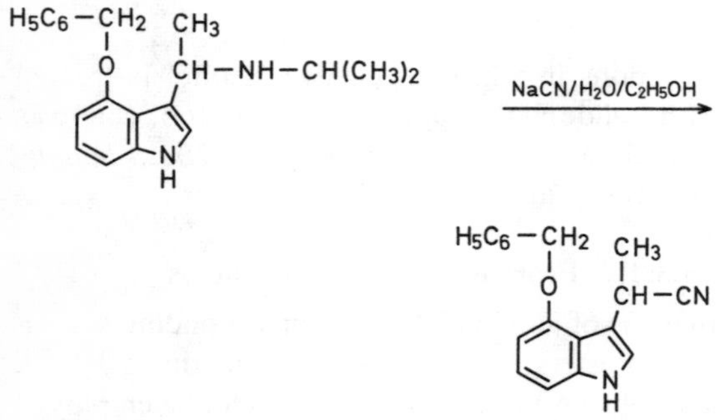

4-(Benzyloxy)-α-
methylindole-3-
acetonitrile [142]; 85%

16.3.1.4 From diazonium salts

$$ArNH_2 \xrightarrow[NaNO_2]{HCl/} ArN_2Cl \xrightarrow{CuCN} ArCN$$

This variant of the Sandmeyer reaction, details of which are reported in two surveys [235, 222], is based on the action of copper(I) cyanide, potassium cyanide-copper powder, potassium copper(I) cyanide, or potassium nickel(II) cyanide on diazonium salts. Correspondingly, only aromatic nitriles can be obtained in this way [143-146]. A drawback is the amount of liquid required ($\sim 1\,l$ solution/10 g nitrile); another the possible strong foaming. Very satisfactory yields are obtained.

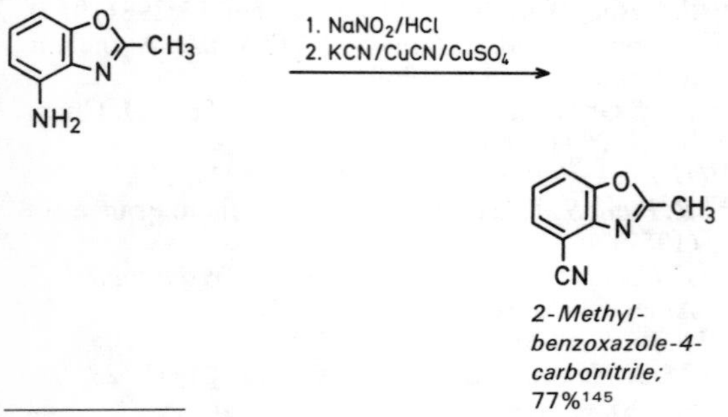

2-Methyl-
benzoxazole-4-
carbonitrile;
77% [145]

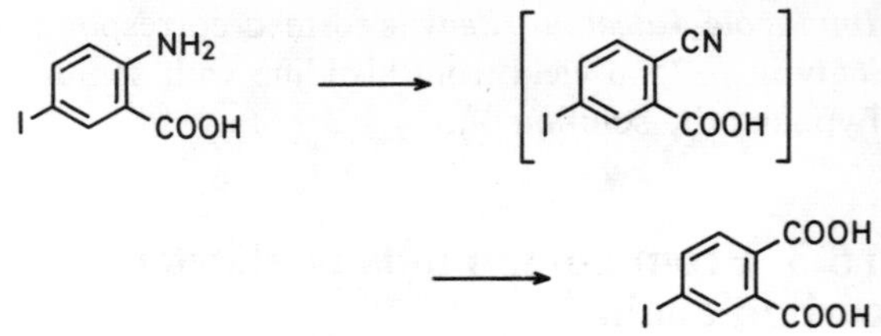

4-Iodophthalic
acid; 91% [144]

16.3.1.5 From carboxylic acid derivatives

Acyl halides [147-150] and *orthoesters* [151] react with metal cyanides or hydrogen cyanide to form *2-keto carbonitriles* or their derivatives. The first reaction is generally performed using copper(I) cyanide [147, 148] or hydrogen cyanide with added pyridine in diethyl ether [149, 150], *e.g.:*

2-Furylgly-
oxylonitrile;
60% [149]

Orthoesters react smoothly with hydrogen cyanide under acid catalytic conditions, preferably with zinc chloride [151], *e.g.:*

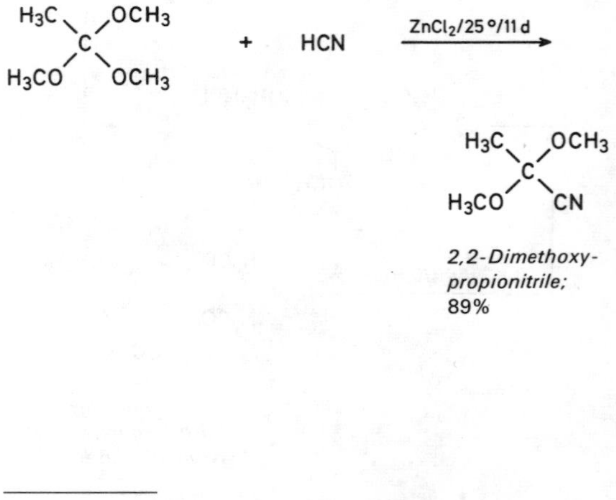

2,2-Dimethoxy-
propionitrile;
89%

[139] *D. Lednicer, C.R. Hauser,* Org. Synth. *40,* 45 (1960).
[140] *T.A. Geissmann, A. Armen,* J. Amer. Chem. Soc. *74,* 3916 (1952).
[141] *H. Hellmann, I. Löschmann, F. Lingens,* Chem. Ber. *87,* 1690 (1954).
[142] *F. Troxler, F. Seemann, A. Hofmann,* Helv. Chim. Acta *42,* 2073 (1959).
[143] *H.H. Hodgson, F. Heyworth,* J. Chem. Soc. 1131 (1949).

[144] *R.W. Higgins, C.L. Hilton, S.D. Deodhar,* J. Org. Chem. *16,* 1275 (1951).
[145] *C. Sannie, H. Lapin,* Bull. Soc. Chim. France 369 (1952).
[146] *H.T. Clarke, R.R. Read,* Org. Synth. Coll. Vol. I, 500 (1932).
[147] *C.H. Hurd, O.E. Edwards, J.R. Roach,* J. Amer. Chem. Soc. *66,* 2013 (1944).
[148] *T.S. Oakwood, C.A. Weisgerber,* Org. Synth. *24,* 14 (1944).
[149] *E. Fisher, F. Brauns,* Ber. *46,* 892 (1913).
[150] *G. Jander, G. Scholz,* Z. Physik. Chem. *192,* 163 (1943).
[151] *J.G. Erickson,* J. Amer. Chem. Soc. *73,* 1338 (1951).

Unlike the examples mentioned above, *lactones* react with alkali metal cyanide to form *substituted carbonitriles*[152], e.g.:

4-Chloro-2-(cyano-methyl)-benzoic acid; 72%

16.3.2 Nitriles by addition of hydrogen cyanide

16.3.2.1 To olefins and acetylenes

Addition to nonactivated $C-C$ multiple bonds is not a generally applicable method, but can be important in individual cases[222]. Thus, acetylene adds on hydrogen cyanide catalytically in the gas phase to form $\sim 60\%$ *succinonitrile* and some *acrylonitrile*[153], while olefins add hydrogen cyanide with collaboration of octacarbonyldicobalt[154]:

$$R-CH=CH_2 \quad + \quad HCN$$

By contrast, addition of hydrogen cyanide to the $C=C$ double bond of α,β-unsaturated carbonyl compounds (Michael reaction) is of general importance[230, 226, 220].

3-Cyclohexanonecarbonitrile forms smoothly from 2-cyclohexen-1-one[155], *3-(2-pyridyl)propionitrile* is formed from 2-vinylpyridine[156].

Free hydrogen cyanide may be replaced by 2-methyllactonitrile (acetone cyanhydrin)[157], *e.g.:*

$H_2C=CH-CN \quad + \quad (H_3C)_2C\begin{smallmatrix}CN\\OH\end{smallmatrix}$

$$\xrightarrow{Na_2CO_3/H_2O} \quad NC-CH_2-CH_2-CN$$

Succinonitrile; 88%

2-Methyl-3-oxocyclohexanecarbonitrile; 73%
2-Methyl-3-oxocyclopentanecarbonitrile; 77%

are obtained analogously.

Alternatively again, diethylaluminum cyanide[158] or a mixture of triethylaluminum and hydrogen cyanide[159] can replace hydrogen cyanide:

$$R^1N=CH-\underset{|}{C}=C\begin{smallmatrix}/\\\backslash\end{smallmatrix} \quad \xrightarrow{HCN/AlR_3}$$

$$R^1NH-\underset{|}{CH}-\underset{|}{CH}-\underset{|}{\overset{|}{C}}-CN$$
$$CN$$

$$\xrightarrow{(COOH)_2} \quad O=CH-\underset{|}{CH}-\underset{|}{\overset{|}{C}}-CN$$

16.3.2.2 From carbonyl compounds and their derivatives

16.3.2.2.1 Cyanhydrin synthesis (α-hydroxynitriles)

The cyanhydrin synthesis

$$\begin{smallmatrix}R^1\\R^2\end{smallmatrix}C=0 \quad \underset{\longleftarrow}{\overset{HCN}{\longrightarrow}} \quad \begin{smallmatrix}R^1\\R^2\end{smallmatrix}C\begin{smallmatrix}OH\\CN\end{smallmatrix}$$

is an equilibrium reaction[230, 220, 235] and is displaced towards cyanhydrin in acid medium. In addition, equilibrium is dependent on the constitution of the compounds and the temperature. The reaction is catalyzed by small quantities of basic substances such as potassium cyanide[222], amines[160], or ion exchangers[161], *e.g.:*

$$\begin{smallmatrix}H_3C\\H_3C\end{smallmatrix}CH-CHO \quad + \quad HCN$$

$$\xrightarrow{Amberlite\ IRA-400/15-20°} \quad \begin{smallmatrix}H_3C\\H_3C\end{smallmatrix}CH-\underset{|}{\overset{OH}{CH}}-CN$$

2-Hydroxy-3-methylbutyronitrile; 72%

---

[152] *J.N. Wells, W.J. Wheeler, L.M. Davisson,* J. Org. Chem. *36*, 1503 (1971).

[153] USP 2 415 414 (1947), Inv.: *W.M. Campbell;* C.A. *41*, 2747 (1947).

[154] *P. Arthur, D.C. England, B.C. Pratt, G.M. Whitman,* J. Amer. Chem. Soc. *76*, 5364 (1954).

[155] *G.A. Berchtold, G.F. Uhlig,* J. Org. Chem. *28*, 1459 (1963).

[156] *V. Boekelheide, W.J. Linn, P.O'Grady, M. Lamborg,* J. Amer. Chem. Soc. *75*, 3243 (1953).

[157] *I.N. Nazarov, S.I. Zavyalov,* Zh. Obshch. Khim. *24*, 466 (1954); C.A. *49*, 6139 (1955).

[158] *W. Nagata, M. Yoshioka, T. Okumura, M. Murakami,* J. Chem. Soc. 2355 (1970).

[159] *W. Nagata, T. Okumura, M. Yoshioka,* J. Chem. Soc. 2347 (1970).

In a further modification a mixture of potassium cyanide and carbonyl compound are acidified [162, 163]

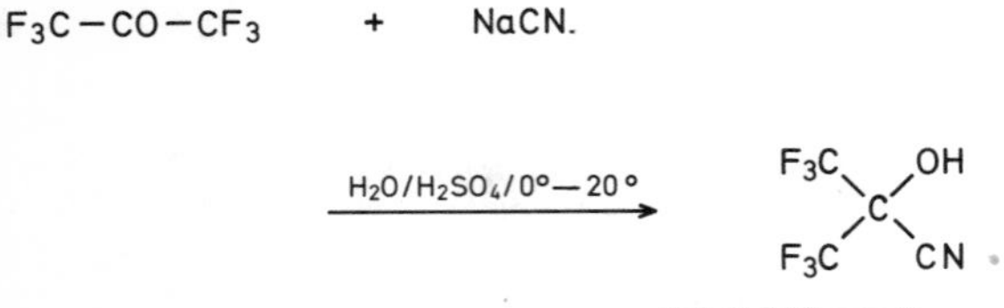

$$F_3C-CO-CF_3 \quad + \quad NaCN.$$

$\xrightarrow{H_2O/H_2SO_4/0°-20°}$

3,3,3-Trifluoro-2-
(trifluoromethyl)
lactonitrile; ~ 100%

or the interaction between aldehyde and ketone bisulfite adducts and aqueous alkali metal cyanide solution is made use of [164], *e.g.*:

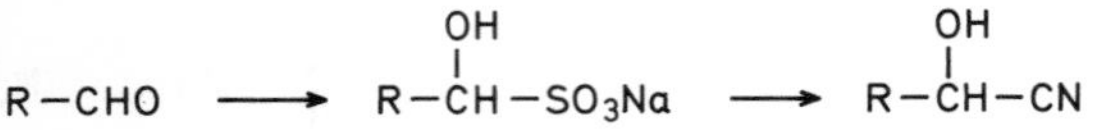

$$R-CHO \longrightarrow R-\overset{OH}{\underset{|}{CH}}-SO_3Na \longrightarrow R-\overset{OH}{\underset{|}{CH}}-CN$$

Often the very selective *transcyanohydrination* is a recommended procedure [165], *e.g.*:

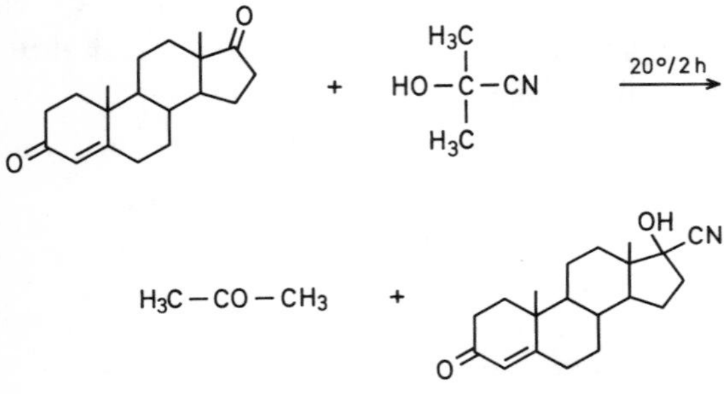

17-Hydroxy-3-oxo-
androst-4-ene-17-
carbonitrile; 96%

With very sluggish ketones *O*-trimethylsilyl enol is made the starting point, *e.g.* [166]:

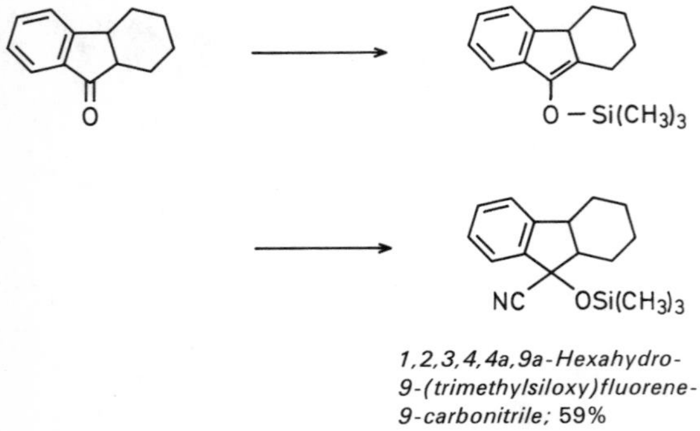

1,2,3,4,4a,9a-Hexahydro-
9-(trimethylsiloxy)fluorene-
9-carbonitrile; 59%

The general method for preparing *oxiranecarbonitriles (epoxynitriles)* [167] is also based on the cyanhydrin synthesis principle, *e.g.*:

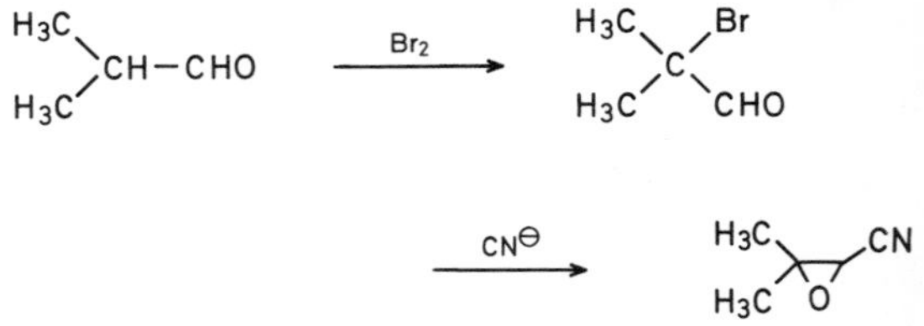

2,3-Epoxy-3-methyl-
butyronitrile
(2,2-Dimethyl-3-oxi-
ranecarbonitrile); ~ 80%

16.3.2.2.2 Strecker and analogous syntheses

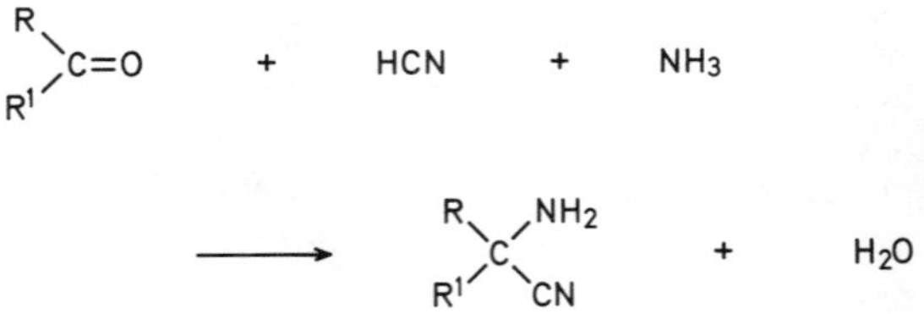

$$\underset{R^1}{\overset{R}{>}}C=O \quad + \quad HCN \quad + \quad NH_3$$

$$\longrightarrow \quad \underset{R^1}{\overset{R}{>}}\overset{NH_2}{\underset{CN}{C}} \quad + \quad H_2O$$

This classic method for preparing α-amino carboxylic acids *via* α-*aminonitriles* has given widely successful results. Hydrazine and hydroxylamine derivatives can be employed in place of ammonia. Aldehydes and ketones are equally good components [230, 220]. The aminonitriles obtained are often saponified to amino carboxylic acids directly [168].

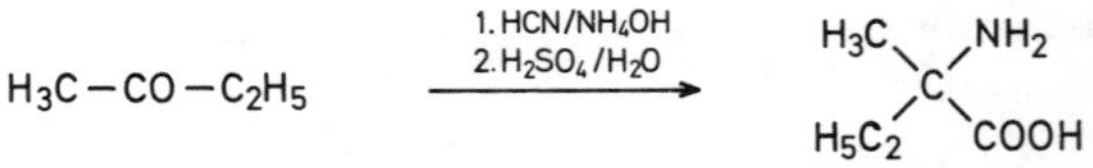

$$H_3C-CO-C_2H_5 \xrightarrow[\substack{1.\,HCN/NH_4OH \\ 2.\,H_2SO_4/H_2O}]{} \underset{H_5C_2}{\overset{H_3C}{>}}\overset{NH_2}{\underset{COOH}{C}}$$

Isovaline;
73.8%

[160] *H. Gudgeon, R. Hill, E. Isaacs*, J. Chem. Soc. 1926 (1951).

[161] *C.J. Schmidle, R.C. Mansfield*, Ind. Eng. Chem. 44, 1388 (1952).

[162] *K.N. Welch, G.R. Clemo*, J. Chem. Soc. 2629 (1928).

[163] *R.A. Darrall, F. Smith, M. Stacey, J.C. Tatlow*, J. Chem. Soc. 2329 (1951).

[164] *D. Biguard*, Compt. Rend. 194, 983 (1932).

[165] *A. Ercoli, P. de Ruggieri*, J. Amer. Chem. Soc. 75, 650 (1953).

[166] *W.E. Parham, C.S. Roosevelt*, Tetrahedron Lett. 923 (1971).

[167] *J. Cantacuzene, J.M. Normant*, Tetrahedron Lett. 2947 (1970).

[168] *W. Cocker, A. Lapworth*, J. Chem. Soc. 1391 (1931).

Using excess ammonia increases the yield of nitrile[168, 169]:

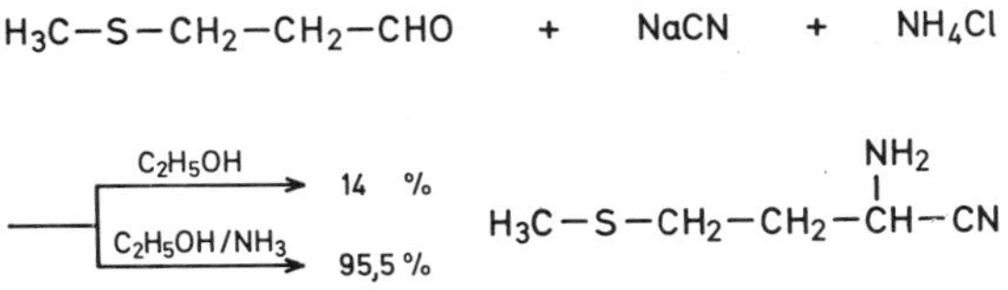

$$H_3C-S-CH_2-CH_2-CHO \quad + \quad NaCN \quad + \quad NH_4Cl$$

2-Amino-4-(methylthio)-butyronitrile

When amines are employed instead of ammonia α-(alkylamino)nitriles and α-(dialkylamino)nitriles[170–172] are obtained.

Alternative preparative methods for aminonitriles are from cyanohydrins and ammonia, e.g., in ethanolic solution[173], or from the bisulfite adducts of the carbonyl compounds with amine and alkali metal cyanide[172, 174, 175]. This modification serves above all for preparing α-dialkylamino nitriles[172, 175].

N-Substituted aminonitriles can be obtained also by adding hydrogen cyanide to aldimines[176–178], hydrazones[179], or oximes[180]:

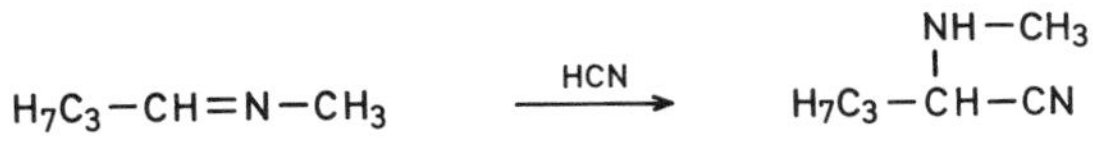

2-(Methyl-amino)valeronitrile; 77%

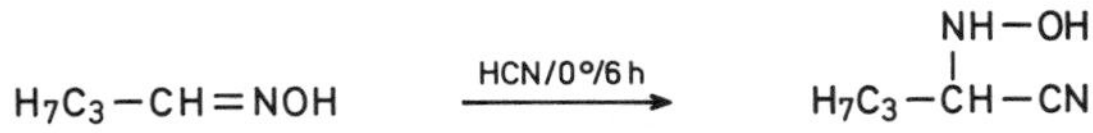

2-(Hydroxy-amino)valeronitrile; 75%

With hydrazine the Strecker synthesis[181–184] leads to addition products that can be converted into vicinal dinitriles[182–184], e.g.:

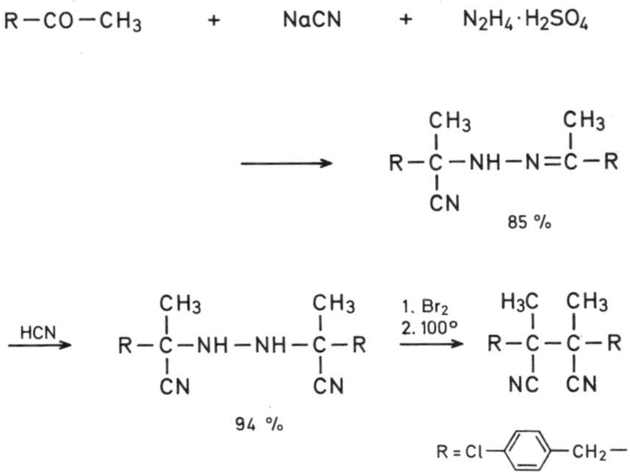

$$R-CO-CH_3 \quad + \quad NaCN \quad + \quad N_2H_4\cdot H_2SO_4$$

85%

94%

$$R = Cl-\underset{}{\bigcirc}-CH_2-$$

2,3-Bis(4-chloro-benzyl)-2,3-dimethylsuccino-dinitrile; 100%

16.3.2.3 From oxiranes

Addition of hydrogen cyanide to oxiranes has given good results for preparing 3-hydroxy carboxylic acid nitriles[185–187]. The reaction proceeds smoothly and much faster than a simple cyanide alkylation[185], e.g.:

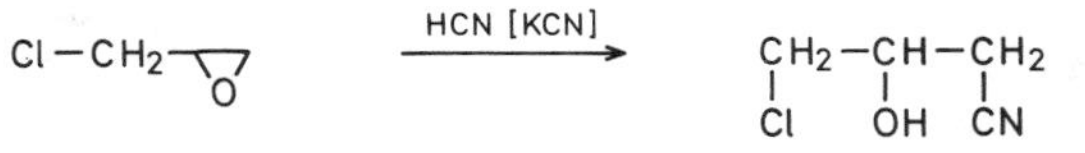

4-Chloro-3-hydroxybutyro-nitrile; 85%

With starting compounds of appropriate make-up oxiranes appear as intermediate products[136]:

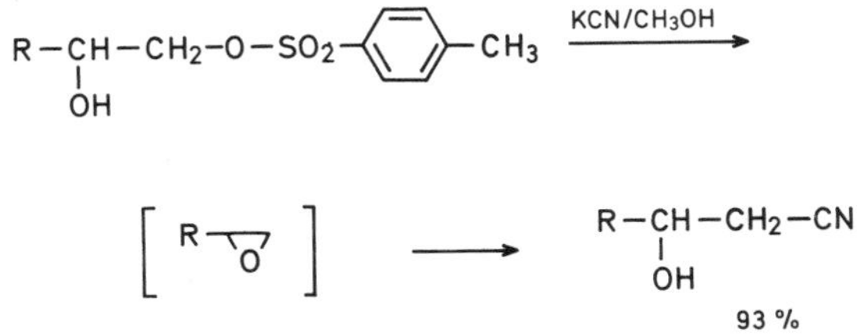

93%

[169] D.O. Holland, J.H.C. Nayler, J. Chem. Soc. 3403 (1952).
[170] H. Zahn, H. Wilhelm, Justus Liebigs Ann. Chem. 579, 1 (1953).
[171] R.A. Langdale-Smith, J. Org. Chem. 36, 226 (1971).
[172] E. Knoevenagel, E. Mercklin, Ber. 37, 4087 (1904).
[173] F. Tiemann, Ber. 13, 381 (1880).
[174] H.R. Henze, M.B. Knowles, J. Org. Chem. 19, 1127 (1954).
[175] D.B. Lutten, J. Org. Chem. 3, 588 (1939).
[176] G.H. Harris, B.R. Harriman, K.W. Wheeler, J. Amer. Chem. Soc. 68, 846 (1946).
[177] G.E.P. Smith, F.W. Bergstrom, J. Amer. Chem. Soc. 56, 2095 (1934).
[178] M.R. Tiollais, Bull. Soc. Chim. France (5) 14, 966 (1947).
[179] H.L. Yale, K. Losse, J. Martins, M. Holsing, F.M. Perry, J. Bernstein, J. Amer. Chem. Soc. 75, 1933 (1953).
[180] C.D. Hurd, J.M. Longfellow, J. Org. Chem. 16, 761 (1951).

[181] T. Bacchetti, Gazz. Chim. Ital. 80, 783 (1950).
[182] C.G. Overberger, T.B. Gibb, S. Chibnik, P. Huang, J.J. Monagle, J. Amer. Chem. Soc. 74, 3290 (1952).
[183] C.G. Overberger, H. Biletch, A.B. Finestone, J. Lilker, J. Herbert, J. Amer. Chem. Soc. 75, 2078 (1953).
[184] C.G. Overberger, H. Biletch, J. Amer. Chem. Soc. 73, 4880 (1951).
[185] M.R. Rambaud, Bull. Soc. Chim. France (5) 3, 138 (1936).
[186] G.N. Walker, D. Alkalay, A.R. Engle, R.J. Kempton, J. Org. Chem. 36, 466 (1971).

For less reactive oxiranes, diethylaluminum cyanide or a mixture of triethylaluminum and hydrogen cyanide has proven best[187]:

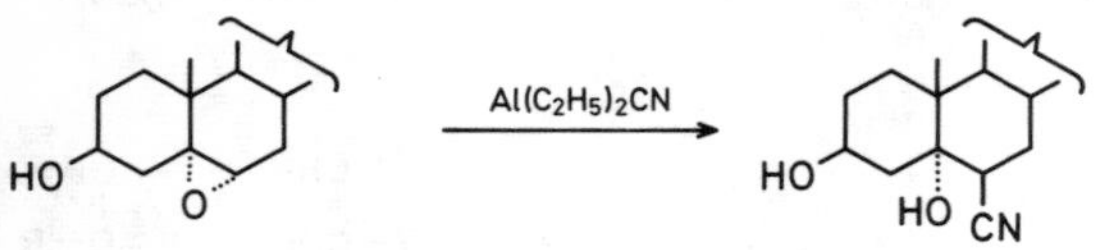

e.g., *3 β, 5 α-Dihydroxy-androstane-6-carbonitrile; 91%*

16.3.3 Introducing the nitrile group with cyanogen, cyanogen halide, cyanic acid, cyanamide, and trichloroacetonitrile

16.3.3.1 Electrophilic substitution of aromatic compounds

Aromatic compounds can be converted into aromatic cyano compounds by means of electrophilic substitution[220]:

$$ArH \quad + \quad XCN \quad \xrightarrow{AlCl_3} \quad ArCN \quad + \quad XH$$

Thus with, for example, cyanogen halide[188, 189], cyanogen[190], or mercury fulminate[191], it is possible to nitrilate benzene[188–191], naphthalene[189], acenaphthene[189], anthracene[188], and thiophene[189] derivatives. The yields obtained vary from moderate to good. For example, *terephthalonitrile* and *isophthalonitrile* are obtained in satisfactory yield from benzonitrile with cyanogen in the gas phase at 745° and a three-second dwell[192].

An interesting variant is the *Houben-Fischer method*[193]:

$$ArH \quad + \quad Cl_3C-CN \quad \xrightarrow[\text{2. NH}_3]{\text{1. AlCl}_3/\text{HCl}} \quad Ar-\underset{\underset{NH}{\|}}{C}-CCl_3$$

$$\xrightarrow{NaOH} \quad ArCN \quad + \quad CHCl_3$$

65 – 87 %

16.3.3.2 Nitrilation of organometallic compounds

The nitrilation of organometallic compounds originally performed with Grignard compounds has gained some importance with the introduction of the nitrile group into aromatic and heteroaromatic compounds *via* organolithium compounds and *N*-methyl-*N*-phenylcyanamide[194].

16.3.4 By other methods
16.3.4.1 Cyanomethylation

An apparently particularly efficient method of cyanomethylation proceeds *via* trialkylboranes (obtained by hydrobaration of olefins by the Braun technique) as intermediates[195, 196]:

$$R_3B \quad + \quad N_2CHCN$$

$$\xrightarrow[\text{2. KOH-H}_2\text{O}/0°-20°/30\,\text{min}]{\text{1. THF}/0°-20°/2,5\,\text{h}} \quad R-CH_2CN$$

90 – 100 %

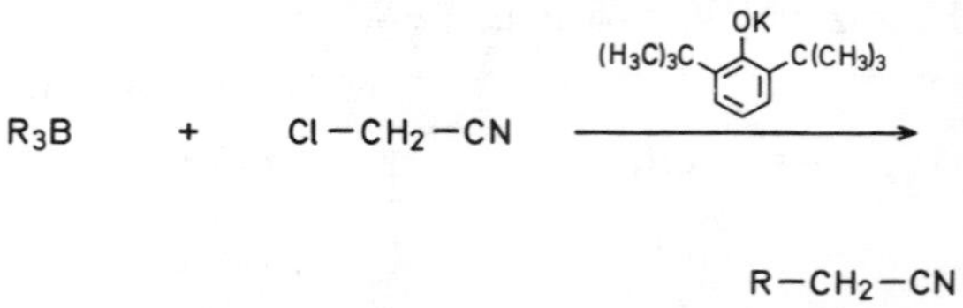

$$R_3B \quad + \quad Cl-CH_2-CN \quad \xrightarrow{\quad}$$

$$R-CH_2-CN$$

R = C₂H₅; *Butyronitrile; 95%*

One alkyl group only is converted into nitrile. When 9-alkyl-9-borabicyclo[3.3.1]nonane derivatives are used only the *B-alkyl group* reacts[196], e.g.:

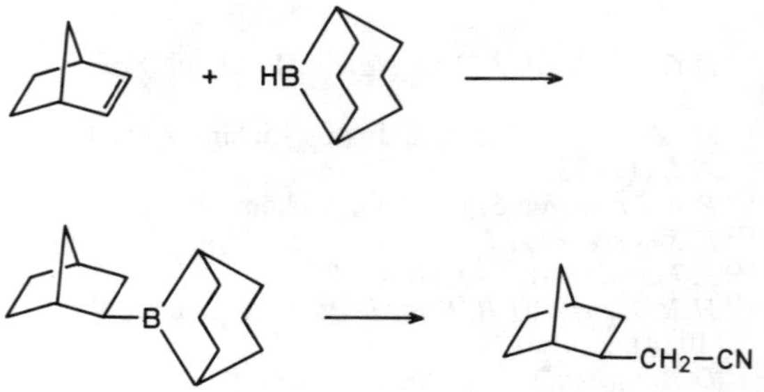

2-Norbornaneacetonitrile; 65%

[187] *W. Nagata, M. Yoshioka, T. Okumura*, J. Chem. Soc. 2365 (1970).

[188] *P. Karrer, E. Zeller*, Helv. Chim. Acta *2*, 482 (1919).

[189] *P. Karrer, A. Rebmann, E. Zeller*, Helv. Chim. Acta *3*, 261 (1920).

[190] *D. Vorländer*, Ber. *44*, 2455 (1911).

[191] *R. Scholl*, Ber. *36*, 10 (1903).

[192] *G. J. Janz*, J. Amer. Chem. Soc. *74*, 4529 (1952).

[193] *J. Houben, W. Fischer*, J. Prakt. Chem. (2) *123*, 313 (1929).

[194] *H. Lettré, P. Jungmann, J. C. Salfeld*, Chem. Ber. *85*, 397 (1952).

[195] *J. Hooz, S. Linke*, J. Amer. Chem. Soc. *90*, 6891 (1968).

[196] *H. C. Brown, H. Nambu, M. M. Rogić*, J. Amer. Chem. Soc. *91*, 6854 (1969).

By using the *Wadsworth-Emmons* technique [198] the carbonyl oxygen can be quite generally replaced by a cyanomethylene group [197], *e.g.:*

$(H_5C_2O)_2PO=CH-CN$ / $H_3C-O-CH_2-OCH_3$ / NaH

Octahydro-8,8a-dimethyl-2-naphthyl-ideneacetonitrile; 79%

16.3.4.2 Cyanoethylation

Cyanoethylation possibly represents the most important method for extending the chain length by three carbon atoms. It is a variant of the base-catalyzed Michael addition, affords excellent yields, and has been described on a number of occasions in surveys [230, 227, 220, 222, 223].

All *CH*-acid hydrogen atoms can be replaced by cyanoethyl groups.

The following, *e.g.,* react in this sense: substituted cyclopentadienes [199], phenylacetonitrile [200], nitroparaffins [201], hydrogen cyanide [202], aldehydes [203], ketones [204], 1,3-diketones [206], β-keto carboxylic acid esters [205], malonic acid diesters [207], derivatives of cyanoacetic acid esters [208], and alcohols [209], hydrogen sulfide [210], amines [221], and many others.

The *base catalysis* ensues with ∼5 molar % alcoholate, alkali metal hydroxide, or, better, with tetraalkylammonium hydroxides or strongly basic ion exchangers. For further details see the surveys in refs. [230, 227].

Addition of *CH*-acid compounds to substituted acrylonitriles succeeds occasionally but, preparatively speaking, is not as versatilely applicable as the cyanoethylation with the unsubstituted acrylonitrile [227, 230].

16.3.4.3 Anodic syntheses

Electrolysis of polymethoxybenzenes and alkali metal cyanides on platinum electrodes in methanol, acetonitrile, or water furnishes aromatic nitriles in high yield; a methoxy group is exchanged against the cyano group. From 1,2-dimethoxybenzene, *e.g.,* more than 95% *2-methoxybenzonitrile* are formed [212]. In *N,N*-dimethylbenzylamine a methyl hydrogen atom is replaced by the cyano group [212] and *N-methylbenzylaminoacetonitrile* is formed.

Furan derivatives are cyanomethoxylated anodically in methanolic solution [213], *e.g.:*

NaCN/CH$_3$OH anodic

2,5-Dihydro-5-methoxy-2,5-dimethyl-2-furonitrile; 30%

16.4 Nitrile syntheses with labeled carbon

As during synthesis of labeled compounds the maximum utilization of the labeled component is more important than the straightforwardness of the synthesis and the overall yield, the synthetic pathways occasionally differ from those chosen normally. The problem of the accessibility of la-

[197] *J.A. Marshall, G.M. Cohen,* J. Org. Chem. *36,* 877 (1971).

[198] *W.S. Wadsworth, W.D. Emmons,* J. Amer. Chem. Soc. *83,* 1733 (1961).

[199] *A. Campbell, S. Tucker,* J. Chem. Soc. 2623 (1949).

[200] *G. Misra, J. Shukla,* J. Indian Chem. Soc. *29,* 455 (1952).

[201] *H. Bruson, T. Rienert,* J. Amer. Chem. Soc. *65,* 23 (1943).

[202] *A. Terentyev, A. Kost,* Zh. Obshch. Khim. *21,* 1867 (1951).

[203] *R. Schreyer,* J. Amer. Chem. Soc. *73,* 3194 (1952).

[204] *A. Terentyev, A. Kost, A. Berlin,* Zh. Obshch. Khim. *25,* 1613 (1955).

[205] *C. Yoho, R. Levine,* J. Amer. Chem. Soc. *74,* 5597 (1952).

[206] *G. Zellars, R. Levine,* J. Org. Chem. *13,* 911 (1948).

[207] *D. Floyd,* J. Amer. Chem. Soc. *71,* 1746 (1949).

[208] *P. Dutta,* J. Indian Chem. Soc. *31,* 875 (1954).

[209] *O. Bayer,* Angew. Chem. *61,* 232 (1949).

[210] *J. MacGregor, C. Pugh,* J. Chem. Soc. 737 (1950).

[211] *D.S. Tarbell,* J. Amer. Chem. Soc. *68,* 1217 (1946).

[212] *S. Andreades, E.W. Zahnow,* J. Amer. Chem. Soc. *91,* 4181 (1969).

[213] *T. Yoshida, T. Fueno,* J. Org. Chem. *36,* 1523 (1971).

beled starting materials, too, plays a role in this connection. The following equations illustrate some examples of the preparation of ^{14}C-labeled nitriles[233, 214, 215]:

$$NC-^{14}CH_2-^{14}CO-NH_2 \xrightarrow[\text{LiCl}]{P_2O_5/} NC-^{14}CH_2-^{14}CN$$

Malonodinitrile-(1,2-^{14}C); 85.7%

$$Cl-^{14}CH_2-^{14}COOH \xrightarrow{KCN/H_2O/1h\ boiling}$$

$$NC-^{14}CH_2-^{14}COOH \longrightarrow HOOC-^{14}CH_2-^{14}COOH$$

Cyanoacetic acid-(1,2-^{14}C); 84%

$$(H_3CO)_2SO_2 \quad + \quad Na^{14}CN$$

$$\xrightarrow{H_2O} H_3C-^{14}CN \quad + \quad H_3C-O-SO_3Na$$

Acetonitrile-(1-^{14}C); 90%

$$^{14}CH_3I + NaCN \xrightarrow[20°/24h]{H_2O/} {}^{14}CH_3CN + NaI$$

Acetonitrile-(2-^{14}C); 70%

Reacting ^{14}C-methanol to labeled acetonitrile-*(2-^{14}C)* is an example of *vacuum line* technique and typical of working with hazardous volatile substances[216]:

$$^{14}CH_3OH \xrightarrow{SO_3/\sim180°\to20°} {}^{14}CH_3-O-SO_3H$$

$$\xrightarrow{KCN/\sim180°\to20°} {}^{14}CH_3-CN$$

96 %

To synthesize *9,10-dihydroxystearic acid-(1-^{13}C)* (85%) potassium cyanide-(^{13}C) is reacted with 8,9-dihydroxyheptadecyl bromide and the nitrile obtained is saponified directly[217].

Preparation of ^{14}C-labeled hydrogen cyanide from barium carbonate is another interesting example[218]:

$$H_5C_6-CH_2-MgCl \xrightarrow{^{14}CO_2} H_5C_6-CH_2-^{14}COOH$$

$$\xrightarrow{NH_3/Silica\ gel/490°} H_5C_6-CH_2-^{14}CN$$

Phenylacetonitrile-(1-^{14}C); 92%

$$H_5C_6-CH_2-^{14}CN \xrightarrow[Na/C_2H_5OH/boiling]{Toluene} [Na^{14}CN]$$

$$\xrightarrow{H^{\oplus}} H^{14}CN$$

~95 %

Mono-^{14}C-benzene has been prepared from α-labeled cyclohexanone built up with labeled hydrogen cyanide *via* a cyanhydrin reaction[219]:

1-Hydroxycyclopen-tanecarbonitrile-(1-^{14}C)

20–25 %

16.5 Transformations and analysis – references

The reactions of nitriles can be divided into two main groups. The first comprises reactions in which the nitrile group disappears, for example, the reaction leading to formation of

Carboxylic acid	Vol. V
Carboxamides	Vol. VI, p. 697
Carboxylic acid imides	Vol. VI, p. 724
Carboxylic acid esters	Vol. V
Amines	Vol. VI, p. 512
Ketones	Vol. V
Alcohols	Vol. V

[214] *E.M. Gal, A.T. Schulgin*, J. Amer. Chem. Soc. *73*, 2938(1951).

[215] *E.V. Brown, E. Cerwonka, R.C. Anderson*, J. Amer. Chem. Soc. *73*, 3735(1951).

[216] *D.N. Hess*, J. Amer. Chem. Soc. *73*, 4038(1951).

[217] *S. Bergström, K. Pääbo, M. Rottenberg*, Acta Chem. Scand. *6*, 1127(1952).

[218] *W.R. Vaughan, D.I. McCane*, J. Amer. Chem. Soc. *76*, 2504(1954).

[219] *H.S. Turner, R.J. Warne*, J. Chem. Soc. 789(1953).

[220] *P. Kurtz, H. Henecka* in *Houben-Weyl*, Methoden der organischen Chemie, Bd. VIII, 4. Aufl., p. 247, 359, Georg Thieme Verlag, Stuttgart 1952.

[221] *R. Schröter, F. Möller* in *Houben-Weyl*, Methoden der organischen Chemie, Bd. XI/1, 4. Aufl., p. 345, 545, Georg Thieme Verlag, Stuttgart 1957.

The second group includes those reactions in which the nitrile group remains intact, *i.e.*, nitrile syntheses starting from other nitriles. Condensations of nitriles with carbonyl compounds[228-230] are of paramount interest here.

The nitrile group can be determined quantitatively by both chemical and physical methods[237, 234]. Hydrolysis is the most important chemical determination, *IR* spectroscopy the most important physical technique[236] (cf. Vol. I, p. 274).

All nitriles are *poisonous*, especially the lower and unsaturated ones. The extreme toxicity of all metal cyanides and of hydrogen cyanide cannot be overemphasized[222] and strictest safety precautions are advisable during any work with them.

[222] *D.T. Mowry*, Chem. Rev. *42*, 189 (1948), The Preparation of Nitriles.

[223] *R.B. Wagner, H.D. Zook*, Synthetic Organic Chemistry, p. 590, John Wiley & Sons, New York 1953.

[224] *H. Wolff*, Org. Reactions, Vol. III, p. 307, John Wiley & Sons, New York 1947 (Schmidt-Reaction).

[225] *L.G. Donaruma, W.Z. Heldt*, Org. Reactions, Vol. XI, p. 1, John Wiley & Sons, New York 1960 (The Beckmann Rearrangement).

[226] *E.D. Bergmann, D. Ginsburg, R. Pappo*, The Michael Reaction, Org. Reactions, Vol. X, p. 179, John Wiley & Sons, New York 1959.

[227] *H.A. Bruson*, Org. Reactions, Vol. V, p. 79, John Wiley & Sons, New York 1949 (Cyanoethylation).

[228] *G. Jones*, Org. Reactions, Vol. XV, p. 204, John Wiley & Sons, New York 1967 (The Knoevenagel Condensation).

[229] *A.C. Cope, H.L. Holmes, H.O. House*, The Alkylation of Esters and Nitriles, Org. Reactions, Vol. IX, p. 107, John Wiley & Sons, New York 1957.

[230] *J. Plešek, A. Zobáčová*, Preparative Reactions in Organic Chemistry, Vol. V, Aldolization and Related Reactions, Czechoslovakian Academy of Sciences Press, Prague 1960 (czech.); p. 38 (cyanhydrines), p. 51 (Strecker reaction), p. 787 (cyanoethylation), p. 749 (nitrile condensation), p. 329, 447 (Knoevenagel type reactions), p. 773 (Michael reactions).

[231] *J. Kovář, M. Hudlický, I. Ernest*, Preparative Reactions in Organic Chemistry, Vol. VIII, Molecular Rearrangements, Czechoslovakian Academy of Sciences Press, Prague 1960 (czech.); p. 749 (Beckmann reaction), p. 549 (Smith reaction, reaction of azides with aldehydes), p. 606 (Hofmann degradation, anomalous reactions).

[232] *K. Bláha*, Preparative Reactions in Organic Chemistry, Vol. VI, Reactions of Organo-metallic Compounds.

[233] *E.H. Rodd, H.R.V. Arnstein*, Chemistry of Carbon Compounds, The Preparation and Use of Isotopically Labelled Organic Compounds, Vol. V, p. 1, Elsevier, Publ. Co., New York 1962.

[234] *R.L. Shriner, R.C. Fuson, D.Y. Curtin*, The Systematic Identification of Organic Compounds, 4. Aufl., p. 154, 178, 256, 321, John Wiley & Sons, New York 1956.

[235] *V. Migrdichian*, Chemistry of Organic Cyanogen Compounds, Reinhold Publ. Co., New York 1947.

[236] *M.F.R. Ashworth*, Analytical Methods for Organic Cyano Groups, Pergamon Press, Oxford 1971.

[237] *F. Kurtz, H. Roth* in *Houben-Weyl*, Methoden der organischen Chemie 4. Aufl., Bd. II, p. 533, Georg Thieme Verlag, Stuttgart 1953.

[238] *P.E. Spoerri, A.S. Du Bois*, The Hoesch Synthesis, Org. Reactions, Vol. V, p. 387, Jonn Wiley & Sons, New York 1949.

17 *N*-Functional carboxylic acid and thiocarboxylic acid derivatives

(Editor: *Gerhard Simchen*)

Contributed by

U. Kraatz
Organisch-Chemisches Institut der Universität Bonn,
Bonn

S. Linke
E. Wehinger
H. Wollweber
Farbenfabriken Bayer AG,
Wuppertal-Elberfeld

G. Simchen
Institut für Organische Chemie der Universität Stuttgart,
Stuttgart

W. Walter
Institut für Organische Chemie und Biochemie,
Universität Hamburg,
Hamburg

17.1 Acyclic and cyclic amides, hydrazides, hydroxyamides, imides, azides of carboxylic acids and of carbonic acid

17.1.1 Direct introduction of the aminocarbonyl and carbonylamino group

17.1.1.1 By reacting isocyanates

17.1.1.1.1 With aromatic compounds

Direct introduction of the carbonylamino group into aromatic compounds was reported as long ago as 1885[1]. Suitably substituted aromatic compounds are reacted with phenyl isocyanate in the presence of aluminum chloride to form the corresponding *anilides:*

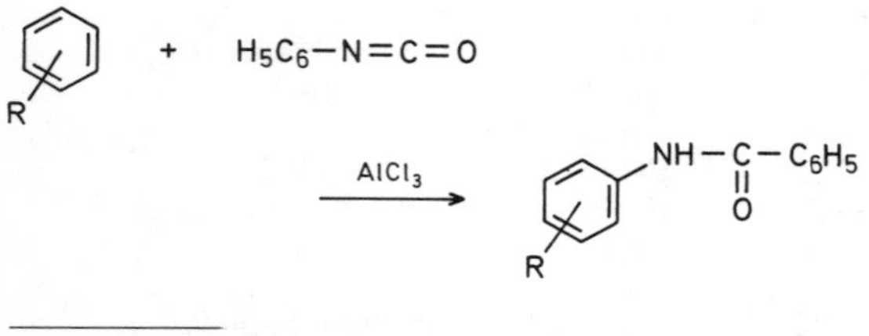

The reaction can also be carried out intramolecularly in order to prepare *9,10-dihydro-10-phenanthridinone*[2] from biphenyl 2-isocyanate or *benz-[cd]indol-2-(1H)ones (8-amino-1-naphthoic acid lactams)* from naphthalene 1-isocyanate[3]:

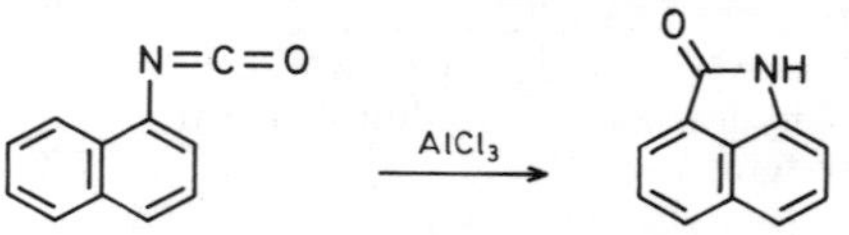

Using this modified Leuckardt amide synthesis, stilbene derivatives have been cyclized to *azocines*[4] and biphenyl ethers to *7-membered ring lactams*[5]:

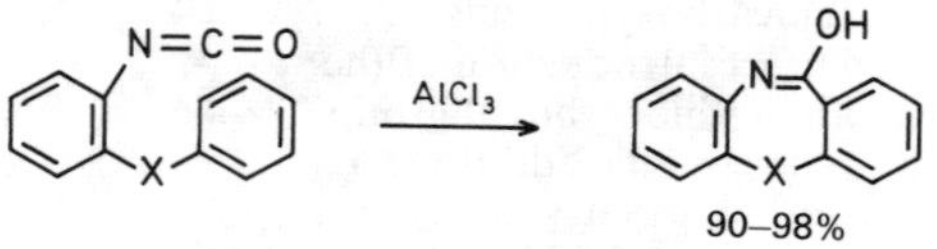

X = O;   *Dibenz[b,f] [1,4]oxazepin-11-ol*
X = S;   *Dibenzo[b,f] [1,4]thiazepin-11-ol*
X = CH₂; *Dibenz[b,e]azepin-11-ol*

[1] R. Leuckart, Ber. 18, 873 (1885);
R. Leuckart, M. Schmidt, Ber. 18, 2338 (1885);
R. Leuckart, J. Prakt. Chem. [2] 41, 301 (1890).

[2] J.M. Butler, J. Amer. Chem. Soc. 71, 2578 (1949).
[3] USP 2628964 (1953), American Cyanamid; Inv.: M. Scalera, B. Hardy, J.J. Leavitt; C.A. 48, 1442 g (1954);
Brit. P. 690307 (1953), American Cyanamid; Inv.: M. Scalera, B. Hardy, J.J. Leavitt; C.A. 48, 4003 i (1954).

Employing aluminum chloride as catalyst here affords much better results than polyphosphoric acid.

While initially only aromatic isocyanates were brought to react with aromatic compounds, success has meanwhile been obtained also with aliphatic isocyanates[6]; because of the reduced electrophilism of the alkali isocyanates only those aromatic compounds react which are more reactive than 3-chlorotoluene. The results obtained to date indicate that the reaction between the isocyanate/aluminum chloride complex and aromatic compounds is a normal electrophilic substitution, which can be described as follows:

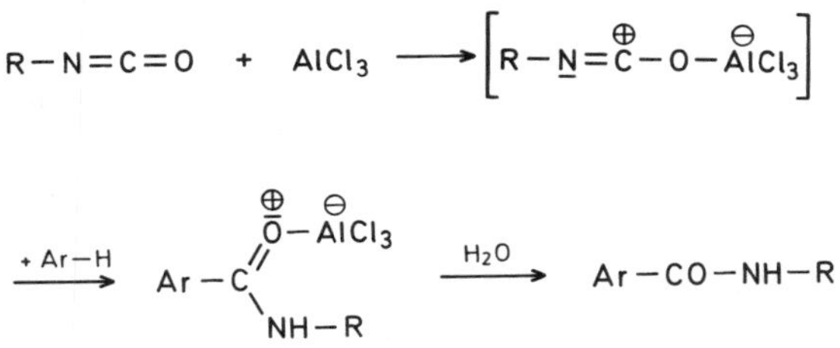

$$R-N=C=O \quad + \quad AlCl_3 \quad \longrightarrow \quad \left[R-\underline{N}=\overset{\oplus}{C}-O-\overset{\ominus}{Al}Cl_3 \right]$$

$$\xrightarrow{+ \; Ar-H} \quad Ar-\overset{\overset{\oplus}{\overset{|}{O}}-\overset{\ominus}{Al}Cl_3}{\underset{NH-R}{C}} \quad \xrightarrow{H_2O} \quad Ar-CO-NH-R$$

R = Aryl-, Alkyl-

Thus the mechanism proposed earlier is refuted[1,7].

Synthesizing anilides by this technique is preparatively quite simple and invariably proceeds in good yield. The following sequence can be written down for the reactivity of the aromatic compound used:

anisole (methoxybenzene) $\simeq$ 1, 3, 5-trimethylbenzene > toluene > benzene $\simeq$ fluorobenzene > chlorobenzene > 1, 3-dichlorobenzene

Using the very reactive 4-tolylsulfonyl isocyanate allows the range of application to be extended to preparing *N-(4-tolylsulfonyl)benzamides*[6]:

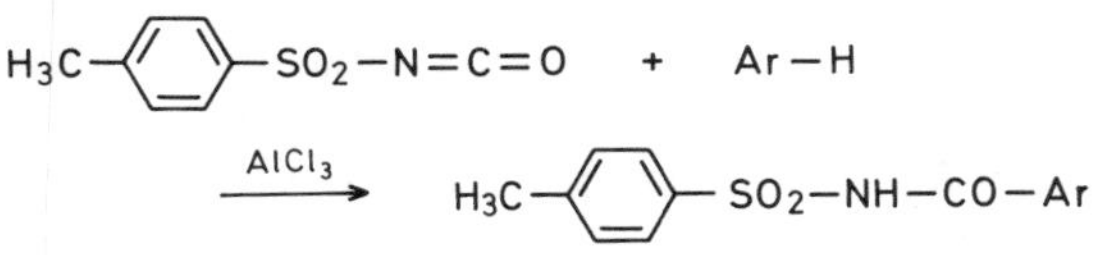

$$H_3C-\langle\!\!\!\bigcirc\!\!\!\rangle-SO_2-N=C=O \quad + \quad Ar-H$$

$$\xrightarrow{AlCl_3} \quad H_3C-\langle\!\!\!\bigcirc\!\!\!\rangle-SO_2-NH-CO-Ar$$

In the absence of a catalyst chlorosulfonyl isocyanate reacts with indene, anthracene, or the heteroaromatic compounds furan and thiophene to form *(inden-2-ylcarbonyl)sulfamoyl chloride, (2-furoyl)sulfamoyl chloride*, and *(2-thenoyl)sulfamoyl chloride* respectively[8]:

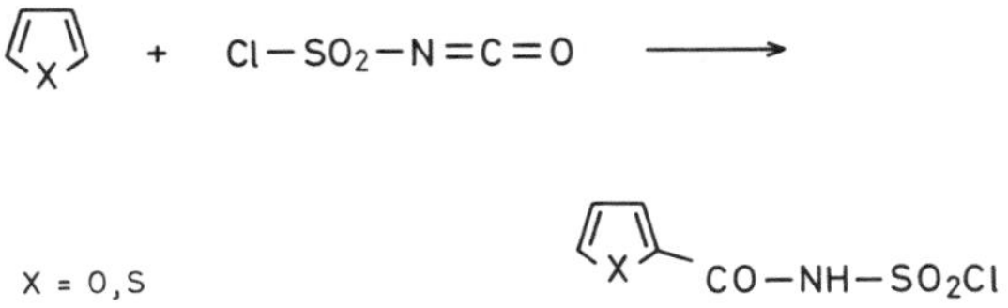

$$\langle\!\!\!\overset{\frown}{\underset{X}{\bigcirc}}\!\!\!\rangle \quad + \quad Cl-SO_2-N=C=O \quad \longrightarrow$$

X = O, S

$$\overset{\frown}{\underset{X}{\bigcirc}}\!\!\!\sim\!\!\!CO-NH-SO_2Cl$$

17.1.1.1.2 Aminocarbonyl or carbonylamino compounds by reaction of isocyanates with olefins

In general alkyl or aryl isocyanates add to olefins only if the *C=C* double bond is activated by alkoxy or amino groups. It is only where the electrophilic character of the *C* atom of the isocyanate group becomes very marked by virtue of electron-abstracting groups, as in sulfuryl chloride isocyanate, that addition to simple olefins, too, is possible[8a]. With this isocyanate the *N-chlorosulfonyl-β-lactams* are formed very rapidly and in exothermic reaction; selective hydrolysis enables the free *β-lactams* to be liberated[8,9,10]:

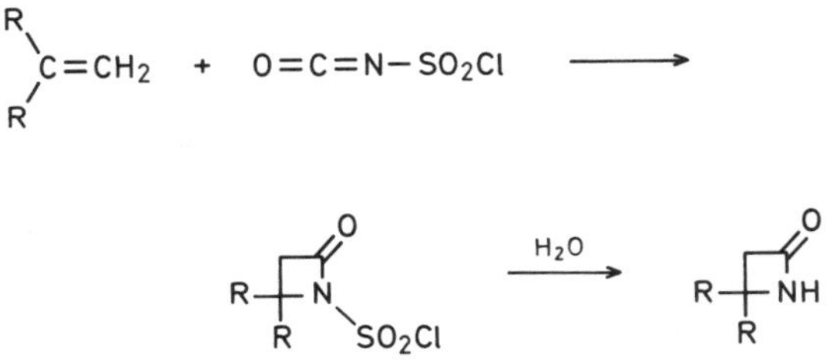

$$\overset{R}{\underset{R}{\diagdown}}C=CH_2 \quad + \quad O=C=N-SO_2Cl \quad \longrightarrow$$

$$R\!-\!\!\!\overset{\overset{O}{\parallel}}{\underset{R}{\boxed{}}}\!\!\!\overset{N}{\underset{SO_2Cl}{\diagdown}} \quad \xrightarrow{H_2O} \quad R\!-\!\!\!\overset{\overset{O}{\parallel}}{\underset{R}{\boxed{}}}\!\!\!NH$$

In addition to this cycloaddition, in which according to Markovnikov's rule the now negative nitrogen of the isocyanate group becomes attached to the *C* atom of the olefin with the fewest hydrogen atoms, substitution to *α,β-unsaturated carboxamides* is an alternative possibility:

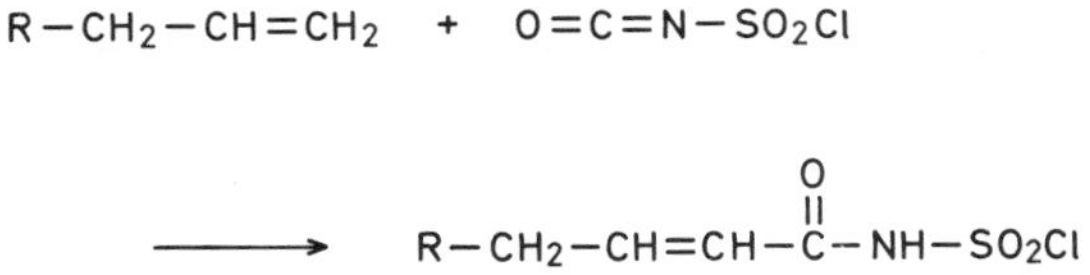

$$R-CH_2-CH=CH_2 \quad + \quad O=C=N-SO_2Cl$$

$$\longrightarrow \quad R-CH_2-CH=CH-\overset{\overset{O}{\parallel}}{C}-NH-SO_2Cl$$

[4] O. Schindler, R. Blaser, F. Hunziker, Helv. Chim. Acta 49, 985 (1966).

[5] J. Schmutz, F. Künzle, F. Hunziker, A. Bürki, Helv. Chim. Acta 48, 336 (1965).

[6] F. Effenberger, R. Gleiter, Chem. Ber. 97, 472 (1964).

[7] G.A. Olah, in G.A. Olah, Friedel-Crafts and Related Reactions, Vol. III, p. 1262, Wiley Intersci., Publ., New York, London, Sydney 1964.

[8] R. Graf, Justus Liebigs Ann. Chem. 661, 111 (1963).

[8a] H. Bestian, Pure Appl. Chem. 27, 611 (1971).

[9] R. Graf, Chem. Ber. 89, 1071 (1956).

[10] R. Graf, Angew. Chem. 80, 179 (1968).

Which of the two products predominates is dependent on the olefin used[10]. Unsymmetrically substituted olefins add sulfuryl chloride isocyanate mainly to form the isomeric β-lactam carrying the longer alkyl group adjacent the N-atom. Sterically speaking, the reaction of the isocyanate with the olefin proceeds by a highly *stereoselective cis* addition to the corresponding *cis* or *trans* substituted β-lactam in dependence on which isomeric olefin is employed[11, 12].

Olefins of the diene type, too, react with sulfuryl chloride isocyanate even at low temperature to form β-lactams by cycloaddition[13, 13a]:

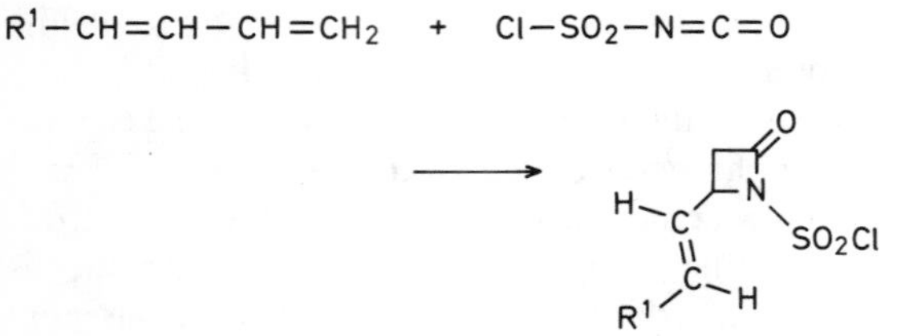

$$R^1-CH=CH-CH=CH_2 \quad + \quad Cl-SO_2-N=C=O$$

R[1] = H;   *N-Chlorosulfonyl-β-vinyl-β-propiolactam*

R[1] = CH$_3$;   *N-Chlorosulfonyl-β-propenyl-β-propiolactam*

R[1] = C$_2$H$_5$;   *N-Chlorosulfonyl-β-(1-butenyl)-β-propiolactam*

The $4+2$ cycloaddition allowed by symmetry was observed in none of the examples cited; invariably, only the Markovnikov-oriented 1,2-cycloadducts were obtained[13]. The addition ensues exclusively to the terminal $C=C$ double bond as, for example, in the nonconjugated 1,4-hexadiene to *N-chlorosulfonyl-β-(2-buten-1-yl)-β-propiolactam*.

17.1.1.1.3 Aminocarbonyl or carbonylamino compounds by reaction of isocyanates with enol ethers

Enol ethers are more nucleophilic than olefins and therefore should react with alkyl or aryl isocyanates more easily. However, a reaction of vinyl ethers with phenyl isocyanate can be achieved only under drastic conditions, the *N*-phenyl substituted β-*lactams* being formed by cycloaddition[14]:

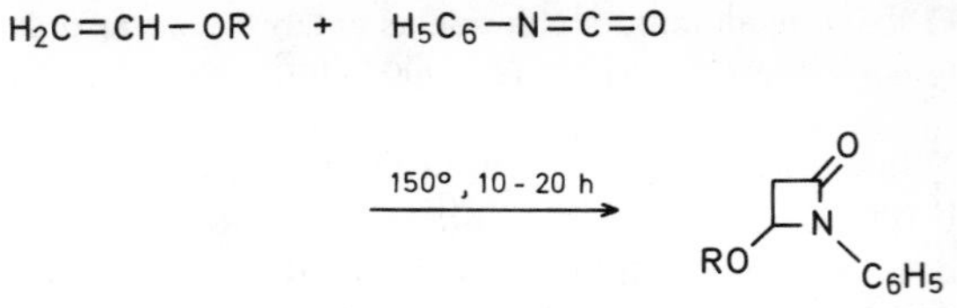

$$H_2C=CH-OR \quad + \quad H_5C_6-N=C=O$$

The yields vary a great deal with the alkyl group R and at best amount to 60%. Substituted vinyl ethers no longer afford cycloaddition products but, instead, β-*alkoxyacrylamides*[14]:

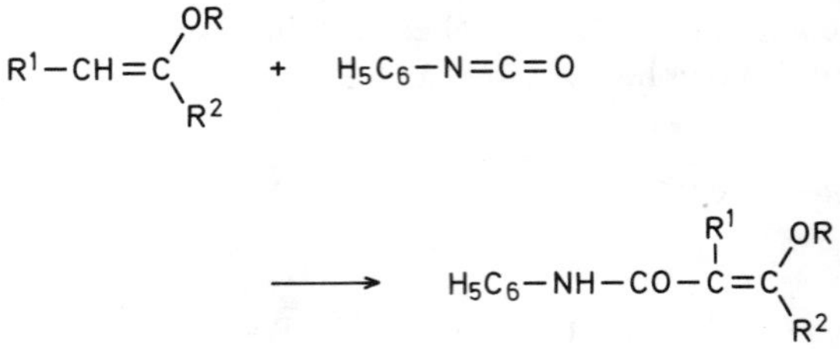

Enol ethers react very readily with the considerably more electrophilic sulfuryl chloride isocyanates. Here, too, two reaction paths are possible, namely, **either cycloaddition or substitution**[15, 16]:

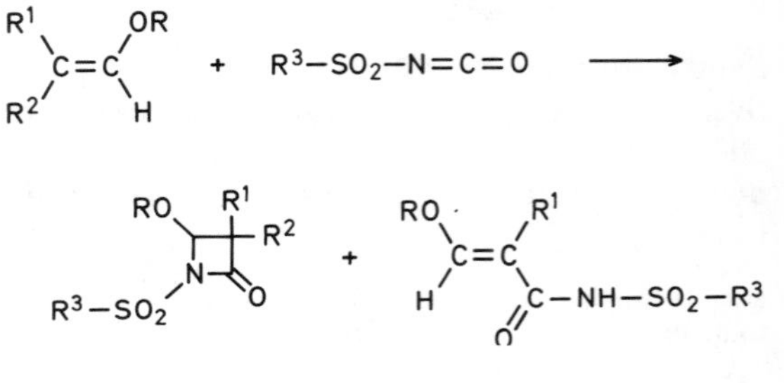

$R^3 = -Cl, p\text{-Tosyl}, 4-Cl-C_6H_4$

The ratio of cycloaddition to substitution is dependent on the substituents, the solvent, and the reaction conditions. Where R^1 and R^2 are alkyl groups β-lactams are formed exclusively, while when $R^2 = H$ lactam formation predominates. In boiling benzene the reaction always affords the thermodynamically more stable β-*alkoxyacrylamides*[15, 17].

[11] *H. Bestian, H. Biener, K. Clauss, H. Heyn,* Justus Liebigs Ann. Chem. *718,* 94 (1968).

[12] *E.J. Moriconi, J.F. Kelly,* Tetrahedron Lett. 1435 (1968).

[13] *E.J. Moriconi, W.C. Meyer,* J. Org. Chem. *36,* 2841 (1971).

[13a] *J.R. Malpass, N.J. Tweddle,* Chem. Commun. 1247 (1972).

[14] *T. Mukaiyama, R. Yoda, J. Kuwajima,* Tetrahedron Lett. 6247 (1966).

[15] *F. Effenberger, R. Gleiter,* Chem. Ber. *97,* 1576 (1964).

[16] *F. Effenberger, P. Fischer, G. Prossel, G. Kiefer,* Chem. Ber. *104,* 1987 (1971).

[17] *F. Effenberger, R. Gleiter,* Chem. Ber. *99,* 3903 (1966).

17.1.1.1.4 Aminocarbonyl or carbonylamino compounds by reaction of isocyanates with enamines

Enamines are more easily attacked electrophilically than enol ethers and this is manifest in their ready reaction with alkyl and aryl isocyanates. Those with at least one H atom in the β-position react with isocyanates to α,β-*unsaturated* β-*aminocarboxamides* from which β-*keto carboxamides* are obtained by hydrolysis[18-20]:

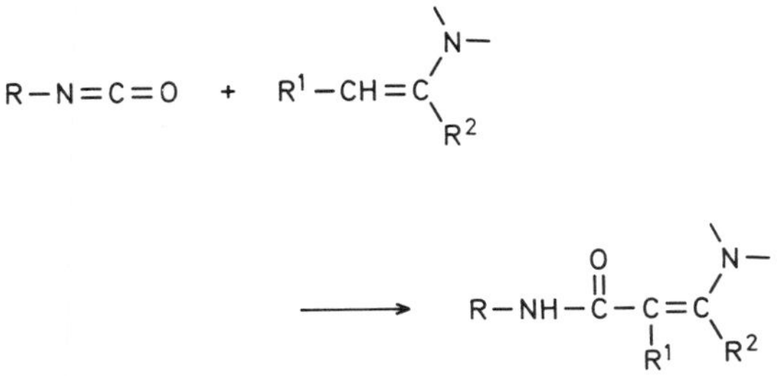

The speed of addition of the isocyanates to enamines increases with their electrophilism in the sequence[19]:

Alkyl- < Aryl- < Aroyl- < Arylsulfonyl-

Ynamines, too, react with phenyl isocyanate to form *4-amino-2-quinolinols* (31–68%)[21]:

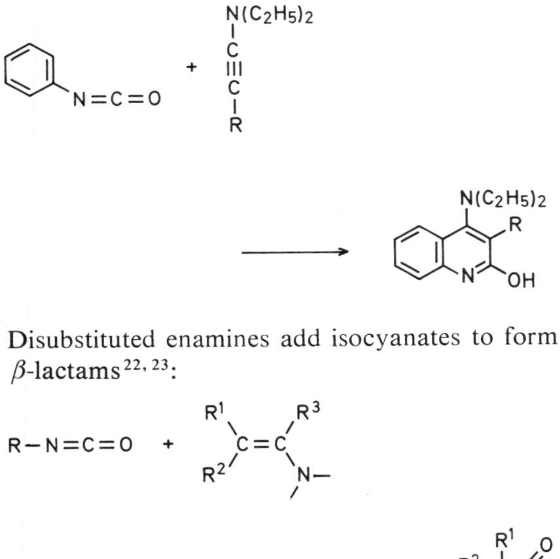

Disubstituted enamines add isocyanates to form β-lactams[22, 23]:

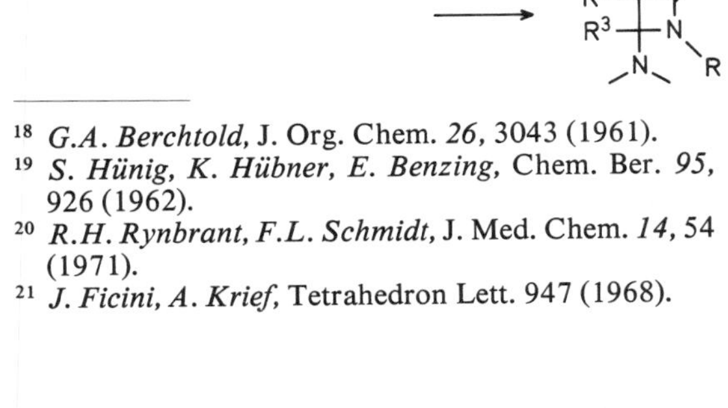

Carrying out the reaction between β-disubstituted enamines with aryl isocyanates at 120–140° makes 2 molecules isocyanate react and leads to *6-aminohexahydro-5,5-dimethyl-2,4-pyrimidinediones* ($\sim 80\%$)[24]:

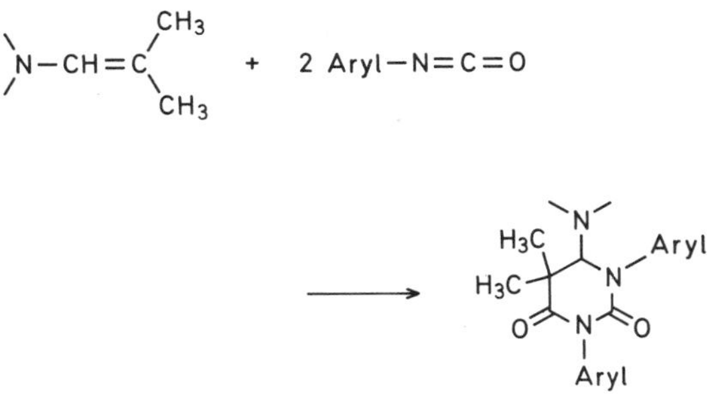

17.1.1.1.5 Aminocarbonyl or carbonylamino compounds by reaction of isocyanates with ketene acetals and analogs

By comparison to enol ethers or enamines the nucleophilic character of ketene acetals, ketene O,N-acetals and ketene N,N-acetals is substantially enhanced, so that these compounds react smoothly with desired isocyanates. In every case *substituted acrylamides* are formed with one molecule isocyanate[25-27]:

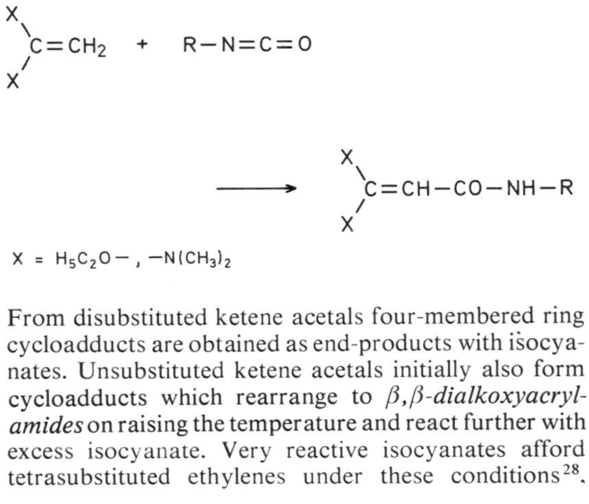

$X = H_5C_2O-$, $-N(CH_3)_2$

From disubstituted ketene acetals four-membered ring cycloadducts are obtained as end-products with isocyanates. Unsubstituted ketene acetals initially also form cycloadducts which rearrange to β,β-*dialkoxyacrylamides* on raising the temperature and react further with excess isocyanate. Very reactive isocyanates afford tetrasubstituted ethylenes under these conditions[28].

[18] *G.A. Berchtold*, J. Org. Chem. 26, 3043 (1961).
[19] *S. Hünig, K. Hübner, E. Benzing*, Chem. Ber. 95, 926 (1962).
[20] *R.H. Rynbrant, F.L. Schmidt*, J. Med. Chem. 14, 54 (1971).
[21] *J. Ficini, A. Krief*, Tetrahedron Lett. 947 (1968).

[22] *G. Opitz, J. Koch*, Angew. Chem. 75, 167 (1963).
[23] *M. Perelman, S.A. Miszak*, J. Amer. Chem. Soc. 84, 4988 (1962).
[24] *A.K. Bose, G. Mina*, J. Org. Chem. 30, 812 (1965).
[25] *R. Scarpati*, Rend. Accad. Sci. Fis. Mat., Soc. Nazl. Sci. Napoli 25, 7 (1958); C.A. 55, 11 423 (1961);
R. Scarpati, G. Delre, T. Maone, Rend. Accad. Sci. Fis. Mat., Soc. Nazl. Sci. Napoli 26, 20 (1959); C.A. 55, 11 423 (1961).
[26] *F. Effenberger, R. Gleiter, G. Kiefer*, Chem. Ber. 99, 3892 (1966).

while less electrophilic isocyanates react to form β,β-*dialkoxyacryloyl carbamoylamides*[28] and not, as was postulated originally, derivatives of 1,3-disubstituted barbituric acids[25, 26]:

$$\begin{array}{c} RO \\ \\ RO \end{array}\!\!C=CH-CO-NH-R^{1(2)}$$

$$R^2-N=C=O \qquad\qquad R^1-N=C=O$$

$$\begin{array}{c} RO \\ \\ RO \end{array}\!\!C=C\begin{array}{c} CO-NH-R^2 \\ \\ CO-NH-R^2 \end{array} \qquad \begin{array}{c} RO \\ \\ RO \end{array}\!\!C=CH-C\begin{array}{c} O \\ \\ N-R^1 \\ R^1-NH-C \\ O \end{array}$$

Ketene N,N-acetals form N-[(dimethylamino)methyl]malonamides in good yield following a further C-acylation[26].

Aryl isocyanates react entirely similarly with cyclic O,N-, O,S-, and S,N-ketene acetals[29].

17.1.1.1.6 Aminocarbonyl or carbonylamino compounds by reaction of isocyanates with *CH*-acid compounds

CH-acid compounds likewise attack isocyanates in a nucleophilic reaction. Substances with the β-dicarbonyl structure or nitro compounds are the compounds with active methylene groups that represent the main interest here. In β-dicarbonyl compounds the *CH*-activity is very important because only very active methylene groups, for instance, those in 2,4-pentanedione (acetylacetone) react with isocyanates in the absence of catalysts. Less active compounds, for example, malonic acid derivatives, require base catalysis to produce success:

$$\begin{array}{c} R \\ \\ O \end{array}\!\!C-CH_2-C\begin{array}{c} O \\ \\ R \end{array} \quad + \quad H_5C_6-N=C=O$$

$$\longrightarrow \quad \begin{array}{c} R \\ \\ O \end{array}\!\!C-CH-CO-R \atop CO-NH-C_6H_5$$

α-Acyl-β-oxocarboxy-anilides

The following β-dicarbonyl compounds have been employed[30, 31]:

Malonic acid derivatives
Malonaldehyde
Acetoacetic acid derivatives
2,4-Pentanedione derivatives
Benzoylacetic acid derivatives
1-Phenyl-1,3-alkanedione derivatives

Aromatic isocyanates[32] and also 1,6-hexyl diisocyanate[33] react to the corresponding carboxamides in addition to phenyl isocyanate, *e.g.*:

$$2 \begin{array}{c} COOR \\ \\ CH_2 \\ \\ COOR \end{array} \quad + \quad \begin{array}{c} N=C=O \\ \\ (CH_2)_6 \\ \\ N=C=O \end{array} \quad \longrightarrow$$

$$ROOC-CH-CO-NH-(CH_2)_6-NH-CO-CH-COOR \atop ROOC COOR$$

Dialkyl N,N'-hexamethylenedimalonamates

With very reactive methylene compounds it is sufficient to mix the two components together with or without a solvent to bring about the reaction. In the other cases sodium methoxide or ethoxytriethyllead[32] initiate the exothermally proceeding addition. Cyclic 1,3-diones, too, react readily with isocyanates without a catalyst[8, 31, 34], *e.g.*:

$$\begin{array}{c} H_3C \\ \\ H_3C \end{array}\!\! \bigcirc\!\!\begin{array}{c} O \\ \\ \\ O \end{array} \quad + \quad R-N=C=O$$

$$\longrightarrow \quad \begin{array}{c} H_3C \\ \\ H_3C \end{array}\!\! \bigcirc\!\!\begin{array}{c} OH \\ \\ CO-NH-R \\ \\ O \end{array}$$

2-Hydroxy-4,4-dimethyl-6-oxo-1-cyclohexene-1-carboxamide

---

[27] *D.H. Clemens, A.J. Bell, J.L. O'Brien*, J. Org. Chem. *29*, 2932 (1964).

[28] *F. Effenberger*, priv. com.

[29] *R. Richter, H. Ulrich*, Justus Liebigs Ann. Chem. *743*, 10 (1971).

[30] *A. Michael*, Ber. *38*, 22 (1905).

[31] *W. Dieckmann, J. Hoppe, R. Stein*, Ber. *37*, 4627 (1904).

[32] *A.G. Davies, R.J. Puddephatt*, J. Chem. Soc. C 1479 (1968).

[33] *S. Petersen*, Justus Liebigs Ann. Chem. *562*, 205 (1947).

[34] *M.M. Shemyakin, Yu.A. Arbuzov, M.N. Kosolov, G.A. Shatenshtein, V.V. Onoprienko, Yu.A. Konnova*, Zh. Obshch. Khim. *30*, 542 (1960); C.A. *54*, 24575 (1960);

M.M. Shemaykin, M.N. Kosolov, Yu.A. Arbuzov, V.V. Onoprienko, Yu-Yuan Hsieh, Zh. Obshch. Khim. *30*, 545 (1960); C.A. *54*, 24576 (1960).

Enolized β-dicarbonyl compounds of the 2H-pyran-2-one or 4-hydroxycoumarin type add isocyanates analogoulsy[35]:

3-Carbamoyl-4-hydroxycoumarins

Very weakly CH-active compounds must first be converted into the carbanion with strong bases such as sodium hydride to make the addition of isocyanates possible. Thus dimethyl sulfoxide very readily reacts with isocyanates[35a, 35b] under these conditions to form *methylsulfonylacetamides* 1 and *methylsulfonylmalonamides* 2[36].

1   2

R = Phenyl-, 1-Naphthyl-, biphenylyl-

CH-active nitro compounds react analogously with aromatic isocyanates to form α-*nitroacetanilides*[37–39]:

$$R-CH_2-NO_2 \ + \ R^1-N=C=O \longrightarrow$$

R = H, H_5C_2-COO-,
$H_5C_6-NH-CO-$, H_5C_6-CO-

$$R^1-NH-CO-\overset{\overset{\displaystyle R}{|}}{CH}-NO_2$$

The reaction is carried out in benzene with potassium carbonate or triethylamine as catalyst.

Addition reactions on the isocyanate group are not limited to *CH*-active compounds; triesters of orthocarboxylic acids, acetals, or phosphinalkylenes[39a, 39b] also react, to form amides[40]. Thus, triethyl orthoformates or formaldehyde diethylacetal add isocyanates in the presence of Lewis acids to form *carbamic acid esters:*

X = RO-, H-
R^1 = Aryl-, Alkyl-

17.1.1.1.7 Aminocarbonyl or carbonylamino compounds by reaction of isocyanates with organometallic compounds

Isocyanates are converted in good yield into *N-substituted carboxamides* by the very reactive organometallic compounds in a nucleophilic reaction. Often Grignard reagents are used[41, 42] to prepare, for instance, *N-substituted phenylmalonamic acids*[43, 44]:

$$H_5C_6-\overset{\overset{\displaystyle |}{CH}}{\underset{\underset{\displaystyle MgCl}{|}}{}}-COOMgCl \ + \ R-N=C=O$$

$$\longrightarrow \ H_5C_6-\overset{\overset{\displaystyle COOH}{|}}{CH}-CO-NH-R$$

[35] Brit. P. 919807 (1963), LIPHA (Lyonnaise Industrielle Pharmaceutique); C.A. *60*, 507a (1964).
[35a] S.B. Kadin, J. Med. Chem. *15*, 551 (1972);
[35b] G. Gref, B. Sabouroult, J. Bourais, Tetrahedron Lett. 1957 (1972).
[36] M. v. Strandtmann, S. Klutchko, D. Connor, J. Shavel jr., J. Org. Chem. *36*, 1742 (1971).
[37] W. Steinkopf, H.M. Daege, Ber. *44*, 497 (1911).
[38] R.N. Boyd, R. Leshin, J. Amer. Chem. Soc. *75*, 2762 (1953).

[39] A. Dornow, A. Müller, S. Lüpfert, Justus Liebigs Ann. Chem. *594*, 191 (1955).
[39a] H.J. Bestmann in W. Foerst, Neuere Methoden der präparativen organischen Chemie, Bd. V, p. 44, Verlag Chemie, Weinheim/Bergstr. 1967.
[39b] H. Saikachi, K. Takei, Yakugaku Zasshi *89*, 1401 (1969); C.A. *72*, 12452y (1970).
[40] H.v. Brachel, R. Merten, Angew. Chem. *74*, 872 (1962).
[41] R.B. Carlin, L.O. Smith jr., J. Amer. Chem. Soc. *69*, 2007 (1947).
[42] H.M. Singleton, W.E. Edwards jr., J. Amer. Chem. Soc. *60*, 540 (1938).
[43] F.F. Blicke, H. Zinnes, J. Amer. Chem. Soc. *77*, 4849 (1955).
[44] F.F. Blicke, H. Zinnes, J. Amer. Chem. Soc. *77*, 5395 (1955).

Benzoyl isocyanate furnishes a variety of addition products with Grignard reagents whose formation is governed by the reaction conditions and the Grignard reagent employed[45]. Other organometallic compounds such as triphenylsodium[46] or benzo[b]thien-2-yllithium[47], too, react fundamentally to anilides with direct introduction of the aminocarbonyl group:

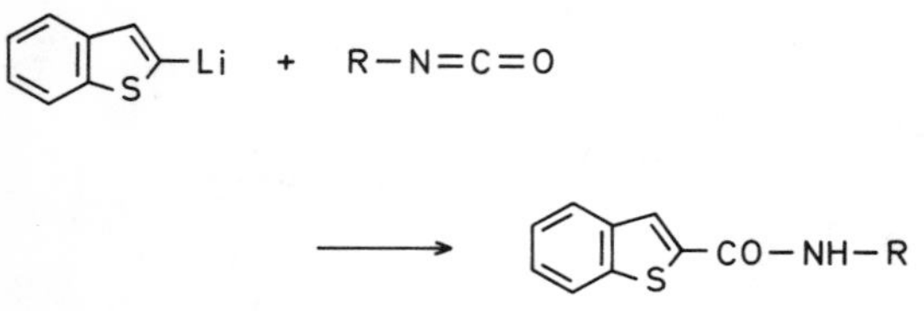

Benzo[b]thiophene-2-carboxanilide (R = C₆H₅)

The smoothly ensuing reactions of triethylaluminum with isocyanates to *N-substituted propionamides*[48, 49] display particular interest:

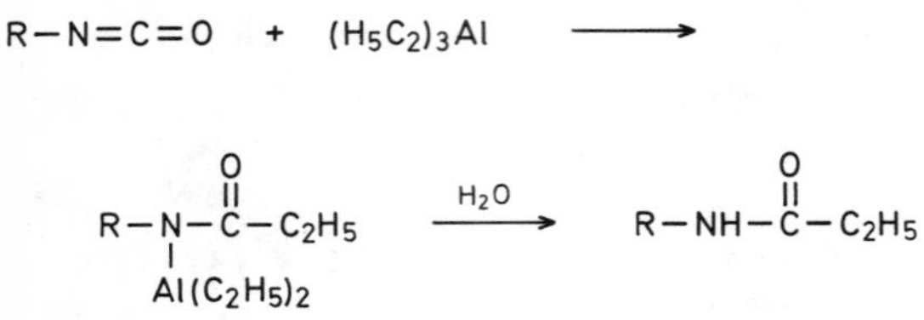

$$R-N=C=O \; + \; (H_5C_2)_3Al \; \longrightarrow$$

Aliphatic isocyanates not do not react as readily, as is shown by the utilization of the triethylaluminum. Better yields are obtained when dichloroethylaluminum is used in place of triethylaluminum.

17.1.1.1.8 Aminocarbonyl or carbonylamino compounds by reaction of isocyanates with amines

The most frequently employed carbamoylation of amines by reacting with isocyanates leads to *N,N'-disubstituted ureas*. Primary and secondary amines attack the electrophilic *C* atom of the isocyanate group nucleophilically. Reaction-promoting factors are, firstly, the amine nucleophilism

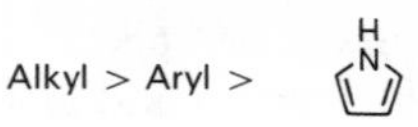

and, secondly, the isocyanate electrophilism[50, 51]

Acyl > Aryl > Alkyl

In general, the reactions, which are usually performed in an inert solvent such as benzene, acetone, 1,4-dioxane, or chloroform, proceed smoothly and afford a high yield of urea derivatives. By choosing the amine and isocyanate component appropriately, ureas substituted in desired manner can be prepared:

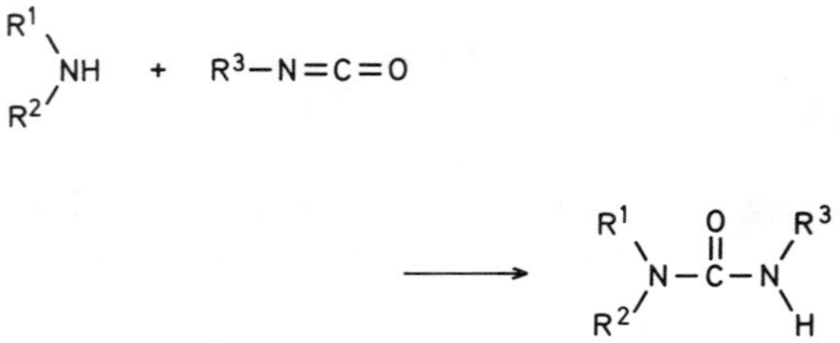

With less basic amines tertiary amines may be added as catalysts[52]:

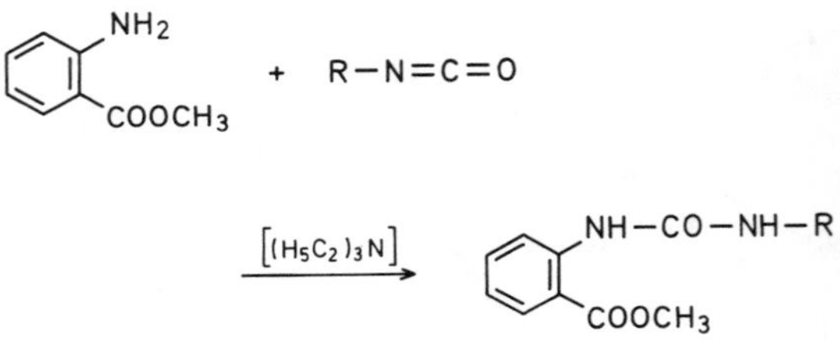

Methyl N-(phenyl-carbamoyl)anthranilate (R = C₆H₅)

Pyrrole needs to be reacted in the form of its potassium salt with phenyl isocyanate to form *N-phenylpyrrole-1-carboxamide*[53]:

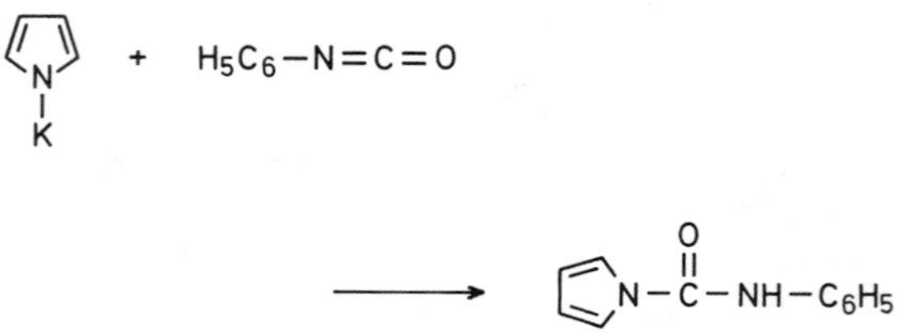

[45] J.W. McFarland, R.L. Harris, J. Org. Chem. *32*, 1273 (1967).
[46] W. Schlenk, E. Bergmann, Justus Liebigs Ann. Chem. *464*, 1 (1928).
[47] D.A. Shirley, M.D. Cameron, J. Amer. Chem. Soc. *74*, 664 (1952).
[48] H. Reinheckel, D. Jahnke, Chem. Ber. *97*, 2661 (1964).
[49] H. Reinheckel, D. Jahnke, G. Kretzschmar, Chem. Ber. *99*, 11 (1966).

[50] S. Petersen in *Houben-Weyl*, Methoden der organischen Chemie, 4. Aufl., Bd. VIII, p. 157, Georg Thieme Verlag, Stuttgart 1952
[51] R.G. Arnold, J.A. Nelson, J.J. Verbanc, Chem. Rev. *57*, 47 (1957);
J. Burkus, J. Org. Chem. *26*, 779 (1961).
[52] B. Taub, J.B. Hino, J. Org. Chem. *26*, 5238 (1961).
[53] E.P. Papadopoulos, H.S. Habiby, J. Org. Chem. *31*, 327 (1966).

Other heteroaromatic amines such as 2-amino-1,3,4-oxadiazoles react straightforwardly with aromatic and aliphatic isocyanates[54]:

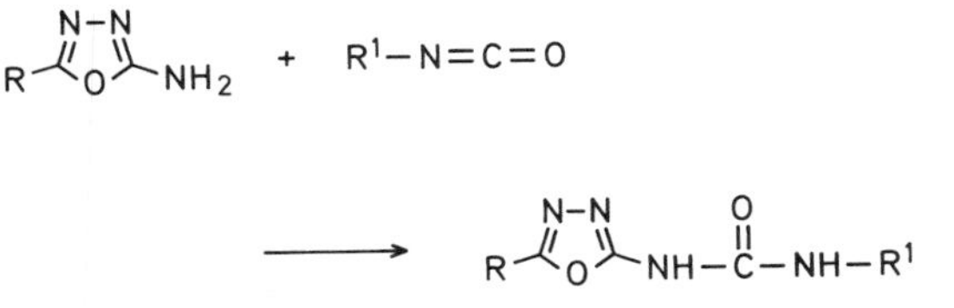

2-Phenylcarbamoyl-1,3,4-oxadiazole ($R^1 = C_6H_5$)

Phenyl isocyanate reacts further with the urea derivative formed to give *uretediones*[54]:

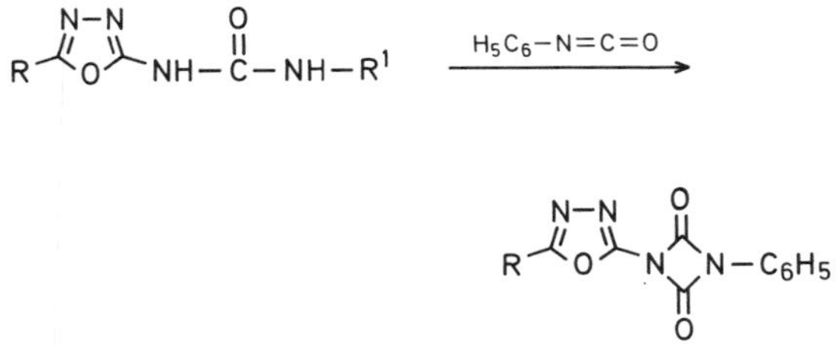

3-(1,3,4-oxadiazol-2-yl)-1-phenyl-2,4-uretidinedione

Enamine esters and related derivatives as amine components, too, react to form ureas[55]. Thus *coumarinylureas* are obtained from 2-amino-4-hydroxycoumarin and isocyanates[56]:

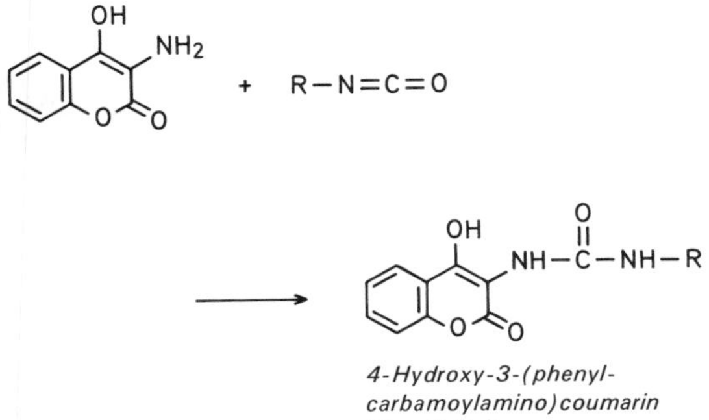

4-Hydroxy-3-(phenyl-carbamoylamino)coumarin ($R = C_6H_5$)

Tertiary amines react only seldom with isocyanates but catalyze isocyanate reactions such as di- and trimerizations and reactions with alcohols, amines, or phenols[51].

By contrast, *N*-substituted aziridines react with isocyanates at 100° to form *1,3-disubstituted 2-imidazolidinones* accompanied by ring expansion[57]:

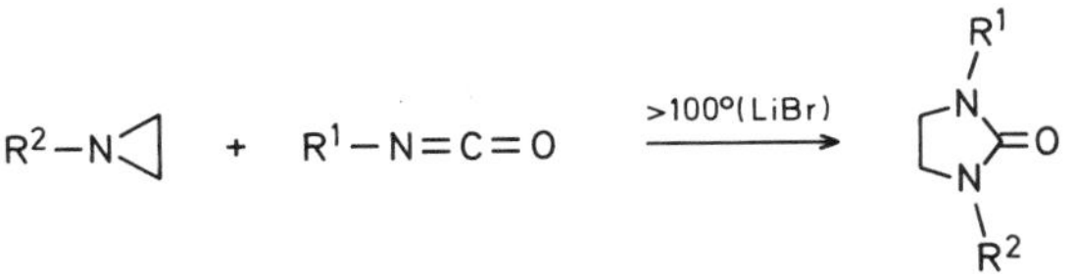

R^1 = Aryl-, Cycloalkyl-, Acyl-
R^2 = Acyl-, Alkyl-

Imidazolines also undergo ring expansion with isocyanates to form *hexahydro-1,3,5-triazepin-2-ones*[58]:

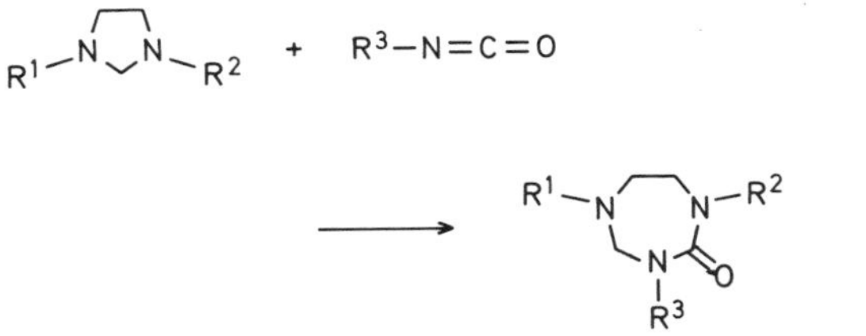

17.1.1.1.9 Aminocarbonyl or carbonylamino compounds by reaction of isocyanates with hydrazines

The very vigorous reaction between isocyanates and unsubstituted hydrazine seldom stops at the aminocarbonylation stage but leads to *biurea (1,2-hydrazinedicarboxamide) derivatives*, as is shown by the reactions of α-isocyanato esters of carboxylic acids[59]:

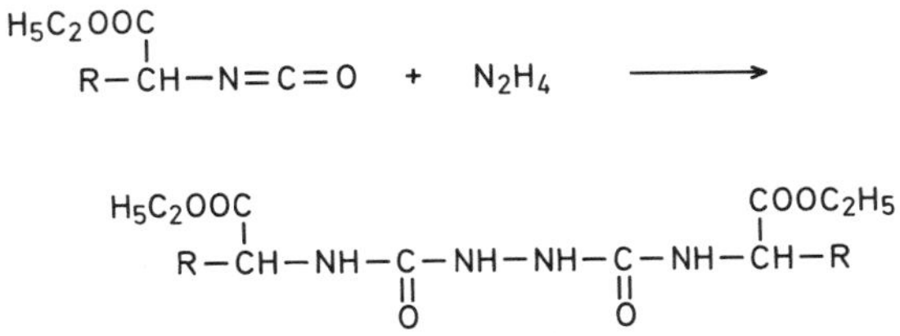

Diethyl 3,3'-(diureylene)dipropionate

With excess hydrazine the ester group is additionally hydrazinolyzed to form *β-(hydrazinocarbonylamino)carboxylic acid hydrazides*[60]:

[54] *H. Gehlen, M. Just*, Justus Liebigs Ann. Chem. *692*, 151 (1966).
[55] *H. Wamhoff*, Chem. Ber. *101*, 3377 (1968).
[56] *L. Reppel, W. Schmollack*, Arch. Pharm. (Weinheim, Ger.) *297*, 45 (1964).

[57] *E. Gulbins, R. Morlock, K. Hamann*, Justus Liebigs Ann. Chem. *698*, 180 (1966).
[58] *H. Böhme, W. Pasche*, Arch. Pharm. (Weinheim, Ger.) *302*, 81 (1969).
[59] *K. Schlögl*, Naturwissenschaften *44*, 466 (1957).

667

$$H_5C_2OOC$$
$$R-CH-N=C=O \quad + \quad 2\ H_2N-NH_2$$

$$\longrightarrow \quad R-CH-CH_2-CO-NH-NH_2$$
$$\qquad\qquad\qquad NH-CO-NH-NH_2$$

Substituted hydrazines, especially phenylhydrazine, react smoothly to *semicarbazides*[60, 61]. Acyl hydrazines react equally readily with isocyanates to afford very good yields of *4-substituted 1-acylsemicarbazides*[62, 63, 63a].

17.1.1.1.10 Aminocarbonyl or carbonylamino compounds by reaction of isocyanates with hydroxylamines

As the hydroxy derivative of ammonia, unsubstituted hydroxylamine reacts very rapidly with isocyanates to form *N'-substituted N-hydroxyureas:*

$$R-N=C=O \quad + \quad H_2N-OH \quad \longrightarrow$$

$$R-NH-C-NH-OH$$
$$\qquad\qquad \overset{\|}{O}$$

However, the addition does not stop here but, with further isocyanate, for example, can lead to *O,N,N-tris(aminocarbonyl)hydroxylamines* via *N,N-bis(aminocarbonyl)hydroxylamine*[64]. Acyl isocyanates react with hydroxylamine practically exclusively to give *3-acyl-1-hydroxyureas:*

$$R-C\overset{O}{\underset{N=C=O}{\diagup}} \quad + \quad H_2N-OH$$

$$\longrightarrow \quad R-C\overset{O}{\underset{NH-C-NH-OH}{\diagup}}$$
$$\qquad\qquad\qquad\qquad \overset{\|}{O}$$

R = Alkyl-, Aryl-

O-Methylhydroxylamine adds two molecules acyl isocyanate to form *N,N-bis(acylaminocarbonyl)-O-methylhydroxylamine*[65]. *O*- and *N*-alkylated hydroxylamines correspond substantially to amines in their reactivity and with isocyanates afford hydroxy- and alkoxy-substituted ureas respectively[66-68]:

$$R^3-N=C=O \quad + \quad HN\overset{R^1}{\underset{OR^2}{\diagdown}} \quad \longrightarrow$$

$$R^3-NH-C-N\overset{R^1}{\underset{OR^2}{\diagdown}}$$
$$\qquad\qquad \overset{\|}{O}$$

R^1 = Alkyl-, H-
R^2 = Alkyl-, H-
R^3 = Alkyl-, Aryl-

Under corresponding conditions *O*-alkylated esters of hydroxamic esters, too, react to form *N-carbamoylhydroxamic acid esters*[69].

17.1.1.1.11 Aminocarbonyl or carbonylamino compounds by reaction of isocyanates with aldehyde imines (azomethines)

Isocyanates can add to the C=N double bond of the aldehyde imines in a variety of ways[70]. Polar cycloaddition to formaldehyde imines affords *hexahydro-1,3,5-triazin-2-ones*[71, 72]

$$2\ R^1-N=CH_2 \quad + \quad R^2-N=C=O$$

$$\longrightarrow$$

R^1 = Alkyl-
R^2 = Alkyl-, Aryl-

[60] *K. Schlögl*, Monatsh. Chem. *89*, 61 (1958).
[61] *U. Kraatz, H. Wamhoff, F. Korte*, Justus Liebigs Ann. Chem. *744*, 33 (1971).
[62] *J. Gante, W. Lautsch*, Chem. Ber. *97*, 983 (1964).
[63] *A.P. Grekov, V.V. Shevchenko*, Zh. Org. Khim. *4*, 2113 (1968); C.A. *70*, 67832m (1969).
[63a] *K.A. Nuridzhanyan, G.V. Kuznetsova*, Zh. Org. Khim. *9*, 1171 (1973).
[64] *G. Zinner, R.O. Weber, W. Ritter*, Arch. Pharm. (Weinheim, Ger.) *298*, 869 (1965).

[65] *G. Zinner, R. Stoffel*, Arch. Pharm. (Weinheim, Ger.) *302*, 838 (1969).
[66] *G. Zinner, R.O. Weber*, Arch. Pharm. (Weinheim, Ger.) *298*, 580 (1965).
[67] *O. Scherer, G. Hörlein, K. Härtel*, Angew. Chem. *75*, 851 (1963).
[68] *G. Zinner, M. Hitze*, Arch. Pharm. (Weinheim, Ger.) *303*, 139 (1970).
[69] *J.H. Cooley, J.R. Throckmorton, W.D. Bills*, J. Org. Chem. *27*, 3131 (1962).
[70] *H. Ulrich*, Accounts Chem. Res. *2*, 186 (1969); *H. Ulrich, R. Richter* in *W. Foerst*, Neuere Methoden der präparativen organischen Chemie, Bd. VI, p. 267, Verlag Chemie, Weinheim/Bergstr. 1970.
[71] *D.H. Clemens, W.D. Emmons*, J. Org. Chem. *26*, 767 (1961).
[72] *W. Bartmann*, Chem. Ber. *100*, 2938 (1967).

while higher aldehyde imines afford *1,3-di-azetidin-2-ones* 3 or *dihydro-1,3,5-triazine-2,4-(1H,3H)-diones* 4[72-75]:

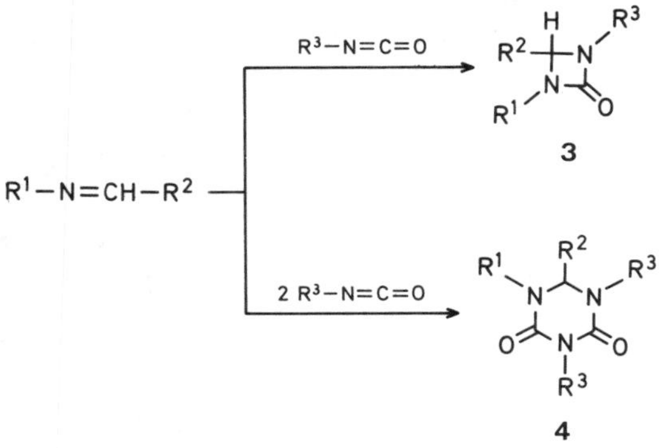

3

4

Which of the two products is formed is dependent on the isocyanate and imine employed as well as on the reaction conditions; sulfonyl isocyanates give mainly 4[72], *C,N*-diaryl substituted imines 3[73]. Benzoyl isocyanate forms *2,3-dihydro-4H-1,3,5-oxadiazin-4-ones* from the initially formed dipolar cycloadduct[74]:

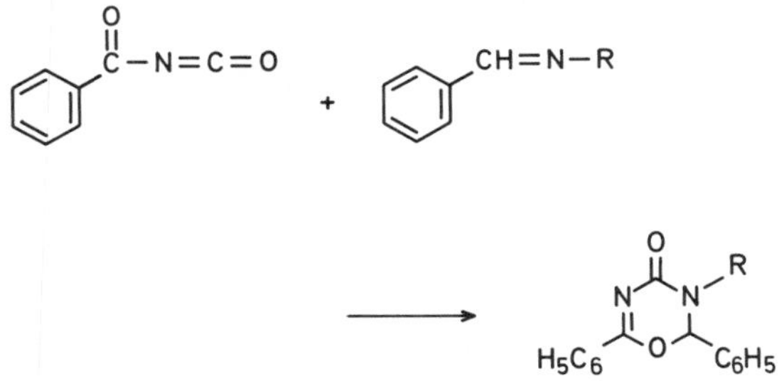

2,3-Dihydro-2,3,6-
triphenyl-4H-1,3,5-
oxadiazin-4-one
(R = C$_6$H$_5$)

Similarly, the addition of isocyanates to guanidines and amidines yields predominantly *dihydro-1,3,5-triazine-2,4(1H,3H)-diones*[76-78]:

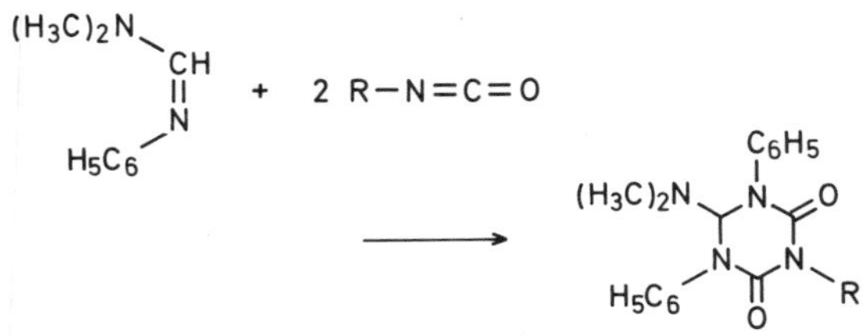

6-(Dimethylamino)-
dihydro-1,3,5-triphenyl-
1,3,5-triazine-2,4(1H,3H)-
dione (R = C$_6$H$_5$)

17.1.1.1.12 Aminocarbonyl or carbonylamino compounds by reaction of isocyanates with isocyanates

From aromatic isocyanates *1,3-diazetidine-2,4-diones*[51,79,80] can be obtained under the catalytic effect of trialkylphosphines or pyridines:

This old-established dimerization cannot be transferred to aliphatic isocyanates, which trimerize to *1,3,5-triazine-2,4,6(1H,3H,5H)-triones* under these conditions.

The trimerizations proceed especially readily in the presence of tertiary amines or metal alkoxides, which cause also aromatic isocyanates to trimerize readily[70]:

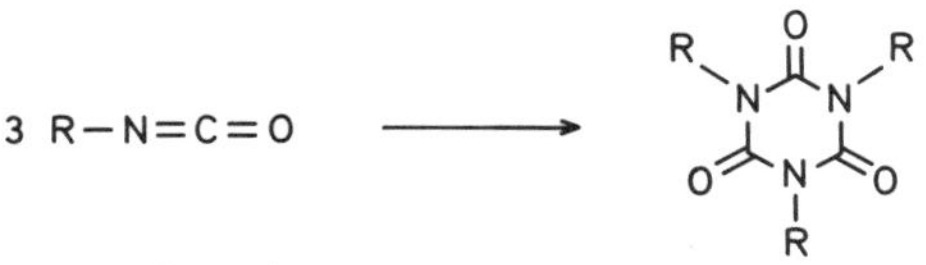

R = Alkyl-, Aryl-

17.1.1.1.13 Aminocarbonyl or carbonylamino compounds by reaction of isocyanates with alcohols, phenols, their thio analogs, and water

Addition of alcohols, phenols, and analogous thio compounds leads to *urethans* and *thiourethans*[33,51,81,81a]:

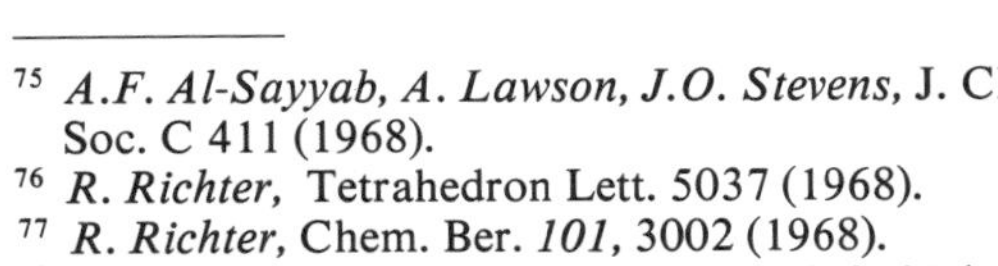

X = O, S
R^1 = Alkyl-, Aryl-

[73] *R. Richter*, Chem. Ber. *102*, 938 (1969).
[74] *R. Neidlein, R. Bottler*, Arch. Pharm. (Weinheim, Ger.) *302*, 306 (1969).

[75] *A.F. Al-Sayyab, A. Lawson, J.O. Stevens*, J. Chem. Soc. C 411 (1968).
[76] *R. Richter*, Tetrahedron Lett. 5037 (1968).
[77] *R. Richter*, Chem. Ber. *101*, 3002 (1968).
[78] *R. Richter, H. Ulrich*, Chem. Ber. *103*, 3525 (1970).
[79] *L.C. Raiford, H.B. Freyermuth*, J. Org. Chem. *8*, 230 (1943).
[80] *H. Ulrich*, Cycloaddition Reactions of Heterocumulenes, p. 122, Academic Press, New York, London 1967.
[81] *S. Petersen* in *Houben-Weyl*, Methoden der organischen Chemie, 4. Aufl., Bd. VIII, p. 130, Georg Thieme Verlag, Stuttgart 1952.
[81a] *S. Ozaki*, Chem. Rev. *72*, 457 (1972).

Simple mixing of the two components or using an inert solvent such as benzene, chloroform, or tetrahydrofuran is sufficient to bring about the reaction. Alcohols react more rapidly than the corresponding mercaptans; this is confirmed by the readier thermolysis of the thiourethans back to the starting compounds[81]. The additions can be catalyzed quite generally by tertiary amines[51, 81] if the alcohols used react sluggishly[82]. This catalysis is essential if phenols or thiophenols are to be reacted with isocyanates.

Water adds to isocyanates with formation of the unstable *carbamic acid* which decomposes at once into carbon dioxide and amine. The amine formed reacts with still remaining isocyanate to *N, N'-disubstituted urea*, the end-product of the reaction between isocyanates and water[81]:

$$2 \ R-N=C=O \ + \ H_2O \ \xrightarrow[-CO_2]{}$$

$$R-NH-\overset{\overset{\displaystyle O}{\|}}{C}-NH-R$$

17.1.1.2 Aminocarbonyl or carbonylamino compounds by carbonylation of amines

Carbonylation of amines with carbon monoxide leads to *formamides* or *ureas*[83, 83a] directly under the catalytic influence of Group 8b metals or their compounds at elevated pressure and temperature:

$$R-NH_2 \ + \ CO \ \longrightarrow$$

$$R-NH-\overset{\overset{\displaystyle O}{\|}}{C}-H \ + \ R-NH-\overset{\overset{\displaystyle O}{\|}}{C}-NH-R^1$$

Tertiary amines likewise react to formamides with loss of an alkyl group. Often the metal catalysts are employed in the form of their carbonyl compounds[84, 85], although some reactions proceed also with sodium methoxide[86] in methanol, which

here result in a quantitative yield of amide. Using platinum as catalyst *oxamides* can be prepared under carbonylation conditions[87]. The reactions of carbon monoxide with *N*-substituted allylamines to *pyrrolidones* involving intramolecular cyclization display some interest[88, 89]

$$H_2C=CH-CH_2-NH-R \ + \ CO \ \longrightarrow$$

while 2-alkenylamines generally lead to an isomeric mixture of *3-substituted pyrrolidones* and *piperidones*

$$R-CH=CH-CH_2-NH_2 \ + \ CO \ \longrightarrow$$

$$R = CH_2-R^1$$

17.1.1.3 Aminocarbonyl or carbonylamino compounds by free-radical amidation with formamide

The free-radical amidation of olefins to *carboxamides* with formamide or *N, N*-disubstituted formamides can be effected both photochemically or with peroxide[90, 91]:

$$R-CH=CH_2 \ + \ HC\overset{\overset{\displaystyle O}{\|}}{\underset{NH_2}{}} \ \longrightarrow$$

$$R-CH_2-CH_2-C\overset{\overset{\displaystyle O}{\|}}{\underset{NH_2}{}}$$

Not only olefinic double bonds add the aminocarbonyl radical formed from formamide but acetylenes, too, react analogously by taking up two radicals[92].

The *field of application* of this amidation procedure with formamide[92a] is not limited to isolated

[82] *L. Capuano, R. Zander*, Chem. Ber. *104*, 2212 (1971).

[83] *W. Reppe*, Justus Liebigs Ann. Chem. *582*, 1 (1953).

[83a] *J.J. Byerky, G.L. Rempel, N. Takebe, B.R. James*, Chem. Commun. 1482 (1971).

[84] *F. Calderazzo*, Inorg. Chem. *4*, 293 (1965).

[85] *W. Hieber, L. Schuster*, Z. Anorg. Allgem. Chem. *287*, 214 (1956).

[86] *A. Schiffers, F. Glaser*, Chemiker-Ztg., *85*, 435 (1961).

[87] *J. Tsuji, N. Iwamoto*, Chem. Commun. 380 (1966).

[88] *J. Falbe, F. Korte*, Chem. Ber. *98*, 1928 (1965).

[89] *J. Falbe* in *W. Foerst*, Neuere Methoden der präparativen organischen Chemie, Bd. VI, p. 181, Verlag Chemie, Weinheim/Bergstr. 1970.

[90] *D. Elad*, Fortschr. Chem. Forsch. 7, 528 (1967).

[91] *H.H. Vogel*, Synthesis 99 (1970).

[92] *G. Friedmann, A. Komen*, Tetrahedron Lett. 3357 (1968).

[92a] *A. Arnone*, Gazz. Chim. Ital. *103*, 13 (1973).

double bonds and can be extended to heteroaromatic bases[93, 94]. Redox systems such as $H_2O_2/Fe^{2\oplus}$ or *tert-butyl hydroperoxide*/$Fe^{2\oplus}$ serve to generate the aminocarbonyl radicals, whose reactivity and orientation properties during substitution reveal a marked nucleophilic character[93], *e.g.*:

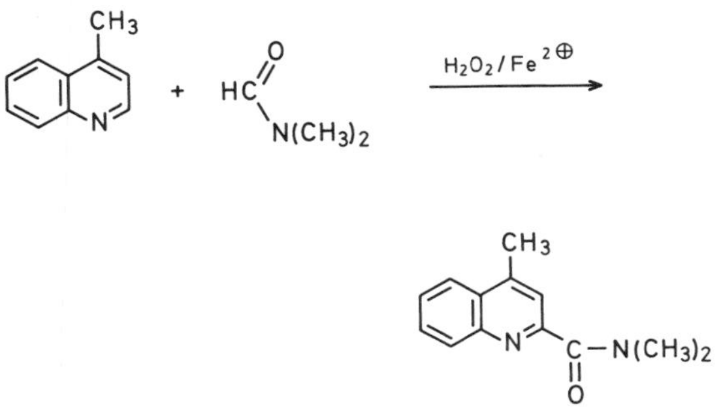

4-Methyl-N,N-dimethyl-
2-quinoline carboxamide

17.1.1.4 Introduction of the aminocarbonyl or carbonylamino group with carbamoyl chlorides

Aminocarbonyl groups can be introduced directly by Friedel-Crafts reaction between aromatic compounds and amides of chlorocarbonic acid[7, 95, 95a]:

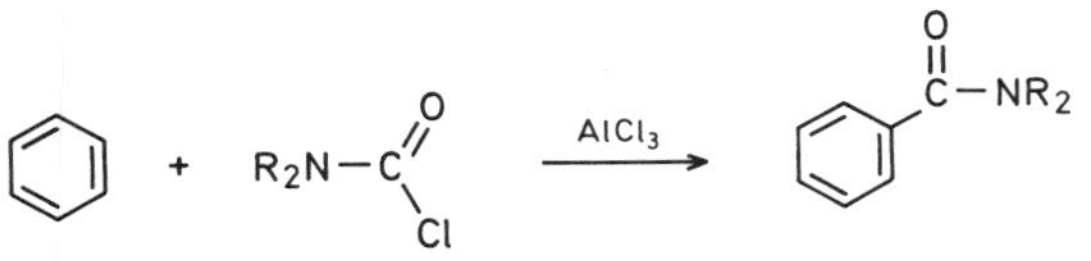

With unsubstituted chlorocarboxamides the yields are substantially higher than with the N-substituted derivatives. Alkylaromatic compounds are acylated particularly rapidly, halobenzenes very poorly or not at all[96].

17.1.2 Acylations with carboxylic acids

17.1.2.1 Acylations of amines and amine derivatives

Formation of amides by acylation of amines with carboxylic acids in general requires quite drastic reaction conditions. To prepare unsubstituted amides the ammonium salts of the relevant carboxylic acid are subjected to dry distillation[97]. With primary or secondary amines, too, the reaction succeeds in quite good yield[98, 99]:

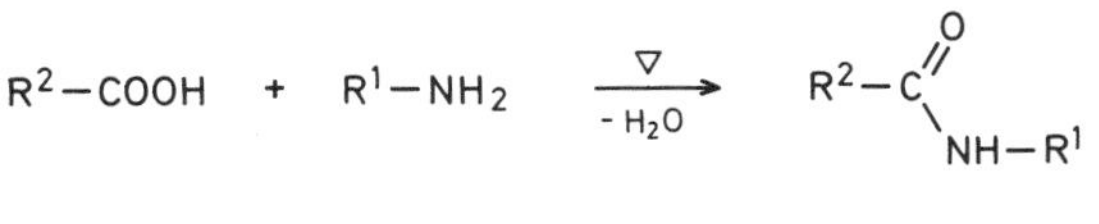

$$R^2-COOH \;+\; R^1-NH_2 \;\xrightarrow[-H_2O]{\nabla}\; R^2-C\!\!\begin{array}{c}O\\ \\NH-R^1\end{array}$$

It is sufficient to heat equimolar amounts of amine and carboxylic acid to 180–200° for some time in order to complete the splitting off of water. The thiocarboxylic acids have a stronger acylating action and the reactions proceed more rapidly with evolution of hydrogen sulfide[100, 100a].

Dehydration of the ammonium salts of carboxylic acids is naturally an equilibrium reaction[101] in which the amide formation can be made to ensue almost quantitatively by removing the water of reaction in a water separation tube. In this way it is possible to prepare *amides* of *formic acid* or *oxalic acid*. With less strongly acid carboxylic acids it is suitable to work in the presence of an acid catalyst, for instance, catalytic amounts of sulfuric acid or Amberlite IR 300[102]. In the same way brief heating of the amines with carboxylic acids in the presence of β-trichloromethyl-β-propiolactone or 3,3,3-trichloropropene-1,2-sulfone as catalyst affords the N-substituted amides in good yield[103]. The reaction proceeds in the same way also with amino acids, hydrazine, and urea deriva-

[93] *F. Minisci, G.P. Gardini, R. Galli, F. Bertini*, Tetrahedron Lett. 15 (1970).

[94] *G.P. Gardini, F. Minisci, G. Palla*, Tetrahedron Lett. 59 (1971).

[95] *H. Henecka* in *Houben-Weyl*, Methoden der organischen Chemie, 4. Aufl., Bd. VIII, p. 380, Georg Thieme Verlag, Stuttgart 1952.

[95a] *A. Yu. Naumov, A.P. Isakowa, A.N. Kost, N.F. Moiseikina, A.A. Stepanova, S.V. Nikerryasova*, Zh. Org. Khim. *9*, 591 (1973).

[96] *J.F.K. Wilshire*, Australian J. Chem. *20*, 575 (1967).

[97] *H. Henecka, P. Kurtz* in *Houben-Weyl*, Methoden der organischen Chemie, 4. Aufl., Bd. VIII, p. 654, Georg Thieme Verlag, Stuttgart 1952.

[98] *F. Möller* in *Houben-Weyl*, Methoden der organischen Chemie, 4. Aufl., Bd. XI/2, p. 4, Georg Thieme Verlag, Stuttgart 1958.

[99] *E.R. Shepard, H.D. Porter, J.F. Moth, C.K. Simmans*, J. Org. Chem. *17*, 568 (1952).

[100] *A. Schöberl, A. Wagner* in *Houben-Weyl*, Methoden der organischen Chemie, 4. Aufl., Bd. IX, p. 747, Georg Thieme Verlag, Stuttgart 1955.

[100a] *R.L. Caswell, J.A. Lanier, M.A. McAdams*, J. Chem. Eng. Data *17*, 269 (1972).

[101] *H. Morawetz, P.S. Otaki*, J. Amer. Chem. Soc. *85*, 463 (1963).

[102] *M. Walter, H. Besendorf, O. Schnider*, Helv. Chim. Acta *44*, 1546 (1961).

[103] *F.I. Luknitskii*, Dokl. Akad. Nauk SSSR *185*, 385 (1969); C.A. *71*, 3057 a (1969).

tives. *Hydrazides* and *hydroxamic acids* have been prepared from the free carboxylic acids with hydrazines and hydroxylamines using the above methods in only very few cases[104]. On heating 1-methyl-2-phenylhydrazine in acetic acid, *acetic acid 1-methyl-2-phenylhydrazide* is obtained in very good yield[105].

$$H_3C-COOH \quad + \quad H_3C-NH-NH-C_6H_5$$

$$\xrightarrow[-H_2O]{}$$

17.1.2.2 Acylations of ureas with carboxylic acids

Reacting ureas with carboxylic acids to form amides succeeds in yields of 70% and more on heating equimolar quantities[106]:

$$R-COOH \quad + \quad$$

$$R-CO-NH_2 \quad + \quad CO_2 \quad + \quad NH_3$$

Dicarboxylic acids furnish *diamides*[106a], while succinic acid is converted into *succinimide*. Employing thiourea gives the same result although the yields are much less favorable[107]. Substituted amides can be obtained similarly from substituted ureas[108].

$$R-COOH \quad +$$

The water formed during amide formation can be bound by adding an additional component as well as be removed by azeotropic distillation. In most cases this component does not react with the water formed initially but, instead, forms a reactive intermediate with the amine or the carboxylic acid, which then reacts further rapidly in its turn. It is unnecessary to isolate these intermediate compounds during the amide synthesis, so that the acylation of the amine can be effected in a *combined reaction* in one vessel.

17.1.2.3 Acylations with carboxylic acids in the presence of dehydrating agents[98]

The water formed during amide formation can be bound by adding an additional component as well as be removed by azeotropic distillation. In most cases this component does not react with the water formed initially but, instead, forms a reactive intermediate with the amine or the carboxylic acid, which then reacts further rapidly in its turn. It is unnecessary to isolate these intermediate compounds during the amide synthesis, so that the acylation of the amine can be effected in a *combined reaction* in one vessel.

17.1.2.3.1 Acylations with carboxylic acids in the presence of N,N'-dicyclohexylcarbodiimide

Dehydration with carbodiimides has found its main application for *synthesizing peptides*[109]. During this preparation of amides from carboxylic acids and amines in the presence of carbodiimides certain conditions need to be maintained in order to suppress the formation of *N*-acylureas as side-products:

$$R-COOH$$

$$+$$

$$R^1-N=C=N-R^1$$

[104] *T. Rabini, G. Vita*, J. Org. Chem. *30*, 2486 (1965).

[105] *K. Kratzl, K.B. Berger*, Monatsh. Chem. *89*, 83 (1958).

[106] *A. Rahman*, Rec. Trav. Chim. Pays-Bas *75*, 164 (1956).

[106a] *G.N. Freidlin, K.A. Solop*, Zh. Org. Khim. *8*, 1162 (1972).

[107] *A. Rahman, M.A. Medrano, O.P. Mittal*, Rec. Trav. Chim. Pays-Bas *79*, 188 (1960).

[108] *E. Cherbuliez, F. Lanolt*, Helv. Chim. Acta *29*, 1438 (1946).

[109] *F. Kurzer, K. Douraghi-Zadeh*, Chem. Rev. *67*, 128 (1967).

The effect of temperature, solvents, and amine are particularly marked in terms of the amide yield, unlike the carboxylic acid. In boiling polar solvents (acetic acid esters, tetrahydrofuran) predominantly *N*-acylureas are formed due to rearrangement of the *O*-acylureas. This side-reaction is restrained by working in nonpolar solvents (dichloromethane, chloroform), using low reaction temperatures, and addition of tertiary amines[109, 110], and the amides are obtained in very good yield.

For synthesizing *hydroxamic acids* according to the following equation

$$R-COOH \quad + \quad H_2N-OR^1 \quad \xrightarrow[-\,H_2O]{\text{Carbodiimid}} \quad R-C\overset{O}{\underset{NH-OR^1}{\big\backslash}}$$

the use of the unsubstituted hydroxylamine is unsuitable, because the corresponding *O*-carbamoylhydroxamic acids are formed. Using *O*-alkylhydroxylamines affords an only average yield of the hydroxamic acid in correspondence with its reduced nucleophilism by comparison to the amines[110, 110a].

By contrast, hydrazines react exothermally with carboxylic acids much more rapidly and better to form *carboxylic acid hydrazides*[111, 112]:

$$R-COOH \quad + \quad H_2N-NR_2^1$$

$$\xrightarrow[-\,H_2O]{\text{Carbodiimid}} \quad R-C\overset{O}{\underset{NH-NR_2}{\big\backslash}}$$

Methylhydrazine furnishes *1-methylhydrazides*. A further acylation of benzoylhydrazine to form *N,N'*-dibenzoylhydrazine with benzoic acid in the presence of carbodiimides is also feasible.

17.1.2.3.2 Acylations with carboxylic acids in the presence of phosphorus compounds

17.1.2.3.2.1 Phosphorus(V) oxide

It is possible to utilize the dehydrating effect of phosphorus(V) oxide to remove the water of reaction during amide formation[98]. Thus *N, N-diethylnicotinamide* is obtained by boiling molar quantities of nicotinic acid with diethylamine and a threefold excess of phosphorus(V) oxide in toluene[113]. An acid anhydride is formed intermediately by virtue of the strongly dehydrating effect of the phosphorus(V) oxide and can then very easily acylate the amine. Under these conditions it is possible also to prepare numerous *N, N-dimethylcarboxamides* using dimethylformamide[114]:

$$2\,R-COOH \quad + \quad HC\overset{O}{\underset{N(CH_3)_2}{\big\backslash}} \quad \xrightarrow[-\,H_2O]{P_2O_5}$$

$$2\,R-C\overset{O}{\underset{N(CH_3)_2}{\big\backslash}} \quad + \quad 2\,CO$$

Here, too, the intermediate formation of acid anhydrides is postulated because the reaction fails with sulfuric acid, phosphoric acid, or phosphoryl chloride. Aromatic *hydroxy acids* furnish particularly good yields of *amides* using this technique; in many cases they exceed 70%. *N*-Unsubstituted carboxamides can be converted into nitriles by a repeated dehydration.

17.1.2.3.2.2 Acylations with carboxylic acids in the presence of phosphorus(III) chloride

In inert solvents amines form phosphorazo compounds with phosphorus(III) chloride with evolution of hydrogen chloride and the products react with carboxylic acids to form amides[115, 116]:

$$PCl_3 \quad + \quad 2\,R-NH_2 \quad \xrightarrow[-\,3\,HCl]{}$$

$$R-N=P-NH-R \quad \xrightarrow[-\,HPO_2]{2\,R^1-COOH} \quad 2\,R^1-CO-NH-R$$

N-(Benzyloxycarbonyl) alanylglycine ethyl ester; 87%

$$R = -CH_2-COOC_2H_5\,; \quad R^1 = -\underset{CH_3}{\underset{|}{CH}}-NH-CO-O-CH_2-C_6H_5$$

[110] *M.T.W. Hearn, A.D. Ward*, Australian J. Chem. *22*, 1731 (1969).

[110a] *D. Geffken, G. Zinner*, Chem. Ber. *106*, 2246 (1973).

[111] *M.L. Hoefle, A. Holmes*, J. Med. Chem. *11*, 974 (1968).

[112] *R.F. Smith, A.C. Bates, A.J. Battisti, P.G. Byrnes, Ch.T. Mroz, Th.J. Smearing, F.X. Albrecht*, J. Org. Chem. *33*, 851 (1968).

[113] *R. Nylander*, Dtsch. Apoth.-Ztg. *93*, 678 (1958).

[114] *H. Schindlbauer*, Monatsh. Chem. *99*, 1799 (1968).

[115] *W. Grassmann, E. Wünsch*, Chem. Ber. *91*, 449 (1958).

[116] *S. Goldschmidt, H.L. Kraus*, Angew. Chem. *67*, 471 (1955).

Pyridine serves as a particularly suitable means for binding the hydrogen chloride and at the same time acts as solvent and reaction accelerator.

17.1.2.3.2.3 Acylations with carboxylic acids in the presence of phosphoryl chloride

Amide formation using phosphoryl chloride proceeds according to a different mechanism than that with phosphorus(III) chloride. The carboxylic acid and the amine are both dissolved in an inert solvent to form the ammonium salt, from which the amide is formed rapidly and in good yield with phosphoryl chloride while giving off hydrogen chloride [117]:

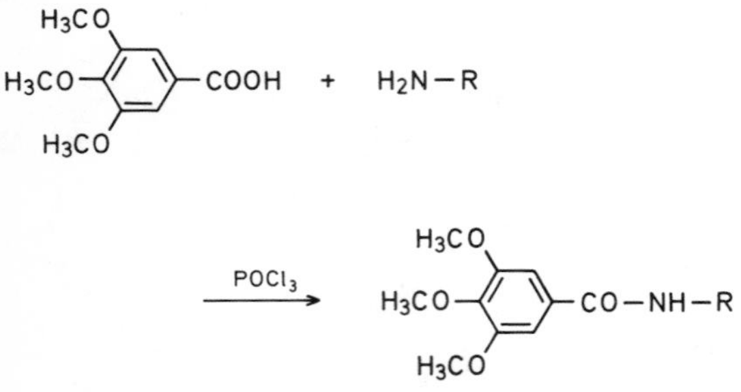

3,4,5-Trimethoxy-N-methyl-benzamide (R = CH$_3$)

Secondary amines as well as primary amines react. With aliphatic amines the yields are somewhat lower than with aromatic ones.

17.1.2.3.2.4 Acylations with carboxylic acids in the presence of triaryl phosphite

The dehydrating action of triaryl phosphites on imidazole to form amides has been used chiefly for peptide syntheses. Excellent yields of amide are obtained by carrying out the reaction in dimethylformamide or 1,4-dioxane as solvent [118]:

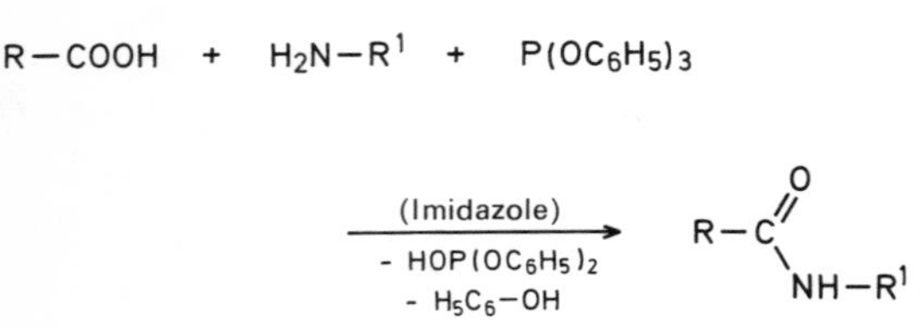

17.1.2.3.2.5 Acylations with carboxylic acids in the presence of triarylphosphine

With triphenylphosphine and carbon tetrachloride carboxylic acids form the phosphonium salts

5, which on addition of amines react smoothly to the amides **6** [119]:

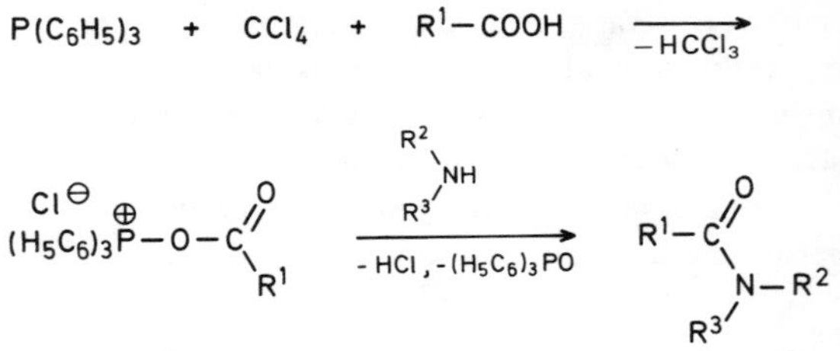

Using bromotrichloromethane in place of carbon tetrachloride enables amides to be prepared directly by this technique, including from aromatic amines.

17.1.2.3.2.6 Acylations with carboxylic acids in the presence of hexachlorocyclotriphosphatriazenes

The great reactivity of hexachlorotriphosphatriazene enables a reactive intermediate stage to be formed rapidly with carboxylic acids; amines or hydrazines readily attack this intermediate compound nucleophilically [120]:

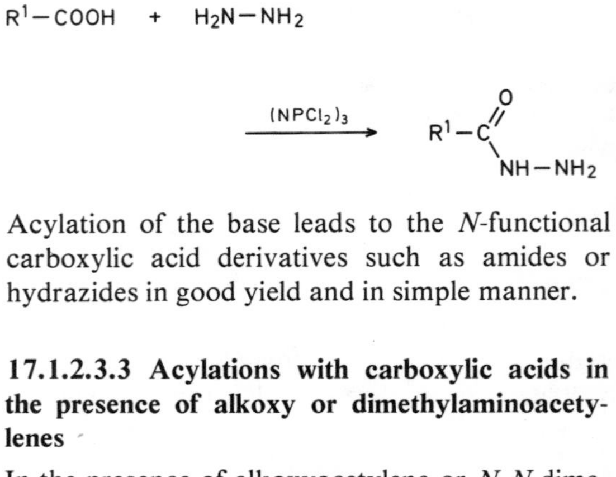

Acylation of the base leads to the *N*-functional carboxylic acid derivatives such as amides or hydrazides in good yield and in simple manner.

17.1.2.3.3 Acylations with carboxylic acids in the presence of alkoxy or dimethylaminoacetylenes

In the presence of alkoxyacetylene or *N,N*-dimethylalkynylamines, which act as anhydride forming agents, carboxamides can be very readily synthesized from amines and carboxylic acids [121]. The reaction proceeds under protective conditions and for this reason can be employed for peptide syntheses [122]:

[119] *L.E. Barstow, V.J. Hruby*, J. Org. Chem. *36*, 1305 (1971).

[120] *C. Caglioti, M. Poloni, G. Rosini*, J. Org. Chem. *33*, 2979 (1968).

[121] *G. Tadema, E. Harryvan, H.J. Panneman, A.J. Arens*, Rec. Trav. Chim. Pays-Bas *83*, 345 (1964).

[122] *H.J. Panneman, A.F. Marx, J.F. Arens*, Rec. Trav. Chim. Pays-Bas *78*, 487 (1959).

[117] *J. Klosa*, J. Prakt. Chem. *19*, 45 (1963).

[118] *Y.V. Mitin, O.V. Glinskaya*, Tetrahedron Lett. 5267 (1969).

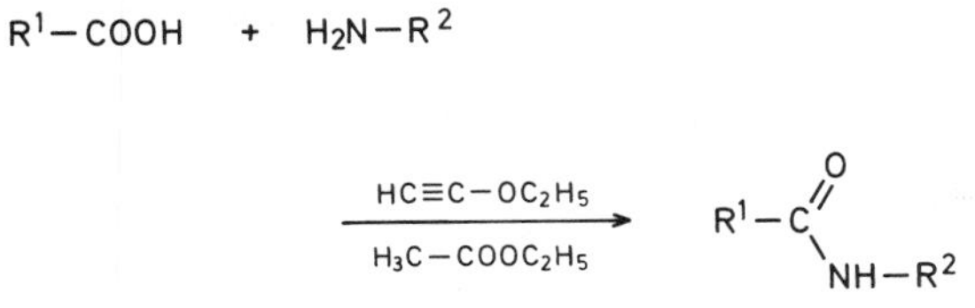

R^1—COOH   +   H$_2$N—R^2

$$\xrightarrow[\text{H}_3\text{C}-\text{COOC}_2\text{H}_5]{\text{HC}\equiv\text{C}-\text{OC}_2\text{H}_5} \quad \text{R}^1-\text{C}\underset{\text{NH}-\text{R}^2}{\overset{\text{O}}{\diagup}}$$

N,N-Dimethylethynylamine can be used for amide syntheses like alkoxyactylene[123, 124], although the basic character of the amine is a drawback that causes a racemization of the peptides formed which does not occur when alkoxyacetylenes are used[125].

17.1.2.3.4 Acylations with carboxylic acids in the presence of silicon(IV) chloride

The strong dehydrating action of silicon(IV) chloride can be made use of for synthesizing amides[126]:

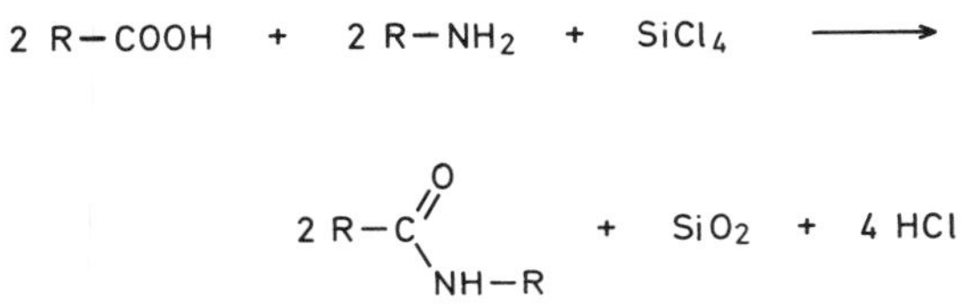

2 R—COOH   +   2 R—NH$_2$ +   SiCl$_4$ $\longrightarrow$

$$2\ \text{R}-\text{C}\underset{\text{NH}-\text{R}}{\overset{\text{O}}{\diagup}} \quad +\ \text{SiO}_2\ +\ 4\ \text{HCl}$$

The reaction is carried out in boiling pyridine as solvent which at the same time binds the hydrogen chloride formed. Both aromatic amines and tertiary *tert*-butylamine and aromatic carboxylic acids afford quite good yields. Salicylic acid alone forms a stable chelate with silicon(IV) chloride and then no longer reacts with amines.

17.1.3 Acylations with acid chlorides

17.1.3.1 Acylation of amines and amine derivatives

The most commonly used acylating agents are acyl halides, which enable *N*-functional carboxylic acid derivatives to be prepared very readily and smoothly from the corresponding amines and amine derivatives[127, 128]:

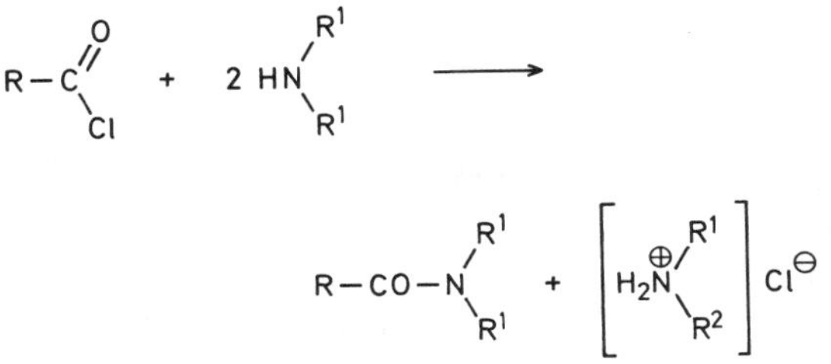

$$\text{R}-\text{C}\underset{\text{Cl}}{\overset{\text{O}}{\diagup}} \quad +\ 2\ \text{HN}\overset{\text{R}^1}{\underset{\text{R}^1}{\diagdown}} \quad \longrightarrow$$

$$\text{R}-\text{CO}-\text{N}\overset{\text{R}^1}{\underset{\text{R}^1}{\diagdown}} \quad +\ \left[\text{H}_2\overset{\oplus}{\text{N}}\overset{\text{R}^1}{\underset{\text{R}^2}{\diagdown}}\right]\text{Cl}^\ominus$$

The reaction proceeds with evolution of hydrogen chloride and in most cases is so exothermic that adequate cooling must be provided. As the reaction equation shows, an equivalent of the amine is not available for acetylation because of loss by ammonium salt formation, and it is normal to add a cheap auxiliary base to bind the hydrogen chloride for economic reasons. Two variants of the acetylation reaction using an auxiliary base can be distinguished:

1 *Schotten-Baumann* technique
2 *Einhorn* technique

In the *Schotten-Baumann* technique the aqueous solution of an amine with an inorganic auxiliary base such as, for example, alkali metal carbonates or hydroxides, is employed and to which the acid chloride is added drop by drop. Naturally, only halides which are difficult to hydrolyze can be employed, above all aromatic and long-chain acid halides. The *Einhorn* procedure enables the acylation to be conducted under anhydrous conditions in an inert solvent such as benzene, ether, chloroform, *etc.* Pyridine, *N*-ethyldiisopropylamine (Hünig base) or other tertiary bases are especially suitable for trapping the hydrogen chloride. They are added to the solution of the amine. Pyridine itself may serve as solvent simultaneously. The *rate of acylation* of an amine is dependent on its nucleophilic properties and on the electrophilism of the acyl halide. In general aromatic primary amines are less basic than aliphatic primary amines. Among secondary amines the difference in basicity is even greater between these two types.

The following diminishing reactivity sequence of the acid halides has been established in terms of the halogen atom[129, 130]:

[123] *R. Buyle, H.G. Viehe,* Angew. Chem. *76*, 572 (1964).
[124] *H.G. Viehe,* Angew. Chem. *79*, 744 (1967).
[125] *A.S. van Mourik, E. Harryvan, J.F. Arens,* Rec. Trav. Chim. Pays-Bas *84*, 1344 (1965).
[126] *T.H. Chan, L.T.L. Wong,* J. Org. Chem. *34*, 2766 (1969).
[127] *F. Möller* in *Houben-Weyl,* Methoden der organischen Chemie, 4. Aufl., Bd. XI/2, p. 7, Georg Thieme Verlag, Stuttgart 1958.

[128] *N.O.V. Sonntag,* Chem. Rev. *52*, 237 (1953).
[129] *D.P.N. Satchell,* J. Chem. Soc. (London) 1752 (1960).
[130] *M.L. Bender, J.M. Jones,* J. Org. Chem. *27*, 3771 (1962).
[131] *T. Fujisawa, S. Sugasawa,* Tetrahedron *7*, 185 (1959).

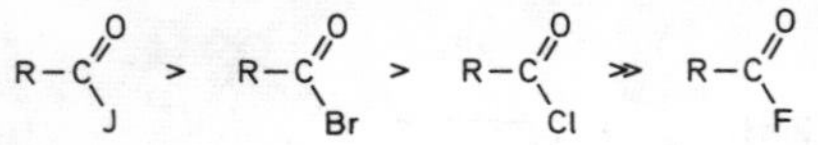

The substituent group R of a particular halide governs the reactivity with its electronic and steric effect; it diminishes in the aliphatic series as the chain length of R grows longer. In aroyl halides the reactivity is determined by the substituents on the aromatic ring by inductive and mesomeric effects in the sense of the Hammett equation[130].

17.1.3.1.1 Acylation of amines with acid chlorides

Unsubstituted carboxamides can be prepared by allowing acyl chlorides to act on excess concentrated ammonia solution[128]:

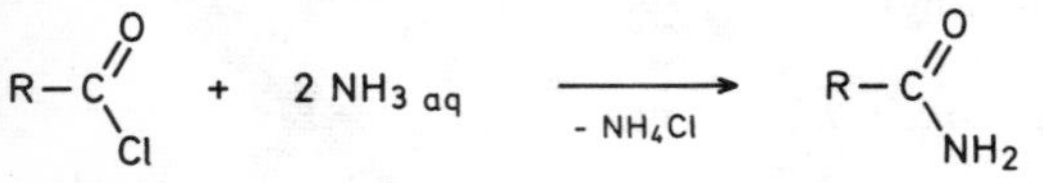

The reaction furnishes very good yields especially with longer chain acyl chlorides[131], *e.g.*:

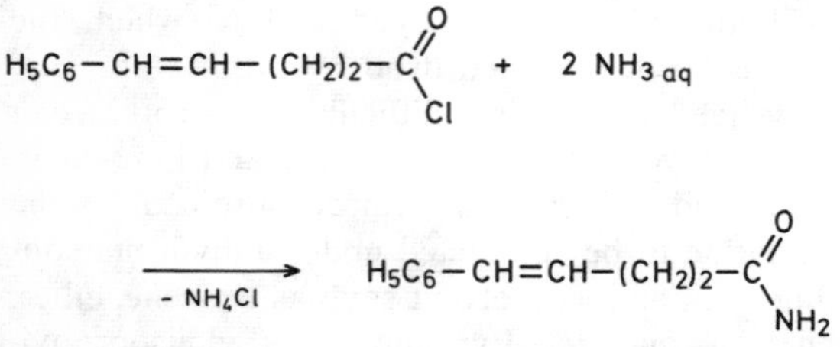

5-Phenyl-4-pentenamide

Lower amines, too, can be acylated in aqueous solution; thus methylamine affords *N-methyloleamide* (90%)[132]. The ready hydrolysis of the lower aliphatic acid chlorides means that their aminolysis must be carried out in anhydrous medium. For this purpose the acyl chloride is added drop by drop to a hot benzene solution through which ammonia streams[133]:

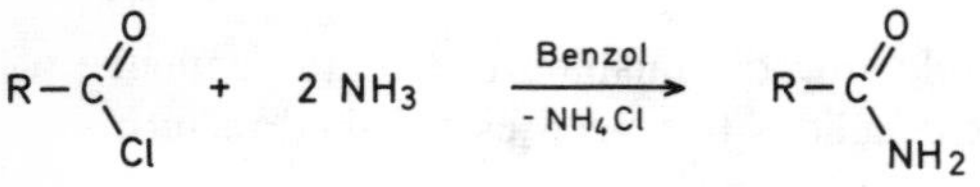

R = Alkyl

By contrast, if ammonia is introduced into the solution of the acid chloride then substantial diacylamine is formed. Under mild conditions amides can be obtained by the action of acyl halides on ammonium acetate in acetone at room-temperature[134]:

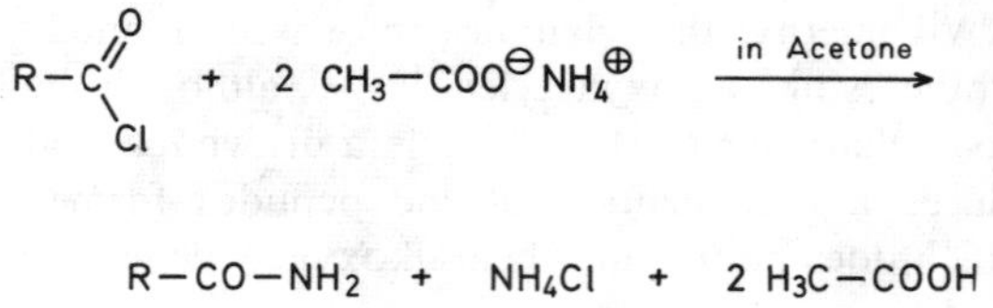

$$R-CO-NH_2 \ + \ NH_4Cl \ + \ 2\ H_3C-COOH$$

R = Aryl

The reaction is considered to be a nucleophilic attack of ammonia molecules formed by dissociation of the ammonium acetate in the solvent.

Many amides can be prepared from acyl halide and ammonium chloride without using a solvent; for this purpose the two components are heated to 120–130° until the evolution of hydrogen chloride ceases[135], *e.g.*:

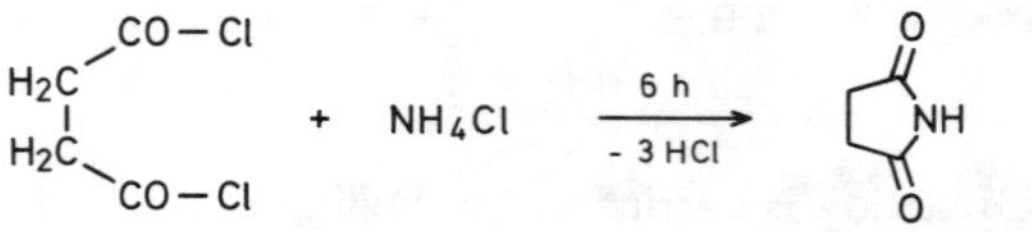

Succinamide; 100%

Succinyl and phthaloyl chlorides and their derivatives afford cyclic imides. The yields are very good.

Hydrochlorides of secondary amines form *carbamoyl chlorides* with phosgene[136, 136a]. The reaction proceeds at elevated temperature in an inert solvent or in the molten state:

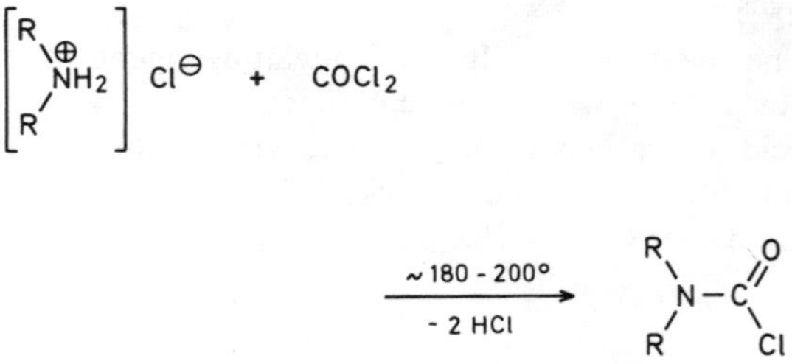

α-Amino acids can be acylated readily with acid chlorides while suspended in ethyl acetate; the *N-acylamino acids* formed dissolve (no auxiliary

[132] *E.T. Roe, J.T. Scanlan, D. Swern,* J. Amer. Chem. Soc. *71,* 2215 (1949).

[133] *G.E. Philbrook,* J. Org. Chem. *19,* 623 (1954).

[134] *P.A. Finan, G.A. Fothergill,* J. Chem. Soc. (London) 2824 (1962).

[135] DBP. 1 222058 (1966), Farbwerke Hoechst; Erf.: *H. Klug, K. Kuclinka; C.A. 66,* 2228 a (1967).

[136] *S. Petersen* in *Houben-Weyl,* Methoden der organischen Chemie, 4. Aufl., Bd. VIII, p. 117, Georg Thieme Verlag, Stuttgart 1952.

[136a] *H. Babad, A.G. Zeiler,* Chem. Rev. *73,* 75 (1973).

base is required because the hydrogen chloride escapes from the boiling solution[137]):

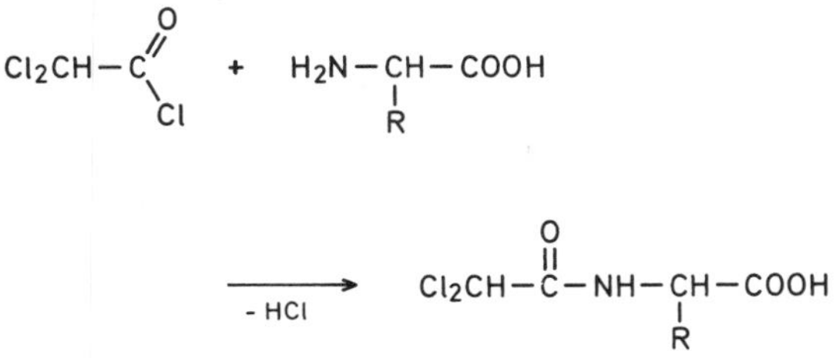

N-(Dichloroacetyl)amino acids

For the preparation of various *N*-benzylcarboxamides the Schotten-Baumann technique can be employed; an excess of either benzylamine or sodium hydroxide is used as auxiliary base[138].

With more reactive acyl halides (*e.g.*, acetyl chloride) the reaction is conducted by adding the reagent drop by drop to a two-phase system[139]. The latter consists of an aqueous solution of the auxiliary base and benzylamine in a water-immiscible solvent.

The Einhorn variant is more generally applicable; it allows amides to be synthesized also from sensitive amines or acyl chlorides. Thus oxaziridines can be readily acylated in ether solution with triethylamine as auxiliary base[140, 141], *e.g.*:

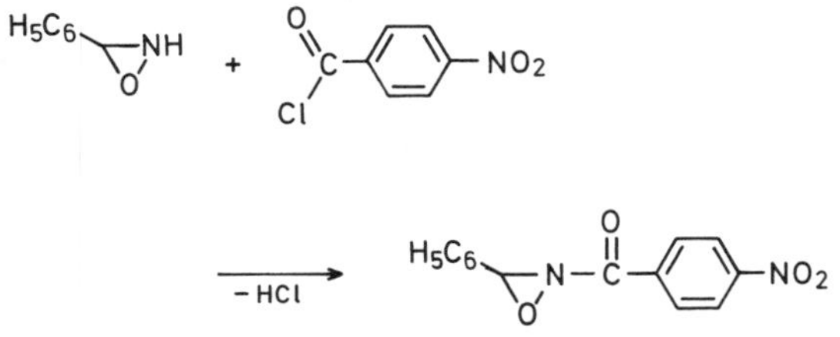

2-(4-Nitrobenzoyl)-3-phenyl-oxaziridine

Under analogous conditions 1-acylaziridines are obtained from aziridines[142, 143], *e.g.*:

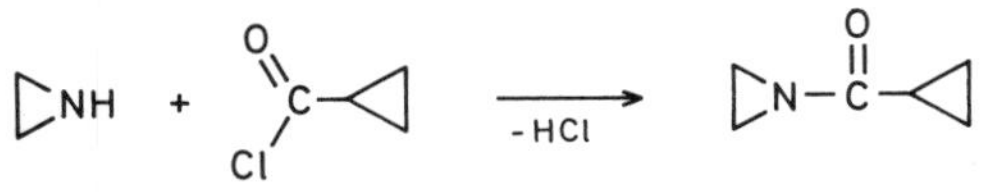

1-(Cyclopropyl-carbonyl)aziridine

Acylations in pyridine as auxiliary base and solvent proceed equally smoothly. Under these conditions less basic amines, too, react further rapidly with the intermediately formed, particularly active 1-acylpyridinium salts. Various *3-aminocoumarins* have been acylated in this way[56, 144]:

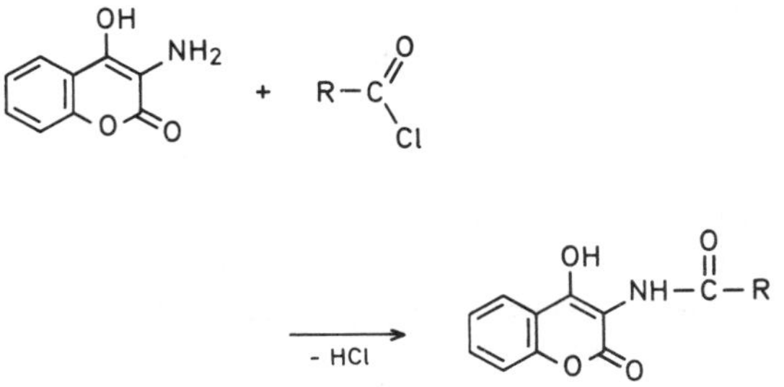

3-Acylamino-4-hydroxycoumarins

With formyl fluoride amines can be converted readily and in very good yield into *formamides* in ether/triethylamine[145, 146]; both aromatic and aliphatic amines undergo the reaction:

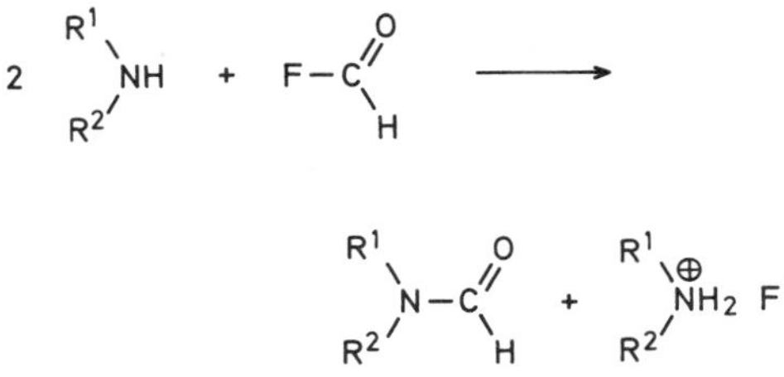

Chlorocarbonic acid derivatives and also vinylogous carboxylic acid chlorides[147] behave like normal acyl halides and correspondingly acylate primary and secondary amines under analogous conditions[148]. Heterocyclic amines such as pyrrole, imidazole, 1,2,4-triazole, or tetrazole, too, can be converted into *N-acyl heterocycles* with acyl halides[149]. The reaction is carried out in ab-

[137] E. Ronwin, J. Org. Chem. *18*, 127, 1546 (1953).
[138] S. Kushner, R.I. Cassell, J. Morton, II, J.-H. Williams, J. Org. Chem. *16*, 1283 (1951).
[139] A.R. Surrey, M.K. Rukwid, J. Amer. Chem. Soc. *77*, 3798 (1955).
[140] E. Schmitz, R. Ohme, S. Schramm, Tetrahedron Lett. 1857 (1965).
[141] E. Schmitz, S. Schramm, Chem. Ber. *100*, 2593 (1967).

[142] H.C. Brown, A. Tsukamoto, J. Amer. Chem. Soc. *83*, 2016 (1961).
[143] H.W. Heine, M.S. Kaplan, J. Org. Chem. *32*, 3069 (1967).
[144] L. Reppel, W. Schmollack, Arch. Pharm. (Weinheim, Ger.) *297*, 711 (1964).
[145] G. Oláh, S. Kuhn, Chem. Ber. *89*, 2211 (1956).
[146] G.A. Oláh, S.J. Kuhn, J. Amer. Chem. Soc. *82*, 2380 (1960).
[147] Z. Arnold, J. Zemlicka, Collect. Czech. Chem. Commun. *24*, 2378, 2385 (1959).
[148] M. Matzner, R.P. Kurkjy, R.J. Cotter, Chem. Rev. *64*, 645 (1964).
[149] H.A. Staab, Chem. Ber. *89*, 1927, 2088 (1956).

solute tetrahydrofuran with 1 molecule excess amine in order to trap the halogen hydride formed. Under these conditions the very reactive phosgene enables *1,1-carbonyl diimidazole* and *1,1'-carbonyl bis-4H-1,2,4-triazole* to be prepared very smoothly[150, 150a]:

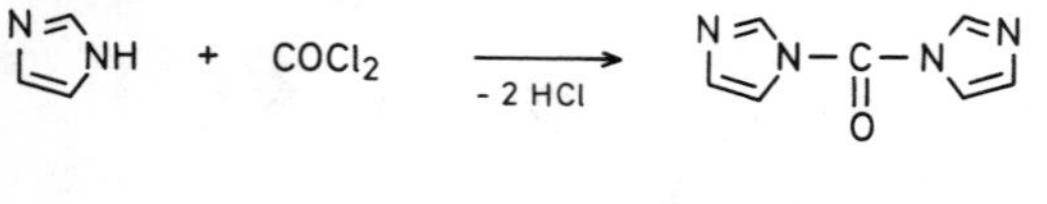

~ 90%

Urethans are formed from chlorocarbonic acid esters and amines including imidazoles and 1,2,4-triazoles[151], while carbamoyl chlorides afford substituted *ureas*[152]:

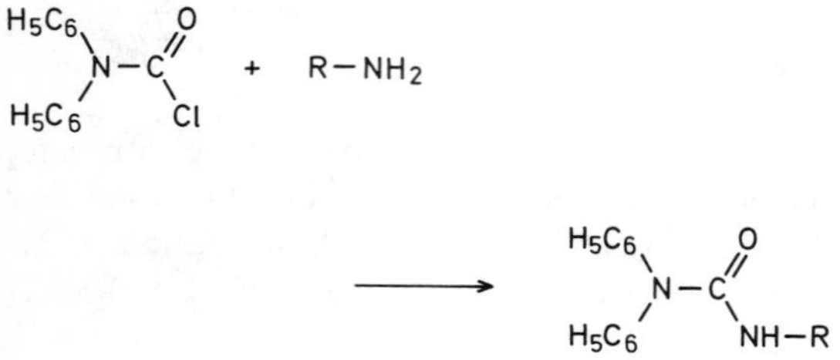

17.1.3.1.2 Acylations of hydrazines with acid chlorides

Acylation of hydrazines with acid chlorides in general ensues very vigorously; the initially formed *monoacylhydrazines* are easily acylated further to *N,N'-diacylhydrazines* on account of their basicity[153, 154]. Thus methylhydrazine and phenylacetyl chloride in equimolar amounts and with potassium carbonate as hydrogen chloride trap react exclusively to *1,2-bis(phenylacetyl)-1-methylhydrazine* and not to the monoacyl product[155]:

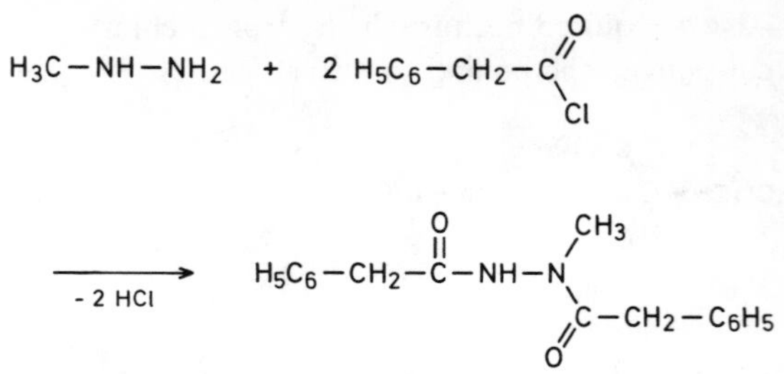

In the same way *dicarbamic acid diesters (1,2-hydrazinedicarboxylic acid diesters)* plus a little monoacylhydrazine are readily accessible by reacting chlorocarbonic acid ester with hydrazine[148]:

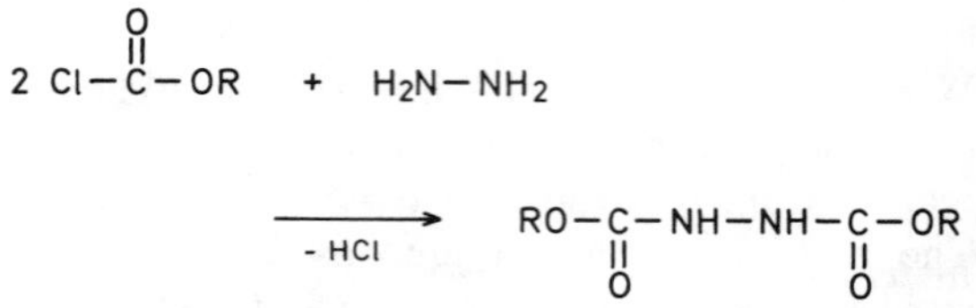

The monoacyl derivative is evidently readily soluble in the nonaqueous phase containing the acid chloride, so that it is acylated much more rapidly than the hydrazine present in the aqueous phase[156].

By contrast, monoacylation of hydrazine succeeds in satisfactory yield in chloroform as solvent[156]:

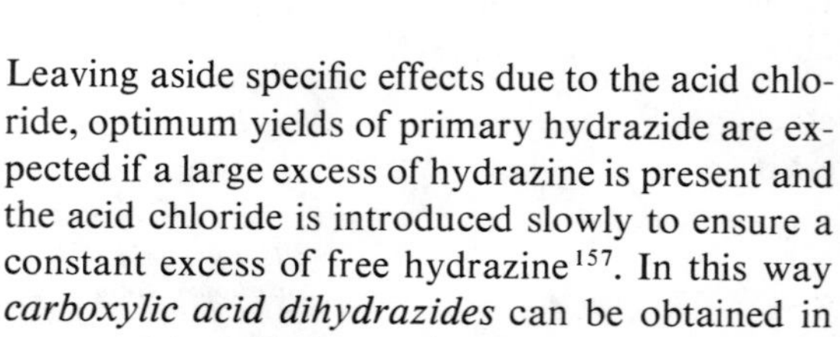

Benzyl carbazate; 60%

Leaving aside specific effects due to the acid chloride, optimum yields of primary hydrazide are expected if a large excess of hydrazine is present and the acid chloride is introduced slowly to ensure a constant excess of free hydrazine[157]. In this way *carboxylic acid dihydrazides* can be obtained in very good yield from phosgene and hydrazine hydrate[158]:

[150] *H.A. Staab*, Justus Liebigs Ann. Chem. *609*, 75 (1957).

[150a] *H.A. Staab, W. Rohr* in *W. Foerst*, Neuere Methoden der präparativen organischen Chemie, Bd. V, p. 53, Verlag Chemie, Weinheim/Bergstr. 1967.

[151] *H.A. Staab*, Justus Liebigs Ann. Chem. *609*, 83 (1957).

[152] *F.L. Scott, M.T. Scott*, J. Amer. Chem. Soc. *79*, 6077 (1957).

[153] *A. Kreuzberger*, J. Org. Chem. *22*, 679 (1957).

[154] *N.O.V. Sonntag*, J. Amer. Oil Chemist's Soc. *45*, 571 (1968).

[155] *W.J. Theuer, J.A. Moore*, J. Org. Chem. *29*, 3734 (1964).

[156] *H. Böshagen, J. Ullrich*, Chem. Ber. *92*, 1478 (1959).

[157] *C. Naegli, G. Stefanovitsch*, Helv. Chim. Acta *11*, 621 (1928).

[158] *T. Lieser, G. Nischk*, Chem. Ber. *82*, 527 (1949).

COCl$_2$ + 2 H$_2$N—NH$_2$

$$\xrightarrow{-\,2\,HCl} \quad H_2N-NH-\underset{\underset{O}{\|}}{C}-NH-NH_2$$

With *N,N*-diarylsubstituted carbamoyl chlorides the reaction leads to *4,4-diaryl*substituted *semicarbazides*

$$\underset{R^1}{\overset{R^2}{\diagdown}}N-\underset{\underset{O}{\|}}{C}-Cl \quad + \quad H_2N-NH_2$$

$$\xrightarrow{-\,HCl} \quad \underset{R^1}{\overset{R^2}{\diagdown}}N-\underset{\underset{O}{\|}}{C}-NH-NH_2$$

R^1 = R^2 = Aryl

while aliphatically substituted carbamoyl chlorides afford only *dialkylbiureas (dialkyl-1,2-hydrazinecarboxamides)* even when used in equimolar proportions[159]:

$$2 \quad \underset{R}{\overset{R}{\diagdown}}N-COCl \quad + \quad H_2N-NH_2$$

$$\longrightarrow \quad \underset{R}{\overset{R}{\diagdown}}N-\underset{\underset{O}{\|}}{C}-NH-NH-\underset{\underset{O}{\|}}{C}-N\underset{R}{\overset{R}{\diagup}}$$

R = Alkyl

In concentrated solution *N,N*-diphenylcarbamoyl chloride, too, reacts to form *diphenylbiurea*. Where a phenyl group has been replaced by the 2-naphthyl group the reaction stops at the semicarbazide stage [giving *4-(2-naphthyl)-4-phenylsemicarbazide*][159].

Unsymmetrically dialkylated hydrazines are readily monoacylated in a nonaqueous solvent such as ether, chloroform, or dichloromethane with triethylamine as auxiliary base[160]:

$$R-\underset{\underset{Cl}{}}{\overset{\overset{O}{\|}}{C}} \quad + \quad H_2N-N\underset{R}{\overset{R}{\diagup}} \quad \xrightarrow{-\,HCl} \quad \underset{R}{\overset{\overset{O}{\|}}{C}}-NH-N\underset{R}{\overset{R}{\diagup}}$$

[159] *W. Ried, H. Hillenbrand, G. Oertel*, Justus Liebigs Ann. Chem. *590*, 123 (1954).
[160] *W. Walter, K.-J. Reubke*, Chem. Ber. *103*, 2197 (1970).

Sodium carbonate[161] or an excess of the hydrazine[162] that is to be acylated are alternative auxiliary bases. The attainable yields are good to very good throughout. Dicarboxylic acid chlorides such as sebacoyl chloride react to *sebacic acid bis-2,2-dimethylhydrazide* with 1,1-dimethylhydrazine[163]. Phosgene reacts with hydrazones to form *carbonic acid dihydrazides* when certain reaction conditions are maintained[164], *e.g.*:

$$2 \quad \underset{H_5C_6}{\overset{H_5C_6}{\diagdown}}C=N-NH_2 \quad + \quad COCl_2 \quad \longrightarrow$$

$$\underset{H_5C_6}{\overset{H_5C_6}{\diagdown}}C=N-NH-\underset{\underset{O}{\|}}{C}-NH-N=C\underset{C_6H_5}{\overset{C_6H_5}{\diagup}}$$

Carbonic acid bis[2-diphenylmethylene-hydrazide]; 94%

Symmetrically substituted hydrazines can serve for preparing *1,2-diazetidine-3,4-diones* if the acylation is performed with oxalyl chloride[165]:

$$\underset{R-NH}{\overset{R-NH}{\underset{|}{}}} \quad + \quad \underset{Cl}{\overset{Cl}{}}\underset{\underset{C}{\|}}{\overset{\overset{C}{\diagup\!\!\!=O}}{}} \quad \xrightarrow{-\,2\,HCl} \quad \text{[1,2-diazetidine-3,4-dione]}$$

R = CH(CH$_3$)$_2$; *1,2-Diisopropyl-1,2-diazetidine-3,4-dione; 20%*
R = C(CH$_3$)$_3$; *1,2-Di-tert-butyl-1,2-diazetidine-3,4-dione; 58%*

17.1.3.1.3 Acylations of hydroxylamines and derivatives with acid chlorides

Hydroxylamines can be acylated by the same methods used for amines. In general the acylation on the nitrogen proceeds first and only then that on the oxygen. Good yields of *hydroxamic acids* are obtained by acylating hydroxylamine by the *Einhorn* technique with triethylamine in an organic solvent[166, 167]

$$\text{[2-methoxybenzoyl chloride]} \quad + \quad H_2N-OH$$

$$\xrightarrow{-\,HCl} \quad \text{[2-methoxybenzohydroxamic acid]}$$

2-Methoxybenzhydroxamic acid; 77%

or in pyridine[168]. *N*-Substituted hydroxylamines can be readily acylated by the *Schotten-Baumann* variant (using aqueous alkali metal carbonate solution as auxiliary base)[168, 169], but *O*-acyl derivatives may be coformed[170]. A stepwise reaction on the nitrogen and oxygen enables different acyl groups to be introduced[170, 171]:

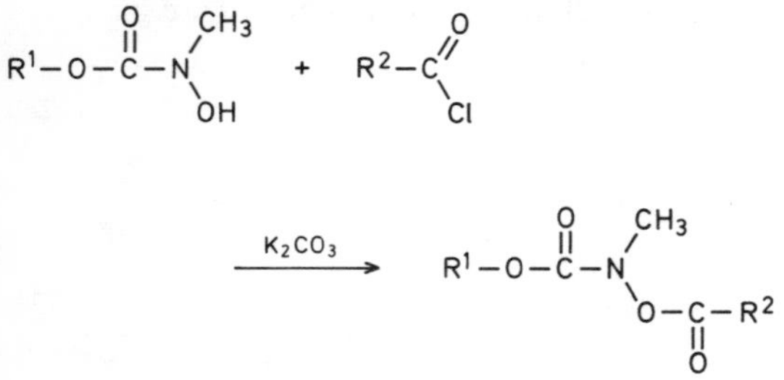

Phthaloyl chloride reacts with formation of *alkyl 1,4-dioxo-2 H-1,2-benzoxazine-3-carboxylate* (~70%)[172]:

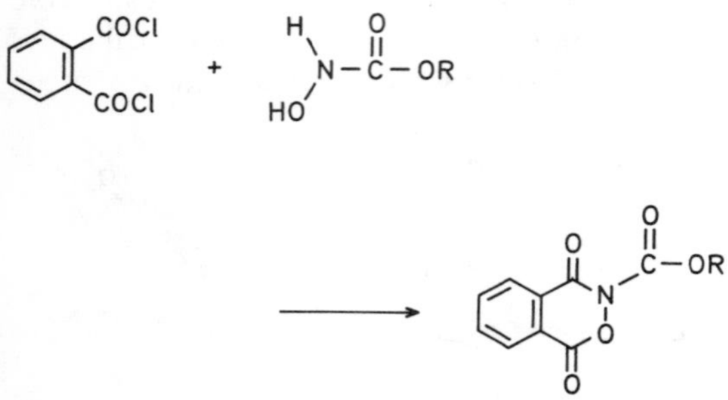

Bulky substituents or polar effects by the group *R* effect the simultaneous or even almost exclusive formation of the *O*-acyl derivative with *N*-substi-

tuted hydroxylamines. This was demonstrated by the reaction between *N-tert*-butylhydroxylamine and 4-nitrobenzoyl chloride[173, 174].

Under Einhorn conditions *O*-alkylhydroxamic acids also are readily very readily acylated. Aliphatic *O*-alkylhydroxamic acids afford only one product *(N-acetyl-N-acyl-O-alkylhydroxylamine)* **7**, while *O*-alkylarylhydroxamic acids in addition form the *N-(α-acetyloxyalkylidene)-O-alkylhydroxylamine* **8**[175].

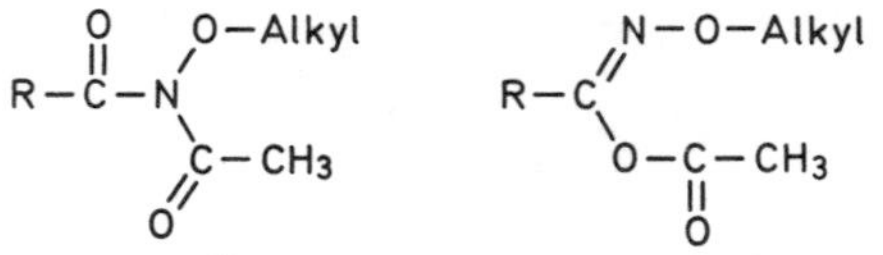

The ratio of the two isomers is dependent on electronic and steric effects. Thus *O*-benzocinnamohydroxamic acid furnishes *N-(1-acetoxy-3-phenylallylidene)-O-benzylhydroxylamine* as main product, a result which appears to be a general one for conjugated *O*-benzylhydroxamic acids[175]:

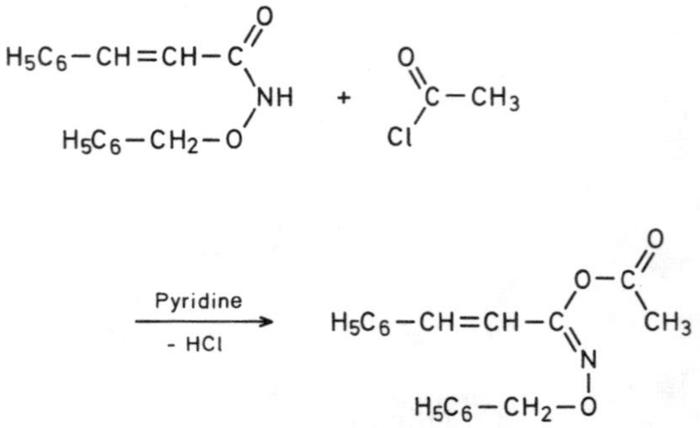

Triacylhydroxylamines are prepared mainly by acylating metal salts of *O*-acylhydroxamic acids. Here the influence of the metal cation governs the direction that the acylation takes[175].

17.1.3.2 Acylations of *C=N* double bonds with acid chlorides

The reaction between acyl halides and aldehyde imines (azomethines) enables substituted *β-lactams* (azetidinones) to be prepared[176, 176a]:

[161] *P. Hope, L.A. Wiles*, J. Chem. Soc. C 2636 (1967).
[162] *R.F. Meyer, B.L. Cumming*, J. Heterocycl. Chem. *1*, 186 (1964).
[163] *W.J. McKillip, L.M. Clemens, R. Haugland*, Can. J. Chem. *45*, 2613 (1967).
[164] *P.W. West, J. Warkentin*, J. Org. Chem. *33*, 2089 (1968).
[165] *J.C. Stowell*, J. Org. Chem. *32*, 2360 (1967).
[166] *M.A. Stollberg, W.A. Mosher, T. Wagner-Jauregg*, J. Amer. Chem. Soc. *79*, 2615 (1957).
[167] *R.E. Plapinger*, J. Org. Chem. *24*, 802 (1959).
[168] *H. Ulrich, A.A.R. Sayigh*, J. Chem. Soc. (London) 1098 (1963).
[169] *D.C. Bhura, S.G. Tandon*, J Chem. Eng. Data 106 (1971).
[170] *L. Horner, H. Steppan*, Justus Liebigs Ann. Chem. *606*, 24 (1957).
[171] *G. Zinner*, Arch. Pharm. (Weinheim, Ger.) *292*, 329 (1959).
[172] *G. Zinner, V. Ruthe, M. Hitze, R. Vollrath*, Synthesis 148 (1971).

[173] *O. Exner, J. Holubek*, Collect. Czech. Chem. Commun. *30*, 940 (1965).
[174] *O. Exner, B. Kakác*, Collect. Czech. Chem. Commun. *25*, 2530 (1960).
[175] *M.T.W. Hearn, A.D. Ward*, Australian J. Chem. *22*, 161 (1969).
[176] *J.C. Sheehan*, Org. Reactions *9*, 398 (1957).

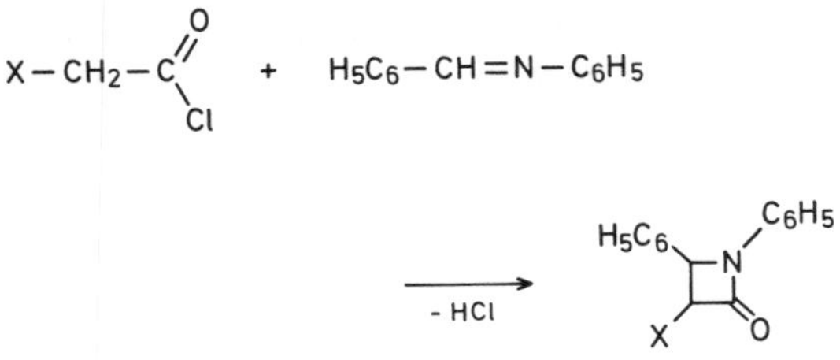

X = Phthalimido; *β,N-Diphenyl-α-phthalimido-β-lactam*

The reaction ensues rapidly in an inert solvent using triethylamine as base and gives good to quantitative yields[177]. 4,5-Dihydrothiazoles[178] or phenylmethanephosphonic acid ester imides[179] serve as alternative azomethines[178, 179]. Where the C=N double bond of the azomethines is part of a conjugate system 1,4-cycloadducts form, e.g.[177]:

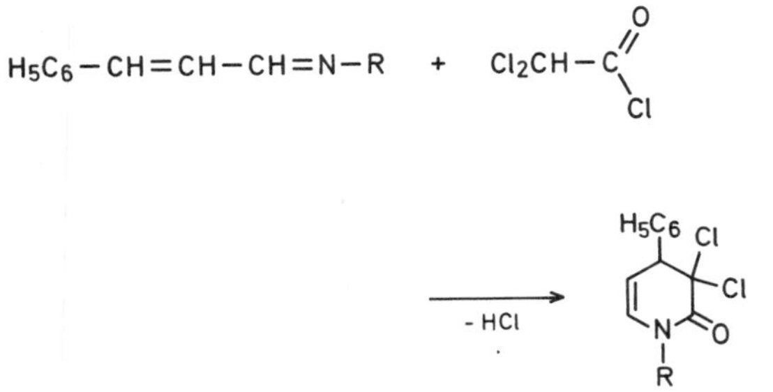

N-Substituted 3,3-dichloro-4-phenyl-3,4-dihydro-2-pyridinones

Imidocarboxylic acid esters behave analogously to azomethines under these conditions and afford 4-ethoxy-2-azetidinones[180]. 2,3-Diphenylazirine adds benzoyl chloride quantitatively in benzene solution to form *N-benzoyl-2-chloro-2,3-diphenylaziridine*[181].

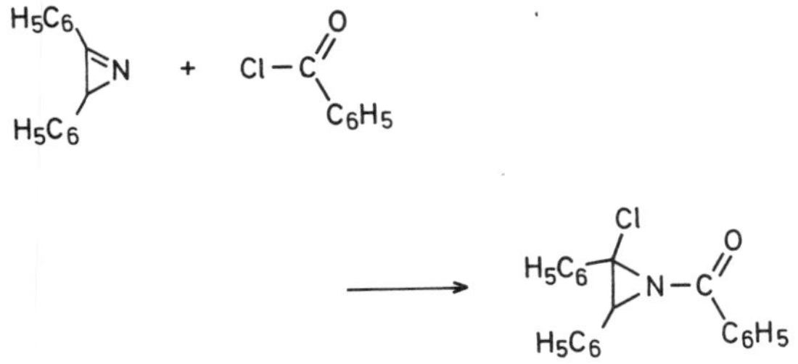

Acetyl chloride reacts with aliphatic carbodiimides in an inert solvent to form the isolatable *N-acetylchloroformamidines*[182]:

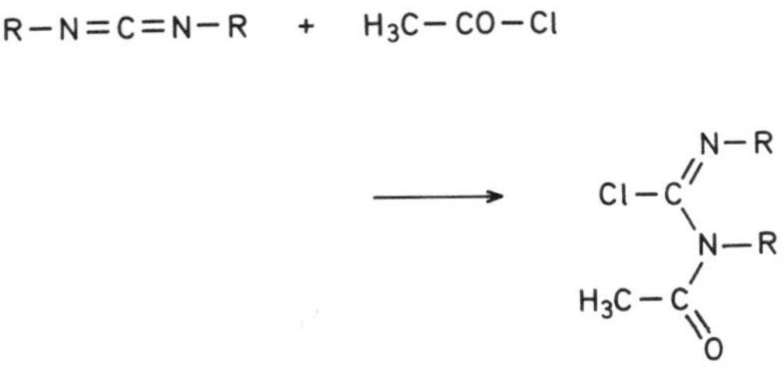

By contrast, the reaction with dicarboxylic acid dihalides takes a different course leading to cyclic products or ureides[183]. Thus, *2,2-dichloroimidazolidine-4,5-diones* are obtained with oxalyl chloride, and *barbituric acid* derivatives with malonyl chloride:

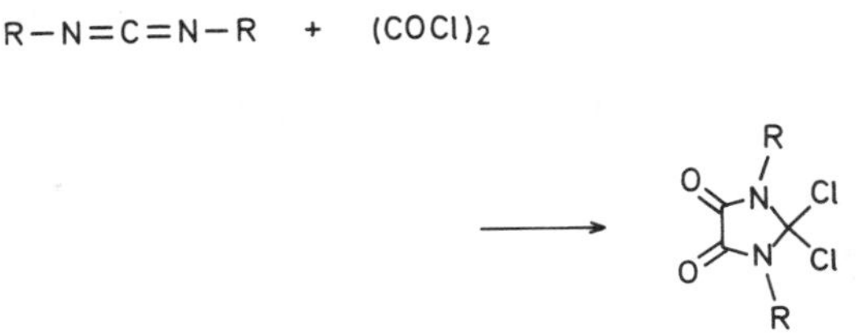

17.1.3.3 Acylations of carboxamides with acid chlorides

The slight nucleophilism of the acylamides means that their acylation succeeds best with acyl chlorides in pyridine; lactams are acylated in good yield in this way[184], e.g.:

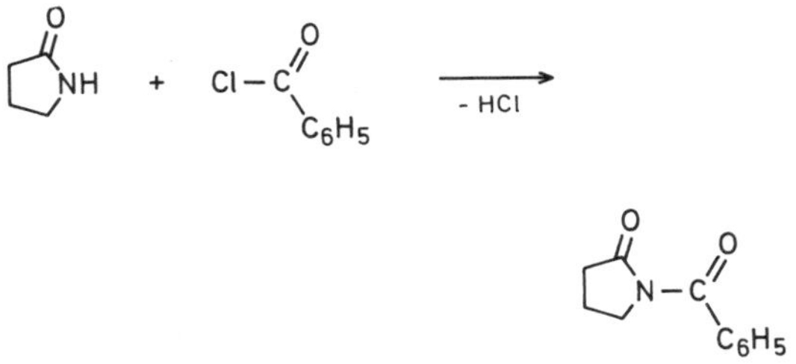

N-Benzoyl-2-pyrrolidinone; 75%

[176a] A.K. Bose, H.P.S. Chawla, B. Dayal, M.S. Manhas, Tetrahedron Lett. 2503 (1973).
[177] F. Duran, L. Ghosez, Tetrahedron Lett. 245 (1970).
[178] L. Paul, P. Polczynski, G. Hilgetag, Chem. Ber. *100*, 2761 (1967).
[179] L. Paul, K. Zieloff, Chem. Ber. *99*, 1431 (1966).
[180] L. Paul, A. Draeger, G. Hilgetag, Chem. Ber. *99*, 1957 (1966).
[181] F.W. Fowler, A. Hassner, J. Amer. Chem. Soc. *90*, 2875 (1968).
[182] K. Hartke, E. Palou, Chem. Ber. *99*, 3155 (1966).
[183] G. Zinner, R. Vollrath, Chem. Ber. *103*, 766 (1970).
[184] F. Korte, H.-J. Schulze-Steinen, Chem. Ber. *95*, 2444 (1962).

The reactions between hexahydro-*2 H*-azepin-2-one (*ε*-caprolactam) with aliphatic and aromatic acyl halides proceed equally smoothly[185]. In few cases only is the sodium salt of the amide reacted with an acyl halide[185]. *Dimethylamides* can be obtained from the reaction between acyl chlorides and dimethylformamide in good yield[186], *e.g.*:

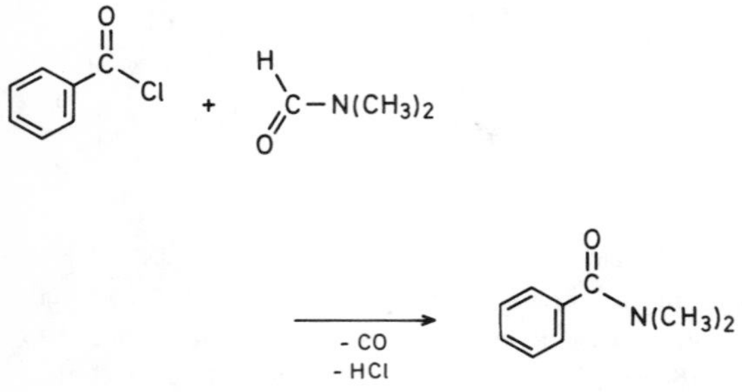

N,N-Dimethyl-benzamide; 97%

The reaction cannot be transferred to all desired acyl halides[187].

Using the system pyridine/acyl halide it is possible to acylate amides to form triacylamines even at low temperature[188]:

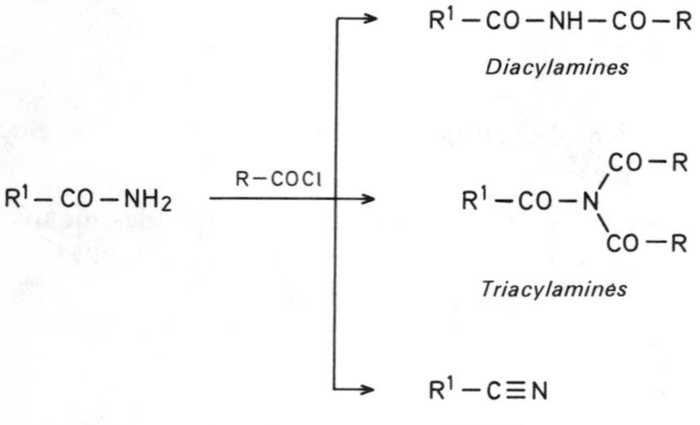

$R^1-CO-NH-CO-R$

Diacylamines

$R^1-CO-N\begin{smallmatrix}CO-R\\CO-R\end{smallmatrix}$

Triacylamines

$R^1-C\equiv N$

Nitriles

In general, aromatic acyl halides furnish *triacylamines*, aliphatic ones only *diarylamines*, while nitrobenzoyl chlorides give *nitriles*[188]. The base used has a substantial influence on the reaction[189]. Thus, with 2,6-dimethylpyridine (2,6-DMP) as auxiliary base, amides can be acylated also with aliphatic acyl halides to form triacylamides; these are not formed in the presence of pyridine, *e.g.*:

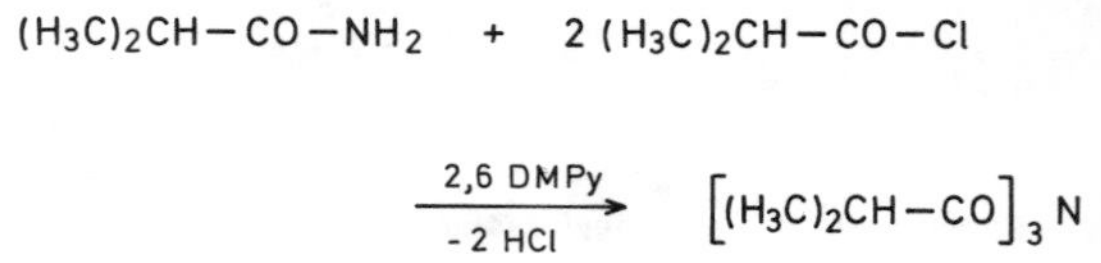

$(H_3C)_2CH-CO-NH_2 \quad + \quad 2\,(H_3C)_2CH-CO-Cl$

Triisobutyramine; 75%

Unsaturated acyl halides, too, furnish a good yield of triacylamine[189].

Reacting acetyl chloride with silver or sodium diformamide and *N*-(trimethylsilyl)diformamide leads to *N,N-diformylacetamide:*

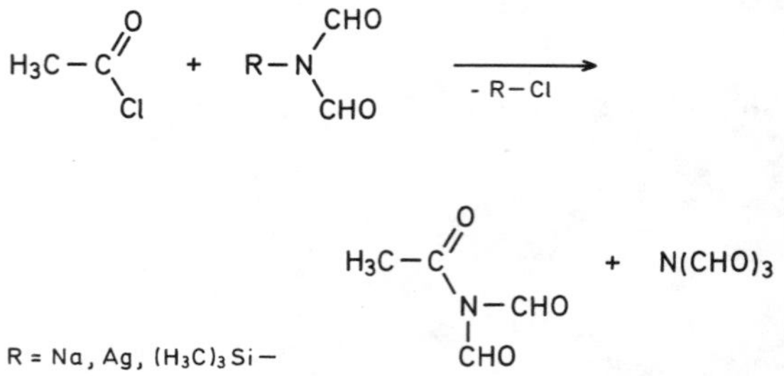

$R = Na, Ag, (H_3C)_3Si-$

Except in the first-named reaction triformamide is formed as by-product[190].

Both triacylamines are very strong formylating agents, so that triformamide can be obtained also by formylating diformamide with *N,N*-diformylacetamide[190].

Urethans are readily acylated to *N-acylurethans* with aliphatic and aromatic acyl halides[191]. No auxiliary base is required and heating the two components until liberation of hydrogen chloride is complete is sufficient:

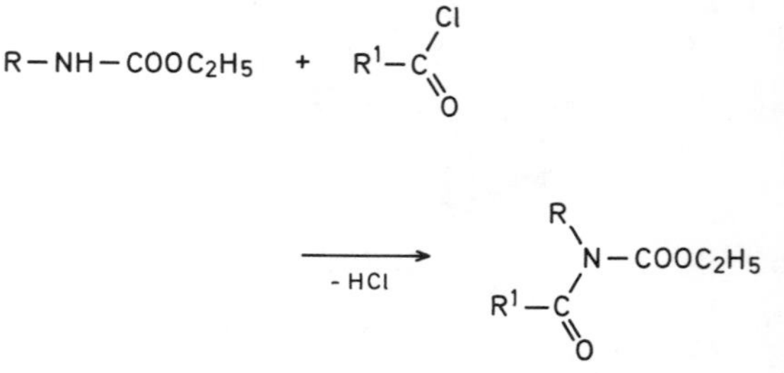

$R-NH-COOC_2H_5 \quad + \quad R^1-C\begin{smallmatrix}Cl\\ \\O\end{smallmatrix}$

[185] *L.G. Donaruma, R.P. Scelia, S.E. Schonfeld*, J. Heterocycl. Chem. *1*, 48 (1964).

[186] *G.M. Coppinger*, J. Amer. Chem. Soc. *76*, 1372 (1954).

[187] *I.L. Knunyants, Y.A. Cheburkov, Y.E. Aranov*, Izv. Akad. Nauk SSSR, Ser. Khim. 1038 (1966); C.A. *65*, 10491 g (1966).

[188] *Q.E. Thompson*, J. Amer. Chem. Soc. *73*, 5841 (1951).

[189] *R.T. LaLonde, C.B. Davis*, J. Org. Chem. *35*, 771 (1970).

[190] *E. Allenstein, V. Beyl, W. Eitel*, Chem. Ber. *102*, 4089 (1969).

[191] *D. Ben-Ishai, E. Katchalski*, J. Org. Chem. *16*, 1025 (1951).

Oxalyl chloride reacts with ethylurethan to form *diethyl N,N'-oxalyldicarbamate*, while phthaloyl chloride affords *ethyl 2-isoindolecarboxylate*[192]:

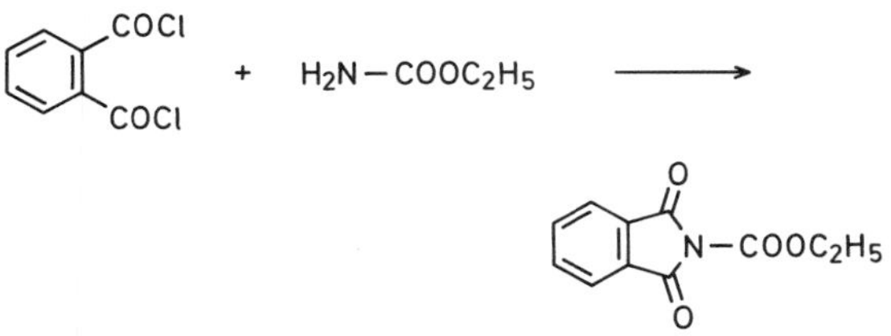

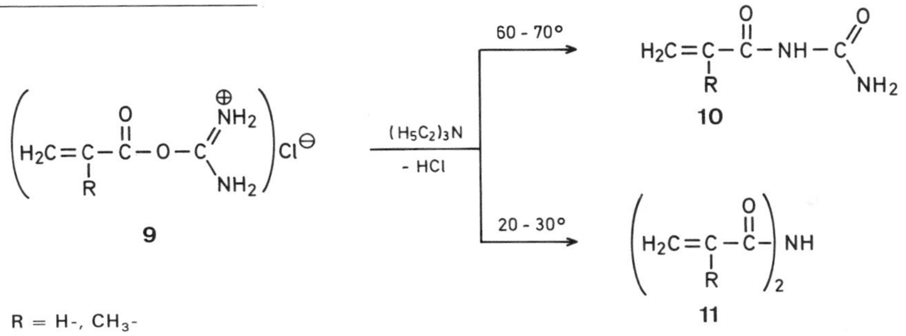

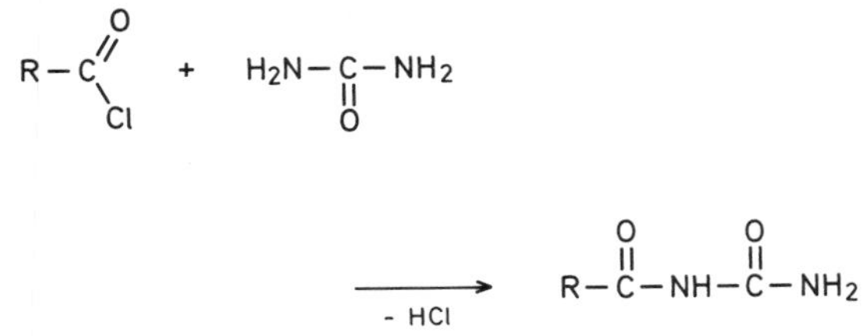

It is not essential to work in solution, because heating the two components together also affords the ureides[194]. *Acryloylurea* is prepared by dehydrohalogenation of adduct **9** of acryloyl chloride and urea with triethylamine or sodium carbonate[195]. In dependence on the reaction temperature either *acyloylurea* or *methacroylurea* **10**, or diacrylamide or dimethacrylamide **11** respectively are obtained as main product

R = H-, CH₃-

By contrast, acetyl bromide reacts with urethans to form acetamides while giving up alkyl bromide[191]:

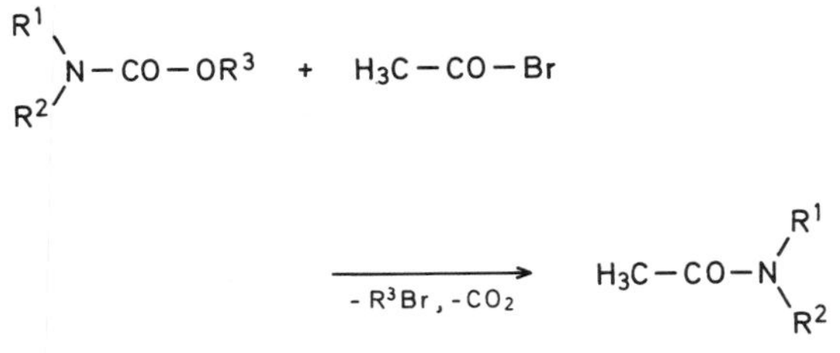

17.1.3.4 Acylations of ureas with acid chlorides

Many aliphatic acid chlorides acylate ureas in benzene solution smoothly to *monoacylureas* under the catalytic action of sulfuric acid[193]:

Unlike during the reaction of urea with acryloyl chloride increasing peracylation to *tris(methacrylamide)* is observed as progressively more methacryloyl chloride is made available. Intramolecular cyclization of allophanyl chlorides accompanied by loss of hydrogen chloride affords 1,3-disubstituted *1,3-diazetidine-2,4-diones* in very good yield[196]:

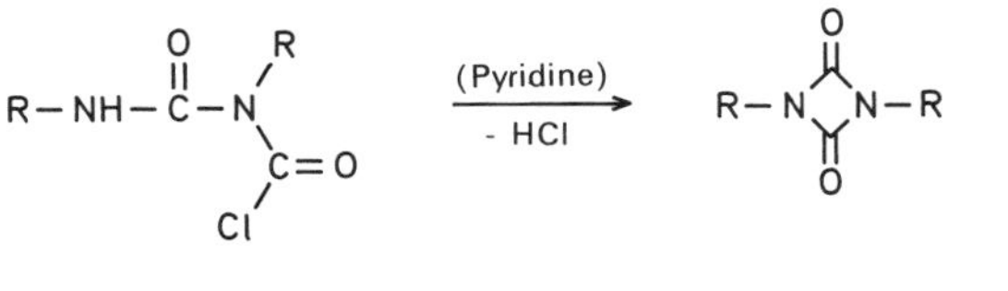

Phthaloyl chloride reacts with ureas to form *1,3-dioxo-2-isoindolecarboxamides:*

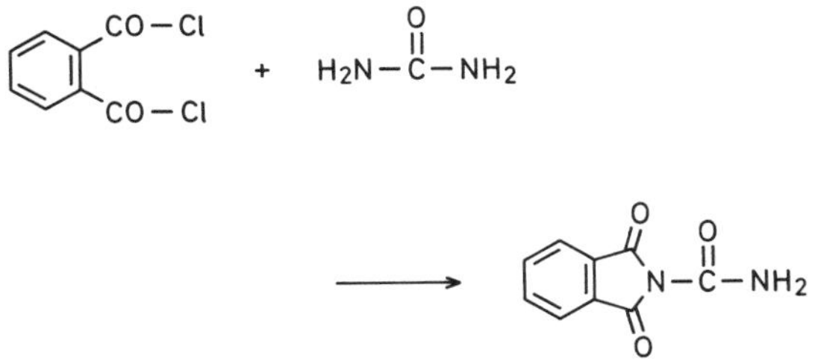

[192] P. Adams, F.A. Baron, Chem. Rev. *65*, 567 (1965).
[193] R.W. Stoughton, J. Org. Chem. *2*, 514 (1938).
[194] I.A. Pearl, W.M. Dehn, J. Amer. Chem. Soc. *61*, 1377 (1939).
[195] F. Merger, Chem. Ber. *101*, 2419 (1968).

[196] H. Helfert, E. Fahr, Angew. Chem. *82*, 362 (1970).

With N, N'-disubstituted ureas cleavage of the urea component ensues with formation of N-substituted phthalimides[197]. By contrast, succinyl chloride furnishes cyclic amides with a 7-membered ring structure [*5,6-dihydro-1 H-1,3-diazepine-2,4,7-(3 H)triones*] in over 90% yield:

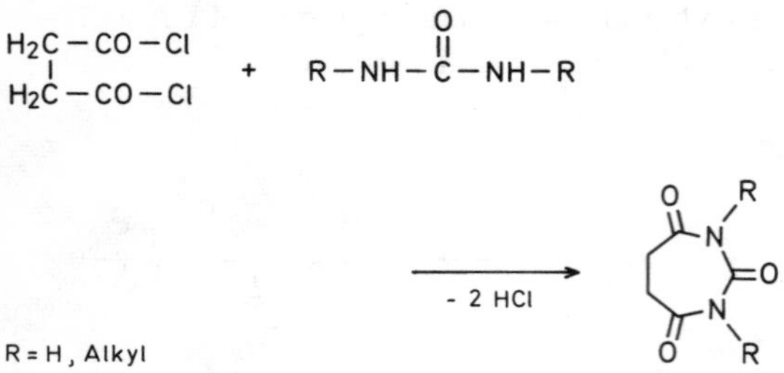

R = H, Alkyl

With R = aryl, cycloalkyl the reaction once more proceeds with formation of imides[183, 197].

With excess acyl chloride ureides react quite generally to N, N'-*diacylureas*. To prepare *mixed* N, N'-diacylureas it is advantageous to first prepare the ureide with the longer R group and then to react this compound with the lower acyl halide[193]:

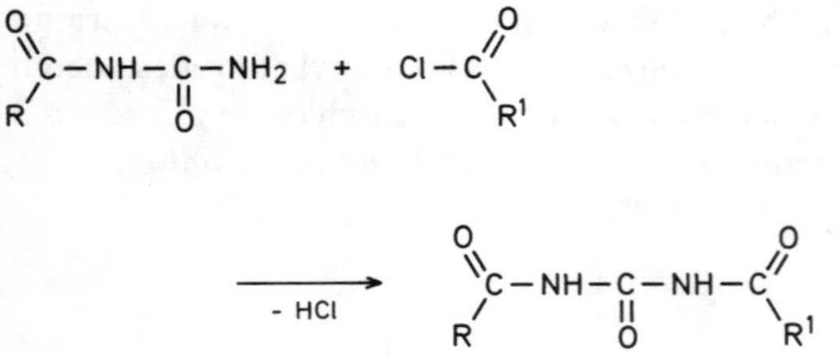

17.1.3.5   Acylations of acylhydrazines with acid chlorides

Acyl halides readily convert monoacylhydrazines into N, N'-diacylhydrazines[198, 199]. The reaction is performed by using either the Schotten-Baumann or Einhorn technique[200] and affords high yields in both cases:

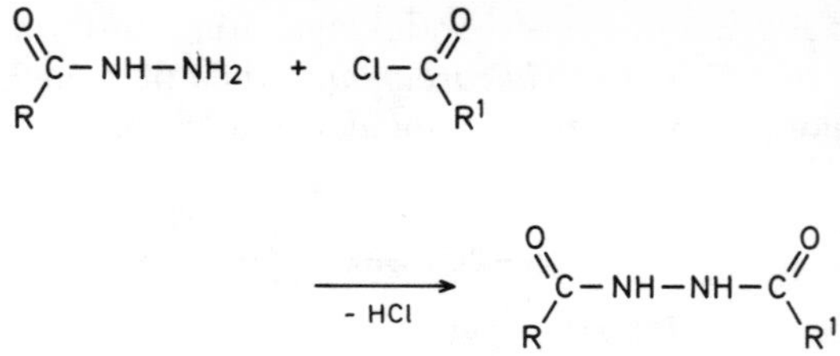

An interesting way of carrying out this reaction is to allow disulfur dichloride to act on monoacyl-hydrazines. The latter are converted into acyl chlorides while losing nitrogen and afford a good yield of the N, N'-*diacylhydrazine* with as yet unreacted acylhydrazine[201]:

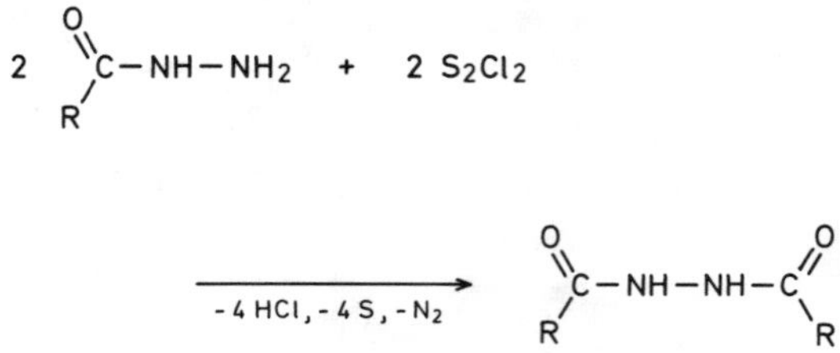

Phosgene reacts with monoacylhydrazines to form *1,3,4-oxadiazol-2-one*[162, 202, 136a] by intramolecular cyclization:

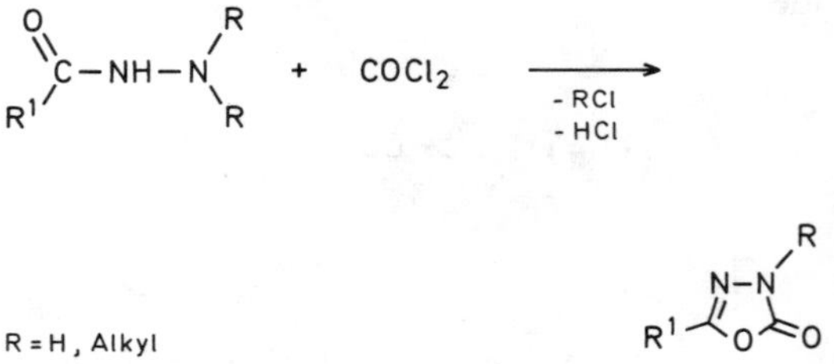

R = H, Alkyl

With additional acyl halide N, N'-diacyl hydrazides can be converted into *triacyl* and *tetraacyl hydrazines*[199]. The Einhorn technique is once more used, but, alternatively, the sodium or mercury(II) salt[203], of the hydrazide can be subjected to the acylation, *e.g.*:

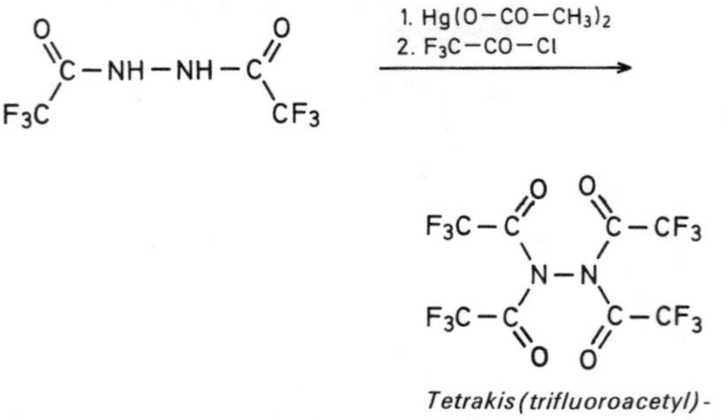

Tetrakis (trifluoroacetyl)-hydrazine

[197] *G. Losse, E. Wottgen, H. Just*, J. Prakt. Chem. *7*, 28 (1958).

[198] *P.A.S. Smith*, Org. Reactions *3*, 366 (1946).

[199] *E. Müller* in *Houben-Weyl*, Methoden der organischen Chemie, 4. Aufl., Bd. X/2, p. 127, Georg Thieme Verlag, Stuttgart 1967.

[200] *A. Winterstein, B. Hegedus, B. Fust, E. Böhni, A. Studer*, Helv. Chim. Acta *39*, 229 (1956).

[201] *B. Hope, L.A. Wiles*, J. Chem. Soc. (London) 5837 (1964).

[202] *A. Dornow, K. Bruncken*, Chem. Ber. *82*, 121 (1949).

[203] *P.A.S. Smith*, Org. Reactions *3*, 373 (1946).

17.1.3.6 Acylations of sodium azide with acid chlorides

Carboxylic acid azides can be prepared by acylating sodium azide with acyl halides in anhydrous medium or aqueous solution[203, 204]:

$$R-CO-Cl \;+\; NaN_3 \;\xrightarrow[-NaCl]{}\; R-CO-N_3$$

Working under anhydrous conditions, *i.e.*, in an inert solvent, is recommended for reactive acyl halides and unstable azides. The reactions with activated sodium azide proceed especially well[203].

17.1.4 Acylations with acid anhydrides

17.1.4.1 With carboxylic acid anhydrides

17.1.4.1.1 Acylations of amines

Like acid chlorides, acid anhydrides are in general very strong acylating agents and convert amines into amides smoothly[205, 206]:

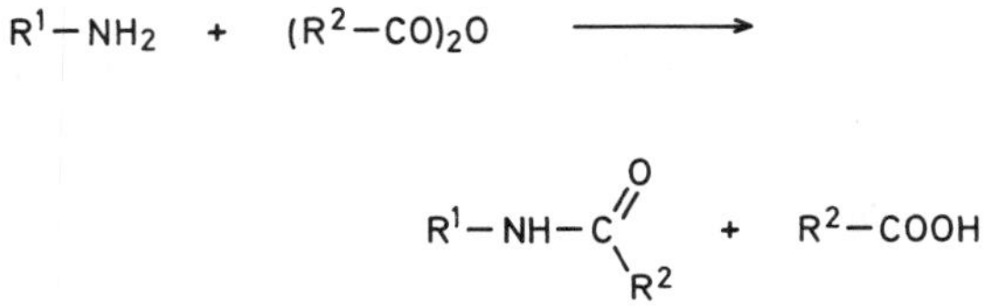

$$R^1-NH_2 \;+\; (R^2-CO)_2O \;\longrightarrow$$

$$R^1-NH-\overset{\displaystyle O}{\underset{\displaystyle R^2}{C}} \;+\; R^2-COOH$$

According to the reaction equation only half the anhydride is consumed for amide formation. Weakly basic amines, too, are acylated; kinetic investigations reveal that acylations are catalyzed by acids[207]. Both acylating carboxylic acid or mineral acid (sulfuric or perchloric acid) can serve as the catalyst, acylation of aromatic amines is accelerated especially markedly. Excess acetic anhydride diacylates the amine in very good yield; in lower aliphatic amines this reaction ensues very easily[208, 209] (brief boiling of the amines in acetic anhydride), *e.g.:*

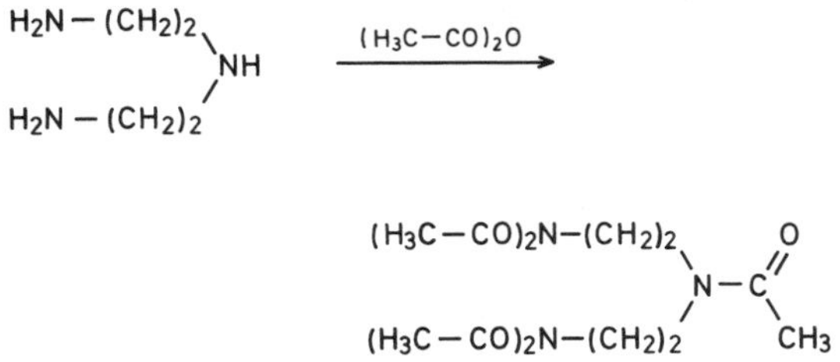

[204] *H. Henecka, P. Kurtz* in *Houben-Weyl,* Methoden der organischen Chemie, 4. Aufl., Bd. VIII, p. 683, Georg Thieme Verlag, Stuttgart 1952.
[205] *H. Henecka, P. Kurtz* in *Houben-Weyl,* Methoden der organischen Chemie, 4. Aufl., Bd. VIII, p. 655, Georg Thieme Verlag, Stuttgart 1952.
[206] *F. Möller* in *Houben-Weyl,* Methoden der organischen Chemie, 4. Aufl., Bd. XI/2, p. 14, Georg Thieme Verlag, Stuttgart 1958.
[207] *P.J. Lillford, D.P.N. Satchell,* J. Chem. Soc. B, 360 (1967).
[208] *C. Nolde, S.-O. Lawesson, J.H. Bowie, R.G. Cooks,* Tetrahedron **24,** 1051 (1968).
[209] *M.B. Frankel, C.H. Tieman, C.R. Vanneman, M.H. Gold,* J. Org. Chem. **25,** 744 (1960).

$$H_2N-(CH_2)_2 \diagdown \atop H_2N-(CH_2)_2 \diagup NH \;\xrightarrow{(H_3C-CO)_2O}$$

$$(H_3C-CO)_2N-(CH_2)_2 \diagdown \atop (H_3C-CO)_2N-(CH_2)_2 \diagup N-\overset{\displaystyle O}{\underset{\displaystyle CH_3}{C}}$$

N,N'-[(Acetylimino)diethylene]-bisdiacetamide

It is not always necessary to employ the free amines; amine hydrochlorides also can be acylated[208]. Comparative acylations of primary amines of different structure show that primary amines of the type $R-CH_2-NH_2$ afford only N,N-diacetylamines *[R—CH$_2$—N(CO—CH$_3$)$_2$]* in good yield. Primary amines of the type

$$R^1 \diagdown \atop R \diagup CH-NH_2$$

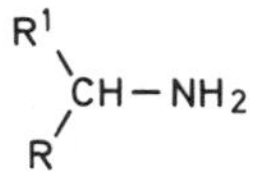

afford a mixture of N-acetylamine and N,N-diacetylamine whose relative proportion is dependent on the nature of the groups R and R^1. With amines of the type

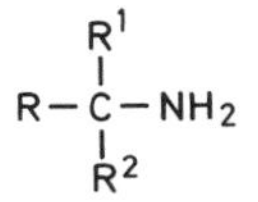

$$R-\underset{\displaystyle R^2}{\overset{\displaystyle R^1}{\underset{\displaystyle |}{\overset{\displaystyle |}{C}}}}-NH_2$$

solely monoacylation takes place. Thus the size of R, R^1, and R^2 governs the predominance of mono- and diacetylation (see Table 1)[210].

Table 1. Acylations with acid anhydrides

	[%] Yield of		
	N-Acetylamine		N,N-Diacetylamine
Cyclopropylamine	0		69 *Cyclopropyldiacetylamine*
Cyclobutylamine	0		77 *Cyclobutyldiacetylamine*
Cyclopentylamine	74 *Acetylcyclopentylamine*		23 *Cyclopentyldiacetylamine*
Propylamine	0		65 *Diacetylpropylamine*
Butylamine	0		90 *Butyldiacetylamine*
Benzylamine	0		70 *Benzyldiacetylamine*
Isopropylamine	15 *Acetylisopropylamine*		65 *Diacetylisopropylamine*
Isobutylamine	0		70 *Diacetylisobutylamine*
*sec-*Butylamine	35 *Acetyl-sec-butylamine*		40 *Diacetyl-sec-butylamine*
*tert-*Butylamine	60 *Acetyl-tert-butylamine*		0

Aromatic amines are likewise rapidly diacylated in boiling anhydrides of carboxylic acids[211].

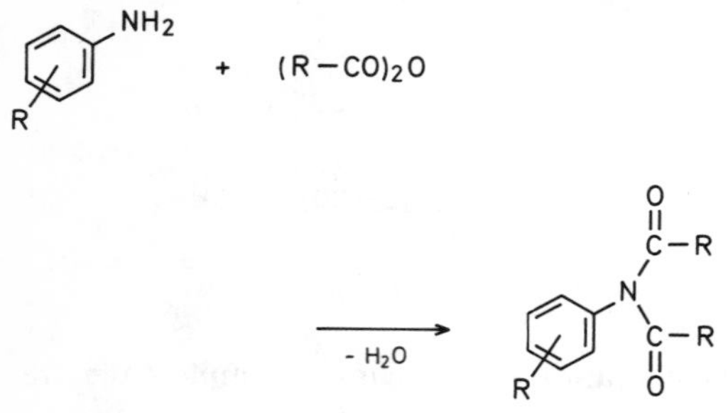

Less basic heteroaromatic compounds such as imidazole and purine derivatives, too, react with excess acetic anhydride[212].

In general, acylations, and especially acetylations, are conducted without solvent, because the anhydride itself or the acid liberated during the reaction acts as solvent. Where, for particular reasons, the acylation is to proceed in a solvent the yield of amide need not suffer as acylations of aromatic amines with trifluoroacetic anhydride in benzene demonstrate[213], e.g.:

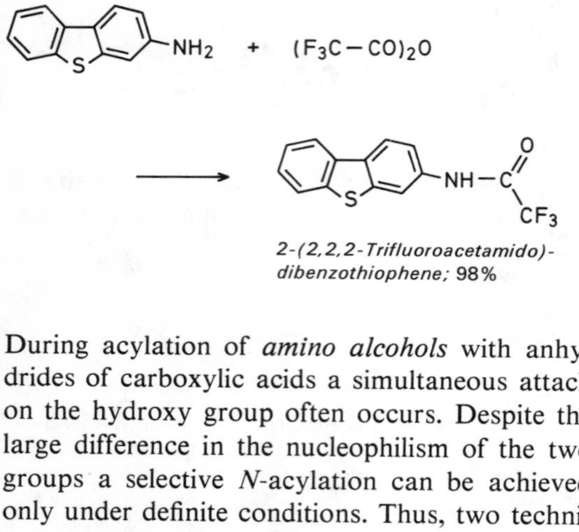

2-(2,2,2-Trifluoroacetamido)-
dibenzothiophene; 98%

During acylation of *amino alcohols* with anhydrides of carboxylic acids a simultaneous attack on the hydroxy group often occurs. Despite the large difference in the nucleophilism of the two groups a selective *N*-acylation can be achieved only under definite conditions. Thus, two techniques have been developed for preparing *N-acylglucosylamines*[214]:

1 Acylation either in methanol or dimethylformamide as solvent at 0° using half a molecule excess anhydride to give yields in excess of 55%. Anhydrides of long-chain carboxylic acids, too, react only with the free amino group without complications.

2 In *cysteine* selective *N-acylation* is achieved by an alternative variant involving working with equimolar anhydride and amino acid in aqueous tetrahydrofuran as solvent[215]:

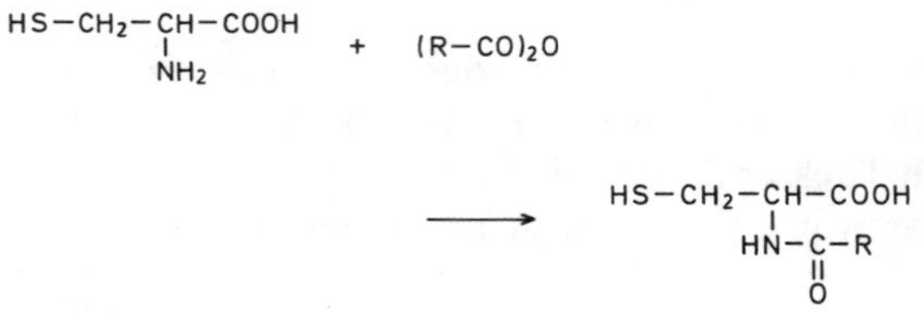

Formation of *S-acyl* derivatives is not observed under these conditions. With α- and β-amino alcohols acyl-migration from the nitrogen to the oxygen can take place following an *N*-monoacylation[216].

Cyclic anhydrides of dicarboxylic acids add to primary or secondary amines at room temperature in solvents such as chloroform, benzene, or ether to *monoamides* of *dicarboxylic acids*, which at higher temperatures[217, 218] or in the presence of water-abstracting agents change into *N*-substituted *imides*[206]:

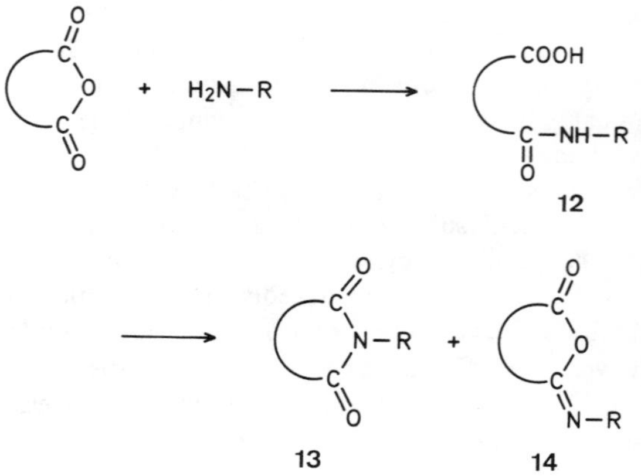

Side by side the imidodicarboxylic acid anhydrides **14** can form[219] in dependence on the conditions employed and the nature of the dicarboxylic acid monoamides

[210] R.P. Mariella, K.H. Brown, J. Org. Chem. *36*, 735 (1971).

[211] S.A. Abbas, W.J. Hickinbottom, J. Chem. Soc. C 1305 (1966).

[212] G.S. Reddy, L. Mandell, J.H. Goldstein, J. Chem. Soc. (London) 1414 (1963).

[213] E. Sawicki, F.E. Ray, J. Amer. Chem. Soc. *75*, 2266 (1953).

[214] K. Onodera, S. Kitaoka, J. Org. Chem. *25*, 1322 (1960).

[215] T.A. Martin, J.R. Corrigan, C.W. Walter, J. Org. Chem. *30*, 2839 (1965).

[216] G. Vodor, V. Bruckner, J. Org. Chem. *14*, 337 (1949).

[217] W. Flitsch, Chem. Ber. *94*, 2494 (1961).

[218] P.O. Tawney, R.H. Snyder, R.P. Conger, K.A. Leibbrand, C.H. Stiteler, A.R. Williams, J. Org. Chem. *26*, 15 (1961).

[219] W.R. Roderick, P.L. Bhatia, J. Org. Chem. *28*, 2018 (1963).

12. Phosphorus(V) oxide, acetic anhydride/sodium acetate, acetyl chloride, or thionyl chloride, *etc.*, are used as dehydrating agent. Certain of these, for instance, trifluoroacetic acid, afford exclusively the derivatives **14**[219, 220] from **12**, while acetic anhydride furnishes the imides **13**[221].

Phthalimide derivatives are readily obtained by melting together the amine component with phthalic anhydride[221, 222]. Boiling under reflux in benzene or acetone, too, effects cyclization of the phthalamic acid **12**, with *R* a long-chain aliphatic group[223]. On refluxing in toluene other cyclic anhydrides, and succinic anhydride on heating with amines, afford the imides **13** in every case[224, 225, 226], *e.g.*:

R² = Aryl

4-Cyclohexene-1,2-dicarboxylic anhydride reacts quantitatively with ammonia to form *4-cyclohexene-1,2-dicarboximide*[227].

In aqueous acetone succinic and phthalic anhydrides react with D-glucosamine[228] to give succinic acid D-glycosimide and phthalic acid D-glycosimide respectively.

Phthalic anhydride reacts with a number of amino acids in nonpolar solvents in the presence of triethylamine to yield that phthalimide derivative directly[229]

which is formed also on boiling the corresponding phthalamic acid with triethylamine[230].

Mixed anhydrides are often used for amide syntheses, mainly in *peptide* chemistry[230a, 231]. Fundamentally, two possible amides can form, *e.g.*[232]:

| **15** | **16** |

Working in water/acetone affords the *amide* **15**, in benzene the *chloroacetamide* **16** as main product. In addition to the nucleophilic nature of the amine the nature of the group *R*, too, determines the direction that the reaction takes[233, 234]. With formic acid/acetic anhydride amines react smoothly to *formamides*[235]; when several amino groups are present a selective reaction can be achieved, *e.g.*:

[220] *W.R. Roderick*, J. Org. Chem. **29**, 745 (1964).

[221] *J.H. Billman, W.F. Harting*, J. Amer. Chem. Soc. **70**, 1473 (1948).

[222] *M. Viscontini, W. Kaiser, H.A. Leidner*, Helv. Chim. Acta **48**, 1221 (1965).

[223] *L. Sterk, J. Hasko, K. Nádor*, Arzneimittel-Forsch. **18**, 798 (1968).

[224] *M.Kh. Gluzman, R.S. Mil'ner*, Izv. Vyssh. Ucheb. Zavedenii, Khim. Tekhnol. **3**, 684 (1960); C.A. **55**, 2557c (1961).

[225] USSR.P. 202921 (1967), All-Union Scientific-Research Institute of Chemicals for Plant. Protection, Inv.: *S.D. Volodkovich, N.N. Mel'nikov, T.K. Yudovskaya*; C.A. **69**, 106009z (1968).

[226] *S.D. Volodkovich, N.N. Mel'nikov, T.K. Yudovskaya*, Zh. Prikl. Khim. (Leningrad) **41**, 1387 (1968); C.A. **69**, 105982z (1968).

[227] *N.N. Mel'nikov, E.M. Sokolova, P.P. Trunov, G.I. Brusenia*, Zh. Prikl. Khim. (Leningrad) **34**, 2550 (1961); C.A. **56**, 10035g (1962).

[228] *H. Rudy, F. Krüger, J. Miksch, L. Bauer, J. Kimmig*, Chem. Ber. **93**, 2851 (1960).

[229] *A.K. Bose, F. Greer, C.C. Price*, J. Org. Chem. **23**, 1335 (1958).

[230] *E. Hoffmann, H. Schiff-Shenhav*, J. Org. Chem. **27**, 4686 (1962).

[230a] *E. Felder, D. Pitre, S. Boveri*, J. Med. Chem. **15**, 210 (1972).

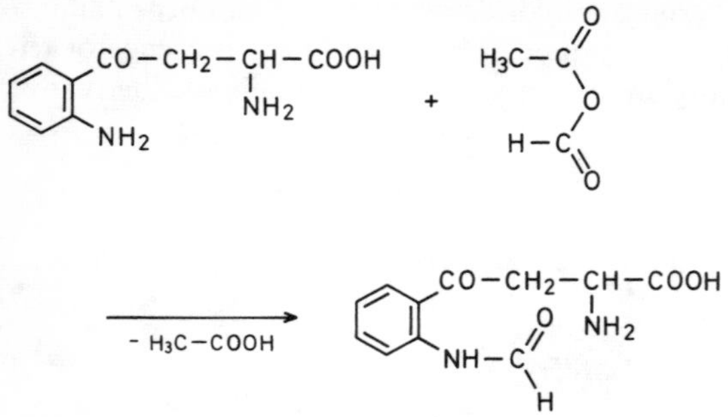

2-Amino-4-(2-formamido-
phenyl)-4-oxobutyric acid

It requires a further molecule of mixed anhydride to cause the α-amino group to by formylated giving *2-formamido-4-(2-formamidophenyl)-4-oxobutyric acid*.

Mixed anhydrides of sulfonic and carboxylic acids form excellent acylating agents. The yields obtained are very high, while no sulfonation is observed[236]. Very often mixed anhydrides of carboxylic acids and carbonic acid are also used; as a rule they react with amines[237] or ammonia[238, 239] to *carboxamides*. *Urethans* are formed only if the carboxylic acid in the mixed anhydride contains bulky or branched substituent groups[240].

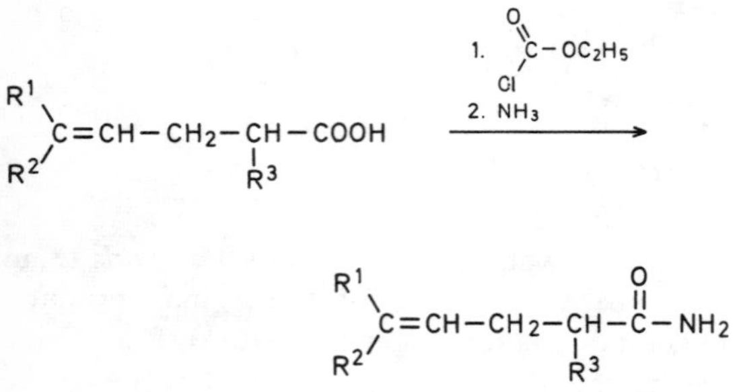

17.1.4.1.2 Acylations of hydrazines with acid anhydrides

Methylhydrazine has been the chief hydrazine representative to have been reacted with anhydrides of carboxylic acids. In every case the *N*-methyl nitrogen was acylated with formation of *1-acyl-1-methylhydrazines*[155, 241]:

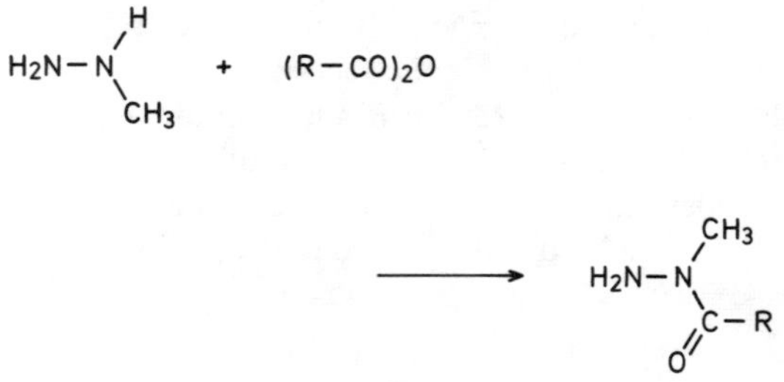

R = CH₃;   N-Acetyl-N-methylhydrazine
R = C₆H₅;   N-Benzoyl-N-methylhydrazine
R = H₅C₆—CH₂;   N-Methyl-N-phenylacetylhydrazine

The reaction can be carried out in benzene or in aqueous solution. With *N, N'*-dimethylhydrazine only monoacylation (giving *N-acetyl-N, N'-dimethylurea*) takes place on boiling in acetic anhydride[242]. Cyclic anhydrides of carboxylic acids easily form *2,6-pyridazinediones* with hydrazine[243], a reaction that takes place also with heteroaromatic anhydrides[244], *e.g.:*

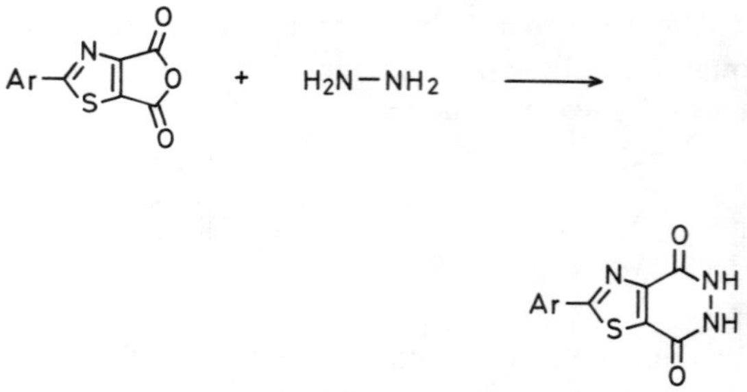

2-Arylthiazolo-
[4,5-d]pyrid-
azine-4,7-
(5H, 6H)-dione

Alkyl-substituted hydrazines give *N*-alkyl-substituted *2,6-pyridazinediones*, while with aryl-substi-

[231] *N.F. Albertson*, Org. Reactions *12*, 157 (1962).
[232] *A.R. Emery, V. Gold*, J. Chem. Soc. (London) 1443 (1950).
[233] *R.P. Steiger, E.B. Miller*, J. Org. Chem. *24*, 1214 (1959).
[234] *G. Losse, E. Demuth*, Chem. Ber. *94*, 1762 (1961).
[235] *C.W. Huffman*, J. Org. Chem. *23*, 727 (1958).
[236] *C.G. Overberger, E. Sarlo*, J. Amer. Chem. Soc. *85*, 2446 (1963).
[237] *W. Ried, K. Marquard*, Justus Liebigs Ann. Chem. *642*, 141 (1961).
[238] *T. Kishi*, Yakugaku Zasshi *81*, 782 (1961); C.A. *55*, 24 718i (1961).
[239] *F.J. McCarty, P.D. Rosenstock, J.P. Paolini, D.D. Micucci, L. Ashton, W.W. Bennetts, F.P. Palopoli*, J. Med. Chem. *11*, 534 (1968).
[240] *M. Nacken, P. Pachaly, F. Zymalkowski*, Arch. Pharm. *303*, 122 (1970).

[241] *R.L. Hinman, D. Fulton*, J. Amer. Chem. Soc. *80*, 1895 (1958).
[242] *R.L. Hinman*, J. Amer. Chem. Soc. *78*, 1645 (1956).
[243] *H. Henecka, P. Kurtz* in *Houben-Weyl*, Methoden der organischen Chemie, 4. Aufl., Bd. VIII, p. 678, Georg Thieme Verlag, Stuttgart 1952.
[244] *M. Robba, Y. Le Guen*, Bull. Soc. Chim. France 4317 (1970).

tuted hydrazines the corresponding *N*-arylamino-imides are formed[245, 245a]:

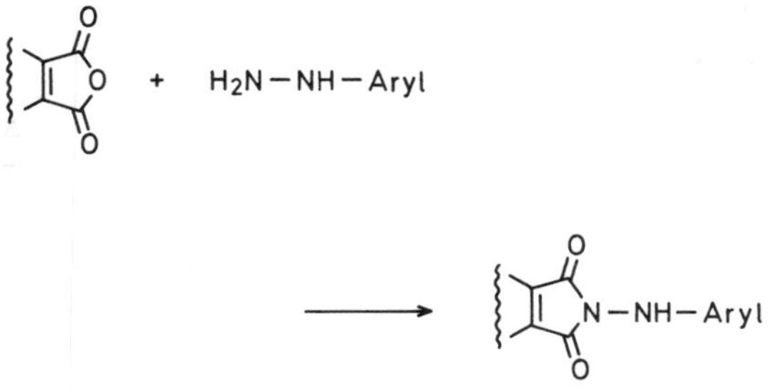

17.1.4.1.3 Acylations of hydroxylamines with acid anhydrides

When anhydrides of carboxylic acids act on hydroxylamines *hydroxamic acids* are formed; with excess anhydride these are readily converted into *N,O-diacylhydroxamic acids*[246, 247]. *O*-Alkyl-hydroxamic acids are acylated on the oxygen or nitrogen in good yield with acetic anhydride in pyridine or a little perchloric acid[175]:

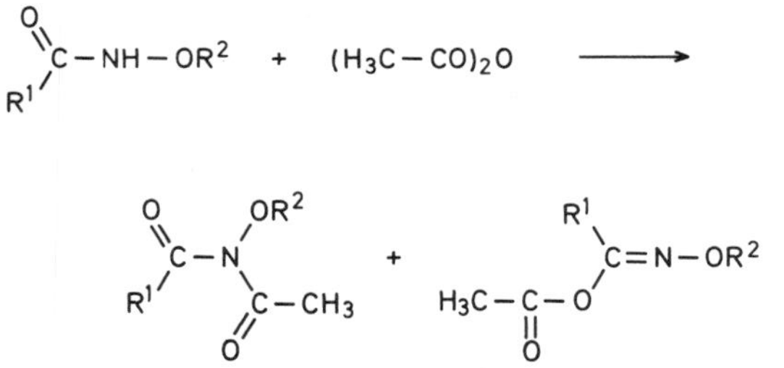

Cyclic anhydrides react with hydroxylamine even in aqueous solution to form *N-hydroxy-imides*[248, 249], *e.g.:*

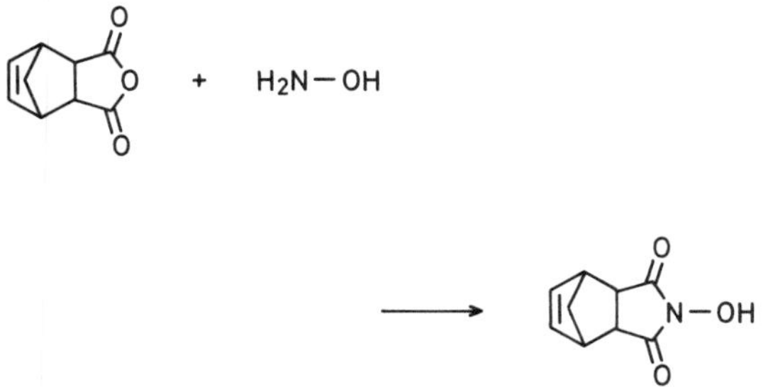

5-Norbornene-
2,3-dicarbox-
imide

17.1.4.1.4 Acylations of amides with acid anhydrides

Acylation of amides with anhydrides of carboxylic acids evidently proceeds uniformly and in good yield to *diacylamines* only when catalyzed by acids (see p. 685). The dehydrating effect of acid anhydrides can also make nitrile formation from certain amides possible[221]. In earlier times hydrochloric acid or acetyl chloride were added to the anhydride as catalyst[251, 252], it required the use of sulfuric acid to catalyze the acylation at 100° to make the technique a general method of preparation[250, 253, 254]:

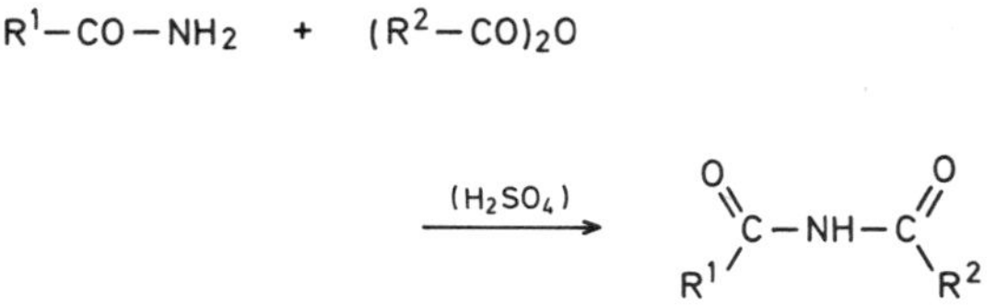

$$R^1\text{—CO—NH}_2 \ + \ (R^2\text{—CO})_2O$$

Replacing the sulfuric acid by perchloric acid reduces the yield of diacylamines again[254].

As *N*-substituted amides, lactams are converted into *N-acetyllactams* by boiling acetic anhydride[255].

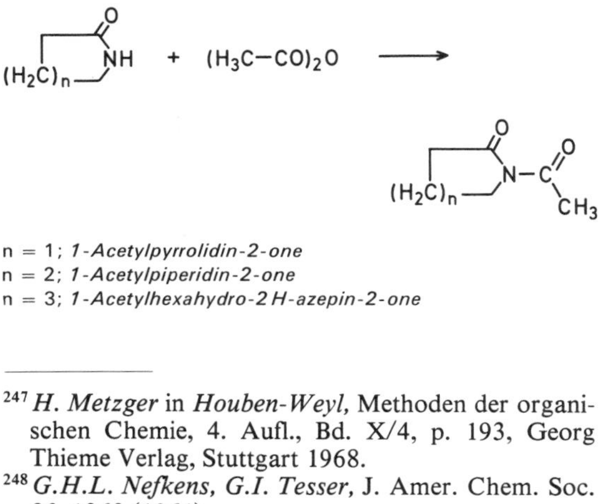

n = 1; *1-Acetylpyrrolidin-2-one*
n = 2; *1-Acetylpiperidin-2-one*
n = 3; *1-Acetylhexahydro-2H-azepin-2-one*

[247] *H. Metzger* in *Houben-Weyl*, Methoden der organischen Chemie, 4. Aufl., Bd. X/4, p. 193, Georg Thieme Verlag, Stuttgart 1968.
[248] *G.H.L. Nefkens, G.I. Tesser*, J. Amer. Chem. Soc. *83*, 1263 (1961).
[249] *L. Bauer, S.V. Miarka*, J. Org. Chem. *24*, 1293 (1959).
[250] *D. Davidson, H. Skovronek*, J. Amer. Chem. Soc. *80*, 376 (1958).
[251] *J.B. Polya, T.M. Spotswood*, Rec. Trav. Chim. Pays-Bas *67*, 927 (1948).
[252] *P. Dunn, E.A. Parkes, J.B. Polya*, Rec. Trav. Chim. Pays-Bas *71*, 676 (1952).
[253] *C.D. Hurd, A.G. Prapas*, J. Org. Chem. *24*, 388 (1959).
[254] *K. Baburao, A.M. Costello, R.C. Petterson, G.E. Sander*, J. Chem. Soc. C 2779 (1968).
[255] *H.K. Hall, M.K. Brandt, R.M. Mason*, J. Amer. Chem. Soc. *80*, 6420 (1958).

[245] *F.G. Baddar, M.F. El-Newaihy, M.R. Salem*, J. Chem. Soc. C 716 (1971).
[245a] *M. Augustin, P. Reinemann*, Z. Chem. *13*, 214 (1973).
[246] *H. Henecka, P. Kurtz* in *Houben-Weyl*, Methoden der organischen Chemie, 4. Aufl., Bd. VIII, p. 688, Georg Thieme Verlag, Stuttgart 1952.

17.1.4.1.5 Acylations of ureas with acid anhydrides

On heating, cyclic acid anhydrides react with ureas to form *dicarboxylic acid monocarbamoylamides,* which cyclize to *imides* in dependence on the conditions[256]. In glacial acetic acid maleic anhydride affords *N-carbamoylmaleamic acids* in very good yield at 60°; the products cyclize to *N-carbamoylmaleimide* in acetic anhydride at 100°[257]:

R = H;   *N-Carbamoylmaleamic* → *N-Carbamoyl-*
                  acid   *maleimide*

R = (CH₃)₃C; *N-(tert-Butylcarbamoyl)-* → *N-(tert-Butyl-*
                  maleamic acid   *carbamoyl)-*
                                          maleimide

R = C₆H₅;   *N-(Phenylcarbamoyl)-* → *N-(Phenyl*
                  maleamic acid   *carbamoyl-*
                                          maleimide)

Carboxylic acid ureides can be prepared by the action of urea on mixed anhydrides of carboxylic and sulfuric acid. For this purpose the relevant carboxylic acid is reacted with chlorosulfonic acid to form the mixed anhydride and urea is added at 70°[258]:

$$R{-}COOH \;+\; ClSO_3H \xrightarrow[-HCl]{}$$

17.1.4.2 Acylations of amines and amine derivatives with ketenes[259]

Acylations of *amines* and amine derivatives with ketenes can be performed very rapidly. With ammonia and primary amines especially the reactions proceed very vigorously and it is generally necessary to work in a solvent. Secondary amines are somewhat less reactive, and especially aromatic amines. During reaction of 2-methyl-2-butyl-3-ynylamines the failure of acetylation to take place when $R = tert\text{-}butyl$ is noteworthy[260]:

R = CH₃;   *N-Methyl-N-(2-methyl-3-butyn-2-yl)-*
                  acetamide

R = C₂H₅;   *N-Ethyl-N-(2-methyl-3-butyn-2-yl)-*
                  acetamide

R = (CH₃)₂CH; *N-Isopropyl-N-(2-methyl-3-butyn-2-yl)-*
                  acetamide

Ketene monoacylates *amides* only under acid catalysis[261]; formamide alone affords *formyldiacetamide.*

$$R{-}CO{-}NH_2 \;+\; H_2C{=}C{=}O$$

Cyclic imides behave strikingly in that they react with ketene only in the presence of sodium acetate[262].

Phenylimines can react with ketenes to form lactams while undergoing cycloaddition[263, 264]. Thus

[256] *F.L. Dunlap,* J. Amer. Chem. Soc. *18,* 332 (1896).

[257] *P.O. Tawney, R.H. Snyder, C.E. Bryan, R.P. Conger, F.S. Dovell, R.J. Kelly, C.H. Stiteler,* J. Org. Chem. *25,* 56 (1960).

[258] *I.G. Khaskin, G.I. Vishnevskaya, O.D. Litvinchuk,* Zh. Prikl. Khim. (Leningrad) *33,* 986 (1960); C.A. *54,* 16384f (1960).

[259] *D. Borrmann* in *Houben-Weyl,* Methoden der organischen Chemie, 4. Aufl., Bd. VII/4, p. 124, Georg Thieme Verlag, Stuttgart 1968.

[260] *N.R. Easton, R.D. Dillard,* J. Org. Chem. *28,* 2465 (1963).

[261] *R.E. Dunbar, G.C. White,* J. Org. Chem. *23,* 915 (1958).

[262] *R.E. Dunbar, W.M. Swenson,* J. Org. Chem. *23,* 1793 (1958).

N-benzylideneanilines add chloroketene to form mixed *cis* and *trans 3-chloro-1,2-diphenyl-4-azetidinones*[265]:

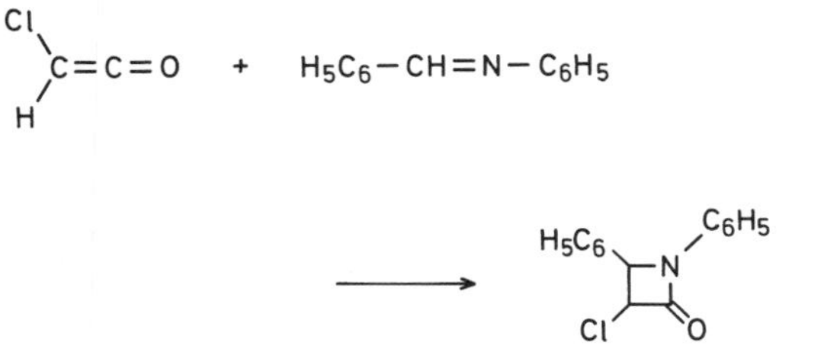

Cycloaddition of ketenes to azo compounds likewise readily leads to *1,2-diazetidin-3-ones*[266]:

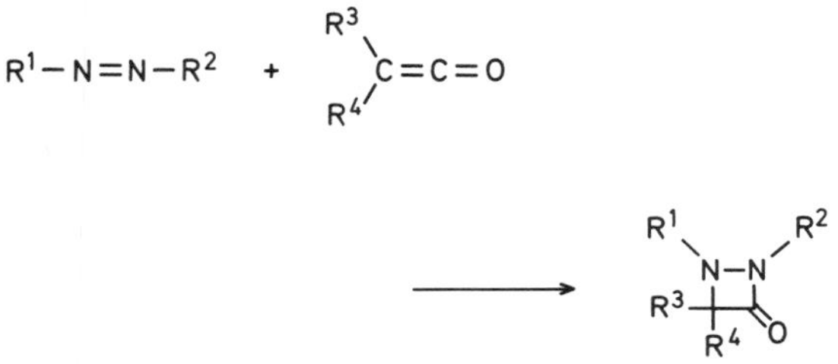

17.1.4.3 Acylations of amines and amine derivatives with carbon suboxide[267]

The inner anhydride of malonic acid, carbon suboxide, reacts smoothly with amines to form *N-substituted malonamides:*

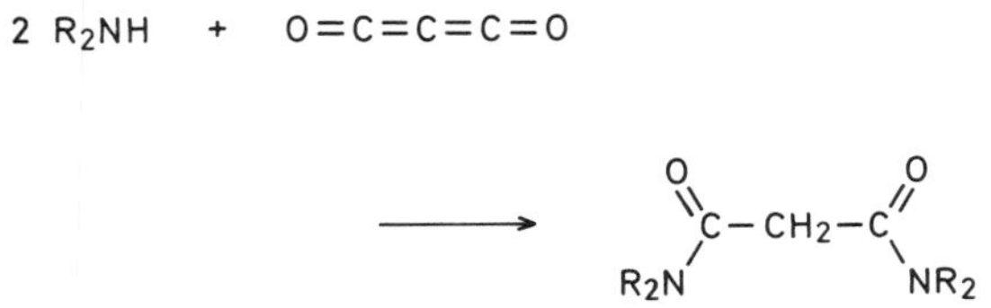

Hydrazine and hydroxylamine derivatives react analogously to the corresponding malonic acid derivatives, while hydrazines form cyclic derivatives:

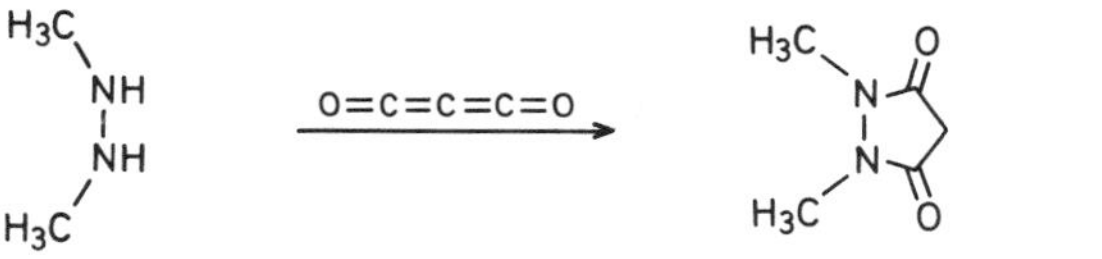

1,2-Dimethyl-3,5-pyrazolidinedione; 62%

This reaction is also observed when carbon suboxide is reacted with other bifunctional amine derivatives such as diamines, ureas, or amidines, *e.g.:*

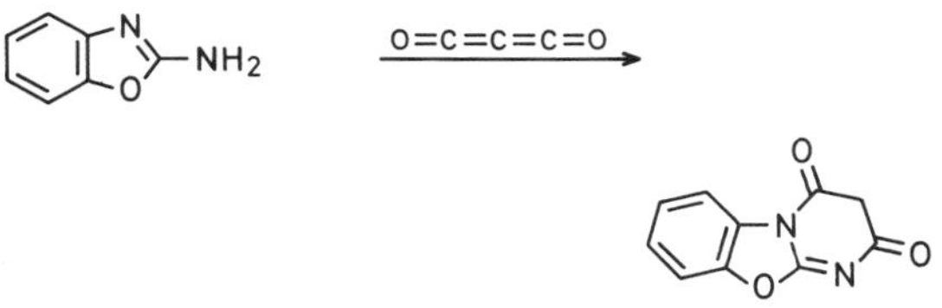

4H-Pyrimido-[2,1-b]benz-oxazole-2,4-dione

17.1.5 Acylations with carboxylic acid esters

17.1.5.1 Acylation of amines

Acylating amines with esters of carboxylic acids represents a widely applicable method for preparing *carboxamides*. The reaction proceeds in dependence on the constitution of the ester and the amine[268] (no amide formation ensues on steric hindrance). *Activated* esters such as cyanomethyl[269] or 4-nitrophenyl[270] esters react very rapidly and in good yield even at room temperature with primary amines and are therefore especially suitable for synthesizing peptides[271], *e.g.*[272]:

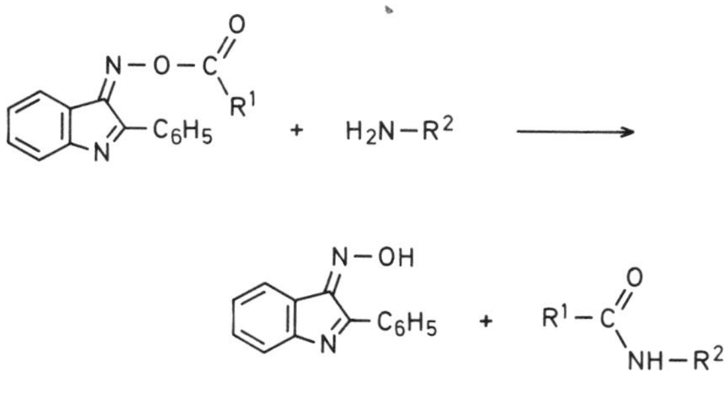

[263] *H. Ulrich*, Cycloaddition Reactions of Heterocumulenes, p. 75, Academic Press, New York, London 1967.

[264] *Th. Haug, F. Lohse, K. Metzger, H. Batzer*, Helv. Chim. Acta *51*, 2069 (1968).

[265] *D.A. Nelson*, Tetrahedron Lett. 2543 (1971).

[266] *H. Ulrich*, Cycloaddition Reactions of Heterocumulenes, p. 84, Academic Press, New York, London 1967.

[267] *D. Borrmann* in *Houben-Weyl*, Methoden der organischen Chemie, 4. Aufl., Bd. VII/4, p. 297, Georg Thieme Verlag, Stuttgart 1968.

[268] *M. Gordon, J.G. Miller, A.R. Day*, J. Amer. Chem. Soc. *71*, 1245 (1949).

[269] *R. Schwyzer, B. Iselin, M. Feurer*, Helv. Chim. Acta *38*, 69 (1955).

[270] *M. Goodman, K.C. Steben*, J. Amer. Chem. Soc. *81*, 3980 (1959).

[271] *E. Schröder, K. Lübke*, The Peptides, Vol. I (1965), Vol. II (1966), Academic Press, New York, London.

[272] *N.C. Bellavista, A. Colonna*, Ann. Chim. (Rome) *59*, 630 (1969); Synthesis 226 (1971).

α-Amino acids react readily to form *N-trifluoroacetyl amino acids* with phenyl trifluoroacetates[273]:

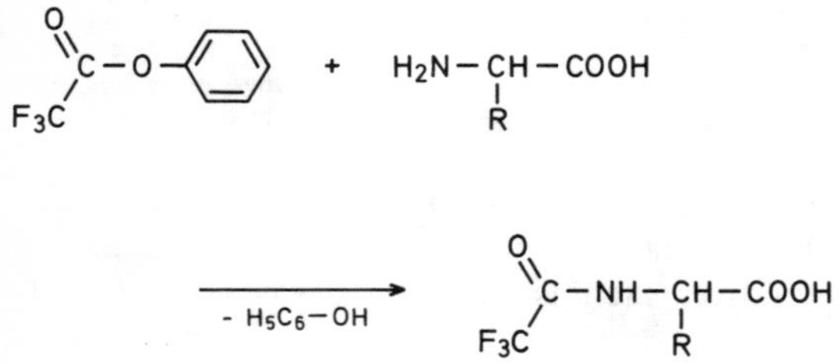

$$\xrightarrow{-\ H_5C_6-OH}$$

The aminolysis of the carboxylic acid esters can be catalyzed by ammonium salts[274] and bases[275]. Thus, a little ammonium chloride is added to the reaction mixture of benzylamine and esters for preparing *N-benzylamides* to speed up the reaction[276]. The solvent used, and also water, likewise accelerate the rate of reaction[268].

With readily reacting esters the amidation is performed with an aqueous solution of the amine[277]; in other cases ethanol is employed as the reaction medium, for example for preparing *N,N-dialkylmalonamides*[278]:

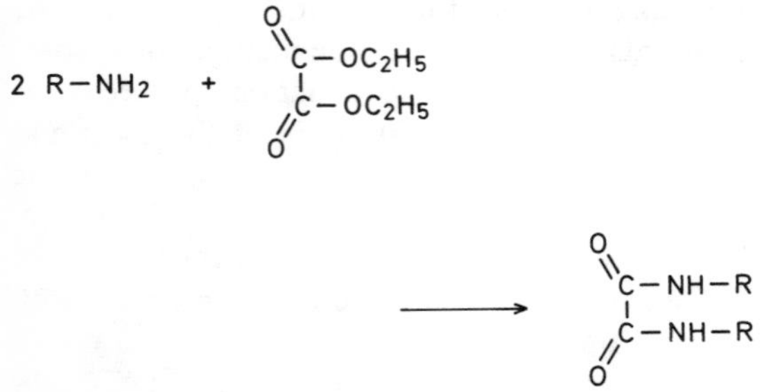

$R = H_3C-(CH_2)_n-$, n = 3 - 18

Secondary amines do not react with diesters of oxalic acid under these conditions.

Should the reactions with concentrated ammonia solution occupy too much time[279],

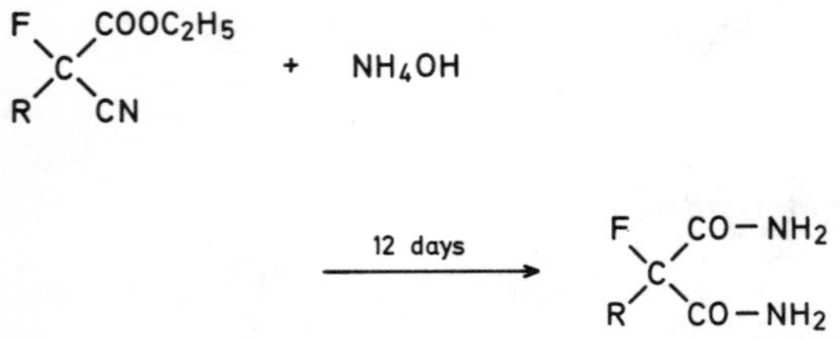

$$\xrightarrow{12\ days}$$

Fluoromalonamide

then the reaction is conducted in liquid ammonia[132, 279] or benzene as solvent[280]:

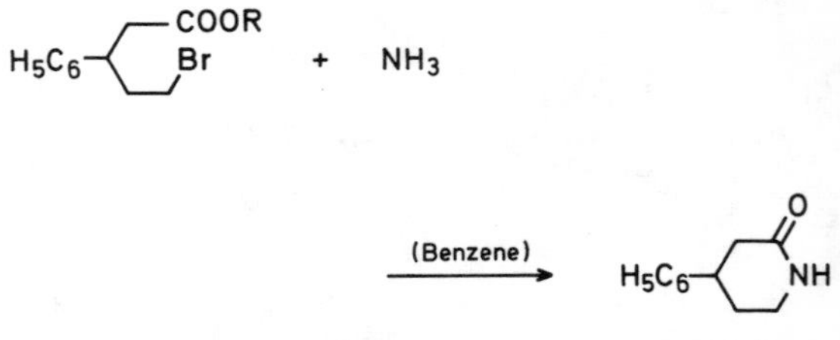

$$\xrightarrow{(Benzene)}$$

4-Phenyl-2-piperidinone

Cyclic carbonates of 1,2-glycols (1,3-dioxolan-2-ones) react with primary and secondary cycloaliphatic amines to form *carbamic acid esters*. In the presence of lithium salts aromatic amines, too, react to *1,3-oxazolidin-2-ones* in over 70% yield[281]:

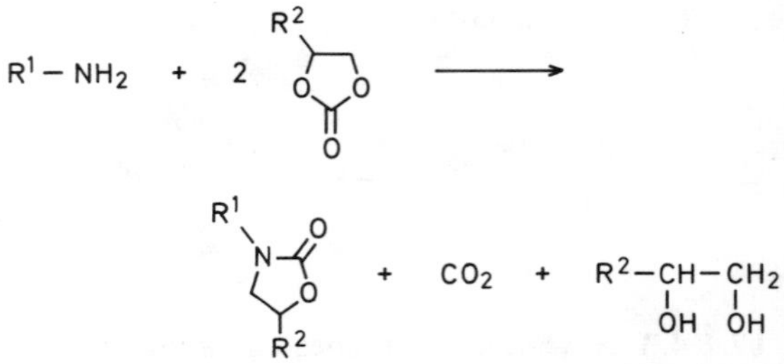

These are obtained also by using isocyanates in place of the amines[282].

In the presence of molar amounts of alcoholate aromatic amines also can be reacted with esters in very good yield[283, 284]:

$$H_3C-CH_2-COOR^1\ +\ R^2-NH_2$$

$$\xrightarrow{RO^\ominus}\ H_3C-CH_2-CO-NH-R^2$$

R^1 = Alkyl
R^2 = Aryl-

[275] *J.F. Bunnett, G.T. Davis*, J. Amer. Chem. Soc. *82*, 665 (1960).
[276] *O.C. Derner, J. King*, J. Org. Chem. *8*, 168 (1943).
[277] *F. Korte, H. Wamhoff*, Chem. Ber. *97*, 1970 (1964).
[278] *L.M. Rice, C.H. Grogan, E.E. Reid*, J. Amer. Chem. Soc. *75*, 242 (1953).
[279] *H. Gershon, S.G. Schulmann, A.D. Spevack*, J. Med. Chem. *10*, 536 (1967).
[280] *A. Burger, A. Hofstetter*, J. Org. Chem. *24*, 1290 (1959).
[281] *E. Gulbins, K. Hamann*, Chem. Ber. *99*, 55 (1966).
[282] *K. Gulbins, G. Benzing, R. Maysenhölder, K. Hamann*, Chem. Ber. *93*, 1975 (1960).
[283] *R.L. Betts, L.P. Hammett*, J. Amer. Chem. Soc. *59*, 1568 (1937).
[284] *R.J. DeFond, P.D. Strickler*, J. Org. Chem. *28*, 2915 (1963).

[273] *F. Weygand, A. Röpsch*, Chem. Ber. *92*, 2095 (1959).
[274] *L.L. Fellinger, L.F. Audrieth*, J. Amer. Chem. Soc. *60*, 579 (1938).

4-Nitro and 2-substituted anilines and methylaniline do not react even under these conditions.

Formylation of secondary aliphatic amines with formic acid esters succeeds very easily with catalytic amounts of sodium in a very strongly exothermic reaction[285], *e.g.:*

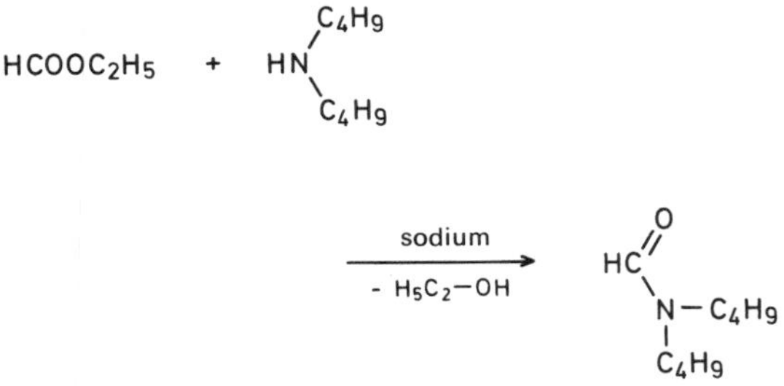

N,N-Dibutyl-
formamide;
98%

For this purpose equimolar quantities of ethyl formate and amine are refluxed for 1–4 hours with catalytic amounts of sodium and the formamide formed is distilled off under vacuum (89–94%).

This elegant technique fails only with sterically hindered secondary amines such as diisopropylamine or dicyclohexylamine, and with aromatic amines[285].

While diesters of malonic acid and α-monoalkyl-substituted derivatives still afford *malonamides* smoothly with concentrated ammonia, success is seldom achieved in the case of α,α-dialkyl-substituted diesters of malonic acid diesters. However, the latter can be prepared very readily with excess formamide and molar quantities of sodium methoxide in methanolic solution[286]:

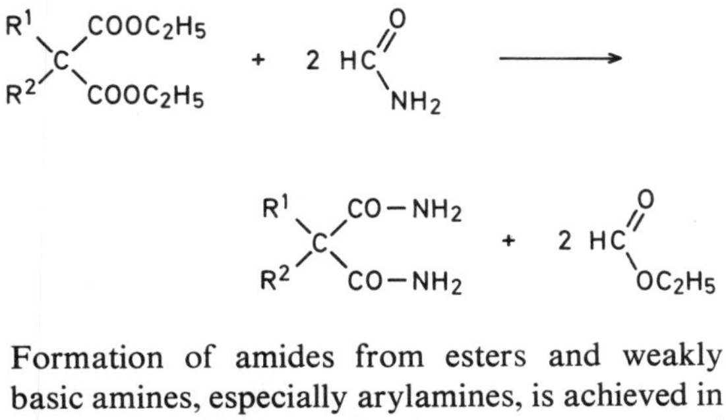

Formation of amides from esters and weakly basic amines, especially arylamines, is achieved in the *Bodroux reaction*. Here amines are converted into magnesium halide amines with Grignard

reagent and these then react with esters to form carboxamides in often quite good yield[287, 288], *e.g.:*

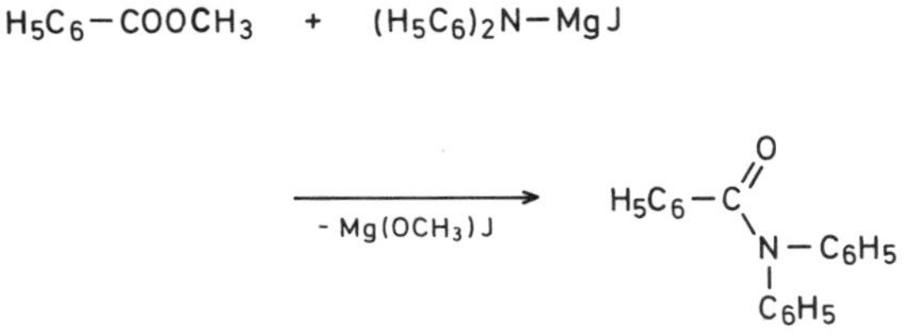

$H_5C_6-COOCH_3$ + $(H_5C_6)_2N-MgJ$

N,N-Diphenylbenz-
amide; 88%

Sometimes alternative metal amides, too, successfully aminolyze esters, for example sodium arylamides[289] or lithium amides[290] (prepared from butyllithium and the relevant amine), the carboxamides are obtained in very good yield. In polyfunctional esters *selective* amidations are feasible[290], *e.g.:*

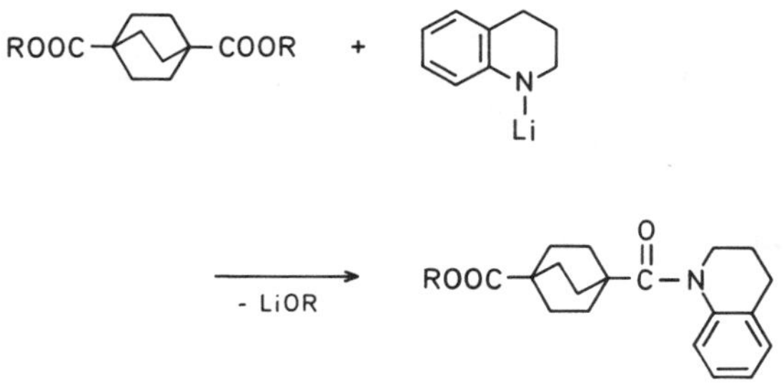

4-Alkoxycarbonyl-1-
(1,2,3,4-tetrahydroquinolino-
carbonyl)bicyclo[2.2.2]-
octane

The base-catalyzed procedure with sodium hydride in dimethyl sulfoxide as solvent is of general value[291]. Equimolar quantities of ester, amine, and sodium hydride are reacted overnight at room temperature; with aniline or cyclohexylamine many esters form the amide in high yield, *e.g.:*

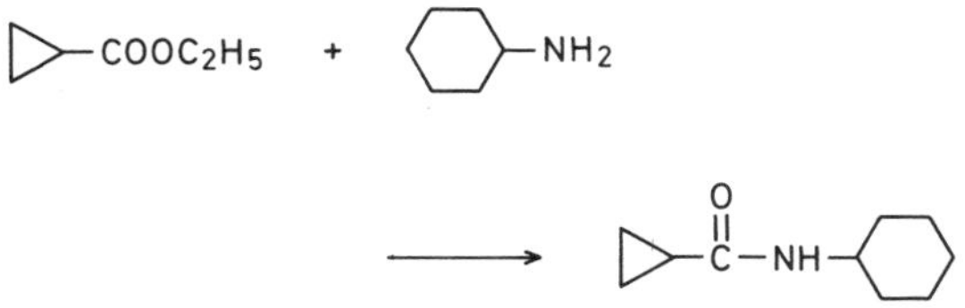

N-Cyclohexylcyclo-
propanecarboxamide;
68%

[287] *H.L. Bassett, C.R. Thomas*, J. Chem. Soc. (London) 1188 (1954).
[288] *R.P. Houghton, C.S. Williams*, Tetrahedron Lett. 3929 (1967).
[289] *E.S. Stern*, Chem. & Ind. (London) 277 (1956).
[290] *K.-W. Yang, J.G. Cannon, J.G. Rose*, Tetrahedron Lett. 1791 (1970).
[291] *B. Singh*, Tetrahedron Lett. 321 (1971).

[285] *W. Kantlehner*, Dissertation, Universität Stuttgart (1968).
[286] *W. Hackbart, M. Hartmann*, J. Prakt. Chem. *14*, 1 (1961).

The bifunctional catalytic action of hydroxylated heterocyclic bases is of interest; it speeds up the aminolysis of esters markedly[292, 293]. 2-Piperidinone appears to be especially effective and in many cases makes the aminolysis possible in addition to increasing the yield[293].

Esters of amino acids cyclize to *lactams* very rapidly and often already during their preparation[293a] from esters of nitro- and cyanoalkanoic acids[294]:

Hexahydro-2,5-pyrazinediones are formed spontaneously from α-amino acid esters[295].

By comparison to open-chain esters lactones react relatively rapidly with amines to form *hydroxycarboxamides* with opening of the ring; at 200° the products lose water intramolecularly with formation of *lactams*[296, 297, 297a]:

4-Hydroxybutyr-amides

While γ and δ-lactones react in the same way, the method is claimed to fail in the majority of cases where the hydroxy acid that forms the basis of the lactone contains a phenolic group[298] (*e.g.*, with coumarins). On the other hand, the action of methylamine on the 2*H*-pyranone derivative **17** at 190° does furnish the corresponding lactam **18**[299, 300]:

R = H;   7,8,9,10-Tetrahydro-1-hydroxy-3-methyl-6(5H)-
            phenanthridinone; 25%

R = C₂H₅; 5-Ethyl-7,8,9,10-tetrahydro-1-hydroxy-3-methyl-
            6(5H)-phenanthridinone; 15%

R = CH₃;   7,8,9,10-Tetrahydro-1-hydroxy-3,5-dimethyl-
            6(5H)-phenanthridinone; 75%

Attempts to extend this reaction to other coumarins revealed that the lactam formation is dependent on the nature of the coumarin and the base[300]. Quite fundamentally a second phenolic group in the 5- or 7-position of the coumarin system is required.

3-Hydroxy-6H-dibenzo[b,d]pyran-6-one reacts only with methylamine to form *3-hydroxy-5-methyl-6(5H)-phenanthridin-6-one* (80%)[300].

β-Lactones react much more easily with ammonia or amines than γ- or δ-lactones to give *3-hydroxypropionamides* or β-*alanine (3-aminopropionic acid)*:

[292] *H.C. Beyerman, W. Maassen van den Brink,* Proc. Chem. Soc. 266 (1963).

[293] *H.T. Openshaw, N. Whittaker,* J. Chem. Soc. C 89 (1969).

[293a] *M. Baues, U. Kraatz, F. Korte,* Justus Liebigs Ann. Chem. 1301 (1973).

[294] *H. Schnell, J. Nentwig* in *Houben-Weyl,* Methoden der organischen Chemie, 4. Aufl., Bd. XI/2, p. 536, 542, Georg Thieme Verlag, Stuttgart 1958.

[295] *M. Augustin,* Chem. Ber. *99,* 1040 (1966).

[296] *H. Kröper* in *Houben-Weyl,* Methoden der organischen Chemie, 4. Aufl., Bd. VI/2, p. 792, Georg Thieme Verlag, Stuttgart 1963.

[297] *C. Lange, H. Wamhoff, F. Korte,* Angew. Chem. *80,* 317 (1968).

[297a] *A.A. Shazhenov, Ch.Sh. Kadyrov, P. Kurbanow,* Khim. Geterotsikl. Soedin. 641 (1972).

[298] *E. Späth, J. Lintner,* Chem. Ber. *69,* 2727 (1936).

[299] *J.F. Hoops, H. Bader, J.H. Biel,* J. Org. Chem. *33,* 2995 (1968).

[300] *U. Kraatz, F. Korte,* Chem. Ber. *106,* 62 (1973).

$$\text{(ketene dimer)} \quad + \quad 2\ HNR_2 \quad \longrightarrow$$

$$\begin{array}{ccc}
CH_2-CH_2 & & CH_2-CH_2 \\
| \quad\quad | & + & | \quad\quad | \\
HO \quad CO-NR_2 & & R_2N \quad COOH
\end{array}$$

Which of the two products is formed predominantly is not as dependent on the basicity of the amine as on the solvent used and the mode of working[301]. β-Lactams cannot be prepared in this way.

Acetoacetamides can be prepared by reacting diketene with primary or secondary amines. Aqueous medium may be used for the reaction because diketene reacts substantially more rapidly with amines than with water[302].

$$\begin{array}{c}
H_3C \\
\quad\quad C-CH_2-CO-NR_2 \\
O
\end{array}$$

17.1.5.2 Acylations of hydrazines with carboxylic acid esters

The readily proceeding hydrazinolysis of esters generally leads to *monoacyl hydrazines (carboxylic acid hydrazides)*. As a rule the rapidly reacting methyl or ethyl esters are reacted with hydrazine hydrate, often in ethanol as solvent[303]. Both aliphatic and aromatic esters form the hydrazides in good yield under these conditions[304-307], *e.g.:*

$$\longrightarrow$$

2-Acetamido-3-hydroxy-3-phenyl-propionic acid hydrazide

The very *active* esters are reacted in chloroform or dichloromethane with hydrazine[308] and are suitable particularly for preparing α,β-*unsaturated* carboxylic acid hydrazides[309]:

$$\begin{array}{c}
R^2 \\
R^1-CH=C \\
\quad\quad\quad COOCH_2-X
\end{array} \quad + \quad H_2N-NH_2 \quad \longrightarrow$$

$$\begin{array}{c}
R^2 \\
R^1-CH=C \\
\quad\quad\quad CO-NH-NH_2
\end{array} \quad + \quad \begin{array}{c} R^1 \quad O \\ R^2 \quad NH \\ N \\ H \end{array}$$

$X = -CN, -OCH_3, -COOC_2H_5, 4-NO_2-C_6H_4-$

In only few cases are 3-pyrazolidinones formed as side-products; generally where $X=H$ or *alkyl*[309].

During hydrazinolysis with methylhydrazine *1-acyl-1-methyl-* and *2-acyl-1-methylhydrazines* are formed:

$$R^1-COOCH_3 \quad + \quad H_2N-NH-CH_3$$

$$R^1-CO-NH-NH-CH_3 \quad\quad\quad \begin{array}{c} R^1-CO \\ N-NH_2 \\ H_3C \end{array}$$

19   **20**

With $R^1 = CH_3-, CH_3CH_2-, C_6H_5-,$ and $(CH_3)_2CH-$ a mixture of **19** and **20** is formed with **19** predominating[241]. As R^1 increases in size both the reaction velocity of the hydrazinolysis and the proportion of **20** in the product mixture decrease[241].

Methyl phenylacetate reacts with methylhydrazine to give *2-methyl-1-phenylacetylhydrazine* (76%)[155]. The following sequence has been set up

[301] *H. Kröber* in *Houben-Weyl*, Methoden der organischen Chemie, 4. Aufl., Bd. VI/2, p. 533, Georg Thieme Verlag, Stuttgart 1963.

[302] *F. Möller* in *Houben-Weyl*, Methoden der organischen Chemie, 4. Aufl., Bd. XI/2, p. 19, Georg Thieme Verlag, Stuttgart 1958.

[303] *H. Henecka, P. Kurtz* in *Houben-Weyl*, Methoden der organischen Chemie, 4. Aufl., Bd. VIII, p. 676, Georg Thieme Verlag, Stuttgart 1952.

[304] *Ng. Ph. Buu-Hoi, Ng.D. Xuong, Ng.H. Nam, F. Binon, R. Royer*, J. Chem. Soc. (London) 1358 (1953).

[305] *H. Zimmer, E. Shaheen*, J. Org. Chem. *24*, 1140 (1959).

[306] *L. Horner, H. Fernekess*, Chem. Ber. *94*, 712 (1961).

[307] *E.D. Nicolaides*, J. Org. Chem. *32*, 1251 (1967).

[308] *M.L. Hoefle, L.T. Blouin, H.F. DeWald, A. Holmes, D. Williams*, J. Med. Chem. *11*, 970 (1968).

[309] *R. Harada, H. Kondo*, Bull. Chem. Soc. Japan *41*, 2521 (1968).

to describe the relative reactivity of *N*-methylhydrazines toward esters[241]:

$$H_2N-NH_2 > H_3C-NH-NH_2 > (CH_3)_2N-NH_2$$

1, 1-Dialkylhydrazine reacts only with esters of formic acid[310] and with diesters of oxalic acid[311] to give *2-acyl-1, 1-dialkylhydrazines, e.g.:*

COOR
|
COOR   + $H_2N-N(C_4H_9)_2$ ⟶

CO—NH—$N(C_4H_9)_2$
|
CO—NH—$N(C_4H_9)_2$

Oxalic acid bis(2,2-dibutylhydrazide)

Hydrazinolysis of methyl benzoate with 1, 1-dimethylhydrazine or 1-aminopiperidine succeeds only when sodium methoxide is used (giving *benzoic acid 2, 2-dimethylhydrazide* and *N-piperidino benzamide;* ~60%)[112]. Esters of formic acid acylate alkylhydrazine on the more nucleophilic, *N*-methyl-substituted nitrogen[312, 314]:

$HCOOCH_3$ + $R-NH-NH_2$ ⟶

$$H_2N-N\begin{smallmatrix}R\\|\\CHO\end{smallmatrix}$$

R = CH_3;   *1-Formyl-1-methylhydrazine*

R = $C_6H_5-CH_2$;   *1-Benzyl-1-formylhydrazine*

R = F_3C;   *1-Formyl-1-trifluoromethylhydrazine*

Diesters of carbonic acid behave like formic acid esters; dibenzyl carbonate alone forms both isomers[314].

Lactones are very easily opened by hydrazines with formation of *hydroxy acid hydrazides*[315]. β-Lactones react particularly rapidly and in exothermic reaction[316]; alkyl substituents in the α-position reduce the reaction velocity. Coumarins likewise react with hydrazine with opening of the ring[315]:

5-(2-Hydroxyphenyl)-3-pyrazolidinones

17.1.5.3 Acylations of hydroxylamines with carboxylic acids esters

A simple *hydroxamic acid synthesis* consists in the action of hydroxylamine on esters of carboxylic acids[317]. The reaction proceeds in good yield with both aliphatic and aromatic esters, especially if the aminolysis is catalyzed with sodium methoxide[318, 319]. Acylamino carboxylic acid esters[320], pyrazinedicarboxylic acid diesters[321], or 1-ethoxycarbonylmethylpyridinium chloride[322] also react with hydroxylamine in methanolic solution in the desired manner, *e.g.:*

Pyridinioacethydroxamoyl chloride

[310] *R.L. Hinman,* J. Amer. Chem. Soc. *78,* 2467 (1956).

[311] *H. Zimmer, L.F. Audrith, M. Zimmer,* Chem. Ber. *89,* 1116 (1956).

[312] *H. Dorn, A. Zubeck, G. Hilgetag,* Chem. Ber. *98,* 3377 (1965).

[313] *H. Dorn, A. Zubeck, K. Walter,* Justus Liebigs Ann. Chem. *707,* 100 (1967).

[314] *C.T. Pedersen,* Acta Chem. Scand. *18,* 2199 (1964).

[315] *C.F.H. Allen, E. Magder,* J. Heterocycl. Chem. *6,* 349 (1969).

[316] *E. Testa, L. Fontanella, G.F. Cristiani, L. Mariani,* Justus Liebigs Ann. Chem. *639,* 166 (1961).

[317] *H. Henecka, P. Kurtz* in *Houben-Weyl,* Methoden der organischen Chemie, 4. Aufl., Bd. VIII, p. 686, Georg Thieme Verlag, Stuttgart 1952.

[318] *S.A. Bernhard, Y. Shalitin, Z.H. Tashjin,* J. Amer. Chem. Soc. *86,* 4406 (1964).

[319] *J.H. Cooley, W.D. Bills, J.R. Throckmorton,* J. Org. Chem. *25,* 1734 (1960).

[320] *E. Hoffmann, I. Faiferman,* J. Org. Chem. *29,* 748 (1964).

[321] *B.E. Hackley jr., R. Plapinger, M. Stolberg, T. Wagner-Jauregg,* J. Amer. Chem. Soc. *77,* 3651 (1955).

[322] *D.G. Coe,* J. Org. Chem. *24,* 882 (1959).

Diesters of dicarboxylic acids such as diethyl phthalates, furnish *N*-hydroxyimides[323] on base-catalyzed reaction with hydroxylamine[323], *e.g. N-hydroxyphthalimide:*

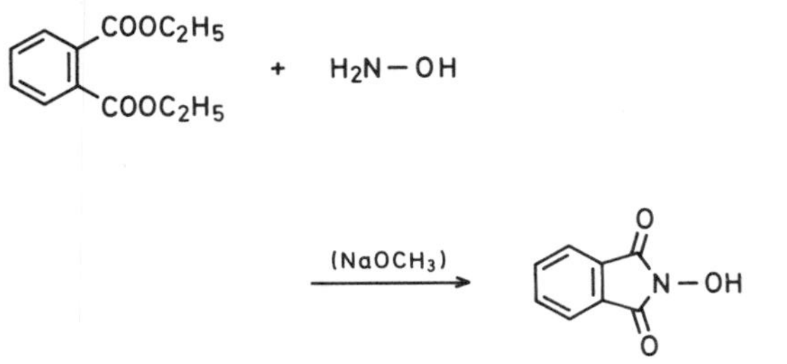

150–175°. Alternatively, the esters can be reacted with sodium amides under conditions which may also lead to ester condensations[327]:

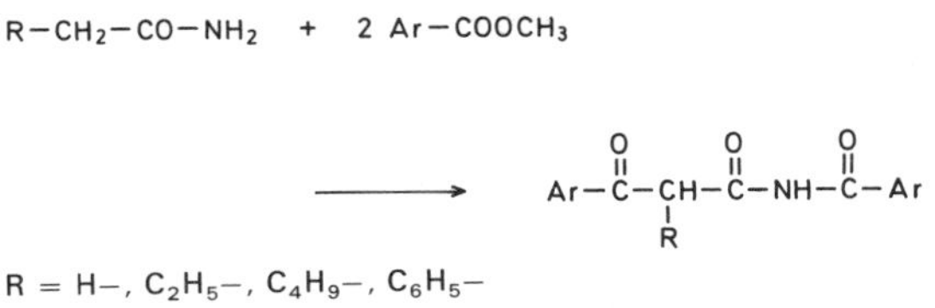

$$R-CH_2-CO-NH_2 \quad + \quad 2\ Ar-COOCH_3$$

$$R = H-,\ C_2H_5-,\ C_4H_9-,\ C_6H_5-$$

Excess sodium hydride serves as base and 1,2-dimethoxyethane as solvent.

17.1.5.4 Acylations of carboxamides with carboxylic acid esters

Esters can acylate carboxamides only under forced conditions. An exception are carboxylic amic acid esters in which the base-catalyzed reaction very easily leads to *imides* with intramolecular loss of alcohol[324, 325]:

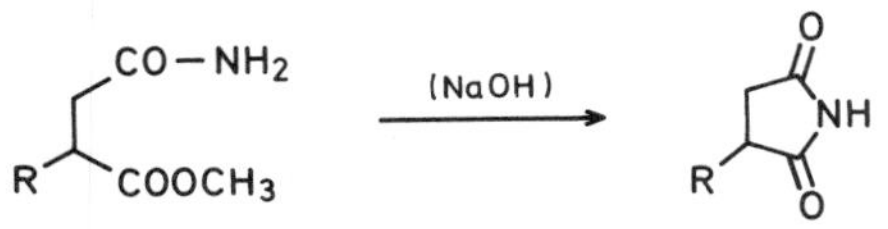

Isopropenyl esters lend themselves in excellent manner for preparing *N-acyl imides* from the imides themselves[326], *e.g.:*

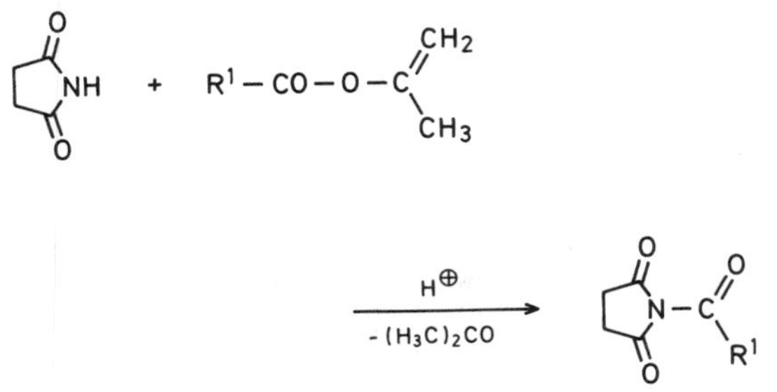

The reaction likewise proceeds to acylamides in high yield using acid catalysis with *p*-toluenesulfonic acid at

17.1.5.5 Acylations of ureas with carboxylic acid esters

Ethylene carbonate (1,3-dioxolan-2-one) reacts with 1,3-diarylureas at 175° in the presence of lithium chloride to form *3-aryl-1,3-oxazolidin-2-ones*[328]:

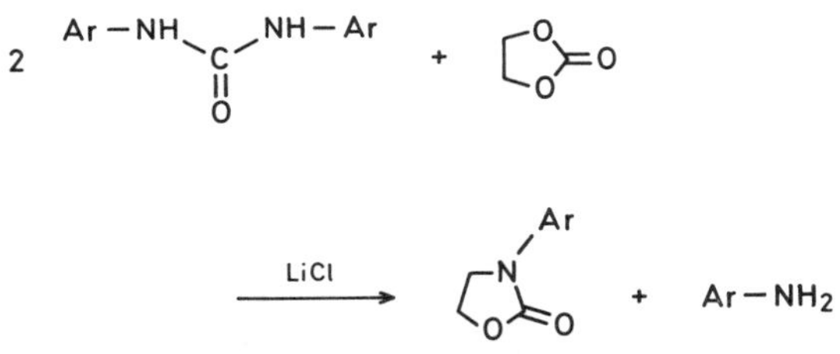

The best conversions are obtained with by using 1 molecule excess urea. Other esters react with ureas to *ureides* only under base catalysis or to *barbituric acids* with diesters of malonic acid[329]:

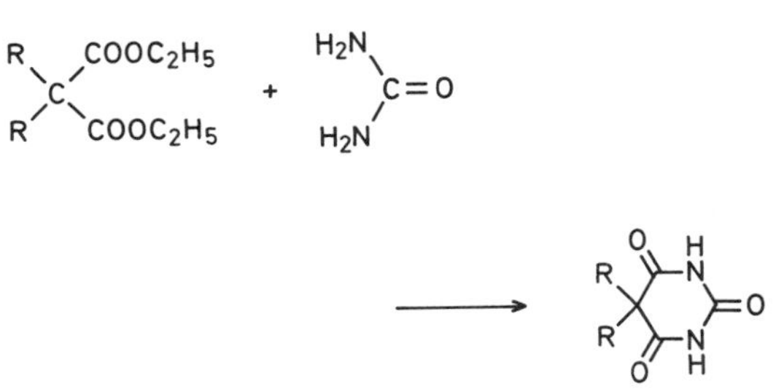

[323] *L. Bauer, S.V. Miarka,* J. Amer. Chem. Soc. *79,* 1983 (1957).

[324] *J.A. Shafer, H. Morawetz,* J. Org. Chem. *28,* 1899 (1963).

[325] *E. Sondheimer, R.W. Holley,* J. Amer. Chem. Soc. *76,* 2467 (1954).

[326] *E.S. Rothman, S. Serota, D. Swern,* J. Org. Chem. *29,* 646 (1964).

[327] *J.F. Wolfe, G.B. Trimitsis,* J. Org. Chem. *33,* 894 (1968).

[328] *F. Gulbins, K. Hamann,* Chem. Ber. *99,* 62 (1966).

[329] *R.A. Jacobson,* J. Amer. Chem. Soc. *58,* 1984 (1936).

N-Substituted ureas afford the corresponding *N*-substituted barbituric acid derivatives with malonic acid diesters[330].

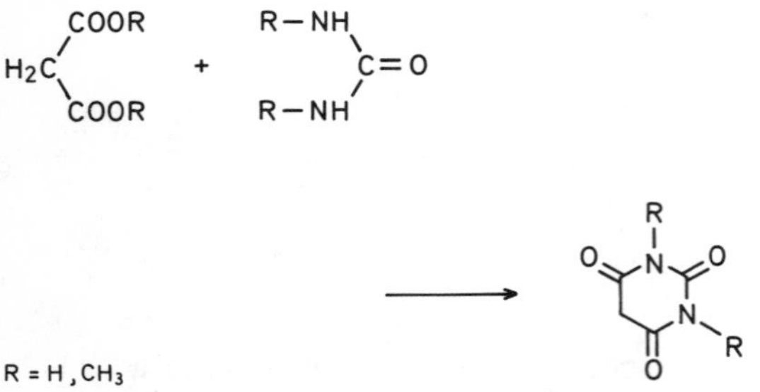

R = H , CH₃

Acylation of ureas is often carried out with their sodium derivatives in acetone or liquid ammonia at room temperature. Under these conditions heteroaromatic esters afford good yields also with substituted ureas[331], *e.g.*:

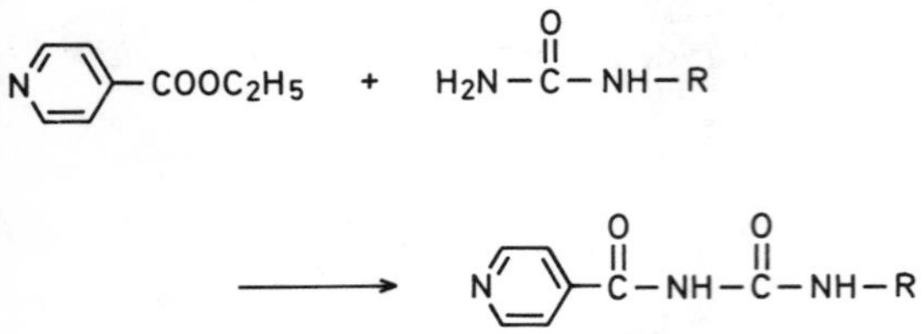

R = H;   *Isonicotinoylurea*

R = (CH₃)₂CH– ; *1-Isonicotinoyl-3-isopropylurea*

N, N'-disubstituted ureas do not react. The synthesis of *ureides* and *diacylureas* from esters and ureas with sodium hydride in 1, 2-dimethoxyethane[332] or dimethylformamide[333] is probably generally applicable; it proceeds smoothly:

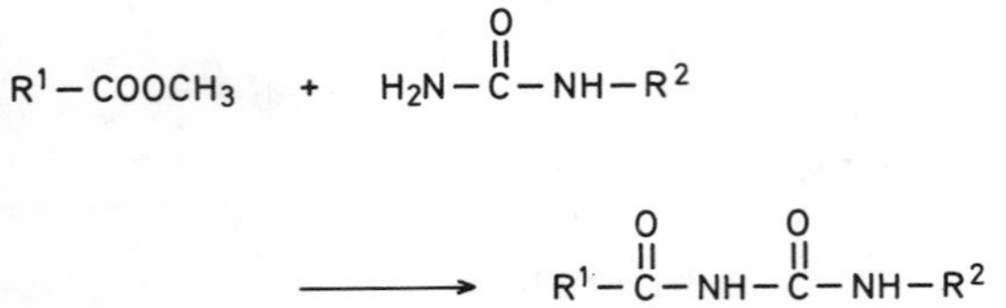

R¹ = Aryl-, -Pyrazyl-

R² = Aroyl-, CH₃–, C₆H₅–, H–

17.1.6 Acylations of amines and hydrazines with carboxamides

Carboxamides display an only weak acylating action on amines, so that these reactions require drastic conditions for success. In most cases astonishingly high yields are obtained by heating the amides with amine hydrochlorides[334-336]:

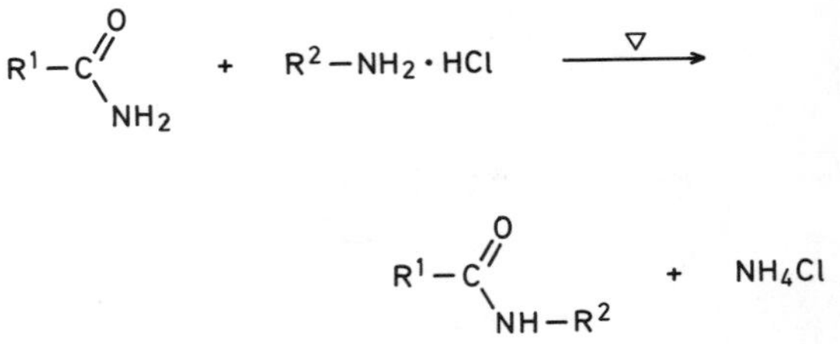

Both aromatic and aliphatic amines and amides can be reacted with one another in this way. Formamide reacts with amines at relatively low temperatures to form *formylamines*[334]. Intramolecular cyclization of ω-dimethylaminocarboxylic acids to *pyrrolidone* and *piperidone* derivatives represents an interesting variant of this reaction[337]; it ensues very rapidly with ω-aminocarboxamides at 200–220°[338].

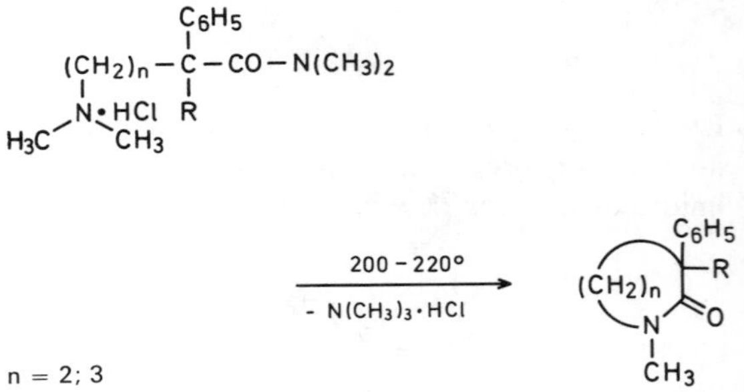

n = 2; 3

Diacylamines react with amines more readily than carboxamides. Cyclic dicarboxylic acid imides react particularly rapidly in aqueous solution[339]:

[330] G. Simchen, priv. com.

[331] A. Stempel, J. Zelauskas, J.E. Aeschlimann, J. Org. Chem. 20, 412 (1955).

[332] J.F. Wolfe, G.B. Trimitsis, Can. J. Chem. 47, 2097 (1969).

[333] J.W. Hanifin, R. Capuzzi, E. Cohen, J. Med. Chem. 12, 1102 (1969).

[334] A. Galat, G. Elion, J. Amer. Chem. Soc. 65, 1566 (1943).

[335] D. Klamann, Monatsh. Chem. 84, 923 (1953).

[336] E.N. Zil'berman, A.E. Kulikova, Zh. Vses, Khim. Obshchest. 5, 107 (1960); C.A. 54, 20923 h (1960).

[337] R.E. Stenseth, F.F. Blicke, J. Org. Chem. 34, 3007 (1969).

[338] C. Berther, Chem. Ber 92, 2616 (1959).

[339] S.P. Janes, J. Chem. Soc. (London) 625 (1945).

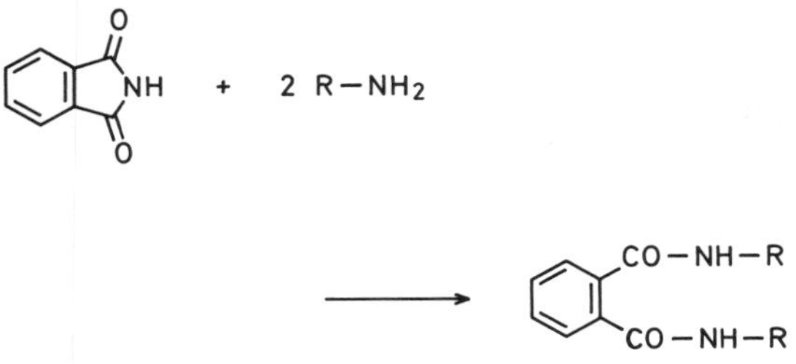

R = H;   *Phthalamide*

R = Alkyl; *N,N-Dialkylphthalamide*

In peptide syntheses the reactions of phthaloyl-carbamates with esters of amino acids are often made use of for introducing the phthaloyl group for protection[340].

Probably the procedure for preparing *carbox-amides, carboxylic acid hydrazides*, and *hydroxyl-amides* that proceeds under the most conservative conditions is the reaction between *N*-acyl-imidazoles and the corresponding nitrogen bases[150a].

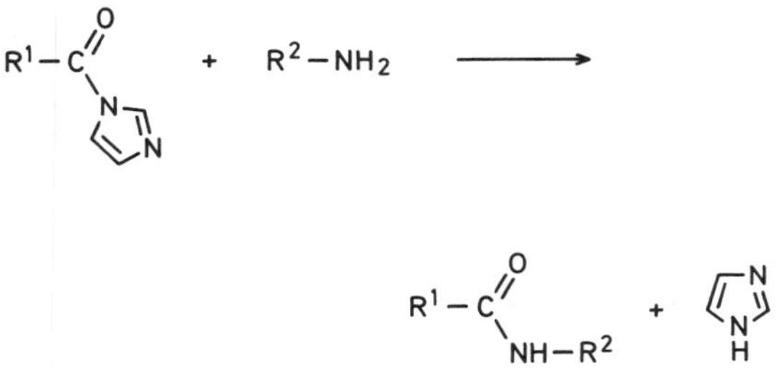

The method is made particularly simple by the fact that, starting from 1,1'-carbonyldiimidazole, the *N*-functional carboxylic acid derivatives are obtained directly by reacting carboxylic acids and amines in a combined reaction in one and the same vessel. The reactions are performed at room temperature in tetrahydrofuran, chloroform, or dimethylformamide[341]. One field where this technique has been applied widely is peptide chemistry:

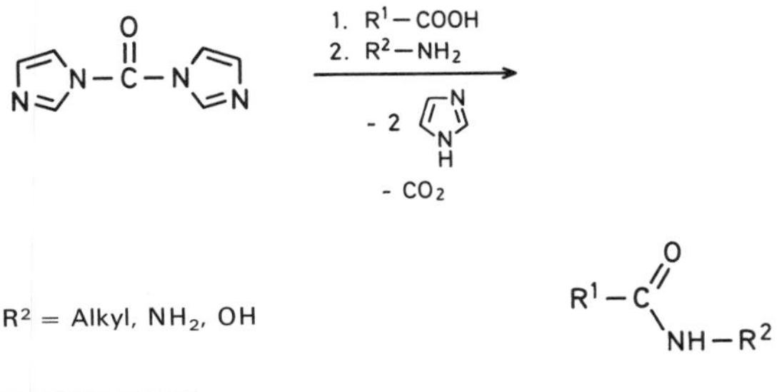

R^2 = Alkyl, NH_2, OH

Hydrazine readily converts cyclic imides into *3,6-pyridazinediones*[342, 343], *e.g.*:

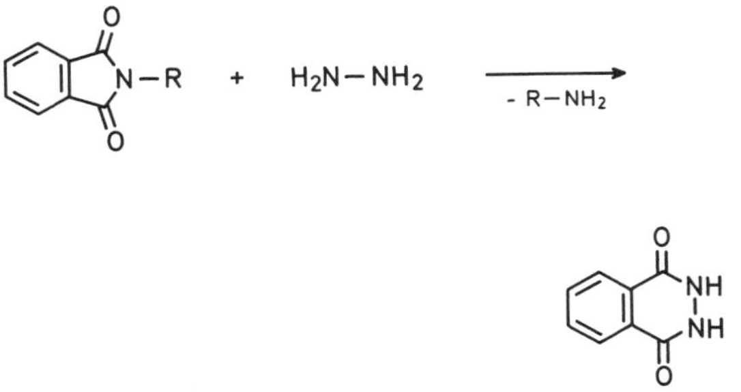

1,2,3,4-Tetrahydro-1,4-phthalazinedione

The reaction proceeds at boiling point in aqueous or alcoholic solution. *N*-Anilinophthalimide is obtained with phenylhydrazine[344]. Carboxamides are formed very readily by aminolysis of acylhydrazones, which are accessible by condensing carboxylic acid hydrazides with chloral hydrate[345]:

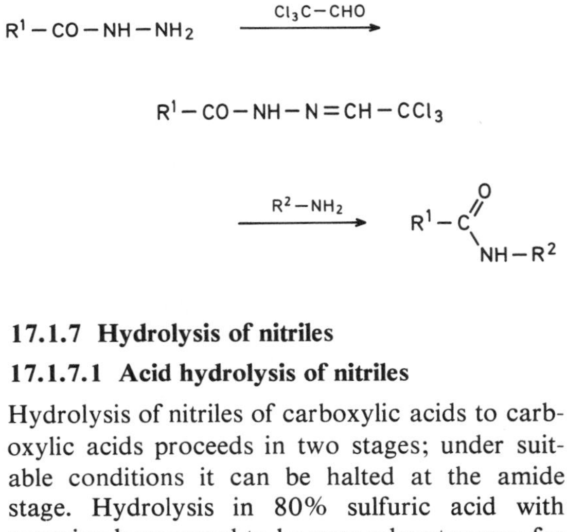

17.1.7 Hydrolysis of nitriles

17.1.7.1 Acid hydrolysis of nitriles

Hydrolysis of nitriles of carboxylic acids to carboxylic acids proceeds in two stages; under suitable conditions it can be halted at the amide stage. Hydrolysis in 80% sulfuric acid with warming has proved to be very advantageous for this purpose as sterically hindered nitriles, too, afford the *acylamines*[346, 347]:

[340] *G.H.L. Nefkens*, Nature 185, 309 (1960).
[341] *H.A. Staab, M. Lüking, F.H. Dürr*, Chem. Ber. *95*, 1275 (1962);
H.A. Staab, Angew. Chem. *74*, 407 (1962).

[342] *J.C. Sheehan, V.S. Frank*, J. Amer. Chem. Soc. *71*, 1856 (1949).
[343] *W. Grassmann, E. Schulte-Uebbing*, Chem. Ber. *83*, 244 (1950).
[344] *I. Schumann, R.A. Boissonnas*, Nature *169*, 154 (1952).
[345] *T. Kametani, O. Umezawa*, Chem. Pharm. Bull. (Tokyo) *14*, 396 (1966).
[346] *N. Sperber, D. Papa, E. Schwenk*, J. Amer. Chem. Soc. *70*, 3091 (1948).
[347] *F.S. Prout, B. Burachinsky, W.T. Brannen jr., H.L. Young*, J. Org. Chem. *25*, 835 (1960).

17.1.7 Hydrolysis of nitriles

$$R-\underset{\underset{R}{|}}{\overset{\overset{R}{|}}{C}}-C\equiv N \quad + \quad H_2O \quad \xrightarrow{H_2SO_4}$$

$$R-\underset{\underset{R}{|}}{\overset{\overset{R}{|}}{C}}-CO-NH_2$$

The yields of amide are very good, even with nitriles that cannot be hydrolyzed with potassium hydroxide in butanol or by the method of Radziszewski[346]. Instead of sulfuric acid the hydration can be carried out with polyphosphoric acid at $110-150°$ [348, 349]. Under these conditions β-keto nitriles form *β-keto carboxamides* just like by hydrolysis in glacial acetic acid/boron trifluoride at $120-140°$ [350]. At room temperature *2,3-diphenyltartramide* is obtained from the corresponding nitrile in hydrogen bromide/glacial acetic acid[351]:

$$\begin{array}{c} H_5C_6-\overset{\overset{OH}{|}}{C}-CN \\ H_5C_6-\underset{\underset{OH}{|}}{C}-CN \end{array} \quad + \quad 2\,H_2O$$

$$\xrightarrow{\hspace{2cm}} \quad \begin{array}{c} H_5C_6-\overset{\overset{OH}{|}}{C}-CO-NH_2 \\ H_5C_6-\underset{\underset{OH}{|}}{C}-CO-NH_2 \end{array}$$

Hydrolysis with formic acid in the absence of solvent can be accomplished very readily to give almost quantitative yields[352]:

$$R-C\equiv N \quad + \quad HCOOH \quad \xrightarrow{180-200°}$$

$$R-CO-NH_2 \quad + \quad CO$$

To bring it about hydrogen chloride or bromide is passed into a formic acid solution of the nitrile[353].

By contrast, treating nitriles in ether solution with hydrogen bromide leads to an initial formation of imidium acid chloride chlorides, which hydrolyze to the corresponding amides in 79–94% yield with water or aqueous hydrochloric acid[354–356]:

$$R-C\equiv N \quad + \quad 2\,HCl \quad \rightleftharpoons$$

$$R-\overset{\overset{Cl}{|}}{\underset{\underset{NH_2}{|}}{C}}{}^{\oplus} \quad Cl^{\ominus} \quad \xrightarrow[-2\,HCl]{H_2O} \quad R-C\overset{O}{\underset{NH_2}{\diagdown}}$$

Hydrolysis of nitriles in the presence of boron halides and one equivalent water probably proceeds similarly[357].

17.1.7.2 Alkaline hydrolysis of nitriles – The Radziszewski method

Alkaline hydrolysis of nitriles to carboxamides does not succeed in every case, so that this technique is less widely used than acid hydrolysis. Alkaline hydrolysis is carried out in special cases only, for instance with pyridinecarbonitriles[358] or dinitriles of dicarboxylic acids[338] using basic ion exchangers. Alkane diacid dinitriles having their functional groups more than one methylene group apart and non*ortho* aromatic dicyano compounds can be hydrolyzed selectively with the aid of ion exchangers[338]. *Cyanocarboxamides* are obtained in good yield in this way side by side with smaller proportions of cyanocarboxylic acids and diamides. The reaction is performed with strongly basic ion exchangers in aqueous solution or in pyridine, *e.g.:*

$$\begin{array}{c} H_2C-CN \\ | \\ H_2C-CN \end{array} \quad + \quad H_2O \quad \xrightarrow{OH^{\ominus}} \quad \begin{array}{c} H_2C-CO-NH_2 \\ | \\ H_2C-CN \end{array}$$

3-Cyanopropion-amide; 36%

[348] H.R. Snyder, C.T. Elston, J. Amer. Chem. Soc. 76, 3039 (1940).

[349] F.D. Popp, W.E. McEwen, Chem. Rev. 58, 390 (1958).

[350] C.R. Hauser, C.J. Eby, J. Amer. Chem. Soc. 79, 725 (1957).

[351] F. Micheel, R. Austrup, A. Striebeck, Chem. Ber. 94, 132 (1961).

[352] F. Becke, J. Gnad, Justus Liebigs Ann. Chem. 713, 212 (1968).

[353] F. Becke, H. Fleig, P. Pässler, Justus Liebigs Ann. Chem. 749, 198 (1971).

[354] E.N. Zil'berman, A.I. Kulkova, N.A. Sazanova, Khim. Nauka i Prom., 4, 135 (1959); C.A. 53, 14992h (1959).

[355] A.E. Kulikova, E.N. Zil'berman, N.A. Sazanova, Zh. Obshch. Khim. 30, 2180 (1960); C.A. 55, 8345g (1961).

[356] E.N. Zil'berman, A.E. Kulikova, Zh. Obshch. Khim. 29, 1694 (1959); C.A. 54, 8629b (1960).

[357] J.R. Blackborrow, J. Chem. Soc. C, 739 (1969).

[358] J.M. Bobbitt, D.A. Scola, J. Org. Chem. 25, 560 (1960).

A much more often successful use of the Radziszewski hydrolysis is the reaction with hydrogen peroxide in the presence of alkali at moderate temperature[359, 360]:

$$R-CN \ + \ 2\ H_2O_2 \longrightarrow$$

$$R-CO-NH_2 \ + \ O_2 \ + \ H_2O$$

The reaction is initiated by the attack of a peroxide anion and is followed by the fast reaction of a further hydrogen peroxide molecule accompanied by splitting off of oxygen[361, 361a]. Certain α,β-unsaturated nitriles suffer epoxidation to *glycidamides (oxiranecarboxamides)* in addition to the hydrolysis[362]:

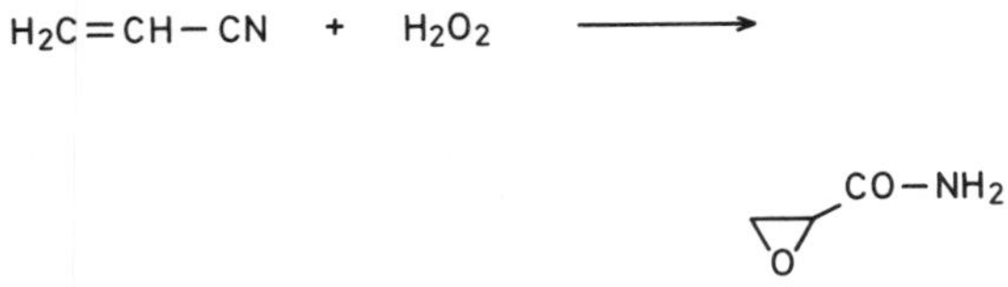

$$H_2C=CH-CN \ + \ H_2O_2 \longrightarrow$$

while cinnamonitrile furnishes *cinnamamide* during the normal reaction[363].

17.1.7.3 The Graf-Ritter reaction

Formation of *N*-substituted amides by addition of prussic acid to olefins in the presence of strongly acid catalysts (first described by R. Graf)[364] can also be transferred to nitriles (Ritter)[365]. These reactions involve a nucleophilic addition of nitriles or hydrogen cyanide to a carbonium ion formed from an olefin with a strong acid, generally sulfuric acid[366]:

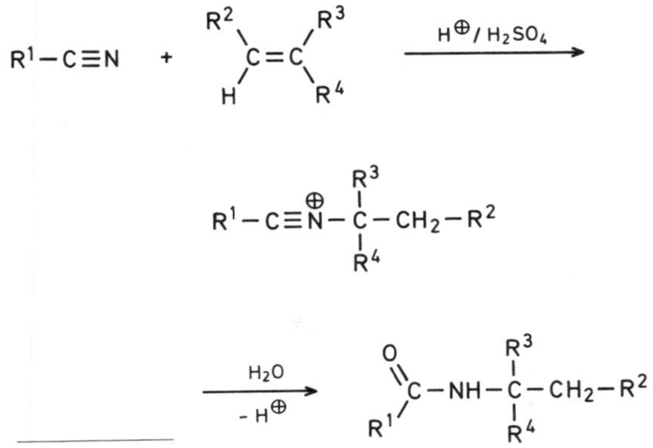

Best results are obtained in 85–90% sulfuric acid, but polyphosphoric acid, formic acid, boron trifluoride, and other Lewis acids (not aluminum chloride!) can also be used[366].

The *scope* of the Graf-Ritter reaction is not limited to olefins; alkanes, alcohols, alkyl halides, or α,β-unsaturated carbonyl compounds, react analogously, *i.e.*, compounds which can form carbonium ions under the action of an acid[366]. A phenyl group or a double bond conjugate with respect to the nitrile group does not affect the reaction much but promotes the addition in the following sequence[366]:

$$H_2C=CH-CN \ > \ H_5C_6-CN \ > \ H_3C-CN$$

Secondary or tertiary alcohols afford *acrylamides* with acrylonitrile in the presence of sulfuric acid, *e.g.*,[367]:

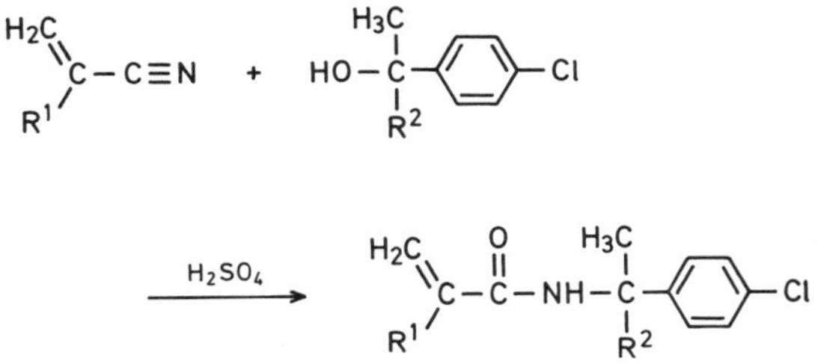

$$R^1 = H, CH_3$$
$$R^2 = H, CH_3, C_2H_5$$

In a variant of the Graf-Ritter reaction proceeding under mild conditions mercury(II) nitrate is used as catalyst and sodium tetrahydroborate as reducing agent[368]:

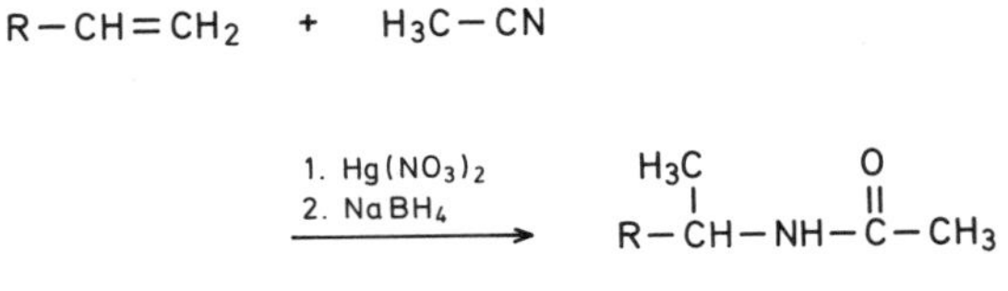

$$R-CH=CH_2 \ + \ H_3C-CN$$

[359] B. Radziszewski, Ber. 18, 355 (1885).
[360] L. McMaster, F.B. Langreck, J. Amer. Chem. Soc. 39, 103 (1917).
[361] K.B. Wiberg, J. Amer. Chem. Soc. 75, 3961 (1953).
[361a] J.E. McIsaac, R.E. Ball, E.J. Behrman, J. Org. Chem. 36, 3048 (1971).
[362] G.B. Payne, P.H. Williams, J. Org. Chem. 26, 651 (1961).
[363] J.V. Murray, C.B. Cloke, J. Amer. Chem. Soc. 56, 2749 (1934).
[364] DRP. 870856 (1940), Farbwerke Hoechst; Erf.: R. Graf; Chem. Zentr. 6341 (1954).
[365] J.J. Ritter, P.P. Mineri, J. Amer. Chem. Soc. 70, 4045 (1948).
[366] L.I. Krimen, D.J. Cota, Org. Reactions 17, 213 (1969).

17.1.7.4 The Passerini-Ugi reaction

Passerini discovered an elegant and universally applicable synthesis of α-*acyloxycarboxamides* in the three-component reaction between carboxylic acids, carbonyl compounds and isonitriles[369]:

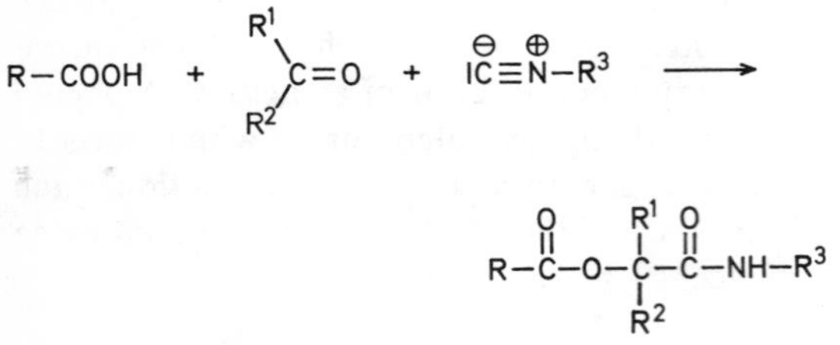

Careful hydrolysis yields the α-*hydroxycarboxamides* that are obtained also directly from isonitriles and ketones on catalysis with mineral acid[370]. If the reaction is conducted in the presence of immonium ions formed from amine and carbonyl compound then α-*aminocarboxamides* are formed[369]:

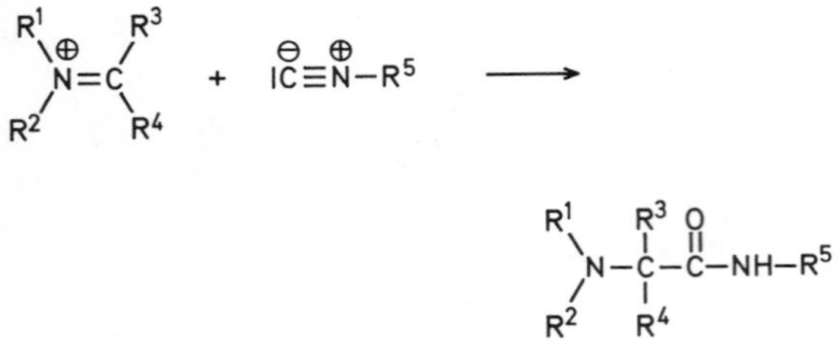

Primary amines occasionally react with both hydrogen atoms, especially if used in less than the theoretical amount. Thus, *2,2'-(butylimino)bis[N-cyclohexylacetamide]* (52%) is obtained from butylamine, formaldehyde, and cyclohexyl isocyanide[369]:

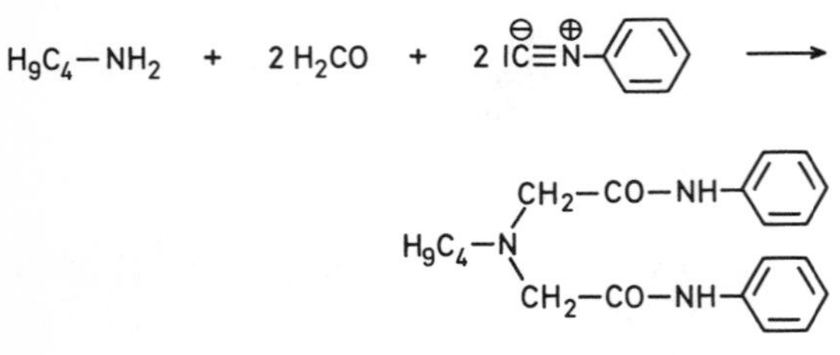

Ammonia can be reacted only in the presence of carboxylic acids, because otherwise no definite products are formed. It is most successful during α-aminoacylation of isonitriles and carboxylic acids, which affords excellent yields of α-*acylaminocarboxamides* in a four-component condensation[369]:

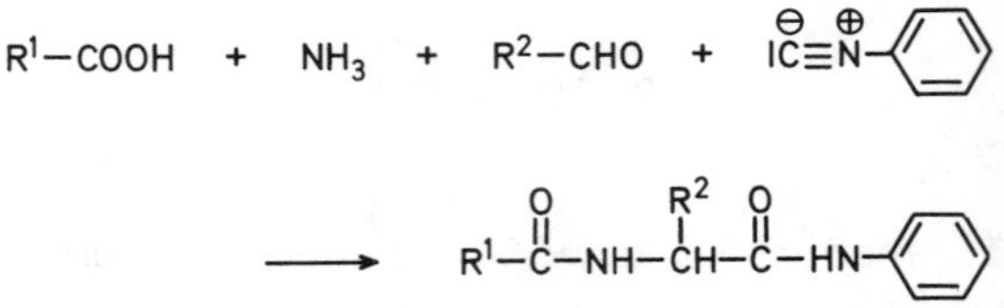

17.1.8 Acyclic and cyclic carboxamide, amides, hydrazides, hydroxyamides, imides, azides, from aldehydes and ketones[371]

17.1.8.1 Beckmann rearrangement of aldoximes

The Beckmann rearrangement[372, 373] of *ketoximes* to amides (p. 564) and of *aldoximes* (p. 639), which leads quite generally to nitriles[372, 374], is described elsewhere in this volume. Under certain reaction conditions *primary carboxamides* can be obtained from aldoximes in the presence of phosphorus(V) chloride, sulfuric acid, or trifluoroacetic acid[375]; substituted formamides arise as side-products in some cases[376].

Hydrolysis of the initially formed nitriles is postulated as the amide formation mechanism[377].

By contrast, using polyphosphoric acid[378-380] or boron trifluoride[380, 381] smoothly affords a good

---

[367] *S. Tanimoto, M. Kimura, M. Okano*, Yuki Gosei Kagaku Kyokai Shi *28*, 1035 (1970).

[368] *H.C. Brown, J.T. Kurek*, J. Amer. Chem. Soc. *91*, 5647 (1969).

[369] *I. Ugi* in *W. Foerst*, Neuere Methoden der präparativen organischen Chemie, Bd. IV, p. 1, Verlag Chemie, Weinheim/Bergstr. 1966.

[370] *I. Hagedorn, U. Eholzer*, Chem. Ber. *98*, 936 (1965).

[371] Re the Schmidt reaction and Wolff rearrangement see pp. 567 and 288; re the Willgerodt reaction see p. 707.

[372] *L.G. Donaruma, W.Z. Heldt*, Org. Reactions *11*, 1 (1960).

[373] *P.A.S. Smith* in *P. de Mayo*, Molecular Rearrangements, Vol. I, p. 457, Interscience Publishers, New York, London 1963.

[374] *A. Hantzsch, A. Lucas*, Ber. *28*, 744 (1895);
A.H. Blatt, Chem. Rev. *12*, 215 (1933);
E.C. Horning, V.L. Stromberg, J. Amer. Chem. Soc. *74*, 5151 (1952);
G.W. Wheland, Advanced Organic Chemistry, p. 449, J. Wiley & Sons, New York 1960.

[375] See references in ref. [378].

[376] *A. Hantzsch, A. Lucas*, Ber. *28*, 744 (1895).

[377] *P.A.S. Smith*, The Chemistry of Open Chain Organic Nitrogen Compounds, Vol. II, p. 29, Benjamin, New York, Amsterdam 1966.

[378] *E.C. Horning, V.L. Stromberg*, J. Amer. Chem. Soc. *74*, 5151 (1952).

[379] *E.C. Horning, V.L. Stromberg, H.A. Lloyd*, J. Amer. Chem. Soc. *74*, 5153 (1952).

[380] *I.P. De Keersmaeker, F. Fontyn*, Ind. Chim. Belge *32*, 1087 (1967).

yield of amides from oximes of aliphatic aldehydes:

$$R-CH=N-OH \xrightarrow{H_3PO_4 \text{ or } BF_3} R-CO-NH_2$$

R = C_3H_7 to C_8H_{17}

With oximes of aromatic aldehydes the Beckmann rearrangement using polyphosphoric acid as catalyst furnishes a uniform yield of amides only in the case of the *anti* configuration[378, 382].

Syn-aldoximes afford either amides, amides admixed with formamides, nitriles, or unidentified products[378, 382]. These results are especially interesting in respect of the mechanism, because during the Beckmann rearrangement that group migrates which stands in the *anti* position to the oxime hydroxy group. In most oximes the exact stereochemical configuration and the *syn* : *anti* ratio are unknown; in addition *syn-anti* rearrangements can ensue under the conditions of the Beckmann rearrangement[372, 373, 383].

An interesting variant is the synthesis of *2-cyanolactamide* (in 82.5–92.9% yield) from pyruvaldehyde 1-oxime and hydrogen cyanide in the presence of a basic catalyst (alkali metal hydroxides, cyanides, organic bases[384], Eq. 1). Formation of *2-phenylglyoxylamide* from 2-phenylglyoxal 1-oxime with sodium bicarbonate represents a similar reaction[385] (Eq. 2):

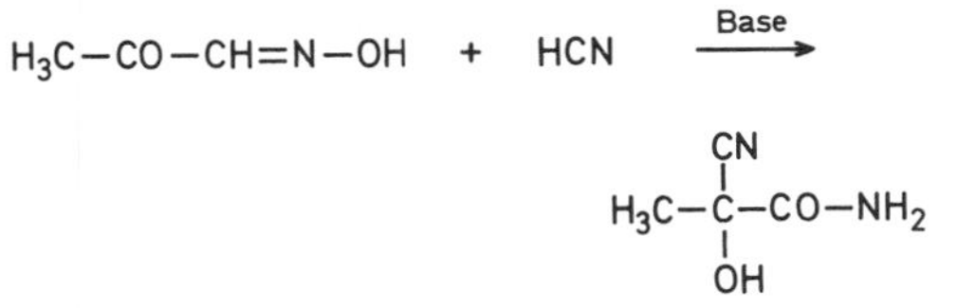

That Raney nickel catalysts and copper(I) halides catalyze the Beckmann rearrangement of aldoximes has been known for a long time[386-388] but no use has been made of this reaction. It is only recently that heavy metal compounds have been reinvestigated as catalysts[389]. Nickel(II) acetate in boiling xylene was found to be particularly suitable for the isomerization of aldoximes of aliphatic and aromatic aldehydes to primary carboxamides (yields around or more than 80%).

Photolytic Beckmann rearrangements of aldoximes are feasible just like with ketoximes[390, 391]; an intermediate oxime is postulated as the mechanism; *e.g.*:

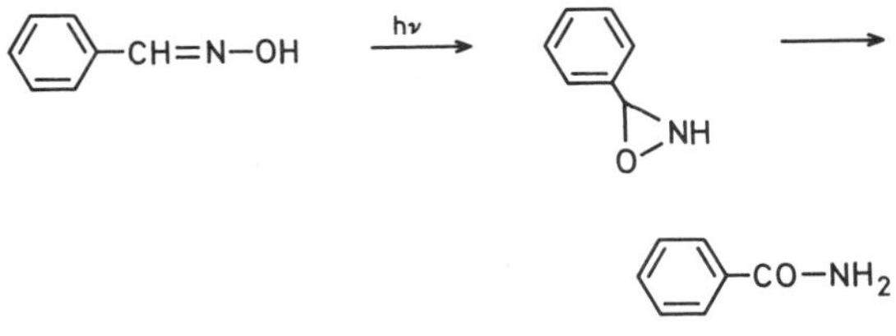

Benzamide

17.1.8.2 Diazotization of hydrazones and semicarbazones

Hydrazones and semicarbazones can be diazotized with sodium nitrite in sulfuric acid[392-394] or polyphosphoric acid[395]. Amides are formed in a rearrangement similar to the Schmidt reaction[396]. Hydrazones and semicarbazones

$$H_3C-CO-CH=N-OH + HCN \xrightarrow{Base}$$

(1)

$$H_5C_6-CO-CH=N-OH \xrightarrow{NaHCO_3}$$

$$H_5C_6-CO-CO-NH_2 \quad (2)$$

[381] *C.R. Hauser, D.S. Hoffenberg*, J. Org. Chem. *20*, 1482 (1955);
C.R. Hauser, D.S. Hoffenberg, J. Org. Chem. *20*, 1491 (1955);
D.S. Hoffenberg, C.R. Hauser, J. Org. Chem. *20*, 1496 (1955).

[382] *M.G. Deshmukh, K.C. Jain*, Indian J. Chem. *6*, 337 (1968); C.A. *69*, 85930a (1968).

[383] *C.G. McCarty* in *S. Patai*, The Chemistry of the Carbon-Nitrogen Double Bond, p. 363, Wiley Intersci. Publ., London, New York, Sydney, Toronto 1970.

[384] USP. 3238244 (1966), Sicedison Societa per Azioni; Inv.: *A. Nenz, L. Marangoni;* Belg. P. 631288 (1963), Sicedison Societa per Azioni; Inv.: *A. Nenz, L. Marangoni;* C.A. *60*, 14395b (1964).

[385] *S. Kodama*, J. Chem. Soc. Japan *44*, 339 (1923); C.A. *17*, 3023 (1923).

[386] *W.J. Comstock*, Amer. Chem. J. *19*, 485 (1897).

[387] *R. Paul*, C.R. Acad. Sci., Paris *204*, 363 (1937).

[388] *R. Paul*, Bull. Soc. Chim. France [5] *4*, 1115 (1937).

[389] *L. Field, P.B. Hughmark, S.H. Shumaker, W.S. Marshall*, J. Amer. Chem. Soc. *83*, 1983 (1961).

[390] *H. Izawa, P. De Mayo, T. Tabata*, Can. J. Chem. *47*, 51 (1969).

[391] *K.H. Grellmann, E. Tauer*, Tetrahedron Lett. 1909 (1967).

[392] *D.E. Pearson, C.M. Greer*, J. Amer. Chem. Soc. *71*, 1895 (1949).

[393] *D.E. Pearson, K.N. Carter, C.M. Greer*, J. Amer. Chem. Soc. *75*, 5905 (1953).

[394] *K.N. Carter*, J. Org. Chem. *23*, 1409 (1958).

[395] *P.T. Lansbury, N.R. Mancuso*, J. Amer. Chem. Soc. *88*, 1205 (1966).

[396] *L.G. Donaruma, W.Z. Heldt*, Org. Reactions *11*, 49 (1960).

of diaryl ketones, and alkyl aryl ketones[392-394], indanones, and tetralones[395] affords good yields:

$$R_2C=N-NH_2 \xrightarrow[-N_2]{HNO_2}$$

$$R_2C=N-NH-CO-NH_2 \xrightarrow[-N_2,\ -CO_2]{HNO_2,\ H_2O}$$

$$\left[R_2C=N^{\oplus}\right] \xrightarrow{H_2O} R-CO-NH-R$$

Hydrazones rearrange to amides in a similar reaction at 300° in the presence of Lewis acids[397]. This reaction is of technical interest for the preparation of *caprolactam (hexahydro-2H-azepin-2-one);* either cyclohexanone semicarbazone is treated with sodium nitrite-acid[398] or acid is added to an aqueous solution of nitrocyclohexane, hydrazine, and sodium nitrite[399].

17.1.8.3 Aminolysis of ketones

17.1.8.3.1 With ammonia and amines

During aminolysis of ketones the first reaction step must be a nucleophilic addition of ammonia or amine to the $O{=}C$ double bond. The *zwitterion* formed undergoes cleavage with formation of an amide cation and a carbanion:

$$R^1-\overset{O}{\overset{\|}{C}}-R^2 + H-N\diagdown \rightleftharpoons R^1-\overset{\overset{\ominus}{|\overset{\cdot\cdot}{O}|}}{\underset{\overset{\oplus}{NH}}{C}}-R^2$$

$$\left[R^1-\overset{O}{\overset{\|}{C}}-\overset{\oplus}{N}H\right]^{\ominus}R^2 \longrightarrow R^1-\overset{O}{\overset{\|}{C}}-N\diagdown + R^2H$$

$$\left[R^2-\overset{O}{\overset{\|}{C}}-\overset{\oplus}{N}H\right]^{\ominus}R^1 \longrightarrow R^2-\overset{O}{\overset{\|}{C}}-N\diagdown + R^1H$$

Formation of amides by aminolysis of carbonyl compounds will therefore be dependent above all on the stabilization of the carbanions $^{\ominus}R^1$ or $^{\ominus}R^2$ and, as a result, succeeds only in special cases.

Thus the *S,S*-diesters of acyldithiomalonic acids can be cleaved with ammonia in ethanol or 1,4-dioxane to form *amides* and *malonamide*[400]:

$$R-CO-CH(CO-S-C_2H_5)_2 \xrightarrow[-2\ HS-C_2H_5]{NH_3}$$

$$R-CO-NH_2 + H_2C(CONH_2)_2$$

R = CH₃; *Acetamide*
R = C₄H₉; *Butyramide*
R = C₆H₅; *Benzamide*

Straight-chain or cyclic β-keto sulfones only sometimes react with pyrrolidine to form enamines; normally cleavage to amides takes place[401], *e.g.:*

$$H_5C_6-CH_2-CO-\overset{R}{\overset{|}{C}}H-SO_2-CH_2-C_6H_5 + HN\diagup$$

$$\xrightarrow[-CH_2-SO_2-CH_2-C_6H_5]{} H_5C_6-CH_2-CO-N\diagup$$

R = H; C₆H₅   *N-Phenylacetylpyrrolidine*

$$H_5C_6-CH_2-SO_2-CH_2-(CH_2)_n-(CH_2)_2-CO-N\diagup$$

n = 2; *5-Benzylsulfonyl pentanoic pyrrolidide*
n = 3; *6-Benzylsulfonyl hexanoic pyrrolidide*

Perhalogenated ketones afford amides even under mild conditions with ammonia and amines in a type of haloform cleavage. Aliphatic, aromatic, and heterocyclic amines can be trichloroacetylat-

[397] *J. Stieglitz, J.K. Senior,* J. Amer. Chem. Soc. *38,* 2727 (1916).
[398] USP. 2763644 (1956), Du Pont; Inv.: *L.G. Donaruma;* C.A. *51,* 5822c (1957).
[399] USP. 2777841 (1957), DuPont; Inv.: *L.G. Donaruma;* C.A. *51,* 10565d (1957).
[400] *L.B. Dashkevich, L.V. Konovalova, V.A. Pechenyuk,* Zh. Org. Khim. *6,* 1511 (1970); engl.: *6,* 1523 (1970).
[401] *J.J. Looker,* J. Org. Chem. *31,* 2714 (1966).

ed with hexachloro-2-propanone[402, 403]. Amino acids and peptides afford *N-trichloroacetyl* derivatives in dimethylformamide at room temperature[404]. Anilines react with 1,3-dibromotetrachloro-2-propanone[405], and ammonia reacts with 2,4-bis(trifluoromethyl)octafluoro-3-pentanone in an entirely analogous manner[406].

$$Cl_3C-CO-CCl_3 \quad + \quad HN\diagdown \quad \xrightarrow[-HCCl_3]{} \quad Cl_3C-CO-N\diagdown$$

Unsymmetrical, perhalogenated ketones ought to be less suitable, but during synthesis of *chloramphenicol* an amino alcohol was selectively dichloroacetylated with pentachloro-2-propanone[407, 408] in good yield:

$$O_2N-\!\!\!\!\bigcirc\!\!\!\!-\!\!\overset{\overset{OH}{|}}{CH}-\!\!\overset{\overset{NH_2}{|}}{CH}-\!CH_2OH \quad \xrightarrow{Cl_3C-CO-CHCl_2}$$

$$O_2N-\!\!\!\!\bigcirc\!\!\!\!-\!\!\overset{\overset{OH}{|}}{CH}-\!\!\overset{\overset{NH-CO-CHCl_2}{|}}{CH}-\!CH_2-\!OH$$

1-(4-Nitrophenyl)-2-(dichloro-acetamido)-1,3-propanediol (Chloramphenicol);

1,1,3,3-Tetramethyl-2,4-cyclobutanedione, the normal dimer of dimethyl ketene, affords with ammonia[404, 410] and with prim. and sec. amines[410] an excellent yield of *2,4,4-trimethyl-3-oxovaleramides* (75–81%)[411]:

$$\begin{array}{c} H_3C \\ H_3C \end{array}\!\!\diagup\!\!\overset{O}{\overset{||}{\square}}\!\!\diagdown\!\!\begin{array}{c} CH_3 \\ CH_3 \end{array} \quad + \quad H_2N-R \quad \longrightarrow$$

$$(H_3C)_2CH-CO-\overset{\overset{CH_3}{|}}{\underset{\underset{CH_3}{|}}{C}}-CO-NH-R$$

[402] *B. Sukornick*, Org. Synth. *40*, 103 (1960).

[403] *V.P. Rudavskii, I.G. Khaskin*, Ukr. Khim. Zh. *33*, 391 (1967); C.A. *67*, 63963v (1967).

[404] *C.A. Panetta, T.G. Casanova*, J. Org. Chem. *35*, 2423 (1970).

[405] USP. 3 405 176 (1968), Allied Chemical Corp., Inv.: *B.S. Farah, E.E. Gilbert*; C.A. *70*, 28670n (1969).

[406] *R.D. Smith, F.S. Fawcett, D.D. Coffman*, J. Amer. Chem. Soc. *84*, 4285 (1962).

[407] *J. Kollonitsch, A. Hajós, V. Gábor, M. Kraut*, Acta Chim. Acad. Sci. Hung. *5*, 13 (1954); C.A. *50*, 230b (1956).

[408] *J. Kollonitsch, A. Hajós, V. Gábor, M. Kraut*, Experientia *10*, 438 (1954).

[409] *R.H. Hasek, E.U. Elam, J.C. Martin*, J. Org. Chem. *26*, 4340 (1961).

17.1.8.3.2 Aminolysis of ketones with alkali metal amides (Haller-Bauer reaction)

Cleavage of nonenolizable ketones with sodium amide[413] in boiling benzene, toluene, or xylene *(Haller-Bauer reaction)* affords amides and hydrocarbons[415–417] (*cf.* the scheme on p. 704).

As the reaction scheme on p. 704 reveals, the cleavage can lead to two different acyl amides if R^1 and R^2 are unlike. The site of the cleavage is dependent on the electronegativity of R^1 and R^2; the group forming the more stable carbanion is split off preferentially.

Where the stability of the carbanions of R^1 and R^2 is similar mixtures of amides are invariably obtained[418, 419].

Large substituent groups R^1-R^3 make the Haller-Bauer reaction difficult. In such cases the less stable group R^4 may be split off[420, 421]. Where $R^1 = R^2 = CH_3$ the cleavage proceeds without

[410] Tetramethyl-1,3-cyclobutanedione, 2,2,4,4-Tetramethyl-1,3-cyclobutanediol, Properties, Reactions, Technical Data Report, Eastman Chemical Products, May 1960.

[411] It has been shown that primary amines form a Schiff base in the first reaction step. The cleavage of the fourmembered ring is caused by the water formed[412].

[412] *G.R. Hansen, R.A. DeMarco*, J. Heterocycl. Chem. *6*, 291 (1969).

[413] Sodium amide is used exclusively in the *Haller-Bauer reaction*. Mixtures of $NaNH_2/KNH_2$ cause a complete elimination of the carboxyl group[414].

[414] *L.Kh. Freidlin, A.A. Balandin, A.I. Lebedeva*, Bull. Acad. Sci. USSR, Div. Chem. Sci. 167 (1941); C.A. *37*, 3749 (1943).

[415] *F.W. Semmler*, Ber. *39*, 2577 (1906).

[416] *A.L.J. Beckwith* in *J. Zabicky*, The Chemistry of Amides, p. 111, Wiley Intersci. Publ., London, New York, Sydney, Toronto 1970;
J.F. Bunnett, B.F. Hrutfiord, J. Org. Chem. *27*, 4152 (1962);
C.L. Bumgardner, K.G. McDaniel, J. Amer. Chem. Soc. *91*, 6821 (1969);
G.W. Kenner, M.J.T. Robinson, C.M.B. Tylor, B.R. Webster, J. Chem. Soc. (London) 1756 (1962).

[417] *K.E. Hamlin, A.W. Weston*, Org. Reactions *9*, 1 (1957).

[418] *T.R. Lea, R. Robinson*, J. Chem. Soc. (London) 2351 (1926).

[419] *P. De Ceuster*, Natuurw. Tijdschr. *14*, No. 3–6, 188 (1932); C.A. *26*, 4323 (1932).

[420] *C.L. Bumgardner, K.G. McDaniel*, J. Amer. Chem. Soc. *91*, 6821 (1969).

[421] *C.L. Carter, S.N. Slater*, J. Chem. Soc. (London) 130 (1946).

difficulty and the group R^3 can have almost any desired size[421].

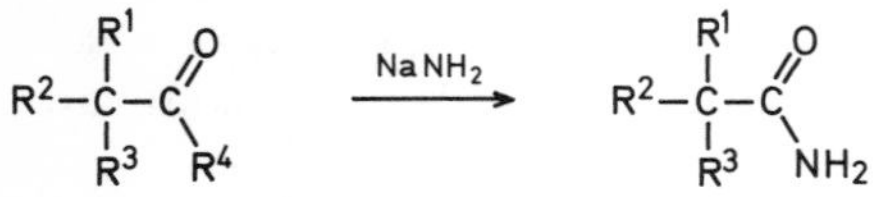

Even replacing one methyl group by an ethyl group makes the reaction more difficult to carry out. As a general rule it may be stated that the Haller-Bauer reaction proceeds readily if the sum of the C atoms of $R^1 + R^2 + R^3 \leqq 10$.

Because of the unambigous direction of the cleavage and the good yields the Haller-Bauer reaction is exceptionally useful for synthesizing *α,α-dialkyl-substituted carboxamides* (Eq. **1**), *α,α,α',α'-tetraalkyl amides* of *dicarboxylic acids* (Eq. **2**), and *1-alkylcycloalkane-1-carboxamides* (Eq. **3**)[417]:

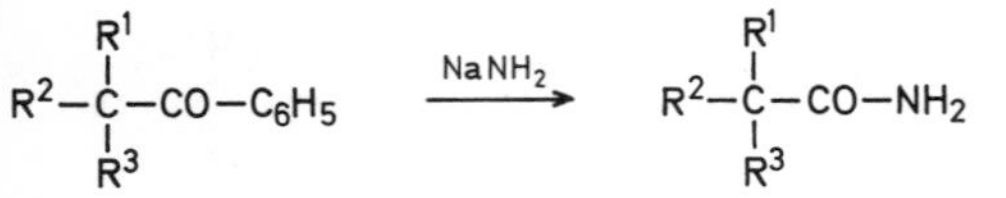

$$ X = \langle benzene \rangle \; ; -(CH_2)_n- $$

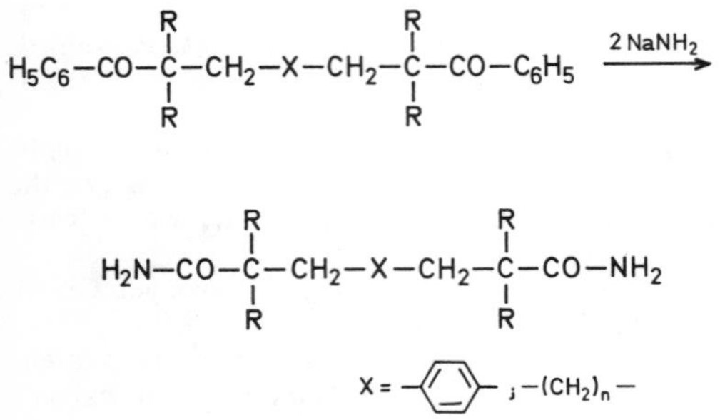

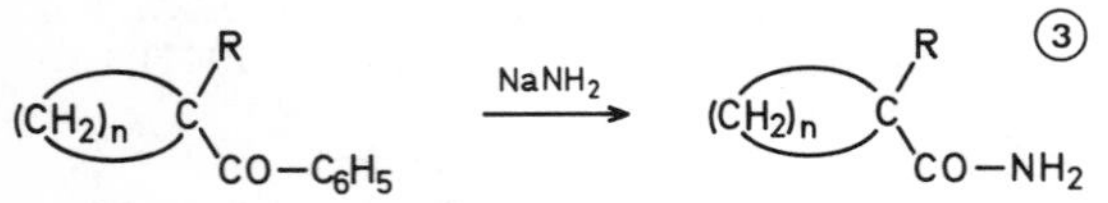

n = 2*, 3, 4

---

* Optically active compounds retain their configuration during the reaction[420].

Re additional applications of the Haller-Bauer reaction for solving synthetic problems reference is made to the original work[422-427].

---

[422] *D. Varech, C. Ouannes, J. Jacques,* Bull. Soc. Chim. France 1662 (1965).

[423] *H.J. Teuber, O. Glosauer,* Chem. Ber. **98**, 2939 (1965).

[424] *A.J. Forlano, C.I. Jarowski, H.F. Hammer, E.G. Merrit,* J. Pharm. Sci. **59**, 121 (1970); C.A. **72**, 125024d (1970).

Of cyclic ketones cyclohexanone is stable toward sodium amide[428], while symmetrically substituted alicyclic ketones afford alkanecarboxamides *via* ring opening (for a survey see ref.[417]).

Applying the sodium amide cleavage to 4-benzoyl-4-alkyl-1-alkenes gives *3,3-dialkyl-5-methyl-2-pyrrolidinones* (the *2,2-dialkyl-4-pentenamides* initially formed by aminolysis cyclize under the basic reactions conditions)[417, 429]:

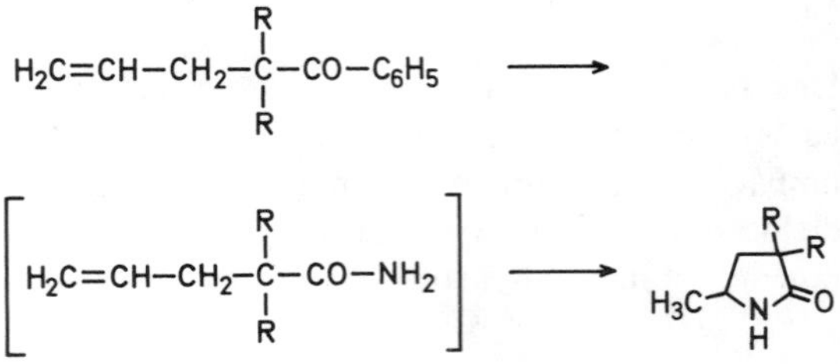

From the following chalcone *4-methoxycinnamamide* is obtained with sodium amide[430]:

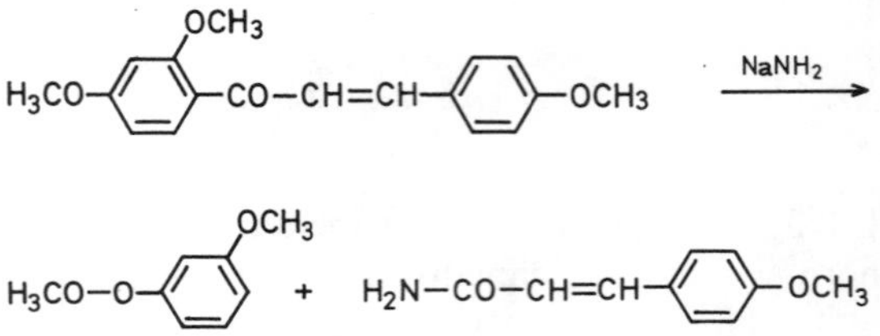

17.1.9 Oxidation reactions

17.1.9.1 Of Amines

The oxidation of amines is a very complex process and only in the case of tertiary amines does it lead to *carboxamides* if certain conditions are maintained. Thus some *N*-alkylanilines can be oxidized to *N-methylformanilides* with a large excess of

---

[425] *A.L.J. Beckwith,* J. Chem. Soc. (London) 2248 (1962).

[426] *K. Winterfeld, H. Meyer,* Arch. Pharm. (Weinheim, Ger.) **294**, 630 (1961).

[427] Tert-carboxamides furnish the corresponding acids and their derivatives. Tertiary amines, which are otherwise difficult to prepare, can be synthesized by Schmidt, Hoffmann, or Curtius degradation (cf. this volume, pp. 562–568).

[428] *A. Haller,* Bull. Soc. Chim. France [4] *31*, 1117 (1922).

[429] *R.F. Brown, N.M. van Gulick,* J. Amer. Chem. Soc. **77**, 1092 (1955).

[430] *R. Calcinari,* Ann. Chim. (Rome) *60*, 405 (1970); C.A. *73*, 87603c (1970).

manganese(IV) oxide[431]. The reaction affords quite good yields at room temperature and succeeds also with aliphatic trialkylamines in cyclohexane as solvent[432]:

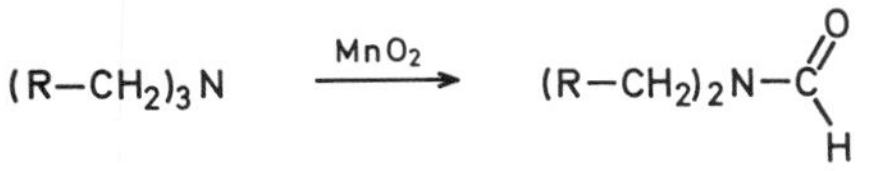

$$(R-CH_2)_3N \xrightarrow{MnO_2} (R-CH_2)_2N-\overset{\displaystyle O}{\underset{\displaystyle H}{C}}$$

Triethylenediamine (1,4-diazabicyclo[2..2.2]octane) affords 1,2-piperazinedicarboxaldehyde (9%) in substantially lower yield under these conditions[433]:

Chromic acid oxidizes tertiary amines in pyridine to *formamides* in substantially better yield, as has been shown for the example of several steroid derivatives[434].

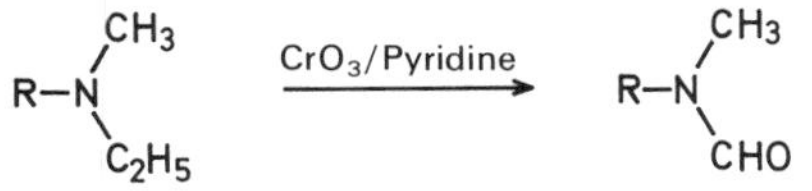

$$R-\overset{\displaystyle CH_3}{\underset{\displaystyle C_2H_5}{N}} \xrightarrow{CrO_3/Pyridine} R-\overset{\displaystyle CH_3}{\underset{\displaystyle CHO}{N}}$$

In *N*-methylpyrrolidines a methylene group adjacent the nitrogen is oxidized to the keto group and the corresponding *1-methyl-2-pyrrolidinone* is obtained[434]. The direct oxidation of tertiary amines with oxygen is of interest[435]. Platinum has been found to be the most efficacious catalyst for this reaction, which proceeds in benzene. *N*-Methyl groups alone are attacked selectively, while *N*-benzyl or *N*-ethyl groups remain untouched. *N-Formylpiperidine* is formed quantitatively from *N*-methylpiperidine[435]:

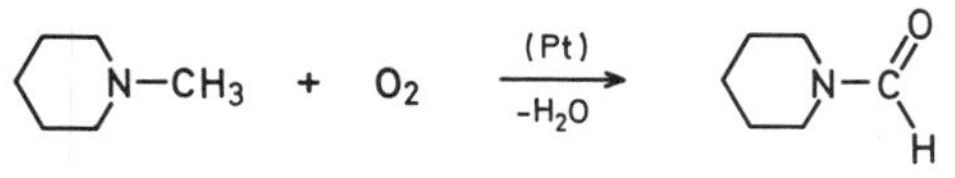

$$\text{(piperidine)}N-CH_3 + O_2 \xrightarrow[-H_2O]{(Pt)} \text{(piperidine)}N-\overset{\displaystyle O}{\underset{\displaystyle H}{C}}$$

17.1.9.2 Oxidation reactions of active methylene compounds by the method of Willgerodt

The *Willgerodt reaction* consists of the oxidation of active methyl or methylene groups with sulfur in the form of ammonium polysulfide[436, 437]. Alkyl aryl ketones are used mainly; on heating for several hours to 150–200° they can be converted into *ω-aryl carboxamides* with just as many carbon atoms as the starting product:

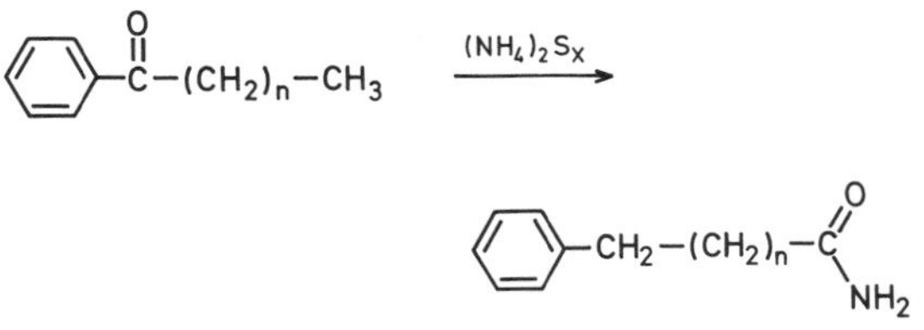

$$\text{(phenyl)}-\overset{\displaystyle O}{C}-(CH_2)_n-CH_3 \xrightarrow{(NH_4)_2S_x}$$

$$\text{(phenyl)}-CH_2-(CH_2)_n-\overset{\displaystyle O}{\underset{\displaystyle NH_2}{C}}$$

Substituents such as alkyl, alkoxy, or halogen on the aromatic ring have little effect on the course of the reaction, nor a branching of the alkyl chain. As the chain length increases the yield of amide falls and side-reactions become more pronouced.

Under these conditions purely aliphatic ketones can be converted into amides in only small yield[438]. Alkylbenzenes[439], olefins[440], and some other compounds[441] as well as ketones react to form carboxamides.

17.1.10 Alkylation reactions

17.1.10.1 Alkylations of *N*-functional carboxylic acid derivatives

On treatment of primary or secondary amides with strong bases salts of amides are formed that can be readily alkylated on the *N*-atom by the action of alkyl halides or alkyl sulfates. A convenient procedure for carrying out the *N*-alkylation enabling long-chain 1-chloroalkanes to be employed as well is to work in dimethyl sulfoxide in the presence of potassium hydroxide[442]:

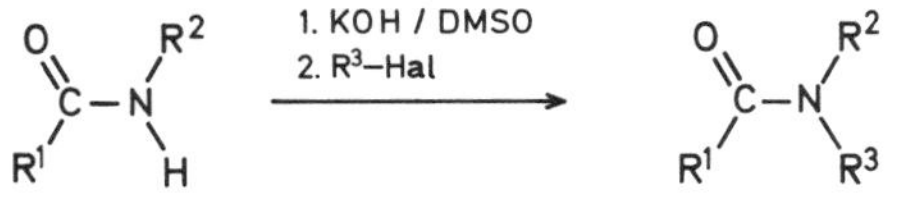

$$\overset{\displaystyle O}{\underset{\displaystyle R^1}{C}}-\overset{\displaystyle R^2}{\underset{\displaystyle H}{N}} \xrightarrow[2.\ R^3-Hal]{1.\ KOH/DMSO} \overset{\displaystyle O}{\underset{\displaystyle R^1}{C}}-\overset{\displaystyle R^2}{\underset{\displaystyle R^3}{N}}$$

[431] *H.B. Henbest, A. Thomas*, J. Chem. Soc. (London) 3032 (1957).
[432] *H.B. Henbest, M.J.W. Stratford*, J. Chem. Soc. C 995 (1966).
[433] *E.F. Curragh, H.B. Henbest, A. Thomas*, J. Chem. Soc. (London) 3559 (1960).
[434] *A. Cave, C. Kan-Fan, P.V. Potier, J. Le Men, M.-M. Janot*, Tetrahedron 23, 4691 (1967).
[435] *G.T. Davis, D.H. Rosenblatt*, Tetrahedron Lett. 4085 (1968).
[436] *M. Carmack, M.A. Spielman*, Org. Reactions 3, 83 (1946).

[437] *R. Wegler, E. Kühle, W. Schäfer* in *W. Foerst*, Neuere Methoden der präparativen organischen Chemie, Bd. III, p. 1, Verlag Chemie, Weinheim/Bergstr., 1961.
[438] *L. Cavalieri, D.B. Pattison, M. Carmack*, J. Amer. Chem. Soc. 67, 1783 (1945).
[439] *W.G. Toland jr., D.L. Hagmann, J.B. Wilkes, F.J. Brutschy*, J. Amer. Chem. Soc. 80, 5423 (1958).
[440] *M.A. Naylor, A.W. Anderson*, J. Amer. Chem. Soc. 75, 5392 (1953).
[441] *V. Franzen*, Chemiker-Ztg. 83, 328 (1959).
[442] *G.I. Isele, A. Lüttringhaus*, Synthesis 266 (1971).

Using stronger bases such as sodium hydride or potassium alkoxides can be eschewed under these conditions. Where sodium hydride is used nonetheless, for instance, for *N*-alkylating carboxamides[443], urethanes[444], or *N*-acyl amino acids[445], xylene, dimethylformamide, or even excess alkyl halide are recommended solvents (alkyl halides do not react with sodium hydride even on boiling)[446]. A particularly interesting case is the intramolecular *N*-alkylation of ω-halocarboxamides to *lactams*:

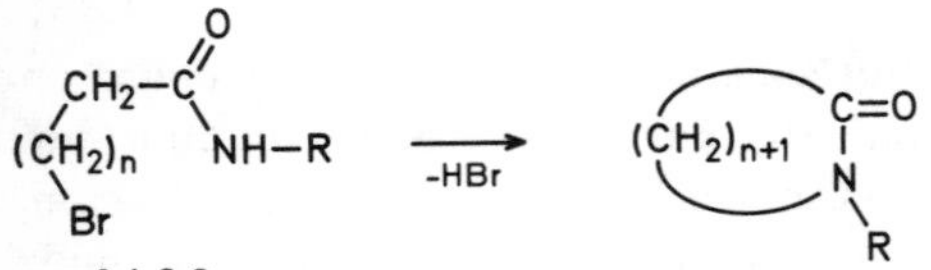

n = 0,1,2,3

However, attempts to prepare a ϵ-caprolactam in this way failed[447]. The following are used as basic condensing agents:

1 Sodium in liquid ammonia[447]
2 Sodium hydride in dimethyl sulfoxide[447] or dimethylformamide[448]
3 Potassium *tert*-butoxide in ether[449] or dimethyl sulfoxide[447]
4 Sodium ethoxide in ethanol[277, 450]

They all allow the lactam to be obtained in more or less the same yield. Potassium *tert*-butoxide is used especially for synthesizing α-*lactams (aziridinones)*[451]. Repeated intramolecular *N*-alkylation of *p*-phenylenebis-(ω-halocarboxamides) with sodium hydride in dimethylformamide leads to *diansa diamides* by using the dilution principle[448], *e.g.:*

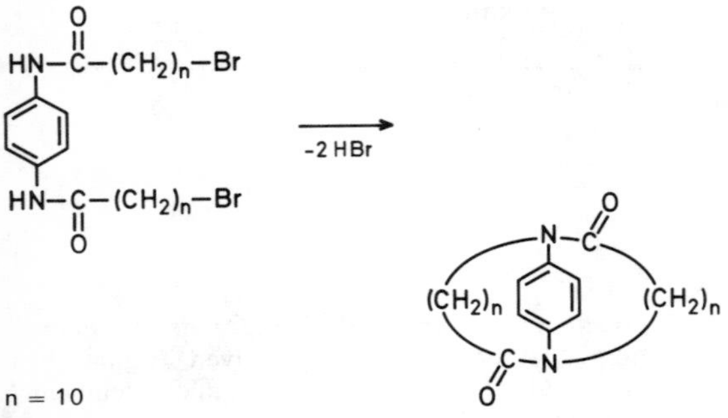

n = 10

As the potassium salt, phthalimide can be alkylated very easily and rapidly with alkyl halides in dimethylformamide to form *N-alkylphthalimides*[452] (re alkylation of the acylhydrazines).
Salts of hydroxamic acids can afford several products on alkylation, as was shown for the example of benzohydroxamic acid[453]:

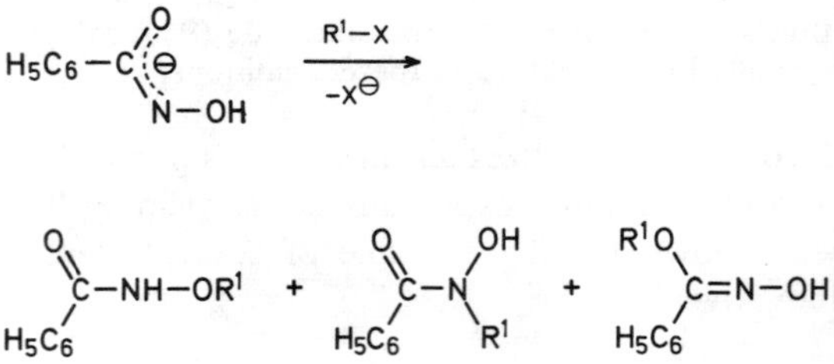

The product distribution is dependent on the alkylation reagent, the solvent, and the hydroxamate cation[453].

While alkylation of ureas always ensues on the carbonyl oxygen atom and that of carbamic acid esters predominantly so[454], *N*-alkoxyureas are alkylated only on the *N*-atom under alkaline conditions[67], *e.g.:*

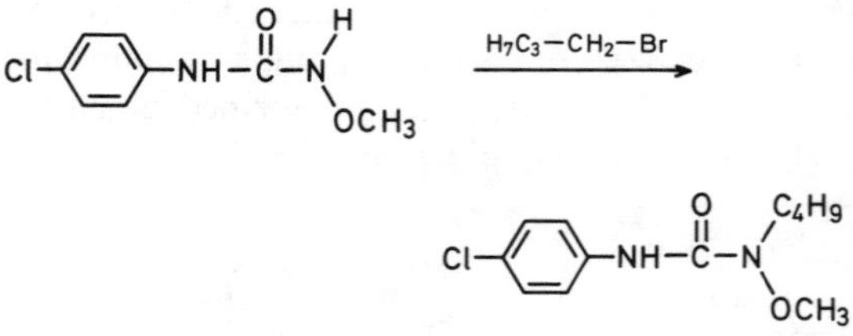

1-Butyl-3-(4-chlorophenyl)-
1-methoxyurea

[443] *W.S. Jones*, J. Org. Chem. *14*, 1099 (1949).
[444] *R.L. Dannley, M. Lukin*, J. Org. Chem. *22*, 268 (1957).
[445] *J.R. Coggins, N.L. Benriton*, Can. J. Chem. *49*, 1968 (1971).
[446] *S.J. Cristol, J.W. Rogsdale, J.S. Meek*, J. Amer. Chem. Soc. *71*, 1863 (1949).
[447] *M.S. Manhas, S.J. Jeng*, J. Org. Chem. *32*, 1246 (1967).
[448] *G. Schill, H. Neubauer, K. Rothmaier, H. Zollenkopf*, Synthesis 436 (1971).
[449] *J.C. Sheehan, I. Lengyel*, J. Amer. Chem. Soc. *86*, 1356 (1964).
[450] *U. Kraatz, W. Hasenbrink, H. Wamhoff, F. Korte*, Chem. Ber. *104*, 2458 (1971).
[451] *E.R. Talaty, C.M. Utermoehlen, K.H. Stekoll*, Synthesis 543 (1971).
[452] *M.S. Gibson, R.W. Bradshaw*, Angew. Chem. *80*, 986 (1968).
[453] *J.E. Johnson, J.R. Springfield, J.Sh. Hwang, L.G. Hayes, W.C. Cunningham, D.L. McClaugherty*, J. Org. Chem. *36*, 284 (1971).
[454] *P. Adams, F.A. Boron*, Chem. Rev. *65*, 594 (1965).

With N-hydroxyureas the oxime O-alkyl derivative is formed first and then the N-alkyl compound[67].

Carboxylic acid ester alkylamides can be converted into *carboxylic acid ester dialkylimidinium salts* on reacting further with alkyl halides[455]:

17.1.10.2 Pinner cleavage

The Pinner cleavage consists in the breakdown of a carboxylic acid ester imide salt into a carboxamide and an alkyl ester; the O-alkyl group of the carboxylic acid ester imide alkylates the anion of the salt-forming acid[456, 457]:

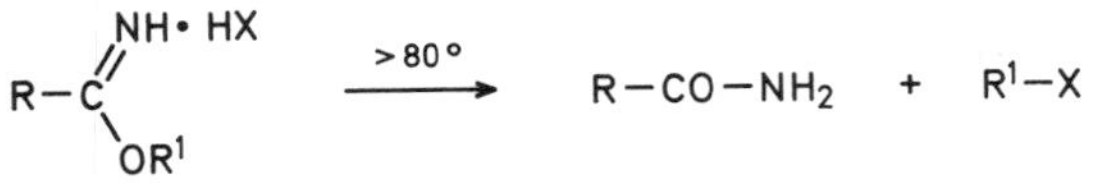

Carboxylic acid ester alkylimides, too, can be cleaved to alkylamides from their salts[458]. The reaction is performed mainly with the halogen hydride salts of the carboxylic acid ester alkylimides and ensues at temperatures of 80° or above in dependence on the acid group X, the alkyl group R^1, and the solvent[456]. It represents a convenient technique for preparing carboxamides that are sometimes difficult to obtain and often proceeds in high yield:

R = Alkyl

2-Chloro- (R² = Cl)[458], 2-Hydroxy- (R = OH)[459]
or 2-Amino- (R = NH₂)[460] alkane acid amides

17.1.10.3 Chapman rearrangement

The Chapman rearrangement (see p. 571) is taken to denote principally the thermal rearrangement of N-arylbenzimidic acid esters to form *N,N-diphenylaroylamides*[461, 462]. However, it succeeds also with carboxylic acid ester alkylimides, but only under the catalytic action of alkyl halides[461], *e.g.*:

N,N-Dimethylbenzamide

In the same way O-alkyl lactim ethers can be readily converted into *N-alkyl lactams*. Traces of dimethyl sulfate, for example, catalyze the rearrangement[461, 461a].

17.1.11 Special methods

17.1.11.1 α-Lactams from N-haloamides

Dehydrohalogenation of N-chloroamides with potassium *tert*-butoxide as base furnishes a moderate yield of α-lactams[463, 463a], *e.g.*:

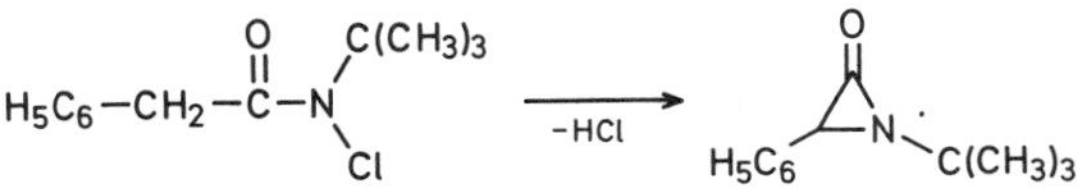

1-tert-Butyl-2-
phenyl-3-azirid-
inone; 24%

Preparatively speaking this method is not as suitable for obtaining α-lactams as that from α-halocarboxamides[463].

17.1.11.2 Rearrangement of oxaziridines

The oxaziridine ring system isomerizes to *carboxamides* on thermal treatment[464]. A clear-cut reaction ensues if the ring carbon carries two like substituents.

[455] *A.E. Arbuzov, V.E. Shishkin*, Dokl. Akad. Nauk SSSR *141*, 81 (1961); C.A. *56*, 10038 (1962).

[456] *F. Cramer, K. Pawelzik, F.W. Lichtenthaler*, Chem. Ber. *91*, 1555 (1958).

[457] *F. Cramer, H.-J. Baldauf*, Chem. Ber. *92*, 370 (1959).

[458] *S.M. McElvain, C.L. Stevens*, J. Amer. Chem. Soc. *69*, 2667 (1947).

[459] *H.E. Johnson, D.G. Crosby*, J. Org. Chem. *28*, 3255 (1963).

[460] *H.E. Johnson, D.G. Crosby*, J. Org. Chem. *27*, 798 (1962).

[461] *J.W. Schulenburg, S. Archer*, Org. Reactions, *14*, 1 (1957).

[461a] *H. Lüssi*, Helv. Chim. Acta *27*, 65 (1973).

[462] *O.H. Wheeler, F. Roman, M.V. Santiago, F. Quiles*, Can. J. Chem. *47*, 503 (1969).

[463] *J. Lengyel, J.C. Sheehan*, Angew. Chem. *80*, 27 (1968).

[463a] *E.R. Talaty, C.M. Utermoehlen*, Tetrahedron. Lett. 3321 (1970).

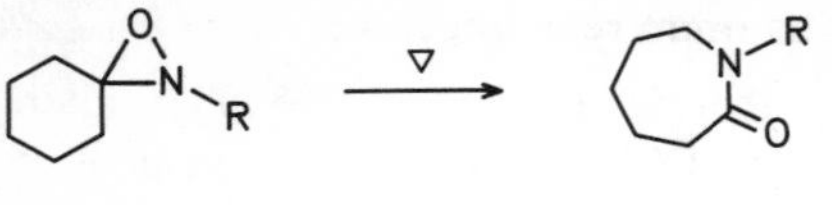

ε-Caprolactams

In other cases a mixture of two isomeric carbox-. amides is obtained, *e.g.*:

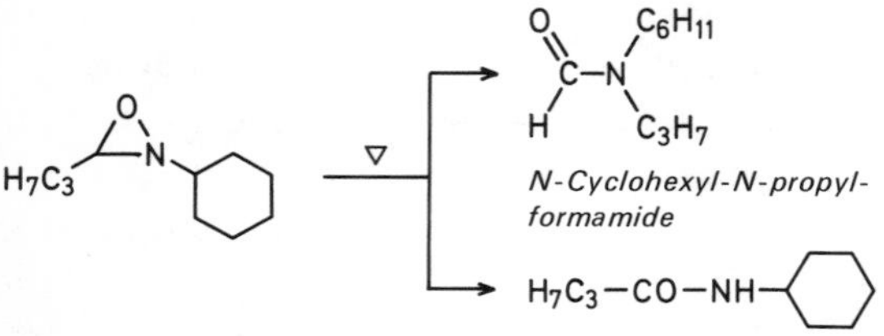

N-Cyclohexyl-N-propyl-formamide

N-Cyclohexylbutyramide

The rearrangement, which with 2-alkyloxaziri-dines requires temperatures of 200–300°, gener-ally proceeds spontaneously at room temperature in the case of the 2-aryloxaziridines. Redox sys-tems such as $Fe^{2\oplus}/Fe^{3\oplus}$ also bring about the isomerization of the oxaziridines[464].

17.1.11.3 Addition of dichlorocarbene to amines

Dichlorocarbene adds to secondary amines to form dichloromethylamines *via* an intermediate stage; the products hydrolyze to *dialkylform-amides* under the reaction conditions[465–467]:

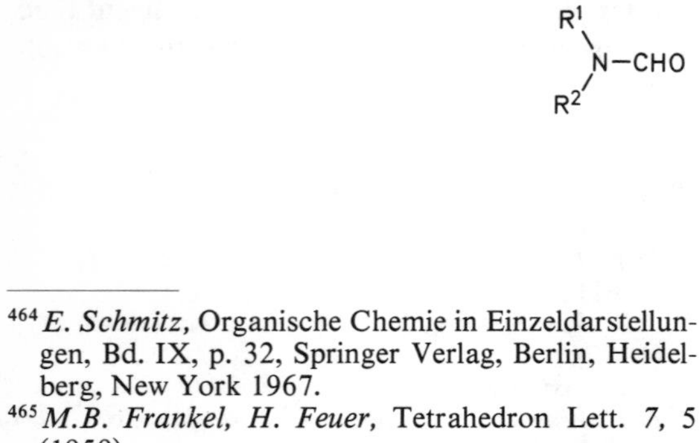

The simplest way to bring about the reaction is to heat a mixture of amine and chloroform to boiling in aqueous methanolic caustic soda[465]. With other procedures for preparing the dichlorocarbene from chloroform potassium *tert*-butoxide in ben-zene is employed[466, 467]; thus triethylamine reacts to a mixture of 15% *diethylformamide* and 12% *2-chloro-N,N-diethylpropionamide*[467].

17.2 Acyclic and cyclic imido, *N*-hy-droxyimido, hydrazido derivatives of carboxylic acids and carbonic acid

17.2.1 Halides, their salts and anhydrides

This section describes the preparation of sub-stance classes derived from the hypothetical imide **21** and the imidium salts **22** by having the hy-droxy groups acylated or substituted by halogen atoms.

21   **22**

The compounds are therefore characterized by the function (*X = Halogen; O*-Acyl)

The carbon and nitrogen atoms can be linked to organ-ic substituent groups and optionally to halogen, ox-ygen, or nitrogen.

Table 2 surveys some typical derivatives of these classes of compounds.

[464] *E. Schmitz*, Organische Chemie in Einzeldarstellun-gen, Bd. IX, p. 32, Springer Verlag, Berlin, Heidel-berg, New York 1967.

[465] *M.B. Frankel, H. Feuer*, Tetrahedron Lett. 7, 5 (1959).

[466] *M. Saunders, R.W. Murray*, Tetrahedron 6, 88 (1959).

[467] *M. Saunders, R.W. Murray*, Tetrahedron 11, 1 (1960).

[468] *H. Ulrich*, The chemistry of Imidoyl Halides, p. 74, Tables 1–3, Plenum Press, New York 1968.

[469] *F. Baumann, B. Bienert, G. Rösch, H. Vollmann, W. Wolf*, Angew. Chem. 68, 133 (1956).

[470] *O. Wallach*, Ber. 9, 1214 (1876).

[471] *J. v. Braun, W. Pinkernelle*, Ber. 67, 1218 (1934).

[472] *W.R. Vaughan, R.D. Carlson*, J. Amer. Chem. Soc. 84, 769 (1962).

[473] *I. Ugi, F. Beck, U. Fetzer*, Chem. Ber. 95, 126 (1962).

Table 2. Carboxylic acid imide and imidium derivatives

Formula	Name
R—C=N—R │ Hal	*Imidoyl halides*
R—C=N—R │ O—Acyl	*O-Acyl imidic acids* (mixed anhydrides of an imidic acid and, especially, a carboxylic acid)
R—C=N[⊕](R)(R) │ O—Acyl	*O-Acyl imidium salts*
R—C=N[⊕](R)(R) │ Hal	*Imidium halide salts* (amide halides)
R—C=N—N(R)(R) │ Hal	*Hydrazoyl halides*
R—C=NOH │ Hal	*Hydroxamoyl halides*
(R)(R)N—C=N—R │ Hal	*Halocarbamidines*
(R)(R)N—C=N[⊕](R)(R) │ Hal	*Halocarbamidinium salts*
RO—C=N[⊕](R)(R) │ Hal	*Halocarbonic acid imidium ester salts*
Acyl—O—C=N—R │ Hal	*O-Acylhalocarbonic acid imides*
R—S—C=N[⊕](R)(R) │ Hal	*Halothioformimidium ester salts*
Hal—C=N—R │ Hal	*Imidocarbonyl dihalides* (isocyanide dihalides, carbonimidic dihalides = new C.A. listing)
Hal—C=N[⊕](R)(R) │ Hal	*Imidiumcarbonyl dihalide salts* (dihalomethylene ammonium salts)

17.2.1.1 Halogenation of *N*-functional carboxylic and carbonic acid derivatives

Imidoyl halides, of which the *chlorides* are the most important representatives, are prepared best by the classical technique from equivalent quantities phosphorus(V) chloride and *N*-monosubstituted carboxamides[468].

Further halogenating agents are phosphorus(III) chloride/chlorine[469], phenyltetrachlorophosphorane[470],

thionyl chloride (especially useful for amides of aromatic carboxylic acids[471-473]), phosgene[473, 474] (preferred for aliphatic carboxamides because here no side-chlorination occurs with careful working[473]), phosphorus(III) bromide[475], and phosphorus(III) bromide/bromine[476] (*imidoyl bromides* are less stable than the corresponding *chlorides*).

The reaction, which can be carried out in inert solvents, leads primarily to imidium halides[477], from which hydrogen halide is split off either thermally or by adding bases[478] (conversely imidium halides can be prepared from imidoyl halides and halogen hydride[479]):

$$R-CO-NH-R^1 \longrightarrow \left[\begin{array}{c} Hal \\ R-C \\ \overset{\oplus}{NH}-R^1 \end{array} \right] Hal^{\ominus}$$

$$\longrightarrow \begin{array}{c} Hal \\ R-C \\ N-R^1 \end{array}$$

Intermediate tetrachlorophosphoranes are occasionally isolated when phosphorus(V) chloride is employed[480, 481]:

$$R-CO-NH-R^1 \xrightarrow{PCl_5} R-C\begin{array}{c} O-PCl_4 \\ N-R^1 \end{array} \xrightarrow{-POCl_3} R-C\begin{array}{c} Cl \\ NH-R^1 \end{array}$$

Phosgene and thionyl chloride are especially useful for halogenating strongly nucleophilic amides (alkylamides of aromatic and aliphatic carboxylic acids), while sluggish amides (diacyl amides, amides of halogen carboxylic acids) are more suitably treated with phosphorus(V) chloride[473].

Imidoyl chlorides derived from aliphatic carboxylic acids are less stable than representatives of the aromatic series, and the following *side-reactions* are often observed during their preparation.

[474] *R. Buyle, H.G. Viehe*, Tetrahedron *24*, 4217 (1968).
[475] *J. v. Braun, C. Müller*, Ber. *39*, 2018 (1906).
[476] *J.A. Arvin, R. Adams*, J. Amer. Chem. Soc. *50*, 1983 (1928).
[477] *H. Eilingsfeld, M. Seefelder, H. Weidinger*, Angew. Chem. *72*, 836 (1962).
[478] *V. Hahn, M. Grdinic*, J. Chem. Eng. Data *11*, 211 (1966).
[479] *M. Grdinic, V. Hahn*, J. Org. Chem. *30*, 2381 (1965).
[480] *H. Ulrich, E. Kober, H.J. Schroeder, R. Rätz, C. Grundmann*, J. Org. Chem. *27*, 2585 (1962).
[481] *W.P. Norris, H.B. Jonassen*, J. Org. Chem. *27*, 1449 (1962).

1 Self-condensation (α-hydrogen atoms need to be present)[482]:

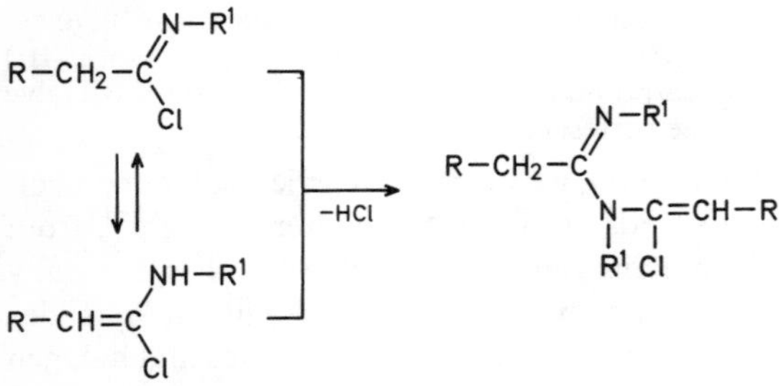

With *R* = phenyl with *ortho* substituents steric hindrance prevents self-condensation[483].

2 α-Halogenation[484], *e.g.*:

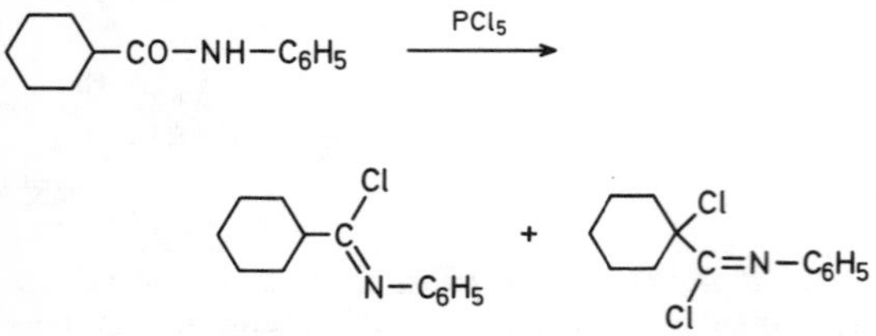

N-Phenylcyclo-
hexanecarbox-
imidoyl chloride;
80%

*1-Chloro-N-phenyl-
cyclohexanecarbox-
imidoyl chloride;*
20%

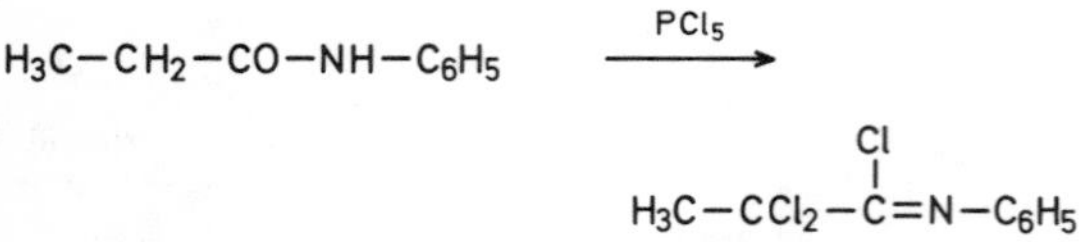

*2,2-Dichloro-N-phenyl-
propionimidoyl chloride;*
100%

3 Cyclization to heterocycles[485]

N-Alkylformimidoyl halides are relatively unstable and therefore can be formed only at lower temperatures (0°) with the aid of phosgene. At 10–40° self-condensation to *N,N'*-dialkyl-*N*-formylformamidinium chlorides occurs; these are rapidly halogenated further with phosgene[486]:

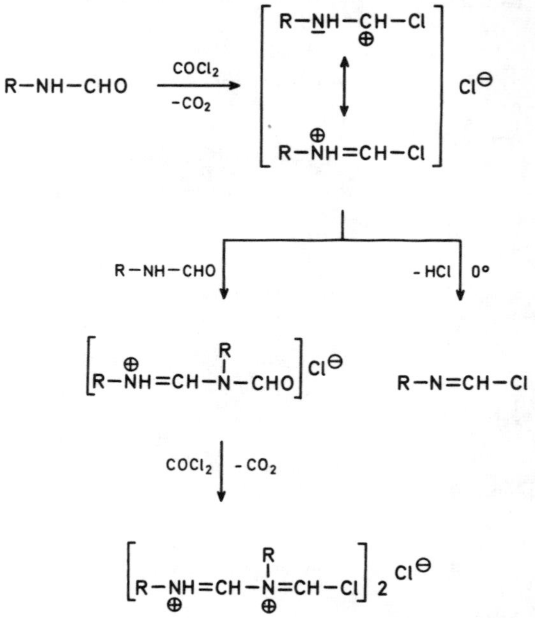

Technical importance accrues to the process using phosphorus(V) chloride for the partial synthesis of 6-aminopenicillanic acid[492].

Phosphorus(V) chloride converts primary amides into nitriles *via* the acyliminotrichlorophosphorane intermediate stage[493], *e.g.*:

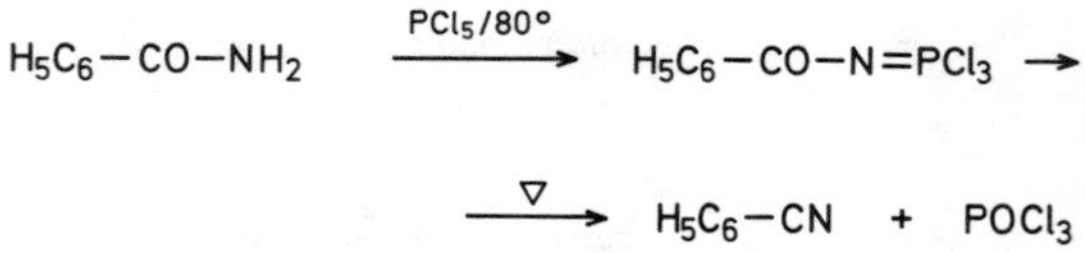

[482] *J. v. Braun, F. Jostes, W. Münch*, Justus Liebigs Ann. Chem. *453*, 113 (1927);
J. v. Braun, F. Jostes, A. Heymons, Ber. *60*, 92 (1927).

[483] *J. v. Braun, A. Silbermann*, Ber. *63*, 498 (1930).

[484] *E.E. Smissman, J.L. Diebold*, J. Org. Chem. *30*, 4002, 4005 (1965);
J. v. Braun, F. Jostes, W. Münch, Justus Liebigs Ann. Chem. *453*, 113 (1927);
H. Holtschmidt, E. Degener, H.G. Schmelzer, Justus Liebigs Ann. Chem. *701*, 107 (1967).

[485] *J. v. Braun, A. Heymons*, Ber. *63*, 3191 (1930);
J. v. Braun, W. Rudolph, Ber. *64*, 2465 (1931);
P.A. Petjunin, M.E. Konsin, J.V. Kozevnikov, Z. Vses. Khim. Obshchest. *12*, 238 (1967); C.A. *67*, 54025 P (1967);
E.F. Godefroy, C.A.M. van der Eycken, P.A. Janssen, J. Org. Chem. *32*, 1259 (1967).

[486] *W. Jentzsch*, Chem. Ber. *97*, 1361 (1964).

[487] *B. Elpern, F.W. Gubitz*, J. Org. Chem. *26*, 5215 (1961).

[488] *A. Sonn, E. Müller*, Ber. *52*, 1927 (1919).

[489] *K.C. Brannock, R.D. Burpitt*, J. Org. Chem. *30*, 2564 (1965).

[490] *P.M. Kocergin, R.M. Paleg*, Zh. Obshch. Khim. *38*, 1132 (1968), engl. 1085;
C.C. Price, B.H. Velzen, J. Org. Chem. *12*, 386 (1947).

[491] *C.H. Bolton, A.M. Dempsey, L. Hough*, Chem. Commun. 658 (1967).

[492] DDR.P. 62838 (1968), Koninklikge Nederlandsche Gisten Spiritusfabriek N.V.;
Belg. P. 643899 (1964), Ciba.

[493] *A.V. Kirsanov*, Izv. Akad. Nauk SSSR 646 (1954), engl. 551; C.A. *49*, 13161 (1955);
A. Lapidot, D. Samuel, J. Chem. Soc. (London) 2110 (1962) (with several references).

Table 3. Imidoyl chlorides from amides and phosphorus(V) chloride

Amide	Imidoyl chloride		Yield [% of th.]	M. p. [°C]	Ref.
$H_5C_6-CO-NH-CH_2-CH_2OH$	$H_5C_6-\underset{\underset{Cl}{\|}}{C}=N-CH_2-CH_2Cl$		*N-(2-Chloroethyl) benzimidoyl chloride*	76	[487]
$H_5C_6-CH=CH-CO-NH-C_6H_5$	$H_5C_6-CH=CH-\underset{\underset{Cl}{\|}}{C}=N-C_6H_5$		*N-Phenylcinnami- midoyl chloride*	~90–100	[488]
$H_3C-CO-NH-CH_2-CH=CH_2$	$H_3C-\underset{\underset{Cl}{\|}}{C}=N-CH_2-CH=CH_2$		*N-Allylacetimidoyl chloride*	good	[489]
$\underset{CO-NH-C_3H_7}{\overset{CO-NH-C_3H_7}{\|}}$	(see structure)		*N,N'-Dipropyl- oxalimidoyl chloride*	50	[490]
(acetylated glucose structure) NH—CO—CH₃	(imidoyl chloride structure) N=C(Cl)CH₃	+ (CCl₃ structure) N=C(Cl)CCl₃	*β-D-2-[(1-Chloro- ethylidene)amino]-2- deoxyglucopyranose 1,3,4,6-tetraacetate + β-D-2-Deoxy-2- [(tetrachloroethyl- idene)amino]gluco- pyranose 1,3,4,6- tetraacetate*	65	[491]
(penicillin structure) R—CO—NH ... CO—OSi(CH₃)₃	(penicillin imidoyl chloride structure)		*Penicillin analogs*	~90–100	[492]

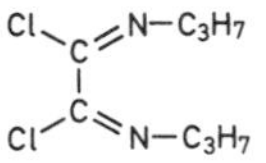

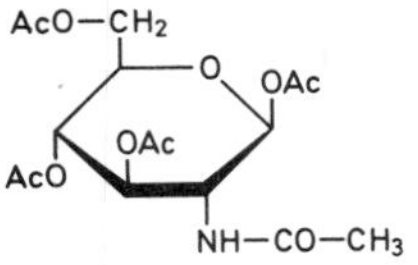

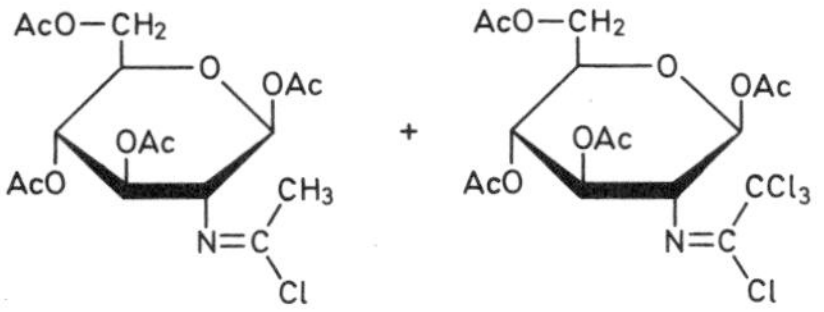

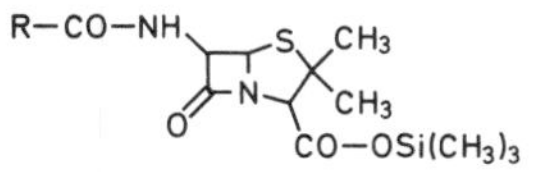

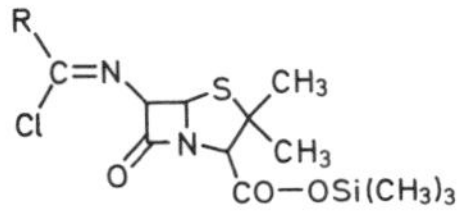

Occasionally, isomerization of the imidoyl halides such as that observed in *N-(4-nitrobenzyl)benzimi- doyl chloride* in the presence of triethylamine, must be anticipated[494]:

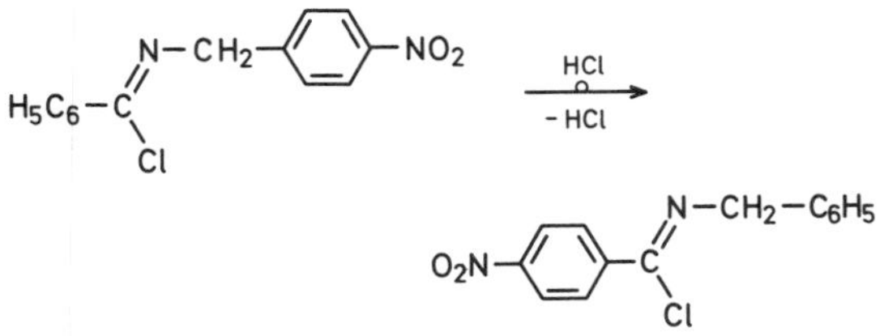

N-(4-Nitrobenzyl)- benzimidoyl chloride

The following techniques represent special variants of the imidoyl chloride synthesis:

1 Phosgenation of acylalkyltrimethylsilylamine **23** leads to acylalkylchlorocarbonyl amines **24** which decompose into imide chlorides (**25**; ~70%) at 100–130° (the latter cleave to nitriles and alkyl chlorides at 200°[495]), *e.g.:*

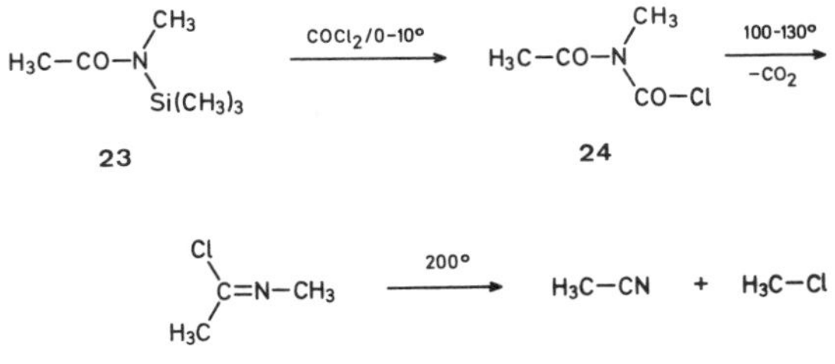

N-Methylacetimidoyl chloride

2 The reactions between triethyl phosphite and *N*-chlo- ro-*N*-methylacetamide[496], and between dichloro- and trichloroacetamides and trialkylphosphines or trial- kyl phosphites[497], proceed in analogy with the Per- kow reaction, *e.g.:*

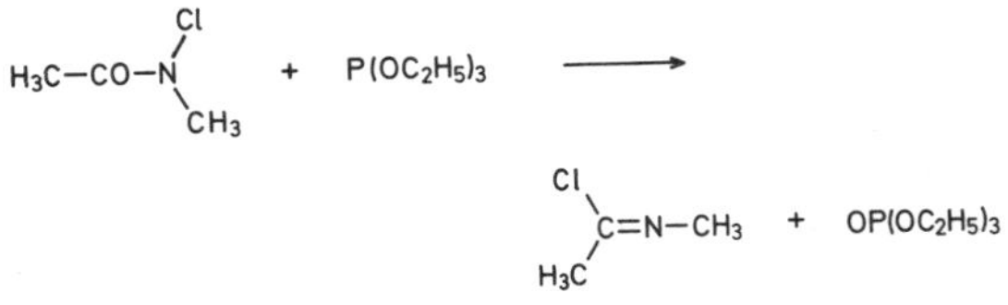

N-Methylacetimidoyl chloride

[494] *R. Huisgen, R. Raab,* Tetrahedron Lett. 649 (1966).
[495] *V.F. Mironov, V.D. Seludjakov, V.P. Kozjukov,* Zh. Obshchest. Khim. *39,* 220 (1969), engl. 208.

[496] *J. v. Mitin, G.P. Vlasov,* Zh. Obshchest. Khim. *35,* 861 (1965), engl. 864 (with several references).

Cl₃C—CO—NH—R + R¹—P (with R¹, R¹) $\xrightarrow[30-45\%]{40°}$ Cl₂CH—C(Cl)=N—R + OP—R¹ (with R¹, R¹)

R = C₆H₅; *N-Phenyldichloroacetimidoyl chloride*

R = C₂H₅; *N-Ethyldichloroacetimidoyl chloride*

R¹ = Alkyl, *O*-Alkyl

Although quite generally applicable, the transformation of thiocarboxamides into imidoyl chlorides with halogenating agents such as chlorine, phosphorus(V) chloride, thionyl chloride, or phosgene has been seldom used, presumably because thiocarboxamides are not so easily accessible [498, 499], *e.g.*:

N-Methylbenz-
imidoyl chloride

Hydrazide halides [500] are readily accessible from carboxylic acid hydrazides only in the aromatic series [501].

With phosphorus(V) chloride a phosphorus-containing intermediate product is formed from the hydrazides which is readily converted into the *hydrazide halide* on adding phenol [502]:

H₅C₆—CO—NH—NH—C₆H₅ $\xrightarrow{PCl_5}$

$\xrightarrow{H_5C_6-OH}$

N-Phenylbenzhydrazidoyl chlorid; 58%

Treating trisubstituted ureas and thioureas with halogenating agents leads to *haloformamidines* (*via* their hydrohalides). *N,N*-Disubstituted thioureas, too, react analogously. *N,N*-Disubstituted ureas yield chloroformamidines if the alkyl groups are secondary or tertiary ones [503]. Phosgene [477, 503, 504] or phosphorus(V) chloride [503, 504, 504a] are the recommended halogenating agents.

H₅C₆—NH—CO—N (with C₆H₅, C₆H₅) $\xrightarrow{PCl_5}$

$\left[H_5C_6-N=\overset{\oplus}{\underset{Cl}{C}}-N\begin{smallmatrix}C_6H_5\\C_6H_5\end{smallmatrix} \right] Cl^{\ominus}$ $\xrightarrow{-HCl}$

H₅C₆—N=C(Cl)—N (with C₆H₅, C₆H₅)

*1-Chloro-N,N',N'-tri-
phenylformamidine*

$\xrightarrow{COCl_2/130°}$

+ ČOS

*1-Chloro-N-(4-chlorobenzene-
sulfonyl)-N'-propylformamidine;
86%*

[497] *A.J. Speziale, R.C. Freeman*, J. Amer. Chem. Soc. *82*, 903 (1960);
A.J. Speziale, L.R. Smith, J. Amer. Chem. Soc. *84*, 1868 (1962);
USP. 3 230 255 (1966), Monsanto; Inv.: *A.J. Speziale, L.R. Smith*; C.A. *64*, 8044 (1966);
E. Kühle, B. Anders, G. Zumach, Angew. Chem. *79*, 663 (1967);
E. Kühle, Angew. Chem. *81*, 18 (1969);
H. Ulrich, The Chemistry of Imidoyl Halides, p. 23, Plenum Press, New York 1968.

[498] *H. Holtschmidt, E. Degener, H.G. Schmelzer*, A. *701*, 107 (1967);
K. Heyns, W. von Bebenburg, Chem. Ber. *89*, 1303 (1956).

[499] DBP. 1 166 771 (1964), BASF; Erf.: *P. Dimroth*.

[500] *H. Ulrich*, The Chemistry of Imidoyl Halides, p. 173, Plenum Press, New York 1968.

[501] *H. v. Pechmann*, Ber. *27*, 320 (1894);
H. v. Pechmann, L. Seeberger, Ber. *27*, 2121 (1894);
R. Huisgen, V. Weberndörfer, Chem. Ber. *100*, 71 (1967).

[502] *R. Huisgen, M. Seidel, G. Wallbillich, H. Knupfer*, Tetrahedron *17*, 3, 18 (1962);
J.K. Stille, F.W. Harris, M.A. Bedford, J. Heterocycl. Chem. *3*, 155 (1966).

[503] *H. Ulrich*, The Chemistry of Imidoylhalides, p. 112, Plenum Press, New York 1968;
H. Eilingsfeld, M. Seefelder, H. Weidinger, Angew. Chem. *72*, 836 (1960).

[504] *H. Ulrich, B. Tucker, A.A.R. Sayigh*, Tetrahedron *22*, 1565 (1966);
H. Ulrich, A.A.R. Sayigh, Angew. Chem. Intern. Ed. Engl. *3*, 781 (1964).

[504a] *A. Steindorff*, Ber. *37*, 963 (1904).

R—NH—C—NH—R $\xrightarrow{COCl_2/20°}$
     ‖
     S

$$\left[R-NH \overset{\oplus}{=} \underset{\underset{S-COCl}{|}}{C} \overset{\ominus}{=} NH-R \right] Cl^{\ominus} \xrightarrow{-COS}$$

$$\left[R-NH \overset{\oplus}{=} \underset{\underset{Cl}{|}}{C} \overset{}{=} NH-R \right] Cl^{\ominus}$$

Cyclic chloroamidines, too, are readily accessible by this method[477, 505], while urethans are generally converted into isocyanates[506, 507].

The *salts* of the *imidoyl halides*[508] are a class of substances which are significant for many syntheses. They are formed on reacting *N,N*-disubstituted carboxamides[477, 509–518, 479], tetrasubstituted ureas, or thioureas[477, 519] with phosgene[477, 479, 511, 515–517], thionyl chloride[517, 518], oxalyl chloride[517], phosphoryl chloride[509–513, 518], or phosphorus(V) chloride[479, 508, 517] in nonpolar solvents. To obtain the imidium·chloride chlorides in the pure state phosgene and thionyl chlorides are especially indicated. The corresponding fluorides are obtained with carbonyl fluoride[520] or from imidinium chloride chlorides with hydrogen fluoride[521].

Bromides[522] and *iodides*[523] are formed analogously from, par example, *N,N*-dimethyl formidinium chloride chloride and hydrogen bromide or iodide. The yields exceed 90% throughout. Initially, all acylation reagents form adducts on the carbonyl oxygen of the *N*-functional carboxylic acid derivative. The *O*-acyl group is then substituted by the halide ion[523–526]; when phosgene or thionyl chloride are used carbon dioxide and sulfur dioxide respectively are splitt off:

$$X-\underset{\underset{R^1}{|}}{\overset{\overset{Y}{\|}}{C}}-N-R^2 \xrightarrow{ZCl_2} \left[X-\underset{\underset{R^1}{|}}{\overset{\overset{Y-Z-Cl}{\|}}{C}}\overset{\oplus}{\underset{N-R^2}{}} \right] Cl^{\ominus}$$

$$\xrightarrow{-YZ} \left[X-\underset{\underset{R^1}{|}}{\overset{\overset{Cl}{|}}{C}}\overset{\oplus}{\underset{N-R^2}{}} \right] Cl^{\ominus}$$

$$R-\underset{\underset{R^1}{|}}{\overset{\overset{O}{\|}}{C}}-N-R^2 \xrightarrow{POCl_3} \left[R-\underset{\underset{R^1}{|}}{\overset{\overset{Cl}{|}}{C}}\overset{\oplus}{\underset{N-R^2}{}} \right] POCl_2^{\ominus}$$

X = Alkyl, Aryl, Dialkylamino
Y = O, S
Z = CO, SO, COCO

Imidium acid anhydrides, which in recent times have achieved some importance for acylation reactions, are obtained by reacting acid halides with *N,N*-disubstituted carboxamides[527–529].

[505] *H. Ulrich*, The Chemistry of Imidoyl Halides, p. 199, Plenum Press, New York 1968.
[506] *H. Ulrich*, Angew. Chem. *79*, 651 (1967).
[507] *E. Kühle*, Angew. Chem. *74*, 861 (1962); *E. Kühle*, The Chemistry of the Sulfonic Acids, p. 55, Georg Thieme Verlag, Stuttgart 1973; *E. Kühle, B. Anders, G. Zumach*, Angew. Chem. *79*, 663 (1967).
[508] *H. Ulrich*, The Chemistry of Imidoyl Halides, p. 62, p. 78, Tabelle 3, Plenum Press, New York 1968.
[509] *H. Bredereck, R. Gompper, H.G. v. Schuh, G. Theilig*, Angew. Chem. *71*, 753 (1959).
[510] *H. Bredereck, R. Gompper, K. Klemm, H. Rempfer*, Chem. Ber. *92*, 837 (1959).
[511] *H. Bredereck, K. Bredereck*, Chem. Ber. *94*, 2278 (1961).
[512] *H. Bredereck, R. Gompper, K. Klemm*, Chem. Ber. *92*, 1456 (1959).
[513] *H. Bredereck, F. Effenberger, G. Simchen*, Chem. Ber. *97*, 1403 (1964).
[514] *H. Eilingsfeld, M. Seefelder, H. Weidinger*, Chem. Ber. *96*, 2671 (1963).
[515] *Z. Arnold, F. Sorm*, Collect. Czech. Chem. Commun. *24*, 452 (1958).
[516] *Z. Arnold*, Collect. Czech. Chem. Commun. *24*, 4048 (1959).
[517] *H.H. Bosshard, R. Mory, M. Schmid, H. Zollinger*, Helv. Chim. Acta *42*, 1653 (1959).
[518] *H.H. Bosshard, H. Zollinger*, Helv. Chim. Acta *42*, 1659 (1959).
[519] *H. Eilingsfeld, G. Neubauer, M. Seefelder, H. Weidinger*, Chem. Ber. *97*, 1232 (1964).
[520] USP. 3 092 637 (1963), Inv.: *M. Brown*; C.A. *59* 12764 Du Pont (1963).

[521] *Z. Arnold*, Collect. Czech. Chem. Commun. *28*, 2047 (1963).
[522] *Z. Arnold, A. Holy*, Collect. Czech. Chem. Commun. *26*, 3059 (1961).
[523] *Z. Arnold, A. Holy*, Collect. Czech. Chem. Commun. *27*, 2886 (1962).
[524] *G.J. Martin, M. Martin*, Bull. Soc. Chim. France 637 (1963).
[525] *M.L. Filleux-Blanchard, M.T. Quemeneur, G.J. Martin*, Chem. Commun. 836 (1968).
[526] *G.J. Martin, S. Poignant, M.L. Filleux, M.T. Quemeneur*, Tetrahedron Lett. 5064 (1970).
[527] *H.K. Hall jr.*, J. Amer. Chem. Soc. *78*, 2717 (1956).
[528] *K. Klemm*, Dissertation Universität Stuttgart (1958).
[529] *D.E. Horning, J.M. Machowski*, Can. J. Chem. *48*, 193 (1970).

Imidocarbonyl chlorides (conventionally called *isocyanide dichlorides*)[508, 521] are prepared from formanilides with chlorine/sulfur dichloride, phosphorus(III) chloride, phosphoryl chloride, or thionyl chloride. One equivalent sulfuryl chloride dissolved in excess thionyl chloride is used advantageously[507, 530]:

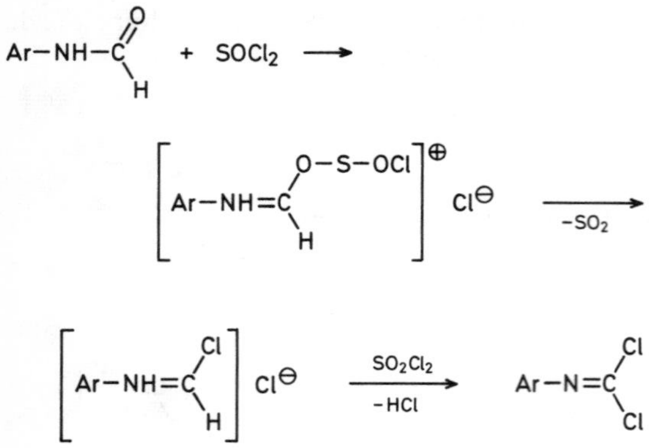

Reference may be made here to the chlorination of tetramethyl thiuram disulfide in dichloromethane to the preparatively important *(dichloromethylene)dimethylammonium chlorides*[531–533], e.g.:

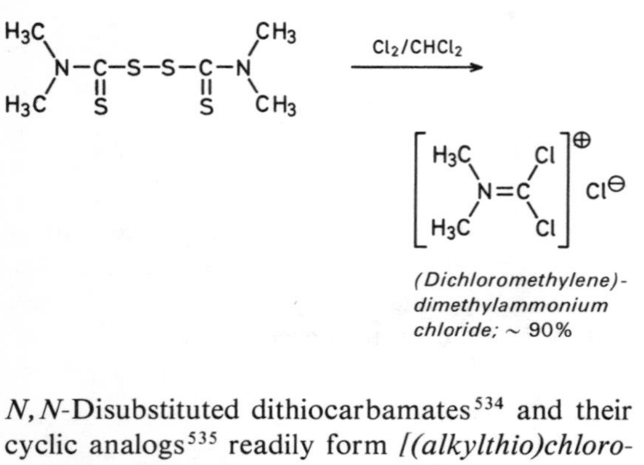

*(Dichloromethylene)-
dimethylammonium
chloride; ~ 90%*

N,N-Disubstituted dithiocarbamates[534] and their cyclic analogs[535] readily form *[(alkylthio)chloromethylene]dialkylammonium chloride* and *2-chloro-dihydro-1,3-thiazolinium chloride* with halogens or phosphorus(V) chloride:

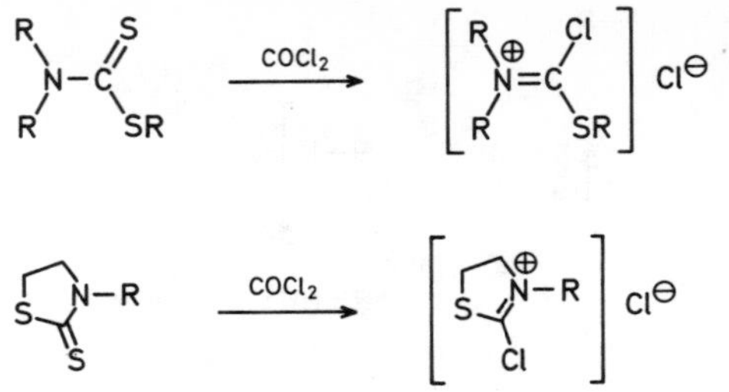

17.2.1.2 Halogenation of heterocumulenes (isocyanates, isothiocyanates)

The classic technique of mustard oil chlorination still possesses substantial preparative significance. In chloroform or carbon tetrachloride imidocarbonyl dichlorides are obtained in a yield of 90%[507]. *N*-Substituted *S-(chlorothio)imidocarbonyl chlorides* can be isolated as intermediate stages[507], e.g.:

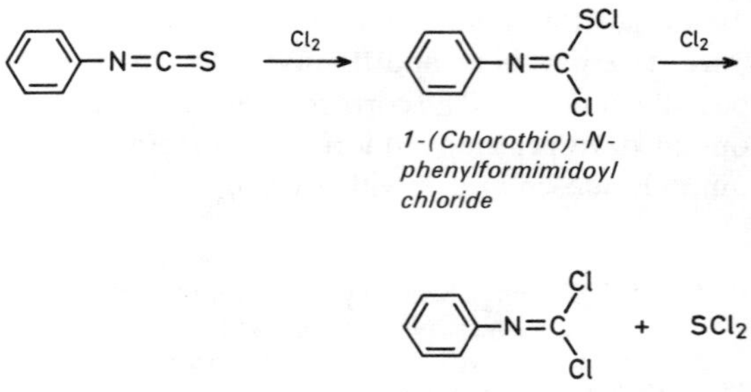

*1-(Chlorothio)-N-
phenylformimidoyl
chloride*

*N-Phenylimidocarbonyl
dichloride; 96%*

In this way aliphatic, aromatic, acyl, phosphoryl and sulfonyl mustard oils can be converted into the corresponding imidocarbonyl dichlorides in simple manner[507].

Chlorination of isocyanates requires the use of phosphorus(V) chloride as halogenating agent[536]. In the aliphatic series the reaction proceeds at 20–50° in moderate to good yield. Aromatic compounds require temperatures of about 100°

[530] *H. Ulrich*, The Chemistry of Imidoyl Halides, p. 13, Plenum Press, New York 1968.
[531] *H.G. Viehe, Z. Janousek*, Angew. Chem. *83*, 614, 615, 616 (1971); *85*, 817 (1973) (review); *A. Senning*, Chem. Rev. *65*, 388 (1965).
[532] *N.N. Jorovenko, A.S. Vasilev*, Zh. Obshch. Khim. *29*, 3786 (1959); *36*, 1309 (1966); C.A. *54*, 19466 (1960); C.A. *65*, 16885f (1966).
[533] *L.M. Yagupol'skii, M.I. Dronkina*, Zh. Obshch. Khim. *36*, 1309 (1966); C.A. *65*, 16885h (1966).

[534] *H. Eilingsfeld, L. Möbius*, Chem. Ber. *98*, 1293 (1965).
[535] Belg. P. 660941 (1965), Badische Anilin & Soda-fabrik; Inv.: *H. Eilingsfeld, L. Moebius*; C.A. *64*, 3364 (1966);
DBP. 1187620 (1965), Badische Anilin & Soda-fabrik; Erf.: *H. Weissauer, D. Weiser*; C.A. *62*, 16254 (1965).
[536] DBP. 1126371 (1962), Farbenfabriken Bayer; Erf.: *E. Kühle, R. Wegler*; C.A. *57*, 11102 (1962);
USP. 3267144 (1966), Olin Mathieson Chem. Corp.; Inv.: *G. Ottmann, H. Hooks*; C.A. *65*, 20029 (1966).

or over, but under these circumstances *carbodi-imides* are formed preferentially as well as a number of other side-products[537]:

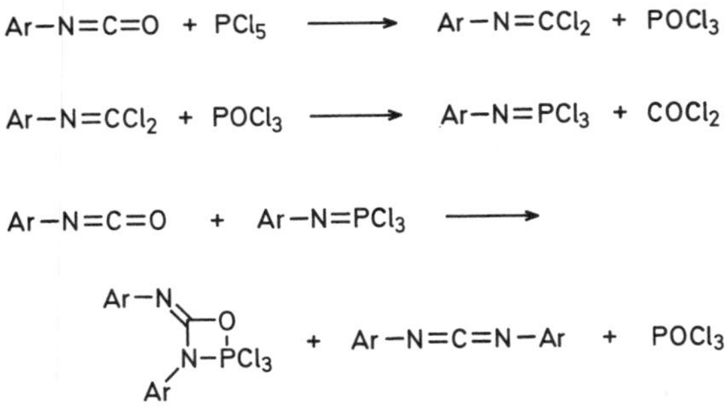

$$Ar-N=C=O + PCl_5 \longrightarrow Ar-N=CCl_2 + POCl_3$$

$$Ar-N=CCl_2 + POCl_3 \longrightarrow Ar-N=PCl_3 + COCl_2$$

$$Ar-N=C=O + Ar-N=PCl_3 \longrightarrow$$

$$+ Ar-N=C=N-Ar + POCl_3$$

17.2.1.3 Addition of hydrogen halides, halogens, and acyl halides to nitriles and heterocumulenes

17.2.1.3.1 Nitriles

Carbonitriles add hydrogen halides in a complex reaction to form thermally labile adducts[538, 539]. Formally speaking, one molecule hydrogen halides adds initially to yield *imidoyl chloride* **26** that can be isolated in exceptional cases only[538]. In a second stage a further molecule hydrogen chloride is added to form an imidium chloride salt **27**:

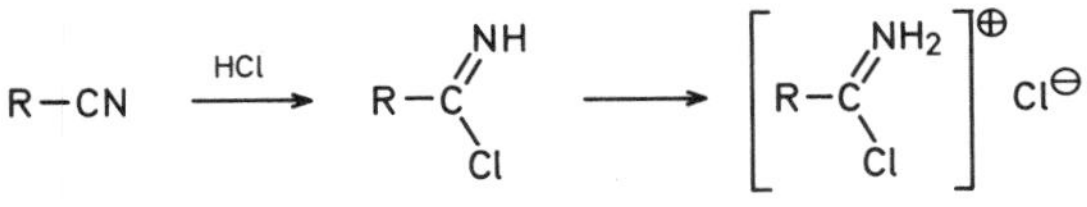

26   **27**

---

[537] *H. Ulrich, A.A.R. Sayigh*, J. Chem. Soc. (London) 5558 (1963).

[538] *R. Bonnet* in *S. Patai*, The Chemistry of the Carbon-Nitrogen Double Bond, p. 609, Wiley Intersci. Publ., London, New York, Sydney, Toronto 1970;
H. Ulrich, The Chemistry of Imidoyl Halides, p. 66, Plenum Press, New York 1968;
E.N. Zilberman, Russian Chem. Reviews *31*, 615 (1962);
F. Klages, R. Rühnau, W. Hauser, Justus Liebigs Ann. Chem. *626*, 60 (1959).

[539] *E. Allenstein, A. Schmidt*, Spectrochim. Acta *20*, 1451 (1964).

[540] *N.O. Brace*, J. Org. Chem. *28*, 3093 (1963);
USP. 2810726 (1957), Du Pont; Inv.: *E.G. Howard, W. Wolf, E. Degener, S. Petersen*, Angew. Chem. *72*, 963 (1960);
F. Johnson, W.A. Nasutavicus, J. Heterocycl. Chem. *2*, 26 (1965);
K. Blaha, O. Cervinka in *A.R. Katritzky, A.J. Boulton*, Advan. Heterocycl. Chem., Academic Press, New York, London 1966;
E. Allenstein, P. Quis, Chem. Ber. *97*, 3162 (1964);
E. Kühle, Angew. Chem. *74*, 861 (1962).

α- or β-alkanedinitriles form heat-stable cyclic imidoyl and imidium halides respectively[540], *e.g.*:

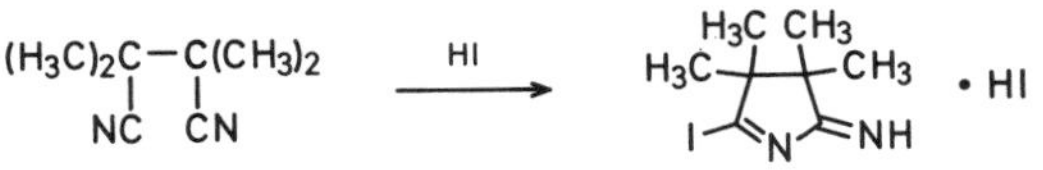

5-Imino-2-iodo-3,3,4,4-tetramethyl-2-pyrroline hydriodide; ~ 95–100%

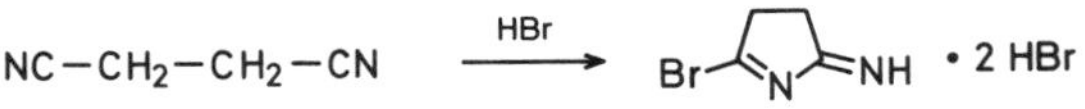

2-Bromo-5-imino-2-pyrroline dihydrobromide; ~ 95–100%

Acid halides add cyanamide to form *N-acyl imid-oyl chlorides*[541]; with phosgene the primary addition product is halogenated further[541]:

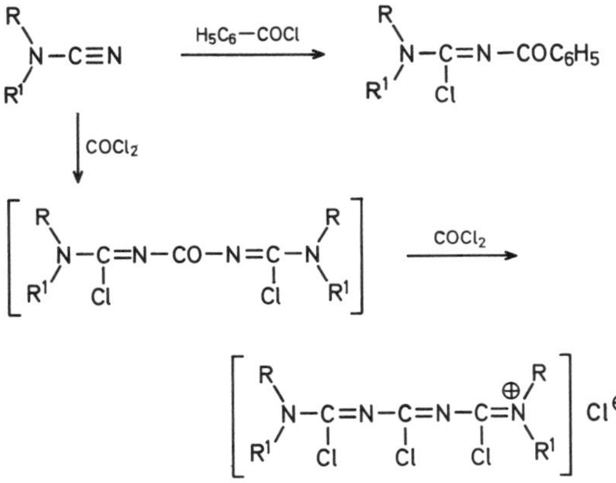

Electrophilic additions of halogen to the $C=C$ double bond in the presence of nucleophilic cyano compounds leads to *N-(β-haloalkyl)imidoyl halides*[542, 543] by virtue of incorporation of the

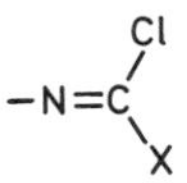

group.

---

[541] *K. Bredereck, R. Richter*, Chem. Ber. *99*, 2454, 2461 (1966).

[542] Niederl. P.-Anm. 6804554 (1968), DBP. 1618401 (1967), Farbenfabriken Bayer; Inv.: *D. Arlt*; C.A. *72*, 110745r (1970);
D. Arlt, Synthesis 20 (1970).

[543] *J. Beger, K. Günther, J. Vogel*, J. Prakt. Chem. *311*, 15 (1969).

Thus, *(3-chloroalkyl)imidocarbonyl dichlorides* **28** and **29**[542, 543] are formed from cyanogen chloride, olefins, and chlorine. Addition to vinyl chlorides requires catalysis with Lewis acids, *e.g.*, iron(III) chloride (to **29**)[542]. The *(β-chloroethyl)-imidothiocarbonyl chlorides* can be isolated[543], while carbonitriles, cyanamides, and cyano acid esters afford only unstable *N*-(β-haloalkyl)imidoyl halides **31–33**, which can be hydrolyzed to *β-haloalkylamides*, *N,N'-dialkyl-N-(β-haloalkyl)ureas,* and *N-(β-haloalkyl)urethanes*[543], for example.

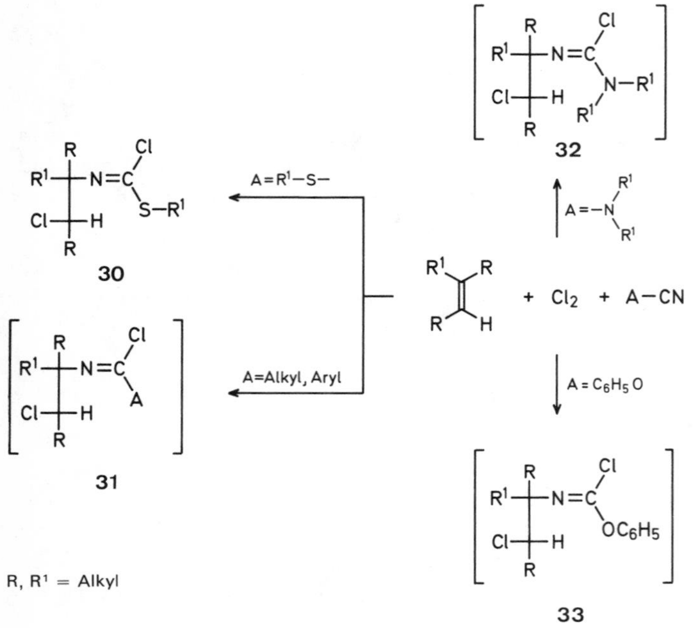

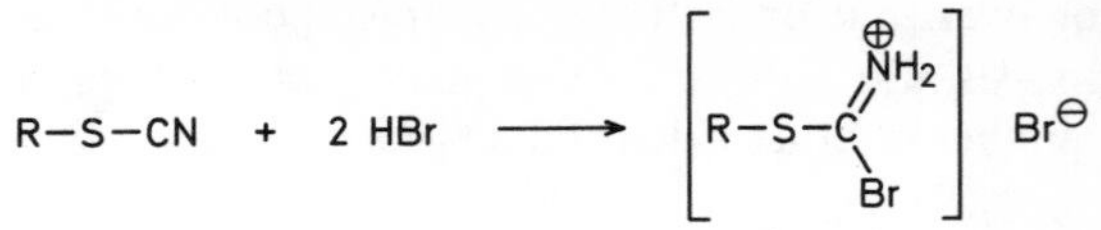

1-Bromothioformimidic
acid S-ester bromide

17.2.1.3.3 Addition of hydrogen halides, halogens, and acyl halides to carbodiimides

On performing the reaction carefully, carbodiimides add 1 molecule hydrogen chloride to form

R, R¹ = Alkyl

17.2.1.3.2 Addition of hydrogen halides, halogens, and acyl halides to thiocyanates

While hydrogen chloride adds to methyl or phenyl thiocyanate to form unstable adducts, hydrogen bromide yields somewhat more stable 2:1 adducts[544]:

chloroformamidines which readily add a further molecule hydrogen chloride to chloroformamidinium halides[545]:

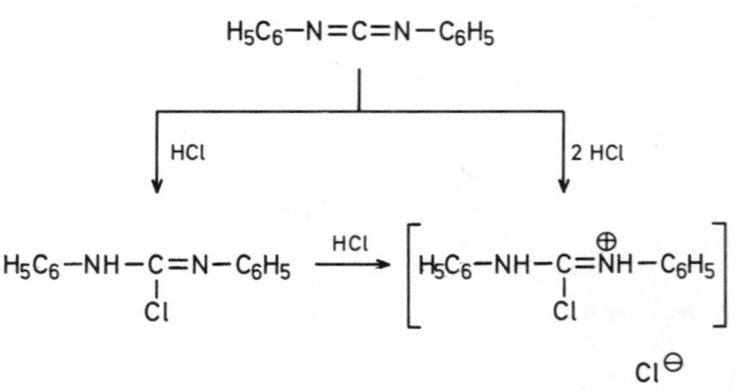

*N,N'-Diphenylchloroform-
amidine; ~ 100%*

*N,N'-Diphenylchloroform-
amidinium chloride;
~ 100%*

[544] *E. Allenstein, P. Quis,* Chem. Ber. *97,* 3162 (1964);
E. Kühle, Angew. Chem. *74,* 861, 865 (1962).

[545] *F. Lengfeld, H. Stieglitz,* Amer. Chem. J. *17,* 98, 108 (1895);
H. Ulrich, J.N. Tilley, A.A.R. Sayigh, J. Org. Chem. *29,* 2401 (1964).

[546] *K. Hartke, J. Bartulin,* Angew. Chem. *74,* 214 (1962).

Acid halides react equally readily with carbodi-imides to form *N-acylchloroformamidines*:

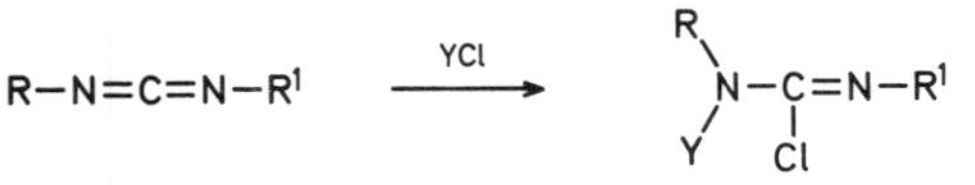

$$R-N=C=N-R^1 \xrightarrow{\text{YCl}} \begin{array}{c} R \\ | \\ N-C=N-R^1 \\ | \quad | \\ Y \quad Cl \end{array}$$

Y = CO–R[2 546]   CO–Cl[546–548, 550]
    CS–Cl[547]   SO–Cl[548]
    POCl[2][548, 549]   PCl[2][548]
    S–Cl[548]   SO[2]Cl[548]

17.2.1.3.4 Addition of hydrogen halides, halogens, and acyl halides to ketenimines

Addition of chlorine to ketene imines leads to α-*chloroimidoyl chlorides*[551], that of halogen hydride to *imidoyl chlorides*[552]:

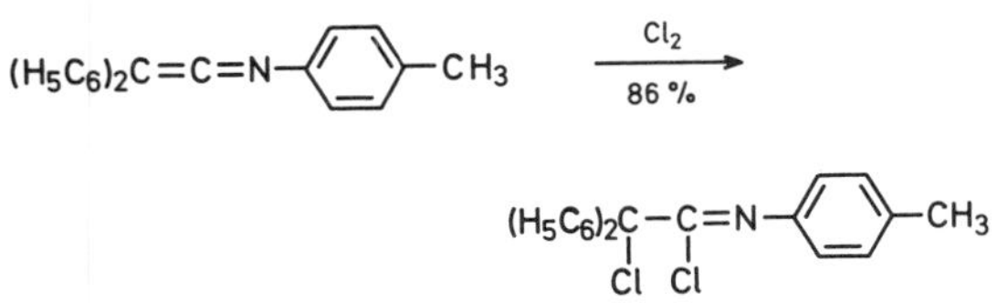

$$(H_5C_6)_2C=C=N-\langle\rangle-CH_3 \xrightarrow[86\%]{Cl_2}$$

$$(H_5C_6)_2C-C=N-\langle\rangle-CH_3$$
$$\qquad\qquad | \quad |$$
$$\qquad\qquad Cl \quad Cl$$

1-Chloro-N-(4-tolylmethyl-phenyl)diphenylacetimidoyl chloride; 86%

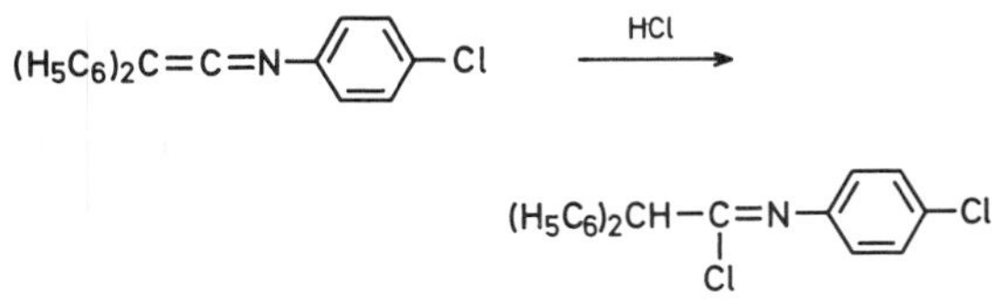

$$(H_5C_6)_2C=C=N-\langle\rangle-Cl \xrightarrow{HCl}$$

$$(H_5C_6)_2CH-C=N-\langle\rangle-Cl$$
$$\qquad\qquad\quad |$$
$$\qquad\qquad\quad Cl$$

N-(4-Chlorophenyl)-diphenylacetimidoyl chloride; ~ 90%

17.2.1.3.5 Addition of halogen hydride, halogens, and acyl halides to nitrile oxides

Addition of halogen hydride to nitrile oxides to form hydroxamoyl halides[553] is particularly suitable for the synthesis of the bromides and iodides. Sulfonyl chlorides are taken up with formation of *hydroxamoyl chloride O-sulfonic acid esters*[554]:

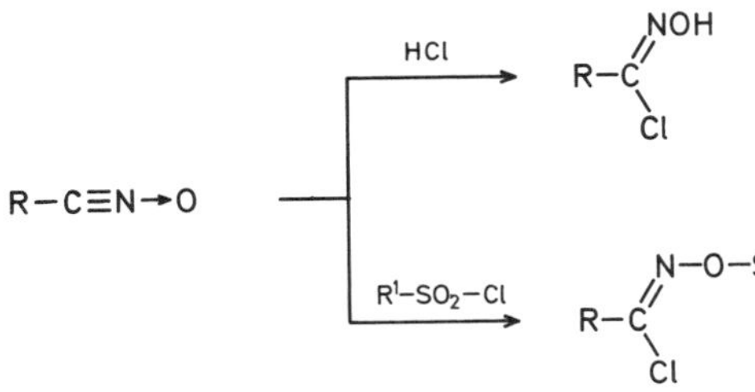

$$R-C\equiv N\rightarrow O$$

with HCl →

$$R-C\begin{array}{c} NOH \\ \| \\ \\ Cl \end{array}$$

with R[1]–SO[2]–Cl →

$$R-C\begin{array}{c} N-O-SO_2-R^1 \\ \| \\ \\ Cl \end{array}$$

17.2.1.4 Addition of halogens and acyl halides to isonitriles

Isonitriles readily add

1 halogen to form *imidocarbonyl dihalides*[555]
2 esters of hypochlorous acid to form *chlorocarbonic acid imidic esters*[556]
3 acid chlorides to form α-*keto imidoyl chlorides*[557]
4 trihalomethanesulfonyl chloride to form *trihaloacetimidoyl chlorides* while losing sulfur dioxide[558].

This technique is of preparative importance especially for synthesizing aliphatic imidocarbonyl dihalides that are difficult to prepare from formamides and halogenating agents by the method described on p. 716:

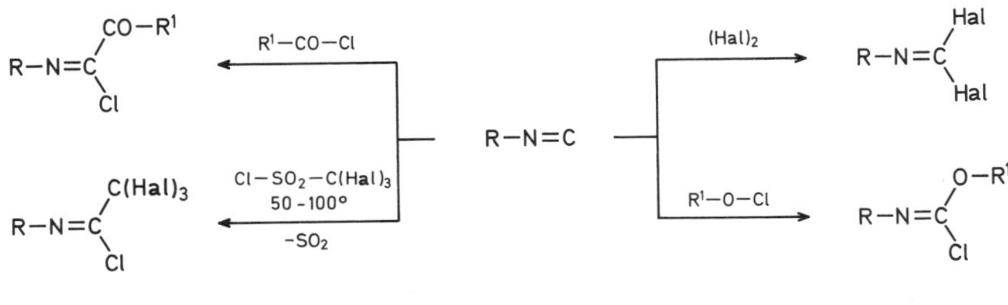

[547] *K. Hartke, E. Palou*, Angew. Chem. *77*, 727 (1965);
    K. Hartke, E. Palou, Chem. Ber. *99*, 3155 (1966).
[548] DBP. 1 131 661 (1962), Farbenfabriken Bayer; Erf.:
    P. Fischer; C.A. *58*, 1401 d (1963).
[549] *H. Ulrich, A.A.R. Sayigh*, J. Org. Chem. *30*, 2779 (1965).
[550] *H. Ulrich, A.A.R. Sayigh*, J. Org. Chem. *28*, 1427 (1963).
[551] *C.L. Stevens, J.C. French*, J. Amer. Chem. Soc. *75*, 657 (1953).

[552] *K. Ichimura, M. Otah*, Bull. Chem. Soc. Japan *40*, 2135 (1967); C.A. *68*, 48897d (1968).
[553] *C. Grundmann, J.M. Dean*, J. Org. Chem. *30*, 2809 (1965);
    C. Grundmann, H.D. Frommeld, J. Org. Chem. *31*, 157 (1966);
    C. Grundmann, P. Grünanger, The Nitrile Oxides, p. 153, Springer Verlag, Berlin, Heidelberg, New York 1971.
[554] *W.E. Truce, A.R. Naik*, Can. J. Chem. *44*, 297 (1966).

With aliphatic compounds it is often advantageous to employ sulfuryl chloride in ether as the chlorinating agent.

17.2.1.5 α-Halogenation of *N*-functional carbonyl derivatives

17.2.1.5.1 Of aldimines

The direct action of halogen on aldehyde imines (azomethines) is difficult to control. Using *tert*-butyl hypochlorite[559] is more suitable; it forms imidoyl chlorides in this reaction which were characterized in the form of their further reaction products with amines or alcohols:

$$R-CH=NH \xrightarrow[\;-(H_3C)_3C-OH\;]{(H_3C)_3C-O-Cl} R-\overset{N-R}{\underset{Cl}{C}}$$

17.2.1.5.2 α-Halogenation of hydrazones and azines

Halogenation (chlorine or bromine) of aromatic hydrazones is a much used method of synthesizing *halogen hydrazones*[560, 561].

Like in many other halogenation reactions it is not always possible to avoid introducing halogen into the aromatic ring bound to the nitrogen. During halogenation of azines investigated hitherto ring substitution does not occur[562]:

$$R^1-CH=N-NH-R^2 \xrightarrow{Br_2} R^1-\overset{N-NH-R^2}{\underset{Br}{C}}$$

$R^1 = C_6H_5$; $R^2 = 4-NO_2-C_6H_4$; *Benzoyl bromide (4-nitrophenyl) hydrazone; 85%*

$R^1 = C_6H_5$; $R^2 = 2,4-(Br)_2-C_6H_3$; *Benzoyl bromide (2,4-dibromophenyl) hydrazone*

$$H_5C_6-CH=N-N=CH-C_6H_5 \xrightarrow{Cl_2}$$

$$\overset{H_5C_6}{\underset{Cl}{}}C=N-N=C\overset{C_6H_5}{\underset{Cl}{}}$$

Benzoyl chloride azine; 74–83%

17.2.1.5.3 α-Halogenation of aldoximes

Chlorination of aldoximes is the classic method for preparing *hydroxamoyl chlorides*[563]. Chlorine in chloroform, ether, or hydrochloric acid, nitrosyl chloride, or *tert*-butyl hypochlorite in methanol are used as the chlorinating agent. Ring or side-chain chlorination are observed occasionally and in these cases the technique of adding hydrogen halides to nitrile oxides (p. 718) is preferred:

$$R-CH=NOH \xrightarrow[60-95\%]{Cl_2/-60°} R-\overset{NOH}{\underset{Cl}{C}}$$

17.2.1.6 Halogenation of acylamides and tertiary amines (high temperature chlorination)

Chlorination of tertiary amines at elevated temperature proceeds accompanied by dealkylation, that of acylamines is accompanied by deacylation. In most cases compounds are formed that contain the imidoyl chloride or the isocyanide dihalide group. In dependence on the reaction conditions secondary chlorinations of various types occur (for details see the literature surveys)[564], *e.g.*:

N-(2,4,6-Trichlorophenylimidocarbonyl dichloride

N-(Pentachlorophenyl)imidocarbonyl dichloride; ~90%

$$(H_3C)_3N \xrightarrow{Cl_2} Cl_3C-N=C\overset{Cl}{\underset{Cl}{}}$$

N-Trichloromethylimidocarbonyl dichloride

[555] E. Kühle, B. Anders, G. Zumach, Angew. Chem. 79, 663 (1967).

[556] J.U. Nef, Justus Liebigs Ann. Chem. 287, 301 (1895).

[557] J.U. Nef, Justus Liebigs Ann. Chem. 270, 267 (1892);
J.U. Nef, Justus Liebigs Ann. Chem. 280, 298 (1894);
I. Ugi, U. Fetzer, Chem. Ber. 94, 1116 (1961);
W. Walter, K.D. Bode, Angew. Chem. 74, 694 (1962).

[558] DBP. 1 163 315 (1964), Farbenfabriken Bayer; Erf.: E. Kühle.

[559] H. Paul, A. Weise, R. Dettmer, Chem. Ber. 98, 1450 (1965).

[560] H. Ulrich, The Chemistry of Imidoyl Halides, p. 176, Plenum Press, New York 1968.

[561] F.L. Scott, J.B. Aylward, Tetrahedron Lett. 841 (1965).

[562] K. Issleib, A. Balszuweit, Chem. Ber. 99, 1316 (1966);
R. Stollé, J. Prakt. Chem. 85, 386 (1911).

N-Trichloromethylimidocarbonyl dichloride; 87%   **1**

N-Chloromethylimidocarbonyl dichloride; 62%   **2**

N-Bis(dichloromethyl)carbamoyl dichloride; (good yield)   **3**

17.2.1.7 Reaction of imidocarbonyl or imidium-carbonyl dihalides with nucleophiles [alcohols, thiols, amines, aromatic compounds, copper(I) cyanide]

The reaction of the imidocarbonyl dihalides with nucleophilic reaction partners ensues in steps. During the first stage it is generally possible to iso-

late the corresponding *imidoyl chloride* derivatives by exchange of a halogen atom (see Table 4 p. 722).

Carbonic acid dialkylimidinium dichlorides (see p. 717) likewise react with nucleophilic partners, for example, amines, to form *chloroformamidines*, or with dialkylamides to *dichlorotrimethiniminium* salts.

Table 4 (p. 722) provides a survey (for details see the cited literature references)[565, 566].

17.2.1.8 Other procedures

17.2.1.8.1 Reaction of diazonium halides with α-halocarbonyl compounds

Coupling diazonium salts with α-halo ketones and carboxylic acids represents a good method for preparing *hydrazidoyl halides*[575]:

Glyoxyloyl acid chlorid (2-phenylhydrazone); ∼ 50%

Glyoxyloyl acid ester chloride (2-phenylhydrazone); ∼ 70%

1 Glyoxyloyl chloride (2-phenylhydrazone); 100%

2 Pyruvoyl chloride (2-phenylhydrazone); 100%

[563] *G. Casnati, A. Ricca,* Tetrahedron Lett. 327 (1967);
A. Werner, H. Buss, Ber. 27, 2193 (1894);
H. Ley, M. Ulrich, Ber. 47, 2941 (1914);
L. Claisen, O. Manasse, Justus Liebigs Ann. Chem. 274, 95 (1893);
H. Rheinboldt, Justus Liebigs Ann. Chem. 451, 161 (1927);
H. Rheinboldt, O. Schmitz-Dumont, Justus Liebigs Ann. Chem. 444, 113 (1925);
H. Ulrich, The Chemistry of Imidoyl Halides, p. 158, Plenum Press, New York 1968.

[564] *H. Holtschmidt,* Angew. Chem. 74, 848 (1962);
H. Holtschmidt, E. Degener, H.-G. Schmelzer, H. Tarnow, W. Zecher, Angew. Chem. 80, 942 (1968);
H. Holtschmidt, E. Degener, H.-G. Schmelzer, Justus Liebigs Ann. Chem. 701, 107 (1967).

[565] *E. Kühle,* Angew. Chem. 81, 18 (1969).

[566] *H. Ulrich,* The Chemistry of Imidoyl-Halides, p. 39, Plenum Press, New York 1968.

[567] DBP. 1 126 380 (1962), Farbenfabriken Bayer; Erf.:
E. Kühle; C.A. 58, 3361 (1963).

[568] DBP. 1 154 089 (1960), Farbenfabriken Bayer; Erf.:
E. Enders, E. Kühle, H. Matz; C.A. 57, 13694 (1962).

[569] DBP. 1 211 163 (1966), Farbenfabriken Bayer; Erf.:
E. Kühle, L. Eue;
Brit. P. 888 646 (1962), Farbenfabriken Bayer; Inv.:
E. Kühle, L. Eue; C.A. 57, 13696 (1962).

[570] DBP. 1 224 306 (1965), Farbenfabriken Bayer; Erf.:
E. Kühle, B. Anders; C.A. 65, 18536 (1966).

[571] *E. Kühle,* Angew. Chem. 81, 31 (1969).

[572] DAS. 1 235 291 (1965), Farbenfabriken Bayer; Erf.:
E. Klauke, E. Kühle;
E. Kühle, B. Anders, G. Zumach, Angew. Chem. 79, 663, 670 (1967).

[573] *H.G. Viehe, Z. Janousek,* Angew. Chem. 83, 614 (1971).

[574] *Z. Janousek, H.G. Viehe,* Angew. Chem. 83, 615 (1971);
H.G. Viehe, Z. Janousek, M.-A. Defrenne, Angew. Chem. 71, 616 (1971).

[575] *H. Ulrich,* The Chemistry of Imidoyl-Halides, p. 175, Plenum Press, New York 1968.

Table 4. Reaction of Imidocarbonyl and imidium carbonyl dihalides with nucleophiles

Dihalide	Nucleophile	Reaction conditions	Reaction product		Ref.
	NaOH, H$_2$O, CH$_3$OH	10–15°, toluene	Cl—C$_6$H$_4$—N=C(Cl)(OCH$_3$)	1	567
Cl—C$_6$H$_4$—N=CCl$_2$	NaS—C$_6$H$_5$	Acetone	Cl—C$_6$H$_4$—N=C(Cl)(S—C$_6$H$_5$)	2	568
	HN(pyrrolidine)	Benzene	Cl—C$_6$H$_4$—N=C(Cl)(pyrrolidinyl)	3	569
	CuCN	180° C$_6$H$_4$Cl$_2$	C$_6$H$_5$—N=C(Cl)(CN)	4	570
C$_6$H$_5$—N=CCl$_2$	H$_5$C$_6$—OCH$_3$	AlCl$_3$	H$_5$C$_6$—N=C(Cl)(C$_6$H$_4$—OCH$_3$)	5	571
	Na$_2$F$_2$	200–240° Tetrahydrothiophene 1,1-dioxide	C$_6$H$_5$—N=CF$_2$	6	572
	H$_5$C$_2$O—OC—C$_6$H$_4$—NH$_2$	40°, CHCl$_3$	H$_5$C$_2$O—OC—C$_6$H$_4$—N=C(Cl)—N(CH$_3$)$_2$	7	573
[H$_3$C)$_2$⊕N=CCl$_2$] Cl$^{\ominus}$	H$_3$C—CO—N(CH$_3$)$_2$	60°, CHCl$_3$	(H$_3$C)$_2$N=C(Cl)—C=C—C(Cl)=N(CH$_3$)$_2$ ⊕ Cl$^{\ominus}$	8	574

1 *Methyl 1-chloro-N-(4-chlorophenyl)formimidate*
2 *S-Phenyl 1-chloro-N-(4-chlorophenyl)thioformimidate*
3 *Pyrrolidinyl 1-chloro-N-(4-chlorophenyl)formimidate*
4 *1-Cyano-N-phenylformimidoyl chloride*
5 *N-Phenyl-4-anisimidoyl chloride*
6 *N-Phenylimidocarbonyl difluoride*
7 *Ethyl 4-[chloro(dimethylamino)methylene]aminobenzoate*
8 *Bis(dihydro) malonic acid bis (dimethylimide)dichloride chloride*

In acetate-buffered solution diazonium salts couple with chloromalonic acid to form formazyl halides while undergoing double decarboxylation[576], *e.g.*:

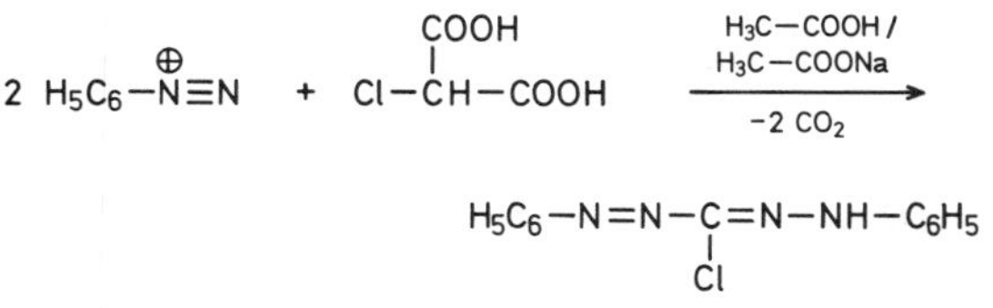

$$2\ H_5C_6-\overset{\oplus}{N}\equiv N\ +\ \underset{COOH}{\underset{|}{Cl-CH-COOH}}\ \xrightarrow[-2\ CO_2]{\underset{H_3C-COONa}{H_3C-COOH/}}$$

$$\underset{Cl}{\underset{|}{H_5C_6-N=N-C=N-NH-C_6H_5}}$$

3-Chloro-1,5-diphenylformazan

17.2.1.8.2 Rearrangement of 1-halo 1-nitrosocycloalkanes with triarylphosphines

1-Halo-1-nitrosocycloalkanes afford imidoyl chlorides with triarylphosphines accompanied by ring expansion[577], *e.g. 2-chloro-3,4,5,6-tetrahydropyridine* and *caprolactim chloride* which are hydrolysed to lactames, *e.g.*:

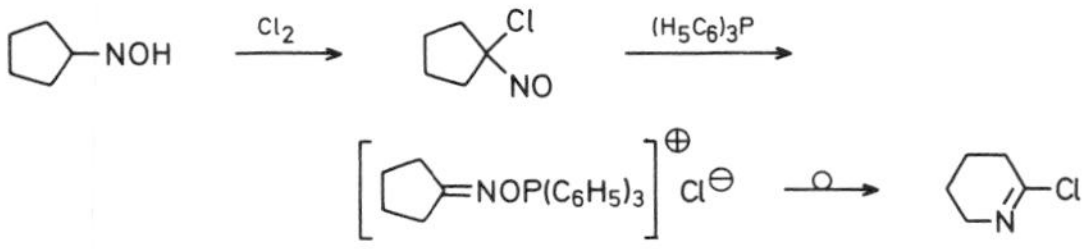

17.2.1.8.3 Reaction of aldehyde imines (azomethines) with dichlorocarbene

Reacting aldehyde imines with dichlorocarbene leads initially to *2,2-dichloroaziridines*[578a]. These are present in the imidium form and are easily rearranged into α-*haloimidoyl chlorides* by heating[578]:

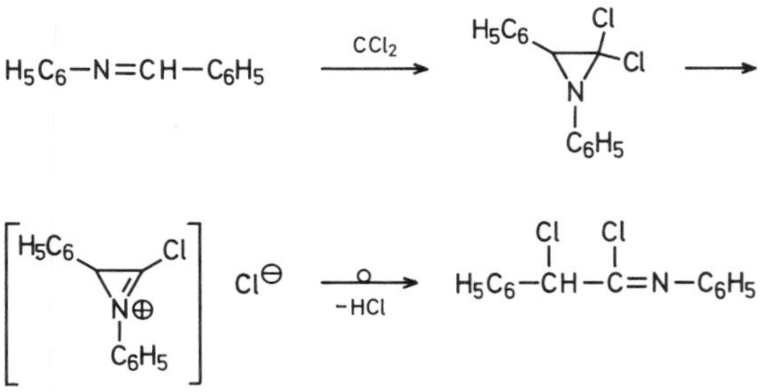

2-Chloro-1,3-diphenyl-3H-azirinium chloride; 55%   *N-Phenyl-2-chloro-2-phenyl-acetimidoyl chloride;* 95%

17.2.1.8.4 Imidoyl halides by Beckmann rearrangement

During Beckmann rearrangement of oximes with halogenating agents[579] imidoyl chlorides arise as isolable intermediates, *e.g.*[580, 581]:

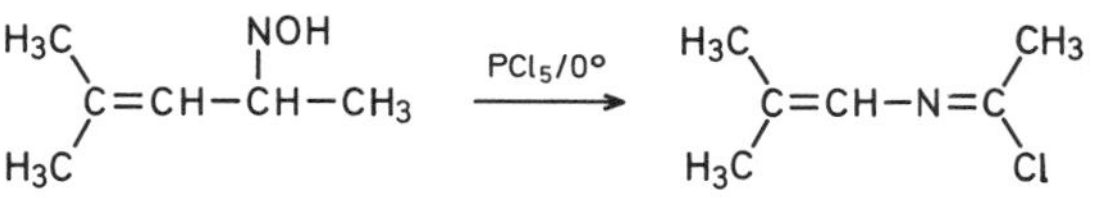

N-(2-Methyl-1-propenyl)acetimidoyl chloride

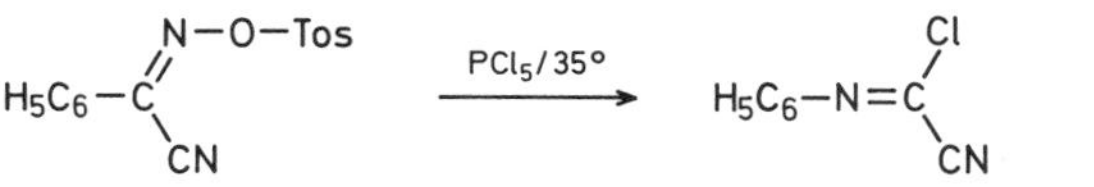

1-Cyano-N-phenyl-formimidoyl chloride; ~ 50%

17.2.2 Imidic acid esters and derivatives

17.2.2.1 By alkylation reactions

17.2.2.1.1 Of carboxamides

Imidic esters can be prepared by *O*-alkylation of carboxamides. Reacting with dialkyl sulfates, trialkyloxonium tetrafluoroborates, or dialkyl fluorosulfates are preferred to the action of alkyl halides on the silver salts of amides as the alkylation means[582]. On account of the more nucleophilic sulfur the alkylation proceeds more readily with thiocarboxamides than with carboxamides. Using alkyl halides in ethanol or aprotic polar solvents or alkyl sulfates in aqueous alkali leads to salts of the *S-alkyl esters* of *thioimidic acids*, which on treatment with potassium carbonate afford the free *S*-esters[583]. Thus thioformamide, unlike formamide, reacts with alkyl halides even in ether solution to form *thioformidium esters halides*[584].

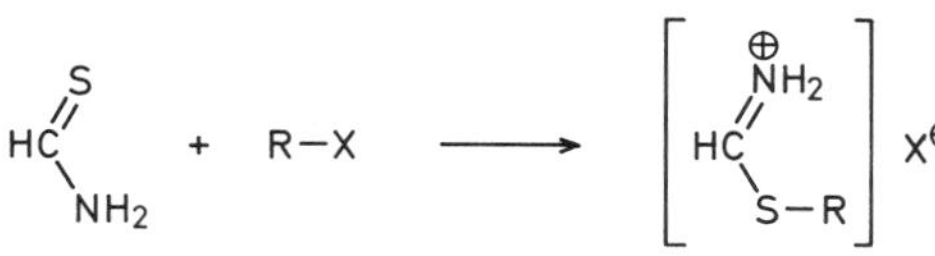

[576] *R. Fusco, R. Romani*, Gazz. Chim. Ital. *76*, 419 (1946);
R. Fusco, R. Romani, Gazz. Chim. Ital. *78*, 332 (1948).
[577] *M. Ohno, J. Sakai*, Tetrahedron Lett. 4541 (1965).
[578] *H.W. Heine, A.B. Smith*, Angew. Chem. *75*, 669 (1963);
R.E. Brooks, J.O. Edwards, G. Levey, F. Smyth, Tetrahedron *22*, 1279 (1966);

K. Ichimura, M. Otah, Tetrahedron Lett. 807 (1966);
K. Ichimura, M. Otah, Bull. Chem. Soc. Japan *40*, 1933 (1967); C.A. *68*, 68759x (1968).
[578a] *E.K. Fields, J.M. Sandri*, Chem. & Ind. (London) 1216 (1959).
[579] see this volume, p. 565.
[580] *J. Wielmann, N. Thoai, H. Poksoon*, Bull. Soc. Chim. France 3920 (1967).
[581] *T.W. Stevens*, J. Org. Chem. *32*, 670 (1967).
[582] *R. Roger, D.G. Neilson*, Chem. Rev. *61*, 181 (1961).

Alkylation of N-(ω-hydroxyalkyl)amides can be accomplished readily with water-abstracting agents (*e.g.*, thionyl chloride, phosgene, sulfonyl chloride), and *cyclic imidic esters* are obtained[585]:

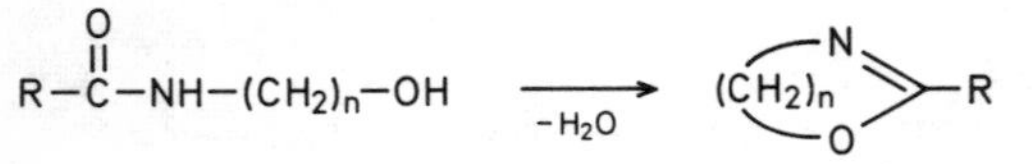

n = 2; 2-Substituted *2-oxazolines*

n = 3; 2-Substituted *5,6-dihydro-4H-1,3-oxazines*

The cyclization is made very much easier if either the C atom adjacent the nitrogen or that adjacent the hydroxy group is disubstituted[585].

Amides and *N*-substituted amides require stronger alkylating agents. With dimethyl sulfate *imidium ester methyl sulfates* are formed initially and with potassium carbonate, triethylamine[586, 587], or sodium cyanide solution[588] afford the free *imido esters*:

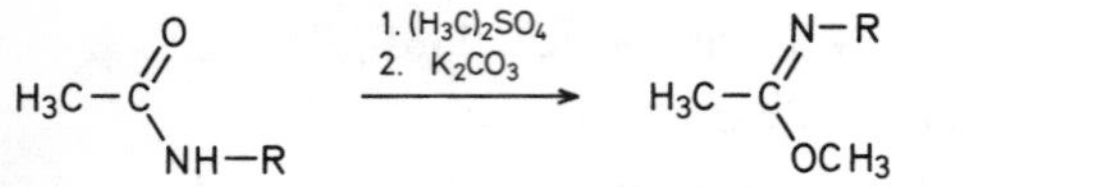

Methyl acetimidates; 57%

N,N-Dialkylamides and their vinylogous compounds[589] react correspondingly with dimethyl sulfate to undergo *O*-alkylation and form *N,N-dialkylimidium esters salts*[590, 591]:

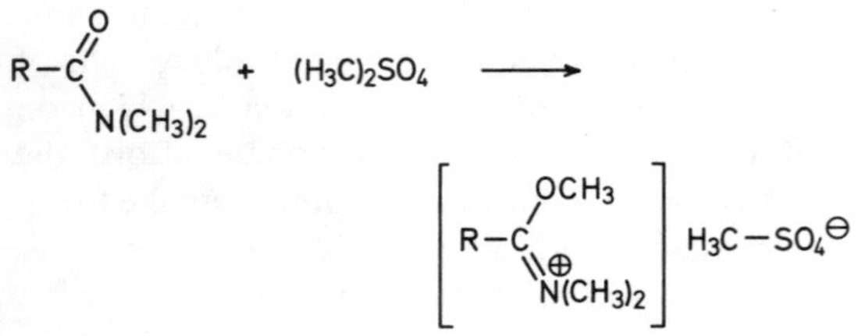

Methyl N,N-dimethyl-imidiniumcarboxylate methyl-sulfate; > 90%

This reaction is the simplest technique for preparing *N,N-dialkylimidium esters.*

It fails with a very few nucleophilic amides containing electron-abstracting groups in the α-position.

Using sultones as alkylating agents affords inner *imidium ester sulfonates*[592]:

X = O, S

O-Alkylations proceed very smoothly and provide good protection for the reactants when the Meerwein reagent trialkyloxonium tetrafluoroborate is employed[593–596]. With protected α-amino carboxamides they afford the corresponding imidoyl esters on brief boiling in dichloromethane[597]:

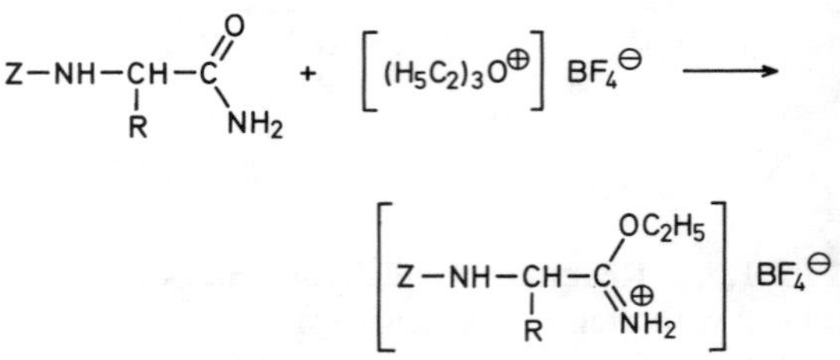

Methyl fluorosulfate (a very strong alkylating agent) produces practically only *O*-alkylation in amides side by side with a little *N*-alkylation[598].

[583] *R.N. Hurd, G. Delamater*, Chem. Rev. *61*, 78 (1961).

[584] *H. Bredereck, R. Gompper, H. Seiz*, Chem. Ber. *90*, 1837 (1957).

[585] *W. Seeliger, E. Aufderhaar, W. Diepers, R. Feinauer, R. Nehring, W. Thier, H. Hellmann*, Angew. Chem. *78*, 913 (1966).

[586] *H. Bühner*, Justus Liebigs Ann. Chem. *333*, 289 (1904).

[587] *H. Bredereck, F. Effenberger, E. Henseleit*, Chem. Ber. *98*, 2754 (1965).

[588] *H. Bredereck, G. Simchen, W. Kantlehner*, Chem. Ber. *104*, 924 (1971).

[589] *H. Bredereck, F. Effenberger, D. Zeyfang, K.H. Hirsch*, Chem. Ber. *101*, 4036 (1968).

[590] *H. Bredereck, F. Effenberger, G. Simchen*, Chem. Ber. *96*, 1350 (1963).

[591] *H. Bredereck, F. Effenberger, H.-P. Beyerlin*, Chem. Ber. *97*, 3076 (1964).

[592] *W. Ried, E. Schmidt*, Justus Liebigs Ann. Chem. *676*, 114 (1964).

[593] *H. Meerwein, E. Battenberg, H. Gold, E. Pfeil, G. Willfang*, J. Prakt. Chem. *154*, 83 (1940).

[594] *H. Meerwein, P. Borner, O. Fuchs, H.J. Sasse, H. Schrodt, J. Spille*, Chem. Ber. *89*, 2060 (1956).

[595] *S. Hanessian*, Tetrahedron Lett. 1549 (1967).

[596] *H. Bredereck, W. Kantlehner, D. Schweizer*, Chem. Ber. *104*, 3475 (1971).

[597] *W. Ried, E. Schmidt*, Justus Liebigs Ann. Chem. *695*, 217 (1966).

[598] *M.G. Ahmed, R.W. Alder, G.H. James, M.L. Sinnott, M.C. Whiting*, Chem. Commun. 1533 (1968).

The simple and high-yielding imidoyl ester synthesis from carboxamides or thiocarboxamides with ethyl carbonate is also likely to gain preparative interest[599]:

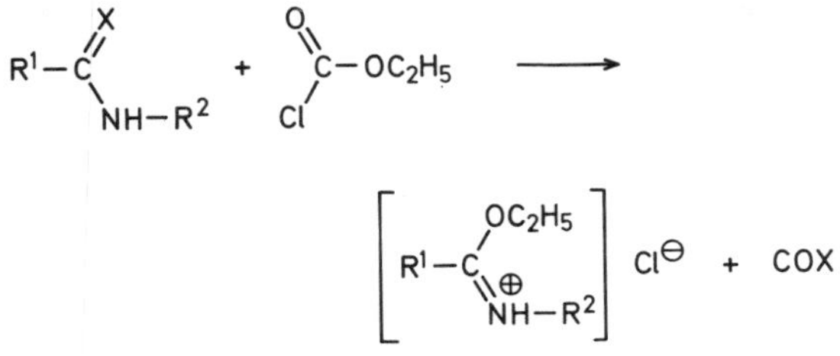

X = O, S

The reaction proceeds very well even with long-chain carboxamides, but not with amides of aromatic carboxylic acids. Some aromatic thiocarboxamides react by virtue of the greater reactivity of the fulfur.

17.2.2.1.2 Imidic acid esters and derivatives by alkylation reactions of lactams

Lactams behave like open carboxamides and form *lactim ethers* with the above alkylating agents. In the majority of cases the *O*-alkylation succeeds with alkyl sulfate in benzene[600, 601] with or without solvent[591, 601, 602, 461a] at moderate temperature to furnish good yields. With azetidinones (4-membered lactones) this method does not always lead to the desired result; by contrast, triethyloxonium tetrafluoroborate affords the expected *2-ethoxyazetines* in every case[603]:

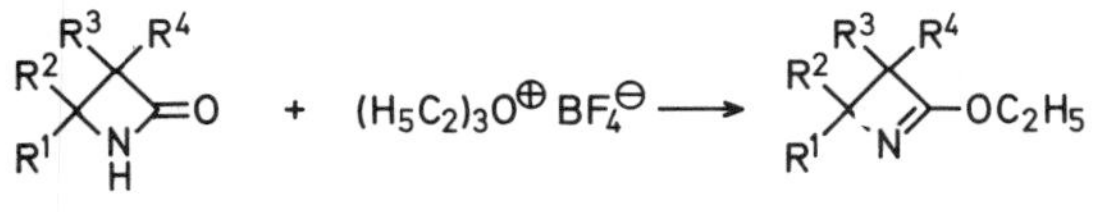

The reaction is accomplished by brief heating in dichloromethane and subsequent liberation of the base with potassium carbonate and affords up to 88% yields. Other lactams such as *N*-methylpyrrolidone[594], piperidinone[604], pyrrolidone deriva-

tives[605], and 7-azabicyclo[4.2.2]deca-2, 4, 9-trien-8-one[606] react with trialkyloxonium tetrafluoroborates even at room temperature:

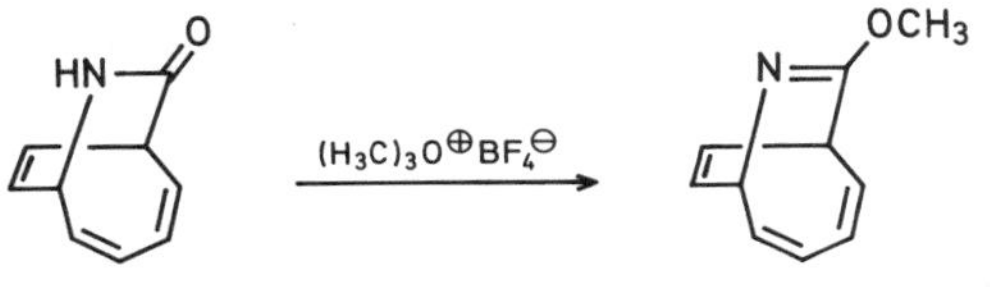

8-Methoxy-7-aza-bicyclo[4.2.2]decatetraene; 81%

The reactions do not proceed as straightforwardly with diazomethane, because either *O*- or *N*-alkylation can occur[607]. 6- and 7-membered lactams react by *O*-alkylation, while pyrrolidone affords only *N*-methylpyrrolidone[608]. Alkylating amides with diazomethane requires alcohol or tetrafluoroboric acid catalysis[608].

17.2.2.1.3 Imidic acid esters and derivatives by alkylation reactions of ureas and thioureas

Dialkyl sulfates, *p*-toluenesulfonamides, acid esters and triethyl phosphate alkylate ureas with formation of *pseudourea salts*[609-611]. Suitably, the reaction is conducted at around 100–150° because higher temperatures contribute to lower yields while lower temperatures slow the reaction down too much[611]:

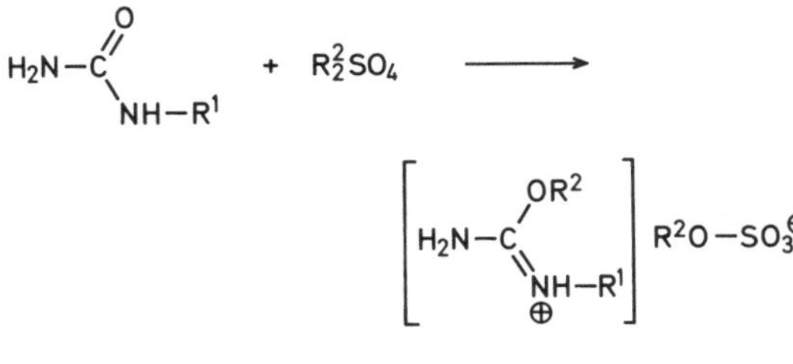

Tetraalkylureas and vinylogs[612], too, can be alkylated on the *O*-atom with dialkyl sulfates[613], thioureas derivatives particularly readily[614].

[599] F.H. Suydham, W.E. Greth, N.R. Langerman, J. Org. Chem. 34, 292 (1969).

[600] R.B. Benson, T.L. Cairns, J. Amer. Chem. Soc. 70, 2115 (1948).

[601] S. Petersen, E. Tietze, Justus Liebigs Ann. Chem. 623, 166 (1959).

[602] J. Körösi, J. Prakt. Chem. 23, 212 (1964).

[603] D. Bormann, Justus Liebigs Ann. Chem. 725, 124 (1969).

[604] W.A. Ayer, K. Piers, Can. J. Chem. 45, 451 (1967).

[605] T. Oishi, M. Nagai, Y. Ban, Tetrahedron Lett. 491 (1968).

[606] L.A. Paquette, T.J. Barton, J. Amer. Chem. Soc. 89, 5480 (1967).

[607] R. Gompper, Chem. Ber. 93, 187 (1960).

[608] J.W. Ralls, J. Org. Chem. 26, 66 (1961).

[609] E.A. Werner, J. Chem. Soc. (London) 105, 923 (1914).

[610] P.A. Ongley, Trans. Proc. Roy. Soc. New Zealand 77, 10 (1948); C.A. 42, 8165c (1948).

[611] J.W. Janus, J. Chem. Soc. (London), 3551 (1955).

[612] H. Bredereck, G. Simchen, B. Funke, Chem. Ber. 104, 2709 (1971).

Thiourea itself reacts very readily with alkyl halides to undergo *S*-alkylation and form *pseudourea halides* or *thiuronium salts*[614]. To obtain *S*-arylthiuronium salts in this way it is necessary to employ the very reactive chloropolynitrobenzenes[615], *e.g.*:

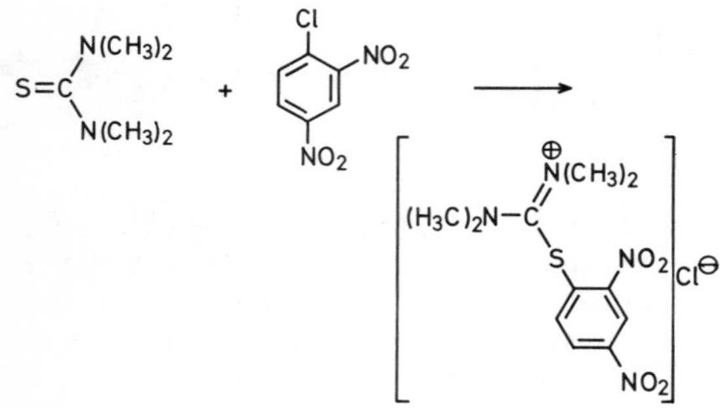

Tetramethyl-2-(2,4-dinitrophenyl)-thiopseudourea chloride

Room temperature *O*-alkylation of urea proceeds with triethyloxonium tetrafluoroborate[593] or diazoalkanes[616, 617], *e.g.*:

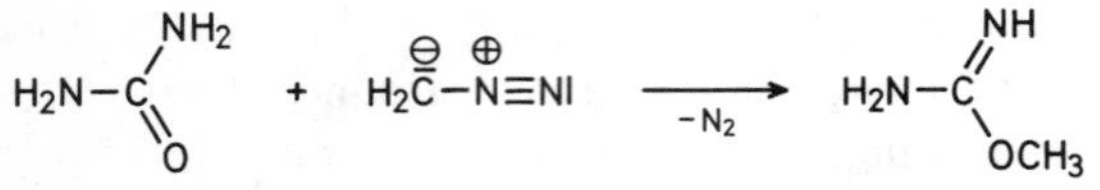

O-Methyl-pseudourea

17.2.2.2 Esters and derivatives by alcoholysis
17.2.2.2.1 Of imidoyl halides

Alcoholysis of imidoyl halides leads to imidic esters. With sodium alkoxides the reaction is generally complete in a few hours at room temperature[618, 619]:

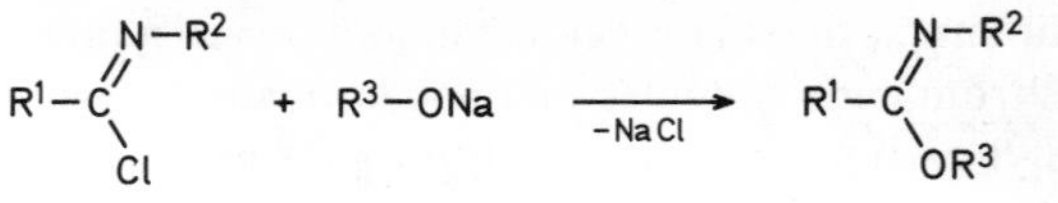

Thiophenoxides[620] and phenoxides[621-623] react analogoulsy, *e.g.*:

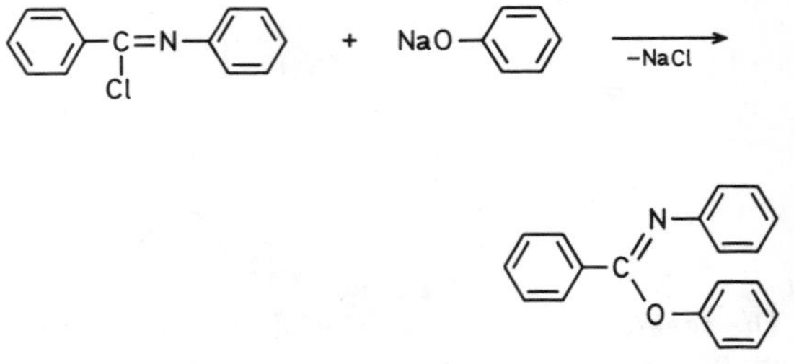

Phenyl benzimidate;
64%

It is advantageous to react the imidoyl halides with the equivalent quantity of phenoxide in a solvent such as *N*-methylpyrrolidone, because in alcoholic solution the yields of the imidic ester fall[624].

With *N*,*N*-dialkyl imidium chlorides alcoholysis or thioalcoholysis leads to *imidium esters* **34** and *thiocarboxylic acid imidium esters* **35** even in the absence of bases[625]

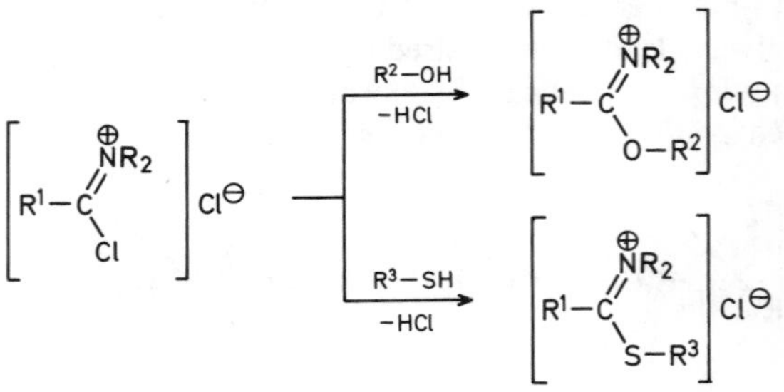

The imidium ester halides **34** are very reactive; with alcohols they form amides and alkyl chlo-

[613] *H. Bredereck, F. Effenberger, H.-P. Beyerlin,* Chem. Ber. *97*, 1834 (1964).

[614] *M. Bögemann, S. Petersen, O.-E. Schultz, H. Söll* in *Houben-Weyl,* Methoden der organischen Chemie, 4. Aufl., Bd. IX, p. 900, Georg Thieme Verlag, Stuttgart 1955.

[615] *H. Kessler, H.-O. Kalinowski, Ch. v. Chamier,* Justus Liebigs Ann. Chem. *727*, 228 (1969).

[616] *Z. Piasek, T. Urbanski,* Tetrahedron Lett. 723 (1962).

[617] *B. Eistert, M. Regitz, G. Heck, H. Schwall* in *Houben-Weyl,* Methoden der organischen Chemie, 4. Aufl., Bd. X/4, p. 739, Georg Thieme Verlag, Stuttgart 1968.

[618] *A.E. Arbuzov, V.E. Shishkin,* Dokl. Akad. Nauk SSSR *141*, 349 (1961); C.A. *56*, 11491h (1962).

[619] *H. Paul, A. Weise, R. Dettmer,* Chem. Ber. *98*, 1450 (1965).

[620] *A.W. Chapman,* J. Chem. Soc. (London), 2296 (1926).

[621] *O. Mumm, H. Hesse, H. Volquartz,* Ber. *48*, 379 (1915).

[622] *K.B. Wiberg, B.I. Rowland,* J. Amer. Chem. Soc. *77*, 2205 (1955).

[623] *E.R.H. Jones, F.G. Mann,* J. Chem. Soc. (London), 786 (1956).

[624] *G. Bock,* Chem. Ber. *100*, 2870 (1967).

[625] *H. Eilingsfeld, M. Seefelder, H. Weidinger,* Chem. Ber. *96*, 2671 (1963).

rides, with bases *orthocarbonic acid diester mono-amides* or *ketene O,N-acetals* (*cf.* p. 766).

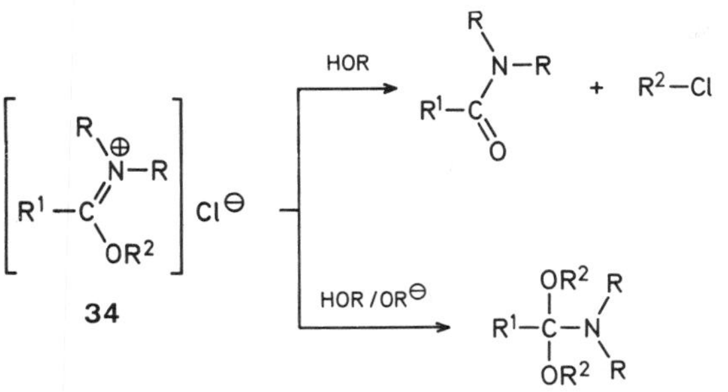

34

17.2.2.2.2 Esters and derivatives by alcoholysis of chlorocarboxamidines

Chloroformamidines react very readily with alkoxides or phenoxides in alcoholic solution to *O-alkyl* or *O-aryl pseudoureas*[626, 627]:

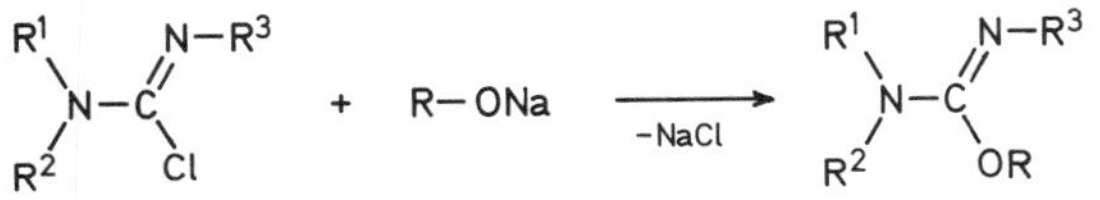

It is not essential to use phenols as the phenoxides; triethylamine in acetone, too, yields excellent results[624] (up to 95%). Thiophenols react analogously with chloroformamidines to *S-arylpseudo-thioureas*. Their reaction with tetramethylchloroformamidinium chloride gives high yields of *S-aryl thiouronium salts*[615], *e.g.*:

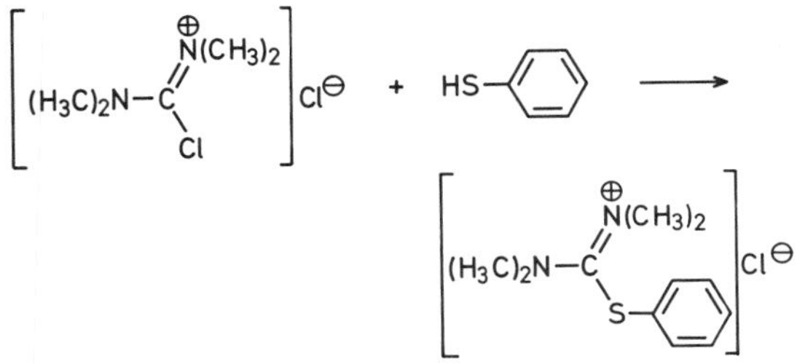

Tetramethyl-2-(phenylthio)-pseudourea chloride

Substituted chloroformamidines and their *N*-sulfonyl derivatives likewise react with alcoholates to form *O-alkylpseudoureas*[628]:

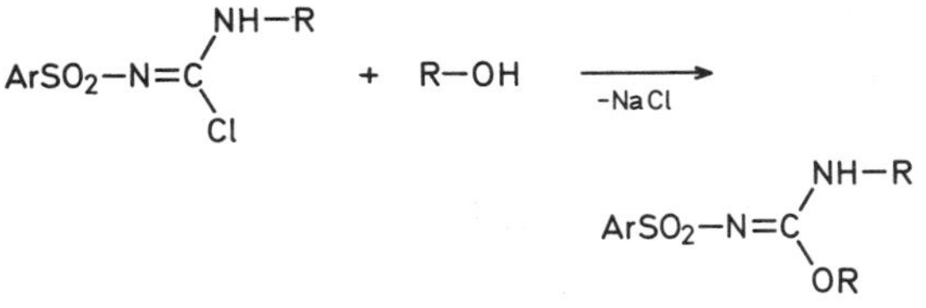

17.2.2.2.3 Esters and derivatives by alcoholysis of hydroxamoyl chlorides

Hydroxamoyl chlorides are generally reacted with alcoholates in the form of their *O-alkyl* derivatives[629-631]. The reaction proceeds to *alkyl O-alkyl-hydroximates* in good yield in every case, *e.g.*:

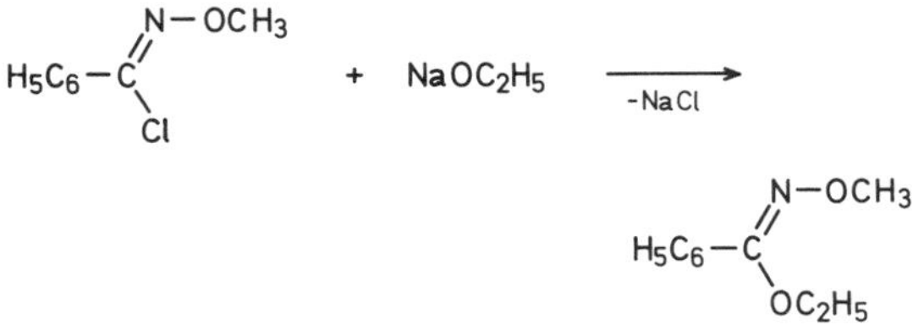

Ethyl O-methylbenz-hydroximate

The cyclic hydroxamoyl chloride (3,5-dichloro-2,1-benzisoxazole) requires 140–160° to react with alkoxide or phenoxide in dimethyl sulfoxide as solvent[632] with formation of 3-alkoxy or 3-aryloxy-5-chloro-2,1-benzisoxazole:

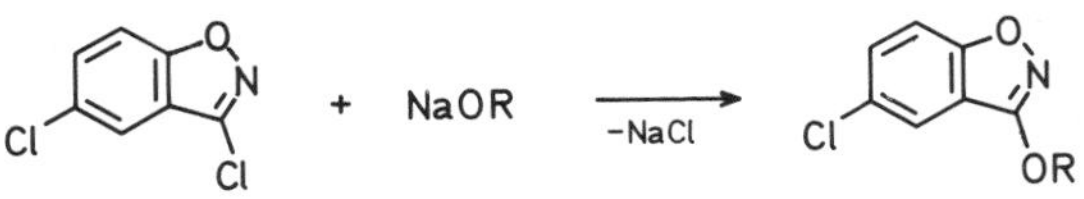

17.2.2.2.4 Esters and derivatives by alcoholysis of hydrazide chlorides

Ester hydrazones can be prepared from acid chloride hydrazones with alcohols, *e.g.*[633]:

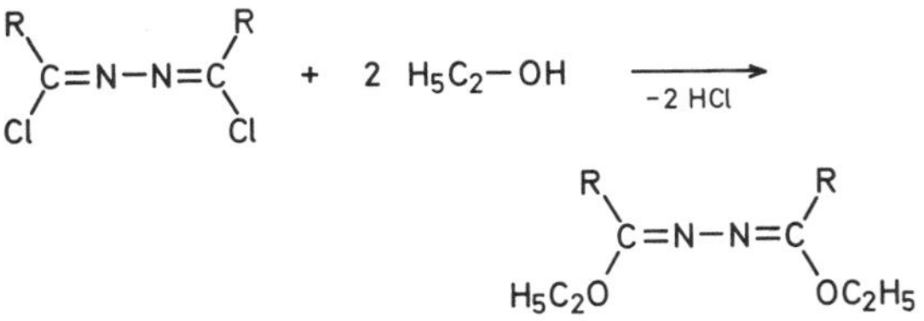

Ethyl carboxylate azine

[626] *S.O. Winthrop, G. Gavin*, J. Org. Chem. *24*, 1936 (1959).
[627] *S.E. Forman, C.A. Erickson, H. Adelman*, J. Org. Chem. *28*, 2653 (1963).
[628] *H. Ulrich, B. Tucker, A.A.R. Sayigh*, Tetrahedron *22*, 1565 (1966).

[629] *W. Lossen*, Justus Liebigs Ann. Chem. *281*, 169 (1895).
[630] *F. Tiemann, P. Krüger*, Ber. *18*, 727 (1885).
[631] *A. Werner, A. Gemeseus*, Ber. *29*, 1161 (1896).
[632] *H. Böshagen*, Chem. Ber. *100*, 3326 (1967).
[633] *R. Stollé*, J. Prakt. Chem. *73*, 277 (1906).

Alcoholysis of acid bromide hydrazones proceeds very readily in the presence of equivalent amounts of sodium acetate on boiling[634]:

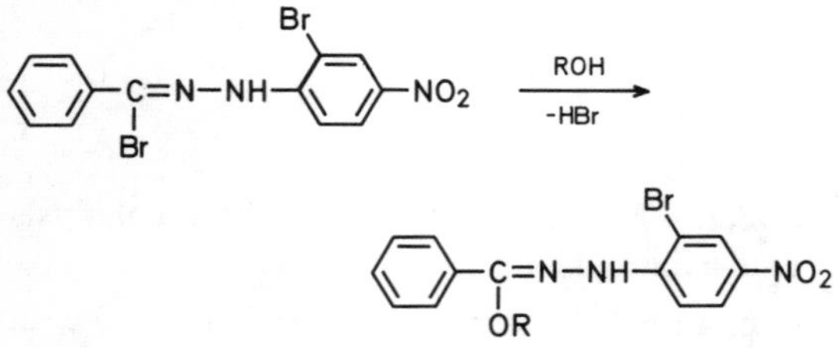

Alkyl benzoate (2-bromo-4-nitro-phenyl)hydrazone

In 6-chloro-3-pyridazinone, however, successful substitution with sodium methoxide requires 130°[635]:

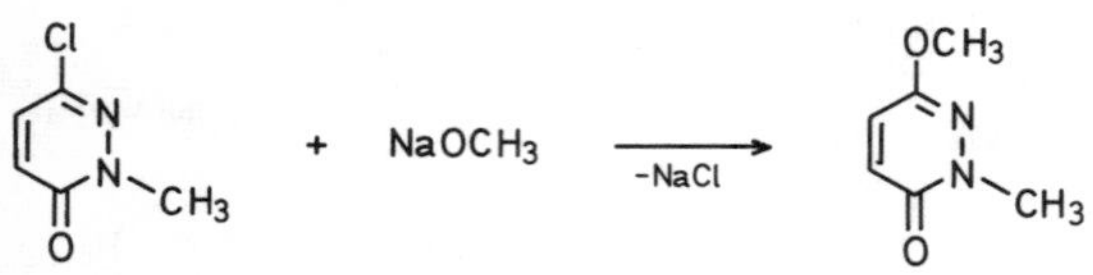

6-Methoxy-2-methyl-3-pyridazinone

17.2.2.2.5 Esters and derivatives by alcoholysis of nitriles

One of the most important pathways for synthesizing imidic esters consists in allowing alcohols or thiols to act on nitriles in the presence of acids such as hydrochloric acid, hydrogen bromide, or sulfuric acid[582, 636]

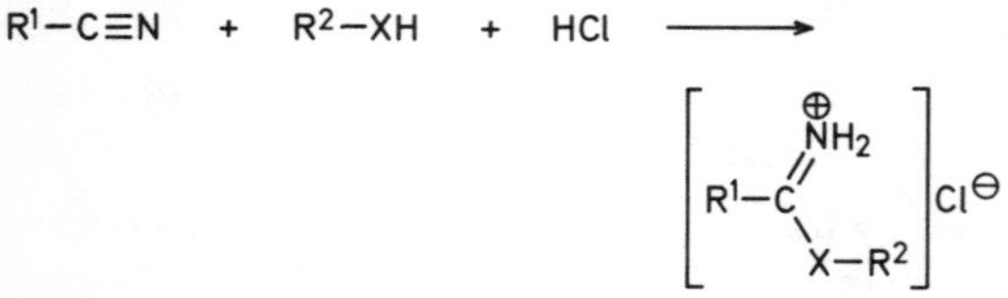

X = O, S

The reaction is conducted under anhydrous conditions with the molar quantity of alcohol in ether, chloroform, or benzene. Higher alcohols, glycols, and the corresponding sulfur compounds, too, react well[582]. *Alkyl formimidate hydrochlorides*, formerly accessible by

this technique only with anhydrous hydrogen cyanide, are obtained more conveniently from formamide by allowing benzoyl chloride and alcohol to react simultaneously[637].

A further elegant method is the base-catalyzed alcoholysis of nitriles with alkoxides in alcoholic solution[638]; the yield of the free imidic esters exceeds 50% in every case.

17.2.2.2.6 Esters and derivatives by alcoholysis of nitrile oxides

Aromatic nitrile oxides add alcohols in the presence of strong bases (*e.g.*, alcoholates) to form *alkyl hydroximates*[639].

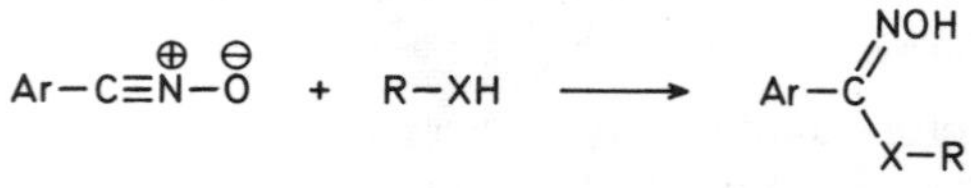

X = O, S

Dilute sulfuric acid is an alternative catalyst. Thiols, which are more nucleophilic than alcohols, react with aliphatic nitrile oxides without catalysis to form *S-alkyl thiohydroximates* even more readily, albeit having triethylamine present is beneficial[639].

17.2.2.2.7 Esters and derivatives by alcoholysis of carbodiimides and cyanamides

Addition of alcohols to carbodiimides to form *O-alkylpseudoureas* is catalyzed by bases, acids, or copper(I) chloride[640]:

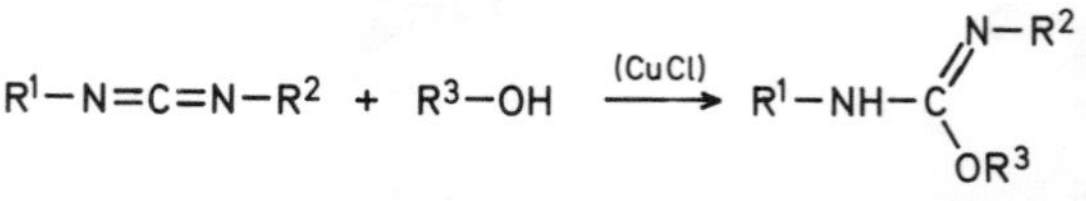

Catalysis with copper(I) chloride is found to be very favorable and affords pseudoureas in 90% yield[641, 642].

[634] *F.L. Scott, J.B. Aylward*, Tetrahedron Lett. 841 (1965).

[635] *K. Eichenberger, A. Staehelin, J. Druey*, Helv. Chim. Acta *37*, 837 (1954).

[636] *H. Henecka, P. Kurtz* in *Houben-Weyl*, Methoden der organischen Chemie, 4. Aufl., Bd. VIII, p. 697, Georg Thieme Verlag, Stuttgart 1952.

[637] *R. Ohme, O. Schmitz*, Angew. Chem. *79*, 531 (1967).

[638] *F.C. Schaefer, G.A. Peters*, J. Org. Chem. *26*, 412 (1961).

[639] *Ch. Grundmann*, Fortschr. Chem. Forsch. *7*, 62 (1966).

[640] *E. Vowinkel*, Chem. Ber. *99*, 42 (1966).

[641] *E. Schmidt, F. Moosmüller*, Justus Liebigs Ann. Chem. *597*, 235 (1955).

[642] *E. Vowinkel*, Chem. Ber. *99*, 1479 (1966).

In the presence of catalytic amounts of sodium alkoxide[627] or potassium *tert*-butoxide[643] disubstituted cyanamides react with primary, secondary, or tertiary alcohols to give *O-alkylpseudoureas* likewise very readily:

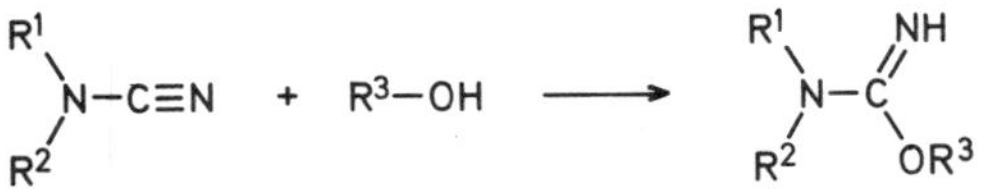

$$R^1R^2N-C{\equiv}N \;+\; R^3-OH \longrightarrow R^1R^2N-C(=NH)OR^3$$

$$HC{\equiv}C-OC_2H_5 \;+\; H_2N-R \longrightarrow \left(H_2C{=}C\!\!\begin{array}{c}OC_2H_5\\NH-R\end{array}\right) \longrightarrow H_3C-C\!\!\begin{array}{c}N-R\\OC_2H_5\end{array}$$

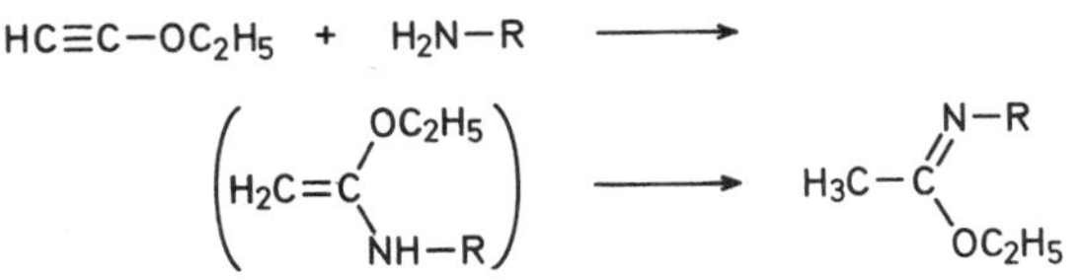

With secondary amines the reaction stops at the ketene *O,N*-acetal stage.

17.2.2.2.8 Esters and derivatives by alcoholysis of esters of cyanic acid

Esters of cyanic acid are characterized by a very reactive nitrile group. In the presence of bases, alcohols, phenols and sulfur analogs it adds on to form imidocarbonic diesters or imidothiocarbonic *O,S*-diesters[644-646]:

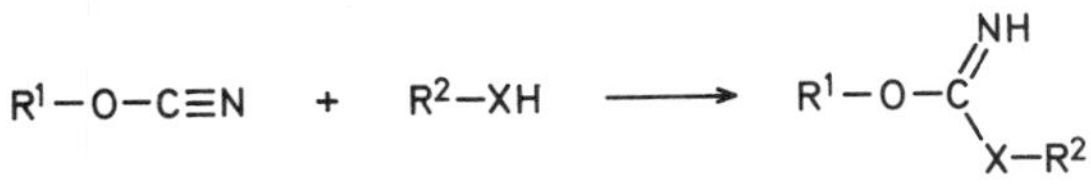

$$R^1-O-C{\equiv}N \;+\; R^2-XH \longrightarrow R^1-O-C(=NH)X-R^2$$

X = O, S

Esters of thiocyanic acid $R-S-C{\equiv}N$ react to form imidothiocarbonic acid diesters under either alkaline[646a] or acid catalysis[647].

17.2.2.3 Esters and derivatives by aminolysis

17.2.2.3.1 Of alkoxyacetylenes

The very reactive alkoxyacetylenes add primary amines to *ethyl N-alkylacetimidate* under anhydrous conditions[648, 649]:

17.2.2.3.2 Esters and derivatives by aminolysis of esters of cyanic acid

Esters of cyanic acid react with ammonia and primary amines to form *bis-(aryloxyimidocarbonyl)-amines*[650],

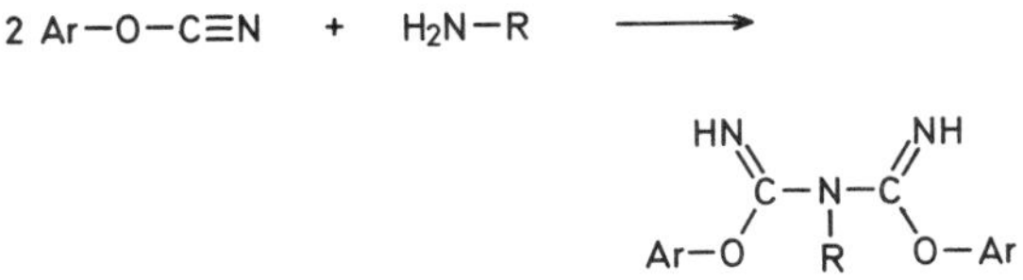

$$2\,Ar-O-C{\equiv}N \;+\; H_2N-R \longrightarrow$$

$$Ar-O-C(=NH)-N(R)-C(=NH)-O-Ar$$

R = H-, Alkyl-, Aryl-

while secondary amines or primary amine hydrochlorides react to form *O-arylpseudoureas*[650, 651].

Hydrazine also adds two molecules aryl cyanate, and even with excess hydrazine present[652], to give *N,N'-bis(aryloxyimidocarbonyl)hydrazines* even at low temperatures:

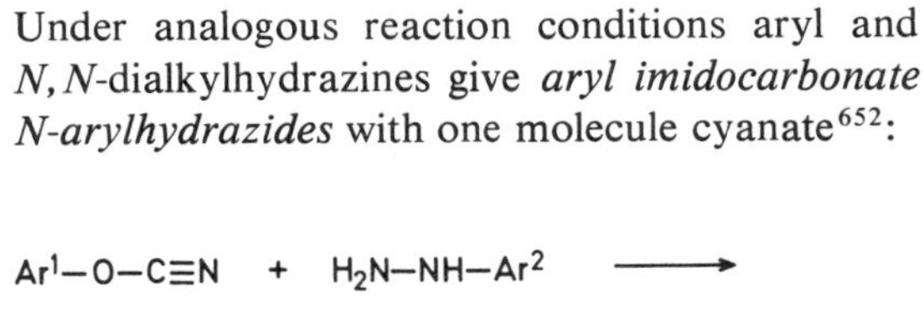

$$2\,Ar-O-C{\equiv}N \;+\; H_2N-NH_2 \longrightarrow$$

$$Ar-O-C(=NH)-NH-NH-C(=NH)-O-Ar$$

Under analogous reaction conditions aryl and *N,N*-dialkylhydrazines give *aryl imidocarbonate N-arylhydrazides* with one molecule cyanate[652]:

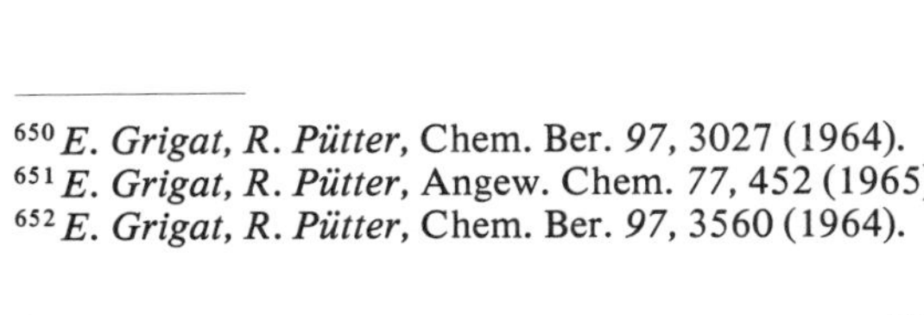

$$Ar^1-O-C{\equiv}N \;+\; H_2N-NH-Ar^2 \longrightarrow$$

$$Ar^1-O-C(=NH)-NH-NH-Ar^2$$

[643] *J.H. Amin, J. Newton, F.L.M. Pattison*, Can. J. Chem. 43, 3172 (1965).

[644] *E. Grigat, R. Pütter*, Chem. Ber. 97, 3018 (1964).

[645] *K.A. Jensen, M. Due, A. Holm, C. Wentrup*, Acta Chem. Scand. 20, 2091 (1966).

[646] *E. Grigat, R. Pütter* in *W. Foerst*, Neuere Methoden der präparativen organischen Chemie, Bd. VI, p. 159, Verlag Chemie, Weinheim/Bergstr. 1970; *E. Grigat, R. Pütter*, Angew. Chem. 79, 219 (1967); *E. Grigat, R. Pütter*, Angew. Chem. Intern. Ed. Engl. 6, 206 (1967).

[646a] *K. Tanaka*, Bull. Chem. Soc. Japan 45, 834 (1972).

[647] *E. Bögemann, S. Petersen, O.-E. Schultz, H. Söll* in *Houben-Weyl*, Methoden der organischen Chemie, 4. Aufl., Bd. IX, p. 835, Georg Thieme Verlag, Stuttgart 1955.

[648] *J.F. Arens* in Advan. Org. Chem. Vol. II, p. 179, Interscience Publishers, New York 1960.

[649] *J.F. Arens, T.R. Rix*, Proc. Koninkl. Ned. Acad. Wetenshap B 57, 270, 275 (1954); C.A. 49, 8798i (1955).

[650] *E. Grigat, R. Pütter*, Chem. Ber. 97, 3027 (1964).

[651] *E. Grigat, R. Pütter*, Angew. Chem. 77, 452 (1965).

[652] *E. Grigat, R. Pütter*, Chem. Ber. 97, 3560 (1964).

Reacting esters of cyanic acid with hydroxylamine hydrochloride smoothly gives the salts of *3-hydroxy-2-arylpseudoureas* even at room temperature in an exothermic reaction

$$Ar-O-C\equiv N \quad + \quad H_2N-OH \cdot HCl \longrightarrow$$

$$\left[Ar O-C \begin{matrix} \overset{\oplus}{N}H_2 \\ \\ NHOH \end{matrix} \right] Cl^{\ominus}$$

Sodium carbonate liberates the bases[653].
Analogously, *N*-arylhydroxylamines can add to esters of cyanic acid[653]:

$$Ar^1-O-C\equiv N \quad + \quad Ar^2-NH-OH \longrightarrow$$

$$Ar^1-O-C \begin{matrix} NH \\ \\ N-Ar^2 \\ HO \end{matrix}$$

Aryl imidocarbonate
N-arylhydroxy-
amides; ~ 86%

17.2.2.3.3 Esters and derivatives by aminolysis of orthocarboxylic acid triesters

In the presence of catalytic amounts of acid, trialkyl orthocarbonates react with aniline to form *alkyl phenylformimidates*[654, 655]:

$$HC(OR)_3 \quad + \quad H_2N-C_6H_5 \xrightarrow[-2\ ROH]{H^{\oplus}} HC \begin{matrix} N-C_6H_5 \\ \\ OR \end{matrix}$$

The reaction proceeds in high yield on heating the components while distilling off the alcohol. Other aromatic amines also react with triesters of orthoformic, orthoacetic, and orthocarbonic acids; the imide ester is formed in more than 85% yield even without acid catalysis[656], but the latter is advantageous because it enables the reaction time to be shortened appreciably. In general a few drops gla-

cial acetic acid are added to the reaction mixture and is followed by heating to 120° and distilling off of the alcohol formed[657]. Under these conditions 2,2-diethoxytetrahydrofuran reacts smoothly with substituted anilines to form *2-(arylimino)-tetrahydrofurans*[658]:

$$\underset{OC_2H_5}{\overset{OC_2H_5}{\diagdown}} + H_2N-C_6H_4-R \xrightarrow{-2\ H_5C_2-OH}$$

$$O \diagup N-C_6H_4-R$$

Not only aromatic amines but also sulfonamides form imidic esters with orthocarboxylic acids[659] and triesters of orthocarbonic acid[660] under corresponding conditions. With *p*-toluenesulfonylhydrazine triesters of orthocarboxylic acids react in a few minutes at room temperature to give *carboxylic acid ester tosyl hydrazides*[661]:

$$H_3C-C_6H_4-SO_2-NH-NH_2 \quad + \quad R^1C(OR)_3 \xrightarrow{-2\ ROH}$$

$$R^1-C \begin{matrix} N-NH-SO_2-C_6H_4-CH_3 \\ \\ OR \end{matrix}$$

$$R^1 = H-, CH_3-, C_2H_5-, C_6H_5-$$

17.2.2.3.4 Esters and derivatives by transimidation of imidic acid esters

Shaking ether solutions of imidic acid esters or imidocarbonic acid diesters with aqueous hydroxylamine hydrochloride solution affords *alkyl hydroximates* with substitution of the imide group[662, 663]:

$$R-C \begin{matrix} NH \\ \\ OC_2H_5 \end{matrix} + H_2NOH \cdot HCl \xrightarrow{-NH_4Cl} R-C \begin{matrix} NOH \\ \\ OC_2H_5 \end{matrix}$$

[653] *E. Grigat, R. Pütter, C. König*, Chem. Ber. **98**, 144 (1965).
[654] *R.M. Roberts*, J. Amer. Chem. Soc. **71**, 3848 (1949).
[655] *R.M. Roberts, T.D. Higgins, P.R. Noyes*, J. Amer. Chem. Soc. **77**, 3801 (1955).
[656] *R.H. De Wolfe*, J. Org. Chem. **27**, 490 (1962).

[657] *E.C. Taylor, W.A. Ehrhart*, J. Org. Chem. **28**, 1108 (1963).
[658] *T. Mukaiyama, K. Sato*, Bull. Chem. Soc. Japan **36**, 99 (1963); C.A. **59**, 5271b (1963).
[659] *B. Loev, M.F. Kormendy*, Can. J. Chem. **42**, 176 (1964).
[660] *R.F. Meyer*, J. Org. Chem. **28**, 2902 (1963).
[661] *R.M. McDonald, R.A. Krueger*, J. Org. Chem. **31**, 488 (1966).
[662] *J. Houben, E. Schmidt*, Ber. **46**, 2447, 3618 (1913).
[663] *J. Houben, E. Pfankuch*, Ber. **59**, 2392, 2397 (1926).

S-Esters of thioformimidic acid react analogously to form *thioformohydroximic acid S-esters*[664]. This exchange technique enables the imide group to be replaced by other nitrogen bases such as amines[665-667], thiosemicarbazides[668], or hydrazines[665]

as well and to obtain *esters* of *alkylimidic acids* and of *carboxylic acid alkylhydrazides* in this way.

17.2.3 Amides, hydroxyamides, hydrazides, amidrazones and related compounds (amidines, amidoximes, amidrazones, hydrazidines, aminoguanidines, formazans, *etc.*)

This section describes compounds with the following structural element:

$$X = N\diagdown \; ; \; -N\diagup^{N\diagdown}_{\diagdown N\diagup} \; ; \; -N\diagup_{OH} \; ; \; -N\diagup_{OR} \quad etc.$$

Table 5 surveys the types of compounds concerned.

Table 5. N-Functional imidic acid and imidocarbonic acid derivatives

Formula	Name	Survey literature
	Amidines	669, 670, 671
	Amidrazones	672, 673, 674

Table 5 (continue)

Formula	Name	Survey literature
	Hydrazidines (hydrazide hydrazones, 1,2-dihydroformazans)	673
	Formazans	675, 676, 677
	Amidoximes	678, 679, 680
	N-Hydroxyamidoximes (hydroxamic acid oximes)	681
	Hydroxamic acid hydrazides	682
	Thioamidoximes	683
	Nitrosolic acids (nitrosoaldoximes; hydroxyimidonitrosides)	684, 685
	Nitrolic acids (nitroaldoximes)	686
	Guanidines	687, 688
	Aminoguanidines	689, 690, 691
	N,N'-Diaminoguanidines	690

[664] *J. Houben, R. Zivadinovitsch*, Ber. *69*, 2352 (1936).
[665] *E. Schmidt*, Ber. *47*, 2545 (1914).
[666] *G. Shaw, R.N. Warrener, D.N. Butler, R.K. Ralph*, J. Chem. Soc. (London) 1648 (1959).
[667] *A. Kjaer*, Acta Chem. Scand. *7*, 1024 (1953).

Table 5 (continue)

Formula	Name	Survey literature
	N,N',N''-Triaminoguanidines	690
	Azoamidines, Azoformamidines	692
	N-Hydroxyguanidine	692
	N,N'-Dihydroxyguanidine	692
	Nitrosoguanidines	693
	Nitroguanidines	694
	Nitrosoamidoximes (nitroso-aminoaldoximes, imidohydrox-amide nitrosides)	692
	Cyanoguanidines (Dicyandiamides)	695
	Biguanides	696, 697, 698

The bonds denote hydrogen or a (possibly functionally substituted) alkyl or aryl group.

17.2.3.1 Aminolysis
17.2.3.1.1 Of amides and thioamides

The reaction of thiocarboxamides with NH_2-group containing compounds possesses significance where the thiocarboxamides are readily accessible. With ammonia or amines *amidines* are formed, with hydroxylamine *amides oximes,* and with hydrazine *amidrazones* (see Table 6 on p. 733)[699, 701]:

[668] *C. Ainsworth*, J. Amer. Chem. Soc. *78*, 1973 (1956).

[669] *H. Henecka* in *Houben-Weyl*, Methoden der organischen Chemie, 4. Aufl., Bd. VIII, p. 702, Georg Thieme Verlag, Stuttgart 1952.

[670] *A. Kreutzberger* in *E. Jucker*, Arzneimittel-Forsch. *11*, 356 (1968).

[671] *R.L. Shriner, F.W. Neumann*, Chem. Rev. *35*, 351 (1944).

[672] *D.G. Neilson, R. Roger, J.W.M. Heatle, L.R. Newlands*, Chem. Rev. *70*, 151 (1970).

[673] *P.A.S. Smith*, The Chemistry of Open Chain Organic Nitrogen Compounds, p. 173, Benjamin, New York, Amsterdam 1966.

[674] *C.C. Clark*, Hydrazine, p. 55, Mathieson Chem. Co., Baltimore Maryland, 1953.

[675] *H. Henecka* in *Houben-Weyl*, Methoden der organischen Chemie, 4. Aufl., Bd. VIII, p. 706, Georg Thieme Verlag, Stuttgart 1952.

[676] *A.W. Nineham*, Chem. Rev. *55*, 355 (1955).

[677] *W. Ried*, Angew. Chem. *64*, 391 (1956).

[678] *P.A.S. Smith*, The Chemistry of Open Chain Organic Nitrogen Compounds, p. 68, Benjamin, New York, Amsterdam 1966.

[679] *F. Eloy, R. Lenaers*, Chem. Rev. *62*, 155 (1962).

[680] *H. Henecka* in *Houben-Weyl*, Methoden der organischen Chemie, 4. Aufl., Bd. VIII, p. 692, Georg Thieme Verlag, Stuttgart 1952;
H. Metzger in *Houben-Weyl*, Methoden der organischen Chemie, 4. Aufl., Bd. X/4, p. 209, Georg Thieme Verlag, Stuttgart 1968.

[681] *H. Henecka* in *Houben-Weyl*, Methoden der organischen Chemie, 4. Aufl., Bd. VIII, p. 695, Georg Thieme Verlag, Stuttgart 1952;
H. Metzger in *Houben-Weyl*, Methoden der organischen Chemie, 4. Aufl., Bd. X/4, p. 212, Georg Thieme Verlag, Stuttgart 1968.

[682] *H. Henecka* in *Houben-Weyl*, Methoden der organischen Chemie, 4. Aufl., Bd. VIII, p. 696, Georg Thieme Verlag, Stuttgart 1952;
H. Metzger in *Houben-Weyl*, Methoden der organischen Chemie, 4. Aufl., Bd. X/4, p. 212, Georg Thieme Verlag, Stuttgart 1968.

[683] *F. Eloy*, Chem. Rev. *62*, 179 (1962).

[684] Rodd's Chemistry of Carbon Compounds, Vol. Ic, p. 192, Elsevier Publ., Co., Amsterdam, London, New York 1965.

Table 6. *N*-Functional imidic acid derivatives from thiocarboxamides

Thiocarboxamide	Amine component	Condensing agent (solvent)	Reaction product		Yield [% of th.]	Ref.
$H_5C_6-CS-NH-C_6H_5$	H_2N-NH_2	H_5C_2-OH	$H_5C_6-C\begin{smallmatrix}N-NH_2\\\\NH-C_6H_5\end{smallmatrix}$	Benzanilide hydrazone	65	707
$HOOC-CS-NH_2$	H_2N-NH_2	H_5C_2-OH, H_2O	$HOOC-C\begin{smallmatrix}N-NH_2\\\\NH_2\end{smallmatrix}$	Oxalamiddrazone	87	708
$H_5C_6-CS-NH_2$	NH_2OH	H_5C_2-OH, H_2O	$H_5C_6-C\begin{smallmatrix}NOH\\\\NH_2\end{smallmatrix}$	Benzamidoxime	good	709
$H_5C_6-NH-CS-NH-C_6H_5$	NH_3	$H_2O, ZnCl_2, NaOH$	$H_5C_6-NH-C\begin{smallmatrix}NH\\\\NH-C_6H_5\end{smallmatrix}$	N,N'-Diphenylguanidine	90	710
$H_5C_6-NH-CS-NH-C_6H_5$	H_2N-NH_2	H_2O, H_5C_2-OH, KOH	$H_5C_6-NH-C\begin{smallmatrix}N-NH_2\\\\NH-C_6H_5\end{smallmatrix}$	N''-Amino-N,N'-diphenylguanidine	75	705
$\begin{smallmatrix}H\\\\C-N(CH_3)_2\\\\S\end{smallmatrix}$	$H_7C_{10}-NH_2$	$NaOCH_3$	$HC\begin{smallmatrix}N-C_7H_{10}\\\\N(CH_3)_2\end{smallmatrix}$	N-Naphthyl-N',N'-dimethyl-formamidine	94	711

Heavy metal salts or heavy metal oxides such as mercury(II) chloride, and oxide, freshly precipitated lead(II) oxide or hydroxide[700], basic lead(II) carbonate[702], and zinc chloride or zinc salts in the presence of alkali[703] are recommended for improving the separation of hydrogen sulfide. The last-named technique is indicated especially for synthesizing *guanidine* derivatives from thioureas.

Unless the hydrogen sulfide is eliminated the aminolysis generally proceeds nonuniformly. Thus transamidation ensues with primary aliphatic amines[704]. *Acid aminolysis* behaves similarly; according to the constitution of the reaction partners the thioxo and/or the amide function are exchanged[699].

[685] *H. Metzger, H. Meier* in *Houben-Weyl,* Methoden der organischen Chemie, 4. Aufl., Bd. X/1, p. 968, Georg Thieme Verlag, Stuttgart 1971.

[686] *H. Metzger* in *Houben-Weyl,* Methoden der organischen Chemie, 4. Aufl., Bd. X/4, p. 47, 48, 87, 101, 173, Georg Thieme Verlag, Stuttgart 1968.

[687] *S. Petersen* in *Houben-Weyl,* Methoden der organischen Chemie, 4. Aufl., Bd. VIII, p. 98, 180, Georg Thieme Verlag, Stuttgart 1952.

[688] *G.J. Durant, A.M. Roe, A.L. Green,* Progr. Medicinal Chemistry 7, p. 124, Butterworth & Co., London 1970.

[689] *S. Petersen* in *Houben-Weyl,* Methoden der organischen Chemie, 4. Aufl., Bd. VIII, p. 190, Georg Thieme Verlag, Stuttgart 1952.

[690] *E. Lieber, G.B.L. Smith,* Chem. Rev. 25, 213 (1939).

[691] *F. Kurzer, L.E.A. Godtry,* Chem. & Ind. (London), 1584 (1962).

[692] Rodd's Chemistry of Carbon Compounds, Vol. Ic, p. 351, Elsevier Publ. Co., Amsterdam, London, New York 1965.

[693] *A.F. McKay,* Chem. Rev. 51, 309 (1952).

[694] *A.F. McKay,* Chem. Rev. 51, 301 (1952).

[695] *H.F. Piepenbrink* in *Houben-Weyl,* Methoden der organischen Chemie, 4. Aufl., Bd. VIII, p. 209, Georg Thieme Verlag, Stuttgart 1952.

[696] *H.F. Piepenbrink* in *Houben-Weyl,* Methoden der organischen Chemie, 4. Aufl., Bd. VIII, p. 215, Georg Thieme Verlag, Stuttgart 1952.

[697] *P. Ray,* Chem. Rev. 61, 313 (1961).

[698] *F. Kurzer, E.D. Pitchfork,* Fortschr. Chem. Forsch. 10, 375 (1968).

[699] *R.N. Hurd, G.M. Delamater,* Chem. Rev. 61, 61ff. (1961).

[700] *J. Aalway, C.E. Vail,* Amer. Chem. J. 28, 158 (1902).

[701] *A. Bernthsen,* Justus Liebigs Ann. Chem. 192, 1, 29 (1878).

[702] DRP. 455 586 (1925), I.G. Farbenindustrie; Friedländer: Fortschritte der Teerfabrikation u. verwandter Industriezweige 16, 453 (1931).

[703] DRP. 481 994 (1926), I.G. Farbenindustrie; Erf.: *H. Meis;* Friedländer: Fortschritte der Teerfabrikation u. verwandter Industriezweige 16, 2516 (1931); DRP. 550 571 (1927), I.G. Farbenindustrie; Erf.: *H. Meis;* Friedländer: Fortschritte der Teerfabrikation u. verwandter Industriezweige 17, 310 (1932).

[704] *O. Wallach,* Justus Liebigs Ann. Chem. 262, 360 (1891).
H.M. Woodburn, C.E. Sroog, J. Org. Chem. 17, 371 (1952).

Condensation of N,N-diarylthioureas with hydrazine succeeds best in the presence of ethanolic potassium hydroxide[705]. An additional special variant is represented by the transamidation of alkyl and aryl amides of thiocarboxylic acids with N-sulfinyl sulfonamides to form *amidines*[706]. With primary thiocarboxamides *1,2,4-thiadiazoles* or *nitriles* are obtained in dependence on the constitution of the starting products employed[706]:

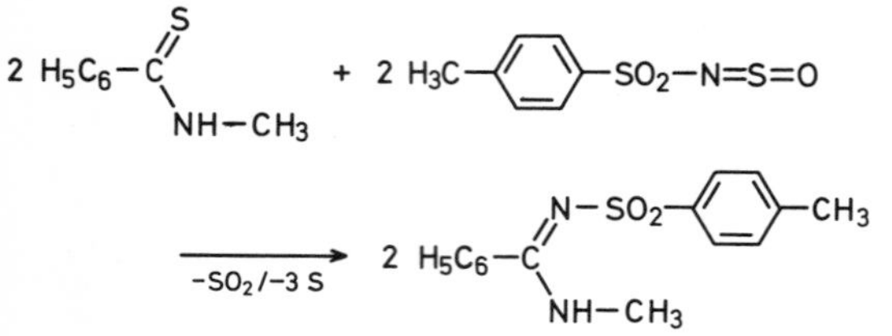

$$2\ H_5C_6-\underset{\underset{NH-CH_3}{|}}{\overset{\overset{S}{||}}{C}} + 2\ H_3C-\langle\ \rangle-SO_2-N=S=O$$

$$\xrightarrow[-SO_2/-3\ S]{} 2\ H_5C_6-\underset{\underset{NH-CH_3}{|}}{\overset{\overset{N-SO_2-\langle\ \rangle-CH_3}{||}}{C}}$$

N-Methyl-N'-tosylbenz-amidine; 42%

Direct replacement of the oxygen in carboxamides and amides of carbonic acid by NH-group containing components has been only seldom accomplished. It is necessary to employ tetrakis-(dimethylamino)titanium and phosphoric acid diamides and triamides (addition of amine hydrochlorides is necessary for better yields in this case) to obtain a transfer of the amino groups to monosubstituted carboxamides above $160°$[712]:

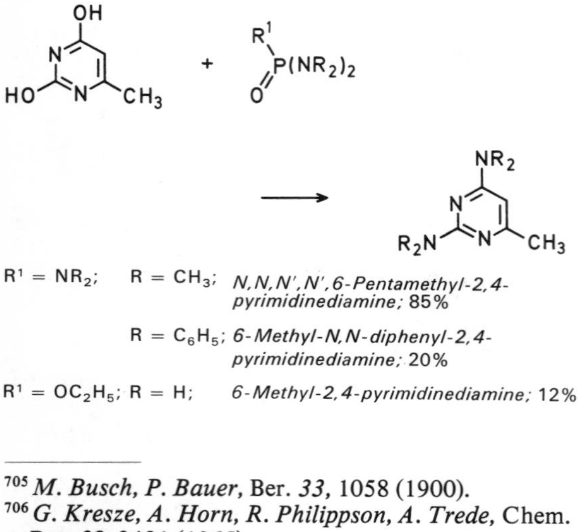

R¹ = NR₂;   R = CH₃; *N,N,N',N',6-Pentamethyl-2,4-pyrimidinediamine; 85%*

R = C₆H₅; *6-Methyl-N,N-diphenyl-2,4-pyrimidinediamine; 20%*

R¹ = OC₂H₅; R = H;   *6-Methyl-2,4-pyrimidinediamine; 12%*

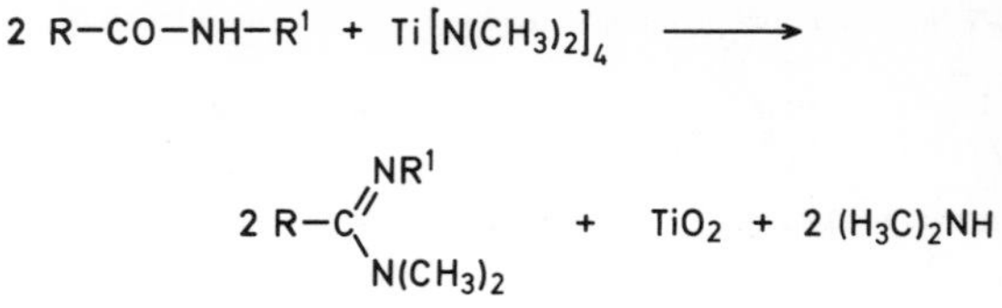

$$2\ R-CO-NH-R^1\ +\ Ti[N(CH_3)_2]_4\ \longrightarrow$$

$$2\ R-\underset{\underset{N(CH_3)_2}{|}}{\overset{\overset{NR^1}{|}}{C}}\ +\ TiO_2\ +\ 2\ (H_3C)_2NH$$

R = H, Alkyl, Aryl

R¹ = Alkyl, Aryl

Yields = 30–90%

A widely applicable method for converting carboxamides into amidines is the reaction between N-monosubstituted and N,N-disubstituted amides with isocyanates *via* 1,2-cycloadducts[713] accompanied by elimination of carbon dioxide. The reaction succeeds with aryl isocyanates at the boiling point[714], with arylsulfonyl isocyanates[715, 716] in an exothermic reaction and, especially well, with chlorosulfonyl isocyanate at $0-5°$ in dichloromethane[717]:

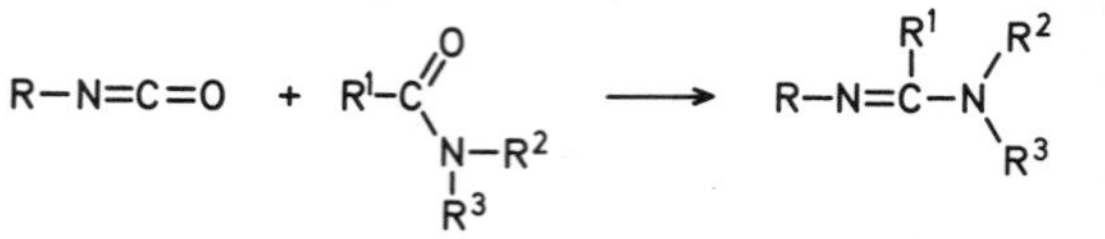

$$R-N=C=O\ +\ R^1-\underset{\underset{R^3}{\overset{|}{N-R^2}}}{\overset{\overset{O}{||}}{C}}\ \longrightarrow\ R-N=\underset{}{\overset{}{C}}-\underset{\underset{R^3}{|}}{N}\overset{R^1}{\underset{R^2}{\diagdown}}$$

R = Aryl, Arylsulfonyl, Chlorosulfonyl

R¹ = H, Alkyl

R² = H, Alkyl, Aryl

R³ = Alkyl, Aryl

17.2.3.1.2 Aminolysis of amide halides, imidium halides and O-acylimidium compounds

This reaction, which proceeds in clear-cut manner, leads to many of the substance classes named in Table 5 (p. 731) and has been described in detail in the literature[718]. The following scheme provides information about the possible syntheses:

---

[705] M. Busch, P. Bauer, Ber. *33*, 1058 (1900).

[706] G. Kresze, A. Horn, R. Philippson, A. Trede, Chem. Ber. *98*, 3401 (1965);
A. Senning, Acta Chem. Scand. *18*, 95 (1964).

[707] A. Spassov, E. Golovinsky, G. Demirov, Chem. Ber. *98*, 933 (1965).

[708] R. Rätz, H. Schroeder, J. Org. Chem. *23*, 1931 (1958).

[709] F. Tiemann, Ber. *19*, 1668 (1886).

[710] DRP. 481994 (1926), I.G. Farbenindustrie AG; Erf.: *H. Meis;* Friedländer: Fortschritte der Teerfabrikation u. verwandter Industriezweige *16*, 2516 (1931).

[711] G.R. Pettit, L.R. Garson, Can. J. Chem. *43*, 2640 (1965).

[712] J.D. Wilson, J.S. Wager, H. Weingarten, J. Org. Chem. *36*, 1613 (1971);
E.A. Arutjunjan, E.A. Gunar, V.I. Zav'jalov, S.I. Zav'jalov, Izvest. Akad. Nauk SSSR Ser. Chim. 904 (1970).

[713] H. Ulrich, Cycloaddition Reactions of Heterocumulenes, p. 162, Academic Press, New York, London 1967.

[714] Belg. P. 629317 (1963), Ciba; C.A. *60*, 14443 (1964).

[715] C. King, J. Org. Chem. *25*, 352 (1960).

[716] W. Logemann, D. Artini, G. Tosolini, Chem. Ber. *92*, 2565 (1958).

[717] DBP. 1144718 (1963), Farbwerke Hoechst; Erf.: R. Graf, D. Günther, H. Jensen, K. Matterstock; C.A. *59*, 6368 (1963).

[718] H. Ulrich, The Chemistry of Imidoyl-halides, p. *43*, 85, 132, 150, 166, 188, Plenum Press, New York 1968.

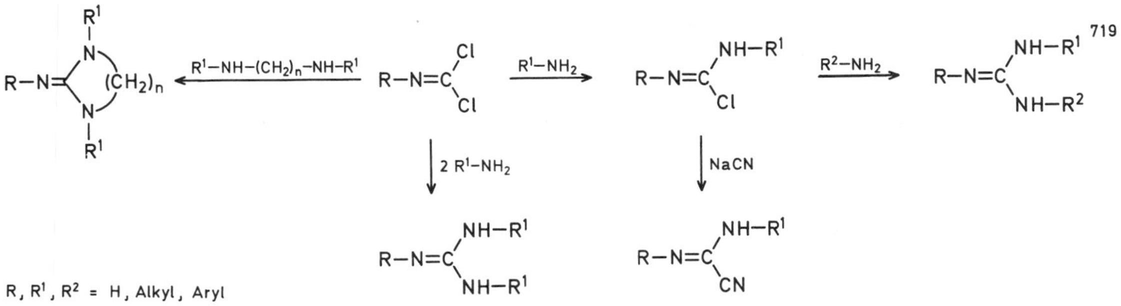

R, R¹, R² = H, Alkyl, Aryl

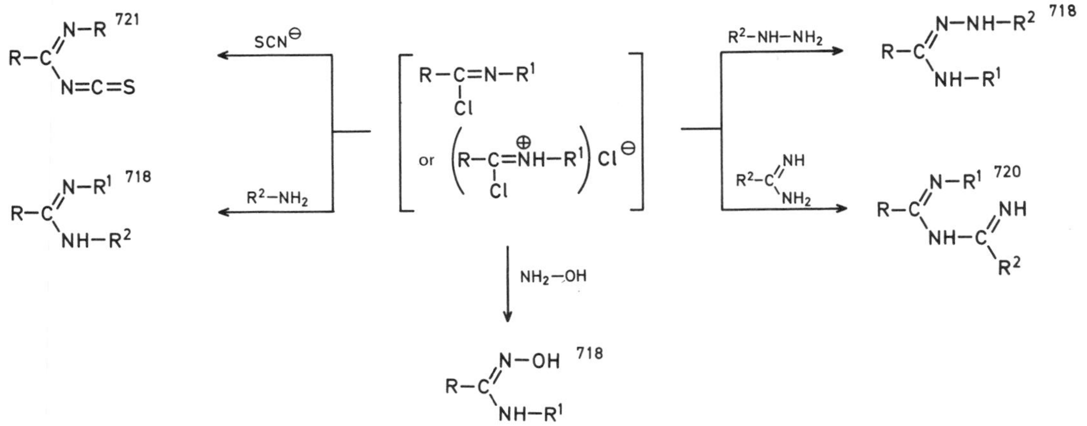

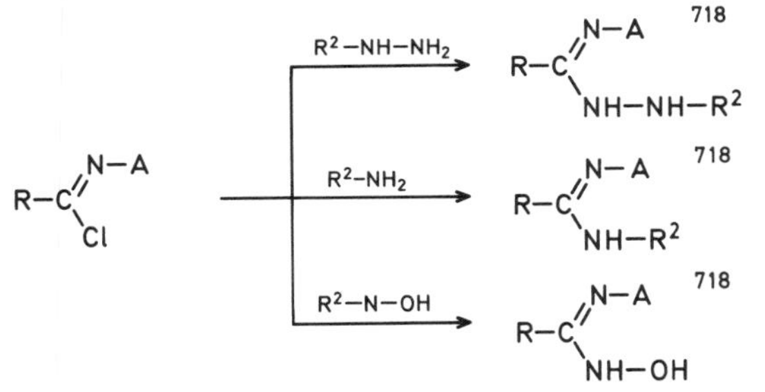

A = —NH—R¹ ; OR¹

R, R¹, R² = H, Alkyl, Aryl

Vinylogous derivatives are obtained by reacting vinylogous imidium chlorides with amines. In general, these reactions are carried out by initially adding the acyl halide (phosphoryl chloride, phosgene, thionyl chloride, *p*-toluenesulfonyl chloride, carboxylic acid chlorides) to the solution of the carboxamide or urea at 0–20° to give complex formation. The further reaction is performed by adding the amine component at 0–20° without isolating the activated amides[722–725] (see table 7, p. 737).

For synthesizing *N, N, N', N'*-tetrasubstituted amidinium salts it is advisable to use *in situ* prepared complexes[726].

Tetramethylformamidinium chloride is obtained from dimethylformamide and chlorodimethylformamide[727] or, particularly simply, by heating a mixture of dimethylformamide and thionyl chloride[728] (in an ∼ 3:1 molar ratio)

[719] E. Kühle, Angew. Chem. *81*, 18 (1969).

[720] H. Ley, F. Müller, Ber. *40*, 2957 (1907);
F. Cooper, M.W. Partridge, W.F. Short, J. Chem. Soc. (London) 391 (1951).

[721] J. Goerdeler, D. Weber, Tetrahedron Lett. 799 (1964).

[722] H. Bredereck, R. Gompper, K. Klemm, H. Rempfer, Chem. Ber. *92*, 837 (1959).

[723] H. Ulrich, The Chemistry of Imidoyl-Halides, p. 132, Plenum Press, New York 1968.

[724] H. Eilingsfeld, M. Seefelder, H. Weidinger, Angew. Chem. *72*, 836 (1960).

[725] H. Bredereck, K. Bredereck, Chem. Ber. *94*, 2278 (1961).

[726] C. Jutz, A. Amschler, Chem. Ber. *96*, 2100 (1963).

[727] Z. Arnold, Collect. Czech. Chem. Commun. *24*, 760 (1959).

[728] W. Kantlehner, P. Speh, Chem. Ber. *104*, 3714 (1971).

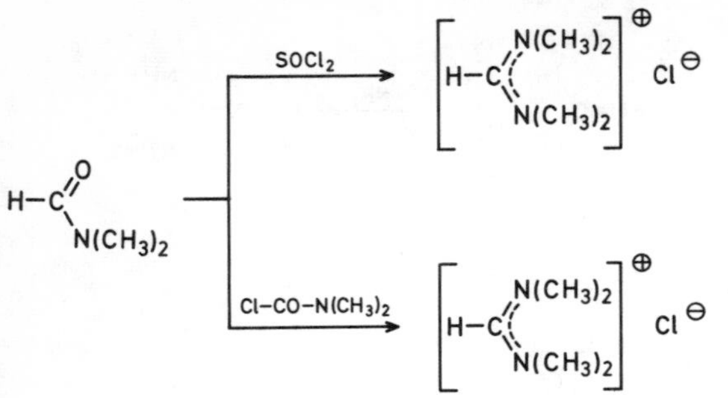

The synthesis of *amidinium* salts from trihalomethanes and amines probably proceeds *via* intermediate imidium halides[737, 738]. It is employed above all for *N,N'-dialkyl-N,N'-diarylformamidinium* salts. Initially, tris-(*N*-alkylanilino)methanes are obtained and are subsequently converted into the amidinium salts with acids, alkylating agents or acylating agents[737]:

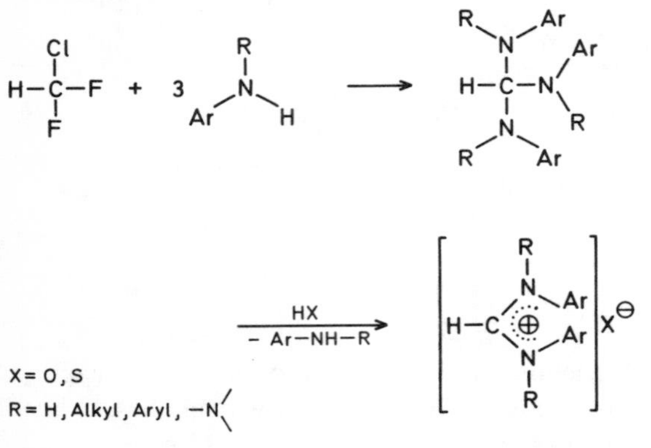

X = O, S
R = H, Alkyl, Aryl, —N⟨

Trichloroacetamidine derivatives react with primary amines in similar fashion[739]:

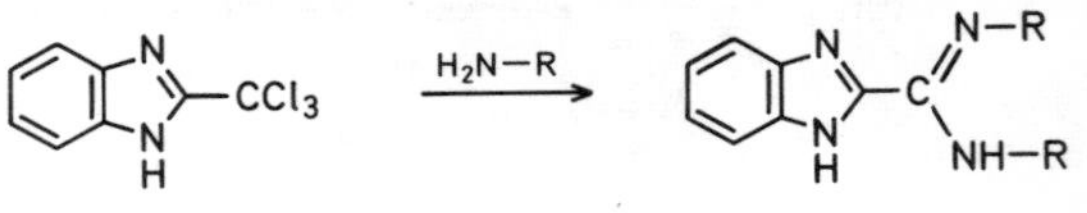

17.2.3.1.3 Aminolysis of thioimidic and thioimidium acid esters

Aminolysis of imidic esters is a frequently used method for preparing amidines, amidoximes, and amidrazones, and is extensively described in the literature[740]. Aminolysis of imidium esters obtained by *O*-alkylation of amides represents the same technique (see p. 738). Esters of thioimidic acids[741–744], thioimidium esters[745], and pseudo-thioureas[746] undergo aminolysis just as readily; the last-named reaction is important for preparing *guanidines, aminoguanidines,* and related compounds:

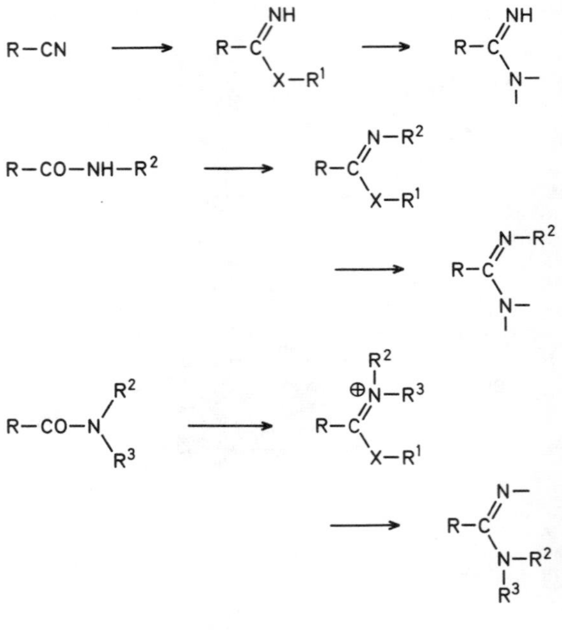

[729] *R. Ohme, E. Schmitz,* Angew. Chem. *79*, 531 (1967).

[730] *H. Finkbeiner,* J. Org. Chem. *30*, 2861 (1965).

[731] *D.E. Horning, J.M. Muchowski,* Can. J. Chem. *45*, 1247 (1967);
A. Holy, Collect. Czech. Chem. Commun. *31*, 2973 (1966);
K. Thinius, W. Lahr, Z. Chem. *6*, 315 (1966).

[732] *Y. Omote, H. Yamamoto, S. Tomioka, N. Sugiyama,* Bull. Chem. Soc. Japan *42*, 2090 (1969).

[733] *B. Föhlisch, R. Braun, K.W. Schultze,* Angew. Chem. *79*, 318 (1967);
R.K. Barthlett, J.R. Humphrey, J. Chem. Soc. C 1664 (1967).

[734] *H. Rapoport, R.M. Bonnes,* J. Amer. Chem. Soc. *72*, 2783 (1950).

[735] *F.L. Scott, J.A. Barry,* Tetrahedron Lett. 2457 (1968);
W. Hoyle, J. Chem. Soc. C 690 (1967).

[736] *H. Dorn, H. Welfle,* Z. Chem. *7*, 277 (1967).

[737] *D.H. Clemens, E.Y. Shropshire, W.D. Emmons,* J. Org. Chem. *27*, 3664 (1962).

[738] *J.W. Scheren, R.J.F. Nivard,* Rec. Trav. Chim. Pays-Bas *88*, 289 (1969).

[739] *G. Holan, E.L. Samuel,* J. Chem. Soc. C 25 (1967);
A. Kreutzberger in *E. Jucker,* Arzneimittel-Forsch. *11*, 376 (1968).

[740] *R. Roger, D.G. Neilson,* Chem. Rev. *61*, 179, 193 (1961);
R.G. Gluschkov, V.G. Granik, Russian Chem. Reviews *38*, 913 (1969).

[741] *P. Chabrier, H. Renard,* Bull. Soc. Chim. France D 13 (1950).

[742] *P. Reynand, R.C. Moreau, T. Gousson,* Compt. Rend. *259*, 4067 (1964).

[743] *S. Hünig, F. Müller,* Justus Liebigs Ann. Chem. *651*, 89 (1962);
D.A. Peak, F. Stansfield, J. Chem. Soc. (London) 4067 (1952).

[744] *T. Mukaiyama, S. Ono,* Tetrahedron Lett. 3569 (1968).

[745] *T. Mukaiyama, T. Yamaguchi, H. Nohira,* Bull. Chem. Soc. Japan *38*, 2107 (1965).

Table 7. *N*-Functional carboxylic acid imide derivatives from amines and condensing agents

Amide	Amine component	Condensing agent	Reaction conditions Solvent	[°C]	[h]	Reaction product	Yield [% of th.]	Ref.
2-pyrrolidinone (N–CH$_3$)	aniline (C$_6$H$_5$–NH$_2$)	POCl$_3$	Benzene	70	8	1-Methyl-2-phenylimino-pyrrolidine	76	725
H$_2$N–CO–H	NH$_3$	H$_5$C$_6$–CO–Cl	Ether	10–20	—	Formamidine	80–95	729
(H$_3$C)$_2$N–CO–H	C$_6$H$_5$–CO–NH$_2$	H$_3$C–CO–Cl	DMF	0–20	18	N'-Benzoyl-N,N-dimethyl-formamidinium chloride	68	730
(H$_3$C)$_2$N–CO–H	H$_2$N–CO–NH$_2$	H$_5$C$_6$–CO–Cl	DMF	0–20	18	N'-Aminocarbonyl-N,N-dimethyl-formamidinium chloride	92	730, 731
H$_3$C–NH–CO–NH–CH$_3$	H$_5$C$_6$–NH$_2$	POCl$_3$	Benzene	70	6	N,N'-Dimethyl-N''-phenylguanidine	60	723, 725
(H$_3$C)$_2$N–CO–H	1,4-dihydroxyphthalazine (H$_2$N, OH)	SOCl$_2$	DMF	—	—	N-(1,4-Dihydroxy-5-phthalazinyl)-N,N-dimethyl-formamidinium chloride		732
(H$_3$C)$_2$N–CO–H	H$_2$N–NH$_2$	COCl$_2$	CHCl$_2$	Reflux	20	N,N-Dimethylformamide azine	90	733
(H$_3$C)$_2$N–CO–H	H$_2$N–N(CH$_3$)–C$_6$H$_5$	POCl$_3$	Benzene	Reflux	5	N,N-Dimethyl-2-phenyl-acetamide methyl(phenyl)hydrazone	40	734
(H$_3$C)$_2$N–CO–H	4-O$_2$N–C$_6$H$_4$–NH–NH$_2$	(H$_3$C)$_2$N–SOCl$_2$	—	–75	1	Dimethylformamidinium chloride N'-(4-nitrophenyl)-hydrazide	81	735
(H$_3$C)$_2$N–CO–H	H$_2$N–N(CH$_2$–CH$_2$–Cl)$_2$	POCl$_3$	Benzene	20	14	N,N-Dimethylformamide bis-(2-chloroethyl)hydrazone	53	736

Imidium ester methyl sulfates are aminolyzed in high yield[747, 748]. With primary and secondary amides either *N, N'*-disubstituted, *N, N, N'*-trisubstituted, or *N, N, N', N'*-tetrasubstituted *amidines* are obtained in dependence on the degree of substitution of the carboxamide.

Potassium hydroxide/potassium carbonate solutions set the disubstituted and trisubstituted amidines free from their salts. Vinylogous amidinium salts and amidines are successfully synthesized in entirely analogous manner[749]:

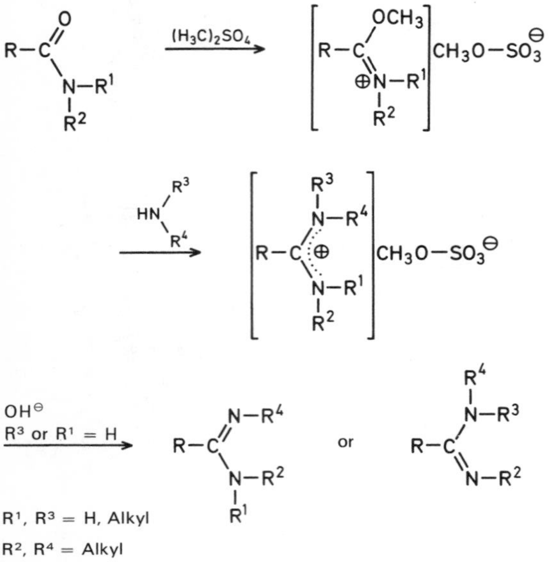

R¹, R³ = H, Alkyl
R², R⁴ = Alkyl

With less nucleophilic carboxamides it is advisable to employ imidium ester tetrafluoroborates[750]. Alkylation of amides with propanesultone leads to imidium esters having a betaine structure and affording a good yield of amidines on aminolysis[751]:

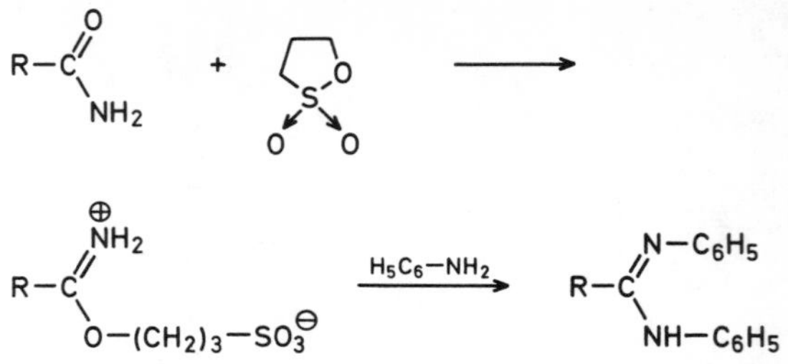

N, N'-Diphenyl-amidines

Other imidium esters, too, behave analogously. The aminolysis can be performed either by reacting with ammonia or primary or secondary amines, or the free imide esters are reacted with ammonium salts[740]. From amino acids *amidinocarboxylic acids* are obtained in this way[752]:

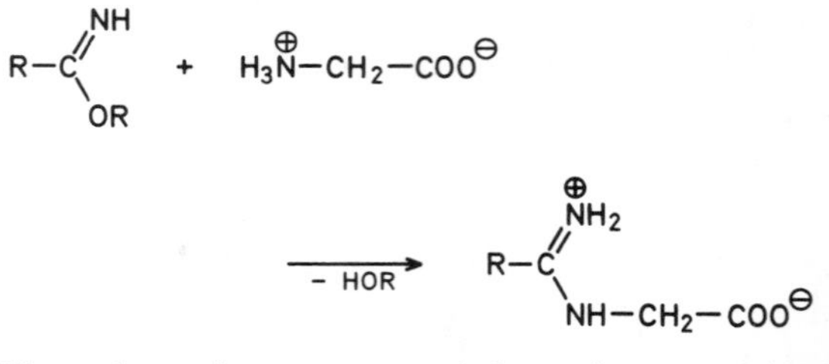

Where the amine reagent and the amino group in the imidic esters are mutually different, the complex nature of the aminolysis reaction finds expression in the formation of different amidines in dependence on the *pH* of the reaction medium[753]:

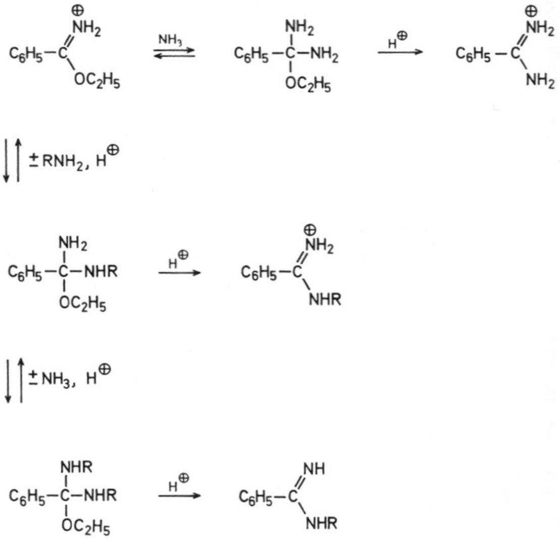

---

[746] *E.E. Reid*, Organic Chemistry of Bivalent Sulfur, Vol. V, p. 33, Chemical Publishing Co., New York 1963;
S. Petersen in *Houben-Weyl*, Methoden der organischen Chemie, 4. Aufl., Bd. VIII, p. 183, 191, Georg Thieme Verlag, Stuttgart 1952.
[747] *H. Bredereck, F. Effenberger, E. Henseleit*, Chem. Ber. *98*, 2754, 2887 (1965);
H. Bredereck, F. Effenberger, G. Simchen, Angew. Chem. *74*, 353 (1962);
H. Bredereck, F. Effenberger, G. Simchen, Chem. Ber. *98*, 1079 (1965).
[748] *H. Bredereck, F. Effenberger, H.P. Beyerlin*, Chem. Ber. *97*, 3081 (1964).
[749] *H. Bredereck, F. Effenberger, K.A. Hirsch, D. Zeyfang*, Chem. Ber. *101*, 4036 (1968).
[750] *L. Weintraub, S.R. Oles, N. Kalish*, J. Org. Chem. *33*, 1679 (1968).
[751] *W. Ried, E. Schmidt*, Justus Liebigs Ann. Chem. *676*, 114 (1964).
[752] *W. Ried, W. von der Emden*, Justus Liebigs Ann. Chem. *661*, 76 (1963);
W. Ried, D. Piechaczek, Justus Liebigs Ann. Chem. *696*, 97 (1966);
F. Weygand, W. Steglich, D. Hoffter, Chem. Ber. *95*, 2264 (1962).

In some cases the amino group may be replaced more rapidly than the alkoxy group[754]. The reactivity of *S*-alkylpseudothioureas, which are important in respect of the synthesis of *guanidines*, can be enhanced further if the *S*-alkyl group contains electron-attracting substituents[755]:

$$R-NH_2 \;+\; HOOC-CH_2-S-C\Big\langle{}^{N-}_{N-}$$

$$\longrightarrow \left[R-NH-C\overset{\overset{\oplus}{N}H_2}{\underset{N-}{\big|}} \right] HS-CH_2-COO^{\ominus}$$

17.2.3.1.4 Aminolysis of orthocarboxylic acid esters and amides and of ketene acetals

Acids, for instance, acetic acid, catalyze the aminolysis of triesters of orthocarboxylic acids[756]. Addition of a few drops boron trifluoride etherate has given particularly good results[757]. Imidic esters are formed intermediately but their isolation succeeded only when molar quantities of aromatic amine were used for the reaction. To obtain good yields the orthocarboxylic acid triester is reacted with twice the molar quantity of a primary amine. For aliphatic amines a molar quantity of acetic acid is added, for aromatic amines catalytic amounts[756]:

$$R-NH_2 \;+\; R^1-C\overset{OC_2H_5}{\underset{OC_2H_5}{\big|}}-OC_2H_5 \longrightarrow$$

$$\left[R^1-C\Big\langle{}^{N-R}_{OC_2H_5} \right] \xrightarrow{R-NH_2} R^1-C\Big\langle{}^{N-R}_{NH-R}$$

R = H, CH$_3$

R^1 = C$_6$H$_{11}$, C$_6$H$_5$CH$_2$, C$_4$H$_9$, Aryl

A synthesis of *N,N,N'-trisubstituted amines* and *amidrazones* that is interesting above all because it is simple to perform and widely applicable consists in reacting orthocarboxylic acid monoamide diesters[758] and diamide monoesters[758] with *NH*-acid compounds. Aromatic amines (including little nucleophilic ones such as 2,4,6-trinitroaniline and 2-aminopyridine), primary carboxamides, ureas, thioureas, urethanes, semicarbazides, *N,N*-dialkylformamidines, and sulfonamides react with orthocarboxylic acid monoamide diesters in some cases at room temperature, in others on heating to split off alcohol[759-766].

While primary aliphatic amines do not react with orthocarboxylic acid amide diesters[759] the reaction with alkylammonium salts should succeed.

Reacting orthoformic acid mono(dialkylamide) diesters with nucleoside bases containing primary amino groups leads to the corresponding *formamidine* derivatives[759].

The considerably more reactive diamide monoesters react analogously; with primary carboxamides further reaction on the α-methylene group occurs at elevated temperature[765]:

$$R-C\overset{OR^1}{\underset{N(R^2)_2}{\overset{|}{\big|}}}-OR^1 \;+\; X-NH_2 \xrightarrow{-2\,HOR^1} R-C\Big\langle{}^{N-X}_{N(R^2)_2}$$

$$\overset{RO}{\underset{H}{\big\rangle}}C[N(CH_3)_2]_2 \xrightarrow{R-CO-NH_2} R-C\Big\langle{}^{O}_{N}$$
$$\overset{}{\underset{H\diagdown{}^{C}\diagup N(CH_3)_2}{}}$$

X = Aryl, R—CO, H$_2$N—CO, RO—CO, NH—Aryl, *etc.*

[753] *E.S. Hand, W.P. Jencks,* J. Amer. Chem. Soc. *84,* 3505 (1962).

[754] *L. Baiochi, G. Palazzo,* Ann. Chim. (Rome) *58,* 608 (1968).

[755] *S. Schütz,* Farbenfabriken Bayer, Elberfeld, unpubl.

[756] *E.C. Taylor, W.A. Ehrhardt,* J. Org. Chem. *28,* 1108 (1963).

[757] *G. Lehmann, H. Seefluth, G. Hilgetag,* Chem. Ber. *97,* 299 (1964).

[758] *H. Bredereck, G. Simchen, S. Rebsdat, W. Kantlehner, P. Horn, R. Wahl, H. Hoffmann, P. Grieshaber,* Chem. Ber. *101,* 41 (1968).

[759] *J. Gloede, L. Haase, H. Groß,* Z. Chem. *9,* 201 (1969).

[760] *H. Meerwein, W. Florian, N. Schön, G. Stopp,* Justus Liebigs Ann. Chem. *641,* 1 (1961).

[761] *C. Feugeas, P. Olschwang,* Bull. Soc. Chim. France 4985 (1968).

[762] *E. Niemers,* Farbenfabriken Bayer, Elberfeld, unpubl.

[763] *H. Bredereck, W. Kantlehner, D. Schweizer,* Chem. Ber. *104,* 3475 (1971).

[764] USP. 3 121 084 (1964), Du Pont; Inv.: *H.E. Winberg;* C.A. *60,* 13 197 (1964).

[765] *H. Bredereck, G. Simchen, B. Funke,* Chem. Ber. *104,* 2709 (1971).

[766] Belg. P. 629 972 (1963), Badische Anilin- & Sodafabrik; Inv.: *H. Weidinger, H. Eilingsfeld;* C.A. *61,* 1803 (1964).

In place of the orthocarboxylic acid amide diesters (amide acetals) the alkoxydimethylaminoaceto-nitriles[767] very easily accessible from imidium ester methyl sulfates and sodium cyanide lend themselves for synthesizing *formamidines* and for *amidrazones*. Aliphatic and aromatic primary amines react in the same way at 0–60° in good yield:

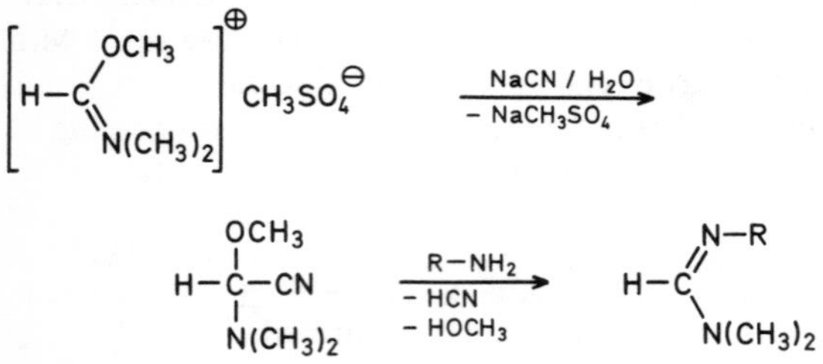

Vinylogous orthocarboxylic acid amide diesters (amide acetals) react with amines to form the *vinylogous amidines*, albeit more slowly. The reaction succeeds in every case with the vinylogous orthocarboxylic acid diamide monoesters (aminal esters) and monoamide diesters (amide aminals) respectively[768].

Dithiocarboxamide *S,S*-diesters[769], ketene *O,O*-acetals[770], and ketene *S,S*-acetals[771] react analogously but more slowly.

17.2.3.2 Addition of nucleophiles

17.2.3.2.1 To nitriles and nitrilium salts[772]

Addition of amines to nitriles in general ensues only at higher temperatures. The reaction is facilitated by adding alkali metals (sodium, potassium, or lithium) or sodium or potassium amide, or magnesium dialkylamide[772, 773], *e.g.*:

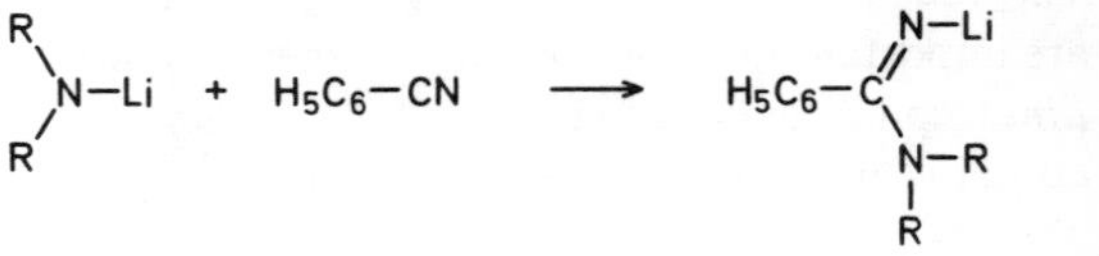

[*α*-(Dimethyl-amino)benzyl-imino]lithium; 60%

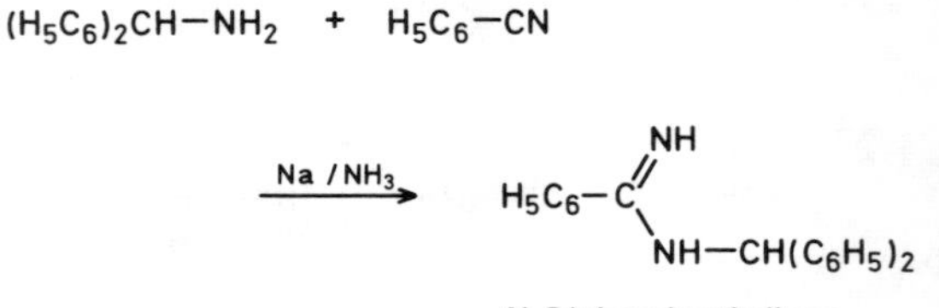

N-Diphenylmethylbenz-amidine; 75%

N,N-Dialkylformamidines (including cyclic ones) can be synthesized without difficulty from hydrogen cyanid and primary amines[774] at ~80° in autoclaves.

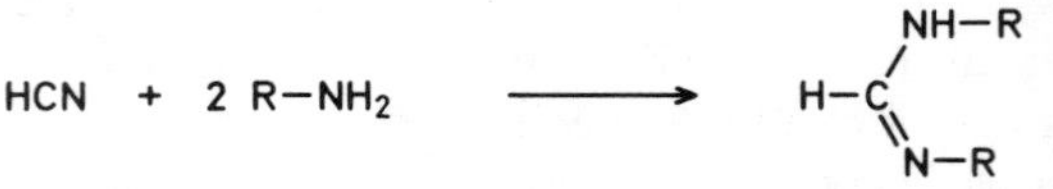

R = C₃H₇; *N,N'*-Dipropylformamidine

R = (CH₃)₂CH–CH₂; *N,N'*-Diisobutylformamidine

No condensing agents are needed for synthesizing amide oximes from nitriles and hydroxylamines[775], but alkali metals or alkali metal hydrazides are added for preparing amide hydrazides from nitriles and hydrazines[776, 777].

Using aluminum chloride as condensing agents (formation of acceptor complexes)[778], both ali-

[767] *H. Bredereck, G. Simchen, W. Kantlehner,* Chem. Ber. *104*, 932 (1971).

[768] *H. Bredereck, F. Effenberger, K.A. Hirsch, D. Zeyfang,* Chem. Ber. *103*, 222 (1970).

[769] *F.M. Stojanovic, B.P. Fedorov, G.M. Andrianova,* Dokl. Akad. Nauk SSSR *145*, 584 (1962);
T. Mukaiyama, T. Yamaguchi, Bull. Chem. Soc. Japan *39*, 2005 (1966);
T. Mukaiyama, T. Yamaguchi, Bull. Chem. Soc. Japan *38*, 2107 (1965).

[770] *D. Borrmann* in *Houben-Weyl,* Methoden der organischen Chemie, 4. Aufl., Bd. VII/4, p. 363, Georg Thieme Verlag, Stuttgart 1968.

[771] *D. Borrmann* in *Houben-Weyl,* Methoden der organischen Chemie, 4. Aufl., Bd. VII/4, p. 421, Georg Thieme Verlag, Stuttgart 1968.

[722] *A. Kreutzberger* in *E. Jucker,* Arzneimittel-Forsch. *11*, 368 (1968);
R.L. Shriner, F.W. Neumann, Chem. Rev. *35*, 351 359, 363 (1944);
F.C. Schaefer in *Z. Rappoport,* The Chemistry of the Cyano Group, p. 270, Wiley Intersci. Publ., London, New York, Sydney, Toronto 1970.

[773] *D.J. Scherer, P. Hornig,* Chem. Ber. *101*, 2533 (1968);
J.A. Gautier, M. Miocque, C. Fauran, A.-Y. Le Cloarec, Bull. Soc. Chim. France 791 (1969);
R.P. Hullin, J. Miller, W.F. Short, J. Chem. Soc. (London) 394 (1947).

[774] *W. Jentzsch, M. Seefelder,* Chem. Ber. *98*, 1342 (1965).

[775] *F. Eloy, R. Lenaers,* Chem. Rev. *62*, 155, 157 (1962).

[776] *D.G. Neilson, R. Roger, J.W. Heatlie, L.R. Newlands,* Chem. Ann. *70*, 151, 152 (1970).

[777] *T. Kauffmann, L. Bán, W. Burkhardt, E. Rauch, J. Sobel,* Angew. Chem. *77*, 1085 (1965);
T. Kauffmann, L. Bán, D. Kühlmann, Angew. Chem. *79*, 243 (1967).

[778] *F.C. Schaefer* in *Z. Rappoport,* The Chemistry of the Cyano Group, p. 250, Wiley Intersci., Publ., London, New York, Sydney, Toronto 1970.

[779] *L. Villa, V. Ferri, E. Grana,* Farmaco Ed. Sci. *22*, 491 (1967).

phatically and aromatically substituted *benzami-dines* have been prepared[779–781], *e.g.*:

$$H_5C_6-CN \quad + \quad H_2N-R$$

$$\xrightarrow{AlCl_3 / 100°} \quad H_5C_6-CH_2-C \begin{array}{c} NH \\ \\ NH-R \end{array}$$

Cyclic amidines (5–7 membered rings) are prepared in simple manner by heating ω-halocarbonitriles with amines[782], *e.g.*:

$$R-NH_2 \quad + \quad X(CH_2)_n-C\equiv N$$

$$\xrightarrow{\nabla} \quad R-NH-(CH_2)_n-CN \cdot HX$$

$$\longrightarrow \quad (CH_2)_n \begin{array}{c} C \\ N \oplus NH_2 \cdot X^\ominus \\ R \end{array}$$

R = H, Alkyl, Aryl

n = 3,4,5

Nitrilium salts, formed by *N*-alkylation or *N*-acylation of nitriles in the presence of complex-forming metal halides[568, 783], add nucleophilic partners, for instance amines, to form amidines. Chlorosulfonic

esters have proven especially useful for preparing nitrilium salts able to react to amidines with amines without having to be isolated[783a].

$$H_5C_6-CN \quad + \quad Cl-SO_2-OCH_3 \xrightarrow{60-70°}$$

$$\left[H_5C_6-C\equiv N-CH_3\right]^\oplus \left[OSO_2Cl\right]^\ominus \xrightarrow{H_7C_3-NH_2}$$

$$H_5C_6-C \begin{array}{c} N-CH_3 \\ \\ NH-C_3H_7 \end{array}$$

17.2.3.2.2 Addition of nucleophiles to nitrile oxides

Addition of ammonia, hydroxylamine, hydrazine, and other nucleophilic partners to nitrile oxides is an entirely general reaction which proceeds well even under mild conditions[784]. The technique is suited particularly for preparing a variety of amidoximes (see Table 8).

17.2.3.2.3 Addition of nucleophiles to heterocumulenes (ketenimines, carbodiimides)

The reactivity of the $C=N$ double bond of ketene imines and carbodiimides has been utilized for numerous addition reactions involving nucleophil-

Table 8. Addition of nucleophilic *NH*-compounds to nitrile oxides

Nitrile oxide	Amine component	Reaction product		Yield [% of th.]	Ref.
$H_5C_6-C\equiv N\rightarrow O$	NH_3	$H_5C_6-C\begin{smallmatrix}NOH\\\\NH_2\end{smallmatrix}$	*Benzamidoxime*	82	785
	$H_2N-NH-C_6H_5$	$H_5C_6-C\begin{smallmatrix}NOH\\\\N=N-C_6H_5\end{smallmatrix}$	*Benzohydroximic acid 2-phenylhydrazide*	20	785
$H_3C-C\equiv N\rightarrow O$	$H_9C_4-NH_2$	$H_3C-C\begin{smallmatrix}NOH\\\\NH-C_4H_9\end{smallmatrix}$	*N-Butylacetamidoxime*	good	786
$(H_3C)_2CH-C\equiv N\rightarrow O$	$HN(C_2H_5)_2$	$(H_3C)_2CH-C\begin{smallmatrix}NOH\\\\N(C_2H_5)_2\end{smallmatrix}$	*N,N-Diethylisobutyramidoxime*	good	786
$R-C\equiv N\rightarrow O$	NR'_3	$R-C\begin{smallmatrix}N-O^\ominus\\\\\oplus NR'_3\end{smallmatrix}$		good	784
$O\leftarrow N\equiv C-\!\!\!\bigcirc\!\!\!-C\equiv N\rightarrow O$	$R-NH_2$	$\begin{smallmatrix}HON\\\\R-HN\end{smallmatrix}C-\!\!\!\bigcirc\!\!\!-C\begin{smallmatrix}NOH\\\\NH-R\end{smallmatrix}$	*N,N-Dialkylterephthal-amidoxime*	—	787
$O\leftarrow N\equiv C-C\equiv N\rightarrow O$	$HN(C_2H_5)_2$	$(C_2H_5)_2N-\underset{\underset{NOH}{\|}}{C}-\underset{\underset{NOH}{\|}}{C}-N(C_2H_5)_2$	*Bis(diethylamino)glyoxime*	78	788

[780] *L.B. Daslevic*, Zh. Obshch. Khim. *32*, 2346 (1962); *P. Oxley, W.F. Short*, J. Chem. Soc. (London) 449 (1949).

[781] *M. Ohosone, T. Tanimonto, R. Oda*, J. Soc. Org. Synth. Chem. Japan *24*, 562 (1966).

ic partners. Ketene imines add ammonia, primary and secondary aliphatic and aromatic amines, and also hydrazine to form *amidines* and *amidrazones* (80–95%)[789, 790]; varying reaction temperatures are required in dependence on the constitution of the ketene imines:

$$F_3C\!\!-\!\!C\!\!=\!\!C\!\!=\!\!N\!-\!C_6H_5 \xrightarrow[\text{Ether}]{NH_3/}$$

Dicyano-*N'*-phenyl-acetamidine

$$(H_5C_6)_2C\!\!=\!\!C\!\!=\!\!N\!-\!\!\bigcirc\!\!-\!CH_3$$

$$\xrightarrow{H_2N-NH_2} (H_5C_6)_2CH\!-\!C$$

N^3-(p-tolyl)diphenyl-acetamidrazon; 97%

Addition of nucleophilic partners to carbodiimides makes numerous types of *N*-functional *carboxylic acid imides* and *carbonic acid imides* available. With *N,N'*-diaryl and *N'*-alkyl-*N*-aryl-carbodiimides the reactions proceed especially rapidly and well. The rate of addition of *N,N'*-dialkylcarbodiimides to various bases in *tert*-butanol is a function of the basicity. The stronger the base is, the more readily does the addition take place[791]:

$$H_{11}C_6\!-\!N\!=\!C\!=\!N\!-\!C_6H_{11} + \begin{array}{c} R^1 \\ N\!-\!H \\ R \end{array}$$

$$\longrightarrow H_{11}C_6\!-\!N\!=\!C\!-\!NH\!-\!C_6H_{11}$$

Reactivity:

Table 9 surveys the synthetically accessible substance groups.

---

[782] *R. Kwok, P. Pranc,* J. Org. Chem. *32,* 738 (1967); *J. Moriconi, A.A. Cevasso,* J. Org. Chem. *32,* 2109 (1968).

[783] *J. Grundnes, P. Klaboe* in *Z. Rappoport,* The Chemistry of the Cyano Group, p. 129, Wiley Intersci. Publ., London, New York, Sydney, Toronto 1970.

[783a] DOS 1 920 499 (1969); Bayer AG; Erf.: *G. Grögeler;* C.A. *74,* 12858 (1971).

[784] *C. Grundmann* in *Z. Rappoport,* The Chemistry of the Cyano Group, p. 824, 829, Wiley Intersci. Publ., London, New York, Sydney, Toronto 1970; *C. Grundmann, P. Grünanger,* The Nitrile Oxides, p. 161, Springer Verlag, Berlin, Heidelberg, New York 1970.

[785] *C. Grundmann, H.D. Frommeld,* J. Org. Chem. *31,* 157 (1966).

[786] *G. Zinner, H. Günther,* Angew. Chem. *76,* 440 (1964).

Table 9. Addition of nucleophiles to carbodiimides giving *N*-functional imide derivatives

Nucleophile	Reaction product	Reference
Ammonia	*N,N'*-disubstituted guanidine	[792]
Amines	*N,N',N''*-trisubstituted and *N,N,N',N''*-tetrasubstituted guanidines	[792]
Guanidines	Biguanidines $$R\!-\!NH\!-\!C\!-\!NH\!-\!C\!-\!NH_2$$	[792]
Hydrazines NH₂–NR¹	*N*-Aminoguanidines $$R\!-\!NH\!-\!C\!-\!NH\!-\!NH\!-\!R$$	[793]
R¹ = H, Aryl, Alkyl	$$R\!-\!NH\!-\!C\!-\!NH\!-\!N\!-\!C\!-\!NH\!-\!R$$	
Aminoguanidines	Biguanidines $$R\!-\!NH\!-\!C\!-\!NH\!-\!NH\!-\!C\!-\!NH_2$$	[792]
Thiosemicarbazides	Guanylthiosemicarbazides $$R\!-\!NH\!-\!C\!-\!NH\!-\!NH\!-\!C\!-\!NH_2$$	[793]
Semicarbazides	Guanylsemicarbazides $$R\!-\!NH\!-\!C\!-\!NH\!-\!NH\!-\!C\!-\!NH_2$$	[793]
Hydroxylamines RHNOH	*N*-Hydroxyguanidines $$R\!-\!NH\!-\!C\!-\!NH\!-\!R$$ $$R\!-\!NH\!-\!C\!=\!N\!-\!R$$ R–NOH	[792]
R = H, Alkyl, Aryl		
Cyanamide	*N*-Cyanoguanidines $$R\!-\!NH\!-\!C\!-\!NH\!-\!R \text{ oder}$$ $$R\!-\!NH\!-\!C\!=\!N\!-\!R$$ NH–CN	[794]
CH-Acid compounds X–CH₂ Y X = Y = COOR	Amidines $$X\!-\!CH\!-\!C\!-\!NH\!-\!R$$	[795]
Grignard compounds R¹–MgHal	*N,N'*-disubstituted amidines $$R\!-\!NH\!-\!C\!=\!N\!-\!R$$ R¹	[796]
Dialkylamino-stannanes (H₃C)₃Sn–N(C₂H₅)₂	Stannylguanidines $$R\!-\!N\!=\!C\!-\!N(C_2H_5)_2$$ R–N–Sn(CH₃)₃	[797]

17.2.3.2.4 Addition of nucleophiles to isonitriles

Isonitriles add primary and secondary amines only in the presence of catalytic amounts of transition metal compounds such as copper(I) chloride[798] (80–95%). Salts of weakly basic amines and salts of strong bases in the presence of excess weak acid likewise add isonitriles to form *disubstituted formamidines*[799]:

$$R-N=Cl \ + \ R^1-NH_2 \ \xrightarrow{\ Cu\,Cl\ }$$

$$R-N=CH-NH-R^1$$

17.2.3.2.5 Addition of nucleophiles to acetylenes

Alkoxyacetylenes, alkylthioacetylenes, or dialkylaminoacetylenes react *regiospecifically* with amines[800, 801]. Hydrazines[801], or hydroxylamines[801] to form *acetamidines, acetamide hydrazides,* and *acetamide oximes,* or their homologs. During the reaction with alkoxy- and alkylthioacetylenes imidic esters are formed intermediately, *e.g.:*

$$R-C\equiv C-OR^1 \ + \ R^2-NH_2 \ \longrightarrow$$

$$R-CH_2-C \overset{NR^2}{\underset{OR^1}{\big\langle}} \ \xrightarrow{\ +\ R^3-NH_2\ } \ R-CH_2-C \overset{N-R^2}{\underset{NH-R^3}{\big\langle}}$$

Using ynediamines makes *N*-persubstituted *aminoacetamidines* accessible in this way[801]:

$$R_2N-C\equiv C-NR_2 \ + \ H_2N-R^1 \ \longrightarrow$$

$$R_2N-CH_2-C \overset{NR_2}{\underset{N-R^1}{\big\langle}}$$

17.2.3.3 Special procedures

17.2.3.3.1 Reduction of *N*-functional imidic acid derivatives

Reduction of *N*-functional imides such as nitroguanidines, nitrazones, formazanes, tetrazolium salts, cyanamides, carbodiimides, and ureas can be performed with a variety of hydrogenating agents in principle. It is advantageous in individual cases where alternative methods of synthesis do not exist (see Table 10).

[787] *F. Eloy,* Bull. Soc. Chim. Belg. *73,* 639 (1964).
[788] *C. Grundmann, V. Mini, J.M. Dean, H.D. Frommeld,* Justus Liebigs Ann. Chem. *687,* 191 (1965).
[789] *N.P. Gambarjan, E.M. Rokhlin, J.V. Zeifman, J.L. Knunjants,* Izv. Akad. Nauk SSSR 749 (1965); C.A. *63,* 2914 b (1965).
[790] *C.L. Stevens, R.C. Freeman, K. Noll,* J. Org. Chem. *30,* 3718 (1965); *R. Dijkstra, H.J. Backer,* Rec. Trav. Chim. Pays-Bas *73,* 695 (1954).
[791] *J.G. Moffatt, H.G. Khorana,* J. Amer. Chem. Soc. *83,* 649 (1961).
[792] *F. Kurzer, K. Dourgahi-Zadeh,* Chem. Rev. *67,* 107 (1967).
[793] *L.E.A. Godfrey, F. Kürzer,* J. Chem. Soc. (London) 3561 (1962).
[794] Brit. P. 626 663 (1949), American Cyanamid; Inv.: *H.Z. Lecher, R.P. Parker, R.S. Long;* C.A. *44,* 4925 (1950); Brit. P. 643 012 (1950), American Cyanamid; Inv.: *H.Z. Lecher, R.P. Parker, R.S. Long;* C.A. *45,* 5180 (1951).
[795] *W. Traube, A. Eyme,* Ber. *32,* 3176 (1899); *W.E. Tischtschenko, N.V. Koshkin,* Zh. Obshch. Khim. *4,* 1021 (1934).
[796] *M. Busch, R. Hobein,* Ber. *40,* 4296 (1907).
[797] *T.A. George, K. Jones, M.F. Lappert,* J. Chem. Soc. (London) 2157 (1965).
[798] *T. Saegusa, Y. Ito, S. Kobayashi, K. Hirota, H. Yoshioka,* Tetrahedron Lett. 6121 (1966).
[799] *J.V. Mitin, V.R. Glusenkova, G.P. Vlasov,* Zh. Obshch. Khim. *32,* 3867 (1962).
[800] *L. Brandsma, H.J.T. Bos, F.J. Arens,* in *H.G. Viehe,* Chemistry of Acetylenes, p. 788, Marcel Dekker, New York 1969.

[801] *H.G. Viehe,* Chemistry of Acetylenes, p. 889, 894, 895, Marcel Dekker, New York 1969; *D.G. Neilson, R. Roger, J.W.M. Heatle, L.R. Newlands,* Chem. Rev. *70,* 151 (1970).
[802] *H.U. Sieveking, W. Lüttke,* Angew. Chem. *81,* 431 (1969).
[803] *E. Ichikawa, K. Odo,* J. Soc. Org. Synth. Chem. Japan, *24,* 1241 (1966).
[804] *A. Larizza, G. Brancaccio, G. Lettieri,* J. Org. Chem. *29,* 3697 (1964).
[805] *Y. Kikugawa, S. Yamada, H. Nagushima, K. Kaji,* Tetrahedron Lett. 699 (1969).
[806] *J.C. Jochims,* Chem. Ber. *98,* 2128 (1965); *J.C. Jochims,* Angew. Chem. *77,* 454 (1965).
[807] *A.F. McKay,* Chem. Rev. *51,* 309 (1952).
[808] *D.G. Neilson, R. Roger, J.W.M. Heatle, L.R. Newlands,* Chem. Rev. *70,* 151, 160 (1970).
[809] *J. Armand, B. Furth, J. Kossanyi, J.P. Morizur,* Bull. Soc. Chim. France 2499 (1968); *J. Armand, G. Champetier,* C.R. Acad. Sci., Paris, Ser. C *262,* 592 (1966).
[810] *A.Q. Nineham,* Chem. Rev. *55,* 412 ff. (1955); USP. 2 375 611 (1945), May & Baker; Inv.: *H.J. Barber, A.D.H. Selt;* Chem. Zentr. II, 1383 (1945).
[811] *H. Ley, M. Ulrich,* Ber. *47,* 2941 (1914).
[812] *P. Oxley, W.F. Short,* J. Chem. Soc. (London) 449 (1949); *P. Oxley, W.F. Short,* J. Chem. Soc. (London) 147 (1946); *M. Seefelder,* Chem. Ber. *99,* 2678 (1966); *V.S. Misra, R.S. Varma,* J. Indian Chem. Soc. *39,* 109 (1962).

Table 10. Reduction of *N*-functional imidic acid derivatives

Starting compound	Reducing agent	Reaction product	Status of method	Ref.
benzene-1,2-dicarbonitrile (ortho-dicyanobenzene)	$LiAlH_4$	3*H*-Isoindolamine; 45%	+	802
R_2N-CN (R = Alkyl)	H_2, Pd, HX	$R_2N-CH=NH \cdot HX$	+ +	803
$H_2N-\overset{\displaystyle NH}{\underset{\displaystyle CN}{C}}$	H_2, Pd, HX or electrolytically	$H_2N-\overset{\displaystyle NH}{\underset{\displaystyle CH=NH \cdot HX}{C}}$ *Iminoacetamidine hydrohalides*	+ +	803
$H_5C_6-NH-CO-N(C_2H_5)_2$	$LiAlH_4$	$H_5C_6-N=CH-N(C_2H_5)_2$ *N,N-Diethyl-N'- phenylformamidine*	+ +	804, 805,
$R-N=C=N-R$ (R = Alkyl, Aryl)	Pd, $BaSO_4$, H_2, or Raney Ni, H_2	$R-N=CH-NH-R$ *N,N'-disubstituted formamidines*	+ + +	806
$O_2N-NH-\overset{\displaystyle NH}{\underset{\displaystyle NH_2}{C}}$ $ON-NH-\overset{\displaystyle NH}{\underset{\displaystyle NH_2}{C}}$	Raney Ni, H_2	$H_2N-NH-\overset{\displaystyle NH}{\underset{\displaystyle NH_2}{C}}$ *Aminoguamidine*	+ + +	807
$R-\overset{\displaystyle N-NH-R}{\underset{\displaystyle NO_2}{C}}$ (R = Alkyl, Aryl)	Raney Ni, H_2; $SnCl_4$ or $(NH_4)_2S$	$R-\overset{\displaystyle N-NH-R}{\underset{\displaystyle NH_2}{C}}$ *Amidrazones*	+ + +	808
$H_3C-\overset{\displaystyle N-NH-C_6H_5}{\underset{\displaystyle NO_2}{C}}$	$(NH_4)_2S$ or electrolytically	$H_3C-\overset{\displaystyle NH-NH-C_6H_5}{\underset{\displaystyle N-OH}{C}}$ *N-Phenyl-acethydroxamid acid* (good yield)	+	809
$R-\overset{\displaystyle N-NH-R}{\underset{\displaystyle N=N-R}{C}}$ $\left[R-\underset{}{}\overset{}{} \right]Cl^{\ominus}$	Pd, $BaSO_4$, H_2; Raney Ni, H_2; $Na_2S_2O_4$	$R-\overset{\displaystyle N-NH-R}{\underset{\displaystyle NH-NH-R}{C}}$ *Hydrazidines* $\longrightarrow$ $R-\overset{\displaystyle N-NH-R}{\underset{\displaystyle NH_2}{C}}$ *Amidrazones*	+ + +	808, 810
$H_5C_6-\overset{\displaystyle N-OH}{\underset{\displaystyle NH-OH}{C}}$	SO_2	$H_5C_6-\overset{\displaystyle N-OH}{\underset{\displaystyle NH_2}{C}}$ *N'-Hydroxybenzamidine* (good yield)	+	811

+ = Usable in some cases
+ + = Can be used in several cases
+ + + = Readily usable

17.2.3.3.2 Aminolysis, hydrazinolysis, and hydroxyaminolysis of amidines and related reactions

On passing in ammonia at 100–160°, N-arylamidinium sulfonates are ammonolyzed almost quantitatively. The reaction is important only where the amidines cannot be readily prepared from nitriles[812]:

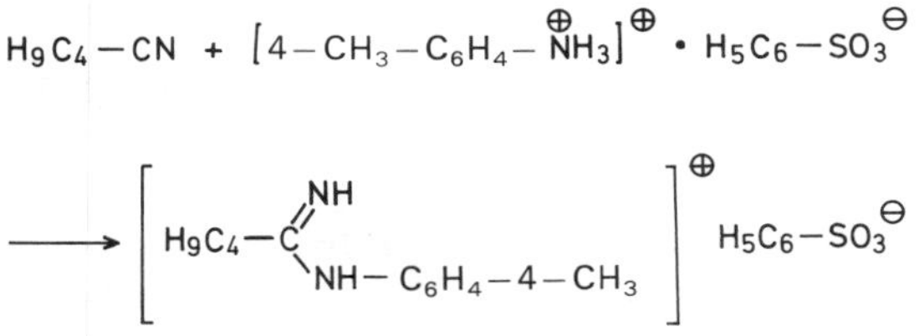

$$H_9C_4-CN \ + \ [4-CH_3-C_6H_4-\overset{\oplus}{N}H_3]^{\oplus} \cdot H_5C_6-SO_3^{\ominus}$$

$$\longrightarrow \left[H_9C_4-\overset{NH}{\underset{NH-C_6H_4-4-CH_3}{C}} \right]^{\oplus} H_5C_6-SO_3^{\ominus}$$

N-(4-Tolyl)valeramidinium benzenesulfonate; 74%

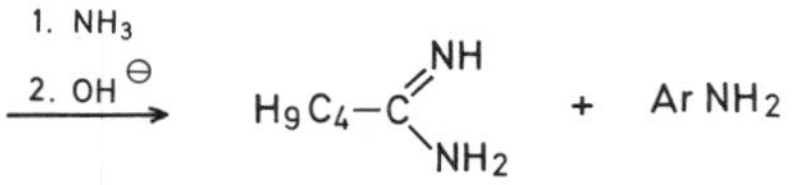

1. NH₃
2. OH⁻

$$H_9C_4-C\overset{NH}{\underset{NH_2}{}} \ + \ Ar\,NH_2$$

Valeramidine; 96%

In principle the method appears to be transferable to alkylamines, as is indicated by the corresponding reaction between N-phenylbenzamidine and cyclohexylammonium chloride[872]. Heating unsubstituted amidinium salts with arylamines to split off ammonia leads to *N-arylamidinium salts* in analogous manner[812]:

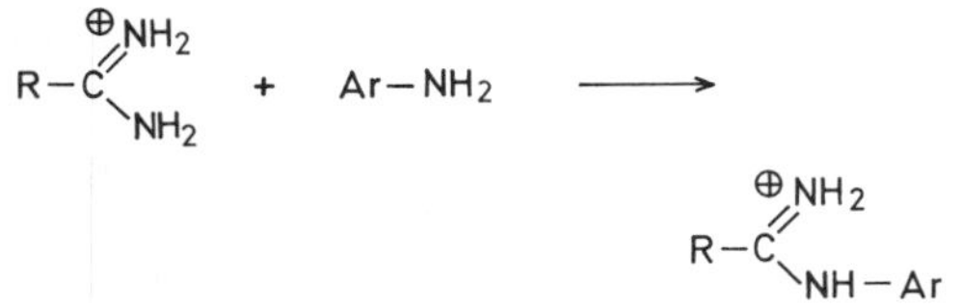

$$R-C\overset{\overset{\oplus}{N}H_2}{\underset{NH_2}{}} \ + \ Ar-NH_2 \ \longrightarrow$$

$$R-C\overset{\overset{\oplus}{N}H_2}{\underset{NH-Ar}{}}$$

Hydroxyaminolyses[813] and hydrazinolyses[814] of amidines are not very important procedurally because the amide oximes and hydrazones that are formed can generally be prepared from nitriles or imidic esters.

Vinylogous formamidinium salts can be substituted by strongly nucleophilic bases. By using N,N'-tetramethylmethylenemethanediamine, for example, *vinylogous guanidinium* and *ω,ω-diaminoacrylamidinium salts* are synthesized accompanied by loss of N-methylaniline[815]:

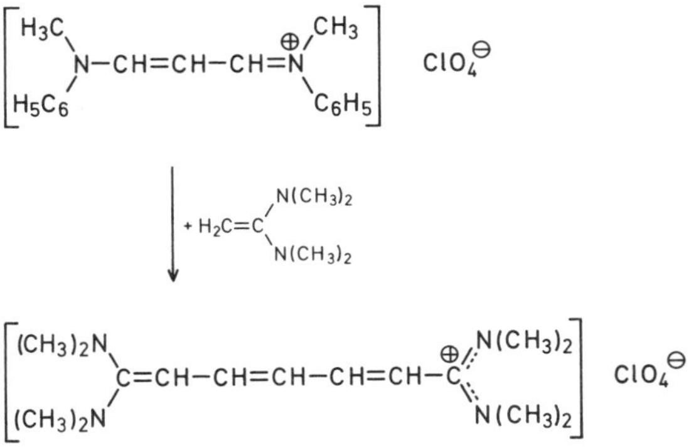

$$\left[\overset{H_3C}{\underset{H_5C_6}{}}N-CH=CH-CH=\overset{\oplus}{N}\overset{CH_3}{\underset{C_6H_5}{}} \right] ClO_4^{\ominus}$$

$$+ H_2C=C\overset{N(CH_3)_2}{\underset{N(CH_3)_2}{}}$$

$$\left[\overset{(CH_3)_2N}{\underset{(CH_3)_2N}{}}C=CH-CH=CH-CH=CH-\overset{\oplus}{C}\overset{N(CH_3)_2}{\underset{N(CH_3)_2}{}} \right] ClO_4^{\ominus}$$

N,N,N',N'-Tetramethyl-7,7-bis-(dimethylamino)-2,4,6-heptatrienecarboxamidinium perchlorate; 94%

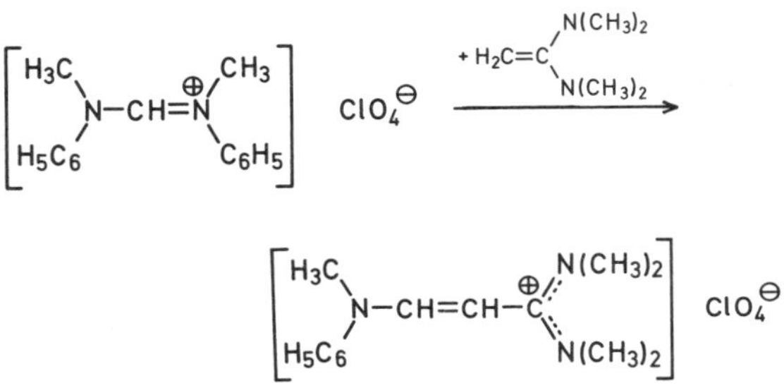

$$\left[\overset{H_3C}{\underset{H_5C_6}{}}N-CH=\overset{\oplus}{N}\overset{CH_3}{\underset{C_6H_5}{}} \right] ClO_4^{\ominus} \xrightarrow{+ H_2C=C\overset{N(CH_3)_2}{\underset{N(CH_3)_2}{}}}$$

$$\left[\overset{H_3C}{\underset{H_5C_6}{}}N-CH=CH-\overset{\oplus}{C}\overset{N(CH_3)_2}{\underset{N(CH_3)_2}{}} \right] ClO_4^{\ominus}$$

N,N,N',N'-Tetramethyl-3-(methylphenylamino)-acrylamidinium perchlorate; 97%

17.2.3.3.3 Addition of azides to enamines, imines, and hydrazones

Enamines of aldehydes initially form 4,5-dihydrotriazole *cycloadducts* with aryl azides which are readily converted into *tetrasubstituted amidines* accompanied by loss of nitrogen and substituent group migration[816]:

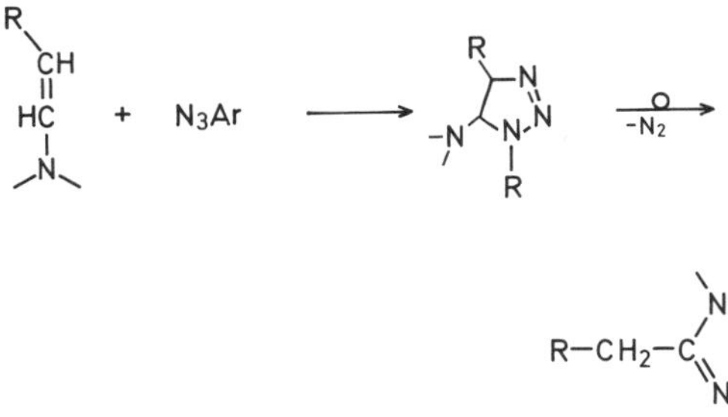

[813] F. Eloy, R. Lenaers, Chem. Rev. *62*, 162 (1962).
[814] D.G. Neilson, R. Roger, J.W.M. Heatle, L.R. Newlands, Chem. Rev. *70*, 151, 158 (1970).
[815] C. Jutz, E. Müller, Angew. Chem. *78*, 747 (1966);
J. Singh, J.L. Boivin, P.E. Gagnon, Can. J. Chem. *40*, 1921 (1962).
[816] R. Fusco, G. Bianchetti, D. Pocar, R. Ugo, Chem. Ber. *96*, 802 (1963);
D. Pocar, G. Bianchetti, P. Dalla Croce, Gazz. Chim. Ital. *95*, 1220 (1965);
R. Fusco, G. Bianchetti, D. Pocar, Gazz. Chim. Ital. *91*, 849, 933 (1961).

[817] G. Bianchetti, P. Dalla Croce, D. Pocar, Tetrahedron Lett. 2043 (1965);
G. Bianchetti, P. Dalla Croce, D. Pocar, G.G. Gallo, Rend. Ist. Lombardo Sci, Pt. I, Classe Sci Mat. e Nat. *99*, 296 (1965);
R.D. Burpitt, V.W. Goodlett, J. Org. Chem. *30*, 4308 (1965).

Imines react from their tautomeric enamine form with, for example, benzoyl azide to form *N-benzoylamidines*[817] (the corresponding conversion of hydrazones furnishes *formazans*[818]):

$$(H_3C)_2CH-C\begin{smallmatrix}N-R\\\\H\end{smallmatrix} + N_3-CO-C_6H_5 \xrightarrow[-N_2]{}$$

$$(H_3C)_2CH-C\begin{smallmatrix}N-R\\\\NH-CO-C_6H_5\end{smallmatrix}$$

N-Benzoyl-N'-ethyl-2-methyl-propionamidine; 60%

17.2.3.3.4 Coupling of diazonium salts with *CH*-acid compounds and arylhydrazones

An important method for preparing *formazans*[675, 676] consists in coupling diazonium salts with *CH*-acid compounds and arylhydrazones; some typical examples will serve to illustrate the possible syntheses:

$$H_5C_2-OOC-CH_2-CO-CH_3 \xrightarrow{+[H_5C_6-\overset{\oplus}{N_2}]X^{\ominus}}$$

$$H_5C_2OOC-\underset{\underset{CO-CH_3}{|}}{CH}-N=N-C_6H_5$$

$$[H_5C_2OOC-\underset{N-NH}{\overset{N=N-C_6H_5}{C}}]\text{—}\underset{R}{\bigcirc}$$

Ethyl 1,5-diphenyl-3-formazancarboxylate; 40–80%

$$H_5C_2OOC-\underset{N-NH-C_6H_5}{\overset{N=N-C_6H_5}{C}}R$$

Ethyl 1-phenyl-5-aryl-3-formazancarboxylate

$$H_3C-NO_2 + [H_5C_6-\overset{\oplus}{N_2}]X^{\ominus} \xrightarrow[H_3C-COONa]{H_3C-COOH}$$

$$O_2N-\underset{N-NH-C_6H_5}{\overset{N=N-C_6H_5}{C}}$$

1,5-Diphenyl-3-nitro formazane

$$H_5C_6-NH-N=CH-C_6H_5 + [H_5C_6-\overset{\oplus}{N_2}]X^{\ominus}$$

$$\xrightarrow{\text{Alkali or Pyridine}} H_5C_6-\underset{N-NH-C_6H_5}{\overset{N=C_6H_5}{C}}$$

1,3,5-Triphenyl formazane

$$H_5C_6-CH=N-NH-C\begin{smallmatrix}NH\\\\NH_2\end{smallmatrix} + [H_5C_6-\overset{\oplus}{N_2}]X^{\ominus}$$

$$\longrightarrow H_5C_6-\underset{N-NH-C}{\overset{N=N-C_6H_5}{C}}\begin{smallmatrix}NH\\\\NH_2\end{smallmatrix}$$

Guanazyls

Acylaminomalonic acids couple with diazonium salts to form N^3-*acylamidrazones*[819], *e.g.*:

$$HOOC-\underset{\underset{NH-CO-CH_3}{|}}{CH}-COONa + [H_5C_6-\overset{\oplus}{N_2}]X^{\ominus}$$

$$\longrightarrow HOOC-\underset{NH-CO-CH_3}{\overset{N-NH-C_6H_5}{C}}$$

17.2.3.3.5 Oxidation of thioureas

A very valuable complementing of methods of synthesizing *trisubstituted formamidines* is the oxidation of trisubstituted thioureas[820] with hydrogen peroxide or peroxyacetic acid in methanol or chloroform at −50 to +20°. The particular value of the procedure lies in that formamidines with bulky alkyl or aryl substituent groups are accessible in good to very good yields. The yields increase as the substituents increase in their spatial requirement; the reason may be the decreasing stability of the intermediately formed thiourea *S,S*-dioxides[820]:

$$R^1-NH-\underset{R^3}{\overset{S}{C}}N-R^2 \xrightarrow{\text{Oxidation}} R^1-NH-\underset{R^3}{\overset{SO_2}{C}}N-R^2$$

$$\longrightarrow R^1-N=CH-\underset{R^3}{\overset{R^2}{N}}$$

R^1 = Alkyl, Aralkyl, 2,6-disubstituted aryl

$R^2 = R^3$ = Alkyl, Aralkyl

17.2.3.3.6 Oxidation of tetrakis(dialkylamino)-ethylenes to octaalkyloxalobis-(amidinium) derivatives

Octaalkylethenetetramines[821] are oxidized to the *octaalkyloxalobis(amidinium) dication*[821] with oxidizing agents (halogens, dinitrogen tetraoxide, copper(I), silver(I), carbon tetrachloride, tetracya-

[818] *W. Ried, W. Kuntel,* Chimia *20,* 279 (1966).
[819] *D.G. Neilson, R. Roger, J.W.M. Heatle, L.R. Newlands,* Chem. Rev. *70,* 160 (1970).
[820] *W. Walter, K.P. Ruess,* Chem. Ber. *102,* 2640 (1969).
[821] *N. Wiberg,* Angew. Chem. *80,* 809 (1968).

noethylene[821]). The product is thermally pretty stable and resistant to water and acids:

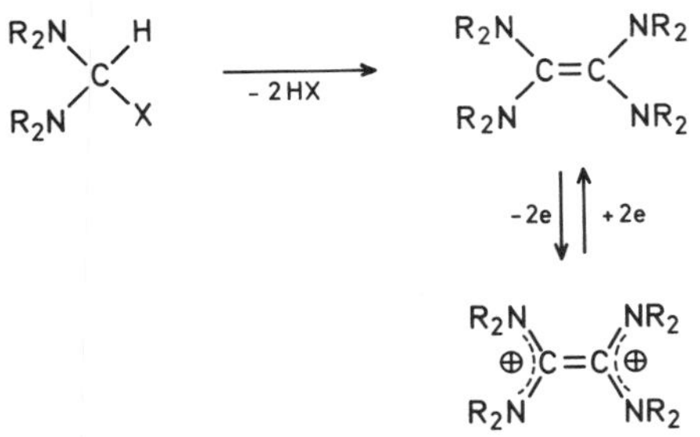

17.2.3.3.7 Amidines by modified Wittig reaction

α-Heterosubstituted phosphonate carbanions react with *C*-nitroso compounds (which are isoelectronic with aldehydes) to *amidines*[822], *e.g.:*

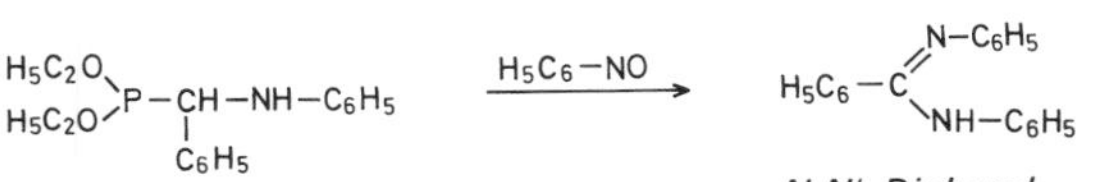

N,N'-Diphenyl-benzamidine; 66%

17.3 Carboxylic acid azides

17.3.1 Scope of the section, nomenclature, hazards of the azides

This section describes the azides of carboxylic acids except those of simple carboxylic acids.

In addition to alkyl and aryl azides, the azides of sulfur and phosphorus acids and similar compounds will not be discussed. Azides of nitrogen heterocycles derived formally from carboxylic acid derivatives such as, for example, 2,4,6-triazido-1,3,5-triazene (cyanuric triazide)[823] are also not included.

Synthesis of azides of simple carboxylic acids by reacting acid halides with the azide anion is dealt with on pp. 685, 749; reference to these here is therefore limited to general synthetic methods. Re preparing azides by diazotizing hydrazides see p. 751 and *Houben-Weyl*, VIII, p. 681.

The nomenclature of the azides is based on that of the corresponding halogen compounds, for example, imidoyl chloride and *imidoyl azide*, chloroformic ester and *azidoformic ester*.

Caution!

Many azides are highly **explosive** substances which may decompose suddenly especially on

heating or through the action of light, impact, or chemical reagents (concentrated sulfuric acid, copper powder). For this reason all work with azides, above all with the pure substances, should be restricted to small quantities and with taking of appropriate precautions[824].

For a detailed survey describing the preparation, properties, and reactions (embracing the literature to 1970) of acyl azides see ref.[825].

17.3.2 On carboxylic acid azide/heterocycle tautomerism

Carboxylic acid azides **36** may be in tautomeric equilibrium with the heteroaromatic compound **37**:

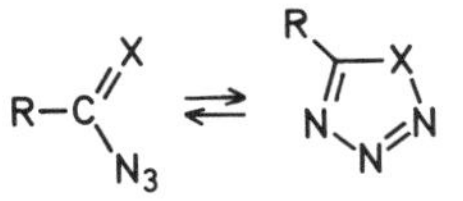

36   **37**

Which of the two structures is present or whether an equilibrium exists is dependent on the nature of X.

To conclude the presence of one of the two tautomers from reactions is not allowable. An unambiguous assignment succeeds with spectroscopic techniques and especially *IR* measurements in that all azides absorb at 2180–2095 and 1260–1210 cm^{-1}[825, 825a, 826]. *IR* and *UV* spectroscopy has shown that compounds with X=O occur *only* in the open-chain form[827], those with X=S exclusively in the cyclic form *(1,2,3,4-thiatriazoles):*

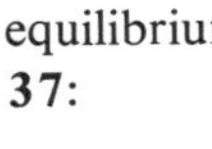

R = Alkyl, Aryl, Heterocycle[828–833]

R = N = [834–841, 856]

R = S = [838, 842–844]

R = O = [845–855]

R = Cl, N$_3$[856]

[822] *H. Zimmer, P.J. Bercz, G.E. Heuer,* Tetrahedron Lett. 171 (1968).

[823] *E. Ott, E. Ohse,* Ber. *54*, 179 (1921).

[824] *J.H. Boyer, R. Moriarty, B. de B. Darwent, P.A.S. Smith,* Chem. Eng. News *42*, No. 31, Aug. 3, 6 (1964).

[825] *W. Lwowski* in *S. Patai,* The Chemistry of the Azido-Group, p. 503, Wiley Intersci. Publ., New York, London, Sydney, Toronto 1971.

[825a] *J.E. Gurst* in *S. Patai,* The Chemistry of the Azido-Group, p. 191, Wiley Intersci. Publ., New York, London, Sydney, Toronto 1971.

[825b] *M.E. Biffin, J. Miller, D.B. Paul* in *S. Patai,* The Chemistry of the Azido-Group, p. 57, Wiley Intersci. Publ., New York, London, Sydney, Toronto 1971.

[826] *E. Lieber, C.N.R. Rao, A.E. Thomas, E. Oftedahl, R.L. Minnis, C.V.N. Nambury,* Spectrochim. Acta *19*, 1135 (1961).

[827] *E. Lieber, R.L. Minnis jr., C.N.R. Rao,* Chem. Rev. *65*, 377 (1965).

[828] *T. Bacchetti, A. Alemagna,* Rend. Ist. Lombardo Sci., Pt. I, Classe Sci. Mat. e Nat., *91*, 617 (1957); C.A. *53*, 6217g (1959).

[829] *W. Kirmse,* Chem. Ber. *93*, 2353 (1960).

[830] *E. Lieber,* U.S. Dept. Com., Office Tech. Serv., P B Rept. *148, 532* (1960); C.A. *58*, 2446g (1963).

In compounds **36/37** with $X=N-R^1$ (R^1=alkyl, aryl, or an alicyclic or heterocyclic group linked to R) azide-heteroaromatic tautomerism has been studied intensively. They can exist both as azide and *1H*-tetrazole; for a survey and a discussion see refs. [825, 825b, 827, 857–860]. The direction in which the equilibrium is shifted is dependent on various factors, but especially the nature of R and R^1. In general, the aromatic character of the *1H*-tetrazole[859, 860] favors the formation of this five-membered ring[861–866]. Azides are obtained if their structure is favored by mesomerism[857–860, 867–870], *e.g.*:

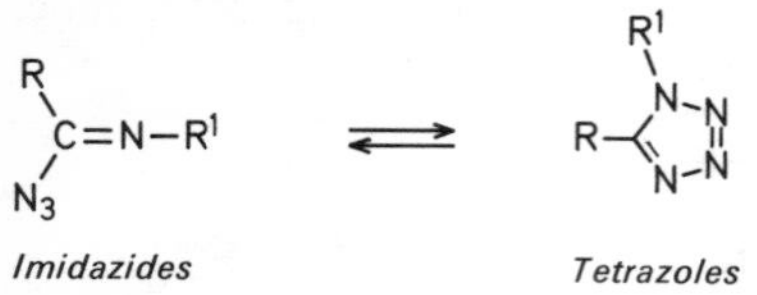

Imidazides   *Tetrazoles*

Thus, the compounds in which R is *aryl* and R^1 are strongly negative groups such as *OH*, *OCH₃*, $OCOC_6H_5$, or *NH-aryl* occur only as azides[867–870]. Compounds with $R=aryl$, $R^1=N=CH-C_6H_5$[861] and $R=CH_2-CO-C_6H_5$, $R^1=CH_3$[862] can be isolated as azides but readily cyclize to tetrazoles[861–863]. 1,5-Diaryltetrazoles exist only in the cyclic form but an imidazide/tetrazole tautomerism was detected[864–866].

Introducing electron-attracting groups in position 1 of 5-amino-1*H*-tetrazole leads to *N*-substituted *azidoamidines (guanyl azides)* accompanied by opening of the ring[871–873]:

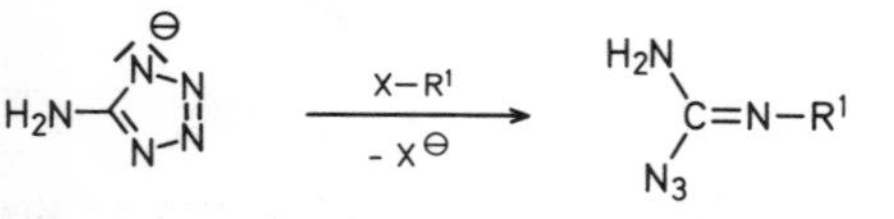

X   = Halogen
R¹ = CN;   *1-Azido-N'-cyanoformamidine*; 20–44%
     –SO₂–Aryl; *N-(1-azidoformimidoyl)arylsulfonamide*; 46–85%

Of the simple azidoformamidines the *1-azido-formamidine (guanyl azide)* can be prepared by diazotizing aminoguanidine in strongly acidic solution[874–878] and isolated as nitrate, chloride, perchlorate, hexachloroantimonate, and azidopentachloroantimonate. In weakly acid or basic solution the nitrate changes into *5-amino-1H-tetrazole*[874]. Other azidoamidines (guanyl azides)[871], carbonic acid cyanoimideamide azides (cyanoguanyl acide)[873], carbonic acid arylsulfonyl imide azide [(acyl sulfonyl)-guanyl azide] and carbonic acid nitroimideamide azide (nitroguanyl azide)[879] likewise isomerize to *1H-tetrazoles* under base catalysis[859].

Carbamoyl azides are stable substances and cannot be cyclized to tetrazoles[827, 880]. N,N'-Diaminoguanidine affords *5-azido-1H-tetrazole* with nitrous acid and not iminomethyl diazide; the same compound results when 5-hydrazino-1*H*-tetrazole is diazotized[881]:

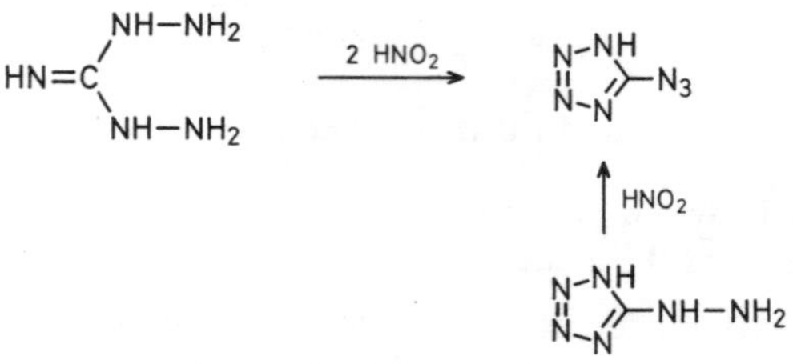

[837] *E. Lieber, C.N.R. Rao, C.N. Pillai, J. Ramachandran, R.D. Hites,* Can.J. Chem. *36*, 801 (1958).
[838] *E. Lieber, J. Ramachandran, C.N.R. Rao, C.N. Pillai,* Can. J. Chem. *37*, 563 (1959).
[839] *M. Kühn, R. Mecke,* Z. Anal. Chem. *181*, 487 (1961).
[840] *E. Lieber, C.N.R. Rao, C.B. Lawyer, J.P. Travedi,* Can. J. Chem. *41*, 1643 (1963).
[841] *E. Lieber, N. Calvanico, C.N.R. Rao,* J. Org. Chem. *28*, 257 (1963).
[842] *E. Lieber, C.N. Pillai, J. Ramachandran, R.D. Hites,* J. Org. Chem. *22*, 1750 (1957).
[843] *E. Lieber, E. Oftedahl, S. Grenda, R.D. Hites,* Chem. & Ind. (London) *893* (1958).
[844] *E. Lieber, E. Oftedahl, C.N.R. Rao,* J. Org. Chem. *28*, 194 (1963).
[845] *D. Martin,* Z. Chem. 7, 123 (1967).
[846] *D. Martin,* Angew. Chem. *76*, 303 (1964).
[847] *D. Martin,* Chem. Ber. *97*, 2689 (1964).
[848] *K.A. Jensen, A. Holm,* Acta Chem. Scand. *18*, 826 (1964).
[849] *D. Martin,* Tetrahedron Lett. 2829 (1964).
[850] *M. Hedayatullah, L. Denivelle, C.R. Acad. Sci., Paris 260*, 2839 (1965).
[851] *P. Reich, D. Martin,* Chem. Ber. *98*, 2063 (1965).
[852] *D. Martin, W. Mucke,* Chem. Ber. *98*, 2059 (1965).
[853] *K.A. Jensen, A. Holm, B. Thorkilsen,* Acta Chem. Scand. *18*, 825 (1964).
[854] *K.A. Jensen, M. Due, A. Holm,* Acta Chem. Scand. *19*, 438 (1965).
[855] *K.A. Jensen, A. Holm, C. Wentrup,* Acta Chem. Scand. *20*, 2107 (1966).

[831] *P.A.S. Smith, D.H. Kenny,* J. Org. Chem. *26*, 5221 (1961).
[832] *E. Lieber, C.N.R. Rao, R.C. Orlowski,* Can. J. Chem. *41*, 926 (1963).
[833] *K.A. Jensen, C. Pedersen,* Acta Chem. Scand. *15*, 1104 (1965).
[834] *E. Lieber, E. Oftedahl, C.N. Pillai, R.D. Hites,* J. Org. Chem. *22*, 441 (1957).
[835] *E. Lieber, C.N. Pillai,* J. Org. Chem. *22*, 1054 (1957).
[836] *E. Lieber, C.N. Pillai, R.D. Hites,* Can. J. Chem. *35*, 832 (1957).

Of the carbonic acid imide diazides the *N*-cyano compound *(dicyanodiazide)* is stable and has been prepared from cyanogen bromide with sodium azide in aqueous solution[871, 882]. *Iminomethyl diazide* itself has been obtained as the *hexachloroantimonate*[878].

17.3.3 Carboxylic acid azides by means of substitution reactions by the azide anion

Except for a few special reactions (see p. 650, 651) the synthesis of the acyl azides being discussed here presents no unusual features. All general azide syntheses can be applied and general guidelines can be deduced from other surveys[825, 825b, 883–890] and from p. 685.

Because of their ready accessibility almost exclusively acid halides are employed as starting compounds for preparing azides by nucleophilic substitution.

Substitution of quaternary ammonium salts[891, 892], toluenesulfonates[893, 894], and thioglycolates[829, 832] with the azide anion is possible in uncomplicated cases but displays no procedural importance. Mixed anhydrides of ethyl carbonate, too, yield carboxylic acid azides[895–897].

$$R-CO-O-CO-OC_2H_5 \xrightarrow[-H_5C_2O-CO-O^\ominus]{N_3^\ominus}$$

$$R-CO-N_3$$

As well as the normally used sodium azide, lithium azide[898, 899], ammonium azide[900, 901], and tetraalkylated ammonium azides[902–904] are employed for synthesizing azides. *Carbonyl azides* have been prepared from tetramethylguanidinium azides, which is soluble in organic solvents (*e.g.* chloroform)[906]. The solution of equimolar hydrazoic acid plus pyridine in toluene has so far been employed only for synthesizing azides of simple carboxylic acids[907].

Advantageous solvents for use are water admixed with organic *solvents* such as acetonitrile, acetone, 1,4-dioxane, alcohols, pyridine, dimethylacetamide, dimethylformamide (as acid complex[908]), or glacial acetic acid, but ether and benzene hydrocarbons are possible alternatives[825, 890]. Alkali metal azides are insoluble in the last-named solvents and freshly prepared *(active)* sodium azide is employed under these conditions[890, 909].

With certain azides of derivatives of carbonic acid water is a suitable solvent from which the azide then separates. The use of perfluorinated hydrocarbons as solvent is described for the synthesis of *azidoformyl fluoride* and *difluorocarbamoyl azide* from the corresponding chlorine compounds with sodium and silver azide[910]. Because of the thermal instability working is generally performed at 0°, but with more stable azides higher temperatures can be employed.

Carbamoyl azides can be synthesized, for example, by boiling carbamoyl chlorides with sodium azide in alcohol[827, 911–915].

Table 11 surveys the acyl azides which have been prepared from halogen compounds with the azide anion.

[856] *E. Lieber, C.B. Lawyer, J.P. Trivedi*, J. Org. Chem. *26*, 1644 (1961).
[857] *M. Tišler*, Synthesis 123 (1973).
[858] *H. Reimlinger*, Chem. Ber. *103*, 1900 (1970).
[859] *F.R. Benson* in *R.C. Elderfield*, Heterocyclic Compounds Vol. VIII, p. 1–104, John Wiley & Sons, New York, London, Sydney 1967.
[860] *R. Huisgen*, Angew. Chem. *72*, 359, 363 (1960).
[861] *R. Stollé, F. Helwerth*, Ber. *47*, 1132 (1914).
[862] *R.B. Woodward, R.A. Olofson*, J. Amer. Chem. Soc. *83*, 1007 (1961).
[863] *R. Stollé, A. Netz*, Ber. *55*, 1297 (1922).
[864] *P.A.S. Smith*, J. Amer. Chem. Soc. *76*, 436 (1954).
[865] *P.A.S. Smith, E. Leon*, J. Amer. Chem. Soc. *80*, 4647 (1958).

[866] *K. v. Fraunberg, R. Huisgen*, Tetrahedron Lett. 2599 (1969).
[867] *F. Eloy*, J. Org. Chem. *26*, 952 (1961).
[868] *M.S. Chang, A.J. Matuszko*, J. Org. Chem. *28*, 2260 (1963).
[869] *A.F. Hegarty, J.B. Aylward, F.L. Scott*, Tetrahedron Lett. 1259 (1967);
A.F. Hegarty, J.B. Aylward, F.L. Scott, J. Chem. Soc. C 2587 (1967).
[870] *C. Grundmann, H.D. Frommeld*, J. Org. Chem. *31*, 157 (1966).
[871] *W.P. Norris, R.A. Henry*, J. Org. Chem. *29*, 650 (1964).
[872] *K.A. Jensen, C. Pedersen*, Acta Chem. Scand. *15*, 991 (1961).
[873] *H.K. Nagy, A.J. Tomson, J.P. Horwitz*, J. Amer. Chem. Soc. *82*, 1609 (1962).
[874] *A. Hantzsch, A. Vagt*, Justus Liebigs Ann. Chem. *314*, 339 (1901).
[875] *A. Thiele*, Justus Liebigs Ann. Chem. *270*, 1, 54 (1892);
A. Thiele, Justus Liebigs Ann. Chem. *303*, 57 (1898).
[876] *K.A. Hofmann, H. Hock, R. Roth*, Ber. *43*, 1087 (1910).
[877] *K.A. Hofmann, R. Roth*, Ber. *43*, 682 (1910).
[878] *A. Schmidt*, Chem. Ber. *100*, 3725 (1967).
[879] *E. Lieber, E. Shermann, R.A. Henry, J. Cohen*, J. Amer. Chem. Soc. *73*, 2327 (1951).
[880] *F.L. Scott, A. Koczorski, J. Reilly*, Nature *170*, 922 (1952).
[881] *E. Lieber, D.R. Levering*, J. Amer. Chem. Soc. *73*, 1313 (1951).
[882] *C.V. Hart*. J. Amer. Chem. Soc. *50*, 1922 (1928).

Table 11. Azides by substitution reactions from acyl halides*

Name	General Formula	Reference
Imidoyl azides (Imide azides, Imido carboxylic acid azides)	$R^1-C\begin{smallmatrix}N-R^2\\ \\N_3\end{smallmatrix}$	825, 862
Azid oximes (Azide hydroxyimides)	$R^1-C\begin{smallmatrix}N-O-R^2\\ \\N_3\end{smallmatrix}$	867, 868
Azidobenzaldehyde azines [(Benzoic acid azide) azine]	$H_5C_6-C\begin{smallmatrix}N-N=C\begin{smallmatrix}C_6H_5\\R\end{smallmatrix}\\ \\N_3\end{smallmatrix}$	863
Azide hydrazones (Hydrazidic azides)	$R^1-C\begin{smallmatrix}N-NH-R\\ \\N_3\end{smallmatrix}$	869
Alkyl 1-azido-*N*-(aryl-sulfonyl)thioformimidates	$R^1-S-C\begin{smallmatrix}N-SO_2-R^2\\ \\N_3\end{smallmatrix}$	916
Azidoformates (Azidocarbonates)	$R^1-O-C\begin{smallmatrix}O\\ \\N_3\end{smallmatrix}$	906, 917–919
Carbamoyl azides (Carbamyl azides; Carbonic acid amide azides)	$\begin{smallmatrix}R^1\\ \\R^2\end{smallmatrix}N-C\begin{smallmatrix}O\\ \\N_3\end{smallmatrix}$	827, 911–915
Dicyanodiazide (*N*-Cyanocarbimidic acid diazide; Carbonic acid cyanoimide diazide)	$N\equiv C-N=C\begin{smallmatrix}N_3\\ \\N_3\end{smallmatrix}$	871, 882
Cyanogen azide	$N\equiv C-N_3$	920

* Re tetrazoles and thiatriazoles which are tautomers or isomers of imidoyl azides or thiocarbonyl azides, see p 747.

[883] *H. Henecka, P. Kurtz* in *Houben-Weyl,* Methoden der organischen Chemie, 4. Aufl., Bd. VIII, p. 680, Georg Thieme Verlag, Stuttgart 1952.

[884] *Ng-Ph. Buu-Hoï, P. Cagniant* in *V. Grignard, G. Dupont, R. Locquin,* Traité de Chimie organique, Vol. XV, p. 715 (1948).

[885] *F. Möller* in *Houben-Weyl,* Methoden der organischen Chemie, 4. Aufl., Bd. XI/1, p. 863, Georg Thieme Verlag, Stuttgart 1957.

[886] *G.H. Boyer, F.C. Canter,* Chem. Rev. *54,* 1 (1954).

[887] *C. Grundmann* in *Houben-Weyl,* Methoden der organischen Chemie, 4. Aufl., Bd. X/3, p. 777, Georg Thieme Verlag, Stuttgart 1965.

[888] *C.G. Overberger, J.-P. Anselme, J.G. Lombardino,* Organic Compounds with Nitrogen-Nitrogen Bonds, p. 99, Ronald Press Co., New York 1966.

[889] *P.A.S. Smith,* The Chemistry of Open Chain Organic Nitrogen Compounds, Vol. II, p. 211, Benjamin, New York, Amsterdam 1966.

[890] *P.A.S. Smith,* Org. Reactions *3,* 337 (1946).

[891] *A.N. Nesmeyanow, M.I. Rubinskaya,* Isv. Akad. Nauk SSSR, Otdel. Khim. Nauk, 816 (1962); Engl. text: 761 (1962).

17.3.4 Carboxylic acid azides by adding hydrazoic acid on to double bond systems

Hydrazoic acid adds to polar multiple bonds and heterocumulenes with formation of azides, but *sequential reactions* take place in many instances. Thus nitriles, cyanates, thiocyanates, isocyanides, cyanamides, carbodiimides, thiocyanates, and iso-thiocyanates react in the above sense; where imide and thiocarbonyl azides are formed initially cyclization to *tetrazoles* and *1,2,3,4-thiatriazoles* invariably occurs (see p. 748). On the other hand, this reaction represents the best technique for preparing *N*-monosubstituted carbamoyl azides side by side with diazotization of the corresponding hydrazides[827]:

$$R-N=C=O \xrightarrow{HN_3} R-NH-CO-N_3$$

R = H, Alkyl, Aryl, CO-Aryl, *p*-Tosyl

From acyl and sulfonyl isocyanates *carbamoyl* and *sulfamoyl azides* respectively are obtained[921,922]; mesitylbenzonitrile *N*-oxide reacts with hydrazoic acid to form *mesitylbenzhydroxamic acid azide*[870].

Methyl ketene furnishes methylcarbamoyl azide with two *molecules* hydrazoic acid *via* a Curtius rearrangement of the initially formed acetyl azide[923]:

$$H_2C=C=O \xrightarrow{HN_3} H_3C-C\begin{smallmatrix}O\\ \\N_3\end{smallmatrix} \xrightarrow{-N_2}$$

$$H_3C-N=C=O \xrightarrow{HN_3} H_3C-NH-C\begin{smallmatrix}O\\ \\N_3\end{smallmatrix}$$

4-Phenyl-1,3-dioxane-5-carbamoyl azide was obtained similarly

[892] *J.P. Horwitz, A.J. Tomson,* J. Org. Chem. *26,* 3392 (1961).

[893] *J.P. Horwitz, A.J. Tomson, J.A. Urbanski, J. Chua,* J. Org. Chem. *27,* 3045 (1962).

[894] *A.K. Bose, J.F. Kistner, L. Farber,* J. Org. Chem. *27,* 2925 (1962).

[895] *J. Weinstock,* J. Org. Chem. *26,* 3511 (1961).

[896] *R.M. Palmere, R.T. Conley,* J. Org. Chem. *35,* 2703 (1970).

[897] *Y.G. Perron, L.B. Crast, J.M. Essery, R.R. Fraser, J.C. Godfrey, C.T. Holdrege, W.F. Minor, M.E. Neubert, R.A. Partyka, L.C. Cheney,* J. Med. Chem. *7,* 483 (1964).

[898] *R. Huisgen, I. Ugi,* Chem. Ber. *90,* 2914 (1957).

17.3.5 Carboxylic acid azides by diazotization of hydrazides with nitrous acid or its derivatives

Hydrazides afford azides on treatment with nitrous acid or its derivatives.

In the majority of cases the hydrazide is employed as the hydrochloride and nitrite solution is added at 0–5°. The azide separates or is taken up in a superior layer of ether. For sparingly soluble hydrazides glacial acetic acid is used as solvent. Other solvent systems are described in survey articles[825, 825b, 883, 890].

Sodium nitrite is employed as the *diazotizing agent* in almost every case but the potassium and ammonium salts, nitrosyl chloride[925], alkyl nitrites[925–929], and oxides of nitrogen oxides (NO[930], N_2O_3[901]), are used.

Acylation of the hydrazide by already formed azide is a possible *side-reaction*, especially at higher *pH* values[890]:

$$R-CO-NH-NH_2 \xrightarrow{N_3-CO-R} $$
$$R-CO-NH-NH-CO-R$$

Table 12 surveys acyl azides that have been synthesized by diazotizing hydrazides (cf. the note in Table 11 on p. 750).

Table 12. Azides by diazotization of hydrazides

Name	General Formula	Reference
Imidoyl azides (Imide azides)	$R^1-C\begin{smallmatrix}N-R^2\\\\N_3\end{smallmatrix}$	825
Azid oximes (Azide hydroxyimides)	$R^1-C\begin{smallmatrix}N-O-R^2\\\\N_3\end{smallmatrix}$	867, 933, 934
1-Azidoformamidines (Guanyl azides; Carbonic acid imide amide azides)	$\begin{smallmatrix}R^1\\R^2\end{smallmatrix}N-C\begin{smallmatrix}N-R^3\\\\N_3\end{smallmatrix}$	931, 932, 938
Azidoformamidinium salts	$\left[\begin{smallmatrix}R^1\\R^2\end{smallmatrix}N-C\begin{smallmatrix}\oplus\\\\N_3\end{smallmatrix}N\begin{smallmatrix}H\\\\R^3\end{smallmatrix}\right]X^\ominus$	874–877, 879, 938
Azidobenzaldehyde azines (Benzaldehyde azide) azine	$H_5C_6-C\begin{smallmatrix}N-N=C\begin{smallmatrix}C_6H_5\\\\R^3\end{smallmatrix}\\\\N_3\end{smallmatrix}$	861, 935
Azidoformates (Azidocarbonates)	$R-O-C\begin{smallmatrix}O\\\\N_3\end{smallmatrix}$	936, 937
Carbamoyl azides (Carbamyl azides; Carbonic acid amide azide)	$\begin{smallmatrix}R^1\\R^2\end{smallmatrix}N-C\begin{smallmatrix}O\\\\N_3\end{smallmatrix}$	827, 901, 938
Carbonyl azide (Carbonic acid diazide)	$N_3-C\begin{smallmatrix}O\\\\N_3\end{smallmatrix}$	827, 939, 940

17.3.6 Carboxylic acid azides from other azides and from tetrazoles

The **highly explosive** oxalyl diazide (obtainable from oxalic acid dihydrazide by diazotization[941]) undergoes Curtius rearrangement to *azidocarbonyl isocyanate*, which reacts with alcohols to form ethyl (azidoformyl)imidocarbonate and with water to give *carbamoyl azide*[941]:

$$N_3-CO-CO-N_3 \longrightarrow N_3-CO-N=C=O$$

$$N_3-CO-NH-C\begin{smallmatrix}OC_2H_5\\\\O\end{smallmatrix} \xleftarrow{C_2H_5OH} \qquad \xrightarrow{HOH} N_3-CO-NH_2$$

1-Azido-N'-cyanoformamidines can be obtained also from dicyanodiazide with ammonia or amines[882]:

$$N\equiv C-N=C\begin{smallmatrix}N_3\\\\N_3\end{smallmatrix} \xrightarrow{H_2N-R} N\equiv C-N=C\begin{smallmatrix}NH-R\\\\N_3\end{smallmatrix}$$

Fluorinated azides have been prepared by fluorination from carbonyl azide (carbonic acid diazide) and carbamoyl azide[910]:

$$N_3-C\begin{smallmatrix}O\\\\N_3\end{smallmatrix} \xrightarrow{F_2} F-C\begin{smallmatrix}O\\\\N_3\end{smallmatrix}$$

Azidoformyl fluoride

$$H_2N-C\begin{smallmatrix}O\\\\N_3\end{smallmatrix} \xrightarrow{F_2} F_2N-C\begin{smallmatrix}O\\\\N_3\end{smallmatrix}$$

Difluorocarbamoyl azide

[899] *M. Brown, R.E. Benson*, J. Org. Chem. *31*, 3849 (1966).

[900] *W.G. Finnegan, R.A. Henry, R. Lofquist*, J. Amer. Chem. Soc. *80*, 3908 (1958).

[901] *T. Curtius, K. Heidenreich*, J. Prakt. Chem. [2] *52*, 454 (1895).

[902] *F. Weygand, M. Reiher*, Chem. Ber. *88*, 26 (1955); *F. Weygand, R. Geiger*, Chem. Ber. *90*, 634 (1957).

[903] *V. Gutmann, G. Hempel, O. Leitmann*, Monatsh. Chem. *95*, 1034 (1964).

[904] *C.G. Swain, M.M. Kreevoy*, J. Amer. Chem. Soc. *77*, 1122 (1955).

[905] *A.J. Papa*, J. Org. Chem. *31*, 1426 (1966).

[906] *K. Sakai, J.-P. Anselme*, J. Org. Chem. *36*, 2387 (1971).

[907] *J.W. van Reijendam, F. Baardman*, Synthesis 413 (1973).

[908] *D.E. Horning, J.M. Muchowski*, Can. J. Chem. *45*, 1247 (1967).

Azidoformimidinium and *azidoformimidinium methyl ester hexachloroantimonates* are accessible in the following way[942] (87% and 74%):

$$2 \left[\begin{array}{c} Y \\ C = \overset{\oplus}{N}H_2 \\ Cl \end{array} \right] SbCl_6^{\ominus} \xrightarrow[- 2\, SbCl_5]{(SbCl_4N_3)_2}$$

$$2 \left[\begin{array}{c} Y \\ C = \overset{\oplus}{N}H_2 \\ N_3 \end{array} \right] SbCl_6^{\ominus} \xrightarrow[- HCl]{ROH}$$

$$2 \left[\begin{array}{c} RO \\ C = \overset{\oplus}{N}H_2 \\ N_3 \end{array} \right] SbCl_6^{\ominus}$$

1-Azidoformidinium hexachloroantimonate; 100%

Azidochloroformidinium hexachloroantimonate; 100%

Azidochloro-, Azidodichloro-, Azidotrichloroacetimidinium hexachloroantimonate; 89–95%

Y = H, Cl, CH₂Cl, CHCl₂, CCl₃

R = H; CH₃

Azidoformamidinium salts (82.3%), too, can be synthesized from chloroformamidinium salts with trimethyl silyl azide[878]:

$$\left[\begin{array}{c} H_2N - C = \overset{\oplus}{N}H_2 \\ Cl \end{array} \right] Cl^{\ominus} + (H_3C)_3SiN_3 \longrightarrow$$

$$\left[\begin{array}{c} H_2N \\ C = \overset{\oplus}{N}H_2 \\ N \end{array} \right] N_3^{\ominus} + (H_3C)_3SiCl$$

17.4 *N*-Functional derivatives of thiocarboxylic acids

17.4.1 Direct introduction of the aminothiocarbonyl group

$$-\overset{\overset{\displaystyle B}{|}}{\underset{\displaystyle \parallel}{C}} - N \overset{\displaystyle B}{\underset{\displaystyle C}{}}$$

17.4.1.1 Nucleophilic addition to isothiocyanates

17.4.1.1.1 Addition of *CH*-acid compounds

The addition of *CH*-acid compounds to isothiocyanates proceeds according to the scheme, with *R = alkyl* or *aryl* (preferentially the latter)

$$R^2 - \overset{\overset{\displaystyle R^1}{|}}{\underset{\displaystyle R^3}{C}}{}^{\ominus} Na^{\oplus} + R - N = C = S \longrightarrow$$

$$\left[R^2 - \overset{\overset{\displaystyle R^1}{|}}{\underset{\displaystyle R^3}{C}} - \overset{\overset{\displaystyle \overset{\ominus}{N} - R}{\|}}{\underset{\displaystyle S}{C}} \right] Na^{\oplus} \xrightarrow{HX} R^2 - \overset{\overset{\displaystyle R^1}{|}}{\underset{\displaystyle R^3}{C}} - \overset{\overset{\displaystyle NH - R}{\|}}{\underset{\displaystyle S}{C}}$$

[909] *J. Nelles*, Ber. *65*, 1345 (1932).
[910] USP 3 418 341 (1968), *DuPont*, Inv.: *R.J. Shozda*.
[911] *R. Stollé, N. Nieland, M. Merkle*, J. Prakt. Chem. [2] *116*, 192 (1927);
    R. Stollé, N. Nieland, M. Merkle, J. Prakt. Chem. [2] *117*, 185 (1927).
[912] *R. Stollé*, Ber. *57*, 1063 (1924).
[913] *C.D. Hurd, L.U. Spence*, J. Amer. Chem. Soc. *49*, 266 (1927).
[914] *T. Kametani, K. Sota, M. Shio*, J. Heterocycl. Chem. *7*, 807 (1970).
[915] *T. Kametami, M. Shio*, J. Heterocycl. Chem. *7*, 831 (1970).
[916] *R. Neidlein, W. Haussmann*, Tetrahedron Lett. 5401 (1966).
[917] *W. Lwowski, T.W. Mattingly jr.*, J. Amer. Chem. Soc. *87*, 1947 (1965).
[917a] *H. Yajima, Y. Kiso*, Chem. Pharm. Bull. (Tokyo) *17*, 1962 (1969).
[918] *G. Smolinsky, B.I. Feuer*, J. Amer. Chem. Soc. *86*, 3085 (1964).

[919] USP 3 324 148 (1967), Union Carbide Co., Inv.: *R.J. Cotter*;
    R.J. Cotter, W.F. Beach, J. Org. Chem. *29*, 751 (1964).
[920] *F.D. Marsh, M.E. Hermes*, J. Amer. Chem. Soc. *86*, 4506 (1964);
    F.D. Marsh, J. Org. Chem. *37*, 2966 (1972).
[921] *R. Neidlein*, Angew. Chem. *78*, 333 (1966).
[922] *R. Neidlein*, Arch. Pharm. (Weinheim, Ger.) *299*, 1003 (1966).
[923] *E. Oliveri-Mandala, E. Calderaro*, Gazz. Chim. Ital. *43*, 538 (1913); Chem. Zentr. III, 668 (1913).
[924] *F.W. Brugman, J.F. Arens*, Rec. Trav. Chim. Pays-Bas *74*, 209 (1955).
[925] *J. Honzl, J. Rudinger*, Collect. Czech. Chem. Commun. *26*, 2333 (1961);
    R.A. Clement, J. Org. Chem. *27*, 1904 (1962).
[926] *R. Pschorr, H. Einbeck, O. Spangenberg*, Ber. *40*, 1998 (1907).
[927] *A. Windaus, W. Vogt*, Ber. *40*, 3691 (1907).
[928] *A. Windaus, H. Opitz*, Ber. *44*, 1721 (1911).
[929] *T.M. Sharp*, J. Chem. Soc. (London) 1234 (1936).
[930] Jap. P. 23 611/68 (1968), Sankyo Kasei Co.
[931] *R.A. Henry, W.G. Finnegan, E. Lieber*, J. Amer. Chem. Soc. *76*, 88 (1954).
[932] *W.G. Finnegan, R.A. Henry, E. Lieber*, J. Org. Chem. *18*, 779 (1953).
[933] *K. Golankiewicz*, Roczniki Chem. *37*, 309 (1963).
[934] *K. Golankiewicz*, Bull. Acad. Polon. Sci., Ser. Sci. Chim. *10*, 417 (1962); C.A. *59*, 1640h (1963);
    K. Golankiewicz, Bull. Acad. Polon. Sci., Ser. Sci. Chim. *12*, 307 (1964); C.A. *61*, 14673b (1964);
    K. Golankiewicz, Bull. Acad. Polon. Sci., Ser. Sci. Chim. *12*, 459 (1964); C.A. *62*, 1659h (1965).
[935] *H. Wieland*, Ber. *42*, 4199 (1909).

When diethyl malonate is the starting material, *diethoxycarbonylthioacetamides* are obtained in yields between 60 and 80% [943]. The possibility of variation in R^1, R^2, and R^3 is large: even aromatic compounds react to give thioamides in the presence of aluminium chloride [944, 945]. Enamines and isothiocyanates form vinylogous *thioureas*, which are readily hydrolyzed to *β-oxothiocarboxamides* [946-948].

Alkylidene phosphoranes and isothiocyanates react *via* a betaine intermediate to give ylides, the Wittig reaction of the ylides gives the *α,β-unsaturated thiocarboxamides* [949].

With sodium tetrahydroborate as the source of hydride ions isothiocyanates give *thioformamides* [950]; this is of preparative significance when the substituent R is large [951].

Grignard reagents also react with isothiocyanates to give *N*-monosubstituted thiocarboxamides in good yields [952, 953] (the preparative value of this method has been recognized only very recently).

17.4.1.1.2 Addition of compounds with amino, hydroxy, or thiol groups

Addition of compounds containing amino, hydroxy, and thiol groups, furnishes *thioureas, thiocarbamic acid esters,* and *dithiocarbamic acid esters* [954-957]:

$$YH = NH_3, R^1-NH_2, R^1_2NH, R^1OH, R^1SH$$

With aliphatic and saturated cyclic amines the reaction proceeds smoothly while the thioureas formed from *N* alkylanilines tend to cleave to the starting materials. The *N'N'*-disubstituted-*N*-(benzo-1,3,2-dioxaphosphol-2-yl) thioureas from 2-isothiocyanatobenzo-1,3,2-dioxaphosphole can even be cleaved by water at the *P–N* bond, and therefore are valuable intermediates for the synthesis of *N,N-disubstituted thioureas* [956, 957].

[936] *L.A. Carpino, C.A. Giza, B.A. Carpino*, J. Amer. Chem. Soc. *81*, 955 (1959).

[937] *A. Hantzsch*, Ber. *36*, 2057 (1903).

[938] *K.A. Jensen, A. Holm, S. Rachlin*, Acta Chem. Scand. *20*, 2795 (1966).

[939] *W. Kesting*, Ber. *57*, 1321 (1924).

[940] *L.E. Chapman, R.F. Robbins*, Chem. & Ind. (London) 1266 (1966).

[941] *H. Roesky, O. Glemser*, Chem. Ber. *97*, 1710 (1964).

[942] *A. Schmidt*, Chem. Ber. *100*, 3319 (1967).

[943] *J. Goerdeler, U. Keuser*, Chem. Ber. *97*, 3106 (1964).

[944] *A. Friedmann, L. Gattermann*, Ber. *25*, 3525 (1892).

[945] *F. Mayer, A. Mombour*, Ber. *62*, 1921 (1929).

[946] *S. Hünig, K. Hübner, E. Benzing*, Chem. Ber. *95*, 926 (1962).

[947] *W. Ried, W. Käppeler*, Justus Liebigs Ann. Chem. *673*, 132 (1964).

[948] *G. Bianchetti, P. dalla Croce, D. Pocar*, Gazz. Chim. Ital. *94*, 606 (1964).

[949] *H.J. Bestmann, S. Pfohl*, Angew. Chem. *81*, 750 (1969).

[950] *S.E. Ellzey jr., C.H. Mack*, J. Org. Chem. *28*, 1600 (1963).

[951] *R.F. Becker*, Dissertation, Universität Hamburg (1971).

[952] *G. Allinger, G.E.P. Smith, E.L. Carr, H.P. Stevens*, J. Org. Chem. *14*, 962 (1949).

[953] *N.B. Mehta, J. Zupicich Strelitz*, J. Org. Chem. *27*, 4412 (1962).

[954] *E. Schmidt, W. Striewsky, M. Seefelder, F. Hilzler*, Justus Liebigs Ann. Chem. *568*, 192 (1950).

[955] *W. Walter, G. Randau*, Justus Liebigs Ann. Chem. *722*, 52 (1969).

[956] *G. Barnikow, T. Gabrio*, Z. Chem. *8*, 143 (1968).

[957] *W. Walter, K.-P. Ruess*, Justus Liebigs Ann. Chem. *743*, 167 (1971).

If *YH* = hydrazine, *thiosemicarbazides*[958] are formed. With substituted hydrazines 2 isomers are formed.

When the reaction is carried out in polar solvents in the cold the thiosemicarbazides **39** are favored. In nonpolar solvents and at higher temperatures the thiosemicarbazide **38** predominates, since it is thermodynamically more stable than **39**[959].

Alcohols and isothiocyanates from *thiocarbamic esters (R¹O–CS–NH–R)*[960–963] in a reaction of wide latitude in respect of conditions and substituents; the reactivity of *meta* and *para* substituted arylisothiocyanates correlates in terms of the Hammett equation[964, 965].

While phenols and isothiocyanates frequently form thiocarboxamides[945], the reaction with thiophenols is a valuable method for the preparation of *dithiocarbamates (R¹S–CS–NH–R)*[966].

17.4.1.1.3 Addition of hydroxylamines

O,N-disubstituted hydroxylamines react readily with isothiocyanates to *N-hydroxythioureas*[967–970].

This is a potentially general reaction of wide scope in respect of conditions.

17.4.1.2 Reaction of thiocarbamoyl chlorides with aromatic compounds

N,N-disubstituted thiocarbamoyl chlorides react with aromatic compounds and aluminum chloride or tin(IV) chloride to give thiocarboxamides in good yield[971].

17.4.2 Thioacylation

17.4.2.1 With thiocarboxylic acids

17.4.2.1.1 With dithiocarboxylic acids

Dithiocarboxylic acids as well as their salts are suitable for the transfer of the thioacyl group to amino groups. The yield of *thioamide hydrazides*[972–974] *(R¹=NR₂, NH–R)*[973, 975, 976] and *thio-*

$R^1 = R^3, NH-R^4, N(R^4)_2, OH$

hydroxyamic acids (R¹=OH)[977, 978] vary and are below 50% in most cases, due to *side-reactions*, of which many types arise in the reaction with hydrazines[976, 979]. Nevertheless, the reaction is of preparative interest, because the dithiocarboxamides or their salts are easily prepared *in situ*, and need not to be isolated for carrying out the reaction[980].

[958] *G. Pulvermacher*, Ber. 27, 613 (1894).

[959] *M. Busch*, Ber. 42, 4596 (1909).

[960] *A.W. Hofmann*, Ber. 3, 770 (1870).

[961] *W. Schneider, D. Clibbens, G. Hüllweck, W. Steibelt*, Ber. 47, 1248 (1914).

[962] *H.P. Kaufmann, K. Lüthje*, Arch. Pharm. (Weinheim, Ger.) 293, 150 (1960).

[963] *D. Göckeritz, R. Pohloudek Fabini*, Pharmazie 17, 679 (1962).

[964] *D. Vlachova, R. Zahradnik, K. Antos, P. Kristian, A. Hulka*, Collect. Czech. Chem. Commun. 27, 2826 (1962).

[965] *A.M. Kardos, J. Volke, P. Kristian*, Collect. Czech. Chem. Commun. 30, 931 (1965).

[966] *W. Walter, K. Wohlers*, Justus Liebigs Ann. Chem. 752, 115 (1971).

[967] *B. Hirsch*, J. Prakt. Chem. [4] 12, 264 (1961).

[968] *L. Capuano, W. Ebner, J. Schrepfer*, Chem. Ber. 103, 82 (1970).

[969] *F. Grambal, J. Mollin, M. Hejsek*, Monatsh. Chem. 101, 120 (1970).

[970] *G. Zinner, R.O. Weber*, Pharmazie 21, 23 (1966).

[971] *H. Viola, S. Scheithauer, R. Mayer*, Chem. Ber. 101, 3517 (1968).

[972] *H. Wuyts, M. Goldstein*, Bull. Soc. Chim. Belg. 40, 497 (1931).

[973] *H.B. König, W. Siefken, A.H. Offe*, Chem. Ber. 87, 825 (1954).

[974] *G. Bähr*, Angew. Chem. 73, 628 (1961).

17.4.2.1.2 Thioacylation with dithiocarbamic acids

Because of the instability of the dithiocarbamates their salts **40**, which can be prepared in aqueous or alcoholic solution, are used almost exclusively for thioacylation[981, 982]:

$$R-NH_2 \ + \ CS_2 \ + \ NaOH \ \longrightarrow$$

$$\left[R-NH-\underset{\underset{S}{\|}}{C}-S^{\ominus} \right] Na^{\oplus} \ + \ H_2O$$

40

$$2\ R-NH_2 \ + \ CS_2 \ \longrightarrow \ \left[R-NH-\underset{\underset{S}{\|}}{C}-S^{\ominus} \right] R-\overset{\oplus}{N}H_3$$

40

On heating, the salts form *1,3-disubstituted thioureas via* isothiocyanates[983, 984]:

$$\left[R-NH-\underset{\underset{S}{\|}}{C}-S^{\ominus} \right] Na^{\oplus}(R-\overset{\oplus}{N}H_3) \ \xrightarrow{-H_2S} \qquad \mathbf{2}$$

$$R-N=C=S \ + \ R-NH_2 \ \longrightarrow \ S=C\overset{\displaystyle NH-R}{\underset{\displaystyle NH-R}{<}}$$

The second step of equation **2** is thus a special case of equation **1** (p. 753).

Often, it may be preferable to prepare first the isothiocyanate by heating the salts **40** with a heavy metal salt and to continue according[985] to equation **1** (p. 753).

By the action of mild oxidants the salts of secondary dithiocarbonic acid amides can be converted to *bis(aminothiocarbonyl) disulfides*[986]; *bis(dimethylaminothiocarbonyl)* and *bis(piperidinothiocarbonyl) disulfide* decompose to *tetramethylthiourea* or *thiocarbonic acid dipiperidide* on heating in vacuum[987].

$$\left[R_2N-\underset{\underset{S}{\|}}{C}-S^{\ominus} \right] Na^{\oplus} \ + \ J_2 \ \xrightarrow{-2\,NaJ}$$

$$\underset{\underset{S}{\|}}{\overset{R_2N}{C}}-S-S-\underset{\underset{S}{\|}}{\overset{NR_2}{C}} \ \xrightarrow[{[R = CH_3 ; -(CH_2)_5-]}]{\nabla;\ i.\,Vacuum}$$

$$\underset{\underset{S}{\|}}{\overset{R_2N}{C}}-NR_2 \ + \ S \ + \ CS_2$$

Hydrazine derivatives react analogously to give *thiocarbohydrazides* **41**[988] or *thiosemicarbazides* **42**[989].

$$\left[Ar-NH-NH-\underset{\underset{S}{\|}}{C}-S^{\ominus} \right] \overset{\oplus}{H_3}N-NH-Ar \ \longrightarrow$$

$$Ar-NH-NH-\underset{\underset{S}{\|}}{C}-NH-NH-Ar$$

41

$$\left[Ar-NH-\underset{\underset{S}{\|}}{C}-S^{\ominus} \right] NH_4^{\oplus} \ \xrightarrow{H_2N-NH_2} \ Ar-NH-\underset{\underset{S}{\|}}{C}-NH-NH_2$$

42

The valuable insertion reaction of carbon disulfide into aminals or Mannich bases for the preparation of *dithiocarbamic acid esters*[990–992] proceeds *via*

[975] *H. Wuyts*, Bull. Soc. Chim. Belg. *46*, 27 (1937).

[976] *K.A. Jensen, C. Pedersen*, Acta Chem. Scand. *15*, 1124 (1961).

[977] *L. Cambi*, Gazz. Chim. Ital. *41*, 166 (1911).

[978] *M.G. Ettlinger, A.J. Lundeen*, J. Amer. Chem. Soc. *79*, 1764 (1957).

[979] *W. Walter, J. Voss* in *J. Zabicky*, The Chemistry of Amides, p. 383, Wiley Intersci. Publ., New York, London, Sydney, Toronto 1970.

[980] *W. Walter, M. Radke*, Justus Liebigs Ann. Chem. *739*, 201 (1970); 636 (1973).

[981] *R. Anschütz*, Justus Liebigs Ann. Chem. *359*, 202 (1908).

[982] USP. 2 693 485 (1954), Du Pont; Inv.: *R.J. Gobeil*; C.A. *49*, 12533 d (1955).

[983] *N.P. Buu-Hoi, N.D. Xuong*, Compt. Rend. *237*, 498 (1953).

[984] *L.C. Raiford, G.M. Mc Nulty*, J. Amer. Chem. Soc. *56*, 680 (1934).

[985] *A.H. Blatt*, Orig. Synth. Coll. Vol. I, 2. Edition, 447 (1948).

[986] *J. v. Braun*, Ber. *35*, 3368 (1902).

[987] *J. v. Braun, K. Weissbach*, Ber. *63*, 2846 (1930).

[988] *E. Fischer*, Justus Liebigs Ann. Chem. *190*, 67, 118 (1882).

[989] *Houben-Weyl*, Methoden der organischen Chemie, 4. Aufl., Bd. IX, p. 910, Georg Thieme Verlag, Stuttgart 1955.

[990] *N. Kreutzkamp, H.Y. Oei*, Arch. Pharm. (Weinheim, Ger.) *299*, 906 (1966).

[991] *H.Y. Oei*, Dissertation, Universität Hamburg (1966).

[992] *A.O. Fitton, A. Rigby, R.J. Hurlock*, J. Chem. Soc. C 230 (1969).

intermediates having the structure of *N*-substituted *inner dithiocarbamate salts*

$$R_2N-CH_2-NR_2 \;+\; CS_2 \longrightarrow$$

$$\underset{S}{\overset{\oplus}{R_2N}}\!\!\begin{array}{c}CH_2-NR_2\\ |\\ N\\ |\\ C-S^{\ominus}\end{array}$$

$$\longrightarrow \quad \underset{S}{R_2N}\!\!>\!C-S-CH_2-NR_2$$

The postulated addition of the morpholinium salt of 4-morpholinecarbodithioic acid to a polarized double bond is well supported by the reaction of acrylonitrile, which is of preparative value[993].

$$H_5C_6-CO-CH_2-CH_2-N\!\!<\!\!\bigcirc O \quad \xrightarrow[-\;O\!\!<\!\!\bigcirc NH]{}$$

$$H_5C_6-CO-CH=CH_2 \quad \xrightarrow{\big[O\!\!<\!\!\bigcirc N-CS-S^{\ominus}\big]\,H_2\overset{\oplus}{N}\!\!<\!\!\bigcirc O}$$

$$H_5C_6-CO-CH_2-CH_2-S-\overset{S}{\overset{\|}{C}}-N\!\!<\!\!\bigcirc O$$

3-Hydroxypropiophenone 4-morpholinecarbodithioate; 58%

$$H_2C=CH-CN \;+\; \big[H_5C_6-CH_2-NH-\overset{S}{\overset{\|}{C}}-S^{\ominus}\big]Na^{\oplus}$$

$$\longrightarrow \quad H_5C_6-CH_2-NH-\overset{S}{\overset{\|}{C}}-S-CH_2-CH_2-CN$$

2-Cyanoethyl benzyl-dithiocarbamate; 83% (R = C₆H₅—CH₂—)

17.4.2.2 Thioacylation with thioacid halides

Of the thioacid halides thiobenzoyl chloride alone has gained importance for the synthesis of certain[980] thiocarboxamides.

$$R-\overset{Cl}{\underset{S}{\overset{|}{C}}} \;+\; H_2N-R^1 \longrightarrow R-\overset{NH-R^1}{\underset{S}{\overset{|}{C}}}$$

In contrast, thiophosgene is important for the preparation of *thiocarbamoyl halides*.

It is advisable to purify thiophosgene by distillation immediately before the reaction.

17.4.2.2.1 Thioacylation with thiocarbamoyl halides

Only *N,N-disubstituted thiocarbamoyl halides* are of interest for synthetic purposes. They may be prepared from secondary amines or amine

hydrochlorides and thiophosgene or by treatment of bis-(aminothiocarbonyl) disulfides with chlorine.

$$Cl-\overset{S}{\underset{Cl}{\overset{\|}{C}}} \;+\; R_2NH$$

$$R_2N-\overset{S}{\overset{\|}{C}}-S-S-\overset{S}{\overset{\|}{C}}-NR_2 \;+\; Cl_2 \quad\Bigg] \longrightarrow R_2N-\overset{S}{\underset{Cl}{\overset{\|}{C}}}$$

Thiocarbamoyl *bromides* and *chlorides* are obtained readily by halogenating the corresponding thioformamides[951, 994].

$$R_2N-\overset{H}{\underset{S}{\overset{|}{C}}} \quad \xrightarrow{X_2} \quad R_2N-\overset{X}{\underset{S}{\overset{|}{C}}} \;+\; HX$$

X = Cl, Br

By using an excess of primary or secondary amine, the reaction may be carried beyond the thiocarboxamide stage in one step, and *thioureas* are obtained[995, 996]. Because of the high cost and unpleasant properties of thiophosgene other thiourea syntheses are preferable (p. 755), but it is almost irreplaceable for the synthesis of aryl-thioureas with strongly electron-attracting substituents[997]. Alcoholates and phenolates react with thiocarbamoyl chlorides to give good yields of thiocarbamic acid esters[998–1000], while *dithiocarbamic acid esters (dithiourethans)* are obtained from thiolates[1001].

$$R_2N-\overset{S}{\underset{Cl}{\overset{/\!\!/}{C}}} \;+\; \big[R^1\!-\!X^{\ominus}\big]Na^{\oplus} \longrightarrow R_2N-\overset{S}{\underset{X-R^1}{\overset{/\!\!/}{C}}}$$

X = O, S

Depending on *R*, *thiocarbamic acid S-esters* can be formed instead of thiocarbamic acid *O*-esters by rearrangement from thiocarbamoyl chloride[1002].

[993] *R. Delaby, R. Damiens, R. Seyden-Penne*, Bull. Soc. Chim. France *26*, 190 (1959).

[994] *W. Walter, R.F. Becker*, Justus Liebigs Ann. Chem. *733*, 195 (1970).

[995] *G.M. Dyson, H.J. George, R.F. Hunter*, J. Chem. Soc. (London) 3041 (1926); *G.M. Dyson, H.J. George, R.F. Hunter*, J. Chem. Soc. (London) 436 (1927).

[996] *J.S. Morley, J.C.E. Simpson*, J. Chem. Soc. (London) 2617 (1952).

[997] *G.M. Dyson, H.J. George*, J. Chem. Soc. (London) *125*, 1702 (1924).

[998] *W. Walter, K.D. Bode*, Justus Liebigs Ann. Chem. *681*, 64 (1965).

[999] Niederl. P. Anm. 287 549 (1965), Japan Soda Co., C.A. *63*, 4227b (1965).

17.4.2.2.2 Thioacylation with thiocarbonic acid O-aryl and S-aryl esters

Under mild conditions (a 2-phase system of chloroform and aqueous alkali[998, 1003]) thiophosgene is reacted with phenols[998, 1003, 1004] or thiols[1005] to form thiocarbonyl chloride O-aryl and S-aryl esters. These afford respectively *thiocarbamic acid* and *dithiocarbamic acid esters* with ammonia and with aliphatic or aromatic primary or secondary amines, often in yields over 80%[998, 967, 1004–1008].

$$Ar-X-\underset{\underset{S}{\|}}{C}-Cl \; + \; R-NH_2 \; \longrightarrow \; Ar-X-\underset{\underset{S}{\|}}{C}-NH-R$$

$$X = O, S$$

O,O-diesters of *thiocarbonic acid* ($R^1O-CS-OR^1$), diesters of *trithiocarbonic acid* ($R^1S-CS-SR^1$), *thioureas*, and *bis-(alkylthio)-thiocarbonylamines* ($R^1S-CS-NR-CS-SR^1$) can be formed especially in the presence of strong bases[1009, 1010].

17.4.2.3 Thioacylation with thiocarboxylic acid esters

In general, amines can be thioacylated according to the equation:

$$R^2-\underset{\underset{S}{\|}}{C}-X-R^1 \; + \; R-NH_2 \; \longrightarrow$$

$$R^2-\underset{\underset{S}{\|}}{C}-NH-R \; + \; R^1-XH$$

$$X = O, S$$

There are some important preparative procedures in this group.

17.4.2.3.1 Thioacylation with thiocarboxylic acid O-esters

On reacting primary amines with thiocarboxylic acid O-esters amidines are also always formed[1011].

$$R^2-\underset{\underset{S}{\|}}{C}-OC_2H_5 \; + \; 2 \, R-NH_2 \; \longrightarrow \; R^2-\underset{\underset{NR}{\|}}{C}-NH-R$$

By replacing the primary amine with its magnesium salt and working in tetrahydrofuran, this unwanted side-product (which naturally does not occur with secondary amines) can be avoided, and at the same time a higher reactivity is obtained[1011]. *O*-Ethyl thioformate[1012] is the reagent of choice for preparing *thioformamides* and *thiohydrazides*[1013–1016]. On reaction with *N*-methyl-*N*-phenylhydrazine some informative side-reactions have been discovered[979].

17.4.2.3.2 Thioacylation with dithiocarboxylic acid esters

Carboxymethyl esters of *dithiocarboxylic acids* are among the most important thioacylating agents, albeit their preparation requires some effort[1017, 1018]. With amines *thiocarboxamides* are formed.

$$\underset{S}{\overset{R^1}{\underset{\|}{C}}}-S-CH_2-COOH \; + \; R-NH_2 \; \longrightarrow$$

$$\underset{S}{\overset{R^1}{\underset{\|}{C}}}-NH-R \; + \; HS-CH_2-COOH$$

During thioacylation of amino acids to *N-thioacylamino* acids configuration at the center of chirality is retained[1019]. With hydrazines thiocarboxylic acid hydrazides are formed[1020–1022],

[1000] *M.S. Newman, H.A. Karnes*, J. Org. Chem. *31*, 3980 (1966).

[1001] *J.J. Ejtingon, N.P. Strelnikova*, Zh. Obshch. Khim. *32*, 3888 (1962); C.A. *58*, 12447c (1963).

[1002] *K.H. Risse, K. Hörlein, W. Wirth, R. Gösswald*, Medizin und Chemie, Bd. VII, p. 171, Verlag Chemie, Weinheim/Bergstr. 1963.

[1003] *P. Reich, D. Martin*, Chem. Ber. *98*, 2063 (1965).

[1004] *D.L. Garmaise, A. Uchiyama, A.F. McKay*, J. Org. Chem. *27*, 4509 (1962).

[1005] *A. Rieche, G. Hilgetag, D. Martin, J. Kreyzi*, Arch. Pharm. (Weinheim, Ger.) *296*, 310 (1963).

[1006] *W. Walter, K.D. Bode*, Justus Liebigs Ann. Chem. *698*, 122 (1966).

[1007] *A. Rieche, D. Martin, W. Schade*, Arch. Pharm. (Weinheim, Ger.) *296*, 770 (1963).

[1008] *A. Rieche, G. Hilgetag, A. Martini, R. Philippson*, Arch. Pharm. (Weinheim, Ger.) *294*, 201 (1961).

[1009] *K. Wohlers*, Dissertation, Universität Hamburg (1970).

[1010] *D.G. Crosby, C. Niemann*, J. Amer. Chem. Soc. *76*, 4458 (1954).

[1011] *P. Reynaud, R.C. Moreau, J.P. Samama*, Bull. Soc. Chim. France 3623, 3628 (1965).

[1012] *R. Mayer, H. Berthold*, Z. Chem. *3*, 310 (1963).

[1013] *R. Mayer, J. Orgis*, Z. Chem. *4*, 457 (1964).

[1014] *W. Walter, E. Schaumann, H. Paulsen*, Justus Liebigs Ann. Chem. *727*, 61 (1969).

[1015] *W. Walter, R.F. Becker*, Justus Liebigs Ann. Chem. *727*, 71 (1969).

[1016] *W. Walter, K.J. Reubke*, Angew. Chem. *79*, 381 (1967).

[1017] *F. Kurzer*, Chem. & Ind. (London) 1333 (1961).

[1018] *K.A. Jensen, C. Pedersen*, Acta Chem. Scand. *15*, 1087 (1961).

[1019] *G.C. Barrett*, J. Chem. Soc. (London) 2825 (1965).

[1020] *K.A. Jensen, C. Pedersen*, Acta Chem. Scand. *15*, 1097 (1961).

$$\underset{S}{\overset{R^1}{\underset{\parallel}{C}}}-S-CH_2-COOH \ + \ R-NH-NH_2 \ \longrightarrow$$

$$\underset{S}{\overset{R^1}{\underset{\parallel}{C}}}-NH-NH-R$$

and with monosubstituted hydrazines the result is a mixture of **43** and **44** (as *R* increases in size **44** is favored)[1023].

$$\underset{S}{\overset{R^1}{\underset{\parallel}{C}}}-S-CH_2-COOH \ + \ R-NH-NH_2 \ \longrightarrow$$

$$\underset{S}{\overset{R^1}{\underset{\parallel}{C}}}-\underset{R}{\overset{NH_2}{\underset{|}{N}}} \ + \ \underset{S}{\overset{R^1}{\underset{\parallel}{C}}}-NH-NH-R$$

$$\qquad\quad \textbf{43} \qquad\qquad \textbf{44}$$

When *N*, *O*-disubstituted hydroxylamines are used *O*, *N*-disubstituted thiohydroxamic acids are obtained[1024, 1025].

$$\underset{S}{\overset{R^1}{\underset{\parallel}{C}}}-S-CH_2-COOH \ + \ R^2-NH-OR \ \longrightarrow$$

$$\underset{S}{\overset{R^1}{\underset{\parallel}{C}}}-\underset{R^2}{\overset{OR}{\underset{|}{N}}}$$

17.4.2.3.3 Thioacylation with dithiocarbonic and trithiocarbonic acid esters

Diesters of dithiocarbonic acids, easily accessible by alkylation of xanthates, are synthetically useful starting materials for preparing aliphatic thiocarbamic esters[1026, 1027]

$$\underset{S}{\overset{R^1O}{\underset{\parallel}{C}}}-SR^2 \ + \ R-NH_2 \ \longrightarrow \ \underset{S}{\overset{R^1O}{\underset{\parallel}{C}}}-NH-R^2$$

Diesters of trithiocarbonic acids are well suited for thioacylating secondary amines[1028],

$$\underset{S}{\overset{R^1S}{\underset{\parallel}{C}}}-SR^2 \ + \ R_2NH \ \longrightarrow \ \underset{S}{\overset{R^1-S}{\underset{\parallel}{C}}}-NR^2$$

but the resulting *dithiocarbamic acid esters* are obtained more easily by alkylating metal salts of dithiocarbamic acid (p. 764).

17.4.2.3.4 Thioacylation with dithiocarbamic acid esters and dithiocarbonic acid hydrazide esters

While for preparing thioureas other thioacylating agents are used preferentially, *thiosemicarbazides* may be advantageously synthesized from dithiocarbamic acid esters[1029] or esters of dithiocarbonic acid hydrazides[1030].

$$Ar-NH-\underset{S}{\overset{S-R}{\underset{\parallel}{C}}} \ + \ H_2N-NH_2 \ \longrightarrow$$

$$Ar-NH-\underset{S}{\overset{NH-NH_2}{\underset{\parallel}{C}}}$$

$$H_3C-S-\underset{S}{\overset{NH-NH_2}{\underset{\parallel}{C}}} \ + \ R_2NH \ \longrightarrow$$

$$R_2N-\underset{S}{\overset{NH-NH_2}{\underset{\parallel}{C}}}$$

17.4.2.4 Thioacylation with thioketenes

Thioketenes are considered to be an intermediate in thioacylating amines with 1-(acetylthio)-1-alkynes[1029, 1030].

$$R^1-C\equiv C-S-CO-CH_3 \ \xrightarrow{R_2NH} \ R^1-CH=C=S$$

$$\xrightarrow{R_2NH} \ R^1-CH_2-\underset{S}{\overset{NR_2}{\underset{\parallel}{C}}}$$

$$\begin{array}{c} (H_3C)_3C-Br \\ R_2NH \end{array} \Bigg\uparrow \begin{array}{c} -\ LiBr \\ -\ CH_2=C(CH_3)_2 \end{array}$$

$$R^1-C\equiv C-Li \ \xrightarrow{S_8} \ R^1-C\equiv C-S-Li$$

[1021] *P.A.S. Smith, D.H. Kenny,* J. Org. Chem. *26*, 5221 (1961).

[1022] *E. Lieber, C.N.R. Rao, R.C. Orlowski,* Can. J. Chem. *41*, 926 (1963).

[1023] *K.-J. Reubke,* Dissertation, Universität Hamburg (1970).

[1024] *W. Walter, J. Curts, H. Pawelzik,* Justus Liebigs Ann. Chem. *643*, 29 (1961).

[1025] *W. Walter, E. Schaumann,* Justus Liebigs Ann. Chem. *743*, 154 (1971).

[1026] *M. Delépine, P. Schving,* Bull. Soc. Chim. France [4] *7*, 894 (1910).

[1027] *B. Holmberg,* Ber. *59*, 1558 (1926).

[1028] *L. Field, H.K. Kim,* J. Org. Chem. *31*, 597 (1966).

[1029] *M.G. Paranjpe, P.H. Deshpande,* Indian J. Chem. *7*, 186 (1969).

[1030] *R.S. McElhinney,* J. Chem. Soc. C 950 (1966).

17.4.2.5 Thioacylation with thiocarboxamides

The amide groups of thioamides may be exchanged in an acid-catalyzed reaction[1032–1034], e.g.:

$$H_2N-\underset{\underset{S}{\parallel}}{C}-\underset{\underset{S}{\parallel}}{C}-NH_2 \quad + \quad 2\ RNH_2 \quad \longrightarrow$$

$$R-NH-\underset{\underset{S}{\parallel}}{C}-\underset{\underset{S}{\parallel}}{C}-NH-R \quad + \quad 2\ NH_3$$

N,N'-Dialkyldithiooxamide; 50–60%

By means of the azolide method thiocarboxamides can be so activated that they act as thioacylating agent without an acid catalyst[980].

$$R^1-\underset{\underset{S}{\parallel}}{C}-N\overset{\frown}{\underset{\smile}{N}}$$

$$\xrightarrow{\ R-NH_2\ } \quad R^1-\underset{\underset{S}{\parallel}}{C}-NH-R \quad + \quad HN\overset{\frown}{\underset{\smile}{N}}$$

Thiocarboxylic acid amide

$$\xrightarrow{\ H_2N-NR_2\ } \quad R^1-\underset{\underset{S}{\parallel}}{C}-NH-NR_2$$

Thiocarboxylic acid hydrazide

17.4.2.6 Thioacylation with thioureas

1,3-Diarylthioureas are capable of analogous exchange reactions[1035, 1036].

$$Ar-NH-\underset{\underset{S}{\parallel}}{C}-NH-Ar \quad + \quad Ar^1-NH_2 \quad \longrightarrow$$

$$Ar-NH-\underset{\underset{S}{\parallel}}{C}-NH-Ar^1 \quad + \quad Ar-NH_2$$

This property can be used for the preparation of *1,4-diarylthiosemicarbazides*[1037, 1038].

$$Ar-NH-\underset{\underset{S}{\parallel}}{C}-NH-Ar \quad + \quad Ar^1-NH-NH_2 \quad \longrightarrow$$

$$Ar-NH-\underset{\underset{S}{\parallel}}{C}-NH-NH-Ar^1$$

17.4.3 Rearrangement of ammonium thiocyanates

When ammonium thiocyanates of primary and secondary amines are heated in water or chlorobenzene, *N-mono-* or *N,N-disubstituted* thioureas are obtained[1039–1042].

$$R_2NH \quad + \quad HCl \quad + \quad \left[NH_4^{\oplus}\right]SCN^{\ominus} \quad \longrightarrow$$

$$\left[R_2NH_2^{\oplus}\right]SCN^{\ominus} \quad \longrightarrow \quad R_2N-\underset{\underset{S}{\parallel}}{C}-NH_2$$

The usefulness of the method is limited by the formation of 1,3-disubstituted thioureas as by-products[1043–1044].

17.4.4 Thiolysis of carboxylic acid derivatives

17.4.4.1 Of imidoyl halides

17.4.4.1.1 Of imidoyl chlorides

The reaction of imidoyl chlorides with hydrogen sulfide is a long-standing method for preparing thiocarboxamides[1045–1050] (with thiols thioformidic acid S-esters are formed, cf. p. 764)[1049, 1050]:

$$R^1-\underset{\underset{Cl}{\mid}}{\overset{\overset{N-R}{\parallel}}{C}} \quad + \quad H_2S \quad \longrightarrow \quad R^1-\underset{\underset{S}{\parallel}}{C}-NH-R$$

[1031] *E.U. Elam, F.H. Rash, J.T. Dougherty, V.W. Goodlett, K.C. Brannock,* J. Org. Chem. *33,* 2738 (1968).

[1032] *M.J. Schlatter,* J. Amer. Chem. Soc. *64,* 2722 (1942).

[1033] USP. 3 354 156 (1967), American Home Products; Inv.: *G.R. Wendt, E. Hertz;* C.A. *68,* 114638 (1968).

[1034] *Y. Otsuji, N. Matsumura, E. Imoto,* Bull. Chem. Soc. Japan *41,* 1485 (1968).

[1035] *R. v. Walther, A. Stenz,* J. Prakt. Chem. [2] *74,* 226 (1906).

[1036] *G. Vasilev,* Farmatsiya [4],*16* [4], *20 (1966); C.A. 66,* 65205 c (1967).

[1037] *L.C. Ralford, W.T. Daddow,* J. Amer. Chem. Soc. *53,* 1556 (1931).

[1038] *U. Anthoni, C. Larsen, P.H. Nielsen,* Acta Chem. Scand. *22,* 1898 (1968).

[1039] *P. de Clermont, E. Wehrlin,* Ber. *9,* 446 (1876); *P. de Clermont, E. Wehrlin,* Ber. *10,* 493 (1877).

[1040] *H. Passing,* J. Prakt. Chem. [2] *153,* 13 (1939).

[1041] *F. Kurzer,* Org. Synth. *31,* 21 (1951).

[1042] *W. König, W. Kleist, J. Götze,* Ber. *64,* 1664 (1931).

[1043] *A. Bertram,* Ber. *25,* 48 (1892).

[1044] *F. Hüter,* Chem. Ber. *80,* 275 (1947).

[1045] *A. Bernthsen,* Ber. *10,* 1238 (1877).

[1046] *H. Leo,* Ber. *10,* 2133 (1877).

[1047] *V.L. Dubina, S.J. Burmistrov,* Zh. Org. Khim. *2,* 1845 (1966); C.A. *66,* 55172 (1967).

[1048] *W. Walter, K.D. Bode,* Angew. Chem. *74,* 694 (1962).

[1049] *A.W. Chapman,* J. Chem. Soc. (London) 2296 (1926).

[1050] *W. Walter, J. Voss, J. Curts,* Justus Liebigs Ann. Chem. *695,* 77 (1966).

Disubstituted amides and phosgene form imidium chlorides[1051], which can be thiolyzed to *N,N*-disubstituted *thiocarboxamides*[1051, 1052].

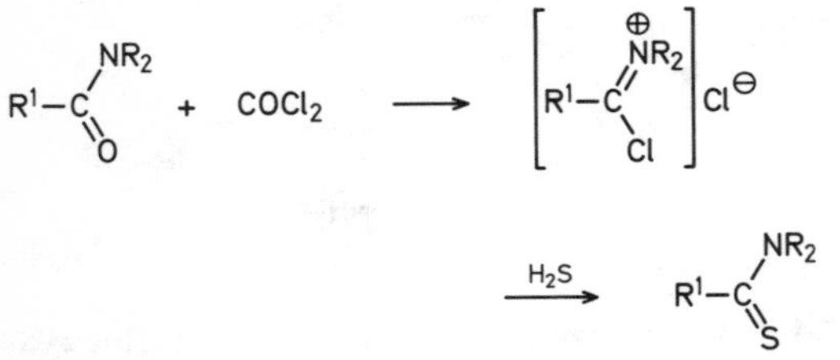

17.4.4.1.2 Thiolysis of chloroformamidines

Available from isocyanide dichlorides[1053, 1054], these are an excellent starting material for preparation of *N*-functional derivatives of thiocarbonic acids. The preparative development is in full swing: the following examples are given as an illustration:

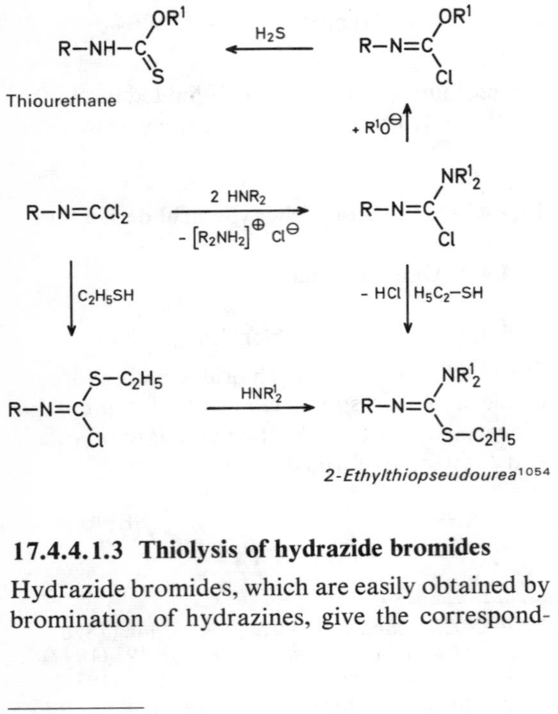

Thiourethane

2-Ethylthiopseudourea[1054]

17.4.4.1.3 Thiolysis of hydrazide bromides

Hydrazide bromides, which are easily obtained by bromination of hydrazines, give the corresponding *thiocarboxylic acid hydrazides* with hydrogen sulfide[1023, 1055–1057].

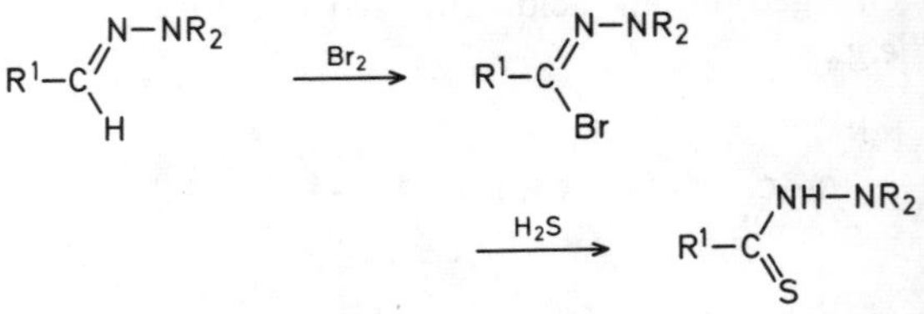

The hydrazide bromides may be employed also in the *in situ* prepared form (for the structure of the thiocarboxylic acid hydrazides *cf.* ref. [1058]).

17.4.4.2 Thiolysis of imidic acid esters, amidines, and amidrazones

When imidates and amidines are synthesized from nitriles, their thiolysis is preparatively important if superior yields are obtained than from thiolysis of the nitriles themselves. This is the case for the preparation of aliphatic *thiocarboxamides* from amidines[1011].

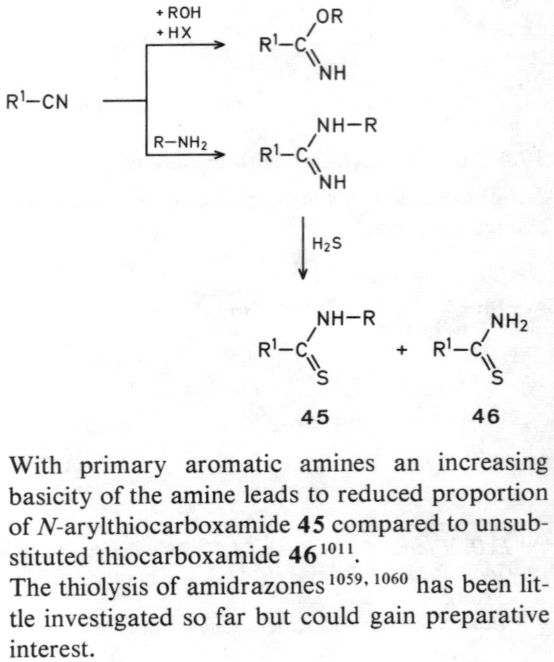

45   46

With primary aromatic amines an increasing basicity of the amine leads to reduced proportion of *N*-arylthiocarboxamide **45** compared to unsubstituted thiocarboxamide **46**[1011].

The thiolysis of amidrazones[1059, 1060] has been little investigated so far but could gain preparative interest.

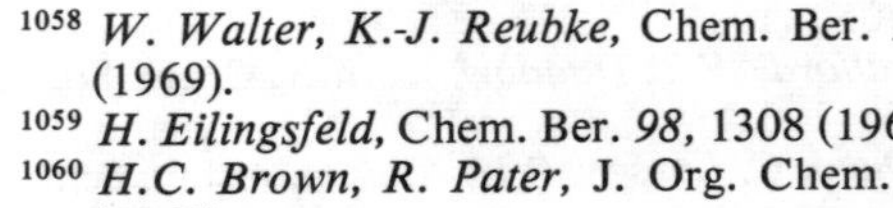

*Thiocarboxylic acid
hydrazides*

[1051] *H. Eilingsfeld, M. Seefelder, H. Weidinger*, Chem. Ber. *96*, 2671 (1963).

[1052] *A.J. Speziale, L.R. Smith*, J. Org. Chem. *28*, 3492 (1963).

[1053] *E. Kühle, B. Anders, G. Zumach*, Angew. Chem. *79*, 663 (1967).

[1054] *E. Kühle, B. Anders, E. Klauke, H. Tarnow, G. Zumach*, Angew. Chem. *81*, 18 (1969).

[1055] *A.F. Hegarty, F.L. Scott*, J. Org. Chem. *33*, 753 (1967).

[1056] *F.L. Scott, F.A. Groeger, A.F. Hegarty*, Tetrahedron Lett. 2463 (1968).

[1057] *W. Walter, K.-J. Reubke*, Tetrahedron Lett. 5973 (1968).

[1058] *W. Walter, K.-J. Reubke*, Chem. Ber. *102*, 2117 (1969).

[1059] *H. Eilingsfeld*, Chem. Ber. *98*, 1308 (1965).

[1060] *H.C. Brown, R. Pater*, J. Org. Chem. *30*, 3739 (1965).

Formazans can be thiolyzed analogously[1061, 1062]

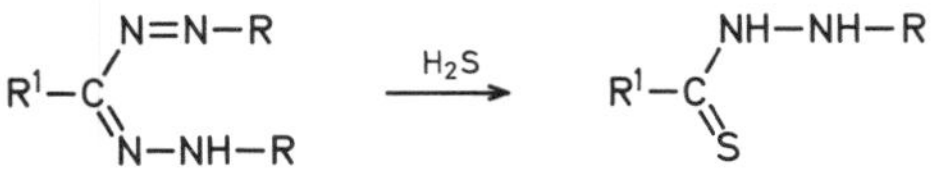

17.4.4.3 Thioamidation

Thioamidation is used here to denote the replacement of the oxygen of the amide group by sulfur. In special cases this substitution can be performed by sulfuration with elemental sulfur[1063], *e.g.:*

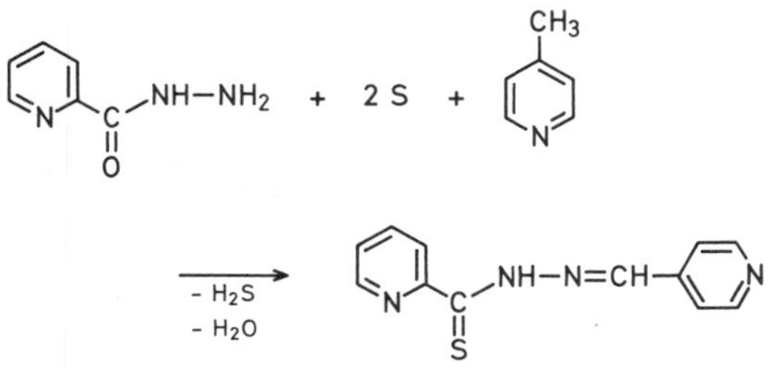

Isonicotinaldehyde thionicotinylhydrazone; 23%

17.4.4.3.1 Reaction with amides

The most important reagent for preparing thiocarboxamides is *phosphorus(V) sulfide* (P_4S_{10}):

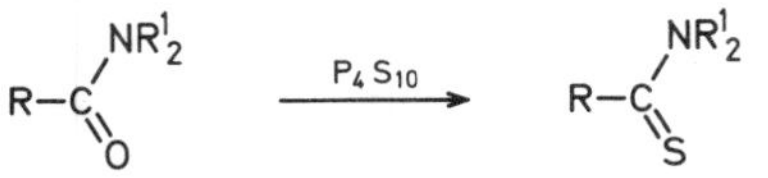

If the amides are stable to heat, the substituents R and R^1 can be varied nearly without limit. In general *N-monosubstituted* amides give the best yields. With *unsubstituted* amides further reaction to nitrile must be anticipated.

The reaction is suitable for the *thioamidation* of lactams also[1064], here, carbon disulfide and thiocarboxylic acids, too, are effective[1065, 1066]. For the reaction, which most often is carried out in a heterogenous system, *solvents* are employed

whose polarity can be varied between aromatic hydrocarbons and pyridine[1067–1070]. The good yields in the latter[1071, 1072] and in diethyleneglycoldiethyl ether[1069] [bis-(2-ethoxyethyl) ether] deserve to be mentioned. In xylene it was possible to thioamidate optically active amides while retaining the configuration[1073]. With pyridine as a solvent it is possible to thionate amides prepared *in situ* from acid chlorides and amines in one step.

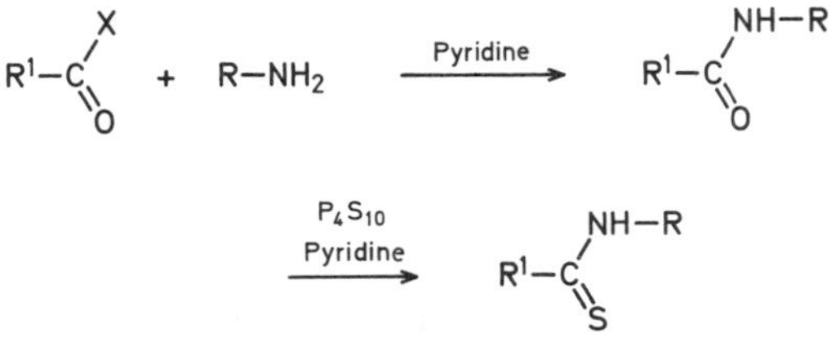

Often the yields are better than in the normal two-step reactions[1074].

Addition of alkali in different stages of the reaction can increase the yield[1075, 1076].

Normally, the ester group will not be attacked by P_4S_{10} but hydroxy groups need to be protected by acylation prior to thioamidation[1077].

17.4.4.3.2 Reaction with hydrazides

Converting hydrazides into *thiocarboxylic acid hydrazides* with P_4S_{10} is much less important[1020, 1078]. By using dichloromethane as solvent and adding triethylamine the results obtainable can be improved[1023], but the method remains far from efficient. On the other hand, the technique is

[1061] *J. Ya. Postowskii, M.J. Ermakowa*, Zh. Obshch. Khim. *29*, 1333 (1959); C.A. *54*, 8796 c (1960).

[1062] *G. Zemplén, L. Mester, A. Messmer*, Chem. Ber. *86*, 697 (1953).

[1063] USP. 2 774 757 (1956), Farbenfabriken Bayer, Inv.: *E. Kühle, R. Wegler;* C.A. *51*, 6705 (1957).

[1064] *H. Schnell, J. Nentwig* in *Houben-Weyl*, Methoden der organischen Chemie, 4. Aufl., Bd. XI/2, p. 574, Georg Thieme Verlag, Stuttgart 1958.

[1065] USP. 2 539 325 (1951), Bata Narodi Podnik; Inv.: *J. Procházka;* C.A. *45*, 4740a (1951).

[1066] DBP. 943 227(1956), BASF, Erf.: *H. Behringer, G.F. Grundwald;* C.A. *53*, 6262 a (1959).

[1067] *L.R. Cerecedo, J.G. Tolpin*, J. Amer. Chem. Soc. *59*, 1660 (1937).

[1068] *S. Gabriel*, Ber. *49*, 1110 (1956).

[1069] *H. Behringer, M. Ruff, R. Wiedenmann*, Chem. Ber. *97*, 1732 (1964).

[1070] *E. Klingsberg, D. Papa*, J. Amer. Chem. Soc. *73*, 4988 (1951).

[1071] *N. Pravdic, V. Hahn*, Croat. Chem. Acta *34*, 85 (1962); C.A. *57*, 12377 (1962).

[1072] *W. Walter, K.-D. Bode*, Justus Liebigs Ann. Chem. *660*, 74 (1962).

[1073] *J.V. Burakevich, C. Djerassi*, J. Amer. Chem. Soc. *87*, 51 (1965).

[1074] *J. Voss, W. Walter*, Justus Liebigs Ann. Chem. *716*, 209 (1968).

[1075] *K. Kindler*, Justus Liebigs Ann. Chem. *431*, 187 (1923).

[1076] *L. Ruzicka, M.W. Goldberg, M. Hürbin, H.A. Boekenoogen*, Helv. Chim. Acta *16*, 1323 (1933).

[1077] *G. Wagner, D. Singer*, Z. Chem. *2*, 306 (1962).

[1078] *H. Bredereck, B. Föhlisch, K. Walz*, Justus Liebigs Ann. Chem. *688*, 93 (1965).

well suited for thioamidating cyclic hydrazides[1079, 1080].

3-Pyridazinethiol; 51%

17.4.4.3.3 Thioamidation of ureas

While cyclic ureas with 6–12 members can be thioamidated to thioureas in yields of 60–70%[1081], the method is limited narrowly with open-chain ureas. Tetrasubstituted ureas alone react to thioureas in good yield, and even with tri-substituted ureas side-reactions are so marked that the technique loses its preparative value[1082].

17.4.5 Thiolysis and ammonolysis of halo hydrocarbons

The simultaneous action of chloroform and hydrogen sulfide on primary or secondary amines affords *thioformamides* in good yield. Probably the reaction proceeds *via* dichlorocarbene as intermediate[1083].

At elevated temperature the thioacyl carbon can be furnished by carbon tetrachloride and other halo hydrocarbons; thus, for example, *dithiooxamides* are formed from hexachloroethane[1084–1086].

17.4.6 Thiolysis of the nitrile group

Addition of hydrogen sulfide to nitriles is a valuable method especially for preparing *unsubstituted thiocarboxamides*.

The reaction may be catalyzed by acids or bases.

17.4.6.1 Base-catalyzed thiolysis

Metal hydroxides, basic ion exchangers[1087], alkoxides, ammonia, and organic bases can be used. Triethylamine and pyridine are especially suitable; the latter can be used in excess to serve as solvent[1088, 1089]. From *aromatic* and *heteroaromatic* nitriles the thiocarboxamides are obtained in yields of between 50 and 90%. With *aliphatic* nitriles it is necessary to employ strong bases such as diethylamine, quaternary ammonium bases, or tetraalkylguanidines in order to produce amide yields of 19–50%. In these cases it is advisable to use dimethylformamide, dimethyl sulfoxide, or sulfolane[1090], but occasionally difficulties are encountered during working up.

Yields of 50–60% are obtained when the thiolysis is conducted at 50° under pressure in alcohol saturated with hydrogen sulfide in the cold and using catalytic amounts of alkoxide[1075].

Nitriles containing a chiralic center adjacent to the nitrile group and having an attached hydrogen atom are totally racemized[1073].

Thioformamides are obtained in good yield by reaction of ammonium cyanide, amines, and hydrogen sulfide[1091, 1092].

With ammonia *thioformamide* is formed in 70% yield[1093].

[1079] *G.F. Duffin, J.D. Kendall*, J. Chem. Soc. (London) 3789 (1959).

[1080] *S. Kakimoto, S. Tonooka*, Bull. Chem. Soc. Japan *40*, 153 (1967).

[1081] *H. Behringer, H. Meier*, Justus Liebigs Ann. Chem. *607*, 67 (1957).

[1082] *J. Voss*, Justus Liebigs Ann. Chem. *746*, 92 (1971).

[1083] *W. Walter, G. Maerten*, Justus Liebigs Ann. Chem. *669*, 66 (1963).

[1084] DBP. 1 227 452 (1966), Dunlop Rubber Co., Erf.: *R.T. Wragg*; C.A. *66*, 10599 (1967).

[1085] DBP. 1 227 451 (1966), Dunlop Rubber Co., Erf.: *R.T. Wragg;*, C.A. *66*, 10601 (1967).

[1086] *F. Becke, H. Hagen*, Chemiker-Ztg. *93*, 474 (1969).

[1087] DDR. P. 50 830 (1966), Erf.: *H.D. Eilhauer, G. Reckling*; C.A. *66*, 55394 (1967).

[1088] *A.E.S. Fairfull, J.L. Lowe, D.A. Peak*, J. Chem. Soc. (London) 742 (1952).

[1089] *H. Möller*, Justus Liebigs Ann. Chem. *749*, 1 (1971).

[1090] *E.E. Gilbert, E.J. Rumanowski, P.E. Newallis*, J. Chem. Eng. Data *13*, 130 (1968).

[1091] *M. Seefelder*, Chem. Ber. *99*, 2678 (1966).

[1092] *P.L. de Benneville, J.S. Strong, V.T. Elkind*, J. Org. Chem. *21*, 772 (1956).

[1093] *R. Tull, L.M. Weinstock*, Angew. Chem. *81*, 291 (1969).

[1094] *E.C. Taylor, J.A. Zoltewicz*, J. Amer. Chem. Soc. *82*, 2656 (1960).

17.4.6.2 Acid-catalyzed thiolysis

The acid-catalyzed thiolysis (*cf.* p. 760) is of value where a sulfur compound other than hydrogen sulfide is used for the thiolysis[1094].

The addition of *O*, *O*-diethyl dithiophosphate, easily obtainable from ethanol and phosphorus(V) sulfide, to nitriles is synthetically significant; the addition product is cleaved to the *thiocarboxamide* by hydrogen chloride[1095].

17.4.6.3 Thiolysis of cyanamides and cyanic acid esters

Addition of hydrogen sulfide to cyanamide is an efficient method for preparing *N, N-disubstituted thioureas*[1096]; with thiols *2-alkylpseudothioureas* are formed[1097, 1098].

The improvements to the method mentioned on p. 762 are applicable to the thiolysis of cyanamides. Thiourea in acid solution once more serves as a

source of hydrogen sulfide[1099]. Under analogous conditions *thiocarbamic acid esters* are obtained from cyanates in good yield[1072, 1100, 1101].

17.4.6.4 Thiolysis of nitrile oxides

From hydroxamoyl chlorides nitrile oxides are formed so rapidly that thiolysis of the former under normal conditions proceeds *via* nitrile oxides[1102].

With hydrogen sulfide nitrile oxides readily for *thiohydroxamic acids*, with thiols *S-aryl* or *S-alkyl thiohydroximic acids*[1103].

17.4.7 Rearrangement of ketoxime benzenesulfonate esters

Ketoximes can be sulfonated with benzenesulfonyl chloride in pyridine. Letting these sulfonic acid esters undergo spontaneous Beckmann rearrangement in the presence of hydrogen sulfide leads to the formation of sulfonic acid imidates, which are 'thiolyzed' to *thiocarboxamides* in 50–90% yield[1104]. In the presence of thiols *S-esters* of *thioimidic acids* are formed[1105].

[1095] DAS. 1 111 172 (1962), Farbenfabriken Bayer, Erf.: *H.-G. Schicke, G. Schrader;* C.A. *56*, 2474a (1962).

[1096] *O. Wallach,* Ber. *32*, 1872 (1899).

[1097] *B. Rathke,* Ber. *18*, 3102 (1885).

[1098] Brit. P. 296 782 (1927), Schering-Kahlbaum; C.A. *23*, 2447 (1929).

[1099] *H. Beyer, R. Giebelmann,* J. Prakt. Chem. [4] *20*, 263 (1963).

[1100] *E. Grigat, R. Pütter,* Chem. Ber. *97*, 3022 (1963).

[1101] *E. Grigat, R. Pütter, E. Mühlbauer,* Chem. Ber. *98*, 3777 (1965).

[1102] *J. Armand,* Bull. Soc. Chim. France 882 (1966).

[1103] *C. Grundmann,* Fortschr. Chem. Forsch. *7*, 62 (1966/67).

The technique lends itself to expansion in respect of the oximes; *thiolactams,* too, can be prepared in very good yield in this way.

17.4.8 Oxidation of activated methyl groups by the method of Willgerodt-Kindler

In the Willgerodt-Kindler procedure alkyl aryl ketones are heated to the boiling point of morpholine, often used as amine, with sulfur [1075]. The original keto group is reduced to a methylene group and the end of the side-chain is oxidized to the thioamide group:

$$\text{Ar--CO--CH}_2\text{--CH}_3 \xrightarrow[130°]{\text{HNR}_2,\ \text{S}_8} \text{Ar--CH}_2\text{--CH}_2\overset{\text{S}}{\underset{\|}{\text{--C}}}\text{--NR}_2$$

Despite much effort the reaction mechanism has not yet been elucidated. The preparative usefulness of the technique is considerable [1106-1109]; with a suitable choice of the starting materials it is possible to work at room temperature [1109].

17.4.9 Alkylation of *N*-functional derivatives of thiocarboxylic acids

17.4.9.1 Dithiocarbamic acids

The anions of dithiocarbonic acid amides mentioned on p. 754 are so nucleophilic that they will react with a large number of alkylating agents to form *dithiocarbamic acid esters* [998, 1005, 1110]:

$$\left[\text{R--NH--C}\overset{\text{S}^{\ominus}}{\underset{\text{S}}{\diagdown}}\right]\text{R--NH}_3^{\oplus}(\text{Na}^{\oplus})\ +\ \text{R}^1\text{X}\ \longrightarrow$$

$$\text{R--NH--C}\overset{\text{SR}^1}{\underset{\text{S}}{\diagdown}}$$

In this way allyl groups [1111], and, with α-halo ethers or amines, alkoxymethyl or aryloxymethyl [1112] and (dialkylamino)methyl or (diarylamino)methyl [1113] groups respectively can be introduced. Olefins, too, are suitable alkylating agents [1005, 1114, 1115], and it is noteworthy that diazonium salts enable the aryl group to be introduced [1005, 1116].

$$\left[\text{R}_2\text{N--C}\overset{\text{S}^{\ominus}}{\underset{\text{S}}{\diagdown}}\right]\text{Na}^{\oplus}\ +\ \left[\text{Ar--N}_2^{\oplus}\right]\text{Cl}^{\ominus}\ \longrightarrow$$

$$\text{R}_2\text{N--C}\overset{\text{S--Ar}}{\underset{\text{S}}{\diagdown}}\ +\ \text{N}_2\ +\ \text{NaCl}$$

17.4.9.2 Alkylation of thiocarboxamides

In general, the electrophilic attack on the ambifunctional thiocarboxamide anions ensues on the sulfur to give *thioimidic acid S-esters* [1117-1119]:

$$\left[\text{R}^1\text{--C}\overset{\overset{\ominus}{\text{N--R}}}{\underset{\text{S}}{\diagdown}}\right]\text{Na}^{\oplus}\ +\ \text{R}^2\text{X}\ \longrightarrow$$

$$\text{R}^1\text{--C}\overset{\text{N--R}}{\underset{\text{S--R}^2}{\diagdown}}\ +\ \text{Na}^{\oplus}\text{X}^{\ominus}$$

The reaction can be carried out also with the thiocarboxamide itself; the hydrohalides are obtained even in the case of the *N,N*-disubstituted thiocarboxamides [1018, 1120-1124]:

$$\text{R}^1\text{--C}\overset{\text{NH}_2}{\underset{\text{S}}{\diagdown}}\ +\ \text{R}^2\text{X}\ \longrightarrow\ \left[\text{R}^1\text{--C}\overset{\text{NH}_2}{\underset{\text{S--R}^2}{\diagdown}}\right]^{\oplus}\text{X}^{\ominus}$$

[1104] *R. Huisgen, J. Witte,* Chem. Ber. *91,* 972, 1129 (1958).

[1105] *A. Kaneda, M. Nagatuska, R. Sudo,* Bull. Chem. Soc. Japan *40,* 2705 (1967).

[1106] *R. Wegler, E. Kühle, W. Schäfer,* Angew. Chem. *70,* 351 (1958).

[1107] *R. Wegler, E. Kühle, W. Schäfer* in *W. Foerst,* Neuere Methoden der präparativen organischen Chemie, Bd. III, Verlag Chemie, Weinheim/Bergstr. 1961.

[1108] *F. Asinger, W. Schäfer, K. Halcour, A. Saus, H. Triem,* Angew. Chem. *75,* 1050 (1963).

[1109] *F. Asinger, H. Offermanns, A. Saus,* Monatsh. Chem. *100,* 724 (1969).

[1110] *J. v. Braun,* Ber. *35,* 3368 (1902); *J. v. Braun,* Ber. *42,* 4568 (1909).

[1111] *J. v. Braun, R. Murjahn,* Ber. *59,* 1202 (1926).

[1112] DBP 1 178 416 (1964), Farbenfabriken Bayer, Erf.: *A. Frank, F. Grewe, H. Kaspers;* C.A. *62,* 16070e (1965).

[1113] *N. Kreutzkamp,* Dtsch. Apoth.-Ztg. *102,* 1286 (1962).

[1114] *J.L. Garraway,* J. Chem. Soc. (London) 4004, 4008 (1964).

[1115] *J.E. Jansen, R.A. Mathes,* J. Amer. Chem. Soc. *77,* 2866 (1955).

[1116] *A.M. Clifford, J.G. Lichty,* J. Amer. Chem. Soc. *54,* 1163 (1932).

[1117] *O. Wallach,* Justus Liebigs Ann. Chem. *262,* 354 (1891).

[1118] *R. Boudet,* Ann. Chim. (Paris) *10,* 178 (1955).

[1119] *P. Reynaud, R.C. Moreau, T. Gousson,* Compt. Rend. *259,* 4067 (1964).

[1120] *D.A. Peak, F. Stansfield,* J. Chem. Soc. (London) 4067 (1952).

[1121] *H. Yamaguchi,* Chem. Pharm. Bull. (Tokyo) *16,* 448 (1968).

[1122] *E.P. Nesynov, M.M. Besprozvannaya, P.S. Pel'kis,* Zh. Org. Khim. *5,* 58 (1969); C.A. *70,* 87236 (1969).

Thiocarboxamides of high kinetic acidity such as thiobenzamide are methylated on the sulfur by diazomethane to give *thiobenzamide S-methyl ester*[1121]. Aryldiazonium ions also attack on the sulfur[1122]. *N*-Alkylation products of the thiocarboxamides are obtained if the alkylating agent is able to form adequately stable cations. In some instances it has been demonstrated that the *N-alkyl* compounds arise by rearrangement from the primarily formed *S*-alkyl thioimidates[1125]:

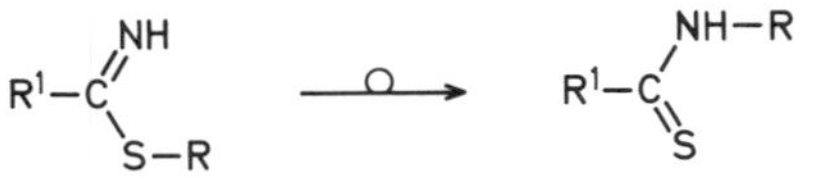

17.4.9.3 Alkylation of thiocarboxylic acid hydrazides

Open-chain thiocarboxylic acid hydrazides are alkylated exclusively on the sulfur atom. The reaction products of unsubstituted thiocarboxylic acid hydrazides are so unstable that they have been isolated in exceptional cases only[1060]. By contrast, the alkylation of N^2-substituted thiocarboxylic acid hydrazides has been described on a number of occasions[1058, 1126, 1127]:

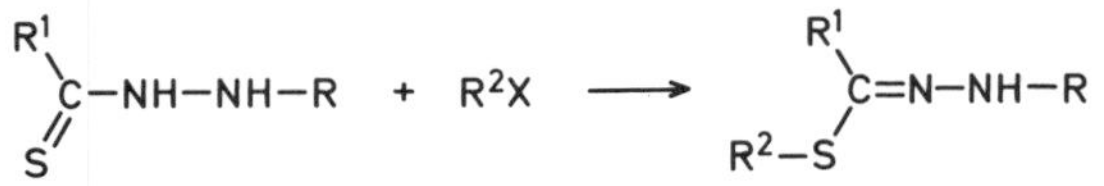

Thiocarboxylic acid
N-alkyl or N-aryl-
hydrazide S-esters

Mixtures of *N*- and *S*-alkylation products have been observed only where cyclic thiocarboxylic acid hydrazides were used[1128].

17.4.9.4 Alkylation of thiohydroxamic acids and *N*-hydroxythioureas

During alkylation of thiohydroxamic acids the attack on the sulfur predominates as well and *S-alkylthiohydroxamic acids* are obtained:

To carry out the reaction the alkali metal salts of the thiohydroxamic acids are reacted in alcohol[977, 978], acetone, or 1,2-dimethoxyethane[1129]. When nickel complex salts are employed *O,S-dialkylthiohydroxamic acids* are obtained side by side with the *S-alkyl* acids[1130].

With diazomethane and *N*-hydroxythioureas a selective *O*-alkylation to *N-methoxythiourea* has been observed[968], while with alkyl halides *2-alkyl-1-hydroxypseudothioureas* are formed[1131].

17.4.9.5 Alkylation of thioureas

With the usual alkylating agents thioureas react to form the corresponding *S-alkylisothiuronium salts*[1132-1134]:

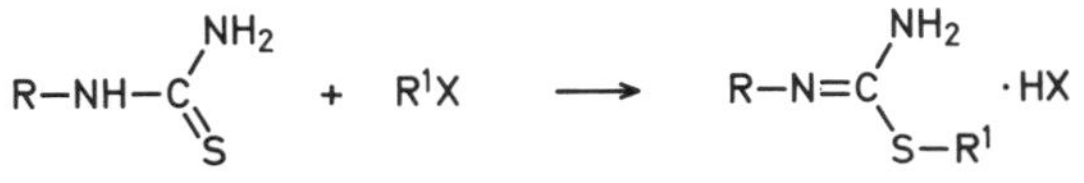

The preparative scope of the reaction is large. Thus, an alkylation is possible under the conditions of the Mannich reaction[1135], *e.g.*:

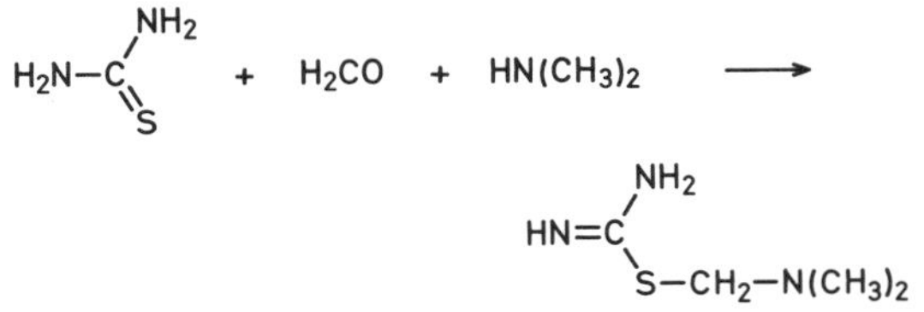

2-[2-(Dimethylamino)-
methyl]-2-pseudothio-
urea; 59%

In these cases rearrangement to *N*-alkylthiourea easily occurs.

17.5 Orthocarboxylic acid amides*

Replacing the carbonyl group in carboxamides or carbonic acid amides by respectively two mono-

[1123] *N. Stojanac, V. Hahn*, Bull. Sci. Conseil Acad. RSF Yugoslavie *11*, 98 (1966); C.A. *65*, 20084g (1966).

[1124] *A. Bernthsen*, Justus Liebigs Ann. Chem. *197*, 341 (1897).

[1125] *W. Walter, J. Krohn*, Chem. Ber. *102*, 3786 (1969).

[1126] *T. Sato, M. Ohta*, Bull. Chem. Soc. Japan *27*, 624 (1954); C.A. *50*, 213d (1956);
T. Sato, M. Ohta, J. Pharm. Soc. Jap. *74*, 821 (1954); C.A. *49*, 9537 (1955).

[1127] *J. Sandström*, Acta Chem. Scand. *16*, 2395 (1962).

[1128] *A. Pollak, M. Tišler*, Monatsh. Chem. *96*, 642 (1965).

* see Lit.[1197], bicyclic orthocarboxylic acid amides.

[1129] *R.E. McClure, A. Ross*, J. Org. Chem. *27*, 304 (1962).

[1130] *K. Nagata*, Chem. Pharm. Bull. (Tokyo) *17*, 653 (1969).

[1131] *C. Kjellin, K.G. Kuylenstjerna*, Justus Liebigs Ann. Chem. *289*, 117 (1897).

[1132] *E. Brand, F.C. Brand*, Org. Synth. *22*, 59 (1942).

[1133] *P.R. Shildneck, W. Windus*, Org. Synth. Coll. Vol. II, 411 (1943).

[1134] *R.L.N. Harris*, Tetrahedron Lett. 4045 (1968).

[1135] *H. Böhme, K. Dietz, K.D. Leidreiter*, Arch. Pharm. (Weinheim, Ger.) *287*, 201 (1954).

valent oxygen, sulfur, or nitrogen functions leads to orthocarboxylic acid amides.

17.5.1 From imidium ester salts

The simplest method for preparing *orthocarboxylic* and *dithiocarboxylic acid monoamide diesters* is nucleophilic addition of alkali metal alkoxides[1136-1145] or thiolates[1146-1148] to imidium ester salts. The latter are obtained by alkylating dialkylamides, their vinylogs, *N*-alkyl lactams, *N,N*-dialkylurethanes, and *N,N,N',N'*-tetraalkylureas with trialkyloxonium tetrafluoroborates, and especially readily with dialkyl sulfates[1138-1143, 1149]. Only in a few nucleophilic amides such as urethans[1150] or diamides of dicarboxylic acids[1145] are the latter unsuitable as alkylating agents. Alkyl halides can be used only for *S*-alkylation of thiocarboxamides[1147, 1148]. Reacting imidium ester tetrafluoroborates[1136, 1137, 1145] (Method A) or alkyl sulfates[1138-1143] (Method B) with alkoxides is carried out most suitably in alcohols, and particularly readily by adding the generally liquid imidium ester alkyl sulfates drop by drop:

$$\left[\begin{array}{c} R^1O \cdots \\ \overset{\oplus}{C} \overset{}{\text{---}}(CH=CH)_n \text{---} N \\ R \end{array} \begin{array}{c} R^2 \\ R^3 \end{array} \right] Y^{\ominus} \xrightarrow[- Y^{\ominus}, R^1\text{---}OH]{R^4\text{---}O^{\ominus} / R^4\text{---}OH}$$

$$R\text{---}\underset{\underset{OR^4}{|}}{\overset{\overset{OR^4}{|}}{C}}\text{---}(CH=CH)_n\text{---}N \begin{array}{c} R^2 \\ R^3 \end{array}$$

n = 0,1

R = Alkyl, substituted Alkyl, Aryl, Dialkylamino, Alkyloxy

R^1, R^2, R^3, R^4 = Alkyl

Y = CH_3SO_4, BF_4, CN

The reaction of 1-dialkylamino-1-alkoxyacetonitriles with alcoholates represents a special case; *orthoformic acid mono(dialkylamide) diesters* are obtained in good yield[1144] (Method C). To furnish good yields the reaction temperature should not exceed 10°; higher temperatures favor nucleophilic substitution:

$$R\text{---}\underset{\underset{R^3}{\nearrow}\overset{|}{\underset{N\text{---}R^2}{}}}{\overset{\overset{OR^1}{\nwarrow}}{\overset{\oplus}{C}}} \xrightarrow{R^4\overline{\underline{O}}^{\ominus}} R\text{---}\underset{\underset{R^3}{\nearrow}\overset{|}{\underset{N\text{---}R^2}{}}}{\overset{\overset{\overline{\underline{O}}|}{\|}}{C}} + R^4\text{---}O\text{---}R^1$$

In general the yields of amide diesters are between 50 and 85%. They are dependent on the stability and the spatial form of the anion R–$O^{\ominus}$, on the electrophilism of the cation, and on the polarizability of the O–R bond in the cation.

Thus, phenoxides, thiophenoxides, and potassium *tert*-butoxide react with the *N,N*-dimethylformamide-dimethyl sulfate adduct by substitution[1146, 1151]. The yield of ~30% *orthocarbonic acid mono(N,N-dialkylamide) diester* using Method A is less than usual.

The alkoxide addition is accompanied by *transacetalization* by the excess alcohol[1139]. On account of dismutation reactions[1152] mixed acetals, except vinylogous ones[1143], cannot be isolated even if the reaction is performed with molar alkoxide in the absence of excess alcohol. *Orthocarbonic acid bis(N,N-dialkylamide) diesters* need to be prepared in aprotic solvents[1137]; with alcohols they react to form *orthocarboxylic acid N,N-dialkylamide triesters*[1141].

Dithioorthoformic acid mono(dialkylamide) diesters are obtained from formidium ester methyl sulfates[1149] and alkali thiolates in nonpolar solvents in the presence of excess thiol[1146]. In place of the thiolates thiols plus tertiary amines are advantageously employed as bases[1146] (Method D). With the less electrophilic, higher homologous imidium ester salts thiolates react by undergoing alkylation[1146], but the corresponding *dithiocarboxylic acid amide diesters* are obtained from thiocarboxylic acid imidium ester salts and thiolates[1147, 1148] (Method E).

[1136] H. Meerwein, P. Borner, O. Fuchs, H.J. Sasse, H. Schrodt, J. Spille, Chem. Ber. **89**, 2060 (1956).

[1137] H. Meerwein, W. Florian, N. Schön, G. Stopp, Justus Liebigs Ann. Chem. **641**, 1 (1961).

[1138] H. Bredereck, F. Effenberger, G. Simchen, Angew. Chem. **73**, 493 (1961).

[1139] H. Bredereck, G. Simchen, S. Rebsdat, W. Kantlehner, P. Horn, R. Wahl, H. Hoffmann, P. Grieshaber, Chem. Ber. **101**, 41 (1968).

[1140] H. Bredereck, F. Effenberger, H.P. Beyerlin, Chem. Ber. **97**, 3081 (1964).

[1141] H. Bredereck, F. Effenberger, H.P. Beyerlin, Chem. Ber. **97**, 1834 (1964).

[1142] H. Bredereck, F.Effenberger, D. Zeyfang, Angew. Chem. **77**, 219 (1965);
H. Bredereck, F. Effenberger, D. Zeyfang, Angew. Chem. Intern. Ed. Engl. **4**, 242 (1965).

[1143] H. Bredereck, F. Effenberger, D. Zeyfang, K.A. Hirsch, Chem. Ber. **101**, 4036 (1968).

[1144] W. Kantlehner, P. Speh, Chem. Ber. **105**, 1340 (1972).

[1145] H. Bredereck, W. Kantlehner, D. Schweizer, Chem. Ber. **104**, 3475 (1971).

[1146] I.A. Ivanova, B.P. Fedorov, F.M. Stoyanovich, Isv. Akad. SSSR 2179 (1965); C.A. **64**, 12538 (1966).

[1147] T. Mukaiyama, T. Yamaguchi, H. Nohira, Bull. Chem. Soc. Japan **38**, 2107 (1965); C.A. **64**, 9618 (1966).

[1148] T. Mukaiyama, T. Yamaguchi, Bull. Chem. Soc. Japan **39**, 2005 (1966); C.A. **66**, 18564 (1967).

One of the few thermally stable orthocarbox-amides with *N*-monosubstituted amino groups is *N,N',N''*-methylidenetrisformamide (orthoformic acid tris[formylamide])[1153, 1154], which is obtained best by reacting dialkyl sulfates with excess form-amide:

$$H-\overset{\displaystyle O}{\underset{\displaystyle NH_2}{C}} \xrightarrow{R_2SO_4} \left[H-\overset{\displaystyle OR}{\underset{\displaystyle NH_2}{\overset{\oplus}{C}}} \right] RSO_4^{\ominus}$$

$$\xrightarrow{HCONH_2} H-\overset{\displaystyle NHCHO}{\underset{\displaystyle NHCHO}{C}}-NHCHO$$

69–73%

17.5.2 Orthocarboxylic acid amides from amidinium salts

The most suitable method for preparing *orthoformic acid bis(dialkylamide) monoesters* and *tris(dialkylamides)*, *thioformic acid bis(dialkylamide) monoesters* as well as their vinylogs is nucleophilic addition of alkali metal alcoholates[1139, 1143, 1155–1158], thiolates[1159], and dialkylamides[1143, 1160, 1161] to *N,N,N',N'*-tetraalkylformamidinium salts (Method F):

$$\left[H-\overset{\displaystyle \overset{R^2}{\underset{}{\overset{\oplus}{N}}-R^1}}{\underset{\displaystyle (CH=CH)_n N \overset{R^2}{\underset{R^1}{}}}{C}} \right] Y^{\ominus}$$

$$\xrightarrow[-Y^{\ominus}]{R^3-X^{\ominus}} H-\overset{\displaystyle \overset{R^1}{\underset{}{N}-R^2}}{\underset{\displaystyle (CH=CH)_n N \overset{R^2}{\underset{R^1}{}}}{C}}-X-R^3$$

n = 0,1
$Y^{\ominus}$ = $H_3C-O-SO_3^{\ominus}$, BF_4, ClO_4, Halide
X = O, S, *N*-Alkyl, *N*-Aryl
R^1, R^2 = Alkyl, Aryl
R^3 = Alkyl

Alcoholates and thiolates are reacted in nonpolar, absolute solvents at 0–50°. The easily available methyl sulfates serve particularly well as the amidinium salts[1139, 1156], alternatives are halides[1144, 1162] or perchlorates[1143].

Synthesis of *orthocarboxylic acid bis(dialkylamide) alkyl esters* from *2,2-bis(dialkylamino)acetonitriles (tetraalkylformamidinium cyanides)* and alkoxides succeeds readily (Method G)[1144]:

$$H-\overset{\displaystyle \overset{R^2}{\underset{}{N}-R^1}}{\underset{\displaystyle \underset{R^1}{\overset{}{N}}-R^2}{C}}-C\equiv N \xrightarrow{R^3-\overline{O}|^{\ominus}} H-\overset{\displaystyle \overset{R^2}{\underset{}{N}-R^1}}{\underset{\displaystyle \underset{R^1}{\overset{}{N}}-R^2}{C}}-OR^3$$

R^1, R^2, R^3 = Alkyl

The *yields* in all the methods are between 60 and 80% independently of the alcoholate ion. Diamide monoesters of orthoformic acid, which are appreciably more reactive than the monoamide diesters, dismutate on heating, especially in the presence of traces of alcohol, to a proportion of *orthoformic amide diester* and *orthoformic acid tris(dialkylamide)*:

$$2\ H-\overset{\displaystyle \overset{R^2}{\underset{}{N}-R^1}}{\underset{\displaystyle \underset{R^1}{\overset{}{N}}-R^2}{C}}-X-R^3 \rightleftharpoons H-\overset{\displaystyle \overset{R^2}{\underset{}{N}-R^1}}{\underset{\displaystyle \underset{R^1}{\overset{}{N}}-R^2}{C}}-N\overset{R^2}{\underset{R^1}{}}$$

$$+\ H-\overset{\displaystyle X-R^3}{\underset{\displaystyle \underset{R^1}{\overset{}{N}}}{C}}-X-R^3$$

[1149] *H. Bredereck, F. Effenberger, G. Simchen*, Chem. Ber. **96**, 1350 (1963).

[1150] *H.P. Beyerlin*, Dissertation Universität Stuttgart (1964).

[1151] *G. Simchen*, unveröffentlicht.

[1152] *G. Simchen, H. Hoffmann, H. Bredereck*, Chem. Ber. **101**, 51 (1968).

[1153] *H. Bredereck, R. Gompper, H. Rempfer, K. Klemm, H. Keck*, Chem. Ber. **92**, 329 (1959).

[1154] *H. Bredereck, R. Gompper, H.G. v. Schuh, G. Theilig*, Angew. Chem. **71**, 753 (1959).

[1155] *H. Bredereck, F. Effenberger, G. Simchen*, Angew. Chem. **74**, 353 (1962).

[1156] *H. Bredereck, F. Effenberger, G. Simchen*, Chem. Ber. **98**, 1078 (1965).

[1157] *H. Bredereck, F. Effenberger, R. Gleiter, K.A. Hirsch*, Angew. Chem. **77**, 1010 (1965);
H. Bredereck, F. Effenberger, R. Gleiter, K.A. Hirsch, Angew. Chem. Intern. Ed. Engl. **4**, 955 (1965).

[1158] *D.H. Clemens, W.D. Emmons*, J. Amer. Chem. Soc. **83**, 2588 (1961);
D.H. Clemens, E.Y. Shropshire, W.D. Emmons, J. Org. Chem. **27**, 3664 (1962).

[1159] *H. Hoffmann*, Dissertation Universität Stuttgart (1968).

[1160] *H. Bredereck, F. Effenberger, Th. Brendle*, Angew. Chem. **78**, 147 (1966);
H. Bredereck, F. Effenberger, Th. Brendle, Angew. Chem. Intern. Ed. Engl. **5**, 132 (1966).

[1161] *H. Bredereck, F. Effenberger, Th. Brendle, H. Muffler*, Chem. Ber. **101**, 1885 (1968).

[1162] *W. Kantlehner*, priv. com.

[1163] DBP. 1 140892 (1961), Farbenfabriken Bayer, Erf.: *H. Gold*; C.A. **59**, 10070 (1963);
H. Gold, Angew. Chem. **72**, 956 (1960).

Where very pure products are to be prepared — which is unnecessary where further reaction are to be performed — alcohols need to be excluded completely[1152].

The likewise extremely reactive *orthoformic acid tris(dialkylamides) [methanetris(dialkylamines)]* and their vinylogs are accessible by addition of (dialkylamino)lithium or -sodium to (vinylogous) formamidinium chlorides[1160, 1161] in absolute ether at −20 to +80°; methyl sulfates are not suitable here. Perchlorates[1143] are alternative compounds for use. As the dialkylamino group increases in size the yields diminish (70–20%). Unlike with amide diesters mixed *orthocarboxylic acid tris(dialkylamides)*, too, can be synthesized[1161].

N,N,N′,N′-Tetrasubstituted formamidinium salts including cyanides, finally, are important for preparing *orthoformic* and *orthothioformic acid amide diesters* with bulky alkoxy and aryl substituted amino groups. The reaction is performed in alcoholic alkoxide solutions*[1144, 1158, 1163, 1164] (Method H); the procedure can be interpreted as an alcoholysis of initially formed orthoformic acid diamide esters:

R¹ = R³ = Alkyl
R² = Alkyl, Aryl

17.5.3 Orthocarboxylic acid amides by transacetalization and transamination

The basis of transacetalization and transamination reactions are the dissociation equilibria of orthocarboxylic acid amides, which adjust themselves rapidly especially in protic solvents[1152]:

X = O, S

The transacetalization of diesters of orthocarboxamides (above all orthoformic acid dimethylamide dimethyl ester) possesses general importance because it leads to very pure end-products. Orthocarboxylic acid amide diesters thus become readily accessible also where the preparation of alcoholate solutions presents difficulties[1137, 1139, 1164–1166].

The procedure consists in heating the orthoacid amide diester with twice to three times the molar quantity of an alcohol while distilling off the lower boiling alcohol derived from the orthoacid amide diester in an efficient column (the same holds for the dithioorthocarboxylic acid amide diester) (Method J). Orthoformic acid dimethylamide di-*tert*-butyl esters[1164] and certain dibenzyl esters[1165, 1166] cannot be obtained by transacetalization; dithioorthocarboxylic amide diesters behave analogously[1146].

In addition, alkylthio groups can be exchanged against alkoxy groups and *vice versa*[1146, 1159] (Method K).

Reacting orthoformic acid monodialkylamide diesters or diamide monoesters with secondary amines[1167, 1168] makes *orthoformic acid tris(dialkylamides)* available (Method L).

The reaction succeeds particularly well with higher boiling, secondary amines (for a reversal of this reaction see p. 767):

R¹ = Alkyl
R² = Alkyl, Aryl

*   In place of the tetramethylformamidinium chloride[1164] the methyl sulfate may be used[1151].

[1164] *Z. Arnold, M. Kornilov*, Collect. Czech. Chem. Commun. *29*, 645 (1964).
[1165] *H. Brechbühler, H. Büchi, E. Hatz, J. Schreiber, A. Eschenmoser*, Angew. Chem. *75*, 296 (1963).
[1166] *H. Brechbühler, H. Büchi, E. Hatz, J. Schreiber, A. Eschenmoser*, Helv. Chim. Acta *48*, 1746 (1965).
[1167] *H. Bredereck, G. Simchen, H.U. Schenck*, Chem. Ber. *101*, 3058 (1968).
[1168] *J.W. Scheeren, R.J.F. Nivard*, Rec. Trav. Chim. Pays-Bas *88*, 289 (1969).

Transamination[1137] of orthocarboxylic acid monoamide diesters, too, is based on the alcoholysis of diamide monoesters. This procedure displays little importance and leads to the desired results only with higher boiling, secondary amines while the alcohol remains in the reaction mixture (Method M):

R¹, R², R³, R⁴ = Alkyl

17.5.4 Orthocarboxylic acid amides from salts of imidium halides

The so-called amide halides, their vinylogs, and the chloramidinium salts lend themselves above all for the reaction with alkali metal alcoholates[1169–1175], thiolates[1176, 1177], and dialkylamides[1178] to form orthocarboxylic acid amides

X = Cl, F
Y = Cl, F, POCl₂
Z = O, S
R = Alkyl, Aryl, Dialkylamino
R¹, R², R³ = Alkyl

In this way *orthocarboxylic acid amide diesters* and *diester lactams*, and also *orthocarbonic acid amide triesters* and *diamide diesters* are made accessible. To prevent formation of alkyl halides from the intermediately formed imidium esters halides, it is necessary to perform the reaction at 0° and to ensure a rapid supply of the alkoxide ions. This is taken care of by adding solutions of the amide halides in nonpolar solvents to the alkoxide solutions. The reaction succeeds also with phenoxides[1170]. Employing triethylamine as base in place of alkoxides is less suitable. *Orthocarbonic acid bis(dialkylamide) diesters* can be obtained from chlorocarboxamidines only if alcohols are completely excluded[1174] (Method N):

R, R¹ = Alkyl

Reacting N,N,N',N'-tetramethylchloroformadinium chloride with (dimethylamino)lithium gives *octamethylmethanetetramine [orthocarbonic acid tetrakis-(dimethylamide)]*[1178]. To date it has not been possible to obtain orthoformic acid tris(dialkylamides) from N,N-dialkylformamide chlorides. By contrast, the reaction of sodium diacylimides to form *orthoformic acid tris(acylamides)* can be accomplished[1179]. Preparation of N,N',N''-methylylidenetrisformamide from formamide and acyl halides is also significant[1153, 1154]. Direct preparation of thiocarboxylic acid imide S-ester halides and of dithiocarb-

---

[1169] H. Bredereck, K. Bredereck, Chem. Ber. *94*, 2278 (1961).

[1170] H. Bosshard, E. Jenny, Hch. Zollinger, Helv. Chim. Acta *44*, 1203 (1961);
Brit. P. 911475 (1962), Ciba; C.A. *58*, 13852 (1963).

[1171] H. Eilingsfeld, M. Seefelder, H. Weidinger, Angew. Chem. *72*, 836 (1960);
DBP. 1119872 (1961), Badische Anilin & Sodafabrik; Erf.: H. Eilingsfeld, M. Seefelder, H. Weidinger; C.A. *56*, 14083 (1962).

[1172] H. Eilingsfeld, M. Seefelder, H. Weidinger, Chem. Ber. *96*, 2671 (1963).

[1173] DBP. 1122936 (1960), Badische Anilin & Sodafabrik; Erf.: H. Weidinger, H. Eilingsfeld.

[1174] H. Eilingsfeld, G. Neubauer, M. Seefelder, H. Weidinger, Chem. Ber. *97*, 1232 (1964).

[1175] USP. 3092637 (1963), DuPont; Inv.: M. Brown; C.A. *59*, 12764 (1963).

[1176] B.P. Fedorov, F.M. Stoyanovich, Izv. Akad. SSSR 1828 (1960); C.A. *55*, 14298 (1961).

[1177] F.M. Stoyanovich, B.P. Fedorov, G.M. Andrianova, Dokl. Akad. Nauk SSSR *145*, 584 (1962); C.A. *58*, 4448 (1963).

[1178] H. Weingarten, W.A. White, J. Amer. Chem. Soc. *88*, 2885 (1966).

[1179] H. Böhme, F. Soldan, Chem. Ber. *94*, 3109 (1961).

oxylic acid monoamide diesters from amide chlorides and mercaptans is made possible by the lesser reactivity of the two former compounds; even the phosphoryl chloride complexes that are unsuitable for preparing orthocarboxylic acid monoamide diesters can be employed with success[1176, 1177] (Method O).

17.5.5 Orthocarboxylic acid amides from orthoformic acid triesters and trihalomethanes

The reaction of orthoformic acid triesters and orthoformic acid trihalides (trihalomethanes) respectively displays some importance, intrinsically in that these compound classes can be reacted successfully with secondary, araliphatic amines in the presence of sodium hydride to yield *orthoformic acid triamides (methanetriamines)*[1158] (Method P):

X, Y = F, Cl, OR
R^1 = Alkyl
R = Aryl, Alkyl

Reacting chloroform with higher boiling, secondary amines in alcoholic alcoholate solution likewise leads to orthoformic acid triamides[1168] (Method R). Ethenetetramines frequently arise as side-products[1168, 1180].

Orthoformic acid dialkylamide diesters can be obtained from chloroform (dichlorocarbene) where the alcohol used boils at a higher temperature than the secondary amine[1168, 1181] (Method S). Trichloroacetic acid esters are an alternative possible source of the dichlorocarbene.

R, R^1, R^2 = Alkyl

Here, difficulties are encountered in preparing the orthoformic acid monoamide diester in the pure state because of simultaneous formation of ortho-

formic acid triesters. Other routes for accomplishing directed syntheses of orthoformic acid triamides are the following (Method T)[1182, 1183]:

X = S–R , OAr

Preparation of *orthoformic acid tris(acylamides)* from orthoformic acid triesters and primary carboxamides using acid catalysis displays greater importance[1184, 1185] (urethans react analogously)[1186]:

R = Alkyl
R^1 = Alkyl, Aryl, Alkyloxy

17.5.6 Orthocarboxylic acid amides by acetalization of carboxamides

Successful *acetalization* of amides is so far limited to few examples. *Orthoformic acid tris(dimethylamide) [methanetris(N,N-dimethylamine)]* is obtained from tetrakis(dimethylamino)titanium and N,N-dimethylformamide[1187] or N,N,N',N'-tetramethyloxalamide[1188]. Similarly trialkyl thioborates and N,N-dimethylformamide afford the corresponding *dithioorthoformic acid dialkylamide diester*[1189] (Method U):

R = Alkyl

[1180] H.U. Schenk, Diplomarbeit Universität Stuttgart (1966).
[1181] DBP. 1 161 285 (1964), Badische Anilin- & Soda-fabrik; Erf. W. Stilz; C.A. 60, 9156 (1964).
[1182] H. Böhme, J. Roehr, Justus Liebigs Ann. Chem. 648, 21 (1961).
[1183] H. Böhme, R. Neidlein, Chem. Ber. 95, 1859 (1962).
[1184] H. Bredereck, R. Gompper, F. Effenberger, H. Keck, H. Heise, Chem. Ber. 93, 1398 (1960).
[1185] H. Bredereck, F. Effenberger, H.J. Treiber, Chem. Ber. 96, 1505 (1963).
[1186] A.V. Stavrovskaya, T.V. Protopopova, A.P. Skoldinov, Zh. Org. Khim. 3, 1749 (1967); C.A. 68, 12448 (1968).
[1187] H. Weingarten, W.A. White, J. Amer. Chem. Soc. 88, 850 (1966);
H. Weingarten, W.A. White, J. Org. Chem. 31, 2874 (1966).

17.5.7 Special procedures

Ethenetetramines (nucleophilic carbenes) react with alcohols to form *orthocarboxylic acid trisamides*[1190]. Isocyanates and orthoformic acid triesters afford *orthoformic acid (alkoxycarbonylalkylamide) diesters*[1191]. Reacting orthoformic acid mono(dialkylamide) diester with isocyanates and isothiocyanates furnishes *O,N*-acetals of *parabanic acid (imidazolinetrione)* and *dithioparabanic acid* respectively[1192, 1193]:

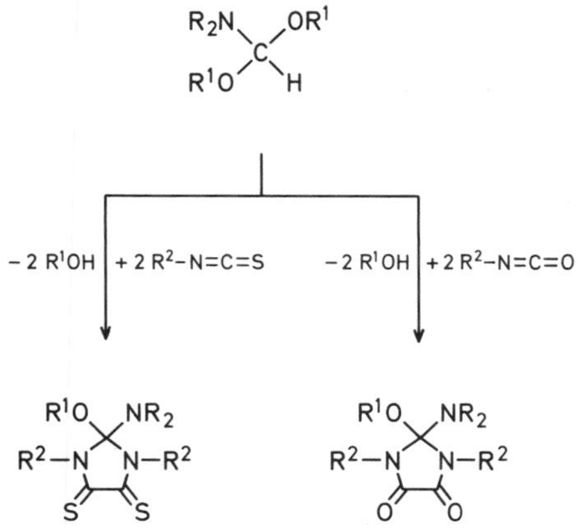

Formation of orthocarboxylic acid amide diesters from cyclic aldehyde acetals and nitrenes seems to display interest[1194]. Because of the reactivity of the formyl proton in orthoformic acid mono(dialkylamide) diesters the corresponding *C-deuteroorthoformic acid mono(dialkylamide) diesters* are readily accessible in simple manner by reacting with *O*-deutero alcohols[1195, 1196].

[1188] *J.D. Wilson, H. Weingarten*, J. Org. Chem. *33*, 1246 (1968).

[1189] *V.A. Dorokhov, B.M. Mikhailov*, Izv. Akad. SSSR 364 (1966); C.A. *64*, 17624 (1966).

[1190] *H.W. Wanzlick*, Angew. Chem. *74*, 129 (1962).

[1191] *H. v. Brachel, R. Merten*, Angew. Chem. *74*, 872 (1962).

[1192] *H. Bredereck, G. Simchen, S. Rebsdat*, Chem. Ber. *101*, 1863 (1968).

[1193] *H. Bredereck, G. Simchen, E. Göknel*, Chem. Ber. *103*, 236 (1970).

[1194] *H. Nozaki, S. Fujita, H. Takaya, R. Noyori*, Tetrahedron 23, 45 (1967).

[1195] *G. Simchen, S. Rebsdat, W. Kantlehner*, Angew. Chem. *79*, 869 (1967); *G. Simchen, S. Rebsdat, W. Kantlehner*, Angew. Chem. Intern. Ed. Engl. *6*, 875 (1967).

[1196] *G. Simchen, W. Kantlehner*, Tetrahedron 28, 3535 (1972).

[1197] *R. Feinauer*, Synthesis 16 (1971).

[1198] *S. Petersen* in *Houben-Weyl*, Methoden der organischen Chemie, 4. Aufl., Bd. VIII, p. 125, Georg Thieme Verlag, Stuttgart 1952.

Table 13. Preparation of orthocarboxylic acid triamides

Orthocarboxylic acid triamide	Method	Yield [% of theory]	Reference
Orthoformic acid dialkylamide dialkyl esters	A	52–72	1136, 1137
	B	50–87	1138, 1139
	C	63–74	1144
	H	30–82	1144, 1163, 1164
	J	57–94	1137, 1139, 1164–1166
	K	50–80	1146, 1159
	M	52–64	1137
	N	27–75	1169–1172, 1175
	S	27–61	1168, 1181
	T	18	1183
Orthocarboxylic acid dialkylamide dialkyl esters and diorthocarboxylic acid bis-(dialkylamide) tetraalkyl esters	A	40–76	1137, 1145
	B	50–75	1140
	J	86	1137
	N	37–65	1170–1172, 1175
ω-Aminoorthocarboxylic acid diester lactams	A	56–87	1136, 1137
	B	73	1140
	N	40–70	1169, 1171, 1172, 1175
Orthocarbonic acid dialkylamide triesters	A	36	1132
	B	14–50	1141
	N	27–73	1173, 1174
Orthocarbonic acid bis-(dialkylamide) diesters	A	56–65	1137
	N	30	1174, 1175
Dithioorthocarboxylic acid dialkylamide diesters	D	46–72	1146
	E	52–63	1147, 1148
	H		1158
	J	60–78	1146
	K	50–80	1159
	M	40–50	1146
	O	11–45	1176, 1177
	T	36	1182
	U	64	1189
Orthoformic acid bis-(dialkylamide) alkyl esters	F	62–79	1139, 1155, 1156
	G	70–80	1144
Orthoformic acid bis-(alkylarylamide) alkyl esters	F	55–67	1158
Thioorthoformic acid bis-(dialkylamide) alkyl esters	F	30–40	1159
Orthoformic acid tris-(dialkylamides) [Methane tris-(dialkylamines)]	F	37–67	1160, 1161
	L	10–80	1167, 1168
	U	83	1187, 1188
Orthoformic acid tris-(alkylarylamides) [Methane-tris(alkylarylamines)]	P	21–58	1158
	R	60	1168
	F	44	1158
	L	10–70	1168
Vinylogous orthocarboxylic acid amide diesters	A	70	1137
	B	46–80	1142, 1143
	N		1170
Vinylogous orthocarboxylic acid diamide esters	F	51–60	1143, 1157
Vinylogous orthocarboxylic acid triamides	F	20–61	1143, 1157

[1199] *M. Bögemann, S. Petersen, O.E. Schultz, H. Söll* in *Houben-Weyl*, Methoden der organischen Chemie, 4. Aufl., Bd. IX, p. 867, Georg Thieme Verlag, Stuttgart 1955.

[1200] *H. Holm, C. Wentrup*, Acta Chem. Scand. *20*, 2123 (1966).

17.6 Isocyanates, isothiocyanates, isoselenocyanates, carbodiimides

17.6.1 Direct introduction of the heterocumulene function

17.6.1.1 Alkylation and silylation of heterocumulene anions NCY$^\ominus$ (Y = O, S, Se, N$^\ominus$)

Alkylation of salts of cyanic acid[1198] and thiocyanic acid[1199] initially often leads to mixtures of *cyanates* and *isocyanates* and the corresponding *sulfur analogs* due to the ambient nature of the NCY$^\ominus$ion[1200–1202].

However, while alkyl cyanates readily isomerize to alkyl isocyanates in an exothermic reaction[1203], and therefore exclusively isocyanates are obtained under the preparative alkylation conditions, mixtures of thiocyanates and isothiocyanates generally

$$R{-}X \;+\; NCY^\ominus \longrightarrow R{-}N{=}C{=}Y \;+\; X^\ominus$$

Y = O, S, Se

require elevated temperature for uniform conversion into isothiocyanates (mustard oils) to take place[1204]. Zinc chloride[1205, 1206], cadmium iodide[1199], or thiocyanate ions[1207, 1208] accelerate the reaction. Salts of selenocyanic acid generally form *alkyl selenocyanates*[1209] during the alkylation[1209] which either do not rearrange into *isoselenocyanates*[1212, 1213] at all[1210] or only under drastic conditions[1211].

Alkylation of alkali metal cyanates is performed especially advantageously in dimethylformamide[1214, 1215]. In addition to alkyl halides dialkyl

sulfates[1198], 4-toluenesulfonates[1198], and trialkyl phosphates[1216] may be employed.

Silylation of salts of cyanic thiocyanic and selenocyanic acid leads to *silyl isocyanates*[1217–1219], *isothiocyanates*[1217, 1220] and *isoselenocyanates*[1221]. A reference to the preparation of *1,3-disilylated carbodiimides* by silylation of silver cyanamide[1222–1224] or calcium cyanamide is particularly indicated. Re the synthesis of *phosphorylated* carbodiimides from silver cyanamide see ref.[1226].

17.6.1.2 Addition of isocyanic acid, isothiocyanic acid and iodoisocyanate to double bonds

Isocyanic acid, isothiocyanic acid, and iodoisocyanates can be added to olefins, α,β-unsaturated ethers, α,β-unsaturated ketones, reactive carbonyl compounds (formaldehyde, chloral, hexafluoro-2-propanone, *etc.*) or isocyanates, sometimes in excellent yield. The resulting products are *alkyl isocyanates*[1227], *alkyl isothiocyanates*[1228, 1229], α-*alkoxy*[1230], α-*hydroxy*[1231], and β-*iodoisocyanates*[1232, 1233], *(aminocarbonyl)isothiocyanates*[1234, 1234a, 1242], and *2-oxoalkyl isothiocyanates*[1235].

[1201] *D.W. Emerson, J.K. Booth*, J. Org. Chem. 30, 2480 (1965).
[1202] *L.A. Spurlock, T.E. Parks*, J. Amer. Chem. Soc. 92, 1279 (1970).
[1203] *K.A. Jensen, M. Due, A. Holm*, Acta Chem. Scand. 19, 438 (1965).
[1204] *P.A.S. Smith, D.W. Emerson*, J. Amer. Chem. Soc. 82, 3076 (1960).
[1205] *W.U. Malik, P.K. Srivastava, S.C. Mehra*, J. Chem. Eng. Data 14, 110 (1969).
[1206] *H. Böhme, U. Scheel*, Arch. Pharm. (Weinheim, Ger.) 300, 326 (1967).
[1207] *A. Fava, A. Iliceto, S. Bresadola*, J. Amer. Chem. Soc. 87, 4791 (1965).
[1208] *L.A. Spurlock, P.E. Newallis*, Tetrahedron Lett. 303 (1966).
[1209] *H. Rheinboldt* in *Houben-Weyl*, Methoden der organischen Chemie, 4. Aufl., Bd. IX, p. 939, Georg Thieme Verlag, Stuttgart 1955.
[1210] *C. Collard-Charon, M. Renson*, Bull. Soc. Chim. Belg. 71, 531 (1962).
[1211] *T. Tarantelli, D. Leonesi*, Ann. Chim. (Rome) 53, 1113 (1963).
[1212] *T. Tarantelli, C. Precibe*, Ann. Chim. (Rome) 52, 75 (1962).

[1213] *C.T. Pedersen*, Acta Chem. Scand. 17, 1459 (1963).
[1214] *W. Gerhardt*, J. Prakt. Chem. 38, 77 (1968).
[1215] USP. 2866801 (1958), Ethyl Corp., Inv.: *C.M. Himel, L.M. Richards.*
[1216] *Th.J. Bieber*, J. Amer. Chem. Soc. 74, 4700 (1952).
[1217] *W. Sundermeyer*, Pure Appl. Chem. 13, 93 (1966).
[1218] *R.M. Pike, E.B. Moynahan*, Inorg. Chem. 6, 168 (1967).
[1219] *W. Airey, G.M. Sheldrick*, J. Chem. Soc. A 2865 (1969).
[1220] *H.H. Anderson*, J. Amer. Chem. Soc. 81, 4785 (1959).
[1221] *H. Bürger, U. Goetze*, J. Organometal. Chem. 10, 380 (1967).
[1222] *L. Birkofer, A. Ritter, P. Richter*, Tetrahedron Lett. 195 (1962).
[1223] *J. Pump, E.G. Rochow, U. Wannagat*, Monatsh. Chem. 94, 588 (1963).
[1224] *E.A.V. Ebsworth, M.J. Mays*, Angew. Chem. 74, 117 (1962).
[1225] *J. Stenzel, W. Sundermeyer*, Chem. Ber. 100, 3368 (1967).
[1226] *A. Weisz, K. Utvary*, Monatsh. Chem. 99, 2498 (1968).
[1227] *F.W. Hoover, H.S. Rothrock*, J. Org. Chem. 29, 143 (1964).
[1228] *L.S. Luskin, G.E. Gantert, W.E. Craig*, J. Amer. Chem. Soc. 78, 4965 (1956).
[1229] *W.R. Diveley, G.A. Buntin, A.D. Lohr*, J. Org. Chem. 34, 616 (1969).
[1230] *F.W. Hoover, H.S. Rothrock*, J. Org. Chem. 28, 2082 (1963).
[1231] *F.W. Hoover, H.B. Stevenson, H.S. Rothrock*, J. Org. Chem. 28, 1825 (1963).

17.6.1.3 Acylation of heterocumulene anions NCY$^\ominus$ (Y = O, S, Se)

Acylation of salts of cyanic, thiocyanic and selenocyanic acid by derivatives of carboxylic, sulfonic, and carbonic acids affords the compounds listed in Table 14 (p. 774), where applicable after thermal isomerization:

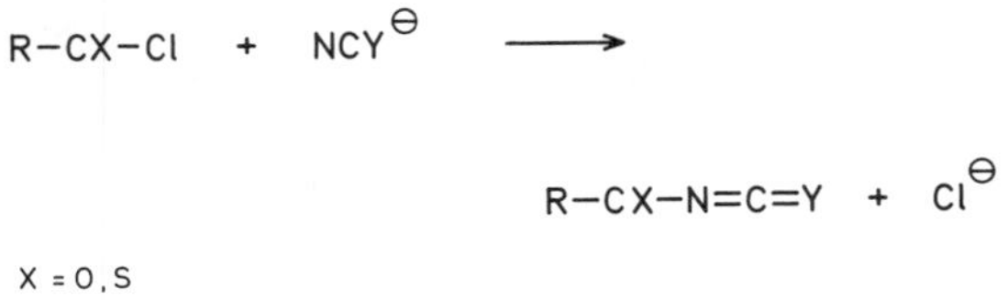

$$R-CX-Cl \;+\; NCY^{\ominus} \longrightarrow$$

$$R-CX-N=C=Y \;+\; Cl^{\ominus}$$

X = O, S
Y = O, S, Se

The classic reaction of acid chlorides with salts of thiocyanic acid is still one of the most important methods for preparing *acyl isothiocyanates*[1236, 1237]. A variant of acylating silver cyanate to *acyl isocyanates*[1238] consists in reacting acyl chlorides with isocyanic acid in the presence of tertiary amines[1239]; the reaction of trifluoroacetic anhydride with isocyanic acid to form *trifluoroacetyl isocyanate* has also been reported[1240].

While carboxylic acid chlorides form exclusively *acyl isothiocyanates* with alkali metal thiocyanates, thiocyanates are often formed as main or side products with carbonic acid derivatives. However, in analogy with the alkyl thiocyanates they can be rearranged to acyl isothiocyanates[1241]. This dualism is exhibited particularly clearly in the two following reactions (in acetonitrile)[1242]:

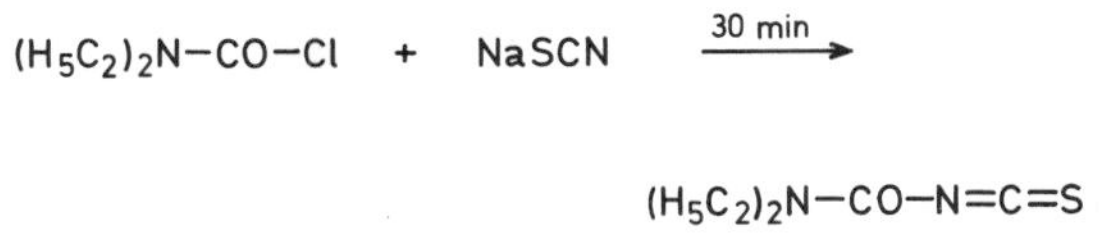

$$(H_5C_2)_2N-CO-Cl \;+\; NaSCN \xrightarrow{\text{30 min}}$$

$$(H_5C_2)_2N-CO-N=C=S$$

Diethylaminocarbonyl isothiocyanate; 45%
No thiocyanate!

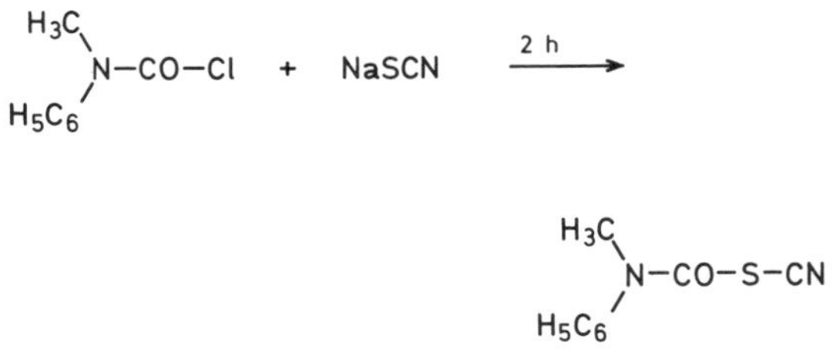

$$\begin{array}{c} H_3C \\ \diagdown \\ N-CO-Cl \\ \diagup \\ H_5C_6 \end{array} \;+\; NaSCN \xrightarrow{\text{2 h}}$$

$$\begin{array}{c} H_3C \\ \diagdown \\ N-CO-S-CN \\ \diagup \\ H_5C_6 \end{array}$$

N-Methylanilinocarbonyl thiocyanate; 89%
No isothiocyanate!

The acylations are carried out either heterogeneously in benzene or toluene with heavy metal thiocyanates or homogeneously with alkali metal thiocyanates in acetone, acetonitrile, or ethyl acetate. In individual cases silicon tetraisocyanate[1243], trimethylsilyl isocyanate[1243] or the corresponding isothiocyanate[1244, 1245] and also dialkyl arsinoisothiocyanate[1246] have been employed.

Acyl isocyanates[1247], *sulfonyl isocyanates*[1248], and *acyl isothiocyanates*[1241] have been described in surveys[1249]. Re the preparation and conversion of *phosphoryl isocyanates* and *phosphoryl isothiocyanates* see refs.[1250, 1251].

[1232] *A. Hassner, M.E. Lorber, C. Heathcock,* J. Org. Chem. *32,* 540 (1967).

[1233] *A. Hassner, R.P. Hoblitt, C. Heathcock, J.E. Kropp, M. Lorber,* J. Amer. Chem. Soc. *92,* 1326 (1970).

[1234] DBP. 1443873 (1968), Farbenfabriken Bayer AG; Erf.: *G. Oertel, H. Holtschmidt.*

[1234a] *L. Birkenbach, K. Kraus,* Ber. *71,* 1492 (1938).

[1235] *O.S. Bhanot, N.K. Ralhan, K.S. Narang,* Indian J. Chem. *2,* 238 (1964).

[1236] *M. Bögemann, S. Petersen, O.E. Schultz, H. Söll* in Houben-Weyl, Methoden der organischen Chemie, 4. Aufl., Bd. IX, p. 878, Georg Thieme Verlag, Stuttgart 1955.

[1237] *R. Pohloudek-Fabini, E. Schröpl,* Pharm. Zentralhalle *107,* 277 (1968).

[1238] *S. Petersen* in Houben-Weyl, Methoden der organischen Chemie, 4. Aufl., Bd. VIII, p. 128, Georg Thieme Verlag, Stuttgart 1952.

[1239] *P.R. Steyermark,* J. Org. Chem. *28,* 586 (1963).

[1240] *W.C. Firth jr.,* J. Org. Chem. *33,* 441 (1968).

[1241] *J. Goerdeler,* Quart. Rep. Sulfur Chem. *5,* 169 (1970).

[1242] *J. Goerdeler, D. Wobig,* Justus Liebigs Ann. Chem. *731,* 120 (1970).

[1243] *O. Glemser, U. Biermann, M. Fild,* Chem. Ber. *100,* 1082 (1967).

[1244] DBP. 1215144 (1966), Farbenfabriken Bayer AG; Erf.: *B. Anders, H. Malz.*

[1245] *K.C. Murdock, R.B. Angier,* J. Org. Chem. *27,* 3317 (1962).

[1246] USSR. P. 215964 (1968), Phytopathology Res. Inst.; Inv.: *A.F. Kolomiets, G.S. Levskaya, N.K. Bliznyuk.*

[1247] *K.A. Nuridzhanyan,* Russ. Chem. Rev. *39,* 130 (1970).

[1248] *H. Ulrich,* Chem. Rev. *65,* 369 (1965).

[1249] *M.O. Lozinskii, P.S. Pel'kis,* Russian Chem. Reviews *37,* 363 (1968).

[1250] *G.I. Derkatsch,* Angew. Chem. *81,* 407 (1969).

[1251] *G.I. Derkatsch, Sh.M. Ivanova,* Z. Chem. *9,* 369 (1969).

Table 14. Acylation of heterocumulene anions NCY$^\ominus$

Formula	Name	Ref.
(Ar)R–CO–NCO	*Aroyl* and *acyl isocyanate*	[1238, 1245]
(Ar)R–SO$_2$–NCO	*Aryl-* and *alkylsulfonyl isocyanate*	[1248]
F–CO–NCO	*Fluorocarbonyl isocyanate*	[1243]
OCN–CO–NCO	*Carbonyl diisocyanate*	[1252]
(Ar)R–CO–NCS	*Aroyl and acyl isothiocyanate*	[1236, 1237] [1253]
F–CO–NCS	*Fluorocarbonyl isothiocyanate*	[1252]
SCN–CO–NCS	*Carbonyl diisothiocyanate*	[1244]
(Ar)R–O–CO–NCS	*Phenoxycarbonyl* and *Alkoxycarbonyl isothiocyanate*	[1244] [1254–1256]
(Ar)R–S–CO–NCS	*Alkyl-* and *Phenylthio-carbonyl isothiocyanates*	[1241, 1254]
(Ar)HNH–CO–NCS	*(Aryl) Aminocarbonyl isothiocyanate*	[1242]
(Ar$_2$)R$_2$N–CO–NCS	*Diphenyl-* and *Dialkyl-carbamoyl isothiocyanate*	[1242, 1257] [1258]
Ar–CS–NCS	*Thioaroyl isothiocyanate*	[1259]
F–CS–NCS	*Fluorothiocarbonyl isothiocyanate*	[1260]
(Ar)R–S–CS–NCS	*Alkyl-* and *Phenylthiothio-carbonyl isothiocyanate*	[1241]
R$_2$N–CS–NCS	*Dialkylaminothiocarbonyl isothiocyanate*	[1261, 1261a]
RO–SO$_2$–NCS	*Alkoxysulfonyl isothiocyanate*	[1262]
(Ar)R–N=C–Ar | NCS	*(Arylimidocarbonyl)-isothiocyanate*	[1263, 1263a]
Ar–N=C–NR$_2$ | NCS	*Amidino-isothiocyanate*	[1263b]
H$_5$C$_6$–CO–NCSe	*Benzoyl isoselenocyanate*	[1210]
Ar–N=C–Ar | NCSe	*(Arylimidocarbonyl)-isoselenocyanate*	[1264]

17.6.2 Isocyanates, isothiocyanates, carbodiimides from amines, imines, amides, amidines, and imidic acid esters

17.6.2.1 Reaction of amines, imines, sulfonamides and amidines with phosgene, thiophosgene and imidocarbonylchlorides

The conversion of aliphatic and aromatic primary amines into *isocyanates*[1265–1267] and *isothiocyanates*[1268] with phosgene and thiophosgene respectively has been known for a long time and is described in detail in the literature. On account of its almost universal applicability to mono- and polyamines[1269] and their hydrochlorides, this phosgenation reaction is among the most important methods of preparing isocyanates. It requires elevated temperatures, is carried out in inert solvents (toluene, xylene, chlorobenzen, *etc.*) and can be understood by the initial formation of a carbamoyl chloride which decomposes into *isocyanate* and hydrogen chloride on heating:

$$R{-}NH_2 \quad + \quad COCl_2 \quad \xrightarrow[-\ HCl]{} $$

$$R{-}NH{-}CO{-}Cl \quad \xrightarrow[-\ HCl]{} \quad R{-}N{=}C{=}O$$

The lower alkyl isocyanates (methyl to propyl) are not readily accessible by this technique, nor short-chain 1,ω-alkyl diisocyanates. Double bonds, halogens, sulfones, ester, cyano, and azo groups are not attacked as a rule. Carboxylic acids[1270] and sulfonic acids are converted into acid chlorides, while aliphatic ether and thioether bonds are often cleaved with formation of chlorinated products. Under certain conditions aziri-

[1252] *W. Verbeek, W. Sundermeyer,* Angew. Chem. *79,* 860 (1967).

[1253] *M. Lipp, F. Dallacker, G. Koenen,* Chem. Ber. *91,* 1660 (1958).

[1254] *A. Takamizawa, K. Hirai, K. Matsui,* Bull. Chem. Soc. Japan *36,* 1214 (1963).

[1255] *J. Goerdeler, J. Neuffer,* Chem. Ber. *104,* 1606 (1971).

[1256] *J. Goerdeler, H. Horn,* Chem. Ber. *96,* 1551 (1963).

[1257] *L.A. Spurlock, P.E. Newallis,* J. Org. Chem. *33,* 2073 (1968).

[1258] *J. Goerdeler, D. Wobig,* Angew. Chem. *79,* 272 (1967).

[1259] *G. Barnikow, T. Gabrio,* Z. Chem. *8,* 142 (1968).

[1260] *A. Haas, W. Klug,* Angew. Chem. *79,* 978 (1967).

[1261] *J. Goerdeler, H. Lüdke,* Tetrahedron Lett. 2455 (1968).

[1261a] *J. Goerdeler, H. Lüdke,* Chem. Ber. *103,* 3393 (1970).

[1262] *W. Ried, B.M. Beck,* Justus Liebigs Ann. Chem. *673,* 131 (1964).

[1263] *J. Goerdeler, D. Weber,* Chem. Ber. *101,* 3475 (1968).

[1263a] *N.A. Kirsanova, G.I. Derkac,* Zh. Org. Khim. *5,* 1285 (1969).

[1263b] *G. Barnikow, W. Abraham,* Z. Chem. *10,* 193 (1970).

[1264] *N.A. Kirsanova, G.I. Derkac,* Ukr. Khim. Zh. *36,* 372 (1970).

[1265] *W. Siefken,* Justus Liebigs Ann. Chem. *562,* 75 (1949).

[1266] *S. Petersen* in *Houben-Weyl,* Methoden der organischen Chemie, 4. Aufl., Bd. VIII, p. 120, Georg Thieme Verlag, Stuttgart 1952.

[1267] *J.H. Saunders, K.C. Frisch,* Polyurethanes, Chemistry and Technology, Part 1, p. 18, Interscience Publishers, New York, London 1962.

[1268] *M. Bögemann, S. Petersen, O.E. Schultz, H. Söll* in *Houben-Weyl,* Methoden der organischen Chemie, 4. Aufl., Bd. IX, p. 875, Georg Thieme Verlag, Stuttgart 1955.

[1269] *W.J. Schnabel, E. Kober,* J. Org. Chem. *34,* 1162 (1969).

[1270] *Y. Iwakura, K. Uno, S. Kang,* J. Org. Chem. *30,* 1158 (1965).

[1271] *C.K. Johnson.* J. Org. Chem. *32,* 1508 (1967).

dines afford *β-chloroalkyl isocyanates*[1271, 1271a] and, with thiophosgene, *isothiocyanates*[1272]. On account of the greater resistance of thiophosgene and mustard oils to hydrolysis amines can be reacted with thiophosgene in aqueous medium; the reaction proceeds entirely analogously *via* thio-carbamoyl chlorides.

To prevent formation of symmetrical ureas during the *phosgenation* the amine hydrochlorides are generally used. In the same way it is necessary to employ the salts

$$R-NH_2 \cdot HCl \quad + \quad Cl_2C=N-R^1 \quad \xrightarrow{-3\,HCl}$$

$$R-N=C=N-R^1$$

for the preparation of *carbodiimides* from imido-carbonyl chlorides and amines *(imidophosgena-tion)*[1273], because the free bases undergo a direct further reaction to guanidines under the normal reaction conditions. While alkyl and aryl imido-carbonyl chlorides require temperatures of be-tween 170 and 180° (1, 2-dichlorobenzene solu-tion)[1274], (sulfonylimido)carbonyl chlorides react even in boiling toluene[1275–1277]. *N-(2-Chloroal-kyl)-N'-arylcarbodiimides* can be obtained from aziridines and (arylimido)carbonyl chlorides[1277a]. In dependence on the nature of the substituents ketimines form *1-alken-1-yl isocyanates* or *N-chlorocarbonyl ketimines (methylenecarbamoyl chlorides)* with phosgene; the latter products are in equilibrium with the α-chloroisocya-nates[1278, 1279]:

[1271a] *D.A. Tomalia, J.N. Paige*, J. Heterocycl. Chem. *4*, 178 (1967).

[1272] *D.A. Tomalia*, J. Heterocycl. Chem. *3*, 384 (1966).

[1273] *E. Kühle*, Angew. Chem. *81*, 18 (1969).

[1274] DBP. 1 149 712 (1961), Farbenfabriken Bayer AG; Erf.: *E. Kühle*.

[1275] *B. Anders, E. Kühle*, Angew. Chem. *77*, 430 (1965).

[1276] *R. Neidlein, E. Heukelbach*, Tetrahedron Lett. 2665 (1965).

[1277] *R. Neidlein, E. Heukelbach*, Arch. Pharm. (Wein-heim, Ger.) *299*, 944 (1966).

[1277a] *D.A. Tomalia, Th.J. Giacobbe, W.A. Sprenger*, J. Org. Chem. *36*, 2142 (1971).

[1278] *L.I. Samaraj, O.W. Wichnewskij, G.I. Derkatsch*, Angew. Chem. *80*, 620 (1968).

H_5C_6 — $C=NH$ + $COCl_2$ $\xrightarrow{-HCl}$

(Diphenylmethylene)-carbamoyl chloride   *(Chlorodiphenyl-methyl) isocyanate*

Silyl derivatives of the amines react with phosgene, thiophosgene[1280, 1281, 1281a], and imidocarbonyl chlorides[1282] like free amines or their hydrochlo-rides.

$$\xrightarrow{-2\,(H_3C)_3SiCl} \quad R-N=C=X$$

X = O, S, N – Ar

The action of phosgene on primary carboxamides normally does not lead to acyl isocyanates but to nitriles with loss of water[1283]. By contrast, direct phosgenation of sulfonamides[1284, 1285] to *sulfonyl isocyanates* can be accomplished successfully, but requires more drastic conditions and as a rule does not afford high yields ($\sim 60\%$). For exam-ple, 4-aminobenzenesulfonamide reacts with phosgene in nitrobenzene to give initially the car-bamoyl chloride, which on further phosgenation at 90° affords *(4-sulfonylaminophenyl) isocyanate*,

[1279] *L.I. Samaraj, V.P. Belaja, O.V. Visnevskij, G.I. Derkac*, Zh. Org. Khim. *4*, 720 (1968).

[1280] *U. Wannagat, H. Bürger, C. Krüger, J. Pump*, Z. Anorg. Allgem. Chem. *321*, 208 (1963).

[1281] *G. Barnikow, W. Abraham*, Z. Chem. *8*, 335 (1968).

[1281a] *V.F. Mironov, V.D. Sheludyakov, V.P. Kozyukov*, Zh. Obshch. Khim. *39*, 2598 (1969); C.A. *72*, 66 300 (1970).

[1282] *K. Itoh, A. Mozawa, Y. Ishii*, Tetrahedron Lett. 1421 (1969).

[1283] *E. Waltmann, E. Wolf*, C.A. *38*, 3990 (1944).

[1284] *F. Effenberger, R. Gleiter*, Chem. Ber. *97*, 1576 (1964).

[1285] *C. King*, J. Org. Chem. *25*, 352 (1960).

[1286] *J. Smith, T.K. Brotherton, J.W. Lynn*, J. Org. Chem. *30*, 1260 (1965).

[1287] *H. Ulrich, B. Tucker, A.A.R. Sayigh*, J. Org. Chem. *31*, 2658 (1966).

[1288] *H. Ulrich, B. Tucker, A.A.R. Sayigh*, Angew. Chem. *79*, 726 (1967).

[1289] *L.I. Samaraj, W.A. Bondar, G.I. Derkatsch*, Angew. Chem. *79*, 897 (1967).

[1290] *L.I. Samaraj, W.A. Bondar, G.I. Derkatsch*, Zh. Org. Khim. *5*, 119 (1969).

but at 150° *(4-isocyanatobenzenesulfonyl) iso-cyanate* is formed [1286]:

SO$_2$—NH$_2$

NH$_2$

↓ + COCl$_2$ / 0°

SO$_2$—NH$_2$

HNCOCl

├ + COCl$_2$ / 90°   ┤ + COCl$_2$ / 150°

SO$_2$—NH$_2$ SO$_2$—N=C=O

N=C=O   N=C=O

The phosgenation of arylsulfonamides was greatly improved by the addition of catalytic amounts of an alkyl isocyanate [1287, 1288], to make a wide variety of substituted *arylsulfonyl isocyanates* accessible in good yield (see p. 779).

Other noteworthy reactions are those between amidines and phosgene [1289, 1240], which in the case of the *N*-arylamidines lead quantitatively to *4-quinazolinol via* the *N*-arylimino isocyanate* intermediate stage, while from *N*-alkyl amidines the relatively stable *N-alkylimino isocyanates* are obtained with phosgene:

$$R-C\overset{NH-R^1}{\underset{NH}{}} + COCl_2 \xrightarrow{-2\,HCl} R-C\overset{N-R^1}{\underset{N=C=O}{}}$$

R = CCl$_3$, C$_6$H$_5$
R^1 = Alkyl

Re the reaction of monosilylated amidines with thiophosgene see ref. [1290a].

17.6.2.2 Reaction of amides, imidic acid esters and heterocyclic amines with oxalyl chloride

A particularly valuable method for preparing *acyl isocyanates* is the reaction between amides and oxalyl chloride. In the particular case of the carb-

oxamides [1291–1293] (X=O, Y=R) the reaction displays the advantage over acylating silver cyanates (*cf.* p. 773) that α-halocarbonyl compounds, too, can be reacted [1294, 1295].

$$Y-\overset{X}{\overset{\|}{C}}-NH_2 + (COCl)_2 \xrightarrow{-2\,HCl}$$

$$\xrightarrow{-CO} Y-\overset{X}{\overset{\|}{C}}-N=C=O$$

X = O, S, N–R
Y = R, OR, SR, NR$_2$
R = Alkyl, Aryl

Excellent results are obtained with amides of aromatic mono [1293]- and dicarboxylic acids [1296, 1297], while aliphatic amides generally react satisfactorily only if they carry electron-attracting substituents in the α-position [1293]. Cyclopropanecarboxamide furnishes *4-chlorobutyryl isocyanate* (87%) in place of the expected cyclopropylcarbonyl isocyanate [1293]. In the same way aromatic thiocarboxamides react to *2-aryl-4,5-thiazolediones* (X=S, Y=Aryl) in absolute acetone at low temperatures [1299]; on heating in an inert solvent these decarbonylate to form the highly reactive monomeric *thioaroyl isocyanates* [1300, 1301].

The reaction between oxalyl chloride and sulfonamides proceeds in less clear-cut manner [1302].

[1290a] *G. Barnikow, H. Ebeling*, Z. Chem. *11*, 103 (1971).

[1291] *A.J. Speziale, L.R. Smith*, J. Org. Chem. *27*, 3742 (1962).

[1292] *A.J. Speziale, L.R. Smith*, J. Org. Chem. *28*, 1805 (1963).

[1293] *A.J. Speziale, L.R. Smith, J.E. Fedder*, J. Org. Chem. *30*, 4306 (1965).

[1294] *A.J. Speziale, L.R. Smith*, Org. Synth. *46*, 16 (1966).

[1295] *R. Lattrell*, Justus Liebigs Ann. Chem. *722*, 142 (1969).

[1296] *R. Neidlein, R. Bottler*, Chem. Ber. *100*, 698 (1967).

[1297] *O. Tsuge, T. Itoh, M. Tashiro*, Tetrahedron *24*, 2583 (1968).

[1298] *G. Zinner, R. Stoffel*, Arch. Pharm. (Weinheim, Ger.) *302*, 691 (1969).

[1299] *J. Goerdeler, H. Horstmann*, Chem. Ber. *93*, 671 (1960).

[1300] *J. Goerdeler, H. Schenk*, Angew. Chem. *75*, 675 (1963).

[1301] *J. Goerdeler, H. Schenk*, Chem. Ber. *98*, 2954 (1965).

[1302] *J.W. McFarland, C.F. Gerber, W.M. McLamore*, J. Med. Chem. *8*, 781 (1965).

Benzenesulfonamide and excess oxalyl chloride afford *(phenylsulfonyl)oxamoyl chloride* (95%), which can be decomposed thermally into *benzenesulfonyl isocyanate*[1303]:

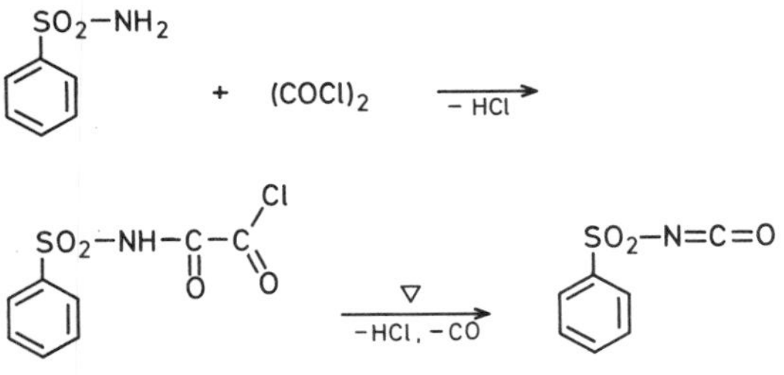

Similarly methyl *N*-chlorooxalyl imidates are formed from imidic esters; their fragmentation leads to *acyl isocyanates*[1304]:

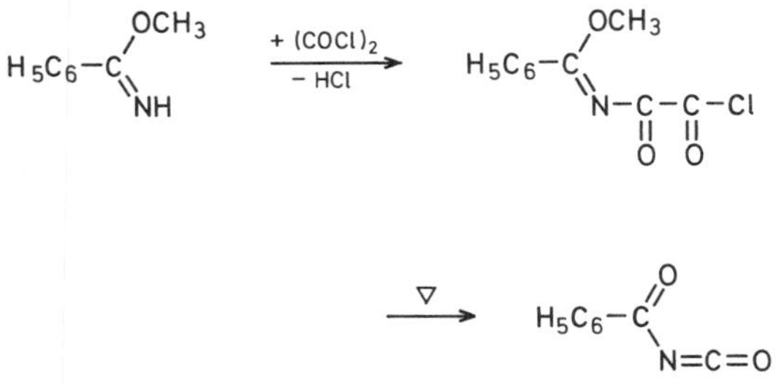

Benzoyl isocyanate

The *1-arylimidazole-4,5-diones (X=N, Y=R)* formed from *N*-arylamidines with oxalyl chloride are converted into 4-quinazolinenols on heating[1305].

In addition, *2-isocyanatopyridines, 2-isocyanatopyrimidines, 2-isocyanatopyrazines*[1306], and *2-isocyanatoquinolines*[1308], and also *(di)isocyanato-1,3,5-triazines*[1307] have been successfully prepared by reacting the corresponding amino derivatives with oxalyl chloride.

Alkyl and phenyl carbamates *(X=O, Y=OR)* form *alkoxy* and *phenoxy carbonylisocyanates* respectively with oxalyl chloride[1293, 1298]. In the same way *N,N*-disubstituted thioureas, *O*-esters of thiocarbamic acid, and esters of dithiocarbamic acid react with oxalyl chloride to form *2-substituted thiazole-4,5-diones (X=S, Y=NR$_2$, OR, SR)* in analogy with the reaction of thiocarboxamides

(see p. 776). Thermolysis of the products opens up a route to the synthesis of *dialkylamino* and *diarylamino*[1301], of *alkylarylamino, alkoxy, phenoxy, alkylthio* and *phenylthio thiocarbonyl isocyanates*[1309, 1310]. However, a few only of these could be obtained in substance, because their preparative isolation is made difficult by the readily ensuing dimerization to *tetrahydro-1,3,5-thiadiazinediones*[1310] and their rearrangement to acyl mustard oils[1309, 1310]:

$$Y-CS-N=C=O \longrightarrow Y-CO-N=C=S$$

Y = NR$_2$, OR, SR
R = Alkyl, Aryl

17.6.3 Isocyanates, isothiocyanates, isoselenocyanates, carbodiimides from ureas, thioureas and thiopseudoureas, aminocarbonylimidazoles, and aminothiocarbonylimidazoles, carbamic, thiocarbamic, dithiocarbamic, diselenocarbamic and imidodithio carbonic acid derivatives

17.6.3.1 By reaction with heavy metal salts

Reacting heavy metal salts or oxides with

1 *N*-monosubstituted salts of dithiocarbamic acid[1311, 1312]
2 *N*-monosubstituted esters of dithiocarbamic acid[1313, 1314]
3 *N*-monosubstituted *S*-esters of thiocarbonic acid amide[1313]
4 1,3-disubstituted thioureas[1315–1318]
5 1,3-disubstituted *S*-alkylpseudoureas[1313]
6 *N*-monosubstituted salts of diselenocarbamic acid[1319]

[1303] *J.E. Franz, C. Osuch*, J. Org. Chem. *29*, 2592 (1964).
[1304] *L.I. Samaraj, V.P. Belaja, G.F. Galenko, G.I. Derkac*, Zh. Org. Khim. *6*, 85 (1970).
[1305] *L.I. Samaraj, W.A. Bondar, G.I. Derkatsch*, Angew. Chem. *79*, 897 (1967).
[1306] *U. v. Gizycki*, Angew. Chem. *83*, 406 (1971).
[1307] *U. v. Gizycki*, Angew. Chem. *83*, 407 (1971).
[1308] *U. v. Gizycki*, Angew. Chem. *83*, 408 (1971).

[1309] *H. Schenk*, Chem. Ber. *99*, 1258 (1966).
[1310] *J. Goerdeler, K. Jonas*, Chem. Ber. *99*, 3572 (1966).
[1311] *M. Bögemann, S. Petersen, O.E. Schultz, H. Söll* in *Houben-Weyl*, Methoden der organischen Chemie, 4. Aufl., Bd. IX, p. 869, Georg Thieme Verlag, Stuttgart 1955.
[1312] *F.B. Dains, R.Q. Brewster, C.P. Olander*, Org. Synth. Coll. Vol. I, 447 (1958).
[1313] *A.F. Ferris, B.A. Schutz*, J. Org. Chem. *28*, 71 (1963).
[1314] *G.J.M. van der Kerk, C.W. Pluygers, G. de Vries*, Rec. Trav. Chim. Pays-Bas *74*, 1262 (1955).
[1315] *F. Kurzer, K. Douraghi-Zadeh*, Chem. Rev. *67*, 107 (1967).
[1316] *G. Losse, H.J. Sirch*, Z. Chem. *7*, 234 (1967).
[1317] *G. Rapi, G. Sbrana*, Chem. Commun. 128 (1968).
[1318] *Y. Iwakura, K. Noguchi*, Bull. Chem. Soc. Japan *40*, 2383 (1967).
[1319] *W.J. Franklin, R.L. Werner*, Tetrahedron Lett. 3006 (1965).

affords in part excellent yields of *isothiocyanates* (**1**+**2**), *isocyanates* (**3**), *carbodiimides* (**4**+**5**), and *isoselenocyanates* (**6**). Emphasis should be given here to the classical method of preparing mono-, bis-, and cyclic *carbodiimides*

$$R-NH-CS-NH-R^1 \quad + \quad HgO \quad \longrightarrow$$

$$R-N=C=N-R^1 \quad + \quad HgS \quad + \quad H_2O$$

by desulfurization of 1,3-disubstituted thioureas with mercury oxide[1315-1318]. *1,3-Disilylated carbodiimides*, too, can be obtained in this way from 1,3-disilylated *thioureas*, but, because of their extreme sensitivity to hydrolysis, it is necessary to employ (imidazolyl)silver in place of mercury oxide[1320].

17.6.3.2 By acidolytic elimination

Ureas, urethans, and their thio analogs, dithiocarbamates and salts of dithiocarbonic acid sulfonylimides can be converted into *isocyanates, isothiocyanates,* and *carbodiimides* by reaction with acid chlorides particularly with phosgene. Thus, dehydrating *N,N'*-dicyclohexylurea with *p*-toluenesulfonyl chloride in pyridine affords dicyclohexylcarbodiimide (82%)[1321]:

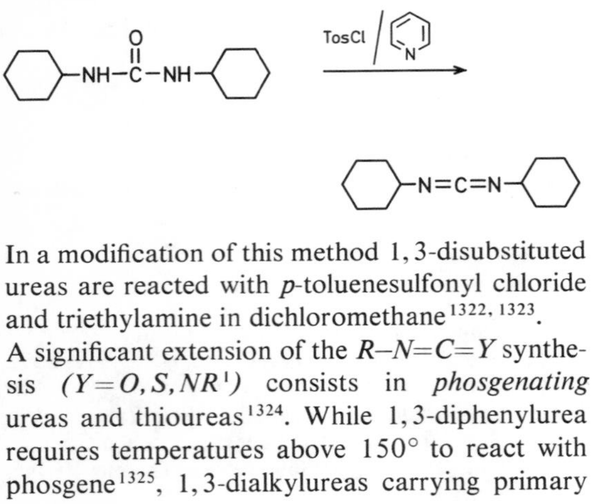

In a modification of this method 1,3-disubstituted ureas are reacted with *p*-toluenesulfonyl chloride and triethylamine in dichloromethane[1322, 1323].

A significant extension of the *R–N=C=Y* synthesis *(Y=O,S,NR¹)* consists in *phosgenating* ureas and thioureas[1324]. While 1,3-diphenylurea requires temperatures above 150° to react with phosgene[1325], 1,3-dialkylureas carrying primary carbon atoms adjacent the nitrogen react at room temperature to give for the most part *allophanoyl chlorides (N-chlorocarbonylureas)*[1326]:

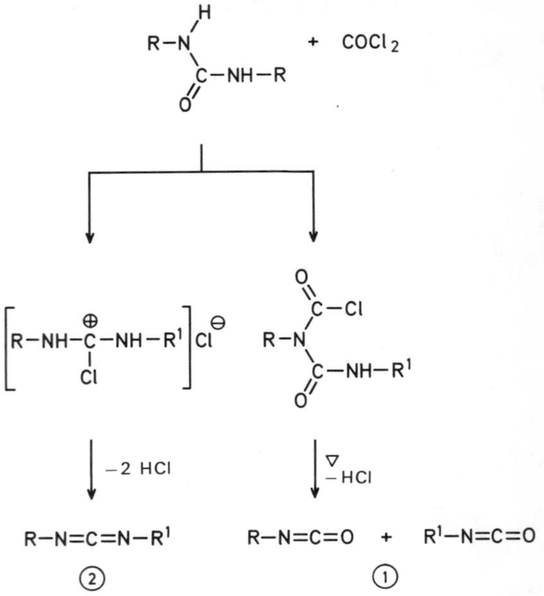

R = R¹ = Alkyl

On heating in an inert solvent at above 100° the allophanoyl chlorides decompose almost quantitatively into *isocyanates* and hydrogen chloride[1327]. By contrast, where the carbon atom adjacent the nitrogen atom is a secondary one almost exclusively chloroformamidinium chlorides are formed[1326, 1328] which dehalogenate to *carbodiimides* in the presence of bases[1328]. Diphenyl carbonate[1329] or thionyl chloride[1324] may be used in place of phosgene.

The transferability of the phosgenation of 1,3-dialkylureas to 1-alkyl-3-arylsulfonylureas is of preparative significance[1330]; *alkyl* and *sulfonyl isocyanates*, easily separated by distillation, are formed:

$$p-H_3C-C_6H_4-SO_2-NH-CO-NH-R \xrightarrow[-2\,HCl]{COCl_2}$$

$$p-H_3C-C_6H_4-SO_2-N=C=O \quad + \quad R-N=C=O$$

p-Tosyl isocyanate

[1320] *L. Birkofer, A. Ritter, P. Richter,* Tetrahedron Lett. 195 (1962).
[1321] *G. Amiard, R. Heymes,* Bull. Soc. Chim. France 1360 (1956).
[1322] *J.C. Sheehan, P.A. Cruickshank, G.L. Boshart,* J. Org. Chem. *26*, 2525 (1961).
[1323] *J.C. Sheehan, P.A. Cruickshank,* Org. Synth. *48*, 83 (1968).
[1324] *H. Ulrich, A.A.R. Sayigh,* Angew. Chem. *78*, 761 (1966).
[1325] *S. Petersen* in *Houben-Weyl*, Methoden der organischen Chemie, 4. Aufl., Bd. VIII, p. 123, Georg Thieme Verlag, Stuttgart 1952.
[1326] *H. Ulrich, A.A.R. Sayigh, J.N. Tilley,* J. Org. Chem. *29*, 2401 (1964).
[1327] *A.A.R. Sayigh, J.N. Tilley, H. Ulrich,* J. Org. Chem. *29*, 3344 (1964).
[1328] *H. Eilingsfeld, M. Seefelder, H. Weidinger,* Angew. Chem. *72*, 836 (1960).
[1329] DBP. 1154090 (1963), Farbenfabriken Bayer AG; Erf.: *H. Morschel, C. Skopalik.*
[1330] *H. Ulrich, B. Tucker, A.A.R. Sayigh,* J. Org. Chem. *31*, 2658 (1966).

This reaction lends itself for the preparation of isocyanates from amines which cannot be phosgenated directly and as a result need to be converted into urea derivatives. *Arylsulfonyl isocyanates* are made readily accessible starting from arylsulfonamides (cf. p. 776) by phosgenating these in the presence of catalytic amounts of an alkyl isocyanate (with intermediate formation of the 1-alkyl-3-arylsulfonylurea and continuous regenerating of the alkyl isocyanate).

The reaction of cyclic ureas with phosgene can be used successfully for preparing short-chain $1,\omega$-*alkyl diisocyanates (cf.* p. 774)[1327]:

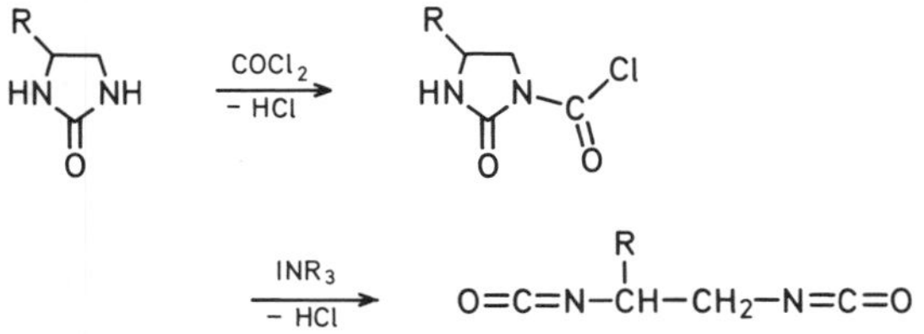

N-monosubstituted (thio)carbamic acid *O*-esters, too, can be converted into isocyanates with phosgene (in the presence of dimethylformamide)[1331] or 2,2,2-trichloro-1,3,2-benzodioxaphosphole[1332].

Like phosgene and thionyl chloride, phosphorus-(V) chloride also reacts in a different way with 1,3-disubstituted ureas depending on the nature of the substituents. If there are primary carbon atoms adjacent the nitrogen the initially formed 1,3,2-diazophosphetanes[1333, 1334] decompose thermally into isocyanates and *N*-alkylphosphorimidic trichlorides, which react further to *carbodiimides* and phosphoryl chloride[1335],

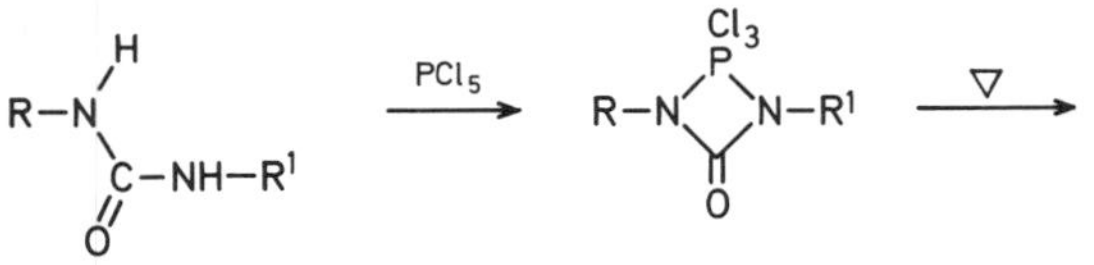

$$R^1N=C=O + R-N=PCl_3 \longrightarrow R-N=C=N-R^1 + POCl_3$$

unlike the correspondingly prepared sulfonylphosphorimidic trichlorides, which do not form carbodiimides with isocyanates[1334]. *Carbodiimides* are obtained under very mild conditions and in high yield also by replacing phosphorus(V) chlo-

ride by triphenylphosphine dibromide[1336] or triphenylphosphine/carbon tetrachloride[1337]:

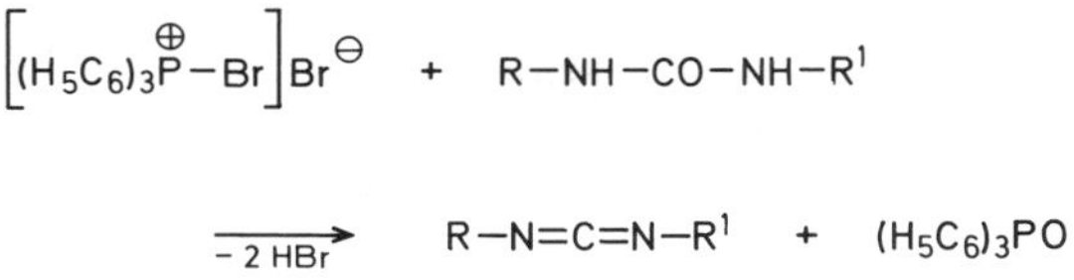

Aromatically and aliphatically 1,3-disubstituted thioureas react with phosgene to form chloroformamidinium chlorides (and with aromatic substitution to some iminouretidinones)[1328]:

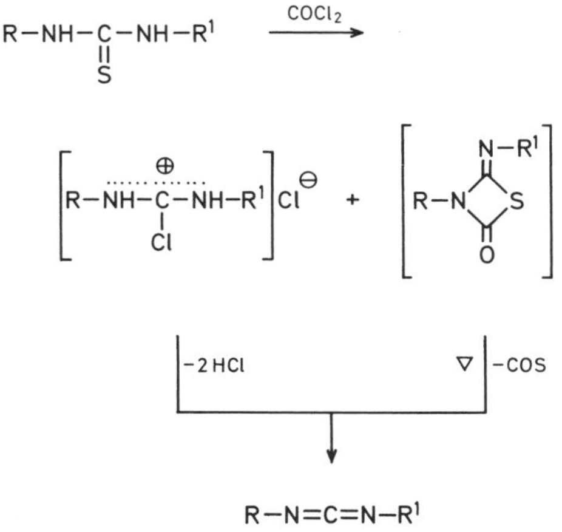

a) R = R¹ = Alkyl, Aryl
b) R = Acyl; R¹ = Alkyl
c) R = Sulfonyl; R¹ = Alkyl

As the basicity of the nitrogen in the aromatic chloroformamidinium chlorides is diminished appreciably, they lose hydrogen chloride even on heating to form *carbodiimides* (83%)[1338].

Treating *N*-acyl-*N*′-alkylthioureas with phosgene makes the extremely reactive *N*-acyl-*N*′-alkylcarbodiimides available[1339, 1340]. In the same way the *N*′-alkyl-*N*-sulfonylcarbodiimides are obtained from 1-alkyl-3-aryl- and 1-alkyl-3-alkylsulfonylthioureas *via* the *N*-sulfonylchloroformamidines[1341, 1342].

[1331] *H. Ulrich, B. Tucker, A.A.R. Sayigh,* Angew. Chem. *79,* 651 (1967).

[1332] *H. Gross, J. Gloede,* Chem. Ber. *96,* 1387 (1963).

[1333] *H. Ulrich, A.A.R. Sayigh,* Angew. Chem. *76,* 647 (1964).

[1334] *H. Ulrich, A.A.R. Sayigh,* J. Org. Chem. *30,* 2779 (1965).

[1335] *H. Ulrich, A.A.R. Sayigh,* Angew. Chem. *74,* 900 (1962).

[1336] *H.J. Bestmann, J. Lienert, L. Mott,* Justus Liebigs Ann. Chem. *718,* 24 (1968).

[1337] *R. Appel, R. Kleinstück, K.D. Ziehn,* Chem. Ber. *104,* 1335 (1971).

[1338] USP. 3301895 (1967), Upjohn Co., Inv.: *A.A.R. Sayigh, H. Ulrich.*

[1339] *R. Neidlein, E. Heukelbach,* Tetrahedron Lett. 149 (1965).

[1340] *R. Neidlein, E. Heukelbach,* Arch. Pharm. (Weinheim, Ger.) *299,* 709 (1966).

[1341] *H. Ulrich, A.A.R. Sayigh,* Angew. Chem. *76,* 781 (1964).

Imino, aminocarbonyl, and *aminothiocarbonyl carbodiimides* were obtained with cyanuric chloride/triethylamine and correspondingly substituted thioureas[1342a].

The reactions of 1,3-disubstituted thioureas with *N*-acylchloroformamidines[1343], aryl cyanates[1344], quinones[1344a], or diethyl azodicarboxylates and triphenylphosphine[1345] represent further means for synthesizing carbodiimides.

Isothiocyanates are obtainable in good yield from *N*-monosubstituted salts of dithiocarbamic acid by the action of phosgene[1346, 1347], ethyl chloroformate[1348, 1349], phosphoryl chloride, or 2,2,2-trichloro-1,3,2-benzodioxaphosphole[1350], dicyclohexylcarbodiimide[1351, 1352], or aryl cyanates[1353]:

$$R{-}NH{-}CS{-}S^{\ominus} \longrightarrow R{-}N{=}C{=}S$$

While the phosgene method is used mainly in the aromatic series, ethyl carbonate, phosphoryl chloride, and dicyclohexylcarbodiimide can be employed for preparing both aliphatic and aromatic isothiocyanates.

Analogously, the salts of dithiocarbamic acid sulfonylimides afford excellent yields of *N-sulfonyl isothiocyanates*[1354, 1355]:

$$R{-}SO_2{-}N{=}C\begin{smallmatrix}SK\\\\SK\end{smallmatrix} + COCl_2 \xrightarrow{-\,COS}$$

$$R{-}SO_2{-}N{=}C{=}S + 2\ KCl$$

4-Nitroanilines and 4-cyanoanilines are not converted into the salts of dithiocarbamic acid with carbon disulfide and base. In this particular case reaction with *N,N*-diethylthiocarbamoyl chloride has proven successful; here the isothiocyanates are formed from the amines in a single-stage synthesis[1356].

17.6.3.3 By basic elimination

1,3-Disubstituted thioureas are converted into carbodiimides by sodium amide, sodium *tert*-butoxide, or sodium hydride in inert solvents[1357]:

$$R{-}NH{-}CS{-}NH{-}R + NaX \longrightarrow$$

$$R{-}N{=}C{=}N{-}R + NaSH + HX$$

X = NH$_2$, OR^1, H

N-Monoalkyl carbamates can be cleaved into isocyanate and alcohol in the presence of strong bases at 170–230°[1358]. The preparative scope of this reaction is limited by the recombination of isocyanate and alcohol at low temperature and by the tendency of the isocyanates to polymerize in alkaline media. Both side-reactions occur least with *tert*-alkyl isocyanates, so that these can be obtained in good yield in some cases:

$$R{-}NH{-}CO{-}O{-}R^1 \longrightarrow R{-}N{=}C{=}O + R^1{-}OH$$

R = *tert*-Alkyl

Ammonium salts of *N*-monosubstituted dithiocarbamic acids and 1,3-disubstituted thioureas can be converted into *isothiocyanates*[1359] and *carbodiimides*[1360] in excellent yields with sodium chlorite or hypochlorite:

$$R{-}NH{-}CS{-}XH + O \longrightarrow$$

$$R{-}N{=}C{=}X + H_2O + S$$

X = S, N–R

[1342] *H. Ulrich, B. Tucker, A.A.R. Sayigh,* Tetrahedron 22, 1565 (1966).
[1342a] *J. Goerdeler, H. Lohmann, R. Losch, S. Raddatz,* Tetrahedron Lett. 2765 (1971) (cf. 1342 b).
[1342b] *S. Furumoto,* Nippon Kagaku Zasshi 92, 1005 (1971).
[1343] *K. Hartke,* Chem. Ber. 99, 3163 (1966).
[1344] *E. Grigat, R. Pütter,* Chem. Ber. 98, 1168 (1965).
[1344a] *S. Furumoto,* Nippon Kagaku Zasshi 92, 357 (1971).
[1345] *O. Mitsunobu, K. Kato, M. Tomari,* Tetrahedron 26, 5731 (1970).
[1346] *M. Bögemann, S. Petersen, O.E. Schultz, H. Söll* in *Houben-Weyl,* Methoden der organischen Chemie, 4. Aufl., Bd. IX, p. 869, Georg Thieme Verlag, Stuttgart 1955.
[1347] USP. 3 404 171 (1968), Upjohn Co.; Inv.: *H. Ulrich.*
[1348] *J.E. Hodgkins, M.G. Ettlinger,* J. Org. Chem. 21, 404 (1956).
[1349] *J.E. Hodgkins, W.P. Reeves,* J. Org. Chem. 29, 3098 (1964).
[1350] *D. Martin, E. Beyer, H. Gross,* Chem. Ber. 98, 2425 (1965).
[1351] *J.C. Jochims, A. Seeliger,* Angew. Chem. 79, 151 (1967).
[1352] *J.C. Jochims,* Chem. Ber. 101, 1746 (1968).
[1353] *D. Martin,* Chem. Ber. 98, 3286 (1965).
[1354] *K. Dickoré, E. Kühle,* Angew. Chem. 77, 429 (1965).
[1355] *K. Hartke,* Arch. Pharm. (Weinheim, Ger.) 299, 174 (1966).
[1356] *A.A.R. Sayigh, H. Ulrich, J.S. Potts,* J. Org. Chem. 30, 2465 (1965).
[1357] *P. Schlack, G. Keil,* Justus Liebigs Ann. Chem. 661, 164 (1963).
[1358] *N. Bortnick, L.S. Luskin, M.D. Hurwitz, A.W. Rytina,* J. Amer. Chem. Soc. 78, 4358 (1956).
[1359] *E. Schmidt, L. Fehr,* Justus Liebigs Ann. Chem. 621, 1 (1959).
[1360] *E. Schmidt, D. Ross, J. Kittl, H.H. von Düsel,* Justus Liebigs Ann. Chem. 612, 11 (1958).

This technique is applicable generally[1361] and is catalyzed by copper salts.

17.6.3.4 By thermal elimination

Under certain conditions thermal fragmentation of substituted ureas, thioureas, and isothioureas represents a good method for preparing *isocyanates*, *isothiocyanates*, and *carbodiimides*. Thus *aryl isothiocyanates*[1362] are obtained from *N*-aryl-thioureas, *methyl isocyanate* (100%)[1363] from 3-methyl-1, 1-diphenylurea. Cleavage of S-methyl-thiopseudoureas allows numerous *mono-* and *bis-carbodiimides* to be synthesized[1364]; their rate of formation is markedly dependent on the space occupied by the groups R and R¹:

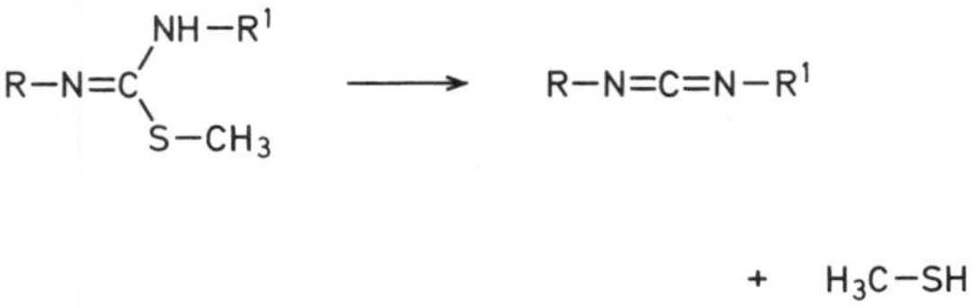

Pyrolysis of 1-alkyl-3-sulfonylureas and 1-aryl-3-sulfonylureas[1365], accessible readily and quantitatively from sulfonyl isocyanates and amines, is a particularly advantageous procedure for preparing *isocyanates* capable of reacting further with hydrogen chloride, *e.g.*, $RO-(CH_2)_n-N=C=O$, and which therefore cannot be obtained by direct phosgenation:

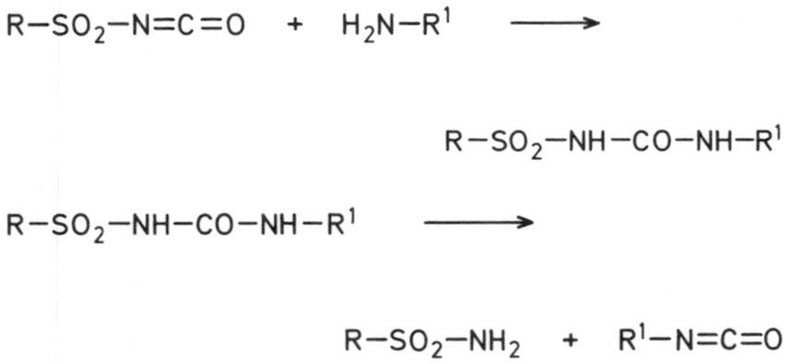

$$R-SO_2-N=C=O \ + \ H_2N-R^1 \longrightarrow$$

$$R-SO_2-NH-CO-NH-R^1$$

$$R-SO_2-NH-CO-NH-R^1 \longrightarrow$$

$$R-SO_2-NH_2 \ + \ R^1-N=C=O$$

As the sulfonylureas need not necessarily be isolated, aliphatic and aromatic amines can be converted into isocyanates directly by reaction with arylsulfonyl isocyanates.

A both simple and elegant isocyanate synthesis relies on the fact that imidazole-1-carboxamides (from carbonyldiimidazole and primary amines) dissociate into isocyanate and imidazole in solution even at room temperature[1366].

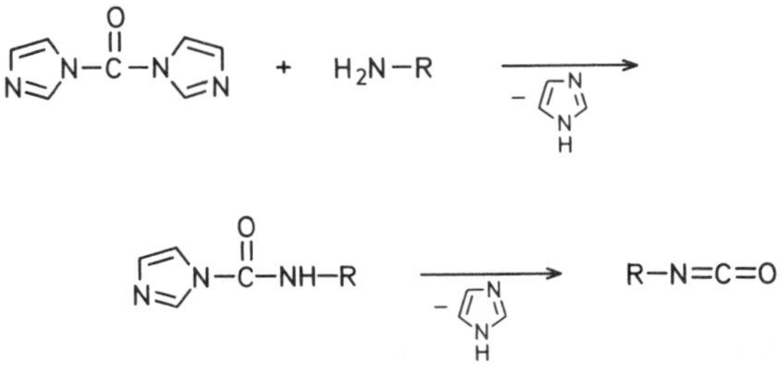

In particular cases this method, too, displays the advantage that it is possible to work in neutral medium without formation of hydrogen chloride[1367–1369]. *Isothiocyanates* are obtained entirely analogously from thiocarbonyldiimidazole and primary amines[1368, 1370, 1371].

As a rule carbamoyl chlorides and esters of carbamic acid afford only poor yields of *isocyanates*, because the temperatures (200–300°) required for fragmentation favor side-reactions. However, the stability of *N, N*-dialkylcarbamoyl chlorides can be reduced drastically by introducing strongly branched α-carbon atoms. Thus *N*-alkyl-*N*-tert-butylcarbamoyl chlorides fragment smoothly to alkyl isocyanates in polar solvents such as nitrobenzene or 1, 2-dichloroethane[1372, 1373]:

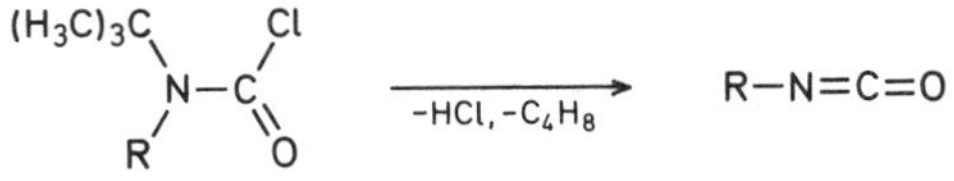

Thermolysis of *N*-monosubstituted carbamoyl chlorides and anhydrides and esters of carbamic acid is facilitated also by their *N*-silylation. The derivatives formed decompose into *isocyanate*

[1361] *H. Stetter, C. Wulff,* Chem. Ber. *95,* 2302 (1962).

[1362] *J.N. Baxter, J. Cymerman-Craig, M. Moyle, R.A. White,* J. Chem. Soc. (London) 659 (1956).

[1363] *S. Petersen* in *Houben-Weyl,* Methoden der organischen Chemie, 4. Aufl., Bd. VIII, p. 127, Georg Thieme Verlag, Stuttgart 1952.

[1364] *P. Schlack, G. Keil,* Justus Liebigs Ann. Chem. *661,* 164 (1963) (several references).

[1365] *H. Ulrich, A.A.R. Sayigh,* Angew. Chem. *78,* 746 (1966).

[1366] *H.A. Staab, W. Benz,* Angew. Chem. *73,* 66 (1961).

[1367] *H.A. Staab, W. Benz,* Justus Liebigs Ann. Chem. *648,* 72 (1961).

[1368] *H.A. Staab,* Angew. Chem. *74,* 407 (1962).

[1369] *H.A. Staab, W. Rohr* in *W. Foerst,* Neuere Methoden der präparativen organischen Chemie, Bd. V, p. 53, Verlag Chemie, Weinheim/Bergstr. 1967.

[1370] *H.A. Staab, G. Walther,* Justus Liebigs Ann. Chem. *657,* 104 (1962).

[1371] *U. Anthoni, Ch. Larsen, P.H. Nielsen,* Acta Chem. Scand. *21,* 2061 (1967).

[1372] *J.N. Tilley, A.A.R. Sayigh,* J. Org. Chem. *28,* 2076 (1963).

[1373] DOS. 1922412 (1970), Badische Anilin- und Soda-Fabrik; Erf.: *K.H. König, H. Kiefer.*

and a silicon compound at temperatures as low as −20 to + 120° [1374, 1374a−c] *(Y = O)*

$$R^1-NH-\underset{X}{\overset{Y}{C}} \xrightarrow{(H_3C)_3SiCl} \underset{(H_3C)_3Si}{\overset{R^1}{N}}-\underset{X}{\overset{Y}{C}}$$

$$\xrightarrow{\triangledown} R^1-N=C=Y + XSi(CH_3)_3$$

$$X=Cl,\ O-\underset{R^2}{\overset{O}{C}},\ OR^2$$

$R^1 = R^2 = $ Alkyl, Cycloalkyl, Aryl, ω-Carboxyalkyl Y = O, S

N-Monosubstituted dithiocarbamic acid esters and *O*-esters and *S*-esters of thiocarbamic acid are converted in good yield into *isocyanates* [1374] and *isothiocyanates* [1375, 1375a] respectively in analogous manner following *N*-silylation.

On being heated to 150−160° *N*-sulfonyldithiocarbamic acid esters cleave into *sulfonylisothiocyanates* and thiol [1376−1378]:

$$R-SO_2-NH-\underset{S}{\overset{\|}{C}}-S-CH_3 \longrightarrow$$

$$R-SO_2-N=C=S + H_3C-SH$$

17.6.4 Isocyanates, isothiocyanates, isoselenocyanates, carbodiimides from heterocumulenes (including carbon dioxide and carbon disulfide)

17.6.4.1 Catalytic conversion of isocyanates into carbodiimides

In the presence of certain catalysts, notably phosphine oxides, aromatic and aliphatic isocyanates are converted into *carbodiimides* in high yield and often under mild conditions [1379−1382]:

$$H_5C_6-N=C=O + R_3PO \xrightarrow{-CO_2}$$

$$H_5C_6-N=PR_3 \xrightarrow{+ H_5C_6-N=C=O}$$

$$H_5C_6-N=C=N-C_6H_5 + R_3PO$$

Diphenylcarbodiimide; 94%

The mechanism [1383] was established by isotopic labeling of the phosphoryl oxygen [1384, 1385]. Structure has a marked effect on the rate of conversion.

Aromatic isocyanates, especially those with electron-attracting substituents, react considerably more readily than aliphatic ones. Reaction in boiling decahydronaphthalene affords higher yields and shorter reaction times with the latter.

Cyclic phosphine oxides (*e.g.* **1**) [1380], cyclic diamides of phosphonic acid (*e.g.* **2**) [1390], or metal carbonyls [1391] are the best catalysts [1386−1389]. Sterically hindered isocyanates (2, 6-dialkylphenyl isocyanates) can be converted into *carbodiimides* by using catalytic amounts of strong bases (alkoxides, *tert*-amines) [1392].

[1374] *G. Greber, H.R. Kricheldorf*, Angew. Chem. *80*, 1028 (1968).
[1374a] *H.R. Kricheldorf*, Synthesis 649 (1970).
[1374b] *V.F. Mironov, V.D. Sheludyakov, V.P. Kozyukov, G.D. Khatuntsev*, Dokl. Akad. Nauk SSSR *181*, 115 (1968); C.A. *70*, 4189 (1969).
[1374c] *H.R. Kricheldorf*, Angew. Chem. *84*, 107 (1972).
[1375] *H.R. Kricheldorf*, Angew. Chem. *82*, 550 (1970).
[1375a] *H.R. Kricheldorf*, Synthesis 539 (1970).
[1376] *R. Gompper, W. Hägele*, Chem. Ber. *99*, 2885 (1966).
[1377] Jap. P. 9329/67 (1967), Toyama Chem. Ind.
[1378] Jap. P. 9054/68 (1968), Toyama Chem. Ind.
[1379] *T.W. Campbell, J.J. Monagle*, J. Amer. Chem. Soc. *84*, 1493 (1962).
[1380] *T.W. Campbell, J.J. Monagle, V.S. Foldi*, J. Amer. Chem. Soc. *84*, 3673 (1962).
[1381] *J.J. Monagle*, J. Org. Chem. *27*, 3851 (1962).
[1382] *T.W. Campbell, J.J. Monagle*, Org. Synth. *43*, 31 (1963).
[1383] *G. Aksnes, P. Froyen*, Acta Chem. Scand. *23*, 2697 (1969).
[1384] *J.J. Monagle, T.W. Campbell, H.F. McShane jr.*, J. Amer. Chem. Soc. *84*, 4288 (1962).
[1385] *J.J. Monagle, J.V. Mengenhauser*, J. Org. Chem. *31*, 2321 (1966).
[1386] *J.O. Appleman, V.J. DeCarlo*, J. Org. Chem. *32*, 1505 (1967).
[1387] *K. Hunger*, Tetrahedron Lett. 5929 (1966).
[1388] *E. Dyer, R.E. Read*, J. Org. Chem. *26*, 4677 (1961).
[1389] *B.A. Arbuzov, A.O. Vizel, K.M. Ivanovskaja*, Dokl. Akad. Nauk SSSR *170*, 585 (1966).
[1390] *H. Ulrich, B. Tucker, A.A.R. Sayigh*, J. Org. Chem. *32*, 1360 (1967).
[1391] *H. Ulrich, B. Tucker, A.A.R. Sayigh*, Tetrahedron Lett. 1731 (1967).
[1392] *W. Neumann, P. Fischer*, Angew. Chem. *74*, 801 (1962).
[1393] *H. Staudinger, E. Hauser*, Helv. Chim. Acta *4*, 861 (1921).

17.6.4.2 Reaction of isocyanates and isothiocyanates, carbon dioxide or carbon disulfide with phosphinimines, phosphoric acid diester amide anions and phosphoric acid imide trichlorides

The conversion of isocyanates into carbodiimides described on p. 782 ultimately rests on the reaction of the isocyanate with the intermediately formed phosphinimine. Alternatively, the latter can be used directly to produce *carbodiimides* from isocyanates and isothiocyanates[1393, 1394]. The latter two are formed from carbon dioxide and carbon disulfide respectively and often react with additional phosphinimine to symmetrical carbodiimides[1393, 1395].

Phosphoric acid diester amide anions display a chemical behavior similar to phosphinimines[1396, 1397].

$$(H_5C_2O)_2\overset{\overset{O}{\|}}{P}-NH-R \quad \xrightarrow{NaH} \quad (H_5C_2O)_2\overset{\overset{O}{\|}}{P}-\overset{\ominus}{N}-R$$

$$\xrightarrow{+R^1-N=C=O} \quad R^1-N=C=N-R \quad + \quad (H_5C_2O)_2\overset{\overset{O}{\|}}{P}-O^{\ominus}$$

They, too, react with carbon dioxide smoothly to afford isocyanates and/or carbodiimides in dependence on the reaction temperature if they carry secondary or tertiary alkyl groups on the nitrogen; carbon disulfide can be *imidated* in the same way[1397]:

It is worth mentioning the reaction between dimeric phosphoric acid trichloride methyl- and phenylimides with phenyl isocyanate at this point which in boiling 1,2-dichlorobenzene leads to *methylphenylcarbodiimide* (51%) and *dibenzylcarbodiimide* (34%) respectively plus phosphoryl chloride[1398] (cf. p. 779):

$$\left[R-N\overset{\overset{PCl_3}{\diagup}}{\underset{\diagdown PCl_3}{}}N-R\right] \rightleftharpoons 2\,R-N=PCl_3 \xrightarrow[-2\,POCl_3]{2\,R^1-N=C=O}$$

$$2\,R-N=C=N-R^1$$

17.6.4.3 Interconversion of isocyanates, isothiocyanates, isoselenocyanates and carbodiimides

Mercury oxide desulfurizes isothiocyanates to *isocyanates*. Conversely isocyanates are converted into *isothiocyanates* on heating with phosphorus(V) sulfide, and into *isoselenocyanates*[1399] with phosphorus(V) selenide[1400]:

$$R-N=C=S \quad \underset{P_4S_{10}}{\overset{HgO}{\rightleftarrows}} \quad R-N=C=O$$

$$R-N=C=O \quad \xrightarrow{P_4Se_{10}} \quad R-N=C=Se$$

The possibility of a mutual conversion is brought out particularly clearly in the finding that at 190–230° isocyanates and isothiocyanates formally exchange sulfur against oxygen[1401, 1402].

The postulated explanation is the intermediate formation of thioxouretidinone (thiono-1,3-diazetanone), $X=S$

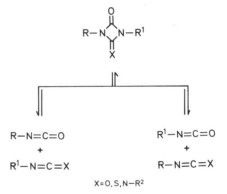

Preparation of a variety of isocyanates (*e.g.*, $R^1 = allyl, methallyl, hexyl$) is achieved by making use of this reaction. The relevant isothiocyanates $(X = S)$ are treated with higher boiling isocyanates (*e.g.*, $R = 3\text{-}chlorophenyl$) and the lowest boiling isocyanate is distilled off[1402].

Reacting an isocyanate with a carbodiimide $(X = N-R^2)$ is an analogous procedure. The initially formed iminouretidinone splits not only in the direction of formation but also in such a way as to form a new isocyanate and a mixed carbodiimide[1392].

Thermolysis of dithiophosphoric acid S,S-dialkyl ester S-aminocarbonyl esters, formed from isocyanates and O,O-diesters of dithiophosphoric

[1394] L. Horner, H. Hoffmann, Angew. Chem. 68, 473 (1956).

[1395] A. Messmer, I. Pintér, F. Szegö, Angew. Chem. 76, 227 (1964).

[1396] W.S. Wadsworth jr., W.D. Emmons, J. Amer. Chem. Soc. 84, 1316 (1962).

[1397] W.S. Wadsworth jr., W.D. Emmons, J. Org. Chem. 29, 2816 (1964).

[1398] H. Ulrich, A.A.R. Sayigh, Angew. Chem. 74, 900 (1962).

[1399] S. Petersen in Houben-Weyl, Methoden der organischen Chemie, 4. Aufl., Bd. VIII, p. 127, Georg Thieme Verlag, Stuttgart 1952.

[1400] C. Collard-Charon, M. Renson, Bull. Soc. Chim. Belg. 71, 537 (1962).

[1401] L.C. Case, Nature 183, 675 (1959).

[1402] W.E. Erner, J. Org. Chem. 29, 2091 (1964).

[1403] G. Ottman, H. Hooks, Angew. Chem. 78, 748 (1966).

acid in inert solvents, afford an almost quantitative yield of *isothiocyanates*[1403, 1404]:

$$R-N=C=O \quad + \quad HS-\overset{\overset{\displaystyle S}{\|}}{P}(OR^1)_2 \quad \xrightarrow{30°}$$

$$R-NH-\overset{\overset{\displaystyle O}{\|}}{C}-\overset{\overset{\displaystyle S}{\|}}{S-P}(OR^1)_2 \quad \xrightarrow{100-160°} \quad R-N=C=S$$

17.6.5 Isocyanates, isothiocyanates, isoselenocyanates, carbodiimides from imodocarbonyl chlorides and isocyanides

17.6.5.1 From imidocarbonyl chlorides

Imidocarbonyl chlorides represent excellent starting substances for preparing *isocyanates, isothiocyanates,* and *carbodiimides* (cf. p. 775)[1405]. The conversion into isocyanate succeeds best with chlorosulfonic acid[1405], methanesulfonic acid, phosphoryl chloride and water[1406], or benzaldoxime[1407].

$$R-N=CCl_2 \quad + \quad H_3C-SO_3H \quad \longrightarrow$$

$$R-N=C=O \quad + \quad HCl \quad + \quad H_3C-SO_2Cl$$

R = Alkyl, Aryl

Isothiocyanates are obtained in high yield in the same way by reacting imidocarbonyl chlorides either with sodium or ammonium sulfide in aqueous acetone or with phosphorus(V) sulfide in inert solvents[1405]:

$$R-N=CCl_2 + Na_2S \longrightarrow R-N=C=S + 2 NaCl$$

$$6 R-N=CCl_2 + P_4S_{10} \longrightarrow 6 R-N=C=S + 4 PSCl_3$$

The latter technique displays the advantage that it can be applied to acylated imidocarbonyl chlorides derived from carboxylic acids, sulfonic acids, and phosphoric acids as well as those with hydrolysis-sensitive groups (*e.g.,* COCl or NCO).

The reaction between phenylimidocarbonyl chloride and sodium selenide furnishes *phenyl isoselenocyanate* correspondingly[1408].

17.6.5.2 From isocyanides

Isocyanides can be oxidized to *isocyanates* with a variety of oxidizing agents including silver and mercury oxides[1409], ozone[1410], peroxybenzoic acid[1411], oxygen (in the presence of transition metal complexes)[1412], or *tert*-butyl hypochlorite[1413]:

$$R-NC \quad + \quad X \quad \longrightarrow \quad R-N=C=X$$
X = O, S, Se, NR¹

When small amounts of halogen are added dimethyl sulfoxide, too, affords isocyanates[1414]:

$$R-NC \quad + \quad Br_2 \quad \longrightarrow \quad R-N=CBr_2$$

$$R-N=CBr_2 \quad + \quad (H_3C)_2SO \quad \longrightarrow \quad R-N=C=O$$

$$+ \quad (H_3C)_2S \quad + \quad Br_2$$

Oxidation with dimethyl sulfoxide in the presence of *p*-toluenesulfonic acid, hydrogen chloride, trityl perchlorate also succeeds to give a 52–92% yield of aliphatic and aromatic isocyanates under relatively mild conditions[1415].

While addition of sulfur to isocyanides has not achieved preparative importance[1416] — re the reaction of isocyanides with thiols see ref.[1417] — addition of elementary selenium represents an important method of preparing aliphatic and aromatic *isoselenocyanates*[1418–1420]. The reaction is generally carried out in petroleum ether, ethanol, or, particularly advantageously, in chloroform[1418].

[1404] USP. 3 409 656 (1968), Olin Mathieson; Inv.: *G. Ottmann, H. Hocks.*

[1405] *E. Kühle,* Angew. Chem. *81,* 18 (1969) (several references).

[1406] *D. Arlt,* Synthesis 20 (1970).

[1407] *T. Mukaiyama, T. Fujisawa, O. Mitsunobu,* Bull. Chem. Soc. Japan *35,* 1104 (1962).

[1408] *C. Collard-Charon, M. Renson,* Bull. Soc. Chim. Belg. *71,* 536 (1962).

[1409] *A. Gautier,* Ann. Chim. (Paris) *17,* 229 (1869).

[1410] *H. Feuer, H. Rubinstein, A.T. Nielsen,* J. Org. Chem. *23,* 1107 (1958).

[1411] *T. Shono, M. Kimura, Y. Ito, K. Nishida, R. Oda,* Bull. Chem. Soc. Japan *37,* 635 (1964).

[1412] *S. Otsuka, A. Nakamura, Y. Tatsuno,* Chem. Commun. 836 (1967).

[1413] *M. Okano, Y. Ito, T. Shono, R. Oda,* Bull. Chem. Soc. Japan *36,* 1314 (1963).

[1414] *H.W. Johnson jr., P.H. Daughhetee jr.,* J. Org. Chem. *29,* 246 (1964).

[1415] *D. Martin, A. Weise,* Angew. Chem. *79,* 145 (1967).

[1416] *J.U. Nef,* Justus Liebigs Ann. Chem. *280,* 291 (1894).

[1417] *T. Saegusa, S. Kobayashi, Y. Ito,* J. Org. Chem. *35,* 2118 (1970).

[1418] *E. Bulka, K.D. Ahlers, E. Tuček,* Chem. Ber. *100,* 1367 (1967).

[1419] *E. Bulka, K.D. Ahlers,* Z. Chem. *3,* 348 (1963).

[1420] *W.J. Franklin, R.L. Werner,* Tetrahedron Lett. 3003 (1965).

Re the preparation of *trialkylsilyl* and *triphenylsilyl isocyanates* and the corresponding *isothiocyanates* and *isoselenocyanates* from the relevant silyl cyanides reference is made to ref. [1421].

In the presence of iron pentacarbonyl isocyanides can be also converted into *carbodiimides* with azides [1421a].

17.6.6 Isocyanates, isothiocyanates, carbodiimides by rearrangement reactions

The degradation reactions involving azides (Curtius), amides (Hofmann), and hydroxamic acids (Lossen), which are employed mainly for synthesizing amines and are thus described in detail elsewhere (*cf.* p. 561 ff), proceed *via* isocyanates as intermediate products. In certain cases the isocyanates can be isolated. However, only the Curtius degradation possesses significance for preparing isocyanates [1422].

A variant of the Hofmann degradation consists in the oxidative rearrangement of carboxamides with lead(IV) acetate which affords partly good yields of isocyanates and their derivatives [1423].

Hydroxamic acids, too, can be converted into *isocyanates* by dehydration in inert solvents with acetic anhydride, phosgene, or thionyl chloride [1424, 1424a–b], or thermally following reaction with diketene to (acetoacetyl)hydroxamates [1425].

The rearrangement of aminimides to isocyanates [1426] is related to the Curtius, Hofmann, and Lossen degradations. It enables *isopropenyl isocyanate* to be prepared in simple manner (86%) [1427]:

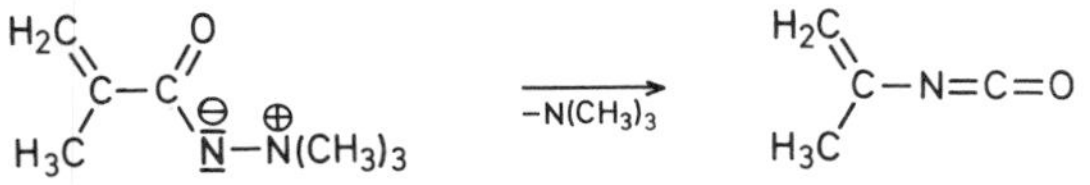

Rearranging 1-bromo-2-azetidinones [1428] in the presence of *CC*-unsaturated compounds and catalytic amounts of free-radical formers leads smoothly to *β-bromoalkyl isocyanates* (up to 93%):

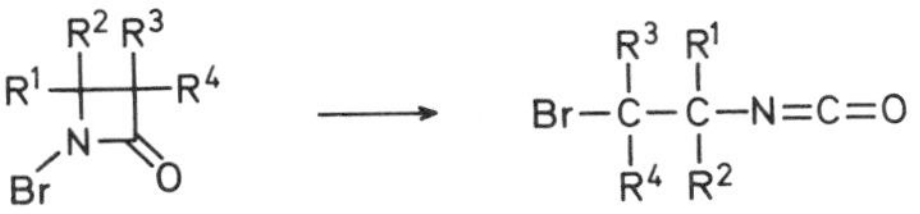

Under similar conditions *N*-bromosuccinimide forms *β-bromopropionyl isocyanate* (64%) [1429].

A further noteworthy reaction is the rearrangement of certain aziridine derivatives which affords 2-substituted alkylisocyanates, alkylisothiocyanates, and carbodiimides [1429a, 1429b].

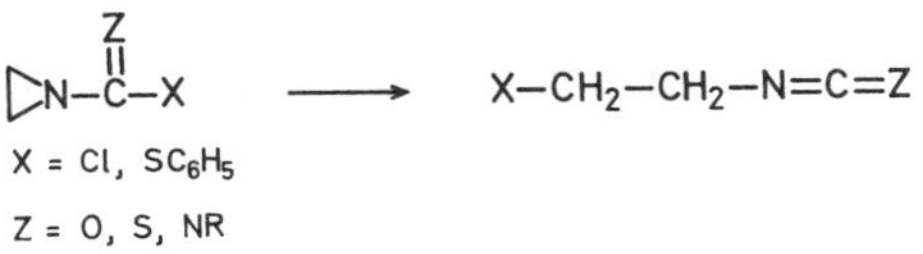

X = Cl, SC$_6$H$_5$

Z = O, S, NR

The Tiemann rearrangement of *N*-hydroxyimidamides (amidoximes) can be utilized successfully for preparing *carbodiimides* [1430]:

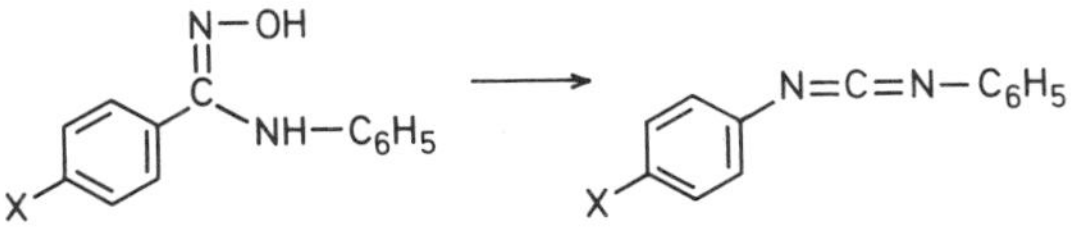

Pyrolytic cleavage of 1,5-diaryl-*1H*-tetrazoles [1431, 1432] to carbodiimides and (little) benzimidazoles can be interpreted as the analogous azide reaction:

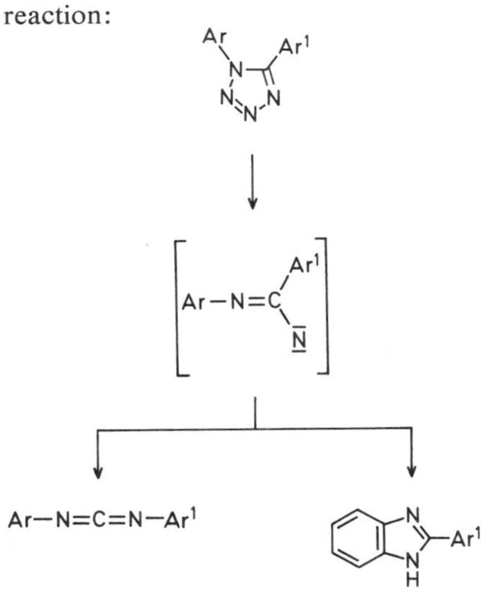

[1421] *J.S. Thayer*, Inorg. Chem. 7, 2599 (1968).

[1421a] *T. Saegusa, Y. Ito, T. Shimizu*, J. Org. Chem. 35, 3995 (1970).

[1422] *J.H. Saunders, K.C. Frisch*, Polyurethanes, Chemistry and Technology, Part 1, p. 22, Interscience Publishers, New York, London 1962.

[1423] *H.E. Baumgarten, A. Staklis*, J. Amer. Chem. Soc. 87, 1141 (1965).

[1424] *S. Petersen* in *Houben-Weyl*, Methoden der organischen Chemie, 4. Aufl., Bd. VIII, p. 125, Georg Thieme Verlag, Stuttgart 1952.

[1424a] *E.H. Burk, D.D. Carlos*, J. Heterocycl. Chem. 7, 177 (1970).

[1424b] *R. Merten, C. Weber*, Synthesis 589 (1970).

[1425] *T. Mukaiyama, H. Nohira*, J. Org. Chem. 26, 782 (1961).

[1426] *W.J. McKillip, L.M. Clemens, R. Haugland*, Can. J. Chem. 45, 2613 (1967).

[1427] *R.C. Slagel, A.E. Bloomquist*, Can. J. Chem. 45, 2625 (1967).

[1428] *K.D. Kampe*, Tetrahedron Lett. 117 (1969).

[1429] *H.W. Johnson, D.E. Bublitz*, J. Amer. Chem. Soc. 80, 3150 (1958).

[1429a] *D.A. Tomalia, Th.J. Giacobbe, W.A. Sprenger*, J. Org. Chem. 36, 2142 (1971).

[1429b] *D.A. Tomalia, D.P. Sheetz, G.E. Ham*, J. Org. Chem. 35, 47 (1970).

[1430] *J. Garapon, B. Sillion, J.M. Bonnier*, Tetrahedron Lett. 4905 (1970).

The relative extent of carbodiimide formation is dependent on the nature of the substituents in Ar^1 [1433].

Fragmentation of 1,4-disubstituted tetrazolium salts in basic medium is suitable for preparing little stable carbodiimides (e.g. methyl vinyl carbodiimide) [1434]:

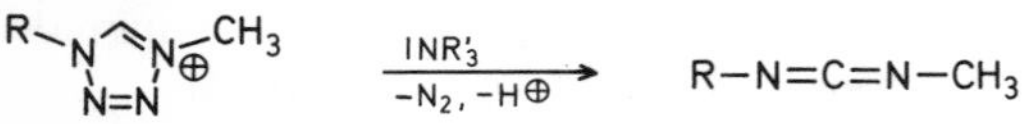

$$R = -CH=CH_2, -N(C_2H_5)_2$$

Thermal cleavage of 1,2,3,5-thiaoxadiazol-1-ones leading almost quantitatively to carbodiimides and sulfur dioxide is a reaction which resembles the pyrolysis of 1,5-diaryl-1H-tetrazoles [1435]:

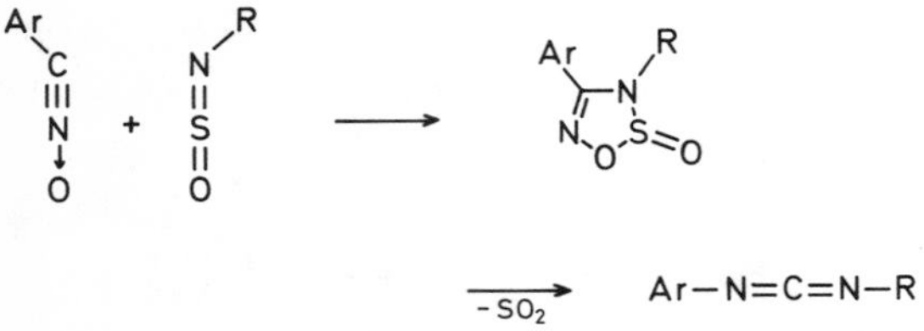

Finally, it may be mentioned that sterically hindered aliphatic and aromatic nitrile oxides isomerize to isocyanates at $\sim 110-140°$ [1436-1438], e.g.:

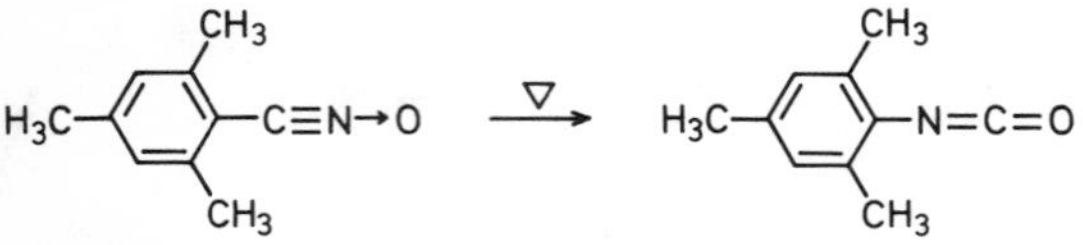

2,4,6-Trimethylphenyl-isocyanate; 68%

17.6.7 Special procedures

Aromatic nitro compounds [1439] and azides [1440] can be converted directly into aryl isocyanates by reaction with carbon monoxide under pressure and at elevated temperature:

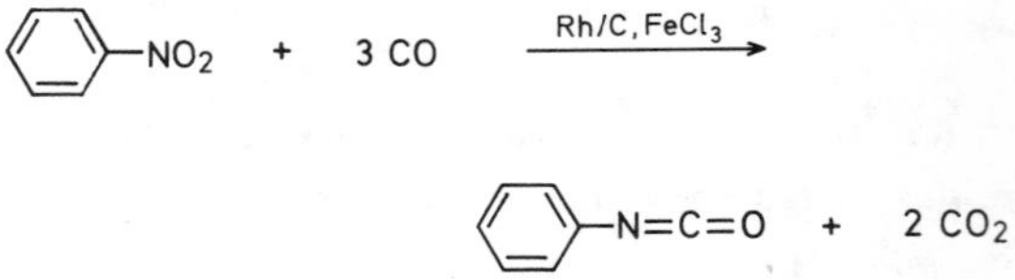

Phenyl isocyanate; 35%

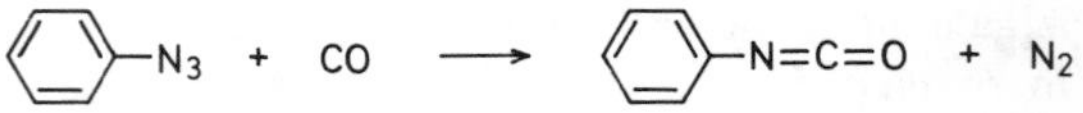

Phenyl isocyanate; 47%

The reaction between alkylamines and arylamines with carbon monoxide in the presence of palladium(II) chloride ensues under much milder conditions [1441].

$$R-NH_2 \;+\; CO \;+\; PdCl_2 \xrightarrow{-2\,HCl}$$

$$R-N=C=O \;+\; Pd$$

Aryl isothiocyanates, too, can be prepared directly from the corresponding nitro compounds with carbon disulfide or carbonyl sulfide in the presence of certain bases [1442].

Phosgenation of N-sulfinylamines and N-sulfinylamides represents a possible means for synthesizing aliphatic and aromatic isocyanates and of sulfonyl isocyanates [1443]:

$$R-N=S=O \;+\; COCl_2 \longrightarrow R-N=C=O \;+\; SOCl_2$$

N-Sulfinylanilines are converted quantitatively into isothiocyanates [1444] with excess carbon disulfide in the bomb tube.

Reference may be made also to the extremely reactive and versatile sulfuryl chloride isocyanate (chlorosulfonyl isocyanate) formed by the action of sulfur trioxide on cyanogen chloride [1445]:

$$SO_3 \;+\; Cl-CN \longrightarrow Cl-SO_2-N=C=O$$

[1431] P.A.S. Smith, E. Leon, J. Amer. Chem. Soc. 80, 4647 (1958).

[1432] P. de Mayo, Molecular Rearrangements, p. 567, Interscience Publishers, New York, London 1963.

[1433] J. Vaughan, P.A.S. Smith, J. Org. Chem. 23, 1909 (1958).

[1434] D.M. Zimmerman, R.A. Olofson, Tetrahedron Lett. 3453 (1970).

[1435] P. Rajagopalan, B.G. Advani, J. Org. Chem. 30, 3369 (1965).

[1436] C. Grundmann, J.M. Dean, J. Org. Chem. 30, 2809 (1965).

[1437] C. Grundmann, P. Kochs, Angew. Chem. 82, 637 (1970).

[1438] C. Grundmann, P. Grünanger, The Nitrile Oxides, p. 62, Springer Verlag, Berlin, Heidelberg, New York 1971.

[1439] W.B. Hardy, R.P. Bennett, Tetrahedron Lett. 961 (1967).

[1440] R.P. Bennett, W.B. Hardy, J. Amer. Chem. Soc. 90, 3295 (1968).

[1441] E.W. Stern, M.L. Spector, J. Org. Chem. 31, 596 (1966).

[1442] G. Ottmann, E.H. Kober, Angew. Chem. 81, 782 (1969).

[1443] H. Ulrich, B. Tucker, A.A.R. Sayigh, J. Org. Chem. 34, 3200 (1969).

[1444] T. Minami, T. Agawa, Tetrahedron Lett. 2651 (1968).

[1445] R. Graf, Angew. Chem. 80, 179 (1968).

17.7 Bibliography

G.A. Olah, Friedel-Craft and Related Reactions, Vol. III, Wiley Intersci. Publ., New York, London, Sydney 1964.

Houben-Weyl, Methoden der organischen Chemie, 4. Aufl., Georg Thieme Verlag, Stuttgart,

Bd. VI/2, 1963

Bd. VII/4, 1968

Bd. VIII, 1952

Bd. IX, 1955

Bd. X/2, 1967

Bd. X/4, 1968

Bd. XI/2, 1958

W. Foerst, Neuere Methoden der präparativen organischen Chemie, Bd. VI, Verlag Chemie, Weinheim/Bergstr. 1970.

H. Ulrich, Cycloaddition Reactions of Heterocumulenes, Academic Press, New York, London 1967.

Organic Reactions, Vol. IX, John Wiley & Sons, New York, London 1957.

Organic Reactions, Vol. III, John Wiley & Sons, New York, London 1946.

Organic Reactions, Vol. XII, John Wiley & Sons, New York, London 1962.

E. Schröder, K. Lübke, The Peptides, Vol. I (1965), Vol. II (1966), Academic Press, New York, London.

W. Foerst, Neuere Methoden der präparativen organischen Chemie, Bd. V, Verlag Chemie, Weinheim/Bergstr., 1967.

Organic Reactions, Vol. XVII, John Wiley & Sons, New York, London 1969.

W. Foerst, Neuere Methoden der präparativen organischen Chemie, Bd. IV, Verlag Chemie, Weinheim/Bergstr. 1966.

W. Foerst, Neuere Methoden der präparativen organischen Chemie, Bd. III, Verlag Chemie, Weinheim/Bergstr. 1961.

Organic Reactions, Vol. XIV, John Wiley & Sons, New York, London 1957.

E. Schmitz, Organische Chemie in Einzeldarstellungen, Bd. IX, Springer Verlag, Berlin, Heidelberg, New York 1967.

Advan. Org. Chem. Vol. II, Interscience Publishers, New York 1960.

Acyclic and cyclic amides, hydrazides, hydroxyamides, imides, azides of carboxylic acids from aldehydes, ketones and derivatives

L.G. Donaruma, W.Z. Heldt, Org. Reactions *11,* 1 (1960).

P.A.S. Smith, The Chemistry of Open Chain Organic Nitrogen Compounds, Vol. I, p. 137, Benjamin, New York, Amsterdam 1965.

P.A.S. Smith in *P. De Mayo,* Molecular Rearrangements, Vol. I, p. 457, Interscience Publishers, New York 1963.

P.A.S. Smith The Chemistry of Open Chain Organic Nitrogen Compounds, Vol. II, p. 29, Benjamin, New York, Amsterdam 1966.

C.G. McCarty in *S. Patai,* The Chemistry of the Carbon-Nitrogen Double Bond, p. 363, Wiley Intersci. Publ., London, New York, Sydney, Toronto 1970.

S.R. Sander, W. Karo, Organic Functional Group Preparations, Vol. I, p. 270, Academic Press, New York, London 1968.

C.A. Buehler, D.E. Pearson, Survey of Organic Synthesis, p. 894, Wiley Intersci. Publ., New York, London, Sydney, Toronto 1970.

K.E. Hamlin, A.W. Weston, Org. Reactions *9,* 1 (1957).

Carboxylic acid azides

W. Lwowski in *S. Patai,* The Chemistry of the Azido-Group, p. 503, Wiley Intersci. Publ., New York, London, Sydney, Toronto 1971.

M.E. Biffin, J. Miller, D.B. Paul in *S. Patai,* The Chemistry of the Azido-Group, p. 57, Wiley Intersci. Publ., New York, London, Sydney, Toronto 1971.

S.R. Sandler, W. Karo, Organic Functional Group Preparations, Vol. II, p. 268, Academic Press, New York, London 1971.

G. Koga, Yuki Gosei Jagaku Kyokai Shi *28,* 996 (1970).

R.F. Benson in *R.C. Elderfield,* Heterocyclic Compounds, Vol. VIII p. 1—104, John Wiley & Sons, New York, London, Sydney 1967.

A.G. Overberger, J.P. Anselme, J.G. Lombardino, Organic Compounds with Nitrogen-Nitrogen Bonds, p. 99, Ronald Press Co., New York 1966.

P.A.S. Smith, The Chemistry of Open Chain Organic Nitrogen Compounds, Vol. II, p. 211, Benjamin, New York, Amsterdam 1966.

E. Lieber, R.L. Minnis, C.N.R. Rao, Chem. Rev. *65,* 377 (1965).

F. Möller in *Houben-Weyl,* Methoden der organischen Chemie, 4. Aufl., Bd. XI/1, p. 863, Georg Thieme Verlag, Stuttgart 1957.

C. Grundmann in *Houben-Weyl,* Methoden der organischen Chemie, 4. Aufl., Bd. X/3, p. 777, Georg Thieme Verlag, Stuttgart 1955.

H. Henecka, P. Kurtz in *Houben-Weyl,* Methoden der organischen Chemie, 4. Aufl., Bd. VIII, p. 680, Georg Thieme Verlag, Stuttgart 1952.

Ng-Ph. Buu-Hoï, P. Cagniant in *V. Grignard, G. Dupont, R. Locquin,* Traité de Chimie organique, Vol. XV, p. 715 (1948).

P.A.S. Smith, Org. Reactions *3,* 337 (1946).

Thiohydrazides

W. Walter, K.J. Reubke in *J. Zabicky,* The Chemistry of Amides, p. 477, Wiley Intersci. Publ., New York, London, Sydney, Toronto 1970.

Thioureas

Houben-Weyl, Methoden der organischen Chemie, 4. Aufl., Bd. IX, Georg Thieme Verlag, Stuttgart 1955.

D.C. Schroeder, Chem. Rev. *55,* 181 (1955).

E.E. Reid, Organic Chemistry of Bivalent Sulfur, Vol. V, Chemical Publishing Co., New York 1963.

Thiocarbamates

W. Walter, K.-D. Bode, Angew. Chem. *79,* 285 (1967).

Dithiocarbamidsäuren

G.D. Thorn, R.A. Ludwig, The Dithiocarbamates and Related Compounds, Elsevier Publ., Co., Amsterdam, London, New York 1962.

Thiohydroxamic acids

W. Walter, E. Schaumann, Synthesis 111 (1971).

Thioamides

P. Chabrier, S.H. Renard, Bull. Soc. Chim. France D 272 (1949).

W. Walter, K.-D. Bode, Angew. Chem. *78,* 517 (1966).

R.N. Hurd, G. de la Mater, Chem. Rev. *61,* 51 (1961).

E.E. Reid, Organic Chemistry of Bivalent Sulfur, Vol. IV, Chemical Publishing Co., New York 1962.

K.A. Petrov, L.N. Andreev, Russian. Chem. Rev. *38,* 41 (1969); engl. 21.

P.L. Compagnon, M. Miocque, Ann. Chim. (Paris) *5,* 11 (1970).

W. Walter, J. Voss in *J. Zabicki,* The Chemistry of Amides, p. 383, Wiley Intersci. Publ., New York, London, Sydney, Toronto 1970.

Orthocarboxylic acid amides

J. Gloede, L. Haase, H. Groß, Z. Chem. *9,* 201 (1969).

R.H. DeWolfe, Carboxylic Orthoacid Derivatives, Academic Press, New York, London 1969.

Isocyanates, isothiocyanates, isoselenocyanates, carbodiimides

S. Petersen in *Houben-Weyl,* Methoden der organischen Chemie, Methoden zur Herstellung und Umwandlung von Isocyanaten, 4. Aufl., Bd. VIII, p. 119, Georg Thieme Verlag, Stuttgart 1952.

J.H. Saunders, K.C. Frisch, Polyurethanes, Chemistry and Technology, Part 1, Interscience Publishers, New York, London 1962.

S.R. Sandler, W. Karo in Organic Functional Group Preparations, Cyanates, Isocyanates, Thiocyanates and Isothiocyanates, Vol. I, p. 301, Academic Press, New York, London 1968.

H. Ulrich, A.A.R. Sayigh in *W. Foerst,* Neuere Methoden der präparativen organischen Chemie, Synthesen von Isocyanaten und Carbodiimiden, Bd. VI, p. 211, Verlag Chemie, Weinheim/Bergstr. 1970.

S. Ozaki, Recent Advances in Isocyanate Chemistry, Chem. Rev. *72,* 457 (1972).

H. Ulrich, The Chemistry of Sulfonyl Isocyanates, Chem. Rev. *65,* 369 (1965).

M.O. Lozinskii, P.S. Pel'kis, The Chemistry of Acyl and Sulfonyl Isocyanates, Isothiocyanates, and Isoselenocyanates, Russian Chem. Reviews *37,* 363 (1968).

K.A. Nuridzhanyan, Acyl Isocyanates, Russian Chem. Reviews *39,* 130 (1970).

G.I. Derkatsch, Phosphorisocyanate, Angew. Chem. *81,* 407 (1969).

M. Bögemann, S. Petersen, O.E. Schultz, H. Söll in *Houben-Weyl,* Methoden der organischen Chemie, Isothiocyansäureester (Senföle), 4. Aufl., Bd. IX, p. 867, Georg Thieme Verlag, Stuttgart 1955.

S.J. Assony in *N. Kharash,* Organic Sulfur Compounds, The Chemistry of Isothiocyanates, Vol. I, p. 326, Pergamon Press, Oxford, New York 1961.

G.I. Derkatsch, Sh. M. Ivanova, Phosphorisothiocyanate, Z. Chem. *9,* 369 (1969).

B.V. Bocharov, Progress in the Chemistry of the Carbodiimides, Russian Chem. Reviews *34,* 212 (1965).

F. Kurzer, K. Douraghi-Zadeh, Advances in the Chemistry of Carbodiimides, Chem. Rev. *67,* 107 (1967).

S.R. Sandler, W. Karo in Organic Funtional Group Preparations, Carbodiimides, Vol. II, p. 205, Academic Press, New York, London 1971.

18 Isonitriles, Cyanates, Thiocyanates

Contributed by

G. Entenmann
Institut für organische Chemie der
Universität Stuttgart,
Stuttgart

18.1 Preparation of isonitriles[1]

18.1.1 Water cleavage from *N*-substituted formamides

The most fruitful synthesis of isonitriles is the cleavage of water from *N*-monosubstituted formamides with acylating agents in the presence of bases:

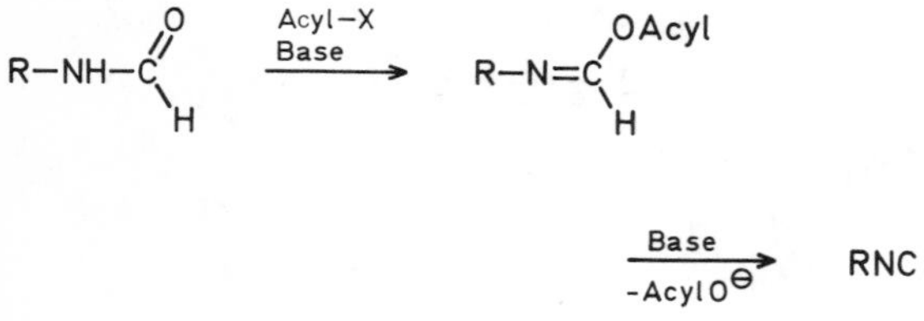

Generally, phosgene in the presence of a base such as trimethylamine, triethylamine, *N,N*-dimethylcyclohexylamine, *N,N*-diethylaniline, pyridine, or quinoline serves as the water-cleaving agent. Triethylamine is used most often. Dichloromethane, and in some instances 1,2-dichlorobenzene or 1,2,4-trichlorobenzene, are suitable solvents.

In a variant of the method ammonia is passed into the reaction mixture for the purpose of working up. The precipitated ammonium chloride is filtered off and the filtrate is distilled. When dichloromethane is employed as solvent, the triethylamine/dichloromethane mixture which distils can be used for analogous reactions without further purification[1].

The method possesses wide scope. Unsubstituted[1], substituted[1], saturated[1], unsaturated[1], aliphatic[1–4], cycloaliphatic[1], bicyclic[5], aromatic[1] and heterocyclic[1,6] isonitriles can be prepared. The procedure fails with compounds containing structural elements which produce decomposition of the isonitrile or further reaction with phosgene. Thus, for preparing *4-aminophenyl isocyanide, 4-isocyanobenzoic acid,* and *4-isocyanobenzenesulfonamide* it is necessary to fall back on the carbylamine reaction (see p. 791)[1].

The phosgene method lends itself well for larger scale work. Where small quantities of isonitrile are required, benzenesulfonyl chloride or *p*-toluenesulfonyl chloride in pyridine are used with advantage (*e.g.,* during syntheses in the *steroid* series[7]).

Phosphoryl chloride in pyridine or in the presence of potassium *tert*-butoxide is the most frequently used reagent after phosgene/triethylamine. Apart from the possibility of its ready application to small-scale work and its simpler manipulation, the use of phosphoryl chloride in place of phosgene offers no benefits.

Using cyanuric chloride in the presence of potassium carbonate[8], or an adduct of triphenylphosphine and carbon tetrachloride together with triethylamine[9] as water-cleaving agent,

$$R{-}NH{-}CHO + (H_5C_6)_3P + CCl_4 + (H_5C_2)_3N \longrightarrow R{-}NC + (H_5C_6)_3PO + CHCl_3 + [(H_5C_2)_3NH]Cl$$

affords good yields of aliphatic and aromatic isonitriles by virtue of the mild reaction conditions (51–76% with cyanuric chloride, 89–91% with triphenylphosphine/carbon tetrachloride).

Triphenylphosphine dibromide also cleaves water from *N*-alkyl- and *N*-arylformamides at elevated temperature in the presence of triethylamine:

$$R{-}NH{-}CHO \xrightarrow[-\,(H_5C_6)_3PO]{(H_5C_6)_3PBr_2\,/\,(H_5C_2)_3N} R{-}NC$$

The corresponding isonitriles are obtained in 56–73% yield[10].

Isonitriles are obtained, further, by reacting thioformamides with cyanogen bromide or 2-chloro-1,3,5-trinitrobenzene in the presence of bases (20–60% yield)[1] and by reacting secondary monoalkylthioformamides with acylated carbodiimide (49–73% yields):

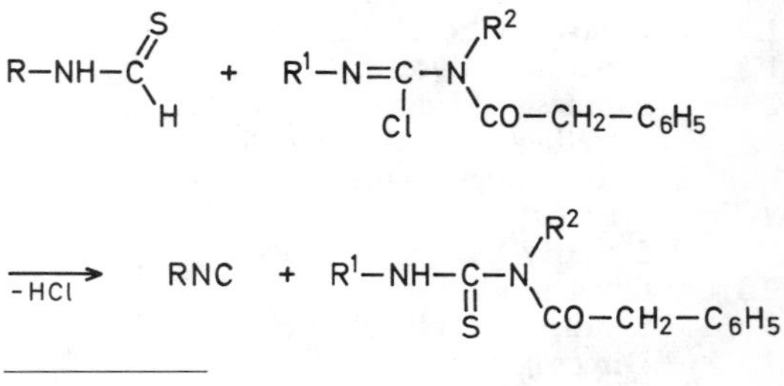

[1] *I. Ugi, U. Fetzer, U. Eholzer, H. Knupfer, K. Offermann,* Angew. Chem. *77,* 492 (1965); Angew. Chem. Intern. Ed. Engl. *4,* 472 (1965).

[2] *D.J. Hoy, E.J. Poziomek,* J. Org. Chem. *33,* 4050 (1968).

[3] *K. Hartke,* Chem. Ber. *99,* 3163 (1966).

[4] Holl. P. 6508902 (1966), Farbenfabr. Bayer AG; C.A. *64,* 19507 (1966).

[5] Brit.P. 1064835 (1967), Farbenfabr. Bayer AG, Inv.: *U. Fetzer, I. Ugi, H. Knupfer, J.A. Renner, F. Grewe;* C.A. *69,* 51732 (1968).

[6] *R. Neidlein,* Arch. Pharm. *299,* 603 (1966).

[7] *R.W. Horobin, N.R. Khan, J. McKenna, B.G. Hutley,* Tetrahedron Lett. 5087 (1966).

[8] *R. Wittmann,* Angew. Chem. *73,* 219 (1961).

[9] *R. Appel, R. Kleinstück, K.-D. Ziehn,* Angew. Chem. *83,* 143 (1971).

On heating N,N'-diaryl-N-hydroxyformamidines isonitriles are formed as side-product[1] and also during the pyrolysis of N-acylated N-alkylform-amidines[2]. Both these reactions are not very suitable for preparative purposes.

18.1.2 Alkylation of cyanide ions

Reacting prussic acid or cyanide ions with alkylating agents leads to a mixture of nitrile and isonitrile, so that this process, like addition of prussic acid to olefins under the action of electrical discharges[1], is seldom used. Reacting prussic acid with olefins in the presence of copper halide furnishes useful results only with olefins forming fairly stable carbonium ions[11].

Alkylation of complex-bound cyanide with alkyl halide or alcohols furnishes good yields[1].

18.1.3 Carbylamine reaction

The reaction of primary amines with chloroform and ethanolic potassium hydroxide or potassium *tert*-butoxide, which has been shown to be a carbene reaction[12], for a long time ranked as the most advantageous method because it affords isonitriles free from nitriles. The yields are moderate. Like the reaction of primary amines with carbon tetrachloride in the presence of copper, or the thermolysis of sodium trichloroacetate plus arylamine, it is being successfully employed in some cases for synthesizing *aromatic isonitriles*[1].

18.1.4 Redox reactions

Isocyanates are reduced by triethyl phosphite, while isothiocyanates are reduced by triethylphosphine, copper, triphenyltin hydride or photolytically. On account of the drastic reaction conditions the yields are moderate[1]. Reacting cyanates with trialkyl phosphite leads to further side-products in addition to a mixture of nitrile and isonitrile and, for this reason, is also unsuitable for preparative purposes[13].

By contrast, the reduction of isocyanates and isothiocyanates (mustard oils) with 3-methyl-2-phenyl-1,3,2-oxaphospholane

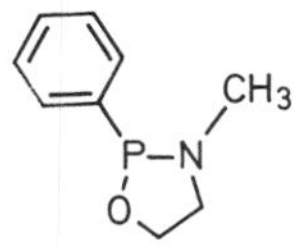

ensues at $5-36°$ and gives good yields ($51-70\%$) of aliphatic and aromatic isonitriles[14].

Finally, there may be mentioned dehalogenation of the now readily accessible *isonitrile dihalides*[15] with tertiary phosphines, potassium iodide, or alkyllithium compounds, which is being used with success in some cases[1].

18.1.5 Other procedures

Other reactions such as conversion of nitrile into isonitrile by gamma radiation[16], the second-order Beckmann rearrangement of oximes and O-tosyl oximes, which is limited to aromatic compounds[1], or ring-opening of heterocyclic compounds[1,17] are generally inferior to other techniques.

Reacting N-monosubstituted α-halo carboxamides with bases leads to α-lactams among other products[18]; these lactams furnish isonitriles on subsequent thermal α-cycloelimination.
The reaction of 2-halo-5,6-dihydro-*4H*-1,3-oxazine with butyllithium in tetrahydrofuran[19], which proceeds in good yield, has not so far met any outside application:

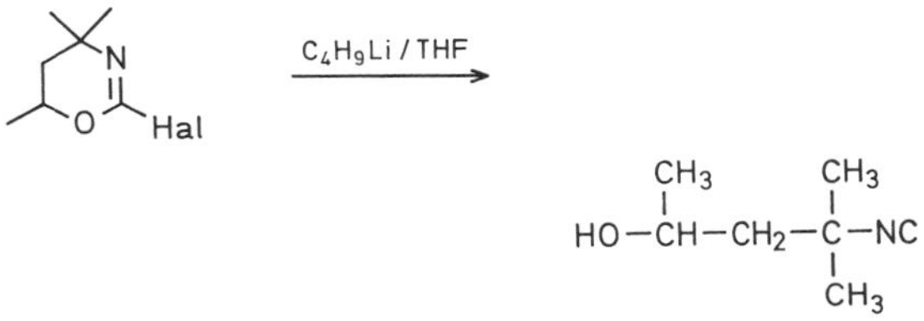

18.2 Preparation of cyanates

18.2.1 Introduction of the cyanate group with cyanogen halide

Sterically hindered alcoholates or phenolates can be reacted to cyanates with cyanogen halide[20-22]:

[10] H.J. Bestmann, J. Lienert, L. Mott, Justus Liebigs Ann. Chem. *718*, 24 (1968).

[11] S. Otsuka, K. Mori, K. Yamagani, J. Org. Chem. *31*, 4170 (1966).

[12] R.C. Whifield, Sci. Rev. *48*, 103 (1966); C.A. *66*, 72301 (1967).

[13] D. Martin, A. Weise, Chem. Ber. *99*, 976 (1966).

[14] T. Mukaiyama, Y. Yokota, Bull. Chem. Soc. Jap. *38*, 858 (1965).

[15] E. Kühle, B. Anders, G. Zumach in W. Foerst, Neuere Methoden der präparativen organischen Chemie, Bd. VI, p. 119, Verlag Chemie, Weinheim/Bergstr. 1970.

[16] M. Talat-Erben, E. Uenseren, N. Seber, Nucl. Sci. Abstr. *19* (24), 5812 (1965); C.A. *65*, 6565 (1966).

[17] S.D. Saraf, Pak. J. Sci. Ind. Res. (13), 48 (1970); C.A. *73*, 109450 (1970).

[18] I. Lengyel, J.C. Sheehan, Angew. Chem. *80*, 27 (1968).

[19] A.I. Meyers, H.W. Adickes, Tetrahedron Lett. 5151 (1969).

[20] E. Grigat, R. Pütter, Angew. Chem. *79*, 219 (1967).

[21] M. Hedayatullah, Bull. Soc. Chim. France 1572 (1968).

With sterically unhindered compounds the primarily formed cyanate reacts with still present alcoholate or phenolate to imidocarbonic acid diester or trimerizes to the corresponding cyanuric acid ester[20]. Under suitable conditions unhindered phenols, too, can be reacted to cyanates with cyanogen halide:

The phenol is dissolved in acetone, cyanogen halide is added (cyanogen bromide allows higher reaction temperatures than cyanogen chloride) and triethylamine is added drop by drop. A 1:1:1 molar ratio of phenol to cyanogen halide to triethylamine is used[20].
Estra-1,3,5(10)-triene-3-carbonitriles[25] are prepared by reacting the corresponding phenols with cyanogen bromide in tetrahydrofuran and slow addition of triethylamine at 0–5°. The molar ratio of phenol to cyanogen halide to triethylamine is 1:3–5:2–5.

Many *aromatic cyanates* can be prepared by such techniques[20, 21, 23, 24, 36], but the method fails with phenols containing strongly electron-withdrawing groups. Consequently, 2,4,6-tribromophenyl cyanate is obtained in a different manner (see below).
The procedure is of limited application for synthesizing aliphatic cyanates. Only the more strongly acid alcohols (*e.g.*, trihaloethanols) react smoothly to form the cyanates[20].

18.2.2 Thermolysis of 5-phenoxy- and 5-alkoxy-1,2,3,4-thiatriazoles

5-Phenoxy- and 5-alkoxy-1,2,3,4-thiatriazoles can be prepared from chlorothioformic acid esters and sodium azide[26] or from thiocarbonic acid hydrazide *O*-ester and nitrous acid at low temperatures. On heating the solution of such a thiatriazole to 20° sulfur and nitrogen are lost and aliphatic and aromatic cyanates are obtained[26, 28, 29].

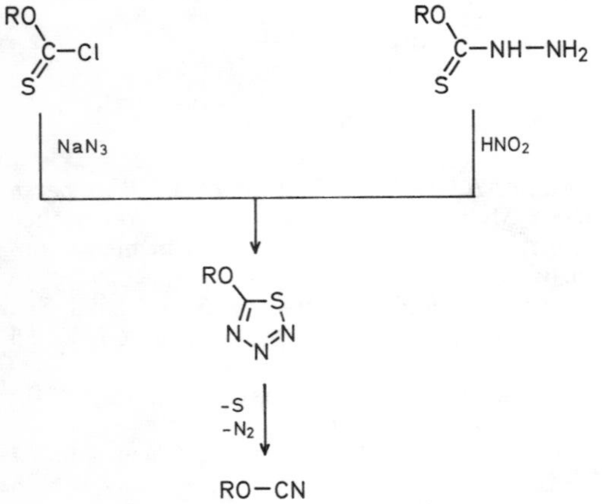

With *aromatic cyanates*[20, 27, 28] the yields are mostly somewhat inferior than with the phenol/cyanogen halide procedure, but the method is used with good success for synthesizing some *biphenylyl cyanates*[29] and aromatic cyanates with electron-withdrawing substituents such as *2,4,6-tribromophenyl cyanate*[20].
With *aliphatic cyanates* the yields are respectively 60%[26] and 75–80%[27]. Unstable cyanates such as *benzyl cyanate* and *phenethyl cyanate*[30] can be isolated as crude products.
1,4-Butyl dicyanate has been detected spectroscopically and by means of trapping reactions; *tert*-butyl cyanate, like cyclohexyl cyanate, readily decomposes into cyanic acid and isobutene and cyclohexane respectively[30].

18.2.3 Other procedures

Small amounts of cyanates are formed during the reaction of alkyl halides with silver cyanate[31] or from diazoalkane and cyanic acid[32]. In contrast, splitting off of sulfur from carbamic acid *O*-esters[33] or of thiol from imidothiocarbonic acid esters with mercury oxide[35] is used with success.
O-Alkyl *N*-hydroxythiocarbamates on reaction with ethyl chloroformate lose sulfur, carbon dioxide and alcohol to give cyanates in good yield[34]:

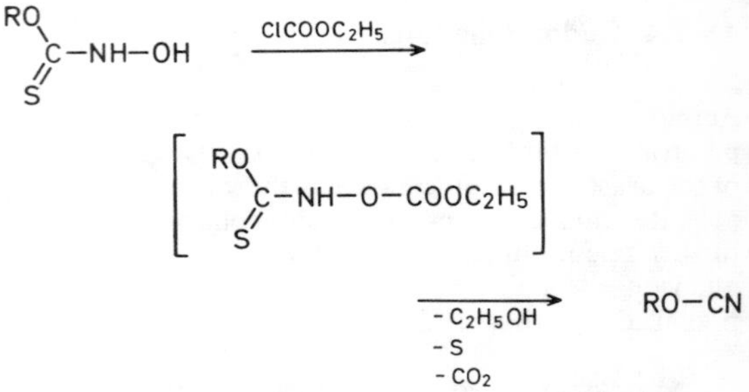

[23] *M. Hedayatullah,* C.R. Acad. Sci., Paris, Ser. C *265* (20), 1124 (1967).
[24] *M. Hedayatullah, J. Pailler, L. Denivelle,* C.R. Acad. Sci., Paris, Ser. C *270* (17), 1464 (1970).
[25] *D. Onken, D. Martin, S. Rackow,* Z. Chem. *9,* 228 (1969).
[26] *D. Martin, W. Mucke,* Chem. Ber. *98,* 2059 (1965).
[27] *K.A. Jensen, M. Due, A. Holm,* Acta Chem. Scand. *19,* 438 (1965).
[28] *D. Martin,* Z. Chem. *7,* 123 (1967).
[29] *M. Hedayatullah,* Bull. Soc. Chim. France 422 (1967).
[30] *K.A. Jensen, A. Holm, J. Wolff-Jensen,* Acta Chem. Scand. *23,* 1567 (1969).
[31] *A. Holm, C. Wentrup,* Acta Chem. Scand. *20,* 2123 (1966).
[32] *N. Groving, A. Holm,* Acta Chem. Scand. *19,* 1768 (1965).

[22] *J.C. Kauer, W.W. Henderson,* J. Amer. Chem. Soc. *86,* 4732 (1964).

Table 1. Aliphatic cyanates

Cyanate	Method 1	2	3	4
Methyl	17[27]			
	60[26]		84[35]	
Ethyl	91[27]			
	60[26]			
Propyl	82[27]	40–57[37]		
	60[26]			
Isopropyl	77[27]			
Butyl	77[27]		76[35]	72[34]
Isobutyl	73[27]		84[35]	75[34]
2-Butyl	77[27]			
tert-Butyl	—[30]			
Pentyl			84[35]	
Hexyl				78[34]
Cyclohexyl	—[30]			
Nonyl				67[34]
Undecyl				30[34]
Benzyl	—[30]			
Phenethyl	—[30]			

Method 1: Pyrolysis of 5-alkoxy-1,2,3,4-thiatriazole
     2: Cleaving sulfur from thiocarbamic acid *O*-esters
     3: Cleaving thiol from imidocarbonic acid esters
     4: Reacting *O*-alkyl *N*-hydroxycarbamates with esters of chloroformic acid

18.3 Preparation of thiocyanates

18.3.1 Direct introduction of the thiocyanate group

18.3.1.1 Reaction of inorganic thiocyanates with halogen compounds

Aliphatic halogen compounds are reacted with ammonium, sodium, potassium, or silver thiocyanate in dimethylformamide, in boiling alcohol or acetone, less often in aqueous acetone, benzene-water mixtures, or in inert solvents such as toluene or xylene to give alkyl thiocyanates. Activated haloaromatic compounds and haloheterocycles react analogously with thiocyanate ions to, for instance, *4-(fluorosulfonyl)-2-nitrophenyl thiocyanate*[36], *4,6,8-trimethylazulenyl thiocyanate*[37], *2-amino-5,6-dimethyl-4-pyrimidinyl thiocyanate*[38], or *4-nitro-3-pyridyl thiocyanate N-oxide*[39]. The halogen in α-halothio ethers (−5° in

ether[55a]) or in α-halo carbonyl compounds, too, is replaced smoothly[40, 56]. However, sequential reactions can occur during reaction of α-halo carboxamides[40, 42]. With potassium or ammonium thiocyanate in acetonitrile or absolute alcohol *N*-halo compounds afford the corresponding *N-amino thiocyanates*[43, 44, 55]:

$$\text{\textbackslash}NCl + SCN^{\ominus} \longrightarrow \text{\textbackslash}N-SCN + Cl^{\ominus}$$

From sulfenyl or selenyl chlorides respectively *sulfenyl* and *selenyl thiocyanates* are obtained with silver, lead(II), or potassium thiocyanate in inert solvents (benzene, chloroform, ethyl acetate) or in bulk[45–50]:

$$RX-Cl + SCN^{\ominus} \longrightarrow RX-SCN + Cl^{\ominus}$$

$$X = S, Se$$

Reacting acid chlorides with thiocyanate ions leads to acyl mustard oils[50, 51]. Only certain aromatic carbamoyl chlorides react with potassium or sodium thiocyanates in acetonitrile to *aminocarbonyl thiocyanates*[52, 53]:

$$\overset{Ar}{\underset{R}{\diagdown}}N-COCl + SCN^{\ominus} \xrightarrow{CH_3CN}$$

$$\overset{Ar}{\underset{R}{\diagdown}}N-C\overset{O}{\underset{SCN}{\diagup}} + Cl^{\ominus}$$

R: Alkyl, Aryl

33 *K.A. Jensen, A. Holm,* Acta Chem. Scand. *18,* 2417 (1964).
34 *A. Holm,* Acta Chem. Scand. *22,* 2019 (1968).
35 *W. Habicher, R. Mayer,* Angew. Chem. *80,* 42 (1968).
36 *M. Hedayatullah, L. Denivelle,* C.R. Acad. Sci., Paris, Ser. C *268* (5), 427 (1969).
36a Jap. P. 6 808 462 (1968), Japan Soda Co., Ltd., Inv.: *S. Kano, T. Noguchi;* C.A. *70,* 3553 (1969).
37 *K. Hafner, H. Patzelt, H. Kaiser,* Justus Liebigs Ann. Chem. *656,* 24 (1962).
38 *J. Kinugawa, M. Ochiai, H. Yamamoto,* Yakugaku Zasshi *83,* 1086 (1963); C.A. *60,* 8026 (1964).

39 *T. Talik, Z. Talik,* Roczniki Chem. *40,* 1675 (1966); C.A. *66,* 94889 (1967).
40 *E. Schröpl, R. Pohloudek-Fabini,* Pharmazie *23* (11), 638 (1968).
41 *M. Renson,* Bull. Soc. Roy. Sci. Liege *29,* 78 (1960); Chem. Zentr. *133,* 1181 (1962).
42 *R. Pohloudek-Fabini, E. Schröpl,* Pharmazie *24* (2), 96 (1969).
43 *H. Bock, K.L. Kompa,* Z. Anorg. Allgem. Chemie *332,* 238 (1964);
J.L. Wood, Org. Reactions *3,* 240 (1946).
44 *E. Allenstein, E. Lattewitz,* Z. Anorg. Allgem. Chemie *333,* 1 (1964).
45 *E.E. Reid,* Organic Chemistry of Bivalent Sulfur, Vol. I, Chemical Publishing, New York 1958.
46 *H. Rheinboldt, M. Perrier,* Bull. Soc. Chim. France 448 (1953).
47 *A. Haas, D.Y. Oh,* Chem. Ber. *100,* 480 (1967).
48 *A. Senning,* Acta Chem. Scand. *17,* 272 (1963).
49 *A.M. Giesbrecht,* Anais Acad. Brasil. Cienc. *30,* 155 (1958); C.A. *57,* 11128 (1962).
50 *W. Ruske, M. Keilert,* Chem. Ber. *94,* 2695 (1961).
51 *R.H. Patton, J.H. Simons,* J. Amer. Chem. Soc. *77,* 2016 (1955).
52 *J. Goerdeler, D. Wobig,* Justus Liebigs Ann. Chem. *731,* 120 (1970).
53 *L.A. Spurlock, P.E. Newallis,* J. Org. Chem. *33,* 2073 (1968).

The yields are 70–93%. Aliphatic aminocarbonyl thiocyanates such as dimethylaminocarbonyl thiocyanate are prepared in alternative manner (p. 799).

With potassium thiocyanate in acetone esters of chloroformic acid furnish a mixture of *alkoxycarbonyl thiocyanate* and alkoxycarbonyl isothiocyanate[54].

Thiocyanates prepared by nucleophilic substitution frequently contain isothiocyanates as side-products. These can arise by both direct reaction under S_N1 conditions and by subsequent isomerization of the initially formed thiocyanate at elevated temperature[41, 55a]. The proportion of isothiocyanate is dependent on various factors. α-Halocarbonyl compounds react with potassium thiocyanate without isocyanate formation[56]. By contrast, in the series benzyl iodide, benzyl bromide, benzyl chloride, 4-methoxybenzyl bromide, isopropyl iodide the ratio of the rate constants $k_{SCN}:k_{NCS}$ falls from 1300 to 85, *i.e.*, the content of isothiocyanate increases[57]. Allyl halides afford mixtures on reaction with thiocyanate ions; in acetonitrile at 100° an equilibrium between isothiocyanate and thiocyanate is established[58]. Working in dimethylformamide or dimethyl sulfoxide at 30°, on the other hand, enables allyl thiocyanates to be obtained in yields of up to 97%. Benzyl compounds give isothiocyanates even under these mild conditions[59].

Substituents other than halogen, for example, methylsulfate or tosylate (in sugars[60, 61] or steroids[62, 63]) and also the hydroxy group[64, 65] can be replaced by the thiocyanate group. Both these procedures are little used.

ω-Thiocyanato carboxylic acids[65] are obtained on reacting lactones with thiocyanate ions.

18.3.1.2   Reaction of thiocyanate ions with diazo compounds

For preparing aromatic thiocyanates the Sandmeyer reaction has lost some importance relative to the direct thiocyanation with thiocyanogen or the reaction of thiocyanate ions in the presence of halogen (p. 796). Isothiocyanates are generally formed as side-products[66, 67], the yields are unsatisfactory in many instances. As a result, the recent literature contains mainly examples relating to the synthesis of *haloaryl thiocyanates, nitroaryl thiocyanates*[68–70], *dithiocyanato* and *trithicyanato phenol ethers*[67] or *3-pyridyl thiocyanate*[71], which are not accessible by direct thiocyanation.

Under suitable conditions aryldiazo ketones also can be reacted with thiocyanic acid[72]:

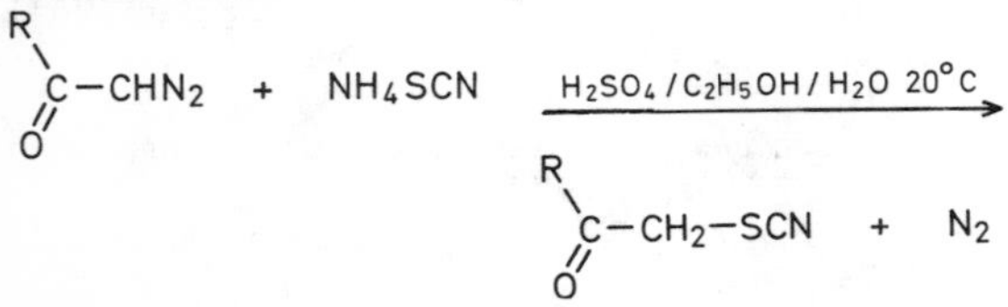

$$\underset{O}{\overset{R}{\underset{\|}{C}}}{-}CHN_2 \;+\; NH_4SCN \quad \xrightarrow{\; H_2SO_4\,/\,C_2H_5OH\,/\,H_2O\ 20°C \;}$$

$$\underset{O}{\overset{R}{\underset{\|}{C}}}{-}CH_2{-}SCN \;+\; N_2$$

but the method hardly offers any advantages over the reaction of α-halo ketones with thiocyanate ions.

18.3.1.3   Reaction of oxiranes with thiocyanic acid

Reacting oxiranes with thiocyanic acid is a reaction technique that is used a great deal in the steroid series[73–79]:

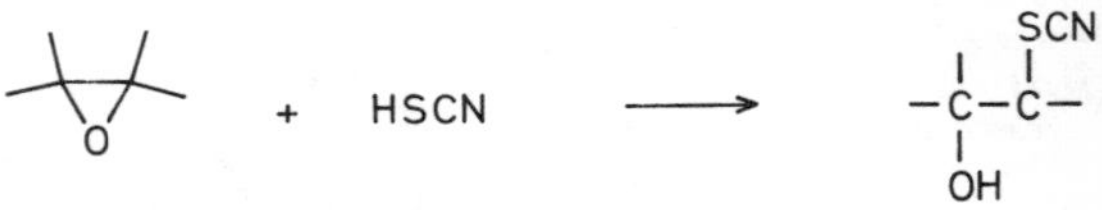

20 *g* potassium thiocyanate are dissolved in water, 40 *ml* ether are added and the whole is shaken with 30 *g* phosphoric acid. The dried ether phase is allowed to act for 12–14 *hours* on ~3–4 *g* of an epoxy steroid. The corresponding *thiocyanato steroid* is obtained in good yield[73].

[54] *A. Takamizawa, K. Hirai, K. Matsui*, Bull. Chem. Soc. Jap. *36*, 1214 (1963); C.A. *60*, 1747 (1964).
[55] *R. Appel, G. Vollmer*, Chem. Ber. *103*, 2555 (1970);
[55a] *H. Boehme, U. Scheel*, Chem. Ber. *100*, 347 (1967).
[56] *G. Barnikow, H. Niclas*, Z. Chem. *6*, 317 (1966).
[57] *U. Tonellato, G. Levorato*, Boll. Sci. Fac. Chim. Ind. Bologna 27 (4), 261 (1969); C.A. *72*, 110525 (1970).
[58] *A. Iliceto, A. Fava, U. Mazzucato*, Tetrahedron Lett. (11), 27 (1960).
[59] *S. Yoneda, H. Kitano, K. Fukui*, Kogyo Kagaku Zasshi *65*, 1816 (1962); C.A. *59*, 2679 (1963).
[60] *J. Defaye, J. Hildesheim*, Carbohyd. Res. *4*, 145 (1967).
[61] *S.D. Gero*, Tetrahedron Lett. 3193 (1966).
[62] *F.O. Bobbio, P.A. Bobbio*, Chem. Ber. *101*, 4241 (1968).
[63] *A. Bertho*, Justus Liebigs Ann. Chem. *714*, 155 (1968).
[64] *E. Ziegler, G. Zigeuner*, Monatsh. Chem. *79*, 316 (1948).
[65] *Yasuo Matsumoto, Toyo Sone*, Kogyo Kagaku Zasshi *63*, 2171 (1960); C.A. *58*, 4496 (1963).
[65a] *C.M. Hendry*, J. Amer. Chem. Soc. *80*, 973 (1958).
[66] *H.P. Kaufmann*, Angew. Chem. *54*, 195 (1941).
[67] *R. Pohloudek-Fabini, K.D. Lüss*, Arch. Pharm. *299*, 866 (1966).
[68] *K. Pilgram, F. Korte*, Tetrahedron *21*, 1999 (1965).
[69] *K. Pilgram, F. Korte*, Tetrahedron *20*, 177 (1964).
[70] *A. Fravolini, C. Salazar*, Ann. Chim. (Rom) *58*, 1155 (1968).
[71] *J.J. Panouse*, C.R. Acad. Sci., Paris, Ser. C *230*, 846 (1950).
[72] *W. Hampel, T. Török*, Z. Chem. *8*, 226 (1968).
[73] *K. Takeda, T. Komeno, J. Kawanami, S. Ishihara, H. Kadokawa, H. Tokura, H. Itani*, Tetrahedron *21*, 329 (1965).

The oxirane ring opens in accord with Barton's rule[80], so that both substituents (hydroxy and thiocyanate group) are axial.

The synthesis of *thiocyanato pyranosides*[81] and *thiocyanato furanosides*[82] and also of open-chain *β-thiocyanato alcohol* derivatives[83–85] can be accomplished analogously.

N-Sulfonylaziridines can be reacted to *N-(2-thiocyanatoalkyl)sulfonamides*[86] but these reactions are of subordinate preparative importance.

18.3.1.4 Addition of thiocyanate to C–C multiple bonds

C=C double bonds readily react readily with a series of thiocyanato compounds:

X: SCN[66, 87]
Cl, J[88–90]
Ar-S[92]
Ar-SO$_2$[93]
H[94]

Thiocyanation of unsaturated compounds with thiocyanogen in glacial acetic acid in general proceeds more selectively than halogenation[87] and, like the reaction of olefins with thiocyanogen iodide and chloride, is applied to analytical problems[66].

The addition ensues in accordance with Markovnikov's rule but not stereospecifically[88, 90]. Thiocyanogen chloride dissociates into Cl$^{\ominus}$ and SCN$^{\oplus}$[91] for the reaction, while thiocyanogen iodide can react in one of two possible ways[89]:

$$JSCN \rightleftharpoons SCN^{\ominus} + J^{\oplus}$$

$$2\,JSCN \rightleftharpoons (SCN)_2 + J_2$$

Alcohol and ester groups have no effect on reactions of olefins with thiocyanogen iodide. Carboxy and oxo groups close to the C=C double bond in part prevent the reaction[95].

The addition of thiocyanic acid to C=C double bonds is being studied mainly in the steroid series and does not display substantial preparative importance[94].

Mention should be made also of the addition of arylsulfonyl thiocyanate (from aryl sulfinate and thiocyanogen) to olefins and acetylenes in the presence of benzoyl peroxide[93].

The nucleophilic addition of thiocyanate to propynal[96], propiolic acid[97], or propiolamide[98]

X: OH[97]
H[96]
NHR[98]

leads to α,β-unsaturated thiocyanates. The predominatly formed *cis* isomers are valuable starting products for synthesizing heterocyclic compounds[96–98]. Preparation of *1,2-ethylene dithiocyanates*[99] from acetylene and thiocyanogen is less important.

[74] *Yo Ueda, E. Mosettig*, Steroids *1*, 361 (1963).

[75] *K. Tori, T. Komeno*, Tetrahedron *21*, 309 (1965).

[76] *K. Takeda, K. Igarashi, M. Narisada*, Steroids *4*, 305 (1964).

[77] *P.D. Klimstra*, J. Med. Chem. *9*, 781 (1966); C.A. *65*, 17018 (1966).

[78] *K. Ponsold, G. Schubert*, Z. Chem. *8*, 465 (1968).

[79] *M.A. Saeed, A. Iftikhav, A.B. Saeed, Zia-ud-Din*, Pak. J. Sci. Ind. Res. *10*, 5 (1967); C.A. *68*, 49869 (1968).

[80] *D.H.R. Barton*, J. Chem. Soc. 1027 (1953).

[81] *J.E. Christensen, L. Goodman*, J. Amer. Chem. Soc. *83*, 3827 (1961).

[82] *L. Goodman*, Chem. Commun. 219 (1968).

[83] USSR. P. 210 154 (1968), Inv.: *A.F. Kolomiets, G.S. Levskaya, N.K. Bliznyuk; *C.A. *69*, 52300 (1969).

[84] Brit. P. 923 582 (1963), Imperial Chem. Industries Ltd., Inv.: *J.M. Pollock;* C.A. *60*, 3011 (1964).

[85] Holl.P. 6 507 964 (1965), Soc. Nat. des Petroles-d'Aquitaine; C.A. *64*, 15753 (1966).

[86] *V.I. Markov, S.I. Burmistrov, O.N. Dolgopolov*, Zh. Obshch. Khim. *35*, 153 (1965); C.A. *62*, 13077 (1965).

[87] *A.A. Bugorkova, L.N. Petrova, V.M. Rodionov*, Zh. Obshch. Khim. *23*, 1808 (1953); C.A. *49*, 237 (1955).

[88] *R.G. Guy, I. Pearson*, Chem. Ind. (London) 1255 (1967).

[89] *P. Mesnard, C. Raby*, Bull. Soc. Pharm. Bordeaux *100*, 209 (1961).

[90] *R.G. Guy, R. Bonnett, D. Lanigan*, Chem. Ind. (London) 47, 1702 (1969).

[91] *A.B. Angus, R.G.R. Bacon*, J. Chem. Soc. (London) 774 (1958).

[92] *N. Kharasch, A.J. Havlik*, J. Amer. Chem. Soc. *75*, 3734 (1953).

[93] DBP. 1 219 924 (1966), Farbenfabr. Bayer AG, Erf.: *J. Goerdeler, P. Rosenthal;* C.A. *65*, 13612 (1965).

[94] *A.A. Akhrem, Z.I. Istomina, C.A. Kogan, A.I. Kuznetsova, A.M. Turnta*, Izv. Akad. Nauk USSSR, Ser. Khim. (11) 2503 (1969); C.A. *72*, 133067 (1970).

[95] *C. Raby*, Ann. Chim. (Paris) *6*, 481 (1961).

[96] *R. Raap*, Can. J. Chem. *44*, 1324 (1966).

[97] *G. Entenmann*, Dissertation, Universität Stuttgart 1971.

[98] *W.D. Crow, N.J. Leonard*, Tetrahedron Lett. 1477 (1964).

18.3.1.5 Replacement of hydrogen by the thiocyanate group

18.3.1.5.1 Reactions with thiocyanogen

Thiocyanogen can be prepared from silver or lead(II) thiocyanate and bromine, from potassium thiocyanate and copper(II) sulfate[100], and electrolytically from alkali metal or ammonium thiocyanate[101] in solution or in bulk[102]. Thiocyanogen in benzene or in carbon tetrachloride reacts with aromatic amines and phenols to *4-aminophenyl* and *4-hydroxyphenyl thiocyanates*[103]. *Mesityl thiocyanate*[104], *1-naphthyl thiocyanate*[104], *2-thienyl thiocyanate*[104, 105], or *pyrrol-2-yl thiocyanate*[104, 106] are prepared analogously in the presence of aluminum chloride or aluminum thiocyanate. *Thiocyanato-1,3-alkanedionatometal derivatives*[107], substituted and unsubstituted *1-azulenyl thiocyanate*[108] and *cyclopenta[c]thiapyranyl-5,7-dithiocyanate*[109], too, can be obtained by this method.

18.3.1.5.2 Reactions with thiocyanate ions in the presence of halogen

Aromatic and heteroaromatic compounds, but also *CH*-acid substances such as diesters of ethylmalonic acid[110], can be reacted in methanol, methyl acetate, or glacial acetic acid with sodium, potassium, or ammonium thiocyanate and bromine and chlorine at temperatures below or around 0° to the corresponding thiocyanates[66]:

$$RH \; + \; SCN^{\ominus} \; + \; Hal_2 \longrightarrow R-SCN$$
$$+ \; Hal^{\ominus} \; + \; HHal$$

Aromatic amines are thiocyanated in the para position[111]. If the para position is occupied, the amine is attacked in the ortho position. However, the yields are then poor because ring closure to 2-aminobenzothiazoles ensues readily[112, 113].

Under suitable conditions, however, a series of 2-aminophenyl thiocyanates can be isolated[113].

Alkyl-substituted phenols and phenol ethers can be reacted smoothly to *4-hydroxyphenyl thiocyanates*[114–116] and *4-thiocyano phenol ethers*[67]. Of polyhydric phenols, catechol but not hydroquinone, resorcinol, or phloroglucinol[66] reacts.

In amino phenol ethers the directing effect of the amino group predominates and *aminohydroxyphenyl thiocyanates* are obtained[67]. Simple phenyl ketones do not react; the corresponding thiocyanates are prepared *via* diazonium salts[117].

For thiocyanating organic compounds with inorganic thiocyanate compounds which readily split off halogen can be used instead of free halogen. Sulfuryl chloride, calcium chloride hypochlorite, and aryl iodide-dichlorides[66] and also haloamides are potential reagents. Systematic investigations of the thiocyanation of aromatic

[99] DBP. 1 924 711 (1969), Am. Cyanamid Co., Erf.: *R.P. Welcher*; C.A. *72*, 42773 (1970).

[100] *A.N. Nesmejanov, E.G. Perevalova, A.O. Nesmejanova*, Dokl. Akad. Nauk. SSSR *119*, 288 (1958).

[101] *G. Canquis, G. Pierre*, C.R. Acad. Sci., Paris, Ser. C *266*, 883 (1968).

[102] *M. Bögemann, S. Petersen, O.-E. Schultz, H. Söll* in *Houben-Weyl*, Methoden der organischen Chemie, Bd. IX, p. 797, 859, Georg Thieme Verlag, Stuttgart 1955;
J. Goerdeler, P. Rosenthal, Tetrahedron Lett. 3665 (1964);
R.T. Major, L.H. Peterson, J. Amer. Chem. Soc. *78*, 6181 (1956).

[103] *A.A. Zelenetskaya, N.N. Nikitina, G.P. Borodina*, Tr. Vses. Nauch-Issled. Inst. Sintetich. Natural Dushist. Veshchestr. *7*, 175 (1965); C.A. *66*, 61629 (1967).

[104] *E. Söderbäck*, Acta. Chem. Scand. *8*, 1851 (1954).

[105] *F.M. Stoyanovich, G.I. Gorushkina, Ya.L. Goldfarb*, Izv. Akad. Nauk SSSR, Ser. Khim. (2), 387 (1969).

[106] *S. Gronowitz, A.B. Hoernfeldt, B. Gestblom, R.A. Hoffmann*, J. Org. Chem. *26*, 2615 (1961).

[107] *P.R. Singh, R. Sahai*, Inorg. Chim. Acta *2*, 102 (1968).

[108] *L.L. Replogle, K. Katsumoto, T.C. Morrill, C.A. Minor*, J. Org. Chem. *33*, 823 (1968).

[109] *A.G. Anderson jr., W.F. Harrison*, J. Amer. Chem. Soc. *86*, 708 (1964).

[110] *F. Eiden, G. Ganglitz*, Arch. Pharm. *302*, 297 (1969).

[111] *K. Papke, R. Pohloudek-Fabini*, Pharmazie *22*, 229 (1967).

[112] *R. Pohloudek-Fabini, M. Schüssler*, Pharm. Zentralhalle *107*, 116 (1968).

[113] *M. Selchau, R. Pohloudek-Fabini*, Arch. Pharm. *302*, 504 (1969).

[114] US. P. 3 202 690 (1965), Consolidation Coal Co., Inv.: *E.P. Previc*; C.A. *63*, 14763 (1966).

[115] *K. Konishi*, Takeda Kenkyusho Nempo *24*, 233 (1965); C.A. *64*, 8076 (1966).

[116] *T. Wieland, E. Bäuerlein*, Chem. Ber. *97*, 2103 (1964).

[117] *K.-D. Lüss, R. Pohloudek-Fabini*, Arch. Pharm. *299*, 878 (1966).

[118] *A. Kaji*, Nippon Kagaku Zasshi *81*, 1776 (1960); C.A. *56*, 2391 (1962).

amines[118, 119], phenols, and naphthols[120] and also various heterocyclic compounds[121] display no substantial advantages over thiocyanation with alkali metal thiocyanate and free halogen[66].

Under certain conditions reacting lead(II) thiocyanate with chlorine affords thiocyanogen chloride and not thiocyanogen[122]:

$$Pb(SCN)_2 + Cl_2 \xrightarrow{H_3C-COOH}$$

$$ClSCN \xrightarrow{Pb(SCN)_2} (SCN)_2$$

Using thiocyanogen chloride, aryl ethers and anilides which do not react with thiocyanogen can be reacted in 65–100% yield to *thiocyanato aryl ethers* and *thiocyanato anilides*. Thiocyanation of phenols and aromatic amines with thiocyanogen iodide[123] reveals no advantages over this procedure.

18.3.1.5.3 Reactions with copper thiocyanate

Copper(II) thiocyanate can be used in thiocyanation reactions[66] in accordance with

$$2\ Cu(SCN)_2 + RH \longrightarrow 2\ CuSCN + RSCN + HSCN$$

Instead of copper(II) thiocyanate a mixture of copper(II) sulfate and alkali metal or ammonium thiocyanate can be used. Although the reaction is simple to carry out, the recent literature gives few examples of its use. Examples are the synthesis of substituted *phenyl thiocyanates*[124], *1,3-azulenyl dithiocyanate*[125], and *5-(diphenylmethylene)-1,3-cyclopentadien-1-yl thiocyanate*[125].

18.3.1.6 Recent techniques

On performing the reaction of olefins (p. 795) or alkylaromatic compounds (p. 796) with thiocyanogen or cyanogen chloride under UV light, the attack occurs in the allyl position or in the side-chain. Thus cholesterol gives *7-cholesteryl thiocyanate*[13], while alkylaromatic compounds give *1-phenyl-1-alkyl thiocyanates*[126, 127]. However, if a secondary alpha-C atom is present isothiocyanates are obtained (*e.g., cumenyl isothiocyanate* from cumene[126]). Aminocarbonylmethylenediphenylsulfan betaines react with thiocyanic acid to give *2-thiocyanatoacetamides*[128]:

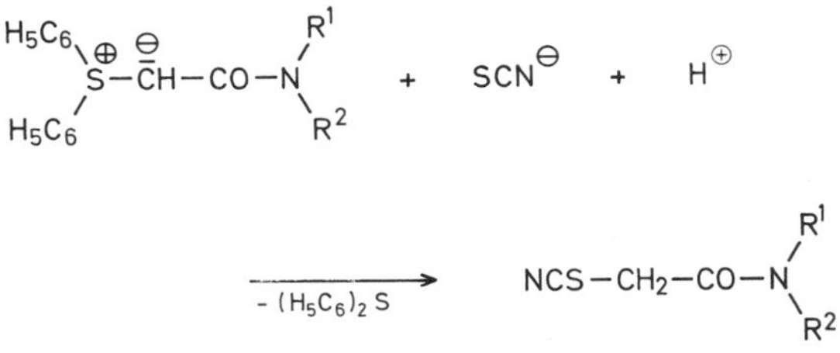

The thermolysis of 5-alkoxy-1,2,3,4-thiatriazoles leading to esters of cyanuric acids (p. 792) can be modified for use for preparing thiocyanates[129]:

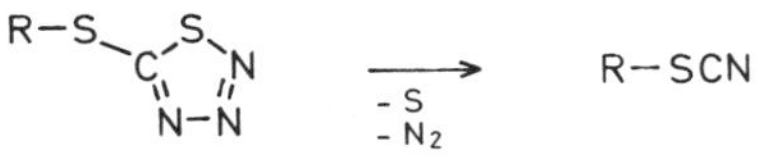

Allene thiocyanates are formed on reacting (β-oxoalkyl)phosphonium salts with lead(IV) acetate,

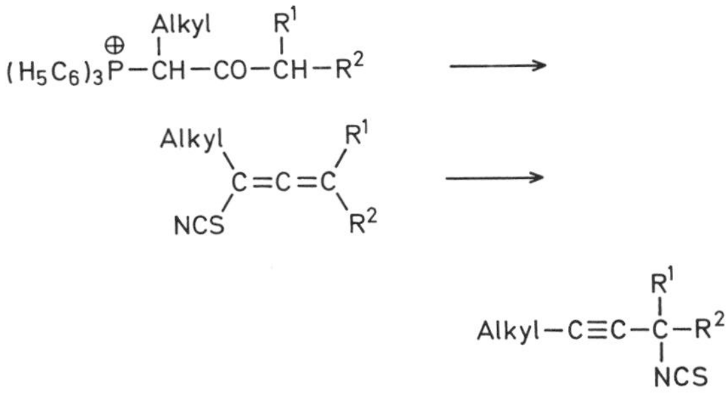

which in certain cases rearrange readily to alkynyl isothiocyanates[130].

On account of the instability of aliphatic diazonium salts the Sandmeyer reaction can be used only for aromatic compounds. Some peroxides reacting similarly to diazonium salts are more

[119] *G. Grandolini*, Boll. Sci. Fac. Chim. Ind. Bologna *20*, 136 (1962); C.A. *59*, 607 (1963).

[120] *A. Kaji*, Nippon Kagaku Zasshi *82*, 382 (1961); C.A. *56*, 10054 (1962).

[121] *A. Kaji*, Nippon Kagaku Zasshi *82*, 502 (1961); C.A. *56*, 10123 (1962).

[122] *R.G.R. Bacon, R.G. Guy*, J. Chem. Soc. (London) 318 (1960).

[123] *P. Mesnard, C. Raby*, Bull. Soc. Pharm. Bordeaux *101*, 33 (1962).

[124] *R.F. Ambartsumova, S.R. Tulyaganov*, Zh. Org. Khim. (6), 1045 (1970); C.A. *73*, 34945 (1970).

[125] *V.A. Nefedov*, Zh. Obshch. Khim. *38*, 2184 (1968); C.A. *70*, 37902 (1969); *E. Frederiksen, S. Liisberg*, Chem. Ber. *88*, 684 (1955).

[126] *R.G.R. Bacon, R.G. Guy*, J. Chem. Soc. (London) 2428 (1961).

[127] *R.G.R. Bacon, R.S. Irwin*, J. Chem. Soc. (London) 2447 (1961).

[128] DBP. 1 280 866 (1968), BASF, Erf.: *H. Koenig, W. Reif;* C.A. *70*, 37467 (1970).

[129] *E. Lieber, E. Oftedahl, C.N.R. Rao*, J. Org. Chem. *28*, 194 (1963).

[130] *E. Zbibal, H. Hengstberger*, Monatsh. Chem. *99*, 412 (1968).

stable. Reacting such peroxides with thiocyanates ions in the presence of copper(I) or iron(I) ions thus enables aliphatic thiocyanates to be prepared:

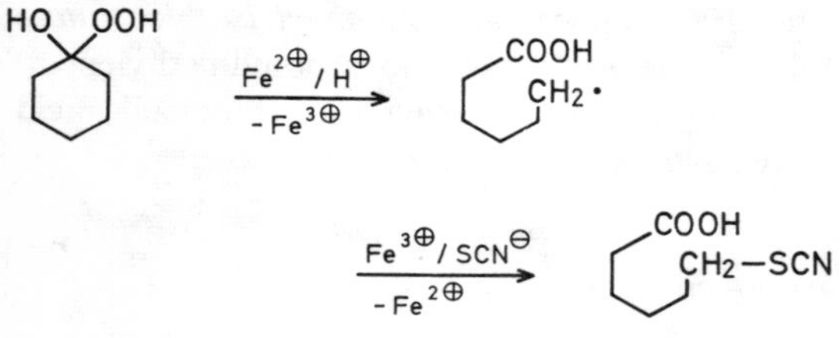

The procedure can be applied to hydroperoxides of various cyclic and open-chain ketones[131, 132]. Ammonium thiocyanate in dimethyl sulfoxide in the presence of hydrogen chloride can be used as an alternative thiocyanation agent[133]:

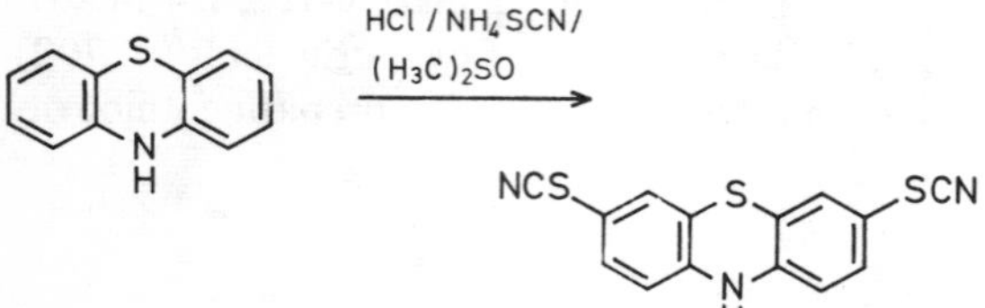

18.3.2 Synthesis of thiocyanates by nitrile group transfer

18.3.2.1 Cleavage of disulfides, thiosulfates, and sulfenyl halides

Cleavage of disulfides with cyanogen bromide formed intermediately from cyanide ions and bromine furnishes thiocyanates in good yield[134]. The reaction with cyanide ions alone has been investigated by several authors with special reference to the reaction mechanism[135–137]. Recent patent applications[138, 139] claim the reaction between organic thiosulfates and sodium cyanate. Like the addition of prussic acid to trithiocarbonates,

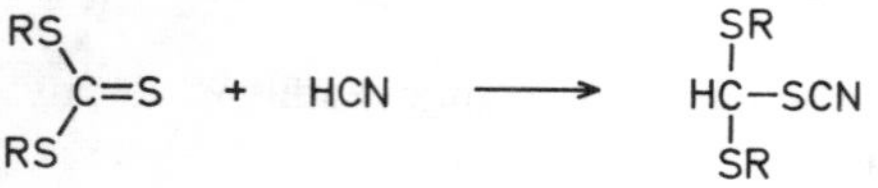

leading to *fluoroalkylthio-substituted thiocyanates*[140], the technique is, however, probably of minor importance for laboratory practise.
Exchanging the halogen in sulfenyl chlorides against cyanide proceeds smoothly[45]:

$$RSCl + KCN \longrightarrow RSCN + KCl$$

The reaction conditions are similar to those during reaction of alkyl halides with thiocyanate ions; as a result, isothiocyanates are obtained as side-products during preparation of sensitive thiocyanates[141]. The reaction is not very often used for preparative purposes, but, for instance, for preparing *thiocyanato-1,3-alkanedionatometal derivatives*[142]; it is suitable for detecting and identifying sulfenyl halides[143, 144].
The preparation of the sulfenyl chloride in a costly, continuously working apparatus[145] is bypassed in a novel technique[144]:
Chlorine is introduced into a suspension of trithiane in thionyl chloride at 10–15°. Formamide is added and the whole is heated to boiling. After distillation *chloromethyl thiocyanate* is obtained:

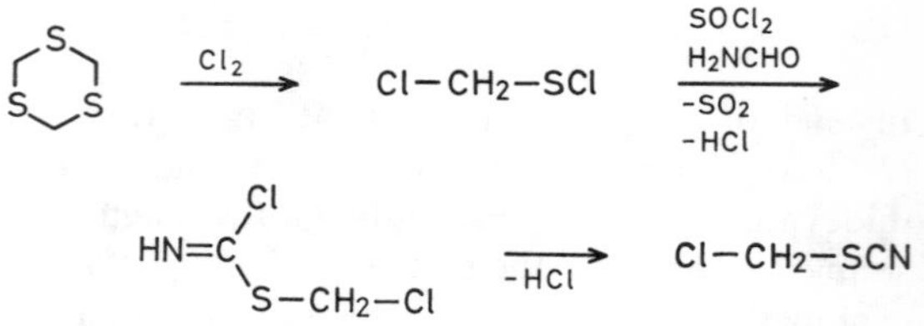

The isolated sulfenyl chlorides, too, can be reacted with formamide in thionyl chloride to, for example, *ethyl, phenyl, pentachlorophenyl, 4-chlorophenyl,* or *1,1,2,2-tetrachloroethyl thiocyanate*[144].

[131] F. Minisci, R. Galli, U. Pallini, Gazz. Chim. Ital. 89, 621, 1910, 1922, 2428 (1959); 91, 1023 (1961).
[132] F. Minisci, M. Cecere, R. Galli, Chem. Ind. (Milano) 50, 225 (1968); C.A. 69, 51500 (1968).
[133] Y. Tsujino, T. Naito, J. Sugita, Nippon Kagaku Zasshi 91 (11), 1075 (1970); Chem. Inform B 10, 362 (1971).
[134] R. Pohloudek-Fabini, K. Kottke, F. Friedrich, Pharmazie 26, 283 (1971).
[135] S. Kawamura, Y. Otsuji, T. Nakabayashi, T. Kitao, J. Tsurugi, J. Org. Chem. 30, 2711 (1965).
[136] R.G. Hiskey, D.N. Harpp, J. Amer. Chem. Soc. 86, 2014 (1964).
[137] R. Pohloudek-Fabini, K. Kottke, F. Friedrich, Pharmazie 24, 433 (1969).

[138] Fr. P. 2008468 (1970), BASF; C.A. 73, 45106 (1970).
[139] Jap. P. 6919524 (1969), Takeda Chemical Ind., Ltd., Inv.: K. Konishi, T. Okuya, T. Aiba; C.A. 73, 45104 (1970).
[140] US. P. 3117985 (1964), E.I. du Pont de Nemours & Co., Inv.: S. Proskow; C.A. 60, 10549 (1964).
[141] U. Miotti, A. Ceccon, Ann. Chim. (Rom) 54, 851 (1964).
[142] R.W. Kluiber, J. Amer. Chem. Soc. 83, 3030 (1961).
[143] E. Müller, E.W. Schmidt, Chem. Ber. 97, 2614 (1964).
[144] DBP. 1157603 (1963), Farbenfabr. Bayer AG, Erf.: E. Kuehle; C.A. 60, 6787 (1964).
[145] H. Brintzinger, M. Langheck, Chem. Ber. 86, 557 (1953).
[146] F.P. Corson, R.G. Pews, J. Org. Chem. 36, 1654 (1971).

Sodium cyanide does not produce sulfonyl cyanides but thiocyanates from sulfonyl chlorides [146]:

$$3 \text{ R}-\text{SO}_2-\text{Cl} \ + \ 5\,\text{NaCN} \ \xrightarrow[\quad]{\overset{\text{CH}_3\text{CN,}}{70^\circ}} \ \text{R}-\text{SCN}$$

$$27,1\%$$

$$2 \text{ R}-\text{SO}_3\text{Na} \ + \ 3\,\text{NaCl} \ + \ 2\,(\text{CN})_2$$

18.3.2.2 Reacting thiolates, thiophenols and thio ethers to thiocyanates

Compounds with thiol groups react with cyanogen halide in alcohol or acetone in the presence of a base [147] (triethylamine or sodium hydroxide):

$$\text{RSH} + \text{XCN} \longrightarrow \text{RSCN} + \text{HX}$$

The reaction can be utilised for preparing aliphatic, aromatic, and heteroaromatic thiocyanates (cf. Table 2) [148–151].
The cyanogen bromide can be generated in the reaction solution itself [159]:

$$\text{RSH} + 2\,\text{NaCN} + \text{Br}_2 \longrightarrow \text{RSCN} + 2\,\text{NaBr} + \text{HCN}$$

The thiol is dissolved or suspended in methanol containing potassium bromide together with sodium cyanide and a methanolic solution of bromine is then added drop by drop whilst cooling.

2- and 3-aminophenyl thiocyanates, which are quite difficult to prepare by other means, can be obtained analogously [137].

Aliphatic and aromatic thiocyanates are prepared from thiols or thiophenols and cyanogen chloride also by using hydrogen chloride as catalyst [152].
In some cases the lead(II) thiolate is used instead of the thiol. Thus, *dimethylaminocarbonyl thiocyanate*[54] is obtained from the corresponding lead(II) salt and cyanogen bromide.

Esters of cyanic acid can be used successfully as nitrile group transfer agents instead of cyanogen halide [153–155]:

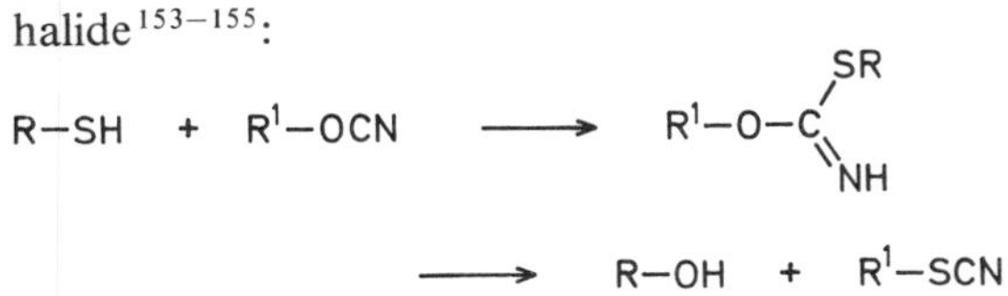

$$\text{R}-\text{SH} \ + \ \text{R}^1-\text{OCN} \ \longrightarrow \ \text{R}^1-\text{O}-\text{C} \underset{\text{NH}}{\overset{\text{SR}}{\big\langle}}$$

$$\longrightarrow \ \text{R}-\text{OH} \ + \ \text{R}^1-\text{SCN}$$

The reaction proceeds smoothly if R^1 is an electron-withdrawing group [155]. The procedure is comparable to the following reaction [156]:

$$\text{RS}-\text{C} \underset{\text{CN}}{\overset{\text{NH}}{\big\langle}} \ + \ \text{Base} \ \xrightarrow[-\text{HCN}]{} \ \text{R}-\text{SCN}$$

$$R = \text{C}_2\text{H}_5,\ \text{C}_3\text{H}_7,\ \text{C}_4\text{H}_9$$

Triethylamine, pyridine, sodium hydroxide, or calcium hydroxide serve as base; alternatively, the reaction can proceed thermally.
The cleavage of thio ethers with cyanogen bromide [157]

$$\underset{R'}{\overset{R}{\big\rangle}}\text{S} \ + \ \text{BrCN} \ \longrightarrow$$

$$\text{R}-\text{SCN} \ + \ \text{RBr}$$

requires elevated temperatures and, as a result, isothiocyanates are obtained side by side with thiocyanates [158]. In some cases, for instance, during preparation of *ferrocenyl thiocyanate*[159], the technique is superior to the direct thiocyanation with thiocyanogen.

Table 2. Heteroarenyl thiocyanates

Heteroaromatic compound	Reaction with thio-cyanogen	Reaction with cyanogen bromide	Replacement of halogen
Thiophene	105, 121		
Pyrrole	106		
Indole	163		
Pyrazole	162	164	
Thiazole	121, 149	150	
1,2,4-Triazole		161, 164, 150	
Imidazole		150, 161	
Pyridine		150	39
Quinoline	71	150	
Isoquinoline	71, 165		
Pyrimidine	121	38	38
Purine		160	

[147] *R. Pohloudek-Fabini, F. Friedrich,* Pharmazie *20,* 114 (1965).
[148] DBP. 1 183 903 (1965), Farbenfabr. Bayer AG, Erf.: *K. Goliasch, E. Grigat, R. Pütter;* C.A. *62,* 7640 (1965).
[149] DBP. 1 195 743 (1965), Farbenfabr. Bayer AG, Erf.: *K. Goliasch, E. Grigat, R. Pütter;* C.A. *64,* 1579 (1966).
[150] *K. Kottke, F. Friedrich, R. Pohloudek-Fabini,* Arch. Pharm. *300,* 583 (1967).
[151] *K. Kottke, F. Friedrich, R. Pohloudek-Fabini,* Pharmazie *24,* 438 (1969).
[152] DBP. 1 270 553 (1968), Deutsche Gold- und Silber-scheideanstalt, Erf.: *C. Kosel;* C.A. *69,* 96212 (1968).
[153] *E. Grigat, R. Pütter,* Chem. Ber. *97,* 3022 (1964).
[154] *K.A. Jensen, A. Holm, C. Wentrup,* Acta Chem. Scand. *20,* 2107 (1966).
[155] *D. Martin,* Z. Chem. *7,* 123 (1967).
[156] DBP. 1 229 075 (1966), Roehm & Haas GmbH, Erf.: *W. Gruber, P. Quis;* C.A. *66,* 28387 (1967).
[157] *D.S. Tarbell, D.P. Harnish,* Chem. Rev. *49,* 1 (1951).
[158] *R.G.R. Bacon, J. Köchling,* J. Chem. Soc. (London) 5594 (1964).
[159] *G.R. Knox, I.G. Morrison, P.L. Pauson,* J. Chem. Soc. C (London) 1842 (1967).

[160] *M. Saneyoshi, G. Chihara,* Chem. Pharm. Bull. (Tokyo) *15*, 909 (1967); C.A. *68*, 49949 (1968).

[161] *J. Kinugawa, M. Ochiai, C. Matsumura, H. Yamamoto,* Chem. Pharm. Bull. (Tokyo) *12*, 433 (1964); C.A. *61*, 4338 (1964).

[162] *M. Guarneri, P. Giori, R. Tomatis,* Ann. Chim. (Rom) *58*, 697 (1968).

[163] *M.S. Grant, H.R. Snyder,* J. Amer. Chem. Soc. *82*, 2742 (1960).

[164] *A.K. Bhat, R.P. Bhamaria, R.A. Bellare, C.V. Deliwala,* Indian J. Chem. *5*, 397 (1967); C.A. *68*, 59501 (1968).

[165] *C.E. Hall, A. Taurins,* Can. J. Chem. *44*, 2465 (1966).

Subject Index

5H-Benzo[b]azepine

~ / 5,5-dimethyl-2-dimethylamino-3-phenyl- 485
Benzo[d]azocine

~ / 4-dimethylamino-5-phenyl-1,2-dihydro- 486
~ / 4-oxo-1,2,3,4,5,6-hexahydro- 568
1H-Benzo[c]azonine

~ / 2-methyl-2,3,4,5,6,7-hexahydro- 571
Benzobicyclo[2.2.1]heptadiene

~ / 7-exo-acetoxy-9-anti-azido 304
Benzobicyclo[2.2.1]heptene / 7-exo-azido-9-anti-iodo-
307
~ / 7-exo, 9-anti-diazido- 304
Benzo[c]cinnoline

~ / 3,8-bis[dimethylamino]- ; 5-oxide 374
~ / 3,8-dimethyl- ; 5-oxide 374
~ / 5-oxide (9-subst.) 371
Benzocyclobutene / (derivatives) 292
~ / 1-cyano- 196
~ / 1-ethoxycarbonyl- 196
Benzocyclohepta[5,6-d]imidazole

~ / 3-hydroxy-2-methyl-2,3,3a,4,5,6-hexahydro- ; 1-oxide
Benzo[d]-1,3,2-diazaborine /

~ / 2-phenyl-1,2-dihydro- ; 3-oxide 367
3H-Benzo[e]-1,4-diazepin

~ / 7-chloro-2-methylamino-5-phenyl- ; 4-oxide 432
Benzo[a,c]difuroxan / 389

Benzo-1,4-dioxin / 2,3-dihydro- 282
11H-Benzo[a]fluorene

~ / 11-diazo- 218
Benzo[b]furan

~ / 5-cyano-2-methyl- 435
~ / 2-(dialkoxymethyl)-3-oxo-2,3-dihydro- 283
~ / 2-(dialkylthiomethyl)-3-oxo-2,3-dihydro- 283
Benzofuro[3,2-c]pyridine

~ / 4a-ethoxy-2-methyl-6-nitro-1,2,3,4,4a,9b-hexahydro-
435
~ / 2-methyl-6-nitro-1,2,3,4-tetrahydro- 435
Benzofuroxan 388

~ / (multiple subst.) 388
~ / 5-acetoxy- 388
~ / 5-acetylamino- 388
~ / 5-amino- 388
~ / 5-azido-6-nitro- 388
~ / 5-carboxy- 388
~ / 5-chloro- 387
~ / 6-chloro-5-methoxy- 388
~ / 5,7-dimethyl- 387
~ / 5-hydroxy- 388
~ / 5-methoxy- 388
~ / 5-methyl- 387
~ / 4,5,6,7-tetrabromo-4,5,6,7-tetrahydro- 382
~ / 4,5,6,7-tetrahydro- 381
Benzoic acid / acylaminonitro- 6
~ / alkoxyamide 708
~ / alkyl- ; amide 671
~ / alkylaminonitro- 6
~ / alkyl ester 2-bromo-4-nitrophenylhydrazonide 728
~ / alkyl ester hydroxyimide 708
~ / N-alkylhydroxyamide 708
~ / alkylimide isocyanate 766
~ / amide 123, 676, 703, 704
~ / amide hydroxyimide 733, 741, 744
~ / amides 671
~ / amidinium salts 739
~ / 2-amino- ; amide 495
~ / 4-(4-aminobenzenesulfonylamino-)- ; amide 578
~ / x-amino-y-diazonio- 167
~ / 4-amino-3,5-dichloro- ; 2-benzylhydrazide 109
~ / 4-amino- ; ethyl ester 501
~ / 4-amino-2-hydroxy- 498
~ / 3-amino- ; methyl ester 579
~ / 2-amino- ; nitrile 495
~ / 3-amino- ; nitrile 579
~ / 4-amino- ; nitrile 579
~ / 2-amino-3-nitro- ; methyl ester 579
~ / -anilide 567, 660
~ / anilide hydrazonide 733
~ / anilide phenylimide 747
~ / 2-anilino- 572
~ / 2-anilinocarbonylamino- ; methyl ester 666
~ / arylmethylamide 587

Benzo[e]indene

Benzo[c,d]indole

Benzo[f]isoquinoline

8H-‹Benzo[c]naphtho[1,8a,8-e,f]-1,2-thiazepine›

Benzo[c]-2,7-naphthyridine

Benzonitrile / see benzoic acid
Benzo-1,2,5-oxadiazole / 1-oxide see benzofuroxan
Benzo[b]-1,4-oxazepine

1H-‹Benzo[d]-1,2-oxazine›

3H-‹Benzo[c]-1,2-oxazine›

2H-‹Benzo[e]-1,3-oxazine›

4H-‹Benzo[d]-1,3-oxazine›

1H-‹Cyclobuta[f]indene›

Cycloheptaimidazole

1H-‹Cyclopenta[c]pyridazine›

2H-‹Cyclopenta[c]pyridazine›

3H-‹Cyclopenta[c]pyridazine›

2H-Cyclopenta[d]pyridazine

1H-‹Cyclopenta[b]quinoxaline›

Cyclopenta[c]thiapyrane

1H-‹Cyclopropa[b]benzofuran›

Naphtho[1,2-d]furoxan / 388

Naphtho[2,1-d]-1,3-oxazole / 349

**3H-‹Naphtho[1',8':4,5,6]pentaleno[1,6a-c:3,3a-c']
  dipyrazole›**

1H-Naphtho[1',8':4,5,6]pentaleno[1,6a-c]pyrazole

3H-Naphtho[1',8':4,5,6]pentaleno[1,6a-c]pyrazole

2H-‹Naphtho[1,8-c,d]-1,2-thiazole›

Naphtho[1,2-e]-1,2,4-triazine

1H-‹Naphtho[1,8-d,e]-1,2,3-triazine› / 168

1H-‹Naphtho[1,2]triazole›

2H-‹Naphtho[2,3]triazole›

Pyrido[3,2]furoxan

Pyrido[2,3,4,5-l,m,n]phenanthridene

Pyrido[1,2-a]pyrazinium

2H-‹Pyrido[1,2-a]pyrimidin

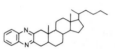

Triazolo[5,1-c]benzo-1,2,4-triazine

1,2,4-Triazolo[4,3-b]pyridazine

1,2,3-Triazolo[3,4-a]pyridine

1,2,4-Triazolo[4,3-a]pyridine

3H-‹Triazolo[4,5-d]-1,2,3-triazine›

1,2,3-Triazolo[4,5-d]-1,2,3-triazole

Tricyclo[2.2.1.0²,⁶]heptane